WITHDRAWN FROM
TSC LIBRARY

The
ENCYCLOPEDIA
of
**BEACHES AND
COASTAL ENVIRONMENTS**

ENCYCLOPEDIA OF EARTH SCIENCES SERIES

Series Editor: Rhodes W. Fairbridge

Volume
- I THE ENCYCLOPEDIA OF OCEANOGRAPHY/*Rhodes W. Fairbridge*
- II THE ENCYCLOPEDIA OF ATMOSPHERIC SCIENCES AND ASTROGEOLOGY/*Rhodes W. Fairbridge*
- III THE ENCYCLOPEDIA OF GEOMORPHOLOGY/*Rhodes W. Fairbridge*
- IVA THE ENCYCLOPEDIA OF GEOCHEMISTRY AND ENVIRONMENTAL SCIENCES/*Rhodes W. Fairbridge*
- IVB THE ENCYCLOPEDIA OF MINERALOGY/*Keith Frye*
- VI THE ENCYLOCPEDIA OF SEDIMENTOLOGY/*Rhodes W. Fairbridge and Joanne Bourgeois*
- VII THE ENCYCLOPEDIA OF PALEONTOLOGY/*Rhodes W. Fairbridge and David Jablonski*
- VIII THE ENCYCLOPEDIA OF WORLD REGIONAL GEOLOGY, PART 1: Western Hemisphere (Including Antarctica and Australia)/*Rhodes W. Fairbridge*
- XII THE ENCYCLOPEDIA OF SOIL SCIENCE, PART 1: Physics, Chemistry, Biology, Fertility, and Technology/*Rhodes W. Fairbridge and Charles W. Finkl, Jnr.*
- XV THE ENCYCLOPEDIA OF BEACHES AND COASTAL ENVIRONMENTS/*Maurice L. Schwartz*

ENCYCLOPEDIA OF EARTH SCIENCES, VOLUME XV

The
ENCYCLOPEDIA
of
BEACHES AND
COASTAL ENVIRONMENTS

EDITED BY

Maurice L. Schwartz
Western Washington University

Hutchinson Ross Publishing Company
Stroudsburg, Pennsylvania

To Stephanie, Phebe, Philip, Howard, and Ivan—
for all the good times that we have had together on beaches

Copyright © 1982 by **Hutchinson Ross Publishing Company**
Library of Congress Catalog Card Number: 81-7250
ISBN: 0-87933-213-1

All rights reserved. No part of this book may be reproduced or transmitted in any form or by any means—graphic, electronic, or mechanical, including photocopying, recording, taping, or information storage and retrieval systems—without written permission of the publisher.

86 85 84 83 82   5 4 3 2 1
Manufactured in the United States of America.

Library of Congress Cataloging in Publication Data
Main entry under title:
The Encyclopedia of beaches and coastal environments.

  (Encyclopedia earth sciences; v. 15)
  Includes index.
  1. Coasts—Dictionaries. 2. Beaches—Dictionaries. 3. Coastal ecology—Dictionaries. I. Schwartz, Maurice L. II. Series.
GB450.4.E53   551.4′57′0321   81-7250
ISBN 0-87933-213-1   AACR2

Distributed worldwide by Van Nostrand Reinhold Company Inc.,
135 W. 50th Street, New York, NY 10020.

PREFACE

The idea that led to the development of this volume was first conceived while the series and volume editors were enjoying a good dinner, fine wine, and a beautiful sunset at a seafood restaurant overlooking Puget Sound. This was a rather auspicious beginning for what turned out to be seven long years of hard, and often frustrating, work.

There was, we reasoned, no single collection of coastal information quite like what we had in mind. Although the AGI *Glossary* contains some coastal terminology, the Corps of Engineers' *Shore Protection Manual* has a useful glossary, the Soviets have recently published a coastal terminology booklet and a four-language geomorphology dictionary, and E. C. F. Bird and I are now working on a forthcoming expanded descriptive version of the world's coastline, there is at present simply no published survey of the field of coastal studies. For this reason we have tried to cover, as best we could in one volume, the geomorphic, biologic, engineering, and human aspects of the world's coast. Omissions of someone's particular topic of interest are inevitable; it is hoped that they are few in number.

How To Use This Encyclopedia

All entries are of course in alphabetical order. The length of the entries (basically short, medium, or long), while somewhat related to the importance of the topic, is, more realistically, an indication of the depth and breadth of treatment. The reader will find that the same topic may be dealt with from various viewpoints in a number of different entries.

To find a particular topic, it is best to look first for that subject in its alphabetical context. Beyond that, the *index* and *cross-references* will provide further information on the subject in question. The comprehensive index at the back of the volume will list for a given term or name every page in the book where that item appears. This is a fine way to find a subject presented in many different contexts. On the other hand, the cross-references at the end of each entry act as a guide to other related entries. To complete this internal system of tying together topics within the volume, the abbreviation q.v. (which see) is used liberally when mention is made in an article of a subject that appears elsewhere in the volume.

To further assist the reader in learning as much as possible about a given topic, *references* are provided at the end of each entry. The number of reference entries usually varies in direct proportion to the length of the entry. *Citations* within the entry are correlated with the references, and these together should provide a further guide to authoritative literature on the subject.

Since the metric system is used throughout this volume, a few basic conversion tables are included to facilitate comparison of data.

Conversion Tables

Metric to English Units—Equivalents of Length

1 micron (μ) = 0.001 millimeter (mm) = 0.00004 inch (in.)
1 mm = 0.1 centimeter (cm) = 0.03937 in.
1000 mm = 100 cm = 1 meter (m) = 39.37 in. = 3.2808 foot (ft)
1 m = 0.001 kilometer (km) = 1.0936 yard (yd)
1000 m = 1 km = 0.62137 mile (mi)
1 in. = 2.54 cm
12 in. = 1 ft = 0.3048 m
1 cm = 0.39370 in. = 0.032808 ft
1 km = 10^5 cm = 0.62137 mile
1 fathom = 6 ft = 1.8288 m
1 nautical mile = 1.85325 km
1 in. = 2.54001 cm
1 ft = 30.480 cm
1 statute mile = 1.60935 km = 5280 ft
1 astronomical unit = 1.496 $\times$ 10^8 km = 92,957,000 miles
1 light year = 9.460 $\times$ 10^{12} km = 5.878 $\times$ 10^{12} miles
1 parsec = 3.085 $\times$ 10^{13} km = 1.917 $\times$ 10^{13} miles

Square Measures

1 square foot = 0.00002295684 acre
1 acre = 43560 ft^2 = 0.0015625 mi^2
1 square yard = 0.836127 m^2
1 hectare = 2.471054 acre
1 square mile (statute) = 640 acres
1 square cm = 0.1550 square in. = 0.0010764 square ft
1 square km = 10^{10} square cm = 0.3861 square mile
1 square in. = 6.452 square cm
1 square ft = 929.0 square cm
1 square mile = 2.5900 square km
1 mm^2 = 0.00155 $in.^2$ 1 $in.^2$ = 6.452 cm^2
1 m^2 = 10.764 ft^2 1 ft^2 = 0.09290 m^2
1 km^2 = 0.3861 mi^2 1 mi^2 = 2.5900 km^2

Cubic Measures

1 gal (UK) = 4.5461 liters = 1.200956 gal (US)
1 liter = 0.219969 gal (UK) = 0.264173 gal (US)
1 gal (US) = 3.7854 liters = 0.832670 gal (UK)

PREFACE

1 cc = 0.0610 cu in. = 0.000035314 cu ft
1 cu in. = 16.387 cc
1 cu ft = 28317 cc
1 mm^3 = 0.000061 in.^{3} 1 in.^{3} = 16.387 cm^3 (cc)
1 cm^3 (cc) = 0.0610 in.^{3} 1 ft^3 = 0.028317 m^3
1 m^3 = 35.315 ft^3 1 mi^3 = 4.1681 km^3
1 km^3 = 0.239911 mi^3

Statute Miles to Nautical Miles to Kilometers

Statute	Nautical	Kilo-meters	Statute	Nautical	Kilo-meters
¼	0.22	0.40	9	7.82	14.48
½	0.43	0.80	10	8.68	16.10
¾	0.65	1.21	20	17.36	32.20
1	0.87	1.61	30	26.05	48.30
2	1.74	3.22	40	34.74	64.35
3	2.61	4.84	50	43.42	80.45
4	3.48	6.45	60	52.10	96.55
5	4.35	8.05	70	61.00	113.00
6	5.22	9.65	80	69.60	129.00
7	6.08	11.27	90	78.16	145.00
8	6.96	12.90	100	87.00	161.00

Fathoms to Feet to Meters

Fathoms	Feet	Meters	Fathoms	Feet	Meters
¼	1.5	0.5	6½	39.0	11.9
½	3.0	0.9	6¾	40.5	12.3
¾	4.5	1.4	7	42.0	12.8
1	6.0	1.8	8	48.0	14.6
1¼	7.5	2.3	9	54.0	16.5
1½	9.0	2.7	10	60.0	18.3
1¾	10.5	3.2	11	66.0	20.1
2	12.0	3.7	12	72.0	21.9
2¼	13.5	4.1	13	78.0	23.8
2½	15.0	4.6	14	84.0	25.6
2¾	16.5	5.0	15	90.0	27.4
3	18.0	5.5	16	96.0	29.3
3¼	19.5	5.9	17	102.0	31.1
3½	21.0	6.4	18	108.0	32.9
3¾	22.5	6.9	19	114.0	34.7
4	24.0	7.3	20	120.0	36.6
4¼	25.5	7.8	30	180.0	54.9
4½	27.0	8.2	40	240.0	73.2
4¾	28.5	8.7	50	300.0	91.4
5	30.0	9.1	60	360.0	109.7
5¼	31.5	9.6	70	420.0	128.0
5½	33.0	10.1	80	480.0	146.3
5¾	34.5	10.5	90	540.0	164.6
6	36.0	11.0	100	600.0	182.9
6¼	37.5	11.4			

Further Comments

While it was originally thought that this volume could be the final arbitrator on technical matters, we now see in fact that the profession is no nearer total agreement on every point than before. The observant reader may still find slight contradictions or differences in point of view expressed in the various entries concerning a particular topic. Although each entry has been contributed by a trained specialist in the field, that fact does not guarantee absolute agreement in every facet of the discipline. The only alternatives would be to have one person write an entire encyclopedia, a virtually impossible and undesirable feat, or to have the editor impose his own views on everything, an even more intolerable situation. It is hoped that the users of this encyclopedia will be able to resolve for themselves any dichotomies that they encounter within the information contained herein.

One liberty taken in the preparation of this volume has been in the use of the English language. Among other things, editors are supposed to correct citations, references, typos, spelling, grammar, and general usage of the language of publication. I have done that, and bear full responsibility for any errors that remain. However, I have never felt it my right or duty to edit out the flavor of those who use the English language in a slightly different manner from my own or of those for whom English is not their native tongue. While maintaining strict discipline standards, the contributors appear here in the personal and colorful ways in which they normally express themselves. On that point, editorial authority has been kept to a minimum.

Acknowledgments

Despite the aforementioned frustrations and hard work, compiling this encyclopedia has been a most rewarding experience. To date, there has been only one really sad event. That was in the passing of Dr. Louis Cutter Wheeler, the eminent botanist who so generously contributed to the volume. It was a pleasure to work with him, and I only wish that he could have seen the publication of his entries.

All the other acknowledgments that are due would fill several pages; just a few very special people can be recognized, and thanked, by name here. There is of course my wife Norma, who has assisted generously in every phase of the project, from compiling lists of potential contributors at the beginning to organizing the index at the end. At my home institution, Western Washington University, vital clerical duties were performed by Patty Combs, Patty Hamilton, Chris Moreland, and Joan Roley; bibliographic-search assistance was given by Evelyn Darrow, Susan Edmonds, Jan Gmeiner Johnson, Jan Moore, Jan Nichols, Wayne Richter, and Priscilla Soucie; artwork was prepared by Joy Dabney. At the production end of the process, invaluable guidance and support were given by Rhodes Fairbridge, series editor; Charles Hutchinson, publisher; Shirley End, managing editor; and Bernice Pettinato, production editor. My heartfelt thanks to each of the above.

Most of all, however, I want to thank the many contributors to this volume for their prized cooperation. Being the editor of an encyclopedia is much like being the conductor of a symphonic orchestra, for whom, without the musicians, there would be no music.

MAURICE L. SCHWARTZ

MAIN ENTRIES

"A-B-C . . ." Model
Accretion Ridge
Aeolianite
Aerosols
Africa, Coastal Ecology
Africa, Coastal Morphology
Air Breakwaters
Aktology
Albufera
Algal Mats and Stromatolites
Alluvial Plain Shoreline
Alveolar Weathering
Amphidromic Systems
Annelida
Antarctica, Coastal Ecology
Antarctica, Coastal Morphology
Apposition Beach
Aquaculture
Aquafacts
Archaeology, Geological Considerations
Archaeology, Methods
Arctic, Coastal Ecology
Arctic, Coastal Morphology
Arthropoda
Artificial Islands
Artificial Shorelines
Aschelminthes
Asia, Eastern, Coastal Ecology
Asia, Eastern, Coastal Morphology
Asia, Middle East, Coastal Ecology
Asia, Middle East, Coastal Morphology: Israel and Sinai
Asia, Middle East, Coastal Morphology: Syria, Lebanon, Red Sea, Gulf of Oman, and Persian Gulf
Attrition
Australia, Coastal Ecology
Australia, Coastal Morphology
Avulsion

Backwash Patterns
Bagnold Dispersive Stress
Ballast
Barrier Beaches
Barrier Flats
Barrier Island Coasts
Barrier Islands
Barrier Islands, Transgressive and Regressive
Bars
Base-line Studies
Beach
Beachcombing
Beach Cusps
Beach Cycles
Beach Firmness
Beach Material
Beach Material, Sorting of
Beach Nourishment
Beach Orientation
Beach Pads
Beach in Plan View
Beach Processes
Beach Processes, Monitoring of
Beach Profiles
Beach Ridge Plain
Beach Ridges and Beach Ridge Coasts
Beaufort Wind Scale
Bed Forms
Bench
Benthos
Biogenous Coasts
Biotic Zonation
Blowholes
Boat Basin Design
Bodden
Bogs
Bombora, Bomby
Bores
Boulder
Boulder Barricades
Brachiopoda
Breakwaters
Bruun Rule

Bryophyta
Bryozoa
Bulkhead
Buoys
Burrows and Borings

Cala and Cala Coast
Calcarenite
Canali
Carolina Bays
Causeway
Cavitation
Central America, Coastal Morphology
Central and South America, Coastal Ecology
Chaetognatha
Chalk Coasts
Chart Datum
Chattermarks
Chenier and Chenier Plain
Chine
Chlorophyta
Chordata
Chrysophyta
Classification
Clay
Cliffed Coast
Cliff Erosion
Climate, Coastal
Coastal Bevel
Coastal Characteristics, Mapping of
Coastal Dunes and Eolian Sedimentation
Coastal Ecology, Research Methods
Coastal Engineering
Coastal Engineering, History of
Coastal Engineering, Research Methods
Coastal Environmental Impact Statements
Coastal Erosion

MAIN ENTRIES

Coastal Erosion, Environmental–Geologic Hazard
Coastal Erosion, Formations
Coastal Fauna
Coastal Flora
Coastal Morphology, History of
Coastal Morphology, Oceanographic Factors
Coastal Morphology, Research Methods
Coastal Plant Ecology, United States, History of
Coastal Reserves
Coastal Sedimentary Facies
Coastal Waters Habitat
Coastal Zone Management
Cobble
Coelenterata
Colk
Computer Applications
Computer Simulation
Contraposed Shoreline
Coral Reef Coasts
Coral Reef Habitat
Corers and Coring Techniques
Coriolis Effect
Cryptophyta
Ctenophora
Current Meters
Currents
Cuspate Foreland
Cuspate Spits
Cut and Fill
Cyanophyceae
Cycadicae (Cycadophyta)

Dams, Effects of
Deflation Phenomena
Deltaic Coasts
Deltas
Demography
Depth of Disturbance
Desert Coasts
Design Wave
Dieback
Dike, Dyke
Dispersion
Dredging

Driftwood
Dune Stabilization

Echinodermata
Echiura
Edge Waves
Effluents
Elevated Shoreline (Strandline)
Embankments
Emergence and Emerged Shoreline
Energy Coefficients
Entrainment
Equilibrium Shoreline
Equisetophyta
Erosion Ramp, Wave Ramp
Estuaries
Estuarine Coasts
Estuarine Delta
Estuarine Habitat
Estuarine Sedimentation
Euglenophyta
Europe, Coastal Ecology
Europe, Coastal Morphology
Euxinic
Evaporites

Fault Coast
Feedback
Feeder Beach
Ferrel's Law
Fiard, Fjärd
Fiord, Fjord,
Firth
Flandrian Transgression
Flocculation
Flotsam and Jetsam
Foam Mark
Food Chain
Foraminifera
Forced Waves
Förde
Fossil Cliff
Fouling
Fourier Analysis
Frost Riving
Froude Number
Fulcrum Effect
Fungi

Geographic Terminology
Geomorphic-Cycle Theory
Ghyben-Herzberg Ratio
Glaciated Coasts
Glaciel
Global Tectonics
Gneticae
Grab Samplers
Gravel
Gravel Ridge and Rampart
Groins
Gyttja

Haff
Halophytes
Harbors
Headland Bay Beach
High-Energy Coast
Highest Coastline
High-Latitude Coasts
Holocene
Human Impact
Humate
Hurricane Effects
Hydraulic Action and Wedging
Hydrogeology of Coasts
Hypersaline Coastal Lakes
Hypsithermal

Ice along the Shore
Ice-Bordered Coasts
Ice Foot
India, Coastal Ecology
India, Coastal Morphology
Inlets and Inlet Migration
Inlets, Marine-Lagoonal and Marine-Fluvial
Insects
Internal Waves
Intertidal Flats
Intertidal Mud Habitat
Intertidal Sand Habitat
Isostatic Adjustment
Isostatically Warped Coasts

Jetties

Kaimoo
Karst Coast
Klint Coast

ix

MAIN ENTRIES

Lagoon and Lagoonal Coasts
Lagoonal Sedimentation
Lagoonal Segmentation
Lakes, Coastal Ecology
Lakes, Coastal Engineering
Lakes, Coastal Morphology
Land Reclamation
Limans and Liman Coasts
Linear Shell Reefs
Links
Littoral Cones
Lobate Coasts
Loose Rock and Stone Habitat
Low-Energy Coast
Low-Latitude Coasts
Low-Tide Deltas
Lycopodiophyta

Magnoliophyta
Major Beach Features
Mangrove Coasts
Marigram and Marigraph
Marine-Deposition Coasts
Marine-Erosion Coasts
Marine Erratics
Marsh Gas
Mathematical Models
Mean Sea Level
Mesozoa
Microseisms
Midden
Mid-Latitude Coasts
Mineral Deposits
Mineral Deposits, Mining of
Minor Beach Features
Moderate-Energy Coast
Mole
Mollusca
Moment Measures
Monoclinal Coast
Monroes
Morro
Mud Lumps
Mud Volcanoes
Muricate Weathering

Navigable Waters
Nearshore Hydrodynamics and Sedimentation
Nearshore Water Characteristics

Nehrung
Nemertina
New Zealand, Coastal Ecology
New Zealand, Coastal Morphology
Niche, Nick, Nip, Notch
Node and Antinode
North America, Coastal Ecology
North America, Coastal Morphology
Nuclear Power Plant Siting
Nutrients

Offset and Overlap
Offshore Platforms
Oil Spills and Pollution
Organism-Sediment Relationship
Oriented Lakes
Outwash Plain Shoreline

Pacific Islands, Coastal Ecology
Pacific Islands, Coastal Morphology
Paleogeography of Coasts
Palvé
Pamet
Paralic
Pebble
Peels
Perched Beach
Periglacial Effects
Phaeophycophyta
Phase Difference
Phoronida
Photic Zone
Photogrammetry
Pier
Pile, Piling
Pinicae
Pinophyta
Platform Beach
Platyhelminthes
Playa
Pleistocene
Polder
Pollutants
Polypodiophyta
Porifera

Postglacial Rebound
Potrero
Present-Day Shoreline Changes, United States
Present-Day Shoreline Changes, Worldwide
Priapulida
Production
Profile of Equilibrium
Profiling of Beaches
Progradation and Prograding Shoreline
Protection of Coasts
Protochordata
Protozoa
Proxigean Spring Tides
Pyrrhophyta

Quarrying Processes
Quay

Radiocarbon Dating
Raised Beach
Ramparts
Rasa
Ravinement
Recession and Retrogression
Reefs, Noncoral
Relict Sediment
Remote Sensing
Restinga
Retrogradation and Retrograding Shoreline
Reynold's Number
Rhodophycophyta
Rhythmic Cuspate Forms
Ria and Ria Coast
Richardson's Number
Ridge and Runnel
Rill Marks
Rip Current
Ripple Marks
Riprap
River-Deposition Coasts
Rock Borers
Rocky Shore Habitat
Rooster Tail

Sabkha
Salcrete

MAIN ENTRIES

Salt Marsh
Salt Marsh Coasts
Salt Weathering
Sample Impregnation
Sand
Sand, Surface Texture
Sand Bypassing
Sand Dune Habitat
Sand Waves and Longshore Sand Waves
Scale Models
Schizomycetes
Schorre
Scour Holes
Sea Caves
Sea Cliffs
Sea Conditions
Sea Level Changes
Sea Level Changes, 1900 to Present
Sea Level Curves
Sea Puss
Sea Slick
Seawall
Sediment Analysis, Statistical Methods
Sediment Budget
Sediment Classification
Sediment Size Classification
Sediment Tracers
Sediment Transport
Seiches
Shelf Sedimentation
Shell Pavement
Shingle and Shingle Beaches
Shoaling Coefficient
Shoestring Sands
Shore Drift Cell
Shoreface and Shoreface Terrace
Shoreline Development Ratio
Shore Platforms
Shore Polygons
Shore Terrace
Silt
Sipuncula
Skerry, Skerry Guard
Slikke
Slope-over-Wall Cliffs
Soils

Solution and Solution Pan
Sounding Sands
South America, Coastal Morphology
Soviet Union, Coastal Ecology
Soviet Union, Coastal Morphology
Spits
Splash and Spray Zones
Spring Pits
Standing Stock
Stokes Theorem
Stone Packing
Stone Reef
Storm Beach
Storm Surge
Storm-Wave Environments
Strand and Strandline
Strandflat
Submarine Springs
Submergence and Submerged Shoreline
Succession
Surf Beat
Surfing
Sweep Zone
Swell and Its Propagation
Swing Mark

Tafone
Tangue
Tar Pollution on Beaches
Tectonic Movements
Thalassotherapy
Thermal Pollution
Thermal Power
Tidal Basin
Tidal Currents
Tidal Deltas
Tidal Flat
Tidal Flushing
Tidal Inlets, Channels, and Rivers
Tidal Power
Tidal Prism
Tidal Range and Variation
Tidal Type, Variation Worldwide
Tide Gauges
Tide Pools

Tides
Tide Tables and Charts
Time Series
Tombolo
Tourism
Trend Surface Analysis
Tsunamis

Vegetation Coasts
Volcano Coasts

Wadden
Washover and Washover Fan
Waste Disposal
Water Layer Weathering
Water Table
Wave-Built Terrace
Wave Climate
Wave-Cut Bench
Wave-Cut Platform
Wave Drag Layer
Wave Energy
Wave Environments
Wave Erosion
Wave Meters
Wave Refraction Diagrams
Waves
Wave Shadow
Wave Statistics
Wave Theories, Oscillatory and Progressive
Wave Work
Weathering and Erosion, Biologic
Weathering and Erosion, Chemical
Weathering and Erosion, Differential
Weathering and Erosion, Mechanical
Wharf
Wind
Wind, Waves, and Currents, Direction of
World Net Sediment Transport

Zero-Energy Coast
Zetaform Bays
Zingg Shape

xi

CONTRIBUTORS

PAUL ADAM, School of Botany, University of South Wales, Kensington, NSW 2033, Australia. *Europe, Coastal Ecology.*

ENAYAT AHMAD, Dept. of Geography, Ranchi University, Ranchi, India. *India, Coastal Ecology; India, Coastal Morphology.*

JAMES R. ALLEN, National Park Service, North Atlantic Region, 15 State Street, Boston, Massachusetts 02109. *Computer Simulation; Feedback; Spits.*

RICHARD J. ANDERSON, Dept. of Geology and Mineralogy, The Ohio State University, 125 South Oval Mall, 2041 North College Road, Columbus, Ohio 43201. *Mineral Deposits; Mineral Deposits, Mining of.*

JOSÉ F. ARAYA-VERGARA, Dept. de Geografia, Universidad de Chile, Av. Larrain 9925, Cas. 10136, Santiago, Chile. *Antarctica, Coastal Morphology; Beach Orientation; Coastal Morphology, Research Methods; Lobate Coasts; Marine Deposition Coasts; Marine-Erosion Coasts; River-Deposition Coasts; Surf Beat; Tectonic Movements; Volcano Coasts.*

SAUL ARONOW, Dept. of Geology, Lamar University, Beaumont, Texas 77710. *Alluvial Plain Shoreline; Avulsion; Lakes, Coastal Morphology; Mud Lumps; Outwash Plain Shoreline; Potrero; Shoestring Sands; Shoreline Development Ratio.*

MICHELE M. BARSON, School of Botany, University of Melbourne, Parksville, Victoria 3052, Australia. *Australia, Coastal Ecology; Vegetation Coasts.*

MARK A. BENEDICT, Dept. of Botany, University of Massachusetts, Amherst, Massachusetts 01003. *Coastal Plant Ecology, United States, History of.*

ROBERT B. BIGGS, College of Marine Studies, University of Delaware, Newark, Delaware 19711. *Estuaries.*

ERIC C. F. BIRD, Dept. of Geography, University of Melbourne, Parksville, Victoria 3052, Australia. *Australia, Coastal Morphology; Chalk Coasts; Coastal Reserves; Colk; Muricate Weathering.*

J. BRIAN BIRD, Dept. of Geography, McGill University, 805 Sherbrooke Street West, Montreal, PQ, Canada H3A 2K6. *Chattermarks; Frost Riving.*

BONNIE BLOESER, Texaco Inc., 3350 Wilshire Boulevard, Los Angeles, California 90051. *Euglenophyta.*

JOANNE BOURGEOIS, Dept. of Geological Sciences, University of Washington, Seattle, Washington 98195. *Accretion Ridge; Attrition; Sediment Classification; Tangue.*

BENNO M. BRENNINKMEYER, Dept. of Geology, Boston College, Chestnut Hill, Massachusetts 02167. *Backwash Patterns; Bagnold Dispersive Stress; Computer Applications; Cut and Fill; Entrainment; Major Beach Features; Minor Beach Features; Moment Measures; Peels; Rill Marks; Rip Current; Sand; Wave Erosion.*

CHARLES L. BRETSCHNEIDER, Dept. of Ocean Engineering, University of Hawaii, Honolulu, Hawaii 96822. *Tsunamis.*

A. C. BROAD, Dept. of Biology, Western Washington University, Bellingham, Washington 98225. *Arctic, Coastal Ecology.*

PER BRUUN, Technical University of Norway, Trondheim, Norway. *Coastal Engineering, History of.*

ALAN PAUL CARR, Institute of Oceanographic Sciences, Taunton, Somerset, England TAI 2DW. *Beach Material, Sorting of; Boulder; Chart Datum; Cobble; Gravel; Pebble; Phase Difference; Sediment Size Classification; Silt.*

RICHARD W. G. CARTER, School of Biological Studies, The New University of Ulster, Northern Ireland BT52 ISA, United Kingdom.

CONTRIBUTORS

Cuspate Foreland; Deflation Phenomena; Depth of Disturbance; Links; Midden; Shell Pavement; Strand and Strandline.

INGEMAR CATO, Marine Geological Laboratory, University of Gothenburg, Fack, S-400 33 Gothenburg, Sweden. *Highest Coastline.*

DAVID M. CHAPMAN, Dept. of Geography, University of Sydney 2006, Sydney, Australia. *Dune Stabilization; Land Reclamation; Zetaform Bays.*

VALENTINE J. CHAPMAN, School of Botany, University of Auckland, Auckland, New Zealand. *Dune Stabilization; New Zealand, Coastal Ecology.*

ROGER H. CHARLIER, Earth Science Dept., Northwestern Illinois University, Bryn Mawr at St. Louis Avenue, Chicago, Illinois 60625. *Canali; Hypersaline Coastal Lakes; Schorre; Slikke; Thalassotherapy; Thermal Power; Tidal Power; Tourism.*

LANNA CHENG, Scripps Institution of Oceanography, University of California, La Jolla, California 92093. *Insects.*

CHARLES B. CHESNUTT, Kingman Building, Coastal Engineering Research Center, Ft. Belvoir, Virginia 22060. *Bulkhead; Causeway; Mole; Pier; Quay; Sand Bypassing; Sea Puss; Wharf.*

H. EDWARD CLIFTON, U.S. Geological Survey, 345 Middlefield Road, Menlo Park, California 94025. *Coastal Sedimentary Facies.*

G. KENT COLBATH, Dept. of Geology, University of Oregon, Eugene, Oregon 97403. *Coastal Ecology, Research Methods.*

DOUGLAS E. CONKLIN, P.O. Box 247, Bodega Bay, California 94923. *Aquaculture.*

LORIN R. CONTESCU, 1410 Chicago Avenue, Evanston, Illinois 60201. *Hypersaline Coastal Lakes.*

DAVID O. COOK, Raytheon Ocean Systems Company, Risho Avenue, Westminster Park, East Providence, Rhode Island 02914. *Effluents; Nuclear Power Plant Siting; Tidal Flushing.*

J. L. DAVIES, School of Earth Sciences, Macquarie University, North Ryde, Australia. *High-Latitude Coasts; Low-Latitude Coasts; Mid-Latitude Coasts; Tidal Type, Variation Worldwide; Wave Environments; World Net Sediment Transport.*

RICHARD A. DAVIS, JR., Dept. of Geology, University of South Florida, Tampa, Florida 33620. *Beach; Beach Cycles; Beach Profiles; Storm Beach; Storm Surge; Time Series; Wind.*

PAUL DAYTON, Scripps Institution of Oceanography, La Jolla, California 92093. *Antarctica, Coastal Ecology.*

CHARLES DILL, Ocean Seismic Survey, Inc., 80 Oak Street, Norwood, New Jersey 07648. *Coastal Engineering, Research Methods.*

JOHN L. DINGLER, U.S. Geological Survey, 345 Middlefield Road, Menlo Park, California 94025. *Wave Shadow.*

JEAN-CLAUDE DIONNE, Dept. of Geography, Université Laval, Québec, Que., Canada G1K 7P4. *Glaciel; Monroes; Mud Volcanoes; Shore Polygons; Stone Packing.*

DONALD DISRAELI, Dept. of Geography, University of California, Santa Barbara, California 93106. *North America, Coastal Ecology.*

D. EISMA, Netherlands Institute for Sea Research, P.O. Box 59, Den Burg, Texel, Netherlands. *Asia, Eastern, Coastal Morphology; Intertidal Sand Habitat; Nearshore Water Characteristics; Polder; Shelf Sedimentation.*

MOHAMED T. EL-ASHRY, Office of Natural Resources, Tennessee Valley Authority, Knoxville, Tennessee 37902. *Photogrammetry; Present Day Shoreline Changes, United States.*

CLIFFORD EMBLETON, Dept. of Geography, King's College, University of London, London W-C-2, England. *Fiard, Fjärd; Fiord, Fjord; Firth; Förde; Glaciated Coasts; Mean Sea Level; Periglacial Effects; Skerry, Skerry Guard.*

LEE D. ENTSMINGER, Dept. of Geology, Florida State University, Tallahassee, Florida 32306. *Beach Pads.*

CONTRIBUTORS

GRAHAM EVANS, Dept. of Geology, Imperial College, London S.W. 7, England. *Intertidal Flats; Sabkha.*

RHODES W. FAIRBRIDGE, Dept. of Geology, Columbia University, New York, New York 10027. *Elevated Shoreline (Strandline); Global Tectonics; Holocene; Hypsithermal; Karst Coast; Marine Erratics; Raised Beach; Submarine Springs; Weathering and Erosion, Chemical.*

CHARLES W. FINKL, JNR., Institute of Coastal Studies, Nova University, Port Everglades, Florida 33004. *Soils.*

JOHN J. FISHER, Dept. of Geology, University of Rhode Island, Kingston, Rhode Island 02881. *Barrier Islands; Coastal Environmental Impact Statements; Fulcrum Effects; Inlets and Inlet Migration; Pamet; Tidal Deltas; Wadden.*

RONALD C. FLEMAL, Dept. of Geology, Northern Illinois University, DeKalb, Illinois 60115. *Pleistocene; Postglacial Rebound; Trend Surface Analysis.*

EDGAR FRANKEL, Dept. of Geology and Geophysics, University of Sydney, Sydney, New South Wales, Australia. *Coral Reef Coasts; Coral Reef Habitat.*

DOREEN LEE FUNDILLER, Dept. of Geology and Geophysics, Yale University, New Haven, Connecticut 06520. *Intertidal Mud Habitat.*

COLIN F. GIBBS, Ministry for Conservation, Marine Science Laboratories, P.O. Box 114, Queenscliff, Victoria, Australia. *Nutrients.*

HANS GÜNTER GIERLOFF-EMDEN, Institut fur Geographie, Luisenstrasse 37, 8000 Munchen 2, West Germany, *Central America, Coastal Morphology.*

EDMUND D. GILL, 1/47 Wattle Valley Road, Canterbury, Victoria 3126, Australia. *Loose Rock and Stone Habitat; Shore Platforms.*

MELINDA M. GODFREY, Dept. of Botany, University of Massachusetts, Amherst, Massachusetts 01002. *Biotic Zonation; Bogs; North America, Coastal Ecology; Rocky Shore Habitat.*

PAUL J. GODFREY, Dept. of Botany, University of Massachusetts, Amherst, Massachusetts 01002. *Biotic Zonation, Bogs; North America, Coastal Ecology.*

VICTOR GOLDSMITH, Dept. of Geology, University of South Carolina, Columbia, South Carolina 29208. *Coastal Dunes and Eolian Sedimentation.*

ABRAHAM GOLIK, Israel Oceanographic and Limnological Research Ltd., POB 1793, Haifa, Israel. *Tar Pollution on Beaches.*

A. N. GOLIKOV, Zoological Institute, Academy of Sciences, Leningrad, U.S.S.R. *Soviet Union, Coastal Ecology.*

ROGER D. GOOS, Dept. of Botany, University of Rhode Island, Kingston, Rhode Island, 02881. *Fungi.*

JACK GREEN, Dept. of Geologic Sciences, California State University, Long Beach, California 90840. *Littoral Cones.*

BRIAN GREENWOOD, Scarborough College, University of Toronto, West Hill, Ontario MIC 1A4, Canada. *Bars; Sand, Surface Texture.*

M. GRANT GROSS, Chesapeake Bay Institute, Johns Hopkins University, Baltimore, Maryland 21218. *Human Impact; Waste Disposal.*

VYTAUTAS GUDELIS, Academy of Sciences, Vilnius 232000, U.S.S.R. *Haff; Nehrung; Palvé.*

GORDON GUNTER, Gulf Coast Research Laboratory, Ocean Springs, Mississippi 39564. *Chordata; Estuarine Habitat.*

JOHN R. HAILS, Manager, Environmental Services, Energy Division, CSR, Knox House, 1 O'Connell Street, Sydney, N.S.W. 2000, Australia. *Aeolianite; Barrier Island Coasts; Beach Nourishment; Beach Processes, Monitoring of; Coastal Zone Management; Grab Samplers; Humate; Ria and Ria Coast; Sediment Analysis, Statistical Methods; Tide Gauges; Washover and Washover Fan.*

JOHN P. HANEY, Ocean Engineering Division, Texas A & M University, College Station, Texas 77843. *Artificial Islands; Bench; Break-*

CONTRIBUTORS

waters; Coastal Engineering; Coastal Erosion; Dike, Dyke; Dredging; Lakes, Coastal Engineering; Navigable Waters; Pile, Piling.

MICHAEL C. HARTMAN, P.O. Box 247, Bodega Bay, California 94923. *Aquaculture.*

ARTHUR HAULOT, Commission of Tourism, Gare Centrale, 1000 Bruxelles, Brussels, Belgium. *Tourism.*

JOEL W. HEDGPETH, 5660 Montecito Avenue, Santa Rosa, California 95404. *Coastal Fauna.*

PAUL HELLER, Dept. of Geology, University of Arizona, Tucson, Arizona 85721. *Splash and Spray Zones.*

JOHN B. HERBICH, Ocean Engineering Division, Texas A & M University, College Station, Texas 77843. *Air Breakwaters; Artificial Islands; Artificial Shorelines; Beach Material; Bench; Breakwaters; Cavitation; Coastal Engineering; Coastal Erosion; Dike, Dyke; Dredging; Groins; Harbors; Lakes, Coastal Engineering; Navigable Waters; Offshore Platforms; Oil Spills and Pollution; Pile, Piling; Riprap; Scour Holes; Seawall; Stokes Theorem; Wave Climate.*

SEWELL H. HOPKINS, Dept. of Biology, Texas A & M University, College Station, Texas 77843. *Platyhelminthes.*

S. A. HSU, Dept. of Marine Sciences, Louisiana State University, Baton Rouge, Louisiana 70803. *Climate, Coastal.*

RALPH E. HUNTER, U.S. Geological Survey, 345 Middlefield Road, Menlo Park, California 94025. *Coastal Sedimentary Facies.*

DAVID JABLONSKI, Dept. of Ecology and Evolutionary Biology, University of Arizona, Tucson, Arizona 85721. *Reefs, Noncoral.*

EDMUND E. JACOBSEN, Dept. of Geology, Western Washington University, Bellingham, Washington 98225. *Perched Beach.*

BARBARA JAVOR, Dept. of Biology, University of Oregon, Eugene, Oregon 97403. *Algal Mats and Stromatolites.*

ROBERT E. JENSEN, Ocean Engineering Division, Texas A & M University, College Station, Texas 77843. *Scour Holes; Seawall.*

IVAN P. JOLLIFFE, Dept. of Geography, Bedford College, University of London, London, N.W.1., England. *Coastal Erosion, Environmental–Geologic Hazard.*

GARRY D. JONES, Union Oil and Gas Division: Western Region, 2323 Knoll Drive, P.O. Box 6176, Ventura, California, 93006. *Euxinic; Microseisms; Niche, Nick, Nip, Notch; Paralic.*

MEREDITH L. JONES, Smithsonian Institution, United States National Museum, Washington, D.C. 20560. *Annelida.*

RAYMOND T. KACZOROWSKI, Dept. of Geological Sciences, University of Wisconsin, Milwaukee, Wisconsin 53201. *Oriented Lakes.*

PAVEL A. KAPLIN, Geographical Faculty, Moscow State University, Moscow, B-234, U.S.S.R. *Flandrian Transgression; Lagoon and Lagoonal Coasts; Paleogeography of Coasts; Shore Terrace.*

CLARENCE KIDSON, Dept. of Geography, University College of Wales, Aberystwyth, Dyfed, SY23 3DB, United Kingdom. *Sediment Transport.*

CUCHLAINE A. M. KING, Dept. of Geograpy, The University, Nottingham, NG7 2RD, England. *Chine; Classification; Rhythmic Cuspate Forms; Ridge and Runnel; Shingle and Shingle Beaches.*

ROBERT J. KING, School of Botany, The University of New South Wales, P.O. Box 1, Kensington, New South Wales, Australia 2033. *Australia, Coastal Ecology.*

ROBERT MILLER KIRK, Dept. of Geography, University of Canterbury, Christchurch, New Zealand. *New Zealand, Coastal Morphology.*

ERNST KIRSTEUER, Dept. of Invertebrates, American Museum of Natural History, Central Park West and 79th Street, New York, New York 10024. *Nemertina.*

DONALD B. KOWALEWSKY, 5908 Topeka Drive, Tarzana, California 91356. *Shoreface and Shoreface Terrace; Wave-Built Terrace.*

JOHN C. KRAFT, Dept. of Geology, University of Delaware, Newark, Delaware 19711. *Archae-*

CONTRIBUTORS

ology, Geological Considerations; Barrier Islands, Transgressive and Regressive.

O. K. LEONTIEV, Geographical Faculty, Moscow State University, Moscow, U.S.S.R. 117234. *Biogenous Coasts; Monoclinal Coast.*

ANTHONY J. LEWIS, ERSAL, Withycombe Hall, Oregon State University, Corvallis, Oregon 97331. *Linear Shell Reefs; Remote Sensing.*

EUGENIE LISITZIN, Skarpskyttegatan 1-C-27, 00130 Helsinki 13, Finland. *Sea Level Changes.*

S. A. LUKJANOVA, Geographical Faculty, Moscow State University, Moscow, V234, U.S.S.R. *Biogenous Coasts; Monoclinal Coast.*

BAYARD H. McCONNAUGHEY, Dept. of Biology, University of Oregon, Eugene, Oregon 97403. *Coastal Flora; Coastal Waters Habitat; Schizomycetes.*

DONALD J. MACINTOSH, Unit of Aquatic Pathobiology, University of Stirling, Stirling, FK9 4LA, Scotland, United Kingdom. *Asia, Eastern, Coastal Ecology.*

ROGER F. McLEAN, University of Auckland, Private Bag, Auckland, New Zealand. *Fault Coast; Gravel Ridge and Rampart; Rock Borers; Weathering and Erosion, Biologic; Zingg Shape.*

REGINALD D. MANWELL, Dept. of Biology, 209 Lyman Hall, Syracuse University, 108 College Place, Syracuse, New York, 13210. *Protozoa.*

MARTEK INSTRUMENTS INC., 17302 Daimier, Irvine, California 92713. *Pollutants.*

THOMAS D. MATHEWS, South Carolina Wildlife & Marine Resources Department, P.O. Box 12559, Charleston, South Carolina 29412. *Radiocarbon Dating.*

JAMES P. MAY, The Citadel, Charleston, South Carolina 29409. *Hydrogeology of Coasts; Lagoonal Segmentation; Moderate Energy Coast; Shoaling Coefficient; Waves.*

MOLLY FRITZ MILLER, Box 6027, Station B, Vanderbilt University, Nashville, Tennessee 37235. *Chaetognatha; Echiura; Priapulida; Sipuncula.*

CHIZUKO MIZOBE, Center for Coastal and Environmental Studies, Rutgers University, New Brunswick, New Jersey 08903. *Central and South America, Coastal Ecology; Mangrove Coasts; South America, Coastal Morphology.*

CARLOS MORAIS, Laboratorio Nacional de Engenharia Civil, Av. Brasil, 1799 Lisboa, CODEX, Portugal. *Design Wave; Wave Meters.*

MARIE MORISAWA, Dept. of Geology, State University of New York, Binghamton, New York 13901. *Geomorphic-Cycle Theory.*

NILS-AXEL MÖRNER, Dept. of Geology, University of Stockholm, Kungstensgatan 45, S113 86 Stockholm, Sweden. *Gyttja; Sea Level Curves.*

JOHN MORTON, Dept. of Zoology, University of Auckland, Auckland, New Zealand. *Pacific Islands, Coastal Ecology.*

FERRUCCIO MOSETTI, Istituto di Geodesia e Geofisica, Universita Degli Studi di Trieste, 34100-Via dell'Universita N. 7, Trieste, Italy. *Amphidromic Systems; Bores; Buoys; Current Meters; Currents; Energy Coefficients; Ferrel's Law; Forced Waves; Fourier Analysis; Froude Number; Marigram and Marigraph; Mathematical Models; Reynold's Number; Richardson's Number; Seiches; Wave Energy.*

R. A. MULLER, Dept. of Geography and Anthropology, Louisiana State University, Baton Rouge, Louisiana 70803. *Climate, Coastal.*

R. S. MURALI, General Development Corporation, 111 South Bayshore Drive, Miami, Florida 33131. *Zero-Energy Coast.*

GEORGE MUSTOE, Dept. of Geology, Western Washington University, Bellingham, Washington 98225. *Alveolar Weathering; Aschelminthes; Brachiopoda; Ctenophora; Magnoliophyta; Mesozoa; Phoronida; Pinicae; Pinophyta; Polypodiophyta; Protochordata; Pyrrhophyta; Tafone.*

LINDA C. NEWBY, Eureka Laboratories, Inc., Sacramento, California 95816. *Dieback; Halophytes; Marsh Gas; Salt Marsh Coasts.*

WALTER S. NEWMAN, Dept. of Earth and Environmental Sciences, Queens College, Flushing, New York 11367. *Isostatic Adjustment; Isostatically Warped Coasts.*

WILLIAM A. NIERING, Dept. of Botany, Connecticut College, New London, Connecticut 06320. *Salt Marsh; Succession.*

YAACOV NIR, Geological Survey of Israel, 30 Malkhe Israel Street, Jerusalem, 95-501 Israel. *Asia, Middle East, Coastal Morphology: Israel and Sinai.*

H. MORGAN NOBLE, 79 West Shore Road, Belvedere, California 94920. *Boat Basin Design.*

SCOTT MORGAN NOBLE, U.S. Army, Corps of Engineers, Portland District, Portland, Oregon 97208. *Boat Basin Design.*

KARL F. NORDSTROM, Marine Sciences Center, Rutgers University, New Brunswick, New Jersey 08903. *Feeder Beach; Tidal Basin; Tidal Prism.*

ALBERT V. NYBERG, Dept. of Geology, University of California, Los Angeles, California 90024. *Cyanophyceae.*

GEORGE F. OERTEL, Dept. of Oceanography, Old Dominion University, Norfolk, Virginia 23508. *Inlets, Marine-Lagoonal and Marine-Fluvial.*

ANTONY R. ORME, Dept. of Geography, University of California, Los Angeles, California 90024. *Africa, Coastal Ecology; Africa, Coastal Morphology.*

G. R. ORME, Dept. of Geology, University of Queensland, St. Lucia, Queensland, Australia. *Calcarenite; Relict Sediment.*

KAAREL ORVIKU, Estonian Academy of Sciences, Tallinn 200101, Estonia, U.S.S.R. *Klint Coast.*

ERVIN G. OTVOS, Geology Division, Gulf Coast Research Laboratory, Ocean Springs, Mississippi 39564. *Barrier Beaches; Barrier Flats; Bruun Rule; Chenier and Chenier Plain; Emergence and Emerged Shoreline; High-Energy Coast; Low-Energy Coast; Strandflat.*

EDWARD H. OWENS, Woodward-Clyde Consultants, 16 Bastion Square, Victoria, B.C., Canada V8W 1H9. *Beaufort Wind Scale; Coastal Characteristics, Mapping of; Ice along the Shore; Ice Foot; Kaimoo; Offset and Overlap; Profiling of Beaches; Sea Conditions; Storm-Wave Environments; Sweep Zone; Tombolo.*

DAVID L. PAWSON, Smithsonian Institution, Washington, D.C. 20560. *Echinodermata.*

D. HOWELL PEREGRINE, Dept. of Mathematics, University of Bristol, Bristol, England BS8 1TW. *Swell and Its Propagation; Tidal Currents; Tide Tables and Charts.*

THOMAS E. PICKETT, Delaware Geological Survey, University of Delaware, Newark, Delaware 19711. *Burrows and Borings; Carolina Bays; Lagoonal Sedimentation.*

MICHAEL R. PLOESSEL, McClelland Engineers, Inc., Geotechnical Consultants, 5450 Ralston Street, Ventura, California 93003. *Ghyben-Herzberg Ratio; Weathering and Erosion, Differential.*

FRANCIS DOV POR, Dept. of Zoology, Hebrew University of Jerusalem, Jerusalem, Israel. *Asia, Middle East, Coastal Ecology.*

W. ARMSTRONG PRICE, 428 Ohio Street, Corpus Christi, Texas 78404. *Beach Ridge Plain; Chenier and Chenier Plain.*

NORBERT P. PSUTY, Rutgers University, Marine Sciences Center, New Brunswick, New Jersey 08903. *Central and South America, Coastal Ecology; Mangrove Coasts; South America, Coastal Morphology.*

MICHAEL R. RAMPINO, Institute for Space Studies, 2880 Broadway, New York, New York 10025. *Apposition Beach; Hydraulic Action and Wedging; Weathering and Erosion, Mechanical.*

DONALD R. REGAN, Dept. of Geology, University of Rhode Island, Kingston, Rhode Island 02881. *Coastal Environmental Impact Statements.*

RUEDEGER REINHARD, Pastoor Buyslaan #64, 2242 Rm, Wassenaar, Netherlands. *Sediment Tracers.*

JOHN A. RIPP, TRC Environmental Consultants, Inc., 125 Silas Deane Highway, Wethers-

CONTRIBUTORS

field, Connecticut 06109. *Corers and Coring Techniques; Wave Refraction Diagrams.*

ANDREW ROBERTSON, GLERL/NOAA, 2300 Washtenaw, Ann Arbor, Michigan 48104. *Lakes, Coastal Ecology.*

PETER S. ROSEN, Dept. of Earth Science, Northeastern University, Boston, Massachusetts 02115. *Boulder Barricades; Cuspate Spits; Water Table.*

CHARLES A. ROSS, Dept. of Geology, Western Washington University, Bellingham, Washington 98225. *Foraminifera.*

JUNE R. P. ROSS, Department of Biology, Western Washington University, Bellingham, Washington 98225. *Bryozoa; Fouling.*

VICENÇ M. ROSSELLÓ, Universidad de Valencia, Facultad de Filosofia Y Letras, Departmento de Geografia, ap. 22005, Valencia, Spain. *Albufera; Cala and Cala Coast; Demography; Morro; Playa; Rasa; Restinga.*

KLAUS RUETZLER, Smithsonian Institution, National Museum of Natural History, Washington, D.C. 20560. *Porifera.*

ASBURY H. SALLENGER, JR., U.S. Geological Survey, 345 Middlefield Road, Menlo Park, California 94025. *Beach Firmness.*

PAUL SANLAVILLE, Maison de l'Orient Mediterraneen, 1 rue Raulin, 69007 Lyon, France. *Asia, Middle East, Coastal Morphology: Syria, Lebanon, Red Sea, Gulf of Oman, and Persian Gulf.*

SHOJI SATO, Marine Hydrodynamics Division, Port and Harbor Research Institute, 3-chome, Nagase, Yokosuka, Japan. *Embankments; Jetties.*

O. A. SCARLATO, Zoological Institute, Academy of Sciences, Leningrad, 199164, U.S.S.R. *Soviet Union, Coastal Ecology.*

J. C. SCHOFIELD, Geological Survey, P.O. Box 61012, Otara, New Zealand. *Pacific Islands, Coastal Morphology; Progradation and Prograding Shoreline; Retrogradation and Retrograding Shoreline.*

J. R. SCHUBEL, Marine Sciences Research Center, State University of New York, Stony Brook, New York 11794. *Estuarine Coasts; Estuarine Sedimentation; Flocculation.*

MAURICE L. SCHWARTZ, Dept. of Geology, Western Washington University, Bellingham, Washington 98225. *Aktology; Amphidromic Systems; Ballast; Beach Cycles; Beach Processes; Contraposed Shoreline; Geographic Terminology; Sample Impregnation; Swing Mark; Wind, Waves, and Currents, Direction of.*

ROBERT B. SETZER, Allan Hancock Foundation, University of Southern California, Los Angeles, California 90007. *Phaeophycophyta; Rhodophycophyta.*

ALAN B. SHAW, Amoco Research Center, P.O. Box 591, Tulsa, Oklahoma 74102. *Evaporites.*

JOSEPH W. SHAW, Dept. of Fine Arts, University of Toronto, Toronto, Canada M5S 1A1. *Archaeology, Methods.*

FRANCIS P. SHEPARD, Geological Research Division, Scripps Institution of Oceanography, La Jolla, California 92093. *North America, Coastal Morphology.*

CHARLES R. C. SHEPPARD, Dept. of Botany, University of Durham, Durham, DH1 3LE, England. *Base-line Studies; Chlorophyta; Food Chain; Photic Zone.*

ROY B. SHILLING, Ocean Engineering Division, Texas A & M University, College Station, Texas 77843. *Artificial Shorelines; Cavitation; Harbors; Oil Spills and Pollution.*

ANDREW D. SHORT, Coastal Studies Unit, Department of Geography, University of Sydney, Sydney, N.S.W. 2006, Australia. *Bombora, Bomby; Cliffed Coast; Cliff Erosion; Erosion Ramp, Wave Ramp; Ice-Bordered Coasts; Platform Beach; Quarrying Processes; Ramparts; Sand Waves and Longshore Sand Waves; Stone Reef; Tide Pools; Water Layer Weathering; Wave-Cut Bench; Wave-Cut Platform.*

YURII D. SHUISKY, Geography Dept., Odessa State University, Odessa, U.S.S.R. *Limans and Liman Coasts; Wave Drag Layer.*

ROBERT T. SIEGFRIED, 1062 Oxford Way, Stockton, California 95204. *Aquafacts; Solution and Solution Pan.*

ELIZABETH J. SIMPSON, Dept. of Geology, University of Rhode Island, Kingston, Rhode Island 02881. *Tidal Deltas.*

ALLISON A. SNOW, Dept. of Botany, University of Massachusetts, Amherst, Massachusetts 01002. *Sand Dune Habitat.*

JOHN SPASARI, Shannon and Wilson, Inc., 1105 N. 38th Street, Seattle, Washington 98103. *Beach Cusps.*

FRANK W. STAPOR, JR., Marine Research Laboratory, P.O. Box 12559, Charleston, South Carolina 29412. *Beach Ridges and Beach Ridge Coasts; Sea Level Changes, 1900 to Present; Sediment Budget; Shore Drift Cell.*

JAMES ALFRED STEERS, Flat 47, Gretton Court, Girton, Cambridge CB3 OQN, United Kingdom. *Europe, Coastal Morphology.*

JAMES E. STEMBRIDGE, JR., Coastal Environmental Resources Institute, 1695 Winter Street S.E., Salem, Oregon 97302. *Beachcombing; Blowholes; Dams, Effects of; Driftwood; Present-day Shoreline Changes, Worldwide; Sea Caves; Sea Cliffs; Tidal Range and Variation.*

DONALD J. P. SWIFT, PRC-G130, Arco Oil and Gas Co., Box 2819, Dallas, Texas 75221. *Nearshore Hydrodynamics and Sedimentation; Ravinement.*

ADA SWINEFORD, 2121 Meadowlark Road, Manhattan, Kansas 66502. *Clay.*

WILLIAM F. TANNER, Dept. of Geology, Florida State University, Tallahassee, Florida 32306. *"A-B-C..." Model; Beach in Plan View; Equilibrium Shoreline; Flotsam and Jetsam; Foam Mark; Hurricane Effects; Low-Tide Deltas; Ripple Marks; Rooster Tail; Spring Pits; Submergence and Submerged Shoreline; Wave Work.*

HJALMAR THIEL, Universität Hamburg, Institut für Hydrobiologie und Fischereiwissenschaft, Hydrobiologische Abteilung, Zeiseweg 9, D2000 Hamburg 50, West Germany. *Benthos; Bodden; Organism-Sediment Relationship; Production; Standing Stock.*

BRUCE G. THOM, University of New South Wales, Royal Military College, Duntroon, A.C.T. 2600, Australia. *Deltaic Coasts; Tidal Flat.*

C. R. TWIDALE, University of Adelaide, Adelaide, South Australia 5001, Australia. *Sounding Sands.*

STEPHEN VONDER HAAR, Dept. of Geology and Geophysics, University of California, Berkeley, California 94720. *Algal Mats and Stromatolites.*

H. J. WALKER, Dept. of Geography and Anthropology, Louisiana State University, Baton Rouge, Louisiana 70803. *Arctic, Coastal Morphology; Climate, Coastal; Coastal Morphology, History of.*

JAMES R. WALKER, Moffatt and Nichol, P.O. Box 7707, Long Beach, California 90807. *Surfing.*

TOM WALTERS, Ocean Engineering Division, Texas A & M University, College Station, Texas 77843. *Air Breakwaters; Beach Material; Groins; Offshore Platforms; Riprap; Stokes Theorem; Wave Climate.*

DETLEF A. WARNKE, Dept. of Earth Sciences, California State University, Hayward, California 94542. *Recession and Retrogression.*

JOHN A. WEST, Dept. of Botany, University of California, Berkeley, California 94720. *Cryptophyta.*

LOUIS CUTTER WHEELER, deceased. *Bryophyta; Cycadicae (Cycadophyta); Equisetophyta; Gneticae; Lycopodiophyta; Magnoliophyta; Pinicae; Pinophyta; Polypodiophyta.*

E. REED WICANDER, Dept. of Geology, Central Michigan University, Mt. Pleasant, Michigan 48859. *Chrysophyta.*

ALAN WOOD, Dept. of Geology, University College of Wales, Aberystwyth, Wales. *Coastal Bevel; Fossil Cliffs; Slope-over-Wall Cliffs.*

CONTRIBUTORS

FERGUS J. WOOD, 3103 Casa Bonita Drive, Bonita, California 92002. *Proxigean Spring Tides; Tides.*

LYNN D. WRIGHT, Dept. of Geography, University of Sydney, Sydney, N.S.W. 2006, Australia. *Coastal Morphology, Oceanographic Factors; Deltas; Estuarine Delta; Internal Waves; Sea Slick; Tidal Inlets, Channels, and Rivers.*

DAN H. YAALON, Dept. of Geology, The Hebrew University, Jerusalem, Israel. *Aerosols; Desert Coasts; Salt Weathering.*

WARREN E. YASSO, Teachers College, Columbia University, New York, New York 10027. *Headland Bay Beach; Salcrete.*

CHRIS F. ZABAWA, Maryland Department of Natural Resources, Tidewater Administration, Tawes State Office Building C-2, Annapolis, Maryland 21401. *Flocculation.*

RYSZARD ZEIDLER, 80-513 Gdansk, Walecznych 14 m 13, Poland. *Bed Forms; Coastal Erosion, Formations; Coriolis Effect; Dispersion; Edge Waves; Node and Antinode; Profile of Equilibrium; Scale Models; Thermal Pollution; Wave Statistics; Wave Theories, Oscillatory and Progressive.*

VSEVOLOD PAVLOVICH ZENKOVICH, Institute of Geography, Akademy of Sciences, Moscow J-17 U.S.S.R. *Protection of Coasts; Soviet Union, Coastal Morphology.*

DONALD J. ZINN, P.O. Box 589, Falmouth, Massachusetts 02541. *Arthropoda; Coelenterata; Mollusca.*

"A-B-C..." MODEL

The "a-b-c..." model is a conceptual device that permits detailed analysis of the history of sediment transport along various shores (Tanner, 1974). The five key points in the model are: (a) the drift divide on the cape or at the updrift end of the transport cell; (b) the point of maximum erosion; (c) the point of maximum transportation; (d) the point of maximum deposition; and (e) the drift divide in the embayment or at the downdrift end of the cell. The *wave energy density E* (dimensions MT^{-2}) decreases systematically from "a" to "e," where it is commonly close to zero. The *littoral component of power* P_L (dimensions MLT^{-3}) increases from zero at "a" to a maximum at "c" and then decreases to zero at "e." The *quantity of sand delivered* per unit time q (dimensions $L^3 T^{-1}$) behaves like P_L.

The *change in load carried* per unit distance along the beach (dq/dx) is maximum (positive and large) at "b" and minimum (negative and large) at "d." These are therefore the points of maximum erosion ("b") and maximum deposition ("d")—that is, the positive first derivative of transport is erosion, and the negative first derivative is deposition.

The *numerical coefficient k*, which relates the immersed mass of q to P_L over a reasonably long period of time and which is one measure of efficiency, is commonly in the range 10^{-2} to 10^{-4}.

The model can be applied under various circumstances (Tanner, 1980), but is particularly clear when applied to beach ridge plains, where the migration with time of some of the key points can be traced. The model also indicates something of the partitioning taking place between *x*-transport (littoral) and *y*-transport (transverse); for example, the model shows that parallel beach ridges were generally built from offshore, not from littoral drift, sand. Where wave climate data are available, the model can be used for reasonably precise numerical studies (May and Tanner, 1973).

The "a-b-c..." model is not to be confused with the scheme of Wyrtki (1953), which was based on a different set of assumptions, including: that the quantity q varies with the amount of sediment present along a coast, that the *angle of wave approach* (β) is of no significance, and that the entire process can be described by a continuous function of some kind. In the application of the "a-b-c..." model, only discontinuous functions are used. The presence of many more or less independent cells, especially along low- to moderate-energy coasts, requires the discontinuous approach.

WILLIAM F. TANNER

References

May, J. P., and Tanner, W. F., 1973. The littoral power gradient and shoreline changes, *in* D. R. Coates, ed., *Coastal Geomorphology.* Binghamton: State University of New York, 43-60.

Tanner, W. F., ed., 1974. *Sediment Transport in the Near-Shore Zone.* Tallahassee: Coastal Research Notes, Florida State University, 147p.

Tanner, W. F., 1980. Variants of the "a-b-c..." model, *in* W. F. Tanner, ed., *Shorelines Past and Present.* Tallahassee: Florida State University, 539-556.

Wyrtki, K., 1953. Die Bilanz des Langstransportes in der Brandungzone, *Deutsch. Hydrograph. Zeitschr.* **6,** 65-76.

Cross-references: Vol. III: *Littoral Processes;* Vol. VI: *Littoral Processes; Sediment Budget; Sediment Transport; Shore Drift Cells; World Net Sediment Transport.*

ABRASION PLATFORM—See WAVE-CUT PLATFORM

ACCRETION RIDGE

In a geomorphological sense, the term *accretion* refers to the addition of sediments to a topographic feature, usually in a lateral sense (*lateral accretion*). Fisk (1959) used the term *accretion ridge* to describe an ancient beach ridge inland of a modern beach on a prograding coastline. Shoreline sediments that could be abandoned by accretion include beach ridges, cheniers, and barriers (Russell, 1968; Todd, 1968; Hoyt, 1969). Ball (1967) used the term *beach accretion bedding* to describe the low-angle bedding produced by an accreting beach face.

Reineck and Singh (1980) review various processes of shoreline accretion. Raised beaches abandoned because of eustatic lowering of sea level or tectonic uplift would not be accretion ridges.

JOANNE BOURGEOIS

References

Ball, M. M., 1967. Carbonate sand bodies of Florida and the Bahamas, *Jour. Sed. Petrology* 37, 556-591.
Fisk, H. N., 1959. Padre Island and the Laguna Madre Flats, coastal south Texas, *Proc. 2d Coastal Geog. Conf., Baton Rouge, La.*, 103-151.
Hoyt, J. H., 1969. Chenier versus barrier, genetic and stratigraphic distinction. *Am. Assoc. Petroleum Geologists Bull.* 53, 299-306.
Reineck, H. -E., and Singh, I. B., 1980. *Depositional Sedimentary Environments.* New York: Springer-Verlag, 549.
Russell, R. J., 1968. Glossary of terms used in fluvial, deltaic, and coastal morphology and processes, *Coastal Studies Inst., La. State Univ. (Baton Rouge), Tech. Rept. 63,* 97p.
Todd, T. W., 1968. Dynamic diversion: influence of longshore current-tidal interaction on chenier and barrier island plains, *Jour. Sed. Petrology* 38, 734-746.

Cross-references: *Beach Ridge and Beach Ridge Coasts; Chenier and Chenier Plain; Progradation and Prograding Shoreline.* Vol. III: *Beach Ridges; Prograding Shoreline;* Vol. VI: *Accretion, Accretion Topography; Littoral Sedimentation.*

AEOLIANITE

Aeolianite, also known as *aeolian calcarenite* or *dune limestone,* is a calcareous dune sand that has been cemented or lithified by percolating rainwater. Large quantities of calcareous sand, composed predominantly of calcium carbonate (90%+) in the form of broken and comminuted shell debris, accumulated at various heights along parts of the Australian coast, for example, as relative sea level changed during Quaternary times (includes Pleistocene and Holocene or Recent epochs) in response to glacial-interglacial periods. Fairbridge (1967) believes that these aeolianites may be associated with *pluvial* phases of the Quaternary that resulted in the calcareous sands being rapidly cemented by percolating rainwater.

Many such aeolianites now form prominent coastal/offshore reefs, shore platforms, and cliffs that may be as high as 60 m. In Australia, aeolianites are particularly prominent along the southern coast, which is exposed to Southern Ocean swell. In contrast, on the east coast of the continent, predominantly quartzose sands are the most common marine deposit. These have been derived from weathered arenaceous sedimentary and igneous rocks and, when leached, remain unconsolidated.

Aeolianite cliffs and shore platforms are common on the coast of Victoria. For example, in the Warrnambool-Port Fairy coastal sector of western Victoria, aeolianite fills a former marine embayment and extends about 4 km inland (Gill, 1967). Elsewhere in this area the lithified calcareous sand forms a cliffed coastline, which in turn truncates the trend of the aeolianite dunes at a slightly oblique angle.

JOHN R. HAILS

References

Fairbridge, R. W., 1967. Coral reefs of the Australian region, *in* J. N. Jennings and J. A. Mabbutt, eds., *Landform Studies from Australia and New Guinea.* New York: Cambridge University Press, 386-417.
Gill, E. D., 1967. Evolution of the Warrnambool-Port Fairy Coast, and the Tower Hill eruption, western Victoria, *in* J. N. Jennings and J. A. Mabbutt, eds., *Landform Studies from Australia and New Guinea.* New York: Cambridge University Press, 340-364.

Cross-references: *Beach Material; Calcarenite; Coastal Dunes and Eolian Sedimentation.* Vol. VI: *Eolianite; Humate; Soils.*

AEROSOLS

A suspension of solid and liquid particulate matter floating in a gaseous medium is called an *aerosol.* In the coastal atmosphere, sea salt particles and spray droplets are the main constituents of the aerosol.

Winds blowing over the ocean pick up a substantial load of sea salt particles. These originate at the sea surface as tiny droplets that form when air bubbles burst at the sea/air interface. Their diameter ranges from 0.02 to 20 μm, with large droplets reaching ~100 μm. The formation of such aerosols is most intensive in the nearshore breaker zone of waves, from where some of the large droplets are carried inland by onshore winds. Ordinary sea winds carry some 1 to 20 $\mu g/m^3$ of sea salts, but above a wind speed of about 9 m/sec the concentration increases rapidly, and storm winds may carry as much as several thousand $\mu g/m^3$ (Monahan, 1968). Vertically, the marine aerosol regime extends to a height of about 2 km.

Onshore winds carry large amounts of the aerosol salts onto the land, where they become intercepted by vegetation, by impinging on some obstacle, or by gravitational deposition on the land surface. The residence time and stability of the aerosol in the atmosphere depend on, in addition to the wind regime, the size distribution and electrical charge of the particles, being from a few hours to a few days long.

The concentration of the larger sea spray droplets ($> 20\ \mu m$), which are essentially of sea water composition, decreases rapidly with distance from the source, and they become deposited within about 1.5 km of the coast, depending on the topographic configuration of the terrain (Yaalon and Lomas, 1970). The concentration of the smaller aerosol particles decreases at a slower, but also exponential rate, to a distance of about 50 km inland. The composition of the aerosol frequently differs from that of ocean salts because of chemical fractionation and interaction with aerosols of continental origin, so that the chemical composition of the precipitation or dry fallout differs considerably from that of sea water (Caddle, 1973; Delany, Pollock, and Shedlovsky, 1973). Man-made pollutants are becoming increasingly important components of these aerosols, with SO_4, Pb, Sn, Zn, and Cu considerably enriched above their crustal average.

The rate of deposition of the coastal aerosols is greatly dependent on terrain characteristics, with exposed sites or positions collecting considerably more than protected (wind-shadow) sites. A direct effect of salt deposition is observed in the zonation of the coastal vegetation. The vegetation communities are directly correlated with the quantity of salt deposited (Malloch, 1972), and frequently in the underlying soil, showing a high content of total and exchangeable sodium and magnesium ions (van der Valk, 1974). In nearshore environments, salty winds also usually cause damage to vegetation, especially to trees, and to buildings, resulting in a more rapid rate of corrosion. In Mediterranean and desert coasts, *tafoni* landforms are ascribed partly to the effect of sea spray.

DAN H. YAALON

References

Caddle, R. D., 1973. Particulate matter in the lower atmosphere, in S. I. Rasool, ed., *Chemistry of the Lower Atmosphere*. New York: Plenum Press, 69-120.

Delany, A. C.; Pollock, W. H.; and Shedlovsky, J. P., 1973. Tropospheric aerosol: the relative contribution of marine and continental components, *Jour. Geophys. Research* 78, 6249-6265.

Malloch, A. J. C., 1972. Salt-spray deposition on the maritime cliffs of the Lizard Peninsula (Cornwall), *Jour. Ecology*, 60, 103-112.

Monahan, E. C., 1968. Sea spray as a function of low elevation wind speed, *Jour. Geophys. Research* 73, 1127-1137.

van der Valk, A. G., 1974. Mineral cycling in coastal foredune plant communities in Cape Hatteras National Seashore, *Ecology* 55, 1349-1355.

Yaalon, D. H., and Lomas, J., 1970. Factors controlling the supply and the chemical composition of aerosols in a near-shore and coastal environment, *Agric. Meteor.* 7, 443-454.

Cross-references: *Alveolar Weathering; Desert Coasts; Salt Weathering; Splash and Spray Zones; Tafoni.*

AFRICA, COASTAL ECOLOGY

The ecology of the African coast is influenced by a complex of environmental variables related to climate, oceanography, hydrology, geomorphology, soils, and interacting biota including human activity. These variables help to explain the nature, behavior, and distribution of the flora and fauna and the occurrence therein of tropical and temperate components and of marine, brackish, and freshwater elements. African coastal ecology is as yet imperfectly understood, and much research remains to be conducted. Nevertheless, it is evident that the terrestrial plants and animals found inland from the coast, although sometimes reaching the shore, offer no clear guide to the nature and variety of organisms and habitats within the coastal zone *sensu stricto*. It is this latter zone, a few hundred meters to many kilometers in width, where oceanic rather than terrestrial influences prevail, that forms the focus of this discussion. The following discussion first outlines some salient environmental considerations and then samples the coastal ecology from the Mediterranean southward down the Atlantic coast to the Cape region and proceeds northward along the Indian Ocean shores to the Red Sea.

Climatic Factors

From Cap Blanc ($37°\ 15'N$) to Cape Agulhas ($34°\ 52'S$) Africa ranges through some $72°$ of latitude. Straddling the equator in this way ensures the predominance of tropical and subtropical conditions over most of the continent. The coastal zone is no exception, although conditions along the Atlantic coast to within $15°$ of the equator are tempered by the cooling effects of relatively cold ocean currents. In terms of the general atmospheric circulation, Africa's coastal climates are much influenced by air-mass movements and concomitant fluctuations of the intertropical convergence zone (ITCZ). During the northern winter the ITCZ shifts southward with the sun, extending from just inland of the Guinea coast southeastward to Madagascar. Apart from the influx of cool, moist maritime polar air from the Atlantic along the coasts of Morocco and the Mediterranean, most of northern Africa is covered by warm, dry outflowing air associated with continental tropical air masses over the Sahara and southwest Asia, the latter promoting a strong northeast monsoon along the east coast. Warm, rainy conditions associated with inflowing oceanic air or local convection cells prevail elsewhere over Africa except along the southwest coast, where the tempering effect of the cold Benguela Current persists. During the northern summer the ITCZ shifts northward to the Sahel zone, and hot, wet conditions prevail over central Africa and the Guinea coast, while arid conditions characterize the Sahara and Mediterranean and, except along a rainy belt around the southern tip, most of southern Africa. A southwest monsoon associated with air flowing toward Asia affects the east coast. The seasonal surface air temperature and rainfall patterns associated with this circulation are shown in Figure 1. Inevitably solar radiation and potential evaporation within the coastal zone are greatest where dry outflowing air and cloudless skies are most persistent, for example, around the Red Sea and the Horn of Africa and off the western Sahara. Values for these variables are lowest along the Guinea coast. Whereas the relationship between these and other climatic parameters and the coastal biota is very broad and at best poorly understood, some correlations are considered important. For example, air temperature appears to determine the limits of mangal or, conversely, salt marsh vegetation. Mangal usually flourishes where the temperature of the coldest month does not fall below 20°C and where the range is around 10°C, although *Avicennia marina* tolerates winter temperatures as low as 15.5°C along the northern Red Sea. And whereas the amount and seasonality of rainfall are clearly important to the character of the terrestrial vegetation reaching toward the shore, even extremely small amounts of moisture carried by onshore winds may be important to coastal cliff communities, for example, along the Namibian coast.

Oceanographic and Hydrologic Factors

The physical and chemical characteristics of the available water supply strongly influence ecologic responses along the African coast. Two qualities are particularly important: water temperature and salinity (Fig. 1). These in turn reflect such variables as solar radiation, air temperature, cloud cover, wind, ocean currents, and freshwater inflows. The surface temperature of coastal waters shows significant differences between the east and west coasts. Along the east coast, the divergence of the warm South Equatorial Current into the south-flowing Mozambique and Agulhas currents; and in the northern summer at least the north-flowing Somali Current disperses warm water along the entire coast. Even the dominant Northeast Monsoon Current of the northern winter is relatively warm. Mean sea-surface temperatures for August, the coolest month south of the equator, are everywhere above 20°C and reach 30°C in the Red Sea, affording ideal thermal conditions for coral organisms and mangal vegetation. Further, the spread of mangrove species along the east coast is attributable to seeds, seedlings, and plant portions carried by these warm ocean currents.

Along the west coast, warm waters are more restricted, mainly to the Guinea coast from Cape Vert to the Cuanza River, Angola, because the cold Canaries and Benguela currents, as they move toward the equator, squeeze the region of warmer water into a narrower zone than would be expected from considerations of latitude alone (Fig. 1). As on the east coast sea-surface temperatures have a significant effect on the nature and distribution of biota. For example, the cold waters and comparable near-surface air temperatures along the coasts of Namibia and Cape Province have long formed an effective barrier to the spread of mangrove species and other Indo-West Pacific organisms from the east coast to the west, and indeed of species from the west coast to the east. Even within the warm waters of the Guinea coast the upwelling of cold water causes a significant temperature drop and brings increased nutrients to the surface from June to August, a fact that helps to explain local differences in coastal biota. Further, inflows of terrestrial waters modify the thermal characteristics of coastal areas, particularly the lagoons along the Ivory Coast and the Bight of Benin and off the Niger delta.

Sea-surface salinity is lowest where solar radiation and potential evaporation are least and where rainfall is highest, notably around the inner angle of the Gulf of Guinea, where August values below 30‰ are recorded (Fig. 1). Conversely, where solar radiation and potential evaporation are highest and rainfall minimal, sea-surface salinities reach 40‰, specifically in the Red Sea. Seasonal differences are important where periodic rainfall and stream discharge effectively dilute the salinities of nearshore waters. This effect is especially pertinent to the biota of coastal lagoons, such as Lake St. Lucia in Natal, where salinities as high as 100‰ periodically cause a significant dieback of the reed-swamp communities during

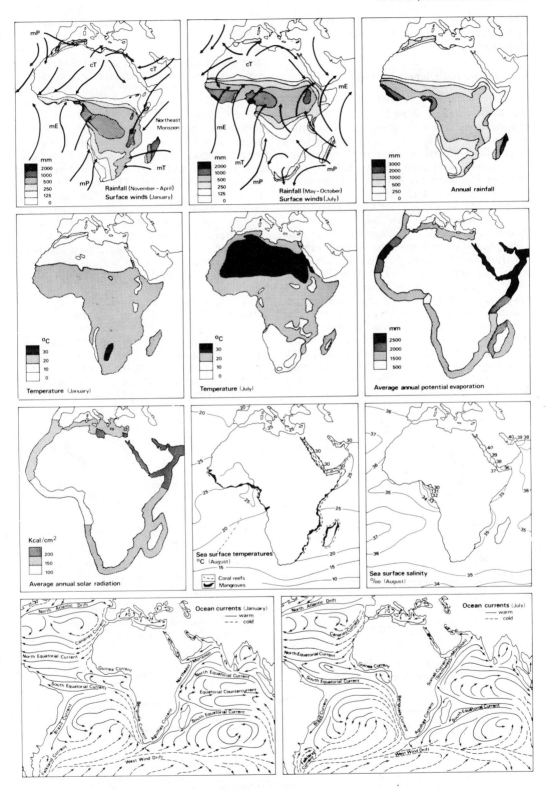

FIGURE 1. Selected environmental factors pertinent to Africa's coastal ecology.

prolonged drought, but where salinities well below seawater values characterize flood periods (Fig. 2).

Geomorphic and Edaphic Factors

Vertical zonation is typical of life within the coastal zone. Thus, along the African coast as elsewhere, we may distinguish between: marine communities that live permanently submerged in seawater; littoral communities that occupy the intertidal zone; strand, dune, and cliff communities that normally live beyond the direct reach of the sea; and those terrestrial communities that to a greater or lesser extent penetrate toward the shore. Such zonation is frequently complex and dynamic in both time and space. For example, the littoral zone is strictly defined as lying between extreme high water and extreme low water of spring tides, but there are transitional zones above and below that blur this definition. Thus marine species such as laminarians may penetrate the littoral zone from below as an infralittoral fringe, while a supralittoral fringe above the upper limit of barnacles may see species such as *Littorina* spp. typical of the littoral zone extend far above extreme high water of spring tides. Seasonal fluctuation in the extent of salt, brackish, and freshwater habitats also has a profound influence on this zonation. The width of these zones and the quality of life therein reflect such physical controls as tidal range, slope, exposure to waves and currents, and the magnitude and frequency of terrestrial stream discharge, including their effect on sediment delivery, water quality, available nutrients, and erosion in estuaries and swamps. For example, steeply shelving coasts will compress the above zones, whereas extensive intertidal shallows may permit broad development of mangal or salt-marsh communities. On the other hand, neither mangal nor salt marsh can develop on exposed coasts where wave action inhibits the establishment of seedlings. Considerations of wind frequency, swell direction, wave height, tidal types, and tidal range have been addressed in the accompanying discussion on Africa: Coastal Morphology (q.v.) and will not be repeated here.

Edaphic conditions vary widely around the African coast and play an important role in determining community composition. Sandy shores predominate, often over long distances in west Africa and Mozambique. Muddy shores are also important, especially in the more sheltered bays, estuaries, and lagoons of the Guinea coast and along the Indian Ocean where mangal

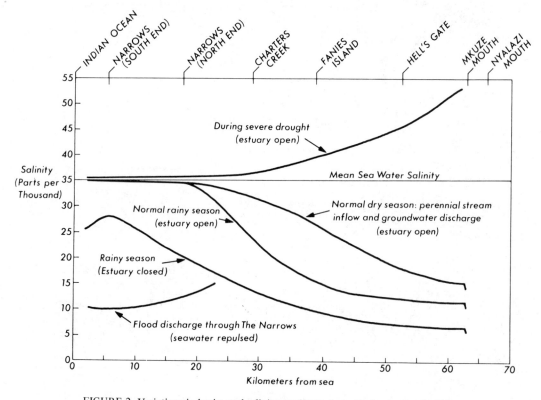

FIGURE 2. Variations in horizontal salinity gradients, Lake St. Lucia, South Africa.

vegetation is well developed, the roots of individual mangrove species encouraging further deposition of mud. Rocky shores are locally significant, but, except in portions of the western Mediterranean, the Red Sea, Somalia, and the southern tip, long stretches of sea cliffs are relatively rare. The intertidal ecology of these rocky shores has been well studied in parts of west Africa and around the Cape region, but the sea cliffs as a whole need more investigation. Substrate texture apart, the role of other edaphic factors such as water chemistry and nutrient availability requires further research before generalized conclusions can be drawn.

Biotic Factors

Although competition and predation among plants and animals inevitably play an important role in African coastal ecology, these forces are poorly understood at this time. More is known about the role of human activities, but in this context it appears that organisms and their habitats have suffered much less from human destruction, pollution, or other forms of interference in Africa than in Europe or North America. There are, however, some notable exceptions. For example, the mangal vegetation of the Indian Ocean coast has suffered from many centuries of timber exploitation, mangrove poles or *boritis* having been transported by dhows from such ports as Lamu, Kenya, to the relatively treeless regions of Arabia and the Persian Gulf for building purposes. Although this particular trade has declined in recent decades, timber harvesting of mangroves continues along both the east and west coasts of Africa and where 25-50 m trees once stood 2-5 m shrubs now prevail. Above the littoral zone plantations of coconuts and other commercial crops have long since replaced the original coastal vegetation. Along the more humid to subhumid parts of the coast increased sediment discharge attributable to the erosion of agricultural lands has in recent times fed abundant sediment into coastal lagoons and estuaries; along the open coast sediment plumes have been dispersed alongshore by waves and currents. This increased sediment budget has not only disrupted existing ecosystems but has created fresh habitats among extensive mud flats and new beach and dune ridges. Along the northern Natal coast, for example, present sediment delivery rates appear to be between two and three times the average for the past 5000 years, attributable at least in part to the accelerated erosion of coastal watersheds resulting from overgrazing and other exhaustive land-use practices (Orme, 1975). Meanwhile nearby coastal dune communities have been severely disrupted by recent mining for ilmenite, rutile, zircon, and other heavy minerals. These are but a few examples of the way in which Africa's coastal vegetation, and implicitly the associated animal life, has been modified by human activities.

The Mediterranean Coast

Ecosystems along Africa's north coast (Fig. 3) are strongly influenced by local geomorphic and edaphic factors as well as by the prevailing temperate climate and modest winter rains. In Egypt, Libya, and south Tunisia, where little-disturbed Cenozoic sediments approach the coast, the most typical habitats along the immediate coast are afforded by barrier beaches, dunes, and low cliffs. The dunes are sometimes succeeded inland by a series of older barrier-dune complexes comprising well-cemented carbonate grains, these in turn merging with the desert. Elsewhere the barrier-dune complexes are backed by salt marshes, salt flats, and shallow lagoons, most notably in the *Barari* (= barren) of the Nile delta. To the west, in north Tunisia, Algeria, and Morocco, the coastal terrain is more rugged, and while barrier beaches, dunes, and salt marshes still occur, they are often smaller and framed within rocky headlands or interspersed between longer stretches of rocky shore.

The Mediterranean coastal land west of Rosetta in the Nile delta is floristically the richest part of Egypt, comprising about 1000 species of flowering plants or 50% of the Egyptian flora (Ayyad, 1973). Dry coastal formations are well exemplified by the sand dunes and inland ridges of the region, comprising mainly coarse carbonate grains fashioned into dunes by predominantly northwesterly winds and then progressively cemented by secondary calcite. Initially the dunes are unstable and only a few plant species are able to grow, specifically those capable of sand binding, resisting burial, and utilizing moisture in coarse substrates. During succession these specialized species give way to those less adaptive to unstable conditions. *Ammophila arenaria* dominates the active, partly stabilized, and stabilized coastal dunes but in the most active dunes is associated with few other species, of which *Euphorbia paralias* is most important. With increasing stability the number of associated species increases, with *Crucianella maritima*, *Launaea resedifolia*, *Agropyron junceum*, and *Echinops spinosissimus* becoming significant, especially on shallow soils and in sand shadows, where they achieve codominance with *Ammophila arenaria*. Because of the destructive effects of man and animals, however, no clear relationships can be established between species distribution and soil characters. Behind the coastal

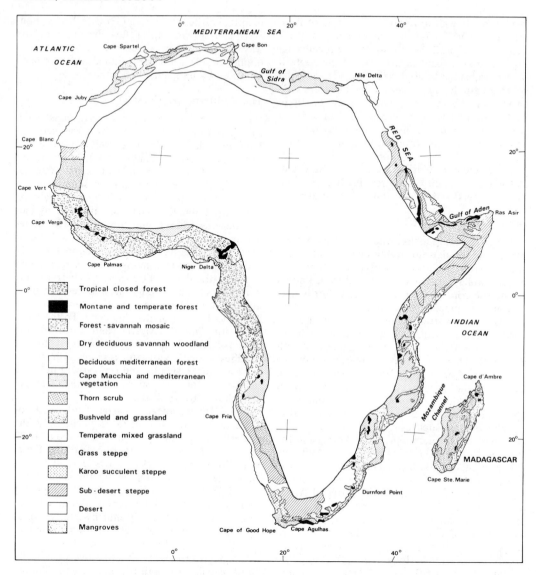

FIGURE 3. Terrestrial vegetation types of Africa's coastal zone *sensu lato*.

dunes lies a series of parallel inland ridges, 20–40 m above sea level, composed of 40–60% oolitic carbonate grains that afford pale brown loamy soils. The vegetation of the shallower soils is codominated by *Thymelaea hirsuta* and *Gymnocarpos decandrum*, whereas the deeper soils are codominated by *Plantago albicans* and *Asphodelus microcarpus* together with several other species (Ayyad and Ammar, 1974).

Wet coastal formations of the Mediterranean shore are typified by salt marshes that are similar from Egypt to Morocco and indeed southward to the Red Sea along the east coast and to Mauritania on the west coast. Frequently it is difficult to determine where salt marsh ends and salt desert commences. Along the coast of Tunisia and Algeria the *Halocnemon strobilaceum* community is a widespread pioneer, collecting among its plants sediment, which is then colonized by *Cutandra memphitica*, *Bassia muricata*, and *Traganum nudatum*. In Tunisia the *Halocnemetum* zone is succeeded by a *Limoniastrum quyonianum* community in which *Zygophyllum album*, *Nitraria retusa*, and *Suaeda vermiculata* occur. Around Oran, Algeria, the pioneer community is a *Salicornietum* dominated by *Salicornia arabica* and *S. herbacea*, succeeded by a floristically rich *Limonietum* dominated by *L. sinuatum* (pan-Mediterranean), *L. sebkarum* (local), and *L.*

spathulatum (farther east). On solonchak soils the primary salt-marsh colonist is either the shrubby perennial *Arthrocnemum glaucum* or *A. macrostachyum*, succeeded landward by a *Salicornia fruticosa* community with either *Centaurium spicatum* (with submergence) or *Monerma cylindrica* (salt desert). *Suaeda fruticosa* is an important landward community in Morocco. At river mouths in Algeria a *Spartina maritima* community dominates in which *Puccinellia palustris*, *P. distans* and *Crypsis aculeata* occur (Chapman, 1977). This *Spartinetum* is essentially an eastern Atlantic community that has spread into the Mediterranean and also reaches South Africa.

The Atlantic Coast

With the Canaries Current offshore, the Atlantic coast of Africa (Fig. 3) as far south as the Senegal River is characterized by relatively cool temperatures and subhumid to arid conditions. Wet formations are restricted to patches of salt marsh similar in composition to those along the Mediterranean coast. Dry formations on the strand and coastal dunes merge inland with typical desert communities. The Cape Vert peninsula marks the effective southern limit of cold-water organisms and cool coastal floras.

From Senegal to Angola annual average air and sea-surface temperatures increase, temperature range decreases, relative humidity rises and precipitation values reach over 3000 mm/yr. Steep, sandy fringing and barrier beaches predominate; muddy estuaries and lagoons are important; but rocky headlands are more restricted, and coastal dunes are rare. Strong swells are common, and since wave amplitudes usually exceed tidal range, wave action is important as an ecological factor. Two features of this coast are particularly worthy of mention: the well-developed mangal vegetation of the muddy estuaries and lagoons, and the flora and fauna of the open rocky shores and sandy beaches.

Six mangrove species are common along this coast. *Rhizophora racemosa* is the common pioneer, and rises from 4 m (where coppiced) to 40 m in height. *R. harrisonii*, which never rises above 7 m, occupies the wetter areas of the main mangrove forest. *R. mangle* grows to about 5 m on the drier, more saline soils to landward. *Avicennia africana* is widely found, and *Laguncularia racemosa* and *Conocarpus erectus* also occur, the latter especially on the landward fringe of the swamps. These species are completely different from those found along the east coast of Africa, showing affinity with mangroves of the western Atlantic shores and the Caribbean. As stated elsewhere, the cold waters of the southern tip and Namibia have formed an effective barrier to migration between the east and west coasts.

Six mangal zones are found along the west coast. First, in northern Senegal the long dry season and low mean winter temperatures (10–15°C) inhibit *Rhozophora* spp., so that *Avicennia africana* is dominant (Fig. 4). Second, from Bathurst at the mouth of the Gambia River to Monrovia *Rhizophora-Avicennia* swamps line the shores of most estuaries and flooding rivers, with *Paspalum vaginatum* as a grassy ground cover. The poorly drained *Avicennia* areas encourage the breeding of the malarial mosquito *Anopheles melas*. Third, beyond the mainly rocky coast between Monrovia and Fresco mangroves are well developed in the extensive lagoons eastward to Lagos. *Rhizophora racemosa*, backed by *R. harrisonii*, typifies those lagoons with continuous open access to the sea, whereas in seasonally closed lagoons the shrubby *Avicennia africana* occurs. Vertical zonation of animal life is demonstrated by a lower zone of red algae *Bostrychia* and *Calaglossa*, a middle zone of the oyster *Ostrea tulipa* in the *Rhizophora* prop roots, and an upper zone of barnacles *Chthamalus rhizophorae* and the snail *Littorina angulifera*. Fourth, the coastal zone from the Niger delta to the Ogooue River, Gabon, sees the most extensive development of mangal in west Africa, extending in riparian bands for as far as 100 km up the Niger. *Rhizophora* species occur in their typical niches, with *Avicennia africana* and *Laguncularia racemosa* growing between the frontal *Rhozophora racemosa* and the *R. mangle* and *Acrostichum aureum* fern on the drier, more landward soils. The transition to fresh water is marked by the invasion of the *R. mangle* zone by the screw pine *Pandanus candelabrum* and the palms *Phoenix spinosa* and *Raphia*. Where protected from felling, which is common, the frontal *Rhizophora racemosa* may reach 50 m, and the *Avicennia africana* 30 m. Animal life is rich, comprising pelicans, terns, sandpipers, fish eagles, crocodile birds, two crocodiles, the Mona monkey, and the mudskipper fish, among many other lesser animals. Fifth, from the Ogooue to the Congo River narrow mangal belts occur similar in zonation to the previous region. Sixth, in Angola mangal occurs only sporadically as fringe vegetation along rivers and estuaries. In the north *Rhizophora* species are primary colonists. South of the Cuanza River, as coastal waters become cooler, *Avicennia africana* (on mud) and *Laguncularia racemosa* (on sand) are primary colonists associated with a ground flora of *Scaevola lobelia* and *Suaeda fruticosa*. Mangroves disappear south of Lobito.

On open rocky shores in west Africa the zone

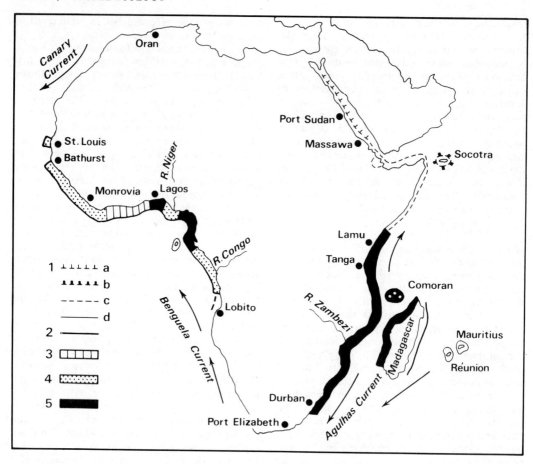

FIGURE 4. Mangal distribution along the African coast (after Grewe, 1941). 1 = Avicennia: (a) single and small groups, (b) closed community, (c) with sporadic Rhizophora, (d) with better developed Rhizophora; 2 = most species present and well developed; 3 = as 2, in lagoons; 4 = complete mangal, well developed; 5 = mangal forest used for commerce.

above high water is usually rather bare except for occasional halophytes such as *Sesuvium portulacastrum*, which grows in soil-filled cracks. Where more soil occurs evergreen coastal shrubs may be found. At the limits of extreme high water of spring tides the supralittoral fringe is characterized by the snails *Littorina punctata* and *Tectarius granosus*, together with such algae as *Entophysalis crustacea* (Lawson, 1966). The main littoral zone is generally divided into an upper belt of barnacles *Chthamalus* spp. together with *Siphonaria pectinata*, *Nerita senegalensis*, *Ostrea cucullata* and such algae as *Enteromorpha* spp. and *Ulva* spp., and a lower belt dominated by pink encrusting sheets of lithothamnia algae together with many other algae and such animals as *Mytilus perna*, *Patella safiana*, *Fissurella nubecula*, and *Thais* spp. In the zone of extreme low water of spring tides the infralittoral fringe is characterized by algae such as *Sargassum* and *Dictyopteris delicatula*, often accompanied by the sea urchin *Echinometra lucunter*, but there is rarely a reliable indicator for this zone throughout the region.

Open sandy beaches are generally bare of plant life, although intertidal sand is sometimes colored green from diatoms *Melosira* spp., and sea grasses such as *Cymodocea* are sometimes exposed during extremely low water. The strand vegetation comprises specialized flowering plants. Steep, coarse sandy beaches are almost devoid of faunal life except for the ghost crab *Ocypoda cursor*. The flatter beaches with finer sand have a much richer fauna, comprising *Ocypoda cursor* and *Ocypoda africana*, the polychaete *Nerine cirratulus*, the isopod *Excirolana latipes*, the small lamellibranch *Donax pulchellus*, the mole crab *Hippa cubensis*, and many others (Lawson, 1956, 1966).

From Lobito south almost to Cape Town negligible rainfall combines with the cold Benguela Current to promote semiarid to arid conditions along the coast, drought being alleviated only by the fog. Along the central Namib coast mean annual rainfall is less than 10 mm, and rain occurs on average only once every two years. Desert biota thus reach down to the shore, which is dominated for long distances by sandy beaches and extensive sand dunes, broken by relatively bare rocky shores. Plants of this region include *Mesembryanthemum salicornioides*, *Hydrodea bossiana*, *Drosanthemum paxianum*, and dense cushions of *Aizoon dinteri* and *Zygophyllum simplex*. Fogs that form over the Benguela Current may be driven as far as 50 km inland by onshore winds, only to dissipate and curl back toward the coast during the day. These fogs are ecologically important. Fog precipitation, or dew, may account to as much as 40–50 mm/yr, assuming 200 foggy days, but this is soon evaporated on level ground by the rising sun and is of little use to plants. Where driven against cliffs, however, fog precipitation may seep into crevices and be more useful to plants. Thus, relatively speaking, sea cliffs tend to have a richer flora than do the open strand and dunes.

The Southern Tip

In general terms, the coastal zone around the southern tip of Africa (Fig. 3) forms a distinct ecological entity, separating the tropical biota of the east and west coasts with a region of temperate forms. The boundaries of this region, however, are less well defined and certainly not with reference to any single criterion. For example, the southern limits of mangal vegetation occur at the Kei River on the east coast and near Lobito on the west coast, whereas tropical faunas, although impoverished south of these limits, extend down the east coast all the way to Cape Town and down the west coast to Walvis Bay. By any criterion, therefore, there is a broad transition zone along both the east and west coasts where tropical or warm water biota are gradually replaced by temperate or cold water forms, the location and width of the transition varying with the life form involved. Certainly the warmth of the Agulhas Current exerts a more pervasive influence over the intertidal fauna bathed in its waters than it does over the wet coastal vegetation, which, although periodically flooded by seawater, is more responsive to the lower winter air temperatures of southern Africa. These two elements—the wetland vegetation and the intertidal fauna—will suffice to demonstrate some aspects of this region's coastal ecology.

South of the Kei River temperatures are too cool for mangal. The wet formations of the south and west coasts of Cape Province are represented by salt marsh. Above an infralittoral fringe of eel grass *Zostera capensis*, the lowest salt-marsh community is dominated by *Spartina maritima*, associated with the red algae *Bostrychia*, the mud prawn *Upogebia africana*, the crabs *Sesarma catenata* and *Cyclograpsus punctatus*, and the barnacles *Balanus elizabethae* and *B. amphitrite* clothing the lower stems of *Spartina* (MacNae, 1957, 1963). At higher elevations the plant *Arthrocnemum perenne* and the mangrove snails *Cerithidea decollata* occur, followed at times of high water by a *Limonietum* (*L. linifolium*), with *Chenolea diffusa*, *Crassula maritima*, *Suaeda fruticosa*, and *S. maritima* among the flora. In muddy areas the zone above the *Limonietum* is dominated by *Arthrocnemum africanum* or *A. pillansii*. In sandy areas *Sporobolus virginicus* dominates this zone. This succession, well seen in the Zwartkops estuary near Port Elizabeth, shows some similarity to the Mediterranean salt marshes.

The intertidal fauna consists of three reasonably distinct populations: the tropical Indo-West Pacific fauna of the southeast coast; the warm, temperate Cape fauna of the south coast; and the temperate or cold-water Namaqua fauna of the southern part of the west coast (Stephenson and Stephenson, 1972). In the warm waters of the southeast coast the supralittoral fringe is dominated by *Littorina africana*, *L. obesa*, and *Tectarius natalensis*. The main littoral zone is characterized in its upper part by barnacles and limpets and in its lower part by algae rather than animals. Typical barnacles include *Chthamalus dentatus*, *Tetraclita serrata*, and *Octomeris angulosa*. The limpets are *Patella granularis*, *P. variabilis*, *Cellana capensis*, and *Siphonaria* spp.. The Indo-West Pacific oyster *Crassostrea cucullata* and the snail *Oxystele tabularis* also occur. The algae comprise a moss-like coating of *Gelidium reptans*, *Caulacanthus ustulatus*, *Gigartina minima*, *Centroceras clavulatum*, and others, replaced toward the infralittoral fringe by the bright green *Hypnea spicifera* and various zoanthids, of which *Palythoa nelliae* is most abundant. Healthy reef-building corals of such genera as *Anamastrea*, *Favia*, *Goniastrea*, *Pocillipora*, *Psammocora*, and *Stylophora* form large colonies in some places but are best developed in rock pools. The infralittoral fringe is usually characterized by a dense sward of small algae exhibiting various shades of red, pink, and purple, in contrast with the vivid green of *Hypnea spicifera* above. The pink *Hypnea rosea*, the deep red *Rhodymenia natalensis*, and various *Sargassum* spp. are typical.

In the Cape fauna of the south coast some warm-water species are still important but never dominant and certain cold-water species are strongly represented. The supralittoral fringe is dominated by *Littorina knysnaensis*, but *L. africana*, although locally common in the east, becomes rare toward Cape Agulhas. The upper part of the littoral zone is again dominated by barnacles and limpets similar to those farther east, but the Natal periwinkle *Oxystele tabularis* is replaced by *O. variegata*. As in Natal, the lower littoral zone has enormous stretches of the gregarious tube-building polychaet *Pomatoleios crosslandi* together with the algae *Gelidium pristoides* and *Colpomenia capensis*. The base of the littoral zone is typified by a mosaic of the limpet *Patella cochlear* with lithothamnia, except in the most exposed localities, where limpets are replaced by mussels, barnacles, and algae. Elsewhere the limpets appear to compete with algae for space, the latter becoming the sole occupants of the most sheltered places. The *Patella cochlear* belt has a distinct lower limit and is replaced in the infralittoral fringe by the leathery ascidian *Pyura stolonifera* and by such algae as the deep red *Gelidium cartaligineum* and various brown or yellow *Sargassum* spp. In contrast with the southeast coast, no reef corals or xeniids appear on the south coast and only one zoanthid occurs. Several species of crabs, snails, barnacles, and algae found in Natal do not occur farther west, their typical niches being commonly occupied by cold-water species from the west.

On the west coast the cold-water component of the fauna outweighs all others, the south-coast element is present but falls off rapidly northward, and few warm-water species remain. The supralittoral fringe is dominated by *Littorina knysnaensis* and by healthy growths of the alga *Porphyra capensis*. The upper littoral zone supports so few barnacles that there is often no balanoid zone, and, except for some *Patella granularis*, the rock is either bare or, in more sheltered localities, characterized by an algal succession from *Porphyra capensis*, through *Chaetangium saccatum*, *C. ornatum*, and *Ulva lactuca*, to *Aeodes orbitosa*, *Iridophycus capensis*, various brown algae, and kelp. The lower littoral zone is characterized by sheets of lithothamnia and the worm *Gunnarea capensis* together with various algae, passing down into a *Patella cochlear–P. argenvillei* belt, the latter limpet being typical of the west coast. The dominant brown *Mytulus perna* of the south and east coasts disappears on the west coast and is replaced by the blue-black *M. meridionalis* and the ribbed *M. crenatus*. Some distinctive algae also occur. The infralittoral fringe represents a transition from the limpet belt above to the giant kelp of the subtidal zone, of which *Ecklonia buccinalis*, *Laminaria pallida*, and *Macrocystis pyrifera* are the most noteworthy inhabitants. The Namaqua fauna and flora of the west coast intertidal zone are distinct from the Cape biota and those plants and animals found farther east, although broad transitions occur between the three regions. Some 42 species (28 algae and 14 animals) typical of the west coast may be found along the south coast for varying distances but do not reach the southeast coast. Conversely, of the warm-water species of the southeast coast that extend farther west along the south coast, 38 species (23 algae and 15 animals) are not found beyond Cape Point.

The Indian Ocean Coast

The shores of the Indian Ocean coast of Africa (Fig. 3) are bathed in warm air and warm water throughout the year. Accordingly mangrove species and reef-building corals thrive under suitable conditions, and plants and animals typical of the shores of the entire Indian Ocean and western Pacific Ocean are frequently found. Between the Kei River and the Gulf of Suez mean winter air temperatures are consistently above $15°C$ and summer temperatures above $25°C$. Most rainfall occurs between November and April, except in northern Kenya and southern Somalia, where some rain between May and October helps to alleviate otherwise semiarid coastal conditions. Sediment yields from nearby watersheds, an important factor in coastal ecology, reflect both this rainfall distribution and the nature and extent of the vegetation cover and land usages found inland. The Horn of Africa is persistently dry, although humidity is often high along the immediate coast. Seasonal discharges of water and sediment are important because they help to dilute nearshore salinities and augment sandy and muddy habitats. Strong surf propelled by relentless southeasterly swells characterizes the entire coast, enhanced by seasonal monsoons, tropical easterlies, and periodic cyclones. The warm waters of the south-flowing Mozambique and Agulhas currents south of $10°S$, of the Equatorial Current off Tanzania, and of the reversing monsoonal flows farther north are of major importance to the maintenance of tropical conditions along this coast.

Although rocky shores occur, notably in Somalia and as coral reefs throughout the region, sandy and muddy habitats predominate. Extensive fringing and barrier beaches topped by massive dunes are found extensively along the northern Natal and Mozambique coasts and also in parts of Somalia where dunes also overlie raised Pleistocene coral reefs. Muddy habitats are found around the lagoons and estuaries between or behind the sandy barriers, notably at the mouths of such rivers as the Zambezi. Systematic studies of the biota have as yet been accomplished only locally, but a sampling of the dry habitats of northern Natal and Kenya and of the wetland communities at intervals along the coast will provide a general picture. More than half the plants found along the coast have wide Indo-West Pacific ranges or are pantropical, reflecting adaptation for long-range dispersal by ocean currents.

Along the Kenya coast the narrow sandy strand is dominated seaward by the sprawling beach vine *Ipomoea pes-caprae* with its 4-m-long stolons, associated successively inland with the grasses *Lepturus repens* and *Sporobolus virginicus* and occasional *Scaevola taccada* bushes. Above the highest tides thickets of *Canavalia maritima, Cyperus maritimus, Colubrina asiatica, Cordia somaliensis*, and others occur (Birch, 1963; Sauer, 1965). The general physiognomy of this seaward zone is one of low succulent plants with occasional small, thick-leaved bushes. Farther back the inland zone along the strand may have a dense cover 1-2 m high of *Grewia glandulosa, Cadaba farinosa, Cleome strigosa, Pemphis acidula*, and *Sideroxylon diospyroides*. Behind the strand are a number of habitats associated with the massive Pleistocene coral reefs that rise to 15 m above sea level. Doum palms (*Hyphaene coriacea*) and other sturdy trees are found where sand overlies coral. A wide range of physiognomic types grows on coral cliffs in the spray zone and is best described as a *Capparis*-evergreen scrub association, with the spiny caper bush *Capparis cartilaginea* forming a dense, low mat along with other shrubs stunted through exposure. Where dunes overlie coral, a rich, dense flora exists that includes trees such as *Afzelia cuanzensis, Ficus tremula*, and *Vitex amboniensis* rising to 5 m or more above a dense undercover. Occasional relict stands of high forest with 30-m-tall evergreen canopy trees and a rich understory are sometimes found on the coral soils, but disturbed shrubland merging into secondary savanna is more common where such soils are shallow and the rainfall inadequate. Human disturbance, especially in almost two centuries of clearance for coconut and *Casuarina* plantations, has played a major role along this coast.

Dry plant communities toward the southern part of this region are well exemplified from northern Natal and Kwa-Zulu. Here the most important pioneer species of the backshore and foredunes are *Scaevola thunbergii* (which usually grows nearest the sea), *Ipomoea biloba*, and the grass *Digitaria erianthia*. These species flourish in shifting sand and help to promote foredune accumulation. Other pioneers include *Mesembryanthemum edule, Tephrosia canescens, Gazania uniflora*, and *Osteospermum moniliferum* (Orme, 1973). Farther back from the shore the pioneer colonies are invaded by *Othonna sarnosa, Gloriosa virescens, Passerina rigida, Anthospermum littoreum*, and various grasses. The mat provided by these plants more or less stabilizes the backshore and enables a low, dense scrub to become established. Under this scrub, which grows to about 2 m high, the outer dunes attain complete stability. Away from the extreme effects of wind, shifting sand, and salt spray, several species of the dune scrub develop into a complex and luxuriant coastal dune forest composed of trees 10-25 m tall. Dominant species include *Mimusops caffra, Diospyros rotundifolia, Brachylaena discolor*, and *Euclea natalensis*. The coastal dune forest does not differ significantly in physiognomy and composition from the evergreen subtropical forest that may once have covered much of the better-drained soils of the Zululand plain northward into Mozambique (Bayer, 1938). These forests afford a rich habitat for many animals, including monkeys and several snake species.

Wet habitats along the Indian Ocean coast are dominated by mangal or reed-swamp vegetation. The mangroves of this coast and the Red Sea are clearly related to the mangal vegetation of India, Malaysia, and Indonesia, comprising part of the Indo-West Pacific realm. From the Bab el Mandeb narrows along the Somali coast south to Lamu, Kenya, mangroves occur only sporadically, mainly in bays and seasonally flooded lagoons, and not in any large quantity. *Avicennia marina* is the dominant species—indeed the only species on Socotra offshore—but on the mainland coast *Rhizophora mucronata, Ceriops tagal*, and *Bruguiera gymnorrhiza* also occur. From Lamu to the Zambezi estuary, climatic, edaphic, and geomorphic conditions combine to promote luxuriant mangrove forests with individual trees reaching 15-25 m high, although long exploitation for timber and tanbark has ensured that most trees are much shorter. The mangal is optimally developed in the Rufiji delta, around Tanga, in the Zambezi delta, and at frequent intervals along the Mozambique coast (Walter and Steiner, 1936). The above-noted four species occur together with the mangrove *Sonneratia alba* and the associated bushes *Lumnitzera racemosa, Xylocarpus granatum*, and *X. moluccensis*. In brackish water the widespread fern *Acrostichum aureum*, the freshwater mangrove *Hibiscus tiliaceus*, and the tree *Heritiera littoralis* are abundant. On salty sand flats behind the mangroves low shrubs of *Suaeda monoica, Arthrocnemum indicum*, and *Sesuvium portulacastrum* and the grass *Sporobolus virginicus* form a ground flora. The wild date palm *Phoenix reclinata* also marks the transition from mangal to terrestrial vegetation.

Zonation among the mangroves seems to be related to exposure and edaphic factors. A *Sonneratia* zone often forms on the exposed outer coast, behind which *Avicennia* occupies the sandier soils while *Rhizophora* prefers muddy soils, especially alongside rivers and their estuaries. However, wherever the seaward substrate is firm and sandy, *Avicennia* tends to replace *Sonneratia* as the pioneer species. Where rainfall is high enough, for instance, in the Zambezi delta, *Bruguiera* forms the main mangrove zone between *Rhizophora* and the landward margin. Elsewhere *Avicennia* forms the most landward zone, the trees becoming smaller and more scattered as the land rises. Crabs (*Uca* spp.) are typical of the mangrove fauna.

From southern Mozambique to the Kei River the mangal vegetation becomes increasingly impoverished, partly because of heavy exploitation for timber. On Inhaca Island *Sonneratia alba* is absent, although present at Inhambane some 3° latitude farther north (MacNae and Kalk, 1962). The primary seaward colonist on the sandy soils is *Avicennia marina*, replaced by *Rhizophora mucronata* on the muddy soils of the creeks. The bulk of the mangal comprises *Ceriops tagal* in the drier areas and *Bruguiera gymnorrhiza* where it is wetter. Where the trees are not too crowded, a ground cover of *Sesuvium portulacastrum*, *Arthrocnemum decumbens*, *A. perenne* and a summer growth of *Salicornia pachystachya* occur. The transition to brackish water is marked by the appearance of *Lumnitzera*, *Xylocarpus*, *Hibiscus tiliaceus*, and *Thespesia acutiloba*. The final transition to freshwater is through a zone of *Juncus krausii* and *Phragmites australis*. Southward *Ceriops tagal* dies out at Kosi Bay, on the Mozambique-Natal border, but the remaining three mangrove species extend as impoverished patches south to the Kei River, about 33°S. In Mozambique the fauna among the mangroves is also ecologically zoned. Fiddler crabs (*Uca* spp.) are found along the more seaward creeks, being replaced by *Sesarma* spp. of crabs and various snails in the dense shade of the *Ceriops-Bruguiera* community. In Natal and the Transkei certain tropical elements of the biota can live in the warmer waters of the coastal lagoons and estuaries, while life on the open shore begins to reflect the occurrence of cooler air and water temperatures from the south.

The east coast of Madagascar (Fig. 3) is not physiographically well suited to mangal development, there being few bays or sheltered estuaries, but there is abundant mangal on the more sheltered and drier west coast (Kiener, 1972; Battistini and Richard-Vindard, 1972). *Sonneratia alba* and *Avicennia marina* are the primary colonists, with *Rhizophora mucronata* and *Bruguiera gymnorrhiza* up the creeks, and *Ceriops tagal* extending landward onto higher ground. *Xylocarpus moluccensis*, rarely found on the mainland coast, and *X. granatum* also occur.

Finally, the character of a changing wetland ecosystem is well shown by Lake St. Lucia, a large coastal lagoon lying behind the massive barrier beach and dune complex of northern Zululand (Fig. 5). Following the Flandrian transgression this lagoon measured 1165 km^2 in area, 112 km long, and up to 40 m deep (Orme, 1973, 1975). Today, after 5000 years of sedimentation, segmentation, and reed-swamp encroachment aided by recent human activity, the lagoon averages only 312 km^2 in area, 40 km long, and less than 2 m deep. Within this context a rich and varied lagoonal ecosystem has developed in which plants and animals characteristic of tropical Africa blend with species from temperate southern Africa. The main plant communities are submerged aquatics, reed swamp, swamp forest, mangrove, and marsh grassland, any changes in which generate responses in the animal life by encouraging migration or disrupting the food chain. Of the aquatics *Potamogeton pusillus* occurs in the less saline areas, being replaced by *Ruppia maritima* and *Zostera capensis* nearer the estuary. Reed swamps are well developed in low-salinity environments, comprising *Phragmites australis* and *Cyperus papyrus*, together with the sedge *Scirpus littoralis* and the rush *Juncus krausii*. These swamps afford excellent habitat for crocodiles (*Crocodylus niloticus*), monitor lizards (*Varanus niloticus*), hippopotamus (*Hippotamus amphibius*) (Fig. 6), and many birds and fishes as well as lesser organisms. Ever-

FIGURE 5. The coastal barrier complex east of Lake St. Lucia, South Africa. The barrier here rises to 160 m above sea level and is clothed in luxuriant coastal dune scrub and forest. (Photo: A. R. Orme)

FIGURE 6. Hippopotamus, vigorous *Phragmites australis* reed swamp, and coastal dune forest, Lake St. Lucia, South Africa. (Photo: A. R. Orme)

FIGURE 7. *Phragmites australis* reed swamp reduced to low stubble by hypersaline conditions, Lake St. Lucia, South Africa. (Photo: A. R. Orme)

green swamp forests up to 25 m tall grow well along watercourses, comprising *Ficus hippopotamus, F. sycamorus, Syzygium cordatum, S. guineense,* Cussonia umbellifera, *Voacanga dregei,* and *Barringtonia racemosa.* This forest also provides excellent protective habitat for a variety of birds, fish, amphibians, and insects. Mud flats toward the estuary are initially colonized by the succulent herbs *Salicornia herbacea, Chenolea diffusa, Dimorphotheca fruticosa,* and *Hydrophylax carnosa,* which prepare the ground for the mangroves *Bruguiera gymnorrhiza* and *Avicennia marina.* The mangroves support a rich fauna of crabs, worms, and snails. In recent times, however, physical changes caused by accelerated sedimentation, dramatic salinity fluctuations, and periodic closures of the estuary by drifting sand have placed much stress on the lagoon's biota. For example, lagoon salinities have varied from 5‰ with heavy influxes of freshwater from inland storms to as much as 100‰ when the estuary has been closed and drought has prevailed. Under hypersaline conditions, the reed swamps die back and come to resemble the stubble of harvested grain fields (Fig. 7). Hippopotamuses are deprived of forage; crocodiles are deprived of their preferred nesting sites and may leave their eggs and hatchlings to be devoured by monitor lizards; the nests of goliath heron and weaver birds are exposed or collapse; and fish that are attracted to the rich food supply generated by the hippo's habit of defacating in the water desert the area, to be followed by predatory birds, such as heron, stork, egret, and spoonbill, that are attracted to the fish. Thus one physical change may generate an entire chain of biotic responses. Toward the estuary excessive sedimentation and human interference have decimated the mangroves and their fauna and have also caused the demise of the *Zostera* beds that afford shelter to young fish, prawns, and shrimps, food for many herbivores, and a habitat for diatom and algae growth. These varied life forms are food for small fish, which are preyed upon by larger carnivorous fish, predatory birds, and crocodiles (Fig. 8). Thus demise of the *Zostera* beds and the mangrove fauna may have serious implications for a wide range of animals. Periodic closure of the estuary by littoral drift inhibits the exchange of water and biota between the lagoon and the sea. Certain plankton, prawns, sharks (*Carcharhinus* spp.), and other fish are all adversely affected by prolonged estuarine closure, which may interrupt feeding and breeding patterns. When trapped within the lagoon and estuary sharks must compete with other carnivorous fish and crocodiles for available food resources. In general terms, the foregoing description of changes in Lake St. Lucia is not untypical of the stresses experienced in many lagoonal ecosystems around Africa, stresses that have been accentuated rather than alleviated by human activity in recent times.

FIGURE 8. A 3 m crocodile basking on a mangal mudflat, St. Lucia, South Africa. (Photo: A. R. Orme)

The Red Sea Coast

Climatically the areas bordering on the Red Sea (Fig. 3) are hot desert. Mean temperatures for the hottest months are 30-35°C, humidity is high along the coast, but rainfall is everywhere deficient, averaging from 3 mm/yr in the north to 150 mm/yr or more in the south, with significant variations from year to year. Further, connection between the waters of the Red Sea and the Indian Ocean is hindered by a shallow submarine sill in the Bab el Mandeb narrows. Accordingly the exchange of water is impaired, and the Red Sea has acquired several distinctive features: warm water, up to 30°C in August, and, with high evaporation rates and negligible terrestrial inflows, mean salinities of 37-43‰. The coast is lined almost continuously with jagged coral reefs between low tide and -100 m in depth. Three main ecosystems comprise the land on both sides of the Red Sea: littoral wetlands, coastal desert plains, and coastal hills, all of which pass from tropical conditions in the south to a more Mediterranean regime in the far north.

The littoral wetlands comprise mangal, reed-swamp, and salt-marsh vegetation types. Except along the Gulf of Suez coast, the mangal vegetation, dominated by *Avicennia marina*, is well developed all along the northern Red Sea Coast. Pure stands of this species vary in extent from limited patches of a few individuals to continuous dense growth several kilometers long, depending largely on shoreline morphology (Zahran, 1977). Three other mangroves, *Rhizophora mucronata*, *Ceriops tagal*, and *Bruguiera gymnorrhiza*, join *Avicennia marina* along the southern Red Sea Coast. The mangal vegetation grows typically in black, sandy muds, rich in organic matter and decaying debris. Reed-swamp vegetation, dominated by *Phragmites australis* and *Typha domingensis*, is found mainly along the creeks at the mouths of large wadis, such as Wadi el Ghweibba. Twenty community types of salt-marsh vegetation occur, variations in the ecological amplitudes of their dominant species causing differences in distribution, density, stratification, zonation, and floristic composition. In general terms, the *Halocnemon strobilaceum* community type, with *Limonium pruinosum* and *Nitraria retusa*, is abundant in the north from Suez to Hurghada. The *Zygophyllum album* community type is widespread along the Egyptian and Sudanese coasts, while the *Arthrocnemum glaucum* community type, with *Halopeplis perfoliata*, *Limonium axillare*, and *Suaeda monoica*, is dominant farther south. Consistent zonation within these salt-marsh communities is not readily apparent.

ANTONY R. ORME

References

Ayyad, M. A., 1973. Vegetation and environment of the western Mediterranean coastal land of Egypt—The habitat of sand dunes, *Jour. Ecology* **61**, 509-523.

Ayyad, M. A., and Ammar, M. Y., 1974. Vegetation and environment of the western Mediterranean coastal land of Egypt—The habitat of inland ridges, *Jour. Ecology* **62**, 439-456.

Battistini, R., and Richard-Vindard, G., 1972. *Biogeography and Ecology in Madagascar*. The Hague: W. Junk, 765p.

Bayer, A. W., 1938. An account of the plant ecology of the coastbelt and midlands of Zululand, *Ann. Natal Mus.* **8**, 371-454.

Birch, W. R., 1963. Observations on the littoral and coral vegetation of the Kenya coast, *Jour. Ecology* **51**, 603-615.

Chapman, V. J., ed., 1977. *Wet Coastal Ecosystems*. New York: Elsevier, 428p.

Grewe, F., 1941. Afrikanische Mangrovelandschaften, Verbreitung und wirtschaftsgeographische Bedeutung, *Wiss. Veroff. Deutsch. Mus. Landerkd.*, N.F. **9**, 105-107.

Kiener, A., 1972. Ecologie, biologie et possibilities de mise en valeur des mangroves malgaches, *Bull. Madagascar* **308**, 49-84.

Lawson, G. W., 1956. Rocky shore zonation on the Gold Coast, *Jour. Ecology* **44**, 153-170.

Lawson, G. W., 1966. The littoral ecology of west Africa, *Oceanogr. Mar. Biol. Ann. Rev.* **4**, 405-448.

Macnae, W., 1957. The ecology of the plants and animals in the intertidal regions of the Zwartkops Estuary area, Port Elizabeth, S. Africa, *Jour. Ecology* **45**, 113-131.

Macnae, W., 1963. Mangrove swamps in South Africa, *Jour. Ecology* **51**, 1-25.

Macnae, W., and Kalk, M., 1962. The ecology of mangrove swamps at Inhaca Island, Mozambique, *Jour. Ecology* **50**, 19-35.

Orme, A. R., 1973. Barrier and lagoon systems along the Zululand coast, South Africa, in D. R. Coates, ed., *Coastal Geomorphology*. Binghamton: State University of New York, 181-217.

Orme, A. R., 1975. Ecologic stress in a subtropical coastal lagoon: Lake St. Lucia, Zululand, in H. J. Walker, ed., *Geoscience and Man*. Baton Rouge: Louisiana State University, 9-22.

Sauer, J. D., 1965. Notes on the seashore vegetation of Kenya, *Ann. Missouri Bot. Gard.* **52**, 438-443.

Stephenson, T. A., and Stephenson, A., 1972. *Life between Tidemarks on Rocky Shores*. San Francisco: Freeman, 425p.

Walter, H., and Steiner, M., 1936. Die Okologie der Ostafrikanischen Mangroven, *Zeitschr. Bot.* **30**, 65-193.

Zahran, M. A., 1977. Africa—A. Wet formations of the African Red Sea Coast, in V. J. Chapman, ed., *Wet Coastal Ecosystems*. New York: Elsevier, 215-231.

Cross-references: *Africa, Coastal Morphology; Asia, Middle East, Coastal Ecology; Australia, Coastal Ecology; Biotic Zonation; Climate, Coastal; Coastal Fauna; Coastal Flora; Europe, Coastal Ecology; India, Coastal Ecology; Mangrove Coasts; Mineral Deposits, Mining of.*

AFRICA, COASTAL MORPHOLOGY

The African continent measures nearly $30 \times 10^6 \text{ km}^2$ and its relatively unbroken coastline is 30,000 km long, compared with the 70,000 km coast of Asia, which is only 1.5 times the area of Africa. Over long distances the African coast is unbroken by sizable inlets, and its major river mouths, except the Congo, are either deltaic or blocked by sand bars. This paucity of approachable and protected natural harbors long hindered the development of the coastal zone and its hinterland. Smoothness in plan is also accompanied by smoothness of profile. Lowland coasts with long, sandy beaches predominate, rugged, mountainous coasts are rare. Offshore Africa's continental shelf covers only $1.28 \times 10^6 \text{ km}^2$, compared with $9.38 \times 10^6 \text{ km}^2$ for Asia and $6.74 \times 10^6 \text{ km}^2$ for North America. Explanations for this morphology are to be found in the geologic history of the coast since the rupture of Gondwanaland during Mesozoic times and in the geomorphic processes that have fashioned the coast, especially during the Quaternary Period.

Coastal Origins and Geologic History

Africa's coastal margins were initially blocked out by the rupture of Gondwanaland and further modified during the subsequent opening of the Atlantic and Indian oceans and the gradual closure of the Tethys Sea around the African plate. The massive slope at the outer edge of Africa's narrow continental shelf marks the approximate margins of the African plate. The continental shelf averages only 25 km wide, its broadest part being the 240-km-wide Agulhas Bank. Beyond the continental slope lie the deep ocean basins, which, beneath a thin cover of mainly Cretaceous and Cenozoic sediments, are floored by volcanic rocks that emerged from the accreting plate margins along the midoceanic ridges (Fig. 1). Geomagnetic lineations and radiometric data from ocean-floor volcanics and coastal rocks show that tensional stresses began affecting the African plate before the close of Paleozoic times, that the subsequent rupture of Gondwanaland did not affect all sections of the coast simultaneously, and that crustal accretion through sea-floor spreading was not achieved at a constant rate. Most expressions of plate tectonic theory assume that over the past 100 m.y. the African plate has remained relatively fixed with respect to the rotational axes of other fragments of Gondwanaland. This should not, however, conceal the facts that the African and Eurasian plates have moved toward one another, that severe crustal warping has locally affected the coastal zone, and that further dislocation of the African plate is now occurring along the east African rift system. Also, compared with the relatively simple opening of the South Atlantic Ocean, the origins and subsequent behavior of the North Atlantic and western Indian oceans continue to pose problems.

The coast of northwest Africa was crudely outlined when North America began separating from Africa in late Triassic or early Jurassic times, 180 m.y. ago or earlier. The subsequent history of this coastal margin was affected by episodic variations in the rate of sea-floor spreading along the Mid-Atlantic Ridge and by variable rates of sediment accumulation on the continental shelf and coastal basins, both of which produced variable epeirogenic seaward subsidence of the coastal margin (Rona, 1973).

The west coast of Africa from Liberia southward was initiated as a tensional rift with the separation of South America from Africa, which, although it may have begun earlier, became important after late Jurassic times, 140 m.y. ago and later (Larson and Ladd, 1973). For instance, the Liberian coast near Cape Palmas formed in response to faulting that attended the formation of three oceanic fracture zones about 140 m.y. ago (Le Pinchon and Hayes, 1971; Behrendt et al., 1974); since then as much as 8000 m of low-density sedimentary rocks have blanketed the shelf and slope. To seaward, between the continental slope and the Mid-Atlantic Ridge, the deep Guinea and Cape basins developed during Cretaceous times to north and south, respectively, of the Walvis Ridge. These features have since continued to influence deep-ocean current systems and sedimentation patterns in the southeast Atlantic (Sclater and McKenzie, 1973). South of the Walvis Ridge the west coast consists of downfaulted continental basement blocks, aligned NNW and overlain by a prograded sediment wedge created by debris brought down by the Orange River and lesser streams that have been eroding the western Great Escarpment since it was first tilted upward in early Cretaceous times (Dingle and Scrutton, 1974).

At the southern tip of Africa plate motion produced a large transform or offset fault that sheared off the Falkland Plateau (formerly south of Durban but subsequently part of the South American plate) from the Outeniqua and other depositional basins at the base of the Hercynian Cape Ranges in late Jurassic to early Cretaceous times, 125-140 m.y. ago. Since then, and notably since the uplift of the southern Great Escarpment in mid-Cretaceous times, terrigenous debris has crossed the narrow continental shelf to blanket the continental slope and neighboring oceanic basins. Farther out the Agulhas Plateau was probably created by supranormal marine volcanism in the fracture zone

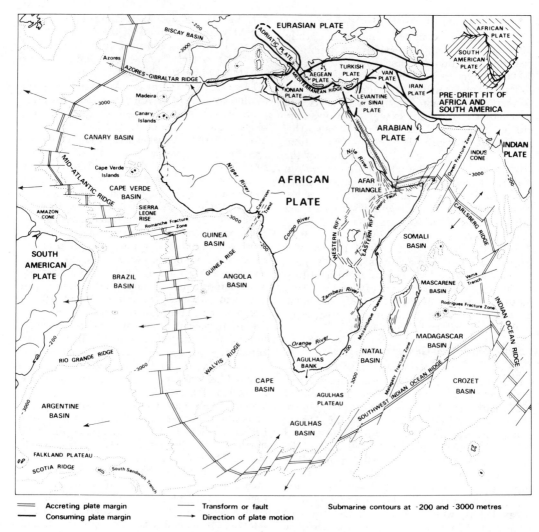

FIGURE 1. Tectonic relations of the African plate.

vacated by the westward-moving Falkland Plateau (Barrett, 1977).

Following Permian tensions and a Triassic saltwater incursion much of Africa's east coast may have been blocked out by early Jurassic times, about 180 m.y. ago (Kent, 1972). In Mozambique, the earliest rift produced a north-south-trending series of large tension-faulted horsts and grabens that were later buried beneath detritus from the Limpopo and Zambezi river systems (Dingle and Scrutton, 1974). The north-south sections of coast south of the Zambezi delta and Limpopo mouth, together with the narrow continental shelf, run parallel with major faults of this graben system. In contrast, the ENE-trending coasts east of these two river mouths reflect right-lateral offsets of the basement rocks, bringing Precambrian rocks close to the coast in northern Mozambique. Offshore more than 6000 m of Cretaceous and Cenozoic sediments, mostly continental, have since accumulated in north-south trending grabens and the massive Zambezi cone.

Madagascar is a large block of continental crust that was intermittently uplifted as a complex horst between two subsiding depressions within the system of NNE-trending fractures of the western Indian Ocean floor. The island's east coast and narrow continental shelf are defined by one such massive fault. Between the east coast fault and a similarly trending continuation of the Malagasy fracture zone lies a 5000-m-deep graben. Madagascar's place in the disruption of Gondwanaland remains elusive. Some authorities believe that it drifted to its present location from a former position off the coast of Kenya and Tanzania. Others believe that it occupies essentially its original place. It has even been suggested, contrary to the simple sea-floor spreading origins for the western Indian Ocean, that Madagascar, India, and the Seychelles formed parts of a Malagasy-Mascarene subcontinent that disintegrated through subsidence, uplift, and lateral motion, with the subsiding portions forming the abyssal deeps of the western Indian Ocean (Kutina, 1975).

Based on the above observations, the Atlantic and Indian Ocean margins of Africa can be considered *trailing-edge coasts* inasmuch as the African plate has moved, relatively, away from the neighboring midoceanic spreading centers (Inman and Nordstrom, 1971). In keeping with this definition, gross coastal form is thus largely dependent on local geology and the nature of erosion before the rupture of Gondwanaland created the coastline. In similar vein, Mitchell and Reading (1969) have defined Africa's margins as Atlantic type on whose continental shelf shallow-water clastic sediments and carbonate reefs typically accumulate. In embryo form, the Red Sea and Gulf of Aden also have trailing-edge coasts in that the Arabian plate and the Horn of Africa, at least over the past 30 m.y., have been diverging from one another and collectively separating from the rest of Africa along extensions of the East African rift system.

In contrast the Mediterranean margin is largely a *collision coast* as Africa and Eurasia have pressed against one another since Jurassic times, swallowing much of the intervening Tethys Sea. The effects of this collision are best seen along the coast of the Alpine zone between Agadir and Gabes. Farther east the situation is less clear, notably because the tectonic history of the Mediterranean region embraces both compressional and extensional interactions between several microplates, activities that continue to yield frequent earthquake and volcanic events (Dewey et al., 1973). In the Aegean subduction zone, for example, the African plate has been computed moving under the Aegean microplate at a rate of 2.7 cm/yr.

Although events during and since the rupture of Gondwanaland most readily explain the broad configurations of Africa's present coastline, an understanding of the more detailed morphology of the coast requires an appreciation of the rocks and other materials reaching the shore. This implies an overview of the continent's history of sedimentation, tectonism, volcanism, and erosion both before and after the breakup of Gondwanaland.

In essence the African continent comprises a widely exposed basement of Cryptozoic or Precambrian rocks, younger mountain chains of limited extent along its northwest and south margins, and widespread platform covers of Phanerozoic or post-Precambrian rocks often tabular but locally folded (Chowlert and Faure-Muret, 1968). Volcanism and salt domes are locally important within the coastal zone, as are major fault zones, the most important of which, the 7000-km-long system embracing the East African Rift Valleys, reaches the shores of the Red Sea and Mozambique Channel.

Precambrian rocks outcrop over 57% of Africa's surface, reflecting several major sedimentary cycles whose deposits were intensely folded and fractured, highly metamorphosed, and often granitized during at least eight orogenic episodes. The oldest of these orogenies occurred over 3500 m.y. ago, whereas the youngest straddled the Precambrian-Cambrian boundary some 570 m.y. ago. Rejuvenation of earlier structures frequently occurred during later Precambrian orogenies and continued episodically during the Phanerozoic, notably during the Hercynian orogeny. Among the wide variety of Precambrian rocks gneiss, schist, quartzite, and migmatite are particularly important, but postorogenic molasse deposits and tabular to strongly folded platform covers of sandstone, limestone, dolomite, and tillite also occur. The development of less-disturbed platform covers and strong volcanism are typical of the later Precambrian. Today these basement rocks reach the coastal zone in the Anti-Atlas of Morocco, along the Guinea coast between Monrovia and Accra, at intervals along the coast of Angola, Namibia, and western Cape Province, and emerge from beneath narrow ribbons of later cover rocks in Mozambique, eastern Madagascar, and along both sides of the Red Sea trough.

Phanerozoic rocks occur mostly as tabular platform covers occupying large basins between swells in the basement complex, but folded covers and younger mountain chains are locally important. In Africa as a whole the great extent of continental deposits and the paucity of marine sediments are noteworthy. Continental rocks include: the *Nubian Sandstones* (Cambrian to Middle Cretaceous) near the Red Sea; the 7000-m-thick Karroo System (Carboniferous to Triassic), whose tillites, shales, sandstones, and basalts locally reach the Indian Ocean coastal zone south of the equator; and thick Cenozoic sandstones within the more arid coastal regions (Furon and Lombard, 1963). For the past 250 m.y. Africa has lain mostly above sea level.

Within the coastal zone, however, marine deposits are generally more important than continental rocks, because, following the rupture of Gondwanaland, the continent's new margins became a series of coastal basins whose dimensions varied in response to tectonism and sedimentation. Further, a remarkable marine transgression of Cenomanian (Upper Cretaceous) age spread across the Sahara to link the proto-Mediterranean with the Gulf of Guinea. After a Danian regression lower Eocene seas again flooded the Sahara region and a major gulf persisted into Miocene times in Libya and Egypt. Since Miocene times marine transgressions have been confined strictly to the immediate littoral zone and the continental shelf, where flights of Quaternary shorelines are locally important above and below present sea level.

There are about a dozen major coastal basins around Africa, all filled to a greater or lesser extent with Cretaceous and Cenozoic deposits. Moving clockwise around Africa from the Red Sea, the first of these is the large Somali Basin, which is over 7000 m deep at the coast. This passes southward into the Kenya Basin, which is over 8000 m deep and in turn merges with the narrower 3000-m-deep Dar-es-Salaam Basin, whose landward margin is partly fault controlled. The 4000-m-deep Mozambique Basin is intensely broken by north-south faults at the southern end of the East African Rift system. Across the Mozambique Channel the 7000-m-thick cover in the Madagascar basin is also severely faulted. Along the west coast of Africa a more or less continuous embayment is represented by the 4000-m-deep Luanda Basin, the 3000-m-deep Cabinda Basin, and the 8000-m-deep Gabon Basin. Both the Luanda (Cuanza) and Gabon basins are folded and characterized by many Cretaceous salt domes. The Niger Basin, the seaward extension of the Benue graben, may contain up to 10,000 m of Cretaceous and Cenozoic deposits beneath the actual delta but becomes considerably more shallow at its eastern and western margins in Cameroon and Dahomey, respectively. The Ivory Coast Basin is narrow but faulted to depths of 4000 m. The Senegal Basin reaches depths of 7000 m along the coast. The Tarfaya (Aaiun) Basin farther north, bounded inland by the great Zemmour Fault, may contain 10,000 m of Mesozoic and Cenozoic deposits at Cape Juby. The Central Saharan Basin reaches the Mediterranean coast as the 5000-m-deep Tripolitanian Basin, which is separated by the Garian horst behind Tripoli and by NNW-SSE faults near Sirte from the extensive Libyan Basin of Cyrenaica and Egypt. This basin also opens toward the Mediterranean and is over 8000 m deep east of Benghazi and 7000 m deep west of Alexandria.

Both the Hercynian (320–280 m.y. ago) and Alpine (30–10 m.y. ago) orogenies intrude locally into the coastal zone, the former in the Cape Ranges of southern Africa and the Anti-Atlas of Morocco, the latter along the coasts of Morocco, Algeria, and Tunisia between Agadir and Gabes. Both orogenies introduce rugged mountainous terrain to the coast, in contrast to the subdued landforms of the basement and platform cover rocks elsewhere along the African coast. However, lacking metamorphism and wide granitization, the Anti-Atlas and Cape ranges are more correctly classified as folded cover rocks than as true Hercynian chains. The main Alpine chains of Morocco and Algeria form mountainous arcs, such as the Rif, opening toward the Mediterranean. Farther south, the High Atlas, Saharan Atlas, and Tunisian Atlas comprise folded forelands flanked by massive molasse deposits in late- and post-tectonic basins.

Coastal Processes

Except for the Mediterranean and northwest coasts, where northwesterly swells predominate, notably in winter, the African coast is influenced by mostly southerly swells generated by storms in the Southern Ocean (Fig. 2). From Cape Vert to Cape Agulhas these swells are mainly southwesterly, the waves decreasing in height from south to north and from southern winter to southern summer. Along the Guinea coast these swells are reinforced by southwesterly winds that flow onshore, especially during the northern summer. Because of the orientation of Africa's west coast, the southwesterly swells produce mainly east-flowing longshore currents in the Gulf of Guinea and mainly north-flowing longshore currents south of Cameroon. This nearshore circulation is reinforced offshore by the Guinea and Benguela currents. Exceptions to this pattern occur, for example, in the lee of major headlands and in the currents that diverge from the nose of the Niger Delta. All along the Atlantic coast onshore winds are stronger and more frequent during the season when swells are highest—from December to March north of Cape Vert, and from June to September farther south (Davies, 1977).

From Cape Agulhas north to the Horn of Africa most swells also originate from the southwest but are strongly refracted to reach the coast from the southeast. The circulation pattern is further complicated by the northeasterly monsoon of the northern winter, by westward-moving cyclones and tropical easterlies off Madagascar and east Africa, and by south-flowing ocean currents off southeast Africa. Thus along the Natal coast, for ex-

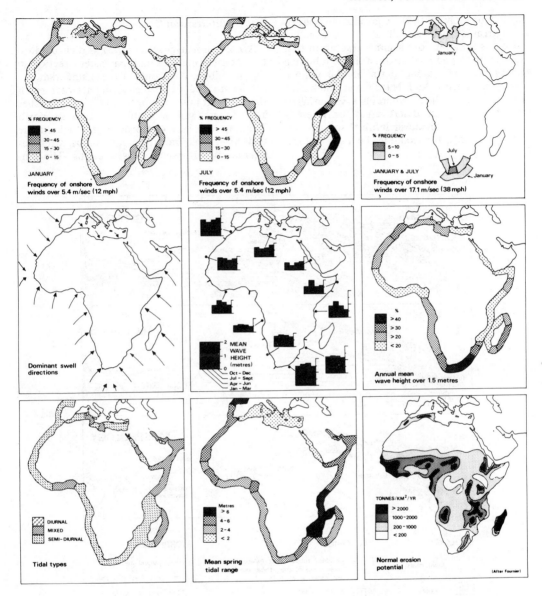

FIGURE 2. Selected processes affecting the African coast (in part after Hayden et al., 1973; Davies, 1977).

ample, the littoral drift of coarser terrigenous sediment is often northward within the surf zone, but the finer sediment that is flushed farther seaward is carried south with the Agulhas Current (Orme, 1973). Onshore winds are stronger and more frequent during the southern winter, when winds in excess of 17 m/s (~60/kph, Beaufort Force 8 and above) may drive strong surf against the continent's southern tip.

Using mean wave height as a surrogate for coastal energy, the highest energy is found at the southern tip, from Cape Point to Durnford Point, and moderately high energy prevails from the Orange to the Limpopo River mouths and along the east coast of Madagascar as well as along the Atlantic coast of Morocco. Relatively low energy conditions occur in the Red Sea, on both sides of the Mozambique Channel, and in sheltered sections of Angola's coast. Semidiurnal tides predominate around Africa, with mixed regimes confined to parts of the Mediterranean, Red Sea, Guinea, Somali, and eastern Madagascar coasts. Mean spring tidal range is less than 2 m throughout the Red Sea and Mediterranean, over 6 m along the Mozambique Channel and north to Kenya, and 2–6 m elsewhere.

Sediment delivery to the coastal zone is greatest in areas of high rainfall, at least seasonally, and of open forest or savanna vegetation that has been modified or destroyed through human activity, notably along parts of the Guinea coast, east Africa, and Madagascar. Sediment yields are least either where heavy rainfall is lacking, notably in desert areas, or where a luxuriant vegetation cover inhibits surface erosion and runoff, for example, beneath tropical forest.

The Mediterranean Coast

Africa's north coast (Fig. 3) is divisible into two contrasting sections: an eastern section in Egypt, Libya, and south Tunisia underlain by little-disturbed Cenozoic sediments that reach the shore in low cliffs or behind barrier beaches and dunes; and a western section in north Tunisia, Algeria, and Morocco in which Alpine tectonism has led to mountainous stretches alternating with small coastal basins. The hin-

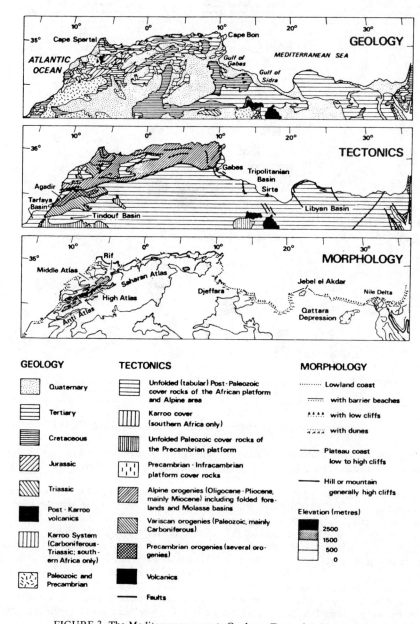

FIGURE 3. The Mediterranean coast: Geology, Tectonics, Morphology.

terlands to each section range from semiarid to true desert so that, with the notable exception of the Nile delta, perennial stream discharge and sediment delivery are largely lacking. Seasonal wadis may introduce sediment, particularly in the Alpine zone. Littoral drift is predominantly eastward.

The 20,000-km^2 Nile delta forms the outlet for a 3-million-km^2 drainage basin whose principal rivers, the 3900-km-long White Nile and the shorter Blue Nile, drain from Lake Victoria and the Ethiopian Highlands, respectively. Although annual precipitation over the vast watershed averages 870 mm/yr, high evaporation rates, locally extensive irrigation, and several major dams, including the High Dam at Aswan, which impounds Lake Nasser, collectively siphon off much of the potential stream flow. Accordingly discharge through the delta averages only 2800 m^3/sec while sediment outflow is correspondingly reduced and, aided by alongshore currents that average 28–40 cm/sec^{-2}, the seacoast of the delta is suffering locally severe erosion (Sharaf el Din, 1974). In early historic times the Nile delta contained six or seven distributaries, but, owing to reduced flows, sedimentation, and human activity, only the Damietta and Rosetta branches, supplemented by a network of irrigation and drainage canals, now function. In physical and human terms, a distinctive littoral zone—the Barari (= barren)—may be distinguished from the rest of the delta. The Barari averages 40 km in width along the 280-km delta coast and comprises marshes, salt flats, and shallow lagoons behind a series of discontinuous barrier beaches. The largest lagoons are Manzala (1450 km^2) between Port Said and Damietta, Burullus (560 km^2), Idku (140 km^2), and Maryut (200 km^2) south of Alexandria. The latter lagoon lies within one of several delta areas that, because of continuing sedimentation, have subsided to below sea level. Elsewhere sedimentation has, until very recently at least, more or less kept pace with subsidence, so that the location of the coastline has not changed significantly during historic times. A cone of faulted later Cenozoic sediments forms a massive submarine extension to the Nile delta.

Eastward from the delta clay-rich and organic sediments are replaced by coarser desert materials, and the 200-km-long coast of Sinai and Gaza comprises a wide, sandy plain with prominent dunes and a major barrier beach fronting Bardawil Lagoon. The ephemeral Wadi el Arish is the only sizable drainage outlet from the limestone terrain of the interior to reach the coast. Westward from the delta a series of parallel sandy ridges, sand dunes fashioned by northwest winds, and saline and nonsaline depressions characterize the coast. Dark Nile sediments decrease in importance, while carbonate grains increase. West of Alexandria the most seaward sandy ridge is 15 m high and 400 m wide, reflecting, like the parallel ridges inland, episodic barrier sedimentation during later Quaternary times.

Between Darnah and Benghazi in Cyrenaica, eastern Libya, the Jebel el Akdar limestone plateau generates higher rainfall and greater surface runoff and reaches the coast in rocky cliffs and narrow terraces that leave little room for coastal plain. South and west from Benghazi, however, the Gulf of Sidra (Sirte) fronts a lowlying barrier-lagoon coast backed by salt flats, sand dunes, and desert. The largest lagoon extends 100 km south from Misratah. Except around Al Khums (Homs), the Tripolitanian coast farther west is similar and extends across the Djeffara agricultural plain into southern Tunisia. There the 150-km^2 Sebkha el Melah affords an example of Holocene evaporitic lagoonal sedimentation in which marine detrital limestones, euxinic beds, magnesian carbonates, gypsum, polyhalite, and halite have successively accumulated over the past 6000 years (Busson and Perthuisot, 1977).

In the Alpine zone structure, neotectonism, eustatic fluctuations, climatic change, and human activity have combined to produce a more complex coast. Neotectonic deformation makes it difficult to correlate, other than locally, between the flights of marine terraces that occur at intervals. In Tunisia, for example, neotectonism is expressed in the subsidence of 400 m of Quaternary littoral sediments beneath the Medjerda delta, in the subsiding shallow Gulf of Gabes (whose 2-m tidal range is greater than anywhere else in the Mediterranean), and in the distinct warping of the *classical* Quaternary shorelines at Monastir (Coque and Jauzein, 1965) and farther north (Crosse, 1969). Marine terraces rise to 200 m above sea level near Bizerte. Farther west, near Algiers, the Calabrian (early Quaternary) marine terrace is warped from over 200 m to below sea level near Tipasa and extends below the coastal plain in the Mitidja depression. Later terraces, including three eolianites, are also somewhat deformed (Cabot and Prénant, 1968). The Tell, the subhumid coastal zone of northern Algeria, reaches the Mediterranean in rugged cliffs, for example, below the Dahra massif, alternating with small, often marshy plains.

The Atlantic Coast

The northwest coast of Africa (Fig. 4) between Cape Spartel and Cape Vert is divided by Cape Dra into two sectors. Northward the Mo-

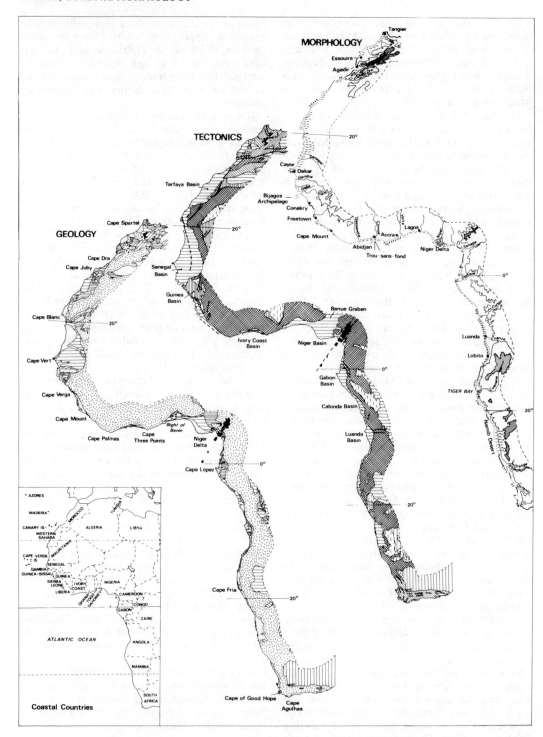

FIGURE 4. The Atlantic coast: Geology, Tectonics, Morphology (see Fig. 3 for key).

roccan coast is framed by mountainous ribs provided by Hercynian and Alpine tectonism, between which nestle coastal basins containing post-Paleozoic sediments. Despite its varied structural framework, the coastal plain between Tangier and Essaouira is uninterrupted and fronted by dunes and marshes. Fossiliferous Quaternary marine terraces rise in flights to over 100 m, as at the seaward end of the Anti-Atlas, and pass inland into well-developed river terraces. Winter rains yield significant runoff and the rivers Sebou, Grou, and Oum er Rbia in the north and the less reliable Sous and Dra in the south are the main arteries of sediment supply to the coast. Strong onshore winds have produced a series of Pliocene and Quaternary dunes, notably near Essaouira and Larache.

South of Cape Dra the coast of southern Morocco and Mauritania comprises the Tarfaya and Senegal structural basins with their great thicknesses of Cretaceous and Cenozoic sediments. The Sahara reaches the coast along a broad front, but desert conditions are ameliorated by moist oceanic air and frequent fogs attributable to the stable atmospheric situation induced by the cool Canaries Current. Lacking both rainfall and a mountain backdrop, water and sediment discharge are negligible except during rare storms, and eolian processes have long dominated, driven by the *alisio*, a persistent onshore wind from the north, and by the occasional *irifi*, a hot easterly wind from the Sahara. Extensive Pliocene and Quaternary dune systems occur, the NE-SW-oriented red dunes of Mauritania extending several hundred kilometers inland. The Cayor coast of Senegal, between the Senegal River and Cape Vert, is characterized by 30-m dunes and interdune swamps and salt flats. Upper Quaternary eolianites occur to at least −50 m off Cape Vert, while the Senegal River mouth has a history of periodic blockage by drifting sand. Human interference in the Cape Vert peninsula has led to rapid dune migration inland (Sall, 1973). Flights of marine terraces up to 100 m are prominent both on the mainland coast and on the off-lying volcanic Canary Islands (Lecointre, Tinkler, and Richards, 1967).

Offshore the continental margin developed by upbuilding and outbuilding on a basement that subsided at least through Eocene times and may still be subsiding along parts of the Moroccan sector. The contrast in runoff regimes and sediment delivery between the northern and southern sectors is reflected in shelf sedimentation. Off the Sahara carbonate and biogenic particles predominate, eolian dust is important near the shore, but terrigenous material is limited at present although *relict* muds off former watercourses reflect Quaternary pluvial conditions.

Off central and northern Morocco and off the perennial Senegal River in the south, terrigenous material is more important. However, lacking much contemporary influx, relict Pleistocene and older sediments characterize much of the shelf and include detrital phosphatic placers derived from the erosion of onshore and offshore phosphorites. The smooth but narrow continental shelf, whose outer edge is only 110–150 m deep, probably owes much of its form to Pleistocene marine abrasion (Tooms, Summerhayes, and McMaster, 1971; Summerhayes et al., 1976).

Along the coast from Cape Vert to the Niger delta hard Precambrian and Paleozoic rocks alternate with post-Paleozoic basinal deposits. Paleozoic rocks emerge from beneath the Senegal Basin along the south coast of Guinea Bissau and at Cape Verga, Guinea, while basic intrusives form the Conakry and Freetown peninsulas. Precambrian granite and gneiss underlie the low, partly swampy, partly cliffed Liberian coast and outcrop as far east as the Sassandra River, Ivory Coast. Coastal morphology is strongly influenced by faults in these rocks. Elsewhere along the coast of Guinea and Sierra Leone mainly Pliocene and Quaternary sediments occur and coastal swamps alternate with tidal creeks. Along this southwest-facing coast tropical rainfall of 2000–3000 mm/yr is accentuated by onshore southwesterly winds and reflected in abundant discharge from the many short coastal rivers. This abundant discharge helps to explain the open creeks and estuaries, the considerable flow of terrigenous sediment through coastal swamps onto the continental shelf, and the paucity of barrier-lagoon systems in this area. Barrier beaches do, however, front the coastal swamps and lagoons in southern Sierra Leone. Offshore the continental shelf ranges from 200 km wide off the Bijagos Archipelago, north of which the Great Geba flat is less than 10 m deep but extends 60 km offshore, to less than 20 km off Cape Palmas. This shelf exhibits many drowned river channels, submarine canyons, active and submerged deltas, barrier spits, barrier-lagoon complexes, and sea cliffs, with a coral 20,750 years old being dredged from a shelf-edge sea cliff. Late Quaternary shorelines occur at −90, −80, −55, −45, −35, and −25 m (McMaster et al., 1971). The surface Canary Current and strong ebbing tidal currents transport fine sediment southeastward over this shelf, while a vigorous bottom countercurrent sweeps material northward (McMaster et al., 1977).

Along the south-facing coast of the Gulf of Guinea, Precambrian granites and schists and Paleozoic sandstones provide a low-cliffed coast from Axim near Cape Three Points nearly to the Volta River, which separates two major sedimentary basins. To the west the subsiding and downfaulted Ivory Coast Basin and to the east the larger Niger Basin are both underlain by great thicknesses of Cretaceous and Cenozoic sediments and characterized at the coast by well-developed barriers and lagoons or, with infilling, by swamp flats. Because of the onshore southwesterly waves and the Guinea Current, a strong eastward littoral drift predominates. Where this drift is partially interrupted, for example, by the moles protecting the Lagos Harbour entrance, downdrift erosion of beach barriers is significant (Usoroh, 1971). The volume of littoral drift is affected by coastal morphology. Immediately east of Cape Palmas, it is comparatively small (200,000 m^3/yr at San Pedro), but it increases eastward (800,000 km^3/yr from Fresco to Port Bouet) and then decreases toward Cape Three Points (Varlet, 1958). There the sequence begins again; there is little transport at Takoradi, more at Sekondi, and a large amount from Accra to Lagos (Martin, 1971). Offshore the continental shelf is relatively narrow (10-70 km), smooth, and featureless. However, noteworthy exceptions are: the remarkable Trou-sans-Fond, a massive submarine canyon that heads at -10m off Abidjan but descends to -2000 m and traps 400,000 m^3/yr of littoral drift; the large Volta River submarine delta; and a series of beach rock and eolianite ridges parallel with the coast that indicate a late Quaternary regression to the outer shelf edge at -110 m. Flandrian deposition of fine terrigenous sediment, though generally less than 20 m thick, has adversely affected existing total faunas off the mouths of larger rivers. Although deeply weathered rocks abound onshore, the luxuriant vegetation cover inhibits the movement of much coarser debris to the coast.

The Niger delta, lying at the seaward end of the Benue Graben, occupies a basin created by the mid-Cretaceous rupture of Gondwanaland. Best-fit reconstructions of Gondwanaland require an overlap of 250 km along the African and South American plate margins that is explained by assuming that the delta postdates the rupture and overlies oceanic crust. This is confirmed by sedimentologic evidence. After a number of Cretaceous and Paleocene marine transgressions, the present delta began forming in Eocene times when the shoreline lay 250 km inland from the modern coast near Onitsha. Later Cenozoic shorelines reflect an intermittent outbuilding, upbuilding, and subsidence of the delta separated by periodic marine transgressions. The late Quaternary shoreline lay 50 km offshore, but the subsequent Flandrian transgression drowned the lower parts of the delta's distributary system before establishing the present coast. The present delta is thought to be in near-isotsatic equilibrium.

The Niger delta is the outlet for a drainage basin covering 1.11×10^6 km^2 whose main artery, the 4460-km Niger River, rises in the Guinea mountains less that 300 km from the Atlantic. This basin acquired its present dimensions only during a later Quaternary pluvial stage when the NE-flowing internal drainage of the Soudanese Niger overflowed near Bourem, Mali, into the SE-flowing Nigerian Niger. The delta covers 28,827 km^2, of which 19,135 km^2 lie above sea level. Approximately 500,000 km^3 of Eocene-Holocene sediment fill the delta to a maximum thickness in excess of 8 km (Hospers, 1971). These sediments are broken by arcuate gravity faults, concave and downthrown toward the gulf. Present yearly discharge of fresh water into the delta is about 200×10^9 m^3 and of sediment about 18×10^6 m^3, including 17×10^6 m^3 of silt and clay but with much sand, especially from the Benue River. Water and sediment are then discharged radially but unequally through eleven major distributaries into the gulf. Depositional environments and corresponding sedimentary facies are distributed concentrically within the delta (Allen, 1970). The deltaic plain comprises upper and lower flood plains, tidal swamps, and barrier beaches. The submerged delta comprises river-mouth bars, delta-front platform, prodelta slope, open shelf, and relict late Quaternary littoral deposits. The prevailing southwesterly winds and waves of the Gulf of Guinea strike symmetrically on the nose of the delta, causing divergent littoral drifts that meet opposing drifts near Lagos and Fernando Poo. Submarine canyons channel about 1 million m^3 of sediment a year from each of these drifts to feed submarine fans on either side of the delta foot.

From the cliffed volcanics of Mount Cameroon (4070 m) south to the Olifants River the west coast of Africa comprises several narrow coastal basins containing Cretaceous and Cenozoic deposits alternating with rocky basement outcrops veneered with Quaternary eolian and marine sediments. Powerful southerly swells, combined in the south with the Benguela Current, promote a predominantly northward littoral drift, reflected in the barriers and spits fronting the basins. The Niger Basin extends through the Cameroon volcanics into the swampy and deltaic Sanaga coastal plain. The Gabon Basin is fronted by several barrier-lagoon complexes and north-oriented spits terminating at Cape Lopez. In contrast, the Cabinda Basin has active sea cliffs north and south of the Congo estuary. South of basement outcrops near Ambriz, the Luanda (Cuanza) Basin is fronted in part by 100-150-m sea cliffs cut in Cretaceous and Cenozoic sediments and in part by *restingas*, northward-oriented spits enclosing bays and lagoons, notably at Luanda and Lobito (Monteiro Marques, 1966). Tiger Bay, north of the Cunene estuary, is enclosed by a restinga 37 km long (Guilcher et al., 1974).

Basement rocks underlie most of the coast of southern Angola, Namibia, and Namaqualand and locally form rugged cliffs. Usually, however, their surface expression is masked by massive Quaternary dune fields, for example, near Mossamedes and in the Namib Desert, where NW-SE-trending dune ridges reach 300 m and extend 30-130 km inland over flights of Quaternary marine terraces. The coastal dunes now effectively bar most ephemeral streams from the coast, but Pleistocene pluvial conditions formerly allowed more runoff and sediment delivery (Kouyoumontzakis and Giresse, 1976). The perennial Cunene and Orange rivers, draining well-watered uplands, are noteworthy exceptions that continue to deliver abundant sediment to the coast. Indeed, it is Orange River debris that has supplied sand to beaches that have in turn fed the Namib coastal dunes to the north. The Cunene River has likewise supplied the sands that feed the dunes of southern Angola. Of special note among the flights of Quaternary shorelines are the diamantiferous marine terraces that range from at least 90 m above to -55 m below sea level along the coast from Kaokoveld, through the Sperrgebied, to Namaqualand (Maree, 1966; Carrington and Kensley, 1969). Tectonism may explain Holocene marine deposits rising to 5 m near Lobito.

Offshore the continental shelf varies from less than 30 km wide off rocky basement coasts, for example, north of Cape Fria, to over 160 km wide off the Orange River, where terrigenous sediment has formed a large submarine delta. South of Cape Fria and the Walvis Ridge two shelves separated by a 50-m submarine cliff occur. Holocene sediments on the shelf comprise terrigenous and shelly sands in the coastal areas, except off the almost rainless Namib Desert, where diatomaceous, organic-rich muds and foraminiferal and skeletal sands occur on the outer shelf. Farther north the Congo River extends through the shelf as a V-shaped submarine canyon with 400-m-high sideslopes to a depth of -2700 m, from which a fan valley with leveed distributaries descends farther to -4900 m. This system was probably produced through erosion and deposition by turbidity currents (Shepard and Emery, 1973). Tidal current scour probably limits the amount of fill in the canyon head.

The Southern Tip

From the Olifants River to the Tugela River strong southerly swells have fashioned a rugged, often cliffed, coast from the Precambrian and Lower Paleozoic rocks of the Cape Ranges and the thick formations of the Karoo System. Small coastal lowlands fronted by sandy barriers (Fig. 5) occur locally, for example, in the Alexandria Basin and similar depressions marginal to the Cape Ranges, but elsewhere mountain ribs and coastal plateaus reach the coast in majestic cliffs. Quaternary marine terraces have been identified to over 60 m, but, although locally important north of Cape Town, they are often poorly preserved. Quaternary dunes also occur onshore, notably in the Cape Flats, but are better expressed as submerged eolianite reefs offshore. Perennial and intermittent rivers introduce much sediment, notably from May to August (winter) rains in Cape Province and after December to March (summer) thunderstorms in Natal. Relict Pleistocene and recent Holocene beach rock forms a broken pavement along much of this coast and extends northward into Mozambique (Siesser, 1974). The beach rock consists mainly of quartz grains and skeletal fragments cemented by micrite or aragonite overlain by a laminated calcrete. Radiometric data indicate that some beach rock along the present shore formed about 25,000 years ago in an intertidal environment during an interstadial high sea level of the last glacial stage, whereas other beach rock has formed within the past 1000 years (Figs. 6 and 7).

Off Cape Agulhas, the continental shelf forms the 240-km-wide Agulhas Bank, whose steep margins toward the Atlantic are cut by several submarine canyons. The bank comprises over 6000 m of Cretaceous and Cenozoic sediments lying on a pre-Mesozoic continental basement. During low Quaternary stillstands,

FIGURE 5. Mtunzini estuary, South Africa, showing double breaker zone, the outer induced by massive offshore bars (Photo: A. R. Orme).

FIGURE 6. Pleistocene raised beach deposit overlying Pleistocene eolianite, Durban Bluff, South Africa (Photo: A. R. Orme).

FIGURE 7. Pleistocene beach rock near Cape Vidal, South Africa (Photo: A. R. Orme).

some 22,500 km² of bank were exposed (Dingle, 1973). Shelf morphology is relatively featureless because of an extensive Quaternary sediment cover except for a continuous nearshore rocky zone. The shelf edge varies from −110 to −380 m.

The Indian Ocean Coast

The mainly low-lying mainland coast from the Tugela estuary to Ras Asir is developed on two sedimentary basin complexes that postdate the rupture of Gondwanaland and contain several thousand meters of Mesozoic and Cenozoic sediments (Fig. 8). In the south, the 4000-m-deep Mozambique Basin is characterized by strong north-south faulting by virtue of its association with both the complex continental margin and the southern end of the East African Rift system. To the north, straddling the equator between $10°N$ and $10°S$, is the 3000–8000-m-deep Dar es Salaam-Kenya-Somali Basin complex. Modern sediment yields are largely a reflection of climate and vegetation cover or land use, diminishing from south to north. The coastal circulation system includes southeasterly swells complicated or enhanced by seasonal monsoons, tropical easterlies, periodic cyclones, the south-flowing Mozambique and Agulhas currents, and well-developed offshore coral reefs.

The Mozambique Basin widens to over 300 km in central Mozambique, covering virtually the entire country, but narrows north of the Zambezi delta and southward in Zululand. It is fronted by long barrier beaches, surmounted in Zululand by sand dunes up to 180 m high, behind which lie shallow lagoons, extensive swamps, old coastal barriers, and extensive eolian cover sands (Orme, 1973). In Zululand the massive coastal barrier, although veneered with Holocene sediments, is clearly superimposed upon remnants of a major Pleistocene barrier-lagoon complex, the Port Durnford Formation, whose nearshore, beach, and swamp facies were probably deposited during the last interglacial time (Hobday and Orme, 1975). Along this coast and northward in Mozambique shore-zone dynamics are now conditioned by prevailing northeasterly and southwesterly winds, by nearshore circulation systems that locally generate alongshore currents over 1m/sec, by reversing coastal currents related to the Agulhas Current that flow parallel to the shore with surface velocities of 0.2 to 0.8m/sec, and by seasonally high water and sediment discharges from nearby watersheds. The Tugela River brings 10.5×10^6 tons of sediment to the coast annually, and its sediment plume, like those of neighboring rivers, may be traced far out into the Agulhas Current. Owing, at least in part, to unwise land use, present sediment delivery rates to the coast, and therefore its beaches and dunes, greatly exceed Pleistocene rates. Beneath the estuaries between the coastal barrier segments, bedrock channels were cut to at least −55 m during Pleistocene low sea levels. These have since been buried beneath complex sequences of marine, lagoonal, and fluvial sediment (Orme, 1974, 1976). Offshore beach rock and eolianites occur to the outer edge of the continental shelf, to as much as 100 m below sea level off Durban, testifying to the magnitude of late Pleistocene drawdown. Farther north the Pleistocene coastal barrier was locally breached by the Flandrian transgression, for example, forming Delagoa Bay behind Inhaca Island, but is now being repaired by Holocene sedimentation. The Zambezi delta covers 7150 km² and forms the outlet for a 1.33×10^6 km² savanna watershed and 3540 km mainstream. Discharge through the delta averages 7000 m³/sec, and sediment that is not trapped within coastal swamps and lagoons or transported as littoral drift into barrier beaches is flushed over

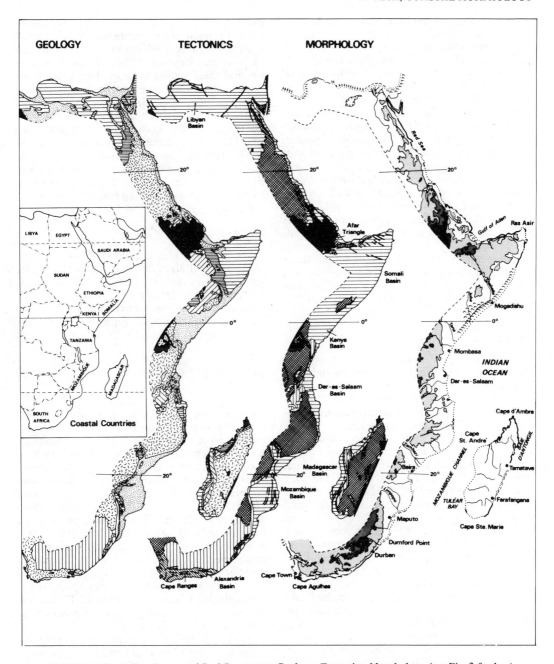

FIGURE 8. The Indian Ocean and Red Sea coasts: Geology, Tectonics, Morphology (see Fig. 3 for key).

the continental shelf into the Zambezi cone and deep ocean basins.

The 15–75-km wide Tanzania coastal zone comprises numerous relict Pleistocene and more active Holocene barrier beaches, offshore bars, and coral reefs, including the basically coralline offshore islands of Zanzibar (1660 km^2), Pemba (984 km^2), and Mafia. Several warped marine terraces with relict coral formations occur along the coast. Three late Quaternary beach ridges occur on the lowest terrace between Dar es Salaam and Tanga, products of an abundant sediment supply that was worked into berm crests by strong swells, modified by wind action, and then stabilized by vegetal growth and tectonic uplift (Alexander, 1968, 1969). Farther north the Kenya coast comprises similar barrier beaches, coral reefs, and mangrove swamps. At least three later Quaternary marine terraces occur up to 140 m, but their boundaries are locally obscured where cliffing was inhibited by the presence of wide lagoons and the continuing growth of coral during marine regressions (Hori, 1970). North of the Tana delta semiarid conditions have long promoted eolian deposition, and Quaternary dunes occur northward to Eil in Somalia. Between Kismayu and Mogadishu, however, the older dunes with their red and yellow soils have been destabilized by overgrazing, although some attempts at dune reclamation have been made in recent years. Coastal dunes are so significant that the Shabelle River, which once reached the shore near Merca, now parallels the coast for almost 400 km before floods sometimes cause it to enter the Juba River. No perennial rivers reach the shore north of the Juba-Shabelle outlet. From Eil around Ras Asir into the Gulf of Aden the coast is more rocky and often cliffed. The peninsula of Ras Hafun is tied to the mainland by a tombolo.

Madagascar

The fault-controlled east coast of Madagascar has its remarkable linearity broken by only one major embayment, the Baie d'Antongil. Basement rocks plunge irregularly from 2000-m highlands onto a narrow coastal plain and continental shelf, underlain by a ribbon of Cretaceous sediments, and steeply down the continental slope. Persistent southeast trade winds, storms in the Southern Ocean, high average wind velocities, and westward-moving cyclones from December to March combine to promote strong wave action along the east coast. In response a series of beach barriers and shallow lagoons have been constructed along 600 km of coast from Tamatave to Farafangana. These linear lagoons, separated from one another by low rises or *pangalanes*, are used by coastal seaborne traffic. In contrast, the west coast is more varied and indented, and the wide coastal lowlands are underlain by Mesozoic and Cenozoic rocks. South from Cape St. André, the coast is low and sandy with vast mangrove areas. Rapid sedimentation reflects the delivery of much fluvial debris to a coast of strong southerly swells and northward littoral drift. Rugged cliffs are found on the resistant igneous rocks towards Cape d'Ambre at the northern tip. Battistini's (1972) investigations of raised beaches, old reefs, eolianites, and paleosols suggest that Quaternary seas rose above the present level only twice, before 500,000 BP and about 100,000 BP. Quaternary reef formations of many types and ages occur: barrier reefs, fringing reefs, inner reefs, coral banks, and coral islets. As in Pleistocene times, Holocene reefs are best developed between major estuaries whose muddy detritus inhibits reef growth, at least in the early stages (Weydert, 1974). Offshore the continental shelf widens to more than 150 km off Cape St. André and on Leven Bank west of Cape d'Ambre. The Îles Glorieuses, northwest of Cape d'Ambre, exhibit coral reefs, 3 m above sea level, that have been dated to 150,000 years BP. The shelf is much narrower off the southwest coast and in Tuléar Bay is notched by massive submarine canyons off the Onilahy and Fiherenana river mouths, the former canyon extending down the continental slope to −2600 m.

The Red Sea Coast

The Red Sea extends 2000 km NNW from the narrow Bab el Mandeb, ranges from 150 to 300 km wide, and deepens to over 2300 m in its central section. This embryo ocean occupies a complex rift system between the African and Arabian plates, whose separation began in mid-Mesozoic times but has increased since mid-Tertiary times. For example, 80 km of left-lateral movement affected the Red Sea rift from late Cretaceous to Eocene time with the opening of the Gulf of Aden, but significant widening has accompanied the 7° counterclockwise rotation of the Arabian block during and since the Neogene. The Red Sea opened to the Indian Ocean in Miocene times, but connection has not been continuous; it remained open to the Mediterranean until the early Quaternary. Thus, with the closure of the Mediterranean connection and eustatic and tectonic changes affecting the Bab el Mandeb narrows, whose sill depth is 125 m below sea level, the Red Sea became isolated during the later Quaternary and, because of intense evaporation, greatly diminished in volume with dramatic impact on shore-zone morphology and ecology. The direction of spreading has been guided by transverse Precambrian fractures trending ENE that extend offshore into transform structures perpendicular to the Red Sea trough.

Owing to the recency of significant plate separation, fault-bounded Precambrian rocks lie close to the coast, and dissected mountain slopes fall steeply toward a narrow ribbon of coastal plain underlain by Neogene and Quaternary littoral deposits. The northern Red Sea coast reflects mainly normal tension faulting of continental margins, as do its extensions: the narrow 10–25-km-wide Gulf of Aqaba, whose faulted sides slope steeply to depths over 1800 m, and the much shallower Gulf of Suez, where massive late Oligocene graben faulting caused Miocene evaporites to accumulate in great thicknesses (Hassan and El Dashlouty, 1970).

South of Ras Banas the Red Sea is structurally more complex. Broad, reef-studded shelves less than 50 m deep fall stepwise toward a narrow central trough over 2000 m deep in which ascending igneous rocks have seemingly engulfed remnants of continental crust.

The African coast of the Red Sea is characterized by several emergent coral reefs of Upper Quaternary age that form wedgelike plates up to 10 m thick, over 1000 m wide, and up to 50 m above sea level (Berry, 1964; Butzer and Hansen, 1968) and up to 6.5 km inland. Recent lateral reef growth in front of now dry early Holocene fluvial deltas has reached several cm/yr (Einsele, Genser, and Werner, 1967). Inland these reefs are usually covered by eolian sand or lagoonal evaporites and, farther inland, by estuarine gravels and alluvial fan and wadi deposits. The coastal plain widens only where major valleys reach the coast—for example, in the Tokar delta, where water and sediment from the swift Baraka River flood the area from June to September each year, or where structural depressions occur, such as the Danakil depression (Dallol basin), a saline depression as much as 116 m below sea level where the Red Sea rift system turns inland south of Massawa.

ANTONY R. ORME

References

Alexander, C. S., 1968. The marine terraces of the northeast coast of Tanganyika, *Zeitschr. f. Geomorph.* **7,** 133-154.

Alexander, C. S., 1969. Beach ridges in northeastern Tanzania, *Geogr. Rev.* **59,** 104-122.

Allen, J. R. L., 1970. Sediments of the modern Niger delta: A summary and review, *in* J. P. Morgan and R. H. Shaver, eds., *Deltaic sedimentation, modern and ancient.* Tulsa, Okla.: Soc. Econ. Palaeont. Miner., Sp. Pub. 15, 138-151.

Barrett, D. M., 1977. Agulhas Plateau off southern Africa: A geophysical study, *Geol. Soc. America Bull.* **88,** 749-763.

Battistini, R., 1972. L'hypothèse de l'absence de hauts stationnements marins quaternaires: Essai d'application à Madagascar et au sud-ouest de l'océan Indien, *Bull. Assoc. Franc. étude de Quaternaire* **31,** 75-81.

Behrendt, J.C.; Schlee, J.; Robb, J. M.; and Silverstein, M. K., 1974. Structure of the continental margin of Liberia, West Africa, *Geol. Soc. America Bull.* **85,** 1143-1158.

Berry, L., 1964. The Red Sea coast of the Sudan, *Sudan Notes and Records* **45,** 148-157.

Busson, G., and Perthuisot, J-P., 1977. Intérêt de la Sebkha el Melah (sudtunesien) pour l'interprétation des séries evaporitiques anciennes, *Sedimentary Geology* **19,** 139-164.

Butzer, K. W., and Hansen, Carl L., 1968, *Desert and River in Nubia.* Madison: University of Wisconsin Press, 562p.

Cabot, J., and Prénant, A., 1968, Observations sur le Quaternaire dans le Sahel d'Alger, *Ann. Algeriennes Géogr.* **3,** 77-92.

Carrington, A. J., and Kensley, B. F., 1969. Pleistocene molluscs from the Namaqualand coast, *Annals S. Afr. Mus.* **52,** 189-223.

Chowlert, G., and Faure-Muret, A., 1968, *International Tectonic Map of Africa (1/5,000,000)* (with 54p. explanatory note). Paris: UNESCO-ASGA.

Coque, R., and Jauzein, A., 1965. Essai d'une carte néotectonique de la Tunisie at 1/1,000,000, *Rev. Géogr. Phys. Géol. Dyn.* **7,** 253-265.

Crosse, M., 1969. *Récherches geomorphologique dans la péninsule de Cap Bon, Tunisie.* Paris: Presses Universitaires de France, 358p.

Davies, J. L., 1977. *Geographical Variation in Coastal Development.* London: Longmans, 204p.

Dewey, J. F.; Pitman, W. C.; Ryan, W. B. F.; and Bonnin, J., 1973. Plate tectonics and the evolution of the Alpine System, *Geol. Soc. America Bull.* **84,** 3137-3180.

Dingle, R. V., 1973. Post-paleozoic stratigraphy of the eastern Agulhas Bank, South Africa continental margin, *Marine Geology* **15,** 1-24.

Dingle, R. V., and Scrutton, R. A., 1974. Continental breakup and the development of post-Paleozoic sedimentary basins around southern Africa, *Geol. Soc. America Bull.* **85,** 1467-1474.

Einsele, G.; Genser, H.; and Werner, F., 1967. Horizontal wachsende Riffplatten am Süd-Ausgang des Roten Meeres, *Senckenbergiana Lethaea* **48,** 359-379.

Furon, R., and Lombard, J., 1963. *Geological Map of Africa* (1/5,000,000) (with 39p. explanatory note). Paris: UNESCO-ASGA.

Guilcher, A.; Medeiros, C. A.; de Matos, J. E.; and de Oliveira, J. T., 1974. Les restingas (fléches littorales) d'Angola, specialement celles du ud et du centre, *Finisterra* **9,** 171-211.

Hassan, F., and El Dashlouty, S., 1970. Miocene evaporites of the Gulf of Suez region and their significance, *Amer. Assoc. Petroleum Geologists Bull.* **54,** 1686-1696.

Hayden, B.; Vincent, M.; Resio, D.; Biscoe, C.; and Dolan, R., 1973. *Classification of the Coastal Environments of the World: Part II–Africa.* Washington: Off. Nav. Res., 46p.

Hobday, D. K., and Orme, A. R., 1975. The Port Durnford Formation: A major Pleistocene barrier-lagoon complex along the Zululand coast, *Trans. Geol. Soc. S. Afr.* **77,** 141-149.

Hori, N., 1970. Raised coral reefs along the southeastern coast of Kenya, east Africa, *Geogr. Repts. Tokyo Metro. Univ.* **5,** 25-47.

Hospers, J., 1971, The geology of the Niger delta area, *in* F. M. Delany, ed., *The Geology of the East Atlantic Continental Margin.* London: ICSU/SCOR, 125-142.

Inman, D. L., and Nordstrom, C. E., 1971. On the tectonic and morphologic classification of coasts, *Jour. Geology* **79,** 1-21.

Kouyoumontzakis, G., and Giresse, P., 1976. L'évolution à la fin du Pleistocene et a l'Holocene du littoral Angolais de Lobito-Benguela et Mossamedes, *Annals S. Afr. Mus.* **71,** 49-67.

Kent, P. E., 1972. Mesozoic history of the east coast of Africa, *Nature* **238,** 147-148.

Kutina, J., 1975. Tectonic development and metallogeny of Madagascar with references to the fracture pattern of the Indian Ocean, *Geol. Soc. America Bull.* 86, 582-592.

Larson, R. L., and Ladd, J. W., 1973. Evidence for the opening of the South Atlantic in the early Cretaceous, *Nature* 246, 209-212.

Lecointre, G.; Tinkler, K. J.; and Richards, H. G.; 1967. The marine Quaternary of the Canary Islands, *Proc. Acad. Nat. Sci. Philadelphia* 119, 325-344.

Le Pinchon, X., and Hayes, D. E., 1971, Marginal offsets, fracture zones, and the early opening of the South Atlantic, *Jour. Geophys. Research* 76, 6283-6293.

McMaster, R. L.; Lachance, T. P.; Ashraf, A.; and de Boer, J., 1971. Geomorphology, structure, and sediment of the continental shelf and upper slope off Portuguese Guinea, Guinea, and Sierra Leone, in F. M. Delany, ed., *The Geology of the East Atlantic Continental Margin*. London: ICSU/SCOR, 109-119.

McMaster, R. L.; Betzer, P. R.; Carder, K. L.; Miller, L.; and Eggimann, D. W., 1977. Suspended particle mineralogy and transport in water masses of the West African Shelf adjacent to Sierra Leone and Liberia, *Deep Sea Research* 24, 651-666.

Maree, B. D., 1966. Die Voorkoms van Diamante op Land en ender die See Langs die Weskus van Suidelike Afrika, *Tegnikon* 15, 149-159.

Martin, L., 1971, The continental margin from Cape Palmas to Lagos: Bottom sediments and submarine morphology, in F. M. Delany, ed., *The Geology of the East Atlantic Continental Margin*. London: ICSU/SCOR, 83-95.

Mitchell, A. H., and Reading, H. G., 1969, Continental margins, geosynclines, and ocean floor spreading, *Jour. Geology* 77, 629-646.

Monteiro Marques, M., 1966. Les grades unités geomorphologiques d'Angola, *Bol. Serv. Geol. Minas. Angola* 13, 13-16.

Orme, A. R., 1973. Barrier and lagoon systems along the Zululand coast, South Africa, in D. R. Coates, ed., *Coastal Geomorphology*. Binghamton: State University of New York, 181-217.

Orme, A. R., 1974. *Estaurine Sedimentation along the Natal Coast, South Africa*. Washington: Off. Nav. Res., Tech. Rept. 5, 53p.

Orme, A. R., 1976. Late Pleistocene channels and Flandrian sediments beneath Natal estuaries, *Annals S. Afr. Mus.* 71, 77-85.

Rona, P. A., 1973. Relations between rates of sediment accumulation on continental shelves, sea-floor spreading, and eustacy inferred from the central North Atlantic, *Geol. Soc. America Bull.* 84, 2851-2872.

Sall, M., 1973. La Côte nord de la presqu'île du Cap Vert. Nouvelle observations de geomorphologie dynamique, *Bull. Ins. Fondamental d'Afrique Noire* 35, 741-763.

Sclater, J. G., and McKenzie, D. P., 1973. "Paleobathymetry of the South Atlantic, *Geol. Soc. America Bull.* 84, 3203-3216.

Sharaf el Din., S. H. 1974. Longshore sand transport in the surf zone along the Mediterranean Egyptian coast, *Limnology and Oceanography* 19, 182-189.

Shepard, F. P., and Emery, K. O., 1973. Congo submarine canyon and fan valley, *Amer. Assoc. Pet. Geol. Bull.* 57, 1679-1691.

Siesser, W. G., 1974. Relict and recent beachrock from southern Africa, *Geol. Soc. America Bull.* 85, 1849-1854.

Summerhayes, C. P.; Milliman, J. D.; Briggs, S. R.; Bee, A. G.; and Hogan, C., 1976. Northwest African shelf-sediments: Influence of climate and sedimentary processes, *Jour. Geology* 87, 277-300.

Tooms, J. S.; Summerhayes, C. P.; and McMaster, R. L., 1971. Marine geological studies on the northwest African margin: Rabat-Dakar, in F. M. Delaney, ed., *The Geology of the East Atlantic Continental Margin*. London: ICSU/SCOR, 13-25.

Usoroh, E. J., 1971. Recent rates of shoreline retreat at Victoria Beach, Lagos, *Nigerian Geogr. Jour.* 14, 49-58.

Varlet, F., 1958. Les traits essentiels de régime côtier de l'Atlantique près d'Abidjan, *Bull. Inst. Fr. Afr. Noire* 20, 1089-1102.

Weydert, P., 1974. Sur l'existence d'une topographie anté-récifale dans la région de Tuléar (Cote sud ouest de Madagascar), *Marine Geology* 16, 39-46.

Cross-references: *Africa, Coastal Ecology; Australia, Coastal Morphology; Antarctica, Coastal Morphology; Asia, Middle East, Coastal Morphology; Australia, Coastal Morphology; Climate, Coastal; Coral Reef Coasts; Desert Coasts; Europe, Coastal Morphology; Flandrian Transgression; Global Tectonics; India, Coastal Morphology, Mangrove Coasts; Restinga; Sabkha; South America, Coastal Morphology.* Vol. VIII, Part 3: *Africa*.

AIR BREAKWATERS

Air breakwaters are a portable, compliant, and feasible method of attenuating certain types of waves. The air breakwater concept (pneumatic) of attenuating waves by forcing compressed air through submerged, perforated pipe was patented in 1907 by Phillip Brasher. Subsequent prototype installation at Million Dollar Pier, Atlantic City (1908), and at El Segundo Pier, California (1915), proved successful. Air breakwaters are now used in harbors and on beach fronts throughout the world.

Current-producing breakwaters are generally divided into two categories: pneumatic and hydraulic. *Pneumatic breakwaters* consist of a simple perforated pipe through which compressed air is allowed to bubble through the water column to the surface. Two surface currents are generated by this process. One opposes the incident waves and is credited with their attenuation. The second, or leeward, current has no appreciable effect on the waves. Thus half of the possible attenuating capacity of pneumatic breakwaters is lost in the leeward current. This waste in attenuating capacity led to the conception of hydraulic breakwaters. *Hydraulic breakwaters* are formed by forcing water through a series of nozzles mounted on a pipe installed perpendicular to the wave direction (Herbich, Zeigler, and Bowers, 1956).

Power requirements of the hydraulic breakwater are similar to those of the pneumatic breakwater, and were found to be primarily dependent upon wave length (L), water depth (d), wave steepness (L/d), jet submergence, spacing, and nozzle diameter. Specifically, it appears that horsepower is fairly constant for wave types with L/d values up to about 2.0 and increases rapidly for L/d values greater than 2.0. True efficiency, or the ratio of attenuated energy for the jet energy, is considerably higher for steep waves than for shallow waves. The zero jet submergence condition is found to be the most efficient. For attenuations of 80 to 100%, the maximum efficiency that can be expected is around 12%.

JOHN B. HERBICH
TOM WALTERS

Reference

Herbich, J. B.; Zeigler, J.; and Bowers, C. E., 1956. *Experimental Studies of Hydraulic Breakwaters*, Project Report No. 51. Minneapolis: University of Minnesota, 137p.

Cross-references: *Boat Basin Design; Breakwaters; Coastal Engineering; Groin; Harbors; Protection of Coasts.*

AKTOLOGY

The term *aktology*—derived from the Greek *akte* and the Latin *acta*, meaning seashore, strand, or promontory—refers to the study of nearshore and shallow-water areas and the life, sediment, and environmental conditions found therein (Jones, 1940; Brown, 1954). As such the word could normally provide a fitting title for this volume. However, it has fallen into such obscurity in the English language that few students of the subject would recognize its meaning (Gary et al., 1972; Lapedes, 1974).

MAURICE L. SCHWARTZ

References

Brown, R. W., 1954. *Composition of Scientific Words: A Manual of Methods and a Lexicon of Materials for the Practice of Logotechnics*. Washington: R. W. Brown, 882p.
Gary, M.; McAfee, R. Jr.; and Wolf, C. L., eds., 1972. *Glossary of Geology*. Washington: American Geological Institute, 805p.
Jones, H. S., 1940. *A Greek-English Lexicon Compiled by Henry George Liddell and Robert Scott*. Oxford: Clarendon Press, 2042p.
Lapedes, D. N., ed., 1974. *McGraw-Hill Dictionary of Scientific Names and Technical Terms*. New York: McGraw-Hill, 1634p.

Cross-references: *Beach Processes; Coastal Engineering; Coastal Fauna; Coastal Flora; Lakes.*

ALBUFERA

Albufera, a Spanish word—from the Arabic *al-buháira*—refers to a shallow, brackish, or salty lagoon lying along the coastline and separated from the open sea by *restingas* or bars with one or more inlets. This phenomenon is found repeatedly in the Mediterranean, for example, Albuferas de Valencia, Mallorca, and Mar Menor de Murcia (Larras, 1964; Rosselló, 1976).

VINCENÇ M. ROSSELLÓ

References

Larras, J., 1964. *Embouchures, Estuaires, Lagunes et Deltas*. Paris: Eyrolles, 171p.
Rosselló, V. M., 1976. Evolution recente de l'Albufera de Valéncia et de ses environs, *Méditerranée* 4, 19-30.

Cross-references: *Barrier Island Coasts; Bars; Deltaic Coast; Estuaries; Haff; Lagoon and Lagoon Coasts; Lido; Liman; Nehrung; Restinga; Salt Marsh; Spits; Tidal Inlet.*

ALGAL MATS AND STROMATOLITES

Many tidal flats of the world today are colonized by a thin film of blue-green algae (cyanophytes), bacteria, and diatoms that coat and bind the fine surface sediments. However, only along a few tropical or desert shorelines (Persian Gulf, Western Australia, Red Sea, Bahamas, Texas Gulf coast, and Baja California, Mexico) do these organisms and sediments accumulate to form a laminated mat more than several mm in thickness. These *algal mats* (or *stromatolites*) not only trap and bind grains but also precipitate carbonate sediments.

Those marine environments where algal mats proliferate are generally characterized by restricted circulation. They are typically hypersaline (20 to 60 or more p.p.t. Cl⁻), and the mats are often adjacent to salt flats or *sabkhas* that contain evaporite minerals including carbonates, gypsum, and halite. The mats usually contain authigenic nonskeletal calcium carbonate granules and traces of lensoid gypsum that range up to 2 or 3 mm in length. Carbonates precipitate as a result of a shift in the carbonate equilibrium due to the photosynthetic removal of CO_2 and the subsequent rise in pH. In most algal mats it is difficult to discern how many of these nonskeletal carbonates are biogenic and how many are products of inorganic precipitation in a hypersaline environment. Where tidal currents or winds transport sediment on the mats, allochthanous grains form laminae with the organic and autochthanous mineral components.

Studies of contemporary stromatolites demonstrate the complex hydrologic, biologic, and sedimentologic interactions that take place in the formation, degradation, and ultimate preservation of coastal algal mats. Intensive analysis has shown that mat morphologies can be related to physicochemical factors such as degree of drainage, degree of desiccation, currents, the nature of the substrate, and biologic factors such as species composition, growth rates, and the roles of grazers and competitors. It has been suggested by a number of researchers that invertebrate grazers may be of paramount importance in limiting algal mat distribution and may have been responsible for the sudden decline in stromatolite diversity and abundance at the end of the Precambrian when invertebrates first evolved.

Examples from Baja California, Mexico

Modern stromatolites are forming in Laguna Mormona (Horodyski, Bloeser, and Vonder Haar, 1977) and Lagunas Guerrero Negro and Ojo de Liebre (Phleger, 1965). These hypersaline lagoons are on the Pacific coast of Baja California, Mexico. A summary of the algal mat types is presented in Table 1. Mats average 2 to 10 cm in total thickness with a maximum size of 35 cm. The accumulation rate is several mm per year with one to several laminae forming annually. The top few mm of these mats constitute the photosynthetic horizon. It is especially apparent in the thick, well-laminated *Microcoleus* mats that the photosynthetic horizon is actually a bilayer of blue-green algae (on top) and purple photosynthetic bacteria (immediately below). The zone inhabited by the photosynthetic bacteria is anaerobic and usually associated with sulfide produced by sulfate-reducing bacteria. Analysis of pigments from mats of different morphologies and dominated by different species of blue-green algae demonstrates that photosynthetic bacterial pigments are present in at least minor amounts in all these mats. Blue-green algae, photosynthetic bacteria, and probably some chemoautotrophic bacteria constitute the primary producers (fixation of CO_2 into organic matter) in these "algal" mats (see Fig. 1).

The organic-rich laminae below the photosynthetic horizon contain anaerobic bacteria and algal and bacterial cells in various stages of decomposition. In general, thicker mats seem to be characterized by slower decomposition rates and are more likely to be preserved and ultimately lithified as stromatolitic sediments.

Several investigators in other parts of the world have noted the limitation of algal mat distribution due to grazing pressure of invertebrates, especially cerithid gastropods. In Laguna Guerrero Negro these organisms graze some mats but appear to be limited by tidal range and salinity. The algal mats tolerate a wider range of environmental conditions. It is interesting to note that blue-green algae also form stromatolites in hot springs above the temperatures normally tolerated by invertebrates. The Baja California algal mats also appear to be locally grazed by small fish that inhabit tidal marsh channels and pools.

Competition by higher algae and vascular plants may also limit algal mat distribution in these marine marshes where both types can coexist. This is especially apparent in environments where vascular plants and algal mats completely colonize the sediment of flat, rather featureless salt marshes. Stands of vascular plants and algal mats may be mixed or may occur as rather pure patches.

The Baja California studies suggest that all intertidal mat material that may be preserved as rock will have horizontal variations in sedimentary and biological features over areas of several square meters and local vertical fluctuations on a scale of 0.5 to 10 cm. Dean, Davis, and Anderson (1975) suggest that even the most regular individual mats interlayered with gypsum sediments in the tidal setting cannot be correlated for more than a few hundred meters. These variations are thus diagnostic and are criteria for separating intertidal-lagoonal algal sediments from the much more uniform postulated deepwater (greater than 30 m) algal or organic-rich sediments. Further, the usefulness of rhythmic laminae in algal mats and adjacent gypsum as cyclic climatic indicators has been challenged (Vonder Haar and Gorsline, 1975).

Literature

Black (1933) was probably the first to recognize and describe algal sediments as incipient algal laminites or stromatolites. Little work on contemporary marine stromatolites was done between 1933 and the early 1960s, when interest was renewed in these "living fossils." Most of the literature on contemporary coastal algal mats has been published within the past fifteen years in geological journals, although a few articles have been published in both geological and biological journals. This is a reflection of the emphasis placed on the sedimentological importance of stromatolites rather than on the

TABLE 1. Algal Mat Types from Hypersaline Lagoons in Baja California, Mexico.

Mat Type	Major Photosynthetic Species	Occurrence
Mammillate *(Entophysalis)*	*Entophysalis* sp. D diatoms A (blooms) other coccoid cyanophytes MR oscillatoriacean cyanophytes R photosynthetic bacteria M	Occurs as a dense encrusting mat in ponded areas of the salt marsh and seaward evaporite flat; requires a firm substrate; hypersaline; permanently submerged; LM
Tufted *(Lyngbya)*	*Lyngbya aestuarii* D cf. *Lyngbya* spp. A-MA *Microcoleus chthonoplastes* MA-MR other oscillatoriacean cyanophytes MR coccoid cyanophytes R diatoms MR-R	Occurs as a surficial mat over better-drained portions of the salt marsh and seaward evaporite flat; hypersaline; commonly intermittently emergent; LM, LGN, LOL
Laminated *(Microcoleus)*	*Microcoleus chthonoplastes* A cf. *Lyngbya* spp. A-MA *Lyngbya aestuarii* MA-R other oscillatoriacean cyanophytes R *Entophysalis* sp. MR diatoms A (blooms) photosynthetic bacteria A	Occurs as a thin layer composed of living cyanophytes underlain by a thick laminated organic deposit; occurs in ponded areas throughout the salt marshes and at the salt-marsh/evaporite flat interface; hypersaline; generally intermittently emergent; LM, LGN, LOL
Enteromorpha	*Enteromorpha* D *Euglena* sp. MR (r) *Lyngbya aestuarii* R other cyanophytes R	Occurs as a diffuse mat in the salt marsh near sites of percolation of seawater into the lagoonal complex; metahaline; permanently submerged; LM, LOL
Black crust *(Rivularia)*	*Rivularia* sp. D oscillatorian cyanophytes MR	A several-mm thick crust with a ropy texture in the high intertidal flats immediately adjacent (but at a slightly higher elevation) to smooth *Rivularia mesenterica* mats; well-drained; LGN and LOL

D = dominant (70–100%)
A = abundant (20–70%)
MA = moderately abundant (5–20%)
MR = moderately rare (1–5%)
R = rare (less than 1%)
Note: Percentages indicated are approximate.

LM = Laguna Mormona
LGN = Laguna Guerrero Negro
LOL = Laguna Ojo de Liebre

biological aspects of algal mats—namely, the physiology and ecology of the microorganisms themselves.

A review of Precambrian paleobiology by Schopf (1975) discusses the contribution of these lithified ancient algal-bacterial mats to the early fossil record. The only comprehensive volume solely dedicated to algal mats is *Stromatolites*, edited by M. R. Walter (1976), in which forty-two contributors, both biologists and geologists, present detailed analyses and discuss many aspects and unsolved problems associated with both ancient and modern stromatolites, including usefulness as biostratigraphic indicators; basin and paleoenvironmental studies; some reflections on mineral resources, including petroleum, in rock sequences that contain algal mat material. A comprehensive bibliography is also presented.

Application of Stromatolite Studies

Holocene sea-level changes on the coast of Mauritania (Einsele, Herm, and Schwarz, 1974) have been determined to be ±0.2 to 1.0 m from mean sea level by noting algal mat formations during the change from lagoonal to sabkha sequences. Indeed the intertidal algal belt facies may be viewed as the final stage of infilling of marine lagoons that have salinities greater than that of ocean water.

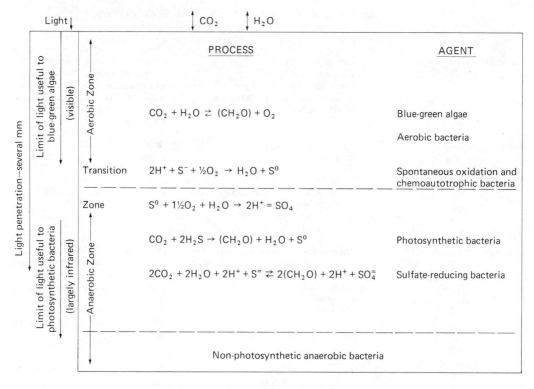

FIGURE 1. Top several mm of an idealized laminated algal mat.

Organogeochemical studies of the insoluble kerogen fractions of contemporary and ancient stromatolites (Philp and Calvin, 1976) may provide information to elucidate aspects of the nature of the early evolution of life chemically locked in ancient sediments. These petrochemical studies of modern mats describe the transformations of various biomolecules as they are broken down and deposited or recycled in the sediments.

Algal mats are buried and preserved in the saline sabkha sediments of Laguna Ojo de Liebre (Baja California, Mexico), where the salt flats are commercially exploited for their sodium chloride. Algal mats buried in the basins of evaporation form an impervious horizon trapping the brines above. In a similar salt-evaporating operation in Brazil algal mats were deliberately introduced in the evaporating pans to prevent the seepage of brines underground (J. I. Bremer, personal communication).

Dalrymple (1965), Friedman et al. (1973), and Horodyski, Bloeser, and Vonder Haar (1977) have studied the petrography of nonskeletal aragonite grains precipitated by algal mats. In the more seaward tidal marshes of Laguna Guerrero Negro (Baja California, Mexico), where the nonalgal mat environment is composed of siliclastic sands, there is generally a positive correlation between mat thickness and carbonate content of the mat and associated sediments. Studies such as these may provide valuable insights into some of the problems of nonskeletal biogenic carbonate precipitation and lithification. For example, the abundance of carbonate stromatolites in the Precambrian sedimentological record may reflect their biological success and their preservability due to early carbonate cementation.

Finally, studies of the microbiology of contemporary stromatolites yield information concerning the physiology and ecology of primitive photosynthetic and nonphotosynthetic microorganisms. It is indeed awesome to think that similar microorganism communities were extant over three billion years ago.

STEPHEN VONDER HAAR
BARBARA JAVOR

References

Black, M., 1933. The algal sediments of Andros Island, Bahamas, *Royal Soc. Lond. Phil. Trans.*, ser. B, **222**, 165-192.

Dalrymple, D. W., 1965. Calcium carbonate deposition associated with blue-green algal mats, Baffin Bay, Texas. *Inst. Mar. Sci. Publ.* **10**, 187-200.

Dean, W.; Davis, G. R.; and Anderson, R., 1975. Sedimentological significance of nodular and laminated anhydrite, *Geology* **3**, 367-372.

Einsele, G.; Herm, D.; and Schwarz, H., 1974. Sea level fluctuations during the past 6000 years at the coast of Mauritania, *Quaternary Res.* **4**, 282-289.

Friedman, G. M.; Amiel, A.; Braun, M.; and Miller, D., 1973. Generation of carbonate particles and laminates in algal mats—Example from sea-marginal hypersaline pool, Gulf of Aqaba, Red Sea, *Amer. Assoc. Petroleum Geologists Bull.* **57** (3), 541-557.

Horodyski, R.; Bloeser, B.; and Vonder Haar, S. P., 1977. Laminated algal mats from a coastal lagoon, Laguna Mormona, Baja California, Mexico, *Jour. Sed. Petrology* **47**, 680-696.

Philp, R., and Calvin, M., 1976. Kerogen structures in recently deposited algal mats at Laguna Mormona. Baja California: A model system for ancient kerogen, in J. Nriagu, ed., *Environmental Biogeochemistry*, vol. 1. Ann Arbor, Mich.: Ann Arbor Sci. Pub., 131-148.

Phleger, F. B., 1965. Sedimentology of Guerrero Negro Lagoon, Baja California, Mexico, *Proceedings of the 17th Symposium of Colston Research Society*. London: Butterworths Scientific Publication, 205-237.

Schopf, J., 1975. Precambrian paleobiology: problems and perspectives, *Annual Rev. Earth and Planetary Sci.* **3**, 213-249.

Vonder Haar, S. P., and Gorsline, P. S., 1975. Flooding frequence of hypersaline coastal environments determined by orbital imagery: Geologic implications, *Science* **190**, 147-149.

Walter, M. R., ed., 1976. *Stromatolites: Developments in Sedimentology*, vol. 20. Amsterdam: Elsevier, 804p.

Cross-references: *Chlorophyta; Chrysophyta; Coastal Flora; Cryophyta; Cyanophyceae; Euglenophyta; Evaporites; Phaeophycophyta; Rhodophycophyta; Sabkha.* Vol. III: *Algae;* Vol. VI: *Algal Reef Sedimentology.*

ALLUVIAL PLAIN SHORELINE

An alluvial plain shoreline is a zone where an alluvial deposit sloping upward into a mountain range contacts ocean or lake waters. The concept is part of D. W. Johnson's (1919) shoreline classification and is a subdivision of his *Neutral Shoreline* category in which the shoreline characteristics are related to neither emergence nor submergence.

A portion of South Island, New Zealand, has an alluvial fan coast "straightened by wave erosion" (Shepard, 1973) that may serve as an example. C. A. M. King (1972) cites the northwest coast of India as example, but E. Ahmad (1972) believes that India's only Neutral Shorelines are deltaic.

SAUL ARONOW

References

Ahmad, E., 1972. *Coastal Geomorphology of India.* New Delhi: Orient Longmans, 222p.

Johnson, D. W., 1919. *Shore Processes and Shoreline Development.* New York: John Wiley and Sons, 584p.

King, C. A. M., 1972. *Beaches and Coasts.* New York: St. Martin's Press, 570p.

Shepard, F. P., 1973. *Submarine Geology.* New York: Harper and Row, 517p.

Cross-references: *Deltaic Coasts; India, Coastal Morphology; New Zealand, Coastal Morphology; Progradation and Prograding Shoreline; Sediment Size Classification.*

ALVEOLAR WEATHERING

Named because of the resemblance to the air sacs (*alveoli*) of the lung, this phenomenon is characterized by extensive networks of small cavities that form *honeycomb patterns* on rock surfaces. These cavities often consist of many shallow depressions, but when well developed they may form deep chambers separated by thin septa of relatively unweathered rock. Individual cavities are typically several centimeters in width and depth, the shape often being controlled by bedding planes, foliation, or other structural features of the rock in which they occur. In many localities the holes occur in association with a hardened surface layer formed when dissolution of ferruginous minerals has been followed by precipitation of ferric hydroxides near the outcrop surface. The thickness of this hardened layer may range from a few millimeters to several centimeters.

Alveolar weathering has been reported in a number of localities, the best-known occurrences being those of Corsica, the Fountainbleau Sandstone of France, and in South Victoria Land, Antarctica. Other sites occur in New Zealand, Australia, Egypt, and the west coast of the United States. The cavities are usually found in sandstone and related sediments, but have also been observed in granite, gneiss, schist, gabbro, and limestone. In some localities the alveolization process is limited to coastal exposures (Fig. 1), but such weathering occurs in a number of inland localities as well.

The mode of formation of these cavities is not well established, and existing hypotheses are based largely on speculation rather than direct evidence (Mustoe, 1981). Since alveolization occurs in widely different outcrops it seems likely that more than one type of origin may be involved. In formulating any explanation it is necessary to account for the extremely selective nature of the erosion. One possibility is that the rock itself possesses varying resistance due to local differences in porosity or cementation. This is difficult to support since alveolar weathering often occurs in rocks having a high degree of homogeneity with no apparent variations in texture or composition. Thus it seems

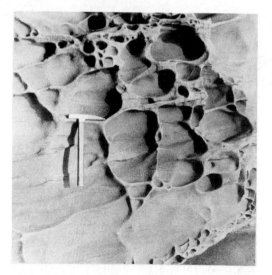

FIGURE 1. Alveolar weathering in Chuckanut Sandstone near Post Point on Puget Sound, Bellingham, Washington (Photo: G. Mustoe).

FIGURE 2. Alveolar weathering in sandstone near Baku on west coast of the Caspian Sea (Photo: M. L. Schwartz).

more likely that attack by the weathering agent occurs in a differential fashion. Alveolar cavities occurring in coastal and arid environments (Fig. 2) have recently been described as products of salt weathering (Evans, 1970), although it is not completely clear how salt weathering can produce such a detailed pattern of differential attack. In any case, alveolization occurs also in environments where salt crystallization seems unlikely. The alveolar weathering of Corsica has been described as a product of selective frost action during the Pleistocene (Cailleux, 1953). Other examples occurring in temperate climates are known to be of post-Pleistocene origin. Wind erosion has been proposed to explain alveolization found in Antarctica, although salt weathering is perhaps a more attractive alternative in this case. Action of microbes seems to offer some promise as an explanation of certain occurrences but is difficult to substantiate with certainty. Although the holes often harbor lichens, algae, and other organisms, it is not possible to assume that these inhabitants were responsible for the initial development of the cavities rather than merely taking advantage of the protective environment that they offer. However, these organisms may play a role in further excavating the depressions by producing organic acids. Future research will, it is hoped, produce additional information, since at present the origin of alveolization is far from clear.

GEORGE MUSTOE

References

Cailleux, A., 1953. Taffonis et érosion alvéolaire, *Cahier Géol. de Thiory* **16-17**, 130-133.

Evans, I. S., 1970. Salt crystallization and rock weathering, a review, *Revue de Géomorphologie Dynamique* 19, 153-177.

Mustoe, G. E., 1981. The origin of honeycomb weathering, *Geol Soc. America Bull.* (in press).

Cross-references: *Abrasion; Frost Riving; Muricate Weathering; Rock Borers; Salt Weathering; Solution and Solution Pan; Tafone; Weathering and Erosion, Biologic; Weathering and Erosion, Chemical; Weathering and Erosion, Differential; Weathering and Erosion, Mechanical.* Vol. III: *Tafoni.*

AMPHIDROMIC SYSTEMS

Tidal-induced resonance in an ocean or marginal basin, acted upon by the *Coriolis Effect* (q.v.), causes successive tidal highs and lows to rotate about a nodal point in what is known as an *amphidromic system.* Progression of the tides is counterclockwise in the Northern Hemisphere and clockwise in the Southern Hemisphere, with the systems in some basins varying slightly in their location with the changing seasons (King, 1972; Bird, 1976).

Cotidal lines, radiating outward from the nodal point, connect equivalent stages—that is, high or low—of the tide. *Corange lines,* concentric around the nodal point, connect locales of equal tidal range or amplitude. At the center the *nodal point* or *amphidromic point,* a tideless condition exists.

FERRUCCIO MOSETTI
MAURICE L. SCHWARTZ

References

Bird, E. C. F., 1976. *Coasts.* Canberra: Australian National University Press, 282p.

King, C. A. M., 1972. *Beaches and Coasts.* New York: St. Martin's Press, 570p.

Cross-references: *Coriolis Effect; Seiches; Tides.* Vol. I: *Amphidromic Point, Region.*

ANNELIDA

Members of the phylum Annelida are wormlike animals that usually possess an extensive internal cavity, extending the length of the body, subdivided into consecutive discrete segments. The segments may be separated from one another internally by septa and indicated externally by outgrowths from the body wall (parapodia) and/or by bundles of bristles (setae).

Three major groups of the phylum Annelida are recognized: The class Polychaeta (polychaetes), almost exclusively marine animals, is characterized by the presence of bundles of many setae on each segment, by heads provided with sensory structures, and by separate sexes. The class Oligochaeta (earthworms) consists mainly of terrestrial and freshwater forms, possessing few setae per segment, lacking sensory structures on their heads, and reproducing as cross-fertilizing hermaphrodites. The class Hirudinea (leeches), primarily found in freshwater and terrestrial habitats (although a few are parasitic on marine fish), lacks setae, sensory structures on the head, and an extensive body cavity, but possesses anterior and posterior suckers; these are also cross-fertilizing hermaphrodites.

The Archiannelida and Myzostomaria are minor classes of the Annelida. The archiannelids are small marine worms, mostly unrelated among themselves and probably representing specialized or aberrant species of polychaetes. The myzostomes are highly modified forms, symbiotic with echinoderms, especially the pelagic crinoids.

The Polychaeta includes by far the most prominent marine annelids and will form the basis of the following discussion. The class is morphologically separable into about 70 families, based on their mode of life (burrowers, tube dwellers, free swimmers in the plankton, or free movers among algae, rocks, or shell debris), and their method of feeding, type of food, and the structure of their feeding organs.

The extensive body cavity (coelom) and the segmentation of the Polychaeta form the basis for the success of the group and are the central features from which much of their evolutionary radiation has proceeded. The fluid-filled coelom is surrounded by longitudinal and circular muscles whose contractions exert pressure on the constant volume of the coelomic fluid. Thus if the longitudinal muscles were contracted while the circular muscles were relaxed, the worm would assume a short, stout shape; if the longitudinal muscles were relaxed while the circular muscles were contracted, the worm would assume a long, narrow shape. In either case, pressure upon the coelomic fluid gives shape to the worm, and the system as a whole can be regarded as a hydrostatic skeleton.

Elements of the nervous system, especially the brain, are found in the head (prostomium), the small segment being located just in front of the mouth opening. From the brain, branches pass posteriorly around the esophagus and unite ventrally to form the main nerve cord, which continues posteriorly. In each segment there is a concentration of nerve cells in the nerve cord (ganglia), from which small nerve branches serve the muscles and other structures of that segment. The muscles of each segment can act independently of all other segments or, by mediation of the brain, can act in concert with other segments or in coordination with them. Thus a wave of contracted circular and relaxed longitudinal muscles can pass forward, followed by a wave of relaxed circular and contracted longitudinal muscles. If segments of the latter wave (short and stout) have purchase on the substratum, then the segments of the former wave (long and narrow) will be thrust forward, where they will remain when they in their turn become short and stout. Contact with the substratum may be improved by paired fleshy parapodia that may bear one or two bundles of setae. By means of the action of internal musculature, again coordinated with the state of the segmental longitudinal and circular muscles, the parapodia may be directed anteriorly or posteriorly and the bundles of setae may be protracted or retracted. These then—the segmented fluid-filled coelom, the longitudinal and circular segmental muscles, the parapodia with their setae, and the brain and segmentally ganglionated nerve cord—are the basis of normal locomotion of the polychaetes (and oligochaetes).

About half of the polychaete families possess an anterior baglike coelomic extension (proboscis) through which the esophagus passes to open to the exterior. By contraction of appropriate muscles in the anterior part of the body, hydrostatic pressure evaginates the proboscis out of the apparent mouth opening, much the same way as air pressure can force the finger tips of a rubber glove. With relaxation of the body muscles and the contraction of retractor muscles inserted on its inner surface, the proboscis can be brought back within the body. The proboscis of some families of polychaetes may be armed with one to several jaws, which serve a grinding or tearing function in feeding (Eunicidae, Onuphidae, Lumbrineridae, Arabellidae, Polynoidae, Polyodontidae, Sigalionidae, Nereidae, Nephtyidae, Glyceridae, Goniadidae);

the surface of the everted proboscis may be provided with adhesive papillae (Phyllodocidae), with minute teeth (Nereidae) or with mucous cells (Arenicolidae, Maldanidae, Capitellidae); rather than being saclike, the proboscis may be digitate (Orbiniidae, Opheliidae, Cossuridae).

In addition to being a feeding mechanism, the proboscis of some polychaete families functions in the process of burrowing (Neredae, Glyceridae, Goniadidae, Magelonidae, Capitellidae, Arenicolidae). In these cases, the sand or mud substratum is agitated, the body is extended, the prostomium is inserted into the substratum, and the proboscis is evaginated. When the quiet substratum has firmed, the longitudinal muscles of the body contract, pulling the worm into the substratum toward the anchor point, which is the everted proboscis. The process is repeated and ultimately the whole body of the worm moves into the substratum; progress through the sand or mud is accomplished in like manner.

Some 16 families of polychaetes live in a fixed tube constructed of solidified mucus (Chaetopteridae, Eunicidae), a mixture of mucus and collected suspended material (Sabellidae), or a mucous tube to which particles from the surrounding substratum may be added (Terebellidae, Maldanidae, Oweniidae, Ampharetidae, Onuphidae, Polyodontidae). Others make tubes of a sand-mucus mixture, thick-walled (Sabellariidae) or thin-walled (Pectinariidae), and still others construct tubes of calcium carbonate (Serpulidae). In all cases the mucus or calcium carbonate is produced from special cells in the anterior region of the body; where sand grains or shell fragments are added during tube construction, tentacular structures are utilized to handle and place this material.

Somewhat over one-third of the families of polychaetes feed upon relatively large particles, utilizing the proboscis and/or jaws. They may be carnivores (preying upon smaller animals), herbivores, (feeding upon algae), or scavengers (subsisting on dead material).

Those families of polychaetes that possess an unarmed proboscis (Orbiniidae, Aphroditidae, Opheliidae, Paraonidae, Maldanidae, Arenicolidae, Capitellidae, Sternaspidae) generally feed directly upon the substratum through which they burrow. Food, in the form of bacteria and detritus mixed with the sand or mud, is digested and the indigestible mineral material is defecated. Such fecal pellets often compromise the major constituent of certain *sands* and *muds*.

Polychaetes that live in tubes or undergo little movement through the substratum are provided with structures to bring food particles to the mouth. Some have a pair of tentacles (Spionidae, Cirratulidae), and others have a multitude of feeding tentacles (Terebellidae). Each of the tentacles bears a ciliated groove along one side, and food (clumps of detritus or organic floc) is selected by the tentacle and is passed down the groove, by means of the cilia, to the mouth.

Certain tube-dwelling polychaetes (Sabellidae, Serpulidae) possess feather duster-like structures (*plumes*) surrounding the mouth. Each *feather* bears a large number of cilia that beat in such a way that a water current passes through the plume. Suspended food particles are filtered from the water current and are carried to the mouth along ciliated grooves on the surface of each *feather*. Another filter-feeding family (Chaetopteridae) usually lives in a U-shaped tube, both of whose openings project above the surface of the substratum. Certain parapodia are modified to function as paddles, establishing a water flow through the tube. Other parapodia are modified so that they form a mucous bag that serves to filter detritus from the water current. After a time the bag is cut loose, rolled into a ball, and passed along a ciliated groove to the mouth. The lug worms (Arenicolidae), although provided with a proboscis and being apparent deposit feeders, are considered to be filter feeders. Living and feeding at the short end of an L-shaped burrow, they create water currents by peristaltic waves of their body. The water moves through the loose, muddy sand upon which they feed, through their burrow, and out its opening. As the water passes through the loose substratum, suspended particles are filtered out and, as sand continuously slumps down, previously filtered organic material becomes available as food. Once again, the considerable indigestible material is defecated as a fecal casting adjacent to the burrow opening.

Relevant literature on this subject can be found in Stephenson (1930), Mann (1962), Dales (1963), Clark (1964), Day (1967), Barnes (1968) and Smith and Carlton (1975).

MEREDITH L. JONES

References

Barnes, R. D., 1968. *Invertebrate Zoology*. Philadelphia: W. B. Saunders Co., 743p.

Clark, R. B., 1964. *Dynamics in Metazoan Evolution*. Oxford: Clarendon Press, 313p.

Dales, R. P., 1963. *Annelids*. London: Hutchinson University Library, 200p.

Day, J., 1967. *Polychaeta of South Africa*, Pt. I, Errantia; Pt. II, Sedentaria. London: London British Museum of Natural History, 878p.

Mann, K. H., ed., 1962. *Leeches (Hirudinea), Their Structure, Physiology, Ecology, and Embryology*. New York: Pergamon Press, 201p.

Smith, R. I., and Carlton, J. T., eds., 1975. *Light's Manual: Intertidal Invertebrates of the Central*

California Coast. Berkeley and Los Angeles: University of California Press, 716 p.

Stephenson, J., 1930. *The Oligochaeta.* New York: Oxford University Press, 978 p.

Cross-references: *Coastal Fauna; Intertidal Flats; Nemertina; Platyhelminthes; Tidal Flat.* Vol. VII: *Trace Fossil.*

ANTARCTICA, COASTAL ECOLOGY

Most Antarctic benthic populations are geographically isolated from other such populations in southern seas by currents associated with the Antarctic Convergence and the Antarctic Circumpolar currents. This biological isolation and cold polar facies of the continent became established approximately 40 million years ago. The continental shelf is relatively narrow, usually about 10 km in width, except in the two embayments of the Weddell and Ross seas. In all areas the continental shelf is unusually deep, ranging from 400 to 800 m. These depths are at least twice those of most shelves. The Antarctic continental slope generally consists of small shelves that fall rather abruptly to the abyssal plains of 3500–5000 m. There are no river systems draining into the Antarctic seas, and the sediments tend to be of glacial origin deposited by ice or wind transport.

The narrow, deep nature of the continental shelf probably contributed to massive extinctions of shallow-water species during the heavy Pliocene and Pleistocene glaciations because there were no shallow-water refugia available. The cold temperatures, the relatively small shelf area, and the oceanographic isolation of the Antarctic shallow benthos have reduced the magnitude of reinvasion of shallow-water benthic populations from other shallow continental shelf habitats such as those of the subantarctic islands, Africa, and South America (Dayton and Oliver, 1977). The result is that the shallow Antarctic benthic populations appear to be composed of those more characteristic of deeper habitats. Another result of the long period of geological isolation and the effective temperature-current barriers to dispersal has been an extremely high degree of endemism (Dell, 1972; Kussakin, 1973). For example, almost 70% of the marine species found here are endemic to Antarctic waters. While this type of calculation is subject to several arbitrary geographic limits and there is high variation between phyla, the endemism is around 90% for some phyla such as benthic fish, Pycnogonida, and Bryozoa. It is clear that the Antarctic benthos is in general a highly coevolved endemic fauna.

Much of the Antarctic Sea is overlaid by permanent ice shelves and annual sea ice. The mean northern limit of the ice cover varies from $55°S$ to $65°S$. Much of this ice breaks up by late summer, and the area covered by pack ice is reduced from approximately 25 to 18 million square kilometers. This ice cover obviously effects the rate of primary *productivity*, which is known to be very high in Antarctic waters in general. However, most productivity data have been collected during the austral summer, when there is a maximum amount of sunlight and a minimum amount of ice cover. Because of the solar inclination and the seasonality of the ice cover, most of the annual productivity takes place in about 125 days per year. For this reason the productivity reaching the benthos is very highly pulsed, a factor of considerable importance to the life-history characteristics of benthic populations.

Despite the difficulties of Antarctic research, the benthic fauna is relatively well described. British and French explorers began working in Antarctic waters near the end of the eighteenth century, and by the beginning of the nineteenth century sealers and whalers from many nations were charting this area. Biological and oceanographic observations were made by James Cook from 1768 to 1771 and in the early 1800s by James Eights and many biologists associated with collections by Bellingshausen, Weddell, D'Urville, Wilkes, Ross, and the *Challenger* expedition of 1872 to 1876. Despite a long history of Antarctic exploration, the most important Antarctic benthic collections were obtained during the first two decades of the twentieth century. Many of these collections were extraordinarily thorough, and, possibly because of an enthusiasm to work on specimens from exotic areas, the taxonomists were reasonably prompt to organize and publish the collections. The later *Discovery* investigations began in 1925 and continued through 1951. The results of these extensive collections are still being published. As a result of this long productive history, most of the benthic species are relatively well described. Unfortunately, much of the literature is widely spread through obscure journals published in many, often difficult, languages. Similarly the type specimens are spread throughout the world and are expensive and difficult to study. Desperately needed are comprehensive monographic studies of particular groups (Hedgpeth, 1971).

All major groups of marine benthic animals are represented in Antarctic waters with the exception of brachyuran crabs. Particularly well represented are siliceous sponges, bryozoans, bivalves, pycnogonids, polychaetes, isopods, amphipods, and echinoderms. Quantitative studies show that some species or genera consistently occur in high densities in many areas. This is especially true of the sea star *Odontaster*, the brittle star *Ophiacantha*, the coelenterates

Edwardsia, Lampra, Urticinopsis, Tubularia, Clavularia, Alcyonium, the isopod *Austrosignum,* the amphipod *Orchomena,* the tanaid *Nototanais,* the gastropods *Subonoba* and *Oviressoa,* the sponge *Rossella,* the ascidian *Cnemidocarpa,* and many polychaetes, especially syllids, terebellids, and polynoids.

A much-discussed characteristic of Antarctic benthic animals is viviparity, or the habit of brooding the young (Arnaud, 1974). This seems particularly well developed in echinoderms, pycnogonids, and molluscs and is relatively well developed in ascidians, polychaetes, nemerteans, and possibly some sponges. The selective value of viviparity almost certainly depends upon the particular group in question, but it may often be an adaptation to accommodate to the predictable summer plankton bloom.

Another seemingly general shallow-water pattern around the continent is a *zonation* of benthic flora and fauna that often appears to be caused by physical disturbance of ice formation (Dayton et al., 1974). Thus the *intertidal habitat* is usually covered with ice and has at best a limited association of lichens, algae, and limpets. With the exception of the Antarctic Peninsula, where there is a shallow zone of algae, the uppermost 10-15 m is usually covered with ice for several months per year and free of long-lived attached organisms. During ice-free months the heavy benthic diatom growth in this shallow zone attracts large populations of the asteroid *Odontaster validus,* the echinoid *Sterechinus neumayeri,* and other motile organisms such as the nemertean *Lineus corrugatus* and the isopod *Glyptonotus antarcticus* as well as several species of gammarid amphipods. In most areas there is a middle zone (15-30 m) most conspicuously inhabited by coelenterates such as the alcyonarian *Alcyonium,* several actinarians such as *Urticinopsis,* and *Artemidactis,* the stoloniferan *Clavularia,* and the hydroids *Lampra, Tubularia,* and *Halecium.* The solitary ascidian *Cnemidocarpa* is conspicuous, as are a few species of sponges, especially *Homaxinella.* This zone (usually 10-30 m) is occasionally subject to physical disturbance from grounded ice floes and especially by the formation and uplift of anchor ice.

The substrata below 33 m are rarely disturbed by anchor ice. This deeper zone is dominated by large, long-lived sponges (especially the family Rossellidae) (see Fig. 1) and bryozoans. The substratum is often composed of a mat of siliceous sponge spicules that has a rich infauna of amphipods, isopods, and polychaetes.

The above zonation of the upper 60 m is representative of most areas where diving research has been done, but it is restricted to relatively clean substrata. Those areas with a mud substratum, or where there is a great deal of sedimentation, have different patterns of infaunal species (Dayton and Oliver, 1977).

FIGURE 1. Scuba diver sitting in large glass sponge *Scolymastra joubini* at 53 m at McMurdo Sound, Antarctica. (Photo courtesy of P. Dayton.)

The deeper (<300 m) rocky substrata around the continent seem similar in that they tend to be dominated by sponges, bryozoans, hydroids, alcyonarians, polychaetes, and bivalves. Several assemblages in the Ross Sea have been described from dredging, bottom grabs, and photography. These assemblages are associated with specific substrata or sediment types. One of the most common is the Deep Shelf Mixed Assemblage, which occurs to 523 m on coarse to fine sediments with erratic boulders. Conspicuous members of this assemblage include tubicolous polychaetes, bryozoans, gorgonaceans, ophiuroids, and crinoids. Another is the Deep Shelf Mud Bottom Assemblage, which occurs somewhat deeper (400-750 m) with similar but muddier substrata. Here also tubicolous polychaetes and ophiuroids are common, as are sipunculids, holothurians, asteroids, and scaphopods. A third is termed the Pennell Bank Assemblage, but actually may be more common as it seems to occur generally along the 400-m shelf break in the Ross Sea. This assemblage is found on mud-sand substrata with rocky outcrops and cliffs, and includes cirripeds, stylasterine corals, ascidea, bryozoans, gorgonaceans, ophiuroids, and pycnogonids. Most authors also include a bathyl assemblage, but there are no good descriptions of bathyl and deeper benthos.

The Antarctic Sea and benthos contrast with the Arctic Ocean in several important aspects. The Arctic Ocean is semienclosed by land, while the Antarctic continent is an island isolated by circumpolar oceanic currents. The Arctic Ocean has broad, shallow shelves subject to much terrigenous runoff from large river systems, while the Antarctic continent has a narrow, deep shelf free of much runoff. The rivers of the Arctic Ocean cause a marked dilution of the surface waters and considerable vertical stratification; these physical parameters contrast with the extremely constant and homogeneous physical patterns of the Antarctic Sea.

In many respects some of the most exciting benthic research remains to be done in the Antarctic. The two large ice shelves have been in continuous existence for millions of years, and research at their edge or below will give us valuable and unique insights into natural processes that have been important through geologic time. Further, the benthos exists in an extremely stable and predictable physical regime, and the populations with their generally deep water affinities have been geographically isolated for 40 million years. These species are uniquely coevolved and yet are reasonably accessible to detailed study. Thus future Antarctic research will undoubtedly make important conceptual contributions to the understanding of evolutionary processes.

PAUL DAYTON

References

Arnaud, P. M., 1974. Contribution a la bionomie marine benthique des régions Antarctiques et subantarctiques, *Tethys* 6(3), 465-656.

Dayton, P. K., and Oliver, J. S., 1977. Antarctic softbottom benthos in oligotrophic and eutrophic environments, *Science* 197, 55-58.

Dayton, P. K.; Robilliard, G. A.; Paine, R. T.; and Dayton, L. B., 1974. Biological accommodation in the benthic community at McMurdo Sound, Antarctica, *Ecol. Monogr.* 44, 105-128.

Dell, R. K., 1972. Antarctic benthos, *Advances Marine Biology* 10, 2-216.

Hedgpeth, J. W., 1971. Perspectives of benthic ecology in Antarctica, *in* L. D. Quam, ed., *Research in the Antarctic.* Washington: American Association for the Advancement of Science, 93-136.

Kussakin, O. G., 1973. Peculiarities of the geographical and vertical distribution of marine isopods and the problem of deep-sea fauna origin, *Marine Biology* 23, 19-34.

Cross-references: *Antarctica, Coastal Morphology; Arctic, Coastal Ecology; Australia, Coastal Ecology; Biotic Zonation; Coastal Fauna; Coastal Flora; Ice along the Shore; Ice-Bordered Coasts; Organism-Sediment Relationship; South America, Coastal Ecology.*

ANTARCTICA, COASTAL MORPHOLOGY

The Antarctic coast may be classified as *glacial tectonic* and *glacial erosional*, because in its genesis the ice action was more important than the marine influence.

Many works have been written on Antarctic geology, but only a few have coastal morphologic importance. Among the first were Cailleux (1963), Adie (1964), and Markov et al. (1968-70); sea-level changes were studied by Adie (1964). An attempt at reconstitution of ancient coastal environments has been made by Araya and Hervé (1966).

Principal Parts of the Antarctic Coasts

Two principal parts are separated by means of a line that follows the Transantarctic Mountains lineament (Fig. 1). This lineament coincides with the eastern coast of the Weddell Sea and the western coast of the Ross Sea. The West Antarctic coast is strongly indented and delimited by the Ross and the Weddell seas. Three parts distinguish this coast: the Antarctic Peninsula is the most prominent and indented part of the coast. The coast of the Amundsen Sea, at the western side of the peninsula, is more straightened, with some well-conformed bays. The coast of the Weddell and Ross seas is more regular than those named above, but with diverse orientation.

The East Antarctic coast is more regular. Straightened sectors alternate with open bays. The only important gulf is that of Christensen Land (Fig. 1).

Geological Background

Geologically Antarctica may be divided into two great provinces: the Western Province and the Eastern Province. The Western Province is orogenic and unstable. Geophysical surveys have determined its morphology and cortical structure, as evidenced by the works of Bentley (1964) and Elliot (1974). The structural units are lineated in the same direction as the Transantarctic Mountains but with a tendency toward arc tectonics, affecting sedimentary and volcanic rocks in discordance with granitic intrusions. The unstable character of the diastrophism shows (Baker, 1974) the relationship between the settling of the volcanic rocks and the internal geodynamics. Thus the lithology of the Western Province is complicated, but the structure permits us to understand the agreement with the coastal configuration.

While the Western Province is orogenic, the Eastern is cratonic. Granitic gneisses and undifferentiated basement appear as a stable and ancient platform. The coastal morphology is simplest here.

Classification

Two principal factors are taken into account in determining coastal morphology in Antarctica: the structural province, and the Quaternary

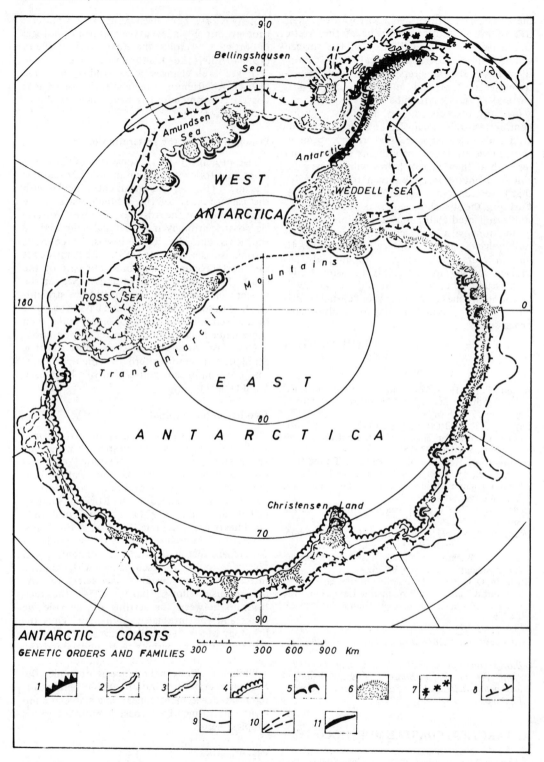

FIGURE 1. 1. Coastal and insular arc. 2. Lobate bay and smooth coast. 3. Great lobated gulfs. 4. Glacial erosional coast. 5. Important fiards and fiords. 6. Ice shelf with barrier. 7. Volcanic ice "calotte" islands. 8. Continental border. 9. Continental slope foot. 10. Submarine canyon. 11. Submarine tectonic trench.

and the actual presence of ice. The sea influence is of minor importance, defining the local forms. Therefore Antarctic coasts are predominantly formed by subaerial and tectonic processes, little changed by the sea. All are also primary dissected (in this case by the ice). The existence of two morphotectonic provinces also shows two subgroups within this group: tectonic in Western Antarctica, and glacial erosional in Eastern Antarctica.

Glacial Tectonic Coasts

In Western Antarctica the folded and faulted structure has directed the glacial erosion and the shorelines are indented, but the indentation is different in the three sectors noted above.

Coastal and Insular Arc (Antarctic Peninsula and Adjacent Islands). *Emerged Part.* The arc structure is defined as matching the general arcing of southern South America. In this structure the primary action of the glaciers has produced embayments of complicated pattern on glaciated peneplaines of hard rock. The embayments have resulted from unguided glacial erosion. In addition to the arced peninsula there is the *Antarctic archipelagos*, with the same peninsular lithology. The minor islands seem to be reefs connected by the recent isostatic uplift. Near the peninsula and the Antarctic archipelagos is the *insular arc* of South Shetland. This is a *volcanic island arc*, defined by González-Ferrán and Katzui (1970) as extinct craters, whose relics show actual volcanic activity. This volcanic chain is strongly worked by the ice (Fig. 1).

Submerged Part. The shelf only appears to be broad. The nautical charts show great depths near the shoreline, especially between the islands. Those parallel to the arcs and orogenic belts are tectonic trenches; the others are in general glaciogenic dissections. Research on the volcanic rocks permits correlation of these trenches with plate tectonics, as evidenced by the works of Elliot (1974) and Baker (1974).

Lobated Bay and Smooth Coast (Bellingshausen and Roald Amundsen Sea Coasts). *Emerged Part.* Above the granitic metamorphic basement lies a discordant pyroclastic plateau. An uplifted block extends onto Marie Byrd Land (González-Ferrán, 1969). From this volcanic area the ice mantle descends to the sea as a piedmont. Therefore the shoreline is usually lobatel with protected bays. Only the Bellingshausen coast is smooth. The hard rocks in the lobatel coasts are generally gneissoidic granites. The bays are covered by *ice shelves*, which undergo distension and contraction, thus creating limiting factors for settlements.

Submerged Part. The shelf is extensive here (100–300 km) and reveals a connection with the emerged morphology as a sedimentary continuation of the piedmont (Fig. 1).

Great Lobated Gulfs (Weddell and Ross Seas). *Emerged Part.* Both coasts have a similar morphostructural placement. Between both great gulfs there is a marginal trough, the continuity of which explains the existence of both seas. The ice drainage, as on piedmonts, also explains the lobate form of the seas. Ice shelves cover these seas and are subject to distension, contraction, and advance at different rates. In the Filchner ice shelf the ice front moves northward at an annual rate of more than 1 m.

Submerged Part. The shelf is wide here (500–900 km) and the continental border is dissected by submarine canyons (Fig. 1).

Glacial Erosional Coast

Emerged Part. The coast of East Antarctica is homogeneous, showing the lithology of the ancient platform with crystalline basement. Open bays reveal little tectonic activity, and the forms seem to reflect glacial erosion. Some bays have an ice shelf. The most important bay is that of Christensen Land, which seems to be an important ancient tectonic range in front of the Kerguelen Ridge.

Submerged Part. A relatively narrow shelf (100–200 km) is heavily dissected by submarine valleys.

Marine Forms

The superficial relief of the bedrock is especially exposed in the coastal margin, where the concept of *terrace* is not necessarily equivalent to *marine terrace*. There are structural, nival, corrasional, glacial and, solifluidal terraces (Markov et al., 1968–70) but only individual cases where the terrace is marine. Voronov found seven high levels between $110°$ and $78°$ E longitude; these are respectively 5–7, 10–15, 25–30, 45, 60, 90, and 120 m high. Marine terraces were found at 15, 40, and 70 m, and prove neotectonic uplift. It is possible to find marine terraces in the fiord-formed valleys, below the leveled surfaces. Japanese researchers found molluscs between 15 and 20 m. In West Antarctica the fauna is thermophilic, and is imputed to the Pliocene. Unfortunately, the works mention the ancient coasts without clear description.

According to Flores-Silva (1970), "the slope of the Antarctic ice cliff is clearly vertical, it reaches heights of 70 and 80 m, and it is developed either on a rocky substratum which is somewhat exposed through pebble beaches or simply on the strand" (Fig. 2). The base frequently has a deep notch, which separates the cliff from the pebble beach. The fissures of the

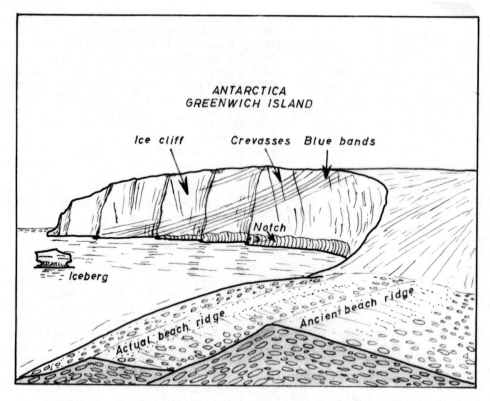

FIGURE 2. Main elements of Antarctic coastal morphology.

ice front, the notch, and mechanical tension produce the collapse of ice blocks onto the beach or into the sea. The movement of the ice mass from the interior and wave action assures the permanent regeneration of the cliff.

It is doubtful whether the beach pebbles are a result of transportation by the Antarctic ice sheet. Flores-Silva (1970) suggests that the beaches may have been derived from piles of clastic forms related to gelifraction and especially to the action of the *ice foot*. But in South Shetland, Araya and Hervé (1966) found an important supply of internal moraines on the beach.

Araya and Hervé (1966) describe the presence of beach ridges, tombolos, littoral banks, comet tails, and patterned gravel beaches with paved structures in South Shetland. The *beach ridges* may be divided into *actual* and *ancient*. The actual outlines the shoreline; its width is from 2 to 8 m, and its height above the high tide level approximately 2 m. The ancient is similar but higher. There are also old tombolos with pebbles and boulders.

The low level in Eastern Antarctica found by Voronov (1964) may be correlated with the ancient beach described in South Shetland by Araya and Hervé (1966). Recent uplift of the islands appears as isostatic response to a decrease in thickness of the ice sheet. The post-glacial emergence is revealed by at least two groups of ancient beach ridges up to 22 or 23 m above the present sea level. This upward movement is correlative with the postglacial retreat of the ice front.

From the top down the principal coastal elements are: ice cliff; ancient and actual strand; intertidal zone with *pack-ice* waters (derivative floes), icebergs, and ice barrier. The major forms that contain these elements are: open and lobate bays; *fiards* or embrional *fiords*, and *ice "calotte" islands*—for example, the South Shetlands (Fig. 3). Finally, an ice-foot occupies a sub-cliff position along the coast (Nichols, 1968; Nielsen, 1979).

Alterations and Opinions

Division of the Antarctic coast has been based on the Zenkovich and Leont'yev classification criterion. It is also possible to use the Shepard Classification, which is the basis of the former.

JOSÉ F. ARAYA-VERGARA

References

Adie, J. R., ed., *Antarctic Geology*, SCAR-IUGS Symposium on Antarctic Geology, Cape Town. Amsterdam: North Holland Publ., 758p.

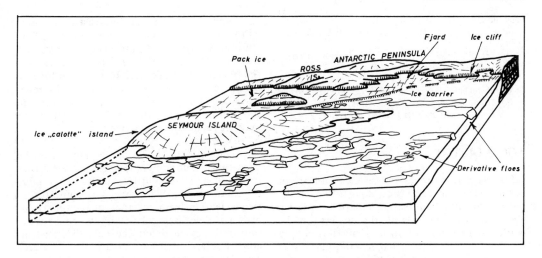

FIGURE 3. Main elements and regional individuals of Antarctic coastal morphology.

Araya, R., and Hervé, F., 1966. *Estudio geomorfológico y geológico en las Islas Shetland del Sur, Antarctica.* Santiago: Inst. Antárct. Chil., Publ. 8, 76p.

Bentley, C. R., 1964. The structure of Antarctica and its ice cover, in H. Idishaw, ed., *Solid Earth and Interface Phenomena.* Cambridge: MIT Press, 335–389.

Baker, P. E., 1974. Volcanism and plate tectonics in the Antarctic Peninsula and Scotia Arc, *I.A.V.C.E.I. Symp. Internac. Volcanol. Santiago Preprint 4,* 6p.

Cailleux, A., 1963. *Géologie de Antarctique.* Paris: S.E.D.E.S., Expéd. Polaires Françaises, No. 242, 201p.

Elliot, D. H., 1974, The tectonic setting of the Jurassic Ferrar Group, Antarctica, *I.A.V.C.E.I. Symp. Internac. Volcanol. Santiago Preprint 27,* 12p.

Flores-Silva, E., 1970. Geomorphological observations and generalizations on the coasts of the South Shetland Islands and Antarctic Peninsula, *SCAR-IUGS Symposium on Antarctic Geology, Oslo, Noruega,* 99–103.

Gonzáles-Ferrán, O., 1969. Cadenas volcánicas de la Tierra María Byrd, *Bol. Inst. Antarct. Chil.* 4, 19–23.

Gonzáles-Ferrán, O., and Katzui, Y., 1970. Estudio integral del volcanismo ceonzoico superior de las Islas Shetland del Sur, Antarctica, Inst. Antarct. Chil. *Ser. Cient* 1(2), 123–174.

Markov, K. K.; Bardin, V. I.; Levedev, V. L.; Orlov, A. I.; and Suetova, I. A., 1968-70. *The Geography of Antarctica.* Jerusalem: Trans. Nat. Sci. Found., Israel Progr. Sci. Trans., 370p.

Nichols, R. L., 1968. Coastal geomorphology, McMurdo Sound, Antarctica, *Jour. Glaciology* 51, 449–478.

Nielsen, N., 1979. Ice-foot processes, *Zeitschr. Geomorphologie, N. F.* 23(2), 233–242.

Voronov, P. S., 1964. Tectonics and neotectonics of Antarctica, in R. J. Adie, ed., *Antarctic Geology,* 681–691.

Cross-references: *Antarctica, Coastal Ecology; Ice along the Shore; Ice-Bordered Coasts; Ice Foot; South America, Coastal Morphology.* Vol. VIII, Part 1: *Antarctica.*

APPOSITION BEACH

An *apposition beach* is one of a series of roughly parallel beaches formed by successive accumulation of material on the seaward side of an older beach. The term was first used by Carey and Oliver (1918), and described as a beach in which "... the materials instead of continuing their course along an existing spit or other beach accumulate in front of it. With the advent of a gale oblique to the line of foreshore, such accumulations may be raised above tidal limits to form a bank parallel to the one previously in contact with the sea. Offshore gales often produce successive closely approximated parallel beaches. From the above causes apposition beaches come into existence and if the process be continued, very extensive ribbed areas of shingle are produced."

An apposition beach, as defined by Carey and Oliver, is thus equivalent to a *beach ridge* (q.v.) of modern usage (Savage, 1959; Friedman and Sanders, 1978). The formation of a series of parallel or semiparallel apposition beaches, or beach ridges, is equivalent to the process of constructional progradation or seaward growth of a beach or spit. These ridges may be produced as storm beaches during *meteorologic* high sea-level conditions. Over longer periods of time, a small rise and fall of sea level will also lead to the building of successive lines of such beaches.

MICHAEL R. RAMPINO

References

Carey, A. E., and Oliver, F. W., 1918. *Tidal Lands: A Study of Shore Problems.* London: Blackie and Sons, Ltd., 284p.

Friedman, G. M., and Sanders, J. E., 1978. *Principles of Sedimentology.* New York: John Wiley & Sons, 792p.

Savage, R. P., 1959. Notes on the formation of beach ridges. *Beach Erosion Board Bull.* **13**, 31-35.

Cross-references: *Accretion Ridge; Beach; Beach Ridge and Beach Ridge Coasts; Contraposed Shoreline; Progradation and Prograding Shoreline.*

AQUACULTURE

Just as agriculture is the science and art of raising terrestrial organisms, *aquaculture* is the science and art of cultivating aquatic organisms. Traditionally man viewed the waters of the world as the source of an inexhaustible supply of food. Occasionally a particular aquatic species was depleted by overfishing, but there were always other species and other waters just over the horizon. An increasing world population forced the development of large, highly mobile fishing fleets to seek out and harvest this vast food supply. From an annual harvest of 20 million tons in 1950, the world catch nearly quadrupled by 1970, but is expected to reach its maximum sustainable yield of around 100 million tons by the 1980s. To surpass this limit, man must learn to farm the water as effectively as he farms the land.

Although the concept of aquaculture probably arose from the practice of harvesting fish trapped in flood basins, its exact origins are not clear. The earliest book on aquaculture, a Chinese treatise from about 500 B.C., dealt with the subject as an already well-established practice.

There are many kinds of aquaculture, ranging from the simplest modification of natural situations to highly engineered technical installations; these two extremes have been classified as *extensive* and *intensive*, respectively (Bardach, Ryther, and McLarney, 1972). However, in practice it is difficult to place specific aquaculture operations into these categories. Many operations combine both approaches: for example, an intensive, totally controlled hatchery program and an extensive grow-out phase in the natural environment.

A good example of extensive aquaculture is the grow-out phase of *ocean ranching*. Hatchery-reared anadromous fish fry are released to forage and mature in the open ocean, and are then captured when the adults return to spawn in the freshwater streams of their origin. Other extensive techniques include the capture and transfer of wild stock to more favorable growing areas, the provision of artificial substrates or habitats to attract and keep organisms in a given location, and the culture of organisms in net or pen enclosures within natural waters. As controls over environmental factors are increased, the aquaculture systems are classified as intensive. Pond culture, which was traditionally an extensive technique, has become an intensive process with the increased manipulation of environmental parameters. Further development of the intensive pond concept has led to tank, raceway, and silo culture where the growing environment may be maintained relatively free of natural influences.

As animals are grown under more crowded conditions, four areas of concern assume critical importance: water purity, nutrition, reproduction, and disease. Water purity or impurity is a two-edged problem for the aquaculturist. Not only must detrimental pollutants from outside sources be avoided but the culturist must not allow his system to become a pollution source. The development of *closed* aquaculture systems where the water is treated and recycled offers obvious advantages: it allows the conservation of heat in those systems where the water temperature is raised; it allows the establishment of an aquaculture installation in a location without a dependable outside supply of water; and it frees the system from the vagaries of nature. Knowledge of the nutritional requirements of the cultured species and the development of inexpensive feeds are essential to any type of intensive culture. Reproductive control and maintenance of breeding stock eliminate dependence on unpredictable wild stocks. Not only does this provide a reliable source of animals but also allows for the genetic improvement of the culture stock through selective breeding. Disease, which is always a problem in animal husbandry, tends to become epidemic under intensive culture systems. Early diagnosis and treatment are necessary, although disease is best prevented through careful management and good water quality.

There are thousands of aquatic species available, but only a relative few have been found suitable for aquaculture (Pillay and Dill, 1979). The ideal aquaculture organism must posses a hardiness to withstand environmental stress, tolerate crowding, reproduce easily, be disease-resistant, grow fast, consume inexpensive food, and have good food or market value.

Single-celled *algae*, the simplest of all aquaculture organisms, are cultivated primarily as a food for cultured shellfish and directly utilized by man only as a health-food commodity. It has not proved to be economically feasible to cultivate and prepare this potentially enormous food source for direct, regular human consumption.

Seaweed culture is most widely practiced in Japan and China. Rope nets are set out to provide a substrate for the spores of certain algae, and other kinds of nets have seaweed cuttings manually attached to them. These nets are then placed in areas that favor algal growth, and the crop is harvested at maturity. Some kelps and red seaweeds have been cultivated for the reestablishment of natural beds, and some are being cultivated for their colloid content. Marine colloids such as alginates, carrageenans, and agar form the basis of a large industry but are supplied mostly by wild crops.

Mollusc cultivation may be the oldest form of marine aquaculture. Records from ancient Chinese and Roman civilizations mention large areas of artificial oyster beds. Oysters, clams, cockles, scallops, mussels, and abalone comprise the majority of cultivated molluscan species. Traditional aquaculture for these animals involves seasonal placement of artificial substrates in areas known for good *spatfalls* and subsequent transfer of this *set* to areas more suitable for growth. Establishment of hatchery techniques has reduced dependence on wild sets and has allowed extensive development of the shellfish industry. Although most cultured molluscs are produced from natural sets or hatchery spat grown in natural waters, they can be readily grown in completely controlled artificial systems.

Crustaceans, such as shrimp, lobster, and crab, command a high market price and are thus particularly attractive for commercial cultivation. Reproductive control and the high cost of feeding have been the main difficulties in crustacean culture. Cannibalistic behavior among lobsters and crabs requires that they be raised in individual compartments, but freshwater and marine shrimp may be reared communally.

Finfish account for approximately three-fourths of the total world aquaculture yield, with the greatest portion coming from relatively simple pond-culture methods (Stickney, 1979). Pond culture involves the construction of basins, dikes, or levees to trap or separate a body of water. Ponds may be fresh, salt, or brackish, and are used for fish that do not require large expanses of open or running water.

The most basic method of pond culture for finfish begins with the fertilization of the water to stimulate abundant aquatic plant life, which is then directly consumed by various herbivorous fish. The species of fish most commonly used in pond aquaculture are carp and *Tilapia* in fresh water and mullet and milkfish in brackish water. Increasing control over factors such as dissolved oxygen and food has allowed fish culturists to economically grow species that would normally not do well in a pond. Examples include catfish, trout, salmon, sturgeon, and baitfish minnows in freshwater; pompano, red sea bream, yellowtail, salmon, flatfish, and sea perch in saltwater; and eels, silversides and sea bass in brackish water. This list comprises only a few major species, and certainly many others will continue to be evaluated for their aquaculture potential. Some fish are able to grow well in ponds, but do significantly better where there is rapid water exchange, such as net enclosures in areas of high tidal flux, impounded streams, flowing water ponds, and in artificial raceways, tanks, and silos. Finfish will always constitute the major portion of world aquaculture outputs. Pond culture of herbivorous fish is viewed as a means of providing high quality *protein* for developing countries. However, as countries become more technologically advanced, aquaculture may be exploited for the development of new luxury-food industries.

The ramifications of aquaculture are as varied as those of agriculture. Although its primary concern is with food and food-related products, aquaculture for nonfood items such as tropical fish, pearls, shells, hides, pharmaceuticals, and industrial products constitutes an area of vital economic importance and of expanding worldwide interest.

DOUGLAS E. CONKLIN
MICHAEL C. HARTMAN

References

Bardach, J. E.; Ryther, J. H.; and McLarney, W. O., 1972. *Aquaculture: The Farming and Husbandry of Freshwater and Marine Organisms.* New York: Wiley-Interscience, 868p.

Pillay, T. V. R., and Dill, W. A., eds., 1979. *Advances in Aquaculture.* Farnham: Fishing News Books Ltd., 651p.

Stickney, R. R., 1979. *Principles of Warmwater Aquaculture.* New York: Wiley and Sons, 375p.

Cross-references: *Algal Mats and Stromatolites; Asia, Eastern, Coastal Ecology; Coastal Ecology, Research Methods; Coastal Fauna; Coastal Flora; Mollusca; Nutrients; Organism-Sediment Relationship; Production; Standing Stock; Waste Disposal.*

AQUAFACTS

Aquafacts are wave-faceted rocks found on beaches. Although their occurrences are not rare, little attention has been given them. The characteristics of the features are similar to those of *ventifacts*, which owe their origin to wind action and abrasion (Keunen, 1947).

The requirements for the formation of aquafacts include: a sand supply sufficient to cause abrasion (pebbles tend to destroy the faceted surface); a boulder that is firmly wedged in place, a small projection of a rocky subsurface, or, most commonly, a rock large enough not to be moved by wave action.

Abrasion occurs on the seaward side of the rock. Sand-sized particles, brought up the beach by swash action, strike the rock, gradually producing the smooth, faceted surface. Backwash is not strong enough to cause abrasion on the leeward side. As the facet is cut into the original surface, a distinct ridge, perpendicular to the onrushing swash, is produced across the crest of the rock.

Aquafacts are not likely to form from rocks that lend a streamlined shape to the oncoming swash. Water will tend to flow around the boulder, and sand grains will produce little effect. The best aquafacts are produced from oblong, angular rocks with their longest axis perpendicular to the uprushing swash.

In addition to *faceting*, *fluting* may also take place. Fluting tends to be the dominant process in rocks of nonhomogeneous composition.

ROBERT T. SIEGFRIED

Reference

Kuenen, Ph. H., 1947. Water-faceted boulders, *Am. Jour. Sci.* 245, 779-783.

Cross-references: *Cobble; Boulder; Sand, Surface Texture; Wave Erosion; Wave Work.* Vol. III: *Ventifacts.*

ARCHAEOLOGY, GEOLOGICAL CONSIDERATIONS

Approximately 15,000 years ago the world was near the last peak of the Wisconsin-Würm glaciation. At that time sea level was approximately 100 meters lower than it is now. For the previous 80,000 years terrestrial erosional and depositional processes had been occurring along the areas of the present continents, continental shelves, and continental margins. Accordingly land forms derived by the climatic, erosional, depositional aspects of that last great Ice Age made an indelible impression upon the topography of the continents and particularly of the coastal zones.

With the waning of the late Wisconsin ice sheets to their present position, sea level, at first rapidly and then at slower rates, rose to its present position. A great deal of argument continues regarding the nature of this rise in sea level and transgression of the marine relative to the land. The nature of the sea level change is still in doubt. We do know that a relatively rapid sea-level rise and transgression of the marine across the world's continental shelves occurred. Approximately 6000 or 7000 years ago this transgression ultimately began to slow down or, in the opinion of some experts, to reach its present level.

The effects on man's occupation of the outer shelf area must have been drastically altered. Thus in Mesolithic and Neolithic times man who occupied the shoreline area must have gradually migrated landward with the rising sea. By the time that civilizations began to develop major cities and ports, approximately 4000 years ago, the effects on man may have reached a catastrophic stage. However, at that time sea level had either reached its present position or was rising at an ever slowing rate relative to land. These considerations must of course include all of the geologic factors involved, such as relative tectonism (uplift and downwarp) climatic changes, variations in rock type, and the nature of the geomorphic surface being transgressed.

Consideration of Mesolithic and Neolithic occupation of the coastal zone, excepting rocky, clifflike areas, is fruitless with present technology. The future holds outstanding promise with new technologies for the study of these cultures on the outer continental shelf. On the other hand, the effects of the latter part of the relative sea-level change and coastal erosion of the Holocene Epoch may be studied now in considerable detail.

First, it is important to establish relative sea level. The local relative sea-level condition and position of the shoreline eroding landward or regression of the sea by accretion are all important in archaeologic considerations and studies. Figure 1 shows an example of the migration of the sea across the Atlantic continental shelf in the eastern United States. The record of man's occupation of this submerged shelf is almost totally lacking. On the other hand, it would be illogical to assume that man was not occupying the coastal zone throughout the Holocene Epoch. Thus paleogeographic maps constructed by sedimentologic, paleontologic, and stratigraphic evidence, supported by radiocarbon dates, may be used to reconstruct the paleogeographies of the waning of the world's ice caps through the late Wisconsin Age and onward through the entire Holocene Epoch to the present. As we approach the present, we can begin to see close associations of American Indian cultures with the shoreline area as opposed to the inland areas. Man occupied the coast because of the marine food resources plus the fact that hunting and agriculture were just as available to him there as inland. Figure 1 shows a sequence of schematic maps of the Delaware Bay–Atlantic coastal area of the paleogeography of the continental shelf, from the peak Wisconsin glaciation at the beginning of the Holocene Epoch, through mid-Holocene time (the Archaic period of the archaeologist) and late Archaic-transitional time to present time (including the late Woodland time of the archaeologist). Relatively complete

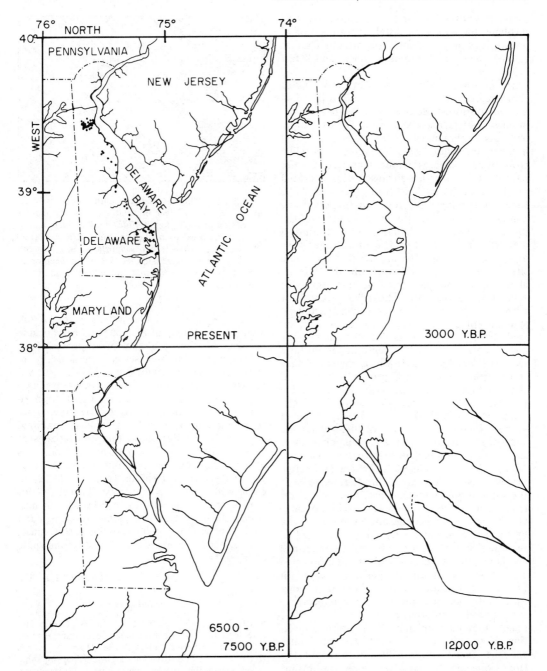

FIGURE 1. A paleogeomorphological sequence of coastal transgression across the mid-Atlantic continental shelf in the area of the mid-Atlantic bight off Delaware Bay. Interpretations are based on radiocarbon analysis of organic specimens from drill holes, sedimentological studies, and geophysical investigations in limited areas. Study of Woodland Period (past 2000 years) occupation of the Delmarva peninsula shows that heavy concentrations of American Indians lived along the tidal and coastal areas (see NW quadrant map). Accordingly it is reasonable to assume that concentrations of archaeological sites once lay along the earlier coastal positions shown. The presence of man in the area to Paleoindian Times (circa 10,000 years before present) has been proved. The possibility of much earlier American man is evident and now being studied.

summaries of the known data and interpretation involving the Atlantic coastal transgression and archaeology may be found in Kraft and Thomas (1976), Kraft and others (1980), and Kraft (1971, 1977).

An intensive study has been made of one of these regressions in the Pamisos River valley area of southwest Messenia by members of the Minnesota-Messenia Expedition (Kraft, Rapp, and Aschenbrenner, 1975; Kraft, Aschenbrenner, and Rapp, 1977). The present shoreline is one of regression. However, subsurface studies have shown clearly that a transgression occurred in this area in late Neolithic, early Helladic times. Accordingly the rounded shapes of the contact between the present alluvium and the pre-Quaternary bedrock in the immediate inland area are erosional landforms formed under wave attack. Interestingly, two important archaeologic sites from early to middle Helladic time lie along this contact line. These include the site of Bouxas to the west and Akovitika, a temple or palace, to the east. Both were constructed on shoreline or coastal sites; both are now far inland. Relative sea level is still changing in the area, whether this is for tectonic or eustatic reasons is not known. However, the subsurface stratigraphic record clearly proves the point and also enables the identification of strandline positions earlier in time. This study by Kraft, Rapp, and Aschenbrenner (1975) may be noted as an example of the use of the subsurface stratigraphic technique in analyzing changes in shoreline position in depositional coastal areas.

With the continuing late Holocene relative sea-level rise in the grabenlike depressions of the eastern Mediterranean in Greece and Turkey, the intensive culture developed by the early Greeks and other peoples and in Mesopotamia by the Sumerians and early Babylonians led to the establishment of maritime-oriented cultures and large cities associated with the sea. Some of these cities, towns, and villages were located on the immediate shoreline in grabens that penetrated into the mainland in deep-water embayments, and were surrounded by *horstlike*, rugged mountain uplands. In addition, alluviation of the valleys continued at a rapid rate, and the embayments infilled throughout the late Holocene Epoch. Eventually important cities such as Miletus, Priene, Heracleia, Pella, Elos, Tiryns from middle Helladic time onward in the Aegean, and Ur and Eridu in Mesopotamia, became abandoned by the sea as a sedimentary regression (*infilling*) continued (Kraft, Kayan, and Erol, 1980; Kraft, Rapp, and Aschenbrenner, 1980).

Many late Holocene (Classical to present) maritime civilizations have come and gone. Figure 2 shows the projected shoreline for Classical-

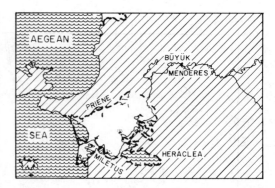

FIGURE 2. The grabenlike embayment of the Büyük Menderes River in Anatolia, showing massive alluviation-progradation (*infilling*) of a deep marine embayment present in Classical times. The important Classical-Hellenistic-Roman seaports of two millennia ago are shown as abandoned inland archaeological sites. Extent of the original embayment is shown by the dashed line.

Hellenistic times and the present in the Büyük Menderes in Ionia, an area of rapid geomorphic change. Whether the geological processes involved had a cause-and-effect relationship on these changes in civilization loci is not yet known. Nevertheless, without question these important cities from Roman times approximately two millennia ago to Helladic times four millennia ago were not able to maintain themselves. Studies show that the citizens of these cultures and cities made numerous attempts to keep their port lines open by dredging canals, as at Ephesus (Miltner, 1958) in Ionia and Tarsus in southeast Anatolia. Ultimately they failed and their cities and civilizations died. Historians, archaeologists, and geologists will long argue as to the reason, cause, and effect of these cultural and geomorphic changes.

JOHN C. KRAFT

References

Kraft, J. C., 1971. Sedimentary facies patterns and geologic history of a Holocene marine transgression, *Geol. Soc. America Bull.* 82, 2131-2158.

Kraft, J. C., 1977. Late Quaternary paleogeographic changes in the coastal environments of Delaware, Middle Atlantic Bight, related to archeological settings, *New York Acad. Sci. Trans.* 288, 35-69.

Kraft, J. C., and Thomas, R., 1976. Early man at Holly Oak, Delaware, *Science* 192, 756-761.

Kraft, J. C.; Aschenbrenner, S. E.; and Rapp, G., Jr., 1977. Paleogeographic reconstructions of coastal Aegean archaeological sites, *Science* 195, 941-947.

Kraft, J. C.; Kayan, I.; and Erol, O., 1980. Geomorphic reconstructions in the environs of ancient Troy, *Science* 209, 776-782.

Kraft, J. C.; Rapp, G., Jr.; and Aschenbrenner, S. E.,

1975. Late Holocene paleogeography of the coastal plain of the Gulf of Messenia, Greece, and its relationships to archaeological settings and coastal change, *Geol. Soc. America Bull.* 86, 1191-1208.

Kraft, J. C.; Rapp, G., Jr.; and Aschenbrenner, S. E., 1980. Late Holocene paleogeomorphic reconstructions in the area of the Bay of Navarino, Sandy Pylos, *Jour. Archaeological Sci.* 7, 187-210.

Kraft, J. C.; Allen, E. A.; Belknap, D. F.; John, C. J.; and Mauermeyer, E. M., 1980. Processes and morphologic evolution of an estuarine and coastal barrier system, in S. P. Leatherman, ed., *Barrier Islands, from the Gulf of St. Lawrence to the Gulf of Mexico.* New York: Academic Press, 149-183.

Miltner, F., 1958. Ephesos, Stadt der Artemis und des Johannes, Ungebung von Ephesos, a map by Captain A. Schindler. Vienna: Verlag Franz Deuticke.

Cross-references: *Archaeology, Methods; Flandrian Transgression; Holocene; Isostatic Adjustment; Midden; Paleogeography of Coasts; Pleistocene; Sea Level Changes; Sea Level Curves; Submergence and Submerged Shoreline.*

ARCHAEOLOGY, METHODS

A variety of techniques have been used for excavating and interpreting the remains of past cultures discovered along littorals. If the materials are exposed above water, or can be isolated by means of *cofferdams,* they can be cleared by conventional methods such as those described by Kenyon (1961) and Wheeler (1961), two archaeologists well known for their careful, systematic approach to excavation. When the remains are underwater, however, the techniques used must be adapted to specific marine environments (e.g., shallow and deep, those with strong currents, heavy underwater growth or silt accumulation, or bare bedrock) as well as to the types of materials being investigated (e.g., wooden or metal shipwrecks, masonry harborworks, submerged dwellings constructed of rubble).

Underwater Equipment

Pioneer work in studying remains underwater was done in the Mediterranean either by inspection from above via sponge divers' viewers, from airplanes, or by means of actual excavation by helmeted divers. Early work concentrated on port facilities once on land (e.g., by Georgiades in Greece, Gunther near Naples, Jondet at Alexandria, Poidebard at Tyre and Sidon) and important work was also done by heroic sponge divers who salvaged statues from sunken ships (at Artemisium, Mahdia, and Antikythera). With the development of the aqualung by Émile Gagnan and Jacques-Yves Cousteau in 1942, however, the new mobility and relative safety made possible by the apparatus enabled archaeologists to extend their sphere of activity (e.g., at Albenga by Lamboglia and at Grand Congloué by Cousteau and Benoît). Using ordinary compressed air and scuba equipment, archaeologists can now work in up to 50 meters of water without resorting to the more dangerous mixed gasses. While certain equipment common to land excavations can still be used (e.g., picks, knives), work in the underwater environment has prompted the adaptation of certain instruments for use underwater (e.g., sealed cameras, simple alidades) as well as the invention of excavation machines that would not work on land and that depend on columns of water being forced through pipes or hoses.

Of these machines, the most important are the water jet, the airlift, and the dredge. The water jet, powered by a pump set up either on land or on a platform floating above the site, has been used successfully to undermine sunken hulls (such as that of the *Vasa* in Stockholm harbor) and make soundings (by McCann at Cosa), but it often clouds the water and can cause havoc with delicate artifacts. The airlift, the most practical tool, uses air pumped into a vertical pipe suspended in the water to create a current of water at the lower end, the result being an underwater vacuum cleaner that can be used for careful work while still maintaining clarity of water in areas where visibility is generally good. The dredge has recently been developed for work in shallow water (0.30 m-5 m depths) where airlifts become inefficient. Since the sides of trenches excavated in soft mud or sand collapse easily, some excavators have employed large pipe sections (as at Cosa), one set upon the other as the depth of the excavation increases.

Search, Survey, and Recording

While the majority of discoveries of ancient wrecks are made casually by amateur sportsmen and by fishermen, relatively recent wrecks have also been located by means of contemporary documents indicating where ships went down. Surveys of underwater areas—by means of divers swimming along grid lines laid on the bottom, by means of towing television cameras or bottom-scanning sonar, or by mud-penetrating geological sounding instruments—have had some success. In shallow areas where visibility is good, low-level aerial observation from manned planes and helicopters or from balloons equipped with remote-control cameras can add significantly to an understanding of the distribution and general shape of remains. Photogrammetry has been used successfully for deep-water wrecks but becomes less useful when camera-to-subject distance is less than three meters.

Complete recording of underwater remains is made possible by means of photography, adequate labeling of objects, the use of spirit levels, and drawings, sections, and notes made on white plastic sheets. Particularly good recording work has been done by Bass at Cape Gelidonia in Turkey, by Katsev at Kyrenia in Cyprus, and by Throckmorton at various sites. Finished drawings can be prepared on land during and after the excavation. Grids are often established over wrecks; when dealing with architectural remains, topographical grids already established on land can be extended into shallow areas by means of buoys and land-based theodolites (Shaw at Kenchreai [Fig. 1]).

Removal and Restoration

Certain artifacts (e.g., gold) retain their original surface and composition when underwater and can therefore be removed and stored in conventional ways without special treatment. Organic remains, however, must be treated by conservators, who often soak them in dense liquids so that interior voids created by rotting are thereby filled. Entire ships, treated piece by piece, have been restored by this method (e.g., the *Vasa* and *Kyrenia* wrecks). Certain metals (e.g., iron) decay completely with time, but their original forms can sometimes be recovered (as at Cape Gelidonia by Bass and Katsev) by sectioning the mass of corrosion found and, after cleaning, filling the voids with plaster or plastic substances. Thus the original form of the metal remains when the outer layers of corrosion are stripped off. Methods of raising material from the bottom vary from site to site, but air-filled balloons (Fig. 1) and hand-held or winch-controlled ropes are most often used.

The Future

Underwater excavation of settlements, port structures, and wrecks, although expensive, should continue to provide new material otherwise unavailable on land. It is hoped that archaeologists will be able to keep pace with the looting of underwater sites by amateurs, however well intentioned. Techniques still to be developed are less expensive means of preserving large masses of organic materials, and it is conceivable that wreck excavations in deep water will sometime be carried out by means of underwater environments set on the bottom near the work areas.

Literature

Further information on this subject may be found in Bass (1966, 1967, 1972), Blackman (1973), Casson (1971), Dumas (1962), Shaw (1967, 1970), Taylor (1966), and Throckmorton (1964).

JOSEPH W. SHAW

FIGURE 1. Many Mediterranean coastal sites, usually harbor towns, are now partially submerged and can be investigated only by divers. In the illustration an archaeologist-diver sketches architectural details of a Roman building at Kenchreai (near Corinth, in Greece), excavated by a group from the University of Chicago and Indiana University. A much-eroded block to his left, no longer *in situ*, awaits lifting by balloon and removal once the area drawing has been completed. Rope grid lines (behind), used to establish the basic axes of the structure, form an outline useful for making a complete topographical plan of all submerged structures. (Photo: Joseph W. Shaw)

References

Bass. G., 1966. *Archaeology under Water.* London: Thames and Hudson, 224p.
Bass, G., 1967. Cape Gelidonya: A Bronze Age shipwreck, *Am. Philos. Soc. Trans.* n.s. 57(8), 1–177.
Bass, G., ed., 1972. *A History of Seafaring Based on Underwater Archaeology.* London: Thames and Hudson, 320p.
Blackman, D., ed., 1973. *Marine Archaeology.* London: Butterworth, 522p.
Casson, L., 1971. *Ships and Seamanship in the Ancient World.* Princeton: Princeton University Press, 441p.
Dumas, F., 1962. *Deep-Water Archaeology.* London: Routledge and Kegan Paul, 71p.
Kenyon, K., 1961. *Beginning in Archaeology.* New York: Praeger, 228p.
Shaw, J., 1967. Shallow-water excavation at Kenchreai (I), *Am. Jour. Archaeology* 71, 223–231.
Shaw, J., 1970. Shallow-water excavation at Kenchreai (II), *Am. Jour. Archaeology* 74, 179–180.
Taylor, J. du P., 1966. *Marine Archaeology.* New York: Thomas Y. Crowell, 208p.

Throckmorton, P., 1964. *The Lost Ships.* Boston: Little Brown, 260p.

Wheeler, Sir M., 1961. *Archaeology from the Earth.* Baltimore: Penguin, 252p.

Cross-references: *Archaeology, Geologic Considerations; Midden; Nearshore Hydrodynamics and Sedimentation.* Vol. 1: *SCUBA as a Scientific Tool.*

ARCTIC, COASTAL ECOLOGY

The term *Arctic coasts* is used here to refer to shores that border the Arctic Ocean and its component seas, including those that have been called Arctic, high Arctic, Subarctic, and Boreo-Arctic.

Arctic shorelines may be precipitous and deeply cut by fjords, as are those of much of the Arctic archipelago, Greenland, Iceland, Spitzbergen, Norway, the Murman coast, Nova Zemlya, and eastern Siberia. The Arctic coastlines of Alaska, western Canada, and most of the U.S.S.R. are, however, of low relief. Arctic coasts are in the tundra biome, and much of this low-relief shoreline is carpeted by treeless meadows of grasses, sedges, mosses, lichens, and flowering plants. Permanently frozen ground (*permafrost*) is found everywhere along Arctic shorelines with a shallow, active (thawed) layer, usually less than 1 m in depth, present during the brief summer. The active layer is waterladen. There are many standing lakes.

The biota of Arctic beaches is impoverished. Indeed, in the high Arctic where the effect of the annual ice is extreme and the tidal amplitude is minimum, there is almost no littoral (intertidal) biota. Owing primarily to the effects of the North Cape, Spitzbergen, and West Greenland Currents, the flora and fauna of beaches of the Murman Peninsula, the Norwegian coast, southern Spitzbergen, and southwestern Greenland are actually more Boreal than Arctic.

Arctic beaches of Alaska and western Canada have no littoral flora. The supralittoral or upper beach zone may have a sparse vegetation of the grass *Puccinellia phryganodes* and the sedges *Carex ursinus, C. ramenskii,* and *C. subspathacea.* *Stellaria humifusa* and *Dupontia fisheri* also are pioneer species. Shores of unconsolidated sediments often have a berm of coarser material that is maintained by waves or ice scouring. Behind this beach berm is a lower region that may contain standing brackish water and often has a characteristic, dense marsh vegetation of *Carex* spp. and *Puccinellia.* Mosses, blue-green algae, and diatoms form mats on the surface of marshes, and small pools of brackish water support populations of cladocerans, copepods, branchiopods, dipteran (chironomid) larvae, and oligochaete worms. Spiders, mites, collembolids, saldid flies, and other insects are terrestrial inhabitants. Geese, shorebirds, some passerine birds, and even caribou feed in these supralittoral, Arctic marshes.

Sandy beaches may have low dunes stabilized by a characteristic vegetation mainly of beach rye *Elymus mollis* and *Mertensia maritima.* Other species of the *Elymus* community are *Honkenya peploides, Lathyrus maritima, Cochlearea officinalis,* and others.

Between the pioneer grasses or the marshes and the upland tundra is a transitional flora of scrub willows (*Salix* spp.), mosses, grasses (*Artemesia* spp., *Poa* spp., *Festuca* spp.), other flowering plants, and *Equisetum* spp.

High Arctic beaches have a sparse but characteristic littoral fauna. In Alaska the region between the shoreline and a depth of about 2 m, which corresponds roughly to the thickness to which the annual, shore-fast ice freezes, harbors probably fewer than fifty species of macrobenthos, of which almost the entire biomass ($3\pm5g/m^2$) consists of enchytraeid (oligochaete) worms, chironomid (midge) larvae, the amphipods *Gammarus setosus* and *Onisimus litoralis,* and the isopod *Saduria entomon.* Three of these are annual migrants from deeper water. Beyond the depth of the annual ice, the enchytraeid worms and chironomid larvae of the littoral are lacking but the sublittoral fauna is richer (biomass of about $30g/m^2$) and more diverse, including as additional major species: the mysiid shrimp *Mysis relicta* and *Neomysis rayii;* polychaete worms *Scolecolepides arctius, Ampharete vega, Prionospio cirrifera, Terebellides stroemi,* and others; two bivalue molluscs *Cyrtodaria kurriana* and *Liocyma fluctuosa;* the priapulid *Halicryptus spinulosis;* the amphipods *Pontoporeia affinis* and *Calliopus laevuisculus;* and the four-horned sculpin *Myoxocephalus quadricornis.* In the southern Chukchi Sea, the blue mussel *Mytilus edulis* is an abundant, sublittoral form.

Benthic macroalgae of the sublittoral region of Alaskan beaches include three species of Chlorophyta (mainly *Enteromorpha* spp.) that may infrequently occupy the littoral zone, fourteen species of Phaeophyta (principally *Laminaria solidungula, L. saccharina,* and *L. spp.*), and eleven species of Rhodophyta (*Rhodomela lycopoides* and others).

Eel grass (*Zostera marina*) occurs subtidally on the southern Chukchi coast of the United States and elsewhere but is not abundant in the Arctic (Broad et al., 1978).

The littoral and sublittoral zones of Alaskan beaches are the principal feeding areas of about thirty species of fish during the summer months. These include, in addition to anadromous salmonids and some truly marine fishes (gadids, pleuronectids, cottids, and others), a few freshwater forms found near river mouths. The principal species are: *Salvelinus alpinus* (Arctic char), *Coregonus autumnalis* (Arctic cisco), *C. sardinella* (least cisco), *C. nasus* (broad whitefish), *Boreogadus saida* (Arctic cod), and *Myoxcephalus quadricornis*. The principal foods of these species are amphipods, myseids, copepods, insect larvae, isopods, small Arctic cod, and four-horned sculpin. Both pink (*Oncorhynchus gorbuscha*) and chum (*O. keta*) salmon are found along the American Arctic coast and may be occasionally abundant.

The shorelines are also important feeding areas for many birds and nesting sites of a few. Glaucous gull, Sabine's gull, Arctic tern, both northern and red pehalarope, golden plover, ruddy turnstone, and sanderling are among the more abundant summer residents. Rock cliffs house nesting colonies of murre, kittiwake puffin, and guillemot. Eider and oldsquaw are especially abundant in coastal regions in late summer.

The principal carnivores of the Arctic are mammals, primarily the seals (ringed, bearded, ribbon, and spotted), the beluga, and the bowhead whale. Seals feed on fish, shrimp, and smaller invertebrates; beluga are piscivorous; and the bowhead whale is a plankton feeder (Weller, et al., 1978).

The native people of the American Arctic are Inuit or Eskimo. They inhabit the coastal region and have traditionally depended to a considerable extent on marine mammals, coastal fishes, and birds for food and clothing.

The biota of the shores of eastern Greenland, northern Spitzbergen, Bear Island, and most of Nova Zemlya is generally similar to that described for Alaskan beaches. There is a sparse algal flora on rock faces, with *Fucus inflata* found in the lower littoral. The littoral fauna is principally oligochaete worms (*Enchytreaus albidus* and *Lumbricella lineatus*), mites (*Molgus littoralis* and *Ameronothrus lineatus*), and two ubiquitous, sublittoral crustaceans *Gammarus wilkitzkii* and *Mysis oculata*. Other littoral or supralittoral animals are collembolids, turbellarians, harpacticoid copepods, menatodes, and chironomid larvae. The vegetation of beaches with unconsolidated sediments and of marshes is similar to that described. A drift line of seaweed is common on Greenland beaches and usually contains enchytraeid worms, collembolids, and mites. The avifauna of Greenland beaches is similar to that of Alaskan (Madsen, 1936).

Zenkevitch (1963) states that there is no littoral biota of the Russian coast of the Chukchi Sea or of the East Siberian or Laptev seas, but this may refer to macrobenthos of large size. Many of the sublittoral forms of the Alaskan and Canadian Arctic are also found in the U.S.S.R. and extend westward into the Kara and Barents seas. However, in the Barents and White seas, along the Arctic coasts of the Murman peninsula and Norway, in northern Iceland and southern Spitzbergen, and on the west shore of Greenland, currents of warm, saline Atlantic water extend the Boreal, littoral biota into the Arctic. Madsen (1936) describes this fauna as Subarctic and characterizes it by the presence in the littoral zone of *Mytilus edulis* and *Balanus balanoides*. Also found here is *Littorina saxatilis* var. *groenlandica*. Both Madsen (1936) and Dunbar (1968) extend the Subarctic zone south along the Labrador coast to Cape Charles.

The Subarctic littoral biota is rich and varied. There are almost 200 species of macroalgae, of which *Pelvetia, Fucus, Ascophyllum, Chorda,* and *Laminaria* are the predominant genera of Phaeophyta; *Enteromorpha* and *Monostroma* the principal greens; and *Rhodomenia, Odonthalia, Ptilota, Delesseria, Phyllophora*, and *Lithothanmium* the major reds. Zonal distribution is marked. The *Ascophyllum* zone is the most extensive. *Zostera marina* occurs in both the Barents and White seas.

There are nearly 2000 species of animals in the littoral and sublittoral beach fauna of the Barents Sea, of which *Mytilus edulis, Balanus balanoides, B. crenatus,* and *Littorina* spp. characterize the upper littoral, which can be further divided into five biocoenoeses: *Balanus balanoides; Mytilus edulis; Ascophyllum* along with *Sertularia* and *Flustrella; Rhodophyta;* and *Sphacellaria* with polychaetes and small molluscs. In the Barents and White seas this littoral Boreal or Subarctic biota coexists with truly Arctic sublittoral and deeper benthic forms. The mussels and barnacles become sublittoral in the eastern Barents and in the other Siberian seas (Zenkevitch, 1963).

The source of energy in Arctic Sublittoral ecosystems is a subject of current research. Photosynthesis during the Arctic spring results in blooms of ice-bound algae, but measurements of this production are few. During the ice-free summer, algal populations and, presumably, photosynthesis are reduced. There are few obviously herbivorous species of animals and even fewer macrophytes. By inference, detritivores and deposit-feeding organisms must be a major element of energy and nutrient flow. The possibility that the shallow Arctic Seas are in part a detritus-based ecosystem driven by terrestrial plant debris from eroding shorelines has been suggested and is under investigation. Other biological research is directed toward extending knowledge of the flora and fauna of less known regions and toward understanding trophic relationships within the community and population dynamics of principal species. Interest in these and other aspects of Arctic ecology has been stimulated by the discovery of petroleum resources in the region.

A. C. BROAD

References

Broad, A. C., Koch, H., Mason, D. T., Petrie, G. M., Schneider, D. E., and Taylor, R. J., 1979. *Reconnaissance, Characterization of Littoral Biota, Beaufort and Chukchi Seas, Environmental Assessment of Principal Investigators for the Year Ending March, 1978.* U. S. Department of Commerce, National Oceanic and Atmospheric Administration, Environmental Research Laboratory, Outer Continental Shelf Environmental Assessment Program, Boulder, Colorado, 86p.

Dunbar, M., 1968. *Ecological Development in Polar Regions.* Englewood Cliffs, N. J.: Prentice Hall, 119p.

Madsen, H., 1936. Investigations on the shore fauna of east Greenland with a survey of the shores of other Arctic regions, *Meddelelser om Gronland* 10, 5-79.

Weller, G., Norton, D., and Johnson, T., 1978. Environmental Impact of OCS Development in Northern Alaska: Revised DRAFT Beaufort Sea Synthesis Report. *Arctic Project Bulletin: Special Bulletin,* U. S. Department of Commerce, National Oceanic and Atmospheric Administration, Center Continental Shelf Environmental Assessment Program. Fairbanks, Alaska, 362p.

Zenkevich, L. A., 1963. *Biology of the Seas of the U.S.S.R.* New York: Interscience, 955p.

Cross-references: *Arctic, Coastal Morphology; Asia, Eastern, Coastal Ecology; Biotic Zonation; Coastal Fauna; Coastal Flora; Europe, Coastal Ecology; Ice Along the Shore; Ice-Bordered Coasts; North America, Coastal Ecology; Soviet Union, Coastal Ecology.* Vol. VIII, Part 2: *Europe and Asia.*

ARCTIC, COASTAL MORPHOLOGY

The Arctic is a complex region that has outer limits that vary with the specific criterion used. From the standpoint of the coastal morphologist, the most extensive delimitation probably results when sea ice is considered the defining factor. Included would be the coasts of Labrador and Newfoundland and those bordering the Sea of Okhotsk and most of the Baltic and Bering Seas—regions normally not considered Arctic and herein considered only incidentally. Concentration is on those coasts directly facing the Arctic Ocean, the islands of the Canadian Archipelago, and Greenland. The length of this coastline (as determined from the American Geographical Society 1:5,000,000 *Map of the*

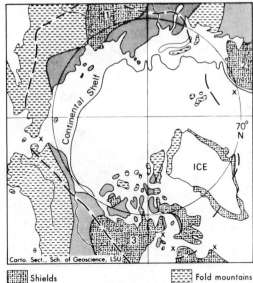

FIGURE 1. Structure and permafrost in the Arctic.

Arctic Region) is 31,000 km for the continental portions, 16,000 km for Greenland, 26,000 km for the islands of the Canadian Archipelago, and 9000 km for other islands in the Arctic Ocean (Fig. 1).

The Geologic Base

Continental structure in the Arctic is dominated by three large, stable shields composed primarily of Precambrian granites and gneisses. Coastal exposures of shield materials (especially on the northern margin of the Angara Shield) are less lengthy than the size of the shields themselves would suggest because of extensive burial. Between shields (including their buried margins) fold mountains are common and extend to the coast in a number of locations (Fig. 1). From the standpoint of tectonic and volcanic activity, the Arctic coastline is quiescent, similar to the embedded coasts surrounding the Atlantic Ocean. Extensive coastal plains and wide continental shelves (Fig. 1) extend from Alaska westward across Siberia (Sater, 1969).

Basic structures and materials along nearly the entire Arctic coast have been directly modified by Pleistocene ice (Fig. 2). The two major exceptions are the east-central Siberian lowlands and the north coast of Alaska. Even these coasts were affected by the change in sea level

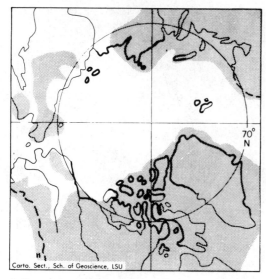

FIGURE 2. Arctic coasts and ice.

Conditions and Processes

The appearance of any coast, the Arctic included, depends on the alterations that have occurred to the geologic base it inherited. Forms may be major or minor, and even ephemeral, but all reflect a number of processes and conditions. In the Arctic permafrost, sea ice, and snow are important, in addition to the more universal factors of relief, structure, sediment type, river discharge, waves, and tides.

Permafrost. Permafrost—perennially frozen ground—is continuous along most of the Arctic coast (Fig. 1) and extends seaward as well as landward along much of the shoreline. Permafrost is important in coastal morphology in several ways (Mackay, 1972). In unconsolidated sediments included water in the form of ice tends to strengthen the ground and helps account for extensive low to moderately high cliffs. Ice content and bonding efficiency vary inversely with texture; on the other hand, the susceptibility to thermal erosion increases with ice content. Near sea level thermoerosional niches are formed and large blocks collapse onto the shore, where further disintegration and reworking occur (Fig. 3). Larger ground-ice slumps are also common along coasts backed by high, ice-rich cliffs.

that accompanied volume changes in glacial ice. Coasts overridden by ice were modified in different ways. Along upland coasts *fiord* development was extensive and is still occurring in Greenland, Baffin Island, Novaya Zemlya, and a few other islands. Pleistocene channeling by glacial ice was intensive on some continental shelves and helped create the present-day coastal configuration within much of the Canadian Archipelago. Low-lying coasts overridden by glaciers show evidence of both glacial scour and glacial deposition with the former tending to dominate. Nonetheless morainal forms of various types are not uncommon along coasts in northern Canada and western Siberia.

The melting of glacial ice exposed new areas to marine action, brought on rebound, and caused a rise in sea level. The rate and intensity with which these three dynamic conditions affected different coasts varied greatly. Some peripheral coasts may have become ice free as much as 13,000 years ago and others only recently (John and Sugden, 1975). The vertical displacement of many Arctic coastlines is evidenced by the occurrence of coastal and marine features at elevations ranging from present-day sea level to over 250 m. In the Canadian Archipelago rebound is continuing along some coasts at rates of as much as 1 m/century (Andrews, 1970).

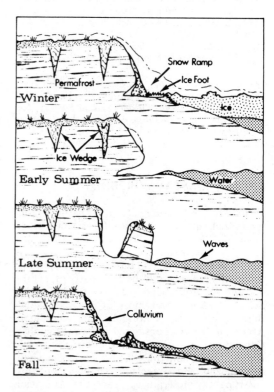

FIGURE 3. Erosion in permafrost-conditioned cliffs.

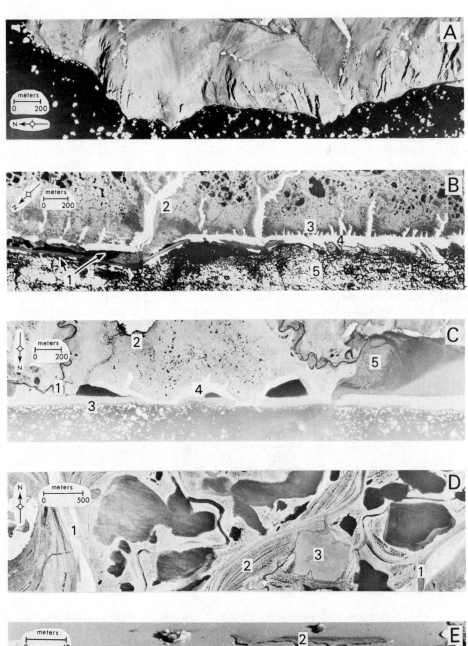

FIGURE 4. Arctic coastal forms.   A. Coastal cliffs at Cape Lisburne, Alaska.   B. Low cliff coast, Chukchi Sea, Alaska. 1. Barrier bar and lagoon. 2. Snow-filled stream channel. 3. Gullies. 4. Snow ramp. 5. Disintegrating sea ice.   C. Barrier bar coast. 1. Bar-deflected stream. 2. Ice-covered lake. 3. Bar. 4. Snow ramp. 5. Lagoonal delta.   D. Deltaic coast. 1. Distributary. 2. Ice-wedge polygons. 3. Tapped lake.   E. Ice-wedge polygons. 1. Thawed ice wedge. 2. Stranded, sediment-laden ice. 3. Polygonal trough.

ARCTIC, COASTAL MORPHOLOGY

Distinctive permafrost forms along many of the shores of the Arctic are *ice-wedge polygons* (Fig. 4E). *Ice wedges* paralleling the shore may serve as a line of weakness aiding block collapse (Fig. 3), whereas those perpendicular to the shore serve as points of more rapid retreat than in the bordering permafrost, resulting in highly gullied shorelines (Fig. 4B).

Sea Ice. Much of the Arctic Ocean is ice-bound during most of the year. Nearshore bottomfast ice effectively isolates the beach from the sea, and shore modification is minimal. Even during summer sea ice shortens fetch and dampens waves, thus reducing the effect of ocean waves alongshore. The duration of the period during which sea ice is a coastal modifier varies greatly. Some coasts are perennially affected, whereas others are affected for short periods of time only (Fig. 2).

Although this passive role is sea ice's most important, it is by no means the only one. Sea ice is also an eroding, transporting, and depositing agent. When it freezes to the shore during fall, sediments are incorporated into and added on top of the ice. During breakup these materials are frequently carried away from shore or, under proper wave conditions, carried farther up the beach. Along many Arctic coasts throughout the summer and along all coasts during breakup, *pack ice* pushes onto beaches or gouges into them and sea cliffs. Numerous mounds, ridges, kettles, and other forms create a highly irregular local relief. Most of these forms are ephemeral but some persist for several years, especially near the back of these beaches when Arctic seas are more or less continuously choaked with ice.

Ice action is not limited to onshore positions. Its effects can be seen at varying depths offshore. In high tidal areas ice-pushed *boulder barricades* parallel the shore to varying depths. Negative relief forms also develop. For example, *bottom scores,* recorded by sidescan sonar (Fig. 5), are made by the keels of drifting ice blocks (Reimnitz and Barnes, 1974).

Snow. Snow is also primarily passive. Accumulating along the shoreline at the time sea ice is forming, it becomes incorporated in the *icefoot* (a combination of snow, froth, seawater and freshwater ice, and organic and inorganic

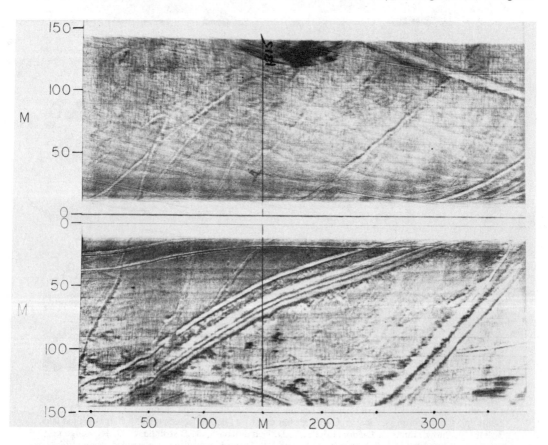

FIGURE 5. Sonograph of ice scores under 8–10 m of water, Colville River delta, Alaska (courtesy Office of Marine Geology, Geological Survey, Menlo Park, California).

sediments) that forms on the beach (McCann and Taylor, 1975). More important, it accumulates in drifts alongshore, creating *snow ramps* (Figs. 3 and 4B) that may last throughout the year in suitable locations.

Rivers. Four of the ten longest rivers of the world (the Ob, Yenisei, Lena, and Mackenzie) and innumerable shorter ones drain across the shore into the Arctic Ocean. As with rivers in other climates, their influence on nearshore environments is manifold (Walker, 1974). The highly seasonal nature of these rivers does not mitigate against their effectiveness as coastal modifiers. During their low- or no-flow period coastal modifiers of other types are also of minimal importance. Some of the world's largest deltas have been formed by such rivers as the Lena and Mackenzie. In addition, smaller deltas, with their many distinctive forms (Fig. 4D), dot the shores of Alaska, Canada, and Siberia. For example, of the 1441 km of coastline between Cape Lisburne in Alaska and the Canadian border, 135 km (9 percent) is deltaic (Wiseman et al., 1973).

Ocean Currents, Tides, and Waves. The Arctic Ocean, relatively isolated in terms of worldwide wind systems and partially or wholly covered with sea ice depending on season, is a body of water in which low tides, slow-moving currents, and low-energy waves predominate. Nonetheless, waves are the most important modifier of Arctic coastlines and during the ice-free period serve as significant eroding, transporting, and depositing agents. Local storms, especially during fall, occasionally result in extensive coastal alteration (Hume and Schalk, 1967).

Conclusions

Many coasts in the Arctic are retreating, some rapidly. Although the retreat in northern Canada and Alaska for ocean-exposed coasts apparently averages about 1 m/year. Lewellen (1970) reports over 10 m/year along parts of the Alaskan coast, and Are (1976) has recorded retreat averaging 50 m/year along some Siberian coasts. Deposition also occurs. In addition to deltaic growth mentioned above, many shorelines possess *barrier islands, offshore bars,* and *onshore spits* (Fig. 4C), which are being continuously modified.

H. J. WALKER

References

Andrews, J. T., 1970. Present and postglacial rates of uplift for glaciated northern and eastern North America derived from postglacial uplift curves, *Canadian Jour. Earth Sci.* 7, 703-715.

Are, F. E., 1976. Thermoabrasion of shores and permafrost, *in Dynamics of Shore Erosion.* Tbilisi: IGU, 170-172.

Hume, J. D., and Schalk, M., 1967. Shoreline processes near Barrow, Alaska—A comparison of the normal and catastrophic, *Arctic* 20, 86-103.

John, B. S., and Sugden, D. E., 1975. Coastal geomorphology of high latitudes, *Progress in Geography* 7, 53-132.

Lewellen, R., 1970, *Permafrost Erosion along the Beaufort Sea Coast.* Denver: University of Denver, 25p.

McCann, S. B., and Taylor, R. B., 1975. Beach freezeup sequence at Radstock Bay, Devon Island, arctic Canada, *Arctic and Alpine Research* 7, 379-386.

Mackay, J. R., 1972. The world of underground ice, *Assoc. Am. Geographers Annals* 62, 1-22.

Reimnitz, E., and Barnes, P. W., 1974. Sea ice as a geologic agent on the Beaufort Sea shelf of Alaska, *in* J. C. Reed and J. E. Sater, eds., *The Coast and Shelf of the Beaufort Sea.* Arlington: Arctic Institute of North America, 301-354.

Sater, J. E., ed., 1969. *The Arctic Basin.* Washington, D.C.: Arctic Institute of North America, 337p.

Walker, H. J., 1974. The Colville River and the Beaufort Sea: Some interactions, *in* J. C. Reed and J. E. Sater, eds., *The Coast and Shelf of the Beaufort Sea.* Arlington: Arctic Institute of North America, 513-540.

Wiseman, W. J.; Coleman, J. M.; Gregory, A.; Hsu, S.-A.; Short, A. D.; Suhayda, J. N.; Walters, C. D.; and Wright, L. D., 1973. *Alaskan Arctic Coastal Processes and Morphology.* Baton Rouge: Coastal Studies Institute, Tech. Rept. 149, 171p.

Cross-references: *Arctic, Coastal Ecology; Europe, Coastal Morphology; Glaciated Coasts; High Latitude Coasts; Ice along the Shore; Ice-Bordered Coasts; Isostatically Warped Coasts; North America, Coastal Morphology; Soviet Union, Coastal Morphology.* Vol. VIII, Part 2: Arctic.

ARTHROPODA

The phylum Arthropoda (Gr. *arthros* joint + *podos* foot) contains segmented animals whose epidermis secretes an exoskeleton of rings. The rings are of varying widths corresponding with the segments and are connected by more or less flexible membranes that act as joints. Many segments bear paired lateral appendages, each with a similar chitinous exoskeleton of jointed rings. The exoskeleton is an organic chemical complex that includes a nitrogenous polysaccharide —chitin—as its chief element.

The phylum Arthropoda includes: the crabs, shrimps, barnacles, lobsters, most members of the zooplankton, and other crustaceans (Class Crustacea [L. *crusta,* a hard shell]), flies, fleas, cockroaches, springtails, bees, grasshoppers, moths, butterflies, lice, beetles, and other insects (Class Insecta [L. *inciso,* into distinct parts]); spiders, scorpions, mites, ticks, harvestmen, and their allies (Class Arachnida [Gr. *arachne,* spider]); centipedes (Class Chilopoda [L. *Chilos + poda,* jaw feet]); millipedes (Class

Diplopoda [L. *diplos* + *poda*, double feet]); and other less familiar and fossil forms.

The body is basically tripartite and includes a head, a thorax, and an abdomen, each composed of several segments that may be fused in a variety of ways. The body musculature, made up of numerous small muscles extending across joints and attaching to the inside of the rings, helps form an intricate mechanism capable of precise and often intricate movements. The exoskeleton not only provides muscle space for muscle attachment and forms a protective cover successful worldwide and in virtually all habitats, but its division into numerous parts makes possible many structural and functional adaptations to particular environments.

In this way the evolutionary potentialities of the arthropodan structural system have adapted them for highly specialized habitats, in some cases so precisely adapted that it has resulted in severely limited ecologic distribution (Edmondson et al., 1959). This is thought to be one of the main explanations for the existence over the years of the enormous number of species of arthropods. The phylum contains about 80% of all known animals, about 1,000,000 species have already been described, many tremendously abundant as individuals; some authorities think that in time the number will exceed 5,000,000 (Gosner, 1971). It is the only major invertebrate phylum with many members adapted for life on land, often far away from moist surroundings; and the insects (q.v.) are the only invertebrates capable of flight.

Arthropods are both cosmopolitan and ubiquitous between the poles; they live at altitudes of over 20,000 feet on mountains, and have been collected from depths of more than 32,000 feet in the sea. A great variety of species are adapted for life in the air, on land, in soil, and in freshwater, brackish water, or saltwater, as well as in association with other animals and with plants as commensals, phoretics, and parasites (Shaller, 1968; Eltringham, 1971). While some are gregarious, several kinds of colonial insects have evolved highly developed social organizations with a division of labor organized among members of different castes. They range in size from microscopic mites to the giant Japanese spider crab, which may measure 10 to 12 feet in its overall dimension.

Many arthropods are important economically. The larger crabs, lobsters, shrimps, and several kinds of insects are eaten by man; small crustaceans are staple foods for commercially important fishes; and insects and spiders are common in the diets of birds and in the food chains of many valuable land vertebrates. Although the largest number of insects are beneficial, others are man's chief competitors, eating his crops, stored foods, household foods, or clothing. Some insects and ticks injure or carry diseases to man, his domestic animals, and his crops as well as to forests and to wildlife.

Morphologic and embryologic evidence indicates that arthropods have evolved from annelid ancestors. The Arthropoda and the Annelida (*segmented roundworms*) are the only invertebrate phyla with both conspicuous external segmentation and segmented muscular and nervous systems. However, the arthropods differ from the annelids in lacking cilia, reduction of the coelom (the body cavity), concentration of excretory organs and gonads, separation of the sexes, compound eyes, and, as already noted, the presence of an exoskeleton and jointed appendages. It is perhaps the thick chitinous covering, preventing the loss of water by evaporation from the body and providing not only skeletal but also muscle support, that has enabled arthropods to have invaded the dry land successfully early during their evolution.

Arthropods have other distinctive structural characteristics as well as complex behavior patterns. Not the least of their beneficial characteristics has been the general presence of striated muscle. Their evolutionary divergence and successful adaptive radiation have depended on many factors, particularly the mechanism of locomotion. A successful invasion of land and air environments must involve methods for travel, and arthropods have succeeded remarkably, for example, in utilizing rapid locomotion when avoiding enemies and in pursuing prey. Inherent in this accomplishment is the basic arthropodan body plan providing a series of segments with muscular movement, each segment with one or more pairs of jointed appendages, both segments and appendages being specialized for many varied functions.

Arthropods often display a range of color patterns in the different groups, particularly in insects and crustaceans. However, many of the smaller planktonic species are virtually translucent when living, eyes, oil droplets, and gut contents bringing the only vestiges of color to otherwise almost transparent bodies. Both true pigments and structural colors are represented through carotenoids and melanins as brilliant yellows, oranges, and reds as well as shades of blues and greens, either in the cuticle or in chromatophores. Some colors, and especially iridescence, is produced primarily by the varied effects of light interference.

The color patterns are a function of the unique five-layered integument (*exoskeleton*) whose 0.1–1.0 micron-thick outermost layer (*epicuticle*) lacks chitin and is often the waxlike substance that provides most of the impermeability of this structure to water. The texture of the cuticle varies from rough to smooth, often being sculptured with pits, spines, or ridges, and perforated for sensory setae (thick hairs) and gland openings. The chitin, characteristic of arthropodan integument, is a polymer (*polysaccharide*) of high molecular weight, similar to cellulose, and is always associated with proteins. Flexible portions of the cuticle called membranes (outermost part of integument) are associated with movable articulations of the exoskeleton and in addition are part of the lining of the foregut, hindgut, trachae, and sections of the reproductive tract.

Throughout the phylum as well as in single species, the appendages, showing a diversity in form unique in the animal kingdom are adapted to entirely different functions, some forming specialized tools that mandate a highly developed nervous system. The limbs and other appendages between sternite and tergite form hollow outgrowths of the body wall and are connected to the body by a variety of well-adapted muscles. The arthropod limb, basically and primitively biramous, may have fixed or movable lateral lobes, called exites and endites. Although the number of somites varies in primitive groups, it is fixed in recent, advanced forms. The body is divided into groups of segments (*tagmata*); a number of fused segments may form a continuous shield, as in the head (*cephalon*), or several similar movable segments (*thorax*, abdomen), and two such regions may in turn be fused (*cephalothorax*).

As might be expected, the often thick and nearly always inflexible arthropodan exoskeleton presents problems both of locomotion and of growth (Barnes, 1980). The first problem is solved by specially adapted jointed and segmented limbs made flexible by articular membranes and sockets as well as by systems of separate muscles. Division of the exoskeleton into hard plates (*sclerites*) connected by the flexible membranes referred to above permits both body and jointed leg movements. Although each body segment ordinarily has a tergum (*dorsal sclerite*), a sternum (*ventral sclerite*), and two pleura (lateral sclerites), there are numerous variations of this arrangement accomplished by fusion, elimination, or other structural modifications. S. M. Manton (1970), foremost student of arthropod locomotion, explained that movement takes place by the exertion of forces proportional to the volume of contractible fibers in the muscles and that speed depends on length of stride, which in turn increases with the length of the leg. She noted that arthropods with high speeds have quick thrusts and that most of the legs are off the ground most of the time.

The second problem is that the jointed but fixed external armor necessarily limits the interior size of the animal and allows little room for expansion. For this reason, it must be shed periodically (molting or *ecdysis*) to permit growth (Hickman, 1973). Just before molting (under hormonal control), most of the inner cuticle is digested away by enzymes, freeing it from the hypodermis. At the same time, a new cuticle is secreted beneath the old by the hypodermis. The actual molting or shedding of the old cuticle now proceeds, aided by swelling induced by absorption of air or water; the cuticle is usually split mid-dorsally first. The animal struggles out of the old cuticle, increases its body size by further absorption, and the new cuticle is soon hardened by the resorbed chitin and the reserve of lime salts. During this period the animal has little protection and usually remains in hiding. The period between molts is called instar; molting occurs most frequently during larval development; most arthropods have from four to seven molts.

While the sizes of many species of arthropods have been restricted by difficulties connected with molting, other arthropodan taxa have been limited by the relatively great weight and bulk of the exoskeleton and the involved mechanics of muscle origins and insertions. For example, the largest living isopod known is *Bathynormus giganteus*, 35 cm in length, and, no living insect exceeds 27.5 cm in wing spread or in length. The tremendous numbers of small species of crustaceans, insects, and mites included in the interstitial fauna worldwide are under 1 mm long (Swedmark, 1964).

Both because the arthropod body is enclosed in a more or less solid exoskeleton and because the joint-legged invertebrates grouped together in this phylum are an extremely numerous and varied assemblage, a great number of fossil forms have been found and described from all geological systems younger than Precambrian (Moore, 1969). No arthropod remains are known with certainty from Precambrian rocks. The arthropod fossils of great antiquity include several kinds of complexly organized *trilobites* found in the lowest fossil-bearing Paleozoic strata, and so the origin of this group surely belongs to some part of Precambrian time. The trilobites appeared near the beginning of the Cambrian, had a very strong development and distribution through the Middle and the Late Cambrian, and then declined and became extinct in the Permian. Both the Crustacea and the primitive Merostomata very probably existed in the Lower (Early) Cambrian. The Eurypterida (ancestors of the insects) became extinct in the Permian, but the Xiphosura still exist in much the same form as they did in the Ordovician period and are good examples of so-called *living fossils*. The Arachnida were well established in the Silurian, and the first Myriapoda appeared near the start of the Devonian. The first wingless insects occurring in the Middle Devonian preceded the strong development of the winged forms from the Early to the Late Carboniferous. Fossils of the Pycnogonida have been found in the Lower Devonian. Today the groups with many species appear to be in a state of relatively rapid evolution or radiation, and the same is considered to have been true in earlier periods.

Arthropods are primarily dioecious, a few are hermaphroditic, and some groups are parthenogenetic. During copulation many use special appendages that in a large number of cases are highly modified. Fertilization is largely internal in terrestrial forms and external in aquatic species (Pennak, 1978). The commonly yolk-rich centrolecithal eggs have modified the usual course of cleavage into what is called superficial cleavage resulting in an egg lacking cell membranes and yolk cleavage but containing a large number of syncytial nuclei within the center. The gonads are nearly always in direct contact with the coelomoducts, and the genital openings are not constant in position: they may open near the posterior end, near the middle of the body, or even toward the head end. During development, metamorphosis may be absent, gradual, or abrupt, and in nearly all cases there are larval stages, the number varying with the family or genus. Parasitic crustaceans and arachnids have generally aberrant patterns of development.

The nervous system arrives at a great degree of complexity in higher arthropods, although the basic plan is similar to that of annelids (Megalitsch, 1972). It usually consists of two ventral nerve cords, a dorsal brain, and a pair of ganglia for each somite. Depending on the arthropod group, the number of ganglia fused together in the head varies with the merging of the head segments. From the ganglia nerves run to the body wall and to the visceral organs. Organs of sense include variously structured internal proprioceptors as well as more or less complex surface receptors sensitive to olfactory, tactile, chemical, and visual (simple and compound eyes) stimuli.

DONALD J. ZINN

References

Barnes, R. D., 1980. *Invertebrate Zoology*. Philadelphia: Saunders College, 1089p.
Edmondson, W. T.; Ward, H. B.; and Whipple, G. C., eds., 1959. *Freshwater Biology*. New York: John Wiley & Sons, 1248p.
Eltringham, S. K., 1971. *Life in Mud and Sand*. New York: Crane, Russak & Co., 218p.
Gosner, K. L., 1971. *Guide to Identification of Marine and Estuarine Invertebrates*. New York: Wiley Interscience, 693p.
Hickman, C. P., 1973. *Biology of the Invertebrates*. St. Louis: C. V. Mosby Co., 757p.
Manton, S. M., 1970. Arthropods: Introduction, in M. Florkin and B. T. Scheer, eds., *Chemical Zoology*, vol. 5(A). New York: Academic Press, 1-34.
Megalitsch, P. A., 1972. *Invertebrate Zoology*. New York: Oxford University Press, 834p.
Moore, R. C., ed., 1969. *Treatise on Invertebrate Paleontology*, Pt. R, Arthropoda 4. Geological Society of America and University of Kansas Press, 651p.
Pennak, R. W., 1978. *Freshwater Invertebrates of the United States*. New York: John Wiley & Sons, 803p.
Schaller, F., 1968. *Soil Animals*. Ann Arbor: University of Michigan Press, 144p.
Swedmark, B., 1964. The interstitial fauna of marine sand, *Biol. Rev.* 39, 1-42.

Cross-references: *Biotic Zonation; Coastal Fauna; Insects; Organism-Sediment Relationship; Succession.*

ARTIFICIAL ISLANDS

Artificial islands, also known as offshore or floating islands, are man-made structures situated some distance away from the shoreline to serve as a part of a transport system or as an isolated industrial, residential, or defense site. As such, they can reduce the risk of collisions of increasingly large cargo ships, can provide a built-in design for handling industrial air, noise, and water pollution, and can offer an aesthetic alternative for residential and recreational purposes. To date only a few permanent artificial islands have been constructed, such as the one in Kobe Harbor, Japan and in the Beaufort Sea. Several were constructed to facilitate drilling as part of exploration for oil and gas offshore. Others are in the research or planning stage.

These offshore islands are being planned for use primarily as a transshipment location—a place where cargo will be unloaded from larger incoming ships and either piped onshore or transferred onto smaller ships and vice versa for outgoing ships. Incoming materials could also be processed on the island before shipment ashore.

Another possible major function is that these islands may serve as industrial locations. The benefits to industry would be: shorter distances to the market area than isolated land sites; islands can be designed to the specifications of industry without problems caused by hills, streams, and highways, as in inland sites; islands can be expanded as necessary, perhaps using solid wastes; fewer environmental constraints; unlimited availability of cooling water; and availability as a deepwater port. The major disadvantages are: labor force and material transportation problems; must be self-supporting and offer fire and police protection, sewage treatment, and electric power; destructive power of waves and storms; jurisdictional problems; and sociological problems caused by the work environment.

The use of offshore islands by industries such as nuclear power and fuel-processing plants may need to be exclusive. Best suited include processing industries, oil-related industries, liquid natural gas (LNG) industries, fossil power plants, coal industries, offshore mining industries, fish farming industries, solid waste disposal industries, and noxious product industries.

Other possible functions include residential and recreational islands surrounded by their own exclusion zones, in which certain industries would not be allowed to locate, thus forming a type of offshore zoning plan. These islands could also serve as military defense sites.

Artificial Island Studies

The *Bos Kalis Westminster* report (North Sea Island Study Group, 1975), which was prepared by a group of dredging companies, resulted in a conceptual design for a tentative North Sea island. As a result, the North Sea Island Study Group was formed representing the Dutch government and some thirty industries in Holland, England, France, Belgium, and Sweden. This group conducted a feasibility study for the construction of a multipurpose industrial island some thirty miles from the coast of Holland. It was found that at the proposed location sand of good quality is available in sufficient quantities for island construction. The total net island area will be about 3330 hectares. It was determined that the construction of the island and the industries upon it will have a considerable beneficial effect on the Dutch national economy. The island is expected to generate between 1 and 2 percent of the total Dutch national income after completion. This island would significantly reduce the sometimes dangerously high traffic levels of ships in the Europort area.

Another study was conducted by the University of Delaware, College of Marine Studies, with participation from Texas A & M University, Gilbert Associates, and Frederic R. Harris, Inc. The study considered the feasibility of artificial islands off the United States east and Gulf coasts for industrial and port purposes. They developed a hypothetical 1200-hectare industrial site plan with space for personnel housing (Anonymous, 1976).

Studies conducted for the Public Service Electric and Gas Company of New Jersey have led to a proposal for the construction of an offshore nuclear power plant.

Other studies were conducted at the University of Hawaii in 1974 considering the construction of floating offshore islands for residential and recreational purposes (St. Denis, 1974). Studies in Japan have led to the construction of an artificial island inside Kobe Harbor.

Construction Methods and Materials

The types of methods and materials proposed for use in the construction of offshore islands are as diverse as the functions of these islands: unprotected beach, polder, sheet-pile cell, caisson, and dike and fill. The unprotected beach method would require constant maintenance and is considered impractical. In the *polder* method, a dike is built around an area whose level is below mean low water. The risk involved in this type of structure is considered to be high, especially during earthquakes or hurricanes. In the sheet-pile cell method, the island area between the sheet piles is filled to a level above mean low water. The caisson-protected island is similar to the sheet-pile structure, except that filled steel caissons are used for protection. The dike-and-fill method is considered to be the most practical. In this method, rubble dikes are constructed and the island area is then filled to a level above mean low water.

The most effective shape for an offshore island is that of a circle, because it provides the maximum area with a minimum perimeter that must be protected. It also limits the possibility of local wave energy being concentrated on the sea defense system.

Floating islands, either semisubmersible or surface floating, are a possibility in locations where wind, wave, and current conditions afford a relatively stable position.

Many factors influence the selection of a specific location for artificial islands including: sea-bottom strength; proximity to shipping lanes; waves, tides, currents; dominant winds; and frequency of occurrence of hurricanes, tornadoes, and earthquakes.

Legal and Environmental Aspects

The legal aspects of the construction of offshore islands within the national territorial waters has been established, but the right to build outside these waters is unclear. Artificial islands are rarely mentioned in the existing written laws of nations. Within the territorial waters it is clear that the nation can govern island activities. Construction in waters outside this zone must be settled through international agreements such as those considered at the Law of the Sea Conferences.

The construction of an offshore island will

definitely cause environmental changes. Some of the effects may be beneficial. The heat of diffusion from industrial processes into the ocean could possibly enhance the area for fishermen. The movement of industries to offshore locations could cause onshore noise abatement, and solid waste disposal from onland sources could be used for island expansion. Materials for the construction of some types of islands will come from the seabed. The result of these changes and any possible effects that the island might have on the coastline must be studied for individual situations. Other effects that must be considered are the effects of the islands on ocean currents. Careful planning and construction practices are required to control turbidity, spills, particulate emissions, water treatment, and waste disposal.

Artificial islands appear to hold great potential for the future. More research is required in all phases of planning and construction, and action is especially necessary in the area of jurisdiction and other legal aspects.

JOHN B. HERBICH
JOHN P. HANEY

References

Anonymous, 1976. *Multi-Purpose Offshore Industrial-Port Islands.* Newark: College of Marine Studies, University of Delaware, 23p.

North Sea Island Study Group, 1975. *Artificial Islands.* Working Plan, April, 1975.

St. Denis, M., 1974. *Hawaii's Floating City—Development Program.* Honolulu: UNIHI-Seagrant-CR-75-01, 93p.

Cross-references: *Coastal Engineering; Coastal Engineering, History of; Coastal Zone Management; Land Reclamation; Nuclear Power Plant, Siting.* Vol. XIII: *Offshore Nuclear Plants; Offshore Platforms.*

ARTIFICIAL SHORELINES

Three conditions may exist along any given beach or shoreline: predominant accretion exceeds erosion rates, and the shore is aggrading; the shoreline is stable, and neither erosion nor accretion predominates; the beach is in a state of erosion. Only this latter condition is usually considered of economic importance, although the erosive accretion may also cause problems to existing facilities that lead to the disappearance of natural protective beaches and shorelines and cause the greatest economic impact on public as well as private concerns.

Artificial shorelines encompass a variety of containment areas and methods. Several types of structures are currently being used to create or recreate new shorelines.

Sea walls and *revetments* are structures placed parallel to the shoreline with the primary purpose of protecting the backshore property from damage by wave forces. *Bulkheads* are used to retain or prevent these lands from sliding into the surf zone.

Groins are shore-protective structures designed to nourish or maintain a beach by trapping littoral drift. These are erected perpendicular to a beach, are relatively narrow in width, and may vary in length from less than 30 meters to several hundred meters.

Jetties are structures, like groins, erected perpendicular to the shore. They are more massive than groins and function primarily to redirect and confine the stream or tidal flow so as to minimize littoral drift from an endangered beach front or to provide protection to a ship channel.

Stabilizing barrier dunes (Savage and Woodhouse, 1969), with various types of ground cover and aligned parallel to existing beach fronts, have also proved successful in trapping wind-blown sand. Sand fences and dune grasses will trap wind-blown sand and create these barrier dunes.

Once significant erosion has taken place, waste material and sanitary landfill techniques are an effective regional coastal-zone planning practice to satisfy the high-volume shoreline nourishment needs and attenuate some of the site-selection problems in land disposal of waste (Williams and Duane, 1975).

JOHN B. HERBICH
ROY B. SHILLING

References

Savage, R. P., and Woodhouse, W. W., Jr., 1969. *Creation and Stabilization of Coastal Barrier Dunes,* Reprint 3-69. Washington, D.C.: U.S. Army, Corps of Engineers, 31p.

Williams, S. J., and Duane, D. B., 1975. *Construction in the Coastal Zone: A Potential Use of Waste Materials.* Reprint 2-75. Washington, D.C.: U.S. Army, Corps of Engineers, 15p.

Cross-references: *Breakwaters; Bulkheads; Coastal Engineering; Groins; Jetties; Protection of Coasts; Riprap.*

ASCHELMINTHES

This group of diverse organisms is sometimes considered to be a single phylum, but more often the term is used to describe a general category containing a number of phyla that share certain similarities. Included are the Rotifera,

Gastrotricha, Kinorhyncha, Nematoda, Nematomorpha, Gnathostomulida, and Acanthocephala. The phyla Priapulida and Entoprocta are sometimes included as well (Hyman, 1951; Hickman, 1963). All are *"pseudocoelomate,"* possessing a central body cavity that contains the internal organs and gives the organism a tube-within-a-tube structure. The Rotifera, Gastrotricha, and Kinorhyncha are microscopic wormlike forms. The latter is strictly a marine bottom dweller, while the former two groups are found attached to solid substrates in marine and aquatic habitats. Nematoda (true *roundworms*) dwell in marine, aquatic, and terrestrial environments as well as being common plant and animal parasites. Thirteen thousand species are known, ranging in size from 1000 microns to 1 mm. Similar to the nematodes, the Nematomorpha (*"hairworms"*) are slender but grow to lengths of up to one meter. Adults may be terrestial, marine, or aquatic, while the larva are parasitic upon insects and other invertebrates. The Gnathostomulida are a recently discovered phylum, first reported in 1928 from the Baltic Sea. Found in fine marine sediments in the intertidal zone, the tiny semitransparent wormlike forms occur in great number in the interstitial spaces between sand grains. Included are some 100 species, with worldwide distribution. Acanthcephala (*"spiny-headed worms"*) are parasites in the digestive tract of vertebrates during adulthood but infect arthropods during the juvenile stage. Over 500 species are known, occurring mostly in fish but also infecting amphibians, reptiles, and mammals (Dougherty, 1963).

The Priapulida have a cylindrical, wart-covered body several centimeters long, and possess an eversible proboscis used to capture polychaetes, small crustaceans, and even other priapulids during feeding. They occur in colder latitudes in both hemispheres at depths from 500 meters to the intertidal zone. Because of their dissimilarity to the other Aschelminthes, some zoologists class the Priapulida among the *"eucoelomate"* phyla such as the Echiurida, Siphunculida, Tardigrada, Pentostomida, and Onychophora. Although the Entoprocta are often classed as "Bryozoa" or "Polyzoa" along with the Ectoprocta, they possess the basic pseudocoelomate structure shared by other phyla of the Aschelminthes group.

GEORGE MUSTOE

References

Dougherty, E. C., 1963. *The Lower Metazoa.* Berkeley and Los Angeles: University of California Press, 478p.

Hickman, C. P., 1973. *The Biology of Invertebrates,* 2d ed. St. Louis: Mosby, 757p.

Hyman, L. H., 1951. *The Invertebrates,* Vol. 3. New York: McGraw-Hill, 572p.

Cross-references: *Annelida; Biotic Zonation; Bryozoa; Chaetognatha; Coastal Fauna; Coastal Waters Habitat; Echiuroidea; Intertidal Sand Habitat; Nemertea; Platyhelminthes; Priapuloidea; Siphuncoloidea.*

ASIA, EASTERN, COASTAL ECOLOGY

Eastern Asia (89° to 129°E.) lies centrally in the Indo-West Pacific biogeographical region (Ekman, 1967), and includes Bangladesh and the eastern Bay of Bengal, continental and insular Southeast Asia, eastern China, and Korea. The Java and Flores seas are the boundaries of the Indian Ocean and West Pacific Ocean influences in the region.

Coastal environments in eastern Asia are the result of climatic, geomorphological, tidal, and biological processes. The Malay-Indonesian province is bathed by shallow, highly productive seas overlying the Sunda Shelf (the drowned peninsula of Pleistocene Sundaland). Here extensive coastal areas lie within the 60-m contour. At the eastern limit of the shelf, the Philippines border deep seas that separate the Sunda Shelf region from the Sahul (Arafura) Shelf of Australasia (Fig. 1).

North of the equatorial belt the NE and SW monsoons dominate the climate and surface ocean circulation. Rainfall becomes increasingly seasonal east and north from the *everwet* Malaysian region. Water currents are reversed seasonally with monsoon changes in wind direction; upwellings are also generated (Wyrtki, 1961, 1962) that bring nutrient enrichment to the surface waters.

Tranter (1974) has reviewed the coastal biology of Southeast Asia. Large areas, particularly the archipelagoes, remain unstudied, although this region contains marine communities of unequaled richness. Mangrove swamps and coral reefs characterize tropical Asian coasts, where these communities attain their greatest diversity. However, even the limited biotas of sandy beach and rocky shore habitats are poorly known in eastern Asia.

Mangrove Swamp Forests

Distribution. Chapman (1970) noted the extreme diversity of the Indo-Pacific mangrove flora (containing 63 of 90 genera worldwide) in comparison with that of equivalent latitudes on the Atlantic shores of Africa and the Americas. Mangrove is the dominant coastal community in tropical Asia (Fig. 2) with the Malay-Indonesian region its center of distribution. Macnae (1968) and Walsh (1974) give excellent accounts of mangrove ecology.

Mangroves flourish on sheltered shores where rainfall is high and aseasonal; average minimum

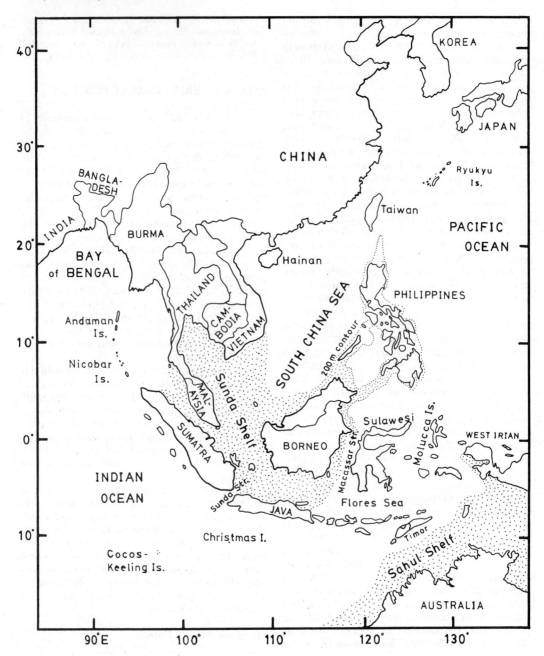

FIGURE 1. General map of eastern Asia; the Sunda and Sahul Shelf regions are shaded.

temperatures must exceed 20°C. Their northern limits in Asia are China (about 25°N.), the Ryukyu Islands (26–27°N.), and the Yaeyama Islands (25°N.), where winter temperatures fall to 17–22°C. Salt marsh replaces mangrove in temperate latitudes. The extensive salt marshes of Korea are probably similar to those of Japan (e.g., Ito and Leu, 1962); the primary community is apparently a *Salicornia* association.

Ideal conditions for mangrove occur along the coasts of the Sunda Shelf, especially within the Malacca Straits (a drowned estuary), which are protected from prevailing winds and ocean currents by Sumatra. This shelter and the shallowness of the coastal shelf allow the accumulation of fine, riverborne sediments into intertidal mud banks that mangroves rapidly colonize (Fig. 3). Extensive mangrove shores in Borneo face the

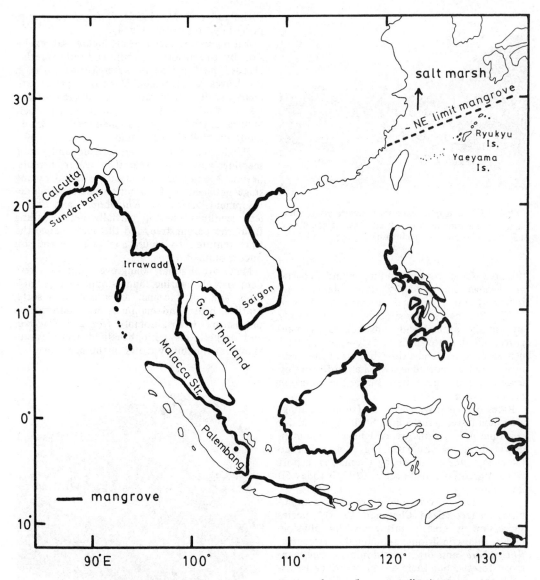

FIGURE 2. Distribution of mangrove swamp forests (heavy coastlines).

open sea, but are sheltered from the main prevailing winds by their direction (Macnae, 1968).

In contrast, river sediments carried to the east coast of Peninsular Malaysia are dispersed offshore by the stronger wave and current forces of the South China Sea. Only heavier marine sands, too unstable and infertile for mangrove, accumulate along the coast. These are thrown up as low dunes (*cheniers*) extending parallel to the shore. Similar dispersive forces exclude mangrove from the Indonesian shores outside the Sunda Shelf region, which are predominantly rocky and sandy.

Elsewhere in eastern Asia mangroves are associated with deltas, notably those of the Ganges and Irrawaddy rivers. The southern Gangetic delta (Sundarbans, 21°45′ to 22°30′N.) borders both India and Bangladesh and contains almost 6000 km^2 of tidal forest, including mangrove, dominated by the important timber tree *Heritiera* (*sundri*). Extensive coastal areas of eastern Bangladesh (Khulna to Chittagong) are also tidally flooded. Mangroves border estuaries and tidal creeks throughout this region; dune vegetation, typically *Casuarina*, fringes the open coastline (Fasberg, 1966). *Casuarina* also prevails along the sand dune coasts of Arakan and Tenasserim in Burma.

About 5200 km^2 of the Irrawaddy coastal zone is inundated by spring high tides. Much of

FIGURE 3. Mangrove seedlings (*Avicennia*) colonizing an estuarine mud flat, west coast of Peninsular Malaysia. (Photo: D. J. Macintosh)

this area is mangrove extending behind a coastal sand ridge colonized by dune plants (e.g., *Ipomaea*) and trees such as *Eugenia* (Stamp, 1925). Sediments deposited by the Irrawaddy are rapidly invaded and consolidated by mangrove, but the rate of delta extension (10 km^2 per year) is relatively slow because some sediments are transported eastward and offshore by coastal currents generated during the monsoon period (Volker, 1966).

Rates of mangrove accretion (shore building) are much faster in the Sunda Shelf region. Major coastline progressions are evident on some mangrove shores of Sumatra, Java, and Borneo (Macnae, 1968). Palembang (Sumatra), a port when Marco Polo visited it in 1292, is today 50 km inland.

Vegetation. Several mangrove plant communities are recognized (Chapman, 1970). Species succession is greatly influenced by physiography, tidal conditions, and climate. *Avicennia*, *Rhizophora*, and *Bruguiera* dominate *true* mangrove forests, but there are many other associated genera that under local conditions may exceed the mangrove in importance. In eastern Asia *Heritiera* (Sundarbans, Irrawaddy), nipah palm (Borneo, Philippines), the fern *Acrostichum* (South Vietnam), and sago palm (Moluccas) occupy extensive areas of tidal swampland. Mangrove floral and faunal diversity decreases northward from the Malay region; eastward mangroves become less luxuriant because rainfall is increasingly seasonal, but their diversity remains high.

Mangrove trees are variously adapted to the littoral (intertidal) environment. Coastal sediments are unstable, saline, and generally anoxic. *Avicennia* trees develop extensive cable roots that bear secondary roots for anchorage and aerial roots (pneumatophores) for respiration (Scholander et al., 1955). *Rhizophora* is supported by prop roots (Fig. 4).

Salt-regulating mechanisms include salt exclusion by the absorptive roots and salt-excreting glands; the function of water-storing tissue in the leaves is uncertain. Mangroves grow optimally in salt concentrations 20–50% that of seawater (Walsh, 1974), which may explain their luxuriance in the Sunda Shelf region, where rainfall is high all year round.

The buoyancy of mangrove fruits and young seedlings aids their dispersal by water currents. Several species are viviparous with no resting stage between seed formation and seedling development. *Avicennia*, *Rhizophora*, and *Bruguiera* seedlings develop partially before falling from the parent tree, and this may reduce the time required to secure anchorage in the soil once grounded.

Mangrove Fauna. Mangroves contain a resident, largely marine fauna dominated by molluscs and crustaceans; a variety of insects, reptiles, birds, and mammals are semiresidents or visitors. Crabs, notably *Uca* and *Sesarma*, and amphibious periophthalmid fishes (mudskippers) inhabit burrows in the mangrove soil

FIGURE 4. *Rhizophora* trees with ramifying prop roots for support. (Photo: D. J. Macintosh)

but emerge onto the surface to feed. The permanent soil fauna includes polychaete worms and sipunculids (peanut worms). Gastropods, bivalves, and barnacles colonize the trees (Sasekumar, 1974).

Mangrove animals are zoned both horizontally (over the shore) and vertically on the vegetation (Berry, 1972) according to their ability to withstand exposure to air and salinity fluctuations caused by rainfall and evaporation. Each species seems well adapted to its particular intertidal location (Berry and Chew, 1973; Macintosh, 1977).

Coral Reefs

Distribution. Coral reefs flourish within a depth of 40 m in tropical, fully marine waters. Strong water circulation is necessary to carry food (zooplankton) and oxygen over the reef

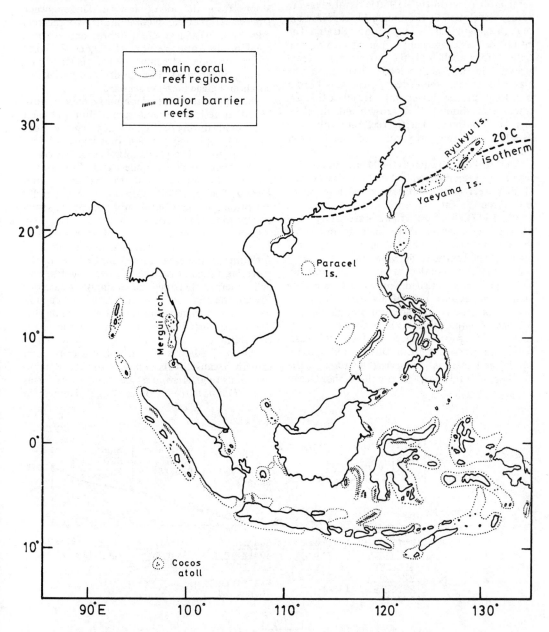

FIGURE 5. Distribution of coral reefs; dotted lines enclose areas with abundant reefs. Major barrier reefs are marked (after Molengraaff, 1930). 20°C isotherm is sea temperature in coldest month of the year.

and to remove waste products. Reef-building corals do not tolerate temperatures less than 20°C, and their northern limit in eastern Asia coincides with the 20°C isotherm (Fig. 5).

Fringing (shore-edge) reefs occur along the Indian Ocean coasts of Sumatra, Java, the Mergui Archipelago (Burma), the Andaman and Nicobar islands, and Christmas Island (10° 30'S., 105° 40'E.). The Cocos Islands of the Cocos-Keeling group (12° 10'S., 96° 55'E.) form a coral atoll (Gibson-Hill, 1950). Coral reefs are poorly developed in the Sunda Shelf region except in marginal areas (e.g., Sunda Strait). The inflow of freshwater and sediment into coastal waters of the shelf is highly detrimental to coral growth. In the archipelagoes east from the Makassar Strait, where deep, clear seas border island coastlines descending steeply offshore, coral reefs abound. The growth and development of Southwest Pacific reefs are enhanced similarly by the proximity of deep water (Maxwell, 1968).

Coral reef distribution in the Malay-Indonesian region has been mapped in detail by Molengraaff (1930). Reef shape is influenced by the strength and direction of the monsoons (Umbgrove, 1930, 1947). Because of the seasonal reversal of water circulation, these reefs do not show the characteristic windward and leeward reef forms of the Pacific trade wind zone.

Biology. The coral community becomes increasingly impoverished outward from the Malay-Indonesian center of diversity. The number of species of familiar coral reef gastropods (*Conus, Strombus, Cypraea*) and fishes (e.g., surgeon fishes: Acanthuridae) decline both eastward and westward from this center, (Kohn, 1967), and there is a westward decrease in the number of coral genera across the Indian Ocean (Wells, 1957).

Unfortunately, recent coral reef studies have concentrated on the Pacific Ocean (Marshall Islands, Great Barrier Reef) and West Indian Ocean (Seychelles, Maldives, Aldabra), and general accounts of reef biology (Mukundan and Gopinadha Pillai, 1972; Jones and Endean, 1973) are noticeably lacking in information from the Malay-Indonesian region.

Two main reef zones—the reef flat and reef slope—are generally recognized. The reef flat slopes from the lower shore or lagoon beach (in barrier reefs and coral atolls) to the reef edge (Fig. 6). Sand, coral debris, and isolated coral heads form the upper flat (Fig. 7); coral heads provide attachment for algae (e.g., *Padina, Sargassum*). The density and variety of corals and algae increase seaward.

The upper reef flat community may be subjected to desiccation and rainfall dilution when uncovered during extremely low spring tides (Fig. 8), and only certain organisms tolerate this exposure. The characteristic reef flat corals in the Malay region include *Favia, Goneastrea, Porites, Goniopora*, and the solitary coral *Fungia* (Searle, 1956). Soft corals (Alcyonaria) are often abundant but do not contribute structurally to the reef flat.

Crevice-dwelling molluscs and anenomes and molluscs that bore directly into coral heads (e.g., *Lithophaga*, the date mussel) are features of the reef flat fauna. Crustaceans, polychaetes, and echinoderms become increasingly numerous toward the reef edge; xanthid crabs, sea cucumbers (holothurians), and the crown urchin *Diadema* are particularly abundant on the lower reef flat.

The reef edge is a zone of vigorous coral growth because of its exposure to surf. Branching corals (e.g., *Acropora*) and foliaceious forms (e.g., *Echinopora*) colonize surge channels in

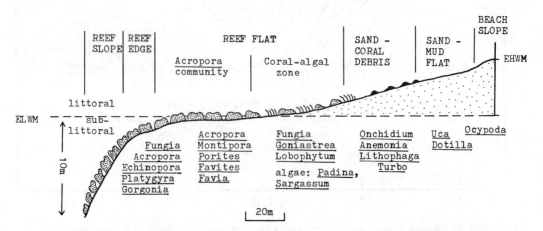

FIGURE 6. Diagrammatic zonation of a fringing coral reef and sandy shore in the Malaysian region. Representative corals and shore organisms are indicated.

FIGURE 7. Coral heads and coral debris on the upper zone of a reef flat. (Photo: D. J. Macintosh)

FIGURE 8. Coral reef flat exposed at extremely low tide, Phuket, Thailand. (Photo: D. J. Macintosh)

the reef edge and slope. Cyclones, which can extensively damage reefs (Wiens, 1962), are rare in the Malay-Indonesian region, and this may account for the dominance of fast-growing, branching corals. Relatively massive corals and alcyonarians become important further down the reef slope.

Community structure and variation in the reefs of the Malay-Indonesian region are probably comparable to those of the better-documented Indian Ocean region (Stoddart, 1972).

Sandy Beaches

Distribution. Sandy beaches occur extensively on the shores of coral islands and are interspersed among other shore formations throughout continental Asia. Steep beaches of coarse sand are built up on ocean-facing coasts exposed to strong surf. Intertidal flats of mixed sediments (sand-silt), with a narrow sandy fringe at high water mark, develop on more protected shores.

Biology. Sandy beach communities of the Malay-Indonesian region are similar to those described in India (Tranter, 1974). Only a restricted fauna tolerates the surf forces and instability of an exposed sandy shore. Tropical organisms are further inhibited by high temperatures and desiccation. Most animals must burrow for protection or limit their surface activity to periods when the sand is moist. Scavenging hermit crabs and ghost crabs (*Ocypoda*) colonize the upper shore; smaller ocypodid crabs (*Uca, Dotilla*), the middle shore. Mole crabs (*Hippa*) and various molluscs burrow at lower levels of the beach; in Malaysia venus shells (*Meretrix, Gafrarium*) and moon snails (*Polynices, Natica*) are the common forms. The middle and lower beach animals are absent from shores with severe wave action.

The fauna of sheltered sandy beaches is much richer by comparison (Berry, 1964; Vohra, 1971). On sand flats containing a proportion of silt, burrowing polychaetes, echinoderms, and coelenterates become important components of the fauna and a seaward zone of the marine herb *Enhalus* is developed.

Large marine turtles nest on the sandy beaches of the Malay-Indonesian region (Hendrickson, 1958). The world's largest concentration of giant leathery turtles (*Dermochelys coriacea*) occurs along the east coast of Peninsular Malaysia; two million turtle eggs are collected for food each year on this coast (Hendrickson and Balasingam, 1966).

Rocky Shores

Distribution. Rocky shores occur on the sea-facing coasts of many Asian islands. The southwest coast of Sumatra and the Pacific coastline of the Philippines and Sulawesi have extensive rocky topographies. Smaller rocky outcrops and boulder formations are ubiquitous above coral reef flats and on headlands bordering sandy bays. Wave erosion of limestone creates sheer or fissured cliffs with little or no beach formation. Limestone cliffs 10–15 m high surround Christmas Island and provide nesting sites for oceanic seabirds (Gibson-Hill, 1947).

Biology. The zonation of organisms on rocky shores in tropical Asia (Fig. 9) conforms to the universal classification of Stephenson and Stephenson (1949) in which three major zones (supra-, mid-, and sublittoral), characterized by *key* organisms (littorinid snails, barnacles, and algae, respectively), are recognized. High surface temperatures and desiccation greatly limit the tropical fauna and flora in comparison with those of temperate rocky shores. Large seaweeds (e.g., fucoids, laminarians) typical of cooler

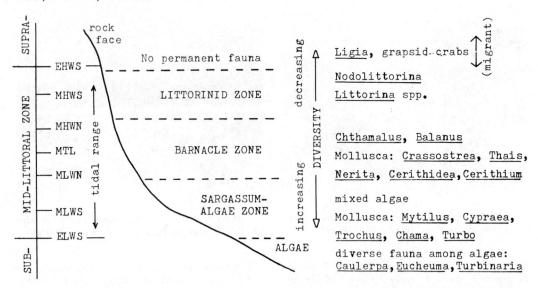

FIGURE 9. Diagrammatic zonation of organisms on rocky shores in tropical eastern Asia. Common genera of Malaysian region shores are shown.

latitudes, and the organisms they support, are absent from tropical rocky shores, and there is a general lowering of the zonation levels toward the equator. A rich assemblage of organisms occurs at the lowest tidal level and in crevices (Berry, 1964; Chuang, 1973), where the environment is more equable.

Tropical rock pools are subject to extreme heating and wide fluctuations in salinity and consequently support a minimal biota.

Economic Exploitation

Fishing and Aquaculture. Some 7.0 million km^2 of shallow seas (depth <200 m) occur in the East Indian Ocean, and a further 4.5 million km^2 surround the Pacific coasts of Asia. Marine organisms are the major protein source for most of the human population in eastern Asia. Inshore fishing, using fixed nets, traps, and hand nets, remains primitive. Yields from coastal trawling are fair (>200 kg/hr) within 60 m depth and poorer in deeper waters. Demersal fish occur all over the shallow Sunda Shelf, with rich areas off South Vietnam and in the Gulf of Thailand.

Deep-water trawling and long-line fishing for tuna are being developed rapidly by several Asian countries. Substantial and increasing catches of mackerel (*Rastrelliger*) are now taken by Thailand and Malaysia from the eastern Bay of Bengal, a previously unexploited area. Potential yields are estimated at twice the present catch in the West Pacific region of Asia (demersal fish) and twice the total fish catch in the Indian Ocean region (Gulland, 1971). The East China Sea is fished heavily by Japan and Korea, and increased yields from this area are unlikely.

Marine aquaculture has been long practiced in Asia (for at least 400 years in the Philippines and Indonesia), and modern developments are now poised to increase the role of coastal farming tremendously (Ling, 1977). The potential area suitable for aquaculture greatly exceeds that in present use (Table 1). More than 20 species each of fish, crustaceans, and molluscs and about 10 species of algae are under cultivation.

Aquaculture (q.v.) is concentrated in sheltered bays and coastal swamplands. By traditional methods prawns or fish are trapped and held in embanked ponds (*tambaks* in Indonesia) until of marketable size (Schuster, 1952). However, the availability of fresh stocks from coastal waters is unpredictable, and this drawback is now aggravated by incidences of pollution.

Modern techniques involving controlled stocking, supplementary feeding, and culture of all stages of the organism (eggs, juveniles, adults) are progressing rapidly. Although cost-intensive, production several times that from traditional methods is attainable by modern culture (Chen, 1976).

Mangrove Forest. Mangrove timber, particularly *Rhizophora*, is used extensively for firewood, poles, building timber, and charcoal production. Forest management is well developed in several countries. In Malaysia trees are felled selectively or cleared along narrow strips (Fig. 10). Felling on a 30-to-40-year rotation is generally followed with trees being replaced by natural seedling regeneration or by replanting.

Mangroves have a number of secondary uses.

TABLE 1. The Status of Coastal Aquaculture in Eastern Asian Countries (after Pillay, 1973)

Country	Area under culture (km²) Developed	Potential	Main organisms cultivated
Philippines	1660	5000	milkfish, prawns, oysters, mussels
Indonesia	1650	>4000	milkfish (*Chanos*), prawns (penaeids)
Taiwan	276	100	milkfish, mullet, clams, oysters
Korea	275	1690	clams, oysters, seaweeds, prawns
Thailand	200	>5000	prawns, cockles, oysters
Malaysia	38	500	cockles, (*Anadara*), prawns
Vietnam	26	1500	prawns, cockles
Cambodia	–	>500	

FIGURE 10. Clear-felling of *Bruguiera* mangrove forest for poles, west coast of Peninsular Malaysia. (Photo: D. J. Macintosh)

Nipah palms are utilized for thatching throughout Southeast Asia. The inflorescence is tapped to obtain sugar and alcohol by fermentation. Mangrove bark contains a poor-quality tannin still used in Bangladesh. Mangrove wood has recently proved usable for chipboard and rayon manufacture. The pharmacological basis of native medicines concocted from mangrove leaves and bark (Watson, 1928) is unknown.

Coastal Malpractices and Problems. Overfishing and pollutants from coastal industries have depleted marine food stocks severely in confined areas such as Manila Bay and Kaohsiung harbor (Taiwan) and have become increasing problems. Malaysia and Indonesia have suffered periodic incidents of oil pollution from tankers using the Malacca Straits. Coral reefs are particularly susceptible: oil pollution and silting have reduced the reef flat corals adjacent to Singapore harbor from 30 to 8 species (Chuang, 1973).

Some 40% of the mangrove in Vietnam was destroyed by the military spraying of herbicides. Forests in the Rung-Sat area (southeast of Saigon) were completely defoliated and have shown poor regeneration even after several years. Soil erosion and cases of high animal mortality have been reported from sprayed areas (Davis, 1974; Ross, 1974). Tidal erosion of coastal soil is a common consequence of excessive mangrove destruction.

A major problem in Asia is to create and implement protective legislation for coastal communities (blast fishing by dynamiting coral reefs is still widely practiced, for example). Coordinated research is urgently needed to develop the region's high potential for coastal aquaculture in order to meet the food requirements of its rapidly increasing human population.

DONALD J. MACINTOSH

References

Berry, A. J., 1964. The natural history of the shore fauna of north Penang, *Mal. Nat. Jour.* **18**, 81-103.

Berry, A. J., 1972. The natural history of West Malaysian mangrove faunas, *Mal. Nat. Jour.* **25**, 135-172.

Berry, A. J., and Chew, E., 1973. Reproductive systems and cyclic release of eggs in *Littorina melanostoma* from Malayan mangrove swamps (Mollusca: Gastropoda), *Zool. Soc. London Jour.* **171**, 333-344.

Chapman, V. J., 1970. Mangrove phytosociology, *Trop. Ecol.* **11**, 1-19.

Chen, T. P., 1976. *Aquaculture Practices in Taiwan.* Farnham, England: Fishing News Books Limited, 162p.

Chuang, S. H., 1973. Life on the seashore, *in* S. H. Chuang, ed., *Animal Life and Nature in Singapore.* Singapore: Singapore University Press, 150-174.

Davis, G. M., 1974. Mollusks as indicators of the effects of herbicides on mangroves in South Vietnam, working paper from *The Effects of Herbicides in South Vietnam.* Washington, D.C.: Nat. Acad. Sci. Nat. Research Council, 29p.

Ekman, S., 1967. *Zoogeography of the Sea.* London: Sidgwick and Jackson, 417p.

Fosberg, F. R., 1966. Vegetation as a geological agent in tropical deltas, *in Scientific Problems of the Humid Tropical Zone Deltas and Their Implications, Proceedings of the Dacca Symposium, Humid Tropics Research.* Paris: UNESCO, 227-233.

Gibson-Hill, C. A., 1947. The nature of the coast, *in* Contributions to the natural history of Christmas Island in the Indian Ocean, *Raffles Mus. Singapore Bull.* **18**, 8-17.

Gibson-Hill, C. A., 1950. A note on the Cocos-Keeling Island, *Raffles Mus. Singapore Bull.* 22, 11-28.

Gulland, J. A., ed., 1971. *The Fish Resources of the Ocean.* London: Fishing News Books Limited, 255p.

Hendrickson, J. R., 1958. The green sea turtle, *Chelonia mydas* (Linn.), in Malaya and Sarawak, *Zool. Soc. Lond. Proc.* 130, 455-535;

Hendrickson, J. R., and Balasingam, E., 1966. Nesting beach preferences of Malayan sea turtles, *Nat. Mus. Singapore Bull.* 33, 69-76.

Ito, K., and Leu, T., 1962. Ecological studies on the salt marsh vegetation in Hokkaido, Japan, *Jour. Ecol.* 12, 17-20.

Jones, O. A., and Endean, R., 1973. *Biology and Geology of Coral Reefs.* 4 vols., *Biology 1, 2; Geology 1, 2.* New York: Academic Press.

Kohn, A. J., 1967. Environmental complexity and species diversity in the gastropod genus Conus on the Indo-West Pacific reef platforms, *Am. Naturalist* 101, 251-259.

Ling, S. W., 1977. *Aquaculture in Southeast Asia.* Seattle: University of Washington Press, Washington Sea Grant Publication, 108p.

Macintosh, D. J., 1977. Some responses of tropical mangrove fiddler crabs (*Uca* spp.) to high environmental temperatures, in *Proceedings 12th European Marine Biology Symposium.* Oxford: Pergamon Press, 49-56.

Macnae, W., 1968. A general account of the fauna and flora of mangrove swamps in the Indo-West-Pacific region, *Advances in Marine Biol.* 6, 72-270.

Maxwell, W. G. H., 1968. *Atlas of the Great Barrier Reef.* New York: Elsevier, 268p.

Molengraaff, G. A. F., 1930. The coral reefs in the East Indian Archipelago, their distribution and mode of development, *Proc. 4th Pacif. Sci. Congr., Java, 1929,* 2A, 55-89.

Mukundan, C., and Gopinadha Pillai, C. S., eds., 1972. *Proceedings of the Symposium on Corals and Coral Reefs.* Cochin: Marine Biol. Assoc. India, 591p.

Pillay, T. V. R., ed., 1973. *Coastal Aquaculture in the Indo-Pacific Region* (Indo-Pacific Fisheries Council Symposium on Coastal Aquaculture, Bangkok, 1970) London: Fishing News Books Limited, 497p.

Ross, P., 1974. The effects of herbicides on the mangrove of South Vietnam, working paper from *The Effects of Herbicides in South Vietnam.* Washington, D.C.: Natl. Acad. Sci. Natl. Research Council, 33p.

Sasekumar, A., 1974. Distribution of macrofauna on a Malayan mangrove shore, *Jour. Anim. Ecol.* 43, 51-69.

Scholander, P. F.; van Dam, L.; and Scholander, S. I., 1955. Gas exchange in the roots of mangroves, *Am. Jour. Botany* 42, 92-98.

Schuster, W. H., 1952. Fish culture in brackish water ponds of Java, *Spec. Publ. Indo-Pacif. Fish. Coun.* No. 1. Rome: F.A.D., 143p.

Searle, A. G., 1956. An illustrated key to Malayan hard corals, *Malay. Nat. Jour.* 11, 1-28.

Stamp, L. D., 1925. The aerial survey of the Irrawaddy delta forests (Burma), *Jour. Ecol.* 13, 262-276.

Stephenson, T. A., and Stephenson, A., 1949. The universal features of zonation between tide marks on rocky coasts, *Jour. Ecol.* 37, 289-305.

Stoddart, D. R., 1972. Regional variation in Indian Ocean coral reefs, in *Proceedings of the Symposium on Corals and Coral Reefs.* Cochin: Marine Biol. Assoc. India, 155-174.

Tranter, D. J., 1974. Marine biology, in *Natural Resources of Humid Tropical Asia, Natural Resources Research.* Paris: UNESCO, 355-393.

Umbgrove, J. H. F., 1930. The influence of monsoons on the geomorphology of coral islands, *Proc. 4th Pacif. Sci. Congr., Java, 1929,* 2A, 49-54.

Umbgrove, J. H. F., 1947. Coral reefs of the East Indies, *Geol. Soc. America Bull.* 58, 729-778.

Vohra, F. C., 1971. Zonation on a tropical sandy shore, *Jour. Anim. Ecol.* 40, 679-708.

Volker, A., 1966. The deltaic area of the Irrawaddy River in Burma, in *Scientific Problems of the Humid Tropical Zone Deltas and Their Implications, Proceedings of the Dacca Symposium, Humid Tropics Research.* Paris: UNESCO, 373-379.

Walsh, G. E., 1974. Mangroves: A review, in R. J. Reimold and W. Queen, eds., *Ecology of Halophytes.* New York: Academic Press, 51-174.

Watson, J. G., 1928., Mangrove forests of the Malay Peninsula, *Malay. Forest. Rec.* 6, 1-275.

Wells, J. W., 1957. Coral reefs, in J. W. Hedgpeth, ed., *Treatise on Marine Ecology and Paleoecology, vol. 1: Ecology, Mem. 67.* New York: Geol. Soc. America, 609-631.

Wiens, J. W., 1962. *Atoll Environment and Ecology.* New Haven: Yale University Press, 532p.

Wyrtki, K., 1961. Physical oceanography of the Southeast Asian waters, *Naga. Rep.* 2, 1-195.

Wyrtki, K., 1962. The upwelling in the region between Java and Australia during the south-east monsoon, *Australian Jour. Marine Freshw. Res.* 13, 217-225.

Cross-references: *Aquaculture; Asia, Eastern, Coastal Morphology; Australia, Coastal Ecology; Biotic Zonation; Coastal Fauna; Coastal Flora; Coral Reefs; India, Coastal Ecology; Mangrove Coasts; Pacific Islands, Coastal Ecology; Soviet Union, Coastal Ecology.* Vol. VIII, Part 2: *Asia.*

ASIA, EASTERN, COASTAL MORPHOLOGY

The shores of eastern Asia, extending from western Pakistan to eastern Siberia and including a large part of the Indonesian archipelago, follow largely the tectonically active zones where the Pacific and Indian Ocean plates collide with the mainland Asia plate (Inman and Nordstrom, 1971). Large stretches of coastal area in Pakistan, Burma, Indonesia, the Philippines, Taiwan, Japan, Sakhalin, Kamchatka, and the East Asian island chains are therefore tectonically unstable. In these areas the structural trends are generally parallel to the coast, which was distinguished by Suess (1892) as the Pacific type. Outside these areas, away from the tectonically active collision zones, the coastal regions are generally more stable and the structural trends are usually not parallel to the coast, which corresponds to Suess' Atlantic type. This type is present in India and along most of the Asian mainland from Thailand to Siberia.

Comparatively straight coasts, situated along mountain chains, sometimes with river deltas and local alluvial foreland, are found mainly in western Sumatra, southern Java, northern Vietnam, and eastern Siberia. A drowned, older topography with an irregular coastline is present on the west coast of India, parts of southern Vietnam, the mainland coast north of the Red River, on the islands of eastern Indonesia, on northern Kalimantan (Borneo), the Philippines, Japan, and southern Sakhalin. These coasts are somewhat remodeled by the sea, with bays containing beaches, spits, and barriers, sometimes being filled up with sediment and partially surrounded by alluvial foreland. The main exceptions are the large river deltas, the west coast of Kamchatka that consists largely of sand barriers, some beach barriers in Japan and on Sakhalin, and some areas like eastern Taiwan where tectonic uplift matches the Holocene sea-level rise. Elsewhere the coast is predominantly depositional, consisting of beaches, spits, barriers, tombolos, mudflats, marshes, mangrove swamps, and coral reefs.

The general direction of beaches, spits, and barriers is related to the direction of the swell; between northern Japan and Indonesia the swell comes chiefly from northern directions, and the beaches and spits face largely E-NE. Where the swell is southerly, as in the Indian Ocean and in the Pacific north of 45°N, the beaches face mainly SE-SW. A major distinction can be made between the coasts north and south of about 40°N (Davies, 1973). North of this latitude stormwaves are predominant and pebbles form an important part of the beach material. During the winter *sea ice* (pack ice) borders the coast as far south as Vladivostok and northern Japan. Further north, along the northern shore of the Sea of Okhotsk and north of Kamchatka peninsula, *permafrost* reaches the sea, and from the eastern Siberian coast ice cliffs have been reported. South of 40°N the ocean swell is more important, and large areas from southern Japan to Malaysia, Bangladesh, western India, and Pakistan, are affected by tropical storms. In this area, especially south of 25°N, sandy beaches are predominant while pebbles are relatively unimportant as beach material.

Coral reefs and *beach rock* are confined to the tropical and subtropical zones—they are not present north of 35°N. *Mangrove* grows south of 31°N (on the Asian mainland coast south of 26°N) and dominates large stretches of coast in Indonesia, Malaysia, Thailand, and Burma, especially in areas that are relatively sheltered from the ocean swell. An important part of the East Asian coastline consists of several large river deltas (those of the Indus, the Ganges-Brahmaputra, the Irrawaddy, Mekong, Yangtse-Kiang, Hoang-Ho) and a large number of smaller deltas often of still appreciable dimensions such as the deltas of the Chao Phraya, the Red, and the Si-Kiang. This preponderance of deltas is due to the high mechanical erosion on the mainland between Pakistan and Korea (>60 ton per km^2 per year), which, associated with the presence of large and numerous rivers, has resulted in extensive deltaic deposition.

Tidal effects are especially important along the mainland coast from Taiwan to Kamchatka, where the tides have an appreciable range. Usually the tidal range in eastern Asia is less than 4 m (and around Japan mostly less than 2 m) but in the northern Sea of Okhotsk, in the Yellow Sea, and along the Chinese coast it reaches more than 6 m. Tidal marshes have been formed locally, as, for example, along the Gulf of Po-Hai and in southern Korea.

A striking feature on many coasts from Japan through southeast Asia to southern India is the evidence of raised shorelines at 0.5-1 m, about 2 m, and 5-6 m above present sea level. These are indicated by flat, sandy terraces, beach ridges, coral reefs, mollusks and/or barnacles, notches, benches, platforms, and beach rock. Similar raised shorelines are known from many other coasts in Asia, Africa, South America, and Australia (for references, see Tjia, 1975). Daly (1934) and many others therefore assumed that during the Holocene, about 6000-4000 years ago, sea level stood about 5 m higher than at present. Others have contended that these features are related to temporary high water during storms or to local uplift of the land. The question is not settled. Tjia (1975) has recently suggested that glacioeustatic uplift of the areas in the Northern Hemisphere that were glaciated during the Pleistocene may account for the absence of any indications for high sea levels during the Holocene. C_{14} dates of raised reefs and mollusk shells from the Pacific and Indian oceans, southeast Asia, and Australia point to a slight Holocene transgression as well as to local uplift (Stoddart and Gopinadha Pillai, 1972; Tjia, 1975).

The general features of the east Asian coasts are summarized in Figure 1, which is based essentially on the world map given in Valentin (1952), but with emphasis on the shoreline characteristics. More detailed knowledge, however, is still lacking for large stretches of coast, especially in eastern Siberia and Southeast Asia.

Japan, Sakhalin

The shores of Sakhalin and Japan consist of large stretches of ria coast alternating with many small and several large stretches of beaches, spits, barriers, sand dunes, lagoons, and river outlets. Especially on northern Sakhalin, but also on Hokkaido and in the north and center of Honshu, large parts of the coast are depositional. Coastal terraces are widely distributed in northern Japan; in the south they occur mainly on the promontories along the Pacific coast. The terraces are predominantly present on the more easily abraded Tertiary and Quaternary formations, whereas the indented ria coasts have been developed mainly in folded Palaeozoic and Mesozoic rock. The highest terraces have been found in southwestern Hokkaido at +585 m, but usually they reach to +150 m (Kosugi, 1971). Below present sea level submerged flat surfaces have been found to depths of −150 and −190 m.

The terraces as well as raised coral reefs (Yabe and Sugiyama, 1935) indicate that the Quaternary eustatic changes in sea level may have had an important effect on coastal development in Japan, but there is also much evidence for differential uplift, tilting, and local subsidence. Consequently in an appreciable number of papers by Japanese authors the combined effects of both eustatic sea-level changes and crustal movements have been discussed (Yoshikawa, Kaizuka, and Oka, 1965; Richards and Fairbridge, 1965; Richards, 1970). The presence of ria coasts is generally seen as evidence of an overall subsidence since the Pliocene, but narrow terraces scattered along the rias suggest a more recent emergence. Some terraces (at +40 to +60 m) are widely distributed along the Japanese coast and are seen as the result of eustatic sea level changes. Coastal terraces, formed at the same sea level, have also been found at different altitudes within short distances, even locally disappearing below present sea level as a result of crustal movements. From this complex situation it has been deduced that the Holocene sea level rise started around 18,000 BP and that between 6000 and 3000 BP sea level was 3-5 m higher than at present, which is indicated by former wave-cut notches, sea caves, benches, small terraces, marine deposits, and neolothic remains (Tada, Nakano, and Iseki, 1952).

A marked feature along the Japanese and Sakhalin coasts is the widespread occurrence of shore platforms (Takahashi, 1974). These are most conspicuous along the coasts of the Pacific and the East China Sea, less on the Sea of Japan coasts, and least along the Seto Inland Sea. Their maximum width varies from 530 m along the Pacific to 80 m on the Sea of Japan coasts, which corresponds roughly to the wave conditions on each coast. The average width is 30-80 m and the maximum length about 4 km; they occur on all types of rock, but tend to be wider on Neogene mudstones, sandstone, and tuff-breccia.

Mainland, Korea to Malaya

The mainland coast from Korea to the Malayan peninsula is largely a drowned or ria-type coast, with abrasion at the headlands, accumulation (with sandy beaches, spits, and beachridges) in the bays and, especially in China and Southern Korea, numerous islands. Large estuaries and river deltas are present where major rivers reach the coast. In West and South Korea tidal flats have been formed in the bays and in the shallower parts between the islands (Guilcher, 1976). Coastal terraces are present along the East Korea coast up to 100-150 m and along the West Korea coast up to 30-60 m, which reflects the emergence of the east coast and the submergence of the west coast. Coastal terraces and old beaches have been reported from the area around Hong Kong at +5 and +14 m (Berry, 1961) and are probably also present elsewhere. At higher levels, up to +70 m, benches are found; below present sea level flat surfaces could be distinguished down to −60 m. Especially the terraces at +5-6 m and +15 m are thought to be caused by eustatic sea-level changes. Flat surfaces at +130 m and +230 m are probably not of marine origin.

Around the Gulf of Po-Hai, between the Liao-Tung and Shan-Tung peninsulas, the coast has been filled with recent alluvium from the Huang-Ho, Luan-Ho, Liao, and a number of smaller rivers. Long stretches of tidal marshland have been formed. Along the northwestern shore, however, up to 80 km of alluvial deposits were eroded again in historical times (von Wissmann, 1940), probably because the Hoang-Ho then had a southerly course and no new sediment from that river reached the Gulf of Po-Hai at that time. The present situation dates from 1853, when the Hoang-Ho again took a northern course. Between 1448 and 1853 it had been reaching the sea south of Shan-Tung peninsula after having joined the Huai-Ho River. While at present the silt from the Huang-Ho moves northward anticlockwise through the Gulf of Po-Hai, the silt flowing out of the southern mouth before 1853 was transported further south and deposited in front of the Yang-tse-Kiang, which was deflected southward. In 400 years the coastline in this area moved 36-62 km eastward.

South of the Yang-tse-Kiang the ria coast ends at the delta of the Si-Kiang, while further south the coast is interrupted by the deltas of the Red, Mekong, Chao Phraya, and other, smaller rivers. South of Hong Kong mangrove swamps are present and fringing coral reefs become increasingly numerous. From Dong Hoi to Cape Varella in Vietnam the coast is formed by a 15–40 km-wide belt of barrier complexes, lagoons, and sand dunes; from Cape Varella to Cape Padaran the coast is again a typical ria coast. Terraces have been found in Vietnam and Cambodia between +2 m and +80 m (Carbonnel, 1964; Saurin, 1965). The terrace at +2 m and raised reefs, mollusks (oysters), and conglomerates at the same level have been found all along the coasts of Vietnam and Cambodia. Other terraces occur more regionally; the +4 m terrace only in Vietnam, the +15 m and +25 m terraces in southern Vietnam and Cambodia, and the terrace at +80 m only near Cape Padaran.

Taiwan, Philippines

Compared to the mainland coast Taiwan is very different. Situated in the tectonically active zone it has been slowly uplifted during the Quaternary. This uplifting continues at present, although there is some evidence for regional temporary subsidence. Eustatic sea-level changes are considered to have had little effect on the coast, since the changes in sea level have been more than balanced by the amount of uplift (Hsu, 1962). On the Pacific side the coast is steep and erosion predominates (as also on the southwestern and northern coasts), but on the west coast, which is sheltered from the ocean swell and the tropical storms as well as from the strong northeastern monsoon, a flat, marshy coast is formed with tidal flats. Coral reefs are present on the Pacific side as well as around the Ryukyu Islands further north and along the Pacific coasts of Kyushu and Shikoku.

The Philippine coasts, situated like those of eastern Taiwan in the tectonically active zone, are comparatively steep but intersected by river outlets, broad valleys, and inlets with beaches, surrounded by coral reefs. Although the area has been drowned during the Holocene transgression, the islands have been rising during the Pleistocene and there is evidence of differential uplift. Numerous and often extensive marine terraces are present; on Sabtan Island and on Luzon the highest are at +180 m, on Bataan at +275, and in northern Mindanao at +360 m (van Bemmelen, 1970), but no correlations have been made.

Malay Peninsula

The Malay Peninsula, together with the eastern part of Sumatra, western and central Kalimantan (Borneo), and the Sunda-Java Sea, belongs to the Sunda landmass, which has probably been tectonically stable since the middle Pleistocene. The presence of beach ridges several meters above present sea level has been interpreted as being the result of recent uplift but can also be explained by Holocene eustatic sea-level changes.

The east coast of Malaya is relatively smooth with shallow bays and few indentations. More than 80% of this coast is formed by sandy beaches interrupted by river outlets and small deltas; cliffs form about 10% (Nossin, 1965; Swan, 1968). Tidal swamps with mangrove are usually found landward of the beaches and sandbars. They are sheltered against the swell of the strong monsoon from the northeast from October to April. During this period there is more erosion than accretion, especially between late October and January, when spring tides are at their highest. Accretion is widespread between February and September, when the tides are less effective and the wind blows mainly offshore. A minor period of erosion occurs from May to July when the spring tides reach secondary maxima.

The west coast of Malaya is covered by mangrove, which continues northward to the Irrawaddy delta and from there to the mouths of the Ganges. Locally this mangrove belt is interrupted by cliffs, river outlets, and beaches, as on the exposed southern Chittagong coast (Rizvi, 1969). The islands around Singapore have been studied in detail by Swan (1971), who has drawn attention to the fact that in this sheltered, low-energy area the coastal forms are very different from those supposed to characterize such coasts in the humid tropics. Mangrove swamps, tidal flats, beaches, and fringing reefs are present as well as cliffs, caves, shore platforms, and beach conglomerate (*ironstone*). This diversity is due to the intense and continuous chemical weathering, which makes possible subaerial and marine erosion by small waves, and to differences in rock type and exposure. Coastal sediment is produced through cliff erosion, which takes place along both sheltered and exposed coasts.

Indonesia

In Indonesia mangrove swamps are present along the coasts of Sumatra, northern Java, and parts of southern Java, along most of Kalimantan (Borneo), along northern Borneo (Sarawak), and at the southern end of Sulawesi (Celebes). All these areas are relatively sheltered against the ocean swell. Cliffs and beaches are locally present (Wall, 1964), but although the rivers of Sumatra and Kalimantan carry large

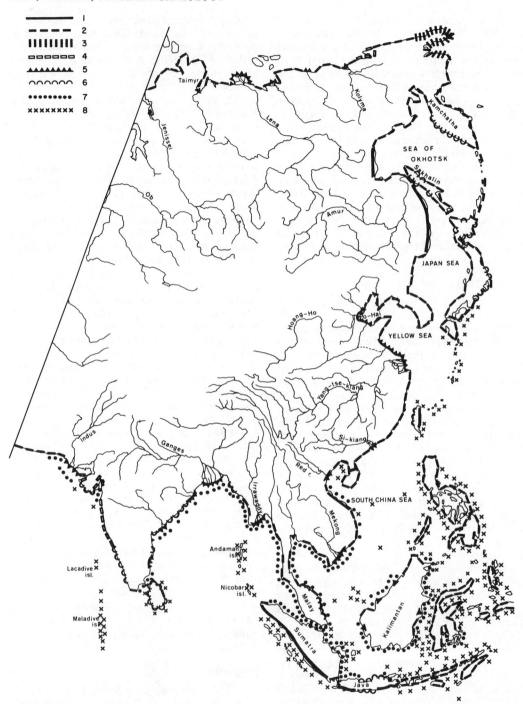

FIGURE 1. Coastal morphology of eastern Asia. 1. Comparatively straight coasts along mountain chains, sometimes with river deltas and local alluvial foreland. 2. Drowned older topography with an irregular coastline, somewhat remodeled by the sea. 3. Fjords. 4. Tableland coast with cliffs. 5. Large river deltas. 6. Predominantly spits, barriers, mud flats, marshes. 7. Mangrove swamp. 8. Coral reefs.

amounts of sediment to the sea, only the Kapuas and Pawan rivers and on the northern coast the Rajang and Baram rivers have built up sizable deltas. The coasts of Java are predominantly beaches interrupted by river outlets and deltas and locally by cliffs. However, along eastern Java, western Sumatra, and on most of the eastern Indonesian islands, the coasts are relatively steep. A few river deltas with associated swamps and alluvial foreland are present on the west coast of Sumatra.

The main feature in western Indonesia is the large, drowned Sundaland, (Molengraaff, 1921). Two large submarine valley systems, being the continuation of the present rivers in Sumatra and Kalimantan, are present on the continental shelf. During periods of low sea level one system drained Sumatra and east Kalimantan, discharging into the South China Sea. The other system, draining Java and south Kalimantan, discharged south of Makassar Strait. A divide crossed the Sunda Sea between Sumatra and Kalimantan across Billiton and the Karimata Islands. The Sundaland plain has been dissected at least twice during the Pleistocene as a consequence of the eustatic lowering of sea level (van Overeem, 1960). In the whole area, including Malaya, eustatic terraces are present from +50 m to −90 m (Tjia, 1970).

On the former east coast of the Sundaland a large barrier reef has been formed, stretching from Balikpapan on Kalimantan to the island of Sumbawa in the south. It is interrupted in many places and shows a large gap of about 100 km wide facing a deep embayment in the former Sundaland. Here the main river outlet was probably situated. The reef began as a late Pleistocene *fringing reef* and has grown upward with the gradual rise in sea level. Locally it reaches the sea surface as separate coral islands. Similar fringing reefs of smaller dimensions are present around Sulawesi and the smaller islands in eastern Indonesia, along the west coast of Sumatra (where they show gaps in front of the river deltas), and further north along the Nicobar and Andaman islands. In the Sunda Sea coral growth is restricted to a number of isolated areas away from muddy river outlets (Kuenen, 1933; Umbgrove, 1947); in eastern Indonesia numerous atolls and barrier reefs have grown upward from gradually subsiding submarine ridges and platforms, rising abruptly and steeply from a depth of 1000-2000 m. The effects of winds and waves on the reefs are conspicuous, especially in eastern Indonesia, where the reefs grow more vigorously on the windward side. Sea currents cause erosion and may shape a whole group of reefs. Solution of coral occurs within the tidal range. On the former northern coast of the Sundaland no barrier reef has been formed. Here the former coast was flat and gradually merged into an extensive sandy and probably muddy shelf. The water in this area was presumably too turbid for coral reef growth.

Ceylon

Ceylon is surrounded by beaches, sand flats, barriers, spits, lagoons, marshes, fringing coral reefs, beach rock, and at a higher level by raised beaches, detrital coral deposits, and raised beach rock (Cooray, 1968; Swan, 1975). On the east coast, which is more sheltered from the southwest monsoon, there is also mangrove. The coastline is interrupted by outlets of rivers and lagoons and in the more exposed southwestern part by tracks of rock, cliffed headlands, rocky inlets, and stacks. During the onshore monsoon the beaches in the latter area are eroded; during the offshore monsoon and the intermonsoonal periods they are built up again. Terraces and platforms are present at +2-3 m, +4-6 m, 15-20 m, and 30-40 m. The present coast was formed by the straightening of a drowned older topography with sandbars and barriers. Many former bays and estuaries have been cut off.

Pakistan

The Pakistan coast west of the large deltaic plain of the Indus River comprises rocky cliffs interspersed with pocket beaches and wide, sandy beaches with river outlets, lagoons, sand dunes, and tidal mud flats with mangrove (Snead, 1967). The region is tectonically active with continuous large-scale uplift of the coastal region even in historical times. Correlating the marine benches and platforms on most of the headlands is difficult because at several locations sea-level changes are thought to have had more influence than the tectonic movements. Because of these movements and the large amounts of sediment brought into the sea, especially by the Indus River, the deltaic parts of the coast have prograded by an average of 10-25 m per year (Wilhelmy, 1968).

D. EISMA

References

Bemmelen, R. W. van, 1970. *The Geology of Indonesia.* The Hague: Martinus Nijhoff, vol. 1, 732p.; vol. 2, 265p.

Berry L., 1961. Erosion surfaces and emerged beaches in Hong Kong, *Geol. Soc. America Bull.* **72**, 1383-1394.

Carbonnel, J. P., 1964. Sur l'existence d'un complexe de terrasses quaternaires dans l'île de Saracen (Cambodge), *Soc. Géol. France Compte Rendu* **9**, 371-373.

Cooray, P. G., 1968. The geomorphology of part of the northwestern coastal plain of Ceylon, *Zeitschr. Geomorphologie,* 7(suppl.), 95-113.

Daly, R. A., 1934. *The Changing World of the Ice Age.* New Haven, Conn.: Yale University Press, 271p.

Davies, J. L., 1973. *Geographical Variation in Coastal Development.* New York: Hafner, 204p.

Guilcher, A., 1976. Les côtes à rias de Corée et leur évolution morphologique, *Annales Géographie* 85, 641-671.

Hsu, T. L., 1962. A study on the coastal geomorphology of Taiwan, *Geol. Soc. China Proc.* 5, 29-45.

Inman, D. L., and Nordstrom, C. E., 1971. On the tectonic and morphologic classification of coasts, *Jour. Geology* 79, 1-21.

Kosugi, K., 1971. Etude analytique des dépots de sable et galets actuels d'origines diverses: Application au processus de formation des surfaces de terrasse littorale dans le Hokkaido (II), *Science Rep. Tôkohu Univ. 7th series* 20, 257-326.

Kuenen, Ph. H., 1933. *Geology of Coral Reefs.* Leiden: E. J. Brill, 125p.

Molengraaff, G. A. F., 1921. Modern deep-sea research in the East Indian archipelago, *Geog. Jour.* 27, 95-118.

Nossin, J. J., 1965. Analysis of younger beach ridge deposits in eastern Malaya, *Zeitschr. Geomorphologie* 9, 186-208.

Overeem, A. J. A. van, 1960. The geology of the cassiterite placers of Billiton, Indonesia, *Geologie en Mijnbouw* 39, 444-457.

Richards, H. G., 1970. *Annotated Bibliography of Quaternary shorelines.* Supplement 1965-1969, VIII INQUA-Congress, Acad. Nat. Sci. Philadelphia Spec. Pub. 10, 240p.

Richards, H. G., and Fairbridge, R. W., 1965. *Annotated bibliography of Quaternary shorelines (1945-1964),* VII INQUA-Congress, Acad. Nat. Sci. Philadelphia Spec. Pub. 6, 280p.

Rizvi, A. I. H., 1969. Morphological changes in the coast of Chittagong, *Oriental Geographer* 13(1), 25-40.

Saurin E., 1965. Terrasses littorales de Son-Hai, *Arch. Géol. de Viet-Nam* 7, 20-24.

Snead, R. E., 1967. Recent morphological changes along the coast of West Pakistan, *Assoc. Am. Geographers Annals* 57, 550-565.

Stoddart, D. R., and Gopinadha Pillai, C. S., 1972. Raised reefs of Ramanathapuram, South India, *Inst. British Geographers Trans.* 56, 111-125.

Suess, E., 1892. *Das Anlitz der Erde,* vol. 1. Vienna: F. Tempsky, 778p.

Swan, S. B. St. C., 1968. Coastal classification with reference to the east coast of Malaya, *Zeitschr. Geomorphologie,* 7(suppl.), 114-132.

Swan, S. B. St. C., 1971. Coastal geomorphology in a humid tropical low energy environment: The islands of Singapore, *Jour. Trop. Geogr.* 33, 43-61.

Swan, S. B. St., C., 1975. A model for investigating the coast erosion hazard in southwest Sri Lanka, *Zeitschr. Geomorphologie,* 22(suppl.), 89-115.

Tada, F.; Nakano, T.; and Iseki, H., 1952. Shoreline development of the Pacific coast of Japan in prehistoric time, *Proc. 17th Int. Congr., Int. Geogr. Union,* 386-391.

Takahashi, T., 1974. Distribution of shore platforms in southwestern Japan, *Science, Rep. Tôhoku Univ. 7th series* 24, 33-45.

Tjia, H. D., 1970. Quaternary shore lines of the Sunda Land, Southeast Asia, *Geologie en Mijnbouw* 49, 135-144.

Tjia, H. D., 1975. Holocene eustatic sea levels and glacio-isostatic rebound, *Zeitschr. Geomorphologie,* 22(suppl.), 57-71.

Umbgrove, J. H. F., 1947. Coral reefs of the East Indies, *Geol. Soc. America Bull.* 58, 729-778.

Valentin, H., 1952. *Die Küsten der Erde,* Petermanns Geog. Mitt. Erg. 246, Gotha, Justus Perthes, 118p.

Wall, J. R. D., 1964. Topography-soil relationships in lowland Sarawak, *Jour. Trop. Geogr.* 18, 192-199.

Wilhelmy, H., 1968. Indusdelta und Rann of Kutch, *Erdkunde* 22, 177-191.

Williams, A. T., 1971. Beach morphology and tidal cyclic fluctuations around Hong Kong Island, *Jour. Trop. Geogr.* 32, 62-68.

Wissmann, H. von, 1940. Südwest-Kiangsu, der Wuhu-Taihu-Kanal und das Problem des Yangdse-Deltas, *Wiss. Veröff. Deutsch. Mus. Landerkd. Leipzig,* N.F. 8, 63-105.

Yabe, H., and Sugiyama, T., 1935. Geological and geographical distribution of reef corals in Japan, *Jour. Paleontology* 9(3), 183-217.

Yoshikawa, T.; Kaizuka, S.; and Ota, Y., 1965. Coastal development of the Japanese islands, *Proc. VII Congr. Int. Assoc. Quaternary Res.* 8, 457-465.

Cross-references: *Asia, Eastern, Coastal Ecology; Australia, Coastal Morphology; Coral Reef Coasts; Flandrian Transgression; Global Tectonics; India, Coastal Morphology; Mangrove Coasts; Pacific Islands, Coastal Morphology; Soviet Union, Coastal Morphology.* Vol. XII, Part 2: *Asia.*

ASIA, MIDDLE EAST, COASTAL ECOLOGY

The living world of the southeastern coasts of the Mediterranean is impoverished qualitatively and quantitatively as compared with that of the western Mediterranean. The quantitative scarceness is a result of the extremely low biological productivity of the Levantine basin. The littoral flora and fauna have a reduced diversity because of a number of factors. The subtropical temperatures and high salinities are suboptimal to the Mediterranean species of Atlantic origins. The consequences of late Pleistocenic brackish (*euxinic*) events are evident (Por, 1964). The poorly developed relief of the coast, devoid of gulfs, with the exception of the Bay of Haifa, is characterized by flat sedimentary shores and rocky bottoms carpeted by Nile sediments. With the exception of the Cenomanic chalk cliffs of Rosh-HaNiqra, there are only low aeolinitic (*kurkar*) cliffs and beach-rock slabs along the Israel shore.

Since the opening of the Suez Canal, over 200 Indo-Pacific migrant species have invaded the Mediterranean from the Red Sea (Por, 1978). There has been a gradual impoverishment in species diversity from north to south with the gradual disappearance of rocky bottoms near the Nile delta. The Indo-Pacific species, a recent addition to the impoverished biota, apparently do not increase in frequency toward the south, but are more or less equally or locally distributed along the coasts of Sinai and Israel.

There have been several taxonomic analyses of the fauna and flora of the Israel Mediterranean coast: algae (Rayss, 1941, 1954, 1955, 1963); foraminiferans (Reiss, Klug, and Merling, 1961); sponges (Tsurnamal, 1968, 1969a, 1969b, 1975); ostracodes (Lerner-Seggev, 1965); polychaetes (Fauvel, 1955, 1957; Tebble, 1959; Ben-Eliahu, 1970, 1972a, 1972b, 1974, 1975-1977); decapod crustaceans (Holthuis and Gottlieb, 1958; Gottlieb, 1959; Lewinsohn and Holthuis, 1964); molluscs (Pallary, 1938; Haas, 1951; Eales, 1970; Barash and Danin, 1971, 1972); and echinoderms (Tortonese, 1953-54, 1966; Achituv, 1969, 1973). Interstitial and meiobenthic crustacea of the shore sediments were analyzed by Por (1964), Masry (1970a, 1970b), and Masry and Por (1970).

There have been relatively few ecological studies on the littoral and shelf biota of the coasts of Israel and Sinai. The basic pattern of the ecological zonation of a kurkar coast has been analyzed by Lipkin and Safriel (1971) at Mikhmoret. The nearshore bottoms of Haifa were studied by Gottlieb (1959) and Tom (1976). The shallow sedimentary bottoms have been analyzed by Wirszubski (1953) and especially by Gilat (1964, 1969, 1974) who analyzed several profiles from north to south (Alexander, Ashdod, El Arish profiles). The associations of sponges in the sublittoral caves and crevices was analyzed by Tsurnamal (1968, 1969a, 1969b, 1975).

The Rocky Intertidal

Following Lipkin and Safriel (1971), a set of characteristic species delimits the different levels of the periodically exposed rocky shore. Despite the lack of a regular and perceptible tide, the zonation of the rocky intertidal is clearly represented though vertically much compressed. If compared with the intertidal of the western Mediterranean, there are several species missing. As an example, two species pairs of gastropoda are represented by only one species: *Patella lusitanica* is absent, and *P. caerulea* alone is present; *Monodonta turbiformis* is absent, and only *M. turbinata* is present. According to Lipkin and Safriel (1971), the strong development of the upper midlittoral belt of the barnacle *Chthamalus stellatus* may be due to the absence of these gastropoda. The only Indo-Pacific additions to the rocky intertidal of Israel that have reached importance are the small mussel *Brachidontes variabilis* and the cerithid *Cerithium scabridum*.

The Infralittoral Reefs

The coast of Israel is characterized by *trottoirs*, biogenic platforms covering the aeolinitic rocky substrate (Safriel, 1974, 1975). These reeflike structures are bioherms of two vermetid gastropods, *Vermetus triquetrus* and *Dendropoma petraeum*, cemented together by the calcareous red alga *Neogoniolithona notarisi*. The vermetids form elevated rims that surround shallow basins several tens of meters broad. The platforms are slightly tilted leeward. Locally, such as at Shiqmona, the vermetid rims form veritable *micro-atolls*. Similar vermetid reefs occur also in Bermuda. Safriel considers the formation as being characteristic "in historically depauperated coral reef situations." In the northwestern Mediterranean the trottoirs are formed exclusively of calcareous red algae.

The Caulerpa Scalpelliformis Meadows

The infralittoral sedimentary bottoms of the Israel coast are devoid of the aquatic flowering plant *Posidonia oceanica* (extremely frequent in the western Mediterranean) and *Zostera* spp. The only widespread macrophytic association is that of the circumtropical green alga *Caulerpa scalpelliformis*, abundantly represented, especially in the Bay of Haifa. A recently immigrated Indo-Pacific brittle-star *Ophiactis parva* is especially abundant among *Caulerpa* (Tom, 1976).

The Circalittoral Coralligenous Ridges

At a depth of 20-30 m, especially along the northern coast of Israel, in the Bay of Haifa and south to Caesarea, a biogenous rocky ridge is found. It is formed mainly by dead bioherms of the coral *Cladocora* previously described as *Dendrophyllia* and the calcareous algae *Lithothamnium* spp. and *Halimeda tuna* (Gottlieb, 1959).

Gilat (1970) reports some local occurrences of coralligenous bottoms also from the more southern shores. However, there is no new report of a continuous sublittoral coralligenous ridge off the Sinai coast as indicated by Gorgy (1966).

The Nearshore Level Bottoms

According to Gilat (1970), the sublittoral sandy shores of Israel and Sinai are inhabited by the Mediterranean community of *Echinocardium cordatum-Venus gallina* with several additional species such as *Aloidis gibba, Mactra corallina,* and others. The dominant, almost exclusive, species in the sandy-shore thanatocoenosis is the shell *Glycimeris violacescens*. However, this species was only rarely found alive in the samples of the past few decades.

Several species of Indo-Pacific origin are widespread on the deeper sandy-muddy bottoms: the cerithid *Cerithium kochi*, the snail *Isanda* cf. *holdsworthiana*, and the swimming crab *Charybdis longicollis*. The penaeid shrimps of the sublittoral of the Israeli coasts, such as *Penaeus japonicus* and others, are chiefly Indo-Pacific species.

The Estuarine Areas

Only seven permanent streams of modest size reach the Mediterranean coast of Israel. Their estuaries support limited accumulations of *Cerastoderma glaucum* (= *Cardium edule*). Typical in the estuaries were the small reeflike structures of the serpulide *Mercierella enigmatica*. Other typical animals were *Balanus amphitrite* and the soft-bodied bryozoan *Victoriella pavida*. Today all the coastal rivers and estuaries, with one or perhaps two exceptions, are heavily polluted.

The Geographical Subdivision of the Israeli and Sinai Shores (from North to South)

The Coast from Rosh-HaNiqra to Nahariya. The white calcareous cliffs of Rosh-HaNiqra are rich in submarine caves, whence Tsurnamal (1968, 1969, 1975) described several typical communities of light-avoiding sponges. In the small rocky lagoons of the area a particularly rich boulder fauna lives. Several species of echinoderms—for example, the black sea urchin *Arbacia lixula*, the sea-star *Asterina wega*, and the big brittle star *Ophioderma longicaudata*—are frequent here but absent or nearly so from more southern shores. The same applies also to the small intertidal chiton *Middendorfia caprearum*, which is rarely found south of Nahariya. The fauna and flora of this region are probably a more or less faithful model of the rocky bottom fauna of Lebanon. The sandy shores in this area are characterized by coarser sand; they harbor the richest interstitial fauna in Israel (Masry and Por, 1970; Masry, 1970a, 1970b).

The Bay of Haifa. The northern part of the bay, near the old port of Acre, is protected and warm. Here a tropical association—banks of the little pearl-oyster *Pinctada radiata* (= *Pteria occa*)—is well developed. Reeflike structures are built by the bryozoan *Schizoporella errata*. Several Indo-Pacific immigrants, in addition to the pearl oyster, are known to have made their first appearance in this sheltered bay. The *Caulerpa* beds and *Dendrophyllia* ridges of the Bay of Haifa have been mentioned above.

The Coast from Cape Carmel to Yavne-Yam. This section of the coast is characterized by sandstone (*kurkar*) cliffs on which vermetid reefs develop. These formations are best represented between Dor and Mikhmoret. The kurkar outcrops along the shores gradually become more rare toward the south as the shoreline straightens out and offers fewer protected environments. The last littoral kurkar ridge—and also the last vertical rocky shore—is that at Yavne-Yam a few kilometers south of Tel Aviv. Toward the south the stretches of bare beach rock and sandy shores become more and more frequent.

In accordance with this trend, the diversity of the littoral flora and fauna decreases gradually (Por, 1964). Tsurnamal (1968), in his study on the littoral sponges, demonstrates the increasing deleterious effect of sand abrasion on the exposed rock fauna of the beach rocks.

Finally, only a limited diversity of littoral flora and fauna lives on the beach-rock slabs: the barnacle *Chthamalus* spp., the limpet *Patella coerulea*, the snail *Monodonta turbinata*, the crab *Pachygrapsus transversus*, and species of such algae as *Enteromorpha* and *Ulva*. Many stretches of this section of the coast are severely polluted by sewage outlets from the growing megalopolis of Tel Aviv.

Tsurnamal (1968) described the environments of underwater caves and channels among the vermetid platforms.

Bardawil Lagoon (Sirbonian Lagoon). This shallow and metahaline lagoon occupies most of the northern Sinai shore and connects with the open sea through three openings. The shallow, 1–2-m-deep bottom of the lagoon is overgrown by the aquatic flowering plant *Ruppia maritima*, which is covered with epiphytic *Cladophora*. The dominant molluscs are the lamellibranchs *Cerastoderma glaucum* (= *Cardium edule*), *Mactra olorina*, and the brine snail *Pirenella conica*. *Mactra* is an Indo-Pacific immigrant. The lagoon serves as a rich feeding ground for several species of fish as well as for the Indo-Pacific shrimp *Metapenaeus stebbingi*.

FRANCIS DOV POR

References

Achituv, Y., 1969. Studies on the reproduction and distribution of *Asterina burtoni* Gray and *A. wega* Perrier (Asteroidea) in the Red Sea and eastern Mediterranean, *Israel Jour. Zool.* 18, 329–324.

Achituv, Y., 1973. On the distribution and variability of the Indo-Pacific sea star *Asterina wega* (Echinodermata: Asteroidea) in the Mediterranean Sea, *Marine Biology* 18, 333–336.

Barash, Al., and Danin, Z., 1971. Opisthobranchia (Mollusca) from the Mediterranean waters of Israel, *Israel Jour. Zool.* **20**, 151-200.

Barash, Al., and Danin, Z., 1972. The Indo-Pacific species of Mollusca in the Mediterranean and notes on a collection from the Suez Canal, *Israel Jour. Zool.* **21**(3-4), 301-374.

Ben-Eliahu, M. N., 1970. The Polychaeta, *in* The Hebrew University-Smithsonian Institution Joint Program Biota of the Red Sea and the Eastern Mediterranean. *Progress Report, Appendix to Research Proposal 1970/71.* Jerusalem: Hebrew University (mimeo).

Ben-Eliahu, M. N., 1972*a*. Studies on the migration of the Polychaeta through the Suez Canal. *Theme No. 3: Les conséquences biologiques des canaux inter-océans,* 17e Congr. Int. Zool. (Monte Carlo, Sept. 1972). 8p.

Ben-Eliahu, M. N., 1972*b*. Polychaeta Errantia of the Suez Canal, *Israel Jour. Zool.* **21**(3-4), 189-237.

Ben-Eliahu, M. N., 1974. Polychaeta from parallel intertidal biotopes from the Eastern Mediterranean and the Gulf of Elat: The Fabriciinae (Polychaeta: Sabellidae), *Israel Jour. Zool.* **23**, 213.

Ben-Eliahu, M. N., 1975*a*. Polychaete cryptofauna from rims of similar intertidal Vermetid reefs on the Mediterranean coast of Israel and in the Gulf of Elat, I: Sabellidae (Polychaeta Sedentaria), *Israel Jour. Zool.* **24**, 54-70.

Ben-Eliahu, M. N., 1975*b*. Polychaete Cryptofauna from rims of similar intertidal Vermetid reefs on the Mediterranean coast of Israel and in the Gulf of Elat, II: Nereidae (Polychaeta Errantia), *Israel Jour. Zool.* **24**, 177-191.

Ben-Eliahu, M. N., 1976*a*. Polychaeta cryptofauna from rims of similar intertidal Vermetid reefs on the Mediterranean coast of Israel and in the Gulf of Elat, III: Serpulidae (Polychaeta Sedenteria), *Israel Jour. Zool.* **25**, 103-119.

Ben-Eliahu, M. N., 1976*b*. Polychaeta cryptofauna from rims of similar intertidal Vermetid reefs on the Mediterranean coast of Israel and in the Gulf of Elat, IV: Sedentaria: rare families, *Israel Jour. Zool.* **25**, 121-155.

Ben-Eliahu, M. N., 1976*c*. Polychaeta cryptofauna from rims of similar intertidal Vermetid reefs on the Mediterranean coast of Israel and in the Gulf of Elat, V: Errantia: rare families, *Israel Jour. Zool.* **25**, 156-177.

Ben-Eliahu, M. N., 1977*a*. Polychaeta cryptofauna from rims of similar intertidal Vermetid reefs on the Mediterranean coast of Israel and in the Gulf of Elat, VI: Syllinae and Eusyllinae (Syllidae: Polychaeta Errantia), *Israel Jour. Zool.* **26**, 1-58.

Ben-Eliahu, M. N., 1977*b*. Polychaeta cryptofauna from rims of similar intertidal Vermetid reefs on the Mediterranean coast of Israel and in the Gulf of Elat, VII: Exogoninae and Autolytinae (Syllidae: Polychaeta Errantia), *Israel Jour. Zool.* **26**, 59-99.

Eales, N. B., 1970. On the migration of tectibranch molluscs from the Red Sea to the eastern Mediterranean, *Malacol. Soc. London Proc.* **39**, 217-220.

Fauvel, P., 1955. Contribution à la faune des Annélides Polychetes des côtes d'Israël, I, *Sea Fish. Res. Stn. Haifa Bull.* **10**, 1-12.

Fauvel, P., 1957. Contribution à la faune des Annélides Polychètes des côtes d'Israël, II, *Res. Counc. Israel Bull.* **6B**(3-4), 213-219.

Gilat, E., 1964. The macrobenthonic invertebrate communities on the Mediterranean continental shelf of Israel, IAEA Radioactivity in the Sea. Publication No. 8, *Inst. Oceanogr. Monaco Bull.* **62**(1290), 1-46.

Gilat, E., 1969. The macrobenthic communities of the level bottom in the Eastern Mediterranean, *in* The Hebrew University-Smithsonian Institution Joint Program Biota of the Red Sea and the Eastern Mediterranean, *Interim Report*. Jerusalem: Hebrew University, 82-89.

Gilat, E., 1970. Macrobenthic communities off the Sinai Peninsula in the Mediterranean, in the Hebrew University-Smithsonian Institution Joint Program Biota of the Red Sea and the Eastern Mediterranean. *Appendix to Research Proposal 1970/71.* Jerusalem: Hebrew University, 19p.

Gilat, E., 1974. Macrobenthic communities off the Sinai Peninsula in the Mediterranean, *Proc. 5th Sci. Conference Israel Ecological Soc.*, B-1-15.

Gorgy, S., 1966. Les Pécheries et le milieu marin dans le secteru Méditerranéen de la Republique Arabe Unie, *Rev. Trav. Inst. Pêches Marit.* **30**(1), 25-80.

Gottlieb, E., 1959. *A Study of the Benthos in Haifa Bay: Ecology and Zoogeography of Invertebrates,* Ph.D. dissertation. Jerusalem: Hebrew University (in Hebrew with English summary).

Haas, G., 1951. Preliminary report on the Molluscs of the Palestine coastal shelf, *Sea Fish. Res. Stn. Haifa Bull.* **1**, 1-20.

Holthuis, L. B., and Gottlieb, E., 1958. An annotated list of the Decapod Crustacea of the Mediterranean coast of Israel, with an appendix listing the Decapoda of the eastern Mediterranean, *Res. Counc. Israel Bull.,* **7B**(1-2), 1-126.

Lerner-Seggev, R., 1965. Preliminary notes on the Ostracoda of the Mediterranean coast of Israel, *Israel Jour. Zool.* **13**(4), 145-176.

Lewinsohn, Ch., and Holthuis, L. B., 1964. New records of Decapoda Crustacea from the Mediterranean coast of Israel and the eastern Mediterranean, *Zool. Meded. Rijks. Mus. Nat. Hist. Leiden* **40**(8), 45-63.

Lipkin, Y., and Safriel, U. N., 1971. Intertidal zonation on rocky shores at Mikhmoret (Mediterranean, Israel), *Jour. Ecology* **59**, 1-30.

Masry, D., 1970*a*. *Microcerberus remanei israelis* n. ssp. (Isopoda) from the Mediterranean shores of Israel, *Crustaceana* **19**, 200-204.

Masry, D., 1970*b*. Ecological study of some sandy beaches along the Israeli Mediterranean coast, with a description of the interstitial harpacticoids (Crustacea, Copepoda), *Cahiers de Biologie Marine* **11**, 229-258.

Masry, D., and Por, F. D., 1970. A new species and a new subspecies of Mystacocarida (Crustacea) from the Mediterranean shores of Israel, *Israel Jour. Zool.* **19**, 95-103.

Pallary, P., 1938. Les mollusques marins de la Syrie, *Jour. Conchyl. Paris* **82**(1), 5-58.

Por, F. D., 1964. A study of the Levantine and Pontic Harpacticoida (Crustacea, Copepoda), *Zool. Verh. Rijks. Mus. Nat. Hist. Leiden* **64**, 1-128.

Por, F. D., 1978. *Lessepsian Migration—The Influx of Red Sea Biota into the Mediterranean by Way of the Suez Canal,* Ecological Studies 23. Berlin: Springer-Verlag, 230p.

Rayss, T., 1941. Sur les Caulerpes de la côte Palestinienne, *Palestine Jour. Botan. Jerusalem Ser.* **2,** 103-124.

Rayss, T., 1954. Les algues tropicales de la Méditerranée orientale et leur origine probable, *8ᵉ Int. Bot. Congr. Paris, Rapp. et Comm. Sect. 17,* 148-149.

Rayss, T., 1955. Les algues marines des côtes Palestiniennes. I. Chlorophycaeae, *Sea Fish. Res. Stn. Haifa Bull.* **9,** 1-35.

Rayss, T., 1963. Sur la présence dans la Méditerranée orientale des algues tropicales de la famille des Soliériacées, *Lucr. Gradinii Bot. Acta Bot. Hort. Bucuresti 1961-1962,* 91-106.

Reiss, Z.; Klug, K.; and Merling, P., 1961. Notes on Foraminifera from Israel, Recent Foraminifera from the Mediterranean and Red Sea coasts of Israel, *Geol. Survey of Israel Bull.* **32,** 27-28.

Safriel, U. N., 1974. Vermetid gastropods and intertidal reefs in Israel and Bermuda, *Science* **186,** 1113-1115.

Safriel, U. N., 1975. The role of vermetid gastropods in the formation of Mediterranean and Atlantic reefs, *Oecologia* **20,** 85-101.

Tebble, N., 1959. On a collection of Polychaetes from the Mediterranean coast of Israel, *Res. Counc. Israel Bull.* **8B**(1), 9-30.

Tom, M., 1976. *The Benthic Faunal Associations of Haifa Bay,* M.S. thesis. Tel Aviv: Tel Aviv University (in Hebrew with English summary).

Tortonese, E., 1953-54. Gli Echinodermi viventi presso le coste dello stato di Israele (Mar di Levante, Golfo di Elath), *1st Zool. Univ. Torino Boll.* **4**(4), 1-35.

Tortonese, E., 1966. Echinoderms from the coast of Lebanon, *Misc. Papers Nat. Sci. Am. Univ. Beirut* **5,** 2-5.

Tsurnamal, M., 1968. *Studies on the Porifera of the Mediterranean Littoral of Israel,* Ph.D. dissertation. Jerusalem: Hebrew University (in Hebrew with English summary).

Tsurnamal, M., 1969a. Sponges of Red Sea origin on the Mediterranean coast of Israel, *Israel Jour. Zool.* **18**(2-3), 149-155.

Tsurnamal, M., 1969b. Four new species of Mediterranean Demospongiae and new data on *Callites lacazii* Schmidt, *Cah. Biol. Mar.* **10**(4), 343-357.

Tsurnamal, M., 1975. The calcareous sponges of rocky habitats along the Mediterranean coast of Israel, *Israel Jour. Zool.* **18,** 149-155.

Wirszubski, A., 1953. On the biology and biotope of the red mullet, *Mullus Barbatus* L, *Sea Fish. Res. Stn. Caesarea Bull.* **7,** 1-20.

Cross-references: *Africa, Coastal Ecology; Asia, Middle East, Coastal Morphology: Israel and Sinai; Asia, Middle East, Coastal Morphology: Syria, Lebanon, Red Sea, Gulf of Oman, and Persian Gulf; Biotic Zonation; Coastal Fauna; Coastal Flora; Europe, Coastal Ecology; India, Coastal Ecology;* Vol. VIII, Part 3, *Middle East; Pollutants.*

ASIA, MIDDLE EAST, COASTAL MORPHOLOGY: ISRAEL AND SINAI

The first modern study of the Sinai and Israel coasts took place at the beginning of the nineteenth century shortly after Napoleon's campaign to Egypt. Jackotin introduced a new series of triangulation base maps of the shorelines in 1823.

Bathymetric maps of the eastern Mediterranean were published by Mansell in 1862, showing in detail some of the shorelines and the nearby sea. These and the Palestine Exploration Fund maps were used as basic maps for the British and Germans during World War I operations in the Middle East. Rosenan (1937) published revised and more detailed bathymetric maps at a 1:100,000 scale, which were later resurveyed by the Israel Oceanographic and Limnological Company and published in 1973.

Avnimelech (1950, 1952, and 1960), Itzhaki (1961), and Issar and Picard (1971) published on Quaternary oscillations of the shoreline and the nature of its sediments.

The first oceanographic and shoreline study of the Israeli shores was published in 1960 by Emery and Neev on the beaches, and by Emery and Bentor on the shelf. Ron (1962), Safra (1962), and Michelson (1970) described the beaches of the Carmel coastal plain, while Schattner (1967) described and classified the northern coasts of Israel. Nir (1973) and Bauman (1972, 1973) described the beaches north of Akko. A detailed study of the shallow-water sand transport was carried on at Ma'agan Mikhael by Eitam, Hecht, and Sass (1978). Other studies on nearshore artificial constructions and their influence on the Israeli beaches were published by Fried (1976), Tauman (1975), Nir (1976), Spar (1976) and Goldsmith and Golik (1978).

The mineralogy of north Sinai beaches was studied by Shukri (1950) and by Shukri and Philip (1960). The tectonic origin of the Mediterranean shores of Israel was discussed by Neev et al. (1973), Neev (1977), Neev and BenAvraham (1977), and Neev et al. (1978). Mazor (1974), however, objects to this theory.

Geographic Background

The Israel and Sinai Mediterranean coasts form the southeastern edge of the Levantine Basin, which by itself forms the extreme eastern part of the Mediterranean. The length of these beaches is about 440 km: 220 km in northern Sinai and 220 km in Israel.

The Israeli coast has a relatively low relief, with only two mountains reaching the seashore: Mount Carmel and the ridge of Rosh Haniqra

(Fig. 1). The coastal plain of northern Sinai has no such mountainous features and presents a smooth, lowland relief with well-developed sand dunes.

Bardawil lagoon, with its low relief and narrow shore bar, forms more than one-half of the northern Sinai shoreline. The lagoon is shallow, 1-2 m in depth on the average, and covers an area of about 600 km^2. Its connection with the sea is maintained via two or three artificial outlets (*boughaz*). This lagoon differs from the Nile delta lagoons (Mariyut, Manzala, and Burulus) in that it is of tectonic origin (Neev, 1967), and not a deltaic lagoon that receives its water from the Nile floods and the sea during winter storms. The only water source at Bardawil is the sea; therefore under natural conditions it has been a large *sabkha* (playa) with seasonal floods coming over the bar from the sea. No other lagoons or inland water bodies are found along the Sinai shore at the present time. In historical times, on the other hand, permanent swamps existed east of the *kurkar ridges* found near the shores of Israel. These ridges, which are a carbonate-cemented quartz sandstone, parallel the shore and influence much of the Israeli shoreline. The above-mentioned swamps lasted until modern settlement took place.

The coastal plain of northern Sinai is poorly defined, since the area between the mountains that form isolated ridges and the shores of northern Sinai is mostly low, covered with active sand dunes. The Israeli coastal plain is different, being bounded by foothills on the east. It consists of Quaternary to Recent sandy to clayey soils (Yaalon and Dan, 1967), with a few kurkar ridges. This kurkar found along the Israeli beaches "fades" in northern Sinai. The coastal plain of Israel is fairly wide in the south, up to 40 km, but narrows toward the north, reaching a width of only a few kilometers and less in the Carmel Plain, where Mount Carmel meets the sea at Haifa. North of Carmel, in Haifa Bay, the plain again widens, reaching an average width of 10 km; but thereafter it narrows northward to Rosh Haniqra, where the mountains again meet the sea. Beaches of similar nature exist north of Rosh Haniqra along the south Lebanon shores.

Climate

The climate along the Israel and Sinai Mediterranean coast is subtropical and typical of the entire Mediterranean area: semiarid in Israel and arid in Sinai, with a winter rainy season. The summers are hot and dry, yet the relative humidity along the coastal plain is moderately high. The weather in the winter season is dominated by cyclones passing in easterly directions. These result in rather unstable conditions, with the most frequent winds occurring from directions between SW and NW. Such a situation results in westerly winds that may generate high waves in the sea off the coast of Israel. Spring and autumn are relatively more quiet and are transitional seasons. Summer winds are mostly light, while in winter and occasionally in spring stronger winds blow, with storm winds reaching speeds of up to 80 km/hour or more. However, periods with no wind at all are also typical for winter.

Average rainfall is 600 mm/year north of Haifa, 500 mm/year in the central region, and about 350 mm/year and less on the southern shores of Israel. Rainfall further decreases westwards toward the north Sinai coasts, where rains are scarce (80-100 mm/year).

Average air temperatures along the coastal plain usually range between 12 and 14°C in January, which is the coldest month (with minima reaching the freezing point), and between 24 and 26°C in August, the warmest month, when the maximum temperatures may reach 45°C.

Waves, Currents, and Tides

Wave activity during the winter, mainly during storms, has a considerable effect on the narrow beaches found along many parts of the Israeli shoreline. The beaches along Haifa Bay, south Israel, and north Sinai are wide, and therefore winter storms do not abrade them much. The Bardawil lagoon bar has a low relief and is narrow along most of its length, so that large quantities of Mediterranean water enter the lagoon during storms. This is the only natural way of refreshing the lagoon waters. According to Levy (1972), the subrecent sediments of the lagoon reflect various sabkha and lagoonal conditions. At present the Bardawil lagoon waters are up to three times more saline than their source, the Mediterranean.

A similar phenomenon occurs west and north of Bardawil lagoon, in the eh-Tineh region. Here too winter storms push huge quantities of water over large areas of the low sabkha.

Winter southwestern storms have the greatest effect on the beaches. Wave heights in these storms may reach 5-7 meters (Israel Port Authority, 1966).

Two main current systems affect the beaches: (1) The anticlockwise east Mediterranean current distributes the Nile sediments eastward and northward from the two Nile mouths at Rosetta and Damietta. The fine Nilotic particles are deposited mostly in quiet waters at about 25 meters and deeper (Nir, 1973). (2) Longshore currents

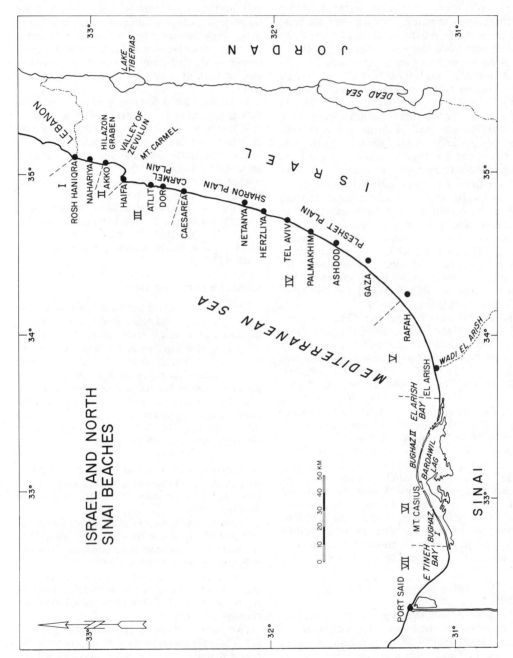

FIGURE 1. Israel and North Sinai beaches.

are locally determined by the existing wave fronts and sea-bottom morphology. The net transport of sand is generally toward the east in northern Sinai and toward the north in southern Israel (Shukri, 1950; Shukri and Philip, 1960; Emery and Neev, 1960; Inman, 1975; Nir, 1976; and Goldsmith and Golik, 1978). The north Sinai longshore sand transport is to a certain extent greater than that along the Israeli beaches because of the relations between shore configuration and storm and wave directions.

Israel Port Authority (1966) and Uziel (1968) data show that the maximum seasonal tide is 80 cm (from −40 to +40 cm), but the amplitude of the daily tides is much smaller, never exceeding 30 cm.

Rivers

The rivers in Israel that drain into the Mediterranean are small, and are mostly of the ephemeral type, having the local name *wadi*. A few springs that flow from the foothills once supplied a few of these wadis, but their exploitation in modern times has changed even these small rivers to dry wadis.

Water discharge during floods is relatively large, but there are actually no active floodplains, since the river channels can in most cases accommodate the water supply. In historical times, on the other hand, a large number of swamps were developed along the lower parts of the coastal plain where dunes and kurkar ridges acted as a barrier for winter floods. Channels and tunnels were therefore dug by the Romans to drain a few of these swamps in the Sharon Plain.

Northern Sinai has a much different drainage scheme. Here at present only one large wadi, Wadi El Arish, drains most of the area. Although floods are rare, when they do occur a few hundred million cubic meters may reach the sea in a few days, producing a small delta. During the following dry winters, these deltaic deposits are transported eastward by the longshore currents.

Bays

The only important bay along the Israel coast is found at Haifa. This is a morphological feature directly related to the Carmel Mountain horst and the Valley of Zevulun graben. Small bays a few tens of meters in length and usually rocky (mostly kurkar rocks) occur locally in central and northern parts of the Israel shore, mostly between Dor and Atlit.

North Sinai has two large bays, that of eh-Tineh in the west, between Bardawil lagoon bar and Port Said, and El Arish east of the Bardawil lagoon.

Offshore Ridges

Submarine kurkar ridges are found in the shallow shelf to depths of about 100 meters (Neev et al., 1976). Some of these ridges are found in shallow water, and their exposed crests were abraded during the post-Würm rise in sea level. Most of the rocky exposures are covered by biogenic activity.

Flat platforms of recent age are found on the rocky beaches, usually at mean sea level (Fig. 2), and the platform is divided into small basins. According to Safriel (1965, 1974), the rock surfaces are incrusted by a dense cover of vermetid shells cemented by coralline algae.

The shallow water region between Tel Aviv and Caesarea (water depths of 2–4 meters) consists of a flat, rocky bottom that extends out from the surf zone to a distance of 200–250 m on the average (Nir, 1973; Bakler, 1976). This rocky bottom is evident in aerial photos and in diving.

Sediments of the Israel and Sinai Beaches

General. The beaches of Israel and Sinai are usually sandy, with a few exceptions where local kurkar or limestone pebbles are found (mainly in central and northern Israel). The sands are light in color, light yellow to very light gray, and are composed mostly of quartz grains. Heavy minerals, generally opaque, are concentrated as laminae, mainly along the beaches of north Sinai. The beaches of northern Israel differ in the bulk mineralogy of the sands; they are of local origin and are composed of broken calcium carbonate skeletons of different pelecypods and calcareous algae. The most abundant pelecypod that forms the bulk of the shells found along the Israel beaches is *Glycimeris violacescens*. The north Sinai beaches, on the other hand, are rich in *Cardium edule*. The transition from one pelecypod type to the other is also seen in the central Bardawil lagoon bar, where the *Glycimeris* start to appear in increasing numbers eastward while the numbers of *Cardium* decrease. The two pelecypods represent sandy and muddy sea bottoms, respectively, and their distribution is also affected by longshore transport.

A few relicts of unconsolidated subrecent terrestrial black clays, appearing in lense-shaped outcrops, are found along a few beaches in northern Israel. These are mostly terrestrial clays of swampy origin deposited along the beaches and in the present shallow shelf area during periods of lower sea level during Würm times (Nir, 1973). These deposits are also found in many shallow offshore drillings, for example, off Hadera (Nir, 1977; Tur Caspa, 1978; Almagor and Nir, 1979; Nir, 1980).

FIGURE 2. Typical abraded rocky platform, Akhziv shoreline, north Israel. (Photo: Y. Nir)

The sandy beaches are relatively wide in regions where cliffs are absent, reaching a width of 60–80 m in some parts of Israel and are even wider in northern Sinai. In the central parts of Israel, where the backshore consists of a kurkar cliff, beaches are narrow, reaching a width of only 10–20 m in summer, and 5–15 m during winter (Fig. 3). Winter storms reach the bases of the cliffs and accelerate erosion. Many high dunes are found at the backshore area in northern Sinai from the eastern parts of the Bardawil lagoon to Rafah. These dunes are distributed and more developed west of the El Arish River (Tso'ar, 1970).

Beach rock has been observed all along the Israel coast with the exception of Haifa Bay and is also widespread along the Sinai beaches. It is usually rich in coarse local components, mostly pelecypod shells. A few beach-rock exposures along the kurkar cliff beaches contain flat kurkar pebbles.

Sources of the Sediments. Mineralogical studies carried out along the Nile River, its delta beaches, northern Sinai, and Israel (Shukri, 1950; Shukri and Philip, 1960; Emery and Neev, 1960; Pomerancblum, 1966), show that the main foreign source of sediments is the Nile River and the Blue Nile originating in Ethiopia. The Ethiopian source was also proved by palynological studies of the shelf and slope sediments of the Israel Mediterranean Sea by Rossignol (1961) and Horowitz (1974), and by mineralogical studies of these sediments (Pomerancblum, 1966; Nir, 1973).

Quartz is the most abundant and significant mineral along the beaches and in the sandy belt along the shallow parts of the continental shelf (Nir, 1973). Calcite, mostly in the form of skeletal grains is second in importance, and heavy minerals are third.

These sediments are transported along the beaches by means of longshore currents and in

FIGURE 3. Typical upper part of the kurkar cliffs, south Netanya. The beach is narrow; therefore winter and spring storms abrade the lower portions of the cliffs. (Photo: Y. Nir)

the shallow shelf waters by other currents, primarily the surface currents of the eastern Mediterranean. According to Inman (1975), until the beginning of the twentieth century the discharge of sediments exceeded the potential of erosion, and the Nile delta built seaward. Man's intervention, beginning with the construction of barriers in the lower reaches of the Rosetta and Damietta branches and of the first Aswan Dam in 1903, changed the pattern to one of erosion. This has been much accelerated after the construction of the Aswan High Dam in 1964. The delta beaches have suffered severe damage and according to Zenkovich (1971) and Inman (1975), erosion has averaged 143 meters per year between 1943 and 1973 east of Ras el Bar.

A few artificial features that have been constructed recently along the Sinai and Israel beaches show a clear and distinct eastward and northward net sediment transport (Nir, 1976; Goldsmith and Golik, 1978). Inman (1975) suggests that the amount of easterly littoral transport of sand in northern Sinai is high and reaches rates as high as 500,000 m^3 per year in Boughaz in the Bardawil bar. He also states that quantitatively the whole subject is not well known, but changes in shoreline orientation and wave approach suggest that the transport rates must generally decrease to the northeast. Still high rates of shore abrasion were recently observed in the apex of the Bardawil bar.

The decrease in easterly and northerly sand transport was also proved by Dornhelm (1972), who shows that the construction of the port of Ashdod, whose main breakwater reaches a water depth of 18 m, has interrupted 80–90% of the annual northward transport of the sand estimated to be only 50,000–100,000 m^3 per year there.

Second in importance to the Nile, a much smaller source of sand is the erosion of the kurkar cliffs. The bases of the cliffs are being abraded, and undercuts (*nips*) are formed: accelerated abrasion conditions are therefore developed. According to Nir (1973), these cliffs were abraded at an average rate of about 3 cm/year during the past 6000 years. Relicts of these eroded kurkar strata are still found in the offshore waters, at depths of 2–8 m.

Erosion of the beach kurkar, by both marine and atmospheric agents, supplies the beaches with clastic sediments: loosely to well-cemented calcareous quartz sandstone, soft *hamra* (loam), and unconsolidated quartz sand. As the abrasion of the kurkar is quick, it supplies relatively large quantities to the nearby beaches. The ratio of imported Nile sediments to local kurkar abrasion sediments is not clear, mostly because of the mineralogical uniformity of the two sources. According to the mean grain sizes that decrease and the CaCO$_3$ content that increases in the kurkar cliff beaches, it is estimated that more than one-half of the present sediments here are derived from local sources, mainly the kurkar.

This source is limited to the Israeli beaches only; in northeastern Sinai there are a few cemented sandy ridges that resemble the kurkar but are different in both their field appearance and areal distribution.

Third in importance for sediment supply to the beaches are those imported by local wadis (dry washes). These are active only during winter floods, and provide only minor amounts of sediments of the course type, since their main load is eroded soils composed mostly of clay-size grains, which, like the small particles derived from the kurkar erosion, are carried out and deposited in much deeper waters.

Wadi el Arish is the only river draining most of central and northern Sinai. Floods here are infrequent and may not occur for years. When a flood does occur, however, it is torrential and may over a few days bring hundreds of millions of cubic meters of water loaded with large quantities of clastic sediments, mostly yellow desert dust, which is also transported to deeper water. Some of the coarse clastic sediments are deposited in a provisional small delta.

The role of sand dunes as a source of sand to the Israeli beaches is small at present (though it has had a significant importance in historical time), as most sand dunes are isolated from the shore by the kurkar ridges. At river outlets and the Haifa Bay beaches, where sand dunes reach the beaches, sand *exchange* can take place. In Sinai, on the other hand, where sand dunes cover enormous areas and are connected with the beaches, the role of beach to backshore sand-dune sediment exchange is much larger.

Artificial changes occur along the Israeli beaches and in the Bardawil outlets. In Israel this was the result of sand quarrying for construction purposes, which lasted until the early sixties. Israeli beaches still suffer a deficit of almost one-third of the total beach sand reserves (Emery, 1963), thus indirectly influencing and accelerating the kurkar cliff abrasion. As a result of this enormous sand mining, quantitative sand transport studies along the Israeli beaches are difficult. At present many beaches are only at their reconstruction stage and they accumulate more sand than the natural deposition at a certain point. It is difficult to foresee how long this phenomenon will last and at what stage of accumulation the deposition conditions will return to normal.

The following qualitative budget of sand along north Sinai and Israel Mediterranean beaches illustrates these sand sources and sinks onshore and offshore.

Offshore Losses
1. Sand quarrying
2. Sand accumulation in harbors and near artificial objects (breakwaters, groins)

Onshore Losses
1. Sand mining (to 1964)
2. Using of beach and tombolo areas (formed at the "shadow" of breakwaters), for construction purposes
3. Dune building

Seaside Contribution
1. Sands originating in the Nile, transported along its delta and north Sinai to the Israeli beaches
2. Destruction of beaches in the Nile delta (since 1965) and north Sinai
3. Organic origin (mostly Pelecypod shells)

Land-side Contribution
1. Erosion and abrasion of the kurkar cliff
2. River alluvium
3. Sand dunes

Grain-size Distribution and CaCO₃ Content of the Beach Sediment. Because the Nile is the main source of sediment for the Sinai and Israel beaches, a clear trend of decrease in grain size from the Nile's eastern mouth at Damietta eastward toward Sinai and Israel beaches might be expected. However, studies of the beach sands from near Port Said to the Lebanese border at Rosh Haniqra show that the decrease in grain size is not significant. The eastern flanks of the Nile delta, the Port Said-Rumani beaches have a mean grain size of 250 microns, the result of the introduction of relatively fine material from the eh-Tineh Bay sediments to the coarser Nile sediments transported along these beaches. The Bardawil bar shows a certain increase in average mean size to about 300 microns, an average maintained along the Sinai beaches to Rafah.

The Gaza region beaches have a higher mean size average of 315 microns, with relatively high peaks of mean grain size (460–480 microns). These sizes and the mean grain sizes from Gaza to Tel Aviv are generally the result of the abrasion of the kurkar cliffs that have a coarser grain size in the area. From Tel Aviv northward the mean sizes sharply decrease, having an average of only 190 microns, a figure that is well in agreement with the kurkar cliffs grain sizes accompanying the beaches here. In the Carmel beaches there is another increase of mean size up to 245 microns. This distinct increase along the Carmel beaches is a result of the introduction of local coarse material, which is high in $CaCO_3$.

The Haifa Bay beaches have a small mean grain size that does not exceed 180 microns. The sediments are poor in $CaCO_3$ and are mostly derived from the recent and subrecent Nilotic sediments that bypass the nose of the Carmel in relatively deep water. Some of these quartz sediments are derived from the nearby sand dunes. There is almost no contribution from the large coarse calcareous sediment lenses found in and around Haifa Bay itself (Nir, 1980). Similar sediments are found southwest of Haifa, around the Cape Carmel beaches, Bat Galim, around Shiqmona, and north of the Carmel beach detached breakwater tombolo.

The northern section, from Akko to Rosh Haniqra, is, as mentioned earlier, a separate sedimentary cell and not connected at present with the supply from the Nile. If there is some supply it takes place in deeper water, and the sediments transported are only fine sand particles, mostly quartz grains. This section is sedimentologically isolated from the northern and the southern beaches. The two ends of this section protrude to the sea and cut sediment transport, and so the sediments are mostly of local organic origin (quartz percentage is low) and are coarse in grain and high in carbonate content.

Table 1 shows the mean grain size of the different beach sections between Port Said and northern Israel.

Because the mineralogy of the Pleistocene coastal plain sediments is similar to that of the recent sediments, it is difficult to differentiate between the present sediments transported along the beaches and those introduced by such local sources as the El Arish River, Israeli wadis, and the sediments resulting from the abrasion of the kurkar cliffs. This introduction of coarser and/or finer material appears to mask the expected trend of decrease in grain size along the 400 km and more of Nile-derived sediments transported along the beaches.

The calcium carbonate content of the sediments of Nilotic origin is low. Therefore the entire area of Nile sedimentation in the Levantine Basin is poor in $CaCO_3$ (Emelyanov, 1968; Anwar and Mohamed, 1970; Nir and Nathan, 1972; Venkatarathnam and Ryan, 1971). The Nilotic sediments and the Israel shelf sediments contain 10% or less of $CaCO_3$ (Nir, 1973).

North Sinai beaches are relatively poor in $CaCO_3$, usually less than 8%, decreasing eastward to only 3–4%. South Israel beaches are

TABLE 1. Mean Grain Size and $CaCO_3$ of the Beach Sand of the eight Different Sections of the Beach (average of total of 92 samples). (The kurkar cliff shores of central Israel are given here as two separate sections.)

Section of beach	Mean grain size (microns)	Average CaCO₃ content (%)	Range (%)
Port Said-Rumani	255[a]		
Bardawil lagoon bar	305	6.37 (10)	1.25-15.21
El Arish-Rafah	300	4.41 (15)	1.51- 7.79
Rafah-Tel Aviv	315	9.07 (19)	3.98-41.24
Tel Aviv-Caesarea	190	11.81 (13)	5.77-20.62
Mt. Carmel beach (Caesarea-Shiqmona)	245	36.14 (13)	8.87-91.31
Haifa Bay (Shiqmona-Akko)	170	15.23 (6)	10.86-21.83
Akko-Rosh Haniqra	400	93.37 (16)	83.15-97.84

[a]Samples analyzed by Beni Sagiv of the Beer Sheva University.

somewhat higher in $CaCO_3$, having about 8% in the southern segment and higher values in the northern segment with its kurkar cliff beaches. The increase in the northern segment is directly related to the local source, the abrasion of the kurkar cliff, which is rich in $CaCO_3$. The Carmel beaches are rich in local material, which is primarily $CaCO_3$ and in a few places the carbonate content reaches 60% or more.

A relatively significant decrease is found in the Haifa Bay beaches, where the carbonate content decreases to an average of about 15%. This value is somewhat higher than the average of the Nilotic, Sinai, and southern Israel beach sediments, and is derived from the offshore sediments, which are rich in carbonate (Neev et al., 1970; Bakler, 1976).

A high carbonate content is typical of most of the northern section, from Akko to Rosh Haniqra. Here, because of the above-mentioned isolation and the local introduction of $CaCO_3$-rich grains, values reach an average of 95%.

The southern Lebanon beaches, which are totally isolated from those of Israel, show an average of about 31% $CaCO_3$ (Emery and George, 1963). Recent studies of these beaches show a resemblance to Rosh Haniqra-Akko beaches of northern Israel.

In general, the beach sediments are well sorted (Table 2). A clear tendency of improvement in sorting from the Nile delta beaches along Sinai and most Israel beaches can be seen. With the exception of a few samples, most of the beach sediments are well sorted. The eh-Tineh Bay, Carmel, and western Galilee beaches are to a certain extent an exception, where sorting is poorer.

TABLE 2. Standard Deviation-Sorting Values of the Different Beach Sections

Beach section	Standard deviation	No. of samples
eh-Tineh Bay	.583	9
Bardawil lagoon bar	.496	9
El Arish Bay	.455	17
Pleshet Plain	.401	19
Sharon Plain	.391	13
Carmel Mountain Plain	.430	9
Haifa Bay	.358	6
Western Galilee	.550	10

Eastward and then northward from eh-Tineh Bay beaches the sediments become better sorted with distance from the Nile mouth, reaching a peak of extremely well-sorted grains in the Sharon Plain and Haifa Bay beaches.

The Carmel Mountain beaches and the western Galilee beaches are much less sorted, mostly because of the introduction of coarse $CaCO_3$ grains of local origin. This newly introduced material can be seen also in the bimodality of grain-size histograms. The sediment at Haifa Bay beaches are the best sorted of all those studied here. This is due to its longest distance of travel from the Nile on one hand, and to the fact of the sediment exchange with the near by sand dunes which have a very well sorted grains on the other hand.

In order to find the grain-size relationship between the Nile-derived and the local sediments, seventeen samples from different locations were twice analyzed: original sediment, and carbonate-free ones (Table 3). The samples containing

TABLE 3. Standard Deviation-Sorting Values, and Mean Size of Selected Samples Having Different $CaCO_3$ Content in Comparison Analyses of Original Versus $CaCO_3$-free Samples

Sample No. Y.N.	%$CaCO_3$	Standard Deviation		Mean Size (microns)	
		original	$CaCO_3$-free	original	$CaCO_3$-free
2469	10	.334	.287	182	171
2465	11	.304	.286	79	158
2466	12	.357	.274	157	148
2468	15	.416	.324	177	169
2464	15	.333	.279	160	152
2484	16	.393	.324	213	188
2467	17	.396	.321	174	163
2473	20	.368	.345	202	196
2463	22	.342	.271	156	152
2485	27	.524	.374	268	210
2483	38	.449	.478	253	216
2470	66	.445	.500	314	252
2482	74	.839	.668	490	253
2481	75	.922	.695	566	228
2478	86	.848	.881	877	191
2477	91	.494	1.167	914	248
2479	93	.422	1.204	1024	199

both Nilotic and local sediments are much less sorted compared with the carbonate-free sediments. The high-carbonate sediments are to a certain extent different, and sorting is improved in the residue; this is probably due to the fact that the Nilotic-origin grains comprise only a minor part of the entire sample (10-20%); therefore the sorting coefficient cannot well represent the original sample. With a few exceptions, there is a clear decrease in mean size of the original sample as compared with the $CaCO_3$-free ones. Here the large differences are in the medium- and high-carbonate content group (27-93% of $CaCO_3$).

The Shoreline

The Mediterranean shore of Israel gently curves from an almost easterly to an almost northerly direction. Northern Sinai shores are also curved, having mostly NW-SE and SW-NE directions. The curved bar of the Bardawil lagoon is its most significant feature. East of Bardawil the shoreline trends almost east-west before it slowly curves to the northeast where it reaches Israel.

Active faulting is responsible for the shape and to a certain extent for the morphology of the coastline of Israel, which is characterized by cliffs in its central parts (Neev et al., 1973). The arc of Bardawil is, according to Neev (1967), also of tectonic origin, since it is basically an emergent structure. As a result, Bardawil's western bar is of tectonic origin while its eastern bar is depositional. The tectonic origin is disputed by some authors, among them Mazor (1974), who explains the central Israel shore cliffs as of abrasional and not of tectonic origin.

The only important bay along the coast of Israel is at Haifa, north of Mount Carmel. The bay is a morphological feature related to the Mount Carmel horst and the Zevulun Valley graben on the north. The bay's northern edge is formed by the Akko ridge and probably by the E-W border fault of the Hilazon graben.

A few small but important bays bounded by relicts of the abraded kurkar ridges exist along the northern parts of the shoreline.

Two large bays are found along the north Sinai shores: eh-Tineh Bay, between the western bar of Bardawil and the fossil delta of the Nile's Pelusaic arm; and the bay east of Bardawil, between its eastern bar and the southern shores of Israel. The two are essentially large embayments.

Relicts of Pleistocene kurkar ridges form small islands and islets adjacent to the northern part of the Israel shores. These islands are small in area, having dimensions of not more than 50 X 200 m. No islands exist between Caesarea and the mouth of the Nile at Damietta. To the north, off the coast of Lebanon and Syria, there are similar islands on which the Phoenicians built the cities of Tyre, Sidon, and Ardos.

There are no deltas protruding along the beaches, since no large rivers drain into the sea here. At El Arish a temporary delta develops after each flood. Such a delta developed after the torrential flood of the winter of 1975. This delta was, however, quickly destroyed by waves and longshore currents that spread the sediments along the shores to the east. In its first stages, this delta covered an area of about 180,000 m^2, while a year and a half later its shape had changed and its area reduced to less than 40,000 m^2. This phenomenon progressed rapidly, and after two years most of the delta had disappeared and the shoreline had returned almost to its original condition.

According to morphological criteria, seven different sections can be identified on the Israeli and Sinai beaches. These differ mostly in their beach and inland features, on the one hand, and in various sedimentary properties of the beach, on the other hand. The different sections are, from north to south and west (see Fig. 1): Rosh Haniqra-Akko, Haifa Bay, Mount Carmel beach, kurkar cliff shores (Sharon and Pleshet beaches), the wide, sandy beaches of the El Arish embayment, Bardawil lagoon bar, eh-Tineh embayment beaches.

Morphology of the Shore Regions. The Nile delta beaches and the Sinai and the Israeli beaches receive most of their sand from the Nile. Hilmy (1951) assumed that west of Alexandria the percentage of Nile-derived sediments decreased sharply with increasing distance from the western Rosetta mouth of the Nile. North of the Rosh Haniqra-Akko-region, the most northern shore region of Israel, no sediment transport can occur to the Lebanese beaches at present because of the steep cliffs of Rosh Haniqra, which present an effective barrier. There is no doubt that the sediments were transported along these shores during lower sea-level stages (e.g., during Würm times).

The large area from the apex of the Nile delta to Akko is actually one large cell of Nile-origin sediments in which each subregion is dependent on the other.

Rosh Haniqra-Akko Region (Western Galilee). As earlier described, this region is exceptional for its sediment content because of its isolation from neighboring Haifa Bay to the south and the Lebanese beaches to the north.

The white chalky cliffs of Cenomanian age at Rosh Haniqra are the only mountain formations in the area that reach the sea at such an elevation (Fig. 4.). These cliffs from the northern border of the region. Directly to the south the shore relief is much lower, and in places rich carbonate kurkar ridges are the only morpho-

FIGURE 4. Rosh Haniqra cliff of white Cenomanian chalk acting as a barrier to longshore sediment transport. (Photo: Y. Nir)

logical features found along the shore. Coarse carbonate grains form the narrow, sandy beaches. These beaches, as described above, are isolated, and there is no supply of sediments other than the addition of minute quantities from the upper Galilee Mountains and from the destruction of some of the kurkar hills. Overquarrying of beach sand during the fifties and the early sixties almost ruined these beaches and exposed huge areas of beach rock along most of the coast. Pebbles are also found, mostly a product of the nearby mountains (Bauman, 1973).

Beach sands here are composed of almost pure, coarse-grained carbonate (90–95%) of local origin. The kurkar found in this region, both on land and offshore, has essentially the same clastic components. The fine grains on the beaches are mostly quartz derived from offshore currents bypassing the Akko ledge. They are of Nilotic origin and resemble the sediments of the Haifa Bay.

Haifa Bay. The Haifa Bay beaches are broad, low in relief, with no cliff-lined areas. Sand dunes of small height border the beach belt, and there is a certain interchange of sediments between the two. The beach sediments are fine-grained and composed mostly of quartz of Nilotic origin. The region is also to some extent isolated by the ledge of Akko in the north and the underwater continuation of the Carmel "nose" in the southwest. The deeper water currents are therefore the only carrier of sand to the bay, a supply that is relatively meager. The Carmel nose and the shape of the beach here divert sediment transport both toward the bay on the east side of the nose and southward toward the northern Carmel beaches.

The shore sediments are high in carbonate content by comparison with those of the regions to the south, but much lower than those of the Haifa Bay offshore sands (Nir, 1980). The beaches have quartz grains of typical Nilotic origin, while the offshore sediments are mostly syngenetic and composed of organic bank debris.

Mount Carmel Beaches. The Carmel coastal plain is flat and narrow—a few hundred meters wide in the north to a few kilometers in the south. One to three inshore kurkar ridges are found here, parallel to the shoreline and at varying distances from the shore. In a few places the westernmost ridge forms a small cliff the base of which acts as a seawall during storms. The area between the kurkar ridges is mostly flat, composed of a dark, black-brown soil that is primarily of swampy origin. Soils of identical origin are also found as lenses along the beaches and as a widespread layer in the shallow water of the Carmel beaches.

The beaches are narrow and are indented in many places between the kurkar rocks, forming small bays. Atlit, the famous Crusaders' fort is located on a relict of kurkar ridge, relatively close to the shoreline, and has a naturally constructed wide tombolo. A few other smaller kurkar relicts form similar structures in the Carmel beaches. A few islets, which are also relicts of former kurkar ridges, are found close to the shore (Fig. 5).

Beach material is of both local and Nilotic origin. Lenses of coarse sandy to pebbly sediments derived from the Carmel Mountain are found along this stretch of coast. Carbonate percentage is therefore high in these quartz sands of otherwise typical Nilotic origin.

The Kurkar Cliff Shores: Sharon and Pleshet Beaches. A fault line described by Neev et al. (1973) is probably a segment of the main fault forming the eastern border of the Mediterranean

FIGURE 5. Islets found mostly in the central and northern parts of the Israeli coast. These are relicts of former kurkar ridges located close to the shore. (Photo: Y. Nir)

that runs from Alexandretta Bay in Turkey to southern Israel (Neev, 1977). This fault line is responsible for the formation of most of the Israeli shore section and its cliffy morphology. Dated terraces of beach rock and seashells (Neev et al., 1973) found at different elevations above present sea level prove the existence of this line in the Israeli section of the Mediterranean.

A 20-50-m-high cliff forms the present coastline, leaving in most cases a narrow beach that often disappears during winter storms. These cliffs are effectively eroded by atmospheric agents and suffer much abrasion at their base during winter storms. As a result the cliffs have retreated landward at a high rate of about 3 cm/year, at least since 6000 years B.P. (Nir, 1973).

The beaches are narrow except at river mouths where there is local widening. In general, widths are only a few meters. Large relicts of kurkar are found at the waterline and in shallow water, acting as natural shore protectors. The Pleshet beaches are to a certain extent wider than the Sharon ones. Here summer beaches may reach widths of 20-40 m; therefore the winter damage to the cliffs is somewhat less.

Beach material is uniform and is derived from two main sources: from the Nile via the Sinai beaches and shallow offshore waters, and from the destruction of the kurkar ridges adjacent to the beaches (which also originated from the Nilotic sediments). Mineralogically, sands of both origins are similar, and the main component is quartz (the kurkar may have some higher carbonate content).

The average grain size is larger in the south than in the north—300 microns in the Pleshet and 200 microns in the Sharon beaches. This variation in grain size is due both to the northerly transport of the sand grains along the beaches and the addition of kurkar-derived, smaller grains mostly Nilotic in origin and aeolian in nature.

Recent beach rock composed of shell fragments and other sandy components is frequently found along these beaches. Some outcrops were formed on the kurkar relicts.

The Wide, Sandy Beaches of the El Arish Embayment. North Sinai beaches are wide and sandy, with high sand dunes on their landward sides. The few artificial structures that were constructed here clearly show that the trend of sand transport along these beaches is toward the Israeli shores. Quantitatively it seems that because of shore direction, the amount of transported sand through longshore currents is greater along the Sinai than along the Israeli beaches. Beach rock with different components of shell fragments and sands is found in patches all along these beaches.

The beach sand is composed mostly of Nile-derived quartz sands with low carbonate content. Because of *sand storms* there certainly appears to be a much more effective exchange of sediment between the well-developed sand dunes and the north Sinai beaches than on the Israeli beaches, where most of the sand dunes are isolated from direct contact with the sea by kurkar cliffs.

The only river (wadi) along this coast is at El Arish, and it floods only every few years. Its sediments differ greatly from those coming from the Nile, but their characteristics are lost among the great bulk of the beach sediments. Most of the sediment load of this river consists of fine-grained particles, which are quickly carried into deep water where they settle and are deposited.

The Bardawil Lagoon Bar. This arched bar, about 75 km long, forms the outer barrier between the Bardawil Lagoon and the open sea. The bar usually has a low relief, which allows seas in winter storms to breach it and penetrate the lagoon and its nearby sabkha areas. This is the only natural way of refreshing the lagoon, which of course during quiet years becomes a huge sabkha.

According to Neev (1967), the western bar is of tectonic origin (resulting from the substructure of the Bardawil, connected with the Pelusium line), while the eastern bar is of depositional origin. This bar widens close to its apex, about 7 km to the SW, where 60-m-high dunes develop. This chain of dunes is a few kilometers long and has two distinct ages—pre-Roman and subrecent—which are easily distinguished and identified by means of well-defined ceramics (as along many beaches in northern Sinai and southern Israel).

At the apex itself the bar is narrow, probably because of the sand-transport regime at this point. Its shape indicates that it is a limited and local western transport toward Mount Casius in addition to being the regional main eastward sand transport; therefore sedimentation conditions here are erosional.

Beach sand grain sizes are uniform to the east and west of the lagoon and average 300-350 microns. However, faunal distribution is much different, for the Bardawil bar forms the contact between the two main pelecypod species found along the Israeli and Sinai beaches: *glycimeris sp.* and the *cardium sp.*, respectively.

Inman (1975) estimated the eastward sand transport to be on the order of 500,000 m^3/year on Boughaz No. 1. No other quantitative estimations were made, but other field studies have showed that the figure is probably somewhat high and is much smaller along the Israeli shores (Dornhelm, 1972).

eh-Tineh Embayment Beaches. This beach

forms the southeastern portion of the large historical delta of the Nile, where the Pelusaic branch once drained into the sea (Sneh and Weissbrod, 1973). Sediments, both inland and offshore, are fine-grained (the name of the region, eh-Tineh means *fine sediments* in Arabic). Sabkhas are the typical terrestrial environment, whereas the marine sediments are muddy sands. The only coarse-grained sediments are found along a narrow belt that comprises the beach. These sands are derived mostly from the Nile.

Here too, because of the shallowness of the beaches and the low relief, water can penetrate the sabkhas during winter floods.

YAACOV NIR

References

Almagor, G., and Nir, Y., 1979. *Detailed Bathymetric and Shallow Seismic Survey Off the Site for the M.D. Electric Station near Hadera.* Geol. Survey Israel Rept. MG/1/79, 14p.

Anwar, Y. M., and Mohamed, A., 1970. The distribution of calcium carbonate in continental shelf sediments of Mediterranean Sea north of Nile delta, *UAR Bull. Inst. Oceanog. and Fishery, Egypt* **1**, 451-459.

Avnimelech, M. A., 1950. *Contribution to the Knowledge of the Quaternary Oscillations of the Shoreline in Palestine.* Firenze: Rev. Ital. Paleon. 13p.

Avnimelech, M. A., 1952. Late Quaternary sediments of the coastal plain of Israel, *Geol. Survey of Israel Publ.* **2**, 51-57.

Avnimelech, M. A., 1960. The latest history of the Mediterranean Coast of Israel, *CIESMM Rapp. et Proc. verbaux* **15**(3), 331-340.

Bakler, N., 1976. Calcareous sandstone and sands of the Israel Mediterranean offshore aggregate reserves, *UN/UNDP-GSI, Offshore Dredging Project*, Rept. 5/76.

Bauman, D., 1972. *Akhziv-Rosh Haniqra, Morphology and Sediments of the Coastal Plain* (in Hebrew). Haifa: Avalon Pub., 54p.

Bauman, D., 1973. Morphology and sediments of the Mediterranean Beach–Akhziv-Rosh Haniqra (in Hebrew), in Flexer and Yedaya, eds., *Studies and Discoveries on the Western Galilee.* Haifa: Avalon Pub., 123-129.

Dornhelm, R. B., 1972. Determination of the longshore sand transport rate in the vicinity of Ashdod port, *Geol. Survey of Israel* (unpublished), 43p.

Eitam, Y.; Hecht, A.; and Sass, E., 1978. Topographic and granulometric variations on the shore of Ma'agan Mikhael, eastern Mediterranean, Israel, *Israel Jour. Earth-Sci.* **27**, 1-13.

Emelyanov, E. M., 1968. Mineralogy of the sand-silt fraction of recent sediments of the Mediterranean sea, *Lithology and Mineral Resources* **2**, 3-21.

Emery, K. O., 1963. Some remarks on the sediments of the Israeli beaches (in Hebrew), *Zifzif Com. Rep., Appendix No. 2.* Jerusalem: Ministry of Development, 2p.

Emery, K. O., and Bentor, Y. K., 1960. The continental shelf of Israel, *Geol. Survey Israel Bull.* **26**, 25-41.

Emery, K. O., and George, C. J., 1963. The shores of Lebanon, *Am. Univ. Beirut Misc. Papers Nat. Sci.* **1**, 1-11. (Woods Hole Oceanog. Contrib. No. 1385.)

Emery, K. O., and Neev, D., 1960. Mediterranean beaches of Israel, *Geol. Survey Israel Bull.* **26**, 1-25.

Fried, I., 1976. Coastal Protection by means of offshore breakwaters, in *Proceedings of the 5th Coastal Engineering Conference, Honolulu.* American Society of Civil Engineers, 19p.

Goldsmith, V., and Golik, A., 1978. *The Israeli Wave Climate and Longshore Sediment Transport Model.* Haifa: Israel Oceanog. and Limnol. Res. Rep. 78/1, 56p.

Goldsmith, V., and Golik, A., 1980. Sediment transport model of the southeastern Mediterranean coast, *Marine Geology* **37**, 147-175.

Hilmy, M. E., 1951. Beach sands of the Mediterranean coasts of Egypt, *Jour. Sed. Petrology* **21**, 109-120.

Horowitz, A., 1974. Preliminary palynological indications as to the climate of Israel during the last 6,000 years, *Paleorient* **2**, 407-413.

Inman, D. L., 1975. Application of nearshore processes to the Nile Delta, in *Proceedings of the UNDP Symposium, Alexandria, Egypt*, 205-255.

Israel Port Authority, 1966. Coast Study Division, Wind and Tide Observations at Ashod; Tide Prediction for Ashod and Eliat (unpublished data).

Issar, A., and Picard, L., 1971. On Pleistocene shorelines in the coastal plain of Israel, *Quaternaria* **15**, 267-272.

Itzhaki, Y., 1961. Contribution to the study of the coastal plain of Israel: Pleistocene shore-lines in the coastal plain of Israel, *Geol. Survey Israel Bull.* **32**, 1-9.

Levy, Y., 1972. Interaction between Brines and Sediments in the Bardawil area–Northern Sinai, Ph.D. dissertation. Jerusalem: Hebrew University, 108p.

Mazor, E., 1974. On the stability of the Mediterranean coast of Israel since Roman times: A discussion, *Israel Jour. Earth-Sci.* **23**, 149-150.

Michelson, H., 1970. *The Geology of the Carmel Plain* (in Hebrew). Tel Aviv: Tahal Rep. No. HG/70/025, 60p.

Neev, D., 1967. Geological observations on the coastal plain of Northern Sinai (in Hebrew), *Geol. Survey Israel Marine Geol. Div. Rep. No. 1/67*, 15p.

Neev, D., 1977. The Pelusium Line–A major transcontinental shear, *Tectonophysics* **38**, T1-T6.

Neev, D., and BenAvraham, Z., 1977. The Israeli coastal plain, in A. E. M. Nairn, W. H. Kanes, and F. G. Stehli eds., *The Ocean Basins and Margins*, vol. 4A, New York: Plenum, 355-377.

Neev, D.; Almagor, G.; Arad, A.; Ginzburg, A.; and Hall, J. K., 1976. The geology of the southeastern Mediterranean Sea, including a note on seismicity by E. Arieh, *Geol. Survey Israel Bull.* **68**, 51p.

Neev, D.; Shachnai, E.; Hall, J. K.; Bakler, N.; and BenAvraham, Z., 1978. The young (post Lower Pliocene) geological history of the Caesarea structure, *Israel Jour. Earth-Sci.* **27**, 43-64.

Neev, D.; Bakler, N.; Moshkovitz, S.; Kaufman, A.; Megeritz, M.; and Gofna, A., 1973. Recent faulting along the Mediterranean coast of Israel, *Nature* **245**, 254-256.

Nir, Y., 1973. Geological history of the recent and subrecent sediments of the Israel Mediterranean shelf and slope, *Geol. Survey Israel Marine Geol. Div. Rep. No. MG/10/73,* 135p.

Nir, Y., 1976. Detached breakwaters, groins, and other artificial offshore structures, and their influence on the Israel Mediterranean beaches (in Hebrew), *Geol. Survey Israel Marine Geol. Div. Rep. No. 2/76,* 34p.

Nir, Y., 1977. Jet-drillings in the sea off the Hadera electric power station, *Geol. Survey Israel Marine Geol. Div. Rept. No. MG/9/77,* 5p.

Nir, Y., 1980. Recent sediments of Haifa Bay, *Geol. Survey Israel Rept. No. MG/11/80,* 8p.

Nir, Y., and Nathan, Y., 1972. Mineral clay assemblages in recent sediments of the Levantine Basin, *Mediterranean Sea Bull. Groupe Français Argiles* **24,** 187-195.

Pomerancblum, M., 1966. The distribution of the heavy minerals and their hydraulic equivalents in sediments of the Mediterranean continental shelf of Israel, *Jour. Sed. Petrology* **36,** 162-174.

Ron, Z., 1962. The morphographic configuration of the southern Carmel coast (in Hebrew), *Israel Explor. Soc. Bull.* **26,** 31-40.

Rosenan, E., 1937. Fishermen's chart for the Palestine offshore, 1:100,000 scale. Gov't of Palestine Dept. of Agric. and Fish. Services. 4 sheets.

Rossignol, M., 1961. Analyse pollinique des sédiments marins quaternaires en Israël. I. Sédiments recents, *Pollen et Spores* **3,** 303-324.

Safra, D., 1962. The abrasion platform and the cliff of the southern Carmel plain (in Hebrew), *Israel Explor. Soc. Bull.* **26,** 15-30.

Safriel, U., 1965. Recent vermetid formation on the Mediterranean shore of Israel, *Malacol. Soc. London Proc.* **37,** 26-34.

Safriel, U. N., 1974. Vermetid gastropods and intertidal reefs in Israel and Bermuda, *Science* **186,** 1113-1115.

Schattner, I., 1967. Geomorphology of the northern coast of Israel, *Geografiska Annaler* **49(A),** 310-320.

Shukri, N. M., 1950. The mineralogy of some Nile sediments, *Geol. Soc. London Quart. Jour.* **105,** 511-534.

Shukri, N. M., and Philip, G., 1960. The mineralogy of some recent deposits in the Arish-Gaza area, *Cairo Univ. Faculty of Science Bull.* **35,** 73-85.

Sneh, A., and Weissbrod, T., 1973. Nile delta: The defunct Pelusian branch identified, *Science* **180,** 59-61.

Spar, M. S., 1976. *Sedimentological Behavior of the Beach Sands in the Vicinity of the Netanya Breakwaters,* M.S. thesis. Jerusalem: Hebrew University, 56p.

Tauman, J., 1976. Enclosing scheme for bathing-beach development, *in Proceedings of the 5th Coastal Engineering Conference, Honolulu.* American Society of Civil Engineers, 1425-1438.

Tso'ar, H., 1970. *The Sand Dunes of the El Arish Region, Northern Sinai* (in Hebrew), M.S. thesis. Jerusalem: Hebrew University, 93p.

Tur Caspa, Y., 1978. *Water-Jet Drillings in the Sea Bottom Off Hadera M.D. Power Station* (in Hebrew). Haifa: Haifa University Center for Maritime Studies, 7p.

Uziel, J., 1968. Sea Level at Ashod and Eilat: Differences between prediction and observations, *Israel Jour. Earth-Sci.* **17,** 137-151.

Venkatarathnam, K., and Ryan, W. B. F., 1971. Dispersal patterns of clay minerals in the sediments of the eastern Mediterranean Sea, *Marine Geology* **11,** 261-282.

Yaalon, D. H., and Dan, Y., 1967. Factors controlling soil formation and distribution in the Mediterranean coastal plain of Israel during Quaternary, *Quaternary Soils Congr., VII, Proc.* **9,** 322-338.

Zenkovich, V. P., 1971. Several fundamentals in the dynamics of the Nile delta (in Russian), *Akad. Nauk SSSR Okeanogr. Kom. Biul.,* 35-49.

Cross-references: *Africa, Coastal Morphology; Asia, Middle East, Coastal Ecology; Asia, Middle East, Coastal Morphology: Syria, Lebanon, Red Sea, Gulf of Oman, and Persian Gulf; Europe, Coastal Morphology; Global Tectonics; India, Coastal Morphology; Sabkha, Sebkha; Sediment Analysis, Statistical Methods; Sediment Transport. Vol. VIII, Part 3: Middle East.*

ASIA, MIDDLE EAST, COASTAL MORPHOLOGY: SYRIA, LEBANON, RED SEA, GULF OF OMAN, AND PERSIAN GULF

Syria and Lebanon

The north and south Syro-Lebanese coast is situated on the western edge of the Arabic plate and runs parallel to the folded coastal mountain range of Ansarieh Jebel in Syria (1562 m) and the Lebanon mountains culminating at 3083 m. In Lebanon, the mountains descend abruptly to a deep sea, where the continental shelf is narrow but cut by erosional platforms of uniform depths (circa 15-20 m and 40-50 m below sea level). Nevertheless, except in the extreme north of Syria and at Ras Chekka in Lebanon, the cliffs are generally small (a few meters to about 30 m), because the slopes have been cut by erosional shelves of the Pleistocene shorelines (Vaumas, 1954).

In the north of Syria ophiolitic rocks outcrop, and the coast is wild and picturesque. Elsewhere carbonaceous rocks are almost exclusively found—mainly Cretaceous or Neogene limestones or dune or marine Quaternary calcarenites (Daubertret, 1966). Numerous *wadis* or rivers run down to the sea through deep and narrow gorges, so that the flows are irregular and the rivers carry considerable sediment to the sea.

The coastal outline is regular, and the coast is open to the western winds and swells. Islands are rare, small, and generally formed by Quaternary dune sandstones—for example, Rouad Island in Syria, Zire islet near Saida, and the Tripoli archipelago. The only noteworthy irregularities of the coast are such peninsulas as Ras el Bassit (Syria) and Ras Beirut. The Ras ibn Hani (Syria), Tripoli, and Tyr (Lebanon) peninsulas are ancient rocky islands recently connected to the mainland by a sandy bar (*tombolo*). The bays are generally small, the most important being Akkar Bay. The best anchorages are on the north shore of the peninsulas (Beirut, Tripoli); but in ancient times towns often had two harbors, one on each side of the peninsula.

In central Lebanon the coastal plain is non-existent or narrow and discontinuous. The coast is ragged, with numerous small coves edged with pebbly beaches and separated by rocky points. Jounieh Bay, north of Beirut, is one of the most beautiful. South of Saida, there begins a coastal plain that progressively widens out southward but it is never more than 3 km wide. Syria has two relatively broad coastal plains: the Latakia-Jable in the north and the Akkar in the south, which extends into Lebanon.

In the rocky areas, which are frequent in Lebanon, the coast is extremely ragged. A *corrosional platform*, the edge of which is built by vermetus and algae, is cut in the carbonaceous rocks, 10-20 cm above sea level (Fig. 1). It is 2-4 m wide in limestone and 3-10 m wide in Quaternary sandstones. A fossil platform, broader than the present one, has been preserved in many places from 60-80 cm above sea level. Behind it the rock is etched into *clints* and *solution pits* (Sanlaville, 1977).

Elsewhere the coast consists of a pebbly or, more frequently, a sandy beach. Sands are siliceous near Beirut where Cretaceous sandstones outcrop and rich in dark minerals in the Akkar because of the presence of lava flows. But everywhere else the sands are mainly biodetrital and generally rich in carbonates (Emery and George, 1963; Kareh, 1970). Dunes are rare and small (Tyr, Beirut, Akkar) and much disturbed by intensive exploitation.

Everywhere the coast is retreating, sometimes rapidly, as in the Akkar. This reteat can be accounted for mainly by the excessive exploitation of the sands and pebbles for use in the building industry or for road purposes. Slabs of *beach rock* often occupy the lower part of the beach (Fig. 2). In some areas the beach is stable, because of sand nourishment (Tyr, Beirut), but none is prograding (Sanlaville, 1977).

The tidal range is small (20-40 cm). The seawater is rather warm (17° in January and 28° in August) and saline (an average of 38.6‰), with a minimum of temperature and salinity toward the end of winter (Engel, 1966; Oren and Engel, 1965). Submarine springs are numerous and sometimes strongly fed. The swells are distinguished by a short period (5-7.5 sec.) and a small wave-length (30-70 m), but the winter storms (from January to March mainly) may be dangerous on this coast exposed to the strong west and southwest winds. From June to the middle of August the *etesian* winds create a diurnal swell that can be dangerous to bathers. On the other hand, in May and from the end of August to the beginning of November, the sea is generally smooth. The longshore currents trend to the north and may be relatively strong.

This coast is now badly polluted; there are four important oil harbors (Banyas, Tartous, Tripoli and Saida), and both towns and manufacturers dispose of their waste directly into the sea.

Red Sea

The Red Sea measures 2100 km from Suez to the Strait of Bab el-Mandeb, which connects it with the Gulf of Aden and the Indian Ocean.

This narrow sea is rather deep—an average of 500 m with a maximum of over 2000 m—but the Strait of Bab el-Mandeb, a narrow passage

FIGURE 1. Corrosion platform on the coast of Lebanon. (Photo: P. Sanlaville)

FIGURE 2. Beach of pebbles and beach rock near Batrun on the Lebanese coast. (Photo: P. Sanlaville)

where the Perim Island stands, is only 110 m deep. This sea is noteworthy for some of the hottest and most saline seawater in the world (Allan and Morelli, 1970). No river or stream of any size flows into the Red Sea, and rainfall is very scanty—an average of 50 mm. But the degree of relative humidity is high in summer. The surface water temperatures range from 20° in the north and 25° in the south in winter, to 27° in the north and more than 30° in the south in summer. The salinity is 40‰ in the north in winter. The sea has a semidiurnal tide of 0.30 to 0.50 m, rising to 1.80 m in the Gulf of Suez. The waves are generally small because this area is sheltered; winds blow mainly from the NW (Maillard, 1972; Neumann and McGill, 1961).

The Red Sea is actually a rift valley, a large fault depression separating two great plates— those of Arabia and Africa (Drake and Girdler, 1964; Quennel, 1958). Plateaus and mountains rise steeply to more than 1000 m above sea level north of Jeddah and 2000 m in the south. The coastal plain, the *tihama,* is from 2-20 km wide, sloping up gently to the east, only to be towered above by the abrupt face of the mountains. The coastal plain is often covered by dune sands. In the mountains the valleys are deeply cut, but streams flowing in the uplands fail to cross the coastal plain to reach the sea. Therefore pebbles are rare along the coast, and sands are essentially organogenic.

The coast is oriented NNW-SSE. The general outline is regular, except for two small salients (each with a large radius of curvature) north of Yanbo and in the Jeddah area.

However, the coastline is irregular when considered in detail because of the presence of *rias* (*sharm*) 1-5 km long, which afford good shelter to small boats.

Coral reefs are numerous everywhere along this coast from the Gulf of Akaba to the Strait of Bab el-Mandeb, especially in the south. These reefs are of different types: Fringing reefs, forming a shallow shelf 50-1500 m broad, run parallel to the shore but allow for a narrow boat channel between the coast and the reef, up to a few meters deep and 10-100 m wide. Platform and patch reefs, on the continental shelf in waters of moderate depths, are generally rounded or ovoid, sometimes forming an atoll-like reef but with only a shallow pool. These reefs have sometimes built an important bank, such as the Farasan Islands (Fig. 3), which are about 600 km long and 100 km wide, south of Saudi Arabia, or the Kamaran Islands along the coast of Yemen. These reefs are of Australian type, often with a small sand island in the middle of the lagoon (*cays,* mere accumulation of loose coral sands). Sometimes, these coral

FIGURE 3. Holocene erosion in fossil coral along shore of Farasan Islands. (Photo: A. Guilcher)

reefs are found beyond the continental platform as patches or pinnacles, dropping abruptly down to 500-900 m (Guilcher; 1955; Macfayden, 1930; Nesteroff, 1955).

Because the Red Sea is narrow and the waves small, the reefs have no *algal rim* with *Lithothamnion.* Between the reefs there are chalky mud deposits. The sand of cays and beaches is organogenic and not of continental origin. Often it has been cemented into *beach rock.* There are numerous traces of Holocene and Pleistocene reefs, with faults and deformations (Deuser, Ross, and Waterman, 1976).

The local harbors are not active, except Jeddah, but since the oil trade began to flourish, the Red Sea has been one of the most heavily traveled waterways in the world. As a consequence, pollution is a real danger.

Gulf of Oman and Persian Gulf

Whereas the Gulf of Oman is deep and largely open to oceanic influences, especially to the swells of the Indian Ocean, the Persian Gulf is shallow (average depth 31 m, seldom more than 80 m) and situated between hot and arid lands. Precipitation, during the winter is generally less than 80 mm, except between Bushehr and

Qeshm (Iran), and the summers are hot. Freshwater is supplied only by the Shatt el Arab and some Persian rivers. As a consequence, the seawater is highly saline and warm; the surface water temperatures range from 32-33° in summer to 22-24° in winter in the south but only 16° in the north. Salinity is about 40‰; higher on the Arabian shores, it falls to 37‰ in the Gulf of Oman. The sea has an average semidiurnal tide of 2 m, rising to 3 m in the northwest, where the tide goes upstream to Basra, 300 km from the sea (Emery, 1956; Purser, 1973). Everywhere broad, sandy or muddy strands are uncovered at low tide, whereas during storms the level of coastal waters rises, causing extensive flooding of the low coastal plains. The currents are counterclockwise, running along the Persian coast and going back southward along the Arabian coast. The warm seawater offers conditions favorable to animal and vegetable life, though relatively few species are to be found. In the southern part of the Persian Gulf the mild winter climate allows the development of coral reefs.

A prolongation of Mesopotamia, the Persian Gulf is situated between two highly different structural areas: (1) the Arabian plate to the west, tilted eastward and covered with Cenozoic and Quaternary sandstone, limestone, and gypsum, sediments that are little deformed and that form large, flat anticlines in which gas and oil have been trapped; (2) the Iranian plate to the east, strongly uplifted, faulted, and folded into NW-SW-aligned foothills and ranges of mountains (Purser, 1973). The consequence is a clear contrast between the sheer eastern coast and the flat, low western coast. But because the waves are small, silty and muddy warping forms prevail everywhere.

Apart from those carried by the Shatt el Arab and some Iranian rivers, continental deposits are now few to be found, especially along the Arabian shores. But wadis and rivers were active during the Quaternary pluvials, and now silts are deposited by winds and dust storms blowing mostly from the NNW (*Shamal*). Today deposits are fine (sands, silts, and muds) and mainly of marine origin with biogenic sands, frequently oolitic, and muds rich in calcium carbonate (Evans, 1966; Emery, 1956; Picha, 1975).

At the northwest end of the Persian Gulf is the vast deltaic plain of the Euphrates, Tigris, and Karun rivers, formed of swamps, sandbars, spits, and islands (Boubian Island, for instance) with fluctuating boundaries. The important silty deposits of these rivers seem to be compensated for today by subsidence and currents and wave erosion, so that the delta does not extend seaward appreciably, but the coast may have been prograding during some wetter Holocene periods or as the result of antropic interference (Brice, 1978; Lees and Falcon, 1952).

The Arabian western coast is generally low, flat, rectilinear and sandy. On the beach the carbonate sediments are frequently cemented into beach rock. Often a sandbar overtopped by dunes isolates large lagoons or *sebkhas* flooded in winter but dry and covered by salt or gypsum the rest of the year and exhibiting many old strandline features accounted for by the progradation of the coastline.

Some areas have a more irregular outline; the bay of Kuwait is some 40 km wide and long. Ras Tanura is a long, sandy hooked spit. The Qatar peninsula is a large limestone platform with cliffs, oriented northward. The broad, shallow shelf of the southwest part of the Persian Gulf displays numerous banks, shoals and islands. The islands are often *salt plugs*—for instance, Halul ou Das Islands, east of Qatar, equipped today for the shipping of oil.

South of the latitude of Bahrain banks, shoals, and islands are surrounded by coral reefs or capped with reefs and small sand cays. These reefs are a danger to navigation; the rocky bottom of the sea is rich in pearl oysters, which were actively collected a few decades ago.

South of the Bahrain archipelago, southeast of the Qatar peninsula (Fig. 4), the boundary between land and sea is rather imprecise. Algal flats as well as intertidal muddy flats spread extensively. The coast of the United Arab Emirates (the Trucial Coast) is characterized by a number of broad, sandy islands enclosing intertidal muddy flats and lagoons and edged with barrier and fringing reefs. The town of Abu Dhabi was built on an island artificially raised and enlarged. Broad, sandy spits isolate lagoons on the shore of Umm el Qawain and Ras el Khaima, and Dubai stands at the mouth of a narrow ria more than 15 km long that from

FIGURE 4. Barchan dunes along coast southeast of Qatar. (Photo: J. P. Perthuisot)

early days has allowed Dubai to be a busy harbour (Purser, 1973).

The mountains of Ras Masendam dip abruptly into the sea, barren and ragged, just opposite the Strait of Hormuz.

The Iranian coasts is a region of extensive continental sedimentation. It is flat and low as far as Bushehr, which is situated near the Hilleh Rud delta. Next there is a Dalmatian type of coast; the rocky shore with cliffs parallel to the ranges of mountains, whereas bays and deltas correspond to synclines and depressions. In front of Ras Masendam, the Iranian coast forms a large recess (Strait of Hormuz), with two main islands—Qeshm and Hormuz. Along the north shore, diagonal to the structure, cliffs and deltaic plains with intertidal mud flats alternate. The eastern shore is muddy and all along the coast of Makran is formed of offshore bars, lagoons, rias, and coastal swamps with tropical mangroves.

South of the Masendam peninsula the coast of Oman is sandy along the Batina plain and rocky south of Mascate, with ragged cliffs as far as Ras el Hadd.

The southern coast of the Arabian peninsula has never been studied in detail. Rocky headlands with cliffs alternate with immense rectilinear shores of fine sands buffeted by oceanic swells and access made difficult by the presence of bars. Some islands, such as Masira Island or the Kuria Muria archipelago, are to be found in Dhufar.

This hot and arid coast, mostly sandy and flat, has been famous for its maritime and trading activities since early times. Today this area is the foremost producer and exporter of oil in the world. Its harbors are among the most important, and maritime traffic is intense. As a consequence, the landscapes have been altered, sometimes beyond recognition, by creation of harbors, artificial islands, and towns. The equilibrium of the sandy shores may be destroyed and oil pollution is a danger to be reckoned with in the Persian Gulf.

PAUL SANLAVILLE

References

Allan, T. D., and Morelli, C., 1970. The Red Sea, in A. E. Maxwell ed., *The Sea*, vol. 4. New York: Wiley-Interscience, 492-542.

Brice, W. C., ed., 1978. *The Environmental History of the Near and Middle East Since the Last Ice Age*. London: Academic Press, 384p.

Deuser, W. G.; Ross, E. H.; and Waterman, L. S., 1976. Glacial and pluvial periods: Their relationship revealed by Pleistocene sediments of the Red Sea and Gulf of Aden, *Science* 191, 1168-1170.

Drake, C. L., and Girdler, R. W., 1964. A geophysical study of the Red Sea, *Jour. Geophysics* 18, 473-495.

Dubertret, L., 1966. Liban, Syrie et bordure des pays voisins: Tableau stratigraphique avec carte géologique au millionième, *Not et Mém. Moy. Orient Paris* 8, 251-283.

Emery, K. O., 1956. Sediments and water of Persian Gulf, *Am. Assoc. Petroleum Geologists Bull.* 40, 2354-2383.

Emery, K. O., and George C. J., 1963. The shores of Lebanon, *Am. Univ. Beirut Misc. Papers Nat. Sci.* 1, 1-10.

Engel, I., 1966. Les températures dans la Méditerranée orientale, *Cahiers Ocean.* 18, 507-514.

Evans, G., 1966. The recent sedimentary facies of the Persian Gulf region, *Royal Soc. London Philos. Trans.* 259A, 291-298.

Guilcher, A., 1955. Géomorphologie de l'extrémité septentrionale du Banc Farsan (Mer Rouge), *Inst. Océanogr. Annales* 30, 55-100.

Houbolt, I. I. H. C., 1957. *Surface Sediments in the Persian Gulf near the Qatar Peninsula.* The Hague: Mouton, 113p.

Kareh, G. el, 1970. *Some Sedimentological Aspects of St. George's Bay,* Ph.D. thesis, American University, Beirut, 120p.

Lees, G. M., and Falcon, N. L., 1952. The geographical history of the Mesopotamian plains, *Geog. Jour.* 118, 24-39.

Macfadyen, W. A., 1930. The geology of the Farsan Islands, Gizan and Kamaran Islands, Red Sea, *Geog. Jour.*, 75, 26-34.

Maillard, C., 1972. Etude hydrologique et dynamique de la Mer Rouge en hiver, *Inst. Océanogr. Annales* 48(2), 113-140.

Nesteroff, W., 1955. Les récifs coralliens du Banc Farsan Nord (Mer Rouge), *Inst. Océanogr. Annales* 30, 7-53.

Neumann, A. C., and McGill, D. A., 1961. Circulation of the Red Sea in early summer, *Deep Sea Research* 8, 223-235.

Oren, O. H., and Engel, I., 1965. Etude hydrologique sommaire du bassin levantin (Méditerranée orientale), *Cahiers Océan.* 17, 457-465.

Picha, F., 1975. Paleoclimatic and tectonic factors in the Pleistocene and Holocene oolithic sedimentation of the Persian Gulf in Kuwait, *Congr. Int. Sediment., Nice, Proc.* 1, 147-151.

Purser, B. H., 1973. *The Persian Gulf: Holocene Carbonate Sedimentation and Diagenesis in a Shallow Epicontinental Sea.* Berlin: Springer-Verlag, 471p.

Quennel, A. M., 1958. The structural and geomorphic evolution of the Dead Sea rift, *Geol. Soc. London Jour.* 114, 1-24.

Sanlaville, P., 1977. *Etude Géomorphologique de la Région Littorale du Liban,* 2 vol. Beyrouth: Public. Univ. Libaniase, Section des Etudes Geogr., 859p.

Vaumas, E. de, 1954. *Le Liban, étude de géographie physique.* Paris: Firmin-Didot, 367p.

Cross-references: *Africa, Coastal Morphology; Asia, Middle East, Coastal Ecology; Asia, Middle East, Coastal Morphology: Israel and Sinai; Coral Reef Coasts; Demography; Europe, Coastal Morphology; Global Tectonics; Human Impact; India, Coastal Morphology; Mangrove Coasts; Pollutants; Sabkha, Sebkha; Soviet Union, Coastal Morphology.* Vol. I: *Persian Gulf;* Vol. VIII, Part 3: *Middle East.*

ATTRITION

Attrition is the act or process of wearing down by friction—the abrasion, impact, and grinding together of clasts in motion, resulting in smaller and usually better-rounded particles. Studies of attrition of beach clasts include experiments, contrived situations, and observations in natural settings. Kuenen (1964) experimentally correlated rate of attrition with relative resistance and median clast size. Zhdanov (1958) placed 2500 pebbles along the Caucasus-Black Sea shore and noted rates of attrition for nearly five years. He calculated an average attrition rate of 5%/yr, from 1.6% for basalt to 8% for marble. Comparisons of natural sediment distributions were done by Marshall (1929) and Landon (1930), but lack of control makes their conclusions about attrition the least reliable—observed diminution of grain size along a coast could be due also to change in source material or to sorting by longshore transport.

JOANNE BOURGEOIS

References

Kuenen, Ph. H., 1964. Experimental abrasion: surf action, *Sedimentology* 3, 29-43.

Landon, R. E., 1930. An analysis of beach pebble abrasion and transportation, *Jour. Geology* 38, 437-446.

Marshall, P., 1929. Beach gravels and sands, *New Zealand Inst. Trans. Proc.* 60, 324-365.

Zhdanov, A. M., 1958. Pebble attrition under surf action (in Russian), *Akad. Nauk SSSR Okeanogr. Kom Biul.,* 1.

Cross-references: *Beach Material; Sand Source; Zingg Shape.* Vol. VI: *Attrition.*

AUSTRALIA, COASTAL ECOLOGY

The Australian coastline is over 19,650 km long, stretching from the tropical waters of northern Australia to the cool temperate waters of Tasmania. The coast is associated with the three major ocean systems of the Southern Hemisphere: the Indian Ocean, the Pacific Ocean, and the Southern Ocean.

The following brief summary of environmental features on the Australian coast is based largely on Rochford (1975). The major influence off eastern Australia is the *East Australian Current.* This current is not a continuous stream but a complex of strong movements to the south near the coast, with an offshore counter-current to the north. At around 34°S the bulk of transport of water by this current moves eastward, although residues of warmer water may sometimes drift southward with anticyclonic eddies. These can on occasion drift southwest and even influence water conditions in eastern Tasmania. Off the west Tasmanian coast there is evidence of a north-flowing current (the *Flinders Current*) in summer, bringing cooler water to western Victoria and southeastern South Australia.

Australia lies north of the subtropical convergence and therefore does not experience cold subantarctic waters. Mean surface temperatures of coastal waters in summer range from below 17°C in southern Tasmania to approximately 30°C in northern Queensland. There is a departure from a strictly latitudinal temperature relationship due to the East Australian Current. The annual surface temperature range is small in the tropical open ocean, but a range is recorded of 6-7°C in the Tasman Sea and 5-7°C in the open southeast Indian Ocean. These temperature ranges are considerably increased in shallow coastal waters and gulfs. Knox (1963) has classified the fauna of coastal waters according to temperature regimes, and his nomenclature is employed in the discussion below.

On the Australian coastline all major tidal phenomena may be observed. The most relevant feature in relation to coastal ecology is the variation in range, from less than 1 m in southwest Australia to over 10 m in northern Western Australia. There is no clear latitudinal gradient, although on most of the southern coastline the range is less than 2 m. Local topography can cause major variations.

Biogeography of Australian Coasts

The Australian coastline has been subdivided into a number of biogeographic provinces. In the initial proposal of Hedley (1904) these regions were named after famous men connected with Australian history, with the exception of the geographical term Adelaidean. This pattern was followed in subsequent work. Stephenson and Stephenson (1972) suggest that such nomenclature is not informative and that these terms obscure the relationships of Australia to ecologically similar regions. If biogeographic provinces are expected to be a reflection of oceanographic conditions, then such terms as *tropical-subtropical region* are preferred. Knox (1963) has equated the biogeographic provinces of Australia directly with sea-temperature regimes.

The marine biogeographic provinces were originally based on mollusc distribution, but data have since accumulated covering a wide range of animal groups and algae. A brief historical survey is given in Womersley (1959) and Knox (1963). Based on the biota of rocky shores the marine biogeographic regions of Australia can be summarized as follows:

Tropical-subtropical Coasts. *Dampierian Province* consists of the tropical and subtropical waters from north of Houtman Abrolhos (28°50'S) to about 25°S on the Queensland coast. The eastern Queensland coast is sometimes recognized as a separate province, the *Solanderian. Great Barrier Reef Province* is the Great Barrier Reef Region.

Warm Temperate Coasts. *Peronian Province* includes southern Queensland, New South Wales, and the easternmost Victorian coast. *Western Australian Province* consists of the Western Australian coasts from south of Houtman Abrolhos to between Esperance and Albany on the south.

Transitional Warm Temperate Coasts. *Flindersian Province* includes the area from east of Albany and Esperance in Western Australia to Robe in southeastern South Australia.

Cold-temperate-mixed-water Coasts (Cool Temperate). *Maugean Province* includes the Victorian, southeastern South Australian west to Robe, and the Tasmanian coastlines. There is a considerable overlap between the biota of this region and that of the *Flindersian*. Womersley (1959) includes the *Maugean* as a subprovince of the *Flindersian*.

Intertidal Ecology of Australian Coasts

Detailed accounts of littoral ecology in Australia generally followed the establishment of the descriptive framework for shore zonation established by Stephenson and Stephenson (1972). In most cases general accounts for different regions were preceded by detailed investigations at specific localities. Reviews are provided by Womersley (1959), Knox (1963), Stephenson and Stephenson (1972), and Clayton and King (1981). Womersley (1959) noted that most studies had been undertaken by zoologists and were largely restricted to open-coast rocky coastlines. A notable exception is the paper by Womersley and Edmonds (1958). These imbalances have been partially corrected in more recent papers. All of the earlier studies paid little attention to the sublittoral zones.

A description of the rocky shore zonation for each region is given below. A general pattern of zonation is found on all rocky coasts with considerable wave action, which can be summarized as follows: The *littoral fringe* is dominated by littorinids and lichens, particularly in the south. The *eulittoral zone* is characterized by barnacles. Most Australian workers subdivide the eulittoral into three subzones. In temperate waters the lower subzone is generally dominated by algae. The *upper sublittoral zone* is dominated by Fucalean algae and, in colder waters, Laminariales. Particularly in southeastern Australia a sublittoral fringe is recognized.

The subtidal ecology of most of Australia is not well known. On the basis of joint work with Shepherd, Womersley (1972) has recognized three sublittoral zones of algae on rough water coasts of southern Australia. These are characterized as: an upper zone of turf-forming red algae including *Corallina* and *Pterocladia;* a broad midzone with large brown algae, such as *Ecklonia, Cystophora,* and *Sargassum;* a lower zone with larger red algae, such as *Plocamium* and *Phacelocarpus*. In calmer waters the upper, turf-forming band is reduced or even absent.

Following are descriptions of zonation patterns around Australia, beginning with South Australia.

Transitional Warm Temperate Coasts: South Australia. The intertidal ecology of South Australia has been described by Womersley and Edmonds (1958) and Womersley and Thomas (1976). The southeastern part of the coastline will be included in the discussion of cold-temperate-mixed waters (Victoria and Tasmania). The basic zonation pattern on open-water rocky shores is as follows.

The littoral fringe is characterized by *Littorina unifasciata* with *L. praetermissa* in moister situations. The blue-green alga *Calothrix fasciculata* occurs in the upper part of this zone. The black lichen *Lichina confinis* may also be conspicuous; colored lichens extend into the spray zone. The isopod *Ligia exotica* and the crab *Leptograpsus variegatus* occur among rocks and boulders at this level. On more sheltered coasts *Littorina* are still dominant, but they are restricted to more sheltered microhabitats.

The eulittoral zone is dominated by barnacles. In the upper eulittoral *Chamaesipho columna* occurs often with *Chthamalus antennatus*. With shelter these species are less abundant. *Littorina* may migrate to this level. In the mid-eulittoral *Catomerus polymerus* is dominant, but the mussel *Brachiodontes rostratus* can be prominent in slightly calmer situations and on gentler slopes. Scattered in the mid-eulittoral are a number of molluscs, including the limpets *Cellana tramoserica, Patelloida alticostata,* the pulmonate *Siphonaria diemenensis,* and *Thais orbita*. With shelter or shade the tube worm *Galeolaria caespitosa* may become common in the lower part of this zone. Seasonal growths of algae such as *Isactis plana, Rivularia firma, Nemalion helminthoides,* and *Splachnidium rugosum* are characteristic. On moderate-wave-action coasts *Catomerus* is absent and the gastropods *Austrocochlea concamerata, Bembi-*

cium nanum, and *Nerita atramentosa* occur with the limpets. *Xenostrobus pulex* occurs on shelving rock. The lower eulittoral in roughest-water localities is an association of *Megabalanus nigrescens* with an algal mat in which coralline algae *(Corallina cuvieri, C. officinalis)* are dominant. With decrease in water action *Megabalanus* disappears. Chitons of the genus *Poneroplax* occur at this level. Where wave action is considerably reduced, *Hormosira banksii* is the commonest alga. The coralline algal mat extends into the upper sublittoral zone, which is dominated by a range of algae including *Caulerpa* species, *Ecklonia radiata*, and fucalean algae such as *Cystophora* and *Sargassum*. In roughest localities there is a distinctive sublittoral fringe of *Cystophora intermedia*.

Together with Victoria and Tasmania, this region of southern Australia has a rich and highly endemic algal flora (Womersley, 1959). The chlorophyta are well represented, particularly the siphonous algae such as *Caulerpa* and *Codium*. In the phaeophyta the Fucales are especially well developed. Almost half of the world's genera of red algae occur in the region—about 800 species, most of which are endemic.

Cold-temperate-mixed Water Coasts: Victoria and Tasmania. No part of the Australian coastline extends far enough south to be truly cold temperate, and so the term *cool temperate* has been coined for these coasts. The intertidal ecology is described in detail in Bennett and Pope (1953, 1960), Dartnall (1974), and King (1972).

The pattern of zonation is essentially the same as in South Australia, but some major differences are given below. There is a progressive thinning out of even the common species in the south; for example, the littorinid *Littorina unifasciata* does not reach the far southern coasts of Tasmania. This trend is most apparent with the barnacles; *Megabalanus nigrescens* does not occur in Tasmania, although it occurs in eastern Bass Strait and Victoria. *Chamaesipho columna* is the only barnacle recorded in southern Tasmania, where it is sparse; on the Victorian coastline it may occur with 100% cover. The reduction in species favoring warmer waters is also apparent on the colder-water coasts of western Victoria.

Dartnall (1974) supports Hedley's earlier (1904) notion that Tasmania had acted as a distribution barrier during periods of lower sea level in the Pleistocene. He also notes that there are relatively few endemic Tasmanian species.

As on South Australian coasts, animals characterize the upper zones, with algae becoming dominant in the lower eulittoral zone. The mussels *Austromytilus* and *Xenostrobus* are better developed than in South Australia. *Galeolaria caespitosa* becomes less conspicuous in southern Tasmania and seems to be replaced by the coralline alga *Pseudolithophyllum hyperellum*, which offers a similar habitat for the associated fauna. Again seasonal algae, such as *Splachnidium rugosum* and *Porphyra columbina*, are important.

The lower eulittoral is well developed especially where broad low-tide shore platforms occur as they do on the Pleistocene dune limestone in Victoria. In these self-sheltering situations *Hormosira* carpets the zone, often completely covering the substrate. In the lower part of this zone *Cystophora torulosa* can form a distinct subzone. On rougher-water coasts *Pyura stolonifera* var. *praeputialis* occurs at this level.

A most obvious feature of the Victorian and Tasmania coastline is the presence of large brown algae in the upper sublittoral zone. *Macrocystis* (*M. angustifolia* on the Victorian coast is replaced by *M. pyrifera* in southern Tasmania) is the common kelp. *Phyllospora comosa* and *Ecklonia radiata* are restricted to areas of local shelter, particularly at the warmer end of the region. The giant fucoid *Durvillaea potatorum* generally forms a distinct sublittoral fringe, although it occasionally extends below this. In Tasmania *Durvillaea* is found at a higher level with respect to tide than it is on the mainland. This underlines the biological nature of the zones and emphasizes the response by particular species to more than just tidal gradients. In more sheltered localities *Durvillaea* is replaced by *Xiphophora gladiata* in Tasmania; in Bass Strait *X. chondrophylla* can occur in this situation. *Cystophora intermedia* is found only at Wilsons Promontory. *Macrocystis*, *Phyllospora*, and *Durvillaea* do not extend west of Robe, South Australia.

Warm Temperate Coasts: New South Wales (Eastern Warm Temperate). A general account of the intertidal ecology of New South Wales is given in Dakin, Bennett, and Pope (1948).

The littoral fringe is characterized by the two littorinids: *Nodilittorina pyramidalis* generally at a higher level than *Littorina unifasciata*. The eulittoral has a typical barnacle population. As in cooler waters, *Chthamalus antennatus* and *Chamaesipho columna* are in the upper eulittoral. In the mid-eulittoral *Catomerus polymerus* occurs, but mixed with *Tesseropora rosea*, and sometimes slightly above. *Galeolaria caespitosa* forms a lower band in this zone and even lower. A range of molluscs occurs at this level, and again they become increasingly important in more sheltered localities. A broad band of *Pyura stolonifera* var. *praeputialis* is often developed in the lower eulittoral, sometimes with the barnacle *Megabalanus nigrescens* and a mixed algal turf. The upper sublittoral zone is dominated by *Ecklonia radiata* and *Phyllospora comosa*. *Sargassum* species and the burrowing echinoderm *Heliocidaris erythrogramma* can be abundant at this level. In the southern region the biota overlaps considerably with that of the cooler waters and even the bull kelp (*Durvillaea*) may be abundant.

Western Australia (Western Warm Temperate). There are no general accounts of the intertidal ecology of south-west Western Australia, although Knox (1963) has summarized data from a number of papers on specific localities. Because this region has often been included within the Flindersian biogeographic region (Womersley, 1959) it is appropriate to compare the zoning pattern with that in South Australia.

In the littoral fringe *Littorina praetermissa* is replaced by the subtropical *Tectarius rugosus*, which occurs with *Littorina unifasciata*. There is only one barnacle in the eulittoral zone (*Megabalanus nigrescens*), and the limpets of the eulittoral zone are mostly endemic. *Galeolaria caespitosa* does not form the characteristic mid-eulittoral band, and *Pyura stolonifera* var. *praeputialis* does not occur as a band in the eulittoral zone. *Hormosira banksii* is absent west of Albany.

The dominant subtidal alga is *Ecklonia radiata*. *Sargassum* species and lithothamnia are relatively more important than they are further east, but *Cystophora* is less so.

Tropical Coasts: (Eastern Tropical and Subtropical) Queensland. The ecology of Queensland mainland rocky shores is covered in Endean, Kenny, and Stephenson (1965). This work is primarily zoological; as the authors indicate, algae are not conspicuous in the zonation. The break with the biota of warm temperate areas is strongly marked, occurring at about 25°S. The zonation pattern south of this point is similar to that described for the New South Wales coast.

On the north Queensland coast the sequence is as follows. There is a littoral fringe with *Nodilittorina pyramidalis* above *Nodilittorina millegrana*. The eulittoral has *Chthamalus malayensis* in the upper part above *Saccostrea* (oyster) and *Tetraclitella squamosa*. A few algae, including lithothamnia, occur but are comparatively unimportant in the zoning pattern.

The ecology of the Great Barrier Reef is described by a number of workers, particularly Stephenson and Stephenson (1972) and in *Biology and Geology of Coral Reefs*, edited by Jones and Endean (1973). The zonation pattern in this region is less clear than on the mainland and is in large part related to the reef structures. Knox (1963) lists a number of differences. Littorinids are not abundant, the upper barnacle (*Chthamalus malayensis*) is absent, and the oyster zone poorly developed. *Tetraclitella squamosa* is replaced by *T. vitata*, and the algal zone replaced by a lithothamnion, zoanthid-coral zone. Other faunistic differences support the separate biological identity of the Great Barrier Reef Province.

Other tropical coasts are poorly known, and for that reason Knox (1963) retains the Dampierian separate from the Solanderian biogeographical province. No generalized accounts exist for this region, although the marine botany and zoology of Arnhemland are briefly described in Womersley (1959) and Stephenson and Stephenson (1972).

Vegetation of the Land Margin

The Australian coastline provides many diverse habitats for vegetation, including cliffs, beaches, dunes, estuaries, and lagoons. Variations in climate, lithology, physiography, and tidal amplitude all contribute to the diversity of plant habitats.

Mangrove and Salt Marsh Vegetation. Mangrove and salt marsh vegetation is best developed in northern and southeastern Australia where tectonically stable land margins have been exposed by a eustatic fall in sea level during Holocene times. These are usually found in the low-wave-energy environments of estuaries, deltas, and embayments and at the head of gulfs and in the lee of barriers, spits, and offshore islands. The biogeography of mangrove and coastal salt marsh vegetation has been summarized by Saenger et al. (1977). Mangrove vegetation is best developed in northeastern Australia, where up to twenty-seven tree species have been recorded and ten species noted as associated liane, epiphyte, or understory plants. Species numbers diminish southward along subtropical sectors of the coast, and *Avicennia marina* is the sole mangrove species in temperate southern Australia.

In the tropical north mangroves commonly form closed forests (Specht, 1970) 10-30 m high; subtropical communities are often low, closed forest (5-10 m); in the temperate zone mangroves occur as low woodlands or scrub. However, the structural development of these communities is often reduced by local conditions. Mangroves of the tropical coast between Broome and Fraser Island occur in distinct zonations, although there is considerable variation with topography.

In contrast, tidal salt marsh communities increase in diversity from tropical to temperate latitudes; only seven species have been recorded from tropical salt marshes, while up to thirty-four species may occur on some sites in temperate Australia. Some of the salt marsh species belong to the family Chenopodiaceae and form a low or open shrubland; the chenopod *Salicornia quinqueflora*, often with other salt-tolerant herbs and grasses, typically forms herbland. Closed sedgelands (*Gahnia, Juncus*) are also present in some temperate salt marshes. Marshes in southern Australia rarely have a vegetation-free zone, although the landward margins of salt marshes in areas where average annual rainfall is less than 500 mm may be only sparsely vegetated. On tropical salt marshes bare flats become extensive in drier regions.

Sand-dune Vegetation. Most of the long, gently curved, sandy beaches of the Australian coastline are backed by beach ridge or dune topography; cliff-top dunes are extensive along some sectors of the coast. Most of these sandy shorelines are eroding, and beaches are backed by cliffed dunes where former vegetation successions have been truncated.

In southern Australia the strandline may be colonized by scattered plants of *Cakile maritima*, *Atriplex cinerea*, or *Festuca littoralis*. Plants of the foredune are initially grasses such as *Spinifex hirsutus* or the introduced *Ammophila arenaria* and *Agropyron junceum*. These may be subsequently invaded and replaced by shrubs, the species depending on the locality. On older, more stable dunes such species form a dense closed scrub that may grade into a woodland of *Eucalyptus, Casuarina, Acacia*, or *Banksia*. On exposed or seasonally waterlogged sites, dune scrub may give way to heath (Barson and Calder, 1981). In humid tropical Australia dunes are generally limited and local in extent, and dune succession is not clearly defined.

The Vegetation of Steep Coasts. The vegetation of steep sectors of the Australian coastline is poorly known. The plant communities vary in response to geology, cliff morphology, and degree of exposure. On plunging cliffs and cliffs with vertical faces, vegetation is restricted to cliff tops, ledges, and crevices. The gentler, more sheltered slopes of coastal ranges and bluffs may support forest communities.

In southeastern Australia closed-tussock grasslands are common on some exposed sites. Heath communities that are usually dominated by members of the Epacridaceae. Casuarinaceae, and Myrtaceae and are often floristically diverse commonly occur on exposed sites on sandy soils that may be seasonally waterlogged. Cliffs cut in dune calcarenite, extensive on the western and southern coasts of Australia, commonly support some of the shrub species of coastal dune systems that may be mixed with species of the communities adjacent to the cliffs. Cliff-top plant communities frequently show a zonation relation to exposure to salt spray; on extremely exposed sites salt marsh species such as *Salicornia* are found.

Transitional Warm Temperate Coasts. The coastal vegetation of South Australia has been described by Specht (1972). Salt marshes (q.v.) are common in bays, gulfs, and estuaries; in Spencer and St. Vincent gulfs and at Streaky Bay they may be fringed by a low *Avicennia marina* woodland. Lower levels of the salt marshes are dominated by *Arthrocneumum arbusculum* shrubland, while *Arthrocnemum halocnemoides* forms low shrubland on areas not subject to regular flooding. Shrubland at the upper limits of the salt marsh is usually dominated by *Atriplex paludosa* with low-growing halophytic understory species. *Nitraria schoberi* may be codominant in some areas. The grass *Sporobolus virginicus* grows on occasionally flooded ground; it is replaced by *Distichlis distichophylla* on higher ground.

Strand species of the calcareous beaches include *Cakile maritima, Cakile edentula,* and *Atriplex cinerea* and *Calocephalus brownii,* dune colonizer, followed by *Scirpus nodosus* and succulent *Carpobrotus rossii.* In more sheltered sites, shrubs such as *Rhagodia baccata, Enchylaena tomentosa, Olearia axillaris, Leucopogon parviflorus,* and *Acacia longifolia* occur. Stabilized dunes support low woodlands of *Casuarina stricta* and/or *Melaleuca lanceolata.*

Cliffs cut in limestones and granites support similar low shrublands. Species tend to be zoned according to their salt tolerance, with *Arthrocnemum halocnemoides* on sites more exposed to salt spray grading into *Disphyma australe–Enchylaena tomentosa* followed by *Atriplex cinerea* and *Calocephalus brownii,* which gives way to a zone of *Atriplex paludosa, Rhagodia crassifolia,* and *Nitraria schoberi,* with *Olearia axillaris* and *Leucopogon parviflorus* in slightly more sheltered sites. *Casuarina stricta* and *Melaleuca lanceolata* form the most landward zone. The limestone cliffs of the Great Australian Bight support a low, sparse shrubland that includes species of saltbush, bluebush, and *Arthrocnemum*, with some areas of scrub dominated by *Eucalyptus, Melaleuca,* or *Acacia* spp.

Cool Temperate Coasts. Tidal marshes are best developed along the central Victorian coast where the tidal range is greater than one meter. To the east and west there are isolated occurrences of salt marsh estuaries. Victorian salt marshes show marked regional variations in the structure and floristics of their plant communities. Common communities include tall shrubland or open shrub of *Avicennia marina, Arthrocnemum* spp. low shrublands, open tussock grasslands of *Stipa teretifolia–Juncus kraussii,* sedgeland and herbfields dominated by *Salicornia quinqueflora*. However, not all these communities may be present in a particular locality. *Melaleuca ericifolia* closed scrub commonly fringes the landward margin of salt marshes east of Port Phillip Bay.

In Tasmania salt marshes border most sheltered waters, and are most extensive on the southeast, northeast, and northwest coasts (Kirkpatrick and Glasby, 1981). *Salicornia quinequeflora* herbland commonly fringes the seaward margin and grades into *Arthrocnemum arbusculum* open heath, which is followed by *Juncus kraussii* rushland and *Gahnia filum-J. kraussii* tussock sedgeland, with *Stipa teretifolia* confined to better-drained sites. *Melaleuca* or *Leptospermum* spp. sometimes form closed scrub at the rear of the marsh, but more commonly there is a direct transition to eucalypt woodland or forest.

The earliest colonizers of sandy shorelines above high tide level are *Cakile* spp. and *Atriplex cinerea*. Foredune species are mainly grasses, *Spinifex hirsutus, Ammophila arenaria, Festuca littoralis*. These grasses may be subsequently invaded and replaced by shrubs, including *Myoporum insulare, Acacia longifolia, Correa alba, O. axillaris, Helichrysum paralium, Leucopogon parviflorus*, and *Rhagodia baccata*. Older ridges support dune scrub, usually dominated by *Leptospermum laevigatum* with emergent *Banksia intergrifolia*. In western Victoria *L. parviflorus* forms low closed scrub on dunes, sometimes with *Melaleuca lanceolata*. The sequence of communities on Tasmanian dune systems is somewhat similar, although *Leptospermum laevigatum* is restricted to the far northeast and northwest.

Cliff top shrublands commonly include species of dune scrubs, and are well developed where wind-blown sand has accumulated as cliff-top dunes. On extremely exposed sites and on heavy basalt-derived soils *Poa* tussock grasslands are common. Heaths occur on sandy, often seasonally waterlogged cliff top sites, and are variously dominated by *Leptospermum juniperimum, L. myrsinoides, Melaleuca squarrosa, Casuarina pusilla* and *C. paludosa*. In the drier parts of Tasmania cliff communities are similar to those in Victoria; in wetter areas scree slopes below cliffs may be occupied by *Bedfordia* closed scrub.

Warm Temperate Western Coasts. Salt marshes are found in estuaries along the coast, but *Avicennia* is absent except at Bunbury. As in Victoria and South Australia, *Salicornia* herbfield and *A. arbusculum* shrublands dominate the lower marsh, with *Arthrocnemum halocnemoides* on rarely inundated sites. The *Arthrocnemum* community is succeeded by a sedge zone (*Gahnia trifida, Juncus kraussii, Scirpus nodosus*), which in less maritime situations may grade into a zone of *Melaleuca* and *Casuarina* woodland (Smith, 1973).

The dune vegetation of Western Australia has been described by Sauer (1965). *Spinifex longifolius* and *Salsola kali* are the main strandline species of the central west region, while the extensive dune systems are covered by open shrubland often less than one meter high. Common species include *Thelkeldia diffusa, O. axillaris, Scaveola crassifolia* and *N. schoeberi*. Further south *Spinifex hirsutus* becomes the dominant grass of the foredunes with *C. maritima* and *Arctotheca nivea*. The dune vegetation here is taller and contains a greater number of species, but the chenopods are less prominent. *Acacia* spp. and *Agonis flexuosa* occur on stable dunes. The foredune and outer dune vegetation of the coast between Perth and Albany is similar to that of the southwest sector, but a large number of new shrubs, including *Adenanthos sericea* and *Leucopogon parviflorus*, are found on the inner dunes. Limestone sea cliffs support species of the mobile dunes and some halophytes.

Tropical-subtropical Coasts. Plant communities on sectors of this coast have been studied by Macnae (1966), Pedley and Isbell (1971), and Specht (1958). Mangroves—including species of *Aegialitis, Aegicerus, Avicennia, Bruguieria, Ceriops, Exocoecaria, Lumnitzera, Osbornia, Rhizophora*, and *Sonneratia*—are common in estuaries and deltas and often extend inland along stream banks for several miles. In high rainfall regions mangroves occupy most of the intertidal zone, but contract seaward where rainfall is lower or seasonally distributed. Here tidal flats landward of the mangrove fringe may be bare or support scattered *Salicornia, Arthrocnemum, Tecticornia* or *Suaeda*, which grades into saline meadows of *Sporobolus virginicus* or *Xerochloa*.

Spinifex longifolius and *Ipomoea pes-caprae* are generally the main colonizers of the foredune, while scattered *Casuarina equisitifolia* and several shrub species are common on the seaward margin of stabilized dunes. In drier areas stabilized dunes may support a grassland of *Triodia* or other species with scattered shrubs such as *Acacia* spp. or herbfields (*Gomphrena canescens-Bulbostylis barbata*). In wetter regions dune scrub develops and may include *Guettarda speciosa, Thespesia populnea, Hibiscus tiliaceus, Calophyllum inophyllum, Messersmidia argentea, Scaevola sericea, Celtis phillippinesis*, and *Clerodendron*.

Warm Temperate Eastern Coasts. Between Fraser Island and the northern New South Wales coast eight species of mangrove are found; in central New South Wales only *Avicenna* and

Aegiceras occur, but up to nineteen salt marsh species have been recorded. On this coast salt marshes and mangroves are not extensive as there are no large areas of suitable habitat (Clarke and Hannon, 1967).

On the northern coast *Spinifex hirsutus*, *Ipomoea pes-caprae*, and *Casuarina equisitifolia* are common foredune species; older dunes support *Eucalypt* forest and *Casuarina* or *Banksia* woodlands or heaths. Further south *Cakile maritima* is a common strand plant, while foredunes support *Spinifex hirsutus* and mats of *Carpobrotus* and *Scaevola*. *Leptospermum laevigatum*, *Correa alba*, *Leucopogon parviflorus*, and *Acacia longifolia* form dense scrubs on landward dunes. Heaths dominated by *Casuarina distyla* and *Banksia serrata* are found on drier exposed sand hills and rocky sites. The coastal heaths of the central east coast of New South Wales, Victoria, and South Australia to southwest Western Australia have a number of species in common, but there are some regional floristic differences.

MICHELE M. BARSON
ROBERT J. KING

References

Barson, M. M., and Calder, D. M., 1981, The vegetation of the Victorian Coast, *Royal Soc. Victoria Proc.* **92**, 55-65.

Bennett, I., and Pope, E. C., 1953. Intertidal zonation of the exposed rocky shores of Victoria, together with a rearrangement of the biogeographical provinces of temperate Australian shores, *Australian Jour. Marine and Freshwater Research* **4**, 105-159.

Bennett, I., and Pope, E. C., 1960. Intertidal zonation of the exposed rocky shores of Tasmania and its relationship with the rest of Australia, *Australian Jour. Marine and Freshwater Research* **11**, 182-221.

Clarke, L. D., and Hannon, N. J., 1967. The mangrove swamp and salt marsh communities of the Sydney district. I. Vegetation, soils, and climate, *Jour. Ecology* **55**, 753-771.

Clayton, M. N., and King, R. J., eds., 1981. *Marine Botany: an Australasian Perspective*. Melbourne: Longman-Cheshire, 468p.

Dakin, W. J.; Bennett, I.; and Pope, E. C., 1948. A study of certain aspects of the ecology of the intertidal zone of New South Wales coast, *Australian Jour. Sci. Research* **B1**, 176-230.

Dartnall, A. J., 1974. Littoral biogeography, *in* W. D. Williams, ed., Biogeography and Ecology in Tasmania, *Mongr. Biol.* **25**, 171-194.

Endean, R.; Kenny, R.; and Stephenson, W., 1965. The ecology and distribution of intertidal organisms on the rocky shores of the Queensland mainland, *Australian Jour. Marine and Freshwater Research* **7**, 88-146.

Hedley, C., 1904. The effect of the Bassian isthmus upon the existing marine fauna: A study in ancient geography, *Linnean Soc. N. S. W. Proc.* **28**, 876-883.

Jones, O. A., and Endean, R., eds., 1973. *Biology and Geology of Coral Reefs*. New York: Academic Press, 2 vols.

King, R. J., 1972. *The Distribution and Zonation of Intertidal Organisms of Rocky Coasts in South Eastern Australia*, Ph.D. dissertation. Melbourne: University of Melbourne, 263p.

Kirkpatrick, J. B., and Glasby, J., 1981. *Salt Marshes in Tasmania: Distribution, Community Composition and Conservation*, Univ. Tasmania Dept. Geog. Occasional Paper 8, 52p.

Knox, G. A., 1963. The biogeography and intertidal ecology of the Australasian coasts, *Oceanogr. Mar. Biol. Ann. Rev.* **1**, 341-404.

Macnae, W., 1966. Mangroves in eastern and southern Australia, *Australian Jour. Bot.* **14**, 67-104.

Pedley, L., and Isbell, R. F., 1971. Plant communities of the Cape York Peninsula, *Royal Soc. Queensland Proc.* **82**, 51-74.

Rochford, D. J., 1975. Oceanography and its role in the management of aquatic ecosystems, *in* H. A. Nix and M. A. Elliot, eds., Managing Aquatic Ecosystems, *Ecol. Soc. Australia Proc.* **8**, 67-83.

Saenger, P.; Specht, M. M.; Specht, R. L.; and Chapman, V. J., 1977. Mangal and coastal salt marsh communities in Australasia, *in* V. J. Chapman, ed., *Wet Coastal Ecosystems*. Amsterdam: Elsevier, 293-345.

Sauer, J. D., 1965. Geographic reconnaissance of Western Australian seashore vegetation, *Australian Jour. Bot.* **13**, 39-69.

Smith, G. G., 1973. *A Guide to the Coastal Flora of South Western Australia*. Perth: Western Australian Naturalists Club, Handbook No. 10, 60p.

Specht, R. L., 1958. The climate, geology, soils, and plant ecology of the northern portion of Arnhem Land, *in* R. L. Specht and C. P. Mountford, eds., *Records of the Australian-American Scientific Expedition to Arnhem. Botany Land, vol. 3: Botany and Plant Ecology*. Melbourne: Melbourne University Press, 333-413.

Specht, R. L., 1970. Vegetation, *in* G. W. Leeper, ed., *The Australian Environment*. Melbourne: C.S.I.R.O. and Melbourne University Press, 47-67.

Specht, R. L., 1972. *The Vegetation of South Australia*. Adelaide: Government Printer, 328p.

Stephenson, T. A., and Stephenson, A., 1972. *Life between Tidemarks on Rocky Shores*. San Francisco: Freeman, 425p.

Womersley, H. B. S., 1959. The marine algae of Australia, *Bot. Rev.* **25**, 545-614.

Womersley, H. B. S., 1972. Aspects of the distribution and ecology of subtidal marine algae on southern Australian coasts, in *Proceedings 7th International Seaweed Symposium (1971)*. Tokyo: University of Tokyo Press, 52-54.

Womersley, H. B. S., and Edmonds, S. J., 1958. A general account of the intertidal ecology of South Australian coasts, *Australian Jour. Marine and Freshwater Research* **9**, 217-260.

Womersley, H. B. S., and Thomas, I. M., 1976. Intertidal ecology, *in* C. R. Twidale, M. J. Tyler, and B. P. Webb, eds., *Natural History of the Adelaide Region*. Adelaide: Royal Society of South Australia, 175-185.

Cross-references: *Antarctica, Coastal Ecology; Asia,*

Eastern, Coastal Ecology; Australia, Coastal Morphology; Cliffed Coasts; Coral Reef Coasts; Mangrove Coasts; New Zealand, Coastal Ecology; Pacific Islands, Coastal Ecology; Salt Marsh. Vol. VIII, Part 1: Australia.

AUSTRALIA, COASTAL MORPHOLOGY

The coastline of Australia (including Tasmania) measures about 20,000 kilometers (Fig. 1) and is notable for its long, gently curving sandy beaches. The sands of the western and southern coasts (from Broome in the northwest to Wilson's Promontory in Victoria and the western coast of Tasmania) are predominantly calcareous; those of the eastern and northern coasts are generally quartzose. Many of the beaches form the seaward margins of depositional sand barriers bearing beach ridge or dune topography (Bird, 1973), and on the western and southern coasts the calcareous sands of the older (Pleistocene) dunes have been partially lithified by secondary carbonate precipitation to form dune calcarenites (calcareous aeolianites). These occur extensively between Broome and Cape Leeuwin on the Western Australian coast, between Streaky Bay in South Australia and Warrnambool in Victoria, and locally around the shores of Bass Strait, including the west coasts of King Island and Flinders Island and the northwest coast of Tasmania. In some sectors these have been cut back to form cliffs and shore platforms (Fig. 2).

Cliffs and rocky shores have also developed where older formations reach the coast: in Tertiary formations near Cape Cuvier, in granites and Palaeozoic and older shield formations between Cape Leeuwin and Albany, and in Tertiary formations at the head of the Great Australian Bight. Pre-Cambrian rocks reach the coast locally on the Eyre and Yorke peninsulas, on the Fleurieu Peninsula south of Adelaide, and on Kangaroo Island, in South Australia. In Victoria there are cliffs cut in various formations, including Cainozoic basalts, Tertiary and Mesozoic sediments, Palaeozoic rocks, and intruded granites. Tasmania has extensive rocky and steep coasts developed on strongly folded pre-Carboniferous rocks and intruded granites, with Mesozoic formations, including Jurassic dolerites, on its southeastern coast. In New South Wales there are cliffed and rocky headlands on Mesozoic and older rock formations, and in Queensland steep coasts, with limited cliffing, occur on similar formations inshore from the Great Barrier Reef (Fig. 3). On the north coast of Australia cliffed and rocky sectors are extensive in Arnhemland and where the Kimberley Ranges reach the sea, both areas of Palaeozoic and older rock formations.

In southwestern and southeastern Australia most of the rivers flow into estuarine lagoons with outlets constricted by sandy barrier formations (Fig. 4). These lagoons show stages of infilling, largely by fluvial sediments, and some have been reclaimed by river deposition as broad alluvial plains. In northern Australia, river mouths are less encumbered by coastal sand deposition, and form estuarine gulfs and deltaic regions (Jennings and Bird, 1967). The shores of estuaries, lagoons, and sheltered embayments are often bordered by marshes and swamps, which have advanced onto tidal sandflats or mudflats. The latter become extensive on sectors of the northern coast where tide range is large (Russell and McIntire, 1966).

Coralline reef formations are extensive off the northern half of the continent, particularly in the Great Barrier Reef off Queensland. Numerous outlying patch reefs and coastal fringing reefs occur off the north and northwest coasts, around to Houtman Abrolhos, off Western Australia (Fairbridge, 1967).

Variations in morphology around the Australian coast can be related to geological, climatic, oceanic, and biological factors that vary regionally around the margins of the continent (Davies, 1972, 1977), taking account of changes in climate and sea level that have occurred on and around the Australian land mass (7.8 million sq. km) and its bordering continental shelves (2.5 million sq. km).

Geological Factors

The western half of the Australian continent is a shield region dominated by Palaeozoic and older formations that produce the hard, rocky coasts of the Kimberley region and southwest Western Australia. However, on much of the western coast these older formations are obscured by Pleistocene dune calcarenites and associated Holocene dunes. On the Eyre Peninsula in South Australia the dune calcarenite cover has been cut back by marine erosion to expose underlying harder Precambrian rocks, which become rocky promontories as bordering cliffs in the less resistant calcarenites recede.

The eastern margin of Australia consists of folded Palaeozoic rock formations, with structural trends roughly parallel with the eastern coastline. These formations have been uplifted and dissected to form the Eastern Highlands, with a watershed varying from 20 to 400 km inland, and a succession of relatively small drainage basins feeding rivers that flow down to the Pacific coast. The structural pattern is complicated by the presence of sedimentary basins occupied by Mesozoic formations: the Sydney Basin, for example, contains the major sandstone

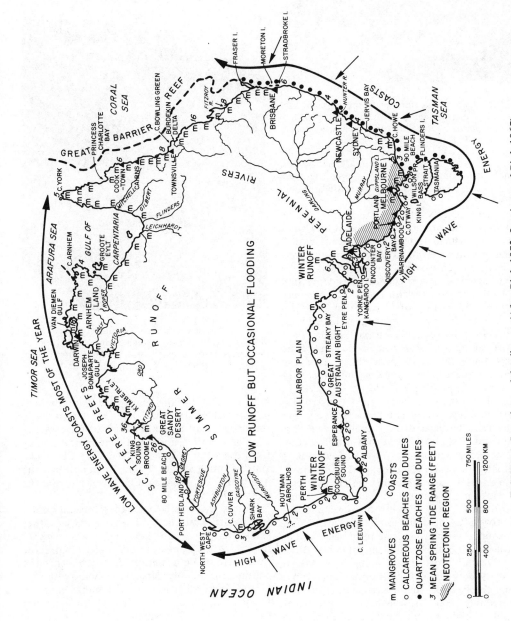

FIGURE 1. Location and detail map. (After Bird, 1973; Jennings and Bird, 1967)

FIGURE 2. Cliffs and shore platforms developed in Pleistocene dune calcarenite on the ocean coast of the Nepean Peninsula, near Melbourne, Victoria, Australia. (Photo: E. C. F. Bird)

formations prominent in cliffs along the New South Wales coast between Newcastle and Jervis Bay.

Between the Western Shield and the Eastern Highlands the continent is dominated by Cainozoic rocks, which occupy a number of basins, several of which extend to and beyond the present coastline. In eastern Victoria the Latrobe Valley syncline passes beneath a depression occupied by the Gippsland Lakes region, where complex sand barriers of Pleistocene and Holocene age are fringed by the Ninety Mile Beach (Bird, 1965). In western Victoria Tertiary rocks dip off the uplifted Mesozoic formations of the Otway ranges and pass beneath an extensive volcanic province, locally active into the early Holocene: the sequence is clearly displayed in cliff sections westward from Cape Otway to Discovery Bay. Bass Strait is underlain by another sedimentary basin, and on a smaller scale Port Phillip and Westernport Bays in Victoria occupy fault-bounded sunklands. Western Tasmania is dominated by Lower Plaeozoic rock formations, but in the southeast these are overlain by Mesozoic rocks in an area where Cainozoic faulting has influenced the evolution of ridges and valleys now partly drowned as rias (Davies, 1965).

In South Australia a sedimentary basin underlies the Lower Murray valley, and the river flows across Tertiary formations to enter a coastal lagoon system (Lakes Albert and Alexandrina) behind the Pleistocene dune calcarenite barriers of Encounter Bay. At the head of the Great Australian Bight the Eucla Basin contains Tertiary formations that outcrop in a long stretch of even-crested cliffs bordering the riverless Nullarbor Plain. Similar cliffed formations mark the Carnarvon Basin at Cape Cuvier, but the Tertiary rocks in the Canning Basin, in northwest Australia, vanish beneath the Great Sandy Desert and the dune calcarenites of the Eighty

FIGURE 3. Steep coast of Macalister Range, north Queensland, Australia, showing lack of cliffing on moderate wave energy coast inshore from Great Barrier Reef. (Photo: E. C. F. Bird)

Mile Beach, and the Bonaparte Gulf Basin is largely a shelf structure. The Carpentaria Basin runs northward under the Gulf of Carpentaria, a graben with Cainozoic formations declining from the western side of Cape York Peninsula, and the much smaller Laura Basin occupies the hinterland of Princess Charlotte Bay on the eastern side.

Despite the existence of these several Cainozoic basins, Australia has been generally a rather stable continent, and there are extensive *planation surfaces,* Cainozoic and older, that show little tectonic deformation. In consequence the continent is notably flat and low-lying, with 80% of the land area below the 500-meter contour. Many of the rivers (the Murray, for example) have low channel gradients as they approach the sea, and carry predominantly fine-grained sediments (silt and clay) to their mouths. Steeper gradients exist in rivers draining the Pacific slopes of the Eastern Highlands, and some of these carry sand and gravel in the load delivered to the sea. Fluvial sediment yields are also influenced by outcrop lithology and

FIGURE 4. Barrier islands and tidal entrances east of Corner Inlet, Victoria, Australia. (Photo: N. J. Rosengren)

past as well as present weathering regimes. Most of the granite outcrops in the Eastern Highlands have been deeply weathered, and rivers draining them derive quartz-rich sands; weathered basalts supply silt and clay, and river gravels are produced by the disintegration of Palaeozoic metasediments. On the coast marine erosion has generally removed weathered mantles, so that the granites, for example, commonly protrude in rocky headlands, as at Wilson's Promontory in Victoria (Fig. 5).

Although the general picture is one of prolonged stability in Australia, marginal neotectonic activity has been recorded locally, especially in Victoria, and the possibility of slight coastal warping within Holocene times cannot be ruled out.

Climatic Factors

The prevailing aridity of the interior and much of western and southern Australia has resulted in few rivers draining to the coast; those few flow intermittently, with occasional brief

FIGURE 5. Granite headland at Wilson's Promontory, Victoria, Australia. (Photo: N. J. Rosengren)

episodes of flooding. In the southwest of Western Australia and the southeast of South Australia winter rainfall nourishes seasonal runoff, but Australia's largest river system, the Murray-Darling, draining a catchment of about a million sq. km, has only a limited discharge at its mouth compared with systems of similar scale in more humid regions. In consequence of hinterland aridity, terrigenous sediment supply to the coasts of much of western and southern Australia is meager; coastal deposition has been nourished primarily from sea floor sources and is dominated by carbonate sands of biogenic origin. The prevalence of such material in shelf sediments is a further consequence of the lack of terrigenous supply, hinterland aridity having evidently prevailed also during Pleistocene phases of lowered sea level.

By contrast, the Eastern Highlands have a higher and more regular rainfall, and the steeper catchments draining to the Pacific coast nourish river systems that deliver a more substantial terrigenous sediment supply. Coastal sediments are dominated by quartzose sands, some of which are supplied directly by rivers such as the Shoalhaven, the Hunter, the Fitzroy, and the Burdekin, particularly during episodes of flooding. Much of the sand deposited in beaches and barriers along the Gippsland and New South Wales coasts has, however, been delivered from the sea floor during successive marine transgressions in Pleistocene and Holocene times by the reworking of shelf deposits of quartzose sand derived partly from fluvial supply during low sea level phases and partly from weathered rock outcrops now submerged. The contrast between the carbonate sands and dune calcarenites of southern and western Australia and the quartzose sands of the southeastern and eastern coast is thus related to a contrast in shelf sedimentation, due in turn to a contrast in hinterland topography and climate through Quaternary times (Bird, 1976).

Along the Queensland coast terrigenous sediment supplied by rivers includes quartzose sands and lithic gravels derived from relatively steep catchments as well as muddy sediment derived from tropical weathering mantles on catchment outcrops. In addition, coastal sedimentation is locally complicated by the presence of fringing and nearshore reefs built by corals and associated organisms in tropical waters. Reef clastics derived from these are added to adjacent beaches, but despite the proximity of a rich coralline province the beaches of Queensland are generally poor in carbonates (Bird, 1971).

In northern Australia, summer rainfall results in substantial runoff from river systems, but catchments are generally larger and less steep than those of the east coast, and fluvial loads are dominated by fine-grained sediment derived from prolonged weathering of catchment outcrops, sandy yields being limited and localized. Muddy sedimentation is extensive on the shores of the Gulf of Carpentaria and in northern embayments and inlets, particularly around Van Diemen Gulf, Joseph Bonaparte Gulf, and King Sound, each of which receives major inputs of river water and sediment.

Other features related to climate include the nature and rate of rock weathering in the shore zone, most active under humid tropical conditions on the northeast coast of Queensland (Fig. 6) and the vigor of wind action, which Jennings (1965) showed to be much stronger in the temperate zone than on the tropical northern coasts, a contrast that helps to explain the more extensive development and greater mobility of coastal dunes in western, southern, and southeastern Australia than in the north and northeast. Davies (1977) has elucidated such contrasts as these within a climatically based sector classification, clockwise around the continent, into warm temperate humid (Fraser Island to Portland), warm temperate arid (Portland to North West Cape), tropical arid (North West Cape to Broome) and tropical humid (Broome to Fraser Island) coastal sectors.

FIGURE 6. Intensive weathering of Paleozoic rocks in shore outcrops on a humid tropical coast south of Cairns, Queensland, Australia. (Photo: E. C. F. Bird)

Oceanic Factors

Swell generated by storms in the Southern Ocean moves onto the western, southern and southeastern coasts of Australia, and its refracted patterns are effective in shaping the long, curved outlines typical of sandy beaches. In New South Wales, for example, refraction of the dominant southeasterly swell is responsible for the asymmetrically curved outlines of beaches between bedrock headlands (Langford-Smith and Thom, 1969). Southern coasts are frequently exposed to storm wave action generated in coastal waters as cyclonic depressions travel through from west to east. With relatively deep water close inshore, these are typical high wave energy coasts, with bold cliffs and headlands (Fig. 7) and smooth depositional outlines; these have also been in receipt of sand carried shoreward from the sea floor during successive marine transgressions and built into massive coastal barrier and dune systems (Fig. 8). Extensive sand deposition has constricted the valley-mouth inlets formed by submergence during the last marine transgression and formed estuarine thresholds of inwashed sand at their entrances (Bird, 1967).

On the east coast north of Fraser Island ocean swell is interrupted by reef structures, and in the lee of the Great Barrier Reef it is largely excluded from coastal waters (Maxwell, 1968). Within these waters the prevailing southeasterly winds generate only moderate wave energy, and consequently the coast has but limited sectors of cliffing, poor development of shore plat-

FIGURE 7. Cliffed coast near Port Campbell, Victoria, Australia, showing intricate joint-guided dissection on a high wave energy coast. (Photo: N. J. Rosengren)

FIGURE 8. The sandy beach of Discovery Bay, Victoria, Australia, backed by extensive dunes moved inland by strong southwesterly winds. (Photo: N. J. Rosengren)

forms, and irregular depositional outlines, with many spits, cuspate forelands, and some protruding deltas, notably the Burdekin (Hopley, 1970); sectors of mud flat and mangrove swamp occur on shores in the lee of headlands and islands. Sandy barriers and dune systems are built on a much smaller scale than on the oceanic coasts, and take the form of beach-ridge plains, often built as cheniers upon mud flats. River mouths are less encumbered by sand deposition, and estuarine lagoons do not occur north of Fraser Island. In the absence of ocean swell, shoreward drifting of sea floor sand has made only a limited contribution to coastal sedimentation on the Queensland coast.

On the north coast of Australia, swell from the Indian Ocean is attenuated by the exceptionally broad continental shelf, and its effects are further diminished by high tidal ranges. In the Timor and Arafura seas and within the Gulf of Carpentaria local wind action generally produces low to moderate wave energy on bordering coasts. Steep sectors show little cliffing and retain the intricate outlines of submerged landscapes of fluvial dissection, while depositional sectors are typically irregular and varied in detail. Storm surges accompanying tropical cyclones accomplish rapid short-term changes, and may be responsible for the emplacement of sandy cheniers on coastal mud flats and alluvial plains, for example, on the depositional lowlands south of Van Diemen Gulf and on the low-lying coasts of the Gulf of Carpentaria, where surges of up to 7 meters have been reported.

The broad patterns of longshore drifting on the Australian coast are northward and eastward from Cape Leeuwin and northward along the east coast, but where coastal outlines are adjusted to the dominant swell patterns, little net drifting occurs (Davies, 1977; Bird, 1978).

In addition to variations in wave energy, there are marked contrasts in tide range around the Australian coast. In the southern part of the continent, from North West Cape around to Fraser Island, mean spring tide ranges are generally less than 2 meters, the exceptions being within the South Australian gulfs and in Western Port Bay and Corner Inlet in Victoria. In southwestern Australia the spring tide range is less than 0.5 meters, and it can be difficult to distinguish tidal oscillations from the meteorological effects of wind stress and barometric pressure that raise or lower the sea surface by greater amounts. On the Queensland coast and in the southeastern part of the Gulf of Carpentaria tide ranges are typically between 2 and 4 meters, but on the northwestern coast between Port Hedland and Darwin the range exceeds 5 meters, with a maximum of 10.5 meters in Collier Bay, near Yampi Sound.

Where tide range is small, wave energy is concentrated within a narrow vertical range; where it is large, wave energy is dispersed. Thus the tidal variation emphasizes the contrast between high wave energy conditions in the southern part of Australia and low wave energy conditions along the northern coast. Large tide ranges also generate strong ebb and flow currents, which impede the deposition of wave-built structures, such as spits and barriers, and lead to the formation of wide, funnel-shaped estuaries, such as that of the Fitzroy, opening into King Sound. Broad areas of sand flat and mud flat are exposed at low tide and, with diminished and intermittent wave action above midtide level, salt marshes and mangrove swamps become more extensive than on microtidal coasts, where the small tide range compresses the biological zonation.

Biological Factors

The effects of organisms on coastal processes are limited by their geographical range, conditioned mainly by climatic and oceanic factors. Reef-building corals live in Australian coastal waters north from Houtman Abrolhos, off the west coast, around to the Capricorn Group at the southern end of the Great Barrier Reef. They are most prolific off the northeast coast, while off the northwest coast they give place to bryozoan and oyster banks, and off the west coast the reefs are of submerged dune calcarenite with only a veneer of corals (Fairbridge, 1967).

Mangroves extend south as far as Corner Inlet, adjacent to Wilson's Promontory in Victoria. In the humid tropical sector of northeast Queensland there are up to 27 species in broad mangrove swamps backed by rain forest vegetation (Fig. 9), but in drier sectors there are fewer species and the seaward mangrove fringe is backed by salt marshes and hypersaline mud flats. The effects of mangroves on sedimentation vary from one species to another, but the white mangrove (*Avicennia marina*), the most widespread of the Australian mangroves and the only one extending as far south as Victoria, has a pneumatophore network that acts as a sediment filter and is consequently an agent of land building alongside estuaries and embayments sheltered from strong wave action (Bird, 1972). Introduced *Spartina anglica* has reshaped the intertidal profiles of Anderson's Inlet, a Victorian estuary, and the Tamar estuary in Tasmania by developing a depositional marshland terrace, but it does not grow in warmer environments farther north.

Modern shelly beaches (as distinct from beaches of shelf-derived biogenic carbonate sand) are confined to environments where shell-bearing organisms live abundantly, notably on rocky shores and reefs and within estuaries and lagoons. These are extensive toward the heads of the South Australian gulfs and on the western shore of Port Phillip Bay, where their dominance reflects the local absence of other types of beach material on a coast formed by the marginal submergence of a basalt plain.

Inherited Features

Australian coastal landforms include various features inherited from past phases of contrasted environment. Some of these are related to changes of sea level during Quaternary times: the submergence of valley mouths and the shoreward drifting of sand from sea floor sources to form beaches, barriers, and coastal dunes have already been noted. In addition, there are coastal features formed during past phases of higher sea level, some of which have persisted from Pleistocene times while others may be referable to a higher Holocene sea level episode. These features include: emerged shore platforms, some with associated fossil beach deposits backed by former cliffs now degraded as bluffs; emerged coralline reefs; emerged sandy thresholds near the mouths of estuaries and lagoons; and beach ridge and barrier features standing higher, in relation to sea level, than they did at the time of their formation.

The interpretation of some of these features has been questioned. It is necessary to distinguish between emerged shelly beaches and the aboriginal *kitchen middens* frequently encountered above high tide level and to be sure that apparently emerged shoreline features and deposits were not formed when storm surges carried wave action briefly to higher levels. Shoreline features that undoubtedly emerged may have been the outcome of tectonic uplift of the land margin, a fall of sea level, or some

FIGURE 9. Mangrove swamps on the shore of Cairns Bay, north Queensland, Australia, showing an outer zone of *Avicennia* backed by *Rhizophora*, with more varied mangrove forest to the rear. (Photo: E. C. F. Bird)

combination of the two; and if the coastal margin has been tectonically stable, *radiocarbon dating* of suitable associated organic materials is needed to decide whether the emerged shoreline developed during a higher sea level phase in Pleistocene times or whether it should be referred to a higher Holocene sea level. Emerged features dating from Pleistocene times often show evidence of dissection by fluvial incision or wind scour extending below present sea level during the last phase of lowered sea level; the inner barrier in the Gippsland Lakes is assigned a Pleistocene age on these grounds (Bird, 1965). Emerged features dating from Holocene times cannot show such dissection. The question of a higher Holocene sea level remains controversial in Australia (Gill and Hopley, 1972; Thom et al., 1972).

On parts of the Australian coast the landforms bear the imprint of weathering under past phases of climate that were warmer, cooler, wetter, or drier than those now prevailing. These include the presence of lateritic weathering profiles in cliff exposures of Tertiary rock in Victoria, inherited from a past phase of warmer and wetter climate, and the presence of periglacial features on cliffed coasts in southernmost Tasmania, inherited from a past phase of colder climate. On the east coast of the Cape York Peninsula near Cape Flattery extensive systems of elongated parabolic dunes, now largely covered by scrub and forest, originated under drier and perhaps windier conditions in the past. Patterns of coastal rock weathering northward and southward of the wettest sector of the humid tropical coast of northeast Queensland indicate that the warm and wet environment was at times more extensive in the past. The high, parabolic dunes on Fraser Island must also have formed under different conditions in the past, for they now carry a cover of rain forest. It is possible that sand mobilization here was also partly related to phases of rising sea level, when sands deposited during earlier phases of stable or falling sea level were trimmed back and exposed to the effects of onshore wind action. The shores of King Sound include narrow, sandy peninsulas and intervening corridor embayments that are features inherited from an arid landscape of sand ridges and swales that has been partly submerged by the last marine transgression.

Shore Platforms

Shore platforms on the Australian coast show a variety of forms, some sloping seaward, others almost horizontal; the former are basically abrasion ramps, but the latter have been influenced by shore weathering processes, which have tended to flatten and maintain platforms originally cut by abrasion. Such horizontal platforms are well developed at about high tide level on cliffed coasts cut in basalt and fine-grained sedimentary formations such as sandstones and shales, which are subject to disintegration by wetting and drying processes (and possibly by salt crystallization) down to a level of saturation by seawater. These are well developed on the coast of New South Wales south of Sydney (Bird and Dent, 1966).

On dune calcarenites similar platforms have developed at a slightly lower level, equivalent to the level attained by fringing reefs built by corals and algae. Their flatness is evidently the outcome of solution processes effective down to a level at which precipitation of carbonates from seawater becomes dominant, indurating the platform surface (Hills, 1971). Where waves are supplied with sand or gravel derived from cliff erosion or washed in from the sea floor, abrasion becomes more effective, and the horizontal platforms give place to benches that slope seaward.

Modern Dynamics

Comparison of the present coastal configuration with evidence from historical sources, notably maps compiled by nineteenth-century surveyors, indicates that few sectors of the Australian coast have advanced seaward (prograded) during the past century. Indeed, much of the sandy barrier coast around Australia has receded over this period, so that most beaches are backed by cliffed dunes and truncated beach ridge systems that had formed during previous progradation. Sectors of advance have been confined to growing portions of spits and cuspate forelands, some minor deltaic additions (mainly near the mouths of northern rivers), and accumulations alongside artificial structures such as harbor breakwaters.

The prevalence of erosion on sandy shorelines is not confined to Australia (Bird, 1976). It could be due to a slight rise of sea level within the past century, to increased storminess in coastal waters, to a diminution in sand supply from rivers modified by dam construction, or to a cessation in the supply of sand from the sea floor in the aftermath of the last major marine transgression (Thom, 1974). In general it appears that beach sand carried away offshore or alongshore by wave erosion, lost landward by deflation to dune systems, and washed into the mouths of estuaries as threshold banks has not been replaced by new inputs from rivers, from cliff erosion, or from the sea floor.

Cliffed coasts have continued to recede, especially on sectors where soft formations confront stormy seas, as on the Tertiary sands and clays of the Port Campbell district in Victoria (Fig. 7). Minor progradation can be detected locally on mangrove-fringed shorelines, as in Cairns Bay in northeast Queensland (Bird, 1972), and some salt marshes have spread seaward, particularly where *Spartina* has been introduced in recent decades. Elsewhere there is evidence of erosion of the mangrove and salt marsh fringe, for example, around the shores of Western Port Bay in Victoria, as a sequel to the clearing and draining of swamps and hinterlands and the construction of boat harbors (Bird and Barson, 1975).

Impact of Man's Activities

A last factor influencing the modern development of coastal morphology in Australia has been the impact of man's activities. In the vicinity of urban centers the coast has been modified by the construction of harbors, the building of sea walls to halt coastal erosion, and the insertion of groins in an attempt to retain beach materials. A substantial part of the formerly cliffed northeast coast of Port Phillip Bay, adjacent to Melbourne, is now lined by such structures, and similar works have been introduced to the Adelaide coast. Where breakwaters interrupt the longshore drift of sand, as at Adelaide Outer Harbour, there has been local accretion. Frequently such accretion is balanced by the onset of erosion downdrift of a breakwater, as at Queenscliff in Port Phillip Bay. Where sea walls have been built (whether of wood, masonry, concrete, or dumped boulders) wave reflection has reduced the adjacent beach. In recent years, some attention has been given to *artificial beach nourishment* as a means of reducing coastal erosion at the same time as ensuring the persistence of a beach for recreational use. As has been noted, the past century has been a period of general retreat of sandy shorelines around Australia, and erosion has been the normal condition even on coasts remote from intensive use and modification by man.

Coastal dune areas near urban centers and seaside resorts show evidence of vegetation loss and initiation or acceleration of blowouts as a consequence of clearing, burning, and tampling dune vegetation. In an era of prevailing sandy shoreline erosion some of this mobilization of dunes would have occurred anyway, and blowouts are indeed developing on remote and little-visited dune coasts, as on Fraser Island. However, in coastal dune areas that are intensively used for recreation the erosion has been accelerated by man's activities, and conservation procedures involving fencing and replanting of vegetation are now being implemented. Such activities as beach nourishment and dune stabilization mark a trend toward management of coastal morphology in Australia with the aim of optimizing conservation values and recreational opportunities in the coastal environment.

ERIC C. F. BIRD

References

Bird, E. C. F., 1965. *A Geomorphological Study of the Gippsland Lakes.* Canberra: Australian National University Press, 101p.

Bird, E. C. F., 1967. Depositional features in estuaries and lagoons on the south coast of New South Wales, *Australian Geogr. Studies* **5**, 113-124.

Bird, E. C. F., 1971. The origin of beach sediments on the north Queensland coast, *Earth Sci. Jour.* **5**, 95-105.

Bird, E. C. F., 1972, Mangroves and coastal morphology in Cairns Bay, north Queensland, *Jour. Trop. Geogr.* **35**, 11-16.

Bird, E. C. F., 1973. Australian coastal barriers, *in* M. L. Schwartz, ed., *Barrier Islands.* Stroudsburg, Pa.: Dowden, Hutchinson, & Ross, 410-426.

Bird, E. C. F., 1976. *Shoreline Changes during the Past Century.* I.G.U. Working Group on the Dynamics of Shoreline Erosion, Department of Geography, University of Melbourne, 54p.

Bird, E. C. F., 1978. Variations in the nature and source of beach materials on the Australian coast, *in* J. L. Davies and M. A. J. Williams, eds., *Landform Evolution in Australia.* Canberra: Australian National University Press, 144-157.

Bird, E. C. F., and Barson, M. M., 1975. Shoreline changes in Westernport Bay, *Proc. Roy. Soc. Victoria* **87**, 15-28.

Bird, E. C. F., and Dent, O. F., 1966. Shore platforms on the south coast of New South Wales, *Australian Geogr.* **10**, 71-80.

Davies, J. L., ed., 1965. *Atlas of Tasmania.* Hobart: Lands and Surveys Dept., 128p.

Davies, J. L., 1972. *Geographical Variation in Coastal Development.* Edinburgh: Oliver and Boyd, 204p.

Davies, J. L., 1977. The coast, *in* D. Jeans, ed., *Australia: A Geography.* Sydney: Sydney University Press, 134-151.

Fairbridge, R. W., 1967. Coral reefs of the Australian region, *in* J. N. Jennings and J. A. Mabbutt, eds., *Landform Studies from Australia and New Guinea.* Canberra: Australian National University Press, 386-417.

Gill, E. D., and Hopley, D., 1972. Holocene sea levels in eastern Australia—A discussion, *Marine Geol.* **12**, 223-233.

Hills, E. S., 1971. A study of cliffy coastal profiles based on examples in Victoria, Australia, *Zeitschr. Geomorphologie* **15**, 137-180.

Hopley, D., 1970. *The Geomorphology of the Burdekin Delta, North Queensland.* Department of Geography, James Cook University, Townsville, 66p.

Jennings, J. N., 1965. Further discussion of factors affecting coastal dune formation in the tropics, *Australian Jour. Sci.* **28**, 166-167.

Jennings, J. N., and Bird, E. C. F., 1967. Regional geomorphological characteristics of some Australian estuaries, *in* G. H. Lauff, ed., *Estuaries.* Wash-

ington, D.C.: American Association for the Advancement of Science, 121-128.
Langford-Smith, T., and Thom, B. G., 1969. New South Wales morphology, *Geol. Soc. Australia Jour.* **16**, 572-580.
Maxwell, W. G. H., 1968. *Atlas of the Great Barrier Reef.* Amsterdam: Elsevier, 258p.
Russell, R. J., and McIntire, W. G., 1966. *Australian Tidal Flats.* Louisiana State University, Coastal Studies 13, 48p.
Thom, B. G., 1974. Coastal erosion in eastern Australia, *Search* **5**, 198-209.
Thom, B. G.; Hails, J. R.; Martin, A. R. H.; and Phipps, C. V. G., 1972. Postglacial sea levels in eastern Australia—A reply, *Marine Geol.* **12**, 233-242.

Cross-references: *Asia, Eastern, Coastal Morphology; Australia, Coastal Ecology; Coral Reefs; Magnoliophyta; Mangrove Coasts; New Zealand, Coastal Morphology.* Vol. VIII, Part 1: *Australia.*

AVULSION

Avulsion in a legal sense refers to the *tearing away* of someone's property or *soil* by stream or wave erosion and its deposition on or adjacent to someone else's or to the division of one's property by the shifting of a river channel, for example, by a neck cutoff.

More recently geologists (Allen, 1965; Pettijohn, Potter, and Siever, 1972) have used the term to refer to the abandonment of a major channel segment (as opposed to a simple neck or chute cutoff) of a stream meandering in an alluvial valley and the development of a new meander belt. This usually occurs by the *crevassing* of a natural levee during a major flood. The segment downstream from the diversion into the lower flood basin or backswamp is then abandoned.

In North America the best-known examples of avulsion are the many changes in the lower reaches of the Mississippi River during the building of the several Holocene deltas (see review in Gould, 1970). Other examples are cited by Allen (1965).

SAUL ARONOW

References

Allen, J. R. L., 1965. A review of the origin and characteristics of recent alluvial sediments, *Sedimentology* **5**, 89-191.
Gould, H. R., 1970. The Mississippi Delta complex, in J. P. Morgan and R. H. Shaver, eds., *Deltaic Sedimentation, Modern and Ancient.* Tulsa, Okla: Soc. Econ. Paleont. Miner. Spec. Pub. 15, 3-30.
Pettijohn, F. J.; Potter, P. E.; and Siever, R., 1972. *Sand and Sandstone.* New York: Springer-Verlag, 618p.

Cross-references: *A-B-C Model Coastal Accretion Forms; Coastal Erosion; Coastal Erosion, Environmental-Geologic Hazard; Coastal Erosion, Formations; Cliff Erosion; Sediment Budget; Sediment Transport.*

B

BACKWASH PATTERNS

Rhombohedral ripples are so common on the lower foreshore of a beach that Johnson (1919) called them *backwash marks*. The flow of the backwash down a beach often results in diamond-shaped rhombohedral ripples of low height that are best recognized by the crisscrossed pattern of intersecting lines of the lee slopes of the ripples. Generally there is a well-developed scour on the seaward side of the ripple rhombs, whereas the landward side of the diamond shape is more gentle.

Rhombohedral ripples seem to have been first described by Williamson (1887), who viewed them as resembling "the overlapping scale leaves of some cycadean stems." Observations by Woodford (1935) and Demarest (1947) show that rhomboid ripples form as a lee wave, radiating seaward from coarser than usual grains or more compact sand or from centers of escaping interstitial water. Rhomboid ripples form only in the water-saturated lower part of beaches that slope between 2 and 10 degrees (Emery, 1960). Rhomboid ripples form in a film of water a few mm thick and never exceeding 1-2 cm (Reineck and Singh, 1973). The angle of the apex of the rhomb becomes progressively smaller as the backwash velocity increases. In general the neighboring rhombs are strongly out of phase.

In some cases the rhomboid ripples seem to grade into forms where only small scours are present, arranged in rhombohedral patterns. In such cases the scours migrate landward depositing laminae dipping landward (Wunderlich, 1972). A similar pattern, only larger in size, will form if shells or gravel lie in the backwash, which will deflect the backwash to form a zone of erosion in front of the obstacle. This zone fans out around and behind the obstacle. Directly behind the obstacle deposition of sand scoured from the eroded zone takes place producing a ridge (Sengupta, 1966). These features are known as *current crescents*. If alternate backwashes come from several directions, more than two ridges, alternating with furrows, may be formed.

BENNO M. BRENNINKMEYER

References

Demarest, D. F., 1947. Rhomboid ripple marks and their relationship to beach slope, *Jour. Sed. Petrology* 17, 18-22.

Emery, K. O., 1960. *The Sea off Southern California*. New York: John Wiley & Sons, 365p.

Johnson, D. W., 1919. *Shore Processes and Shoreline Development*. New York: Hafner Publishing Co., 584p.

Reineck, H.-E., and Singh, I. B., 1973. *Depositional Sedimentary Environments*. New York: Springer-Verlag, 439p.

Sengupta, S., 1966. Studies of orientation and imbrication of pebbles with respect to cross-stratification, *Jour. Sed. Petrology* 36, 362-369.

Williamson, W. C., 1887. On some undescribed tracks of invertebrate animals from Yoredale rocks, and some inorganic phenomena produced on tidal shores, simulating plant remains, *Manchester Lit. and Philo. Soc. Mem.* 10, 19-29.

Woodford, A. O., 1935. Rhomboid ripple marks, *Am. Jour. Sci.* 29, 518-525.

Wunderlich, F., 1972. Beach dynamics and beach development Georgia coastal region, Sapelo Island, U.S.A., *Senckenbergiana Maritima* 4, 47-79.

Cross-references: *Beach Processes; Entrainment; Minor Beach Features; Rill Marks; Ripple Marks; Sediment Transport; Swash Mark*.

BACTERIA—See SCHIZOMYCTES

BAGNOLD DISPERSIVE STRESS

The best-known theory of the movement of high sediment concentrations, in water is the one proposed by Bagnold (1954, 1956, 1966a, 1966b, 1973). According to his theory, at sediment concentrations as low as 9% by volume the interactions between the grains themselves is much more important in determining the types of movement than the impact of the fluid on the particles. A layer of grains cannot be moved without dilatation, during which the individual grains are moved apart from each other and upward against gravity. He argued that if sediment is moving the dispersion of grains is maintained by mutual encounters between the grains, which push against each other as they are driven along. Thus there comes into existence a dispersion of the grain momenta and a *dispersive pressure*.

The fluid forces, which are composed of the *drag force* (F_D) and the *fluid lift force* (F_L), drive the grains forward and upward while gravity pulls them down. Bagnold hypothesizes that: the shearing grains generate a *dispersive pressure* (P) that is directed upward and perpendicular to the bed; there is a retarding *tangential shear stress* (T_t) of the grains due to grain to grain encounters; and there exists a *resisting shear stress* (T_r) of the intergranular shearing fluid. He showed that T_t could exceed more than 100 times T_r. Moreover, the fluid stress does not become important unless the sediment concentration drops below 9% by volume. So for high bed load concentrations T_r can be ignored, so that $T_O = T_t + T_r = P \tan\phi = (\rho_s - \rho_f)/\rho_s (g\,M\tan\phi)$ for a flat bed and $T_O = P\tan\phi - t = (\rho_s - \rho_f)/\rho_s(g\,M[\cos a \tan\phi - \sin a])$ for a sloping bed, where t is the gravity-generated tangential stress of the sloping bed of angle of inclination a and M is the mass of the moving grains. Finally, he showed that the dispersive pressure (P) is spatially related to the tangential grain collision shear stress T_t by $T_t/\rho = \tan\phi$, where ϕ is the friction angle corresponding to the equilibrium grain orientation in the plane of transport. ϕ can be looked upon as the pivot angle through which the grain in question has to be moved to rise above the surrounding grains.

Two different regimes were distinguished by Bagnold: a *viscous regime*, in which the sand grains do not actually touch each other because of the viscous shear stresses set up by the near approach to each other by grains moving at different speeds; and an *inertial regime*, in which the actual particle collisions are important, so that the fluid viscosity is no longer significant.

Originally Bagnold stated that $\tan\phi = 0.75$ in the viscous regime but could be as low as 0.32 in the inertial regime. More recently (1973) he corrected the inertial regime figure for "natural sand" grains to 0.63, which is approximately the same as the mass angle of repose of sand. Thus ϕ can be thought of as a dynamic analog of the static friction coefficient.

Bagnold's theory of dispersive pressure is also applicable to *saltation*. As grains rise in the fluid they become subject to fluid drag, where the thrust of the fluid is transferred to the grains. When the saltating grains fall, the thrust is transformed by grain collision into an upward force strong enough to lift the influenced grains. As Moss (1972) has observed, when bed load movement increases, a stage is reached wherein collisions become important and thereafter the whole load proceeds as a dense mass of colliding particles buoyed up by dispersive pressure generated by the collisions and held to the bed by gravity. This moving mass behaves as a viscous fluid with a sharp upper boundary. Its onset coincides with the transition phase of Simons, Richardson, and Nordin (1965) while *plane bed* sedimentation represents its full development.

The concept of grain flows has also been derived from the dispersive pressure. For the viscous regime the dispersive pressure is produced by viscous force due to the near approaches of the grains to each other, so that the viscosity of the interstitial fluid is paramount but the density of the grains is not. Dzulinski and Sanders (1962) termed this the *Bagnold effect*. More recent work (Middleton, 1970; Lowe, 1976), however, questions whether dispersive pressure is the primary mechanism supporting the grains in grain flow.

The dispersive pressure in particles finer than 0.2 mm varies as the square of the grain diameter. Therefore on a beach where grains of various sizes are sheared together the larger grains will tend to migrate upward to zones of low shear, whereas the finer grains will tend to migrate toward the bed where the shear is at a maximum. The deposits left behind by this process have inverse grading. Inman, Ewing, and Corliss (1966) have termed this process shear sorting. It is one of the means of obtaining inverse grading on beach lamination by the backwash shearing the sand as it returns to the sea. Others, however, (Middleton, 1970; Bluck, 1967) believe that the sorting is produced by the smaller particles falling between the larger and thereby displacing the larger particles to the surface.

BENNO M. BRENNINKMEYER

References

Bagnold, R. A., 1954. Experiments on a gravity-free dispersion of large solid spheres in a Newtonian fluid under shear, *Royal Soc. (London) Proc.*, Ser. A **225**, 49-63.

Bagnold, R. A., 1956. The flow of cohesionless grains in fluids, *Royal Soc. (London) Philos. Trans*, Ser. A **249**, 235-297.

Bagnold, R. A., 1966a. *An Approach to the Sediment Transport Problem from General Physics*, U.S. Geol. Survey Prof. Paper 422-I, 37p.

Bagnold, R. A., 1966b. The shearing and dilatation of dry sand, *Royal Soc. (London) Proc.*, Ser. A **295**, 219-232.

Bagnold, R. A., 1973. The nature of saltation and of "bedload" transport in water, *Royal Soc. (London) Proc.*, Ser. A **332**, 473-504.

Bluck, B. J., 1967, Sedimentation of beach gravels: Examples from South Wales, *Jour. Sed. Petrology* **37**, 128-156.

Dzulinski, S., and Sanders, J. E., 1962. Current marks on firm mud bottoms, *Connecticut Acad. Arts Sci. Trans.* **42**, 57-96.

Inman, D. L.; Ewing, G. C.; and Corliss, J. B., 1966. Coastal sand dunes of Guerrero Negro, Baja California, Mexico, *Geol. Soc. America Bull.* **77**, 787-802.
Lowe, D. R., 1976. Grain flow and grain flow deposits, *Jour. Sed. Petrology* **46**, 188-199.
Middleton, G. V., 1970. Experimental studies related to problems of flysch sedimentation, *Geol. Assoc. Canada Spec. Paper 7*, 253-272.
Moss, A. J., 1972. Bed-load sediments, *Sedimentology* **18**, 159-219.
Simons, D. B.; Richardson, E. V.; and Nordin, C. F., 1965. Sedimentary structures generated by flow in alluvial channels, *Soc. Economic Paleontologists and Mineralogists Spec. Publ.* **12**, 34-52.

Cross-references: *Beach Material, Sorting of; Beach Processes; Nearshore Hydrodynamics and Sedimentation; Ripple Marks; Sediment Transport.* Vol. VI: *Flow Regimes.*

BALLAST

The term *ballast* has two meanings applicable to matters relating to the coastal environment. One refers to heavy material put into the hold of a ship, instead of cargo, to steady the vessel while underway; the other denotes the broken stone laid down to form a firm base for further construction, as in the case of a foundation for a sea wall. It is the former meaning that will be expanded here.

Particularly in the early days of sailing, when a vessel had to travel between ports without its usual complement of cargo, local supplies of cobbles and boulders were loaded into the hull to make the ship ride deeper in the water and thus be more stable while at sea. Upon reaching a harbor where cargo was to be taken on, the expendable ballast was simply jettisoned over the side into the water. Such *erratics* then became a part of the local marine sediment, and in some cases have been transported by wave action to the shore. If not identified as glacial erratics, ice-rafted material, or relict sediment, anomalous cobbles and boulders on beaches and in the nearshore zone should be considered as possibly of a transported ballast origin.

Reid (1892) has stated, "There has always been some uncertainty as to which of the far transported blocks found on the Sussex coast were genuine erratics, and which had been brought in ballast, or had been derived from wrecks." In 1975, while on a submarine dive in the Corinthian Gulf searching for evidence of the submerged ancient coastal city of Helice, Jacques Cousteau and Harold Edgerton observed anomalous mounds of gravel and cobble on the shallow marine substrate (Edgerton, personal communication). One plausible explanation is that these mounds were composed of ballast from trading ships that called on ports in the area.

MAURICE L. SCHWARTZ

Reference

Reid, C., 1892. The Pleistocene deposits of the Sussex coast, and their equivalents in other districts, *Geol. Soc. London Quart. Jour.* **48**, 344-363.

Cross-references: *Boulders; Cobbles; Nearshore Hydrodynamics and Sedimentation; Sediment Transport.* Vol. VI: *Transportation.*

BARRIER BEACHES

The term *barrier beaches* is frequently employed as a synonym for *barrier islands*, especially narrow ones, with a few or single sand ridge. Because beaches on barriers represent only a relatively small portion of the total barrier body, this use is not recommended (Price, 1951). The term is accurate only for beaches of barrier islands and barrier spits. For additional details see Gary, McAfee, and Wolf (1972), Komar (1976), Shepard (1972), and Zenkovich (1967).

ERVIN G. OTVOS

References

Gary, M.; McAfee, R., Jr.; and Wolf, C. L., eds., 1972. *Glossary of Geology.* Washington, D.C.: American Geological Institute, 805p.
Komar, P. D., 1976. *Beach Processes and Sedimentation.* Englewood Cliffs, N.J.: Prentice-Hall, 429p.
Price, W. A., 1951. Barrier island, not "offshore bar," *Science* **113**, 487-488.
Shepard, F. P., 1972. *Submarine Geology.* New York: Harper and Row, 517p.
Zenkovich, V. P., 1967. *Processes of Coastal Development.* New York: Wiley Interscience, 738p.

Cross-references: *Barrier Flats; Barrier Island Coasts; Barrier Islands; Barrier Islands, Transgressive and Regressive; Beach.*

BARRIER FLATS

Barrier flats are often lobate or fan-shaped, marshy and/or sandy plains, situated slightly above sea level on the mainland-facing side of certain *barrier islands* and *barrier spits*, lagoonward of the islands' and spits' central dune or beach ridge complex. A multitude of round and oval lakes and interconnecting tidal channels dot several Texas barrier island flats (Matagorda Island, for instance,) with arcuate spits along the lagoonal shores. Sandy deposits dominate in flats, strongly affected by eolian and/or storm washover sediment supply from the open marine shores. The sand ranges from structureless to well laminated. Elsewhere the mud content of the sediments is higher, reaching 20% in central Texas barrier island flats (Shepard, 1960; Weise and White, 1980). Plant remains are common. In areas with extensive evaporation of the interstitial water, calcareous aggregates and veinlets and even *gypsum* accumulate. The Laguna Madre Flats, landward of Padre Island of southern Texas (Fisk, 1959; Weise and White, 1980), represent mostly a feature in transition toward real barrier flats, caused by large-scale infilling of an arid-zone lagoon. Barrier flats are absent from lagoonal-sound shores with higher energy conditions and/or with limited sediment supply.

ERVIN G. OTVOS

References

Fisk, H. N., 1959. Padre Island and the Laguna Madre Flats, coastal south Texas, *Second Coastal Geography Conference, Washington, D. C.,* 103–152.

Shepard, F. P., 1960. Gulf coast barriers, *in* F. P. Shepard et al., eds., *Recent Sediments, Northwest Gulf of Mexico.* Tulsa, Okla.: Am. Assoc. Petroleum Geologists Bull., 197–220.

Weise, B. R., and White, W. A., 1980. *Padre Island National Seashore, Guidebook No. 17.* Austin: Texas Bureau of Economic Geology, 94p.

Cross-references: *Barrier Island Coasts; Barrier Islands; Cuspate Spits; Estuarine Sedimentation; Evaporites; Lagoonal Sedimentation; Washover and Washover Fans.*

BARRIER ISLAND COASTS

Barriers are accumulations of sand or shingle (water-worn pebbles and cobbles known as gravel) that stand permanently above high-tide level along coasts. According to their physiographic features, barriers can be termed *barrier beaches, barrier spits,* or *barrier islands.* There is no general agreement about the evolution of barrier island coasts. Various theories have been advanced since the early work of Elie de Beaumont in 1845 by, for example, Gilbert, Shepard, Curray, Leont'yev and Nikiforov, Hoyt, and Hails, Cooke, Fisher, Swift, and Otvos (see references in King, 1972; Schwartz, 1973). Barrier islands are believed to develop during either emergence or submergence or when sea level is relatively stable, provided that there is an adequate supply of material to be transported from offshore or alongshore or from a combination of sources and that the coast has a low offshore gradient.

Although these are distributed widely throughout the world, the most continuous and best-developed barrier islands are in lower latitudes, and in zones of low to moderate tidal range (see map in King, 1972, based on the work of Gierloff-Emden).

Agreement on the origin and development of barrier islands appears to be complicated by conflicting conclusions on Quaternary (Pleistocene and Holocene) changes of sea level. D. W. Johnson's theory of barrier island development from offshore bars and the view of Leont'yev and Nikiforov that barriers have formed during a general lowering of sea level in late Holocene time do not adequately account for the universal distribution of Pleistocene and Holocene barriers along coastal plains. The majority of the world's barrier island coastlines are submerged, even those that are tectonically active or recovering isostatically during the postglacial period. If barrier islands form from offshore bars, various stages of their evolution should be seen along modern shorelines. However, these have been reported only as short-lived features. Further, open ocean beaches and neritic sediments (found in water along continental shelves at depths ranging from that of low tide to about 200 meters) should have developed landward of the bars. Recent sedimentological studies of barrier islands and lagoons in various regions of the world indicate that open marine conditions did not prevail in the lagoons during the development of major barrier island chains.

JOHN R. HAILS

References

King, C. A. M., 1972. *Beaches and Coasts.* London: Edward Arnold, 570p.

Leatherman, S. P., ed., 1979. *Barrier Islands from the Gulf of St. Lawrence to the Gulf of Mexico.* New York: Academic Press, 336p.

Schwartz, M. L., ed., 1973. *Barrier Islands.* Stroudsburg, Pa.: Dowden, Hutchinson & Ross, 451p.

Cross-references: *Barrier Beaches; Barrier Flats; Barrier Islands; Emergence and Emerged Shoreline; Sea-Level Curves; Submergence and Submerged Shoreline.* Vol. III: *Barriers—Beaches and Islands.*

BARRIER ISLANDS

Although barrier islands are the most common coastal landform in the United States, they make up only 10–15% of the world's shorelines. Different theories on the origin of barrier islands have been a common topic of discussion for more than 100 years. Barrier island origin, development, geomorphology, sedimentation, and stratigraphy have been the subject of special publications (c.f., Schwartz, 1973) and geomorphology symposiums (c.f., Coates, 1973;

Leatherman, 1979a) in recent years. Several important economic aspects of barrier islands are the reason for this widespread and long-term interest. Subsurface former barrier islands are often important petroleum reservoirs, while present-day surface barrier island shorelines are among the most important recreational and residential coastlines. In recent years man's impact on these fragile barrier island shorelines has led to the need for studies of barrier origin and development for efficient coastal management.

Barrier Terminology and Features

Barrier islands proper are one of a series of barrier type coastal landforms, some of which are often spatially and, therefore probably, genetically related (Fig. 1). They include the following specific forms (Johnson, 1919; Shepard, 1952, 1960): *barrier*, a general term for an offshore elongated ridge usually parallelling the shore, composed of sand to cobble-size material, rising above high tide level, separating an open water body (usually marine) from an enclosed or partially enclosed water body (usually brackish) referred to as a lagoon, often with one or more inlets connecting this lagoon to the open water body; *barrier beach*, a barrier composed of a single ridge sometimes with a single line of low foredunes (in part, *offshore bar* of Johnson, 1919, and *bay barriers* of others); *barrier island*, a complex barrier, often composed of multiple sand ridges, foredunes, dunefields, washover fans, and tidal marshes (*offshore bar* of Johnson, 1919); *barrier spit*, a barrier connected to the mainland at one end, but inlet openings may transform the barrier spit to a barrier island (a barrier spit differs from the ordinary spit in that it parallels the coast and encloses a lagoonal water body); *barrier chain*, a series of interconnected barrier spits, barrier islands, and barrier beaches, often in sequence from the mainland and extending along a considerable length of coast (Fig. 2); *barrier complex*, barrier sedimentary environments consisting of a shoreface, beach, dunes, sand flat or barrier flat, and a lagoon beach (Curray, 1969); and *bay barrier*, a barrier attached at both ends to the mainland, limited by some to such barriers within the lagoon's mainland shoreline, but for a bay on any shoreline it is the *baymouth bar* of Johnson (1919) and *baymouth barrier* of others.

Earlier descriptive terms for barrier islands have included *offshore bars, barrier bars, sand reefs, sand bars,* and *sand banks,* but these are no longer in active use (Price, 1951). A *barrier reef,* which may be linear and enclose a lagoon, is a biogenic coralline form always below high tide level. Louis Agassiz, the naturalist, erroneously suggested (1869) that the barrier islands of the southern United States developed on earlier coral reefs.

Certain other shore forms are often associated with barrier islands and sometimes are even mistaken for them in the geologic record. A *longshore bar* is a linear sand or gravel ridge seaward

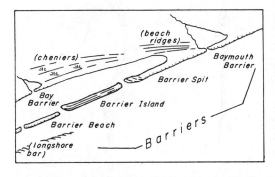

FIGURE 1. Diagram illustrating different barrier island types including: barrier, barrier beach, barrier island, barrier spit, barrier chain, bay barrier, and baymouth barrier. Features similar to barriers such as longshore bars, beach ridges, and cheniers are also diagramed.

FIGURE 2. Outer Banks barrier chain of North Carolina's Cape Hatteras coast, illustrating various barrier features: barrier spit, barrier island, barrier beach, lagoon, and inlet (Photo: NASA).

and parallel to a beach and submerged by high tide. Although smaller than barrier forms, it is considered by some as a precursor to a barrier island (*ball* of Johnson, 1919; *offshore bar* or *barrier bar* of others). A *beach ridge* is an elongated low sand or gravel ridge paralleling the shoreline and usually, occurring in multiple sets on both barrier islands and coastal plains. A *chenier* is a similar elongated low sand or shell ridge usually occurring only in a coastal plain marsh environment.

Distribution of Barrier Islands

Extensive barrier chains occur along the African, Australian, and South American coasts, the east Indian, Ceylon, and Arctic north slope coasts, as well as along the Baltic, Mediterranean, Black, and Caspian Seas. Barriers are associated with the Rhone and Nile river deltas, extend along the North Sea coast of the Netherlands, Germany, and Denmark (c.f., *Wadden*), and protect the historic city of Venice from the open Adriatic Sea. The most extensive barrier chain coastlines are along the Atlantic and Gulf coasts of the United States, where they extend along 5000 km of shoreline.

Although barriers are worldwide in distribution, King (1972) pointed out that the most continuous and well-developed barrier islands are in the lower latitudes, have low to intermediate tidal ranges, and occur off sandy coastal plains. However, in terms of plate tectonics, barrier island coastlines are most common on the trailing edge of a migrating continental plate (Inman and Nordstrom, 1971). This type of plate boundary is usually nonmountainous with wide coastal plains and continental shelfs and is generally nontectonic, although it exhibits Quarternary sea level changes.

Along the United States Atlantic and Gulf barrier coasts all of the features *commonly* referred to as *barrier islands* are not morphogenically similar and differ in several respects (Fisher, 1968). For example, barrier island chain shorelines along the Middle Atlantic coast of Long Island, New Jersey, Delaware, Maryland, Virginia, and the North Carolina *Outer Banks* (Fig. 3A) are long, linear, and associated with major embayments such as Delaware and Chesapeake bays; the *Sea Island* barrier of Georgia and South Carolina (Fig. 3B) are shorter, more equidimensional in shape, with numerous tidal inlets. In contrast, along the Florida east coast (Fig. 3C) the barriers are again long and linear, with extensive linear lagoons. Along the Gulf Coast of Texas and Mexico (Fig. 3D) the barrier islands are long and linear, with few tidal inlets, and are associated with the deltas of major rivers such as the Mississippi and Rio

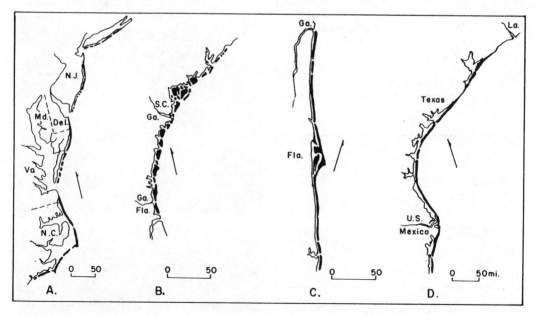

FIGURE 3. Different barrier island coastlines of the United States: A. Middle Atlantic states, long linear barriers with large bays; B. South Carolina and Georgia Sea Islands, shorter barriers with numerous inlets; C. Florida east coast, long linear barriers with linear lagoons; and D. Texas and Mexico coast, long linear barriers with few inlets but with river deltas (Fisher, 1968).

Grande. Smaller barrier island chains are found in the United States along the Massachusetts, Rhode Island, Louisiana, Alabama, and west Florida coasts, in Canada along the Prince Edward Island coast, and along the Gulf and Yucatán Peninsula coasts of Mexico.

Origin of Barrier Islands

Mechanisms for the origin and development of barrier islands have been discussed by many researchers for more than 100 years (c.f., Schwartz, 1973), producing conflicting theories. This suggests that there is probably more than one type of barrier island (as indicated by the previous discussion of Atlantic and Gulf barriers) with different barriers undergoing different types of development because of differences in controlling mechanisms such as: regional coastal topography, source, type, and rate of sediment supply, rate and direction of sea change, tidal range, meteorological conditions, climatic conditions, and stage of coastal development. However, most theories suggest a controlling process that appears to dominate in forming different types of barriers in different regions. Figure 4 outlines the primary mechanism for some of these different hypotheses.

Initially, based on European barriers, Elie de Beaumont (1845) suggested origin by wave action depositing sediment offshore to form an emerging barrier that would be in equilibrium with this wave action. Later relict shorelines of Lake Bonneville, Utah, indicated to Gilbert (1885, 1890) that spitlike longshore deposition in the breaker zone on a gentle offshore slope could build a barrier above water level. Sea level changes were introduced with Merrill (1890) suggesting that elevating the sea bottom brought submarine bars above sea level to form a barrier, while McGee (1890) claimed that coastal subsidence of *shore ridges* formed the Gulf of Mexico barriers (called *keys* by McGee). Thus by the end of the nineteenth century there were the following four theories for barrier island origin: formation from offshore deposition (Elie de Beaumont's *emergent bar*); formation by longshore deposition (Gilbert's *longshore spit*); formation by coastal uplift (Merrill's *emergent offshore bar*); formation by coastal subsidence (McGee's *submergent shore ridges*). Throughout the twentieth century each of these concepts would be applied to explain barrier island origin.

Johnson's (1919) worldwide analysis of barrier island offshore profiles suggested that Elie de Beaumont's concept was correct and that wave energy on an emerging shoreline would build a barrier upward to maintain an equilibrium profile (Fig. 4). This theory was accepted for decades, with barrier islands considered as features of a *shoreline of emergence* developed by a falling sea. However, Evans (1942) observed that bars in Michigan lakes were limited by wave action from building above the water's surface, as suggested earlier by Johnson, and Evans felt that perhaps barriers developed as spits. Similarly, along the English East Anglia coast Steers (1946) showed that several shingle barrier islands developed as spits, although recently Steers (1969) has suggested an initial offshore bar forming the spit. In contrast, the South Carolina and Georgia Sea Island barriers (Zeigler, 1959) were of three types: eroded remnants of a coastal plain cuesta, longshore deposition as beach ridges from these remnants, and tidal marsh islands. For those barriers formed by a combination of longshore deposition at both ends of the remnant islands, Hayes (1976) has used the descriptive term *drumstick barriers*. During this time stratigraphic studies on Galveston Island on the Texas coast (LeBlanc and Hodgson, 1959) indicated that these barrier chains developed by seaward beach ridge progradation during sea level stillstand (Fig. 5) about 5000 years B.P. with sediment sources offshore from the shelf and longshore from the Rio Grande, Brazos, and Colorado rivers. Shepard (1960) classified Gulf of Mexico barriers into four possible *genetic* types: long, straight, or curved barriers (e.g., Texas coast); segmented island chains (e.g., Mississippi-Alabama coast); cuspate islands and spit (Florida panhandle coast); and lobate or crescentric islands (Florida west coast). He felt that all of these barriers developed with a rising sea level and that thus Johnson's earlier theory was not valid. Later Kwon's geomorphic study (1969) indicated that these gulf barriers (Fig. 5) developed in response to sediment sources such as deltas (e.g., *interdeltaic Texas barrier arc* or Chandeleur and Grand Isle barriers of the Mississippi delta) and downdrift from mainland coasts (e.g., Bolivar, Matagorda, and Mississippi Sound islands). Weide (1968) could therefore suggest the following factors as *optimal* for gulf barrier development: low relief of coastal plain and continental shelf; slow regional subsidence; sediment sources, especially deltas; and longshore currents to form barriers from spits and increase their size.

In Europe during the 1960s several barrier concepts appeared. Relict barrier islands inland from the Netherlands coast were found to have developed on a rising sea level (van Straaten, 1965) with an offshore source for the lower barrier deposits but a longshore source from the Rhine delta for the upper deposits. In the Soviet Union transgressive Black Sea subsurface off-

HYPOTHESIS	MAJOR PROPONENT	COASTAL REGION	PRIMARY MECHANISM	SEDIMENT SOURCE
EMERGENT BAR	DeBeaumont (1845)	Northern Europe	Bar emergence/ onshore sediment movement	Seaward of the breaker zone sediments
SHORELINE SPIT	Gilbert (1890)	Induced from Utah & Michigan Lakes	Longshore drift causing spit extention	Cliff erosion/ updrift from fluvial sediments
EMERGENT OFFSHORE BAR	Johnson (1919)	Atlantic	Shoreline emergence onshore sediment movement	Offshore sediments
SUBMERGENT BAR	Shepard (1960)	Gulf Coast	Submerging shoreline	Offshore and fluvial deposits
RIDGE EMBAYMENT	McIntire and Morgan (1963)	Massachusetts	Holocene transgression/longshore drift	Existing Pleistocene/ updrift sediments
RIDGE EMBAYMENT	Hoyt (1967,1968)	Georgia	Holocene Transgression	Existing Pleistocene sediments
COMPLEX SPIT	Fisher (1968)	Middle Atlantic	Longshore drift/ wave refraction	Updrift sediments
EMERGENT SHOAL	Otvos (1970)	Gulf of Mexico	Shoal emergence/ longshore drift	Updrift sediments
RIDGE EMBAYMENT SPIT GENERATION	Pierce and Colquhoun (1970)	North Carolina	Either Holocene transgression or spit development	Existing Pleistocene sediments
COMPLEX SPIT/ DUNAL MIGRATION	Jones (1977)	Massachusetts	Longshore drift/ high energy storm conditions	Updrift/offshore sediments

FIGURE 4. Diagram illustrating different hypotheses for barrier island development, including major proponent and date theory proposed, coastal region where field evidence exists for that theory, primary mechanism for origin with necessary sea level changes and dominant littoral processes, and sediment source for barrier origin (Jones, 1977).

shore barrier deposits indicated to Zenkovich (1962), however, only an offshore rather than a longshore sediment source. Leont'yev and Nikiforov (1965), studying the same coast, suggested that higher sea level (Holocene Flandrian transgression) submarine offshore bars became sub-aereal barrier islands when sea level dropped after that transgression. Evidence for this Holocene higher sea level stand, however, is not always present worldwide, and Zenkovich later (1969) suggested that regardless of sea level changes barriers will form whenever there is a gently offshore slope. However, in the United States in the 1960s discussions between several researchers of Atlantic coast barriers (Hoyt, Fisher, Cooke, and others) revived inter-

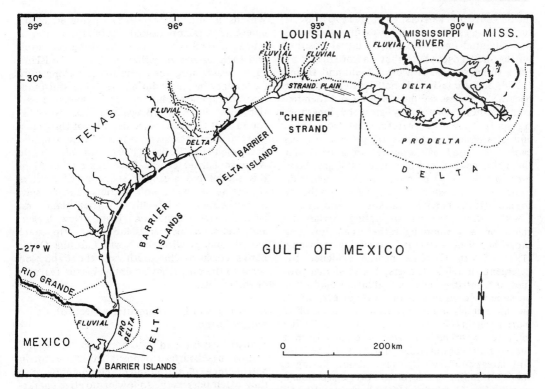

FIGURE 5. Diagram of different barrier islands of Texas, Louisiana, and Mississippi Gulf coast illustrating: long, straight or curved barriers of the Texas coast relative to former and present fluvial deltas (interdeltaic barrier arc of Kwon, 1969); the chenier strand plain of the Louisiana coast; and delta-margin barriers of the Mississippi River delta coast (after A. J. Scott and W. L. Fisher, 1969).

est in theories for barrier origin. McIntire and Morgan's Plum Island, Massachusetts, stratigraphic study (1963) indicated that the island originated offshore, migrated landward as sea level rose, and grew by longshore transport as the rate of sea level rise decreased, although Jones and Cameron (1977) later showed that landward dune migration has also occurred during the present stable sea level (Fig. 4). Hoyt (1967), citing the absence of open marine sediments behind the Georgia, Texas, and Netherlands barriers, suggested that barriers did not develop according to Johnson's theory but by submergence of mainland coastal ridges (Fig. 4), similar in concept to McGee's 1890 theory. However, Cooke (1968), discussing Hoyt's theory, claimed that the North Carolina, Georgia, and Texas barriers originated initially as offshore shoals and migrated landward, while Fisher (1968) claimed that morphology and dating of beach ridges indicated that the Atlantic and Texas barriers developed as complex segmented spits (Fig. 4) in a manner after Gilbert's 1885 theory. In reply Hoyt (1968a,b) pointed out that both Cooke's and Fisher's theories still required the presence of back-barrier marine sediments, which were absent. However, Otvos (1970) claimed that open marine sediments might be absent landward of barriers forming along the Mississippi Sound from emerging shoals (Fig. 4) because of low sedimentation rates and brackish rather than marine salinities. Thus Otvos required terrestrial sediments, which are missing, below a barrier to accept Hoyt's theory, while Hoyt required marine sediments landward of a barrier, also missing, to accept either Fisher's or Otvos' theory. Both Hoyt and Otvos agreed that landward migration or seaward progradation of a barrier could eliminate the true subsurface record of either origin. This is a likely possibility since a slowly rising sea allows barriers to migrate continuously landward (*roll-over* of Dillon, 1970), while rapidly rising sea level causes barriers to *jump* intermittently landward (Sanders and Kumar, 1975). In addition, inlet migration on present barriers also erodes the earlier subsurface deposits that might explain the barrier's origin (Susman and Heron, 1979). Investigation of the Atlantic continental shelf for evidence of these earlier offshore barriers indicated only extensive sand sheets overlying lagoonal sediments but no barriers (Swift, 1975). Swift further suggested that the present-day barriers had migrated onshore from an initial position on the shelf edge where they had originated by ridge submergence. However, Field and Duane (1976) felt these initial shelf barriers originate because of excess sediment supply and then migrate onshore where they grow by longshore transport. They claim therefore that geomorphic and stratigraphic features of present barriers cannot be used as evidence for a particular theory since barriers originate offshore; however, Otvos (1977) pointed out that when such evidence does exist in support of a certain theory for a particular barrier it cannot be dismissed.

With various theories suggesting barriers develop on a submerging rather than emerging shoreline, Fisher (1973), reexamining Johnson's (1919) North Carolina profile evidence for emergent shoreline barriers, found similar profiles on nonbarrier coasts, a situation that is not possible if Johnson's theory was correct. Stratigraphic studies along this same North Carolina barrier chain (Pierce and Colquhoun, 1970) indicated a *primary barrier* developed from a submergent coastal ridge during sea level rise and later modified by wave action during a stable or falling sea level to form a *secondary barrier* (Fig. 4). Cooke (1971), however, claimed that this North Carolina primary barrier was actually a much earlier Pleistocene coastal terrace, while Hails (1971) questioned the subsurface paleosol evidence for this primary barrier. Reviewing this work and other recent theories, Schwartz (1971) proposed the following barrier island classification suggesting multiple modes of origin:

I. Primary barriers
 1. Engulfed beach ridges (c.f. Hoyt)
II. Secondary barriers
 1. Breached spits (c.f. Fisher)
 2. Emergent offshore bars (c.f. Otvos)
  a. Sea level rise (c.f. Zenkovich)
  b. Sea level fall (c.f. Leont'yev)
III. Composite barriers—any combinations of above (c.f. Pierce and Colquhoun)

Applications of Schwartz's classification may be seen in Orme's (1973) South African study where the present (secondary?) barriers are superimposed by littoral drift on fragments of an earlier Pleistocene (primary?) barrier; while Bird's (1973) Australian study indicates present-day longshore developing barriers (secondary?) incorporating earlier consolidated calcareous barriers (primary?). Similarly, along the Texas coast stratigraphic studies indicate that Matagorda Island and Peninsula (Wilkinson, 1975; Wilkinson and Basse, 1978) originated as an emerging offshore shoal (primary barrier?), migrated landward during the rising sea level, and then, as shown earlier for Galveston Island to the east, prograded seaward and alongshore (secondary barrier?) during the recent stillstand.

Stratigraphic studies along the Delmarva barrier coast by Kraft, Biggs, and Halsey (1973) found that the subsurface deposits of this transgressive barrier system fitted in part a model developed by Fisher (1967) for all Middle Atlantic barrier coasts. Each coast (Fig. 6) in this model has a similar sequence of barriers consisting of baymouth barriers, barrier spit, barrier island, and finally a barrier island and barrier beach chain. Long Island, New Jersey, and Delmarva barrier island coasts are exact model analogs with the North Carolina Outer Banks coast to the south and the Cape Cod coast to the north, having similar barrier features of the model.

Economic and Environmental Aspects of Barrier Islands

Identification and interpretation of barrier islands in subsurface deposits are important in petroleum exploration and development because they often possess the following positive characteristics (Weide, 1968): ideal reservoir properties with clean, well-sorted sands, ideal source-bed relationship with marine shales down dip, and excellent trap geometry with adjacent and overlying finer sediments. These characteristics will vary significantly if the barrier forms on a transgressive or regressive shoreline; hence sea level position is of importance in discussions on theories of barrier origin. Because barrier island shorelines—such as Atlantic City, Miami Beach, and the Lido, Venice—are among the world's most important recreational beaches, the economic aspects of careful coastal planning and management are significant. For example, man-made tall barrier frontal dunes built to protect the highways and commercial development along Cape Hatteras, North Carolina, barrier island actually concentrate wave action and increase shoreline erosion (Dolan, 1972); while on Core Banks, a barrier beach to the south, where there are only low natural frontal dunes, dune overwash is allowing upward growth of the island as sea level rises (Godfrey and Godfrey, 1973). On barrier islands where shoreline erosion is rapid, this overwash is more common than on less erosive shores (Leatherman, 1979b). On the worldwide scale, all barrier beaches are undergoing erosion and the reason may relate to sea level changes associated with their origin. If barriers form as shoals at higher sea level stands (Leont'yev and Nikiforov, 1965), with a subsequent sea level fall, than on the present rising

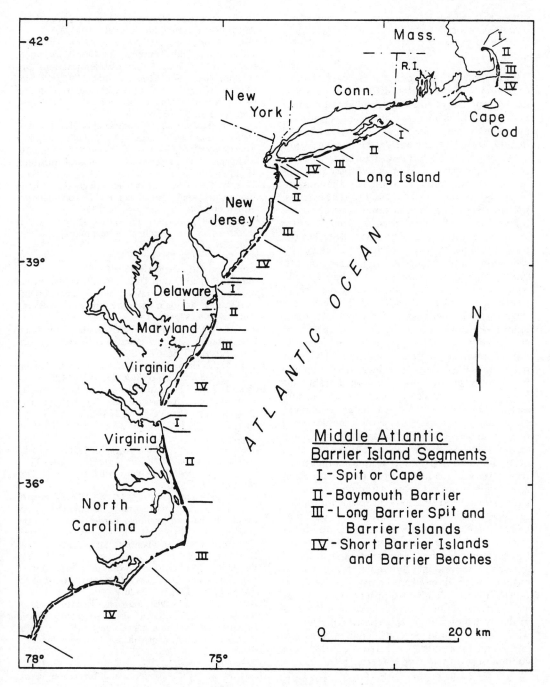

FIGURE 6. Middle Atlantic barrier island coastlines, showing similar sequence of barriers for each coast consisting of: I. spit or cape, II. baymouth barriers along mainland coast, III. barrier spit and long linear barrier island with few inlets, and IV. short barrier island and barrier beach chain with many inlets. Adjacent North Carolina Outer Banks coast and Cape Cod coast show similar features (after J. Fisher in Swift, 1969).

sea level, these barriers are not in hydrodynamic balance (*profile of equilibrium*) with the present coast and are eroding. However, barrier island shoreline erosion is only part of a worldwide trend of erosion of all shorelines. Other causes not directly related to barrier origin include equilibrium readjustment to sea level rise, increased storm activity, manmade shore construction activities, and cessation of onshore sediment movement (Fisher, 1977). Some of these causes may also relate to specific barrier island shoreline erosion.

Summary

Barrier islands are common features of most coastlines, but they often differ significantly in size, morphology, and stratigraphy in a manner that suggests differences in types of littoral processes, magnitude of these processes, and the stage of development of the barrier, if not in its origin itself. Various theories for barrier island origin have been proposed based on different geologic interpretations of the morphology and stratigraphy of these different barriers, similar barriers, or, even in some cases, the same barrier islands. Differing theories usually center on interpretations of evidence as to whether the littoral sediment movement forming the barrier was dominantly alongshore, onshore, or offshore; whether the source of the barrier sediments was from alongshore, offshore, relict terrestrial, or relict submarine; and whether the stage of sea level necessary for origin was rising, falling, or stable. For each of these factors or combination of factors, it seems that a barrier can be found to fit those conditions. Thus barrier islands have been considered to originate from: emergent offshore bars on a stable sea level, emergent shoals on a falling sea level, submergent bars on a rising sea level, submergent dune ridges on a rising sea level, complex spits on a rising sea level, emergent shoals on a rising sea level, migration onshore on a rising sea level, or a combination of these. It appears that most recent theories see most barrier islands as originating on a rising sea level. Although other barriers probably do originate on stable and falling sea levels, these would exhibit a different morphology. In a similar manner, the dominant littoral process is a continuous spectrum for the different barriers consisting exclusively of longshore movement dynamics for some to onshore wave dynamics for others. Classification of barrier islands into primary and secondary barriers is an improvement, but there is probably a continuous spectrum of barriers of which these should be considered only as the initial and final stages for barriers developed in conjunction with a rising sea level.

JOHN J. FISHER

References

Agassiz, L., 1869. On the relation of geological and zoological researches to general interests, in the development of coast features, *U.S. Coast Survey Report for 1867*, 183-186.

Beaumont, L. E. de, 1845. Septième lecon, *in Lecons de Géologie Practique*. Paris: P. Bertrand, 221-252.

Bird, E. C. F., 1973. Australian coastal barriers, *in* M. L. Schwartz, ed., *Barrier Islands*. Stroudsburg, Pa.: Dowden, Hutchinson & Ross, 410-426.

Coates, D. R., ed., 1973. *Coastal Geomorphology*. Binghamton: State University of New York, 404p.

Cooke, C. W., 1968. Barrier island formation: Discussion, *Geol. Soc. America Bull.*, **79**, 945-946.

Cooke, C. W., 1971. Holocene evolution of a portion of the North Carolina coast: Discussion, *Geol. Soc. America Bull.* **82**, 2369-2370.

Curray, J. R., 1969. Shore zone sand bodies: Barriers, cheniers, and beach ridges, *in* D. J. Stanley, ed., *The New Concepts of Continental Margin Sedimentation*. Washington, D.C.: American Geological Institute, JC-11-1-JC-11-16.

Dillon, W. P., 1970. Submergence effects on a Rhode Island barrier and lagoon and inferences on migration of barriers, *Jour. Geology* **78**, 94-106.

Dolan, R., 1972. Barrier dune system along the Outer Banks of North Carolina: a reappraisal, *Science* **176**, 286-288.

Evans, O. F., 1942. The origin of spits, bars, and related structures, *Jour. Geology* **50**, 846-863.

Field, M. E., and Duane, D. B., 1976. Post-Pleistocene history of the United States' inner continental shelf: Significance to origin of barrier islands, *Geol. Soc. America Bull.* **87**, 691-702.

Fisher, J. J., 1967. Origin of barrier island chain shorelines, Middle Atlantic states, *Geol. Soc. America Spec. Paper 115*, 66-67.

Fisher, J. J., 1968. Barrier island formation: Discussion, *Geol. Soc. America Bull.* **79**, 1421-1426.

Fisher, J. J., 1973. Barrier island formation, *in* M. L. Schwartz, ed., *Barrier Islands*. Stroudsburg, Pa.: Dowden, Hutchinson & Ross, 278-282.

Fisher, J. J., 1977. Coastal erosion revealed, *Geotimes* **22**, 24-25.

Gilbert, G. K., 1885. The topographic features of lake shores, *U.S. Geol. Survey 5th Ann. Rept.*, 69-123.

Gilbert, G.K., 1890. Lake Bonneville, *U.S. Geol. Survey Mono. 1*, 23-65.

Godfrey, P. J., and Godfrey, M. M., 1973. Comparison of ecological and geomorphic interactions between altered and unaltered barrier island systems in North Carolina, *in* D. R. Coates, ed., *Coastal Geomorphology*, Binghamton: State University of New York, 239-258.

Hails, J. R., 1971. Holocene evolution of a portion of the North Carolina coast: Discussion, *Geol. Soc. America Bull.* **82**, 3525-3526.

Hayes, M. O., 1976. Beaches-barrier islands, *in* M. O. Hayes and T. W. Kana, eds., *Terrigenous Clastic Depositional Environments*. Columbia, S.C.: Coastal Research Division, Department of Geography, University of South Carolina Tech. Rept. 11-CRD, I-79-I-108.

Hoyt, J. H., 1967. Barrier island formation, *Geol. Soc. America Bull.* **78**, 1125-1136.

Hoyt, J. H., 1968a. Barrier island formation, Reply, *Geol. Soc. America Bull.* **79**, 947.

Hoyt, J. H., 1968b. Barrier island formation, Reply, *Geol. Soc. America Bull.* **79**, 1427-1432.

Inman, D. L., and Nordstrom, C. E., 1971. On the tectonic and morphologic classification of coasts, *Jour. Geology* **79**, 1-21.

Johnson, D. W., 1919, *Shore Processes and Shoreline Development*. New York: John Wiley & Sons, 584p.

Jones, J. R., 1977. *Alternative Hypothesis for Barrier*

Island Migration, Boston University, Ph.D. dissertation, 210p.
Jones, J. R., and Cameron, B., 1977. Landward migration of barrier island sands under stable sea level conditions: Plum Island, Massachusetts, *Jour. Sed. Petrology* 47, 1475-1483.
King, C. A. M., 1972. *Beaches and Coasts*. New York: St. Martin's Press, 570p.
Kraft, J. C.; Biggs, R. B.; and Halsey, S. D., 1973. Morphology and vertical sedimentary sequence models in Holocene transgressive barrier systems, *in* D. R. Coates, ed., *Coastal Geomorphology*. Binghamton: State University of New York, 321-354.
Kwon, H. J., 1969. *Barrier Islands of the Northern Gulf of Mexico Coast, Sediment Source and Development*. Baton Rouge: Louisiana State University Press, Coastal Studies Series No. 25., 51p.
Leatherman, S. P., ed., 1979a. *Barrier Islands from the Gulf of St. Lawrence to the Gulf of Mexico*. New York: Academic Press, 325p.
Leatherman, S. P., 1979b. Migration of Assateague Island, Maryland, by inlet and overwash processes, *Geology* 7, 104-107.
LeBlanc, R. J., Sr., and Hodgson, W. D., 1959. Origin and development of the Texas shoreline, *Gulf Coast. Assoc. Geol. Socs. Trans.* 9, 197-220.
Leont'yev, O. K., and Nikiforov, L. G., 1965. Reasons for the world-wide occurrence of barrier beaches, *Oceanology* 5, 61-67.
McGee, W. J., 1890. Encroachment of the sea, *Forum* 9, 437-449.
McIntire, W. G., and Morgan, J. P., 1963. *Recent Geomorphic History of Plum Island, Massachusetts, and Adjacent Coasts*. Baton Rouge: Louisiana State University Press, Coastal Studies Series No. 8, 44p.
Merrill, F. J. H., 1890. Barrier beaches of the Atlantic coast, *Pop. Sci. Monthly* 37, 744.
Orme, A. R., 1973. Barrier and lagoonal systems along the Zululand coast, South Africa, *in* D. R. Coates, ed., *Coastal Geomorphology*. Binghamton: State University of New York, 181-217.
Otvos, E. G., 1970. Development and migration of barrier islands, northern Gulf of Mexico, *Geol. Soc. America Bull.* 81, 241-246.
Otvos, E. G., 1977. Post-Pleistocene history of the United States inner continental shelf: Significance to origin of barrier islands: Discussion, *Geol. Soc. America Bull.* 88, 691-702.
Pierce, J. W., and Colquhoun, D. J., 1970. Holocene evolution of a portion of the North Carolina coast, *Geol. Soc. America Bull.* 81, 3697-3714.
Price, W. A., 1951. Barrier island, not "offshore bar," *Science* 113, 487-488.
Sanders, J. E., and Kumar, N., 1975. Evidence of shoreface retreat and in-place "drowning" during Holocene submergence of barriers, shelf off Fire Island, New York, *Geol. Soc. America Bull.* 86, 65-76.
Schwartz, M. L., 1971. The multiple causality of barrier islands, *Jour. Geology* 79, 91-94.
Schwartz, M. L., ed., 1973. *Barrier Islands*. Stroudsburg, Pa.: Dowden, Hutchinson & Ross, 451p.
Scott, A. J., and Fisher, W. L., 1969. Delta systems and deltaic deposition, *in* W. L. Fisher, L. F. Brown, Jr., A. J. Scott, and J. H. McGowen, eds., *Delta Systems in the Exploration for Oil and Gas*. Austin: Texas Bureau of Economic Geology, 10-29.

Shepard, F. P., 1952. Revised nomenclature for depositional coastal features, *Am. Assoc. Petroleum Geologists Bull.* 36, 1902-1912.
Shepard, F. P., 1960. Gulf Coast barriers, *in* F. P. Shepard et al., eds., *Recent Marine Sediments, Northwest Gulf of Mexico*. Tulsa, Okla.: American Association of Petroleum Geologists, 338-344.
Steers, J. A., 1946, *The Coastline of England and Wales*. Cambridge: Cambridge University Press, 644p.
Steers, J. A., 1969. *The Coastline of England and Wales* (reprinted with additions). Cambridge: Cambridge University Press, 762p.
Susman, K. P., and Heron, S. D., Jr., 1979. Evolution of a barrier island, Shakleford Banks, Carteret County, North Carolina, *Geol. Soc. America Bull.* 90, 205-215.
Swift, D. J. P., 1969. Inner shelf sedimentation: Process and products, *in* D. J. Stanley, ed., *The New Concepts of Continental Margin Sedimentation*. Washington, D.C.: American Geological Institute, 4-1-4-46.
Swift, D. J. P., 1975. Barrier island genesis: Evidence from the central Atlantic shelf, eastern U.S.A., *Sed. Geology* 14, 1-43.
van Straaten, L. M. J. U., 1965. Coastal barrier deposits in south and north Holland, *Geol. Stichting Med.* N.S., 17, 41-87.
Weide, A. E., 1968. Bar and barrier island sands, *Gulf Coast Assoc. Geol. Socs. Trans.* 18, 405-415.
Wilkinson, B. H., 1975. Matagorda Island, Texas: The evolution of a Gulf coast barrier complex, *Geol. Soc. America Bull.* 86, 959-967.
Wilkinson, B. H., and Basse, R. Z., 1978. Late Holocene history of the central Texas coast from Galveston Island to Pass Cavallo, *Geol. Soc. America Bull.* 89, 1592-1600.
Zeigler, J. M., 1959. Origin of the sea islands of the southeastern United States, *Geograph. Rev.* 49, 222-237.
Zenkovich, V. P., 1962. Some new exploration results about sand shores development during the sea transgression, *De Ingenieur, No. 17, Bouw en Waterbouwkunde* 9, 113-121.
Zenkovich, V. P., 1969. Origin of barrier beaches and lagoon coasts, *in Lagunas Costeras, Un Simposio,* Mem. Simp. Internat. Lagunas Costeras, UNAM-UNESCO, 27-37.

Cross-references: *Barrier Beaches; Barrier Flats; Barrier Island Coasts; Barrier Islands, Transgressive and Regressive; Bars; Beach Processes; Coastal Morphology, History of; Coastal Morphology, Research Methods; Geographic Terminology; Global Tectonics; Major Beach Features; Sea Level Curves; Sediment Transport*. Vol. III: *Barriers-Beaches and Islands*.

BARRIER ISLANDS, TRANSGRESSIVE AND REGRESSIVE

Intensive discussions of *barrier islands* are presented elsewhere in this encyclopedia. However, it is obvious to many that barrier islands are considered to be of relatively simple structure and, more important, of single origin. This is definitely not the case. Over one-half of the world's barrier island coasts are transgressive with the sea moving landward, whereas the remainder are the so-called *classical* barrier islands so familiar to all of us with their repetitive lines of beach ridges. Only a study of the subsurface stratigraphy of barrier islands and their internal structure can determine whether a barrier island is transgressive or regressive. Short-term map studies will not solve the problem, because local conditions fluctuate and frequently reverse, sometimes for great lengths of time. Figure 1 shows two simple schematic cross sections illustrating transgressive and regressive concepts of barrier islands. These are from the work of the author (Kraft, 1971) on the Atlantic coast and from Bernard and LeBlanc (1965) on the Gulf of Mexico shoreline in Texas.

Walther's Law of Sedimentary Facies applies. Walther stated: "The various deposits of the same facies-areas and similarly the same of the rocks of different facies-areas are formed beside each other in space, though in cross-section we see them lying on top of each other. As with biotopes, it is a basic statement of far reaching significance that only those facies and facies-areas can be superimposed primarily which can be observed beside each other at the present time." (Middleton, 1973.) Thus the present geographic distribution of adjacent coastal sedimentary environments should be repeated in the vertical sequence of sedimentary deposits that remain as coastal environments migrate. Coastal sedimentary environmental lithosomes, the sedimentary deposit seen in three dimensions, and in their sequence of origin vertically, will immediately inform the geologist whether a barrier is transgressing or regressing. Obviously a barrier system located over a lagoon, over a basal salt marsh, overriding a highland is *transgressing;* whereas a nearly identical surficial geology might lead to a stratigraphic sequence that vertically includes the younger barrier on top of older and seaward barrier elements ultimately overlying shallow, nearshore marine coastal sediments, which obviously indicates a *regression* of the sea. Ultimately in a regression the barrier system may be overridden by coastal lagoonal and fluvial sediments. Preservation of the vertical stratigraphic record of the sedimentary environmental lithosomes is a happenstance of the tectonics of the region and eustasy combined into resultant relative sea level change. The ancient geological stratigraphic record contains many examples of both transgressive and regressive barrier islands. However, to date there has been a tendency of geologists to identify nearly all barrier islands in the ancient record as being regressive. Much restudy of vertical sedimentary sequences in the light of Walther's Law of Sedimentary Facies is needed in both modern and ancient sediments.

JOHN C. KRAFT

References

Bernard, H. A., and LeBlanc, R. J., Sr., 1965. Résumé of Quaternary geology of the Northwestern Gulf of Mexico province, *in* H. E. Wright, Jr., and D. G. Frey, eds., *The Quarternary of the United States*, VII Congr. Internat. Assoc. Quaternary Research. Princeton: Princeton University Press, 137-185.

Bernard, H. A.; Major, C. F., Jr.; Parrott, B. S.; and LeBlanc, R. J., Sr., 1970. *Recent Sediments of Southwest Texas: A Field Guide to the Brazos Alluvial and Delta Plains in the Galveston Barrier Island Complex*. Austin, Texas: Bureau of Economic Geology, 138p.

Kraft, J. C., 1971. Sedimentary facies patterns and geologic history of a Holocene marine transgression, *Geol. Soc. America Bull.* 82, 2131-2158.

Kraft, J. C., 1977. Late Quaternary paleogeographic

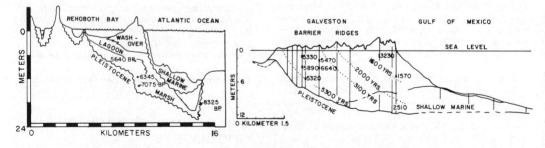

FIGURE 1. A schematic presentation of two cross sections of a transgressive and regressive barrier island. Note that the vertical sequence of sedimentary environmental lithosomes in motion landward or seaward, according to Walther's Law, clearly define the nature of a transgressive or regressive barrier island. In the transgressive case, the barrier island is overriding landward environments. In the regressive case, the barrier island is overlying more marine environments and a tendency may develop for lagoonal and fluvial environments to override the barrier (after Kraft, 1971, 1977; Bernard and LeBlanc, 1965; Bernard et al., 1970).

changes in the coastal environments of Delaware, Middle Atlantic Bight, related to archaelogical settings, *N.Y. Acad. Sci. Annals* **288**, 35-69.

Middleton, G. V., 1973. Johannes Walther's Law of the Correlation of Facies, *Geol. Soc. America Bull.* **84**, 979-988.

Cross-references: *Barrier Beaches; Barrier Flats; Barrier Island Coasts; Barrier Islands; Beach Stratigraphy; Coastal Sedimentary Facies; Nearshore Hydrodynamics and Sedimentation.* Vol. III: *Barriers— Beaches and Islands.*

BARS

Sedimentary ridges, generally larger than bedforms, that characterize the beach face and nearshore zones dominated by wave activity are termed *wave-formed bars*. These were recognized as early as 1845 on the marine coasts of Europe (Elie de Beaumont), by 1851 in the Great Lakes of North America (Desor) and subsequently on marine and lacustrine coasts worldwide (see extensive references in Zenkovich, 1967; King, 1972; Greenwood and Davidson-Arnott, 1975). Confusion still surrounds the term *bar*, however, because of its use for ridges with a wide range of size, morphology, location, and orientation relative to the shoreline and that occur in a wide variety of environments, from subaerial to those dominated by tidal currents or river currents (see papers in Schwartz, 1972). Further, the origins and dynamics of *wave-formed bars* are still poorly defined.

Shepard (1950) called shore-parallel ridges and troughs *longshore bars* and *troughs*, equating them with the terms *ball* and *low* of Evans (1940), and clearly showed their relationship to plunging wave activity. He also emphasized the seasonality of such bars on the west coast of the United States, and subsequently terms such as *winter* and *summer, storm* and *normal,* and *storm* and *swell* have been applied to beach profiles to denote the presence or absence of bars. Although a correlation between profile type and storm waves or season may exist in some localities, it is not universal; both *barred* and *nonbarred* profiles occur at all times in some areas, while in others only one type may persist throughout the year. Bars therefore represent *the equilibrium morphology for many coastal environments, controlled by the incident wave field, secondary waves, and currents modulated by the local slope, grain size, and tidal regime.*

Morphology and Sedimentology

Bars are most clearly identified as symmetrical or asymmetrical undulations in the beach profile (Fig. 1) and may range in numbers from one to over thirty. They form continuous or compartmented linear, sinuous, or crescentic patterns in plan view, and range from shore parallel to shore normal in orientation, often producing periodic or rhythmic topography alongshore. The vast majority of bars are made of sand-sized materials, although they have been recorded in small gravels. Bar sediments are generally better sorted than those of the associated trough, but may be finer or coarser depending upon whether the latter are erosional lags or suspension deposits of quiescent periods (Greenwood and Davidson-Arnott, 1972).

The structural characteristics of wave-formed bars have been described from a number of regions by Davis et al., Davidson-Arnott and Greenwood, Howard and Reineck, Reineck and Singh, and others; summaries of this work are found in Elliot (1978), Davidson-Arnott and Greenwood (1976), and Reineck and Singh (1973). Bars are aggradational features, and contain primary sedimentary structures reflecting bedforms generated by symmetric and asymmetric oscillatory currents associated with wave shoaling and breaking and also unidirectional longshore and rip currents. Stratification ranges from small-scale chevron or trough cross-lamination, to medium-scale trough cross-bedding, to low-angle parallel bedding. Cross-stratification dips are predominantly landward, but parallel lamination may dip both into and away from the direction of wave approach according to the position on the seaward (lakeward) or landward slopes. In highly mobile bars medium- to large-scale sets of tabular cross-bedding are produced by avalanching as the bar front migrates. Trough sediments contain parallel bedding and small to medium-scale trough cross-bedding: dips are highly varied as both waves and longshore currents are active in this zone. In rip channels and on the crests of some bars seaward (lakeward)- dipping medium-scale units of tabular and trough cross-bedding occur in association with flows away from the surf

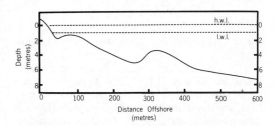

FIGURE 1. Typical barred profile: example from a microtidal, moderate wave energy environment in the southern Gulf of St. Lawrence, Canada, illustrating two bars.

zone. Bioturbation is controlled largely by the degree to which bar sediments are continuously reworked by the hydraulic regime and, where more than one bar is present, facies are repeated across the shoreface (Davidson-Arnott and Greenwood, 1976).

Classification

A classification of *wave-formed bars* based on morphology and the environmental constraints on bar formation is illustrated in Table 1. The group names are those in common use, and the definitive paper describing each type is referenced in the source paper.

Classic *ridge and runnel* topography is found on low-angle foreshore slopes dominated by surf action and foreshore drainage associated with tidal fluctuations in macrotidal environments. Although low in amplitude these bars are usually stable in form and position. In contrast, the *cusp or bar-type sand waves* are often destroyed by storm waves and regenerated by the following low, long period waves as a result of surf-swash action near the toe of the swash slope: further, they migrate relatively rapidly both alongshore and onshore, and in the latter case may *weld* to the foreshore. The *multiple parallel* and *transverse bars* are both limited to low-angle slopes and low to moderate wave energy conditions coupled with limited water level shifts. The height and spacing of the multiple bars are nearly uniform in the offshore direction, and bar form is near symmetrical in contrast to the *nearshore type II* group. The division of *nearshore bars* into two groups is based upon size, stability, and the controlling wave form. *Type I bars* are associated with high plunging breakers, which produce narrow, low amplitude ridges on relatively steep slopes: they lack a well-defined asymmetry and are essentially unstable modifications of nonbarred nearshore profiles. *Type II bars*, in contrast, are large, stable configurations formed seaward of the low water level; distance offshore, depth of water over the crest, and bar height all increase offshore in a regular manner.

Bar Genesis

The conditions necessary for bar formation have been studied theoretically, experimentally, and in the field; reviews of this work are provided by Komar (1976) and Greenwood and Davidson-Arnott (1979). In general, barred profiles are associated with high values of wave steepness and wave height to grain size ratios, but it is apparent that the way in which wave energy is either reflected or dissipated is the important control on nearshore morphology (Wright, et al., 1979). Although cause and effect are far from clear, it is evident that equilibrium bar profiles can exist only where the time-averaged sediment transport is zero everywhere on that profile.

A number of mechanisms have been proposed for sediment transport leading to bar formation. Among the most important are: the seaward transport of sediment by vortices under plunging breakers (Miller, 1976; Fig. 2a); alternating scour and deposition by mass transport velocities near the bed generated by standing waves, resulting from the interaction of reflected and incident waves (Carter, Liu, and Mei, 1973; Fig. 2b) or from reflection in the infragravity range (Suhayda, 1974); sediment movement to null positions in the drift velocity field of standing edge waves, which are periodic both alongshore and offshore (Bowen and Inman, 1971; Fig. 2c); and alongshore and offshore sediment movement under meandering or cellular nearshore circulations produced by the instability of longshore flows (Hino, 1974; Fig. 2d).

The vortex mechanism is most applicable to *nearshore type I bars* of the American west coast and justifies the early correlations of bar formation with wave steepness, but it does not explain the many barred coasts dominated by spilling breakers. Mass transport velocities under reflected standing waves would explain the formation of *multiple parallel bars,* but simple reflection of the incident waves could not be the cause since the length scales of the bars would require much longer periods. The theoretical convergence patterns of drift velocities under standing edge waves provide strong support for their role in forming crescentic *nearshore type II bars;* it is possible that progressive edge waves generate linear bars of the same group. Further, the edge wave periods necessary to produce the length scales found in nature is of the same order as the well-known *surf beat*. However, the generation of these trapped modes of oscillation still remains ill defined, even though field observations of low-frequency peaks in the nearshore energy spectrum have been made on barred coasts and related to the presence of edge waves (Huntley, 1980). Barred topography has long been associated with the occurrence of longshore and rip current circulations (Shepard, Emery, and LaFond, 1941), and the work of Hino (1974) proposes instability of the fluid sediment interface as a means of generating variations in sediment transport resulting in sinuous or crescentic undulations of the surf-zone bed. Certainly the role of rip-cell circulation in bar dynamics has been well documented for *bar-type* and *cusp-type sand waves* (Bowen and Inman, 1969; Davis and Fox, 1972; Greenwood and Davidson-Arnott,

TABLE 1. Bar Classification (after Greenwood and Davidson-Arnott, 1979)

Name	MORPHOLOGY						ENVIRONMENTAL CONSTRAINTS				
	Definitive Description	Size	Planform	Profile	Number Seaward	Location	Wave Energy	Wave Processes	Tidal Range	Nearshore Slope	Intertidal Slope
Ridge and runnel	King and Williams (1949)	$h \sim 0.2\text{-}1.5$ m $l \sim 10^3$ m	Straight, shore parallel	Asymmetric landward	1-4	Intertidal	Low to moderate	Surf-swash, beach drainage	Ma-Me	–	<0.01
Cusp-or bar-type sand wave	Sonu (1968) Davis and Fox (1972)	$h \sim 0.2\text{-}1.5$ m $l \sim 10^2$ m	Straight to spit shaped, shore parallel	Asymmetric landward	1-2	Intertidal and low tide terrace	Moderate	Breakers, surf-swash	Me-Mi	–	>0.01
Multiple parallel	Zenkovich (1967)	$h \sim 0.2\text{-}0.75$ m $l \sim 10^3$ m	Straight to sinuous, shore parallel	Near symmetric	4-30 or more	Nearshore and intertidal	Low to moderate	Spilling breakers, surf	Mi-Me	<0.01	–
Transverse	Niederoda and Tanner	$h \sim 0.2\text{-}0.75$ m $l \sim 10^2$ m	Straight, shore normal	Symmetric and asymmetric landward	1	Nearshore and intertidal	Low	Surf-swash, spilling breakers	Mi	<0.01	–
Nearshore I	Shepard (1950)	$h \sim 0.25\text{-}1.0$ m $l \sim 10^2$ m(?)	Straight, shore parallel	Asymmetric landward	1-2	Nearshore	Moderate to high	Plunging breakers	Mi-Me	<0.1	–
Nearshore II	Evans (1940) Hom-ma and Sonu	$h \sim 0.25\text{-}3.0$ m $l \sim 10^3$ m	Straight, sinuous to crescentic, shore parallel	Asymmetric landward	1-4	Nearshore	Moderate	Spilling breakers	Mi	<0.01	–

h = bar height; l = bar length; Ma = macro; Me = Meso; Mi = Micro

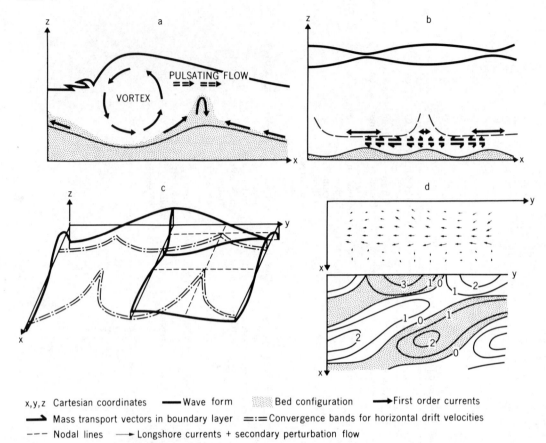

x,y,z Cartesian coordinates — Wave form ░ Bed configuration ➤ First order currents
➥ Mass transport vectors in boundary layer ═╪═ Convergence bands for horizontal drift velocities
--- Nodal lines ⟶ Longshore currents + secondary perturbation flow

FIGURE 2. Velocity fields and the driving mechanisms proposed for bar formation: a, vortex generation by plunging breakers; b, mass transport in the boundary layer of a strongly reflected wave; c, drift velocities generated by standing edge waves; and d, longshore current and secondary perturbation flow resulting from hydrodynamic instability.

1975; Sonu 1973), for *transverse bars* (Niedoroda, 1973) and for *nearshore type II bars*, both crescentic and straight (Greenwood and Davidson-Arnott, 1979; Greenwood and Mittler, 1979). However, it is also possible that the regularity in nearshore circulations is controlled by the presence of edge waves.

BRIAN GREENWOOD

References

Bowen, A. J., and Inman, D. L., 1969. Rip currents, 2: Laboratory and field observations, *Jour. Geophys. Res.* 23, 5479-5490.

Bowen, A. J., and Inman, D. L., 1971. Edge waves and crescentic bars, *Jour. Geophys. Res.* 76, 8662-8671.

Carter, T. G.; and Liu, P. L. F.; and Mie, C. C., 1973. Mass transport by waves and offshore sand bedforms, *Am. Soc. Civil Engineers Proc., Jour. Waterways, Harbors and Coastal Eng. Div.* 99, 165-184.

Davidson-Arnott, R. G. D., and Greenwood, B., 1976. Facies relationships on a barred coast, Kouchibouguac Bay, New Brunswick, Canada, *in* R. A. Davis, Jr., and R. L. Ethington, eds., *Beach and Nearshore Sedimentation.* Tulsa, Okla.: Soc. Econ. Paleont. Miner. Spec. Pub. 24, 140-168.

Davis, R. A., Jr., and Fox W. T., 1972. Coastal processes and nearshore sand bars, *Jour. Sed. Petrology* 42, 401-412.

Elliott, T., 1978. Clastic shorelines, *in* H. G. Reading, ed., *Sedimentary Environments and Facies.* New York: Elsevier, 143-177.

Evans, O. F., 1940. The low and ball of the east shore of Lake Michigan. *Jour. Geology* 48, 467-511.

Greenwood, B., and Davidson-Arnott, R. G. D., 1972. Textural variation in the sub-environments of the shallow-water wave zone, Kouchibouguac Bay, New Brunswick, Canada, *Jour. Earth Sci.* 9, 679-688.

Greenwood, B., and Davidson-Arnott, R. G. D., 1975. Marine bars and nearshore sedimentary processes, Kouchibouguac Bay, New Brunswick, Canada, *in* J. Hails and A. Carr, eds., *Nearshore Sediment Dynamics and Sedimentation.* London: John Wiley, 123-150.

Greenwood, B., and Davidson-Arnott, R. G. D., 1979.

Sedimentation and equilibrium in wave-formed bars: A review and case study, *Canadian Jour. Earth Sci.* 16, 312-332.
Greenwood, B., and Mittler, P. R., 1979. Structural indices of sediment transport in a straight, wave-formed, nearshore bar, *Marine Geology* 32, 191-203.
Hino, M., 1974. Theory on formation of rip-current and cuspidal coast, *Coastal Eng. Japan* 17, 23-38.
Huntley, D. A., 1980. Edge waves in a crescentic bar system, *in* S. B. McCann, ed., *Coastline of Canada: Littoral Processes and Shore Morphology.* Geological Survey of Canada, Paper 80-10, Report 2, 111-121.
King, C. A. M., 1972. *Beaches and Coasts.* London: Edward Arnold, 470p.
Komar, P. D., 1976. *Beach Processes and Sedimentation.* Englewood Cliffs, N.J.: Prentice-Hall, 429p.
Miller, R. L., 1976. Role of vortices in surf zone prediction: Sedimentation and wave forces, *in* R. A. Davis, Jr., and R. L. Ethington, eds., *Beach and Nearshore Sedimentation.* Tulsa, Okla.: Soc. Econ. Paleont. Miner. Spec. Pub. 24, 92-114.
Niedoroda, A. W., 1973. Sand bars along low energy beaches Part 2: Transverse bars, *in* D. R. Coates, ed., *Coastal Geomorphology.* Binghamton: State University of New York, 103-113.
Reineck, H.-E., and Singh, I. B., 1973. *Depositional Sedimentary Environments.* New York: Springer-Verlag, 439p.
Schwartz, M. L., ed., 1972. *Spits and Bars.* Stroudsburg, Pa.: Dowden, Hutchinson, & Ross, 452p.
Shepard, F. P., 1950. *Longshore Bars and Longshore Troughs.* Beach Erosion Board, Tech. Memo 15, 32p.
Shepard, F. P.; Emery, K. O.; and LaFond, E. C., 1941. Rip currents: a process of geological importance, *Jour. Geology* 49, 337-369.
Sonu, C. J., 1973. Three dimensional beach changes, *Jour. Geology* 81, 42-64.
Suhayda, J. N., 1974. Standing waves on beaches. *Jour. Geophys. Res.* 79, 3065-3071.
Wright, L. D.; Chappell, J.; Thom, B. G.; Bradshaw, M. P.; and Cowell, P., 1979. Morphodynamics of reflective and dissipative beach and inshore systems, southeastern Australia, *Marine Geology,* 32, 105-140.
Zenkovich, V. P., 1967. *Processes of Coastal Development.* London: Oliver and Boyd, 738p.

Cross-references: *Beach; Beach Processes; Beach Stratigraphy; Major Beach Features; Minor Beach Features; Ripple Marks.* Vol. I: *Bars.*

BASE-LINE STUDIES

Base-line studies, also called *impact studies,* are surveys designed to provide ecological information about an area that may be used for monitoring man-induced changes in the environment.

There are two basic variants, *Before and after* studies, in which the *before* study is the base-line study and is used as a reference or norm with which to compare the *after* study—after a coastal development program has been completed, for example; when environmental impact has already begun or is suspected, in which case the base-line information must be taken from an equivalent but unaffected site nearby.

In several parts of the world, notably in the United States, base-line studies are often required by law when coastal development threatens the marine environment. Such legislation is far from universal, however, and is often lacking where it is probably most needed, as in the countries bordering the Mediterranean, for example.

A typical approach in a base-line survey might be to lay a series of transect lines from which samples are collected at regular intervals throughout the period of study. The samples will include comprehensive collections of the flora and fauna. These are identified and counted, and from the data obtained various indexes, such as diversity indexes, can be formulated that are felt to describe in general terms the condition and health of that region. In addition to the general biological survey, selected organisms, sediment cores, or samples of water are collected for the purpose of pollutant analysis. Heavy metal concentrations, pesticides, nutrient- and sewage-related compounds, and bacteria levels are determined, so that any subsequent changes in these may be detected.

In general, information is collected on those physical and biological aspects of the area that may be changed by man's subsequent activity. Many typical examples of base-line studies can be found in Ruivo (1972), in which the parameters analyzed or investigated by the authors generally fall into three categories: *physical parameters* including radiation monitoring, water clarity, suspended solids, temperature, and direction and velocity of currents; *chemical parameters* which include pH, dissolved gasses (especially oxygen), biochemical oxygen demand, nutrient elements such as nitrogen and phosphorous, heavy metals, trace toxic organic substances, and oil fractions; and *biological parameters,* often the largest group, including the collection, identification, and interpretation of the biota. Within this last group interpretation of the data has often proved to be the most contentious aspect of base-line studies or of any environmental monitoring. The numerous approaches used very often rely on certain statistical properties of populations and assume that the communities have a known theoretical basis, which is not always a valid assumption. Several examples are provided by Gray (1976), who argues against the usefulness of many base-line studies. The natural variability of marine environments has always made ecological interpretation difficult.

Many of the approaches to interpretation are contained in Olson and Burgess (1967). The problems with base-line studies lie not in obtaining information but in attributing measured changes to pollution outfalls and other industrial activity, in predicting changes in the environment, and in formulating policy for coastal development and allowable discharge levels.

CHARLES R. C. SHEPPARD

References

Gray, J., 1976. Are marine base-line surveys worthwhile? *New Scientist* 70, 219-221.

Olson, T. A., and Burgess, F. J., eds., 1967. *Pollution and Marine Ecology*. New York: Wiley Interscience, 364p.

Ruivo, M., ed., 1972. *Marine Pollution and Sea Life*. West Byfleet: Fishing News, Ltd., 624p.

Cross-references: *Aquaculture; Coastal Ecology, Research Methods; Coastal Fauna; Coastal Flora; Human Impact; Nutrients; Oil Spills and Pollution; Pollutants; Waste Disposal* Vol. XIII: *Environmental Impact Statements*.

BEACH

Although the term *beach* is in everyone's vocabulary, there is little consistency among definitions that are commonly used or presented in textbooks (Bascom, 1980; Davis and Ethington, 1976; Ingle, 1966; Komar, 1976; U.S. Army Corps of Engineers, 1973). For purposes of discussion, the best working definition of a beach is the accumulation of unconsolidated sediment that is limited by low tide on the seaward margin and by the limit of storm wave action on the landward side. In the case of the landward limit, there is also typically an abrupt change in slope as a new geomorphic feature is encountered such as a dune, bluff, or even a man-made structure. This landward limit is rather consistently used as a boundary throughout the literature, but a wide range exists in designation of the seaward limit. Some authors include the surf zone or the bar and trough topography; at least one extends the beach to the zone of significant bottom sediment movement by waves (King, 1972). Although processes and topography across this broad zone are closely related, the completely subtidal area (i.e., below low tide) is best separated from the beach proper.

Beaches are found throughout the world where there is sufficient material available and where the proper setting exists for the accumulation of that material. Some beaches are limited in extent and are termed *pocket beaches,* whereas others stretch for hundreds of kilometers such as those along the Carolina Outer Banks or the Texas gulf coast. Beaches are usually comprised of well-sorted, medium-grained sand primarily of quartz with some feldspar, carbonate (shell material) and traces of accessory minerals. That may describe many—perhaps even most—beaches, but there is a great variety in terms of both grain size and mineral composition. Generally however, beaches are characterized by good sorting regardless of grain size and somewhat less typically by well-rounded grains. The composition of a beach is most often a reflection of the sediment that is available in the vicinity and shows infinite mineralogic variety. It may come from bedrock or unconsolidated sediments, from longshore transport from up or down coast, from offshore, or from biogenic sources, such as shells or coral fragments.

The geometry of the *beach profile* tends to be similar throughout the world within some predictable variations. The beach can be subdivided into two major parts, the *foreshore* and the *backshore* (Fig. 1). The foreshore is the seaward portion and is typically narrow with respect to the backshore. It is generally inclined seaward at a slope of 2-10°, the slope being controlled largely by grain size (i.e., coarse sediment slopes are steeper than fine ones). The foreshore is sometimes equated with the *swash* zone and encompasses the normal intertidal part of the beach. There may be a minor topographic *step* near the low water mark that is the *plunge step* (also called *foreshore step* and *beach scarp*). On gently sloping, fine-sand beaches this position

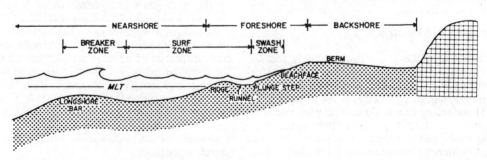

FIGURE 1. Generalized profile of beach and adjacent environments showing primary subdivisions and zones.

may be marked only by a linear concentration of shells and/or gravel. The foreshore may contain *swash marks, air holes, current crescents,* or *rill marks.*

The backshore zone is commonly much broader than the foreshore and slopes slightly landward (Fig. 1). The change in slope between the two major zones is abrupt and is commonly called the *berm crest.* The *berm* is a nearly flat, terracelike area that is the seaward portion of the backshore. Some beaches may have multiple berms. In normal conditions the backshore is subaerial. It is only in times of high wave energy that this portion of the beach is awash. As a normally subaerial environment wind may exert an important influence on the backshore by transporting sediment, generally landward, and thereby nourishing the *foredunes.Wind shadow dunes, sand shadows, ripples,* and burrows are common on the backshore (Davis, 1978).

The beach is one of the most dynamic of all coastal environments and is constantly experiencing change, except in high latitudes during times of shore ice. In normal conditions the changes are modest and the general profile configuration remains as shown in Figure 1. When storms and their associated high wave energy conditions attack the beach, there is generally much erosion. The result is a *storm beach* that displays a distinctly different profile. The removal of much sediment results in a more extensive foreshore, and in extreme cases the foreshore may constitute the entire beach.

Although scientists have long visited and described beaches as well as collected their sediment, we are just now beginning to understand the complex process and response mechanisms which are operating in this dynamic environment.

RICHARD A. DAVIS, Jr.

References

Bascom, W., 1980. *Waves and Beaches.* New York: Doubleday & Company, 366p.

Davis, R. A., Jr., and Ethington, R. L., eds., 1976. *Beach and Nearshore Sedimentation.* Tulsa, Okla.: Soc. Econ. Paleont. Miner. Spec. Pub. 24, 187p.

Davis, R. A., Jr., 1978. Beach and nearshore zone, *in* R. A. Davis, ed., *Coastal Sedimentary Environments.* New York: Springer-Verlag, 237-285.

Ingle, J. C., Jr., 1966. *The Movement of Beach Sand.* New York: Elsevier, 221p.

King, C. A. M., 1972. *Beaches and Coasts.* London: Edward Arnold, 570p.

Komar, P. D., 1976. *Beach Processes and Sedimentation.* Englewood Cliffs, N.J.: Prentice-Hall, 429p.

Meyer, R. E., ed., 1972. *Waves on Beaches.* New York: Academic Press, 462p.

U.S. Army Corps of Engineers, 1973. *Shore Protection Manual,* vol. 1. Ft. Belvoir, Va.: Coastal Engineering Research Center.

Cross-references: *Beach Cusps; Beach Cycles; Beach Firmness; Beach Material; Beach Material, Sorting of; Beach Nourishment; Beach Orientation; Beach Pads; Beach in Plan View; Beach Processes; Beach Profiles; Beach Ridge Plain; Beach Ridges and Beach Ridge Coasts; Sediment Transport.*

BEACHCOMBING

Beachcombing, the act of searching along the shore for objects of value, is a favorite activity of coastal visitors, especially after storms drive floating debris ashore.

The eroded wintertime beach is a particularly good place to find fine specimens of shell and stone. The high water and onshore winds of storms in any season can bring treasures of fishing floats, netting, bottles, shipwreck debris, or driftwood. *Ambergris,* once among the most valued of beached treasures, has reportedly become nearly worthless. Originating in the intestines of sperm whales, it has been largely replaced by synthetic ingredients in the manufacture of perfumes.

One treasure that still brings searchers to the beach in all seasons is the kind buried by pirates, or its modern equivalent, the coins and jewelry dropped by summer tourists. Modern beachcombers utilize sophisticated electronic equipment in their search for metallic riches.

Beachcombing was the main source of income on the Outer Banks of North Carolina during the late eighteenth century, because of the numerous shipwrecks along Cape Hatteras beaches.

A common meaning of beachcombing, *to loaf at the seacoast* (referring especially to Caucasian males in the South Pacific), probably derives from the efforts of *beachcombers* to rake together enough income to finance their loafing— not unlike the contributors to this volume, who spend a great deal of their time at beaches.

JAMES E. STEMBRIDGE, Jr.

Cross-references: *Coastal Reserves; Driftwood; Flotsam and Jetsam; Thalassotherapy; Tourism.*

BEACH CONCENTRATES—See MINERAL DEPOSITS

BEACH CUSPS

Beach cusps are rhythmically spaced, crescentic beach forms that consist of cusp-shaped ridges or mounds, often referred to as *horns,* separated by concave troughs or *bays.* They occur in the foreshore zone of a beach and are most commonly found at or near the high tide mark. Beach cusps seem to form best on beaches consisting of a mixture of sand and shingle, but have been seen on beaches that range in sediment size from fine sand to shingle (Evans, 1938; King, 1972). It has also been noted that the cusp horns generally consist of coarser material than the bays. Beach cusps are called cusplets if the spacing between the tips of the horns is less than 1.5 m or cusps if the spacing is between 1.5 m and 25 m (Dolan and Ferm, 1968).

The origin of beach cusps is still not well understood. Several theories have been proposed to explain their origin; some of these stress an erosional origin. Evans (1938) attributed the origin of beach cusps to erosion of beach ridges, while Johnson (1919) attributed their origin to erosion of beach depressions. Other theories stress an erosional-depositional origin. Kuenen (1948) held that wave swash moves erosively up the center of the embayments, and backwash returns, depositing material along the sides of the horns. On the other hand, Bagnold (1940) and Russell and McIntire (1965) stated that swash converges on the horns and divides along each side depositing coarser sediment, and the backwash flows erosively through the embayments.

There are also theories that attempt to explain the origin and rhythmic character of beach cusps by wave action. Dalrymple and Lanan (1976) describe possible beach cusp formation by *intersecting waves* of the same wavelength. Bowen and Inman (1971), Huntley and Bowen (1975), Guza and Inman (1975), and Komar (1973) consider that *edge waves* form beach cusps. Gorycki (1973) attributes the origin of beach cusps to crescent-shaped *sheetflood wave swash*. Cloud (1966) attributes their origin to the rhythmic collapse of the breaking wave itself.

Topographic fluctuations in the foreshore zone bed have also been considered to form rhythmic beach cusps (Dubois, 1978). The nature of these fluctuations has been discussed by Bruun (1954), Schwartz (1967), Zenkovich (1967) and Sonu (1968). They believe that shore drift can move in the form of *sand waves*. Schwartz (1972) writes that beach cusps are the shoreline extensions of topographic fluctuations in the littoral zone shore drift.

JOHN SPASARI

References

Bagnold, R. A., 1940. Beach formation by waves: Some model-experiments in a wave tank, *Inst. Civil Engineers Jour.* **15**, 27-52.

Bowen, A. J., and Inman, D. L., 1971. Edge waves and crescentic bars, *Jour. Geophys. Research* **76** (36), 8662-8671.

Bruun, P. M., 1954. Migrating sand waves or sand humps, with special reference to investigations carried out on the Danish North Sea coast, *Conf. Coastal Eng., 5th, Grenoble, 1954, Proc.*, 269-295.

Cloud, P. E., Jr., 1966. Beach cusps: Response to Plateau's rule? *Science* **154**, 890-891.

Dalrymple, R. A., and Lanan, G. A., 1976. Beach cusps formed by intersecting waves, *Geol. Soc. America Bull.* **87**, 57-60.

Dolan, R., and Ferm, J. C., 1968. Crescentic landforms along the Atlantic coast of the United States, *Science* **159**, 627-629.

Dubois, R. N., 1978. Beach topography and beach cusps, *Geol. Soc. America Bull.* **89**, 1133-1139.

Evans, O. F., 1938. The classification and origin of beach cusps, *Jour. Geology* **46**, 615-627.

Gorycki, M. A., 1973. Sheetflood structure: Mechanism for beach cusp formation, *Jour. Geology* **81**, 109-117.

Guza, R. T., and Inman, D. L., 1975. Edge waves and beach cusps, *Jour. Geophys. Research* **80**, 2997-3012.

Huntley, D. A., and Bowen, A. J., 1975. Field observations of edge waves and their effect on beach material. *Geol. Soc. London Jour.* **131**, 69-81.

Johnson, D. W., 1919. *Shore Processes and Shoreline Development.* New York: John Wiley & Sons, 584p.

King, C. A. M., 1972. *Beaches and Coasts.* New York: St. Martin's Press, 570p.

Komar, P. D., 1973. Observations of beach cusps at Mono Lake, California, *Geol. Soc. America Bull.* **84**, 3593-3600.

Kuenen, Ph. H., 1948. The formation of beach cusps, *Jour. Geology* **56**, 34-40.

Russell, R. J., and McIntire, W. G., 1965. Beach cusps, *Geol. Soc. America Bull.* **76**, 307-320.

Schwartz, M. L., 1967. The Bruun theory of sea level rise as a cause of shore erosion, *Jour. Geology* **75**, 76-92.

Schwartz, M. L., 1972. Theoretical approach to the origin of beach cusps, *Geol. Soc. America Bull.* **83**, 1115-1116.

Sonu, C. J., 1968. Collective movement of sand in littoral environment, *Conf. Coastal Eng., 11th, London, 1968, Proc.*, 373-400.

FIGURE 1. Beach cusps in the process of development, Rosario Beach, Fidalgo Island, Washington (Photo: R. S. Babcock).

Zenkovich, V. P., 1967. *Processes of Coastal Development.* New York: Wiley Interscience, 739p.

Cross-references: *Beach Pads; Beach Processes; Coastal Morphology, Research Methods; Minor Beach Features; Sediment Transport.*

BEACH CYCLES

Beach sedimentation and *profiles* exhibit distinct patterns that are commonly referred to as *beach cycles*. Such cycles are developed around the erosion and deposition (accretion) that takes place at a given coastal location. There is considerable difference from place to place in both the scale of change incorporated in these cycles and also the rate at which these changes take place. Some of the fundamental variables that must be considered are orientation of coast relative to prevailing winds and passage of storm systems, nature and intensity of storms, sediment texture, wave climate, and nearshore and shoreface topography (Hayes, 1972).

At most locations there are cycles with quite different periods, one superimposed upon the other. The fundamental cycles include the storm and its associated erosion plus the post-storm recovery period and tidal cycles. These cycles may have a period of between one-half day and several weeks. The above cycles are superimposed upon one of considerably longer period, usually associated with seasons. The seasonal cycle is based primarily on annual variation in wave energy as the result of seasonal changes in weather patterns.

Storm Beach Cycle

Regardless of the location of the beach or the absolute intensity of a particular storm, the result is one of relatively high energy conditions imparted on the coast that cause significant removal of sediment from the beach. This typically takes place in only a day or so, although some slow-moving storms may exert their influence for a week. The range in the severity of the storm and its damage is almost infinite, from a typical midlatitude low-pressure system to a severe northeaster of the New England states or a hurricane in the Gulf of Mexico.

The result is a *storm beach profile* (Fig. 1) with its characteristic uniform slope and veneer of lag concentrates of heavy minerals. Near the low tide position a sand bar (ridge) is deposited. This poststorm profile has the same basic geometry regardless of location, but the scale may range markedly (Davis et al., 1972). During the low energy period following the storm, it is possible for the beach to recover much of the sediment lost by erosion during the storm. As

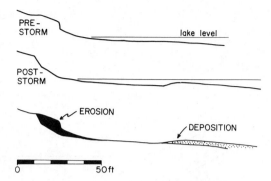

FIGURE 1. Pre- and post-storm profiles showing erosion of beach and foredunes and deposition of a bar or ridge.

the small waves pass over the ridge, a change in ridge configuration takes place, from a nearly symmetrical profile to one resembling a large fluvial bedform with a steep landward face and a gentle seaward slope (Fig. 2). Continued low energy conditions cause the landward migration of this ridge as small waves generate landward moving currents over the ridge. These currents cause sediment to move over the gently seaward slope and tumble down the steep landward slipface. Eventually the ridge migrates over the shallow trough (runnel) between it and the beach and is welded to the storm beach (Fig. 2).

This cycle takes place most rapidly along coasts where there is little or no tidal fluctuation in water level. In such circumstances the bar is almost continuously subjected to the above-mentioned processes. The greater the tidal range, however, the greater the period of time during which the ridge is exposed subaerially and thus not migrating landward. For comparison, on Lake Michigan this welding process occurs in 7-10 days (Fig. 2), whereas it takes several weeks on the Massachusetts coast where tides are 3 m (Davis, et al., 1972).

Tidal Cycles

The first of the tidal cycles affecting the beach profile is that of the semidiurnal or diurnal tide. During flood tide the beach profile is eroded down, and during the ebb tide the beach profile is built up again. Thus in a period somewhat in excess of 12 or 24 hours, respectively, the profile undergoes erosion and deposition of an envelope that averages 12 cm (Strahler, 1966; Schwartz, 1967, 1968). The second of the tidal cycles is related to the neap-spring cycle, when during a 14-day period spring tides cut down the slope by an average of 1 m and neap tides restore it by the same amount (Schwartz, 1968). In both flood tide and spring tide, wave height and reach are relatively greater, analagous to

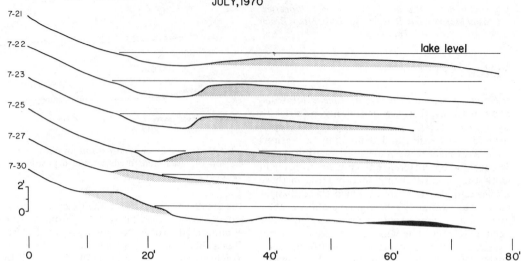

FIGURE 2. Sequence of profiles during recovery phase on a Lake Michigan beach.

the aforementioned storm conditions, whereas ebb tide and neap tide act to restore the beach in the manner of the *summer* or *accretion* beach phase. (The vertical dimensions cited here are averages of observations reported in the literature.)

Seasonal Cycles

The storm beach cycle is the most noticeable beach cycle because of the abrupt changes associated with it. Seasonal variations in the strength and occurrence of storms cause a long-term beach cycle. At many, perhaps most, coastal locations winter storms are more intense and more frequent than during the summer. This results in prolonged erosion during the winter and extended periods of accretion during the low energy summer months. Because of this relationship the term *winter beach* has been equated with *erosion* and *summer beach* with *accretion*—which is a great oversimplification for a variety of reasons. First of all, some regions experience less coastal energy during the winter than during the summer (e.g., Queensland, Australia). Further, it is not at all uncommon to have an accretional profile in winter and an erosional or storm profile during the summer. Typically storms are closely spaced during the winter and therefore do not generally provide the length of time necessary between storms for recovery to take place. Similarly, summer storms are generally more widely separated in time and less intense than in winter, thereby commonly providing time for recovery. The terms *winter beach* and *summer beach* can be misleading, and should be replaced with *storm beach* and *accretional beach*, respectively.

RICHARD A. DAVIS, Jr.
MAURICE L. SCHWARTZ

References

Davis, R. A., Jr.; Fox, W. T.; Hayes, M. O.; and Boothroyd, J. C., 1972. Comparison of ridge-and-runnel systems in tidal and non-tidal environments, *Jour. Sed. Petrology* 42, 413–421.

Hayes, M. O., 1972. Forms of sediment accumulation in the beach zone, *in* R. E. Meyer, *Waves on Beaches*. New York: Academic Press, 297–356.

Schwartz, M. L., 1967. Littoral zone tidal-cycle sedimentation, *Jour. Sed. Petrology* 37, 677–683.

Schwartz, M. L., 1968. The scale of shore erosion, *Jour. Geology* 76, 508–517.

Strahler, A. N., 1966. Tidal cycle of changes in an equilibrium beach, Sandy Hook, New Jersey, *Jour. Geology* 74, 247–268.

Cross-references: *Beach Processes; Beach Profiles; Coastal Morphology, Research Methods; Cut and Fill; Profiling of Beaches; Sweep Zone; Time Series.*

BEACH DRIFT—See SEDIMENT TRANSPORT

BEACH FIRMNESS

Different portions of a beach often exhibit varying degrees of firmness. The *foreshore,* that part of the beach extending from the low tide line shoreward to the highest part of the beach or berm crest, tends to be firm in its lower parts and soft on its upper margin. The *backshore*, that part of the beach shoreward of the foreshore, is generally softer than the lower foreshore but often firmer than the upper foreshore.

The difference between lower foreshore and backshore deposits can be explained by differences in moisture content (Kindle, 1936). Lower foreshore deposits are saturated with water during most stages of the tide, and sandy lower foreshores generally retain significant fluid at low tide when the deposit is exposed. Backshore deposits are less frequently subjected to wave swash and consequently tend to be dry. The capillary effect of water in the voids between grains on the exposed lower foreshore tends to act as a cohesive agent binding the grains together. Consequently lower foreshore deposits tend to be firmer than backshore deposits. The difference in firmness would, however, be less where the grain size of the beach sediment is large because the capillary effect of water is greatly diminished where the void spaces between grains are large.

Other factors may be of significance in explaining the difference in firmness of lower foreshore and backshore deposits. Trefethen (1941) reasoned that lower foreshore deposits are closely packed and firm because they are exposed to the constant dynamic activity of wave and swash forces. Backshore deposits are more exposed to processes such as the burrowing of organisms and transport by wind that cause a more open packing of grains.

The soft nature of some upper foreshore deposits is due to a different process according to Kindle, Emery, Harris, and Hoyt and Henry (Harris, 1974). Interstitial water tends to drain from the upper foreshore between high tides much more completely than from the lower foreshore. Consequently a flooding tide will tend to trap air in the upper foreshore deposit, and as the sediment approaches saturation the air will tend to collect in bubbles, forming vesicules in the sediment. This sponge-cake type of structure in the sand represents an open grain packing and consequently the sediment will be exceedingly soft. Vesicular sand, according to Emery (Harris, 1974), is common in upper foreshore fine to coarse sands that are relatively protected from heavy surf.

ASBURY H. SALLENGER, Jr.

References

Harris, C. R., 1974. A quantitative study of compaction with depth in a vesicular sand layer in an intertidal area of Dublin Bay, Ireland, *Jour. Sed. Petrology* 44, 1024-1028.

Kindle, E. M., 1936. Dominant factors in the formation of firm and soft sand beaches, *Jour. Sed. Petrology* 6, 16-22.

Trefethen, J. M., 1941. Dominant factors in the formation of firm and soft sand beaches–A discussion, *Jour. Sed. Petrology* 11, 42-43.

Cross-references: *Beach Material; Beach Material, Sorting of; Beach Processes; Beach Stratigraphy; Intertidal Sand Habitat; Organism-Sediment Relationship; Sand.* Vol. III: *Beach.*

BEACH MATERIAL

The size and character of beach material are directly related to the type of material available (from offshore, updrift, land, and so on) and also to forces to which the beach area is subjected (waves, winds). "Most beaches are composed of fine or coarse sand and in some areas of small stones called shingle or gravel" (Coastal Engineering Research Center, 1973:1-9). This material can be deposited by streams or it can be eroded material from adjacent parts of the coast, from onshore movement of material, and also from cliffs behind the beach. A broad classification of beaches was presented by Trask (1952), consisting of three divisions: sand, gravel, shingle, and cobble beaches; muddy, silty, or clayey beaches; and bedrock and reef beaches.

JOHN B. HERBICH
TOM WALTERS

References

Coastal Engineering Research Center, 1973. *Shore Protection Manual.* Washington, D.C.: U.S. Army Corps of Engineers, 750p.

Trask, P. D., 1952. Source of beach sand at Santa Barbara, California as indicated by mineral grain studies, *U.S. Army Corps of Engineers Beach Erosion Tech. Memo. No. 28*, 24p.

Cross-references: *Beach; Clay; Gravel; Pebble; Sand; Sediment Size Classification; Shingle; Silt.*

BEACH MATERIAL, SORTING OF

Sorting of sand and coarser material is a response to the prevailing physical beach processes. However, sorting of the silt and clay fractions is dominated more by chemical and biological factors (e.g., flocculation) rather than by any direct physical, process-response relationship. Even within the range of noncohesive sediments, sorting is not adequately understood and results for sand-size particles do not appear consistent with those for pebbles and cobbles. In part these difficulties may reflect problems in sampling and in analytical techniques.

Approaches have been via theory, laboratory modeling, and field observation and experiment. Further, workers have shown a distinct difference in emphasis concentrating either on sedimentological or hydrodynamic data acquisition and interpretation.

Experimental results, particularly in the field, are complicated by the range of factors that may influence sorting and selection. Sediments differ according to specific gravity, size, shape, angularity, and susceptibility to abrasion of differing rock types. Availability and gross topography provide limiting factors. Physical processes include currents, whether wave-generated or tidal; wave characteristics (especially wave height); lag effects, such as settling at slack water; edge waves; and, in the intertidal zone, the effect of winnowing through wind transport. Movement can be in suspension by saltation or as bed load, and different particle sizes may be affected by different mechanisms at the same time.

Sorting occurs either parallel with or normal to the coastline, and can vary with depth; it may occur on various scales, temporal or spatial. For example, progressive longshore grading may be evident over a distance of many kilometers, as on the pebble foreshore at Chesil Beach, England, where consistent size grading extends over a length of some 23 km. Alternatively, sorting may be observed on many beaches from time to time, as in cusps spaced as little as 10 m apart.

Movement Normal to the Shore

Ippen and Eagleson (1955) provide a useful, albeit brief, summary of knowledge of sand movement prior to that date. They quote Inman, who attributed zoning parallel to the beach to the effect of standing waves or surf beat, seaward transportation of sediment by diffusion of suspended material, or net onshore/offshore transportation along the bottom due to the difference between onshore and offshore orbital velocity components. Transport would be selective, depending on particle size, wave character, and depth of water. Inman observed that the best sorting occurred on the shelf and beach, the poorest in the surf zone. In general, grading of sediment became progressively finer toward the sea (as energy became less).

More quantitative studies by other workers have shown the importance of specific variables. Laboratory experiments on a plane beach by Ippen and Eagleson (1955) showed that hydrodynamic drag and sediment particle weight were the primary forces involved in net bed sediment motion by wave action. Imbalance of hydrodynamic forces always resulted in a shoreward component of sediment transport. Net sediment transport depended upon the balance between this and offshore transport because of gravity. A *null point* (where net sediment particle velocity equaled zero) existed at a particular depth for each sediment size under a given incident wave. Further work has elaborated this concept. Model studies, using a steady flow, have helped to simplify basic problems of sediment movement and shown that fall velocity is the significant sediment parameter in suspended load. The Hydraulics Research Station (1969) in the United Kingdom has used a pulsating water tunnel to simulate the action of waves over a shingle bed at full scale to predict pebble movement. In general, hydraulic models using a mobile sand bed and thus generating sand ripples have, however, not been particularly useful in classifying relationships.

King (1972) quotes the work of Bagnold, who examined the sorting of larger particles, 0.5-0.9 cm, in a wave tank. He found that grading appeared to be confined to the surface layers. Coarse material was at or near the beach crest, with mean size material from there to the step near low water mark. The coarsest material occurred below this step, with finest material further seaward again. This is broadly typical of pebble and cobble beaches, such as Chesil, where Gleason and Hardcastle (1973) showed that transverse sorting on the exposed beach correlated well with wave frequency and the square root of significant wave height. For the same beach Neate (Carr, 1969) had demonstrated that the sorting process was not effective below low water mark.

Humbert (1968) has observed that shape selection does not appear to take place across a tideless beach; however, this observation seems reasonable only where beach processes are confined to a narrow zone. Moss (1962, 1963) explained vertical grading of beach sediments through the rejection of atypical size particles from a given lamina.

Movement Alongshore

The processes responsible for sorting normal to the beach are also important in longshore grading. In addition, the angle of wave approach and longshore current so generated become relevant. While on self-contained granule and pebble beaches in a high energy environment, an element of grading is rapidly achieved following changes in wave direction, height, or period; such grading tends to be restricted to the surface or even to bands along the beach that are within the zone of hydraulic processes at the particular time. Wave parameters rarely remain sufficiently constant for equilibrium to be attained across the whole foreshore or to any depth.

Hails (1974) has noted that in a series of papers, co-workers Inman and Komar (Inman, Komar, and Bowen, 1968; Komar and Inman, 1970) have shown that longshore transport of sand is directly proportional to the longshore component of wave power regardless of grain size. Nevertheless Komar and Miller (1973) have pointed out that the threshold for sediment movement in fine and medium sand (particles less than 0.5 mm) falls in laminar flow, whereas for coarser sand, flow is turbulent.

Bajournas found that sediment transport increased with the square root of sediment size in the fine to medium sand grade, while Larras and Bonnefille showed in their wave tank experiments that maximum transport occurred with sand having a median value of 0.8 mm (Castanho, 1970). Castanho argued that the conclusions reached depended upon the proportion of suspended load and bed load. Where suspended load dominates, the net longshore movement of sand is faster for finer grades of material, and this fact has resulted in a policy of *nourishing* beaches with sand slightly larger than the grade that originally occurred at the site.

The frequent link between shape, angularity, and particle size presents an analytical problem. Although shape has been used as a separate diagnostic factor in studies of sand grade particles (e.g., Winkelmolen, 1969; Winkelmolen and Veenstra, 1974), Humbert (1968) correctly observed that particle shape is more critical with larger material. Data for pebble and cobble beaches emphasize some of the complications that arise in interpretation. For these grades of material individual particles are examined, unlike in the sand range, where the weight of particular ϕ classes is obtained. Thus there is an emphasis on frequency by number rather than on frequency by weight. Flemming (1965) observed that there were 13 dimensions that could be measured on a pebble. Studies have tended to concentrate on long diameter, weight, and shape ratios: e.g., b/a; c/a; $(a+b)/2c$; $\sqrt[3]{bc/a^2}$; $\sqrt[3]{c^2/ab}$; where a, b, and c are long, intermediate, and short axes, respectively (Humbert, 1968).

Carr (1969) showed that on Chesil Beach grading was best described by the pebble thickness (c). Subsequent experimental work on the same beach using geological tracers (Carr, 1975) showed that within limits the largest material traveled farthest. At a less exposed site—Slapton Beach, Devon, England—where indigenous material was finer (-1 to -4 ϕ), the smallest material traveled farthest. Equally important is the fact that for Slapton as well as a third beach, grading was sometimes best described by linear dimensions, and for others by shape ratios. Most studies with pebbles have concentrated on longshore variablility in shape.

Various statistical approaches to defining and analyzing sorting have been made. Sorting is most often described by the first four moment measures of a population; because particles appear to be distributed log-normally (at least in the sand grades and finer), the ϕ scale (Pettijohn, 1975) is generally used to represent this. Values are obtained for *mean* (plus *median* and *mode*), *standard deviation* (dispersion), *skewness* (asymmetry of distribution), and *kurtosis* (peakedness). Skewness and kurtosis apply only to normal distributions. These types of analysis, either on their own, modified in some fashion (e.g., as *sorting coefficients*), or in conjunction with one another, have been used to identify specific environments with a degree of success (Bigarella et al., 1969). Folk (1966) gives a summary of the many attempts that have been made.

Bimodal populations are common, and are usually taken to imply two processes at work or a lack of equilibrium with the environmental processes prevailing.

A considerable body of data has been obtained by using tracer material, generally fluorescent, to represent the movement and relative rates of transport of different sizes of natural beach sediment. Useful results have been obtained in this direction, but initial transport has tended to be excessive because of exposure of tracer on the beach surface rather than in equilibrium with the mobile surface layer; subsequent results have been complicated by variability in the direction of wave approach, by tracer being cast outside the zone of wave action, by burial, and by pulsing of new tracer from a reservoir of buried material at the injection site. Transport of sediment may occasionally be more apparent than real because of the diffusion process.

In spite of the extensive effort that has been put into research on sorting of beach material, neither the precise mechanism of the hydrodynamics, the most relevant sedimentary parameters, nor the relationships between detailed process and result are adequately understood. This difficulty is compounded in the interpretation of paleoenvironments.

ALAN PAUL CARR

References

Bigarella, J. J.; Alessi, A. H.; Becker, R. D.; and Duarte, G. M., 1969. Textural characteristics of the coastal dune, sand ridge, and beach sediments, *Parana, Brazil, Univ., Paranaense Geociencias Bol.* 27, 15-80.

Carr, A. P., 1969. Size grading along a pebble beach:

Chesil Beach, England, *Jour. Sed. Petrology* **39**, 297-311.
Carr, A. P., 1975. Differential movement of coarse sediment particles, *Conf. Coastal Eng., 14th, Copenhagen, Am. Soc. Civil Engineers, Proc.* **2**, 851-870.
Castahano, J., 1970. Influence of grain size on littoral drift, *Conf. Coastal Eng., 12th, Washington, D.C., Am. Soc. Civil Engineers, Proc.* **2**, 891-898.
Flemming, N. C., 1965. Form and function of sedimentary particles, *Jour. Sed. Petrology* **35**, 381-390.
Folk, R. L., 1966. A review of grain size parameters, *Sedimentology* **6**, 73-93.
Gleason, R., and Hardcastle, P. J., 1973. The significance of wave parameters in the sorting of beach pebbles, *Estuarine and Coastal Marine Sci.* **1**, 11-18.
Hails, J. R., 1974. A review of some current trends in nearshore research, *Earth-Sci. Rev.* **10**, 171-202.
Humbert, F. L., 1968. *Selection and Wear of Pebbles on Gravel Beaches*. Groningen, The Netherlands, Geological Institute Pub. No. 190, 144p.
Hydraulics Research Station, 1969. Threshold movement of shingle subject to wave action, *Hydrol. Research Station Notes No. 15*, 6-7.
Inman, D. L.; Komar, P. D.; and Bowen, A. J., 1968. Longshore transport of sand, *Conf. Coastal Eng., 11th, London, Am. Soc. Civil Engineers, Proc.* **1**, 298-306.
Ippen, A. T., and Eagleson, P. S., 1955. *A Study of Sediment Sorting by Waves Shoaling on a Plane Beach*. Washington, D.C.: U.S. Beach Erosion Board Tech. Memo. 63, 83p.
King, C. A. M., 1972. *Beaches and Coasts*. London: Arnold, 570p.
Komar, P. D., and Inman, D. L., 1970. Longshore sand transport on beaches, *Jour. Geophys. Research* **75**, 5914-5927.
Komar, P. D., and Miller, M. C., 1973. The threshold of sedimentary movement under oscillatory water waves, *Jour. Sed. Petrology* **43**, 1101-1110.
Moss, A. J., 1962. The physical nature of common sandy and pebbly deposits, Part 1, *Am. Jour. Sci.* **260**, 337-373.
Moss, A. J., 1963. The physical nature of common sandy and pebbly deposits, Part 2, *Am. Jour. Sci.* **261**, 297-343.
Pettijohn, F. J., 1975. *Sedimentary Rocks*. New York: Harper and Row, 628p.
Winkelmolen, A. M., 1969. *Experimental Rollability and Natural Shape Sorting of Sand*. Groningen, The Netherlands: Geological Institute Pub. No. 191, 141p.
Winkelmolen, A. M., and Veenstra, H. J., 1974. Size and shape sorting in a Dutch tidal inlet, *Sedimentology* **21**, 107-126.

Cross-references: *Beach Cusps; Beach Material; Beach Nourishment; Beach Processes; Classification; Research Methods; Coastal Morphology, Flocculation; Near-Shore Hydrodynamics and Sedimentation; Sediment Analysis, Statistical Methods; Zingg Shape.* Vol. VI: *Sorting.*

BEACH NOURISHMENT

The artificial supply of beach sand to restore a deficiency caused by erosion is known as *beach nourishment*. Most beaches afford some protection for the backshore or contiguous frontal dune system along a coast because they dissipate wave energy. Therefore in this sense they serve the same purpose as man-made structures like sea walls, revetments, and bulkheads. On the other hand, groins and jetties are constructed either to retard or to arrest the movement of material alongshore and are generally used in conjunction with sea walls, beach replenishment schemes, or both.

Many coastal engineering establishments such as the Coastal Engineering Research Center (CERC) in the United States and the Hydraulics Research Station in England are attempting to solve coastal erosion problems these days by beach nourishment rather than by conventional man-made structures. Thus the artificial replenishment of beaches is now a fairly standard practice whereby sand, deposited within a few meters of the shoreline, is pumped ashore along the length of a beach. If conditions are particularly favorable for such nourishment, long reaches of the shoreline may be protected.

One of the advantages of the method is that the main cause of coastal erosion, a deficiency in natural sand supply, is remedied. Following beach nourishment it is possible to evaluate the magnitude and location of changes in the beach profile. This was done, for example, at Atlantic City, New Jersey, where beach replenishment in 1963 remained effective for six years.

As a result of successful experiments at Bournemouth, England, where 110,000 m^3 of sand were deposited within 200 m of the shore by a trailing hopper suction dredger and pumped ashore, and at Portobello near Edinburgh, Scotland, where 200,000 m^3 of sand were pumped ashore, Willis and Price (1975) conclude that ". . . it would seem realistic to look to the future when perhaps many more of our coastal erosion problems will be resolved by specially designed dredgers carrying out 'topping up' operations of beaches on a national scale." Many engineers are now convinced that beach nourishment, at ten-year intervals, is still more economical than constructing a series of groins backed by massive sea walls.

Despite these facts, it is still debatable whether beach nourishment is the complete answer or only a short-term solution to coastal erosion problems. It is not always possible to predict how frequently beaches need to be renourished, and some schemes could therefore prove to be expensive commitments over the long term.

Therefore it is essential that systematic monitoring and interdisciplinary studies are undertaken in order to determine the cause(s) of erosion, particularly when some beaches are eroded continuously despite beach nourishment.

Ideally, the size of the *artificial* material should, as far as possible, approximate that of the natural beach but, depending on its source, the sediment used for beach nourishment is usually smaller or larger in diameter.

JOHN R. HAILS

Reference

Willis, D. H., and Price, W. A., 1975. Trends in the application of research to solve coastal engineering problems, *in* J. R. Hails and A. P. Carr, eds., *Nearshore Sediment Dynamics and Sedimentation.* New York: John Wiley & Sons, 316p.

Cross-references: *Beach Processes; Coastal Engineering; Coastal Erosion, Environmental-Geologic Hazard; Present-Day Shoreline Changes.* Vol. III: *Beach Erosion and Coastal Protection;* Vol. XIII: *Beach Engineering.*

BEACH ORIENTATION

Beach orientation is the beach direction with respect to oceanographic factors and with reference to coastal forms.

Classical morphology does not expressly indicate a corresponding statement for the expression. Only modern geomorphology uses the term *orientation* with reference to the oceanographic factors. W. V. Lewis and A. Schou (Tanner, 1962) established two important laws concerning beach orientation.

Lewis established that beaches have a tendency to be perpendicular to the principal sea and swell. Schou found a very close correspondence between the resultant of the wind regime and the evolution of the windward coastline. The beach tends to form in a direction at a right angle to the wind resultant, but the final orientation of the beach depends, likewise, on the fetch. Four cases can occur: equal fetch in all directions—orientation of beach is perpendicular to the resultant; fetch coincides with wind resultant—beach is also perpendicular to the resultant; fetch distinct from resultant—beach orientation is perpendicular to a line oriented between fetch and resultant; and resultant parallel to beach orientation—beach ridges follow as bars and spits (Guilcher, 1954).

From this point of view, a beach oriented in agreement with the oceanographic factors is an *equilibrium beach*. The beach curvature depends on the combined action between the wave direction and the planimetric structure of the coast. Beaches tend to be concave, following the Schou and Lewis laws. Araya-Vergara (1967) found that the curvature may state itself as $C = a/Rd$, where C is the curvature index, Rd is one radian, and a the subtended angle of the curvature arc. The index oscillates between 0 and 3.14. A beach with radius equal to beach length has curvature 1.

Lewis and Schou laws indicate only the tendency of evolution. However, lithologic and structural factors sometimes control the actual orientation. There are three cases of beach orientation with regard to the common swell direction (Fig. 1): *transverse beaches, oblique beaches,* and *longitudinal or parallel beaches.* When the beaches are independent from the hard rocks, these terms may be used: *equilibrium beach,* in agreement with the fetch and wind resultant; and *unbalanced beach,* with orientation independent of the oceanographic factors. Transverse beaches are built by swash alignment. Longitudinal and oblique beaches are built by drift alignment (Davies, 1977).

The Schou theory is based on the idea of local wind-sea attack on the beach, but the common important waves are produced by swell. For a more correct explanation of beach orientation it would seem better to plot the swell orthogonals (Fig. 1).

JOSÉ F. ARAYA-VERGARA

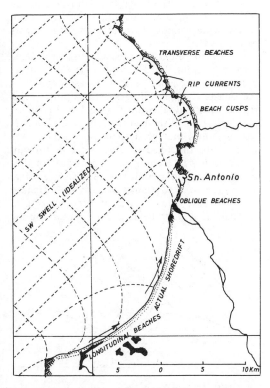

FIGURE 1. Types of beach orientation and curvature along the central coast of Chile. Curvature index between 0.4 and 1.2.

References

Araya-Vergara, J. F., 1967. Morfometría de la curvatura de las playas entre Punta de Talca y Punta Toro (Chile Central), *Inform. Geog. Chile* **17**, 5–30.

Davies, J. L., 1977. *Geographical Variation in Coastal Development.* London: Longman, 204p.

Guilcher, A., 1954. *Morphologie Littorale et Sous-Marine.* Paris: Press Univ. France, 216p.

Tanner, W. F., 1962. Reorientation of convex shores, *Am. Jour. Sci.* **260**, 37–43.

Cross-references: *A-B-C Model; Beach in Plan View; Headland Bay Beach; Orientation of Shorelines; Sediment Transport; Waves; Wind; Zetaform Bays.*

BEACH PADS

Beach pad is a term for a roughly triangular sand body that has been reported several times under various descriptive names, including the catchall *sand wave* (Bruun, 1954). Beach pad was proposed in an effort to simplify the nomenclature of an important and distinctive beach feature that relates beach morphology to sand transport. The first available reference (Tanner, 1975) used the term beach pad to describe an asymmetrical roughly triangular body of sand on the beach face landward of the surf which migrated downdrift, but the term was in wide colloquial use at least fifteen years earlier.

The beach pad is not a beach cusp or a transverse bar (Sonu, 1973). In plan view its outline is roughly triangular, with the base parallel to the coast, the next longest side updrift, and the shortest side downdrift (see Fig. 1). Beach pads alternate with deeper water embayments that at times of high activity allow the waves to cut into the adjacent dune ridge or upland. From each pad a bar extends seaward roughly parallel to the long updrift side. This bar offers an avenue of sand transport into the offshore bar complex. The volume of sand moving in the pad as it migrates is the minimum volume of sand transported in the downdrift littoral direction, because additional sand is passing through the system at a velocity exceeding that of the pad.

Minimum migration rates for 145-m spaced pads in Lake Michigan were reported as 51 m for the 8-month ice-free season (Tanner, 1975). Migration rates for beach pads on St. Joseph Peninsula, Florida, with a spacing of 200–450 m, averaged 40 m per month (1.3 m/day) for the interval October to April.

Field observations have shown that the beach pad is an excellent mechanism of sand transport that is associated primarily with eroding beaches. The presence of beach pads indicates beaches with strongly unidirectional littoral drift systems such as those on spits or barrier islands experiencing lateral growth.

LEE D. ENTSMINGER

References

Bruun, P. M., 1954. Migrating sand waves and sand humps, with special reference to investigations carried on the Danish north seacoast, *Conf. Coastal Eng., 5th Grenoble, Council Wave Research, Proc.*, 269–295.

Sonu, C. J., 1973. Three-dimension beach changes, *Jour. Geology* **81**, 42–64.

Tanner, W. F., 1975. Beach processes, Berrien County, Michigan, *Jour. Great Lakes Res.* **1**, 171–178.

Cross-references: *Beach Cusps; Beach Processes; Coastal Accretion Forms; Coastal Erosion, Formations; Edge Waves; Minor Beach Features; Rhythmic Cuspate Forms; Ripple Marks; Sediment Transport.*

BEACH IN PLAN VIEW

A *plan view* approach to the study of a beach has the advantage that the obvious requirement (map outline) can be met from maps and/or air photos. It has the disadvantage that the three-dimensional geometry, and time changes in that geometry are ignored. Nevertheless certain useful results can be obtained from a study of the plan view.

The Cell Concept

Along a sandy coast, the ideal cell can be defined as a single area of erosion, a single area of deposition, and the transport path between. Cells are bounded along the coast in various ways: rocky headlands, sandy capes, bay heads, wide inlets, estuary mouths, deltas, submarine canyons, marine marsh or swamp, and others. For two cells to be isolated from each other there must be little or no leakage of sand past their common boundary. In nature some cells are well isolated, and others are poorly isolated,

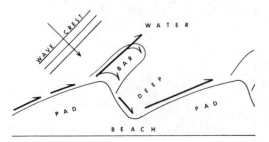

FIGURE 1. Two beach pads, with a migration direction of left to right. Arrows indicate the direction of approach of the incident waves and the resulting avenues of sand transport.

commonly gaining or losing significant amounts of sand from or to adjacent cells; these last are better considered as subcells.

Maturity in such a system is characterized by a few large cells with few or poorly developed subcells. Submaturity, on the other hand, is characterized by many small cells and subcells. The rate of evolution toward maturity is a function of energy level; high energy coasts tend to develop rather quickly toward a chain of a few large cells, whereas *zero* energy shores may have many small or poorly defined cells.

A circular lake having sandy shores shaped by one effective prevailing wind direction exhibits two concave cells, one on each side, extending from the upwind side to the downwind side. A sandy island exposed to wave attack from one direction will develop two convex cells: one extending from the drift divide near the center of the exposed beach downdrift to the right, and the other extending to the left.

Key points in one cell are conveniently identified with the letters *a* through *e*, the first being located at the updrift drift divide, and the latter at the downdrift drift divide. Identification of key points for subcells is difficult, because what may be point *a* for one wave approach may well be point *b* or *c* for some other wave approach or no key point at all.

The Half Heart

In 1960 Silvester (1974) published what various workers had earlier felt—that there is a certain geometry (in plan view) of sandy beaches that can be recognized in many places. He described this as the *half heart* shape, using as his reference the outline of a geometrical figure commonly printed in red ink on Valentine greeting cards. The *Valentine heart* is similar to the geometrical figure known as the *cardioid* except that the *heart* has a point opposite the indentation whereas the cardioid does not. The *half heart* comes closer than the half cardioid to matching the plan geometry of many beaches, but is not satisfactory because the shape is not really defined.

The Log Spiral

Yasso (1965) provided a rigorously defined standard when he proposed that the *log spiral* be adopted to represent what Silvester had called the half heart. The equation for the log spiral can be expressed in polar coordinates, as $r = ae^{k\phi}$, where r is the radius vector, ϕ the vector angle, and a and k numerical coefficients to be determined for any specific log spiral. The log spiral is a true spiral and turns through much more than 360°. Therefore only part of the log spiral can be fitted to a beach outline.

Even though a reasonably precise match can be found, the limitations are numerous. First, the center (or origin) of the log spiral is not known a priori and cannot be found by inspection of the map pattern of a beach. As a general rule, it does not lie on the beach itself but on the land or in the water to one side or the other. Second, the position of the beach pattern along the log spiral cannot be specified from inspection except in a general way (tightly curved segments must of course be closer to the center). Third, many beaches, because of their sub cell nature, cannot be fitted to any part of the log spiral unless they are subdivided into many small segments. Fourth, successful application of the log spiral provides no physical insight.

It is likely that the log spiral is a first approximation to a more involved relationship, having to do with the interplay of littoral power ($P_L = fn$ of wave height and breaker angle) and distance along the beach from *a* toward *e*, but this has not been developed.

"a–b–c . . ." Patterns

The *"a–b–c . . ." model* is a graphic (or mathematical) device that can be applied immediately and directly to many beaches. A series of stages in coastal evolution may provide a great deal of information about the history of the area particularly if the "a–b–c . . ." patterns are clear. This ordinarily means that old beach features, such as beach ridges, must be preserved for analysis, as on aerial photos. If the beach has been eroding over a long period of time, then historical records will have to be consulted as far back as they go.

The most obvious elements of the "a–b–c . . ." pattern are parallelism (or nonparallelism) of successive beach positions and littoral displacement of key points in the model (such as *c*). If old beach ridges, for example, are parallel, whether curved or straight, they were built essentially from offshore sand driven by wave action up onto the beach; littoral transport was not a significant factor (for step-by-step development of this concept, see Tanner, 1974: 120–122). If they depart from parallelism, then littoral transport was significant, and the greater the departure from parallelism, the greater the part played by littoral transport. (It is not possible, however, to conclude that the widest spacing of the ridges is located near the updrift end; examples of the opposite configuration are known. Other procedures must be used to resolve this dilemma [Murali, 1973].)

The point *c* in the model separates the eroding updrift part of the cell from the aggrading downdrift portion. The histories of many coasts show a littoral transfer of the c point with time. Where c has been migrating downdrift, erosion has been increasing relative to deposition; the "lost" material may have been blown ashore into dunes or washed seaward into deeper water. Where c has been migrating updrift, deposition has been increasing relative to erosion; the excess material may have been derived from offshore. The c point may behave as if it were a moving fulcrum about which the shoreline has been rotating in the map plane.

Lewis (1938) discussed reorientation of the shoreline due to a change in wind direction; in this case also the key points in the model shift position with time.

A corollary of the "$a-b-c\ldots$" model is that as the importance of littoral drift increases there is less and less tendency to build beach ridges. This is another way of saying that as the coast matures cells tend to be integrated into larger and larger units with less and less local deposition.

Equilibrium Curvature

The idea of an equilibrium geometry is not new. It has been expressed for vertical profiles taken at right angles to the beach outline in map view for many years, and has been integrated more recently into a three-dimensional model (Tanner, 1958).

In such formulations it is appropriate to assume that there is an equilibrium curvature (as seen in plan view) but to specify or imply that the details of the curvature cannot be known in advance—that is, there is no fixed or "standard" curvature against which we can then measure all beaches. This position is attractive if we think that beach geometry is the result of an interaction between processes and materials (see *Equilibrium Shoreline*), and that different interactions (different energy levels, for instance) will produce different geometries.

It is possible to ask whether there are curvature characteristics, such as radius and arc of curvature, that represent some ultimate equilibrium form. In general the answer seems to be no. The radius of curvature appears to be partly a function of differences between the cell constraints at a and e, whereas the arc of curvature is probably determined largely by the degree of maturity of the system, which is partly determined by the energy level, the latter commonly dependent on the fetch.

However, from the success one has in matching many beaches with parts of the log spiral, it should be clear that in many instances even a single coastal segment has both arc and radius that vary from one point to another. Only a static geometry could have fixed arc and radius.

The "high energy but zero littoral power" coast would fit into such a category, but at present there is no way to predict for any one segment what the arc and radius will be.

The Segmented Lagoon

One of the most interesting problems associated with map analysis of the coast arises from the so-called segmented lagoon. In special cases this lagoon in map view looks somewhat like a string of sausages. The segmentation arises from the scalloped outline of the two shores; where the opposite points match (which is not common), the sausage outline is seen (Fig. 1).

Because of the presence of a submerged sand deposit downdrift from each such point and because of the smooth curvature of the beach updrift from each point, it can be inferred that these features have a hydrodynamic origin. A critical factor is that these are formed along beaches where the fetch is short (e.g., in lagoons) and hence where the growth rate of wave energy density per meter of fetch is high ($\partial E/\partial F$ is large). Any wave train proceeding along the main axis of the lagoon must be continuously subject to refraction (on both sides), and small changes in bottom roughness or relief will have the effect of focusing the refracted waves on selected parts of one or both of the two beaches. Such differential refraction should provide for oscillations in the change in littoral power in the downdrift direction and hence in littoral transport in the same direction (dg/dx).

Any deposit initiated in this manner will tend to produce a curved, concave outline (deposit), perhaps having a log spiral nature and exhibit-

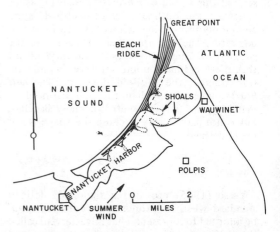

FIGURE 1. Nantucket Harbor, Massachusetts, showing the alternation of embayments on the convex northwestern shore. The southeastern edge of the old beach ridge complex permits recognition of eroded areas (concavities) and deposited features (spits and shoals) (from Tanner, 1962).

ing the widest part of the deposit near the downdrift end.

Sand moves through such a system, tending to close the lagoon where two points are essentially opposite each other. This tendency will be offset, however, by either currents in the lagoon, or changes in the wind direction, or both. The submerged bar, which forms downdrift from each point, shows a reversed curvature, providing an overall S shape to the shallow isobaths on any one side.

The radius of curvature of the beach within one arc is roughly one-fourth of the fetch. The arc of curvature varies from a maximum of 180° to a minimum of roughly 50°, with the larger arc appearing in smaller systems.

Irregularities

Despite all efforts to systematize thinking about the map outline of the coast, there are still irregularities that must be explained on an individual basis. These can be grouped in three major classes: zero energy coasts, submature coasts, and eroding coasts (nonequilibrium). In the first class, where average breaker heights are less than about 4 cm, irregularities in coastal geometry are commonly due to vegetation in saltwater swamp (such as mangrove) or saltwater marsh. In the second class, irregularities may be left over from any previous history, whether glacial, volcanic, or some other. Contemporary nonmarine work (such as new lava flows) would also fit into this category, in which coasts have not yet achieved a marine equilibrium. In the third class, irregularities may represent the etching out of less resistant rocks by wave action as the shore migrates slowly inland.

WILLIAM F. TANNER

References

Lewis, W. V., 1938. The evolution of shoreline curves, *Geol. Assoc. London Proc.* 49, 107-127.
Murali, R. S., 1973. Wave-power gradient: An approach to Holocene depositional history, *Gulf Coast Assoc. Geol. Socs. Trans.* 23, 364-367.
Silvester, Richard, 1974. *Coastal Engineering*, Vol. II. Amsterdam: Elsevier Scientific Publishing Co., 338p.
Tanner, W. F., 1958. The equilibrium beach, *Am. Geophys. Union Trans.* 39, 889-891.
Tanner, W. F., 1962. Reorientation of convex shores, *Am. Jour. Sci.* 260, 37-43.
Tanner, W. F., ed., 1974. *Sediment Transport in the Near-Shore Zone*. Tallahassee: Geology Department, Florida State University, 147p.
Yasso, W. E., 1965. Plan geometry of headland-bay beaches, *Jour. Geology* 73, 702-714.

Cross-references: *A-B-C Model; Beach Orientation; Coastal Accretion Forms; Coastal Erosion, Formations; Cuspate Spits; Equilibrium Shoreline; Headland Bay Beach; Lagoonal Sedimentation; Lakes, Coastal Morphology; Zero Energy Coasts; Zetaform Bays.* Vol. III: *Beach; Cuspate Foreland or Spits.*

BEACH PLATFORM—See WAVE-CUT BEACH

BEACH PROCESSES

Beaches

A *beach* is the noncohesive material accumulated in the region affected by wave action along the perimeter of a body of water. The material is commonly sand (0.06-2.0 mm), but may range through gravel (2-60 mm), cobbles (60-200 mm), and boulders (over 200 mm). The *shore* is the zone between low tide limit and the maximum upper swash limit, while the *coast* includes both the shore and landward areas affected by it.

The beach zone is subdivided into *nearshore*, *foreshore*, and *backshore*; its boundaries may be translated by cyclic migrations. The *nearshore* is the submerged portion of the beach seaward of the low tide limit in the region of shoaling waves. The *foreshore* extends from low tide limit to the high tide upper swash limit. The *backshore* lies landward from the high tide upper swash limit to the upper limit of storm swash, and may include cliffs, dunes, or marshes. The crest of the backshore zone, above the uppermost limit of normal swash, is referred to as a *berm* or *ridge*.

Standard texts dealing with this subject are those by Bird (1976), Davis (1978), Guilcher (1958), Johnson (1919), King (1972), Komar (1976), Shepard (1973), and Zenkovich (1967).

Wave Action

When a wave reaches relatively shallow coastal water, bottom friction causes a decrease in velocity. Where wave approach is at an angle to the coast, the wave is turned toward the shore by *refraction* so as to tend to approach in a parallel front. In the process of refraction a *longshore current* that flows parallel to the shore may be generated landward of the breakers and in the breaker zone. A second effect of the friction is to shorten wave length and increase height and steepness. As the wave continues to steepen it becomes unstable and finally collapses, forming *breakers* that plunge, spill, or surge. The water thus propelled forward moves up the foreshore slope as *swash*, first as turbulent flow, then, impeded by friction and gravity and thinned by infiltration, as laminar flow. The return of the water down the slope reversing the flow regime and direction is known as *backwash*. Unusually strong backwash is commonly called *undertow*. The water piled up at the shore by wave action may under certain conditions of wind and bottom topography return to greater depths by means of concentrated surface flows called *rip currents*.

Sediment Transport Normal to the Beach

In the breaker zone sediment transported down the foreshore by backwash is mixed with that transported shoreward by waves of translation and that churned up by the breaking wave. Caught up in the turbulence, only the larger particles can settle out; these form the *step* that is found under the breaker zone. A small part of the finer particles diffuse to the seaward side; the balance, in suspension, is carried forward with the swash.

As a single swash reaches its upper limit with decreasing velocity, caused by frictional drag and the pull of gravity, its sediment load is gradually deposited. At the upper limit, deposition may be facilitated by percolation of the water into the interstices between the particles of which the beach is composed. The degree of percolation is dependent on the grain size and beach water table (Duncan, 1964). The water returned in the backwash entrains larger particles as its velocity increases, and transports the sediment seaward until it meets, and is incorporated by, the next breaker.

The mechanics of the swash-backwash zone result in initial deposition within the inner limit of swash, scour in the zone of swash and backwash, and deposition of the step under the breaker zone. These features move up the foreshore slope with advancing flood tide and down the foreshore slope with receding ebb tide. In the latter, since the inner swash limit depositional feature is being deposited in a scour zone, the beach slope is restored essentially to its original form instead of being aggraded to form a microberm as in flood tide (Schwartz, 1966; Strahler, 1966).

On the seaward side of the breaker zone, landward pulses accompanying orbital movement of the shoaling waves transport particles toward the step. According to the nature of the particle, prevailing depth, and wave parameters, a *null point* will be reached where net movement along a line normal to the wave direction is zero. The results of this and the foregoing swash-backwash process are a decrease in median grain size and a decrease in sorting encountered as one progresses seaward and landward of the breaker zone step (Zeigler and Tuttle, 1961).

Wave steepness is of considerable importance in beach sediment transport. The *steepness ratio* is defined as deep-water height divided by deep-water wavelength (Ho/Lo). Change in beach sedimentation occurs at a critical ratio of about .025. On sand beaches, waves with steepness of .03 or greater erode while waves of .02 or lower aggrade.

Steep short-period waves erode the beach through a combination of processes. When these waves are of large height, they impart more energy upon breaking and project the swash farther up the foreshore slope, thus extending the upper limits of erosion. Then too, the greater turbulence keeps more material in suspension and transport, the increased saturation of the beach facilitates a larger percent of backwash and particle entrainment, and the volume of backwash is necessarily enlarged by both the greater swash and reduced percolation. Mass transport is rapid, tending to build a hydraulic head and intensify seaward return flow along the bottom. Transport to deeper water is often facilitated by rip currents.

The effect of low waves of low steepness is to return sediment to the beach. These are usually swells of long period and low height as well. Not only are the erosive characteristics of waves of low steepness (i.e., reach, turbulence, saturation, and backwash) less than waves of greater steepness, but the landward drag upon particles by the orbital motion under the shoaling wave zone transports sediment back toward the beach.

Sediment Transport along the Beach

Two forms of sediment transport along the beach depend on wave action.

Although a wave is refracted so as to approach parallel to the shore, the transformation is not complete and there remains a slight obliquity, generating an impulse in the direction in which the wave originally traveled. Therefore the ensuing swash moves obliquely up the foreshore slope toward a point somewhat along the beach in that direction. The return backwash flows down the slope, carrying through the same impetus along the beach. The path of an individual sand grain traced through the transport of succeeding swashes and backwash is a series of arched (arcuate) trajectories. The term applied to this type of sediment transport is *beach drift*.

The longshore current developed in the area of shoaling waves in and landward of the breaker zone flows parallel to the shore in the same direction as the beach drift. Sediment is transported by the current principally as bed load and to a lesser degree in suspension in what is called *longshore drift*.

The combined action of beach drift and longshore drift is known as *shore drift*. Shore drift is the process involved in beach depletion or accretion along the coast, inlet filling, and construction of bars, spits, and tombolos.

Another form of sand transport on the beach

is wind (eolian) transport in the upper, dry areas. This may be in any direction, across or along the beach, and may result in improved sorting. Finer grains are removed, leaving a surface layer of coarse sand or fine gravel.

Beach Cycles

Various cycles affect the beach, depending essentially on changes in wave steepness and effective sea level.

In regions of tidal action diurnal or semidiurnal tides translate the swash-backwash and breaker zones, with their accompanying phases of deposition and scour, landward and seaward across the foreshore slope with the flood and ebb tide. The range of the tides and the foreshore slope determine the distance over which this translation will take place. With other parameters remaining constant, breaker height will be greater at high tide than at low. At high tide the prevailing waves approach less hindered over the relatively deeper water of the nearshore, whereas at low tide approaching waves are modified by the shoaler depths of the gently sloping nearshore bottom. Under this changing regime, scour may be slightly increased at high tide.

Neap-spring tides generated by the changing positions of the sun and moon relative to the earth are the cause of a near semimonthly cycle, reinforced or cancelled by the perigean-apogean effect. At approximately 7½-day intervals the tidal range changes from minimum (neap) to maximum (spring). Upper foreshore erosion with lower foreshore and nearshore deposition at spring tides accompanied by a reversal at neap tides is well documented in the literature (LaFond, 1938; Inman and Rusnak, 1956; Emery, 1960; Inman, 1960; Inman and Filloux, 1960; U.S. Navy Department, 1960). LaFond and Rao (1954) postulate the net transfer during spring tides to be the accumulative results of higher effective sea level, higher waves, and time lag in replacement against gravity during low spring tide, with restoration in the more limited neap tide beach zone under low waves.

Seasonal cycles generally follow summer-winter storm patterns, with deposition in the former and erosion in the latter. Onshore winter storms raise sea level by piling up water against the coast, particularly during storm surges, and greatly increase wave height and steepness, both of which may increase shore erosion enormously. In summer low waves of low steepness tend to restore material to the beaches.

Long-range changes in relative sea level, such as those that accompany glacial advance and retreat or isostatic adjustments, translate the shore to new positions where beach development proceeds anew. Recent studies have introduced the concept of *surf base* or *surge base* (Dietz, 1963, 1964), formerly termed *wave base* (Bradley, 1958), the depth, approximately 10 m, at which normal wave erosion is terminated. A *profile of equilibrium*, a transverse beach profile that maintains its form over long periods with only minor cyclic changes may develop in the sediment lens that forms the beach zone.

Minor Beach Features

Sediment features in the shoal water areas of the nearshore zone may take the form of ripples or migrating humps. Symmetrical *oscillation ripples* develop in the shoaling wave zone normal to the direction of wave advance. Asymmetric *current ripples* form in the sediment being transported by longshore current, tidal currents, and backwash. The *migrating humps* or *sand waves* are expressions of relatively large sediment transport. Occasionally intersecting sets of ripples cause gridlike patterns called *tadpole nests*.

At the upper swash limit deposition of debris forms scalloped marginal lines known as *swash marks*. Backwash flowing around small obstacles forms diagonal *backwash patterns*, and in some cases rhombic patterns develop as a result of the minor currents generated in the backwash. Seepage of interstitial water down the foreshore slope at low tide cuts miniature channels termed *rill marks*, and the escaping of entrapped air in this zone forms small *sand domes*.

Beach cusps, varying greatly in height and length, are crescent-shaped forms concave to the swash-backwash zone. Although their origin is not generally agreed upon, they appear to be initiated by slight irregularities in the beach acted upon by wave approach, normal or oblique, and their length is a function of the wave height.

Major Beach Forms

Aside from the major subdivisions of the beach outlined earlier, there exist a number of other recognizable beach elements (Thompson, 1937). Of great importance is the *slope* of the foreshore. The foreshore slope, through numerous beach cycles, is a function of particle size (Bascom, 1964) and local wave parameters. Thus coarse material beaches are steeper than those composed of fine material. Similarly, eroding beaches flatten slightly, while aggrading beaches become slightly steeper.

Shore drift transports large quantities of sand along the beach in the form of *longshore sand waves*. These sediment concentrations generally occur as long-wavelength, low-amplitude, migrating waves or humps in the region of the berm, foreshore, and shoaling wave zone. Longshore sand waves may be delineated by contouring the beach or plotting a series of transverse profiles at regular intervals along the beach during low diurnal or neap tides.

Seaward of the breaker zone a *submarine bar* may form, particularly accompanying the cutting back of the beach or berm by steep waves (storm or seasonal). When similar forms are found at low tide in the lower foreshore of low-sloped beaches, they are called *ridges and runnels* as well as *low and ball;* the "low" and "runnel" referring to the hollow on the landward side, which is sometimes known as a *trough.*

One of the most striking of contemporary beach features is the *berm* or *ridge.* Known as *boulder ridge* or *cobble ridge* if composed of particles of these sizes, the berm or ridge is built of material transported beyond the normal upper swash limit. Guilcher (1958) cites an exceptional crest at 13 m above high water level at Chesil, England. Impressive cobble ridges, formed by material eroded from glacial debris headlands, may be seen on the beaches of Nova Scotia (Fig. 1).

MAURICE L. SCHWARTZ

FIGURE 1. Cobble ridge at Smith Cove, Guysborough County, Nova Scotia. The stepped appearance of the ridge results from a series of storms of decreasing energy (Photo: M. L. Schwartz).

References

Bascom, W. N., 1964. *Waves and Beaches.* Garden City, N.Y.: Doubleday Book Co., 267p.

Bird, E. C. F., 1976. *Coasts.* Canberra: Australian National University Press, 282p.

Bradley, W. C., 1958. Submarine abrasion and wave-cut platforms, *Geol. Soc. America Bull.* 69, 967-974.

Davis, R. A., Jr., ed., 1978. *Coastal Sedimentary Environments.* New York: Springer-Verlag, 420p.

Dietz, R. S., 1963. Wave base, marine profile of equilibrium, and wave built terraces: A critical appraisal, *Geol. Soc. America Bull.* 74, 971-990.

Dietz, R. S., 1964. Wave base, marine profile of equilibrium, and wave built terraces: Reply, *Geol. Soc. America Bull.* 75, 1275-1281.

Duncan, J. R., 1964. The effects of water table and tide cycle on swash-backwash sediment distribution and beach profile development, *Marine Geology* 2, 186-197.

Emery, K. O., 1960. *The Sea off Southern California.* New York: John Wiley & Sons, 366p.

Guilcher, A., 1958. *Coastal and Submarine Morphology.* London: Methuen and Co., 274p.

Inman, D. L., 1960. Shore processes, in *McGraw-Hill Encyclopedia of Science and Technology.* New York: McGraw-Hill, 299-306.

Inman, D. L., and Filloux, J., 1960. Beach cycles related to tide and local wind wave regimes, *Jour. Geology* 68, 225-231.

Inman, D. L., and Rusnak, G., 1956. *Changes in Sand Level on the Beach and Shelf at La Jolla, California.* Washington, D.C.: U.S. Beach Erosion Board Tech. Memo. No. 82. 30p.

Johnson, D. W., 1919. *Shore Processes and Shoreline Development.* New York: John Wiley & Sons, 584p.

King, C. A. M., 1972. *Beaches and Coasts.* New York: St. Martin's Press, 570p.

Komar, P. D., 1976. *Beach Processes and Sedimentation.* Englewood Cliffs, N.J.: Prentice-Hall, 429p.

LaFond, E. C., 1938. Relationship between mean sea level and sand movements, *Science* 88, 112-113.

LaFond, E. C., and Rao, R. P., 1954. Beach erosion cycles near Waltair on Bay of Bengal, *Andhra Univ. Mem. Oceanog.* 1, 63-77.

Schwartz, M. L., 1966. Littoral zone tidal-cycle sedimentation, *Jour. Sed. Petrology* 37, 677-683.

Shepard, F. P., 1973. *Submarine Geology.* New York: Harper and Row, 517p.

Strahler, A. N., 1966. Tidal cycle of changes in an equilibrium beach, Sandy Hook, New Jersey, *Jour. Geology* 74, 247-268.

Thompson, W. O., 1937. Original structures of beaches, bars, and dunes, *Geol. Soc. America Bull.* 48, 723-751.

U.S. Navy Department, 1960. *Oceanography for the Navy Meteorologist.* Norfolk, Va., U.S. Navy Weather Research Facility, 128p.

Zeigler, J. M., and Tuttle, S. D., 1961. Beach changes based on daily measurements of four Cape Cod beaches, *Jour. Geology* 69, 583-599.

Zenkovich, V. P., 1967. *Processes of Coastal Development.* New York: Wiley Interscience, 738p.

Cross-references: *Beach Cusps; Major Beach Features;*

Minor Beach Features; Sediment Transport; Tides; Waves. Vol. III: *Beach.*

BEACH PROCESSES, MONITORING OF

It is virtually impossible to predict accurately rates of coastal erosion and accretion and to forecast the effects of man-made structures on shoreline equilibrium (which occurs when the rate of sediment supply to a beach balances, or is equal to, the rate at which sediment is removed from a beach) unless beach processes have been monitored for some considerable time in order to analyze *extreme* rather than average conditions. Bearing in mind such events as the 50- or 100-year storm, one realizes that this is often impracticable.

Consequently, many researchers have resorted to theoretical studies of beach processes that have also involved laboratory wave-tank investigation of sediment transport and sorting by oscillatory waves, but with only a limited number of field studies to verify the results of such work (Hails, 1974). It is not the intention here to discuss the obvious constraints of laboratory experiments, but to mention in passing the severe limitations imposed on in situ measurements by heavy swell and wave turbulence in the breaker zone as well as the fact that increasing public use of formerly less-frequented beaches is limiting, to some degree, the type of instruments that can be installed and left safely for long-term data collection.

Of course it is essential to collect relevant data, so that they can be used meaningfully; otherwise both monitoring and compiling statistics are uneconomical exercises. This consideration, in turn, necessitates some systematic and standardized method of data collection. What is more fundamental, however, in the case of beach processes, is the type of monitoring that may be required from one locality to another along a coast (Morisawa and King, 1974). In other words, what should be monitored, and for how long? Sometimes these questions cannot be answered until crude data are available. Further, the two questions are even more difficult to answer when existing monitoring programs in the coastal zone are often unrelated to clearly defined aims and objectives.

Beach Processes and Monitoring Techniques

Although there are several definitions of the word *beach*, for the purpose of this article the word is defined as the zone that extends from the extreme upper limit of wave action (excluding catastrophic waves, such as *tidal waves*, *tsunamis*, and storm tides that accompany hurricanes or cyclones) to offshore beyond low-water mark, where bottom material is subject to movement by ordinary wave action (Bascom, 1964; Shepard, 1964; King, 1972; Hails and Carr, 1975).

Several interrelated processes regulate *beach profiles* and *sediment budgets* and ultimately determine whether or not a beach will erode. *Sediment transport* is an important process that is difficult to monitor. Despite qualitative and quantitative studies of the rate of littoral drift, there is a need for concurrent measurements and monitoring of bulk transport, bulk dispersion, and overall sediment transport paths by means of relatively sophisticated instrumentation.

The magnitude and frequency of variations in the sediment budget of a beach are controlled not only by the long-term sources of sand supply but also by the composition, size, and shape of the material comprising the beach; these properties determine its mobility. Coarse material that is too large to be transported by waves will remain near the source area, while fine-grained material is entrained as suspended sediment by wave turbulence and transported by currents into deeper water, where it is eventually deposited. Thus the sand fraction is the main supply for nourishing beaches naturally. In the case of *shingle* beaches (composed mainly of granules, pebbles, and small cobbles, which are often referred to as *beach gravel*) the degree of sorting parallel with, normal to, and vertically through a beach reflects the relationship between particle size (mean phi (ϕ)) and such parameters as wave height and direction of wave approach ($\Theta°$). Wave energy indicators, such as significant wave height and period, can be evaluated from data obtained from wave recorders located a short distance offshore. Pressure-type wave recorders are reasonably reliable for monitoring nearshore wave conditions, while shore-based radar is also proving useful in the routine measurement of wave direction. A Shore and Shore System (SAS System) developed by Lowe, Inman, and Brush (1972) is now able to provide continuous wave climate data, including two-dimensional wave spectra. It is anticipated that storms may be instrumented in the immediate future so as to afford a comprehensive and unifying study of the hydrodynamics of beaches and the nearshore zone.

The seasonal occurrence of storm and swell waves is manifest in winter *cut* and summer *fill*. Regular surveys of beach profiles are one way in which beaches have been monitored in the past in order to determine the amount of material that has been moved offshore-onshore or alongshore. Now such changes can be monitored more accurately by aerial remote sensing and by sequential photographs from satellites and aircraft. The Earth Satellite Corporation, for example, has employed the technique of coordinating aerial and orbital imagery to calculate rates of beach change along parts of the North Carolina and New Jersey coasts (see references in Hails, 1974).

Experiments have been conducted to measure the velocity fields on beaches of widely different slopes, under both plunging and spilling breakers, by using a two-component electromagnetic flowmeter mounted 15 cm above the seabed and oriented so as to measure the horizontal velocity field in turbulent nearshore water (Huntley and Bowen, 1975). The influence of beach slope on nearshore hydrodynamics can be determined by this instrument, and in due course it should be possible to monitor the interaction between water motion and beach profiles.

Some progress has been made toward quantifying and monitoring the relationship that exists between *swash* (*uprush* of water on the beach face following the final breaking of each wave on the shore; the swash is only a film of water < 6 mm deep at its upper edge) percolation, ground-water flow, and beach dynamics. Wells for monitoring the level of the water table and pipe stations for recording changes in beach height have been positioned on some beaches (see references in Hails, 1974). The significance of beach permeability in controlling wave energy dissipation, beach slope, and the phase difference between swash and wave period has been studied intermittently in the field, but so far no continuous monitoring programs have been put into operation.

Almometers, consisting of a high intensity light source and a series of photoelectric cells encased in watertight acrylic cylinders, have been used to measure suspended sediment concentration in water across the beach face. At present the potential use of almometers for studying wave-generated sediment motion across the beach and into the surf zone is unknown because the size distribution and composition of sediment must be known before this type of instrument is installed and calibrated. In view of the engineering problem of arresting the loss of beach sand to the backshore and adjacent dune system of some coasts, all-purpose wind-recording systems are being developed to provide the surface shear stress coefficients of sea breezes. Hsu (1970) has measured the temperature and wind profiles in the surface boundary layer at an experimental site on the Gulf coast near Fort Walton Beach, Florida. However, because these systems are only at an early stage of development and because more information is required on synoptic and subsynoptic wind systems, it may be several years before monitoring programs are fully operational on beaches.

Conclusion

So far in coastal research studies, field data have been obtained mainly at infrequent intervals over extended periods of time, and therefore may not necessarily relate to the most significant natural events that are required to evaluate laboratory investigations. Thus the main problem is simply to collect data for a sufficiently long period in order to gain a representative picture of the changes that are occurring in the coastal zone, particularly on beaches. Technological innovation and the development of sophisticated instrumentation to monitor coastal processes are the main solutions to this problem. Already some progress has been achieved in monitoring beach processes, but it may be several years before continuous monitoring programs are implemented to record the subtle interaction between the various processes that occur from sector to sector along a coastline (Inman and Brush, 1973).

JOHN R. HAILS

References

Bascom, W. N., 1964. *Waves and Beaches.* Garden City, N.Y.: Doubleday and Company, 267p.

Hails, J. R., 1974. A review of some current trends in nearshore research, *Earth-Sci. Rev.* **10**, 171-202.

Hails, J. R., and Carr, A. P., eds., 1975. *Nearshore Sediment Dynamics and Sedimentation.* London: John Wiley & Sons, 316p.

Hsu, S.-A, 1970. The shear stress of sea breeze on a swash zone, *Conf. Coastal Eng., 12th, Am. Soc. Civil Engineers, Washington, D.C., Proc.* **1**, 243-255.

Huntley, D. A., and Bowen, A. J., 1975. Comparison of the hydrodynamics of steep and shallow beaches, in J. R. Hails and A. P. Carr, eds., *Nearshore Sediment Dynamics and Sedimentation.* London: John Wiley & Sons, 316p.

Inman, D. L., and Brush, B. M., 1973. The coastal challenge, *Science* **181**, 20-32.

King, C. A. M., 1972. *Beaches and Coasts.* London: Edward Arnold, 570p.

Lowe, R. L.; Inman, D. L.; and Brush, B. M., 1972. Simultaneous data system for instrumenting the shelf, *Conf. Coastal Eng., 13th, Am. Soc. Civil Engineers, Vancouver, Proc.* **3**, 2027-2043.

Morisawa, M., and King, C. A. M., 1974. Monitoring the coastal environment, *Geology* **2**, 385-388.

Shepard, F. P., 1964. *The Earth beneath the Sea.* New York: Atheneum, 275p.

Cross-references: *Beach; Beach Material; Beach Processes; Coastal Engineering; Coastal Morphology, Research Methods; Computer Simulation; Mathematical Models; Sediment Analysis Methods; Sediment Budget; Sediment Transport; Waves.*

BEACH PROFILES

The beach is constantly being changed by waves and currents. Although these changes are commonly rather subtle on a day-to-day basis, they may be prominent on a seasonal basis. The most marked changes to the beach are the result of storms, whereby prominent changes may occur in only a day or so. One of the best methods for displaying these changes is through a representation of the *beach profile*. Although there are commonly changes in beach geometry along the coast, the most significant changes are those that take place at a given location through time. These changes are those that reflect the interaction of the beach with wave and current energy directed against it from the sea.

It is possible to generalize by placing beach profiles in two categories—an erosional profile (*storm profile*) and an *accretional profile* (Coastal Engineering Research Center, 1973). Shortly after storm conditions the beach is rather flat and uniformly seaward sloping with a slight concave-upward configuration. It is evident that much sediment has been removed and a significant quantity has been deposited just seaward of the low tide level in the form of a shadow sand bar. A veneer of heavy mineral concentration commonly covers the upper storm beach, although it may blanket the entire beach (Fig. 1). This veneer is the result of a lag concentration of minerals such as magnetite and garnet that are commonly present as accessory minerals in the beach and dune sediments. When erosion occurs, the dominant and less dense grains of quartz, feldspar, and carbonate are carried seaward by storm swash, leaving the more dense grains behind (Davis, 1978).

Accretional beach profiles show configurations considerably different from those of erosional profiles. The foreshore is generally rather steep

FIGURE 2. Accretional beach showing relatively steep foreshore and nearly horizontal backshore (Photo: R. A. Davis, Jr.).

(5-15°) and is essentially equated with the swash zone. A distinct break occurs at the landward limit of the foreshore where the backshore begins. This zone slopes 1-2° landward and is unaffected by normal wave action. It is commonly the site of wind shadow dunes (Fig. 2).

Early literature made common reference to *summer beach profiles* (accretional) and *winter beach profiles* (erosional). Although it is true that in most places this seasonal relationship holds, the genetic terminology is preferred and does not allow for misinterpretation. It is not uncommon for pronounced erosional profiles to exist in the summer, especially in the Gulf of Mexico or the southeastern United States where tropical storms are frequent.

RICHARD A. DAVIS, Jr.

References

Coastal Engineering Research Center, 1973. *Shore Protection Manual*, vol. 1. Washington, D.C.: U.S. Army Corps of Engineers.

Davis, R.A., Jr., 1978. Beach and nearshore zone, *in* R. A. Davis, ed., *Coastal Sedimentary Environments*. New York: Springer-Verlag, 237-285.

Cross-references: *Beach Cycles; Coastal Characteristics, Mapping of; Coastal Morphology, Research Methods; Sweep Zone; Time Series.*

FIGURE 1. Storm beach profile showing uniformly sloping beach (Photo: R. A. Davis, Jr.).

BEACH RIDGE PLAIN

A beach ridge plain is a progradational strand plain composed of a series of parallel *beach ridges*. The ridges are commonly closely spaced, the bottoms of the interridge swales commonly being at about the normal elevation of the back of the beach before the prograding ridge was formed in front of it (Stapor, 1975).

Changes in the trend commonly affect sandy

shorelines. These cause diagonal, curving, or end truncation of the beach plain, with a new progradational series forming parallel with the changed shoreline.

The ridge may originate immediately back of the active beach as a flood-level ridge commonly of the coarser beach materials, or it may form as an eolian accumulation caught in the vegetation immediately back of the beach proper. Many beach ridges are water-laid below and eolian above, or there may be alternations of the two types, as in the plain at the mouth of Sabine Lake (bay), Texas. Where excess amounts of sand are blown to it from the beach, the ridge may become hummocky, deserving to be called dunes, with the ridge becoming a *foredune*.

Beach ridge plains occur as *cuspate forelands, deltaic forelands,* and at the mouths of incoming streams that break through a *chenier plain* depositing sand in the line of a dominantly clay plain (Otvos and Price, 1979).

The swales are commonly dotted with rain ponds and may be dark with organic materials. Locally swales may be partly filled with sand from the ridges.

W. ARMSTRONG PRICE

References

Otvos, E. G., and Price, W. A., 1979. Problems of chenier genesis and terminology—an overview, *Marine Geology* 31, 251-263.

Stapor, F. W., 1975. Holocene beach ridge plain development, northwest Florida, *Zeitschr. Geomorphologie* 22, 116-144.

Cross-references: *Beach; Beach Material; Beach Ridges and Beach Ridge Coasts; Chenier and Chenier Plain; Coastal Dunes and Eolian Sedimentation; Cuspate Foreland.*

BEACH RIDGES AND BEACH RIDGE COASTS

Beach ridges are linear, mound-shaped sand deposits roughly paralleling a coast (Fig. 1). Ridge crests have elevations well above mean high tide, and the bottoms of the adjacent troughs or swales have elevations not far from mean low tide. Their constituent clasts range in size from very fine sand to pebbles. The mineralogic composition of the clasts is dependent on the nature of local materials, although quartzo-feldspathic sand is the most common. Most ridges have relief measured in meters, widths measured in tens of meters, and lengths measured in kilometers. Ul'st (1957) and Tanner and Stapor (1972) reported a complex internal structure of beach face or swash bedding oriented generally perpendicular to ridge crests.

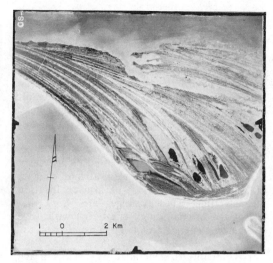

FIGURE 1. Holocene beach ridges of St. Vincent Island, Franklin County, Florida (Photo: F. W. Stapor).

Psuty reported a complex structure of washover bedding generally perpendicular to ridge crests, and Bigarella reported an internal structure composed of master sets of beach-face bedding perpendicular to ridge crests interbedded with wash-over cross beds parallel to ridge crests. Ul'st, Tanner, Stapor, Psuty, Bigarella, Curray and Moore, and Nossin emphasized the predominance of marine processes in the formation of beach ridges; Davies and King and Barnes recognized a significant eolian contribution to

FIGURE 2. Barshageudd, southern Gotland, Sweden. In 1741 Linnaeus counted 77 ridges here; there are now approximately 130, representing about 6500 years of beach ridge development (Photo: Arne Philip).

beach ridge genesis and preservation in addition to a fundamental marine process (see references in Stapor, 1975).

Beach ridges typically occur in fairly distinct groups, the grouping being a function of topographic position, physical size, and/or geographic pattern (Fig. 2). One (or more) beach ridge group defines a beach ridge plain. Beach ridges and beach ridge plains comprise the bulk of many sandy, *barrier islands,* and hence are common along sandy coasts that have experienced Holocene and late Pleistocene deposition.

FRANK W. STAPOR, Jr.

References

Stapor, F. W., 1975. Holocene beach ridge plain development, northwest Florida, *Zeitschr. Geomorphologie* **22**, 116-144.

Tanner, W. F., and Stapor, F. W., 1972. Precise control of wave run-up in beach ridge construction, *Zeitschr. Geomorphologie* **16**, 393-399.

Ul'st, V. G., 1957. *Morphology and Developmental History of the Region of Marine Accumulation at the Head of Riga Bay.* Akademiya Nauk Latvia SSR, 179p.

Cross-references: *Apposition Beach; Barrier Islands; Beach Ridge Plain; Chenier and Chenier Plain; Coastal Accretion Forms; Coastal Dunes and Eolian Sedimentation.* Vol. III: *Beach Ridges.*

BEAUFORT WIND SCALE

Originally the Beaufort wind scale was used to classify the force of the wind at sea from force 0 (calm) to force 12 (hurricane) (Kinsman, 1968). Introduced by Sir Francis Beaufort of the Royal Navy in 1805, the scale defined the strength of the wind by its effects on an early nineteenth-century frigate "in chase." The scale was adopted by the Royal Navy in 1838 and subsequently used to record wind speed at sea and on land as systematic weather observations were introduced in the late nineteenth century. Several different tables of equivalent wind speeds were produced, and these were standardized by the International Meteorological Committee in 1926. Since its inception, major additions to the Beaufort wind scale include: wind speed equivalents; an extension of the scale to force 17 by the U.S. Weather Bureau in 1955 (Table 1); descriptive parameters for the visible effects on the sea and on the land of the corresponding wind forces; and estimates of wave height corresponding to wind forces (D.M.A.H.C., 1977; Russell and MacMillan, 1952). The wind force scale was introduced before direct measurements of wind velocity were possible and is now rarely used except by mariners.

EDWARD H. OWENS

References

Defense Mapping Agency Hydrographic Center, 1977. *American Practical Navigator.* Washington D.C.: Government Printing Office, H. O. Pub. No. 9, 1386p.

Kinsman, B., 1968. *An Exploration of the Origin and Persistence of the Beaufort Wind Scale.* Baltimore,

TABLE 1. Beaufort Wind Scale

Beaufort Number	Wind Speed				
	Knots	Mi/hr	M/sec	Km/hr	
0	Under 1	Under 1	0.0- 0.2	Under 1	Calm
1	1-3	1-3	0.3- 1.5	1-5	Light air
2	4-6	4-7	1.6- 3.3	6-11	Light breeze
3	7-10	8-12	3.4- 5.4	12-19	Gentle breeze
4	11-16	13-18	5.5- 7.9	20-28	Moderate breeze
5	17-21	19-24	8.0-10.7	29-38	Fresh breeze
6	22-27	25-31	10.8-13.8	39-49	Strong breeze
7	28-33	32-38	13.9-17.1	50-61	Moderate gale
8	34-40	39-46	17.2-20.7	62-74	Fresh gale
9	41-47	47-54	20.8-24.4	75-88	Strong gale
10	48-55	55-63	24.5-28.4	89-102	Whole gale
11	56-63	64-72	28.5-32.6	103-117	Storm
12	64-71	73-82	32.7-36.9	118-133	
13	72-80	83-92	37.0-41.4	134-149	
14	81-89	93-103	41.5-46.1	150-166	Hurricane
15	90-99	104-114	46.2-50.9	167-183	
16	100-108	115-125	51.0-56.0	184-201	
17	109-118	126-136	56.1-61.2	202-220	

Source: Modified from Pub. No. 9, *American Practical Navigator* (Bowditch), vol. I, 1977 ed. courtesy of the Defense Mapping Agency Hydrographic Center.

Md.: The Johns Hopkins University, Chesapeake Bay Institute, Tech. Rep. 39, 55p.

Russell, R. C. H., and MacMillan D. H., 1952. *Waves and Tides.* London: Hutchinson, 348p.

Cross-references: *Sea Conditions; Waves; Wind.* Vol. I: *Sea State; Wind Principles;* Vol. II: *Marine Climatology; Wind.*

BED FORMS

Forms of sediment accumulation in the beach zone include ridge-and-runnel systems, berms, a multiplicity of types of near-shore bars, cusp-type sand waves and crescentic bars (rhythmic topography), complex sand bodies affiliated with ebb-tidal deltas, and an ordered system of minor features that correlate with *flow-regime conditions.* It is the latter features that are generally referred to as *bed forms.* The sequence of bed forms that occur with increasing flow-regime conditions is shown in Figure 1.

The American Society of Civil Engineers (1966) proposed the following nomenclature for bed forms in alluvial channels, which can be accepted equally well for coastal conditions:

1. *Bed Configuration.* Any array of bed forms, or absence thereof, generated on the bed by the flow.

2. *Flat Bed.* A bed surface devoid of bed forms.

3. *Bed Form.* Any deviation from a flat bed that is readily detectable by eye, or higher than the largest sediment size present in the parent bed material, generated on the bed by the flow.

4. *Ripples.* Small bed forms with wavelengths less than approximately .3 m and heights less than approximately 3 cm. Ripples occur at flow velocities slightly higher than those required for initation of sediment motion but at lower velocities than flat bed or antidunes. Parts of the upstream slopes of dunes and bars are occupied by ripples under some flow conditions. In plan view, a ripple configuration can vary from an irregular array of three-dimensional peaks and pockets to a regular array of continuous, parallel crests and troughs transverse to the direction of flow. In longitudinal section, ripple profiles vary from triangular, with long, gentle upstream slopes and downstream slopes approximately equal to the angle of repose of the bed material, to symmetrical, nearly sinusoidal shapes. The triangular ripples commonly observed in alluvial channels move downstream with low velocities compared to the mean velocity of the generating flow. These are observed to occur only rarely in sediments coarser than approximately 0.6 mm.

5. *Bars.* Bed forms having heights comparable to the mean depth of the generating flow and the heights of the generating waves. In longitudinal sections, bars are triangular, with long, gentle upstream slopes and short downstream slopes that are approximately the same as the angle of repose of the bed material. Parts of the upstream slopes of bars are often covered with ripples or dunes.

6. *Dunes.* Bed forms smaller than bars but larger than ripples that are out of phase with any water-surface gravity waves that accompany them. Dunes generally occur at higher velocities and sediment transport rates than ripples, but lower velocities and transport rates than antidunes; however, ripples do occur on the upstream slopes of dunes at the lower velocities in the dune regime. The length of dune crests are usually of the same order as the wavelength. The longitudinal profiles of dunes are triangular, with fairly gentle upstream slopes and downstream slopes that are approximately equal to the angle of repose of the bed material. Dunes move slowly downstream with low velocities compared to the mean velocity of the generating flow. The large lee eddies that occur in dune troughs often cause surface boils of intense

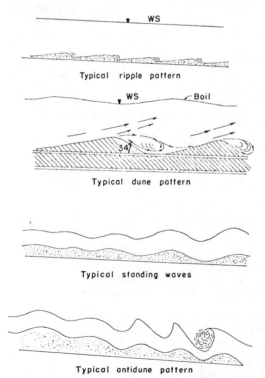

FIGURE 1. Development of bed forms. The top tier represents the lower flow regime and the bottom tier, the upper flow regime (from D. B. Simons and E. V. Richardson, 1962, Resistance to flow in alluvial channels, *Am. Soc. Civil Engineers Trans.* **127**, 927–1006).

turbulence and high sediment concentration above and slightly downstream from dune crests.

7. *Transition.* A bed configuration consisting of a heterogeneous array of bed forms, primarily low-amplitude ripples or dunes and flat areas. The transition bed is a configuration generated by flow conditions intermediate to those producing dunes and flat beds. In laboratory flumes, the transition configuration has been observed in some cases to consist of dunes or ripples over a reach covering part of the channel length and a flat bed over the remainder. The flow in the flat-bed reach is shallower than in the reach with ripples or dunes, and the discontinuity in the bed between the two reaches moves slowly downstream. In other instances, the bed configuration has been observed to consist of small, widely separated ripples or dunes over the entire bed.

8. *Antidunes.* Bed forms that occur in trains and that are in phase with and strongly interact with water-surface gravity waves. Antidunes can move upstream or downstream or can remain stationary, depending on the properties of the flow and sediment. The free-surface waves have larger amplitudes than the antidunes. At higher velocities and *Froude numbers* (q.v.), the surface waves usually grow until they become unstable and break in the upstream direction. The agitation accompanying breaking obliterates the antidunes, and the process of antidune initiation and growth is then repeated. At lower velocities, the antidunes will grow and then diminish in amplitude without the surface waves ever breaking. In longitudinal sections, antidune profiles vary with flow and sediment properties from triangular to sinusoidal, the latter occurring at larger Froude numbers than the former. However, the sharp-crested, triangular-shaped antidunes have been observed only in laboratory flumes and always more downstream. The crest lengths of antidunes are usually of the same order as wavelength.

9. *Chutes and Pools.* A sediment bed configuration occurring at relatively large slopes and sediment discharges that consists of large mounds of sediment that form chutes on which the flow is supercritical, connected by pools in which the flow may be supercritical or subcritical. A hydraulic jump often forms at the downstream end of each chute where it joins a pool or at the downstream end of the pools where the flow approaches the next mound. The chutes move slowly upstream. Chutes and pools are infrequently observed in the field.

Most of the terms defined herein have one to several synonyms in common use in the literature and among workers in the field. The following summary lists several of the more widely used synonyms:

Bed configuration: bed form, bed geometry, bed irregularities, bed phase, bed regime, bed shape, forms of bed roughness, sand waves.
Flat bed: plane bed, smooth bed.
Bed form: bed feature, bed irregularity, bed wave, dune, gravel bar, ripple, sand bar, sand wave.
Ripples: current ripples, dunes, ripple marks, sand waves.
Bars: banks, deltas, sand banks, sand waves, slipoff points for point bars.
Dunes: ripples, sand bars, sand waves.
Transition: sand waves, washed-out dunes.
Antidunes: antiripples, standing waves, stand waves (also applied to the accompanying water waves).
Chutes and pools: violent antidunes.

The beach zone contains all the variations of these bed forms; however, because of the shallow depths and high velocities that normally occur in the beach zone, Froude numbers greater than 1.0 are frequently produced and bed forms in the upper flow regime predominate.

The ridge-and-runnel (q.v.) beach profile is a common type. The following conditions prevail over the different portions of this profile, according to Hayes (1972):

Low-tide terrace. At near low-water stages, upper flow regime conditions are prevalent. Antidunes and plane beds are commonplace. At high tide, megaripples (bedforms with wave lengths between .6 and 6.6 m; *dunes* in some classification) and ripples occur because of the increase in water depth.
Ridge. As waves break across the ridge at high tide, upper flow regime conditions prevail. Antidunes are common, but plane bed predominates. Flow separation occurs along the top of the slip face, and sediment avalanches down the slip face, producing the landward migration of the ridge.
Runnel. Almost exclusively lower flow regime conditions because of the deeper water in the runnel.
Incipient berm. This zone is dominated by shallow, sheet flow at high tide (not covered at low tide); hence upper flow regime conditions predominate.

The physical processes of the formation of bed forms have been analyzed by numerous scientists (Kennedy, 1969). Although the study of sand dunes, including megadunes generated by tidal flows, is becoming increasingly important (McCullagh, Hardy, and Lockman, 1972; Yalin and Price, 1974), much attention has been paid to sand ripples under flow and water waves (Carstens, Neilson, and Altinvilek, 1967;

Shulyak, 1971; Mogridge and Kamphuis, 1972; Sleath, 1975).

RYSZARD ZEIDLER

References

American Society of Civil Engineers Task Force on Bed Forms in Alluvial Channel, 1966. Nomenclature for bed forms in alluvial channels, *Am. Soc. Civil Engineers Proc., Jour. Hydraulics Div.* **92**, (HY3).

Carstens, M. R.; Neilson, F. M.; and Altinvilek, H. D., 1967. An analytical and experimental study of the bed forms under water waves, *Georgia Inst. Tech. Eng. Exp. Station, Final Rept., Proj. A-798.*

Hayes, M. O., 1972. Forms of sediment accumulation in the beach zone, *in* R. E. Meyer, ed., *Waves on Beaches,* New York: Academic Press, 297-356.

Kennedy, J. F., 1969. The formation of sediment ripples, dunes, and antidunes, *Rev. Fluid Mechanics Ann.* **1**.

McCullagh, M. J.; Hardy, N. E.; Lockman, W. O., 1972. Formation and migration of sand dunes: A simulation of their effect in the sedimentary environment, *in* D. F. Merriam, ed., *Mathematical Models of Sedimentary Processes.* New York: Plenum Press.

Mogridge, G. Rm., and Kamphuis, J. W., 1972. Experiments on bed form generation by wave action, *Conf. Coastal Eng. 13th, Vancouver, Am. Soc. Civil Engineers, Proc.,* 1123-1142.

Shulyak, A., 1971. *Fizika voln na poverkhnosti sipuchei sredi i zhidkosti* (The Physics of Waves on the Interface of a Loose Medium and a Liquid). Moscow: Nauka.

Simons, D. B., and Richardson, E. V., 1962. Resistance to flow in alluvial channels, *Am. Soc. Civil Engineers Trans.* **127**, 927-1006.

Sleath, J. F. A., 1975. A contribution to the study of vortex ripples, *Internat. Assoc. Hydrology Research Jour.* **13**(3).

Yalin, M. S., and Price, W. A., 1974. Formation of dunes by tidal flows, *Conf. Coastal Eng., 14th, Copenhagen, Proc.,* 369-372.

Cross-references: *Bars; Beach Processes; Currents; Major Beach Features; Minor Beach Features; Ripple Marks; Ridge and Runnel.* Vol. VI: *Primary Sedimentary Structures.*

BENCH

A *bench* is a coastal feature formed because of the unequal solubility of various sediments. Waves and currents create these steplike structures in which the more soluble layers form treads. Benches can be developed on layers of durable sandstone that alternate with beds of readily eroded shale. Small shells are found as high as 18 m on benches, where they have been transported by spray. The downslope flow of spray has sufficient force to remove rock fragments from most bench surfaces. Erosional bench inclination depends upon the original rock bedding. Nearly horizontal benches can be formed when rock bedding is flat. Some benches extend to positions below sea level (Coastal Engineering Research Center, 1973).

JOHN B. HERBICH
JOHN P. HANEY

Reference

Coastal Engineering Research Center, 1973. *Shore Protection Manual.* Washington, D.C.: U.S. Army Corps of Engineers, 750p.

Cross-references: *Water-Layer Weathering; Wave-Cut Bench; Wave-Cut Terrace.*

BENTHOS

All organisms living in or on bottom substrates are defined as the *benthos* and sometimes as the *benthon*. The biome of the benthic or benthonic species is termed *benthal*. The benthic organisms are subdivided into groups according to organism size, methods of investigation, and taxonomic unit.

Organism Size Groups

Macrofauna are those animals larger than 1 mm (or, in some cases, larger than 0.5 or 0.4 mm) (Thorson, 1957). The term benthos is restricted to this animal size group in many publications. Important taxa of the macrofauna are polychaetes, bivalves, gastropods, crustaceans, echinoderms, ascidians, sponges, and coelenterates. Samples are taken with trawls and dredges for qualitative research and with grabs or suction devices for quantitative results. The meiofauna are the invertebrates smaller than macrofauna, with the lower size limit of 40 μm. In most sandy and muddy substrates the nematodes make up 90% or more of the multicellular meiofauna, while copepods and other crustaceans, polychaetes, and minor groups (tardigrades, gastrotrichs, turbellarians, halacarids) account for only a few percent. The unicellular foraminiferans may be included with the meiofauna, since they are sampled from the environment with the same small diameter corers or as subsamples from grabs. Elutriation and stereoscopic examination of the samples are used to isolate this benthic component from the substrate. Larvae and juveniles of macrofaunal species fall in the size group of the meiofauna. They are termed temporary meiofauna as opposed to the permanent meiofauna spending its total life span in this size order. Geologists use ostracod and foraminiferan shells for stratigraphy and count these as microfauna, while the biologists include them in the meiofauna. The microorganisms constitute the part of the benthos that cannot be properly studied without live observation and cultures (Fenchel, 1969). Protozoa such as ciliates and amoeboids may be counted while preserved, with some hesitation. Fungi and bacteria must be baited with nutrient media, and colonies are counted as infectious units after a few days. Culture conditions can never copy the natural habitat and its nutrient situation. Therefore total numbers of bacteria and of the different physiological groups encountered in the cultures may be lower than the standing stock. In some cases direct counts of viable bacteria under the flourescent light are preferable, and tracer methods are essential to study their ecological functions. Within the microorganisms heterotrophic and autotrophic nutrition is found. The microalgae are unicellular plants living on all substrates, which include mud, sand, rocks, shell, animals, and plants. Most important are the diatoms and the blue-green algae. Diatoms can cover the substrates totally, for example, giving mud flats a yellow to brown surface layer. The green, red, or brown macroalgae are multicellular, bushy, or with leaves, and can reach considerable size, as in the *Laminaria*. Algae settle on many kinds of stable substrates, on rocks and stones, on shells, and on harbor constructions. In tropical and subtropical regions, green and red algae incrusting $CaCO_3$ may be of importance in shallow-water assemblages and in the formation of sandy sediments. Mud and sand flats are often inhabited by eelgrasses belonging to the flowering plant.

Zonation

According to ecological conditions, mainly water level and tidal exposure time, the coastal regions are subdivided into horizontal zones on a worldwide scale. Table 1 gives the zonation of the littoral zone as presented by various authors. The mid-littoral zone—*étage mediolittoral* or *eulittoral*—is the zone between high water and low water tide marks (Friedrich, 1969).

There are differences in tide levels from place to place, resulting in a wide variety of ecological conditions and organism assemblages. In narrow straits the water level may change by 14 m (English Channel), while changes ranging from zero to a few centimeters are measured in other areas (Baltic). The reaches exposed during low tide may be small where steep rocky coasts prevail, or may cover many km^2 of muddy and sandy flats (North Sea). Organisms in this tidal belt are exposed to changing water level, currents, and wave action, and to a great variability in temperature and salinity conditions depending on the type of weather during low tide (Table 2).

Organisms living in this zone are able to withstand these severe conditions. The stress is stronger in the upper than in the lower reaches of the intertidal zone. Coverage by algae can protect many animals from dessication and allow them to settle in otherwise unsuitable habitats. Higher up the coast is the *supralittoral* or splash zone—the *étage supralittoral*. This region is not washed by the tides. Only occasional events, such as the splash of breaking waves and wind-driven spray, bring marine water into this zone. Otherwise organisms are fully exposed to terrestial conditions. Out of the coastal domain, and deeper than where tidal influences can determine faunal and phytal assemblages, is the *infralittoral* zone—*étage circalittoral* or *sublittoral* border. This zone is constantly covered with water, and conditions for life are less severe. Between the three zones transitional belts may be found that vary in width and formation according to tidal conditions.

TABLE 1. Some proposals for the vertical zonation of shallow marine benthal (after Friedrich, 1969).

Stephenson, 1949	Giordani-Soika, 1950	Yonge	Pérès, 1957	Ercegovic, 1957	Pérès & Piccard, 1958	Friedrich, 1969
Supralittoral zone		Splash zone	Étage supralittoral	Étage holophotique	Étage supralittoral	Supralittoral
Supralittoral fringe	Zone intercotidale supérieure	Upper shore				Übergangszone
Midlittoral	Zone intercotidale moyenne	Middle shore	Étage mesolittoral	Étage talanto-photique	Étage mediolittoral	Eulittoral
Infralittoral fringe	Zone intercotidale inférieure	Lower shore	Étage infralittoral	Étage mégaphotique Étage métriophotoque	Étage infralittoral	Übergangszone
Infralittoral zone						Sublittoral

TABLE 2. Physical and Chemical Measurements in a Wadden Pool (3–5 m across, 20 cm deep) in the Estuary of the Elbe River. The pool was inhabited by diatoms, barnacles, amphipods, and polychaetes (after Kühl, 1952).

Time	t°C	S‰	pH	Alkalinity cm³ n/10 HCl	O_2 mg/l	O_2 % sat.
0807	18.0	22.56	8.9	2.0	25.7	301
0907	21.7	22.79	9.4	1.6	24.1	311
0920	22.9	22.37	9.7	1.7	23.1	293
0945	24.0	22.43	9.9	1.5	22.2	283
1020	25.2	22.34	10.0	1.4	21.6	282
1030	25.9	22.43	9.1	1.4	20.5	274
1035	20.8	22.85	8.7	2.0	11.19	137
1040	19.8	17.60	8.1	2.3	9.57	114
1045	18.6	13.60	8.0	2.2	8.84	95
upcoming tide water:						
1005	17.0	12.41	7.8	2.2	–	–
river water:						
0915	17.1	13.09	7.9	2.2	–	–

Substrates and Ecological Groups

Benthic organisms are found in communities, associations, or assemblages throughout the world and are altered by the geographical distribution of species (Thorson, 1957). The associations are defined by typically dominant and abundant animal and plant species. Important descriptive factors are the types of substrate: rocks, boulders, gravel, sand, and mud; and faunal and floral components. Sand and mud themselves are indications of current conditions and water percolation through the sediment. This in turn governs the transport of oxygen into the sediment. Sandy sediments are normally well supplied with oxygen by water transport through the grains. The sedimentation rate of organic matter is low; it is used as food or is chemically oxidized. Muddy sediments develop under low current conditions, and the content of organic matter is high because of sedimentation. Oxygen replacement for the degradation of organic substances in the sediment is low, and it is zero in sediment depth of a few centimeters or even millimeters and hydrogen sulfide replaces the oxygen. Between the upper oxidized and the lower reduced sediments, a more or less steep discontinuity layer is developed in which the redox potential (Eh) changes from positive to negative values and other chemical alterations occur (Halberg, 1973). Macrofaunal animals can rarely survive for a long time in the presence of hydrogen sulfide unless they create a breathing current with palps, antennae, or legs. A few of the smaller faunal components can live in the sulfide biome (Fenchel and Riedl, 1970; Powell, Crenshaw, and Rieger, 1979; Felbeck, Childress, and Somero, 1981). Some species of bacteria can reduce nitrate, nitrite, or sulfate in the anaerobic decomposition of organic matter and thus are capable of living in unoxygenated sediments. Macroalgae require hard substrates; they constitute an important structural component for the fauna and act as food for herbivores. Microalgae grow on all surfaces, but under favorable conditions diatoms may move into the sediments. Some species, living on the surface of mud flats, give the flats a brown or yellow cover during low tide, but disappear among the grains when the tide is rising.

The macrofauna can be divided into the two subgroups of infauna and epifauna. The infauna live in the sediment, in more or less permanent burrows or creep in the sediment, which causes bioturbation. Feeding types are sediment or deposit feeders inside the sediment or on the sediment surface. Others project their palps (bivalvia), tentacles (polychaetes), or antennae and legs (crustacea) into the water above the bottom to catch floating food particles. The epifauna live on the substrate and developed special mechanisms for fastening to rocks, wood, or in the sediment. These organisms, such as sponges, actinians, hydrozoans, bryozoans, polychaetes, bivalves, crustaceans, and ascidians, are active or passive filter feeders or suspension feeders. Similar to algae, they act as substrate for other animals. In tropical regions reef corals, epifauna themselves constitute an important substrate for algae, epifauna, and infauna that live in the crevices. Most of the epifauna, and some infauna are exposed to predation by other benthic invetrabrates (echinoderms, polychaetes, crustaceans, gastropods, cephalopods) and by bottom and demersal fish.

The meiofauna live partly epifaunisticly on live and dead substrates, but most are found in the sediment. In muddy bottoms the animals dig through the grains, shifting them aside as they pass. In well-sorted sands having an intersitital space, the meiofauna may move between and around the grains in the interstitium. These meiofaunal species are grouped together as interstitial fauna or mesopsammon.

Recruitment and Densities

Recruitment of assemblages may occur through migration, but is mainly the result of reproduction and spat fall. Many benthic metazoans have planktic larvae that live in the water from a few hours to several months. These are transported by currents and are capable of selecting a suitable biotope. Larval time can be prolonged and metamorphosis shifted to a later time when no suitable environment is found. Aggregate species (oysters and barnacles) are attracted by already settled specimens. *Standing stock*, number of organisms per unit of area, range from a few to many thousands per square meter in the macrofauna, 10^4-10^7 in meiofauna, 10^6-10^9 in the ciliates, and up to 10^{12} in bacteria and in diatoms (Fenchel, 1969). In most environments *biomass* is made up mainly of the macrofauna, and ranges from a few grams to several kilograms per square meter. Abundance is determined by food availability, where chemical or physical conditions are not severe. Stable environments with a high predictability allow higher faunal and phytal diversity—i.e., number of species per number of specimens and unit of area. Diversity decreases under strongly changing environmental conditions. Production, increase in biomass per unit of time, is found between 10 and 10^2 gram per square meter (wet weight) per year, and may reach 50 kg in natural mussel beds.

HJALMAR THIEL

References

Felbeck, H.; Childress, J. J.; and Somero, G. N., 1981. Calvin-Benson cycle and sulfide oxidation enzymes in animals from sulfide rich habitats, *Nature* **293**, 291-293.

Fenchel, T., 1969. The ecology of marine microbenthos. IV. Structure and function of the benthic ecosystem: Its chemical and physical factors and the microfauna communities, with special references to the ciliated Protozoa, *Ophelia* **6**, 1-182.

Fenchel, T., and Riedl, R., 1970. The sulfide system: A new biotic community underneath the oxidized layer of marine sand bottoms. *Marine Biology* **7**, 255-268.

Friedrich, H., 1969. *Marine Biology: Introduction to Its Problems and Results*. London: Sidwick and Jackson Ltd., 474p.

Hallberg, R. O., 1973. The microbiological C-N-S cycles in sediments and their effect on the ecology of the sediment-water interface, *Oikos* **15**, 51-62.

Kuhl, H., 1952. Über die Hydrographie von Wattenpfützen, *Wiss. Meeresunters. Helgolander* **4**, 101-106.

Powell, E. N.; Crenshaw, M. A.; and Rieger, R. M., 1979. Adaptions to sulfide in the meiofauna of the sulfide system. 1. 35 S-sulfide accumulation and the presence of a sulfide detoxification system, *Jour. Exp. Marine Biol. Ecol.* **37**, 57-76.

Thorson, G., 1957. Bottom communities (sublittoral or shallow shelf.) *Geol. Soc. America Mem.* **67**, 461-534.

Cross-references: *Biotic Zonation; Coastal Ecology, Research Methods; Coastal Fauna; Coastal Flora; Corers and Coring Techniques; Grab Samplers; Organism-Sediment Relationship; Production; Spray and Splash Zone; Standing Stock.*

BIOGENOUS COASTS

Biogenous, or *biogenetic,* coasts are those that are developed mainly by the growth of plants (*phytogenetic coasts*) or living organisms (*zoogenetic coasts*). Mangrove coasts (q.v.) of tropical seas and reed coasts of shallow seas are of the phytogenetic type (Morskaja geomorphologia, 1980). Shelter from intensive open sea waves and frequent alternating submergence by seawater, combined with runoff from the land, are of the utmost importance for the growth of plant organisms forming the phytogenetic coasts. For this reason, phytogenetic shores are shallow shores exposed to tides and/or storm surges. Since even a small range of tide or storm surge results in the flooding of a wide belt of the shore with such a gently sloping surface, the shallowness of the shore is of special importance.

Mangrove Coasts

Shallow coasts of tropical seas develop swampy landscapes of *mangroves*—i.e., evergreen halophyte amphibian bushes and trees primarily related to the *Rhizophora, Avicennia,* and *Bruguiera genera*. Growths of these plants are usually dispersed zonally in a certain sequence. Thus on Pacific shores *Rhizophora* make up the forefront, being replaced by *Bruguiera* in the landward direction, and then *Avicennia,* sheltered by the latter or behind plantless sandy, silty spits (*Laguncularia racemosa* on west African and South America shores). *Bruguiera* is frequently replaced by *Sonnenatia alba* on east African shores.

A typical feature of all mangrove species is a stilted root system, including respiratory roots or pneumatophoras, holding the trees and bushes in a silty and often diluted bedrock. All mangrove species are capable of evaporating water in large amounts because of high osmotic pressure (30-60 atmospheres) in the cells of their roots and leaves (Alekhin, Kudryashov, Govorukhin, 1961). They thus obtain needed salts from seawater and gradually drain the coastal mangrove swamps. Consequently the mangrove front slowly moves toward the sea and the drained area is rapidly colonized by ordinary tropical plants. Therefore, mangroves generally promote stabilization and drainage of sandy, silty shore and expand the land. There are theories, however, that mangroves could precipitate suspended particles and therefore *actively* affect the accumulation of sediments and the shallowing process of the coastal sea. These theories are inaccurate as it has been pointed out that mangroves of the Colombian Pacific coast settle only in those places where they are protected by spits or shoals against direct sea wave influence or in coastal sites of broad drainage. Zenkovich (1962) described the same situation for the mangrove shores of the South China Sea. Therefore shallowness promoting an increase of shoreline accumulation is not only a consequence of mangrove settlement, but it is also a prerequisite.

Mangrove coasts are distinguished by an intensive accumulation of organic substances in the coastal deposits due to the gradual burying of an increasing amount of plant remnants.

Reed Coasts

Reed coasts are extensive in moderate wave zones, but they are especially noted on the Aral Sea. They occur on the northern coast of the Caspian Sea and are usually accompanied by deltaic shores. Like mangroves, coastal reeds evolve only on shallow seashores protected against a constant direct influence of waves. They are mainly composed of several genera of reed (*Phragmites* and the more wide-spread *Ph. australis*) and "rogoz" (*Thipha*), the latter being more wide-spread in Europe and Asia (*Th. latifolia* or "chakan"). The plant formations of the Caspian Sea long-grass coastal marshes are frequently called "chakan." The chakan has a width of 6-8 km at some sites on the northern Caspian coast but more often only a few hundred meters. Sedimentary material is accumulated in these growths, settling out from turbid waters during the chakan flood periods caused by storm surges and atmospheric dust storms. This deposition includes large amounts of plant residue. As with mangroves, coastal reeds promote drainage of the coastal swampy belt by intensive transpiration. All this benefits land expansion.

Reeds and "rogozas" as well as mangroves hold sandy, silty deposits because of the development of a root network, promote the soil layer formation, and thus protect loose coastal soils from washing out at unusually high water levels during storm tides or a collaboration of storm surges with spring tides.

Coral Coasts

The coasts of tropical seas are distinguished by coral reef constructions (see Coral Reefs)— i.e., a coastal (framing) barrier and circlelike atolls. The main morphodynamic elements of the coral shore are coral reefs, lagoons, and beaches if we speak of a barrier reef or atoll (Fairbridge, 1950). The backward, lower part of a reef flat looks like an elongate depression and is called a *boat channel* on shores with a fringing reef rather than a lagoon. The reef is a principal supplier of sediment-producing material for the coral coast, paving some sections of the reef flat and accumulating sand, gravel, and pebble for beaches (and on circlelike reefs and islands as well). Coarse material, such as boulders and blocks, substitutes for beaches when piled up at windward sites on some atolls of the Pacific Ocean; more often, however, the beach of a coral shore is composed of fine coral sand (Emery, Tracey, and Ladd, 1954). In general, coral coasts are accumulative, where accumulation is original and evolved by means of an interaction of biogenetic and wave factors.

Under certain conditions coral coasts can be subjected to destruction by waves and tides— for example, when a strong storm occurred at Funafuti Atoll in 1973. This phenomenon is approximately of the same order of magnitude as a catastrophic erosion of accumulative shores. Elevated coral limestone could be subjected to abrasion as it is on the uplifted coasts of ancient Nauru Atoll. However, this is not a coral shore but an ordinary abrasive shore composed of limestones.

O. K. LEONTIEV
S. A. LUKJANOVA

References

Alekhin, V. V.; Kudryashov, L. V.; and Govorukhin, V. S., 1961. *Geografiya Rastenii*. 532p.

Emery, K. O.; Tracey, J. I.; and Ladd, H. S., 1954. Geology of Bikini and Nearby Atolls, *U.S. Geol. Survey Prof. Paper 260-A,* 265p.

Fairbridge, R. W., 1950. Recent and Pleistocene coral reefs of Australia, *Jour. Geology,* 58, 340-401.

Morskaja geomorphologia, 1980. *Terminologicheskji spravochnic.* Mysl, 279p.

Zenkovich, V. P., 1962. *Osnovy Ucheniya o Razvitii Morskih Beregov.* 710p.

Cross-references: *Coastal Flora; Coral Reef Coasts; Halophytes; Mangrove Coasts; Vegetation Coasts.* Vol. III: *Coral Reefs.*

BIOMASS—See STANDING STOCK

BIOTIC ZONATION

The arrangement of groups of organisms having similar ecological tolerances and requirements into bands perpendicular to environmental gradients can be called *biotic zonation*. Where gradients are short and changes abrupt, these groups tend to be discrete and take the appearance of well-defined communities. Biotic interactions such as competition, predation, or grazing may force a particular species to occupy a narrower band than its ecological requirements alone would dictate.

The transition from the marine to the terrestrial environment gives rise to especially clear examples of zonation. Some environmental influences that produce distinct gradients in the coastal zone include degree of exposure to air, salinity (of water, soil, and aerosols), temperature, light, oxygen availability, and the presence of various nutrients.

Barrier structures provide a paradigm of zonation in the coastal zone. Figure 1 shows a general pattern of plant communities to be found across a typical barrier island of the United States east coast. Each physical region contains a group of organisms adapted to the conditions of that region. In the nearshore zone are marine organisms that can survive in an environment of high wave energy and shifting sand. Burrowing pelecypods such as *Spisula*, *Mactra*, and *Mercenaria* are major components of this community.

The intertidal foreshore is an especially difficult habitat since it is exposed alternately to air (frequently accompanied by hot sun or bitter cold) and then to pounding surf. Both marine and terrestrial predators can exact their toll. Larger organisms are often rapidly burrowing filter feeders, typified by hippid crabs and clams of the genus *Donax*. Interstitially, one finds large numbers of bacteria, single-celled algae, protozoans, and meiofaunal invertebrates.

The backshore, or beach berm, is exposed to the atmosphere for long periods, punctuated by violent storm attack and flooding. Most of the surface is bare sand, and the organisms that survive here do so as transients. Storms regularly level small sand dunes and wash away rooted plants, but these same events also leave drift lines containing seeds and fragments of dune and beach plants that can renew the cycle of growth and destruction in the next season. Rotting organic matter in the drift material releases nutrients to the plants and also supports a community of microorganisms and arthropods. Shorebirds forage here, and some species, such as terns and skimmers, prefer the backshore for nesting. On warm shores, ghost crabs (*Ocypode*) burrow in the sand, hunting and scavenging from the dunes to the low tide mark, and sea turtles nest in this zone.

The dunes behind the berm are built by salt-spray-tolerant perennial grasses (*Elymus, Ammophila, Uniola, Spartina patens*) that trap sand blowing along the beach. These plants grow upward through the accumulating sand, gradually stabilizing the dune. Salt-spray-resistant herbaceous plants, including legumi-

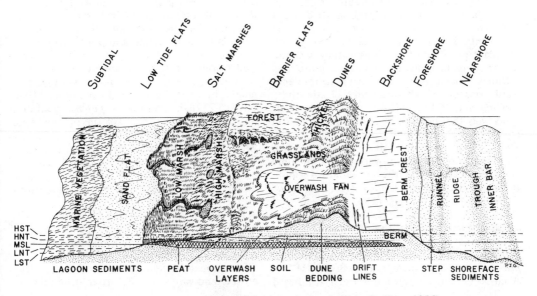

FIGURE 1. Zonation on a generalized barrier beach (from Godfrey, 1976).

nous nitrogen fixers (*Lathyrus, Strophostyles*), invade as well, and the stability and diversity of the community increase together as long as the dune is spared storm damage.

Rainwater falling on the building dune percolates down to lower layers of sand, to create a saturated zone, a fresh-water lens "floating" on top of seawater-saturated sands beneath. Between this lens and the uppermost dry layer (10–25 cm thick) is a zone of sand containing about 2% water, from which most dune plants draw their moisture. These plants cannot tolerate seawater around their roots while actively growing.

The special adaptations of dune grasses are numerous. These plants can reproduce vegetatively by rapid rhizomatous growth and by germination of water-borne fragments; their seeds are transported by the sea as well. Specialized leaves confer resistance to both salt spray and drought: stomates are located on the inner walls of deep parallel grooves along the undersides of these leaves and are thus protected from salt entry. During a dry period the leaves curl ever more tightly, narrowing the grooves and conserving moisture. The upper sides of the grass leaves have a hard, waxy cuticle that both retains water and keeps out salt. The photosynthetic chemistry of dune grasses from southern latitudes (*Uniola, Spartina*) features intermediates containing four carbon atoms ("C–4 metabolism"), a physiological adaptation to hot, dry conditions; more northern species (*Ammophila, Elymus*) are "C–3" plants.

Herbaceous plants of the dunes may have woolly leaf surfaces, thick cuticles, and other adaptations that prevent salt entry and water loss. The succulent, fleshy nature of some of these plants is not an adaptation to drought but a response to chloride ions absorbed by the plant. By retaining a great deal of water, the plant can dilute these ions as they accumulate in its tissues and thus avoid salt poisoning.

Dunes provide a physical barrier to the landward transport of salt spray. In a classic study, Oosting and Billings (1942) showed that salt spray was the major factor dictating zonation of plants on a North Carolina barrier island. They demonstrated that salt spray accumulation decreases as one goes landward across the dune zone, correlating with the appearance of additional species such as *Andropogon scoparius*. Experiments by Oosting (1945) showed that those plants found furthest from the foredunes are indeed least resistant to salt spray.

The actual mechanism of aerosol damage to plants and the role of salt spray in forming *pruned* trees were elucidated by Boyce (1954). Wind-borne spray strikes the tree's windward branches, leaves, and buds. Salt enters through tiny lesions and gradually kills the tissue; in time all the windward portions of the tree are killed. However, little salt lands on the leeward branches, which continue to grow normally. (Little or no salt is translocated horizontally within the plant.) The result is a wind-shaped appearance with all the branches sweeping away from the salt source. Many evergreen members of the maritime woodlands of southeastern United States have small, stiff, waxy leaves on short petioles. This structure resists twisting and turning in the wind, with concomitant cracking and salt entry.

Protection provided by seaward dunes allows development of shrub thickets and woodlands behind the main dune system. The most important members of the shrub community in eastern United States are oaks, hollies, cedars, myrtles, and woody members of the rose family. Where there is enough shelter the shrubland may grade into a substantial forest, usually dominated by oaks and pines with hollies, myrtles, red bay, cabbage palm, and various vines forming the understory; the species mix depends of course on the latitude. The tops of the woody plants will be level with the highest seaward dune ridge, or will slope gradually down to whatever protecting dune is present, as exposed twigs are killed by salt spray.

Another factor affecting zonation of plants on certain barrier structures is oceanic overwash. Where the shore is regularly subjected to storm-driven floods, the natural succession from grassland to shrubland is held back. Instead, a grassland vegetation containing species adapted to overwash persists indefinitely. *Spartina patens* is usually the dominant. Like other dune builders it can tolerate considerable burial, sending up vigorous rhizomes through the new sand and developing new culms and roots. But unlike *Ammophila* or *Uniola*, it can withstand occasional seawater flooding and so survive burial by water- as well as wind-borne sand. Experiments by the authors and colleagues (1970, 1976, 1979) showed that *S. patens* could push up through more than 0.5 m of new sand. Nearly complete recovery can occur within one year following overwash deposition on a dense *S. patens* stand. If overwash should be shut off by a solid dune line of natural or artificial origin, natural succession progresses and shrubs invade the grassland.

The fauna of these terrestrial coastal communities will usually be a depauperate subset of the animal life to be found inland at the same latitude.

Plant zonation mediated by tidal cycles occurs on the landward sides of barrier structures and on protected mainland shores. This is the salt marsh community, with its bands of vegetation directly related to frequency of seawater flooding. The upper limit of the highest spring tides coincides with the upper limit of species that can tolerate some seawater flooding. From that point to the upper level of the highest neap tides, one finds the *high salt marsh,* dominated in eastern United States by *Spartina patens* and *Distichlis spicata,* along with other graminoids and various succulent herbs. The elevations below the upper level of neap tides are flooded during each high tide and constitute the *low salt marsh. Spartina alterniflora* is widely dominant in this zone, and grows considerably taller close to standing water than it does in the upper low marsh. It has been shown that this height difference is not genetic, but is related to environmental conditions (Shea, Warren, and Niering, 1975).

The salt marsh is home to many insects, mites, spiders, snails of the genera *Melampus* and *Littorina,* amphipods, isopids, burrowing fiddler crabs (*Uca* spp.), the mussel *Modiolus demissus,* and certain specialized rodents.

The lower limit of *Spartina alterniflora* is a little above mean sea level in the north and slightly below in the south. There is evidence that this limit is set by the amount and duration of light reaching the plants during high tide. Between the edge of the marsh and the lowest spring tide line are mud and sand flats crowded with crustaceans, mollusks, and annelids. Below the spring tide line one finds underwater beds of marine angiosperms: *Zostera marina* and *Ruppia maritima* in the northeast and *Thalassia, Diplanthera,* and *Ruppia* to the south. These plants can tolerate little exposure to air at their upper limits, and their lower limits seem to be set by sunlight penetration, as with *Spartina.* Fields of *Zostera* and *Thalassia* teem with marine life of all kinds.

Another factor influencing zonation is salinity. Extremes occur in depressions in the marsh surface called salt pannes, where salt concentrations may surpass 60‰. In the saltiest pannes and in hypersaline lagoons, such as Laguna Madre in Texas, no higher plants can exist. The highly salt-resistant species of *Salicornia* are able to grow in only slightly less extreme conditions; they may even need some salt for optimum growth. *Spartina alterniflora* can persist in pannes where salinity is somewhat above that of sea water, but its best growth occurs at about 20‰. Other marsh plants, such as *Typha angustifolia* and *Spartina pectinata,* can survive only at the highest level of the marsh where the soil water is nearly fresh. *Spartina patens* can survive anything from fresh to nearly full-strength seawater, but its intolerance of regular flooding excludes it from the low marsh.

On tropical shores, intertidal mangrove swamps replace salt marshes in low energy locations. Like marsh plants, these trees are zoned according to tide height and salinity. In Florida, *Rhizophora mangle* dominates the lowest zone and also invades bare intertidal mud. Above it one finds *Avicennia germinans,* and then, bordering on the terrestrial vegetation, *Conocarpus erectus. Laguncularia racemosa* is found within the *Avicennia* zone and into the upper part of the *Rhizophora* area. The mangrove forest serves as a habitat for specialized forms of marine life and for many birds.

Many other examples of zonation as a function of environmental or biotic factors can be found in the coastal zone.

PAUL J. GODFREY
MELINDA M. GODFREY

References

Boyce, S. G., 1954. The salt spray community, *Ecol. Monogr.* 24, 29-67.

Godfrey, P. J., 1970. *Oceanic Overwash and Its Ecological Implications on the Outer Banks of North Carolina.* U.S. National Park Service, 37p.

Godfrey, P. J., 1976. Barrier beaches of the east coast, *Oceanus* 19, 27-40.

Godfrey, P. J., and Godfrey, M. M., 1976. *Barrier Island Ecology of Cape Lookout National Seashore and Vicinity, North Carolina.* Washington, D.C.: National Park Service Sci. Monogr. Series No. 9, 160p.

Godfrey, P. J.; Leatherman, S. P.; and Zaremba, R., 1979. A geobotanical approach to classification of barrier beach systems, *in* S. P. Leatherman, ed., *Barrier Islands.* New York: Academic Press, 99-126.

Oosting, H. J., 1945. Tolerance to salt spray of plants of coastal dunes, *Ecology* 26, 85-89.

Oosting, H. J., and Billings, W. D., 1942. Factors affecting vegetation zonation on coastal dunes, *Ecology* 23, 131-142.

Shea, M. L.; Warren, R. S.; and Niering, W. A., 1975. Biochemical and transplantation studies of the growth forms of *Spartina alterniflora* on Connecticut salt marshes, *Ecology* 56, 461-466.

Cross-references: *Aerosols; Bogs; Coastal Fauna; Coastal Flora; Halophites; Rocky Shore Habitat; Salt Marsh; Washover and Washover Fan.*

BLOWHOLES

A *blowhole* is a puncture in the roof of a *sea cave,* sometimes called *spouting horn.* Both names derive from the water and spray that are forced through the roof cracks high into the air whenever the proper conditions of wave-created pressure exist within the cave. Blowholes, sometimes located several meters away from the surf zone, occur along high energy shores composed of extensively jointed yet highly resistant rock. The same pressure that creates the blowhole and its occasional fountain may continue to enlarge the hole and may eventually cause collapse of the cave.

The Lawai blowhole on the south shore of Kauai, Hawaii, has been known to spout over 30 m into the air, while in the same vicinity an even larger blowhole was destroyed by dynamite because of the damage it was causing nearby sugar cane fields (Macdonald and Abbott, 1970).

JAMES E. STEMBRIDGE, Jr.

Reference

Macdonald, G. A., and Abbott, A. T., 1970. *Volcanoes in the Sea: The Geology of Hawaii.* Honolulu: University of Hawaii Press 441p.

Cross-references: *Sea Caves; Sea Cliffs.*

BLUE-GREEN ALGAE—See CYANOPHYCEAE

BLUFF—See SEA CLIFFS

BOAT BASIN DESIGN

There are many textbooks and reports that cover the details of boat basin design. We present here important factors that a property owner should consider when contemplating developing a boat basin. He can follow up the guidelines given herein to secure engineering expertise to carry out the details, which usually involve knowledge of all or part of the following disciplines: soils and dredging, which can involve either mechanical or hydraulic means of excavation; coastal and harbor installations involving analyses of basin exposure to waves and currents and design of floating and pier structures; engineering of buildings and road construction; and consideration of environmental factors that could affect the surrounding area and the water quality of the boat basin (Noble, 1977).

In the past few years it has become highly important to consider the socioeconomic impact brought about by the development of a boat basin. This involves analysis of: water quality; increased automobile traffic because of boat basin; parking requirements; increased boat navigation in the waterway; effect on business activities; effect on shoreline processes of erosion and accretion; and changes in noise patterns.

Another important consideration is to check the economic feasibility wherein an estimate is made of annual income versus annual expenses, which should include amortization of the project, unless it is to be performed under a tax-supported system or other governmental support. It is often also important to secure assistance in developing a program for obtaining the financial backing of the project either by institutional funding or by government program support.

The important details to consider in boat basin design are grouped below into different planning categories.

Soils and Dredging (Geotechnical Services)

Soils along waterfronts and tidelands can vary extremely in characteristics. Sandy and noncohesive soils are much easier to dredge and present far fewer problems and lower resulting costs than do silty and muddy soils. Rock can of course be expensive to excavate. In silty and muddy areas, basin slopes must be flatter and fills will cause subsidence, adding to costs and affecting harbor aesthetics. Placement of sand fills by hydraulic dredging is easier and faster and results in fewer foundation problems than with handling silty and muddy soil.

If the site is in an active seismic area, sandy soil must be checked for the occurrence of liquefaction wherein loose nongraded sands can become *quick* and lose bearing capacity during an earthquake. Soft mud can similarly present problems because lateral support is weak and can fail during an earthquake.

Protection of the Boat Basin

Protection of a boat basin from wave action and currents is most important and usually costly unless nature provides a good degree of natural protection. Breakwater protection from wave action and offshore jetties to guide currents are not as likely to directly produce income as would land fill, berthing, and building areas. Bulkheads and stone riprap to protect shorelines from erosion are much more expensive when built in mud and clay than in sand because of foundation and slope stability problems. The coarser the sand, the more erosion-resistant the shoreline.

Design of Harbor Structures

Design of floating structures should be varied to meet wave exposure conditions. Waves inside a marina should usually not exceed .3 m in height, and surge and swell should be eliminated. Special attention should be given to joint connections, especially where finger berths attach to the main head walks. Pontoons should be securely attached to superstructure, to prevent displacement or stress from waves.

Dense reinforced concrete becomes stronger when placed in an aqueous environment unless tensile or torsional forces are not compensated for in the design. Prestressed or posttensioned concrete can be used advantageously in the more exposed sites to avoid maintenance problems. Laminated timber, if preserved with salt treatments or creosote, is useful in such stressed areas so that loads can be distributed. Preservative treatments should cover any bored holes for connections since these areas are most susceptible to sea worm attack. Care should be used in handling treated lumber as any damage to the protective cover will open the member to attack.

Building and Road Construction

Such construction should take into account foundation conditions. Often marina basins are located in soil areas poorly suited for construction such as marshes and tidelands consisting of clays and bay muds. The most common method to overcome these conditions is to place buildings on pilings to secure solid foundations. However, when the soil around the building starts to consolidate from the weight of fill, subsidence of the surrounding ground lowers the elevation so that buildings are left high in the air. Subsidence also places stresses on pipeline connections that can crack the pipe at the buildings or between joints if they are rigid.

Some of the ways to overcome these problems are: the use of flexible connections; the use of light-weight fill or none at all if the area is suitable for traffic and any rising waters can be diked off from entering the building area; the use of floating, structurally-sound building mat foundations that will subside with the surrounding area (care should be taken to avoid differential settlement). If soils consist of loose sands and a high water table and the area is seismically active, liquefaction should be investigated. Liquefaction occurs when sands become *quick* and nonsupporting. To avoid liquefaction, loose soils should be compacted to densify them.

Where clays and silts exist, good drainage systems must be designed to remove moisture from under pavements. Rock base and the use of filter cloth over such soft material are used to eliminate soft spots that result in pavement settlement. Flexible asphaltic pavements can help eliminate cracks that could develop in rigid pavements.

Water Quality

Maintenance of acceptable water quality has become an integral part of marina design. In this respect engineers find themselves faced with the problem of presenting a balanced design that will meet the requirements of protecting the marina from rough water, providing the desired moorage capacity, and ensuring sufficient exchange of water. While this is not an insurmountable problem it is a relatively recent one and consequently planning techniques are in a stage of infancy.

Characteristically, marinas are sub-basins of a main body of water used to moor boats. To ensure protection from waves and currents, small boat basins are either located behind headlands, in indentations in the main body of water, or behind breakwaters or other engineering structures. Inherently, all of these regions provide areas of restricted water flow that may act as a trap for incoming pollutants (Noble, 1977). Sources of incoming pollutants may be associated with marina activities (gasoline spillage, boat bottom cleaning and painting, cleaning of fish, and so on) or from pollutants in the main body of water that may flow into the marina basin. Most of the water quality problems associated with marinas will be due at least indirectly to the entrapment of materials introduced into the system.

Conditions that are most often considered pollutant problems are: high levels of organic material; algae blooms; low dissolved oxygen concentration; high concentration of heavy metals and pesticides; high grease content; high degree of turbidity; high water temperature; and, from an aesthetic point of view, general uncleanliness of the marina waters. Many of these problems are interrelated. For example, a high organic content often may be related to low dissolved oxygen levels due to a high degree of bacterial decomposition of the organic material (a biochemical process that will utilize oxygen). Low dissolved oxygen can be a direct cause of fish kills inside marina areas.

Basically then, the more a marina acts as a trap, or is indicative of a low water exchange area, the higher is its potential to accumulate pollutants. The degree to which it accumulates pollutants is highly dependent on the influx from the main body of water and from the marina area itself. Thus it becomes obvious that if a marina is designed to the extreme where the water is very calm, it may be good for boat safety but also counter-productive environmentally. Thus once a decision is made to build a marina, a compromise or balance must be reached between a safe calm harbor and a sufficiently flushed, environmentally safe harbor.

To attain the necessary balance, two basic approaches can be used: to provide means to create circulation inside the marina, such as a pumping system to be used in times of low circulation; or to attempt to create a design that will provide natural flushing action. The latter is probably the way of the future. Recent research efforts have suggested that physical characteristics of marinas can be controlled to enhance flushing. While this approach is in its infancy, the mechanical approach can provide needed circulation for marinas that experience water stagnation and possible pollution problems.

The focus so far has been on operational water quality problems. Pollution problems that may be associated with the construction of marinas are related to dredging and filling activities. Both activities may disrupt benthic organisms and/or natural feeding areas; or they may cause turbidity problems, or introduction of toxicants into the water column.

Water quality as well as other environmental problems of marina construction and operation may be eased greatly by proper design techniques and the cooperation of the people using the facilities.

Property owners planning a boat basin design are advised not to neglect consideration of boat basin maintenance. Shoaling, pollution, and maintenance of protective breakwaters cause extreme problems if proper engineering is not followed.

H. MORGAN NOBLE
SCOTT MORGAN NOBLE

Reference

Noble, S. M., 1977. *Use of Benthic Sediments as Indicators of Marina Flushing*. Corvallis: Oregon State University, Master's thesis.

Cross-references: *Breakwater; Bulkheads; Coastal Engineering; Coastal Reserves; Human Impact; Jetties; Nearshore Water Characteristics; Pollutants; Soils; Thalassotherapy.*

BODDEN

Bodden originates in a north German dialect and means *bottom*. It characterizes a wide, shallow sea bay; however, by definition, bodden is restricted to bays of postglacial origin. During the rise of the sealevel 7000-4000 years ago (*Litorina transgression* in the Baltic), lowlands between ground moraines were flooded and became shallow marine regions. Bays of this type are found in Denmark and in Germany on the Baltic south coast (Hurtig, 1954). Only in Germany are these bays termed bodden, being used locally as part of a proper name, for example, Greifswalder Bodden, the shallow bay of the Baltic near the town of Greifswald. As bodden are part of a moraine-molded area, sand transport by prevailing currents parallel to the coast may lead to the formation of sand barriers (*Nehrung*) and lagoons (*Haff*).

HJALMAR THIEL

Reference

Hurtig, T., 1954. *Die mecklenburgische Boddenlandschaft und ihre entwicklungsgeschichtlichen Probleme*. Berlin: Deutscher Verlag der Wissenschaften, 148p.

Cross-references: *Europe, Coastal Morphology; Flandrian Transgression; Förde; Geographic Terminology; Haff; Nehrung; Sea Level Curves.*

BOGS

A *bog* is a wetland with acidic water, poor drainage, and a distinctive vegetation of sphagnum moss, insectivorous plants, and members of the Ericaceae (heath family) (Deevey, 1958). Coastal bogs have been formed in various ways. On coasts made up of glacial deposits, kettle holes are common. These were initiated when large blocks of ice were surrounded by outwash sediments as the glaciers diminished. When the blocks finally melted, ponds appeared in any of the resulting depressions that were deep enough to intersect the water table. If drainage was limited, organic acids accumulated in these ponds. The resulting low pH inhibited decay of organic matter, which gradually built up in the pond. Lack of oxygen and minerals soon limited growth of most plants, but sphagnum moss and certain tolerant shrubs such as swamp loosestrife (*Decodon verticillatus*) began to build a floating mat of vegetation around the edges of the pond. This mat eventually covered the whole pond; organic matter continued to accumulate, and in time the whole basin was filled.

Among the ericaceous plants that invade such a growing mat, leatherleaf (*Chamaedaphne calyculata*) is the most important. Others include sheep laurel (*Kalmia angustifolia*), bog huckleberry (*Gaylussacia dumosa*), and the cranberries (*V. macrocarpon* and *V. oxycoccos*). Insectivorous species such as the sundews (*Drosera* spp.) and pitcher plants (*Sarracenia purpurea*) can derive nitrogen from the insects they trap and digest with the help of specially adapted leaves. Although insects are not essential to their survival, evidence suggests that these plants maintain a competitive edge in a nitrogen-poor environment by their carnivority. Around the edges of most bogs may be found a characteristic zone of taller shrubs such as white swamp azalea (*Rhododendron viscosum*), highbush blueberry (*Vaccinium corymbosum*), poison sumac (*Rhus vernix*), and stunted red maples (*Acer rubrum*). Bogs are excellent places to demonstrate plant succession.

In eastern Canada and the Arctic bogs can be found anywhere on the coast where freshwater accumulates. Reflecting the slow decay of organic matter in these cold climates, some are even raised above the surrrounding landscape rather than confined to depressions. The vegetation includes most of the species of New England coastal bogs, along with boreal plants such as Labrador tea (*Ledum groenlandicum*), bog rosemary (*Andromeda glaucophylla*), and black spruce (*Picea mariana*) (Dansereau and Segada-Vianna, 1952; Tanner, 1944).

Bogs may also be found in shallow depressions throughout the flat coastal plain of southeastern United States. Like all bogs, they owe their existence to poor drainage. In the acid waters that accumulate in the depressions, sphagnum moss is again the most characteristic plant. Growing in the moss are several kinds of pitcher plants and sundews and other insectivorous plants such as Venus flytrap (*Dionaea muscipula*), confined to coastal bogs of southern North Carolina and northern South Carolina. Other members of the diverse southern bog flora are various blueberries and huckleberries, sweet bay (*Magnolia virginiana*), loblolly bay (*Gordonia lasianthus*), titi (*Cyrilla racemiflora*), fetterbush (*Pieris nitida*), bog dog laurel (*Leucothoe axillaris*), inkberry, bamboo brier (*Smilax laurifolia*), honeycup (*Zenobia* spp.), leatherleaf, wax myrtle (*Myrica cerifera*), pepperbush, redbay (*Persea borbonia*), and sheep laurel (Wells, 1967). A number of rare plants, especially orchids, are found only in these southern bogs. The general appearance of the vegetation is evergreen with small, leathery leaves.

On barrier spits and islands, wetlands often occur in interdune depressions. In southern latitudes these are marsh communities, but in cooler climates bog vegetation may be found here. Boglike communities among the dunes of Cape Cod's Province Lands are dominated by sphagnum moss and cranberries; indeed, the commercial cranberry bog is more like an interdune wetland than a kettle hole bog. Cranberry growers simulate the conditions of the interdune wetland by flooding their bogs in winter to protect the cranberry vines from frost and by regularly adding fresh sand to stimulate the growth of the plants. A few interdune ponds develop a true bog mat, with most of the typical bog plants present. An example is Shank Painter Pond, near Provincetown, Massachusetts, in which rare orchids have been found.

Although bogs are common in the far north, they become increasingly restricted as one goes south. Many contain unusual plants, such as the orchids of Shank Painter Pond and the southern *pocosin* bogs, and the Venus flytrap with its extremely limited range. Southern bogs are especially threatened by drainage programs likely to destroy the habitats of plants found nowhere else on earth. In addition to their unique flora, bogs are important reservoirs of water and stored carbon. And perhaps most important, the cold, acid waters of bogs have preserved good chronological pollen records of past climates and vegetation. For all these reasons, temperate zone bogs are appropriate objects of conservationists' concern.

PAUL J. GODFREY
MELINDA M. GODFREY

References

Dansereau, R., and Segada-Vianna, F., 1952. Ecological study of the peat bogs of North America, *Canadian Jour. Botany* **30**, 490-520.

Deevey, E. S., Jr., 1958. Bogs, *Sci. American* **199**, 114-122.

Tanner, V., 1944. Outlines of the geography, life, and customs of Newfoundland-Labrador, *Acta Geographica* **8**.

Wells, B. W., 1967. *The Natural Gardens of North Carolina*. Chapel Hill: University of North Carolina Press, 458p.

Cross-references: *Biotic Zonation; Coastal Flora; Estuaries.*

BOILER—See BLOWHOLES

BOMBORA, BOMBY

Bombora (bomby) is an Australian aboriginal term for a dangerous stretch of water where waves break over a submerged reef of rocks. This meaning is maintained in Australia and is often more widely applied to include any relatively small area of submerged rocks or coral reef lying off the coast over which waves break either continuously or only during storms or large swell conditions (Fig. 1). Bomboras are

FIGURE 1. Two-meter high waves breaking on submerged (background) and emerged (foreground) bomboras off Long Reef, New South Wales, Australia (Photo: A. D. Short).

therefore prevalent off *cliffed coasts.* Locally the term may be applied to exposed or semi-submerged rocks over which waves break. In some cases where waves are ridden by surfers, the term applies to the breaking wave itself.

ANDREW D. SHORT

Cross-references: *Cliffed Coasts; Skerry and Skerry Guard; Surfing.*

BORES

In some shallow estuaries particular phenomena can be produced by a special interaction between river flow and tide oscillation. The rise of tide occurs, with some difficulty, against the flow; seawater can penetrate into the estuary only during high tide and in some cases only during spring tide. Thus a long wave appears (a special kind of *solitary wave*, deformed by a great slope in the advancing front), running against the current for some hundred kilometers.

The speed of such a wave $c = \sqrt{g(H + h)}$ (where g is gravity, h river depth, H wave height) can be remarkable—more than 10 m/sec. This speed can increase in narrow channels to values higher than the current speed. Such a phenomenon is called a *bore* (*mascaret* in French; *pororoca* by the Indians in South America).

Bores can be responsible for severe damage at the edges of rivers and in shipping or fluvial harbors. Damage can be comparable to that caused by a *storm surge* or *tsunami.* Small and almost inappreciable bores can appear in various estuaries and rivers, as in the Severn and the Gironde (Europe). Great and dangerous bores can be produced in rivers of other regions, as in Argentina, Brazil, and China. If damage is to be avoided, navigation must be regulated according to the rhythm of the tide or to the maximal tides. South American Indians hang their small boats in trees when a bore is coming. Of major importance, however, is the effect of the bore on erosion at the river edge and also on the dissolution of sediment accumulation.

FERRUCCIO MOSETTI

Cross-references: *Coastal Engineering; Estuaries; Protection of Coasts; Seiches; Tides; Waves.*

BOULDER

A *boulder* is a rock detached from the parent body (Gary, McAfee, and Wolf, 1972). Characteristically, some degree of rounding has taken place through abrasion during transport. The size ranges from 256 mm to several meters in diameter (greater than $-8\ \phi$). Smaller boulders are usually more rounded.

Boulders may be of any kind of rock or even of clay or peat (Twenhofel, 1950), and in general there is a wider range of composition than in *cobbles, pebbles,* and finer clastics. Weathering *in situ* is dealt with by qualifying the type of boulder—hence *boulder of exfoliation.*

ALAN PAUL CARR

References

Gary, M., McAfee, R., Jr.; and Wolf, C. L., 1972. *Glossary of Geology.* Washington, D.C.: American Geological Institute, 805p.

Twenhofel, W. H., 1950. *Principles of Sedimentation.* New York: McGraw-Hill, 673p.

Cross-references: *Beach Material; Beach Material, Sorting of; Clay; Cobble; Gravel; Pebble; Sand; Sediment Size Classification Shingle and Shingle Beach; Silt; Soils.*

BOULDER BARRICADES

Boulder barricades are elongate rows of boulders that flank the coastline, separated from the shore by a low gradient nearshore zone (Fig. 1). They are the result of ice transport and therefore occur only in Arctic or sub-Arctic regions. In North America these forms have been reported in Labrador (Daly, 1902; Rosen, 1979, 1980); Hudson Strait, east Foxe Basin, Baffin Island (Bird, 1964); and the St. Lawrence River (Brochu, 1957); in other areas they have been reported in the Baltic Sea and Fennoscandia (Lyell, 1862; Tanner, 1939). Løken (1962) utilized uplifted boulder barricades in northern Labrador as an accurate sea level indicator to delineate the Holocene regression.

Numerous hypotheses have been offered to explain these accumulations. Tanner (1939) suggested the diminution of barricade development in Lake Melville, Labrador, was due to the concomitant reduction in tidal range. However, boulder barricades in the Baltic Sea occur in areas with astronomic tide ranges—as low as 2 cm.

Model of Boulder Barricade Formation

Two process systems must be considered to explain the deposition of boulder barricades: lifting of boulders onto ice rafts during the winter, and transportation during spring breakup.

In tidal regions, such as Labrador, intertidal ice freezes downward with increased freezing at each high tide. Boulders are frozen in the ice at the intertidal bottom and lifted as ice melts

FIGURE 1. Boulder barricade in Makkovik Bay, Labrador, at low tide (Photo: P. Rosen).

from the surface. Observations on a broad intertidal flat indicate that more boulders are lifted from the upper intertidal zone, so apparently the less frequent lifting of the ice during spring tides was more effective at encasing boulders than the diurnal lifting from the lower intertidal zone. High melting rates occur from the ice surface in late winter, so the continued freeze-down and surface melt result in the transportation of boulders up through the ice.

In the Baltic Sea off Tallinn, Estonia S.S.R., where astronomic tides are minimal, the infrequent lifting of the ice necessary for encasement and freeze-down may be due to meteorological tides that are the major cause of sea level fluctuations in the region.

By spring, when snow cover is melted, boulders are commonly observed sitting on the surface of small intertidal ice pans. The intertidal zone breaks up first, because of numerous tidal cracks and the decreased albedo of the mud-laden nearshore ice. Shore leads up to 1 km. wide serve as thoroughfares for these wind-transported ice rafts. The boulders are randomly deposited as *boulder flats*. This commonly occurs in Lake Melville, Labrador, and on the deltaic flats at the heads of embayments in Labrador. However, many of the sedimentary intertidal zones in central Labrador consist of steeply sloping uplifted marine clays with the top surface planed off to a low gradient by contemporary wave and ice processes. This results in a slope break near the low water line (Fig. 2). Since the ice thickness is comparable to the tide range, the highest probability for ice rafts to ground is at this slope break. Accumulation of boulders at this point over successive seasons results in an intermittent barricade, which further serves to trap ice rafts during breakup. Landward of this slope-break/boulder-barricade position, random *boulder flats* are also common. These result from ice rafts that are grounded during higher tidal stages.

At Tallinn (Fig. 3), the barricades form in a morphologically similar setting. In this case, the nearshore is a rock-cut bench, and the barricades accumulate at the seaward limit of this bench (M. Schwartz, personal communication).

Boulder barricades are the result of grounding of boulder-laden ice rafts in the nearshore zone. The requisite conditions for the formation of boulder barricades are: a rocky coastal setting, to serve as a source of boulders; sufficient winter ice and water level fluctuations to entrain boulders in ice rafts; and a distinct slope break in the nearshore zone. Without the third condition, boulders will be deposited as boulder flats.

The terminology of ice-formed shore structures has been confusing in the past. Boulder barricades are distinct from *boulder ramparts* and *ice-push ridges* in that they are located seaward of the strandline because of the conditions cited above.

PETER S. ROSEN

References

Bird, J. B., 1964. *The Physiography of Arctic Canada.* Baltimore, Md.: Johns Hopkins University Press, 220p.

Brochu, M., 1957. *Déplacement de Blocs par la Glace le long du Saint-Laurent.* Ottawa: Ministère des Mines et des Relevés Techniques, Étude Géogr. No. 30, 27p.

Daly, R. A., 1902. Geology of the northeast coast of

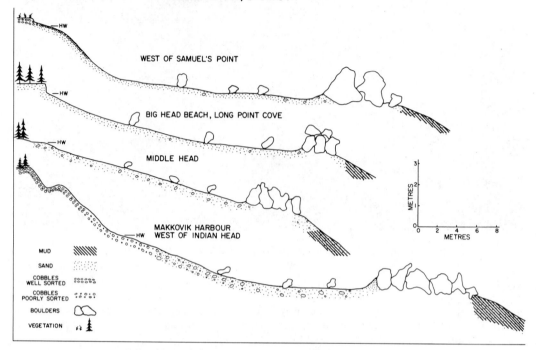

FIGURE 2. Nearshore profiles at selected sites in Makkovik Bay, Labrador. Random boulders are common in the intertidal zone. Note steep drop offs seaward of boulder barricades.

FIGURE 3. Boulder barricade west of Tallinn, Estonia S.S.R. (Photo: M. L. Schwartz).

Labrador, *Harvard Univ. Mus. Comp. Zool.* **38**, 203–270.
Løken, O., 1962. The late-glacial and post-glacial emergence and the deglaciation of northernmost Labrador, *Geog. Bull.* **17**, 23–56.
Lyell, C., 1862. *Principles of Geology.* New York: Appleton, 834p.
Rosen, P. S., 1979. Boulder barricades in central Labrador, *Jour. Sed. Petrology* **49**, 1113–1124.
Rosen, P. S., 1980. Coastal environments of the Makkovik region, Labrador, *in* S. B. McCann, ed., *The Coastline of Canada.* Ottawa: Geological Survey of Canada Paper 80-10, 267–280.

Tanner, V., 1939. Om de blockrika strandgördlarna (Boulder barricades) vid subarktiska oceankustar, förekomstsatt og upkomst, *Terra* **51**, 157–165.

Cross references: *Boulder; Cliff Erosion; Ice along the Shore; Ice-Bordered Coasts; Ramparts; Wave-Cut Platform.*

BRACHIOPODA

One of the dominant invertebrates of the Paleozoic Era, the phyllum Brachiopoda is represented by more than 25,000 fossil species. At present only about 250 species survive, occurring throughout the world but abundant only in such locales as the Sea of Japan and Australian waters. Brachiopods have a bivalve shell superficially resembling the Pelecypoda, but the internal anatomy is strikingly different. The organism feeds by means of a pair of organs called *brachia* (known collectively as the *lophophore*), consisting of loops or spirals of cilia-covered tissue. The lophophore is similar to the feeding apparatus of the Ectoprocta and Phoronida, suggesting that these phyla have a common ancestry (Hickman, 1973). The name *brachiopod* (Greek for arm-footed) was proposed by Dumeril in 1806, who assumed that the brachia were used for locomotion. Internal organs include muscles, a digestive tract, several glands, and a simple nervous system. A fleshy foot, the *pedicle*, extends between the valves or through an opening in one valve. Members of the class Inarticulata bear a long, flexible pedicle used for burrowing. The Inarticulata are also marked by the absence of a hinge, the valves being held together only by muscles. A second class, Articulata, possess an interlocking hinge and a short, rigid pedicle used to attach the organism to solid substrates. The Articulata have calcareous shells, while the relatively primitive Inarticulata secrete a mixture of chitin and calcium phosphate. Both classes produce free-swimming, ciliated larvae that persist for several weeks before settling to the bottom.

The Brachiopoda are highly adaptive, inhabiting depths from 5000 m to the intertidal zone, although the majority of species dwell in shallow water (Ager, 1967). Appearing in the Precambrian, brachiopods rapidly evolved into diverse forms and occupied many environments, making them invaluable as index fossils for paleontologists (Moore, 1965). By the late Paleozoic brachiopods were in a state of rapid decline, their habitats being taken over by the Mollusca. Members of both classes may still be found at extremely low tides and include *Terebratalia* from rocky beaches of the North Pacific and *Lingula*, an ancient burrowing species that inhabits sandy areas. The latter species is particularly abundant near southern Japan and the Philippine Islands, where it is boiled and eaten by the local residents.

GEORGE MUSTOE

References

Ager, D. V., 1967. Brachiopod paleoecology, *Earth Sci. Rev.* **3**, 157-179.
Hickman, C. P., 1973. *Biology of Invertebrates.* St. Louis, Mo.: C. V. Mosby Co., 757p.
Moore, R. C., ed., 1965. *Treatise on Invertebrate Paleontology, Part H, Brachiopoda.* New York: Geological Society of America, 523-927.

Cross references: *Asia, Eastern, Coastal Ecology; Australia, Coastal Ecology; Coastal Fauna; Mollusca; Organism-Sediment Relationship; Phoronida.*

BREAKERS—See WAVES

BREAKWATERS

Breakwaters are structures that protect beaches, harbors, anchorages, and basins from waves by decreasing wave energy or wave heights through wave reflection and energy dissipation (Coastal Engineering Research Center, 1973). Breakwater types include rubble mounds, impermeable, mobile or floating, submerged, hydraulic, and pneumatic. They can be either shore-connected or detached. The choice of type depends upon economics, available materials, position and size of the breakwater, wave characteristics, and the desired degree of wave attenuation. Breakwaters are constructed mainly for navigation purposes, and can have a detrimental effect on downstream beaches because the sediment transport is interrupted by the presence of the breakwater—in the case of shore connected breakwaters—or by the absence of wave action—as in the case of offshore (detached) breakwaters. In these cases, it may be necessary to pump the sand, using stationary dredge units, from upstream accreted side to the eroded downstream side. These are often referred to as *sand bypassing* plants.

JOHN B. HERBICH
JOHN P. HANEY

Reference

Coastal Engineering Research Center, 1973. *Shore Protection Manual.* Washington, D.C.: U.S. Army Corps of Engineers, 750p.

Cross references: *Air Breakwaters; Coastal Engineering; Waves; Wave Shadow.*

BROWN ALGAE—See PHAEOPHYCOPHYTA

BRUUN RULE

Originally a theory by P. Bruun (1962), which was later elevated to the level of a rule by Schwartz (1967), the *Bruun Rule* applies to low coastal and adjacent nearshore zones, composed of unconsolidated sediments (Fig. 1). Assuming an *equilibrium profile* as sea level rises, material

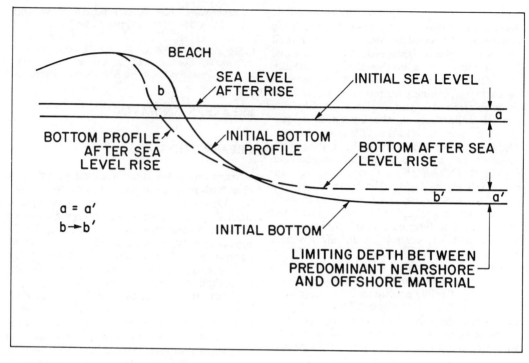

FIGURE 1. Shore erosion following a rise in sea level according to the Bruun Rule (after Schwartz, 1967).

eroded from the beach foreshore and backshore is deposited on the shoreface down to a limiting depth: there is a parallel shoreward beach profile displacement during the erosion of the upper beach zone; the material eroded equals the volume of material deposited on the nearshore bottom; the rise of the nearshore bottom level due to the deposition equals the sea level rise. This maintains a constant water depth in that area and elevates the nearshore profile proportionally to the rising sea level (Dubois, 1975). The theory was tested on stream tables (Schwartz, 1967) and short-term marine and lake beach-nearshore profile field surveys (Schwartz, 1967; Dubois, 1975, 1976; Rosen, 1977, 1978; Schwartz and Fisher, 1980). Swift et al. (1972) pointed out that the concept is applicable primarily to mainland beaches and to the substructures of coastal barriers, and that, taken literally, it requires that longshore transport must be in a steady-state process with no net change of the sediment volume at a given area. In other circumstances, including barrier islands, as well as general coastal retreat involving shoreface erosion and simultaneous seafloor aggradation (Swift et al., 1972), other (accumulational and erosional) factors modify the rule.

ERVIN G. OTVOS

References

Bruun, P. M., 1962. Sea level rise as a cause of shore erosion, *Am. Soc. Civil Engineers Proc., Jour. Waterways and Harbors Div.* 88, 117-130.

Dubois, R. N., 1975. Support and refinement of the Bruun Rule on beach erosion, *Jour. Geology* 83, 651-656.

Dubois, R. N., 1976. Nearshore evidence in support of the Bruun Rule of beach erosion, *Jour. Geology* 84, 485-491.

Rosen, P. S., 1977. Nearshore evidence in support of the Bruun Rule of beach erosion: Discussion, *Jour. Geology* 85, 491-492.

Rosen, P. S., 1978. A regional test of the Bruun Rule on shoreline erosion, *Marine Geology* 26, M7-M16.

Schwartz, M. L., 1967. The Bruun theory of sea-level rise as a cause of shore erosion, *Jour. Geology* 75, 76-92.

Schwartz, M. L., and Fisher, J. J., eds., 1980. *Proceedings of the Per Bruun Symposium, Newport, Rhode Island, November 1979.* Bellingham, Wash.: Bureau for Faculty Research, Western Washington University, 83p.

Swift, D. J. P.; Kofoed, J. W.; Salisbury, F. P.; and Sears, P., 1972. Holocene evolution of the shelf surface, central and southern Atlantic shelf of North America, *in* D. J. Swift, D. B. Duane, and O. H. Pilkey, eds., *Shelf Sediment Transport: Process and Pattern.* Stroudsburg, Pa.: Dowden, Hutchinson & Ross, 499-574.

Cross-references: *Beach Profiles; Coastal Ecology, Research Methods; Coastal Engineering; Coastal Engineering, Research Methods; Coastal Morphology, Research Methods; Equilibrium Shoreline; Profile of Equilibrium; Profiling of Beaches; Shore Defense; Sweep Zone.*

BRYOPHYTA

Bryophytes are green plants, mostly small. The plant body ranges from a pad of soft, fleshy tissue with little external differentiation to erect, trailing or pendulous forms resembling higher plants by possession of a stemlike axis (*caulidium*) bearing leaflike appendages (*phyllidia*) (Sharp, 1974a, 1974b; Watson, 1964). The two phases of the life history are obvious. The dominant gametophytic (*haploid*) sexual phase produces eggs and sperms, and bears the resultant dependent sporophytic (*diploid*) phase that bears the capsule (*spore case*). Specialized conducting tissues and roots are lacking in both generations; water and mineral nutrients are absorbed through the surface of the plants. Reproduction is by both spores and fragmentation.

The geographic range is cosmopolitan. Greatest abundance is in moist to wet sites with some kinds in fresh water. Special abundance occurs in tropical rain- and cloud-forests, with few in the deserts. Occurrence in saline habitats is rare.

The two classes of bryophytes are mosses and liverworts. Mosses (Bryopsida or Musci), are distinguished by possession of a gradually elongating stalk of the capsule, opening of the capsule usually by a definite cap usually exposing a ring of minute delicate teeth and absence of elaters. Sphagnum (peat moss) is the only member of appreciable commercial importance. Liverworts (Heptopsida or Hepaticae) are distinguished by being thelloid, or, if "leafy," with the stalk of the capsule elongating, if ever, only at maturity, and capsule opening irregularly or by longitudinal valves. The two subclasses are: Hepatidae (Hepaticae or liverworts in a narrow sense), and Anthocerotidae (Anthocerotes or hornworts or horned liverworts).

LOUIS CUTTER WHEELER*

References

Sharp, A. J., 1974a. Hepatopsida, *in Encyclopaedia Britannica*, 15th ed., Macropaedia 8. Chicago: Encyclopaedia Britannica, 779-781.

Sharp, A. J., 1974b. Bryopsida, *in Encyclopaedia Britannica*, 15th ed., Macropaedia 3. Chicago: Encyclopaedia Britannica, 351-354.

*Deceased

Watson, E. V., 1964. *The Structure and Life of Bryophytes.* London: Hutchinson & Co., 192p.

Cross-references: *Biotic Zonation; Coastal Ecology, Research Methods: Coastal Flora; Coastal Plant Ecology; United States, History of.*

BRYOZOA

Bryozoa (also known as Ectoprocta, sea mats, or sea mosses) are aquatic, benthic animals. The majority are marine, and only one class (Phylactolaemata) has a few genera that live in freshwater. Bryozoans are filter-feeding, invertebrate organisms having sessile colonies that range in size from a few millimeters to large, ramose growths of several meters in height (Fig. 1), and are composed of calcium carbonate and/or hardened chitin. These colonies are often mistaken for those of hydroids or corals. An individual animal (zooid) of a colony is generally 0.05-0.20 mm in diameter; the ciliated crown of tentacles of a zooid extends above the surface of the colony when feeding. Colonies are hermaphroditic, but not all zooids are. A colony develops by metamorphosis of a sexually produced, free-swimming larva that settles, attaches, and then undergoes asexual budding to produce extensions of the colony. The anatomy of bryozoans, as well as extensive documentation of the phylum, is presented in Brien (1960), Hyman (1959), and Ryland (1970).

Most species flourish best on stable bottoms in shallow waters less than 100 m deep on the continental shelves, however bryozoans occur from within the intertidal zone to the bathyal depths. The colony is usually encrusting or attached to a firm substrate, such as rock, shell, or algae. Bryozoans are common members of fouling communities. Temperature, wave action, salinity, turbidity, water currents, and substrate are important factors affecting the distribution of species. Where the substrate is an alga, the bryozoan epophyte community appears commonly to be species specific. A bryozoan colony may be short-lived and survive only one or two seasons of a year or for one whole year. Others are perennial species that survive for several years. Many epiphytic bryozoans have a short life cycle with two or more overlapping generations in a year. In short-lived species, a great number of zooids may produce embryos in rapid succession. Some species have two generations each year. One generation in the fall, in which the larvae overwinters, forms a colony in the spring, which grows rapidly, produces larvae in the summer, and then dies. The other generation grows and matures quickly in the

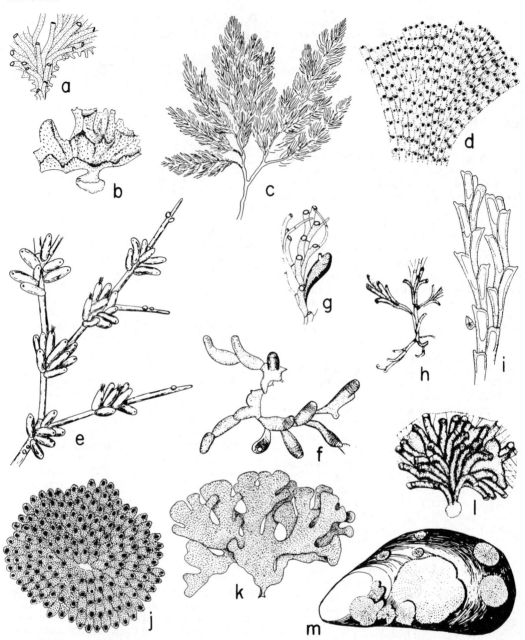

FIGURE 1. Colony forms. Order Cyclostomata: a, *Tubulipora*, several long, tubular zooecia on each branch; g, *Bicrisia*, biserial branches with chitinous joints between zooecia; h, *Bicrisia*, clusters of zooecia on colony branch; l. *Tubulipora*, fan-shaped colony. Order Cheilostomata: b, *Sertella*, reteporid colony with tiny windows between rows; c, *Bugula*, highly branched colony; d, *Schizoporella*, encrusting, fan-shaped colony; i, *Bugula*, portion of colony showing two rows of zooecia; j, *Cryptosula*, radiating colony; k, *Flustra*, ramose colony; m, several encrusting cheilostomes on bivalve shell. Order Ctenostomata: e, *Bowerbankia*, clusters of zooecia along branching stolon; f, *Sundanella*, contiguous zooecia in single file (Hyman, 1959; Ross, 1970).

summer, produces the larvae for the fall generation, and dies before the onset of winter.

The class Gymnolaemata includes the two orders Cheilostomata and Ctenostomata. Cheilostomes (Fig. 1: b, c, d, i, j, k, m) are the dominant group of living bryozoans, and most are characterized by boxlike zooecia (calcareous exoskeleton containing the zooid), hood-shaped brood chambers, and two or three different kinds of zooids that perform specialized functions.

The ctenostomes have only a soft, cuticular skeleton and commonly have a series of stolons connecting individual zooids. The class Stenolaemata has one living order, the Cyclostomata (Fig. 1: a, g, h, l), which are characterized by long, tubular zooecia.

JUNE R. P. ROSS

References

Brien, P., 1960. Bryozoaires, in Traité de Zoologie, vol. 5, fasc. 2. Paris: Masson et Cie, 1054-1335.
Hyman, L. H., 1959. The Invertebrates, vol. 5, Smaller Coelomate Groups. New York: McGraw-Hill, 783p.
Ross, J. R. P., 1970. Keys to the Cyclostome Ectoprocta of marine waters of northwest Washington State, Northwest Science 44, 154-169.
Ryland, J. S., 1970. Bryozoans. London: Hutchinson University Library, 175p.

Cross-references: *Algal Mats and Stromatolites; Coastal Fauna; Fouling; Organism-Sediment Relationship.*

BUILT PLATFORM OR TERRACE—See WAVE-BUILT TERRACE

BULKHEAD

A bulkhead (or *quay wall*) is a wall-like structure, generally constructed of wood, concrete, or steel pilings, separating a land mass from a body of water (Coastal Engineering Research Center, 1977). Its primary purpose is to retain or support the ground in back of it, but it can also protect the upland from erosion by wave action.

Bulkheads are generally used to maintain a depth of water along the shoreline in a harbor or river, as for a wharf, or support development in an advanced position relative to the adjacent shoreline (Quinn, 1961). However, for ocean-exposed locations, bulkheads do not provide a long-term protection against erosion; a more substantial wall is required because the beach continues to recede and larger waves reach the structure.

CHARLES B. CHESTNUTT

References

Coastal Engineering Research Center, 1977. *Shore Protection Manual.* Washington, D.C.: U.S. Army Corps of Engineers, 1264p.
Quinn, A. F., 1961. *Design and Construction of Ports and Marine Structures.* New York: McGraw-Hill Book Co., 531p.

Cross-references: *Coastal Engineering; Protection of Coasts; Quay.*

BULLER—See BLOWHOLES

BUOYS

Buoys are devices for oceanographic measurements that permit *in situ* surveys, including those extended in the time. The form and size of the buoy differ according to use (Myers, Holm, and McAllister, 1969). For example, if a buoy is used for wave recording, its dimension must be such as to have a floating period different (i.e., larger) from the wave period (about 1-20 seconds). Buoys can accumulate the data or broadcast in real time to a sampling center, which records the data. Power for data sampling and/or transmission is obtained from electric batteries on board, but replacement or substitution of the batteries is a problem involving expensive service. The first experiments for self-powered buoys, by means of electric power furnished by oceanic waves, are now being carried out.

At present, surveys by buoys replace or complement surveys by ship (cruises). Real-time transmission of data is more useful especially for meteorologic and oceanographic forecasting (e.g., surges, hurricanes). According to the purpose, buoys can sample both meteorological and oceanographic parameters. The most common meteorological parameters are air pressure, temperature, humidity, and wind speed and direction. In special cases, even radiation (from sun and sky), light absorption, and polarization can be measured. Oceanographic measurements commonly involve current speed and direction, salinity, and temperature. All these parameters are sampled at different depths (i.e., pressures). Sea level is measured in order to provide information on tides, seiches, and other long-range oscillations or on waves produced by wind. In special cases, buoys measure optical (turbidity), acoustical, and electrical properties or some special chemical constituent (nutrients, for example) of seawater. Thus, they are useful in providing pollution information.

Modern use of buoys is possible because various (physical or chemical) parameters can be transduced into electrical signals owing to the progress in data transmission and computerization.

FERRUCCIO MOSETTI

Reference

Myers, J. J.; Holm, C. H.; and McAllister, R. F., 1969. *Handbook of Ocean and Underwater Engineering.* New York: McGraw-Hill, 1100p.

Cross references: *Coastal Ecology, Research Methods; Coastal Engineering; Coastal Engineering, Research Methods; Coastal Morphology; Coastal Morphology,*

Research Methods; Current Meters; Marigram and Marigraph; Protection of Coasts; Tide Gauges; Tides; Waves; Wind.

BURROWS AND BORINGS

Burrows are tubes or holes of varying diameter, length, configuration, and composition excavated or constructed by organisms, chiefly worms, crustaceans, or molluscs, along a bedding plane or penetrating the bedding. If a lining is present, it may be preserved as a fossil (*trace fossil*), or only the tube filling of sediment may be preserved. Burrows are made into soft sediment; a similar tube or bioerosion into hard material (shell, wood, rock) is a *boring*.

Seilacher (1953) classified burrows, borings, and other trace fossils according to behavior (feeding, dwelling, or locomotion).

The main contribution of burrows and borings has been to paleoecology, behavioral studies, and environmental reconstructions (Frey, 1975). They are also good paleobathymetric indicators.

Burrows and borings are usually found even when other fossils are absent, thus providing some environmental information from an otherwise barren formation. However, caution should be used to study them in conjunction with fossils of the organisms that made them, whenever possible, and not to overinterpret from the burrows and borings alone.

THOMAS E. PICKETT

References

Frey, R. W., 1975. *The Study of Trace Fossils.* New York: Springer-Verlag, 562p.

Seilacher, A., 1953. Studien zur Palichnologie. I. Über die Methoden der Palichnologie, *Neues Jahrb. Geologie u. Paläontologie Abh.* **96,** 421-452.

Cross-references: *Benthos; Coastal Fauna; Intertidal Mud Habitat; Intertidal Sand Habitat; Organism-Sediment Relationship.*

BYPASSING—See SAND BYPASSING

C

CALA AND CALA COAST

Sometimes considered as a particular case of *ría coasts*, the word *cala* better suits a narrow and drowned valley with steep sides, peculiar to the limestone coasts of southeast Majorca. The Balearic Islands, Catalonia, Provence, Corsica, and the Portuguese and Italian coasts have the typical cala or *calanque* that—although in some cases match with granitic lands (Catalonia, Provence) or schists—generally corresponds to calcareous coasts such as the *sherum* (sherm) of the Red Sea. The cala is normally associated with a *rambla*, arroyo, or dry gulch with subvertical edges in a subarid climate, distinct from the *ría* in Galicia or the *aber* in Bretagne. The bottom gradients in calas vary between 0.5 and 4%; while the short arroyos are more inclined than the long ones and often remain hung without forming calas. Other types are karstic canyons, not related to surficial flow, located in high, retreating coasts.

The relevant literature on this topic includes: Barbaza (1971), Butzer (1962), Drooger (1973), Penck (1894), and Rossello (1964, 1975).

VICENÇ M. ROSSELLÓ

References

Barbaza, Y., 1971. Morphologie des secteurs rocheux du littoral catalan septentrional, *Service Doc. Cartograph. Géogr. Mem. et Documents (Paris), 1970* **11,** 152p.

Butzer, K. W., 1962. Coastal geomorphology of Majorca, *Assoc. Am. Geographers Annals* **52,** 191-212.

Drooger, C. W., 1973. *Messinian Events on the Mediterranean.* Amsterdam: North-Holland Publishing Co., 272p.

Penck, A., 1894. *Morphologie der Erdoberfläche,* 2 vols. Stuttgart: Englehorn.

Rosselló, V. M., 1964. *Mallorca. El sur y sureste* (Thesis, University of Valencia). Palma de Mallorca: Cámara de Comercio, Ind. y Naveg., 553p.

Rosselló, V. M., 1975. El litoral de Mallorca. Assaig de genètica i classificació, *Mayurqa* **14,** 5-19.

Cross-references: *Canali; Karst Coast; Ria and Ria Coasts.*

CALCARENITE

Calcarenite is a clastic *limestone* consisting predominantly of sand-grade (1/16 to 2mm in diameter) calcitic or aragonitic particles (allochems, Folk, 1959); it is a consolidated carbonate sand. Grabau (1904) introduced this term as *calcarenyte* in his genetic classification of sedimentary rocks to convey the basic calcareous composition, arenite grain size, and exogenetic or clastic nature of the rock.

Calcarenite, like the related terms *calcirudite* and *calcilutite* that describe the coarser-grained and the finer-grained limestones, respectively, has survived as a general term that is particularly useful for field descriptions. It may be qualified by terms that indicate the dominant clastic components—e.g., brachiopod calcarenite, crinoid calcarenite, oolitic calcarenite. In the strict sense, the term should not be applied to autochthonous accumulations such as reef rock or coquinoid limestones (oyster banks), but should be employed only to describe limestones that have a mechanically produced texture. Thus the term calcarenite not only indicates mineral composition and texture but also carries a genetic connotation.

The clastic carbonate grains of calcarenites are of subaqueous origin, being largely biogenic (skeletal debris, fecal pellets), chemical (ooids), or consisting of fragments of reworked carbonate material (intraclasts, Folk, 1959). Typically calcarenites are *grain stones* (grain-supported framework, Dunham, 1962), but where the proportion of sand-grade particles to matrix is reduced, they grade to *wackestone* (carbonate-mud-supported framework). Although the majority of calcarenites are water-laid (hydroclastic), locally wind-transported grains give rise to eolian calcarenites—e.g., the Pleistocene eolinites of Bermuda and the Bahamas.

Impure varieties occur, such as argillaceous calcarenite, carbonaceous calcarenite, siliceous calcarenite; detrital components, such as quartz, felspar, and rock fragments, may occur as ooid nuclei in oolitic calcarenite. Other noncarbonate impurities may be authigenic—e.g., chalcedony, lutecite, felspar, and glauconite.

Current bedding, ripple marks, and scour-and-fill structures are common features of calcarenites that provide clues to depositional conditions (transportation media, current strength). Information regarding the nature of the environment is also provided by the types of clastic grains in the deposit. However, in modern limestone classifications much significance is placed on the presence or absence of fine carbonate "ooze" matrix (micrite, Folk, 1959) as an index of energy levels prevailing at the time of deposition of the carbonate sediment. Thus calcarenites may be "-sparites," or "-micrites." In the former, clastic grains are enclosed by a crystalline cement, this texture being the consequence of deposition under the influence of current agitation or wave and surf action, which winnow away fine particles, leaving behind a grain-supported sediment with voids for subsequent occupation by cement. On the other hand, the "-micrites," represent deposits accumulated under low energy conditions and consist of angular grains enclosed by a matrix of carbonate mud.

Therefore, in more precise work carbonate petrologists have replaced calcarenite by terms that more accurately describe texture and composition and that have narrower genetic implications—e.g., oosparite, intrasparite, biosparite (well-sorted deposits accumulated in higher energy conditions), biomicrite, pelsparite (less well-sorted deposits formed in low energy environments: see Folk, 1959).

The extensive calcarenites of the Paleozoic, Mesozoic, and Tertiary constitute quantitatively the most important limestone type. Holocene examples of calcarenites are provided by most beach rocks, and in the contemporary coral-reef environment the most abundant sediment type will form calcarenite after lithification. For further references see Pettijohn (1975).

G. R. ORME

References

Dunham, R. J., 1962. Classification of carbonate rocks according to depositional texture, *in* W. E. Ham, ed., *Classification of Carbonate Rocks.* Tulsa, Okla.: American Association of Petroleum Geologists, 108-121.
Folk, R. L., 1959. Practical petrographic classification of limestones, *Am. Assoc. Petroleum Geologists Bull.* **43**, 1-38.
Grabau, A. W., 1904. On the classification of sedimentary rocks, *Am. Geol.* **33**, 228-247.
Pettijohn, F. J., 1975. *Sedimentary Rocks,* 3d ed., New York: Harper and Row, 628p.

Cross-references: *Coral Reefs; Dalmatian Coastline; Humate; Karst Coast; Reefs; Ripple Marks; Sand; Sand Source; Sediment Analysis, Statistical Methods.* Vol. VI: *Calcarenite.*

CANALI

The term *canali* is common among French and southern European geomorphologists. It has sometimes been equated with *fjord*. It is considered by many to be a constituent part of the *Dalmation coast* type. Schmidt (1923) considered them to be similar to Red Sea sherum. The *maremme* (French) or *maremma* (Italian), which has been identified by some researchers as similar to calanques and canali, is a low, marshy tract along the coast.

Canali are the regional equivalent of *calanques*, common along the Provence Mediterranean coast of France, and may be considered synonymous with them. The head of canali and calanques is usually a small beach. Both are dry valleys that have been excavated in limestone relief by the normal processes of erosion during a wet period. The dry valley has subsequently been submerged by a rise in sea level. Deep calanques, separating peninsulas, are common in the region near Marseille, and deep canali provide excellent harbors along the Yugoslav coast.

The term *ría* has also been proposed to cover the above types of inlets and embayments. Gulliver (1899) suggested using it to designate all types of subaerially carved troughs, including *fjords, dalmans,* and *limans* (hypersaline coastal lakes). Such generalization may be too broad since the ría more specifically designates estuary bays that widen and deepen seaward and extend far into land—e.g., the rías of the Río de Vigo and the Río de la Coruna (in northwestern Spain).

ROGER H. CHARLIER

References

Gulliver, F. P., 1899. Shoreline topography, *Am. Acad. Arts Sci. Proc.* **34**, 149-258.
Schmidt, W., 1923. Die Scherms an der Rotmeerküste von el-Hedschas, *Petermann's Geog. Mitt.* **69**, 118-121.

Cross-references: *Cala and Cala Coast; Dalmatian Coast; Fiord, Fjord, Fyord; Geographic Terminology; Karsts Coast; Liman and Liman Coast; Ria and Ria Coasts.*

CARBON 14—See RADIOCARBON DATING

CAROLINA BAYS

Carolina bays are shallow, elliptical, usually sand-rimmed, poorly drained, closed depressions along the Atlantic coastal plain from northern

Florida to New Jersey. There may be over one-half million such features (Prouty, 1952). In some areas they are so abundant that they intersect. The long axis of most Carolina bays is oriented NW-SE. Most are only marshy, but others are deep enough to have ponds or even lakes in them. They range in diameter from about 50 m to many km. Sedimentation and vegetation in Carolina bays are usually distinct from those of surrounding areas. They are best developed in the Carolinas (Cooke, 1940). Somewhat similar forms are found in Arctic regions.

The origin of Carolina bays is much debated. They have been ascribed to: meteorite impacts, springs, sand bar dams in drained drowned valleys, giant sand ripples, submarine scour, segmentation of lagoons, several kinds of solution processes, erosion by eddy currents, fish nests, eolian blowouts, or oriented depressions in interridge swales of former beach plains and dune fields (Price, 1968).

THOMAS E. PICKETT

References

Cooke, C. W., 1940. Elliptical bays in South Carolina, and the shape of eddies, *Jour. Geology* **48**, 205-211.
Price, W. A., 1968. Carolina bays, *in* R. W. Fairbridge, ed., *The Encyclopedia of Geomorphology*. New York: Reinhold Book Corp., 102-109.
Prouty, W. F., 1952. Carolina bays and their origin, *Geol. Soc. America Bull* **63**, 167-224.

Cross-references: *Geographic Terminology; Lakes, Coastal Morphology; Oriented Lakes; Shoreline Development Ratio.* Vol. III: *Carolina Bays.*

CAUSEWAY

A *causeway* is a road built across wetlands or water. It can be constructed as a raised embankment or as a bridge (California Division of Highways, 1960; Coastal Engineering Research Center, 1977).

CHARLES B. CHESTNUTT

References

California Division of Highways, 1960. *Bank and Shore Protection in California Highway Practice.* Sacramento, Calif.: State of California Department of Public Works, 423p.
Coastal Engineering Research Center, 1977. *Shore Protection Manual.* Washington, D.C.: U.S. Army Corps of Engineers, 1264p.

Cross-references: *Coastal Engineering; Protection of Coasts.*

CAVE—See SEA CAVES

CAVITATION

Cavitation is difficult to define precisely. A brief description of the main features and the fundamental processes can be best illustrated by the following comments. If a mass of liquid is heated isobarically, or if its pressure is reduced isothermally by static or dynamic means, a state is ultimately reached at which vapor or gas- and vapor-filled bubbles or cavities become physically apparent and begin to grow. This bubble growth, if by diffusion of dissolved gas into the cavity, may be at a normal rate. The bubble growth will be explosive, however, if it is primarily a result of vaporization into the cavity. This latter condition is commonly called *boiling* if it is caused by a temperature rise and *cavitation* if it is caused by an isothermal dynamic-pressure reduction (Knapp, Hammitt, and Daily, 1970).

Cavitation is, then, a liquid phenomenon and does not occur in any normal circumstances in either a solid or a gas. Cavitation is the direct result of pressure reductions in a liquid, and thus it can presumably be controlled by controlling the amount of reduction. Cavitation is a dynamic phenomenon, since by definition it is concerned with the growth and collapse of cavities. However, cavitation can occur in a liquid whether the liquid is in motion or at rest.

There are four main types of cavitation. *Traveling cavitation* is composed of individual transient cavities or bubbles that form in a liquid and move with it as the cavities or bubbles expand, shrink, and collapse. *Fixed cavitation* occurs when a liquid flow detaches from the rigid boundary of an immersed body or a flow passage to form a cavity attached to the boundary. *Vortex cavitation* occurs in the cores of cavities that form in zones of high fluid shear. This cavitation may appear as traveling or as fixed cavities. *Vibratory cavitation* is caused by a continuous series of high-amplitude, high-frequency pressure pulsations in the liquid. No cavities will be formed unless the amplitude of the pressure variation is great enough to cause the pressure to drop below the vapor pressure of the liquid.

JOHN B. HERBICH
ROY B. SHILLING

References

Knapp, R. T.; Hammitt; F. G., and Daily; J. W., 1970. *Cavitation.* New York: McGraw-Hill, 578p.

Cross-references: *Cliff Erosion; Coastal Engineering; Coastal Erosion; Wave Erosion.*

CENTRAL AMERICA, COASTAL ECOLOGY–See CENTRAL AND SOUTH AMERICA, COASTAL ECOLOGY

CENTRAL AMERICA, COASTAL MORPHOLOGY

Central America is the land-bridge between North and South America. The distance from northwest to southeast is about 2200 km. The narrowest part is the Isthmus of Panama, where the width is about 50 km. The widest part is in the Honduras-Nicaragua border region, where the width is about 500 km. The west coast opens on the Pacific Ocean; the east coast, the Caribbean Sea and the Atlantic Ocean.

Morphology

The Pacific coast is more crenulated than the Caribbean coast. The Pacific coast has peninsulas, gulfs, and bays, while the Caribbean coast has some large-scale gulfs (Honduras, Mosquito, Darien) and some small-scale lagoons. Thus the coastline is represented partly by continental land and partly by coastal barriers.

Profiles of the coasts of Central America show a variety from flat coasts, such as *wadden* areas (q.v.) and large littoral plains, to steep coasts with cliffs.

The submarine profile is generally gently sloped nearshore, with shelves of varying width (10–50 km). On the Mosquito Bank of Honduras, in the Carribean Sea, the width is up to 200 km; in the Gulf of Panama, on the Pacific Ocean, it is up to 150 km. On the Pacific Coast, from Nicaragua to Guatemala, the offshore area slopes from the shelf to the Middle America Deep Sea Trench (more than 6000 m deep).

Large areas are of intertidal character; wadden, bays, and mud flats are developed in large embayments and lagoons. Barriers and lagoons are developed along the Caribbean and the Pacific coasts. The barriers are of 1–2 m in height, and differ from small scale (10 m wide and 100 m long) up to 10–50 km long and 1–3 km wide. The lagoons on the Pacific coast are in general 1–3 m deep and in tidal channels to 10 m or more, while on the Caribbean coast lagoons are mostly flat, 0.5–1.5 m.

Dunes on the coast of Central America are scattered, mostly flat, with heights of 1–2 m.

Cliffs are not developed along the Pacific coast of Guatemala. However, they are partly developed along the coast of El Salvador, the Gulf of Fonseca, Nicaragua, and Panama, and are intensively developed along the coast of Costa Rica. Along the Caribbean coast, cliffs are present along only some parts of coastal Honduras, Nicaragua, and Costa Rica, with somewhat more on the coast of Panama. Ancient raised terraces are developed where cliffs occur along the Pacific coast of El Salvador and Costa Rica.

Along the Pacific coast there are islands only off Panama, while on the Caribbean coast there are small islands, called *cayos*, scattered along the Mosquito Bank.

Tectonics

In the context of plate tectonics, Central America is bordered on the Pacific Ocean by the Cocos Plate, with the subduction zone underthrusting the coast along the Middle American Trench. The main seismic belt is situated along this zone. Central America is bordered on the Caribbean Sea by the Caribbean Plate, which is connected to the continental crust.

One large transverse fault system crosses Central America between Guatemala and Honduras in an east-west direction. This is the Comayagrea Graben, continued to the east in the Caribbean Sea as the Cayman Graben (7000 m deep). There is also the Fonseca transverse fault system, on the Pacific Coast, where the Gulf of Fonseca originates. Plate tectonic and smaller-scale movements are active and seismic activity occurs along faults and trenches.

Volcanoes, some active, are close to the Pacific coast from Guatemala, through El Salvador, to Nicaragua (Coseguina Volcano), and are farther inland in Costa Rica. The volcanoes erupt ash over the coastal plains, and volcanic rocks occur on cliff sites (Weyl, 1961).

Vertical movements and uplift have occurred since Tertiary times. Lake Managua was once part of the Pacific Ocean, and ocean connections were developed between the Pacific and Caribbean in earlier geologic time.

Such movements still occur on the Pacific coast of Central America. Many coastlines with cliffs and terraces were formed by this geotectonic mechanism. Central Panama was an area of crustal movement in recent geological times. The uplifting of the land bridge above sea level began during the late lower Miocene in east Panama, and later in west Panama. In east Panama, valleys and estuaries were flooded during the *Flandrian transgression* (q.v.). On parts of the Pacific coast and on most of the Caribbean coast, sedimentation and alluvial coastlines prevail.

Environment

The Pacific coast of Central America is attacked generally by waves of local wind systems, mostly moderate, except during tropical storms. This coast is south of the region of cyclones or hurricanes, which do occur in the Gulf of Tehuantepec. So-called allochtone-generated waves attack the Pacific coast of Central America, mainly in the area from Guatemala to El Salvador and Nicaragua. These waves are generated far out in the Pacific Ocean by cyclones, and travel as swell over a long distance. In the coastal region these transform to *surf* up to 2 m high. Beach ridges are built up by these waves, while erosion is going on along cliffs and barrier islands. The swell may block harbors for some weeks.

The Caribbean coast is not attacked by swell but, rather, for short intervals, by heavy waves during hurricanes. The coast of Honduras has been hit catastrophically at times.

Strong ocean currents are not developed directly along the coasts of Central America. Coastal currents, instead, are local depending on wind and waves, and vary during the year. Along the Pacific coast, prevailing currents are from NW to SE during February and March and from SE to NW during August and September (except in the Gulf of Panama). The equatorial countercurrent reaches the coast of Panama during August and September. On the Caribbean coast, currents are influenced by trade winds.

The tidal range along the Pacific coast of Central America varies in the north (Guatemala) from 1.5 to 2 m; in the middle (El Salvador, Nicaragua) from 2 to 3 m; in the Gulf of Fonesca 4 m; in the Gulf of Panama from 4.5 to 6 m. All are semidiurnal. The tidal range along the Caribbean coast of Central America varies between 0.5 and 0.6 m, mostly diurnal and mixed diurnal, semidiurnal. The greatest tidal range difference between the Pacific Ocean and the Caribbean Sea occurs at the Isthmus of Panama; Balboa has a 6 m tidal range on the south side of the Panama Canal; Cristobal has less than 1 m on the north side. The canal could not have been constructed without *locks* to change the level of the waterway in steps or increments.

The discharge of rivers is important for sedimentation along the coast. On the Pacific coast of Central America there are short rivers with significant water and suspended-matter discharge during the rainy season from May to September. Soil erosion is heavy then, and has increased in the last few decades because of the enormous growth in population and density of agriculture, even on steep slopes and volcanic soils. Thus there has been acceleration of suspended material accumulated along the coast. On the Caribbean coast (Honduras and Nicaragua) there are more longer and larger rivers, with lower inclinations. The discharge of these rivers is more uniform throughout the year, because on the Caribbean coast there is no significant dry season. Precipitation varies from 2000 to 5000 mm (at Bluefields), so that there is a considerable discharge of water; but because of less slope inclination, denser vegetation, and a lower population density, the suspended material is relatively less.

Small, windblown flats have developed at some locations. *Beach rock* (q.v.) has developed in some places because of exposure to the sun and the hot climate.

Common plants on the coasts of Central America are palms; on the Caribbean coast of Honduras, pines. *Mangrove* (q.v.) occurs on the beaches of bays, in lagoons, and in estuaries on both coastlines. A widely distributed plant on sandy beaches is *Ipomea*. Coral reefs have developed near the Caribbean coast on the cayos of Mosquito Bank and along the Honduras coast, and in the Gulf of Panama on the Pacific coast.

Large Pacific Coast Gulfs

The Gulf of Fonseca, on the Pacific coast of Central America, was discovered in 1522. The gulf, about 70 km wide, extends inland about 50 km, and is bordered by Nicaragua, Honduras, and El Salvador. It is restricted on the west by Punta Amapala and on the east by the northernmost peninsula of Nicaragua. The border river between El Salvador and Honduras, the Goascorán, is the largest river emptying into the gulf, and its delta is covered with mangroves. Volcanoes bordering the gulf are Coseguina, in Nicaragua, and Conchagua. A group of small islands are situated in the gulf, some of the larger ones being 5–10 km in diameter and of volcanic origin. Among these are Meanguera, Conchaquita, and El Tigre. The average depth of the gulf is 10 m in the central section and 20 m on the seaward side. The best natural harbor on the Pacific coast of Central America is in the gulf at La Unión (Cutuco), El Salvador.

The Gulf of Panama forms the eastern half of the Pacific coast of Panama. The gulf extends from Punta Mala, on the Azuero Peninsula, on the west to Punta Pinas 180 km to the east, and inland 140 km. Depths range from 50 to 100 m in the middle of the gulf. Most of the gulf coast is flat land with small river mouths, swamps, and mangroves. West of the Panama Canal there are extensive tidal flats. On the eastern coast, along the Peninsula of Darien, there are predominantly marshes and mangrove swamps. The Gulf of San Miguel here, with its many estuaries, divides the region into a narrow northwest segment, the Isthmus of Panama, and a wider mountainous southeasterly segment extending to the Colombian border.

Islands off The Coasts of Central America

Coiba, formed by volcanic eruptions from the Cretaceous through the Eocene, is an island off the Pacific coast of southwest Panama. It measures 24 by 40 km, reaches 400 m above sea level, and is bordered by a mangrove coast on the east.

Otóque, with a population of 900 in 1950, is in the inner Gulf of Panama 40 km south of the terminus of the Panama Canal. An important navigational point, it is 2 km in diameter and reaches an altitude of 192 m.

The Pearl Islands (Archipelago of Pearls), also known as Islas del Istmo or Islas del Rey, in the Gulf of Panama consist of one large, three midsized, and over 180 smaller islands discovered in 1513 by Balboa. The main resource is the cultivation of pearls, from which the islands got their name. For the most part the islands are wooded and have steep coasts. Volcanic eruptions in the Cretaceous and Tertiary account for their formation.

About 40 km off the Caribbean coast of Honduras are the Islas de la Bahía, chief of which is Roatán or Rootan Island, about 50 km long and 5 km wide.

The islands in the Gulf of Fonseca were described above.

The Mosquito Coast as a Special Type of Central American Coastal Landscape

The Mosquito Coast is a flat coastal landscape encompassing 16,000 km^2 of savanna, 6000 km^2 in northeastern Honduras and 10,000 km^2 in eastern Nicaragua. Only 1000 people live in this area of 500 by 175 km, located between the Honduras Río Patuca and the Nicaraguan lagoons of northern Bluefields. Elevations are generally less than 100 m, and the slope of the coastal region averages less than 1 m/km.

Because of precipitation in excess of 2000 mm/yr., a high water table, and considerable runoff from nearby mountains, the drainage system is closely spaced and interconnected. The mouths of several large rivers (Río Grande, Río Escondido, Río Prinzapolca, Río Coco, and Río Patuca) are along the Mosquito Coast, as are the mouths of many lowland streams called *criques*, which are deep and carry clear water. Because of the high amount of sediment in the large rivers that come from the mountains, and because of shore drift caused by waves, many lagoons and spits have been formed and these constantly change their shape.

Cape Gracias a Dios forms the northeast tip of the Mosquito Coast on the border between Nicaragua and Honduras, and also forms the northern bank of the large Río Coco delta. The cape is about 55 km wide and extends about 40 km seaward. The coastal plain consists of alternating small, flat plateaus and swampy basins. The rises between the basins are called *ocotales,* after the native name *ocote,* and are covered with a standing growth of pine (*Pinus caribaea*). Because of logging in the wooded areas and erosion of the soil, the greater part of the Mosquito Coast has the appearance of a pine- savanna or steppe. The climate is hot and damp, with noonday temperatures of 29–35°C and morning temperature of about 25°C. The weather fluctuates, however, between stormy and calm, wet and dry. Occasionally hurricanes make their way to the Mosquito Coast. Because of the many bodies of freshwater, insects are especially numerous; however, the name Mosquito Coast is derived not from the insect of the same name but from the original inhabitants, the Mosquito (Miskitto) Indians.

The Río Coco, the largest river on the Caribbean coast of Central America, is some 725 km long, with its source in headwaters no more than 100 km from the Gulf of Fonseca on the Pacific side of Central America. There are many rapids in the mid and upper sections, while portions of the lower stretch are used for shipping. The discharge of the river is 500 m^3/sec. In its lower region there are many meanders and, at its mouth, the aforementioned delta comprising the Cape Gracias a Dios.

Mosquito Gulf, about 130 km long and 16 km wide, is part of the Caribbean Sea surrounded by the coast of west Panama. In the gulf are the Escudo de Veraguas Islands. Mosquito Shoal is a coastal shelf up to 200 km wide off the east coast of Honduras and Nicaragua. Here coral reefs, sandbanks, and cays rise from the shallow waters.

Harbors

There are several harbors along the coast of Central America, mostly of local or special character such as banana harbors. Among these are the following: on the Pacific coast, Champerico (Guatemala); Acajutla, La Libertad, La Unión (El Salvador); Corinto, Puerto Somoza, San Juan del Sur (Nicaragua); and Punta Arenas and Golfito (Costa Rica). On the Caribbean Sea there are: Puerto Barrios, Puerto Motaquia (Guatemala); Puerto Cortés (Honduras); Puerto Cabeza, Puerta Isabela, El Bluff (Nicaragua);

Puerto Limón (Costa Rica); and Cristobal (Panama).

Literature

While there is considerable literature available on this subject, a few noteworthy items that should be mentioned here are: Allen (1956), Gierloff-Emden (1974, 1976, 1977), Lewis and MacDonald (1972), Parsons (1959), Sauer (1975), Termer (1936) and the U.S. Navy Hydrographic Office (1937).

HANS GÜNTER GIERLOFF-EMDEN

References

Allen, P. H., 1956. *The Rain Forests of Golfo Dulce.* Gainesville, Fla.: University of Florida Press, 417p.
Gierloff-Emden, H. G., 1974. Anwendung von Multispektralaufnahmen des ERTS-Satelliten zur kleinmasstäbigen Kartierung der Stockwerke amphibischer Küstenräume am Beispiel der Küste von El Salvador, *Kartogr. Nachr.* **24,** 35-76.
Gierloff-Emden, H. G., 1976. *La costa El Salvador.* San Salvador: Ministerio de Educación, Dirección de Publicaciones, Monografía morfologica-oceanografica.
Gierloff-Emden, H. G., 1977. *Orbital Remote Sensing of Coastal and Offshore Environments: A Manual of Interpretation.* Berlin: Walter de Gruyter, 179p.
Lewis, A. J., and MacDonald, H. C., 1972. Mapping of mangrove and perpendicular-oriented shell reefs in southeastern Panama with side looking radar, *Photogrammetria* **28,** 187-199.
Parsons, J. J., 1959. The miskito pine savanna of Nicaragua and Honduras, *Assoc. Am. Geographers Annals* **45,** 36-63.
Sauer, J. D., 1975. Remnant seashore vegetation of northwest Costa Rica, *Madroño* **23,** *174.*
Termer, F., 1936. Geographie der Republik Guatemala. Beiträge zur physischen Geographie von Mittel- und Südguatemala. *Geog. Gesell. Hamburg Mitt.* **44,** 89-275.
U.S. Navy Hydrographic Office, 1937. *Naval Air Pilot, Central America.* Washington, D.C.: Hydrographic Office Publ. No. 195, 262p.
Weyl, R., 1961. *Die Geologie Mittelamerikas.* Berlin: Borntraeger, 226p.

Cross-references: *Central America and South America, Coastal Ecology; Global Tectonics; Mangrove Coasts; North America, Coastal Morphology.* Vol. VIII, Part 1: *Central America, Regional Overview.*

CENTRAL AND SOUTH AMERICA, COASTAL ECOLOGY

Central (Middle) and South America offer almost every type of environment possible, from the driest deserts to the wettest rain forests. Many plant and animal species of these different environments are the most diverse and distinctive of any continent. This is especially true for South America, where the animals evolved during a time of isolation (Bates, 1964). The major ecological systems of Latin America are represented within the coastal regions (see Figs. 1 and 2).

The occurrence of the different vegetative communities are controlled by many factors such as precipitation (total amount and distribution throughout the year), temperature, humidity, elevation, and soil types, but human activities are responsible for the most rapid and numerous changes. Agricultural practices, exploitation of native species, overgrazing by domesticated farm animals, clear-cutting, burning practices, and air and water polution have greatly damaged, altered, or destroyed many areas of natural vegetation. Changes, whether natural or man-induced, are still occurring, and thus the ecological systems remain dynamic.

Central America—Atlantic Coastal Ecological Communities

Along the Atlantic coastline of Central America four types of coastal communities exist: coastal dunes and beach vegetation, mangrove swamps, salt marsh, and coral reefs (Fig. 1).

Coastal dune and beach vegetation is found along most of Mexico and from Honduras to northeastern Panama on open shorelines and barrier islands. The vegetation is specialized for edaphic aridity and can withstand salt spray, wind action, a low nutrient supply, and, on the dunes, accumulation of sand (Eltringham, 1971; Howard, 1973). Beach plants are mostly creeping vines such as *Ipomoea pes-caprae* and *Canavalia maritima,* and frequently one or more erect plants—e.g., *Uniola paniculata, Scaveola keonigii, Sesuvium portlaccastrum, Sporobolus virginicus,* and *Distichlis spicata* are scattered among or in place of the vines (Standley, 1937; Sauer, 1967; Porter, 1973; Gómez-Pompa, 1973). *Chrysobalanus icaco, Coccoloba uvifera, Croton punctatus,* and *Hippmane mancinella* are some of the characteristic low trees and shrubs that occur on the dunes or areas farther inland (Sauer, 1967; Poggie, 1963).

Over 190 plant species have been recorded on these sandy beaches and dunes, but because of individual requirements not all species are distributed along the Atlantic coast (Sauer, 1967). For example, *Chrysobalanus icaco* is not found in arid regions, *Coccoloba uvifera* is absent from temperate regions, *Uniola paniculata* is confined to quartz sand beaches, and the more common species are missing from the calcareous shorelines of the Yucatán Peninsula.

The faunal inhabitants of beach and dune areas are scant. At the water/land interface, the most predominant burrowing invertebrates are

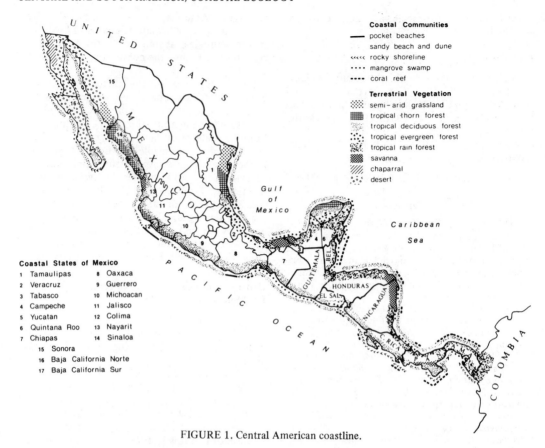

FIGURE 1. Central American coastline.

llamellibranchs, polychaetes, and crustaceans, whereas the microfauna include harpacticoids, copepods, rotifers, and nematodes (Moore, 1966). The distribution of these small creatures depends on grain size and wind and wave action (Tait and DeSanto, 1975). On the dunes dwell insects, reptiles, a few small mammals, and coastal birds (Henderson, 1967). Some animals, such as turtles, also use the beach as migration routes.

In areas sheltered from wind and wave action, such as bays, estuaries, and the leeward side of barrier islands, *mangrove swamps* thrive. Most coastlines between latitudes 25°N and 25°S support mangrove formation (Aubert de la Rue, Bourliere, and Harroy, 1957). The most solid occurrence is along the western coast of Yucatán Peninsula and Laguna de Terminos, Campeche, Mexico, and Belize (Vermeer, 1959; West, 1956). *Rhizophora mangle* (red mangrove), *Avicennia nitida* (black mangrove), *Laguncularia racemosa* (white mangrove), and *Conocarpus erectus* (buttonwood) are the dominant trees. Within the mangrove forest there is a species zonation (in the sequence noted above with *Rhizophora* next to the open water) related to the extent and the duration of tidal flooding, sedimentary action, and composition of the soil (Vann, 1959; Daubenmire, 1978). *Rhizophora* is renowned for its well-developed tangle of adventitious roots that arch outward from the trunk and penetrate into the mud. Because of the anaerobic condition and soft substrate in which they grow, the roots play an important role for support and for conduction of gases between the air and the submerged parts of the root system (Aubert de la Rue, Bourliere, and Harroy, 1957).

Mangrove swamps are vital for the life cycle of many marine animals and also harbor a complicated community of animals. The roots of the trees support many sessile invertebrates such as barnacles, bivalves, serpulid worms, and tunicates (Tait and DeSanto, 1975). Numerous fish, nonsessile mollusks, and crustaceans find shelter and food among the roots (West, 1976). Contained in the mud are a number of burrowing crabs, mollusks, and small fish, while in the treetops thrive a multitude of insects, numerous snakes, lizards, and birds with a variety of colors and shapes (Henderson, 1967).

Salt marsh formation in tropical environments occurs on the margins or within mangrove formations as: a pioneer community on new mud flats and shores of open coast, estuaries, and

FIGURE 2 : SOUTH AMERICAN COASTLINE
FIGURE 2. South American coastline.

tidal channels; a halophytic community; and a secondary formation on disturbed areas (West, 1977). The extent of salt marsh in tropical climate is thought to be limited because of competition with mangrove swamps (Chapman, 1960). In pioneer situations the dominant plants are *Spartina* spp. These salt marsh grasses occur with *Sporobolus virginicus, Scripus maritima, Paspalum vaginatum, Salicornia perennis,* and *Batis maritima* as a halophytic community behind mangroves (West, 1977; Gómez-Pompa, 1973). The third situation has not been encountered in Middle America, but occurrences have been recorded in South America. These species include *Spartina brasillensis, Acrostichum aureum*, and *Paspalum vaginatum*.

Coral reef formations are found in several locations along the Atlantic coast of Middle America, usually 24-48 km offshore. These exist off: Chetumal Peninsula, Mexico, to southern Belize; northwest corner of Honduras and Nicaragua; southern Nicaragua; northwest Panama, near Chiriqui Lagoon; and paralleling the Archipélago de las Mulatas, Panama (Dolan et al., 1972; Vermeer, 1959; Moser, 1975). Coral reefs are the most complex, diverse, and delicately balanced of all marine communities (Curtis, 1975). They grow in warm, well-lighted surface waters, in which the temperature rarely falls below 21°C. The reef itself is constructed of coelenterates that secrete a calcium-containing skeleton of new generations building on top of the old (Hannau, 1974). Symbiotic algae occur on and/or in the polyps, and are believed to contribute oxygen and organic compounds and remove waste products for the animals. Studies have concluded that the polyps can survive without the algae, but their presence is necessary for the rate of reef development (Curtis, 1975; Moore, 1966). The reef furnishes both food and shelter for numerous invertebrates and fish species.

Coastal waters are abundant with sea life, much of which is important for sport and commercial fisheries, including shrimp, bass, mackerel, pompano, red snapper, tarpon, and mullet (Hubbs and Roden, 1964; Henderson, 1967). Many species are often found concentrated above upwelling of fresh water originating from the ground water beneath the limestone surface of the Yucatán Peninsula (Henderson, 1967). Marine animals such as the manatee and the West Indian wolf seal exist in Atlantic waters. Manatee can sometimes be seen grazing on flowering marine plants in lagoons and embayments. The range and number of these sirenians have been greatly reduced by man. Once hunted for their flesh, they are now endangered by collisions with powerboats (Moser, 1975; Bourliere, 1955).

Central America—Atlantic Vegetation Cover Types

The vegetation inland along the Atlantic coast of Central America usually reflects the climate of an area (Fig. 1). Wet and dry seasons occur in the northern and central regions, resulting in grasslands and seasonal formations (tropical thorn, deciduous, and evergreen forests and savannas). These formations exhibit a seasonal rhythm in phenology—e.g., losing leaves in the dry season. A tropical rain forest occurs in the southeastern areas where the dry seasons are short to nonexistent.

Tamaulipas, Mexico, is dry and hot with an annual average rainfall of 50-75 cm/year (Henderson, 1967). Even though this is twice the precipitation of deserts, it is still not sufficient for tree development. Thus, in northern Mexico, behind the dune vegetation, a semiarid grassland exists. This vegetation cover is the southern extent of the temperate grasslands of Texas, and like all grasslands is typified by short, perennial grasses (*Andropogon* spp., *Cenchrus, Chloris*, and *Paspalum*). Because of the low precipitation, desert species also abound. Xerophilous species include low spiny shrubs (*Prosopis* and *Acacia*) and cacti (*Opuntia*) (Henderson, 1967; Daubenmire, 1978). During the pronounced dry season the grasses usually estivate and are therefore highly flammable.

In southern Tamaulipas the rainfall increases, and a *tropical thorn forest* begins to develop. This formation extends to northern Veracruz, and is also found on the tip of the Yucatán Peninsula in an east-west band (Henderson, 1967; Beard, 1944). Common thorny woody plants include *Acacia, Prosopis, Fiscus, Randia, Caesalpinia, Cordia*, and *Colubrina*. *Cephalocerus, Pachycereus, Pterocereus*, and *Opuntia* are some of the cacti, whereas the shrub layer consists of *Bromelia, Agave*, and *Yucca* (Daubenmire, 1978; Gómez-Pompa, 1973; Cloudsley-Thompson, 1975). Perennial and annual herbs appear abundantly in the brief rainy season, but grasses are essentially lacking.

The inhabitants of the dry regions (the thorn forest and semiarid grassland) include lizards, snakes, rodents, rabbits, insects, and birds (Henderson, 1967; Cloudsley-Thompson, 1975). Many of the animals estivate, whereas others are migratory becuase of the hot and dry climate.

Tropical deciduous forests are located south of the thorn forests in northern Veracruz and in an east-west band of the Yucatán Peninsula (Henderson, 1967; West and Augelli, 1976). This formation can be thought of as a transition zone between the thorn forest and the tropical evergreen forest (which occurs south of this formation), for it has constituents of both. The majority of trees are deciduous, however, and include *Bursera, Cecropia, Ceiba, Coccoloba, Enterolobium, Guazuma, Hymenaea, Pisonia, Pterocarpus, Spondias, Tabebuia, Trema*, and many others (Daubenmire, 1978). Other characteristic features, besides leaf loss during the dry season, are the closed canopy usually consisting of two stories, the dominant story without thorns and sparseness of ground vegetation (Beard, 1944; Wagner, 1964).

Along central Veracruz and on the Yucatán

Peninsula in an east-west band, south of the deciduous forest, a *tropical evergreen forest* exists (Henderson, 1967). This forest resembles a rain forest in many aspects, such as its high species diversity, similar species, canopy consisting of several layers, and similar pollinating agents. But unlike a true rain forest, this formation has a definite rhythm of flowering and leaf renewal timed to the seasons, even in regions where seasonal distinctions are subtle (Daubenmire, 1978).

Sandwiched between forest and coastal vegetation are *savannas* (tropical grasslands). These communities are found in southeastern Veracruz; around Laguna de Terminos, Campeche; and on the corner of northeastern Honduras and Nicaragua (Henderson, 1967; West and Augelli, 1976; Wagner, 1964). Unlike the semiarid grassland, savannas are usually moist enough to sustain tree growth, even though not all savannas have trees intermixed with the grasses. The composition of the native grasses vary from place to place, but all the major species are of the Graminae family. *Curatella americana* and *Byrsonima crassifolia* are two common trees that are found scattered among the grasses. Along the Caribbean coast of Nicaragua and Honduras, *Pinus caribaea* is incorporated with them (Wagner, 1964).

Unlike grasslands, savannas have poor, shallow soils that lead to saturation during the rainy season and desiccation during the dry period. Fires therefore become a major problem, as in the grasslands. Many genera are preadapted to periodic fires, and both *Curatella* and *Byrsonima* have the ability to sprout from the roots should the trunk be destroyed (Daubenmire, 1978).

Tropical rain forests extend from southern Veracruz up into the Yucatán Peninsula and southeastward into Panama (Standley, 1937; Porter, 1973; Gómez-Pampa 1973). This formation occurs in areas of high humidity, little or no temperature or rainfall variation, and freedom from frost (Daubenmire, 1978). Because of these optimum environmental conditions, the rain forest has the highest number and diversity of both plant and animal species in comparison with other terrestrial communities. Diversity is so high that usually no two trees of the same species occur next to each other.

True rain forest is often confused with tropical evergreen formation. However, flora of a rain forest do not have cycles timed to the seasons as do seasonal formations. There are always some species blooming or losing leaves throughout the year. There is no one dominant plant, but evergreen trees are most abundant. Characteristic lowland rain forest generally include *Andira, Brosimum, Calophyllum, Castilla, Gwettarda, Manilkara, Miconia, Pithecolobium, Pouteria, Terminalia,* and *Vochysia* (Daubenmire, 1978). Other characteristic plants are woody vines, lianas, ferns, orchids, epiphytes, mosses, and lichens (Aubert de la Rue, Bourliere, and Harroy, 1957). Because of the density of the forest, wind pollination is virtually impossible. Therefore insects, birds, and bats become the major disseminating agents (Daubenmire, 1978).

The forests (deciduous, evergreen and rain) have many common species, and every niche from the treetops down to the ground level is filled. Insects and birds, of a variety of shapes and colors, are found at all levels; reptiles and amphibians inhabit the ground, trees, and waters; and mammals are usually nocturnal and aboreal (Cloudsley-Thompson, 1975).

Central America—Pacific Coastal Ecological Communities

The Pacific coastline of Central America is different from its Atlantic counterpart. Rocky shores and cliffs are common, as are sandy beaches and mangroves. Most of the Mexican state of Baja California and the countries of Costa Rica and Panama have rocky shorelines. Between the Mexican state of Sonora and the country of Honduras, sandy beach and dune communities are interrupted in many places by rocky shorelines with pocket beaches. Beach and dune biota consist mainly of the same species as on the Atlantic coast.

Rocky shorelines are a difficult environment. Both plants and animals are exposed to pounding waves, periods of desiccation, and changes in temperature due to tidal action and insolation. Insolation is the major factor in determining local and latitudinal distribution in the intertidal zone (Moore, 1966). For example, temperate regions have an abundance of algae growth when compared with tropical localities. In the tropics the intense rays of the sun causes overheating, actinic, and desiccation problems. Therefore growth of algae in the tropics is confined to the sublittoral zone (extreme low or high water of spring tides). Red and green algae are more suited for tropical and subtropical waters, whereas browns attain optimum growth in cooler waters (Stephenson and Stephenson, 1972).

The rocky coasts supply a variety of habitats, such as exposed rock faces, sheltered overhangs, crevices, and shallow pools. Life on the rocky shores is usually zoned depending on species tolerance to the factors discussed above and also the type of substrate they inhabit. In areas where the wave action is not strong, algae dominate. Zonation occurs among algae species and can be observed by the various belts of color attached to the surface (Tait and DeSanto, 1975).

Barnacles are usually on more exposed surfaces, where wave action is too disruptive for plants (Daubenmire, 1978). Other common animals among the rocks are mussels, limpets, periwinkles, crabs and chitons (Tait and DeSanto, 1972; Moore, 1966). Cliffs and rocky offshore islands also provide breeding and brooding grounds for innumerable sea birds, while in the surrounding waters some fish rely on attached vegetation for breeding grounds (Daubenmire, 1978).

Mangrove swamps are also numerous along the sheltered regions of the Pacific coast, and even the aridity of Baja California and Sonora, Mexico, does not stop this vegetation type from developing (Aubert de la Rue, Bourliere, and Harroy, 1957; Johnson, 1972). The species are the same as found on the Atlantic coastline, but *Pelliciera rhizophporae* becomes another important constituent (Daubenmire, 1978). The most extensive mangrove formations occur along the coasts of Chiapas, Mexico, and Guatemala (Steyermark, 1950; Breedlove, 1973). Mangrove swamps along the Pacific coast also support less fauna. This is because of the large tidal range, which leaves the roots dry for long periods of time during low tide (West, 1977).

Salt marsh formations are not very extensive, nor have they been well investigated. They occur in northwest Mexico in deeply indented bays along the coasts of the states of Baja California and Sonora. *Salicornia pacifica, Allenriofea occidentalis, Batis maritima,* and *Monmonanthochloe littoralis* grow on lower ground, whereas in higher, less saline areas, shrubs such as *Sueda, Maytenus,* and *Frankenia* are found. Mud flats at the mouth of Colorado River, however, are virtually devoid of vegetation because of the high tidal range (West, 1977).

Salt marshes support a variety of life because of the magnitude of primary productivity. They are an important natural resource serving as a nursery and feeding ground for many species of fish. They also harbor shore and wading birds by providing feeding, resting, and nesting areas. Burrowers in the mud, such as bivalves, crustaceans, and annelids, are consumed by coastal birds that scavenge in the soft substrate. The salt marsh plants usually have numerous gastropods and insects browsing on them.

Coral reefs of the Pacific are not nearly so extensive as along Atlantic coasts. The main reefs are found in waters off Panama, south of Azuero Peninsula and Coiba Island (Dolan et al., 1972).

The Pacific waters of Central America are nutritive and hold an abundance of microscopic plankton, fish, and several types of turtles. The most productive waters are around Baja California, which are teeming with life because of the upwelling and cooler waters. A number of sea mammals have found sanctuary along the northwest coast of Mexico. The gray whale, Guadalupe fur seal, and the northern elephant seal have escaped the peril of extinction in this area (Johnson, 1972). Other animals which can be seen in this area include the green sea turtles, sea otters, and humpback, sperm, and finback whales (Henderson, 1967; Johnson, 1972).

Pacific Central America—
Vegetation Cover

Inland along the Pacific coast the vegetation cover is affected by the pronounced long dry season (Fig. 1). Most of coastal Central America is predominately seasonal forest; in the northwest a desert exists. Proceeding south, one reaches eastern Panama before a rain forest begins to dominate.

Baja California and Sonora, Mexico, are mainly coastal deserts (Petrov, 1976). The region is arid, receiving no more than 12.5 cm of rain/year with mild winters and hot summers, with average summer temperatures of 29-35°C (Henderson, 1967). Because of this climate, vegetation is sparse, but many types are unique. For example, Baja California contains approximately 110 cactus species, of which 80 are endemic (Johnson, 1972). Characteristic plants include the boojum tree, cardon cactus, creosote bush, burrow weed, organ-pipe cactus, and mesquite. Depending on the geographic distribution of moisture, some plants will predominate over others. Creosote bush (*Larrea divaricata*) is the major species existing in moister areas such as Sierra Vizcaino, which is covered with fog. Farther south around Magdalena Plain, where it is drier, cardon cactus becomes dominant (Johnson, 1972).

Fauna are adapted physically and physiologically to the arid climate. Soil protozoans, insects, reptiles, amphibians, small mammals, and desert birds all find a way to exist. Many inhabitants have life cycles timed to periods of favorable rainfall and temperature; others go into estivation. Animals that must remain active each day avoid the daytime heat by finding shelter in the shade or in the ground, and later become active during the night (Brown, 1968; Cloudsley-Thompson, 1975). Reptiles, birds, and mammals of Baja California are marked by endemism and confined to the southern areas (Neill, 1970).

Interruptions in the desert vegetation are the *chaparral* in northwestern Baja California, and the deciduous forest (which is similar to the Atlantic side) located in southern Baja California (Henderson, 1967; Johnson, 1972; West and Augelli, 1976). The chaparral formation extends south to about the city of Rosario, and differs from the desert in both climate and vegetation. The climate is moist enough to be classified as Mediterranean with its mild summers and moist winters. The precipitation in the area averages about 20-30 cm/year (Wagner, 1964). The flora is distinctive, with short trees and shrubs that are evergreen, sclerophyllous, and fire-sensitive. The most characteristic species *Artemisia california*, is joined by other common plants such as *Encelia californica, Eriogonum fasciculatum, Haplopappus sugarrosus, Eriodictyon, Lotus, Rhus,* and *Salvia* (Daubenmire, 1978; Ochoterena, 1945).

In central Sonora the desert receives more moisture (12.5-25 cm/year), and the vegetation cover becomes similar to the semiarid grassland in northeastern Mexico. This area is a transition zone between the desert and the thorn forest that slowly starts to predominate in southern Sonora (Henderson, 1967). In northern Sinaloa, the thorn forest is established and extends to about northwest Guerrero.

Tropical deciduous forest becomes the predominant vegetation cover from Guerrero to about central Panama, because the region still has noticeable wet and dry seasons. In western Panama the climate is suitable for tropical rain forest (West and Augelli, 1976).

Numerous savanna formations are scattered inland along the coast, but the major expanses are found: from southern Sinaloa to northern Nayarit; around Laguna Superior in Oaxaca and Chiapas; in Costa Rica north of Nicoya Peninsula; and in Panama extending west-east, approximately from David to the Gulf of Panama (Wagner, 1964; Henderson, 1967; Beard, 1944; Breedlove, 1973; Porter, 1973). The vegetation is similar to that of the Atlantic coast, but some of the savannas have seasonal swamp species in areas that are inundated for periods of time. Plants associated in these wet areas are *Crescentia* spp., *Fiscus,* and several species of palms.

South America—Vegetation Cover

The tropical rain forest continues from Panama into South America, extending to northern Ecuador. Between northern Ecuador and northern Peru is a transition zone known as a *bushland*. Off the coast of this area the cold Peruvian current comes in contact with the warm waters of the Gulf of Panama. The meeting immediately causes a decrease in temperature, absolute humidity, and precipitation. Farther to the south the combination of the Peruvian current and the Andes creates an arid coastal environment that has a high incidence of fog (Fig. 2).

The region known as bushland is difficult to classify, and many investigators have put it in the same group as the tropical thorn forest. But unlike the thorn forest, grasses are also common (*Anthephora, Chloris* and *Cenchrus*) but not dominant enough for this area to be called a semiarid grassland. The characteristic plants are xerophilous cushion-shaped woody shrubs *Acacia, Prosopis, Capparis, Pithecolobium,* and *Cryptocarpus.* These provide the basis for describing these regions as bushland (Jones, 1930; Svenson, 1945; Brown, 1968).

Coastal deserts extend from northern Peru to northern Chile, latitude 7-31°S (Petrov, 1976). Even though most areas have sparse to nonexistent vegetation cover, there are some areas where plants can be found. On stony and gravely mountain slopes, cacti, succulents, and spurges survive (Petrov, 1976). *Lomas* are areas covered with luxuriant, seasonal vegetation because of heavy sea fogs occurring at altitudes between 300 and 1000 m (Penaharrera del Aguila, 1969). Lomas occur in both Peru (latitude 8-18°S) and Chile (20-24°S) abounding with simple algae, lichens, herbs, shrubs, cacti, succulents, and grasses (Brown, 1968; Petrov, 1976; Penaharrera del Aguila, 1969). Many of these species are endemic. Because of human interference, over 40% of the flora are now endangered (Ferreyra, 1977). This coastal desert supports a limited number of animals. The most characteristic is the lizard of the genus *Dicrodon.* Other animals such as toads find shelter during the day and at dusk feed on insects. Birds (oven birds, cactus wrens, flycatchers) and mammals are confined to areas close to sources of freshwater (Neill, 1970).

South of the desert in middle Chile, the climate and the vegetation (even though of different species) are similar to those of the chaparral in Baja California. This region, referred to as the *Chilean mattaral,* has species of both the coastal desert and the temperate forest that occur south of this region. Characteristic plants are cacti, acacias, cypresses, laurels, and beeches (Jones, 1930).

Further south, rainfall steadily increases and plants common to the dry regions are no longer found but are replaced by trees. A *temperate mesophytic forest* (often called *temperate rain forest*) extends from central Chile to Tierra del Fuego (Jones, 1930; Daubenmire, 1978). In the northern areas, rainfall of 150-200 cm/year is favorable for a dense forest of the evergreen beeches (*Nothofagus dombeyi*), deciduous beeches (*N. obliqua*), and Chilean pines (*Araucaria imbricata*). *Libocedrus chilensis* and *Fitzroya patagonica* occur in the swamps or bog areas. Around the Gulf of Ancud two small beeches become dominant, *Nothofagus antarctica* and *N. betaloides.* Southward toward Tierra del Fuego increasing rains (250 cm/year), strong winds, and decreasing temperatures cause a reduction in physical size and distribution of vegetation. *N. antarctica, N. betaloides,* and *Pilgerodendron uviferum* occur as stunted growths, and in areas of snow and ice fields there is no vegetation (Jones, 1930; Stephenson and Stephenson, 1972; Goodspeed, 1945).

On the Atlantic side, grasslands are found in two regions: from Tierra del Fuego to approximately the state of Santa Cruz, Argentina, and from the state of Buenos Aires, Argentina, to the state of Rio Grande, Brazil. The latter region is locally referred to as pampas. *Stipa,* a hard bunch grass, is the major constituent, accompanied in the extreme southern districts of the southern grasslands by mosses and lichens (Jones, 1930; Daubenmire, 1978).

Between these two grasslands another bushland occurs. This region is also known as the Patagonia steppe or cold coastal desert. The cushion-shaped plants include *Prosopis, Larrea,* and *Gourliea* and are joined by grasses and cacti (Brown, 1968; Petrov, 1976; Jones 1930). The southern areas are cold, whereas the northern reaches have a warmer temperature and are covered predominantly by desert scrub. Both regions, however, have low precipitation (12.5–25 cm/year).

Characteristic fauna of the pampas and the bushlands (of the Atlantic side) include Patagonia weasel, pampas deer, rodents (mara and viscacha), burrowing owl, and cavy, a type of guinea pig (Neill, 1970).

North of the pampas, rainfall increases and the area becomes forested with deciduous and evergreen trees. The seasonal forests extend to approximately southern São Paulo, Brazil. There a rain forest begins to form and extends to Rio Grande do Norte (Jones, 1930; Dolan et al., 1972).

In the states of Rio Grande do Norte and Ceará, Brazil, a thorn forest exists that is locally known as *caatinga* (Daubenmire, 1978). The species are similar to those of other thorn forests, but the dominant plants belong to the Mimosae and Bombacaceae.

West of the caatinga, in the states of Piaui and Maranhão, Brazil, the porous soils and six rainy months support a *palm forest* (Daubenmire, 1978; Vann, 1959). Common palms are *Acoelorraphae, Copernicia, Mauritia, Orbignya, Raphia, Bactris,* and *Scheelia,* and there may be a scattering of evergreen dicot trees. The ground is usually flooded for such a long period of time that almost no shrubs or herbs can survive.

Rain forest predominates from the palm forest of Brazil to the eastern edge of Venezuela (Dolan et al., 1972; Aubert de la Rue, Bourliere, and Harroy, 1957). Between the rain forest and coastal ecological communities occur savannas and many types of swamps. In the savannas *Bowdichia virgiloides* is an additional codominant (Daubenmire, 1978). Depending on the duration of inundation, there will exist a mosaic pattern of swamps with numerous vegetation transition zones. In several types of herbaceous swamps, trees are lacking and the dominant plants are herbs (*Cyperus* and *Typha*). Where an area is inundated most of the year, swamp forest predominates with trees such as *Annona, Andira, Pterocarpus, Chrysolablanus,* and *Fisvu.* Regions where there are fluctuations in the water table have species such as *Bombax auqaticum, Pterocarpus officinalis* and several types of palms (Vann, 1959; Daubenmire, 1978).

From the rain forest of eastern Venezuela there is a thorn forest that extends into the Guajira Peninsula, Colombia. But around Lake Maracaibo the vegetation cover is characteristic of wet and humid climates. A rain forest exists around the southern edges of the lake; in the northern regions, it is surrounded by a deciduous forest. From the Guajira Peninsula to about the Gulf of Morrosquillo a deciduous forest exists. Here a rain forest occurs and combines with the rain forest of Panama and southern Colombia (Dolan et al., 1972). However, most of the natural vegetation in northern Colombia has been harvested, and only isolated forest remnants remain.

South America—Coastal Ecological Communities

The Pacific coast of Colombia consists mainly of two coastal ecological communities: rocky shores and mangrove swamps (Fig. 2). The flora and fauna of both areas are similar to those on the Pacific tropical coast of Central America. A steep, mountainous coast occurs in the northern third of the country, extending from southeastern Panama to the vicinity of Cabo Corrientes. From here to northeastern Ecuador there is a narrow discontinuous sand beach with mangrove swamps forming inland (West, 1956). These swamps do not, however, form a continuous strip. At the Bay of Buenaventura, Las Tortugas, and Ensenada de Tumaco, there are outcroppings of Tertiary material forming high cliffs extending to the sea.

The mangrove forest of northwestern South America is distinguished from all other mangrove communities of Latin America. This coastline has a *high mangrove forest* (West, 1956). *Rhizophora* attains heights up to 30 m, whereas it normally grows to 9–12 m.

The mangrove swamps give way to a sandy beach and dune community in northern Ecuador, a little south of Santiago River. This sandy community continues to the Gulf of Guayaquil, where once more mangrove swamps predominate. This ecological community extends to Tumbes, Peru, which is also the southern extent of mangrove swamps. This is a high mangrove forest as in Colombia, but has no barrier sand beaches (West, 1956).

The warm offshore waters hold some small coral formations. Larger ones are found off south central Colombia and central Ecuador (Snead, 1972).

Between Tumbes, Peru, and Trujillo, Peru, sandy beaches and dunes become prevalent. At Trujillo the beaches are crescent-shaped and interrupted by rocky headlands. This pocket beach formation continues to approximately Arica, in northern Chile (Dolan et al., 1972).

The beaches of Peru do not have much vegetative growth and consist mainly of active dunes. In some areas along the coast the saline soils hold a halophytic association of *Distichlis spicata, Sporobalus virginicus, Sesuvium portulocastrum, Cressa truxillensis, Spilanthes urens,* and many other species (McGinnies, 1968).

Rocky shoreline extends from Arica, Chile, to Valparaiso, in central Chile. The rocky shoreline and headlands of both Peru and northern Chile are within the tropics but display no tropical elements. Mangroves are not present, and even salt marshes are scarce. The lack of tidal marshes is primarily the result of the absence of lagoons and river deltas along this stretch of coastline. The rocky habitats are abundant with biota.

Along the supralittoral fringe a littorina belt exists. In some areas this fringe is bare because of the trampling of sea birds and guano effects, and in other places their absence is the product of strong wave action (Stephenson and Stephenson, 1972). Seaward, both numbers and diversity of species increase. Barnacles (*Chthamalus cirratus*) span the littoral zone. Mussels (*Brachyodontes* and *Mytilus*) are found in the midlittoral zone except around Antofagasta, where *Pyura chilensis* replace them. Other characteristic fauna include *Chiton* spp., *Siphonaria* spp., *Fissurella* spp., *Tegula, Leptograpsus,* and *Concholepas.* Belts of algae are conspicuous in the midlittoral zone but vary with location. Common algae include *Porphyra, Ulva, Endarachne, Enteromorpha,* and *Colpomenia* (Guiler, 1959a).

A pocket beach community extends from Valparaiso, Chile, to Chiloe Island, Chile. Along the southern tip of South America from Chiloe Island to an area a little south of Rio Chubut, Argentina, rocky shore biota is found. South of central Chile the vegetation and animal species and/or their distribution gradually changes. South of Chiloe the supralittoral fringe is lined with black lichens and dark red algae. Littorinids are not present in this or any other zone. The midlittoral zone consists of barnacles, mussels, and siphonarids (*Kerguelenella*). These species do not extend the whole midlittoral zone as in northern Chile, for in the lower midlittoral zone species of barnacles (*Elminius* and *Balanus*) prevail over *Chthamalus* and *Mytilus edulis* replace mussels and siphonarids. Chief algae are *Isidaea* spp., *Adenocystis utricularis, Caepidium antarcticum, Ceramium rubrum, Porphyra umbilicalis, Ulva lactuca,* and many others. The lower midlittoral rocks are also covered by corallines and lithothamnia, forming thick pink coverings (Guiler, 1959b).

There is also an extensive kelp bed found in the intralittoral fringe from Peru south to Tierra del Fuego and extending north to Río Chubut in Argentina (Snead, 1972). The kelps consist of *Lessonai nigrescens, Durvillea antarctica, Macrocystis pyrifera,* and *Melanophseal* spp. (Snead, 1972; Stephenson and Stephenson; 1972).

In southern Chile and along the Patagonian coast there are a few salt marsh communities (West, 1977). The tidal marshes in south Chile are confined to small disjunct inlets, with *Spartina densiflora* and *Salicornia*. Salt marshes are found at mouths of small rivers in Patagonia with salt shrubs (*Salicornia, Atriplex, Swalda, Frankenia,* and *Lepidophyllum*) being the most common.

From Río Chubut, Argentina, north to Río Paraná, in northeast Brazil, the coastal community is mainly sandy beach and dune. This formation occurs with or is interrupted by salt marsh, mangroves, or rocky communities. Common plants on the beaches are *Sesuvium portulacostrum, Salicornia guadichaudiana,* and *Spartina* spp. The dunes are stablized by *S. portulacostrum, Cotula, Alternathera, Heliotropium, Panicum, Fiscus* and *Hyptis* (Delaney, 1966).

The southernmost salt marshes interrupting the sandy beach community occur along the Gulf of San Matías. Other marshes are found in areas of Río Negro, Río Colorado, and the Río de la Plata estuary, along the southern shore and the shore east of Montevideo, Uruguay (Dolan et al., 1972). The marshes are dominated by *Spartina, Distichlis, Juncus* and *Salicornia* (West, 1977). However, in some areas of the La Plata estuary the vegetative zonation is reversed from normal. The outer marsh is flooded with brackish water and composed of grasses and sedges normally found in freshwater and brackish water. These include *Paspalum, Scripus,* and *Eleocharis.* Inland the ground is higher in salinity because of evaporation and capillarity. Thus the halophytes are found inland instead of along the coast.

Pocket beaches are found along the states of Santa Catarina and São Paulo, Brazil. Mangrove swamps also begin in this proximity. North from 25° latitude south to Santiago mangroves are found behind the sandy beaches. (Snead, 1972; Dolan, et al., 1972). Flora and fauna are similar to those of the Atlantic coast of Central America.

There are many large and small coral reef formations offshore of northeast Brazil. The largest one extends between the states of Alagoas and Rio Grande do Norte (Dolan et al., 1972).

In the state of Maranhão, Brazil, sandy beach and dunes gradually decrease and mangrove swamps become prevalent. The mangroves extend to the Orinoco delta in Venezuela. The flora and fauna are similar to those of the Atlantic Central America coast. However, along the shores of the Guianas, *Avicennia* is found seaward of *Rhizophora*. This reversed situation is believed to be caused by deposition of sand coupled with the ability of *Avicennia* to grow in sandy soil (Vann, 1959).

The Venezuelan and Colombian coasts have a variety of ecological communities. (For the characteristic flora and fauna of the communities for Venezuela and Colombia see Central America–Atlantic Coastal Ecologic Communities.) Rocky shore communities are found along the coasts of northern Trinidad and Paria Peninsula, Venezuela. Westward from Paria Peninsula to Paraguaná Peninsula occur pocket beaches with sandy beach and dune formations. From Paraguaná Peninsula, Venezuela, and northern Atlantic Colombia, a sandy beach and dune community predominates. Between central and southern Colombia, mangrove swamps are present in sheltered areas.

The waters off South America hold an abundance of life. In the tropical and temperate waters some of the more important fish include swordfish, tuna, tarpon, shark, snook, bluefish, black bass, mackerel, and mullet; crustaceans include various lobsters, crabs, and shrimps. The cooler waters are, however, more productive. Waters off Tierra del Fuego have a *standing crop* of life four to seven times that of tropical waters. These cold waters abound with krill and plankton and sustain seals, whales, penguins, and petrels. Dusky dolphins are often seen in bays and estuaries of southern Argentina, where they seek shelter from killer whales. The cool current and associated upwelling along Peru and Chile create ideal conditions for plankton. The population of plankton, in turn, is important for supporting anchovy fisheries and dense colonies of seabirds (cormorants, gulls, pelicans, and boobies). Because of the cool current, penguins and southern fur seals inhabit waters north to the Galápagos Islands.

Conclusion

Central America and South America have a diversity of marine and terrestrial environments. Climate is the major factor in determining the vegetation cover, which includes coastal desert, grassland, savanna, bushland, chaparral, seasonal forest (thorn, deciduous, and evergreen), and rain forest. Additional environmental parameters affect specific coastal ecological communities. Insolation, tidal extent and range, salinity, temperature, and siltation are all important in determining particular distribution of salt marshes, mangrove swamps, sandy beach and dune, rocky shores and coral reefs.

NORBERT P. PSUTY
CHIZUKO MIZOBE

References

Aubert de la Rue, E.; Bourliere, F.; and Harroy, J. P., 1957. *The Tropics.* New York: Alfred A. Knopf, 208p.

Bates, M., 1964. *The Land and Wildlife of South America.* New York: Time-Life Books, 200p.

Beard, J. S., 1944. Climax vegetation in tropical America, *Ecology* 25, 127-157.

Bourliere, F., 1955. *Mammals of the World: Their Life and Habitats.* New York: Alfred A. Knopf, 223p.

Breedlove, D. E., 1973. The phytogeography and vegetation of Chiapas, Mexico, *in* A. Graham, ed., *Vegetation and Vegetation History of Northern Latin America.* New York: Elsevier Scientific Publishing Co., 149-165.

Brown, G. W., Jr., 1968. *Desert Biology.* New York: Academic Press, 635p.

Chapman, V. J., 1960. *Salt Marshes and Salt Deserts of the World.* London: Hill, 392p.

Cloudsley-Thompson, J., 1975. *Terrestrial Environments.* New York: John Wiley & Sons, 253p.

Curtis, H., 1975. *Biology.* New York: Worth Publishers, 1065p.

Daubenmire, R., 1978. *Plant Geography: With Special Reference to North America.* New York: Academic Press, 338p.

Delaney, P. J. V., 1966. *Geology and Geomorphology of the Coastal Plain of Rio Grande do Sul, Brazil, and Northern Uruguay.* Baton Rouge: Louisiana State University Press, Coastal Studies Series No. 15, 58p.

Dolan, R.; Hayden, B.; Hornberger, G.; Zieman, J.; and Vincent, M., 1972. *Classification of the Coastal Environments of the World. Part 1, The Americas.* Charlottesville: University of Virginia, Department of Environmental Science Tech. Rept. No. 1, 163p.

Eltringham, S. K., 1971. *Life in Mud and Sand.* New York: Crane, Russak, and Company, 218p.

Ferreyra, R., 1977. Endangered species and plant communities in Andean and coastal Peru, *in* G. T. Prance, ed., *Extinction Is Forever.* New York: New York Botanical Gardens, 150-157.

Gómez-Pompa, A., 1973. Ecology of the vegetation of Veracruz, in A. Graham, ed., *Vegetation and Vegetation History of Northern Latin America*. New York: Elsevier Scientific Publishing Co., 73-148.

Goodspeed, T. H., 1945. Notes on the vegetation and plant resources of Chile, in F. Verdoon, ed., *Plants and Plant Science in Latin America*. Waltham, Mass.: Chronica Botanica Company, 145-149.

Guiler, E. R., 1959a. Intertidal belt-forming species on rocky coasts of Northern Chile, *Royal Soc. Tasmania Proc.* 93, 33-58.

Guiler, E. R., 1959b. The intertidal ecology of Monteman area, Chile, *Royal Soc. Tasmania Proc.* 96, 164-183.

Hannau, H. W., 1974. *In the Coral Reefs of the Caribbean, Bahamas, Florida, and Bermuda*. Garden City: Doubleday and Company, 135p.

Henderson, D. A., 1967. Land, man, and time, in R. D. Ewing, ed., *Six Faces of Mexico*. Tucson: University of Arizona Press, 103-160.

Howard, R. A., 1973. The vegetation of the Antilles, in A. Graham, ed., *Vegetation and Vegetation History of Northern Latin America*. New York: Elsevier Scientific Publishing Co., 1-38.

Hubbs, C. L., and Roden, G. I., 1964. Oceanography and marine life along the Pacific coast of Middle America, in R. Wauchope, ed., *Handbook of Middle American Indians. Vol. I, Natural Environment and Early Cultures*. Austin: University of Texas Press, 143-186.

Johnson, W. W., 1972. *Baja California*. New York: Time-Life Books, 184p.

Jones, C. F., 1930. *South America*. New York: Henry Holt and Company, 798p.

McGinnies, W. G., 1968. Appraisal of research on vegetation of desert environments, in W. G. McGinnies, B. J. Goldman, and P. Paylore, eds., *Deserts of the World*. Tucson: University of Arizona Press, 381-566.

Moore, H. B., 1966. *Marine Ecology*. New York: John Wiley & Sons, 495p.

Moser, D., 1975. *Central American Jungles*. New York: Time-Life Books, 184p.

Neill, W. T., 1970. *The Geography of Life*. New York: Columbia University Press, 480p.

Ochoterena, I., 1945. Outline of the geographic distribution of plants in Mexico, in F. Verdoon, ed., *Plants and Plant Science in Mexico*. Waltham, Mass.: Chronica Botanica Company, 261-265.

Penaherrera de Aguila, C., 1969. *Geografía General de Peru, Tomo I/Aspectos Fisicos, Síntesis*. Lima: Institute of Pan-American Geography and History, 119p.

Petrov, M. V., 1976. *Deserts of the World*. New York: John Wiley & Sons, 447p.

Poggie, J. J., Jr., 1963. *Coastal Pioneer Plants and Habitat in the Tampico Region, Mexico*. Baton Rouge: Louisiana State University Press, Coastal Studies Series No. 6, 62p.

Porter, D. M., 1973. The vegetation of Panama: A review, in A. Graham, ed., *Vegetation and Vegetation History of Northern Latin America*. New York: Elsevier Scientific Publishing Co., 167-201.

Sauer, J. D., 1967. *Geographic Reconnaissance of Seashore Vegetation along the Mexican Gulf Coast*. Baton Rouge: Louisiana State University Press, Coastal Studies Series No. 21, 59p.

Snead, R. E., 1972. *Atlas of World Physical Features*. New York: John Wiley & Sons, 158p.

Standley, P. C., 1937. *Flora of Costa Rica, Part I*. Chicago: Field Museum of Natural History, 398p.

Stephenson, T. A., and Stephenson, A., 1972. *Life between Tidemarks on Rocky Shores*. San Francisco: W. H. Freeman and Co., 425p.

Steyermark, J. A., 1950. Flora of Guatemala, *Ecology* 31, 368-372.

Svenson, H. K., 1945. The vegetation of Ecuador, a brief review, in F. Verdoon, ed., *Plants and Plant Science in Latin America*. Waltham, Mass.: Chronica Botanica Company, 304-306.

Tait, R. V., and DeSanto, R. S., 1975. *Elements of Marine Ecology*. New York: Springer-Verlag, 327p.

Vann, J. H., 1959. *The Physical Geography of the Lower Coastal Plain of the Guiana Coast*. Baton Rouge: Louisiana State University Press, 91p.

Vermeer, D. E., 1959. *The Cays of British Honduras*. Berkeley and Los Angeles: University of California Press, 71p.

Wagner, P. L., 1964. Natural vegetation of Middle America, in R. Wauchope, ed., *Handbook of Middle American Indians. Vol. 1, Natural Environments and Early Cultures*. Austin: University of Texas Press, 216-263.

West, R. C., 1956. Mangrove swamps of the Pacific coast of Colombia, *Assoc. Am. Geographers Annals*, 46, 98-121.

West, R. C., 1976. Conservation of coastal marine environments, *Rev. Biol. Trop.* 24, 187-209.

West, R. C., 1977. Tidal salt marsh and mangal formations of Middle and South America, in V. J. Chapman, ed., *Wet Coastal Ecosystems*. New York: Elsevier Scientific Publishing Co., 193-213.

West, R. C., and Augelli, J. P., 1976. *Middle America: Its Lands and People*. Englewood Cliffs, N.J.: Prentice Hall, 494p.

Cross-references: *Antarctica, Coastal Ecology; Biotic Zonation; Central America, Coastal Morphology; Coastal Fauna; Coastal Flora; Coral Reef Coasts; Halophytes; Insects; Mangrove Coasts; North America, Coastal Ecology; Rocky Shores Habitat; Salt Marsh Coasts; South America, Coastal Morphology; Standing Stock*. Vol. VIII, Part 1: *Central America, Regional Review; South America*.

CHAETOGNATHA

Chaetognaths, also called *arrowworms,* are common carnivorous, planktonic animals resembling darts. The small body (usually about 3 cm long) is divided into a head, trunk, and tail. The head contains eyes and other sensory structures and the mouth region, which is surrounded by numerous spines and teeth (the entire head can be covered by a hood). Externally the trunk and tail have lateral fins and a tail fin, and internally they contain the straight digestive system and both male and female reproductive organs. Chaetognaths dart toward prey (including euphausids, copepods, small fish, and other chaetognaths), capture, and hold it with the spines and teeth, and push it to the posterior part of the digestive tract, where it can be digested in as little as 40 minutes. Adults lack a peritoneum surrounding the coelom, and thus resemble Aschelminthes (including nematodes), but the embryology indicates that chaetognaths are deuterostomes, as are echinoderms and chordates (Barnes, 1974).

All but one genus in the phylum float in the water, from the surface to depths greater than 1000 m; they are, however, most common in the upper 200 m (Alvarino, 1965). They occur in both coastal and oceanic waters from the tropics to the poles. Although some of the approximately 50 species are widely distributed, others are restricted to waters of particular temperatures and salinities (Hyman, 1959). Chaetognaths are therefore useful for identification of water masses; their presence or absence has also been used for predicting fish catches.

MOLLY FRITZ MILLER

References

Alvarino, A., 1965. Chaetognaths, *Oceanog. Mar. Biol. Annual Rev.* **3**, 115-194.

Barnes, R. D., 1974. *Invertebrate Zoology*, 3d ed., Philadelphia, Pa.: W. B. Saunders Co., 870p.

Hyman, L. H., 1959. *The Invertebrates: Smaller Coelomate Groups*, Vol. V, New York: McGraw-Hill Book Co., 783p.

Cross-references: *Annelida; Aschelminthes; Biotic Zonation; Chordata; Coastal Fauna; Echinodermata; Nemertea; Platyhelminthes.*

CHALK COASTS

Cliffed coasts are strongly influenced by lithology, the features of limestone, granite, or basalt outcrops being distinctive on the world coastline. Analysis of the effects of other factors, such as structure, weathering, exposure to wave attack, and variations in tidal range, can be usefully made in terms of comparative studies of features within one type of rock formation outcropping at various places along a coast. The *Chalk coasts* of northwest Europe (Fig. 1) show a distinctive range of cliff and shore-platform topography with variations that can be related to other factors.

The Chalk is an Upper Cretaceous formation consisting of stratified white limestone with associated nodular flint layers, especially along the bedding planes. It is divided into Upper Middle, and Lower Chalk, the latter being more varied in lithology, with marly horizons and fewer flints. The extensive Chalk cliffs of Kent and Sussex and of northern France are cut mainly in gently dipping Upper Chalk, and show vertical cliffs fronted by fallen boulders and narrow beaches of flint shingle, with broad shore platforms exposed at low tide (Fig. 2). The width and gradient of these platforms vary in relation to wave exposure and tide range.

The cliff crest undulations are due to intersection of late Pleistocene dry valleys that dissect the coastal slopes, and are matched by reentrants in the shore platform where the valley mouths passed below present sea level (Fig. 3). Cliff erosion is due to basal undercutting by storm waves, followed by collapse of the cliff face, especially when freeze and thaw occur in cold winters; the resulting rock fans are gradually dispersed and consumed by abrasion and solution over ensuing decades (So, 1965; May, 1971). Wave erosion is facilitated by the availability of hard flint nodules weathered out of the Chalk. These are used as ammunition by storm waves, and after a storm freshly abraded white chalk is exposed at the cliff base and on the shore platform. Foreshore dissection yields a microtopography of mesas and cuestas of flint-capped chalk, and the platform is grooved at right angles to the cliff line by abrasion as waves move flints to and fro, especially along joints. In addition, there is evidence of solution by sea spray and rainwater, notably on the upper surfaces of fallen boulders beyond the reach of wave attack and also in the pits and pools etched out on the shore platform. A varied shore fauna (*Littorina, Patella, Balanus* spp.) contributes to bioerosion of the intertidal zone, and at lower levels a carpet of marine plants (*Fucus, Ascophyllum, Pelvetia, Laminaria* spp.) protects the platform from abrasion by flint-laden waves. Where a broad shingle beach has accumulated at the cliff base, as at Walmer north of Dover, wave attack is impeded and cliffs become degraded into vegetated bluffs. On the other hand, local absence of flint shingle, as at Seaford Head between Brighton and Eastbourne, results in diminished abrasion and a dominance of solution and bioerosion features. A classic account of the morphology of the Chalk cliffs and platforms of northern France was presented by Prêcheur (1960).

Laterally, sectors of Chalk coast show transitions to features of overlying Tertiary-Quaternary or underlying older Cretaceous sands and clays. Near Eastbourne and Le Havre the vertical cliffs give place to landslips behind a more irregular and complex shore topography as the marly Lower Chalk and underlying Greensands and Gault clay rise to the shore level. In Yorkshire the Chalk cliffs of Flamborough have a capping of glacial drift, and between Sheringham and Cromer the north Norfolk coast shows a shore platform cut in

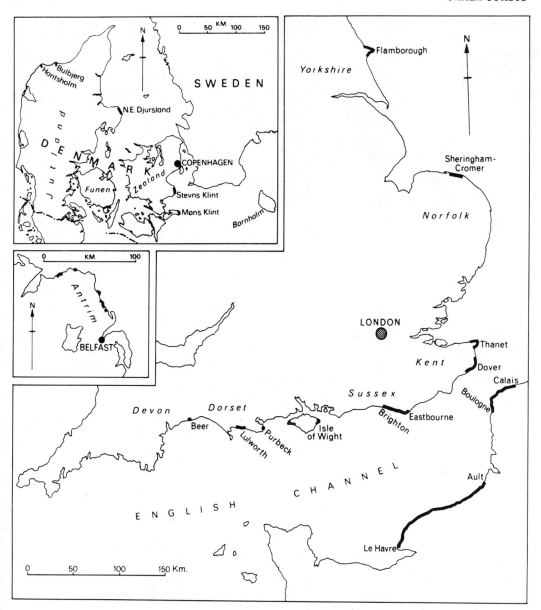

FIGURE 1. Locations of chalk coasts (in black) in northwest Europe.

flinty chalk (yielding shingle to the beach and to Blakeney Point to the west) backed by unstable slopes where glacial drift largely conceals the Chalk hinterland.

On the Isle of Wight and parts of the Dorset coast, where the Chalk is steeply dipping and locally faulted on the flanks of strong anticlinal folds, the vertical cliffs give place to coastal escarpments (Ballard Down, Culver Cliff) where the dip is landward and steep dip slopes (Alum Bay, Whitecliff Bay) where it is seaward. Shore platforms become narrow and irregular, with reefs of chalk hardened by compressional folding, as at Culver Cliff at the eastern end of the Isle of Wight, or are absent altogether, as at Ballard Point on Purbeck, which is geologically and locationally similar to Culver Cliff—a contrast that has not been elucidated.

In Northern Ireland the Antrim coast shows Chalk with much variation in dip and strike penetrated by volcanic necks and dikes associated with the overlying Tertiary basalt, and the cliffs and platforms are correspondingly intermittent. In Denmark the Upper Chalk outcrops locally on the coast in the Djursland Peninsula and at Stevns Klint and Møns Klint (Fig. 4);

CHALK COASTS

FIGURE 2. Chalk coast near Eastbourne in southeast England (Photo: C. T. Bird).

because the Baltic is here almost tideless the chalk cliffs are fronted by only narrow shore platforms, a feature seen also on the microtidal sector of the Purbeck coast in southern England (Fig. 5). Planed-off Chalk nevertheless extends out over the adjacent sea floors, a feature well developed off the Hantsholm-Bulbjerg coast in northern Jutland.

The typical association of vertical cliff and broad intertidal shore platform (Fig. 2) is best developed on relatively resistant Upper Chalk where flints are abundant and the dip is gently seaward on shores exposed to moderate or strong wave action. Shore platforms are less well developed on the marly Lower Chalk, where flints are rare, where the Chalk is steeply dipping, or where tidal range is small or wave attack weak (Figs. 4 and 5).

ERIC C. F. BIRD

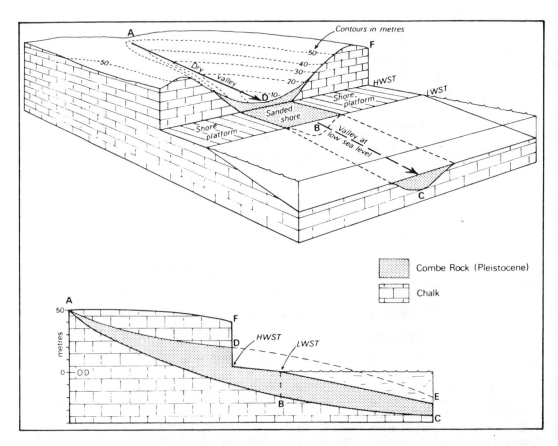

FIGURE 3. Where chalk cliffs are intersected by dry valleys (Fig. 2) the shore platform is often interrupted by a sandy sector. This occurrence is a result of the valleys having been incised during Pleistocene glacial low sea level phases (ABC) and then partly infilled with periglacial drift (Coombe Rock) to give an aggraded profile (ADE) into which the present cliff and shore platform have been cut. The diagram is based on features at Birling Gap, Sussex (seen in the distance in Fig. 2).

204

FIGURE 4. Chalk coast at Møns Klint, Denmark (Photo: A. Schou).

FIGURE 5. Chalk coast near Ballard Point, Purbeck, England, at low tide on a microtidal shore (spring tide range about 1 m). Contrast with Figure 2, where the spring tide range is more than 5 m. Note cloudy white chalk being removed in suspension from fallen rock area (Photo: E. C. F. Bird).

References

May, V. J., 1971. The retreat of chalk cliffs, *Geog. Jour.* **137**, 203-206.

Prêcheur, C., 1960. *Le littoral de la Manche, de Ste. Adresse à Ault: Étude morphologique.* Poitiers: Norois, 138p.

So, C. L., 1965. Coastal platforms of the Isle of Thanet, Kent, *Inst. British Geographers Trans.* **37**, 147-156.

Cross-references: *Cliff Erosion; Cliffed Coast; Klint Coast; Splash and Spray Zone; Weathering and Erosion, Biologic.*

CHANNEL SAND—See SHOESTRING SANDS

CHART DATUM

Soundings on hydrographic surveys must be reduced to a common plane that is low enough to ensure that a tide level will seldom (preferably never) fall below it. However, datum should not be so low that the resulting chart gives unduly shallow values for navigation purposes (U.K. Hydrographic Office, 1969). Soundings must be in close agreement where different datums join. Greatest problems occur in estuaries where there is a rapid spatial variation in tide height and time of low water.

Various datums have been used on different historical charts. At one time *Mean Low Water Spring* tides was adopted, but *Lowest Astronomical Tide* (L.A.T.) is now preferred. L.A.T. is defined as the lowest level that occurs under average meteorological conditions and is based on observations, ideally over an 18-year period. Because of difficulties in obtaining these data, a margin of 0.1 m error is accepted in British Admiralty surveys.

Chart datum for nontidal waters, as in most land surveys, is based on mean sea level.

ALAN PAUL CARR

Reference

U.K. Hydrographic Office, 1969. *Admiralty Manual of Tides,* vol. 2, chap. 2 (published as individual chapters).

Cross-references: *Coastal Characteristics, Mapping of; Mean Sea Level; Reference Plane; Tidal Range and Variation; Tidal Type, Variation Worldwide; Worldwide; Tide Gauge; Tides.*

CHATTERMARKS

Chattermarks is the name given to *fractures* and *gouges* preserved on some rock surfaces after the passage of glacier ice. The marks are crescentic in shape and vary in length from microscopic to typically 5-10 cm across. They occur spaced 2-4 cm apart, with the long axis of the series parallel to the motion of the ice. Chattermarks are best preserved on brittle rocks such as quartzites, and were first described by Chamberlin (1888), who noted their resemblance to scars produced by the vibration of a cutting tool when forced across a metal surface. Several varieties of these cracks have been described (Harris, 1943) with curvature both concave upstream and downstream. Although the term chattermarks has been restricted to crescentic (or straight) cracks *within* large grooves (Flint, 1971), it is frequently used for all similar kinds of cracks.

Passage of debris-laden sea and lake ice over shoreline rock surfaces rarely produces well-developed chattermarks but irregular, *shock marks* may be left on soft rocks (limestone, siltstone), and infrequently on crystalline rocks, that may be confused with glacially derived features (Dionne, 1973).

J. BRIAN BIRD

References

Chamberlin, T. C., 1888. The rock-scorings of the great ice invasions, *U.S. Geol. Survey 7th Annual Rept, 1885-86,* 147-248.
Dionne, J.-C., 1973. Distinction entre stries glacielles et stries glaciaires, *Rev. Géographie Montréal* **27,** 185-190.
Flint, R. F., 1971. *Glacial and Quaternary Geology.* New York: John Wiley & Sons, 892p.
Harris, S. E., 1943. Friction cracks and the direction of glacial movement, *Jour. Geology* **51,** 244-258.

Cross-references: *Antarctic, Coastal Morphology; Aquafact; Arctic, Coastal Morphology; Frost Riving; Glacial Drift Ice; High-Latitude Coasts; Ice along the Shore; Ice-Bordered Coasts; Ice Foot; Ice Thrusting.* Vol. I: *Sea Ice;* Vol. III: *Chatter Marks.*

CHENIER AND CHENIER PLAIN

Russell and Howe (1935) and Howe, Russell, and McGuirt (1935) were the first to use the term *chenier* (Anglicized pron., shanear) in the geological literature for long (sometimes 40-60 km), narrow, low, fossil beach ridges in Cameron and Vermilion parishes, southwestern Louisiana. This local Acadian French expression comes from *chêne* ('oak tree'), live oak (*Quercus virginiana*) trees being common on the ridges. Another form, *cheniere,* not adopted in geology, can refer both to the fossil beach ridges and to higher river levees in the marsh and swamplands. The cheniers are composed of sand and, in areas where shells are abundant near shore, only of shells. A 22% shell content was found in Louisiana cheniers (Byrne, LeRoy, and Riley, 1959). Although greater thicknesses have been observed in Guiana, these are usually less than 4 m thick. Cheniers rest on muddy, nearshore, intertidal deposits, sometimes on marsh sediments, and are entirely surrounded by mud flats and marshes. The ridges often branch out with a multitude of arms in a feathery, fanlike fashion and bend landward, outlining shore positions of earlier coastal indentations. Where a river mouth is the sand source, fans open down current.

Chenier plains (Price, 1955) are progradational belts composed of chenier ridges and the intervening muddy, marshy sediment zones of marine origin. Beach ridges that front chenier plains cannot be considered cheniers until individuals become inactivated by intertidal and supratidal mud flats that develop on their seaward side and isolate them from the sea. In a superficially similar facies setting, subsiding barrier beach plains, composed entirely of sand ridges (sometimes even fossil nearshore barrier islands), may be largely covered by salt marshes near sea level and could give the appearance of chenier plains. Only by careful drilling can the actual identity of such plains be identified. Beach ridge fans lying within the area of a delta are typically separated by sand and local deltaic clay (Otvos and Price, 1979).

Chenier plains develop on shores that receive large supplies of mud and also some sand and periodically undergo erosion. The ridges form partly from the sand fraction of eroding mud flats, marsh sediments, and partly from sand carried in longshore transport from nearby sediment sources. Low energy and microtidal conditions are not prerequisites for chenier plain development; maximum tidal ranges of 3 m and wave heights of 2 m have been described from the Guiana coast as well as maximum tidal range of 10 m from the Gulf of California (Diephuis, 1966; Thompson, 1968; cited in Cook and Polach, 1973). Shores of the 1600-km-long Guiana coast, where the world's largest chenier plains are located, receive huge amounts of mud from onshore-moving currents that incorporate suspended material from Amazon River discharge to the east (Gibbs, 1976). Mud bank-shoals form at regular distance intervals, and their steady westward migration along the shore creates cyclic alternations of mud flat accretion and erosion, sand-shell ridge formation (Diephuis, 1966; cited in Allersma, 1971). The different chenier generations in Louisiana were attributed by Gould and McFarlan (1959) to sediment influx variations, regulated by shifting Holocene Mississippi subdelta positions to the east. Steadily increasing mud volumes from the Atchafalaya River in recent decades are credited with mud flat development in front of several southwest Louisiana beach segments. Short-term, climatic-hydrographic regimen changes may alone be responsible locally for shifts from mud flat to beach ridge accumulations and vice versa (Vann, 1959; cited in Allersma, 1971).

Mud flat stabilization in Louisiana is largely accomplished by *oyster grass* colonization, while *mangrove* growth, leading to marsh development, is essential in Guiana in this respect.

The presence of reworked older shell material and other types of contamination often causes difficulties in absolute chenier age determination, but it appears that most of the known chenier plains started to prograde at the time the worldwide sea level had nearly occupied its present semistillstand position. Well-documented chenier plains have also been described from the Colorado River delta, Gulf of California (Thomson, 1968), Broad Sound, Australia (Cook and Polach, 1973), the Firth of Thames, New Zealand (Schoefield, 1960, cited in Cook and Polach); and a few other worldwide locations.

ERVIN G. OTVOS
W. ARMSTRONG PRICE

References

Allersma, E., 1971. Mud on the oceanic shelf off Guiana, in *Symposium on Investigation and Resources of the Caribbean Sea and Adjacent Regions.* Paris: UNESCO, CICAR, 192-203.

Byrne, J. V.; LeRoy, D. O.; and Riley, Ch. M., 1959. The chenier plain and its stratigraphy, southwestern Louisiana, *Gulf Coast Assoc. Geol. Socs. Trans.* 9, 237-260.

Cook, P. J., and Polach, H. A., 1973. A chenier sequence at Broad Sound, Queensland, and evidence against a Holocene high sea level, *Marine Geology* 14, 253-268.

Gibbs, R. J., 1976. Amazon River sediment transport in the Atlantic Ocean, *Geology* 4, 45-48.

Gould, H. R., and McFarlan, E., Jr., 1959. Geologic history of the chenier plain, southwestern Louisiana, *Gulf Coast Assoc. Geol. Socs. Trans.* 9, 261-270.

Howe, H. V.; Russell, R. J.; and McGuirt, J. H., 1935. Physiography of coastal southwest Louisiana, *Louisiana Geol. Survey Geol. Bull.* 6, 1-72.

Otvos, E. G., and Price, W. A., 1979. Problems of chenier genesis and terminology—an overview, *Marine Geology* 31, 251-263.

Price, W. A., 1955. Environment and formation of the chenier plain, *Quaternaria* 2, 75-86.

Russell, R. J., and Howe, H. V., 1935. Cheniers of southwest Louisiana, *Geog. Rev.* 25, 449-461.

Cross-references: *Beach Plain; Beach Ridge and Beach Ridge Coasts; Beach Stratigraphy; Deltas; Sea Level Curves.* Vol. III: *Beach Ridges;* Vol. VIII, Part 1: *French Guiana.*

CHIMNEY—See MARINE-EROSION COASTS

CHINE

In the Bournemouth area on the south coast of England, the term chine refers to a steep, narrow valley, about 18 m deep, cut into an older, broader valley. The chine is cut in the soft Eocene and Oligocene sediments that fill the Hampshire Basin. The broader, upper valley was probably cut during a cold phase when frost would render the strata more resistant, and snowmelt would have provided adequate flood water seasonally (Bury, 1920). The upper valleys were graded to the Solent River then running about 2-3 km from the present coast. The inner valleys—the chines—were incised under present climatic conditions and eventually graded steeply down to the modern coast. The small power of the present streams is reflected in the steep, narrow character of the chines through which they flow. Marine erosion is cutting the coast back more quickly than the chines are being deepened and lengthened by headward erosion, so that in time they will be eroded away or remain as notches on the cliffs. Similar features are called bunnies locally near Christchurch.

CUCHLAINE A. M. KING

Reference

Bury, H., 1920. The chines and cliffs of Bournemouth, *Geol. Mag.* 57, 71-76.

Cross-references: *Geographic Terminology,* Vol. III: *Valley Evolution.*

CHLOROPHYTA

The Chlorophyta are a division (phylum) of the *algae*. Green in color, their photosynthetic pigments are similar to those of flowering plants—namely a complex of chlorophylls a and b, xanthophylls, and carotenes contained in well-defined chloroplasts (Dawson, 1966).

The Chlorophyta are found in a wide range of environments from freshwater, where they are the dominant form of algae, through brackish and estuarine conditions to marine situations. In the latter, they are usually far less conspicuous; especially in cold and temperate seas, they are far less prevalent than the reds (Rhodophyta) and the browns (Phaeophyta). Their sizes range from planktonic, where they occupy an important place, to large macrophytic forms, the largest recorded size being 8 m, attained by *Codium magnum*.

Although they rarely dominate the coast, in some places such as quiet lagoons or near sewage outlets their biomass can be considerable and their locally overwhelming presence can exclude almost all other vegetation. Some members of the Ulvaceae in particular can show high local productivity.

CHARLES R. C. SHEPPARD

Reference

Dawson, E. Y., 1966. *Marine Botany: An Introduction.* New York: Holt, Rinehart and Winston, 371p.

Cross-references: *Algal Mats and Stromatolites; Chrysophyta; Coastal Flora; Coral Reefs; Cryptyphyta; Cyanophyceae; Euglenophyta; Phaeophycophyta; Reefs; Rhodophycophyta.* Vol. III: *Algae;* Vol. VI: *Algal Reef Sedimentology.*

CHORDATA

The ancestors of all chordates doubtless arose as small, sluglike organisms in the shallow sediments of pre-Cambrian seas, where they probably settled down from tornarian larvae similar to those of their deuterostome relatives, the starfishes.

These creatures gave rise to descendants existing today, of which the most primitive are the Protochordata. They have an ectodermal tubular nerve cord, a pharynx with slits, and as ciliary-mucous feeders they have a groove along the pharynx—the endostyle, also found in the more advanced Amphioxus and Urochorda (*tunicates*). There are two groups. The Pterobranchia are antipodal, sometimes colonial, deep and cool water habitants coming into the shallows only in the East Indies. They have no sign of a notochord or anything that could become one. The Enteropneusta (*acorn worms*) are weakly moving buried worms a few cm to 2.5 m long, the large ones being so fragile that they cannot be picked up without tearing apart. They are common in muddy soft-bottomed estuaries. They are noted for their iodoform odor, and apparently serve as food for many bottom feeders. The Enteropneusta have a short diverticulum extending from the anterior digestive tract into the proboscis, which has been interpreted as being a prenotochord. This organ is now thought to be part of an excretory complex.

The Urochorda or Tunicata are respectable chordates as larvae, with a cerebral vesicle and a functional tail containing the notochord. But all this degenerates, and in sessile species the animals attach at the head end and become mostly a large sac around the digestive system and the reproductive elements. There are many floating or nektonic species on the high seas, single and colonial. Some of these do not regress as adults. Near the coasts and especially in the estuaries a few sessile species abide as singles and in small numbers. The tunicates are saliniphilous, and their influence is not strong in estuaries. They are not euryhaline, as has been stated. Atlantic North American material has been summarized by Plough (1978).

The Cephalochordata, the lancelets or amphioxus, are the only complete nonlarval chordates in a sense. The notochord lasts throughout life; it is the only element of the axial skeleton, and there are no vertebrae. The notochord and the dorsal tubular nerve cord run into the anterior region, where the nerve ends in a slight swelling, too small to call a brain. There are no eyes. The mouth is at the bottom of a small buccal cavity surrounded by cirri. There is an anterior end but no proper head. The name Euchordata would be better for this group than Cephalochordata, which is something of a misnomer. Instead, however, that name is used for the vertebrata in which the notochord is whole only in early embryonic life. The cephalochords have dorsal, ventral, and tail fins supported by short bits of connective tissue.

The larval Tunicata seem to be more complicated or advanced as larvae than the adult Cephalochordata, but the latter do not regress. The body is made of a long series of V-shaped muscle segments, and there is a special chamber called the atrium for the outflow of water through the pharyngeal slits. Otherwise, the body arrangement is fish-shaped, but there are no scales. Amphioxus lives buried and seems to prefer mixtures of mud and sand, into which it can dive in from either end. It is abundant in estuaries and shallow seas, and various species are found over much of the world except the polar zones. In China it is abundant enough to serve as food for man. In Mississippi Sound they have been recorded in the billions and are sometimes killed in catastrophic numbers by freshwater floods (Boschung and Gunter, 1962; Dawson, 1965).

The other Chordata are all vertebrates. The Cyclostomata or Agnatha, hagfishes and lampreys, have round mouths with chitinous teeth and a piston like tongue. They rasp the flesh and suck the blood of fishes. They are thought of as parasitic and loathesome, but actually they are predators with poor equipment. The axial skeleton consists of the notochord, with somewhat crude vertebrae and a cartilaginous branchial basket and a partially enclosed brain case. The brain is small and primitive but has the same parts as the advanced vertebrates.

The hagfishes live in pure saltwater and into

the depths, while the lampreys are largely anadromous, with some freshwater species.

The cyclostomes have the ecological effects of prey species in estuaries and marine coastal waters and are known to be destructive in certain enclosed waters.

The other Chordata, all gnathostomes, are the Pisces (Elasmobranchii and Osteichthyes), Amphibia, Reptilia, Aves, and Mammalia.

The Pisces, or fishes, are divided into two groups. The Elasmobranchii, which are the sharks, rays, sawfishes, and their relatives, have no true bone and are thus called *cartilaginous* fishes. They are common in low salinity waters, and many species are euryhaline. Many of them are large and are predators on other estuarine and marine organisms. They are an important complex in the life of the seas, being exceeded only by the Osteichthyes, the bony fishes that live in all oceans and estuaries except the cold polar seas. Their breeding and feeding habits and habitats are so diverse that it may be said they are the most dominant, influential, and important of all aquatic organisms.

Amphibia have never been known to be denizens of the seas, apparently because of their inability to live in waters hyperosmotic to their blood. Today a few frogs are found on the shore of bays and sounds following freshwater floods, but their sojourn there is temporary.

The Reptilia were abundant in the Mesozoic seas, and there were large, wide-ranging species such as Ichthyosaurs and Pleisiosaurs. There were even the flying reptiles, Pteranodon, up to 9 meters in wingspread, which apparently lived over shallow seas and fed on marine animals. There were also some large turtles in the ancient seas, but the marine reptiles faded away as they did on land; and now only some marine serpents exist in the tropical Pacific, and a few species of marine turtles are found in the tropic and subtropic oceans. An essentially terrestrial iguana of the Galápagos Islands has learned to dive and feed on marine algae several feet under the water. The crocodilians have been in existence for about a quarter of a billion years. A few of them still swim into the oceans where the marshes and jungle swamps meet the sea.

The birds, Aves, are abundant on seashores and estuaries. Next to the fishes, the vertebrates of this class are probably the most abundant and have the greatest ecological effect of any group of the Chordata on the ecology of the bays and shores of the ocean. Hundreds of species have adapted for feeding in the waters or probing the shores and have also adjusted anatomically and physiologically for migrating with the seasons toward or away from the poles.

They pursue and catch fishes or eat the plant food of saltwater marshes. Unfortunately, they have been easily influenced and destroyed by man and his activities. The great waterfowl era of the recent past seems to be over, especially in North America, where no one will ever again see flocks of a million geese in the air at one time, unless climatic and geological changes occur that man cannot control.

The various species of whales, walruses, seals, sea otters, and sea cows are also disappearing at an alarming rate. The sea cows probably had little influence, but the seals consumed vast amounts of fishes, and even the sea otters are reputed to have reduced the populations of abalones and certain sea urchins and clams along the shores where they live. But as a whole, the probabilities are that these mammals will decrease and never again dominate the shores and shallow seas as they have in some areas in the recent past.

The Vertebrata treated here are all members of the Chordata. They develop a notochord in early stages, which is invaded and taken over in part by the vertebral bones, which also surround the nerve tissue of the spinal cord. In this complex the notochord remains as pads of soft resilient tissue between the vertebrae, the whole forming the backbone.

GORDON GUNTER

References

Boschung, H. T., and Gunter, G., 1962. Distribution and variation of *Branchiostoma caribaeum* in Mississippi Sound, *Tulane Studies in Zoology* 9, 245-257.

Dawson, C. E., 1965. Rainstorm induced mortality of lancelets, *Branchiostoma,* in Mississippi Sound, *Copeia* 4, 505-506.

Plough, H. H., 1978. *Sea Squirts of the Atlantic Continental Shelf from Maine to Texas.* Baltimore: Johns Hopkins University Press, 118p.

Cross-references: *Biotic Zonation; Coastal Fauna; Organism-Sediment Relationship; Succession.*

CHRYSOPHYTA

The Chrysophyta or *golden brown algae* are microscopic, generally unicellular organisms, although there are some colonial and a few filamentous forms. Chrysophytes are generally planktonic or epiphytic, numerous, and are found in a variety of environments, mostly freshwater. The silicoflagellates (Dictyochaceae) and coccolithophorids (Coccolithaceae) are, however, important elements of the modern marine phytoplankton as well as being useful fossils for geologic age determinations.

Chrysophytes are characterized by photosynthetic pigments chlorophyll *a* and *c* and carotenoids fucoxanthin and diadinoxanthin. The principal reserve products are oil and chrysolaminarin (leucosin). The main sterols are fucosterol and porifasterol. The majority of species are naked and lack a cellulose wall, while many show a high degree of silicification (e.g., silicoflagellates) or calcification (e.g., coccolithophorids). The production of endogenous cysts (statospores) is also common. Finally, chrysophytes are either uniflagellate, biflagellate (the flagella either equal or unequal in length and similar or dissimilar in nature), or biflagellate with a third filiform or anchorage appendage, called a haptonema.

Chrysophyte reproduction is commonly by cell division, by zoospores, or by statospores. Sexual reproduction, although rare, has been reported and is isogamous.

The division Chrysophyta is generally accepted as being divided into two classes, the Chrysophyceae and the Haptophyceae.

Taxonomic treatment of the Chrysophyceae has varied widely with anywhere from five to twelve orders proposed. The classification of Chapman and Chapman (1973) recognizes five orders. The Chrysomonodales are the simplest types, and include unicells (*Ochromonas, Mallomonas*), colonial sheathed forms (*Dinobryon, Synura*), or organisms with an internal siliceous skeleton (silicoflagellates). The Rhizochrysidales include pseudopodial amoeboid forms that can be naked (*Rhizochrysis*), ensheathed (*Lagynion*), or plasmoidal (*Myxochrysis*). The Chrysocapsales include colonial forms with the cells contained in a gelatinous matrix (*Chrysocapsa*). The Chrysosphaerales are nonflagellate coccoid or colonial cells with a distinct cell wall (*Chrysosphaera*); the fifth order, the Phaeothamniales, contain the most highly evolved members, which are simple or branched, erect or prostrate threads (*Phaeodermatium, Phaeothamnion*).

The only Chrysophyceae group with a fossil record are the silicoflagellates, which are found from the Cretaceous to the Recent.

The Haptophyceae include two orders, the Isochrysidales and the Prymnesiales. The Prymnesiales are the more important of the two orders, and include the family Coccolithaceae (*Coccolithus*). These phytoplankton are characterized by the presence of a haptonema and the production of calcareous plates that cover the external cell (*Hymenomonas*). These calcareous plates (coccoliths) have been reported in the fossil record from the Pennsylvanian, Permian, and Jurassic to the Recent.

E. REED WICANDER

Reference

Chapman, V. J., and Chapman, D. J., 1973. *The Algae*. London: Macmillan and Co., 497p.

Cross-references: *Algal Mats and Stromatolites; Chlorophyta; Chrysophyta; Coastal Flora; Cryptophyta; Cyanophyceae; Euglenophyta; Phaeophycophyta; Rhodophycophyta*. Vol. VII: *Algae*.

CLASSIFICATION

Classification is a useful means of summarizing knowledge in a systematic way. The type of information used in the classification of coasts depends on a number of factors, such as the purpose of the classification, the scale of the area under consideration, and the amount of data available. Data concerning both the present coast and its development through time can be taken into account. Classifications range from those applicable to the coasts of the whole earth to those of small local areas. Some of them are land-based, examining the coast from the point of view of the character of the land against which the sea rests; others concentrate on sea conditions and their effect on the coast.

Here the classifications to be considered can be subdivided according to the main criteria used for initial classification, and then within each group the classification depends on the size of the area covered. The first group depends on the *structural* characteristics of the land; the second on the *morphological* attributes of the coastal zone, a group that can be considered on a wide range of scales. The third group includes many of the well-known *genetic* classifications, based on the *processes* that have influenced the coast and given it its present character. The fourth and fifth groups of classification are based on the *dynamic* character of the area in terms of the changing *morphology* and the *energy* of the marine system. The sixth group is also *dynamic*, and is based mainly on the *wave* characteristics.

Structural Classification

One of the earliest attempts to differentiate coasts of different types was that of Suess (1888), who recognized the striking contrasts between many coasts in the Pacific Ocean and those in the Atlantic, using these terms to distinguish his two major types. In the Atlantic type the general trend of the coast runs at a high angle to the structures, a type well exemplified in the coast of southwestern Britain, with its long peninsulas and deep ria embayments. The Pacific type, on the other hand, is typified by coastal mountain ranges running parallel to the coast, as exemplified by the coast of western North and South America, where the Rocky Mountains and Andes run nearly parallel to the coast for a great distance. The island arcs of the western Pacific ocean represent another form of this coastal type.

The dichotomy is generalized, applies only to extensive stretches of coastline, and is not strictly applicable throughout the two oceans. It does, nevertheless, draw attention to significant differences in the structural nature of the coastal margin over great lengths of the world's coast. It is essentially a land-oriented type of classification. Recent work in tectonic geology has given renewed interest to this type of classification.

The new approach to coastal classification by Inman and Nordstrom (1971) builds on the structural element of Suess' classification, and is based on global tectonics and sea floor spreading. Three size scales associated with coasts are recognized. The first is associated with moving plates and has linear dimensions of about 1000 km along the coast, 100 km normal to it with a vertical range of 10 km. The second-order scale brings in erosional and depositional modifications of the first order features, and has dimensions of about 100 km, 10 km, and 1 km for the three directions, respectively. The third-order scale is dependent on wave action, type of material and similar variables. Its dimensions are 1–100 km along the coast, 1 km normal to the coast direction, and it includes such features as beach face, berm and bars. The term *coastal zone* is used to refer to the first two scales, and the *shore zone* to the third. The shore zone includes the foreshore, surf zone, and nearshore zone, while the coast zone includes the coastal plain and continental shelf to the shelf edge, often taken conventionally to a depth of 200 m.

The first-order features fall into three tectonic classes: collision edge coasts, trailing edge coasts, and continental coast of marginal seas behind island arcs. The collision coasts are further subdivided according to the presence of two continents, one continent and one ocean, or no continents; the last is the island arc and trench system. If the continent is part of a stable plate, the potential for erosion and deposition will be reduced compared with a collision continent. The rivers Congo and Nile illustrate the first type, and the Amazon and Mississippi the second. The ratio of drainage area to suspended sediment load illustrates the importance of the distinction. The Congo and Nile are second and fourth in drainage area, but 21st and 15th in sediment load, while the Amazon and Mississippi are first and third in drainage area and third and sixth in suspended sediment load. Further subdivision of the trailing edge coasts into three types is based on this distinction, giving the neo-

TABLE 1. Comparison of First-Order Tectonic Coastal Classification, Second-Order Process, and Shelf Character (after Inman and Nordstrom, 1971).

Second Order	First Order					% World Coast
	1	2a	2b	2c	3	
Wave erosion	47.9	5.6	3.8	4.9	37.8	44.7
Wave deposition	15.5	12.4	21.4	33.3	17.4	11.6
River deposition	5.7	6.4	9.9	62.4	15.6	3.2
Wind deposition	1.9	18.8	79.3	0	0	1.2
Glaciated	6.4	0	7.2	86.4	0	30.7
Biogenous	36.1	21.0	11.3	15.8	15.8	3.0
% World	39.1	4.3	6.8	35.4	8.8	94.4*
Shelf type %						
tectonic dam	86.2	1.7	0	0	74.2	31.4
salt dome dam	0	0	0	0	13.4	1.6
reef dam	10.5	81.5	4.5	0	12.4	10.9
progradational	3.3	16.8	95.5	100.0	0	56.1
Total	100	100	100	100	100	100
Shelf width %						
narrow <50 km	23.9	6.5	11.2	12.6	3.3	57.5
wide >50 km	1.4	0.1	3.6	29.1	8.3	42.5
Total	25.3	6.6	14.8	41.7	11.6	100

*excluding Antarctica, 5.6%.

First order: 1, collision coasts; 2a, trailing edge–Neo; 2b, trailing edge–Afro; 2c, trailing edge–Amero; 3, marginal sea coasts.

CLASSIFICATION

trailing edge, the Afro-trailing edge, and the Amero-trailing edge.

Some relationships between the first- and second-order features are given in Table 1. The first part of the table relates to the major process operating on the different primary divisions, while the second part relates the type of shelf to the nature of the primary coast. The importance of deltas on the Amero-trailing edge coasts is related to the heavy sediment load of rivers discharging along these coasts, while wave erosion is dominant on the collision coasts, which also have exceptionally narrow shelves, compared with the much wider ones associated with the Amero-trailing edge coasts. Figure 1 illustrates the geographical distribution of the different coastal and shelf types, and provides a useful summary of the coastal character in terms of modern global tectonics. For some purposes, however, a morphological classification is more suitable; Inman and Nordstrom (1971) give a classification in these terms also.

Morphological Classification

The morphological classification of Inman and Nordstrom (1971) is related to their tectonic classification, but stresses the major morphological features of the tectonic types. Table 2 relates the two types of classification. There are six main coastal morphological types: mountainous coasts, narrow shelf coasts, wide shelf coasts, deltaic coasts, reef coasts, and glaciated coasts. The last two types are related to climatically controlled features and therefore show a latitudinal distribution—the former in low latitudes, and the latter in high ones. The mountainous coasts are essentially collision coasts, while the wide shelves off plain coasts are mainly the Amero-trailing edge type. The neo-trailing edge coasts are mainly narrow shelf, hilly coasts; the marginal coasts are predominantly hilly coasts with wide shelves. These relationships provide a useful link between the purely morphological and the tectonic coastal types, as illustrated in Figures 1 and 2.

Alexander (1966) attempted to map part of the coast of east Africa, using a morphological symbolism to represent different coastal features. His scheme can be applied empirically on a variety of scales, the detail increasing as the scale is enlarged. The symbols apply to the shore zone, and include both the nature of the land margin and the nearshore characteristics of marine-formed features. Cliffed and non-cliffed coasts, smooth and crenulate coasts are distinguished, and the shape of bays indicated. Further details such as rock type and cliff height can be added where data are available. The marine features include various types of bars, barriers, reefs, and salt flats. Further details that can be added when scale and information allow include beach slope and vegetation.

Shore accumulation forms are noted for their

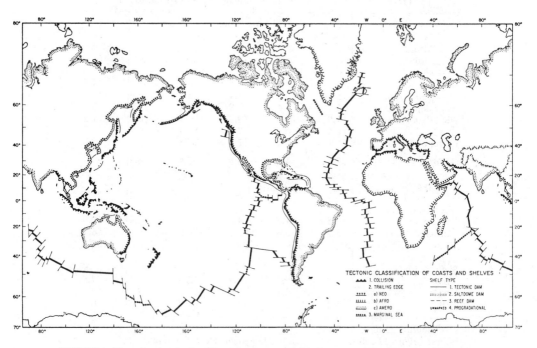

FIGURE 1. Map of tectonic classification of coasts (after Inman and Nordstrom, 1971).

TABLE 2. Comparison of First-Order Tectonic Classification and Morphological Classification (after Inman and Nordstrom, 1971).

Morphological class	First-order tectonic class				
	1. Collision coast	2. Trailing edge coast			3. Marginal coast
		a. Neo	b. Afro	c. Amero	
Mountainous coast (1)	97.2	8.0	–	–	2.5
Narrow shelf, hilly coast (2a)	–	75.1	14.1	–	5.6
Narrow shelf, plains coasts (2b)	–	15.9	46.2	1.5	–
Wide shelf, plains coast (3a)	–	–	4.0	89.3	3.1
Wide shelf, hilly coast (3b)	–	–	–	2.2	77.4
Deltaic coast (4)	–	1.0	3.4	1.3	5.8
Reef coast (5)	–	–	3.0	1.9	5.6
Glaciated coast (6)	2.8	–	29.3	3.8	–
% World coast line (excluding Antarctic)	39.0	4.6	7.5	35.2	8.1

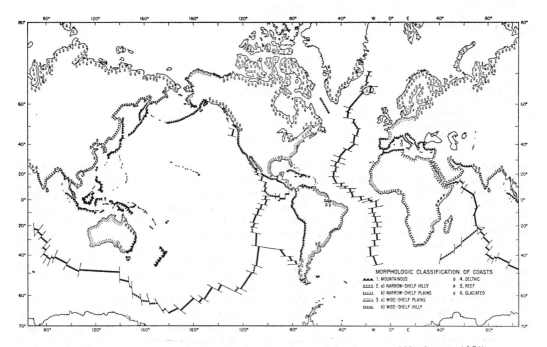

FIGURE 2. Map of morphological classification of coasts (after Inman and Nordstrom, 1971).

great range and variety, so that a classification of these forms is of value. One has been given by Zenkovich (1967). The classification applies to a scale on which individual features can be shown. Table 3 gives the range of forms classified, which include attached forms, free forms, barriers, and looped forms, each group being further subdivided according to its morphology and degree of complexity. The classification of features is useful in that it provides evidence of the nature of coastal accretion. Information concerning longshore movement of material, which is an essential process in the building of most accumulation forms, can also be inferred.

These forms can be recognized on large-scale maps, but they show up particularly well on clear vertical air photographs, on which the individual elements of the features can often be distinguished, such as shingle beach ridges, enabling the stages of development of the feature to be elucidated. The major distinction in shape is between rounded and cuspate forms, which depends on the nature of wave attack; the pointed forms are largely restricted to bays or sheltered localities behind islands, while rounded forms indicate greater wave energy and more effective wave refraction of longer waves.

When the area under consideration is small

TABLE 3. Classification of Accumulation Forms of Zenkovich (1967).

I ATTACHED FORMS
   1 *Beaches*
   2 *Above-water terraces (berms):*
      a) On smooth and projecting sectors of coast
      b) Filling indentations: Simple (unilateral), or double
   3 *Forelands, relative length to width less than unity:*
      a) Simple (unilateral, symmetrical), rounded, cuspate
      b) Double (symmetrical), rounded, cuspate
II FREE FORMS (relative length greater than unity)
   4 *Spits (simple forms fed from one direction only):*
      a) Continuing line of the original coast or curving into bays
      b) Trending seawards from the coast
   5 *Spits (fed from both sides, usually symmetrical relative to the original coast)*
III BARRIERS (extending from the original coast and enclosing a lagoon)
   6 *Looped barrier (symmetrical rounded form)*
   7 *Double fringing barrier*
   8 *Bracket-shaped barrier*
   9 *Double fringing spit*
IV LOOPED FORMS (both ends connected to the shore)
   10 *Barrier beaches (enclosing a bay)*
   11 *Tombolos linking island to mainland:*
      a) Asymmetrical
      b) Symmetrical
   12 *Tombolos between islands*
V DETACHED FORMS
   13 *Barrier beaches (not in contact with the coast for long distances)*
   14 *Accumulation islands:*
      a) Relict forms
      b) In the course of formation

and a great deal of data are available, the classification can best be indicated on a geomorphological map, on which the individual elements of the coastal system can be differentiated. An example is provided by the barrier island system. This system can be classified into separate *units* that occur both *normal* to and *parallel* with the coastline. The units normal to the coastline include the mainland shore of the lagoon, the lagoon, and its barrier island shoreline. The barrier island itself involves washover fans, dunes, and berms from land to sea. The foreshore covers the intertidal section of the beach, followed seaward by the swash zone, the surf zone, and the nearshore zone. Parallel to the coastline, tidal inlets and the tidal deltas and bars associated with them alternate with the barrier island and its terminal spits. In dealing with a system of this type it is essential that all the units are classified and recognized and their interrelationships studied.

Genetic classifications

The oldest and best-known classifications are the genetic ones, including those of Johnson, Cotton, and Shepard. Johnson's (1919) classification was for a long time a standard one that grew out of earlier work by Davis, Gulliver, and Richthofen (1886). Dana (1849) was one of the first to note the effect of subsidence on the coastal form in the deeply embayed coast of Tahiti. All these workers had recognized the importance of subsidence in explaining the characteristics of many coasts, and this element formed the major dichotomy of Johnson's classification. He divides coasts into four groups: submerged coasts, emerged coasts, neutral coasts, and compound coasts. The last group consists of coasts showing a combination of the first three types.

Johnson's classification is given in more detail in Table 4. Submerged coasts are divided on the basis of fluvial and glacial erosion, which determine the nature of the drowned land. One of the major drawbacks of this classification is re-

TABLE 4. Coastal Classification of Johnson (1919).

1. Submergence coasts	i Ria coast
	ii Fjord coast
2. Emergence coasts (with barriers)	
3. Neutral coasts	i Delta coasts
	ii Alluvial plain coasts
	iii Outwash plain coasts
	iv Volcano coasts
	v Coral reef coasts
	vi Fault coasts
4. Compound coasts: any combination of the first three types	

lated to Johnson's class of emergent coasts. He associated emergent coasts with barrier islands, assuming that the emerging land would be gently sloping, a characteristic that he considered necessary for the formation of barrier islands. He failed to consider the instance of emergence on a steep coast or the possibility that barrier islands are in areas that have undergone rapid submergence of an exposed continental shelf (Putnam, 1937).

The neutral category in the classification includes coasts dominated by processes other than submergence or emergence, such as volcanic action, glacial outwash, deltas, coral reefs, and faults. All coasts have been influenced by the repeated and rapid changes of sea level characteristic of the glacial period. The dominance of coasts of submergence results from the last major swing of sea level, the rapid rise in the Flandrian transgression during the period from about 15,000 to about 4000 years ago. Sea level has risen over 100 m since 20,000 years ago. It is only in the areas of rapid postglacial isostatic recovery and in some tectonically unstable areas that the land has risen relative to the sea. In practice nearly all coasts should strictly be placed in Johnson's fourth category, compound coasts, since nearly all coasts show elements of two or more of his first three categories. The classification of Johnson was an advance in that it attempted to consider the agents genetically responsible for the coastal type, thus implying an understanding of the processes involved.

Cotton (1952, 1954a) advanced a rather different dichotomy in his classification of coasts, which also falls within the genetic type. His major divisions were between coasts of stable regions and coasts of mobile regions, thus anticipating some elements of Inman and Nordstrom's classification based on the global tectonics. Within his major groups, however, Cotton returns to submergence and emergence for his secondary categories, especially in the coasts of stable regions, which he recognized as being influenced by the recent swings of sea level, particularly the Flandrian transgression. He also includes a miscellaneous group, in which fjord coasts are placed because these need not be affected by submergence, owing to glacial overdeepening. His classification, however, follows the tradition of Johnson, and is related to the cyclic concept of coastal development.

Another much used coastal classification is that of Shepard (1937, 1963), whose major dichotomy is based on the degree to which the coast has been altered by marine agencies. Thus he divides coasts into two major groups: primary coasts and secondary coasts. The first group are those shaped mainly by subaerial agencies, while the second group includes those shaped by marine agencies. Table 5 gives details of the subdivisions within the two major categories. The primary coasts, which Shepard recognizes as youthful, include four major subdivisions: those shaped by land erosion, those shaped by subaerial deposition, volcanic coasts, and those shaped by diastrophic movement. These groups are further subdivided as shown in the table, providing a comprehensive classification. The second major category is divided into three classes: coasts shaped by wave erosion, coasts resulting from marine deposition, and coasts built by organisms. Further subdivision is given in the table.

Shepard in his classification recognizes the youth of many of the world's coasts, owing to the short time that sea level has been at or near its present height. He also implicitly recognizes the fact that the tempo of coastal change varies, in that some coasts have been sufficiently modified to come under the second category, while others are still in the first. Shepard has completely eliminated the emergent category, although he does recognize drowned forms resulting from submergence. Some coasts have emerged, so that it is not desirable to eliminate this group altogether. Another problem associated with the application of this classification is to establish when the coast has been modified enough to transfer from the first to the second major category.

A classification proposed by Leontyev, Nikiforov, and Safyanov (1975) is also based on the cyclic theory. Coasts of normal development are discussed in terms of the relationship between the type of coast at different stages of development and the genetic groups, as indicated in Table 6. This classification is based essentially on the cycle of coastal development, as put forward by Davis and followed by Johnson, although it does also take into account the contrast between coasts of erosion, coasts of accretion, and those that show a combination of these states.

Dynamic Morphology

The youthful nature of many coasts and the lack of modification that many of them have undergone are indicated in Shepard's classification. This characteristic is the result of the recent rapid changes in sea level. Coastal modification is fast at first as the coast adapts to a new base level. The tempo of coastal change is still rapid in many areas, but it also varies widely. This important characteristic of coastal geomorphology is basic to the dynamic morphological classification of Valentin and Bloom, who introduced the time element and the element of change.

TABLE 5. Coastal Classification of Shepard (1963).

I PRIMARY (YOUTHFUL) COASTS
   A. *Land erosion coasts*
      1 Ria coasts (drowned river valleys)   a Dendritic type
                                                          b Trellis type
      2 Drowned glacial erosion coasts   a Fjord coasts
                                                           b Glacial troughs
      3 Drowned karst topography

   B. *Subaerial deposition coasts*
      1 River deposition coasts   a Deltaic coasts   i Digitate (birdfoot)
                                                                                     ii Lobate
                                                                                     iii Arcuate
                                                                                     iv Cuspate
                                                            b Alluvial plain coasts
      2 Glacial deposition coasts   a Partially submerged moraines
                                                           b Partially submerged drumlins
                                                           c Partially submerged drift features
      3 Wind deposition coasts   a Dunes
                                                            b Fossil dunes
                                                            c Sand flats
      4 Landslide coasts

   C. *Volcanic coasts*
      1 Lava flow coasts
      2 Tephra coasts
      3 Volcanic collapse or explosion coasts

   D. *Shaped by diastrophic movements*
      1 Fault coasts   a Fault scarp coast
                                             b Fault trough type
                                             c Overthrust type
      2 Fold coasts
      3 Sedimentary extrusions   a Salt domes
                                              b Mud lumps

II SECONDARY COASTS
   A. *Wave erosion coasts*
      1 Wave straightened coasts   a Cut in homogeneous materials
                                                          b Hogback strike coasts
                                                          c Fault line coasts
      2 Made irregular by wave erosion   a Dip coasts
                                                           b Heterogeneous formation coasts

   B. *Marine deposition coasts*
      1 Barrier coasts   a Barrier beaches
                                           b Barrier islands
                                           c Barrier spits
                                           d Bay barriers
                                           e Overwash fans
      2 Cuspate forelands
      3 Beach plains
      4 Mud flats or salt marshes

   C. *Coasts built by organisms*
      1 Coral reef coasts
      2 Serpulid reef coasts
      3 Oyster reef coasts
      4 Mangrove coasts
      5 Marsh grass coasts

Valentin (1952, 1969) recognized not only the importance of sea level changes but also the significance of accretion and erosion in his major dichotomy. He divides coasts into those that have advanced and those that have retreated, while also distinguishing the effects of emergence and submergence. Table 7 gives his full classification, while Figure 3 diagrammatically illustrates four possible situations. A line ZZ' can be drawn through the diagram to represent

TABLE 6. Coastal Classification of Leontyev, Nikiforov, and Safyanov (1975).

Genetic Group	Type of Coast			
	Initial stage	Youth	Maturity	Old age
Coasts not changed by the sea	Smooth or indented, smooth-faulted, indented-fjord, ria, tectonically indented	–	–	–
Abrasion coasts	–	Marine ingression, scoured with 2nd stage crenulation	Smoothed by abrasion	Dead cliffs protected by attached, fringing accumulation forms
Abrasion-Accumulation coasts	–	Sea ingression and scoured. Varied free accumulation forms and bays	Smoothed by abrasion and accumulation	Dead cliffs with accumulation forms washed away
Accumulation coasts	–	Ingressive accumulation	Smooth with bars or other accumulation forms	Eroded

TABLE 7. Coastal Classification of Valentin (1952).

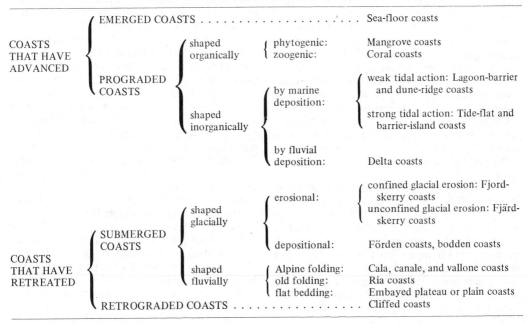

the conditions under which the coast remains stationary owing to the balance between accretion and submergence on one side, and on the other between erosion and emergence. One side of the line ZZ' represents advancing coasts, and the other retreating coasts; the farther from ZZ', the greater the change in coastal position. Any coast can be placed in the appropriate position on the diagram provided the changes of sea level and morphology are known.

Within the classification, provision is made for a wide range of coastal types. In the advanced coasts category a distinction is made between emerged sea floor, typical of areas of isostatic uplift, and prograded coasts, which can be shaped organically or nonorganically. In

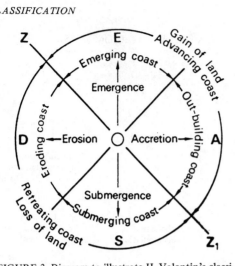

FIGURE 3. Diagram to illustrate H. Valentin's classification of coasts (after Valentin, 1952).

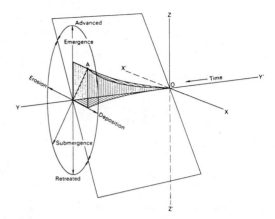

FIGURE 4. Diagram to illustrate A. L. Bloom's classification of coasts (after Bloom, 1965).

the second major category of coasts that have retreated are those showing cliffing, while submerged coasts are divided according to the nature of the land that has been submerged.

A useful attribute of this classification is that it allows for a continuous change in sea level and does not imply a cyclic development of coasts. To be applied successfully, however, it does presuppose a detailed knowledge of the coastal processes and sea level changes, which are not known everywhere as yet. Valentin has applied his classification to the coasts of the world, although on the scale used a considerable amount of generalization is necessary. This classification has a great deal to recommend it, and where it can be applied in detail and accurately it provides a valuable statement of the nature of the coast and the processes operating on it.

A development of this classification is that of Bloom (1965). He adds a third dimension to the classification, as indicated diagrammatically in Figure 4. He emphasizes the dynamic nature of the coastal processes by adding a third dimension, which represents time. The time axis is placed orthogonal to the other two axes, indicating emergence and submergence, and erosion and accretion, respectively. Provided sufficient data are available, it is possible on a diagram of this type to illustrate the history of a particular coast through time. This method of coastal classification and description avoids the problem mentioned in connection with Shepard's classification, in that the stages of change the coast has undergone can be indicated. The scheme also incorporates emergence and submergence.

Dynamic Energy

Process-based classifications provide useful information concerning currently acting forces. It is sometimes possible to relate the processes and their intensity to the morphological characteristics of the coast. Tanner and Armstrong-Price have provided useful ideas on this dynamic aspect of coastal classification in their suggestions concerning the energy criterion.

The energy classification of Armstrong-Price (1954a, b; 1955) has recently been updated and applied to the coast of the Gulf of Mexico in a rather more elaborate form than the earlier energy classification, which divided coasts into those of low energy, moderate energy, and high energy. The original classification was based on the ramp angle, the categories being equal to or less than 28.6 cm/km for low energy coasts, 28.6 cm/km to 47.6 cm/km for moderate energy coasts, and greater than 47.6 cm/km for high energy coasts. Breaker heights can be used in a similar way, with breaks at 10 and 50 cm. Coasts of zero energy have negligible breakers.

Armstrong-Price shows that a number of variables change together along the coast of the Gulf of Mexico counterclockwise from Florida. These include both static and dynamic elements. In the first category are shelf width, delta size, offshore shelf gradient, and climate. The first two decrease counterclockwise, and the third increases, and the last becomes drier. In the second category are equilibrium profile, shoreface height, ramp width, shoreline smoothness, perfection of barrier chain unit alignment, onshore wave energy, and sand transport calculated by an engineering criterion. All six variables increase in value counterclockwise. These can all

TABLE 8. Coastal Classification of Tanner (1960).

Subequilibrium	Equilibrium	Nonequilibrium	
"Zero" drift and "zero" energy	Stable equilibrium	Uplift balances wave erosion, but equilibrium form not present	Shoreline not shifting
Equilibrium still developing	Retrograded: 1. Downshore, increase in energy level 2. Loss of materials downshore 3. General imbalance	Eroded (shape depends on exposed materials): 1. Downshore increase in energy level 2. Loss of materials downshore 3. General imbalance	Landward
Water level or tectonic instability precludes equilibrium development	Prograded: 1. Deltaic 2. Glacial 3. Volcanic etc.	Constructed: 1. Deltaic 2. Glacial 3. Volcanic etc.	Nonmarine agencies
	Prograded: 1. From offshore 2. Downshore decrease in energy level 3. Biogenous a. Coral b. Marsh c. Swamp	Constructed: 1. From offshore 2. Downshore decrease in energy level 3. Biogenous a. Coral b. Marsh c. Swamp	Marine agencies

Nonmarine agencies + Marine agencies = Seaward

Landward + Seaward = Shifting Shoreline

be related to the increase of wave energy in this direction, the first through an increase in material size associated with increasing energy. The shelf steepness is a factor that affects the amount of energy reaching the shore. In this area the offshore relief is an important element in determining the amount of incoming energy that reaches the coast. On more open oceanic coasts wind waves play a greater part, and their variation determines the energy reaching the shore more directly.

Tanner (1960a, b) added the concept of equilibrium to the energy concept in his coastal classification, which depends on exposure and offshore character. To establish the equilibrium conditions, it is necessary to know both the energy level and the average rate of littoral drift past a point. Table 8 gives details of the coastal classification in terms of equilibrium. It is related to Valentin's classification in the stress laid on the movement of the shoreline, although Tanner emphasizes the progradation and erosion aspects rather than the sea level changes, which are implied by assuming sea level stability over the time span considered. The factors considered in the classification include energy level, littoral drift rate, equilibrium, downshore variability, tectonic stability, sea level stability, type of material, and nonmarine agencies, such as rivers. These variables are used in the table of classification, although the energy level and drift rates are implied rather than specified in the equilibrium concept. Subequilibrium shores are immature. The type of material is a variable that has not been previously considered in coastal classification. It is one that should certainly be taken into account, because it exerts an important effect on the character of the coast, providing such distinctive features as coral coasts, salt marshes, sand beaches and dunes, shingle ridges, and mangrove swamps. This classification, although less easy to apply than some, does indicate that a wide range of variables should be taken into account. It also stresses the dynamic character of the coast, recognizing the great importance of longshore transport of sediment in the building of coastal accretional features and accounting for zones of coastal erosion.

Dynamic Processes

Wave energy was one of the criteria used in the previous classifications of Tanner and Armstrong-Price. This aspect is further developed in the dynamic classification of Davies (1964, 1980). His classification is based on the nature of the waves reaching the coast and their relationship to the coastal forms. He recognizes four major types of wave activity: storm waves, west coast swells, east coast swells and trade wind waves, and low energy wave environments. The distribution of these major types is shown in Figure 5.

The low energy environment occurs where coasts are protected from the full force of the waves. Protection may be provided by short sea fetches, resulting from the distribution of land, offshore sea ice in polar regions, or coral reefs in tropical seas. The characteristics of the low energy shores depend on the shore waves, the only ones that can be generated in sheltered waters. These waves approach the shore from oblique angles, and result in complex shore accumulation forms.

The storm wave environment covers the high latitudes dominated by the west winds. These strong winds generate large storm waves in the North Atlantic, North Pacific, and in the belt of open water in high southern latitudes. The waves are high, relatively short, and steep, and thus destructive in their action on the coast. Coasts subject to these show evidence of erosion in the form of cliffs and abrasion platforms, although beaches can accumulate in the more sheltered bays. Shingle structures are built by these storm waves and face their direction of approach. The Northern Hemisphere storm wave belt varies more with the seasons than the southern one, migrating south in the northern winter.

The swell wave environment covers lower latitude coasts on which long, low swells are dominant. These waves have traveled far from their generating areas in the stormy west wind belt, and arrive as swells of 14 second period and 300 m length. They are higher on west coasts. East coasts are dominated by lower swell waves and waves generated by the steady trade winds. The low latitude coasts are usually predominantly sandy, and the long swells are constructive, building up berms and offshore barrier islands. The form of the beach in plan is adjusted to conform to the approach pattern of the long, refracted swells, which depends on their direction of arrival and the offshore relief.

The tide is another important process factor in coastal geomorphology. It can also be classified in terms of the range of the tide and the tidal type. The range is more significant from the geomorphological point of view, and a division of coasts into macro-, meso-, and microtidal environments is of value. The macrotidal environment has ranges over 6 m, and covers relatively small areas, including parts of the coast of Britain, northeast Canada, southeast South America, northwest Australia, and other small areas. The microtidal environment is more extensive, covering, for example, south Australia, the Gulf of Mexico, the Mediterranean, the Baltic, southwest Africa, and parts of western South America.

Another type of dynamic process classification

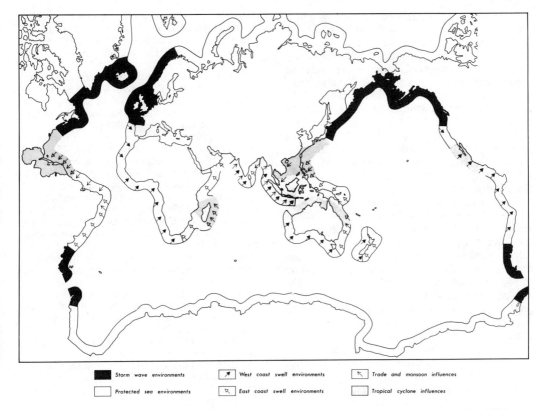

FIGURE 5. Map of dynamic classification of coasts in relation to wave environments (from Davies, 1980).

is based on the influence of the subaerial climate on coastal processes. An early attempt by Aufrère (1936) recognized six classes: the zone of permanent ice cover, in which development was virtually stopped; seasonal ice cover, in which seasonal activity acts on material supplied largely by glacial action; temperate humid zone, in which *normal* processes are dominant; hot-, wet areas, in which coral can thrive and constructional forms tend to develop in the wet season; deserts, in which estuaries and deltas are absent and littoral sediment is of marine origin; semiarid areas, in which seasonal continental influence is strong and lagoons become *sebkhas*. A classification of this type stresses processes depending on climate, including features related to ice on the shore in polar latitudes, coral and its associated features in tropical latitudes, beach rock in suitable climates, and hurricane activity in low latitude areas.

More recent studies of coastal classification in terms of climate, oceanography, biology, and materials include the work of Putnam et al. (1960) and Dolan et al. (1972). The latter work covers North and South America. Data are presented on maps, and are classified first on atmospheric regimes and second on oceanic subregimes. The latter are related to interface units, which are composed in turn of elemental units. The climatic regimes of the Americas are divided into 15 types, while five wave classes are recognized in the oceanic subregimes. The dominant shoreline processes are divided into four groups: marine, fluvial, mass wasting, and biological. Maps are provided showing landform types for the different regimes, on which five categories of geology are depicted with six types of relief; the shoreline character is indicated by seven symbols, which cover: sand beach or barrier island, sand beach and rocky headland, pocket beach, mud flat or marsh, swamp or mangrove, coral, and rock. The elemental units are also shown symbolically on maps. Six marine processes are distinguished, three fluvial ones, two biological types and mass wasting. Of a possible 1800 different combinations of types, taking into account the 15 different regimes, 10 different subregimes, and 12 interface types, only 116 were actually found, indicating a close coupling of the different aspects; thus 94% of the possible combinations did not occur.

Conclusion

A number of different coastal classifications have been briefly introduced. They cover a considerable range of approach, some of which are based on the nature of the land, others on the nature of the processes that operate to modify the land/sea interface. Some are purely descriptive or empirical, while others refer to the current state of the coast. Scale varies from worldwide classifications to those that can be applied to a short stretch of coast, giving in considerable detail the processes and sediments as well as the morphology of the area. They all provide useful information that enables the wide range of coastal features to be ordered more meaningfully into related groups according to the criteria used.

As more becomes known about coasts, it may be possible to improve some of the classifications already in use, while others may be devised based on other dichotomies. Criteria that might be of value, particularly in terms of coastal engineering, include the amount of longshore transport of beach sediment, the nature of the coastal material, and the current changes in sea level. All these variables have an important bearing on the stability of the coast, while the present practice of beach replenishment, which is probably the best means of coastal defense and conservation, requires a detailed knowledge of the natural beach material. Process, material, and morphology should ideally all be included in a successful coastal classification.

CUCHLAINE A. M. KING

References

Alexander, C. S., 1966. A method of descriptive shore classification and mapping as applied to the northeast coast of Tanganyika, *Assoc. Am. Geographers Annals* 56, 128-140.

Aufrère, L., 1936. Le rôle du climat dans l'activité morphologique littorale, *Congr. Internat. Geogr., 14th, 1934, Warsaw, Comptes Rendus* 2.

Bloom, A. L., 1965. The explanatory description of coasts, *Zeitschr. Geomorphologie*, N. F., 9, 422-436.

Cotton, C. A., 1952. Criteria for the classification of coasts, *Congr. Internat. Geogr., 17th, 1952, Washington, Abstract of Papers* 15.

Cotton, C. A., 1954a. Deductive morphology and the genetic classification of coasts, *Sci. Monthly* 78, 163-181.

Cotton, C. A., 1954b. Tests of a German non-cyclic theory and classification of coasts, *Geog. Jour.* 120, 353-361.

Dana, J. D., 1849. Geology, in *U.S. Exploring Expedition*. Philadelphia, Pa.: C. Sherman.

Davies, J. L., 1964. A morphogenetic approach to world shorelines, *Zeitschr. Geomorphologie*, Sp. No., 8, 127*-143*.

Davies, J. L., 1980. *Geographical Variation in Coastal Development*. Edinburgh: Oliver and Boyd, 212p.

Dolan, R.; Hayden, B. P.; Hornberger, G.; Zieman, J.; and Vincent, M. K. 1972. *Classification of the Coastal Environments of the World. Part I. The Americas.* Office of Naval Research, O.N.R. N00014-69-A00060-0006, NR 389-158 Tech Rept., 161p.

Inman, D. L., and Nordstrom, C. E., 1971. On the tectonic and morphological classification of coasts, *Jour. Geology* 79, 1-21.

Johnson, D. W., 1919. *Shore Processes and Shoreline Development*. New York: John Wiley & Sons, 584p.

Leontyev, O. K.; Nikiforov, L. G.; and Safyanov, G. A., 1975. *The Geomorphology of Sea Coasts* (in Russian). Moscow: Moscow University, 336p.

Price, W., 1954a. Shorelines and coasts of the Gulf of Mexico, in *Gulf of Mexico Fishery Bull.* 55, 39-65.

Price, W., 1954b. Correlation of shoreline type with offshore conditions in the Gulf of Mexico, in *Proceedings of the Coastal Geography Conference, 1954*. Washington, D.C.: Office of Naval Research, 11-30.

Price, W., 1955. *Correlation of Shoreline Types with Offshore Bottom Conditions*. Austin, Texas: A & M College of Texas, Department of Oceanography, Project 63, 2p.

Putnam, W. C., 1937. The marine cycle of erosion for a steeply sloping shoreline of emergence, *Jour. Geology* 45, 844-850.

Putnam, W. C.; Axelrod, D. I.; Bailey, H. P.; and McGill, J. T., 1960. *Natural Coastal Environments of the World*. Berkeley and Los Angeles: University of California Press, 140p.

Richthofen, F. von, 1886. *Führer für Forschungsreisende*. Hanover: Janecke, 734p.

Shepard, F. P., 1937. Revised classification of marine shorelines, *Jour. Geology* 45, 602-624.

Shepard, F. P., 1963. *Submarine Geology*, 2d ed. New York: Harper and Row, 557p.

Suess, E., 1888. *The Faces of the Earth*, vol. II (English translation in 1906 by H. B. Sollas). London: Oxford University Press, 5 vols.

Tanner, W. F., 1960a. Florida coastal classification, *Gulf Coast Assoc. Geol. Socs. Trans.* 10, 259-266.

Tanner, W. F., 1960b. Bases of coastal classification, *Southeastern Geology* 2, 13-22.

Valentin, H., 1952. *Die Küsten der Erde*, Petermanns Geog. Mitt. Erg. 246. Gotha: Justus Perthes, 118p.

Valentin, H., 1969. Principles of a handbook on regional coastal geomorphology of the world, *Zeitschr. Geomorphologie*, N. F., 13, 124-129.

Zenkovich, V. P., 1967. *Processes of Coastal Development*. Edinburgh: Oliver and Boyd, 738p.

Cross-references: *Coastal Characteristics, Mapping of; Emergence and Emerged Shoreline; Flandrian Transgression; Global Tectonics; Isostatic Adjustments; Isostatic Uplift Coasts; Postglacial Rebound; Sea Level Changes; Sea Level Curves; Submergence and Submerged Coasts; Tectonic Movements.* Vol. III: *Coastal Geomorphology.*

CLAY

Clay is used as a rock term and also as a particle-size term. It implies a natural, earthy, fine-grained material composed essentially of silicates (Grim, 1968). The upper limit of particle size of clay is generally regarded as 2μ.

Clay material is defined by Grim (1968) as any fine-grained, natural, earthy, argillaceous material.

Clay minerals are generally finely crystalline hydrous silicates essentially of aluminum, magnesium, and iron. Most are phyllosilicates, but some are chain structures or inosilicates. The major clay mineral groups include kaolin, serpentine, and kindred minerals; smectite or montmorillonite minerals; the mica clay minerals (chiefly illite); chlorites; and the inosilicates sepiolite and palygorskite. Another major group consists of various clay minerals interstratified with each other, such as illite-montmorillonite.

Nearshore argillaceous deposits are common in tidal flats, lagoons, and estuaries. The clay mineralogy seems to be a function of the clay mineralogy of the source area (Weaver, 1958; Griffin, 1962; Griffin and Ingram, 1955; Pevear, 1972). Generally there is no clear evidence of alteration in the shallow marine environment other than absorption or exchange of cations from seawater. There is some evidence to suggest, however, that sepiolote and palygorskite may form in magnesium-rich hypersaline lagoons (Demangeon and Salvayre, 1961; Wollast, Mackenzie, and Bricker, 1968) and that iron-rich chlorite may form in the reducing environment of normal lagoons.

Quick clays are generally marine-marginal glacial or postglacial silty clays that represent fine-grained material produced by abrasion of bedrock by an ice sheet. Most are composed of illite, chlorite, and silt-size quartz grains (Rosenqvist, 1958). The term "quick" refers to their tendency to change to highly fluid mud when disturbed.

ADA SWINEFORD

References

Demangeon, P., and Salvayre, H., 1961. Sur la genese de palygorskite dans un calcaire dolomitique, *Soc. Française Minéralogie et Cristallographie Bull.* **84**, 201-202.

Griffin, G. M., 1962. Regional clay-mineral facies; Products of weathering intensity and current distribution in the northeastern Gulf of Mexico, *Geol. Soc. America Bull.* **73**, 737-768.

Griffin, G. M., and Ingram, R. L., 1955. Clay minerals of the Neuse River estuary, *Jour. Sed. Petrology* **25**, 194-200.

Grim, R. E., 1968. *Clay Mineralogy.* New York: McGraw-Hill, 596p.

Pevear, D. R., 1972. Source of recent nearshore marine clays, southeastern United States, *Geol. Soc. America Mem.* **133**, 317-335.

Rosenqvist, I. Th., 1960. Marine clays and quick clay slides, *Norges Geol. Undersökelse* **208**, 463-471.

Weaver, C. E., 1958. Geologic interpretation of argillaceous sediments. Part I. Origin and significance of clay minerals in sedimentary rocks, *Am. Assoc. Petroleum Geologists Bull.* **42**, 254-271.

Wollast, R.; Mackenzie, F. T.; and Bricker, O. P., 1968. Experimental precipitation and genesis of sepiolite at earth-surface conditions, *Am. Mineralogist* **53**, 1645-1662.

Cross-references: *Beach Material; Classification; Cobbles; Gravel; Pebbles; Sand; Silt.* Vol. VI: *Clay as a Sediment.*

CLIFF—See SEA CLIFFS

CLIFFED COAST

A *cliffed coast* is a section of coast that has been cliffed by marine erosion or drowning. Cliffed coasts are prevalant along mountainous and hilly coasts where the offshore slope is steep and little sediment is available for coastal progradation. In particular, they occur along collision, neo- and Afro-trailing, and marginal seacoasts that occupy approximately 50% of the worlds coastline (Inman and Nordstrom, 1971). Drowning of vertical coastal slopes produces

FIGURE 1. Vertical cliff 70 m high in sandstone-shale complex at Warriwood, New South Wales, Australia. Note the abundant debris at the base generated by rock falls. To the left enough debris has accumulated to permit soil and vegetation to develop and protect the backing cliff until the debris is removed (Photo: A. D. Short).

FIGURE 2. Steep, tundra-covered cliffs fronted by narrow beaches around Corwin Bluffs (center), north Alaska. These cliffs, generated by drowning, have undergone little recent erosion, as evidenced by the vegetation cover. Despite their steepness and height (170 m in foreground), slumping is the principal means of erosion on this low energy, generally ice-bordered coast (Photo: A. D. Short).

plunging cliffs. On gentler slopes, wave erosion at or just above sea level notches the slope, producing such features as *wave-cut platforms*, benches, boulder debris, and cliffs. The cliff itself is also denuded by the processes of mass movement, in particular rock falls, landslides, and slumps. Hills (1971) found that the base level of cliff planation caused by wave action is affected by lithology, rock weathering, degree of exposure, and duration of platform development. The height of a cliff depends upon the elevation of the backing coast, the slope on the nature and strength of the material forming the cliff, and the rate of retreat on atmospheric and marine erosion. Cliffs tend to be more vertical and clear of debris in more resistant rocks, such as sandstone and granite and where wave attack is intense (Fig. 1). They tend to deteriorate and be covered by debris, soil, and vegetation in less resistant rock and where marine attack diminishes because protection afforded by beaches, boulder debris, and wide platforms (Fig. 2).

ANDREW D. SHORT

References

Hills, E. S., 1971. A study of cliffy coastal profiles based on examples in Victoria, Australia, *Zeitschr. Geomorphologie* 15, 137-180.

Inman, D. L., and Nordstrom, C. E., 1971. On the tectonic and morphologic classification of coasts, *Jour. Geology* 79, 1-21.

Cross-references: *Cliff Erosion; Coastal Bevel; Fossil Cliffs; Sea Cliffs; Slope-over-Wall Cliffs; Wave-Cut Bench; Wave-Cut Platform.* Vol. III: *Platform, Wave-Cut.*

CLIFF EROSION

Cliff erosion is generated by two processes: *notching* at the base of the cliff by marine processes, and collapse and denudation of the entire cliff face by a combination of atmospheric and marine processes. Concentration of wave attack at the cliff base accelerates erosion and notches the base, permitting collapse or movement of the overlying material. This debris is subsequently removed by wave action, and the marine attack continues at the base.

The level of planation of the cliff base lies immediately above the level of permanent rock saturation, where continued wetting and drying weakens the rock, allowing accelerated wave erosion. This level will, however, vary depending on wave energy, rock lithology and structure, platform width, and accompanying atmospheric processes.

The cliff face itself is denuded by processes related to marine attack, including *water-layer weathering*, honeycombing, spray and abrasion, and the atmospheric processes of physical and chemical weathering plus the effect of gravity. These processes combine to produce all forms of mass movement in delivering material to the cliff base for its subsequent removal. The rate of cliff retreat, combined with rock type and structure, climate (especially rainfall), and vegetation will determine the nature of mass movement down the cliff face. Along the Oregon coast landsliding peaks during December and January, when precipitation is highest (Byrne, 1963). On mudstone cliffs in New Zealand, slumping is dominant and greatest in areas of wave convergence behind actively extending *wave-cut platforms* (McLean and Davidson, 1968). Sunamura (1973), in an extensive examination of Japanese cliff erosion, found that long-term cliff retreat is logarithmically related to rock strength and unrelated to cliff height.

ANDREW D. SHORT

References

Byrne, J. V., 1963. Coastal erosion, northern Oregon, in T. Clements, ed., *Essays in Marine Geology in Honor of K. O. Emery.* Los Angeles, Calif.: University of Southern California Press, 11-33.

McLean, R. F., and Davidson, C. F., 1968. The role of mass-movement in shore platform development along the Gosborne coastline, New Zealand, *Earth Sci. Jour.* 2, 15-25.

Sunamura, T., 1973. Coastal cliff erosion due to waves—Field investigations and laboratory experiments, *Tokyo Univ. Fac. Eng. Jour.* **32**, 1–86.

Cross-references: *Cliffed Coasts; Coastal Bevel; Niche, Nick, Nip, Notch; Sea Cliffs; Wave-Cut Bench; Wave-Cut Platform; Wave Erosion; Weathering and Erosion, Biologic; Weathering and Erosion, Chemical; Weathering and Erosion, Mechanical.* Vol. III: *Platform, Wave-Cut.*

CLIMATE, COASTAL

For some distance on both sides of the 450,000-km-long world shoreline, meteorological processes and resulting climates differ from those farther seaward and landward. The coast is in essence a transitional zone between marine on one side and terrestrial on the other, a transitional zone or band that varies in total width as well as in the relative width of its two parts, land and water. The distinctiveness of this band varies with the amount and rate of meteorologic and climatic change across it. Although most of its characteristics are a blend of marine and terrestrial phenomena, there are a few characteristics, such as land and sea breezes and coastal fogs, that can be truly labeled *coastal*.

The climatic characteristics of the coastal zone have most often been considered a landward invasion of adjacent sea air and have variously been referred to as oceanic, marine, maritime, littoral, and coastal. The first three designations have been loosely used, and the last two appear only rarely in the literature. *Maritime* (see Fairbridge, 1967) is the term that comes closest to our usage here of *coastal*. We believe that with the rapid evolution of man's concept of the coastal zone, and especially because of the recent tendency to include the seaward segment of the coastal zone as one of its integral parts, the terms *coastal meteorology* and *coastal climatology* are now more appropriate than any of the others.

Climatic Controls

Three of the most important climatic controls on earth are variation of insolation with latitude, distribution of land and water, and surface configuration. Coastlines cross virtually every parallel on earth except those immediately surrounding the two poles. Although coasts do not occur south of 78°S latitude because of the presence of Antarctica and its ice shelves, and they are not present north of 82°N latitude because of the presence of the Arctic Ocean, they nonetheless have a wide range of latitudinally influenced characteristics.

Whereas the juxtaposition of any two environments (e.g., forest/grassland, mountain/plain, and city/country) give rise to climatic modification, the land/water combination that is the most distinctive on a worldwide basis. There are marked variations in the heat and moisture budgets of these two surfaces because of contrasts in albedo, evaporation rate, transparency, surface mobility, heat capacity, and sensible heat flux. Moreover, there are important modifications in circulation patterns because of variations in surface irregularities across the interface, including variations in the actual relief as well as in the mobility and spatial characteristics of roughness forms. Possibly the two most important differences are in wetness and roughness—the sea is almost always wetter and smoother than the land.

These controls are especially important in affecting atmospheric motion in the coastal zone. Pertinent motions have a great range in scale both spatially and temporally (Fig. 1), and although all of these phenomena have some coastal expression, only a few have actually been considered from that standpoint. Studies to date have most often dealt with the impact of these levels of atmospheric motion (especially turbulence, sea breezes, and hurricanes) on the coastal environment rather than with the motion itself as a coastal phenomenon. Such an emphasis reflects the fact that coastal meteorology and climatology are integral parts of the total systems approach to the study of coastal environments.

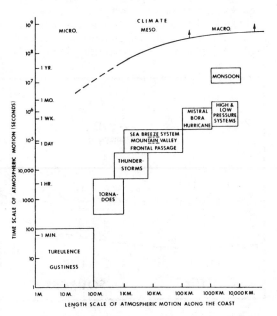

FIGURE 1. Time-length scales of atmospheric motions.

Boundary-layer Meteorology
(Micro- and Meso-Scale Meteorology)

The atmospheric boundary layer is the region of the atmosphere that is directly affected by friction caused by interaction with the earth's surface. Transport in this layer in nearshore and estuarine environments is important from several points of view. For example, the wind stress or momentum flux is one of the most essential driving forces in shallow-water circulation. Heat and convection are the origins of some localized coastal weather systems. Sensible heat and water vapor fluxes are necessary elements in radiation and heat budget considerations, including the computation of salt flux for a given estuarine system. Experiments in these environments ranging from the tropics to the Arctic have produced a large number of conflicting drag and bulk transfer coefficients for both deep and shallow waters. There are several reasons for the discrepancies, but the most important is that early methods did not take into account the simultaneous contribution of wind, dominant waves, and atmospheric stability. This difficulty was recently removed by the development of a wind/wave interaction method of determining wind stress from commonly available wind and wave parameters (Hsu, 1976).

Aerosol Transport. Atmospheric particles, particularly sea salts, have become an increasingly important subject for investigation in recent years. Sea-salt aerosols are a significant source of condensation nuclei. The quantity and quality of sea-salt particles deposited on land may be important in determining the physical and chemical characteristics of coastal soils and plants. The generation of aerosols depends upon many meteorological and oceanographic factors. Among those in the coastal region are wind speed, direction, duration, fetch (which govern *sea state* and whitecap distribution), and subaqueous bathymetry (which controls the breaking wave condition in the surf zone). Aerosol contribution from these point sources (*whitecaps*) and line sources (*surf zone*) can be considered as originating from an area source (Fig. 2). Experiments measuring the vertical distribution of sea salt in the atmospheric boundary layer over beaches have shown that on a coast influenced by synoptic onshore winds, such as a beach-dune complex, mixing depths are less thick than they are under sea breeze conditions (Hsu, 1977).

Sand Dune and Ice Ridge Air Flow. When air passes over an obstacle, it must adjust to a new set of boundary conditions. Major characteristics associated with this adjustment are the displacement zone, the wake, and the cavity. Measurements of the wind field in the region of coastal sand dunes in the tropics and ice ridges in the Arctic were combined with data from laboratory experiments in the preparation of a basic flow model. This model (Fig. 3) is aiding the study of the modification of the structure of airflow over selected coastal topographies, and is furthering the understanding of low-altitude atmospheric dispersion characteristics in the coastal zone.

Sea Breeze. The sea breeze, a coastal local wind that blows from sea to land, occurs when the temperature of the sea surface is lower than that of the adjacent land (Defant, 1951; Hsu, 1970). It usually blows on relatively calm, sunny, summer days, and alternates with an oppositely directed, usually weaker, nighttime land breeze. As a sea breeze regime progresses,

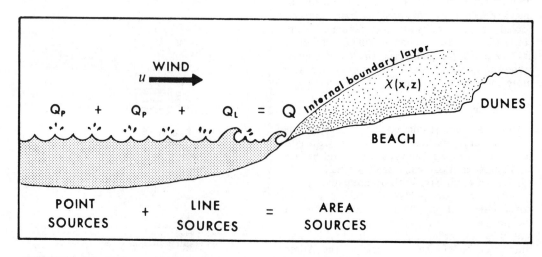

FIGURE 2. Schematic representation of point and line sources for sea salt and the effect of the internal boundary layer (from Hsu, 1977).

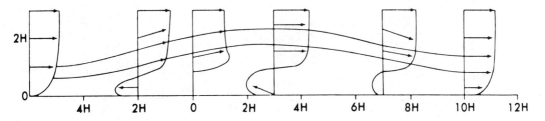

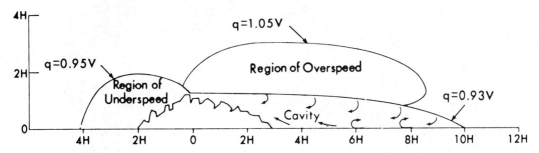

FIGURE 3. A basic flow model and a similarly shaped ice ridge. Q is the local resultant mean velocity, and V is the reference velocity in the uniform stream (from Hsu, 1977).

coriolis deflection causes the development of a wind component parallel to the coast. Coastal air circulations of this kind have been known and recorded since the time of Aristotle, about 350 B.C., and have been studied more or less scientifically since the seventeenth century. Of all coastal meteorological phenomena, these have received by far the greatest amount of attention and are continuing to do so. For example, during the First Conference on Coastal Meteorology (see *Bulletin of the American Meteorological Society* 57, 747-759, June 1976), about 40 percent of the papers dealt in some way with the *sea breeze*.

Synoptic Meteorology and Its Coastal Impact (Macrometeorology)

Associated with the readjustment of airflow at the sea/land interface are other parameters such as temperature and humidity. As considered above at the micro- and meso-scale, these are quite distinctive. Although their effect is less obvious at the larger scale, they nonetheless have a significant impact on synoptic and planetary circulation systems.

Data support for the conduct of studies at this scale is scarce. Few meteorological data are routinely obtained at coastline stations; the major exception is at the all too few United States Coast Guard stations where observations of temperature, precipitation, and wind are made. The most useful data, however, come from coastal airports, even though these are usually located several kilometers inland.

Such data have been used in the analysis of the weather input and environmental response in the coastal zone in Louisiana (Muller, 1977). This approach classified Louisiana coastal weather into eight inclusive synoptic types: Pacific High (PH), Continental High (CH), Frontal Overrunning (FOR), Coastal Return (CR), Gulf Return (GR), Frontal Gulf Return (FGR), Gulf Tropical Disturbance (GTD), and Gulf High (GH). Because each type can be identified on weather maps, climatic calendars were devised that are useful in the analysis of mean properties by months and seasons, climatic variation from year to year, and geographical comparisons of one coastal region with another (Muller and Wax, 1977).

Synoptic weather type changes have been found to be related to water level and salinity changes in the coastal marshes and estuaries of Louisiana (Borengasser, Muller, and Wax, 1978). When the effects of astronomical tides were "filtered" from water level data, it was found that water levels tended to drop during the 24-hour periods associated with a progression from the Frontal Gulf Return type to the Pacific High, Continental High, or Frontal Overrunning types. In contrast, water levels were shown to rise during other types of progression. Such analyses are only beginning but appear to hold promise for other nontropical coastal areas as well.

Coastal Climates

Whereas the atmospheric processes thus far considered are measured in intervals of

TABLE 1. Climatic Characteristics Related to Coastal Landscapes.

Climatic Type and Percent Of World Coastline	Typical Locality	Climatic Conditions*					
		Temperature (°C.)			Precipitation		
		Mean Maximum Warmest Month (tm)	Mean Minimum Coldest Month (tm)	Mean Annual Depth, cm (P)	Mean Annual No. Days ≥0.004 cm (FP)	Winter Concentration of Precipitation** (R)	
Rainy tropical (20%)	Inner tropics	30–32	19–23	198–310	134–185	14–49	
Subhumid tropical (10%)	Border tropics	30–33	15–21	104–140	61–114	17–38	
Warm semiarid (2%)	Tamaulipas, Venezuela	32–34	13–19	53–71	42–60	9–40	
Warm arid (5%)	Horn of Africa, Sonora	33–37	11–20	13–25	10–32	36–94	
Hyperarid (4%)	Cool-water coasts of subtropics	24–34	9–14	<5	1–4	42–100	
Rainy subtropical (6%)	East coasts, lat. 20–35°	29–32	6–9	114–147	93–142	29–49	
Summer-dry subtropical (7%)	Mediterranean	27–31	6–9	43–69	54–103	74–87	
Rainy marine (1%)	W. coasts, Tasmania, New Zealand	17–20	4–7	109–206	166–187	51–69	
Wet-winter temperate (2%)	Oregon, Washington	17–22	0–6	99–170	120–198	67–78	
Rainy temperate (9%)	NE United States, W. Europe	20–27	−7–2	66–112	127–188	41–54	
Cool semiarid (1%)	Bahia Blanca	19–31	−3–9	30–53	45–87	37–52	
Cool arid (2%)	Patagonia	21–26	1–7	10–15	24–41	54–88	
Subpolar (6%)	Gulfs of Alaska, Bothnia	15–23	−23–−13	46–104	106–184	32–50	
Polar (25%)	Arctic Sea border	9–13	−34–−13	18–66	91–131	30–49	

Source: Bailey, 1976
*Each pair of numerals in the body of the table refers to data from climatic stations at the 25th and 75th percentiles of the frequency distribution appropriate to the climatic type and element. As only long-period means have been entered into the frequency distributions, the data in the table above show the spread in average conditions of climate in the most representative parts of the several climatic regions. Because approximately equal spacing was employed in the station network, it is also true that the data illustrate, for a given climatic type, conditions in about 50% of the aggregate length of coastline affected by that climatic type.
**The winter concentration of precipitation is defined as the percentage of the mean annual total that falls in the winter half-year. October through March in the Northern Hemisphere, April through September in the Southern Hemisphere. The computation was not carried out for those places where the difference between the mean monthly temperatures of the warmest and coldest month was less than 3°C.

seconds to weeks, there remain longer time periods to be considered—periods representative of climate (Fig. 1). Climate, as a synthesis of atmospheric events, is unique for each location on earth. Yet vast areas have climates that are sufficiently similar to justify consideration together. In most classifications (Wilson, 1967) continental types are extended directly to the shore, so that their coastal ramifications are indistinguishable.

The major exception to date is the work of Bailey (1960, 1976), who developed a classification of coastal regions through the utilization of data from coastal stations. Defining his climatic types so that they closely approximated the distribution of coastal vegetation Bailey (1976) arrived at 14 types (Table 1). The relative length of coastline represented by each type varies from 1 and 2% for 5 types to 20 and 25% for rainy tropical and polar climates respectively. Only about one-fourth of the coastline of the world has a temperate climate.

The longitudinal gradients of the 14 coastal climates in Bailey's classification are generally weak. In contrast, the climatic gradients across the coastal zone are frequently steep. That there is a distinction between oceanic, coastal, and continental conditions has been demonstrated by Bailey (1976) through comparison of the data in Table 1 with that from non-coastal areas. When the mean annual range of temperature is related to mean annual temperature, the intermediate nature of the coastal zone is evident (Fig. 4). There are decided interhemispheric contrasts in the curves, contrasts that are especially evident in the seasonal swings (A of Fig. 4) of temperature.

Bailey's classification of coastal climates, like most climatic classifications, is based on temperature and precipitation. Dolan et al., (1972), however, utilized air mass climatology as their basis for separating the coastal climates of the Americas. The air mass characteristics they considered significant are: air mass seasonality, air mass source region, nature of the surface over which the air mass moves, and the confluence of airstreams at the coast. Further conditions strengthening this approach are: each air mass possesses a set of secondary characteristics, including the traditionally utilized meteorological elements; the air mass is modified upon crossing the coast; and the boundary between air masses (i.e., a front) is distinct, and separates natural climatic complexes. Although the two approaches are different, the mapped distributions of coastal climates are nonetheless similar in both.

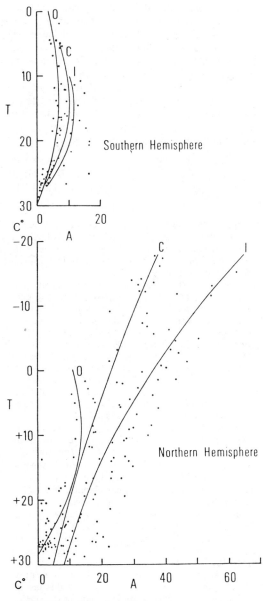

FIGURE 4. The relation between mean annual range of temperature (A) and mean annual temperature (T) in each hemisphere for oceanic (O), coastal (C), and interior (I) places (from Bailey, 1976).

The Future

Coastal meteorology and *coastal climatology* are in their infancy as research disciplines, but the future for both looks bright. Many research organizations are beginning to place greater emphasis on the meteorology and climatology of coasts, albeit an emphasis that is still related mostly to the demand for data by other coastal sciences. In addition to the topics briefly discussed in this paper, research is underway on storm surges, upwelling, tidal stirring, coastal fog, coastal dunes, and coastal frontogenesis, among others. The organization in 1976 of the first Conference on Coastal Meteorology by the American Meteorological Society is perhaps the most encouraging sign to date of what the future holds for the developing sciences of coastal meteorology and climatology.

H. J. WALKER
S. A. HSU
R. A. MULLER

References

Bailey, H. P., 1960. Climates of coastal regions, in W. C. Putnam, D. I. Axelrod, H. P. Bailey, and J. T. McGill, eds., *Natural Coastal Environments of the World*. Berkeley and Los Angeles: University of California Press, 59–77.
Bailey, H. P., 1976. Coastal climates in global perspective, in H. J. Walker, ed., *Geoscience and Man*, vol. 15, *Coastal Research*. Baton Rouge: Louisiana State University School of Geoscience, 65–72.
Borengasser, M.; Muller, R. A.; and Wax, C. L., 1978. *Barataria Basin: Synoptic Weather Types and Environmental Responses*. Baton Rouge: Louisiana State University Center for Wetland Resources, Sea Grant Pub. No. LSU-7-78-001, 60p.
Defant, F., 1951. Local winds, in T. F. Malone, ed., *Compendium of Meteorology*. Boston: American Meteorological Society, 655–672.
Dolan R.; Hayden, B. P.; Hornberger, G.; Zieman, J.; and Vincent, M. K., 1972. *Classification of the Coastal Environments of the World. I. The Americas*. Charlottesville, Va.: Department of Environmental Sciences, University of Virginia, Tech. Rept. 1, 163p.
Fairbridge, R. W., 1967. Maritime climate: Oceanicity, in R. W. Fairbridge, ed., *Encyclopedia of Atmospheric Sciences and Astrogeology*. New York: Reinhold Publishing Corp., 546–549.
Hsu, S. A., 1970. Coastal air circulation system: Observations and empirical model, *Monthly Weather Rev.* 10, 187–194.
Hsu, S.-A., 1976. Determination of the momentum flux at the air-sea interface under variable meteorological and oceanographic conditions: Further applications of the wind-wave interaction method, *Boundary-Layer Meteorology* 10, 221–226.
Hsu, S.-A., 1977. Boundary-layer meteorological research in the coastal zone, in H. J. Walker, ed., *Geoscience and Man, Research Techniques in the Coastal Environment*, vol. 18, Baton Rouge: Louisiana State University School of Geoscience, 99–111.
Muller, R. A., 1977. A synoptic climatology for environmental baseline analysis, New Orleans, *Jour. Appl. Meteorology* 16, 20–33.
Muller, R. A., and Wax, C. L., 1977. A comparative synoptic climatic baseline for coastal Louisiana, in H. J. Walker, ed., *Geoscience and Man*, vol. 18, *Research Techniques in the Coastal Environment*. Baton Rouge: Louisiana State University School of Geoscience, 121–129.
Wilson, L., 1967. Climatic classification, in R. W. Fairbridge, ed., *Encyclopedia of Atmospheric Sciences and Astrogeology*. New York: Reinhold Publishing Corp., 171–193.

Cross-references: *Classification; Coastal Morphology, Oceanographic Factors; Coastal Morphology, Research Methods; Storm Wave Environments; Wind; Wind Incidence on Coasts*. Vol. II: *Maritime Climate*.

CLIMATIC OPTIMUM— See HYPSITHERMAL

CLUB MOSSES—See LYCOPODIOPHYTA

COASTAL BEVEL

A *coastal bevel* is a seaward slope, commonly covered by rough grass or brushwood, generally abruptly terminated by a cliff. This is a common feature along coastlines cut in resistant rock, particularly in once glaciated or periglacial regions (s.w. England, Brittany, Nova Scotia). The appearance may be identical with *slope over wall cliffs* (q. v.), but the origin is quite different.

This feature is so frequently seen that it has often been referred to as *the coastal slope* and regarded as a natural consequence of subaerial weathering above a cliffed coastline. Indeed, most diagrams of the sequence of cliff forms begin with a coast sloping seaward, into which the first cliff is cut. The origin of the coastal bevel is nevertheless complex. In its simplest form, it represents an older cliff degraded by subaerial weathering and solifluction. The two-cycle coastline of Cotton (1951) shows a coastal bevel overlying a steep, recently cut cliff. Wood (1959, 1974) and Savigear (1962) have independently suggested that the coastal bevel is polygenetic, formed during a long period during which sea level rose and fell repeatedly, but was unable at any level to erode deeply into the resistant rocks of the coastline. It has recently been suggested that the original cliffed coastline that was repeatedly freshened during the Pleistocene was cut by an early Pliocene (post Messinian) transgression.

Often the coastal bevel, prolonged seaward, meets the outer edge of the abrasion platform bordering the modern cliff, this being interpreted as proving that the outer edge of the abrasion platform is a *fossil cliff* cut during a period of lowered sea level.

A *hogs-back cliff* is a steep, uniform seaward slope, not formed by present-day marine erosion, commonly truncated by the modern cliff. This is a special case of the *coastal bevel* where landward-dip of the rocks encourages the formation of a uniform slope (Arber, 1911).

ALAN WOOD

References

Arber, E. A. N., 1911. *The Coast Scenery of North Devon.* London: Dent and Sons, 269p.
Cotton, C. A., 1951. Atlantic gulfs, estuaries and cliffs, *Geol. Mag.* **88**.
Savigear, R. A. G., 1962. Some observations on slope development in North Devon and North Cornwall, *Inst. British Geographers Trans.* **18**, 23-42.
Wood, A., 1959. The erosional history of the cliffs around Aberystwyth, *Liverpool and Manchester Geol. Jour.* **2**, 271-287.
Wood, A., 1974. Submerged platform of marine abrasion around the coasts of south-western Britain, *Nature* **252**, 563.

Cross references: *Fossil Cliffs; Sea Cliffs; Slope-Over-Wall Cliffs; Wave-Cut Platform.*

COASTAL CHARACTERISTICS, MAPPING OF

The results of a systematic survey of coastal environments are frequently presented in map form to illustrate the variability of the processes or morphology. The method of data collection and the mapping techniques depend primarily on the scale of the area involved and on the objectives of the survey. The *spatial variability* of coastal characteristics results from interactions between the dynamic processes (geophysical factors: winds, waves, tides, currents) and the resistance factors (geological factors: structure, lithology, topography). These characteristics may be recorded in terms of a generalized classification of shoreline types or a symbolic representation of the actual processes and morphologic features. The degree of *generalization*, or scale, is controlled by the dimensions of the area that is studied and by the level of accuracy that is required. At global scale, the degree of complexity is low, and only a few topics can be shown (e.g., McGill, 1958). At the local level, detailed morphologic maps can be used as a base for illustrating variations in sediment size, beach slope, nearshore bar types, and so on. As the scale increases, so the level of generalization decreases, less information is filtered out, and the final map more closely represents the real world situation.

The initial selection of data or parameters to be recorded varies with the objectives of the map maker. Maps of individual environmental parameters or characteristics, such as tidal range or storm wave environments (Davis, 1964), require the collection of only a small volume of data. More complex projects that involve, for example, plotting the location and size of barrier islands and inlets or the distribution of vegetation types require the selection and processing of field or remote-sensing data. One of the critical elements during the planning of the data compilation or collection is to determine the scale transformations of the real world distributions. This involves selection of those parameters to be recorded and choice of the cartographic symbolization to display the data.

There are a great variety of approaches to the mapping of coastal characteristics; examples of studies at different scales with different objectives will be used to indicate some commonly used methods.

Small-Scale Mapping

At small scales (1:1,000,000 or less) the level of generalization is necessarily high, and only major elements can be shown. In order to present information in a meaningful way, most approaches use a classification or categorization that may be genetic, descriptive, or quantitative to systematize coastal characteristics or features. The information given at these scales is usually based on secondary data sources, such as topographic maps, hydrographic charts, or published literature. As the level of generalization is high, no field mapping or checking of the information is undertaken.

A map of coastal landforms of the world prepared by McGill (1958) at a scale of 1:25,000,000 emphasizes geologic-geomorphic features. The map units, of which there are approximately 50, are descriptive elements based on shore features, landform classes, and the principal agents that shape the coast. In addition, data are included to show tidal range, the approximate limits of Wisconsin and present-day glaciation, permafrost, sea ice, and coral. At this scale only major coastal landforms and generalized data can be shown, but the use of multicolor printing permits the inclusion of relatively detailed information. By contrast, Dolan, Hayden, and Vincent (1975) mapped the coastal landforms of North and South America at scales of 1:20,000,000 and 1:25,000,000, respectively, using black and white symbols. Only 18 landform units are defined, and these are based on geology, relief, and shoreline location (Fig. 1). This classification of the coastal landforms of the Americas expanded a system developed by Dolan (1970) for a multicolor coastal landform map of the United States at a scale of 1:7,500,000 that utilized 16 units. A further set of maps of the Americas (Dolan et al., 1972) that give supplementary information

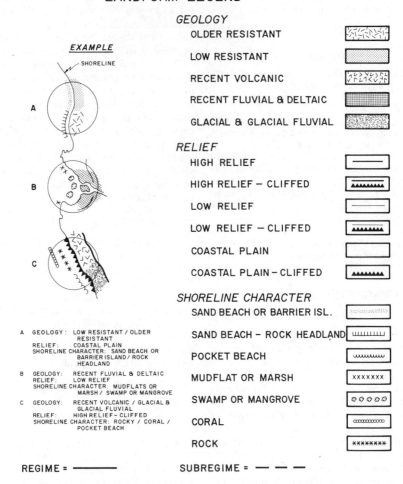

FIGURE 1. Legend used by Dolan, Hayden, and Vincent (1975) for small-scale mapping of landforms.

on coastal landforms present data on climatic regimes (15 units), coastal water-mass characteristics (19 units), terrestrial vegetation (13 units), and the dominant coastal processes (16 units) (Fig. 2). This classification, using a series of black and white maps, permits definition of a maximum of 1800 coastal environments, although only 116 of these were found to occur in the Americas.

Alexander (1962, 1966) developed a descriptive approach, based on shore form, to map the characteristics of the northeast coast of Tanganyika. Using 27 symbols (Fig. 3), Alexander produced a set of maps at scales of 1:100,000, 1:500,000 and 1:1,000,000 that illustrate different levels of generalization from the same data base. This approach was subsequently refined by Dolan, Hayden, and Vincent (1975), who derived maps of the coastal landforms of the Americas from larger-scale data. Base-line information was plotted at 1:5,000,000, and regional patterns were then established from maps at 1:7,500,000 to 1:10,000,000.

Medium-Scale Mapping

At scales between 1:100,000 and 1:1,000,000, more data are considered in the information selection and reduction stages than are required at smaller scales. Some form of classification or categorization is still necessary, however, in order to present the information in map form. As the scale increases, so the variety and accuracy of the data sources increase (Brown and Fisher, 1974; Owens et al., 1981). Frequently, field data collection or field checking of remote-sensing data is necessary where existing data or information is not available. The complexity

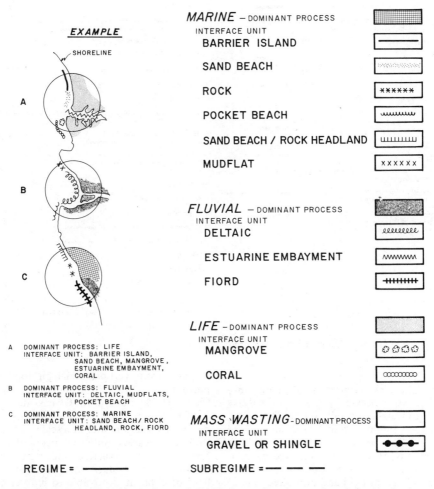

FIGURE 2. Legend used by Dolan et al. (1972) for small-scale mapping of elemental units.

of the mapping or data collection program is determined by the purpose of the map. The two examples discussed below present contrasting approaches: one dealing with a single major parameter (coastal morphology), the other with a multipurpose set of maps that attempt to depict a complete regional analysis and inventory of a section of coast.

An example of a single-purpose map is given by Swan (1968), who classified the coastal morphology of Malaya and presented his results in a series of 17 maps at a scale of 1:350,000. Only four map units were used to describe the morphology, but the physiographic nature of the coastal zone is given by five shaded relative-relief categories. The descriptive text that accompanies the maps emphasizes the coastal history (prograding or retrograding), contemporary trends, and the plan form of the shoreline (straight, convex, or concave beaches). By contrast, the multipurpose *Environmental Geologic Atlas of the Texas Coastal Zone*, which covers 7 map areas, has the objective of presenting "accurate maps of physical and biological environments, landforms, areas of significant processes, sedimentary or substrate units, and manmade features" (Brown and Fisher, 1974). To achieve this objective, 8 major systems comprising a total of approximately 130 individual units have been defined and are presented at a map scale of 1:125,000. This information is a synthesis of field, remote-sensing, and published data that is initially plotted at a scale of 1:24,000. To supplement the geologic data, a set of 8 special-use environmental maps is included in the atlas. These maps, at 1:250,000, cover: Physical Properties; Environments and Biologic Assemblages;

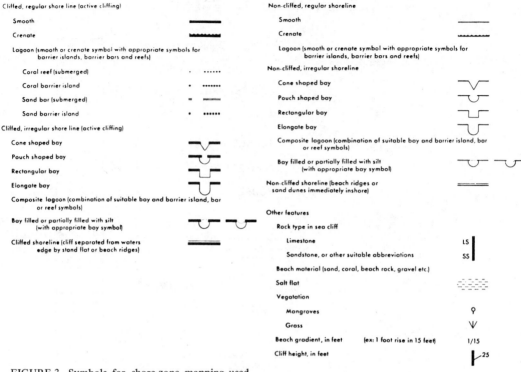

FIGURE 3. Symbols for shore-zone mapping used by Alexander (1966).

Current Land Use; Mineral and Energy Resources; Active Processes; Man-Made Features and Water Systems; Rainfall, Stream Discharge, and Surface Salinity; and Topography and Bathymetry. This set of special-purpose maps contains about 150 units. The maps in the atlas provide a comprehensive and detailed synthesis of a large volume of field and published data of the Texas coast.

Large-Scale Mapping

Detailed studies of the characteristics of small sections of coast, presented at scales greater than 1:100,000, usually require mapping from field and/or remote-sensing data. Topographic maps at scales of 1:50,000 or 1:63,360 provide basic information on relief and form. These maps also contain information on the general nature of the coast (rocky, sand shorelines, marshes, jetties, and the like), which can be used as a basis for smaller-scale morphological mapping. At larger scales, the information presented on topographic maps is frequently too general, and the sheets are used only in planning of a field-mapping program or as a base for plotting data.

The geomorphological features of the barrier beaches of the Magdalen Islands have been mapped from vertical aerial photography and plotted on 1:63,360 topographic base maps (Paul, 1974; quoted in Owens and McCann, 1980). These maps, subsequently reproduced by Owens and McCann at a scale of 1:136,000, show the location and distribution of individual features such as active and relict washover channels, dune ridges, and blowouts. At this level of detail it is possible to plot, either from remote sensing or field mapping, such features as nearshore bar morphology, inlet morphology, and vegetation distribution. The information that can be shown on large-scale morphological maps relates to specific features rather than to units based on a generalization of the character of the coastal zone.

An example of detailed mapping of coastal morphology is a study of sections of the northern coast of Scotland (e.g., Ritchie and Mather, 1969; Mather, Smith, and Ritchie, 1974). These maps are intended as a resource inventory of the nature and use of the coastal zone with emphasis on the physical features of the shoreline. The geomorphological maps, at a scale of 1:5,280, are based largely on field mapping of the beaches and the beach material "to the landward limit of marine activity" (Ritchie and Mather, 1969). The data are presented in black and white, using 48 symbols, with additional information on bedrock geology printed directly on the maps. Local relief is given by contours.

In a recent series of studies the use of integrated coding sheets and maps has been applied to detailed biophysical coastal mapping, in which the scale can be varied. To date, this approach has been used for both detailed (1:5,000) and regional (1:250,000) mapping (Robilliard and Owens, 1981; Owens et. al., 1981).

Maps of coastal features or coastal environments can be used to display data or information on spatial variations in numerous ways. At small scales the characteristics of a section of coast are presented in terms of some form of classification that relates to more than one element of the coast. Only as the scale increases is it possible to map individual features or parameters. Consequently, there is no uniform system of mapping or plotting that can be applied to this wide range of scales.

EDWARD H. OWENS

References

Alexander, C. S., 1962. A descriptive classification of shorelines, *California Geographer* 3, 131-136.

Alexander, C. S., 1966. A method of descriptive classification and mapping as applied to the northeast coast of Tanganyika, *Assoc. Am. Geographers Annals* 56, 128-140.

Brown, L. F., and Fisher, W. L., 1974. Environmental geologic atlas, Texas coastal zone: The role of geology in land-use planning, *Gulf Coast Assoc. Geol. Socs. Trans.* 24, 4-24.

Daves, J. L., 1964. A morphogenic approach to world shorelines, *Zeitschr. Geomorphologie* 8, 127-142.

Dolan, R., 1970. Coastal landforms and bathymetry, in *National Atlas of the United States*. Washington, D.C.: U.S. Department of the Interior, 78-79.

Dolan, R.; Hayden, B. P.; Hornberger, G., Zieman, J.; and Vincent, M. K., 1972. *Classification of the Coastal Environments of the World. I. The Americas.* Charlottesville, Va.: Department of Environmental Sciences, University of Virginia, Tech. Rept. 1, 163p.

Dolan, R.; Hayden, B. P.; and Vincent, M. K., 1975. Classification of coastal landforms of the Americas. *Zeitschr. Geomorphologie* 22, 72-88.

McGill, J. T., 1958. Map of coastal landforms of the world, *Geog. Rev.* 48, 402-405.

Mather, A. S.; Smith, J. S.; and Ritchie, W., 1974. *Beaches of Orkney*. Aberdeen: Department of Geography, University of Aberdeen, 168p.

Owens, E. H., and McCann, S. B., 1980. The coastal geomorphology of the Magdalen Islands, *Geol. Survey Canada Paper 80-10*, 51-72.

Owens, E. H.; Taylor, R. B.; Miles, M.; and Forbes, D., 1981. Coastal geology mapping: An example from Svedrup Basin, N.W.T., *Geol. Survey Canada Paper 81-1B*, 39-48.

Ritchie, W., and Mather, A. S., 1969. *The Beaches of Sutherland*. Aberdeen: Department of Geography, University of Aberdeen, 84p.

Robilliard, G. A., and Owens, E. H., 1981. An integrated biological and physical shoreline classification applicable to oil spill countermeasure mapping, *Arctic Marine Oilspill Program, Environmental Canada, Ottawa, Proc.*, 517-535.

Swan, S. B. St. C., 1968. Classification with reference to the east coast of Malaya, *Zeitschr. Geomorphologie* 7, 114-132.

Cross-references: *Beach Processes, Monitoring of; Classification; Coastal Ecology, Research Methods; Coastal Engineering, Research Methods; Coastal Morphology, Research Methods; Geographic Terminology; Photogrammetry; Remote Sensing; Tidal Variation; Wave Environments.* Vol. III: *Geomorphic Maps;* Vol. XII: *Map, Maps.*

COASTAL DUNES AND EOLIAN SEDIMENTATION

The importance of *eolian deposition* in coastal areas is clearly demonstrated by the size and bulk of *coastal sand dunes* in many areas. The coastal dune fields of Coos Bay, Oregon; Washington; California; Texas Gulf coast; Truro, Cape Cod, Massachusetts; the southern end of Lake Michigan; Currituck Sound, North Carolina; Guerrero Negro, Baja, California; Veracruz, Mexico; Peru; Paraná, Brazil; Lincolnshire, England; Gascogny, France; Israel; West Indies; Diamond Coast, South-West Africa; Baltic Sea, Black Sea, and Caspian Sea coasts of the U.S.S.R.; Sheveningen and Ijmuiden, Netherlands; Akita, Japan; Tasmania; and Auckland, New Zealand are just a few outstanding examples (Goldsmith, 1978).

Smaller sand dune accumulations are an integral part of almost all depositional coasts. More subtle forms of eolian deposition on beaches, marshes, intertidal sand beaches, and in shallow bays and estuaries may not be as noticeable but are also important. Clearly, eolian deposits form an imposing percentage of total sediment accumulations on depositional coasts.

Sand dunes may occur where there is a large supply of sand, a wind to move it, and a place in which it will accumulate. The coastline is an ideal location for these criteria to be met. Longshore drift supplies the sand, and waves accumulate the sand on a beach. Differential cooling and warming between land and sea assures an onshore wind at least some of the time, regardless of the general wind circulation pattern. Climate (q.v.) is definitely not a criteria for accumulations of eolian deposits along the coast, since large dunes form in arid climates (e.g., Guerrero Negro, Mexico; Diamond Coast, South-West Africa), temperate climates (Cape Cod; Gascogny, France), and in humid areas with up to 300 cm of annual rainfall (Oregon, Washington, and southeast Alaskan coasts, and Veracruz, Mexico). The only climatic condition lacking extensive coastal dunes appears to be humid-tropical. This may be related to a lack of sand supply due to intense chemical weathering (q.v.) higher shear stress required to move almost continuously wet sand, and dense vegetation adjacent to the beach. Large-scale migrating dunes or transverse dune ridges, which form through high-angle, slip-face deposition, appear to be the most common type of eolian deposition on arid coasts, as in deserts. However, such dunes are often present on humid coasts as well (e.g., the medaños on the Outer Banks of North Carolina; Cape Cod, Massachusetts, and Coos Bay, Oregon; and the Gascogny and Netherlands coasts of Europe). Coasts with abundant rainfall commonly contain eolian deposits largely fixed in place from the beginning by the abundant and well-adapted coastal vegetation. Such eolian deposits take the form of interconnected dune ridges with gently undulating upper surfaces and as distinct parabolic dunes. These *vegetated dunes* grow upward largely in place and are characterized by low-angle dipping beds or a combination of low-angle and high-angle beds. A detailed review of dune geography and internal geometry is given in Goldsmith (1973, 1978).

The occurrence of dunes on coasts appears to be unrelated to present climate, but is directly related to sand supply and a favorable wind regime. The original sand source could be glaciofluvial sediments as on the Baltic, Cape Cod, and Lake Michigan, or abundant fluvial sediments, as on the west coast of the United States (Cooper, 1958) and Australia. In any case, these sediments are moved to the shoreline and deposited on the beach by the longshore currents and waves. The sand grains are then picked up and moved by the winds at low tide. Therefore, all other factors being the same, a coast with a large tidal range should promote sand dune development, because the sand deposition is spread over a much larger intertidal area and is thus made more accessible to the work of the wind. However, as will be discussed below, there is much back-and-forth eolian transport of sand, and much of the sand now residing in dunes appears to have come directly from the landward side of the coast.

Wind Regime

There is no simple relationship between the directional distribution of wind velocity and sand dune development; low velocity winds are unable to move the sand, while extreme velocity winds may tend to destroy dunes. The dominant or highest velocity, less frequent winds may move more sand per unit time than lower-velocity, more frequent prevailing winds. However, because of the lower frequency, the dominant winds may not be as important in dune development as the prevailing winds. Further, the orientation of the shoreline with respect to both dominant and prevailing winds is also critical. The best way of ascertaining the characteristics of wind regime is to draw vector diagrams that take into account the relative frequency of occurrence of winds of different velocities (King, 1972). The sand-moving power of the wind depends on the cube of the wind velocity above 16 km/hr.

Maximum efficiency in dune growth occurs when the resultant wind vector is oriented normal to the coast. When the resultant wind vector is not normal to the coast, dunes with crests oblique to the adjacent shoreline may form, as on the Baltic coast (Zenkovich, 1967).

Dune Classification

The two principal types of coastal sand dunes are: vegetated dunes (i.e., fixed), and transverse dune ridges (i.e., generally migrating and bare of vegetation).

Vegetated Dunes. The most common types of coastal dunes are vegetated ones (Bigarella, 1972). These are generally in the form of ridges, with flat to undulating upper surfaces and continuous but irregular crests often punctuated by blowouts (i.e., low places in the dune crest through which eolian transport occurs), and washover sluice channels (i.e., low places in the dunes through which water transport occurs during storms). Vegetated dune ridges are commonly made up of stabilized parabolic or *upsiloidal dunes*, with the ends anchored by vegetation and the centers recessed back from the beach.

Commonly there occurs a series of dune ridges, usually but not always parallel to each other and the coastline. reflecting the accretional history of the coastline much as do growth rings on a tree. Elsewhere they tend to reflect the wind regime rather than the coastal outline.

Formation of Vegetated Dunes. Coastal sand dunes are initiated on accretional coasts above the spring high tide line. Sand accumulation usually begins behind some obstacle or roughness element on the beach, such as flotsam washed up on the beach during storm conditions, or behind vegetation. King (1972) notes that the growth of incipient dunes on the Lincolnshire coast is promoted by the collection of *flustra* at the high tide level.

The roughness element deflects the airstream

(i.e., wind) around it. In the *shadow zone* downwind of the obstacle the airflow is in the form of swirls and vortices, and the net forward velocity of this air is much less than that of the airstream outside the shadow zone. Sand grains moved by the wind do not follow the exact flow lines of the wind, since the bulk of the sand is moved by creep and saltation (defined and discussed below). Eventually many grains come to rest inside this relatively stagnant shadow zone, accumulating in a growing heap with the slopes standing at the *angle of repose*. The angle of repose for dry sand varies between 32 and 34° depending on the grain size, with coarser sand lying at steeper angles.

An excellent example of incipient dune growth aided by vegetation is described for the Texas coast by McBride and Hayes (Goldsmith, 1978). who tell of the formation of *pyramidal dune wind shadows* behind large, isolated clumps of grass on the supratidal beach. These beds then dip away from the dune crest in two oblique directions, with the dune crest being parallel to the prevailing wind direction. The direction of dip of the resulting beds, the azimuth, is bimodal, and the two dip directions of the beds are bisected by the prevailing wind direction.

Incipient dune growth was monitored for three years on the accretional portions of Monomoy Island and Nauset Spit, Cape Cod, Massachusetts, and on Currituck Spit, North Carolina-Virginia. These observations indicate that the dunes undergo a growth of 0.3-0.5 m per year, even at heights up to 10 m above m.s.l. The dunes begin around small isolated hummocks of dune vegetation. Wind shadows also form behind logs, peat blocks, and other material deposited by storms at the former storm high tide line.

Dune Vegetation. The incipient dune area, generally in the supratidal zone, is rapidly colonized by plants. On Cape Cod the first plants are usually marram grass (*Ammophila arenaria*) and saltwort (*Salsola kali*); on Cape Hatteras, where articifial plantings are common, these are American beach grass (*Ammophila breviligulata*) and sea oats (*Uniola panicilata*). In Britain the first colonizers are marram grass, sand couch grass (*Agropyron junciforme*), saltwort (*Salsola kali*), and *Cakile maritina*. On the Baltic Coast the first plants are lyme grass (*Elymus arenarius*), sea sandwort (*Honkenya peploides*) and sea rocket (*Cakile maritima*); on the coast of the North Sea and the Atlantic coast of Europe the first plants are usually *Agropyrum junciforme* and *Ammophila arenaria* (Zenkovich, 1967). On the northwest coast of the United States dune vegetation is not widespread despite the high rainfall. This may be due in part to the fact that the usually abundant marram grass is not native to this coast and was introduced to San Francisco Bay only in 1869 and to Coos Bay, Oregon, in 1910.

All these plants are characterized by high salt tolerance and long, elaborate root systems that reach down to the freshwater table and have additional rhizomes that grow parallel to the upper dune surface. Maximum growth and branching of the root system are achieved when the marram grass is covered by abundant sand. Thus the effect of these plants on stabilization of the dune is immense. Within three years the areas between the many isolated wind shadows fill in as the grass spreads along the rhizomes that grow out from the initial plant. After small heaps of sand reach a height of 2-3 m, the rate of growth decreases because sand is not so easily transported to the upper dune surface. With continued growth of the dunes, the roots continue to reach toward the water table. Low-angle dipping beds are therefore abundant. At the top of the highest dunes (generally up to 10-15 m) the surface is gently undulatory as a result of the upward growth of the grass coincident with the vertical growth of the dunes. An average of 1.5 m of sand accumulated in six years on the foredunes at Gibraltar Point, Lincolnshire coast, Britain at dune elevations of 4-7 m.

Because dune vegetation is sensitive to salt spray and elevation above the water table, its distribution is distinctive and changes in a regular floral succession with distance inland from the beach. The second and third lines of dunes have a greater variety of plants than the foredunes, including sea rocket (*Eatule adentula*), bayberry bushes (*Myrica pensylvanica*), goldenrod salt spray rose (*Rosa virginiana*), bayonet grass (*Scirpus paludosus*), and, in England, sea buckthorn (*Hippophae rhamnoides*). The presence of these latter plants in the first dune line is indicative of severe beach erosion with the original foredune ridge having been removed. This vegetation succession in dunes is described in more detail in Chapman (1964).

Thus the effect of plants and their root systems on stabilization and growth of vegetated (i.e., fixed) dunes is immense, and just as important as the steel girder framework of tall buildings. The vertical growth of these dunes, coincident with little horizontal migration, results in a distinctive internal dune geometry. Despite their high salt tolerance and growth on the highly exposed barrier islands, many of these plants are fragile. The most important contribution that man can make toward preservation of barrier islands is to prevent damage to the dune vegetation.

Internal Geometry of Vegetated Coastal Sand Dunes. Data from five widely scattered localities (Cape Cod [Fig. 1], Texas, Georgia, Brazil, and Israel) show that vegetated coastal sand dunes have a distinctive internal geometry (Goldsmith, 1973, 1978). The azimuth distributions correlate closely with the prevailing wind directions rather than with the dominant winds, except where the two coincide, as in Israel. The crossbeds may dip with the same orientation as the prevailing mean wind vectors (e.g., Monomoy Island, Cape Cod; Sapelo Island, Georgia; Israel; and Porto Nova and Guairamar, Brazil) or may dip in two directions. The bi-

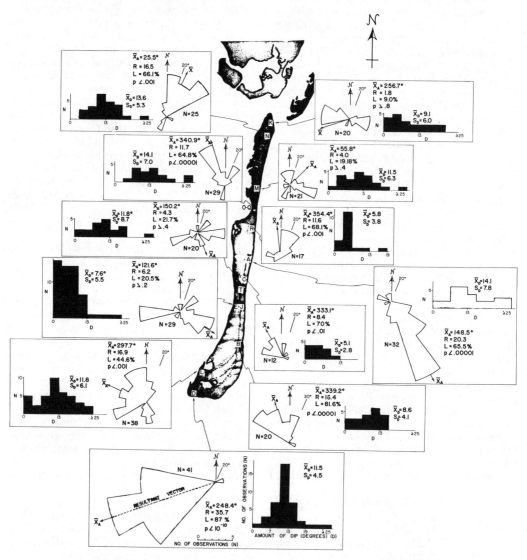

FIGURE 1. Dune crossbed dips and azimuths at each of 15 sample locations on Monomoy Island, Massachusetts. Note the low dips and the large variation in azimuths, with the beds tending to dip seaward around the island.

Key: $\bar{X}_D$ = mean dip, S_D = standard deviation of the dip, $\bar{X}_A$ = azimuth of the resultant vector, R = magnitude of the resultant vector, L = magnitude of resultant vector in percent, and P = probability that the azimuth distribution is due to chance.

modal azimuth distributions are bisected by the mean wind vector (e.g., Mustang Island, Texas; Jardim Sao Pedro, Brazil; and Israel).

The dip distributions from all five coastal dune areas contain an abundance of low-angle crossbeds. These crossbeds form as sand accumulates around the dune vegetation that acts as a baffle trapping the wind-blown sand, as observed in the Monomoy coastal dunes. Thus the vegetation anchors and stabilizes the dunes, preventing dune migration and encouraging the formation of low-angle crossbeds. It follows then that there should be a relation between vegetation density and proportion of low-angle beds (as suggested by D. Yaalon).

The higher-angle dipping dune beds probably form as pyramidal wind-shadow dunes with sand accumulating as slip-face deposition. A bimodal azimuth distribution (e.g., Brazil and Israel) is highly suggestive of a pyramidal wind-shadow dune origin as postulated for the Texas coast.

Transverse Dune Ridges. Transverse dune ridges or *migrating dunes*, characterized by a lack of anchoring vegetation, move generally in response to the prevailing winds, either landward or along the long axis of the barrier. They contain a well-defined slip face at, or close to, the angle of repose and stand in a single, large, distinct feature. The large transverse ridges either have a single straight to sinuous-shaped crest up to 1 km long or take the form of isolated *barchan dunes* that migrate inland at rates of 10–30 m annually, as in Mexico and Peru. They are as much as 30–50 m in height, and may or may not be associated with vegetated dunes. Transverse dune ridges therefore fit better than vegetated dunes into Bagnold's definition of a true dune, which is ". . . a mound or hill of sand which rises to a single summit."

Precipitation dunes, a form of transverse ridges, deposit sand in front of, and often migrate over, houses, roads, streams, forests, and even the marshes and estuaries behind barrier islands. The origin of the name is in the process whereby the sand precipitates down on whatever is in its way. The precipitation ridges of the west coast of the United States are spectacular examples of this process. Uninterrupted growth and migration of dunes may cut off streams flowing into the sea and may cause accumulation of water in the troughs between dune ridges, especially if the ground water table is high enough. This aspect is promoted by deep eolian deflation in the interdune lows. Thus these immense mounds of sand, largely bare of vegetation, are often associated with small bodies of water.

Formation of Transverse Dunes. A sand dune is an accumulation of loose sand that acts as an obstacle to the wind but also is subject to deformation by the wind. Sand particles start to move when the shear stress exerted by the wind exceeds a critical value. The value of the shear stress varies from point to point on the obstacle because the fluid flow is not uniform, as indicated by the streamlines. The airflow velocity and therefore the shear stress increase on the windward face and decrease on the lee face. Where the airflow is accelerated, as indicated by the converging flow lines, the shear stress increases; where the flow is being retarded by the diverging flow on the lee side of the crest, the shear stress decreases. Because the rate of sand movement is directly related to the shear stress, which in turn is related to the airflow velocity, the result is that sand is removed and transported from the windward face to the lee face, where it is deposited (Bagnold, 1954).

The streamlines, and hence the dune shapes, also vary with the velocity of the wind. With low velocity winds the shear stress does not reach the critical value for sand removal and transport except near the crest of the dune. Sand is then removed from the windward side of the crest of the dune and deposited in the lee of the crest where the streamlines diverge. Therefore low velocity winds have a tendency to flatten and lengthen dune profiles. During increasing wind velocity the location of the critical value of shear stress (i.e., the convergence of flow lines) moves down the windward dune slope toward the toe. High velocity winds will thus steepen and increase the height of the dune. During extremely high wind velocities the wind may arrive at the dune already charged with sand, much of which may be deposited on the lee slope. When this occurs, the dune grows wider and higher as it advances.

Deposition of sand on the lee slope occurs a certain distance downwind of the airflow divergence due to a lag effect. The location of the flow line divergence on the lee slope exhibits less variation in location with variation in wind velocities than the location of the flow line convergence on the windward slope. Therefore maximum deposition occurs more or less a set distance downwind from the dune crest for a large range of wind velocities. As the size of the dune increases, the location of maximum deposition increases in elevation and moves closer to the dune crest. This results in the upper part of the lee face receiving more deposition than the lower part, eventually causing oversteeping of the lee face and the formation of a slip face when the angle of repose is exceeded. Once formed, this slip face acts as a sediment trap for all the sand moved over the crest by the wind. The rate of advance of such a dune is thus directly proportional to the quantity of sand moved over the crest and inversely proportional to the height of the slip face (Bagnold, 1954).

A *blowout* will form where there is a lowering of the dune crest. Wind will be "funneled" across the lowered crest, resulting in an increased velocity and sand removal from the crest, which in turn increases the size of the gap at the dune crest. Eventually this sand is transported beyond the base of the original dune, and may accumulate at a new location, resulting in the formation of a new dune. Blowouts also occur on vegetated foredune ridges, and are commonly associated with parabolic dunes. Blowouts result in large amounts of windblown sand being transported onto the beach.

Internal Geometry of Transverse Dunes. Local exposures of portions of these dunes have indicated the dominance of uniformly dipping high angle bedding.

Eolianites

Under tropical conditions a sand composed of calcium carbonate rather than quartz is common. Calcium carbonate sand, acted on by the wind, will produce a distinct type of coastal dune, referred to as *eolianite* or eolian calcarenite, commonly known as *dune limestone*. This type of dune forms where eolian accumulations of carlcareous sand have become lithified. The processes of dune formation and growth for calcium carbonate sand dunes are similar to the processes for quartz sand dunes. Whereas a calcium carbonate dune is permanently lithified, a quartz sand dune fixed by vegetation may suddenly start to move again if the vegetation cover is destroyed. Eolianite may show most of the common sand dune forms. In southeast South Australia, for example, there are several ridges of dune limestone parallel to the shore that were formed during emergence of the coast resulting from both eustatic lowering of sea level and uplift of the land. These parallel eolianite ridges are now separated by broad troughs containing lakes and swamps. Occasionally *parabolic dune* forms are found preserved. Elsewhere in South and Western Australia extensive eolianite has been noted along the coast where the dunes have been partially planed off by an advancing sea. They have even been found underwater along this coast, testifying to their ability to preserve their form intact and to a lower Pleistocene sea level at the time of their formation.

Entrainment and Transport of Sediment by Wind (Physics of Grain Movement)

Sand transport by wind occurs by *saltation*, *creep*, and *suspension* (in order of importance). Because of the large differences of density between sand grains and air, transportation by suspension is relatively unimportant in coastal dunes.

Basically, a stationary particle will begin to move when the shear stress (τ_0) at the grain surface from the wind exceeds a certain critical value. The particle can be entrained in one of two ways: either directly by the wind, or by being struck by another particle already in motion (i.e., impact), The critical shear velocity is dependent on the relative density of the particle and its size, as follows (Bagnold, 1954, p. 101):

$$U_{*c} = A\sqrt{\frac{\tau_{oc}}{\rho}} = A\sqrt{\frac{\sigma\rho}{\rho}gD} \quad (1)$$

where U_{*c} = critical shear velocity; τ_{oc} = critical shear stress; ρ = fluid density (i.e., wind); σ = grain density; D = grain diameter; A = constant; and g = gravity acceleration.

Saltating grains (bounding motion initiated by impact) take a path with an initial, almost vertical, rise followed by a long, low trajectory almost parallel to the ground and a velocity close to that of the wind. The loss of momentum from the wind thus gained by this particle is transferred to one or more other particles upon impact, or may result in the same original particle bounding back up into the airstream.

Experimental data also show that during transport the coarsest grains are concentrated at two levels: at the bed, and several centimeters or more upward. This complex vertical distribution is explained by the fact that when the coarse grains undergo saltation they "get more bounce to the ounce," i.e., the bigger grains bounce higher.

Coincident with saltation, *surface creep* is occurring by other grains or by the same grains alternating between saltation and creep. Creeping particles travel slowly and irregularly along the surface under the influence of the bombarding grains or directly under that of the fluid.

The movement of these grains is dependent on the critical shear velocity (U_{*c}), which in turn is directly proportional to the rate of increase of the wind velocity with the log height. Using the Prandtl log height equation

$$U = \frac{U_{*c}}{K}\log\frac{Z}{Z_0} \quad (2)$$

where U = wind velocity at elevation Z; K =

Karman constant (0.40); and Z_o = roughness at the surface.

This equation is of great advantage because it relates the fluid velocity at any height with both the surface roughness and the critical shear stress at the surface. However, once U_{*c} is exceeded and sand particles are entrained, the wind velocity profile is affected by the sand movement. With increasing wind velocities, the layer affected increases in height because of the increased height of sand trajectories (Fig. 2 and 3).

For the same size sand, however, all the wind velocity distributions plotted on semilog paper against height pass through a common point, called the focal point. This means that no matter how strong the wind blows, no matter how great the velocity gradient, the wind velocity at a certain height (about 3 mm) remains almost the same (Bagnold, 1954, p. 58, Fig. 17). The reason for this phenomenon is unclear, but it allows for the simplification of equation (2) to

$$U = 5.75 U_* \log_{10} \frac{Z}{Z'} + U' \qquad (3)$$

where, according to Zingg (Goldsmith, 1978), $Z' = 10D$ (in mm) at the focal point and $U' = 20D$ (in km/hr) at the focal point.

Of great interest is the total amount of sand transported. Two empirical solutions of note are by Bagnold and by Kawamura (discussed in Goldsmith, 1978).

Eolian Ripples

Eolian ripples are common on all parts of the dunes and beach. The following commonly occurring characteristics make wind-formed ripples distinct from water-formed ones: Eolian ripples fall into two distinct types: *sand ripples*, composed of well-sorted medium to fine-grained sand; and *granule ripples*, isolated ripple forms, often parabolic-shaped, composed entirely of coarse sand or granule-size particles. (The next three characteristics refer primarily to sand ripples.) Sand ripples often occur on the sides of dunes with their long axes parallel to the slope. There is a tendency for the coarsest grains and/or heavy mineral grains to accumulate on the crests of the sand ripples. Eolian ripples tend to have large (>17) ripple indexes (ripple wave length/height) and symmetry indexes between 2.0 and 4.0, although there are many exceptions.

The sharp distinction between sand ripples and granule ripples is recognized not only for the coastal zone but also for most desert environments as well (Bagnold, 1954). Granule ripples are usually asymmetrical, whereas sand ripples are either symmetrical or asymmetrical.

Artificial Insemination of Coastal Sand Dunes

The eolian history of many coastal areas in Europe and North America is similar in that once stable forest-covered dunes have been

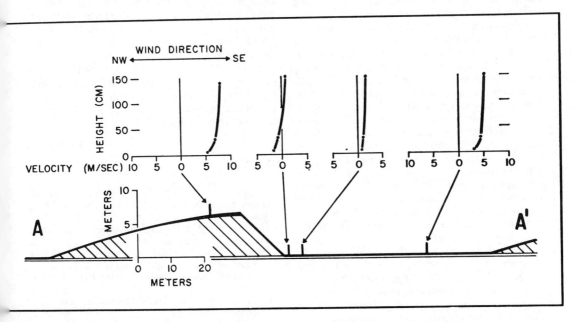

FIGURE 2. Wind velocity profile over a dune. Note the reversal of wind direction at the toe of the slip face (from Inman, Ewing, and Corliss, 1966).

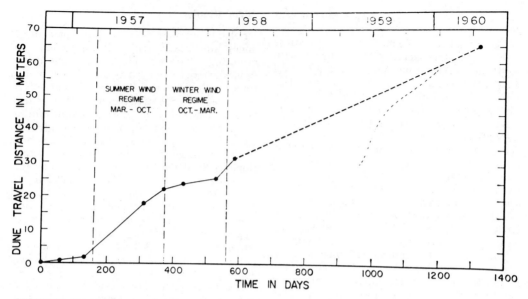

FIGURE 3. Travel distance of a dune; total travel distance of 65 m in 1322 days or 18 m/yr (from Inman, Ewing, and Corliss, 1966).

denuded by man's activities. This has had a definite effect on dune processes and coastal geomorphology in many areas. On the northeast coast of the United States the first European settlers, in the seventeenth century, adversely affected dune vegetation almost immediately through overgrazing. Thus dune rebuilding and stabilization has had a long history and is definitely not a new problem.

The two principal methods for creating and promoting stabilized coastal barrier dunes are through *artificial* plantings and sand fencing, aimed at promoting *natural* eolian sand accumulations. Alternative methods, such as using junk cars on the beaches at Galveston, Texas, have been attempted without much success. Much literature on this subject is in U. S. Army Corps of Engineers publications or results from corps-sponsored projects, inasmuch as the corps has a vested interest in stable barrier islands.

Implications. The great success of the dune rebuilding program since the 1930s by means of fencing in the coastal areas of Virginia and North Carolina, between Chesapeake Bay and Ocracoke Inlet, has resulted in some changes that require a reassessment of the program. Godfrey and Godfrey (1973) have discussed the extensive botanical changes in the dune ecosystem and the disturbing implications, resulting from the development of the "artificial" ecological system on Cape Hatteras. Dolan (1972) has suggested that a coarsening, narrowing, and steepening of the beach profiles in front of the stabilized Cape Hatteras dunes can be directly attributed to the success of this dune fencing program, which results in the inability of the fixed dune line to respond to the natural processes of a dynamic system. Also, the sand eroded by waves from the foredune during storms that had formerly reached the backdune area is now prevented from doing so. This storm wave overwash deposition is at least as important a process in some coastal dunes as wind deposition. Moreover, storm overwash onto the marshes and bays behind the barrier islands may be a critical process in maintaining the existence of these barrier islands. Leatherman (1979) has suggested that much of the overwash sediments on Assateague Island, Maryland and Virginia, are blown back onto the beach and into the dunes, causing the dunes to grow from the backside.

VICTOR GOLDSMITH

References

Bagnold, R. A., 1954. *The Physics of Blown Sand and Desert Dunes.* New York: William Morrow and Co., 256p.

Bigarella, J. J., 1972. Eolian environments–Their characteristics, recognition, and importance, *in* J. K. Rigby and W. K. Hamblin, eds., *Recognition of Ancient Sedimentary Environments.* Tulsa, Okla.: Soc. Econ. Paleont. Miner. Spec. Pub. No. 16, 12–62.

Chapman, V. J., 1964. *Coastal Vegetation.* New York: Macmillan Co., 245p.

Cooper, W. S., 1958. *Coastal Sand Dunes of Oregon and Washington.* New York: Geol. Soc. America Mem. 72, 169p.

Dolan, R., 1972. Barrier dune system along the Outer

Banks of North Carolina: A reappraisal, *Science* **176**, 286-288.
Godfrey, P. J., and Godfrey, M. M., 1973. Comparison of ecological and geomorphic interactions between altered and unaltered barrier island systems in North Carolina, in D. R. Coates, ed., *Coastal Geomorphology*. Binghamton: State University of New York, 239-258.
Goldsmith, V., 1973. Internal geometry and origin of vegetated coastal sand dunes, *Jour. Sed. Petrology* **43**, 1128-1142.
Goldsmith, V., 1978. Coastal dunes, in R. A. Davis, Jr., ed., *Coastal Sedimentary Environments*. New York: Springer-Verlag, 171-235.
Inman, D. L.; Ewing, G. C.; and Corliss, J. B., 1966. Coastal sand dunes of Guerrero Negro, Baja California, Mexico, *Geol. Soc. America Bull.* **77**, 787-802.
King, C. A. M., 1972. *Beaches and Coasts*. London: Edward Arnold, 570p.
Leatherman, S. P., 1979. Migration of Assateague Island, Maryland, by inlet and overwash processes, *Geology* **7**, 104-107.
Zenkovich, V. P., 1967. *Processes of Coastal Development*. New York: Wiley Interscience, 738p.

Cross-references: *Beach; Beach Orientation; Beach Processes; Climate, Coastal; Coastal Flora; Coastal Reserves; Deflation Phenomena; Dune Stabilization; Major Beach Features; Ripple Marks; Sediment Transport; Swing Mark; Washover and Washover Fan; Wind; Wind-Deposition Coast.* Vol. III: *Sand Dunes.*

COASTAL ECOLOGY, RESEARCH METHODS

The methods employed for the study of coastal ecology are almost as numerous as are investigators in this field. To a large degree the procedures used will be dictated by the type of information that is desired. Methodology will vary greatly, depending upon whether the research is descriptive or experimental, whether the substrate is hard or soft, and what temporal and spatial parameters are considered important.

Hard Substrate Benthos

Of all the elements of the coastal biota, the inhabitants of the rocky intertidal environment have been most thoroughly investigated. Here a great deal of valuable information can be obtained by careful observation and note taking. A more detailed approach involves laying out transects perpendicular to the coastline and counting the studied organisms within a given distance or counting the organisms within random quadrants of an appropriate size. Rocky environments are commonly highly heterogenous; thus it may not always be possible to obtain a representative sample of the microhabitats present that is strictly quantitative.

An experienced observer will be able to identify many of the larger organisms in the field, but it may be necessary to remove some of the smaller forms by scraping them off the rock or by removing cobbles or boulders for laboratory study. In many cases an estimate of *biomass* will be desired, requiring removal of a representative group of organisms that are weighed either wet or dry. The values obtained can then be used to estimate *standing crop* and, when integrated through time, to estimate *productivity*. Growth rates can also be measured in the laboratory or by marking individuals in the field (Holme and McIntyre, 1971; see also papers by Doty, Frank, and Kohn in Nybakken, 1971).

The use of scuba gear has allowed many of the same methods to be used in subtidal areas of resonable visibility. Tropical reefs in particular have been amenable to scuba investigation (Odum and Odum, in Nybakken, 1971). Technological advances in still and motion photography and in television have made the use of these tools increasingly widespread, and they are often used in conjunction with diving. The study of fast-moving or rare organisms in particular is facilitated by such means (Holme and McIntyre, 1971).

Soft Substrate Benthos

Because of the increased importance of the infauna in soft-bottomed communities, direct observation loses much of its usefulness, and removal of substrate samples is often required. One of the simplest devices used in the intertidal and shallow subtidal zones is a coffee can with the bottom cut out of it. Used to take shallow cores, it provides a sample of constant surface area and convenient size for the study of meiobenthic (.062-.5 mm median diameter) and small macrobenthic (greater than .5 mm median diameter) organisms. More elaborate glass or metal corers are often used to sample the meiobenthos where it is considered important to retain the laminations in the sediment (Holme and McIntyre, 1971).

Mangrove swamps and salt marshes are somewhat special cases of the soft-bottomed intertidal zone, for they commonly require the integration of techniques used by terrestrial and aquatic ecologists. *Insects* (q.v.) may be an important part of the biota, and these are commonly censused using a net or other trapping devices. Birds and mammals may also be important, and these are commonly censused by observational methods. Aerial photos may also prove of considerable value for the estimation of the amount of vegetation present (see papers by Golley et. al. and Teal, in Nybakken, 1971).

In deeper water many types of grabs, dredges, and coring devices are used. The use of dredges for scientific purposes dates back to the middle of the eighteenth century, and they are still in common use where qualitative results are adequate. The first systematic, quantitative survey of the soft-bottomed benthos was conducted by Petersen in the early part of this century using a *grab sampler* (Petersen and Boysen Jensen, 1911). Many technical improvements have been made since that time, but the most popular devices still follow the same principle. A more recent development is the use of diver-operated suction samplers, which have the added advantage of the direct observation of any possible sources of error (Holme and McIntyre, 1971).

Once the sample is obtained, there are a number of techniques by which the organisms can be extracted. Many of these involve washing the sample over a mesh of variable size in order to remove fine particulate matter. The material can then be examined live or fixed with buffered formalin or alcohol. The sorting of preserved material can be facilitated by staining the protoplasm with Rose Bengal. Living meiofauna can sometimes be extracted using temperature and salinity gradients. Small algae can be trapped by placing lens tissue over a fresh sample and allowing the algae to migrate up into it. Pigments can also be extracted directly from preserved material with no sorting at all, although remnants of cells in detritus may bias the results (Holme and McIntyre, 1971).

Nekton

Methods for sampling the nekton date back to the first fishermen, and even today a great deal of valuable information is obtained from commercial catches. Trawls, gill nets, and lines are among the most commonly used devices Evaluating sample bias is a particular problem with such data, for not only must mesh size and sampling area be taken into account but behavioral aspects of the fish also become important (Harrison, 1967).

Since the first successful recovery of Atlantic salmon in 1873, *tagging* has become an important method employed by fisheries biologists for the study of migration, growth rate, survivorship, and related phenomena. A wide variety of inserted and collar tags are available, and dyes, radio transmitters, and fin clipping are just a few of the other devices used to tag fish. Some of these have also been applied to crustaceans (Everhart, Eipper, and Youngs, 1975).

Plankton

The adequate sampling of planktonic organisms has long been a problem. For the phytoplankton, nets are inadequate for quantitative purposes because they miss many of the important small forms. The most acceptable methods in current use involve either bottles that close at a specified depth or pumps with measured flow rates. Biomass can then be estimated by measuring pigment density using spectrophotometry or fluorescence. A direct estimate of photosynthesis can be obtained by measuring the difference in carbon 14 uptake between paired light and dark bottles (Schlieper, 1968).

The zooplankton are sampled with a net. No single mesh size provides an adequate sample, because fine nets clog too quickly to provide a good sample of large forms, while coarse nets allow many of the small animals to pass through. This difficulty can be somewhat ameliorated, by nesting several sizes of nets in a row. For quantitative surveys, a current meter is placed in the mouth of the net to estimate the area sampled. Bacteria must be sampled separately using sterilized containers (Ahlstrom, 1969).

Physical and Chemical Parameters

The measurement of the physical and chemical aspects of the environment forms an important part of most studies of coastal ecology. Measurements of important chemical factors, such as salinity, and concentrations of oxygen, nitrogen, phosphorus, and organic carbon, are carried on routinely (Strickland and Parsons, 1968). For other substances, more sophisticated techniques may be necessary, such as the use of gas chromatography for studying the effects of oil pollution (Blumer, 1973). Important physical parameters include temperature, pressure, light, current velocity, turbidity, wave shock and tidal range. In the soft-bottom, additional measurements may be desired as to grain size, compaction, and permeability (Holme and McIntyre, 1971).

Experimental Methods

A vast amount of laboratory experimentation has been done on plants and animals from coastal areas, and much important physiological and biochemical data in particular have been obtained in this way (Mariscal, 1974). In recent years increasing attention has been paid to experimentation in the field. As explained by Connell (in Mariscal, 1974), there is a major difference in the design of field versus laboratory experiments. In the lab an attempt is made to keep all untested variables constant, while in the field these are allowed to vary naturally. In this way the field experimentalist hopes to obtain results more directly applicable to the natural environment. Excluding predators by the use of cages, shading, and transporting or removing organisms are a few of the types of experiments that are commonly attempted. One of the most important problems to be addressed during such experiments is the establishment of a control. For instance, a cage constructed to exclude predators might also alter the flow characteristics of the water in the immediate area. In this case, Connell suggests the use of a cage with the sides removed as a control that will allow entrance of predators but also alter the flow regime.

Procedural Design

A great deal of thought must often go into determining the procedure that will provide the information desired by the ecologist. In descriptive studies, sufficient stations should be sampled to adequately represent the community under study, while taking duplicate samples at a given station will provide some feeling for error in the methods. The experimentalist may need to conduct simultaneous runs in different parts of an organism's range. In addition to such problems of spatial arrangement, the temporal setting is also important. An investigator interested in long-term community changes may sample at yearly intervals; one interested in life cycles of the organisms might want quarterly samples; another studying the effects of an oil spill on the biota might require samples spaced weeks or days apart. An experimentalist might find it desirable to repeat his or her procedures during different seasons to detect possible variations among results.

Practical considerations such as availability of equipment and accessibility of the study area are important in designing a research project. A frequently overlooked limitation concerns the amount of trained help available for sorting samples and identifying specimens, for this process requires a considerable amount of time.

The methods employed for studying coastal ecology are numerous, and what procedures are used can have a significant effect on the data that are generated. It is important that the scientist remain aware of this, and any conclusions reached should be tempered accordingly.

G. KENT COLBATH

References

Ahlstrom, E., 1969. *Recommended Procedures for Measuring the Productivity of Plankton Standing Stock and Related Oceanic Properties.* Washington, D.C.: National Academy of Sciences, 59p.

Blumer, M., 1973. *Interaction between Marine Organisms and Oil Pollution.* Washington, D.C.: Environmental Protection Agency, Ecol. Res. Series, R3-73-042, 97p.

Everhart, W. H.; Eipper, A. W.; and Youngs, W. D., 1975. *Principles of Fishery Science.* Ithaca, N.Y.: Comstock Pub. Associates, 288p.

Harrison, C. M. H., 1967. On methods for sampling mesopelagic fishes, in N. B. Marshall, ed., *Aspects of Marine Zoology.* London: Zoological Society of London, Symp. 19, 71-126.

Holme, N. A., and McIntyre, A. D., eds., 1971. *Methods for the Study of Marine Benthos, IBP Handbook 16.* Oxford: Blackwell, 334p.

Mariscal, R. N., ed., 1974. *Experimental Marine Biology.* New York: Academic Press, 373p.

Nybakken, J. W., ed., 1971. *Readings in Marine Ecology.* New York: Harper and Row, 544p.

Petersen, C. G. J., and Boysen Jensen, P., 1911. *Valuation of the Sea. I. Animal Life of the Sea-Bottom, Its Food and Quantity.* Danish Biological Station Rept. 20, 81p.

Schlieper, C., ed., 1968. *Methoden der Meeresbiologischen Forschung.* Jena: Gustav Fischer Verlag, 322p.

Strickland, J. D. H., and Parsons, T. R., 1968. *A Practical Handbook of Sea Water Analysis.* Fisheries Research Board of Canada Bull. 167, 311p.

Cross-references: *Base-line Studies; Benthos; Coastal Fauna; Coastal Flora; Coastal Morphology, Research Methods; Coastal Waters Habitat; Coastal Environmental Impact Statements; Grab Samplers; Intertidal Mud Habitat; Intertidal Sand Habitat; Nearshore Water Characteristics; Organism-Sediment Relationship; Pollutants; Production; Rocky Shores Habitat.*

COASTAL ENGINEERING

Coastal engineering is the practical application of diverse scientific knowledge in and along coastal waters for the maximum economic benefit of mankind and with regard for overall shoreline processes and respect for the tremendous destructive forces of the sea. The term *coastal waters* refers to waters covering the continental shelves, estuarine waters, and large bodies of inland waters. Although coastal engineering itself has been practiced for centuries, the term coastal engineering appears to have been first used in conjunction with the first conference on coastal engineering in 1950 (Johnson, 1951). It was then defined as "a branch of Civil Engineering which leans heavily on the sciences of oceanography, meteorology, fluid mechanics, electronics, structural mechanics and others."

Coastal Engineering Structures

Most of the structures related to coastal engineering are those that provide shore protection. The coastal engineer is faced with a problem, such as excessive erosion along a beach, and he or she must determine the most economically feasible and effective structure or procedure to halt or reverse the problem. Structures that cause the interception of longshore transport can severely disrupt the delicate balance between erosion and deposition at downdrift locations.

In locations where the beach is eroding and protection is needed for adjacent buildings and highways to prevent undermining, sea walls, bulkheads, and revetments have been constructed. *Sea walls,* such as the one shown in Figure 1 are massive concrete structures designed to stop storm waves. The downward rush of water from the sea wall caused by striking waves can cause severe scour problems unless methods such as the stone apron shown in Figure 1 are used. A *bulkhead* is a structure composed of a series of concrete, steel, or timber pilings, as shown in Figure 2, that are driven into the ground for storm wave protection. Bulkheads may be only a temporary solution; as larger waves reach the receding beach, a sea wall may be necessary. *Revetments,* as shown in Figure 3, are constructed by armoring the sloping face of a dune or bluff with one or more layers of rock, concrete, or asphalt (Coastal Engineering Research Center, 1973).

Other structures are constructed to control or modify sand movement. These include groins and jetties. *Groins* are structures built to increase the width of a beach by either natural means or artificial nourishment. The groin acts as a dam to block the flow of sediments, thereby building up the beach. Downdrift beaches will erode until the sand builds up behind the groin to the point where it can continue around the groin. To minimize downstream property damage, a groin should be designed with the top profile equal to that of a beach of reasonable dimensions. *Artificial nourishment* is a process in which sand is pumped or barged to

FIGURE 2. Two types of bulkheads: Virginia Beach, Virginia (top); Plymouth, Massachusetts (bottom) (from Coastal Engineering Research Center, 1964).

FIGURE 1. Sea wall at San Francisco (from Coastal Engineering Research Center, 1964).

the desired location and spread out along the shore. This sand may be obtained from offshore sites, inland sites, or from locations where sand has been excessively deposited. Artificial nourishment in the area before the groin would speed up the filling process and decrease the amount of time allowed for downshore erosion to occur.

FIGURE 3. Stone revetment, Johns Pass, Treasure Island, Florida (from Coastal Engineering Research Center, 1964).

FIGURE 5. Offshore breakwater at Channel Islands Harbor, California (from Coastal Engineering Research Center, 1964).

A *jetty* is also a structure that interrupts the longshore flow of sediments and causes a build-up on the upstream beach and starvation on the downstream end, as shown in Figure 4. However, this is not 'the purpose of the jetty. Its purpose is to keep a navigation channel clear of sediment build-up. Jetties are usually much larger than groins, and extend to a depth of water equal to that of the navigational channel. To be effective, it must be high enough to completely intercept the flow of sand. One means of controlling erosion on the downstream beach is to dredge the sand from behind the upstream jetty and to pump it though a pipeline to behind the downstream jetty. This technique is called inlet or sand *bypassing*.

FIGURE 4. Jetties at Cold Spring Inlet entrance to Cape May Harbor, New Jersey. Net sediment transport is from right to left in the photo. Note accretion of the beach at the right and erosion of the beach at the left (from Coastal Engineering Research Center, 1964).

Breakwaters are coastal structures constructed to dissipate and reflect wave energy to provide safe navigation into a harbor. A shore-connected breakwater acts as a jetty in cutting the sand stream. Methods such as those used with jetties can be used to control downstream erosion. Offshore breakwaters, such as the one shown in Figure 5, also reduce wave energy. However, offshore breakwaters can in some cases still allow the continuation of sediment transport. Such breakwaters are used in some locations instead of jetties because they can cut off the longshore transport when they are of sufficient length and are close enough to the shore. The sediment that is deposited in such a system is then dredged and pumped to the beach downdrift from the navigation channel.

Coastal engineering also includes the construction of harbors, offshore islands, and marinas. Figure 6 summarizes the general problems involved in the construction of coastal engineering structures along with the problems that necessitate the construction and some of the considerations that must be taken into account.

Waves: Description, Generation, Prediction, and Theories

As shown by Figure 6, the coastal engineer must consider many factors when planning for construction in the coastal zone. He must ask questions such as: What will be the maximum wave that the structure can expect to encounter during its life? What forces will various waves induce in the structure? What effect will the structure have on its environment? These are just a few of the thousands of questions that must be considered. This section will deal with maximum wave height.

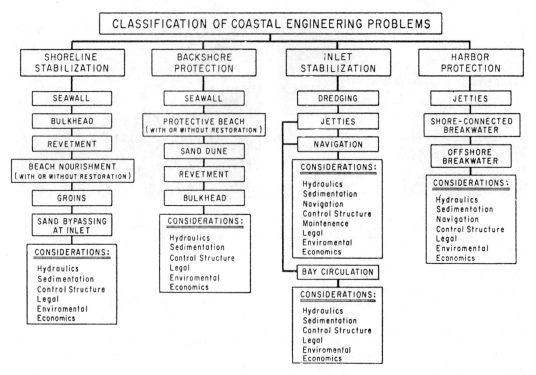

FIGURE 6. General classification and scheme for solution of coastal engineering problems (from Coastal Engineering Research Center, 1973).

Waves are defined as surface undulations between fluids of varying densities. For coastal engineering, the two fluids are water, either fresh or salt (usually the latter), and the atmosphere above it. Waves are divided into two main categories: *storm waves (sea)* and *swell*. Storm waves are those that are still being generated and are within the *fetch* (length over which the wind is blowing) or storm area. Swell applies to waves that have left the storm area and are dispersing across the water (Kinsman, 1965).

Figure 7 illustrates the various characteristics of waves, such as *wave height* (H), *wave length* (L), and crest and trough positions. In addition to these, there are several other important definitions concerning waves. *Wave amplitude* is taken as half the wave height and is measured from the mean water level (m.w.l.) to the wave crest or trough. The *significant wave height*, $H_{1/3}$ is the average height of the highest one-third of the waves, while H_{avg} is the *average height* of all the waves. The *wave steepness* is the ratio of a wave's height to its length. The

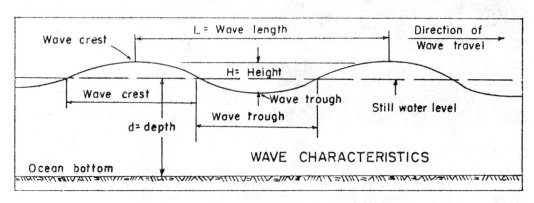

FIGURE 7. Wave characteristics (from Coastal Engineering Research Center, 1964).

mean water level is defined as the halfway mark between crest and trough, and is not necessarily coincident with the *still water level* (s.w.l.). The *wave period* (T) is the time it takes for two successive crests or troughs to pass one point on the water surface. The period is usually expressed in seconds, and is the only characteristic of a wave that remains constant at all times. The *wave celerity* (C) is the speed at which a wave travels across the surface and it is defined mathematically by:

$$C = L/T \qquad (1)$$

where L = wave length and T = wave period.

As stated above, waves are generated by the wind blowing over the surface. For a given wind velocity, there is a certain maximum wave energy that can be produced. This stage is known as *fully arisen sea* (FAS), and occurs when the wind energy being transferred to the sea is being dissipated at the same rate through wave breaking and other forms of losses. Waves can be fetch-limited or duration-limited. In areas where the wind does not have enough area over which to generate FAS conditions, the wave energy will not reach its maximum (Pierson, Neuman, and James, 1955).

Swell waves travel over the decay, or dispersal, area where the waves disperse longitudinally and traversely. There is a reduction in height caused by the spreading over an increasing area of ocean. Conditions outside the fetch are highly variable as the wave energy is reduced.

To determine the characteristics of waves that a given area will encounter, it is necessary to use meteorological data from the past. From these data a prediction is made as to the maximum winds that can be produced. Hurricanes and tornadoes must be especially considered. This method is known as *hindcasting*. The coastal engineer must decide whether the available data are sufficient and accurate enough to make a forecast of future maximum winds. The use of satellites has tremendously increased the availability of weather data for many parts of the world. From the maximum winds an energy spectrum must be derived for a given area.

To be able to determine the forces that these waves will exert, it is necessary to be able to determine the actual wave motion. Some of the major theories for the description of waves includes Airy or Stokes I (Airy, 1845; Ippen, 1966; Silvester, 1974), Stokes II (Coastal Engineering Research Center, 1973), Stokes III (Skjelbreia, 1959), Stokes V (Skjelbreia and Hendrickson, 1962), Stokes VII, Cnoidal (Masch and Wiegel, 1961), Solitary (Ippen, 1966), and Stream Function (Dean, 1965).

The theory that should be used depends on factors such as wave height, wave period, and water depth (Le Méhauté, 1969). The most basic theory is that derived by Airy in 1845, referred to as small-amplitude, linear, or Stokes I wave theory. It is easy to apply and is applicable over a wide range of waves.

Wave Forces on Structures

After assessing the types of waves a structure will probably encounter on a statistical basis, the coastal engineer must determine the effect that these waves will have upon the structure.

For cylindrical structures, such as piles (Agerschou and Eden, 1965), an equation known as the Morison equation can be used (Morison, 1950).

$$F_T = F_I + F_D = \frac{1}{2} C_D \rho D u |u| + C_M \frac{\pi D^2}{4} \rho \dot{u} \qquad (2)$$

where F_T = total force per unit length; F_D = drag force per unit length; F_I = inertia force per unit length; C_D = drag coefficient, typically (0.5-1.2); ρ = fluid density; D = diameter of member; u = water particle velocity; $|u|$ = absolute value of u; C_M = mass (or inertia) coefficient, typically (1.5-2.0); and $\dot{u}$ = particle acceleration. This equation is used to compute forces for both uniform flow ($F_I = 0$) and for wave motion.

Wave forces on other types of structures require special procedures and design criteria. Sometimes this involves only a slight modification of equation 2. Many tables and graphs have been prepared directly relating wave heights, water depths, and other factors to the size of the members required to withstand the forces created by these waves. Other forces that the coastal engineer must consider are winds, currents, and possibly ice and/or earthquakes, depending upon the area. Other methods in the development stage for nonsymmetrically shaped structures include the "diffraction theory" and "marker and cell" methods.

Diffraction theory is applied in circumstances where viscous and separation effects are negligible. The problem is set up in terms of a velocity potential, and the solution is sought that satisfies the conditions relating to the size of the object and the free surface effects. The dynamic pressure distribution is determined from the velocity potential using Bernoulli's equation. The forces and moments are then determined by surface integration (Rao and Garrison, 1971).

The *marker-and-cell* (MAC) method uses a numerical approximation of the two-dimensional Navier-Stokes equations with free surface boundary conditions. This method allows the fluid to move through large amplitude contortions in several space dimensions. From the approximate solution of the Navier-Stokes equations, the load vector and consequently the resultant forces on the structure can be determined.

Special Design Practices

Structures designed for construction in the coastal zone are subjected to the strong *corrosive* forces of seawater. Steel should be protected by using cathodic means or coatings of concrete, corrosion-resistant metals, or organic and inorganic paints (Watkins, 1969). Concrete in the tidal zone must have a high cement content, a low water-cement ratio, and admixtures to increase durability. Timber piles should be treated with creosote oil with a high phenolic content or encased in a gunite armor sealed at the top. It is usually more economical to design for round members, because they have the smallest surface areas and the best flow characteristics that result in longer usage.

Coastal engineering is one of the most truly multidisciplinary fields of engineering that exist today, as witnessed by the numerous and variable factors that must be considered when designing a coastal structure. The role of coastal engineer will continue to grow in the future as the cost of coastal protection and the demand for residential, recreational, and industrial areas increase. Careful planning and sound coastal engineering are the keys to solving these problems.

JOHN B. HERBICH
JOHN P. HANEY

References

Agershou, H. A., and Eden, J. J., 1965. Fifth and first order wave force coefficients for cylindrical piles, in *Coastal Engineering, Santa Barbara Specialty Conference*, Am. Soc. Civil Engineers, 219-248.

Airy, G. B., 1845. Tides and Waves, *in* E. Smedley et al., eds., *Encyclopedia Metropolitana*. London: B. Fellowes, 241-396.

Coastal Engineering Research Center, 1964. *Land against the Sea*. Washington, D.C.: U.S. Army Corps of Engineers, 43p.

Coastal Engineering Research Center, 1973. *Shore Protection Manual*. Washington, D.C.: U.S. Army Corps of Engineers, 750p.

Dean, R. G., 1965. Stream function representation of nonlinear ocean waves, *Jour. Geophys. Research* 70, 4561-4572.

Ippen, A. T., 1966. *Estuary and Coastline Hydrodynamics*. New York: McGraw-Hill, 744p.

Johnson, J. W., ed., 1951. *First Conference on Coastal Engineering, Proceedings*. Berkeley, Calif.: Council on Wave Research, The Engineering Foundation, University of California.

Kinsman, B., 1965. *Wind Waves*. Englewood Cliffs, N.J.: Prentice-Hall, 676p.

Le Méhauté, B., 1969. *An Introduction to Hydrodynamics and Water Waves*, 2 vols. U.S. Dept. Commerce, ESSA Tech. Rept., 503p. and 725p.

Masch, F. D., and Wiegel, R. L., 1961. *Cnoidal Waves. Tables of Functions*. Richmond, Calif.: Council on Wave Research, The Engineering Foundation, 129p.

Morison, J. R., 1950. The force exerted by surface waves on piles, *Petroleum Transactions* 189, 149-154.

Pierson, W. J., Jr.; Neumann, G.; and James, R. W., 1955. *Practical Methods for Observing and Forecasting Ocean Waves by Means of Wave Spectra and Statistics*. Washington, D.C.: U.S. Navy Hydrographic Office Pub. No. 603, 284p.

Rao, V. S., and Garrison, C. J., 1971. *Interaction of a Train of Regular Waves with a Submerged Ellipsoid*. College Station, Texas: Texas A&M University, Sea Grant Pub. No. TAMU-SG-71-209, 156p.

Silvester, R., 1974. *Coastal Engineering*, vol. I. Amsterdam: Elsevier Scientific Publishing Company, 457p.

Skjelbreia, L., 1959. *Gravity Waves. Stokes' Third Order Approximation. Tables of Functions*. Berkeley, Calif.: Council on Wave Research, The Engineering Foundation, University of California, 337p.

Skjelbreia, L., and Hendrickson, J. A., 1962. *Fifth Order Gravity Wave Theory and Tables of Functions*. Pasadena, Calif.: National Engineering Science Company, 424p.

Watkins, L. L., 1969. *Corrosion and Protection of Steel Piling in Seawater*. Washington, D.C.: U.S. Army Corps of Engineers, Coastal Engineering Research Center TM-27, 100p.

Cross-references: *Air Breakwaters; Artificial Islands; Boat Basin Design; Coastal Engineering, History of; Coastal Engineering, Research Methods; Lakes, Coastal Engineering; Offshore Platforms; Protection of Coasts; Stokes Theorem; Wave Shadow; Waves*. Vol. I: *Wave Theory*.

COASTAL ENGINEERING, HISTORY OF

Initial developments in coastal engineering were probably related to the establishment of ports in sheltered waters. Observations concerned what was immediately visible: winds, tides, currents, and perhaps local waves. Sediment transport was probably also known to occur but hardly understood. Characteristically, many of the ancient ports, like the famous naval port on the Tiber River at Ostia near Rome, drowned in sediments from the river flow that was allowed to pass through the harbor basin. The same was also true of the magnificent Viking naval bases in Denmark.

Ancient ports were built on rivers, estuaries, and fjords protected against wave action. This was true of the ancient Egyptian ports as well as the Viking ports in Scandinavia. It was still true of most of the medieval ports, like London, Rotterdam, and Hamburg, located in estuaries with considerable tidal flow that maintain deep channels, until the development of vessels of much deeper draft—mainly during the past few decades—imposed further requirements of increased depths. Attempts to establish ports on the open, exposed seacoast, facing its littoral drift problems, were few. We only know of one major attempt. Some 5000 years ago the Phoenicians attempted, and succeeded, in establishing a port on the open seacoast at Tyre, in what is now Lebanon. It was built of heavy blocks locked together with copper dowels. Apparently it was known by experience that unless the blocks were keyed together the breakwater had no chance of survival. We do not know for how long a period this port was functional and under what circumstances it vanished, leaving only a few traces behind.

That studies in coastal engineering are of recent origin is true if "studies" refers only to *instrumented studies* as we define modern instruments. Depths, currents, and waves were observed in the early times, although analyses of data were not undertaken in the sense that they are today. If any particular country was the cradle of coastal engineering, it must have been the Netherlands, and studies there were all related to coastal protection and reclamation (Bruun, 1973).

Netherlands

Although a great many training and irrigation wells, dams, or dykes were built in the Far and Middle East, coastal protection per se probably first developed in the low countries of Europe where rivers poured soft materials, mainly clay and silt, out into the ocean. These soft materials settled by means of a slow consolidation, while at the same time the sea level was rising. To avoid loss of land by flooding and to protect themselves from drowning, the Frisians and the Dutch first built earth dykes, starting about the year 1000. By the thirteenth century, the Dutch had accomplished major coastal protection and reclamation works, particularly in the Dordrecht area. Dr. van Veen (1962, pp. 10, 11, 14, 15, 16, 17, 23) writes:

The earliest written records about the Frisians (or Coastal Dutch) describe them as water-men and mud-workers. The Romans found in the North of the country the artificial hillocks upon which the inhabitants, already called "Frisii", made a living. We shall follow their history, because written records are available about the early reclamation works they made. One and the same race, now called the Dutch, took, held and made the low country.

Pliny, who saw these mound-dwelling tribes in the year 47 A.D. described them as a poor people. He apparently exaggerated when he wrote that they had not cattle at all. Or did he see some much-exposed mounds near the outer shores where the sea had swallowed every bit of marshland? At stormtide, Pliny said, the Frisians resembled groups of miserable shipwrecked sailors, marooned on the top of their self-made mounds in the midst of a waste of water. It was impossible to say whether the country belonged to the land or to the sea. "They try to warm their frozen bowels by burning mud, dug with their hands out of the earth and dried to some extent in the wind more than in the sun, which one hardly ever sees".

No doubt the mud Pliny refers to was the peat which was found in the "wolds", or swamps, some distance south of the clay marshes, where the artificial mounds had been made.

In all they built 1260 of these mounds in the northeastern part of the Netherlands, an area of a mere 60 × 12 miles. Further East there are more of them in East Friesland. The areas of the mounds themselves vary from 5 to 40 acres; they rise sometimes to a height of 30 feet above normal sea level. The contents of a single mound may be up to a million cubic yards.

They built their mounds on the shores of the creeks in which the tide ebbed and flowed. In their scows they went [in their language in which the roots of so many English words can be found]: "uth mitha ebbe, up mitha flood"—out with the ebb, up with the flood. The tide bore them towards the peat regions, or perhaps to the woods still farther inland and then brought them back. Or they went out with the ebb in the morning towards the sea, where they gathered their food, and returned in the evening with the incoming tide.

The Coastal Dutch have now lived 24 centuries in their marshes and of these the first 20 or 21 were spent in peril. It was not until 1600 or 1700 that some reasonable security from flooding was achieved. During these long treacherous centuries the artificial mounds made their survival possible.

It was a work which might be compared with the building of the pyramids. The pyramid of Cheops has a content of 3,500,000 cubic yards, that of Chephren 3,000,000 and that of Mycenium 400,000 cubic yards. The amount of clay carried into the mounds of the northeastern part of the Netherlands can be estimated at 100,000,000 cubic yards.

In Egypt it was a great and very powerful nation which built the pyramids throughout a series of dynasties. The aim was to glorify the Pharaohs. With us it was a struggling people, very small in number and often decimated, patiently lifting their race above the dangers of the sea, creating large monuments, not in stone, but in native clay.

In [the] Lex Frisionum of 802 there is not yet any mention of seawalls, but the first attempts at dike building must have been made shortly afterwards. Frisian manuscripts still extant, dating from the early Middle Ages, deal chiefly with the following three points: First, the right of the people to freedom, all of them, "the bern and the unbern". Secondly, the "wild Norsemen" whose invasions took place roughly from 800 to 1000, and thirdly: the Zeeburgh or Seawall.

This novel means of defence against the sea by means of a continuous clay wall was called a Burgh, or stronghold. The people were apparently very proud of this seaburgh, because they described in in poetical language as "the Golden Hop", the Golden Hoop. "This is also the Right of the Land to make and maintain a Golden Hoop that lies all around our country where the salt sea swells both by day and by night".

The spade, the hand barrow, and the fork were the instruments used for diking, the fork presumably for the grass turfs which were used to heighten the dykes and make them stronger.

Despite the tremendous efforts the sea was the strongest. This was due partly to our insufficient technical skill and partly to lack of cooperation. For a single night, Dec. 14th 1287, the officials and priests estimated that 50,000 people had been drowned in the coastal district between Stavoren and the Ems. This is a large number considering that this was the area where so many dwelling mounds could be used as places of refuge.

The advances and successes have been tied to a few names. Says van Veen (1962, pp. 68, 69):

We often wondered who was the master engineer who created the marvellous Great Holland Poulder, south of Dordrecht, the work which had included the damming off of the tidal mouth of the river Maas, and the leading of that river into the Rhine. This proved to be William I. He had already finished that gigantic undertaking by 1213. The polder was destroyed in 1421 by the St. Elisabeth's flood. . . . William was a man of great conceptions. He surrounded the entire area of Holland-Proper with strong dikes and made several canals intended to drain the vast moors. They also served as a splendid network of shipping canals. It is likely that he made the dikes around the Zeeland islands Walcheren and Schouwen too, and that he established the still-existing administrations for the upkeep of these islands. The other part of his clever and amazing reclamation and construction programme cannot be described here, but it is very clear that he knew the geography of his country by heart. No maps as yet existed!

The earliest reference to the art of accelerating the natural rate of accretion is the manuscript *Tractaet van Dijckagie* (*Treatise on Dikebuilding*), written by the Dutch dykemaster Andries Vierlingh, between 1576 and 1579. Vierlingh discusses the construction of *cross-dams* on mud-flats that are not yet dry at low water. In this connection he also advised that old ships be sunk and earth dumped on top of them to make artificial islands or flats to hold back the silt and sand suspended in the water. These islands were subsequently to be connected with low dams. Although this method was never commonly used, it is known that shipwrecks had been used in numerous places to close dyke breaches. These wrecks formed the basis for the fill material that was secured with mats or brushwood. Vierlingh, however, was much against closing of dyke breaches with shipwrecks because of the nonhomogeneity they created in the dyke structure. Nevertheless, this method was widely used not only in Holland but in the Schleswig-Holstein (at that time Danish) over a long period of time.

Of Vierlingh, van Veen writes (1962, p. 33):

[He] was found to be a real master of the dikes and waters, a man of great ability and spirit—one of the greatest of his kind. Luckily the greater part of his manuscript has survived. Its ancient picturesque style is a joy to every hydraulic engineer. This remarkable book already shows the special vocabulary of the Dutch diking people in all its present-day richness. In some ways it is even richer.

His advice is simple and sound. The leading thought is: *Water will not be compelled by any "fortse" (force), or it will return that fortse onto you.*

This is the principle of streamlines. Sudden changes in curves or cross-sections must be avoided. It is the law of action and reaction. And truly, this fundamental law of hydraulics must be thoroughly absorbed by any one who wants to be a master of tidal rivers.

The work by the dykemasters and farmers to protect and to gain land has been remarkable. Dykes were built up gradually by adding a layer of silt, or silt and sand, shell, willow mattresses, and the like on top of each other. Remains of old ships, brick walls, and pile walls were also used. No less than two-thirds of the lower part of the Netherlands is man-made, while the other third is just "natural" sea marsh or moorish swamp. Since about the year 1200 the following areas have been gained according to van Veen (1962, p. 55):

On the sea shores	940,000 acres
By pumping lakes dry	345,000 acres
By pumping the Zuiderzee dry	550,000 acres
	(partly future)
In all	1,835,000 acres

With respect to the distribution, the 100,000,000 cubic yards of earth that the Dutch carried to their artificial hills were made only in a small area, covering roughly eight percent of the country. Van Veen (1962, p. 61) writes:

The sea walls or dikes were our second work. In 1860, that is just before the advent of steam dredging, we had about 1750 miles of them, containing about 200,000,000 cubic yards of material. Moreover, there were many old deserted dikes, whose contents may be estimated at 50,000,000 cubic yards. Those 250,000,000 cubic yards were practically all transported by handbarrows, wheel-barrows and horse-drawn carts.

The third great work was the digging of the ditches and canals. In the lower half of the country about 800,000,000 cubic yards of earth have been removed, in order to drain the land and separate the fields. Of shipping canals there are about 4800 miles in Holland, for which a figure of 200,000,000 cubic yards would be a fair estimate.

The fourth and greatest task was the digging of peat. This digging served a double purpose: the provision of fuel and creation of lakes which, when drained, gave more fertile land than the original moors themselves.

In total we have dug according to this rough estimate the enormous volume of some 10,000,000,000 cubic yards. This includes the making of lakes as well as the digging of moors in the higher eastern regions of the Netherlands.

Compare this figure with the dredging of the Suez Canal. We constructed about 100 Suez Canals of the size made by De Lesseps. All this *was done by hand,* whereas De Lesseps used 60 steam dredges.

But the work would never have been completed without the dykemasters, their foremen and *polderboys,* who often were the farmers themselves.

Weed dykes were a special kind of dyke-building. Construction was limited to West Friesland and the Zuiderzee area, where seaweed or seagrass was found in ample quantities along the coast. The West Frisian sea dykes for a long time were reinforced with seaweed, and so were some of the Wieringen dykes. It is not known with certainty how old the weed dykes are, but it is known that they were constructed as early as the eighth century. A weed dyke was built at the northernmost point of the Island of Schokland in the sixteenth or seventeenth century, and another one was built in 1734 in the northern part of Noord Holland. Seagrass was collected offshore in the Zuiderzee and the Wadden areas. Following drying, a broad, tough layer was placed on the sea side of the dike.

As man's ambition grew, dykes also grew. Moving them still closer to the dangers, it became necessary to reinforce the dykes by hard surfaces like basalt blocks and/or other structures parallel as well as perpendicular to shore. These reinforcing or supporting structures developed as experience and exposure increased. The gradual reinforcement by structures such as seawalls and groynes may have contributed to a not fully justified sense of security. It seems that dykes were not raised rapidly enough in step with the sinking of the land and the rise of sea level, and that they were not subjected to a thorough enough investigation of their structural soundness.

On February 3-4, 1953, a spring tide whipped up by a raging gale overwhelmed the sea defences and made tremendous breaches in the dykes. Most of the islands in the southwest were inundated; 1850 people lost their lives. Available material and manpower were mobilized and within a year all the gaps in the dykes were closed and the flooded areas reclaimed. On November 5, 1957, the *Delta Bill* was enacted which contains plans for closing the tidal entrances in the southwest. When this project is completed the Dutch coast will have been shortened by 700 kilometers. The Delta project provided for the closure by massive dams of four broad, deep sea inlets; namely, the Haringvliet in 1968, the Veerse Gat in 1961, the Brouwershavense Gat in 1972, and the Eastern Scheldt in 1978. It also provided for the building of secondary dams in the Zandkreek, the Grevelingen, and the Volkerak. The Rotterdam Waterway and the Western Scheldt will be left open since they provide access to the ports of Rotterdam and Antwerp, respectively. This sequence was chosen after due consideration because the transition from small to large sea arms enables experience gained to be profitably used in the larger projects. Another reason for the sequence was the desire to achieve the highest degree of safety for the largest possible area in the shortest possible time. This—the world's largest coastal protection project—is thoroughly described in a number of publications, including the Dutch periodical *Deltawerken.* The Veerse Gat and the Haringvliet were closed according to schedule; two sections of the Brouwershavense Gat were closed in the spring of 1971 and the work was to be completed by 1972. The southern gap was closed by *telpher* (concrete blocks dumped from cable cars), the northern one by means of 14 caissons. The closing of the last gap meant that tidal currents involving the movement of 360 million cubic meters of water into and out of the inlet (each movement taking about 6 hours) ceased to flow. There remained the dam that would close off the Eastern Scheldt. This would be about 9 kilometers long and would stop tidal currents involving the movement of 1,100 million cubic meters of water into and out of the inlet every six hours. The construction of three artificial islands was needed to build the dam: the first was completed in 1969, the second in 1970, and the third in 1971. This dam, the last and largest to be constructed (it fills up channels as deep as 35 meters), was completed in 1978.

The construction of the large sluices presented enormous problems. Protection of the bottom was obtained by placement of large *Zinkstükken* (willow mattresses) as was done 1000 years ago. Although many tools and construction practices have changed, willow mattresses are still used although in some cases they may have been replaced by mattresses of asphalt. The cost of the Delta project was estimated to be 3,500 million guilders ($1.1 billion). It is an expensive project, but it will ensure greater safety for the entire southwest of the Netherlands, it will reduce the cost of dyke maintenance due to the shortening of coastline by nearly 700 km, it will open up a whole series of islands, it will reduce silting, it will offer fast traffic links across the dams, and it will improve control of the supply of fresh water in almost all of Holland. In addition, it will provide new recreational possibilities for the vast population in the southwest urban areas with unique aquatic sports areas.

Development of Dutch Groins. The first Dutch groins were probably built at the beginning of the sixteenth century, but groinlike structures may have been built much earlier. Their exact appearance is unknown, but the history of their development during the last 100 to 150 years is known and represents a continuous line of development of a streamlined structure exposing itself as little as possible to the force of currents and waves.

Although groins have grown in size, the principles of construction are the same: stone pitching on gravel on mattress in the middle and stones on mattresses on the sides with two or more pile walls as supports (Bruun, 1973). The length is usually approximately 200 meters with the space between them of the same order as is described in more detail in a later section. Offshore elevations are at about mean sea level. Occasionally groins are provided with piggybacks to break the longshore currents. Analyses by Bakker and Joustra (1970) have demonstrated that the Dutch groins have not only decreased or stopped erosion in certain areas but have even caused accretion. This is probably due to the fact that (tidal) currents combined with swell action provided the shore with material from offshore so that the groins did not suffer starvation as is normally the case.

The groins, however, are not the cornerstone of Holland's protection; the dykes are. But foreigners seeing the results of the Dutch groins, sometimes misinterpreted the situation. The massive Danish North Sea groins, which gradually increased in length to several hundred meters at the Thyborøn Barriers due to continued shore recession, can be attributed to such a misinterpretation. Enormous quantities of material were sacrificed because of insufficient understanding of the mechanism involved. Such misinterpretation also found its way to the New World, with the groins at Miami Beach being one of the most startling examples of their inadequacy as coastal protection. On the other hand, the Long Island Atlantic shore groins live up to the Dutch example.

The Dutch also carried out many dyke and drainage projects in France, Germany, Poland, and Russia. Along the Molotschna, there were 46 Dutch villages in 1836; the district Chortitza had at that time 20 such villages. In Poland, there were about 2,000 villages inhabited by the descendants of the Dutch immigrants; in Posen there were 830 villages. The first great canal in the United States, the Erie Canal, was financed in 1772 by the Dutch and its locks were devised by Dutch engineers.

England

Coastal protection also has a long history in England because of continuous erosion of strategic areas on the south coast, in Lincolnshire, in South Yorkshire, and in many estuaries. There is clear evidence of reclamation works by construction of *walls* (dykes) in the Dungeness area during the Roman occupation, the Rhee wall being the best known example. Historical evidence gives a consistent picture of the incursion of the sea along the Lincolnshire coast, by references to loss of land and damage to *sea banks,* which had been a necessary defense since the thirteenth century. In 1335, according to records, the waves breached the sea banks at Mablethorpe and the land was flooded. By 1430 the seawall again needed repair. Erosion has continued and the history of this area has been one tough fight against the sea.

As in Holland, the first measures against erosion were sea banks, the design being modified to serve as seawalls according to the local situation. Some were just earth dams, others were fascine or pilewalls (Matthews, 1934). Later, vertical bulkheads were developed. On the English shingle beaches, abrasion presented a severe problem and called for the application of flint, basalt, or other suitable materials (backed by concrete) to resist abrasion. The block walls at Pett Level (Dungeness) and Walland are examples of modern sloping walls, providing flexibility rather than rigidity and low reflection of wave energy thereby being more considerate to the beach in front than vertical or slightly curved structures (Thorn, 1960).

Groins were used as an additional protective measure. They were put into use in early times, probably as a result of observations of the effect of headlands protruding from the shore. This was likely to have been the case at Hornsea, South Yorkshire, where, during an inquisition held in 1609 concerning heavy losses by erosion, it was stated that "there was a peere at Hornsea Beach, during the continuance whereof the decay was very little" (Pickwell, 1877–1878). In 1864 six groins were built on the heavily eroding Spurn Head, South Yorkshire, at the entrance to the River Humber, where nature's forces were assisted by man's removal of shingle from the beach. The groins were of the King Pile type with horizontal boards that could be adjusted, similar to the Withernsea Groynes erected in the 1870s. They were strutted at the downdrift side to resist the pressure of the accumulating beach on the updrift side. Sheet pile groins were also tested, but the result was less satisfactory; they were too rigid and lacked any means of adjustment.

Some enthusiasm seems to have been generated by the groin construction works of limited length along the shore, but observations were also made about ill effects in the form of downdrift erosion. In a discussion of an article by J. Murray on "Sunderland Docks," published in the *Proceedings of the Institute of Civil Engineers,* Rennie and Walker (1849), referring to a coastal protection report of 1832, admit

that groins were, under certain circumstances the best defense for a coast, for wherever the waves brought the sand and shingle in quantities, the seaward side filled up while on the lee side it was generally scooped out, but by a judicious distribution of these groins, such an accumulation of material might be produced, as would effectually protect a shore, or any sea works.

Inexpensive types were devised. Murray (1847) discusses the design of groins and says that they

might be formed with stones, timber, or fascines, either of the two first-named materials lasted well, but in cases where the deposit was rapid, and of such nature as to entirely fill up interstices, and prevent decay, the latter material would be sufficiently durable for all ordinary purposes.

The entire situation with respect to sea protection works was reviewed by a Royal Commission on Coast Erosion *etc.* (1911). One of the most significant statements in this report is that sea walls, unless properly constructed are "agents of their own destruction." In particular, the report refers to scour at the toe and the necessity of constructing a special toe, apron, or groin protection in front of the sea wall to prevent undercutting.

The commission fully realized the advantages and disadvantages of groins.

The evidence laid before us goes to show that in many cases on the coast of the United Kingdom groins have been constructed of a greater height than was necessary to fulfil the required conditions, with the result that they have so unduly interfered with the travel of the shingle as to lead to impoverishment of the beach to leeward, causing in many districts serious injury to the coast.

The commission also discussed the length of groins and the distance between them; 1 to 1 ratios are common but "satisfactory results were also obtained by 1 to 2 ratios." Alignment at right angles to the shore was found to work best and provision for adjustment by adding or removing planks was found to be preferable as low groins often proved to more more efficient than high ones, being less adverse to downdrift beaches at the same time.

It is regrettable that the wisdom contained in this 70-year-old British document was realized so late elsewhere and that designs as contradictory as possible to the century-old British experience were advocated for a long period of time and to some extent still are being promoted. The difference between British and Dutch practices in groin design is related to the grain size of the material that groins are expected to accumulate. A good many English beaches are of shingle, and some of them experience high rates of beach drift and significant fluctuations in beach profiles. A high (but adjustable) groin may therefore be practical. Energy loss along the stem is of less importance due to the coarseness of the material. Conversely, all beaches in Holland consist of fine to medium sand that moves easily and fluctuations of beach profiles are of relatively small magnitude. Smooth streamlined cross-sectional geometry causing little turbulence is therefore best for such conditions and groins should be low to conform with relatively gentle sand slopes. Groins having high vertical walls would result in scour and lowering of he beach on either side of the groin.

Denmark

In Denmark coastal protection started on the North Sea coast in 1840 with a government project to increase the height of dunes on the Lime Fiord Barriers (Bruun, 1954). In the 1870s experimental groins were built on the west coast using a Dutch design that soon proved too weak to withstand the violent wave action on that shore. Reinforcements were added to the design and over the next 50 to 60 years almost 100 massive groins, ranging in length from 100 to 400 meters, were built in this general area of approximately 50 km in length. These groins were built of concrete blocks that ranged from 4 to 8 tons and often had granite side slopes of 2 to 8 tons. These blocks are put in place with specially designed cranes. Erosion continued outside the extreme end of the groins and the outer parts were not maintained. The land ends were extended gradually as the dunes and dikes were withdrawn (Bruun, 1954). Artificial nourishment from bay or offshore sources has, surprisingly enough, not been applied yet but is urgently needed, particularly on the Lime Fiord Barriers.

North America

Before 1930, federal interest in shore problems was limited to protection of federal property and improvements for navigation. An advisory board on sand movement and beach erosion, appointed by the Chief of Engineers, was the principal instrument of the federal government in this field. In 1930, Congress assumed a broader role in shore protection by authorizing creation of the Beach Erosion Board. Four of the seven members of the board were Corps of Engineers officers and the other three were from state agencies. The board was empowered to make studies of beach erosion problems at the request of, and in cooperation with, cities, counties, or states. The federal government funded up to half of the cost of each study but did not bear any of the construction costs unless federally owned property was involved.

This important first step was followed by a series of improvements, in 1945, 1955, 1962, 1965, and 1968; demonstrating an ever-increasing interest and involvement in the matter by federal authorities. Several states created their own beach erosion and shore development agencies that established cooperation with the local U.S. Army Corps of Engineers district and division offices and with the office of the Chief of Engineers. A great number of studies of actual beach erosion problems, led to reports by the Secretary of the Army to Congress, which authorized federal contributions to actual improvements. These efforts were supported by research projects of the Beach Erosion Board and, from 1963, by the Coastal Engineering Research Center (CERC). A number of special projects were handled by model tests at the Waterways Experiment Station of the USCE and CERC.

Structurally speaking, coastal protection suffered shortcomings compared to the low countries in Europe. Patented, more or less useless, coastal protection devices, for example, permeable groins, have been better accepted by U.S. business than elsewhere; but the newest and most effective measure—artificial nourishment—although not discovered in the United States, was developed there and so far has been most successful.

The differences between the European low country and the U.S. coastal protection practices lie in scale and in degree of involvement. In Europe coastal protection is high, short, and often complex. In America it is long, relatively low (excluding hurricane protections), and relatively simple. In Europe coastal protection is tough and silent, but in America it is flexible and noisy because it not only provides protection but also acts as a *nourishing machine*.

A summary of combined experience, gained through years of struggling, may be expressed briefly as:

1. Whatever is done, avoid waves and currents turning their full force onto the protection structure (Vierlingh, c. 1570).
2. Don't be nearsighted, think large if possible. It is better to solve problems of kilometers or miles than only of meters or feet.
3. Look oceanward, landward, and up and down the shore and evaluate carefully how the coastal protection structure may be influenced by or may influence the surrounding areas of land and water.
4. Coastal protection need not be just coastal defense. Old Dutch experience and military tradition seem to favor defense by attack.

An American version of this experience may be expressed as "the best protection for real estate is plenty of real estate in front of the real estate you want to protect."

PER BRUUN

Note: This entry has been largely extracted from Bruun, 1973.

References

Bakker, W. T., and Joustra, D. Sj., 1970. The history of the Dutch coast in the last century, *Conf. Coastal Eng., 12th, Am. Soc. Civil Engineers, Washington, D.C., Proc.*

Bruun, P. M., 1954. *Coastal Stability*. Copenhagen: Danish Society of Civil Engineers Press, 7 pamphlets.

Bruun, P. M., 1973. The history and philosophy of coastal protection, *Conf. Coastal Eng., 13th, Am. Soc. Civil Engineers, Vancouver, Proc.*, 33-74.

Coastal Engineering Research Center, 1970. *Shore Protection Program*. Washington, D.C.: U.S. Army Corps of Engineers, 16p.

Coastal Engineering Research Center, 1971. *Shore Protection Guidelines*. Washington, D.C.: U.S. Army Corps of Engineers, 59p.

Matthews, E. R., 1934. *Coast Erosion and Protection*. London: Charles Griffin and Co., 228p.

Murray, J., 1847. Design of groins, *Inst. Civil Engineers Proc.*

Murray, J., 1849. Sunderland Docks, *Inst. Civil Engineers Proc.*

Pickwell, R., 1877-1878. The encroachment of the sea from Spurn Point to Flamboro Head, and the works executed to prevent the loss of land, *Inst. Civil Engineers Proc.*

Royal Commission on Coast Erosion, 1911. *Report on Coast Erosion.*

Thorn, R. B., 1960. *The Design of Sea Defense Works.* London: Butterworths Scientific Publications, 106p.

van Veen, J., 1962. *Dredge, Drain, Reclaim, The Art of a Nation.* The Hague: Martinus Nijhoff, 200p.

Cross-references: *Beach Processes; Coastal Engineering, Research Methods; Coastal Morphology, History of; Dike, Dyke; Europe, Coastal Morphology; Groins; North America, Coastal Morphology; Polder.* Vol. XIII: *Coastal Engineering.*

COASTAL ENGINEERING, RESEARCH METHODS

Coastal engineering research provides the engineer with data on many geological and oceanographic parameters. Those parameters of major interest vary, depending on the type of project involved, but usually include the following: local surface and subsurface geology, the geotechnical properties of the geological layers present, wave height and direction, tidal range, current velocities, storm magnitude and frequency, and salinity and temperature ranges of the water.

Beach Erosion Control

Beach replenishment is the process of placing sand on the beach to replace what has been eroded away. The prime consideration is the grain-size distribution of the eroding beach sand versus the grain size-distribution and volume of sand available for placement on the beach. Grain-size analysis by sieving of a large number of surficial grab samples, taken by scooping into the upper few inches of the beach sand, will tell the distribution of sand grain sizes across the eroding beach. Sand that is coarser and better sorted than the old beach sand is best for replenishment purposes. The U.S. Army Coastal Engineering Research Center (1973) *Shore Protection Manual* provides details on beach replenishment and on all the research methods involved in this and most other types of coastal engineering.

The horizontal and vertical extent of potential areas from which replenishment sand may be taken must be known. Continuous reflection profiling, using high-resolution seismic systems, shows the extent of any sediment layering present in the subbottom of the potential *borrow* area. Vibratory coring techniques with free-standing pneumatic vibratory hammers are used to take continuous cores of most unconsolidated sediments. Using this method, samples have been taken and recovered from over 12 m below the ocean bottom. The cores are compared with the seismic data to determine the nature of the sediment in each layer, and the extent of the layers containing the best replenishment sand is thus outlined.

The placement of sand on the beach is governed by the direction and rate of littoral drift, which is the surf-zone current created by waves breaking at an angle to the shoreline. The velocity of this drift is measured by one of several methods. *Drogues,* each consisting of a small reference float with a set of neutrally buoyant vanes hung below to cause the unit to move with the water, are placed in the surf zone, where they are carried along by the drift. The location of the drogue float is mapped every few minutes, and the velocity of the current is thus mapped. Dye patches from small, discrete batches of dye placed in the water are monitored in the same manner.

Another method involves taking samples of sand from the beach and "tagging" the sand for *tracing.* The sand is dried, stained with dye, and placed back in the surf zone at a central location. The sand movement is monitored by sampling the sand in the surf zone at fixed distances from the place where the tagged sand was released. This gives the direction and rate of movement of the various sand sizes. Colored glass beads can be used instead of tagged sand, but these are usually of only one size.

Repetitive measurements of the shape of the beach—i.e., its cross-sectional *profile*—at fixed locations over time, tell the rate of erosion or deposition of sand at various locations along the beach. These data are used, along with the traced sand movements, to give a more complete understanding of sand movement along the beach.

Once the average direction and rate of sand movement have been determined, the best beach erosion prevention method must be determined on a case-by-case basis. Besides beach replenishment, other methods include groins, sea walls, breakwaters, among others, all of which are designed to reduce erosion by slowing the movement of sand away from an area. The method used depends on the wave energy and tidal range of the area under study. Wave energy is a function of the shape of the nearshore ocean bottom, the local wind velocity, the frequency and magnitude of storms in the vicinity, and the frequency of large swells from distant storms.

Hindcasting, the mathematical development of probable wave heights and wave periods from known wind patterns, is used to determine historical wave patterns for a given stretch of shoreline. Much of the wave data are gathered in this way from historical weather records, supplemented by monitoring present-day weather patterns. Present-day wave height, periods, and direction of travel are measured directly by groups of sensors placed near the shoreline, either on moored buoys or on fixed structures such as piers.

Designing Large Coastal Structures

The major problem with new coastal structures is that they tend to interfere in some way with the preexisting environment. The structure must be designed to create minimum interference, and, if possible, to correct any interference that does occur where it is unavoidable. Many of the parameters needed for this type of engineering are the same as those used for beach erosion control, but the emphasis is different. Wave activity and sand motion must be known to ensure that the structure will not be undermined by rapid erosion or, conversely, buried by rapid deposition of sand. The methods for determining beach sand movement are widely used in the latter type of problem.

The nature of the local geology is important for the foundation design of the proposed structure. The structural foundation data are obtained by the coring and high-resolution seismic methods as described above. The coring information must be supplemented by more sophisticated tests of the geotechnical parameters of the geological layers present. These include in-place *vane shear tests* and *cone penetrometer tests,* done by instruments either free-standing on the ocean bottom, or introduced down through the pipe of a small drilling platform placed on site for these tests. The drilling methods conducted from small platforms are more sophisticated than the vibratory hammers used for continuous coring and allow deeper penetration and rock sampling.

The emphasis on high-resolution seismic profiling is to determine the presence of any subbottom structures such as faults or sharp erosional valleys in the bedrock that might be hazardous to the proposed structure. Depth contours of the ocean bottom are also determined by the profiling study.

Large-diameter intake and outfall pipes are required for *nuclear power plants* and *sewage treatment plants* being built today. In these cases, *water quality testing* becomes necessary in addition to the foundation parameters needed for construction of the pipe. In most areas temperature of the discharge, measured as an increase above the ambient temperature, is regulated by law. Therefore long-term temperature monitoring of both discharge and surrounding water is necessary for power plant discharge pipe projects. This monitoring is done by taking a series of readings at known depths and fixed locations for at least a year before the pipe is put into operation. These measurements can be done from a boat or with a series of sensors mounted on buoys. The data are relayed either by radio or cable from the mooring buoy to an onshore recording station. This monitoring of ambient and outfall temperatures is continued after the pipe is in use as a check on compliance with the law.

Sewage effluent, usually a secondary treatment product, has the potential of polluting beaches. Modern outfall pipes are being made longer, up to 4 km from the shore, with their discharge ends often in water over 15 m deep. Water circulation characteristics near the end of the proposed pipe must be known, to determine where the effluent will go upon discharge. If there are strong currents directly toward the beach, then the proposed pipe location may have to be changed.

The direction of drift is determined in many ways. Both drogues and dye patches can be used, as described for determining littoral drift velocities. In the deeper water at the ends of outfalls, the tides are more important than wave direction in creating currents. Therefore measurements must be continued over at least one tidal cycle. When drogues are used, several, set at different depths, are released simultaneously at one place. Their movements relative to each other give an indication of the dispersion of the effluent water.

An alternative method is to mount a series of recording *current meters* at various depths in the area of interest. These meters record current direction and velocity. The data are stored on either tape or film in the meter, read out through a cable to a small boat, or relayed by cable to an onshore recorder. These meters are often left in place for more than one year.

Effluent water will rise from the discharge point until the density of the effluent equals that of the ambient water. Density is a function of temperature and salinity. Therefore the salinity and temperature of the ambient water must be known for the whole year, since they will probably vary more over that period than the effluent. By knowing the density of the ambient water and making assumptions as to the density of the effluent, assumptions can also be made as to the dispersion of the effluent. Salinity and temperature measurements, taken either from boats or from moored instrument arrays, are therefore extremely important in coastal engineering design of large effluent pipes.

Dispersion can be estimated by conducting dye studies. The dye is mixed with oceanwater of a known temperature similar to that at the release depth, and the mix is released continuously at a fixed known rate and concentration. The area is then sampled in a grid pattern at several depths, and the concentration of the dye present in the water is determined by a *fluorometer*. The patterns of dispersion are mapped over at least one tidal cycle. The process should be repeated several times during the year to determine seasonal variations in the dispersion pattern.

Conclusion

Certain types of information, such as the effects of extreme storms, cannot be obtained in the field. Based on the normal field data *models*—either *physical scale models* or-*mathematical computer* models—are constructed. The effects of changing wave height or tide height can then be predicted from the model. For example, changes in the shoreline due to extreme wave activity can be predicted.

All of the above research methods are concerned with the research needed to design structures that counteract the effects of the environment. Recently increased ecological awareness has made it necessary for the engineer to consider the effects of the structure on the environment (D'Appolonia, 1971; Ling, 1972; Task Committee, 1974). The *environmental impact statement* that is required for any large construction project emphasizes, among other things, the effect of the project on the biological community in the area. Such considerations include the human, fish, wildlife, and flora populations of the proposed site.

CHARLES DILL

References

Coastal Engineering Research Center, 1973. *Shore Protection Manual.* Washington, D.C.: U.S. Army Corps of Engineers, 3 vols., 1160p.

D'Appolonia, E., ed., 1971. *Underwater Soil Sampling, Testing, and Construction Control.* Philadelphia, Pa.: American Society for Testing and Materials, Spec. Tech. Pub. 501, 241p.

Ling, S. C., 1972. *State of the Art of Marine Soil Mechanics and Foundation Engineering.* Vicksburg, Va.: U.S. Army Engineer Waterways Experiment Station Tech. Rept. S–72–11, 153p.

Task Committee, 1974. Ocean engineering–Selected references, *Am. Soc. Civil Engineers Proc., Jour. Waterways and Harbors Div.* 100, 241–248.

Cross-references: *Beach Nourishment; Coastal Engineering; Coastal Morphology, Research Methods; Corers and Coring Techniques; Nuclear Power Plant Siting; Pollutants; Scale Models; Sediment Analysis, Methods; Sediment Tracers; Thermal Pollution; Waste Disposal; Wave Refraction Diagrams.* Vol. XIII: *Coastal Engineering.*

COASTAL ENVIRONMENTAL IMPACT STATEMENTS

United States federal law and many state laws require environmental impact studies for any federal- or state-funded projects that will modify or interact with the natural resources of the land or sea. The basis for the federal environmental impact studies is found in the federal National Environmental Policy Act (NEPA) of 1969. Because there is often confusion as to what are the limits of this law, the following outline of policy is excerpted directly from Title I, Section 101 of this law (National Environmental Policy Act, 1969):

a. Congress, recognizing the impact of man's activity on the natural environment and the importance of restoring and maintaining environmental quality, established the policy in a manner calculated to promote the general welfare, to create and maintain conditions under which man and nature exist in harmony and to fulfill social, economic requirements of present and future generations.
b. To carry out this policy, the Federal Government, consistent with national policy will:
  1. be responsible for succeeding generations;
  2. assure safe, healthful, productive, esthetically and culturally pleasing surroundings;
  3. attain the widest range of beneficial uses of the environment without degradation, health or safety risk, or other undesirable consequences;
  4. preserve important historic, cultural and natural aspects of our heritage and wherever possible an environment of diversity and variety of individual choice;
  5. achieve balance between population and resource use;
  6. enhance renewable resources and recycle depletable resources;
  7. recognize that each person should enjoy a healthful environment and each person is responsible to preserve and enhance the environment.

As can be seen from the above, this policy does not call for a preservation of the present environment, but provides for a "balance" of activities "consistent with a national policy" of man's necessary activities and his environment. The procedures for the enactment of this policy to be followed by all federal agencies, also set forth in the same act, are as follows:

a. Utilize an interdisciplinary approach in the natural and social sciences and "environmental design arts" in planning and decision making;
b. That unquantified environmental amenities be considered in decisions along with economic and technical considerations;
c. Include in major Federal actions affecting the environment, a statement on the environmental impact of proposed action.

In 1975 the Environmental Protection Agency revised its basic act (U.S. Environmental Protection Agency, 1975) to include the need for environmental impact statements on all federal projects in coastal zones and coastal waters as well as wild and scenic rivers, prime agricultural land, and wildlife habitat. The Coastal Management Act of 1972 (Public Law 92-583) was cited as indicating the need to add coverage of these "environmentally sensitive" areas to the environmental protection law.

Terminology and Procedures

The following definitions relate primarily to the various aspects of the federal environmental policy acts and the procedures required for review by federal agencies (see Fig. 1 for flow sheet). While some terms may be used synonymously in general conversation, federal regulations require the exact terminology of the enabling act. Often accepted abbreviations are used in government reports (e.g., EIS for Environmental Impact Statement) rather than the entire term.

Environmental Impact. This term implies the effect of man's activity (planned and unplanned, although usually thought of only as unplanned, such as coastal oil spill) on natural environment, both organic (flora, fauna) and inorganic (water resources, sediment erosion), usually only when of a negative nature. The original Federal National Environment Policy Act refers to impact on "human environment" although all impacts could be shown to be related to man. Unless stated as human environment impact, *environmental impact* perhaps should refer to inorganic aspects and *ecological impact* to organic aspects.

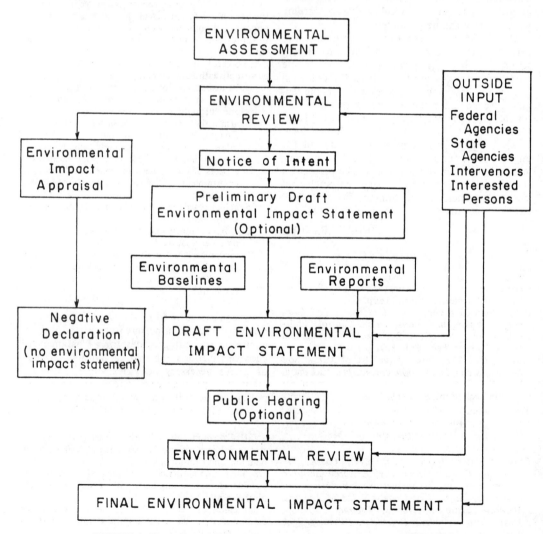

FIGURE 1. Flow chart of Federal Environmental Impact Statement Procedure.

Environmental Impact Statement (EIS). This report identifies and analyzes in detail: objectives of the proposed action; proposed action; environmental impacts both beneficial and detrimental; feasible alternatives; short- and long-term productivity; commitment of natural resources; and costs of all these actions, including alternatives. This report is required for all projects, legislation, funding (complete or partial), and other major federal actions by the National Environmental Policy Act of 1969. Such a report shall consist of an integrated, interdisciplinary, natural- and social-science approach.

Environmental (Impact) Assessment (EA or EIA). This can be a written analysis submitted to the U.S. Environmental Protection Agency by its grantees or contractors describing the environmental impact of proposed action and alternatives as input to determine the need for an environmental impact statement. It is also a general term for scientific analysis of effect of human activities on the environment as a result of development of ecological awareness and environmental problems of the 1970s.

Environmental Review. This is a formal evaluation by federal agencies, including input of public meetings and written comments at the interagency, federal, state, and general public levels to determine whether a proposed action may have a significant impact on the environment. This review, together with the environmental assessment, is used to: determine if significant impacts are anticipated; whether changes can be made to eliminate significant adverse impacts; and decide if an environmental impact statement is required.

Notice of Intent. When the environmental review indicates that a significant environmental impact may occur and that these impacts cannot be eliminated by making changes in the project, a notice of intent will announce to federal, state, local agencies, interested persons, and intervenors that a draft environmental impact statement will be prepared.

Negative Declaration. When the environmental review indicates that no significant adverse impacts are anticipated or when the project is changed to eliminate these impacts, a written announcement will state that the federal agency will not prepare an environmental impact statement but will prepare an environmental impact appraisal.

Environmental Impact Appraisal. This report, based on an environmental review supporting a negative declaration, describes a proposed action, its expected environmental impacts, alternatives, and the basis for the conclusion that no significant impact is anticipated.

Interested Persons. These are individuals, citizen groups, conservation organizations, corporation, or other nongovernment units who are interested in, affected by, or technically competent to comment on the environmental impacts of the proposed action.

Interveners. These individuals, corporations, state or federal agencies request to testify or submit comments on the basis that they are directly affected or impacted by the specific proposed action.

Preliminary Draft Environmental Impact Statement. This initial EIS report may be circulated internally for input from certain federal, state, and local agencies with special expertise or jurisdiction by law of the specific project.

Draft Environmental Impact Statement (Draft EIS). This first required public report is prepared after the "notice of intent." A public hearing to solicit comment on this Draft EIS by "interested persons" and "interveners" may be held if the project is considered controversial. No administrative action can be taken until 90 calendar days after distribution of the Draft EIS.

Final Environmental Impact Statement (Final EIS). This final report, in addition to including proposed action, major environmental impacts, alternatives, and costs, shall also respond to all substantive comments raised during the review of the Draft EIS. Special care will be taken to respond fully to comments disagreeing with the federal position. No administrative action can be taken until only 30 calendar days after the Final EIS has been made public.

Environmental Overview Statement (EOS). This optional report, less detailed than an environmental impact statement, usually directs itself to one aspect of the project that has indicated need for additional review, study, or comment after the Final EIS is completed.

Environmental Impact Report (EIR). This environmental study is similar to an environmental impact statement but is usually required at the state level (Warden and Dagodag, 1976), as, for example, by California Environmental Quality Act of 1969. Often these are not as specific as to methods of analysis as federal EIS, and often the federal EIS can be substituted, but some state-funded projects will require a state EIR.

Environmental Report (ER). This constitutes preliminary studies, usually of field or baseline nature, to obtain needed data on a specific project for the final EIS. Often only the conclusion and major points may be discussed in the final EIS, although all data, and calculations must be available and filed.

Environmental Baseline Studies. These inventory, usually field, studies for specific projects indicate base values and the range of values on a certain parameter for the environmental impact study. These parameters may be either natural science, social, or economic. For values that may vary seasonally (e. g., meteorology, current, or river flow), the baseline study may extend over a one-year period.

Coastal Environmental Impact Studies

Nuclear Power Plants. Perhaps the greatest expenditure in time and money for coastal environmental impact statements has been for nuclear power plants, with perhaps the most detailed and extensive field studies. Since power plants, especially nuclear power plants, require water-cooling systems, location of sites is often adjacent to rivers or coasts, even if cooling towers are to be used. For coastal-sited plants, information required formerly on plant flood-protection design (U.S. Atomic Energy Commission, 1974) relative to flood levels has been taken to also imply coastal flooding, and specifically the 100-year or "design" hurricane water levels as well as possible maximum tsunami flooding and significant (33 1/3%) and maximum (1%) wave heights (U.S. Nuclear Regulatory Commission, 1975b). Hydrology impacts, including coastal hydrology, especially for systems with "once through cooling systems", require site-specific baseline data on both current speed and direction through all tidal stages collected by on-site measurements, vertical and areal, to evaluate interaction of proposed thermal releases with the receiving marine waters in the form of *distributional isopleths* (U.S. Nuclear Regulatory Commission, 1975a). Environmental baseline studies extending over a one-year period that relate to *thermal discharge* in either protected waters or oceanic outfall would include the following: literature survey; tidal data; bathymetry; hydrography of offshore, nearshore, and protected waters; and littoral processes, including materials, methods, direction, and sediment budget of adjacent beaches (Grumman Ecosystems, 1974).

Similar baseline studies would be developed for water quality, aquatic ecology, and terrestrial ecology. All important marine species need to be inventoried, and the mortality of organisms passing through the cooling condenser and pumps must be estimated. In addition, increased heating of marine waters by the "once-through" cooling systems cause direct biological impact on both marine fauna and flora, while the necessary water intake and outfall pipes cause impact on bottom and surface currents, bottom and suspended sediment movement, and adjacent beach erosion. In addition, the amount of water available and its rate of movement or *turnover* need to be determined or controlled dilution and disposal of the cooling water if it becomes contaminated by short-lived radioactivity.

Nuclear power plants also require a *Safety Analysis Report* for plant integrity under Nuclear Regulatory Commission Law (U.S. Nuclear Regulatory Commission, 1975b) as well as the *Environmental Impact Statement* under the National Environmental Policy Act (U.S. Nuclear Regulatory Commission, 1975a). The Safety Analysis Report, prepared first, evaluates the geologic site-specific impact of seismicity and surface faulting (U.S. Nuclear Regulatory Commission, 1975c), but some material, including coastal flooding, may be common to both.

Offshore Nuclear Power Plants. Unique coastal Environmental Impact Statement problems had to be considered for the first offshore "floating" nuclear power plant proposed for the New Jersey coast and anchored in 60 feet of water with a protective fixed breakwater (Public Service Electric and Gas Company, 1976). Principal aspects studied in the field included: an oceanographic survey, a meteorological program, a thermal prediction evaluation, an ecological survey, an environmental radiation survey, and a geologic survey program (Roney, 1973). In addition, the study had to determine what effect the plant would have on the waves, currents, ocean bottom, and adjacent coastline. The extent of erosion, scouring, and sedimentation were projected from an analysis of wave refraction, bottom currents, underwater stake field surveys and historical shoreline changes (Fischer and Fox, 1973). Even the siting of the underground transmission lines from offshore to onshore adjacent to an inlet required determining inlet migration rates (Fischer, Watson, and Salomone, 1973).

Shoreline Studies. Federal funding, even if only partial, for coastal management shoreline erosion protection projects, hurricane protection programs, construction of groins, jetties, breakwater, and sea walls requires EIS studies. The engineering implications of these techniques and structures to the coastal environment are common knowledge (U.S. Army Corps of Engineers, 1971), although the environmental implications in actual coastal applications of these techniques are often ignored (McHarg, 1969). In most cases, even careful development of a coastal region produces negative as well as positive impacts (Price, 1971).

In a unique well-known situation, a coastal Environmental Impact Statement was required to support a proposed nonaction. For the Cape Hatteras Seashore Park of North Carolina, the

National Park Service prepared an environmental assessment (U.S. Department of the Interior, 1974) to evaluate whether to continue the building and maintenance of the frontal barrier dunes along the shoreline in the park. Studies had shown that these dunes, while protecting the land behind them, might actually contribute to increased shoreline erosion (Dolan, 1972). Of several alternative management objectives, the proposed one, which would not maintain the dunes, would have as its impact the restoration of the natural environment (positive) but with loss of highways, village properties, and economic development (negative), but with no direct federal costs. A second alternative continued present practice at an estimated $1 million annually; while a third would stabilize the shoreline by both structural and dune building, with the impacts being destruction of the natural setting and resource base (negative) and development of villages into urban complexes (negative or positive, depending on point of view) but with an initial construction cost of $40-$56 million and an annual maintenance of $3.2-$6.4 million. Intervenor input against the proposed action indicated the need for a complete EIS.

Federally funded shore protection structures as well as harbor and docking facilities along both marine and lake shores will require Environmental Impact Statements. Projects such as those in the newly developing National Lakeshore Parks along the Great Lakes especially need such statements, since these are both federally funded and are within a federal park (Rittschof, Dickas, and Fisher, 1976). Although not dramatic, municipal or regional sewage treatment plants along coastal and lakeshore regions are partially federally funded and require EIS coastal studies primarily in the siting of the outfall pipes in the nearshore region and the passage of these pipes across the fragile beach and tidal marsh systems (Grumman Ecosystems, 1975).

Even the application for a federal permit may require an EIS. Permission is required from the U.S. Army Corps of Engineers to alter, dredge, or build structures in tidal and navigable waters; although this regulation is subject to various interpretations (Friedman, Donnellan, and Nickerson, 1976), it has been assumed that the Corps of Engineers can request an environmental impact statement from the applicant if the project would affect the quality of the human environment.

Organization of Environmental Impact Statements

Federal regulations indicate specifically the exact nature and content of an Environmental Impact Statement (U.S. Environmental Protection Agency, 1975). These specifications need to be followed exactly in order for the report to be legally reviewed and approved. Preparation of a federal EIS therefore requires the following: cover sheet indicating type EIS (draft or final), project name, number, responsible office, date, and responsible official; summary sheet (note: the summary of the report *precedes* the body of the EIS); and an EIS, which should be clear, concise, and include all facts to evaluate benefits and adverse environmental effects of proposed action and alternative actions. To the extent possible, it should not be in a style requiring extensive scientific or technical expertise to comprehend or evaluate. It should be divided into the following sections: background and description of the proposed action; alternatives to the proposed action; environmental impacts of the proposed action; adverse impacts that cannot be avoided and steps to minimize environmental harm; relationship between local short-term uses of man's environment and the maintenance and enhancement of long-term productivity; irreversible and irretrievable commitments of resources to the proposed action if it is implemented; and problems and objections raised by other federal, state, and local agencies and by interested persons in the review process.

Environmental impacts of the proposed action (section 3c) should address itself to these specific aspects: positive and negative impacts, both beneficial and adverse; primary impacts directly related to proposed action; secondary impacts of indirect or induced changes; socio-economic changes related to the action; and land-use changes related to the action. All documents, research correspondence, and field studies used as references should be included in the bibliography. While federal statutes usually do not require a public hearing, one can be held during the Draft EIS stage if it may facilitate resolution of conflicts or significant public action. Evaluation of six years' experience by seventy federal agencies in preparing Draft Environmental Impact Statements (Peterson, 1976) indicates the following average preparation time for an EIS: simple projects prepared by experienced personnel, 1-6 months; complex projects prepared by experienced personnel or simple ones prepared by inexperienced personnel, 2-12 months; complex projects prepared by inexperienced personnel, 18 months. While environmental impact assessments can often be made on the basis of inventories of existing information, the Draft Environmental Impact Statement usually necessitates specific project research. Coastal Environmental Impact Studies, because of the dynamic processes involved, will almost always require field research. This research will usually consist of on-site measurements and seasonal environmental baseline inventories, and may require as much as 20-50% of available research time.

Evaluation of Environmental Impact Studies

An evaluation of about 100 typical Environmental Impact Statements with respect to their geologic content indicated that many often supply the necessary information in too generalized a form, if not erroneous or even omitted (Lessing and Smosna, 1975).

A general matrix analysis system has been proposed by the U.S. Geological Survey (Leopold et al., 1971) to evaluate environmental impact statements. This analysis weights the relative magnitude and importance of the action within a matrix, plotting actions that cause an environmental impact (horizontal axis of the work sheet) against existing environmental conditions (vertical axis) that might be affected. The listed 100 possible actions against a possible 88 conditions of the Leopold et al. (1971) matrix give a total of 8800 possible interactions, but only a few would be of such magnitude and importance as to require additional study. Long-term impacts should be differentiated from short-term impact (less than 1 year) as well as the situation in which initial lesser impacts may produce long-term secondary impacts. Matrix analysis is now common in many land-use planning decisions. While providing a visual inventory checklist, statistical techniques are available to quantitatively express the relationships. The reasoning that assigned the specific numerical values for the magnitude and importance should be set forth in the report. Specifically discussed should be: row and columns with high total computed values; those individual boxes with large numerical magnitude and importance values; those columns indicating a large number of actions regardless of numerical value; and environment elements (rows) with large numbers of marked boxes.

However, the environmental impact study should not include superfluous operating detail of the proposed project or descriptions, quantitative or qualitative, of the environment, but should include only material that the matrix analysis indicates is needed for evaluating the environmental impact.

JOHN J. FISHER
DONALD R. REGAN

References

Dolan, R., 1972. Barrier dune system along the Outer Banks of North Carolina: A reappraisal, *Science* 176, 286-288.

Fischer, J. A., and Fox, F. L., 1973. Siting constraints for an offshore nuclear plant, *Engineering Bull.* 42, 3-6.

Fischer, J. A.; Watson, I.; and Salomone, L. A., 1973. Considerations in planning transmission lines to offshore nuclear plants, *Engineering Bull.* 42, 19-28.

Friedman, J. M.; Donnellan, R. A.; and Nickerson, G. A., 1976. *Regulation of Harbors and Ponds of Martha's Vineyard: Final Report–WHOI-76-56.* Woods Hole, Mass.: Woods Hole Oceanographic Institution, 55p.

Grumman Ecosystems Corp., 1974. *A Proposal for the Environmental Baseline Study–Charlestown, R.I., site-hydrography.* Bethpage, N.Y.: Grumman Ecosystems Corp., 97p.

Grumman Ecosystems Corp., 1975. *Proposal for Environmental Impact Statement for Outfall of Sewer District, Suffolk, N.Y., RFP No. WA 76X-024,* Bethpage, N.Y.: Grumman Ecosystems Corp., 178p.

Leopold, L. B.; Clarke, F. E.; Hanshaw, B. B.; and Baisley; J. B., 1971. *A Procedure for Evaluating Environmental Impact.* Washington, D.C.: U.S. Geological Survey Circ. 645, 13p.

Lessing, P., and Smosna, R. A., 1975. Environmental Impact Statements–worthwhile or worthless?, *Geology* 3, 241-242.

McHarg, I. L., 1969. *Design with Nature.* Garden City, N.Y.: Natural History Press, 197p.

National Environmental Policy Act, 1969. An act to establish a national policy for the environment, to provide for the establishment of a Council on Environmental Quality, and for other purposes, approved January 1, 1970 (Public Law 91-190; 91 STAT. 852; 42 U.S.C. 4321-4347).

Peterson, R. W., 1976. *Environmental Impact Statements–Analysis by Federal Agencies, Report.* Washington, D.C.: Council on Environmental Quality, Executive Office of the President, 65p.

Price, W. A., 1971. Environmental impact of the Padre Isles development, *Shore and Beach* 39, 4-10.

Public Service Electric and Gas Company, 1976. *Draft Environmental Statement Relating to Construction of Atlantic Generating Station Units 1 and 2, Docket Nos. STN 50-477 and STN 50-478.* Washington, D.C.: U.S. Nuclear Regulatory Commission, Office of Nuclear Reactor Regulation, 329p.

Rittschof, W. F.; Dickas, A. B.; and Fisher, J. J., 1976. Coastal parameters relative to harbor design (abstract), in *First Conference on Scientific Research in National Parks.* Arlington, Va.: National Park Service and American Institute of Biological Sciences, 94.

Roney, J. R., 1973. Environmental studies for an offshore nuclear power plant, *Engineering Bull.* 42, 29-31.

U.S. Army Corps of Engineers, 1971. *Shore Protection Guidelines: National Shoreline Study.* Washington, D.C.: U.S. Army Corps of Engineers, 59p.

U.S. Atomic Energy Commission, 1974. *Additional Information, Water Level (Flood) Design for Nuclear Power Plants.* Regulatory Guide 1.70.5, May, 1974, pp. 1.70.5-1-1.70.5-2.

U.S. Department of the Interior, 1974. *Environmental Assessment–Cape Hatteras Shoreline Erosion Policy Statement, NPS 826.* Washington, D.C. Denver Service Center, National Park Service, 150p.

U.S. Environmental Protection Agency, 1975. Title

40—Protection of environment, Chapter 1—Environmental Protection Agency (FRL 327-5); Part 6—preparation of environmental impact statements, final regulations, *Federal Register* **40**, no. 72, Monday, April 14, 1975.

U.S. Nuclear Regulatory Commission, 1975a. *Preparation of Environmental Reports for Nuclear Power Stations*. Washington, D.C.: Office of Standards Development, Regulatory Guide 4.2, January, 1975, Revision 1, 77p.

U.S. Nuclear Regulatory Commission, 1975b. *Standard Format and Content of Safety Analysis Reports for Nuclear Power Plants*. Washington, D.C.: Office of Standards Development, Regulatory Guide 1.70, September, 1975, Revision 2, 275p.

U.S. Nuclear Regulatory Commission, 1975c. *General Site Suitability Criteria for Nuclear Power Stations*. Washington, D.C.: Office of Standards Development, Regulatory Guide 4.7, November, 1975, Revision 1, 30p.

Warden, R. E., and Dagodag, W. T., 1976. *A Guide to the Preparation and Review of Environmental Impact Reports*. Los Angeles, Calif.: Security World Publishing Co., 138p.

Cross-references: *Coastal Engineering; Coastal Reserves; Coastal Zone Management; Dredging; Human Impact; Land Reclamation; Mineral Deposits, Mining of; Nuclear Power Plant Siting; Protection of Coasts; Tourism.*

COASTAL EROSION

Natural forces, including winds, waves, currents, and tides, interact with sediments to create sediment transport. The sediments are carried until the net force causing the movement becomes less than the forces tending to stop the motion, such as particle weight. When this happens, the sediments are deposited. Coastal erosion is the process that occurs as a result of a net loss of sediment, meaning that more sediments are carried away from an area than are deposited.

Natural Forces Causing Coastal Erosion

Waves. Water waves are the most dominant forces causing coastal erosion. They affect sediment motion by initiating sediment movement and by driving current systems that transport the sediment once motion is initiated. Waves induce an orbital motion in the fluid beneath them. These orbits are not closed, so that the fluid undergoes a small wave-induced drift, known as *mass transport*. The amount and direction of this mass transport are functions of elevation above the seabed, various wave parameters, and temperature gradients. As waves approach breaking, the bottom motion induced by the waves becomes more intense and there is a greater effect upon the sediment. Breaking waves move sediments by creating intense local currents. *Steep waves* (those with lengths 7-20 times the wave height) cause the beach to erode by taking the material from the area just above the water level and carrying it into deeper water, thus tending to tear the beach down. *Low steepness waves* (length 30-500 times the height), which are usually generated from distant storms, bring material in from the shallow water portion of the nearshore zone and deposit it just above the still-water level on the beach, thereby building up the beach.

In most areas there is a constant change in the beaches caused by the alternation of the two types of wave steepnesses. During the winter season the beaches tend to be worn away by the high frequency of storm waves, which does not allow enough time for rebuilding. During the summer season the opposite action occurs (Coastal Engineering Research Center, 1973).

Most wave crests do not approach parallel to the shoreline. This means that there is usually a longshore (parallel to the shoreline) component of momentum in the breaking waves. This longshore component of momentum is the major cause of longshore currents that flow parallel to the shoreline within the surf zone. The movement of sediments by these currents is known as *longshore transport* or sometimes as *littoral drift,* although this latter term applies also to movement perpendicular to the shore.

Tides. Tides, which are actually long waves, can also produce currents that can effect the shore. Tide-induced currents are especially strong near entrances to bays, lagoons, and in regions of large tidal range. Tides are important in transporting large piles of sand known as *shoals* that can greatly endanger surface navigation, especially around entrances to bays and estuaries. Further, the change in water level caused by the tides affects sediment transport, because with a higher water level, waves can attack a greater range of elevations on the beach (Bruun, 1962).

Winds. Winds can have a significant effect upon coastal erosion by both direct and indirect means. The direct method of blowing sand off beaches is known as *deflation*. Deflation usually removes only the finer-grained material, leaving behind coarser sediment and shell fragments. Sand blown seaward usually falls in the surf zone, where it is introduced into the littoral transport system. Sand blown landward may form dunes or be deposited in lagoons behind barrier islands. Sand dunes immediately landward of the beach, known as *foredunes,* are formed by the beach grass, which traps some of the landward blown sand. Sand dunes are the last zone of defense in absorbing the energy of storm waves. The foredunes act as a barrier to prevent waves and high water from moving inland and as a supply of sand to replenish the nearshore area during severe erosion.

The indirect effect of wind is in producing surface currents that, when directed toward the shore, can cause significant bottom return flows that can transport sediments seaward. Offshore-directed winds have just the opposite effects.

Amounts of Erosion

Beaches change in plan view as well as in cross-sectional distribution as a result of coastal

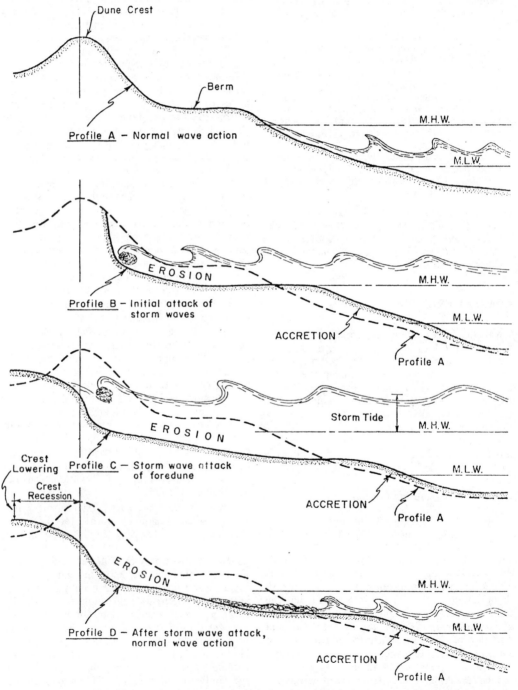

FIGURE 1. Schematic diagram of storm wave attack on beach and dune (from Coastal Engineering Research Center, 1964).

erosion. Figure 1 shows how the cross-sectional area of a beach changes under the attack of storm waves (Coastal Engineering Research Center, 1964). Comparison of beach profiles during and after storms has indicated erosion of the beach above mean sea level of 13,000-61,000 cubic m/km of shoreline during storms expected to occur about once a year, with the actual amount depending on the locality. It also depends on the wave conditions, *storm surge* (the rise above normal water level caused by the action of wind stress on the water surface), the stage of the tide, and storm duration. The amount of horizontal change in shoreline position can be as much as 20-30 m or more for a moderate storm.

The volume of erosion caused by storms is much smaller than the amounts of net longshore transport rates along some ocean beaches. The *net longshore transport rate* refers to the summation of the amounts of sediment passing a given point during a certain amount of time, usually a year, with a positive quantity applied to transport in one direction and a negative quantity in the opposite direction. Zero net transport refers to an equilibrium position of a beach where the amount of material eroded is balanced by the amount supplied. The rates of longshore transport are typically 80,000-380,000 cubic m/year, but may vary from 0-800,000 cubic m/year.

Long-term changes in the shoreline position are a result of a net longshore transport. Long-term erosion rates are rarely more than a few m per year in horizontal motion of the shoreline, except in coastal areas particularly exposed to erosion, such as near inlets or capes. It has been estimated that beach and cliff erosion along all coasts of the world totals about 120 million cubic m/year.

JOHN B. HERBICH
JOHN P. HANEY

References

Bruun, P., 1962. Sea-level rise as a cause of shore erosion, *Am. Soc. Civil Engineers Proc., Jour. Waterways and Harbors Division* 88, 117-130.
Coastal Engineering Research Center, 1964. *Land against the Sea.* Washington, D.C.: U.S. Army Corps of Engineers, MP4-64, 43p.
Coastal Engineering Research Center, 1973. *Shore Protection Manual.* Washington, D.C.: U.S. Army Corps of Engineers, 750p.

Cross-references: *Beach Processes; Coastal Dunes and Eolian Sedimentation; Coastal Erosion, Environmental-Geological Hazard; Coastal Erosion, Formations; Nearshore Hydrodynamics and Sedimentation; Sediment Transport; Wave Erosion. Vol. III: Beach Erosion and Coastal Protection.*

COASTAL EROSION, ENVIRONMENTAL-GEOLOGIC HAZARD

The Nature of Coastal Erosion

The coast is a dynamic zone of interaction: between land and sea; between lithosphere, hydrosphere, atmosphere, and biosphere; between terrestrial, marine, and subaerial processes. The term *coastal erosion* applies, therefore, not only to the coast itself but also to the land strip and a strip of sea floor immediately bordering on the coast. Erosion can affect cliffs and dunes behind high water mark, beaches and the beach substrate between extreme tidal levels, and the sea floor below lowest tide level. In the case of lacustrine environments, which are essentially tideless, erosion again affects the zone between extreme high and extreme low water level as well as the zones immediately above and below these levels.

The erosion of cliffs, the beach substrate, and the *solid* parts of the sea floor is progressive in the sense that the materials removed are, except perhaps in a long-term geologic sense, permanently removed. Cliff erosion in hard rocks may be almost imperceptible except, perhaps, over several decades; but in soft rocks it may be clearly apparent from storm to storm or even after a period of heavy rain. The spatial manifestation may be of a sporadic nature.

By contrast, erosion of dunes, beaches, and the unconsolidated sediments below low tide level may be of a transient or fluctuatory nature in the sense that cut-and-fill processes freely operate in marine and lacustrine environments. For instance, the combing down of a beach in rough weather is usually followed by the rebuilding of that beach in ensuing calmer weather. Thus a state of dynamic equilibrium may obtain. In rather the same manner, dunes may experience both deformation and reformation (although not necessarily in the same position). In the case of beaches, changes in the rate of littoral drift necessarily imply either erosion or accretion; on the other hand, beaches with strong rates of littoral drift do not necessarily erode (Jolliffe, 1978).

It is important to point out that erosion, transportation, and deposition operate conjointly, although the relative importance of erosion as compared with accretion will vary spatially and temporally. Indeed, the history of numerous coastal areas testifies to significant changes in the incidence of coastal erosion and accretion; moreover, it is often possible to relate events occurring in one place to those that take place elsewhere. One example is the gradual migration of *ness* features on the east coast of England, apparently largely because erosion on one flank is more or less matched by accretion on the other. Another example is the east Kent coast, off which the Goodwin Sands are slowly rotating counterclockwise; in response, there have been changes in the rates of littoral drift and related changes in the incidence of coastal erosion and accretion.

Factors Causing Coastal Erosion

The factors and processes involved in coastal erosion are numerous, complex, and often superimposed, virtually precluding the possibility of reliable extrapolation of past trends of coastal behavior to an accurate prediction of future events. On the other hand, long-established erosion patterns can usually be expected to continue, provided there is no human intervention. An example would be the Holderness coast of Yorkshire, England, where the coast has been receding on average some 1.6 m per century, at least since Roman times.

Basic factors in coastal erosion include the action of winds, waves, tides, and subaerial processes. Geology and climate are also important. Chemical, physicochemical, biochemical, mechanical, and biological factors may also be important in particular environmental situations. We can thus recognize so-called dominant, as opposed to generally subordinate, factors and processes.

Winds raise surface waves that can influence the sea floor as well as the intertidal zone; and they may elevate water level, *wind set-up,* thereby enabling waves to operate at higher levels. Onshore winds, as well as raising surface waves, cause surface water to move shoreward and the lower layers of water to move slowly seaward; in this way some of the material placed into suspension within the breaker zone may be transported seaward. Winds may deflate beaches and dunes by the removal and transportation of sediment particles. Waves in shallow water generate currents that can erode both the submarine slope and the shoreface, these currents being the wave-induced oscillatory current at the seabed, the shoreward-directed mass-transport current, the seaward-directed rip current, and the longshore current. Moreover, the effect of plunging waves at wave breakpoint is to comb down a beach. However, such currents are not necessarily erosive, for they may promote deposition, or both. Tidal flow tends generally to be rectilinear, i.e., to flow roughly parallel to the shoreline; within the breaker zone, tidal currents usually play a subordinate role to the wave-induced currents in affecting beach behavior. Tidal currents tend to be negligible over the beach face, but assume increasing importance seaward of the breaker zone. In shallow bays and estuaries, tidal scour of the familiar expanses of generally cohesive silty and muddy deposits occurs mainly because of side scour caused by migrating low water channels.

From empirical observations of beach behavior, it has been shown that beaches generally *erode* when waves become higher and shorter, when winds are onshore, when water level is raised, and when the littoral drift is inhibited more successfully in the updrift direction than on the beach in question. Coastal erosion is closely related to wave energy levels at the coast, which in turn are influenced by the wave refraction patterns near the coast. Where the wave orthogonals converge, energy levels tend to increase and erosion rates in these segments may also increase. Indeed, one can distinguish between so-called macro-, meso-, and micro-energy environments.

Much erosion is caused by subaerial processes, including surface runoff and subsurface seepage, sometimes leading (depending on the lithology and structure of the coastal rocks) to slips, slumps, and flows. Wet and incompetent strata, such as clays, may be squeezed out seaward by the superincumbent weight of overlaying strata. Again, a whole range of generally subordinate processes, including freezing and thawing, heating and cooling, and chemical solution effects, may accelerate or generally assist the main erosive agents—for example, by rotting and otherwise weakening rocks, thereby making them more susceptible to hydraulic action. Many microerosion forms develop in this way. Interesting and by no means inconsiderable factors are the roles performed by marine animals and plants—for example, the role seaweeds exert in removing clastic sediments from parts of the sea floor (Jolliffe, 1976a) and the action of marine borers such as *Clione* and *Patella* on rocks, including limestones and some sandstones.

A potent factor in coastal erosion is human intervention. The dredging of offshore sediments may inadvertently lead to erosion at the coast, while deliberate removal of beach material clearly erodes the local beaches (Fig. 1) and possibly in time, beaches some distance away (Jolliffe, 1974, 1976b). Groining and walling, while perhaps locally beneficial, often induce erosion at one or both ends of the wall or groin system, giving rise to the so-called *terminal scour* problem. Structures such as pipes or cables laid across a beach or the sea floor may induce scour around them. The bulldozing of dunes along parts of the barrier coast of eastern United States, to inhibit *washover* from storm

FIGURE 1. West Bay, Dorset, southern England. An eroding coast with both natural and man-modified cliffs and beach mining for gravel (photo: I. P. Jolliffe).

waves, has reduced sand supplies to the back-beach zone, with ensuing erosion.

Coast Erosion Problems

Coastal erosion presents mankind with a wide array of practical problems; the remedial measures taken to combat them are often costly and difficult. Some erosion problems are virtually intractable—for example, dealing with a high landslip coast. Erosion may mean serious loss of coastal land, including buildings, roadways, and coastal footpaths; while any diminution in the size of beaches may nowadays be viewed with considerable alarm because beaches often provide protection to land behind the beach and also valuable amenity-recreational resources. Along low-lying coasts, erosion may be accompanied by serious flooding—for example, the effect of the 1953 storm surge that affected much of the southeast coast of England and the northwest coast of Europe. High water levels often mean increased erosion. Lowered beach levels may result in the failure of walls and other structures, especially if foundations are exposed to direct wave attack. Sewer outfalls, pipelines, and other structures placed on the sea bed are often vulnerable to scouring action. However, serious though erosion problems may be, erosion has some beneficial aspects: it is partly because of erosion that coastal configuration is what it is, and bays and coves provide shelter, sites for recreational and other resource development, and an integral part of scenery that makes the coastal environment so attractive. Erosion exposes new geologic sections that in themselves may be of considerable scientific interest and value. Especially, erosion produces raw materials that are destined in the long term for new rock formation and in the short term for the nourishment of beaches. For instance, although some cliffs provide little beach material (the rapidly eroding Holderness cliffs of eastern England are said to contribute only some 3% of their material to local beaches), there are other areas, such as the Selsey Peninsula in Sussex, where cliff protection has seriously reduced the natural replenishment of local beach material. In this context *cost-benefit analysis* has shown that in many cases the cost of protecting a shoreline segment against erosion can barely be justified in terms of the benefits accruing from such protection.

Erosion Abatement Measures

In those cases where man has to try to reduce erosion, a whole variety of methods have been devised, although these are by no means always practicable or successful. Deflationary losses from dune areas may be reduced by the use of brushwood fences or other obstacles or by planting suitable vegetation. Low cliffs may be walled, and the toe of a high cliff may be protected by walls or some form of *riprap*. However, cliff erosion is often as much the result of subaerial weathering and ground-water flow, as much as of marine scour, so that remedial measures often necessitate cliff drainage and even cliff trimming.

Localized beach erosion has been tackled in various ways—for example, by using flexible concrete mattresses. Usually, however, attempts are made to inhibit the littoral drift by use of groins made of timber or concrete. Isolated, or groups of, groins (if successful) seriously inhibit the littoral drift, although they rarely prevent cut and fill from occurring. Unfortunately, both individual and groups of groins usually cause some scouring, especially at the downdrift end of the system (Fig. 2). In England, an increasingly used erosion abatement measure is that of artificial recharge or nourishment, involving the tipping of suitable material from trucks or from shallow-draught boats onto many beaches. At Bournemouth, Hampshire, sand dredged from offshore is pumped ashore as a slurry. At Dungeness, Kent, controlled beach dredging in areas of surplus is matched by tipping of the same material in areas of paucity, resulting in a man-improved coastal sediment budget. At Lake Worth Inlet, Florida, sand bypassing of a tidal inlet is undertaken to assist in both erosion abatement and navigation.

Erosion of the nearshore seabed is difficult to combat, especially if the seafloor is cut in clay. A principal cause of such erosion is the filelike action of the mobile sediment layer resident on the sea floor substrate. Attempts have been made in some areas to accumulate sea floor and beach sediments using synthetic seaweed grids (Jolliffe, 1976a). In parts of the Netherlands, this method has proved successful, as has the reduction of scour along the flanks of dredged navigation channels.

Conclusion

Thus, while coastal erosion can be beneficial, it is more usually regarded as an environmental

FIGURE 2. Middleton, Sussex, southern England. An example of man-induced terminal scour. The direction of shore drift is from lower left to upper right (photo: I. P. Jolliffe).

problem. Its control calls for ongoing research leading to an improved knowledge of coastal processes, natural and seminatural (man-modified) coastal systems, and erosion abatement measures. The use of scale hydraulic and mathematical models has provided much valuable data, while another fruitful line of research relates to the monitoring of coastal *sediment budget* utilizing aerial stereo-photogrammetric techniques. Erosion abatement schemes are increasingly subject to critical *cost-benefit analysis*.

IVAN P. JOLLIFFE

References

Jolliffe, I. P., 1974. Beach-offshore dredging: Some environmental consequences. *Offshore Technology Conference, 6th, Houston, Proc.* **2**, 257-268.
Jolliffe, I. P., 1976a. Roles performed by seaweed in the marine environment, *Offshore Technology Conference, 8th, Houston, Proc.* **1**, 993-1004.
Jolliffe, I. P., 1976b. Man's impact on the coastal environment, *Geographuca Polonica* **34**, 73-90.
Jolliffe, I. P., 1978. Coastal and offshore transport, *Prog. Phys. Geog.* **2**, 264-308.

Cross-references: *A-B-C Model; Beach Nourishment; Beach Processes; Coastal Dunes; Coastal Engineering; Coastal Erosion; Coastal Erosion, Formations; Currents; Lakes, Coastal Morphology; Nearshore Hydrodynamics and Sedimentation; Rock Borers; Sediment Transport; Tidal Currents; Waves.* Vol. III: *Beach Erosion and Coastal Protection;* Vol. XIII: *Geologic Hazards.*

COASTAL EROSION, FORMATIONS

Abrasion, the mechanical action of the ocean against its rocky bed, results in many formations, some of which are controlled primarily by the dynamics of ocean waters, while others are subject to a variety of mechanical (ice, floating materials), physical (thermal effects, weathering), physicochemical (some types of weathering in the swash zone), chemicobiological (bacteria), and biological factors. Many abrasion formations, such as niches, caves, platforms, residual fingers, and chimneys, are illustrated by Zenkovich (1967). *Erosion platforms* and *ramps* are defined in respective entries in this volume.

Coastal rock composition, hardness, porosity, and texture determine the kinds of formations that will result from abrasion. While harder and consolidated rocks are ground gradually, less resistant rocks are crushed at a more rapid rate. Both processes result in continuous *corrasion*—direct attack by waves with sand grains and pebbles—of the coast.

Cliffs, important elements of the coastal profile, are produced by coastal erosion. During the initial stages of cliff development, the shore retreats because of the alternate formation of erosional caves and niches and successive cliff falls. Because cliffs are vulnerable to attack by rain and frost, poor drainage is blamed for many cliff falls. The process of breaking up pieces of rock fallen from the cliff by undercutting or other means is referred to as *attrition* (q.v.). Cliff erosion may be accelerated by the transport of material into deep water or its lateral transport by waves breaking obliquely on the shore. These processes are accompanied by beach expansion and the resulting formation of underwater accretion terraces. Waves dissipate their energy on the gradually widening beach so that the process of cliff retreat decelerates, finally making the cliff stable. Among other cliff-related features, *scarps* are often formed when steep waves erode the beach. Types and morphology of cliffs are discussed in detail by Zenkovich (1967).

Rhythmic forms along the open coast as well as within embayments are among the coastal forms that may be attributed partially to the erosion of loose materials. These forms occur on a wide range of scales, from small *beach cusps* (q.v.), through larger protuberances, to *major cuspate features.* Their origin is still a matter of speculation, but their importance in controlling localized zones of erosion and accretion is generally agreed upon.

According to Bowen and Inman (1971), two types of rhythmic forms exist: *cuspate* and *crescentic.* The former occur at the shoreline, are concave seaward, and vary in size. The latter are convex seaward and occur in bays and on straight shorelines with normally low tidal ranges. A crescentic, sometimes referred to as a *lunate bar,* is essentially a submerged sand bar that is concave shorewards.

Both cuspates and crescentics have wave lengths of 100-1000 m, but primarily 200-300 m. They may be generated by *edge wave* formation and are also associated with *rip currents.* The edge waves, which move parallel to the beach, set up *nodes* resulting from mutual interference. The pattern of nodes determines the pattern of water movement, which in turn determines the position at which deposition takes place—hence the pattern of the crescentic bars. The bars have half the wavelengths of the edge waves.

The motion associated with edge waves could also account for the circulating type of pattern that results in beach cusp formation on the foreshore. In order to produce crescentic bars on long, straight beaches, standing edge waves of a 30-60 sec. period with a significant amplitude would be required. The larger features, which have been called *sand waves* but are more accurately called *protuberances* or *nesses,* on the shoreline at Cape Hatteras generally have wavelengths of 500-600 m. They move rapidly along the coast, particularly under stormy conditions, and cause local zones of intensified erosion, or temporary respite from erosion, according to the position of the embayment or the horns.

Cuspated forms may also be produced by irregular shoreline configuration. Headlands may cause the diversion of beach material, as indeed may harbors, piers, jetties, and groins. The erosional effect of such structures is felt for distances of several km (Silvester, 1974).

Erosion patterns and equilibrium shapes of bay physiographic units are discussed by Silvester (1974). The formation of erosion bays and embayments—with intermediate festoons, teeth, and similar features—is analyzed by Zenkovich (1967).

RYSZARD ZEIDLER

References

Bowen, A. J., and Inman, D. L., 1971. Edge waves and crescentic bars, *Jour. Geophys. Research* 76, 8662-8671.

Silvester, R., 1974. *Coastal Engineering,* vol. 2. Amsterdam: Elsevier, 338p.

Zenkovich, V. P., 1967. *Processes of Coastal Development.* New York: Wiley Interscience, 738p.

Cross-references: *Accretion Ridge; Bars; Beach Cusps; Cliffed Coast; Cliff Erosion; Coastal Erosion; Coastal Erosion, Environmental-Geologic Hazard; Edge Waves; Erosion Ramp, Wave Ramp; Platform Beach; Rhythmic Cuspate Forms; Sea Cliffs; Wave-Cut Platform; Weathering and Erosion, Biologic; Weathering and Erosion, Chemical; Weathering and Erosion, Mechanical.*

COASTAL FAUNA

The coastal zone is a prime interface of physicochemical and biological interactions, and harbors the richest and most varied assemblages on earth per comparable area, especially in rocky intertidal regions. Almost every popular article or introductory work on the sea is illustrated with composite drawings or diagrams to show the typical fauna of the shore: seastars, crabs, clams, and snails on the beach, whether the shore be of rock or sand, a seal or sea lion basking in the sun, and gulls and terns flying overhead. The fauna of the coastal zone, however, includes much more than the animals seen on a beach, for there are often hordes of fishes near shore, great aggregations of birds, and, in some parts of the world, large herds of seals, sea lions, and elephant seals. Even some of the Cetacea are characteristically coastal in habitat, the harbor porpoise and the California gray whale in particular. In numbers and variety, however, the predominant members of the coastal fauna at or near the shore interface are invertebrates.

Yet this abundant development of massive populations of animals depends for the most part on comparatively short-term life spans of plants; most of the larger plants, the sea weeds, surf and sea grasses, and colonial diatoms are annual or die back to rootstocks or holdfasts each year, and the neritic phytoplankton usually have several generations within a growing season. In contrast to this, many of the conspicuous invertebrates of the shore environment have life spans of several years: clams, seastars, urchins, and anemones may live 5-15 years, and it is suspected that some herbivorous gastropods may live as long as 25 years. Sea anemones may live for decades in aquaria. These life spans are characteristic of middle and higher latitudes; in the tropics life spans are probably shorter. The exception to the pattern of short-lived plants and longer-lived animals (the reverse of most terrestrial systems except grasslands) is that of the tropical *mangrove coast* (q.v.), where the development of sediment-holding, maritime trees and consequent development toward a tropical savanna is characteristic of ecological succession in upland regions. Here, as in temperate latitudes, however, the bulk of the food supply for the detritus-, deposit-, and filter-feeding invertebrates is derived from the reworked algae and animal remains, recirculated, and made accessible by the mixing processes of tidal action. These processes are best developed, and support the largest populations of such organisms as clams, worms, detritus-feeding snails, and the like, in *Estuaries* (q.v.), which by their location and nature are nutrient traps. Much of the nutrient resource in estuaries is derived from the breakdown and recycling of marsh grasses and submerged sea grasses, which are important components of coastal ecosystems (Chapman, 1977; Phillips and McRoy, 1980).

Although dependent to a large extent on the inorganic nutrients derived from land, the fauna of the coastal zone is predominantly marine in origin. Several of the most common groups of organisms in the coastal zone are representatives of exclusively marine phyla or orders—e.g., echinoderms, scyphomedusae, sea anemones, corals, cephalopods, and chitons. Several of the other major groups of marine animals have limited freshwater representation; these include sponges, bryozoa, hydroids, polychaete worms, and bivalves. Among the groups more successful in freshwater (and on land) are the gastropods and crustacea. On the other hand, predominantly freshwater or terrestrial organisms have not fared well in the sea, the coastal zone, or estuaries. While there are substantial populations of maritime flies, estuarine and mangrove mosquitoes, and an array of beetles in shore environments, the total representation of this largest and most varied group of organisms on earth is relatively insignificant from the systematic point of view. No *insect* (q.v.) is known to be completely marine, independent of the need for atmospheric oxygen throughout its life. Among the arthropods, only a group of mites (often numerous in the coastal zone) do not depend on atmospheric air. A few other arthropods—e.g., spiders, pseudoscorpions, and centipedes—occur sparingly at the edge of the sea in various parts of the world. This terrestrially derived arthropod fauna, especially the insects, is summarized by Cheng (1976).

The two major groups of marine mammals, the Pinnipedia and the Cetacea, are considered to be derived from terrestrial stock. One obviously more recent invasion of coastal waters from land is that of the comparatively unmodified mustelid (weasel family), the sea otter of North Pacific shores.

The fauna of coastal regions includes the animals living near shores in shallow bottoms and that of the intertidal regions. In a sense these are part of the same complex, for many of the intertidal animals also occur below the low tide line, and the intertidal zone represents the uppermost extreme of the edge of the sea colonized by marine life. In both regions the primary dependence for food and for an environment in which reproduction can occur and dispersion of young is on the sea.

Although the fauna of rocky intertidal shores was first observed in detail by naturalists (see below), the extensive areas of shallow bottom have received perhaps more intensive notice because of their relationship to coastal fisheries and culture of marine animals. Another type of faunal association is that of the coral reefs, themselves formed primarily through the agency of colonial animals in tropical regions where the average water temperature does not fall much below 20°C and in localities where sediment from rivers and offshore drainage is reduced; this association is not strictly coastal, but may form its own coasts around tropical islands; the development of *coral reefs* (q.v.) is not entirely dependent on the corals, but also involves calcareous algae.

Shallow Seas

The study of the fauna of the shallow sea bottom was stimulated by the decline of the oyster beds of the German Waddensee, which were studied by Karl Moebius (1825-1908) about a hundred years ago. In his 1877 report, *"The Oyster and Oyster Culture"*, Moebius first stated the concept of an interrelated system of organisms and environmental conditions, which he termed the *biocoenosis*. The study of these bottom associations was placed on a quantitative basis by the landmark studies of C. G. Johannes Petersen, in a series of studies of Danish waters, who counted, weighed, and measured the animals according to a unit area system based on samples taken with a small, specialized clam-shell dredge, since known as the *Petersen Grab* (Petersen, 1918). Petersen noticed recurrent groupings or patterns of associations of bivalves, ophiuroids, worms, and other burrowing organisms living with the sediments (the *infauna*). From this work a large body of knowledge has been developed concerning these associations of the shallow bottoms because of their significance as the food supply of commercial stocks of fishes; much of this work was summarized by Gunnar Thorson (1971), who carried on in Denmark the study and concern for "level bottom communities" (Fig. 1). Thorson developed a concept of parallel bottom communities on the basis of his work in the Arctic and north European seas. According to his idea, these shallow bottom communities are similar, with different species of the same or similar genera in comparable environments throughout the shallow seas of the world. Tropical shallow bottoms, however, did not conform to this generalization, but Thorson, who realized his idea was not universally applicable, was unable to reexamine the problem before his death. Although considerable attention has been given to counting and weighing, much of the research on marine bottom communities has been descriptive or cataloging, and attention has since been turned to analyzing the data for internal consistencies and degrees of difference between similar communities or in seasonal variations.

The fauna of estuaries is similar to that of shallow oceanic bottoms; most of the communities studied in Danish waters are actually estuarine. They are somewhat more restricted, lacking echinoderms and some smaller groups of invertebrates, but include the productive oyster and mussel beds of commerce as well as the clam and scallop beds found in many parts of the world. Thus they are communities of filter- and deposit-feeding organisms, obviously dependent on the rich fare provided by the estuarine nutrient trap and, to a less understood extent, on the influx of freshwater into estuaries and the consequent fluctuation of salinity.

COASTAL FAUNA

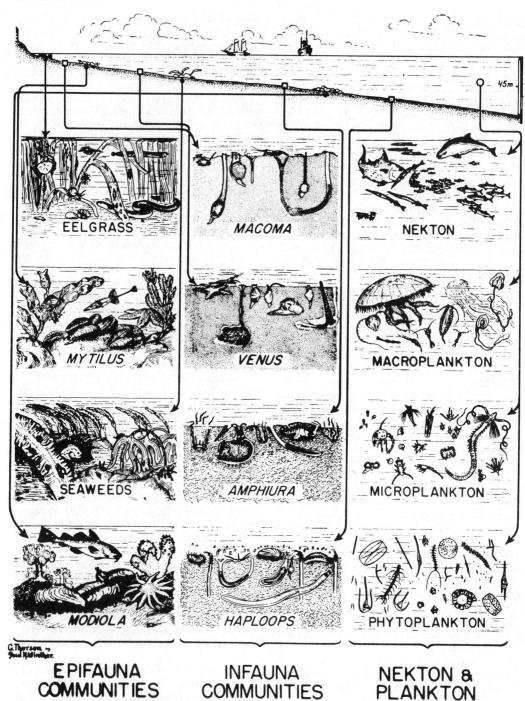

FIGURE 1. The communities of the shallow Danish seas, where the study of marine communities began (by G. Thorson and P. H. Winther; from Hedgpeth, 1957).

Intertidal Regions

The coastal environments between tidal or storm extremes are of three basic kinds: rock, sand, and mud. In all these environments, but most obviously on rocks, animal life is abundantly developed, and in all these environments different kinds of organisms are more abundant at different levels, roughly associated with tidal fluctuations, although it is not possible to equate the zonal distribution on rocks directly with that of sand or mud (Fig. 2). A good account of the life in sand and mud flats is provided by Eltringham (1971).

Rocky Intertidal. The most abundant animals of rocky intertidal shores are various invertebrates, such as barnacles, mussels, ascidians, sea anemones and hydroids, sea urchins, starfish, and worms, often occurring in dense epifaunal aggregations interspersed with seaweed and forming discernible bands or zones related to tidal level but modified by degree of wave action. This banding or zonation has been observed and studied by European biologists since about 1830. Characteristically the animals of the uppermost level are small herbivorous snails (littorines and limpets) and wandering shore crabs and isopods. Below this somewhat bare region is the barnacle zone, followed by patches or bands of rockweed. At about mean sea level there may be aggregations of mussels, as on the Pacific coast of North America, or large, solitary ascidians, as in several shores of the Southern Hemisphere. Below this the zonation becomes less regular, with more seaweed. At approximately mean low tide, there are often dense growths of Laminaria and similar kelps. These zones and associations of plants and animals have been described for most of the rocky shores of the world in the temperate zone in many books and research papers; a useful summary will be found in Stephenson and Stephenson (1972); Lewis (1964) provides exhaustive details for British shores.

Zonation is usually considered to be caused in some way by tidal levels, although Stephenson and Stephenson considered it a universal but nontidal phenomenon. Recent studies suggest that zonation is the result of complex interaction of physical and biological causes—that is, the levels of optimum or massive occurrence of the organisms depend on both adaptation to tidal levels and interactions between predators and prey and grazers upon the seaweed. The problem with relating zonation to tidal level per se is that most tidal records are from gauges maintained in harbors, and do not reflect fluctuations some distance away on open rocky coasts, where it is also easily apparent that the zones tend to expand vertically with the degree of wave exposure. Another factor of significance is the abrupt changes in exposure and/or immersion time associated with the rise and fall of the tide, a phenomenon known as *critical levels*. The intertidal region of rocky shores is a rich interface zone as attested by the dense accumulation of organisms; it is a mistake to think of these organisms as tolerant of the ex-

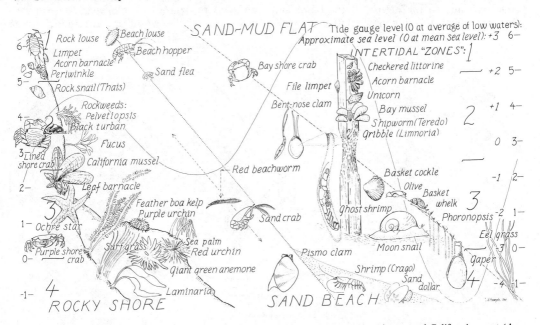

FIGURE 2. Zonal arrangements of intertidal biota on various substrates on the central California coast (drawing by J. W. Hedgpeth, courtesy of the University of California Press).

tremes of exposure to air and water; they are adapted to this changing environment, perhaps in many cases requiring the fluctuations for maximum development. Herbivorous gastropods and chitons, for example, may require the warmth of air at low tide for digestive enzymes to function.

In this crowded region between the tides, where the major need of many sedentary or attached organisms is for space to settle and remain, the lower limits of some of these may be controlled by the upward limits of predators, such as sea stars, whereas the upper limits are related to the tides and other physiographic factors. These matters of resource partition, interspecific interactions, and competition for space are at present the principal interest of students of intertidal ecology. For a recent summary, see Hedgpeth (1976) and Price, Irvine, and Farnham (1980).

Zoogeographical Patterns

The fauna of the world's coastal regions is distributed in space and time according to prevailing temperature patterns, locally modified by upwelling, salinity (i.e., runoff from land), and sediment and, in polar regions, by ice conditions. We recognize in a general way arctic and antarctic faunas, tropical faunas and various intergrades of temperate faunas, especially in the shallow seas and coastal regions. These various divisions of the fauna are often considered provinces, although the term itself has never been clearly defined. This difficulty is in large part due to the nature of the fauna and the consequent differing viewpoints of the special interest of the biogeographer. Patterns of distribution of mollusks, for example, may be somewhat different from those of crustacea. Nevertheless, the best general summary for the arrangement of marine littoral (or shallow water and shore) provinces is that of Ekman (1953):

Arctic
Boreal
  European Atlantic
  North American Atlantic
Temperate Pacific
Warm Water
  Indo-West Pacific
  Atlantic-East Pacific
    Tropical and subtropical America
    Tropical and subtropical West Africa
  Mediterranean Atlantic
  Sarmatic
Warm temperate of Southern Hemisphere
  South coast Africa
  Southwest Africa
  Southern Australia and Tasmania
  New Zealand
  Peru and northern Chile
Antiboreal
  South America
  Oceanic islands
  Kerguelen
Antarctic

The major divisions of Ekman's scheme are illustrated in figure 3. Many of the subsidiary divisions recognized by biogeographers in various parts of the world are considered to be

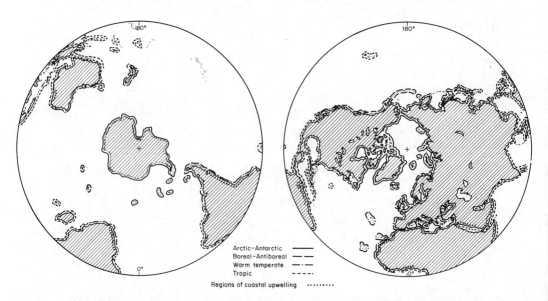

FIGURE 3. Littoral provinces of the world, compiled principally from Ekman (1953) (from Hedgpeth, 1957).

provinces, and are given names of local implication; as long as one is in the middle of one of these regions, it seems clear enough but toward the limits of such well-defined zones there is uncertainty. For that reason many biogeographers prefer to work within the physical limits of temperature, as indicated by the implied definitions in figure 4. However, temperature, especially in the coastal region, acts in different ways. Some species may be limited by the temperatures favorable for reproduction in the north, and those that permit survival in the south, or vice versa. These patterns of reproduction and survival were recognized by Hutchins (1947), and are illustrated, along with the various isotherms and provincial categories for the Atlantic coast of North America in, figure 5. The phenomena illustrated here indicate a comparatively wide rage of tolerance for environmental variations in coastal regions, but do not take into account the well-known limit of tolerance for high temperatures in tropical species—that is, many if not most tropical invertebrates live near the upper limit of their tolerance, and are more susceptible to increased temperatures than temperate species are to comparable increases.

This classification recognizes the approximate prevailing temperature limits illustrated in figure 4. The kinds of animals living between these various limits in comparable parts of the world are similar, although there are few cosmopolitan species in temperate latitudes separated by the tropical regions, and even fewer similar systematically related species at the polar extremes. Tropical faunas are best developed on the eastern sides of the oceans; along upwelling coasts reef-forming corals do not thrive, and some representatives of the temperate fauna occur far away from the major populations of temperate regions in pockets of upwelling. It is usually considered that tropical coasts, especially coral reefs, have the highest diversity of animal life, but this remains to be demonstrated on a comparable unit basis with other seas. In shallow Antarctic waters there is a luxuriant development of epifaunal animals that may be as diverse as the fauna of a coral atoll. The temperate faunas of the world are somewhat similar in having conspicuous echinoderms (sea stars and urchins), abundant attached animals including barnacles, mussels, ascidians, hydroids, and sponges, on rocky shores, and large populations of various arthropods such as isopods, amphipods, and crabs. In the polar regions intertidal fauna is reduced because of ice action, so that there are no barnacles or other attached organisms, few limpets or other snails, and a conspicuous reduction in decapod crustacea. There are no shallow water crabs in the high latitudes, and in both Arctic and Antarctic seas the major shallow water arthropod scavenger/predator is a large isopod. There is a tendency toward shortened larval stages, brooding of young, and viviparity rather than reproduction by free release of gametes that is the major pattern of reproduction in tropical and temperate invertebrates.

The seas of the higher latitudes are rich in fish and crustaceans (large anomuran crabs in the sub-Arctic and euphausiids in the Antarctic), and there is an abundant vertebrate fauna, both birds and mammals. Seals and sea lions are not characteristic of tropical regions, and such birds as auks, murres, and penguins form massive populations in Arctic and Antarctic waters.

The most obvious contrast between the fauna of the Arctic and that of the Antarctic is that there are no resident vertebrates on Antarctic shores—no land carnivores such as the polar bear of the Arctic. Both are essentially sea-based natural economies, depending on the resources of the sea. The penguins, for example, are marine animals, feeding on fish and euphausiids, that nest on land. Most of the seals of the Antarctic are more abundant on sub-Antarctic island shores or the pack ice; only the Weddell seal hauls out on the Antarctic ice with any frequency. In the Arctic there is a similar distribution exemplified by the herds of the Pribilof Islands. The exception to this vertebrate pattern is the Galápagos, where the cool westward current from South America results in water temperatures that are temperate rather then tropical at the equator. Thus a few Galápagos penguins are actually in the Northern Hemisphere.

Systematic Composition

The variety and complexity of life on the seashore and in the shallow seas are major attractions of this environment for both the casual amateur and the dedicated zoologist; there is no

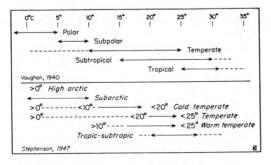

FIGURE 4. Approximate temperature limits of various biogeographic zones (from Hedgpeth, 1957).

COASTAL FAUNA

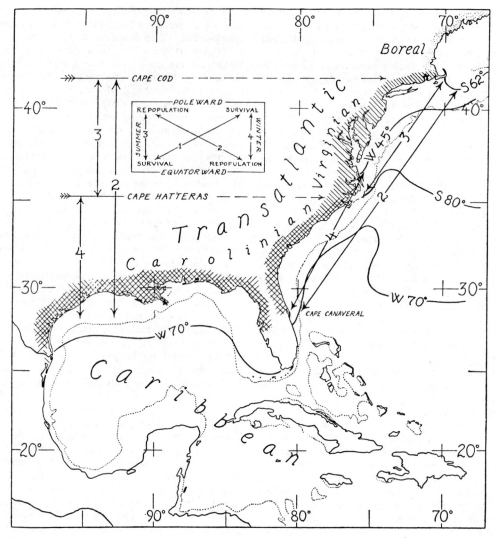

FIGURE 5. Biogeographic regions of the Atlantic and Gulf coasts of North America (from Hedgpeth, 1953).

other environment where representatives of all the major phyla of animals may be found. Such representation does not occur everywhere, but the interest and appeal of every type of seashore and coastal locality are attested to by the presence of marine laboratories in every major maritime country and by the numerous guidebooks to the littoral fauna of all the major regions of the world.

The following synopsis is an attempt to list, by class or order (as appropriate), the characteristic animals of the coastal regions of the sea.

Phylum Protozoa

Class Mastigophora. Of the nineteen or so orders, three are characteristically marine: Silicoflagellida, Coccolithophorida and Dinoflagellata. Among the dinoflagellates many are coastal, and such genera as *Gymnodinium, Peridinium,* and *Gonyaulax* often occur in massive *red tide* populations. *Noctiluca* is sporadically abundant on shores, where, as its name implies, it glows in the dark.

Class Sarcodina. The two important marine orders are Foraminifera and Radiolaria. There are many near-shore species of foraminifers, some of them occurring in the more saline parts of bays, but representatives of this order, like the radiolarians, are more characteristically pelagic.

Class Ciliata. Ciliates are ubiquitous, on the outside and inside of animals (and plants), amongst crud and bottom detritus, and in the surface film; one order, Tintinnida, is characteristically marine and pelagic. Orders represented along the shore include Gymnostomatida, Chonotrichida (epizoic on crustacea), Apostomatida, Astomatida, Hymenostomatida, Thigmotrichida, Peritrichida, Oligotrichida, Tintinnida, and Hypotrichida.

The Sporozoans, variously classified as a major class or as subphyla, are all parasitic, mostly in terrestrial vertebrates. Representatives of the Class Myxosporidea occur in fishes and annelids.

Phylum Mesozoa
Mesozoans are organisms of obscure affinities. The class Dicyema are parasites in the nephridia of cephalopods; the class Orthonectida are parasites of various marine invertebrates. All are of course marine.

Phylum Porifera
All sponges are marine except the family Spongillidae (class Demospongia, order Haplosclerina); many coastal species are conspicuous components of the intertidal fauna, forming bright red, yellow, or purple crustlike growths at or slightly below mean low water. The class Hexactinellida or glass sponges are confined to deeper water, from 100 m to abyssal depths.

Phylum Coelenterata
Jellyfish, sea anemones, and corals are familiar creatures of the coast, often occurring in enormous aggregations.

Hydrozoa are the hydroids and small medusae; many hydroids have no medusa stage; all are marine except the small solitary freshwater *Hydra* and *Chlorohydra*, the colonial estuarine to freshwater *Cordylophora*, and the small freshwater jellyfish *Craspedacusta*.

Scyphozoa are large jellyfish and are exclusively marine. Representatives of the small order Stauromedusae are sessile, found near shore and in bays, attached to seaweeds and eel grass.

Anthozoa are the coelenterates of the shore—sea anemones and various kinds of corals. The reef-forming corals belong to the order Madreporaria or Scleractinia.

Phylum Ctenophora
Comb jellies, sea walnuts; sometimes included with coelenterates as phylum Cnidaria; all marine; many in coastal waters.

Phylum Platyhelminthes
Many members of the classes Trematoda and Cestoda (flukes and tapeworms) are parasitic in marine fishes (all members of both classes are parasitic); free-living flatworms belong to the Turbellaria and are marine except for a few representatives of the orders Rhabdocoela, Allcoela, and Tricladida. Littoral turbellarians of the order Polycladida are often large, spectacularly colored animals; *Stylochus* is a predator on young oysters.

Phylum Nemertea (Ribbon worms)
Nemerteans are often encountered in intertidal waters, looking like damp string hanging across rocks and attached organisms. Some are dull green; others bright orange. Some may be several feet long. A few are small, commensal in bivalves. The representatives of the order Hoplonemertini include: a freshwater genus, *Prostoma;* a terrestrial genus on marsh shores, *Geonemertes;* and a number of bathypelagic species occuring at depths of 1000 m or more.

Phylum Aschelminthes
Many zoologists consider the *aschelminth complex* to be a gathering of small organisms deserving separate phyletic rank; others prefer to regard these groups as phyla. Most are small, difficult to see and observe, but often abundant.

Class (or phylum) Rotifera. Most characteristically freshwater organisms, but there are many marine and estuarine forms, mostly epizoic on larger animals.

Class (or phylum) Gastrotricha. Uncommon, more conspicous in fresh-water.

Class (or phylum) Kinorhyncha (Echinodera). Exclusively marine, small to minute.

Class (or phylum) Nematoda. Nematodes, or threadworms, are among the more numerous animals of the earth, living in soil, freshwater, and the sea. Many are parasitic of animals or plants. They are often abundant in the bottom detritus in tide pools.

Class (or phylum) Nematomorpha. Horsehair worms are characteristically found in freshwater; the larval stage is parasitic in arthropods. There is one marine genus, *Nectonema*, parasitic in crustacea.

Class (or phylum) Acanthocephala. Parasites of vertebrates; some species occur in marine mammals.

Class (or phylum) Priapuloidea. All marine; the most common genus, *Priapulus*, occurs in the muds of marine bays.

Phylum Gnathostomulida
These small worms, living obscurely in anaerobic sands or muds as well as in cleaner sediments, were only recently recognized as a separate phylum. They are apparently abundant and are characteristic of coastal environments.

COASTAL FAUNA

Phylum Sipuncula
Peanut worms. An exclusively marine phylum, with many coastal representatives. Most are nestlers in cavities, empty shells, or among algal holdfasts and Zostera rootstocks.

Phylum Echiura
Another exclusively marine, predominantly nestling, group, although some species maintain permanent burrows in sheltered tidal flats. Some occur in the deep sea.

Phylum Annelida
Class Oligochaeta. Not all oligochaetes are earthworms; there are many small freshwater species and an important group of marine and estuarine species, often abundant in foul and anaerobic situations, These include representatives of the families Enchytraedae, Naididae, and Tubificidae.

Class Hirudinea. Leeches are predominantly freshwater animals, but there are some marine species, ectoparasites of sharks, fishes, and some crustacea.

Class Polychaeta. Polychaetes are marine worms par excellence, and are part of the characteristic fauna of the shore. Errant or wandering forms are found crawling almost everywhere; a few species may attain lengths of more than 1 m. Sedentary polychaetes form tubes of sand grains or calcareous encrusting tubes on rocks and larger animals, or burrow within the sediments. Some have elaborate feeding fans of tentacles with branches armed with cilia (*feather duster worms*). There are a few freshwater polychaetes.

Phylum Arthropoda
Most of the animals known in the world, or least those given names, are arthropods, and most of these are insects, an overwhelmingly terrestrial group (many with freshwater larval stages). Inevitably arthropods are also a conspicuous and significant element of the coastal and intertidal fauna. In order to cope with this vast assemblage of at least a million species, a somewhat more detailed hierarchy of subphyla, classes, and orders is required.

Subphylum Onychophora. Terrestrial
Subphylum Trilobita. Marine, but—alas—extinct.
Subphylum Chelicerata
Class Merostomata. Consists of the extinct aquatic Eurypterids and the extant Xiphosura, the horseshoe crabs. The most eminent of these survivors of Paleozoic days, *Limulus polyphemus*, is a common animal of north Atlantic shores. Other species are found in East India and southeast Asia.

Class Arachnida. Predominantly terrestrial, with nine orders, of which only the Trombidiform mites have significant representation in the sea; these marine mites (Halacaridae) are often abundant. There are a few species of Pseudoscorpions in tidal marshes and rocky intertidal shores, an occasional maritime scorpion, and one or two marine diving-bell spiders.
Subphylum Pycnogonida
Exclusively marine, sometimes in high salinity bays, pycnogonids are found from intertidal situations to depths of several thousand m. They are most abundant in colder waters, expecially the Antarctic, but there are many small species on all shores rich in algae. Their appropriate hosts are mostly coelenterates and other soft-bodied invertebrates.
Subphylum Mandibulata
Class Crustacea
Subclass Cephalocarida: small crustacea living between sand grains (interstitial). They have been overlooked until recent years but may be widespread albeit elusive. One species was found in San Francisco Bay some years ago but has not been collected since.

Subclass Branchiopoda: phyllopods or fairy shrimp. Most of these are freshwater or inland saline animals. The brine shrimp *Artemia* occurs worldwide in marine salt evaporation works; it can live in seawater but seldom does. The widespread freshwater order Cladocera includes a few coastal species.

Subclass Ostracoda: small, bivalved crustaceans looking like beans or seeds, abundant in both freshwater and seawater. They are often found in bottom accumulations of detritus and crud.

Subclass Copepoda: the key animals of pelagic systems, occurring in vast populations. As the prime herbivores, they convert the productivity of the diatoms to food for fishes. There are seven orders of Copepods, several of which are specialized parasites (often on fishes) or commensals of invertebrates. The common orders represented in coastal environments are: Calanoida, which includes most of the actively swimming coastal and estuarine species; Cyclopoida, mostly small commensal species in coastal environments; and Harpacticoida, of which there are thousands of small benthic, detritus-living species. The orders Caligoida and Lernaepoida are fish parasites, usually visible only as paired egg sacs hanging from the fish.

Subclass Branchiura: related to copepods, but easily visible flattened ectoparasites of both freshwater and coastal fishes.

Subclass Cirripedia: exclusively marine (a few may survive for a while in freshwater but cannot reproduce) and are among the most characteristic of seashore animals, forming conspicuous intertidal bands. These are the acorn and stalked barnacles of the order Thoracica. There are two orders, Acrothoracica and Ascothoracia, of minute creatures that bore into shells, corals, and limestone, and are especially numerous in such calcareous environments as coral reefs. The order Rhizocephala are parasites so specialized that they appear only as sacs of reproductive tissue on various species of crabs.

Subclass Malacostraca: crabs, shrimp, lobsters, pill bugs, beach hoppers, and so on.

Nebaliacea: a small order of bivalved marine crustacea, often found on tidal flats in an aerobic situation under seaweed.

Mysidacea: abundant near shore and in bays. One genus is a freshwater glacial relict. *Neomysis mercedis* is apparently the key forage animal of the San Francisco Bay *null zone.*

Cumacea: an order of small crustaceans living for the most part in marine sediments.

Tanaidacea: small, living in tubes or in sediments from shore to several thousand m.

Isopoda: one of the larger crustacean orders, including the terrestrial familiar garden sow bugs, the high intertidal rock louse, many freshwater species, and hundreds of marine species, including wood borers, scavengers, and fish parasites.

Amphipoda: another large order, consisting of beach hoppers, found on every sandy shore, especially under wrack and flotsam (there is a terrestrial species as well), hundreds of marine species in shallow waters, and a pelagic group. One peculiar group, the Caprellida, are semisedentary, attenuated inhabitants of crowded intertidal situations. One family, cyamidae, are external parasites of whales, cetacean "crab lice." There are several important tubiculous species living in bays and a large contingent of freshwater species.

Stomatopoda: mantis shrimps are medium (8 cm) to large-sized burrowing crustaceans, characteristic of warm to tropical shallow waters.

Euphausiacea: planktonic, occurring in vast numbers; the krill *Euphausia superba* and its nearer shore relative *E. crystallorophias* are the mainstay of penguins and crabeater seals, creatures of the coastal environment of Antarctica.

Decapoda: crabs and shrimps, mostly marine. There are a few isolated freshwater shrimps, a group of freshwater crabs, and the familiar freshwater crayfish. There are "terrestrial" hermit crabs—e.g., *Birgo,* the coconut crab—but they must release their eggs in the ocean. Crabs, shrimp, and lobsters are bottom feeders, and their occurrence in the accessible shallow waters of the coastal regions is a significant food resource.

Class Insecta. Of the more than 20 orders of insects, marine representation is restricted to the following:

Collembola: *Anurida* is a genus of intertidal or littoral springtails, often locally abundant. Other collembolans occur in wrack and marshes.

Thysanura: a few species of maritime springtails (*Petrobius*) all on northern European coasts, are known.

Dermaptera: a maritime earwig on European shores, now naturalized on the Atlantic coast of North America.

Mallophaga: biting lice found on marine birds.

Anoplura: sucking lice found on seals, sea lions, walrus, and the like.

Hemiptera: some corixids (water boatmen) occur in bays. The marine water striders (*Gerridae, Halobates*) are the only insects living on the surface of the sea, often far from land, but most species are usually not distant from shore. There are 40 or more species of closely related bugs found on marine and bay shores in various parts of the world mostly of the families Veliidae, Mesoveliidae, and Harmatobatidae. There are many species of shore bugs (*Saldidae*) on temperate shores.

Trichoptera: of the 5000 species of caddis flies, only two can live in saltwater, a species found in salt-marsh pools in England, and one in tide-pools in New Zealand and southeast Australia.

Diptera

Culicidae: tideland mosquitoes of salt marshes and mangrove swamps are proverbially numerous and annoying. At least 9 genera are represented.

Ceratopogonidae: biting midges are numerous almost everywhere except in polar latitudes. There are 3 genera with species considered marine by virtue of larval habitat: *Culicioides, Leptoconops,* and *Dasyhelea.*

Chironomidae: among the nonbiting midges there are 65 or more species considered marine, including the Antarctic *Belgica antarctica* and several other species of the sub-Antarctic islands. Many of these midges spend most of their lives in green turflike algal growth (enteromorpha) of the high intertidal regions. In some species the females are wingless.

Flies: among the flies, the Tabanidae (horseflies and deer flies) are well represented on coasts and marshlands; the Coelopodidae (seaweed flies) live in seaweed wrack and may be pestiferously abundant, but the most widespread and abundant maritime species are those of the Ephydridae (brine flies). Some species withstand high salinities and occur in salt evaporation ponds.

Coleoptera: most marine beetles are representatives of the Staphylindae (rove beetles). There are at least 56 genera with marine species, occurring in beach wrack and under flotsam, on sandy shores and bay shores; some of those occurring on rocky shores live deep in crevices or in intertidal sponges. At least 9 other families of beetles occur in marine or maritime situations, notably the carabid beetles from northern Europe to the Antarctic (Crozet Islands), some diving and whirligig beetles and some phytophagous beetles, including a weevil with an intertidal larval stage in the Kerguelen Islands. In view of the predominantly substrate and benthic habitat of adult beetles, this representation is not surprising.

Phylum Mollusca

The mollusks are predominantly marine animals. Of the major groups, the Polyplacophora (chitons) are exclusively marine, in shallow to intertidal waters, primarily of rocky shores, where they are more abundant and diverse in temperate Pacific waters than on other shores, and the Cephalopoda (squids and octopus), also marine, are represented in nearshore waters by octopus on rocky shores and squid and cuttlefish in the shallows, but most species of squid are pelagic or bathypelagic.

Class Gastropoda. Snails, slugs (univalves). A shore or bay without some kind of snail, limpet, or slug is inconceivable. They are everywhere, often in great abundance. Of this great array of 40,000 species or more, only one group, the subclass Pulmonata, representing perhaps 16,000 species, is characteristically nonmarine, living in freshwater and on land. Even some of these have returned to the seashore as specialized limpets. A few relatives of the more characteristically marine littorines are also in freshwater.

Class Bivalvia. The clams, mussels, oysters, and wood-boring shipworms are even more marine as a group than the gastropods. While some species of oysters, mussels, and clams may do well in upper bays, none can survive long in freshwater. Only one group, the order Schizodonta, the freshwater mussels, has colonized streams and lakes.

Phylum Bryozoa

The moss animals are predominantly shallow water and coastal in habitat. They form lacy encrustations on rocks, seaweed, other animals, or small bushy growths. In some shallow subtropical waters, enormous massive clumps of a gelatinous bryozoan, *Zoobotryon,* may occur. The Gymnolaemata, with 5 orders, are marine; The Phylactolaemata is a small class of freshwater bryozoa.

Phylum Brachiopoda

The lamp shells are curious, mollusklike animals, once abundant in shallow Paleozoic seas. The survivors of this once thriving host are few in species and numbers. *Lingula* and *Glottidia* are sometimes found in intertidal mud or sand; the other species of brachiopods are mostly attached to hard substrates.

Phylum Phoronida

There are only 2 genera and half a dozen species of these worms; they are burrowers nestlers, and tube formers.

Phylum Chaetognatha

Arrowworms. One genus, *Spadella,* lives near shore and clings to rocks and seaweed; the others (*Sagitta* and so on) are predaceous, actively swimming creatures of the plankton, near shore and pelagic.

Phylum Echinodermata

There are no freshwater echinoderms, and none tolerant of water with salinity less than about half that of seawater. Of all the living orders, only the Crinoids are deep-water forms, yet the shallow-water representatives, *Antedon* (feather stars) are coastal, shallow-water forms, especially in the tropics. Sea stars, sea urchins, brittle stars, and sea cucumbers (there is a good representation of these in the deepest oceans) are all part of the shore and nearshore scene.

Phylum Hemichordata

A small group of somewhat enigmatic wormlike creatures that, like chordates, feed by taking food through a mouth and involving the anterior end of the respiratory system in sorting and selecting it. This pharyngeometry suggests affiliation with chordates. They are all marine, often abundant in patches on intertidal flats (*Balanoglossus*).

Phylum Chordata

The chordates are divided into two divisions: Acrania and Craniata—or those without a brain case or well-defined head; and those with a brain case and usually with a vertebral column. Thus, by the broadest definition, the Chordates are an invertebrate phylum.

Subphylum Tunicata

All marine. There are three classes, the pelagic Larvacea and Thaliacea and the benthic, mostly shallow-water and intertidal Ascidiacea, the sea squirts, the filter feeders par excellence of the sea.

Subphylum Cephalochordata

Lancets or *Amphioxus* of shallow, sandy bottoms.

Subphylum Agnatha. Jawless vertebrates.

Class Cyclostomata. Eellike, jawless fishes. Lampreys are anadromous or freshwater; the hagfishes are marine, in deeper waters.

Class Chondrichthyes. Sharks and rays belong to the subclass Elasmobranchii, with 4 living orders. All of them are marine, found in coastal waters, some resident, others vagrant from the open sea. Many of the rays are bay and estuarine species, active predators of mollusks and worms. A few sharks have invaded freshwater. Holocephalii (Chimaeras) are the other subclass, a small group of deeper-water species.

Class Osteichthyes. The bony fishes are the largest group of vertebrates, with about 30,000 species in 42 orders.

Crossopterygii: the coelacanth *Latimeria* lives near shore in relatively shallow depths in the Comoro Islands. Its nearest relatives thrived in Mesozoic times.

Acipenseriformes: some sturgeons are anadromous, and may be found in estuaries during their spawning migration.

Elopiformes: tarpons. There are two species, both frequenting coastal waters and estuaries in warm waters and spawn near shore.

Anguilliformes: in addition to the catadromous eels (living in freshwater and spawning in the sea), there are the moray eels, common in the tropics, especially in coral reefs.

Clupeiformes: there are about 300 species of herrings, sardines, anchovies, and menhaden; they are fishes of coastal regions and the shelf seas and are the most important to commercial fisheries.

Salmoniformes: most salmon are anadromous, living in the sea, often within the coastal region; trout, char, and grayling are freshwater fish.

Siluriformes: marine catfish, living in shallow coastal seas, are of the family Ariidae.

Batrachoidiformes: toadfish and midshipmen, usually in tropical or subtropical seas; the midshipman *Porichthys* is common in bays and deposits its eggs under rocks.

Gobiescoiformes: clingfish adhere to rocks and shells in shore waters.

Atheriniformes

Cyprinodontoidei: killifish and mosquitofish are mostly freshwater; some species of *Fundulus* live in bays and can withstand high salinities.

The atherinid *Leuresthes tenuis,* the California grunion, spawns high on intertidal beaches at night during new or full moon.

Gasterosteiformes: includes sticklebacks, pipefishes and sea horses, all shallow-water species; sea horses occur in tropical waters.

Scorpaeniformes: scorpion fish and cottids (bullheads) are fishes of rocky shores and coastal shallows.

Perciformes: the largest order of fishes; most of them are freshwater; but the Embiotocidae, surf perch, are viviparous fishes of the North Pacific from California to Japan; the Mugilidae, mullets, are important tropical and warm-water coastal and bay species. This order also includes barracuda, gobies (mostly shallow-water and shore species), mackerel, tuna, and the remora. Another important suborder is the Blennioidea (Blennies), which include many tide-pool species.

Pleuronectiformes: flatfishes, especially flounders, are common in bays, and many species—e.g., halibut, plaice, turbot—are important fishes of the shallow seas and fishing banks.

Tetraodontiformes: puffers, trunkfishes, and the like. Fishes of warm shallow seas.

Class Amphibia. There are no marine amphibians.

Class Reptilia. Once in Mesozoic seas there were numerous large marine reptiles, some of them similar in aspect to whales and porpoises, others like nothing now on earth unless the Loch Ness monster is a relict Plesiosaur.

Chelonia: the turtles and tortoises, include the principal marine reptiles of these days; there are two families of sea turtles: Chelonidae and Dermochelyidae. They live in tropical and warm temperate seas, and crawl ashore on isolated beaches to dig nests and deposit their eggs.

Of the hosts of lizards, only the Galápagos iguana (*Amblyrhynchus cristatus*) can be considered marine. It feeds on seaweeds, basks in the sun for its digestive juices to work, and buries its eggs in the warm beach sand.

There are several maritime snakes, especially those of the genus *Natrix,* that live in salt marshes, and the poisonous sea snakes of the family Hydrophiidae (related to cobras and coral snakes). There are about 50 species of these live-bearing snakes in tropical seas, usually in coastal regions.

Among the crocodiles (Crocodilia), the saltwater crocodile of northern Australia and Indonesia, *Crocodylus porosus,* frequents nearshore waters and estuaries, and is considered the most dangerous of the crocodiles to man.

Class Aves. There are 8600 species of birds in 30 orders, but representatives of only 10 orders are marine or coastal.

Order Sphenisciformes: penguins are maritime birds that spend much of their lives feeding like porpoises. They nest (or brood) on islands and continental shores, from about 74°S to the equator (at the Galápagos).

Podicipediformes: grebes are mostly freshwater birds, but some are often seen on bays or near shores.

Procellariformes: the albatrosses, fulmars, and petrels are birds of the open seas and stormy coasts, nesting on islands, mostly in the Southern Hemisphere.

Pelecaniformes: includes not only the pelicans but also cormorants, boobies, gannets, and man-of-war birds. These birds are especially abundant in tropical and warmer waters, and most of them are part of the nearshore scene.

Ciconiformes: some herons and ibises are common in bays and along shores; flamingos nest in salt lakes.

Falconiformes: Of the eagles, falcons, and vultures, the best-known coastal species are the almost cosmopolitan osprey (*Pandion haliaëtus*), often seen along seacoasts, the marsh hawk (*Circus cyaneus*), a marine and coastal bird as well as one of inland marshes, and the North American–Siberian Bald eagle (*Haliaetus leucocpehalus*), a bird primarily of coastal environments near the sea, as its generic name implies.

Anseriformes: of this order, the most frequently encountered in bays and on coasts are the diving ducks; brant (a small goose) feed on eel grass in bays.

Gruiformes: cranes and coots; cranes are often seen on salt marshes; the whooping crane (*Grus americana*) winters on salt marshes along a restricted part of the Texas coast. The coot, or "tourist duck," is found almost everywhere in America, including bays.

Charadriiformes: these are the gulls and shorebirds. Gulls and terns are in the family Lariidae. The Alcidae—auks, murres, and puffins—are birds of rocky coasts; the extinct great auk was the Northern Hemisphere counterpart of penguins.

Passeriformes: most birds—5000 species—belong in this order, which includes sparrows, warblers, and all the other hosts of little brown birds. Some are characteristic of salt marshes, but crows and ravens are the principal frequenters of seashores among this large groups of birds.

Class Mammalia. There are more than 4000 species of mammals, but the aquatic marine species are represented in 3 comparatively small orders.

Cetacea: whales, dolphins, and porpoises. Whales from the high seas are often seen near shore; others live near shore. The harbor porpoise is a coastal and bay species; the largest of the porpoises, the killer whale (*Orcinus*), is often encountered near shore. There are freshwater dolphins in China, India, and South America. Among the whalebone whales, the California gray whale is a species of coastal waters, and enters coastal lagoons to give birth.

Pinnipedia: seals and sea lions. This group includes the elephant seals of the North Pacific and subAntarctic, the Arctic walrus, as well as the familiar harbor seal of the Northern Hemisphere. They often occur in large bands or herds and are a familiar trademark of temperate to boreal Arctic shores. Several species live near the subAntarctic islands, and the large Weddell seal is the southernmost mammal, spending the cool summers in the southern part of the Ross Sea.

Sirenia: the extant manatees are found mostly in rivers of west Africa, South Africa, Florida, and Indo-Pacific coasts, but also in bays and marine shoals. They are inoffensive vegetarians. Once the great Steller's sea cow, the most marine of manatees, flourished in the North Pacific, and is generally thought to have been exterminated in 1768, but some people would like to think otherwise.

Carnivora and other coastal mammals: along Pacific and Aleutian shores the sea otter *Enhydra lutris,* a large member of the weasel family (Mustelidae), is flourishing after a severe decline from exploitation for its splendid pelt. This animal spends most of its life on the sea. The European river otter often takes up residence along the rocky shores of northern Britain and Scotland. In Arctic seas the polar bear is maritime to marine, living on seals and fishes.

That characteristic American raccoon, *Procyon lotor,* is, wherever occasion warrants, a coastal animal of marshes and seashores, leaving its sign of cracked crabs and ravaged shellfish along with its tracks on the beach the morning after. It is now becoming naturalized in Europe.

The largest order of mammals, the rodents, has no marine representative, although muskrats may occur in brackish or low-salinity bays, and even ground squirrels have been observed to feed on fish eggs (and probably turtle eggs) on sandy shores. Some rats, mice and voles are native to salt marshes, and the ubiquitous Norway and black rats occur wherever they can find food.

Primates: there are no marine primates, but the most ubiquitous, omnivorous, and eurytopic of the primates *Homo sapiens* frequents many seashore and coastal sites, and, by his preemptive and destructive occupancy, alters the environments and endangers the continued existence of the fauna of the marginal seas, shores, bays, and estuaries of the coastal regions.

JOEL W. HEDGPETH

References

Chapman, V. J., 1977. *Ecosystems of the World,* vol. 1, *Wet Coastal Ecosystems.* Amsterdam: Elsevier Scientific Publishing Company, 428p.
Cheng, L., ed., 1976. *Marine Insects.* Amsterdam: North Holland Publishing Co., 581p.
Ekman, S., 1953. *Zoogeography of the Sea.* London: Sidgwick and Jackson, 417p.
Eltringham, S. K., 1971. *Life in Mud and Sand.* London: English Universities Press, 218p.
Hedgpeth, J. W., 1953. An introduction to the zoogeography of the northwestern Gulf of Mexico with reference to the invertebrate fauna, *Inst. Marine Sci. Pub.* 3, 102-224.
Hedgpeth, J.W., ed., 1957. *Treatise on Marine Ecology and Paleoecology,* vol. I, *Ecology.* New York: Geological Society of America Mem. 67, 1296p.
Hedgpeth, J. W., 1976. The living edge. *Geoscience and Man* 14, 17-51.
Hutchins, L. W., 1947. The bases for temperature zonation in zoogeographical distribution, *Ecol. Monogr.* 17, 325-335.
Lewis, J. R., 1964. *The Ecology of Rocky Shores.* London: English Universities Press, 323p.
Moebius, K., 1877. *Die Auster und die Austernwirthschaft.* Berlin: von Wiegandt, Hempel & Parey, 126p. (An English translation appears in *1880 Rept. U.S. Fish and Fisheries Commission,* 1883, 683-751.)
Perkins, E. J., 1974. *The Biology of Estuaries and Coastal Waters.* London: Academic Press, 678p.
Petersen, C. G. J., 1918. The sea bottom and its production of fish-food, *Danish Biol. Station Rept.* 18, 1-62.
Phillips, R. C., and McRoy, C. P., 1980. *Handbook of Seagrass Biology: An Ecosystem Perspective.* New York: Garland STPM Press, 353p.
Price, J. H.; Irvine, D. E. G.; and Farnham, W. F., 1980. *The Shore Environment,* vol. 1, *Methods;* vol. 2, *Ecosystems.* London: Academic Press.
Stephenson, T. A., and Stephenson, A., 1972. *Life between Tidemarks on Rocky Shores.* San Francisco: W. H. Freeman, 437p.
Thorson, G., 1971. *Life in the Sea.* New York: McGraw-Hill, 256p.

Cross-references: *Antarctica, Coastal Ecology; Aquaculture; Arctic, Coastal Ecology; Biotic Zonation; Coastal Flora; Coastal Waters Habitat; Coral Reef Habitat; Dieback; Estuaries; Estuarine Habitat; Grab Samplers; Intertidal Mud Habitat; Intertidal Sand Habitat; Insects; Loose Rock and Stone Habitat; Rocky Shore Habitat; Sand Dune Habitat; Thalassotherapy.*

COASTAL FLORA

Coastlines are among the most fascinating places in the world for a biologist. Here we find numerous biotic communities compressed into relatively narrow belts paralleling the coast, often with surprisingly sharp boundaries or *ecotones* between them. A wide variety of habitats, from strictly terrestrial to completely marine, can be observed in the same area. The factors that determine their location, composition, limits, productivity, interactions, and change with time can be readily studied. The marine plant communities are discussed in a general way in McConnaughey (1978); the terrestrial maritime communities, by Chapman (1976).

Transition: Terrestrial to Marine

On rocky coasts a dark band of microscopic plants, the *black zone,* occurs at or just above the level of the extreme high tides, marking the transition from terrestrial to marine-dominated habitats.

On other types of coastline there is no black zone, but the lower boundaries of salt marshes, mangrove swamps, or beach vegetation form well-marked, sometimes abrupt transitional zones.

Marine Communities

Zonation of marine communities below the boundaries mentioned above is determined largely by the tides. There are commonly two high tides and two low tides in each 24-hour day, each reaching a different level. The tidal cycle shifts nearly half an hour each day in keeping with the lunar cycle. There is also an approximately fortnightly rhythm in tidal amplitude, a week of high high tides and low low tides (the spring tides) being followed by a week of tides of intermediate amplitude (the neap tides). Although tides are the dominant factor determining the level of intertidal communities, they are not the only factor. Exposure to wave action, sun, rain, wind, and interactions between species also exert influences. It is more helpful to describe zonation in terms of the biotic communities than in terms of strictly defined tide levels.

Usually a division into upper, middle, lower, and sublittoral zones is evident. The upper littoral comprises the levels exposed most of the time, covered only by high spring tides. The middle littoral is the range of the neap tides, covered and uncovered by most successive tides. The lower littoral is covered most of the time, exposed only during low spring tides. In areas where there is considerable tidal amplitude subdivisions within these zones may be evident. The sublittoral lies below the extreme low tides but is still strongly influenced by wave action, turbulence, storms, and coastal runoff. The type of community that develops is determined by the character of the substrate.

Sharp boundaries between intertidal communities may indicate critical tide levels—levels at which the greatest average change in degree of exposure and submergence occurs over the least vertical distance. Interactions between species may also be important in determining upper or lower levels of particular communities.

The Marine Flora. The two most important elements of the marine flora are not evident to the casual observer. These are the phytoplankton and the thin algal film growing on all intertidal and subtidal surfaces.

The Phytoplankton. Phytoplankton consists of mostly microscopic floating or drifting plant cells—diatoms, dinoflagellates, and others—found throughout the euphotic zone of the world's oceans. These are the producers in the seas and the primary source of food for animals. Their abundance and importance are attested to by the fact that most dominant intertidal animals are sessile plankton feeders—barnacles, bivalve molluscs, tube worms, tunicates, bryozoa—to mention a few.

The Algal Film. The second most abundant food source in the intertidal is the thin film of algal cells, diatoms, starts of larger algae, bacteria, and microscopic animals coating the rock surfaces. The second greatest animal component of the rocky intertidal consists of the gastropods and chitons adapted to scraping this film from the rocks. In and on soft substrates, detritus and substrate ingesters represent this component.

The Larger Marine Plants. Seaweeds and sea grasses comprise the readily apparent flora of coastal waters. Seaweeds are algae of the classes Phaeophyceae (brown algae), Rhodophyceae (red algae), and Chlorophyceae (green algae).

Sea grasses are flowering plants, monocots of the order Niadales, which includes most submerged freshwater angiosperms as well. Some, such as *Ruppia* and *Althenia,* are found only in brackish water. Eel grass (*Zostera*) forms extensive beds in shallow, protected waters, and is an annual. It contributes heavily to the productivity of such waters. Surf grass (*Phyllospadix*) prefers somewhat more open surge channels. In shallow tropical flats, turtle grass (*Thalassia*) and others such as *Halodule* form extensive beds and are the major food of the green sea turtle.

The Rocky Intertidal

Upper Littoral Zone. Few obvious plants occur here. The green alga *Cladophora* may form mats in favorable spots, and the blackish tar spot alga (*Ralfsia*) is also sometimes found. The rocks are mostly barren or inhabited by barnacles, periwinkles, and a few limpets.

Middle Littoral Zone. The uppermost bands of seaweed are commonly fucoids. Genera such as *Pelvetia* or *Pelvetiopsis* may form the uppermost band, with *Fucus* or *Ascophyllum* slightly lower. Below these are the thin flat thalli of sea lettuce, *Ulva, Enteromorpha,* or *Porphyra.* The tops of rocks in the middle littoral often have the erect bladderlike *Halosaccion* or the wrinkled *Leasthesia.* The sides of the rocks below these may be coated with a mat of the small, much-branched *Odonthalia, Microcladia,* or *Rhodomela.*

The Lower Littoral Zone. In the upper part of the lower littoral zone there are somewhat larger algae with thicker thalli—the shiny *Iridea* and the rough *Gigartina.* Occasional growths of the coenocytic green algae *Bryopsis* and *Codium* may be found here. Colorful coralline algae such as the nodose erect *Corallina* or the flat encrusting *Lithothamnion* add to the beauty of tide pools, rock surfaces, and even some mollusc shells.

In the lower part of the lower littoral are a number of larger kelps *Cystoseira, Sargassum, Alaria, Egregia, Laminaria,* and beds of surf grass. The giant kelps *Macrocystis, Nereocystis,* and in the Southern Hemisphere, *Pelagophycus* extend from the lower margin of the lower littoral out into the sublittoral, often forming large offshore kelp beds. The curious sea palm (*Postelsia*) is found on the most exposed rocks subjected to the maximum surf, where few other organisms can maintain themselves.

Many tropical areas around coral islands lack larger marine plants in habitats that would seem favorable for them. This is because herbivorous fish (e.g., parrot fish and tang) scrape off the starts as fast as they appear. Exclusion of these fish from sample areas results in a good growth of algae (Bakus, 1966).

Terrestrial Maritime Communities

We shall consider here those coastal communities dominated by vascular plants other than sea grasses.

Priseres. In contrast to the stable conditions found in inter- and subtidal rocky coasts, where long-lasting equilibria or climax communities occur, the most prominent coastal terrestrial communities—salt marshes, mangrove swamps, and sand dunes—are in a continuous state of succession, beginning with colonization of bare ground at the seaward side and progressing toward more stable terrestrial communities on the landward side.

In such communities a transect at right angles to the shoreline will pass through a series of successional stages, such as might be observed over a period of many years if one spot at the seaward edge was chosen for annual observation.

Salt Marshes. First there are bare mud flats deposited by and covered by the high tides. The second stage begins with colonization of the upper margins by plants able to tolerate flooding by saltwater. These plants cause faster silt deposition plus accumulation of organic debris from plant bodies. The roots stabilize the mud.

This raises the level of the substrate, reduces exposure to saltwater, and enriches the soil, making possible the establishment of other species. As more species establish themselves, several may become codominant in the marsh. These will compete with each other for plant nutrients. For any plant the maximum demand for nutrients comes at the period of flowering. The marsh can support more species if their flowering times are staggered. Just such a sequence is found in the main periods of flowering in well-developed salt marshes.

The final stage is stabilization after the substrate has been built up beyond the reach of high tides, and succession has led to a terrestrial community dominated by one or a few species—a climax type community.

Salt marshes form along the edges of protected waters, in estuaries, or where sheltered by spits or barrier islands. At the seaward edge, raised areas are often covered with pickleweed (*Salicornia*) and small unattached fucoids. In the channels between these raised areas or around their bases *Ruppia* may occur. The *Salicornia* becomes mixed with, and succeeded by, sedges such as *Eleocharis, Scirpus,* or *Carex,* the particular species depending on the climate and on whether the substrate is sandy or silty.

As soil builds up and becomes less subject to tidal flooding, though still waterlogged, the surface water becomes more brackish and finally fresh from rain or entrapped surface runoff. The sedges begin to give way to stands of rushes (*Juncus*), arrow grass (*Triglochin*), cord grass (*Spartina*), salt grass (*Distichlis*), *Deschampsia, Puccinellia,* and others. In the upper parts of the marsh numerous other plants establish themselves, including *Cochlearia* (Brassicaceae), *Suaeda* or sea blite (Chenopodiaceae), *Potentilla* (Rosaceae), *Glaux* (Primulaceae), plantains (Plantaginaceae), *Archanglica* (Apiaceae), *Aster* and *Jaumea* (Asteraceae). Willows, alders, or other trees may occur near the upper border, or the marsh may abut abruptly against the edge of coniferous forests on surrounding hills.

Salt marshes are mong the most productive areas known in terms of the synthesis of new organic matter. Their contribution to marine food cycles through the detritus pathways is only now beginning to be properly understood. Too often they are regarded as worthless, filled in to create cheap land for development, diked off to create pastures for grazing, or used as dumps for urban trash or dredge spoils. Neuenschwander, Thorsted, and Vogl (1970) describe a relatively pristine salt marsh in Baja California; Gallagher, Reimold, and Read (1979) analyze the dynamics of salt marshes; and Welsh (1980) points out the interactions and nutrient cycling between a salt marsh and the adjacent tidal mud flat.

Mangrove Swamps. Mangrove swamps are tidal woodlands. Like salt marshes they are priseres, and most of their production is cycled through the detritus pathways. They are limited to the humid tropics or subtropics. The best-developed mangrove swamps are in the Florida-Caribbean region and the Indo-Pacific.

The most characteristic trees are species of the Rhizophoraceae. Their odd prop and accessory roots growing down from the branches enable them to stand in and grow out over tidal flats vegetatively. They also have long pendulous seedlings, some of which penetrate the substrate and grow where they drop while others are carried off by currents and establish new colonies if they happen to lodge in suitable shallow places.

As the swamp develops, other trees become part of the complex. In Atlantic mangrove communities these are mostly *Laguncularia* (Combretaceae) and *Avicennia* (Verbeniaceae). In the Indo-Pacific, *Rhizophora* and other genera of Rhizophoraceae (*Ceriops, Kandellia,* and *Bruguiera*) as well as various other trees occur in addition to Combretaceae and Verbeniaceae. Some of the families represented are Lythraceae, Meliaceae, Myrsiniaceae, Acanthaceae, and palms (Macnae, 1968).

Sand Dunes. Coastal dunes form where low lying land extends for some distance back of extensive sand beaches subjected to onshore winds. Sand blown inland forms dunes, which are gradually colonized by pioneer plants and later by more permanent vegetation.

The earliest colonization is by plants such as couch grass (*Agropyron*), saltwort (*Salsola*), sea rocket (*Cakile*), marram grass (*Amophila*), sandwort (*Honkenya*), and others. The ability of couch grass to tolerate seawater allows it to dominate foredunes, while the power of plants like marram grass to grow up through covering sand allows them to dominate the dunes beyond the reach of the sea—the so-called *yellow dune phase,* where there is still mostly bare sand.

Most of the thirty or so plants to be found in the yellow dune phase are characteristically dune plants, seldom found elsewhere. With increasing age, farther from the sea other plants become more and more abundant until eventually the dunes are fully vegetated and no longer regarded as sand dunes.

Dunes are unstable hills and ridges with depressions or slacks between them that sometimes extend below the water table forming temporary ponds or marshy areas, colonized by a different set of plants. The interaction of dune and slack vegetation makes the situation very complex.

In the Oregon Dunes, European beach grass introduced for dune stabilization, has spread along the upper edges of the beaches forming a foredune. The grass grows up through the sand as fast as it is covered, causing the foredune to become higher and more efficient in trapping sand.

The sand back of the foredune is gradually blown farther inland until a low, flat deflation plane develops at the level of the water table. This becomes colonized by marsh plants, willows, and conifers.

The main dunes, no longer fed by a supply of new sand, continue to move farther inland until finally stabilized by vegetation.

Shingle Beaches. Shingle beaches are covered with loose stones derived from glacial deposition, cliff erosion, or brought down by rivers. Wave action is the most important factor in their formation. Usually the stones or pebbles are moved about enough so that vegetation does not become established. However, if sand, silt, or beach wrack becomes incorporated, some vegetation may establish itself, at least temporarily. Four kinds of shingle beaches are recognized, depending on the presence and nature of the matrix material: *no matrix,* only lichens occur on the rocks; *sandy matrix,* some grasses and other plants, including species similar to those found on dunes, *Agropyron, Festuca, Sedum, Poa; silt,* often from an adjacent salt marsh, some species extending onto the upper margin of the beach, *Glaux, Artemisia, Spergularia, Halimione, Puccinellia;* and *wrack,* seaweed debris abundant, a richer flora develops, depending on drainage and salinity, and the chenopod *Atriplex* is often prominent.

Coastal Cliffs. The vegetation above high water on coastal cliffs are the least studied of all maritime vegetation systems, partly because it does not seem to form clearly demarcated or unique communities, and partly because of the difficulty of access.

Where there are large sea bird colonies or rookeries, the vegetation is much altered by trampling and excrement. The degree of exposure to sun, wind, rain, spray, and storms also exerts strong influences. As in salt marshes and dunes, many of the plants are hemicryptophytes and therophytes.

A few of the plants characteristically found on coastal cliffs are: spleenwort (*Asplenium*), stock (*Matthiola*), fennel (*Foeniculum*), samphire (*Crithium*), sea kale (*Crama*), sea beetroot (*Beta*), scurby grass (*Cochlearia*), sea pink (*Armeria*), plantains (*Plantago*), chickweed (*Cerastium*), and sea campion (*Silene*).

Conclusion

Coastlines present sharper gradients of a greater number of environmental factors than do any other major environments. The interactions of these factors and of the components of the biota result in a kaleidoscopic panorama of complexly interrelated biotic communities such as can be found nowhere else. The complexity and beauty of these communities and of their setting make coastlines a perennial source of inspiration and challenge both for professional biologists and for those who come to enjoy the aesthetic and recreational opportunities they afford.

BAYARD H. McCONNAUGHEY

References

Bakus, G. J. 1966. Some relationships of fishes to benthic organisms (algae, invertebrates) on coral reefs, *Nature* **210**, 280-284.

Chapman, V. J., 1976. *Coastal Vegetation.* Elmsford, N.Y.: Pergamon Press, 292p.

Gallagher, J. L.; Reimold, R. J.; Linthurst, R. A.; and Pfeiffer, W. J., 1980. Aerial production, mortality and mineral accumulation and export dynamics in *Spartina alterniflora* and *Juncus roemerianus* plant stands in a Georgia salt marsh, *Ecology* **61**, 303-312.

Macnae, W., 1968. The fauna and flora of mangrove

swamps of the Indo West-Pacific, *Advances in Marine Biol.* **6**, 73-269.

McConnaughey, B. H., 1978. *Introduction to Marine Biology.* St. Louis, Mo.: C. V. Mosby Co., 624p.

Neuenschwander, L. F.; Thorsted, T. H.; and Vogl, R. J., 1979. The salt marsh and transitional vegetation of Bahia de San Quentin, Baja California, Mexico, *Calif. Acad. Sci. Bull.* **78**, 163-182.

Welsh, B. L., 1980. Comparative nutrient dynamics of a marsh-mudflat ecosystem. *Estuarine and Coastal Marine Sci.* **10**, 143-165.

Welsh, B. L.; Bessette, D.; Herring, J. P.; and Read, L., 1979. Mechanisms for detrital cycling in near shore waters at Bermuda, *Marine Sci. Bull.* **29**, 125-139.

Cross-references: *Biotic Zonation; Coastal Dunes and Eolian Sedimentation; Coastal Ecology, Research Methods; Coastal Fauna; Coastal Plant Ecology, United States, History of; Halophytes; Mangrove Coasts; Organism-Sediment Relationship; Salt Marsh; Tides.*

COASTAL MORPHOLOGY, HISTORY OF

I am only too painfully aware how increasingly difficult it is to find time for a careful study of the work of our predecessors . . . (Geike, 1897)

The coast has been of primary concern to man since he first set foot upon it, and will undoubtedly continue to be so long into the future. It must have been "studied" by the earliest dwellers in their quest for food and also by later inhabitants for other reasons, such as a need of harbors or for defense. Men learned early about cliffs, rocky shores, and sandy beaches; they must also have discovered the operation and importance of tides, currents, and waves upon the shore. In light of man's long-continued interest in the coastal zone, it is surprising that its scientific study lagged so far behind that of many of the earth's other environments.

One explanation may well be that the shoreline, like so many of nature's other boundaries, failed to attract the attention of scientists. Scientifically, it was long a no man's land—neither "ocean nor land." Oceanographers were reluctant to tread where their ships would not go, and geologists were equally reluctant to tackle the sea near the shore.

Early Observations and Theories

Although the actual beginnings of coastal study are shrouded in the haze of history past, some of the speculations of the early Greeks and Romans have been preserved. They were aided by having a keen practical interest in things coastal; as early as the fourth century B.C. they were well acquainted with Mediterranean and Black Sea littorals. Their knowledge of coasts was advanced greatly by such men of action as Alexander the Great. The anonymous Greek *Periplus of the Erythraean Sea,* which described the coasts of the western Indian Ocean and even, albeit hazily, the coast east of India, was used by sailors during the days of Pliny. Although based on direct observation and unhindered by religious teachings, early theories about coasts were nonetheless tempered by the superstitions, legends, and myths that were in the heritage of all Mediterranean peoples at the time.

The presence of marine fossils far from the sea led Aristotle (384-322 B.C.) and others to conclude that the sea had previously occupied higher levels. Strabo (54 B.C.-25 A.D.) even went so far as to write that it frequently changed levels—rising at times, falling at others. The role of rivers in altering the landscape also attracted the attention of these natural philosophers, and, although they had several ideas about where and how river water originates, there seems to have been little doubt in their minds about what happens when it enters the sea. Aristotle, Strabo, and others recognized deltaic and alluvial plain deposits. They knew that river-carried sediment first shallows the sea, then converts it into marshland, and changes it eventually into dry, farmable land. Erosion as a coastal phenomenon was not ignored. Strabo, for example, noted that the ebb and flow of tides prevent deltas from advancing continuously outward into the sea. The role of man as coastal agent was also studied. Strabo cited an example of the attempt to improve the harbor of Ephesus in the second century B.C.:

The mouth of the harbor was made narrow by engineers, but they, along with the king who ordered it, were deceived. . . . He thought the entrance would be deep enough for merchant vessels. . . . But the result was the opposite, for the silt, thus hemmed in, made the whole of the harbor . . . more shallow. Before this time the ebb and flow of the tides would carry away the silt and draw it to the sea outside (Russell, 1967a: 299).

Such was the state of knowledge and theory during those Greek and Roman times at about the beginning of the Christian era. Although these notions persisted in Europe only until the fall of the Roman empire, they were nonetheless preserved and nurtured in the Arab world during much of the time leading up to the European Renaissance.

The Renaissance

The translation of Ptolemy's *Geography* in the early fifteenth century was a major step toward the great discoveries of the fifteenth through the eighteenth centuries. Much of the actual knowledge held by the Greeks, Romans, and Arabs about coasts came from navigators, and so it was to continue for many centuries. Many of the voyages during this period (such as those of the Cabots, Verrazano, and Gómez along the northeast coast of North America) were strictly coastal; there was no attempt at colonization or at exploration of the interior, Penrose (1955: 147) wrote: "A fair coastal survey was the sole result—nothing more...."

Such surveys added rapidly to the body of knowledge that was beginning to accumulate, and influenced those who would put their minds to coastal problems. Most of the material was pure description, and just what influence it actually had on the development of coastal morphology as a science is uncertain. Nonetheless it was available for such thinkers as Leonardo da Vinci (1452-1519), who recognized terraces for what they are, and for Steno (1631-1687), who established a depositional sequence in explanation of the strata he observed.

Bernhard Varenius (1622-1650), because he successfully merged description with theory, has been credited with laying the foundations for geography as a science (Mather and Mason, 1939). These foundations are recorded in his *Geographia Generalis,* a book that was used as a text in British universities for a century by such notables as Isaac Newton. On the topic of coastal morphology, Varenius wrote, "The ocean in some places forsakes the shores, so that it becomes dry land where it was formerly sea" (Mather and Mason, 1939: 25). He reasoned that many factors may be involved in such a change, including those of erosion and deposition, ocean currents and tides, texture and structure, river flow, and wind. Process was very real in his science.

The Influence of Scripture

Nearly all "scientists" of the seventeenth, eighteenth, and much of the nineteenth centuries were influenced to greater or lesser degree by the Bible and the Church. In some cases this acceptance was probably not unlike that of the Greek philosophers, influenced unknowingly as they were by mythology. The biblical quotations of greatest significance, as found in the King James version of the Bible are:

Let the waters under the heaven be gathered together unto one place, and let the dry land appear... (Genesis, I, 9). Let the waters bring forth abundantly the moving creature... (Genesis, I, 20). And... the flood was forty days upon the earth; ... and the waters prevailed exceedingly upon the earth; and all the high hills, that were under the whole heaven, were covered (Genesis, 7, 17, 19).

Because, according to the Bible, the separation of land from sea occurred two days before animal life in the sea was created, the utilization of *fossils* in explaining the history of the earth, as the Greeks had done, was heresy. However, the Bible offers the universal flood of Noah's time as a possible out, one that was used in many explanations. A major difficulty was time. Because of the strict application of scripture, time was unavailable as a basis for applying observable processes to earth history. The major challenge was how to present the facts as offered by landscape observations without running counter to theological precepts.

Theories of Landscape Development and Coastal Morphology

Within such a Church-dominated intellectual environment, it is not surprising that a number of theories were proposed to explain the earth and its surface forms. During the eighteenth and nineteenth centuries many were proposed, accepted, modified, merged, and abandoned. Contemporary and subsequent history has labeled many of these theories by their predominant theme, and some also by the name of their principal proponent. Examples include neptunism (Wernerism), plutonism, diluvialism, and fluvialism (Huttonism). These theories and their major advocates are treated at length in volume one of *The History of the Study of Landforms,* by Chorley, Dunn, and Beckinsale (1964). Basically most of the earlier theories were catastrophic in approach, reflecting a high degree of biblical influence; later ones tended toward uniformitarianism, although there was much overlap. In any event, each theory in its own way has played a role in the development of coastal geomorphology.

The neptunists, of whom Werner (1749-1817) was the principal spokesman, invoked a primeval universal ocean in accounting for both stratigraphic sequence and landform types. The earth's rocks were formed, according to the neptunists, through the successive accumulation of chemical precipitates within the ocean. Although some surface forms were caused by submarine deposition and erosion during the presence of the universal ocean, most forms resulted from erosion during a rapid recession of oceanwater. From this viewpoint, every landform originally was coastal—at least, in the sense that it was formed by turbulent oceanic currents and waves.

The diluvialists, as exemplified by Buckland (1784-1856), held a theory that was similar in some ways to that of the neptunists. The catastrophy they invoked, however, was Noah's flood. In the pure form of diluvialism, the flood was more a destructive than constructive agent. Again, it was rapidly receding water that carved the landscape. Some of the many types of otherwise unexplained phenomena that were accounted for by the flood were erratics, underfit streams, and terraces both river and marine.

Most of the catastrophists recognized that coastal and riverine erosion and deposition did occur but at such a slow pace that they had no place in earthly or "Godly" schemes. Nonetheless, even before and especially during the dominance of catastrophism, uniformitarian ideas surfaced and by the mid-nineteenth century were dominant.

The uniformitarians were united on the concept that currently operating processes were capable of creating present-day landscapes. They were not united, however, as to which processes were most important. In general, there were two camps: one, represented by Hutton (1727-1797) and Playfair (1747-1819), believed that the river was the most important agent in landscape development; the other, represented by Lyell (1797-1875) and Ramsay (1814-1891), advocated marine abrasion as dominant.

Form and Process:
Seventeenth-Nineteenth Centuries

Although it is basically correct to conclude that the dominance of the catastrophic schools during their heyday retarded the study of geomorphology, the study of *coastal forms and processes* actually progressed to some extent. Because such forms and processes were considered insignificant from the standpoint of the overall scheme of landscape development, they could be looked at without fear of countering religious dogma if one did not try to conclude too much. Coastal cliff erosion and deposition are so conspicuous, especially in parts of the British Isles, that they did not escape the consideration of layman, geologist, and engineer alike. Hutton and Playfair, although mainly fluvialists, recognized coastal processes. Hutton, for example, wrote, ". . . we never see a storm upon the coast, but that we are informed of the hostile attack of the sea upon our country" (1788; quoted in Chorley, Dunn, and Beckinsale, 1964: 39). Playfair, in the same vein, emphasized the obviousness of coastal erosion when he wrote, "If the coast is bold and rocky, it speaks a language easy to be interpreted." He also noted that once fragments of rock are detached they "become instruments of further destruction . . ." (1802; quoted in Chorley, Dunn, and Beckinsale, 1964: 60). Thus throughout the period of, and subsequent to the Renaissance, statements appeared that indicated some relatively advanced thoughts about coasts, some of which did overstep the bounds of Church dogma.

For example, John Ray (1627-1705), a keen observer, went so far as to propose that the combination of subaerial erosion and coastal cliff retreat would eventually reduce all land to a level below the sea. Guettard (1715-1786), famous for his geological maps, believed the sea to be the major agent in land erosion, and that cliff coasts were the remnants of former extensive hill systems. He observed that sediments brought to the sea by rivers mixed with material eroded from adjacent cliffs and submerged rocks. However, he tempered this view by noting that the action of the waves would have little effect beneath the surface of the sea. Lavoisier (1743-1794) adopted the Guettard idea that littoral beds are composed of materials from varied sources, but went a step further and noted that the coarsest materials are highest on the shore, and are followed downslope by coarse sands, fine sands, and clay. The width of each band, he maintained, varies with the steepness of the slope.

One of the most perceptive of the natural historians of this period was the little-recognized John Walker (1731-1803). He was the first effective teacher of geology at the University of Edinburgh where he held the chair of natural history from 1779 to 1803. His students included the geologists Playfair, Hall, and Jameson. His lecture notes—not published until 1966—contained many advanced notions that must have guided much of the thinking of the geologists of the early part of the nineteenth century. Some of Walker's notions, of relevance to coastal geomorphology, dealt with: continental drift—". . . why not America from Europe and Asia and indeed every one continent from another"; coral reefs—he was apparently the first geologist actually to describe the growth of coral reefs; subsidence—he not only described the processes of alluviation but wrote that sediments ". . . are found in great quantity and to considerable depth, being the sediment of rivers . . ." (Walker, 1966: 178, 183) anticipating R. J. Russell's work of 150 years later. Unlike most geologists of the time, Walker believed that sea sand was formed by the weathering (his term) of rocks rather than by chemical precipitation from the sea (Walker, 1966).

Marine Planation

With Ray's seventeenth-century and Guettard's eighteenth-century views of marine erosion, coupled with the fact that even the most dedicated of fluvialists placed great importance on marine processes, it is not surprising that some men in the nineteenth century considered the sea as more powerful than the river. Lyell was one of them, although he had not always been so. During his career he gave increasing importance to marine erosion, finally considering it to be the major modifier of the landscape. His ideas were modified by Ramsay, and eventually evolved into the theory of marine planation. Ramsay's concepts included two ideas: one, that the sea is capable of planing surfaces over which it moves, regardless of rock composition; the other, that unequal hardness in cliffs will result in differential erosion and the creation of an irregular coastline. He also maintained that marine planation accompanies shifts in sea level, using as evidence the presence of plains at different elevations above sea level. The escarpments between these levels he explained as old sea cliffs.

The importance of this theory is emphasized by Chorley, Dunn, and Beckinsale (1964: 313): "Even when the idea of universal marine erosion had been discredited, the planation part of the theory lived on in Davis's cycle of erosion and in the writings of mid-twentieth-century geomorphologists."

As far as the coastline is concerned, Ramsay and others emphasized increasing irregularities because of differential erosion, a view that was not difficult to accept in the British Isles. However, Dana (1813–1895), a confirmed fluvialist, disagreed. He believed that: ". . . waves tend rather to fill up the bays and remove by degradation the prominent capes, thus rendering the coast more even, and at the same time, accumulating beaches that protect it from wear" (1849; quoted in Chorley, Dunn, and Beckinsale, 1964: 363). These conclusions were based on observations made during Dana's four-year voyage in the Pacific Ocean.

During this period, thought was being given to a number of agents previously little considered. Hutton, it is true, had recognized the importance of chemical action in soil formation, and others had discussed the transport of matter being carried to the sea in solution. Nonetheless, it was von Richthofen (1833–1905), a staunch follower of Ramsay, who applied such ideas to coasts. He wrote:

The weathering and loosening of rock by sea salts, carbonic acid, the formation of ozone, and the gripping of plants and animals—to which must be added the action of frost in higher latitudes—aids the mechanical action of the striking billows (1882; quoted in Mather and Mason, 1939: 515–516).

Gilbert and Lake Bonneville

It is somewhat curious that the western explorations in the United States, especially those during the last half of the nineteenth century, should be important from the standpoint of the history of the study of coastal morphology. This importance is even more surprising when one realizes that much of the western field work combined with the studies being made on the Ganges (Everest, 1793–1874) and the Mississippi (Humphreys, 1810–1883) rivers and in the heavy rainfall areas of the tropics in helping to reestablish the notion that the river is the dominant agent in geomorphology.

Especially significant are the coastal concepts presented by Gilbert (1843–1918) in his *Lake Bonneville* (1890) and *The Topographic Features of Lake Shores* (1885). He treated a variety of shore-related topics, including beaches, cliffs, terraces, barriers, lagoons, waves, currents, undertow, backwash, and sorting, among many others. His deductions, based on intensive fieldwork, were lucidly presented. The main limitation to their usefulness was that, having been derived from work on lakes, they are not always applicable to oceanic situations. A major case in point is the concept of the bottomset, foreset, and topset bed composition of delta terraces, a concept that has little value when dealing with major oceanic deltas. Nonetheless, by describing coastal landforms in terms of physical processes, Gilbert set the style for present-day research in coastal geomorphology.

Gilbert, like most of the other geologists involved in the western explorations, was not tradition-bound and thus was better able to distinguish between the relative importance of subaerial and marine processes, as he did for Lake Bonneville.

Possibly his most important contribution in the development of geomorphology is his concept of grade, a concept that he used in the development of his ideas on beach equilibrium. He wrote: ". . . in order that the local process be transportation only, and involve neither erosion nor deposition, a certain equilibrium must exist between the quantity of the shore drift on the one hand and the power of the waves and currents on the other" (1885: 101).

Gilbert utilized lake shorelines and lake deposits as indicators of past climates and tectonic history. For example, he was able to correlate lake-terrace width with rock type in a lake's discharge channel and the lack of horizontality and parallelism of shorelines with orogenic movement.

Davis, Gulliver, and Johnson

The role of William Morris Davis (1850-1934) in the development of geomorphology has been analyzed many times (most thoroughly by Chorley, Beckinsale, and Dunn, 1973). Davis influenced in some way nearly every aspect of geomorphology, including coastal geomorphology. This influence has been realized in several closely linked forms, including his own research and publications on coastal topics, the wide adoption of his concepts of the cycle of erosion, and the work of his students (especially Gulliver and Johnson), who emphasized the study of coasts.

Much of Davis's research on coastal problems was related to reestablishing support for Darwin's subsidence theory of coral-reef growth. By writing some three-quarters of a century after Darwin, Davis was able to incorporate in his writings data about sea level changes during the glacial period. He believed that the only way to properly understand coral reefs was through an examination of the ". . . physiographic features of the coasts, either insular or continental, that are bordered by fringing reefs or fronted by barrier reefs . . ." (1928; quoted in Chorley, Beckinsale, and Dunn, 1973: 592) and not from the reefs themselves.

Another example of his contributions in coastal geomorphology is his "The Outline of Cape Cod" (1896). It illustrates an attempt at geomorphic reconstruction: "Let the activities of the sea be resolved into two components: one acting on and off shore, the other along shore; and let the effects of the first of these components be now examined alone, . . ." (Davis, 1896; 700).

Davis incorporated the ideas of Gilbert's beach equilibrium within his cyclic concepts: "Here the sea is able to do more work than it has to do. Its action is like that of a young river. . . . When a graded profile is attained, the adolescent stage of shore development is reached" (Davis, 1896: 701). This paper also presents numerous diagrams, a Davisian hallmark, illustrating the development of graded profiles (both normal and longitudinal) and of bars and spits.

The cyclic concepts presented in the Cape Cod paper preceded by three years the publication of his most famous and influential paper, "The Geographical Cycle" (Davis, 1899). The Cape Cod paper is only one example of the fact that the cyclic concept had been in Davis's mind for many years before the turn of the century. The influence of this concept in coastal geomorphology is further evidenced in the dissertation by Gulliver (1865-1919) that was entitled simply "Shoreline Topography" (1899) and was published in the same year as "The Geographical Cycle."

Davis was continuously working with the cycle, and considered that it is ". . . not arbitrary or rigid, but elastic and adaptable. . . ." For example: "Like the processes of surface carving, the processes of shore-line development are subject to variation with climate, from the work of the ice foot in polar regions to the work of coral reefs and mangrove swamps in the torrid zone" (Davis, 1905: 290).

The most influential of Davis's disciples was Douglas Johnson (1878-1944). Despite Gulliver's early coastal work under Davis's direction, it remained for Johnson to publish the first inclusive book dealing with coastal morphology, a book that Zenkovich (1967) considers to be the most complete theoretical study of coasts available. Entitled *Shore Processes and Shoreline Development* (Johnson, 1919), it was aimed at presenting an analysis of the forces operating along the shore together with a systematic discussion of the cycles of shoreline development. This book had a major influence on the study of coasts for at least 40 years, an influence that must be considered to have been detrimental in some regards. Johnson's emphasis on the importance of submergence and especially emergence in shoreline development, as well as the incorporation of these aspects of shore profile development in his classification scheme, delayed more meaningful approaches to coastal understanding for several decades. Nonetheless, much of his material is still useful, and should be consulted by any serious student.

1920-1950

Although the Davisian and Johnsonian evolutionary and qualitative approaches to geomorphology dominated coastal research from World War I until about 1950, there were nonetheless a number of important developments that occurred during this thirty-year period. One of the most significant of these developments was that coastal research became a respectable research endeavor after Johnson's book was published. Earlier most coastal research was of a practical nature—harbor construction, shoreline defense, coastal mining, and the like. Few university scholars studied coasts for their own sake. Geologists, for example, often resorted to coasts, but only because coastal cliffs provided them with good exposures of the rocks they were studying, not because of any interest in their existence as coastal forms.

A second and concurrent development was the rapid rate at which human utilization of the coast developed. The increase resulted directly from increases in coastal populations and indirectly from man's increased mobility, desire for coastal vacations, and use of coastal resources. Fortunately, this increased utilization of the coast was accompanied by an increase in its study, although not to the extent that might have been desirable. Group research and publication were initiated, research laboratories were developed, new techniques and equipment such as aerial photography and sediment-coring devices were devised, coastal maps became more detailed, and long-term research was initiated.

The sponsorship by the United States National Research Council of separate studies on tides and sea level changes in the 1920s, the creation of the Commission of Coastal Studies in the U.S.S.R. and the Coastal Engineering Research Center in the United States, and the publication of such volumes as *Recent Marine Sediments* (Trask, 1939) are examples of the organized endeavors that began between the two world wars.

Despite all of the technical advances suggested above, there were surprisingly few actual substantive developments. In a very real sense, this period of time was transitional and set the stage for the rapid rise in coastal research that began during the 1950s and has continued at an ever increasing pace since. The way in which this transition proceeded might be illustrated by considering the way in which depositional landforms came to be recognized as significant elements of the landscape. One of the pioneers of depositional geomorphology was R. J. Russell (1895-1971). Trained in the Davisian School with its emphasis on erosion, Russell had to develop a completely new perspective in order to understand the depositional landforms he found when he moved to Louisiana. He utilized the thousands of cores taken during the drilling for oil in the Mississippi River delta, sets of detailed topographic maps (the drafting of which were prompted by a severe flood in 1927), and aerial photography (Russell, 1936). This research of Russell and his co-workers over a twenty-year period led to the acceptance of the importance of three-dimensional studies in geomorphology, a clarification of the different types of subsidence, a reevaluation of Johnson's concepts of emergence and submergence and of Davis's concept of old age, and to new notions about sea level fluctuations.

The work of Axel Schou in Denmark, J. A. Steers in England, André Guilcher in France, Veselod Zenkovich in the U.S.S.R., and R. J. Russell in the United States during the 1940s set the stage for the most recent period of coastal research.

Present-Day Research in Coastal Geomorphology

The most significant of the recent trends in coastal morphologic research is based on the concept of process and response in geomorphology. Practitioners in general follow the procedure advocated by Strahler for fluvial morphologists: ". . . dynamic-quantitative studies require, first, a thorough morphological analysis in order that the form elements of a landscape may be separated, quantitatively described, and compared from region to region" (Strahler, 1952: 1118). Such a procedure is especially valuable for morphologic research in the coastal zone because of the great number of forms and processes present there and because of the complex nature of the interrelationships that exist between them. The process-response models that result from such studies are not only of value in the prediction of specific types of coastal behavior but they are possibly even more valuable because they can lead to clearer understanding of the integrated nature of the coastal system.

Today such varied techniques as radioactive tracers, aqualung diving, satellite imagery, the electron microscope, and high-speed computers are providing data and analyses about coastal forms and processes at scales both larger and smaller than was possible only a few years ago.

Although the scientific study of coasts has traditionally been in the hands of western Europeans and Americans, during the past three decades it has truly become international. Evidence of this development is indicated by: the frequency with which international symposia (such as the International Geographical Union-sponsored symposium on the Dynamics of Shoreline Erosion held on the Black Sea coast of the U.S.S.R. in 1976) are being convened; the increasing numbers of research papers from non-Western countries such as Australia, Japan, and the U.S.S.R. (Walker, 1976); and the increasing frequency of research along Arctic, desert, and tropical coasts.

In 1968 R. J. Russell wrote that he regarded " . . the subscience of coastal morphology as one in relative infancy. . . ." In the intervening years a great deal of progress has been made, and if book publication rate is a valid indicator of the growth rate of a discipline, it can be stated that coastal morphology is rapidly approaching maturity—a concept that Davis might have liked. During this period the number of relevant books has mushroomed. Some of these books, such as Zenkovich's *Processes of Coastal Development* (1967), Russell's *River Plains and Sea Coasts* (1967b), Bird's *Coasts* (1969), King's *Beaches and Coasts* (1972), and Davies's *Geographical Variation in Coastal Development* (1973), are broad and inclusive in coverage. Others, such as Ippen's *Estuary and Coastline Hydrodynamics* (1966), Ingle's *The Movement of Sea Sand* (1966), Shepard and Wanless's *Our Changing Coastlines* (1971), and Komar's *Beach Processes and Sedimentation* (1976), deal with special topics.

There is yet another kind of volume that is becoming especially common and apparently popular. In such a volume a number of articles by different authors are published together. They may be special issues of standard periodicals, such: as Longinov's *Dynamics and Morphology of Sea Coasts* (1969), volume 48 of the *Transactions of the Institute of Oceanology;* Tedrow and Deelman's *Soil Science* (1975), devoted to soil formation in sediments under water; and Fairbridge's *Contributions to Coastal Geomorphology* (1975) and Kaiser's *Küstengeomorphologie* (1968), both special issues of *Zeitschrift für Geomorphologie*. They may also result from symposia such as: *Estuaries*, edited by Lauff (1967); *Waves on Beaches and Resulting Sediment Transport*, edited by Meyer (1972); *Coastal Geomorphology*, edited by Coates (1973); *Nearshore Sediment Dynamics and Sedimentation*, edited by Hails and Carr (1975); and *Research Techniques in Coastal Environments* edited by Walker (1977). A third type of compilation, and one that is especially valuable from the standpoint of the development of coastal concepts, is that in which the key papers of the past are brought together. A prime example is the Benchmark Papers in Geology series, of which the two volumes edited by Schwartz, *Spits and Bars* (1972) and *Barrier Islands* (1973), are especially apropos.

R. J. Russell (1968) wrote that coastal morphology should be regarded as being in relative infancy, he went on to state that it is a subscience ". . . offering opportunities for exciting discussions for many, many years to come." Certainly the list of books above, especially when thought of in combination with the vast number of recently published research papers, suggests that opportunities were present and, more important seized by coastal morphologists. It is unlikely that such a manifested interest is only a passing fad; the future of coastal morphology appears brighter than ever before.

H. J. WALKER

References

Bird, E. C. F., 1969. *Coasts.* Cambridge, Mass.: M.I.T. Press, 246p.

Chorley, R. J.; Dunn, A. J.; and Beckinsale, R. P., 1964. *The History of the Study of Landforms, Vol. 1: Geomorphology before Davis.* London: Methuen and Co., 678p.

Chorley, R. J.; Beckinsale, R. P.; and Dunn, A. J., 1973. *The History of the Study of Landforms, Vol. 2: The Life and Work of William Morris Davis.* London: Methuen and Co., 874p.

Coates, D. R., 1973. *Coastal Geomorphology.* Binghamton: State University of New York, 404p.

Davies, J. L., 1973. *Geographical Variation in Coastal Development.* New York: Hafner Publishing Co., 204p.

Davis, W. M., 1896. The outline of Cape Cod, in *Geographical Essays.* New York: Dover Publications, 1954 reprint, 690-724.

Davis, W. M., 1899. The geographical cycle, *Geog. Jour.* 14, 481-504.

Davis, W. M., 1905. Complications of the geographical cycle, in International Geographical Congress, 8th, Washington, 1904, Report, 150-163. (Reprinted in W. M. Davis, 1954, *Geographical Essays.* New York: Dover Publications, 279-295.)

Fairbridge, R. W., ed., 1975. Contributions to Coastal Geomorphology, *Zeitschr. Geomorphologie* suppl. 22, 170p.

Geike, A., 1897. *The Founders of Geology.* New York: Dover Publications, 1962 reprint, 486p.

Gilbert, G. K., 1885. The topographic features of lake shores, *U.S. Geol. Survey Annual Rept.* 5, 65-123.

Gilbert, G. K., 1890. *Lake Bonneville.* Washington, D.C.: U.S. Geological Survey Monograph I, 438p.

Gulliver, F. P., 1899. Shoreline topography, *Am. Acad. Arts Sci. Proc.* 34, 149-258.

Hails, J. R., and Carr, A. P., eds., 1975. *Nearshore Sediment Dynamics and Sedimentation.* New York: John Wiley & Sons, 316p.

Ingle, J. C., Jr., 1966. *The Movement of Beach Sand.* Amsterdam: Elsevier, 221p.

Ippen, A. T., 1966. *Estuary and Coastline Hydrodynamics.* New York: McGraw-Hill, 744p.

Johnson, D. W., 1919. *Shore Processes and Shoreline Development.* New York: John Wiley & Sons, 584p.

Kaiser, K., ed., 1968. Küstengeomorphologie, *Zeitschr. Geomorphologie* suppl. 7, 199p.

King, C. A. M., 1972. *Beaches and Coasts.* New York: St. Martin's Press, 570p.

Komar, P. D., 1976. *Beach Processes and Sedimentation.* Englewood Cliffs, N.J.: Prentice-Hall, 429p.

Lauff, G. H., ed., 1967. *Estuaries.* Washington, D.C.: American Association for the Advancement of Science, 757p.

Longinov, V. V., ed., 1969. *Dynamics and Morphology of Sea Coasts,* Inst. Oceanology (Moscow) Trans. 48 (translated from the Russian). Washington, D.C.: U.S. Department of Commerce, 372p.

Mather, K. F., and Mason, S. L., 1939. *A Source Book in Geology 1400-1900.* Cambridge, Mass.: Harvard University Press, 702p.

Meyer, R. E., ed., 1972. *Waves on Beaches and Resulting Sediment Transport.* New York: Academic Press, 462p.

Penrose, B., 1955. *Travel and Discovery in the Renaissance.* Cambridge, Mass.: Harvard University Press, 377p.

Russell, R. J., 1936. *Physiography of Lower Mississippi River Delta.* Baton Rouge: Louisiana Dept. Conservation, Geol. Survey Bull. 8, 199p.

Russell, R. J., 1967a. Aspects of coastal morphology, *Geog. Annaler,* ser. A, **2-4,** 299-309.

Russell, R. J., 1967b. *River Plains and Sea Coasts.* Berkeley and Los Angleles: University of California Press, 173p.

Russell, R. J., 1968. Foreword, in Kaiser, 1968, v-vii.

Schwartz, M. L., ed., 1972. *Spits and Bars.* Stroudsburg, Pa.: Dowden, Hutchinson & Ross, 452p.

Schwartz, M. L., ed., 1973. *Barrier Islands.* Stroudsburg, Pa.: Dowden, Hutchinson & Ross, 451p.

Shepard, F. P., and Wanless, H. R., 1971. *Our Changing Coastlines.* New York: McGraw-Hill, 579p.

Strahler, A. N., 1952. Hyposemetric (area-altitude) analysis of erosional topography, *Geol. Soc. America Bull.* **63,** 1117-1142.

Tedrow, J. C. F., and Deelman, J. C., eds., 1975. *Soil Science,* vol. 119, no. 1. Baltimore, Md.: Williams and Wilkins, 112p.

Trask, P. D., ed., 1939. *Recent Marine Sediments.* New York: Dover Publications, 1968 reprint, 736p.

Walker, H. J., ed., 1976. *Geoscience and Man, Vol. XIV: Coastal Research.* Baton Rouge: Louisiana State University, 153p.

Walker, H. J., ed., 1977. *Geoscience and Man, Vol. XVIII: Research Techniques in Coastal Environments.* Baton Rouge: Louisiana State University, 320p.

Walker, J., 1966. *Lectures on Geology,* H. W. Scott, ed. Chicago, Ill.: University of Chicago Press, 280p.

Zenkovich, V. P., 1967. *Processes of Coastal Development.* New York: Wiley Interscience, 738p.

Cross-references: *Beach; Beach Processes; Coastal Characteristics, Mapping of; Coastal Ecology, Research Methods; Coastal Engineering, History of; Coastal Engineering, Research Methods; Coastal Morphology, Research Methods; Geomorphic-Cycle Theory.* Vol. III: *Geomorphology, History of.*

COASTAL MORPHOLOGY, OCEANOGRAPHIC FACTORS

By definition the coastal environment is dominated by the interactions between atmospheric, terrestial, and marine factors. Spatial and temporal variations in the relative intensities of various coastal processses and in associated depositional patterns are related to variations in the relative contributions of the different factors. Although the ultimate sources of all of the earth's energy are astronomical in the form of solar radiation and gravitational attractions, the sea is the immediate source of most of the energy for transporting coastal sediments and driving coastal processes along the world's coasts as a whole. There are numerous *oceanographic factors* that play major roles in determining the dynamic regime and morphology of the coastal zone. These factors are the fundamental inputs from deep water into the coastal system. The higher-order processes that produce coastal change and morphology result from the interactions of these factors with each other and with the solid boundaries of the coast and inner shelf.

The oceanographic factors that influence coastal morphology may be broadly grouped into two general categories: primary factors that directly affect morphologic development through the immediate input of energy or mass into the coastal system; and secondary factors that influence coastal morphology indirectly by their ability to affect the primary factors. These factors are discussed individually in more detail elsewhere in this encyclopedia.

Primary Factors

The manner and extent of influence by primary oceanographic factors on coastal morphology depend on the frequency and intensity of the individual factors. The intensities of these factors and their ability to perform work on the coastal landscapes vary considerably with location and with time; Davies (1973) has described the geographic variability of some of the more important factors. However, in terms of world averages, the three most important factors (ranked in descending order of power expended at the shoreline) are *waves, tides,* and *ocean currents.* Other significant oceanographic sources of coastal energy include *internal waves, shelf seiche,* and *tsunamis.* Inman and Brush (1973) cite estimates of the rates of dissipation of mechanical energy in the coastal waters of the world for each of the primary factors. In kw $\times$ 10^9 these are: breaking ocean waves, 2.5; tidal currents, 2.2; large-scale ocean currents, 0.2; internal waves, .01; shelf seiche, .01; tsunamis (average rate), .0001.

Waves. Ocean waves are the most important single source of energy for the morphologic development of the world's coasts as a whole. Excellent up-to-date discussions of the theory of wave generation and propagation and wave behavior in coastal waters may be found in LeMéhauté (1976) and Komar (1976); mechanisms of sand transport by waves and beach response to varying wave conditions are discussed in detail by Komar (1976). In addition to their well-known and obvious role in causing beach erosion and accretion, waves are responsible for inshore *setup,* for generating longshore currents and rips, and are the sources of energy for phenomena such as *edge waves.*

The intensity of wave processes and their relative dominance of coastal morphologic development vary considerably on both regional and local scales. The heights and periods of the waves that enter the nearshore and inshore zones depend on the deepwater wave climate; inshore wave height also depends on the shallow subaqueous topography. Global variations in deepwater wave climate reflect differences in both wind-wave and swell regime and are described in general terms by Davies (1973). Wind-wave climate depends directly on local wind regime and on the fetch to which the coast is exposed.

The topography of the inner continental shelf and nearshore zones may exert a significant influence on the power of the waves that reach the shore, primarily through the mechanisms of refraction and frictional dissipation. Although wave attenuation by bottom friction is relatively minor over steep subaqueous profiles, friction may cause appreciable reduction of wave height when the offshore profile is broad and flat (Wright and Coleman, 1973; Wright, 1976).

Tides. Astronomical tides generated by the gravitational attractions of the moon and sun on the sea surface are second in importance to waves in terms of their relative influences on the morphology of the world's coasts as a whole. However, in many parts of the world, tides may dominate over waves by orders of magnitude as the primary form-molding and sediment-transporting force. Tides affect coastal morphologic development by means of: rotary and bidirectional shallow-water tidal currents that redistribute sediments alongshore; tidal flows and associated sediment transport in estuaries and tidal inlets; and horizontally and vertically extending the region of land-sea interaction.

The primary determinant of relative tidal influence is the tidal range near the coast; tidal type (i.e., semidiurnal, diurnal, mixed) is of significance but secondary importance. On the basis of tidal range, Davies (1964) classified the world's tide regimes as *microtidal* (range less than 2 m), *Mesotidal* (2–4m range), and *macrotidal* (range greater than 4 m). Easton (1970) added a fourth category: *tideless* (range less than 1 m). In general, tides are of greatest morphologic significance in macrotidal environments and are least significant in tideless environments. Tidal ranges in the deep ocean are universally low, averaging only about 50 cm (Komar, 1976). The large tidal amplitudes of coastal waters in many parts of the world are produced by the shallow water transformations of the tidal wave as it crosses the continental shelf or enters constricted gulfs or embayments. The largest tidal ranges are normally associated with wide continental shelves and narrow straits or semiconfined seaways.

Ocean Currents. In comparison with relatively powerful wave- and tide-induced coastal currents, oceanic currents intruding into coastal waters exert a minor influence on coastal morphology in most cases. An exception is found along the coasts of northern Brazil and the Guianas, where the swift, northerly-setting Guiana current impinges on the inner continental shelf, sweeping fine-grained sediments debouched by the Amazon River hundreds of kilometres to the northwest. Although early theories attributed the Carolina Cape systems to large-scale eddies off the Gulf Stream, oceanographic observations have failed to reveal the existence of the eddies (Komar, 1976).

It appears that the major direct effect of oceanic currents on coastal morphology is the transport and subaqueous deposition of fine, suspended sediments parallel to continental margins. Perhaps more important in many cases are the indirect influences of ocean currents on wave climate and wind regime related to the effects of the currents on coastal climate.

Internal Waves. Internal gravity waves that propagate along thermoclines or density gradients beneath the sea surface have been the subjects of considerable oceanographic research in recent years (Garrett and Munk, 1975; Wunsch, 1975). However, the coastal morphologic roles of oceanic internal waves entering coastal waters have received scant attention. Recent studies suggest that shoaling internal waves may cause movement of bottom sediments at depths well below the reach of normal surface waves (Cacchione and Southard, 1974). Higher-frequency shear-generated internal waves at stratified river mouths may be highly significant to river-mouth bar formation (Wiseman et al, 1976).

Shelf Seiche and Tsunamis. Low-frequency *shelf seiches* with periods on the order of 2 minutes to 1.5 hours and amplitudes of less than 1 cm can produce long, low-standing oscillations along a coast (Neumann and Pierson, 1966). The energy of these oscillations is generally exceedingly low; however, under certain circumstances the oscillations may serve to focus large-scale nearshore circulation systems that can, in turn, significantly influence beach morphology (Inman and Brush, 1973).

Tsunamis generated by seismic disturbances have periods of 15 to 20 minutes (Neumann and Pierson, 1966). Although the heights of tsunamis are low in the deep ocean, shoaling of the tsunami waves as they enter shallow water can produce extreme heights of up to 30 m, often with dramatic consequences. Because of the infrequency of their occurrence, tsunamis are of comparatively little significance as coastal geomorphic agents. Inman and Brush (1973) estimate that large tsunamis with high energy content occur only about 5 times each century. However, when tsunamis do occur they can produce long-lasting geomorphic manifestations in the form of elevated scarps and debris accumulations (Shepard, 1973). Tsunamis are most prevalent around the tectonically active margins of the Pacific.

Secondary Factors

Important secondary oceanographic factors that influence coastal morphology by way of their effects on the primary factors include seawater density structure and sea-surface temperature.

Density Structures. Horizontal and vertical gradients in the density of seawater result from contrasts in salinity and temperature. In the deep ocean the effects of these density gradients are superimposed on those of wind shear in generating ocean currents and circulations (Neumann and Pierson, 1966). In coastal waters density gradients are often appreciably strengthened by the discharge of fresh water from rivers. These strong density gradients can significantly influence currents over the inner continental shelf and nearshore zone. Murray (1972) found that density gradients in the vicinity of the Mississippi delta may alter or reverse the wind-induced velocity field. The depths and steepnesses of vertical *pycnoclines* and *thermoclines* are also important in determining the depths and periods of internal waves.

Sea Surface Temperature. The temperature of surface waters off coasts can play an important role in influencing the coastal climate and wind regime. Dramatic examples occur along the coasts of northwest Africa, southwest Africa, and Peru, where abnormally cool sea surfaces produced by coastal *upwelling* create foggy but otherwise arid coastal climates. The coastal environment of Peru is particularly sensitive to variations in offshore surface temperatures: the occasional southward intrusion of warm equatorial waters brings massive mortality of marine organisms as well as precipitation. Less spectacular, but of greater consequence to the world's coasts as a whole are the influences of temperature differences between land and sea on coastal sea breezes. Recent analyses of long-term sea surface temperature records off the southeast coast of Australia (B. G. Thom, personal communication) suggest that high temperature anomalies are associated with storms and beach erosion related to the intensification of lows centered over the coast and immediately offshore.

LYNN D. WRIGHT

References

Cacchione, D. A., and Southard, J. B., 1974. Incipient sediment movement by internal gravity waves, *Jour. Geophys. Research* 79, 2237-2242.

Davies, J. L., 1964. A morphogenic approach to world shorelines, *Zeitschr. Geomorphologie* 8, 127-142.

Davies, J. L., 1973. *Geographical Variation in Coastal Development.* New York: Hafner, 204p.

Easton, A. K., 1970. *The Tides of the Continent of Australia.* Adelaide: Horace Lamb Centre, Flinders Univ., Research Paper 37, 326p.

Garrett, C., and Munk, W., 1975. Space-time scales of internal waves. A progress report, *Jour. Geophys. Research* 80, 291-297.

Inman, D. L., and Brush, B. M., 1973. The coastal challenge, *Science* 181, 20-32.

Komar, P. D., 1976. *Beach Processes and Sedimentation.* Englewood Cliffs, N.J.: Prentice-Hall, 429p.

LeMéhauté, B., 1976. *An Introduction to Hydrodynamics and Water Waves.* New York: Springer-Verlag, 315p.

Murray, S. P., 1972. Observations on wind, tide, and density driven currents in the vicinity of the Mississippi River Delta, *in* D. J. P. Swift, D. B. Duane, and O. H. Pilkey, eds., *Shelf Sediment Transport.* Stroudsburg, Pa.: Dowden, Hutchinson, & Ross., 127-142.

Neumann, G., and Pierson, W. J., 1966. *Principles of Physical Oceanography.* Englewood Cliffs, N.J.: Prentice-Hall, 545p.

Shepard, F. P., 1973. *Submarine Geology.* New York: Harper and Row 517p.

Wiseman, W. J.; Wright, L. D.; Rouse, L. J.; and Coleman, J. M., 1976. Periodic phenomena at the mouth of the Missippi River, *Contrib. Marine Science* 20, 11-32.

Wright, L. D., 1976. Nearshore wave-power dissipation and the energy regime of the New South Wales Coast: A global comparison, *Australian Jour. Marine and Freshwater Research* 27, 633-640.

Wright, L. D., and Coleman, J. M., 1973. Variations in the morphology of major river deltas as functions of the ocean wave and river discharge regimes with seven examples, *Am. Assoc. Petroleum Geologists Bull.* 52, 370-398.

Wunsch, C., 1975. Deep ocean internal waves: What do we really know? *Jour. Geophys. Research* 80, 339-343.

Cross-references: *Aktology; Climate, Coastal; Coastal Morphology, Research Methods; Internal Waves; Seiche; Tidal Range and Variation; Tidal Type Variation, Worldwide; Tides; Tsunami; Waves.* Vol. I: *Oceanography, Nearshore.*

COASTAL MORPHOLOGY, RESEARCH METHODS

One of the first attempts to describe the research methods used in the study of coastal morphology was that of A. Schou in 1945, but the first complete book on the subject was that of V. I. Budanov published in 1964. V. P. Zenkovich published a paper in 1965 on the aims and lines of research on the shore zone of the sea. Also, the U.S. Beach Erosion Board published various memoranda and technical reports, constituting a set of methodological aspects, on topics such as wave study, coast erosion, and sand movement.

Lithology and Structure

Submarine Zone. Echo-sounding may be used to make *deep measurements*. A good nautical chart with detailed notations will give reasonable results. *Profiles* should be made normal to the shoreline and referred to a known level as much as possible. It is often difficult to coordinate the surveys below and above sea level. Bedrock *samples* are collected by standard geological methods. Some superficial samples of submerged reefs can be obtained through diving practices (Aqua-lung), which serve also for collecting samples of bottom sediments. For other conditions, piston corers are necessary. A special device is needed to obtain undisturbed cores with sediment lamination and to take thick oriented samples. To study the *tectonic elements*, a good bottom survey or pre-existing bottom charts are needed. Cross-profiles in different directions are useful, because it is necessary to note the condition of rock bedding. In ideal circumstances a sub-bottom profiler can be used to show thickness, stratification, and dislocations of sedimentary beds.

Sub-aerial Zone. Topographical and nautical charts are indispensable to identify the rocks exposed to abrasion on the cliffs, reefs, or bottoms. Aerial photography is an excellent instrument for this work. Standard geological techniques are used for *sample* collection after good areal selection. The most interesting rocks are those of the cliffs, reefs, and other forms attacked by the waves. Topographic charts, nautical charts, and air photographs are important for studying the *structural and tectonic elements*. Fault lines can be located by relief studies that will also indicate the conditions of rock bedding and contacts between different hardnesses or beddings. For regional studies, satellite images permit one to find large structural patterns. Detailed heights survey and geochronological research are necessary to detect vertical displacements of marine terraces.

Oceanographic Factors

Wave Action Studies. One should distinguish between *sea* and *swell*, and their differential effects in shore dynamics. The use of wave recorders is desirable, but wind data and the use of *swell forecasting or hindcasting* methods can supply satisfactory results. The *refraction, reflection,* and *diffraction* of waves are of interest in shore dynamics. Wave charts with *orthogonals* and *crest lines,* and aerial photographs are good research instruments. Some experiments in model wave tanks, if possible, are useful. In studying *littoral drift* (shore drift), the incidence angle of common swell must be determined. The orientation of shorelines and the distribution of coastal forms give good references. It is also useful to observe the existence of beach cusps as a result of swash alignment.

Tide and Tsunami Studies. Monthly *mareograms* (q.v.) obtained for the nearest port permit one to establish the *tidal regime,* and its influence on the physical, biological, and chemical phenomena of the strand. The observations of *tidal currents* in estuaries are very useful. Flow-meters, drifters, and current-markers complement observations on movement of sediment. The influence of *tsunamis* on coastal platforms permit one to see them as factors of evolution. The relative position of headlands with respect to the tsunamis can also be an element of analysis.

Longshore and Rip Currents. Both types of currents may be observed directly and indirectly; directly by means of flow-meters, drifters, and current markers; indirectly by means of sand movement as beach ridges and spits or bars showing, respectively, the longshore current and the rip current. In order to make a good correlation between currents and their causes, it is necessary to use meteorological data.

Fluvial Factors

River Discharge. Hydrologic stations furnish records of fluvial discharge. It is also necessary to know the sediment discharge. Alluvial banks must be studied for gravelly and sandy samples, their granulometry and petrography, and compared with the detrital deposits of the shore. The bottom and suspended load (clays and silts) may also be studied and compared, in turbidity samples with bottles and in bottom samples with dredges.

Deltaic and Anastomosis Processes. In studying *deltaic processes,* one must observe the evolution of distributary channels—their banks, levees, and erosion zones. The comparative study of aerial photographs and charts is helpful. The development of marine forms at delta shores must be determined, as well as the existence of distributary fingers. In studying *anastomosis processes,* aerial photographs, charts, and periodic visual observations record the growth, degradation, or equilibrium of lenses and banks near and in the river mouth. Difluence and anastomosed channels appear as a result of this evolution.

Bar and Embankment Processes. The growth, degradation, or equilibrium of river mouth bars and beach ridges near the bars may be measured by visual observations, aerial photographs, and charts. Comparative studies of sediments also furnish useful results.

Ancient Evolution

Submarine Zone. When studying the ancient evolution of submerged shorelines it is necessary to use the plotting of bottom profiles in addition to deep measurement methods. To distinguish between forms of recent origin, those of the Holocene age, and those that are ancient, one may use geochronological methods such as paleontological analysis (micropaleontological, if possible), radioisotopic analysis, and the study of cores.

Sub-aerial Zone. The analysis of topographic charts and aerial photographs permits one to find the principal pattern of evolution. An important complement is the geomorphological observation of terraces, as well as their general sedimentology. One can also compare old and modern shorelines and distinguish recent, Holocene, and Pleistocene forms. Conventional methods of dating, relative or absolute, are naturally useful in dating the high stands of Quaternary sea levels.

Correlations. The comparison of sea levels is the first approximation of long distance correlation; but only the description of landform is insufficient. One must compare sedimentary bodies. An accurate study of the bibliography on this topic is obligatory.

Present-Day Processes

In studying submarine zones, make repeated soundings, sediment samplings, and diving operations. Compare repeated surveys and charts. Experiment, if possible, with radioactive or luminescent tracers in sediments. Determine the zone of wave erosion and extend the research to where the depth is equal to half the length of average storm waves. Coordinate, if possible, the heights above and the sounding below sea level. Employ the interpretation of historical and archaeological data. Follow the movement of offshore bars and sediment deposits. Investigate the existence of submarine canyons and their importance in sediment loss. Some remote sensing images furnish useful information. Thus, infrared and radar images, as well as thermal scannings, give excellent results. Successive surveys with infrared photographs and thermal scanning are usually available.

In studying sub-aerial zones, compare successive aerial photographs and charts. Study thermal images and infrared photographs from aircraft. Marine abrasion and sediment formation must be observed in reefs, abrasional platforms, cliffs, and other adjacent forms. A system of standard symbols for characterizing the abrasional system and the formation of sediments is desirable. The rock type, weathering type, and the principal patterns of chemical and physical wasting must be noted, as well as the biological activity. Observe the evolution of the beach, beach ridges, and dunes. Carry out studies on the sediments. Gravels and soils of similar grain size are commonly studied in the field; sand, silts, and clays, in the laboratory. For sediment analysis, refer to the work of Krumbein and Pettijohn (1938). In order to study the present-day evolution, only some sedimentologic methods are useful. Classifying gravels and sands is most important. The shape and petrography of gravels permit their identification. For sands it is important to determine heavy minerals, calcareous content, and insoluble residue (quartz and feldspars). These three properties define the sands and their *provenance*. Dune sands are well defined by their granulometry. It is important to determine the dune type and evolution. Relative vertical movements, actual and neotectonic, can be found by tide gauge records and repeated levelings. These should be taken especially, if possible, after earthquakes and tsunamis. Finally, the study of cataclysmic events, such as the occurrence of surges and tsunamis, permits one to determine their influence on coastal evolution.

Systems Analysis and Models. Structural and genetic relationships may be studied by means of morphologic and process-response systems theory, and by using conceptual, mathematical-deterministic, statistical, and simulation models (King, 1972).

Geomorphological Maps. The main principles of geomorphological maps are described by Tricart (1965). The basic concept is for readable and instructive detailed maps giving morphographic, morphometric, morphogenetic, and morphochronological information.

Individual *morphographic forms* with adequate identity should be given. Contours are the basis of morphography and morphometry. Groups of *slope gradient* must be expressed. The origin of the different *morphogenetic forms* must be indicated, in color if possible, with the marine forms in blue. For *morphochronological* information, according to the

International Geographical Union Subcommission on Geomorphological Mapping (Gellert, 1967), a differentiation of full colors for landforms will permit identification of the ages of the morphogenesis. The older forms have lighter tints and the more recent and better preserved forms, darker tints. For example, dark blue for Holocene (Q1) and Upper Pleistocene (Q2); medium blue for Middle Pleistocene (Q3); light blue for Lower Pleistocene (Q4); and celestial blue for Pliocene (N1) and Miocene (N2) formations.

Small-scale geomorphological coastal maps are described by Araya-Vergara (1972), who proposes that they include a *morphogenetic context* defining regional coastal types; *primary or continental factors* defining the local types and individuals; *secondary or marine factors* also defining local types and individuals; and *submarine morphology* correlating the submarine factors. To obtain this information it is necessary to study aerial photographs, charts, and field reality. Satellite images are also available.

JOSE F. ARAYA-VERGARA

References

Araya-Vergara, J. F., 1972. Proyecto del mapa geomorfológico de las costas de Chile, in *1st Symposium Cartográfico Nacional Instituto Geográfico Militar, Santiago,* 336-340.
Budanov, V. I., 1964. *Research Methods on the Littoral Zone* (in Russian). Moscow: Nauka.
Gellert, J. F., 1967. Further works on the unification of signs and signatures of geomorphological detail maps, *Zeitschr. Geomorphologie N.F.* 11, 506-509.
King, C. A. M., 1972. *Beaches and Coasts,* 2d ed. London: Arnold, 570p.
Krumbein, W. C., and Pettijohn, F. J., 1938. *Manual of Sedimentary Petrography.* New York: Appleton, 549p.
Schou, A., 1945. *Det Marine Forland.* Kobenhaven: Folia geographica danica 4, 236p.
Tricart, J., 1965. *Principes et Méthodes de la Géomorphologie.* Paris: Masson, 496p.
Zenkovich, V. P., 1965. Buts et principaux axes de recherches des études sur les zones maritimes littorales, *Cahiers Oceanogr.* 17, 605-623.

Cross-references: *Coastal Characteristics, Mapping of; Coastal Ecology, Research Methods; Coastal Engineering; Coastal Engineering, Research Methods; Profiling of Beaches; Protection of Coasts; Sediment Analysis, Statistical Methods; Sediment Tracers; Tides; Waves.*

COASTAL PLANT ECOLOGY, UNITED STATES, HISTORY OF

The science of coastal plant ecology, with beginnings extending back to the 1860s, started to develop around the turn of the twentieth century and has been expanding ever since. By definition, coastal plant ecology is the ecological study of plants growing in the coastal zone at the land/sea interface. These plants live under the continuous influence of the physical oceanic environment, manifested in tides, salt spray, sand movement, and storms. The coastal zone is a region of constant physical change in which all facies of the environment are controlled primarily by the sea.

The seaward boundary of the coastal zone, and therefore of coastal ecology, is the lowest limit of the low spring tides. Coastal plants are distinguished from marine plants in that the former are exposed at some times, however infrequently and briefly, to the air. The terrestrial boundary of the coastal zone is the landward limit of regular physical oceanic influence, salt spray and tides being the factors that extend farthest inland.

The coastal zone has long been recognized as an ideal place to study the fundamental concepts of ecology.

Descriptive Coastal Plant Ecology

Period I—Early (Classic) Studies Without Quantitative Sampling. Around the turn of the century there was a profusion of descriptive coastal plant ecology papers. These papers gave the first indication of the evolution of this new unified branch of plant ecology in the United States. Each author employed the same descriptive techniques and referred to the other coastal scientists who had published similar papers. Though none of these papers employed quantitative sampling techniques, their discussion of coastal plant classification, of the major environmental conditions under which coastal plants live, and of anatomical and physiological adaptations to these conditions laid the foundations for many of the modern concepts of coastal plant ecology.

In 1899 Henry Chandler Cowles published a classic series of papers on the ecological relations of the vegetation on the sand dunes of Lake Michigan. Although these papers did not deal with plants living under the influence of the oceanic environment, Cowles could still be considered the father of modern coastal plant ecology because he was the first to suggest many of the concepts essential to the development of this field. In his papers Cowles (1899) discussed "(1) the plant formations, their characteristic species, the progressive changes that take place, and the environmental factors which cause these changes; and (2) the adaptations of plants, both gross adaptations of plant bodies and organs and others dealing with the anatomical structure of the plant tissues, to their dune environment." In addition, his classification scheme for beach vegetation is frequently referred to in other early descriptive papers.

Between 1899 and 1905 a series of papers appeared on the vegetation of the Atlantic and Gulf coasts of the United States. Excluding those that contained only brief lists of coastal plants as parts of larger floristic studies, the major papers that dealt in detail with the coastal vegetation, the coastal environment, and the adaptations of plants to that environment were: Webber, 1898 (Florida); Kearney, 1900 (North Carolina); Johnson, 1900 (North Carolina); Harshberger, 1900 (New Jersey); Kearney, 1901 (North Carolina, Virginia); Lloyd and Tracy, 1901 (Louisiana, Mississippi); Snow, 1902 (Delaware), Ganong, 1903 (New Brunswick); and Coker, 1905 (South Carolina); the three most significant being those by Kearney (1900), Harshberger, and Ganong. However, all of these authors could be considered the first coastal plant ecologists of the east coast of the United States.

Kearney's and Harshberger's 1900 publications were detailed accounts of the coastal vegetation of North Carolina and New Jersey, respectively. For easy comparison of these two areas, Harshberger followed Kearney's format and nomenclature. Both of these extensive publications contained sections on climate, physiography, geology and soils, plant formations—their composition and physiognomy, ecological forms—adaptations to the environment, and phytogeographical affinity of the flora. In addition, Kearney described and diagrammed the anatomy of 32 coastal plants.

Although it discusses only marsh vegetation, Ganong's 1903 publication ranks with Kearney's and Harshberger's because of the thoroughness with which he treated marsh formations, the marsh soils and environment, and the ecological adaptations of the marsh plants to this habitat. In addition, Ganong goes even deeper than Kearney or Harshberger into the application of basic ecological concepts to marsh studies. These three works are unique because, unlike more recent publications in coastal plant ecology, they contained information on the coastal habitats (climate, soils, environmental factors) and coastal plants (ecology and anatomy).

Following this initial burst of publications on coastal plant ecology, a few more descriptive accounts of east coast vegetation appeared between 1909 and 1930, including those of: Transeau, 1909 (Nova Scotia); Lewis, 1917 (North Carolina); Nichols, 1920a and b (Connecticut); and Chrysler, 1930 (New Jersey). George Nichols's work most closely resembled the earlier efforts by Kearney, Harshberger, and Ganong. In 1920, as part of a survey of the vegetation of Connecticut, Nichols published two accounts of coastal vegetation, one on the plants of eroding coasts, the other on plants of depositing areas. He described the major habitat factors along the seacoast, the plant formations and their classification, and the ecological adaptations and anatomy of common coastal plants.

A number of descriptive papers on the coastal vegetation of the Pacific states were published between 1902 and 1905. These include Davy, 1902 (northwest California coast); Olsson-Seffer, 1910 (San Francisco, Monterey Bay, Santa Monica, and Santa Barbara); House, 1914 (Coos Bay, Oregon); Kellogg, 1915 (northern California and southern Oregon); Ramalay, 1918 (San Francisco); and Cooper, 1919, 1922 (Monterey Peninsula). All were short descriptions of dune and beach vegetation, with no discussion of environmental conditions, ecological adaptations, or classification of plant communities. In 1909 Olsson-Seffer published an extensive discussion of the environmental condititions under which plants live in the coastal zone. This work encompassed observations on coastal vegetation around the world. None of these authors acknowledged the work of other west coast scientists, although Ramalay did refer specifically to Cowles's work, while Olsson-Seffer referred to Cowles, Harshberger, Webber, Kearney, Ganong, Lloyd and Tracy, and a few other east coast researchers.

Period II—Descriptive Studies with Quantitative Sampling of Plant Communities. In the mid-1930s descriptive coastal plant ecology took on a new dimension. In addition to the traditional classification of plant communities and discussion of environmental factors, publications began to contain quantitative data on the makeup of coastal vegetation. These data were usually obtained by placing sampling quadrats in major plant communities, often along transects laid out perpendicular to the coast. In addition to quantifying the species composition, most of the authors measured some of the important environmental factors influencing plant growth and distribution.

The most important descriptive coastal plant ecology studies to appear between 1934 and 1959 were those by: Penfound and O'Neill, 1934 (Mississippi); Conrad, 1935 (Long Island); Penfound and Hathaway, 1938 (Louisiana); Kurz, 1942 (Florida); Dexter, 1947 (Massachusetts); Penfound, 1952 (southern swamps and marshes); Bordeau and Oosting, 1959 (North Carolina); and Martin, 1959 (New Jersey). These papers all dealt with Atlantic and Gulf coast vegetation; there were no similarly significant publications on west coast vegetation. The few west coast plant ecologists working at this time concentrated on experimental studies and autecological investigations of major species.

Of all the descriptive coastal plant ecology papers published prior to 1960, Martin's (1959) is the most complete. The techniques, thoroughness, and results are on a par with any of the more recent publications; the paper can therefore be considered transitional to modern coastal plant ecology and serves as the cutoff point for this article. Martin's study had two objectives: to map, classify, describe, and sample the vegetation of Island Beach; and to determine the relationship of plant distribution to topography and environmental patterns. In fulfilling his first objective, Martin included a complete classification of all plant communities present, along with a detailed vegetation map. Vegetation was measured quantitatively, using relative cover, abundance, density, and frequency parameters obtained from quadrats located on several cross-island transects. For each community he also quantified topography, elevation, edaphic factors, and exposure to wind. Related to the second objective, Martin conducted four sets of experiments. To study the relationship of sand dune erosion rates to the presence or absence of *Ammophila,* he placed stakes every 10 m along three transects and measured the distance to the ground throughout the year. On his transects, he related plant-distribution data to topography. Using salt spray traps, he determined the distribution of salt spray across Island Beach. To determine the tolerance of plants to salt spray, he periodically sprayed 50 species of plants with a saltwater solution.

Modern Concepts in Coastal Plant Ecology

Many of the principal concepts of modern coastal plant ecology (e.g., the importance of salt spray in plant distribution) originated in the minds of early workers in the coastal zone. Because these concepts are essential to our understanding of plant growth and survival in the coastal environment, it is important to trace their origin and development.

Coastal Geobotany. Cowles (1899) was the first ecologist to stress the importance of plants in controlling physiography. He stated that in their interactions with common physical forces, plants play a major role in creating sand dunes and in determining their structure. He concluded that since different plants have different growth patterns, the form of a dune will relate to the species that stabilizes it. To emphasize the importance of such interactions, this concept has been termed *coastal geobotany.*

This idea was referred to briefly in a few early east coast descriptive plant ecology papers.

Webber (1898) stated it the most directly by saying that the interaction of plants with processes yields form. In a footnote Harshberger (1900) stated, "The student of the dynamics of dunes who desires to study the modification of vegetable organisms is referred to the papers by Dr. Cowles." Snow referred to this concept indirectly in her 1902 and 1913 papers. In his 1942 paper on the dune and scrub vegetation of Florida, Kurz discussed in some detail dune patterns as determined by different coastal plants. Olson's 1952 paper, on Lake Michigan Dune Development, addressed the role and importance of plants in dune establishment and growth. As he stated at the end, "Thus vegetation not only initiates and regulates dune growth but also provides an autobiographical record which can be useful to the geomorphologist who would translate its record."

After this paper, the idea of coastal geobotany faded until the present. The coastal plant ecology group at the University of Massachusetts under the leadership of Dr. Paul Godfrey is currently emphasizing this concept in their research on the barrier beaches of the United States Atlantic and Gulf coasts.

Characterization of Coastal Zone Environmental Factors and the Adaptation of Plants to These Factors. The coastal literature from 1898 to 1920 is replete with information on the coastal environment. The earliest papers, published by Cowles (1899), Kearney (1900 and 1901), and Harshberger (1900) contained general discussions of the climate, geology, and soils of the areas described. Later contributions by Ganong (1903), Harshberger (1908, 1909*a* and *b*), Olsson-Seffer (1909*a* and *b*), Transeau (1913), and Nichols (1920*a* and *b*) were much more specific, presenting the major environmental factors responsible for plant distribution on the coast. These included atmospheric, hydrodynamic, edaphic, and historic forces and other more specific influences, such as tides, salinity, and erosional and depositional patterns.

The most complete early study on the coastal environment appeared as two papers by Olsson-Seffer in 1909. In his publication "Hydrodynamic Factors Influencing Plant Life on Sandy Sea Shores" (1909*b*), he dealt with all aspects of the soil-water relationships (e.g., evaporation layer, percolation, salinity) in coastal areas. In the longer work, "Relation of Soil and Vegetation on Sandy Sea Shores" (1909*a*), he discussed in detail all factors influencing plant life on the coast, but concentrated on environmental factors relating to the soils. This significant paper provided a tremendous amount of information on the coastal environments of plant growth.

Starting in the 1930s a few ecologists began to make quantitative measurements of environmental conditions along transects set out to sample vegetation. Most notable was the work of Penfound and O'Neill (1934) and Penfound and Hathaway (1938), who measured depth to water table, soil water content, soil water salinity, percent organic matter, and pH. However, it was not until the start of the modern period, with Martin's 1959 paper, that it became common to make quantitative environmental measurements along sampling transects.

One environmental concept of particular importance is that dealing with the availability of water in coastal soils. Early ecologists disagreed on the extent of water available. Johnson (1900), Harshberger (1900, 1908, 1909a and b), Ramalay (1918), and Whitfield (1932) believed that plants growing on the strand lived under xeric conditions. They stated that the lack of water in the dune soil was the major environmental factor influencing the distribution of coastal plants. In his early experimental paper on the osmotic concentrations of chaparral, coastal sagebrush and dune species of southern California, Whitfield stated that a single species indicated "that the dunes were also more xeric than the coastal chaparral" and "with only a single exception, all values [osmotic concentration] were higher in the dunes than in the coastal sagebrush."

In contrast, Kearney (1900, 1901) and Olsson-Seffer (1909a and b) believed that conditions on the dunes are not xeric, and that water availability does not play a major role in determining what plants grow there. Kearney (1900) stated "There is no lack of moisture in this sandy substratum. Even in the dryest looking beach sand, water usually stands at a depth of only 15 to 30 centimeters from the surface"; and (1901) "Despite the excessive permeability the soil is here rarely dry, except at and very near the surface. Even on dunes one can easily reach moist sand with the hands . . . it is probable that only the smallest plants of the strand formation ever have difficulty in reaching a sufficient water supply with their roots." In the long run, Kearney's and Olsson-Seffer's belief has proved correct; however, even today one still hears statements such as those originally proposed by Johnson, Harshberger, and Ramalay.

Along with their discussion of environmental conditions, some early ecologists also described the adaptations of coastal plants to major environmental factors. Three scientists talked about adaptations in general. Cowles (1899) discussed adaptations of plants to the dune and beach environment. In his 1900 and 1901 papers, Kearney mentioned two types of ecological adaptations: changes of form to cope with the mechanical action of the wind and shifting sand; and modifications of morphology to help plants increase or conserve water supply against excessive transpiration. In his 1903 paper, Ganong described the particular adaptations of major marsh species to their environment.

A common topic of early coastal literature was the xeromorphic nature of coastal plants and its causes. Kearney (1901) stated that coastal plants, particularly their leaves, have xeromorphic characteristics in order to reduce water loss by transpiration. However, this is not due to soil drought. He went on to say that in salt marsh plants, "it is chiefly the presence of comparatively high percentages of NaCl in the soil water which necessitates a xeromorphic character." Harshberger (1900) attributed the "extreme xeromorphic character of succulence" in middle beach plants to their existence in "a porous soil of drifting sand and within the influence of salt spray." He then explained the presence of nonsucculent *Euphorbia* on the beach by its possession of latex, "which is probably instrumental in reducing transpiration."

Associated with the concept of coastal xeromorphy there was a debate on whether or not plants of coastal beaches and dunes are true halophytes and whether salt is the cause of the common xeromorphic characteristic. Some early ecologists stated that dune and beach sand contained a large amount of NaCl, while others said the salt content was low. In his 1904 paper "Are Plants of Sea Beaches and Dunes True Halophytes?" Kearney described an experimental investigation of this problem. He showed that dune and beach soils from Woods Hole, Massachusetts, and Norfolk, Virginia, contained only low concentrations of NaCl, compared with the highly saline soils of the marshes. He stated that neither the sand nor the water under the beach and dunes is saline, and therefore beach and dune plants are not true halophytes. These plants' xeromorphic character is caused, rather, by the strong winds, intense light, and excessive transpiration of the beach and dune environment.

As a part of their description of plant adaptation to the coastal environment, a number of early ecologists described in great detail the anatomical and morphological characteristics of common coastal species. In his 1900 paper "The Plant Covering of Ocracoke Island," Kearney diagramed and described the anatomy of 32 sand strand and salt marsh plants. This was followed by a detailed discussion listing many anatomical adaptations widely recognized today. In 1904 Chrysler published "Anatomical Notes on Certain Strand Plants," in which he described a comparative study of the leaf anatomy of certain plants from Woods Hole and Lake Michigan. He listed nine anatomical differences between plants growing in saline coastal soils and plants growing inland. In 1908 and 1909 Harshberger published two papers on the comparative structures of the sand dune plants of Bermuda and New Jersey, respectively. In both he not only diagramed and described in detail the anatomy of strand plants but also discussed the environmental factors that he believed responsible for these characteristics. Throughout he separated the strand and salt marsh plants, and at the end of his 1909 publication he compared the structural characteristics of these two groups. Purer, in her 1936 studies of important dune plants of southern California, presented the most morphological, anatomical, and ecological information in common dune species. Her study included: investigation of the gross morphology of the roots, stems, leaves, flowers, and fruits; measurement and charting of subterranean and aerial portions of the plants; anatomical study of stained cross sections of leaves, stem, and roots. She combined this information with environmental and plant distribution data to produce an extremely detailed and informative work. These detailed anatomical and morphologic papers on coastal plants reflect the close ties between early coastal plant ecology and traditional botany. Unfortunately, in succeeding years the disciplines separated, so that today one rarely sees anatomical analysis in coastal plant ecology literature.

Salt in the Coastal Zone. One of the principal factors that make plant life in the coastal zone different from everywhere else is salt in the air, soil, and ground water. The adverse effect of salt on coastal plants was first mentioned by Warming and Schimper near the turn of the century, but a full understanding of the effect of salt on plant life and a knowledge of the tolerances of different plants to salinity were not attained until the late 1950s.

Salt in the air, in the form of salt spray, was not initially recognized as a causative factor in the pruned, krumholtz-shaped, coastal vegetation form or as an important factor in determining the distribution of plants along the shore. Harshberger (1900, 1908, 1909a and b) and Olsson-Seffer (1909a) did mention the presence of salt spray in the coastal zone but did not acknowledge its importance. Many early ecologists, Kearney (1900, 1901), Snow (1902), and Ganong (1903), omitted the effect of salt spray entirely from their papers and attributed the pruned forms of coastal vegetation to mechanical damage caused by the wind. As Kearney (1900) states: "The shrubs and trees of the islands show the effect of much exposure to high winds, in their short gnarled branches and in often onesided positions of their crown foliage . . . here however we have to do rather with direct mechanical effect of the wind than with a protective modification." Other scientists state that this form was due to the drying of plant parts by wind. As expressed in *Plant Ecology* by Weaver and Clements (1938): ". . . on wind swept coasts . . . excessive water loss results in a stunted and gnarled growth. This is however partly due to the mechanical effect of wind." Other suggestions for this distinct form included the effect of sand blasting or substrate salinity.

The first to suggest salt spray as a major environmental factor in the coastal zone was Shreve in 1910. In 1920 Boodle hypothesized that salt spray played an active role in pruning. In papers published in 1937 and 1938 Wells and Shunk proved that the distinct coastal shrub form was caused by salt spray, not wind. As they stated (1937): ". . . the young tender, exposed shoots are so severely injured by the spray that only the protected lateral and leeward terminals develop, resulting in the characteristic compact repressed stooping form" and (1938) ". . . the injury may thus be interpreted as necrosis due to excessive water loss from the young tissue resulting from the osmotic action of high salt concentration on the unprotected surface." In 1941 J. K. Doutt published an article entitled "Wind Pruning and Salt Spray as Factors in Ecology." He compared pruned trees in the Arctic and the southeast, and found no apparent differences in form. He therefore concluded that the pruned form must be due to salt spray and not wind.

In papers published by Oosting and Billings (1942) and Oosting (1945), the salt spray tolerances of coastal plants were determined experimentally. Oosting and Billings potted *Uniola, Andropogon,* and *Spartina patens* and applied the following daily watering regimes: freshwater, saltwater, freshwater every other day, freshwater with daily sprayings of seawater. This test established that the distributions of these three important North Carolina coastal plants were related to their variation in salt spray tolerance. In 1945 Oosting conducted a similar experiment with common coastal plants growing in the wild. He used the above treatments with 1, 2, 3, or 4 applications daily for nine days. His results showed a large variation in tolerance of coastal plants to salt spray.

Boyce, in contributions published in 1952 and 1954, investigated the source of salt spray, the transportation and deposition of salt particles, the relation of salt spray to zonation and growth, the variation in tolerance of the species to salt spray, and the response of different species to salt spray. Boyce conducted his investigation using laboratory and greenhouse studies, the conclusions from which were tested in the field. He demonstrated that the active toxic agent in salt spray is chloride. "Anatomical studies of leaves of plants exposed to different intensities of salt spray showed that chloride induced hypertrophy attained before coagulation of the protoplasm was correlated with the degree of tolerance of the species to the chloride salt." Boyce's experiments conclusively documented the important role of salt spray in the coastal environment.

Once the deleterious effect of salt spray on coastal plants was acknowledged, researchers started to look into it as a causative factor in coastal plant distribution. This causation was first implied in a paper by Bowman (1918). Wells and Shunk (1938) stated, "It seems probable that the distinctive composition of the dune community is based primarily upon the species adaptations to the spray factor." Wells (1939) in fact hypothesized that live oak, *Quercus virginiana,* dominated mature forests of coastal North Carolina because its relative resistance to salt enabled it to survive where other potential dominants could not. This forest type they termed the *salt spray climax.*

In 1942 Oosting and Billings clearly demonstrated the role of salt spray in the distribution of the coastal vegetation of North Carolina. In this classic study they sampled and mapped plant distribution over a transect running inland perpendicular to the beach. At five stations along the transect they measured soil moisture, soil temperature, soil pH, air temperature, relative humidity, evaporation, salt content of soils, and amount of wind-borne salt. The only clear differences between stations were in salt spray intensity. Correlating salt spray pattern, plant distribution data, and the known salt spray tolerances of dominant plants, Oosting and Billings proved that plant distribution on the beaches and dunes of North Carolina is controlled by salt spray. Oosting (1954) presents an extremely detailed and complete review of all salt spray literature to that date.

Although it has not received as much attention as salt spray, the amount of salt in the soil, or in the water in the case of marsh plants, is also important in plant distribution. Papers by Olsson-Seffer (1909a and b), Harshberger (1911), and Penfound and Hathaway (1938) dealt with the tolerances of various plants to these forms of salinity. To investigate the tolerances of coastal plants to soil salinity, Olsson-Seffer grew common coastal species in sand to which different amounts of salt had been added. Harshberger conducted an extremely detailed set of experiments on the tolerance ranges of coastal wetland species to water salinities. Using a hydrometer to measure water salinity in various marsh zones and specifically at the locations of a number of kinds of marsh plants, Harshberger showed how the several species differed in their "ability to withstand degrees of salinity of water." He then ranked all species in order of their salinity tolerance levels and ranges. In 1938 Penfound and Hathaway conducted a similar hydrometric study of the tolerance of common wetland plants in southeastern Louisiana.

Algal Zonation Studies. Although they transcend the boundaries of coastal plant ecology, zonation studies of marine algae provided tremendous amounts of information on factors influencing the distribution of plants in the intertidal zone. Most work on this topic was from areas other than North America, but four notable papers were published on the zonation of United States algae.

The earliest paper describing the vertical distribution of United States intertidal and subtidal algae and therefore perhaps the first true paper in North American coastal plant ecology, was Kemp's "On the Shore Zones and Limits of Marine Plants on the Northeastern Coast of the United States" (1862). While on a trip to Casco Bay, Maine, in 1861, Kemp described in detail the six zones (drift, *Ulva, Fucus, Laminaria, Chondrus,* and deep-sea) characteristic of the northeastern rocky shore. He discussed the species composition of each zone, but did not specifically mention the environmental factors responsible for the zones and gave only a rough idea of their vertical limits.

In the words of Henry Conard (1935): "The most serious study of the plants of this region [Long Island] is *The Relation of Plants to Tide Levels* by D. S. Johnson and H. L. York.... Johnson and York have presented in great detail the vegetation of the inner harbor below the level of the flood tides. They have shown a clear relation of the plants to the tide levels, and interpreted their findings as due to the daily period of exposure of the plant to salt water or the air." Their extensive treatise on this area covered the zonation of both intertidal and subtidal algae as well as the seed plants of the salt marsh. Nichols (1920a and b) listed the five major factors responsible for the algal zonation: salinity, the tides, illumination at different depths, temperature of saltwater, and erosional and depositional patterns. As with Johnson and York, in Nichols' paper the distinct algae zones were given precise vertical definition.

A series of papers published in 1928 by Johnson and Skutch presented the results of a detailed four-year research project "to determine the precise limits of distribution, vertically and horizontally, of each littoral plant association found on a high rocky point of Mt. Desert Island, Maine." Following a thorough review of world literature on algal zonation, they classified each zone and looked specifically at the environmental factors characterizing it. In addition, using reference points painted on the rocks, they surveyed the elevation of the zone boundaries. In their paper they correlated "the observed vertical and horizontal distribution with the character of the substratum and with all these other environmental factors such as light, temperature, salinity, acidity and evaporation which are affected by the alternating submergence and exposure of the plants by the tide."

Salt Marsh Ecology. Today salt marshes are receiving a tremendous amount of attention in the literature. Although they have been intensively studied only in the past ten years, research on salt marshes goes back to the beginning of coastal plant ecology. The first papers were purely descriptive, but they were soon followed by studies of the factors responsible for the distinct salt marsh zonation and of salt marsh successional patterns. A complete discussion of salt marsh research to 1938 is presented in Penfound and Hathaway (1938).

The earliest paper on North American salt marshes was published by D. M. Nesbit in 1885. Works by Eaton in 1893 and Shaler in 1895, though written from a geological standpoint, provided additional general information on the salt marshes of the Atlantic coast. Kearney, in his 1901 survey of the Dismal Swamp region, presented the first detailed ecological description of salt marshes. Not only did he characterize the major environmental factors and the recognizable salt marsh zones and their plants, he also related these zones to the frequency and duration of tidal flooding. Ganong's 1903 study of the Bay of Fundy salt marshes was devoted entirely to the ecology of the vegetation. He discusses in great detail the major factors determining the ecological features of marsh vegetation and the specific adaptations of salt marsh plants to these factors.

Between 1902 and 1915 the major salt marsh zones and their causative factors were discussed by Harshberger (1902, 1909a and b), Smith (1907), Davis (1910), and Johnson and York (1915). Harshberger (1909b) divided the New Jersey salt marshes into three zones and clearly related each to the levels of the tide: "Three kinds of salt marshes may be distinguished in New Jersey. The first and the smallest in area is that which is covered at every neap high tide. The second in area is rarely covered at ordinary tides, but is so little above mean high water that even the slightest rise due to wind, storm or moon changes results in a partial covering with water. The third type of marsh is that above mean high water and more or less completely covered with vegetation. All three kinds of marshes may be covered at the time of spring and fall tides." In addition, Harshberger discussed marsh establishment by *Spartina alterniflora* and the successional change of the marsh from low marsh to high marsh vegetation. Charles Davis (1910), a geologist studying the salt marshes near Boston, described zonation in terms of differences in the tolerance of common marsh plants to saltwater:

That all of these few types are not equally able to endure exposure to and submergence by the tidal water or are equally fond of it, for by watching the rise of a single tide, it will be found that some species are covered long before others, while those parts of the surface of the marsh on which the greatest number of species grow will not be reached at all by even the crest of the average tidal wave. In other words the plants of a typical salt marsh will be found to be arranged in well-defined zones, the limits of which are marked apparently by the length of time the area occupied by each is covered periodically by the tides.

Johnson and York's 1915 paper gave the most detailed account of marsh zonation to that date. Their results clearly showed a definite relation between plant distribution and tide levels.

In Johnson's 1925 geology text *The New England Acadian Shoreline,* a number of pages were devoted to the description of the three Atlantic Coast marsh types (New England, Fundy, and Coastal Plain) that he recognized. These extensive chapters included a detailed description of each type with its common plants and their zones and a general discussion of the marsh type's distribution.

Many other papers deal with salt marsh ecology, the plants, zones, and major environmental factors. The most notable of the early and mid 1900s include: Transeau, 1913 (New York); Nichols, 1920b (Connecticut); Conard, 1935 (New York); Chapman, 1938 a and b, 1939, 1940a and b, 1941, 1960 (many locations); Penfound and Hathaway, 1938 (Louisiana); Taylor, 1938 (New York); Dexter, 1947 (Massachusetts); Miller and Egler, 1950 (Connecticut); Penfound, 1952 (southeastern United States); Hinde, 1954 (California); Bordeau and Adams, 1956 (North Carolina); and Butler, 1959 (Massachusetts). A large number of these include quantitative, vegetative, and environmental analysis. Extensive salt marsh literature lists are included in Harshberger (1911), Nichols

(1920a and b), and Johnson (1925). V. J. Chapman is noted as a worldwide expert on salt marshes, having published extensively on both European and North American marshes. In 1960 he published *Salt Marshes and Salt Deserts of the World*, the most comprehensive treatment of salt marshes to date.

Salt marsh ecology has drawn and is continuing to draw a tremendous amount of research attention, more than any other area of coastal plant ecology. Initially it was a perfect place to study plant distribution with respect to environmental factors, and in addition to look at the adaptation of plants to coastal stress. Most recently salt marshes have been emphasized because of their great productivity and ability to export some of the excess output to larger but less productive ecosystems.

Experimental Coastal Plant Ecology

Modern coastal plant ecologists use a number of common experimental techniques in studying plants and ecosystems. Since the development of these techniques was an important part of the growth of this field, it is instructive to look at the origin of some of them. The earliest papers describing the use of these techniques are listed as follows. For techniques involving *environmental analysis:* Ganong (1903), Kearney (1904), and Olsson-Seffer (1909a and b) utilized analysis of soil characteristics (e.g., minerals, moisture content); Smith (1907) and Harshberger (1909a and b), the chemical analysis of salt marsh peat; Olsson-Seffer (1909a and b) and Oosting and Billings (1942), salt spray traps on beach and dunes; Cooper (1919), Penfound and O'Neill (1934), Puere (1936), Penfound and Hathaway (1938), and Taylor (1938), quantitative sampling of coastal environmental characteristics; Johnson and Skutch (1928), surveying. For techniques involving *plant analysis:* Wells and Shunk (1937, 1938), Oosting and Billings (1942), Oosting (1945), and Boyce (1952, 1954) did experimental tests of salt tolerance in air-simulated salt spray; Olsson-Seffer (1909b), in soil-simulated salt in soil; and Harshberger (1911) and Penfound and Hathaway (1938), in water-hydrometer tests. Quantative vegetation sampling by quadrat and/or transect was done by Couch (1914), Cooper (1919), Penfound and O'Neill (1934), Conard (1935), and Penfound and Hathaway (1938). Cooper (1919) utilized vegetation mapping and greenhouse studies, and compiled a photographic record of permanent quadrats. Cores of salt marsh peat to determine vegetation succession were take by Smith (1907), Davis (1910), Harshberger (1911), and Johnson (1925). Whitfield (1932) undertook the measurement of osmotic concentration in physiological studies; Purer (1936) did a detailed autoecological investigation of coastal plants; and Burkholder (1959) did nutritive studies of coastal plants.

Although many new methods have come into use in coastal plant ecology, these early workers pioneered many of the techniques upon which the science has been built.

Summary

Several review papers have been written on coastal plant ecology. Strand vegetation was reviewed in the greatest detail by Oosting's 1954 paper giving a detailed discussion of coastal process, vegetation, and environmental factors of the southeast. Much of his discussion is easily applied to coastal areas and plants not specifically discussed in his review. As his title implies, however, Oosting dealt primarily with the sand strand vegetation and factors and little with salt marsh vegetation. Nichols (1920a) presents a brief review of papers published on the strand vegetation of the North Atlantic.

Salt marsh research has been reviewed by Harshberger (1911), Johnson (1925), and Penfound and Hathaway (1938). The later review is the most complete and detailed in that it includes a short discussion of all previous papers involving salt marshes. Harshberger's and Johnson's works did not discuss previous authors, but did include extensive bibliographies of the salt marshes of the east coast. Johnson and Skutch (1928) gave a worldwide review of research on algal zonation with a short discussion of the major papers on North American algal zonation.

Since 1959 the study of coastal plant ecology has been expanding rapidly. Coastal research is still concentrated on the Atlantic and Gulf coasts, but the plant ecology of the Pacific coast is now being investigated by an increasing number of scientists. Because of their great productivity and importance in the estuarine ecosystem, salt marshes are currently receiving the greatest attention. Research on the ecology of east coast barrier islands, encompassing studies of salt marsh, grassland, dune and beach vegetation, is secondary to salt marshes in terms of increasing research attention.

Like most other fields of ecology, coastal plant ecology is moving away from traditional descriptive studies. Research emphasis is now on physiological ecology and population biology of major coastal species. In addition, a large number of investigations are being conducted into the dynamics, productivity, energetics, and modeling of these coastal ecosystems. There has also been a rebirth of the idea of coastal geobotany. In the studies of barrier islands the importance of the interaction of plants with the physical process in forming common coastal structures is often emphasized.

MARK A. BENEDICT

References

Boodle, L. A., 1920. The scorching of foliage by sea winds, *Jour. Ministry Agric. Great Britain* 27, 479-486.

Bordeau, P. F., and Adams, D. A., 1956. Factors in vegetational zonation of salt marshes near Southport, North Carolina, *Ecol. Soc. America Bull.* 37, 68.

Bordeau, P. F., and Oosting, H. J., 1959. The maritime live-oak forest in North Carolina, *Ecology* 40, 148-152.

Bowman, H. H. M., 1918. Botanical ecology of the Dry Tortugas, *Dept. Marine Biol. Paper 12, Carnegie Inst. Washington Pub.* 252, 109-138.

Boyce, S. G., 1952. An ecological study of coastal plants on Cape Cod, *Biol. Bull.* 103, 296-297.

Boyce, S. G., 1954. The salt spray community, *Ecol. Monogr.* 24, 29-67.

Burkholder, P. R., 1959. Studies on the nutritive value of *Spartina* grass grown in the marsh areas of coastal Georgia, *Torrey Bot. Club Bull.* 83, 237-334.

Butler, P., 1959. Palynological studies of the Barnstable marsh, Cape Cod, Massachusetts, *Ecology* 40, 735-737.

Chapman, V. J., 1938a. Coastal movement and the development of some New England salt marshes, *Geol. Assoc. Proc.* 49, 373-384.

Chapman, V. J., 1938b. Studies in salt-marsh ecology, Sections I-III, *Jour. Ecol.* 26, 144-179.

Chapman, V. J., 1939. Studies in salt-marsh ecology, Sections IV and V, *Jour. Ecol.* 27, 160-201.

Chapman, V. J., 1940a. Studies in salt-marsh ecology, Sections VI and VII, *Jour. Ecol.* 28, 118-152.

Chapman, V. J., 1940b. Succession on the New England salt marshes, *Ecology* 21, 279-282.

Chapman, V. J., 1941. Studies in salt-marsh ecology, Section VIII, *Jour. Ecol.* 29, 69-82.

Chapman, V. J., 1960. *Salt Marshes and Salt Deserts of the World*. London: Leonard Hill Books Ltd., 392p.

Chrysler, M. A., 1904. Anatomical notes on certain strand plants, *Bot. Gaz.* 37, 461-464.

Chrysler, M. A., 1930. The origin and development of the vegetation at Sandy Hook, *Torrey Bot. Club. Bull.* 57, 163-176.

Coker, W. C., 1905. Observations on the flora of the Isle of Palms, Charleston, South Carolina, *Torreya* 5, 135-145.

Conard, H. S., 1935. The plant associations of central Long Island, *Am. Midland Naturalist* 16, 433-516.

Cooper, W. S., 1919. Ecology of strand vegetation of the Pacific coast of North America: A geographical study, *Carnegie Inst. Washington Year Book*, 18, 96-99.

Cooper, W. S., 1922. Strand vegetation of the Pacific Coast, *Carnegie Inst. Washington Year Book*, 21, 74-75.

Couch, E. B., 1914. Notes on the ecology of sand dune plants, *Plant World* 17, 204-208.

Cowles, H. C., 1899. The ecological relations of the vegetation on the sand dunes of Lake Michigan, *Bot. Gaz.* 27, 95-117, 167-202, 281-308, 361-391.

Davis, C. A., 1910. Salt marsh vegetation near Boston and its geological significance, *Econ. Geology* 5, 623-639.

Davy, J. B., 1902. *Stock Ranges of Northwestern California*. Washington, D.C.: U.S. Dept. Agric. Bureau of Plant Industry Bull. 12, 81p.

Dexter, R. W., 1947. The marine communities of a tidal inlet at Cape Ann, Massachusetts, *Ecol. Monogr.* 17, 261-294.

Eaton, F. H., 1893. The Bay of Fundy tides and marshes, *Pop. Sci. Monthly* 43, 250-256.

Ganong, W. F., 1903. The vegetation of the Bay of Fundy salt and diked marshes: An ecological study, *Bot. Gaz.* 36, 161-186, 280-302, 349-367, 429-455.

Harshberger, J. W., 1900. An ecological study of the New Jersey strand flora, *Acad. Nat. Sci. Philadelphia Proc.* 52, 623-671.

Harshberger, J. W., 1902. Additional observations on the strand flora of New Jersey, *Acad. Nat. Sci. Philadelphia Proc.* 54, 642-699.

Harshberger, J. W., 1908. The comparative leaf structure of the sand dune plants of Bermuda, *Am. Philos. Soc. Proc.* 47, 97-110.

Harshberger, J. W., 1909a. The comparative leaf structure of the strand plants of New Jersey, *Am. Philos. Soc. Proc.* 48, 72-89.

Harshberger, J. W., 1909b. The vegetation of the salt marshes and of the salt and fresh water pools of northern coastal New Jersey, *Acad. Nat. Sci. Philadelphia Proc.* 61, 373-400.

Harshberger, J. W., 1911. An hydrometric investigation of the influence of sea water on the distribution of salt marsh and estuarine plants, *Am. Philos. Soc. Proc.* 50, 457-496.

Hinde, H. P., 1954. The vertical distribution of salt marsh phanerograms in relation to tide levels, *Ecol. Monogr.* 24, 209-225.

House, H. D., 1914. The sand dunes of Coos Bay, Oregon, *Plant World* 17, 238-243.

Johnson, D. S., 1900. Notes on the flora of the banks and sounds at Beaufort, North Carolina, *Bot. Gaz.* 30, 405-410.

Johnson, D. S., and Skutch, A. F., 1928. Littoral vegetation on a headland of Mt. Desert Island, Maine. Section III, Adlittoral or non-submersible region, *Ecology* 9, 429-448.

Johnson, D. S., and York, H. H., 1915. *The Relation of Plants to Tide-levels: A Study of Factors Affecting the Distribution of Marine Plants*. Washington, D.C.: Carnegie Inst. Washington Pub. No. 206, 161p.

Johnson, D. W., 1925. *The New England Acadian Shoreline*. New York: John Wiley & Sons, 608p.

Kearney, T. H., 1900. The plant covering of Ocracoke Island: A study in the ecology of the North Carolina strand vegetation, *Contr. U.S. Natl. Herb.* 5, 261-319.

Kearney, T. H., 1901. Report on a botanical survey of the Dismal Swamp region, *Contr. U.S. Natl. Herb.* 5, 321-585.

Kearney, T. H., 1904. Are plants of sea beaches and dunes true halophytes? *Bot. Gaz.* 37, 424-436.

Kellogg, F. B., 1915. Sand-dune reclamation of the

coast of northern California and southern Oregon, *Soc. Am. Foresters Proc.* **1**, 41-64.
Kemp, A. F., 1862. On the shore zones and limits of marine plants on the northeastern coast of the United States, *Canadian Nat.* **7**, 20-34.
Kurz, H., 1942. *Florida Dunes and Scrub, Vegetation and Geology.* Tallahassee, Fla.: Florida Geological Survey Bull. No. 23, 154p.
Lewis, I. F., 1917. *The Vegetation of Shackleford Bank.* Raleigh, N.C.: North Carolina Geological and Economic Survey Econ. Paper No. 46, 32p.
Lloyd, F. E., and Tracy, S. M., 1901. The insular flora of Mississippi and Louisiana, *Torrey Bot. Club Bull.* **28**, 71-101.
Martin, W., 1959. The vegetation of Island Beach State Park, New Jersey, *Ecol. Monogr.* **29**, 1-46.
Miller, W. R., and Egler, F. E., 1950. Vegetation of the Wequetequock-Pawcatuck tidal marshes, Connecticut, *Ecol. Monogr.* **20**, 143-172.
Nesbit, D. W., 1885. *Tide Marshes of the United States,* U.S.D.A. Misc. Spec. Rept. 7, 259p.
Nichols, G. E., 1920a. The vegetation of Connecticut. VI. The plant associations of eroding areas along the seacoast, *Torrey Bot. Club. Bull.* **47**, 89-117.
Nichols, G. E., 1920b. The vegetation of Connecticut. VII. The plant associations of depositing areas along the coast, *Torrey Bot. Club Bull.* **47**, 511-548.
Olson, J. S., 1958. Lake Michigan dune development. 2. Plants as agents and tools in geomorphology, *Jour. Geology* **66**, 345-351.
Olsson-Seffer, P., 1909a. Relation of soil and vegetation on sandy sea shore, *Bot. Gaz.* **47**, 85-126.
Olsson-Seffer, P., 1909b. Hydrodynamic factors influencing plant life on sandy sea shores, *New Phytologist* **8**, 37-51.
Olsson-Seffer, P., 1910. *The Sand Strand Flora of Marine Coasts.* Rock Island, Ill.: Augustana Library Pub. 7, 183p.
Oosting, H. J., 1945. Tolerance to salt spray of plants of coastal dunes, *Ecology* **26**, 85-89.
Oosting, H. J., 1954. Ecological processes and vegetation of the maritime strand in the southeast United States, *Bot. Rev.* **20**, 226-262.
Oosting, H. J., and Billings, W. D., 1942. Factors affecting vegetation zonation on coastal dunes, *Ecology* **23**, 131-142.
Penfound, W. T., 1952. Southern swamps and marshes, *Bot. Rev.* **18**, 413-446.
Penfound, W. T., and Hathaway, E. S., 1938. Plant communities in the marshlands of south-eastern Louisiana, *Ecol. Monogr.* **8**, 1-56.
Penfound, W. T., and O'Neill, M. E., 1934. The vegetation of Cat Island, Mississippi, *Ecology* **15**, 1-16.
Purer, E., 1936. Studies of certain coastal plants of southern California, *Ecol. Monogr.* **6**, 1-87.
Ramalay, F., 1918, Notes on dune vegetation at San Francisco, California, *Plant World* **21**, 191-201.
Shaler, N. S., 1895. Beaches and tidal marshes of the Atlantic coast, *Natl. Geogr. Monogr.* **1**, 137-168.
Shreve, F., 1910. The ecological plant geography of Maryland, coastal zone, eastern shore district, *Md. Weather Service Spec. Pub.* **3**, 101-148.
Smith, J. B., 1907. *The New Jersey Salt Marsh and Its Improvement.* New Brunswick, N.J.: N.J. Agric. Exp. Station Bull. 207, 24p.
Snow, L. M., 1902. Some notes on the ecology of the Delaware coast, *Bot. Gaz.* **34**, 284-306.
Snow, L. M., 1913. Progressive and retrogressive changes in the plant associations of the Delaware coast, *Bot. Gaz.* **55**, 45-55.
Taylor, N., 1938. A preliminary report on the salt marsh vegetation of Long Island, New York, *N.Y. State Mus. Bull.* **316**, 21-84.
Transeau, E. N., 1909. Successional relations of the vegetation about Yarmouth, Nova Scotia, *Plant World* **12**, 271-281.
Transeau, E. N., 1913. The vegetation of Cold Spring Harbor, Long Island. I. The littoral succession, *Plant World* **16**, 189-209.
Weaver, J. E., and Clements, F. E., 1938. *Plant Ecology.* New York: McGraw-Hill, 601p.
Webber, H. J., 1898. Notes on the Strand Flora of Florida. *Science N.S.* **8**, 658p.
Wells, B. W., 1939. A new forest climax: The salt spray climax of Smith Island, North Carolina, *Torrey Bot. Club. Bull.* **66**, 629-634.
Wells, B. W., and Shunk, I. V., 1937. Seaside shrubs: Wind forms vs. spray forms, *Science* **85**, 499.
Wells, B. W., and Shunk, I. V., 1938. Salt spray: An important factor in coastal ecology, *Torrey Bot. Club. Bull.* **65**, 485-492.
Whitfield, C. J., 1932. Osmotic concentrations of chaparral, coastal sagebrush, and dune species of southern California, *Ecology* **13**, 279-285.

Cross-references: *Biotic Zonation; Climate, Coastal; Coastal Ecology, Research Methods; Coastal Flora; Halophytes; Salt Marsh; Splash and Spray Zone; Tides.*

COASTAL RESERVES

Coasts typically show a variety of morphological and biological features within neighboring terrestrial, intertidal, and nearshore zones. They are also subject to relatively rapid physiographic and ecological changes and interactions, both natural and man-induced. Diversity of coastal environments contributes to the aesthetic attraction of coastal scenery, while the dynamic nature of coasts stimulates scientific research and educational use.

Coasts offer a range of opportunities for beach and water-borne recreation, and are thus attractive for holiday making, resorts, and residential development. Coastal land is also in demand for mining, quarrying, engineering works, power station sites, waste disposal plants, and military installations. Ports, based on natural or man-made harbors, have been both a cause and a consequence of coastal urbanization and industrial growth. Thus there are many competing claims for the utilization of coastal land and water, and in many countries attempts are being made to resolve conflicts by means of coastal zone management, requiring the deliberate planning of coastal resource use based on environmental, economic, and sociological research.

Coastal zone management is now operating in several European countries, notably in Britain and Scandinavia; it has been, or is being, introduced to parts of the United States, Canada, Soviet Union, Australia, New Zealand, Japan, and South Africa; and it is in the discussion stage elsewhere. Essential to coastal zone management is the division of a coast into sectors for specific usage, some sectors being declared *coastal reserves* to maintain features of scenic, scientific, or historical value, or to preserve a pleasant setting for coastal recreation. Within such coastal reserves it is necessary to prevent uses and developments that would damage or destroy the features of interest. Some manmade features may enhance the interest of a coast—lighthouses and fishing villages, for example—but in many cases it has been necessary to modify or remove man-made features that spoil or detract from the values for which the coastal reserve was declared.

In practice, reserves are generally located within sectors of coast that have remained in a natural or near-natural condition or are subject only to rural land use. They come into existence as the result of surveys and campaigns by groups interested in the conservation of scenery and wildlife, in the maintenance of sites of special scientific or historical interest, and in various forms of coastal recreation including water-borne activities. Coastal reserves are delineated, and put under the control of organizations, usually governmental, that become responsible for their maintenance and management.

Some types of coastal reserve are designed to be exclusive. It is necessary to restrict public access and exclude off-road vehicles, Hovercraft, and similar environmentally disruptive contraptions from areas where the aim is to conserve fragile ecosystems, such as dune and marshland habitats, to manage wildlife sanctuaries, or to carry out scientific research and experimental work in an undisturbed environment. On the other hand, there are coastal reserves explicitly intended for public enjoyment of coastal scenery and for recreational activities. Within these there are often difficulties over the provision of such features as roads, parking areas, motels, boat-launching ramps, marinas, and other structures needed to facilitate visitor usage. Inevitably such structures intrude upon coastal scenery; and while much can be achieved with the aid of careful siting and landscaping, there is a finite limit to the visitor population that can be accommodated in a coastal reserve without destroying the aesthetic values and recreational experience that the visitors came to seek.

Where recreation, conservation, and scientific requirements are incompatible, it may be possible (in a large coastal reserve) to separate these activities by internal use zoning. Public access, especially by motor vehicle, remains a key problem. In many countries the building of roads that follow the coastline is now considered bad management. The maintenance of landscape diversity and variety of visitor experience is often better served when roads branch from an inland highway to reach the coast at various locations with contrasting features, including resorts and recreation centers as well as different kinds of coastal reserve (Fig. 1).

Another problem is posed by the dynamic nature of many coastal features: mobile dunes, accreting marshlands, eroding cliffs, and silting estuaries. Where the changes are due to natural physiographic and ecological processes there may be no case for interfering, but often the changes are at least partly man-induced. In Australia, for example, the extensive mobile dune landscape at Discovery Bay (see *Australia, Coastal Morphology*) is now a coastal reserve under the management of the Victorian National Park Service. It is thought that dune mobilization here was largely due to destruction of a natural bushland cover by grazing and burning in the nineteenth century. However, the active dunes are of scenic as well as scientific interest, and a decision has to be made on whether to leave them as they are or to attempt the restoration of the nineteenth-century landscape. On the other hand, the widespread devastation of dune-terrain-carrying relict rain forest and woodland communities on the large sand islands (Moreton, Stradbrokes, and Fraser) off Queensland by open-cast mineral sands dredging is regarded as incompatible with conservation in a coastal reserve.

Variety of form and function in coastal reserves is illustrated by the range of terms applied to such features in various countries. The most widely used terms are as follows.

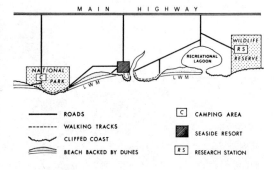

FIGURE 1. Scheme of coastal access avoiding the use of roads along the coastline.

COASTAL RESERVES

Beach Park (Beach Reserve)

This term is usually applied to a public beach area of outstanding recreational value, often with an immediate hinterland also in public ownership, managed by local authorities. As a rule, the only structures permitted in these areas are those directly serving the needs of beach users (e.g., life-saving facilities, showers, and toilets). In Australia some states (notably Victoria) have a coastal fringe of crown land at least 30 m wide, declared a reserve for public purposes a century ago, and parts of this are now managed as beach parks (Fig. 2). Such reserves are of especial value in countries (New England, United States, for instance) where public access to the coast is confined to limited sectors that have not passed into private ownership.

FIGURE 2. Beach Park reserve on the coast of Port Phillip Bay, Australia, ten miles from the center of the city of Melbourne (photo: E. C. F. Bird).

Heritage Coast

This term is used in Britain to describe a sector of high quality scenery subject to the rigorous planning control needed to maintain its appearance and also permit a limited range of recreational activities without excessive concentrations of people and vehicles. In 1970 the British Countryside Commission recommended 34 sectors of Heritage Coast extending along some 730 miles (26.6%) of the coastline of England and Wales. These are mainly cliffed sectors with a rural hinterland (Fig. 3), but in north Norfolk the Heritage Coast consists of shingle spits, dunes, and tidal marshlands, while Chesil Beach Heritage Coast is a shingle barrier 15 miles long. The concept of a Heritage Coast has traveled to Australia, where in 1974 the National Estate Committee of Inquiry recommended the setting up of Coastal Heritage Parks to preserve wilderness sectors of the Australian coast.

FIGURE 3. Heritage Coast at Seven Sisters, on the south coast of England (photo: E. C. F. Bird).

Marine Park

This refers to an area of coastal or offshore water, together with the submerged sea floor, coral reefs, and intertidal features, such as mangrove swamps. Marine parks have been declared off Florida, Bahamas, Virgin Islands, Kenya, Sri Lanka, and Japan to conserve submarine environments and marine ecosystems. They involve restrictions on such activities as diving, fishing, collecting, and power boating. Glass-bottomed boats, submarine vessels, and underwater viewing facilities are among the features used to enable visitors to appreciate the marine park environment. In Australia there are plans to declare a large marine park in the Great Barrier Reefs province off Queensland.

National Park

This term usually refers to an area of scenic (and often wildlife) value that has been kept clear of modification and development, except that permitted to enable visitors to see and explore the area. Rangers are employed to manage national parks and to interpret them for visitors with the aid of such facilities as visitor centers and guided trails. In many countries national parks are surviving regions of natural or near-natural landscape, but in Britain they include farmland, villages, and even small towns. In the United States, Everglades National Park is one of the few that extends to the coastline; in Britain the Pembrokeshire Coast National Park is a narrow fringe of mainly rocky coastline in southwest Wales, while three other national parks (Lake District, Exmoor, North York Moors) reach the coast of England. In Australia, Wilson's Promontory (Fig. 4) is one of several coastal National Parks.

FIGURE 4. National Park at Wilson's Promontory, Victorian coast, Australia (photo: E. C. F. Bird).

National Seashore

This term was devised in the United States for large coastal reserves under the control of the National Parks Service, in which beach and water-borne recreation is of major significance. They include Cape Cod and Cape Hatteras on the east coast and Point Reyes in California.

Nature Reserves

This term is used for areas of physiographic and biological interest, managed primarily for the conservation of plant and animal species and communities of particular scientific importance, for research work, and, to some extent, for educational use (Countryside Commission, U.K., 1969). In Britain, many nature reserves have been established as "living museums and outdoor laboratories," managed by warden staff, and interpreted with the aid of nature trails and field study centers (Fig. 5). Designated by the Nature Conservancy Council and other organizations, including local authorities and conservation societies such as the National Trust, they include coastal reserves on island, estuary, marsh, dune, spit, cliff, and landslip habitats. Some have access restricted to permit holders.

The Nature Conservancy Council has also listed over 2000 *Sites of Special Scientific Interest,* many of which are on the coast. These are referred to local authorities for inclusion as coastal reserves in planning schemes. Some are in private ownership, in which case the Nature Conservancy Council seeks agreements with landowners to manage the site for conservation purposes. On the British coast there are also some *Wildfowl Refuges,* wildlife resources for controlled shooting, and *Bird Sanctuaries,* established under the Protection of Birds legislation.

Similar reserves have been declared on coasts elsewhere, notably in Europe; the Coto Doñana is a large dune and wetland area at the mouth of the Guadalquivir River, Spain; La Camargue is a wetland island reserve in the Rhone delta, France; the Dutch island of Terschelling is a dune and marshland reserve. Each of these is of major importance for the conservation of sea birds. In Australia, state authorities have declared *Wildlife Reserves* in coastal regions, and in 1976 a survey of coastal sites of special scientific interest by the Town and Country Planning Board in the state of Victoria located and described over 200 geological, geomorphological, biological, and archeological sites, some of which are already within National Parks and Wildlife Reserves, while others await designation as coastal reserves in the context of a state coastal management scheme.

FIGURE 5. Nature Reserve at Slapton Ley, Devonshire coast, England. A shingle barrier separates a freshwater lagoon from the sea, producing an area of varied physiographic and ecological features, much studied by groups using the nearby Slapton Field Studies Center (photo: E. C. F. Bird).

Recognition, delimitation, and management of coastal reserves is still at an early stage on much of the world's coastline. Reserves are needed to protect both unique areas (such as the habitat of a rare sea bird) and typical samples of more widespread features (such as mangrove associations in tropical estuaries) that are threatened by modification in the course of coastal development. Planning is more acceptable in some countries than others, but disfigurement of high quality coastal scenery, damage to biological communities in the coastal zone, and destruction of coastal features of special scientific and historical interest all represent serious losses to man's coastal heritage. Such losses can be prevented only by instituting some form of control of coastal use, and by securing coastal reserves to safeguard an adequate and representative coverage of scenic and scientific features around the world's coastline.

ERIC C. F. BIRD

References

Countryside Commission (U.K.), 1969. *Nature Conservation at the Coast.* London: Her Majesty's Stationery Office, 97p.
Countryside Commission (U.K.), 1970. *The Coastal Heritage.* London: Her Majesty's Stationery Office, 99p.

Cross-references: *Coastal Zone Management; Conservation; Human Impact; Surfing; Thalassotherapy; Tourism.* Vol. XIII: *Coastal Zone Management.*

COASTAL SEDIMENTARY FACIES

Sedimentary facies is a term used by Moore (1949) to describe an areally restricted, lithologically or paleontologically distinctive component of any body of genetically related sedimentary deposits. The dimensional scope of the term *facies* ranges from the broad (*coastal facies of a sedimentary basin*) to the narrow (*upper foreshore facies of a beach*). In this discussion, we consider three levels of complexity, termed here *facies elements, facies systems,* and *facies complexes.*

Facies elements are the smallest facies units that can be assigned to an areal setting or set of processes. Smaller stratigraphic entities result from repetition or change in magnitude of processes in the same setting. We consider facies elements here only in their contribution to the broader categories.

Facies systems comprise a group of contiguous, related sedimentary facies elements. Each system is generally formed by several processes operating in a relatively specific geomorphic setting. A sedimentary facies system can be considered as a veneer of sediment that is controlled largely by processes operative in the immediate vicinity. Our analysis concentrates on those systems that are common to most parts of the world.

Facies complexes are a higher-order entity and consist of a three-dimensional array of related facies systems. For example, in simple progradation (seaward movement of the shoreline due to sediment accretion), facies systems of deeper water systematically underlie those of shallower water. Our analysis includes only those complexes that occur widely in space and time.

Criteria used to delineate specific facies elements, systems, and complexes are commonly subjective and arbitrary. Moreover, the boundaries between contiguous facies are difficult to define precisely. A place near the boundary of two facies elements, systems, or complexes may be part of one under certain conditions and part of the other under different conditions.

The term facies can also be in the sense of *facies models*—i.e., groups of deposits that bear a common set of distinctive, genetically related characteristics (Blatt, Middleton, and Murray, 1972). The facies systems and facies elements described here may also be considered as facies models for the sets of deposits discussed. Coastal sedimentary facies have been discussed in great detail by Hayes and Kana (1976); Reineck and Singh (1973); Selley (1970); Pettijohn, Potter and Siever (1972); Blatt, Middleton, and Murray (1972); Reading (1978); Walker (1979); and Heward (1981).

Controls of Coastal Sedimentary Facies

An extremely complicated interaction of various influences determines the character of coastal sediments (Fig. 1). Fundamentally, however, the controlling factors are of two basic types: the material available and the processes. The texture and composition of the available materials depend on the geologic nature of the source area and on a number of processes acting in the source area and in subsequent sediment transport. Processes near and at the depositional site act as a final control on the nature of the material and as the main control on the structure of the sediment. Processes may be classed as physical, biological, or chemical. Physical processes are primarily those involving the movement of sediment by fluids, especially water currents. Biological processes may be dynamic (the movement of sediment or the baffling of currents), or they may affect the dynamic properties of the sediment by binding or cementing it. Chemical and some biological processes involve the precipitation or solution of materials within either the water column or the sediment.

Coastal sedimentary facies depend directly on

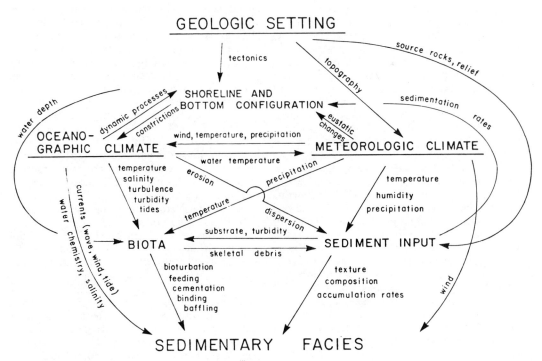

FIGURE 1. Schematic diagram showing interrelation of environmental factors and their control of coastal sedimentary facies.

oceanic climate, meteorologic climate, biota, and sediment input. These, in turn, are influenced by one another or by external controls related to the geologic setting (Fig. 1). Particularly important is the shoreline configuration produced by the interaction of geologic, oceanic and meteorologic conditions. Although the configuration may not directly determine the character of the sediment, it profoundly affects the processes that do.

Facies Systems

The following capsular summaries indicate the geomorphic setting of each system considered, the dominant processes operating therein, and the assemblage of the sedimentary textures, compositions, and surface and internal structures that characterizes the system.

Inner Shelf System. Setting: open sea, in water depths at which the bottom is stressed by shoaling surface waves before they begin rapidly to deform near the surf zone. Processes: oscillatory currents induced by surface waves; currents induced by wind stress or thermohaline structure of the water column; tidal currents where the shelf is embayed; bioturbation; particulate settling from suspension; biogenic carbonate accretion. Textures: predominantly fine sand, silt, and clay under low wave energy or near river influences; coarser relic or biogenic sediment; relatively coarse storm layers. Compositions: siliciclastic; carbonate where terrigenous clastic supply is low; locally glauconitic. Surface structures: ripple marks, tracks, trails, and burrow entrances. Internal structures: bioturbate structures, planar lamination, low-angle hummocky crossbedding, ripple lamination, high-angle crossbedding only in medium- to coarse-grained sand. References: Goldring and Bridges (1973), Howard and Reineck (1972).

Nearshore Systems. Setting: open very close to shoreline; area of rapid wave deformation. Processes: oscillatory currents of shoaling waves; longshore and seaward (rip) unidirectional currents (commonly channelized); minor biological activity. Textures: mostly sand and/or gravel; silt or clay rare. Compositions: siliciclastic or carbonate, depending on source. Surface structures: ripples, megaripples, and plane beds. Internal structures: crossbedding of diverse orientation; planar lamination; ripple lamination; bioturbate structures generally minor. References: Clifton, Hunter, and Phillips (1971), Davidson-Arnott and Greenwood (1974).

Beach Systems. Setting: shorelines intermittently exposed to the air and at least intermittently subject to intense wave action. Processes: dominantly wave swash and backwash; less influenced by eolian processes and locally by small streams. Textures: mostly sand and/or gravel; silt and clay rare. Compositions: siliciclastic and/or carbonate detritus, heavy minerals commonly concentrated on upper beach. Surface structures: plane beds dominant; foreshore antidunes common on lower foreshore; swash and rill marks; rhomboid ripples; wind ripples and small dunes may occur on backshore; current ripples, megaripples, and antidunes in stream channels or runnels. Internal structures: predominantly gently inclined planar lamination; laminae commonly inversely graded on seaward slopes, may be normally graded on landward slopes; ripple bedding and crossbedding (landward facing where overwash spills into ponded water) may occur; pebble imbrication well defined, direction variable but generally inclined seaward; small-scale or scattered bioturbate structures. References: Thompson (1937), McKee (1957), Logvinenko (1970).

Storm-overwash Channel and Fan Systems. Setting: low barrier spits or islands. Processes: erosion of channels by water passing over barrier during storms; deposition of splay fans on landward side of barrier as currents spread out and enter lagoon. Textures: sand or gravel. Compositions: siliciclastic and/or carbonate, depending on composition of barrier material. Surface structures: shallow fan channels; large rhomboid bedforms, megaripples, ripples, and plane beds; slip faces where splay fans terminate in lagoon. Internal structures: landward-facing large- and medium-scale crossbedding, ripple lamination, and planar lamination. References: Andrews (1970), Pierce (1970).

Coastal Dune System. Setting: a belt immediately landward of beach. Processes: transport of sand from beach by onshore winds; growth of vegetation, with consequent baffling of wind and binding sediment; minor bioturbation. Textures: well-sorted sand, mostly fine- to medium-grained; thin mud beds in interdune areas. Surface structures: nonmigratory, vegetated foredune ridges and coppice dunes; depressions and partially vegetated parabolic dunes; wind-shadow drifts behind coppice dunes; small, migratory, unvegetated dunes. Internal structures: low- and high-angle, large- and medium-scale cross stratification, the dip directions commonly symmetrically bimodal around a vector pointing in the landward semicircle; bedding commonly irregular and faint because of deposition on vegetated surfaces and disturbance by roots; minor small- and large-scale bioturbate structures. References: Bigarella, Bicker, and Duarte (1969), Goldsmith (1973).

Tidal Channel Systems. Setting: associated with semienclosed bodies of water, at inlet or extending through the water body; topographically higher portions may be intertidal. Processes: tidal currents; bioturbation; minor wave effects. Textures: Highly variable—gravel, sand, silt, or clay; coarsest material generally on channel floor; sediment generally finer-grained away from ocean, but may coarsen near river mouths. Compositions: Siliciclastic and/or carbonate; lag deposits common (mud clasts, shells, and/or wood fragments); on channel floor local concentrations of carbonaceous detritus. Surface structures: dependent on texture—generally ripples or megaripples in sand; mud bottoms planar with numerous tracks, trails, and burrows. Internal structures: dependent on texture and location in channel; channel floor deposits generally intensely bioturbated; channel margins ripple bedded or crossbedded (sand) or evenly laminated (mud); bioturbate structures less common on channel margins. References: Green (1975), Klein (1970), Smith (1969), Terwindt (1971), Dalrymple, Knight, and Middleton (1975).

Tidal Flat Systems. Setting: relatively protected shores, particularly in semienclosed bodies of water; areas of low topographic relief within the intertidal (inundated by astronomic tides) and supratidal (inundated by a combination of astronomic and meteorologic tides) environments. Processes: faunal mixing; relatively weak tidal currents; small waves; binding of sediment and baffling of currents by vegetation. Textures: strongly dependent on wave energy; characteristically sand, silt, and/or clay; gravel uncommon; grain size may progressively decrease shoreward. Compositions: siliciclastic and/or carbonate detritus, carbonaceous debris may abound in temperate climates; evaporite minerals and dolomite in arid climates. Surface structures: symmetric and asymmetric ripples on sand; planar surfaces on mud; desiccation cracks; abundant burrow openings, tracks, and trails; algae mats; small runoff channels common; growing vegetation (salt-marsh grasses, mangroves) common in upper subtidal to supratidal areas. Internal structures: ripple lamination and horizontal lamination; bioturbate structures abundant in intertidal, and less pronounced in supratidal areas, algae stromatolites; nodular evaporites. References: Evans et al. (1969), Reineck and Singh (1973), Shinn, Lloyd, and Ginsburg (1969).

Tidal Delta Systems. Setting: shoal areas on landward side (flood delta) or seaward side (ebb delta) of inlet into semienclosed body of water; shoals are traversed by tidal channels. Processes: Complex interaction of ebb and flood tidal currents and wave-generated currents; intense sediment transport and deposition during storms; minor faunal activity; possible sediment binding by algae. Textures: predominantly clean sand; rarely mud. Compositions: siliciclastic and/or carbonate detritus. Surface structures: giant relief forms, megaripples, ripples, and plane beds. Internal structures: crossbedding and ripple bedding of variable orientation; planar lamination; thin mud drapes and beds; minor bioturbation. References: Kumar and Sanders (1974), Ludwick (1974).

Reef Systems. Setting: in the photic zone at the margins of continents or islands where supply rate of terrigenous sediment is low, particularly in tropical climates. Processes: intense sediment production and cementation by faunal activity; wave-generated and tidal currents. Textures: where clastic, commonly coarse. Compositions: predominantly biogenic carbonate. Surface structures: mostly living or dead calcareous colonial organisms; channels floored by carbonate sand. Internal structures: cemented mass of skeletal material; local beds of biogenic sand or gravel. References: Jones and Endean (1973), Laporte (1974), Milliman (1967, 1974).

Bay Floor Systems. Setting: broad subtidal basins within semienclosed bodies of water; may be of diverse origin (i.e., interdistributory bays of deltas, lagoons, or drowned river valleys without well-developed tidal channels). Processes: abundant infaunal activity; generally feeble dynamic processes (small waves, relatively weak tidal currents); sediment binding by vegetation; chemical precipitation in arid climates. Textures: predominantly fine-grained, except for biogenic clasts. Compositions: siliciclastic and/or carbonate detritus, evaporites in arid climate, possible abundant carbonaceous material in temperate climates. Surface structures: burrows, tracks, and trails; ripples; algal mats; living grasses; oyster reefs. Internal structures: intensely bioturbated; horizontal lamination and ripple bedding; root structures. References: Castañares and Phleger (1969), Elliott (1974).

River Mouth Systems. Setting: interface of fresh and saline waters, either an open coast or within semienclosed body of water. Processes: river discharge combined with effects of waves or tides. Textures: variable, depends on stream sediment load; fine sediment may accumulate in submarine levees where discharged into an open, low energy environment; commonly sandy on river mouth bar. Compositions: predominantly siliciclastic. Surface structures: ripples, megaripples, and plane beds; some burrows, tracks, and trails. Internal structures: crossbedding, ripple bedding, and planar lamination; minor bioturbate structures. References: Clifton, Phillips, and Hunter (1973), Coleman and Gagliano (1965).

Facies Complexes

Our analysis includes both regressive or progradational complexes and transgressive complexes. Regressive or progradational complexes imply a seaward movement of the strandline, caused primarily by the accumulation of sediment. Relative sea level at such a time may be unchanging, falling, or rising too slowly to overcome the rate of sediment accumulation. Transgressive complexes imply a landward movement of the shoreline, caused by a rise in relative sea level that exceeds the rate of sediment accumulation. Transgressive complexes are in general thinner and less abundant in the geologic record than regressive complexes.

The following brief summaries give the geomorphic and tectonic setting for each complex considered, the three-dimensional form of preserved deposits, the character of the geometric array of facies systems that compose the complex, and the variability of the complex under different environmental conditions.

Deltaic Complexes (Coleman and Gagliano, 1965; Fisher and McGowen, 1969; Fisher et al., 1970; Morgan, 1970; Shirley and Ragsdale, 1966). Commonly major centers of deposition along coasts, deltaic complexes result from relatively local accumulation where the volume of sediment discharged from a river exceeds the dispersal capability of oceanic processes. Accordingly, deltaic deposits generally consist of a lobe of sediment associated with a river mouth. The character of a deltaic complex depends on the rate of sediment supply, the texture of that sediment, the nature of the oceanic processes (particularly waves and tides), the rate of relative sea level change due to compaction, tectonism, or eustatic events, the bottom configuration of the body of water into which the delta builds, and the climate. Under most combinations of subsidence and sedimentation, deltas are progradational and build outward into deeper water. In the open sea, these deeper water deposits will likely include an inner shelf facies system (commonly referred to in part as *delta front*) as well as even deeper water deposits (*prodelta*). Deltas building into a semienclosed body of water, in contrast, are likely to extend laterally over a bay floor system.

The upper part of a deltaic complex incorporates fluvial facies not described here; its nature depends on a variety of factors. A fine-grained delta that progrades into relatively quiet, deep water such as the Mississippi River delta, tends to have a digitate *bird's foot* form, each distributary forming a distinct finger. River-mouth facies systems are well developed and grade laterally into nearshore and beach systems. Interdistributary bays initially contain bay-floor facies systems, which grade up into crevasse-splay and tidal-flat (salt-marsh) systems as the bay fills. Such deltas commonly subside because of compaction of underlying fine sediment, and the depositional lobes shift position as the delta progrades. The shallowing-upward sequence that generally characterizes deltas may be greatly complicated by upward transport of deep-water facies in mud diapirs (*mud lumps*) on the delta front.

Other types of deltas occur where waves or tides vigorously redistribute the sediment. Wave-dominated deltas tend to have straighter shorelines than other types. Beach and nearshore systems may be more in evidence than river-mouth systems, and interdistributary bay-floor deposits are likely to be subordinate to both. Tide-dominated deltas tend to have irregular shorelines that are transected by numerous tidal channels. Tidal-channel and tidal-flat systems, accordingly, are major components of the delta complex.

A special type of delta develops where an alluvial fan progrades into the marine environment. *Fan deltas* tend to be relatively small features with steep delta-front slopes. Grain sizes are generally coarse relative to other deltas, and the delta is likely to be capped by intermittent stream deposits.

Chenier-plain Complexes (Curray, 1969; Hoyt, 1969). Chenier plains (q.v.) are the product of progradation along coasts where the mud supply is so great and the wave energy is so low that tidal flats take the place of beaches along the open coast most of the time. The progradation of tidal-flat deposits is repeatedly interrupted, however, by the building of cheniers, which are beach ridges composed of sand or shells eroded from the tidal-flat deposits. A chenier results from an increase in wave energy, such as a storm, or a decrease in mud supply, such as might occur when a delta lobe is abandoned. The decrease in mud supply that allows a chenier to form may be a local phenomenon associated with the alongshore migration of broad, low mud waves oriented transversely or obliquely to shore. Chenier plains occur òn or marginally to deltas or marginally to the mouths of large rivers that do not form deltas because of rapid alongshore dispersal of sediment.

A chenier plain complex is rather simple in vertical section, the muddy inner-shelf system grading up into a muddy tidal-flat system, the uppermost facies of which is a salt marsh or mangrove swamp if the climate is humid. Sandy or shelly beach-ridge deposits locally overlie the marsh or swamp deposits and extend as seaward-dipping tongues into the intertidal and subtidal deposits.

Strand Plain Complexes (Curray, Emmel, and Crampton, 1969; Davies, Ethridge, and Berg, 1971; Evans et al., 1973; Hoyt and Henry, 1967; Otvos, 1970; Straaten, 1965). Strand plain complexes develop where the general shoreline shifts seaward in response to sediment accumulation. The sediment may come from numerous, relatively small streams, from a single large river, or from older deposits undergoing wave erosion along the coast or on the shelf. As a strand plain complex progrades, its individual facies systems shift laterally over previously deposited more seaward systems, producing a shallowing-upward vertical sequence. The completeness of the vertical sequence depends on topographic complexity of the strand plain and the balance between rates of sedimentation and relative sea level change. If the strand plain and related shelf surface slope continuously seaward, all the facies systems are likely to be preserved. The balance between rates of sediment accumulation and relative subsidence will dictate the thickness of the facies complex.

On topographically complex strand plains—for example, those with barrier islands and lagoons—the topographically low components (i.e., lagoons) may be filled and left behind as the rest of the complex builds in a seaward direction. A complete vertical sequence of facies systems will be preserved only if the land subsides substantially relative to the sea (Fig. 2).

The facies systems present in any prograding strand plain complex differ depending on the environment. The simplest complexes on most coasts consist of inner-shelf, nearshore, beach, coastal dune and/or alluvial plain systems. Strand plain complexes with barrier islands and lagoons also include bay floor and possible tidal delta, tidal channel, and tidal flat facies systems as well as storm overwash channel and fan systems.

Restricted Embayment Complexes (de Raaf and Boersma, 1971; Lauff, 1967; Ginsburg, 1975; Nelson, 1972). These complexes occur in embayments along coasts where communication with the open sea is restricted by barrier islands, spits, or other topographic features. The physical character of such embayments is highly variable. They may be diluted by freshwater discharge (estuaries) or be hypersaline. Tidal effects are commonly pronounced, particularly near inlets, and wind stresses are important in some embayments.

The facies systems developed within a restricted embayment depend primarily on the magnitude of tidal effects, the discharge of streams entering the embayment, and the nature of the connection with the open sea. Filling of the embayment produces a shallowing-upward vertical sequence. Embayments may have an initial deposit of the bay-floor facies system. If the embayment is dominated by tides, the bay-floor system will be progressively succeeded by tidal channel and tidal-flat facies systems. These facies, in turn, may merge laterally near the inlet with a tidal delta system and near any rivers with a river-mouth system and related fluvial facies. Beach and nearshore facies systems are generally limited, as wave effects are relatively unimportant in most semi-enclosed embayments. Storm-overwash fan systems may be an important component on the seaward side of embayments isolated from the sea by a low, sandy barrier.

Transgressive Complexes (Swift, 1968; Kraft, 1978). Transgressive facies complexes tend in general to be thin; moreover, coastal facies are especially subject to erosion during transgression, and thus tend to be incompletely represented or missing (Fig. 2). The most incomplete transgressive complex possible would consist of shelf or deeper marine facies resting unconformably on older rocks. With more rapid deposition and/or weaker wave and current action, other facies may be preserved. An inner-shelf sheet sand may be preserved as a basal part of the shelf deposits, and fluvial deposits may be preserved as a layer separated by erosional surfaces from the underlying older rocks and overlying shelf deposits.

With still more complete preservation, coastal facies that were deposited farther inland or in topographically low areas may be preserved as lenses or tongues beneath the shelf deposits. Among coastal facies, the most important are those typical of estuarine and lagoonal complexes. Coastal dune, beach, and nearshore facies systems are least likely to be preserved in transgressive complexes. The precise nature of the complex will of course vary, depending on whether the coast is a mainland sandy beach, a barrier and lagoon, a reef and lagoon, a mainland tidal flat, a coast indented by estuaries of the drowned river valley type, or a deltaic coast.

Reef-lagoon Complexes (Logan et al., 1974; Milliman, 1974; Friedman, 1969; Purser, 1973). Reef-lagoon complexes are most common in warm climates and in areas where the contribution of terrigenous sediment is limited. A reef, constructed of a skeletal framework of coral, algae, or other organisms, forms a barrier to the open sea; its seaward margin is commonly

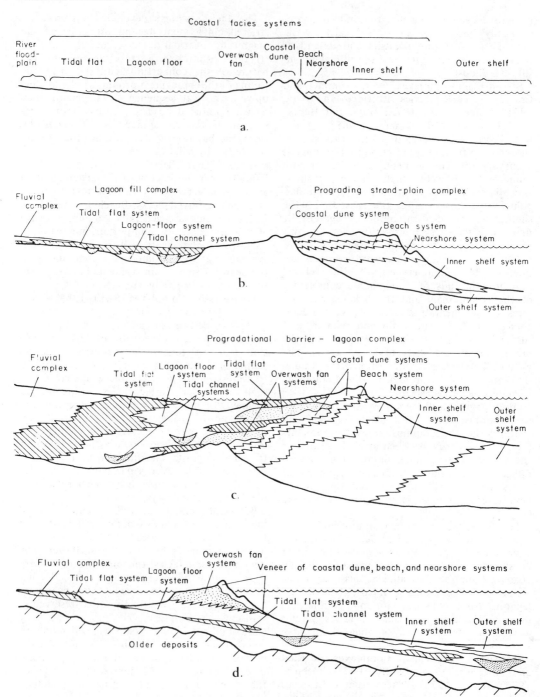

FIGURE 2. Schematic cross sections showing some of the more common types of facies complexes produced by deposition of a barrier-lagoon coast. *a*: distribution of surficial facies systems at a moment in time; *b*: facies complexes produced by progradation and lagoon fill during a stillstand of sea level on a tectonically stable coast; *c*: facies complexes produced by progradation during a rise of sea level relative to land (sea level rising eustatically or land subsiding); *d*: facies complexes produced by transgression during a rise of sea level relative to land.

marked by a submarine cliff with a talus slope at its base. The reef itself is composed of the reef-facies system.

For a variety of reasons—restricted turbulence, changes in temperature or salinity, or an excess of terrigenous detritus—the reef-forming organisms do not commonly thrive on the landward side of the reef where a lagoon develops. Such lagoons resemble the semi-enclosed embayments discussed above, but contain more carbonate sediment than do most embayments. The carbonate sediment may be in the form of mud or coarser detritus (skeletal clasts, oolites, fecal pellets). The character of the clasts commonly differs systematically across the lagoon; sediment on the seaward side may be dominated by skeletal debris derived from the reef during storms, whereas sediment on the landward side may show a greater abundance of terrigenous detritus. Small patch reefs may exist within the lagoon. In areas of marked tidal range, a carbonate tidal flat is likely to occur along the landward margin of the lagoon, and tidal channels may transect the reef.

Reef-lagoon complexes are likely to achieve substantial thickness only under conditions of relative subsidence or if the reef builds laterally into relatively deep water. The complex may be either transgressive or regressive, depending on the balance of relative subsidence and lateral growth of the reef.

H. EDWARD CLIFTON
RALPH E. HUNTER

References

Andrews, P. B., 1970. *Facies and Genesis of a Hurricane-Washover Fan, St. Joseph Island, Central Texas Coast*. Austin: University of Texas Bureau of Economic Geology Rept. Inv. RI 67, 147p.

Bigarella, J. J.; Bicker, R. D.; and Duarte, G. M., 1969. Coastal dune structures from Paraña (Brazil), *Marine Geology* 7, 5-55.

Blatt, H.; Middleton, G.; and Murray, R., 1972. *Origin of Sedimentary Rocks*. Englewood Cliffs, N.J.: Prentice-Hall, 634p.

Castañares, A. A., and Phleger, F. B., eds., 1969. *Coastal Lagoons: A Symposium*. Mexico City: Universidad Nacional Autónoma, 686p.

Clifton, H. E.; Hunter, R. E.; and Phillips, R. L., 1971. Depositional Structures and processes in the non-barred high-energy nearshore, *Jour. Sed. Petrology* 41, 651-670.

Clifton, H. E.; Phillips, R. L.; and Hunter, R. E., 1973. Depositional structures and processes in the mouths of small coastal streams, southwestern Oregon, in D. R. Coates, ed., *Coastal Geomorphology*. Binghamton: State University of New York, 115-140.

Coleman, J. M., and Gagliano, S. M., 1965. Sedimentary structures: Mississippi River deltaic plain, in G. V. Middleton, ed., *Primary Sedimentary Structures and Their Hydrodynamic Interpretations*. Tulsa, Okla.: Soc. Econ. Paleont. Miner. Spec. Pub. 12, 133-148.

Curray, J. R., 1969. Shore zone sand bodies: Barriers, cheniers, and beach ridges, in D. J. Stanley, ed., *The New Concepts of Continental Margin Sedimentation:* Washington: American Geological Institute, JC-II-1-18.

Curray, J. R.; Emmel, F. J.; and Crampton, P. J. S., 1969. Holocene history of a strand plain, lagoonal coast, Nayarit, Mexico, in A. A. Castañares, and F. B. Phleger, eds., *Coastal Lagoons: A Symposium*. Mexico City: Universidad Nacional Autónoma, 63-100.

Dalrymple, R. W.; Knight, R. J.; and Middleton, G. V., 1975. Intertidal sand bars in Cobequid Bay (Bay of Fundy), in L. E. Cronin, ed., *Estuarine Research*, vol. II. New York: Academic Press, 293-307.

Davidson-Arnott, R. G. D., and Greenwood, B., 1974. Bedforms and structures associated with bar topography in the shallow-water wave environment, Kouchibouguac Bay, New Brunswick, Canada, *Jour. Sed. Petrology* 44, 698-704.

Davies, D. K.; Ethridge, F. G.; and Berg, R. R., 1971. Recognition of barrier environments, *Am. Assoc. Petroleum Geologists Bull.* 55, 550-565.

de Raaf, J. F. M., and Boersma, J. R., 1971. Tidal deposits and their sedimentary structures, *Geologie en Mijnbouw* 50, 479-504.

Elliott, T., 1974. Interdistributary bay sequences and their genesis, *Sedimentology* 21, 611-622.

Evans, G.; Murray, J. W.; Biggs, H. E. J.; Bate, R.; and Bush, P. R., 1973. The oceanography, ecology, sedimentology, and geomorphology of parts of the Trucial Coast barrier island complex, Persian Gulf, in B. H. Purser, ed., *The Persian Gulf.* New York: Springer-Verlag, 233-277.

Evans, G.; Schmidt, V.; Bush, P.; and Nelson, M., 1969. Stratigraphy and geologic history of the sabhka, Abu Dhabi, Persian Gulf, *Sedimentology* 12, 145-159.

Fisher, W. L., and McGowen, J. H., 1969. Depositional systems in Wilcox Group (Eocene) of Texas and their relation to occurrence of oil and gas, *Am. Assoc. Petroleum Geologists Bull.* 53, 30-54.

Fisher, W. L.; Proctor, C. V., Jr.; Galloway, W. E.; and Nagle, J. S., 1970. Depositional systems in the Jackson Group of Texas—Their relationship to oil, gas, and Uranium, *Gulf Coast Assoc. Geol. Soc. Trans.* 20, 234-261.

Friedman, G. M., ed., 1969. *Depositional Environments in Carbonate Rocks*. Tulsa, Okla.: Soc. Econ. Paleont. Mineral. Spec. Pub. 14, 209p.

Ginsburg, R. N., ed., 1975. *Tidal Deposits—A Casebook of Recent Examples and Fossil Counterparts*. New York: Springer-Verlag, 428p.

Goldring, R., and Bridges, P., 1973. Sublittoral sheet sandstones, *Jour. Sed. Petrology* 43, 736-747.

Goldsmith, V., 1973. Internal geometry and origin of vegetated coastal sand dunes, *Jour. Sed. Petrology* 43, 1128-1142.

Green, C. D., 1975. A study of hydraulics and bed-forms at the mouth of the Tay Estuary, Scotland, in L. E. Cronin, ed., *Estuarine Research*, vol. II. New York: Academic Press, 323-344.

Hayes, M. O., and Kana, T. W., eds., 1976. *Terrigenous*

Clastic Depositional Environments. Columbia, S.C.: Coastal Research Division Department of Geology, University of South Carolina Tech. Rept. 11-CRD, 171p.

Heward, A. P., 1981. A review of wave-dominated clastic shoreline deposits, *Earth-Sci. Rev.* 17, 223-276.

Howard, J. D., and Reineck, H.-E., 1972. Georgia coastal region, Sapelo Island, U.S.A.: Sedimentology and biology. IV. Physical and biogenic sedimentary structures of the nearshore shelf, *Senckenbergiana Maritima* 4, 81-123.

Hoyt, J. H., 1969. Chenier versus barrier, genetic and stratigraphic distinction, *Am. Assoc. Petroleum Geologists Bull.* 53, 299-306.

Hoyt, J. H., and Henry, V. J., Jr., 1967. Influence of island migration on barrier-island sedimentation, *Geol. Soc. America Bull.* 78, 77-86.

Jones, O. A., and Endean, R., eds., 1973. *Biology and Geology of Coral Reefs, vol. I, Geology.* New York: Academic Press, 410p.

Klein, G. deV., 1970. Depositional and dispersal dynamics of intertidal sand bars, *Jour. Sed. Petrology* 40, 1095-1127.

Kraft, J. C., 1978, Coastal stratigraphic sequences, in R. A. Davis, Jr., ed., *Coastal Sedimentary Environments.* New York: Springer-Verlag, 361-383.

Kumar, N., and Sanders, J. E., 1974. Inlet sequence: A vertical sequence of sedimentary structures and textures created by the lateral migration of tidal inlets, *Sedimentology* 21, 491-532.

Laporte, L. F., ed., 1974. *Reefs in Time and Space.* Tulsa, Okla.: Soc. Econ. Paleont. Mineral. Spec. Pub. 18, 256p.

Lauff, G. H., ed., 1967. *Estuaries.* Washington, D.C.: Am. Assoc. Adv. Sci. Publ. 83, 757p.

Logan, B. W.; Read, J. F.; Hagan, G. M.; Hoffman, P.; Brown, R. G.; Woods, P. J.; and Gebelein, C. D., 1974. *Evolution and Diagenesis of Quaternary Carbonate Sequences, Shark Bay, Western Australia.* Tulsa, Okla.: Am. Assoc. Petroleum Geologists Mem. 22, 358p.

Logvinenko, N. V., 1970. A description of the sandy beach facies, *Acad. Sci. U.S.S.R. Dokl., Earth Sci. Sect.* 188, 180-182.

Ludwick, J. C., 1974. Tidal currents and zig-zag sand shoals in a wide estuary entrance, *Geol. Soc. America Bull.* 85, 717-726.

McGowen, J. H., 1971. *Gum Hollow Fan Delta, Nueces Bay, Texas.* Austin: University of Texas Bureau of Economic Geology Rept. Inv. RI 69, 91p.

McKee, E. D., 1957. Primary structures in some recent sediments, *Am. Assoc. Petroleum Geologists Bull.* 41, 1704-1747.

Milliman, J. D., 1967. Carbonate sedimentation in Hogsty Reef, a Bahamian atoll, *Jour. Sed. Petrology* 37, 658-676.

Milliman, J. D., 1974. *Recent Sedimentary Carbonates. Part 1, Marine Carbonates.* New York: Springer-Verlag, 375p.

Moore, R. C., 1949. The meaning of facies, *Geol. Soc. America Mem. 39,* 1-23.

Morgan, J. P., and Shaver, R. H., eds., 1970. *Deltaic Sedimentation—Modern and Ancient.* Tulsa, Okla.: Soc. Econ. Paleont. Mineral. Spec. Publ. 15, 312p.

Nelson, B. W., ed., 1972. *Environmental Framework of Coastal Plain Estuaries.* Boulder, Colo.: Geological Society of America Mem. 133, 619p.

Otvos, E. G., 1970. Development and migration of barrier islands, northern Gulf of Mexico, *Geol. Soc. America Bull.* 81, 241-246.

Pettijohn, F. J.; Potter, P. E.; and Siever, R., 1972. *Sand and Sandstone.* New York: Springer-Verlag, 618p.

Pierce, J. W., 1970. Tidal inlets and washover fans, *Jour. Geology* 78, 230-233.

Purser, B. H., ed., 1973. *The Persian Gulf: Holocene Carbonate Sedimentation and Diagenesis in a Shallow Epicontinental Sea.* New York: Springer-Verlag, 471p.

Reading, H. G., ed., 1978. *Sedimentary Environments and Facies.* New York: Elsevier, 557p.

Reineck, H.-E., and Singh, I. B., 1973. *Depositional Sedimentary Environments.* New York: Springer-Verlag, 439p.

Selley, R. C., 1970. *Ancient Sedimentary Environments: A Brief Survey.* Ithaca, N.Y.: Cornell University Press, 237p.

Shinn, E. A.; Lloyd, R. M., and Ginsburg, R. N., 1969. Anatomy of a modern carbonate tidal flat, Andros Island, Bahamas, *Jour. Sed. Petrology* 39, 1202-1228.

Shirley, M. L., and Ragsdale, J. A., eds., 1966. *Deltas in Their Geologic Framework.* Houston, Texas: Houston Geological Society, 251p.

Smith, J. D., 1969. Geomorphology of a sand ridge, *Jour. Geology* 77, 39-55.

Straaten, L. M. J. U. van, 1965. Coastal barrier deposits in South- and North-Holland in particular in the areas around Scheveningen and Ljmuiden, *Geol. Stichting Med.* 17, 41-75.

Swift, D. J. P., 1968. Coastal erosion and transgressive stratigraphy, *Jour. Geology* 76, 444-456.

Terwindt, J. H. J., 1971. Lithofacies of inshore estuarine and tidal inlet deposits, *Geologie en Mijnbouw* 50, 515-526.

Thompson, W. O., 1937. Original structures of beaches, bars, and dunes, *Geol. Soc. America Bull.* 48, 723-752.

Walker, R. G., ed., 1979. *Facies Models.* Geoscience Canada Reprint Series 1, 211p.

Cross-references: *Beach Processes; Chenier and Chenier Plain; Coastal Dunes and Eolian Sedimentation; Coastal Flora; Coral Reefs Coasts; Deltas; Global Tectonics; Minor Beach Features; Nearshore Hydrodynamics and Sedimentation; Organism-Sediment Relationship; Ripple Marks; Sediment Transport; Soils; Tidal Flat.* Vol. VI: *Littoral Sedimentation.*

COASTAL WATERS HABITAT

The ocean basins are in most places bordered by shallow, marginal shelf regions, the *continental shelves.* These shelf areas include the areas extending from the low water line out to the depth at which there is a marked increase in slope to much greater depths. Conventionally this is regarded as being at about the 200 m (100 fathom) depth. Actually it varies from as little as 20 m to as great a depth as 550 m, with the average approximately 135 m.

The width of the shelf zone ranges from zero to about 1500 km, averaging about 78 km. Continental shelves occupy about 7.5% of the ocean area. The most extensive shelf areas are found in the northern boreal and Arctic regions, along the eastern coasts of Asia, in the Malay Archipelago between Sumatra, Java, and Borneo, and along the eastern coast of southern South America. The shelves around Africa, the eastern coast of India, the western coast of South America, and the western coast of Central America south of the Gulf of California are relatively narrow.

Coastal waters above the continental shelves are termed *neritic*. They differ in important respects both from the waters of the open ocean and from those of bays and estuaries.

In the first place, circulation of neritic waters tends to be more active and complex than that in the open ocean. The major planetary currents or gyres tend to create some *upwelling* along their continental borders, and eddies from them contribute to the circulation and turbulence of the shelf waters. Tidal currents are in evidence along the outer edges of the shelves and wherever the bottom topography funnels the waters. Offshore winds tend to drive surface waters seaward and cause intermediate depth water to move shoreward and upwell near the coast. Onshore winds have the opposite effect. The runoff from rivers and streams dilutes the neritic waters in most areas, making them slightly less saline and also complicating the patterns of water movement. The bottom and shoreline topography also exercises strong effects on water movements.

The seaward borders of neritic waters can often be determined as that distance from shore at which there is no further increase in salinity of the surface waters as one proceeds from land toward the open sea. In a general way this corresponds to the width of the continental shelf, but of course weather conditions or alteration in patterns of circulation may alter this locally at times, either causing oceanic water to sweep in over the shelves or carrying neritic water farther out to sea than usual. In a few places, far from the influx of freshwater from rivers, and subject to high rates of evaporation, neritic water may at times become more saline than oceanic water.

Neritic waters also tend to be richer in dissolved plant nutrients than are oceanic waters, and hence more productive biologically. There are many reasons for this. The upwelling along the continental edges of the major ocean currents contributes to it, as does the upwelling caused by offshore trade winds in regions such as northern South America. Where these currents also bring cold, nutrient-rich Antarctic or Arctic water as they do along the western shores of continents, especially the Americas and Africa, this contributes heavily to the enrichment of the regional waters. Rivers and streams, in addition to somewhat diluting neritic waters, also add dissolved calcium, phosphate, nitrate, and silica, the very nutrients that are most commonly depleted in oceanic surface waters because of their uptake by phytoplankton and their loss to deeper water due to the sinking of organic residues before their decomposition is completed. The sediments of continental shelves are mostly of terriginous origin and richer in organic material than are truly oceanic sediments. They are also in or near the euphotic zone, so that any nutrients dissolved out of them are in the productive part of the water column and tend to become well distributed in the water by the generally more active water movements of neritic areas. Further, loss of nutrients to deep water is prevented by the shallowness of the water, and the nutrients are recycled with very little loss.

Most of the water column and much of the sea floor of neritic areas lie within the euphotic zone, so that the entire body of water is productive, and in most shelf areas there is also a rich benthos (organisms living on or burrowed into the bottom) in addition to the plankton and nekton in the water itself. This benthos supports large populations of demersal fish and other bottom-feeding animals in addition to those characteristic of upper ocean waters generally. The benthic organisms also contribute heavily to the plankton in the form of swarms of larvae of all groups, which adds to the food for animals at all levels in the water. Further, numerous animals living at or near the bottom by day enter the plankton at night. Animals that are in the plankton for only part of their life cycle, or only at certain times of the day, or seasons of the year, are termed *meroplankton*, in contrast to the *holoplankton*, planktonic organisms that spend their entire life in the water. Abundant and diverse meroplankton is one of the characteristic features of neritic waters compared with those of the open ocean.

Another feature of neritic waters is the much greater diversity of available habitats or ecological niches, because of the sea floor and the continental margins. The greater diversity of habitats and of available foods makes for both richer and more diverse biota than is found in the open sea.

Neritic waters differ from those of bays and estuaries in that they are much less influenced by freshwater influx and by tides. The water of bays and estuaries is usually even less saline than that of adjacent neritic waters, shows much greater fluctuation in salinity, water chemistry, and temperature, and commonly either is strongly stratified with respect to salinity or shows a marked gradient of salinity from the lower to the upper reaches of the estuary.

Many neritic animals do, however, utilize bays and estuaries as breeding grounds and nurseries for their young, both because there is even more food available for their young there and because of the protection such waters offer from some of the predators found offshore. Changes in the character of a major bay or estuary are often reflected in changes in the regional offshore fisheries as well as in those of the bay or estuary itself.

Because of the greater abundance of plankton and of suspended particles swept in from land, neritic waters are generally more turbid than oceanic waters and commonly have a greenish cast rather than being clear blue.

Almost all of the world's great fisheries are in neritic waters, because of the greater richness and diversity of foods available there and the diversity of habitats. Next to bays and estuaries, neritic waters are the most productive of any marine waters. They are also the most vulnerable to depletion, contamination, and other alterations resulting from the activities of man.

Our technology has just reached the point at which not only all the living resources but also the oil, mineral, sand, gravel, and other resources of the shelf regions can be exploited. The level of our industrial activities and weapons developments has also reached such heights that vast areas can be poisoned either by accident or design. Major oil spills and the drainage and dumping of industrial wastes, pesticides, and radioactive nuclides, and other poisons have an impact directly upon the waters of bays and estuaries and the neritic waters of the shelf regions. Increased erosion from deforestation and agricultural activities also greatly increases the sediment load reaching these waters.

For the next few decades at least, we can expect increasing pressures for change and degradation of the world's neritic waters and their biota as part of the general destruction and simplification of the world ecosystem currently in progress.

General discussions of basic oceanography and productivity can be found in Raymont (1963), Russell-Hunter (1970), and McConnaughey (1978). Rounsfell (1975) exemplifies the application of this knowledge to the management of marine fisheries. The impact of man on marine environments has been discussed widely in recent years and several articles related to this topic now appear annually in *Ocean Yearbook,* edited by Borgese and Ginsburg (1980).

BAYARD H. McCONNAUGHEY

References

Borgese, E. M., and Ginsburg, N., eds., 1980. *Ocean Yearbook.* Chicago, Ill.: University of Chicago Press, 713p.

McConnaughey, B. H., 1978. *Introduction to Marine Biology.* St. Louis, Mo.: C. V. Mosby Co., 624p.

Raymont, J. E. G., 1963. *Plankton and Productivity in the Oceans.* Oxford: Pergamon Press, 660p.

Rounsfell, G. A., 1975. *Ecology, Utilization, and Management of Marine Fisheries.* St. Louis, Mo.: C. V. Mosby Co., 516p.

Russell-Hunter, W. D., 1970. *Aquatic Productivity.* New York: Macmillan, 306p.

Cross-references: *Biotic Zonation; Coastal Fauna; Coastal Flora; Nearshore Water Characteristics; Organism-Sediment Relationship; Pollutants; Production; Standing Stock.*

COASTAL ZONE MANAGEMENT

There are many different interpretations of the term *coastal zone.* Some definitions refer to narrow areas of land and wetland immediately adjacent to the shoreline, the line of intersection of the sea with the land. On the other hand, the Resource Agency of California, for example, refers to: "the area extending inland approximately ½ mile (ca. 1 km) from the mean low tide and seaward to the outermost limit of the state boundaries." The main difficulty, however, is to define the inland limit of the coastal zone. Some planners view it as a complete natural system that should not be defined according to political boundaries. Following these brief comments, it is proposed here to define the coastal zone as "an area of variable width that extends seaward to the edge of the continental shelf but that has no set landward demarcation." It is within this zone that man's activities can interrupt or destroy natural systems (ecosystems) and natural processes whether they are biological, chemical, or physical.

Resources of the Coastal Zone

Increasing demands for land within the coastal zone by industrialists, land developers, commercial organizations, and people providing recreational facilities are making it increasingly necessary to plan short- and long-term coastal zone management programs in order to minimize the adverse effects of man's activities. Although the coastal zone has both finite and renewable resources, legislation to control the use of these is either vague or nonexistent.

The need to compile inventories of coastal resources and to record changing land use patterns is readily apparent. In Queensland, Australia, for example, twelve landform units have been delineated along the coast between Coolangatta and Fraser Island. It is believed that there may be competition for the use of this coastal zone between housing, industrial development, forestry, grazing, sand mining, and recreation. In addition, a projected increase in population will eventually promote further demands for water supply, drainage, sewage disposal, and sanitary fill, as well as vegetable and horticultural production, but little is known about the physical and chemical processes in this area and its diverse resources (Hails, 1977).

Many coastal states in the United States are attempting to describe and to record their resources, possible threats to these resources, the conflicts that may arise through multiple use, and finally the objectives for rational and effective management. According to Ketchum (1972), California and Florida have already produced and published comprehensive surveys and plans.

For economic as well as for practical reasons, effective management in the future will depend largely upon the results of feasibility studies and systematic monitoring in addition to ongoing, long-term research programs. Collectively, these investigations will create data banks that will furnish, in turn, information on natural and modified coastal systems, demographic patterns, land use, industrial location, and socioeconomic requirements. A great deal of information already exists, and the rate of data collection is being accelerated at local, state, and national levels in the United States. But there are two key questions that must be posed at this juncture: How relevant or meaningful are the data? How were the data-collection programs devised? The long-term benefit to be gained from data collection and monitoring programs is related largely to co-ordinated activities, but in the United States, as elsewhere, various agencies are still working independently of each other. Further, however scientifically data are collected, decision-making is invariably based upon value judgments.

Environmental Quality and Pollution

It is essential to understand what such terms as "quality of the environment" really mean with regard to long-term planning objectives. This can be illustrated by an example cited in Ketchum (1972): "The effects of sudden catastrophic events, such as a violent storm or a large accidental oil spill in the coastal zone, are there for all to see in the form of land erosion, damaged buildings, or dead fish and shellfish. It does not need much analysis to determine cause and effect and to assess the damages accurately.... [But] Much more disturbing are the chronic, relatively low-level changes created by man which are not obvious to the eye but which are steadily degrading the quality of the environment." One only has to refer to the dangers of persistent toxins such as DDT (dichlorodiphenol trichloroethene) and PCBs (polychlorinated biphenyls) and heavy metals such as cadmium, lead, and mercury. In coastal areas without adequate facilities, water supply and waste disposal pose complex problems because a preponderance of septic tanks and poor land disposal cause soil and surface water pollution. Also, the conflicts that arise from the multiple use of the coastal zone are manifest in industrial pollution that can severely curtail or even preclude recreational activities. These are only two examples of pollution that present difficult management problems because they involve priorities; these, in turn, hinge on subjective deduction and reasoning.

Beach Mining

In the view of many conservationists and planners, sand mining companies operating in the coastal zone have either caused or contributed to coastal erosion because of the indiscriminate way *placer deposits* have been exploited in the recent past. Beach placers are surficial heavy mineral deposits that are concentrated by the winnowing action of waves on the foreshore. Typical examples are ilmenite, magnetite, monzonite, rutile, tourmaline, and zircon (see references in Hails, 1976). Mining for placers has continued along the east coast of Australia, for example, since the early 1930s. During the rapid expansion of the industry in the late 1940s and mid-1950s many operators left large tracts of unstabilized sand after mining without attempting to restore dunal systems to their original state. As larger areas of the coast were disturbed by new mining methods, the need for reclamation schemes became more apparent, particularly the importance of restoring frontal dunes. Even today, despite sophisticated remedial measures, the success of stabilization programs after mining depends upon the rate at which a coast is actually eroding. This, in turn, can be determined only by obtaining accurate data by means of well-planned monitoring programs.

In some areas valuable heavy mineral deposits have been lost to housing development. On the other hand, some previously mined areas have been restabilized and restored to their original form and used for residential development. From the planning viewpoint, this example shows that where there is conflict between sand mining and property development, the problem can often be resolved by giving mining companies the option to extract minerals first.

Credit must be given to those mining companies that have provided new roads, recreational amenities, and parks, as in the Kingscliff-Cudgen area of the far north coast of New South Wales, Australia (Hails, 1977). The main consideration is to incorporate if possible the plans of the sand mining companies into development programs in order to reduce costs and to avoid conflicting priorities.

Hurricanes and Cyclones

Because hurricanes and cyclones often cause severe damage, amounting to many millions of dollars, in densely populated and highly industrialized coastal areas, it is imperative that management programs be responsible for improving forecasting systems and warning services. These should not only furnish information on the direction, intensity, position, and rate of movement of tropical storms but also predict *storm surges* (wind-driven tides) and the strength of wind and amount of rainfall expected. The value of such services is exemplified by the case of hurricane Camille which struck the Gulf coast of the United States near New Orleans in August 1969. Because of advance warnings, most people living less than 6 m above sea level were evacuated safely, and an estimated 50,000 lives were saved.

Recent statistics show that deaths caused by hurricanes are declining but that damage to property is increasing because development along the coast has increased. Of course the cost of emergency storm preparations for major urban centres is also increasing. For example, five years ago the average cost of a hurricane alert on the Florida coast was $2 million. Protection from hurricane surges is both complicated and costly, because many municipal and industrial developments are located in areas subject to inundation from such surges. In localities where hurricane surge protection barriers are required to provide economic protection for property along the shore of an estuary, it may be essential to construct a barrier across the estuary. The engineer, however, must design the barrier with tidal openings so that navigation and tidal movement within the estuary are not impeded, thus ensuring that salinity intrusion, sediment movement, water circulation, water level, and the flushing of pollutants are not adversely affected. There is no doubt that increased public pressure and interest to preserve the coastal zone will escalate the cost of protection against natural disasters in the immediate future. In retrospect, the damage caused by hurricane Carla along the Texas coast in 1961, which totaled more than $400 million, and by Betsy along the Louisiana coast in 1965, totalling $372 million, was probably modest compared with likely costs in the mid- to late-1970s (see references in Tanner, 1973).

Coast Protection Schemes

About 80,000 km of the United States coast are vulnerable to erosion, particularly in southern California, Florida, New Jersey, and Texas, but it is often difficult to distinguish between man-induced and natural erosion. A *coastline* (technically, the line that forms the boundary between the coast and the shore) can be modified considerably by man-made structures like breakwaters, causeways, groins, jetties, levees, and seawalls, and by reclamation for housing and industrial development, beach and dune stabilization projects, hurricane protection barriers and bulkheading.

Because beaches are basically in a state of dynamic equilibrium, it is important to understand the processes contributing to that equilibrium before man attempts to arrest coastal erosion by constructing sea defenses. It is now generally accepted that engineering structures along one sector of a coast are likely to affect adjacent areas to leeward, and that many coast protection structures are totally unrelated to the hydrodynamics of a site or to other neighboring schemes. Despite these facts, the literature abounds with well-documented case histories of man's activities affecting shoreline equilibrium (see references in Hails, 1977). For example, reduction in sediment supply to the Florida coast resulting from farmers adopting contour plowing, antierosion measures, and the construction of dams on streams and their tributaries have indirectly caused beach erosion several kilometres south of the St. Johns River. These activities demonstrate that coastal zone management often has its roots in coastal hinterlands and river catchments as well as in the nearshore zone. Interference with natural processes by the construction of harbor works is also recorded widely in the literature, particularly in the United States. At Lake Worth, Florida, Manasquan Inlet, New Jersey, Tillamook Bay, Oregon, Santa Barbara and Port Hueneme, California, shorelines have eroded over a distance of 32 km from these respective harbors with resultant damage totaling millions of dollars.

After Port Hueneme harbor was constructed in 1940, the shoreline downcoast lost about 920,000 m^3 of material annually because the north jetty of the harbor diverted material moving alongshore into the Hueneme canyon. This rapid rate of erosion was partly arrested in 1960-61 by the construction of Channel Islands Harbor as a sand bypass system, 2 km upcoast. The harbor is now fronted by an offshore breakwater, located on the 10 m depth contour, that serves a dual function of sheltering the harbor entrance and acting as a littoral sand trap.

The Queensland Gold Coast beaches in Australia, in addition to being eroded periodically by cyclonic storms, have undergone extensive changes since the training walls of the Tweed River, on the New South Wales–Queensland border, were built in 1962-64. It has been predicted that the alongshore movement of material on the southeast Queensland coast will not be fully reestablished for at least another decade and that by this time about 1,500,000 m^3 of sand will have been lost from the beaches. In the interim, sand nourishment has been introduced in an attempt to arrest long-term erosion and the problems caused by man-made structures. Despite the adverse effects of constructing the training walls at the entrance to the Tweed River, a groin was built at Kirra Beach on the Queensland Gold Coast in 1972. Although this beach has been restored, erosion is now occurring downdrift of the groin.

At present in the United Kingdom and Australia there is no national body that is responsible for monitoring coastal processes and coastal changes. In fact, in England one local authority (usually a Water Authority responsible for a certain sector of the coast) will undertake measures to protect its *designated coast* without necessarily consulting neighboring authorities. In this way, for the past few years, erosion problems have migrated along the coast instead of being resolved on a regional or national basis. Obviously this ad hoc, unplanned, uncontrolled, and unregulated modus operandi cannot continue with so many demands on the coastal zone.

Future Potential of the Coastal Zone

The social, economic, and political processes through which societal decisions are made are unlikely to curtail the continuing and increasing use of the coastal zone. Thus its capacity to meet future demands will depend upon technological innovation combined with management strategies. Following is a list of some recommendations for coastal zone management based on Ketchum (1972):

1. Compilation of coastal resources, both finite and renewable.
2. Development of predictive models to assess the effect of activities and structures upon the coastal zone.
3. Improved environmental impact statements.
4. On-going short- and long-term biological, chemical, and physical research directed toward the following problems: monitoring and surveillance of input levels of pollutants, particularly chlorinated hydrocarbons, petroleum, heavy metals, industrial effluents, detergents, and toxins; effects of solid waste disposal and sewage outfalls; effects of contaminants on organisms and ecosystems, factors affecting stability, diversity, and productivity of coastal zone ecosystems; and epidemiologic and virologic studies.
5. Research in the economic, legal, political, and social aspects of the coastal zone directed toward the following problems: cost benefit analysis of different uses of the coastal zone; resource evaluation; siting, construction, and operation of coastal and offshore power plants and ports; dredging and dumping of spoil; economic models for policy guidance in calculating inputs and outputs; group interests and priorities; decision-making processes at the local, state and national levels; statutory guidelines for shoreline development; and administration and judicial enactment of legislation.
6. Establishment of regional and national monitoring systems in order to establish reliable data banks.
7. Training and educational schemes.
8. National coastal zone management programs.
9. Preservation of those areas noted for their unique scientific interest and ecological character.
10. Protection of existing ecological systems; and protection from environmental degradation of coastal wetlands and estuaries that are highly productive habitats for aquatic life.

There seems little doubt that the construction of offshore facilities will play a significant role in future development. According to Ketchum (1972), the concept of *floating cities* or *satellite communities offshore* is being explored as a means to accommodate the population pressures in high-density coastal areas. But, as stated by Ketchum (1972), proposals to build offshore ports, energy plants, or cities will result in keen debate and argument about the use of the coast in the immediate future.

The availability of comprehensive aerial and space photography now provide useful data for observing changes in beaches, sand dunes, and tidal inlets, and may resolve in the long-term how best the coastal zone may be used and managed. Further relevant treatment of this topic may be found in Australian UNESCO Committee (1974), Brahtz (1972), Duane, (1968, 1969), Hardin (1968), Herron and Harris (1966), Inman and Brush (1973), Ippen (1970), Lewis (1973), Matthews, Smith, and Goldberg

(1971), Moss (1967), Silvester (1974), Smith (1968), and Snodgrass (1964).

JOHN R. HAILS

References

Australian UNESCO Committee for Man and the Biosphere (MAB), 1974. *The Impact of Human Activities on Coastal Zones,* Pub. No. 1. Canberra: Australian Government Publishing Service, 189p.
Brahtz, J. F., 1972 *Coastal Zone Management: Multiple use with Conservation.* New York: John Wiley & Sons, 352p.
Duane, D. B., 1968. Sand inventory program in Florida, *Shore and Beach* 36, 12-15.
Duane, D. B., 1969. Sand inventory program: A study of New Jersey and northern New England coastal waters, *Shore and Beach* 37, 12-16.
Hails, J. R., 1976. Placer deposits, *in* K. Wolf, ed., *Handbook of Strata-Bound and Stratiform Ore Deposits,* vol. 3. Amsterdam: Elsevier, 213-244.
Hails, J. R., 1977. Applied geomorphology in coastal-zone planning and management, *in* J. R. Hails, ed., *Applied Geomorphology.* Amsterdam: Elsevier, 317-362.
Hardin, G., 1968. The tragedy of the commons, *Science* 162, 1243-1248.
Herron, W. J., and Harris, R. L., 1966. Littoral by-passing and beach restoration in the vicinity of Port Hueneme, California, *Conf. Coastal Eng., 10th, Am. Soc. Civil Engineers, Proc.* 1, 651-765.
Inman, D. L., and Brush, B. M., 1973. The coastal challenge, *Science* 181, 20-32.
Ippen, A. T., ed., 1970. *The Water Environment and Human Needs.* Cambridge, Mass.: MIT Press, 365p.
Ketchum, B. H., 1972. *The Water's Edge: Critical Problems of the Coastal Zone.* Cambridge, Mass.: MIT Press, 393p.
Lewis, T. E., 1973. Coastal zone management—the Florida experience, *Shore and Beach* 41, 12-14.
Matthews, W. H.; Smith, F. E.; and Goldberg, E. D., eds., 1971. *Man's Impact on Terrestrial and Oceanic Ecosystems.* Cambridge, Mass.: MIT Press, 540p.
Moss, F. E., 1967. *The Water Crisis.* New York: Praeger Publishers, 303p.
Silvester, R., 1974. *Coastal Engineering,* vol. II. Amsterdam: Elsevier, 338p.
Smith, J. E., ed., 1968. *"Torrey Canyon" Pollution and Marine Life.* New York: Cambridge Press, 196p.
Snodgrass, F., 1964. How to tell breakwaters from elephants, *Seahorse* 1, 1-4.
Tanner, S. H., 1973. Hurricane barrier environmental planning in Texas, *Am. Soc. Civil Engineers Proc., Jour. Waterways and Harbors Div.* 99, 459-470.

Cross-references: *Beach Processes, Monitoring of; Coastal Engineering, History of; Coastal Engineering, Research Methods; Coastal Erosion, Environmental-Geologic Hazard; Coastal Reserves; Demography; Hurricane Effects; Mineral Deposits, Mining of; Pollutants; Sediment Transport; Waste Disposal.* Vol. XIII: *Coastal Zone Management.*

COBBLE

A *cobble* is a subangular or rounded rock fragment larger than a *pebble* but smaller than a *boulder,* hence 64-256 mm in diameter (-6 to -8 ϕ). Rounding is generally through abrasion during transport, but may be by weathering in place. Because cobbles generally have traveled farther than boulders, they tend to be composed of hard, resistant rocks (Gary, McAfee, and Wolf, 1972).

A cobble can also be a rock or mineral fragment in soil between 20-200 mm in diameter. In the United States, the term is used for soil particles 75-250 mm; in engineering, for particles above 76 mm. In the United Kingdom, the British Standards Institution (1967), which covers soil testing for civil engineering purposes, defines cobble size range as 60-200 mm.

A cobble is sometimes also called a *cobblestone* because of former use in paving.

ALAN PAUL CARR

References

British Standards Institution, 1967. *Methods of Testing Soils for Civil Engineering Purposes, B.S. 1377.* London, 234p.
Gary, M.; McAfee, R., Jr.; and Wolfe, C. L., eds., 1972. *Glossary of Geology.* Washington, D.C.: American Geological Institute, 805p.

Cross-references: *Beach Material; Beach Material, Sorting of; Boulder; Clay; Gravel; Pebble; Sand; Sediment Analysis, Statistical Methods; Shingle and Shingle Beach; Silt; Soils.*

COELENTERATA

Originally called Zoophyta or plant animals, the *coelenterates* (Greek *koilos,* hollow + *enteron,* intestine) are generally considered to be the evolutionarily lowest phylum of animals constructed of well-defined tissues (Eumetazoa), with distinct form and symmetry and with a digestive tube (coelenteron) lined with epithelium derived from "entoderm." Present throughout the epidermis are characteristic specialized cells containing stinging structures (nematocysts) called cnidocytes, whence the often used phyletic name for this group, Cnidaria. The radially symmetrical and basically tentaculate coelenterates or cnidarians include a variety of diverse forms that are actually variations on a simple body plan: the familiar, relatively simply constructed freshwater hydras as well as the more complex siphonophores, jellyfish, sea fans, sea anemones, sea pens, corals, and their close relatives.

In this phylum the digestive tract has a mouth but no anus. The body has a continuous surface and limited organ development producing simple digestive, muscular, nervous, and sensory systems built from epithelial, muscular, and connective tissues. However, there is no second body cavity (coelom), no cephalization, no trace of a centralized nervous system, no definite respiratory, circulatory, or excretory systems, and the modified genital system has only sex cells that are often localized into gonads. Coelenterates have a body wall consisting of two layers of cells of many different types in which there is an outer epithelium ("ectoderm") and an inner gastrodermis ("entoderm"), sandwiching a variably proportioned jellylike substance, the *mesoglea,* that ranges from a thin, noncellular membrane to a thick, fibrous or gelatinous and mucoid material sometimes containing wandering cells (amoebocytes) (Hyman, 1940). The radial symmetry of these animals is arranged around the oral-aboral axis, and the number of units is 4, 6, or a multiple of these.

The coelenterates characteristically have two basically similar types of individuals differing in structural details, called the *polyp* and the *medusa*. The polyp is sessile. Its body is tubular or cylindrical, with the free end (oral) bearing mouth and tentacles directed upward, and the opposite end or base (aboral) with pedal disc, rootlike stolons, or calcareous skeletal structures attached to the substratum. The tentacles are most often covered with clusters or batteries of nematocysts, and are thus adapted for predation, defense, food ingestion, and sometimes reproduction. Polyps reproduce asexually.

The medusa is the free-living form. It is ordinarily a bowl or umbrella-shaped bell of stiff mesogleal jelly composed of more than 95% water (the jellyfish) that is adapted for swimming and, in polymorphic species, for sexual reproduction (mostly dioecious). The medusa has a four-part or tetramerous symmetry, and although it has the same basic structure it is morphologically more advanced than the polyp. In nature, the convex side of the bell is uppermost, the mouth is centered in the concave undersurface, while the tentacles hang from the rim of the bell. Like polyps, medusae are carnivorous on all kinds of animals from zooplankton to fish. Some coelenterates are entirely polypoid, others are solely medusoid, and many pass through both forms alternately during their life cycles (alternation of generations or metagenesis).

In general, fertilization of medusan eggs and sperm may be external in the sea or internal in the gonad. Cleavage is complete and is followed by blastula, gastrula, and finally a free-swimming planktonic planula larva that settles to the bottom, either becoming a creeping actinula or metamorphosing into the polypoid form. The polypoid form will mature and bud off medusae, thus completing the life cycle. However, there are many variations in the sequence of development including both suppression and modification of entire stages in the life cycles of various coelenterate groups.

Except for the hydras and a few other freshwater hydrozoans, coelenterates are all marine. Although most species live subtidally in shallow waters, often as members of the local zooplankton, many are sessile, inhabiting rocky coasts and tropical coral formations, as members of hard-bottom benthic communities and as interstitial inhabitants of subtidal sandy bottoms.

In some forms, brilliant coloring combined with the unique radial symmetry often creates an unusual kind of beauty. Of greatest direct importance to most people are the coelenterates of the several groups with nematocysts capable of pentrating human skin and inflicting painful stings. Probably best known are the hydrozoan siphonophores, of which the beautiful *Portuguese man-of-war* is properly feared by swimmers and beachcombers in tropical and even sometimes in temperate waters. The scyphozoan sea nettles, called by Aristotle "acalephae," are also recognized along the Atlantic and Gulf coasts as dangerously stinging coelenterates. Among these is the giant jellyfish known as the lion's mane, made famous in a Sherlock Holmes story, whose lens-shaped disc may be 2½ m in diameter and whose 800 fully expanded tentacles may reach 60 m in length. In the same class, there is also a group of virulent cubomedusan man-killers in Australian waters, the sea wasps. At the other end of the scale are the far less effective (to man) nematocysts of attached hydroids whose delicate branching colonies growing on wharves, docks, pilings, ship's bottoms, and algae were originally thought to be marine plants and not recognized as animals until well into the eighteenth century.

Although coelenterates are of little direct economic value to man, they are particularly useful in biomedical, developmental, physiological, geological, and evolutionary research. They go back in geological time as far as the sponges. The bodies of jellyfish and particularly the calcareous skeletons of corals have been found as fossils as far back as the Precambrian rocks of Australia. In the Silurian and Ordovician periods, corals formed reefs that became extremely favorable sites for the collection of petroleum deposits and so are now of great interest to the oil industry. Coral limestone from these ancient reefs is quarried in the Florida Keys, and is used as a building stone in areas where other kinds of rock are not readily available.

The phylum Coelenterata contains three classes: (1) The Hydrozoa are typically colonial with an alternation of generations in which the polyps (hydroid stage) give rise by asexual budding to free-living or sessile dioecious medusae that in turn reproduce sexually, producing a larval ciliated planula that eventually metamorphoses back into the polyp; the mesoglea is never cellular; the gastrodermis lacks nematocysts, and the gonads are usually epidermal. (2) In the Scyphozoa, the sessile hydroid stage is minimized and usually becomes insignificant, while the free-swimming medusoid stage is highly developed, as in the larger jellyfish (true jellyfishes). The mesoglea contains wandering amoeboid cells, and the gonads are gastrodermal. (3) The Anthozoa, which include the sea anemones, corals, and gorgonians, have no free-swimming medusoid stage. The hydroid stage is more fully developed, with the lining of the polyp cavity being thrown into folds called mesenterial ridges; the mesoglea contains fibers as well as amoebocytes, part of the gastrodermis has nematocysts, and the gonads are in the gastrodermis of all or only certain mesenteries.

Hydrozoa (Greek hydro + *zoon,* animal)

The hydroid or polyp is the familiar form in this class. The freshwater hydras, with one exception, lack a medusoid stage and are cylindrical solitary polyps up to 2.5 cm long and 1 mm in diameter. The oral end is cone-shaped (hypostome), with 5-12 terminal tentacles and a terminal mouth, while aborally a basal disc attaches the animal to the substratum. They reproduce asexually by budding in the summer, and often by sexual reproduction (usually dioecious) in the fall, the encapsulated embryo wintering on the bottom and developing directly into a young hydra as the water warms in the spring. Additional but less well-known freshwater forms include a species of colonial hydroid (colonial polyp) and a medusoid species, the so-called fresh-water jellyfish (Pennak, 1978). All of these hydroids live in ponds and lakes, while those attached to mollusk shells may be distributed through entire drainage systems. There is one species of hydroid, 2-5 mm long, that lives in brackish estuaries, swamps, creeks, and intertidal sand, that lacks tentacles, and that reproduces both sexually (dioecious) and asexually (transverse and longitudinal fission).

The marine hydroids, except for the rare protohydroid from brackish coastal waters, are usually colonial with tubular stolons forming a supporting network for the branches bearing terminal hydranths (polyps), and with small transparent medusae 0.5-6 cm in diameter. Hydranths vary structurally with or without a supporting chitinous cylinder and with different arrangements of nematocyst-armed tentacles. The colonies have two to several types of polypoid individuals or zooids (polymorphism) concerned with reproduction, feeding, protection, and predation. Solitary marine hydroids are usually large, and ordinarily each individual is adapted to carry on all of its life functions. The small, free-swimming, bell-shaped hydromedusae are tetramously arranged with general sense organs, ocelli, statocysts, as well as tentacles along the bell margin. They are isotonic with seawater. A strongly muscled thin flap projecting inward from the margin of the bell (the velum) helps control the rhythmic pulsations that are part of the vertical swimming movement. Horizontal movement depends mostly on water currents. The appearance of hydroid colonies and hydromedusae is seasonal and temperature-related.

The tropical and semitropical stylasters and the so-called stinging corals or millepores are two small orders of colonial polypoid hydrozoans that secrete an external calcareous skeleton resulting in encrusting or in upright growth forms from sheetlike crusts to antlerlike branches (Barnes, 1980). The more delicately branched stylasters may be white, pink, red, yellow, or purple, while the millepores are green to mustard-colored with a white, furry appearance when the polyps are expanded from their pores. Millepores are common on coral reefs in the West Indies.

Mong the best-known and most unusual hydrozoans are the planktonic or pelagic tropical and semitropical siphonophores. In members of this order, which includes the well-known *Physalia* or Portuguese man-of-war, polymorphism is more highly developed than in any other group of coelenterates. The several kinds of specialized medusoid and hydroid members of the colony bud off from special zones of a supportive stalk that arises directly from the planula larva. The polypoid individuals adapted for feeding, reproduction, predation, swimming, flotation, and protection are suspended in most forms from a pneumatophore that acts either as a float or as a hydrostatic organ. The beautifully blue to pink iridescent float of every *Physalia,* a gas-filled bladder with an erectile crest up to 30 cm long, curves gently either to the left or to the right, an adaptation well suited to effecting wide distribution while sailing before the wind. The 15 m long tentacles of *Physalia* attached to the specialized feeding polyps are abundantly supplied with knoblike or spirally arranged batteries of nematocysts. Contact with these nematocysts in the water or on the beach is always painful and can be dangerous. *Physalia* sometimes appears on the North Atlantic coast following late summer storms.

Several marine, subtidal, interstitial, solitary polypoid forms have been discovered recently on both sides of the Atlantic. Although they have been placed in a separate order, Actinulida, the lack of planuloid and medusoid stages and the presence of ciliation on the tentacles and sometimes on the body have raised questions concerning their primitiveness (Russell-Hunter, 1979; Rees 1966).

Coelenterates as a whole are of little economic importance. Seldom used as food by man, they are eagerly devoured by certain fishes and by invertebrates such as the *crown-of thorns starfish*. Certain corals used as ornaments and for the manufacture of jewelry have become rare because of exploitation. A few common species of hydrozoans, including the freshwater hydras, have been used extensively and successfully in regeneration studies in explaining the roles of bioletric fields, dominance, and control during development and organismal growth.

Scyphozoa (Greek *skyphos,* cup
+ *zoon,* animal)

The dominant medusoid phase in this class is commonly 15–30 cm larger than the hydromedusae; all lack a velum. Their tetramerous symmetry is also seen in the reduced inconspicuous polypoid stage, the scyphistoma. Through the transparent cartilaginouslike mesoglea of the domelike to shallow, saucer-shaped umbrella can be seen the often striking coloration of the gonads and other internal organs. The margin of the bell is notched, producing scalloped lobes called lappets, and is fringed with variouslv shaped tentacles in multiples of four. Several kinds of nematocysts are on both umbrella surfaces and on the tentacles. Between the lappets are club-shaped rhophalia that contain sensory pits, a statocyst, and sometimes one or more ocelli that may be involved in phototactic reactions of the medusae. Although locomotion is accomplished in a manner similar to that in hydromedusae, the gastrovascular canal system is often more branched and more generally complex.

One order, the Stauromedusae, contains two boreal genera of unique, permanently polypoid, green or brown, sessile scyphozoans that attach to a substratum (usually marine plants) by a narrow stalk. They cannot swim or change locations, and their development is direct without production of a free-swimming stage.

Nearly all scyphozoans are dioecious with four readily visible gastrodermal gonads. Fertilization precedes development to a planula that becomes a polypoid, hydralike larva, called a scyphistoma, after settling to the bottom. In general, it may produce new scyphistomae by asexual budding or form immature medusae (ephyrae) by transverse fission of the oral end (strobilization). The ephyrae mature to sexually reproducing medusae within two years.

The more than 200 described species of scyphozoan jellyfishes may occur singly or in great schools in all seas. Although they float quietly and can swim feebly by rhythmic contractions of the bell, the majority inhabit coastal waters. Great numbers of them may be cast up on shore during storms because of their being largely at the mercy of currents and waves. The order Cubomedusae contains the transparent tropical and subtropical box jellies or sea wasps responsible for many documented cases of fatal jellyfish stings. The sea nettle of our eastern coast, belonging to another order, may be troublesome, but its stings are not nearly so serious. Adult scyphozoans feed on all types of small animals, including ctenophores and other medusae, and the members of several species are additionally suspension feeders on plankton. One genus, *Cassiopeia,* has zooxanthellae (symbiotic algae) in its tissues, and so can live on the products of algal photosynthesis. However, there are several kinds of larval fish that use the subumbrella space and the oral arms of certain species of scyphozoans for protection.

Anthozoa (Greek *Anthos,* flower
+ *zoon,* animal)

This is the largest class of coelenterates, with more than 6000 living species. Anthozoans are solitary or colonial marine polyps, somewhat flowerlike, of small to large size and rather firm texture, with a tendency toward biradial symmetry in arrangement of the gullet and the internal radially arranged membranes (mesenteries or septa). Many anthozoans secrete a calcareous skeleton known as coral. They are divided into two subclasses: the Octocorallia or Alcyonaria, with octomerous symmetry, mesogleal calcareous spicules, and epidermal horny skeleton, as in the gorgonians; and the Hexacorallia or Zoantharia, with hexamerous symmetry, with or without skeletal support, and with a great diversity of nematocysts. In addition to sea anemones and stony corals, this class includes the soft, horny, and black corals, the colonial sea pens and sea pansies and their close relatives, all of which lack a medusa stage.

Anthozoan polyps differ from hydrozoan polyps in being larger and squatter, in having a flattened oral end (oral disc) surrounded by one to several circles of hollow nematocyst loaded tentacles, in the possession of a more specialized gastrovascular cavity, in being more strongly and diversely muscled, and in having a thick and richly cellular mesoglea. The supporting element for solitary or colonial aggregations of polyps may be calcareous spicules secreted by mesogleal cells, a horny ectodermal secretion, a calcareous ectodermal secretion around the polyp bases (coral), or the incorporation of sand grains or sponge spicules into the column (body) wall. However, certain forms, the anemones, lack a skeleton and have body walls that are strongly contractile.

The Octocorallia may be massive or creeping colonies, such as the fleshy, leathery, lobed, soft corals, the milliporelike blue corals, and the reef-dwelling organ-pipe corals; colonial branched forms including the precious red coral of commerce (jewelry), and the yellow or red featherlike sea fans and sea whips (gorgonians); or bilaterally symmetrical colonies as the stalk-anchored sea pens and sea feathers, and the white, featheryy polyped pink or violet sea pansies. The octocorallians are all colonial with small polyps connected by a mesogleal tissue called coenenchyme and an internal skeleton. The familiar yellow, orange, or lavender gorgonians or horny corals, with their supportive central axial rods of gorgonin, are common and conspicuous members of reef faunas, while yellow brown reef forms are the result of the presence of symbiotic zooanthellae.

The Hexacorallia are simple or colonial, rarely dimorphic, with or without a skeleton, with hollow pentamerous or hexamerous tentacles. Included in this group are: the well-known, colorful, inshore, solitary, skeletonless, variously sized sea anemones; the usually colonial, rapidly asexually reproducing, mostly reef-building, stony or scleractinian corals; the endangered colonial, branching or whiplike (7 m), horny, black corals; and the large, solitary, aborally rounded, tubiculous, skeletonless burrowing sea anemones. Members of this subclass are as diverse morphologically as the octocorallians are homogenous.

Stony corals (Madreporaria) are the principal reef builders, incorporating into their masses other coelenterates and inverbrates as well as marine algae. Together with the whole spectrum of animals and plants that live in and on them, they form the extensive and complex reef communities on eastern continental shores worldwide except for the western Atlantic Ocean between latitudes $30°N$ and $30°S$. In these communities, many coelenterates are involved in phoretic, communal, symbiotic, and other symbiotic associations. Probably the best-known coral reef is the Great Barrier Reef off northeastern Australia, 160-2400 km offshore and more than 19000 km long.

Reef corals require undiluted, sediment-free, clear seawater at $20°C$ or warmer. They live normally from the surface to depths of about 30 m because of their dependence on the light necessary for the photosynthesis of the algae living symbiotically in their tissues. Reef corals may form: a fringing reef extending up to half a kilometer from shore; a barrier reef separated from shore by a lagoon of varying width and depth; or an atoll or circular reef that encircles an islandless lagoon (Dakin, 1950). At present, reefs grow 5-200 mm a year, keeping pace with rising postglacial sea levels. It is estimated that all existing reefs could have developed in from 10,000 to 30,000 years.

DONALD J. ZINN

References

Barnes, R. D., 1980. *Invertebrate Zoology,* 4th ed. Philadelphia: W. B. Saunders Co., 1089p.

Dakin, W. J., 1950. *Great Barrier Reef.* Melbourne: Australian National Travel Association, 135p.

Hyman, L. H., 1940. *The Invertebrates: Protozoa through Ctenophora,* vol. 1. New York: McGraw-Hill, 726p.

Pennak, R. W., 1978. *Freshwater Invertebrates of the United States,* 2d ed. New York: Wiley Interscience, 803p.

Rees, W. J., ed., 1966. *The Cnidaria and Their Evolution.* New York: Academic Press, 449p.

Russell-Hunter, W. D., 1979. *A Life of Invertebrates.* New York: Macmillan, 605p.

Cross-references: *Australia, Coastal Ecology; Biotic Zonation; Coastal Fauna; Coastal Waters Habitat; Coral Reef Coasts; Coral Reef Habitat; Organism-Sediment Relationship; Thalassotherapy.* Vol. VI: *Coral Reef Sedimentology.*

COFFEE ROCK—See HUMATE

COLK

A *colk* is a relatively deep depression on the sea floor excavated or kept clear of sedimentation by locally strong current action. Such holes are usually found where current flow has been intensified in passing through a narrow strait or tidal entrance or in the vicinity of a sharp curve in an estuarine channel. They are also known as *tidal colks* and *scour holes.*

A major example of a colk occurs in the narrow strait between Shikoku and Kyushu at the entrance to the Japanese Inland Sea. It extends to a depth of 390 m more than 300 m below the general depth of the strait floor. Colks occur at the entrances to San Francisco Bay, (110 m deep) and Port Phillip Bay, Australia (65 m deep), both in areas of strong tidal current scour. In the Bristol Channel and off the Dorset coast in England, sand moved by tidal currents has excavated colks in softer rock outcrops (Donovan and Stride, 1961), but off the Brittany coast the colks of the Goulet of the Rade de Brest (50 m deep) and Morbihan (31 m deep) are in hard rock, and it is probable that these are maintained by strong currents that prevent infilling by sedimentation. Colks exist where strong currents flow between islands in Bass Strait and Torres Strait, Australia. In some cases there is a single colk within a constriction; in others (as in Otago Harbour, New Zealand) there are paired colks on either side of a constriction (Benson and Raeside, 1963).

Colks are common in soft sediments of estuary floors. The Scheldt estuary in the Netherlands has a colk 59 m deep off the Zieriksee promontory (Van Veen, 1950), and there are similar features in the Ord estuary in northern Australia.

ERIC C. F. BIRD

References

Benson, W. N., and Reaside, J. D., 1963. Tidal colk in Australia and New Zealand, *New Zealand Jour. Geology and Geophysics* 6, 634-640.
Donovan, D. T., and Stride, A. H., 1961. Erosion of a rock floor by tidal streams, *Geol. Mag.* 98, 393-398.
Van Veen, J., 1950. Eb-en vloedschaar systemen in de Nederlandse getijwateren, *Koninkl. Nederlandsch Aardrijksk. Genoot. Tijdschr. (Waddensymposium)* 42-65.

Cross-references; *Scour Holes; Tidal Currents.*

COMPOSITE COAST—See CLASSIFICATION

COMPOUND SHORELINE—See CLASSIFICATION

COMPUTER APPLICATIONS

Any time large data sets are collected, their manipulation and interpretation become laborious and tedious if done by hand. On the other hand, *data analysis* by computers can be done in pico-seconds, and most statistical analyses or iterations take only several seconds.

The investigation of a coastal process such as longshore fluctuation in energy and currents requires a knowledge of the energy distribution in waves as they reach the nearshore zone. This energy is equal to that of deep-water waves if the dissipation loss due to gravitational surface tension and frictional forces is subtracted. Moreover, deep-water wave energy may be distributed by *refraction, reflection,* or *diffraction* as the wave progresses to the shore. By using the simplifying assumption of classical wave theory, the resulting equations are still unsolvable since they are nonlinear. Perturbation techniques are used to linearize the group of equations over small distances.

When a wave moves into shallow water, its velocity will change, the speed of each part of the wave front depending on the water depth beneath it. If the depth is not constant, then the wave front must bend. The theoretical paths of waves over a given seabed may be predicted by refraction diagrams. This is usually done by drawing the paths of the wave rays (lines in the direction of travel of the waves, perpendicular to the wave crest).

These diagrams may be drawn by hand (Munk and Traylor, 1947), but this is both tedious and time-consuming especially if a spectrum of wave periods (see Time Series, below) and directions is included. The main problem in a computer solution is the determination of the curvature of a wave ray as it changes velocity due to changes in depth. Stokes's velocity formula for waves at depth D and wave period T is

$$C = \frac{gT}{\sqrt{2\pi}} \tanh \frac{2\pi D}{TC}.$$

Solution for this requires an iterative process since C appears on both sides of the equation. Therefore, an initial approximation must be supplied. The curvature of the ray may be represented in Cartesian coordinates by

$$K = \left(\frac{1}{C} \sin \alpha \, \frac{2C}{2X} - \cos \alpha \, \frac{2C}{2Y}\right),$$

where α is the approach angle relative to the X axis. Given this curvature, a new position of the wave ray a short distance closer to shore can be calculated by iteration.

Several programs have been published that solve these equations (Dobson, 1967; Hardy, 1968; Goldsmith, 1976).

Time Series

Just as data can be interpolated and smoothed in the space domain, so it can be in the time domain. Time can be the independent variable, with wind velocity, wave height, current velocity, and amount of erosion or deposition being the dependent variables. Several techniques are available for separating the trend component from the oscillating and random variations in a time series. One of the most popular is the moving average (Vistelius, 1961). Successive averages are computed for a series by dropping the earliest data point and adding the next value in the sequence. The number of data points used at one time in computing the average ranges up to 21 terms.

If the data, whether temporal, spatial, or periodic, are cyclic in nature, such as wave height and period, or longshore current velocities, or ripple size and spacing, it can best be described by *Fourier analysis* (Gaskell, 1958; Wylie, 1960). An objective in Fourier series is to determine whether variations in some natural feature can be represented by a series of oscillatory functions. By this method, a complicated curve can be broken down into an aggregate of simple wave forms that can be described by the amplitude of a series of sine and cosine curves. Although the total function $Z = f(X)$ usually is not known, the coefficients of the cosine and sine terms may be determined (Harbaugh and Merriam, 1968) by numerical integration (at each point in time X_i and Z_i are known) using the following equation in which $n = 0, 1, 2, \ldots k/2$:

$$Z_i = F(X_i) = \frac{a_o}{2} + \sum_{n=1}^{N} a_n \cos\frac{n\pi X_i}{L} - b_n \sin\frac{n\pi X_i}{L}$$

where

$$a_n = \frac{2}{K}\left\{\frac{Z_o + Z_k}{2} + \sum_{i=1}^{K-1} Z_i \cos\frac{n\pi X_i}{L}\right\}$$

and

$$b_n = \frac{2}{K} \sum_{i=1}^{K-1} Z_i \sin\frac{n\pi X_i}{L}.$$

The amplitude (α_i) of the ith harmonic can be determined directly by taking the square root of the power or variance spectrum, which is

$$\tfrac{1}{2}\sqrt{a_n^2 + b_n^2}.$$

The coefficients are generally expressed as the power or variance given by

$$S_n^2 = a_n^2 + b_n^2.$$

Since each Fourier component is a discrete harmonic of the observed data, the Fourier components are theoretically independent. Then by least squares technique, the total variance explained by each harmonic can be calculated. Usually only a few of the basic harmonics account for most of the variation.

Fourier techniques are frequently applied in water and wave studies (Longuet-Higgins, 1952). Fourier analysis also forms the basis for edge wave studies (Bowen, 1969) and their influence on cusp development and rip currents (Komar, 1971). Brenninkmeyer (1975) used spectral analysis to show the diverse periodicity of sand movement within the surf zone.

Two different time series can also be compared. In cross-spectral analysis, the variance in the different frequency bands is correlated separately, so that a high coherence value appears only if both time series have a similar periodicity. By this method King and Mathew (1972) showed the importance of both semidiurnal and diurnal tidal cycles on the beach slope.

In a similar way, Trenhaile (1973) showed that water depth is the main hydrologic factor for both ripple height and steepness changes of low frequency, whereas bed slope is more important in determining higher frequency changes.

Statistical Inference

Most studies of the nearshore start out with a qualitative conceptual model that is meant to express this segment of the world in idealized form. It represents the essential features with extraneous details omitted. One such model is a *process-response model*. The scientist organizes his observations according to the relevance to the problem and builds up a framework within which the processes and their results are satisfactorily explained in their relationship to each other. If this is done correctly, the model can be used as a predictive tool. Krumbein (1964) has listed the components of such a model for a beach. Because there are so many interrelationships, it is difficult to separate the variables that are pure causes from those that are pure effect. Statistical analysis may serve to identify and delineate the more important variables.

Computers are frequently used to provide the inferential statistics such as regression, trend surfaces, harmonic analysis discriminant, factor and Markov analysis. These provide many methods of analysis that give greater precision to statements of relationships among the variables. Statistical models play an important part in making coastal studies more rigorous and quantitative, so that the results can be expressed within the framework of probability theory. By building upon these statistical and mathematical models, simulations of the environment are made possible.

Several methods are available for analyzing data that involve interrelationships among variables. *Regression analysis* is one. If the foreshore slope is assumed to be linearly dependent on wave height, then by fitting a least squares straight line to a scatter diagram of beach slope against wave height, the straight line has the general form of: Slope = $A + B$ (wave height). The degree of relationship between the two variables can be evaluated by the reduction of the sum of squares of the slope produced by the regression function. The sum of squares (SS) is the sum of the squares of the difference between individual measurements and the mean of all measurements:

$$SS_{slope} = \Sigma(\text{slope} - \text{mean of slope})^2.$$

If this value is small, nonlinear regression can be used. Or, since the independent variables may interrelate, sequential multiregression may be used to estimate the simultaneous influence of two or more independent variables on the foreshore slope (Krumbein, 1961); Slope = $A + B$ (wave height) + C (wave period) + D (angle of wave approach) + E (currents).

This has been done for more variables by Harrison (1969), while allowing a time lag for finding the greatest influence of the process variables to the response. The best predictor equation for beach slope is: Slope = 0.345 + 0.351 (hydraulic head/breaker distance run up) (9 hr.) − 0.374 (breaker height/gravity breaker period^2) (3 hr.) + 0.238 (hydraulic head/breaker distance run up) (0 hr.) + 0.032 (angle of wave approach) (9 hr.) − 0.11 (grain diameter/mean trough-to-bottom distance at breaker) (0 hr.).

Trend maps are a special application of multiple regressions. *Trend analysis* deals with the recognition, isolation, and measurement of trends that can be represented by lines, surfaces, or hypersurfaces. It seeks to separate broad-scale variation or trends from local variations. The simplest type of trend of a variable over a geographical region is a plane. Least squares methods are used to fit a plane to the data. Where irregularities are marked, further refinements lead to fitting surfaces of higher degree to the data until a sufficiently close fit is obtained. Such a surface can be expressed analytically, and points for contouring computed from the resulting equation (Fox and Davis, 1971a). Miller (1956) has found a good correlation between wave energy and sorting distribution.

Discriminant analysis is also related to multiple regression, and is frequently used for classification. A discriminant function categorizes a sample into two or more groups of samples that are known to be different. It can be used to determine which of the two or more groups an unknown sample belongs to. The analysis consists of finding a transform that gives the minimum ratio of the difference between a pair of group multivariate means to the multivariate variance within the groups. Or if the measurements on the two groups consist of two clusters of data points in multivariate space, there is only one orientation along which the two distributions have the greatest separation and at the same time the least inflation. The coordinates of the axis of orientation form the linear discriminant function (Krumbein and Graybill, 1965; Davis, 1973).

Sahu (1964) is one among many who utilized discriminant analysis to derive equations to distinguish several nearshore environments based on the moment measure of sediments. To separate beach from dune deposits, $Y = 3.5688 \, M + 3.701\sigma^2 + -2.0766 \, SK + 3.1135 \, K$. If Y is above −2.7411, it indicates a beach; below that, a dune environment. To distinguish beach from shallow marine deposits, $Y = 5.6534 \, M + 65.7091\sigma^2 + 18.107 \, SK + 18.5043 \, K$. If Y is above 65.3650, it is a beach; below that, a marine environment. The discriminant function can also be extended to hypersurfaces. Using this, Wright (1977) obtained separation and 78% correct classification between dune, intertidal, upper beach, lower beach, channel, and flood tidal delta deposits.

If the interrelationship of the measured variables is not known, *factor analysis* can be of great help. The purpose of factor analysis (Davis, 1973; Harbaugh and Merriam, 1968; Joreskog, Klovan, and Reyment, 1976) is to interpret the structure within the variance-covariance matrix of multivariate data collection. The technique extracts the eigen values and eigen vectors from the matrix of correlations of standardized covariances. Elements in a M × M matrix can be regarded as defining points lying on an m-dimensional ellipsoid. The eigen vectors of the matrix yield the mutually perpendicular axes of the ellipsoid, and the eigen values represent the lengths of those axes.

A graphical representation can be made for bivariate data. If the variance of x is plotted along the x axis and the covariance of x with y is plotted from the end of the variance line parallel to the y axis, and the same is done for the variable Y, then the variance covariance structure of the data can be represented by vectors. If these vectors lie on the boundary of an ellipse that just encloses the points, then the lengths of the major and minor axis of the ellipse (eigen values) can be calculated.

Factor analysis is concerned with finding these axes and measuring their magnitude, because these principal axes also represent the total variance of the data set. Further, each axis accounts for an amount of the total variance equal to the eigen values divided by the race of the matrix (sum of the eigen values). If the original series of data points are projected onto the principal axis, a new data set is constructed that maximizes the variance on this set of axes and minimizes them on all others (since the axes are mutually perpendicular the correlation between them is 0). Thus the purpose of factor analysis is to provide a method of sorting out a large number of variables into a smaller number of composite factors that best account for the variability of the original data.

This type of analysis has been used by Winant, Inman, and Nordstrom (1975) to show variation with time of beach profiles.

Dal Cin (1976) used grain-size analysis to predict whether coasts are accreting or eroding. Accretion of beaches is marked by fine and very fine sands. Eroding beaches are composed mainly of medium sands. Orford (1975) used this means to discriminate pebble zonation on a beach.

Markov Chains

Frequently in temporal or spatial data, the tendency is to change from one state to another. If the interest is in the nature of the transition from one state to another, then *Markov chains* can be used. Markov chains represent the intermediate point between wholly deterministic models in which past events completely control future events and independent event models in which the past has no influence whatever on future events.

A discrete-time, discrete-state, one-step memory Markov chain can be expressed as:

$$P = [t_{n+1} = j | t_n = i] P_{ij},$$

in which P_{ij} is the conditional probability of the process being in state j at time t_{n+1} given that it was in state i at time t_n. If there are three states or processes, a transition matrix can be constructed:

For state i		to state j	
	A	B	C
A	P_{AA}	P_{AB}	P_{AC}
B	P_{BA}	P_{BB}	P_{BC}
C	P_{CA}	P_{CB}	P_{CC}

where P_{AA} is the probability that the process will remain in that state; P_{AB}, that the system will move from state A to state B at the chosen time or space interval; P_{AC}, that it will change to state C. If these are the only three possibilities, then the probabilities must add up to 1.

Transition probability matrixes can be generated from any succession of events, but this does not guarantee that the process is Markovian. The most widely used test is that of Anderson and Goodman (1957), which tests the hypothesis of an independent events model against the alternative—that is, the system possesses a "memory" or is Markovian.

A Markovian model has been used by Sonu and James (1973) to predict changes in beach profiles through six stages and by Krumbein (1964) to study transgressive-regressive sequences.

Simulation

In order to get a full grasp of the interrelationship between the many variables affecting a process, mathematical models are appropriate. *Simulation* employs computers to perform calculations and logic operations to simulate geologic processes. The usefulness of simulation lies in the ability to explore the interactions between different variables, many of which may not be known. But if a series of assumptions are made to form a multidimensional continuum within which there is an infinity of possible combinations, some combinations are more plausible than others. The assumption incorporated into a model may be in error, and the results of the model will be influenced by these errors. The model cannot be used to verify the assumptions. The purpose of the model is to help formulate hypotheses that in turn may be rigorously tested.

One of the most complete programs for nearshore sedimentation simulation is that by Harbaugh, Carter, and Merrill (1971). It is a dynamic deterministic model that takes into account: quantity of material supplied, from one to five sediment size fractions; initial geometry of the basin; tectonic warping through time and space; equilibrium surface or base level for each sediment size.

Fox and Davis (1971b) have developed an empirical mathematical model for the response of beach and nearshore topography to the effects of wind, waves, and longshore currents.

King and McCullagh (1971) simulated, with remarkable success, the complex recurved spit at Hurst Castle in England, based on changes in wave state and direction in addition to a depth and refraction factor. Theirs was a probabilistic model. Storm waves from the southwest caused the spit, while northeast waves formed the recurves. By proportioning a series of random numbers allocated to different wave directions, they adjusted the simulated spit pattern to fit the actual morphology. Thereby it is possible to come to some conclusion about which wave directions are responsible for the growth of the actual spit.

BENNO M. BRENNINKMEYER

References

Anderson, T. W., and Goodman, L. A., 1957. Statistical inference about Markov chains, *Annals Math. Statistics* 28, 89-110.

Bowen, A. J., 1969. Rip currents, *Jour. Geophys. Research* 74, 5467-5478.

Brenninkmeyer, B. M., 1975. Mode and period of sand transport in the surf zone, *Conf. Coastal Eng., 14th, Am. Soc. Civil Engineers, Proc.* 2, 813-827.

Dal Cin, R., 1976. The use of factor analysis in determinining beach erosion and accretion from grain-size data, *Marine Geology* 20, 95-116.

Davis, J. C., 1973. *Statistics and Data Analysis in Geology.* New York: John Wiley & Sons, 550p.

Dobson, R. S., 1967. Some applications of a digital computer to hydraulic engineering problems, *Stanford Univ. Dept. Civil Engineering Tech. Rept.* 80, 7-35.

Fox, W. T., and Davis, R. A., Jr., 1971a. *Computer Simulation Model of Coastal Processes in Eastern Lake Michigan.* Williamstown, Mass.: Williams College, O.N.R. Tech. Rept. 5, 114p.

Fox, W. T., and Davis, R. A., Jr., 1971b. *Fourier Analysis of Weather and Wave Data from Holland, Michigan, July, 1970.* Williamstown, Mass.: Williams College, O.N.R. Tech. Rept. 3, 79p.

Gaskell, R. E., 1958. *Engineering Mathematics.* New York: Henry Holt, 195p.

Goldsmith, V., 1976. Wave climate models for the continental shelf: Critical links between shelf hydraulics and shoreline processes, *in* R. A. Davis and R. L. Ethington, eds., *Beach and Nearshore Sedimentation.* Tulsa, Okla.: Soc. Econ. Paleont. Mineral. Spec. Pub. 24, 24-47.

Harbaugh, J. W., and Merriam, D. F., 1968. *Computer Applications in Stratigraphic Analysis.* New York: John Wiley & Sons, 282p.

Harbaugh, J. W.; Carter, G. B.; and Merrill, W. M., 1971. *Programs for Computer Simulations in Geology.* Stanford, Calif.: Stanford Univ. Dept. Geology Tech. Rept. 389-154, 389p.

Hardy, J. R., 1968. Computation of wave refraction diagrams and wave energy balance along a coast line, *British Geomorph. Res. Group Occasional Paper* 6, 73-85.

Joreskog, K. G.; Klovan, J. E.; and Reyment, R. A., 1976. *Geological Factor Analysis.* Amsterdam: Elsevier, 178p.

King, C. A. M., and McCullagh, M. J., 1971. A simulation model of a complex recurved spit, *Jour. Geology* 79, 22-36.

King, C. A. M., and Mathew, P. M., 1972. Spectral analysis applied to the study of time series from the beach environment, *Marine Geology* 13, 123-142.

Komar, P. D., 1971. Nearshore cell circulation and distribution of giant cusps, *Geol. Soc. America Bull.* 82, 2643-2650.

Krumbein, W. C., 1961. *The Analysis of Observational Data from Natural Beaches.* Washington, D.C.: U.S. Army Corps of Engineers, Beach Erosion Board, Tech. Memo. 130, 59p.

Krumbein, W. C., 1964. Geological process response model for analysis of beach phenomena, *U.S. Army Corps of Engineers, Beach Erosion Board, Annual Bull.* 17, 1-15.

Krumbein, W. C., and Graybill, F. A., 1965. *An Introduction to Statistical Models in Geology.* New York: McGraw-Hill, 475p.

Longuet-Higgins, M. S., 1952. On the statistical distribution of the height of sea waves, *Jour. Marine Research* 11, 245-266.

Miller, R. L., 1956. Trend surfaces: Their application to analysis and description of environments of sedimentation, I. The relation of sediment size parameters to current-wave systems and physiography, *Jour. Geology* 64, 425-466.

Munk, W. H., and Traylor, M. A., 1947. Refraction of ocean waves: A process linking underwater topography to beach erosion, *Jour. Geology* 55, 1-26.

Orford, J. D., 1975. Discrimination of particle zonation on a pebble beach, *Sedimentology* 22, 441-463.

Sahu, B. K., 1964. Depositional mechanisms from the size analysis of clastic sediments, *Jour. Sed. Petrology* 34, 73-83.

Sonu, C. J., and James, W. R., 1973. A Markov model for beach profile changes, *Jour. Geophys. Research* 78, 1462-1471.

Trenhaile, A. S., 1973. Near-shore ripples: Some hydraulic relationships, *Jour. Sed. Petrology* 43, 558-568.

Vistelius, A. B., 1961. Sedimentation time trend functions and their application for correlation of sedimentary deposits, *Jour. Geology* 69, 703-728.

Winant, C. D.; Inman, D. L.; and Nordstrom, C. E., 1975. Description of seasonal beach changes using empirical eigen functions, *Jour. Geophys. Research* 80, 1979-1986.

Wright, A., 1977. *Statistical Verification of Geomorphic Changes for a Breached Spit.* Boston College, unpublished thesis, 147p.

Wylie, C. R., 1960. *Advanced Engineering Mathematics.* New York: McGraw-Hill, 640p.

Cross-references: *Computer Simulation; Fourier Analysis; Mathematical Models; Trend Surface Analysis; Wave Refraction Diagrams.*

COMPUTER SIMULATION

The essential idea in simulation is to use a computer system as the raw material for constructing a process model to be studied. The computer program thus embodies a theory that can be compared with the behavior of what is modeled by these comparisons and subjected to conditions that may be impossible in the real world. Recent models involve the mathematicization of a physical system regarding input, analysis, and conclusions. As such, these are deterministic because of the reliance on mathematical statements of wave refraction, sediment transport, tidal currents, and so on (Komar, 1976). Other simulations have been of a stochastic nature, recognizing that exactly predictable relationships of cause and effect are not now possible (King and McCullagh, 1971). The latter models are generally of the Markov type, because any particular state is partially dependent on the previous state.

Computer simulations are being used at an increasing rate to study coastal interactions. If the simulation model replicates the real situation, one may assume that the processes and responses identified in the model are also important in reality. By isolating the action of variables and changing the frequencies of interaction, the relative contributions of time and process can be evaluated. One of the chief advantages of computer simulation over other types of modeling is the speed with which the results can be obtained and variable combinations tested.

Many different problems have been studied using this method. Goldsmith et al. (1974) present the history and applications of wave refraction studies of wave energies to wave forecasting, shoreline evolution, and environmental quality. Through shoreline compartmentalization and assumptions of continuity in sediment transport relative to energy flux, Willis and Price (1975) have been able to simulate the amount of beach erosion/deposition as a function of the wave climate. Others are trying to correlate complex coastal processes phenomena with weather data (see Fox and Davis, 1973). Another example is in the study of the hydrodynamics of semienclosed bodies. Current structures within inlets and estuaries have implications to storm surges, salinity and temperature distributions, water quality, and biologic dynamics.

A rapid evolution has occurred from the generalized first-generation studies of planimetric form and sedimentary stratigraphy to more precise second-generation programs that correlate real data inputs with relatively accurate responses in a two-dimensional framework. Three-dimensional modeling lies in the near future, and this will allow even more sophisticated replication, analysis, and understanding of coastal dynamics.

JAMES R. ALLEN

References

Fox, W. T., and Davis, R. A., Jr., 1973. Simulation model for storm cycles and beach erosion on Lake Michigan, *Geol. Soc. America Bull.* **84,** 1769-1790.

Goldsmith, V.; Morris, W. D.; Byrne, R. J.; and Whitlock, C. H., 1974. *Wave Climate Model of the Mid-Atlantic Shelf and Shoreline (Virginian Sea).* Washington, D.C.: Natl. Aeronaut. Space Admin. Spec. Paper 358, 146p.

King, C. A. M., and McCullagh, M. J., 1971. A simulation model of a complex recurved spit, *Jour. Geology* **79,** 22-37.

Komar, P. D., 1976. *Beach Processes and Sedimentation.* Englewood Cliffs, N.J.: Prentice-Hall, 429p.

Willis, D. H., and Price, W. A., 1975. Trends in the application of research to solve coastal engineering problems, *in* J. Hails and A. Carr, eds., *Nearshore Sediment Dynamics and Sedimentation.* New York: John Wiley & Sons, 111-121.

Cross-references: *Coastal Engineering, History of; Coastal Engineering, Research Methods; Coastal Morphology, Research Methods; Computer Applications; Mathematical Models; Scale Models.*

CONIFERS—See PINICAE

CONTINENTAL FLEXURE—See TECTONIC MOVEMENTS

CONTRAPOSED SHORELINE

In 1913 Clapp introduced the term *contraposed shoreline* (contraposed meaning *placed against*) to denote a shoreline that has been cut through soft mantle until it reaches underlying hard rock (Bates and Jackson, 1980). Thus the shoreline changes character rapidly as it evolves from an initial nearly straight and cliffed form to an irregular one following the shape of the hard rock surface. As an example, Clapp cited the coast in the vicinity of Victoria, British Columbia, where the shoreline has been developed in a "drift-covered crystalline rock."

In a footnote to the article (Clapp, 1913) the author acknowledged Professors W. M. Davis and D. W. Johnson, who suggested that he devise a name for this type of shoreline.

MAURICE L. SCHWARTZ

References

Bates, R. L., and Jackson, J. A., 1980. *Glossary of Geology.* Falls Church: American Geological Institute, 749p.

Clapp, C. H., 1913. Contraposed shorelines, *Jour. Geology* **2,** 537-540.

Cross-references: *Classification; Coastal Morphology, History of; Geomorphic-Cycle Theory.*

CORAL—See COELENTERATA

CORAL REEF COASTS

Coral reefs grow in the warm, shallow waters along many tropical coastlines. Such coasts are classified as *Zoogenic* in Valentin's (1952) scheme, and a close relationship exists between the nature of the shoreline and the reef types developed adjacent to it (Bird, 1976).

Fringing reefs occur where hard, stable substrates immediately front the shore—for example, around islands or on rocky headlands. In general these reefs consist of a platform at low tide level similar to the low tide platforms found on cliffed coasts. However, unlike the latter, which are erosional features, fringing reefs prograde, continuously increasing their width by coral growth at the seaward margin.

Barrier reefs grow in areas where suitable foundations exist on the continental shelf some distance from the coast. In general the protected shore behind barrier reefs is prograding and characterized by mangrove swamps. Such low-lying muddy coast lines are not suited to reef coral growth; where inshore and fringing reefs do manage to become established, they are usually depauperate and swamped by terrigenous sediments. Fringing reefs around islands in the lee of barrier reefs are often luxuriant, although the effects of freshwater runoff, for example, may have devastating consequences (Hedley, 1925).

Many types of shoreline, ranging from cliffed, through rocky and sandy deposit, to low and muddy with associated swamps, may occur in the lee of a barrier reef. While both the barrier and fringing reefs in a region afford protection from marine erosion if the fetch in the lagoon behind the former is relatively long, as, for example, in the Great Barrier Reef Province (Maxwell, 1968), considerable coastal erosion may still occur during storms. This is more pronounced on exposed sandy shores; muddy mangrove environments are generally well protected in embayments and stabilized by the vegetation.

EDGAR FRANKEL

References

Bird, E. C. F., 1976. *Coasts.* Canberra: Australian National University Press, 282p.
Hedley, C., 1925. The natural destruction of a coral reef, *Great Barrier Reef Commission Rept.* **1**, 35-40.
Maxwell, W. G. H., 1968. *Atlas of the Great Barrier Reef.* Amsterdam: Elsevier, 258p.
Valentin, H., 1952. *Die Küsten der Erde,* Petermanns Geog. Mitt. Erg. 246, Gotha, Justus Perthes, 118p.

Cross-references: *Algal Mats and Stromatolites; Australia, Coastal Ecology; Biotic Zonation; Coastal Fauna; Coastal Flora; Coastal Waters Habitat; Coral Reef Habitat; Mangrove Coasts; Organism-Sediment Relationship.* Vol. III: *Coral Reefs;* Vol. VI: *Coral Reef Sedimentology.*

CORAL REEF HABITAT

Habitat is defined as *the natural home of a plant or animal.* Within a coral reef ecosystem a large number and variety of habitats are available, and an ecological hierarchy defining these *places to live* can be established.

Each habitat has its own characteristic species and substrates, which change with changing environmental factors such as light, diurnal temperature variations, oxygen content of the water, sedimentation rate, wave energy, and a number of others, including the biota and their effects.

The difficulties of describing such a complex situation are manifest, and in coral reefs the word habitat is commonly used in the restricted sense of topography and substrate.

Three basic *environmental systems* affect coral reefs: submarine, intertidal, and supratidal. In each of these, *major environments* are defined. Further division leads to *general habitats.* Specific habitats within the general scheme may then be identified. This scheme is illustrated in Table 1.

Reef Substrates

The calcium carbonate substrates of coral reefs are of two distinct types: reef framework, and unconsolidated sediment.

The *reef framework* is composed principally of the skeletons of hermatypic corals cemented together by encrusting coralline algae into a rigid, consolidated, albeit porous mass. Skeletal material from numerous other organisms supplement the framework, but their importance is generally minor.

Numerous organisms burrow and bore into the reef rock, and secondary voids created by these are coated by epibenthos and filled with lime mud. This process, in addition to inorganic cements, creates a massive rock, markedly different from the original coral-coralline algal mass—called *reef rock.*

The *unconsolidated sediment* substrates of coral reefs generally fall within the sand and gravel classes. Silt and clay-sized material usually constitutes less than about 5% of the sediment, and collects only in the more protected environments such as lagoons.

The sediments are composed principally of coral, coralline algae, *Halimeda,* and foramini-

CORAL REEF HABITAT

TABLE 1. Coral Reef Habitats

Environmental System	Major Environment	General Habitat
Submarine	Interreef Channel (off reef floor)	Unconsolidated substrate (sand/mud) Isolated coral pinnacles
	Fore Reef	Live coral Reef rock (dead coral) Unconsolidated sediment in pockets
	Lagoon	Lagoonal ("Patch") reef Unconsolidated substrate (poorly sorted) "Grass" beds
Intertidal	Reef Crest	Live coral Reef rock (dead coral, boulders) Pools
	Reef Flat (outer and inner)	Live coral Reef rock Pools Unconsolidated sediment in pockets
	Beach	Unconsolidated sediment Beach rock Reef rock of "raised reefs"
	Mangrove Swamp	Unconsolidated sediment Trees
Supratidal	Cay	Sand Shingle "Broken" beach rock Terrestrial vegetation
	Raised Reef	Reef rock

feral and molluscan debris. Material from other organisms generally forms a minor component, although it may occasionally become significant locally, while the overall composition of the sediment varies depending upon the depositional environment.

Reef Environments

In an ideal coral reef the following major environments are recognized (Fig. 1, Table 1): interreef channel, forereef, reef crest, reef flat, lagoon, cay (and raised reef), beach, mangrove swamp. Some of the more important characteristics of topography, substrata, inhabitants and the relationships between them of the major environments, are as follows.

Interreef Channel (Off Reef Floor). The substrate is ripple-marked, poorly sorted unconsolidated sediment and occasional coral pinnacles that sometimes reach considerable size. The unconsolidated materials are generally biogenic, except in areas adjacent to continents or large islands, where a considerable proportion of terrigenous material may be present.

The fauna that inhabits the unconsolidated substrates is variable, but echinoderms are generally dominant. Sponges and soft corals, particularly gorgonians, are often found in considerable numbers, and a large infauna of crustaceans, molluscs, and worms is always present. The coral pinnacles usually support luxuriant coral growth, although siltation becomes a problem in areas where terrigenous sediments occur. Large, diverse populations of both demersal and pelagic fish inhabit the waters in the interreef areas.

Forereef (Reef Front). This extends from the windward edge of the reef flat to the lower limit of coral growth (approximately 40 m). An important feature is the spur-and-groove system that consists of closely spaced, steep-sided highs and lows that extend perpendicularly from the reef flat. This system apparently aids in dissipating wave energy.

Corals grow luxuriantly in this environment, a great variety of fish inhabit crevices and caves, and a host of other organisms and plants live on and in both the exposed reef rock and the unconsolidated sediments that collect in shallow pockets.

Reef Crest. Exposed at low water and exposed to wave action at most times, the reef

CORAL REEF HABITAT

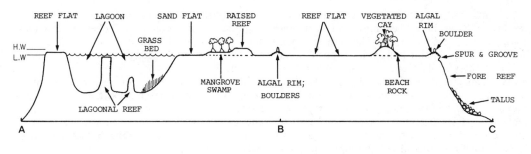

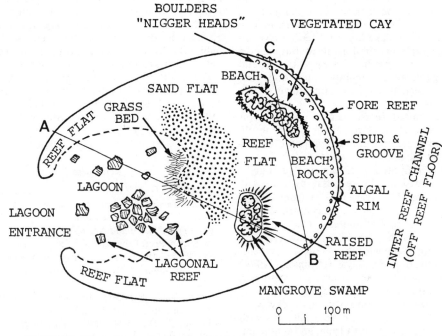

FIGURE 1. Plan and cross section of an ideal coral reef, showing the major environments.

crest appears bare, but the area supports a considerable algal population (mainly calcareous), which binds the reef framework together. Many molluscs inhabit this environment, and crustaceans and echinoderms shelter under large boulders and loose rubble. Small colonies of coral are sometimes common.

Reef Flat. On the *outer flat,* deep pools and flourishing corals are predominant. Further from the reef edge, on the *inner flat,* pools are shallower and the area is characterized by small coral colonies surrounded by coarse sediments. Holothurians are the most conspicuous inhabitants, although a considerable infauna, particularly molluscs and crustaceans and occasionally fish, inhabits the unconsolidated substrate. The large foraminifera *Marginopora vertebralis, Calcarina hispida, Baculogypsina sphaerulata,* and several others are notable inhabitants of the reef flat environment in the Indo-Pacific region.

Lagoon. Lagoons are shallow environments bounded on at least one side by a topographic high. In coral reefs, lagoon slopes connect the peripheral reef flats to the lagoon proper. These slopes may be either gentle or steep. Patch reefs occasionally grow on the slopes of deeper lagoons, and *Halimeda* thickets are a notable feature of many slopes.

The lagoon floor is the part of the lagoon not directly covered by patch (lagoonal) reefs. The substrate is variable depending on local conditions and may contain a considerable quantity of mud-sized material. The environment generally contains large populations of holothurians, echinoids, browsing gastropods, pelecypods, benthonic foraminifera, and often supports luxuriant green algae.

Currents and/or grass and algal mats play important roles in the ecology and sedimentology of lagoonal sediments. The former, when slow, allow fine-grained material to settle, and infaunal filter feeders are common. When currents are of greater velocity, sediments tend to be coarser and the organisms are often epifaunal. Grass bands and algal mats trap fine material and stabilize the bottom, causing the areas covered by them to rise above the general level. Numerous benthic and epiphytic organisms and plants inhabit these areas.

Lagoonal reefs are coral reefs that rise above the lagoon floor. These may be small and low-lying (*knolls*) or be represented by large reefs, both dispersed and reticulate, that extend to the water surface. The lagoonal reefs are generally depauperate in coralline algae, but support luxuriant coral growth. These are the habitat for a large variety of mainly epibenthic organisms and usually support large, diverse fish populations.

Cay (and Raised Reef). Islands on coral reefs are of two types: cays consisting of unconsolidated sediments, and islands composed of pre-Recent reef rock.

Cays are composed of either sand or shingle, and are formed on the reef flat by wave and current action. They vary in size from tens to hundreds of meters, and in height from a few centimeters to several meters. The long axis usually parallels the reef front, and lengths are about 3-9 times the widths.

A classification of cays (Milliman, 1974) recognizes several types: submerged sand bar, sand or shingle cays, low cay with pioneer strand-line plant community, high cay with well-developed vegetation cover, and cays with mangrove swamps. A sixth type, an island composed of pre-Recent emerged reef rock, is also recognized.

Sand cay sediments are generally well sorted, and contain little or no mud-sized material. Vegetation and beach rock slow, but do not prevent, gradual migration across the reef flat and possible eventual disappearance into the lagoon. A notable feature of many sand cays is the presence of a *Ghyben-Herzberg* freshwater lens that helps support the terrestrial vegetation.

A terrestrial fauna, predominantly birds that nest both in trees and other vegetation or in burrows in the unconsolidated substrate, inhabits the sand cays. Other organisms, the majority of which are introduced, include rodents, lizards, and pulmonate molluscs. In addition, turtles lay their eggs in the sediments of many cays.

Beach. The composition of beach sediments is a function of sediment size and source, and beaches on cays range in size from sand to gravel (shingle). Fauna are dependent on the grain size: finer-grained beaches have a considerable fauna of molluscs, crustaceans, and echinoderms, while shingle beaches are almost barren.

The intertidal zone of beaches is often cemented to form beach rock. Large algal populations grow on the beach rock both in pools and on exposed surfaces. Numerous fish and molluscs, particularly chitons, browse and graze over these areas.

Mangrove Swamp. Mangrove swamps are developed in the intertidal of some cays. The mangrove species vary from place to place. In general these environments are characterized by finer, organic-rich substrates. Large mollusc and crustacean populations are the predominant inhabitants, living both on and in the unconsolidated substrate and on the vegetation.

For additional literature on this topic, the reader is referred to Bathurst (1971); Emery, Tracey, and Ladd (1954); Folk (1967); Maxwell (1968); Scoffin (1970); Stoddart (1969); and Wells (1957).

EDGAR FRANKEL

References

Bathurst, R. G. C., 1971. *Carbonate Sediments and Their Diagenesis.* Amsterdam: Elsevier, 620p.

Emery, K. O.; Tracey, J. I.; and Ladd, H. S., 1954. *Geology of Bikini and Nearby Atolls.* U.S. Geological Survey Prof. Paper 260-A, 265p.

Folk, R. L., 1967. The sand cays of Alacran Reef, Yucatán, Mexico: Morphology, *Jour. Geology* **75**, 412-437.

Maxwell, W. G. H., 1968. *Atlas of the Great Barrier Reef.* Amsterdam: Elsevier, 258p.

Milliman, J. D., 1974. *Marine Carbonates.* Berlin: Springer-Verlag, 375p.

Scoffin, T. P., 1970. The trapping and binding of subtidal carbonate sediments by marine vegetation in Bimini lagoon, Bahamas, *Jour. Sed. Petrology* **40**, 249-273.

Stoddart, D. R., 1969. Ecology and morphology of recent coral reefs, *Biol. Rev.* **44**, 433-498.

Wells, J. W., 1957. Coral reefs, *in* J. Hedgpeth, ed., *Treatise on Marine Ecology,* Geological Society of America Mem. 67, 609-631.

Cross-references: *Algal Mats and Stromatolites; Biotic Zonation; Coastal Fauna; Coastal Flora; Coastal Waters Habitat; Coral Reef Coasts; Ghyben-Herzberg Ratio; Mangrove Coasts; Organism-Sediment Relationship.* Vol. III: *Coral Reefs;* Vol. VI: *Coral Reef Sedimentology.*

CORERS AND CORING TECHNIQUES

Corer is a general name for any cylindrical or box-shaped device that takes soil samples, whether underwater or surface, from which pertinent geologic information is derived. A cylindrical corer usually consists of a barrel-shaped pipe, a removable plastic liner, a cutting edge on the lower end, a check valve to allow air to escape, and a retainer to hold the sample once taken. A box corer is used to take an undisturbed sample of the ocean bottom from a few centimeters up to a meter thick. It operates by lowering the open-floor box until it is embedded into the ocean bottom. Once completely embedded, a spring or mechanically triggered scoop is released that closes the bottom of the box.

Coring techniques involve refinements and adaptations to the basic cylindrical corer. Among them are such corers as gravity corers, piston corers, vibracorers, and Shelby corers (Tirey, 1972; U.S. Army Corps of Engineers, 1972).

Gravity Corer

The gravity corer is identical to the cylindrical corer described above. It is usually allowed to free fall to the ocean floor where gravity forces the corer into the bottom. Stabilizing fins and a leaded pipe end allow for vertical stability and deeper penetration, respectively.

Piston Corer

Friction between soil and liner is greatly reduced in a cylindrical corer by the presence of a piston. The piston is situated inside the liner so that as the corer strikes the sea floor the piston remains on the surface while the pipe penetrates into the bottom. The piston provides a vacuum, so that it is directly connected to the hoisting cable on return. Penetration by a piston corer is greater than by the gravity corer, but the sediment will be more disturbed in the former.

Vibracorer

Vibracorers (Fig. 1) are bottom-rest systems that either pneumatically or hydraulically vibrate a 3-12 m pipe with liner into the ocean floor. Umbilical cords deliver the air or hydraulic fluid to the vibrator. The entire coring unit is lowered and hoisted to the ship via cable through an "A" frame or davit. Vibracorers are extensively used in coastal work in water 60 m deep or less.

Shelby Corer

The Shelby or *Split-Spoon* corer is used on inland waters, lakes and beaches. It hydraulically presses into the soil a thin walled tube which can be separated for sample retrieval upon return.

FIGURE 1. Vibracorer being lowered from a barge (photo courtesy of Alpine Geophysical Associates, Inc.).

Other Coring Techniques

Other techniques range from simple, hammer-driven corers to elaborate, diver-operated vibratory corers to rocket-fuel samplers, such as the one developed by the Norwegian Geotechnical Institute.

JOHN A. RIPP

References

Tirey, G. B., 1972. Recent trends in underwater soil sampling methods, *in Underwater Soil Sampling, Testing, and Construction Control*. Philadelphia: Am. Soc. Testing and Materials, STP 501, 42-54.

U.S. Army Corps of Engineers, 1972. *State-of-the-Art of Marine Soil Mechanics and Foundation Engineering*. U.S. Army Corps Engineers, Waterways Experiment Station Tech. Rept. S-72-11, 153 p.

Cross-references: *Coastal Engineering; Coastal Morphology, Research Methods; Coastal Sedimentary Facies; Grab Samplers.* Vol. I: *Sediment Coring.*

CORIOLIS EFFECT

The *Coriolis effect* is applied to every fluid particle in a stationary system of coordinates on the earth, revolving around the earth's axis

(Wiegel, 1964). The total acceleration of this element moving with velocity $\vec{v}$ depends mostly on the double product of $\vec{v}$ and the earth's rotational speed $\vec{\omega}$, called the *Coriolis acceleration*. If the positive direction of the z axis is oriented along the earth's radius outward, the y axis points northward, and the x axis eastward, then the components of the Coriolis acceleration are:

$$a_{cx} = 2(v \cdot \omega \cdot \sin\phi - w \cdot \omega \cdot \cos\phi),$$
$$a_{cy} = 2(u \cdot \omega \cdot \cos\phi),$$

and

$$a_{cz} = 2(u \cdot \omega \cdot \cos\phi)$$

where u, v, and w are x, y, and z components of the velocity $\vec{v}$, and ϕ is the latitude of a given point.

The Coriolis effect is important only for these motions where the horizontal components of $\vec{v}$ are much greater than w, which yields

$$a_{cx} = 2(v \cdot \omega \cdot \sin\phi) = v\Omega$$

and

$$a_{cy} = -u\Omega.$$

The Coriolis effect is sometimes negligible in the wide spectrum of oceanic motions, but the respective terms should be examined for significance in specific applications of the equations of motion. Except near boundaries of the ocean, where sharp horizontal and vertical gradients of velocity can be developed, the opposition of the pressure gradient and Coriolis effect is the major dynamical restriction to be satisfied by the steady flow. In regions removed from solid boundaries and the surface of the ocean, water masses move with velocity $\vec{v}$ and turn permanently to encircle an area of the radius of inertia $r_c = |v| \cdot 2\omega \cdot \sin\phi$ during one-half of pendulum day, $T_c = \pi/\omega \sin\phi$. Since Ω changes in these large regions, in theoretical models of oceanic circulation it is necessary to introduce the so-called β-parameter and β-plane. By this means the changing Coriolis parameter Ω can be found responsible for the western intensification of currents in the Northern Hemisphere.

RYSZARD ZEIDLER

Reference

Wiegel, R. L., 1964. *Oceanographical Engineering*. Englewood Cliffs, N.J.: Prentice-Hall, 532p.

Cross-reference: *Amphidromic Systems; Currents; Swell and Its Propagation; Tides; Waves; Wind.* Vol. I: *Coriolis Force*.

CORRASION—See WEATHERING AND EROSION, MECHANICAL

CORROSION—See WEATHERING AND EROSION, MECHANICAL

CRYPTOPHYTA

The Cryptophyta is a division of algae comprised principally of unicellular photosynthetic flagellates. About 23 genera and 150 species are divided among 7 families of one order (Butcher, 1967).

They are free-swimming in both freshwater and marine environments, principally in more eutrophic habitats. Little is known of marine cryptomonad ecology in nature. They are frequently present in shallow coastal waters, estuaries, and salt marshes with a salinity of 5–35‰. Their role in primary productivity of coastal waters has not been assessed, but is probably minor relative to other unicellular algae such as *diatoms* and *dinoflagellates*. The photosynthetic forms are easily cultured, and often "bloom" in mixed cultures after diatoms decline in abundance.

The bean-shaped, laterally compressed cell has an outer periplast of helically arranged plates closely bound to the outer cell membrane. The anterior end of the cell is obliquely truncate, and contains a canal or furrow in which two mastigoneme-bearing, equal or subequal, heterodynamic flagella are inserted (Fig. 1). The photosynthetic forms generally have a single H-shaped chloroplast with a central pyrenoid partially enclosed by starch granules ($\alpha,1,4$ glucan). The principal photosynthetic pigments are chlorophyll a and c and the phycobiliproteins phycoerythrin and phycocyanin, with minor pigments including α-carotene, alloxanthin, and other carotenes and xanthophylls. The cells are red, blue-green, yellow, or brown, depending on the species and the physiological state. The nucleus is usually located in the cell center and nearby is a nucleuslike structure, the nucleomorph, unique to the cryptomonads (Gantt, 1980).

The biliproteins are localized in the photosynthetic thylakoids. The chromophore and the protein moieties of the biliproteins are chemically similar to those of the *blue-green* and *red algae*. Most forms have a conspicuous red eye-

pounds). *Cyanophora* and 4 other genera contain what are possibly endosymbiotic blue-green algae (cyanelles) instead of chloroplasts. The marine ciliate *Mesodinium rubrum* contains a photosynthetic endosymbiont believed to be a cryptomonad, and occasionally causes extensive *red tides*. The free living, photosynthetic, chloroplast-containing, marine cryptomonads require vitamin B_{12} and thiamine; one requires biotin in addition. In laboratory culture some photosynthetic species can utilize glycerol and other organic compounds, including urea, in the dark as a carbon source, but the roles of these compounds in the natural habitats is not understood.

JOHN A. WEST

References

Butcher, R. W., 1967. *An Introductory Account of the Smaller Algae of British Coastal Waters. Part IV. Cryptophyceae.* London: Fish. Investigations Ser. IV. Ministry Agric. Fish and Food, 54p.

Dodge, J. D., 1969. The ultrastructure of *Chroomonas mesostigmatica* Butcher (Cryptophyceae), *Arch. Mikrobiol.* **69**, 266-280.

Gantt, E., 1980. Photosynthetic cryptophytes, *in* E. Cox, ed., *Phytoflagellates.* New York: Elsevier, 381-405.

Cross-references: Algal Mats and Stromatolites; Chlorophyta; Chrysophyta; Coastal Flora; Coral Reef Habitat; Cyanophyceae; Euglenophyta; Organism-Sediment Relationship; Phaeophycophyta; Reefs, Noncoral; Red Tides; Rhodophycophyta. Vol. III: Algae. Vol. VI: Algal Reef Sedimentology.

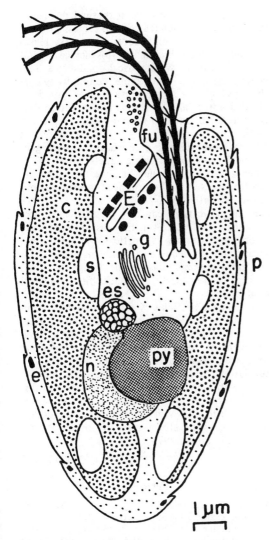

FIGURE 1. A drawing to illustrate a typical cryptomonad (*Chroomonas*) and the arrangement of major organelles. Key: *c*, chloroplast; *E*, large ejectosome; *e*, small ejectosome; *es*, eyespot; *fu*, flagellar furrow; *g*, Golgi body; *h*, flagellar mastigonemes; *n*, nucleus; *p*, periplast; *py*, pyrenoid; *s*, starch granules (after Dodge, 1969).

spot in the chloroplast near the end of the cell. Unique ejectile organelles, called ejectosomes or trichocysts, are located in the canal and beneath the cell surface. They are discharged when the cell is irritated, and have no known function.

One colorless freshwater genus (*Cyathomonas*) is phagotrophic, and ingests particles through a special cytostome located near the canal. Another freshwater genus (*Chilomonas*) is osmotrophic (absorbing soluble organic com-

CTENOPHORA

Although members of this phylum have worldwide distribution, they are seldom recognized because of their strong superficial resemblance to jellyfish. Known as *comb jellies* or *sea gooseberries,* ctenophores are usually characterized by a transparent globular body having a diameter of less than 1 cm. A few specialized genera have ribbon or wormlike shapes. Most ctenophores are active swimmers, locomotion being achieved by eight rows of cilia called *ctenes.* Two classes are recognized: Tentaculata, whose members possess a pair of long, feathery tentacles used to collect plankton during feeding; and Nuda, lacking tentacles. Unlike jellyfish, ctenophores almost always lack stinging cells. Strictly marine, they are generally pelagic in habitat, although a few modified forms are creepers that live on the ocean floor. Many species are strongly bioluminescent. The Ctenophora are important predators since they feed upon the planktonic larva of molluscs and other marine organisms, sometimes congregating in great numbers. Their relationship with other phyla is uncertain, although their structure bears some similarity to both the Coelenterata and Turbellaria (Hyman, 1940; Konai, 1963).

GEORGE MUSTOE

References

Hyman, L. H., 1940. *The Invertebrates, Vol. I.: Protozoa through Ctenophora.* New York: McGraw-Hill, 726p.

Konai, T., 1963. A note on the phylogeny of the Ctenophora, *in* E. C. Dougherty, ed., *The Lower Metazoa.* Berkeley and Los Angeles: University of California Press, 181-188.

Cross-references: *Coastal Fauna; Coelenterata; Organism-Sediment Relationship.*

CURRENT METERS

According to the two essential ways of describing an oceanic current (*Lagrangian* and *Eulerian* criteria), there are two kinds of current meters.

Lagrangian criterion regards the motion of a single water particle (or of a small body of water) in its spatial and temporal evolution. Eulerian criterion consists of defining the motion in a fixed point (the description of velocity field) and the subsequent time evolution.

Lagrangian current meters are floating or submersible bodies having different buoyancies. The motion of such a current meter can be observed from ships by radar (if the current meter is floating) or by ultrasonic location (if the current meter is submerged). In an extreme case, the floating body is reduced to a spot of labeled matter (a color, for example) that is located by means of aerial photography. The location of Lagrangian current meters is made continuously or from time to time. A well-known current meter of this kind is the *parachute*. Lagrangian current meters are used in pollution studies, in sediment transport studies, or for understanding ice drift.

Eulerian current meters are fixed instruments (at surface, at various depths, at the sea bottom) that gather or broadcast data (speed and direction of the current). The sampling can be continuous or not, according to the purpose and the sensors.

Mechanical recording instruments (*propeller current meters*) cannot give continuous records because the measurement is related to turns of the propeller. Acoustical (*Doppler's effect*) or electrical (*Faraday's effect*) devices can give continuous records. There are also other kinds of current meters: based on thermal diffusion, the slope of a suspended rod, and so on.

The Faraday current meter (Von Arx, 1962) is useful. The principle of this device is that oceanic currents are electrical conductors (seawater denotes a remarkable conductivity). If the conductor is moving in a magnetic field, an electric potential V develops in a direction that is orthogonal both to the magnetic vector and to the current vector: $V = K\ H\ L\ v$ (where V is electric potential; K, a constant involving geometrical and electrical characteristics; H, a magnetic field; L, the distance between the electrodes used in the potential measuring; and v, the current speed).

A magnetic field can also be artificially produced; in this case, even the least speeds (if the electrode distance is sufficiently small) are measurable by using an intense magnetic field. In more general cases, the magnetic field is the weaker earth magnetic field (vertical component).

By using a sufficiently large distance L (usually 100 m), oceanic currents are measurable. Two electrodes can be dragged by a ship, fixed between the coast and a point in the open sea, or placed between the edges of a channel. In the latter, the water budget through a channel can be obtained. Recently the author used this method successfully in determining the water budget into the Venice lagoon in connection with forecasting storm surges (Mosetti, Accerboni, and Manca, 1974).

FERRUCCIO MOSETTI

References

Mosetti, F.; Accerboni, E.; and Manca, B., 1974. Marine currents measurements by means of electromagnetic methods in lagoon mouths, for forecasting the storm surges in Venice, *Accad. Naz. Lincei Internat. Conf. Physics of the Seas* I, 105-107.

Von Arx, W. S., 1962. *An Introduction to Physical Oceanography.* Reading, Mass.: Addison-Wesley Publishing Co., 442p.

Cross-references: *Coastal Engineering; Coastal Morphology, Research Methods; Currents; Protection of Coasts; Remote Sensing.*

CURRENTS

Sea currents consist of bodies of water in motion, caused by different factors (internal or external) and affected by both friction and the rotation of the earth. In the ideal case of frictionless water, and without the rotation of the earth, the motion of the currents would be different from reality.

Currents in the ocean are responsible for the fundamental transport of energy, in the form of heat, from the hot equatorial zones to the cold polar zones and from the frozen zones to the lower latitudes. Warm water is transported by surface currents, cold water by abyssal ones.

Currents also transport substances such as salts, oxygen, nutrients, organic substances, animal or vegetable populations, and sediment. Currents recycle water through the oceans and, to some extent, between ocean and atmosphere. Near the shore, currents are an important component in the erosion and transport of sediments.

Oceanographic studies deal principally with currents that develop in the horizontal (or almost horizontal) plane, but vertical motion (*upwelling*, for example) is interesting in special cases.

Horizontal currents are not only surficial or abyssal; some intermediate currents (such as *compensation currents*) are present in the oceans. All oceans are set in motion and mixed by currents: the motion and the exchange of energy or substance may have a different velocity or intensity in different zones of the ocean or at different times in the same zone.

Oceanographers describe different kinds of currents, based on their causes. By contrast, early geographers divided the currents according to some property (temperature and/or salinity). Such a distinction has little importance in modern times. The distinction between the different causes is generally only an heuristic approximation, even if in some cases one cause is dominant over the other. The general condition is the coexistence of many causes. Different forces interact, so that each point of the ocean is forced to move by a complex of causes. These are studied both experimentally, by means of oceanographic instrumentations, and theoretically, by means of mathematical models. The application of these studies includes the transport of energy (with attendant climatic implications), the transport and budget of substances, and economic implications (productivity, fisheries, deposits of minerals on beaches and coasts, and the protection of beaches and harbors).

In the past, simple analytical models of currents were acceptable, these models being only a first approximation of the real problem. At the present time, the true conditions of motion, based on the experimental observation of some fundamental parameters, are calculated by computers, utilizing convenient mathematical models and by resolving the motion equations. Since the causes can be various and interacting, the equations do not have a simple form but contain a number of terms corresponding to the causes. Their solutions are not always simple.

Classification

A first classification of the different kinds of currents is possible by considering each cause of motion separately. The approximation is not always sufficient, but is useful for a preliminary knowledge of the meaning of some of the basics of physical oceanography and hydrodynamics. To begin, currents are observed and measured. The measurement of currents is made with *current meters* (q.v.) or by indirect methods. Interesting information about the motion of water bodies is obtained from the mixing of different waters. Salinity, temperature, oxygen content, or other similar parameters are useful indexes for this purpose (Mosetti, Accerboni, and Lavenia, 1972; Grancini, Lavenia, and Mosetti, 1972).

Density Currents. An important cause of motion is the density difference between two or more coexisting waters. Density currents are present at every depth, even if the best-known currents are the relatively shallow surface ones: i.e., the Gulf Stream and the Kuroshio Current. Because of density differences, the hydraulic pressure distribution is not coincident with the level surfaces; at the sea surface, for example, the true level is lower in conjunction with the greater density and higher at the lesser density. Water must flow from higher to lower levels, but the *Coriolis effect,* generated by the rotating earth, affects this simple condition. The resulting current is directed along isobaric lines; thus the current veers to the right in the Northern Hemisphere and to the left in the Southern Hemisphere.

Turbidity Currents. The density of water varies not only with different values of salinity and temperature (as above) but also according to the degree of turbidity. Turbidity currents occur in the presence of soft and unconsolidated sediments along the major slopes of the ocean basins. Turbidity currents can erode the bottom in the form of *submarine canyons* or produce submarine deposits as, for example, *submarine cones* and *fans.*

Wind-driven Currents. A typical surface current, not present at depth, is the wind-driven current. This current is generated by the drag of the wind on the ocean surface. Since in this case the Coriolis effect is also interacting, the current does not flow in the same direction as the blowing wind, but deviates (to the right in the Northern Hemisphere and to the left in the Southern Hemisphere) in a uniform and infinitely deep ocean. Theoretically, if the depth is infinite, the angular deviation is $45°$. In a limited basin, or in the presence of stratified water, the deviation is different. It then becomes less than 45°, to 0°, at minimal depths. Also, some interacting factors, principally the tides, could greatly modify this simple model.

At the surface the relationship between wind and current velocity is not always well described, since complex actions between wind and water turbulence diversify the conditions. But, on the average, the current velocity at the surface is 1-3% of the wind speed. At depth the direction and velocity of this current are progressively deviated and slowed down. At a certain depth D, whose value is strongly related to the turbulence, the current is in the opposite direction to the surface current and its velocity practically vanishes. This depth D is variable with the condition of turbulence; it can be only 20-30 m or as much as 200-300 m.

The great oceanic currents—i.e. northern and southern equatorial currents or, at the middle latitudes, the western (driven) currents—are all of this kind. These currents are not independent of the density currents, but are often interconnected with them. Both wind-driven currents and density currents make up, for example, the surface circulation of the oceans. This circulation is based on a great system of circulating flows directed counterclockwise in the Southern Hemisphere and clockwise in the Northern Hemisphere. The sector of the circulation disposed along the meridians is prevalently density current; the sector disposed along the parallels is wind-driven current.

Wind currents occur not only in the great systems of oceanic circulation but also near the coasts in marginal seas. Here the currents are very important in the transport of terrigenous sediments and in modifying the beaches. Wind blowing against the coast develops not only currents, but also raises the level of the sea in connection with the transport of water against the closed extremity of the sea. This phenomenon is often responsible for severe coastal damage and also for countercurrents that produce considerable removal of the coastal waters.

Tidal Currents. Another fundamental kind of current is the tidal current. Wind-driven currents and density currents are both stationary (or nearly so) currents or occasional currents. By contrast, tidal currents are a typically alternative motion of water bodies acting everywhere in the seas. Tidal currents force the water to move, independently of depth, and the deepest waters also may move under the action of tidal forces. The stratification of water and friction with the sea bottom or margins modifies the structure of tidal currents and their distribution according to depth. This distribution is not simple, and can be modified by changing conditions. Tidal currents are superimposed on other currents, so that the cumulative effect can be complicated. In the open ocean, tidal currents do not have strong intensities, but the velocity can increase greatly in channels, on coastal platforms, and in some estuaries.

The strongest tidal currents can modify, by erosion and transport, the coastline or the sea bottom. For example, in a part of the bottom of the Straits of Messina (Italy), sediment cannot be deposited because of the strong tidal currents that are superimposed on wind-driven currents. Current velocities there can reach values of 5-7 knots, about the maximal values known for sea currents.

Seiches. Seiche currents are similar to tidal currents. In some seas, seiches are well developed, appearing frequently and intensely. While tidal currents are a persistent occurrence, most seiche currents are, on the contrary, occasional phenomena. Even when these currents do occur frequently, in almost all seas their intensity is not constant. Such currents, like tidal currents, act both at the surface and at depth.

Wave-induced Currents. There are other currents caused by wave motion. A wave consists not only of a transfer of energy but also of mass. As waves are surficial phenomena, so the currents connected with waves are typical surface currents. These are important near some coasts—i.e., *longshore currents*—for their effect in the transport of sand or other terrigenous materials (Winant, 1979). In the open ocean, currents related to waves are not of much importance and are often confused with wind-driven currents, which occur for the same reason (both wind-driven and wave-induced currents are caused by the wind).

Inertia Currents. Inertia currents (Henderschott, 1973) appear after another cause of motion (the density difference of water or the wind) has disappeared. While the cause of motion soon ceases, the motion continues for some time because of *inertia*. Since the Coriolis effect and frictional forces always act on marine waters, the inertia motion is periodical and progressively smoothed. The period of this current is given by $T = \pi/\Omega \sin \phi$. This current can frequently appear in marginal seas, depending on different meteorologic or climatic conditions. Since the dampening effect is not very strong, some long series of oscillations can persist for many days, and often, before the current disappears, new oscillations are produced by renewed causes.

Sea Currents. Some currents are formed near the mouths of rivers. The water flowing downward in the river continues its motion in the sea as a *sea current*. The freshwater of the river, with its lesser density, floats on the more saline seawater. The buoyancy persists differentially; in fact turbulence mixes the waters, and so the current vanishes. Turbidity currents, because of solid materials transported by rivers, are often related to sea currents.

FERRUCCIO MOSETTI

References

Grancini, G.; Lavenia, A.; and Mosetti, F., 1972. A contribution to the hydrology of the Strait of Sicily, in *SACLANTCEN Conf. Proc., La Specia, Italy,* 68-81.
Hendershott, M. C., 1973. Inertial oscillation of tidal period, in B. Warren, ed., *Progress in Oceanography,* vol. 6. New York: Pergamon Press, 1-27.
Mosetti, F.; Accerboni, E.; and Lavenia, A., 1972. Recherches océanographiques en Méditerranée orientale (août 1967), *Comm. Internat. Mer. Médit. Rapp.* 20, 623-625.
Winant, C. D., 1979. Coastal current observations, *Rev. Geophysics and Space Physics* 17, 89-98.

Cross-references: *Climate, Coastal; Coastal Engineering; Coriolis Effect; Current Meters; Inlet and Inlet Migration; Mineral Deposits; Nearshore Hydrodynamics and Sedimentation; Nutrients; Protection of Coasts; Seiche; Tides; Waves; Wind.* Vol. I: *Dynamics of Ocean Currents.*

CUSPATE FORELAND

Seaward-projecting accumulations of loose marine sand or gravel, when bounded on both sides by active wave-dominated shorelines, have been termed *cuspate forelands* by Gulliver (1896). Cuspate forelands showing a wide range of form and occurrence have been described on accretion coasts from many parts of the world (Davies 1972; King 1972).

Basically a cuspate foreland is thought to form where a long-term balance, or near balance, exists between two or more opposing, but constructional, coastal forces. These forces may be either alongshore currents–drift-aligned features (Zenkovich 1959, 1967); or onshore wave action–swash-aligned (Davies 1972). Alternatively, a combination of the two may exist. The size and growth rate of a foreland will depend both upon the magnitude and frequency of formative processes and also upon the availability and input rate of sediment.

Ideal sites for the formation of cuspate forelands include those locations where a major change in shoreline direction occurs–at the entrances to bays, inlets, or fjords; or in wave shadow areas–behind headlands, offshore, or barrier islands, or even other spits (Zenkovich, 1959). Some forelands show progressive alongshore movements (hence *traveling forelands*); often these features are asymmetrical in plan form, and have distinct erosion and accretion coasts. This type of foreland is caused by a dominant process from one direction. A symmetrical foreland is most likely a stationary form reflecting a dynamic equilibrium between opposing forces and expanding or contracting in size only in response to sediment availability.

Some cuspate forelands may have originated as looped bars (Evans 1942) and have been modified by minor falls in sea level. Still another theory holds that some coastal plain *capes,* such as those of the southeast coast of the United States, were formed by Flandrian transgression reworking of deltas developed previously at a lower stand of sea level (Hoyt and Henry, 1971).

RICHARD W. G. CARTER

References

Davies, J. L., 1972. *Geographical Variation in Coastal Development.* Edinburgh: Oliver and Boyd, 204p.
Evans, O. F., 1942. The origin of spits, bars, and related structures, *Jour. Geology* 50, 846-865.
Gulliver, F. P., 1896. Cuspate Forelands, *Geol. Soc. America Bull.* 7, 399-422.
Hoyt, J. H., and Henry, V. J., Jr., 1971. Origin of capes and shoals along the southeastern coast of the United States, *Geol. Soc. America Bull.* 82, 59-66.
King, C. A. M., 1972. *Beaches and Coasts,* 2d ed. London: Edward Arnold, 570p.
Zenkovich, V. P., 1959. On the genesis of cuspate spits along lagoon shores, *Jour. Geology* 67, 269-277.
Zenkovich, V. P., 1967. *Processes of Coastal Development.* Edinburgh: Oliver and Boyd, 738p.

Cross-references: *Cuspate Spits; Deltas; Lagoonal Segmentation; Rhythmic Cuspate Forms; Spits.* Vol. III: *Cuspate Forelands or Spits.*

CUSPATE SPITS

Cuspate spits are projections of a beach into an enclosed or semienclosed lagoon. Genetically, cuspate spits are a wave-induced phenomenon representing the reorientation of a shoreline into dominant wave-approach directions. Cuspate spits are one of a family of shoreline reorientation features that includes cusplike structures, giant cusps, looped spits, and cuspate forelands. These forms are generated by longshore processes as the result of dominant waves approaching a shoreline at a high angle. A shoreline will tend to orient into dominant wave-approach directions as a result of net longshore transport.

Cuspate spits commonly form in elongate lagoons where the basin shape limits fetch, so as to act as a selective filter on the wave climate. The result is a dominant bidirectional wave field parallel to the long axis of the basin, regardless of the wind rose.

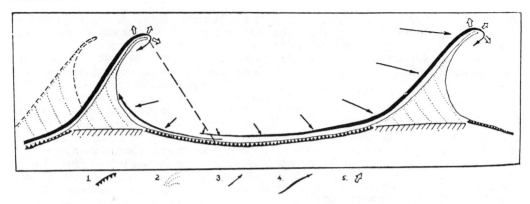

FIGURE 1. Processes of cuspate spit development (after Zenkovich, 1959): 1. abraded shore; 2. dune ridges; 3. relative magnitude and orientation of local wave resultants; 4. longshore sediment streams, with width of arrows showing the relative sediment capacity; 5. portion of the sediment deposited as a bar.

In an elongate lagoon, where dominant wave approach is at a high angle to the shore, there will be a point where longshore sediment transport is constant. If the sediment load is obstructed in any way, the longshore capacity will decrease because of the decreased wave-approach angle (Fig. 1), causing deposition. A spit will begin to accrete out from the shore at an angle of about 45°. Waves from the opposite direction will act similarly on the opposite side of the spit, resulting in the cuspate form. The growing spit will form a wave shadow, and the process can be repeated along the lagoon shore. As the process continues, spits in favored positions with relation to the wave regime will grow at the expense of others (Zenkovich, 1959).

When a series of cuspate spits have formed, sediment is removed from the shoreline concavities between spits by longshore drift (because of the wave shadow effect), and is deposited as a bar projecting from the ends of the spits, under both dominant wave directions.

Early investigations of the origin of cuspate spits by Gilbert, Shaler, Gulliver, Johnson, and Jones (see references in Rosen, 1975) concluded that such spits are the result of scour and deposition from eddies of tidal currents forming circulation cells in a lagoon. Field studies (Rosen, 1975) have shown that tidal currents in lagoons serve to erode the ends of cuspate spits and straighten the shoreline. For this reason, cuspate spits are found only in zero or low tidal-range settings.

There are two major types of cuspate spits: *accumulative* and *abrasional*. On accumulative spits (Fig. 2A), the cuspate form is composed primarily of redeposited material; on abrasional cuspate spits (Fig. 2B), the form of the spit has been eroded from previously deposited material. No cuspate spit can be purely accumulative or abrasional, since the formative process involves both erosion in the concavities and deposition

FIGURE 2. *A*: Abrasional cuspate spit (Bass Point, Nantucket, Massachusetts). Note truncated dune ridges. The spit formed by erosion of the preexisting landform. Deposition of this material is a bar extending over 1 km from the end of the spit. *B*: Accumulative cuspate spit (Cove Point, Maryland). Dune ridges outline accretional form of spit. Older ridges on north side of spit are being submerged by high local subsidence.

TABLE 1. Survey of Cuspate Spit Shorelines

Area	Mean Tidal Range (Meters)	Shoreline Trend	Major Wind Components	Harbor Segmentation	Form	Rhythmicity	References (see Rosen, 1975)
Nantucket Harbor, Massachusetts	1	NE-SW	NE-SW	No	Abrasional	Yes	Rosen (1972)
Waquoit Bay, Massachusetts	1	N-S	NE-SW	No	Abrasional	No	–
Pond, South end Monomoy Island, Massachusetts	0	NE-SW	NE-SW	Yes	Accumulative	No	Goldsmith (1972)
Chappaquidick Island, Marthas Vinyard, Massachusetts	0	N-S	NE-SW	No	Accumulative	No	–
Sengenkontacket Pond, Martha's Vinyard, Massachusetts	1	NW-SE	NE-SW	No	Abrasional	No	–
Chuktotsky Peninsula, U.S.S.R.	0	–	–	Yes	Accumulative	Yes	Zenkovich (1959)
St. Lawrence Island, Alaska	0	NW-SE and E-W	N-E-SW	Yes	Accumulative	Yes	Fisher (1955)
Coskata Pond, Nantucket, Massachusetts	1	NE-SW	NE-SW	No	Abrasional	No	–
Zululand Coast, South Africa (Kosi Bay, Lake Sibayi, Lake St. Lucia, Richards Bay)	0	NE-SW	NE-SW	Yes	Accumulative	No	Orme (1972)
Chilka Lake, India	0.8	NE-SW	NE-SW	Yes	Accumulative	No	Venkatarathnam (1970)
Mobile Bay, Alabama	0.6	E-W	N-SW	Yes	Mixed	Yes	Shepard & Wanless (1971)

at the spit ends. Accumulative spits commonly form in nontidal settings, where growth of the bar at the end of the spit is not diminished by tidal scour (Table 1).

The end product of the erosional-depositional system in the cuspate spit process is segmentation of the lagoon by the bar accreting from the end of the spit. Accumulative cuspate spits often occur in opposing pairs. As the bar from an initial spit extends across the lagoon, it can form an irregularity on the opposite shore sufficient to initiate the cuspate spit-longshore process. If the two spits join, segmentation is the result. Scour from tidal currents serves to prevent segmentation. As a pair of cuspate spits segment, the volume of tidal flow is reduced in the lagoon, thereby promoting segmentation by other spits. The resultant shore is a series of circular ponds.

Cuspate spits occur either as a single spit or in a series. When several spits occur, there is often a rhythmicity to their spacing. The most symmetrical and most evenly spaced cuspate spit system in the literature is in Nantucket Harbor, Massachusetts (Rosen, 1975). The long axis of this basin is in exact alignment with a highly bidirectional, opposing wind field. There has been no significant shoreline change in over 200 years. The action of the tidal scour eroding the ends of the spits approximately equals the competency of the longshore processes building out the spit ends. The rhythmic form is not fully understood, but the spacing appears to be a function of available fetch (Tanner, 1962).

Cuspate forelands appear to be genetically identical to cuspate spits, except that the basin shape does not act as a selective filter on the wave spectrum. The two major cuspate foreland systems in North America—the Carolina Capes on the southeastern coast of the United States and the Icy Cape, Point Franklin, Point Barrow system on the northwest coast of Alaska—occur on shorelines parallel to a strongly bidirectional, opposing wind regime.

PETER S. ROSEN

References

Rosen, P. S., 1975. Origin and processes of cuspate spit shorelines, in L. E. Cronin, ed., *Estuarine Research*, vol. II. New York: Academic Press, 77-92.

Tanner, W. F., 1962. Reorientation of convex shores, *Am. Jour. Sci.* 260, 37-47.

Zenkovich, V. P., 1959. On the genesis of cuspate spits along lagoon shores, *Jour. Geology* 67, 269-277.

Cross-references: *Beach in Plan View;\Coastal Erosion, Formations; Headland Bay Beach; Lagoonal, Sedimentation; Lakes, Coastal Morphology; Spits; Zetaform Bays.* Vol. III: *Cuspate Foreland or Spits.*

CUSPS—See BEACH CUSPS

CUT AND FILL

A cyclical erosion-depositional couplet exists on almost all beaches. This cycle is present in two spatial and temporal scales: one is *tidal*, with relatively small amounts of cut and fill; while the other is *seasonal* or caused by *storms* in which large quantities of sand are moved from the beach to the offshore zone and back again after the winter months or the storm has passed.

The works of Duncan (1964), Otvos (1965), Strahler (1966) Schwartz (1967), and Brennink-meyer (1973) have shown that there are three phases of beach changes that can be recognized within each tidal cycle. First, there is an initial phase of deposition of up to 5 cm, as the swash zone on a rising tide acts upon a section of the beach. This is followed by a period of scour of up to 20 cm, as the area is under the turbulent zone of the bore-backwash interaction. As the tide rises still higher, while the area is under the surf zone, some of the sediment transported from bore-backwash interaction and the breaker zone is deposited. This depositional wedge is on the order of tens of centimeters. Finally, under the breaker zone, there is scour. This is also on the order of tens of centimeters. The depth of erosion ranges from 3% of the breaker height (King, 1951) to 40% and more (Williams, 1971). On the ebb tide this erosional-depositional sequence is reversed as an area on the beach again goes through the surf, transition, and swash zones. The erosion in the transition zone is due to the high turbulence.

Saville (1950) and Scott (1954) have shown that wave steepness is one of the most important variables in determining the beach slope. Many authors follow Saville, who divides the beach profile into two categories: a *storm profile*, produced by waves with a steepness (H/L) greater than 0.023, and a *summer profile* formed by waves less than 0.025 steep.

When waves grow steeper, during storms or in the winter months, sand is moved from the beach to the offshore. When waves decay after the passage of a storm or are long when compared to their height during the prevalent summer swell, the offshore deposit gradually moves onshore, and the sand is spread out over the beach. This forms the so-called winter or concave-up profile or the wide concave-down bermed summer profile of the beaches so common on the Pacific coast (Shepard, 1950). By way of contrast, the Atlantic coast of the United States is subject to frequent variations in wave energy caused by the passage of storms.

These may overshadow any seasonal variations in wave climate (Hayes and Boothroyd, 1969).

Sonu and Van Beek (1971) have made the observation that it is the length of time of either growing or decaying waves that is important in the estimation of deposition or erosion from the beach face rather than the actual wave steepness. This may be due to the formation of the offshore bar. If the foreshore slope is less than 4° the offshore deposit tends to accumulate in a bar (Shepard, 1950). This bar will grow until it is high enough to steepen and force the early breaking of storm waves. Since the waves break farther from shore, most of the erosion of the beach face stops. Erosion usually reaches its peak within several hours after the onset of a storm. Conversely, the first of the winter storms will cause the largest and most spectacular beach erosion, since the offshore deposit or bar still has to be built. After a storm or during the summer, the bar begins to migrate landward. If it continues long enough, the bar is *welded* onto the beach (Davis et al., 1972), resulting in a beach with a wide backshore.

This seasonal cut and fill can in part be explained by the ratio of swash to breaker periods. In the summer, when the long period swell is common, this ratio is less than one (Rockwell and Schweizer, 1956). This implies that the swash-bore interaction is at a minimum. The incoming wave has a greater probability of transporting sand up to the beach to be deposited. In the winter or during storms, the ratio is greater than one. The resultant is a greater number of collisions of the bore with the backwash, retarding the shoreward motion of the sediment carried in the bore and allowing the backwash to carry sediment further through the surf zone.

Moreover, during storms when waves are steep, more water is thrown onto the beach. The beach soon becomes saturated, and the water table becomes almost coincident with the beach face. Because of this, little of the swash percolates through the sand, so that the backwash volume is almost equal to the uprush. This downflow drags sediment from the beach face and increases the turbulence within the scour zone where the backwash meets the incoming bore. At this mean water line, the ground water flow returns to the sea. Its effluent out of the sediment provides a quicksand effect that sets the stage for more rapid removal of sediment by the backwash (Silvester, 1974).

BENNO M. BRENNINKMEYER

References

Brenninkmeyer, B. M., 1973. *Synoptic Surf Zone Sedimentation Patterns*. Ph.D. dissertation, University of Southern California, 274p.

Davis, R. A., Jr.; Fox, W. T.; Hayes, M. O.; and Boothroyd, J. C., 1972. Comparison of ridge and runnel systems in tidal and non-tidal environments, *Jour. Sed. Petrology* 42, 413-421.

Duncan, J. R., 1964. The effects of water table and tidal cycle on swash-backwash sediment distribution and beach profile development, *Marine Geology* 2, 186-197.

Hayes, M. O., and Boothroyd, J. C., 1969. Storms as modifying agents in coastal environments, *in* M. O. Hayes, ed., *Coastal Environments, Northeast Massachusetts and New Hampshire*. Amherst: University of Massachusetts, Coastal Research Group Contribution No. 1, 245-265.

King, C. A. M., 1951. Depth of disturbance of sand on sea beaches by waves, *Jour. Sed. Petrology* 21, 131-140.

Otvos, E. G., 1965. Sedimentation-erosion cycles of single tidal periods on Long Island Sound beaches, *Jour. Sed. Petrology* 35, 604-609.

Rockwell, F. G., and Schweizer, J. E., 1956. *A Study of Beach Sediment Motion in the Breaker Region*. Ph.D. dissertation, Massachusetts Institute of Technology, 76p.

Saville, T., 1950. Model studies of sand transport along an infinitely straight beach, *Am. Geophys. Union Trans.* 31, 555-556.

Schwartz, M. L., 1967. Littoral zone tidal-cycle sedimentation, *Jour. Sed. Petrology* 37, 677-683.

Scott, T., 1954. *Sand Movement by Waves*. Washington, D.C.: U.S. Army Corps of Engineers, Beach Erosion Board Tech. Memo 48, 37p.

Shephard, F. P., 1950. *Longshore Bars and Longshore Troughs*. Washington, D.C.: U.S. Army, Corps of Engineers, Beach Erosion Board Tech. Memo 15, 32p.

Silvester, R., 1974. *Coastal Engineering*. Amsterdam: Elsevier, vol. 1, 475p.; vol. 2, 338p.

Sonu, C. J., and Van Beek, J. L., 1971. Systematic beach changes on the Outer Banks, North Carolina, *Jour. Geology* 79, 416-425.

Strahler, A. N., 1966. Tidal cycle changes in an equilibrium beach, Sandy Hook, New Jersey, *Jour. Geology* 74, 247-268.

Williams, A. T., 1971. An analysis of some factors involved in the depth of disturbance of beach sand by waves, *Marine Geology* 11, 145-158.

Cross-references: *Beach Cycles; Beach Processes; Beach Profiles; Profiling of Beaches; Sweep Zone; Water Table.*

CUT PLATFORM—See WAVE-CUT PLATFORM

CUT TERRACE—See WAVE-CUT PLATFORM

CYANOPHYCEAE

The Cyanophyceae (Cyanobacteria or *blue-green algae*) are a group of photosynthetic microorganisms that are more closely related to the bacteria than to higher (eukaryotic) algae. Cyanophyceans and bacteria differ from other organisms in having a prokaryotic cellular organization—that is, their cells lack nuclei, mitochondria, chloroplasts, and other specialized organelles. This primitive cellular organization reflects the early evolution of blue-green algae and bacteria; *stromatolites* (laminated structures of biogenic origin) have recently been discovered in 3.5 billion-year-old sediments from Australia (Walter, Buick, and Dunlop, 1980; Awramik, 1981). While the evidence is not conclusive, it appears that *true* blue-green algae (which, unlike bacteria, produce oxygen during photosynthesis) evolved considerably later than bacteria, perhaps 2.5 to 2.0 billion years ago. The production of oxygen by these early cyanophyceans was almost certainly the major cause of the oxygenation of the initially anoxic Precambrian atmosphere (Schopf, 1975). Numerous cyanophycean microfossils and stromatolites have been reported from Middle and Late Precambrian sediments, a time when the group dominated the majority of marine communities. The post-Precambrian history of blue-green algae has been one of decline and restriction to specialized habitats not well suited to higher (eukaryotic) life forms (Schopf, 1975).

Of the 2,000 extant cyanophycean species, a few live as single cells of microscopic size, while others are aggregated in colonies but live essentially independently. Most Cyanophyceae are filamentous, however. It is in this form that the blue-green algae are most prominent, visible as mosslike growths in moist areas on land or as mucilaginous masses in the water (Fogg et al., 1973). Reproduction in all species is asexual.

The Cyanophyceae are widely distributed, often living in habitats where no other organisms can survive. They live in fresh and hypersaline water, in hot springs and the Antarctic, in moist soils and the Atacama Desert, and in parasitic and symbiotic association with other organisms. The largest number of species are found in freshwater. Although Cyanophyceae are widespread in the marine environment, they are seldom dominant and are usually ignored by the casual observer. In the supralittoral and littoral zones of the rocky shore, there is often a distinct growth of epilithic and endolithic cyanophytes (Fogg et al., 1973). Endolithic cyanophytes dwell within tubes that they bore in calcareous rocks and shells. They may play an important role in the erosion of limestone and marble coasts. Blue-green algae are usually abundant in estuaries and salt marshes, habitats where the remarkable capacity of the group to withstand extremes of dessication, salinity, and redox potentials facilitates their survival. There are few genera of planktonic marine cyanophytes, although *Trichodesmium* blooms are rather common in tropical seas.

ALBERT V. NYBERG

References

Awramik, S. M., 1981. The Pre-Phanerozoic biotic crisis—Three billion years of crises and opportunities, in M. H. Nitecki, ed., *Biotic Crises in Ecological and Evolutionary Time*. New York: Academic Press, 83–102.

Fogg, G. E.; Stewart, W. D. P.; Fay, P.; and Walsby, A. E., 1973. *The Blue-Green Algae*. New York: Academic Press, 459p.

Schopf, J. W., 1975. Precambrian paleobiology: Problems and perspectives, *Annual Rev. Earth Planetary Sci.* **3**, 213–249.

Walter, M. R.; Buick, R.; and Dunlop, J., 1980. Stromatolites 3400–3500 Myr-old from the North Pole area, Western Australia, *Nature* **284**, 443–445.

Cross-references: *Algal Mats and Stromatolites; Chlorophyta; Chrysophyta; Coastal Flora; Cryptophyta; Estuaries; Euglenophyta; Phaeophycophyta; Rhodophycophyta; Rock Borers; Salt Marsh.* Vol. III: *Algae;* Vol. VI: *Algal Reef Sedimentology.*

CYCADICAE (CYCADOPHYTA)

This class of the gymnosperms (Pinophyta) includes some fossil orders in addition to the only living order, the Cycadales, which is treated here. It includes only one family, the Cycadaceae or cycads, which includes 9 living genera and about 65 species.

The cycads are mostly more or less arborescent, palmlike, woody plants with usually unbranched, columnar stems 1–18 m tall or, less frequently, with shorter, often partially subterranean stems with a branch or two. The internodes are usually suppressed in the arborescent kinds, and the trunk is covered with the persistent leaf bases.

The conductive tissues of the stem are arranged in a central cylinder somewhat as in other gymnosperms, but the central pith is much larger, and there are often successive concentric cylinders of mostly *secondary* vascular tissue separated by parenchyma. The cortex, or parenchyma outside the vascular cylinder, is also usually thick. In contrast, the vascular tissues of the palms (members of the Magnoliophyta (angiosperms) class Liliopsida (monocotyledons)), which often superficially resemble cycads in form, are scattered through a mass of parenchyma as small strands and are all *primary* tissue.

The leaves of cycads, which resemble the leaves of feather palms, are evergreen, spirally arranged, and borne in a terminal crown. They are evergreen—i.e., survive more than one year.

All are once-pinnately compound except in the genus *Bowenia,* in which they are bipinnate. In length, they range from 5 cm–3 m. They are persistent after their ultimate death, but eventually fall, leaving their persistent bases covering the trunk. The plants are unisexual. The female strobili range from loosely organized clusters of long carpels (or megasporophylls) borne terminally on the trunk in *Cycas,* a primitive genus, to closely aggregated cones each with its own axis bearing short megasporophylls. In primitive genera, there is only the one strobilus borne at the apex of the stem; in more advanced genera, there may be one, two, or three female cones borne obviously laterally when there is more than one. In size the female cones range from 7–60 cm long and up to the unusual extreme of 40 kg in weight, thereby far surpassing the carpellate cones of all other gymnosperms in both weight and length.

The numerous carpels composing the strobili range in form from elongate, upwardly curving structures with the distal portion resembling a foliage leaf and the proximal portion bearing several ovules marginally with their micropyles pointing obliquely toward the tip of the carpel or at right angles to the axis of the carpel in primitive genera, such as *Cycas,* through a reductional series in various genera, to short wholly nonleaflike apically truncate carpels borne at right angles to the axis of the cone, bearing beneath their apices a pair of pendulous ovules parallel with the stalks of the carpels with their micropyles pointing toward the axis of the cone in advanced genera such as *Encephalartos* and *Zamia.*

Archegonia are present in the ovules. The seeds range from 1½–6 cm in length. The outer seed coat is often both fleshy and colored; the inner seed coat is hard. The endosperm, the haploid female gametophyte, the food supply in which the embryo is embedded, is abundant and starchy. The embryo has two cotyledons, except in *Ceratozamia* in which there is only one.

The male cones are compact and lateral, but borne, as the cones in the female plants, above the crown of leaves. In length the male cones rival the female, but they are smaller in diameter and much lighter than the female cones full of mature seeds. There may be one or a few male cones per stem. The numerous spirally arranged microsporophylls of each male cone bear from a couple of dozen to about a thousand pollen-producing microsporangia on the abaxial surface. At maturity of the pollen the axis of the male cone elongates, thereby separating the microsporoplylls and thus allowing release of the pollen to be wind-borne, or perhaps insect-borne in some cases, to the pollination droplets at the apices of the ovules in the female cones. Resorption of the pollination droplet carries the pollen grains into the pollen chamber of the ovules. Eventually each pollen grain sends out a pollen tube that delivers the *ciliated motile* sperms to the archegonia, where one sperm fuses with one egg to form the zygote that eventually grows into the embryo of the seed.

The wide geographic range of the cycads reflects their antiquity. In the New World, they range from Florida, the islands of the Caribbean, and southern Mexico south to $20°$ S latitude in South America. In the Old World, they range from the Ryukyu Islands and Taiwan south to Australia, westward across southern Asia to Ceylon (Sri Lanka), to Madagascar, and in Africa from $16°$ N latitude south to the Cape of Good Hope.

Economically the cycads are of little importance. Some are cultivated as ornamentals on a limited scale, often under the misleading name of sago palm. *Cycas revoluta* is often grown out-of-doors in subtropical climates. *Cycas circinalis* is a popular greenhouse subject. Occasionally the starch from the parenchyma of the stem of cycads is used for food, but some cycads are poisonous. One in Guam that has been used during famine to avoid death from starvation contains a mutagenic carcinogen that may result in cancer years later. In some areas, especially in Australia and Mexico, certain cycads are important livestock-poisoning plants.

Additional details on the subject may be found in Chamberlain (1935), Foster and Gifford (1974), Pilger (1926), and Schuster (1932).

LOUIS CUTTER WHEELER*

References

Chamberlain, C. J., 1935. *Gymnosperms Structure and Evolution.* Chicago: University of Chicago Press, 484p.

Foster, A. S., and Gifford, E. M., Jr., 1974. *Comparative Morphology of Vascular Plants.* San Francisco: W. H. Freeman, 749p.

Pilger, 1926. Cycadaceae, *in* A. Engler and K. Prantl, eds., *Die Naturlichen Pflanzenfamilian,* 2d ed., vol. 13. Leipzig: Verlag von Wilhelm Englemann, 44–82.

Schuster, J., 1932. Cycadaceae, *in* A. Engler, ed., *Das Pflanzenreich,* No. 99. Weinheim: Verlag von Wilhelm Engelmann, 168p.

Cross-references: *Biotic Zonation; Coastal Ecology, Research Methods; Coastal Flora; Coastal Plant Ecology, United States, History of; Pinophyta.*

CYCLE OF EROSION—See GEOMORPHIC-CYCLE THEORY

*Deceased

D

DAMS, EFFECTS OF

The construction of dams across rivers greatly restricts the downstream flow of sediments, a restriction that is especially important where river-originating sand comprises a large proportion of the total beach sand supply.

Dams trap sand-sized sediments reaching their reservoirs from upstream, and also greatly reduce the flood flows that are the main forces moving sand toward the seacoast. This reduction in sand supply creates a long-run beach disequilibrium in which normal sand losses are not offset by new supplies. The consequences are shrinking beaches and coastal erosion.

After the blockage of nearby fluvial sand sources, shore erosion may develop rapidly, or may not begin to appear for many years. Accelerated erosion (up to 40 m/yr) of the Nile delta began soon after the completion of the Aswan High Dam in 1964 (Nielson, 1973). The effects of dams have only recently been recognized as an increasingly important factor contributing to shrinking beaches in the northwestern United States, although dams have been in place for decades. Sediment bypasses reportedly built into the new dams along the Hwang Ho may help reduce the sand-trap problem in northern China.

In southern California, where the situation is perhaps most severe, reservoirs must be periodically drained and dredged to maintain their flood-control and water-holding capacities, but the trapped and recovered sediments are removed from the river system and never reach the beach. Severe coastal erosion has been averted only by extensive harbor dredging and expensive artificial *beach nourishment* (Norris, 1964).

Worldwide, as more dams are completed and sand supplies continue to dwindle, coastal erosion is likely to grow in both severity and distribution.

JAMES E. STEMBRIDGE, JR.

References

Nielsen, E., 1973. Coastal erosion in the Nile delta, *Nature and Resources* 9, 14-18.
Norris, R. M., 1964. Dams and beach-sand supply in southern California, *in* R. L. Miller, ed., *Papers in Marine Geology*. New York: Macmillan, 154-171.

Cross-references: *Beach Nourishment; Coastal Engineering; Coastal Erosion; Sand.*

DEFLATION PHENOMENA

In many coastal areas beach sediments are transported by wind processes, sand often accumulating as *dunes*. Normally the water-lain sediments exhibit considerable spatial variations in texture and composition that, under aeolian processes, result in differential rates of *deflation*. During deflation a number of ephemeral primary sedimentary structures are formed, including: asymmetrical structures (Fig. 1A) caused by variations in sand moisture content and non-uniform compaction over the beach (Berry, 1973; Vortisch and Lindstrom, 1980); pebble or shell pedestals (Fig. 1B and 1C) in mixed coarse/fine deposits, where only exposed fine material is removed, leaving coarser material raised on pedestals (Carter, 1978) often with a distinct downwind tail; salt-crust structures (Fig. 1D), where a higher salt content increases cohesion of the surface layers, causing peripheral wind scour; and beach plains or pavements, where flat, wide surfaces (Fig. 1E and 1F) occupy positions on upper beach, berm, or overwash areas when deflation has almost ceased because of the presence of wind-resistant surface lag deposits. (*Note:* The figure appears on the facing page.)

RICHARD W. G. CARTER

References

Berry, R. W., 1973. A note on asymmetrical structures caused by differential wind erosion on a damp sandy forebeach, *Jour. Sed. Petrology* 43, 205-206.
Carter, R. W. G., 1978. Ephemeral sedimentary structures formed during the aeolian deflation of beaches, *Geol. Mag.* 115, 379-382.
Vortisch, W., and Lindstrom, M., 1980. Surface structures formed by wind activity on a sandy beach, *Geol. Mag.* 117, 491-496.

Cross-references: *Beach Processes; Dunes and Eolian Sedimentation; Salcrete; Sediment Transport; Shell Pavement.* Vol. III: *Deflation.*

DELTAIC COASTS

Deltaic depositional features result from interacting geomorphic processes, such as wave energy, tidal currents, and river discharge. The

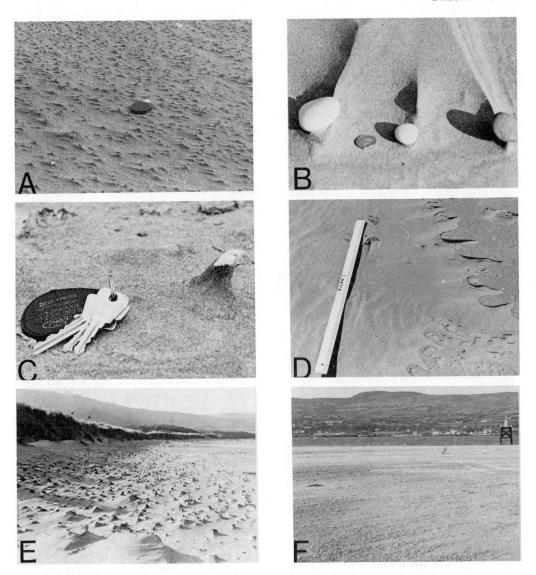

FIGURE 1. Beach deflation phenomena: *A*. asymmetrical structures; *B*. pebble pedestals with downwind tails (coin 29 mm); *C*. shell pedestal (key 60 mm); *D*. salt-crust scouring; *E*. pebble pavement during deflation; *F*. shell pavement forming a wind-resistant surface on the beach berm (photos: R. W. G. Carter).

plan form of deltas; the location, orientation, and geometry of individual landforms; and the three-dimensional facies relationships of sediments within a deltaic depocenter—all express the degree to which coastal processes operating within a given tidal, climatic, and tectonic setting can modify and disperse riverborne clastics.

In the coastal zone the tendency for equilibrium adjustment between wave forces and morphology is continuously upset near the mouths of rivers carrying significant sediment loads. Depositional patterns depend on the spatiotemporal distribution of wave forces, where the latter is a function of subaqueous and shoreline topography (Wright and Coleman, 1972). The degree of interaction between wave forces and the river's ability to supply sediments varies between deltas, within a deltaic area, and over time. The result is a spectrum of configurations and landform combinations (Morgan, 1970; Wright and Coleman, 1972, 1973; Coleman and Wright, 1975; Wright, 1978). Where wave activity is dominant, sediments deposited at river mouths are rapidly dispersed by shore-normal and alongshore littoral processes, producing barriers, spits, and beach ridges. At the other extreme, digitate distributaries are well developed along coasts, or in lagoons and lakes, where wave forces are insignificant. Wright and Coleman (1972) have shown that there is a progressive decrease in abundance, magnitude, and relative dominance of river-built features (digital distributaries, open bays, marshes, crevasses) and an increase in the development of wave-built features, especially beach ridges, as wave energy in the nearshore zone increases. Of primary importance to delta morphology is the role of frictional attenuation in reducing wave power in the nearshore. Attenuation increases with decreasing offshore slope. Hence deltaic coasts with flat to convex offshore profiles (e.g. Mississippi) exhibit more river control on delta form and landform composition, while deltaic coasts with wave-dominated landforms have steeper, concave offshore profiles (e.g., Nile).

Besides influencing the incident wave regime and river discharges, climatic factors affect the type and rate of sedimentation in the various depositional environments of a deltaic coast (compare the tropical rain forest-mangrove deltas of west Malayasia—Coleman, Gagliano, and Smith, 1970—with the vast bare tidal flats of the Ord River—Thom, Wright, and Coleman, 1975). Tidal forces significantly modify deltaic topography. For instance, high-tidal deltas are characterized by linear tidal ridges off the mouth of the river (Coleman and Wright, 1975; Wright, 1977). Finally, the tectonic setting is important. Coasts undergoing subsidence for whatever reason form complexes of overlapping and laterally displaced sediment lobes, as in the Mississippi. With emergence, deltaic sediments form terrace complexes, but a rigid basement forces distributary streams to build the deltaic plain upward and outward during progradation (e.g., Mekong; see Morgan, 1970).

BRUCE G. THOM

References

Coleman, J. M.; Gagliano, S. M.; and Smith, W. G., 1970. Sedimentation in a Malaysian high tide tropical delta, in J. P. Morgan and R. H. Shaver, eds., *Deltaic Sedimentation, Modern and Ancient.* Tulsa, Okla.: Soc. Econ. Paleont. and Miner., Spec. Pub. 15, 185-197.

Coleman, J. M., and Wright, L. D., 1975. Modern river deltas: Variability of processes and sand bodies, in M. L. S. Broussard, ed., *Deltas: Models for Exploration.* Houston, Texas: Houston Geological Society, 99-149.

Morgan, J. P., 1970. Depositional processes and products in the deltaic environment, in J. P. Morgan and R. H. Shaver, eds., *Deltaic Sedimentation, Modern and Ancient.* Tulsa, Okla.: Soc. Econ. Paleont. and Miner., Spec. Pub. 15, 31-47.

Thom, B. G.; Wright, L. D.; and Coleman, J. M., 1975. Mangrove ecology and deltaic-estuarine geomorphology: Cambridge Gulf-Ord River, Western Australia, *Jour. Ecology* 63, 203-232.

Wright, L. D., 1977. Sediment transport and deposition of river mouths: A synthesis, *Geol. Soc. America Bull.* 88, 857-868.

Wright, L. D., 1978. River deltas, in R. A. Davis, Jr., ed., *Coastal Sedimentary Environments.* New York: Springer-Verlag, 5-68.

Wright, L. D., and Coleman, J. M., 1972. River delta morphology: Wave climate and the role of the subaqueous profile, *Science* 176, 282-284.

Wright, L. D., and Coleman, J. M., 1973. Variations in morphology of major river deltas as functions of ocean wave and river discharge regimes, *Am. Assoc. Petroleum Geologists Bull.* 57, 370-398.

Cross-references: *Barrier Islands; Currents; Deltas; Global Tectonics; Sediment Transport; Tides; Waves.*

DELTAS

Definition and General Characteristics

Deltas are subaerial and subaqueous accumulations of river-derived sediment deposited at the *coast* when a stream decelerates by entering and interacting with a larger receiving body of water. The receiving body of water may be an ocean, gulf, lagoon, estuary, or lake, while the source stream may be virtually any size. The term *delta* originated with Herodotus' (c. 450 B.C.) description of the triangular plain of deltaic alluvium at the mouths of the Nile River. For this reason the Nile delta is one of the traditional deltaic models. More recently the Mississippi delta, with its extended lobes, digitate distributaries, and *bird's-foot* configuration has become the most widely accepted model of deltaic sedimentation. However, although both the Nile and the Mississippi are representatives of important delta types, a coastal accumulation of river-derived sediment need not conform to either model in order to be classed as a delta. Indeed, the majority of the world's modern deltas differ from these models in major respects.

Deltas, particularly those associated with large river systems, exhibit appreciable variability of morphology and depositional patterns, reflecting global variations in such controlling factors as river discharge regime, marine energy conditions, and structural controls (Coleman and Wright, 1975; Wright, 1978). Because deltas are subject to numerous interacting coastal processes, the large deltas cannot simply be regarded as being individual landforms but, rather, are *complexes* of many coastal depositional features, including those that have been secondarily molded by waves, tides, currents, and winds. Some of the more common features that may make up a deltaic complex include river mouths, distributary channels, bays, flats, swamps, beaches and beach ridges, dunes and dune fields, but the particular combination of forms varies immensely, depending on the coastal environment. In spite of this broad range of deltaic variability, however, all actively forming deltas share at least one important attribute: a river supplies sediments that accumulate more rapidly than can be removed by marine processes.

The Occurrence of Deltas

A delta may occur whenever a stream enters a receiving body of water. Consequently deltas of various sizes may be found at virtually all latitudes and on the shores of all seas. However, owing to the fact that major deltas require large river systems to supply the necessary quantities of sediment, they are relatively restricted in their occurrence: the existence of a major river system requires extensive catchment areas that cannot occur in all parts of the world. With respect to their tectonic classification of coasts, Inman and Nordstrom (1971) found that of 58 major rivers having drainage areas greater than 10^5 km^2, 46.6% entered the sea along Amero-trailing-edge coasts, 34.5% along marginal-sea coasts, 8.6% along Afro-trailing edge coasts, 1.7% along neo-trailing-edge coasts, and only 8.6% along tectonically active collision coasts, where drainage divides are normally close to the sea. Table 1 lists some of the largest and most significant river deltas together with their locations and areas.

Major Delta-Molding Factors

The areal extent, configuration, and characteristic depositional sequence of a delta depend on numerous dynamic factors, geological controls, and interacting processes. Global variability of deltaic landscapes results because of corresponding variations in the relative roles played by these different factors. Some of the more important determinants of delta development include the discharge regime and sediment load of the contributing river, climate, the nature of the river-mouth effluent diffusion processes, the wave climate of the receiving basin, the coastal wind regime, the tidal range, the strength and direction of coastal currents, the width and

TABLE 1. Locations and Areas of 33 Major Deltas

River	Continent	Receiving Body of Water	Deltaic Area km^2 × 10^3
Amazon	South America	Atlantic Ocean	467.1
Burdekin	Australia	Coral Sea	2.1
Chao Phraya	Asia	Gulf of Siam	11.3
Colville	North America	Beaufort Sea	1.7
Danube	Europe	Black Sea	2.7
Dneiper	Asia	Black Sea	—
Ebro	Europe	Mediterranean Sea	0.6
Ganges-Brahmaputra	Asia	Bay of Bengal	105.6
Grijalva	North America	Gulf of Mexico	17.0
Hwang Ho	Asia	Yellow Sea	36.3
Indus	Asia	Arabian Sea	29.5
Irrawaddy	Asia	Bay of Bengal	20.6
Klang	Asia	Straits of Malacca	1.8
Lena	Asia	Laptev Sea	43.6
Mackenzie	North America	Beaufort Sea	8.5
Magdalena	South America	Caribbean Sea	1.7
Mekong	Asia	South China Sea	93.8
Mississippi	North America	Gulf of Mexico	28.6
Niger	Africa	Gulf of Guinea	19.1
Nile	Africa	Mediterranean Sea	12.5
Ord	Australia	Timor Sea	3.9
Orinoco	South America	Atlantic Ocean	20.6
Paraná	South America	Atlantic Ocean	5.4
Pechora	Europe	Barents Sea	—
Po	Europe	Adriatic Sea	13.4
Red	Asia	Gulf of Tonkin	11.9
Sagavanirktok	North America	Beaufort Sea	1.2
São Francisco	South America	Atlantic Ocean	0.7
Senegal	Africa	Atlantic Ocean	4.3
Shatt-al-Arab	Asia	Persian Gulf	18.5
Tana	Africa	Indian Ocean	3.7
Volga	Europe	Caspian Sea	27.2
Yangtze-Kiang	Asia	East China Sea	66.7

slope of the continental shelf, and the tectonics and geometry of the receiving basin (Coleman and Wright, 1975; Silvester and LaCruz, 1970; Galloway, 1975).

Fluvial Regime. The discharge regime and sediment yield of the river depend on the size, climate, geology, and relief of the drainage basin (catchment). In addition to determining deltaic volume, discharge and sediment yield also play important roles in affecting the morphology of the delta. For example, the Mississippi, with a comparatively regular discharge and a fine-grained sediment load, exhibits straight, stable, and well-channelized distributaries in its delta, in contrast to the numerous shallow, migrating, sand-filled distributary channels that are frequently associated with erratic discharge and coarse-grained sediment loads. The rates at which the effluent from a river mouth spreads and decelerates on entering the receiving basin control the river-mouth morphology and have an important effect on the number and nature of the distributaries (Wright and Coleman, 1974).

Wave Regime. Marine and atmospheric energy inputs in the form of waves, winds, tides, and currents influence deltaic morphology by affecting the initial dispersal and deposition patterns near the river mouth and by reworking and redistributing deltaic sediments following initial deposition. Along delta coasts as along most other coastal types, ocean waves constitute one of the most important sources of marine energy. High wave energy increases the rate of mixing and momentum exchange between river effluents and ambient marine water, thereby increasing outflow deceleration and causing rapid deposition near the outlet. More important, powerful waves sort, redistribute, and remold deltaic sediments into beaches and beach ridges, thereby straightening the delta shoreline and reducing the extent of delta protrusions. In a comparative analysis of deltaic morphologic variability in terms of wave power and river discharge regime, Wright and Coleman (1973) found that deltas that experience high wave power relative to river discharge exhibit straight delta shorelines with *beach ridges* occupying interdistributary regions. The additional input of strong onshore winds commonly leads to the development of extensive dune sheets in the delta plain.

Tide Regime. Tides influence the rate of mixing of riverine and marine waters within as well as seaward of the delta distributaries, and play a major role in distributing the river-derived sediments. Many of the world's major river deltas occur in macrotidal environments where tidal range equals or exceeds 4 m (e.g., the Ord of Australia, the Shatt-al-Arab of Iraq, the Yangtze of China, the Klang of Malaysia). In such cases it is not uncommon for the tidal prism in the lower reaches of the delta to exceed substantially river discharge, with the result that tidal currents dominate the transport and deposition of clastics (e.g., Wright, Coleman, and Thom, 1973, 1975). Upstream transport of bed load by flood-tide currents leads to sand-filled channels, while linear tidal ridges paralleling the direction of flow develop seaward of the river mouth. Tidal flats and tidal creeks tend to prevail along distributary margins and in interdistributary regions.

Ocean Current Regime. Persistent unidirectional coastal currents occasionally transport riverine sediments, particularly fines, considerable distances along the adjacent coasts. The swift Guiana Current, which flows northwestward over the front of the Amazon delta, has inhibited the seaward progradation of the delta by sweeping the sediments debouched by the river hundreds of kilometers along the coast (Metcalf, 1968).

Shelf Configuration. The width and slopes of the continental shelf fronting a delta have an important bearing on the rate and extent of deltaic progradation: deltas can prograde faster and farther over flat, shallow shelves. Shelf configuration also exerts pronounced indirect influences on deltaic environments through its modification of tides and waves. Broad, flat shelves frequently cause amplification of tides and attenuation of incident wave power.

Receiving-basin Geometry and Tectonics. The overall large-scale configuration of deltas and the gross three-dimensional geometry of deltaic accumulations are dependent to a significant extent on the tectonics and geometry of the receiving basin. Vertically thick deltaic deposits are favored by rapid subsidence of the underlying basement (e.g., the Mississippi). The size and configuration of the receiving basin and the direction of delta growth relative to the structural axis of the basin often impose severe limits on the lateral or longitudinal extent of a delta (Coleman and Wright, 1975). The deltas of the Shatt-al-Arab and Chao Phraya, for example, have accumulated at the closed ends of narrow, elongate gulfs (the Persian Gulf and the Gulf of Siam) that have restricted lateral growth but permitted rapid progradation along the gulf axes. By contrast, the Klang delta of Malaysia (Coleman, Gagliano, and Smith, 1970) has advanced into the narrow Malacca Straits from the side, with the result that lateral spread has generally exceeded seaward progradation.

Delta Components

In spite of the high variability of deltaic environments and morphologic suites, the majority of the world's deltas may be broadly subdivided into the basic components shown diagramatically in Figure 1. It should be noted, however, that the configurations and relative dimensions of these components may vary appreciably between deltas.

All deltas consist of both *subaqueous* and *subaerial* zones, which lie below and above low tide level, respectively. The subaqueous delta is the foundation of a delta, and is normally much thicker vertically than the subaerial delta but will vary considerably in relative area. The subaqueous delta is usually characterized by seaward fining of sediments, and can be further subdivided into a basal *prodelta* unit, consisting primarily of fine silts and clays deposited from suspension near the outer reaches of the river effluent, overridden by a *delta front* unit composed of coarser-grained material deposited at least in part from bed-load transport. Rivermouth bar sands, tidal ridges, and, in the case of wave-dominated deltas, regressive beach and shoreface deposits may comprise the delta front. On the scale of the major deltas, the prodelta and delta front are analogous, respectively, to the bottomset and foreset deposits of Gilbert's (1884) classic delta model; however, the bedding patterns and sequence of "real" deltas are far more complex than in the simple model.

The subaerial delta that caps the subaqueous delta is subject to a wider range of processes and energy conditions. The subaerial delta consists of a *lower delta plain* and an *upper delta plain*. The lower delta plain (as defined by Coleman and Wright, 1971) extends inland to the effective limit of tidal influence; it is consequently widest in low-gradient macrotidal deltas and narrowest in microtidal deltas. Saltwater vegetation, such as *mangroves* and *saltmarsh*, together with estuarine processes tend to prevail in the less river-dominated portions of the lower delta plain. The upper delta plain lies above the limit of tidal influence, and is dominated largely by riverine processes. It is basically a seaward continuation of the riverine alluvial valley, and is the older portion of the subaerial delta.

Both the subaqueous and subaerial deltas can also be separated into abandoned and active zones. Significant regions of many deltas, including the Mississippi, have been abandoned by the river and its distributaries, owing either to course changes in the river or to delta lobe switching. While these abandoned regions may continue to prograde, they are dominated by marine and estuarine rather than riverine processes. The active delta, on the other hand, is that part of the delta under the immediate influence of the riverine distributary system, and receives sediment directly from the river.

At a higher level of resolution, every active deltaic system consists of several basic elements: one or more river-mouth systems with associated subaqueous and subaerial features; distributary channels and networks; interdistributary deposits with a variety of delta-plain features, and a delta shoreline.

The River Mouth

The river mouth is the point at which the riverine sediments that compose the delta are initially dispersed and deposited and as such it is the most basic and important single element of any deltaic system. Depositional patterns at river mouths are nearly as varied as deltas themselves, and are closely dependent on several primary processes that control the diffusion and deceleration of river effluents and on the modifying roles played by marine forces. Investigations in contrasting river-mouth environments (Wright and Coleman, 1974; Wright, Coleman, and Thom, 1973) indicate that effluent expansion and deceleration rates and consequent depositional patterns are governed by: outflow inertia; turbulent bed friction seaward of the mouth; and effluent buoyancy arising from density contrasts between issuing and ambient water. In addition to these primary influences, tide- and wave-induced processes interact with the effluent to modify the dispersion patterns and cause penecontemperaneous reworking of river sediments. Figure 2*A–C* illustrates the idealized depositional patterns associated

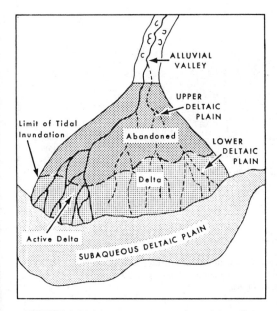

FIGURE 1. Major components of a delta (from Coleman and Wright, 1971).

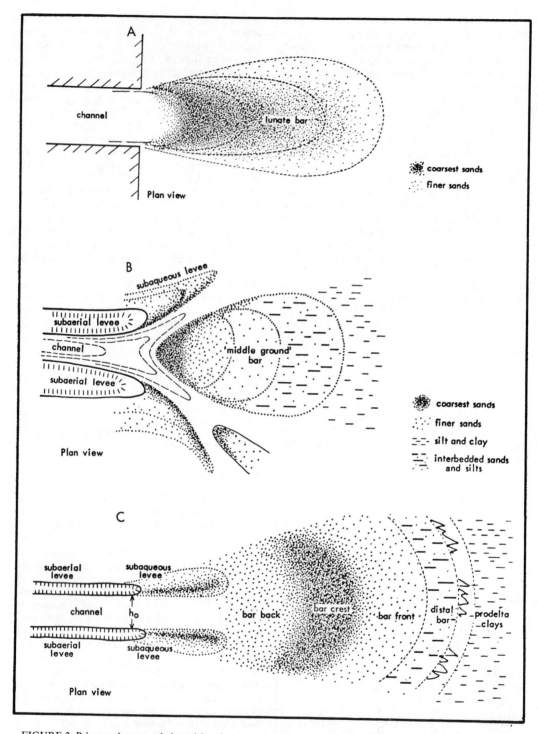

FIGURE 2. Primary river-mouth depositional patterns. *A*: narrow lunate bar formed by fully turbulent (inertia-dominated) effluent diffusion. *B*: middle-ground bar and bifurcating channels associated with friction-dominated effluent. *C*: depositional pattern associated with buoyant effluents.

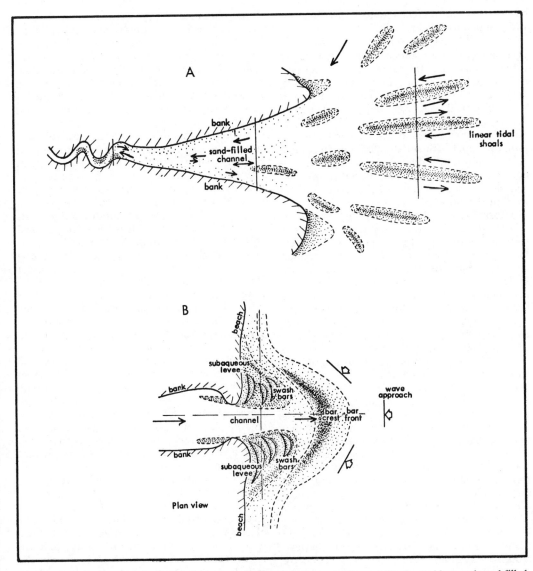

FIGURE 3. Marine-dominated river-mouth depositional patterns. *A*: linear delta-front ridges and sand-filled channels associated with tide dominance. *B*: depositional patterns associated with wave-dominated river mouths.

with the primary effluent processes; Figure 3 shows common marine-dominated river-mouth patterns.

Primary River-mouth Processes. The effect of inertia on effluent behavior is to cause the generation of turbulent eddies, which exchange momentum between the outflow and the receiving basin water. Inertia-dominated effluents exhibit fully turbulent jet diffusion similar to that described by Bates (1953) with low, lateral-spreading angles, and produce narrow, elongate bar deposits (Fig. 2*A*). Shallow water depths seaward of the mouth result in adding the effects of turbulent bed friction to lateral turbulent diffusion, causing a marked increase in expansion and deceleration rates (Borichansky and Mikhailov, 1966). This results in "middle-ground" type bars separating incipient bifurcating channels (Fig. 2*B*). Deep distributary mouths in low tide-range environments favor the intrusion of discrete salt wedges into the channels, resulting in buoyant effluents. Buoyancy tends to suppress turbulent diffusion but to promote buoyant expansion of the lighter effluent above denser seawater. The mouths of the Mississippi normally display this type of effluent (Wright and Coleman, 1974). Buoyant effluents commonly produce laterally restricted distributary-mouth bars and narrow elongate distributaries with parallel banks (Fig. 2C).

Effects of Tides and Waves. At river mouths in macrotidal environments, tidal currents are often stronger than river flow, and bidirectional currents redistribute the river-derived sediments, creating sand-filled, funnel-shaped distributaries with linear tidal ridges replacing the distributary mouth bar as the major delta-front feature (Fig. 3A). Powerful waves attacking a river outlet promote rapid effluent deceleration and transport river-mouth sediments back onshore, producing constricted river mouths with wide bars and shoals accumulating at short distances seaward of the entrance (Fig. 3B).

Deformational Processes. Following their original deposition, the deposits at many river mouths, particularly the mouths of the rivers that debouch large quantities of fine-grained material (the Mississippi, for example), are subject to penecontemporaneous subaqueous deformation. Some of these processes have been described by Morgan, Coleman, and Gagliano (1968), Shepard (1955, 1973), and most recently by Coleman et al. (1974). Subaqueous deformation occurs in response to rapid deposition near the river mouth combined with differential loading of bar sands onto prodelta clays. The subaqueous Mississippi delta exhibits several types of deformation: rotational slumping of oversteepened delta-front sands downslope into the prodelta; radial tensional faulting that creates delta-front gullies radiating from the locii of maximum deposition rates; mass wasting induced by degassing of fine sediments; deep-seated clay flowage; and clay diapirism (Coleman et al., 1974). Clay diapirism, described by Morgan, Coleman, and Gagliano (1968), is responsible for producing the well-known *mudlumps* (q.v.) that occur around the peripheries of the Mississippi's distributary mouth bars. Compression of prodelta clays by the overlying bar sands causes the former to flow toward locii of least pressure. Stress is ultimately relieved by diapiric intrusion, anticlinal folding, and faulting. The resulting diapirs may intrude upward through bar sands from depths as great as 200 m to assume surface expression as mudlump islands.

The Delta Plain

As a delta progrades, the delta shorelines and river mouths advance seaward and a surficial delta plain is produced surmounting delta-front deposits. Most delta plains are highly complex and encompass numerous subenvironments and depositional features. A delta plain is made up of at least three broad physiographic subdivisions: the distributary network; distributary-margin deposits; and interdistributary regions.

Distributary Networks. Delta distributary networks are produced by river-mouth progradation and associated branching processes; network patterns are consequently dependent on river-mouth environment. When the channel is highly stable or confined or when bifurcation is inhibited, for example, by high energy waves attacking the mouth, the river may follow a single channel all the way to the coast. Flat, shallow offshore profiles tend to favor frequent bifurcations that lead to a progressive seaward decrease in channel efficiency but increase the width of the subaerial delta plain (Coleman and Wright, 1975). Networks produced by rivers with erratic discharge regimes and coarse-grained sediment loads are often characterized by alternate channel bifurcation and rejoining, creating a braiding pattern. Quantitative descriptions of various types of distributary networks can be found in Smart and Moruzzi (1972).

Distributary Margins. Distributary margin features include natural levees, overbank crevasse splays, and minor secondary channels. In low-gradient deltas such as the Mississippi, the distributary natural levees are low, subtle, and relatively wide. It appears that these levees form in large part from the coalescence of *crevasse splays*. Crevasse splays are small mini-deltas of sediment debouched overbank into adjacent interdistributary bays. The nature and role of Mississippi Delta crevasse splays have been discussed by Arndorfer (1973) and Wright and Coleman (1974).

Interdistributary Features and Environments. The interdistributary regions (the regions between distributaries) account for most of the area of the deltaic plain and display a multitude of forms; the interdistributary morphologic suite is closely dependent on the dynamic environment and on the relative dominance of the various processes previously described. Rapid distributary progradation combined with subsidence and compaction of clays in a low wave-energy environment often yields shallow open or closed interdistributary bays, such as those that occupy the Mississippi's active deltaic plain. Slower subsidence rates allow the interdistributary bays to become infilled with marsh, as in the Danube delta. In Arctic deltas, freshwater thaw lakes are a common interdistributary feature. Extensive barren evaporite-crusted or mud-cracked tidal flats commonly make up the interdistributary surfaces of deltas in macrotidal arid or semiarid environments (e.g., the Ord delta of Western Australia; see Thom, Wright, and Coleman, 1975, or the Shatt-al-Arab of Iraq). In moister macrotidal environments densely vegetated mud flats and intricate networks of tidal creeks replace the barren flats of drier climates (e.g., the Ganges-Brahmaputra delta).

Alternations between periodic high wave energy and rapid shoreline progradation by continuous addition of river-borne fines produces interdistributary plains of widely spaced *chenier ridges* separated by tidal flats or marsh (e.g., the Burdekin delta of northeastern Australia). More powerful and persistent waves form continuous sandy plains of successive beach ridges or dunes (e.g., the São Francisco of Brazil and the Senegal of west Africa). High wave energy also produces straighter and more regular delta shorelines. Many of the *Holocene* (q.v.) deltaic plains along the wave-dominated, sediment-deficient Australian east coast have the form of flood plains filling deeply indented estuarine embayments that have been impounded by barrier formations (e.g., the Shoalhaven delta of southeastern Australia).

Long-term Evolution of Delta Plains. Over long periods of time extensive deltaic plains may be produced by abrupt changes in, or lateral migrations of, the loci of active river-mouth deposition. In many cases where a river is relatively unconfined in its lower course, a delta lobe may prograde in one direction until the channel becomes overextended and inefficient and then switch to a new position that offers better gradient advantage and a shorter route to the sea. During the Holocene, the Mississippi delta has changed position at least seven times (Kolb and Van Lopik, 1966) and possibly as many as sixteen times (Frazier, 1967). Coastal retreat and coastline smoothing under the influence of wave reworking and subsidence followed the abandonment of each lobe.

Deltaic Sediments and Sedimentary Structures

The composition and grain size of deltaic sediments may vary appreciably depending on the sediment yield and geology of the river catchment. Deltas may be composed of mud, sand, or gravel. However, the range of grain sizes debouched by most rivers is normally rather large, and significant quantities of both suspended silt and clays and sandy bed load may be supplied to the coast. The Mississippi transports 4.97×10^{11} kg of sediment to the Gulf of Mexico annually, of which 45% is clay, 36% is silt, and 19% is very fine to fine sand (Everett, 1971). In contrast, rivers that exhibit steeper gradients along their lower courses or that experience erratic or peaked discharge regimes may debouch higher relative amounts of coarse material. The Burdekin River of northeastern Australia, for example, has a highly erratic discharge, and arises from a granitic drainage basin in relatively close proximity to the delta. Consequently, considerable quantities of coarse sand reach the river mouth together with a high suspended load of silt and clay (Coleman and Wright, 1975). Whatever the absolute size of the delta sediments, the coarsest material tends to accumulate nearest the river mouth or in the high energy regions, while the finest material is deposited farthest from the outlets or in regions of lowest energy. Environmental differentiation of grain sizes is accompanied by differentiation in diagnostic suites of primary sedimentary structures. Some descriptions of sediments and sedimentary structures of various deltas include the following: Mississippi: Coleman and Gagliano (1964, 1965), Coleman, Gagliano, and Webb (1964), Gould (1970); Niger: Allen (1970); Rhone: Russell (1942), Oomkens (1970); Klang: Coleman, Gagliano, and Smith (1970); Colorado: Meckel (1975); Ebro: Maldonado (1975). Other examples are discussed by Coleman and Wright (1975).

Prodelta Sediments. In the majority of deltaic sequences, the prodelta, which is the basal unit of a delta, is the most laterally extensive unit and consists of the finest sediments. Characteristically the prodelta is a blanket of clays deposited from suspension. Parallel laminations are the dominant primary structure of the prodelta, and faunal content is usually high.

Delta-front Sediments. Progressing upward through the delta-front, lateral continuity of depositional units decreases, and variability both within and between deltas increases. Delta-front sediments are typically coarser than prodelta sediments and tend to coarsen and become better sorted upward. Current- and wave-induced structures, such as crossbedding, scour-and-fill structures, planar bedding, ripple marks, and flasers, also increase in abundance with decreasing depth, while faunal content decreases. The actual depositional features that prevail in different regions of the delta front depend on the process environment of the individual delta and may include distal bar, distributary mouth bar, tidal ridge, or shoreface deposits.

Deltaic Plain Sediments. Deposits associated with distributary channels, channel margins, and interdistributary environments exhibit extreme variability in terms of texture, composition, and primary structures. In macrotidal environments, distributary channels tend to fill with sands exhibiting large-scale bidirectional crossbedding; however, significant sand fill is commonly precluded in highly stratified distributaries, such as those of the Mississippi, that normally fill with fines upon abandonment. Channel levees are commonly characterized by multidirectional crossbedded sands capped by thin zones of intense burrowing. In many deltas the levee deposits are laterally continuous with wedge-shaped crevasse splay deposits consisting of thin stringers of alternating sands, silts, and clays. Interdistributary sediments may range from laminated or bioturbated silts, clays, and sands, in cases where bays are the dominant interdistributary feature, to well-sorted planar, laminated sands in deltas occupied by beach ridge plains or dune fields.

Delta Variability and Delta "Models"

Global variations in the relative intensities of the delta-molding forces produce a wide range of deltaic morphologies and depositional sequences. For examples of studies of contrasting deltas and deltaic environments, the reader is referred to Broussard (1975) and to the references listed by LeBlanc (1975) and Coleman and Wright (1975). In detail, each individual delta is unique; however, deltas exhibit certain common responses to various dominating processes or process combinations, and can be grouped into a finite number of more or less discrete deltaic types or "models." Depending on the relative dominance of fluvial, tidal, or wave processes, deltaic morphologies and sand-body geometries pass through a spectrum of types ranging from the *birds-foot* Mississippi pattern, to flaring, shoal-filled estuaries, to highly regular cuspate or straight beach ridge plains. Figure 4 shows six common deltaic sand-distribution patterns. The stratigraphic sequences of each of these types are described by Coleman and Wright (1975).

Type 1 is represented by the Mississippi, and is associated with fluvial-dominated deltaic environments having low tide range, low wave energy, flat offshore slopes, and relatively fine-grained sediment loads. The deltaic shoreline is highly crenulate, and bays, lakes, and marshes occupy the interdistributary regions. The dominant sand bodies are elongate, distributary-mouth bar deposits such as those described by Fisk (1961).

Type 2 occurs in association with extreme tidal ranges, which cause redistribution of sediments by means of strong bidirectional tidal currents. The Ord delta of Western Australia, the Fly delta, the Shatt-al-Arab delta, and the Colorado Delta are examples of this type. Seaward-flaring, sand-filled estuarine channels fronted by linear tidal shoals and flanked by tidal flats or heavily vegetated mudflats are the typical features. Deltas of this type are often referred to as estuarine deltas (Galloway, 1975) because the major sand accumulations assume the form of shoals filling broad estuaries.

Type 3 results from the combination of moderate tides and intermediate wave energy. Sand-filled channels are produced by tidal currents, while the higher waves redistribute sands parallel to the coastline trend, remolding them into beach ridges and chenier plains. Examples include the Burdekin (Australia), the Irrawaddy, the Mekong, and the Red (North Vietnam) deltas.

Type 4 is best represented by the Appalachicola, the Sagavanirktok, Horton, and Brazos deltas. The combination of intermediate wave energy, flat offshore slopes, and relatively small tides favors the occurrence of this type. The Niger delta (Allen, 1964, 1970) and Nile delta are probably transitional between Types 2 and 4.

Types 5 and 6 are both dominated by persistent high wave energy acting over relatively steep offshore slopes. Type 5 is represented by the São Francisco delta (Brazil), the Godavari Delta (India), the Tana (Kenya), and the Grijalva (Mexico). A cuspate deltaic plain occupied primarily by beach ridges or dune fields and a minimum number of distributaries are characteristic of Type 5. Type 6 is distinguished from Type 5 by the prevalence of strong unidirectional littoral drift, which causes longshore deflection of the river mouth. This produces a straight, nonprotruding delta shoreline. The Senegal (Senegal) and Marowijne (Surinam) Deltas are prominent examples.

Summary

Deltas are highly complex depositional landscapes that may be comprised of virtually the full range of individual coastal features in various combinations. Delta configuration, morphology, and stratigraphy may vary appreciably with geographic location and with time, and depend on the prevailing combination of fluvial regime, wave regime, tide regime, current regime, shelf width and slope, and basin shape and tectonics. There is no single model to which a sedimentary accumulation must conform in order to be classified as a delta; there is a wide spectrum of greatly contrasting delta types. Whatever their morphology and environment of deposition, however, all deltas by definition owe their existence to the presence of a sediment-supplying stream and to river-mouth processes that initially disseminate and deposit sediments at the coast.

LYNN D. WRIGHT

References

Allen, J. R. L., 1964. Sedimentation in the modern delta of the river Niger, West Africa, *in* M. J. U. Van Straaten, ed., *Deltaic and Shallow Marine Sediments*. Amsterdam: Elsevier, 26–34.

Allen, J. R. L., 1970. Sediments of the modern Niger

1. Low wave energy; low littoral drift; high suspended load.

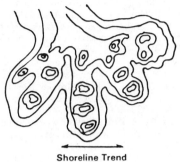

Shoreline Trend

2. Low wave energy; low littoral drift; high tide

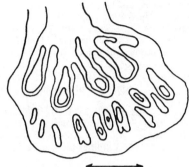

3. Intermediate wave energy; high tide; low littoral drift.

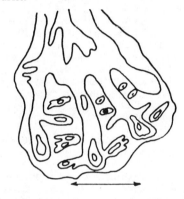

4. Intermediate wave energy; low tide

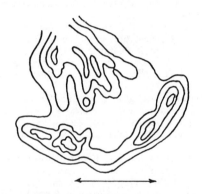

5. High wave energy; low littoral drift; steep offshore slope

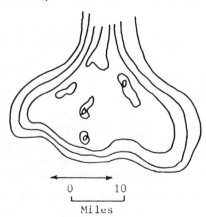

0   10
Miles

6. High wave energy; high littoral drift; steep offshore slope

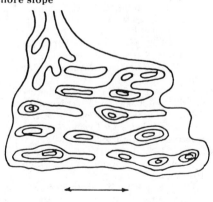

FIGURE 4. Six models of deltaic sand-body geometry (from Coleman and Wright, 1975).

Delta: A summary and review, *in* J. P. Morgan and R. H. Shaver, eds., *Deltaic Sedimentation, Modern and Ancient*. Tulsa, Okla.: Soc. Econ. Paleont. and Miner., Spec. Pub. 15, 138–151.

Arndorfer, D., 1973. Discharge patterns in two crevasses of the Mississippi River delta, *Marine Geology* 15, 269–287.

Bates, C. C., 1953. Rational theory of delta formation, *Am. Assoc. Petroleum Geologists Bull.*, 37 2119–2161.

Borichansky, L. S., and Mikhailov, V. N., 1966. Interaction of rivers and sea water in the absence of tides, *in Scientific Problems of the Humid Tropical Deltas and Their Implications*. Dacca: UNESCO, 175–180.

Broussard, M. L. S., ed., 1975. *Deltas: Models for Exploration.* Houston, Texas: Houston Geological Society, 555p.

Coleman, J. M., and Gagliano, S. M., 1964. Cyclic sedimentation in the Mississippi River deltaic plain, *Gulf Coast Assoc. Geol. Socs. Trans.* **14,** 67-80.

Coleman, J. M., and Gagliano, S. M., 1965. Sedimentary structures—Mississippi River deltaic plain, *in* G. V. Middleton, ed., *Primary Sedimentary Structures and Their Hydrodynamic Interpretation.* Tulsa, Okla.: Soc. Econ. Paleont. and Miner., Spec. Pub. 12, 133-148.

Coleman, J. M., and Wright, L. D., 1971. *Analysis of Major River Systems and Their Deltas: Procedures and Rationale with Two Examples.* Louisiana State University Coastal Studies Institute Tech. Rept. 95, 125p.

Coleman, J. M., and Wright, L. D., 1975. Modern river deltas: Variability of processes and sand bodies, *in* M. L. S. Broussard, ed., *Deltas: Models for Exploration.* Houston, Texas: Houston Geological Society, 99-149.

Coleman, J. M.; Gagliano, S. M.; and Smith, W. G., 1970. Sedimentation in a Malaysian high-tide tropical delta, *in* J. P. Morgan and R. H. Shaver, eds., *Deltaic Sedimentation, Modern and Ancient.* Tulsa, Okla.: Soc. Econ. Paleont. and Miner., Spec. Pub. 15, 185-197.

Coleman, J. M.; Gagliano, S. M.; and Webb, J. E., 1964. Minor sedimentary structures in a prograding distributary, *Marine Geology* **1,** 240-258.

Coleman, J. M.; Suhayda, J. N.; Whelan, T.; and Wright, L. D., 1974. Mass movement of Mississippi River delta sediments, *Gulf Coast Assoc. Geol. Socs. Trans.* **24,** 49-68.

Everett, D. K., 1971. *Hydrologic and Quality Characteristics of the Lower Mississippi River.* New Orleans: Louisiana Dept. Public Works, 48p.

Fisk, H. N., 1961. Bar finger sands of the Mississippi Delta, *in* J. Peterson and J. C. Ormond, eds., *Geometry of Sandstone Bodies.* Tulsa, Okla.: American Association of Petroleum Geologists, 29-52.

Frazier, D. E., 1967. Recent deltaic deposits of the Mississippi River: Their development and chronology, *Gulf Coast Assoc. Geol. Socs. Trans.* **17,** 287-315.

Galloway, W. E., 1975. Process framework for describing the morphologic and stratigraphic evolution of deltaic depositional systems, *in* M. L. S. Broussard, ed., *Deltas: Models for Exploration.* Houston, Texas: Houston Geological Society, 87-98.

Gilbert, G. K., 1884. The topographical features of lake shores, *U.S. Geol. Survey Annual Rept.* **5,** 104-108.

Gould, H. R., 1970. The Mississippi delta complex, *in* J. P. Morgan and R. H. Shaver, eds., *Deltaic Sedimentation, Modern and Ancient.* Tulsa, Okla.: Soc. Econ. Paleont. and Miner., Spec. Pub. 15, 3-30.

Inman, D. L., and Nordstrom, C. E., 1971. On the tectonic and morphologic classification of coasts, *Jour. Geology* **97,** 1-21.

Kolb, C. R., and van Lopik, J. R., 1966. Depositional environments of Mississippi River deltaic plain, southeastern Louisiana, *in* M. L. Shirley, ed., *Deltas in Their Geological Framework.* Houston, Texas: Houston Geological Society, 17-61.

LeBlanc, R. J., 1975. Significant studies of modern and ancient deltaic sediments, *in* M. L. S. Broussard, ed., *Deltas: Models for Exploration.* Houston, Texas: Houston Geological Society, 13-85.

Maldonado, A., 1975. Sedimentation, stratigraphy, and development of the Ebro Delta, Spain, *in* M. L. S. Broussard, ed., *Deltas: Models for Exploration.* Houston, Texas: Houston Geological Society, 311-338.

Meckel, L. D., 1975. Holocene sand bodies in the Colorado Delta, Salton Sea, Imperial County, California, *in* M. L. S. Broussard, ed., *Deltas: Models for Exploration.* Houston, Texas: Houston Geological Society, 239-266.

Metcalf, W. G., 1968. Shallow water currents along the northeastern coast of South America, *Jour. Marine Research* **26,** 232-243.

Morgan, J. P.; Coleman, J. M.; and Gagliano, S. M., 1968. Mudlumps: Diapiric structure and Mississippi Delta sediments, *in* J. Braunstein and G. D. O'Brien, eds., *Diapirism and Diapirs.* Tulsa, Okla.: Am. Assoc. Petroleum Geologists Mem. 8, 145-161.

Oomkens, E., 1970. Depositional sequences and sand distribution in the postglacial Rhone delta complex, *in* J. P. Morgan and R. H. Shaver, eds., *Deltaic Sedimentation, Modern and Ancient.* Tulsa, Okla.: Soc. Econ. Paleont. and Miner., Spec. Pub. 15, 198-212.

Russell, R. J., 1942. Geomorphology of the Rhone Delta, *Assoc. Am. Geographers Annals* **32,** 149-254.

Shepard, F. P., 1955. Delta front valleys bordering the Mississippi distributaries, *Geol. Soc. America Bull.* **66,** 1489-1498.

Shepard, F. P., 1973. Sea floor off Magdalena and Santa María area, Colombia, *Geol. Soc. America Bull.* **84,** 1955-1971.

Silvester, R., and LaCruz, C., 1970. Pattern forming processes in deltas, *Am. Soc. Civil Engineers Proc. Jour. Waterways and Harbors Div.* **96,** 201-217.

Smart, J. S., and Moruzzi, U. L., 1972. Quantitative properties of delta channel networks, *Zeitschr Geomorphologie* **16,** 268-282.

Thom, B. G.; Wright, L. D.; and Coleman, J. M., 1975. Mangrove ecology and deltaic estuarine geomorphology: Cambridge Gulf-Ord River, Western Australia, *Jour. Ecology* **63,** 203-232.

Wright, L. D., 1978. River deltas, *in* R. A. Davis, ed., *Coastal Sedimentary Environments.* New York: Springer-Verlag, 5-68.

Wright, L. D., and Coleman, J. M., 1973. Variations in morphology of major river deltas as functions of ocean wave and river discharge regimes, *Am. Assoc. Petroleum Geologists Bull.* **57,** 370-398.

Wright, L. D., and Coleman, J. M., 1974. Mississippi River mouth processes: Effluent dynamics and morphologic development, *Jour. Geology* **82,** 751-778.

Wright, L. D.; Coleman, J. M.; and Thom, G. B., 1973. Processes of channel development in a high-tide-range environment: Cambridge Gulf-Ord River delta, *Jour. Geology* **81,** 15-41.

Wright, L. D.; Coleman, J. M.; and Thom, G. B., 1975. Sediment transport and deposition in a macrotidal river channel, Ord River, Western Australia, *Estuarine Research* **2,** 309-321.

Cross-references: *Bars; Beach Ridge and Beach Ridge Coasts; Chenier and Chenier Plain; Currents; Deltaic*

Coasts; Global Tectonics; Mangrove Coasts; Salt Marsh; Sediment Size Classification; Tides; Waves. Vol. III: *Delta Dynamics; Deltaic Evolution.*

DEMOGRAPHY

Seacoasts are fringes of land 100 km wide or less that are adjacent to a sea or ocean. Three great blocks of the world's population inhabit seacoasts: Southeast Asia, with more than 1500 million people; Western Europe, with more than 600 million; and the eastern United States, with more than 250 million. The factors that influence the settlement contrast of continental inland areas also apply to seacoasts (Beaujeu-Garnier, 1965).

Factors Influencing Littoral Settlement

Physical Factors. Except for special instances of settlement for scientific research, men are usually repelled by polar or montainous regions as well as by tropical forest and desert areas. For example, the humid heat of the tropics has discouraged human settlement because of the epidemics and parasites (malaria, tsetse fly, bilarziosis, intestinal parasites) that attack the inhabitants and their animals (Hawley, 1962). *Hydric deficit,* more than thermal extremes, is responsible for the sparse population of tropical areas in western South America, western South Africa, western Australia, and both sides of the Red Sea. The amount of population in these areas contrasts sharply with that in the same latitudes on the Gulf of Mexico, India, and Southeast Asia, where water for farming is more readily available. This disymmetry is increased by the leeward/windward variations in rainfall; for example, in Sumatra, Brunei, Sunda, Canaries, and Caribbean islands. Finally, two factors determine the habitability of some coastal areas: the presence of littoral deserts caused by cold longitudinal currents (e.g., Perú, Benguela), and the mitigation of thermal extremes by relatively high humidity or warm or cold currents (e.g., the Gulf Stream, Kuroshio, and Oia Shio). With the exception of the western coast of Europe, most coastlines above 50° latitude have smaller populations.

Historical and Economic Factors. History shows that harsh physical circumstances of coastline are often balanced by advantages as large platforms, good currents, and rich fishing grounds. The *polders* or English *fens* are an important example of man's refusal to give up to the sea. The littoral dunes have been occupied since 2000 B.C. About A.D. 500 the Frisian lived on *terpen,* large mounds that provided refuge from flooding. The first collective defenses were built in the tenth and eleventh centuries with the beginning of urbanism. From the fifteenth to the nineteenth century, Holland led the maritime movement. Subsequently the state took control of the dikes in this country in which the population density was already more than 200 h/km^2 on littoral land mostly reclaimed from the sea. Engineers from the Low Countries later took part in bonification projects throughout the world.

Mediterranean historians have written about thalassocracy of Greece, Syria, Egypt, Crete, and other nations since the time of the Bronze Age. In the Iron Age the Mediterranean area was dominated by the Phoenicians. Carthage and then Rome later held power. Byzantium controlled the Mediterranean Sea from the sixth to the eleventh century, followed by the maritime powers of Genoa, Flanders, and Venice, which were influential from the eleventh to the thirteenth century. The Hanseatic League gained power in the Mediterranean during the thirteenth and fourteenth centuries.

Spain neglected the area during the following two centuries and built a colonial empire with settlements that were, naturally, started on the coasts. The maritime powers of Great Britain, Holland, and France followed this pattern of colonization of new areas. Whoever controlled the coastlines had an important effect on the practices of the day, including piracy, slavery, and smuggling.

The littoral settlement of black Africa has been affected by the intrusion of Europeans seeking land, rubber, ivory, gold, and slaves, as well as by the geographic disadvantages of unfavorable climate and difficult land access (Landry, 1945).

Today littoral areas in France, 30% of the nation's land, support 35% of the people. In addition, the littoral townships, with an area of only 3.8% of the land, contain more than 10% of the French population with an average density of 280 h/km^2 (Michaud, 1976). Because of rapid urban development, 60% of the people of the Iberian peninsula live on the littoral.

Fishing, agriculture, and commerce have always played major roles in the development of coastline settlements. Fishing provides work and food and was the principal means of livelihood for some ancient cultures, such as the people of the Maglemose cultures. When the climate improved on the western coast of Denmark, the Maglemosians were able to occupy the area and increase their fishing activity.

The tradition of agriculture depends on the depth and fertility of alluvial soils and on abundant irrigation such as is found in southeast Asia and in the deltas of the Nile and Ganges. Before the IV millenium B.C., the Nile delta had dikes and cultivated fields based on the regulation of flood water. The long history of human settlement in such areas is responsible for the density of settlement. In East Mekong, long a farming area, the population density is about 200 h/km^2 while the marshy West Mekong, more recently occupied, has population densities between 20 and 50 h/km^2. The Irrawaddi delta does not exceed 60 h/km^2 but the Red River delta exceeds 500 h/km^2 with 7.5 million inhabitants in 1936. In Djawa, high productivity of rice and healthful conditions are responsible for the extraordinary population densities of 500-800 h/km^2. The enrichment of coastal land with sand and other marine elements (*goëmen* in Brittany and *Possidonia* in the Mediterranean) greatly contribute to successful littoral farming.

Because of the natural accessibility to sea routes and the existence of labor and markets in coastal cities, industry also is often found on coastal strips. Tourism, an important activity in littoral areas, requires only sunshine and sandy beaches. The coastlines of low and middle latitudes with beaches open all year are especially appealing to tourists. The exotic foods and atmosphere and the native folklore are also enticing.

Littoral Population Concentrations

Continents and Climatic Zones. In Europe the seacoasts are populated more homogeneously than on any other continent. Although latitude has an important influence, there are inhabitants at latitude 70°. Oslo and Stockholm are at the same parallel as less-inhabited areas of Canada.

In Brittany the coastal population density, 150-180 h/km^2, contrasts sharply with that of the inland density, 40-80 h/km^2, because of the ports, fishing, intensive agriculture, and industry that exist on the coast. Greater inland-than coastal-density is exemplified in Ireland, Jersey, and the Hebrides where raising sheep is a major industry. Gascony was mainly uninhabited until the advent of oyster-culture and tourism. The steep coastline in the west of Scotland with its adverse climate remains nearly empty (Nonn, 1972).

The variation in littoral population densities of the Mediterranean coastlines has been related to periods of greater or lesser cultural and mercantile interchange and of maritime problems such as piracy, malaria, and so on.

The Soviet coasts of the Black Sea have a population density similar to that of the area around Vladivostock on the Pacific. In subdesert areas of the Middle East (Lebanon, Israel) people settle on the coastlines and riverbanks. In eastern Asia (Japan, Taiwan, Vietnam) most of the population is located on deltas and islands and there are great contrasts, such as overpopulated Djawa and sparsely populated Sumatra and Brunei. In Japan, a maritime country, the great cities and harbors are clustered on coastal plains on only 18% of the nation's land area, with rural densities of 500-100 h/km^2.

The demographic symmetry of Africa is a reflection of its climatic symmetry. The northern flank on the Biafra Gulf is irregularly populated with densities of about 25 h/km^2. The northern end of southwest Africa constitutes a deserted littoral. However, the seacoast of southwest Africa has a larger population than the inland area.

In the western hemisphere there are few inhabitants of the arctic area of North America and of the forests of equatorial South America; the Eskimos—35,000 human beings on 8,000 km of littoral land—and a handful of Fueguians fight for survival in a barely habitable environment. The great estuaries and ramified bays in the eastern United States from Boston to Washington, D.C. support the most densely populated area in the world, a *megalopolis* of 45 million people. The eastern coast of South America is populated mainly between the estuary of the Amazon and the fortieth latitude. On the western coast, the population density is very low in the tropics but is higher in the temperate latitudes of Chile. Other coastal clusters of population in South America are in Río de la Plata, São Paulo, and Río de Janeiro. Australia is sparsely populated with the main concentration on the southeast coast (Sydney-Newcastle and Melbourne).

In the temperate zone, contrasts in population density are due to natural or historical causes.

In the intertropical latitudes the environments are diverse. The lagoons and coasts of the Benin Gulf and Ivory Coast are not so much hostile as they are inaccessible and the population densities there range from 25-200 h/km^2. In the low Casamance, where rain is frequent, the cultivation of rice supports some population despite the presence of malaria. The average population in Liberia does not reach 15 h/km^2 but the densities in Nigeria are far greater.

Islands. Many Pacific islands can support settlers on coconut, fish, molluscs, and birds. In Melanesia, the average population density is 15 h/km^2 on the coasts as compared with 1.5 h/km^2 inland. In Micronesia and south Polynesia there are as many as 51 persons per km^2 inhabiting the coast compared to one person per km^2 inland. The coastal populations of islands under the influence of the trade winds dwindles greatly on the leeward side of the island with its desert-like climate.

Cities. The tendency toward urban communities is typified in littoral environments all over the world (Schwarz, 1966). In 1968, half of the French population was clustered in seven main agglomerations of more than 250,000 inhabitants that comprised 13% of the littoral townships. Of 70 of the world's greatest littoral cities, 40—not including political capitals—are centers of large metropolitan areas or dense population clusters. Included among those 40 are: San Francisco, Baltimore, Río de Janeiro, Hamburg, Rotterdam, Barcelona, Naples, Bombay, Calcutta, Shanghai, Osaka, Yokohama, Djakarta, and Alexandria.

Political Capitals. Older capitals, such as Havana, Lima, Callao, Montevideo, Buenos Aires, Copenhagen, Lisbon, Athens, London, Colombo, Seoul, Tokyo, and Manila, traditionally have been centers of finance, commerce, and industry and have thus attracted large populations. The population densities of newer capitals are being regulated by government restrictions (e.g., Brasilia, Canberra).

Ports and Their Facilities. The existence of a good port gives rise to activities and services that encourage the growth of the adjacent city. Shipping and commerce generate industry that may reach large proportions. Before the onset of air travel and air freight, most of the communication between continents went through the ports. Consequently, port cities such as New York, Río de Janeiro, Rotterdam, Singapore, Casablanca, and Sydney have been sites of large littoral agglomerations.

"Mushroom" Cities. The recent growth of some littoral cities has increased so suddenly as to approach geometric progression. It is difficult to find a common denominator for Hong Kong, Los Angeles, Río de Janeiro, Bombay, Calcutta, Singapore, and Shanghai but some of the growth factors include the presence of industry and the desirability of the residential aspect. In all of these cities, immigrants have increasingly provided the labor force. The annexation of quarters, villages, or districts has provided for the total urbanization of large areas (Fig. 1).

New and Old Cities. Ancient seacoast cities such as Istanbul or Alexandria achieved a status that more recent cities reached in one or two centuries. The urban rush continues and few countries are capable of regulating or assimilating it. Old littoral cities with ancient structures adapt with great difficulty to modern transportation and industry. Today's cities, even those newly planned, can rarely escape the material and social disadvantages of rapid growth. The Peoples' Republic of China is promoting a return to the land and villages from the cities more actively than any other nation in an effort to slow the progressive agglomeration in cities and it still has been unable to stop Shanghai from becoming the world's third largest metropolis.

Extreme Densities. Although a long tradition of inhabitation is important to the continued development of an area, more important is the general burst of polymorphic urbanism. To the original city are added more dwellings, industry, port facilities, tourist attractions, and tourist accommodations. It is the urbanism that provides an environment that draws large numbers of people. In Djakarta, the population reaches 7,930 h/km^2; in Washington, D.C., 4,298; in greater Athens, 5,865; in greater London, 4,623. The densities of the populations of these cities exceed the approximately 2,000 h/km^2 of Montevideo, Hamburg, Singapore, Hong Kong, and Alexandria.

Islands are often highly populated because of their limited areas. In the West Indies, overcrowded Haiti, Santo Domingo, and Puerto Rico reach a population density of 315 h/km^2; Martinique reaches 312; and Barbados, 545. In the Atlantic, Bermuda has 1,056 h/km^2 and Madeira, 322. In the Mediterranean, Malta has a population density of 1,009 h/km^2. Other African (Cape Verde and Mauritius) and Asiatic (Bahrein, Okinawa, Bali) islands exemplify the same phenomena in diverse circumstances. Older industrial areas—such as Liguria, West Flanders, the French north, Euzkadi, Catalunya, and Honshu—often exceed 300 h/km^2 and some exceed 400 or 500 h/km^2 (Lancashire, Lanark, Middlothian). However, long-inhabited areas with an agrarian base show demographic structures that are more regular and similar; for example, in Europe: northern Holland, 859 h/km^2, southern Holland, 1,052; in Asia: Bangla Desh, Jiang Su, Taiwan, Kerala, West Bengal, and Sri Lanka, 861 h/km^2, Djawa (an island), 500-800; and in Africa: Damietta, 721 h/km^2 and Beheira, 431 h/km^2.

Extremely low densities of population (fewer than 10 h/km^2 according to 1968-1970 statistics, see Fig. 2) are found in warm desert areas such as Queensland (1 h/km^2), Saudi Arabia (3 h/km^2), Sahara (0.3 h/km^2), Somalia (4 h/km^2), southwestern Australia (1 h/km^2), southwestern Africa (Moçamedes having 1 h/km^2), and Pacific America (Antofagasta, Coquimbo, and Arequipa with fewer than 8 h/km^2). Conditions in certain leeward areas are similar to those of the warm desert—for example, Madagascar has an average of 15 h/km^2 windward and 6 h/km^2 leeward.

The cold littorals in high latitudes are often nearly empty. The average population density

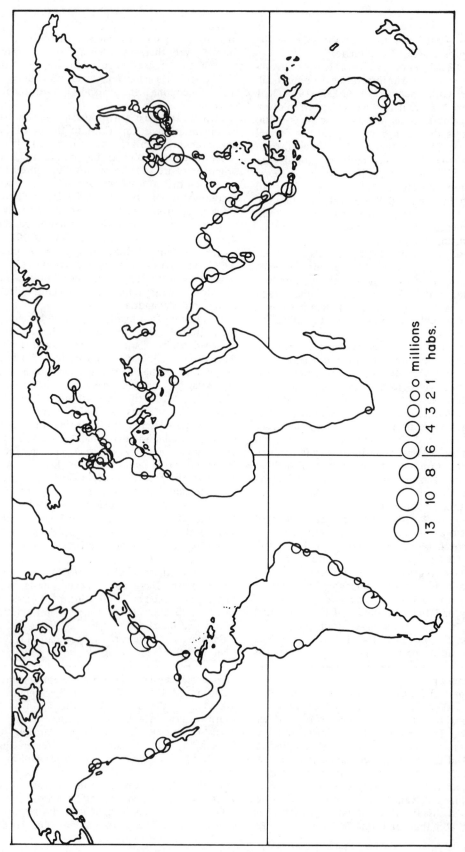

FIGURE 1. The seventy largest littoral cities of the world. These are coastal cities that during 1968–70 exceeded one million inhabitants. The greatest concentration is in the Far East followed by that in northwestern Europe (around the North Sea).

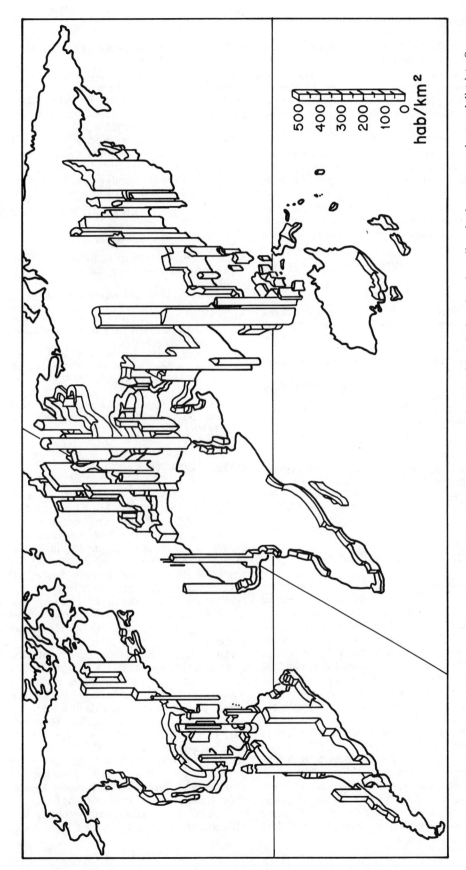

FIGURE 2. Littoral densities in 1968–70. A uniform conventional coastal band has been adopted, although there is a deep disparity between provinces and districts. Overpopulation is seen in the eastern United States, Brazil, and countries of La Plata as well as around the Mediterranean and in Western Europe. Most islands stand out because of their relatively high populations.

in such areas in the Americas is 9 h/km² in New Brunswick, 2 in British Columbia, 0 in the Yukon and Alaska as well as in Río Negro, 1 in Chubut, and 0.7 in Tierra del Fuego. In Europe, Iceland has 2 h/km² and Lapin, Inverness, Finmark, Norrland, Vasterbotten, and Norbotten have 2-8 h/km². In Asia, Kamchatka has 0.6 and Skahalin, 7.

For various reasons, equatorial forests also are nearly uninhabited. In Central America, Colon has 7 h/km²; Campeche, 4; and Quintana Roo, 2. Densities of 2-7 h/km² can be found in Pará, Amapa, Guyana, Marahano, Piaui, and Surinam in Amazonia; in Papua, West Irian, East Kalimantan, and Sarawak in Indonesia; and in Koilu in Africa.

VINCENÇ M. ROSSELLÓ

References

Beaujeu-Garnier, J., 1965. *Trois Milliards d'Hommes. Traité de Démographie.* Paris: Hachette, 402 p.
Hawley, A. H., 1962. *Human Ecology.* New York: Ronald Press, 433p.
Landry, A., 1945. *Traité de Démographie.* Paris: Payot, 651p.
Michaud, J.-L., 1976. *Manifeste pour le Littoral.* Paris: Berger-Levrault, 306p.
Nonn, H., 1972. *Géographie des Littoraux.* Paris: Presses Universitaires de France, 238p.
Schwarz, G., 1966. *Allgemeine Siedlungsgeographie.* Berlin: Walter de Gruyter, 751p.

Cross-references: *Climate, Coastal; Coastal Engineering, History of; Coastal Environmental Impact Statement; Coastal Reserves; Coastal Zone Management; Desert Coasts; Human Impact; Pollutants; Thalassotherapy; Tourism.*

DEPTH OF DISTURBANCE

The term *depth of disturbance* was introduced by King (1951) to describe the vertical limit of erosion and subsequent accretion experienced by a sand beach during a tidal cycle. King (1951, 1966) proposed a simple technique for measuring disturbance depth by recording, against fixed reference markers, the cut and fill of a column of dyed sand. Williams (1971) has modified the method, introducing more accurate location and height control by using a metal detector.

For a number of British beaches, King (1951) has shown that the expected depth of disturbance is on the order of 4% of wave height (e.g., waves 40 cm high will disturb the top 1 cm of beach), but experiments by Otvos (1965) and Williams (1971) indicate far higher ratios (up to 40%) can occur. Both Otvos and Williams describe the deposition of various types of dual and complex sediment units caused by specific wave and tide processes.

King (1951) suggested that the disturbance depth was related to wave height and beach sediment size, an increase in either allowing greater disturbance. However, Williams (1971) infers that a more complex relationship must exist, depending not only on wave height but on wave period, beach slope, and position on the foreshore. Interestingly, neither Williams nor Otvos (1965) found a significant correlation between disturbance and sand size.

RICHARD W. G. CARTER

References

King, C. A. M., 1951. Depth of disturbance of sand on sea beaches by waves, *Jour. Sed. Petrology* 21, 131-140.
King, C. A. M., 1966. *Techniques in Geomorphology.* London: Arnold, 342p.
Otvos, E. G., 1965. Sedimentation: Erosion cycles of single tidal periods on Long Island Sound beaches, *Jour. Sed. Petrology* 35, 604-609.
Williams, A. T., 1971. An analysis of some factors involved in the depth of disturbance of beach sand by waves, *Marine Geology* 11, 145-158.

Cross-references: *Beach Processes; Beach Profile; Coastal Morphology, Research Methods; Water Table; Wave Energy; Wave Erosion.*

DESERT COASTS

Desert regions meet the seacoast for a total length of about 30,000 km of coastline—i.e., only some 7% of the total world shoreline length. Since deserts occupy one-third of the earth's land surface, frequently in the rain shadow of nondesert coastal mountains, a relatively smaller portion of coasts has a desert environment as compared with its share of the land surface. Where the desert reaches down to the sea, the coasts are less incised and with fewer embayments than, for example, glaciated coasts.

The longest continuous stretch of desert coast is in southwest Asia, extending from the Gulf of Suez to the Savarashtra coast of the Kathiawar Peninsula in India. In Africa, desert coasts stretch from Tunis via the Red Sea to south of Mogadisho in Somalia. The western part of the Sahara comes down to the coast from Ogadir in Morocco to Mauritania, a distance of almost 2000 km. In southwestern Africa, an even longer stretch—2800 km—along the Namib desert from Luanda in Angola to Saint Helena Bay in South Africa, is a continuous desert coast. The southwestern coast of Madagascar, about 400 km long, is also a desert coast. In Europe, only the southern Spanish coast, stretching for some 300 km west of Alicante, can be classified as a desert coast.

In the Americas, mainly the western coasts are desertic. In North America, about 3200 km of the coastline of Baja California are desert. In South America, a 3700-km-long stretch bordering the Pacific from Peru to the Atacama Desert in Chile is a coastal desert. In Patagonia, the 1600 km from 39°S to 53°S latitude are a desert coast.

In Australia, with its extensive interior desert, only the western and southern littorals have desert coasts. In Western Australia it stretches from about 18°S to 27°S latitude, bordering the Great Sandy Desert. In South Australia, it stretches for about 1000 km along the littoral of the Nullarbor Plain.

Climate

Precise definitions of deserts use various climatic or ecological parameters for their delimitation, but a simple definition as "dry places supporting few life-forms" is frequently adequate. Because of the proximity of the oceans, desert coasts frequently differ in some of these specific parameters from inland deserts. They are especially distinguished from the adjacent inland deserts by a different temperature regime, with a much smaller diurnal temperature range than the interior, the consequence of shore winds (Fig. 1).

In general, humidity, fogs, cloudiness, and amounts of dew are also greater in desert coasts than in the interior desert. The chemical climate—i.e., the amount and rate of deposition of airborne salts—is high at the shore and generally decreases exponentially with the distance from the coast for the first few kilometers, then at a much slower rate farther inland. These airborne salts frequently have a profound effect on the landform features and soils of the desert coasts. The amount of rainfall, which by definition is small and irregular, is usually not related to the distance from the coast but may be controlled by orographic factors (Meigs, 1966).

Landforms

Desert coasts exhibit a great variety of physiographic features. Although sea cliffs do occur, constructional coastal landforms are more frequent than erosional. Relatively narrow littoral sedimentary plains at the foot of mountains or plateaus are the commonest landforms. Some alluvial fans may reach down to the shore, but frequently long stretches of desert coasts are only occasionally breached by gravelly or sandy wadi beds, carrying the episodic floods with their huge sediment load of mixed, but mainly coarse, detritus down to the coast. Some of the

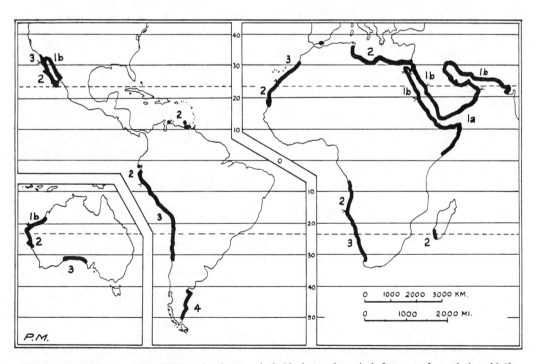

FIGURE 1. Desert seacoast climates. 1a: hot tropical; 1b: hot, subtropical; 2: warm; 3: cool; 4: cold (from Meigs, 1969; copyright © 1969 by The [University of] Arizona Board of Regents).

largest deltas (Nile, Mesopotamia, Indus) occur on desert coasts. However, their fine-grained alluvium has been transported over long distances from nondesert regions and is thus not a characteristic of desert deltas.

Material for the construction of the desert beach is frequently derived from clastic sediments transported by longshore currents and supplemented by local biogenic material, and salts precipitated in lagoons and *sebkhas*. Where longshore bars and barrier islands develop, the high evaporation rate of the shallow coastal and lagoonal waters turns the area into a sebkha environment, which is probably the most characteristic desert coastal landform.

Beach rock, commonly cemented by aragonite, is also known on desert coasts. In tectonically active areas, uplifted marine terraces, unless covered by sand dunes, are usually well preserved, especially where there is some cementation of the coastal sediments by $CaCO_3$.

Coastal sand dunes, with their sand either derived from inland or blown onto the beach from offshore, are another characteristic of desert coasts. If noncalcareous, the dunes remain mobile, covering extensive coastal plains. Transverse ridges are the common dune type. Where the sand is calcareous, and subject to intermittent wetting and drying, rapid cementation may occur, producing *eolianites*. These eolianites are frequently aligned in a number of parallel ridges along the coast. The presence of $CaCO_3$ also produces pisolitic or massive calcrete, frequently capped by a laminated calcrete crust.

Fringing and barrier reefs are common on desert coasts, but are more characteristic of tropical rather than desert conditions. Tidal flat sediments are less frequent along desert coasts than in other environments.

Nearshore shelf sediments in desert regions are frequently highly calcareous (Purser, 1973; Logan, 1970, 1974). The intensive modern studies of sebkha and carbonate sedimentation, which have changed considerably our understanding of the origin and distribution of such chemical sediments, are essentially studies of coastal desert environments (Schreiber, 1967).

DAN H. YAALON

References

Logan, B. W., ed., 1970. *Carbonate Sedimentation and Environment, Shark Bay, Western Australia.* Tulsa, Okla.: Am. Assoc. Petroleum Geologists Mem. 13, 223p.

Logan, B. W., ed., 1974. *Evolution and Diagenesis of Quaternary Carbonate Sequences, Shark Bay, Western Australia.* Tulsa, Okla.: Am. Assoc. Petroleum Geologists Mem. 22, 358p.

Meigs, P., 1966. *Geography of Coastal Deserts.* Paris: UNESCO, 158p.

Purser, B. H., ed., 1973. *The Persian Gulf: Holocene Carbonate Sedimentation and Diagenesis in a Shallow Epicontinental Sea.* New York: Springer-Verlag, 471p.

Schreiber, J. F., Jr., 1967. Desert coastal zones, in W. G. McGinnies, B. J. Goldman, and P. Paylore, eds., *Deserts of the World.* Tucson: University of Arizona Press, 647-724.

Cross-references: *Aerosols; Africa, Coastal Morphology; Asia, Middle East, Coastal Morphology: Israel and Sinai; Asia, Middle East, Coastal Morphology: Syria, Lebanon, Red Sea, Gulf of Oman, and Persian Gulf; Australia, Coastal Morphology; Climate, Coastal; Deltas; Europe, Coastal Morphology; North America, Coastal Morphology; Sabkha; Salt Weathering; South America, Coastal Morphology; Tafoni.*

DESIGN WAVE

The design of a maritime structure under wave attack must be the result of a compromise between excessive strength (conservative design) and allowance for expensive damages. Considering the height of the wave (H), the problem is to find a value H_{proj}, height of the *design wave*, derived by optimization. Several factors influence the selection of H_{proj}: type of structure (a breakwater or an oil platform require distinct H_{proj}), type of phenomenon (different H_{proj} if we are studying overtopping or the stability of a breakwater), available wave data, financial aspects, and so on.

The study of H_{proj} always requires a *statistical study of extreme values*. Let us consider some of the random variables involved: H (wave height) follows Rayleigh distribution under the assumption of fully arisen sea and narrow spectrum; $H_M(t)$ (maximum wave height for time interval t); $H_{1:m}$ (extreme value of H during m years); T_R (length of time record); I (record interval or time between the beginning of two consecutive records); and H_{max} (value of $H_M[t] = T_R$).

It is necessary to solve two distinct statistical problems: to extrapolate H_{max} from $t = T_R$ to $t = I$; and to extrapolate $H_M(t = I)$ for long periods (10, 100, 1000 years).

To solve the first problem, the assumption is made that the sea state is stationary during time I. Under the assumption that H follows Rayleigh distribution, the value $\mu(H_M[t])$ of the *most probable maximum wave height* during the time interval I is computed.

For the second problem, let us define *encounter probability* (p) as the probability that within n years there is at least one year in which a given value of H is exceeded; and *return period* ($m = E[N] = 1/p$) as the mean value (or expectation) of N (random variable representing the number of years elapsed from a given moment until the year in which there occurs an exceedance of a given value of H). Let $H_{1:100}$ be a given extreme value of H, exceeded on the average once in 100 years. The return period is $m = 100$, and the encounter probability in one year is $p = 1/100 = 1\%$. It is easy to calculate the encounter probability of this wave height for different periods of time: 1 year (1%), 10 years (10%), 100 years (64%), 500 years (99%), and 600 years (100%).

To estimate values for the random variable $H_{1:m}$ (the ten year wave, the one hundred year wave, the thousand year wave, and so on), it is necessary to find well-fitted extreme distributions. Commonly used are the *Gumbel distribution,* the *logarithmic-normal distribution,* and the *Weibull distribution* (with adaptation by Baatjes).

There are physical limitations to the growth of a wave (breaking, fetch, and so on), and we have to keep this in mind when we estimate design wave characteristics in order to get actual values.

H is the most widely used characteristic to define a design wave for studies of stability of a breakwater, workability of a port, behavior of a beach, and so on. In dynamic studies, other parameters may have to be considered in order to extrapolate extreme values: orbital velocity, acceleration, pressure, and so on.

For further readings on this subject see Brahtz (1968), Bretschneider (1969), Bruun (1976), and McCormick (1973).

CARLOS MORAIS

References

Brahtz, J. F., ed., 1968. *Ocean Engineering.* New York: John Wiley & Sons, 720p.
Bretschneider, C. L., ed., 1969. *Topics in Ocean Engineering,* vol. 1. Houston, Texas: Gulf Publishing Co., 420p.
Bruun, P., 1976. *Port Engineering.* Houston, Texas: Gulf Publishing Co., 586p.
McCormick, M. C., 1973. *Ocean Engineering Wave Mechanics.* New York: John Wiley & Sons, 179p.

Cross-references: *Boat Basin Design; Breakwaters; Coastal Engineering; Coastal Engineering, Research Methods; Waves.*

DIEBACK

Dieback represents patches of dying or dead vegetation or *rotten spots* in an otherwise healthy salt marsh. Although dieback is most frequently associated with *Spartina anglica* (a polyploid of *Spartina townsendii*) in English marshes, a similar phenomena has been studied in North American marshes (see references in Smith, 1970). Dieback is commonly limited to two locations within a salt marsh: along tidal creeks (*channel dieback*) and in small pans or areas where drainage is impeded (*pan dieback*) (Chapman, 1960). Both types of dieback sites exhibit similar characteristics, including fine-sized sediment, poor drainage, low oxygen content, high organic content and microbial activity, high salinities, and high concentrations of reduced ions, particularly sulfides.

Dieback is usually an ephemeral feature. The disintegration and revegetation of *rotten spots* occur in several stages, beginning with poorly drained vegetated pools, death of the vegetation, recolonization by dwarfed vegetation, and, finally, the establishment of plants that are normal in size and abundance. In the initial stages of dieback in a *Spartina anglica* marsh, plants begin to yellow and produce fewer tillers and rhizomes. Those rhizomes produced are long and soft, rotted at the apexes, and have a tendency to run horizontally rather than to turn upward as in healthy sites. Up to 90% of the elongate stems do not survive the winter, and the remaining 10% do not flower (Goodman and Williams, 1963).

There have been numerous observations but few satisfactory explanations of dieback in marshes. Earlier studies concentrated on pan dieback and attributed its formation to high salinity values (see references in Chapman, 1960). Water that is retained in small depressions after spring tides evaporates, thus increasing salinities to toxic levels. Although hypersalinity kills are known to occur, this theory does not explain the occurrence of dieback along the channels where salinity values approximate those of seawater. Chapman (1960) suggests that accumulation of debris may be responsible for the creation of bare spots. In places where debris and trash remain after snow melt or are washed up from creeks, the growth of vegetation is retarded. This area of lower growth becomes susceptible to the accumulation of more trash, and the vegetation eventually dies.

Dieback may be associated with some aspect of waterlogging such as anaerobiosis or toxicity of reduced ions (see references in Goodman and Williams, 1963). Analysis of plants from a dieback site in Lymington marsh, southern England, have shown that *Spartina anglica* is able to obtain sufficient oxygen to perform life processes. This suggests that rhizome rot may be due to a toxic inorganic ion. However, other researchers have found that sulfide concentrations are not toxic and suggest anaerobiosis as the major cause of dieback.

The effects of fungi and bacteria on apical rot in dieback areas have also been investigated as possible causes of dieback (see references in Sivanesan and Manners, 1972). The number and type of fungi are not appreciably different between healthy and dieback sites. However, the number of anaerobic and aerobic bacteria are less in the areas of dying plants. Because of the overall anaerobic nature of the dieback sites, the bacteria in their metabolic processes reduce ions to obtain oxygen, thus increasing the anaerobic state. Although not the initial cause of the rot, bacterial activity appears important in accentuating the problem.

It is evident that further investigations are needed to explain the development of dieback in coastal marshes, and it may be that no one theory is applicable to all sites.

LINDA C. NEWBY

References

Chapman, V. J., 1960. *Salt Marshes and Salt Deserts of the World*. London: Leonard Hill Limited, 392p.
Goodman, P. J., and Williams, W. T., 1963. Investigations into "Die-back" in *Spartina townsendii* agg., III: Physiological correlates of "Die-back" *Jour. Ecology* 49, 391-398.
Sivanesan, A., and Manners, J. G., 1972. Bacteria of muds colonized by *Spartina townsendii* and their possible role in *Spartina* die-back, *Plant and Soil* 36, 349-361.
Smith, W. G., 1970. *Spartina* "Die-back" in Louisiana marshes, *Coastal Studies Bull.* 5, 89-95.

Cross-references: *Coastal Flora; Halophytes; Salt Marsh; Salt Marsh Coasts; Vegetation Coast.*

DIFFRACTION—See WAVES

DIKE, DYKE

A *dike* is a wall, embankment, mound, or dam built around a low-lying area to prevent flooding by sea or by a river or stream (Griswold, 1976). Dikes built along rivers are known as *levees*, and were constructed as early as 2000 B.C. along the Nile River. A dike is usually constructed of soil from nearby land placed so that the top of the bank is above the maximum expected water level. The seaward side of the dike is protected against erosion by waves and currents by using various types of protective mats, including vegetation, concrete, and stone (Coastal Engineering Research Center, 1973). The most extensive dike system in the world exists in the Netherlands, where they protect that low-lying country. The most extensive levee system is along the Mississippi River, with 2545 km of mainline river levees and 3195 km of backwater, tributary, and floodway levees.

JOHN B. HERBICH
JOHN P. HANEY

References

Coastal Engineering Research Center, 1973. *Shore Protection Manual*. Washington, D.C.: U.S. Army Corps of Engineers, 750p.
Griswold, A. H., 1976. Levees, *in Encyclopedia Americana*, vol. 17. New York: Americana Corp., 261.

Cross-references: *Coastal Engineering; Coastal Engineering, History of.*

DISPERSION

Dispersion is the process or result of spreading, by active migration or of passive transfer, of organisms and physical properties, usually coupled with the scattering of the respective values of frequency distributions from their average. In physics, it is the separating of a beam of white light; in chemistry, the scattering of fine particles of one substance throughout another substance. The latter meaning has much in common with the term *dispersion* in oceanography, coastal engineering (q.v.), and related fields. Examples can be found in the spreading of wastes and other pollutants, dissipation of heat discharged from power plants, migration of plankton, scattering of sediment particles, and random transfer of chemical agents. All these seaborne quantities are subject to the action of numerous physical, chemical, and biological factors, such as currents, waves, radiation, chemical, and biochemical reactions. Since the latter are random, the former not only participate in the more or less regular processes (e.g., transfer in a certain direction by currents, long-wave back radiation of thermal plume, biochemical decay of bacteria) but also undergo random scattering to different places. Accordingly, dispersion can be observed in a fluid at macroscopic rest, owing to molecular and turbulent diffusion—i.e., random fluctuations of fluid particles. For further elaboration of these principles, see Hinze (1959), Ippen (1966), and Monin and Yaglom (1965, 1967).

RYSZARD ZEIDLER

References

Hinze, J. O., 1959. *Turbulence: An Introduction to Its Mechanism and Theory*. New York: McGraw-Hill, 586p.

Ippen, A. T., 1966. *Estuary and Coastline Hydrodynamics.* New York: McGraw-Hill, 744p.

Monin, A. S., and Yaglom, A. M., 1965. *Statisticheskaya Gidromekhanika,* vol. 1. Moscow: Nauka, 639p. (English translation, 1972. *Statistical Hydromechanics.* Cambridge, Mass.: M.I.T. Press, 605p.)

Monin, A. S., and Yaglom, A. M., 1967. *Statisticheskaya Gidromekhanika,* vol. 2. Moscow: Nauka, 720p.

Cross-references: *Coastal Engineering; Nearshore Water Characteristics; Nuclear Power Plant Siting; Thermal Pollution; Waste Disposal.*

DOUGLAS SCALE—See SEA CONDITIONS

DREDGING

Dredging is the process by means of which soil and/or minerals are removed from underwater locations. The dredges used to perform these operations vary with the type of deposit, the size of the deposit, the water depth, and the economics of the situation. The main purpose of dredging is to keep harbors and waterways navigable. There are two basic categories of dredges: mechanically operated and hydraulically operated (Herbich, 1975). Mechanical dredges include *dipper dredges* and *bucket ladder dredges.* The hydraulic type is the most efficient and economical dredge. It digs the material and transports it through a floating pipeline to the dredged material disposal area or stores it in hoppers that are moved to the disposal area and dumped. In the hydraulic type dredge, the material is first loosened by cutterheads or by agitation with water jets and is then transported as a fluid. Hydraulic or airlift dredging also shows promise as a means of removing deep ocean manganese nodule deposits (Herbich, 1975).

JOHN B. HERBICH
JOHN P. HANEY

Reference

Herbich, J. B., 1975. *Coastal and Deep Ocean Dredging.* Houston, Texas: Gulf Publishing Co., 622p.

Cross-references: *Coastal Engineering; Human Impact; Mineral Deposits; Protection of Coasts.*

DRIFTWOOD

Driftwood is logs, lumber, wreckage, and other bouyant, vegetative debris floated into the shore zone by currents and waves. Where deposited in large volumes by storm waves, driftwood may anchor sand deposits, thus contributing to accretion or cliff stability. Along cliffed coasts, however, individual logs and boards can become tool material of the crashing surf, thus contributing to erosion and shoreline retreat. Such logs have been known to demolish seashore buildings during storms.

Driftwood can be found on virtually any oceanic coast, but is significant as a geomorphic agent mainly in heavily forested coastal environments. Along the northwestern coast of North America, for example, floating timber has long been an environmental factor, judging from deposits found buried in centuries-old dunes. Driftwood accumulation has been greatly accelerated in the present day, however, by the inland activities of logging and land clearance. In recent decades, huge piles of driftwood, several hectares in area, have accumulated there, helping some beaches to prograde more than 100 m (Stembridge, 1979).

Driftwood collecting is a favorite activity of *beachcombers,* but logs rolling in heavy surf have killed many coastal visitors. Driftwood piles are considered hazards in some areas, and are intentionally burned at some resorts because they are considered unattractive.

JAMES E. STEMBRIDGE, JR.

Reference

Stembridge, J. E., 1979. Beach protection properties of accumulated driftwood, *in Proceedings of the Speciality Conference on Coastal Structures 1979.* Alexandria, Va.: American Society of Civil Engineers, 1052-1068.

Cross-references: *Beachcombing; Cliff Erosion; Flotsam and Jetsam; Human Impact; Tourism.*

DUNE STABILIZATION

Dune stabilization involves arresting the movement of entire subaerial *sand dunes* or parts thereof that have become mobilized by wind action following weakening or destruction of the vegetative cover.

Reshaping of the dune into an aerodynamically stable form is sometimes necessary prior to stabilization, and in certain circumstances, such as in the case of badly eroded frontal dunes, complete reforming of the dune with earth-moving equipment or by means of sand traps may be necessary.

The most popular method of dune stabilization is encouragement of a *psammosere;* plantings of *Ammophila* spp. have been carried out in Europe for this purpose since the fourteenth century. Sometimes mechanical stabilization of the sand surface and perhaps fertilizing are all that is required for initiation of a psammosere, but usually seeding or planting (in the case of species that are propagated vegetatively) of suitable species is required (Ranwell, 1972).

Secondary and tertiary colonization may then be left to natural processes or aided by plantings/seedings of suitable species. Usually some leguminous species are included in plantings of secondary colonizers, and woody species in the case of tertiary colonizers (Mitchell, 1974; Tsuriell, 1974; Wendelken, 1974).

Such colonization may also be directed toward the production of a particular type of cover or habitat, as in the case where the dune area is destined for use as, for example, recreation area, golf course, pine plantation, or wildlife refuge.

Although various types of artificial coatings, such as bitumen or latex sprays (with or without additives such as chopped straw), matting, or brush surfacing, have been used to stabilize dune surfaces, the principal value of such treatments seems to lie in their use as temporary stabilizing agents pending the establishment of vegetation in difficult situations.

In some cases, protection from wind or cutting off the sand supply may be feasible and adequate.

DAVID M. CHAPMAN

References

Mitchell, A., 1974. Plants and techniques for sand dune reclamation in Australia, *Internat. Jour. Biometeor.* **18**, 168-173.
Ranwell, D. S., 1972. *Ecology of Salt Marshes and Sand Dunes.* London: Chapman & Hall, 258p.
Tsuriell, D. E., 1974. Sand dune stabilization in Israel, *Internat. Jour. Biometeor.* **18**, 89-93.
Wendelken, W. J., 1974. New Zealand experience in stabilization and afforestation of coastal sands, *Internat. Jour. Biometeor.* **18**, 145-158.

Cross-references: *Coastal Dunes; Coastal Flora; Conservation; Deflation Phenomena; Major Beach Features; Sand Dune Habitat; Washover and Washover Fan.* Vol. III: *Sand Dunes.*

DYNAMIC PROCESSES, MEASUREMENT OF —See BEACH PROCESSES, MONITORING OF

E

EAGRE—See BORES

EARTHQUAKE SEA WAVE—See TSUNAMI

ECHINODERMATA

The echinoderms are a phylum of exclusively marine invertebrate animals that includes the familiar starfish (asteroids, Class Asteroidea), brittle stars or serpent stars (ophiuroids, Class Ophiuroidea), sea urchins (echinoids, Class Echinoidea), sea cucumbers (holothurians, Class Holothuroidea), and sea lilies and feather stars (crinoids, Class Crinoidea). All living forms display a more or less conspicuous five-part (*pentamerous*) symmetry. The dermal skeleton is composed of numerous ossicles of calcium carbonate in the form of calcite; thus the body is generally *bony* to the touch. Part of the body cavity or coelom is converted into a complex system of fluid-filled vessels, termed the water-vascular system. Some of these vessels project through the body surface as tube feet and tentacles, and assist in feeding, locomotion, and respiration. Excellent overviews of the phylum are provided by Hyman (1955) and Nichols (1969).

The five living classes comprise approximately 6000 species, which display a great variety of forms and habits. Most echinoderms are small, usually a few centimeters in diameter, but some sea cucumbers may attain a length of 2 m, and some starfish exceed 1 m in diameter. Echinoderms occur in all seas and can be found at the greatest recorded depths. In deep abyssal and hadal zones, holothurians may comprise more that 90% of the total biomass present. Tropical shallow-water forms usually have a wide distribution. Temperate forms tend to have more restricted distribution patterns. Arctic and Antarctic species are frequently circumpolar. Dispersal is achieved by migration of adults, by passive transport of adults or juveniles in floating seaweed or other flotsam, or by transport in ocean currents of planktonic larval stages.

The sexes are usually separate, and usually echinoderms shed eggs and sperm into surrounding seawater, where fertilization takes place. The fertilized egg may develop into a distinctive type of planktotrophic larval stage, or if a large amount of yolk material is present, the zygote may pass through a variety of nonfeeding developmental stages to form a juvenile. Some echinoderms retain broods of young until they develop into juveniles; this form of reproduction is common in Antarctica.

In addition to the five classes that exist today, another 16 classes (approximately 13,000 species) are extinct, all of these having died off during the Paleozoic era. Fossil echinoderms frequently occur in vast beds, and obviously they were important contributors to sea floor sediments during the Paleozoic. Similar vast aggregations can be found today; these usually occur in response to availability of suitable food material.

Classes

Asteroidea (Fig. 1A, B). Asteroids or sea stars typically have five approximately equal hollow arms, although forms with more than five arms are common. They usually live with the mouth facing the substrate. In shallow water, asteroids tend to live on hard substrates, where they can cling and *walk* with the aid of their suctorial tube feet. Those that have pointed nonsuctorial tube feet live in sand, where they can conceal themselves by burrowing. Feeding habits can vary considerably. Some deeper-water forms ingest mud and extract organic matter from it; others are selective detritus feeders; still others are opportunistic scavengers. The best-known feeding method is active predation upon bivalved mollusks, such as clams, oysters, and scallops. By means of a steady pull of the tube feet, the starfish can cause the mollusk's valves to open slightly. The starfish then everts its stomach into the mollusk shell and digests the animal within.

Seven orders of starfish are recognized. These include: the order Phanerozonida, with conspicuous marginal plates (Fig. 1*A*); the order Spinulosida, with reduced marginal plates, with the skeleton of the upper body appearing as a reticulated mesh, and with the submicroscopic pincer organs (pedicellariae) absent or very simple; the order Forcipulatida, with reduced marginal plates, and the pedicellariae forcipulate well developed as defensive structures (Fig. 1*B*).

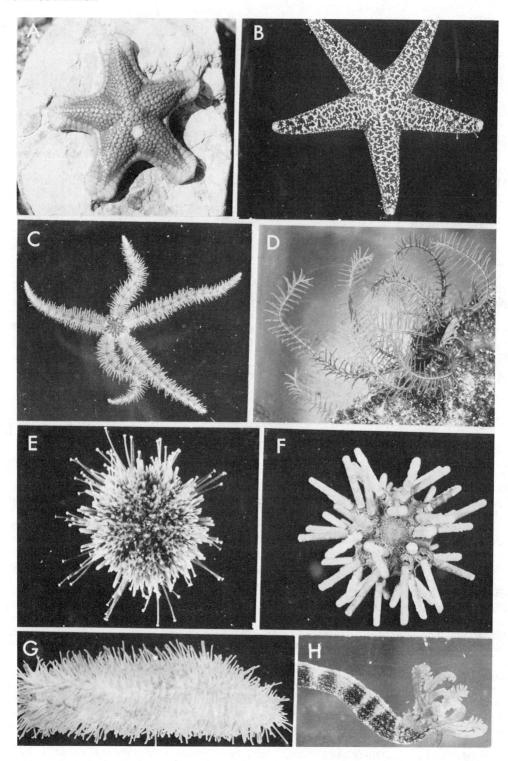

FIGURE 1. *A*. Starfish, *Asterodon*. *B*. Starfish, *Piaster*. *C*. Brittle star, *Ophiothrix*. *D*. Feather star, *Nemaster*. *E*. Regular sea urchin, *Strongylocentrotus*. *F*. Regular sea urchin, *Eucidaris*. *G*. Sea cucumber, *Cucumaria*. *H*. Sea cucumber, *Euapta* (photos: *A*, *F* by D. L. Pawson; all others by K. Sandved).

Ophiuroidea (Fig. 1 C). Ophiuroids or brittle stars have five solid, serpentine arms attached to a circular central disc. The internal organs, apart from branches of the nervous and water-vascular systems are all restricted to the disc. Brittle stars are so named because of a tendency for the arms to break into several pieces when the animal is disturbed. New arms are easily regenerated, the power of extensive regeneration being well developed in the echinoderms. While most ophiuroids have five unbranched arms, representatives of one group have arms that branch repeatedly, giving rise to a mass of tendrils that form a feeding net for the capture of small planktonic organisms. Ophiuroids are usually secretive, living under rocks, in crevices, or in burrows. In deeper water, they frequently form vast aggregations on the substrate surface. Some ophiuroids are selective detritus feeders, while others capture small organisms on their sticky tube feet and pass them to the mouth.

Two orders of ophiuroids exist today. The order Phrynophiurida has the body skeleton covered by soft integument, and the arms are capable of movement in all directions. Several members have branching arms. The order Ophiurida comprises forms whose skeleton is not covered by soft integument; the arms are unbranched, and generally capable of lateral movement only.

Crinoidea (Fig. 1D). Crinoids occur in two basic forms. Feather stars are common in tropical areas, where they are conspicuous members of coral reef biotopes. As in the ophiuroids, most organs are contained within a central chamber, here termed a calyx. On the underside of the calyx, several short cirri are used to hold the animal to a hard substrate. The upper surface of the calyx gives rise to several feather-like arms used for the capture of food. Sea lilies differ from feather stars in having a stem in place of the cirri. The calyx and arms are raised above the sea floor on the stem, which is anchored in the mud by means of a holdfast or root. Crinoids are plankton feeders, capturing minute drifting organisms in mucus secreted by the tube feet and passing them to the mouth.

Living crinoids are classified into the single order Articulata.

Echinoidea (Figs. 1E, F; 2A-D). Echinoids also occur in two basic forms. *Regular* echinoids

FIGURE 2. *A*. Portion of a large population of the sea urchin *Echinometra* that has excavated shallow burrows in a rocky shore at high tide level. *B*. Oral area of *Echinometra*, showing mouth with five pointed teeth. *C*. *Echinometra* in a deep burrow in a coral reef. *D*. Irregular sea urchin, *Brissopsis*, in process of burrowing in a sandy substrate (photos: *A* by R. B. Manning; *B* by K. Sandved; *C*, *D* by P. M. Kier).

are usually approximately spherical, with mouth and anus on opposite sides of the body. Five meridional bands of tube feet (ambulacra or ambs) run from mouth to the anal area; spaces between the ambs are termed interambs. Large and small movable spines are carried on hemispherical tubercles that are most common in the interambs. The mouth is equipped with five sharp teeth, which are used to chew and grind food material. Regular echinoids are most common on hard substrates, although in deeper water then can be numerous on mud bottoms. *Irregular* echinoids tend to be elongate, bilaterally symmetrical. On the underside of the body the mouth tends to be placed anteriorly. On the upper surface the anus is no longer centrally placed; it may lie toward or at the posterior end of the body or even on the oral surface. Distinctive petal-shaped areas of pores for tube feet radiate from the sieve plate or madreporite on the upper surface. Irregular echinoids are usually rather fragile, and carry a dense coat of short spines. In some forms considerably longer spines may be scattered among the short ones. These echinoids are well adapted to life as burrowers, and most live in sandy to muddy substrates, often burrowing to depths of 20 cm or more. Most lack teeth and feed by selectively or nonselectively ingesting detritus. Regular sea urchins usually graze upon algae.

There are approximately 15 orders of echinoids alive today; their classification is rather specialized. The primitive order Cidaroida (Fig. 1*F*) is placed in the subclass Perischoechinoidea. All other living echinoids are placed in the subclass Euechinoidea. The regular echinoids comprise the superorders Diadematacea and Echinacea; all possess teeth throughout their lives The irregular echinoids comprise the superorder Gnathostomata, which have teeth throughout life, and the superorder Atelostomata, which lack teeth.

Holothuroidea (Fig. 1*G, H*). The generally soft-bodied sea cucumbers or holothurians are more or less cylindrical, with mouth and anus at opposite ends of the body. In four of the six orders, five rows of tube feet traverse the body from mouth to anus, defining five ambulacra or radii. Two orders lack tube feet, locomotion being effected by muscular contractions of the body wall. The mouth is surrounded by a ring of modified tube feet, usually termed tentacles, which are used for feeding. While most echinoderms have a hard body made up of a large number of skeletal elements, in the holothurians the elements are reduced in size to form microscopic ossicles embedded in the body wall.

Sea cucumbers may capture small organisms with their sticky tentacles and pass them into the mouth, or they may use shovellike motions of the tentacles to pass quantities of the sediment into their mouth. In the intestine, organic matter is extracted and unused sediment is passed from the anus in the form of castings. Some holothurians are the most important reworkers of sediments in many areas of the world (see below).

The six orders include the order Aspidochirotida, with 10 to 30 shield-shaped tentacles, numerous tube feet, and a thick, leathery body wall. Most conspicuous tropical shallow-water forms belong to this order. The order Dendrochirotida comprises mostly inconspicuous small forms with 10 to 30 richly branched tentacles, numerous tube feet, and a thin body wall. Members of the order Apodida are wormlike, fragile, lacking tube feet and possessing featherlike tentacles. The order Molpadida comprises fusiform holothurians with the posterior end produced to form a tail; tube feet are absent; the tentacles are more or less clawshaped, with few branches.

Effects of Echinoderms on Coastal Environments

In rocky intertidal areas, regular echinoids can dramatically affect the environment in a variety of ways. By using spines and teeth, many species can excavate shallow to deep burrows in rock, causing the gradual breakdown of this substrate (Fig. 2*A-C*). Regular echinoids usually graze upon algae and are capable of completely denuding vast areas of coastline, rendering such areas inhospitable for many kinds of marine organisms. Relationships between these sea urchins and algae are discussed in a recent review by Lawrence (1975). Off western North America, the complex interactions between echinoids, the giant kelp *Macrocystis,* sea otters, abalones, and other predators and vegetarians have been the subject of numerous recent studies (see, for example, Lawrence, 1975). Irregular urchins actively rework subsurface sediments during their burrowing activities in sand and mud (Fig. 2*D*). They often form huge subsurface communities. Bromley and Asgaard (1975) have studied fossil and recent workings of these echinoids.

The burrowing starfish and brittle stars also disturb sediments to varying degrees. Crinoids do not significantly affect sediments directly.

Holothurians are the prime reworkers of mud, sand, and ooze substrates at all depths. Several groups of holothurians actively ingest large quantities of the substrate and pass it through their intestine (see review by Pawson in Boolootian, 1966). Rhoads and Young (1971) intensively studied the holothurian *Molpadia* in Cape Cod Bay, and found that burrowing and feeding activities produced extensive vertical sediment sorting, high sediment water content, and altered topographic relief of the sea floor.

Echinoderms in general, because of their well-developed calcite skeletons, contribute significantly to the formation of carbonate sediments. The contributions of echinoderms to sediments throughout geologic time are reviewed by Lowenstam (1963).

DAVID L. PAWSON

References

Boolootian, R. A., ed., 1966. *Physiology of Echinodermata.* New York: John Wiley & Sons, 822p.

Bromley, R. G., and Asgaard, U., 1975. Sediment structures produced by a spatangoid echinoid: A problem of preservation. *Geol. Soc. Denmark Bull.* 24, 261-281.

Hyman, L. H., 1955. *The Invertebrates, Vol. IV: Echinodermata.* New York: McGraw-Hill, 763p.

Lawrence, J. M., 1975. On the relationships between marine plants and sea urchins, *Oceanog. Marine Biol. Annual Rev.* 13, 213-286.

Lowenstam, H. A., 1963. Biologic problems relating to the composition and diagenesis of sediments, *in* T. W. Donnelly, ed., *The Earth Sciences: Problems and Progress in Current Research.* Chicago: University of Chicago Press, 137-195.

Nichols, D., 1969. *Echinoderms,* 4th ed. London: Hutchinson, 192p.

Rhoads, D. C., and Young, D. K., 1971. Animal-sediment relations in Cape Cod Bay, Massachusetts. II. Reworking by *Molpadia oolitica* (Holothuroidea), *Marine Biology* 11, 255-261.

Cross-references: *Algal Mats and Stromatolites; Biotic Zonation; Coastal Fauna; Coastal Waters Habitat; Coral Reef Habitat; Organism-Sediment Relationship; Rock Borers; Weathering and Erosion, Biologic.*

ECHIURA

Echiurans are *unsegmented worms* that range in size from 7-470 mm. The body is divided into two parts: a nonretractable anterior proboscis, and a posterior trunk. The proboscis is folded to form a ventral ciliated gutter; it is variable in length (up to 1.5 m), and in some species is highly extensible. Although the trunk appears segmented because of rows of papillae, internally it is a single sac of coelomic fluid, which serves as a hydrostatic skeleton and holds the internal organs, including the highly convoluted digestive system extending from the mouth at the anterior tip of the trunk to the anus at the posterior. Unusual structures in the echiurans are anal sacs that function as excretory organs like nephridia but empty into the anus. These and the presence of setae and a circulatory system serve to differentiate the phylum Echiura from the phylum Sipuncula. The nervous system, circulatory system, setae, and aspects of the excretory system and of the early development of the echiurans are similar to those of the annelids. The Echiura appear to lack, however, the segmentation characteristic of the Annelida, although thickenings of the mesodermal bands in the larva have been interpreted as incipient segmentation (Stephens and Edmonds, 1972).

Most echiurans feed by extending the proboscis over the substrate and passing diatoms, bacteria, algae, and so on down the ventral groove to the mouth (Kaestner, 1967). The Pacific Coast form, *Urechis caupo,* however, constructs a mucous net across its burrow, pumps water, and ingests the net and entrapped particles. Some echiurans live on rocks and in crevices, but many live in burrows as deep as 40-60 cm below the sediment/water interface. *Echiurus* excavates a U-shaped tube with hardened walls, and is also capable of burrowing through sediment using peristaltic motion of the trunk and the anterior and posterior setae (Schäfer, 1972). Of the 100 or so species, some occur at depths to 10,000 m, but most live in shallow water, sometimes in great density.

MOLLY FRITZ MILLER

References

Kaestner, A., 1967. *Invertebrate Zoology,* Vol. 1. New York: Wiley Interscience, 597p.

Schäfer, W., 1972. *Ecology and Paleoecology of Marine Environments,* trans. I. Oertel, ed. G. Y. Craig. Edinburgh: Oliver and Boyd, 568p.

Stephens, A. C. and Edmonds, S. J., 1972. *The Phyla Sipuncula and Echiura.* London: British Museum (Natural History), 528p.

Cross-references: *Annelida; Aschelminthes; Biotic Zonation; Coastal Fauna; Nemertea; Organism-Sediment Relationship; Sipuncula.*

EDGE WAVES

Edge waves are free-wave motions introduced by a coast in its interaction with surges (the latter being understood as disturbances of atmospheric origin with dominant periods ranging roughly from 10^3 to 10^5 sec, thus falling between the *tsunamis* and the lower-frequency *astronomical tides*) or lower-period oscillations (Bowen and Inman, 1971). The simplest mathematical way of introducing a coast into the theoretical models of surges is to take an infinitely long vertical wall, in which case the edge waves can be obtained by simple reflection of the ordinary sinusoidal waves at sea.

A more complicated model, but one much more satisfying from the standpoint of similarity with a real coast, is an infinitely long beach of constant slope. Eckart (1951), using shallow water theory, showed that an edge wave of longshore wave number $\lambda = 2/$wavelength and radian frequency $\omega = 2\pi/$(wave period) had a velocity potential Φ, which is given by

$$\Phi = \frac{ga_n}{\omega} \cos\lambda y \cdot L_n(2\lambda x) e^{-\lambda z} \cos \omega t$$

where $\omega^2 = g\lambda(2n+1)\tan \beta$; x and y are the horizontal coordinates in the offshore and longshore directions; $L_n(2\lambda x)$ is the Laguerre polynomial of order n; and a_n is the wave amplitude at the shoreline ($x = 0$). For higher modes n, the curves are more oscillatory, and their amplitude decreases seaward more slowly. The longshore variation is given by $\cos \lambda y$. The solution has a further condition—that the dispersion relation between δ and λ is given by

$$\delta = g\lambda(2n+1)\tan \beta$$

where $\tan \beta$ is the beach slope. For any given frequency, there are a set of modes of modal number n and amplitude a_n. The surface elevation has *nodes*, where the surface displacement is always zero, when $\cos \lambda y = 0$ and $L_n(2\lambda x) = 0$. The surface behavior, particularly for the higher modes, can be rather complex. Although the incoming wave may interact with all the possible edge wave modes of the same frequency, Bowen and Inman (1969) found that the interaction with one particular mode is often dominant.

Bowen and Inman (1971) have shown that standing edge waves provide a satisfactory explanation of crescentic bars in regions of small tidal range, the bars having a longshore wavelength of one-half that of the edge waves. In the absence of large incoming surface waves, the edge waves may also form cuspate features on the beach face.

The effect of the earth's rotation becomes important only for waves whose period is appreciable compared to that of half pendulum day.

RYSZARD ZEIDLER

References

Bowen, A. J., and Inman, D. L., 1969. Rip currents, *Jour. Geophys. Research* 74, 5479-5490.

Bowen, A. J., and Inman, D. L., 1971. Edge waves and crescentic bars, *Jour. Geophys. Research* 76, 8662-8671.

Eckart, C., 1951. *Surface Waves in Water of Variable Depth*. La Jolla, Calif.: Scripps Institution of Oceanography, Wave Rept. 100, S10Ref.S1-12, 99p.

Cross-references: *Bars; Beach Cusps; Coastal Engineering; Coriolis Effect; Nearshore Hydrodynamics and Sedimentation; Waves.*

EFFLUENTS

In the context of coastal environments, the term *effluent* may refer to river water being discharged into the sea or, alternatively, to waste water undergoing disposal in coastal waters.

A *river effluent*, by virtue of its low salinity, is initially confined to the surface of the water column, and forms a seaward-extending plume. This plume can be recognized in aerial photographs or satellite imagery because its suspended solids content is normally greater than that of adjacent coastal waters. River discharge (which may vary tidally as well as seasonally), water density differentials, coastal currents, and wave action influence the areal configuration of the plume.

Four types of *waste water effluents* are commonly released in coastal waters: *thermal effluents*, which consist of ambient water that has received an addition of heat (Δt) in the process of electrical power generation; *sewage effluents*, which contain suspended solids, nutrients, and bacteria and are subject to differing degrees of prior treatment; *chemical effluents*, which encompass a variety of constituents resulting from such activities as petrochemical refining, industrial processing, and paper manufacturing (Kraft-Mill effluent); and *dredge spoil effluent*, which consists of a sediment slurry resulting from navigational dredging or offshore mining. These effluents may be introduced into receiving waters by means of shoreline discharges, submerged outfalls or diffusers, or dumping from barges. After being released into coastal waters, waste water effluents also assume a plume configuration, but these plumes tend to be long and narrow in comparison with most river plumes.

The postdischarge behavior of both river and waste water effluents, which is a function of the same advective and diffusive processes, has recently been the subject of extensive field studies, laboratory simulations, and mathematical modeling. Dispersion of river effluent is an important mechanism for suspended sediment transport in the coastal zone, and dispersion of waste water effluents bears heavily upon coastal water quality (Palmer and Gross, 1979).

DAVID O. COOK

Reference

Palmer H. D., and Gross, M. G., 1979. *Ocean Dumping and Marine Pollution.* Stroudsburg, Pa.: Dowden, Hutchinson & Ross, 268p.

Cross-references: *Dredging; Estuaries; Human Impact; Pollutants; Nuclear Power Plant Siting; Thermal Pollution; Waste Disposal.*

ELEVATED SHORELINE (STRANDLINE)

Elevated shorelines or *strandlines* are traces of former coastlines that have become emerged because of some combination of eustatic and crustal processes. The nature of the former shoreline may be in the form of a rock platform (wave cut or eroded by sea ice), sea caves, raised beaches, lagoon deposits, organic reefs, and/or related features (Charlesworth, 1957; Gill, 1972; Gill and Hopley, 1972; Guilcher, 1969; Nansen, 1905; Vita-Finzi, 1973; Zeuner, 1959).

While there are many secondary complications, the causes of the emergence of the old shoreline are generally some combination of the following three factors.

Glacioisostatic Uplift. There are many examples of this phenomenon in Scandinavia (Eronen, 1974), Scotland, Northern Ireland, Svalbard, Greenland, Canada, New England, and the state of Washington, as well as some in the Southern Hemisphere, notably in Patagonia, southern Chile, the Palmer Peninsula, many of the subantarctic islands, and the main shoreline of Antarctica (Andrews, 1974). A classic example of discontinuously emerged lines of raised beaches and abrasional shore platforms in Scotland is the "Parallel Roads of Glen Roy," a former glacial lake.

Glacioeustatic Oscillation. Since 1600 B.P. sea level has oscillated through ±3-4 m about the present mean and with a periodicity of about 500-1500 yr, but progressively decreasing in height. In uplift areas this produces a "staircase" of raised shorelines (Fairbridge, 1961; Hillaire-Marcel and Fairbridge, 1978). Multiple beach lines occur in those areas where there is a generous supply of unconsolidated debris as at river mouths, in tropical regions with deep saprolite, or in glaciated regions with extensive till or outwash sands. The rate of strandline material supply may be exceedingly high. For example at Point Peron, Western Australia, there are approximately 100 successive *beach ridges* (subparallel lines of raised beaches) belonging to the interval 6000 B.P. to present. Comparable sheaves of ridges are described from many parts of the world. Generally 3-5 distinctive sheaves, separated by a truncated boundary (erosional episodes), reflect the eustatic oscillations of the middle to late Holocene. The beach ridges often carry a bank of littoral dunes superposed on them. Coastal lagoons and sabkhas are frequently lined, on the *inner side,* by a raised beach containing shells of open sea facies; from this it is evident that the barrier island cutting off the lagoon or sabkha is a subsequent product (see Vol. VI: *Sabkha Sedimentology*).

Elevated shorelines of eustatic origin are known also from pre-Wisconsinian times (around 100,000 B.P. and earlier). However, it is unusual to find shells in them, on account of long-extended leaching by rainwater solution, and in many of the older examples the loose sand becomes degraded or blown by wind, so that its origin as a raised beach deposit must be deduced by other means (e.g., electron microscopy of sand grains, close analysis of heavy mineral patterns in the bedding). Where such ancient beaches have been capped or protected by a thick cover, the older shells may be preserved. In southwest England there are ancient shelly beaches protected by glacial colluvium ("head"). In northwest France similar beaches are protected by a cover of loess.

Tectonic Uplift. Raised beaches and related strandlines along the coast of Chile were first noted by Charles Darwin (1838) during the voyage of the *Beagle;* those in the south are partly involved in isostatic uplift, and all are partly eustatic, but extensive sectors of the coast have been subject to strong tectonic motions. Some Chilean beaches elevated to over 25 m contain artifacts of mid-Holocene man. Tilted beach lines can be followed to disappear below sea level. Classic examples are known in Greece and Italy. The same features may be seen on parts of the California coast, in parts of Japan, Indonesia, southern Iran, and parts of New Zealand. The islands of Hawaii have also been subjected to both submergence and emergence of tectonic origin. The interactions of eustacy and tectonics have been skillfully analyzed in New Guinea by Chappell (1974a, b).

Resolution of Multiple Causes

The history of the problem was first presented in book form by Chambers (1848). Von Zittel (1901) reviewed and referenced the early literature. The discontinuous step-by-step emergence of Scandinavia was noted as long ago as 1702 by Hjärne. In 1743 Celsius proposed that it was due to a secular loss of volume of the world ocean (i.e., a eustatic cause), and this led to an ill-fated hypothesis—the Desiccation

Theory, which held that eventually, by evaporation and loss to outer space, the ocean would dry up. Linnaeus supported the Celsius view, but it was opposed by others—for example, E. D. Runeberg, who submitted that the crust was rising because of earthquakes; that is, tectonically (references in Von Zittel, 1901).

In 1820-1821 the Swedish Academy reported on the findings of a special commission to study this question; although the field evidence was confirmed, the causes were not resolved. In 1835 J. Berzelius, the famous chemist, suggested that it was the cooling of the earth and shrinkage of its crust that led to the drop of sea level. In 1840 Adhemar suggested that the asymmetric ice loading during glaciations could displace the earth's center of gravity; later Croll, Pratt, Fischer, and, most recently, Munk and Macdonald (1960) have considered this problem. The eustatic concept was generated in 1842 by MacLaren, although it was not until 1885 that Suess (transl. 1906) actually coined the term.

The historical ambivalence is still in effect today. There is evidently no simple solution, but the problem is usually approached by considering *rates of change* and their duration.

RHODES W. FAIRBRIDGE

References

Andrews, J. T., 1974. *Glacial Isostasy*. Stroudsburg, Pa.: Dowden, Hutchinson & Ross, 491p.
Chambers, R., 1848. *Ancient Sea Margins*. Edinburgh: Chambers, Orr, and Co., 388p.
Chappell, J., 1974a. Geology of coral terraces, Huon Peninsula, New Guinea: A study of Quaternary tectonic movements and sea-level changes, *Geol. Soc. America Bull.* **85**, 555-570 (see also *Geol. Soc. America Bull.* **87**, 235).
Chappell, J., 1974b. Late Quaternary glacio- and hydro-isostasy on a layered earth, *Quaternary Research* **4**, 405-428.
Charlesworth, J. K., 1957. *The Quaternary Era*. London: Edward Arnold, 2 vols., 1700p.
Darwin, C., 1838. On certain areas of elevation and subsidence in the Pacific and Indian oceans, as deduced from the study of coral formations, *Geol. Soc. London Proc.* **2**, 552-554.
Eronen, M., 1974. The history of the Litorina Sea, *Fennia* **44**, 79-195.
Fairbridge, R. W., 1961. Eustatic changes in sea level, in *Physics and Chemistry of the Earth*, vol. 4. New York: Pergamon Press, 99-185.
Gill, E. D., 1972. The relationship of present shore platforms to past sealevels, *Boreas* **1**, 1-25.
Gill, E. D., and Hopley, D., 1972. Holocene sea levels in eastern Australia—A discussion, *Marine Geology* **12**, 223-243.
Guilcher, A., 1969. Pleistocene and Holocene sea level changes, *Earth-Sci. Rev.* **5**, 69-97.
Hillaire-Marcel, C., and Fairbridge, R. W., 1978. Isostasy and eustasy of Hudson Bay, *Geology* **8**, 117-122.

MacLaren, C., 1842. The glacial theory of Prof. Agassiz, *Am. Jour. Sci.* **42**, 346-365.
Munk, W. H., and MacDonald, G. J. F., 1960. *The Rotation of the Earth*. Cambridge: Cambridge University Press, 323p.
Nansen, F., 1922. *The Strandflat and Isostasy*. Videnskabs-selskabet i Kristiania, Matematisk-naturvidenskapellig klasse, Skrifter No. 11 (1921).
Suess, E., 1904-1924. *The Face of the Earth*, 5 vols. Oxford: Clarendon.
Vita-Finzi, C., 1973. *Recent Earth History*. London: Macmillan, 138p.
Von Zittel, K., 1901. *History of Geology and Palaeontology*. London: W. Scott, 562p.
Zeuner, F. E., 1959. *The Pleistocene Period*, 2nd ed. London: Hutchinson, 447p.

Cross-references: *Beach Ridges and Beach Ridge Coasts; Emergence and Emerged Shoreline; Global Tectonics; Isostatic Adjustment; Post-Glacial Rebound; Sea Level Changes; Tectonic Movements.*

EMBANKMENTS

Embankments are *artificial banks*, such as mounds or dikes, made of silt and sand (Fig. 1). In coastal engineering, embankments are constructed to protect the lower land against intrusion of seawater or waves, to carry a roadway along the shore, or to prevent some enclosed body of water from flowing out into the sea. Both sloping surfaces may be protected by grass, stones, or concrete. The seaward surfaces are often covered by sea walls, bulkheads, or revetments to prevent erosion or destruction due to wave action or currents. Such covered embankments are usually called sea or coastal *dikes* rather than embankments (Coastal Engineering Research Center, 1973).

SHOJI SATO

Reference

Coastal Engineering Research Center, 1973. *Shore Protection Manual*. Washington, D.C.: U.S. Army Corps of Engineers, 750p.

Cross-references: *Breakwaters; Coastal Engineering; Dikes; Groins; Jetties; Seawalls.* Vol. III: *Beach Erosion and Coastal Protection.*

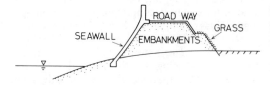

FIGURE 1. Cross-section of an embankment with a sea wall.

EMERGENCE AND EMERGED SHORELINE

Emergence is the rise of land areas, relative to sea level, changing the position of fossil shore zone features in the process. These land areas are called *emerged, raised, uplifted,* or *stranded shorelines.* The emergence may result from localized or regional *tectonic* movements, from *isostatic* uplift and warping, due to the unloading of ice sheets and seawater that covered continental shelves during deglaciation, and/or from the *eustatic* drop of sea level during glaciation. The various factors are often superimposed on each other, sometimes resulting in serious problems in the interpretation of specific emerged shoreline segments and in the evaluation of Quaternary worldwide sea level positions. Tectonic uplift has been most pronounced along Pacific coasts (e.g., Alaska, California, Chile, Japan, New Guinea, New Zealand), while isostatic readjustment has dominated in Scandinavia and large areas of Canada. Early Holocene shorelines have been uplifted to 220 m above present sea level in the past 10,300 years (Andersen, 1965, Fig. 22) near Oslo, Norway, and to 270 m elevation southeast of Hudson Bay in approximately the past 8000 years (Lee, 1968). A stairway of hundreds of successive standlines formed near Hudson Bay during the Tyrrell Sea Holocene regression.

Depositional Indicators of Emerged Shoreline

Direct Sea Level Indicators. Beach dune ridges with occasional associated lagoonal deposits are the most widespread fossil shoreline indicators. The increasing age of the littoral ridges may be shown by their increasing distance from the present shore, by increasing absolute elevations, by their more weathered state and thicker soil cover. Such *coastwise marine terraces,* as they are called, often terminate with coastward scarps. With the partial exception of late Holocene *cheniers,* they rarely contain accurately datable vegetable or shell material. Unless the depositional conditions of the littoral ridge units are verified by sedimentological and paleontological methods, misinterpretation of nonmarine (e.g., alluvial-fluvial) surfaces or interfluve ridges as marine terraces and littoral ridges is always a distinct possiblity (Otvos, 1972). Because littoral ridges are often accentuated by thick dune deposits, the surface of associated lagoonal deposits offers a better reference point for the sea level position. Radiocarbon dating of intertidal salt marsh peat that directly overlies bedrock or noncompacting sediments provides the most reliable absolute age determination of late pleistocene and Holocene sea level positions. Intertidal and supratidal flat deposits may also help in identifying emerged fossil shorelines.

Emerged fossil coral reef terraces in tropical climatic zones also offer reference points for approximate sea level positions, dating shore uplift rates or eustatic sea level changes (Bloom et al., 1974; Broecker and Thurber, 1965). Uranium series dating methods are usually employed, and diagenetically unaffected (nonrecrystallized), mechanically uncontaminated parts of the reefs, preferably in growth position, are sampled for the purpose.

Indirect Sea Level Indicators. Other sediments with known depositional locations, above or below sea level may also be useful in establishing related shoreline positions. Subfacies of fluvial deltas, submarine bars, shoals and shoreface deposits fall into this category.

Erosional Indicators of Emerged Shoreline

Uplifted coastal features, such as cliffs; scarps; high-, inter-, and sub-tidal platforms; intertidal pools; solution pits; visors; notches; sea caves; and other forms are also helpful in establishing ancient shoreline positions if the marine nature of the features can firmly be established. A flight of tectonically uplifted bedrock terraces in the Palos Verdes Hills, southern California (Emery, 1960), occasionally veneered by a few feet of round cobbles and shallow marine shells, is an example of the combination of erosional and depositional shoreline indicators.

ERVIN G. OTVOS

References

Andersen, B. G., 1965. The Quaternary of Norway, *in* K. Rankama, ed., *The Quaternary,* Vol. 1. New York: Wiley Interscience, 91-138.

Bloom, A. L.; Broecker, W. S.; Chappell, J. M. A.; Matthews, R. K.; and Mesolella, K. J., 1974. Quaternary sea level fluctuations on a tectonic coast; new ^{230}Th/^{234}U dates from the Huon Peninsula, New Guinea, *Quaternary Research* 4, 185-205.

Broecker, W., and Thurber, D., 1965. Uranium-series dating of corals and oolites from Bahamas and Florida Key limestones, *Science* 149, 58-60.

Emery, K. O., 1960. *The Sea off California.* New York: John Wiley & Sons, 366p.

Lee, H. A., 1968. Quaternary geology, *in* C. S. Beals, ed., *Science, History, and Hudson Bay,* vol. 2. Ottawa: Department of Energy, Mines, and Resources, 503-543.

Otvos, E. G., Jr., 1972. Pre-Sangamon beach ridges along the northeastern Gulf Coast—Fact or fiction, *Gulf Coast Assoc. Geol. Socs. Trans.* 22, 223-228.

Cross-references: *Classification; Global Tectonics; Isostatic Adjustment; Isostatic Uplift Coast; Post-*

Glacial Rebound; Progradation and Prograding Shoreline; Sea Level Changes; Sea Level Curves; Submergence and Submerged Shoreline; Tectonic Movements. Vol. III: Coastal Stability.

ENERGY COEFFICIENTS

The energy, E, of a wave (q.v.) is related to the height, H, by $E = 1/8(g\rho H^2)$, where g is acceleration of gravity and ρ is water density. The energy is considered, per unit length, in the direction of the propagation. The energy flows parallel to the wave crest and so, per unit length, E is related to a unit area of the sea surface.

Considering the energy related to a wave length λ, the following formula is available: $E = 1/8(g\rho H^2 \lambda)$.

The energy is furnished by the wind; in consequence of the origin of the waves, their energy can be related to the wind intensity. For a completely developed sea we can assume that the unity energy in erg/cm^2 is $E = AU^4/4Bg^2$, where U is the wind speed in cm/sec, and A and B are two coefficients whose values are 1.62×10^{-2} and 0.74, respectively (Wiegel, 1969; Kinsman, 1965).

FERRUCCIO MOSETTI

References

Kinsman, B., 1965. *Wind Waves.* Englewood Cliffs, N.J.: Prentice-Hall, 676p.
Wiegel, R. L., 1964. *Oceanographical Engineering.* Englewood Cliffs, N.J.: Prentice-Hall, 532p.

Cross-references: *Wave Energy; Waves; Wind.*

ENTRAINMENT

Sediment grains begin to move when the lift and drag forces of the water exceed the force of gravity that holds the grain in place (Johnson and Eagleson, 1966; Graf, 1971; Middleton and Southard, 1977). The gravity force acts through the center of the grain. The fluid forces usually act through some point above the center of the grain, for the grain must be lifted above a pivot point. Therefore only that component of the water acting parallel to the easiest path of motion need be considered, the component of gravity acting in the opposite direction. This path of easiest motion will vary widely from grain to grain. It is dependent on the packing arrangement of the grains and the amount of exposure of the grain to the water flow. Not only that, but even in "steady" flow, the water motion fluctuates because of turbulence. Initiation of a particular sand grain's entrainment cannot be completely determined. Rather, it is a stochastic process; only an average grain can be described.

The gravity force is equal to the volume of the grain $(\P/6) D^3$ times the submerged specific gravity $(\rho - \rho_f)g$ or

$$F_G = a_1 \P/6\, D^3 (\rho - \rho_f)g.$$

The part of gravity that opposes the grain movement is $F_G \sin \alpha$ where α is the angle of that direction and the horizontal. A usual assumption is that α equals the mass angle of repose for submerged grains, which for sand is approximately 35°. The vertical distance of the gravity component from the pivot axis is a_1. The gravity force may be written as $F_g = c_1 D^3 (\rho - \rho_f)g$, where c_1 is a coefficient that also takes into account the shape factor.

The drag force F_D is the result of the fluid shear stress (tangential force per unit area) and fluid pressure (normal force per unit area). The important variables in determining F are: fluid velocity, U; fluid density, ρ; fluid viscosity, μ, for it determines the shear force with a given rate of shearing of the fluid; grain area, $\pi D^2/4$; and shear stress, $\tau_0 = \mu(du/dy)$. Keeping in mind that by definition $\mu/\rho = 1/\nu$, the four parameters can be replaced by a dimensionless variable, the Reynolds number $uD/\nu = \text{RE}$, which for given conditions of flow will be constant c_2. So, $F_D = a_2 c_2 D^2 \tau_0 \cos \alpha$ or $a_1 c_1 D^3 (\rho - \rho_f)g \sin \alpha = a_2 c_2 D^2 \tau_0 \cos \alpha$, where $\tau_0 = (a_1 c_1)/(a_2 c_2) \tan \alpha\, D(\rho - \rho_f)g$.

The Shields parameter $\tau_0/D(\rho - \rho_f)g$ is approximately equal to τ_0/F_G, which can be seen as the ratio between two forces per unit area. Results obtained by Shields (1936), with a diagram showing the shear stress required to move an average flat bed sediment under unidirectional currents, have been accepted almost universally, although modified and extended several times by Vanoni (1946) and White (1970). The analysis would not change significantly if lift forces were also considered, for there is a proportionality between the two forces (Jeffreys, 1929) that depends only on grain geometry and boundary Reynolds number. Lift is directly proportional to the shear velocity (Einstein and El Samni, 1949).

When the shear water velocity increases past the critical point, sand grains begin to hop, roll, and slide; but this movement is neither continuous nor uniform. These are short bursts of movement that affect an area—the grains move a short distance and stop and then move again

(Grass, 1970). For sand finer than 0.8 mm, movement of the grains leads to the formation of ripples, which disappear if the velocity increases. In coarser sand, ripples do not form; rather, sand waves and dunes form if the velocity increases.

There have been fewer studies of oscillatory flow and agreement is not always good. Silvester and Magridge (1971) presented thirteen different equations for sediment movement under waves. Komar and Miller (1973, 1975) found that for medium- and finer-grained sands the threshold for movement is best established by the left side of the following equation. If grains are coarser, the right side is a better approximation.

$$0.21(d/D)^{1/2} \leqslant \rho u/(\rho - \rho_f)g \leqslant 0.46\pi(d/D)^{1/4}$$

In this equation, u and d are the orbital velocity and orbital diameter of wave motion obtained from linear Airy wave theory, and $u = \pi d/T = \pi H/T \sin(2\pi h/L)$.

BENNO M. BRENNINKMEYER

References

Einstein, H. A., and El Samni, E. A., 1949. Hydrodynamic forces on a rough wall, *Rev. Modern Physics* 21, 520-524.
Graf, W. H., 1971. *Hydraulics of Sediment Transport*. New York: McGraw-Hill, 513p.
Grass, A. J., 1970. Initial instability of fine bed sand, *Am. Soc. Civil Engineers Proc., Jour. Hydraulics Div.* 96, 619-632.
Jeffreys, H., 1929. On the transportation of sediment by streams, *Philos. Soc. Cambridge Proc.* 25, 272-276.
Johnson, J. W., and Eagleson, P. S., 1966. Coastal processes, *in* A. T. Ippen, ed., *Estuary and Coastline Hydrodynamics*. New York: McGraw-Hill, 404-492.
Komar, P. D., and Miller, M. C., 1973. The threshold of sediment movement under oscillatory water waves, *Jour. Sed. Petrology* 43, 1101-1110.
Komar, P. D., and Miller, M. C., 1975. Sediment threshold under oscillatory waves, *Conf. Coastal Eng., 14th, Am. Soc. Civil Engineers, Proc.* 2, 756-775.
Middleton, G. V., and Southard, J. B., 1977. *Mechanics of Sediment Movement*. Tulsa, Okla.: Soc. Econ. Paleont. Miner., Lecture Notes for Short Course No. 3, 191p.
Shields, A., 1936. Anwendung der Ahnlichkeits Mechanik und der Turbulenz ofrshung auf die Geschiebe Bewegung, Preuss, Versuchanst. Wasserbau Schiffbau Mitt. 26, 20p.
Silvester, R., and Mogridge, F. R., 1971. Reach of waves to the bed of the continental shelf, *Conf. Coastal Eng., 12th, Am. Soc. Civil Engineers, Proc.* 2, 651-667.
Vanoni, V. A., 1946. Transportation of suspended sediment by water, *Am. Soc. Civil Engineers Trans.* 8, 67-133.
White, S. J., 1970. Plane bed thresholds of fine grained sediments, *Nature* 228, 152-153.

Cross-references: *Beach Firmness; Beach Processes; Reynolds Number; Ripple Marks; Sediment Transport.*

EOLIANITE—See AEOLIANITE

EQUILIBRIUM SHORELINE

The equilibrium shoreline is a hypothetical state that actual shorelines may or may not approximate. Sandy beaches tend to approach the ideal closely, whereas rocky or marsh coasts tend to depart from it. Without tectonic, eustatic, or other interference, all shorelines should in due time pass through one or more equilibrium stages until a terminal static (*not* equilibrium) condition is reached.

For many persons, the equilibrium concept applies only to the two-dimensional geometry of the transverse profile, extending seaward from the coast. The more general concept, however, is three-dimensional, and must include the plan view of the shore under study.

Equilibrium is a dynamic state, in which energy and materials are being transported; a certain geometry results. The geometry reflects the balance obtained between *materials, process*, and *climate*. By materials is meant mineralogy and particle size. By process is meant energy type: wave, tide, direct wind, or other. By climate is meant energy level.

The ideal equilibrium beach has curvature and sand-prism characteristics (geometry) adjusted so closely that the energy available (process, climate) transports the detritus supplied, over a period to be measured in years rather than months, days, or seconds (Tanner, 1958). Sand prism characteristics include beach face steepness, surface features (such as cusps, pads, and bars), and roughness.

Many beaches fail to achieve perfect equilibrium, but maintain *equilibrium geometry*—that is, if there is a deficiency of sediment delivered to the system, a given amount will be removed from the beach. (Tanner and Stapor, 1972). The result is an equilibrium form that shifts slowly landward. Similarly, an excess of sediment can produce a shifting equilibrium that moves slowly seaward.

The *zero energy coast* must be placed in the nonequilibrium category, because no significant amount of energy is being expended and no significant changes are taking place. Certain shorelines may fit into the subequilibrium category: they are being changed toward an equilibrium form, but have not yet achieved it. Even spectacularly eroding or aggrading beaches may, however, have the equilibrium form. A detailed wave-climate study is necessary to determine whether or not an irregular beach has equilibrium characteristics; the irregularities may reflect uneven concentration of littoral wave power, owing to bathymetric complications offshore, in which case the beach may well be essentially in equilibrium.

A change in wave climate may cause the beach to change from one equilibrium geometry to another. Such changes, in areas of moderate to high wave energy, may occur as quickly as a new long-term wave climate can be established.

WILLIAM F. TANNER

References

Tanner, W. F., 1958. The equilibrium beach, *Am. Geophys. Union Trans.* **39**, 889-891.

Tanner, W. F., and Stapor, F. W., 1972. Accelerating crisis in beach erosion, *Internat. Geography* **2**, 1020-1021.

Cross-references: *Beach Cycles; Beach Processes; Beach Profiles; Sediment Transport; Wave Climate.* Vol. III: *Littoral Processes;* Vol. VI: *Littoral Processes.*

EQUISETOPHYTA

The Equisetophyta (Sphenopsida, Calamophyta, Arthrophyta, or Equisetopsida) are seedless, photosynthetic, vascular plants with roots, jointed stems, and whorled leaves. Although abundant as trees belonging to several genera in the Carboniferous coal measures, the surviving members are herbaceous plants devoid of secondary growth mostly 0.3-2 m tall, all of which belong to the genus *Equisetum,* including about 20 species of worldwide distribution except for Australia and New Zealand.

The plants have two generations in their life cycle, both photosynthetic. The haploid gametophytic (sexual) stage consists of a flat, green pad or irregular, thin-lobed body, usually not over a centimeter across. It grows from the spore and produces the egg and sperm, which fuse to form the diploid zygote that is the first cell of the diploid sporophytic stage. The gametophyte is rarely seen, and even then may be mistaken for a liverwort.

The sporophyte, which is soon independent of the gametophyte, consists of green, jointed, hollow, erect aerial stems arising from solid, horizontal, underground stems known as rhizomes. The center of the stems both aerial and underground is pith, but as the aerial stems mature, the pith shrinks, leaving these stems hollow except for a transverse septum at each node. The aerial stems of some species are all green and of one sort, and bear the spore-producing cone at the tip. Most of these species with one kind of aerial stem are unbranched and have the common name of *scouring rush.* The aerial stems of the other group of species are of two types: a nongreen, unbranched, short-lived stem bearing a terminal cone; and green, much-branched sterile stems that vegetate through the whole growing season. This latter group has the common name of *horsetail* or *horsetail rush.* The stems, especially of the scouring rush type, are minutely roughened by bits of amorphous, hydrous silica embedded in the epidermis, whose hard roughness suits them for scouring—hence the common name.

The leaves of *Equisetum* are borne in whorls of several on the main stem or as few as three on the branches of some kinds. The leaves are united by their adjacent edges to form a sheath, with their free tips usually projecting as slender, tapering points. On the green stems they are usually not over a centimeter long, but on the nongreen fertile stems of species with dimorphic aerial stems they may be longer.

The cones in which the spores are produced are borne at the tips of the stems, and are from one to several centimeters long. These cones consist of a solid central axis bearing several to many whorls of several sporangiophores in each whorl. These usually hexagonal-headed sporangiophores have their axes at right angles to the axis of the cone. From the proximal faces of their heads 5-10 sporangia project parallel with the axis of the sporangiophore. When the spores are mature, the axis of the cone elongates, separating the heads of the sporangiophores. Then the sporangia open and release the haploid spores, the first cell of the gametophytic generation. Each of the globose spores has four threads affixed to the spore at a common point. These elaters are wound closely around the spore while moist inside the sporangium but, on drying as the sporangium opens, they squirm and help release the spores from the sporangium and also increase the area of the spores and so aid in their dispersal by wind.

Spores that land in a favorable moist site may germinate and produce a gametophyte.

The usual habitat of *Equisetum* is wet to moist, although some kinds grow in sites that may be moist only seasonally. Their tolerance to salinity is low, hence they are not to be expected on the beach or in salt marshes.

For further reading, see Bierhorst (1971), Smith (1955), and Sporne (1962),

LOUIS CUTTER WHEELER*

*Deceased

References

Bierhorst, D. W., 1971. *Morphology of Vascular Plants.* New York: Macmillan, 560p.

Smith, G. M., 1955. *Cryptogamic Botany, vol. 2, Bryophtes and Pteridophytes.* New York: McGraw-Hill, 399p.

Sporne, K. R., 1962. *The Morphology of Pteridophytes: The Structure of Ferns and Allied Plants.* London: Hutchinson Library, 192p.

Cross-references: *Biotic Zonation; Coastal Ecology, Research Methods; Coastal Plant Ecology, United States, History of; Coastal Flora.*

EROSION RAMP, WAVE RAMP

An *erosion ramp, wave ramp,* or *abrasion ramp* is a relatively smooth, seaward-dipping surface extending from the cliff base or notch at the cliff base to the horizontal surface of a shore platform (Fig. 1). In the absence of a platform the ramp may extend seaward below the low-water mark (Hills, 1971). Hills (1972) also noted the difference between this second type of wave ramp and a *shore platform*—the ramp extending to below low water and more dominant in bays, and the platform terminating in a cliff above low water and more prominent on headlands. The location of the cliffbase wave ramp immediately above the platform surface and above the level of permanent rock saturation places them in the optimum zone of erosion associated with wetting and drying, water-layer weathering, and abrasion. Collapse of the cliff face or notch deposits debris on the ramp. Wave action removes ramp and cliff debris and in so doing, abrades the ramp itself. Wave action may also deposit sand and boulders on the ramp, temporarily protecting it from further erosion.

ANDREW D. SHORT

References

Hills, E. S., 1971. A study of cliffy coastal profiles based on examples in Victoria, Australia, *Zeitschr. Geomorphologie* 15, 137-180.

Hills, E. S., 1972. Shore platforms and wave ramps, *Geol. Mag.* 109, 81-99.

Cross-references: *Cliff Erosion; Depth of Disturbance; Quarrying Processes; Ramparts; Shore Platforms; Solution and Solution Pan; Water Layer Weathering; Wave-Cut Bench; Wave-Cut Platform; Wave Erosion; Weathering and Erosion, Mechanical.* Vol. III: *Platforms—Wave-Cut.*

FIGURE 1. Erosion ramp (center) cut in limestone complex at the foot of Diamond Head, Oahu, Hawaii. Seaward of the ramp lies a deteriorating shore platform, being eroded primarily by solution. The ramp is backed by the tuffaceous material of Diamond Head (photo: A. D. Short).

ESTUARIES

An *estuary* is a semi-enclosed coastal body of water that has a free connection with the open sea and within which seawater is measurably diluted with freshwater derived from land drainage (Pritchard, 1967a). The landward limit of the estuary has been defined as the area where the chlorinity falls below 0.01 ‰ and the ratios of the major dissolved ions change radically from the ratios in seawater. The dilution of seawater in the estuary is variable, reflecting the variability of fresh-water runoff. Thus, from an oceanographic point of view, the estuary is an unstable environment. Organisms that live in the estuary are adapted to large changes in the salt concentration and, in temperate regions, to large changes in temperature.

There are almost 900 individual estuaries along all the coasts of the United States, comprising an area of about 68,000 km². As the United States was settled, the regions around these natural harbors were the first to be developed. Cities sprang up in areas near the head of navigation. Seafood was harvested from protected estuarine waters. As the size of ships increased, dredging of previously adequate natural channels was necessary to preserve ports. As population increased, the use of the estuary for industrial and domestic *waste disposal* increased, affecting commercial seafood production. The population living around estuaries (approximately 60 million people live within 90 km of the coast of the United States) have placed high demands on estuaries for recreation. As a result of these competing demands for use of the estuary, such as for transportation, as a waste disposal facility, a food source,

ESTUARIES

and a recreational outlet, it is one of the best-studied entities in oceanography.

Physical oceanographers classify estuaries by the processes that mix the freshwater from upland drainage with seawater. Geologists classify estuaries by their mode of origin or by the range of the tides that shape the features of the estuary.

Classification by Origin

Coastal plain estuaries, typified by the Chesapeake Bay, are the result of flooding of previously created river valleys by the recent Holocene sea level rise. *Barrier beaches* have enclosed large coastal areas to form coastal lagoons, such as Biscayne Bay, Florida, and the major estuaries of the Gulf Coast.

Fjords, created by the action of glaciation, are prevalent along the rocky northwest coast of Northern Hemisphere continents, and represent yet another estuary type. Finally, *fault block estuaries,* such as San Francisco Bay, are formed and controlled by local or regional geological structure.

Classification by Mixing Processes

The densities of freshwater and saltwater masses are typically 1.00 gm/cm^3 and 1.025 gm/cm^3. Because of the density contrast between them, these two fluids tend to maintain themselves as separate water masses, with the denser saltwater overlain by the freshwater. In the absence of currents, mixing of the two water masses would be limited to an upward flux of salt by diffusion processes and vertical advection of saline water by upward-breaking *internal waves* at the sharp boundary between the two water masses. In the presence of *currents*, the mixing processes are much more efficient. Advection processes produce a mass flux of water and salt, resulting in a diffuse boundary layer and a freshwater layer that becomes progressively saltier in a seaward direction. Pritchard's (1967a) classification of *estuarine circulation systems* is based on the relative importance of advective and diffusion processes acting within the water mass.

Type A Estuary. In the type A, or *salt wedge*, estuary described by Pritchard (1967b), the circulation pattern is dominated by river discharge, and tidal effects play a negligible role in the circulation (Fig. 1). The freshwater from the river, because of its lower density, floats on top of the denser saltwater, spreading out in a layer that becomes progressively thinner toward the sea. Underlying the freshwater layer is a wedge-shaped body of seawater that thins in an up-river direction. The interface between the two water masses is characterized by an abrupt

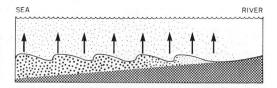

FIGURE 1. Type A, or salt-wedge estuary in which river discharge dominates circulation, producing a density-stratified system. Vertical advection is the primary mechanism for mixing across the freshwater/saltwater interface (from Schubel, 1971).

density contrast and internal waves that break upward carrying saline water into the top layer, where mixing occurs. The primary process for mixing across the interface is vertical advection, and involves the mass transfer of saltwater into the upper layer, which becomes more saline toward the sea. To replenish the water lost to the upper layer, there is a slow landward movement of saltwater up the bottom of the estuary.

Type B Estuary. In a type B, or *partially mixed*, estuary, the tidal influence is increased to the point where river discharge does not dominate the circulation (Fig. 2). Mixing caused by increased turbulence results in the mass transfer of saltwater and freshwater in both directions across the density boundary, which becomes gradational and much less distinct than the salt-wedge boundary. The differences in salinity between surface and bottom waters is approximately constant over much of the estuary. Chesapeake Bay has been classified as a partially mixed estuary.

Type C Estuary. In the type C, or *vertically homogeneous*, estuary, tidal current velocities are large enough and the estuary is wide enough so that there is sufficient mixing to erase the saltwater-freshwater boundary (Fig. 3). The water salinity is constant vertically and decreases longitudinally in an upstream direction. Pritchard (1967b) considers the wider reaches of the Delaware and Raritan estuaries to be examples of type C estuaries.

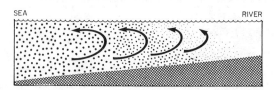

FIGURE 2. Type B, or partially mixed, estuary in which tidal currents are strong enough to prevent the river from dominating the circulation pattern. Mixing is due both to vertical advection and turbulence associated with the tidal currents (from Schubel, 1971).

ESTUARIES

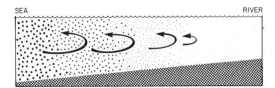

FIGURE 3. Type C, or vertically homogeneous, estuary in which tidal mixing eliminates vertical density stratification. Water is homogeneous vertically, and shows a longitudinal salinity gradient with salinity increasing toward the sea (from Schubel, 1971).

Classification by Tidal Range

Davies (1964) recognized the importance of tidal range (the difference in elevation between high water and low water) to coastal morphology, and proposed the following classification of tides: microtidal—tidal range 0–2 m; mesotidal—tidal range 2–4 m; macrotidal—tidal range greater than 4 m.

Hayes (1975) developed an illustration of the generalized relation between tidal range and the variation of coastal plain shorelines (Fig. 4).

Microtidal Estuaries. A representation of a model microtidal estuary is given in Figure 5. The dominant forces moving sediments in a microtidal estuary are wind- and storm-generated waves and currents. Sandy sediments are found in tidal deltas (near inlets), in washover fans, and in spits. Finer-grained sediments (silts and clays) accumulate in the deeper, more central portions of the estuary and near the river deltas. Most gulf coast estuaries are microtidal.

Mesotidal Estuaries. As tidal range increases, current speed increases. Mesotidal estuaries usually show sediment distribution characteristics of tidal currents. Tidal deltas, meandering tidal creeks, and point bars dominate the sand-sized sediments (Fig. 6). Fine grained silts and clays are found on mud flats and in salt marshes. Most New England estuaries and those of the Wadden Sea, the Netherlands, are mesotidal.

Macrotidal Estuaries. Tidal currents dominate the flow of macrotidal estuaries. Macrotidal estuaries are usually funnel shaped and broad mouthed. Silts and clays are formed near shore on mud flats and in salt marshes (Fig. 7). Sandy sediments are found near the center of the estuary, usually as long bars oriented parallel to the dominant current directions.

Sediments in Estuaries

Materials in estuaries move in suspension or as part of the bed load. *Suspended load* is the

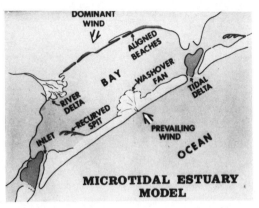

FIGURE 5. Microtidal estuary model (reprinted from M. O. Hayes, 1975, Morphology of sand accumulation in estuaries, *in* L. E. Cronin, ed., *Estuarine Research*. New York: Academic Press).

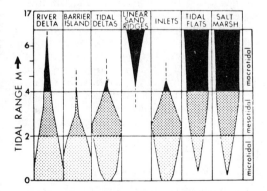

FIGURE 4. Variation of morphology of coastal plain shorelines with respect to differences in tidal height (reprinted from M. O. Hayes, 1975, Morphology of sand accumulation in estuaries, *in* L. E. Cronin, ed., *Estuarine Research*. New York: Academic Press).

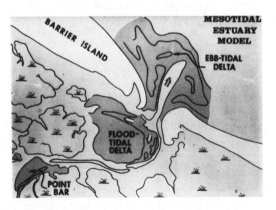

FIGURE 6. Mesotidal estuary model (reprinted from M. O. Hayes, 1975, Morphology of sand accumulation is estuaries, *in* L. E. Cronin, ed., *Estuarine Research*. New York: Academic Press).

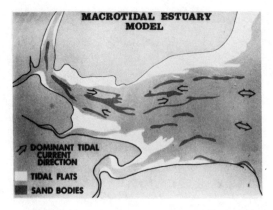

FIGURE 7. Macrotidal estuary model (reprinted from M. O. Hayes, 1975, Morphology of sand accumulation in estuaries, *in* L. E. Cronin, ed., *Estuarine Research*. New York: Academic Press).

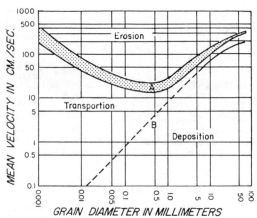

FIGURE 8. Hjulström's diagram of the relation between water velocity and particle size (from Hjulström, 1939).

material moving in suspension in the fluid, being supported by the upward component of turbulence or by colloidal suspension. *Bed load* is a term used to designate coarser material moving on or near the bed. In a strict sense, it is difficult to place a well-defined boundary between bed load and suspended load because of the transitional nature of the dynamics between the two states. Graf (1971) notes that "there exists an active interchange between the suspended load and bed load, but also between the bed load and the bed itself. . . ." and that "suspended load is always accompanied by bed load. . . ." It is therefore important to examine how sediment grains as part of the bed load become part of the suspended load.

As flow velocity increases, pressure differences develop on the upstream and downstream sides of grains. Because the velocity of the fluid increases with increasing distance above the bed, the upper half of the upstream face of the grain is subject to a greater fluid force than the lower half of the grain. The resultant of these forces acts parallel to the bed and tends to roll or slide the grain along the bottom as soon as a high enough velocity has been reached. The minimum current velocity required to initiate movement of a given grain size is called the *critical erosion velocity*.

The relationship between the minimum current velocities required to erode, transport, and deposit a particle of a given diameter has been summarized in a series of diagrams pioneered by Hjulström (1939) and refined by Allen (1965), Sundborg (1967), and Postma (1967). Hjulström's diagram (Fig. 8) is applicable to material of uniform size on a bed of loose material of the same size and "clear water." The velocity axis is the average velocity across a transverse profile of a river. The upper portion of the broad curve *A* represents the critical erosion velocity for different grain sizes. Fine sand (300–600 μ) is more easily eroded than coarse sands, gravels, or component silts and clays. Once eroded, a particle of a given size can be transported at velocities lower than those required to initiate movement. When the velocity drops to curve *B,* the particle can no longer be transported, and begins to settle towards the bottom.

Particle radius, changes in the viscosity of the fluid that may be due to changes of temperature, changes in dissolved ion concentration, or changes in suspended particulate concentration, and a change in the resistance to fall caused by varying particle shape or roughness—all can affect the fall velocity.

Estuarine Sedimentation Rates. Rates of sedimentation within estuaries and the loss of suspended materials to sea are both extremely variable. Average deposition rate varies from almost 4 mm/yr to less than 0.7 mm/yr. Some estuaries (notably Delaware Bay) exhibit a net contribution of sediments from the ocean, while others contribute quantities of sediment to the oceans. Most long estuaries trap sediment within their boundaries. Those that have filled with sediment, notably the Amazon and the Mississippi, discharge large quantities of suspended material into the sea.

Estuaries in the Geologic Record

Examples of estuarine deposits in the geologic record are the Lower Pennsylvania coal measures of the Allegheny plateau (Pottsville and Allegheny groups); the Mississippian Berea Sandstone of north and central Ohio, which has been compared to sediments of the Haringvliet in Holland; the Lower Devonian Holdgate Sandstone of England; the Mid-Jurassic Great Estuarine Series of western Scotland, which has faunal similarities with the lagoons along the Texas coast; and pyrite-bearing Permian and Precambrian shales, which may have been deposited in an environment comparable to Chesapeake Bay (Rusnak, 1967; Biggs, 1967).

ROBERT B. BIGGS

References

Allen, J. R. L., 1965. A review of the origin and characteristics of recent alluvial sediments, *Sedimentology* 5, 89-191.

Biggs, R. B., 1967. The sediments of the Chesapeake Bay, *in* G. H. Lauff, ed., *Estuaries.* Washington, D.C.: Am. Assoc. Advancement Sci. Pub. 83, 239-260.

Davies, J. L., 1964. A morphogenetic approach to world shorelines, *Zeitschr. Geomorphologie* 8, 127-142.

Graf, W. H., 1971. *Hydraulics of Sediment Transport.* New York: McGraw-Hill, 513p.

Hayes, M. O., 1975. Morphology of sand accumulation in estuaries, *in* L. E. Cronin, ed., *Estuarine Research.* New York: Academic Press, 3-22.

Hjulström, F., 1939. Transportation of detritus by moving water, *in* P. D. Trask, ed., *Recent Marine Sediments.* Tulsa, Okla.: Soc. Econ. Paleont. Miner. Spec. Pub. 4, 5-31.

Postma, H., 1967. Sediment transport and sedimentation in the estuarine environment, *in* G. H. Lauff, ed., *Estuaries.* Washington, D.C.: Am. Assoc. Advancement Sci. Pub. 83, 158-179.

Pritchard, D. W., 1967a. What is an estuary: Physical viewpoint, *in* G. H. Lauff, ed., *Estuaries.* Washington, D.C.: Am. Assoc. Advancement Sci. Pub., 83, 3-5.

Pritchard, D. W., 1967b. Observations of circulation in coastal plain estuaries, *in* G. H. Lauff, ed., *Estuaries.* Washington, D.C.: Am. Assoc. Advancement Sci. Pub. 83, 37-44.

Rusnak, G. A., 1967. Rates of sediment accumulation in modern estuaries, *in* G. H. Lauff, *Estuaries.* Washington, D.C.: Am. Assoc. Advancement Sci. Pub. 83, 180-190.

Schubel, J. R., 1971. *Estuarine Circulation and Sedimentation.* Washington, D.C.: Am. Geol. Inst., Lecture Notes, Oct. 1971. VI-1-VI-7.

Sundborg, A., 1967. Some aspects on fluvial sediments and fluvial morphology: General views and graphic methods, *Geog. Annaler* 49A, 333-343.

Cross-references: *Currents; Estuarine Habitats; Human Impact; Lagoon and Lagoon Coasts; Lagoonal Sedimentation; Sediment Transport; Tides; Waste Disposal.* Vol. III: *Estuary;* Vol. VI: *Estuarine Sedimentation.*

ESTUARINE COASTS

The most recent rise in sea level drowned many semienclosed coastal embayments producing *estuaries* in those that received sufficient freshwater to measurably dilute the encroaching seawater. Most of the basins were former river valleys carved during the previous low stand of sea level. Some scientists restrict the term estuary to submerged river valleys, but other coastal embayments that exhibit similar biological, chemical, geological, and physical processes are appropriately considered to be estuaries. These basins include fjords, bar-built embayments, and basins formed by tectonic processes (Schubel, 1971).

Submerged river valley estuaries are also known as *ria coasts.* The basins are commonly V-shaped in cross section, and the depth increases more or less uniformly in a seaward direction. Their configurations are largely inherited, produced by the rising sea coming to rest against landforms shaped by fluvial processes. They may be of either dendritic or trellis type. The Chesapeake Bay is a classic example of a submerged river valley estuary. Other examples in the United States include: Delaware Bay, the Hudson, the Mississippi, and many others. Submerged river valley estuaries are found throughout the world, including the Thames (England), Ems (Germany), Seine (France), Si-Kiang (Hong Kong), and Murray (Australia) (Schubel, 1971).

Fjord estuaries are characteristically narrow, steep-sided, and relatively straight compared with submerged river valley estuaries. They are commonly U-shaped in cross section and, as a class, are the deepest of all estuaries, with depths of greater than 1200 m recorded in Norway and Greenland. Many fjords have shallow sills near their mouths. Fjord estuaries are found in British Columbia, Alaska, Greenland, New Zealand, the Scandinavian countries, and along formerly glaciated coasts in many other countries.

Bar-built estuaries occupy basins produced by the formation of barriers across reentrants in the coastline. The barriers must be broken by one or more inlets to provide a free connection to the ocean to maintain the basins' estuarine character. Frequently more than one river enters the basin, but the total drainage basin is generally small and the freshwater input low. Since the lower reaches of the entering rivers have generally been drowned, bar-built estuaries have a composite origin. Bar-built estuaries are usually shallow and occur along barrier coasts. Albemarle and Pamlico sounds on the North Carolina coasts are examples.

Some estuaries occupy basins produced primarily by tectonic processes. San Francisco Bay, or at least a part of it, is an example. The upper reaches of the San Francisco Bay estuary were formed by the drowning of the lower San Joaquin-Sacramento River systems.

Estuaries are ephemeral features on a geologic time scale being rapidly filled with sediments. Lifetimes are typically measured in thousands of years, occasionally in a few tens of thousands of years (Schubel, 1971).

J. R. SCHUBEL

References

Schubel, J. R., ed., 1971. *The Estuarine Environment: Estuaries and Estuarine Sedimentation.* Washington, D.C.: Am. Geol. Inst. Short Course Lecture Notes, 324p.

Cross-references: *Classification; Estuaries; Estuarine Delta; Estuarine Sedimentation; Fiard, Fjärd; Fiord, Fjord; Ria and Ria Coasts.* Vol. VI: *Estuarine Sedimentation.*

ESTUARINE DELTA

The term *estuarine delta* refers to subaqueous and subaerial deltaic deposits that have accumulated within the semiconfined and protected environment of an estuary. At least two genetically and morphologically distinct types of estuarine delta may be recognized: deltas wholly or partially filling open-ended, funnel-shaped estuaries; and deltas wholly or partially filling barrier lagoons or estuaries impounded by coastal barrier formations. Estuarine deltas of the first type are best developed in macrotidal environments where bidirectional tidal currents redistribute river-derived sediments. Deposits assume the form of elongate, subaqueous tidal ridges, subaerial midchannel islands, and shoals and tidal flats. Examples include the Ord, Fly, Colorado, and Amazon deltas (Wright, 1978; Wright, Coleman, and Thom, 1975).

Estuarine deltas of the second type are typically fluvially dominated accumulations formed within shallow estuaries that are sheltered from marine forces by wave-built coastal barriers. In the early stages of their evolution, deltas of this type prograde across estuarine sediments as elongate, digitate protrusions. A well-documented example is the Mitchell delta in the Gippsland Lakes, Australia (Bird, 1962). In cases where the supply of river-borne sediment is large relative to the depth of the estuary, the subaerial deltaic plain may completely fill the estuary extending to the inner edge of the impounding barrier (e.g., the Shoalhaven Delta, New South Wales, Australia).

LYNN D. WRIGHT

References

Bird, E. C. F., 1962. The river deltas of the Gippsland Lakes, *Royal Soc. Victoria Proc.* 75, 65-74.

Wright, L. D., 1978. River deltas, *in* R. A. Davis, Jr., ed., *Coastal Sedimentary Environments.* New York: Springer-Verlag, 5-68.

Wright, L. D.; Coleman, J. M.: and Thom, B. G., 1975. Sediment transport and deposition in a macrotidal river channel, Ord River, Western Australia, *in* L. E. Cronin, ed., *Estuarine Research,* vol. 2. New York: Academic Press, 309-321.

Cross-references: *Deltas; Estuaries; Estuarine Coasts; Estuarine Sedimentation; Tidal Deltas.*

ESTUARINE HABITAT

Estuarine conditions are those where mixtures of seawater and freshwater, within varying limits, are maintained along sea coasts. Near the mouths of rivers, oceanic currents, tectonic forces, climatic and meteorological conditions, tides, sediments transported by rivers, and even biological growths cause an enclosed geographical area to form separated from the sea incompletely by spits, islands, or banks. The whole geographic entity is called a *bay* or a *sound* or even an *estuary*, and is essentially an estuary according to Pritchard (1967).

But the estuarine condition of the water along with a soft, fine sediment bottom are not necessarily enclosed. Thus the offshore waters of the Louisiana coast, where oyster reefs grow off the mouth of the Atchafalaya River (a distributary of the Mississippi) is estuarine, and has been characterized as such by Percy Viosca (personal communication). This was most fully documented by Barrett et al. (1978), who showed that the salinities from the beach to 16 km offshore ranged from 17.7-20.5‰ over a two-year period, 1974-75. This area covers approximately 7740 square km of water beyond the beaches. A similar area in the Atlantic extends from Long Island to Cape Hatteras, and both correspond to Ketchum's (1951) definition of an estuary as a body of water where seawater is measurably diluted by freshwater.

Estuaries are difficult to define purely in geographic terms or shapes. Some large rivers run directly into the sea, such as the Columbia, which flows directly into the Pacific Ocean with virtually no change in size. The water dissipates quickly in the deep, vast sea without much dilution of the seawater. In fact, the Mississippi flows directly into the Gulf of Mexico, except where drainage from distributaries causes low-salinity bay waters to interdigitate with the separated river channels of the *crowfoot delta,* which forms a large estuarine area over the shallow continental shelf.

The predominant feature of an estuary is its variable intermediate salinity, which determines the type of life living in it. The limit varies theoretically from 0.5-36.0‰ saline, but more commonly is 5-30‰. On dry coasts, such as that of Texas, these limits may vary downward from 64.0‰ saline during droughts to completely fresh for a few weeks following a hurricane. Most estuaries have periods of low salinities following irregular periods of heavy rains or floods. Dawson (1965) gave an example where sudden rains killed millions of *Branchiostoma floridae* in Mississippi Sound, a large estuary. Even Chesapeake Bay has been affected following hurricane floods, as after Camille in 1969, when freshwater extended more than 56 miles in the Atlantic off the Chesapeake (Adrian Lawler, pers. comm. 1978; Elder, 1971).

Estuaries are continuously filling with silt brought down by rivers and are generally shallow and turbid; their temperatures are similar to that of the surrounding land, lagging behind the atmosphere, and even more slowly behind the temperatures of nearby shallow ocean waters (Collier and Hedgpeth, 1950).

In addition to freshwater and sediments, rivers also bring mineral salts of phosphates, silicates, nitrogen, and organic materials into the estuaries. Some of this material, such as fertilizer salts, is used quickly by marsh plants and the photosynthetic microplankton, and thus quickly enters the food chain.

Estuaries are turbid from the fine sediments brought down by the rivers, and the bottoms are soft and anoxic a few mm beneath the surface. Being shallow and with sedimenting bottoms, the shores are often bordered with marsh plants of both freshwater and saltwater variety (Eleuterius, 1972). Farther out, submerged seagrasses cover hundreds of acres where the waters are less turbid, both areas being particularly noted as nursery areas for marine animals.

Estuaries have been described as mixing bowls, which they are in a sense, but actually they are way stations for sediments and salts on their way to the bottom of the deep sea. For instance, it has been calculated that during the time it takes to fill the eastern and southern estuaries in the United States (the lifetime of a bay) with sediment, several thousand times the bay volume of the water passes through in 6000 to 12,000 years. At the same time, up to 200 times the volume of sediment needed to fill a bay passes through (Gunter, Mackin, and Ingle, 1964).

The soft bottoms of estuaries permit large numbers of annelids, crustaceans, and gastropods, to burrow or simply bury in the bottom. Thousands of invertebrate larvae also settle on the bottom. As Dean (1892) pointed out, it takes only one rare *hard point* in the mud to start self-perpetuating oyster reefs that as relicts may equal in volume anything in the sea. Sessile mussels, barnacles, and other nonmotile species live in the same habitat, sometimes in great abundance.

Estuaries have become recognized through biological studies (Gunter, 1938, 1945, 1950; Weymouth, Lindner, and Anderson, 1933; and Hedgpeth, 1957) and fishery analyses (Gunter, 1967; Rounsefell, 1975) as nursery grounds for the chief commercial fishery species of the Atlantic and Gulf coasts of the United States. Estimates range from 95-98% of fishery species as being estuarine-raised. Their general life history shows a shallow oceanic spawning with a migration of the young toward shore, obviously by current riding, a rapid growth in low salinity waters and a return to the sea at maturity. And so this cycle has been called *semicatadromous*, and is more like that of the eels than the salmons. This process has also been referred to as the *estuarine-shelf dependency*.

The importance of the estuaries as nursery grounds has been recognized during the past 40 years. Gunter (1945) said:

Several species of fishes spawn in the Gulf near the passes. This also holds true for the common shrimp. The blue crab spawns in the Gulf or lower bay. The speckled trout, the two marine catfishes, the menhaden, *Brevoortia* sp., the anchovy, *A. m. diaphana*, and the silverside, *M. b. peninsulae*, are the chief species that spawn in the bays. The young of many fishes that spawn in the shallow Gulf work into the bays in the winter and spring. The bays act as a nursery ground for these fishes and for the common shrimp, *Penaeus setiferus*, and blue crab, *Callinectes sapidus*, as well as for those species spawned within their limits. These animals remain in the bays and grow during the spring and summer. Most of the fishes that leave the bays when cool weather begins are the young which have been growing in this nursery area during the warm months. They are mostly from a few months to almost a year old. The other numerous components of the seaward fall migrants are the spawners going to the Gulf to spawn.

Estuaries not only are nursery grounds but they have indigenous species. Because of the nutrient salts brought in by the rivers, estuaries are extremely fertile. The richest fishery area of the North American continent lies around the mouth of the Mississippi (Gunter, 1963). This is said to be basically because of the surrounding marshes and the vast amount of detritus that fertilizes the open waters. On the other hand, the nannoplankton seems to be abundant and a more likely food of the water strainers.

Estuaries are shallow, generally warmer in winter than the nearby sea, rich in nutrients, and sometimes have protective vegetation. They are also less turbid than the seashore and have a milder surf.

Because the numbers of species decline with a falling salinity gradient, a matter known to zoogeographers (Hesse and Doflein, 1910) of more than two generations ago, the numbers of infections of organisms, parasites, and predators decline with the lowering salinities of estuaries along a gradient from seawater to freshwater. Also, the young organisms seem to prefer lower salinities. The optimum salinity for growth and survival of young brown shrimp of the Gulf is about 8–18‰ salt (Venkataramiah, Lakshmi, and Gunter, 1973).

It is also known that little fishes and crustaceans go into the lowest salinities of the species range they can find. Figure 1 (facing page) shows the total length frequencies of *Penaeus fluviatilis* Say, the North white shrimp, taken in Texas bay waters by personnel of the Parks and Wildlife Commission at given salinity ranges taken in all large Texas bay systems. There is a direct correlation between total lengths and salinity of the water at the 95% level. If the highest salinity is excluded, the correlation is at the 99.9% level. This is because the smallest shrimp come in from the sea in the plankton and settle to the bottom at their smallest sizes in the highest salinity area in the bays. This works against the correlation until the little shrimp scatter themselves over the bays at preferred salinities. The same relationship of length frequency to salinity could be demonstrated with many other fishes and crustaceans, but generally we do not have such large amounts of data as shown here for the shrimp.

For further readings see also Biggs (1978), Gunter (1967), and Hedgpeth (1957).

GORDON GUNTER

References

Barrett, B. B.; Merrell, J. L.; Morrison, T. P.; Gillespie, M. C.; Ralph E. J.; and Burdon, J. F., 1978. *A Study of Louisiana's Major Estuaries and Adjacent Offshore Waters.* New Orleans: Louisiana Department of Wildlife and Fisheries, Tech. Bull. No. 27, 197p.

Biggs, R. B., 1978. Coastal bays, *in* R. A. Davis, Jr., ed., *Coastal Sedimentary Environments.* New York: Springer-Verlag, 69–99.

Collier, A., and Hedgpeth, J. W., 1950. An introduction to the hydrography of tidal waters of Texas, *Inst. Marine Sci. Pub.* 1, 121–194.

Dawson, C. E., 1965. Rainstorm induced mortality of lancelets, *Branchiostoma,* in Mississippi Sound, *Copeia* No. 4, 505–506.

Dean, B., 1892. The physical and biological characteristics of natural oyster grounds of South Carolina, *U.S. Fish Comm. Bull.* 10, 335–362.

Elder, R. B., 1971. The Effect of Run-off from Hurricane Camille on the Continental Shelf Waters of the Chesapeake Bight. Unpublished thesis, College of William and Mary, 80p.

Eleuterius, L. N., 1972. The marshes of Mississippi, *Castanea* 37, 153–168.

Gunter, G., 1938. Seasonal variations in abundance of certain estuarine and marine fishes in Louisiana, with particular reference to life histories, *Ecol. Mon.* 8, 313–346.

Gunter, G., 1945. Studies on marine fishes of Texas, *Inst. Mar. Sci. Pub.* 1, 1–190.

Gunter, G., 1950. Seasonal population changes and distributions as related to salinity, of certain invertebrates of the Texas coast, including the commercial shrimp. *Inst. Mar. Sci. Pub.* 1, 89–101.

Gunter, G., 1963. The fertile fisheries crescent, *Mississippi Acad. Sci. Jour.* 9, 286–290.

Gunter, G., 1967. Some relationships of estuaries to the fisheries of the Gulf of Mexico, *in* G. H. Lauff, ed., *Estuaries.* Washington, D.C.: Am. Assoc. Advancement Sci. Pub. 83, 621–638.

Gunter, G.; Mackin, J. G.; and Ingle, R. M., 1964. *A Report to the District Engineer of the Effect of the Disposal of Spoil from the Inland Waterway, Chesapeake and Delaware Canal, in Upper Chesapeake Bay.* Philadelphia: U.S. Army Corps of Engineers, 51p.

Hedgpeth, J. W., 1957. Estuaries and lagoons, *in* J. W. Hedgpeth, ed., *Treatise on Marine Ecology and Paleoecology,* vol. 1. New York: Geol. Soc. America Mem. 67, 693–729.

Hesse, R., and Doflein, F., 1910. *Tierbau und Tierleben in ihrem Zuzamenhang betrachtet.* Berlin: Teubner, 789p.

Ketchum, B. H., 1951. The exchange of fresh and salt water in estuaries, *Jour. Marine Research* 10, 18–38.

Pritchard, D. W., 1967. What is an estuary: A physical viewpoint, *in* G. H. Lauff, ed., *Estuaries.* Washington, D.C.: Am. Assoc. Advancement Sci. Pub. 83, 3–5.

Rounsefell, G. A., 1975. *Ecology, Utilization, and Management of Marine Fisheries.* St. Louis: C. V. Mosby Company, 516p.

Venkataramiah, A.; Lakshim, G. J.; and Gunter, G., 1973. The effects of salinity and feeding levels on the growth rate and food conversion efficiency of the shrimp *Penaeus aztecus, World Mariculture Soc. Proc.* 3, 267–283.

Weymouth, F. W.; Lindner, M. J.; and Anderson, W. W., 1933. Preliminary report on the life history of the common shrimp, *Penaeus setiferus* (Linn.), *U.S. Dept. Commerce Bur. Fish. Bull.* 48, 1–25.

Cross-references: *Biotic Zonation; Coastal Ecology, Research Methods; Coastal Fauna; Coastal Flora; Estuaries; Estuarine Sedimentation; Geographic Terminology; Nutrients; Standing Stock.*

ESTUARINE SEDIMENTATION

Estuaries are major sites for accumulation of sediment along our coastline. Sediments are added to estuaries by rivers, by shore erosion, by primary production, by the sea, and by the atmosphere. Typically, estuaries fill from their heads and their margins. An estuarine delta generally forms near the head of the estuary. The delta grows seaward, extending the realm of the river and expelling the intruding sea from the

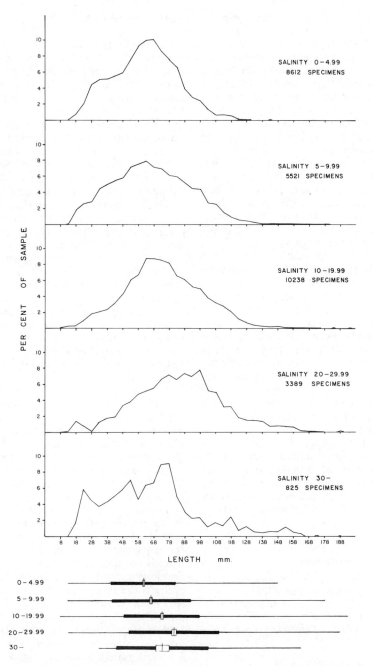

FIGURE 1. Total length frequencies of 28,585 white shrimp taken in Texas bays from 1960–1964 by personnel of the Texas Game and Fish Commission. Salinities were collected at the same time. This covered all large Texas bay systems—Galveston, Matagorda, and Copano-Aransas. The data were worked up at the Gulf Coast Research Laboratory by J. Y. Christmas. (*Note*: Refer to Estuarine Habitat, page 400.)

coastal basin. Lateral accretion by marshes may also play a major role, and in some estuaries deposition of marine sediments near the mouths is important. As a result of these processes, the estuarine basin is converted back into a river valley. Finally, the river reaches the sea through a depositional plain, and the transformation is complete. All modern estuaries were formed during the most recent rise in sea level, and today we find estuaries around the world in various stages of this evolutionary process (Lauff, 1967; Schubel, 1971).

Because of their characteristic circulation processes, estuaries are effective sediment traps. The tidal circulation is important in the formation of channels, tidal flats, and tidal deltas, but it is the net (gravitational) circulation that is of primary importance in determining the rates and patterns of filling of most estuaries.

In most estuaries fluvial sediment sources dominate, and much of this sediment is introduced during events. A flooding river commonly discharges as much sediment in a few days as in many years of average flow. In one week following tropical storm Agnes (June 1972) the Susquehanna River discharged about 25 times as much sediment into Chesapeake Bay as it had during the previous year.

The prevailing mode of sediment transport in most estuaries is as suspended load. Concentrations of suspended matter are relatively high, typically ranging from a few mg/ℓ to a few tens of mg/ℓ under average conditions, and reaching values of hundreds or even thousands of mg/ℓ during floods and hurricanes. In partially mixed estuaries, the most common type, concentrations of suspended matter are higher near the head of the estuary than either farther downstream in the estuary or farther upstream in the source river. Such features, called turbidity maxima, are created and maintained by the net estuarine circulation.

While sedimentation rates in most estuaries are naturally high, man's activities have greatly accelerated the rate of infilling, and the increased concentrations of suspended sediment have seriously degraded the water quality of many estuaries (Lauff, 1967; Schubel, 1971).

J. R. SCHUBEL

References

Lauff, G. H., ed., 1967. *Estuaries.* Washington, D.C.: Am. Assoc. Advancement Sci. Pub. 83, 757p.

Schubel, J. R., ed., 1971. *The Estuarine Environment: Estuaries and Estuarine Sedimentation.* Washington, D.C.: Am. Geol. Inst. Short Course Lecture Notes, 324p.

Cross-references: *Bars; Estuaries; Estuarine Coasts; Estuarine Delta; Inlets, Marine-Lagoonal and Marine-Fluvial; Lagoonal Sedimentation; Lagoonal Segmentation; Tidal Currents; Tidal Deltas; Tidal Inlet, Channel, River.* Vol. III: *Tidal Inlet;* Vol. VI: *Estuarine Sedimentation.*

ETCHED POTHOLE—See SOLUTION AND SOLUTION PAN

EUGLENOPHYTA

Euglenophyta, a division of highly differentiated algal flagellates were probably first described by Antony van Leeuwenhoek, the pioneer in the study of *protists* in the seventeenth century, but it was not until 1838 when Ehrenberg described *Euglena,* for which the division is named and whose features characterize the attributes of Euglenophyta, was there a detailed account of the euglenoids.

Predominately inhabitants of the freshwater environment, these unicellular flagellates are cylindrical, ovoid to fusiform, microscopic plants of eukaryotic organization (possessing a true nucleus and other membrane-bound organelles) with usually two flagella for locomotion, an undifferentiated cell wall, and chloroplasts; colorless varieties are known (Fig. 1). The flagellar arrangement is atypical in that there is a longer emergent flagella of the tinsel variety (pantonematic), but the mastigonemes (hairs) are located on only one side of the flagella; a condition referred to as *"stichonematic"* in the older literature, in addition to a second, shorter, nonemergent stubby flagella, both eminating from the gullet, an invagination at the anterior pole of the cell.

Moreover, the euglenoid cell does not possess a cell wall, but is, instead, bounded by a periplast or pellicle; being either firm or plastic. This cell membrane, when plastic and flexible, gives rise to the condition of metaboly whereby the organism changes its shape while in motion.

Euglenoids, like all plants, photosynthesize, but the storage product, the carbohydrate paramylum ($\alpha\beta$-1:3 linked glucan similar to laminarin, the food reserve of the Phaeophytes) is the reserve product instead of starch, as is typical of most other algal groups. Photosynthetic pigments of euglenophytes are characteristically chlorophyll *a* and *b* in addition to accessory carotenoids. A unique feature of this group is the possession of an eye spot (*stigma*) located adjacent to the base of the flagella, independent of the chloroplasts, that is a collection site of carotenoid pigments, and thus it appears orange-red in color. The most widely accepted function of the stigma is as a light-absorbing shield that prevents light from reaching the paraflagellar body, the true photoreceptor. Reproduction is normally by longitudinal division of the motile cell; sexual reproduction has not as yet been observed, and the formation of cysts does occur in some species.

Nutrition is diverse among the euglenoids, and may be either photoautrophic, saprophytic, holozoic, mixotrophic, or parasitic.

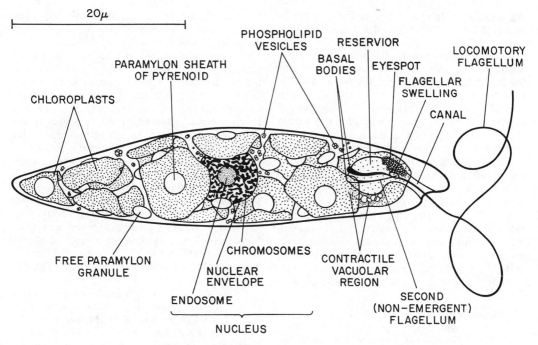

FIGURE 1. A living cell of *Euglena gracilis* (after Leedale, 1967).

Finally, it should be noted that there exists uncertainty about the taxonomic relations of this group; some investigators classify euglenoids with the *protozoa,* others with the plants, and still others in the kingdom Protista. Many biologists believe that this group may be related to those organisms from which both the plant and animal kingdoms developed.

At present there is much active euglenoid research in such fields as chloroplast physiology, genetics, and environmental studies. For additional literature on this subject, the reader is referred to Beutow (1968), Bold (1970), Chapman and Chapman (1973), Fritsch (1965), Fuller and Carothers (1963), Gojdics (1953), Morris (1967), Prescott (1968), Rosowski and Parker (1971), and Wolken (1961).

BONNIE BLOESER

References

Beutow, D. E., 1968. *The Biology of Euglena.* New York: Academic Press, 417p.
Bold, H. C., 1970. *The Plant Kingdom.* Englewood Cliffs, N.J.: Prentice-Hall, 190p.
Chapman, V. J., and Chapman, D. J., 1973. *The Algae.* London: Macmillan, 497p.
Fritsch, F. E., 1965. *The Structure and Reproduction of the Algae.* London: Cambridge University Press, 791p.
Fuller, H. J., and Carothers, Z. B., 1963. *The Plant World.* New York: Holt, Rinehart and Winston, 564p.
Gojdics, M., 1953. *The Genus Euglena.* Madison: University of Wisconsin Press, 268p.
Leedale, G. F., 1967. *Euglenoid Flagellates.* Englewood Cliffs, N.J.: Prentice-Hall, 242p.
Morris, I., 1967. *An Introduction to the Algae.* London: Hutchinson, 189p.
Prescott, G. W., 1968. *The Algae: A Review.* Boston: Houghton Mifflin, 436p.
Rosowski, J. W., and Parker, B. C., eds., 1971. *Selected Papers in Phycology.* Lincoln: University of Nebraska Press, 876p.
Wolken, J. J., 1961. *Euglena–An Experimental Organism for Biochemical and Biophysical Studies.* New Brunswick, N. J.: Rutgers Univ., Inst. Microbiology, 173p.

Cross-references: *Algal Mats and Stromatolites; Chlorophyta; Chrysophyta; Coastal Flora; Coral Reefs; Cryptophyta; Cyanophyceae; Organism-Sediment Relationship; Phaeophycophyta; Reef Habitat; Rhodophycophyta.* Vol. III: *Algae;* Vol. VI: *Algal Reef Sedimentology.*

EUROPE, COASTAL ECOLOGY

Between the Arctic and the Mediterranean the physiographically diverse European coastline provides a range of environmental conditions and habitats. There is a strong north-south climatic gradient, although the climate within the coastal zone is considerably modified and ameliorated in comparison with that of inland regions. The tidal range varies from approximately 12 m in Brittany and the Bristol Channel to the minimal tidal regimes of the Baltic and the Mediterranean. Salinities vary from about 35.0‰ in most of the coastal waters to the much lower salinities of the Baltic and the complex and changing patterns found in estuaries. The major habitats of the coastal zone are determined largely by physiography, but within each habitat type the biota is strongly influenced by the physical factors outlined above.

The boundaries between the major habitats are to some extent arbitary and a number of different classifications are possible. By emphasizing the biological component of these habitats, they can be divided in those in which the biota is predominantly marine: algal communities of rocky shores, mud and sand flats, and sea grass communities; and those in which the biota is essentially terrestrial: saltmarsh, seacliff, driftline, shingle beach, and sand dune.

Algal Communities of Rocky Shores

Extensive stretches of the European coast are rocky, and support communities dominated by algae.

The most striking feature of these communities is their zonation, which is further emphasized by the different colors of the dominant species concerned.

The communities of the rocky shore can be divided into two major zones, each of which can be further subdivided. These two zones are: the *littoral,* occupied by organisms that are adapted to periods of exposure to air; and the *sublittoral.* Although the boundary between these zones is determined by environmental factors, principally tidal, it is normally defined in biological terms owing to the difficulty of choosing one environmental factor from within an interacting complex to make the division. For much of the European coast the sublittoral/littoral boundary can be set at the top of the zone dominated by members of the Laminariales, normally of the genus *Laminaria.*

The upper limit of the littoral is more difficult to define, but is usually taken as the upper edge of a zone characterized by black lichens of the genus *Verrucaria* and the marine gastropod *Littorina*. Within the littoral, zones are normally defined in terms of the dominant species of algae and sessile animals such as barnacles, although the distribution of more mobile animals frequently corresponds with zone boundaries so defined. As in the sublittoral, the dominant algae are brown (Phaeophyceae), but are members of the Fucales, often in the genus *Fucus.*

The vertical extent of the zones varies with tidal amplitude, but is also influenced by wave action. On extremely sheltered shores, the littoral/sublittoral boundary is close to extreme low water spring-tides, so that the laminarian zone is rarely exposed to the air. On exposed coasts the laminarian zones may extend considerably above extreme low water spring-tides as even during periods of exposure there will be continuous splashing by waves. Similarly, on sheltered shores the entire littoral zone, including the uppermost *Verrucaria,* is within the tidal range, but on exposed sites the littoral zone may extend far above the tidal limit but is nevertheless subjected to heavy spray deposition. As well as changes in zone boundaries, there are often changes in the abundance of individual species between exposed and sheltered coasts—for example, on exposed shores the algal cover in the littoral zone may be low and the zone dominated by barnacles (*Balanus* spp.)

Within the sublittoral, zonation of species will be controlled primarily by the availability of light that will also determine the lower limit of the whole sublittoral zone.

Although the broad pattern of zonation can be interpreted in terms of tidal range and exposure to wave action, many local factors will influence the fine pattern at any particular site. In particular the availability of pools and clefts in the rock face that provide habitats for many more restricted species, will be governed by the geology of the rock.

Although the dominant algae of the rocky shore are Phaeophyceae, the habitat is also the most important one for the Rhodophyceae (red algae), which are found in abundance in the lower littoral and sublittoral both as epiphytes on the larger brown algae and on the rocks themselves.

The Phaeophyceae and even more so the Rhodophyceae, are relatively intolerant of varying salinities, and with the exception of a few fucoids (notably *Fucus ceranoides*) do not penetrate far up estuaries. However, rocky shores in the low, but relatively invariant, salinity environment of the Baltic support abundant fucoids that often exhibit growth forms different from the same species growing at high salinities.

A detailed account of the ecology of rocky shores applicable to northern Europe is given by Lewis (1964).

Mud and Sand Flats

Extensive areas of intertidal flats occur in estuaries and embayments and also on the landward side of offshore barrier islands, as in the Wadden Zee.

These flats provide an unstable habitat with patterns of drainage and microrelief changing frequently. This substrate instability, coupled with the turbidity of the covering water during flooding tides, limits the algal populations supported by the habitat, and for the most part the primary productivity of intertidal flats is low. However, there may be considerable heterogeneity in both space and time of algal populations, and locally high primary productivities may occur. The majority of algae in this habitat are unicellular (mainly diatoms), and are either *epipelic* (free-living on the surface) or occur within the upper few millimeters of the sediment. Populations of macroalgae tend to fluctuate in area and abundance. Among the most important of these macroalgae are species of the genus *Enteromorpha* (which locally might be favored by the eutrophic conditions that have developed in many estuaries).

In spite of the low primary productivity, the secondary productivity of intertidal flats is high. In most flats there is a large invertebrate infauna in the upper layers of the sediment. Although the total fauna is large, the majority of the *biomass* is made up of relatively few species. Many of the infauna are filter feeders dependent on particulate organic matter in the estuarine water. The origin of this detritus is still subject to debate; one possible source is from plant material washed out of the salt marshes that fringe many intertidal flats.

The distribution of the species that make up the majority of the biomass is largely controlled by the nature of the substrate, mud supporting a different range of organisms than sand.

The infauna is relatively less affected by minor changes in microrelief than the mud-flat algae, but large channel changes can cause mass mortality. One factor of extreme importance in controlling the infauna in flats in northern Europe is frost. Exposure to the air during the winter can lead to heavy mortality: many populations of the cockle, *Cardium edule* were much reduced during the winter of 1962-3.

A spectacular consequence of the large infauna is the large flocks of waders and wildfowl that gather in the estuaries of northern Europe to overwinter. The majority of species involved spend the summer at higher latitudes. The main concentrations of birds occur in relatively few mud-flat areas, mainly in the British Isles and the Low Countries. Probably the most important single overwintering site is Morecambe Bay in northern England, where over 250,000 waders are regularly recorded.

Sea Grass Communities

Sea grass communities do not fit easily into a classification of coastal habitats.

In northern Europe sea grasses are represented by *Zostera* spp., which are found locally on intertidal mud flats but also occur in the sublittoral. In the 1930's *Zostera* in the Northern Hemisphere was afflicted by a *wasting disease,* the causative agent of which was never conclusively determined. Many European populations, particularly of *Zostera marina,* did not recover from this disease, and the present limited distribution is much reduced compared with the predisease period.

Zostera beds support a rich epiphytic alga flora and a diverse fauna. *Zostera* itself is an important food resource for wildfowl, with, for example, the *Zostera* beds of southeast England and the Wadden Zee providing the overwintering habitat for the majority of the world's dark-bellied Brant geese (*Branta bernicla bernicla*).

In the Mediterranean, *Zostera* is rare, and the predominant sea grass is *Posidonia oceanica,* which is found in exposed to moderately sheltered sites on open shores and the head of bays. It is exclusively sublittoral and occurs down to depths of at least 40 m. Below c. 10 m *Posidonia* beds are often dense and homogeneous, but water and substrate movement in shallower water causes local erosion and reduction in cover. These dense *Posidonia* beds provide a substrate for a large number of epiphytic algae, and also provide food and shelter for a diverse and abundant fauna.

Cymodocea nodosa is also widespread in the Mediterranean, but is absent from areas occupied by dense *Posidonia*. It has been suggested that *C. nodosa* is a pioneer species and that in time it is often replaced by *Posidonia.*

In the Aegean, *Halophila stipulacea* occurs, having migrated into the Mediterranean through the Suez Canal from the Red Sea.

In the brackish conditions of the Baltic, there are extensive sublittoral communities dominated by angiosperms. *Zostera marina* is locally abundant and extends as far east as approximately the 3‰ isohaline. Other important species in the Baltic include *Ruppia maritima, Naias marina, Zannichellia palustris, Potamogeton pectinatus,* and *P. filiformis.* Similar communities are also found in brackish coastal lagoons, a relatively rare habitat in Europe.

Salt Marsh

Coastal salt marsh can be defined as areas dominated by vascular plant communities (excluding sea grasses) and subjected to direct tidal- or weather-affected inundation by more or less dilute seawater.

Salt marsh is found in a number of different physiographic situations, but the most extensive areas occur in estuaries or behind offshore barrier islands or spits.

The plant communities are often clearly zoned, and there is a temptation to interpret the relatively simple zonation pattern as a response to some uniform environmental gradient. However, the salt marsh environment is extremely heterogeneous, both spatially and temporally, and each species has an individual response to the environmental complex. The major factors influencing the zonation can be related to the overriding effects of periodic flooding. In the lower salt marsh, with regular inundation and relatively constant but high soil salinity, the major influences on the plants are the mechanical effects of the tidal currents, the direct effects of submergence (i.e., reduction in photosynthesis) and the effects of repeated flooding on the physical and chemical nature of the substrate. In the upper marsh, with infrequent submergence, the major factors are probably the timing and range of fluctuations in soil salinity. The maximum salinities in the mid and upper marsh are often in excess of those in the lower marsh owing to evaporation during long periods of exposure in the summer. Even in the Baltic or toward the head of estuaries where the flooding water may be of low salinity, high salinities may be recorded in the upper marsh.

The salt marsh zonation is often interpreted as a spatial representation of the succession of communities in time. While salt marsh is a dynamic habitat, change in the environmental conditions during the developmental history of a marsh make the simple reconstruction of succession from zonation a dangerous practice.

Succession is normally driven by the input of tide-borne sediment, although wind-borne material may also be important at some sites. Accretion has been studied on a number of north European marshes; these studies show the same pattern of accretion rate declining with decrease in tidal inundation, but the absolute rates of accretion vary greatly from site to site. Incorporation of appreciable quantities of autochthonous organic material in the soil is rare in European salt marshes; it occurs in brackish reed swamps in lagoons and upper estuaries and also in salt marshes in western Ireland and Scotland, where the substrate throughout the marsh zonation is often peaty. The factors causing the development of these salt marsh peats are unknown, but the small input of allochthonous material and the high regional rainfall in Scotland and Ireland are probably important. In the Baltic, with minimal tidal influence, the development of the salt marshes (generally called sea meadows) is controlled by the continuing high rate of *isostatic uplift*.

There is considerable variation in the vertical extent of salt marsh development. The lower limit of salt marsh growth is normally between mean high water and mean high water neaptides; the factors determining the limit at any one site are complex, although the degree of exposure to wave action is important.

Within Europe three major marsh types have been recognized: Arctic, north European, and Mediterranean.

The marshes of Arctic Europe are similar to those elsewhere in the Arctic, being dominated by grasses and sedges. Climatic conditions probably determine the nature of the flora, but these marshes are often strongly influenced by spring meltwater producing brackish conditions in the early part of the growing season. Elements of the Arctic flora are found in salt marshes as far south as southern Norway, and marshes related to the Arctic type also occur toward the head of the Gulf of Bothnia.

The salt marshes of northern Europe are diverse, and a number of geographical subdivisions have been recognized by some workers. At the broadest scale, there are two major subdivisions: marshes that are principally grass-dominated, and those in which grasses are less important. Grass-dominated marshes are almost always heavily grazed by sheep and/or cattle, and this grazing pressure may be necessary for the development and maintenance of many examples of this marsh type. These marshes are most often found on a sandy substrate, and are widespread in northern Europe. Grass-dominated marshes also occur in the Baltic, but the particular hydrological and salinity conditions of this area allow a greater development of reed swamp communities.

The other major marsh type tends to occur on muddy substrates, and is normally ungrazed. The characteristic plants are *halophytes* (plants restricted to the salt marsh habitat; the balance of present evidence suggests that the majority of these plants do not require saline conditions) and would include *Limonium* spp. and the shrub *Halimione portulacoides*. This marsh type is widespread, but in general its distribution is more southerly than the grass-dominated type.

The variation of marsh type can be correlated with a number of factors, notably grazing, substrate, and climate. Unfortunately, the complex land use history of most north European marshes makes it difficult to assess which are causal.

An important recent change in north European salt marshes has been the spread of *Spartina townsendii* s.l. This species was first recorded in 1870 in Southampton Water, England, and is believed to be a natural hybrid between the native *Spartina maritima* and *S. alterniflora*, an American species accidentally introduced into Britain. *S. townsendii* is a perennial species of low-marsh habitats more effective in promoting marsh development than the previous most important low-marsh species, annual *Salicornia* spp. *S. townsendii* has been spread by deliberate planting and natural means, and is now the dominant of low marsh communities in Channel and southern North Sea marshes and more sporadically elsewhere. Susceptibility to frost damage is likely to limit its spread northward.

In the Mediterranean region, the infrequency of submergence and the high evaporation/precipitation ratio leads to long periods of high soil salinity. In consequence the vegetation tends to be fairly open and dominated by shrubby members of the Chenopodiaceae family, although richer plant communities develop if the soil salinity is ameliorated (as around permanent brackish lagoons).

The fauna of salt marshes may be divided into two components: one primarily marine, similar to that of mud flats, which occupies creek bottoms and pans; the other, essentially terrestrial, which is found on the main marsh surface. The terrestrial invertebrate fauna of European salt marshes has not been subject to extensive study, but recent work has revealed considerable diversity and many species with special adaptations to the rigors of the environment. Salt marshes provide roosting and feeding grounds for many birds, and during the summer the upper marsh may support a high density of ground-nesting birds, such as the skylark (*Alauda arvensis*).

Although salt marshes are dominated by vascular plants, they also support a large algal flora. This probably contributes little to the total productivity of the marshes, but the role of blue-green algae in nitrogen fixation may be important in salt marsh functioning.

In comparison with the eastern United States, there have been relatively few studies on productivity and nutrient cycling on European marshes. Available studies suggest that the extremely high productivities recorded on some American marshes are not matched in Europe, but that productivity figures for Europe fall well within the range of those from America. However, the majority of European marshes have a higher species diversity of vascular plants than do those in the eastern United States.

The ecology of European salt marshes is discussed by Chapman (1974) and Beeftink (1977); aspects of the geographical variation in the vegetation, by Adam (1978).

Sea Cliff

Long stretches of the European coastline are formed of cliffs. The vegetation of sea cliffs is influenced by three major factors: geology and geomorphology, the amount of salt spray, and the presence or absence of large numbers of nesting sea birds.

The geology and geomorphology will determine the availability of sites for plant colonization (ledges, crevices, and the like) and the nature of the soil in these sites. The amount of salt spray will depend on many features that in aggregate determine the degree of exposure of a site. Sea birds have a profound effect on vegetation partly through physical disturbance but most particularly through the eutrophication caused by their droppings. The vegetation of bird cliffs is often extremely lush but of low diversity, comprising plants that are often weeds in inland habitats. The suitability of any particular site for birds is governed partly by geology (insofar as it influences availability of nest sites) and partly by the food resources of the adjacent seas. The major sea bird cliffs in Europe are in the northwest (northern Britain and Scandinavia); single cliffs may support hundreds of thousands of birds.

The grasslands and heathlands above sea cliffs are also influenced by sea spray, and often form a distinctive floristic facies of more widespread inland communities.

A number of plants reach their northernmost limits of distribution in sea cliff sites. This probably reflects the fact that such sites provide a frost-free, humid-equable microclimate. Inaccessibility of many ledges also allows grazing-intolerant species to find refugia on sea cliffs. There have been few detailed studies on cliff vegetation, but Malloch (1971) provides a comprehensive account of the west Cornish cliffs.

Drift Line

The drift line is an unstable ephemeral habitat. Accumulations of tidally deposited organic material are found in salt marshes, in front of dunes, and on shingle beaches. There is variation in the flora between these different habitats and also between different substrates. However, the drift line has been little studied, and it would be difficult to quantify the variation.

There is great variation in the density of vegetation cover on the drift line from extremely open to dense monodominant stands. The majority of the flora are representatives of three families (Chenopodiaceae, Crucifereae, and Polygonaceae), but the total flora is diverse.

The instability of the habitat leads to great fluctuations in the population size of drift line plants. Superimposed on these fluctuations appears to be a general decline in a number of species that are becoming extremely rare. The causes of this decline (even if genuine) are poorly understood although often ascribed to man.

The fauna of drift litter is often abundant but has been little studied.

Shingle Beach

On a world scale, extensive areas of vegetated coastal shingle are a rare habitat. Within Europe, the largest areas of this habitat are found in the British Isles, where the three outstanding examples are Chesil Beach (Dorset), Dungeness (Kent), and Orfordness-Shingle Street (Suffolk).

Many shingle beaches support drift line communities, but development of vegetation above the tidal limit is rare. There are probably few species of plant or animal restricted to the shingle habitat, but several of the communities may be unique to the environment. The nature of the vegetation cover is influenced by the fine fraction in the substrate; on those beaches where the fine fraction is absent, vegetation development is minimal. The major communities that form are either some type of maritime grassland or coastal heath. Scrub communities occur locally, and at Dungeness a holly (*Ilex aquifolium*) woodland, which is probably unique, has developed.

Shingle beaches provide nesting sites for a number of birds, particularly for species of tern (*Sterna* spp.) that nest colonially.

Sand Dune

Like other coastal communities, sand dunes are dynamic systems subject to considerable temporal variation.

There is considerable similarity in the outer developing part of sand dune systems (the yellow dune stage) throughout Europe. The principal colonizer and foredune former is the grass *Agropyron (Elytrigia) junceiforme*. This species can tolerate limited tidal submergence, but is restricted in its ability to keep pace with accretion. After the foredunes have accreted above the tidal limit, *Agropyron* is replaced either by *Ammophila arenaria* (the marram grass) or (chiefly in northern Europe) by *Elymus arenarius*. These species can tolerate considerable accretion and contribute to the building and stabilization of the main dune ridges.

In time the rate of accretion falls, colonization between the tussocks of *Ammophila* or *Elymus* occurs, and the original grasses become moribund. Further community development is dependent on the pattern of soil development, and two major series can be recognized, although each encompasses a range of variation related to climatological and historical factors.

The two major series are influenced by the nature of the parent sand. On dunes with little calcium carbonate in the parent material, the development of vegetation and soil causes a rapid drop in soil pH and the accumulation of humus. Succession leads to the development of heath vegetation (dominated by *Calluna vulgaris* or *Empetrum nigrum*) and possibly in time to the formation of birch (*Betula*) and then pine (*Pinus*) woodland. On calcareous systems the rate and extent of soil acidification are much reduced. Succession leads through a species-rich calcareous grassland community to a scrub in which a number of species may occur, although one of the most important is *Hippophaë rhamnoides*. The end point of succession is probably mixed deciduous woodland, although because of disturbance few dunes reach this stage.

On the European mainland, calcareous and noncalcareous systems are geographically distinct. South of the Gironde, in the Bay of Biscay, dunes are generally noncalcareous; between the Gironde and southern Holland, they are calcareous, while farther north the lime content is again low. Within the British Isles, the distribution of calcareous and noncalcareous dune systems does not follow as clear a pattern, although the majority of calcareous dunes occur in the west. The most calcareous dunes are the *machair* systems of northern and western Scotland, many of which are composed predominantly of shell sand with a $CaCO_3$ content in excess of 75%.

Between the dune ridges are dune valleys (*slacks*) characterized by a varying soil water table. In some, permanent or seasonal lakes form; in others, the water table may never reach the soil surface. In the majority of dune systems, there is little tidal influence on the water table, which is normally fresh. These slacks often support a rich flora and a distinctive range of communities, for which the major controlling factors are the fluctuations in the water table and soil nutrient status.

The invertebrate fauna of sand dunes, particularly calcareous systems, is extremely rich. Many dunes have been influenced in the past by the rabbit (*Oryctolagus cuniculus*), which formerly occurred in large numbers on many systems. Sand dune ecology is discussed in detail by Salisbury (1952) and Ranwell (1972).

Coastal Floras

The vascular plant flora of the *terrestrial* coastal communities is composed of species restricted to the various habitats and species of wider ecological range that occur in coastal as well as inland habitats. The only habitat totally occupied by specialist species is the lower salt marsh, but the drift line contains a high proportion of species restricted to that environment. Many of the inland species found in coastal ecosystems occur as distinct coastal ecotypes. A large number of the inland species in the coast fall in the category of weeds; such plants are often shade-intolerant, and their occurrence at the coast reflects the fact that coastal communities are naturally treeless and would have provided refugia during the period of the postglacial forest maximum.

More detailed accounts of the coastal ecology of Europe may be found in Chapman (1976), Barnes (1977), Ratcliffe (1977), and Jeffries and Davy (1979).

PAUL ADAM

References

Adam, P., 1978. Geographical variation in British salt-marsh vegetation, *Jour. Ecol.* **66**, 339-366.

Barnes, R. S. K., ed., 1977. *The Coastline.* London and N.Y.: John Wiley & Sons, 356p.

Beeftink, W. G., 1977. The coastal salt marshes of western and northern Europe: An ecological and phytosociological approach, *in* V. J. Chapman, ed., *Wet Coastal Ecosystems.* Amsterdam: Elsevier, 109-155.

Chapman, V. J., 1974. *Salt Marshes and Salt Deserts of the World.* London: Leonard Hill, 292p.

Chapman, V. J., 1976. *Coastal Vegetation.* Oxford: Pergamon, 292p.

Jeffries, R. L., and Davy, A. J., eds., 1979. *Ecological Processes in Coastal Environments.* Oxford: Blackwell, 684p.

Lewis, J. R., 1964. *The Ecology of Rocky Shores.* London: Hodder and Stoughton, 323p.

Malloch, A. J. C., 1971. Vegetation of the maritime cliff tops of the Lizard and Land's End peninsulas, west Cornwall, *New Phytol.* **70**, 1155-1197.

Ranwell, D. S., 1972. *Ecology of Salt Marshes and Sand Dunes.* London: Chapman and Hall, 258p.

Ratcliffe, D. A., ed., 1977. *A Nature Conservation Review,* vol. 1. London: Cambridge University Press, 399p.

Salisbury, E. J., 1952. *Downs and Dunes: Their Plant Life and Its Environment.* London: Bell, 328p.

Cross-references: *Africa, Coastal Ecology; Arctic, Coastal Ecology; Asia, Middle East, Coastal Ecology; Biotic Zonation; Climate, Coastal; Coastal Fauna; Coastal Flora; Europe, Coastal Morphology; Halophytes; Intertidal Mud Habitat; Intertidal Sand Habitat; Loose Rock and Stone Habitat; Organism-Sediment Relationship; Phaeophyceae; Rhodophyceae; Rocky Shores Habitat; Salt Marsh; Sand Dune Habitat.* Vol. VIII, Part 2: *Europe.*

EUROPE, COASTAL MORPHOLOGY

It is difficult to know where to start writing a general account of the coasts of a continent. To give a brief description of each country does not offer a proper perspective, and may give the impression that the coast is peculiar to that country. It therefore seems better to begin in much earlier geological times, well before the coasts as we see them now existed but nevertheless when their main outlines began to evolve.

In Europe, the structural framework has been controlled since Precambrian times by three rigid masses or shields: the *Baltic Shield,* which is best seen in and around the Baltic Sea, although it extends much farther under newer rocks; the *North African-Arabian Shield* in the south; and a much larger one in the northwest, only fragments of which remain in the *Hebridean* region. In the course of geological time, these great shields have moved relative to one another, and in so doing they have compressed the more mobile crust between them and given rise to the three great orogenic zones of Europe: the *Caledonian* chain in the early part of the Paleozoic; the *Hercynian* in the Upper Paleozoic; and the *Alpine* in the Tertiary, although preliminary movement began in the Cretaceous.

The picture of rigid masses and orogenic zones is an old one, and was fully discussed by Suess (1904) in *Das Antlitz der Erde.* Since then many theories have been put forward to explain how the rigid masses were formed and why they have moved in such a way as to cause the formation of mountains. In 1914 Wegener promulgated his theory of continental drift (Wegener, 1929), which, after World War I, was widely discussed, and provoked many other theories. It was not until the late 1950s or early 1960s that a theory began to be based on a much surer foundation of facts. The great ocean basins were investigated, and much information about oceanic floor structure became available. Today we think in terms of plate tectonics and sea floor spreading. We assume that the earth's surface is made of a number of rigid, but relatively thin (97-161 km) plates. They interlock and are believed to be in continuous motion with relation to one another. Each plate shows an upper (24-32 km thick) and a lower (80-113 km thick) unit; they float on a mobile layer—the mantle or asthenosphere.

It is not relevant to discuss how the movements of these plates were detected and measured, but we must attempt to show how they have affected Europe. Let us turn first to the origin of the Atlantic Ocean. The central Atlantic ridge is a part of a continuous ridge that encircles the globe, but as a mountain range it is different from those of the continents. It marks a line of separation, a great crack, from which the two sides of the Atlantic have moved away as the crack itself was refilled with molten material from below. Europe and North America were not originally joined in the sense that their present coastlines coalesced, but there was a nearly continuous fit along the margins of their continental shelves, and this only some 200 million years ago, in Late Jurassic times. Since then the west European coasts have evolved in detail, and the Rockall Plateau may have separated, along the line of the Rockall Trough, from the European continental plate only some 90-60 million years ago.

We can, then, suppose that the Caledonian mountains of Scandinavia, Scotland, and Ireland were formed between two ancient plates that later moved apart as the Atlantic opened out. But we meet a difficulty in the Hercynian ranges. Today fragments of them are found in Iberia, Brittany, Auvergne, Ardenne, Bohemia, Corsica, Sardinia, and folds of the same orogeny are present in south Britain. But, as Hallam points out, no one has yet satisfactorily accounted for the later Paleozoic orogeny in western Europe. Perhaps at one time Hercynian folds in western Europe connected with similar folds in North America, and we may thus visualize the sudden ending of the east-west ridges of the ria coasts of southern Ireland, Brittany, and Galicia. We can regard the Alpine ranges as being produced by the advance of Africa-Arabia to the Baltic Shield; but in the Mediterranean, movements have been complex, and a possible evolution is discussed below.

The European coasts, as we see them on a large-scale map are infinitely complex. We must bear in mind movements of sea level, particularly in the Quaternary; the ways in which structures caused by earth movement, folds, and faults affect the nature of a coast; the constant changes brought about by erosion and accretion (marine, fluvial, and glacial); delta growth; the filling up of shallow inlets; recent volcanicity; and many other factors. It is appropriate to consider these matters regionally.

Scandinavia and Northern U.K.

Norway and Sweden and much of Scotland and northern Ireland are Caledonian in structure. The details of their coasts depend largely upon two major factors; the high ground on their western sides, and the glaciations that have been the main agents in making the *fiords*. Along much of Norway there is, slightly above sea level, a broad rock platform, often cut up into numerous islands (the skjaergoard), that is known as the *strandflat*. Behind it there is a slope, often steep, to the upper surface of the mountains that makes a dissected plateau on which the last vestiges of former ice caps rest. Nearly all the fiords lie between these plateaus and the strandflat. Before the advent of the ice, rivers flowed down the slope, cut valleys, and deposited much material on the strandflat. The slope may be the product of more than one cycle of denudation. Whatever the previous history, the fall in temperature caused the ice caps to form on the plateaus. Tongues of ice spread to the upper parts of the valleys and scoured them. The ice spread over the strandflat, removed much loose material, and also had an effective planing action. When the ice finally melted to its present position, it left the fiords free, so that rivers cascaded into the upper parts of the fiords. Somewhat similar conditions existed in Scotland. There the sea *lochs* are on the west coast, and the ice gathered on the high ground behind them. Scotland is a much smaller country than Norway or Sweden; it is also lower, and no ice cap remains. To the east, in both Scandinavia and Scotland, the slope is gradual. Along nearly all the eastern coast of Scotland there is a fringe of Old Red Sandstone that gives rise to fine cliff scenery, and in the Moray Firth and Fifeshire there are extensive sand flats. The great dunes of Culbin now cover much ground that was farmland up to the end of the seventeenth century. The Baltic rests on the crystalline rocks of the shield that extends over nearly all Sweden and Finland. It is also a shallow and narrow sea; cliffs are generally absent; and, as in eastern Scotland, true fiords are missing. The indented coast of Sweden is largely the result of submergence of a low-lying, heavily glaciated coast of resistant crystalline rocks. Submergence, whether wholly the result of fluctuations of sea level or partly as a result of tectonic movements, has given rise to the drowned nature of the whole peninsula. The depths in the Sogne fiord reach more than 1000 m, and in many others exceed several hundred meters. Loch Morar (Scotland) is 305 m deep. But the Swedish inlets and the *firths* on the east coast of Scotland are relatively shallow. The Lofoten and Vesteralen islands may be compared with the Outer Hebrides. The Hebrides, although their higher parts carried small local ice caps, were invaded by the main Scottish ice. The higher latitude of the Lofotens produced conditions in which island and mainland ice coalesced into one major sheet.

The Caledonian trend covers all Scotland, but south of a line from Stonehaven to near Glasgow the crystalline basement no longer appears and, in the central valley, rocks of Old Red Sandstone and Carboniferous ages, with many intrusives, give rise to lower ground and, except for the Old Red cliffs, less spectacular coasts. South of the central valley, the Ordovician and Silurian rocks, on both east and west coasts, show the effects of marine erosion on sedimentary rocks of varying types, locally much folded and occasionally penetrated by intrusions. In Ireland, the crystalline rocks extend to Donegal and County Mayo. But both the west Scottish and Irish coasts are made more interesting as a result of the great outpourings of basalt and intrusions of dioritic rocks that give such character to Skye (the Cuillins), Rhum, Eigg, Mull, Staffa (Fingal's Cave), Antrim (Giant's Causeway), and parts of western Donegal. In all this area, oscillations of sea level have occurred, and narrow sounds— e.g., between Skye and Mull and the mainland— are parts of former valleys. *Raised beaches* are abundant at different levels. In the Firth of Forth some are now below sea level. Since both isostatic and eustatic movements have taken place, and are still active, the beaches are not necessarily horizontal, and it is misleading to refer to a beach by its height, which may vary considerably in 30–50 km or more.

The North Sea and much of the eastern side of the North Atlantic are parts of the continental shelf. Rocks that outcrop on land to east and west of the North Sea can be traced under it. We have recently learned much of the shelf geology as a result of prospecting for oil and gas.

Baltic and North Sea Coasts

The coast of West Germany extends only to Lübeck and is similar to that of east Denmark— low lying and broken by Kiel Bay and Lübeck Bay. Travemünde is the outport of Lübeck; to the east the land rises steeply to 91 m in places, and low cliffs face the sea. Offshore are many rocky reefs. Occasional low cliffs occur as far as Warnemünde, and between that town and Rügen the coast is diversified and long spits, culminating in the Darss, a sand and shingle *foreland,* extending to Pramort. The enclosed waters, *bodden,* are divided by sand and marsh into distinct basins. The irregularity of Rügen follows from its origin as a number of small islands, submerged postglacially, and now united by sand spits and dunes. There are a few outcrops of chalk: at Aikona it forms cliffs 46 m high; inland at Königstuhl it reaches 128 m. The chief feature in the west is the Zalew Szczecinski at the mouth of the Oder. The islands Usedom and Wollin, rising locally to about 61 m, enclose the haffs, and a ship channel is maintained at Swinoujsche. The mainland as far as Danzig (Gdansk) is simple in outline; several rivers are ponded back into lagoons and there are short lines of low, glacial cliffs. At Danzig Bay (Zatoka Gdansk) the great spits begin—Pólwysep Hel, Frische Nehrung, and the Kurische Nehrung (Kurshyo Neriya).

Denmark's present relief is almost wholly produced by glacial deposits resting on Cretaceous and newer rocks. Locally the chalk forms steep cliffs, as at Möns Klint (128 m); Stevns Klint, also in Sjaelland, is formed of limestone overlying chalk. These face the Baltic, and other lines of limestone cliff occur on the eastern coast and also along the northwest of Jylland. Locally they give way to glacial cliffs. However, the most striking features of the west coast north of Esbjaerg are the fine sand dunes, beaches and long spits, such as Blaavands Huk and those that enclose Ringkobing, Nissum, and Lim fiords. South of Esbjaerg the coast is best considered with that of Germany and Holland. The eastern coast of Denmark, with its numerous islands, is a submerged coast of mainly glacial deposits. The many long, narrow, inlets—Mariager Fiord, Randers Fiord, Horsens Fiord, Veile Fiord, and several others—are *föhrden.* They are unlike true fiords, and were probably cut by subglacial streams in soft deposits. The narrow channels between the islands may be of similar origin; they are certainly drowned valleys, and their irregular form is consistent with the nature of the glacial deposits in which they are cut.

The Cretaceous rocks, which are for the most part covered by glacial deposits in Denmark, reappear around Malmo and Kristianstad in Sweden. Apart from a narrow strip of Jurassics, they soon give way to gneisses and granites. On the west coast, there is a low coastal plain reaching the Oslo Fiord and continuing as a narrower belt as far as Kristiansand, even Stavanger, in Norway. On the Baltic coast, the long island of Öland and the adjacent strip of mainland are low and formed of rocks of Cambro-Silurian age. Bornholm is mainly granite except for a narrow strip of sediments (Cambrian, Jurassic, and Cretaceous) on the south coast.

The coast from Esbjaerg to Belgium may be treated as a unit. It consists of a long series of elongated offshore islands in front of a low coast, subject, under natural conditions, to much change. Fano and Romo are the largest of the Danish islands. They and some smaller ones are separated by channels called *dybs.* The mainland is made of fine sands—the *geest;* where outwash plains reach the coast, there are stretches of marsh. The islands are similar— dunes on the west, heathland grading down to mud flats on the east. Esbjaerg is the only port of consequence, but Ribe is a beautiful old city. The frontier runs between Romo and Sylt, which is similar in structure. Sylt is joined to the mainland by a dam that has had considerable effect on sedimentation. As far as the Elbe the coast is irregular—sandy banks, dry at low water, are cut by deeper channels that frequently change course. Everywhere the land slopes almost imperceptibly to the sea. In the thirteenth century, submergence led to the formation of inlets like the Dollart, and the former sandy dune belt was converted into a chain of islands, the North and East Frisian islands. They are similar to those off the Dutch coast, and are separated from the mainland by sand and mud, covered at high water, to form the Watten or Wadden Zee. On the mainland there are more dunes, and since the thirteenth century much reclamation has taken place, so that rich *polder* lands have been created. This process has been partly natural—sediment brought down by rivers, growth of salt marsh, the result of inning—and partly because of the building of dykes.

The mouths of the Elbe, Weser, and Ems rivers make the major breaks in the coast. They are all encumbered with sandbanks. Hamburg and Bremen are now many kilometers from the sea and have outports at Cuxhaven and Bremerhaven. Dredging is constant. On the mainland between the Weser and the Ems there are several small ports used by yachts and fishing boats. A sea wall is continuous along this coast. Offshore is the isolated island of Heligoland, a mass of Triassic sandstone steeply cliffed by marine erosion.

The eastern Frisian Islands are continued westward by five large and five small islands off the Dutch coast. They are of the same nature as those off Germany.

Throughout history Holland has had to protect itself from the sea. The earliest refuges—*terpen*—were mounds built in the marshes on which farms were often sited. Later, walls were made along the coast and main rivers. Reclamation has been a continuous process, and great meres—e.g., the Haarlemmermeer—were drained several centuries ago. The Zuyder Zee dam was completed in 1932, and a freshwater lake, the Ijssel Meer, with many new polders, has replaced a saltwater gulf. It is possible that in the future dams will join the Frisian Islands, and that the Wadden Zee will also be divided by dams between some of the islands and the mainland, so that eventually all of it may be reclaimed. There are strong reasons against this from the point of view of nature conservation. The storm of 1953, which caused so much loss of life and damage, has hastened the Delta Plan. Dams now join the islands of Walcheren, North Beveland, Schouwen, and Goeree to the mainland at Hellevoet. These shorten the coastline and make it easier to protect a large area. Access to Rotterdam and Antwerp is ensured. The other major work will be the enclosure of the Lauwersee a little to the northwest of Groningen. The Dutch coast between Den Helder and the delta is a reinforced dune coast. The whole coast of the Netherlands is a most remarkable example of man's ingenuity in sea defense work.

U.K. Coast

The connection between the European and British Caledonides has been set out. Now we must turn to the southern parts of the U.K. and despite what may appear to be a lack of continuity, we must first look at the west coast.

The sweep of Cardigan Bay and the line of the Lleyn Peninsula in Wales continue the roughly northeast and southwest trend that characterizes Scotland, although between north Wales and the Southern Uplands, the Caledonian rocks—so noticeable in the Lake District—do not reach the coast, along which there is a fringe of Carboniferous and Triassic rocks often covered by glacial deposits and raised beaches. On the Lancashire coast also, the older rocks are similarly hidden by Triassic beds and recent deposits. The Caledonian and Hercynian trends meet in Pembrokeshire, where they both run directly seaward and form a *ria coast*. In southeastern Ireland, a somewhat similar condition exists, but the southwest shows the effect of Hercynian folding to perfection in the long inlets following the Carboniferous limestone separated by ridges of Old Red Sandstone. Sandstones and limestones of Carboniferous age make up the coast between the Shannon and the last outcrops of the crystalline to the north. In this strip are the remarkably fine limestone cliffs of the Burren, Galway Bay, with its contrasted south (limestone) and north (granite) shores, and the spectacular drumlins in Clew Bay. The Irish Sea covers part of the continental shelf; the rocks on either side of it are continuous, although the floor of the sea is divided into separate basins by a ridge running between the Lleyn Peninsula and Wexford in southeast Ireland. The southern coast of Wales owes much of its shape and beauty to the Hercynian folding and the nature of the rocks; the Tenby and Gower peninsulas are noteworthy.

The Alpine folds in southern England followed much the same trends as did the Hercynian. Devon and Cornwall, primarily formed of Palaeozoic and locally crystalline rocks, were folded mainly in the Hercynian orogeny, but the contrast between the north and south coasts of the peninsula is largely the result of the greater exposure of the north coast to the Atlantic. Farther east, from Dorset to the Thames, the coast is formed of Mesozoic and Tertiary rocks; the anticlines of the Isles of Purbeck and Wight and of the Weald and the synclines of the Hampshire and London basins have given rise to a coast of great interest, but large parts of it are urbanized. The chalk cliffs are well known. It must not, however, be forgotten that movements of sea level relative to the land have increased the beauty of all the British coasts. The Bristol Channel and the Thames estuary are the results of major folding and submergence, although this is a grossly oversimplified statement. The small, picturesque inlets of Devon and Cornwall; the separation of the Isle of Wight; and the channel, now silted up, that once surrounded Thanet—all owe their origin to submergence. Raised beaches of interglacial and postglacial age are conspicuous in many places, and the Chesil Beach is a unique feature not only in Britain but in all Europe on account of the remarkable grading of the pebbles of which it is composed.

The east coast of England is comparable to the continental coasts facing the southern North Sea. The London Clay, faced by mud flats and marshes in Essex; the low cliffs cut in easily eroded Neogene and glacial deposits between Harwich and Sheringham; the fine drowned valleys of Suffolk and Norfolk; the Broads, except for Breydon Water—all are ancient peat diggings in alluvium-filled valleys. The magnificent development of salt marshes between Sheringham and the small outcrop of Cretaceous rocks at Hunstanton; the Wash, which is all that is left of a great gulf that in early historical times reached almost to Cambridge and Peterborough; the reclaimed marshland and wide beaches now backed by seawalls of Lincolnshire; the Humber mouth and the extensive fenlands of the lower Trent and Yorkshire Ouse; and the soft-boulder clay cliffs of Holderness—all add up to make one of the most interesting coasts in Europe not only because of the physical variety but also because of the changes that have taken place, many of which can be connected with historical events. Flamborough Head is a chalk *cape;* to the north it is followed by Jurassic rocks in which are some of the highest cliffs in England. In Durham, the magnesian limestone (Permian) makes fine natural cliffs but, as in south Northumberland, often spoiled by coal mining activities inland. Finally, rocks of Carboniferous age, interspersed with long sandy bays and the outcrops of the Whin Sill (dolerite), make up the coast as far as the border and the Southern Uplands.

The British Isles are structurally part of Europe, and are the higher parts of the continental shelf. Because rocks of almost all periods are represented, folding and faulting associated with the main orogenies are well developed. Igneous rocks of many types and ages are prominent, and phenomena associated with low coasts—spits, drowned valleys, marshes—are conspicuous, as are raised beaches and associated features. The coasts of these islands are more varied than those of any other part of the world.

Northern France

The Mesozoic rocks of Kent reappear in the Boulonnais, and formerly made a bridge. We may assume that streams flowed north and south from this ridge. In glacial times, the southern North Sea may have been occupied by a lake held between the ridge in the south and the ice to the north. It is possible that overflows from the lake across the ridge cut down one or more of these valleys, and later, as a result of the postglacial rise in sea level, the straits began to assume their present form. On the other hand, the straits may have been formed simply by a rise in sea level and the widening of the valleys.

Apart from the folding in the Boulonnais that on the coast involves oolitic (Jurassic) rocks and some lower Cretaceous beds, the whole coast from the Rhine delta to the Seine estuary is formed of newer Tertiary and Quaternary beds and chalk. The maritime plain of Belgium and the low coast between approximately the Canche and the Somme is fringed with dunes, and shingle spits partly close the river mouths. South of the Somme, chalk cliffs of great beauty reach almost to the Seine. It is characteristic of such cliffs that they are usually vertical, possibly indicating a balance between marine and subaerial erosion. Beyond the Seine a greater variety of rocks reaches the coast— Jurassic in Calvados, and Carboniferous strata and outcrops of granite at Barfleur and C. de la Hague in the Cotentin.

Near Coutances the crystalline and granitic rocks of Brittany outcrop and, apart from the Lower Carboniferous of the Crozon Peninsula and occasional areas of recent deposits, make up the cliffs almost to Les Sables d'Olonne. The Channel Islands are formed of similar rocks, and all the islands are cliffed. The Brittany coast is in some respects comparable to that of Cornwall, mainly in the sense that the trend of the rocks is generally westward, so that near Brest and Douarnenez there is a ria coast, whereas on the north and south coasts the peninsula is interrupted by numerous small inlets separated by lines of rugged cliffs along which, as in Cornwall, many headlands are formed of igneous rocks. The Brittany coast is a fine stretch of coast, and on the south includes the interesting scenery around Quimper, Lorient, and particularly the Baie de Quiberon and Morbihan. The estuary of the Loire and the Baie de Bourgneuf, enclosed by the Ile de Noirmoutier and the Marais Breton, is also part of it. The change to Jurassic and later Mesozoic rocks takes place near Les Sables d'Olonne, and they extend to the north side of the Gironde and include the islands of Ré and Oleron, behind which are the extensive Marais Poitevin north of La Rochelle and the marshes of the Charente to the south.

The remainder of the west coast almost to the Spanish frontier is low and fringed by the fine dunes of the Landes, the last stage of infilling of the Aquitaine gulf. The dunes pond back several lagoons into which small streams drain. The Bassin d'Arcachon is the largest and has a natural entrance from the sea. The sand that forms the dunes is derived from the marine sand of a short-lived extension of the sea during the Quaternary. The dunes are aligned in subparallel crests, and in a few places reach 100 m; they are the largest moving dunes in Europe. Plantations of conifers and other trees have helped in stabilizing them. The mouth of the Adour is associated with a deep submarine valley, the Fosse de Cap Breton.

Spain and Portugal

The northern coast of Spain is subject to strong wave attack. It is a cliffed coast with many small indentations and one important harbor, Santander. From Santander to Gijon there is often a platform 30-91 m high and about 1000 m wide along the littoral. It is cut by rivers, but there are no good harbors. Seaward the bottom slopes steeply. East of Santander folded Mesozoics make the coast; in Asturias, Hercynian folding is present and trends with the coast, but turns north-south at the Galician border.

Galicia is mainly a Precambrian granite mass that has been much fractured and has long been a peneplain. It is cut by a number of rivers that, as a result of submergence and the structure of the area, reach the ocean in rias that are deep and make good harbors. Between the inlets the coast is steep. Cape Finisterre marks the approximate separation between those on the north coast, the trend of which is generally north-south, except for the beautiful and much-branched inlet of Corunna and El Ferrol, and the rias bajas to the south, which trend roughly northeast to the southwest.

In Portugal, the Precambrian core and Hercynian folds of the Meseta spread over much of the country from Oporto to Cape de Sines. They are most conspicuous north of the Tagus. A major fault line runs from just south of Oporto, passes east of Coimbra, and reaches the Tagus some miles above Santarem. This fault brings Mesozoic and Tertiary rocks to the coast, except between Cape Carvoeiro and the mouth of the Tagus, and again from Cape de Sines to Cape St. Vincent. The northern part of the Portuguese coast geologically resembles Galicia, even if it is almost rectilinear and faultlike in appearance. The Mesozoic hill country is largely calcareous, but the coast south of the Douro is sandy with salt marshes and lagoons; at Mondego the dunes are about four miles wide, and the coastal plain at Coimbra, about twelve miles. The coast is similar to the south, but at Nazaré a rocky headland marks the end of the low coast. There is a large lagoon near Peniche, a tied island similar to those offshore, the Berlenga and Farilhoes islands. Steep cliffs fringe the coast to Cape Raso. There is a wide area of low ground around the lower Tagus and Sado, but the short Sa de Arrábida (Mesozoic) between Setubal and Cape de Espichel (152 m) form a prominent north shore to Setubal Bay. A dune coast runs from the mouth of the Sado to Cape de Sines, from which point to Cape St. Vincent long lines of cliff formed of Paleozoic rocks are interrupted by a number of sandy beaches. From about Cape St. Vincent to the frontier, the Algarve mountains (Mesozoic, calcareous) are near the coast, which in places is steep, rocky, and level-topped. Farther east, between Faro and the mouth of the Guadiana, it is low, sandy, and marshy.

This type of coast continues to Huelva, the Guadalquivir, Cadiz, and Cape Trafalgar, where the first of several spurs between the cape and Tarifa reach the sea and so break the monotony of the low coast. From Tarifa to Gibraltar, a tied island, the coast is rocky and exposed. The long stretch of coast from Gibraltar to Cape de Gata is closely bordered by high mountains behind a narrow coastal plain along which are many fine beaches and shallow, but exposed, bays. Near and beyond Malaga the coastal strip is narrow, and the mountains rise steeply behind it; apart from the lowland at Motril and the Llanos de Almeria, this is generally true of the coast as far as Mazarron, where there is a rugged headland. Near Cape de Palos the coast is backed by hills, and small ridges run more or less parallel to the shore and end in cliffs. At Cartagena there is a break and a good natural harbor. However, from there to Alicante the coast is lower with sandy beaches and large lagoons (e.g., Mar Menor and those near Torrvieja). Between Alicante and Denia, the Betic Cordillera runs out to sea, so that the coast is more broken and picturesque, with sandy bays and high points, including Cape de la Nao, Cape de San Martín, and Cape de Santa Antonia. The continuation of the cordillera is found in the Balearic Islands. In Majorca, folded Jurassic and Triassic rocks build northeast to southwest ridges on the steep north coast; the lower south and east coast is composed mainly of Cretaceous and Eocene rocks. In Minorca, the southern half of the island is a limestone plateau of Miocene age, whereas on the north coast a much indented strip of Jurassic and Triassic separates two masses of Devonian rock to the west and east. Iviza is largely limestone; Formentura is floored by Tertiaries. Nearly all the coasts of the islands are much embayed and picturesque.

The coast from Denia to Cape Oropesa is low and sandy except at Cape Cullera and the hill of Sagunto. There are lagoons south of Valencia, and the coastal plain varies from about 16–24 km in width. The Sierra de Irta interrupts the low coast between Cape Oropesa and Benicarlo. On the south side of the Ebro delta the Sierra de Montsia dominates the extensive low ground of the delta from which a long spit running to the south encloses the Puerto de los Alfaques. On the north side there is a somewhat similar formation. The delta continues to extend seaward and encloses many lagoons.

The Gulf of San Jorge is mainly faced by a steep and rocky coast; the harbor of Tarragona is man-made. Sandy beaches and local cliffs characterize the coast as far as Barcelona; the Costas de Garraf is bold and even precipitous a few miles south of the city. From Barcelona to the mouth of the Tordera, the coast is straight. Crystalline rocks, apart from a patch of Mesozoics just north of the mouth of the Ter and some alluvium behind the Gulf of Rosas, extend to the frontier. The Costa Brava begins north of Blanes. The natural, untouched coast is a beautiful, rocky, and indented coast and includes the small ports of San Feliu and Palamos.

The appearance of much of the Mediterranean coast of Spain, especially the Costa del Sol and the Costa Brava, has been made less attractive by ill-sited buildings.

Southern France

The Mediterranean coast of France presents many points of interest. The high Pyrenees do not reach the sea, so that there is a narrow belt of low ground between the Spanish frontier and Collioure. This, however, is part of the Pyrenean fold belt, and around Llansa and Port Venres there is a ria coast. From Collioure up to and including the Rhone delta, the coast in some ways resembles that of the Bay of Biscay south of the Charente. A belt of dunes ponds back a series of lagoons (*étangs*), and in this way a double coast is formed. The old coast, before the formation of the barriers, was irregular and interrupted by headlands or islands; the Clape (limestone) massif near Narbonne; Cape d'Agde, a basaltic headland that divides the bays of Narbonne and Aigues Mortes; and at Sète, where there is a limestone headland. The étangs have been much commercialized for tourist purposes and industries in recent years, and the coast has lost much of its natural beauty. Narbonne and Aigues Mortes were formerly ports, and are now connected to the sea by canals. The Rhone delta continues to advance seaward at an average rate of about 40 m a year. In the mid-nineteenth century the water reached the sea by six channels, but its natural mouths still present difficulties on account of silting, and the Canal Marseille joins Arles to Port de Bouc. The Carmargue on the west of the delta is a nature reserve of worldwide significance. The growth of the delta and sand bars have enclosed several étangs, the two largest of which are the E. de Vaccarés and the E. de Berre.

At Marseille the nature of the coast changes. Between that city and Toulon, the limestone is much cut up by small inlets called *calanques.* the cliffs are high and often nearly vertical. The crystalline Côte de Maures, between the Gapeau and the Argens, is fringed by the Iles d'Hyères. At Giens a former island is tied to the coast by a double tombolo. The Esterel, also crystalline, gives rise to a much-indented coast with small inlets and fronted by many rocky islets. The Gulf of Fréjus is filled with alluvium. Between Cannes and Nice, the coast is lower, with open bays broken only by Cap d'Antibes. Along the Riviera (Nice to Menton) the mountains plunge steeply down to the sea and result in a natural coast of great beauty.

Italy

The Riviera di Ponente continues that of France and is similar in nature. Mountains rise steeply behind the much-indented coast. The Riviera di Levante is backed by the northern Apennines. The sandstone promontory of Porto Fino, and the Gulf of Rapallo are noteworthy. A succession of headlands and small bays distinguishes the coast to the end of the peninsula enclosing the Gulf of Spezia, both sides of which are rugged. South of the gulf, the nature of the coast changes. As far as Leghorn, dunes and pinewoods line the shore and the plain widens out at the Arno mouth. From Leghorn, to Civitavecchia, wide open bays and prominent headlands and islands make an interesting coast. Elba, situated just off the Piombino peninsula is much embayed; the western part is granite separated by a sedimentary belt from a small outcrop of crystalline rocks in the southeastern peninsula (cf. Corsica and Sardinia). Monte Argentario is a former island joined by two tombolos to the mainland; a third, intermediate, tombolo did not reach the island, but is now connected to it by a road. The island is mainly limestone. Several small islands nearby form the Tuscan archipelago. A sweeping sand beach runs from Monte Argentario to Civitavecchia, where there are low cliffs. Then follows a long stretch of low coast, including the Tiber delta, to Anzio, where there are more cliffs. Another stretch of low coast, enclosing long lagoons and the Pontine Marshes, extends to the former island of Monte Circeo, a local interruption in the arcuate beaches which reach Terracina, where the coast is steep, and thence to the higher ground at Sperlonga-Gaeta. There is more low coast south of Gaeta, straight and often dune-fringed apart from the delta of the Volturno, and enclosing Lago di Patria. Monte di Procida is an old, partly submerged, volcano off which are the islands of Procida and Ischia. The Phlegraean fields lie between Monte di Procida and Naples; the gulf is dominated by Vesuvius, lava flows from which reach the sea. The Sorrento peninsula (limestone) and the island of Capri enclose it on the south, and also make up the rocky and picturesque (e.g., Amalfi) north coast of the Gulf of Salerno, the east side of which is low and sandy and includes the Sele delta. At Agropoli, a long, rocky block of Mesozoics intervenes between the gulfs of Salerno and Policastro. From the southern end of the block almost to Reggio, the coastal strip is narrow except in the gulfs of Eufemia and Gioia. Road and railway hug the shore; the plain is often only about 402 m wide, and locally cliffs rise directly from the sea. The slopes are often terraced. The Poro headland is cliffed, but a railway runs around its seaward margin. Many of the cliffs in this part of Italy are out of reach of the sea, and the plain is, in effect, a raised beach.

The Strait of Messina is about 3.2 km wide; along the mainland coast there is a narrow beach with dissected ridges behind it. Landslides are frequent; small changes of sea level as a result of earthquakes occur all along the coast from Naples southward. Apart from the Crotone peninsula, road and rail continue to run close to the sea, which in many places is edged by pebble beaches. Some towns, such as Roccella, are built on steep slopes. The plain widens in the Rio Crati lowlands, but is narrow again between them and the northeastern part of the Gulf of Taranto. Megapontum, a well-known site, stands on a spur of an 18 m terrace. Beyond Taranto the plain is wider, higher, and often fronted by low cliffs; farther southeast, it is low and rocky, and there are many minor inlets. From Cape San Maria di Leuca (white limestone, c. 76 m) to the north side of the Gargano peninsula the outline is simple. Gargano is a limestone plateau; locally cliffs rise directly from the sea; elsewhere there is a narrow beach plain. On the northern side it is less steep, and two large lagoons are ponded by bars of sand and dunes. The promontory is separated from the higher mountains by the plain of Tavoliere.

The Adriatic coast is simple in outline. As far as Rimini, it is fronted by a narrow beach; occasionally there are low, crumbly cliffs. The most pronounced cliffs are between Vasto and Pescara; where rivers debouch, there are small deltas. Steep cliffs near Ancona give way to a narrow beach that reaches Pesaro and Cattolica. There the mountains retreat inland, and the great plain and delta of the Po and other rivers build the coast. Between Rimini and Monfalcone there is the big lagoon of Comacchio plus many smaller ones in the delta and those around Venice. The flat low coast is crossed by many streams between Venice and Tagliamento, by the Laguna di Marano, and by the delta of the Isonzo. The lagoons are contained by sand bars and spits, broken in many places for access. The Gulf of Trieste is entirely different; the coast is steep, often inaccessible, and may be regarded physically as part of Istria and the Yugoslav coast.

The fact that so much of the coast of Italy is followed by roads and railways means that many miles of it are urbanized and its natural beauty is often marred.

Yugoslavia and Albania

The coast of Yugoslavia from the Gulf of Rijeka (Fiume) to Albania is unique in Europe. The coastal area, including Istria, is almost entirely formed of limestone, although flysch (Tertiary) may spread over considerable areas. Nevertheless, the scenery is karstic. The whole coastal area has been much folded. The direction of folding is toward the southwest and the overthrust masses (nappes) increase in age inland. The coast itself has been submerged, so that in Istria there are a number of drowned valleys—e.g. the Canale di Lerne and the Assa estuary, which are unlike the long, narrow channels, islands, and island chains that run parallel to the coast south of Rijeka. There are often several parallel lines of islands. The islands themselves are limestone; the *canali* have been cut in the flysch. A common feature of nearly all the islands is the absence of cliffing, partly to be explained by recent submergence and partly to limited fetch of waves. Yet wave action is locally severe; the Bora blows fiercely on this coast, and the crossing of even narrow straits may be impossible for days on end. The Bora often reaches the Italian coast of the Adriatic, but is much less violent than on the Yugoslav coast. There is little tidal action in the Adriatic, but even so the influence of the Bora can build up strong currents in some straits. Few rivers reach the coast. The most important river is the Neretva, which has built a delta and a large area of marshy land. The river is now controlled; originally its mouth was well to the north of its present one. At Sibenik, the Krka and Lake Prokljan afford an interesting example of submerged relief; two broad basins parallel to the Dinaric grain have been linked by two narrow valleys to give an inlet shaped rather like a double L. Near Kotor there is a remarkable inlet, a series of deep depressions joined by narrow straits. The main *lakes* are deep and broad and enclosed by high land. They are probably formed in limestones, although faulting has played an important role. The straits were probably river-cut, and the present appearance of the whole area is the result of submergence. The direct distance from the Italian to the Albanian frontier is about 560 km; the mainland coast is some 1520 km long, and since there are 61 major islands and 540 islets, the total length of the coast is approximately 5120 km. Locally high mountains fall steeply to the coast; in other places the coast may be relatively low and fronted by islands and gulfs. The limestones are nearly everywhere pure white. Locally they are forested, and terraced slopes may be cultivated. In one sense the coast may be monotonous; it certainly does not have the colors of the Greek coasts, but it is beautiful and greener in the south. There are several small ports, but Dubrovnik (Ragusa) is the only large town.

In Albania the nature of the coast is different from that to the north and south. The interior is mountainous; the coastlands are lower and are divided into a number of bays by low spurs, usually of limestone, running from east to west. Behind the bays are alluvial flats and, in a sense, the coast may be regarded as transverse and not longitudinal.

Greece

To try to describe the coasts of Greece in limited space is almost impossible. There is a great difference between the rocks and structures of western and northeastern Greece. The Pindhos folds maintain more or less the same trend, and on the west coast Mesozoic limestones are predominant and often marmorized. In the Peloppónnisos, the trend lines spread out and form the south-pointing peninsulas, thus providing a ria type of coast. The Ionian Islands—Kérkira (Corfu), Kefallinía, and Zákinthos—follow a similar trend, and again limestone plays a significant role in their structures. The west coast of the Peloppónnisos is lower and smoother, and for long distances there are dunes and lagoons. Faulting, however, has played an important part throughout the country. Many of the main bays are fracture-controlled; the Gulf of Patras (Patraikós Kólpos) and Kólpos Korinthiakos, the three south-facing gulfs (Messiniakós, Lakonikós, Argolikós), the channels between Evvoia (Euboea) and the mainland, and the *fingers* of the broad peninsula of Khalkidhiki are all of this nature. East of this peninsula the coast is different. Broad, open bays with smooth shores succeed one another, thus giving low-lying ground traversed by sluggish rivers. But inland, and locally reaching the coast, are crystalline plateaus that rise steeply from the coastal plains. Here also faulting is important; the islands of Thásos and Samothráki have been separated from the mainland by vertical movements. Wave action has smoothed this east side of the Gulf of Kavalla. The same type of coast extends to the peninsula that makes the west side of the Dardanelles. This strait and the Bosporus are drowned valleys.

Greece cannot be considered without reference to the islands to the south and east of it. The Aegean Sea is a foundered area; part of an old rigid mass has subsided, and there are three deep basins in the sea: the northern one runs northeast from Larisa to the Saros Körfezi; the central one, between the southern part of Evvoia and between Khíos on the north and Ándros, Tínos, and Ikaría on the south; and the southern one includes the Mirtoan Sea and the Sea of Crete. They are all steep-sided.

The Thracian Islands are Thásos, Samothráki, Límnos, and Áyios Eustrátios. The first is mainly marble and schist; Samothráki is largely volcanic. The other two are the visible remnants of a ridge running northeastward from the smaller islands of the Northern Sporades (Voríai Sporadhes). The northern chain, Skiáthos to Yioúra, is mainly crystalline. Skíros and many small islets lie to the south; Skíros is famed for its marble.

The mountains of south Evvoia continue through Ándros, Tínos, Míkonos, and Yioúra (not the island mentioned above); Síros and Dhílos are more southerly fragments of the same arc. The mountains of the peninsula on which Athinai (Athens) is situated run on to Kéa, Kíthnos, Sérifos, Sífnos, Amorgós, and so to the Sporádhes (Dodecanese). Páros and Náxos on the north, and Folegandros, Sikínos, Íos, and Anáfi on the south belong to this same chain. Finally, there is a volcanic arc—Méthana (in the Saronic Gulf), Mílos, and the Kaimeni

Islands in the explosion crater of Thíra (Santorin). Crete is the largest island of Greece, and in structure and landscape is closely related to western Greece. It is largely limestone; the south coast is high and without harbors, the north coast is lower and indented. The limestones rest on older rocks that are exposed in limited areas on the east and west of the island.

Corsica

Depths of more than 2000 m separate the islands from the French and Italian Rivieras, and 500-1000 m are found between Corsica and Italy. It is rugged, and cliffs surround nearly all except for a long stretch on the east coast south of Bastia and on either side of Aléria. In both these areas there are large lagoons. The island is divided structurally by a line running from the west side of the Gulf de St. Florent to the east coast near Solaro. To the west is a great and rugged crystalline mass; to the east, and locally around Bonifacio in the south, the rocks are Triassic and later. The west coast is much embayed; small beaches occur at the heads of some bays. The limestone cliffs of Bonifacio are imposing. In many ways western Corsica and Sardinia resemble the Maures and Esterel massifs of Provence. The separation between the crystalline west and the later rocks of the east seems to have resulted from a great thrust that was directed to the west. The old view that Corsica and Sardinia are remnants of a foundered continent needs to be revised in the light of recent work on plate tectonics. We may probably assume that fracturing and sinking led to the formation of the Straits of Bonifacio.

Sardinia

Sardinia is largely crystalline, and its surface is a peneplain. It rises steeply, especially on its eastern side, and is isolated by deep seas. The crystalline rocks are much tilted and fractured, and locally Mesozoic (as in the Gulf of Orosei) and Tertiary sediments imply tectonic disturbances; so also do the large spreads of volcanic rocks, especially in the northwest. In general the coast is irregular, with many bays, small inlets, and islands, and has the characteristics of a submerged rocky coast. The Tirso River reaches the sea about the middle of the west coast and during the Fascist regime marshes and lagoons were reclaimed at its seaward end. The southern part of the lowland leads to Campidano, which rises to a rolling area of Quaternary deposits in its midparts and falls to sea level immediately west of Cagliari.

Sicily

Tectonically Sicily is regarded as part of Africa, but along its northern coast the Apennines are continued in the Monte Peloritani, Nebrodi, and Le Madonie. The first are mainly gneisses and schists; the second, sandstones and clays; the third, rugged Mesozoic limestones standing out from the surrounding Tertiaries. Farther west the *range* disappears, and the main relief behind Palermo and Castellammare is formed of Jurassic and Triassic rocks. Much of the center and south is built of Tertiary clays and sandstones that give moderately high ground. As in Italy, road and railway follow a narrow coastal strip almost all the way from Catania to Messina (excluding M. Ciccia) and thence to Palermo and Castellammare. The main areas of low ground on the coast are the Piano di Catania and expanses of varying width, locally interrupted by low cliffs, between Catania, Siracuse, Pozzallo, Gela, and Licata. The most impressive feature is Etna, which dominates the northeast. The far western coast between Trapani and beyond Marsala is low.

The Lipari Islands are volcanic, especially Vulcano and Stromboli, which is in almost constant activity.

To the south the Maltese Islands stand on a shelf extending southward from Sicily. They are largely built of permeable Miocene limestones. There are fine cliffs, often with marked ledges along the coasts of the islands and many sheltered bays and coves, but only Grand Harbor (Valletta) is of major consequence.

The Iberian, Italian, and Balkan peninsulas and the large islands of Corsica, Sardinia, and Sicily present a strange appearance, and explanations based on earlier views of the nature of the Hercynian and Alpine orogenies are inadequate. Recent ideas on plate tectonics offer a more convincing picture, although many difficulties remain. A. G. Smith (1971) has presented these modern views in a comprehensive paper in which references to all recent work are given.

Smith believes that Spain, Corsica-Sardinia, and Yugoslavia-Greece-Turkey must at times have belonged to a plate or plates independent of Africa and Eurasia. In Permo-Triassic times there almost certainly was a large land area between Africa and Eurasia. Corsica and Sardinia are isolated Hercynian massifs, and are surrounded by areas with an oceanic seismic structure. But great uncertainties remain—what happened to the Tethys? Since sea-floor spreading is now assumed in all oceans, Smith argues that there is no good reason to suppose it did not take place in the Mediterranean. Sicily is definitely regarded as part of Africa. The

Balearic Islands have been fitted to Spain; Corsica and Sardinia to Italy; Corsica-Sardinia-Italy to France-Balearics-Africa; Greece and Turkey to Italy-Sicily-Africa. This reassembly is regarded as a possible final product of the Hercynian orogeny, and it may have remained a unit until the Atlantic Ocean began to open. It is not possible to discuss what happened to the Tethys—it is sufficient to quote Smith's (1971) suggestion that "there must have existed during the interval between the beginning of the opening of the central Atlantic and the present day, compressional plate margins along which the Tethys was swallowed up. The only possible site for such margins is the northern branch of the Alpine chain"—the Alps, Caucasus, and beyond.

The breakup or opening of the Mediterranean *continent* took place in stages, and has given rise to the present disposition of land and seas. This is not the place to account for the dismemberment or for any of the other major problems involved in the hypothesis. It has, however, been advantageous to indicate how modern views go so much farther than did earlier ones to explain the curious outline of the northern shores of the Mediterranean, which control much of the present phenomena—winds, waves, currents—at work on the coastlines.

This account of the European coasts is so brief that details of great interest have had to be omitted. Some general points must, however, be made. Although there are many examples of fine mountain and cliffed coasts, it is noteworthy that many hundreds of kilometers are low and flat. These may have less appeal to many people, but they are often the most interesting because natural changes take place in them relatively quickly. Spits and bars of sand or pebbles have been formed by wave action and partly by currents. They grow, and sooner or later a storm destroys part or all of them; then, later, growth recommences. Behind them there may be lagoons or marshes; lagoons tend to silt up and evolve into marshes; they are usually interlaced with intricate creek systems, and some marshes (e.g., Norfolk) show magnificent spreads of color in summer—pink (*Armeria* spp.), mauve (*Limonium* spp.), or yellow (*Aster*). In many places the growth of spits and the silting of marshes can be related historically to the decay of ports—Aigues Mortes, Ribe, Orford. Elsewhere dunes may form and advance inland—Landes, Culbin Sands. The railways and roads that run close to the sea—e.g., in Italy—suggest recent slight changes in the relative levels of land and sea, so that a narrow coastal plain or raised beach platform has been formed. Elsewhere major rivers—Ebro, Rhone, Rhine, Po—build large deltas that, left to the undisturbed regime of natural phenomena, continuously change form. Tides often play an important part; in the Severn and Mont St. Michel, their range may be 13 m or more in spring; in the Mediterranean, they are almost negligible. Nevertheless the small *tide* or *seiche* at the head of the Adriatic increases the man-made problems confronting Venice. The continental shelf off northwest Europe is wide; the North Sea and Baltic are shallow pans; the Mediterranean contains several deep basins. Cliffed coasts do not change so quickly, but careful inspection of any range of cliffs will show how the sea works along lines of weakness to produce *stacks* and *arches* and off-lying rocks. Perhaps the most striking changes have been the result of events in the Ice Age. Sea level may well have been 120 or more meters below the present, but the weight of the ice depressed the land on which it rested; this and its subsequent recovery (still in process) are slow processes. Along the lower Thames, for example, Bronze age, Iron age, and Roman remains are found well below water level, indicating a constant rise of the sea on the land. In fact, this is but a continuation of a much longer-term geological change in southeast England and the Low Countries. Some of the most picturesque coasts have resulted from glacial erosion and later submergence, but we cannot assume that fiords, often thousands of feet deep, have been formed in the same way as the drowned valleys of Brittany, Galicia, and Dalmatia. The coast, left to itself, is ever changing; if man interferes, he may, even for long periods of time, save the coast from destruction. Holland illustrates this constant warfare. But to try to save the glacial cliffs of Yorkshire or Norfolk is, in the long run, not worthwhile. Finally, man has all too often ruined the natural coast; conservation is now effectively practiced in many countries, but anyone who knows the coasts of many parts of the globe can but be depressed at the changes man has brought about so often in the interests of tourism, and money!

Further Readings

A complete bibliography would take up more space than the text of this entry. The following publications are intended for general guidance; most of them contain many important references, but geographical, geological, and ecological journals published in *each* country should be consulted for detailed reading. The series of *Geographical Handbooks* (Naval Intelligence Division, British Admiralty, 1942–current) gives detailed *descriptions* of all coasts in western and southern Europe. These publications are not analytical, but if used with the *Pilot* books (Hydrographic Department, British Admiralty), a clear picture of the nature of every part of the coast can be obtained. The European volumes of *Géographie Universelle* are also helpful, as are the articles are on separate countries in the *Encyclopedia Britannica* (Encyclopedia Britannica, Inc.). For particular regions the following books or journals are essential: *Das Anlitz der Erde* (Suess, 1904); *The Origin of the Continents and Oceans* (Wegener, 1929); *Die Küste*, Archiv-fur Forschung und Technik and der Nord und Ostee (1952 onward) (West Holsteinische Verlagsanstelt Boyens & Co.), which contains many important papers on the German and neighboring coasts; the several volumes of *Meddelelser fra Skalling Laborotoriet* (1953 onward) are highly informative about the Danish area; a short book by J. S. Lingsma, *Holland and the Delta Plan* (1963) gives a good account of the Netherlands coast; G. D. Hubbard (1934) gives a good summary of the physiographic history of Norway; the coasts of England, Wales and Scotland are analyzed by J. A. Steers in the *Coastline of England and Wales* (1969), *The Coastline of Scotland (1973), and Coastal Features of England and Wales* (1980), and by M. J. Tooley (1979) in *Sea-Level Changes: North-West England During the Flandrian Stage*; J. B. Whittow, in *Geology and Scenery of Ireland* (1974), gives excellent descriptions and references to the coast; A. Guilcher, who has written important coastal papers, gives many references to France and other countries in *Coastal and Submarine Morphology* (1958). The Mediterranean countries are discussed in the general books already mentioned, but details must be sought in the references given in those books and in national scientific journals. For recent views on the general structures of European coasts see *Geology of the North-West European Continental Shelf* (Naylor and Mounteney, 1975, Pegrum and Rees, 1975); the article by A. G. Smith (1971) on areas of the Tethys, Mediterranean, and Atlantic; and a valuable paper by R. J. N. Devoy (1979) on recent changes in sea level in northwestern Europe, particularly in different parts of Britain.

JAMES ALFRED STEERS

References

Devoy, R. J. N., 1979. Flandrian sea level changes and vegetation history of the lower Thames estuary, *Royal Soc. London Philos. Trans.* **285B**, 355-407.
Guilcher, A., 1958. *Coastal and Submarine Morphology*. New York: Methuen, 274p.
Hubbard, G. D., 1934. Unity of physiographic history of Norway, *Geol. Soc. Soc. America* **45**, 637-654.
Lingsma, J. S., 1963. *Holland and the Delta Plan*. Rotterdam: N. V. Vitgeverij Nijgsh & Van Dittmer, 199p.
Naylor, D., and Mounteney, N., 1975. *Geology of the North-West European Continental Shelf, vol. 1, The West British Shelf.* London: Graham & Trotman, 162p.
Pegrum, R., and Rees, G., 1975. *Geology of the North-West European Continental Shelf, vol. 2, The North Sea.* London: Graham & Trotman, 224p.
Smith, A. G., 1971. Alpine deformation and the oceanic areas of the Tethys, Mediterranean and Atlantic, *Geol. Soc. America Bull.* **82**, 2039-2070.
Steers, J. A., 1969. *The Coastline of England and Wales.* Cambridge: Cambridge University Press, 750p.
Steers, J. A., 1973. *The Coastline of Scotland.* Cambridge: Cambridge University Press, 353p.
Steers, J. A., 1980. *Coastal Features of England and Wales: Eight Essays.* New York: Oleander Press, 240p.
Suess, E., 1904. *The Face of the Earth* (Das Anlitz der Erde). Oxford: Clarendon Press, 5 vols.
Tooley, M. J., 1979. *Sea-Level Changes: North-West England During the Flandrian Stage.* London: Oxford University Press, 232p.
Wegener, A., 1929. *The Origin of the Continents and Oceans*, trans. J. Biram. London: Methuen, 212p.
Whittow, J. B., 1974. *Geology and Scenery in Ireland.* Middlesex, England: Penguin Books, 301p.

Cross-references: *Africa, Coastal Morphology; Asia, Middle East, Coastal Morphology; Europe, Coastal Ecology; Global Tectonics.* Vol. VIII, Part 2: *Europe.*

EUSTATIC—See SEA LEVEL CHANGES

EUXINIC

Euxinic sediments are deposited in an anaerobic-reducing environment and are characterized by their black color, high organic content, presence of hydrogen sulfide, and lack of a bottom fauna (Pettijohn, 1957).

A euxinic depositional environment occurs when biological, physiographic, or tectonic influences cause isolation of parts of the sea in which free access to the open ocean is restricted. Typically, the climate is humid and evaporation does not exceed inflow. Instead, the restricted sea retains its volume, and varying degrees of stagnation of the bottom waters ensues (Krumbein and Sloss, 1963).

Stagnant conditions resulting from submerged rock sills in the Black Sea and Norwegian fjords are well known (Pettijohn, 1957). In both cases, density stratification of seawater due to salinity and temperature differences results in an upper and lower circulation that inhibits vertical exchange. The lesser salinity and density of the surface waters are due to a freshwater contribution from rivers. At the same time, a lower inflow into the restricted basin of more saline, and therefore more dense, seawater occurs. The surface waters are normally oxygenated and support typical marine life. At depth, however, the stagnant and heavier salt water becomes so depleted of oxygen that a highly reducing environment results—the euxinic environment. The toxicity of the water prevents normal marine life, and only anaerobic bacteria can survive. Pelagic organisms, settling slowly but without interruption to the deeper zone, are only partially decomposed by the bacteria, allowing for preservation of the organic material. The bacteria also reduce the sulfates, generating important concentrations of hydrogen sulfide. The hydrogen sulfide, in turn, causes precipitation of ferrous sulfide that blackens the organic-rich accumulations (Pettijohn, 1957).

GARRY D. JONES

References

Krumbein, W. C., and Sloss, L. L., 1963. *Stratigraphy and Sedimentation*. San Francisco: W. H. Freeman and Co., 660p.

Pettijohn, F. J., 1957. *Sedimentary Rocks*. New York: Harper and Row, 718p.

Cross-references: *Fiord, Fjord; Sediment Size Classification*. Vol. VI: *Euxinic Facies*.

EVAPORITES

Marine evaporites are those sedimentary rocks formed as a result of the evaporation of seawater. The use of the word *evaporite* in geologic literature is inconsistent and sometimes refers to chemical precipitates not the result of evaporation, but if the mode of origin is the main criterion, then evaporites can be separated from chemical precipitates created by the mixture of magmatic fluids or chemically charged terrestrial runoff with seawater.

Marine evaporites precipitate in a succession of mineral species whose broad outlines were first made known by Usiglio (summarized by Stewart, 1963). The Usiglio succession is shown in figure 1. The complex of potash-rich salts that forms as seawater approaches dryness is of great economic importance and has been intensively studied. Yet the details of the formation of these *bittern salts* are still incompletely known (see Borchert and Muir, 1964, for discussion).

Marine evaporites are widespread in rocks that have been buried and not subjected to subsequent exposure on the surface. Except in arid or desert areas, they are not preserved in surface outcrop, because they are quickly redissolved; however, many exposures record the former presence of evaporites in the form of breccias developed when the evaporites dissolved and the enclosing rocks collapsed as rubble into the resulting voids.

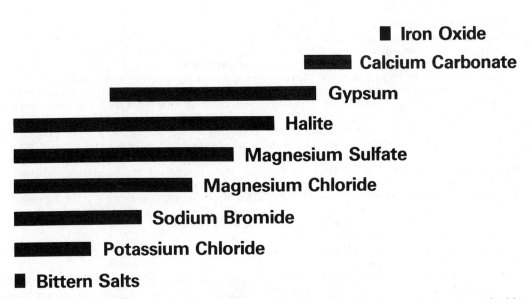

FIGURE 1. Schematic diagram showing the order of mineral precipitation from seawater as determined by Usiglio.

Marine evaporites occur in two basic modes. The commercially significant deposit is an areally widespread layer or series of layers of evaporite minerals either interbedded or in an essentially pure state. Undisturbed evaporites often shown a rhythmic banding that reflects the alternation of pairs of mineral species in seemingly endless succession. This banding has been mistaken for an annual phenomenon and called *varves* or *Jahresringe* (Udden, 1924; Richter-Bernberg, 1960), but Braitsch (1963) early expressed doubt that all such bands could be annual, and Shaw (1977) has pointed out that many bands are too thick to have been created by a single year's evaporation. Most commercial salt (*halite*) and potash minerals are recovered from such widespread deposits by drilling and subsequent solution underground. Such widespread deposits of evaporites often serve as impermeable seals above porous rocks that can be reservoirs for hydrocarbon accumulation and thus have an important indirect economic impact.

The second, less obvious, occurrence of evaporites is as cement, matrix, or infilling in rocks that are not dominantly evaporitic. These evaporites record the existence of waters of elevated salinity either in seas that mainly deposited other sediment types (e.g., the selenite—crystalline gypsum—commonly found in the thick Cretaceous shales of the western United States) or in seas that modified earlier deposited rocks (e.g., the anhydrite- or halite-plugged Silurian reefs in the Michigan Basin). Evaporite plugging has destroyed many potential hydrocarbon reservoirs with a consequent adverse economic impact.

While the basic succession of evaporite minerals is recognizable in the geologic record, the proportions of those minerals in the actual rocks is never the same as those found by Usiglio. For example, in rock sections containing high percentages of anhydrite, there will be a deficiency of carbonate and halite. If halite dominates, then carbonate, anhydrite, and the posthalite mineral species are deficient. The pattern is: whatever mineral is dominant in a particular evaporite deposit, the mineral species that appear both before it and after it in the Usiglio succession will be underrepresented. This geologic observation showed investigators early that evaporites could not have formed simply by drying up an isolated body of seawater. Most of the literature dealing with the problem of evaporite origin has therefore dealt with various proposed mechanisms to account for the "improper" proportions of the various species. Early attempts to account for these anomalies are summarized by Grabau (1920). They included such unlikely mechanisms as accumulation by wind or a series of isolated evaporating pans that emptied from one into another at the critical moment when salinities had risen by evaporation to the level at which one mineral species replaced another in the succession.

Following World War II, the problem was reexamined by a series of investigators (Adams, 1944; King, 1947; Briggs, 1957, 1958; Landes, 1963; Braitsch, 1963; Schmalz, 1966; American Association of Petroleum Geologists, 1971), who generally came to the conclusion that evaporites accumulate at the bottom of deep basins partly cut off from the sea into which access from the open ocean is restricted to such a degree that evaporation can cause elevated salinities. This idea has become the commonly accepted concept (Borchert and Muir, 1964). The papers of King (1947) and Briggs (1958) spell out the details. Recently, Shaw (1977) has attempted to show that the deepwater model requires impossible hydrodynamic conditions within the basin and that diffusion is more effective in reducing salinities than evaporation is in raising them, so that evaporites cannot form in deepwater that retains any connection to the ocean.

The solution to this dispute has important implications for the tectonic history of those regions containing evaporites. If the evaporites are of deepwater origin, then recurrent movements of the sea floor through hundreds or thousands of feet must have occurred. If the evaporites formed only in shallow water, then only trivial tectonic movements need be postulated.

The choice between models also has economic implications because evaporites commonly act as seals preventing the escape of hydrocarbons trapped in reservoirs below the evaporites. Exploration strategy is affected strongly, depending on whether one believes that the evaporite seal was formed under environmental circumstances that were basically a continuation of those that created the underlying reservoirs or, on the contrary, were formed in a wholly new and different environment prior to the development of which the reservoir was exposed, permitting any migrating hydrocarbons to escape. The choice of hypothesis would also affect in much the same way the manner in which one would explore for the evaporite minerals themselves.

ALAN B. SHAW

References

Adams, J. E., 1944. Upper Permian Ochoa Series of Delaware Basin, west Texas and southeastern New Mexico, *Am. Assoc. Petroleum Geologists Bull.* **28**, 1596-1625.

American Association of Petroleum Geologists, 1971. *Origin of Evaporites*. Tulsa, Okla: Am. Assoc. Petroleum Geologists Reprint Series No. 2, 208p.

Borchert, H., and Muir, R. O., 1964. *Salt Deposits*. London: Van Nostrand, 338p.

Braitsch, O., 1963. Die Entstehung der Schichtung in rhythmisch geschichteten Evaporiten, *Geol. Rundschau* 52, 405-417.

Briggs, L. I., 1957. Quantitative aspects of evaporite deposition, *Michigan Acad. Sci., Arts, and Letters Papers* 42, 115-123.

Briggs, L. I., 1958. Evaporite facies, *Jour. Sed. Petrology* 28, 45-56.

Grabau, A. W., 1920. *Geology of the Nonmetallic Minerals Other than Silicates, vol. I: Principles of Salt Deposition*. New York: McGraw-Hill, 435p.

King, R. H., 1947. Sedimentation in Castile Sea, *Am. Assoc. Petroleum Geologists Bull.* 31, 470-477.

Landes, K. K., 1963. Origin of salt deposits, in *Symposium on Salt*. Cleveland: Northern Ohio Geological Society, 3-9.

Richter-Bernberg, G., 1960. Zeitmessung geologischer Vorgange nach Warven-Korrelationen im Zechstein, *Geol. Rundschau* 49, 132-148.

Schmalz, R. F., 1966. Environments of marine evaporite deposition, *Mineral Industries* 35, 1-7.

Shaw, A. B., 1977. A review of some aspects of evaporite deposition, *Mtn. Geologist* 14, 1-16.

Stewart, F. H., 1963. *Data of Geochemistry*, 6th ed., U.S. Geol. Survey Prof. Paper 440-Y, 53p.

Udden, J. A., 1924. Laminated anhydrite in Texas, *Geol. Soc. America Bull.* 35, 347-354.

Cross-references: *Beach Rock; Global Tectonics; Humate*. Vol. VI: *Evaporite Facies*.

EXSUDATION—See SALT WEATHERING

F

FAULT COAST

The essential feature of a fault coast is a *fault scarp* separating a higher-standing earth block—which, after faulting, forms the land—from a lower-lying block—which, after faulting, is depressed below sea level. The prefaulting surface may have any form. This cumbersome definition was first proposed by Cotton (1916), who deduced a series of initial and sequential forms that fault coasts were expected to exhibit, and illustrated some of them from the coasts of New Zealand. The great systematist Douglas Johnson introduced a number of further terms including *fault-line-scarp shoreline,* which differed from a *fault shoreline* in that the feature partially submerged was caused by differential erosion and not by faulting. Such shorelines fall into the neutral or compound categories of Johnson's four-part genetic classification. Shepard (1963) also made the distinction between fault scarp and fault-line scarp coasts, the former being included in his primary class and the latter in his secondary class.

Fault coasts are generally steep-to coasts possessing a simple linear or rectilinear plan outline. Examples include the northeastern side of San Clemente Island, California; Wellington and Thames, New Zealand; Murmansk, eastern Kamchatka, and southern Crimea, USSR; and the Red Sea coast.

Closely related to fault coasts in origin, and frequently placed with them in a single tectonic or diastrophic category, are those initiated by sharp monoclinal flexure, termed *monoclinal coasts* (Cotton) or *fold coasts* (Shepard). Extensively developed tectonic coasts (fault or monoclinal) that are still in their first cycle cannot be common, for there has been little time for the formation of great faults of flexures since the postglacial rise of sea level. However, Cotton argues that the diagnosis of a fault coast is not necessarily a claim for such an origin within the *current cycle.* Instead, he believes that the deductive theory of the development of fault coasts is of value for the study of coasts of remote fault origin. Thus, if some coastlines originated as tectonic scarps in an earlier though possibly rather remote cycle, they are still fault coasts, but in Johnson's compound category also.

Clearly, the term *fault coast* is now utilized in a broader sense than initially defined, and encompasses a suite of coastal forms and histories which have a tectonic component. The development of these changes in usage is best exemplified in the series of papers in *Bold Coasts* (Collins, 1974).

ROGER F. McLEAN

References

Collins, B. W., ed., 1974. *Bold Coasts: Annotated Reprints of Selected Papers on Coastal Geomorphology, 1916-1969.* Wellington, New Zealand: Reed, 354p.

Cotton, C. A., 1916. Fault Coasts in New Zealand, *Geog. Rev.* **1,** 20-47.

Shepard, F. P., 1963. *Submarine Geology.* New York: Harper, 557p.

Cross-references: *Classification; Cliffed Coasts; Geomorphic-Cycle Theory; Sea Level Curves; Submergence and Submerged Shoreline; Tectonic Movements.* Vol. III: *Fault-Line Scarp.*

FEEDBACK

Feedback describes the interdependence of variables, a mutual interaction of the many processes and responses that take place in the coastal zone. This correlation may occur with either positive (reinforcing) or negative (regulating) results (King, 1970).

Systems Analysis

The term *feedback* has been borrowed from the nomenclature of general systems theory, where it is defined as the portion of the output that is monitored back (recycled) as input to a given system. As such, it acts as a control upon the system and indicates three different criteria for systems analysis: preestablished causal trains that are linear and unidirectional (as in most process-response models); the free and random interplay of forces (that often lead towards the dynamic equilibrium state); typically, a certain amount of "closure" in the open, coastal systems (implying entropic tendencies of the systems).

Examples

A simple form of feedback is the decline of *swash* activity with increasing *wave steepness*. As steepening waves saturate more of the foreshore, a decline in swash percolation results in a larger volume of *backwash*. Often this increased backwash will oppose and therefore lessen wave run-up. This is negative feedback. Another example is given within the cyclic model of beach dynamics. A depositional beach is relatively steep; when erosional storm waves act on the beach, a flatter slope evolves because material is shifted from the upper profile (*berm*) to the lower profile—the offshore component of sediment migration. The flatter nearshore causes the breaking waves to shoal over a greater area, and the diffusion of wave work results in decreased erosion rates by the storm waves.

A longer-term, positive feedback is shown by the *eustatic* relationship between glacial development and coastal transgression. As glaciers grow, sea level decreases and thus exposes coastal land. The increased continental area and low relief encourage glacial expansion into the coastal zone.

Time

In general, coastal processes are dynamic and thus of short-term periodicity. Most of these involve negative feedback relationships. The fewer long-term cycles often suggest either positive feedback or no feedback in coastal systems that are unidirectional in change (King, 1972).

Implications

Feedback in the complex coastal environment can be either self-regulating or self-enhancing. This nontrivial complexity demands consideration in the study of beaches and coasts. Any modeling of process-response chains or physical systems must then relax certain assumptions of independence among variables. Hayden and Dolan (1974) have shown how feedback principles must be incorporated in the development of strategies for coastal land use management.

JAMES R. ALLEN

References

Hayden, R., and Dolan, R., 1974. Management of highly dynamic coastal areas of the National Park Service, *Coastal Zone Management Jour.* 1, 133-139.

King, C. A. M., 1970. Feedback relationships in geomorphology, *Geog. Annaler* 52A, 147-159.

King, C. A. M., 1972. *Beaches and Coasts.* New York: St. Martin's Press, 570p.

Cross-references: *Beach Cycles; Beach Processes; Computer Simulation; Mathematical Models; Sediment Transport.* Vol. III: *Littoral Processes;* Vol. VI: *Littoral Processes.*

FEEDER BEACH

A feeder beach is an artificially widened beach that serves to *nourish* downdrift beaches by natural littoral processes. The location and dimensions of a feeder beach are governed by economic considerations related to the methods employed in mining, transporting, and depositing the fill material (Coastal Engineering Research Center, 1973).

If the area to be nourished is part of a continuous and unobstructed beach, the stockpile of fill is located at the updrift end of the problem area. Ideally, stockpiles should be placed at several points along the problem area, especially on shorelines where the rate of sand transported alongshore increases in the downdrift direction. The construction of several feeder beaches within a problem area also decreases the time interval required for complete nourishment of the area. The construction of several stockpiles may also be required on beaches where *beach protection structures* interfere with the natural transport rate, since it is often impractical to modify the structures to increase sand transport.

KARL F. NORDSTROM

Reference

Coastal Engineering Research Center, 1973. *Shore Protection Manual.* Washington, D.C.: U.S. Army Corps of Engineers, 3 vols.

Cross-references: *Coastal Engineering; Coastal Engineering, History of; Conservation; Nourishment; Sediment Transport.*

FERNS—See POLYPODIOPHYTA

FERREL'S LAW

Ferrel's law involves the deflection of a particle (water, air, ice, or the like) in motion of the *Coriolis effect.* The American meteorologist W. Ferrel was the first to describe the Coriolis effect on moving particles. As a result of this action, winds, oceanic currents, and drift ice are deflected rightward (with reference to their original motion) in the Northern Hemisphere and leftward in the Southern Hemisphere. The Coriolis effect, per unit mass, is $F = v \times 2\Omega \sin \phi$ (in scalar notation), Ω is angular speed of the earth's rotation ($\neq 0.729 \times 10^{-4} s^{-1}$), ϕ the latitude and v the speed of the particle. The Coriolis effect does not exist on bodies at rest (McCormack and Crane, 1973).

Although this force is small with respect to gravity (if we assume $v = 10$ cm/sec, $\phi = 45°$, $F \cong 1 \times 10^{-3} cm/s^{-2}$, i.e., $\cong 10^{-6} g$), its effect is very important in horizontal motions.

In oceanography, the Coriolis effect could be balanced by other forces; the gradient of the hydrostatic pressure, the wind drag, the tide, among others. Several important oceanic currents are regulated by such a balancing. If the Coriolis effect is balanced only by friction, then inertia currents are developed.

FERRUCCIO MOSETTI

Reference

McCormack, P. D., and Crane, L., 1973. *Physical Fluid Dynamics.* New York: Academic Press, 487p.

Cross-references: *Coriolis Effect; Currents; Fourier Analysis; Froude Number; Mathematical Models; Reynold's Number; Richardson's Number; Waves; Wind.*

FIARD, FJÄRD

Linguistically, *fjärd* (Swedish) and its anglicized version *fiard* (Gregory, 1913) are derived from the same root as *fiord* (q.v.). The terms are used both in a general sense to describe marine inlets, especially in southern Sweden, and in a more particular geomorphological sense to describe glacially eroded marine inlets on a lowland coast (Werth, 1909). They are distinguished from *rias* (q.v.) of nonglacial origin by the presence of shallow, glacially eroded rock basins in their floors, by their association with other features of glacial erosion, and often by the lack of large rivers draining into them. Werth showed how they are linked in south Sweden with a system of inland valleys and ribbon lakes (Fig. 1) that radiate from the interior

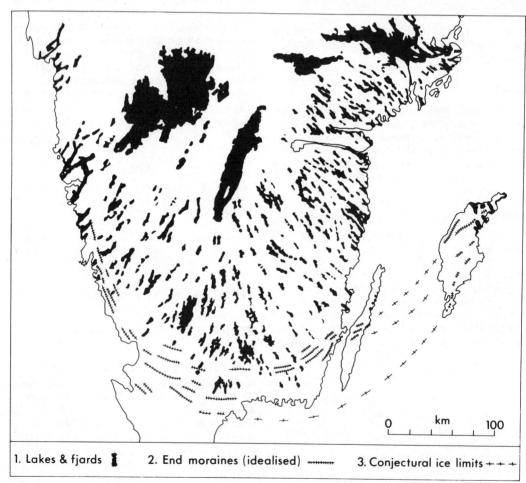

FIGURE 1. Ribbon lakes and fiards in southern Sweden: 1. Lakes and fiards; 2. End moraines (diagrammatic); 3. Conjectural ice margins (from Werth, 1914).

to the coast in a pattern accordant with the former directions of ice movement. The fiards and inland valleys may also have been modified by the erosion of subglacial streams that followed these routeways in escaping from beneath the thicker ice of the interior to reach the edge of the ice and regions of lesser ice pressure (Embleton and King, 1975). The ice, thought to have been the main agent that sculptured the fiards, was guided by preglacial valley systems in most cases (cf. *fjords*), and was able to erode the fiard floors without respect to base level. The present state of submergence of the fiards is therefore, as in the case of fjords, an accident, contingent on the amount of glacial erosion, the amount of postglacial sea level change, and the degree of postglacial isostatic recovery.

CLIFFORD EMBLETON

References

Embleton, C., and King, C. A. M., 1975. *Glacial Geomorphology*. London: Edward Arnold, 573p.

Gregory, J. W., 1913. *The Nature and Origin of Fiords*. London: John Murray, 542p.

Werth, E., 1909. Fjorde, Fjärde, und Föhrden, *Zeitschr. Geomorphologie* 3, 346-358.

Werth, E., 1914. Zur Oberflächengestaltung der südschwedischen Habinsel, *Zeitschr. Geomorphologie* 8, 345.

Cross-references: *Europe, Coastal Morphology; Fiord, Fjord; Firth; Förde; Glaciated Coasts; Ria and Ria Coasts*. Vol. III: *Fjärd, Fjord*.

FIORD, FJORD

The terms *fiord* (Danish and Norwegian), *fjärd* (Swedish), *Förde* (German) (q.v.), *fjardur* (Icelandic) and *firth* (Scottish) (q.v.) are all linguistic variants of the same word, and refer to coastal inlets of various forms and origins possessing the common characteristic of relatively great length compared with width. In Denmark and Norway, *fjord* is a common element of place names belonging to coastal inlets, but in worldwide geomorphological usage the term has taken on a special meaning, referring to the lower portions of glacial troughs partly submerged by the sea. It includes some in Antarctica and Greenland that would be invaded by the sea if glacial ice were not still occupying them.

Fiords, in a geomorphological sense, are best developed on the coasts of British Columbia, southern Alaska, southern Chile, eastern Canada (especially eastern Baffin Island), Norway, Iceland, Greenland, Spitsbergen, and South Island, New Zealand. They thus occur mostly in higher latitudes (more than 45°) and are usually backed by rugged and glacially dissected highlands. They have the features associated with glacially eroded troughs (Fig. 1): cross profiles of catenary form in the deeper parts of the trough; steep, rocky sides above which flatter shoulders may be present; relatively flat floors in the broader parts of the fiord where sediment infilling has occurred; and long profiles characterized by rock basins and by a threshold at the lower end that may rise near to, or even partly above, sea level. The deepest parts of the submerged rock basins descend to 1308 m in the Sogne Fiord (Norway), 1450 m in Scoresby Sound (Greenland), and 2287 m in Vanderford (Antarctica), for example; the principal Sogne Fiord rock basin is separated from the sea at its lower end by a rock threshold that rises to within 200 m of the sea surface, and is marked by numerous rocky islets or *skerries* (q.v.). The patterns of many fiords often show strong relationships to lines of fracture in the bedrock or to other zones of weakness that were selectively exploited by the eroding glaciers.

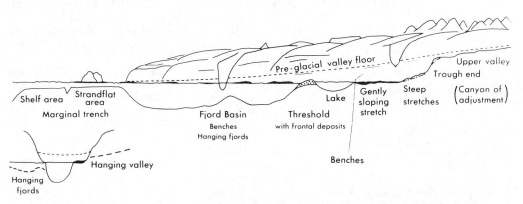

FIGURE 1. Characteristic features of fiords and fiord valleys, represented by long profile and cross section (from Gjessing, 1966).

The drowned appearance of fiords bears no necessary relation to any change of sea level. The postglacial rise in sea level, of the order of 100 m, is small in comparison with most fiord depths. The extent to which the sea penetrates is mainly a function of the depth of glacial excavation of the trough floor, for glacial erosion, unlike fluvial erosion, is not controlled by a *base level*, although sea level partly controls the point at which ice in a fiord begins to float.

The questions as to why the deepest basins in fiords so often occur in their middle and lower sections, and why a submerged rock threshold commonly marks the fiord entrance, have been extensively discussed, and many hypotheses have been advanced (Charlesworth, 1957; Embleton and King, 1975). Variations in rock resistance to erosion along the length of the fiord have been postulated, but by no means all basins are excavated in less resistant rocks. Others associate the threshold with the end of the fiord glacier melting rapidly in contact with saltwater or able to spread itself laterally at the fiord mouth, thus suffering reduced velocity. The deep basins have been supposed to represent the localized action of preglacial weathering or ground freezing, breaking up the bedrock ready for glacial transport. Another general theory relates the seaward shallowing of fiords to the point at which the glacier begins to float (and therefore ceases to erode the floor). This point will vary in position according to changes in both water depth (sea level) and glacier thickness.

An entirely different view of fiord formation is that fiords represent glacially modified raised submarine canyons (Winslow, 1966). Fiords are said to have acquired many of their distinctive characteristics as *submarine canyons* (q.v.) before glaciation and to have been eroded initially by deep submarine processes, such as the action of turbidity currents.

CLIFFORD EMBLETON

References

Charlesworth, J. K., 1957. *The Quaternary Era.* London: Edward Arnold, 2 vols., 1700p.
Embleton, C., and King, C. A. M., 1975. *Glacial Geomorphology.* London: Edward Arnold, 573p.
Gjessing, J., 1966. Some effects of ice erosion on the development of Norwegian valleys and fjords, *Norsk Geogr. Tidsskr.* **20**, 273-299.
Winslow, J. H., 1966. Raised submarine canyons: An exploratory hypothesis, *Assoc. Am. Geographers Annals* **56**, 634-672.

Cross-references: *Europe, Coastal Morphology; Fiard, Fjärd; Firth; Förde; Glaciated Coasts.* Vol. III: *Fjärd, Fjord.*

FIRTH

This is an essentially Scottish word with no precise geomorphological meaning. Etymologically, it is derived from Old Norse and closely related to *fiord* (q.v.) and other similar words. Physiographically, it describes a sea inlet, an estuary, or a strait in Scotland (Steers, 1952). Thus the Firth of Tay is the seaward extension of the Tay estuary; the Pentland Firth is a sea strait; while the Firth of Lorn is a submerged, glacially eroded channel similar to a fiord.

CLIFFORD EMBLETON

Reference

Steers, J. A., 1952. The coastline of Scotland, *Geog. Jour.* **118**, 180-190.

Cross-references: *Europe, Coastal Morphology; Fiard, Fjärd; Fiord, Fjord; Förde; Glaciated Coasts.* Vol. III: *Fjärd, Fjord.*

FLANDRIAN TRANSGRESSION

During the last Würmian (Wisconsin) glaciation about 35 million km^3 of ocean water was locked up in ice sheets. Consequently the world ocean level was 100 m (130 m, according to some authors) below the present level. With the melting and retreat of the European and American ice caps 18,000-17,000 years B.P., the sea level began to rise. The late postglacial transgression, taking place ever since then up to the present has been named the *Flandrian*. The term was originally coined for the Atlantic shore, but lately has been applied to the entire late postglacial transgression.

The *Flandrian transgression* is divided into two stages: the *Upper Pleistocene* (17,000-6,000 B.P.), a period of rapid rise of sea level at the rate of 9 m in a thousand years; and the *Holocene* (from 6000 B.P. to the present), marked by a progressive decline in the rate of sea level rise, from 4-1 m in a thousand years or by sea level fluctuations close to the present level.

Many radiocarbon datings on shelf deposits in areas around the world indicate that the ocean level in 18,000 B.P. was at about 100 m in depth; in 15,000 B.P. at about 80 m; in 10,000 B.P., at 30m; in 8000 B.P., at 20 m; and in 6000 B.P., at 6-0m.

Most of the researchers are certain about the nature of the transgression during the first stage; proceeding at a headlong pace, it was inundating the upper part of the present shelf. Possibly there were stages of faster and of slower rise of ocean level.

Opinions are divided over ocean level changes in the past 6000 years, and can be reduced to three principal views (Fig. 1), as summarized by Curray (1961):

1. Sea level first attained its present position 6000 B.P., and since then has undergone minor positive and negative fluctuations. Fairbridge (1961) thinks that the ocean level 5000 and 3700 B.P. exceeded the present level by 3-4 m. These peaks of transgression have been named the old and the young Peron stage, respectively. Following some insignificant oscillations, 2300 and 1200 B.P., sea level once again rose to +1.5 m elevation (respectively, Abrolhos and Rottnest stage).

2. Sea level attained its present position 3000 to 5000 years ago and has remained so until now (Fisk).

3. Sea level has been slowly and continuously rising, asymptotically approaching the present position that it did not attain until recently (Shepard).

There are geological materials and radiocarbon datings to support each of the three points of view and their numerous variations. Yet the effects of recent tectonic deformations, the difficulty of accurate determination (within ±1 m) of the Holocene coastline, and the specifics of local natural conditions (high tides, gale surges, material dispersion in storms, and so on) are such as to allow no definitive conclusion about a sea level change of 3-5m; so none of the opinions can be given preference thus far. This remains a controversial subject of debate in present-day Quaternary geology.

The *Flandrian transgression* exercised an immense influence on the evolution of shelves and coastlines. Since the coastal zone in the period of transgression was migrating landward up the shelf, subaerial and coastal units were trapped in the mass of shelf sediments. The present coastline did not begin to take shape before the ocean level came close to its present elevations. In this sense, the present coastline may be reasonably considered a 6000-year old product of the *Flandrian transgression*.

Other relevant literature on this topic includes Jelgersma (1961), Kaplin (1973), and Shepard (1961).

PAVEL A. KAPLIN

References

Curray, J. R., 1961. Late Quaternary sea level: A discussion, *Geol. Soc. America Bull.* 72, 1707-1712.

Fairbridge, R. W., 1961. Eustatic changes in sea level, in *Physics and Chemistry of the Earth*, vol. 4. New York: Pergamon Press, 99-185.

Jelgersma, S., 1961. Holocene sea level changes in the Netherlands, *Geol. Stichting Med.*, ser. C, 6, 100p.

Kaplin, P. A., 1973. *Recent History of the World Ocean Coast*. Moscow: Moscow State University, 264p.

Shepard, F. P., 1961. Sea level rise during the last 20,000 years, *Zeitschr. Geomorphologie* 5, 30-35.

Cross-references: *Coastal Morphology, Research Methods; Global Tectonics; Holocene; Isostatic Adjustment; Pleistocene; Present-Day Shoreline Changes; Sea Level Changes; Sea Level Curves*. Vol. I: *Coastal Stability*.

FLOCCULATION

The concept of *flocculation* has generally been reserved to describe the clumping of colloidal and fine, suspended particles by electrochemical forces. Many investigators have, however, used the terms *flocculation, coagulation,* and *agglomeration* interchangeably and referred to any loosely bound composite particles as *flocculates* or *flocs*. This imprecision in terminology to describe composite particles has been partially responsible for the failure of researchers to assess the relative importance of the several mechanisms of agglomeration in controlling sedimentation in the coastal environment. All flocs are loosely bound composite, particles—agglomerates; but not all agglomerates are flocs—only those formed and bound by electrochemical forces (Schubel, 1971; Vold and Vold, 1964).

Electrochemical flocculation depends upon collision and cohesion of fine, suspended and colloidal particles. The most important factors that control flocculation are: the composition and concentration of the particles; their size distribution and shapes; and the salinity and intensity of the mixing of the water. Flocculation is probably important only with colloidal and fine, suspended particles, particles less than

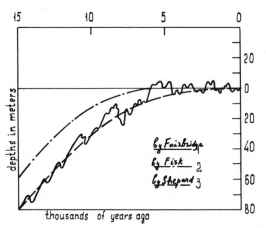

FIGURE 1. Proposed Flandrian transgression curves.

a few microns in diameter (Van Olphen, 1963). Increasing the concentration of particles increases the probability of collision; flocculation is probably important only when the concentration of particulate matter exceeds several hundreds of mg/ℓ, and the critical concentration may be several thousands of mg/ℓ. Increasing the mixing of the waters up to a certain threshold promotes flocculation by increasing the probability of collision; at great mixing intensities the internal shear limits the formation of larger flocs. The threshold is not well documented, but may be in the range 0.5–1.0 dyne/cm^2.

Laboratory experiments have demonstrated that monomineralic suspensions of different clays settle at different rates over a range in salinities. This observation has been widely interpreted to mean that flocculation results in the sequential deposition of clay minerals (first, illite then, kaolinite, then, montmorillonite) with increasing salinity along the axis of an estuary.

Flocculation had been ascribed a major role, particularly by civil engineers, in controlling estuarine and coastal sedimentation. While there is little question that flocculation can promote sedimentation in the laboratory and in the settling ponds of water-treatment plants, the field evidence for electrochemical flocculation in coastal environments is much less convincing (Meade, 1972; Schubel, 1971). Some investigations indicate that the agglomeration of fine particles by filter-feeding phytoplankton and by microbiological organisms may be more important than flocculation in promoting the sedimentation of fine-grained particulate matter in estuarine and coastal environments (Schubel, 1971).

J. R. SCHUBEL
C. F. ZABAWA

References

Meade, R. H., 1972. Transport and deposition of sediments in estuaries, *in* B. W. Nelson, ed., *Environmental Framework of Coastal Plain Estuaries*. Boulder, Colo.: Geol. Soc. America Mem. 133, 91–103.

Schubel, J. R., ed., 1971. *The Estuarine Environment: Estuaries and Estuarine Sedimentation*. Washington, D.C.: Am. Geol. Inst. Short Course Lecture Notes, 324p.

Van Olphen, H., 1963. *Introduction to Clay Colloid Chemistry*. New York: Wiley Interscience, 301p.

Vold, M. J., and Vold, R. D., 1964. *Colloid Chemistry: The Sciences of Large Molecules, Small Particles, and Surfaces*. New York: Van Nostrand Reinhold, 118p.

Cross-references: *Clay; Estuarine Sedimentation, Lagoonal Sedimentation; Nearshore Hydrodynamics and Sedimentation; Nearshore Water Characteristics; Protozoa.* Vol. VI: *Flocculation;* Vol. XIII: *Flocculation.*

FLOTSAM AND JETSAM

The first of these two words means the floating wreckage of a ship or its cargo, in whole or in part, and therefore, by extension, generally unimportant miscellaneous material. The second refers to parts of a ship, or its load, that have been thrown overboard, in the strictest sense to lighten the vessel in time of difficulty such as a severe storm.

In loose usage, the two terms are taken together (*flotsam and jetsam*) to indicate all floating debris, whether or not derived from a ship. Such material commonly washes ashore, and may be found especially along certain beaches and rocky coasts. Tree trunks or logs, tree limbs, planks, wood fragments, bottles of various kinds (both glass and plastic), floats used by fishermen, and many other items are included in the more general usage.

Items found on the beach that had been thrown into the sea with the intent of recovery, fall into the category of *lagan*. The most widely used items in this category are floating cards, designed to be returned by the finders to a central office, so that *surface currents* can be traced. Formerly bottles were used for this purpose.

Where flotsam and jetsam accumulate spectacularly at the top of the youngest ridge or berm, one can generally conclude that long-term accretion is taking place (Tanner and Stapor, 1972). Eroding beach ridges may contain interesting finds, within the sand, but do not collect flotsam and jetsam along the crest lines. Useful items inside eroding beach ridges include boards, some with nails in them, and bottles of a wide variety (Stapor, 1975). The latter especially, because of differences in making the lip, neck, bead, and bottom, provide clues as to the date of accumulation. In the Western Hemisphere, sawed boards place an upper limit on age (not likely to be more than about 300 years old) and nail types (round, square) further subdivide this interval. "Bottle chronology" of beach ridges is commonly internally consistent and in many instances permits an uncertainty of no more than 10–20 years. (Frank Stapor, Marine Resources Division, Charleston, S.C., 29412, had developed this technique, but has not published his results.)

WILLIAM F. TANNER

References

Stapor, F. W., 1975. Holocene beach ridge plain development, Northwest Florida, *Zeitschr. Geomorphologie* 22, 116-144.

Tanner, W. F., and Stapor, F. W., 1972. Precise control of wave run-up in beach ridge construction, *Zeitschr. Geomorphologie* 16, 393-399.

Cross-references: *Beachcombing; Beach Ridge and Beach Ridge Coasts; Driftwood; Foam Mark; Pollutants; Swash Mark.*

FLOWERING PLANTS—See MAGNOLIOPHYTA

FOAM MARK

With a stiff sea breeze, a great deal of foam can be formed in the *surf zone* and blown up onto the beach, in some instances covering the beach completely to a thickness of tens of centimeters. Where less foam is present, individual bubbles may leave faint circular marks on the beach surface, and where bits of foam are driven by the wind across the beach, subtle drag marks may be left (Allen, 1967).

Foam occurring in irregular bands, transverse to the wind direction, may travel downwind in a jerky motion, leaving behind a system of subtle anastomosing ridges likewise transverse to the wind direction. The spacing of the ridges is roughly the bubble diameter (2-5 cm). The foam bands migrate in a manner related to the movement of water-current *ripple marks*, but, in the beach case, represent a three-component system: air, foam, sand. Migration of the foam patches may blur or obliterate previous sand surface markings.

The extent of preservation of the various kinds of foam marks is unknown. However, *swash marks* and other subtle features are known from the rock record, and therefore it is possible that foam marks may be present in well-lithified sand.

WILLIAM F. TANNER

Reference

Allen, J. R. L., 1967. A beach structure due to wind-driven foam, *Jour. Sed. Petrology* 37, 691-692.

Cross-references: *Flotsam and Jetsam; Ripple Marks; Swash Mark.* Vol. VI: *Primary Sedimentary Structures.*

FOOD CHAIN

A *food chain* is a trophic relationship between species or between a species and another food source. In its simplest form, it is a predator-prey or herbivore-plant reaction. More complex ones involve several steps: phytoplankton-zooplankton-fish-predatory fish-man. In most cases such relationships are greatly complicated because each species may have several predators and feed on more than one prey species. A *food web* thus exists, and a second definition of a food chain may be: any linear or continuous component of a food web.

The energy retention at each step of a food chain is low, on the order of 10%. In the longest food chains, the efficiency of energy transfers up to the top predators is lower than in short chains. However, evidence exists suggesting that the ultimate yield of a top carnivore is sometimes greater on a longer chain because of the lower metabolic expenditure required when feeding on large prey (Steele, 1970)—a point of importance in fish culture.

Although in the global sense all food chains are ultimately based on primary producers, in some marine and other ecosystems they may be based more immediately on other trophic levels —e.g., the polluted water chain: suspended sewage solids-bivalves-starfish; or the detritus chain: organic sediment-bacteria-nematodes.

CHARLES R. C. SHEPPARD

Reference

Steele, J. H., ed., 1970. *Marine Food Chains.* Edinburgh: Oliver and Boyd, 552p.

Cross-references: *Biotic Zonation; Coastal Ecology, Research Methods; Coastal Fauna; Coastal Flora; Organism-Sediment Relationship; Photic Zone; Production.*

FORAMINIFERA

Foraminifera, an order of the protozoan subphylum Sarcodina, are single-celled organisms that secrete a shell (or *test*) and possess elongate, linear extensions of the protoplasm that form bifurcating and anastomosing pseudopodia. Modern Foraminifera are assigned to four suborders: Allogromiina, Textulariina, Miliolina, and Rotaliina (Fig. 1), based on the construction of their shell wall (Loeblich and Tappan, 1964).

The Allogromiina has the simplest shell, with only one or a few poorly defined chambers and a shell wall composed of membranous or pseudochitinous material. This wall may occasionally have ferruginous encrustations or include small quantities of agglutinated material, such as fine sand grains. Inclusion of the Allogroniima in the Foraminifera has been questioned; they are, hoever, generally con-

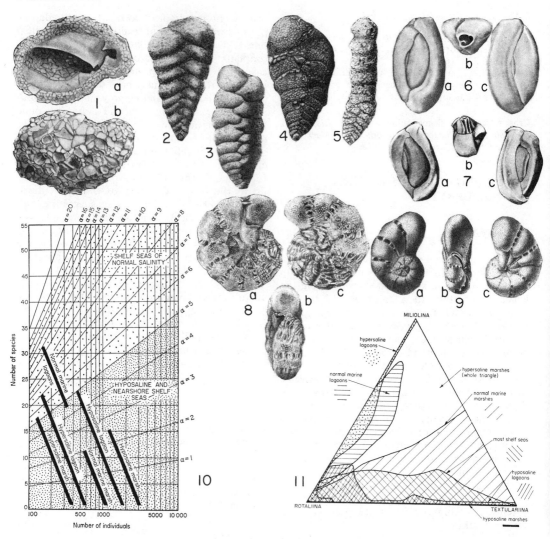

FIGURE 1. Representatives of the four living suborders of modern Foraminifera and their relative diversity and distribution in coastal environments. Allogromiina: 1a, b, *Iridia* × 40. Textulariina: 2, *Olssonia* × 40; 3, *Textularia* × 20; 4, *Textularia* × 50; 5, *Bigeneria* × 30. Miliolina: 6a, b, c, *Quinqueloculina* × 30; 7a, b, c, *Quinqueloculina* × 25. Rotaliina: 8a, b, c, *Epistomaroides* × 20; 9a, b, c, *Epistomaroides* × 20. 10. Range of species diversity in different coastal environments (α is Fisher's index of diversity); 11. Triangular plot of abundance of suborders in different coastal environments (1 through 9 from A. R. Loeblich and H. Tappan, 1964, *Treatise on Invertebrate Paleontology*, courtesy of the Geological Society of America and the University of Kansas; 10 and 11 from Murray, 1973).

sidered a minor group of primitive Foraminifera. Members of the Allogromiina live mainly in marine and brackish environments, and some are able to survive in waters with low salinity approaching freshwater. The presence of Allogromiina and a lack of other Foraminifera generally suggest prevailing freshwater conditions in a particular coastal environment.

The Textulariina has a wall composed of sand-size grains, sponge spicules, and fragments of other shells or other broken material that are cemented together by pseudochitinous material to form an agglutinated shell. In some species, a secondary calcareous lining or cement may also be present. Textulariina includes a wide range of morphological shapes, including those with multiple apertures and tubelike extensions without chambers as well as those with many, regularly formed chambers having a high degree of geometrical configuration. Many similar, well-developed geometrical arrangements are also present in most Miliolina and Rotaliina; however, the agglutinated construction of the Textulariina shell wall is considered distinctive for this suborder. Textulariina is commonly the dominant suborder of Foraminifera in most hyposaline marshes where salinities fluctuate between 6-30 parts per thousand. In normal to high salinity marshes, species of Textulariina usually form a large percentage of the Foraminiferal fauna, and become a gradually smaller proportion of the fauna in environments that have more exchange with open ocean water.

The Miliolina has a wall composed of more or less randomly arranged minute platelets and needles of calcite set in a pseudochitinous matrix. The wall generally lacks pores. Some have multiple chamberlets with multiple apertures. The shell when dry looks porcelaneous in reflected light, and when wet is amber in transmitted light. Representatives of the Miliolina are most common in lagoons having normal marine salinity and hypersalinity, and may comprise more than 50% of the Foraminiferal fauna and occasionally as much as 90% or more. In shallow tropical and subtropical coastal waters in carbonate depositional provinces, this suborder may have an extremely high species diversity and dominate the Foraminiferal fauna.

The Rotaliina has a pseudochitinous and calcareous, crystalline wall of calcite, or rarely aragonite, that appears glassy (or *hyaline*) in reflected light. The microstructure of the wall is complex, and includes many small pores, passageways, and commonly a number of layers. This is the most diverse of the modern suborders, and contains a wide variety of shapes, sizes, and geometric patterns. Rotaliina shares lagoonal environments with Miliolina; however, members of the Rotaliina dominate the Foraminiferal faunas in lagoons in noncarbonate depositional provinces, particularly where strong seasonal fluctuations of salinities occur. The Rotaliins also dominate Foraminiferal populations of most marine parts of deltas and neritic margins of open oceans and seas that form continental shelves and shelf seas.

A number of ecological factors are important in determining the diversity and distribution of species and genera of Foraminifera (Bandy, 1964; Phleger, 1960; Walton, 1964). Species diversity gradually increases from slightly brackish water marshes and lagoons to normal marine lagoons and neritic seas, but decreases in hypersaline lagoons (Murray, 1973). Species diversity markedly decreases from warm tropical and subtropical, through temperate, and into cold Arctic waters when comparing similar environments. Further, the average size of shallow-water Foraminifera decreases from tropical and subtropical into temperate and cold water (Boltovskoy and Wright, 1976). The availability of nutrients and related factors, such as symbionts, are also important and are interrelated to sunlight, temperature, currents, wave energy, and other organisms in the ecosystem. The grain size of the sediment of the substrate affects distributional patterns; many Textulariins are highly selective in the grains that make up their shells. Grain size and current strength influence the distribution of other Foraminiferal groups. Sandy beaches and detrital sands on the shelf commonly have low diversity of living species, although dead, empty shells may be common. Some of these empty shells may have been recently transported to their site of deposition, but many are assemblages of Quaternary age that, at the time of lower sea levels, thrived in a shallower water environment and now have been submerged beyond their depth limits but have remained uncovered by a lack of sedimentation. Some Rotaliin genera are found almost exclusively on mud or muddy substrates, and this relationship extends from neritic into bathyal regions.

In coastal waters, Foraminiferal distribution is influenced by sewage pollution, particularly near large outfall areas from major cities. In general, the number of living individuals per unit surface area adjacent to the point of discharge is low but increases gradually to form an aureole of higher abundance at a short distance from the point of discharge, depending on current direction and dilution. The substrate pH is commonly high in these situations, and the empty shells of calcareous Foraminifera may be rapidly dissolved, so that dead agglutinated shells appear abundant. Away from the auereole of abundance, the number of individuals decreases by about half. The outfall apparently supplies increased nutrients for phytoplankton, which are the major food source for the Foraminifera, and this results in a 5- to 1000-fold increase in the number of individuals of certain species.

CHARLES A. ROSS

References

Bandy, O. L., 1964. General correlation of foraminiferal structure with environment, *in* J. Imbrie and N. L. Newell, eds., *Approaches to Paleoecology*. New York: John Wiley & Sons, 75-90.

Boltovskoy, E., and Wright, R., 1976. *Recent Foraminifera*. The Hague: W. Junk, 515p.

Loeblich, A. R., and Tappan, H., 1964. *Treatise on Invertebrate Paleontology, Part C, Protista 2*. Lawrence, Kansas: Geological Society of America and University of Kansas Press, 900p.

Murray, J. W., 1973. *Distribution and Ecology of Living Benthic Foraminiferids*. New York: Crane, Russak & Co., 274p.

Phleger, F. B., 1960. *Ecology and Distribution of Re-*

cent *Foraminifera.* Baltimore, Md.: Johns Hopkins University Press, 297p.

Walton, W. R., 1964. Recent foraminiferal ecology and paleoecology, *in* J. Imbrie and N. D. Newell, eds., *Approaches to Paleoecology.* New York: John Wiley & Sons, 151-237.

Cross-references: *Biotic Zonation; Coastal Fauna; Organism-Sediment Relationship; Protozoa; Waste Disposal.*

FORCED WAVES

Forced waves involve some problems concerning long oceanic waves, and are considered less interesting than short wind waves (which are *forced* by wind). They are often thought to have the same, or longer, period interval as tides and seiches.

A well-known forced wave is the *induced tide* (tide oscillation), which develops in semiclosed basins by the induction of the true tides (*autonomous tides*) acting in a nearby open sea. The eigenperiods in oscillating basins can be close to tide periods (diurnal, semidiurnal, or others); therefore the induced tide can be reinforced by resonance. Tide-forced waves have many properties similar to the tide-wave properties; but other forced waves also appear in oceans (Henderschott, 1973).

Forced waves are induced by wind or by atmospheric pressure differences; they appear only if wind or pressure differences occur periodically. For example, in some cases low pressure systems can pass over a sea about every 3 days or every 5 days; in consequence of the passing low pressure, the sea level is induced, first, to rise and, then, after the low pressure level has passed to fall. In this manner a sea level fluctuation is produced. This fluctuation can be almost periodic or composed of some superimposed periodicities.

Neither period nor amplitude of the forced oscillation is related to the geometrical characteristics of the moving water body; on the contrary, both depend only on the intensity of the cause. This is a principal difference between proper (free) and forced oscillations in a basin.

Meteorological conditions over the sea and the general climatic situation of some regions can contribute to establish different periodicities that are characteristic of some sea or of some time. So, for example, the Adriatic is particularly affected in winter by forced oscillations of about 30-60 hours (Manca, Mosetti, and Zennaro, 1974) that are certainly not *seiches.* In the eastern Mediterranean Sea, Gulf of Kalamata (Greece), distinct waves of about 12-day periods appear frequently, often persisting and also mixed with shorter waves (6-day period).

On the southeast coast of Australia (for example, Port Phillip Bay) waves with a period of 2-7 (often 5-7) days are well developed, principally in the southern winter (July-August). These waves can be large (50-70 cm) and appreciable on the *marigram* (q.v.). A current, which can be a component in renewing coastal waters, is related to the rising and falling of sea level.

FERRUCCIO MOSETTI

References

Hendershott, M. C., 1973. Inertial oscillation of tidal period, *in* B. Warren, ed., *Progress in Oceanography,* vol. 6. New York: Pergamon Press, 1-27.

Manca, B.; Mosetti, F.; and Zennaro, P., 1974. Analisi spettrale delle Sesse dell 'Adriatico, *Boll. Geof. Teorica e Appl. (Trieste)* **16,** 51-60.

Cross-references: *Seiches; Tides; Waves; Winds.*

FÖRDE

Förde (or *Föhrde*) is a term used in north Germany to describe a marine inlet of relatively narrow form (e.g., Flensburg Förde) akin to those named *fjord* in Denmark (e.g., Vejle Fjord, Limfjord). Like *fiard* (q.v.), Förde stems from the same root as *fiord* (q.v.), but there have been attempts to differentiate these terms geomorphologically. The Förden of north Germany have been regarded as the drowned lower portions of *tunnel valleys* that were eroded by subglacial streams escaping under pressure beneath the Baltic ice to its margins and ice-free ground. A similar hypothesis has been advocated for many of the Danish *fjords.* The shapes of Förden are variable, but they usually narrow inland from wide mouths and cannot be distinguished morphologically from normal river estuaries. Their banks are low, composed of glacial deposits, and their depths are usually less than 20 m. The subglacial hypothesis of their origin rests on their association with other glacial features and on their pattern in relation to the distribution of Pleistocene ice (Embleton and King, 1975). The longest tunnel valleys can be traced from their drowned portions (Förden, fjords) at the Baltic or Kattegat coasts to the ice limits in Schleswig-Holstein or Jylland at the borders of outwash plains or marginal meltwater drainage valleys (*Urstromtäler*), where they often end abruptly. Similar features have been described in eastern England (Woodland, 1970): some of the estuaries in East Anglia may be analogous to

Förden. The relatively great widths of many Förden or tunnel valleys makes it difficult to regard them as single subglacial stream courses, for the ice roof would not be strong enough to bridge such widths. The larger ones were probably formed over considerable periods of time by the lateral migration of smaller ice tunnels; small eskers found on their floors represent infillings of the final ice tunnel and have also been detected in the submerged portions.

A different hypothesis (Woldstedt, 1961; Hansen, 1971) returns to an older view that the tunnel valleys and Förden of Denmark and north Germany are simply the result of selective erosion by ice tongues pushing up existing valleys during the last glaciation.

CLIFFORD EMBLETON

References

Embleton, C., and King, C. A. M., 1975. *Glacial Geomorphology*. London: Edward Arnold, 573p.

Hansen, K., 1971. Tunnel valleys in Denmark and northern Germany, *Dansk Geol. Foren. Medd.* 20, 295-306.

Woldstedt, P., 1961. *Das Eiszeitalter*, vol. 1. Stuttgart: F. Enke, 374p.

Woodland, A. W., 1970. The buried tunnel valleys of East Anglia. *Yorkshire Geol. Soc. Proc.* 37, 521-578.

Cross-references: *Europe, Coastal Morphology; Fiard, Fjärd; Fiord, Fjord; Firth; Glaciated Coasts*. Vol. III: *Fjärd, Fjord*.

FOSSIL CLIFF

A cliff that has been abandoned by a fall in sea level, covered or partially covered by glacial, solifluction, or alluvial deposits, and now is being reexcavated in consequence of the postglacial rise in sea level (Wood, 1959).

Since the term *fossil* is used, the fact of burial or partial burial is important. A *sea cliff* abandoned by fall of sea level or rise of the land but not covered by later deposits may be termed a *dead cliff*, an *old cliff* or an *abandoned cliff*—all terms so colloquially used that the earliest usage is unimportant.

An abrasion platform cut in solid rock is often missing below a recently exhumed fossil cliff. This is good evidence that the cliff was cut by a sea whose level was below that of the present day.

ALAN WOOD

Reference

Wood, A., 1959. The erosional history of the cliffs around Aberystwyth, *Liverpool and Manchester Geol. Jour.* 2, 271-287.

Cross-references: *Coastal Bevel; Sea Cliffs; Slope-over-Wall Cliffs*.

FOULING

Marine fouling is the growth of marine microorganisms, plants, and animals on submerged surfaces including ships, fixed and floating structures and seawater conduits. Fouling commonly affects the longevity and efficiency of underwater structures, and is a complex problem that is solved by using engineering, chemical, and biological techniques (Woods Hole Institute of Oceanography, 1952).

Organisms that form fouling communities are bacteria, diatoms, algae (seaweeds), protozoans, sponges, hydroids, bryozoans (sea mats or sea mosses), barnacles, bivalves (particularly mussels and oysters), and ascidians (sea squirts). Natural fouling communities can be grouped according to the physical exposure of the site—open sea coast, sheltered marine environment, and estuarine environment, each of which includes a characteristic group of organisms. On open, rocky shores, fouling intensity is high. Sheltered marine environments, such as tidal estuaries or sounds, support the greatest variety of fouling organisms. The estuarine environment is frequently subject to extremes of salinity and turbidity. Most fouling organisms have a definite vertical zonation. Algae are confined to the photic zone, which varies with the turbidity of the water. Marine fouling organisms also vary in geographical distribution, some being confined to warm or to cool waters, others being more widespread.

Most fouling organisms have an early motile (free-swimming) stage, after which they settle to a substrate to develop into an adult stage. When the organisms attempt to attach to the substrate, they are most sensitive to environmental conditions, and at this stage control measures are commonly applied. Control of fouling can be effected by creating unfavorable conditions for larval settlement by altering the temperature, salinity, light, water movement, supply of nutrients, or by means of poisons, such as copper and mercury paints. Fouling intensity and the actual kinds of organisms in a fouling community will vary with a number of factors, including the season, nature of the substrate, the amount of pollution, predation, and species competition, as well as the factors noted above in controlling fouling (Sutherland and Karlson, 1977). Because of the great variation in these many factors, the fouling intensity and the composition of the fouling communities in a particular locality cannot be predicted with any degree of accuracy unless observations have been made over several years.

Temperature plays an important part in the growth and reproduction of most marine organisms, but such factors as seasonal rainfall, current movements, or tidal amplitude may also affect the release of ova and sperm. In most nontropical waters, fouling is less in winter, but in some places settlement occurs throughout the year.

Normally the formation of a primary film by microorganisms is the first event in the fouling of a submerged surface, although barnacles and other animals can attach directly on a suitable surface. Bacteria, diatoms, and algal spores are important in making up the primary film. Other fouling organisms become established, depending on time of spawning. Barnacles (Cirripedia) may be the only organisms on some heavily fouled surfaces.

JUNE R. P. ROSS

References

Sutherland, J. P., and Karlson, R. H., 1977. Development and stability of the fouling community at Beaufort, North Carolina, *Ecol. Monogr.* 47, 425–446.

Woods Hole Oceanographic Institute, 1952. *Marine Fouling and Its Prevention*. Annapolis, Md.: U.S. Naval Institute, 388p.

Cross-references: *Algal Mats and Stromatolites; Biotic Zonation; Coastal Engineering; Coastal Fauna; Coastal Flora; Offshore Platforms.*

FOURIER ANALYSIS

Fourier analysis is very useful in research involving oscillating phenomena. This analysis can be applied to both temporal (time series) and to spatial functions. An exact application of Fourier analysis in its original form can be performed on fluctuations composed of a set of components whose periods are in harmonic succession. However, more generalized forms permit other applications. On the basis of Fourier analysis, whatever the components may be, one can calculate special mathematical filters available for the separation of the components. In some cases an important extension of the Fourier analysis is the *Fourier transform*.

Fourier analysis and transform are applicable to both unidimensional and bidimensional problems (tridimensional space oscillating functions have still not been introduced in oceanography).

The basic concepts of Fourier analysis establish that a function $f(t)$, satisfying the so-called *Dirichelet conditions*, can be represented as:

$$f(t) = \frac{a_0}{2} + \sum_{n=1}^{\infty} a_n \cos nt + \sum_{n=1}^{\infty} b_n \sin nt$$

(some functions could be represented only as a cosine or as a sine series).

The coefficients a_n and b_n, where $n = 1, 2, 3, \ldots$, are determinable by means of:

$$a_n = \frac{1}{\pi} \int_0^{2\pi} f(t) \cos nt \, dt$$

and

$$b_n = \frac{1}{\pi} \int_0^{2\pi} f(t) \sin nt \, dt.$$

These calculations are commonly made in the interval in which $f(t)$ is given. According to the form of $f(t)$ (more or less complicated) the number of the coefficients a and b necessary for the best fit of $f(t)$ could be different. But often such a decomposition is valid only in the examined interval; it is not possible to forecast the function of $f(t)$ beyond the analyzed interval. In such a circumstance, $f(t)$ can be supposed as:

$$f(t) = A + \sum_{n=1}^{\infty} h_n \cos \omega_n t$$

(ω_n not necessarily 1, 2, 3).

The various components could be separated by means of filtration—that is, by multiplying $f(t)$ with a special filter-function $M(\omega)$, of which the most common form is:

$$M(\omega) = \begin{cases} 0 & \omega \leqslant \omega_1 \\ 1 & \omega_1 < \omega < \omega_2 \\ 0 & \omega \geqslant \omega_2 \end{cases}$$

So, by developing the function $M(\omega)$ in a cosine Fourier series, one can obtain the coefficient of $M(\omega)$. To multiply $f(t)$ by M is equivalent to performing a linear combination among the several coefficients of $M(\omega)$ and the successive values of $f(t)$. In such a manner, $f(t)$ becomes filtered. The complete decomposition is obtained by testing $f(t)$ by different $M(\omega)$—that is, for $M(\omega) = 1$, related to different intervals $\omega_1 - \omega_2$. This method, or similar ones, is usefully employed in tide or seiche analyses or in long period level changes.

Fourier transform (Goldberg, 1961) is a kind of *integral transform*. When a function $f(t)$ is given, its Fourier transform is:

$$C(\gamma) = \frac{1}{\sqrt{2\pi}} \int_{-\infty}^{+\infty} f(t) \exp(i\gamma t)\, dt.$$

There are also simple sine or cosine transforms:

$$F(\gamma) = \sqrt{\frac{2}{\pi}} \int_0^\infty f(t) \sin \gamma t\, dt$$

and

$$G(\gamma) = \sqrt{\frac{2}{\pi}} \int_0^\infty f(t) \cos \gamma t\, dt.$$

Fourier transform could be usefully applied in wave analysis or in other phenomena where temporal evolution is not interesting, or not significant, but spectral composition is required.

If $f(t)$ represents a wave record extended from 0 to T, its autocorrelation function is:

$$R(\tau) = \lim_{T \to \infty} \frac{1}{T} \int_0^T f(t) f(t+\tau)\, dt.$$

Spectral density $E(\omega)$ of $f(t)$ can be considered as the Fourier cosine transform of $R(\tau)$:

$$E(\omega) = \frac{2}{\pi} \int_0^\infty R(\tau) \cos \omega \tau\, d\tau.$$

FERRUCCIO MOSETTI

Reference

Goldberg, R. R., 1961. *Fourier Transforms*. London: Cambridge University Press, 76p.

Cross-references: *Mathematical Models; Time Series; Waves; Wave Statistics.*

FROST RIVING

Frost riving, known also as *frost shattering*, *frost splitting*, *frost wedging*, and *congelifraction*, is the term applied to the periglacial process in which repeated freezing and thawing of water in pores, crevices, joints, and along bedding planes in rock lead to the disintegration of the surface. The weathered mantle of coarse, angular rock debris observed in many polar and alpine areas, and which in extreme cases may be several meters deep, has long been described as resulting from this process.

The pressure generated by freezing of water in rock is not likely to approach the theoretical maximum of 2100 kg cm^{-2} at $-22°$C, (French, 1976) but it has been shown to be adequate under controlled laboratory conditions to disrupt the rock, and both microsplitting (into flakes) and macrosplitting (into blocks) have been achieved. Factors that influence the intensity of frost riving include the quantity of moisture in the rock, the degree, frequency, and number of freeze-thaw cycles, as well as the physical properties of the rock (Potts, 1970). The most prominent riving is observed in thin-bedded sedimentary rocks, while it is rarely conspicuous in coarse crystalline, igneous rocks. Felsemeer (block fields), talus slopes, and rock glaciers may be an end product of strong frost shattering, but their origins are often complex. Frost riving is pronounced in the most susceptible rocks on lake shores in periglacial environments. Although it has been described in boulders on marine beaches, it is generally not strong in the intertidal zone and becomes prominent only above high water.

J. BRIAN BIRD

References

French, H. M., 1976. *The Periglacial Environment*. London: Longmans, 309p.
Potts, A. S., 1970. Frost action in rocks: Some experimental data, *Inst. British Geographers Trans.* 49, 109-124.

Cross-references: *Antarctic, Coastal Morphology; Aquafact; Arctic, Coastal Morphology; Chattermarks; High-Latitude Coasts; Ice along the Shore; Ice-Bordered Coasts; Shore Polygons; Weathering and Erosion, Mechanical.* Vol. III: *Frost Action.*

FROUDE NUMBER

The *Froude number*, an important parameter for the study of liquids moving in a free surface (e.g., surface wave motion), is defined as $F_r = v^2/gL$ or $F_r = \rho v^2 L^2 /\rho g L^3$. Because $\rho L^3 = m$ (mass), one can also have $F_r = \rho v^2 L^2/mg$. Since ρv^2 is the inertial force per unit area, the numerator of this last formula represents the total inertial force of the fluid. Obviously mg is the gravity force. Like other *dimensionless numbers* of fluid dynamics (see *Reynold's Number*), the Froude number gives a ratio between two forces acting with the fluid in motion. (These numbers are also employed in physical models of motion in similarity problems.)

In a channel where depth is L, $\sqrt{gL}$ represents the velocity of a long *gravity wave* coming into the channel. The first form of the Froude number given above can be interpreted as the ratio between the square of the fluid speed and the square of the speed of a coexisting *long wave*. If $F_r < 1$, the motion is defined as subcritical; if $F_r > 1$, the motion is supercritical (McCormack and Crane, 1973).

FERRUCCIO MOSETTI

Reference

McCormack, P. D., and Crane, L., 1973. *Physical Fluid Dynamics*. New York: Academic Press, 487p.

Cross-references: *Ferrel's Law; Fourier Analysis; Mathematical Models; Reynold's Number; Richardson's Number.*

FULCRUM EFFECT

In his classical study on the development of the northern Cape Cod, Massachusetts, shoreline, W. M. Davis (1896) was the first to point out that during the growth of the Provincelands spit and while the cliff is retreating, there must be a "neutral point of no change," referred to as a *fulcrum*. With continuing cliff erosion, this fulcrum would shift or migrate in the direction of the distal or downdrift end of the spit. The point where the present spit shoreline intersects the cliffed shoreline, while now a fulcrum, is not the original point of attachment of spit and cliff, which was seaward of the present shoreline. Based on present rates of cliff retreat, Davis estimated that the original shoreline was at the initial fulcrum point about 1200 B.P.

At later stages of development, the fulcrum, as it migrates, may be located along the spit shoreline at the intersection of an earlier spit shoreline with that of later spit development. In some cases, the term *fulcrum* is limited to indicate this intersection of a recurved spit with the next stage of spit development producing a "compound recurved spit" (Lobeck, 1939). Sandy Hook spit, at the north end of the New Jersey coastline, is such a spit developing in a manner analogous to the Provincelands spit (Davis, 1896). Development of a spit through a fulcrum point as an extension of a cliff or mainland shoreline under the action of longshore transport was considered by Davis as a *graded shoreline* and the spit as a *tangent bar or spit*. Neither of these terms is now commonly in use.

Fulcrum zones appear to be significant areas in short-term beach erosion shoreline changes. In the Cape Cod National Seashore, the southern fulcrum zone experienced increasing beach and dune erosion for several years (Fisher, 1976). Finally, during the record February 1978 blizzard, the area was overwashed and all frontal barrier dunes along 400 m of shoreline were completely destroyed. For Sandy Hook, the fulcrum zone is in an area the National Park Service considers a critical erosion zone of particular concern (Nordstrom, 1977).

A fulcrum point should not be confused with a shoreline *nodal point* or *zone*, which is an area in which the predominant direction of longshore transport changes (Coastal Engineering Research Center, 1973). These diverging longshore currents may be due to wave refraction or current eddies. Nodal zones may also result from seasonal changes in the longshore transport direction. For example, along the Cape Cod coast, the dominant (strongest) northeasterly winter winds produce a southerly sediment movement, while the prevailing (commonest) southerly summer winds move sediment to the north. The net annual sediment movement is as if there were a nodal zone along this shore.

JOHN J. FISHER

References

Coastal Engineering Research Center, 1973. *Shore Protection Manual*. Washington, D.C.: U.S. Army Corps of Engineers, 750p.

Davis, W. M., 1896. The outline of Cape Cod, *Am. Acad. Sci. Proc.* **31**, 303-332.

Fisher, J. J., 1976. Coastal geology and geomorphology of Cape Cod—A mesoscale view, *in* B. Cameron, ed., *New England Intercollegiate Geological Conference Guide Book*. Boston: Boston University Press, 239-255.

Lobeck, A. K., 1939. *Geomorphology*. New York: McGraw-Hill, 731p.

Nordstrom, K. F., 1977. *Coastal Geomorphology of New Jersey*, vol. 1. New Brunswick, N.J.: Rutgers University Center for Coastal and Environmental Studies Tech. Rept. 77-1, 39p.

Cross-references: *Bars; Beach in Plan View; Beach Processes; Coastal Erosion; Major Beach Features; Paleogeography of Coasts; Present-Day Shoreline Changes; Sediment Transport; Spits.*

FUNGI

Fungi are eukaryotic organisms, traditionally classified with plants, but in contemporary five kingdom systems (Whittaker, 1969) they are placed in the kingdom Fungi. In the great majority of fungi, the thallus is constructed of tubelike filaments termed hyphae, which en masse make up the mycelium. Hyphae may coalesce and fuse to form complex pseudotissues, such as those found in the fruiting bodies of *mushrooms*, or they may remain discrete as they ramify within the substrate upon which the fungus grows. Unicellular forms, such as the *yeasts*, are also found; and in some primitive aquatic forms, the thallus consists of little more than a reproductive structure with a few hypha-like attachment rhizoids.

Nutritionally the fungi are heterotrophic for carbon and dependent on preformed organic matter. Thus they are found growing on dead organic matter as saprobes or as parasites on living hosts. Many fungi are severe pathogens of plants, causing extensive crop damage, while others are pathogenic for man and animals. Some, such as the mycorrhizal and lichen fungi, enter into highly balanced symbiotic relationships, in which the fungus contributes to the nutritional needs of the host while at the same time extracting its own sustenance. The great majority of fungi are saprobes, performing in the ecosystem the essential role of the decomposer. Because of their ability to convert plant and animal remains into detritus and microbial biomass, fungi and other decomposers comprise an important and essential link in the detrital food chains of aquatic habitats.

An important consideration in the nutrition of fungi is the fact that digestion occurs outside the thallus. Enzymes secreted by the fungus into the substrate result in the breakdown of complex organic matter, permitting the fungus to take up simpler molecules by absorption. This mode of nutrition has been aptly termed *osmotrophic*.

Fungi are classified primarily on the basis of their sexual states, although asexual reproduction is common and in many species may actually be the most important method of propagation and dispersal. The critical events of sexual process (nuclear fusion and meiosis) occur in the specialized sexual structures that characterize each major group, and the process culminates in the production of haploid spores, minute masses of cytoplasm surrounded by a cell wall or membrane. Asexual reproduction may occur by several means, including fragmentation of the hyphae, budding, fission, or as is most common, by the formation of spores or conidia. Formation of asexual spores is preceded by mitotic divisions of the nuclei, and it is convenient to refer to them as mitospores to distinguish them from the sexually produced spores, which are then termed *meiospores*.

An outline of the major groups of the fungi, as given by Ainsworth (1973) follows:

KINGDOM FUNGI

Division Myxomycota (plasmodial or pseudoplasmodial forms)
    Class Acrasiomycetes
    Class Myxomycetes
    Class Plasmodiophoromycetes
Also included: Order Labyrinthuales

Division Eumycota
  Subdivision Mastigomycotina (flagellated cells present in the life cycle)
    Class Chytridiomycetes (zoospore with single posterior flagellum)
    Class Hyphochytridiomycetes (zoospore with single anterior flagellum)
    Class Oomycetes (zoospores biflagellate)
  Subdivision Zygomycotina (Zygomycetes)
    Class Zygomycetes
    Class Trichomycetes
  Subdivision Ascomycotina (Ascomycetes)
    Class Hemiascomycetes
    Class Loculoascomycetes
    Class Plectomycetes
    Class Laboulbeniomycetes
    Class Pyrenomycetes
    Class Discomycetes
  Subdivision Basidiomycotina (Basidiomycetes)
    Class Teliomycetes
    Class Hymenomycetes
    Class Gasteromycetes
  Subdivision Deuteromycotina (Fungi Imperfecti)
    Class Blastomycetes
    Class Hyphomycetes
    Class Coelomycetes

It should be noted that Olive (1975) presents a radically different arrangement for those groups included in the Myxomycota, placing them in the kingdom Protista under the subphylum Mycetozoa. The order Labyrinthuales, a group of considerable interest to aquatic biologists, are raised by Olive to the level of a subphylum. Olive's classification is well conceived, and merits thoughtful consideration.

In Ainsworth's classification, the groups Mastigomycotina and Zygomycotina constitute the Phycomycetes of earlier authors. The order Thraustochytriales, a group of marine biflagellates formerly assigned to the Oomycetes, are placed by Olive (1975) with the Labyrinthuales in the subphylum Labyrinthulina of the Mycetozoa, on the basis of their similar ultrastructure. Yeasts and yeastlike fungi are assigned to three widely separated classes: those forming ascospores (ascosporogenous) are placed in the class Hemiascomycetes; those forming teliospores (basidiomycetous) are placed in the class Teliomycetes; those lacking a sexual state are assigned to the class Blastomycetes of the subdivision Deuteromycotina.

By conservative estimate, there are between 50,000 and 100,000 species of fungi, with many species probably still awaiting description. Of these, about 300 species are known to be restricted to the marine environment. Certain fungi that are characteristic inhabitants of soil and litter in terrestrial habitats are regularly isolated from bottom sediments and decaying litter collected in marine sites, but because it has not been shown that this group carries out an active role in marine sites, they are usually excluded from lists of *marine fungi*. Recognizing the problems of rigidly defining a marine fungus, Kohlmeyer (1974) proposed the following definition: "*Obligate marine fungi* are those that grow and sporulate exclusively in a marine or estuarine (brackish water) habitat; *facultative marine* are fungi from freshwater or terrestrial areas able to grow also in the natural marine environment."

Species defined as marine are distributed in the major taxonomic groups approximately as follows: Phycomycetes (includes subdivision Mastigomycotina and Zygomycotina and the order Thraustochytriales), 61; Subdivision Ascomycotina, 135; Subdivision Deuteromycotina, 51; Basidiomycotina (filamentous forms), 2; Yeasts (includes ascomycetous, basidiomycetous and asporogenous forms), 67.

Substrates in the marine or aquatic environment available for colonization by fungi include wood and other plant residues originating from terrestrial environments; algae; animals; dissolved organic matter that are produced in the habitat; and soil, sediments, and mud contiguous to aquatic environments (see Table 1). Phycomycetous fungi attack primarily plankton, algae, and animals, while the Ascomycetes and Deuteromycetes are the major groups involved in the decomposition of wood and other cellulosic plant residues. Yeasts occur primarily as free-floating organisms, although they have been reported to be associated in high numbers with algae (Seshadri and Sieburth, 1975) and with the salt marsh cord-grass, *Spartina alterniflora* (Meyers, 1974.)

The study of marine and estuarine fungi is still a relatively young science. Although pioneer studies were completed before the beginning of the present century, the first definitive work in the United States was the study by Baarghorn and Linder, completed in 1944. By 1961 sufficient information had accumulated to enable Johnson and Sparrow to compile a comprehensive monograph on marine and estuarine fungi. This work provided a great stimulus to the field, and laid a sound foundation for research. The past two decades have seen substantial activity, and the thrust of research has moved from the descriptive phase to one of ecological and biological investigation. Good taxonomic treatments of marine fungi are provided by Johnson and Sparrow (1961) and by Kohlmeyer and Kohlmeyer (1971), while Jones (1975) and Hughes (1975) provide good reviews of the biological and ecological aspects. Kohlmeyer and Kohlmeyer (1979) provide an excellent summation of current knowledge of marine fungi.

ROGER D. GOOS

References

Ainsworth, G. C., 1973. Introduction and keys to higher taxa *in* G. C. Ainsworth and A. S. Sussman, eds., *The Fungi: An Advanced Treatise*, vol. IV-A. New York: Academic Press, 1-7.

Baarghorn, E. S., and Linder, D. H., 1944. Marine fungi: Their taxonomy and biology, *Farlowia* 1, 459-467.

Hughes, G. C., 1975. Studies of fungi in oceans and estuaries since 1961. I. Lignicolous, caulicolous and foliicolous species, *Oceanog. Marine Biol. Annual Rev.* **12,** 69-180.

Johnson, T. W., Jr., and Sparrow, F. K., Jr., 1961. *Fungi in Oceans and Estuaries*. Weinheim: J. Cramer, 668p.

Jones, E. B. G., 1975. *Recent Advances in Aquatic Mycology*. New York: John Wiley & Sons, 749p.

Kohlmeyer, J., 1974. On the definition and taxonomy of higher marine fungi, *Inst. Meeresforsch. Bremerhaven Veroffentl.* 5(suppl.), 263-286.

Kohlmeyer, J., and Kohlmeyer, E., 1971. *Synoptic Plates of Higher Marine Fungi*. Lehrte: J. Cramer, 87p.

Kohlmeyer, J., and Kohlmeyer, E., 1979. *Marine Mycology*. New York: Academic Press, 690p.

TABLE 1. Occurrence of Major Groups of Fungi on Selected Substrates (from Jones, 1975).

	Phycomycetes	Ascomycetes	Basidiomycetes	Deuteromycetes
Lignicolous (wood)	3	91	2	36
Free-floating	10	50 (primarily yeasts)	8 (yeasts)	-
Sediments, soil, mud	35	20	-	180
Algae	70	50	-	14
Animals	19	5	-	5
Decaying leaves of terrestrial plants	3	2	-	9
Decaying leaves of marine plants	2	6	1	30

Meyers, S. P., 1974. Contribution of fungi to biodegradation of *Spartina* and other brackish marshland vegetation, *Inst. Meeresforsch. Bremerhaven Veroffentl.* 5(suppl.), 357-375.

Olive, S. L., 1975. *The Mycetozoans.* New York: Academic Press, 293p.

Seshadri, R., and Sieburth, J., 1975. Seaweeds as a reservoir of *Candida* yeasts in nearshore waters, *Marine Biol.* 30, 105-117.

Whittaker, R. H., 1969. New concepts of kingdoms of organisms, *Science* 163, 150-160.

Cross-references: *Biotic Zonation; Coastal Fauna; Organism-Sediment Relationship; Soils.*

G

GEOGRAPHIC TERMINOLOGY

In the course of compiling this volume, a great many coastal terms were turned up—too many for each to have a separate entry—that were essentially geographic or geomorphic in nature. A number of these have therefore been collected here under the heading of *Geographic Terminology*. While they resist rigid classification, these terms are grouped into such categories as beaches, bays, channels and inlets, reefs, islands, sedimentary deposition, regional. Where possible, the country where used, language, or etymology is given. Chief sources for this compilation have been the *Glossary of Geology* (Gary, McAfee, and Wolf, 1972) and the *Shore Protection Manual* (Coastal Engineering Research Center, 1973), although some condensation or amplification has occasionally been made.

Beaches

Various terms commonly used to denote beaches are the Scottish *air* (etymol.: Old Norse *eyrr*, "gravelly bank"); Italian *lido*, for barrier beach or bathing beach; and French *plage*. The latter two generally bear the connotation of being located at resort areas. A special case is the *false beach* (Veatch and Humphrys, 1966), a subaerial bar situated just offshore.

Two types of beaches that are crescent-shaped and convex toward the sea are *crescent beaches* and *pocket beaches*. Both are located between prominences of some sort, generally along hilly or mountainous coasts, that hinder shore drift to some degree. The term crescent beach implies a more open shape in plan, with limited sediment transport into or out of the feature; while the term pocket beach implies a deeper plan configuration, with virtually no sediment transport to or from the adjacent coast (see *cove beach, bayhead beach*, below).

Location designation is seen in such terms as: *bayside beach*, situated along the side of a bay; *bayhead beach*, at the landward end of a bay; *cove beach*, in a cove; and *headland beach*, at the base of a cliffed headland.

Bays and Embayments

A *bay, embayment*, or *sinus*, is an extension of the sea into a recess or indentation of the coast. If the recess is formed by a long, gentle curve, the term *bight* may be applied. A small embayment is referred to as a *cove*. Other terms for small bays or inlets are *fleet* (England), *hole, hope* (Scotland), basin, or haven—the latter two implying boat anchorage capacity. If long, narrow, and extending into the land, the embayment may be called an *arm, reach*, or *flow* (Scottish). In the western United States a small coastal inlet, used in loading lumber, is called a *doghole* (Gove, 1967). Still another term for a small inlet (Stamp, 1966) or sandy cove in a cliff (Robson and Nance, 1959) is the English *zawn*.

Special cases are: *front bay*, a large, shallow, irregular bay, opening to the sea through a barrier island inlet; *back bay*, a smaller type of front bay, into which coastal streams flow; *backwater*, an arm of the sea or interconnected lagoons, behind a strip of land and parallel to the coast, but opening to the sea over a sill or bar; and *calanque* (French), a small cove or inlet, formed by sea level rise inundation of a former valley in limestone terrain.

The names of some embayments are descriptive of their configuration. Thus, a *funnel sea* is a bay that is wide and deep at its mouth and narrow and shallow at its head; a *hooked bay* is like a bight with a headland at one end; a *bottleneck bay* is narrowed at its entrance by other than barrier islands; a *closed bay* has a narrow pass tying it indirectly to the sea; and a *branching bay* is one having a dendritic pattern. The Scottish terms *geo, gia, gja*, and *goe* (etymol.: Old Norse *gja*, "chasm") connote a long, narrow, deep inlet between parallel rocky cliffs, eroded by marine agents along weak planes—that is, joints. The term *gully* is used similarly; and compared with all of these, an *open bay* is like a large bight, between two headlands or capes, in which the wave environment at its center is similar to that farther up and down the open coast.

On a larger scale than a bay, and partially enclosed by an extension of the land, is the body of water known as a *gulf*. As compared with this, a *sound* is a large, long body of water between two others or between the mainland and an island or peninsula.

A *road* or *roadstead* is a semiprotected anchorage site behind an island, barrier, or reef—

none of which are, strictly speaking, bays—or within a shallow embayment.

Channels and Inlets

Straits and *channels* are elongate connections between larger bodies of water; but the term channel also denotes the deeper course of the water moving in a bay or strait. A *pass* is a narrow opening between two islands, a channel through a barrier reef, barrier island, or delta; or a connection between a bay and the sea. If the channel is kept open for navigation, and so indicated, it may be called a *fairway*.

Other variations are: the Scottish term *kyle* for a narrow channel between one island and another or the mainland, also called a *gap* in the latter case; *gat* or *gate*, a channel between islands, shoals, promontories, or cliffs; *canal*, a long, narrow arm of the sea or channel, of uniform width, connecting two larger bodies of water; and *gut*, a narrow passage similar to the above.

Channel terminology related in some way to tidal phenomena includes: *slough*, a sluggish body of water in a coastal marshland; *tidal creek*, a small tidal estuary or inlet; *gut*, a creek in tide flat or marsh; and *race* or *euripus*, a constricted channel where the tidal flow is fast and turbulent.

Five other terms should be mentioned. Another usage of the term *gully* is to denote a small channel cut into a mud flat below the high water mark (Schieferdecker, 1959). A *cul-de-sac*, in its coastal context, is the single small opening of an inlet. A *sluiceway* is an overflow channel. A ditch cut through the Louisiana marsh to provide access for trapping and fishing is known locally as a *trainasse* (Davis, 1976). The Malaysian term *sungei* or *soengei* refers to a marine channel, caused by a Quaternary antecedent stream, across the Aroe Islands off the coast of New Guinea.

Reefs

The various terms associated with coral reefs seem to be almost limitless. A few regional terms that do not appear elsewhere in this work follow.

In Brazil, two such terms derived from the Portuguese are: *chapeiro* (etymol.: *chapeirao*, "broad-brimmed hat"), a tower or mushroom-shaped isolated coral reef occurring in groups; and *albrolhos*, a widely spreading, near-surface, barrier reef of mushroom shape. From the Polynesian, *motu* is a small vegetated coral island, and *makatea* is the raised rim of a coral reef or a wide uplifted coral reef around an island.

In Morocco, the Arabic term *kess-kess* is used to denote a reef knoll separated by erosion. Many small atolls constituting a large atoll are called an *atollon* in the Maldive Islands area.

Two somewhat similar, but distinctly different, features are the *karang* (etymol.: Malay, "reef," "coral reef"), old, fringing-reef material now forming an emerged terrace (Termier and Termier, 1963); and *trottoir* (etymol.: French, "sidewalk"), coral, or coral over rock, forming a narrow border between the sea and the shoreline, analogous to a sidewalk between street and building.

Islands

Although not an island in the strictest sense, a *land-tied island* is an island connected by a tombolo to the mainland; a *tied island* or *tombolo island* is similarly connected to another island or to the mainland. Schofield (1920) described a *dumbbell* or *dumbbell island* as two areas of land higher than the subaerial narrow, sandy isthmus that connects them. In the same *non-island* category are *floating islands*, free-floating vegetation mats or masses.

An island group, and the sea area itself, may be called an *archipelago*. The term *havsband* is Swedish for the outskirts of an archipelago (Eric Olausson, personal communication) or skerry-guard (qv.) (Stamp, 1961).

A stack or chimney or resistant igneous rock is called a *stac* in the St. Kilda Islands off Scotland; an isolated, offshore rock mass in the region of Scotland is called a *carr* or *carrig*.

Lakes, Lagoons, and Marshes

Coastal lakes are formed by shore processes, and include a great many different features, some of which overlap with those covered previously in the section on Bays and Embayments. Special cases are: *beach pools*, small temporal pools between or behind beaches or beach ridges; *fleet*, a slightly saline lagoon behind a long, coast-parallel, shingle or sand bank (Monkhouse, 1965); and *étang* (French), the ponding among sand dunes and behind beach material of inland drainage. An artificially formed (it is believed by former removal of peat) wide, shallow body of freshwater near a river estuary is the British *broad*.

Terms used in connection with marshy areas are: *telmaro* (etymol.: Greek, *telma*) for a river crossing a peat marsh (Veatch and Humphrys, 1966); *corcagh* or *corcass* (Ireland) for a low marsh along a tidal riverbank; and the Italian *maremma* for a low, marshy coastal area.

Related to the whole salt marsh (qv.) regime are such terms as: *salt pond*, denoting the body of water in a coastal marsh; *salt pan*, the shallow depression and accumulated water between drainage channels, on the marsh surface; *marine salina*, a saline body of water behind a sand or gravel bank, on an arid coast, with only a seawater contribution; and *salting* (British), the grassy, higher portion of a salt marsh lying above all but spring tide inundation. The latter term also has a more limited use in Great Britain to denote intertidal land that is not a salt marsh.

Sedimentary Deposits

Similar to the bayhead beach, the *bayhead bar*, *bayhead barrier*, and *bayhead delta* describe such features lying at the upper reaches of a bay. A *bay delta* may be located elsewhere in a bay.

A *bank* or *embankment* is a narrow deposit of material built out into a body of water, above or below the water surface, by shore processes (see Bars and Spits); the term *shoal* is used to indicate the shallow condition or the accompanying loose sediment deposit.

Spit- or bar-like terms include: *pendant terrace*, for a strip of sand connecting an offshore rock to the mainland; and *barrier bank, trailing spit*, or *tail*, for a barrier or bar in the lee of a skerry (q.v.) or small island. A coastal dune (q.v.) is called a *médano* in Spanish, and formerly was referred to as a *down*.

Special terminology for deltaic or tidal features range from *middle ground* and *dwip* (Strickland, 1940), both akin to a tidal delta (q.v.) to *char, chur* (etymol.: Hindu) or *diara*, a bank of sand and silt developed by, and exposed after, the flood stage of a deltaic river.

Cliffs and Headlands

A *headland* or *promontory* is defined as a high, usually rocky, projection of the land into the sea. Other synonymous terms are: *nab, nase, ness, naze, nore* (all British), *mull* (Scottish), *peak, bold coast, beak, cobb, reach, nook*, and, in special cases, *cliffed headland*. The British *brig* or *brigg* refers to a near-intertidal rocky headland (Stamp, 1961). In the geographic sense, a *horn* is a body of either water or land with that distinctive shape.

If more extensive than a promontory or headland, and whether of low or high relief, the seaward-projecting landform may be called a *foreland* (see Cuspate Foreland) and, if rounded, a *cape*. Low, tapering, and somewhat smaller than a cape is the *point* or *tongue*. A *strand plain* is that form of foreland prograded seaward by shore processes (Cotton, 1958).

Where wasted cliffs are present, the *cliff line* is defined as an imaginary line following the cliff base. In Sweden and Denmark, a steep cliff along the Baltic Sea is called a *klint*. In France, an old, low, marine cliff, in an emerging region and reunited with the sea, is referred to as a *falaise*. Coastal *hanging valleys* are those in which rapid cliff retreat leaves the mouths of streams elevated at the shore.

Peninsulas, sometimes called *half-islands*, are elongated stretches of land extending out into a body of water. The narrow isthmus that sometimes connects a peninsula to the mainland is called a *neck*, the term being applied also to a narrow cape, promontory, or *hook*. Peninsula forms that are nearly islands are: the *presque isle* (etymol.: French, *almost*), a lacustrine peninsula connected to the land by a low neck or isthmus just above water level; and the *bridge islet*, which is an island at high tide and a peninsula at low tide.

Regional

Shores may be characterized by their: *composition*, a *hard coast* being made up of sand, gravel, and cobbles, a *soft coast* of marsh vegetation, peat, muck, mud, or soft marl; *wind climate*, a *lee shore* being protected from waves, since the wind direction is from land to sea, and a *weather shore* (also called an *open coast*), exposed to the full brunt of onshore winds and waves; and *direction, upcoast* being usually to the north, and *down coast* to the south (Coastal Engineering Research Center, 1973).

Over the years many terms have been used to describe various types of shorelines and coastlines. One term, virtually unused at present, is the *cheirographic* (Swayne, 1956) or *cheiragratic* (Stamp, 1961) *coast*, meaning an irregular or *broken shoreline* with many headlands, islands, and peninsulas in a folded or faulted region. More limited are: the *fold coast*, where the irregularities are due to folded structure alone; and the *crenulate shoreline*, highly irregular because of differential erosion during an early stage of submergence. Submergence of a concordant coast, where mountain ranges and intermontaine valleys are parallel to the coast, develops a *Dalmatian coastline*, with offshore islands and inner straits on sounds along the new coast.

The Scottish term *carse* is used for a raised beach or marine terrace, now low, level, and fertile, along an estuary. Somewhat more variable is the term *wash*, denoting a marsh or bog, the shallowest part of an embayment, or the intertidal area of banks of mud or sand.

Cultural Terms

Two terms related to human occupation are included here in closing. An embankment, upon which there is a roadway built along the shore, is called a *bund*. The term *riviera* is used to designate a coastal tourist resort, having a warm, sunny climate, and attractive beaches. Thus, while the term is well known as applied to such areas in France and Italy, the resort area along the Black Sea coast near Sochi is known as the *Caucasian Riviera*.

MAURICE L. SCHWARTZ

References

Coastal Engineering Research Center, 1973. *Shore Protection Manual.* Washington, D.C.: U.S. Army Corps of Engineers, 750p.
Cotton, C. A., 1958. *Geomorphology: An Introduction To The Study of Landforms.* Christchurch, N.Z.: Whitcombe and Tombs, 505p.
Davis, D. W., 1976. Trainasse, *Assoc. Am. Geographers Annals* 66, 349-359.
Gary, M.; McAfee, R. Jr.; and Wolf, C. L., eds., 1972. *Glossary of Geology.* Washington, D.C.: American Geological Institute, 805p.
Gove, P. B., ed., 1967. *Webster's Third New International Dictionary of the English Language, Unabridged.* Springfield, Mass.: Merriam, 2662p.
Monkhouse, F. J., 1965. *A Dictionary of Geography.* Chicago: Aldine, 344p.
Robson, J., and Nance, R. M., 1959. Geological terms used in S.W. England, *Royal Geol. Soc. Cornwall Trans. 1955-1956* 19, 33-41.
Schieferdecker, A. A. G., ed., 1959. *Geological Nomenclature.* Gorinchem: Royal Geological and Mining Society of the Netherlands, 523p.
Schofield, W., 1920. Dumb-bell islands and peninsulas on the coast of South China, *Liverpool Geol. Soc. Proc.* 13, 45-51.
Stamp, L. D., ed., 1961. *A Glossary of Geographical Terms.* New York: John Wiley & Sons, 539p.
Stamp, L. D., ed., 1966. *A Glossary of Geographical Terms.* London: Longmans, Green, 539p.
Strickland, C., 1940. *Deltaic Formation with Special Reference to The Hydrographic Processes of the Ganges and the Brahmaputra.* New York: Longmans, Green, 157p.
Swayne, J. C., 1956. *A Concise Glossary of Geographical Terms.* London: George Philip & Son, 164p.
Termier, H., and Termier, G., 1963. *Erosion and Sedimentation.* New York: Van Nostrand, 433p.
Veatch, J. O., and Humphrys, C. R., 1966. *Water and Water Use Terminology.* Kaukauna, Wis.: Thomas Co., 381p.

Cross-references: *Aktology; Bars; Coastal Dunes and Eolian Sedimentation; Coral Reef Coasts; Cuspate Forelands; Deltas; Salt Marsh; Skerry and Skerry Guard; Spits; Tidal Delta; Tombolo.*

GEOMORPHIC-CYCLE THEORY

This theory concerns the progressive development of a shoreline in terms similar to the fluvial cycle concept. It was explained by Davis (1896) as a change in the coastal configuration from a youthful stage, where excess wave energy erodes the shore, to a graded condition of maturity where energy and work are in *equilibrium*.

Johnson (1919) redefined the cycle concept, discussing the development of *submerged* and *emerged shorelines*. The initial, irregular submerged coast is subjected to vigorous erosion and concomitant deposition during youth. Cliffs, benches, and terraces result from the erosion, while deposition occurs at the heads of bays as spits and bars, and within the bays as beaches. In this stage, headlands act as barriers to sand transport. In the submature stage, bays are nearly shut off by bars and spits, and headlands are significantly eroded back. Maturity is reached when the coast has both a seaward *profile of equilibrium* and a straight alongshore view, with movement of sediment along the whole coast.

On an emergent shoreline, Johnson perceived that the initial gentle offshore profile would prevent erosion. In youth, an offshore bar appears, and is built up as a barrier island, with tidal inlets and a lagoon on the landward side. In time, the barrier island moves landward, and lagoons and marshes are filled. The shoreline is then mature. Strahler (1969) gives excellent diagrams illustrating Johnson's cycle of emergent and submergent shores.

Putnam (1937) added the example of emergent, steep-sloped coasts, as are found in the tectonically active regions of California. Here no barrier islands develop in youth. Instead, the youthful stage is characterized by uplifted cliffs and terraces that are being incised by streams. The eroding rivers deposit the materials removed at their mouths, building fans. The sea cliffs are progressively eroded back, as are the alluvial fans developed along the sea cliffs. Maturity is reached when the shore is retrograded to the position it had before uplift, and the upraised terrace is completely removed.

The theoretical cycle of development, as outlined by these scientists, is seldom realized because of the dynamic nature of the earth and changing environmental conditions along the coast. Seldom are shorelines simply emergent or submergent. Moreover, Johnson's theory that barrier islands are related to youthful emergent coasts is no longer considered valid (Fisher, 1967). So, although the cycle concept is a good pedagogic device, it is not now applied seriously in coastal studies.

MARIE MORISAWA

References

Davis, W. M., 1896. The outline of Cape Cod, *Am. Acad. Arts Sci. Proc.* 31, 303-332.
Fisher, J. J., 1967. Origin of barrier island chain shore-

lines, Middle Atlantic states, *Geol. Soc. America Spec. Paper 115,* 66-67.
Johnson, D. W., 1919. *Shore Processes and Shoreline Development.* New York: John Wiley & Sons, 584p.
Putnam, W. C., 1937. The marine cycle of erosion for a steeply sloping shoreline of emergence, *Jour. Geology* 45, 844-850.
Strahler, A. N., 1969. *Physical Geography.* New York: John Wiley & Sons, 733p.

Cross-references: *Classification; Coastal Morphology, History of; Emergence and Emerged Shoreline; Equilibrium Shoreline; Profile of Equilibrium; Submergence and Submerged Shoreline.* Vol. III: *Geomorphology.*

GHYBEN-HERZBERG RATIO

Fresh ground water floats on denser, salty ground water within islands and along coasts consisting of permeable earth materials (Fig. 1). This principle was developed independently by Ghyben (1889) and Herzberg (1901). The column of freshwater $(a+b)$ is balanced by a column of saltwater (equal to b). The Ghyben-Herzberg ratio or equation is thus: $bg_s = (a+b)g_f$, where a is elevation of water table above sea level, b is the depth of freshwater below sea level, and g_s and g_f are the specific gravity of seawater and freshwater, respectively. If $g_f = 1$ and $g_s = 1.027$, then $b = 37a$. The relationship is dynamic, owing to replenishment, evaporation, and discharge of freshwater (De Wiest, 1965). Any lowering of the water table will result in a much greater intrusion of saltwater. Typically, if the water table (a) is lowered by one meter, then the freshwater lens below sea level (b) is reduced by about 37 m (Tolman, 1937).

MICHAEL R. PLOESSEL

References

De Wiest, R. J. M., 1965. *Geohydrology.* New York: John Wiley & Sons, 366p.
Ghyben, B. W., 1889. Nota in verband met de voorgenomen put boring nabij, *Amsterdam Kononkl. Inst. Ing. Tijdschr. 1888-1889,* 8-22.
Herzberg, B., 1901. Die Wasserversorgung einiger Nordseebäder, *Jour. Gasbeleuchtung Wasserversorg.* 44, 815-844.
Tolman, C. F., 1937. *Ground Water.* New York: McGraw-Hill, 593p.

Cross-references: *Coastal Engineering, Research Methods; Land Reclamation; Protection of Coasts; Water Table.* Vol. XVIII: *Geohydrology.*

GLACIATED COASTS

On certain coastlines in the world, the effects of glaciation are dominant and a distinctive environment results from the juxtaposition of glacial landforms and the sea.

Coastlines Dominated by Glacial Erosion

The features encountered on such coasts vary according to bedrock geology and preglacial

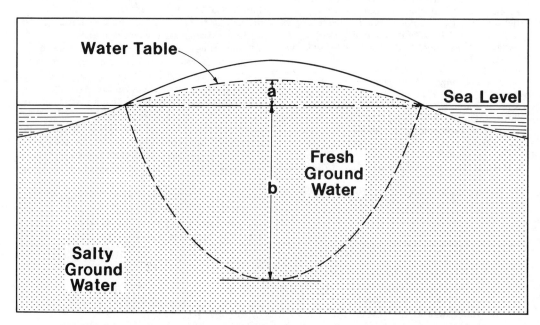

FIGURE 1. Elevation of water table above sea level (a) and depth of freshwater below sea level (b).

relief. The latter played a major part in affecting patterns of ice movement and the type of glacierization (Embleton and King, 1975). *Fjords* (q.v.), associated with dissected hardrock coasts, represent glacial troughs partly occupied by the sea, for example, in western Norway. The glaciation of lowland hard-rock coasts has given rise to a *fiard* (q.v.) coastline, as in southern Sweden. Other features include the partially submerged platforms known as *strandflats* (q.v.), with their accompanying rocky archipelagoes known as the *skerry guard* (q.v.), in whose formation both marine and glacial action may have been involved. Softrock coasts subjected to glacial erosion, as in Denmark or north Germany, bear less distinctive features attributable to glaciation, although the inlets known as *Förden* (q.v.) have been held to represent erosion by subglacial meltwater.

Coastlines Dominated by Glacial Deposition

Along most coastlines built out of glacial deposits, the latter simply provide the material on which marine processes act (Guilcher, 1958). For instance, along much of the coast of East Anglia in Britain or the coast of New England north of Cape Cod, glacial drift has been cut into unstable low cliffs or moved alongshore to form spits and bars. In southern Iceland, bars backed by lagoons fringe the fluvioglacial outwash plains. On a few coasts, the glacial inheritance is more direct: for example, in northeast Ireland, partially submerged drumlins form a distinctive coastal landscape.

Effect of Land- and Sea-level Changes

Such changes comprise: eustatic sea-level change in the postglacial period, causing submergence both on stable coasts and on those where the rate of eustatic rise has outpaced any local uplift; and glacioisostatic rebound since the beginning of deglaciation. Features of emergence and submergence are therefore often intimately associated with glaciated coasts (Valentin, 1952).

CLIFFORD EMBLETON

References

Embleton, C., and King, C. A. M., 1975. *Glacial Geomorphology.* London: Edward Arnold, 573p.

Guilcher, A., 1958. *Coastal and Submarine Morphology.* London: Methuen & Co., 274p.

Valentin, H., 1952. *Die Küsten der Erde,* Petermanns Geog. Mitt. Erg. 246. Gotha: Justus Perthes, 118p.

Cross-references: *Emergence and Emerged Shoreline; Fiard, Fjärd; Fiord, Fjord; Firth; Förde; Ice along the Shore; Sea Level Changes; Sea Level Curves; Submergence and Submerged Shoreline.* Vol. III: *Fjärd, Fjord.*

GLACIEL

Glaciel, a French term created by Hamelin (1959), refers to all processes, sediments, and features related to the action of drift ice and icebergs in fluvial, lacustrine, and marine (including littoral and estuarine) environments. The term refers also to glaciological, hydrological, morphological, sedimentological, biological, and human aspects of drift or floating ice.

The term is widely used in French and in other languages. During the past decade, numerous studies referring to glaciel processes, sediments, and features have been published (Dionne, 1974), and many expressions and composite words have been derived from the term glaciel (Dionne, 1972). The first international symposium entirely devoted to the geological aspects of glaciel was held in Québec City in 1974.

The oldest known reference to the geological effect of glaciel appeared in scientific literature more than 150 years ago (Petros, 1822), and referred to the displacement of boulders by lake ice in Connecticut. In 1854, Lyell emphasized the role of drift ice in sedimentation and made many valuable observations in North America, but he went too far and confused glacial sediments with glaciel sediments, for which he was severely criticized. For this reason the geological action of drift ice became a controversial subject and was largely ignored for more than a century.

Glaciel processes are worldwide and of great importance locally. Today ice action occurs along about 145,000 km of marine littoral in the Northern Hemisphere, and along about 30,000 km in the Southern Hemisphere. Many recent studies have shown the morphosedimentological importance of glaciel in Recent and during Pleistocene times. Because it is a process

FIGURE 1. Glaciel sediments left by an ice floe upon melting at the surface of a mud and clay tidal flat along the Saguenay Fjord, Québec (photo: J. C. Dionne).

of cold regions from high to middle latitudes, it should be classified with periglacial rather than with glacial processes, although icebergs are glacier ice. Conditions being equal, it seems that glacial processes are more active today in subarctic regions of North America, even though the geological action of icebergs is more important at high latitudes.

Further relevant discussion of this subject may be found in Dionne (1968, 1970, 1976) and Tarr (1897).

JEAN-CLAUDE DIONNE

References

Dionne, J. C., 1968. Morphologie et sédimentologie glacielles, littoral sud du Saint-Laurent, *Zeitschr. Geomorphologie* 7(suppl.), 56–84.
Dionne, J. C., 1970. *Aspects morpho-sédimentologiques du glaciel, en particulier des côtes du Saint-Laurent.* Paris: Universite Paris-Sorbonne, Ph.D. thesis, 412p. (Also Quebec: Environment Canada, Inform. Rept. Q-F-X-9, 324p.)
Dionne, J. C., 1972. *Vocabulaire du glaciel (Drift Ice Terminology).* Québec: Environment Canada, Inform. Rept. Q-F-X-34, 47p.
Dionne, J. C., 1974. *Bibliographie annotée sur les aspects géologiques du glaciel.* Québec: Environment Canada, Inform. Rept. LAU-X-9, 122p.
Dionne, J. C., ed., 1976. Le Glaciel (Proceedings of the First International Symposium on the Geological Action of Drift Ice), *Rev. Géographie Montreal* 30, 236p.
Hamelin, L. E., 1959. Dictionnaire français-anglais des glaces flottantes. Québec: Université Laval, Institute Geographie, unpub. manuscript, 83p.
Lyell, C., 1854. *Principles of Geology.* New York: Appleton, 834p.
"Petros," 1822. On certain rocks supposed to move without any apparent cause, *Am. Jour. Sci.* 5, 34–37.
Tarr, R. S., 1897. The Arctic sea ice as a geological agent, *Am. Jour. Sci.,* ser. 4, 3, 223–229.

Cross-references: *Ballast; High-Latitude Coasts; Ice along the Shore; Ice-Bordered Coasts; Ice Foot; Kaimoo; Monroes; Mud Volcanoes; Shore Polygons; Stone Packing.*

GLACIOEUSTATIC—See SEA LEVEL CHANGES

GLOBAL TECTONICS

The form and character of coastlines are strongly influenced by the gelogical behavior of the earth's crust. Highly contrasted, regional variations are observed in the patterns of this behavior. Some regions are relatively stable; others show a rising or sinking tendency (Davies, 1972).

The direction of crustal motion, up (positive) or down (negative), is expressed in terms of *relative* to sea level; thus one refers to the former as an *emergent* or *emerging* coast and the latter as *submergent, submerging,* or *drowned* coast. There is both a terminological and a practical problem here because sea level is also known to be constantly changing; during the past 6000 years it has oscillated through a range of ±3–4 m about the present mean with a rough periodicity of around 500–1500 years. From about 17,000 to 6000 years ago there was a progressive (but discontinuous) eustatic rise of sea level of some 135 m, which means that the more stable coastlines appear to be drowned. However, in some areas the positive motion of the earth's crust during this interval has outweighed the eustatic effect, and the coast is an emergent one (e.g., in Hudson Bay or in the Baltic). On the other hand, in regions of crustal subsidence (e.g., the Mississippi delta or the southern North Sea), the rising eustatic effect is amplified and the drowned features are even more prominent (Fairbridge, 1961).

Coastal features that have developed over the course of the past 10,000 or 20,000 years are long-term as compared with those dating from 100–200 years, but are intermediate as compared with long-term global tectonic events, which have cycles of the order of 1–100 million years. It is these global events that condition the deep-seated skeletal framework of a coastline.

Theory of Plate Tectonics

In the decade beginning about 1960, there emerged an integration of geological and geophysical data that provide us with a comprehensive theory of global dynamics. It had begun with the theory of continental drift of Wegener (1915). Worldwide plots of seismic (earthquake) epicenters disclose a segmentation of the earth's crust into twelve major *plates,* the margins of which are constantly in a state of relative motion, jostling each other, converging, stretching apart, or sliding past one another. Along some of the junction areas there are also numerous microplates, segments and slivers of crust that have been torn off one side or the other during their long history of jostling. The plates themselves are relatively rigid sectors of lithosphere (the earth's rocky outer skin) that averages about 100–200 km in thickness and "floats" in a hydrodynamic equilibrium on a dense but plastic layer of the earth's mantle, the so-called *asthenosphere,* the layer of no strength.

The plates have been compared with ice floes on water; a skater on thin ice will set up a load that depresses an area, surrounded by a peripheral bulge that conserves the volume of the underlying water. Remove the load, and the ice will return to its original plane. This compensatory behavior observed also in the earth's crust is termed *isostasy*. In the delta region of a great river, for example, there is constant subsidence of the crust, because of the progressively increased loading by sediment dumped by the river.

During a glacial interval in the geologic past the loading by the ice sheet, up to 3 km thick, would depress the lithosphere by 1 km; the density of the glacier ice is close to 1, and that of the lithosphere close to 3. In contrast, around the periphery of a major ice sheet (such as that of North America), there will be a *marginal bulge* that reaches out to about 1000 km but rises only to a maximum of about 10 m in elevation; when the ice melts, the bulge collapses (Walcott, 1975). The rate of crustal reaction to loading and reloading is rapid (within a few hundred years), but the curve is exponential and the crust continues to recover several thousand years after the ice has all melted. For example, in parts of Scandinavia and Hudson Bay, the crust is still rising at 8-10 mm/year, although the last ice disappeared 6000 years ago. Parts of the marginal bulge in the eastern United States or in the southern Baltic are still subsiding at rates that may exceed 1 mm/year.

Generation and Growth of Plates

Studies of marine geology and geophysics during the period 1950-1970 disclosed a pattern of *stripes* recorded paleomagnetically in the remanent magnetism of the oceanic crust. The stripes correspond to periods when the earth's magnetic field successively alternated from north-pointing to south-pointing and back again. The polarized evidence of the former magnetic field becomes "locked" into the minerals of submarine lavas that crystallize and the sediments that settle on the sea floor. By dating the rocks concerned, it is possible to demonstrate that the midocean ridges are the youngest zones in each of the world's oceans, and that the geomagnetically tagged stripes that parallel the ridge symmetrically on either side become progressively older away from the ridge (Dietz and Holden, 1970).

The midocean ridge has been known as a topographic feature for more than a century, but its continuity and peculiar attributes have become known only since the 1950s. The ridge is about 60,000 km in length, forming approximately equidistant locus of symmetry between the 2000-m depth contours near the base of the continental rises, and traceable in this way around the earth, except in the North Pacific, where it veers under the margin of the North American plate in the Gulf of California. The ridge can be traced into continental areas in other areas: for example, in the Arctic Ocean, it passes into the Verkhoyansk Range of eastern Siberia (which is the western limit of the North American plate); and in the northwest Indian Ocean, it splits (at a *triple junction*) into the Red Sea rift and the Ethiopian-East African Rift (Burke and Dewey, 1973).

The midocean ridge has thus become recognized as continuous with continental zones of tension (Red Sea-East African and Gulf of California Basin Range) or zones of compression (Verkhoyansk Range). Heat-flow studies along the ridge and these continental zones show anomalously high values, up to ten times the global average. The elevated nature of the ridge, usually 2000-3000 m higher than the normal deep-sea floor (about 5000 m), can therefore be explained as a result of an unusual regional heating of the earth's crust. By precise echo sounding, by dredge hauls from oceanographic ships, and by visual inspection from deep-diving miniature submarines (e.g., on the expedition in 1974), it is now known that the central axis of the ridge is a rift, or complex graben structure, the sides of which are progressively separating, permitting the upwelling and rapid cooling of basic lavas of basaltic type. Numerous pillow lavas have been photographed, proving that actual welling up of lava streams are rapidly quenched by the cold seawater. Coming up with the lavas are also high concentrations of metals (e.g., manganese, copper, zinc) in hydrothermal solution from the deep mantle; in places they are precipitated in what can be economically interesting concentrations. Where ancient midocean ridge sites can now be identified on land, they provide exciting clues for geologists exploring for commercial ore deposits.

Measurements of the rate of rift opening have been made both at sea and where the rift crosses land areas (e.g., in the volcanic area of Iceland). Opening at rates of 1-5 cm a year, the area of sea floor enlargement is of the order of 3-4 km^2/yr. The rate of opening is not constant through time, and there seem to be phases of acceleration and deceleration, with a main periodicity of about 85 million years, but minor variations and discontinuities are also observed. The maximum rates of opening are measured in equatorial to middle latitudes, while in the Arctic the rate drops to nearly zero. The geometry of opening is thus in some way related to the rotation of the earth, inasmuch as angular momentum, centrifugal force, and other parameters decrease in value from the equator to the poles.

Subduction and Consumption of Plates

From the global point of view, it is evident that if every oceanic basin is growing along the "seam" of the midoceanic ridges, then either the earth must be expanding in volume or parts of the oceanic crust must be removed and "recycled" (White et al., 1970). A very small and slow expansion and contraction of the earth are envisaged by some theoreticians to have taken place during its 4.6 billion-year history (possibly in the 300 million-year rhythm of the solar system's swirl around our galaxy), but, be that as it may, this would not compensate for the rapid growth of plates along the midocean ridges.

Seismic monitoring of earthquakes began in a systematic way in the early part of the present century, and by the 1930s it was recognized that certain belts were marked by a high concentration of shocks, the hypocenters of which could be traced down to depths of 500-700 km. With more and closely spaced seismograph stations, it became possible to focus the precise situations of these shocks. These are found in every case to correspond to an obliquely downward-dipping plane that is initiated at the surface corresponding to the site of a deep-sea trench; such trenches, unless filled or masked by heavy deposition of sediment from the land, are commonly up to 10,000 m in depth, thus twice as deep as normal ocean floor.

The sloping planes of seismic activity are called *Benioff zones*, after a seismologist who had been active in their study. The plane is found to slope down usually at 40-45° *away* from the ocean and underneath the marginal belt of island arcs or cordilleran type mountains. These Benioff zones are recognized in a belt around the western Pacific, beneath the Andes, around the Antillean and Scotian arcs, the Indonesian arcs (Sunda-Banda arcs), and in the Mediterranean in the Hellenic arc and Ionian Sea.

Interpreted as the sites where oceanic crust disappears and is "recycled," these Benioff zones are not by any means symmetrical to the growth centers in the midocean ridges. Nevertheless, they correspond to the margins of relatively stable, rigid plates, and careful plotting of the geometry of those plates on large plastic globes with computer-aided precision shows that the rates of crustal growth calculated for the midocean ridges are exactly matched by the rates of consumption at the Benioff zones.

The fact that segments of the earth's crust were forced down obliquely under other segments, in what today are the world's great mountain chains (*orogenic belts*), was recognized already in a 1906 paper by the Austrian geologist Ampferer (1906, 1939). The process was named in the 1940s by a Swiss geologist Amstutz with the term *subduction*.

Volcanic Arcs

If the angle of depression of the descending oceanic crustal slab along a Benioff zone is plotted against the gradient of heat flow, it is observed that at around 50-100 km depth (and at a similar distance back from the deep-sea trench) the oceanic crustal material should be reaching its melting point, somewhat over 1000°C. Thermal conduction through rock is slow, however, and some of the cold descending limb of oceanic crust can actually reach 700 km depth. Melting begins in the most volatile material, and accordingly there are upwelling and liquid magma (molten rock).

These magmatic bodies rise partly by buoyancy, being lighter (density approximately 2.7-3.0) than the surrounding mantle (3.3 or more) in the manner of a salt dome or diapir. This diapirism is aided by the appearance of extensive fracture zones that appear in the crust 100-200 km to the landward side of the deep-sea trench. Some of the magmatic diapirs break through to the surface and form volcanoes, most typically with andesitic lavas, named appropriately after the Andes, that correspond to a clastic Benioff subduction zone. These lavas are of intermediate type and are not simply remelted oceanic crust, but are contaminated by light, upper crustal materials that have evidently been melted and incorporated as the magma rose. It is their intermediate nature, with high volatile content, that makes the circum-Pacific lavas so explosive and dangerous to man. Intra-Pacific lavas, such as appear in Hawaii, are more basic and erupt in a more peaceful manner.

Subduction and the associated melting, magmatism, and volcanism thus account for the removal and recycling of oceanic crust. What is generated at the midocean ridges is gradually displaced landward through time in a symmetrical way in the manner of a pair of conveyor belts, an analogy proposed by Dietz (Dietz and Holden, 1970). The ocean floor, by absolute dating and fossils in its associated cover of pelagic sediments, is thus recognized to become progressively older and older, away from the ridge.

Theory of Continental Drift

Early in the present century, Taylor in North America and Wegener in central Europe were both struck by the similarities of the geology along "matching" coastlines and by dissimilarities on opposing sides of major mountain chains. The continents, acting as semirigid plates, were visualized as "floating" on oceanic crust and drifting across its surface. A complete, integrated theory of continental drift was developed by Wegener (1915), but it lacked a convincing dynamic mechanism and was widely rejected. Today the geological data behind the continental drift theory have been largely confirmed on a worldwide basis. The new global tectonicists, however, define their plates on the basis of seismically defined boundaries; these plates are segments of the lithosphere averaging 100 km in thickness, whereas Wegener's plates were strictly continental and were 30-50 km thick.

Wegener's masterful contribution to modern tectonics was his systematic collection of data and reconstruction of evidence that suggested a splitting of a former mega-continental mass called Pangaea that lay surrounded by Panthalassa (the ancestor of the present Pacific Ocean). This rifting of Pangaea was followed by the progressive opening of the youthful oceans (Atlantic, Indian, Arctic, and Southern).

Pangaea first broke apart into *Gondwanaland* —the southern continents—and *Laurasia*—the northern ones-and these in turn became further segmented. Later, bits of Gondwanaland—e.g., India—came to collide with the southern border of Laurasia, and the Himalyas would represent the collision site. Today it is recognized that subduction occurred here. When the intervening oceanic crust was entirely consumed, the continental block of India collided with that of Tibet, resulting in a gigantic orogenic belt. It was not a simple closure because repeated tectonic pulses and collisions are documented, notably in the Alps (Staub, 1928) and around the Pacific (Kobayashi, 1949).

Thanks to the concept of continental drift, it is now possible also to clearly recognize the rifted coasts, so-called *Atlantic-type* continental margins. A first stage is seen today in the Red Sea; its bordering escarpments are still high (2000-3000 m), because the rifting was relatively recent (about 30 million years ago). Eventually the rift will widen, and the "shoulders" (border scarps) will become lower, as along the opposite side of the South Atlantic. During the initial stretching phase, there is a sinking of the crust, so that there is an inflow of sediment and shallow marine waters in the prerift stage. After rifting, the sedimentary basins are split in two, and are progressively carried apart (Kinsman, 1975). But the fossils and distinctive minerals and structures in them can still be matched on opposite sides of the rifted type of ocean—for example, in basins of eastern Brazil and Gabon (Burk and Drake, 1974).

The nature and history of each plate margin in the world profoundly affect its role as the bedrock of any given coastline type. Those plate boundaries should therefore be closely studied. First, it must be remembered that there are two distinctive categories of crust that make up the lithosphere: sialic, continental crust, 30-50 km thick and containing multiple belts of ancient orogenic belts, dating back to possibly 4 billion years, having been added to the continental nuclei (the oldest spots) by successive accretion; and simatic, oceanic crust, 1-5 km thick, consisting of consolidated lavas and basic to ultrabasic intrusions with a veneer of marine sediments, dating back to no more than 200 million years. Each plate may contain some area of each crustal type. Thus, during subduction of a plate, a sector of this oceanic crust easily slides under the leading edge of the next plate. However, where a continent approaches this subduction zone, the result can be a continent-continent collision, which leads to orogeny or mountain building. Other plate boundaries can involve continent-island arc collision, arc-arc collision, ocean crust-ocean crust collision, and so on. Wherever these situations involve shorelines, particular histories and structures are involved, which must be clearly recognized in any rational analysis of shorelines.

Convection Hypothesis

The fact that there is a strong thermal gradient from the earth's interior (probably more than $6000°C$) to the surface introduces the probability of some sort of convective heat transfer in addition to conduction and radiation. Models of a pair of symmetric conveyor-beltlike convective cells beneath the oceans were sketched out by Arthur Holmes a half century ago and later embellished by Hess and by Dietz. But the models were not carried through in a complete global, three-dimensional form. It is now observed that the plates are of very different sizes, from around 10,000 km in width down to microplates, perhaps 100 km in width. Subduction is not symmetrical around the margins of all oceans. Thus, although the geometry of plate tectonics is now well established, knowledge of the mechanism is still in an unsatisfactory state.

Clearly, there is rising heat flow beneath the midocean ridges. Jason Morgan, Tuzo Wilson and others favor the idea of *hot spots* in the mantle, ascending planes of hot magma that spread out in the upper asthenosphere in a mushroom form to provide a horizontal motive source. The horizontally moving flow in the asthenosphere is coupled to the base of the lithosphere by a zone of friction in which upward cooling is reflected by increasing rigidity. Where a sector of the lithosphere remains steady over a mantle plume, the result may be a large complex of volcanic eruptions, as in Iceland. Where the lithosphere is drifting across the mantle hot spot, the result should be a sequence of volcanic centers, as in the Hawaii chain.

Classification of Coastal Types by Global Tectonics

Disregarding insular coasts and those recently involved in glaciation, there are distinctive coastal categories directly related to the style and history of their plate motions and continental borders (Gregory, 1912; Fairbridge, 1968; Inman and Nordstrom, 1971).

Atlantic Type Continental Margins. These are described as preorogenic—that is, their rocks have not been mobilized or undergone orogeny during the last crustal cycle. They have been stretched, rifted, and torn apart. The trends of former orogenic belts have been abruptly truncated by faulting. They were first identified and called *Atlantic type* by Suess (1885) and *discordant* by Supan (1930). There are two principal subtypes.

Red Sea (Erythrean) Type. Its borders are Precambrian, and stretching began about 40 million years ago, followed by complete rifting. The two sides are now about 100 km apart. The facing shorelines still have the general form of the rift, but they are rarely in direct contact with the Precambrian basement, being eroded in the superficial alluvial fan and salt basin deposits that mark the foot of the scarps. Apart from eustatic phenomena and the like, the general tectonic sign for this settling today is *negative*, because of the slow cooling of the formerly heated and uplifted rift zone.

North Atlantic Type. Two sea-floor-spreading cycles earlier (2 × 85 m/yr—i.e., early Jurassic), the North Atlantic went through a stretching and rupturing comparable with the Red Sea in the contemporary cycle. Hercynian and Caledonian orogenic belts are abruptly ruptured between Newfoundland and Ireland. Overlapping the front of the ruptured belts in many places there are Mesozoic-Cenozoic coastal plain deposits up to several thousand meters thick. During the drifting apart there has been progressive cooling of the basement with subsidence accompanied by progressive accumulation of mainly *miogeoclinal* (miogeosynclinal) type sediments. In places the basins are broader, and paraliageosynclinal-type sedimentation occurs. The tectonic status is approaching stablity, but minor gentle undulations may be identified (e.g., by geodetic relevelling).

Bahama Platform Type. The basement of the Bahamas is a crustal remnant of southern Red Sea type, overgrown by a 7-km thickness of coral reefs and carbonate sediments of shallow-water types (oolites, eolianites, and so on). The age of these sediments goes back to around 180 million years ago, and the tectonic trend has been one of very slow subsidence, averaging 4 cm/1000 yr. The tectonic sign today is still slightly negative, but subsidence is so unimportant that it would scarcely be noticed in Holocene shore features.

Pacific or Cordilleran Type Continental Margins. These are mainly *concordant* types—that is, the trends of the associated mountain belts are more or less paralleled by the principal coastal trends. They are related to the timing of the orogeny in three distinctive ways: synorogenic, postorogenic, and preorogenic. Three corresponding coastal subtypes are as follows.

Aleutian Trench Type. This is the classic synorogenic orthogeosyncline; it is a site of active subduction today, with twin island arcs, an inner belt that is marked by explosive volcanoes, and an outer one (largely submerged) marked by the rising thrusts of the subduction zone. The tectonic sign is mainly positive, but the amplitude variable, so that recent shorelines are found to be strongly deformed, warped, and tilted. Many varieties of this synorogenic type are found, depending partly on the sedimentation rate (e.g., high in the Gulf of Alaska, low in the western Aleutians), latitude and climate (e.g., cliffed, reef-free coast in the wet-tropical regions, such as Java, or reef-lined in drier regions, such as Timor or the Tonga Islands).

California Borderland Type. This is a postorogenic setting. The main orogeny here was in late Jurassic and early Cretaceous time, 150–100 million years ago. Subsequently there has been extensive taphrogeny, block faulting that has segmented the entire area into uplift and downdropped blocks or splinters. The tectonic signs are variable, showing still active subsidence alternating with emergent trends. Coasts are partly concordant, partly discordant.

Gulf of California (Reactivation) Type. This is the belt where a midocean ridge, the East Pacific Rise, joins the North American plate, splitting it away from Baja California (part of the Farallon Plate of the northeast Pacific). It is comparable to the Red Sea situation, except that the sides are not Precambrian; the last orogeny was Mesozoic, as with the California Borderland, and the new rift followed the old trends. However, there are numerous oblique-running fracture zones (transforms), so that the coasts are partly concordant, partly discordant.

Western Pacific (Back-arc) Types. A second type of crustal reactivation is found in the Western Pacific marginal basins and the Mediterranean-Tethyan belt, a type described long ago by Stille as *quasicratonic*. These are regions of multiple microplates and microcontinents, separated by basins with small sea-floor spreading systems, active or dormant today. In general, these are regions where the coasts are discordant, although a few are concordant and correspond to the preorogenic phase. The tectonic sign alternates abruptly, with areas of tremendous uplift side by side with areas of dramatic subsidence. In Indonesia, some areas of late Quaternary coral reef coasts are uplifted to over 1000 m, uplift rates exceeding 1 cm/yr.

RHODES W. FAIRBRIDGE

References

Ampferer, O., 1906. Ueber die Bewegungsbild der Faltengebirge, *Jahrb. Geol. Bundesanst.* 61, 539-622.
Ampferer, O., 1939. Grundlagen und Aussagen der geologischen Unterströmungslehre, *Natur u. Volk* 69, 337-349.
Burk, C. A., and Drake, C. L., eds., 1974. *The Geology of Continental Margins.* New York: Springer-Verlag, 1009p.
Burke, K., and Dewey, J. F., 1973. Plume-generated triple junctions: Key indicators in applying plate tectonics to old rocks. *Jour. Geology* 81, 406-433.
Davies, J. L., 1972. *Geographical Variation in Coastal Development.* Edinburgh: Oliver and Boyd, 204p.
Dietz, R. S., and Holden, J. C., 1970. Reconstruction of Pangaea: Breakup and dispersion of continents, Permian to present, *Jour. Geophys. Research* 75, 4939-4956.
Fairbridge, R. W., 1961. Eustatic changes in sea level, in *Physics and Chemistry of Earth*, vol. 4. New York: Pergamon Press, 99-185.
Fairbridge, R. W., 1968. Atlantic and Pacific type coasts, in R. W. Fairbridge, ed., *Encyclopedia of Geomorphology.* New York: Reinhold, 34-35.
Gregory, J. W., 1912. The structural and petrographical classification of coast-types, *Scientia* 2, 36-63.
Inman, D. L., and Nordstrom, C. E., 1971. On the tectonic and morphologic classification of coasts, *Jour. Geology* 79, 1-21.
Kinsman, D. J. J., 1975. Rift valley basins and sedimentary history of trailing continental margins, in A. G. Fischer, ed., *Petroleum and Global Tectonics.* Princeton, N.J.: Princeton University Press, 83-126.
Kobayashi, T., 1949. The Akiyoshi and Sakawa orogeneses on the southwestern side of the Pacific basin, *Japanese Jour. Geology* 21, 75-90.
Staub, R., 1928. *Der Bewegungsmechanismus der Erde.* Berlin: Borntraeger, 270p.
Suess, E., 1885-1909. *Das Antlitz der Erde.* Vienna: F. Temsky (English translation: *The Face of the Earth*, Oxford: Clarendon, 1902-1924, 5 vols.).
Supan, A., 1930. *Grundzüge der physischen Erdkunde*, vol. 2, part 1, *Das Land (Allgemeine Geomorphologie).* Berlin: De Gruyter, 551p.
Walcott, R. I., 1975. Recent and late Quaternary changes in water level, *EOS (Am. Geophys. Union Trans.)* 56, 62-72.
Wegener, A. L., 1915. *Die Entstehung der Knotinente und Ozeane* (English translation from 4th German ed.: *The Origin of Continents and Oceans.* New York: Dover, 1966, 246p.).
White, D. A.; Roeder, D. H.; Nelson, T. H.; and Crowell, J. C., 1970. Subduction, *Geol. Soc. America Bull.* 81, 3431-3432.

Cross-references: *Classification; Emergence and Emerged Shoreline; Isostatic Adjustment; Isostatically Warped Coasts; Postglacial Rebound; Sea Level Changes; Submergence and Submerged Shoreline; Tectonic Movement.*

GLOUP—See BLOWHOLES

GNETICAE

The Gneticae (Gnetophyta, Chlamydospermae) are vascular, seed-bearing, photosynthetic plants that are trees, shrubs, or vines (Bierhorst, 1971). The group includes plants of diverse aspect, which, however, share the following combination of characters that distinguishes them from other gymnosperms: The leaves are simple in morphological composition and opposite or whorled in arrangement; resin canals are absent; the secondary xylem includes vessels; both male and female strobili are compound and are usually borne on separate plants; the ovules have two integuments, and at least the inner forms a long micropylar tube; the sperms are nonmotile; the embryo has two cotyledons.

The following three orders are included in the subdivision. The Ephedrales (class Ephedropsida of some authors) include only the one family Ephedraceae and the single genus *Ephedra*, which has about 42 species. The genus is known by various common names: *joint fir* and *joint pine*, in allusion to the conspicuously jointed character of the slender twigs; *Mexican tea*, *Mormon tea*, and *desert tea*, in allusion to its use for making an aqueous infusion that, by virtue of its tannin content, resembles black tea in flavor; and simply by its generic name of *ephedra*.

The plants are shrubs from 0.5-2 m. tall with abundant, jointed, long-internoded, slender, light green, yellowish green or gray-green twigs. The leaves, borne in opposite pairs or whorls of 3, being small and withering early, are scarcely noticeable.

The strobili are conelike, small (ordinarily not over 1 cm) and borne at the joints of the stem. They consist of few to several pairs of scales ranging in texture from thin to fleshy. At the time of release of pollen, the male cones may give the male plants a yellowish color by the abundance of exserted pollen-producing microsporangia. The young female cones are green. Their slightly exserted micropylar tubes bear a droplet of fluid at the tip to which the pollen grains, transported either by wind or insects, adhere and are withdrawn into the pollen chamber as the liquid is resorbed. Archegonia are present. At maturity of the one or two seeds in each female cone, the scales of the cone may be dry and membranous, firm, or, in some species, fleshy and reddish.

The usual habitat of *Ephedra* is desert. The geographic distribution is wide: the genus occurs in Eurasia, North Africa, and North and South America, but occurrence is sporadic.

Ephedrine, the alkaloidal nasal decongestant, is extracted from Asiatic species.

The second order, the Gnetales (class Gnetopsida, and division Gnetophyta, in part, of some authors) includes only the one family Gnetaceae and the single genus *Gnetum*, which has about 35 species. These are mostly woody, reclining vines, but some are trees or shrubs. The leaves, which are opposite in arrangement, are unusual among gymnosperms in that they resemble those of dicotyledonous angiosperms in that they are broad and net-veined.

The pollen- and ovule-bearing structures are borne in discrete whorls on unconelike, short lateral branchlets. Archegonia are absent. The seeds are borne entirely unprotected. The outer seed coat is fleshy and often red; the inner seed coat is hard.

Members of the genus *Gnetum* are found in tropical South America, Africa, and southeastern Asia.

The third order, the Welwitschiales (class Gnetopsida, in part, of some authors) includes only the single family Welwitschiaceae and the single genus *Welwitschia*, which includes the single species *W. mirabilis*.

This plant is of unusual form; its short stem is obconical, and usually only the two leaf-bearing lobes of its summit extend above the ground level. Only two leaves are produced. They arch upward and outward, then sprawl over the ground. Their growth is indefinite from their bases at the apices of the lobes of the stem. Their length is limited only by their gradually wearing away at their tips. Venation is parallel, and the leaves become more or less split into longitudinal segments.

The strobili are conelike and borne on short branchlets at the summits of the stem. Pollination is by insects. Archegonia are absent. The seeds are conspicuously winged, hence samaralike. The two cotyledons gradually wither as the two leaves of the seedling grow and assume the photosynthetic function.

Welwitschia is restricted geographically; it occurs only near the coast in the extremely arid desert of South-West Africa (Namibia) and adjacent Angola.

LOUIS CUTTER WHEELER*

Reference

Bierhorst, D. W., 1971. *Morphology of Vascular Plants*. New York: Macmillan, 560p.

*Deceased

Cross-references: *Africa, Coastal Ecology; Biotic Zonation; Coastal Flora; Desert Coasts; Pinophyta.*

GRAB SAMPLERS

Grab samplers are used to obtain bottom samples from the seabed. Several types are available, the most common of which are the *Shipek, van Veen,* and *Petersen* sediment samplers. Grabs are better for collecting fine-grained cohesive sediments, such as silt and clay, than noncohesive sands, comminuted shell, and gravel (waterworn pebbles and cobbles).

They have one distinct disadvantage, compared with corers, in that they considerably disturb bottom sediments. This disturbance can result in an unrepresentative sample being collected at a sampling station. Another problem with bottom sampling has to do with preventing material from being lost from the grab as a result of *washout*, which invariably occurs when the edges of the jaws are kept open by granules or pebbles. Further, fine-grained material can be lost by washing out as grab samplers are raised from the seabed.

Sly (1969) has conducted tests and trials on six grab samplers in order to evaluate their efficiency and suitability for both biological and geological sampling. These are the *Franklin-Anderson* grab, the *Dietz-LaFond* sampler, the *Birge-Ekman* dredge, the *Foerst Petersen* grab sampler, the *Ponar* grab, and the *Shipek* bottom sampler. It appears from the results of Sly's investigations that the reliability of the trigger system and the design, shape, and cut of the jaws are probably the two most important factors that determine the success or failure of grab samplers. The Franklin-Anderson, Shipek, and Ponar grabs are useful for collecting biological material in sandy sediments; the Shipek and Ponar grabs, for obtaining biological material successfully from gravel bottoms.

Although grabs differ slightly in their modus operandi, they all have jaws that are forced to close, on contact with the seabed, by springs, weights, or lever arms.

As stated in Ingham (1975), the winch system on a boat is also important in grab sampling operations. This should run free at all times and be installed in such a way as to provide maximum safety to those persons using the samplers.

JOHN R. HAILS

References

Ingham, A., ed., 1975. *Sea Surveying*. New York: John Wiley & Sons, 306p.

Sly, P. G., 1969. Bottom sediment sampling, in *Proceedings of the 12th Conference on Great Lakes Research*. Buffalo, N.Y.: International Association for Great Lakes Research, 883-898.

Cross-references: *Classification; Coastal Morphology, Research Methods; Corers and Coring Techniques; Peels; Sediment Analysis, Statistical Methods; Sediment Size Classification.*

GRADED SHORELINE—See EQUILIBRIUM SHORELINE

GRAVEL

Gravel is a term widely used for unconsolidated accumulation of rock fragments, particularly detrital material associated with streams and beaches. Characteristically, subangular to rounded *granules* and *pebbles* are mixed with 50-70% *sand*, by weight. The term is also used for unconsolidated and ill-sorted accumulation of eroded rock fragments ranging from 2 (occasionally 4) mm up to *boulder* size, as well as for unconsolidated conglomerate. In soil science and civil engineering, and (occasionally) sedimentation, the term is used for rock or mineral particles instead of the term "pebbles" and size is then variously defined as between either 2 or 4 mm and 20, 60, 64, or 75 mm (British Standards Institution, 1967; Gary, McAfee, and Wolf, 1972).

Gravels may be composed of any kind of rock and some artificial materials, such as brick or concrete fragments. Rate of wear tends to be proportional to weight of particles in water and distance traveled. In general, gravels are hard and resistant to wear (Pettijohn, 1975).

ALAN PAUL CARR

References

Gary, M.; McAfee, R.; and Wolf, C. L., eds., 1972. *Glossary of Geology.* Washington, D.C.: American Geological Institute, 895p.

British Standards Institution, 1967. *Methods of Testing Soils for Civil Engineering Purposes,* B.S. 1377. London, 234p.

Pettijohn, F. J., 1975. *Sedimentary Rocks.* New York: Harper and Row, 628p.

Cross-references: *Beach Material; Beach Material, Sorting of; Boulder; Clay; Cobble; Pebble; Sand; Sediment Analysis, Methods; Sediment Size Classification; Shingle and Shingle Beach; Silt; Soils.*

GRAVEL RIDGE AND RAMPART

Gravel ridges are one type of *beach ridge* (q.v.). They are distinguished from sand ridges by: the caliber of material; elevation above high water mark; and mode of formation. The terms *gravel ridge* and *shingle ridge* are synonymous. As their name implies, gravel ridges are composed predominantly of unconsolidated particles coarser than sand, either boulders, cobbles, or pebbles, or any combination of these. Generally, at any one site, particles consist of a single rock type, and invariably clasts are well rounded, well sorted for size and shape, and loosely packed. Morphologically, gravel ridges are distinctive. They are narrow, elongate banks rising above the reach of normal spring tides and wave action being built to this high level during episodic heavy seas and storms. Although of marine origin, the crests of ridges are subaerial features. In profile, crests may be sharp or flat and either symmetrical or asymmetrical. Slopes, from crest to base, are generally steep; narrow berms or steps may interrupt the regular seaward slope. Only rarely do gravel ridges occur singly as discrete entities. More commonly they are arranged as a series of roughly parallel linear or curving bands congruent with and behind the present shorelines. Each separate ridge marks a successive position of an advancing shoreline. Some suites of ridges, separated by low swales, extend a considerable distance back from the shore, indicating large-scale progradation in Holocene time. Variations in ridge size, spacing, and orientation indicate temporal changes in shoreline geometry, sediment supply of coarse rock particles, and wave energy sufficient to move such material. These limiting conditions occur especially in higher latitudes and in those areas subjected to repeated Pleistocene glacial and periglacial conditions. Spectacular spits, barriers, and forelands composed of gravel ridges are found along Alaskan, Canadian, Russian, and British coasts in the Northern Hemisphere, and in New Zealand and Chile in the Southern Hemisphere. The Russian literature on gravel beaches is particularly voluminous (Zenkovich, 1967), and is matched only by that from Britain (Steers, 1964) where some long-studied, classic shingle landforms such as Dungeness, Orfordness, Chesil Beach, and Blakeney Point are found.

Ramparts are also shingle or gravel ridges found in coral reef seas. Although many American workers use rampart as synonymous with beach ridge, the former term should be retained exclusively for those loosely compacted coral rubble accumulations on the reef flat that may or may not be wholly emergent at high water. A rampart may thus be a subaerial or intertidal feature. Ramparts are storm deposits; sometimes they have been termed *hurricane beaches*. Typhoon or hurricane waves pounding on the reef edge may break and dislodge coral colonies and dump the resulting rubble (coral shingle or gravel) on the reef flat as a linear bank some distance in from the reef edge. If an island is present to landward, a tidal moat may be formed between it and the rampart. Many ramparts associated with particular storms have been described, most recently from Funafuti Atoll, Tuvalu, central Pacific (Baines and McLean, 1976). On October 21, 1972, an enormous rubble rampart, 19 km in length and up to 4 m high, was deposited there on the southeast facing ocean reef flat during Hurricane Bebe. In this case, and generally, the initial storm rampart was a discrete isolated feature, asymmetric in profile, consisting of clean, white, abraded coral shingle. Poststorm shoreface erosion and washover result in inward migration of the rampart that may in time either weld against an adjacent island as a gravel ridge or remain as an individual rampart if another develops seaward of it. Many reef islands, atoll motu, and, low wooded islands possess multiple ramparts, separated by moats or swamps, and/or multiple ridges lapping against one another. Each ridge or rampart represents a particular storm event and its subsequent poststorm movement. Islands on the windward side of reefs are frequently made up of shingle ridges or ramparts whose relative age can be distinguished by the color, degree of weathering, and diagenesis of clasts, and extent of colonization by vegetation and relative location.

ROGER F. McLEAN

References

Baines, B. G. K., and McLean, R. F., 1976. Sequential studies of hurricane deposit evolution at Funafuti Atoll, *Marine Geology* 21, M1-M8.
Steers, J. A., 1964. *The Coastline of England and Wales*. London: Cambridge University Press, 750p.
Zenkovich, V. P., 1967. *Processes of Coastal Development*. Edinburgh: Oliver and Boyd, 738p.

Cross-references: *Beach Ridge and Beach Ridge Coasts; Boulder Barricade; Coral Reef Coasts; Major Beach Features; Pacific Islands, Coastal Morphology.*

GREEN ALGAE—See CHLOROPHYTA

GROINS

Groins are structures whose principal purposes are to trap littoral drift, to inhibit the erosion of a beach, to widen a beach, or to provide a beach where none had previously existed (Coastal Engineering Research Center, 1973; Wiegel, 1964). They are usually constructed perpendicular to the shoreline or in a way in which they are most effective. Materials for construction include timber, concrete, rock, and steel.

There are many types of groins, based upon their plan form. These include corner, inclined, T. Z. angular, and standard groins.

A groin is typically classified as high or low, long or short, permeable or impermeable, or fixed or adjustable. High groins tend to initially intercept all of the longshore moving sand until the areal pattern or surface profile of the accumulated sand allows sand to pass around the seaward end of the groin. Low groins, where the height of the groin is as high as the desired beach height, allow for the accumulation of sand up to the height of the groin, at which point sand passes over and downdrift of the groin. Permeable groins allow for passage of some of the sand and wave energy through the structure, while impermeable groins are, as the name implies, with respect to downdrift flow.

JOHN B. HERBICH
TOM WALTERS

References

Coastal Engineering Research Center, 1973. *Shore Protection Manual*. Washington, D.C.: U.S. Army Corps of Engineers, 750p.
Wiegel, R. L., 1964. *Oceanographical Engineering*. Englewood Cliffs, N.J.: Prentice-Hall, 532p.

Cross-references: *Beach Nourishment; Coastal Engineering; Protection of Coasts.*

GROUNDWATER—See WATER TABLE

GYTTJA

Gyttja is a Swedish word for fine-grained organic matter in sediments (von Post, 1862; Lundqvist, 1924; von Post and Granlund, 1926). One speaks about coarse detrital gyttja (with remaining cellular structures) and fine detrital gyttja (fine organic particles without remaining cellular structures). There is a continuous genetic transition from allochthonous (alluvial) peat to coarse detrital and fine detrital gyttja. Boiled in an alkaline solution, gyttja results in a transparent solution (as compared with humus, sapropel, and dy that render a brown color). Natural fine-grained sediments are usually a mixture of clay and gytta. According to the Swedish nomenclature: a gyttja contains more than 12% organic matter; a clayey gyttja, 6-12% organic matter; and a gyttja clay (clay with gyttja), 3-6% organic matter. Lime gyttja is a calcareous gyttja. Gyttja has (during the past decades) become an international term.

NILS-AXEL MÖRNER

References

Lundqvist, G., 1924. Utvecklingshistoriska insjöstudier i Sydsverige, *Sveriges Geol. Undersökning Årsb. C-330*, 129p.

von Post, H., 1862. Studier öfver nutidens koprogena jordbildningar; gyttja, dy och mull, *Kgl. Svenska Vetenskapsakad. Handl.* 4, 1-59.

von Post, L., and Granlund, E., 1926. Södra Sveriges torvtillgängar, *Sveriges Geol. Undersökning Arsb. C-335*, 127p.

Cross-references: *Sediment Analysis, Methods; Sediment Size Classification;* Soils. Vol. VI: *Gyttja, Dy.*

H

HAFF

Haff is a term widely used in German and Scandinavian geomorphological literature to designate shallow, freshwater lagoons of the Baltic separated by sand spits (*coastal barriers*) from the open sea. The word "haff" is of Dutch-Swedish origin (hav = sea). The largest haffs of the Baltic are the Kurisches Haff (Kuršiu marios in Lithuanian) and the Frisches Haff (Aistmarės in Lithuanian, Zalew wiślany in Polish). In the mouth area of the Odra (Oder) River there is the so-called Oder Haff (Zalew Ordžański in Polish).

The surface area of the Kuršių marios lagoon is about 1610 km^2. The maximum depth in the southern part of the basin reaches 5–6 m, whereas the mean depth does not exceed 3.7 m. The water volume of the lagoon is approximately 5.9 km^3. The lagoon shores are mainly lowlands of alluvial origin. In some parts of the southern and northeastern shores low cliffs are cut in the glacial deposits. The Nemunas (Njemen) River delta comprises the eastern shores of the lagoon. Among the recent bottom sediments, organic oozes and various sand-gravel facies predominate (Pratje, 1931). Areas where glacial deposits outcrop at the bottom are limited. The thickness of the lagoonal (haff) sediments accumulated during the late- and post-glacial times reaches to 40–50 m. The lower part of the sediment sequence consists of a thick series of varved and homogeneous clays and silts as well as sands. The late-glacial sediments are bedded on the glacial drift of the last glaciation. The Kuršių marios lagoon is situated in an area where slow crustal subsidence is prevailing. The annual river load discharge into the lagoon is estimated to be about 635,000 tons. The so-called Klaipėda Strait is the only channel for inflow and outflow. The Kuršių marios lagoon is a freshwater basin (salinity does not exceed 1–2‰). Nevertheless, the lagoon is rich in fish.

The Zalew wiślany lagoon (Frisches Haff) is a little smaller, with a surface area of 838 km^2, mean depth of 2.6 m, and water volume of 2.3 km^3. The physicogeographical and geological conditions are similar to the lagoon described above.

Both Baltic lagoons originated in late-Litorina (late-Atlantic) time, about 6000 years ago, when the transgressing sea waters intruded into the lagoon areas and then formed coastal barriers that cut them off from the open sea.

In the development of the Kuršių marios lagoon, three periods are to be distinguished: the period of ice-dammed and relict lakes (13,000–10,000 B.P.); the period of dry land regime (drainage of coastal lakes, peat bog formation, 10,000–6000 B.P.); and the period of proper lagoon development (6000 B.P. until the present).

VYTAUTAS GUDELIS

Reference

Pratje, O., 1931. Die Sedimente des Kurischen Haffes, *Fortschr. Geologie und Paläontologie* **10**, 176.

Cross-references: *Bodden; Europe, Coastal Morphology; Geographic Terminology; Lagoon and Lagoonal Coasts; Nehrung; Soviet Union, Coastal Morphology.*

HALOPHYTES

Halophytes are defined as plants capable of growing and reproducing in a saline environment—that is, where NaCl comprises greater than 0.5% soil salts. For nonhalophytes, growth in a saline soil results in absorption of toxic levels of salts, ionic imbalances, and eventual death of the plant. Halophytes, however, have adapted special mechanisms to tolerate the adverse condition of high soil salinity.

Ion selectivity, the most advanced mechanism for salt tolerance, allows plants to absorb essential nutrients according to the proportions required by the plant. It appears that all plants have a dual mechanism for the absorption of nutrients (Epstein, 1966). At low concentrations in the external medium, only the first mechanism operates, absorbing increasing amounts with increasing external concentrations. The second mechanism begins to function at high external concentrations but is independent of the actual amount present. In saline environments, many necessary nutrients occur in low concentrations. Thus halophytes must be able to absorb sufficient amounts of nutrients but exclude chemically related, toxic ions (such as sodium) that exist in high concentrations in the external medium. Selectivity of ions is not complete, and all halophytes absorb some sodium into their tissues. Once absorbed, sodium and other excess salts may be removed in a number of ways. Salt glands, which act as transit cells, move ions against a concentration gradient, and pump the excess salts out of the plant. Salts may also be removed by a process called salt leaching, in which salts are moved to the surface by the transpiration stream and are washed away by rain or fog (Waisel, 1972). Although little is known about the leaching of halophytic plants, it does occur in some tropical mangrove and coastal halophytes.

Succulence is another important adaptation that enables plants to tolerate high salinities. The increased volume in the plant tissues dilutes the toxic ion content to levels that the plant can tolerate. A second method by which succulent plants cope with high concentrations of salt is to shed the leaves and cortex where salts are stored. Nonsucculent plants, such as *Juncus maritimus* or *J. gerardii,* also shed leaves to rid themselves of salts. Accumulation of salts away from metabolic sites or retransportation of salts back to the external medium through the roots are additional means of coping with high salinities.

Numerous attempts have been made to classify halophytes based on their tolerance ranges and adaptations (Waisel, 1972). Such classifications are complicated by the difficulties of determining salinity limits for each species. Salinity alone does not inhibit the growth of a plant, and other factors such as aeration, competition, or edaphic characteristics may limit or prohibit a plant's growth. As a result, laboratory attempts to delineate tolerance ranges fail to duplicate salinity ranges in the field (see references in Phleger, 1971). *Spartina alterniflora* becomes chlorotic when grown in solutions with salinities greater than that of freshwater. In other tests, *Spartina alterniflora* as well as *S. patens* and *S. foliosa* grow most rapidly in freshwater.

There is an increasing amount of literature on the effects of nutrients on a plant's ability to tolerate saline conditions (see references in Ranwell, 1972). Calcium enables certain plants such as *Avicennia, Agropyron,* and *Phaseolus* (not a marsh plant) to tolerate high salinities. *Phaseolus* displays massive breakthroughs of sodium to the leaves when the plant is grown in a calcium sulfate solution with less than 0.01% calcium. It has been suggested that calcium is important to maintain the selective ion mechanism. Iron and potassium may also be limiting factors in the growth of halophytes.

LINDA C. NEWBY

References

Epstein, E., 1966. Dual pattern of ion absorption by plant cells and by plants, *Nature* **212**, 1324-1327.

Phleger, C. P., 1971. Effect of salinity on the growth of a salt marsh grass, *Ecology* **52**, 908-911.

Ranwell, D. S., 1972. *Ecology of Salt Marshes and Sand Dunes.* London: Chapman and Hall, 258p.

Waisel, Y., 1972. *Biology of Halophytes.* New York: Academic Press, 395p.

Cross-references: *Coastal Flora; Dieback; Marsh Gas; Salt Marsh; Salt Marsh Coasts; Soils; Vegetation Coast.*

HARBORS

Harbor site selection and features are dictated usually by the size, shape, meterological environment, and hydraulic characteristics of the body of navigable water to which access by a boat or ship is to be provided. Of course the size and draft of vessels the harbor must service are also important parameters. Harbors along open seacoasts require protection far in excess of that required for harbors naturally protected in inland waters. Harbors in areas of extremely high tidal variation or large seasonal water level fluctuations must be designed differently from those on lakes and along seacoasts where water fluctuations are only a few feet. In short, since no two sites are alike in size, shape, meterological environment, and hydraulic characteristics, each harbor is different and requires separate important site considerations.

In selecting a harbor site, two basic needs must be fulfilled. First, the site must provide safe navigation access to cruising waters. Second, the site must have adequate land access for private or commercial needs. Other important considerations are: enough protected water area that can be excavated to navigable depth for future expansion; adequate land perimeter for future expansion; and utility service to site such as electric power, sewage, gas.

Site selection is also dependent upon what type of harbor is needed at a particular location. There are three major types of harbors: commercial, recreational, and harbor of refuge. Commercial facilities usually require a complex site evaluation including both physical, economic, ecological, geological, archeological, and hydraulic analysis of the waters and adjoining land areas. Planning criteria must also be adopted if small crafts are to use the harbor as well, to reduce the collision hazard to a minimum without curtailing the activities of either class more than is essential for navigational safety. Recreational harbors are usually designed for small craft, and require significantly less site evaluation and analysis. Finally, harbors of refuge are an important but often neglected part of current harbor planning. These harbors need only supply the essentials, such as fuel, food, shelter, and medical care, that may be required in time of emergency.

Relevant literature on this topic includes Chaney (1961), Dunham and Finn (1974), Treadwell (1969), and the U.S. Department of Transportation (1975).

JOHN B. HERBICH
ROY B. SHILLING

References

Chaney, C. A., 1961. *Recommendations for Design, Construction and Maintenance.* New York: National Association of Engine and Boat Manufacturers, 247p.

Dunham, J. W., and Finn, A. A., 1974. *Small-Craft Harbors: Design, Construction, and Operation.* Washington, D.C.: U.S. Army Corps of Engineers Coastal Engineering Research Center, 11-26.

Treadwell, G. T., 1969. *Small Craft Harbors.* New York: American Society of Civil Engineers, Manuals and Reports in Engineering Practice No. 50, 139p.

U.S. Department of Transportation, 1975. *Code of Federal Regulations, No. 33—Navigation and Navigable Waters.* Washington, D.C.: U.S. Government Printing Office, 596-635.

Cross-references: *Boat Basin Design; Coastal Engineering; Human Impact.*

HARMONIC ANALYSIS—See FOURIER ANALYSIS

HEADLAND BAY BEACH

A *headland bay beach* is a coastal embayment formed by wave erosion in the dominant downdrift (lee) direction immediately adjacent to a single prominent headland (Yasso, 1965). Headlands include, but are not limited to: masses of erosion-resistant glacial till, as along the western shore of Cape Cod Bay; rock promontories, as along the California coast; and widely spaced groins, as along many sandy coasts. Such headlands block direct dominant wave attack against the downdrift beach. However, waves are *refracted* around the headland to cause retrogradation of the lee shoreline. *Diffracted* waves and eddies from coastal or tidal currents may play a small, but as yet undetermined, role in evolution of the headland bay beach.

Headland bay beaches examined to date all have a logarithmic spiral geometry in plan view (Yasso, 1965). Logarithmic spiral center of curvature for each headland bay beach is at the headland or in reasonable proximity to it. Limiting cases of logarithmic spirals include the straight line and arc of a circle. Between are an infinite number of more typically shaped log spirals. Examples of recent computer-curve-fitting of headland bay beach form to confirm spiral shape and examine log-spiral parameters include Bodega Bay and Bolinas Bay along the California coast north of San Francisco. Bodega Bay has a log-spiral plan shape with spiral angle, of 42.43° (Fig. 1). Bolinas Bay is also a log-spiral with log-spiral center at 37°52'16"W and

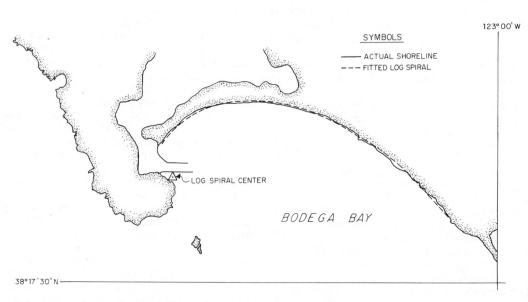

FIGURE 1. Log-spiral fit to the Bodega Bay shoreline. Map represents a portion of the Bodega Head Quadrangle, California (7½-min. series, U.S.G.S., 1942 edition; map drafted by J. J. Pepenella).

spiral angle of 64.99°. Reasons for differences in spiral angle between adjacent headland bay beaches are not yet understood. Nor is it known whether the spiral angle remains constant as the embayment grows over time.

Apparent examples of log-spiral-shaped headland and bay beaches are ubiquitous (Yasso, 1979). In addition, some raised beaches may retain a log-spiral form inherited from an earlier, relative sea level position (Schwartz and Grabert, 1973).

WARREN E. YASSO

References

Schwartz, M. L., and Grabert, G. F., 1973. Coastal processes and prehistoric maritime cultures, in D. R. Coates, ed., *Coastal Geomorphology*. Binghamton: State University of New York, 303-320.
Yasso, W. E., 1965. Plan geometry of headland-bay beaches, *Jour. Geology* 73, 702-714.
Yasso, W. E., 1979. Headland-bay beaches along the western shoreline of Cape Cod Bay, Massachusetts, in G. Halasi-Kun and K. Widmer, eds., *Proceedings of the Columbia University Seminar on Pollution and Water Resources*. Trenton, N. J.: New Jersey Department of Environmental Protection, F1-F6.

Cross-references: *Beach in Plan View; Coastal Erosion, Formations; Cuspate Spits; Zetaform Bays*. Vol. III: *Cuspate Foreland or Spit*.

HIGH-ENERGY COAST

High-energy coasts are those that are exposed to strong, steady, zonal winds and fronts with high wave energies in the lee of high-latitude storm waves and low latitude swells (Davies, 1973), unprotected by shallow offshore topography, and receive the highest energy. The high-energy regime characterizes exposed Atlantic shores of Norway, Iceland, Greenland, Maritime Canada, northern New England, as well as the northern Pacific coast of Alaska, Canada, Washington, Oregon, and the southern tip of South America. On bedrock coasts cliffs, stacks, abrasional platforms, and gravelly, sandy pocket beaches are typical; while on sandy clayey coastal plains, wide sand beaches and broadly arcuate or straight delta fronts predominate.

Price (1954), who developed the coastal energy concept and applied it to the coasts of the Gulf of Mexico, noted the highest shorefaces and the steepest shelf-ramp slopes in the *high* relative wave energy zones of the Gulf. He described fully smoothed, well-aligned shorelines with nearly continuous barrier island chains and bar-obstructed river mouths. Based on Price, Tanner (1960) also published a coastal energy zone classification, using 50 cm mean breaker heights as a cutoff valve between moderate- and high-energy Florida and other south Atlantic states shores.

ERVIN G. OTVOS

References

Davies, J. L., 1973. *Geographical Variation in Coastal Development*. New York: Hafner Publishing Co., 204p.
Price, W. A., 1954. Dynamic environments: Reconnaissance mapping, geologic and geomorphic, of continental shelf of Gulf of Mexico, *Gulf Coast Assoc. Geol. Socs. Trans.* 4, 75-107.
Tanner, W. F., 1960. Florida coastal classification, *Gulf Coast Assoc. Geol. Socs. Trans.* 10, 259-266.

Cross-references: *Low-Energy Coast; Mid-Latitude Coasts; Moderate-Energy Coast; Storm Beach; Storm Wave Environments; Wave Environments*.

HIGHEST COASTLINE

In connection with the recession of the ice sheet during the last deglaciation, the highest coastline was developed as a metachronous level, uplifted by isostatic movements (still in progress). These highest ancient shore marks have previously in Swedish literature been generally named *högsta marina gränsen* or merely *marina gränsen* (De Geer, 1888) often abbreviated to the symbol MG (in English: the highest marine limit, abbreviated to HM or ML). However, as this level, especially within certain parts of the Baltic Sea, was developed by freshwater, Halden (1933) proposed the term *the highest coastline* or *HK* (abbreviation of the Swedish term *högsta kustlinjen*) which now, because of its general applicability, replaces the older terms.

The *HK* has been developed differently, and can, depending on several factors, occur as small deltas, marginal deltas, abrasion or accumulation terraces, wave-cut benches, cobblezones, zones of free washed bedrocks, and so on. For practical reasons the level of the *HK* is defined, where possible, as the upper limit of the zone where the drift sheet has been abraded by wave action. In this sense the HK is identical with the term *limit of wave washing* (Bergström, 1963)—that is, the boundary between nonwashed till and wave-washed till. Where the wash limits are insufficient for a determination of the *HK*, the features previously mentioned and/or the lowest level of supraaquatic features (e.g., lateral channels) are used as indications of the *HK* level. The wash limits and the deltas give the maximum and the minimum levels of the *HK*, respectively (Cato and Lindén, 1973).

In Fennoscandia, the *HK* is developed in the western parts by the ocean, around the southern parts of the Baltic basin by the Baltic Ice-Lake and the Yoldia Sea, and in the northern part of the Baltic basin by the Ancylus Lake (Lundqvist, 1965). Along the Gulf of Bothnia in the northern central part of Sweden, the *HK* has a maximum altitude of about 285 m, which drops peripherally to just above the present-day coastline in the northern and western parts of Fennoscandia. The altitude is about zero in central Denmark.

Emerged marine features, related to glacial-isostatic movements, have also been reported from other previously glaciated parts of the world—for example, North America (especially the eastern parts), Greenland, Iceland, Svalbard, the British Isles, along the coast of Siberia, and in the Southern Hemisphere from Patagonia, Argentina (Flint, 1971).

INGEMAR CATO

References

Bergström, R., 1963. Högsta kustlinjen i norra Gästrikland och södra Hälsingland (The highest shoreline in the northern part of Gästrikland and the southern part of Hälsingland), *Sveriges Geol. Undersökning C-591*, 26p.

Cato, I., and Lindén, A., 1973. The highest shore line at Törnevik, southern Östergötland, *Geol. Fören. Stockholm Förh.* 95, 25-35.

De Geer, G., 1888. Om Skandinaviens nivåförändringar under qvartäperioden, *Geol. Fören. Stockholm Förh.* 10, 366-379.

Flint, R. F., 1971. *Glacial and Quarternary Geology*. New York: John Wiley & Sons, 892p.

Halden, B. E., 1933. Högsta kustlinjen—Ett nytt namn på ett gammalt begrepp, *Geol. Fören. Stockholm Förh.* 55, 429-430.

Lundqvist, I., 1965. The Quaternary of Sweden, *in* K. Rankama, ed., *The Quaternary*, New York: John Wiley & Sons, 139-198.

Cross-references: *Glaciated Coasts; Ice along the Shore; Sea Level Changes; Sea Level Curves; Submergence and Submerged Shoreline.*

HIGH-LATITUDE COASTS

High-latitude coasts may be defined as those where landfast sea ice exists for at least part of the year (Davies, 1980; John and Sugden, 1975). On some sections, notably extensive parts of the Antarctic coast, such ice is perennial and the coast is permanently frozen; on others there is an ice-free season of varying length. On the Antarctic coast, there are also sections where glacier ice extends across the bedrock coastline to form shelf ice. In contrast, Arctic coasts are often characterized by the occurrence of ground ice or permafrost outcropping in sea cliffs.

A prominent feature of high-latitude coasts everywhere is the feature known as *ice foot*. This is a rim of ice mantling the seaward edge of the shore for varying periods of the year, produced by the successive adfreezing of seawater and freshwater on beach, rock, and ice surfaces in a variety of ways to form a variety of types.

Wave energy levels are low because of the comparative rarity of gale-force winds and the relatively short stretches of ice-free water over which waves may be generated. Swell produced in mid-latitude storms tends to be deflected toward the equator, and all waves are frequently damped down by floating ice. Wave climates are characterized by long periods of relative calm, interrupted by infrequent high-energy events, and these tend to determine the nature of coastal evolution. With the exception of the Sea of Okhotsk and the Hudson Bay-Hudson Strait area, tidal range is low and tidal effects do not figure largely in the coastal system.

Because of the way in which glaciers have recently crossed all the Antarctic coasts and still cross much of it, there are few shore sediments in the south; but northern high-latitude coasts display an abundance of sediment, much of it coarse and of pebble size. Some of it is delivered by rivers and glacial meltwater streams. However, along tundra shores large quantities are produced from the erosion of older sands and gravels frozen by ground ice. By comparison with higher-energy coasts, shore sediments can be expected to be relatively poorly sorted and unrounded, but there is not much statistical evidence as yet. Large ice-rafted boulders may be deposited on all types of shores.

The abundance of sediment and the low tidal range are two factors conducive to the development of coastal barrier systems, so that barrier beaches are found extensively in many parts of the Arctic. They are particularly prevalent in northern Alaska and eastern Siberia. The beaches themselves differ from beaches elsewhere because of modifications resulting from ice action. These can be divided broadly into: predominantly ridgelike features produced by the pushing action of sea ice as it is driven onshore by winds and waves; and predominantly hummocky features produced as ice of varying origin becomes temporarily incorporated into the beach and subsequently melts. The beach itself is normally totally frozen for part of the year, and its substratum may be permanently frozen. As a result, beach water tables tend to be notably different from those in warmer regions with wave infiltration much reduced.

Because of the infrequency of strong onshore winds and the common presence of pebbles on beach surfaces, coastal dune development is limited, but there are notable local exceptions where sand is in particularly good supply. At least some high-latitude coasts, for instance in Labrador, are backed by older, vegetated, transgressive dunes, formed presumably when the availability of sand was much greater and perhaps when the larger Pleistocene ice sheets were finally melting.

Whereas Antarctica is notable for the almost complete absence of rivers, the Arctic basin has some of the world's largest rivers flowing into it. These rivers tend to be strongly seasonal in flow and the breakup flood is often of particular significance in determining estuarine form and the supply of coastal sediments. Estuaries are often deltaic. Both here and in coastal lagoons behind barriers, the extent of salt marsh is strictly limited by the unfavorable climatic environment, impoverished flora, and low tidal range. Where they occur, Arctic salt marshes are small and grassy.

Sea cliffs and shore platforms in high latitudes may be particularly inherited from earlier Pleistocene interglacial periods when there was less ice and more wave energy. Certainly in places, these have been described as only now emerging from beneath an ice cover. Little is known of modern processes of bedrock reduction. The prevailing low wave energy suggests that rates of erosion should be low, but it has been suggested, on the other hand, that the effects of freezing sea spray or meltwater should be significant and that the effects of salt crystallization should be particularly prevalent in these relatively arid environments. It has also been pointed out that solution of limestones may be more rapid because of the lower concentrations of lime in high-latitude seawater. In any event, true sea cliffs are relatively rare, and the majority of rocky coasts are glacially eroded surfaces that have been little modified. Fjord coasts are the most spectacular examples. Where cliffing does occur, as in frozen unconsolidated sediments or rock especially susceptible to frost shattering, the rate of debris supply from slumping and solifluction is commonly greater than the rate of wave removal. Talus aprons and slump blocks may fringe the shore.

A particular feature of high latitude coasts is the *strandflat,* an undulating but generally subhorizontal rocky lowland extending for many kilometers inland. Its origin remains unresolved. On the one hand, it has been considered to be essentially a subaerially formed feature, caused by stream planation or to the past occurrence of piedmont ice; on the other, it has been thought of as resulting from prolonged marine erosion at many different times in the Quaternary. It will probably prove to be of composite origin and the result of the juxtaposition of a number of processes in space and time.

Almost all high-latitude coasts show evidence of relatively recent *sea level change* because of the effects of glacioisostasy. As Pleistocene glacier masses dwindled and often disappeared, compensatory upward movement of the earth's crust has gone on into the Holocene and on some coasts, such as those of Hudson Bay and the Gulf of Bothnia, is still discernible. Recently formed shore features may be found stranded in the coastal zone.

J. L. DAVIES

References

Davies, J. L., 1980. *Geographical Variation in Coastal Development,* 2d ed., London: Longman, 212p.

John, B. S., and Sugden, D. E., 1975. Coastal geomorphology of high latitudes, *Prog. Geog.* 7, 53-132.

Cross-references: *Antarctic, Coastal Morphology; Arctic, Coastal Morphology; Beach Material; Glaciated Coasts; Ice along the Shore; Ice-Bordered Coasts; Ice Foot; Low-Energy Coasts; Low-Latitude Coasts; Mid-Latitude Coasts; Salt Weathering; Strandflat.*

HOLOCENE

The youngest epoch in the geologic time scale, the *Holocene,* began about 10,000 years ago (Fairbridge, 1974). This same interval is also sometimes known as the *Postglacial* or the *Recent,* but such terms are rather imprecise and no longer approved in scientific works. The term Holocene was proposed by the French geologist Gervais in 1867 and adopted by the American Stratigraphic Commission in 1967 (Fairbridge, 1968, 1976c). In the deep sea, far from land, the Holocene deposits may be less than 30 cm thick, but in lakes, deltas, lagoons, and nearshore formations (Hageman, 1969), the sedimentation rate may exceed 1 m/1000 years (1 mm/yr).

Geological time within the Holocene is measured by several techniques; radiocarbon dating (q.v.), varve counting, tree ring counting, and historicoarcheological methods (see, for example, Fritts, 1977).

At the beginning of the Holocene, the continental ice borders in North America lay in the latitude of the Great Lakes; in Europe, along a line passing just south of Stockholm. As the glaciers withdrew, the meltwater raised sea level progressively, so that the coastline was gradually moved inward; from near the edge of the continental shelf during the last glacial maximum of the Pleistocene (about -135 m at 17,000 B.P.), it had reached -40 m by 10,000 B.P. and present level by 6000 B.P. Since then it has fluctuated somewhat, gradually dropping in the past few thousand years.

In the formerly glaciated areas, isostatic uplift of the earth's crust followed the retreat of the ice, so that in central Canada and northern Sweden, areas that were below sea level in the early Holocene have now risen to 300 m and more (Andrews, 1974). In contrast, the regions that lay south of the former ice margin in Europe and North America have experienced a sinking crust. At New York and London, for example, the crust is still subsiding at 1-2 mm/yr. In the tropical regions, the crust is generally stable or tends to rise. Thus the Holocene beachlines are preserved like a flight of stairs in the Arctic and former glaciated regions (Hillaire-Marcel and Fairbridge, 1978; Tooley, 1974), whereas in the tropics they are spread out laterally, becoming a few meters lower near the present shore (Fairbridge, 1976a). Because of the slow subsidence, most of the North American mid-Holocene shorelines are slightly below present sea level.

The climatic history of the Holocene has been extraordinarily varied as shown, for example, by ^{18}O fluctuations in Greenland ice cores (Dansgaard, Johnson, and Clausen, 1971). Environments have changed so rapidly that many major mammals have become extinct; but no doubt, hunting by man has also played a role in some cases (Martin and Wright, 1967). Following the peak of the last glaciation (c. 17,000 B.P.), the tropics began to warm first. In place of a cloudless but dusty, hyperarid climate, heavy rains began around 13,000 B.P., and by the early Holocene there was a general filling of lakes and reforestation along all the rivers. By 8000 B.P. savanna or rain forests spread over much of the globe from $30°N$ to $30°S$, beyond which mixed temperate forests took over. In the high latitudes, the postglacial warming was retarded by the presence of glacial ice, patches of which lasted until about 6000 B.P. The warm phase or *climatic optimum* is thus said to be diachronous; it started in the low latitudes and gradually shifted toward the poles. The very high latitude oceans also were retarded in warming, and, in turn, the cold Arctic waters made the land slow to recover from the glaciation.

Since the mid-Holocene mild typical (*interglacial*) period, there has been a global cooling trend. The subtropics have become progressively more affected by droughts. Regions such as the Sahara, that were covered by savanna grasslands and light forest when the first Egyptian dynasties were established along the Nile, have gradually reverted to deserts (Wendorf and Schild, 1976; Fairbridge, 1976b). In the high latitudes, the tree line, for example in northern Canada, has retreated 500 km in the past 5000 years.

In mountainous country, such as the Alps or Alaska, where there are small glaciers, the Holocene chronicle has been punctuated by brief but dramatic readvances of the ice. Glacial surges are suspected. These are known as *neoglacial* or *little ice age* events. In most cases the glaciers would grow rapidly and advance down-valley several kilometers for a few decades or a century and then retreat once more (Denton and Karlén, 1973). The frequency of these advances has been increasing during the past millennia. The level of the sea has reacted eustatically to such growth or retreat of glaciers, so that mean sea level has fluctuated several meters at times. Thus the coastal geography is far from constant, but is in a continuous state of dynamic change. Along the rocky coasts of more stable regions, however, the Holocene shore is often no more than 50 m from the shoreline of 2 million years ago, but wherever the coastal formations are of loose, easily eroded materials, rapid changes of 1000 m or more may be expected.

A special feature of the late Holocene has been the gradual appearance of *anthropogenic sediments* and related phenomena. Man's domestication of animals (cattle, goats, and so on), notably in the Neolithic, followed by general agriculture (deforestation, plowing, and so on) has led to greatly accelerated erosion and increased sedimentation downstream. Flooding and beach erosion are notable consequences of man's environmental impact. Natural processes, such as climatic fluctuations and wind vector changes, are equally important.

RHODES W. FAIRBRIDGE

References

Andrews, J. T., 1974. *Glacial Isostasy*. Stroudsburg, Pa.: Dowden, Hutchinson & Ross, 727p.

Dansgaard, W.; Johnson, S. J.; and Clausen, H. B., 1971. Climatic record revealed by the Camp Century Ice Core, *in* K. K. Turekian, ed., *The Late Cenozoic Glacial Ages*. New Haven, Conn.: Yale University Press, 37-56.

Denton, G. H., and Karlén, W., 1973. Holocene climatic variations: Their pattern and possible cause, *Quaternary Research* **3**, 155-205.

Fairbridge, R. W., 1968. Holocene, postglacial or recent epoch, *in* R. W. Fairbridge, ed., *Encyclopedia of Geomorphology*. New York: Reinhold, 525-536.

Fairbridge, R. W., 1974. Holocene epoch, *in Encyclopedia Britannica*, 15th ed., 998-1007.

Fairbridge, R. W., 1976*a*. Shellfish-eating preceramic Indians in coastal Brazil, *Science* **191**, 353-359.

Fairbridge, R. W., 1976*b*. Effects of Holocene climatic change on some tropical geomorphic processes, *Quaternary Research* **6**, 529-556.

Fairbridge, R. W., 1976*c*. The search for a boundary stratotype of the Holocene, *Boreas* **5**, 194-197.

Fritts, H. C., 1977. *Tree Rings and Climate.* New York: Academic Press, 567p.

Hageman, B. P., 1969. Development of the western part of the Netherlands during the Holocene, *Geologie en Mijnbouw* **48**, 373-388.

Hillaire-Marcel, C., and Fairbridge, R. W., 1978. Isostasy and eustasy of Hudson Bay, *Geology* **6**, 117-122.

Martin, P. S., and Wright, H. E., Jr., eds., 1967. *Holocene Pleistocene Extinctions.* New Haven, Conn.: Yale University Press, 453p.

Mörner, N. -A., 1976. Eustasy and geoid changes, *Jour. Geology* **84**, 123-151.

Tooley, M. J., 1974. Sea-level changes during the last 9000 years in northwest England, *Geog. Jour.* **140**, 18-42.

Wendorf, F., and Schild, R., 1976. *Prehistory of the Nile Valley.* New York: Academic Press, 404p.

Cross-references: *Climate, Coastal; Global Tectonics; Hypsithermal; Radiocarbon Dating; Sea Level Curves.* Vol. III: *Holocene, Postglacial or Recent Epoch.*

HONEYCOMB WEATHERING—See ALVEOLAR WEATHERING

HORSETAIL—See EQUISETOPHYTA

HUMAN IMPACT

Human activities alter shapes and locations of shorelines, the nature of coastal ocean bottoms, and the quality of ocean waters in coastal environments. Some of these alterations are transitory, small in scale, and of little importance geologically. Others are widespread and involve large areas, so that locally man has become a major geological factor. For example, in the New York metropolitan region waste disposal operations are the largest source of sediment reaching the United States Atlantic continental shelf between Cape Cod and Cape Hatteras (Gross, 1972). The number and size of such large scale alterations are likely to increase in the future.

Agricultural Effects

Erosion and Sedimentation. The most extensive alterations of coastal environments have occurred because of *agricultural* practices. Agriculture—especially primitive agriculture— greatly increased erosion of farmed lands, quickly destroying their fertility. Eroded soils were carried by streams to be deposited in estuaries and coastal ocean areas. Many formerly thriving tobacco ports on Chesapeake Bay are now separated from the bay by several kilometers of wetlands.

In 4500 years the delta of the Tigris-Euphrates rivers has prograded about 300 km into the Persian Gulf. At times the rate of progradation has been about 16 m/year. The delta of the Po River has also built out into the Adriatic. Likewise, relatively high erosion rates in China have been attributed to agricultural practices, especially to the extensive deforestation of the hills (Davis, 1956).

Land Reclamation. Increased populations have created demands for agricultural land in coastal regions. In densely populated regions, governments have *reclaimed* land from the shallow continental shelf. Between 1200 and 1950, the Netherlands reclaimed about 6300 sq km, and reclamation of new areas continues. This land is reclaimed from shallow nearshore areas by the use of dikes.

In the 1930s the Zuider Zee, a large, shallow embayment in the Netherlands, was cut off by a dike, changing it from a brackish estuary to a freshwater lake, Ijssel Meer. Land reclamation has also been done in low-lying areas of Belgium and Great Britain (van Veen, 1962).

Industrial Facilities

Offshore facilities, typically *artificial islands*, have been built to provide space for waste disposal operations (especially dredge spoil disposal), sites for large electrical power-generating facilities, deep-water port facilities for deep-draft tankers and other vessels.

Offshore port facilities have been used for loading and unloading supertankers carrying hundreds of thousands of tons of petroleum and drawing up to 25-35 m of water. Savings in transportation costs provided by larger vessels has made them more attractive.

Locating the facility offshore reduces the amount of *dredging*. Offshore port and storage facilities have been used in the North Sea and the Persian Gulf and are planned for United States coastal waters.

Shortages of suitable sites for electrical power plants on land and requirements for large volumes of cooling waters have made offshore sites for large power plants especially attractive. Distance from shore reduces residents' objections to the disruptive appearance of a power plant on the coastal landscape, and also reduces the exposure of the coastal population to potential radiation hazards caused by accidents at *nuclear power plants*.

Mining

Mineral and fuel production from coastal lands and ocean bottoms alter the marine environment. Sand and gravel are prime examples of the causes of alterations resulting from mineral exploitation. Sand and gravel are produced in large volume. Being relatively low-cost, high-bulk commodities, they are typically produced near the market, since shipping costs are high. Near coastal cities, sites become scarce as the urbanization expands, so that sand and gravel deposits on the shallow continental shelf are exploited. This has long been the practice in the United Kingdom, Denmark, Netherlands, Japan, and is increasing in the United States.

Sand and gravel are typically produced from the ocean bottom by dredging. The dredge may remain stationary, in which case it can dredge a relatively large, deep hole (up to 20 m deep and 100 m across) at a single location before moving to a new spot. A moving dredge removes only a layer about 1 m thick. The primary impact of this activity is to alter the bottom topography and to remove the bottom-dwelling organisms. The results of changing bottom topography may be to alter the wave regime on the nearby beaches and perhaps to alter currents. The removal of the bottom organisms may result in recolonization by similar organisms if the bottom is not greatly altered and the current regimes remain comparable. If large holes remain, the waters become commonly stagnant, so that dissolved oxygen is removed in summer, causing odors and preventing the reestablishment of normal bottom-dwelling fauna. Dredging of the deposits and processing them for market can release suspended sediment to the waters.

Tin, phosphorite, carbonate sands, and heavy minerals have also been mined by dredging continental shelves and shallowly submerged banks such as the Bahamas. Tin has been produced by dredging the shallow southeast Asian continental shelf off Indonesia and Thailand. Phosphorite, used for making fertilizer, will likely come from the submerged deposits as the presently exploited deposits on land are depleted and prices increase. The techniques are similar to strip mining on land and are likely to significantly alter the bottom topography and sediment characteristics.

M. GRANT GROSS

References

Davis, J. H., 1956. Influences of man upon coastlines, in W. L. Thomas, Jr., ed., *Man's Role in Changing the Face of the Earth*. Chicago: University of Chicago Press, 504-521.

Gross, M. G., 1972. *Oceanography: A View of the Earth*, 2d ed. Englewood Cliffs, N.J.: Prentice-Hall, 581p.

van Veen, J., 1962. *Dredge, Drain, Reclaim: The Art of a Nation*, 5th ed. The Hague: Martinus Nijhoff, 200p.

Cross-references: *Artificial Islands; Conservation; Coastal Reserves; Dredging; Land Reclamation; Mineral Deposits; Mineral Deposits, Mining of; Nuclear Power Plant Siting; Pollutants; Thalassotherapy; Thermal Pollution; Tourism; Waste Disposal.*

HUMATE

Humate (coffee rock, sandrock) is a dark brown to black, water-soluble organic substance that cements or impregnates layers of dune and beach sand. It also accumulates in and beneath marsh deposits, near ground water seepages, in subsurface soil layers and as a type of organic sediment in bodies of brackish or saline water. According to Swanson and Palacas (1965), it is derived by leaching from decaying plant material or humus when soluble and colloidally dispersed humic substances are transported by surface and subsurface waters to brackish or saline water bodies, as well as to subsurface sands, where flocculation or precipitation of humate takes place by various physical-chemical mechanisms.

In coastal environments in Australia, it is mainly confined to deeply podsolized Pleistocene sands and to lagoonal-swamp environments, and has been termed *coffee rock* by Bird, *Waterloo rock* by Griffin, *coastal sandrock* by Coaldrake and McGarity, and *indurated sand* (Hails, 1968). The thickness of humate varies from a few centimeters to several meters, and it extends below present sea level. Near Evans Head, in northern New South Wales, Australia, humate forms coastal cliffs, which stand about 6 m above present mean sea level, and is overlain by fixed podsolized dunes.

From the results of their work on humate in Florida, Swanson and Palacas (1965) conclude that it can be compared with hydrocarbons in that both have a biologic origin and both migrate or are transported with water through sediments. However, they are quite distinct chemically in that humate is a high-oxygen low-hydrogen substance, is coallike in composition, is insoluble in benzene, and is transported in a soluble or colloidal form in water.

The modes and sites of accumulation of the two in a sediment also differ because most hydrocarbons, being immiscible in water, are transported, physically separated, and concentrated as distinct gaseous or fluid substances in a structural or stratigraphic trap. Therefore humate does not appear to be a progenitor of petroleum hydrocarbons.

Humate is generally deposited when it combines with waters containing dissolved ions, such as aluminum, iron, and magnesium, on complexing with clay colloids, or on combining with waters of pH <5. Nevertheless, many queries about the origin and geochemistry of humate remain to be answered (Thom, 1967).

JOHN R. HAILS

References

Hails, J. R., 1968. The Late Quaternary history of part of the mid-north coast, New South Wales, Australia, *Inst. British Geographers Trans.* **44**, 133-149.

Swanson, V. W., and Palacas, J. G., 1965. Humate in coastal sands of northwest Florida, *U.S. Geol. Survey Bull.* **1214-B**, B1-B29.

Thom, B. G., 1967. Humate and coastal geomorphology, *Louisiana State Univ. Coastal Studies Bull.* **1**, 15-17.

Cross-references: *Beachrock; Calcarenite; Flocculation; Soils.* Vol. VI: *Humic Matter in Sediments.*

HURRICANE EFFECTS

Hurricanes modify the coastal zone in several different ways: through storm wave activity, because of storm tides (the *storm surge*), and as a result of the tremendous amount of rainfall (and hence runoff) that may be dumped on the area quickly.

The storm tide (in some cases coupled with high *astronomical tides*) may raise the water level 3 or more meters. The probability of getting a specific storm tide height, exclusive of the astronomic tide, in a given region in a given year can be estimated by "extreme value" methods, using historical data as a base (Gumbel, 1958). Such estimates cannot be extrapolated from one area to another, except in the loosest sense.

Storm waves in deep water can be estimated from standard wave charts, using weather map data (most obviously, wind direction, velocity, and duration) as a base. Such waves can be tracked ashore by computer methods, using the temporary water level rather than MSL, to produce surf zone estimates. Breakers, under these conditions, tend to be spread over a wide zone parallel with the shore. Large storm waves break farther offshore, in deeper water, whereas smaller storm waves may break landward of the fair-weather surf (because of the high water level). Storm waves are steep (as opposed to swell), hence storm wave energy expenditure on the bottom (for one *wave train*) tends to be concentrated in a narrower band than does the expenditure of swell energy. The distribution of dE/dy, from the point where wave motion first touches bottom to the beach cannot, however, be represented by a simple curve.

A rise in water level of 1 m may increase the littoral component of wave power—with no increase in wave energy levels—by a factor of 10^2 or more (Berquist and Tanner, 1974); therefore storm waves can be highly damaging. Along extremely *high energy coasts*, this may not represent a great departure from the ordinary; on *"zero" energy coasts*, the various factors that normally attenuate wave energy may also minimize storm waves, so that little damage is done. Hence, it is inferred that hurricane wave damage will be most spectacular in regions having intermediate long-term wave energy levels.

Wave energy levels and propagation directions depend on the circulation around the *storm eye* (counterclockwise in the Northern Hemisphere). For a hurricane moving northward across the central Gulf of Mexico, maximum waves tend to be located to the northeast and east of the eye. Such waves, sweeping into a coastal reentrant, coupled with the storm tide enhanced in that reentrant, may be extremely damaging. Such appears to have been the case with Hurricane Camille (1969) (Tanner, 1970).

Most sediment moved from beaches during hurricanes is swept offshore, from where part, or all, of it may perhaps be recovered at a later date, under the influence of *swell*. Experience in the past few decades indicates that beach ridges in general have not been built by hurricanes. Narrow barrier islands may be moved shoreward during hurricanes by the washover process.

The effects of hurricane rainfall include dilution of near-shore salinity, which may be important over the short term in coastal marshes and swamps, and soil erosion and flooding in the adjacent uplands.

WILLIAM F. TANNER

References

Berquist, C. R., and Tanner, W. F., 1974. Analysis of water-level rise effects on littoral transport, *Gulf Coast Assoc. Geol. Socs. Trans.* **24**, 255-256.

Gumbel, E. J., 1958. *Statistics of Extremes.* New York: Columbia University Press, 375p.

Tanner, W. F., 1970. Significance of Camille, *Southeastern Geology* **12**, 95-103.

Cross-references: *Climate, Coastal; Storm Surge; Storm Wave Environments; Waves; Winds.* Vol. VII: *Hurricanes.*

HURRICANE TIDE OR WAVE—See STORM SURGE

HYDRAULIC ACTION AND WEDGING

Wave impact can exert great pressures and can be a highly effective agent of erosion, especially upon jointed and fractured rock. The pressures produced by breaking waves have been measured by dynamometers, and the values agree well with theoretical estimates derived from a formula of Gaillard (1904),

$$1.31(C + V_{max})^2 \frac{\rho}{2g},$$

where C is the wave celerity and V_{max} is the orbital velocity at the crest of the wave. When the waves are long and their celerities are correspondingly high, the pressures generated may be quite substantial; pressures as high as 250 kg/m^2 (6000 lb/ft^2) have been reported.

When waves break against a cliff, water is driven into every crack, and air within the cracks is greatly compressed. The compressed air acts as a wedge driven into the cracks, widening the fractures and loosening blocks of rock. As the wave recedes, the compressed air expands with explosive force, effectively *quarrying* out large blocks (Johnson, 1919; King, 1972). Such quarrying action has provided enough energy to move blocks weighing many tons. While the precise process is not isolated as a rule, there are many reports of the movement of large blocks by wave action. A 2500-ton block of concrete at Wick (Scotland) was thus displaced, and at Bilbao (Spain) a section of a breakwater weighing 1700 tons was completely overturned (Matthews, 1934).

MICHAEL R. RAMPINO

References

Gaillard, D. D., 1904. *Wave Action in Relation to Engineering Structures.* Washington, D.C.: U.S. Army Corps of Engineers, 232p.

Johnson, D. W., 1919. *Shore Processes and Shoreline Development.* New York: John Wiley & Sons, 583p.

King, C. A. M., 1972. *Beaches and Coasts.* New York: St. Martin's Press, 570p.

Matthews, E. R., 1934. *Coast Erosion and Protection.* London: C. Griffin and Co., 228p.

Cross-references: *Cliff Erosion; High-Energy Coasts; Quarrying Processes; Wave Erosion; Weathering and Erosion, Mechanical.*

HYDROGEOLOGY OF COASTS

The term *hydrogeology* was originally defined (Mead, 1950) as the study of "the laws of the occurrence and movement of subterranean waters"—hence a branch of hydrology. Gary, McAfee, and Wolf (1972) define the term as "the science that deals with subsurface waters and related geologic aspects of surface waters. Also used in the more restricted sense of groundwater geology only." The definition *groundwater geology* is the one most frequently encountered in the literature.

The term *coast* generally refers to a linear zone of finite width that includes the shoreline—that is, the line of contact between a surface water body, the atmosphere, and dry land. In the present discussion, however, it is necessary to distinguish between saltwater and freshwater coasts, as the hydrogeologic situation in the two types of coasts is dissimilar.

In most saltwater coastal regions, there is a discharge of freshwater from the aquifer to the sea. The surface of contact between fresh and saltwater is the *freshwater/saltwater interface* (*FSI*). A first approximation to the geometry of the *FSI* (Walton, 1970) can be determined by assuming that the two fluids are immiscible and static (Fig. 1). The depth of the *FSI* below the water table measured vertically below point A can be determined by assuming hydrostatic equilibrium between the two fluids—that is, the hydrostatic pressure at point B is identical to that at point C. Therefore we can set:

$$P_C = P_B \quad (1)$$

$$d_4 \rho_2 g = (d_2 + d_3) \rho_1 g \quad (2)$$

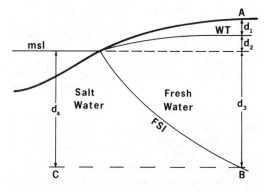

FIGURE 1. The geometry of the water table (*WT*) and the freshwater/saltwater interface (*FSI*) and their relation to mean sea level (*msl*) according to the Ghyben-Herzberg model.

where ρ_2 and ρ_1 are densities of saltwater and freshwater, respectively, and g is the acceleration of gravity.

From Figure 1, note that $d_4 = d_3$; therefore we can write:

$$d_4 \rho_2 = (d_2 + d_4)\rho_1$$

whence

$$d_4 = \frac{\rho_1}{\rho_2 - \rho_1} d_2 \quad (3)$$

Equation (3) is called the *Ghyben-Herzberg relation*.

Assuming that ρ_2 is 1.025 gm/cm^3 and ρ_1 is 1.000 gm/cm^3:

$$d_4 = 40 \, d_2 \quad (4)$$

The total depth to the *FSI* measured vertically below point A is

$$D = d_1 + d_2 + d_3$$
$$= d_1 + 41 \, d_2 \quad (5)$$

Expressed differently, the subsea elevation of the *FSI* measured vertically below point A is:

$$-d_3 = -40 \, d_2 \quad (6)$$

The limitation of the Ghyben-Herzberg approach is seen easily by consideration of *Darcy's law*, which relates specific discharge (q = discharge/area) to aquifer hydraulic conductivity (K) and hydraulic gradient (Darcy, 1856):

$$q = K \, dh/ds \quad (7)$$

where both q and K have the dimensions (LT^{-1}). The slope of the water table (or potentiometric surface) can be written (Fig. 2):

$$dh/ds = \sin a = q/K \quad (8)$$

The slope of the *FSI* is:

$$d/ds \left(\frac{\rho_1}{\rho_2 - \rho_1} h \right) = \sin b$$

$$= \frac{\rho_1}{\rho_2 - \rho_1} \cdot q/K \sim 40 \, q/K \quad (9)$$

Since the Ghyben-Herzberg relation assumes static conditions—that is, q equals zero, both of these equations equal zero; hence both the water table and the FSI are horizontal, contrary to fact.

Hubbert (1969) assumed that the fluids are not static but, rather, that both fluids have nonzero velocities, and derived a general equation for the slope of the *FSI*:

$$\sin b = \left(-\frac{\rho_2}{\rho_2 - \rho_1} \cdot \frac{q_{2s}}{K_2} + \frac{\rho_1}{\rho_2 - \rho_1} \cdot \frac{q_{1s}}{K_1} \right) \quad (10)$$

where q_{1s} is the component of specific discharge of the ρ_1 fluid in the s direction.

Note that in the case where the saltwater body is static ($q_{2s} = 0$), equation (10) reduces to (9); hence the latter does provide a reasonably good description of the *FSI* under natural conditions. This case is illustrated in Figure 2, which shows the velocity potential and stream lines. It is also shown that the freshwater discharges through a zone of finite width extending from above to below mean sea level. The stream lines illustrate the velocity vectors of freshwater flow. The velocity potential lines indicate the magnitude of flow, their spacing being inversely proportional to flow velocity.

Conservation of mass requires that the discharge through any aquifer cross section be constant. It is obvious that the flow velocity must increase, from right to left in the illustration, as the cross-sectional area of flow decreases. From equation (9) the slope of the *FSI* must increase proportionately. It is further to be noted that the depth to the *FSI* according to the Ghyben-Herzberg relation is valid only if the distance is measured down the velocity potential line (to point D) and not vertically to point A. The significance of this discrepancy depends on the slope of the *FSI*,

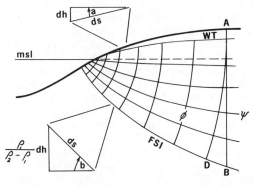

FIGURE 2. Schematic section of a coastal freshwater aquifer, showing the water table (*WT*), freshwater/saltwater interface (*FSI*), velocity potential lines (ϕ), stream lines (ψ), and the freshwater discharge zone. Also shown are the slopes of the *WT* and the *FSI*.

being negligible where the slope is only a few degrees, as is the case throughout much of the eastern United States.

The assumption of immiscible fluids is obviously inaccurate in the case of freshwater and saltwater. The *FSI* is actually a zone across which the fluids mix. An estimate of the thickness of the mixing zone can be made from the data of Cooper (1959) from the Biscayne aquifer near Miami. If one defines freshwater by ppm (parts per million) chlorides less than 200, and saltwater by ppm greater than 10,000, the zone is from 3-4 m thick. Using 15,000 ppm Cl as a criterion for seawater, the thickness increases by 1-2 m. At a distance of about 500 m inland from the shoreline, the thickness of the mixing zone appears to thicken abruptly. Similar data are presented by Lusczynski (1960), which also show a thickening of the mixing zone landward from the coast. The thickness of the mixing zone is apparently inversely related to the specific discharge (De Wiest, 1965).

In the vicinity of freshwater coasts, a different situation exists. Where there is no density contrast between the groundwater and the surface water body, there is no basis for an interface. Except for ephemeral conditions in arid regions, there is generally a flow of groundwater toward the coast because of the same effect that causes groundwater to flow toward a saltwater coast—that is, the water flows down the pressure gradient, which is generally a reflection of the topographic gradient. This water must be removed from the surface water body by surface flow and/or evaporation in order to maintain a nonzero hydraulic gradient.

These natural conditions are frequently altered by artificial removal of groundwater in coastal regions. Pumping water from a well produces a cone of depression around the well. If this cone lowers the water table or potentiometric surface to a position near or below sea level in a coastal region, the position of the FSI beneath the well is proportionately displaced upward toward sea level; hence the well may soon produce saltwater. Because specific discharge through an aquifer is generally low (measured in meters to tens of meters per year), saltwater encroachment problems are not easily remedied (Kashef, 1977; Andrews, 1981). It may take many years, even decades, to flush the salt from the contaminated aquifer and restore it to its original, usable condition.

JAMES P. MAY

References

Andrews, R. W., 1981. Salt-water intrusion in the Costa de Hermosillo, Mexico: A numerical analysis of water management proposals, *Ground Water* **19**, 635-647.

Cooper, H. H., Jr., 1959. A hypothesis concerning the dynamic balance of fresh water and salt water in a coastal aquifer, *Jour. Geophys. Research* **64**, 461-467.

Darcy, H., 1856. *Les Fontaines publiques de la Ville de Dijon.* Paris: Victor Dalmont, 647p.

De Wiest, R. J. M., 1965. *Geohydrology.* New York: John Wiley & Sons, 366p.

Gary, M.; McAfee, R., Jr.; and Wolf, C. L., eds. 1972. *Glossary of Geology.* Washington, D.C.: American Geological Institute, 805p.

Hubbert, M. K., 1969. *The Theory of Ground-water Motion and Related Papers.* New York: Hafner Publishing Co., 311p.

Kashef, A. I., 1977. Management and control of saltwater intrusion in coastal aquifers, *CRC Crit. Rev. Environ. Control* **7**, 217-275.

Lusczynski, N. J., 1960. Head and flow of ground water of variable density, *Jour. Geophys. Res.* **66**, 4247-4256.

Mead, D. W., 1950. *Hydrology.* New York: McGraw-Hill, 728p.

Walton, W. C., 1970. *Groundwater Resource Evaluation.* New York: McGraw-Hill, 664p.

Cross-references: *Beach Firmness; Beach Processes; Coastal Engineering, Research Methods; Ghyben-Herzberg Ratio; Land Reclamation.* Vol. XIII: *Hydrogeology.*

HYPERSALINE COASTAL LAKES

Lakes with an unusual proportion of dissolved salts are called *brackish* or salty; there is no clear-cut limit between freshwater lakes and brackish lakes or between these and salt lakes. Although generally shallow, some are real seas: Caspian, Aral, Dead. F. W. Clarke (1916) classified salt lakes in nine categories, based upon the predominance or concurrent presence of chlorides, sulfates, carbonates, and magnesium salts. He did not account for lakes rich in boron or nitrates.

According to the type of salt contained in the waters, geographers have distinguished among *saline* lakes—salt lakes saturated with precipitated sodium chloride; *bitter* lakes (sodium sulfates), as in Egypt; *alkali* lakes (sodium and potassium carbonate); and *borax* lakes (borax and related borate minerals). The term "brackish" is used to indicate a salinity generally less than that of seawater (15-30‰); such terms as oligohaline, miohaline, mesohaline, pliohaline, and polyhaline indicate different degrees of salinity, placed here in increasing order.

All saline lakes are not coastal lakes by any means; the great salt lakes of the world lie in basins of internal drainage, such as Lake Eyre in Australia and Great Salt Lake in Utah.

Lake water analysis may provide clues to the probable origin of salt lakes, some of which are definitely continental. G. Bertrand (1938) contends that the Dead Sea's salinity is due to material brought to it by the Jordan River, a position already taken by Lortet (1884). Both the Dead Sea and the Great Salt Lake are hypersaline.

Many coastal lakes, some of which are hypersaline, are, however, definitely of marine origin. This is particularly evident in Eastern Europe. The Kara Bogaz, on the east shore of the Caspian Sea, has a salinity of about 282‰, while the nearby Caspian waters reach only 12.8‰; the former is fed by a continuous inflow of waters of the Caspian, which, because of a high degree of evaporation due to the dry climate, are regularly renewed. As a result, important salt deposits accumulate. Salt deposits in lagoons are exclusively of marine origin, and their order of succession as well as their proportions are similar to those of seawater.

Behind spits that build along coasts, marine saltwater lakes are formed. They are common: for instance, along the Mediterranean coast of France, Gulf of the Lion and the coasts of Languedoc and of Landes; in northern Germany, *haffs* are the area of the sea gradually cut off as a bar or spit grows and eventually forms a lagoon that, in turn, becomes a lake when the bar connects with the land at both ends. The lagoon itself could be in communication with the sea by the way of a narrow channel; elsewhere, however, such channels do not exist, and the lagoon is separated by a natural levee from the adjoining sea. In such cases infiltration of sea water occurs through the levee's permeable rocks; again, if climate conditions permit, hypersalinity of lagoon or lake water ensues.

There are in fact two genetic types of such lakes: the marine lagoons, and the fluviomarine *limans*. Both lagoons and limans can be filled with either freshwater or saltwater. In some cases, the water is brackish; in others, hypersaline.

The lagoons are generally associated with low, sandy coasts, and they were in the past marine gulfs or bays. Because of active offshore currents carrying sand and mud, the mouth of such bays was gradually silted, and the water behind the newly built bars was insulated from the marine body of water.

The spits isolating the lagoons from the open sea are stable as long as: the tides are low, and thus have not enough energy to wash away the accumulating detritus; the rivers emptying into the lagoon are small, and discharge only insignificant amounts of water that are not powerful enough to flush seaward the sand bars closing the bay. Marine transgressions and regressions as well as tectonic processes (rise of ridges or formation of horsts) are also sometimes involved in forming lagoons.

The limans have a somewhat similar origin—the damming by baymouth bars of a restricted mass of water, although the mass was not situated in a former gulf, but represented the mouth of a river valley subsequently flooded by a transgressive sea (a process similar to the formation of *ria coasts*) and then severed by the accumulation of detritus building the bar (Banu, 1970). During the time of marine transgression, wave and current erosion reshaped the valley banks, which consequently became much steeper. The limans, often situated along cliffy, rocky coasts, are long, narrow bodies of water, and their longitudinal cross section is typical. Because they are nothing but drowned river valleys with their initial longitudinal profiles dipping seaward, the limans' maximum depths are attained at the inner wall of the sand bar (seaward extremity of the lake). Limans usually have a muddy bottom, and are characteristic of the Black Sea and its rivers.

The *pan* or *panne* is a salt or brackish lake, usually shallow. Pannes are ephemeral, and become part of the coast; the city of De La Panne along the Belgian North Sea coast lies in an area of numerous dried pannes. The depth of water in both lagoons and limans is usually slight (1–2 m up to 12 m, seldom more).

Whether the water in such lakes is fresh, brackish, salty, or hypersaline depends on a series of factors and on their mutual relationships: the ratio between the rate of precipitation, the rate of runoff, and the rate of evaporation; presence (or absence) of underground springs and of a karstic morphology; permeability of the lake bottom; degree of insulation from the sea; type and amount of water discharged by the rivers (if any) emptying into the lake.

Along the shores of the Caspian and the Black seas, the evaporation rate far exceeds the precipitation rate. The ratio between them is highest on the east coast of the Caspian Sea, and reaches especially high values along the west coast of the Black Sea. Here, for instance, during a ten-year period (1958–1968), the evaporation mean value was 960 mm, while that of precipitation was less, only 460 mm.

It is thus obvious that, given a certain amount of time, the waters in insulated lagoons or limans will become increasingly mineralized. Moreover, it appears probable that, at least in some cases, the initial water was salty—relic seawater left from the last marine regression, known in the Dobrudja as the Histrian regression (Cotet, 1970).

If strong underground springs are present, their discharge of freshwater on the lake's bottom can eventually drastically change the salinity of the lake waters, which will become either fresh or, more probably, brackish. Such situations are possible in areas where the rocks flooring the lake are limestones or dolomites and extend at the surface beyond the perimeter of the body of water. A karstic water network will be gradually built, and frequent infiltrations will feed underground springs, some of them situated under the lagoon or the liman. In such cases, the degree of permeability of the lacustrine sediments (if present) is of course important (Breier, 1970a, 1970b). The existence of large subterranean discharges of freshwater is well illustrated in the case of Lake Siutghiol (north of the city of Constanţa, Romania). In this lake the net surplus of water (the difference between the evaporated water and the influx of groundwater) is an impressive 765 l/sec., despite the fact that the annual rate of evaporation is much higher than the corresponding rate of precipitation (Nicolae, 1970).

Another important factor is the amount of water discharged by streams into the limans (since the latter are nothing but the drowned lower portion of their valleys). When such rivers carry a sizable quantity of water (the Dniester, for example), the lakes are filled with freshwater. Where only small creeks empty into the liman, the water will be brackish (the Sinoe-Razelm lakes).

Finally, the degree of insulation of the lagoons or limans from the adjacent sea also plays a role in determining the chemistry of the water. Communication through narrow inlets and sluiceways will obviously allow a certain amount of seawater to penetrate into the lagoon.

When the lake is completely severed from the sea, then the salinity will depend on climate, groundwater, geology, and rivers.

The best-known and the largest hypersaline lagoon in the world is the Kara Bogaz Ghiol, on the east coast of the Caspian Sea. It lies 31 m below the level of the planetary ocean and 3 m below the level of the Caspian Sea (which is itself at -28 m). The concentration of salts is remarkably high (282 g/l), almost eight times the normal salinity of seawater.

Historically, the concentration has increased from 164 g/l in 1894 to 280 g/l in 1938 as a result of the drop in the level of the Caspian Sea. The lowering by 155 cm of the Caspian level in the past 75 years has drastically reduced the amount of less salty water (concentration 12.73 g/l) moving from the sea into the lagoon, and consequently there is less dilution of the hypersaline solution and a steady increase in concentration. The rate of precipitation from the suprasaturated solution is extremely high, and the lagoon's bottom is covered with thick layers of halite, especially mirabilite ($Na_2SO_4 \cdot 10H_2O$).

Laguna Madre on the Texas coast, except for two small inlets, is separated from the Gulf of Mexico by 160-km long Padre Island, a sandy *barrier island* (Rusnak, 1960). Laguna Madre is extremely shallow, its mean depth being 0.9 m. The highest salinity recorded was 110‰; the mean salinity is about 50‰. Storms and heavy rains may reduce the salinity drastically. Padre Island was once a remarkable fishing ground, but the catch has steadily declined because of pollution by industrial and domestic waste waters and agricultural land runoff.

Coastal salt lakes are numerous along the Romanian shores of the Black Sea. The waters of Lake Tekirghiol are a muddy dark green; Lake Sinoe's water leaves circles of light brown salt around the white marble vestiges of Histria, an ancient Greek settlement (700 B.C.).

Lake Tekirghiol is a typical hypersaline lake along the Black Sea shore. Its area is 1,070 hectares; its depth does not exceed 10 m; and it is the only lake on the Romanian coast to lie below sea level (-1.50 m). It originated as a fluviomarine liman that was subsequently totally isolated from the Black Sea and from the feeding river. Its water has an average salinity of 95.5‰, several times that of the Black Sea, whose coastal waters here, up to a depth of 10 m, never exceed 15‰. This difference in salinity, is, however, partially the result of north winds that push the lower-salinity waters of the Danube delta southward. The lake bottom is covered with a thick layer of dark olive and black mud that has therapeutic properties and is extensively used in resort towns along the Romanian Black Sea coast (Pricăjan and Opran, 1970).

Romanian researchers stress the special character of the 732-km² lagoonal complex Razelm, an ancient marine gulf (Halmyris) situated between the Black Sea and the Dobrudja Plateau. Now separated from the sea and the Danube delta, it is compartmented by sand banks into a series of lakes (Sinoe, Smeica, Babadag, Golovitza, Agighiol, and Razelmul Mare) connected with one another. Some (Razelmul Mare) are also connected with the Danube delta; others (Sinoe, Golovitza), with the sea. Salinity is subject to periodic changes because of discharge variations of the Danube, of common breaching of natural coastal dikes (spits), and of the cutoff of communication with the Danube. Normal salt concentration varies from 1 g/l in the Razelmul Mare to 6 g/l in Lake Sinoe; at times it has reached in these lakes, respectively, 6.5 g/l and 22 g/l. The temporary salinity increase favored, after 1950, the development of new species of fish that are prized for their nutrients (*e.g., Rithrohaspanopeus harrissii tridentatus*).

It should be mentioned that several coastal bodies of water here are freshwater lakes fed by rivers and springs—e.g., Lake Siutghiol (Fig. 1).

ROGER H. CHARLIER
LORIN R. CONTESCU

References

Banu, A. C., 1970. Du nouveau sur la genèse et l'age des limans fluviaux situés dans la région du cours inférieur de Danube, *in Géographie des Lacs, Travaux du Colloque National de Limnologie Physique*. Bucharest: State Publishing House, 45-50.

Bertrand, G., 1938. Sur la quantité de zinc contenne dans l'eau de mer, *Acad. Sci. Comptes Rendus (Paris)* **207**, 1137.

Breier, A., 1970*a*. Raportul dintre caracteristicile morfometrice si morfografice ale lacurilor de pre litoralul românesc al Mării Negre, *Studii si Cercetari Geologie, Geografisica și Geografia–serie Geografia* **17**, 187-194.

Breier, A., 1970*b*. Le régime des niveaux dans le complexe Tasaul–Suit–ghiol, *in Géographie des Lacs, Travaux du Colloque National de Limnologie Physique*. Bucharest: State Publishing House, 135-149.

Clarke, F. W., 1916. The data of geochemistry, *U.S. Geol. Survey Bull. 616*, 821p.

Cotet, P., 1970. Las lacs littoraux dobrogéens et leurs rapports génétiques avec les changements de niveau de la Mer Noire, *in Géographie des Lacs, Travaux du Colloque National de Limnologie Physique*. Bucharest: State Publishing House, 27-45.

Lortet, L. C., 1884. *Le Syrie d'aujourd'hui, voyages dans la Phénicie, le Liban et la Judée, 1875-1880*. Paris: Hachette, 675p.

Nicolae, T., 1970. Quelques considérations concernant l'utilisation des eaux du lac Suitghiol, *in Géographie des Lacs, Travaux du Colloque National de Limnologie Physique*. Bucharest: State Publishing House, 297-305.

FIGURE 1. Mamaia, Romania, with the Black Sea on the right and Lake Siutghiol on the left (photo: R. H. Charlier and L. R. Contescu).

Pricăjan, A., and Opran. C., 1970. La protection et l'exploitation rationelle des lacs et des boues thérapeutiques, in *Géographie des Lacs, Travaux du Colloque National de Limnologie Physique.* Bucharest: State Publishing 307-311.

Rusnak, G. A., 1960. Sediments of Laguna Madre, Texas, in F. P. Shepard, F. B. Phleger, and Tj. H. van Andel, eds., *Recent Sediments, Northwest Gulf of Mexico.* Tulsa, Okla.: American Association of Petroleum Geologists, 82-108.

Cross-references: *Bars; Haff; Human Impact; Hydrogeology of Coasts; Liman and Liman Coasts; Nehrung; Ria and Ria Coasts; Spits; Submarine Springs; Thalassotherapy.* Vol. III: *Kara Bogaz, Coastal Morphology.*

HYPSITHERMAL

A climatic phase in the early to middle part of the Holocene (q.v.)—lasting several thousand years—when conditions were appreciably warmer than today, is called *hypsithermal*. First recognized in Scandinavia as the Climatic Optimum, paleobotanical studies showed that the warm annual temperature was around $2.5°C$ higher than today (de Geer, 1912; Faegri and Iverson, 1950; Nilsson, 1965). It followed a peak phase in the solar radiation cycles of Milankovitch (1941) and was recognized somewhat later in North America (Deevey and Flint, 1957). Some authorities use the term in a chronostratigraphic sense, but this procedure is satisfactory only in specific regions. In light of subsequent work, it was discovered that the maximum warm-up after the glaciation began many thousands of years earlier in the subtropics and equatorial regions and was strongly retarded in high latitudes near the ice margins (Fairbridge, 1972). In northern Europe, the peak of this "milk and honey" phase was marked by the Bronze Age of man. In northern Canada, early Indian migrants made settlements in the high Arctic, from which they later retreated, to be replaced by the Eskimos (Wright and Frey, 1965).

Deevey and Flint (1957) defined the hypsithermal as "the time represented by four pollen zones, V through VIII in the Danish system," spanning approximately the interval 9000-250 B.P. (by radiocarbon). Other authorities define it according to the record in their various regions. Antevs (1948) defined the middle part as *altithermal*. It would seem wise for scientists to use the term only in a relative sense—that is, an early- to mid-Holocene phase that on the average was warmer than today. It must be stressed that throughout the Holocene there have been *neoglacial* episodes of decades up to centuries when catastrophic cooling occurred (Fairbridge, 1976). In low latitudes these times were marked by disastrous droughts. Those cool episodes occurred about every 1000 years, but in the post-hypsithermal time have occurred more frequently.

RHODES W. FAIRBRIDGE

References

Antevs, E., 1948. The Great Basin, with emphasis on glacial and post-glacial times. III. Climatic changes and pre-white man, *Utah Univ. Bull.* **38**, 168-191.

Deevey, E. S., and Flint, R. F., 1957. The post-glacial Hypsi-thermal interval, *Science* **125**, 182-184.

De Geer, G., 1912. A geochronology of the last 12,000 years, *Internat. Geol. Cong., 11th, Stockholm, 1910, Compte Rendu* **1**, 241-258.

Faegri, K., and Iverson, J., 1950. *Textbook of Modern Pollen Analysis.* Copenhagen: Munksgaard, 168p.

Fairbridge, R. W., 1972. Climatology of a glacial cycle, *Quaternary Research* **2**, 283-302.

Fairbridge, R. W., 1973. Effects of Holocene climatic change on some tropical geomorphic processes, *Quaternary Research* **6**, 529-556.

Milankovitch, M., 1941. Canon of insolation and the ice-age problem, *Roy. Serbian Acad., Belgrade, Sci. Math. Nat.* **133**, 33. (Translated from German in *Israel Progr. Sci. Transl.,* Jerusalem, 1969.)

Nilsson, T., 1965. The Pleistocene-Holocene boundary and the subdivision of the late Quaternary in southern Sweden, *Internat. Cong. Quat., 6th, Warsaw, 1961, Rept.,* 479-494.

Wright, E., and Frey, D. G., 1965. *The Quaternary of the United States.* Princeton, N.J.: Princeton University Press, 901p.

Cross-references: *Climate, Coastal; Holocene; Sea Level Curves.*

I

ICE ALONG THE SHORE

On high-latitude or cold-climate coasts, *ice* plays an important role in altering normal shoreline processes. The presence of ice on the shore results from: grounded floes or frozen wave swash and snow in the littoral zone; freezing of the sea or lake surface; the presence of a frost table or permafrost in the littoral zone; or glaciers or ice caps that extend into the shore zone. The influence of these distinctly different ice types depends primarily on seasonal variations in air and water temperatures. The intensity of ice effects increases with latitude. Locally, the action of ice is related to the tidal range, the wave environment, and the nature of the shore zone. The primary effect of ice is to reduce energy levels by limiting wave generation or by absorbing wave energy in the littoral zone. Usually the presence of ice has a negative effect inasmuch as sediment transport is reduced to zero, but redistribution of littoral sediments can result from ice push or ice rafting.

Ice on Beaches

The effect of ice on limiting wave-induced littoral processes varies with the length of time the beach is ice-free. In mid-latitudes—for example, the Great Lakes—ice is present either on the beach or adjacent to the shore for periods of up to 4 months each year. The length of the open-water season decreases in higher latitudes, so that on high Arctic beaches open water and an ice-free beach may be limited to only a few weeks each year.

Freeze-up. Ice begins to form on a beach with the onset of subzero daily maximum air temperature as wave spray or swash run-up freezes. As the interstitial water freezes, all topographic features of the beach are preserved. Accumulation of ice occurs by the deposition and consolidation of: ice slush, stranded ice floes or cakes, wave swash or spray, and snow, or by the formation of a pressure ridge several meters high on the beach that results from the onshore movement of sea ice (Fig. 1). The net effect is the development of an *ice foot* (q.v.), which is an immobile, solid ice mass attached to the beach. The form of the ice foot may vary, depending on the beach slope, the tidal range, and the wave conditions during freeze-up (Wiseman, Owens, and Kahn, 1981). On tidal beaches, the ice foot is separated from the sea ice by a series of tidal cracks. The ice within this hinge zone grounds at low tide and is frequently compressed by shearing and/or pressure, associated with movement of the sea ice, to form an ice ridge. Seaward of the tidal cracks, the sea ice rises and falls with the tides. In nontidal or lacustrine environments, the ice may grow seaward from the ice foot, and often a series of offshore ice foot features develop.

FIGURE 1. *a*. Pressure-ridge ice foot prior to breakup at Cape Ricketts, Devon Island, Canada. *b*. Following breakup, the pressure ridge remains on the beach until removed by wave erosion or by ablation.

These are formed by the accumulation of ice on the margin of the fast ice as it grows seaward before freeze-up is complete.

Relatively small amounts of sediment are incorporated within the ice foot, though deposition of sediment, interbedded with ice, may occur during the early stages of freeze-up. A storm-ice foot composed of interlayered ice and sediment that has a flat surface is sometimes referred to as a *kaimoo* (q.v.).

Breakup. Rates of ice ablation during the spring thaw depend on the air temperatures and on the volumes of ice and sediment that accumulated during the previous freeze-up period and during the winter snowfall (Short and Wiseman, 1975). Prior to breakup and removal of the sea ice, all melting produces in situ deposition on the beach and the formation of a variety of small-scale topographic features that result directly from the presence of ice. The most common types of microrelief features are kettles (Fig. 2), dirt cones, and ridges (Greene, 1970; Dionne and Laverdière, 1972). On the back beach, microfans may be deposited as meltwater from the backshore flows onto the frozen beach (Davis, 1973). During the thaw period, the beach has an irregular surface owing to differential ablation or melting of the ice. The ice foot is usually the last to thaw, although buried lenses of ice on the back shore, protected by a cover of sediment, may persist longer.

The ice melt features are rarely preserved, inasmuch as storm-wave activity during the open-water season rapidly redistributes the beach sediments. Ice push ridges (Fig. 3), which

FIGURE 2. Ice kettles forming in the beach. The ice blocks became stranded and buried before freeze-up. As an indication of scale, the photograph covers approximately 2 m of beach in the lower foreground (photo: J. M. Coleman).

FIGURE 3. Ice push scar resulting from the grounding of a floe on a gravel beach, Cornwallis Island, Canada (from Owens and McCann, 1970).

result from the grounding of wind-driven ice floes after melting of the ice foot and before the subsequent freeze-up, can be preserved if the scar extends beyond the limit of storm-wave activity (Hume and Schalk, 1964; Owens and McCann, 1970; Alestalo and Häikiö, 1979; Harpes and Owens, 1981).

Frost Table. Freezing of interstitial water beneath the beach surface produces frozen ground. The depth of the active layer to the frost table varies geographically and seasonally (Owens and Harpes, 1977; Taylor, 1980); in polar regions, the frost table is present throughout the year. In the Canadian Arctic, north of 72°, the depth of the active layer rarely exceeds 75 cm (Taylor and McCann, 1974). The frost table acts as a lower limit for sediment reworking by wave action or by ice push.

Effects of Ice on a Beach. The presence of ice in the intertidal zone reduces the amount of energy that can be transmitted to the beach from wave-generated processes. This has a direct effect on reducing annual sediment transport rates. The extent to which ice is a limiting factor in reducing energy levels is dependent on the relationship between the length of time that the ice is present and the processes that would otherwise be operating. In a mid-latitude storm wave environment, in which storm frequency and wind velocities are at a maximum during winter months, ice is a major factor in reducing energy levels. On high-latitude beaches, which have a short open-water season, there is a large annual variation in rates of sediment transport. The amount of variation depends on the coincidence of an open fetch, the generation of waves by winds blowing over that fetch, and an ice-free beach (McCann, 1972). During a single storm at Barrow, Alaska, in 1963, the volume of sediment transport was equivalent to what would normally be transported in that area in 20 years (Hume and Schalk, 1967).

The positive effects of ice are few. Depositional features are usually destroyed rapidly by wave action following melting of the ice foot. Only those features formed above the limit of storm-wave activity, such as ice-push ridges, are likely to be preserved. Sediments that are incorporated in or deposited on ice in the littoral zone may be removed by ice-rafting during breakup (Fig. 4), but these volumes are small compared with normal rates of littoral zone sediment transport.

Ice on Tidal Flats and Marshes

The continuous rise and fall of the water level in tidal estuaries and marshes prevent formation of a continuous ice cover in the intertidal zone. During breakup, the large-scale movement of ice floes plays a major role locally in the formation of sedimentary features, the redistribution of sediments, and the redistribution of vegetation. As ice grounds at low tide, material adheres to the base of floes. With the subsequent high tide and movement of floes, large volumes of sediment, including boulders, and vegetation can be displaced and rafted offshore or deposited on adjacent tidal flats (Rosen, 1979). On a vegetated tidal flat, this freezing and rafting process can cause creation of a

FIGURE 4. A rockfall has led to deposition of talus on the sea ice. This material will be rafted from the area of the cliff base during breakup. The cliff (on Griffith Island, Canada) is approximately 200 m high.

pitted surface characterized by numerous irregular water-filled pans (Dionne, 1968). Deposition of rafted blocks on a bare surface can contribute to the extension of a marsh inasmuch as the vegetation takes root once a floe becomes stranded and melts. The gouging action of floes on a tidal flat produces ice-push scars, grooves, and striations that, in association with ice-rafted features and *monroes* (q.v.), give a variety of microrelief features that may be preserved in the sediments.

Nearshore Ice

The primary effect of ice on the water seaward of the fast ice is to prevent wave generation or to dampen incoming waves. Ice also affects bottom sediments by the gouging action of grounded floes, the constriction of bottom currents, and scouring of the bottom below drainage holes off river mouths. The grounding of floes or of the keels of pressure ridges in the nearshore zone produces small-scale ice-gouge relief features on the bottom. A secondary effect is the erosion of depressions around the grounded ice by current scour or by turbulence, which may lead to resuspension and transport of bottom sediments (Reimnitz and Barnes, 1974). These ice gouges or scour depressions are rarely preserved in the nearshore zone, since the sediments are readily reworked by wave action during the open-water season. Where ice is present off a river mouth during spring breakup, freshwater floods over the ice surface and drains through holes and cracks. This overflow drainage can cause scour depressions (*strudel scour*) on the bottom up to 4 m deep and 20 m in diameter (Reimnitz, Rodeick, and Wolf, 1974).

The formation of pressure ridges in the nearshore zone results from wind- or current-driven movement of the ice. In lakes, the same effect can be achieved also by thermal expansion or contraction of the continuous ice cover (Pessl, 1969; Alestalo and Häikiö, 1979).

Terrestrial Ice on the Coast

Ice that is formed on the land may crop out as permafrost in coastal cliffs or may extend into the shore zone in the form of glaciers or ice-cap margins. Permafrost is present beneath polar beaches, and acts as a lower limit for sediment redistribution, or can crop out in eroding cliffs. On unresistant cliffed coasts, such as the tundra of northern Alaska and the Yukon or the taiga of eastern Siberia, the exposed permafrost is subject to both thermal and mechanical erosion (Are, 1972; Harpes, 1978). Cliff retreat occurs as a result of either: undercutting by the mechanical or thermal energy of waves during the open-water season that results in block slumping (Fig. 5); or thawing of the permafrost or ground ice by thermal energy of the air or solar radiation, which results in flow slumping of the unconsolidated sediments. If patterned ground is present along the cliff section, melting of the ice wedges will frequently produce gullies that give the cliff a saw-toothed appearance (Fig. 6).

Where a glacier or an ice cap extends into the shore zone, the interface is usually one of ice and water. Occasionally a beach may form on the ice margin as glacial debris is reworked by littoral processes (Nichols, 1961).

EDWARD H. OWENS

References

Alestalo, J., and Häikiö, J., 1979. Forms created by the thermal movement of lake ice in Finland in winter 1972-73, *Fennia* 157, 51-92.

Are, F., 1972. The reworking of shores in the permafrost zone, in W. P. Adams and F. M. Helleiner, eds., *International Geography*. Toronto: University of Toronto Press, 78-79.

Davis, R. A., 1973. Coastal ice formation and its effect on beach sedimentation, *Shore and Beach* 41, 3-9.

Dionne, J. -C., 1968. Schorre morphology on the south shore of the St. Lawrence Estuary, *Am. Jour. Sci.* 266, 380-388.

Dionne, J. -C., and Laverdière, C., 1972. Ice formed beach features from Lake St. Jean, Quebec, *Canadian Jour. Earth Sci.* 9, 979-990.

Greene, H. G., 1970. Microrelief of an arctic beach, *Jour. Sed. Petrology* 40, 419-427.

Harpes, J. R., 1978. Coastal erosion rates along the Chukchi Sea Coast near Barrow, Alaska, *Arctic* 31, 428-433.

Harpes, J. R., and Owens, E. H., 1981. Analysis of ice-override potential along the Beaufort Sea coast of Alaska, *1981 POAC Conf., Quebec City, Proc.* 2, 974-984.

Hume, J. D., and Schalk, M., 1964. The effects of ice-push on arctic beaches, *Am. Jour. Sci.* 262, 267-273.

FIGURE 5. Erosion of a tundra on the Alaskan Chukchi Sea coast. The cliff is 3 m high (photo: J. M. Coleman).

FIGURE 6. Sawtoothed dissection of tundra cliffs (approximately 25 m high) associated with ice wedge thaw; Skull Cliffs near Barrow, Alaska (photo: L. D. Wright).

Hume, J. D., and Schalk, M., 1967. Shoreline processes near Barrow, Alaska: A comparison of the normal and catastrophic, *Arctic* 20, 86-103.

McCann, S. B., 1972. Beach processes in an arctic environment, in D. R. Coates, ed., *Coastal Geomorphology*. Binghamton: State University of New York, 141-155.

Nichols, R. L., 1961. Characteristics of beaches formed in polar climates, *Am. Jour. Sci.* 259, 694-708.

Owens, E. H., and Harpes, J. R., 1977. Frost table and thaw depth in the littoral zone near Pearl Bay, Alaska, *Arctic* 30, 154-168.

Owens, E. H., and McCann, S. B., 1970. The role of ice in an arctic beach environment, with special reference to Cape Ricketts, Southwest Devon Island, Northwest Territories, Canada, *Am. Jour. Sci.* 268, 397-414.

Pessl, F., 1969. *Formation of a Modern Ice-push Ridge by Thermal Expansion of Lake Ice in Southeastern Connecticut*. Hanover, N.H.: U.S. Army Cold Regions Research and Engineering Laboratory, Research Rept. 259, 13p.

Reimnitz, E., and Barnes, P. W., 1974. Sea ice as a geologic agent on the Beaufort Sea shelf of Alaska, in J. C. Reed and J. E. Sater, *The Coast and Shelf of the Beaufort Sea*. Arlington, Va.: Arctic Institute of North America, 301-353.

Reimnitz, E.; Rodeick, C. A.; and Wolf, S. C., 1974. Strudel scour: A unique arctic marine geologic phenomenon, *Jour. Sed. Petrology* 44, 409-420.

Rosen, P. R., 1979. Boulder barricades in central Labrador, *Jour. Sed. Petrology* 49, 1113-1124.

Rudowski, S., 1972. Influence of freeze on active processes in shore zone and on beach structure under moderate climatic conditions, *Acad. Polonaise Sci. Bull.* 20, 139-144.

Short, A. D., and Wiseman, W. J., Jr., 1975. Coastal breakup in the Alaskan Arctic, *Geol. Soc. America Bull.* 86, 199-202.

Taylor, R. B., 1980. Beach thaw depth and the effect of ice-bonded sediment on beach stability, Canadian

Arctic islands, *in Proceedings of the Canadian Coastal Conference.* Ottawa: National Research Council, 103-121.

Taylor, R. B., and McCann, S. B., 1974. Depth of "frost table" on beaches in the Canadian Arctic Archipelago, *Jour. Galciology* **13**, 321-322.

Wiseman, W. J., Jr.; Owens, E. H.; and Kahn, J., 1981. Temporal and spatial variability of ice-foot morphology, *Geog. Annaler* **63A**, 69-80.

Cross-references: *Antarctic, Coastal Morphology; Arctic, Coastal Morphology; Chattermarks; Frost Riving; Glaciel; High-Latitude Coasts; Ice-Bordered Coasts; Ice Foot; Kaimoo; Monroes; Mud Volcanoes; Shore Polygons; Stone Packing.* Vol. I: *Sea Ice;* Vol. III: *Ice Thrusting.*

ICE-BORDERED COASTS

Coasts that for all or part of each year have sea ice frozen to the shore are termed *ice-bordered coasts*. They occur in the high latitudes of the Northern Hemisphere and in the Southern Hemisphere around Antarctica. The period of ice bordering increases with latitude, from four months in the southern Gulf of St. Lawrence to eleven months in the high Canadian Archipelago. Coasts fronted by glaciers and ice sheets are perennially ice-bordered.

Along seasonally ice-bordered coasts ice begins forming with the arrival of below 0°C air and water temperatures, and is termed *freeze-up*. Water initially freezes in more brackish areas; then along the shoreline, possibly forming a *kaimoo* or *ice foot* as a result of freezing swash (Short and Wiseman, 1974). As freeze-up continues, the water surface itself freezes, and wave action ceases at the shoreline. During the following winter, the new ice thickens to approximately 2 m, generating shore-fast ice to this depth along the shoreline. In regions of significant tidal action, the daily up and down movement of the floating ice, immediately seaward of the shore-fast ice, may generate *ball and point* couplings (Owens, 1976) and ice boulder ridges over the boundary. If a storm occurs during freeze-up, ice ridges up to 10 m high may be piled up on the shoreline, over shore-fast ice, or over an offshore obstruction, such as a bar or shoal. The ice may also be pushed onshore plowing the beach and generating ice push features.

Movement of the sea ice during winter tends to severely fracture the ice, generating ice ridges and an irregular surface. However, fast ice, or ice lying inside a grounded ice ridge, is protected from such lateral movement, and will often remain unfractured with a relatively smooth surface (Fig. 1).

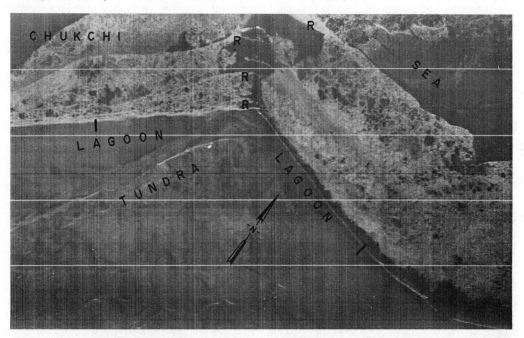

FIGURE 1. Radar image of Icy Cape, north Alaska, showing winter ice conditions May 1, 1974. The lower left portion of the image consists of relatively flat snow- and ice-covered tundra and Kasegaluk Lagoon. Bordering the lagoon to the north is a low barrier (thin white line) with smooth ice lying immediately seaward. The smooth (dark) ice lies inside a prominent offshore bar; seaward of the bar, the ice is broken (white) 1-2 m high. Seaward of the cape (center top), four higher ice ridges are prominent (R). To the south of the cape, the ice is higher along the beach as a result of a fall storm. Lines are 5 km apart.

With the arrival of above 0°C temperatures, breakup of the ice begins (Short and Wiseman, 1975). Snow and ice begin to melt, and offshore fractures are reactivated. Breakup is fastest when the floating ice is free to move seaward, slowest when it grounds and must melt in place.

In marginal ice-bordered coasts, the formation of sea ice and its deposition on and attachment to the coast occurs only during winter. However, in higher latitudes, such as the Arctic Ocean, sea ice derived from the permanent pack ice may be driven onshore during major summer storms (Short, 1976).

ANDREW D. SHORT

References

Owens, E. H., 1976. The effects of ice on the littoral zone at Richibucto Head, eastern New Brunswick, *Rev. Géographie Montréal* **30**, 95-104.

Short, A. D., 1976. Observations of ice deposited by waves on Alaskan Arctic beaches, *Rev. Géographie Montréal* **30**, 115-122.

Short, A. D., and Wiseman, W. J., Jr., 1974. Freezeup processes on Arctic beaches, *Arctic* **27**, 215-224.

Short, A. D., and Wiseman, W. J., Jr., 1975. Coastal breakup in the Alaskan Arctic, *Geol. Soc. America Bull.* **86**, 199-202.

Cross-references: *Antarctica, Coastal Morphology; Arctic, Coastal Morphology; Glaciel; High-Latitude Coasts; Ice along the Shore; Ice Foot; Kaimoo.* Vol. I: *Sea Ice;* Vol. III: *Ice Thrusting.*

ICE FOOT

Ice foot is a term derived from the Danish word *isfod* (Dionne, 1973) to describe a feature that is formed on and attached to beaches in cold climates. Usually the ice foot is formed during winter months and melts in the spring, but in high latitudes it may persist throughout the year. On tidal beaches, the ice foot forms at the high-water mark (Fig. 1), and grows vertically and horizontally into the intertidal zone (Fig. 2), so that it may extend to the low-water mark. The ice foot is separated from the *sea ice*, which rises and falls with the tide, by a hinge zone of tidal cracks (Fig. 3) (Owens, 1976; Nielsen, 1979). The ice foot develops by the accumulation of ice that results from the freezing of wave spray and swash. The size of the ice foot may be increased by: the incorporation of ice slush or floes stranded on the beach or deposited above the high-water mark by waves; the development of a *pressure ridge* on the beach by the action of sea ice; or the accumulation and compaction of snow on the ice foot (McCann and Taylor, 1975).

The horizontal extent of the ice foot is controlled by the tidal range and by the slope of the beach (McCann and Carlisle, 1972), and the vertical extent is dependent on the level of wave activity during freeze-up. On exposed coasts, where storm-wave activity occurs during freeze-up, the ice foot can grow to heights in excess of 5 m above the beach surface. In addition, two distinct ice foots can develop, one at the limit of storm waves, and the second at the limit of normal waves. In nontidal and lacustrine environments, a series of ice foot features may form parallel to the beach as the fast ice grows seaward before freeze-up is complete (Marsh, Marsh, and Dozier, 1973).

A variety of different ice foot types have

FIGURE 1. Initial stage of ice foot development at the limit of wave activity. The scale is graduated at 10-cm intervals.

FIGURE 2. Growth of an ice foot at the high-water mark. Note that frozen swash is covering snow deposits on the landward margin of the ice foot and that the intertidal zone is frozen (from Owens, 1976).

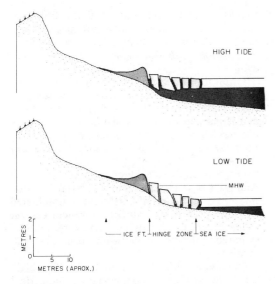

FIGURE 3. Location of the ice foot in relation to the hinge zone of tidal cracks and to the sea ice at high tide and low tide in a microtidal (range 1.0 m) environment (from Owens, 1976).

been defined; these include: tidal platform ice foot, storm ice foot, drift ice foot, pressure ice foot, stranded floe ice foot, false ice foot, wash-and-strain ice foot, shore ice foot, offshore ice foot, upper strand ice foot, and lower strand ice foot. This wide range of descriptive terms is indicative of the variability in form and processes associated with the accumulation of ice in the littoral zone (Levenson and Cohn, 1979; Wiseman, Owens, and Kahn, 1981). In its strictest sense, the definition does not include ice accumulation features that extend seaward or landward of the intertidal zone.

Usually little sediment is incorporated into the ice foot inasmuch as the intertidal beach is often frozen during ice foot formation (Fig. 2). If material is thrown onto the ice foot or onto the landward side of the ice foot and becomes interbedded with ice, this is referred to as a *kaimoo* (q.v.) in cases where a relatively flat rampart is formed.

EDWARD H. OWENS

References

Dionne, J.-C., 1973. La motion de pied de glace (Icefoot), en particulier dans l'estuaire du Saint-Laurent, *Cahiers Géographie Québec* **17**, 221-250.

Levenson, E. B., and Cohn, B. P., 1979. The ice foot: Its morphology, formation and role in sediment transport and shoreline protection, *Zeitschr. Geomorphologie* **23**, 58-75.

McCann, S. B., and Carlisle, R. J., 1972. The nature of the ice foot on the beaches of Radstock Bay, southwest Devon Island, N.W.T., Canada in the spring and summer of 1970, *Inst. British Geographers Spec. Pub.* **4**, 175-186.

McCann, S. B., and Taylor, R. B., 1975. Beach freeze-up sequence at Radstock Bay, Devon Island, Arctic Canada, *Arctic and Alpine Research* **7**, 379-386.

Marsh, W. M.; Marsh, D. B.; and Dozier, J., 1973. Formation, structure, and geomorphic influence of Lake Superior icefoots, *Am. Jour. Sci.* **273**, 48-64.

Nielsen, N., 1979. Ice-foot processes. Observations of erosion on a rocky coast, Disko, West Greenland, *Zeitschr. Geomorphologie* **23**, 321-331.

Owens, E. H., 1976. The effects of ice on the littoral zone at Richibucto Head, eastern New Brunswick, Canada, *Rev. Géographie Montréal* **30**, 95-104.

Wiseman, W. J.; Owens, E. H.; and Kahn, J., 1981. Temporal and spatial variability of ice-foot morphology, *Geog. Annaler* **63A**, 69-80.

Cross-references: *Antarctica, Coastal Morphology; Arctic, Coastal Morphology; Glaciel; High-Latitude Coasts; Ice along the Shore; Ice-Bordered Coasts; Kaimoo.* Vol. I: *Sea Ice;* Vol. III: *Ice Thrusting.*

INDIA, COASTAL ECOLOGY

The word "coast" carries an inevitable vagueness in its meaning regarding the seaward and landward extent of the coastal zone (Ahmad, 1972). The coastal zone is a prominent ecotone —i.e., a contact zone between the terrestrial and marine communities. The Indian coast lies roughly between 24° N (Cutch) and 7° N (Nicobars); with a mean January temperature of 20°-25°C and July temperatures of 30°C in the north and 27.5°C in the south. The low thermal range produces relatively uniform ecological conditions. Salinity distribution is relatively uniform throughout the year. In February, it ranges from 36.5‰ near Cutch to 32‰ around Nicobars. The east coast waters are somewhat less saline than those on the west coast. Annual precipitation ranges from about 250 mm at Cutch to about 4000 mm on the Western Ghats, to under 600 mm in the Gulf of Manaar and 2000 mm in the Sundarbans. The west coast has onshore winds from March to October, and offshore winds from December to February. The east coast has onshore winds from October to February (northeast monsoon), offshore winds from May to September, and south, parallel-to-coast winds from March to April (Heezen, Takashi, and Tharp, 1966). Cloud cover varies from two-tenths from December to March, except on the east coast, to seven during the southwest monsoon from June to October. Tropical cyclones before and after the southwest monsoon periods seriously affect coastal environment and life.

The coastal currents (mostly parallel) check the movement of planktonic (small, microscopic) organisms, which move only feebly against currents. During the southwest monsoon the currents are strong and more persistent on the west than on the east coast. During winter their direction is reversed. Tides cause short-lived transference of organisms above and below the shoreline and aid the formation of tidal marshes. During spring tides, the mean range varies from 5.7 m at Bombay to 1.1 m at Cochin; it is about 1 m on the east coast, but about 4.25-5.2 m in the shallower waters of the Ganges delta.

Waves affect the tranquility of the coastal waters by their magnitude and consequently the habitat of smaller organisms, particularly in shallow water. In the areas of the northeast monsoons, only 5% of the waves are 3.6-6.1 m high, and 80% under 1.2 m. In southwest monsoon areas, 10% of the waves are 6.1 m high; 60%, 1.2-6.1 m high.

Pebbles are common shore deposits between Bombay and Karwar. Sandy beach dominates the eastern coast. Silt flats are seen in the Cambay region. Fine sediments, rich in organic matter, occur in the lagoons.

The deltas are often mangrove-covered swamps. On the beach, sand and dunes (at places the dune belt is 4-6 km wide) of the inter-delta coast foster palm and casuarina. Cliffed coasts, between Ratnagiri and Karwar or near Cape Comorin, plunge steeply into the sea and there is no coastal plain. The lagoons, so characteristic of Kerala, provide remarkable contact zones between marine and terrestrial environments.

Plant Ecology

Terrestrial. Tropical, moist, semievergreen vegetation is found between the Tapti and Cape Comorin and on the Orissa and Bengal coast outside the deltas (Champion, 1936). Coconut palm, date palm (*Phoenix sylvestris*), palmyra palm, and casuarina dominate the sandy beaches and higher coastal interior. On the muddy lagoons, marshy vegetation is dominant. The Konkan coastal landscape is dominated by soft white sand beneath overhanging palm and gray-green casuarina, red-rocked islets, and the flaming flower of the gold mohur tree in the hot season. The herbaceous flora of the Orissa coast consists of plants of both dry and moist regions.

Because of less rainfall (about 1000 mm, most of which occurs between October and January), the region between the Vaigai and the Penner has tropical, dry, evergreen vegetation. The scrub flora of the region is both persistent and invading, the typical species being *Carissa carandas, Maba buxifolia, Randia dumelorum, Albizzia amara,* and *Memecylon edule.*

Tropical, dry, deciduous vegetation of the Rayalaseema and Manaar Coast is believed to be a regression of the moist deciduous type, and is marked by transitional facies. The region has less than 1000 mm of rain during a short rainy season followed by a long intense dry period.

Tropical, moist, deciduous vegetation of the Andhra Coast (annual rain greater than 1000 mm) east of the Godavari delta is dominated by the gregarious, quickly regenerating *sal* (*Shorea robusta*), with dense undergrowth of evergreen shrubs along with bamboos and climbers.

The delta vegetation contains dense, high-tidal forests of *Rhizophora* mangroves along tidal channels. Elsewhere the forest is low and open. Many plants of this vegetation have pneumatophores or breathing roots sticking out of the mud like a field of tent pegs driven upside down. Mangrove dominates the delta shores. In Bengal, the mangrove—called *sundri* (*Heritiera fomes*), a pneumatophore—forms the Sundarban forests, growing about 30 m high. Other species include the stemless gregarious palms (*Phoenix farinifera*) on the Coromandel coast and *Nannorrhops richieana* in the extreme west. Further inland, at higher levels, grow screwpines, palms (e.g., the beautiful *Phoenix paludosa*), and canes. The creeks may be lined with the stemless gregarious palm *Nipa fruticans* having 5 m leaves (Puri, 1960).

The tropical thorn forest (a peneclimatic stage in equilibrium with biotic factors) of Kathiawar and Cutch coastal tracts (less than 700 mm annual rainfall) has xerophytic bush at the moister end, degenerating into complete desert in the northwest. Low (6-9 m), widely dispersed species—ensuring free jurisdiction to the radially spreading roots, including acacias, euphorbias, and wild dates—form the typical vegetation.

The tropical rain forests of the Andamans and Nicobars, with subequatorial climate, has teak (*Tectona grandis*), adauk, and gurjan as the important species, covering the islands from the shore to the hilltop.

Among the littoral sand dune plants are the gregarious *Phoenix farinifera* and Spinifex, *Ipomoea bilboa, Vigna lutea,* and *Conavalia lineata.* Saline water and tides make the number of estuarine species larger than in the other botanical regions of India.

One of the botanical glories of coastal India is the economically important coconut, which prefers moist soil and some sea salt, found in areas extending up to 25° N, penetrating 150 km inland from the eastern shore, 75-120 km from the western coast, and 250 km into Karnataka. It is abundant in the Andamans and Nicobars with the densest cultivation being in Kerala, the Godavari district, and the Cauvery delta.

Aquatic. The marine plants are not as fixed as terrestrial vegetation. The diatoms form the bulk, followed by *Dinophyceae, Cyanophyceae,* and *Silicoflagellaceae*. Studies of the phytoplankton (important as fish food) have shown peak production during the southwest monsoons. The fat content among the west coast planktons is said to be higher than on the east coast. In Indian waters it is the physiological state of the floral elements that finally decides the magnitude of plankton production on which fish production ultimately depends.

The tropical conditions have made the seaweeds or the "algal garlands" at Dwarka, Rameswaram, and elsewhere, look at low tide like a carpet laid over with methodical designs.

Animal Ecology

In contrast with the plant life, which is static, particularly on land, animal life is mobile and migratory. Consequently the species inhabiting the coastal tracts of India may belong to vast zoogeographical regions of even subcontinental or continental extent (Benton, 1964a). It is therefore not feasible to distinguish species that are particularly coastal in their habitat. No doubt because of the great geographical range, the Indian region is rich in fauna, with 401 species of mammals, 1617 species of birds, 534 species of reptiles, 130 species of batrachians, and 1418 species of fishes.

Terrestrial. Among the mammals of the Indian coast land regions are found *Edentata,* which includes the Indian pangolin or *banrohu* (*M. pentadactyla*), and among the birds, Crag Martins are inhabitants of the coastal cliffs. *Ansiodactyli* include the bee-eaters and hornbills of the Malabar coast. The large imperial pigeons of the Malabar coastlands and the pied imperial pigeons of Andamans and Nicobars are among the *Carpophanginae* found. *Limicolae* include stone plover (*E. magnirotris*) of the Adaman shores and *Herodiones* include the reef herons that stay along the coasts (*Imperial Gazetteer of India,* 1907).

Aquatic. The mammals characteristic of the Indian marine coastal regions include *Cetacea* (whales and porpoises) and *Sirenia* (including *Halicore dugong*). Among reptiles are crocodiles inhabiting the estuaries and occasionally seen in coastal waters, the herbivorous green turtles of coastal waters, and the sea snakes seen near the coast and in tidal streams. The batrachians (frogs) are common in the moister regions of deltas, estuaries, and marshy tracts of the coast.

The fishes are by far the most important animals of the coast and adjoining seas. Of the approximate 1418 species, some 361 species are freshwater varieties. Another large group inhabits the brackish water of lagoons and estuaries, but it is difficult to distinguish the truly marine from the estuarine, and the latter from fluviatile groups. The number of migratory species in Indian waters is not large.

All the common tropical sharks and rays occur along the Indian coasts. Most of the freshwater and marine fishes of the Indian coasts are bony fishes (*Teleostii*). They include five orders: Within *Physostomi* are the eels, some of attractively colored species occurring on rocky coasts; the catfishes; the herring family, containing the most important migratory species of Indian water—e.g., *hilsa* of the Ganges delta, Sable fish of south India, oil sardine of the Malabar coast, and Bombay duck plentiful in some parts of the Indian coast. The order Acanthopterygi includes estuarine begti (*Lates calarifer*) of Bengal; beautifully colored sea perches; the prized mullets, *Nandidae*, with several families, ascending tidal waters of Bengal during the southwest monsoons; the swordfish common on the Coromandel coast; the prized pomfrets; the mackerels; and the brilliantly colored marine fishes in the coral coasts of Andamans and Nicobars. The mackerels are commercially the most important; their chief habitat being between Bombay and Quilon. *Plectognatha* include the lancelet or the fishes without head or brain and the other two orders consist of deep-sea species (Chandy, 1970).

With a 9000 km long coastline and 260,000 km^2 of continental shelf, Indian coastal waters are rich in variety and quantity of fish. The west coast, however, contributes over 75% of the total fish catch. The reasons are the wider western continental shelf, stronger southwestern monsoons causing pronounced upwelling of waters and better circulation, and a higher proportion of phosphates and nitrates causing greater production of fish food in the form of plankton.

The estuarine fisheries of the lagoons are also important. After the harvesting of a paddy, the banks of these depressions are raised to arrest seawater and fish at high tide. In this embanked brackish water milkfish (*Chanos*), mullets (*Tialpai mossambica*), and prawns live and grow for six months.

Human Ecology

The importance to man of the coastal environment is related to its physical character, natural endowments, and the stage of material civilization (Benton, 1964b). The coastal lowlands of Kerala and Bengal are among the most densely populated areas of the world because of great soil fertility and moisture leading to abundant production of crops and considerable industrialization. The rest of the coastal lowlands, particularly of the east coast excepting Rayalasesma and the Cambay regions, have only slightly less population density than Kerala and west Bengal. The coastal settlements have the additional advantages of estuarine and marine fisheries, salt industry, and coastal trade. In most of the coastal areas the proportion of urbanization is higher than the national average, particularly in the Kathiawar and Cambay areas in spite of their relative aridity. *Ecocide*, ecological deterioriation and pollution, is not yet serious and is confined to the Hooghlyside, greater Bombay, Ahmadabad, Madras, and other urbanized pockets of coastal Andhra, Tamilnadu, and Gujarat.

ENAYAT AHMAD

References

Ahmad, E., 1972. *Coastal Geomorphology of India*. Delhi: Orient Longman, 222p.

Champion, H. G., 1936. A preliminary survey of the forest types of India and Burma, *Indian Forest Records n.s. Silviculture (Delhi)* 1, 286p.

Chandy, M., 1970. Fishes, in *India—The Land and People*. New Delhi: National Book Trust, 166p.

Benton, W., 1964a. Ecology, in *Encyclopedia Britannica*, vol. 7. London: Encyclopedia Britannica Ltd., 912-923.

Benton, W., 1964b. Ecology, human, in *Encyclopedia Britannica*, vol. 18. London: Enclopedia Britannica Ltd., 235-241.

Heezen, B. C.; Takashi, I.; and Tharp, M., 1966. Indian Ocean, *in* R. W. Fairbridge, ed., *Encyclopedia of Oceanography*. New York: Reinhold, 370-402.

Imperial Gazetteer of India, 1907, vol. 1, new ed. London: Oxford Book Company. (Reprinted in 1960, New Delhi: Today and Tomorrow Printers and Publishers, 568p.)

Puri, G. S., 1960. *Indian Forest Ecology*, 2 vols. New Delhi: Oxford Book Company, 710p.

Cross-references: *Asia, Middle East, Coastal Ecology: Asia, Eastern, Coastal Ecology: Coastal Fauna; Coastal Flora: India, Coastal Morphology; Mangrove Coasts*. Vol. I: *Indian Ocean;* Vol. VIII, Part 2: *India*.

INDIA, COASTAL MORPHOLOGY

India has a remarkably symmetrical and unindented coast. On a map the peninsula appears like a pillar dividing the northern shore of the Indian Ocean into two arches (the Arabian Sea and the Bay of Bengal). The Indian coast is generally an emergent one (Ahmad, 1972). In contrast to the symmetry of the coast, the continental shelf is strongly asymmetrical. It is broad off the Ganges delta and unusually narrow (30-35 km) opposite the other major Indian deltas. It is widest (400 km) off the Gulf of Cambay.

The Indian coastal zone may be regarded as including: *offshore features*, consisting of the continental shelf and offshore islands; *shore features*, consisting of beach, barriers, lagoons, delta and estuary margins, cay sandstone and shore alignment; and *coastal features*—i.e., morphology of the land immediately behind the shoreline. *Coastal morphology* is related to the genesis, morphology, classification, and evolution of these constituents of the Indian coast.

Offshore Features

The continental shelf around the Indian peninsula is roughly delimited by the 100-fathom submarine contour. Its area is about that of the Ganges plain. The eastern shelf is about one-third the width of the western. The most frequent slope of the shelf nearer shore is 5-7'. The most common slope on the eastern shelf as a whole is about 21'. On the western shelf, the slope varies from about 1' in the Cambay region to 10' near Cape Comorin (the average slope of the Ganges plain is 1'). While the eastern shelf appears to continue the slope of the Tamilnadu plain, the extremely gentle slope of the western shelf is unrelated to the high relief of the western coastal interior. The shelf may be classed genetically into five types: that caused by deltaic sedimentation; downfaulted plateau off the lava region; submerged seaboard of a folded mountain (Andamans) (Gee, 1926); coral built (Laccadive) (Sewell, 1932); and that caused by normal sediment filling (east coast).

The western shelf is generally sand-covered, but mud occasionally occurs on the shore. On the east coast (especially on the Bay of Bengal shelf), the sequence from the shore outward is sand up to 3 km, followed by clay up to 20 km, succeeded by shells up to 70 km from the shore, and then oolitic concretions up to the edge of the shelf. The largest submarine canyon near the Indian coast is the northeast-southwest trough-shaped Ganges mouth canyon, not aligned along the trend of the delta. Several other canyons have been discovered on the Andhra coast. It is believed that these canyons perform the important role of channeling the sediments carried by rivers (Heezen, Takashi, and Tharp, 1966).

The Indian coast is notably devoid of islands, but the few islands that do occur are important in a genetic study of coastal morphology. These are of five types: deltaic; coralline; submergent, ancient, crystalline eminences; submerged lava eminences and submerged remnants of Tertiary mountains. Geological identity between the islands and the mainland indicates former connection. The notable absence of islands in the offshore zone appears to indicate dominance of recent uplift.

Shore Features

Beaches—composed of debris coarser than mud largely derived from the sea floor by the action of waves that are particularly under the influence of the monsoons or currents—are characterized by cycles of building, destruction, and rebuilding. They may take the form of shapeless sand accumulation or ridges, dunes, barriers enclosing lagoons, pocket beaches, and so on. Beaches are more extensive on the low eastern (Cushing, 1913) and Kerala coasts and absent from the swampy Indian deltas. They occupy about 55% of the Indian shore. Between Malvan and Cutch on the west, there is a type of *fossil beach* called *littoral concrete* (cay sandstone or beach rock) in Indian geology. It is an agglutinated shelly grit, composed of shells, corals, pebbles and sand, cemented together more or less thoroughly by carbonate of lime, and slightly above sea level (Pascoe, 1964).

The offshore bars and their variants on the Indian coast—spits or tombolos—are largely the result of monsoon-generated waves. On the Kerala coast, they enclose long *kayals* or lagoons, e.g., Vembanad Lake (approximately 75 km long) (Wadia, 1961).

Deltas and estuaries are significant among the shore features of India. The size of the Indian deltas is generally proportional to the basins of the rivers concerned. The large lobate Ganges delta is associated with a basin containing the highest mountains, heaviest showers, unusually intense erosion in the catchment, and heavy deposition at the shore. The north-south banks on the Hooghly mouth indicate the manner in which the sediments are being laid down there.

An important feature of a shoreline is its progradation or recession in recent times. The Indian deltas are generally prograding, despite sedimentation subsidence, at the annual rate of 60 m for the Ganges-Brahmaputra delta; 10 m at Mahanadi; 15 m at both Krishna and Godavari, and 5 m at the Cauvery delta. The alluvial shore on the east coast of the Gulf of Cambay is also believed to be actively prograding.

Locally the role of deltaic prograding is powerful, so much so that the trimming effect of longshore drifting has been overcome by deltaic protrusion, for example, the Krishna and Godavari deltas. As a whole, about one-fourth of the Indian shore is in some sort of progradation. The deposition rate at Madras, Vishakhapatnam, and Godavari Point is one million tons to 400,000 yards^3/year.

Recession characterizes considerable stretches of alluvial or beach shore in Karnataka or Kerala; waves, especially where the winds are orthogonal to the shore, are sufficient cause of coastal recession except at the deltas dominated by sedimentation and land building or where the lay of the land affords protection against the fury of the waves.

Cliffed coast occurs in Maharashtra, southeast Kathiawar, near Cape Comorin, and Vishakhapatnam. Fossil cliffs occur near Balasore and in the northern part of the Rann of Cutch. Coastal terraces, sea caves, stacks, natural bridges, and promontories are prominent between Mangalore and Bombay (Arunachalam, 1964). *Rias* or river systems partly flooded by the sea are prominent between Cape Comorin and Bombay; estuaries are particularly notable in the Cambay region and are believed to be the result of postglacial (Pleistocene) transgression and silting.

About half the length of the Indian shore (east coast, Kerala coast, the southwest Kathiawar coast) is straight due to emergence, longshore drift, or development of barriers. About 430 km of the western shoreline (mostly lava region) between Karwar and Janjira and that of the Andaman and Nicobar Islands (submerged remnant of the Tertiary mountain chain) are crenulate (Tipper, 1911). The deltaic shores are protuberant. Other shores are moderately straight. Shore alignment is significant because it influences intervisibility, navigation, protection from storms, tidal range, force and direction of currents, and the nature of waves.

The evolution of the shore features of India has been influenced by the nature of the continental shelf, inland terrain, winds, fetch, currents, drifts, tides, eustatic changes, rise or fall of the adjoining land, deltaic or estuarine sedimentation, barriers, and biological factors such as corals and vegetation.

There are six types of Indian shoreline: deltaic; barrierless shore of emergent aspect between the Ganges and Godavari; shore of emergent aspect with offshore bars between the Krishna and Cape Comorin; rocky shore of submergent aspect in Karnataka and Maharashtra; marshy lowland shore of submergent aspect in the Cambay region; and compound shore of Kerala and parts of Kathiawar with a straight seaward shore (of offshore bars) and an irregular landward shore of lagoons.

Coastal Features

The land immediately behind the shore may be regarded as the *coastal zone,* which is narrow in areas of high relief and relatively broad in lowlands. If the 50 m contour is taken as the inner limit of the coastal zone, then the coastal zone ranges from a few meters from the shore at Cape Comorin and several points on the Maharashtra coast to 50 km in the Ganges delta and 100 km in the Godavari-Krishna deltas. It is dominated by depositional features—deltaic sedimentation, estuarine marshes, sand dunes, river alluvium, and even low-level laterites.

There appears to be considerable evidence of coastal emergence in the Indian region as indicated by the Dunkirkian, Monastirian, Tyrrhenian, Milazzian, and Sicilian strandlines. The raised beach at Bombay, the marine shells between Madras and Tinnevelly, and the raised coral reefs in Andaman and Nicobar islands provide evidence of the Dunkirkian strandline (3-5 m above the present sea level); beach deposits in North Arcot and a clifflike feature in Balasore, of the Monastirian (20 m); old beach deposits in south Dalbhum (Bihar), of the Tyrrhenian (30 m); an old marine terrace (55-61 m) on the coast, of the Milazzian (c. 60 m); and the inner edge of Pliocene marine deposits, of the Sicilian strandline (100 m). However, there are stretches of the Indian coast where evidence of submergence coexists with that of uplift: the cliffs and indentations on the Konkan coast provide geomorphic evidence; near Calcutta and Bombay there is evidence from boring; and elsewhere there are the remains of submerged forests.

On the basis of the dominant physical characteristics, the Indian coastal zones can be classified into seven types: the Andaman and Nicobar type that is marked by a dominantly rugged terrain and north-south folded structures; a deltaic type where deltaic sedimentation and hydrography override other physical conditions; a depositional plain type of the east coast; the Malabar type with highly irregular terrain, the structural grain of ancient crystalline rocks being parallel to the coast and that of the Tertiaries transverse to it; a lava-dominated Maharashtra type between Goa and Gujarat where terrestrial erosion is well marked (the lower slopes of the Western Ghats, thought to be caused by faulting, dominate this zone); the Gujarat coast zone of the Cambay Gulf that is marked by intense gully erosion; and the Kathiawar coast, suspected to be marked by an east-west fault in the south (Spate and Learmonth, 1967). A ribbon of coastal Tertiaries and Pleistocene windblown sediments have shaped the coastal landforms.

Concerning the overall evolution of the Indian littoral zone, the truth of the classic concept of Johnson is partially (because cliffs and barriers are missing) borne out from Cape Comorin to Karwar, where there are an irregular inner shore of lagoons, caused partly by the submergence of a dissected land, and a relatively straight seaward shore formed by barriers. This area resembles a premature shoreline of submergence. The classic concept of the mature stage of an emergent coast is partially borne out along the entire eastern coast, where beach deposits occur but cliffs are absent. The Maharashtra littoral appears to represent an intermediate stage between the initial crenulate and subsequent youth of submerged irregular terrain.

There is no evidence in the Indian littoral to negate the existence of the *Flandrian transgression* of about 20 cm per century since the peak of Wurm glaciation; the submergent aspect of the western coast bears it out. In the wide, level eastern coastal lowland, minor transgression would still leave the coast with an emergent aspect.

ENAYAT AHMAD

References

Ahmad, E., 1972. *Coastal Geomorphology of India.* Delhi: Orient Longman, 222p.

Arunachalam, B., 1964. Coastal features in the vicinity of Ratnagiri town, *Bombay Geog. Mag.* 12, 33-39.

Cushing, S. W., 1913. The east coast of India, *Geog. Soc. Bull.* 45, 81-92.

Heezen, B. C.; Takashi, I.; and Tharp, M., 1966. Indian Ocean, *in* R. W. Fairbridge, ed., *Encyclopedia of Oceanography.* New York: Reinhold, 370-402.

Gee, E. R., 1926. The geology of the Andaman and Nicobar islands, with special reference to Middle Andaman Island, *Geol. Survey India Rec.* 59, Pt. II, 208-232.

Pascoe, E. H., 1964. *A Manual of the Geology of India and Burma,* III. Delhi: Manager of Publications, 1345-2130.

Sewell, R. B. S., 1932. The coral coasts of India, *Geog. Jour.* 79, Pt. I, 449-465.

Spate, O. H. K., and Learmonth, A. T. A., 1967. *India and Pakistan.* London: Methuen, 877p.

Tipper, G. H., 1911. The geology of the Andaman Islands with reference to the Nicobars, *Geol. Survey India Mem.* 35, Pt. 4, 195-216.

Wadia, D. N., 1961. *Geology of India.* London: Macmillan, 536p.

Cross-references: *Asia, Eastern, Coastal Morphology; Asia, Middle East, Coastal Morphology; India, Coastal Ecology.* Vol. I: *Indian Ocean;* Vol. VIII, Part 2: *India.*

INLETS AND INLET MIGRATION

An interruption in a barrier beach, maintained by tidal flow, is commonly referred to as a *tidal inlet*. Such an inlet usually consists of the following elements (Fig. 1): the *gorge* or main

FIGURE 1. Morphologic features of a tidal inlet (Oregon Inlet, N. C.) and geomorphic features indicative of a migrating inlet, including relict flood delta, relict channel, and relict inlet pond developed behind a relict inlet ridge (photo: U.S. Soil Conservation Service; inlet features labeled by J. Fisher).

tidal channel, which may be straight or slightly sinuous; flanking shallow shoulders or shoals; adjacent barrier beaches, which under littoral drift develop as updrift and downdrift points; flood delta on the lagoon or bay side; and ebb delta on the ocean side. Both the flood and ebb deltas develop distinctive morphology in response to tidal currents (q.v.). The geometry of the inlet points, which may be symmetrical, offset, or overlapping, can be related to: magnitude of the littoral current, quantity of longshore drift; magnitude of the tidal current; or geometry of the bay or lagoon (Bruun and Gerristen, 1958). Inlets may originate as original initial interruptions in a developing barrier beach, as openings remaining as a baymouth spit develops, or by breakthroughs by storm or hurricane waves. Inlets opened by such storms are often temporary and close quickly, while permanent inlets are in a dynamic equilibrium between littoral and tidal currents and are considered stable even though they may migrate.

Inlet migration along a barrier beach shoreline may be due to a predominant (strongest) unidirectional longshore current (Atlantic Coast) or displacement by ebb delta sedimentation (Gulf Coast, W. A. Price, personal communication). Direction of inlet migration and littoral drift is often indicated by the shape, offset, or overlap of the inlet points. The rate of migration depends on the quantity of littoral drift and the velocity of the tidal currents. A pattern of migrating inlets along a barrier island chain is that: some inlets open simultaneously and remain stationary; some inlets open successively and remain stationary; some inlets open simultaneously and migrate rapidly; and some inlets open successively and migrate rapidly (Lucke, 1934). Implications of migrating inlets in the geologic record is that barrier beach deposits will be replaced with inlet deposits down to the depth of the inlet channel, which often reaches to beyond the base of the barrier deposits (Johnson, 1919).

Tidal inlets have both environmental and engineering implications in that they provide navigation and waterway passages through barrier beaches; they affect longshore drifts, whether jettied or left in a natural state; and shoaling of channels occurs in both the flood and ebb deltas. Field, model, and computer studies have been conducted to study these applied aspects of tidal inlets in general and for specific inlets (Barwis, 1976). To stabilize and minimize channel-shoaling effects in an inlet, three operations are common: dredging, sand-bypass pumping, and installing jetties. In the case of jetties, excessive material will accumulate on the updrift side of the inlet with subsequent erosion of the beach on the downdrift side (Fig. 2). Migrating inlets, if stabilized by jetties, often also exhibit channel shoaling, in this case because of an unbalanced sedimentation system.

JOHN J. FISHER

References

Barwis, J. E., 1976. *Annotated Bibliography on the Geologic, Hydraulic, and Engineering Aspects of Inlets.* Vicksburg, Miss.: U.S. Army Corps of Engineers, Waterways Experiment Station, General Investigation of Tidal Inlets Rept. 4, 333p.

Bruun, P., and Gerritsen, F., 1958. Stability of coastal inlets, *Am. Soc. Civil Engineers Proc., Jour. Waterways and Harbors Div.* 3, 1-19.

Johnson, D. W., 1919. *Shore Processes and Shoreline Development.* New York: John Wiley & Sons, 584p.

Lucke, J. B., 1934. Tidal inlets: A theory of evolution of lagoon deposits on shorelines of emergence, *Jour. Geology* 42, 561-584.

Cross-references: *Barrier Islands; Coastal Engineering; Dredging; Jetties; Protection of Coasts; Sand Bypassing; Tidal Deltas.*

FIGURE 2. Jettied tidal inlet (Charlestown Inlet, R.I.) adversely affecting longshore drift that moves from left to right. The updrift (left) beach has accreted, while the downdrift (right) beach has eroded (photo: J. Fisher).

INLETS, MARINE-LAGOONAL AND MARINE-FLUVIAL

Although the term *marine inlet* connotes flow in one direction, most function as orifices of bidirectional flow between the ocean and coastal reentrants, such as bays, sounds, and rivers. Inlet form is controlled largely by the patterns of water and sediment exchanged through the inlet. Early work on inlet dynamics and form was done by Bruun and O'Brien (see references in Oertel, 1975; Swift, Duane, and Pilkey, 1972; and Cronin, 1975). In sandy areas the mobile substrate permits a dynamic equilibrium between inlet size and water flow characteristics.

Because the patterns and characteristics of flow determine the size and shape of the inlet, fluvial or river inlets have distinctly different forms from those that exchange water between the ocean and lagoons. Density differences between freshwater (river) and seawater often causes stratification of the water column in which the more dense seawater underlies the freshwater in the form of a salt wedge. Salt wedges in fluvial estuaries generally form landward of the inlet throat. A net landward flow is produced in the lower portion of the channel, while river water drains seaward in the upper portion of the channel. At the toe of the salt wedge, there is a *nodal point* where residual ebb and flood currents meet. The circulation pattern at the nodal point produces channel shoaling, and is often responsible for the formation of midchannel bars and islands that are characteristic of many river mouths (Oertel, 1979). The result is that inlets with upland sources of freshwater have high width-to-depth ratios at the inlet throat (Oertel, 1974).

Lagoonal inlets exchange water between basins with minor freshwater sources (lagoons) and the sea. The driving mechanism for this exchange are changing water levels produced by the tides. In areas such as the Frisian coast of the Netherlands and the Sea Islands of the United States, the volumes of the lagoons and the thicknesses of the tidal prisms are relatively large (Nummedal et al., 1977). The exchanges of water across the inlet throat are also large and well mixed. Lack of stratification at the throat allows high-velocity tidal currents to scour sediment from the channel bottoms. The distribution and accumulation of this material are controlled by the interaction of inlet and coastal currents. In areas where forces produced by tidal currents dominate over that produced by onshore forces of coastal currents, large ebb deltas form, and the width-depth ratio at the inlet throat is distinctly lower than at the throat of fluvial inlets. In areas where tidal currents are weaker than the onshore forces of wave and coastal currents, midchannel bars and flood deltas form, and width-depth ratios at the inlet throat are proportional to, or higher than, those at the throats of fluvial inlets.

GEORGE F. OERTEL

References

Cronin, L. E., 1975. *Estuarine Research,* vol. 2. New York: Academic Press, 381p.

Nummedal, D.; Oertel, G. F.; Hubbard, D. K.; and Hine, A. C., 1977. Tidal inlet variability, Cape Hatteras to Cape Canaveral, in *Coastal Sediments '77.* New York: American Society of Civil Engineers.

Oertel, G. F., 1974. *Hydrographic Survey of the Doboy Sound Estuary and Surveys of the Other Tidal Inlets along the Coast of Georgia.* Skidaway Island, Ga.: Georgia Marine Science Center, Tech. Rept. Series 74-4.

Oertel, G. F., 1975. Post Pleistocene island and inlet adjustment along the Georgia coast. *Jour. Sed. Petrology* 45, 150-159.

Oertel, G. F., 1979. The Altahama delta system, in *Estuary-Shelf Sediment Exchange (ESSEX) 1979.* Norfolk, Va.: Old Dominion University Center for Marine Studies, 31-39.

Swift, D. J. P.; Duane, D. B.; and Pilkey, O. H., 1972. *Shelf Sediment Transport.* Stroudsburg, Pa.: Dowden, Hutchinson & Ross, 656p.

Cross-references: *Bars; Estuaries; Estuarine Circulation; Estuarine Coasts; Estuarine Delta; Estuarine Sedimentation; Tidal Currents; Tidal Deltas; Tidal Inlets, Channels, and Rivers.* Vol. III: *Tidal Inlets;* Vol. VI: *Estuarine Sedimentation.*

INSECTS

Insects are the most abundant animals on land. About a million species have been described, constituting almost three-quarters of all known animal species. Terrestrial insects are found in almost all conceivable habitats on earth. About 3%, or 25,000-30,000, of insect species are characteristically found in aquatic habitats, but of these only a fraction are associated with the sea. Perhaps for this reason, *marine insects* have been rarely mentioned in standard textbooks on entomology or marine biology. However, information on many marine insects has recently been collected, and is now available in one volume (Cheng, 1976). Much of the present article has been abstracted from that book, to which the reader is referred for further details.

Although all kinds of insects may be blown out to sea or stranded on beaches, sometimes in large numbers (Cheng and Birch, 1978), there are many insects that are specifically adapted to such marine habitats as salt marshes, lagoons, bays, rock pools, intertidal reefs, sandy and rocky beaches, mangrove swamps, and even the open ocean. At least 1400 species, representing some 300 genera, 100 families, and not less than 14 orders of the Insecta, are known to live or to spend at least a part of their life histories in a marine environment. The most important or most interesting marine insects are in the following orders and families:

- Collembola: springtails (Hypogastruridae, Neanuridae, Isotomidae).
- Thysanura: silverfish (Machilidae).
- Hemiptera: water striders (Gerridae, Veliidae, Hermatobatidae, etc.), shore bugs (Saldidae, Omaniidae, etc.), water boatmen (Corixidae), aphids (Aphididae).
- Trichoptera: caddis flies (Philanisidae).
- Diptera: mosquitoes (Culicidae), biting midges (Ceratopogonidae), seashore midges (Chironomidae), horseflies (Tabanidae), seaweed flies (Coelopidae, Anthomyiidae, Dryomyzidae, etc.) shore flies (Ephydridae, Canaceidae, etc.), and other beach flies (Tipulidae, Dolichopodidae, etc.).
- Coleoptera: rove beetles (Staphylinidae), ground beetles (Carabidae, Tenebrionidae, etc.), water beetles (Dytiscidae, Hydrophilidae, Limnichidae, etc.).
- Mallophaga: bird lice (Amblycera, Ischnocera).
- Anoplura: sucking lice (Echinophthiriidae).

The most commonly encountered insects on the seashore are the seaweed flies, houseflylike insects belonging to one of at least five families of the Diptera. The eggs are laid on decomposing cast seaweed or wrack, and when hatched the larvae feed on a mixture of bacteria and other microorganisms associated with the decomposed weed. *Coelopa frigida* Fabricius (Coelopidae) is the best-known and most widely distributed seaweed fly, which breeds wherever wrack is found on European as well as North American coasts. Its life cycle is well synchronized with the tidal cycle. The eggs are laid when spring high tides cast seaweeds high on the shore. The insects complete their larval stages in the next two weeks, emerging as adult flies before the weeds are washed away by the next series of spring tide. During summer months, adult densities reaching several hunred per meter stretch of beach are not uncommon in southern California. Other insects commonly found associated with wrack include springtails (Collembola) and various beetles, most of which belong to the family Staphylinidae. The majority of these intertidal staphylinids are predaceous, being presumably attracted to the wrack by the abundance of other invertebrates that feed and breed there.

The beetles themselves generally do not breed in the wrack, and probably leave this habitat before it becomes fully submerged at high tide.

The algae that cover intertidal rocks provide food and shelter for members of several families of beach flies, but the Chironomidae is the only well-studied family. It has at least 50 marine species. The larvae of *Paraclunio*, a widely distributed genus on the west coast of North America, build cases in intertidal algal mats and feed on common marine algae such as *Porphyra, Enteromorpha,* and *Ulva.* In certain areas, larval densities may reach several hundred per m^2, and so these herbivorous insects could constitute an important factor in the ecology of such intertidal algae.

Tiny midges of another genus, *Pontomyia,* were erroneously reputed to be submarine throughout their lives. In fact, although the larval stages live submerged among fronds of vegetation (*Halophila, Boodlea,* etc.), the mature midges are air-breathing and spend their brief adult lives at the sea surface (Cheng and Collins, 1980). The males have wings modified like oars, by which they scull over the sea surface in search of the females, which are wingless. Mating takes place within a few minutes after emergence, and for both sexes the adult life lasts only an hour or two. The eggs are laid in coils embedded in a jellylike substance that quickly imbibes water; they soon sink to the bottom, where they tend to become entangled in the vegetation. These and other adaptations of intertidal chironomids to marine environments have been reviewed by Neumann (1976).

Confined to the coasts of southern New Zealand and southeastern Australia is a family of marine caddis flies (Chathamiidae) (Riek, 1976). The eggs of *Philanisus plebeius* Walker are laid in the coelomic cavity of the common intertidal starfish *Patiriella exigua* (Anderson, Fletcher, and Lawson-Kerr, 1976). The larvae presumably chew their way out of the host, and then build cases among the coralline algae on which they feed.

Insects that can tolerate changing salinities are commonly found in coastal salt marshes, lagoons, and salt ponds. The water boatman *Trichocorixa reticulata* (Guerin-Meneville) and the brine fly *Ephydra cinerea* Jones can even live in saturated brine, which has a salinity ten times that of seawater. Several species of horseflies (*Tabanus* and *Chrysops*-Tabanidae) breed in *Spartina* marshes along the Atlantic seacoast of North America. Adult females are aggressive biters of man and livestock, and may also carry such diseases as *loa loa* and trypanosomiasis. Salt marshes tend to have a rich variety of insect faunas: in five salt marshes studied, in various parts of the world, over 300 insect species were recorded, including, in order of abundance, species of Diptera, Coleoptera, Hemiptera, Hymenoptera, Orthoptera, Collembola, Thysanoptera, Odonata, and Lepidoptera. Several intertidal ants have recently been studied in Mexico (Yensen, Yensen, and Yensen, 1980). The aphid *Pemphigus trehernei* Foster is restricted to the roots of *Aster tripolium,* a halophyte of European salt marshes. Detailed studies on the biology, distribution, and ecology of this aphid have been published by Foster and Treherne (1976). The salt marsh collembolan *Hypogastrura viatica* (Tullbg.) can sometimes be found in extremely high densities, reaching $400/cm^2$, after flooding of its habitat by seawater or rains. Certain springtails are also common in intertidal rock crevices as well as on sandy and muddy beaches.

From economic and medical standpoints, the most important marine insects are undoubtedly the biting midges (Ceratopogonidae) and the mosquitoes (Culicidae). At least three genera of blood-sucking ceratopogonids (*Culicoides, Leptoconops,* and *Dasyhelea*), with some 50-60 species, breed in coastal salt marshes in the tropics and subtropics. In Florida alone, thousands of acres of coastal land are sometimes rendered almost unfit for human habitation because of pestilent flies, and millions of dollars have been spent in search of effective measures to control these biting insects, so far without great success.

At least 9 genera of mosquitoes, including *Aedes, Anopheles, Culex, Deinocerites,* and *Opifex,* are known to breed in brackish waters. Tideland mosquitoes, especially those known to transmit malaria and filariasis, are important pests of man and domestic animals. As in the case of the biting midges, large sums of money are spent every year on attempts to control tideland mosquitoes.

Several interesting insects can be found around tropical coral reefs. The commonest forms include collembolans (*Anurida*), veliids (*Halovelia*), reef bugs (*Hermatobates*), and several genera of saldids and beetles. These insects do not as a rule possess special adaptations for underwater respiration, although they all have water-repellent cuticles that may also act as plastrons. In general they hide in crevices at high tide and entrap air bubbles that supply them with oxygen during submergence. If they become short of oxygen, they may go into an anoxic coma, from which they recover rapidly when oxygen becomes available again. On rocky shores in temperate zones, such habitats are occupied by different genera of collembolans, saldids, and beetles (*Halovelia* and *Hermatobates* are confined strictly to tropical zones). On European shores, the carabid *Aeposis robini* (Laboulbene) and the saldid *Aepophilus bonnairei* Signoret apparently spend much of their life cycles within crevices in rocks that are uncovered only during the lowest tides. In Australia, similar habitats are occupied by the saldid *Corallocoris marksae* (Woodward) and the limnichid beetle *Hyphalus insularis* Britton.

In many tropical bays, lagoons, and mangrove swamps, water striders belonging to the Veliidae and Gerridae abound. Several genera in these families (*Halovelia, Trochopus, Asclepios, Halobates,* etc.) are exclusively marine. The sea skater *Halobates* is the best known of these, with 42 described species. Only 5 species are widely distributed around the world in the open ocean between $40°N$ and $40°S$ (Cheng and Shulenberger, 1980), the rest being more restricted in their distribution and generally confined to nearshore lagoons, bays, or estuaries (Cheng, 1979).

The Mallophaga and Anoplura are lice parasitic on birds and mammals, respectively. A large variety of sea birds, including penguins, albatrosses, petrels, pelicans, terns, and gulls may be infested by members of the Mallophaga. About 30 genera are known. Anoplura are found only on mammals; 11 spp. have been recorded from seals. Little is known of the host-parasite relationships of these insects, except in the case of the southern elephant seal, *Mirounga leonina,* and its sucking louse *Lepidophthirus macrorhini* Enderlein, the physiology of which has been studied in some detail.

Finally, it should be mentioned that a large number of other air-breathing arthropods—spiders, mites, pseudo-scorpions—are associated with marine coastal environments. Many are predators of marine insects or other invertebrates; together with the insects, they constitute an important component of the marine intertidal community.

LANNA CHENG

References

Anderson, D. T.; Fletcher, M. J.; and Lawson-Kerr, C., 1976. A marine caddis fly, *Philanisus plebeius,* ovipositing in a starfish, *Patriella exigua, Search* **1**, 483-484.

Cheng, L., ed., 1976. *Marine Insects.* Amsterdam: North-Holland Publishing Co., 581p.

Cheng, L., 1979. *Halobates* and other little-known marine insects of the south Pacific region, *Internat. Symp. Marine Biogeog. Evol. So. Hemisphere: Auckland, N.Z., Proc., D.S.I.R. Inf. Series 137* **2**, 583-591.

Cheng, L., and Birch, M. C., 1978. Insect flotsam: An unstudied marine resource, *Ecol. Entomol.* **3**, 87-97.

Cheng, L., and Collins, J. D., 1980. Observations on behavior, emergence and reproduction of the marine midges *Pontomyia* (Diptera: Chironomidae), *Marine Biol.* 58, 1-5.

Cheng, L., and Shulenberger, E., 1980. Distribution and abundance of *Halobates* species (Insecta: Heteroptera) in the eastern tropical Pacific, *Fishery Bull.* 78, 579-591.

Foster, W. A., and Treherne, J. E., 1976. The effects of tidal submergence on an intertidal aphid, Pemphigus trehernei Foster, *Jour. Anim. Ecol.* 45, 291-301.

Neumann, D., 1976. Adaptations of chironomids to intertidal environments, *Annual Rev. Entomol.* 21, 387-414.

Riek, E. F., 1976. The marine caddisfly family Chathamiidae (Trichoptera), *Australian Entomol. Soc. Jour.*, 405-419.

Yensen, N.; Yensen, E.; and Yensen, D., 1980. Intertidal ants from the Gulf of California, Mexico, *Entomol. Soc. America Annals* 73, 266-269.

Cross-references: *Arthropoda; Biotic Zonation; Coastal Fauna; Coastal Flora; Coral Reef Habitat; Estuarine Habitat; Flotsam and Jetsam; Mangrove Coasts; Organism-Sediment Relationship; Salt Marsh; Swash Mark.*

INTERNAL WAVES

Internal waves are gravity waves that propagate beneath the sea surface along pycnoclines or thermoclines and within regions of weak or strong vertical density gradients. Internal waves occur in the deep sea, over the continental shelf, in coastal waters, or in estuaries. They may be generated in the ocean by any of a variety of mechanisms, including: traveling atmospheric pressure fields; interactions of currents with bottom topography; tidal oscillations; or second-order interactions between surface waves of different frequencies (Wunsch, 1975; Thorpe, 1975). High-frequency internal waves are also generated at the pycnocline or density *interface* in stratified estuaries or at river mouths by the shear of swiftly flowing surface layers.

Internal waves occupy a broad spectrum of frequencies ranging from inertial frequency ($\sim$ 1 cycle day^{-1}) to the Brunt-Vaisala frequency, N, approximated by

$$N = \left[\frac{g}{\rho_0} \frac{\partial \bar{\rho}}{\partial z} \right]^{1/2} \qquad (1)$$

where g = acceleration of gravity, ρ_0 = ambient surface density, $\bar{\rho}$ = local mean density, and z is the vertical coordinate (Phillips, 1966). Because oceanic density gradients are relatively weak, N values associated with ocean thermoclines tend to be low, ranging from 10^{-3} to 10^{-2} radians sec^{-1}. Hence, ocean internal wave periods less than 10 minutes are normally precluded. However, in estuaries and at river mouths, strong density gradients induced by salinity contrasts often result in Brunt-Vaisala frequencies on the order of .3-.5 rad sec^{-1}, thereby permitting the occurrence of internal waves with periods as short as 12-20 seconds.

The maximum phase speed, Ci, that an internal wave can attain is limited by the fractional density difference between the two layers of a stratified column and the depths of the layers such that

$$Ci < \left[g \gamma h_1 h_2 / (h_1 + h_2) \right]^{1/2} \qquad (2)$$

where $\gamma = (\rho_2 - \rho_1)/\rho_2$, h is the depth of vertical thickness of a layer, and the subscripts 1 and 2 designate, respectively, the upper and lower layers. Internal wave amplitudes are highly variable, but amplitudes in the range of 1-15 m are common.

Internal waves may play an important role in moving sediment on the continental shelf and nearshore bed. Recent theoretical estimations by Cacchione and Southard (1974) suggest that amplified near-bottom velocities associated with shoaling internal gravity waves may exceed the threshold criterion for incipient sediment motion on the continental shelf and possibly on the upper continental slope. High-frequency internal waves generated at the density interface of stratified river-mouth effluents contribute significantly to effluent deceleration and mixing.

Problems relating to the occurrence, behavior, and dynamics of internal waves are major topics of current frontier research by physical oceanographers and geophysical fluid dynamicists (Garrett and Munk, 1975).

LYNN D. WRIGHT

References

Cacchione, D. A., and Southard, J. B., 1974. Incipient sediment movement by shoaling interval gravity waves, *Jour. Geophys. Research* 79, 2237-2242.

Garrett, C., and Munk, W., 1975. Space-time scales of internal waves: A progress report, *Jour. Geophys. Research* 80, 291-297.

Phillips, O. M., 1966. *The Dynamics of the Upper Ocean*. Cambridge: Cambridge University Press, 261p.

Thorpe, S. A., 1975. The excitation, dissipation, and interaction of internal waves in the deep ocean, *Jour. Geophys. Research* 80, 328-338.

Wunsch, C., 1975. Deep ocean internal waves: What do we really know?, *Jour. Geophys. Research* 80, 339-343.

Cross-references: *Currents; Estuarine Sedimentation; Nearshore Hydrodynamics and Sedimentation; Sea Conditions; Sea Slick; Tidal Range and Variation; Tides; Waves.* Vol. I: *Internal Waves.*

INTERTIDAL FLATS

Intertidal flats are flat or gently inclined, relatively featureless, muddy and sandy areas exposed between tide levels, often cut by meandering creeks and partly vegetated on their inner margins to form coastal marshes or swamps. The intertidal flats may be 10-20 km wide, extend for hundreds of kilometers along the shoreline, and have vertical ranges up to approximately 17 meters.

The term should be restricted to areas exposed between the high- and low-water marks of astronomically induced tides and not include those that are flooded by marine waters during abnormal meteorological conditions, for which the term *supratidal flat* should be used. (See papers in Ginsburg [1975] for correct usage.)

Intertidal flats are best developed in sheltered environments, such as estuaries, lagoons, and coastal embayments, but are also found on exposed coastlines, where they sometimes form a *low-tide terrace* seaward of the beach (van Straaten, 1961).

They pass landward on sheltered coastlines into marshes and peat swamps in temperate and subtropical latitudes, into *mangrove* swamps in some tropical latitudes, and into low, saline flats (*sabkha*) in arid areas. On exposed coastlines, they pass landward into beaches and dunes.

They are composed mainly of either siliciclastic or calcareous sand, silt and clay, but gravel other than skeletal debris is rare. The sediment is derived either from rivers, erosion of neighboring coastlines, or from the adjacent sea floor. Much of the calcareous material may be produced by in situ precipitation of mud or by breakdown of skeletal remains. Both siliciclastic and calcareous mud is extensively pelleted into sand-grade material. Gravel is produced mainly from local skeletal debris. Diagenetically produced gypsum and some other minerals sometimes form an important component of intertidal sediments (Ginsburg, 1975).

In the sheltered areas in temperate and subtropical latitudes, the landward parts are colonized by *halophytic* plants to form coastal marshes. Blue-green *algal mats* are developed, but are best seen in tropical areas; elsewhere mangrove swamps develop.

Intertidal flats support an abundant but poorly diversified infauna and epifauna.

The microtopography of the flats results from the development of a rich variety of structures produced by a combination of wave and tidal current action as well as biological activity; other features result from desiccation during coincidence of high temperature and low tide and also result from freeze-thaw and ice action in high latitudes (Reineck, 1972).

In areas rich in calcareous sediment, lithification of surface sediment produces crusts analogous to *beach rock* that affect the types of fauna and flora.

Where there is an abundant supply of sediment, progradation of the coastline has formed extensive coastal plains that have often been reclaimed, in such areas as that surrounding the North Sea, to form valuable agricultural land.

GRAHAM EVANS

References

Ginsburg, R. N., 1975. *Tidal Deposits: A Casebook of Recent Examples and Fossil Counterparts.* New York: Springer-Verlag, 428p.

Reineck, H. E., 1972. Tidal flats, *in* J. K. Rigby and W. K. Hamblin, *Recognition of Ancient Sedimentary Environments.* Tulsa, Okla.: Soc. Econ. Paleont. Mineral., Spec. Pub. 16, 146-159.

van Straaten, L. M. J. U., 1961. Sedimentation in tidal flats areas, *Alberta Soc. Petroleum Geologists Jour.* 9, 204-226.

Cross-references: *Algal Mats and Stromatolites; Estuaries; Mangrove Coasts; Sabkha; Salt Marsh; Terraces; Tidal Flat; Tides.*

INTERTIDAL MUD HABITAT

Intertidal muds represent shoreline deposits of fine-grained muddy or silty material that accumulate in areas of restricted wave and current activity along rivers, in estuaries, and in areas adjacent to open ocean. These generally broad expanses of sediment are characterized by low topography and are drained to varying degrees during ebb tides.

Intertidal muds cannot exist where there is much wave action, and therefore the mudflat habitat is limited in its extent by certain environmental factors. Muds are normally deposited where water movement is minimal or where currents are laden with a heavy suspension of fine material. Broad mud flats do not commonly develop in areas of significant shoreline relief, because cliff face erosion adds a considerable coarse-grained component to the local sediment and the high wave energy along these areas of exposure winnows away fine particles.

The intertidal mud habitat consists of an assemblage of subenvironments, including tidal pools, tidal creeks, local sandy patches, and areas of marsh or swamp. In poorly drained areas of the flats, shallow tidal pools can form. The water remaining in these pools is subject to wide salinity fluctuations. Heavy rains can cause a rapid decrease in salinity, while during dry periods excessive evaporation results in increased salinity. In more evenly drained areas, runoff on the flats is diverted into a series of tidal creeks. Sandy patches accumulate during storms when waves carrying coarser-grained material from subtidal areas deposit it onto the flats. Marshes and swamps occur in the higher intertidal areas, and can trap sediments brought to these areas during high tides.

Muds tend to accumulate in sheltered places, and are frequently associated with a rich supply of organic material. Because of the mud's fine texture, there is little circulation of water within the sediment. Thus the wide salinity and temperature fluctuations in surface waters of the flat do not occur with depth in the sediment. The limited interchange of interstitial waters, combined with the high organic content, often produces conditions in which oxygen is deficient and hence unsuitable for many forms of life. The shortage of oxygen in many muds is indicated by the presence of a black layer a short distance below the surface. *Hydrogen sulfide* is usually present in this layer and produces a characteristic "stinking" smell.

The muddy substrate is soft, and unless shells or plant roots bind the sediment, there will be few holdfasts for attached forms of life. The softness of the surface also eliminates those animals requiring a hard substratum on which to crawl. On the other hand, the texture makes burrowing easy, resulting in a high proportion of burrowing (infaunal) to surface (epifaunal) forms. The high organic content affords food for a large population of deposit feeders.

The intertidal mud habitat is a difficult one in which to live, and the only plants and animals that can inhabit it successfully are those adapted to withstand low oxygen levels, turbid water, seasonal temperature variations, and salinity changes. The frequent shortage of oxygen necessitates that the fauna either be capable of tolerating anaerobic conditions for a considerable time, as meiofaunal nematodes do, or that it be able to maintain a connection with the surface to obtain oxygen, as many burrowing forms do, or that it be limited in its vertical distribution to the extreme surface layers (Moore, 1958). Turbid water can be hazardous to filter-feeding organisms whose gills tend to be clogged by fine silt particles. Conditions of varying temperature and salinity provide some limitation on the species that occupy the mud habitat, but the fluctuations are offset to some extent by the buffering action of the mud and the protection it affords (from predation and physical stress) to burrowing forms.

The surface of the muddy shore supports a varied crustacean population, made up principally of isopods and amphipods, although crabs of various species are also common. One typically abundant form is the small burrowing amphipod, *Corophium,* which sits in its tube collecting detritus from the bottom or sieving it out of the water. Gastropod molluscs such as *Hydrobia,* which ingest the sediments and their associated bacteria, and the salt-marsh periwinkle, *littorina irrorata,* which grazes along the surface, may be plentiful where there is an abundance of plant detritus. The scavenging mud snails of the species *Nassarius* also move over the surface and often occur in extremely dense populations.

Among the burrowing fauna, bivalve molluscs and polychaete worms predominate, although in some areas, often where the eelgrass *Zostera* occurs, a burrowing sea anemone is fairly common. This small burrowing anemone, *Peachia hastata,* occupies a burrow nearly a foot deep (Yonge, 1966), and has twelve long tentacles that extend up above the surface during feeding.

Deeply burrowing clams are equipped with long retractile siphons that enable them to maintain contact with the surface for food and oxygen. The soft-shelled clam, *Mya arenaria,* an infaunal filter feeder may be dug from depths of up to 30 cm. Deposit-feeding bivalves of the species *Macoma balthica* often congregate in immense numbers. The indrawing siphon can be seen continuously collecting large quantities of the superficial deposits, while a mound of sediment forms near the exhalant siphon.

A varied population of worms live in and upon the muddy sand; some species wander over the surface seeking prey, and others inhabit a variety of tubes embedded in the mud. The ragworm, *Nereis,* a common species that eats smaller invertebrates and filters food particles from the water, lives in a U-shaped tube but can also move out of the tube in search of food. Burrowing worms, such as the common peacock worm, *Sabella pavonina,* whose widely spread ring of colorful plumelike tentacles collect food from the water, live in muddy tubes. *Sabella* may be so common that the tubes, which project about 12.5 cm above the surface, form a miniature forest. A third species, *Myxicola infundibulum,* which has a shorter, more complex ring of tentacles, is distinguished by its translucent gelatinous tube. The flattened ribbon worms occur, as do the permanently tube-dwelling phoronid worms.

Populations of the fiddler crab, *Uca,* and the marsh crab, *Sesarma,* live on the borders of the tidal pools. Closer to the land are burrows of the "land" crabs, *Cardiosoma* and *Ucides.* These crabs are largely herbivorous, but some species may feed on microorganisms in the substrate.

A special type of mud flat is found in the tropics along stretches of the coasts of Indonesia, Africa, and South and Central America. These mud flats contain a dense growth of mangrove scrub, *Rhizophora* being the dominant type. A great variety of land, freshwater, and marine animals live among the aerial stilt-like roots of the mangroves. Oysters cemented to the extensive root system of the mangroves, species of the common mussel, viviparous periwinkles found high up in the branches and stems, and a variety of crabs are the most conspicuous marine animals associated with the mangroves. (See Vermeij, 1973, for an extensive study of the ecology of gastropod molluscs living in the mangrove habitat.) The most common crabs in the mangrove swamps are fiddler and marsh crabs. The male fiddler can be readily recognized by the great difference in the size of its claws, one being much larger and more highly colored than the other and used in fighting and courtship displays. Fiddler crabs emerge to feed when the tide is out, and withdraw to their burrows when the tide floods.

The *climbing perch* or mudskipper, *Periopthalamus,* has evolved the ability to live both in and out of the water. It leaves the water when the tide is out to search for the insects, worms, and crabs on which it feeds, but must return to the water periodically to moisten its gills. The ventral and pectoral fins function as primitive legs, and when the animal is moving, the tail assists in propulsion in the same manner as that of a lizard.

The intertidal mud habitat is a heterogeneous environment, and supports a great diversity of living forms. Organisms living in this environment are adapted not only to physical parameters, such as temperature, salinity, and oxygen levels, but to biotic parameters as well. Plants and animals provide food, attachment sites, microhabitats, and protection for other living organisms. This delicately balanced system is maintained by nature's processes, but is extremely vulnerable to man's perturbations.

Other relevant literature on this topic can be found in Fenchel, Kofoed, and Lappalainen (1975), Lucas and Critch (1974), Russell and Yonge (1975), and Thorson (1957, 1971).

DOREEN LEE FUNDILLER

References

Fenchel, T.; Kofoed, L. H.; and Lappalainen, A., 1975. Particle size selection of two deposit feeders: The amphipod *Corophium volutator* and the prosobranch *Hydrobia ulvae, Marine Biol.* **30,** 119-128.
Lucas, J., and Critch, P., 1974. *Life in the Oceans.* New York: E. P. Dutton, 216p.
Moore, H. B., 1958. *Marine Ecology.* New York: John Wiley & Sons, 493p.
Russell, F. S., and Yonge, C. M., 1975. *The Seas,* 4th ed. London: Frederick Warne and Co., 283p.
Thorson, G., 1957. Bottom communities, *Geol. Soc. America Mem.* **67,** 401-534.
Thorson, G., 1971. *Life in the Sea.* New York: McGraw-Hill, 256p.
Vermeij, G. J., 1973. Molluscs in mangrove swamps: Physiognomy, diversity, and regional differences, *Syst. Zool.* **22,** 609-624.
Yonge, C. M., 1966. *The Sea Shore.* London: Collins Clear-Type Press, 311p.

Cross-references: *Biotic Zonation; Coastal Fauna; Coastal Flora; Coastal Waters Habitat; Coral Reef Habitat; Estuarine Habitat; Intertidal Flats; Intertidal Sand Habitat; Mangrove Coasts; Organism-Sediment Relationship; Rocky Shores Habitat; Sand Dune Habitat; Tidal Flat.*

INTERTIDAL SAND HABITAT

Intertidal sands are found on beaches, sand flats, and—more sheltered—in shallow bays, estuaries, and wadden areas, where they gradually merge into mud flats. Grain size ranges from 60-2000 μm, but there may be a small admixture, sometimes only temporary, of finer and (or) coarser material. Apart from tidal fluctuations, and tidal currents, intertidal sands are usually also affected by surface waves. During onshore storms, the water level may be raised to more than double the tidal range.

The surface of the sands is regularly disturbed; silt and clay normally do not settle; and the sand tends to be sorted out. On exposed beaches, wave action prevents the fine sand sizes from settling near low tide level, whereas at high tide level the fine sand is usually winnowed out, either by returning water or by the wind. Thus there is normally some size sorting on beaches in zones parallel to the coast (Hedgpeth, 1957).

On the more sheltered tidal flats, size sorting is related to the velocity distribution of the tidal currents and to local wave action. A predominant wind direction is usually reflected in the presence of coarser sands on that side of the tidal flat. The surface of intertidal sands is seldom entirely flat, but usually covered with some sort of relief: ripples, megaripples, cusps, bars, scour holes, small gullies, ridges and runnels. On exposed beaches the sand is coarser than on sheltered beaches and estuarine tidal flats. Wave characteristics, average particle size, and surface slope are related; with greater particle sizes, the slope is *steeper*.

During each tide the surface layers of intertidal sands fall dry, and air partially replaces the interstitial water. During the oncoming flood the flow over the surface is usually more rapid than through the sands. So it can happen, especially when flooding occurs in a relatively short time and when the sands have become very dry (as often happens on a beach), that the surface water flows over sands still filled with air. Escaping air bubbles form holes, small domes, and miniature craters that look much like animal burrows. Sometimes air remains in the sand in small pockets which are remarkably stable.

The water content of intertidal sands is greatly influenced by their *porosity* and *permeability*. Porosity is the volume of the space between the grains expressed as a percentage of the total volume of the deposit. In natural sands, porosity varies between 20–38% depending on grain size, grain shape, the structure of the deposit, compaction, and solidifaction. Within these margins, porosity is determined by complicated relationships. Greater angularity of the grains may increase or decrease porosity, depending on the packing of the grains. Also the presence of small particles may either enhance or reduce porosity depending on whether they fill the pore space between the larger grains. The presence of disc-shaped grains greatly diminishes porosity.

Permeability is greatly influenced by porosity and grain size. A small admixture (of only a small percentage) of silt or clay may appreciably decrease the permeability because of adsorption of water on the grain surfaces and the filling of pores. Porosity and permeability strongly influence the flow of water and air through the sands as well as the capillary movement of water. The latter process is important for wetting the sands during the periods they fall dry. Water in coarse sands drains and evaporates easily. Finer sands are not so easily drained, and the capillary rise of water is more pronounced. Evaporation is also less in fine sands because of the larger surface area on which water is adsorbed. Coarse sands are not usually able to retain any water, whereas sands of 150 μm retain about 14% of all the water they hold when saturated. The retention of water is highly important for the sands around or directly above high tide level, as it keeps them damp. On dessication, such sands can be eroded by the wind down to the level where the sands are moist.

Some intertidal sands behave as fluids: the density, viscosity, and amount of water as well as the grain size of the sand determine the percentage of grains that have no direct contact with other grains. In *quicksands* a large proportion of grains are floating. Such sands offer little or no resistance to penetration. When, however, all grains are in contact with other grains, resistance against penetration is large. The *firmness* of intertidal sands is related to whether the resistance is reduced with increasing shear (*thixotropy*) or increased (*dilatancy*). Usually intertidal sands are a mixture of thixotropic and dilatant sands. The firmness may vary during the tidal cycle or during the seasons.

Temperature variations, either during the tidal cycle, during the day, or during the seasons, are largest at the surface. When insolation of surface cooling is strong, large temperature gradients develop within the top few centimeters of the sands. Salinity variations are also highest in the top layers because of overflowing water of higher or lower salinity, rainwater, freshwater runoff, or evaporation. Even in a temperate climate, salinity may rise considerably on a tidal flat during the summer because of evaporation at low tide: an increase of 4.5‰/hour has been measured. The general composition of the interstitial water is influenced by chemical processes in the sand. Trace metal content tends to increase because of solution and desorption from particles; silicon may be dissolved from diatom frustules and phosphates; and nitrates may be released from decomposing organic material. Oxygen content usually becomes low at some depth below the surface, and a dark anaerobic layer is an almost universal phenomenon. In sheltered bays, wadden, and estuaries, the dark layer is present only a few centimeters below the surface; but on exposed beaches with heavy surf, it is present at a depth of 50 cm or more or may be absent. The dark bluish-black color is due to H_2S formed by sulfate-reducing bacteria and the formation of iron sulfide and iron hydroxide present on the grains. This process starts when all the oxygen in the interstitial water has been used, usually by decomposition of organic material. The dark layer therefore indicates stagnant conditions, and is limited to the zone that is not regularly disturbed by waves and currents. In temperate regions, the dark layer is closer to the surface during the summer (when water turbulence—wave action—is minimal) than during the winter. Also at higher temperatures, oxygen is less soluble, whereas the decomposition of organic material proceeds more rapidly. The overall effect is that during the summer anaerobic conditions set in at an earlier stage. On decomposition of organic matter, CO_2 or HCO_3 is produced, which is buffered by the calcium carbonate usually present in the sands as shell fragments. The pH therefore remains about that of seawater.

The surface of intertidal sands in temperate regions is not favorable for plant or animal life, because of large fluctuations in temperature and salinity and of disturbance by waves and currents (Moore, 1958; Green, 1968). Only some algae inhabit the surface. Below a depth of about 5 cm, conditions are more stable, while the top layer offers protection. Intertidal sands are therefore mainly inhabited by burrowing organisms or by temporary visitors, such as predatory birds and crabs, reproducing turtles, and swimming or crawling organisms that move in with the tide. Not all intertidal sands, however, are favorable even for burrowing fauna. Coarse, angular volcanic sands are difficult to penetrate and therefore poor in organisms. The same is probably true for coarse calcareous sands consisting of broken shell, echinoid fragments, and coral debris. Coarse sands on exposed beaches at temperate and high latitudes are deeply disturbed, or become frozen during the winter, when the temperature is low, so that life is not possible during that period.

Fine sands are generally more suitable for a rich animal life to develop. These sands occur in more sheltered areas and are less disturbed. Their surface area is larger, so that bacteria and other organisms that live attached to sand grains are favored. Fine sands usually also contain more organic matter, so that more organic carbon and nitrogen are available. Nevertheless, because of the rather extreme conditions, fine-grained, intertidal sands are mostly inhabited by specialized types occurring in large numbers but fluctuating widely in numbers from year to year.

Many organisms migrate during the tidal cycle. Some burrowing organisms stay in their burrows during low tide but rise to the surface when the tide rises, and; the reverse also happens. Many swimming or crawling organisms move up and down with the tide. Tube-building organisms are typical of sheltered flats, since on an exposed beach there is little chance for tubes to remain intact. On more exposed sites, some mollusks burrow deep and reach the surface water through a long siphon. Burrowing mollusks tend to have solid shells that are smooth or have a reduced ornamentation. Echinoderms tend to be flat, with reduced spines. There is usually a zonation of species living at a certain level with regard to the tide. This is probably related to the availability of food and the ability to withstand desiccation. Also for a number of species there is a zonation according to size or sex. Besides macrofauna there is usually a well-developed interstitial microfauna that shows a marked zonation, which is probably related to the porosity and the permeability of the sands.

D. EISMA

References

Green, J., 1968. *The Biology of Estuarine Animals.* London: Sidgwick & Jackson, 401p.

Hedgpeth, J. W., 1957, Sandy beaches, *Geol. Soc. America Mem.* 67 1, 587-608.

Moore, H. B., 1958. *Marine Ecology.* New York: John Wiley & Sons, 494p.

Cross-references: *Biotic Zonation; Coastal Fauna; Coastal Flora; Coastal Waters Habitat; Coral Reef Habitat; Estuarine Habitat; Intertidal Flats; Intertidal Mud Habitat; Loose Rock and Stone Habitat; Nearshore Water Characteristics; Organism-Sediment Relationship; Rocky Shores Habitat; Sand Dune Habitat; Sediment Size Classification; Wadden.*

ISOSTATIC ADJUSTMENT

Isostatic adjustment refers to the transient (10^2-10^4 years) or long term ($> 10^5$ years) nonelastic response of the earth's lithosphere to loading and unloading due to erosion, deposition, water loading, desiccation, ice accumulation, and deglaciation. Isostasy is essentially the Archimedian principle of hydrostatic balance between floating bodies. The term *isostasy* was proposed by C. E. Dutton in 1889 to define a suggestion made by Sir George Airy in 1855 that the earth's crust is supported by underlying denser material, and that the weight of mountains is balanced by light material extending as roots into the denser mantle. Perhaps the best example of isostasy is the contrast in levels between the earth's major first-order physiographic features: the continental platforms and the ocean basins. The mean elevation of the continents is about 1 km, while the average depth of the oceans is near 4 km. The 5-km difference is best explained by the differing densities of the rocks underlying these features. As a first approximation, a unit cross-section 40-km-deep column composed of continental rock having a mean density of 2.85 would be just about balanced by a 35-km-deep oceanic rock column of the same cross-section having a density of 3.25. Because the second-order physiographic features of both the continents and the ocean basins differ appreciably in relief, the depth of compensation is variable.

The response of the earth to transient loads of ice or water provides the most explicit information available concerning these nonelastic properties of the upper asthenosphere (Cathles, 1975). Isostatic adjustment or compensation due to static disequilibria appears to be accomplished by lateral flow in the upper asthenosphere's low-velocity and low-viscosity channel.

The relaxation time for postglacial uplift, defined as the time taken to recover 1/e (say,

65%) from glacial loading is on the order of 10^3-10^4 years. In other words, half of the potential isostatic adjustment to the disappearance of an ice cap will occur on the order of 1000 years. The inherent strength of the earth's crust is such that the minimum stress change required to induce flow in the low-viscosity channel is on the order of 10 bars. Further, the asthenosphere seems insensitive to loads less than about 100 km wide, which are largely taken up by elastic processes within the lithosphere. Thus isostatic adjustment is a regional phenomenon involving large-scale transfer of load stress on the order of 10 bars or more (a minimum of about 100 m of water, 110 m of ice, or about 36 m of continental rock).

WALTER S. NEWMAN

Reference

Cathles, L. M., III, 1975. *The Viscosity of the Earth's Mantle.* Princeton, N.J.: Princeton University Press, 386p.

Cross-references: *Global Tectonics; Isostatically Warped Coasts; Sea Level Changes.*

ISOSTATICALLY WARPED COASTS

At the close of the last ice age, climatic changes caused vast loads to shift on the earth's surface. Starting about 15,000 years ago, the 1-3 km of ice that had covered northern North America, northwestern Europe, and several areas in Siberia quickly melted and dumped their meltwaters into the world ocean, increasing the average depth of the sea by about 100 m. The earth's response to this load redistribution was one of viscous flow in the upper asthenosphere (Cathles, 1975). As a result, ancient shorelines (shorelines out of reach of present wave action) have been differentially warped by *isostatic adjustment* to positions both above and below present sea level (Andrews, 1974). The deformed shorelines can be visualized as bounding an imaginary water surface extending throughout a basin, although this former surface can be observed only where former shorelines remain, usually around the rim of a basin. In addition to glaciers, water, and sediment, loading and unloading are believed to be major causes of isostatic deformation.

On the North American plate, shorelines about 12,000 years old are found at elevations up to several hundred meters above present sea level within the formerly glaciated area, while shorelines of similar age have been detected to depths of as much as 200 m below sea level off the east coast of the United States and beneath the Mississippi delta. Ancient Lake Bonneville in the Salt Lake Basin of Utah, desiccated around 12,000 years ago, reacted by an up to 64-m domical arching of its once horizontal shorelines.

WALTER S. NEWMAN

References

Andrews, J. T., ed., 1974. *Glacial Isostasy.* Stroudsburg, Pa.: Dowden, Hutchinson & Ross, 491p.
Cathles, L. M., III, 1975. *The Viscosity of the Earth's Mantle.* Princeton, N.J.: Princeton University Press, 386p.

Cross-references: *Elevated Shoreline; Emergence and Emerged Shoreline; Global Tectonics; Isostatic Adjustment; Raised Beach; Sea Level Changes; Submergence and Submerged Shoreline.*

J

JETTIES

A *jetty* is a structure extending into a body of water from the land, designed to direct and confine the stream and tidal flow and to prevent or reduce shoaling of a channel by littoral materials (Coastal Engineering Research Center, 1973). The various types of jetties resemble small *breakwaters* and may consist of rubble-mound; concrete caisson; cellular steel sheet-pile; and cribs using timber, steel, or concrete. In sheltered areas, a single row of braces and tied timber piling and steel sheet-piling have also been used. The term jetty is also improperly used to denote a narrow projecting pier for the berthing of ships in harbors and places where harbor facilities are not available; spurs extending at intervals from the banks of rivers along a wide channel so as to effect a concentration of the current; or a single long groin extending beyond the surf zone to prevent accretion or shoaling in the downdrift area by littoral drift.

Jetties located at the entrance of a bay, lagoon, or river are usually constructed in pairs, either parallel or converging toward the seaward ends, sometimes extending seaward to the position of the contour equivalent to project depth of the channel. They also serve to protect the entrance from wave action and cross currents, which is the reason why jetties are often confused with breakwaters, especially in Japan (Figs. 1 and 2).

A jetty at the shoreline interposes a total barrier in the part of the littoral zone between the seaward end of the structure and the limit of wave uprush on the beach. Accretion takes place updrift of the structure and erosion, downdrift. The quantity of the accumulation and erosion depends on the length of the structure, the longshore transport rate of littoral drift, and the angle at which the resultant of the natural forces, such as winds, waves, and currents, strikes the shore.

SHOJI SATO

FIGURE 2. Aerial view of the entrance to Tosa Bay, Shikoku, Japan, in 1966 showing jetty construction. Broken lines indicate the shorelines in 1947.

FIGURE 1. Aerial view of Miyazaki Port, Kyushu, Japan, in 1971, showing jetties at the mouth of the Cyodo River. Broken lines indicate shorelines in 1947, before construction of the jetties.

Reference

Coastal Engineering Research Center, 1973. *Shore Protection Manual.* Fort Belvoir, Va.: U.S. Army Corps of Engineers, vol. 2, 5-46-5-49; vol. 3, A-19.

Cross-references: *Breakwaters; Coastal Engineering; Embankments; Groins; Inlets and Inlet Migration; Sediment Transport.* Vol. III: *Beach Erosion and Coastal Protection.*

K

KAIMOO

Kaimoo is an Eskimo word referring to a feature of cold-region beaches described by Moore (1966) as "a flat rampart of alternating layers of beach sediment and ice" (Fig. 1). This ice-sediment deposit often develops on the upper part of a beach by wave action before formation of the *sea ice* or of an *ice foot* (q.v.). During swash run-up, a thin layer of ice forms on the upper beach; this may be followed by deposition of a layer of sand or gravel. Repetition of this sequence can form a bed of interlayered ice and sediment several meters thick (Short and Wiseman, 1974). The horizontal extent of a kaimoo is controlled by the length of the wave run-up, and the height is determined by the length of time available for accretion before the ice foot or the sea ice develops.

A kaimoo is one category of ice foot features, and some confusion has arisen from attempts to define a kaimoo as a separate ice form. In general, an ice foot may contain sediment, is often convex upward in profile, has an irregular surface, and has a small scarp at the seaward margin (Rex, 1964). A kaimoo is more specifically defined as an ice foot or storm ice foot that has a flat surface and contains interlayered ice and sediment.

Melting of the kaimoo in spring or summer results in a pitted or corrugated surface that will persist until the sediments are redistributed by wave action. Greene (1970) describes a kaimoo ridge that may form following ablation of the ice.

EDWARD H. OWENS

References

Greene, H. G., 1970. Microrelief of an arctic beach, *Jour. Sed. Petrology* 40, 419-427.

Moore, G. W., 1966. Arctic beach sedimentation, in N. J. Wilimovsky and J. N. Wolfe, eds., *Environment of the Cape Thompson Region, Alaska*. U.S. Atomic Energy Commission Rept. PNE-481, 587-608.

Rex, R. W., 1964. Arctic beaches, Barrow, Alaska, in R. L. Miller, ed., *Papers in Marine Geology*. New York: Macmillan, 384-400.

Short, A. D., and Wiseman, W. J., Jr., 1974. Freezeup processes on arctic beaches, *Arctic* 27, 215-224.

Cross-references: *Antarctica, Coastal Morphology; Arctic, Coastal Morphology; Glaciel; High-Latitude Coasts; Ice along the Shore; Ice-Bordered Coasts; Ice Foot.*

FIGURE 1. A cross-section through a kaimoo, facing landward, at Point Lay, Alaska (photo: W. J. Wiseman, Jr.).

KARST COAST

A *karst coast* (*littoral karst*) is characterized by limestones affected by both subaerial and marine weathering and erosion in a littoral situation. The result is often a development of extremely jagged and bizarre erosion forms. While normal karst is affected by solution only from rainwater and related groundwaters, the coastal karst is influenced in addition by seawater and sea spray (seawater carried by splash momentum and/or wind). In addition to primarily downward-directed erosion, it shows important horizontal activity, involving undercuts, fretting, and honeycomb effects.

The physiographic characteristics of the karst coast are related to the interactions and rates of erosion of two agencies. The horizontal effect of marine activity tends to develop an *undercut* or *notch*, the maximum undercut relating to mean sea level (or, in sites of heavy swell, somewhat higher, at mean swash limit) and correspondingly a more or less flat, intertidal erosional *platform* or *bench* in front of the undercut, with an *overhang* or *visor* above the undercut (Macfadyen, 1930; Kuenen, 1933). The undercut is commonly affected not only by wave action but, more important, by biologic attack (e.g., boring molluscs—such as *Pholas* and *Lithodomus*—boring algae, boring sponges, browsing crabs and gastropods). Above the visor there is a transition to the supratidal splash and rainwater zone, where the marine influence decreases proportionately with its distance from the shore but varies in its development according to relief, exposure to onshore winds, and various climatic factors; in this *littoral karst zone* are a great variety of *lapies, pits, pans* and *pipes* that develop an extremely jagged microrelief. There is a highly sensitive biologic zonation (see, for example, Schneider, 1976, who refers to the biologic etching as *biokarst*).

The physiographic expression of the coastal karst varies also according to rock type. Paleozoic and Mesozoic limestones and dolomites, commonly developed in the Mediterranean and western Europe, are usually fine-grained or are less responsive to the various erosional agencies than are the more porous Cenozoic limestones that include especially Quaternary coral limestones, eolianites, and beach rocks. These Cenozoic rock types are typically developed around many of the tropical and subtropical coasts of the present day, but are scarcely ever encountered poleward of $45°$N or S.

Coastal karst has been called *halokarst* (German: *Salzwasserkarst*, according to Kelletat, 1974) to recognize the saltwater interactions. Most coastal karsts contain evidence of at least two phases of *karstification:* a stage of subaerial processes involving freshwater solution and nonhalophytic plant activity (termed *phytokarst* by Folk, Roberts, and Moore, 1973), the evidence for which is often inherited, and may be partly effaced or overprinted (*palimpsest*) by the halokarst; and a succeeding stage of combined marine and freshwater weathering—the contemporary halokarst. The phytokarst activity is generated mainly by unicellular boring algae, which tend to die under exposure to hot solar radiation, so that the surface is black in color. Where wave action is vigorous, phytokarst is destroyed, and the rock surfaces tend to be whitened.

Many coastal karsts were exposed to Pleistocene lowered sea levels, and were transected by multiple vertical solution pipes during the lowered base level of freshwater solution. In the present erosion cycle, these pipes (which became lined by highly resistant travertine) may now become reexposed under halokarst conditions, with an inversion of preservation, to form isolated cylinders, so-called *organ pipes*. Partly exposed examples in the Bahamas are called *Banana holes*. Large sink holes developed during a preceeding erosion cycle may become drowned and truncated by the present-day erosion; examples in the Bahamas and in the Yucatán-Belize reefs are known as *blue holes,* and have been explored by scuba divers to depths of over 100 m. Multiple examples in the Houtman's Abrolhos Islands, Western Australia, are illustrated in aerial photographs reproduced in Fairbridge (1948). In tropical oceanic situations, the last glacial fall of sea level generated saucer-shaped karstic islands that, after subsequent rise of sea level, led to the evolution of the typical ring form of the Pacific insular atolls, according to the theories of Darwin and Daly, subsequently modified by MacNeil (1954). The same eustatic fall (in some cases modified by crustal movements) led to the submergence of aquifers, and today freshwater springs are found along certain karst coasts (e.g., Florida, Persian Gulf, Adriatic) (Herak and Stringfield, 1972). Drowned karst caves are known—e.g., the famous Blue Grotto on Capri (Italy).

The halokarst sequence may be reversed in the case of an emerged or uplifted coast—e.g., on the north coast of Jamaica, in Barbados, New Guinea, Timor, Taiwan, the halokarst that was associated with a marine-erosional undercut (*wave-cut notch*), but is now physically removed from the coastal area and is in the process of being effaced by purely subaerial karstification.

The precise nature of the halokarst processes has been subject to some controversy. The writer and others have carried out systematic salinity and pH readings in halokarst pools on the coasts of Western Australia and on Pacific coral reefs (Fairbridge, 1948; Revelle and Emery, 1957). During 24-hour sessions of continuous monitoring, it was found that there are regular day-night oscillations, reflecting several diurnal changes: daytime solar warming and evaporation, followed by nightly cooling, which tend to lower CO_2 partial pressure by day and raise the pH, leading to precipitation of $CaCO_3$, $CaSO_4 \cdot 2H_2O$ and $NaCl$ within the pools; daytime algal activity, leading to photosynthetic liberation of CO_2; browsing (at various times) by invertebrate organisms, notably crabs and gastropods of *Littorina* type, the role of which is both to raise CO_2 in the water and mechanically to loosen rock particles, aggravating further chemical attack. The above variables are further modified by two changes affecting the supply of water to the tide pools and pockets in the halokarst. These pools may be refilled at any time of the day or night by: seawater at high tide or due to increased storminess, or by freshwater during rain showers. Thus the salinity exhibits an astonishing range of values: from almost pure rainwater at times, varying to brackish and to normal seawater, up to hypersaline brines that in certain cases evaporate leaving a salt crust. Correspondingly, the pH range is also remarkable: pools rich in algae will show a fall to pH 5 (under the algal mats during daylight hours), but after refilling and particularly after heavy rain, the reading will rise to over pH 10.

These biochemical observations confirm what could be learned from a petrographic microscope study of the rock surfaces in the halokarst. The rock is intensively bored to depths of 1-3 mm by unicellular algae, but the rock remaining behind and at greater depths shows an intense filling of pore spaces by secondary $CaCO_3$. In hot climates, this material crystallizes as aragonite, or in the presence of much Mg-rich calcitic algal debris, the cement will be a high magnesium calcite. Both the aragonite and high Mg cements are unstable under long-continued freshwater exposure, and tend to invert to calcite. In any case, the halokarst develops a *case hardening* or *surface induration*, paradoxically at the same time as its surface becomes pitted and dissected by solution. The physiographic effect is an extremely sharp and jagged texture.

For coastal zone management, the karst presents several virtues and several problems. Of advantage for human requirements are: its stability—its erosion susceptibility is of the order of 1 mm/yr.; and its useful building materials—beach rock and eolianite are best for housing construction, whereas coral or algal limestones are best for foundation rubble. The limestones may also be quarried for cement burning and in rare cases (e.g., Nauru, Christmas Island) are associated with valuable rock phosphate deposits. Also an advantage, there is usually a *perched* aquifer above the saltwater wedge, which makes for a small but valuable potable water supply, while large springs may be found along continental coasts. Disadvantages include the fact that the coast is usually dangerous, abrupt, and unsuitable for boat landing, except in specific inlets or estuaries; and that for tourists the coast is generally unsuitable for swimming, although it may be visually spectacular; deeply eroded halokarst may be impossible to traverse by vehicle, even extremely difficult on foot.

RHODES W. FAIRBRIDGE

References

Fairbridge, R. W., 1948. Notes on the geomorphology of the Pelsart Group of Houtman's Abrolhos Islands, *Royal Soc. Western Australia Jour.* **33**, 1-43.

Folk, R. L.; Roberts, H. H.; and Moore, C. H., 1973. Black phytokarst from Hell, Cayman Islands, British West Indies, *Geol. Soc. America Bull.* **84**, 2351-2360.

Herak, M., and Stringfield, V. T., eds., 1972. *Karst: Important Karst Regions of the Northern Hemisphere.* Amsterdam: Elsevier, 551p.

Kelletat, D., 1974. Beiträge zur regionalen Küstenmorphologie des Mittelmeerraumes, *Zeitschr. Geomorphologie* 19(suppl.), 1-161.

Kuenen, P. H., 1933. Geology of coral, in *Snellius Expedition.* Leiden: E. J. Brill.

Macfadyen, W. A., 1930. The undercutting of coral reef limestone on the coasts of some islands of the Red Sea, *Geog. Jour.* **75**, 27-37.

MacNeil, F. S., 1954. The shape of atolls: An inheritance from subaerial erosion forms, *Am. Jour. Sci.* **252**, 402-427.

Revelle, R., and Emery, K. O., 1957. Chemical erosion of beach rock and exposed reef rock. *U.S. Geol. Survey Prof. Paper 260-T*, 699-709.

Schneider, J., 1976. *Biological and Inorganic Factors in the Destruction of Limestone Coasts.* Stuttgart: Schweizerbart, Contributions to Sedimentology, vol. 6, 112p.

Cross-references: *Cliff Erosion; Rock Borers; Splash and Spray Zone; Tide Pools; Weathering and Erosion, Biologic; Weathering and Erosion, Chemical.* Vol. III: *Biological Erosion of Limestone Coasts; Karst; Limestone Coastal Weathering.*

KLINT COAST

The term *klint* (Swedish and Danish in origin) denotes a long, high cliff; particularly, along the Baltic Coast, a sea cliff formed by wave erosion (Gary, McAfee, and Wolf, 1972).

The North Estonian klint (*escarpment*) is a part of the Baltic klint remaining in the boundaries of Estonian S.S.R. (Tammekann, 1940). The Baltic klint begins south of Ladoga Lake, proceeds along the southern coast of the Gulf of Finland, continues as a scarp to the west on the south shore of the Baltic Sea, and appears as a klint again on the western coast of the island of Öland near the eastern coast of Sweden. The North Estonia klint is present in the area between the Narva River and the island of Osmussaar as an escarpment that has a complicated structure and origin and is dissected by river valleys. Its lower part consists of Cambrian and Ordovician clastic deposits (clays, silt, and sandstones), the upper part, of Lower and Middle Ordovician stratified limestones. In places the klint is present in the form of two scarps. The maximum relative height of the klint is 56 m (Ontika, in the northeastern part of Estonia); in its western part (island of Osmussaar), its height reaches only a few meters. In the western part, the klint is strongly indented by alternation of klint peninsulas and bays.

In Estonia the klint acts as the northern

FIGURE 1. The klint in the northeastern part of Estonia (photo: K. Orviku).

boundary of the North Estonian plateau (Fig. 1). The klint formed as a result of denudation processes before the Quaternary period, and was subjected to the action of continental ice during the Pleistocene. Its present form results from the abrasion of the Baltic Sea in conditions of slow uplift of the earth's crust during the past 11,000 years. During Recent time, the North Estonian klint has been subjected to abrasion in only a few places (cliffs at Päite, Türisalu, Pakri, Osmussaar); in other places, the cliff is inactive and lies farther from the present shoreline.

KAAREL ORVIKU

References

Gary, M.; McAfee, R., Jr.; and Wolf, C. L., eds., 1972. *Glossary of Geology.* Washington, D.C.: American Geological Insitute, 805p.

Tammekann, A., 1940. The Baltic klint: A geomorphological study, *Acta ad Res Naturae Estonicae Perscrutsndas 11,* 103p.

Cross-references: *Chalk Coast; Cliffed Coast; Cliff Erosion; Europe, Coastal Morphology.*

LAGOON AND LAGOONAL COASTS

Lagoons are water areas separated from the sea by narrow and low-lying land strips and usually differing from the sea in salinity. Lagoons are diversified both in water area configuration and in morphology and origin of the land parting. Lagoons are usually separated from the sea by coral reefs, including circular ones (*atolls*), or by coastal aggradation features. In the latter case, the parting may separate the sea from the estuary of a flat-country river (*liman-lagoons*), and inundated trough valley (*fiord-lagoons*), or some other low-lying terrain feature.

In many regions, lagoons arise at the edge of shoaly low-lying shores, and form a string of narrow water areas extending along the coastline. Lagoonal coasts are widespread at present. According to Leontiev (1961), they account for about 13% of the total continental coastline. Most widespread are lagoonal coasts in the United States (Atlantic, Gulf of Mexico), Mexico (Pacific), Central America, Soviet Far East (Chukotka, Sakhalin).

Aggradation features cutting off lagoons are formed of material either supplied by coastal drift or coming up from the sea floor in transverse motion. Most typical are lagoons cut off by *bars*—i.e., features formed by transverse motion of detrital material. In 1845, Beaumont suggested that offshore bars may be formed of material brought up from the sea floor by waves.

Johnson (1919) believed that offshore bars separating lagoons from the sea emerged on rising shores, where flat former sea floor is exposed to wave action.

In recent years it has been shown that lagoons and their aggradation partings are a natural outcome of the development of shoaly aggradation shores under certain physicogeographical conditions. They may arise during both regression and transgression, but the great number of lagoons on present coasts warrants linking their formation with the postglacial *Flandrian transgression*. On the evidence of radiocarbon analysis, many of the lagoon-separating aggradation features are dated at the time when transgression attained a hypsometric level close to the present coastline—i.e., 6000-5000 B.P.

Zenkovich (1967) attributed the formation of bars mainly to sinking of nearly flat plains and the related reshaping of the *equilibrium profile* on a shoaly underwater slope. Here waves forcibly wash out the near-edge portion of the underwater coastal slope, making it steeper and throwing out drift above the water edge where an aggradation feature—i.e., bar—is formed. In further evolvement of transgression, the sea-facing side of the aggradation feature is washed out, while its rear side is being built up. As its body is built up, the bar is progressively shifting landward, the flat plain behind the bar becomes inundated, and a lagoon is formed.

Leontiev, Nikiforov, and Safianov (1975) assume bars to originate during transgression on underwater slopes and during a landward motion of drift within the initial wave-breaking zone. But at constant sea level, an underwater bar thus formed cannot rise above the surface, since with a decreasing depth above the crest of an aggradation feature, wave motion will have a higher rate and wash out the sea floor. Thus an underwater bar cannot emerge above water and form a lagoon, unless sea level falls at least slightly. Leontiev and his colleagues therefore date the formation of most of lagoonal bars by the late-Holocene regression of the world ocean. Subsequent rising of ocean level was unfavorable for the existence of lagoonal bars, and, indeed, many of them are being washed out.

Lagoons are notable for specific hydrologic and hydrochemical regimes and sedimentation processes. In arid regions, lagoonal waters are of a higher salinity than seawater. At the bottom of such lagoons, chemogenic material (mirabilite, astrakhanite, epsolite) is being deposited, which provides valuable chemical raw material. In humid climates, lagoons are mostly freshwater, the bottom sediments being composed of well-sorted and fine high-organic ooze.

Owing to their clear-cut stratification, the presence of organic remains, facies variability, and dependence on natural environment, lagoonal bottom sediments provide a highly valuable material for paleogeographic studies of various coastal zones.

PAVEL A. KAPLIN

References

Beaumont, E. de, 1845. *Lecons de Géologie Pratique.* Paris: P. Bertrand, 557p.

Johnson, D. W., 1919. *Shore Processes and Shoreline Development.* New York: John Wiley & Sons, 583p.

Leontiev, O. K., 1961. *The Principles of Shoreline Geomorphology.* Moscow: Moscow State University, 416p.

Leontiev, O. K.; Nikiforov, L. G.; and Safianov, G. A., 1975. *Geomorphology of Shorelines.* Moscow: Moscow State University, 335p.

Zenkovich, V. P., 1967. *Processes of Coastal Development.* London: Oliver and Boyd, 738p.

Cross-references: *Accretion Ridge; Barrier Islands; Bars; Coral Reef Coasts; Estuaries; Fiord, Fjord; Haff; Lagoonal Sedimentation; Lagoonal Segmentation; Liman and Liman Coast.* Vol. III: *Lagoon.*

LAGOONAL SEDIMENTATION

Lagoons are ephemeral features in geologic time that are gradually filled in with available sediment. The ultimate source of most of the sediments in nontropical lagoons is the continental shelf. This sediment has been either washed over the barrier island into the lagoon, blown into the lagoon from the barrier island, or introduced by tidal currents transporting it through the inlets. The other source of lagoonal sediments is the mainland. Sediment may be transported by rivers or estuaries or directly eroded from the mainland shore.

Pamlico Sound, a large east coast United States lagoon, is a model for sedimentation in other nontropical lagoons (Pickett and Ingram, 1969). The mainland side has muddy, organic-rich fine to very fine sands; clayey silts are in the deeper centers of the basins or river mouths; cleaner fine sands are found on the barrier island (ocean) side of the lagoon. Evidence of biogenic activity is characteristic of lagoonal sediments.

The distribution of sediments in lagoons is controlled largely by wave-generated currents, water depth, salinity, and amount of available sediment (Folger, 1972).

THOMAS E. PICKETT

References

Folger, D. W., 1972. Characteristics of estuarine sediments in the United States, *U.S. Geol. Survey Prof. Paper 742*, 94p.

Pickett, T. E., and Ingram, R. L., 1969. The modern sediments of Pamlico Sound, N.C., *Southeastern Geology* 11, 53-83.

Cross-references: *Coastal Dunes and Eolian Sedimentation; Estuaries; Estuarine Sedimentation; Lagoon and Lagoonal Coasts; Tidal Inlets, Channels, Rivers; Washover and Washover Fans.* Vol. VI: *Lagoonal Sedimentation.*

LAGOONAL SEGMENTATION

A common feature of elongate basins, such as *coastal lagoons,* is the existence of spits that partially or wholly divide the basin into several subbasins of more circular shape.

Gulliver (1896) ascribed these forms to large-scale eddies due to tidal or other currents. This type of current, however, is not generally observed. The existence of segmented lagoons in tideless water bodies further diminishes the importance of tidal currents as a mechanism to account for their presence.

Fisher (Zenkovich, 1967), in describing the segmented lagoons of St. Lawrence Island (Alaska), proposed storm washover of the barrier as an initiating mechanism. Zenkovich objected on the basis of the rather constant spacing of the spits.

Price and Wilson (Zenkovich, 1967) suggest that the deposits might result from resonance of seiche oscillations produced by the wind but largely controlled by basin geometry. Zenkovich argues that the presence of fine sediment between opposing spits indicated that the coastal forms develop before the bottom forms, contrary to the predictions of resonance oscillation effects.

Zenkovich states that the spits that effect lagoonal segmentation result from shore-parallel drift due to wave action on the lagoon shoreline. The most effective wind direction is approximately parallel to the long axis of the lagoon. Waves are developed and become sufficiently large, downwind through a minimum distance of fetch, to move material along the shorelines. Additional fetch distance allows the waves to develop to a maximum height, limited by water depth or wind speed (or wind duration), hence to a maximum shore-parallel drift capacity. Beyond the point of maximum wave height development, there is no increase in drift capacity (assuming a straight coastline), a situation that Zenkovich calls "saturated flow." He claims that any shoreline oriented parallel with the wind (wave) direction is unstable, and that therefore any protuberance on the shoreline will be the locus of spit building. This spit will be oriented roughly 45° to the wave direction. As it grows, it will limit fetch downwind causing another distance of minimum fetch before the waves are capable of moving material along the shoreline. Waves from the opposite direction can operate on the sheltered shore in the lee of the spit; hence drift in the opposite direction tends to infill behind the spit, giving it a more symmetrical *capelike* appearance. Apparently the first protuberances are accidental in location, but continued development of spits and minimum fetch distances, accompanied with downdrift migration of the spits, eventually results in a periodic segmentation of the lagoon.

The process of cape development and shoreline reorientation through erosion and deposition can be comprehended easily by reference to the shore-parallel wave power gradient (May and Tanner, 1973) and the resulting modification of shoreline geometry. According to the a-b-c- model (q.v.), the shoreline tends to be reoriented to a position approaching 90° (rather than the 45° modification of Zenkovich). Uniformity of lagoonal bathymetry and sediment type control the positioning of spits for a specified wind/wave condition. Constancy of these factors promotes a regular, symmetrical development. Variability promotes irregular development, a more frequent situation.

JAMES P. MAY

References

Gulliver, F. P., 1896. Cuspate forelands, *Geol. Soc. America Bull.* 7, 399-422.

May, J. P., and Tanner, W. F., 1973. The littoral power gradient and shoreline changes, in D. R. Coates, ed., *Coastal Geomorphology*. Binghamton: State University of New York, 43-60.

Zenkovich, V. P., 1967. *Processes of Coastal Development*. New York: Wiley Interscience, 738p.

Cross-references: *A-B-C Model; Beach in Plan View; Beach Orientation; Beach Processes; Cuspate Spits; Lagoonal Sedimentation; Lagoon and Lagoonal Coasts; Sediment Transport; Spits.* Vol. III: *Coastal Lagoon Dynamics.*

LAKES, COASTAL ECOLOGY

The coastal zones of lakes possess a distinctive community of organisms usually referred to as the *littoral community*. This community has evolved in response to the unique set of environmental conditions that prevail in this interface between air, water, and land.

In certain ways the conditions in this zone are highly favorable for the development of a rich and varied assemblage of organisms. Light and nutrient conditions are usually conducive to photosynthesis and thus abundant plant growth. The presence of concentrations of organic matter and adequate supplies of oxygen make the zone one of potentially favorable conditions for animals.

However, the actual community of organisms found in the coastal zone varies greatly among lakes and even within the same lake. The overriding determinants of the type and richness of the community present are the degree of exposure to waves and, related to this, the composition of the bottom. Nutrient concentrations, temperature, ice conditions, and other environmental factors also play important roles.

Ecology of Exposed Coasts

Areas periodically subjected to strong wave action, such as the exposed shores of large lakes, usually have few macroscopic organisms present. The force of the waves causes the coarse materials commonly found on these shores to grind together, fragmenting any but the most sturdy macroscopic forms. Thus, in the most exposed situations, the biota is largely limited to organisms small enough to avoid the grinding action. Such forms can, however, reach large numbers, as these areas often have high concentrations of the nutrients needed for plant growth as well as a constantly replenished supply of organic matter that can support large populations of bacteria and detrital-feeding animals. The algae of such areas, especially diatoms and blue-greens, grow on the surface of the sand or other materials during periods of reduced wave action. Animals and bacteria are also present on these surfaces, but, as they do not require light, they can inhabit the interstices of the sedimentary material for an appreciable distance into the bottom as well. Actual studies of the biota of a sandy shore revealed numbers of bacteria reaching millions per 10 cc as well as large numbers of protozoans, rotifers, small crustaceans, and other microscopic animals (Welch, 1952). Interestingly, this community of interstial bacteria and animals is not limited by the edge of the lake. On most sandy shores, only the top sand dries out. The deeper material is constantly surrounded by water that has a direct continuity with the lake, and so is inhabited by some of the same organisms as found between the sand grains offshore.

Coastlines of solid bedrock or large boulders can provide an exception to the observation that exposed shores have few macroscopic organisms. In these situations, there are no sand grains or stones to grind together, so large organisms can thrive if they can maintain themselves in the face of the violent turbulent action of the waves. Algae adapted for clinging to rock faces often grow abundantly in such areas (Ruttner, 1963). In fact, growths on such solid surfaces can reach nuisance proportions if other conditions are especially favorable. Such is the case in certain parts of the Great Lakes, where green algae of the genus *Cladophora* form thick mats in response to nutrient enrichment from sewage and other sources.

Macroscopic animals also can inhabit such rocky shores. A variety of forms have evolved special mechanisms by which they can stay in such areas in the face of the strong water movements (Welch, 1952). The mechanisms include sucking disks and strong grasping claws with which to hold to the surface of the rock or the algae as well as modifications in shape to forms that can flatten tightly against a surface or squeeze into narrow cracks and crevices. Other holding mechanisms include the ability to build cases or tubes that are strongly attached to the surfaces or to spin threads of material to anchor the animal in place.

These mechanisms are the same as those used to maintain animals under conditions of strong currents in streams and rivers. In fact, the organisms of rocky exposed shores bear a strong resemblance to those of rapid streams, with many species being found in both environments. This similarity also applies to shores of stones or sand. On such shores, where wave conditions are reduced enough so that some macroscopic animals can survive, the animals found are often ones that also inhabit riverine environments. On sandy shores, these commonly include forms, such as large clams and certain mayflies, that can burrow into the sand to escape strong wave action.

Ecology of Protected Coasts

The ecological conditions found on the shores of ponds and small lakes and in protected areas of large lakes differ strongly from those of exposed coasts. Here wave action is weak, and a great variety of organisms can often be found. In many cases the most obvious characteristic of such situations is the presence of thick growths of large aquatic plants, known collectively as *aquatic macrophytes*. This group is composed mostly of representatives of the angiosperms or higher plants, but also includes some mosses and macroalgae.

Shores with these plants often possess three rather distinct ecological areas that form concentric zones around the shore, with one group replacing another as the water gets deeper (Wetzel, 1975). The shallowest zone is that of the *emergent macrophytes*—that is, plants that are rooted to the bottom but that project above the water surface for some appreciable distance. This group includes the cattails, bulrushes, reeds, and a variety of other forms. This zone provides a link between land and water and is favored by many water birds and amphibious animals. It usually extends from shore to a water depth of about 1.5 m.

Farther from shore is the second zone—that of the *floating-leaved macrophytes*—in which the plants are rooted but have long flexible stems that allow the leaves to float on the surface. This zone occurs in depths from about 0.5-3 m, and includes such forms as waterlilies.

The deepest zone is that of the *submerged macrophytes* that grow completely underwater. these plants can occur at any depth that has sufficient light, but are most abundant below 3 m down to about 10 m. These plants include a number of forms, but, in the temperate part of the world, members of the genus *Potamogeton*, the pond weeds, often predominate.

While the macrophytes are the most obvious component of the protected shores, they are by no means the only biotic element that thrives in such situations. Microscopic algae grow abundantly on the stems of the larger plants as well as on other available surfaces, including those of the bottom sediment. Water movements often pick up these algae and at times they are found in large numbers in the waters between the macrophytes, where they mix with phytoplankton from the open lake.

Animals also find favorable conditions on such protected shores. High levels of production of organic matter by the plants provide a ready source of food for herbivorous organisms (Odum, 1971). Some forms such as snails, protozoans, certain insects, and various worms graze on the algae and detritus found on the plants and rocks or in the sediments. Other types, such as freshwater sponges and hydra, use the plant stems or other surfaces for support and extract organic particles from the water.

Farther up the food chain, a variety of invertebrate carnivores feed on these herbivores or on each other. Such forms as large protozoans, leeches, and especially a number of types of insects (q.v.) are included in this group.

Finally, such protected areas are favorable habitats for many kinds of fish. The plants help provide secluded areas for breeding and for protecting the eggs and young from predators. On the other hand, the various types of invertebrates provide an abundant source of food.

ANDREW ROBERTSON

References

Odum, E. P., 1971. *Fundamentals of Ecology*. Philadelphia: W. B. Saunders Co., 574p.

Ruttner, F., 1963. *Fundamentals of Limnology*, trans. D. G. Frey and F. E. J. Fry. Toronto: University of Toronto Press, 295p.

Welch, P. S., 1952. *Limnology*. New York: McGraw-Hill, 538p.

Wetzel, R. G., 1975. *Limnology*. Philadelphia: W. B. Saunders Co., 843p.

Cross-references: *Algal Mats and Stromatolites; Coastal Fauna; Coastal Flora; Insects; Lakes,*

Coastal Engineering; Lakes, Coastal Morphology. Vol. III: *Lakes*.

LAKES, COASTAL ENGINEERING

Coastal engineering along lakes poses some unique circumstances not frequently encountered along ocean coasts. Lakes are different from oceans in several ways, the most obvious being the fact that they are smaller. Lakes are tideless, and are shorter-lived than oceans. Some lakes have freshwater, while others have saltwater, which in some instances has a salinity much greater than that of the normal ocean.

A lake is defined as a body of standing water occupying a depression in the earth's surface. They are classified according to their origin in the following groups: glacial lakes; floodplain lakes; coastal or deltaic lakes; deflationary lakes; lakes of volcanic origin; lakes of diastrophic origin; and artificial lakes. Each type has certain characteristics, of which some are important in engineering application.

Important Lake Characteristics

Glacial Lakes. These lakes are situated in depressions developed by glaciation. They are usually shallow, with common shore features, other than rims of boulders or pebbles. The lake regions of Wisconsin and Minnesota are almost entirely of glacial origin. Their bottom deposits include sand or silt, soft, dark mud, and marl (a calcareous sand), overlying the original foundation of stone clay (glacial till) or bedded sand and silt.

Floodplain Lakes. Normal stream processes in valleys cause the formation of floodplain lakes. *Oxbow lakes,* produced by river meander cutoff, are the most common. Other variations of this type of lake are situated along the floodplains of wide valleys, particularly in locations where the river is higher than the floodplain and is confined behind a levee. Floodplain lakes are mostly small and short-lived. They tend to be occupied by vegetation and to become swamps and marshes. They seldom show shore features, and the bottom deposits are mostly organic-rich mud, lying on the foundation of normal floodplain deposits.

Coastal or Deltaic Lakes. Coastal lakes are formed when barrier beaches, bars, spits, or migration of dunes isolate bodies of water from the sea. Deltaic lakes are similar, being formed by changes in river channels or by combinations of stream and coastal processes. Both of these types of lakes are usually shallow, and may sometimes become stagnant. Bottom deposits vary from clean sand to dark, organic-rich mud. Streams from inland supply most of their sediments.

Deflationary Lakes. The life of deflationary lakes is usually short-lived. They are formed in arid regions where the wind has scoured shallow basins. They are highly subject to evaporation, and seldom have a true littoral zone, owing to the wide range in their shrinking and expanding areas.

Volcanic Lakes. Lakes of volcanic origin are usually small to moderate in size, and may have steep sides and considerable depth. They may be formed by the damming of drainage lines by lava flow, such as Lake Tahoe on the California-Nevada line, or by the development of calderas from volcanic explosions, such as Crater Lake in Oregon. Bottom deposits are reworked volcanic debris.

Lakes of Diastrophic Origin. Many of the largest lakes in the world fall into this category. These lakes are formed by large earth movements, such as abrupt dislocation of rock strata during earthquakes, which can cause depressions or alter natural drainage. Examples of lakes of this type are along the San Andreas fault in California. Another cause of this type of lake is the slow downwarping of the earth's surface due to long-term isostatic adjustment. The depths of lakes of diastrophic origin range from shallow to deep. Their deposits are similar to those in the ocean in terms of texture, composition, and structure.

Artificial Lakes. This category includes all man-made lakes, ranging in size from small ponds to major bodies of water. Deposits include sand and silt brought down by the original stream that may spread a blanket of fine material over the lake bottom.

Special Engineering Conditions in Lakes

As a result of their positions and origins, lakes have several unique conditions from a point of view of coastal engineering.

One of the major problems associated with freshwater lakes is that they will freeze, while oceans at the same latitude will not freeze because of their salinity and constant movement. Structures in frozen freshwater lakes will be subjected to ice loadings and abrasion, which must be considered in designing and planning the structures. High-salinity lakes can cause special corrosion problems. Another problem is that lakes are subject to filling by sediment, particularly artificial lakes.

Lake coastal structures are usually not subjected to the large wave forces encountered by structures along ocean coasts, because of the shallowness and limited *fetch* length (distance over which the wind can blow to create waves) of lakes. However, large lakes may attain the magnitude of shore conditions along the oceans.

Although lakes have insignificant tidal variations, they are subject to seasonal and annual hydrologic changes in water level, along with water level changes caused by wind setup, barometric pressure variations, and seiches. Some lakes also undergo occasional water level changes caused by regulatory control works.

In locations where the temperature stays below freezing for significant periods of time, such as the Great Lakes area, the water level falls during the winter months and rises during the warmer months when melting and the associated increase in runoff occur. This seasonal variation is as much as 2 m in the Great Lakes. Structures along the coast must be designed to operate within the entire range of water levels to be encountered.

Lakes are also subject to occasional *seiches* (rise in water level) in irregular amounts and durations. This can result from a resonant coupling that occurs when the propagation speed of an atmospheric disturbance is nearly equal to the speed of free waves on the lake. Lakes are also affected by wind stresses that can raise the water level at one end and lower it at the other. All these factors must be considered in determining the total range of water levels that a structure might encounter.

The soil deposits of lakes vary widely with origin, as previously explained. The soil along the coast of lakes is usually less dense than along oceans, and may therefore be more susceptible to erosion. Sediment transport along lake coasts, if present, is usually less than on ocean sites, which adds to the erosion problem.

In general, coastal engineering structures, such as breakwaters, seawalls, bulkheads, revetments, jetties, and groins to be constructed in lakes may be designed smaller than their ocean counterparts. Also, structures such as floating breakwaters, rather than rubble-mound breakwaters, may be more feasible in lakes.

Coastal engineering along lakes, although similar to that along oceans, poses some special conditions that must be evaluated in the design and construction of structures along with the normal problems and hazards encountered in coastal environments.

For further recommended reading see: Harris (1953), Hellstrom (1941), Krumbein (1950), and Platzman and Rao (1963).

JOHN B. HERBICH
JOHN P. HANEY

References

Harris, D. L., 1953. Wind, tide, and seiches in the Great Lakes, *Conf. Coastal Eng., 4th, Am. Soc. Civil Engineers, Proc.,* 25-51.

Hellstrom, B., 1941. Wind effects on lakes and rivers, *Ing. Vetensk. Akad. Stockholm Handl. 158,* 191p.

Krumbein, W. C., 1950. Littoral processes in lakes, *Conf. Coastal Eng., 1st, Long Beach, Calif., Am. Soc. Civil Engineers, Proc.,* 155-160.

Platzman, G. W., and Rao, D. B., 1963. Spectra of Lake Erie water levels, *Jour. Geophys. Research* 69, 2525-2535.

Cross-references: *Beach Processes; Coastal Engineering; Coastal Erosion; Ice along the Shore; Lakes, Coastal Morphology.*

LAKES, COASTAL MORPHOLOGY

The coastal morphology of lakes, both large and small, is similar to that of the oceans. The larger the lake and the smaller the tidal range along an ocean shore, the more the coastal features compare in size and kind. Almost from the beginning of coastal studies, terms and concepts pertaining to coastal forms and processes have been interchanged between lacustrine and marine studies. One of the early important papers by Gilbert (1885), based largely on his work in the basin of the Pleistocene Lake Bonneville in Utah and, to a lesser extent, on the Great Lakes, is still cited in discussions of *barrier islands* (a comparatively rare lake feature) and both marine and lacustrine deltas. Likewise, the literature on the segmentation and development of cuspate spits in lagoons overlaps that on elongate lakes.

Among the major sources of differences in coastal processes and forms between lakes and the oceans are: the absence of tides in lakes (although some seas have small tidal ranges); the smaller fetch of the wind over many lakes and the resultant low level of wave energy; the similarity of the density of many lake waters to that of their influent streams, thus producing the classic *Gilbert type* delta; the confining of an expansive ice cover and the formation of *ice ramparts;* the opportunity for the removal of shoreline-derived material along some marine coasts to deeper water, beyond the influence of waves, thus not contributing to shoaling; and the apparent rarity or absence of *rip currents* in lakes.

The topographic depressions occupied by lakes have a variety of origins, exhaustively classified by Hutchinson (1957). The shorelines upon initial filling would then be "given" by whatever agency or process fashioned the depression. Such shorelines would fall into Johnson's (1919) *shoreline of submergence* category. At this hypothetical initial stage, shorelines would also belong to Shepard's (1973) category of *primary coasts.* Upon later modification by waves, currents, and delta infilling, they advance to his category of *secondary shorelines*. With merely delta infilling, Johnson's *neutral shoreline* category would apply. Tectonic basins caused by faulting would also have *neutral shorelines*. The application of other shoreline classifications, such as those of Valentin, Bloom, Price, and Davies (see references and summaries in King, 1972) is of considerable interest but cannot be pursued here (see *Classification*). Thinking in the dynamic terms of these classifications would certainly be fruitful in future studies of lacustrine shoreline processes.

Lakes are doomed to disappear through: infilling by sediments of fluvial, shoreline (wave erosion and mass wasting), biological, chemical, and eolian origin; draining by the downcutting of outlets or by tectonic deformation (e.g., tipping); climatic changes causing desiccation by evaporation or by the diminution of discharge of inflowing streams; and the lowering of a local water table, even under a stable climate (Zumberge, 1952).

The relatively short lifetime of lakes, and the oscillations and develeling of the water plane controlling the production of shoreline features precludes, except perhaps in the very largest of lakes, the progressing of shorelines very far in a shoreline cycle in the sense used by Johnson (1919). Shoreline features of lakes face burial, removal by stream erosion and mass wasting, and continuing modifications by the ups and downs of lake levels, as dramatically outlined by Gilbert (1885).

Inventories or catalogs of the coastal forms of lakes and descriptions of specific lakes in the United States can be found in many regional studies (e.g., Fenneman, 1902; Juday, 1914; Scott, 1921; Davis, 1933). The most useful compilations for limnological purposes are probably those of Gilbert (1885) and Zumberge (1952). Details of coastal processes, albeit marine, are given in many more recent works (e.g., Bird, 1969; King, 1972; Shepard, 1973).

The net areal effect of the production of the shoreline features other than deltas described below is to simplify or rectify the shoreline in the sense of removing crenulations and leaving it with a smoothly curving outline. Headlands are eroded away and reentrants, filled or separated from the main body of the lake. This may also result in the segmentation or even a major geographic reorientation of the lake.

The Sea Cliff and the Wave-Cut Bench

Along the steeper margins of a lake basin, wind-generated waves cut, steepen, and undermine a cliff by processes such as abrasion, chemical solution, impact of large rock fragments, and hydraulic action (compression of air between a wave and shoreline materials). The vertical and horizontal geometry of the cliff and its rate of retreat will depend on, among other factors, the initial coastal areal configuration, its steepness, the nature of the shoreline materials and their mode of failure, and the size, kind, and duration of waves from a given direction.

As the cliff retreats and material derived from erosion of the cliff is moved lakeward or laterally by the beach drifting effect and longshore currents, a flat platform, the *wave-cut bench* underlain by the original shoreline materials extends itself inland (Fig. 1). The depth of the bench is determined by the depth of *wave base* (the effective depth of movement of debris by waves).

Both the *sea cliff* and the wave-cut bench can be carved from hard bedrock, such as granite or sandstone, as well as from unlithified materials, such as Pleistocene glacial till or volcanic ash. The coarser debris from erosion and mass-wasting along the shoreline often accumulates on the surface of the wave-cut bench, particularly boulders and cobbles from glacial till, or the larger fragments from jointed bedrock.

The Wave-Built Terrace (Littoral Shelf or Subaqueous Terrace)

The *wave-built terrace,* if present and not removed by longshore currents, is the flat-surfaced wedge of material derived from the erosion of the sea cliff and transported across the wave-cut bench either directly from the adjacent shore or moved in by currents.

The concept of a wave-built terrace seems to have originated in relation to lacustrine shorelines (Gilbert, 1885), but reference to it is largely absent from many modern discussions of marine coastal processes (e.g., King, 1972). Along the shores of larger lakes and seas, it is chiefly the undefined site of offshore bars.

The often reedy and floating-plant-infested shoaling areas (not necessarily immediately adjacent to cliffs) around shallow lakes can also be subsumed under wave-built terraces. Because the many shallow depositional areas around the edges of lakes may not be due directly to the action of waves, the terms *littoral shelf* and *subaqueous terrace* (Zumberge, 1952) have also been used.

Beaches

Along the gently sloping, low-angle shores of many lakes, material derived from the erosion of the shoreline is transported laterally and may locally accumulate as a *beach ridge* at and above the water line. Beach material is almost

LAKES, COASTAL MORPHOLOGY

FIGURE 1. Abandoned sea cliff and wave-cut bench along the shore of Devil's Lake, eastern North Dakota, cut in glacial till. Note residual boulders on the surface of bench (reprinted from S. Aronow, 1957. On the postglacial history of the Devil's Lake region, North Dakota, *Jour. Geology* 65, Plate IB; copyright © 1957 by the University of Chicago).

continuously in transit by the beach-drifting effect—the swash-and-backwash motion of waves obliquely approaching the shore. Made up of sand, gravel (the shingle of British usage), or cobbles, a wide beach may be the above-the-water extension of a wave-built terrace—a narrow one, simply the sparse accumulation along the landward edge of a wave-cut bench.

The slope of beaches varies from about 1° for fine sand to about 25° for cobbles (Shepard, 1973; King, 1972). The grain size largely determines the permeability, which, in turn, governs the swash-backwash ratio; the larger the grain size and permeability, the smaller the volume of returning backwash.

The lowering of levels, for example, in lakes undergoing desiccation or changes in outlets may leave a succession of beach ridges (and other coastal forms) marking the temporary maintenance of a water plane (Fig. 2). Examples include the downhill sequences of beaches around proglacial lakes (e.g., Lake Agassiz and the precursors of the Great Lakes) and so-called pluvial lakes (e.g., Lake Bonneville) in the western United States and elsewhere.

Offshore Bars (Submarine Bars or Longshore Bars)

In modern usage, a *bar* along a marine coast is a subaqueous ridge that remains below sea level at low tide; along a lake shore, it is a ridge that remains below lake level. *Offshore bars* parallel the shoreline, and may be built upon or in fact constitute the wave-built terrace.

They form at the break point of waves moving shoreward, and are essentially within or at

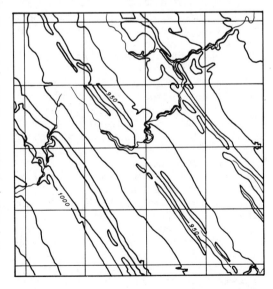

FIGURE 2. Beach ridges, dry bed of proglacial Lake Agassiz, west of Grand Forks, North Dakota, from the Emerado 15-minute quadrangle. Sections are roughly a mile (1.61 km) on a side; contour interval, 10 ft (3.408 m).

the outer edge of the surf zone. They may be ephemeral, mobile features whose presence number, location, and height seem to depend on the direction of wave movement and the wave spectrum. If present in groups of two or three and separated by troughs, usually with a relief of less than 2 m, they may be the product of several different wave spectra affecting different water depths. They are best developed in lakes and seas with low tidal ranges (see discussion in King, 1972).

In the Great Lakes, where they parallel the shores for many miles, the troughs separating the bars and the bars themselves have been called the *low and ball,* respectively, by Evans (1940). See also related papers by Shepard (1950) and Keulegan (1942).

Spits: Simple, Recurved (Hooks), Cuspate, V-Shaped and Looped

A *spit* is a depositional ridge, standing above lake level, attached at one end and composed of sand or gravel (shingle). Many spits, regardless of ground plan, have subaqueous extensions or platforms into deeper water that may be the precursors of a subsequent longer spit. Of interest here are papers by Evans (1942), Kidson (1963), and Meistrell (1966).

A *simple spit* is more or less straight and typically the extension of a beach at the base of a straightened, cliffed, eroded headland. *Double spits* extend from adjacent headlands and partly enclose a bay. Single and double spits may also emerge from the shores of bays, and are called *mid-bay spits.* Should single or double spits completely enclose a bay at its mouth or in mid-bay position, they are referred to, respectively, as *bay-mouth* and *mid-bay barriers;* when near the heads of bays, *bay-head barriers.*

Recurved spits or *hooks* curve shoreward, owing to waves moving in directions other than those causing the original longshore drift of the adjacent beach, or to waves refracting around the end of the hook. *Compound hooks* may develop as the original feature extends itself in successive, joined hooks.

A *cuspate spit* is the pointed depositional extension of the intersection of two arcuate sea cliffs or otherwise eroding shorelines. They are often called simply *points.* They are "fed" by beach drifting or currents from two directions, and usually extend subaqueously lakeward in the form of a bar. In many elongate lakes and lagoons with low tidal range, they form in opposing pairs, and may ultimately segment these water bodies into separate circular or elliptical bodies. For discussion and literature concerning processes of segmentation, see Shepard (1973) and Zumberge (1952).

Similar in ground plan to the cuspate spit is the *V-shaped spit* of Gilbert (1885) and the *looped spit* of Evans (1942). This is a triangular spit, land-tied at both ends with the point of the "V" extending lakeward and enclosing a shallow lagoon. The looped spit is more U-shaped. Gilbert found the V-shaped ones in the dry bed of Lake Bonneville, and thought them rare in existing lakes and deposited in lakes with rising water levels. Evans found some examples of looped spits in recent lakes, and believed them simply spits recurved sufficiently to rejoin the land area.

Tombolos

A tombolo is a spit connecting an island with the shore. It may be composed of sand or coarser material and may occur singly or be paired with the island at the apex of a "V" with inward curving (cuspate) sides that enclose a lagoon. (For a recent discussion, see Farquhar, 1972).

Barriers or Barrier Islands

Barriers are offshore, elongate, depositional features, usually composed of sand, and which may be attached at one end. In form they may merge into compound recurved spits, but differ from spits mainly in that they partly or complete enclose a shallow lagoon. They are comparatively rare lake features, occur along the Black and Caspian seas (see map on p. 526 in King, 1972), and are best developed along coasts with low to moderate tidal ranges. Whether changes of lake or sea level are necessary for their formation, and whence the source of sediment, have been the source of heated debate since their description by de Beaumont in 1845, and later by Gilbert (1885). (See Schwartz, 1973, for these and other relevant papers.)

Deltas

Deltas are accumulations of solid materials carried by streams into lakes, when the stream velocity is checked by lake waters. They are the most obvious and perceptible mode of lake filling.

The low to moderate wave and current energy in most lakes leads to the formation of arcuate, lobate, and digitate deltas rather than the cuspate form or to no delta at all (Hutchinson, 1957; Bird, 1969).

Lacustrine deltas are probably mostly of the *Gilbert type,* so called because of the early description by Gilbert (1885) of their characteristic topset, foreset, and bottomset beds, which he examined in the dry bed of Lake

Bonneville. Some authorities (e.g., Morgan, 1970) believe that this type of delta can be found principally in the rather restricted environment of lakes and in the stream tables and tanks of experimenters.

According to Bates (1953), the lack of density differences between lakes waters and their influent streams (*homopycnal flow*) promotes three-dimensional mixing and an axial jet pattern of flow that results in the three forms of bedding.

Ice Ramparts

In regions where lakes freeze in the winter, the expansion of the ice on the lake surface may cause the shoreward overriding and bulldozing of shoreline materials. The ridges thus produced are called *ice ramparts*. They are particularly persistent and durable when the ice impinges on bouldery glacial till or coarse bedrock-derived rubble. Successive ridges may unite to form an *ice-push terrace*. After the partial melting and breakup of the ice cover, rafts and slabs of the lake ice may be driven against the shoreline by the wind (*ice jams*) and produce similar features, especially on lakes with a large fetch for the wind (Hutchinson, 1957; Zumberge, 1952).

Similar ice-cored, wind-generated ridges and mounds form temporarily along the shores of the Arctic Ocean (King, 1972).

Marshy and Boggy Shorelines

As lakes fill and subaqueous terraces with emergent and floating hydrophytes prograde into lakes, wave action is damped and current velocities decrease. These areas become the sites of deposition of fine-grained clastic and coarse, organic debris (peat), thus producing a *marshy shoreline*. The marine analogy to this might be a mangrove swamp, a mudflat, or a salt marsh. Unlike the marine analogy, the whole body of water ultimately becomes a marsh or swamp.

In some lakes, *bog lakes,* a floating mat of sedges comprises the shoreline. A *quaking bog* results when the mat covers the whole lake.

SAUL ARONOW

References

Aronow, S., 1957. On the postglacial history of the Devils Lake region, North Dakota, *Jour. Geology* 65, 410-427.

Bates, C. C., 1953. Rational theory of delta formation, *Am. Assoc. Petroleum Geologists Bull.* 37, 2119-2162.

Bird, E. C. F., 1969. *Coasts.* Cambridge, Mass.: M.I.T. Press, 246p.

Davis, W. M., 1933. The lakes of California, *California Jour. Mines and Geology* 29, 175-236.

Evans, O. F., 1940. The low and ball of the eastern shore of Lake Michigan, *Jour. Geology* 48, 476-511.

Evans, O. F., 1942. The origin of spits, bars, and related structures, *Jour. Geology* 50, 846-865.

Farquhar, O. C., 1967. Stages in island linking. *Oceanography and Marine Biology* 5, 119-139.

Fenneman, N. M., 1902. On the lakes of southeastern Wisconsin, *Wisconsin Geol. and Nat. History Survey Bull. No. 8,* 178p.

Gilbert, G. K., 1885. The topographic features of lake shores, *U.S. Geol. Survey Annual Rept.* 5, 69-123.

Hutchinson, G. E., 1957. *A Treatise on Limnology,* vol. 1. New York: John Wiley & Sons, 1015p.

Johnson, D. W., 1919. *Shore Processes and Shoreline Development.* New York: John Wiley & Sons, 583p.

Juday, C., 1914. The inland lakes of Wisconsin: The hydrography and morphometry of the lakes, *Wisconsin Geol. and Nat. History Survey Bull. No. 27,* 137p.

Keulegan, G. H., 1948. An experimental study of submarine sand bars, *U.S. Army Corps Engineers Beach Erosion Board Tech. Rept. No. 3,* 40p.

Kidson, C., 1963. The growth of sand and shingle spits across estuaries, *Zeitschr. Geomorphologie* 7, 1-22.

King, C. A. M., 1972. *Beaches and Coasts.* New York: St. Martin's Press, 570p.

Meistrell, F. J., 1966. *The Spit-Platform Concept: Laboratory Observation of Spit Development.* Edmonton, Alberta: University of Alberta, master's thesis, 46p.

Morgan, J. P., 1970. Deltas: A résumé, *Jour. Geol. Educ.* 18, 107-117.

Schwartz, M. L., ed., 1972. *Spits and Bars.* Stroudsburg, Pa.: Dowden, Hutchinson & Ross, 452p.

Schwartz, M. L., ed., 1973. *Barrier Islands.* Stroudsburg, Pa.: Dowden, Hutchinson & Ross, 451p.

Scott, I. D., 1921. Inland lakes of Michigan, *Michigan Geol. and Biol. Survey Pub. No. 30,* 383p.

Shepard, F. P., 1950. Longshore-bars and longshore-troughs, *U.S. Army Corps Engineers Beach Erosion Board Tech. Memo. No. 15,* 32p.

Shepard, F. P., 1973. *Submarine Geology.* New York: Harper & Row, 517p.

Zumberge, J. H., 1952. The lakes of Minnesota: Their origin and classification, *Minnesota Geol. Survey Bull. No. 35,* 99p.

Cross-references: *Barrier Islands; Bars; Beach Processes; Classification; Cliff Erosion; Cuspate Spits; Deltas; Geomorphic-Cycle Theory; Ice along the Shore; Lagoonal Segmentation; Lakes, Coastal Ecology; Lakes, Coastal Engineering; Oriented Lakes; Spits; Wave-Cut Platform.* Vol. III: *Lakes;* Vol. VI: *Lacustrine Sedimentation.*

LAND RECLAMATION

Land reclamation is used in general to describe the process of making lands suitable for more intensive use by such means as cultivation, drainage, revegetation, irrigation, chemical or physical modification, or the like. Reclamation efforts may be concerned: with rehabilitation of areas strip mined or areas of coastal dunes mined for heavy mineral content; with improvement of rainfall-deficient areas by irrigation; or with removal of detrimental constituents from alkaline or saline lands, and so on. But in the coastal zone, reclamation usually signifies exclusion of marine or estuarine waters from littoral or riparian lands formerly periodically or permanently inundated.

Reclamation of land in the coastal zone may involve: raising the level of land above the highest anticipated water level, or exclusion of water by *diking* or *barrages* (or the like) and draining.

Raising Land Level

Reclamation may be accomplished by dumping on the site such materials as quarry waste, excavation spoil, building demolition spoil, or by emplacement of dredged material acquired specifically for the purpose or as a byproduct of dredging operations elsewhere. A variant on the latter method is involved in the construction of *canal estates* (otherwise called waterfront developments or marina developments), where canal cutting and land reclamation are accomplished simultaneously, the spoil from the canals providing the fill for land reclamation. Areas of filled land are usually protected from wave action by sea walls, riprap, sheet piling, or the like.

Tidal lands in proximity to urban areas have also frequently been used for the dumping of refuse, a practice that at one time produced reclaimed land almost incidentally but is usually now carried out with the objective of reclamation. An example is found in Sydney Harbour, Australia, a ria with numerous bayhead intertidal deposits commonly occupied by mangrove (*Avicennia*) communities at the time of European colonization in the late eighteenth century. As European settlement spread around the foreshores and slopes of Sydney Harbour, the mangrove swamps became convenient dumping grounds for refuse, which gradually accumulated to above HWM. Late in the nineteenth century a demand developed for large, level sites to be used for recreation grounds or industrial sites. Some such areas existed on the ridge tops of the natural topography surrounding Sydney Harbour, but most of these had been claimed as prized residential sites, so that attention became focused on the bayhead mangrove swamps/rubbish dumps as potential level sites, to the extent that reclamation of bayhead swamps to provide public parks became regarded as a meritorious local government activity. Today, a map of bayhead parks in lower Sydney Harbour or of riparian industrial areas in the upper reaches is virtually a map of former intertidal wetlands, a situation repeated in many other estuaries of the world.

In recent years demand has grown for large level areas of land close to large and/or densely settled urban areas for the purpose of providing industrial sites or airports/airstrip extensions. The strength of this demand has made feasible expansion of landfill techniques beyond intertidal lands into areas formerly permanently occupied by the waters. Many industrial sites in Japan have been created in this way, as was much of the airport at Hong Kong, while occasionally the technique may enable creation of an artificial island where this is required for a special purpose, such as location of oil-well facilities as on Rincon Island, California (Keith and Skjei, 1974).

Draining of Lands

Numerous drainage techniques have been used to reclaim waterlogged or submerged lands in the coastal zone.

The simplest method, often employed in deltas and barrier-enclosed swamplands where the general level of the land is above sea level, involves facilitating the passage of water to the sea or estuary by means of tile drains, ripping, open cut drains, or the like. The free passage for water thus created locally lowers the water table and allows more intensive use to be made of the land. Where the desired water table level is below HWM but above or at the level of LWM, a *barrage* may be placed in the main drainage line and fitted with one-way valves that allow free passage of water from the land at low tide but prevent entry of water up the channel as the tide rises.

Where the general level of the land is at or below the level of the sea, exclusion of marine or estuarine waters and expulsion of accumulated waters to reduce the water table to the desired level must be effected. Exclusion of waters by means of *dikes* (i.e., barriers, usually of earth material, thrown up around the area to be reclaimed) is commonly practiced and sometimes a sediment-laden stream may be diverted into the impounded area in order that the sediments may be deposited therein and so raise the level of the land. The diked area (known as a *polder* in the Netherlands) must then be drained by means of ditches, tile drains, or the like, which are usually connected in a dendritic type of arrangement to a major drainage canal. This canal may empty to the sea via a system of one-way valves as described above, but more commonly the drained water must be pumped from the polder.

Reclamation of intertidal lands is sometimes initiated by accelerating natural sedimentation processes, as in the case of the Waddenzee coast (Netherlands), where the *Schleswig-Holstein* method of reclamation has long been practiced. The system uses osier breakwaters at about 400-m. intervals parallel and perpendicular to the coast, which enclose 16-ha. *bezinkvelden* (sedimentation fields) within which earthen levees divide the area into 1-ha. units. Gaps allow easy penetration of the tides, and runnels are made at right angles to the coast to speed drainage. When sediments have accumulated to the desired depth, the area is enclosed by a dike and united to the mainland.

Inhabitants of the Netherlands have for centuries been actively involved in land reclamation (Marsh, 1864; Sherlock, 1922; Lambert, 1971). About half the modern area of the country lies below HWM, and six out of every ten Dutchmen live and work below the level of the sea. In the pre-Christian era dwellers in what is now the province of Friesland constructed *terpen* (refuge mounds) to elevate dwellings and other structures above permanently or periodically waterlogged ground. The terpen (some became large enough to support entire villages) fulfilled their role until about A.D. 1000, by which time the whole of Friesland had been surrounded by dikes.

The natives of these parts may have originated the concept of a dike by observing the effects of Roman causeways and road embankments. In any case, the first dikes were simply low earth banks joining together groups of terpen to enclose areas to which the sea no longer had access. As the reclaimed land became more valuable and the consequences of floods more and more serious, the inhabitants of Friesland set themselves the task of building dikes along the whole length of coastline of their territory, a task that produced a virtually continuous sea defense by c. A.D. 1300.

Impoldering technology at this time principally allowed only for evacuation of waters by gravity flow, using the simple sluice invented in the tenth century. Although excess waters could be evacuated by using human or animal power to operate simple scoops and water wheels, such methods had rather restricted application for reclamation of lands where water tables were to be lowered beyond LWM. Large-scale reclamation of such lands awaited development of water mills.

The first wind-driven water mill was deveoped and put in operation in Alkmaar in 1408, but such machines were not widely used until the seventeenth century, when reclamation, which had come almost to a halt in the Middle Ages, again reached a peak. Early *schepradmolens* (scoop-wheel mills) could lift water only 1.5-2 m, so that deeper water had to be raised by a gang of two to four mills, but more efficient *vijzelmolens,* working on the principle of the Archimedean screw and able to lift water up to 4 m at each stage, helped make possible the great Dutch reclamations of the seventeenth century, of which the *droogmakerijen* (lake reclamation schemes) were possibly the most spectacular.

No less than 27 large lakes north of Amsterdam (some as deep as 4 m) were pumped dry before 1640. The characteristic feature of the *droogmakerijen* was a ring dike and encircling canal that acted as a reservoir into which the windmills discharged surplus water. When the polder had been pumped dry, the peats were cut and sold as fuel, leaving the valuable underlying pasture soil exposed at the surface.

In 1667 Hendrik Stevin envisaged the possibility of one day draining the Zuyder Zee, a shallow inlet of the North Sea in the north and central Netherlands. Once a lake, recorded in earliest times as Flevo Lacus, the Zuyder Zee was formed in the thirteenth century by a storm that breached the land barrier, leaving the chain known as the West Frisian Islands. Piecemeal reclamation of the Zuyder Zee commenced in the seventeenth century, but not until 1923 was a major reclamation project begun with the construction of a dam across the mouth.

The Zuyder Zee became a lake in 1932 with the closure of the dam; the first polder (Wieringermeer) of 20,000 ha. had been drained two years previously. In 1942 a further 48,000 ha. were added with the draining of the Noordoostpolder, followed by Oostelijk Flevoland (54,000 ha.), Zuidelijk Flevoland (43,000 ha.), and Markerwaard (60,000 ha.) in postwar years. The 225,000 ha. of agricultural land thus obtained represents about 10% increase in the area of Dutch farmland.

A freshwater lake of 120,000 ha., the Ijsselmeer, remains to supply industry and agriculture, to check infiltration of saltwater into the subsoil, and to provide a recreational facility.

Lands newly reclaimed from the sea are commonly saline, and require drainage and flushing with freshwater to remove excess sodium. Amelioration by adding calcium, usually in the form of gypsum, is often required. Some marshlands are high in sulfides which are stable if air is excluded. Reclamation, followed by drainage and aeration, can result in oxidation of the sulfides to sulfates; if the soil is low in basic constituents free sulfuric acid will be produced, resulting in conditions unsuitable for crop growth. Amelioration may not be economical.

In the period following World War II, concern

was expressed by biologists and conservationists that widespread wetland reclamation, especially in Western countries, would result in the loss of ecosystems valuable to man, especially estuarine wetlands valued for high primary productivity and as nursery grounds for valuable estuarine-dependent commercial fisheries (Keefe, 1972; Pope and Gosselink, 1973; Walker, 1973; Odum and Skjei, 1974). Considerable evidence has been advanced linking wetlands primary production to aquatic secondary production and suggesting that wetlands reclamation results in lowered production of estuarine organisms of interest to man.

DAVID M. CHAPMAN

References

Keefe, C. W., 1972. Marsh production: A summary of the literature, *Contr. Marine Sci.* 16, 163-181.

Keith, J. M., and Skjei, R. E., 1974. Engineering and ecological evaluation of artificial-island design. Rincon Island, Punta Gorda, California, *U.S. Army Corps Engineers Coastal Engineering Research Center Tech. Memo. No. 43*, 76p.

Lambert, A. M., 1971. *The Making of the Dutch Landscape.* London: Seminar Press, 412p.

Marsh, G. P., 1864. *Man and Nature.* (Reprinted in 1965, Cambridge, Mass.: Belknap Press of Harvard University Press, 472p.)

Odum, W. E., and Skjei, S. S., 1974. The issue of wetlands preservation and management: A second view, *Coastal Zone Management Jour.* 1, 151-163.

Pope, R. M., and Gosselink, J. G., 1973. A tool for use in making land management decisions involving tidal marshland, *Coastal Zone Management Jour.* 1, 65-74.

Sherlock, R. L., 1922. *Man as a Geological Agent.* London: Witherby, 322p.

Walker, R. A., 1973. Wetlands preservation and management on Chesapeake Bay: The role of science in natural resources policy, *Coastal Zone Management Jour.* 1, 75-101.

Cross-references: *Coastal Engineering; Coastal Engineering, History of; Dike, Dyke; Dredging; Ghyben-Herzberg Ratio; Human Impact; Planning; Polders; Salt Marsh.* Vol. VI: *Swamp, Marsh, Bog.*

LIMANS AND LIMAN COASTS

Limans—narrow bays cut deep inland—are formed along previously existing river valleys as a result of coastal plain submergence by transgressional seawater (Fig. 1). *Liman coasts* are peculiar to recent platform troughs subjected to neotectonic settling. The main morphological peculiarities of liman coasts are: alternation of active cliffs and barrier beaches along the shoreline; relatively deep submarine slope; and straightening of the shoreline. The most strongly pronounced liman type coasts are found along the Black, Azov, Chuckchi, and some other seas.

Depending on their degree of development and the influence of coastal processes, three types of limans can be distinguished: those openly connected to the sea; those completely separated from the sea; and those periodically connected to the sea. According to the type of liman, there exist different hydrological, hydrochemical, and hydrobiological conditions, as well as different conditions of bottom deposit formation. The rate and direction of development of these conditions are different within arctic, temperate, subtropical, and tropical zones of the earth.

The problem of liman origin is closely related to liman basin formation and barrier beach genesis. A liman basin is always of erosional origin. The development of the liman water regime and accumulation of deposits are influenced by: the basin area, overdeepening relative to recent control of erosion, overland flow, and rate of exchange with the sea. Most liman barrier beaches are formed by longshore drift, frequently supplied by submarine abrasion drift. However, some are formed by transverse drift migration.

Liman coast dynamics are characterized by a predominance of abrasion at seacoast sections between limans; washout of barrier beaches and their retreat together with adjacent abrasion shore sections; retardation of seacoast retreat; and accumulation of coastal deposits. Liman basins are also traps for terrigenous drift washed from the land. According to dynamic indicators, 6 types of liman barrier beaches are distinguished (Fig. 2).

Sea, eolian, and *liman* landscapes are recognized on the barrier beach surfaces and are subjected to natural processes such as the prevailing action of sea waves and eolian activity. The seaside (beach) of the barrier beaches is affected by nearshore wind, wave, and tidal currents; while the liman side (swamped barrier flats) is exposed to liman waves and currents and fluctuations of liman water level. Within the boundaries of each landscape, the predominant action of these strictly defined natural factors is observed and it is the interaction of these factors that defines liman barrier beach morphology and dynamics (Leontiev, 1960; Shuisky, 1970; Shuisky and Shevchenko, 1975).

Limans and their barrier beaches are favored by birds during the nesting season, and provide the necessary conditions for unique flora. Frequently, limans are used as sources of salt, saltwater, and curative mud. They provide basins where fish and shellfish may thrive and be raised commercially. The barrier beaches serve as convenient sites for the construction of

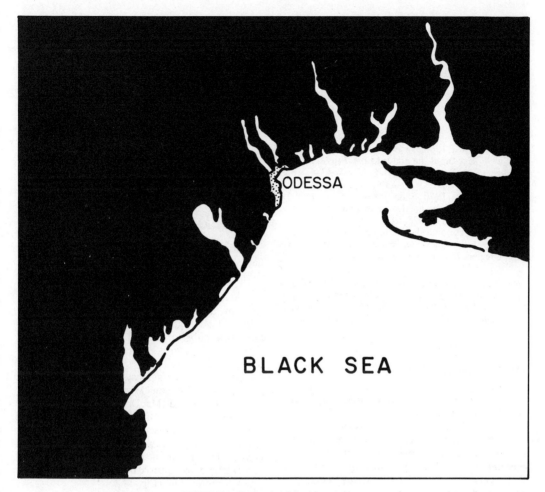

FIGURE 1. Example of liman-type coast.

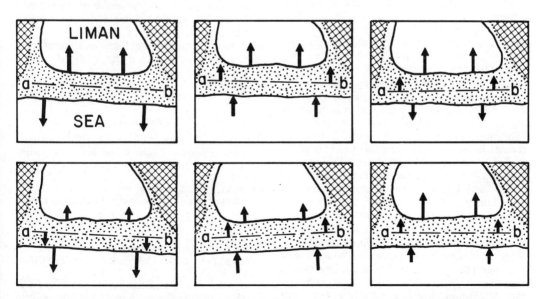

FIGURE 2. Dynamic diagram of liman barrier beach development. The arrows show the direction of shoreline movement and longitudinal axes of barrier beaches, the length of an arrow indicating broadening or narrowing of a barrier beach.

health resorts and of roads. The sand taken from the barrier beaches is used for construction as well as for the extraction of useful heavy minerals.

YURII D. SHUISKY

References

Leontiev, O. K., 1960. Types and formation of seashores of lagoons at the present time, *Internat. Geol. Congr., Rep. Soviet Geologists, 21st Session*, 188-196.
Shuisky, Yu. D., 1970. Some features of the modern development of the northwestern Black Sea shores, *Oceanology* 10, 117-125.
Shuisky, Yu. D., and Shevchenko, V. Ya., 1975. Dynamics of the Black Sea coast within the Burnas Cape region. *Jour. Geomorphology* 4, 98-104.

Cross-references: *Barrier Islands; Bars; Mineral Deposits, Mining of; Ria and Ria Coasts; Spits; Submergence and Submerged Coast; Thalassotherapy; Tourism.* Vol. III: *Ria, Rias Coast and Related Forms.*

LINEAR SHELL REEFS

Linear shell reefs, as first noted by Grave (1901, 1905) along the coast of North Carolina, are narrow, linear accumulations of oyster and clam shell detritus oriented perpendicular or at a high angle to the shoreline. In the *Treatise on Invertebrate Paleontology* (Moore, 1971), linear shell reefs are referred to as *string reefs*. It is also stated that this type of reef is found only in brackish water and is never built by euhaline oysters.

Grave hypothesized that the unique growth pattern was in response to longshore currents paralleling the shoreline. These currents provided favorable growing conditions by supplying nutrients to the reef complex and keeping the growing edge of the reef clean of sediment. The resultant growth of the reef was at a right angle to the direction of current flow, with bifurcation occurring occasionally. Subsequent reports have identified perpendicularly oriented oyster (*Crassostrea virginica*) reefs in Louisiana (Coleman, 1966), Texas (Parker, 1960; Price, 1954), and Upstart Bay, Australia (Coleman, pers. comm., 1972). Although the size of the reefs varies considerably, many in Louisiana and Texas are over 2 km long, with several reaching lengths of 8 km. Shell depths of over 7 m have also been reported (Price, 1954). Coleman (1966) proposed that the position of abandoned distributaries of ancient deltas of the Mississippi River provided a firm substratum for the shellfish to attach to, and was one of the main controlling factors in the location of linear shell reefs in Louisiana. Price (1954) and Parker (1960) postulated that both firm substratum and nearshore circulation patterns were important in linear shell reef development.

FIGURE 1. Oblique aerial photograph of exposed linear shell reefs on tidal flat in San Miguel Bay, Panama (from Lewis and MacDonald, 1972).

Linear shell reefs have also been described in the Gulf of San Miguel, Panama (Lewis, 1971; Lewis and MacDonald, 1972). Unlike the previously reported reefs, the reefs in Panama are composed mostly of small clams (*Anomalocardia subragosa* and *Protothaca grata*). The reefs in Panama are exposed at low tide and are approximately 100 m wide and one kilometer long (Fig. 1).

The practical importance of these linear reefs is two-fold: they are natural groins (groynes) and thereby serve to deter shoreline erosion; and they exhibit many of the characteristics of seed bars that generate high spat counts and may provide insight into the artificial cultivation of oyster beds.

ANTHONY J. LEWIS

References

Coleman, J. M., 1966. *Recent Coastal Sedimentation: Central Louisiana Coast.* Baton Rouge: Louisiana State University Press, Coastal Studies Tech. Rep. No. 29, 73p.
Grave, C., 1901. The oyster reefs of North Carolina: A geological and economic study, *Johns Hopkins University Circular No. 151*, 50-53.
Grave, C., 1905. Investigations for the promotion of the oyster industry of North Carolina, in *Report, U.S. Commissioner of Fish and Fisheries for 1903.* Washington, D.C.: Government Printing Office, 247-341.
Lewis, A. J., 1971. *Geomorphic Evaluation of Radar Imagery of Southeastern Panama and Northwestern Colombia.* Lawrence: University of Kansas, CRES Tech. Rep. 133-18, 164p.

Lewis, A. J., and MacDonald, H. C., 1972. Mapping of mangrove and perpendicular oriented shell reefs in southeastern Panama with side-looking radar, *Photogrammetria* 28, 187-199.
Moore, R. C., 1971. *Treatise on Invertebrate Paleontology* (Part N, Volume 3, Mollusca 6). Boulder, Colo.: Geological Society of America; and Lawrence, Kans.: University of Kansas, N1045-N1048.
Parker, R. H., 1960. Ecology and distributional patterns of marine macro-invertebrates, northern Gulf of Mexico, *in* F. P. Shepard, ed., *Recent Sediments Northwest Gulf of Mexico.* Tulsa, Okla.: American Petroleum Institute, 302-333.
Price, W. A., 1954. Oyster reefs of the Gulf of Mexico, *in Gulf of Mexico: Its Origin, Waters, and Marine Life.* Washington, D.C.: Government Printing Office, Fish. Bull. 89, 491p.

Cross-references: *Biotic Zonation; Coastal Fauna; Coral Reef Coasts; Mollusca; Organism-Sediment Relationship; Reefs, Non-Coral.* Vol. III: *Oyster Reefs.*

LINKS

The word *links* is derived from the Old English *hlinc* meaning a ridge or bank, but today it is normally used to describe areas of sandy, undulating topography on the east coast of Scotland. Precisely, links are found inland of higher, more mobile coastal dunes, and are characterized by a wide variety of grasses, legumes, and mosses growing on well-developed humic soils (Gimingham, 1964). However, in some places, links refer to the total dune areas—e.g. Preston Links, Foveron Links.

Many links have been used for cattle- and sheep-grazing, but most are now developed for recreational activities, especially golf, the playing of which has been recorded since 1625 on Scottish links (hence golf links). Ranwell (1972) considers that golf courses are important in conserving many dunes and links.

RICHARD W. G. CARTER

References

Gimingham, C. H., 1964. The maritime zone, *in* J. H. Burnett, ed., *The Vegetation of Scotland.* Edinburgh: Oliver and Boyd, 67-143.
Ranwell, D. S. 1972. *Ecology of Salt Marshes and Sand Dunes.* London: Chapman and Hall, 258p.

Cross-references: *Coastal Dunes and Eolian Sedimentation; Coastal Flora; Coastal Reserves; Thalassotherapy; Tourism.*

LITTORAL CONES

These volcanic structures are usually half cones formed by ventless nearshore steam explosions at, or slightly below, water level where confined *lava flows* enter the sea or lakes. The half-cone pattern results from the washing away of the radially ejected *tephra* on the seaward or lakeward side. Generally littoral cones consist of crescentic ridges breached by later lava flows; rarely are they perfect cones when lava traverses boglands. The ridge represents the low rim enclosing a shallow crater bordering its seaward or lakeward side; subsequent lava flows can degrade this rim to a mound.

Most littoral cones are less than 75 m high and 800 m wide at the base. Flank slope angles are less than 25°. These cones are often at one side of a feeder lava stream and sometimes on both sides. With bifurcating lava streams, the explosive growth of the cones can assume a complicated pattern with radial ejection of tephra from multiple and often migrating explosion centers. Overlap of tephra from each explosion site produces slump structures and unconformities that are further complicated by penecontemporaneous and subsequent lava streams breaching the deposits.

The characteristic mantle bedding is poorly sorted, and agglutinated layers are often produced near the explosion centers; the lower the viscosity of the feeder flow, the more likely can agglutinating materials form. The tephra may show layering from bottom to top, reflecting the changing chemistry of the early to late lava flows.

Littoral cones do not grow from a vent. They are caused by relatively unconfined steam explosions resulting from the contact of molten lava with water. The explosion centers are confined to small areas into which lava is continuously and rapidly delivered. This concentration of explosive activity is controlled by topographic irregularities that serve to confine the lava channels at the point of entry into the water. Obviously, the larger the surface area of the sole, top, and interior of a flow exposed to water, the greater will be the explosivity. Aa lava moves fast enough and is fractured enough to yield high steam explosivity. Blocky lava, although fractured, moves more slowly than aa and therefore has a smaller volume of steam-generating potential, since less hot rock is in contact with water per unit time. Pahoehoe lava is generally insulated by a continuous skin, and, except for toe protrusions, generates minor steam explosivity. Calculations by Fisher (1968) are in accord with those of Moore and Ault (1965) on the amount of comminution of the aa and pahoehoe flows from which the Puu Hou littoral cones in Hawaii were formed. Five to 6% of the feeder flows were comminuted.

It may be difficult to distinguish between a

littoral cone and a cone formed above a true vent. In general, according to Macdonald (1972), the vitric ash formed by littoral hydroexplosions is less vesicular and denser than vitric ash formed by true magmatic explosions. This is because vesicles are mostly produced by expansion of gas originally contained in the melt in contrast to the gas (steam) formed outside the ash in littoral eruptions. However, the vesiculation of littoral cone tephra is also a function of the distance the feeder flow has traveled. Perhaps a better criterion of littoral (and hydromagmatic) explosivity versus volcanic explosions created under relatively *dry* conditions is the predominance of sideromelane over tachylite in littoral tephra. The sideromelane is often granulated, dehydrated, and oxidized by the steam explosions. Another criterion of littoral explosions may be red-centered bombs with thick rinds in the agglutinated layers as described by Fisher (1968).

JACK GREEN

References

Fisher, R. V., 1968. Puu Hou littoral cones, Hawaii, *Geol. Rundschau* **57**, 837-864.
Macdonald, G. A., 1972. *Volcanoes*. Englewood Cliffs, N.J.: Prentice-Hall, 510p.
Moore, J. C., and Ault, W. V., 1965. Historic littoral cones in Hawaii, *Pacific Sci.* **19**, 3-11.

Cross-references: *Global Tectonics; Lobate Coasts; Pacific Islands, Coastal Morphology*. Vol. III: *Volcanic Landscapes;* Vol. XIII: *Volcanic Hazards*.

LITTORAL DRIFT—See SEDIMENT TRANSPORT

LOBATE COASTS

Lobate coasts have bays with great lobes, sometimes reniform or horseshoe-shaped. Their curvature is strong but smooth. Penck (1894) used the expression *"gelappten Küsten"* (lobate coasts) for coasts with different genetic origins.

Lobate coasts may have various origins. Araya-Vergara (1974) studied the glaciogenic coastal piedmonts of southern South America, which have reniform and lobate bays, as in Chiloé and Strait of Magellan. The lobes correspond to glacial lobes of Alaskian glaciation in the Upper Würm. In middle and low latitudes, marine regularization bays are horseshoe bays, especially the bay head beaches developed by shore-drift or swash regularizing processes, with the building of important beach ridges. Blanche Bay, near Rabaul, New Guinea, occupies an oval-shaped caldera (Ollier and Brown, 1971) (see *Volcano Coasts*).

Lobate coasts of glaciogenic origin are primarily erosional; those of volcanic origin, primarily structural, and those of regularization, secondary (abrasional-accumulative bays).

JOSÉ F. ARAYA-VERGARA

References

Araya-Vergara, J. F., 1974. Relaciones entre los piedmonts glaciogénicos y las formas marinas litorales en el Sur de Sudámerica, *Rev. Geog. Panamer.* **81**, 115-138.
Ollier, C. D., and Brown, N. J. F., 1971. Erosion of a young volcano in New Guinea, *Zeitschr. Geomorphologie* **15**, 12-38.
Penck, A., 1894. *Morphologie der Erdoverfläche*, vol. 2. Stuttgart: Engelhorm, 696p.

Cross-references: *Antarctica, Coastal Morphology; Volcano Coasts.*

LOG-SPIRAL BEACH—See HEADLAND BAY BEACH

LONGSHORE DRIFT—See SEDIMENT TRANSPORT

LOOSE ROCK AND STONE HABITAT

Coastal deposits of this category consist of: *angular rocks,* such as from cliff falls, slides of talus or colluvium onto the shore, and rocks quarried by the sea; *rounded rocks,* such as boulders (256-4096 mm), cobbles (64-256 mm), pebbles (4-64 mm), and granules (2-4 mm). The classification used is that of Folk (1974).

One problem with the terms involved is that they are used in different senses in different countries, and so communication is impeded. *Shingle,* for example, in the United Kingdom means granules to boulders, whereas in New Zealand it means pebbles. Most rounded rocks on coasts come from rivers, and their lithology and size depend largely on the bedrock of the river basin. Some rounded rocks are generated on the coast.

Almost all rounded rocks begin as angular ones, and the rounding is a function of the kinetic energy of moving water. Sand and water constitute an effective *cutting compound,* and the rocks grind against one another and the bedrock. In the United Kingdom and France, great numbers of flint pebbles and cobbles are present on the beaches. Indeed the word *beach* is said to be derived from the Middle English *bayche,* which means a pebbly shore. These coarse flint sediments that are characteristic of parts of Western Europe are a function of geological events: a Cretaceous event when an extensive chalk formation was emplaced that included abundant sponge spicules and such forms of opaline silica—these in time were converted to hard, tough flint nodules; or a Pleistocene event when repeated ice transport and solifluction processes spread them widely across the land and the present sea floor.

Origins

Produced by Rivers. The potential energy of a river depends on the height above sea level of its source and on its volume. The kind of stones in the river and their volume depend on the nature of the country rock. The majority of the sediment produced by a river has its origin in the mountainous part of its course, and therefore their lithology will reflect the bedrock of that terrain. Most of the pebbles found on the coast are produced by rivers, and there are commonly higher concentrations around their mouths than elsewhere. However, river-produced pebbles reach the sea only if there is sufficient energy in the lower reaches of the river to transport them to the coast. When sea level was lower, the rivers were rejuvenated, and had the energy to transport heavier sediments. The *Flandrian Transgression* has betrunked the streams, forming estuaries, which are traps for heavy sediments. Rivers from high mountains may form braided streams across the coastal plain that deliver immense loads of pebbles (*shingle*) to the sea. The surf zone sorts these sediments and further abrades them.

On horsts the rivers are often shorter and faster, delivering pebbles, cobbles, and boulders directly to the coast without traversing any coastal plain or estuary. In the midplate continent of Australia (the world's flattest and driest), comparatively little heavy sediment is delivered to the sea. Thus its coasts are characterized by thousands of kilometers of sandy beaches, pebbles to boulders being delivered only where uplifted areas are close to the coast.

Produced by the Sea. Undercutting of the cliffs by the sea produces rock falls, while landslips bring angular stones and/or pebbles from an earlier transgression under the influence of the waves. Marine quarrying produces angular masses from shoreline rocky outcrops. High cliffs produce a greater rockfall yield per meter than do low cliffs. For example, at Moher in County Claire, Eire, cliffs nearly 210 m high extend for 8 km along the coast. Waves of high energy quarry more than those of low energy. Microtidal coasts concentrate the wave energy in a narrow band, and are thus more effective in undermining the cliffs than are the waves on a macrotidal shore. Thus the supply of primary angular rock to the shore varies greatly. These angular masses are rounded by the energy of the waves. The surf zone is the power tool that turns angular shoreline rocks into rounded ones. The impulsive loading of breaking waves can smash rocks, but the horizontal shear of the surf bowls them over and over, abrades one against the other, and sweeps tool-laden (sand, shells, and so on) waters over and around them.

The fundamental distribution of energy in the ocean is a function of the distribution of land and sea, especially in the higher, stormy latitudes. In the northern land hemisphere the continents divide the ocean, breaking down the fetch of the winds. In the southern ocean hemisphere, the circumpolar fetch is almost unlimited, so that the Southern Ocean is the most energetic per km^2 in the world, and its powerful swell reaches even as far as Alaska (Snodgrass et al., 1966).

In the higher latitudes, ice transports large quantities of rock that is mostly angular. Storm waters round these and aggregate them into shore ramps and spits. Islands can be enlarged by the accretion of storm terraces and spits of such rocks.

Loose Coarse Sediments and the Biosphere

Rocks in constant motion, as in a channel in a shore platform, suffer too much abrasion to carry much life, if any. However, rocks having intermittent motion, such as those on a shore platform, carry life. Algae, on which littoral mollusca graze, are common. Both the algae and the mollusca are zoned. Supratidal platforms carry fine algae on their surfaces on which the littorinids graze. Observations on such a shore platform in southeast Australia showed that during a time of exceptional wave attack the algae were ground off the surface of the platform and loose rocks, so that the littorinids were left virtually without pasture for a time. However, littorinids survived when kept in the laboratory for long periods without food or being wet by the sea. They are thus adjusted to these changes. After the storm there were fewer littorinids, large numbers having apparently been swept away; thus there was a smaller population to survive the time of food shortage.

On the same coast, the mobile rocks in platform channels were smooth and uninhabited, but large blocks quarried from the walls of the channels that were too large to be moved were well populated. The most populated angular and rounded rocks were those that had been immobilized below low water springs or in potholes no longer active. Thus clearly even a small change of sea level can change the relationships of the biosphere with these rocks. *Potholes* are now active a short distance above high water, where the maximum energy of storm waves is exercised. Those below mean sea level on this coast are inactive, the sides of the potholes and the grinders in them being heavily covered with both plant and animal species. Potholes can thus be used for following small changes of sea level. How good the evidence is depends largely on the rate of retrogression of the coast.

EDMUND D. GILL

References

Folk, R. L., 1974. *Petrology of Sedimentary Rocks.* Austin, Tex.: Hemphill, 182p.

Snodgrass, F. E.; Groves, G. W.; Hasselmann, K. F.; Miller, G. R.; Munk, W. H.; and Powers, W. H., 1966. Propagation of swell across the Pacific, *Royal Soc. London Philos. Trans.* A259, 431-497.

Cross-references: *Biotic Zonation; Coastal Fauna; Coastal Flora; Coastal Waters Habitat; Estuarine Habitat; Intertidal Mud Habitat; Intertidal Sand Habitat; Organism-Sediment Relationship; Rocky Shores Habitat; Sand Dune Habitat; Sediment Size Classification.*

LOW AND BALL—See RIDGE AND RUNNEL

LOW-ENERGY COAST

Low-energy coasts are coasts sheltered from storms and swells by adjacent topographical features (barrier island, reef, embayment, shoal, headland), by their position with respect to prevailing wind direction, by their position in a climatic belt, by gentle offshore topography, or by a combination of these factors. Broad and gently sloping shelves (e.g., west of Florida) can effectively dissipate the energy of waves crossing them. Polar coasts, ice-bound most of the year, represent a special low-energy coast category. In Price's Gulf coast relative wave-energy classification (1954), low shoreface heights and gentle shelf-ramp slopes characterize such coasts. He noted the scarcity of longshore-transported sediments and of barriers and active beaches, and the prominence of tidal channels in shore reentrants. The delta shores are strongly protuberant and, because of low wave energies, coastal marshes-swamps prominent. Recent shore modification is minimal. This environment is transitional toward *zero relative wave energy* shores. Tanner (1960) used the 10-cm mean breaker height value as a cutoff between moderate- and low-energy Florida coasts.

ERVIN G. OTVOS

References

Price, W. A., 1954. Dynamic environments: Reconnaissance mapping, geologic and geomorphic, of continental shelf of Gulf of Mexico, *Gulf Coast Assoc. Geol. Socs. Trans.* 4, 75-107.

Tanner, W. F., 1960. Florida coastal classification, *Gulf Coast Assoc. Geol. Socs. Trans.* 10, 259-266.

Cross-references: *High-Energy Coast; High-Latitude Coasts; Low-Latitude Coasts; Moderate-Energy Coast; Wave Environments; Wave Shadow; Wind Incidence on Coasts; Zero-Energy Coasts.*

LOW-LATITUDE COASTS

Although *low-latitude coasts* have often been thought of as tropical or *biogenous* and largely defined by the presence of *coral reefs* and *mangrove swamps,* the most meaningful boundaries, drawn on the basis of the total coastal environment, appear to lie at about $40°$N and S (Davies, 1980). Low-latitude coasts occur between the northern and southern temperate storm belts. Swell waves emanating from these belts travel toward them, and are especially important on west-facing shores. Aggregate wave energy varies generally with distance toward the equator. It is high on the poleward periphery, where swell is relatively unattenuated and occasional storms penetrate from high latitudes; it is low near the equator, especially on sheltered coasts such as those of the Gulf of Panama, Sierra Leone, and northern Australia. At any one point, swell of mid-latitude origin tends to be consistent in direction, although energy levels may change seasonally, especially in the Northern Hemisphere. The trade winds of low latitudes also tend to produce waves of consistent direction, with monsoonal effects again introducing seasonal changes in energy. Big storm waves are expectable only along the poleward fringe with the incursion of mid-latitude depressions and with the seasonal development of tropical cyclones on some coasts, especially in the west Pacific.

Low-latitude shores display an abundance of fine sediments, tending to be sandy in drier and higher-energy regions but muddy in wetter and lower-energy regions. On long stretches of coast, these sediments form an apron separating the bedrock of the hinterland from the sea and incorporating extensive coastal barriers (Tricart and Cailleux, 1972). On sandy, high-energy shores, the barriers are massive features isolating coastal lagoons of varying size; on muddy, low-energy shores, they are commonly of *chenier* form and are incorporated within extensive tidal plains and lagoons. The prevailing consistency of wave direction over large areas gives great regularity of orientation, especially to the bigger barrier systems. Where possible they turn to face toward wave sources in higher latitudes, and repetitive asymmetrical *zetaform* plans are frequent.

Pebbles are particularly scarce on low-latitude coasts, being found on a few arid or semiarid shores and in coral reefs.

The relatively low onshore wind velocities, especially in the tropics, mean that coastal dunes are only moderately developed, except where there has been a favorable sand supply for a long period of time. Humid tropical shores are notably deficient in coastal dunes, even where, as in west Africa, Ceylon, and Malaya, sand is abundant. Wind-blown sand may be virtually absent or accumulate in low-amorphous platforms characteristically covered in trees.

On drier coasts with higher wind velocities, dunes may be much more massive, particularly where sand has been trapped along the shore for long-continued periods. On semiarid coasts, such accumulations may be extensively lithified by alternate solution and precipitation of constituent calcium carbonate, so that Pleistocene dunes persist as limestone hills, cut into cliffs and platforms and truncated offshore as reefs.

Another form of cementation may occur on beaches. Beach rocks are produced within the beach mass as a result of the precipitation of calcium carbonate from seawater or groundwater and are exposed by erosion. Whereas dune rocks are favored by a semiarid subaerial climate, beach rocks are favored by high water temperatures. Their distribution is therefore not always coincident in detail, but their combined world distribution follows closely the low-latitude zone as it is defined here.

Dry climates favor dune rock formation not only because they allow the optimum amount of lime circulation but also because the relative absence of rivers means that quartz is supplied to the shore in only small quantities and carbonate sands are prevalent. The situation is well exemplified in Australia, where the wetter north and east coasts are reached by almost all the large rivers and quartz sands are normal, while the drier south and west coasts have few large rivers and are characterized by carbonate sands.

The comparative absence of big rivers along drier coasts means that there are relatively few estuaries and deltas where tidal plains may accumulate. In estuarine and deltaic situations on wetter coasts, mangrove woodland may develop, sometimes as a fringe to patches of salt marsh and sometimes as extensive tidal forests. By contrast, the tidal plains of drier coasts carry only limited, mangrove shrubbery and much more extensive tracts of salt marsh or bare, saline sediment.

An important feature of many low-latitude coasts is the presence of coral reefs, especially in the Indian and west Pacific oceans. The reefs are predominantly island features, and continental fringing reefs are few. However, where they occur they not only provide massive shore structures but also provide sources of sediment and modify the wave climate of nearby coasts. Perhaps even more important as biological agents on low-latitude coasts are the calcareous algae, which are not only significant constituents of coral reefs but modify shore platforms and form reefs on their own account in regions where coral is largely absent. Although lime-secreting algae are worldwide in distribution, they are particularly effective geomorphologically in low-latitude seas highly supersaturated with calcium carbonate.

Because bedrock reaches the shore relatively rarely and because wave energies are moderate, cliffs cut into pre-Quaternary rocks are few and usually poorly developed. However, spectacular cliffs do occur, often eroded from recent volcanic rocks or from lithified dune limestones. Because of the prevalence of recently formed dune rocks, beach rocks, coral rocks, and algal rocks, limestones are especially important in low latitudes. They often form walled cliffs of vastly varying height, and are eroded into horizontal platforms at about low tide mark. A major part in the erosion of these platforms is played by the boring activity of species of *blue-green algae.*

Other horizontal shore platforms lying close to high tide mark are produced in volcanic rocks and in more porous noncalcareous sedimentaries. The major process here has been called *water-layer weathering,* and results from frequent wetting and drying of the rocks near the shore. The processes of water-layer weathering and the high horizontal platforms they produce have so far been described only from low-latitude coasts.

The prevalence of horizontal shore platforms may be taken as an important characteristic of the low-latitude zone. It is the result not only of the favorable climatic conditions that encourage formational processes but also of the relatively low level of wave erosion resulting from reduced wave energy. The sloping platforms so typical of higher latitudes, and that result directly from wave quarrying and abrasion, tend to be absent except where coastal rocks are unusually susceptible to wave attack.

J. L. DAVIES

References

Davies, J. L., 1980. *Geographical Variation in Coastal Development*, 2d ed. London: Longman, 212p.

Tricart, J., and Cailleux, A., 1972. *The Landforms of the Humid Tropics: Forests and Savannas*. London: Longmans, 306p.

Cross-references: *Algal Mats and Stromatolites; Beach Material; Chenier and Chenier Plain; Coral Reef Coasts; High-Energy Coasts; Low-Energy Coasts; Mangrove Coasts; Mid-Latitude Coasts; Moderate-Energy Coasts; Shore Platforms; Water-Layer Weathering; Waves.*

LOW-TIDE DELTAS

When wave action is minimal, and the water level drops to or below the step at the toe of the beach, water flowing out of *spring pits* (q.v.) higher on the beach may transport sand to the step, where small *deltas* or *fans* are built (Tanner, 1959). Such deposits are spread out over the *ripple mark* field that almost invariably extends seaward from the step. A rise in water level or an increase in wave action commonly redistributes the sand, thus eliminating the deltas and fans.

Water, draining from the spring pits, cuts tiny, narrow, parallel V-shaped valleys. Sand obtained from the spring pits and/or the cutting of these valleys builds the low-tide deltas and fans. In one set of examples typical of a low-energy sand beach, the valleys were about 2 cm deep, 60-70 cm long, and 2-3 m apart; low-tide deltas were 25-50 cm wide; and the top of each delta or fan was marked by a well-developed distributary pattern.

Ripple mark troughs adjacent to the deltas may be partly filled with sand introduced across the deltas, producing distinctive flat-bottomed ripple marks, which grade into obscure ripple marks even closer to each delta or fan.

WILLIAM F. TANNER

Reference

Tanner, W. F., 1959. Near-shore studies in sedimentology and morphology along the Florida Panhandle coast, *Jour. Sed. Petrology* 29, 564-574.

Cross-references: *Beach; Minor Beach Features; Ripple Marks; Spring Pits.* Vol. III: *Littoral Processes;* Vol. VI: *Littoral Processes.*

LYCOPODIOPHYTA

The lycopods (Lycophyta, Lycopsida, Lepidophyta) of today are small, often creeping plants in contrast to some of the extinct tree-sized kinds that formed the bulk of the carboniferous coal measures (Ferre, 1963; Reimers, 1954; Sporne, 1962). All lycopods have true roots, stems, and leaves in the conspicuous sporophytic stage. They are perennial, seedless, vascular plants, with small gametophytes (haploid gamete-producing generation) sometimes ± photosynthetic, at first dominant, but the sporophytes (diploid, spore-producing generation) photosynthetic and soon independent. The sporophytic stage is the conspicuous and leafy stage.

The leaves are microphyllous (with a single unbranched vein, and the leaf traces not subtending gaps in the stem stele). In arrangement they may be alternate, opposite, or rarely whorled. The sporangia are either axillary or borne in the upper (adaxial) surface of the leaf. The spores may be either of one kind (homosporous) or of two kinds (heterosporous).

There are two present-day classes of the division Lycopodiophyta: Lycopodiopsida and Isoetopsida. The Lycopodiopsida stems are elongate and densely leafy. The growth is primary only—i.e., the stems and roots once differentiated add no more tissue. The leaves are small, usually under a centimeter in length, and are alternate, opposite, or rarely whorled. The sporangia are axillary and borne in terminal, usually cone-like strobili. There are two present day orders of the Lycopodiopsida. The Lycopodiales (or class Aglossopsida), including the single family Lycopodiaceae, which includes the genera *Lycopodium* (the club mosses), of about 600 species and *Phylloglossum,* of only one species. The leaves lack ligules, and only one kind of spore is produced. The spore-bearing branches are usually erect, of limited growth, and differentiated from the usually creeping stems.

The gametophyte develops outside the spore, and is either ± photosynthetic or is associated with a fungus that digests the humus of the soil and transfers this food to the gametophyte. *Lycopodium* is abundant in moist tropical regions and often frequent in moist temperate regions. The plants, although mostly terrestrial, are often also epiphytes (growing on other plants). Economically the group is of little importance. Because it is intolerant of salinity, it is not to be expected as a component of the coastal maritime flora.

The Selaginellales (or class Glossopsida) includes the single family Selaginellaceae, which includes only the single genus *Selaginella,* the little club mosses or spike mosses, of about 500 species. The leaves possess ligules, minute appendages at the base on the adaxial side. Two kinds of spores are produced; often both kinds are borne within one strobilus but never both kinds within one sporangium.

The gametophytes are endosporic—i.e., are produced within the spore. The smaller spores produce the ♂ gametophytes that produce the sperms; the larger spores produce the ♀ gametophytes that produce the eggs. Both sexes of gametophytes are minute, and depend on the food stored in the spore from which they develop. *Selaginella* is abundant in moist tropical regions and occasional in temperate regions, even in hot dry deserts. The plants are often epiphytic. Some kinds are cultivated as ornamentals especially in greenhouses.

The class Isoetopsida includes a single living order—the Isoetales—with a single family, Isoetaceae, including a single genus *Isoetes*, the quillworts, of about 50 species. The stems are short, rarely exceeding 4 cm in length, vertically oriented, 2- or 3-lobed, and have secondary growth. The elongate, linear, ligulate leaves are borne in a dense terminal, spirally arranged cluster, and range in size from 5–50 cm. They bear the sporangia embedded in the adaxial side of their broadened bases. The spores, of two different sizes, are borne in different leaves on the same plant. The gametophytes are endosporic.

The quillworts are of no economic importance. They grow in nonsaline sites ranging from meadows wet only in winter and spring to permanent ponds and lakes. The plants may be mistaken for grasses with which they may grow.

LOUIS CUTTER WHEELER*

References

Ferre, Y. de, 1963. Les Lycopodinées, *in* H. des Abbeyes et al., eds., *Botanique Anatomie—Cycles Evolutifs Systematique*. Paris: Masson et Cie Éditeurs, 512–522.

Reimers, H., 1954. Pteridophyta, Farnpflanzen, *in* H. Melchoir and E. Werdermann, eds., *A Engler's Syllabus der Pflanzemfamilien,* 12th ed., vol. 1. Berlin-Nikolassee: Gebrüder Borntrager. 269-311.

Sporne, K. R., 1962. *The Morphology of Pteridophytes: The Structure of Ferns and Allied Plants*. London: Hutchinson University Library, 192p.

Cross-references: *Biotic Zonation; Coastal Ecology, Research Methods; Coastal Flora; Coastal Plant Ecology, United States, History of.*

*Deceased

M

MADE LAND—See LAND RECLAMATION

MAGNOLIOPHYTA

Seed-bearing plants can be separated into two divisions, Magnoliophyta (Anthophyta, Angiospermae) and Pinophyta (Gymnospermae). The Magnoliophyta include more than 200,000 species, comprising approximately two-thirds of the known plant species (Burns, 1974; Cronquist, 1979) and have dominated the land flora as the largest group of plant species since the middle Cretaceous. They appear quite suddenly in the fossil record, perhaps evolving from Paleozoic seed ferns. Charles Darwin described the appearance of flowering plants as an "abominable mystery," and the mystery remains unsolved. Though the earliest angiosperm fossils date back only to the Jurassic period, numerous families of modern flowering plants were already in existence by the mid-Cretaceous. Except in certain regions dominated by conifer forests, the angiosperms are the most conspicuous form of terrestrial life. Most are readily recognizable as plants by their form and color, and the plant body typically consists of roots, stem, and leaves. These plants exhibit a considerable range in size. Some are small, low herbs (nonwoody plants), while others are shrubs, vines, and even magnificent trees. The smallest are diminutive green, aquatic globules about 1–1.5 mm long lacking both roots and leaves, belonging to the genus *Wolffia* (water meal) of the Lemnaceae (duckweed) family.

Length of life of the vegetative plant ranges from that of small herbs, which may complete their life cycle from seed to seed in a few weeks, to trees and shrubs that may survive for centuries. Longevity of seeds ranges from a few days to rarely as long as a century and a half, typically being a decade or two.

The one universal characteristic of the Magnoliophyta is the phenomenon of double fertilization, a process not found in any other plant group. During reproduction, two sperms fuse with a nucleus from the female. One sperm is used to form the plant embryo, while the other initiates development of a food storage tissue, endosperm. The two structures are combined to produce a seed. Magnoliophyta produce egg-containing *ovules* (future seeds) that are completely encased within the maternal tissue, so that fertilization requires growth of a pollen tube to transport sperms to the ovule. In contrast, Pinophyta bear ovules on the surface of cone scales or at the end of a stalk, where they are exposed to air.

Fruits are composed of the ripened ovaries, their contents, and associated structures. Fruits protect the seeds during their development and distribution and aid in seed distribution either by attracting distributors, such as birds or mammals, or by floating in water or being suspended in the air. In some cases, fruits regulate the time of germination and provide water, which assists in germination and the establishment of seedlings. Coconuts (*Cocos nucifera*) and the bottle gourd (*Lagenaria vulgaris*) are fruits noted for long-distance transport by ocean currents because of their buoyancy. Certain mangroves—e.g., the common or red mangrove, *Rhizophora mangle*—retain the one-seeded fruit on the parent plant until the seed has germinated and the root has grown down out of the fruit; then it falls, and the root penetrates the muddy shore, where the seed has an added advantage for survival by already being germinated when planted. Fruits specialized for transport by wind have expanded surfaces such as wings or hairs that increase the ease with which the wind carries them.

The angiospermous vegetation of oceanic islands results from seeds that were airborn, bird-borne, perhaps (rarely) bat-borne, floated either as the bare seed or in the fruit that produced them, or attached to floating debris. Seeds ingested by birds or sticking to their feet or feathers may be transported long distances to oceanic islands. For example, the circumboreal English sundew (*Drosera anglica*) occurs in high-altitude bogs on Mt. Waialeale, Kauai Island, Hawaiian Islands, where it presumably was carried by the Pacific golden plover that nests in Alaska and winters from August to March in the Hawaiian Islands, among other places in the Pacific basin. Some seeds are adapted for flotation by having impervious coats combined with internal cavities. The seeds of *Entada scandens* (gogo vine), a

West Indian pea vine, float on the Gulf Stream to Norway. There they may germinate on the beach (but fail to survive the winter). The seeds of the mangrove, *Excoecaria agallocha* (blinding tree), although small, survive local transport floating on seawater.

Flowers contain some of the most important characteristics used for classifying the Magnoliophyta, and usually include the following structures: *sepals*, the green, outermost floral structures collectively composing the *calyx*; *petals*, which form a larger structure, the *corolla* (together, the calyx and corolla form the *perianth*, usually the showy part of the flower); *stamens*, male structures inside the corolla that produce pollen; and *pistil(s)*, the female reproductive organ composed of one or more *carpals*, each containing *ovules*. There are many possible combinations of these structures. While both stamens and pistil(s) are usually present, flowers having only stamens or only pistil(s) are not uncommon. A few species produce sterile flowers.

Floral structures reflect the type of pollination utilized by the plant. During pollination, pollen produced by the stamens is transported to the stigma, the receptive apex of the pistil. Pollen germinates in the stigma, and triggers formation of a pollen tube extending through the tissue of the pistil to reach the ovule. A combination of showy perianth, scent, nectar, and abundant pollen is attractive to animal pollinators, which are most commonly insects but may also be birds and, less frequently, bats, that utilize the nectar and pollen for food. Families of coastal plants that are pollinated by these biotic agents include: Caryophyllaceae, pink or carnation; Convolvulaceae, morning glory; Fabaceae (Leguminosae), legume or pea; Plumbaginaceae, leadwort; Verbenaceae, verbena; and Brassicaceae (Cruciferae), mustard.

Correspondingly, flowers pollinated by abiotic agents—wind and water—lack nectar, scent, and showy parts since they need not be attractive to pollinators. Wind-pollinated species often have broadened or elongated, feathery stigmas and light, dry pollen. They include: Arecaceae (Palmae), palm; Asteraceae (Compositae) part, sunflower; Chenopodiaceae, saltbush; Cyperaceae, sedge; Juncaceae, rush; Plantaginaceae, plantain, and Poaceae (Gramineae), grass. Water-pollinated types, having elongated pollen grains that are readily wetted, include Hydrocharitaceae and Potamogetonaceae, whose marine members are known as sea grasses.

Near the end of the 17th century, John Ray established a classification system that still remains in use, dividing the flowering plants into two classes, monocotyledons (Liliopsida) and dicotyledons (Magnoliopsida). These groups are distinguished from each other by the number of *cotyledons*, or embryonic leaves, as well as by differences in vascular tissue, leaf veination, root structure, and flower type. Monocotyledons include the grasses, lily family, and palms. The dicotyledons are a highly diverse group, the most primitive members perhaps being catkin-bearing plants such as the willows. These forms may have evolved from gnetalian, or conifer-like, gymnosperm ancestors (Porter, 1967; Sporne, 1975).

In tropical, subtropical, and most temperate regions the conspicuous vegetation is commonly Magnoliophyta. In forested cold regions the bulk of vegetation may be composed of Pinophyta (gymnosperms), but Magnoliophyta is likely to comprise the greatest number of species even in areas of gymnospermous forests. Aside from the gymnospermous forests in boreal regions, the seed plant cover of the tundra is composed almost entirely of Magnoliophyta.

Numerous Magnoliophyta are adapted to the saline conditions of the littoral and vicinity, but Pinophyta rarely are tolerant of saline soils, though a few survive within reach of ocean spray. Several groups of Magnoliophyta have adapted to unusual coastal habitats. The sea grasses (phanerogams) consist of about 45 species of marine flowering plants belonging to two monocotyledonous families, the Potamogetonaceae and Hydrocharitaceae (Dawson, 1967; Hartog, 1970). These plants live wholly submerged, and are most often observed as debris washed ashore. Eel grass (*Zostera marina*) occurs on the Pacific coast from Alaska to Mexico, and is an important food source for marine life, as well as sea brant, Canada geese, and black ducks. The majority of sea grass species are strictly tropical. Particularly important areas of occurrence are Indo-Malaysia and the Caribbean Sea.

About 30 species of mangroves can be found in quiet tropical lagoons and estuaries. Plants are supported by an intricate system of roots that extends 1–2 m above the mud bottom. Mangrove swamps are important habitats for a wide variety of plants and animals. One member of the palm family *Nypa fructicans*, also forms offshore marshes. *Nypa* thrives only with its base at least partially submerged in brackish water, and grows only on tide lands. The low, shrubby plant consists of a short, thin underwater rootstalk instead of a trunk, and bears pinnate leaves 3–10 m long. *Nypa* marshes occur along the coasts of the Phillipines, India, and Malaysia (McCurrach, 1970).

Tidal areas have a great diversity of flowering plants. Monocotyledons include Poaceae (gramineae), grass family; Cyperaceae, sedge family; Arecaceae, palm family; and Juncaceae, rush family. Dicotyledons flourishing in coastal environments include Chenopodiaceae, saltbush family; Asteraceae (Compositae), sunflower family; Plumbaginaceae, leadwort family; and Verbenaceae, verbena family including the principal mangroves.

The Magnoliophyta have great economic importance, yielding food, fiber, building materials, industrial raw materials, medicines, and ornamental plants and flowers. The grass family (Poaceae or Gramineae) produces cereal grains important as food for humans and animals, as well as fodder and sugar. The legume family (Fabaceae or Leguminosae) produces peas, beans, lentils, and other nutritious seeds, while the spurge family (Euphorbiaceae) yields major amounts of starch known variously as cassava, manioc, and tapioca in tropical regions. The spurge family is also a major source of natural rubber, from the Para rubber tree (*Hevea brasiliensis*), and castor oil, from the castor bean (*Ricinus communis*). A wide variety of fruits and vegetables are produced by flowering plants, and dicotyledonous trees provide large quantities of hardwood lumber, as well as pulp used for the manufacture of paper.

Though flowering plants represent a tremendous economic resource, their role in nature extends far beyond their importance in the marketplace. In the coastal environment, these plants offer habitats for many other organisms, particularly in the cases where entire communities develop around particular plant species, as in mangrove and *Nypa* marshes. Offshore sea grass beds provide habitats for marine organisms, and shrubs and grasses play an important role in stabilizing dunes along sandy coasts. These plants require the unique ecological conditions found near the shore, and they in turn participate in the biological and geological processes that occur in this environment.

LOUIS CUTTER WHEELER*
GEORGE MUSTOE

References

Burns, G. W., 1974. *The Plant Kingdom.* New York: Macmillan, 540p.
Cronquist, A., 1979. *How to Know the Seed Plants.* Dubuque, Iowa: W. C. Brown Co., 472p.
Dawson, E. Y., 1967. *Marine Botany.* New York: Holt, Rinehart & Winston, 371p.
Hartog, C. den, 1970. *The Sea-grasses of the World.* Amsterdam: North Holland Publishing Co., 275p.
McCurrach, J., 1970. *Palms of the World.* New York: Harper Brothers, 290p.
Porter, C. L., 1967. *Taxonomy of Flowering Plants.* San Francisco: W. H. Freeman, 472p.
Sporne, K. R., 1975. *The Morphology of Angiosperms:*

*Deceased

The Structure and Evolution of Flowering Plants. New York: St. Martin's Press, 207p.

Cross-references: *Biotic Zonation; Coastal Flora; Coastal Plant Ecology, United States, History of; Insects; Mangrove Coasts; Vegetation Coast.*

MAJOR BEACH FEATURES

A beach can be divided into three regions: *nearshore, foreshore,* and *backshore.* The nearshore is always underwater, while the foreshore is that part of the beach extending from the mean low water line to the highest elevation reached by waves at normal high tide. The backshore encompasses the area landward from the water's reach at normal high tide to the maximum uprush during storms. Each of these zones is subject to different hydraulic conditions that are reflected in the geomorphological features present. The nearshore zone possesses asymmetric and lunate *ripples* and/or longshore *bars.* The foreshore is the most active part of the beach and usually has a steep slope. In the transition between foreshore and backshore, *cusps* may be present. The major forms of the backshore are *berms, beach ridges, wind ripples,* and *washover fans.*

Nearshore

Trask (1955) divides the nearshore zone into three regions: below 20 m in depth, a passive zone with little or no sediment movement under normal conditions; from 20-10 m, an intermediate zone in which sand will move only at intervals; an active zone from 10 m to the shore. In the outer two regions, asymmetric ripples frequently mark the surface (Inman, 1957). The ripples are always present on sandy bottoms if the significant water velocity associated with waves is between 10 cm and 1 m/second. The ripple asymmetry is often landward, but can be seaward, depending on the dominant drift that crosses them. The internal structure of the asymmetric ripples usually consists of thin (generally less than 4 cm thick) shoreward-inclined ripple cross lamination, sometimes suggestive of climbing ripples. Closer to shore, sand movement is predominantly onshore and lunate megaripples are common (Clifton, Hunter, and Phillips, 1971). These ripples migrate slowly landward and exhibit cross-stratification with foresets dipping steeply landward.

Still closer to shore, longshore bars may be present. These are either solitary or in parallel series of two or three. Bars are common only off gently sloping beaches. Shepard (1950) found that if the foreshore slope exceeds $4°$, bars are usually not formed. In general, the lower the wave energy, the fewer the number of longshore bars on the submerged beach. High, steep waves scour the beach eroding the foreshore and even parts of the backshore into a concave-up profile. Some of this material eroded from the beach is deposited in the nearshore zone, where it may be reworked into a bar. Immediately seaward of the breaker zone, an acceleration of shoreward water velocity takes place, while a seaward acceleration occurs in the trough developed below the *plunge point* of the breaker (Morrison and Crooke, 1953). This water motion may produce a bar that accretes from both sides. The crest of the bar never grows above the water level, but reaches a maximum equilibrium height for a given wave condition. If waves increase in size and break further from shore, sand will move from the bar and migrate to a new location seaward of its old position. During lower waves, some of the sand from the storm-formed bar together with sand derived from the foreshore may be piled into a new bar. If the storm-formed bar is, however, in relatively deep water, it may be out of reach of the onshore surges associated with the breaker zone of weaker waves and so retain its shape and position.

Typical bar morphology shows a rather gentle sloping face toward the sea, a rounded crest, and a steeper slope landward. This asymmetry of form is reflected internally by the presence of two sets of laminae. Seaward of the crest, the laminae dip gently seaward, while steeply landward-dipping laminae occur shoreward of the crest (McKee and Sterrett, 1961).

Foreshore

The foreshore is the zone through which sediment moves from a winter, storm, or eroded profile (often with longshore bars) to a summer or normal bermed profile (which may be without bars). A profile of equilibrium (Tanner, 1958) will form if no changes of sea state occur. This profile is characterized by a reduced transport rate and smaller exchange of sediment between the various zones in the littoral zone.

After a storm, the nearshore bar deposit will migrate slowly to the beach, usually in the form of transverse bars, crescent bars, sand waves, ball and ridges, or rhythmic topography (Evans, 1942; Homma and Sonu, 1963; Hayes, 1972; Sonu, 1973). These are some of the names given to the onshore- or longshore-migrating sand deposits. The size, rate of migration, and orientation of the bars is a function of barometric pressure, sea state, angle of wave approach, the strength of the longshore current, and the interference pattern of incoming waves with standing waves. In the landward migration of the bars, sand is moved by bed load over the flat seaward side and avalanches over the landward slip face of the bar. If this continues long enough, the bar is "welded" to the beach (Davis et al., 1972). This results in a beach with a wide backshore and an accretion profile that is concave downward.

Before the migrating bars become part of the backshore, they may have associated lows (*runnels*) on the landward side. Water spilling over the bar becomes trapped within the runnel and can move only parallel to shore. This current moving parallel to shore generates current ripples that are superimposed on and at right angles to the wave ripples, forming ladder-backed ripples (Hayes, Anan, and Bozeman, 1969). The water trapped in the runnels returns seaward through low spots in the bar in a strong, narrow current. These rip currents can scour out a channel 1-3 m deeper than the bar (Cook, 1970). The scoured channel is floored by a rippled lag deposit of coarse sediment, heavy minerals, and shell fragments. The ripples have large-scale crossbedding with foreset laminae sloping seaward.

Constant reworking by the waves makes the foreshore the steepest part of the beach. The slope of the beach increases as the grain size increases (Bascom, 1951; McLean and Kirk, 1969). This overall tendency is modified by different wave conditions and by the porosity and permeability of the beach sediment (Savage, 1958). The slope of the beach increases with decreasing wave height and wave steepness.

The main type of bedding in the foreshore is 1-15-cm thick beds of evenly laminated, seaward-dipping sand with low-angle discontinuities (Thompson, 1937). Back-and-forth movement by the swash and backwash often leads to reverse grading within each lamina (Clifton, 1969), with the finer grained, light colored sand or heavy mineral layer near the base and then grading upward into coarser grained or heavy mineral deficient layers.

In the transition from the foreshore to the backshore, cusps may develop (Fig. 1). *Beach cusps* (q.v.) are cuspate deposits of coarser sediment extending seaward as tongues or horns separated by bow-shaped depressions of finer-grained sediment. They can be formed in all types of sediment, but form most readily when there is a vertical stratification of material. If present, cusps are regularly spaced. In bays and sheltered areas, cusps are close together (3-20 m). On beaches facing an open sea, cusps are higher and further apart (usually more than 30 m). The regularity of the spacings suggest an interference pattern between progressive waves and standing waves as a possible origin for cusps (Bowen, 1969). Cusps usually

FIGURE 1. Beach cusps in the process of formation at West Beach, Whidbey Island, Washington (photo: M. L. Schwartz).

appear shortly after a storm, when wave energy decreases. The base of the cusp deposit shows an erosional contact with the underlying sediment (Thompson, 1937). Cusps are first noticeable as patches a few centimeters thick of gravel, shells, and other coarse material (Russell and MacIntire, 1965). These are due to the swash, which, because of the interference patterns set up by the waves, is not of equal volume everywhere and therefore can move larger-size sediment if the volume is larger. Then the following backwash, instead of returning in dispersed form, moves away from the patches and returns in a channel. This process is repeated and intensified until the backwash flow attains such momentum that the next swash cannot proceed against it and the swash is projected onto the horns. There the coarsest sediment is deposited as the water swings into the adjacent bays without stopping (Bagnold, 1940). The entrained sediment forms trough-shaped laminae (Mii, 1958). The horn deposit is not uniform in thickness, but possesses a shape similar to a Gilbert delta, thinning in a seaward wedge composed of steeply dipping foreset beds. Gravel and heavy mineral concentrations are usually exposed at the surface, and pinch out below the surface in a landward direction.

Backshore

Characteristic of the backshore is the berm, a nearly horizontal to slightly landward-sloping flat portion of the beach. The seaward limit of the berm, the berm crest or edge, may be steep to form a beach scarp. Usually the coarser the size of the sediment, the steeper the berm crest. On the cobble-and-shingle Chesil Beach in southern England, the scarp reaches up to 13 m above high water (Guilcher, 1958). However, the break in slope is usually less dramatic. The upward growth of the berm depends mainly on the largest waves reaching the beach. Sand is carried upward as far as the swash extends, but only the largest waves can pass over the crest and deposit the bulk of their sand load on top of the berm. More than one berm (up to 5 or 6) may be present at any one time. Each of these represents deposition during spring tide high waters or welded longshore bars.

Just landward of the crest at the front edge of the berm, a beach ridge may develop. A beach ridge is a continuous linear mound of the coarsest sediment found on the beach (Fig. 2). These sand, gravel, and shell deposits are heaped up by waves during storm tides. The uprush of large storm waves add sand to the top of the berm even while the beach face is eroded and the berm extent is reduced. On a progradational shore, many beach ridges may be formed one after another. These beach ridges may be up to several meters high, tens of meters in width, and hundreds of meters in length.

The backshore is usually dry, except under storm conditions, and is acted upon primarily by wind. Wind removes the finer material and piles it up in ripples and dunes, leaving behind deflation surfaces where convex-side-up shells are common. During storms, the backshore is flooded by rather shallow water in which surges are active. This destroys the wind-formed features, and leaves behind layers of almost horizontally laminated sand (Andrews and van der Lingen, 1969). The sorting of sand by the swash overrun may produce heavy mineral placers

FIGURE 2. Cobble beach ridge, Smith Cove, Nova Scotia. The steplike character of the ridge is the result of a series of decreasing storms (photo: M. L. Schwartz).

(Inman and Filloux, 1960). Subsequent wind action is important in concentrating these mineral placers, so that in many localities they are mined.

During severe storms, much of the foreshore and even some of the backshore may be eroded. A great part of this eroded sand is deposited offshore, but some of it is carried landward and deposited in *washover fans* (q.v.) on the landward side of the beach, on top of the salt marsh, or in depressions among dunes (Schwartz, 1975). On St. Joseph Island in Texas, a washover fan formed by hurricanes extends over a 7-km area and attains a thickness of 125 cm (Andrews, 1970). In general, washover fans are composed of a series of superimposed sandy units. Each unit starts with a shell-rich layer overlying an erosion surface. The shells of the invertebrates commonly have several biocoenoses mixed together. The top of the unit consists of a sandy layer with well-developed, evenly laminated sand.

BENNO BRENNINKMEYER

References

Andrews, P. B., 1970. Facies and genesis of a hurricane washover fan, St. Joseph Island, central Texas coast, *Texas Univ. Bur. Econ. Geology Rept. Invest.* 67 1-147

Andrews, P. B., and van der Lingen, G. J., 1969. Environmentally significant characteristic of beach sands, *New Zealand Jour. Geology and Geophysics* 12, 119-137.

Bagnold, R. A., 1940. Beach formation by waves: Some model experiments in a wave tank, *Inst. Civil Engineers Jour.* 15, 27-52.

Bascom, W. N., 1951. The relationship between sand size and beach face slope, *Am. Geophys. Union Trans.* 32, 866-874.

Bowen, A. J., 1969. Rip currents, *Jour. Geophys. Research* 74, 5467-5478.

Clifton, H. E., 1969. Beach lamination—Nature and origin, *Marine Geology* 7, 553-559.

Clifton, H. E.; Hunter, R. E.; and Phillips, R. L., 1971. Depositional structures and processes in the non-barred high-energy nearshore, *Jour. Sed. Petrology* 41, 651-670.

Cook, D. O., 1970. The occurrence and geologic work of rip currents off southern California, *Marine Geology* 9, 173-186.

Davis, R. A.; Fox, W. T.; Hayes, M. O.; and Boothroyd, J. C., 1972. Comparison of ridge and runnel systems in tidal and non-tidal environments, *Jour. Sed. Petrology* 42, 413-421.

Evans, O. F., 1942. A discussion of the use and meaning of the term low and ball, *Jour. Geology* 50, 213-215.

Guilcher, A., 1958. *Coastal and Submarine Morphology.* London: Methuen and Co., 274p.

Hayes, M. O., 1972. Forms of sediment accumulation in the beach zone, *in* R. E. Meyers, ed., *Waves on Beaches.* New York: Academic Press, 297-356.

Hayes, M. O.; Anan, F. S.; and Bozeman, R. N., 1969. Sediment dispersal trends in the littoral zone: A problem in paleogeographic reconstruction, *in* M. O. Hayes, ed., *Coastal Environments of Northeastern Massachusetts and New Hampshire.* Amherst, Mass.: University of Massachusetts, Geology Dept. Contrib. 1-CRG, 290-315.

Homma, M., and Sonu, C. J., 1963. Rhythmic patterns of longshore bars and related sediment characteristics, *Conf. Coastal Eng., 8th, Am. Soc. Civil Engineers, Proc.,* 248-278.

Inman, D. L., 1957. Wave generated ripples in nearshore sands, *U.S. Army Corps of Engineers Beach Erosion Board Tech. Memo. 100,* 42p.

Inman, D. L., and Filloux, J., 1960. Beach cycles related to tide and local wind wave regime, *Jour. Geology* 68, 225-231.

McKee, E. D., and Sterrett, T. S., 1961. Laboratory experiments on form and structure of longshore bars and beaches, *in* J. A. Peterson and J. C. Osmond, *Geometry of Sandstone Bodies.* Tulsa, Okla.: American Association of Petroleum Geologists, 13-28.

McLean, R. F., and Kirk, R. M., 1969. Relationship between grain size, size sorting and foreshore slope on mixed sand-shingle beaches, *New Zealand Jour. Geology and Geophysics* 12, 138-155.

Mii, H., 1958. Beach cusps on the Pacific coast of Japan, *Tohoku Univ. Sci. Rep.* 29, 77-107.

Morrison, J. R., and Crooke, R. C., 1953. The mechanics of deep water, shallow water and breaking waves, *U.S. Army Corps of Engineers Beach Erosion Board Tech. Memo. 40,* 14p.

Newton, R. S., 1968. Internal structure of wave formed ripple marks in the near shore zone, *Sedimentology* 11, 275-292.

Reineck, H.-E., and Singh, I. B., 1973. *Depositional Sedimentary Environments.* New York: Springer-Verlag, 439p.

Russell, R. J., and MacIntire, W., 1965. Beach cusps, *Geol. Soc. America Bull.* 76, 307-320.

Savage, R. P., 1958. Wave run up on roughened and permeable slopes, *Am. Soc. Civil Engineers Proc., Jour. Waterways and Harbors Div.* 84, 3.

Schwartz, R. K., 1975. Nature and genesis of some storm washover deposits, *U.S. Army Corps of Engineers Coastal Engineering Research Center Tech. Memo. 61,* 69p.

Shepard, F. P., 1950. Beach cycles in Southern California, *U.S. Army Corps of Engineers Beach Erosion Board Tech. Memo. 20,* 26p.

Sonu, C. J., 1973. Three dimensional beach changes, *Jour. Geology* 81, 42-64.

Tanner, W. F., 1958. The equilibrium beach, *Am. Geophys. Union Trans.* 39, 889-891.

Thompson, W. O., 1937. Original structures of beaches, bars, and dunes, *Geol. Soc. America Bull.* 48, 723-752.

Trask, P. D., 1955. Movement of sand around southern California promentories, *U.S. Army Corps of Engineers Beach Erosion Board Tech. Memo. 76,* 60p.

Cross-references: *Bars; Beach; Beach Cusps; Beach Processes; Beach Ridge and Beach Ridge Coast; Bed Forms; Coastal Dunes and Eolian Sedimentation; Ridge and Runnel; Ripple Marks; Washover and Washover Fan.*

MANGROVE COASTS

Mangrove formations are a common occurrence associated with the low alluvial coasts of the tropics and subtropics. The vegetation, consisting of several species of trees and a few shrubs and vines, grows between extreme high tide and a level close to but above mean sea level (Macnae, 1968). Mangal formations are easily recognized by pneumatophores (which are root adaptations) and a dense tangle of prop roots. The soft and prevailing anaerobic environment is also characteristic of these communities.

Mangal communities grow on mud, sand, rock, or coral primarily in areas free from strong waves (Chapman, 1976). Therefore this vegetation type is found on the shores of bays, lagoons, estuaries, leeward side of barrier islands, and open beaches where the mainland is fronted with coral reefs that break the incoming waves.

Other factors important to the development of mangrove swamps include temperature, rainfall, tidal extent, salinity, water table, and drainage (West, 1956, 1977; Waisel, 1972; Chapman, 1976). The flora prefers regions where the average temperature of the coldest month exceeds 20°C and the seasonal temperature range is not greater than 5°C. The saline environment limits competition, and the farther the tide extends inland, the wider the area of growth. Abundant rainfall distributed evenly throughout the year is also considered to be advantageous. However, mangrove formations exist on the arid shores of the Red Sea, Persian Gulf, and Gulf of California (Snead, 1972).

Most mangrove formations occur between latitudes 25°N and 25°S, but are known to extend to regions out of the tropical zone, in those areas where temperatures remain fairly constant. In the Americas, the northern limit is the Bahamas, whereas the southern limit is the state of Santa Catarina, Brazil (West, 1956, 1977). The southernmost mangroves in Africa occur at the mouth of the Umnagazana River, 39°50'S (Macnae, 1963). In Asia, the poleward limit is extreme southern Japan and Quelpart Island, south of Korea (Putnam et al., 1960). In Australia, they exist as far south as Corner Inlet, Victoria, and in New Zealand, the poleward limit is Chatham Island, 44°S (Clark and Hannon, 1967; West, 1956).

Size of the vegetation varies from 1 m to over 31 m in both height and areal extent. The most luxuriant and developed mangrove vegetation is in latitudes close to or on the equator (West, 1956). An exception is the tall mangroves at Ten Thousand Islands, Florida. Well-developed mangrove forests are found along the Pacific in South America, in parts of Columbia and Ecuador; in Africa, on the shores of the Cameroons and eastern Nigeria (Atlantic) and on the coast of Tanzania (Pacific); and in southeast Asia, along the northern coast of Sumatra and most of Borneo.

Composition and zonation of the flora vary around the world. In the tropical Americas and west Africa, the major genera include *Rhizophora, Avicennia, Laguncularia,* and *Conocarpus. Pelliciera rhizophora* is found on the Pacific coasts of southern Central America and northwest South America (West, 1956). Along the coasts of the Indian Ocean and Southeast Asia, the flora includes at least 15 genera and more than 20 species (Snead, 1972). The major families are Avicenniaceae, Rhizophoraceae, Combretaceae, Meliacae, Myrsinaceae, Plumbaginaceae, and Sonneratiaceae (Waisel, 1972).

The plants become zoned, from sea to land, depending on the environmental factors discussed above. However, toward their latitudinal limits, species drop out, and only one or two form an association. *Avicennia* sp. is usually the pioneer plant and is found at the poleward limits, owing to its ability to tolerate high salinity and high sand content in the substrate (Vann, 1959; Macnae, 1968). *Rhizophora* and *Sonneratia* prefer substrates of organic muck, and are found in the seaward margins of the mangrove community in most areas of Latin America and northern Australia, respectively. Plants such as *Laguncularia, Bruguiera, Ceriops,* and *Conocarpus* occur behind the pioneer plants, and thrive best in clay soil, lower saline environments, and fast-drainage areas.

Mangrove formations are important to both terrestrial and marine life by providing habitat in the canopy, roots, mud, and water channels (Macnae, 1968). Most of the wildlife is not associated with mangroves but comes from neighboring terrestrial communities. Resident species of mangroves are those on the prop roots and in the mud. Insects and birds are most numerous while amphibians and reptiles (frogs, snakes, crocodiles, and/or water monitors) are distributed among all mangroves.

Birds, mammals, and insects that come mainly from adjacent forest can be found in the tree canopies. The prop roots usually harbor a number of sessile invertebrates, including barnacles, oysters, tunicates, mussels, and sepulid worms (Tait and DeSanto, 1975). However, mangal formations in areas of large tidal ranges, such as along Pacific water, support fewer animals because the roots are left dry for extended periods of time during low tide. The mud environment contains burrowing crabs, mud skippers, mud lobsters, prawns, snails, and bivalves. Water channels contain numerous fish species while wading birds can be found along the banks.

Mangrove forests are also utilized by man. The biota (fish, crustaceans, mollusks, and pneumatophores) are collected for food. A major use is for the extraction of tannin from the bark. The mangrove trees are also used for fuel and the production of charcoal. Other minor applications include lumber, pulp, and railroad ties. Some mangroves have been exploited, such as in east Africa and parts of southeastern Asia, causing destruction of mangal formations.

CHIZUKO MIZOBE
NORBERT P. PSUTY

References

Chapman, V. J., 1976. *Mangrove Vegetation.* Vaduz, Austria: J. Cramer, 447 p.
Clark, L., and Hannon, N. J., 1967. The mangrove swamps and salt marsh communities of the Sydney district. I. Vegetation, soils, and climate, *Jour. Ecology* 55, 753-771.
Macnae, W., 1963. Mangrove swamps in South Africa, *Jour. Ecology* 51, 1-25.
Macnae, W., 1968. A general account of the flora and fauna of mangrove swamps and forests in the Indo-West-Pacific region, *Advances in Marine Biology* 6, 73-270.
Putnam, W. C.; Axelrod, D. I.; Bailey, H. P.; and McGill, J. T., 1960. *Natural Coastal Environments of the World.* Los Angeles: University of California, Office of Naval Research Geography Branch, Contract Nonr-233(06), NR 388-013, 140p.
Snead, R. E., 1972. *Atlas of World Physical Features.* New York: John Wiley & Sons, 158p.
Tait, R. V., and DeSanto, R. S., 1975. *Elements of Marine Ecology.* New York: Springer-Verlag, 327p.
Vann, J. H., 1959. *The Physical Geography of the Lower Coastal Plain of Guiana Coast.* Baton Rouge, La,: Louisiana State University Tech. Rept. No. 1, Proj. NR 388-028, 91p.
Waisel, Y., 1972. *Biology of Halophytes.* New York: Academic Press, 395p.
West, R. C., 1956. Mangrove swamps of the Pacific coast of Columbia, *Assoc. Am. Geographers Annals* 46, 98-121.
West, R. C., 1977. Tidal salt marsh and mangal formation of Middle and South America, *in* V. J. Chapman, ed., *Wet Coastal Ecosystems.* Amsterdam: Elsevier, 193-213.

Cross-references: *Africa, Coastal Ecology; Asia, Eastern, Coastal Ecology; Australia, Coastal Ecology; Biotic Zonation; Central and South America, Coastal Ecology; Coastal Flora; Coral Reef Coasts; New Zealand, Coastal Ecology; Pacific Islands, Coastal Ecology.*

MARIGRAM AND MARIGRAPH

A *marigram* is a record of the sea level; since sea level is not constant, the marigram is generally fluctuating. Not only tides but also seiches, forced waves, and long-term fluctuations (stagional or secular) are recorded. Old marigrams were paper strips, with an inked curve; modern marigrams are frequently magnetic or perforated tapes containing a digital record. Paper-strip marigrams could be directly read; taped marigrams must be used with a computer.

A *marigraph* is a sea-level measuring instrument. The simplest marigraph is a graduated rod, involving a reading by sight only. More sophisticated instruments are recorders, based on various principles (Bruns, 1958).

There are floating marigraphs, pressurometers, ultrasonic (bathymetric ultrasonic survey in reverse, from the stable bottom to the variable surface). At great depths, differential pressurometers are used, in which the constant pressure disappears and only the fluctuation remains. At the present time, surveys by satellite are also able to record some level variations (i.e., the largest ones).

Marigraphs are used for both oceanographic and geodetic purposes. In oceanography, even the shortest level variations are interesting. (A special kind of marigraph is the *wave meter,* a device with great sensitivity for short periods and maximal recording speed.) By comparison, for geodetic purposes, short or middle periodicities are not interesting. The marigraph is filtered in such a way as to record only the long periods.

FERRUCCIO MOSETTI

Reference

Bruns, E., 1958. *Ozeanologie.* Berlin: Deutscher Verlag der Wissenschaften, 365p.

Cross-references: *Coastal Engineering; Coastal Morphology, Research Methods; Protection of Coasts; Sea Level Changes; Sea Level Curves; Seiches; Tide Gauge; Tides; Waves.*

MARIN GRÄNS—See HIGHEST COASTLINE

MARINE ABRASION—See WAVE EROSION

MARINE-BUILT TERRACE—See WAVE-BUILT TERRACE

MARINE-CUT PLATFORM—See WAVE-CUT PLATFORM

MARINE-DEPOSITION COASTS

Marine-deposition coasts are those formed by accumulation of sediments by wave action. Classically, F. P. Gulliver distinguished between coasts of *initial form* and *subsequent form*. More recently, this viewpoint was used by F. P. Shepard (1948), distinguishing between *primary* and *secondary coasts*. The subsequent coasts of Gulliver (1899) and the secondary of Shepard are formed by wave action, the marine-deposition coasts being a subgroup of the secondary coasts.

Marine-deposition coasts are accompanied by the following principal elements: *beach ridges, bars, spits, lagoons, limans,* and *tombolos*. Many beaches, bars, tombolos, and forelands are similar to various beach ridges built by the waves and sometimes modified by winds. All of these forms consist of sand, gravel, shell detritus, or a mixture of them. They are interconnected, with respect to their genesis: the submarine continuation of the beach ridges is the bar; spits appear when the beach ridge is projected into the sea in subaerial form; beach ridges may contain the lagoons. Shore drift is the phylogenetic relationship between these forms, whose principal element is the beach ridge.

According to the Zenkovich and Leontiev (1964) classification, marine-deposition coasts belong to the group of shorelines formed by wave action. They are distributed in two sub-groups: those in the process of being straightened, and those that have been straightened. In the first sub-group family is the genus *abrasional accumulative bay* (the *bayhead beaches* of Johnson [1919]). In the second sub-group are the following: *straightened abrasional accumulative, lagoon and liman,* and *shorelines of marine and alluvial marine plains*.

The expression marine-deposition coast may be used to define a coastal environment, but is not a genetic expression. Sometimes it is difficult to decide if a coast is erosional or depositional. However, coasts where the elements named here predominate are generally depositional. Likewise, the expression has a significance in time. A coast may be *depositional* in the paleogeographic sense but may actually be eroding now, and vice versa.

JOSÉ F. ARAYA-VERGARA

References

Gulliver, F. P., 1899. Shoreline topography, *Am. Acad. Arts Sci. Proc.* **34**, 151-258.
Johnson, D. W., 1919. *Shore Processes and Shoreline Development*. New York: John Wiley & Sons, 584p.
Shepard, F. P., 1948. *Submarine Geology*. New York: Harper, 348p.
Zenkovich, V. P., and Leontiev, O. K., 1964. Physical-geographic atlas of the world, *Soviet Geog.* **6**, 1-403.

Cross-references: *Bars; Classification; Coastal Accretion Forms; Geographic Terminology; Lagoon and Lagoonal Coasts; Limans and Liman Coasts; Major Beach Features; Spits.*

MARINE-EROSION COASTS

Marine erosion coasts are those whose forms are products of marine erosion. These coasts were called *subsequent* by Gulliver (Johnson, 1919) and *secondary* by Shepard (1948). A. Penck (1894) distinguished between *consequent cliffed coasts* (*konsequente Steilküste*) and *consequent plain coasts* (*konsequente Flachküste*), the cliffed coast being the most typically erosional.

The principal elements of a marine-erosion coast are the *cliff* and the *abrasion bench*. Sometimes the cliffs are smooth, but frequently they occur in an irregular coast, called a *crenulate coast* by Johnson (1919), with small bays, points, capes, and promontories. Various features are characteristic of the cliff and wave-cut bench: *pinnacles, chimneys or stacks, arches,* and *sea caves*. All these forms have genetic relationships. Pinnacles, stacks, and arches are residual parts of the rock body from which the cliff and the abrasional bench are derived.

According to the classification of Zenkovich and Leontiev (1964) marine-erosion coasts belong to the *group formed by wave action*, each with one sub-group; *in process of being straightened*, with subgroup: *abrasional bay shoreline; straightened,* with sub-group: *straightened abrasional;* and *dissected*, with sub-group: *abrasional bay*.

Classic as well as modern authors commonly dispense with the expression *erosion coast,* because this is not a taxonomic expression. Only small coastal fragments may be called erosional or depositional. Commonly the coastal sub-groups are complicated combinations of both phenomena; even more so if sea level changes are considered. For example, in Constitución (south central Chile), there is a coast with typical abrasional elements (pyramids, arches, and stacks), but they resulted from a sea level higher than the present. Today deposition seems to be the essential process. On many coasts, the common foreshore beaches next to the cliffs complicate the phenomenon.

JOSÉ F. ARAYA-VERGARA

References

Johnson, D. W., 1919. *Shore Processes and Shoreline Development.* New York: John Wiley & Sons, 584p.
Penck, A., 1894. *Morphologie der Erdoberfläche.* Stuttgart: Engelhorm, 696p.
Shepard, F. P., 1948. *Submarine Geology.* New York: Harper, 348p.
Zenkovich, V. P., and Leontiev, O. K., 1964. Physical-geographic atlas of the world, *Soviet Geog.* 6, 1–403.

Cross-references: *Cliff Erosion; Coastal Erosion, Formations; Geographic Terminology; Sea Caves; Sea Cliffs; Wave-Cut Bench; Wave Erosion.*

MARINE ERRATICS

A *marine* or ice-borne *erratic* is a boulder of glacial origin that has been transported by floating ice on the ocean. The floating ice is commonly carried by currents, winds, and tides to ground on or near the coastline, usually in the spring, and to melt there, depositing its load of boulders and finer glacial debris. The erratic boulder is usually recognized first by its anomalous size and exotic nature: it is of a rock type that is not seen commonly cropping out (in situ) on the present-day coast. Sometimes it bears facets and striations characteristic of ice dynamics, but commonly this surface texture is abraded by nearshore wave action and probably less than 1% of the erratics carry distinctive glacial *signatures*. The rock itself may also contain distinctive indicators of its distant origin: a unique lithology or a noticeable assemblage of fossils. The American Geological Institute's *Glossary of Geology* (Bates and Jackson, 1980) includes plants and animals in the definition of erratic transport agents. However, inasmuch as *erratics* normally imply glacial transport, it might be better to identify biologically transported clasts with a non-genetic term such as *dropstone, dropped block, rafted block* (Pettijohn, 1975; Conybeare and Crook, 1968). Biological agencies may range from floating tree-trunks, seaweeds such as kelp, or marine animals—e.g., in the form of sea lion gastroliths (Emery, 1963). The term *glaciel* (in French) has been proposed for all sea-ice transported debris but the term has found no acceptance in Anglo-Saxon circles (Hamelin and Clibbon, 1962).

Marine erratics may be transported by one or more (because of recycling) of several different types of floating ice (Fairbridge, 1966): (a) icebergs of glacier origin (the erratics have then been glacier transported from continental or island interiors to the coast and then, after calving, floated to the point of grounding); (b) icebergs of shelf ice origin (in this case, the boulders are incorporated in the shelf ice by freezing on to its base); (c) *ice-foot floes* (freezing along the shoreline incorporates fragments of bedrock, beach gravels, sands and recycled erratics into the ice foot, which, with rise and fall of the tide or changing winds and currents, tends to break off and drift along the coast); (d) river-borne ice floes (erratics and other debris are incorporated by freezing upstream and carried to the ocean by river currents and there redistributed by marine agencies).

By these multiple transport media, a great variety of exotic littoral debris can be distributed along a coast. Multiple recycling may occur. An erratic block left in a mountain valley after an earlier glaciation can thus become incorporated and carried downstream by river ice; it can then be melted, to be later reincorporated in an ice-foot floe (Dionne, 1973).

Historical Observations

River-borne erratics have been noted in the literature since the early part of the past century. The power of freezing water to dislodge bedrock was first described in New England by Adams (1825) and by Wood (1825). Lyell (1854) mentioned a case of recycling reported on the River Niemen (in Lithuania), where an erratic of Finnish granite was being carried downstream to the Baltic. The role of floating ice in the tidal estuary of the St. Lawrence was first studied in the field by Lyell (1845). Recent analyses by Dionne (1973) have shown that more than 90% of the erratics on the south shore of this estuary (an area of Ordovician limestones) are in fact Precambrian crystalline rocks from the shield area to the north that have been transported partly by continental glacier ice and partly by the ice-foot floes.

Iceberg erratics have also been repeatedly recorded since about the turn of the nineteenth century, particularly by whaling captains whose hunt took them into high latitudes of both hemispheres (Tarr, 1897; Sverdrup, 1931). It was the sighting of granite erratics on icebergs from Antarctica that suggested to the early explorers that there was a great southern continent there, although as yet it was undiscovered (granite is not formed in oceanic islands). The analysis of large volumes of erratics sampled along the coasts of Antarctica by the various expeditions led by Sir Douglas Mawson demonstrated the existence of an immense and varied Precambrian basement for much of East Antarctica. Under West Antarctica, in contrast, the rocks are mainly Paleozoic and younger—again a fact also recognized first from studies of the coastal erratics.

Floating ice and ice-borne erratics played an interesting and curiously misleading role in the early studies of glaciation. The discoveries of large erratics, especially in southern Scandinavia and across the North German Plain attracted the attention of the early geologists; some of the blocks (called *Findlinge*—i.e., lost or abandoned children) were of immense size, 5–10 m and more length, and thus defying biological, human, or stream transport in low relief terrain (Charlesworth, 1957). In the first decades of the past century, the absolute truth of the Flood of Noah was not questioned, but the evidence of an ice age was becoming apparent from the discoveries of erratic-bearing tills associated with layers containing fossil reindeer, mammoth, and the like. An ingenious theory was developed that could explain both phenomena at the same time: the *Drift Theory* (described in the 6th edition of Lyell, 1840), which claimed a former high sea level condition due to the Flood, on which there was a great deal of floating ice—"drift ice"—that carried with it erratics. As the Flood subsided, the ice grounded and melted, leaving the erratics behind. Alas, the beautiful theory was a failure, although the term *drift* has remained in the literature of English and all other European languages, to mean any deposit related directly to glaciation (till, erratics, outwash sands, and so on).

Erosional Effects of Marine Erratics

Ice-borne erratics in the ice-foot floes are observed to produce plow marks, and a large variety of scour and other disturbances (e.g., recycled Precambrian erratics working on the soft estuarine muds of the St. Lawrence: Dionne, 1968). On hard-rock shores, the role of the ice is more often that of plucking and generating new erratics, although preexisting erratics carried by the floating ice may have a sandpapering effect, especially where there is a marked difference in hardness between the erratic and the bedrock.

One of the most universal features of coasts in high latitudes is the *strandflat* (so named by Reusch, 1894). In a detailed study by Nansen (1922), a preliminary rock-loosening role was attributed to subaerial frost action, following which sea ice would freeze onto the debris and transport it away. In this way broad littoral flats are cut into hard bedrock. The strandflats are not by any means limited to contemporary sea ice action, but in part seem to go back to the early Quaternary times, having been repeatedly deformed and attacked during the successive cold cycles. It has been submitted by the writer (Fairbridge, 1971, 1976) that the widespread development of so-called *wave-cut rock platforms* in the middle to high latitudes and their limited and small development in the lower latitudes suggest an ice-foot erosional history during earlier glacial stages.

Marine Erratics of the Glacial Stages

The initiation of bottom sampling in connection with cable laying in the North Atlantic during the last century led to the discovery of veritable submarine moraines of erratics dumped near the edge of the continental shelf (Peach, 1912), which suggests the grounding of seaborne ice (glaciers and floes) during the last glaciation when sea level dropped to about 135 m.

From the beginning of the Quaternary to the present, there appears to have been a progressive drop of sea level, modulated by the relatively short-term and rapid oscillations of the glacial stages (Zeuner, 1959; Fairbridge, 1961). In this 2-million-year period, the glacial oscillations (amplitudes of about 100 m) have occurred at intervals of around 100,000 and 40,000 years, suggestive of the well-known "Milankovitch" planetary orbital cycles. The progressive secular drop of 100–200 m is probably related to the slowing and cooling phase of a long-term plate tectonic cycle. Whatever the cause, there is evidence of warm interglacial sea levels up to 200 m higher in the early part of the Quaternary. If the interglacial levels were so high, then the corresponding glacial stage levels should have been much lower but still above present sea level. Large marine erratics have in fact been discovered at elevations of 3–10 m above present sea level in a number of places well beyond the suspected limits of the Pleistocene ice sheets. For example, at Brighton in the south of England erratics were found that underlie a Tyrrhenian raised beach containing warm- to subtropical-type shells (Mantell, 1822). Later, many other examples were reported (Reid, 1892; Hodgson, 1964). Similarly, in northern France, which was never glaciated, erratics at various levels have been found from the Pas de Calais to Brittany (Velain, 1866; Dubois, 1923; de Heinzelin, 1964). Here too they have been found beneath raised beaches with warm interglacial shells, and in turn these were capped by multiple layers of periglacial solifluction, colluvium, and loess, proving their origin well before the last glaciation. The writer interprets these erratics as having been transported by sea ice during the early Pleistocene stages when the lowest glacial sea level stands were in fact higher than present sea level.

Problems

Charles Darwin, who had personally observed ice-borne erratics off Tierra del Fuego while on the voyage of the *Beagle*, remarked that if one discovered erratic boulders on hillsides thousands of feet above sea level, no amount of sea level or crustal change could be reasonably assumed to interpret them as having been carried by marine drift ice (Darwin, 1846). Only the glacial theory would explain their presence. However, between about 300 m and present sea level, one may encounter erratics that have been glacier-borne, others that came by means of river ice, and still others carried by sea ice. How does one differentiate them? The recycling problem has already been noted. The answer is by the associated evidence—raised beach shells, fluvial sands, nonmarine till. The physical form of littoral rock platforms is exceedingly distinctive, and if the erratics are reposing thereon, the evidence is reasonably presumptive.

RHODES W. FAIRBRIDGE

References

Adams, J., 1825. Remarks on the movement of rocks by the expansive power of freezing water, *Am. Jour. Sci.* 9, 136-143.

Bates, R. L., and Jackson, J. A., eds., 1980. *Glossary of Geology*. Falls Church, Va.: American Geological Institute, 749p.

Charlesworth, J. K., 1957. *The Quaternary Era*, 2 vols. London: Edward Arnold, 1700p.

Conybeare, C. E. B., and Crook, K. A. W., 1968. *Manual of Sedimentary Structures*. Canberra, Aust.: Bureau of Mineral Resources, Geology, and Geophysics, 327p.

Darwin, C., 1846. *Geological Observations on South America*. London: Smith, Elder and Co., 279p.

de Heinzelin, J., 1964. Cailloutis de Wissant, capture de Marquise, et percée de Warcove, *Soc. Belge Géologie Bull.* 73, 146-161.

Dionne, J.-C., 1968. Morphologie et sedimentologie glacielles, littoral sud du Saint-Laurent, *Zeitschr. Geomorphologie*, 7(suppl.), 56-84

Dionne, J.-C., 1973. La notion de pied de glace (Icefoot) en particulier dans l'estuaire du Saint-Laurent, *Cahiers de Géographie de Québec* 17, 221-249.

Dubois, G., 1923. Repartition et origine des galets dans les formations quaternaires marines du Nord de la France, *Soc. Géol. Nord Annales 48*, 188p.

Emery, K. O., 1963. Organic transportation of marine sediments, in M. N. Hill, ed., *The Sea*, vol. 3. New York: Wiley Interscience, 776-793.

Fairbridge, R. W., 1961. Eustatic changes in sea level, in *Physics and Chemistry of the Earth*, vol. 4. New York: Pergamon Press, 99-185.

Fairbridge, R. W., 1966. See ice transportation, in R. W. Fairbridge, ed., *The Encyclopedia of Oceanography*. New York: Van Nostrand Reinhold, 781-782.

Fairbridge, R. W., 1971. Quaternary shoreline problem at INQUA, 1969, *Quaternaria* 15, 1-18.

Fairbridge, R. W., 1976. Pleistocene marine shore platforms eroded by drift-ice, in *Symposium on the Geological Action of Drift Ice, Quebec City, April 1974* (unpublished manuscript).

Hodgson, J. M., 1964. The low-level Pleistocene marine sands and gravels of the West Sussex coastal plain, *Geol. Assoc. Proc.* 75, 547-561.

Lyell, C., 1840. *Principles of Geology*, 6th ed., 3 vols. London: John Murray.

Lyell, C., 1845. *Travels in North America in the Years 1841-2*, vol. 2. New York: Wiley and Putnam, 221p.

Lyell, C., 1854. *Principles of Geology*. New York: Appleton, 834p.

Mantell, G. A., 1822. *The Fossils of the South Downs or Illustrations of the Geology of Sussex*. London: L. Relfe, 337p.

Nansen, F., 1922. The strandflat and isostasy, *Norske Vidensk.-Akad. Oslo Skr., Mat.-Naturv. Kl. 11*, 2 vols., 313p.

Peach, B. N., 1912. Report on rock specimens dredged by the 'Michael Stars" in 1910, *Roy. Soc. Edinburgh Proc.* 32, 262-291.

Pettijohn, F. J., 1975. *Sedimentary Rocks*. New York: Harper and Row, 628p.

Reid, C., 1892. The Pleistocene deposits of the Sussex coast and their equivalents in other districts. *Quart. Jour. Geol. Soc. London* 48, 344-364.

Reusch, H., 1894. Strandfladen, et nyt traek i Norges geografi, *Norges Geol. Undersökelse* 14, 1-14.

Sverdrup, H. U., 1931. Drift-ice and ice-drift, *Geog. Annaler* 12, 121-141.

Tarr, R. S., 1897. The Arctic sea as a geological agent, *Am. Jour. Sci.* 4, 223-229.

Velain, C., 1886. Note sur l'existence d'une rangée de blocs erratiques sur la côte normande, *Soc. Géol. France Bull.* 3, 569-575.

Wood, J., 1825. Remarks on the moving of rocks by ice, *Am. Jour. Sci.* 9, 144-145.

Zeuner, F. E., 1959. *The Pleistocene Period*. London: Hutchinson, 447p.

Cross-references: *Glaciel; Ice along the Shore; Ice-Bordered Coasts; Ice Foot; Strandflat; Wave-Cut Platform.*

MARINE SWAMP—See PARALIC

MARINE TERRACE—See WAVE BUILT TERRACE

MARSH—See SALT MARSH

MARSH GAS

Marsh gas or methane (CH_4) is an end product of the decomposition of organic matter, and is particularly important because of its relationship to carbonate precipitation and its effect on sediment instability. Marsh gas may result directly from the fermentation of organic material or from bacterial reduction of preformed CO_2 subsequent to sulfate reduction. Abundant organic matter, low sulfate concentrations, and anaerobic conditions typically found in swamps, bogs, fresh and saline marshes, and at depth in marine sediments are ideal conditions for methane production.

Methane production is dependent on the rate of microbial activity, which in turn is temperature-dependent. Consequently the highest methane production rates in salt marshes are generally found during the summer season (Atkinson and Hall, 1976). The major portion of methane in sediment exists as bubbles, and actual concentrations are dependent on the production and migration of the gas into the adjacent water. This migration is demonstrated by the fact that the adjacent waters have been shown to have methane concentrations 200 to 300 times that of normal sea water.

In environments where sulfate is abundant, as in saline marshes and marine sediments, three biochemical zones are apparent in the sediment. There is an upper aerobic zone where oxygen is present and two lower anaerobic zones. After oxygen has been depleted, sulfate reduction becomes the dominant process. When sulfate becomes low or is absent, carbonate reduction and methane formation begin. These zones are characterized by less efficient modes of respiratory metabolism, and represent geochemical consequences of species-induced environmental changes (Claypool and Kaplan, 1974). The depth of each zone is dependent on the abundance of oxygen, sulfate, and carbonate and the rate of metabolic processes. In general, where sulfate concentrations are low, methane occurs near the surface of the sediment. It appears that the hydrogen needed by the methane bacteria as an energy source is not available for CO_2 reduction in the presence of dissolved sulfate.

Sulfate reduction may not always provide sufficient CO_2 for the quantities of methane produced in selected areas of the Louisiana marsh (Whelan, 1974). Additional CO_2 is derived from the decomposition of organic matter. This is evident by the kinetic isotope effect associated with CO_2 reduction: $CO_2 + 4H_2 \rightarrow CH_4 + 2H_2O$. The degradation of organic matter results in CO_2 with $\delta C^{13}/C^{12}$ ratios equal to those of the original organic matter (i.e., enriched in C^{12}). The δC^{13} values* of the resulting methane are also isotopically light and parallel the trends in the δC^{13} of CO_2. In most cases, methane-producing environments, such as the Louisiana marshes, are characterized by isotopically light CO_2, indicative of its biogenic sources. This is because of the rapid decomposition of organic matter and the replenishment of C^{12} as CO_2 is being used for methane production.

*$\delta C^{13}(‰) = 1000 \times \overline{\delta C^{13}/C^{12} \text{ of sample} - \delta C^{13}/C^{12} \text{ of standard}}/\delta C^{13}/C^{12}$ of standard

In several studies, however, the kinetic isotope effect has yielded CO_2 enriched in δC^{13} (see references in Claypool and Kaplan, 1974). The concentration of δC^{13} initially increases with depth, then decreases. This initial increase is similar to the behavior of δC^{13} in a closed system of CO_2 reduction (i.e., when a fixed amount of dissolved carbonate can be converted to methane by the Rayleigh distillation process). In the production of methane, C^{12} is used at a greater rate than the δC^{13}; consequently methane produced from the reduction of CO_2 subsequent to sulfate reduction becomes enriched in C^{12} relative to CO_2. The residual unreduced CO_2 becomes enriched in δC^{13}. Continued reduction of the residual CO_2 results in a progressive increase in the δC^{13} of the methane. The point of reversal (i.e., the decrease in δC^{13} values) in nature occurs where the decomposition of organic matter and the replenishment of C^{12} exceeds the use of CO_2 utilization. CO_2 may be removed from the sediment by methane production or by carbonate precipitation. Where the quantities of biogenic CO_2 are low, methane production results in an increase in pH, which favors carbonate precipitation.

LINDA C. NEWBY

References

Atkinson, L. P., and Hall, J. R., 1976. Methane distribution and production in the Georgia salt marsh, *Estuarine and Coastal Marine Sci.* 4, 677-686.

Claypool, G. E., and Kaplan, I. R., 1974. The origin and distribution of methane in marine sediment, in I. R. Kaplan, ed., *Natural Gases in Marine Sediments.* New York: Plenum Press, 99-139.

Whelan, T., 1974. Methane, carbon dioxide, and dissolved sulfate from interstitial water of coastal marsh sediments, *Estuarine and Coastal Marine Sci.* 2, 407-415.

Cross-references: *Coastal Flora; Halophytes; Salt Marsh; Salt Marsh Coasts; Soils; Vegetation Coast.*

MATHEMATICAL MODELS

The general meaning of *mathematical model* is the mathematical approach to a physical (or biological, or other) phenomenon. For example, the wave equation $y = a \sin(\omega t + \phi)$ can describe the profile of a wave; the exponential equation $n = n \exp(-kt)$ can describe radioactive disintegration and the like. These equations are simple mathematical models. In oceanographic sciences, the concept is actually more restrictive, so that *mathematical models* are often considered only the numerical solutions of some differential equations—i.e., principally the *hydrodynamical equation, continuity equation,* and *diffusion equation*. Such a restriction is not always acceptable. In fact the mathematical model should be in any case a mathematical device for the fitting of oceanographic phenomena. In this case, other relations represent a model; for example, an empirical relation between wind and wave, or between the sea rise (during a storm surge) and the blowing wind or the atmospheric pressure gradient. According to the most frequent meaning of mathematical models—that is, numerical solutions of differential equations—we examine some application of numerical solutions.

Equations (or systems of equations) such as the hydrodynamical ones, give by integration the solution of any motion problem; the velocities of currents or areal level modifications are principally furnished. Analytical solution of hydrodynamical differential equations is possible only for unbounded basins (being similar to infinitely extended seas) or for simple geometrical forms of bounded basins (for example, a rectangular basin with constant depth). In this case, although simple geometrical forms do not exist in the real ocean, analytical solutions are only a first approximation of the real condition of the motion. In all other cases (water vertically stratified, limited basin, and so on), only numerical solutions are possible. The modern use of computers makes it possible to solve practically every problem (Hansen, 1962).

The solutions are possible by converting the differential equations into a finite difference among the various terms; initial conditions (at the time $t = 0$) and boundary conditions (at the border of the basin or into the water if this is not homogeneous) must be fixed. X and t axes in monodimensional problems, or x, y, and t axes in bidimensional problems (obviously x, y, z, t, for three dimensions) are divided into a mesh grid having given dimensions (often interconnected for reasons of calculus stability.) A program must be performed, so that any difference (at the knots of the grid or in other points) in the investigated function must be successively added, starting from the initial value. The sums are performed along the t and x direction (or other). If the sum in t direction for each point x (or x, y, z) is truncated at a certain time, t^*, it is possible to give the distribution of the investigated function in the space (for the given t^*); analogously for the spatial summations.

At this point it is convenient to give some of the principal criteria of translation between differential and finite difference equations in oceanography. On the basis of these examples, it is not difficult to extend for analogy the calculation for other conversions in more complicated cases or, if other terms are added, in the equations. Our examples include bidimensional *continuity equations,* monodimensional *hydrodynamic equations,* in the presence of forcing terms, monodimensional and bidimensional *equations of diffusion* in a sea at rest. The solutions are applicable to important studies of currents, seiches, and tides with regard to hydrodynamic equations.

For mixing of waters or pollutants, regarding the diffusion equation, the continuity is a condition that must be added in various problems for finding the true physical validity (O'Brien and Hurlburt, 1971). The *salinity balance* equation often must also be added.

General Criteria for Bidimensional (or Monodimensional) Problems

The *volume x, y, t* is limited by

$$X_o \leqslant x \leqslant X$$
$$Y_o \leqslant y \leqslant Y$$
$$T_o \leqslant t \leqslant T$$

and divided in intervals of the grid h_x, h_y, h_t. A general interval is

$$x_i = x_o + i h_x \quad (1)$$
$$y_j = y_o + j h_y$$
$$t_k = t_o + k h_t$$

We must indicate the step of time and the step of space for defining each difference; time can be assumed each $2h_t$; X each $2h_x$, and so on.

Continuity Bidemensional Equation

If u and w were the velocity components along x and z, the bidimensional equation of continuity can generally be written as

$$\frac{\partial u}{\partial x} + \frac{\partial w}{\partial y} = 0 \quad (2)$$

Before the transition into finite difference, the general equation (2), after some calculations, can become

$$\frac{\partial \eta}{\partial t} = -\frac{1}{b} \frac{\partial (A \bar{u})}{\partial x} \quad (3)$$

with

$$\bar{u} = \frac{1}{A} \int_A u \, dy \, dz,$$

where η is the vertical displacement of level, A is a cross section of the basin, and b is the width of the basin in correspondence with A.

Into finite difference $\partial\eta/\partial t$ becomes $\Delta\eta/\Delta t$ and

$$\frac{\Delta\eta}{\Delta t} = \frac{\eta^{k-1} - \eta^{k+1}}{2h_t}$$

where k is the time index, according to equation (1).

If S represents the superficial area of the basin (orthogonal to A), Δx could be written as

$$\Delta x = \frac{\Delta S_i + \Delta S_{i-1}}{b_i}$$

(the basin is divided into bands ΔS). So $\Delta/\Delta x$ is translated into

$$\frac{b_i}{\Delta S_i + \Delta S_{i-1}} \Delta.$$

$\Delta(A\bar{u})$ is considered at time k (centered between $k-1$ and $k+1$), but for different points along the x axis (derivation with respect to x); these points are labeled with index $i+1$ and $i-1$ (A does not vary with time). Then $1/b_i \, \Delta(A\bar{u})/\Delta x$ becomes

$$\frac{1}{b_i} \frac{b_i}{\Delta S_i + \Delta S_{i-1}} (A_{i+1}\bar{u}_{i+1}^k - A_{i-1}\bar{u}_{i-1}^k)$$

and equation (3) is translated into

$$\frac{\eta^{k-1} - \eta^{k+1}}{2h_t} = -\frac{1}{\Delta S_i + \Delta S_{i-1}} \quad (4)$$

$$(A_{i+1}\bar{u}_{i+1}^k - A_{i-1}\bar{u}_{i-1}^k).$$

The successive sums with respect to t and x indicated by equation (4) are performed by means of computers.

The hydrodynamic equation, which describes the motion in a channel (approaching a narrow sea), is a useful equation in monodimensional problems. If we consider the action of an *external force F*, the equation is

$$\frac{\partial\bar{u}}{\partial t} = -g\frac{\partial\eta}{\partial x} + F \quad (5)$$

where, as in equation (3), η is the displacement of level, $\bar{u}$ is the mean speed of the current, x is the axis of the basin, t is time, and g is the acceleration of gravity.

In some cases, F does not represent a single force but a sum of each external force acting in the problem. The term $-g(\partial\eta/\partial x)$ corresponds to a hydrostatic force (these forces are always pointed to the mass unit). The term $\partial\bar{u}/\partial t$ is the resultant force (acceleration).

If we introduce the air pressure gradient over the basin (acting on the water motion), or the friction of the wind (likewise acting in the motion at the surface), or, also, the bottom friction, the term F can be specialized. In this case F consists of the sum of pressure gradient $(-1/\rho \, \partial p_a/\partial x)$, wind friction $(b/\rho A \, T_\eta)$, and bottom friction $(-b/\rho A \, T_H)$; b and A are the dimension of the basin (see equation 2), ρ is the water density, p_a the air pressure. The friction terms T_η and T_H are related, respectively, to wind (U) and current (u) speed. In practical work the following empirical relations are commonly used: $T_\eta = T = 3{,}2 \cdot 10^{-6} U^2$ and $T_H = R\bar{u} = 2{,}5 \cdot 10^{-3} \bar{u}$.

With these conditions equation (5) becomes more complicated as

$$\frac{\partial\bar{u}}{\partial t} = -g\frac{\partial\eta}{\partial x} - \frac{1}{\rho}\frac{\partial p_a}{\partial x} - \frac{b}{\rho A}R\bar{u} + \frac{b}{\rho A}T \quad (6)$$

showing an analogy to equations (3) and (4); the formula, equation (5), is translated into finite difference, as

$$\bar{u}_i^{k+1} - \bar{u}_i^{k-1} = 2h_x \left[\frac{1}{\rho} \frac{b_i}{\Delta S_i + \Delta S_{i-1}} \right.$$

$$\left(p_{a_{i+1}}^k - p_{a_{i-1}}^k \right) - \frac{gb_i}{\Delta S_i + \Delta S_{i-1}}$$

$$(\eta_{i+1}^k - \eta_{i-1}^k) - \frac{b_i}{\rho A_i} R\bar{u}_i^k$$

$$\left. + \frac{b_i}{\rho A_i} T_i^k \right] \quad (7)$$

With this relation the calculations are introduced into the computer.

If the problem is bidimensional we need to use two equations; these are converted to space and time by means of special grids. Numerical stability must be reached by introducing smoothing factors.

According to Shamir and Harleman (1967), with D representing diffusivity and c, the concentration of the diffusing substance, a simple diffusion equation in the direction x can be given as

$$D\frac{\partial^2 c}{\partial x^2} = \frac{\partial c}{t} \quad (8)$$

Obviously in the finite difference, this becomes

$$\frac{D}{h_x^2}(c_{i+1}^k + c_{i-1}^k - 2c_i^k) = \frac{1}{h_t}(c_i^{k+1} - c_i^k) \quad (9)$$

The bidimensional diffusion equation could be

$$D\left(\frac{\partial^2 c}{\partial x^2} + \frac{\partial^2 c}{\partial y^2}\right) = \frac{\partial c}{\partial t} \quad (10)$$

This is translated, by means of a convenient assumption of the grid, as

$$\frac{D}{h_x^2}(c_{i+1,j}^k + c_{i-1,j}^k - 2c_{i,j}^k) + \frac{D}{h^2 y}(c_{i,j+1}^k$$

$$+ c_{i,j-1}^k - 2c_{i,j}^k) = \frac{1}{h_t}(c_{i,j}^{k+1} - c_{i,j}^k) \quad (11)$$

Numerical stability is assured if $D(1/h_x^2 + 1/h_y^2) \leq 1/2h_t$. In the particular case where $h_x = h_y = h$ and with $h_t = Dmh^2$, equation (11) becomes

$$c_{i,j}^{k+1} = m(c_{i+1,j}^k + c_{i-1,j}^k + c_{i,j-1}^k + c_{i,j+1}^k)$$

$$= (1 - hm)c_{i,j}^k \quad (12)$$

that is convenient for resolution, in each point and time, by means of computers.

FERRUCCIO MOSETTI

References

Hansen, W., 1962. The reproduction of the motion in the sea by means of hydrodynamical-numerical methods, *Hamburg Univ. Inst. Meereskunde Mitt.* 25, 1-57.
O'Brien, J. J., and Hurlburt, H. E., 1972. A numerical model of coastal upwelling, *Jour. Phys. Oceanog.* 2, 19-26.
Shamir, U. Y., and Harleman, D. R. F., 1967. Numerical solutions for dispersion in porous mediums, *Water Res. Research* 3, 557-581.

Cross-references: *Coastal Morphology, Research Methods; Computer Applications; Computer Simulation; Scale Models.*

MEAN SEA LEVEL

To provide a datum for heights shown on topographical maps, a reference point for many coastal investigations and engineering works, and for a variety of other purposes, knowledge of *mean sea level* is required (Doodson, 1960). The level of the sea along a coast varies over time. There are the regular fluctuations of marine *tides* (q.v.) with, most often, a semidiurnal periodicity, on which are superimposed fortnightly and monthly terms; there are fluctuations of an annual nature related to changes of wind and atmospheric pressure; and there are local and irregular variations caused by storms (onshore winds will temporarily raise local sea level), sea surges, or flooding in river estuaries. Additionally, there may be longer-term changes in the relative levels of land and sea, of eustatic or isostatic origin, acting over thousands of years. Mean sea level is determined by a *tide gauge*, in which the movement of a float is either mechanically recorded by pencil on a revolving drum or digitally registered on magnetic tape. Analysis of the curve so recorded can then yield mean sea level over any desired period of time. An approximate value will be obtained from hourly water heights observed over one month, but such a value may be misleading because of abnormal weather conditions and because it ignores the annual tide. If a high degree of accuracy is needed, several years' observations must be obtained. The siting of a tide gauge is an important consideration: both estuaries and highly exposed coasts should be avoided, and the tide gauge must be constructed on the most stable foundations possible, ideally hard bedrock. Long-term tide-gauge records can provide evidence of changes in relative land and sea level (Fairbridge, 1960; Rossiter, 1967): mean sea level computed annually at a given place may reveal a consistent positive or negative shift. For example, the tide gauge at Felixstowe in eastern England showed a relative rise of sea level averaging 1.7 mm/year between 1918 and 1950; the reverse tendency has been recorded at Aberdeen, where relative sea level fell at an average rate of 0.5 mm/year between 1862 and 1913 (Valentin, 1952, 1953). The longest record of mean sea level in Britain is that given by the Newlyn tide gauge in Cornwall that commenced operation in 1916 and provides the present datum for the Ordnance Survey. Much longer, however, is the record of the Amsterdam tide gauge that operated from 1682 to 1930, when it was interrupted by closure of the Zuider Zee. Until 1930 the record showed subsidence of Amsterdam relative to sea level averaging 0.7 mm/year. Other long records in Europe include those from the Baltic, where approximate measurements of water level span three centuries; more precise observations have been kept since 1839. North American records date mostly from this century (Gutenberg, 1941).

CLIFFORD EMBLETON

References

Doodson, A. T., 1960. Mean sea level and geodesy, *Bull. géodésique* **55**, 69-77.
Fairbridge, R. W., 1960. The changing level of the sea, *Sci. Am.* **202**, 70-79.
Gutenberg, B., 1941. Changes in sea level, post-glacial uplift, and mobility of the earth's interior, *Geol. Soc. America Bull.* **52**, 721-772.
Rossiter, J. R., 1967. An analysis of annual sea-level variations in European waters, *Royal Astron. Soc. Geophys. Jour.* **12**, 259-299.
Valentin, H., 1952. *Die Küsten der Erde,* Petermanns Geog. Mitt. Erg. 246. Gotha: Justus Perthes, 118p.
Valentin, H., 1953. Present-day vertical movements of the British Isles, *Geog. Jour.* **119**, 229-305.

Cross-references: *Chart Datum; Sea Level Changes; Sea Level Curves; Tide Gauge.* Vol. III: *Mean Sea Level.*

MESOZOA

Members of this phylum represent the simplest of all multicellular animals, and are characterized by tiny, wormlike, parasitic organisms having a single layer of ciliated, epithelial cells that enclose one or more reproductive cells (Stunkard, 1954; McConnaughey, 1963). Genetic relationships are uncertain, although the Mesozoa are sometimes considered to be allied to the *flatworms*. The life cycle consists of alternate sexual and asexual generations. Two orders are recognized: the Dicemida infect the kidneys of cephalopods; the Orthonectida occur as parasites in ophiurids, nemertines, Turbellaria, pelecypods, and other marine invertebrates. Among the Dicemida, the nematogen phase asexually produces larvae that stay within the host organism. The following rhombogen stage is characterized by hermaphroditic sexual reproduction, giving rise to free-swimming larvae that leave the host, their fate still remaining unknown. The Orthonectida occur in parasitic form as multinucleate ameboid masses (*syncytia*) that multiply by fragmentation. Asexual reproduction also gives rise to free-living males and females, which sexually reproduce to yield the ciliated larvae that enters the body cavities of the new host. Once inside, the larvae give rise to new syncytia.

GEORGE MUSTOE

References

McConnaughey, B. H., 1963. The mesozoa, *in* E. C. Dougherty, ed., *The Lower Metazoa.* Berkeley and Los Angeles: University of California Press, 151-177.
Stunkard, H. W., 1954. The life history and systematic relations of the Mesozoa, *Quart. Rev. Biology* **29**, 230-244.

Cross-references: *Coastal Fauna; Mollusca; Platyhelminthes.*

METEOROLOGY—See CLIMATE, COASTAL

MICROSEISMS

The interference of two opposing wave trains of swell of equal frequency can give rise to secondary standing waves, called *microseisms.* Unlike normal wind waves, microseisms transmit their energy to the bottom via pressure fluctuations that travel as waves through the earth's crust and are recorded on seismographs. The opposing wave trains can result when waves are reflected from a steep coast or when winds are blowing in opposite directions on either side of a small, intense storm (King, 1972).

Observations have shown that microseisms are long waves, do not travel along great circle paths, and have periods equal to half that of the generating swell. Microseisms cannot propagate across geological shadow zones, and are refracted by depth and geological structures. This makes the recording and interpretation of microseisms complex, although they have been used to give warning of approaching storm swells (King, 1972).

GARRY JONES

Reference

King, C. A. M., 1972. *Beaches and Coasts.* New York: St. Martin's Press, 570p.

Cross-references: *Internal Waves; Storm Wave Environments: Swell and Its Propagation; Wave Climate; Waves.* Vol. I: *Microseisms.*

MIDDEN

Kitchen *midden* is an archeological term referring to heaps or mounds of refuse remaining after the human exploitation of a food resource. The majority of recorded middens have a coastal location and they are widely distributed throughout the world (Shackleton, 1969; Butzer, 1971). Of the coastal middens, many are formed predominantly of marine molluscan valves (hence *shell midden* and *shell mound*). This is because: molluscans have a very high ratio (20:1) of skeletal to edible parts compared with other animals; and the immobile populations are easily exploited by food gatherers throughout the year. Meighan (1969) reports that the earliest middens date from around 7000–8000 B.P., and that they have been formed ever since by hunter-gatherer societies.

Extensive studies of kitchen middens provide information not only for the archeologist on past population pressure and social, economic, and domestic habits, but also many data of relevance to the palaeoenvironmentalist, particularly on the ecology and climatology of the midden areas. Such an integrated approach is especially important if a precise chronology is available (Brothwell and Higgs, 1969). However care must be taken to establish the true origin of middenlike deposits, since natural shell and bone mounds can occur (Craig and Psuty, 1971; Carter, 1975) causing confusion and misinterpretation, particularly of archeological data.

RICHARD W. G. CARTER

References

Brothwell, D., and Higgs, E., eds., 1969. *Science in Archeology,* 2d ed. London: Thames and Hudson, 720p.

Butzer, K. W., 1971. *Environment and Archaeology.* Chicago: Aldine-Atherton, Inc., 703p.

Carter, R. W. G., 1975. The origin of the Magilligan Shell Mounds, *Irish Naturalists' Jour.* 18, 184–187.

Craig, A. K., and Psuty, N. P., 1971. Palaeoecology of shell mounds at Otuma, Peru, *Geog. Rev.* 61, 125–132.

Meighan, C. W., 1969. Molluscs as food remains in archeological sites, *in* D. Brothwell and E. Higgs, eds., *Science in Archeology,* 2d ed. London: Thames and Hudson, 415–427.

Shackleton, N. J., 1969. Marine molluscs in archeology, *in* D. Brothwell and E. Higgs, eds., *Science in Archeology,* 2d ed. London: Thames and Hudson, 407–414.

Cross-references: *Archaeology, Geological Considerations; Archaeology, Methods.*

MIDDLE AMERICA—See CENTRAL AMERICA

MID-LATITUDE COASTS

In line with the definitions given elsewhere in this volume for *high-latitude* and *low-latitude coasts,* the coasts of mid-latitudes can be thought of as those that lie to poleward of about latitudes 40°N and S and that are free of coast-fast sea ice at all times (Davies, 1980). They are limited in the Southern Hemisphere, where only the Patagonian coast fits in this category, but are particularly extensive in western Europe and northwestern North America.

Mid-latitude coasts lie in the zones of maximum wave generation under the influence of the temperate frontal systems and their associated depressions. In the Southern Hemisphere, the seas around latitude 55°S are the stormiest in the world, and storm activity varies little between summer and winter. They are a major source of swell penetrating long distances into the low-latitude zone. However, only the coasts of Patagonia lie directly within this zone. The storm areas of the temperate North Atlantic and North Pacific are sources of frequent high-energy waves in winter, but, in contrast to the Southern Hemisphere, there is considerable diminution of activity in summer. In every area, the *west*-facing coasts are stormier than the *east*-facing coasts, and the west coast of Patagonia is the stormiest coast of all.

Beaches occupy a smaller proportion of the coast than they do in low latitudes, and barrier systems are smaller and less extensive, being well developed only in lower-energy sections such as eastern United States and Japan. In higher-energy areas, beaches are rectilinear or concave in profile and have gentle gradients. They rarely carry berms of sand. Instead, the upper beach is usually composed mostly or entirely of pebbles. Pebble beaches are especially frequent in those areas that were glaciated in the Pleistocene or where periglacial processes of mass movement extended down to present sea level. Here the pebbles may be derived directly from glacial or periglacial gravels; in other instances, the frequent occurrence of big storm waves gives maximum opportunity for wave quarrying of rocky shores and for pebble manufacture by the sea itself.

On sandy shores, the relatively common occurrence of high onshore wind velocities means that the rate of deflation of sand from beaches is potentially high. The major limiting factor is sand supply, but where, as in northwestern United States, there is a continuing source of shore sand, the coastal dunes that result are massive indeed.

In barrier lagoons and estuaries, tidal plains are covered by salt marshes, more continuous and more varied floristically than elsewhere in the world. Particularly in western Europe, extensive areas of these marshes have been used for stock grazing or have been wholly or partly reclaimed for a variety of agricultural purposes.

In the mid-latitude zone as a whole, bedrock shores are more extensive than on other coasts. The notable exceptions, such as eastern United States, where sand barriers are unusually well developed, have already been mentioned. However, the *fjord* coasts of Alaska, British Columbia, Norway, Patagonia, and southern New Zealand lie at the opposite extreme. They still owe much of their form to oversteepening of slopes by glaciers in the Pleistocene, but, wherever it can occur, genuine cliffing of bedrock by waves is at a maximum in mid-latitudes. Relatively frequent coastal storms, with their associated big waves, provide optimum conditions for quarrying and abrasion by the sea and rapid removal of resulting debris. Cliffs, such as some in western Ireland claimed to be among the highest in the world, lie within this zone.

High wave energy, conducive to strong quarrying and abrasion, produces the typical sloping shore platform that is virtually universal in mid-latitudes. Horizontal platforms created by weathering and biological activity appear to be completely absent. This is the result partly of the great potential strength of the wave erosion forces, but also partly of the unfavourable environmental conditions for *water-layer weathering* and for biological action. Compared with low-latitude coasts, mid-latitude coasts owe proportionately more to physical processes and less to biological processes.

Because the early work in coastal geomorphology was carried out in the mid-latitude zone, notably in western Europe and northeastern United States, it has been thought of implicitly as the norm from which low-latitude and high-latitude coasts deviate. However, in the past thirty years or so, a great deal of work has been done on low-latitude coasts, which, in terms of extent, are considerably the longest.

J. L. DAVIES

Reference

Davies, J. L., 1980. *Geographical Variations in Coastal Development,* 2d ed. London: Longman, 212p.

Cross-references: *Barrier Islands; Beach Material; Cliff Erosion; Coastal Flora; High-Energy Coasts; High-Latitude Coasts; Low-Energy Coasts; Low-Latitude Coasts; Moderate-Energy Coasts; Wave-Cut Platform; Wave Erosion; Waves.*

MIGRATING HUMPS—See SAND WAVES and LONGSHORE SAND WAVES

MINERAL DEPOSITS

There are three principal categories of marine mineral deposits: *beach and close offshore, shallow continental shelf,* and *deep sea floor.* We are concerned here only with the former category. The term *mineral* is used here in the strict definition of a material of uniform composition, readily identifiable by color, specific gravity, and hardness. The term *deposit* is used to describe a concentration of a given mineral or a group of minerals occurring in quantities large enough to be exploited at a profit.

Beach and Close Offshore Deposits

Along the edge of every continent, where streams dump their loads of sediment in the sea, a *sorting* process is continuously going on. Surf, longshore currents, tidal scour, and wind action combine to carry away the lighter and smaller mineral fragments into deeper water, leaving the heavier and coarser-sized materials close to or on the beach. Thus the beaches of the world have become concentrations of *sand-*sized mineral grains. In special circumstances, beaches may contain valuable mineral deposits.

The principal factors that combine to make possible a valuable beach mineral deposit are: a mineral-rich hinterland, in which a prolonged period of weathering has occurred, and decomposition of rocks in the hinterland permit the liberation of valuable minerals, freeing them for stream erosion and transport; a stream drainage with sufficient hydraulic force to transport mineral fragments to the coast; a gently sloping offshore shelf; detrital minerals hard enough to endure considerable abrasion in the surf; and detrital minerals having a specific gravity higher than average.

Where these requirements are met, valuable mineral deposits have been found. Typical examples are: the diamond-bearing gravels near the mouth of the Orange River in southwest Africa; the ilmenite-rich beaches of Kerala, near the southern tip of the Indian peninsula; and the alluvial tin deposits along the coast of the Malay Peninsula.

The minerals most commonly recovered from beach and near-shore deposits are listed in Table 1. This list does not include the most common constituents of beach sand, such as: quartz; or, in the case of beaches in the vicinity of coral reefs, made up of calcium carbonate (shell or coral) fragments; or beaches along volcanic island coasts, which are often dark in color, consisting of basalt and other volcanic rock fragments.

A complete list of world occurrences of valuable beach and near-shore mineral deposits will not be attempted here. Some of the more noteworthy are discussed briefly below.

Australia. The beaches of northern New South Wales, southern Queensland, and Western Australia contribute major shares of the world supplies of ilmenite, rutile, and zircon (Lynd and Lefond, 1975).

India. The coast of Kerala was once an important supplier of ilmenite. The ilmenite content of some beaches is annually renewed by

TABLE 1. Valuable Minerals Found in Beach Deposits

Mineral	Chemical Composition	Specific Gravity	Hardness
Cassiterite	SnO_2	6.8-7.1	6.0-7.0
Chromite	$(MgFe)O(Cr,Al,Fe)_2O_3$	4.1-4.7	5.5-6.5
Columbite	$(Fe,Mn)Nb_2O_6$	5.2	6
Diamond	C	3.5	10
Gold	Au	19.3	2.5-3.0
Ilmenite	$FeTiO_3$	4.5-5.5	5.0-6.0
Magnetite	Fe_3O_4	5.2	5.5-6.5
Monazite	$(Ce,La,Y,Th)PO_4$	4.6-5.4	5.0-5.5
Platinum	Pt	14-19	4.0-4.5
Rutile	TiO_2	4.25	6.0-6.5
Scheelite	$CaWO_4$	6.12	4.5-5.0
Tantalite	$(Fe,Mn)Ta_2O_6$	7.95	6.0-6.5
Wolframite	$(Fe,Mn)WO_4$	7.1-7.5	4.0-4.5
Zircon	$ZrSiO_4$	4.6-4.7	7.5

storm waves during the monsoon season. This area is believed to have substantial reserves of ilmenite and monazite (Gillson, 1960).

Malaysia. The recovery of tin minerals from the near-shore area along the Malay Peninsula and from the beaches and coasts of nearby islands in the Java Sea has progressed from river placer deposits to the beaches and thence offshore, as the tin rich gravels are traced to drowned stream valleys, now submerged by the sea. This region is one of the world's largest sources of tin.

Sri Lanka (Formerly Ceylon). One of the world's most extensive beach deposits of ilmenite and rutile has been worked for several years along the east coast of Sri Lanka. The black beach sands contain over 70% ilmenite.

New Zealand. The west coasts of both the North and South islands of New Zealand are rich in ilmenite and magnetite. The latter is now used as a source of iron for New Zealand's first steel mill.

RICHARD J. ANDERSON

References

Gillson, J. L., 1960. *Industrial Minerals and Rocks*, 2d ed. New York: American Institute of Mining, Metallurgical, and Petroleum Engineers, 1053-1061.

Lynd, L. E., and Lefond, S. J., 1975. *Industrial Minerals and Rocks*, 4th ed. New York: American Institute of Mining, Metallurgical, and Petroleum Engineers, 1173-1180.

Cross-references: *Africa, Coastal Morphology; Asia Eastern, Coastal Morphology; Australia, Coastal Morphology; Beach Processes; India, Coastal Morphology; Mineral Deposits, Mining of.* Vol. VI: *Beach Sands.*

MINERAL DEPOSITS, MINING OF

The task of extracting valuable materials from the earth is often a most difficult and, from the standpoint of modern technology, a costly enterprise. This is because shafts must be sunk or tunnels driven, and equipment must be transported over great distances—all before mining can begin. The inherent advantage of detrital mineral deposits contained in beaches and near offshore areas is the *already mined, already crushed and ground,* and *already sized* nature of the valuable minerals.

The principal task, then, in recovering minerals of value from a beach deposit is to separate the good mineral grains from the bulk of beach material. This is accomplished by a variety of concentration techniques, including separation by gravity, by magnetic differences, by electrostatic devices, and by flotation.

The initial gathering or *mining* of the beach sand was once accomplished by hand labor. Today this is done with mechanical equipment. Since water is always close by, many beach-mining operations are carried out by floating dredges, which progressively excavate the pond in which they float and *fill in* where they have just been. The sand is picked up either by means of suction or by an endless chain of buckets. Removal of low-value sand overburden or *beach rock* precedes the dreding operation, usually carried out by various earth-moving machines.

Since a somewhat higher specific gravity is a common characteristic of the so-called *heavy* minerals, gravity concentration methods are commonly used to recover the valuable fraction in beach sands. In recent years, a wonderfully simple device, known as the *Humphrey spiral,* has been instrumental in the recovery process. Here a mixture of sand and water is poured down a spiral metal trough, equipped at the outer edges with numerous *ports* or exits. As the swirling water moves down the trough, the heavier minerals tend to move to the outer edge as centrifugal forces come into play. When the heavy minerals are *high* enough on the outer wall of the trough, they find an exit, and, *voilà,* separation has been accomplished.

Magnetic separation is carried on in machines equipped with magnetized cylinders or *rolls.* Electrostatic separation is effected by first exposing the sand mix to an electrostatic field and then catching the desirable minerals on a roll or sheet containing an opposite charge (Lynd and Lefond, 1975).

Another form of gravity separation can be made on a vibrating table, across which a thin stream of sand and water is moving. The heavier minerals are gradually shunted to one side of the table, where they fall into the designated trough. Lighter-weight material then proceeds to the waste dump.

In some cases, especially where the grain size is extremely small, flotation is used. Here a slurry of minerals and water is mixed with a foaming agent, and the slurry then goes to a flotation *cell,* where the mixture is agitated. A foam is produced and, as the bubbles rise through the cell, mineral particles cling to the bubble walls and are thus *floated* to the surface and are swept safely off—just as the bartender removes foam from the top of a stein of beer.

Once separated, the mineral products are sized, graded, and packed for shipment.

RICHARD J. ANDERSON

Reference

Lynd, L. E., and Lefond, S. J., 1975. *Industrial Minerals and Rocks,* 4th ed. New York: American Institute of Mining, Metallurgical, and Petroleum Engineers, 1192-1197.

Cross-references: *Beach Material; Coastal Engineering; Human Impact; Mineral Deposits.*

MINOR BEACH FEATURES

If water flows over a sandy bottom, many different kinds of surface features will form. When water flow increases to the point that it can move sand, ripples will begin to form on the bed. A further increase in the water flow will result in the disappearance of the ripples and the formation of plane bed. Although devoid of large-scale features, there is a distinct lineation of the sediment particles parallel to the flow direction in this plane-bed flow regime. At *Froude numbers* greater than 0.8, antidunes may form. These are deposits roughly sinusoidal in outline.

Ripple Marks

Ripple marks are undulations formed on a noncohesive surface. Their origin may be either currents, waves, or wind. Ripples are common on beaches. They are especially well developed on the extensive tidal flats sometimes formed close to low tide and in troughs that run parallel to the coast higher up on the beach or in troughs that cross the beach and mark the location of *rip currents.* Ripples can occur in many sizes and shapes, often the smaller superimposed on the larger. Sixty cm marks the demarcation between ripples and *megaripples* (Allen, 1968). *Giant ripples* are greater than 30 m in length.

Current ripples (Fig. 1) are asymmetrical in form, and generally do not have a length greater than 60 cm. The height varies from 0.3-6 cm. Based on the nature of the crest, current ripples may be divided into straight, undulatory, or lingoid ripples. These will change from one to the other as the water velocity increases. At relatively low water velocities, straight-crested ripples form with planar foreset beds. The

FIGURE 1. Flood-tide current ripples, Point Roberts, Washington. The current flowed from right to left (photo: F. N. Sausa).

straight ripples have a ripple index (L/H) greater than 5, with most values falling between 8 and 15. If the water velocity increases, the crests become undulatory or wavy. A single crest can be followed for a long distance, but neighboring crests are often out of phase. Internally these ripples are weakly festoon-shaped. At still higher velocities, the ripple crests become discontinuous, lingoid ripples. The centers migrate in the current direction to form tongue- or lobelike forms, so that the acute angle always points downstream. The tongues are always out of phase. The internal crossbedded units are festoon-shaped.

Wave ripples are classified into two groups, based on their shape: asymmetrical, and symmetrical. Symmetrical wave ripples have symmetrical, pointed crests and rounded troughs, and internally consist of superimposed chevronlike laminations (Boersma, 1970). They are formed by an almost equally strong back-and-forth movement of the water. Symmetrical ripples range in length from 0.9-200 cm and in height from 0.3-23 cm (Reineck and Singh, 1973). If one direction of water movement predominates, then the symmetry will be destroyed, so that the ripples will have a steep lee side and a gentle stoss side. The asymmetric ripples range in size from 1.5-105 cm in length and 0.3-20 cm in height with a ripple index from 5 to 16 mainly between 6 and 8. Internally, asymmetric ripples may be concordant or discordant. If concordant, then there are wave laminae dipping in a single direction that are truncated by the stoss shape. If discordant, the ripple is made up of laminae of earlier formed ripples, and the outer shape or the ripple is not related to its internal structure. Discordant ripples are common seaward of the breaker zone.

Small asymmetric ripples can be formed by both currents and waves. The crests of the asymmetric (also symmetric) wave ripples repeatedly bifurcate, whereas straight current ripples do not bifurcate; rather, one crest may be replaced by another some distance away. Frequently, especially in troughs on a beach, ripple marks show a combination of wave and current ripples trending roughly at right angles to each other. These are known as ladder-back ripples or *tadpole nests*. The wave ripples are formed parallel to the shoreline, while the current ripples are generated by the longshore current or by the runoff of the tide.

Parting or primary current lineation is produced by higher water velocities found within a wave bore and especially the backwash. This structure consists of en echelon ridges and hollows a few grain diameter in height, a few mm to 1 cm apart, and 20-30 cm in length (Potter and Mast, 1963). The ridge spacing is directly proportional to the grain size, with the coarser grains in the ridges and the finer grains in the rounded troughs. The grains align themselves with the long axes parallel to the ridges (McBride and Yeakel, 1963). Parting lineation is formed on the flat foreshore as well as on the stoss side of ripples and on the more strongly carved surfaces of antidunes (Allen, 1964). They are formed only in the so called "upper flow" regime with Froude numbers greater than 0.3 (Simons, Richardson, and Norton, 1965).

Antidunes are trains of symmetrical low-relief ripplelike bedforms produced by a free surface flow in or near the supercritical state ($F = 1.0$ or greater) (Allen, 1968). Two dimensional antidunes are long-crested bedforms of almost perfect sinusoidal longitudinal profile where crests are perpendicular but in phase to the line of flow. Three dimensional antidunes are short-crested bed waves that in plan make a pattern of en echelon, ellipsoidal mounds and hollows. Three-dimensional antidunes are common in almost all runnels, whereas two dimensional antidunes are common on the foreshore of many relatively flat beaches. These may be formed by bores or by the hydraulic jump formed either by the incoming bore or by the backwash. Antidunes formed by the backwash are also called *backwash ripples* (Komar, 1976). In these the laminae are inclined at low angles both onshore and offshore, with lenticular and wedge-shaped sets of laminae being common (Middleton, 1965). In Florida these antidunes have a wavelength of 20-50 cm and are of low relief, so that the ripple index changes from 30 to 100, much larger than any other wave formed undulation (Tanner, 1965).

Other Minor Forms

Each hydraulic regime caused by the tide, waves, backwash, or wind leaves its characteristic depositional or erosional features on a beach. If waves coming in to a beach are uniform (i.e., predominantly of the same height and length) and if the beach is relatively steep, a *step* will form at the breaking point of the waves. A step is an abrupt downward inflection of the beach slope. This area is scoured out by the breakers. At the base of this inflection is a relatively flat area that is covered with coarse sediment (Miller and Zeigler, 1958). This is a lag deposit of the particles so coarse that they cannot be transported by the waves, whereas the finer particles are winnowed out (Schiffman, 1965).

As the tide rises, the swash will reach higher and higher up the beach. When the swash reaches above the groundwater table, it will start to deposit sand in a *swash bar* that migrates up the beach with the tide. This lens of sand is deposited since much of the swash water filters into the beach leaving its sediment load behind. These swash bars may reach up to 15 cm in height and have a long wavelength (Duncan, 1964). When the tide ebbs, the swash bar begins to broaden and thin out, for now the backwash is stronger than the swash because there is less infiltration and to the backwash is added the effluent from the groundwater.

During the rapid sedimentation in the swash, many air bubbles can be trapped in the sediment. This imparts a spongy texture to the sediment. The bubbles are a few mm in cross section and more or less ovoid in shape (Reineck and Singh, 1973). If the beach has alternating laminae of coarse and fine sand, the water drives the air ahead of it into the coarser and more permeable layers. This pressure may be sufficient to fold the beach sediment (Mii, 1956). The air may be so compressed that it is able to lift several cm of the overlying sand to form a *sand dome*, with a diameter of 5-15 cm (Emery, 1961). If this dome is later truncated by the waves, the dark laminae leave circular rings on the sand surface.

If the bubbles of foam are left on the sand surface, they will leave behind hemispherical pits with smooth walls without any rims. They generally occur in clusters of varying sizes and look like smallpox scars. If foam is pushed landward by the swash or seaward by the backwash, the pits are elongated into small furrows.

On the upper foreshore, the lobate fronts of the farthest swash advance leave behind imbricated sand ridges. These convex landward ridges, called *swash marks* by Johnson (1919), are 1-2 mm high deposits of sand grains usually rich in foraminifera, micas, and flat shell hash that, because of their low specific gravity or angularity, the swash front, by surface tension, can float the farthest up the beach. Coarser, but hydrodynamically lighter, material, such as sticks and seaweed, accumulate at the swash marks. Swash marks are present only above the water table. They are more closely spaced (±1 m apart) on steep beaches than on gentle ones (3-6 m) (Emery and Galem, 1951).

Just as the rising tide has certain characteristic deposits, so does the falling tide. As the backwash starts to gather momentum down the beach, several features may be left behind. These include: *rhomboid marks*, also called *rhomboid ripples* or *rhomboid rills;* and *current marks*, also called *current crescents*. Rhomboid marks (Otvos, 1965) are diamond-shaped or rhombohedral patterns formed by the backwash (Fig. 2). Each scalelike tongue has its acute angle pointing in the direction of

FIGURE 2. Rhomboid marks on beach at Ocean Shores, Washington. Backwash flow was from top to bottom (photo: W. F. Johnson).

flow. The marks range in size from 10-50 cm in length, with a length to width ratio from 2:1 to 5:1 (Hoyt and Henry, 1963). The steeper the beach, the greater the ratio. They develop under high water velocities in extremely shallow water (1-2 cm), but their origin is uncertain. Most authors require some mechanism to deflect the sheet flow. Among these are areas of compact sand or coarser particles (Johnson, 1919), areas of groundwater seepage (Demarest, 1947), or interference caused by groundwater seepage (Woodford, 1935).

A similar feature, a current mark, is formed by deflection of the backwash around an obstruction such as a pebble or a shell (Fig. 3). The obstruction deflects the backwash and thereby increases its velocity around the front

FIGURE 3. Current marks on Whidbey Island, Washington, beach. The backwash flowed from bottom to top (photo: M. L. Schwartz).

and sides, causing erosion. This increased velocity continues downstream, producing an erosional V. The sand that is eroded is redeposited behind the obstruction forming a depositional V inside of the erosional V (Sengupta, 1966).

As the tide ebbs even lower, there comes a time when the groundwater level becomes higher than the tidal elevation. Groundwater escaping from the beach surface develops scour marks called *rill marks*. The escaping groundwater aggregates into small rills of dendritic pattern (Fig. 4). These rills are originally 2-10 mm wide and 0.5 m or more long and about 1 mm deep. The dendritic pattern may be half as wide as it is long (Twenhofel, 1926). The groundwater does not seep out uniformly, but is spaced irregularly owing to grain orientation and permeability inhomogeneities in the sand. Elongate grains align themselves with their long axis to the water movement. Because the swash moves more or less perpendicular to the beach, the grains align themselves at right angles to the beach trend (Curray, 1956) and imbricate themselves in a shoreward direction. Measurements of permeability (Emery, 1961) show that the lowest values are vertical owing to horizontal laminations. The greatest permeability is horizontal and parallel to the beach. Therefore only after the intermediate values of permeability perpendicular to the beach have been surpassed will the groundwater escape.

At the contact of a sloping sediment surface with a standing body of water, a small step can be formed. When the water level falls discontinuously owing to local wind stresses, small waves have a chance to erode a series of small scarps known as *water level marks* (Hantzchel, 1939). These water level marks are common on protected beaches, the slopes of channels, runnels, and even ripples.

On the back beach, wind is an active transport agent, piling sand into ripples. *Wind ripples* have straight, long, parallel crests and are asymmetrical with high ripple indexes (30-40). The ripple length is dependent on the wind strength (Bagnold, 1954), and increases as the wind speed and grain size increase (Sharp, 1963). In wind ripples, the coarsest grains collect on the ripple crests (Twenhofel, 1926). Few internal structures exist in a wind ripple.

Wind also produces *adhesion warts* and *adhesion ripples*. Where dry sand is blown over a moist sand surface, it will adhere on the spot at which it lands. By capillary action, the dry sand grains will be moistened and capable of receiving new sand grains, thereby producing tiny ridges of sand. Adhesion warts and ripples have crests that are asymmetrical in cross section with a windward side steeper than the lee side. In fact, it may be so steep as to be vertical or overhanging (van Straaten, 1953; Hunter, 1969).

Certain beach features are climatically controlled. In cold climates, the back beach is deformed and cracked by frost action. This produces V-shaped fissures that may be up to several meters deep. These are later filled in, but leave a characteristic V-shaped discontinuity (Rex, 1964). On the foreshore, when the temperature falls persistently below $0°C$, the swash run-up on the beach freezes in place and is contiguous with any deposition of sediment. This forms a *kaimoo* (q.v.), a thick bed of ice interlayered with thin beds of gravel and sand. If pack ice reaches the shore, ice-pushed irregular mounds and ridges (up to 10 m in size and 1 m deep) are common. Summer thawing results in collapse features and extremely soft zones on the beach. The subsidence due to the melting ice causes any continuous layers of sediment to be increasingly deformed toward the surface, producing irregular, mottled, and complex bedding (Short and Wiseman, 1973), and leaves the surface pitted with small *kettles* and sand and gravel *cones* (Greene, 1970).

BENNO BRENNINKMEYER

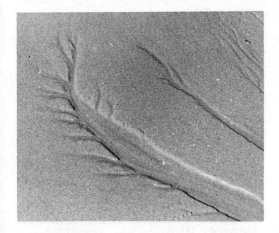

FIGURE 4. Rill marks in the beach at Point Roberts, Washington. Escaping groundwater flowed from upper left to lower right (photo: I. C. Schwartz).

References

Allen, J. R., 1964. Primary current lineation in the lower Old Red Sandstone Devonian, Anglo-Welsh Basin, *Sedimentology* 3, 98-108.

Allen, J. R., 1968. *Current Ripples.* Amsterdam: North Holland, 433p.

Bagnold, R. A., 1954. *Physics of Blown Sand and Desert Dunes.* London: Methuen and Co., 265p.

Boersma, J. R., 1970. *Distinguishing Features of Wave-Ripple Cross Stratification and Morphology.* Unpublished dissertation, University of Utrecht, 65p.

Curray, J. R., 1956. Dimensional grain orientation studies of recent coastal sands, *Am. Assoc. Petroleum Geologists Bull.* 40, 2440-2456.

Demarest, D. F., 1947. Rhomboid ripple marks and their relationship to beach slope, *Jour. Sed. Petrology* 17, 18-22.

Duncan, J. R., 1964. The effects of water table and tidal cycle on swash-backwash sediment distortion and beach profile development. *Marine Geology* 2, 186-197.

Emery, K. O., 1961. *The Sea off Southern California.* New York: John Wiley & Sons, 366p.

Emery, K. O., and Galem, J. F., 1951. Swash and swash marks, *Am. Geophys. Union Trans.* 32, 31-36.

Greene, H. G., 1970. Microrelief of an Arctic beach, *Jour. Sed. Petrology* 40, 419-427.

Hantzchel, W., 1939. Brandungwalle, Rippeln und Fliessfiguren am Strande von Wangeroog, *Natur Volk* 69, 40-48.

Hoyt, J. H., and Henry, V. J., 1963. Rhomboid ripple marks indicator of current direction and environment, *Jour. Sed. Petrology* 33, 604-608.

Hunter, R. E., 1969. Eolian microridges on modern beaches and possible ancient examples, *Jour. Sed. Petrology* 39, 1573-1578.

Johnson, D. W., 1919. *Shore Processes and Shoreline Development.* New York: Hafner Publishing Co., 584p.

Komar, P. D., 1976. *Beach Processes and Sedimentation.* Englewood Cliffs, N.J.: Prentice-Hall, 429p.

McBride, E. F., and Yeakel, L. S., 1963. Relationship between parting lineation and rock fabric, *Jour. Sed. Petrology* 33, 779-782.

Middleton, G. V., 1965. Antidune cross bedding in a large flume, *Jour. Sed. Petrology* 35, 922-927.

Mii, H., 1956. Folded structures observed in beach sand at Fukanuma, Sendai City, Miyagi Prefecture, Japan, *Geol. Soc. Japan Jour.* 62, 189-191.

Miller, R. L., and Zeigler, J. M., 1958. A model relating dynamics and sediment pattern in equilibrium in the region of shoaling waves, breakers, and foreshore, *Jour. Geology* 66, 417-441.

Otvos, E. G., 1965. Types of rhomboid beach patterns, *Am. Jour. Sci.* 261, 271-276.

Potter, P. E., and Mast, R. F., 1963. Sedimentary structures, sand shape, fabric, and permeability, *Jour. Geology* 71, 441-471.

Reineck, H. E., and Singh, I. B., 1973. *Depositional Sedimentary Environments.* New York: Springer-Verlag, 439p.

Rex, R. W., 1964. Arctic beacher, Barrow Alaska, *in* R. L. Miller, ed., *Papers in Marine Geology.* New York: Macmillan, 384-399.

Schiffman, A., 1965. Energy measurements of the swash-surf system, *Limnology and Oceanography* 10, 255-260.

Sengupta, S., 1966. Studies on orientation and imbrication with respect to cross stratification, *Jour. Sed. Petrology* 36, 362-369.

Sharp, R. P., 1963. Wind ripples, *Jour. Geology* 71, 617-636.

Short, A. D., and Wiseman, W. J., 1973. Freezing effects on arctic beaches, *Louisiana State University, Coastal Studies Institute, Tech. Rept. 128*, 9p.

Simons, D. B., Richardson, E. V., and Nortin, C. F., 1965. Sedimentary structures generated by flow in alluvial channels, *in* G. V. Middleton, ed., *Primary Sedimentary Structures and Their Hydrodynamic Interpretation.* Tulsa, Okla.: Soc. Econ. Paleont. Miner. Spec. Pub. 12, 34-52.

Tanner, W. F., 1965. High index ripple marks in the swash zone, *Jour. Sed. Petrology* 35, 968.

Twenhofel, W. H., 1926. *Treatise on Sedimentation.* Baltimore: Williams and Wilkins, 661p.

van Straaten, L. M. J. U., 1953. Rhythmic patterns on Dutch North Sea beaches, *Geologie en Mijnbouw* 15, 31-43.

Woodford, A. O., 1935. Rhomboid ripple marks, *Am. Jour. Sci.* 29, 518-525.

Cross-references: *Backwash Patterns; Beach; Beach Firmness; Beach Processes; Bed Forms; Coastal Dunes and Eolian Sedimentation; Ice along the Shore; Kaimoo; Mud Volcanoes; Rill Marks; Ripple Marks; Shore Polygons; Swash Mark; Swing Mark; Water Table.*

MODERATE-ENERGY COAST

The classification of coasts based on wave energy was first developed by Price (1955). He defined the *moderate-energy coast* as one characterized by a *ramp slope* (offshore bottom slope) of 1.5-2.5 feet per mile, a *shoreface slope* (nearshore bottom slope) of 10-15 feet per mile, and "low" breakers. This method of classification requires the existence of bathymetric charts or field measurements. It does not account for the width of the shelf, the depth to the foot of the shoreface, or the orientation of the coast; hence it is not rigorous.

Tanner (1960) defined the moderate-energy coast as one characterized by mean breaker heights of 10-50 cm. This method of classification avoids some of the difficulties associated with Price's method; however, field measurements are still required.

Moderate-energy coasts are characterized by sand beaches, barrier islands, spits, bars, shoals, and compartmentalized sediment drift systems. The typical delta form is lobate, becoming digitate only with excessive sediment load.

Some examples of moderate-energy coasts are most of the northern Gulf of Mexico, the Atlantic coasts of Georgia and South Carolina, and the coasts of much of the Great Lakes of North America.

JAMES P. MAY

References

Price, W. A., 1955. Correlation of shoreline type with offshore bottom conditions, *Department of Oceanography, A & M College of Texas, Project 63*, 2p.

Tanner, W. F., 1960. Florida coastal classification, *Gulf Coast Assoc. Geol. Socs. Trans.* 10, 259-266.

Cross-references: *High-Energy Coast; Low-Energy Coast; Mid-Latitude Coasts; Zero-Energy Coast.*

MOLD—See FUNGI

MOLE

A mole is a massive, solid-fill structure extending from the shore toward deep water (Gary, McAfee, and Wolf, 1972). It may be constructed of masonry, large stones, or revetment-protected earth, sand, or gravel. Mole is sometimes used synonymously with *pier* or *breakwater,* but it is more precisely used to describe a structure serving both purposes—that is, a mole provides protection from waves along one side, as does a breakwater, and, when provided with a broad superstructure equipped for commercial operations, it functions as a pier (Coastal Engineering Research Center, 1977; Cornick, 1958).

CHARLES B. CHESTNUTT

References

Coastal Engineering Research Center, 1977. *Shore Protection Manual.* Washington, D.C.: U.S. Army Corps of Engineers, 1264p.
Cornick, H. F., 1958. *Dock and Harbor Engineering,* vol. 1. London: Charles Griffin & Co. Ltd., 316p.
Gary, M.; McAfee, R., Jr.; and Wolf, C. L., 1972. *Glossary of Geology,* Washington, D.C.: American Geological Institute, 805p.

Cross-references: *Bulkhead; Coastal Engineering; Pier; Protection of Coasts.*

MOLLUSCA

Mollusca is a phylum of bilaterally symmetrical, unsegmented invertebrate animals in which the body is usually inside of, or contains remnants of, a calcareous shell, and has four parts: anterior head, dorsal visceral mass, circum-visceral mantle and ventral foot. The phylum Mollusca (Latin *Mollis,* soft) is divided into seven main classes of diverse appearance (Newell, 1969; Barnes, 1980): the recently discovered Monoplacophora; the Polyplacophora (Amphineura) or *chitons* and their relatives; the Aplacophora or *solenogasters;* the Gastropoda or *snails;* the Pelecypoda (Bivalvia or Lamellibranchia) or *bivalves;* the Scaphopoda or *tusk* and *tooth shells;* and the Cephalopoda or *nautili, cuttlefish, squids,* and *octopods.*

The mollusks are one of the most important groups in the animal kingdom if for no other reason then that, next to the insects, they include more known species than any other animal subdivision, over 80,000 living species having been described. In addition, more than 35,000 fossil species are known, since paleontologically they are one of the oldest groups, representatives being abundant among the fossils of the Lower Cambrian strata laid down 600,000,000 years ago. Their abundance as fossils is not surprising, the mineral shell possessed by most of these animals having increased the chances for preservation (Newell, 1969). It is thought that the presence of great numbers of kinds of fossil mollusks indicates existence of the group for many millennia, perhaps as naked soft-bodied forms lacking the hard parts necessary to form traces in the rocks.

Although mollusks have been employed variously in literature, art, and commerce from earliest times, and were carefully treated by Aristotle in his *Historia Animalium,* it was not until the time of Carolus Linnaeus (1707-1775) and Georges Cuvier (1769-1832) that the foundation for modern classification of these animals was laid (Hyman, 1967).

However, the first students of mollusks in this country were the early physician-naturalists and the serious collectors who were endeavoring to build large so-called *cabinets* of shells. The earliest scientific descriptions of importance were apparently published in 1817 by Thomas Say, who described a large number of species before his death in 1834. Constantine S. Rafinesque, during nearly the same period, contributed descriptions of large numbers of terrestrial and freshwater mollusks. Together with Timothy A. Conrad (1803-1877), Isaac Lea (1792-1866), Augustus A. Gould (1805-1866), William Stimpson (1832-1872), William G. Binney (1833-1909), Addison E. Verrill (1839-1945), these men laid the foundations of American conchology (Hyman, 1967). In 1878, George W. Tryon (1838-1888) started and, in 1888 Henry A. Pilsbry (1862-1951), continued the publication of the *Manual of Conchology* by the Academy of Natural Sciences in Philadelphia, whose many volumes of monographs cover the entire field. In this century, investigations into all aspects of the biology and the socioeconomic importance of mollusks have been pursued by a constantly increasing number of specialists, whose publications appear in many popular and professional books as well as in a variety of journals and monographs. At the present time, there are several journals, such as *Nautilus* and *Johnsonia,* devoted entirely to investigations of mollusks.

Mollusks, particularly bivalves and cephalopods, have always been an important part of man's sociological environment. The art of the classical world is strewn with mollusk shells—

Roman funerary monuments, designs on coins, domestic utensils, Greek and Christian religious motifs (the badge of St. James) in the decorative arts as well as in classical architecture, sculpture, painting, armorial bearings—as of course is modern art (Morton, 1967).

Perhaps of more pragmatic consideration is the economic relationship of mollusks to man. In the heyday of the wooden clippers, one of the best-known mollusks was the teredo or *shipworm* (a bivalve), whose boring destroyed hundreds of thousands of dollars worth of wooden ships and wharves annually. Many of the bivalves, gastropods, squid, and octopuses have furnished an enormous food resource worldwide from the dawn of history. In this country, there is no doubt that shellfish were an early article of diet for all of the early coastal peoples when the vast shell deposits comprising most of their natural rubbish heaps or middens are examined. Today the oyster, clam, scallop, and squid fisheries are by far the most important for income. Millions of dollars are invested in their development including mariculture, and thousands of men are employed in canneries, packing plants, markets and on the high seas in great numbers of vessels. With growing populations and a concomitant growing demand for proteins; mussels, cockles, razor shells, limpets, periwinkles, whelks, conchs, abalone, and other lesser-known mollusks are increasingly collected for market. Other commercial uses of mollusks include the manufacture of jewelry. In this connection, mention should be made of: the pearl oyster, which produces the precious pearl; and the freshwater mussels, abalones, top shells, and turban shells whose valves are lined with the mother of-pearl used in manufacturing buttons, knife handles, inlays, and ornaments. The American Indians used the shell of hard clams for money (*wampum*), and certain cowries were employed similarly in Micronesia and elsewhere in the Far East. Purple snails were crushed by coastal aborigines on both sides of the Atlantic to obtain purple dye. Shells of common bivalves are still ground to make roads and kilned to produce lime. Shell collecting has long held widespread attention, and in many parts of the world has become big business. Several of the most common mollusks—oysters, clams, slugs, and squid prominent among them—are in demand by physiologists, geneticists, ecologists, embryologists, biochemists, and their colleagues in related fields, for work in basic, applied, and medical research.

Most mollusks are marine; many live in freshwater; and some species have become terrestrial, although a majority of these ordinarily require a moist environment. The marine forms are greatest in numbers of species, individuals, distribution, and variety of habitat, ranging from the giant squids of the North Atlantic (16 m) to the minute subtidal, interstitial, meiofaunal sea slugs (less than 2 mm), and from representation in the oceanic zooplankton worldwide (heteropods and pteropods) to the large, strikingly colored, evolutionarily important land snails of the Society Islands. Although the majority of mollusks with a radula feed on flowering plants or on algae, many gastropods and all cephalopods are predacious; the bivalves are filter feeders, using a mechanism that centers on their ciliated gills. A few species of gastropods are parasitic; others are species-specific hosts for helminthic diseases of man and of other animals; several are possible index species for indicating polluted waters.

General Anatomy

Mollusks are bilaterally symmetrical (except for coiled viscera in the gastropods and in some cephalopods), possess three germ layers, are unsegmented (except possibly in monoplacophorans), and have a single-layered mostly ciliated epithelium with mucous glands. Their body is usually short and enclosed in a thin dorsal mantle that secretes a shell most often of one, two, or eight parts. However, the shell may be entirely lacking, as in slugs, reduced, as in shipworms, or internal, as in squid. The head region is developed, except in the Pelecypoda and the Scaphopoda. The ventral muscular foot may be variously modified for adhesion, gliding, crawling, burrowing, or swimming.

The mouth leads into an often coiled or V-shaped, complete digestive tract. In the mouths of all mollusks except pelecypods is the ribbon-shaped, muscularly controlled radula bearing diagnostic, transverse rows of minute, chitinous teeth that may be used to scrape algae, grasp food, or bore holes. There are often salivary glands, always a large digestive gland or "liver," and an anus that has an opening in the mantle cavity.

The circulatory system includes a dorsal heart with one or two auricles and a ventricle, usually enclosed in a pericardial cavity from which extends an anterior aorta leading to other vessels. Respiration is most often accomplished by one to many gills or ctenidia that are paired in the Monoplacophora. However, this function may be carried out by a "lung" in the mantle cavity, by the mantle itself, or by the epidermis. Excretion is the responsibility of the single, one, two, or six pairs (Monoplacophora) of kidneys or nephridia that connect to the pericardial cavity and the veins. In these animals, the coelom is reduced to the cavities of the nephridia, the gonads, and the pericardium.

Typically, the nervous system has three pairs of ganglia: cerebral, pedal, and visceral. The ganglia are joined by longitudinal and cross-connective nerves. Special sense organs include structures for taste, smell (aesthetes), and touch (tentacles); eyes or eyespots (ocelli), osphradia or chemoreceptors, and statocysts for maintaining equilibrium. Although the sexes are usually separate, hermaphroditism and protandry are not uncommon, and fertilization may be internal or external. Gonads may be single or paired, with oviparity being the rule. During development, egg cleavage is determinate, unequal, and total (discoidal in Cephalopoda), with eventual production of a veliger (trochophore) larva or a glochidium (a parasitic stage in freshwater clams or mussels). Direct development takes place in the pulmonate gastropods and in the cephalopods. Asexual reproduction has never been observed in mollusks.

Monoplacophora. This class of primitive mollusks, established in 1940, is similar to the polyplacophorans and gastropods in possessing a flat, creeping foot. Until 1952, monoplacophorans were believed to be represented only by Cambrian and Devonian fossils (Newell, 1969). At that time, ten living specimens of the limpetlike *Neopalina* were dredged from a deep ocean bench off the Pacific coast of Costa Rica, and a description was published in 1957. Since then, seven additional species of these "living fossils" have been collected from deep water (to 7000 m) in different parts of the world. As the name indicates, monoplacophorans have a single symmetrical shell whose under surface displays three to eight muscle scars. They have a posterior anus and six pairs of segmentally arranged excretory organs (nephridia) and gills (ctenidia), a condition strengthening the concept of a close phylogenetic connection between annelids and mollusks (Barnes, 1980). The survival of *Neopalina* is correlated with their adaptation for life at great depths.

Polyplacophora. The widely distributed, drably colored polyplacophorans are amphineurans, commonly known as chitons. Some features of their structure and embryogeny are primitive, their broad, flat, adhesive foot (eaten in the West Indies as *sea beef*) enables them to cling to smooth stones or to creep sluggishly along shallow, rocky bottoms. They are elliptical, dorsoventrally flattened, and have a characteristic series of eight dorsal transverse overlapping calcareous plates (shell) that allow them to bend their body and roll up into a ball when disturbed. The 600 species vary in size from 3 cm to more than 30 cm. The mantle (girdle in chitons) is heavy, and its groove contains from four to eighty pairs of ctenidia. The head is greatly reduced, and lacks eyes and tentacles. Polyplacophorans feed on algae and other organisms scraped from substratal rock or shell surfaces with their radulae.

Aplacophora. Members of this class are known also as solenogasters. They comprise 130 species of worm-shaped mollusks that through either specialization or primitiveness have lost the typical molluskan shell, mantle, and foot (Hyman, 1967). In addition, the radula may be reduced or absent, the posterior or mantle cavity houses a pair of gills, and the integument contains layers of embedded calcareous spicules. Aplacophorans occur worldwide to depths of 9000 m; the burrowers are scavengers and carnivores, while the creepers live and feed on corals and hydroids, neither group exceeding a length of 5 cm.

Gastropoda (Greek *gaster*, belly + *podos*, foot). There are more than 35,000 living and 15,000 fossil species of snails, slugs, limpets, whelks, and conchs in this largest and most successful (because of extensive adaptive radiation) class of mollusks. Marine species have adapted successfully to pelagic, benthic, and interstitial habitats; they have invaded freshwaters, and pulmonate snails have conquered the land (Hyman, 1967). The diverse and varied development of radulae, modified jaws, proboscises, feet, teeth, and digestive tracts have led to many different food habits. Gastropods may be herbivorous, carnivorous, scavengers, deposit feeders, filter feeders, parasites, or involved in parasite life cycles; some have become serious agricultural pests (Hyman, 1967). They are utilized as food by many higher invertebrates and by all classes of vertebrates, and are of further economic importance to man in destroying agricultural and marine crops, serving as intermediate hosts for trematode parasites, and in their extensive use in genetic and medical research. Structurally, gastropods usually have an asymmetrical, spirally coiled, sinistral or dextral, one-piece shell (which may be reduced or lacking), lined with a mantle that in turn contains much of the viscera. The foot is well developed and flattened, and the head has one or two pairs of tentacles, one pair of eyes, and a variously developed radula. They are mostly oviparous, and may be hermaphroditic or dioecious (one gonad) with variable development.

Pelecypoda (Greek *pelekys*, hatchet + *podos*, foot). Pelecypods include more than 10,000 species of marine and freshwater mollusks with two shells or valves (the bivalves), and include common, often commercially valuable forms as scallops, edible and pearl oysters, clams, and mussels. Members of this totally aquatic class are bilaterally symmetrical and laterally compressed. The headless soft body is enclosed in a bilobed mantle that secretes a two-part rigid shell and contains the wedge-shaped foot. The gills on either side of the mantle cavity are thin and platelike, whence the earlier class name, Lamellibranchiata. The gills are covered with cilia that help carry minute particles of food to the mouth as well as create a current of respiratory water through the mantle cavity. These structural characteristics have enabled bivalves, with few exceptions, to become more or less specialized bottom burrowers.

About 20% of this class lives in freshwater, mostly in lakes and ponds (Pennak, 1978). Marine forms are commonest intertidally and subtidally, often occur in amazingly large quantities in great "beds," and occasionally live to a depth of about 6000 m. They either creep along the bottom or burrow in sand while extending their mantle-derived tubular siphons in maintaining a flow of respiratory microorganism-laden water (Morton, 1967). Notable exceptions are: the marine mussel that temporarily attaches first to one hard substratum and then to another by secreted byssus threads; edible oysters that attach permanently to rocks or shells, pholads that burrow into shell, coral, peat, hard clay, or soft rocks; teredos or shipworms that use their shells to burrow in seawater-wetted wooden ships, pilings, and wharves; scallops that swim erratically by forcefully clapping their shells together, and a large variety of forms that are phoretic, symbiotic, commensal, or parasitic in or on members of other invertebrate groups (Morton, 1967).

Pelecypod shells have a tremendous range of shapes, sizes, surface sculpturings, and colors, varying from species less than 2 mm in length to those more than 1 m long with a weight of over 1100 kg. Shells are usually symmetrical with dorsal hinge and ligament, and are closed by one or two adductor muscles (Abbott, 1954). There are no jaws or radulae, and the mouth has labial palps that guide food into the gut. The majority of bivalves are dioecious, some are hermaphrodites, others are protandric hermaphrodites, and a few of the latter can change from female to male. Freshwater bivalves exhibit modified development compared with marine forms that go through a free-swimming trochophore succeeded by a larval veliger stage before metamorphosing into the adult on the substratum (Pennak, 1978).

Scaphoda (Greek *skaphe,* boat + *podos,* foot). The tooth or tusk shells are all marine, with more than 200 living species and nearly 300 fossil forms. The body is slenderly elongate dorsoventrally, and is surrounded by a mantle that secretes the typically slightly curved and tapered tubular shell. The 3-15cm shell is open at both ends; the small, pointed, conical burrowing foot and the head protrude through the buried (wider) lower end, while the inhalant and exhalant water currents pass through the narrow, upper (posterior), water-immersed end.

Scaphopods are apparently relatively numerous and live subtidally to depths of over 2000 m, partly obliquely buried in mixtures of mud and sand, and feeding on microscopic organisms, especially foraminifera. Tusk shells on strings were the money of the coastal Indians from California to Alaska.

Although the head of tusk shells has many tentacles and a radula, the animals have only a rudimentary circulatory system. They are dioecious, and development includes planktonic trochophore and veliger larval stages.

Cephalopoda (Greek, *kephale,* head + *podos,* foot). This class contains the most specialized and highly organized of all mollusks. Although the octopus and its immediate relatives have secondarily become less active benthic forms, the class is adapted for a swimming existence and has developed aggressive carnivorous habits. The head is well developed, with a "brain" in a cartilage-like covering, a radula, and highly developed large, often complex eyes. Part of the foot has grown around the head, with the mouth in the center of the anterior mobile, sucker-laden, prehensile tentacles (arms). The other part forms the muscular tunnel or siphon through which water is jetted from the mantle cavity, propelling the animal rapidly backward. Squids attain the greatest swimming speeds of any aquatic invertebrates. Cephalopods with characteristically coiled chambered shells live in the terminal and largest chamber. However, most living cephalopods lack a shell or have one that is either reduced, internal, or both. Evolutionarily, cephalopods are a dying group, with about 650 living species and more than 7500 fossil forms, many of which are important index fossils for the geologist.

Included in this group are the largest invertebrates, the giant squids, reaching 16 m in length including tentacles, with a circumference of 4m. They are eaten by sperm whales. Most cephalopods range from 6-70cm. Small squids sometimes occur in enormous schools, where they may be preyed on by fishes and marine mammals. They are economically important as bait by fishermen and as food for man, and in recent years they have also become invaluable for physiological and biomedical research. The Atlantic cuttlefish (*Sepia*), with its retractile tentacles, has an internal calcareous shell. The cuttlebone has been used as a bill sharpener for caged birds, while over the years its ink has provided pigment for artists. In most cephalopods, a large intestine-bordered ink sac contains a brown or black fluid that is released forming a cloud when the animal is alarmed. Giant octopods or devilfishes are fictitious; octopods rarely exceed 36 cm, although in several species their slender arms may reach a length of 5 m. Although octopods ordinarily crawl about rocks and tide pools, when disturbed they can swim by using their siphon. Certain species of deep-sea squids have luminous organs with symbiotic luminous bacteria; others are bioluminescent with luminous photophores of different colors in a variety of body patterns. The amazing, often continuously changing color of cephalopods, other than in the completely shelled *Nautilus,* is created by the presence of yellow, orange, blue, red, and black nerve and hormone-controlled, muscled chromatophores (Barnes, 1980).

Cephalopods are mostly dioecious. Following copulation, in which in some forms the heterocotylus arm of the male is an intromittent organ, fertilization takes place inside or outside the mantle cavity. Eggs attached in gelatinous coverings are deposited singly or in large clusters; development is direct.

DONALD J. ZINN

References

Abbott, R. T., 1954, *American Sea Shells.* Princeton, N.J.: Van Nostrand, 541p.
Barnes, R. D., 1980. *Invertebrate Zoology,* 3rd ed. Philadelphia: W. B. Saunders, 870p.
Hyman, L. H., 1967. *The Invertebrates,* vol. 6, *Mollusca.* New York: McGraw-Hill, 792p.
Morton, J. E., 1967. *Molluscs,* 4th ed. London: Hutchinson University Library, 252p.
Newell, N. D., 1969. Classification of Bivalvia, *in* R. C. Moore, ed., *Treatise on Invertebrate Paleontology,* vol. N, pt. 1, *Mollusca.* Boulder, Colo. and Lawrence, Kans.: Geological Society of America and University of Kansas, N205-N224.
Pennak, R. W., 1978. *Fresh-water Invertebrates of the United States,* 2d ed. New York: John Wiley & Sons, 803p.

Cross-references: *Archaeology, Methods; Biotic Zonation; Coastal Fauna; Coastal Waters Habitat; Intertidal Flats; Intertidal Sand Habitat; Midden; Organism-Sediment Relationship; Thallasotherapy.*

MOMENT MEASURES

There are two approaches to adequately describe a sample sediment size frequency distribution in terms of four descriptive parameters (Krumbein and Pettijohn, 1938). These are used to distinguish either the mode of transportation or the environment of deposition of a sediment. The most mathematically elegant method of describing the complete frequency distribution is by the statistical moment measurements. Since most grain size distributions are approximately normal (or *Gaussian*) when their ϕ size is plotted on an arithmetic scale, conventional moment statistics can be used to characterize an individual sample.

The mathematical expectation of a variable $[g(x)]$ related to x is: $E(x) = \Sigma g(x)f(x)$, where $f(x)$ is the probability function at x and the summation extends over the entire range of $g(x)$ (Kreyzig, 1970). If $g(x) = x^r$ or $E(x^r) = \Sigma x^r f(x) = m_r$, this is called the rth moment about the origin of the distribution of the random variable x. The term *moment* is borrowed from physics. If the values of $f(x)$ are point masses acting perpendicular to the x axis at a distance x from the origin and if $r = 1$, then m_1 would be the x coordinate of the center of gravity. Similarly, if $r = 2$ then m_2 would be the moment of inertia.

In statistics, the first moment, called the mean ($\bar{x}$), is the mathematical expectation of x itself. It is a measure of the central tendency of the frequency distribution curve. The higher moments are usually calculated not from the origin but around the mean, $g(x) = (x - \bar{x})^r$. Thereby, the second moment, the variance, is a measure of the spread of the variable around the mean. The third moment or skewness is a measure of asymmetry of the distribution. It will be 0 for normal or symmetric distribution, positive for a distribution with a right tail, and negative for a left-tailed distribution. Kurtosis, the fourth moment, is sometimes used as a measure of the peakedness of a distribution, but this parameter really does not have a visual analog (Baker, 1968).

In sediment size analysis, it is more convenient to compute the moments about an arbitrary mean, usually the size class midpoints. Moreover, in order to preserve linearity of the moments, it is conventional to take the square root of the variance and to transform the third and fourth moments by dividing them by the appropriate power of the variance.

The moments and their derived parameters then become (McBride, 1971):

$$\text{Mean } \bar{x}_\phi = \frac{\Sigma f(m)}{n}$$

$$\text{Standard deviation } \sigma_\phi = \sqrt{\frac{\Sigma f(m - \bar{x}_\phi)^2}{100}}$$

$$\text{Skewness } Sk_\phi = \frac{\Sigma f(m - \bar{x}_\phi)^3}{100 \sigma_\phi^3}$$

$$\text{Kurtosis } K_\phi = \frac{\Sigma f(m - \bar{x}_\phi)^4}{100 \sigma_\phi^4}$$

where f = wt % (frequency) in each grain-size class present, m = midpoint of each grain-size class in ϕ values, and n = total number in sample, which is 100 when f is in %.

Another approach used to compare sediment characteristics is to use a graphical percentile-intercept method. The first step in this process is to draw a cumulative size percentage curve of the sample on probability paper. Then, from the graph is read the grain size value (in ϕ) that corresponds to the percentile value needed to compute each of the graphic statistics of which many have been prepared. McCammon (1962) has rated some of these in their efficiency in approximating the moment measures (Table 1). Some of these graphic values are also given verbal scales by several authors (Table 2).

These measurements are used to differentiate environments of deposition. For instance, Mason and Folk (1958) found that a plot of skewness versus kurtosis best separated beach, dune, and aeolian flat environments. This does

TABLE 1. Formulas for Computing "Graphical" Parameters

Parameter	Formula	Efficiency (%)
Mean		
Trask (1930)	Median, $\phi 50$	64
Inman (1952)	$M\phi = (\phi 16 + \phi 84)/2$	74
Folk and Ward (1957)	$Mz = (\phi 16 + \phi 50 + \phi 84)/3$	88
McCammon (1962)	$(\phi 10 + \phi 30 + \phi 50 + \phi 70 + \phi 90)/5$	93
McCammon (1962)	$(\phi 5 + \phi 15 + \phi 25 \ldots + \phi 85 + \phi 95)/10$	97
Sorting		
Krumbein (1934)	$QD\phi = (\phi 75 - \phi 25)/1.35$	37
Inman (1952)	$(\phi 84 - \phi 16)/2$	54
Folk and Ward (1957)	$(\phi 84 - \phi 16)/4 + (\phi 95 - \phi 5)/6.6$	79
McCammon (1962)	$(\phi 85 + \phi 95 - \phi 5 - \phi 15)/5.4$	79
McCammon (1962)	$(\phi 70 + \phi 80 + \phi 90 + \phi 97 - \phi 3 - \phi 10 - \phi 20 - \phi 30)/9.1$	87
Skewness		
Krumbein and Pettijohn (1938)	$Skq\phi = [\phi 25 + \phi 75 - 2(\phi 50)]/2$	
Inman (1952)	$a_{1\phi} = \dfrac{\phi 16 + \phi 84 - 2(\phi 50)}{\phi 84 - \phi 16}$	
	$a_{2\phi} = \dfrac{\phi 5 + \phi 95 - 2(\phi 50)}{\phi 84 - \phi 16}$	
Folk and Ward (1957)	$Sk_I = \dfrac{\phi 84 + \phi 16 - 2\phi 50}{2(\phi 84 - \phi 16)} + \dfrac{\phi 95 + \phi 5 - 2\phi 50}{2(\phi 90 - \phi 5)}$	
Kurtosis		
Krumbein and Pettijohn (1938)	$K_{qa} = \dfrac{\phi 75 - \phi 25}{2(\phi 90 - \phi 10)}$	
Inman (1952)	$\phi = \dfrac{(\phi 95 - \phi 5) - (\phi 84 - \phi 16)}{\phi 84 - \phi 16}$	
Folk and Ward (1957)	$K_G = \dfrac{\phi 95 - \phi 5}{2.44(\phi 75 - \phi 25)}$	

TABLE 2. Descriptive Scales of Graphical Moment Measures

Term	Boundary Values (ϕ units)	
Sorting	Folk and Ward, 1957	Friedman, 1962
very well sorted	.35	.35
well sorted	.50	.50
moderately well sorted	.71	.80
moderately sorted	1.00	1.40
poorly sorted	2.00	2.00
very poorly sorted	4.00	2.60
extremely poorly sorted		
Skewness	Folk, 1968	
strongly fine skewed	+.30	
fine skewed	+.10	
near symmetrical	−.10	
coarse skewed	−.30	
strongly coarse skewed		
Kurtosis	Folk, 1968	
very platykurtic	.67	
platykurtic	.90	
mesokurtic	1.11	
leptokurtic	1.50	
very leptokurtic	3.00	
extremely leptokurtic		

not effectively separate river from beach deposits, which is done by plots of skewness versus standard deviation (Friedman, 1967).

BENNO M. BRENNINKMEYER

References

Baker, R. A., 1968. Kurtosis and peakedness, *Jour. Sed. Petrology* 38, 679-680.
Folk, R. L., 1968. *Petrology of Sedimentary Rocks.* Austin: Hempill's, 170p.
Folk, R. L., and Ward, W. C., 1957. Brazos River bar: A study in the significance of grain size parameters, *Jour. Sed. Petrology* 27, 3-26.
Friedman, G. M., 1962. On sorting, sorting coefficient, and the lognormality of the grain size distribution of sandstones, *Jour. Geology* 70, 737-756.
Friedman, G. M., 1967. Dynamic processes and statistical parameters composed for size frequency distribution of beach and river sands, *Jour. Sed. Petrology* 37, 327-354.
Inman, D. L., 1952. Measures for describing the size distribution of sediments, *Jour. Sed. Petrology* 22, 125-145.
Kreyzig, E., 1970. *Introductory Mathematical Statistics.* New York: Wiley & Sons, 470p.
Krumbein, W. C., and Pettijohn, F. J., 1938. *Manual of Sedimentary Petrology.* New York: Appleton-Century Crofts, 549p.
McBride, E., 1971. Mathematical treatment of size distribution data, *in* R. E. Carver, *Procedures in Sedimentary Petrology.* New York: Wiley & Sons, 109-127.
McCammon, R. B., 1962. Efficiency of percentile measures for describing the mean size and sorting of sedimentary particles, *Jour. Geology* 70, 453-465.
Mason, C.C., and Folk, R. L., 1958. Differentiation of beach, dunes, and aeolian flat environments by size analysis, Mustang Island, Texas, *Jour. Sed. Petrology* 28, 211-226.
Trask, P. D., 1930. Mechanical analysis of sediments by centrifuge. *Econ. Geology* 25, 581-599.

Cross-references: *Beach Material; Beach Material, Sorting of; Coastal Morphology, Research Methods; Sediment Analysis, Statistical Methods; Sediment Size Classification.* Vol. VI: *Grain Size Parameters.*

MONOCLINAL COAST

A *monoclinal coast* is a comparatively rare example of a shore developed by monoclinal sedimentary strata descending into the sea (Cotton, 1958), presenting a succession of differentially erodable layers, such as limestone, sandstone, or clay (Gary, McAfee, and Wolfe, 1972). Under conditions of subaerial denudation, these take the shape of asymmetrical limestone or sandstone ridges with steep slopes where the *heads* of layers outcrop facing the land, and the seaward slope corresponds to the *dip* of the bedding plane (Cotton, 1958) (Fig. 1). If this latter slope has an angle of within 7-20° from the horizontal where it crosses the shoreline, the surf run-up along this surface is the same as on shores that have been cemented or paved for protection. A cliff is not developed, and erosion proceeds very slowly, with retreat approximately along the dip of the

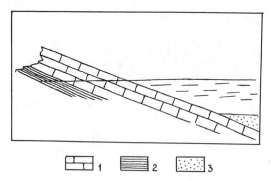

FIGURE 1. Schematic profile of the monoclinal coast: 1. limestone; 2. clays; 3. sand.

bedding plane. Erosion along this surface is evidenced by microrelief structures, usually carved into minor roughness, small holes, little ravinements, and the like. Thus we see a paradoxical example of an abrasive shore without a cliff being developed. This occurs where the dipping strata are not less than 5-7° (i.e., practically horizontal layering) or more than 20-25°, with all the other conditions that are conducive to abrasion.

Typical monoclinal coasts can be observed at a number of places along the Caspian Sea shore, where they are composed of Middle- or Upper-Sarmat limestones, Apsheron coquinas, or Middle-Pliocene sandstones. In those cases where a monoclinal coast is composed of poor limestone (for instance, the chalky limestone of the Albanian shore of the Adriatic and Ionian seas), microstructures of *marine karst*, small ravines and ridges, develop on the monoclinal slope as a result of the dilutive effect of sea water on the limestone surface.

Structural ridges, owing to the erosion of rocky layers, are also typical of the underwater portion of monoclinal coast slopes. O. K. Leontiev (1961), V. F. Solovjev (1954), and V. V. Sharkov (1964) have used the presence of these ridges along the western Caspian Sea shore to conduct marine geological surveys. By tracing the orientation of these ridges, a set of submerged structural folds and fractures were located that, in turn, led to the discovery of some oil and gas fields on the coastal shelf.

O. K. LEONTIEV
S. A. LUKJANOVA

References

Cotton, C. A., 1958. *Geomorphology; An Introduction to The Study of Landforms.* Christchurch: Whitcombe and Tombs, 505p.
Gary, M.; McAfee, R., Jr.; and Wolf, C. L., eds., 1972. *Glossary of Geology.* Washington: American Geological Institute, 805p.
Leontiev, O. K., 1961. *The Basement of Sea Shore Geomorphology* (in Russian). Moscow: Moscow State University, 418p.
Sharkov, V. V., 1964. *Geology of the Underwater Slope of the Western Caspian Sea Coast* (in Russian). Moscow, 430p.
Solovjev, V. F., 1954. Relief and structure of the Apsheron threshold (in Russian), *Akad. Nauk SSSR Izv. Ser. Geol.* **5**, 127-138.

Cross-references: *Differential Erosion; Karst Coast; Protection of Coasts.*

MONROES

Monroes are small mud mounds, 5-20 cm high and 5-25 cm in diameter, having a rounded conical shape, commonly with a small prominence at the top. They often occur in groups, on tidal flats of temperate cold regions (Fig. 1). Monroes form under the *ice foot* by upward extrusion of fluidized mud, and most probably result from downward ice pressure on mud layers (Dionne, 1973). They are a type of so-called *mud volcanoes* that have been observed in mud tidal flats along the St. Lawrence estuary.

JEAN-CLAUDE DIONNE

Reference

Dionne, J.-C., 1973. Monroes: A type of so-called mud volcanoes in tidal flats, *Jour. Sed. Petrology* **43**, 848-856.

Cross-references: *Glaciel; High-Latitude Coasts; Ice along the Shore; Ice-Bordered Coasts; Ice Foot; Kaimoo; Mud Volcanoes; Shore Polygons; Stone Packing.*

FIGURE 1. Monroe in the Montmagny mud tidal flat, St. Lawrence estuary, Quebec (photo: J.-C. Dionne).

MORRO

The Romanic languages from the Iberian Peninsula (and Latin America) as well as the Languedoc, French, and Sardinian, use similar vocabulary—morrito, morra, morro, morrone—to designate a blunt point of high land jutting into the sea (Gary, McAfee, and Wolf, 1972). Typically, these formations are fairly high cliffs, generally lacking an abrasional platform owing to a plunging profile. Most often they are cut or projected points with gradients of higher than 40% when not vertical. The generic word *morro* is often followed by determiners, usually referring to easily distinguished features observed by sailors as bearing points; for example, Morro de Bonifacio (Chile), Morro de Puercos (Panama), Morro Vermelha (Brazil), and la Morra (Spain). Limestone landscapes are particularly effective in producing and preserving the coastal forms called morros.

VINCENÇ M. ROSSELLÓ

Reference

Gary, M.; McAfee, R., Jr.; and Wolf, C. L., 1972. *Glossary of Geology*. Washington, D.C.: American Geological Institute, 805p.

Cross-references: *Cliffed Coasts; Coastal Erosion, Formations; Geographic Terminology.*

MUD FLAT—See TIDAL FLAT

MUD LUMPS

Mud lumps are islands of diapiric or intrusive clay in an area peripheral to the major passes or distributaries of the modern Mississippi delta. They may be unique to this region. The islands are generally elongate, less than a kilometer in length, and exhibit steeply dipping, reverse, and normally faulted clay strata. After elevation above sea level, they erode rapidly and form local shoals.

They form, as indicated by many studies summarized in Morgan, Coleman, and Gagliano, (1968), during the seaward advance of heavier bar *finger sands*, deposited at the mouths of distributaries, over older, previously deposited weaker, more plastic delta front and prodelta clays and silts. The weight of the bar sediments causes the diapiric intrusions, although pressure produced by methane gas generation may play a role (Hedberg, 1974). *Sea floor mounds* (not islands) offshore from the Magdelena delta of Colombia, according to Shepard (1973), may have an origin similar to the mud lumps of the Mississippi delta.

SAUL ARONOW

References

Hedberg, H. D., 1974. Relation of methane generation to undercompacted shales, shale diapirs, and mud volcanoes, *Am. Assoc. Petroleum Geologists Bull.* 58, 661–673.

Morgan, J. P.; Coleman, J. M.; and Gagliano, S. M., 1968. Diapiric structures in Mississippi delta sediments, *in* J. Braunstein and G. D. O'Brien, eds., *Diapirism and Diapirs*. Tulsa, Okla.: American Association of Petroleum Geologists Mem. 8, 145–161.

Shepard, F. P., 1973. Sea floor off Magdalena delta and Santa María area, Colombia, *Geol. Soc. America Bull.* 84, 1955-1972.

Cross-references: *Deltas; Mud Volcanoes; Shoestring Sands.* Vol. III: *Islands.*

MUD VOLCANOES

Mud volcanoes are miniature volcano-shaped features, from a few to many centimeters high and less than 100 cm in diameter, having a tiny crater or a small calderalike depression at the apex. They occur in tidal flats, on the bottoms of seasonally dry lakes, or on river bars (Williams and Rust, 1969). They result from the upward injections of liquefied clay, silt, or fine sand. Various factors could cause liquefaction of sediments that are forced upward: overloading by sediments in a reverse density gradient system, downward pressures by ice floes or ice foot, melting of segregation or buried shore ice, dewatering, seismic shocks, and slumping. Mud volcanoes frequently result from upward expulsion of water and occasionally of air or gas trapped in the surficial layered sediments. Only a few examples of modern mud and sand volcanoes are known (Neumann-Mahlkau, 1976), of which three sites are from cold regions (Bondesen, 1966; Williams and Rust, 1969; Dionne, 1976). Most occurrences in consolidated rocks have been attributed to subaqueous slumping (Burne, 1970; Gill and Keunen, 1958; Harris and Schenk, 1975; Neville, 1957; Rust, 1965; Smith, 1971; Williamsom, 1960).

In James Bay tidal flats (Fig. 1), miniature mud volcanoes are 7-15 cm high and 15-34 cm in diameter. Some volcanoes have at the apex a minute crater, 30-50 mm in diameter, that is the end of a central vent through which liquefied sediments flow; others have a calderalike depression at the top up to 9 cm in diameter and 35 mm deep. The James Bay mud vol-

FIGURE 1. A miniature mud volcano in a James Bay tidal flat, 15 cm high and 50 cm in diameter at the base, showing a calderalike depression 10 cm in diameter and a recent mud flow on the front slope (from J.-C. Dionne, 1976, Miniature mud volcanoes and other injection features in the tidal flats, James Bay, Quebec, *Canadian Jour. Earth Sci.* 3, 424. Reproduced by permission of the National Research Council of Canada).

canoes are related to the melting of segregation or buried shore ice in the layered sediments that produces liquefaction upon melting. The upward injection of sediments is probably induced by pressures of ice floes or *ice foot* (Dionne, 1976). Recent observations have shown that miniature mud volcanoes occur in various environments, including tidal flats.

JEAN-CLAUDE DIONNE

References

Bondesen, E., 1966. Observations on recent sand volcanoes, *Dansk Geol. Foren. Medd.* 16, 195-198.

Burne, R. V., 1970. The origin and significance of sand volcanoes in the Bude Formation (Cornwall), *Sedimentology* 15, 211-228.

Dionne, J.-C., 1976. Miniature mud volcanoes and other injection features in tidal flats, James Bay, Quebec, *Canadian Jour. Earth Sci.* 13, 422-428.

Gill, W. D., and Kuenen, Ph. H., 1958. Sand volcanoes on slumps in the Carboniferous of County Clare, Ireland, *Geol. Soc. London Quart. Jour.* 113, 441-460.

Harris, I. M., and Schenk, P. E., 1975. The Meguma Group, *Maritime Sed.* 11, 25-46.

Neumann-Mahlklau, P., 1976. Recent sand volcanoes in the sand of a dike under construction, *Sedimentology* 23, 421-425.

Neville, W. E., 1957. Sand volcanoes, sheet slumps, and stratigraphy of part of the Slieveardagh Coalfield, County Tipperary, *Royal Soc. Dublin Sci. Proc.* 27, 313-324.

Rust, B. R., 1965. The sedimentology and diagenesis of Silurian turbidites in south-east Wigtownshire, Scotland, *Scottish Jour. Geology* 1, 231-246.

Smith, J. L., 1971. Sand-blows, *Geotimes* 16, 1.

Williams, P. F., and Rust, B. R., 1969. The sedimentology of a braided river, *Jour. Sed. Petrology* 39, 649-679.

Williamson, I. A., 1960. A spring pit with mound structures, *Geol. Assoc. Canada Proc.* 71, 312-315.

Cross-references: *Glaciel; High-Latitude Coasts; Ice along the Shore; Ice-Bordered Coasts; Ice Foot; Kaimoo; Monroes; Shore Polygons; Stone Packing.*

MURICATE WEATHERING

Muricate weathering is the name given to the intricate pitting that develops in the spray zone on rocky shores, especially in limestones, calcareous sandstones, and dune calcarenites. As the name implies, the rock surface is roughened and rendered sharp and spiky as the result of this form of weathering. In addition to the removal of rock material to form the pits, there is secondary hardening of the intervening walls by carbonate precipitation. The result is a microtopography of pits typically 1-5 mm in depth and diameter. Processes involved include the physical and chemical effects of spray (aerated seawater), such as corrosion and salt crystallization, and the biophysical and biochemical effects of the colonizing microflors (notably algae) and browsing shelly organisms (Bird, 1976).

ERIC C. F. BIRD

Reference

Bird, E. C. F., 1976. *Coasts,* 2d ed. Cambridge, Mass.: M.I.T. Press, 246p.

Cross-references: *Alveolar Weathering; Frost Riving; Salt Weathering; Solution and Solution Pan; Tafoni; Weathering and Erosion, Biologic; Weathering and Erosion, Chemical; Weathering and Erosion, Differential; Weathering and Erosion, Mechanical.* Vol. III: *Tafoni.*

MUSHROOM—See FUNGI

N

NAVIGABLE WATERS

In the legal context, the phrase *navigable waters* refers to those waters that are now, or have been in the past, or may be in the future used for purposes of commerce. To be navigable waters of the United States, there must be either a past, present, or potential presence of interstate or foreign commerce (Black, 1976).

Navigable waters may also be a body of water where artificial aids have been or may be used to make the water suitable for navigation. A body of water retains its character as navigable in law even though it is not at present used for commerce or is now incapable of such use because of changed conditions of the presence of obstructions. Also included under the title navigable waters of the United States are all coastal and ocean waters that extend seaward from the coastline a distance of three nautical miles (200 miles for economic purposes). Also included are any bodies of water and their beds subject to tidal action, such as bays, estuaries, marshlands, and saltwater wetlands.

JOHN B. HERBICH
JOHN P. HANEY

Reference

Black, W. L., 1976. Waters of the United States, *in Proceedings of the Speciality Conference on Dredging and Its Environmental Effects,* Mobile, Alabama. New York: American Society of Civil Engineers, 10-38.

Cross-references: *Beach; Coastal Engineering; Coastal Zone Management; Geographic Terminology.*

NEARSHORE HYDRODYNAMICS AND SEDIMENTATION

For the purposes of this discussion, the *nearshore zone* is defined as a zone extending seaward from the high tide line on open coasts to an indefinite outer boundary where coastal flow, and therefore coastal sedimentation, is no longer appreciably affected by the proximity of the coast. The effects of coastal proximity may be diagnosed by grain size distribution patterns, bed form patterns, or other criteria. The transition commonly occurs at the 20-m isobath. The nearshore zone includes two major morphologic provinces. The *shoreface* (Barrell, 1912) is a more steeply dipping sector of the shelf floor, extending from the high tide line to a depth of 10-20 m, some 5-10 km from the beach, where it joins the more gently dipping *inner shelf floor.* On unconsolidated coasts, the bathymetric profile through the two provinces tends to take the form of an exponential curve, with the steeper segment being the shoreface. On rocky coasts, the two provinces tend to be less well developed and may be difficult or impossible to distinguish. The exponentially curved shoreface surface is generally thought to be an equilibrium response to the coastal hydraulic climate, and in particular to the regime of shoaling waves (Fenneman, 1902; Johnson, 1919; Zenkovich, 1967; Wright and Coleman, 1972), but the mechanisms are poorly understood.

Nearshore Hydraulic Climate

The two morphologic provinces described above interact with the fluid motions of the overlying water column in such a way as to result in several hydraulic zones (Fig. 1). A *zone of wave-dominated flow* extends from the water line out to a depth of perhaps 10 m; the precise location of the outer limit depends on the wave climate. In such shallow water, movements of water in response to regional pressure gradients are damped by bottom friction. The dominant fluid motion experienced by the sea floor is oscillatory wave surge, and most other important categories, such as longshore and rip currents, are driven partially or wholly by the momentum flux gradients associated with shoaling and breaking waves (Bowen, 1969a, b). Water motions resulting in bottom shear stress sufficient to transport sediment in this zone are essentially continuous.

A *zone of friction-dominated flow* extends seaward from the outer margin of the wave-dominated zone. Here low-frequency fluid motions induced by the passage of the semidiurnal tidal wave or by periods of prolonged and intense wind stress are more important than high-frequency wave motions in the energy budget of the sea floor, although wave

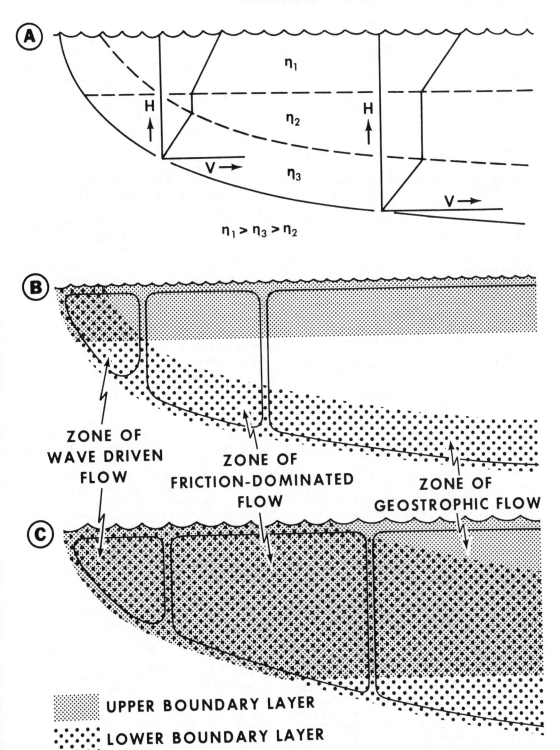

FIGURE 1. Velocity structure of the shoreface and inner shelf. A: general form of velocity profiles through the upper boundary layer, core flow, and lower boundary layer, and relative values of eddy viscosity. B: velocity structure during a period of relatively mild flow. C: velocity structure during peak flow (from Swift, 1976).

surge is still an important subordinate influence. In the equation of motion governing flow in this region, a frictional term is balanced primarily by direct wind stress or by the regional pressure gradient associated with the semidiurnal tidal wave.

Flow events in this zone last for hours or days and, except on coasts with high tide ranges, tend to be separated by weeks of quiescence, in which flow does not exert a bottom shear stress sufficient to entrain sediment. Flow tends to be *Couette-like*, in that there is a top-to-bottom velocity gradient. Flow is predominantly coast-parallel, but an onshore or offshore component of bottom flow may result as a consequence of wind-induced set-up or set-down of water against the coast. Flows in the friction-dominated zone are generally coast-parallel, for continuity reasons. However, they may respond to winds that blow normal or obliquely to the coast with two- or three-layer flow patterns. Onshore winds cause onshore movement of the surface layer downwelling and an offshore movement of the bottom layer; or if the water column is stratified, top and bottom layers may move landward, while the central layer moves seaward (Murray, 1975; Scott and Csanady, 1976). Offshore winds cause offshore movement of surface water, coastal upwelling, and landward movement of bottom water (Lavelle et al., 1976). When winds blow obliquely to the beach, as is more commonly the case, coast-normal flow components are superimposed on a a stronger coast-parallel component.

Seaward of the zone of friction-dominated flow is a *zone of geostrophic* flow. In the equation of motion governing geostrophic flow, a pressure term is balanced primarily by the coriolis term and only secondarily by the friction term. Ideally, a three-layered flow system prevails (Ekman, 1905; Neumann and Pierson, 1966). An upper wind-stressed and wave-stirred boundry layer experiences a vertical velocity gradient and the spiral Ekman velocity structure, in which the trajectory of each deeper, more slowly moving, layer is deviated to the right (in the Northern Hemisphere) of the layer above. A central layer moves in slablike fashion (no vertical velocity gradient) parallel to the beach in a downwind direction. It overrides a basal boundary layer, whose fluid is sheared against the sea floor. Here frictional retardation results in a left-handed spiral velocity structure. Thus a wind blowing equatorward along a shelf with the beach on the right will drive surface water toward the beach and bottom water offshore (east coast of North America), while a wind blowing equatorward along a shelf with the beach on the left will drive surface water seaward and bottom water landward (west coast of North America). The boundary between the friction-dominated and geostrophic zones may be described by a characteristic *Ekman number,* the dimensionless ratio of the coriolis and frictional terms in the equation of motion.

It seems probable that during peak flow events when a critical *Reynolds number* is exceeded, the boundary layers of the geostrophic zone thicken and overlap, and top-to-bottom instability results, in which flow-parallel lanes of downwelling water alternate with lanes of upwelling water (Faller, 1971). A recent theoretical study suggests that the zone of friction-dominated flow may experience a similar instability during peak flow events (Gammelsrod, 1975). Such a flow structure, resulting in an alternation of bottom convergences and divergences, may be responsible for the patterns of sand ribbons and erosional furrows found on continental shelves (Newton, Seibold, and Werner, 1973; McKinney, Stubblefield, and Swift, 1974).

Erosional Retreat of the Coastal Equilibrium Profile

It has been noted above that the shoreface tends to be an exponentially curved concave-upward surface, and is generally inferred to be an equilibrium response to the coastal hydraulic regime. The tacit understanding is that the sandy sea floor erodes (or aggrades) to an ideal surface, whose depth at every point is that at which average wave orbital velocity is just sufficient to stir the available grade of sand for a given time-averaged horizontal gradient of sand discharge. To date a fully satisfactory theoretical foundation for this hypothesis has not been advanced.

The ability of the shoreface surface to respond to the hydraulic regime requires that one portion of the profile be able to erode while another portion be able to aggrade. Thus, if a given unit volume of the coast is a closed system, there must be a mechanism for shifting sediment onshore or offshore; or if it is an open system, then a change in the surface can be accomplished by means of a longshore gradient in sand discharge, so that more sand is introduced into the coastal box than is removed (or vice versa) by the steady stream of littoral drift and by the more intermittent tide- or storm-driven sand flux in the zone of friction-dominated flow seaward of the surf (Fig. 2). The available evidence indicates that both onshore-offshore and coast-parallel sediment transport occurs in both zones, although the latter dominates, so that long-term transport vectors make small angles with the coast.

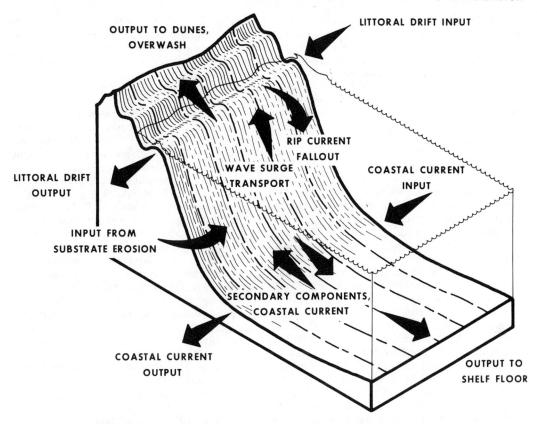

FIGURE 2. The unit volume of the coastal water mass and its substrate. Arrows indicate components of sediment and water transport. Coast-parallel water movements are more intense than coast-normal movements; hence resultant transport vectors tend to make low angles with the coast (from Swift, 1975b).

Onshore-offshore sand transport tends to be cyclic in nature. It has long been known that the high, steep waves of storms tend to strip sand off the beach and move it offshore into a storm-wave-built bar, which in turn tends to move slowly onshore during the ensuing period of fair-weather swells (Davis and Fox, 1972; Sonu and Van Beek, 1971). In many areas, an annual cycle results, in which sand storage in the beach prism is at a maximum during the quiet summer months and at a minimum during the winter period of frequent storms.

This annual cycle of onshore-offshore beach transport tends to interact with a longer-term cycle involving the entire shoreface. A significant portion of the sand removed from the beach during storms is transported beyond the bar by rip currents, to rain out on the upper shoreface. As a consequence, the shoreface tends with time to aggrade toward the ideal wave-graded profile. Moody (1964) has described this process on the Delaware coast. He notes that the steepening is not continuous, but varies with the frequency of storms and duration of intervening fair-weather periods. The slope of the Delaware shoreface steepened from 1:40 to 1:25 between 1929 and 1954, but erosion on the upper shoreface between 1954 and 1961 regraded the slope to 1:40. The steepening process is terminated by a major storm, during which time the gradient is reduced and a significant landward translation of the shoreline occurs. Moody (1964) describes the great Ash Wednesday storm of 1962, bracketed within his time series, as having stalled for 72 hours off the central Atlantic coast. Its storm surge raised the surf into the dunes for six successive high tides. The shoreline receded 18–75 m during the storm. While much of the sand was transported over the barrier to build *washover fans* over 1 m thick, much more was swept back onto the sea floor by rip currents (Moody, 1964) and perhaps by obliquely seaward bottom flow on the shoreface, as described in the preceding section.

The geometry of a retreating shoreface during a period of rising sea level has been examined by Fischer (1961) and Bruun (1962). The profile translates landward and upward in response to sea level rise by a process of episodic upper shoreface erosion and aggradation of the adjacent sea floor. (Fig. 3). If there is no coastwise gradient of sand discharge, so that sand input from upcoast equals sand output downcoast, and if the coast is two-dimensional in the sense that successive downcoast profiles are identical, then the transfer of sand from the upper shoreface to the adjacent shelf floor would be an equal volume process. However, these conditions are rarely met.

Sedimentary Regimes

Erosional shoreface retreat is a special case of a more general mechanism whereby the coastal profile may either retreat landward or prograde seaward, depending on the rate of sediment supply and the rate and sense of sea level displacement (Curray, 1964). The interaction of these two variables gives rise to two basic sedimentary regimes. *Autochthonous sedimentation* (Naumann, 1858; Swift, Stanley, and Curray, 1971) occurs during erosional transgression (term from Curray, 1964) when the rise in sea level is rapid relative to the rate of sediment supply. In these conditions, estuaries deepen in response to sea level rise faster than they can be filled by river sediment, and become efficient traps for river sediment (Fig. 4). The major source of coastal sediment is the erosion of the shoreface as it translates landward and upward in response to rising sea level (Fig. 3A). The debris thus generated tends to accumulate in shoals and spits at the downcurrent ends of littoral drift cells (Swift and Sears, 1974) and also as a transgressive shelf sand sheet, seaward of the shoreface, that unconformably overlies eroded back barrier or lagoonal deposits.

In such a regime, the shoreface tends to consist of two distinct textural provinces: an upper shoreface province of fine, seaward fining sand, and a lower shoreface province of more variable and generally coarser sand (Howard and Reineck, 1972). The fine sand of the upper shoreface is interpreted as veneer of rip current fallout that advances down the shoreface during fair weather, and retreats or is stripped off altogether during major storms. The coarser sand of the lower shoreface is interpreted as a thin lag deposit resulting from storm current erosion of the underlying strata.

The shelf floor sand sheet consists of extremely fine to extremely coarse sand, but coarser material predominates. Fine sand has been swept over the barrier into the lagoon and has been overridden by the retreating barrier, or has been swept downcoast and deposited in shoals (Swift and Sears, 1974) or in low areas on the shelf surface. The sand sheet may exhibit up to 10 m of relief in such areas as the Southern Bight of the North Sea (Swift, 1975a) or the Middle Atlantic Bight of the North American Atlantic shelf, (Swift, Duane, and McKinney, 1973). Relief takes the form of a sand ridge topography that is initiated at the transgressive sand sheet's leading (nearshore) edge, and is maintained by shelf currents as the

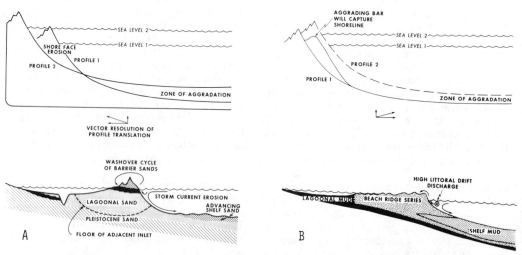

FIGURE 3. Dynamic and stratigraphic models for (A) a retrograding and (B) a prograding coast during a rise in sea level. If coastwise sand imports are balanced by, or are less than, coastwise sand exports, the hydraulically maintained coastal profile must translate upward and landward by a process of shoreface erosion and concomitant aggradation of the adjacent sea floor (Bruun, 1962). If coastwise sand imports exceed exports, as in the case for deltaic coasts, then the profile must translate seaward and upward (based on Curray, Emmel, and Crampton, 1969; from Swift, 1976).

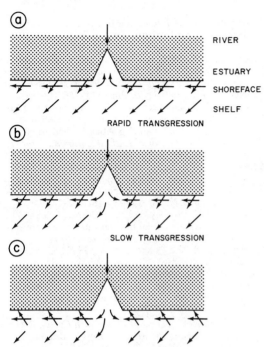

FIGURE 4. Sense of net sediment transport for (*a*) rapid transgression, (*b*) slow transgression, and (*c*) regression. Offshore component of transport is exaggerated (from Swift, 1976).

shoreline continues its landward translation, and the water column deepens.

Autochthonous sedimentation is characteristic of most coasts in the present period of relatively rapid glacio-eustatic sea level rise.

Allochthonous sedimentation occurs under conditions of depositional transgression or regression (Curray, 1964) where the rate of sediment supply is high relative to the rate of sea level rise. Estuaries can close down to equilibrium channels, whose cross-sectional areas are determined by tidal discharge. Such estuaries bypass sand to the lunate river mouth shoal during periods of flooding or spring high tide, and the sand is entrained into the littoral drift system. If the rate at which sediment is delivered to the coast is not adequate to stabilize it, then it will continue to retreat by means of shoreface erosion.

The distribution of textural provinces under such conditions of depositional transgression differs from the case of erosional transgression as described above. The sediment delivered by the eroding hinterland to the river systems undergoes preferential deposition of the coarser size fraction, and the load delivered at the river mouth is primarily fine to extremely fine sand, silt, and clay (Allen, 1965). This fine, more mobile size fraction is introduced into the littoral drift in vast quantities, and is steadily released by rip currents to the inner shelf, where it undergoes periodic resuspension and dispersal in response to peak tidal currents or storm currents. As a consequence, the blanket of rip current fallout continues seaward down the shoreface and across the inner shelf, becoming finer-grained and thicker in this direction. This material is stripped off the upper shoreface during major storms when erosional shoreface retreat occurs, but the seaward portion is too thick to be completely removed, and it becomes slowly thicker as the water column deepens and the shoreline continues to move landward.

On such coasts, extremely fine sand may pass seaward into mud at depths of 10 m or less. The coastal profile is exceedingly gentle as a consequence of the low effective angle of repose of such fine sediment; the 10-m isobath may be as far as 10 km from the beach.

Allochthonous sedimentation under conditions of depositional transgression, in which equilibrium estuaries bypass sand but the shoreface continues to undergo erosional retreat, was probably more characteristic of the slow transgressions of the geologic record than of the rapid post-glacial sea level rise. Modern examples of allochthonous sedimentation usually occur where the supply of sediment is so voluminous as to reverse the sense of shoreface translation, so that the coastline progrades (the sea undergoes regression; Fig. 3B), creating a *chenier* or *beach ridge* plain. The coastal sedimentary regime is similar to that of the depositional regression case, except that sediment is supplied entirely by rivers and there is no coastal erosion.

DONALD J. P. SWIFT

References

Allen, J. R. L., 1965. Late Quaternary Niger Delta, and adjacent areas: Sedimentary environments and lithofacies, *Am. Assoc. Petroleum Geologists Bull.* 49, 547-600.
Barrell, J., 1912. Criteria for the recognition of ancient delta deposits, *Geol. Soc. America Bull.* 23, 377-446.
Bowen, A. J., 1969a. Rip currents, 1. Theoretical investigations, *Jour. Geophys. Res.* 74, 5467-5478.
Bowen, A. J., 1969b. The generation of longshore currents on a plane beach, *Jour. Marine Research* 27, 206-215.
Bruun, P., 1962. Sea level rise as a cause of shore erosion, *Am. Soc. Civ. Engineers Proc., Jour. Waterways and Harbors Div.* 88, 117-130.
Curray, J. R., 1964. Transgressions and regressions, in R. L. Miller, ed., *Papers in Marine Geology: Shepard Commemorative Volume.* New York: Macmillan, 179-203.

Curray, J. R.; Emmel, F. J.; and Crampton, P. J. S. 1969. Holocene history of a strand plain lagoonal coast, Mayarit, Mexico, in *Lagunas Costeras, Un Simposio*, Mem. Simp. Internat. Lagunas Costeras, UNAM-UNESCO, 1967, Mexico, 63-100.

Davis, R. A., Jr., and Fox, W. T., 1972. Coastal processes and nearshore sand bars, *Jour. Sed. Petrology* 42, 401-412.

Ekman, V. W., 1905. On the influence of the earth's rotation on ocean currents, *Arkiv Matematik, Astron., Fysik* 2, 1-53.

Faller, A. J., 1971. Oceanic turbulence and the Langmuir circulations, *Ann. Rev. Ecology and Systematics* 2, 201-235.

Fenneman, N. M., 1902. Development of the profile of equilibrium of the subaqueous shore terrace, *Jour. Geology* 10, 1-32.

Fischer, A. G., 1961. Stratigraphic record of transgressing seas in the light of sedimentation on the Atlantic coast of New Jersey, *Am. Assoc. Petroleum Geologists Bull.* 45, 1656-1660.

Gammelsrod T., 1975. Instability of Couette flow in a rotating fluid and origin of Langmuir circulations, *Jour. Geophys. Res.* 80, 5069-5075.

Howard, J. D., and Reineck, H. E., 1972. Georgia coastal region, Sapelo Island, U.S.A.: Sedimentology and biology. IV. Physical and biogenic sedimentary structure of the nearshore shelf, *Senckenbergiana Maritima* 4, 81-123.

Johnson, D. W., 1919. *Shore Process and Shoreline Development*. New York: John Wiley & Sons, 584p.

Lavelle, J. W.; Brashear, H. R.; Case, F. N.; Charnell, R. L.; Gadd, P. E.; Haff, K. W.; Han, G. A.; Kunselman, C. A.; Mayer, D. A.; Stubblefield, W. L.; and Swift, D. J. P., 1976. Preliminary results of coincident current meter and sediment transport observations for wintertime conditions Long Island inner shelf, *Geophys. Research Letters* 3, 97-100.

McKinney, T. F.; Stubblefield, W. L.; and Swift, D. J. P., 1974. Large scale current lineations on the Great Egg Shoal retreat massif: Investigation by sidescan sonar, *Marine Geology* 17, 79-102.

Moody, D. W., 1964. *Coastal Morphology and Processes in Relation to the Development of Submarine and Sand Ridges off Bethany Beach, Delaware*, Ph.D. dissertation, Johns Hopkins University, 167p.

Murray, S. P., 1975. Trajectories and speeds of wind-driven currents near the coast, *Jour. Phys. Oceanogr.* 5, 347-360.

Naumann, C. F., 1858. *Lehrbuch der Geognosie*, Bd. 1, 2d ed., Liepzig: Wilhelm Engelmann, 960p.

Neumann, G., and Pierson, W. J., Jr., 1966. *Principles of Physical Oceanography*. Englewood Cliffs, N.J.: Prentice-Hall, 545p.

Newton, R.S.; Seibold, E.; and Werner, F., 1973. Facies distribution patterns on the Spanish Sahara continental shelf mapped with sidescan sonar, *Meteor. Forsch. Ergebnisse* 15, 55-77.

Scott, J. T., and Csanady, G. T., 1976. Nearshore currents off Long Island, *Jour. Geophys. Research* 81, 5401-5409.

Sonu, C. J., and Van Beek, J. L., 1971. Systematic beach changes on the Outer Banks, North Carolina, *Jour. Geology* 79, 416-425.

Swift, D. J. P., 1975a. Tidal sand ridges and shoal-retreat massifs, *Marine Geology* 18, 105-134.

Swift, D. J. P., 1975b. Barrier island genesis: Evidence from the central Atlantic shelf, eastern U.S.A., *Sed. Geology* 14, 1-43.

Swift, D. J. P., 1976. Continental shelf sedimentation, in D. J. Stanley and D. J. P. Swift, eds., *Marine Sediment Transport and Environmental Management*. New York: John Wiley & Sons, 311-350.

Swift, D. J. P., and Sears, P., 1974. Estuarine and littoral despositional patterns in the surficial sand sheet, central and southern Atlantic shelf of North America, *Inst. Géologie Bassin d'Aquitaine Mem.* 7, 171-189.

Swift, D. J. P., Duane, D. B.; and McKinney, T. F., 1973. Ridge and swale topography of the Middle Atlantic Bight: Secular response to Holocene hydraulic regime, *Marine Geology* 15, 227-247.

Swift, D. J. P.; Stanley, D.J.; and Curray, J. R., 1971. Relict sediments on continental shelves: A reconsideration, *Jour. Geology* 79, 322-346.

Wright, L. D., and Coleman, J. M., 1972. River delta morphology: Wave climate and the role of the subaqueous profile, *Science* 176, 282-284.

Zenkovich, V. P., 1967. *Processes of Coastal Development*. New York: Wiley Interscience, 738p.

Cross-references: *Barrier Islands; Beach Cycles; Beach Material, Sorting of; Beach Processes; Bruun Rule; Currents; Equilibrium Shoreline; Profile of Equilibrium; Ravinement; Sediment Transport; Tides; Waves.* Vol. VI: *Littoral Processes.*

NEARSHORE WATER CHARACTERISTICS

The characteristics of nearshore waters that distinguish them from the open ocean are determined primarily by the proximity of the land. Usually there is an admixture of fresh water runoff, but this is often hardly noticeable along rocky shores where runoff is small. In areas where evaporation exceeds precipitation (mainly in the subtropics), *salinity* tends to increase toward the shore, especially in inlets and lagoons. Generally insolation and cooling have more effect in nearshore waters because of the shallow depth, resulting in higher or lower *temperatures* than farther offshore. Wave action and tidal effects are concentrated in nearshore waters because the coast forms a barrier and energy is released in the form of breaking waves and higher current velocities. Erosion products of the land, in solution or in particulate form, are present, depending on conditions on the land; but concentrations of land-derived substances such as *nutrients, trace metals,* and *suspended material,* are higher nearshore than in the open ocean. *Organic productivity* is usually higher near the coast because of the inflow of nutrients from the land, the higher water temperatures, and the shallow depth, which enable bottom fauna to get its share of organic matter produced in the surface water. A high *turbidity* of the nearshore water, however, reduces productivity, but when sufficient light is able to penetrate the surface water and to reach the sea floor, a rich algal vegetation or coral growth is possible. As most *pollution* is derived from the land, pollution is concentrated mainly in nearshore waters.

D. EISMA

Cross-references: *Coastal Waters Habitat; Nutrients; Pollutants; Production; Standing Stock; Thermal Pollution.*

NEHRUNG

Nehrung, a German term derived from the Baltic (Old-Prussian) word "nerija," means a submerging land, and defines coastal barriers or spits isolating the lagoons (*firths*) from the open seas. The largest nehrungs of the Baltic are the Kurische (Kuršiu) and the Frische (Vistula). The length of the Kurische is 98 km. Only the narrow Klaipeda Strait separates it from the mainland. The width of the spit varies from 0.4-3.8 km. The sea shoreline of the spit is concave but even. On the other hand, the lagoon shoreline is cuspated. The bedrock of the Kurische spit is represented by pre-Quaternary deposits of Cretaceous and Jurassic ages, covered by a thick layer of glacial drift consisting of till banks with intercallated glaciofluvial and glaciolacustrine sediments. Holocene sediments (marine, lagoonal, and aeolian) make up an overlying complex, reaching a thickness of up to 100 m. On the Kurische there are well-developed dunes, the highest coastal dunes in Europe. Almost half of these mobile sand dunes reach a height of 50-60 m above m.s.l. The maximum height of the forested (fixed) dunes is approximately 64 m above m.s.l. (Gudelis, 1970).

Sea beaches of the spit are predominantly sandy. In some places the presence of gravel and pebble admixture may be noticed. Along the seashore a prominent foredune ridge is seen, formed mainly by human intervention. Between the foredunes and the longitudinal ridge of the high dunes lies a deflation plain, the so-called (in Lithuanian) palvé (q.v.). The lee slopes of the dune ridge contact directly with the lagoon beach.

In many places on the spit, ancient lagoonal and lake sediments consisting of *gyttja*, peat, and silty sands were found beneath the marine layers of the Litorina Sea. This evidence proves that the spit as a whole has migrated eastward upon the former lagoon basin area (Wichdorff, 1919).

The Frische nehrung is smaller, and the geological structure is similar to that of Kurische. Its length is 60 km; the width varies from 1.2-2 km. The dune ridge of this spit is lower (30-40 m in average) and covered by forests. The Frische spit is separated by the narrow Pillau (Baltijsk) Strait from the Sambian Peninsula. In the historical past there were several such channels that were later closed by the shore drift.

Accumulated sand beaches prevail on both of these Baltic spits. Erosional shores are limited and attributed to the root parts of the spits. Along the Kurische spit there has been considerable longshore debris transport, which dates from the late-Litorina period.

The formation of both Baltic spits began in the late-Atlantic period. In initial stages underwater accumulative sand bodies were formed. After the Litorina Sea regression these forms appeared above sea level in the form of bars. The intensive sand transport and accumulation caused the growth of the initial spits and their broadening. The ancient dunes originated by their migration toward the lagoon, which led to the advance of the spit on the formerly lagoon area.

The ancient dunes of the spit were parabolic. The cuspated shoreline of the lagoonal shore of the Kurische spit was created by cusps, formed by blown dune sand intruding into the lagoon. The development of the lagoon shoreline in some places is still intensive. From late Litorina time the Kurische spit has shifted eastward at least 2-4 km in its southern and middle parts.

VYTAUTAS GUDELIS

References

Gudelis, V., 1970. *Kuršiu nerija.* Vilnius: Leidykla Mintis, 75p.

Wichdorff, H. von, 1919. *Geologie der Kurischen Nehrung.* Berlin: Abhandlungen der Preussichen Geologischen Landesanstalt, 196p.

Cross-references: *Bars; Europe, Coastal Morphology; Geographic Terminology; Gyttja; Haff; Palve; Soviet Union, Coastal Morphology; Spits.*

NEMERTINA

The invertebrate phylum Nemertina (Nemertea, Rhynchocoela) comprises bilaterally symmetrical, acoelomate, unsegmented, vermiform animals that are closely related to *turbellarian flatworms* (phylum Platyhelminthes) but exhibit a higher level of organization by possessing a blood vascular system and a definitive anus at the posterior end of the body (Gontcharoff, 1961; Gibson, 1972). The phylum includes the class Anopla with the orders Palaeonemertini and Heteronemertini and the class Enopla with the orders Hoplonemertini and Bdellonemertini. Approximately 800 species are known.

Nemerteans, also called *ribbon worms,* vary in length from 1 mm to about 30 m. Some species are white or grayish and transparent, whereas others are brightly colored in shades of red, orange, yellow, green, and brown, and patterns formed by longitudinal stripes and/or transverse bands in contrasting colors are not uncommon. The usually slender, soft, and strongly contractile body is covered by a ciliated glandular epidermis. A body cavity is lacking, and the space between the body wall musculature and the alimentary canal is filled with connective tissue (parenchyma) in which the internal organs are embedded. A proboscis, contained in a special chamber (rhynchocoel) dorsal to the gut and eversible through a pore near the anterior end of the body, is characteristic of all nemerteans. It is employed for defense and also for the capture of prey, consisting primarily of worms and small crustaceans.

Apart from a few species that have invaded freshwater and humid, terrestrial habitats, nemerteans are marine organisms and are found in all oceans and at all latitudes. They inhabit a wide range of niches from the high tide level down to abyssal depths. A number of species have acquired a bathypelagic mode of life, but the majority of nemerteans are bottom-dwellers and are particularly abundant in coastal environments, where they either burrow in sediments or live among plant tangles and crusts of sessile invertebrates and in crevices on rocky shores and coral reefs. Some species occur as commensals in sponges, bivalved mollusks, and ascidians, and a few are ectoparasitic on crabs.

ERNST KIRSTEUER

References

Gibson, R., 1972. *Nemerteans.* London: Hutchinson & Co., 224p.

Gontcharoff, M., 1961. Némertiens, in P. P. Grasse, ed., *Traité de Zoologie,* vol. 4. Paris: Masson et Campagnie, 783-886.

Cross-references: *Annelida; Coastal Fauna; Intertidal Mud Habitat; Intertidal Sand Habitat; Organism-Sediment Relationship; Platyhelminthes.*

NEUTRAL SHORELINE— See CLASSIFICATION

NEW ZEALAND, COASTAL ECOLOGY

New Zealand has an extensive coastline with every type of habitat represented. Since the country extends over many degrees of latitude, there is considerable variation in the composition of the intertidal flora and fauna. The following habitats can be recognized (Morton and Miller, 1968): exposed and protected *hard shore lines* in both the northern and southern regions, *caves, boulder beaches, harbors and estuaries,* open and protected *sand beaches, mud flats, mangrove swamp,* and *salt marshes.*

Hard Shore Lines

Exposed. Typical zonations on the exposed north Aukland east coast shores are given in Figure 1. Zonation A represents a slightly more protected area than zonation B. These zonations are by no means universal, and variations do occur. Thus, below the *Chamaesipho brunnea* zone, one may find a belt of *Gigartina alveata,* or *C. brunnea* may be replaced by *C. columna,* followed below by *Modiolus neozelanicus, Elminius plicatus, Pachymenia himantophora,* a *Corallina officinalis* turf, and *Perna canaliculus.*

On the exposed west coast of the North Island, one finds a zonation more characteristic of exposed southern shores, a zonation dominated by the giant bull kelp, *Durvillea antarctica.* A typical zonation is given in Figure 2.

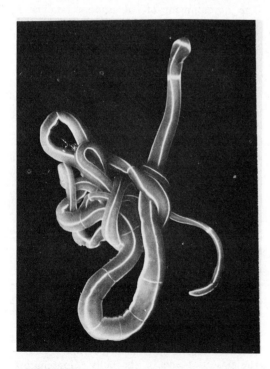

FIGURE 1. *Tubulanus annulatus,* a palaeonemertean of red color with white stripes and bands (photo: E. Kirsteuer).

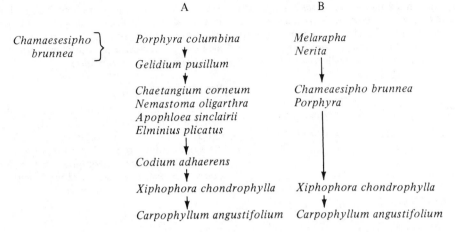

FIGURE 1. Typical zonations on the exposed north Auckland east coast shores.

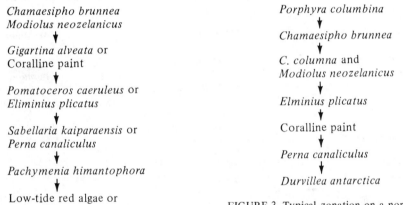

FIGURE 2. Typical zonation on the exposed west coast of North Island.

FIGURE 3. Typical zonation on a northern portion of South Island.

Further south on the northern portion of the South Island, a typical zonation is as given in Figure 3.

On the east coast, the zoning may be (from upper to lower beach) *Modiolus neozelanicus-Mytilus edulis-Eliminius plicatus* or *Aulacomya maoriana→Durvillea* and *Perna canaliculus*. In southern areas (e.g., Otago peninsula), the red algae *Bostrychia arbuscula* is associated with the *Champaesipho,* and the coralline paint also has the associated red alga *Pachymenia lusoria,* while below *D. antarctica* is the other bull kelp, *D. willana*.

Protected. On the sheltered northland coastline, a typical zonation from Great Barrier Island is as follows (Morton and Miller, 1968): *Eliminius modestus → Crassostrea glomerata → Hormosira banksii* and *Corallina* turf → *Ecklonia radiata* and *Sargassum sinclairii*. Reef terraces on sheltered or semisheltered shores are characterized by abundant *Hormosira banksii* in the pools and a mixture of large brown algae (*Carpophyllum* spp., *Cystophora retroflexa, Halopteris hordacea*), and red algal turf species at the reef edge. A comparable zonation from the South Island (Otago) is as given in Figure 4. Southern protected coasts from Cook Strait southward are characterized by the sub-tidal large brown alga *Macrocystis pyrifera*. Zonations that have been recorded are given in Figure 5.

Caves

Zonation in caves depends on the light intensity and seepage moisture. Dellow and Cassie (1955) illustrated the variation that occurs in a typical cave situation (Fig. 6).

Apart from caves, other modifying factors on hard seashores are represented by sand accumulation, where, if it occurs up to the midlittoral, it can result in a zone of *Sabellaria kaiparaensis* on the low side with *Pomatoceros caeruleus* on the exposed side. Freshwater stream runoffs inevitably introduce *Enteromorpha* species. In deep gulleys with vertical walls, where light and evaporation are much reduced and there is heavy surge, algae are largely replaced by brightly colored sponges. In the shade there also occur anemones and ascidians with the soft coral *Alcyonium aurantiacum*.

Boulder Beaches

These occur in a variety of places, and the fauna and flora vary between the lee side and exposed side and also on the underside. Molluscs are common on this type of shoreline. Among the algae, *Apophloea sinclairii* and *Codium adhaerens* can be found with large brown fucoids at low tide mark.

Harbors and Estuaries

In both these physiographic areas, one can find rocky shores and reefs that generally have a fine coating of mud or sand. Despite this,

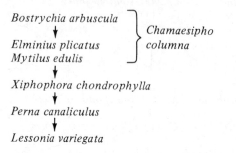

FIGURE 4. Zonation on the protected coastline of South Island.

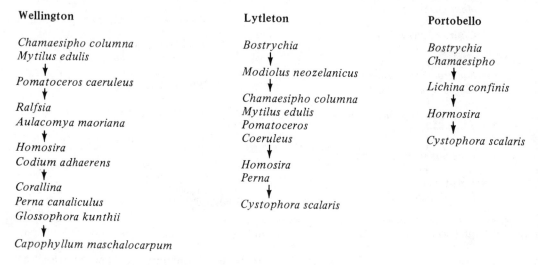

FIGURE 5. Zonations on the southern protected coasts of Cook Strait.

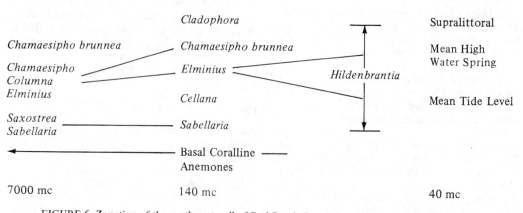

FIGURE 6. Zonation of the northwest wall of Red Beach Cave (after Dellow and Cassie, 1955).

they do have a more or less characteristic fauna and flora. Some brown algae can extend long distances into the harbor; but green algae become more prominent with decreasing salinity. Figure 7 illustrates zonations on reefs in harbors in the north of New Zealand.

As an example of what can be found in estuaries, one may cite the zonation from Riverhead (Morton and Miller, 1968), which starts on the concrete pile of a bridge and terminates on a flat rock shore: *Rhizoclonium*→*Oscillatoria nigroviridis*→*Elminius modestus* and *Caloglossa leprieurii*→*Modiolus fluviatilis* and *Gelidium pusillum*→*Wittrockiella salina* (Chlorophyceae). This zonation extends essentially as far up the estuary as the tide reaches.

Sand Beaches

Open. Because of the instability of the habitat, exposed sand beaches do not carry any vegetation until the drift-line zone is reached at extreme high tide mark. There is, however, an extensive and significant animal population that varies depending on the height above extreme low water mark. Thus the upper beach is occupied by the sea slater, the sand louse, and the sand-hopper. The middle beach is characterized by isopods, while the lower beach provides a home for amphipods, crabs, and shrimps. The mollusca are significant features, especially on particular beaches. Thus the toheroa (*Amphidesma ventricosum*) on North Island beaches is an important food animal, and its collection is regulated by the government. There are also a number of polychaete worms to be found.

Protected. Like the exposed sandy beaches, these are characterized by their crustacea. The upper beach does not possess an abundance of isopods, though the sand-hopper (*Talorchestia quoyana*) is abundant. The middle beach does have two abundant isopods, while the lower beach has an abundance of sand shrimps and burrowing shrimp. Other Crustacea include Ostracoda, Nebaliacea, and Cumacea. Among the mollusca, the pipi (*Amphidesma australe*) is characteristic and much sought after for food. At low tide level, there is the abundant cockle *Chione stutchburyi*. On the protected beaches, sand-burrowing and surface-dwelling gastropods are especially abundant, and the polychaete worms also increase in variety, *Platynereis australis* being associated particularly with the middle beach. On northern sand flats, the burrowing acorn worm, *Balanoglossus australiensis*, is common. There is also a common echinoderm, *Trochodota dendyi*, which resembles a wormlike sea cucumber.

Mud Flats

These areas are found characteristically in harbors and estuaries where there is a mingling of freshwater with seawater so that there can be, especially at the upper reaches, substantial changes in salinity. Some of the flats have a high proportion of sand, so that the substrate is firm and consolidated. Morton and Miller (1968) have divided the lower harbor flats into neap and spring flats, the latter having a much softer substrate than the former. Five bivalves (pipi, cockle, nutshell, the estuarine trough shell, and *Macomona*) are common on all harbor flats, and there is also a small typical group of gastropods, including *Amphibola crenata*, which also occurs in the salt marshes. These flats also have a high population of burrowing

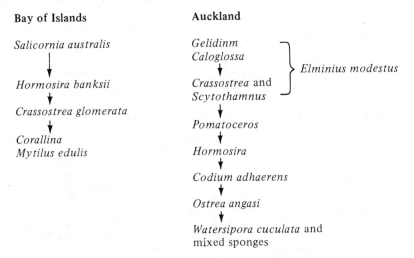

FIGURE 7. Zonations on reefs in harbors in northern New Zealand.

crabs and sea anemones, but these are outnumbered by the burrowing worms, the nereids being particularly important.

In estuarine flats, the animals fall into three groups: truly marine species; estuarine animals tolerating regular, widely fluctuating salinity; and freshwater species that can tolerate some salinity. This division is also reflected in the salt marsh plants that border such mud flats.

Mangrove Swamps

Only one species of mangrove occurs in protected estuaries and harbors of New Zealand, and that forms swamps between North Cape and Raglan on the west coast and Ohiwa on the east coast (Chapman, 1979). The mangrove *Avicennia marina* var. *resinifera* forms the front vegetation zone and is usually succeeded by a *Juncus–Leptocarpus* salt marsh in which isolated mangrove can be found. In front of the mangrove there may be beds of eelgrass (*Zostera*). On the mangrove pneumatophores there is either an algal turf of *Bostrychia harveyi*, *Catenella nipae* and *Caloglossa leprieurii* or an animal population of *Saccrostrea cucullata*, *Elminius plicatus*, and *Modiolus neozelanicus*. The fiddler crab *Helice crassa* and the molluscs *Amphibola crenata* and *Zeacumantus lutulentus* crawl over the mud and up the lower parts of the trees. On the surface mud within the mangrove forest, one can find two algae: the Rhodophyta *Gracilaria secundata* var. *pseudoflagellifera*, and the free-living form of the fucoid *Hormosira banksii*. The mangrove are tall in the north, but decrease in size as they approach their southern boundary, until finally they form a low scrub. This is regarded as a result of frosts killing off the terminal buds of erect shoots (Chapman, 1976).

Salt Marsh

While high salt marsh (rush type) occurs behind mangrove in the north of New Zealand, once the mangrove has disappeared, salt marsh replaces it in the intertidal zone in protected areas to the extreme south (Chapman, 1970). Such salt marsh develops mainly in muddy areas, and frequently there are eelgrass meadows at extreme low tide, but usually with a bare mud zone between them and pioneer salt marsh of *Salicornia australiensis*.

Two species of *Zostera* (*Z. nana* and *Z. tasmanica*) form the eelgrass beds, and contain a rich fauna of marine worms, gastropods, sea hare, crustacea, and small invertebrates (brittle star, marine caddis). The rich fauna and flora of the eelgrass beds make the zone a valuable one for fish, especially flounder and mullet.

Behind the *Salicornia* zone in salt marshes is the *Juncus–Leptocarpus* rush community, succeeded at higher levels by *Leptocarpus–Stipa* associated with *Muehlenbeckia complexa* and *Plagianthus divaricatus*. The primary saline colonist on shingly marshes is *Samolus repens*, and depressed areas generally carry a mixed sea meadow vegetation. At the highest levels of the salt marsh there is commonly a *Scirpus maritimus–Baumea juncea* zone, while in areas of freshwater inflow a transition to freshwater marsh takes place by way of *Cladium ustulatum* and *Typha*.

As may be expected, there is a wide range of marine animals in these plant communities, especially marine amphipods (e.g., *Orchestia chilensis*), burrowing crabs, and nereid worms. These marshes are also a favorite habitat for numerous migratory birds.

Northern and Southern Islands

The Kermadec Islands in the far north at latitude $30°S$ belong to New Zealand, and possess an almost tropical shore pattern. The zonation on a moderately exposed rocky shore is as follows (Morton and Miller, 1968): *Nerita melanotragus* or *Balanus nigrescens* → *Cellana creticulata* → *Sargassum fissifolium* and *Leathesia* → *Penepatella kermadecensis* and *Lithothamnia*.

The subantarctic islands in the far south, as their generalized name implies, have a strong subantarctic component in both fauna and flora. A typical moderately exposed rocky shore intertidal zonation from Campbell Island (Morton and Miller, 1968) is as follows: Mixed lichens → *Lichina confinis* → *Porphyra columbina* → *Bostrychia arbuscula* → midlittoral algal carpet → *Mytilus edulis–Aulacomya* → *Xiphophora* → *Durvillea antarctica*.

It should be noted that a number of mainland algae and marine animals are not to be found on these extreme southern shores.

VALENTINE J. CHAPMAN

References

Chapman, V. J., 1962. *The Algae*. London: Macmillan, 472p.

Chapman, V. J., 1970. *Salt Marshes and Salt Deserts of the World*. Lehre: Cramer, 392p.

Chapman, V. J., 1976. *Mangrove Vegetation*. Lehre: Cramer, 447p.

Chapman, V. J., 1979. *Mangrove Vegetation of the South Auckland District*. New Zealand: Lands and Survey Department, in press.

Dellow, V., and Cassie, R. M., 1955. Littoral zonation in two caves in Auckland District, *Royal Soc. New Zealand Trans*. 83, 321-331.

Morton, J. E., and Miller, M., 1968. *The New Zealand Sea Shore*. London, Auckland: Collins, 638p.

Cross-references: *Antarctica, Coastal Ecology; Australia, Coastal Ecology; Biotic Zonation; Coastal Fauna; Coastal Flora; Halophytes; Mangrove Coasts; New Zealand, Coastal Morphology; Pacific Islands, Coastal Ecology.* Vol. VIII, Part 1: *New Zealand.*

NEW ZEALAND, COASTAL MORPHOLOGY

One consequence of the elongated shape of New Zealand is that it has a long coastline (approximately 10,000 km) and a large amount of continental shelf (about 249,000 km^2) in proportion to its land area (270,000 km^2). No part of the country is more than 130 km from the sea, and a wide variety of rocks are exposed at the coast, so that the coastline displays a great variety of forms.

Since more than three-quarters of New Zealand's three million people live in the larger urban areas near the coast, there is thus a large, coastal resource per capita that is available for recreation but that is subject to increasing pressures of economic development. Geomorphologists have been active in studying the evolution and characteristics of the coast for the past 50 years, but much of the early work was of a broad, descriptive nature. Detailed investigations of contemporary processes of sedimentation have been a feature of the past 15 years, and the tempo of work is increasing. However, it is still possible only to identify the major processes and level of stability for about 10% of the total shoreline. Apart from scientific studies in both universities and government agencies, there is an increasing demand for knowledge on coastal dynamics for a wide range of practical applications including: the downstream effects of modifying catchments for flood and erosion control, for irrigation and hydroelectric power generation; for environmental impact of port, residential, and recreational developments; for seabed prospecting and mining; and for oil and gas exploration. Most of the sites for large new industrial and power-generating plants are also on the coast, so that New Zealanders, accustomed to thinking in terms of a national income earned mainly from farming, are becoming more aware of the coast as urbanization and development proceed.

The New Zealand Coast

Management and understanding of the coastal zone are complicated by both the quantity of the resource and its diversity. New Zealand is a hilly and mountainous land of great internal diversity in which there are major changes of rock type, relief, and landforms within small distances, a feature that is especially marked at the coast. Thus, there are large tracts of Holocene sand beaches facing the open oceans and in quieter waters. Examples of these are the beaches fronting the Canterbury Plains, the Nelson-Golden Bay coast, the Manawatu coast, and the Ninety Mile Beach of Northland (Figure 1). Many of these have large areas of associated prograded ridges and dune sequences both Pleistocene and Holocene in age.

In addition, there are large areas of hard rock erosional shore and embayed or fragmented coastlines such as the Bay of Islands. There are glaciated shores exhibiting typical *fiord* forms in South Westland and Fiordland, although there are no glaciers that now reach sea level; and there are typical *ria* type shores, most notably developed in the Marlborough Sounds, but also present in Northland and elsewhere. There are volcanic coasts, such as those found on the Tertiary shield complexes of Banks and Otago Peninsulas, and shores developed in fluvial sands and gravels deposited as outwash during the Quaternary. These occur predominantly along the east coast of South Island and to a lesser extent on the east coast of North Island. Both the central east and west coasts of North Island have coastal landforms developed in a wide range of volcanigenic materials, including lavas, ignimbrites, and tuffs, that have been frequently mantled in ashes of a variety of ages thus providing a means of dating. In addition, New Zealand has 7500 named rivers, so that there are a large number of tidal estuaries, lagoons, and littoral-drift-barred river mouths throughout the coastal zone.

Morphogenetic Environment

It should be apparent from the above that almost every type of coastal morphology is to be found somewhere in New Zealand, the major exceptions being active directly glaciated shores and coral coasts. Owing to its large latitudinal extent ($34\frac{1}{2}$–$47\frac{1}{2}°S$) New Zealand's climate ranges from subtropical in the north to the verges of the subantarctic in the south. Hence, it is not surprising that there have been abundant deeply weathered and podsolized sands supplied to the coasts of the far north, and even the development of lithified beach and dune deposits; while the shores of the far south have been nourished by glaciers and by broad, braided snow-fed river systems carrying large quantities of angular sands and gravels.

New Zealand is a mountainous land. The

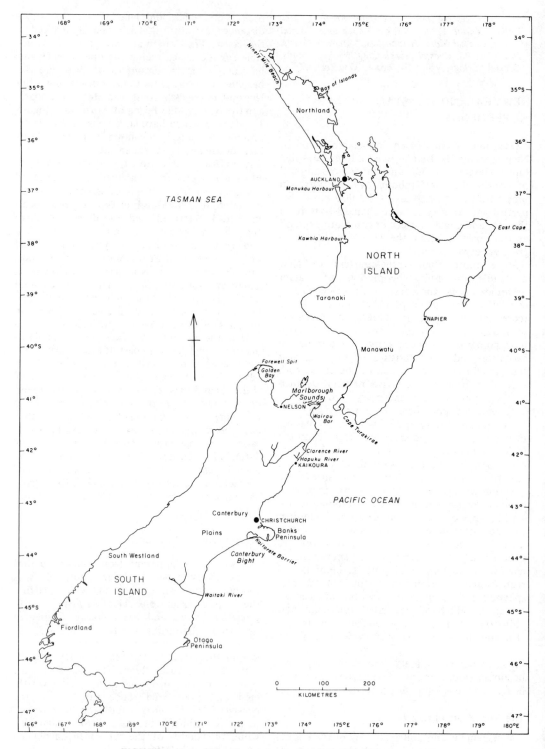

FIGURE 1. Map of New Zealand and places mentioned in the text.

main ranges are aligned SW-NE running the length of the South Island, and continuing through the lower half of the North Island toward East Cape. Because the country lies in the zone of westerly wind circulation this imposes a strong west-east gradient in rainfall particularly in South Island. The west coast is fed by abundant short, steep, fast rivers nourished by up to 3000 mm of rainfall per year, while the drier east coast has fewer longer rivers with much larger catchments. The rivers may flood at any time of year.

Because of the prevailing wind directions the major storm centers lie to the west and south of New Zealand, and owing to isolation in the mass of the Pacific, all of the ocean shores are of a high-energy nature. On both the windward (western) and the leeward (eastern) shores the dominant wave is the southerly, and so longshore transport is predominantly northward on both coasts of the main islands. Prime examples of the results of this process are Farewell Spit, the Kaitorete Barrier system, and the Wairau Bar complex. However, there are important counterdrifts thought to emanate from the northwest (on the west coast); from lesser storm centers to the east and northeast (on the east coast). The western and northern coasts of both islands are occasionally profoundly influenced by heavy seas produced by southward tracking tropical cyclones in the north Tasman Sea, and storm surge damage has been reported from most major ports over the past century.

There is no particularly strong seasonal pattern to wave activity around New Zealand, at least so far as can be discerned from the available short and fragmentary records. Rather, it seems that periods of high-energy input are correlated with the long term average for passage of frontal systems through the region (8-10 days), although there is a tendency for the frequency of occurrence of southerlies to increase in the southern winter (April-August).

Southerly storms produce waves 1-4m in height at the shore and periods of from 8-20 seconds that spill during shoaling and then plunge at the beach. These are capable of large amounts of erosion. Swell waves from a variety of directions including the south have heights of from 0.2-2.0m and periods in the range 5-12 seconds. In keeping with the open ocean nature of the coast, it is common to record a mixture of heights, periods, and directions received at the shore, and there is often little correlation between weather and surf conditions at the coast.

Tsunamis

Because it is part of the tectonically active zone bordering the Pacific Ocean and because it is exposed to the north and east, New Zealand has received numerous *tsunamis* from central Pacific and South American epicenters. However, their effects on coastal settlement and morphology have been mainly small, probably because the wide continental shelf exerts a strong shoaling effect on incoming waves.

Tides

Most of the New Zealand coast is mesotidal (tide range 2-4 m), so that the primary determinants of width and height of development in costal landforms are grain-size, sediment supply, and wave characteristics, including setup and storm surge.

Quaternary History

As elsewhere in the world, the Quaternary of New Zealand was a time of great geomorphic changes. Much of South Island was subjected to at least four major glaciations and associated interglacial periods. A great proportion of North Island was simultaneously affected by volcanism emanating from numerous centers, principally in the central plateau. Associated with climatic change were several major fluctuations of sea level both above and below present level, so that the continental shelf carries numerous relict coastal and nearshore features from previous low stands of the sea. Also, many areas of New Zealand display raised and tilted remnants of coastal landforms dating from higher stands than the present.

An additional complicating factor is widespread tectonism. New Zealand lies at the junction of two major continental plates, and can be thought of structurally as the buckled and sheared zone where they meet. Hence a major feature of the geology is a 480 km transcurrent fault running SW-NE through South Island's Southern Alps. The major movement has been northward on the western side of the fault. However, it is generally believed that the Quaternary was characterized by mountain building (the Kaikoura Orogeny), which gave much of the land its present relief. This episode, which may still be continuing, was accompanied by widespread block faulting that has broken the landscape into what Cotton (1942) aptly described as a "concourse of earth blocks." It is this that is responsible for much of the structural detail of the coast, and, taken together with sea level fluctuations over the very broad continental shelf, it is clear that there have been large changes in the form and position of the coast through the Quaternary.

Tectonism and seismicity are still active forces in the landscape. Pavoni (1971) quotes maximum vertical displacement rates of 1.1 cm/yr^{-1} for the Alpine Fault and rates of from 0.031–0.046 cm/yr^{-1} for the Kaikoura Mountains. Cotton (1954) records an apparent 1.5 m drop in a series of prograding beach ridges in the Wairau Valley following an earthquake in 1848; the Napier earthquakes of 1931 produced uplift of the adjacent coast of about 2 m. Stevens (1974) details an active raised beach system at Cape Turakirae dating to 5000 yrs B.P.

Because of tectonism and shelf loading, the pattern of response to Holocene rising sea levels varies from place to place, but the broad trend was a rapid rise from ca. 18,000–ca. 7000 yrs B.P. that rapidly drowned the shelf. This has been followed by a less rapid rise, possibly with small excursions above present level, to the present position. Schofield (1967) suggests that sea-level is still rising at Auckland at rates up to 20 cm per century. Unfortunately there are no data on hand for other ports.

A final noteworthy feature of the Quaternary is the work of rivers that particularly during glacial intervals, contributed vast amounts of angular sand and gravel to the coast, particularly in South Island. The Canterbury Plains contain more than 1000 m of fluvial outwash that is extensively reworked through beach systems. The Holocene appears to be a time of declining river competence, so that this major source for beach nourishment may be declining in importance. This process is accentuated by engineering works for flood control, hydroelectricity, irrigation, and gravel extraction in the catchments, so that coastal stability adjacent to some river mouths is now a cause for concern.

Sand Beaches

New Zealand has extensive tracts of sand beach, particularly along the west coasts of both islands, in the far north of North Island, and in the Nelson-Golden Bay area. In composition they are predominantly quartz and feldspars, with minor carbonate and heavy mineral components. However, the beaches of the central west coast of South Island and of the west coast of North Island from Taranaki to Kawhia Harbour contain a high proportion of dark titanomagnetite and other iron minerals. The North Island beach placers are derived from the Egmont volcanic complex in Taranaki, and are mined for export.

Elsewhere throughout New Zealand, there are abundant pocket sand beaches of a variety of textures and compositions reflecting local conditions. Much of these sandy shores are erosional but there are some areas of spectacular progradation. For example, Williams and Cochrane (1974) detail progradation of Whatipu Beach at North Manakau Head amounting to more than 1 km in the past 50 years.

Mixed Sand and Gravel Beaches

Along the east coast of South Island, and to a lesser extent on the east coast of North Island, are extensive mixed sand and gravel beaches. These are a rarity among beach types at world scale, and have peculiar dynamics. They are typically steep (6–12°) and plunge steeply into 10 m of water immediately seaward of the breakpoint; and so they have a single, narrow high-energy surf line under all wave conditions, and the subaerial beach is dominated by runup and backwash (Kirk, 1975). In the Canterbury Bight these beaches front a 25 m high cliff of fluvial sands and gravels that is eroding at 1.5–2.0 m/yr^{-1} over a 78 km length of coast. North of this section lies the 30 km of the Kaitorete Barrier complex comprising numerous Holocene beach ridges developed from the longshore-drifted erosion products. This complex is now stable, but has undergone considerable reorientation during the past 5000 years.

Gravel Beaches

There are numerous small gravel and boulder beaches around the coast, particularly on headlands and in small bays on promontories, but there are also some large gravel complexes. Along the east coast of South Island at the Waitaki, Hapuku, and Clarence rivers, there are extensive gravel deltas produced in Quaternary outwash. These features relate to earlier phases of activity, and are being eroded under present conditions.

Hardrock Shores

Much of the New Zealand coast is developed in hard rock, so that there is a wide range of cliff and platform features to be studied. A great deal of early work was done by Bartrum (1916) on the control of shore platform development imposed by water table and saturation levels in coastal materials, and by Cotton on the cycle of coastal development. Cotton (1949) also drew attention to volcanic coasts having *plunging cliffs* with little shore platform development, most notably at Banks Peninsula.

As elsewhere in the world, the study of shore platform processes is not well advanced, but investigations now underway at Kaikoura on the east coast of South Island, where the author has recorded intertidal surface lowering rates of up to 1.5 mm/yr^{-1} on limestones and mudstones. Rates of cliff retreat up to 24 cm/yr^{-1} have been measured.

Problems for Future Study

From the above it should be clear just how variable and dynamic the shores of New Zealand are, and also how important it is, from a variety of scientific and applied viewpoints, to quantify the main changes and delineate principal processes. Because of the diversity of conditions affecting wave processes, sediment type, and supply, and tectonic history, it is necessary to carry out numerous local investigations. It is correspondingly difficult to generalize results from a particular study to other parts of the country.

In an isolated, open ocean location such as New Zealand's, it is necessary to study in some detail the characteristics of the wave environment, because the conventional distinctions between constructive *swell* and destructive *storm* waves may require modification on continuously high-energy shores that are usually subject to mixed wave trains of varying ages, heights, and so on that approach from more than one direction. In addition, the nature of the *time series* of beach changes is in doubt. It may be that some beaches fluctuate seasonally, while others alter in response to a synoptic weather cycle having about a fortnightly recurrence interval. Finally, the nature of river inputs of sediment to the coast is a matter for urgent study, especially in relation to proposed land use changes upstream and their ultimate impact on beach nourishment and stability.

ROBERT MILLER KIRK

References

Bartrum, J. A., 1916. High water rock platforms: A phase of shoreline erosion, *New Zealand Inst. Trans.* 48, 132-134.

Cotton, C. A., 1942. *Geomorphology.* Christchurch: Whitcombe and Tombs Ltd., 505p.

Cotton, C. A., 1949. Plunging cliffs, Lyttleton Harbour, *New Zealand Geographer* 5,130-136.

Cotton, C. A., 1954. Submergence in the lower Wairau Valley, *New Zealand Jour. Sci. and Technology* 35, 364-369.

Kirk, R. M., 1975. Aspects of surf and runup processes on mixed sand and gravel beaches, *Geog. Annaler* 57A, 117-133.

Pavoni, N., 1971. Recent and late Cenozoic movements of the earth's crust, *Royal Soc. New Zealand Bull.* 9, 7-17.

Schofield, J. C., 1967. Sand movement at Mangatawhiri Spit and Little Omaha Bay, *New Zealand Jour. Geology and Geophysics* 10, 697-722.

Stevens, G. R., 1974. *Rugged Landscape: The Geology of Central New Zealand.* Wellington: Reed, 286p.

Williams, P. W., and Cochrane, G. R., 1974. Auckland's physical environment, in W. Moran, ed., *Auckland and the Central North Island,* International Geographical Union Regional Conference 1974, Palmerston North, 17-34.

Cross-references: *Asia, Eastern, Coastal Morphology; Australia, Coastal Morphology; Global Tectonics; Holocene; Pacific Islands, Coastal Morphology; New Zealand, Coastal Ecology; Sea Level Curves; Tidal Type, Variation Worldwide; Tsunami; Waves; World Net Sediment Transport.* Vol. VIII, Part 1: *New Zealand.*

NICHE, NICK, NIP, NOTCH

Since the rise of sea level after the last major glaciation, many cliffed coasts have been actively eroded according to rock type, rock structure, exposure to wave attack, availability of abrasive material, climatic conditions, geomorphic history, and other factors (Russell, 1963). In the traditional concept, a retreating cliff face that eventually results in a *wave-cut platform* is initially scored laterally into an undercut *notch* (Russell, 1963). This notch is also commonly referred to as a *niche, nick,* or *nip.* Neumann (1966) restricts the term notch to a subtidal indentation with a flat roof (*visor*), and the term nip to an intertidal identation with a sloping roof. In general, the best notch and visor cutting is to be found on microtidal coasts (Bird, 1969; Davies, 1973). Exposed coasts tend to have flat notch floors and visors inclined up to 45°. Sheltered coasts tend to produce horizontal cuts with flat visors. The distance between notch floor and visor is seen to increase with tidal range and exposure to storm waves (Russell, 1963). Reports on rates of notch development range from 1 mm/yr. to rates exceeding 10 mm/yr. (Neumann, 1968).

Two basic types of notches exist: *abraded* or *wave quarried* forms occur when storm waves cause abrasion on the cliff face forming the notch (Russell, 1963); and *Solutional-biochemical-biophysical-bioerosional* forms are largely restricted to intertidal, spray, and shallow subtidal zones of tropical limestone coasts. The notch originates and is deepened by an intricate system of poorly understood, interrelated factors, including: direct solution of limestone by seawater; biochemical attack involving metabolic products of the organic community at the rock/water interface; physical removal of biologically loosened materials by waves and currents; and bioerosion—the destruction and removal of consolidated material or lithic substrate by the direct action of boring, browsing, and burrowing organisms (Neumann, 1966, 1968).

There is much disagreement as to the levels where notches form. Russell (1963) reports the following investigations: Fairbridge (1947) believes the maximum undercut below the notch roof occurs at mean sea level (mesolittoral zone); Verstappen (1960) believes the notch forms just below high tide level; Wentworth (1938) notes *nips* produced by *solution benching* as high as 1.6 m above sea level; Kaye (1959) states that notches extend below low-tide, wave-trough level and mean-high-wave level. Kuenen (1950) finds wave attack to abrade notches most quickly just above high tide level. Hodgkin (1964) believes that maximum solution of coastal calcarenites takes place just above mean sea level where the notch develops. Neumann (1966) describes extensive notches that most closely coincide with the level of extreme low tides and are for the most part subtidal in origin.

Much work remains to be done on the origin of notches and the role they play in cliff erosion, especially of limestone coasts.

GARRY JONES

References

Bird, E. C. F., 1969. *Coasts*. Canberra: Australian National University Press, 246p.

Davies, J. L., 1973. *Geographical Variation in Coastal Development*. New York: Hafner Publishing, 204p.

Fairbridge, R. W., 1947. Notes on the geomorphology of the Pelsart Group of the Houtman's Abrolhos Islands, *Royal Soc. Western Australia Jour.* **33**, 1-43.

Hodgkin, E. P., 1964. Rate of erosion of intertidal limestone, *Zeitschr. Geomorphologie* **8**, 385-392.

Kaye, C. A., 1959. Shoreline features and Quaternary shoreline changes, Puerto Rico, *U.S. Geol. Survey Prof. Paper 317-B*, 49-140.

Kuenen, Ph. H., 1950. *Marine Geology*. New York: John Wiley & Sons, 568p.

Neumann, A. C., 1966. Observations on coastal erosion in Bermuda and measurements of the boring rate of the sponge, *Cliona lampa*, *Limnology and Oceanography* **11**, 92-108.

Neumann, A. C., 1968. Biological erosion of limestone coasts, *in* R. W. Fairbridge, ed., *Encyclopedia of Geomorphology*. Stroudsburg, Pa.: Dowden, Hutchinson & Ross, 75-81.

Russell, R. J., 1963. Recent recession of tropical cliffy coasts, *Science* **139**, 9-14.

Verstappen, H. T., 1960. On the geomorphology of raised coral reefs and its tectonic significance, *Zeitschr. Geomorphologie* **4**, 1-28.

Wentworth, C. K., 1938. Marine bench-forming processes, water-level weathering, *Jour. Geomorphology* **1**, 6-32.

Cross-references: *Cliff Erosion; Tidal Range and Variation; Wave-Cut Platform; Wave Environments; Wave Erosion.*

NODE AND ANTINODE

Nodes are points at which the displacement of oscillating water surface is always zero, while the amplitude of oscillations in *antinodes* is maximum. Both features result from the superposition of two or more progressive waves propagating in opposite directions in a body of water. Quasinodes and antinodes appear in partially reflected waves. If a rectangular basin oscillates in both its longitudinal and transverse directions, there appear nodal and antinodal *lines* where m and n are integers of possible values of 0, 1, 2 . . . defining the modes of longitudinal and transverse oscillations, respectively ($m = 0$ and $n = 0$ for the first harmonics, i.e., fundamental modes, and so on). Nodes and antinodes can also be found in many other forms of water motion, as in tides. Standing *edge waves* (q.v.) provide a satisfactory explanation for the formation of crescentic bars (in regions of small tidal range) according to Bowen and Inman (1971). By analogy, the nodal point of coastal sediment determines the location from which sand moves in both directions, upcoast and downcoast (see Wiegel, 1964).

RYSZARD ZEIDLER

References

Bowen, A. J., and Inman, D. L., 1971. Edge waves and crescentic bars, *Jour. Geophys. Research* **76**, 8662-8671.

Wiegel, R. L., 1964. *Oceanographical Engineering*. Englewood Cliffs, N. J.: Prentice-Hall, 532p.

Cross references: *A-B-C Model; Coastal Morphology, Research Methods; Scale Models; Waves; Wave Theories, Oscillatory and Progressive.*

NORTH AMERICA, COASTAL ECOLOGY

Arctic Regions

The eastern Arctic extends from above 80°N in the Canadian Arctic Archipelago to southern Labrador. Depending on latitude, the average frost-free period is between 40 and 100 days, although frost and snow can occur in any month. Winters are extremely cold; summers are cool, but the days are long. An arctic desert climate prevails in the far north, with precipitation of less than 250 mm/yr (little of which evaporates), but southern Labrador receives about 1000 mm/yr. Inland Labrador has a boreal climate and vegetation, but arctic conditions extend far southward down the coast owing to the cold Labrador current. This current keeps the air temperature low, enshrouds the shoreline in fog more than half the days of the year, and delivers much of the area's precipitation in the form of cold drizzle. Arctic shores are icebound 7–9 months out of the year, with profound consequences for plant life.

These severe conditions result in tundra plant communities, treeless and dominated by low mosses, lichens, grasses, sedges, herbs, and shrubs. This vegetation can be subdivided into the *polar desert*, the *moss-heath-shrub*, and *strand* types that include dune, beach, and salt-marsh communities.

The high arctic beach ridges of Devon Island support a sparse polar desert vegetation of *Dryas integrifolia, Saxifraga oppositifolia, Cassiope tetragona, Carex nardina*, and *C. misandra*, along with mosses and lichens. Far northern salt marshes are poorly developed, mainly because of destruction by moving sea ice. Plants of these depauperate marshes include *Puccinellia phryganodes, Stellaria humifusa*, and *Cochlearia officinalis*, and at slightly higher levels, *Primula borealis* and *Carex ursina*. In the brackish marshes, *Carex ramenski, Arctophila fulva, Dupontia fisheri*, and *Hippurus tetraphylla* are most important, while *Elymus arenarius* spp. *mollis, Armeria maritima, Juncus arcticus, Plantago eripoda* and *Arenaria peploides* dominate sand and gravel beaches. Indeed *Elymus* (wild lyme grass) and *Arenaria* (sea chickweed) are the most important sand beach plants throughout the Arctic.

The coastal tundra of Labrador is considerably more diverse than that to the north, featuring the high arctic plants listed above and also species of more southern affinities able to live here where arctic conditions are slightly moderated. A tundra made up of *Empetrum nigrum, Betula nana, Arctostaphylos alpina, Phyllodoce coerulea, Vaccinium uliginosum, Ledum groenlandicum*, the lichens *Cetraria islandica, C. nivalis, Cladonia alpestris*, and *Stereocaulon* spp., and various mosses may be found on wind-swept sites. Where the shore is more protected and has a clay soil, one may see salt meadows containing *Glyceria* spp., *Stellaria* spp., *Juncus* spp., *Carex* spp., *Potentilla anserina, Iris setosa, Plantago maritima, Ranunculus reptans,* and *R. hyperboreus*.

Dunes and beaches of Byron Bay in eastern Labrador are dominated by *Elymus* and *Arenaria*, sprouting from seeds and fragments buried in drift lines. Once these pioneers have begun to build dunes, *Lathyrus japonicus* and *Senecio pseudo-arnica* appear. With increasing stability, a lush grass land of *Elymus, Lathyrus, Senecio, Festuca rubra, Achillea millifolium, Epilobium angustifolium, Heracleum maximum*, and *Stellaria longifolia* appears.

Along the back sides of these stable dunes is a heath tundra containing *Empetrum nigrum, Vaccinium vitis-idaea, V. uliginosum, Actostaphylos alpina, Potentilla tridentata, Rubus pubescens, R. chamaemorus, Pedicularis* sp., *Cornus canadensis*, and *Salix* sp., with scattered stands of *Picea mariana* in well-sheltered locations. The boundary between this community and the strand grassland is sharp, suggesting mutual exclusion by the dominant species of each.

At the water's edge behind these barrier dunes are *Carex maritima, Potentilla anserina, Triglochin* spp., and *Poa eminens*. Most of the intertidal area has no rooted plants, but the algae *Fucus vesiculosus* and *Ascophyllum nodosum* occur where they can escape ice and wave damage.

Other plants that have been reported from Labrador shores include *Mertensia maritima, Aster* spp., *Senecio frigidus, Rumex salicifolius, Angelica atropurpurea, Cochlearia officinalis, Stellaria humifusa, Galium palustre, Parnassia palustris, Atriplex* spp., *Tofieldia borealis, Salix waghornei, Triglochin maritima, T. palustris, Juncus filiformis, J. biglumis, J. triglumis, Carex maritima,* and *C. glareosa*.

Boreal Region

The boreal zone of the eastern North American coast is confined essentially to eastern Newfoundland. The Labrador Current passes along this coast, maintaining arctic conditions as far south as the Strait of Belle Isle, while Gulf Stream eddies and the shallow waters of the Gulf of St. Lawrence tend to warm southwestern Newfoundland and the other Maritime Provinces. The east coast of Newfoundland is squeezed between these two influences. The climate is damp, stormy, and foggy, with cool summers and cold but not extreme winters. The frost-free period averages 100–120 days. The Strait of Belle Isle is frozen for 6–7 months every winter, but St. John's harbor, on the Avalon Peninsula, is ice-free most of the year.

Occasional salt marshes may be seen where fine sediments have accumulated within estuaries or behind barrier beaches. Gravel and cobble beaches support few plants, but more sandy shores are vegetated. In the Little Bay area, there are an open *Festuca rubra* grassland and then a closed *Festuca* turf, with *Rubus, Achillea, Juniperus communis,* lichens, and mosses. Nearby beaches have patches of *Elymus, Lathyrus,* and *Mertensia maritima.* Dunes are few, owing to lack of sand.

Most of the shores of the many bays and fjords are rocky, and here the boreal forest of upland Newfoundland comes almost down to the high tide mark. *Picea mariana, P. glauca, Larix laricina, Abies balsamea, Betula papyrifera, B. alleghaniensis,* and *Populus tremuloides* are the prominent components. In the intertidal zone one finds *Fucus* and *Ascophyllum,* with *Laminaria* subtidally.

Acadian Region

This zone includes the southerly shores of the Gulf of St. Lawrence, the western coast of Newfoundland, and the coasts of Nova Scotia and Maine. Summers are warm and winters cold, although not as frigid as on the boreal coast. Precipitation is adequate and well distributed through the year. Storms and fog are significant, although not as severe as farther east and north; warm ocean currents and Gulf of St. Lawrence waters exert a moderating influence generally. The average frost-free period ranges from 120–170 days. Much of the gulf freezes over in the dead of winter, but farther south, along the northern New England coast, ice occurs mostly in protected bays.

The Acadian upland forest is a mixture of conifers and hardwoods, including *Picea glauca, P. mariana, P. rubens, Abies, Larix, Thuja occidentalis, Acer saccharum, Betula* spp., *Quercus borealis,* and *Fagus grandifolia. Pinus strobus,* and *P. banksiana* are prominent in the northern areas, and *P. rigida* and *Tsuga canadensis* appear toward the south. On rocky shores, these forest trees come almost down to the high tide mark. The coastal forest of eastern Maine is mostly of conifers, reflecting the cool, damp, foggy local climate. In northern New Brunswick, huge raised peat bogs occur just behind barrier islands (see Bogs).

The Acadian coastal vegetation is distinguished from that of the boreal and arctic zones by the appearance of new dominant grasses. American beach grass (*Ammophila breviligulata*) replaces *Elymus* on the dunes and strand, and the cordgrasses (*Spartina alterniflora, S. patens,* and *S. pectinata*) replace the arctic and boreal salt marsh plants. *Ammophila* and the *Spartinas* all reach their northern limits on the southwestern shores of Newfoundland. *Solidago sempervirens* and *Cakile edentula* make their first appearances in the Acadian area, and extend far south along with the *Spartinas*. Members of the Rosaceae and the Ericaceae assume considerable importance in the Acadian and northeastern coasts.

The shore vegetation can be divided into beach and dune communities, heathlands, shrublands and thickets, maritime forest, salt marshes, freshwater marshes and ponds, and intertidal algae. The beach and dune strand may be pure *Ammophila* or a mixture of beachgrass and herbs. *Elymus* and *Arenaria* occur in isolated patches as holdovers from the arctic, but *Cakile* and *Salsola kali* are more important as beach plants. Along with *Ammophila*, one finds *Lathyrus japonicus, Solidago sempervirens, Artemisia stelleriana, Festuca rubra, Carex silicea, Achillea millefolium, Rubus idaeus, Angelica atropupurea,* and *Ribes hirtellum* as important dune plants.

On stable dunes a tundralike heathland may occur, dominated by *Hudsonia tomentosa*. This low shrub often grows in nearly exclusive stands, sharply bounded from neighboring *Ammophila* communities. Among its few associates are *Lechea maritima, Aster linarifolius,* and *Cladonia* lichens. *Vaccinium macrocarpon* may form a dense cover in damp interdune depressions. *Potentilla tridentata,* an arctic species, can be found in heathlands as far south as Cape Cod.

Shrublands and thickets on stable dunes and coastal uplands are dominated by *Myrica pensylvanica, Rosa virginiana, Spiraea latifolia, Rubus* spp., and *Rhus toxicodendron*. Spruce-fir forests develop on long-stable dunes in Canada and northern New England, joined by *Pinus banksiana* in Canada, and in New England by *Acer rubrum, Quercus borealis, Betula papyrifera, Populus tremuloides, Alnus rugosa, Ilex verticillata,* and *Amelanchier* spp. In southern Maine, *Pinus rigida* joins this community.

The salt marshes of eastern North America are remarkably similar from the Acadian region to the Gulf of Mexico. *Spartina alterniflora* dominates the intertidal marsh, while *Distichlis spicata* and *Spartina patens* characterize the high marsh, along with species of *Salicornia*, the latter especially in salt pannes. This general pattern is recognizable all the way to the subtropics. In the Acadian zone, the *Salicornia* is *S. europaea. Puccinellia maritima, Limonium nashii, Glaux maritima, Potentilla egedii, Juncus gerardii, Triglochin maritima, T. palustris, Hordeum jubatum, Plantago maritima, P. juncoides, P. oliganthos, Chenopodium rubrum, C. album, Atriplex hastata, Spergularia marina, Suaeda maritima,* and *Scirpus americanus* may also be found in the high marsh. In some places, the *Spartina patens-Distichlis* zone is actually dominated by *Plantago* and *Triglochin*. As a rule the northern salt marshes are more diverse than their southern counterparts. *Spartina pectinata, Typha angustifolia, Juncus* spp., and *Myrica gale* may be found along the edges of salt marshes and in fresh marshes.

The North Central Region

This section of the coast extends from New Hampshire to Delaware Bay. It can be subdivided into three parts: northeastern Massachusetts, Cape Cod to New York City, and New Jersey. One of this region's distinctive characteristics is a frost-free period considerably longer than that of the Acadian zone: 180 to 220 days, the maximum being reached on Nantucket and Block Island, and Rhode Island. The climate is one of moderately cold winters and warm summers, with well-distributed precipitation. Fog occurs throughout, but maximally in the Nantucket area. Winter northeasters are a regular occurrence, and fully developed tropical hurricanes occasionally visit the area. Eastern Massachusetts bears the brunt of major storms, receiving high winds and huge seas. Winter icing is limited and local toward the southern end of the region, but Cape Cod Bay freezes over in hard winters.

The vegetational transition between the Acadian and North Central regions is rather gradual. Spruce and fir drop out of the coastal forest in southern Maine, and *Pinus rigida* becomes important. *Prunus maritima* makes its appearance on the dunes, and two southern tree species, *Chamaecyparis thyoides* and *Nyssa sylvatica*, begin to occur in coastal swamps. Ericaceous and rosaceous plants are prominent as in the Acadian region.

The *Ammophila*-dominated flora of beaches and young dunes is similar to that of the Acadian area. On older dunes one finds *Andropogon scoparius, Deschampsia flexuosa, Carex silicea, Festuca rubra, Artemisia caudata, Parthenocissus quinquefolia,* and *Rhus toxicodendron.* (These stable dunes are only a marginal habitat for *Ammophila*, which requires a constant supply of fresh sand.)

The most important plants of shrublands and thickets are *Myrica pensylvanica* and *Prunus maritima.* These are accompanied by *Rosa rugosa, Amelanchier* spp. *Spiraea* spp., *Quercus ilicifolia, Gaylussacia baccata, Prunus serotina, Vaccinium corymbosum,* and *V. angustifolium.* Some thickets are primarily *Rhus typhina, R. copallina,* or *Lonicera morrowi. Juniperus virginiana* is commonly present in early stages of succession. *Aronia melanocarpa, Kalmia angustifolia,* and *Rubus* spp. may invade wetter sites.

The heathlands of this region are mostly the result of human disturbance. Salt spray and natural fires tend to maintain this vegetation type, but have not prevented it from giving way to shrubland and forest in most of the area. Cape Cod, Martha's Vineyard, and Nantucket have much of the heathland that remains today. *Hudsonia tomentosa* and *Arctostaphylos uva-ursi* are the most important ground covers. With them grow *Corema conradii, Lechea maritima, Chrysopsis falcata, Aster linariifolius, Gaylussacia baccata, Vaccinium angustifolium,* and *V. vacillans.* Scattered heathlands occur in New York and New Jersey, but *Hudsonia* and *Lechea* are the only major components that extend south into the Virginian region.

Except for *Pinus rigida* and *Juniperus virginiana*, North Central maritime woodlands are primarily deciduous. *Quercus velutina* and *Q. alba* are the dominant hardwoods. Many of the accompanying species in this region are rosaceous: *Prunus maritima, P. serotina, P. virginiana, P. pensylvanica, Amelanchier canadensis,* and *A. laevis* are all highly conspicuous when they flower in the spring.

On large stable dunes in the Province Lands of Cape Cod, one may find what appears to be a remnant of northern mesophytic climax forest. *Fagus grandifolia* dominates, and with it are found *Acer rubrum, Sassafras albidum, Quercus alba, Q. velutina, Betula papyrifera, B. populifolia, Amelanchier* spp., *Viburnum dentatun,* and *Pinus rigida.* The ground cover of *Epifagus virginiana, Maianthemum canadense, Gaultheria procumbens, Epigaea repens,* ferns, and mosses also reminds one of an inland forest community.

Ilex opaca, one of the components of the evergreen broadleaf forest to be found farther south, makes its first appearance in the maritime forests of southern New England. It dominates these woodlands on parts of Fire Island, New York and Sandy Hook, New Jersey. Associated with it are *Sassafras, Amelanchier, Quercus velutina, Q. stellata,* and *Prunus serotina.*

Fresh swamps and marshes show a mixture of northern and southern plants. Wooded swamps are dominated by *Nyssa sylvatica* and *Acer rubrum,* and occasionally by relict stands of *Chamaecyparis thyoides,* which used to be much more abundant. Ponds and marshes contain species of *Typha, Scirpus, Juncus, Carex,* and increasing stands of *Phragmites communis.* These communities can occur in kettle holes, interdune depressions, and former tidal wetlands now cut off from saltwater.

North Central salt marshes are much the same as those of the Acadian region. The importance of *Triglochin* and *Plantago* decreases as one goes south, while *Salicornia bigelovii* and *S. virginica* become important, along with the southern high marsh shrub *Iva frutescens*. Other salt-marsh plants of this area include *Limonium carolinianum, Atriplex hastata* var. *patula, Suaeda maritima, S. linearis, Glaux maritima, Chenopodium rubrum,* and *Festuca rubra*.

Virginian Region

The Delmarva Peninsula, between Delaware Bay and the mouth of Chesapeake Bay, is an area of marked transition, both in vegetation and in coastal physiography. On the Atlantic side of the peninsula, we first see the low, wide barrier islands, with extensive flatlands and open dune lines, that are so characteristic of southeastern United States. The west side of this peninsula is a region of many shallow bays and tidal streams but no major freshwater input; large rivers, however, do enter Chesapeake Bay from the west, making the bay as a whole the world's largest estuary.

The Virginian Region is warmer and less snowy than the North Central Region, although the frost-free period is about the same. Many of the coastal storms and hurricanes pass to the eastward of the area. The bay side of the peninsula, however, is often subject to wave attack and erosion during strong westerly winds.

On the bay side, the shoreline consists of eroding uplands, with farms and some remaining forests coming almost to the bay edge. Coastal vegetation may be found in salt marshes and on the few scattered barrier spits; it is much more extensive on the numerous barrier structures and associated marshes of the Atlantic side.

Cakile edentula, Salsola kali, and *Euphorbia polygonifolia* are the important beach plants. In the dunes *Ammophila* is still dominant in the northern part of the peninsula, but on the southern end of Assateague Island *Uniola paniculata* makes its appearance. A short distance to the south it becomes the dominant dune grass.

Another notable change in this area involves *Spartina patens*. North of Cape Cod this grass grows in decumbent fashion in high salt marshes or weakly upright where conditions permit. The form *S. patens* var. *monogyna* makes its first appearance on Cape Cod, growing upright on low dunes bordering the high marsh. It is taller and more robust than the decumbent form, has more aggressive rhizomes and a larger flowering spike, flowers later, and can tolerate much more sand burial. As one goes south, the upright form becomes steadily more important in the dune and grassland flora, until in the southern Virginian Region it may dominate the low dunes. At the same time the decumbent variety drops out as a salt-marsh plant.

Other species of these dunes include *Strophostyles helvola* (replacing *Lathyrus japonicus* of northern areas), *Panicum amarum, P. amarulum,* and *Cenchrus tribuloides*. *Solidago sempervirens* is conspicuous, but the Artemisias are not to be seen.

It is in the Virginian Region that we first see the extensive barrier flats so typical of the North Carolina coast. These grasslands begin behind the dune zone, and grade into the salt marshes, growing on old inlet sites or overwash surfaces. *S. patens, Andropogon scoparius,* and *A. virginicus* are the dominant grasses; two southern herbs, *Diodia virginiana* and *Lippia nodiflora,* make their first appearances here. In wet depressions one finds *Typha angustifolia, Kosteletskya virginica, Hibiscus moscheutos, Juncus effusus, J. acuminatus,* and six other species of *Juncus*.

The dominant forest tree here and everywhere on the southeast coast is *Pinus taeda*, replacing *P. rigida* as the coastal pine and growing to great size on some of the barrier islands. Most of the other important trees are also primarily southern in their distribution; these include *Querus falcata, Q. nigra,* and *Liquidambar styraciflua,* with *Magnolia virginiana* and *Nyssa sylvatica* appearing on wetter sites.

In the shrublands, thickets, and forest understory, we see an interesting mix of northern and southern plants. Woody species familiar from more northern coasts include *Prunus serotina, Amelanchier* spp., *Myrica pensylvanica, Rosa rugosa, R. virginiana,* and *Pyrus melanocarpa*. Many southern shrubs and small trees are to be found, some making their first appearance in the coastal flora, and others seen to the north as well but becoming abundant in the Virginian Region. The evergreen character of the forests is enhanced as *Persea borbonia* makes its appearance and *Ilex glabra* and *I. opaca* become increasingly common. *Myrica cerifera* begins to displace its northern congener *M. pensylvanica*. Other southern woody plants include *Xanthoxylum clava-hercules, Aralia spinosa, Diospyros virginiana, Baccharis halimifolia,* and *Opuntia humifusa*. Southern forest vines, both woody and herbaceous, include *Vitis rotundifolia, Mikania scandens, Berchemia scandens, Campsis radicans,* and *Smilax* spp. Certain other shrubs and understory trees found in this region are widely distributed along the whole coast; these include *Acer rubrum, Sassafras albidum, Juniperus virginiana, Rhus copallina, Gaylussacia baccata,* and *Vaccinium corymbosum* (along with their varieties). The same is true of the vines *Rhus toxicodendron* and *Parthenocissus quinquefolia*.

Vaccinium macrocarpon, *Hudsonia tomentosa*, and *Lechea maritima* approach their southern limits in the Virginian Region, and the typical heathlands of more northern shores do not develop here.

Virginian salt marshes, however, are well developed. They resemble their counterparts to the south more than those to the north. *Spartina alterniflora* dominates large areas of low marsh, with *S. patens*, *Distichlis*, and *Scirpus robustus* in the high marsh. Two southern graminoids that make their first appearance in the high marsh are *Juncus roemerianus* (in large, exclusive stands) and *Fimbristylis castanea* (the typical sedge of southern marshes and salt flats). Of the succulent dicots of northern marshes, only *Salicornia* spp. and *Limonium* spp. are conspicuously present. Southern dicots significant in the high marsh include *Borrishia frutescens*, *Gerardia purpurea*, *Sabatia stellaris*, and *Pluchea purpurascens*. The shrubs *Iva frutescens* and *Baccharis halimifolia* occur in the marsh borders.

Carolinian Region

This large area begins at Cape Henry, Virginia, extends down the coast to Jacksonville, Florida; begins again in the Florida panhandle, and stretches around the Gulf of Mexico to Aransas, Texas. Natural subdivisions are: the Outer Banks; Cape Henry to Beaufort Inlet, North Carolina; Beaufort Inlet to Cape Romain, South Carolina; the Sea Islands, Cape Romain to Jacksonville; and the Gulf coast.

The climate is warm temperate, with frost-free periods ranging from 230 to 310 days. Annual snowfall averages from 152 mm/yr at Norfolk, Virginia, to mere traces farther south. In long cold snaps, ice forms in lagoons and bays of the Carolinas. Rainfall is abundant throughout, much of it being delivered by convective thunderstorms, tropical storms, and winter northeasters. The first two storm types are a little more prevalent in the southern parts, causing precipitation to peak in summer and fall. In the more northern sections, northeasters are more common, and total rainfall is therefore more evenly distributed through the year.

The flora of the Carolinian Region has many more affinities with the tropics and subtropics than with more northern coasts. On the beach berm, however, we find several species familiar from the north: *Cakile edentula*, *Euphorbia polygonifolia*, *Salsola kali*. These are joined by *Sesuvium portulacastrum* and *Croton punctatus*. The dune strand contains all these plants but is dominated by grasses, especially *Uniola paniculata* and *Spartina patens* var. *monogyna*. Populations of *Ammophila* can be found as far south as Cape Fear, with the help of artificial planting, but they are not prominent. Other dune grasses are *Panicum amarum*, *Cenchrus tribuloides*, *Triplasis purpurea*, *Andropogon virginicus* var. *littoralis*, *Eragrostis pilosa*, *Muhlenbergia capillaris*, and (especially in southernmost sections) *Sporobolus virginicus*. This diversity of grasses is matched by the variety of herbaceous dicots: the ubiquitous *Solidago sempervirens* is joined by *Diodea virginica*, *Oenothera humifusa*, *Erigeron pusillus*, *Hydrocotyle bonariensis*, *Physalis maritima*, *Heterotheca subaxillaris*, and *Strophostyles helvola*, along with the herbs from the beach berm. In the far south, *Euphorbia dentata*, *E. ammannioides*, *Ipomoea sagittata*, and *I. stonifera* make their appearances. Two woody vines of the dunes are *Parthenocissus* and *Ampelopsis arborea*. The shrub *Iva imbricata* acts as a dune builder; as the dunes are stabilized, other shrubs invade, especially *Myrica cerifera*, with *Ilex vomitoria*, *Quercus virginiana*, *Juniperus virginiana*, *Xanthoxylum clava-hercules*, and *Baccharis halimifolia*. From South Carolina southward, *Serenoa repens*, *Sabal palmetto*, *Yucca aloifolia*, and *Opuntia drummondii* increase in importance in the shrub community. Far to the south, we find *Solidago pauciflosculosa*, *Helianthemum arenicola*, *Satureja coccineum*, and *Ceratiola ericoides* among the dune shrubs.

On barrier islands subject to frequent overwash and inlet changes, a special grassland community occurs on the flats between the dune zone and the high salt marsh. This community is especially conspicuous on the Outer Banks and other islands of the northern Carolinian Region. Its species are mostly adapted to overwash, and indeed may require it for their persistence; the community may be termed an *overwash subclimax*. *Spartina patens* var. *monogyna* dominates, along with *Solidago*, *Eragrostis*, *Muhlenbergia*, *Hydrocotyle*, and *Andropogon*. Other associates are *Scirpus americanus*, *Chloris petrea*, *Cynodon dactylon*, *Cynanchum palustre*, and (especially on slightly drier sites) any of the herbs from the dune zone.

The Carolinian maritime forest becomes increasingly dominated by evergreen broadleaf species as one proceeds southward. *Quercus virginiana* is the dominant and most characteristic species. *Quercus laurifolia*, *Q. nigra*, and *Q. phellos* are present as well; the latter two oaks tend to be deciduous in the northern part of the region and evergreen in the south. Other evergreen broadleaf trees and shrubs included *Persea borbonia*, *Ilex opaca*, *Magnolia virginiana*, *Osmanthus americanus*, *Prunus caroliniana*, *Bumelia tenax*, and (especially toward the south) *Magnolia grandiflora*.

These plants form most of the forest on some parts of the coast, but in other areas conifers assume importance. *Pinus taeda* is the dominant pine of the southern Atlantic Coast, to be joined by *P. palustris* and *P. elliottii* in South Carolina and Georgia. On the Gulf coast, *Pinus elliottii* takes precedence, forming a savannalike landscape. *Juniperus virginiana* mixes with the broad-leaved species as far south as South Carolina.

North of Florida the evergreen dicots tend to be confined to a fringe along the coast, but the deciduous species that are found in the understory also grow far inland. These include *Cornus florida, Carpinus caroliniana, Xanthoxylum clava-hercules, Aralia spinosa,* and *Vaccinium arboreum.* On some islands *Celtis* spp. dominate the forest, and on others they are entirely absent.

Stable islands close to the mainland, such as Bogue Banks, North Carolina, may have areas of diverse and well-developed forest similar to those found inland. These forests grow on and between old dune ridges, well back from the beach and protected from salt spray and seawater flooding. *Liquidambar styraciflua* and *Ulmus americana* grow on the drier sites; *Fraxinus pensylvanica* and *Nyssa sylvatica,* in interdune swamps.

Adding to the tropical appearance of these coastal forests are many entangling vines, some of which are also seen on the dunes. Some species are *Parthenocissus quinquefolia, Smilax bona-nox, S. auriculata, S. laurifolia, Ampelopsis arborea, Rhus toxicodendron, Vitis aestivalis, V. rotundifolia, Berchemia scandens, Mikania scandens, Passiflora incarnata, P. lutea, Gelsemium sempervirens, Lonicera sempervirens, L. japonica,* and *Campsis radicans.* Epiphytes, especially *Tillandsia usneoides, Polypodium polypodioides,* and numerous mosses and lichens, are also conspicuous on the forest trees.

From South Carolina to the tropics, palms become increasingly important in the coastal forests, with *Sabal palmetto* growing farthest north.

Freshwater wetlands on coastal islands can form in interdune depressions, in dune-field blowouts, or where saltwater bays have been closed off by sand drift. Parallel or arcuate dune systems especially favor wetland development. The flora varies from depauperate to rich and diverse, depending on the age, stability, and water regime of the wetland. *Typha angustifolia* or *Juncus roemerianus* may form large exclusive stands. *Cladium jamaicense* increases in importance southward. Other freshwater plants include *Typha latifolia, Kosteletzka virginica, Hibiscus* spp., *Dichromena colorata, Fimbristylis spadicea, Scirpus americanus, Hydrocotyle* spp., *Centella asiatica, Sagittaria latifolia, Boehmeria cylindrica, Pluchea* spp., *Ludwigia* spp., *Cicuta maculata, Pteridium aquilinum, Asplenium platyneuron, Lippia nodiflora, Bacopa monnieri, Carex* spp., *and more species of Juncus.*

Carolinian salt marshes are among the best developed in the world, being especially extensive and productive along the coasts of South Carolina and Georgia. However, where the barrier islands are far out to sea, as are some in the Gulf of Mexico, the marshes may be confined to the interior of the islands, protected by small spits on the landward side.

Behind the low and dynamic Outer Banks, salt marshes owe their existence to the opening and closing of tidal inlets, and secondarily to overwash deposition in backbarrier lagoons. The huge marshes behind more stable islands in Georgia and South Carolina are little affected by inlet changes or overwash, but have, instead, been formed as the sea slowly encroached on the mainland, with help from river sedimentation and other processes.

The low marsh is relatively more extensive here than farther north. It is dominated by *Spartina alterniflora;* at the edge of the low marsh and in salt pannes may be found *Distichlis spicata, Salicornia virginica, S. bigelovii,* and (in South Carolina and Georgia) *Batis maritima.* The high marsh is limited to a narrow fringe at the edge of the higher ground, dominated by *Spartina patens* and *Juncus roemerianus. Spartina cynosuroides* takes on increasing importance in this habitat toward the south. Other high-marsh species include *Iva frutescens, Borrichia frutescens, Gerardia maritima, Aster tenuifolius, Atriplex hastata* var. *patula, Fimbristylis castanea, Limonium carolinianum,* and *Pluchea purpurascens.*

Caribbean Region (Peninsular Florida)

The Florida Peninsula extends south from the Georgia border about 708 km, occupying the latitudes between 31° and 25°N. The northern part falls into the warm temperate (Carolinian) zone, while south Florida enjoys a climate closely approximating that of nearby tropical regions. Between these extremes there is a gradient of climatic change that is reflected in the vegetation. Thus this peninsula is of great interest botanically; it possesses a variety of temperate, subtropical, and tropical vegetation types, much diversity within types, and many species found nowhere else in the continental United States. Indeed, some communities, such as the famous Everglades, are unique in the world. Unfortunately, human activities in recent years have fragmented the natural communities.

Most of the coastline not included in parks has been urbanized, especially the east coast and the Miami and Broward County area, with serious consequences to native plant life.

Technically, all of Florida lies outside the tropics proper. However, proximity to the Gulf Stream, plus other factors, confer on the coasts of southern Florida a climate so mild that many tropical species appear in the coastal flora, while the interior of the peninsula has a more temperate climate and vegetation at a given latitude. Snow and frost are virtually unknown south of Cape Canaveral ($28°30'N$) on the east coast and Cedar Key on the west coast, and these points are more or less the northern limits of the occurrence of tropical coastal plants. The entire peninsula is subject to cold snaps, if not to frost, during severe winters, and these may exclude some tropical plants. It should be noted that summers on the south Florida coast are not extremely hot compared with those of inland southeastern United States.

Annual freshwater input is substantial, but tends to peak in the summer and to be scanty in winter. Most comes from convective summer thunderstorms that build up over the peninsular land mass and move eastward, conferring an annual average of 1567 mm of rain on West Palm Beach but only 1016 mm at Key West. Large inputs often come from tropical hurricanes, to which south Florida is markedly susceptible, but these are concentrated in summer and fall, contributing to the annual cycle of drought and deluge.

Southern Florida's coastal substrate is highly variable. Most of the peninsula is formed from marine limestones, sands, and coralline deposits that became exposed during the Pleistocene epoch. Sandy beaches and barrier structures are common on the Atlantic side, being especially well developed at Cape Canaveral. Barrier islands extend south to Miami, but these have been altered almost beyond recognition. Extreme south Florida is notable for coralline islands and beaches, and offshore reefs, the only such tropical marine features on the coast of continental United States. Huge marshes and mangrove swamps extend from the tip of Florida north to Naples on the Gulf coast. In some locations limestone outcrops along this shoreline. On the central Gulf coast there are sandy barriers from Naples north to Clearwater. North of this area, barrier beaches are again absent; marshes, swamps, and mud flats characterize this low-energy shore, and barriers begin once more as the coastline turns westward along the Florida panhandle. In general, the peninsula is low and flat, with little relief, making for gradual changes in plant distribution. In some cases, however, slight changes in elevation can mark a transition between such habitats as dry limestone ridges and freshwater swamps.

The coastal landforms are closely related to oceanic conditions, which are in turn influenced by the orientation of the shoreline. Wave energies are relatively high on the whole Atlantic coast, low from the Keys to Naples, moderate from Naples to Clearwater, and low from Tarpon Springs to Chasshowitzka. From there to Apalachee Bay wave energies drop to zero and then become low to moderate westward along the panhandle. Barrier structures are maintained on moderate- and high-energy shores, and mangrove swamps and marshes occur on low- or zero-energy coasts. Tide ranges around the peninsula vary from 2.1 m at the Georgia border to 0.3 m in the Keys, but are all less than 1 m in southern Florida.

Prevailing winds are from the east or northeast in winter, shifting to west or northwest with the passage of a cold front. Summer winds are mostly easterly and southeasterly.

We have noted a trend toward evergreenness in the coastal vegetation as we move south along the United States east coast. This trend culminates in south Florida, where coastal woodlands are as fully evergreen as are those of the Caribbean in general. Of the 82 tree species found in southern Florida, only 4 are deciduous, and 77 are found nowhere else in continental United States. Southern Florida supports 41 families of woody plants, of which no single family is most important, as is the case in more northern regions. If any one group of plants can be identified as giving the tropics a special appearance, it is the palm family, with 6 species in Florida.

A complete vascular plant listing for Cayo Costa Island, a Gulf coast barrier near Fort Myers shows both temperate and tropical influences as well the lack of dominance by particular families. Listed are 93 families and 309 species. Only 17 families have 5 or more species on the island, and these are mostly cosmopolitan families with intrinsically numerous members. The largest are the Poaceae (36 species), Asteraceae (35), Fabaceae (19), Cyperaceae (13), Rubiaceae (12), Euphorbiaceae (12). The other large families are nearly all strongly tropical: Convolvulaceae (8 species), Bromeliaceae (7), Scrophulariaceae (17), Malvaceae (6), Arecaceae (5), Agavaceae (5), Amaranthaceae (5), Rutaceae (5), Cactaceae (5), and Asclepiadaceae (5). Many of the remaining 76 families are also tropical groups, but have only 1–4 species at the latitude of Cayo Costa Island.

The great diversity of the species, combined

with the variability of the shoreline, makes it difficult to generalize about the vegetation of particular habitats, as one can easily do farther north. It is only in a broad sense that the Floridian coastal vegetation can be grouped into the following community types: beach and dune strand, coastal forest, tropical hammock, marshes and mangrove swamps, salt flats, coral keys, and freshwater wetlands.

Texan Region

West of the Mississippi delta the northern coast of the Gulf of Mexico is characterized by relatively low wave energies. In these conditions lines of barrier islands, such as those of the eastern Gulf, cannot be maintained. Instead, one sees large, low marshes and marsh islands, dominated by *Spartina alterniflora, S. cynosuroides* and *S. bakeri,* with *Phragmites* and *Typha* on more brackish sites. The marsh vegetation usually grades into a muddy beach or a narrow, sandy beach without dunes. In the vicinity of Galveston, with the coast facing more easterly, wave energies are again sufficient for barrier island formation.

To the southwest of the Port Lavaca-Matagordo Island area, there is a steady decrease in rainfall, and the vegetation changes from the Carolinian pine and live oak to that of the central Texas prairies; the *Texan* region can be said to extend from here into northern Mexico. The climate is semiarid, featuring mostly mild winters, and hot, dry summers. The frost-free period exceeds 300 days, being interrupted when major cold fronts sweep through Texas to the coast, accompanied by strong northwest winds. Summer thunderstorms provide most of the moisture, with large but irregular additions from tropical storms. Normal tide range is less than 1 m, and normal waves are small; except during hurricanes, this is only a moderate-energy coast.

The islands are low, wide, and long, and their few small inlets tend to be long-lived. Some washover-inlet features persist for centuries. Continuous onshore (southeasterly) winds favor formation of dune ridges close to the beach.

Vegetation is primarily a grassland: an analysis of South Padre Island reveals the presence of 204 species in 47 families, of which the Compositae (26 spp.), Cyperaceae (10), Poaceae (46), Fabaceae (8), and Scrophulariaceae (8) were most important. The supratidal beach and primary dunes of Padre Island are dominated by *Sesuvium portulacastrum* and *Uniola paniculata,* with *Ipomea stolonifera* and occasionally *Sporobolus virginicus.* Other dune species include *Andropogon scoparius, Croton punctatus, Cassia fasciculata, Heterotheca subaxillaris, Oenothera drummondii, Rhynchosia minima, Paspalum monostachum, Panicum marulum,* and *Fimbristylis castanea.* Behind the primary dunes are secondary dunes and extensive flats dominated by *Andropogon scoparius;* these look much like the barrier flats of the Outer Banks of North Carolina. Where the elevation is low enough for the freshwater lens to be near the surface there may be a 100% cover of vegetation, primarily *Fimbristylis castanea, Eleocharis obtusa, Scirpus americanus, Spartina patens,* and *Dichromena colorata.* The only nonherbaceous species commonly found on the flats are the shrub *Sophora tomentosa* and the cactus *Opuntia lindheimeri.*

Washover sites are fairly common, but are barren except for scattered clumps of *Sesuvium, Uniola,* and *Andropogon* invading the borders. In the dune zone these sites stand out, but they tend to grade into the flats as the vegetation advances over them.

Trees are conspicuously lacking on the Texas barriers. Except for planted exotics, the only trees on Padre Island are a grove of 5 live-oaks (*Quercus virginiana*), 2 of which are dead. These isolated trees are surrounded by grassland and low, migrating dunes. The lack of trees has been attributed to severe grazing in the past, but it is unlikely that forests were ever substantial considering the low rainfall and frequent hurricane flooding. On the Mexican barriers *Prosopis glandulosa* (mesquite) is abundant and attains the size of a small tree.

Salt marshes behind Padre Island are also noticeably absent owing to hypersaline conditions in Laguna Madre and the lack of inlets. The lagoon side of the island features extensive parallel, unvegetated dunes. Where conditions are not extreme on the tidal flats, halophytes such as *Salicornia virginica, Machaeranthera phyllocephala, Suaeda linearis,* and *Borrichia frutescens* occur in scattered patches with low cover values. Other species occasionally found include *Spartina patens, Limonium nashii, Fimbristylis castanea, Fuirena simplex, Eleocharis obtusa, Bacopa monniera, Monanthochloe littoralis, Philoxerus vermicularis, Andropogon scoparius,* and *Sporobolus virginicus.* Vast *Spartina alterniflora* marshes are not to be seen.

A few stands of *Cladium jamaicense* are present in this area. *Avicennia germinans,* the black mangrove, occurs at the southern end of Laguna Madre, and becomes increasingly important behind the barrier islands of Mexico. Tropical scrub and low trees appear on the uplands in addition to the grassland species listed above. Diverse tropical wet forests begin along the central Mexican coast and extend to the Yucatán peninsula.

West Coast Region

Beach Flora. Brecken and Barbour (1974) and Barbour, DeJong, and Johnson (1975) offer excellent reviews of the beach flora on the Pacific coast of North America. They group the 49 species characteristic of the beach flora into 9 distributional types and 7 ecofloristic units.

The circumarctic distributional type consists of *Festuca rubra, Honckenya peploides, Lathyrus japonicus, Ligusticum scoticum,* and *Mertensia maritima*. These species have widespread arctic or circumpolar distributions.

The species that comprise the Beringian-eastern North American type range across Beringia. Their extensions include areas to the north and south. *Angelica lucida, Conioselinum chinense, Elymus mollis, Poa eminens,* and *Senecio pseudo-arnica* are the species that make up this taxa.

The Beringian type consists of *Carex macrocephala* and *Glehnia littoralis*.

Four species make up the Widespread Inland-Maritime unit. *Allenrolfea occidentalis, Distichlis spicata, Heliotropium curassavicum* and *Atriplex patula* have a widespread distribution, occupying beaches on more than one ocean or continent.

The Temperate Beach distribution is characterized by 4 species: *Calystegia soldanella, Fragaria chiloensis, Mesembryanthemum cheilense,* and *Ammophila breviligulata*. This group is made up of those species found on temperate beaches, but not inland, on more than one ocean or continent. The lack of similar species indicates a large degree of isolation from the migrating routes of the temperate oceans.

The Tropical Beach species for the purpose of this review are limited to the tip of Baja California. *Ipomoea brasiliensis, Ipomoea stolonifera, Scaevola plumeri, Sesuvium portulacastrum,* and *Sporobolus virginicus* comprise the flora.

Fourteen species comprise the Maritime-Endemic distributional type. These species are all endemic to the maritime Pacific coast of North America. The flora in this group can be characterized by their affinities to the other distributional types: Arctic or Beringian affinities; *Polygonum paronychia* and *Tanacetum douglasii;* inland taxa of Mesic Temperate areas; *Agrostis pallens;* inland taxa of arid areas: *Abronia latifolia, Abronia maritima, Ambrosia chamissonis,* and *Camissonia cheiranthigolia:* taxa of no apparent affinity; *Chamaesyce leucophylla, Lathyrus littoralis;* and *Poa douglasii;* and species added and not categorized by Barbour, DeJong, and Johnson (1975), *Atriplex leucophylla, Elymus vancouverensis, Malacothrix incana,* and *Diodia crassifolia*.

Species that are endemic to both maritime and inland habitats of western North America are included in the Endemic Inland-Maritime affinity. Six species—*Atriplex barclayana, Arthrocnemum subterminale, Astragalus magdalenae, Frankenia palmeri, Lycium brevipes,* and *Agoseris apargiodes*—make up this taxa.

Isolation of the Pacific coast is apparent by the high percentage of endemism. Whether or not these species arrived by overland routes is not known. Brecken and Barbour (1974) argue that, because of present-day distributions and taxonomic affinities, in situ evolution is most likely.

There are 4 introduced species: *Ammophila arenaria, Cakile edentula, Cakile maritima,* and *Mesembryanthemum edule*. All but *M. edule* have become dominants since their introduction.

Analysis of the coastal flora for species associations identifies 7 ecofloristic zones (Breckon and Barbour, 1974). The Arctic zone extends from 71–60°. The Circumpolar and Beringian-Eastern North America distributional units occur here. The Subarctic zone consists of the same two groups as well as the Beringian taxa. It ranges from 60–54°. The temperate zone is composed primarily of endemics. It is also the widest zone—54–34°30'. The Dry Mediterranean zone is similar to the Subarctic zone because it does not have a unique flora other than *Camissonia cheiranthifolia* spp. *suffruticosa*. It extends from 34°30'–30°. Inland-Maritime species dominate the Northern Arid zone (30–28°) and the Southern Arid zone (28–24°). The Tropical zone is made up of the Tropical Beach flora and lies south of 24°.

Dune Flora. Sand dunes represent 45% of the northern California, Oregon, southern Washington coast. Sand dunes are not well developed elsewhere along the Pacific Coast, therefore the flora of these areas is excluded. Since the flora of these northern dunes is primarily restricted to such a small geographic area, the flora is discussed along with dune vegetation elsewhere.

Coastal Marsh Flora. Chapman (1960) divided the flora of the Pacific coast of North America into two seres: the Arctic sere and the Pacific American sere. He indicates that the identifying plant for the southern sere is *Salicornia virginica*. However, because of the work of Jefferson (1975), the Pacific coast most likely comprises 3 major floristic regions.

The southern marshes (Baja California to northern California) have a large number of annuals, which is typical of marshes in warm temperate areas. *Salicornia virginica* is a prominent species along with *Spartina foliosa*. Five other common species are known only within this range: *Batis maritima, Suaeda californica, Frankenia palmeri, Salicornia bigelowii,* and *Limonium commune*.

A transition occurs in northern California. These taxa include species that are found only in northern California or range from northern California to southern Canada: *Hordeum brachyantherum, Cuscuta salina, Salicornia ambigua, Grindelia cueifolia, Glaux maritima,* and *Scirpus americanus*.

A group of 13 species have been listed that are typical of Oregon and Washington coastal marshes (the central floristic unit). These have been subsequently verified by the work of other investigators in that area (Jefferson, 1975; Burg, Rosenberg, and Tripp, 1976; Disraeli and Fonda, 1979). These species are *Deschampsia caespitosa, Trifolium wormskjoldii, Grindelia integrifolia, Lilaeopsis occidentalis, Juncus lesueurii, Agrostis alba, Eleocharis parvula, Scirpus americanus, Puccinelia pumila, Plantago maritima, Glaux maritima, Potentilla pacifica,* and *Carex lyngbyei*.

Hanson (1951) reports that *Puccinelia phryganodes* is dominant on the Alaskan marshes and that southward *Carex ursina* becomes more dominant. Important species in this northern sere are *Puccinelia paupeala, Salicornia herbica, Puccinelia trifolia, Cochlearia officinalis, Carex cryptocarpa,* and *Carex ramenski*. The transition between Arctic marshes and Washington/Oregon marshes occurs at the southern range of *Carex ursina* and the northern extension of *Carex lyngbyei*.

Several marsh species have been grouped into one unit because of their wide-ranging distribution, Point Barrow, Alaska to Cabo San Lucas, Baja California: *Distichlis spicata, Salicornia virginica, Jaumea carnosa, Cuscuta salina, Triglochin maritimum, Cotula corenopifolia, Spergularia marina,* and *Scirpus maritimus*.

Beach Vegetation. The vegetated beaches of California, Oregon, and Washington exhibit considerable variation in total length and steepness. The beach communities, typical of harsh environments, are low in numbers of species and cover. Thirty-two species are encountered on the beach over this entire coastline; 20 of these are typical beach species. In most areas the vegetation consists of only 5 species, and usually only 1 or 2 of these contribute a significant amount of cover.

Barbour, DeJong, and Johnson (1976) are able to identify five vegetative groups by their dominant species: group A, *Ammophila arenaria;* group B, *Elymus mollis;* group C, *Ambrosia chamissonis;* group D, *Mesembryanthemum chilense;* and group E, *Abronia maritima* or *Cakile maritima*.

The distribution of groups A, B, and C does not correlate well with latitude except that they primarily occur north of 37°. Groups D and E have a narrow range and can be identified by latitude; group D lies between 37–35°, and group E lies between 35°–32°30' (which is the southern limit of this review). The change from grass-dominated communities to the north of 37° to the forb-dominated communities southward is shown in the group shift here. This may correlate with the January, 10° and July, 15° isotherms that are thought to be critical in the transplantation of *Ammophila arenaria*. The 35° group shift appears to correlate with the shift from a Temperate Mediterranean climate to a Dry Mediterranean climate. Group B differs from groups A and C in substrate texture; group B averages 55% fine sand, and the other two average only 10%.

The macroclimatic gradients do not seem to be important in the determination of community structure, since the overall vegetative composition correlates poorly with latitude. Barbour, DeJong, and Johnson (1976) hypothesize that local features, such as salt spray, substrate mobility, presence of introduced species, and storm wave protection, are most important to community structure.

Dune Vegetation. The dunes of northern California, Oregon, and southern Washington comprise 45% of that region's coastal area. Dunes along the rest of the coast are not extensive. Only the dune vegetation of Oregon will be discussed here.

There are 4 principal communities on the sand dunes; behind the dunes the deflation plains comprise 5 community types. *Ammophila arenaria,* the only species found on migrating dunes, was introduced in 1930 and is also associated with the origin of the foredune. *Abronia latifolia* and *Franseria chammisonis* are the species that make up the community found in blowout areas. These 2 communities are not true pioneer communities. The pioneer species (communities) will become established in these 2 areas or on unvegetated dunes, stable or unstable. The principal pioneers are *Poa macrantha, Carex macrocephala,* and *Lathyrus littoralis* on shifting dunes and *Festuca rubra* and *Solidago spathulata* on stable dunes. As the associated microclimate changes, *Tanacetum camphoratum* and *Lupinus littoralis* invade the dunes. These are subsequently followed by the shrubs *Gaultheria shallon* and *Vaccinium ovatum* and the trees *Pinus contorta* and *Pseudotsuga menziesii*.

The deflation plain consists of meadow species. The dry meadow community is found on the edge of the plain, where the sand is the driest. It usually has the fewest number of species, and the water table is greater than 1 m from the surface in the summer and does not stand on the surface in the winter. The principal species are *Lupinus littoralis, Ammophila arenaria,* and *Poa macrantha*. The meadow is the next zone on the plain. It is generally on level ground. It is more moist than the dry meadow, since the water table is .5-1 m from the surface in the summer and stands on the surface for part of the winter. The dominant species are *Festuca rubra* and *Aira praecox*. The rush meadow community has a dense growth of many species. The sand is moist, the water table being only .3-.5 m deep in the summer and standing on the surface 3-4 months of the winter. Characteristic species are *Juncus phaeocephalis, Juncus falcutus,* and *Trifolium wormskjoldii*. The marsh community is characterized by three species: *Carex obnupta, Potentilla pacifica,* and *Ranunculus flamella*. The water table stands on the surface most of the year or at most .2 m below the surface. Shrub and tree species may colonize the drier sites; the driest sites are colonized by *Gaultheria shallon* and *Vaccinium ovatum* and less dry sites by *Salix hookeriana* and *Myrica californica*. *Pinus contorta* and *Picea sitchensis* can colonize both sites.

Marsh Vegetation. In North America the coastal marshes of the Pacific coast are not as numerous or as developed as those on the Atlantic coast. The Pacific coast is characterized by a paucity of marsh-type environments. Where they do occur, the coastal marshes are found primarily in estuaries, protected bays, and lagoons.

Relatively little work has been done on the marsh vegetation on the Pacific coast. The marshes of California have received the most attention (Purer, 1942; Stevenson, 1954; Warme, 1971; and MacDonald, 1967). The northern marshes have only recently been investigated (Jefferson, 1975; Eilers, 1975; Burg, Rosenberg, and Tripp, 1976; Disraeli, 1979; del Moral and Watson, 1978). Most of this work has dealt with vegetation patterns and not the dynamics of marsh ecosystems as has been the case on the east coast.

Chapman (1960) believes that, because of a common sand substrate and flora, the marshes here are closely related units. However, there is a marked difference in marsh environments of the west coast. South of central California the climate tends to be much more arid, the estuaries smaller, lagoons more common, and marshes dominated by *Spartina foliosa* (which has a northern, natural limit, around Humboldt Bay, California). Northward, precipitation and runoff becomes increasingly important as does the occurrence of large rivers and estuaries.

The entire spectrum of California coastal marshes is too lengthy for a complete discussion here. However, the superficial resemblance of these marshes to one another is remarkable. The lower intertidal zone is dominated mostly by *Zostera marina*. This plant acts to decrease wave action and collect sediment and other debris. *Spartina foliosa* dominates the lowest zone of the marsh proper from Baja California to Humboldt Bay, except in southern California, where it is succeeded by *Salicornia bigelowii* and *Batis maritima*. It is usually found in the zone that is inundated by most tides. Occasionally *Salicornia virginica* is found in this zone. The next zone, the intermediate marsh, is still influenced by the tides. The dominant plant is *Salicornia virginica*. The major vegetative difference in the California marshes is observed on the high marsh. To the north (primarily around Humboldt Bay) *Salicornia virginica* remains dominant, but is often codominant with *Distichlis spicata* and *Jaumea carnosa*. Southward, *Salicornia virginica* may be found in pure stands, but is often found mixed with *Limonium californiicum* and *Suaeda californica* along with many other less dominant species.

Jefferson (1975) investigated succession and vegetation patterns on the Oregon marshes. She believes that the marshes of Oregon (and Washington) are transitional between subarctic and temperate marshes. She found 6 saline-brackish marsh types and four freshwater marsh types.

The low-sand-marsh type occurs on sandy substrates on the island side of baymouth spits and islands of sandy bays. This marsh is elevated only slightly above the tideflat and has a slight slope. All high tides flood this marsh; the tidal drainage is diffuse. The dominant vascular plant is *Salicornia virginica, Scirpus americanus* or *Puccinellia pumula*. At higher elevations the dominants are mainly *Distichlis spicata, Jaumea carnosa,* and *Plantago maritima*.

The low-salt marsh is usually located on a silt or mud substrate. Typical species are *Salicornia virginica, Triglochin maritimum, Spergularia marina, Cotula coronopifolia, Eleocharis parvula, Carex lyngbyei, Deschampsia caespitosa* and *Scirpus maritimus*. This marsh is flooded by nearly all high tides. The runoff is diffuse, except where small channels have developed and around the islands of vegetation. The variety of communities (hence species) is large and not well interrelated because of variations in tidal exposure, past usage, and freshwater runoff.

The sedge marsh is normally found on a silt substrate between the low-salt marsh and the more mature marshes on the edges of islands, deltas, and dikes. The surface may be level and elevated 30-100 cm above the tideflat. Most high tides inundate this marsh; tidal runoff is diffuse or well contained in the more mature sections. *Carex lyngbyei* is essentially the only vascular dominant.

The immature high-marsh type is observed on soils high in organics and silt, usually inland of sedge and low-salt marshes. Again, the surface is relatively level, but it is interrupted with salt pans, rotten spots, and well-defined drainage areas. The elevation varies from 40 cm to higher above the tideflat. The vascular plants associated with this marsh are *Deschampsia caespitosa, Distichlis spicata, Carex lyngbyei, Salicornia virginica, Potentilla pacifica, Trifolium wormskjoldii*, and *Juncus lesueurii*.

Points and deltas high in organic matter overlying clay usually support the mature high-marsh type. The surface is level but interrupted, as in the immature high marsh. *Deschampsia caespitosa, Juncus lesueurii*, and *Agrostis alba* are the dominant plants. Other species that help to identify this marsh are *Grindellia integrifolia* and *Potentilla pacifica*.

The bulrush- and sedge-marsh type is found along tidal creeks and ditches and on islands where freshwater dilutes the saltwater substantially. As the salt content decreases upstream, the bulrush-marsh type is formed. This may mature into a complex intertidal freshwater high marsh that is usually 75 cm above the tideflat and characterized by *Deschampsia caespitosa, Phragmites communis, Equisetum spp.*, and *Lilaeopsis occidentalis*.

The intertidal gravel-marsh type is found only in the mouth of the Rogue and Chetco rivers. These two rivers are the only areas where a high volume gravel bed exists. *Eleocharis bella* and *Eleocharis palustris* are covered by high tides of slightly brackish water.

The dikes salt-marsh type is usually a mature high marsh. It has all the same characteristics, but it is normally older and therefore consists of more upland species than does the normal high marsh.

The successional trends have been worked out for only the Oregon marshes. Jefferson (1975) has worked out a rough scheme of succession from three years of observation.

The Washington coast is markedly different from those of Oregon and California because, in addition to the open coast, Puget Sound provides 2240 km of interior coastline. (All published work on marshes in Washington has been done on Puget Sound.) All literature surveyed, with minor exceptions, can be fit into Jefferson's classification scheme of marsh types on the Oregon coast (Burg, Rosenberg, and Tripp, 1976; Disraeli and Fonda, 1979). Of the studies in Washington, only Disraeli and Fonda (1979) correlates vegetation patterns with environmental gradients.

The paucity of information about coastal marshes north of Washington is immediately apparent. All information about these marshes comes from the descriptions of southern Alaskan marshes by Cooper (1931) and Hanson (1951), the study of mosquito breeding sites by Frohne (1953) and the vegetation study on the Stikine Flats, Alaska, by del Moral and Watson (1978).

Three zones are broadly recognized in southern Alaska. The lower zone is dominated by *Puccinellia spp.* upon a mostly open mud flat. *Glaux maritima* and *Puccinellia triflora* occur as a thick mat of vegetation at intermediate levels but still subject to inundation. The upper marsh zone is dominated by *Plantago maritima*, with *Triglochin maritimum, Hordeum boreale*, and *Potentilla pacifica* as frequent members. As in Oregon and Washington, a monoculture of *Carex lyngbyei* occurs in brackish marshes that are rarely flooded by full-strength seawater.

For additional relevant literature on North American coastal ecology see Clovis (1968), Godfrey (1976); Jeffries (1977); Judd, Lonard, and Sides (1977), Martin (1959), Moul (1969), Ogden (1961), and Wagner (1964).

PAUL J. GODFREY
MELINDA M. GODFREY
DONALD DISRAELI

References

Barbour, M. G.; De Jong, T. M.; and Johnson, A. F., 1975. Additions and corrections to a review of North American Pacific coast beach vegetation, *Madroño* 23, 130-134.

Barbour, M. G.; De Jong, T. M.; and Johnson, A. F., 1976. Synecology of beach vegetation along the Pacific coast of the United States of America: A first approximation, *Jour. Biogeog.* 3, 55-69.

Breckon, C. J., and Barbour, M. G., 1974. Review of North American Pacific coast beach vegetation, *Madroño* 22, 333-360.

Burg, M. E.; Rosenberg, E. S.; and Tripp, D. R., 1976. Vegetation associations and primary productivity of the Nisqually salt marsh, *in* S. G. Herman and A. M. Wiedemann, eds., *Contributions to the Natural History of the Southern Puget Sound Region*,

Washington. Olympia, Wash.: Evergreen State College, 109-144.
Chapman, V. J., 1960. *Salt Marshes and Salt Deserts of the World.* New York: Wiley Interscience, 392p.
Clovis, J. F., 1968. The vegetation of Smith Island, Virginia, *Castanea* 33, 115-120.
Cooper, W. S., 1931. A third expedition to Glacier Bay, Alaska, *Ecology* 12, 61-95.
del Moral, R., and Watson, A. F., 1978. Vegetation on the Stikine Flats, Southeast, Alaska, *Northwest Sci.* 52, 137-150.
Disraeli, D. J., and Fonda, R. W., 1979. Gradient analysis of the vegetation in a brackish marsh, Bellingham Bay, Washington, *Canadian Jour. Botany* 57, 465-475.
Eilers, H. P., 1975. *Plants, Plant Communities, New Production, and Tide Levels: The Ecological Biogeography of the Nehalem Salt Marshes, Tillamook County, Oregon.* Ph.D. dissertation. Corvallis; Oregon State University, 368p.
Frohne, W. D., 1953. Mosquito breeding in Alaskan salt marshes, with special reference to *Aedes punctodes* Dyar, *Mosquito News* 13, 96-103.
Godfrey, P. J., 1976. Barrier islands of the east coast, *Oceanus* 19, 27-40.
Hanson, H. C., 1951. Characteristics of some grassland, marsh, and other plant communities in western Alaska, *Ecol. Monogr.* 21, 317-375.
Jefferson, C., 1975. *Plant Communities and Succession in Oregon Coastal Salt Marshes.* Ph.D. dissertation. Corvallis: Oregon State University, 192p.
Jeffries, R. L., 1977. The vegetation of salt marshes at some coastal sites in Arctic North America, *Jour. Ecology* 65, 661-672.
Judd, F. W.; Lonard, R. I.; and Sides, S. L., 1977. The vegetation of south Padre Island, Texas, in relation to topography, *Southwestern Naturalist* 22, 31-48.
MacDonald, K. B., 1967. *Quantitative Studies of the Salt Marsh Mollusk Faunas from the North American Pacific Coast.* Ph.D. dissertation. San Diego: University of California, 316p.
Martin, W. E., 1959. The vegetation of Island Beach State Park, N.J., *Ecol. Monogr.* 29, 1-46.
Moul, E. T., 1969. Flora of Monomoy Island, Massachusetts, *Rhodora* 71, 18-28.
Ogden, J. G., 1961. Forest history of Martha's Vineyard, Massachusetts, I. Modern and pre-colonial forests, *Am. Midland Naturalist* 66, 417-430.
Purer, E. A., 1942. Plant ecology of the coastal salt marshlands of San Diego County, California, *Ecol. Monogr.* 12, 81-111.
Stevenson, R. E., 1954. *The Marshlands at Newport Bay (California).* Ph.D. dissertation. Los Angeles: University of Southern California, 109p.
Wagner, R. H., 1964. Ecology of *Uniola paniculata* in the dune-strand habitat of North Carolina, *Ecol. Monogr.* 34, 79-96.
Warme, J. E., 1971. Paleoecological aspects of a modern coastal lagoon, *California Univ. Pubs. Geol. Sci.* 87, 1-131.

Cross-references: *Biotic Zonation; Central and South America, Coastal Ecology; Coastal Flora; Coastal Plant Ecology, United States, History of; Mangrove Coasts; North America, Coastal Morphology; Salt Marsh Coasts; Washover and Washover Fans.* Vol. VIII, Part 1: *North America.*

NORTH AMERICA, COASTAL MORPHOLOGY

The North American coastal geomorphology is extremely varied, including all types listed in the writer's classification of coasts (Shepard, 1973). About one-third of the coasts including most of Canada and southern Alaska are characterized by glacial erosion forms including *fiords* and *troughs* along with numerous rocky islands. Another third is *barrier coasts*, notably along eastern and southern United States, with some of the longest *barrier islands* in the world and most of the world's *cuspate forelands*. *Ria coasts*, many inside barrier islands, are very common, and *deltaic coasts* are well represented with perhaps the longest group of *compound deltas* in the world along the north coast of Alaska and adjacent Canada. *Mud flats*, *mangroves*, and other types of *marshes* are found along the protected coasts of the tropics and subtropics, particularly along the United States Gulf coast and parts of eastern Central America.

Wave erosion coasts, mostly with *sea cliffs*, are common along the western coasts of North America, south of the glaciated territory. Erosion is important where the cliffs are of alluvium or soft rocks.

Glaciated East and North Coasts

Coasts that have been glaciated are generally so distinct from the other coasts of the world that almost any map or chart should make it clear to a geomorphologist that glaciers had formerly covered the area. Most of these coasts are deeply indented like the Gulf of St. Lawrence, and they are usually margined by numerous rocky islands. All the glaciated north and east coasts of Canada have these characteristics; the embayments have deep water and usually have straight sides as the result of the erosion by the glaciers. These characteristics can be traced south into the United States, where we see the many islands and fiordlike bays of the northern Maine coast.

The low coasts to the north around Hudson Bay differ from the glaciated mountain coasts of Labrador, where we find many fiords. Aside from the barrenness due to the cold climate, the Labrador coast closely resembles that of Norway, British Columbia, and southern Chile, where glacial erosion has also been intense. Uplifted terraces are common all along these coasts that were formerly weighted down by the ice.

The coastal character changes in southern Maine, where glacial deposition and particularly glacial outwash develop a dominant role with only a few rocky areas, like Cape Ann in Massachusetts, where some of the glacial erosion forms can be found along with moraines and sand dunes that are the result of the reworking of the outwash plains.

In Boston harbor and along the nearby coast, we find that most of the relief is related to glacial deposition, with the hills and islands consisting largely of drumlins. Starting at Cape Cod, which is tied to the land by glacial outwash deposits, we have a series of relatively low islands consisting of glacial outwash and morainal material, the latter partly deformed by overriding ice, as at Gay Head in Martha's Vineyard, where the beds are greatly disturbed by ice shove. Most of the coast from southern Maine to New York City has low hills and numerous estuaries partly blocked by barriers. Long Island was originally glacial moraines and outwash plains, but has been greatly modified by formation of barrier islands and by extensive dunes built from the beach sands that have blown inland.

Barrier and Drowned Valley Coastal Lowlands, New Jersey to Florida

South of the glacial boundary we find the end of the rolling, hilly coasts and their outlying glacial deposition islands. Here begins a coastal plain with relatively continuous barriers that are interrupted by small inlets and, farther south, by large embayments with dendritic drowned river valleys. This coastal plain represents the start of the extensive barrier islands that are said to include over 10,000 km of North American coasts and 33% of all barrier coasts of the world (Berryhill, Dickinson, and Holmes, 1969). Much of this coast has extensive swamps and marshes, where the lagoons inside the barriers have been partly filled with sediments and marsh vegetation. As a result, much of the habitation is on the barriers, where there are long, broad beaches used for summer resorts. Near New York the marshes have been largely filled artificially. The principal larger communities are well up in the embayments, particularly at the head of Delaware and Chesapeake bays.

South of Chesapeake Bay we encounter a series of cuspate forelands (Fig. 1), the best developed of any in the world. These include Capes Hatteras, Lookout, Fear, and Romain. All are unstable barriers that are backed by dunes and have wide lagoons that separate them from the mainland. South of Cape Fear the barriers are connected to the mainland in the Myrtle Beach area, and the coastal plain with its many beach ridges is cut by a series of oval lakes called the Carolina Bays. These are all elongate to the northwest, and have been variously interpreted as ancient meteorite craters (Melton and Schriever, 1933), segmented lagoons due to coastal processes (Johnson, 1942), or the result of permafrost during glacial maxima (Prouty, 1967). These are found also on the coastal plain of northwest Alaska.

South of Cape Romain the coast is much straighter, and the only cuspate foreland is in Florida at Cape Canaveral. Barriers and drowned valleys continue along the South Carolina, Georgia, and northern Florida coasts, but are mostly short, stubby sand islands separated by estuaries. Here also old barrier islands, probably formed in interglacial epochs, are found inland separated by low marshy country.

The lagoons inside the barriers are mostly narrow, and many would have been entirely filled by vegetation if it were not for the digging and maintaining of the Intracoastal Canal.

Coral Keys of Southern Florida

South of Miami the barrier islands lose their source of quartz sand and gradually merge into coral keys with slightly raised old coral reefs, the remnants of higher sea level of the last interglacial stage. Growing reefs are found along the east side of these raised reefs. On the west side of the Florida tip, the shallow Florida Bay is studded with small mangrove islands, and the coast is marshy.

Alternating White Sand Barriers and Swamps From West Florida to Mississippi

The low west coast of Florida contains many long barrier islands with almost pure quartz sand, but there are also extensive areas with marshes and swamps. The Everglades just north of Cape Sable have interlocking channels separated by red mangrove swamps. Many of the channels are navigable in shallow-draft boats. The water varies from brackish to almost fresh during the rainy season. To the north the coast is called Ten Thousand Islands, and consists of myriads of small mangrove masses with some small, quartz sand beaches. Low waves characterize the entire coast under normal conditions.

The barrier sand islands begin at Cape Romano and extend north as far as Cedar Keys. This includes the populated barriers around Sarasota, St. Petersburg, Clearwater, and Tarpon Springs. The barriers generally enclose bays with abundance of mangrove islands (Fig. 2). The coast is all low with many lakes and marshes.

North of Cedar Keys the barriers come to an end. Here the rocks of the interior are all lime-

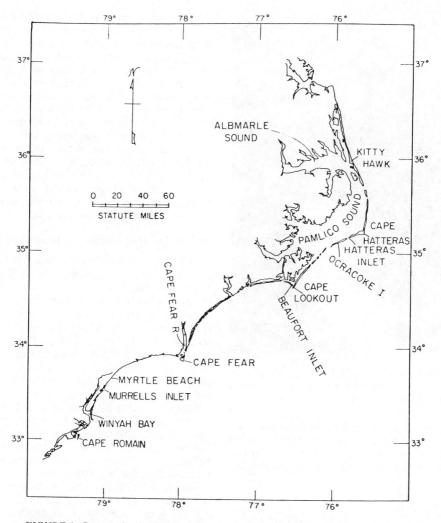

FIGURE 1. Cuspate forelands along the coasts of the Carolinas. These are the best developed of any cuspate forelands in the world. Note large lagoons inside Cape Hatteras.

stone, and the low gradient of the rivers and the numerous ponds, which are mostly sinkholes, account for the absence of quartz sand along the seacoast, leaving a vast marsh with numerous, small, vegetated islands. Mangroves are less common than farther south because of the cooler climate. The sinkholes are merely the result of solution during the glacial stages of lowered sea level. The oyster reefs lying off the coast are some of the largest in the world. Here, except for rare hurricanes, the wave energy is almost zero, partly because of the shallow water along the coast. Several relatively large rivers enter the Gulf coast, including the well-known Suwannee.

The swamp coast comes to an end some 128 km northwest of the Suwannee River, where the coastal trend changes direction from north-south to east-west. Here Ochlockonee Bay has drainage from the southern Appalachians, so that quartz sand is available, and waves are more pronounced, allowing redevelopment of the barriers. These sandy islands, with one large cuspate foreland—Cape San Blas—continue westward as far as the Mississippi delta. Various openings through the barriers give access of navigation to the lowland port cities, such as Panama City, Pensacola, and Mobile; all three are located on large bays that are protected by discontinuous barriers. The bays are a combination of lagoons and estuaries. The Appalachicola River, which has a small delta, provides sand that drifts westward to form the San Blas cuspate foreland. A more rapidly advancing delta is building into the head of Mobile Bay. The barrier islands along the Florida panhandle and

FIGURE 2. Aerial view of Blind Pass, Sanibel Island, west Florida, showing barrier islands and mangrove islands in the lagoons (photo: D. L. Inman).

the coasts of Alabama and Mississippi are subject to frequent hurricanes that have caused great destruction of property and some loss of lives when the water has risen over the low coasts.

Mississippi Delta

Although one thinks of the Mississippi delta as the actively building birdfoot passes southeast of New Orleans, the delta actually includes a series of old passes that extend for 300 km along the coast. The well-known history of the delta, determined by Harold Fisk (1944), Richard Russell (1936), C. R. Kolb (1958), and their many associates at Louisiana State University and the Mississippi River Commission, shows that the river has built a series of deltas into the Gulf of Mexico during postglacial times of which the Balize Delta (Birdfoot) is only the latest, with an age of about 1500 years. The old passes have been greatly eroded and have also sunk to a great extent below the surface, leaving barrier islands as remnants, including Chandeleur Islands on the east and Grand Isle, the great petroleum center, on the west.

A large part of the coast of the greater Mississippi delta is marshland and mud flats, with numerous shallow lakes and intertwining channels. The natural levees of the principal river channels rise about a meter above the water level, and many of these have been built up artificially to protect the towns and cities from floods. The marshland is covered largely with aquatic plants and is remarkable for its huge population of aquatic birds. In the portions representing old delta lobes, the general sinking of the area has often left only the levees rising above the water level. A few hills are the result of subsurface salt domes, and directly outside the main passes there are *mud lump islands*, which are raised above the surface as a type of mud diapir. They form rapidly and are soon cut away by wave erosion.

Barrier Coasts of Western Louisiana and Texas

West of the Mississippi Delta, the barrier islands become the dominant coastal feature. Near the delta the barriers, called *cheniers* because of the live oak trees that grow on them, constitute a series of ridges separated by marshland and extending inland for several kilometers (Gould and McFarlan, 1959). These formed mostly during the past two or three thousand years while the sea level has been relatively stable. They provide high ground for many of the roads in the coastal area.

Along the Texas coast are some of the longest barrier islands in the world, notably Padre Island, which, combined with Mustang Island, extends for 208 km almost to the Mexican border. These barriers have extensive dune tracts inside the broad beaches. Also, marshy washover deltas have extended the barriers into the lagoons. The highest parts of the barriers are the dunes, although these rarely rise more than 10 m.

The lagoons and estuaries in general decrease in depth to the southwest. Laguna Madre is almost filled, so that a large part of it is inundated only during flood periods or when the north winds carry water onto the flats from Corpus Christi Bay. Most of the bays to the north are filling rapidly (Shepard, 1973). The deltas are responsible for much of this fill, but sediment is also settling on the floors, making large differences in the depths of recent charts from those of 100 years ago. Also, oyster reefs are filling some of the bays. Two deltas, those of the Brazos and the Rio Grande, have filled their estuaries and are built slightly seaward into the gulf.

Landward of the lagoons and estuaries the broad plains are interrupted by many ancient barriers (Le Blanc and Hodgson, 1959). These are related mostly to higher stands of sea level in interglacial stages.

Coast of Eastern Mexico

Not much study has been made of the Mexican coast south of the Rio Grande. Certainly the northern portion represents a continuation of the barrier islands and shallow lagoons to the north. Although the coastal plain continues farther south, the lagoons have been largely filled, except at Cape Rojo, a cuspate foreland with a 20-km wide lagoon on the inside. In the Gulf of Campeche there are more lagoons and broad barrier reefs. On the east side the Laguna de la Términos has a lagoon 50 km wide with a narrow barrier that nearly separates it from the gulf. The north side of the Yucatán Peninsula has a long, narrow lagoon, Río de la Cienega, along its entire length. Thus we find that the entire Gulf of Mexico is ringed with barrier islands and lagoons.

On the east side of Yucatán is the beginning of coral reefs, which are partly elevated along the island coasts although best developed seaward of the barrier islands. One of the largest barrier reefs in the world lies a short distance outside a series of sand barrier islands, and continues south past the deltaic lowland coast of Honduras. Mangrove shores predominate in this area, with many passageways along the low coasts of Belize. The city of Belize is built on a delta. Several other deltas are found farther south on the Honduras coast, but elsewhere the coast has drowned valleys and barrier islands. Beyond the bend of the Gulf of Honduras the coast becomes straight, and low mountains extend along it over a considerable distance on the east side of the gulf. Near the next southerly bend, the coastal plain and deltaic encroachments are redeveloped. These extend south past the relatively straight coasts of Nicaragua with a large deltaic bulge near the Costa Rica border. The coast again becomes mountainous approaching Panama.

Western Central America

In general the west coast of Central America is more mountainous than the east coast. Several large embayments are found, including the Gulf of Panama and the Gulf of Fonesca in Nicaragua. Narrow coastal plains with intermittent barriers exist at the base of the mountains in El Salvador and Nicaragua. Some of the barriers are of black sand, as in Nicaragua, where streams have carried the volcanic sediments to the coast.

Western Mexico

The coastal lowlands continue along the Mexican coast as far as the west side of the Gulf of Tehuantepec with barriers and marsh-filled lagoons that become broad bodies of water east of Tehuantepec. West of this lowland town the mountains extend down to the coast, and there are few barriers until approaching Punta Galero 150 km east of the large resort of Acapulco. The coast is generally straight or gently curving in this mountainous area, and continues mountainous with occasional delta plains to the northwest past the picturesque Zihuatanejo with its scallop-shaped bay and rugged hillsides that may make it another Acapulco in the future.

When the Río Balsas, the second largest river on the west coast of Mexico, comes to the coast 250 km northwest of Acapulco, the largest of the Mexican deltas has built out into the Pacific (Shepard and Reimnitz, 1981). Farther northwest the mountainous coast continues, extending past the compound embayments of Manzanilla. Near here a number of coastal islands with their white guano deposits rise above the sea.

At Cape Corrientes the coastal trend changes, and a large fault trough with mountainous walls extends into the city of Puerto Vallarta. Beyond this gulf only a small stretch of mountainous coast continues, and then a northwesterly trend begins the long but narrow coastal lowland with resort cities like Tuxpan and Mazatlán. The peninsula of Baja California exerts its influence on wave conditions along this stretch, and there are many deltas built into the sea and many shallow embayments of the ria type.

The coast again becomes mountainous at Guaymas, where rocky promontories extend into the sea and the harbor has several arms, giving it good protection from the waves. Northwest of Guaymas we encounter several large islands.

At the head of the Gulf of California a large delta has been formed by the Colorado River, and the extensive lowland extends up into Imperial Valley in the United States. Along the west side of the gulf there are extensive fault scarps (Fig. 3), notably that inside the long Ángel de la Guarda Island. Almost the entire coast is rugged and relatively straight. High islands rise as fault blocks off this coast, and long embayments extend into the coast.

The west coast of Baja California is also rugged except for the lowland on the east side of Magdalena Bay and a rather extensive area south and inside the large Vizcaino Bay. These extend southeast of the bay to form the amazing Guerro Negro and Scammons lagoons where whales come in annually for breeding. Here salt flats and broad belts of sand dunes extend along the bay shores (Fig. 4). Northwest of Vizcaino Bay the mountainous coast continues with many rocky points. Lowlands exist inside some

NORTH AMERICA, COASTAL MORPHOLOGY

FIGURE 3. Fault coast along the west side of the Gulf of California, the Giganta Range north of Santa Rosalia (photo: F. P. Shepard).

FIGURE 4. Aerial photo of the great mass of sand dunes at the mouth of Guerro Negro Lagoon on the south side of Vizcaino Bay, Baja California (photo: N. Marshall).

of the bays where rivers have built deltas into estuaries.

Southern California

The coast of southern California has mostly low sea cliffs bordered with terraces and a few coastal plains and deltas, as in the Los Angeles area and near Ventura. Several ridges rise above the coast, such as Point Loma and Soledad Mountain to the south and San Pedro Hills in the Los Angeles area. These blocks all form projections into the sea, and all have a series of raised terraces (Fig. 5). Beyond the Los Angeles

FIGURE 5. The terraces along the seaward side of Palos Verdes Hills, one of the recently raised blocks along the coast of southern California (photo: R. C. Frampton).

plain, the Santa Monica Mountains follow the coast. A highway has been maintained along the mountain front despite landslides that threaten to carry it into the sea on occasions.

Many of the sea cliffs are being eroded actively, particularly where they have been cut into alluvium. Just how unstable some of these cliffs are has not been well recognized, notably in the area north of San Diego, where old maps show that entire blocks of property on the terraces adjoining the cliffs have disappeared since some of the nineteenth-century land surveys (Kuhn and Shepard, 1981).

Central and Northern California

With the turn in the coast at Point Conception a somewhat different coastal geomorphology is encountered. Despite the entrance of a series of diagonal northwesterly trending mountain ranges that cut across the north-northwest trend of the coast, the rugged central and northern California coast is one of the straightest in the world. The mountainous coast is interrupted by the entrance of a few broad river valleys. At these valleys we have some unusually large dune tracts, the result of river sediments being returned by waves to the beaches and thence being carried inland by westerly winds. One of the most scenic drives in the world extends north from San Simeon along the mountain wall of the Santa Lucia range to Carmel.

The two largest embayments, Monterey and San Francisco, are found, respectively, at the mouths of the Salinas and the San Joaquin-Sacramento rivers, the latter draining the great Central Valley of California. North of San Francisco some of the smaller bays, including Bolinas, Bodega, and Tomales, follow the San Andreas Fault, which finally goes to sea just south of Cape Mendocino, where the coast changes trend to the north.

Most of the central and northern California coast has high cliffs, and many greatly elevated marine terraces are found along them.

Storm-Exposed North-Trending Coasts of Oregon and Washington

The next coastal subdivision actually includes the northernmost portion of California and, aside from rocky points, trends almost directly north to where it ends at the entrance to the Strait of Juan de Fuca. This north-trending coast alternates between lowland valleys at the mouths of a series of large rivers and relatively low mountainous tracts of short length. Barriers, largely missing to the south, again become important at Eel River and Humboldt Bay, both in California. Farther north they are found in many places along the Oregon and southern Washington coasts. Locally they have large dune tracts inside the barriers, as at Coos Bay. Many of the rivers, including the giant Columbia, enter estuaries, indicating that the rivers have not been able to fill the drowned valleys that resulted from the rise of sea level when the great glaciers melted. It is perhaps no coincidence that the long estuaries of southern Washington are comparable to the equally long estuaries south of New York, both areas being just south of the glaciated land where upbulging of the coast probably occurred during the time of ice weighting to the north, so that sinking would be expected when the ice melted.

Glaciated Coast of Northern Washington to Alaska

The Strait of Juan de Fuca is just north of the limits of continental glaciation, and a great difference of coastal type is indicated even more clearly than along the glacial boundary of the east coast. The strait is a typical glacial trough with straight sides and deep basins. To the east there are other troughs that trend in various directions along the courses of the old glaciers, notably Puget Sound, which extends south along the west side of Seattle. To the north numerous islands and passageways exist allowing inside sailing along the coast almost all the way to Alaska. The glacial troughs in the Puget Sound area are bordered mostly by glacial deposits, but to the north, beginning with the San Juan Islands, typical glaciated rock islands with many passageways are found. The coast of British Columbia is mountainous, and has numerous fiords extending deep into the land; the same condition continues into southeastern Alaska, but there we find many of the glaciers coming down to sea level (Fig. 6). Beyond where the Alaskan coast bends to the west we encounter the Malaspina Glacier, the largest glacier in North America. When first seen by Vancouver, this ice mass extended well out into the Gulf of Alaska but has now shrunk considerably. The mountains in this area include the St. Elias Range, which, next to Mt. McKinley, has the highest peaks on the continent.

The most densely populated area of Alaska is along the fiords west of the St. Elias Range. Here the topography is partly the result of recent faulting. Farther west volcanic peaks become important, and the topography is a combination of vulcanism, diastrophism, and glacial erosion. The volcanic Aleutian Islands extend out from the southwest corner of Alaska for 2200 km.

Bering Sea and Arctic Alaska

Around the corner beyond the Aleutian Peninsula we find a great change in the character of the coast. Extensive coastal plains are found along the Bering Sea with numerous lakes and winding stream channels. A few mountain ranges extend as points into the sea. A large delta has been formed by the Yukon that includes many old lobes forming a vast plain connecting small, elevated tracts. The oldest arm of the delta developed at the now drowned mouth of the Kuskokwim River. Aside from the scarcity of mountains along the Bering Sea, the coastal area differs also from southern Alaska because it was largely ice-free during the Pleistocene. On the other hand, permafrost becomes increasingly important to the north. This

FIGURE 6. Columbia Glacier entering Prince William Sound in south-central Alaska. Note the calving ice front and the medial moraines on the glaciers.

process greatly increases the abundance of depressions in the surface when melting occurs in summer, causing *thaw lakes* that dot the surface.

North of the Yukon delta, Norton Sound is deeply indented into the coast. On its north side the city of Nome is located on such a straight coast that no harbor has been feasible. The city developed because of the gold found in the series of beaches at various elevations. Farther north, mountains of over 1000-m elevation constitute the Seward Peninsula, with Cape Prince of Wales as the nearest point to Siberia.

North to Seward Peninsula there is 240 km of barrier coast extending to Kotzebue Sound. Farther north much of the coast is bordered by other barriers including the unusual cuspate foreland of Point Hope (Fig. 7), and three more cuspate forelands, terminating with Point Barrow, the extreme northern point of Alaska. Except for Point Hope, these are rather similar to the cuspate forelands extending south of Cape Hatteras. Curiously, some of the coastal plain south of Point Barrow has oriented lakes similar to the Carolina Bays of the east coast, although the Alaskan lakes lie on a plain above a 20-m-high sea cliff.

We finally complete our tour around North America with one of the most continuous delta coasts in the world (Shepard and Wanless, 1971). These deltas include the Colville and the Kupuruk on the north slope built out by rivers draining the Brooks Range and farther east by the Mackenzie draining the northern Canadian Rockies. Extensive barrier islands are found wherever the deltas are not building actively into the sea. The rivers in this cold area flow only for a short period each year.

To the east of the deltaic coast, we encounter the glaciated territory and the deep channels and islands described earlier.

FRANCIS P. SHEPARD

References

Berryhill, H. L.; Dickinson, K. A.; and Holmes, C. W., 1969. Criteria for recognizing ancient barrier coastlines, *Am. Assoc. Petroleum Geologists Bull.* 53, 706-707.

Fisk, H. N., 1944. *Geological Investigation of the Alluvial Valley of the Lower Mississippi River.*

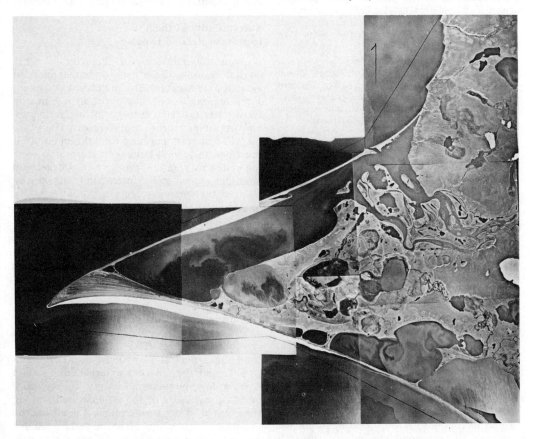

FIGURE 7. Aerial view of Point Hope, a cuspate foreland built into the Chukchi Sea beyond the Kupuruk Delta. The sea is frozen seven months of the year (photo: U.S. Geological Survey).

Vicksburg, Miss.: Mississippi River Commission, U.S. Army Corps of Engineers.
Gould, H. R., and McFarlan, E., Jr., 1959. Geologic history of the chenier plain, southwestern Louisiana, *Gulf Coast Assoc. Geol. Socs. Trans.* 9, 1-10.
Johnson, D. W., 1942. *Origin of the Carolina Bays.* New York: Columbia University Press, 341p.
Kolb, C. R., 1958. *Geological Investigations of the Mississippi River–Gulf Outlet Channel.* Waterway Experimental Station, U.S. Army Corps of Engineers Tech. Rept. No. 3.
Kuhn, G. G., and Shepard, F. P., 1981. Should Southern California build defenses against violent storms resulting in lowland flooding as discovered in records of past century? *Shore and Beach* 49, 3-10.
LeBlanc, R. J., and Hodgson, W. D., 1959. Origin and development of the Texas shoreline, *Gulf Coast Assoc. Geol. Socs. Trans.* 9, 197-220.
Melton, F. A., and Schriever, W., 1933. The Carolina "Bays"–Are they meteorite scars? *Jour. Geology* 41, 52-66.
Prouty, W. F., 1962. Carolina Bays and their origin, *Geol. Soc. America Bull.* 63, 167-224.
Russell, R. J., 1936. Physiography of lower Mississippi River delta–Louisiana, *U.S. Geol. Survey Bull.* 8, 3-193.
Shepard, F. P., and Wanless, H. R., 1971. *Our Changing Coastlines.* New York: McGraw-Hill, 579p.
Shepard, F. P., and Reimnitz, E., 1981. Sedimentation bordering the Rio Balsas delta and canyons, western Mexico, *Geol. Soc. America Bull.* 92(part 1), 395-403.

Cross-references: *Arctic, Coastal Morphology; Barrier Islands; Carolina Bays; Deltaic Coasts; Glaciated Coasts; North America, Coastal Ecology; South America, Coastal Morphology; Soviet Union, Coastal Morphology.* Vol. VIII, Part 1: *North America.*

NUCLEAR POWER PLANT SITING

With fossil fuels rising in cost and alternate sources of power (solar, sea, geothermal) not yet feasible on a large scale, nuclear power is playing an increasingly important role in providing energy to the industrialized world. As of 1978, 581 nuclear plants were operating, under construction, or planned in 35 countries; the United States alone accounted for over one-third of this total (Anonymous, 1978). Nuclear plants are frequently located in the coastal zone, because of the needs for cooling water, proximity to electrical load centers, and accessibility to transportation. In reflection of public safety concerns and the advanced technology associated with nuclear power, siting of these plants is often controversial, and is given careful attention by governments and the utility industry.

Those factors bearing on the siting of nuclear plants can be categorized as follows: susceptibility of the site to natural phenomena; susceptibility of the site to human-related events; potential effects of the plant on the region; and economic feasibility of the site. These factors are discussed below.

Susceptibility of the Site to Natural Phenomena

A primary engineering criteria of the site is the potential for disturbance by natural phenomena, relating to ground stability, water levels, and meteorology. Ground stability can be adversely affected by landslides, rockslides, surface faulting, earthquakes, subsidence or collapse, and soil liquifaction. Of these items, displacement from surface faulting and vibratory motion induced by earthquakes are given special attention because they are generally difficult to predict. Unusually high water levels can result from runoff, hurricanes, typhoons, tsunamis, and seiches; drawdown associated with tsunamis and seiches is also important regarding cooling water source. Adverse meteorological phenomena include high winds from hurricanes, typhoons, and tornadoes and heavy precipitation in the form of rain, hail, or snow.

Susceptibility of the Site to Human-Related Events

Given the industrialized character of many coastal regions, the effects of human-induced catastrophic events on the site must be assessed. Fires and explosions may occur at such nearby facilities as arsenals, chemical plants, refineries, and gas or petroleum storage areas. The possibility of aircraft crashing into the plant itself must be considered. Long-term removal of cooling water may result from river blockage, ship collisions, or oil spills. Interaction with nearby nuclear reactors represents a potential safety hazard.

Effects of the Plant on the Region

Construction and operation of a nuclear plant may affect the adjacent region from the standpoints of public safety, socioeconomics, environmental disruption, and aesthetic considerations. Of these, safety factors are paramount, and dispersion of radioactive materials in the air, surface waters, and ground waters must be carefully analyzed. The immediate area surrounding a nuclear plant must be strictly controlled, and the adjacent external zone should have a low population density to minimize potential exposure of the public to radiation in the event of a reactor accident. Aside from the radiation hazard, the plant may have human implications in terms of modifying the demography, increasing employment (particularly during construction), and raising the tax base of the local community.

The potential for environmental disruption by a nuclear plant arises through use of natural waters for cooling purposes and rejection of waste heat. In the first place, the continuous intake of water raises the possibility that significant quantities of plankton or fish larvae could be impinged on protective screening or entrained into the plant-cooling water system. Release of waste heat into natural waters causes a temperature rise that may be detrimental to the biota; also, supersaturation of gases in the discharge may cause local mass mortality of fish from gas bubble disease. Should waste heat be released into the air by means of a cooling tower, drift or salt could have detrimental effects on downwind vegetation. The impacts on the biota of routine releases of radiation must be addressed. Environmental issues may be intensified by the presence of any rare or important aquatic or avian species at the proposed plant site and in adjacent waters.

Aesthetic factors include effects of the plant's presence on scenery, destruction of historical or archeological sites, and impingement on unique land types such as marshes or dunes. Aesthetic considerations must extend to the power transmission corridor as well as the plant site iteself.

Economic Feasibility

Aside from safety and envrionmental factors in site selection, the location of the site usually has important economic ramifications. Proximity to a cooling water source translates into costs for pipelines; similarly, proximity to feed-in points on the existing power grid governs the length of the transmission corridor. Access to the site may require construction of new roads, and labor and construction material must be available. Design of the plant to accommodate any site-related safety or environmental problems, such as extensive foundation work or building a cooling tower, can entail substantial additional costs.

Deciding on the acceptability of a proposed nuclear plant site represents a complex interplay of the above siting criteria. While these criteria are recognized by most nations, they are given differing priorities in various parts of the world. In general, environmental/aesthetic issues and community acceptance take on considerable importance in affluent countries, whereas basic safety and engineering considerations are dominant in less developed nations. Siting decisions are in most cases approved by an arm of the federal government, be it a regulatory agency such as the United States Nuclear Regulatory Commission or a nationalized electricity-generating body such as Great Britain's Ministry of Power. A supranational group, the International Atomic Energy Agency, has summarized nuclear plant siting practices for various countries (Anonymous, 1975).

DAVID O. COOK

References

Anonymous, 1975. *Siting of Nuclear Facilities.* Vienna: International Atomic Energy Agency, 625p.

Anonymous, 1978. *Power and Research Reactors in Member States.* Vienna: International Atomic Energy Agency, 123p.

Cross-references: *Demography; Effluents; Human Impact; Hurricane Effects; Oil Spills and Pollution; Thermal Pollution; Tsunami; Waste Disposal.*

NUTRIENTS

Nutrients are chemicals required for the growth and multiplication of living organisms. Strictly, these include major constitutents of seawater such as carbon, in the form of bicarbonate for photosynthesis and organic carbon for heterotrophic organisms, as well as trace metals (including iron, manganese, cobalt, magnesium) and vitamins. However, in marine chemistry, the term is usually applied to the most important *'micronutrients';* phosphorus, inorganic nitrogen, and silicon. Phosphorus and nitrogen are required by all organisms; and silicon, by those important phytoplankton that have a siliceous covering.

Because concentrations of these three elements in surface seawater are usually low in proportion to plant requirements, one or more of them often limits biological productivity—that is, phytoplankton growth almost ceases when one nutrient is exhausted, and continues only as fast as this nutrient is replenished. Nitrogen to phosphorus ratios in plankton and in seawater are similar (about 15:1 on a molar basis), although variations occur in some seawater masses. The most important forms of nitrogen are nitrate and ammonia. Phosphorus exists in many forms, of which orthophosphate is most important. Dissolved silicon probably occurs mainly as unionized silicic acid (Spencer, 1975).

Input of nutrients into coastal waters is largely by river runoff. Regeneration of nutrients by autolysis and bacterial decomposition of dead organisms is also vital in maintaining a supply of these elements in an inorganic form suitable for phytoplankton. This can involve complex seasonal cycles requiring vertical mixing.

In the interstitial waters of beaches and sediments, nutrient concentrations are higher than in seawater (Pugh et al., 1974). Darkness precludes photosynthetic growth that otherwise provides the main demand for nutrients; then too, large surface areas promote growth of bacteria that degrade particulate and dissolved organic matter, regenerating inorganic nitrogen and phosphorus. Owing to these microbiological processes and the limited gas exchange, the supply of oxygen may become depleted. This is more marked in less porous beaches, resulting in inorganic nitrogen being present as ammonia rather than as nitrate. Silicic acid may dissolve directly from the silica in the sediment.

Although subtidal sediments have high nutrient concentrations relative to the overlying water, intertidal beaches are more likely to interact with seawater. Depending on particle size and tide and wave conditions, considerable volumes of seawater can be filtered through beaches (Reidl, 1971), resulting in decomposition of its organic matter and reintroduction of nutrients to inshore waters. Thus sandy beaches can form an interesting site of regeneration of the essential nutrients that limit primary productivity in surface waters.

COLIN F. GIBBS

References

Pugh, K. B., Andrews, A. R., Gibbs, C. F., Davis, S. J., and Floodgate, G. D., 1974. Some physical, chemical, and microbiological characteristics of two beaches of Anglesey, *Jour. Exp. Marine Biology and Ecology* **15**, 305-333.

Reidl, R. J., 1971. How much seawater passes through sandy beaches? *Internat. Rev. Gesamten Hydrobiolie* **56**, 923.

Spencer, C. P., 1975. The micronutrient elements, in J. P. Riley and G. Skirrow, eds., *Chemical Oceanography*, 2d ed. New York: Academic Press, 245-300.

Cross-references: *Aquaculture; Nearshore Water Characteristics; Photic Zone; Production; Standing Stock.* Vol. I: *Chemical Oceanography; Nutrients in the Sea.*

OFFSET AND OVERLAP

These are terms that refer to the plan form of a tidal inlet in a barrier island system with respect to the shape of the barrier shorelines adjacent to the inlet. Galvin (1971) suggests that the factors which control inlet form are: the availability of sediment in the longshore transport system; the relative strength of tidal and longshore transport rates; and the ratio of net to gross longshore transport rates. On this basis he classifies offset inlets into four groups (Fig. 1): *Overlapping inlet,* in which group there is an adequate supply of littoral drift and waves approach from one direction. Therefore longshore sediment transport rates are high, and the direction of transport is constant. Tidal flow is sufficient to maintain the inlet channel. *Updrift offset,* in which the supply of littoral drift is adequate and waves approach from both (updrift and downdrift) directions, causes reversals in the direction of longshore sediment and preventing development of an overlap. *Negligible offset,* in which waves approach equally from all directions, results in a ratio between net and gross longshore transport rates that is low. *Downdrift offset,* in which the sediment supply is limited and waves approach from both directions, causes erosion on the updrift barrier. The offset is accentuated by deposition on the downdrift barrier that results from refraction around the ebb-tidal delta at the inlet entrance (Hayes, Goldsmith, and Hobbs, 1971; Goldsmith et al., 1975).

EDWARD H. OWENS

References

Coastal Engineering Research Center, 1973. *Shore Protection Manual.* Washington, D.C.: U.S. Army Corps of Engineers, 750p.

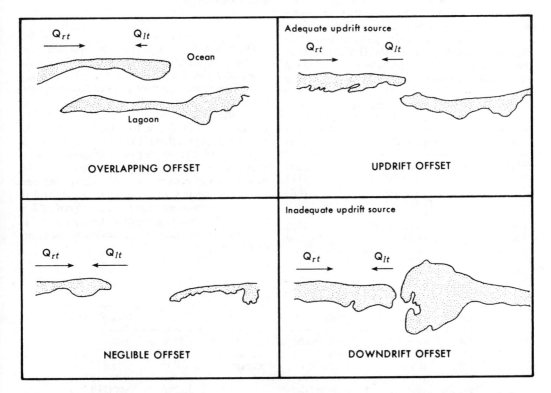

FIGURE 1. Four types of barrier island offset, proposed by Galvin (1971). The length of the arrows indicates the relative magnitude of longshore sediment transport rates (from Coastal Engineering Research Center, 1973).

Galvin, C. J., Jr., 1971. Wave climate and coastal processes, *in* A. T. Ippen, ed., *Water Environments and Human Needs.* Cambridge, Mass.: M.I.T. Parsons Laboratory for Water Resources and Hydrodynamics, 44-78.

Goldsmith, V., Byrne, R. J., Sallenger, A. H., and Drucker, D. M., 1975. The influence of waves on the origin and development of the offset coastal inlets of the southern Delmarva Peninsula, Virginia, *in* L. E. Cronin, ed., *Estuarine Research,* vol. 2. New York: Academic Press, 183-200.

Hayes, M. O., Goldsmith, V., and Hobbs, C. H., III, 1971. Offset coastal inlets, *Conf. Coastal Eng., 12th, Am. Soc. Civil Engineers, Washington, D.C., Proc.,* 1187-1200.

Cross-references: *Barrier Island Coasts; Barrier Islands; Inlets and Inlet Migration; Tidal Inlets, Channels, Rivers.*

OFFSHORE PLATFORMS

Offshore platforms are supporting structures placed in a body of water, and are independent of the accompanying coastal areas. *Man-made islands* is another name sometimes given to these structures because of their independence from adjacent coasts. Platforms as we know them today more or less had their origin in the Gulf of Mexico. The first platform was installed there in 1947, and since then they have increased in size and in depths of water into which they are being placed. Table 1 shows this increase for drilling rigs in the gulf throughout the years. The oil industry alone has installed over 2800 fixed platforms in the gulf. Another high-density area for offshore platforms is the North Sea. This area, with its hostile environment, has brought about the development of many assorted platform designs (see Table 2). As for future projections, Figure 1 graphically describes the giant increases in technology expected in the area of offshore platforms.

Construction

In the past, offshore platforms were basically onshore structures moved into the offshore environment. With the need to extend to greater water depths, these structures had to be redesigned with their own set of procedures to withstand the harsh conditions of the open seas. Therefore designers of platforms must be concerned with: environmental conditions; environmental effects; "a kinematic theory to translate the environmental severity parameter into motion parameters"; and "a force theory to translate the motion parameter into force parameters on the system" (Geer, 1975).

Two materials predominate in the construction of offshore platforms: concrete and steel. Steel structures are usually constructed totally or in sections on land, shipped to the installation site, and assembled on site. Concrete structures, however, are usually constructed and assembled on site.

Types

There are two basic types of offshore platforms available at the present time. They are the *fixed* and the *mobile* offshore platform. Each has its own special advantages and disadvantages.

A *fixed offshore platform* is one that is built onto the sea bottom, designed for only one location, and intended to remain at that location for its operational life (Geer, 1975). These units are usually employed in shallower waters than mobile structures owing to the structural problems and economics of the giant members needed in deep water. Fixed platforms will in the future, however, extend to depths of 287 m and beyond with increasing technology. See Table 1 for typical Gulf of Mexico tower structure data.

TABLE 1. Gulf of Mexico
(29 Years of Platform Development)

Year	Feet of Water	Tons
1947	20(6.1m)	1200
1955	100(30.5m)	2430
1959	206(62.8m)	1520
1965	285(86.9m)	5000
1967	340(103.6m)	6510
1970	373(113.7m)	7000
1980	1050(320.0m)	33,000

TABLE 2. North Sea (9 Years of Platform Development)

Year	Location	Type		Tons	Feet of Water
1966	Leman Bank	(template)	12 pile	2500	100(30.5m)
1972	Ekofisk	(template)	12 pile	6000	220(67.1m)
1973	Ekofisk	(gravity)	storage	225,000	220(67.1m)
1974	(BP) Forties	(template)		37,000	420(128.0m)
1975	Brent A	(template)		33,000	460(140.2m)
1975	Brent B	(gravity)		350,000	460(140.2m)

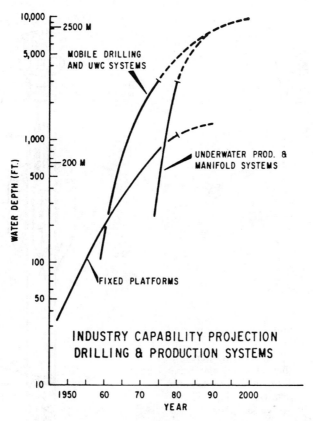

FIGURE 1. Future projections for offshore platforms.

With the moving of platforms to hostile areas, a new type of fixed structure was needed. This is the concrete gravity structure shown in Figure 2. This type of structure actually approaches a true man-made island. It is relatively simple in construction, and has a much lower installation cost than that of previous tower structures. Owing to the harshness of the Arctic areas, this design is being given much consideration for these areas. The stability when subjected to such loadings as ice loadings is excellent.

Certain factors must be considered in fixed platform design, including forces, environmental and man-made, foundation suitability, and material selection and behavior. The forces acting on offshore platforms are almost always irregular, and include earthquakes, ice, winds, waves, and currents. Man-made forces include installation and ship-ramming forces (impact forces). The foundation upon which the structure sits (all fixed platforms are bottom-supported) is important. It must not only hold the weight of the platform, but it must not scour away so that the structure shifts from its original position. Also, materials placed in offshore environments must be able to resist corrosion, and must have high fatigue limits owing to the movements to which they are subjected (from the constant force of winds, waves, and so on).

Many different types of fixed platforms exist other than the aforementioned tower and concrete gravity platforms (Antonakis, 1975a, b; Hancock and Peevey, 1975). The articulated column (Fig. 3), the tension leg platform, and the compliant tower (Fig. 4), are others that have received strong support and are now being designed for protection.

There are two main types of *mobile offshore platforms:* column stabilized and self-elevating. Column-stabilized platforms are supported "by either lower displacement-type hulls by means of columns as by large caissons with or without bottom footings" (Det Norske Veritas, 1976). A unit of this type that is designed to float is termed a *semisubmersible,* while one designed to operate when it is supported by the seabed is termed a *submersible.* "A self-elevating unit is defined as a unit which in the normal operating conditions rests on the sea bottom by means

OFFSHORE PLATFORMS

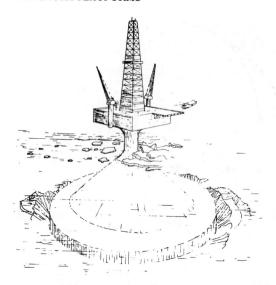

FIGURE 2. Ballasted concrete cone.

of legs with the main hull clear of the water" (Det Norske Veritas, 1976).

The number of mobile platforms in operation is greater than the number of fixed platforms because of: their versatility and mobility, and their decreased initial capital outlay. Therefore they are usually used for smaller jobs where larger fixed platforms are uneconomical. Figure 1 shows the relative difference in numbers between the two.

Uses and Needs

With the increased need for coastal development and the advent of environmental protection, much of our coastal needs will have to be satisfied by offshore platforms (McAleer, 1974). These platforms can be used as: offshore drilling rigs, offshore construction headquarters, fishing and recreation locations, power and energy plants, and scientific observations. As the costs of platforms decrease with increased technology advances and as the cost of coastal land increases, platforms will definitely become viable alternatives to waterfront land areas.

The national needs for artificial islands and platforms are, according to McAleer (1975):

Artificial islands and platforms can provide space for development in strategic locations to relieve pressures on crowded cities and preserve the environmental quality of coastal regions.

Economical offshore opportunities already exist, but seaward advancement is evolving slowly because advantages of offshore opportunities are recognized and the difficulties of on-land development and their total costs (social, environmental, and economic) have not been evaluated or passed on to land developers.

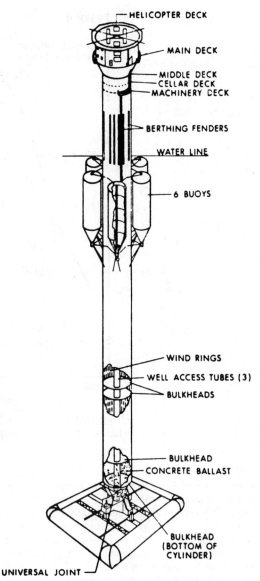

FIGURE 3. Articulator column.

The national needs to be served by offshore platforms are clear; the questions are: where, what kind, and how rapidly will offshore platforms develop? The answers will depend largely on national policies and cooperative technological efforts.

JOHN B. HERBICH
TOM WALTERS

References

Antonakis, J., 1975a. Steel versus concrete platforms, part I, *Petroleum Engineer* 47, 12-16.
Antonakis, J., 1975b. Steel versus concrete platforms, part II, *Petroleum Engineer* 47, 11-14.

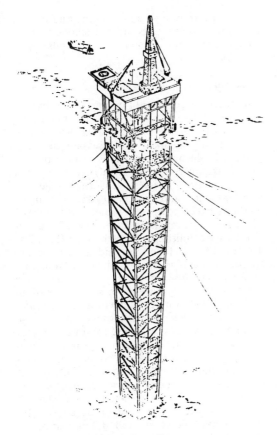

FIGURE 4. Compliant tower.

Det Norske Veritas, 1975. *Rules for the Construction and Classification of Mobile Offshore Units.*
Geer, R. L., 1975. National needs, current capabilities, and engineering research requirements offshore petroleum industry, *in Background Papers on Seafloor Engineering,* I. Washington, D.C.: National Academy of Science, National Science Foundation, and the National Research Council, 1-71.
Hancock, J. L., and Peevey, R. M., 1975. New concepts in offshore platforms, *Seventh Annual Offshore Tech. Conf. Proc.* 2, 235-242.
McAleer, J. B., 1974. Multiple-use potential of offshore facilities – Artificial islands and platforms in bays and estuaries, *Marine Tech. Soc. Proc.* 697-714.
McAleer, J. B., 1975. National needs, current capabilities, and engineering research requirements for artificial islands and platforms, *in Background Papers on Seafloor Engineering,* I. Washington, D.C.: National Academy of Science, National Science Foundation, and National Research Council, 287-332.

Cross-references: *Artificial Islands; Coastal Engineering; Coastal Engineering, History of; Coastal Engineering, Research Methods; Coastal Zone Management; Demography; Human Impact; Oil Spills and Pollution.*

OIL SPILLS AND POLLUTION

In the rapidly advancing technological societies prevalent in leading nations around the world today, the assurance of a constant, plentiful supply of energy has taken on an incalculable political and economic importance. With the depletion of land-based fuel oil reserves, man has turned increasingly to the sea to fill his energy requirements. With this boom of offshore oil production has come the burden of environmental maintenance and strict pollution control.

While presently it is difficult to predict accurately the relationship of offshore production with specific ecological changes, there is some increase in chronic oil pollution in production areas. The repeated exposure of the biosphere to these pollutants is a problem requiring further research. Likewise, the potential for insiduous long-term effects on the biosphere due to short-term sublethal toxic levels is little understood and deserves more attention. Both these situations take on added importance in the context that the application of chemical dispersants and absorbants is being recommended by the oil industry for both exceptional and chronic oil spills.

To understand the complex environmental impact of oil pollution, a working knowledge of the physical properties of an oil spill is of the utmost importance. The principal physical properties of an oil spill that have a direct bearing on its eventual fate are: velocity, specific gravity, volatility, and flash point. In conjunction with the influence of the natural environment in which the spill occurs, these properties will govern the extent of a spill and the degree to which it may damage the marine ecosphere.

Oil Spill Dispersion

The tendency of oil to spread, undeterred by environmental effects, is governed by the physical forces of surface tension and gravity. The downward pull of gravity actually causes the horizontal dispersion of a slick of significant thickness. As the slick spreads and becomes thinner, the force of gravity naturally decreases, and surface tension, which is not dependent upon film thickness, becomes the dominating force. At the same time, viscosity and inertia tend to hinder the spread of oil. Coupling these effects with environmental loads, the movement and behavior of escaped oil are found to be influenced most significantly by the following factors. The wind direction, strength, and duration are the dominating parameters determining the direction and distance of travel of the surface and subsurface oil plumes. The actual processes involved are the drag of the wind on the surface, sea surface roughness (waves), wave celerity, and the formulation of *wind rows*. The drift direction of a spill has been found to be the same as the wind; and the speed, approximately 3.5% of the wind speed. Major freshwater/saltwater interfaces, known as *rips*, will in some cases act as effective barriers to surface oil driven away from the spill area. Turbidity boundaries, when present, have been known to block the oil suspended in the subsurface (brackish) layer. However, the oil plume floating on the surface will usually cross this barrier with little or no distortion. The net outflow of water from rivers and estuaries near the spill area tends to flush these areas and in most cases prevents major pollution problems. Finally, tidal flows and oceanographic currents also offer a mechanism by which oil slicks may be transported. However, these effects are usually secondary to wind-induced flow.

Physical Characteristics of the Oil Spill

As the oil moves away from the spill area under the action of the above-mentioned forces, three easily distinguishable physical states are usually observed. The first is a narrow, reddish-brown surface string, typically 2–10 m in width, extending off the slick's body in a ropelike fashion. This thick oil layer is thought to be a water-in-oil emulsion, and is commonly referred to as *mousse*. The next state is the oil slick itself. This widening surface plume is readily observable with characteristic oil film colors, depending upon local thickness, as follows: a barely visible gray (approximately 25 gallons per square mile); a silvery sheen (approximately 50 gallons per square mile); a slight trace of colors (100 gallons per square mile); bright, banded colors (200 gallons per square mile); a dull brown color (600 gallons per square mile); and a darker brown color (over 1300 gallons per square mile). The final state is a widening, creamy yellow subsurface plume. This yellow plume is in the upper water column and is thought to be an oil-in-water emulsion formed by chemical dispersion of the oil.

Where liability needs to be established, these macroscopic, physically observable characteristics of oil spills are unfortunately too uniform to make any sort of source identification. To identify particular oil spills and their origin, several sample identification methods have been developed. The simplest, most rapid, and most reliable methods of crude oil spill identification are: gas liquid chromatography (GLC), atomic absorption, infrared spectroscopy, and Bomb Sulfur and Kjeldahl Nitrogen tests. Of these, infrared spectroscopy and Bomb Sulfur and Kjeldahl Nitrogen tests are the simplest, most inexpensive, and quickest methods now available.

Clean Up

Whether source identification is positive or negative, there always remains the responsibility to clean up the polluted area. Several mechanical and chemical processes are currently practiced. Mechanical methods include the use of oil skimmers employing dynamic inclined planes, suction nozzles, separating columns, floating weirs, and rotating disks. Mechanical methods have shown real promise with overall efficiency performances exceeding 97% under optimum conditions (Battelle, 1971; EPA, 1971a). Chemical herding and absorption are the other major means by which spill oil is recovered. The use of polyurethane foam is a practical recovery system for spilled oils of all types in a wide variety of environmental conditions. Unlike the mechanical methods, the use of foam is equally efficient for thick or thin surface layers, and actually improves in efficiency in deteriorated environmental conditions. Gelling of crude oil is a containment method that can result in a reduction of marine pollution caused by accidental discharge of crude oil (EPA, 1971b). This procedure is still experimental, however, and determination of what gel flow properties must be in relation to environmental loads must be made before any reliable gelling method can be employed (EPA, 1970a, b; EPA, 1971c-f).

JOHN B. HERBICH
ROY B. SHILLING

References

Battelle, 1971. *Concept Development of Hydraulic Skimmers System for Recovery of Floating Oil.* Richland, Wash.: Battelle Memorial Institute, 1–4.

EPA, 1970a. *Testing and Evaluation of Oil Spill Recovery Equipment.* Portland, Maine: Maine Port Authority.

EPA, 1970b. *Vortex Separation for Oil Spill Recovery Systems.* Panama City, Fla.: American Process Equipment Corporation, 1–5.

EPA, 1971a. *Floating Oil Recovery Device.* Las Cruces, N.M.: New Mexico State University Physical Science Laboratory, 1–5.

EPA, 1971b. *Gelling Crude Oils to Reduce Marine Pollution from Tanker Spills.* Richardson, Tex.: Western Company Research Division, v–viii.

EPA, 1971c. *Oil Pollution Control Training Manual.* Edison, N.J.: Edison Water Quality Research Laboratory, 1–1, 12–1.

EPA, 1971d. *Oil Pollution Incident Platform Charlie, Main Pan Block 41 Field, Louisiana.* Norwood, N.J.: Alpine Geophysical Associates, 15–64.

EPA, 1971e. *Oil Spills Control Manual for Fire Departments.* Edison, N.J.: Edison Water Quality Laboratory, 1-1, 12-1.

EPA, 1971f. *Recovery of Floating Oil Rotating Disc Skimmer.* Costa Mesa, Calif.: Atlantic Research Systems Division, Marine Systems, 1-9.

Cross-references: *Artificial Islands; Human Impact; Offshore Platforms; Pollutants; Tar Pollution on Beaches; Thermal Pollution; Waste Disposal.* Vol. XIII: *Pollution, Geochemical Aspects.*

ORGANIC WEATHERING—See WEATHERING AND EROSION, BIOLOGIC

ORGANISM-SEDIMENT RELATIONSHIP

Biotic and abiotic conditions determine the establishment of faunal assemblages in the seabed, resulting in the ecological zonation of the sea floor. In turn, all organisms influence their environment by altering it to some extent. Many kinds of interactions are observed between organisms and sediments that are important in understanding coastal dynamics, biological groupings, and diagenetic processes.

Grain Size and Organism Abundance

The grain size of sediments is often regarded the most prominent factor ruling the distribution of the benthic fauna that live on and in the sand and mud bottom (Gray, 1974). However, this is correct only for the mesopsammic meiofauna, organisms 0.04-0.5 or 1 mm in size that depend on the space between sand grains. Macrofaunal species are related to an integrated factorial complex, in which grain size is only one of the most easily recognized and determined factors. Grain size distribution itself depends predominantly on sedimentary conditions, which also determine the food particle transport and the food input into the sediments and, in turn, sediment chemistry and faunal assemblages. In another, indirect interdependence, organisms are related to grain size. According to the ratio of surface area to volume, adsorption is higher in the smaller particles. Together with the smaller, hydrodynamically lighter grains, most of the organic particles settle according to their equivalent grain size diameter, while fewer sedimentate with the heavier sand grains. Thus, organic matter is higher in muds where it supports higher population densities of bacteria, meiofauna, and macrofauna than in sands.

Sediment and Feeding Type

The most obvious interrelations exist between sandy sediments and filter feeders, while deposit feeders predominantly live in bottoms with muddy sediments. Filter feeders take particulate food from the near bottom waters. They may be active by beating their appendages (Cirripedia, Amphipoda), by movement of cilia (Ascidia, Porifera, Bryozoa, Brachiopoda, Bivalvia). Others are inactive, spreading their arms or tentacles out into the water, waiting for occasional contact with food items (Hydrozoa, Anthozoa, Polychaeta, Crinoida, Ophiuroidea). Hydrozoa and Gorgonaria, growing in a vertical plane, may spread perpendicular to the predominent current, while Pennatularia and Crinoidea orientate themselves perpendicular to actual current for optimal filter efficiency. Whenever currents transport particles, there is a good availability of food for filter feeders. This is the case on most rocky shores and near many secondary hard substrates (harbor and beach constructions) but also where sandy sediments are found. Particles are transported and do not settle owing to water movements. In irregular current conditions, particles that have already been sedimented may be resuspended. Filter feeders cement themselves to hard substrates (Porifera, Cirripedia, Bryozoa, Anthozoa, Hydrozoa, Polychaeta, Ascidia); they may be supported by their own tube constructions (Polychaeta, Amphipoda, Isopoda); or they may develop rootlike holdfasts projecting into the sediment (Porifera, Hydrozoa, Anthozoa, Bryozoa, Ascidia).

Deposit feeders depend basically on the same food source as the filter feeders—that is, on organic matter previously transported by the water but finally incorporated into the sediment. Deposit feeders live on or in the sediment and feed selectively on discrete particles or swallow indiscriminately all particles of suitable size, including living smaller organisms (Polychaeta, Sipunculida, Holothuroidea).

Sediment Reworking and Bioturbation

Feeding processes change the sediment characteristics in many instances. Most prominent is the reworking of the sediment by deposit feeders. Ingestion and casting of fecal material can occur in different sediment layers. The lugworm *Arenicola marina,* which lives in a U-shaped tube, collects surface sediment with the aid of a sand funnel and feeds on it. After the material has passed out of the gut of the animal, it is decast in long curls to the sediment surface. In contrast, the holothuria *Molpadia oolitica* is positioned with its mouth in a depth of 10-20 cm, while the anus is located near the surface. Within the sediment column, material slowly moves downward to the feeding level. It is taken up, transported through the gut, and discharged together with respiratory water above the sediment/water interface. A sediment cone around the anal opening is produced by the accumulation of feces, while a circular depression around the cone is filled with unconsolidated fecal material. The cone shows a defined sediment structure with feeding voids, vertical mixing of the deeper darker and the lighter surface materials, and a loose upper zone, which is inhabited by the suspension-feeding polychaete *Euchone incolor* stabilizing the cone with its tube meshwork against erosion (Rhoads and Young, 1971). In the depressions around the cones, a few polychaetes may be found in the fecal rich mud.

The sea urchin *Echinocardium cordatum* gives another example of bioturbation. It slowly moves horizontally in a sediment depth down to 10 cm. The funnel-building tube feet produce a slime-stabilized canal through which a breathing current is created. While feeding, the sea urchin moves through the sediment by transporting grains in front of him to the back with the spatulate spines that compress the sediment and mix it with slime into vertically standing saucerlike structures (Reineck, Gutmann, and Hertweck, 1967).

Reworking intensity depends on population size and temperature. Rhoads (1963) gives 257 and 377 ml/organism/year of reworked sediment for the bivalve *Yoldia limatula* (14.2 mm in length) and calculates 23 $1/m^2$/year and 52 $1/m^2$/year for different stations. Discarding of fecal material above the sediment surface increases resuspension, transport of particles, and sedimentation in other areas. Bioturbation to a lesser extent originates from the creation of breathing current, from movement on the sediment surface, and from burrowing into deeper layers. Fossil and recent sediments show tubes and burrows filled with surface sediment. Gas bubbles originating from microbial processes also disturb sedimentary structures.

Sediment Stabilization

The influence of filter feeders on the structure of the sediment is mainly a result of the production of fecal and pseudofecal pellets.

Particles are combined with larger particles, exhibiting higher sinking and lower transportation rates. In some areas the sediment surface layer is dominated by fecal pellets, and their shape and structure often allow the determination of which animal species produced them. Filtration of unsuitable or too many particles may result in the production of pseudopellets, which are rejected without having passed through the alimentary canal. These contribute in large amounts to the sediment. In turn, the different types of pellets, determining the sediment structure, are of nutritional importance for other species because of undigested, non-refractory organic matter and bacteria content.

The production of pellets and their sedimentation may cause an alteration in grain size distribution. Pellets are much larger than the original sediment particles, changing the sediment type from predominantly silty to sandy. Grain size analysis has to be done with great caution, in order not to break the pellets to obtain the effective grain size distribution. Together with grain size, other characteristics of the sediment are altered by the reworking of the fauna. Sediment porosity and water content increase through bioturbation, while cohesion and compaction decrease, enabling weaker currents to resuspend and transport sediments. In such areas, bottom surface structures are temporary and the sea bed is more or less flat.

Other organisms stabilize the surface of the sediment. Slime production by actinians, polychaetes, bivalves, and gastropods results in a binding of particles, and the same is observed through the intrusion of the protoplasmatic pseudopodia of foraminifera into the sediment, by which an area much larger than their shell may be covered. The extent of this process is largely unknown, but the importance of particle accumulation through organisms is well known. Agglutinating foraminifera and certain polychaetes and crustaceans collect grains for the construction of shells, spheres, and tubes, resulting in their restricted transport through currents. Some species use grains of similar size, while others are less selective. Tube building in in the sediment, the rhizoids of hydrozoans, sponges, bryozoans, foraminifers, and of macro-algae, as well as the roots of eel grasses stabilize the upper sediment layer. Microorganisms may stabilize the sediment surface. Filamentous and coccoid green, red, and blue-green algae, diatoms, and bacteria constitute mats, and grains are also held together by the rhizoids of the fungi and by the fimbriae of bacteria (Rheinheimer, 1974).

Oxygenation and Recycling

Together with the physical properties of the sediments, chemical conditions are influenced by the organisms. Metabolic activities of organisms tend to acidify their environment and lower the pH where water percolation through the sediment is low. PH changes may cause the deposition of calcium carbonate. Bioturbation is of importance for the distribution of oxygen and hydrogen sulfide in the sediment. Several polychaetes and holothurians feed in or below the O_2/H_2S discontinuity layer. These transport reduced materials to the sediment/water interface, into the oxygenated zone, while oxygen needed for respiration is transported down with the bottom water. Dense populations of these organisms, conveyor belt species (Rhoads, 1974), keep the discontinuity layer deeper down in the sediment and allow other organisms to inhabit this zone. Through this process, biological and chemical degradation of organic substances and the fluxes of dissolved compounds are enhanced at the interface (Aller and Yingst, 1978). The same effects are obtained through currents created by animals for respiration and food gathering.

The amphipod *Corophium* lies in the sediment surface and with its appendages creates a water current for respiration and food transport, while other particles are resuspended and the exchange of interstitial water and near-bottom water is increased. Without the activity of organisms, transport through the interface would be restricted to the slow process of physical diffusion. The *effective transport rate* is strongly dependent on the inhabiting species and on population densities.

Organic matter sinking to the sediment surface as well as that produced in the sediments is degraded through feeding activities of the fauna and through microbial decomposition. Above the redox-discontinuity layer, microbial decomposition is aerobic and complex organics and oxidized chemical compounds are found (Berner, 1976). In the deeper zone, which can be recognized by its black iron sulfide color, anaerobic decomposition and reduced low molecular compounds determine the chemical processes. The interaction of the two layers through the activity of the fauna is of importance in the biogeochemical cycles, for the permanent deposition of organic matter, and rules the chemical environment of diagenetic processes. In addition, the recycling of nutrients into the water column is essential for primary production in the ocean surface layers. The recycling of minerals is enhanced and controlled to a large extent by the activity of the infauna moving through the aerobic and anaerobic layers of the sediment.

Dynamic Aspects

Organism-sediment relationships have been described so far as stable situations or steady-state processes in a defined habitat. This is true for short time intervals but not for long periods. In higher latitudes the physical conditions are governed by seasonal changes. Temperature and light and food availability exhibit strong variations influencing the activity of organisms. Age and size structure of animal and plant assemblages vary within the year, and between years the species composition and abundance are not the same in a single locality. With species fluctuations and population successions, the interactions between the organisms and sediments exhibit related fluctuations and successions influencing the sediment structure and the chemical regime. Compaction, water content, porosity, permeability, grain size, depth of the redox-discontinuity and the effective transport rate, along with recycling intensity, vary with time and type of inhabitants. The life cycle of a species is marked by spatfall, metamorphosis, population establishment, growth, reproduction, and mortality. The larvae have the ability to choose a suitable sediment type, recognizing grain size, bacterial coating of grains, sediment structure, and chemical components. In some species a preconditioning of the habitat by the adults attracts the larvae. Metamorphosis and growth are followed by gradually changing behavior and activity, introducing new influences on the sediment. Reproduction may lead to the new settlement of larvae and higher population densities, whereas mortality brings about the return to earlier conditions, allowing another species with other influences to settle on the sediments. Mills (1969) gave a good example of the fluctuations in species and related sediment structures. The first year the gastropod *Nassarius obsoletus* dominated a large sandflat, and by grazing and mechanically disturbing the sediment surface some infaunal species disappeared. The following year, the amphipod *Ampelisca abdita* was able to settle in this sandflat and apparently, with the tubes of the amphipods projecting vertically, the sediment surface was newly structured and stabilized, resulting in a larger area for the growth of flagellates, diatoms, infaunal polychaetes, and other species. However, the dense population of amphipods brought the finer grains back into suspension, finally creating a more sandy environment suitable for a recolonization through *Nassarius*.

Beaches and coastal environments are highly variable owing to seasonal and long-term influences on the interrelated physical, chemical, and biological successions. The complex dynamic processes in the sediment surface layer and the sediment/water interface must be carefully considered in coastal research and engineering.

HJALMAR THIEL

References

Aller, R. C., and Yingst, J. Y., 1978. Biogeochemistry of tube dwellings: A study of the sedentary polychaete Amphitrite ornata (Leidy), *Jour. Marine Research* **36**, 201-254.

Berner, R. A., 1976. The benthic boundary layer from the viewpoint of a geochemist, *in* I. N. McCave, ed., *The Benthic Boundary Layer*. New York: Plenum Press, 33-55.

Gary, J. S., 1974. Animal-sediment relationship, *Oceanog. Mar. Biol. Ann. Rev.* **12**, 223-262.

Mills, E. L., 1969. The community concept in marine zoology, with comments on continua and instability in some marine communities: A review, *Jour. Fish. Res. Board Can.* **26**, 1415-1428.

Reineck, H.-E.; Gutmann, W. F.; and Hertweck, G., 1967. Das Schlickgebiet südlich Helgoland als Beispiel rezenter Schelfablagerung, *Senckenbergiana Lethaea* **48**, 219-274.

Rheinheimer, G., 1974. *Aquatic Microbiology*. New York: John Wiley & Sons, 184p.

Rhoads, D. C., 1963. Rates of sediment reworking by Yoldia Limatula in Buzzards Bay, Massachusetts, and Long Island Sound *Jour. Sed. Petrology* **33**, 723-727.

Rhoads, D. C., 1974. Organism-sediment relations on the muddy sea floor, *Oceanog. Mar. Biol. Ann. Rev.* **12**, 263-300.

Rhoads, D. C., and Young, D. K., 1971. Animal sediment relations in Cape Cod Bay, Massachusetts. II. Reworking by Molpadia oolitica (Holothuroidea), *Marine Biology* **11**, 255-261.

Cross-references: *Algal Mats and Stromatolites; Benthos; Biotic Zonation; Coastal Ecology, Research Methods; Coastal Fauna; Coastal Flora; Coastal Waters Habitat; Coral Reef Habitat; Estuarine Habitat; Intertidal Mud Habitat; Intertidal Sand Habitat; Loose Rock and Stone Habitat; Nutrients; Production; Rocky Shore Habitat; Sediment Analysis, Statistical Methods; Sediment Size Classification; Standing Stock.*

ORIENTED LAKES

Oriented lakes differ from nonoriented lakes in that they are elliptical or elongate in plan view and demonstrate a preferred long axis orientation. The shape and orientation result from surficial processes associated with wind activity and wave action. Varying degrees of lake orientation and ellipticity exist in different geographic areas as a function of wind regime, wave climate, and geologic or environmental conditions. However, because the terms *oriented* and *aligned* have been used interchangeably in the past it has been proposed (Kaczorowski, 1977) that the term aligned be applied to those lakes that appear to be related to some type of structural or topographic control (e.g., development along joints or faults or in dune swales). Oriented lakes, on the other hand, show no clear relationship to any structural or topographic control as described above, and consistently have their long axes perpendicular to prevailing winds. Aligned lakes may or may not be elongate, but in some cases they may show weak shoreline orientations. Theoretically, lake basins could be both oriented and aligned, and although some examples do exist (Kaczorowski, 1977), this is not considered to be common.

Whether the initial depression where an oriented lake develops is the result of subsurface or surficial processes is of little consequence because the morphology of the subsequent oriented lake is the product of a different set of processes. Once a lake basin has developed (or contemporaneous with development), opposing or unidirectional prevailing winds produce a wave climate that induces lake shore erosion in zones perpendicular to wind direction where wave approach angles are high (Fig. 1). Theoretical models (Bruun, 1953; Rex, 1961) as well as actual field observations, measurements, (Carson and Hussey, 1962; Kaczorowski, 1977) and model studies (Kaczorowski, 1977) support this mechanism of lake orientation. When quasi-stability is achieved, a noticeable beach/dune complex may develop around the lake shore.

The best-developed and best-known examples of oriented lakes are the Carolina Bays (q.v.) of the Atlantic coastal plain, previously thought to be unique geomorphic features resulting from meteorite impact or subsurface processes. Other similar modern oriented lakes occur on the

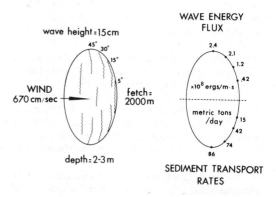

FIGURE 1. Theoretical average oriented lake model based on field measurements and observations. Wave approach angles are noted to be highest in the lake ends and approaching zero in areas perpendicular to wave approach. Calculations of wave energy flux and sediment transport volumes are shown for the various wave approach angles. (Note: The sediment transport rates assume a constant wind regime for twenty-four hours, one day.)

northern Alaskan coastal plain, in Siberia, on Tierra del Fuego, Chile, in Bolivia, and South Africa.

RAYMOND T. KACZOROWSKI

References

Bruun, P., 1953. Forms of equilibrium coasts with a littoral drift, *University of California Inst. Engineering Research,* ser. 3, issue 347, 1-17.

Carson, C. E., and Hussey, K. M., 1962. The oriented lakes of Arctic Alaska, *Jour. Geology* 70, 417-439.

Kaczorowski, R. T., 1977. *The Carolina Bays: A Comparison with Modern Oriented Lakes.* Columbia: University of South Carolina, Geology Department, Coastal Research Division Tech. Rept. 12, 124p.

Rex, R. W., 1961. Hydrodynamic analysis of circulation and orientation of lakes in northern Alaska, *in* G. O. Raasch, ed., *Geology of the Arctic,* Proceedings of the First International Symposium on Arctic Geology. Toronto: University of Toronto Press, 1021-1043.

Cross-references: *Carolina Bays; Geographic Terminology; Lakes, Coastal Morphology; Shoreline Development Ratio.*

OUTWASH PLAIN SHORELINE

An *outwash plain shoreline* is a zone where ocean or lake waters impinge directly upon a recent or Holocene deposit of galcial outwash. The concept is a subidivision of D. W. Johnson's (1919) *neutral shoreline* category, in which the shoreline characteristics are related to neither submergence nor emergence. In the Johnson classification, outwash from Pleistocene glaciers along ocean shores would have been drowned by the postglacial rise in sea level, and would thus fall into his category of *shorelines of submergence.*

Examples may be found: on the southern coast of Alaska (Shepard and Wanless, 1971), where outwash merges into coastal deltas and associated marshes; and along the southeast coast of Iceland, where the shoreline, *sensu stricto,* is actually located on barrier beaches (King, 1972). Probably the term *outwash plain coast* would be more appropriate for most examples within these areas.

SAUL ARONOW

References

Johnson, D. W., 1919. *Shore Processes and Shoreline Development.* New York: John Wiley & Sons, 584p.

King, C. A. M., 1972. *Beaches and Coasts.* New York: St. Martin's Press, 570p.

Shepard, F. P., and Wanless, H. R., 1971. *Our Changing Coastlines.* New York: McGraw-Hill Book Co., 731p.

Cross-references: *Beach Processes; Classification; Coastal Accretion Forms; Glaciated Coasts; Progradation and Prograding Shoreline.*

P

PACIFIC ISLANDS, COASTAL ECOLOGY

Tropical Pacific shorelines are diverse and widely scattered. This article draws upon features of the Solomons, New Hebrides, New Caledonia, Fiji, Samoa, Cook, and Hawaii to bring out their common pattern rather than local distinctions (Dawson, 1971; Gibbs, Stoddart, and Vevers, 1971; Morton, 1973; Morton and Challis, 1969).

Temperate coasts can be sharply divided into those of *erosion* (hard shores) and of *deposition* (beaches and soft flats), supporting different biota. The richest habitats of tropical shores, the *coral reefs* (q.v.), partake of both divisions. They maintain an equilibrium between organic accretion and physicobiological erosion. At the outward growing edge they are in contact with waves and surge; behind the reef front they are a habitat-mosaic incorporating an immense diversity of sheltered substrata, including soft flats.

The different types of reefs include *barrier reefs, fringing reefs, atoll systems,* and *patch reefs* or *sand cays*. The reef shoreline may be *subsiding,* as in the majority of small island reefs, or *emerging,* as in the reefs and limestone terraces of the Solomons and New Hebrides. But the community composition is affected chiefly by degree of wave action, salinity, and turbidity, rather than directly by history and geomorphology.

Closest to the temperate pattern of shore zonation are the simpler *volcanic coasts* (Fig. 1), with organisms zoned on hard rock in the absence of a coral reef system. On wave-exposed shores, the littoral fringe shows grapsid crabs, littorinid and neritid snails, onchidiid slugs, and siphonariid limpets. The eulittoral has pink crustose *Lithophyllum,* the barnacle *Tetraclita squamosa,* and sometimes zoanthid anemones. The sublittoral fringe generally has a short turf of brown algae (*Sargassum, Turbinaria, Chnoospora*), and *Pterocladia,* corallines and *Jania*. On more sheltered shores occur barnacles (*Chthamalus*), the rock oyster *Crassostrea,* mussels (*Modiolus, Septifer*) and sometimes the vermetid *Dendropoma*. Brown algae include *Dictyota, Padina,* and more luxuriant fucoids.

Coral Reefs

The strong characteristic of a coral shore is the seaward prolongation of the rich, briefly emersed sublittoral fringe, represented in temperate zonation only by a narrow strip. With their requirements of shallow depth, good illumination, and high temperature, hermatypic corals extend horizontally many hundreds of meters seaward. They construct biotic reefs, supporting the richest and most diverse associ-

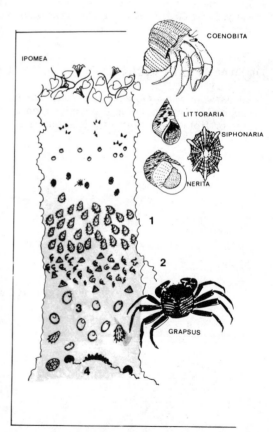

FIGURE 1. Simple zonation of an intertidal shore on volcanic rock: 1. zone of oysters, *Crassostrea;* 2. vermetids (*Dendropoma*), with barnacles (*Tetraclita squamosa*); 3. bivalves *Chama* and *Spondylus;* 4. beginning of scleractinian coral zone.

ated faunas of any marine habitat (Morton, 1973; Stoddart, 1969; Wells, 1957).

Fringing Coral Reefs. The simplest reefs present a wave-washed seaward crest, reached by crossing a reef flat, extensively emersed at low spring tide. Of drab color compared with the reef crest, the flat is a highly populated habitat. The bottom may be: a firm coral limestone platform, silt-strewn and scoured with small urchins *Echinometra mathaei*, or sheltering the arm-waving snake star *Ophiocomina scolopendrina*, with crevice-concealed crustacea, molluscs, and worms abounding; covered with bleached coral rubble, tinged by inconspicuous green algae (*Boodleya, Struvea, Dictyosphaeria, Valonia*) or festooned with *Halimeda*; clad with sheltered living corals, in a few inches of standing water (finger species of *Montipora, Porites lutea*); or sand-silt, with a green sward of monocotyledonous sea grasses, including *Syringodium* (turtle grass) *Halodule*, or (in the west Pacific) *Thalassia*, with sparser *Enhalus acoroides* in greater depths, and with green algae *Halimeda, Caulerpa, Avrainvillea,* and *Udotea* often common.

Tepid and turbid shallow stretches support sheltered brown algae (*Padina, Dictyota, Turbinaria, Hydroclathrus*) as well as sponges (funnel-shaped *Phyllospongia* and *Hippospongia, Heteronema, Dactylia*) and large cucumbers (*Holothuria, Stichopus, Thelonotus, Actinopypga*). *Synapta* and the pipefish *Siphonops* are common among sea grass. Colorful pistol shrimps (Alpheidae) and mantis shrimps (Squillidae) abound. The urchins *Tripneustes* and *Toxopneustes* are typically found in moats.

Dead Reef. The dead reef is formed by a stretch of old coral heads and slabs, frequently lying behind the surf-washed reef crest. Barren-looking at the surface, it has a diverse cryptofauna. Under boulders and coral heads are some of the most colorful species of the whole shore. Encrusting animals include sponges of many shapes and hues: yellow to brown halichondriids, scarlet microcionids, "golf balls" (*Tethya*), black elastic cylinders (*Heteronema*), small calcareous tubes and vases (syconids and asconids). In smaller numbers are compound ascidians (*Didemnum, Aplidium*), sertulariid hydroids, Bryozoa jointed, clathrate (*Reteporella*) or crustose, tubeworms (sabellids, terebellids and serpulids, especially *Filograna*), and sessile Foraminifera (red *Homotrema, Miniacina*). The most obvious mobile animals are crabs: the fast-moving grapsid *Percnon planissimum*; the highly numerous xanthids (the shawl crabs) *Atergatis floridus*; red-eyed *Eriphia*; large, blood-spotted *Carpillus maculatus*; poisonous *Zozimus*; and small, sculptured *Liomera* or hairy *Pilumnus*. Scuttling porcellanid crabs are common and sometimes colorful, as are spider and decorating crabs (Maiidae) and sponge-masking Dromiidae. Mantis and pistol shrimps are abundant in rubble and under boulders. Synalpheid prawns live in many commensal associations.

Each class of echinoderms is well represented: crinoids (Comatulidae) at the open margins of boulders; urchins underneath (black-bristling *Diadema*, and *Echinothrix calamarius*); starfish (blue *Linckia*, and bright-colored *Nardoa, Henricia, Fromia, Oreaster*), with the bulky *Culcita* and sometimes *Acanthaster*, living among boulders in shallow moats. Brittle stars include fragile, long-armed *Ophiomastix* and bright-colored species of *Ophiarachna, Ophiarthrum,* and *Ophiolepis*.

Gastropods under boulders range through trochoids (*Tectus* and *Trochus*), turbans (*Turbo* and *Lunella*), ormers (*Haliotis*), and fissurellids (*Scutus, Diodora,* and *Fissurella*), to the small and choice cowries grazing on sponges and other sessile animals. The commonest larger cowry is frequently *Cypraea arabica*. The smaller *Monetaria annulus* and *Monetaria moneta* are among the most abundant reef gastropods. Bright-colored, carnivorous opisthobranchs include: sponge grazers (Doridacea); ascidian and sponge feeders (pleurobranchs); dendronotaceans, feeding chiefly on alcyonarians: and aeoliids predating hydroids and anemones.

Byssus-attached bivalves are diverse and common: *Isognomon, Avicula, Lima, Chlamys, Pedum, Arca,* and *Barbatia*. *Chama* and *Spondylus* cement firmly to the rock.

Still more modified bivalves are those excavating and tunneling the coral limestone, such as *Tridacna*, the giant clam (farming symbiotic algae, Zooxanthellae, in its pigmented mantle) lodged on the upper sides of boulders. Powerful bioerosive agents include the vast numbers of rock boring mussels (*Lithophaga* and *Botula*), the bivalve *Gastrochaena*, the excavating barnacle *Lithotrya*, and, above all, the boring worms. Sipunculids (*Phascolosoma, Aspidosiphon, Cloeosiphon, Lithacrion*) are the most numerous, followed by polychaetes, especially Euniciidae (with *Eunice* (*Palolo*) *viridis* and many others). Terebellids, sabellids, serpulids, and nereids take up lodging in vacated galleries or they bore. The chitons *Acanthozostera* and *Cryptochiton* also lodge in excavations. A widely important eroding species is the yellow sponge *Cliona celata*.

The Outer Reef. Backward from the reef margin, in well-circulating water, waist-deep at low tide, are large scleractinian corals, with several habit forms: microatolls of Poritidae (*Porites lutea, P. fusca, P. andrewsi Goniopora*), also of Mussidae (*Lobophyllia* and *Symphyllia*) and of Faviidae. Numerous branched *Acropora* species are distinguished by bright colors of the terminal calices. Flexible and rubbery to the touch are the soft corals: smooth or undulating cushions of short-branched *Sarcophyton;* longer-branched *Sinularia;* and *Nephthya,* fleshy stemmed with clustered terminal zooids.

As the water shallows to seaward, the reef crest ascends. Coral entities multiply immensely in this clean, well-washed zone, still sheltered from the heaviest effects of surf and surge. Deeper moats and trenches with good circulation may be crowded with convex *Porites* heads or with faviid mounds: large, bright-cupped *Favia* and *Favites,* smaller-cupped *Goniastrea,* or the specialized *Hydnophora,* as well as the distinctive meandrine brain corals *Platygyra* and *Leptoria.*

Branched *Acropora* becomes increasingly important, with their bright tips, blue, mauve, yellow or pink according to species. Some species (e.g., *A. hyacinthus*) form low, corymbose heads or broad spreads of short, clustered branches. Others are woven together in wide, concave bowls or flat bracket shapes (*A. reticulata*). Branched corals common in local shelter include brittle and sharp-pointed *Seriatipora hystrix*, heavier *Stylophora mordax* or *S. pistillata*, and *Pocillopora damicornis*. Dull gray-brown or green *Montipora* with inconspicuous calices also grow here, forming crusts with erect flanges and branches, fragile scrolls, and horizontal brackets.

Free-lying mushroomlike corals (Fungiidae) are to be found on the sandy bottom: smaller *Fungia* spp. or their larger colonial relatives *Herpolithon, Polyphyllia,* and *Halomitra.*

The red organ-pipe coral *Tubipora* is common, interspersed on flat surfaces with the soft zoanthideans *Zoanthus* and *Palythoa* and with the delicate, naked polypes of *Xenia.*

Fragile Reefs. In some situations (lagoons of barrier reefs or enclosed bays), luxuriant groves of branched corals can extend into shelter as far as increasing turbidity and diminished illumination and turbidity will ultimately permit. Lacking resistance to currents and wave shock, these delicate corals are periodically vulnerable to hurricanes and excessive land runoff.

These can sometimes be reached by snorkeling, but are generally best observed by scuba. Such reefs deserve careful protection from wanton damage. Generally predominant are the long-branched species of stagshorn and bottlebrush *Acropora.*

Brackets and delicate branch systems may also spread horizontally. The most sheltered *Acropora* growth forms are the fragile bottlebrushes (e.g., *A. hystrix*). Thin and foliose life forms include the explanate scrolls of *Merulina ampliata, Echinopora,* and *Montipora foliosa,* often curled into funnels and scrolls.

Fragile corals grow with their living tips spreading upward and reaching for light. Forests may be built up of close-packed vertical columns of *Goniopora* species or branching and thickets of *Porites.* Near the tops these may be draped with *Halimeda* and other green algae. Their deeper levels, among the dead growth, present an increased wealth of habitat space. In strong shade is the red alga *Rhodymenia* and brittle sheets of pink *Peyssonelia*, with many encrusting animals. Tubeworms, bryozoans, and sessile Foraminifera occur at the darkest parts.

Mobile animals (brittle stars, shrimps, crabs) abound, while crinoids (feather stars or sea lilies) appear upon the canopy in a wide range of colors: black, maroon, lime green, and black and white cross-striped.

As well as corals, surge-swept places show *Zoanthus* and *Palythoa,* sponges (*Adocia*), and yellow or brown corrugated sheets of the soft coral *Lobophytum.*

The Wave Front. Calcareous red algae reach their maximum importance at the wave front: rough veneers or rigid branches of *Lithophyllum,* crusts of *Porolithon,* heavy-jointed *Amphiroa,* and the more delicate *Jania* and *Cheilosporum.* The encrusted rock surface forms a rich substrate, harboring many molluscs, particularly gastropods: large *Trochus* and *Tectus* spp., *Turbo argyrostoma,* and *Turbo petholatus; Drupa* species (*morum, violacea, ricinus, rubus-caesius*); and the surge zone cowries (*Cypraea caput-serpentis* and *Cypraea mauritiana*). Urchins include the slate-pencil (*Heterocentrotus mammillatus*) and the smaller and ubiquitous *Echinometra mathaei.*

As well as coralline reds, other algae are important on the reef crest. This is the principal haunt of the short-growing, crisp *Sargassum* turfs, especially *S. cristaefolium,* and of *Turbinaria* (*T. turbinata, T. murrayana*). More diverse are the smaller green algae: succulent *Caulerpa* spp. (*racemosa, peltata, serrulata, ashmeadii*); *Chlorodesmis comosa* in bright green tresses; thick filaments of *Chaetomorpha; Dictyosphaeria* in stiff, large-celled pads; and the strangely large, single-celled thalli of *Valonia.* Suctorial opisthobranch slugs (Sacoglossa) abound among green algae.

On the briefly glimpsed walls of surge channels, numerous ahermatypic corals grow: bright hydrocorals (pink *Stylaster* and mauve to apricot *Distichopora*) as well as cups of *Tubastrea* and caryophlliids. The hydrocoral *Millepora* is common in strong surge, while shade algae include *Amphiroa, Cheilosporum, Peyssonnelia,* and *Rhodymenia. Palythoa* and *Zoanthus* as well as the soft alcyonacean *Lobophytum* are also common. In the farthest depths of shade are yellow and scarlet sponges, tubeworms *Salmacina*, and sessile forams (*Homotrema*).

The Subtidal. Beyond the low water mark, illumination falls off with depth and even faster with increasing turbidity. On the open reef front, the stresses of wave action largely cease below 15 m. Down to 50m in clear water, a number of reef-building (hermatypic) corals, with symbiotic algae, still appear. But ahermatypic species now tend to predominate: immediately beyond low water, the shade forms *Stylaster*, *Distichopora*, and *Tubastrea* increase, along with cup corals (*Caryophyllia*, *Balanophyllia*, *Culicia*, and *Flabellum*) and brightly colored branching corals with few and large cups (*Dendrophyllia*).

Two much modified scleractinian families are almost wholly subtidal: the Agariciidae, with mushroomlike coralla, (*Leptoseris* and *Pachyseris*) (with one genus, *Pavona*, between tides); and the Pectiniidae (*Pectinia*, *Echinophyllia*).

On subtidal vertical faces, other Anthozoa abound: first, below the corals, the sea fans (*Gorgonia*, *Eunicella*, and other genera); and at greatest depth, the black corals (Antipatharia).

Numerous carnivorous gastropods live subtidally on soft corals: Cypraeida (*Ovula* and *Calpurnus*); and on Gorgonacea (*Primovula* and *Volva*). *Architectonica* grazes living corallites; *Heliacus* lives on zoanthids. Of the coral-feeding Magilidae, *Coralliobia violacea* grazes on *Porites*, and *Leptoconchus* and *Magilus* deeply penetrate the heads of faviids. *Rapa* is a predator of soft corals.

Soft Shores

Open Beaches. The wave-exposed open beaches are the most mobile. These form rather narrow slopes, often steeply ramped, girding islands and sand cays, and composed of coarse-textured white sand (foraminiferan tests, comminuted coral rubble, and shell fragments). There are also *gray beaches*, generally girding mainland volcanic coasts, with finer-textured ferromagnesian and quartz grains. Both kinds of open beach come frequently under wave attack at high tides. Their sands are clean and well sorted, and undergo constant shifting. They slope from a high berm, through upper beach and middle beach to about low water neap, where they give place, by a *beach step*, to a low tidal flat.

Above the beach there generally grows a rich coastal forest or scrub. Characteristic trees are *Calophyllum*, *Barringtonia*, *Terminalia*, *Messerschmidtia*, *Pandanus*, and the ubiquitous *Cocos*. In an *adlittoral* forezone grow the shrub *Scaevola* and trailing *Ipomea*, *Vigna*, *Canavalia*, and *Cassytha*. The litter fauna of the upper beach includes *Melampus* and other ellobiid snails, terrestrial insects, lizards, and coenobitid hermits. Running swiftly or burrowing into clean sand are the ghost crabs *Ocypode cordimanus* and *O. ceratophthalmus*.

In the middle beach, reaching down to the water table, *Hippa pacifica* is, in the West Pacific, the most abundant crustacean. Over the same stretches, squadrons of soldier crabs (*Dotilla*) are common. The most prolific bivalves of clean mobile sand are *Donax cuneatus* and *Donax faber*. In coarse white sand, *Atactodea* is highly numerous. The commonest burrowing gastropods at this level are *Oliva* and *Olivella*.

At the beach step, the coarse sand is well drained and oxygenated by wave runoff. The extensive interstitial spaces carry a rich fauna of miniaturised metazoans: turbellarians, archiannelids, syllids, harpacticoid copepods, ostracods, tanaids, amphipods, hydracarines, tardigrades. Tiny interstitial gastropods, many feeding on minute polypes (*Halammohydra* and *Psammohydra*) include the opisthobranchs *Microphiline*, *Philinoglossa*, and *Acochlidium*. The prosobranch *Caecum* is also typical.

Protected Flats. These are level and extensive intertidal stretches lying beyond the beach step. Well immersed for most of the tide cycle, they are protected from wave attack at low tides by outlying reef barriers or their placement in sheltered estuaries. Their fine-textured sand includes much silt, with a gray, anoxic sulfide layer at about 10 cm depth. Sheets of standing water generally remain at low tide, and green swards of sea grass are conspicuous and typical, with similar plant species to reef flats.

The burrowing fauna is richer and more diverse than on any other soft shore. Characteristic bivalve families are: suspension-feeding Cardiidae (*Laevicardium*, *Vasticardium*, and *Fragum*), Veneridae (*Periglypta*, *Lioconcha*, *Antigona*, *Pitar*, *Paphia*), Lucinacea (*Fimbria*, *Codakia*, *Anodontia*); and deposit feeding Tellinidae (*Tellinella*, *Pinguitellina*, *Arcopagia*), Asaphidae (*Soletellina*). Fan mussels *Pinna* and *Atrina*, are common. Even more diverse are the gastropods: browsing Cerithiidae (*Cerithium* and *Rhinoclavis*); deposit-grazing Strombidae (*Oostrombus gibberulus*, *Strombus*, and *Labiostrombus* spp., spider shells *Lambis*.) Burrowing carnivorous snails include: Naticidae (*Polinices*, *Natica*, *Uber*, *Sinum*), Cymatiidae, Cassididae, Tonnidae (feeding on echinoderms and other molluscs); Mitridae, Volutidae, and Harpidae, predators of bivalves; and, finally, active Conidae, stalking polychaetes, opisthobranchs, and other snails, and even fishes. Individually most numerous are the bright olives and terebrids and the small teemingly abundant scavenging whelks Nassariidae.

Crustacea abound on sand-silt flats: active portunid crabs (*Thalamita, Ovalipes, Portunus, Charybdis*); box crabs *Calappa*, feeding on hermits; *Matuta*, a predator of soldier crabs. The commonest anomurans are *Callianassa* and *Albunea;* decapod shrimps include Penaeidae and Alpheidae; stomatopods are represented by large *Lysiosquilla* spp.

The most ubiquitous sand asteroids are *Archaster typicus*, sometimes with the less common *Luidia*. Common irregular urchins are *Laganum* and *Spatangus*. Burrowing holothurioids include *Leptosynapta* and *Cucumaria* spp., *Holothuria scabra*, and many others.

Polychaete worms of protected flats are represented chiefly by Glyceridae, Nephtyidae, Ariciidae, Capitellidae, Euniciidae, and Amphinomiidae (*Eurythoe*). Nonsegmented worms include *Sipunculus, Siphonosoma,* and the echiuroids *Urechis* and *Ochetostoma*. Constant elements of tropical soft flats are burrowing anemones *Edwardsia* and *Cerianthus;* also large pedunculate brachiopods *Lingula*, and yellow proboscis worms *Balanoglossus*.

Enclosed Soft Shores. Conspicuously different from protected flats, these are soft and unstable muddy expanses in the shelter of estuaries, harbors, and wide river mouths. Completely secluded from wave action, they form deep accumulations of terrigenous silt and clay, unwalkable except where consolidated by mangroves. Such shifting as these mud flats undergo is the work of river currents and estuarine tidal flow. The sea grasses of protected flats are lacking. The distinctive life form here is the *mangrove*. Toward the west Pacific and Australo-Malaysian coasts, these tidal swamps (the *mangal*) increase greatly in productivity and species number.

Mangroves (Fig. 2) are amphibious trees or shrubs, forming a unique supratidal or intertidal forest. According to species, they may be regularly, occasionally, or never tidally submerged. The mangrove life form has been produced in several different families, mostly with convergent adaptations. Living in high salt concentrations, mangroves are *halophytes* specialized for physiological drought, with heavy leaf cuticle, leaf waxing, and reduced or sunken stomata. Roots are shallow, scarcely penetrating the black, anoxic sulfide layer. As well as for nutrient absorption, roots are also respiratory: buttresses, strut roots, and twig or stumplike

FIGURE 2. Mangrove of different genera, with details of fruits, foliage, and roots: 1. *Avicennia;* 2. *Sonneratia;* 3. *Bruguiera;* 4. *Rhizophora;* 5. *Xylocarpus.*

pneumatophores, with much internal aerenchyma. Reproductive specializations include viviparous embryos, as with *Avicennia, Aegiceras*, and the torpedo-shaped *droppers* of *Rhizophora* and *Bruguiera*.

The fully zoned mangal of northern Australia and Malaysia comprises three belts: on the seaward edge, to various extents tidally covered, on soft muds strut-rooted *Rhizophora* spp., and on sandier stretches *Avicennia* and *Sonneratia;* taller and denser forest of *Bruguiera* spp., on better-drained supratidal soils, with high humic content; and a greater diversity of back-shore mangroves, including *Excoecaria, Ceriops, Xylocarpus, Aegiceras, Lumnitzera* and such mangrove associates as *Pandanus* and *Acanthus ilicifolius*.

The most widespread Pacific genus is *Rhizophora*, with *Rh. mucronata* reaching only to western Melanesia, and *Rh. apiculata* and *Rh. stylosa* extending farther east. *Rh. stylosa* has been introduced into the Society Islands, but no mangroves reach the Cook Islands. *Rh. mangle*, from Pacific Central America, has found its way to Fiji and Hawaii. Of the other important genera, *Bruguiera* is found in Indo-Malaysia, Queensland, Melanesia, and reaches east through Fiji and Samoa; *Lumnitzera* has a similar range; *Sonneratia* and *Ceriops* do not spread east of Melanesia; but *Avicennia*, stopping short of Fiji, reaches south to New South Wales and northern New Zealand.

The fauna of the mangal is rich and characteristic. Trunk and branches generally support a *hard shore* zonation, with *Littorina scabra, Nerita exuvia*, a barnacle (*Chthamalus* sp.), oysters (*Lopha* or *Crassostrea*), *Modiolus* spp., and sometimes *Sabellaria* (worm tubes). Typical fishes are the mudskippers, one or more species of the Periophthalmidae. Common gastropods of the mud surface are Neritidae (*Vittina, Clithon, Septaria*), trailing cerithiids (*Terebralia, Pyrazus*, and *Cerithidea*); ellobiid pulmonates (*Ellobium, Melampus, Cassidula, Pythia*).

Burrowing bivalves of the mangal include most commonly *Asaphis, Geloina, Quidnipagus, Anodontia, Gafrarium*, and *Anadara*. Crustacea are diverse and prominent. Largest of the burrowing crabs are generally the heavy portunid *Scylla serrata*, with species of *Sesarma, Metopograpsus, Helice, Heloecius*, and *Macrophthalmus*. Common terrestrial crabs include *Geograpsus* spp. and *Cardisoma carnifex*. The large anomuran mud lobster *Thalassina anomala* forms burrow exits in large conical mud piles. It is in other localities replaced by its smaller relative *Laomedia*. The most ornamental mud crabs of all are the highly colorful fiddlers (*Uca*). Common among many others are *U. lactea* and *U. vocans* on sandier, more consolidated stretches and *U. marionis* and *U. dussumierii* in soft muds (for soft shores, see Morton and Raj, 1982).

JOHN MORTON

References

Dawson, E.W., 1971. Marine biology in Tonga—Vava'u and the Western Islands, *Royal Soc. New Zealand Bull.* 8, 108-120.

Gibbs, F. E., Stoddart, D. R., and Vevers, H. G., 1971. Coral reefs and associated communities in the Cook Islands, *Royal Soc. New Zealand Bull.* 8, 107-120.

Morton, J. E., 1973. Intertidal ecology of British Solomon Islands. I. Weather coasts, *Royal Soc. London Philos. Trans.* 265, 491-542.

Morton, J. E., and Challis, D.A., 1969. Biomorphology of Solomon Islands shores. *Royal Soc. London Philos. Trans.* 255, 459-516.

Morton, J. E., and Raj, U., 1982. *Shore Ecology of Fiji.* Suva, Fiji: University of the South Pacific Marine Science Institute.

Stoddart, D. R., 1969. Ecology and morphology of recent coral reefs, *Biol. Rev.* 44, 433-498.

Wells, J. W., 1957. Coral reefs, *in* J. Hedgpeth, ed., Treatise on Marine Ecology and Paleoecology. Boulder, Colo.: Geological Society of America, Mem. 67, 609-632.

Cross-references: *Asia, Eastern, Coastal Ecology; Australia, Coastal Ecology; Biotic Zonation; Coastal Fauna; Coastal Flora; Coral Reef Coasts; Halophytes; Littoral Cones; Mangrove Coasts; New Zealand, Coastal Ecology; Organism-Sediment Relationship; Pacific Islands, Coastal Morphology.* Vol. VIII, Part 1: *Oceania.*

PACIFIC ISLANDS, COASTAL MORPHOLOGY

The coasts described below are limited to those of islands in the Pacific Basin that rest on relatively thin crust and are either volcanoes or coralline limestone on a probable volcanic base. In the east (Fig. 1) the Pacific Basin is bounded by the American continent and in the west by an almost continuous active volcanic-arc system where uplift and tilting complicate the local coastlines (New Zealand and eastern Asia). Such regions and the circumscribing continental coasts are not included.

Besides the two relatively simple coastal forms found on active volcanoes (the most youthful coastal form) and *atolls* (which commonly represent the end point of coastal development), there are a number of intermediate forms. These are associated with extinct volcanoes in various degrees of erosion, some with caldera harbors and many with living *fringing* or *barrier reefs*.

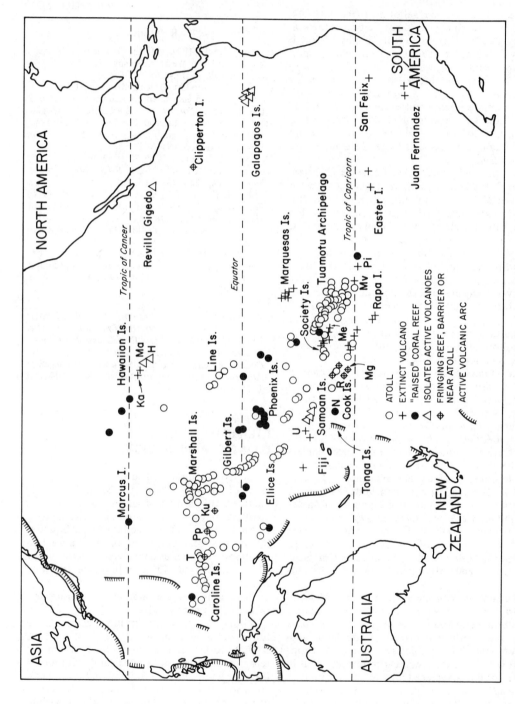

FIGURE 1. Map of the Pacific Basin. Individual islands mentioned in the text are Hawaii (H), Kauai (Ka), Kusaie (Ku), Maui (Ma), Mehetia (Me), Mangaia (Mg), Mangareva (Mv), Niue (N), Pitcairn (Pi), Ponape (Po), Rarotonga (R), Truk (T), and Uvea (U).

High-level, dead, coral reefs that may or may not have been eustatically stranded are referred to as *raised* reefs and include raised atolls. It is impossible here to mention other than a few islands; for further details, see Fairbridge (1975).

Active Volcanoes

Inside the marginal volcanic arcs, active volcanoes are confined to four isolated regions: the islands of Hawaii and Maui at the southeast end of the Hawaiian chain of islands; Revilla Gigedo, west of Mexico; the Galápagos Islands, near the equator on the eastern margin of the Pacific Basin, where Darwin considered there could be 2000 craters; and at the northwest end of the Samoan chain, where again there are a remarkable number of craters. Others, likely to be only dormant, are Uvea in the Wallis Islands northeast of Fiji and Mehetia at the southeast end of the chain of Society Islands.

None of these have coasts like the youthful, shoal-stage coasts of Falcon Island in the nearby Tongan Volcanic Arc—the island that, because of erosion and subsequent rebuilding, is sometimes there and sometimes not. Instead, their coasts reflect their long histories of development. Most of the coasts are steep and/or rocky. Some of the older coasts, particularly those exposed to the prevailing winds, have been cut back to form high cliffs, commonly more than 300 m high. Where coastal retreat has been faster than stream downcutting, the cliffs are featured by hanging valleys and waterfalls, as on the northeast coast of Hawaii (Shepard and Wanless, 1971). Some steep, straight or curvilinear coasts may be the result of faulting or slumping, as along the southeast coast of Hawaii; the southeast coast of Upolu in Samoa may have a similar origin (Kear and Wood, 1959). Where relatively young lava flows project beyond the older parts of the coast, they form protective headlands.

The newness of the coast and the easily eroded terrigenous sediments inhibit coral reef growth. However, coral, foraminiferal, and molluscan fragments are important constituents in some beaches, such as those on the leeward side of Hawaii. Other beaches, where coastal erosion is active or where streams export large quantities of sediment, are dark with volcanic fragments. The steep offshore slope carries eroded material out of the coastal system, and thus, except for minor examples in special conditions, *prograded coasts* are absent. As pointed out by Shepard and Wanless (1971), "The beaches are subject to serious erosion. The best known of these, Waikiki Beach, is at present largely artificial, and it has been a serious problem to find enough sources of sand . . . to replace the losses that occur during storm periods. . . ." and on the windward side of Maui Island, "The beaches and the calcareous sand dunes . . . retreat each winter, and summer fill does not make up for the erosion."

Coral-Fringed Volcanoes

Coral reefs (q.v.) depend on suitable conditions of light, food supply, oxygen supply, and temperature, which restrict them to a zone mainly between the Tropics of Cancer and Capricorn. Lack of coral reefs around the Marquesas Islands, despite their low latitudes, could be the result of an occasional influx of the cold Humboldt Current, but may reflect the eastward dwindling in the number of coral species (Vaughan and Wells, 1943). Nevertheless, the near-atoll island of Clipperton lies farther east still and is much more isolated than the Marquesas.

Living coral reefs that encircle volcanic islands have two forms: a *barrier reef* separated from the island by a lagoon; and a *fringing reef* attached to the island except at the mouths of rivers. A fringing reef forms an extension of a low-tide, *wave-cut platform*, whereas an island with a barrier reef has in effect a triple coastline: an outer reef coast exposed to the full force of the ocean, particularly boisterous on the windward side; an inner, lagoonal reef coast; and an innermost lagoonal volcanic coastline. Both barrier and fringing reefs provide a living and hence renewable bulwark against coastal erosion, softening even the effects of *tsunamis* and hurricane-force winds.

The reef coasts are described more fully in the sections on *atolls* (see below) and coral reef coasts (q.v.). The usually tranquil volcanic coast on the inside of a barrier lagoon is of particular interest. As early as 1849 Dana recognized that its commonly irregularly embayed outline is due to drowning of stream valleys; such embayment could not be produced by sea action, but must ultimately be destroyed by it. Davis (1928) made full use of this principle to support Darwin's subsidence theory for the origin of barrier reefs and atolls and to argue against the part of Daly's (1915) glacial-control theory that stated that reefs were killed during a glacioeustatic low sea level, allowing part planation of the unprotected island. Purdy (1974) writes of Davis:

Like Daly he supposed that wave abrasion could only have been effective in producing cliffed headlands if the protective influence of a surrounding reef were completely removed. . . . In fact Davis found that cliffed headlands were singularly absent within the barrier reef lagoons of the central coral sea but become increasingly apparent toward the northern and southern limits of present day coral growth. He therefore concluded that the limits of reef growth were merely narrowed by colder water temperatures during the glacial low stand of sea level and that consequently Daly's hypothesis was only applicable to this marginal belt.

Fringing coral reefs and barrier lagoons are an important and usually almost exclusive source of coastal sand. For example, not only are the beaches around the high volcanic island of Rarotonga formed of carbonate sands, but these continue inland as a continuous vegetated strip of carbonate sand up to a height of 8 m and commonly 200-500-m wide (Wood and Hay, 1970). Similar relationships are shown on the geological maps for Samoa (Kear and Wood, 1959). Another important example is the western coast of Kauai, which "is unique in the Hawaiian Islands" (Shepard and Wanless, 1971) in consisting of a 1.5 to 3 km wide coastal plain of beach ridges and dunes formed from carbonate sand. This plain lies in front of a "row of definite cliffs that were cut at least 33 m below present sea level" (Shepard and Wanless, 1971). The cliffs were indeed probably cut during rising sea level, but it does not seem necessary that "there must have been a diminution of westerly storms that allowed the waves to build out the plain as a combined effect of the currents flowing north-west from the south coast and south-west from the Napali Coast" (Shepard and Wanless, 1971). In fact these authors state that volcanic sediments from these latter coasts are not important constituents of the coastal plain. On the other hand, the plain is associated with an equally and unusually extensive, shallow (ca. 20 m deep) offshore region of coral sand from which sand could have been derived as a result of a late Holocene fall in sea level (see *islets* below).

Atoll and Near-Atoll Islands

The atoll, that "most extraordinary of natural structures" (Davis, 1928), consists of a coral reef surrounding and protecting a usually tranquil lagoon from a usually restless ocean, and supporting small vegetated islets 3-4 m high. The opalescent light greens and blues of the lagoon are in marked contrast to the startling white beaches, palm-green islands, lithothamnion-red outer reef, and deep, clear, blue ocean. "Yet these low, insignificant coral islets stand and are victorious. . . . Let the hurricane tear up its thousand huge fragments. . . . What will this tell against the accumulative labour of myriads of architects at work night and day" (Darwin, 1842).

Near atolls retain small vestiges of the mainly submerged volcanic base as an island or several islands.

Passes. Passes into the lagoon may be absent or as deep as the maximum depth of the lagoon. They are thought to be the result of the rapid rise in postglacial sea level. When the rates of sea level rise relative to the average coral growth (Wiens, 1962) ensured that reef growth could not be maintained up to low-tide level in some regions. Further, conditions on the leeward side of an atoll are less conducive to coral growth, and it is here that passes have most often developed. Less than ideal conditions would also account for some drowned atolls and for the incomplete development along the leeward sides of many Gilbert Islands atolls.

Reef Width and Lagoonal Depth. Equatorial regions provide optimum conditions for coral reef growth and dependent atoll reef widths (Fig. 2). Maximum lagoonal depth varies inversely with the average lagoonal radius (Fig. 3), but as well as atoll perimeter, growth rates of lagoonal coral and algae such as *Halimeda* (Agadzanin et al., 1973) must also be important factors in determining how quickly a lagoon is infilled. This effect of organic growth rate is shown in the family of curves for atolls of Gilbert, Ellice, and Lau Islands (eastern Fiji group) that lie in different latitudes (Fig. 3). There are, however, other factors such as are required by the steeper curve for the Cook Islands (Fig. 3) and the near absence of corals in the Marquesas.

Islets. Islets owe their preservation from all but the greatest storms to the presence of cemented boulder ramparts on their windward sides (Fig. 4), partly aided by *beach rock*. In some islets multiple parallel ramparts may form much of their width. In others their greater

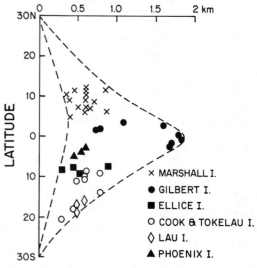

FIGURE 2. Average coral reef width relative to latitude.

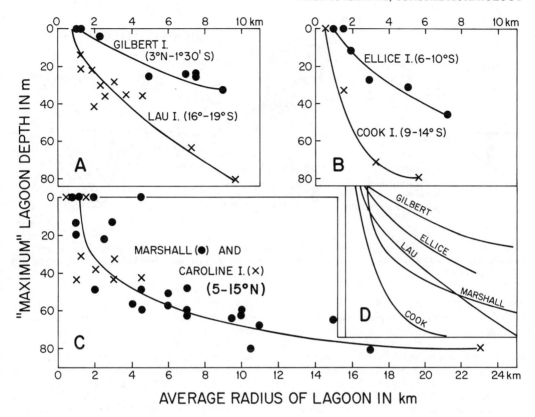

FIGURE 3. Relationship of maximum lagoonal depth with average radius of lagoon. The Marshall and Caroline curve is after Emery, Tracey, and Ladd (1954). Part D shows a family of curves dependent mainly on latitudinal ranges. However, some other factor is necessary to explain the Cook curve in relation to the Ellice and Lau curves.

width consists of unconsolidated sand on the leeward side of the rampart. The predominance of islets on the windward reefs is probably the result of a more common construction of boulder ramparts there.

Skeletal soils and radiocarbon dates suggest that the islets are late Holocene in age. Thus if the doubt raised by the *Carmarsel* expedition (Shepard et al., 1967; Newell and Bloom, 1970; Bloom, 1970)—that there has been no postglacial sea level above the present in the West Pacific—was correct, there could be no relationship between sea-level fall and islet formation. This doubt is challenged by six well-dated, second-order transgressions for the same West Pacific region that correlate remarkably well in time with those on the Firth of Thames, New Zealand, sea-level curve, and that demonstrate that the *Flandrian Transgression* reached a maximum of 2.4 m for the Gilbert Islands almost 1000 B.C. These six transgressions (Schofield, 1977a) are supported by uncommon high-level remnants of biohermal reef rock, but are based mainly on subtidal, back-reef breccias that were not differentiated from reef rubble by the *Carmarsel* expedition. Quantitative and mineralogical studies of prograded coasts in New Zealand show that 95% of the coastal sand has been derived from the sea floor as a result of the 2-m fall in late Holocene time. This is also probably true for atoll islets. Even those on the windward reef are built from lagoonal sand (Schofield, 1977b), but this is to be expected since the wide, shallow, marginal lagoonal floors are the only suitable source from which sand could be derived as a result of sea-level fall. The preservation of more or less permanent islets has almost certainly been favored by the net fall in sea level since the maximum of the Flandrian Transgression, which took place locally about 1000 B.C. It may be no coincidence that this is also approximately the earliest known Polynesian settlement date.

Low-Tide Platform. The outermost zone is a 1/2–1-m-high algal ridge, alive in constant spray at the breaker line. Across it there are deep grooves that impart a saw-toothed appearance to the edge of the reef. Inside the ridge there is in some places the so-called *boat channel* or string of deep pools that are persistent even at

PACIFIC ISLANDS, COASTAL MORPHOLOGY

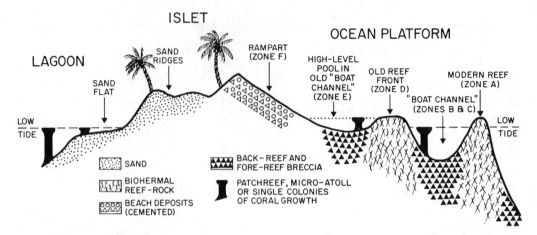

FIGURE 4. Diagrammatic representation of some important reef zones. These may not always be present or, if present, may not always be readily noticeable. The so-called boat-channels may seem absent though being infilled with back-reef breccia.

low tide and that are interconnected with the ocean and support live corals. Inland, the platform proper may consist of an older reef and its outer algal ridge, which have been abraded flat but in which the saw-toothed nature is still recognizable from the air, presumably since the algal ridge was abraded the cross grooves have been infilled. In the Gilbert Islands this abraded part of the platform lies about 0.3 m above low spring tide, and forms a barrier to another more inland set of shallow pools that support *Heliopora* and *Porites* and that are probably remnants of an older, incompletely infilled *boat channel* (Fig. 4). The older reef probably lived during some slightly higher sea level and died when sea level fell. At the same time a new breaker line was developed farther offshore where a new algal ridge grew and continued to grow upward to its present position with the present rise in sea level. (See also *Coral Reef Coasts*.)

Raised Reefs and Pacific Basin Stability

Raised reefs and their coastal terracings are dotted throughout the Pacific Basin and provide some of the most interesting and least investigated coastlines. Yet their origin is inextricably woven with the stability or instability of the Pacific Basin. "The assertion is commonly made that the island areas of the South Pacific represent . . . parallel folds of the ocean floor" (Williams, 1933) but Williams concluded that despite local tilting such as in the Hawaiian Islands (which is "only proper to volcanic regions") "the islands of the South-central Pacific as a whole seem to indicate a vast region of comparative stability." More recently, drilling in the Marshall Islands (Emery, Tracey, and Ladd, 1954) has shown that the average rate of oceanic sinking for that part of the Pacific has been about 0.025 m/1000 years during the Cenozoic and perhaps twice that rate for the past 5 million years. If this is the order of crustal movement within the Pacific Basin, there could be areas rising at 25-50 m/million years. This could explain much about the raised reefs. Nevertheless, unless there has been epeirogenic uplift over a distance of 1300 km it does not explain the close correlation of nontilted, raised reefs and coastal terrace sequences of 70 m, 55 m, 34-39 m, 23 m, 12-15 m, and 4-6 m at Niue and Mangaia and at other islands in the Cook Group (Wood and Hay, 1970; Schofield, 1967). Other raised reefs within the Pacific Basin have not received the same detailed study, but it may be significant that the few levels recorded (Fairbridge, 1975; Davis, 1928) lie closely within this framework. Hence, in view of their nontilted and possible widespread nature, these raised levels could be truly eustatic.

Whatever their origin, raised reefs usually present a steep coastline, in some places with plunging cliffs and in others fringed with a living reef, but everywhere rugged with karrenfeld solution features. A wave-cut or solution notch is common at the base of the cliff, and in some cliffs is preserved at levels of higher oceanic stillstands. Solution has lowered the 23-m and 70-m strandlines at Niue by 3 m and perhaps 15 m, respectively (Schofield, 1959), but is unlikely to have constructed the interior basin of this raised atoll. It contains uncemented carbonate sand with plio-Pleistocene lagoonal faunas that probably lived well below tide mark. The theory that such basins have been formed solely through solution (Purdy, 1974) depends on the comparison with solution effects on blocks of

626

nonpervious limestone. In fact raised reefs are extremely permeable, and there is no differential runoff of regional significance so necessary for such a theory. The rain soaks straight into the ground, and any lowering of the surface is likely to be fairly general with perhaps some more cemented parts such as beach rock and cemented boulder ramparts being preferentially preserved as relict or *shadow* structures (Schofield, 1959). However, solution at the contact between a fringing reef and less impermeable weathered volcanic rock, from which there is runoff, results in the development of swampy hollows that are much favored for taro growing. Such might even be the origin of the so-called 70-m high raised barrier reef at Mangaia which may, before taro-swamp solution, have been a fringing reef (Wood and Hay, 1970).

Notwithstanding the probability that raised reefs may have their origin in high eustatic sea levels of Quaternary age, crustal movement, particularly subsidence within the Pacific Basin, seems to have been important in the long term. This overall tendency for subsidence has been illustrated by Chubb (1957) for a number of island chains in which "vulcanicity began at one end and moved progressively along the chain, erecting a series of volcanoes, each of which passed through a succession of stages, ending as an atoll." Chubb cited a number of atolls at the northwest end of the Society Islands. Southeastward from these there are: Maupiti, a near atoll; then Bora-Bora, a deeply dissected and embayed volcano with a barrier reef and broad lagoon; thence progressively "less dissected and less embayed volcanoes, Tahaa, Raiatea, Huahine, Moorea and Tahiti, the last consisting of two volcanoes, both still showing their original volcanic form and having an almost unindented shoreline. All these islands have barrier-reefs which on Tahaa extend to nearly two miles from shore, on the succeeding islands generally to about a mile, and around parts of Tahiti to only half a mile, becoming a fringing reef in places. Finally, at the eastern end of the chain, lies Mehetia, a bold volcanic cone, hardly dissected and unembayed, with a narrow fringing-reef and a well preserved crater from which many recent lava flows have issued, though there is no record of activity within human memory" (Chubb, 1957). Other examples given by Chubb include: the Hawaiian chain, the increasing northwest age being confirmed by K/Ar dates (McDougall, 1964); the Samoan chain, which gets older southeastward; and the older Tuamotu (Mangareva to the younger Pitcairn) and eastern Caroline chains (Truk, Ponape, to the youngest Kusaie).

J. C. SCHOFIELD

References

Agadzanin, A. K., Voronov, A. G., Ignatiev, G. M., Kaplin, P. A., Leontiev, O. K., Medvedev, V. C., and Nikiforov, L. G., 1973. *Geography of Atolls in South-West Pacific Ocean.* Moscow: Oceanographic Commission of the U.S.S.R., Academy of Science, 140p.

Bloom, A. L., 1970. Paludal stratigraphy of Truk, Ponape, and Kusaie, Eastern Caroline Islands, *Geol. Soc. America Bull.* 81, 1895-1904.

Chubb, L. J., 1957. The pattern of some Pacific island chains, *Geol. Mag.* 94, 221-228.

Daly, R. A., 1915. The glacial-control theory of coral reefs, *American Acad. Arts Sci. Proc.* 51, 155-241.

Darwin, C. R., 1842. *The Structure and Distribution of Coral Reefs.* London: Smith, Elder, and Co., 214p.

Davis, W. M., 1928. *The Coral Reef Problem.* New York: American Geographical Society, Spec. Pub. No. 9, 596p.

Emery, K. O., Tracey, J. I., Jr., and Ladd, H. S., 1954. *Geology of Bikini and Nearby Atolls, Part 1, Geology,* U.S. Geol. Survey Prof. Paper 260-A, 264p.

Fairbridge, R. W., ed., 1975. *The Encyclopedia of World Regional Geology, Part 1, Western Hemisphere.* Stroudsburg, Pa.: Dowden, Hutchinson & Ross, Inc., 704p.

Kear, D., and Wood, B. L., 1959. The geology and hydrology of Western Samoa, *New Zealand Geol. Survey Bull.* 63, 92p.

McDougall, I., 1964. Potassium-argon ages from lavas of the Hawaiian Islands, *Geol. Soc. America Bull.* 75, 107-128.

Newell, N. D., and Bloom, A. L., 1970. The reef flat and "two metre eustatic terrace" of some Pacific atolls, *Geol. Soc. America Bull.* 81, 1881-1894.

Purdy, E. G., 1974. Reef configurations: Cause and effect, *in* L. F. Laporte, ed., *Reefs in Time and Space.* Tulsa, Okla.: Soc. Econ. Paleont. Miner., Spec. Pub. 18, 9-76.

Schofield, J. C., 1959. The geology and hydrology of Niue Island, South Pacific, *New Zealand Geol. Survey Bull.* 62, 28p.

Schofield, J. C., 1967. 1. Post-glacial sea-level maxima a function of salinity? 2. Pleistocene sea-level evidence from Cook Islands, *Jour. Geosci. (Osaka City Univ.)* 10, 115-119.

Schofield, J. C., 1977a. Late Holocene sea-level and atoll development: Gilbert and Ellice Islands, West Central Pacific Ocean, *New Zealand Jour. Geology and Geophysics* 20, 503-530.

Schofield, J. C., 1977b. Effect of Late Holocene sea-level fall on atoll development, *New Zealand Jour. Geology and Geophysics* 20, 531-536.

Shepard, F. P., and Wanless, H. R., 1971. *Our Changing Coastlines.* New York: McGraw-Hill Book Co., 579p.

Shepard, F. P.; Curray, J. R.; Newman, W. A.; Bloom, A. L.; Newell, N. D.; Tracey, J. I., Jr.; and Veeh, H. H., 1967. Holocene changes in sea level: Evidence in Micronesia, *Science* 157, 542-544.

Vaughan, T. W., and Wells, J. W., 1943. *Revision of the Suborders, Families, and Genera of the Scleractinia.* Boulder, Colo.: Geological Society of America, Spec. Paper No. 44, 363p.

Wiens, H. J., 1962. *Atoll Environment and Ecology.* New Haven: Yale University Press, 532p.

Williams, H., 1933. *Geology of Tahiti, Moorea, and Maiao,* Bernice P. Bishop Museum Bull. **105**, 89p.

Wood, B. L., and Hay, R. F., 1970. Geology of the Cook Islands, *New Zealand Geol. Survey Bull.* **82**, 103p.

Cross-references: *Asia, Eastern, Coastal Morphology; Australia, Coastal Morphology; Coral Reef Coasts; Coral Reef Habitat; Global Tectonics; New Zealand, Coastal Morphology; Pacific Islands, Coastal Ecology.* Vol. VIII, Part 1: *Oceania.*

PALEOGEOGRAPHY OF COASTS

The world ocean's coasts took shape in the process of development of ocean basins, and their outlines and structure have been determined by potent tectonic processes of a planetary nature. However, the pre-Quaternary outlines of oceanic basins and the distribution of major structural elements were just initial conditions in shaping the relief of the oceanic coastal zone. The present coasts were formed during the Pleistocene by repeated glacio-eustatic oscillations of the ocean level. During glaciation, the ocean subsided 100–130 m below its present level, detrital material was supplied into the shore zone, piling up atop the present shelf in considerable accumulations, which, during transgressions, would be processed into specific shelf and shore sediments. During interglacials, the ocean level would rise to present elevations or even somewhat higher. Repeated migrations of the coastline during glaciations and interglacials left their imprint on the coasts, having formed complexes of coastal features at different levels such as coastal terraces and old coastlines. Glacio-eustatic oscillations of the ocean level would superimpose themselves on isostatic uplifts of ancient and recent glaciation regions and on regional tectonic movements of continental structures. Further evidence of these movements is furnished by numerous *shore terraces,* widespread around the world.

Study of shore terraces and their deposits (Kaplin, 1973; Zenkovich, 1967) reveals changes in paleogeographic environment, the rate and duration of transgressions, as well as the pattern of tectonic movements over long periods. In this sense, shore terraces are a major criterion in coastal paleogeography. Especially important are terrace complexes forming a *terrace series* or *terrace ladder.* A case in point is the terraces of Africa's Mediterranean coast, from the highest pre-Pleistocene (+200 m) one down to the most recent Holocene terrace a few meters above sea level. Terrace series contribute to reconstructions of coastal Pleistocene paleogeography and primarily of the history of sea level changes.

The paleogeographical conditions in the Pleistocene are studied by analysis of offshore deposits of the underwater coastal slope, aggradation features, and lagoonal water bodies. Deposits are recovered by drilling with various core tubes and, when investigated by lithological and paleontological methods, permit reconstruction of the coastal relief.

For the formation of the present coastal zone, the latest post-Würmian, *Flandrian Transgression* of the world ocean was most important; the present coastal relief has developed during this transgression. At the time of the Flandrian Transgression sea level approached its present level about 6000 years ago and afterward never changed in excess of 10 m—that is, it remained within the limits of the present coastal zone. This is definitive for the conclusion that the present coastal zone is 6000 years old.

Within this brief geological time, the coastal zone in certain areas has evolved from shores completely unaltered by the sea to extinct forms now in a stage of geomorphological senility. The intensity of coastal zone evolution depended on specific physicogeographical conditions in each area and on factors of the latitudinal zonality inherent in the coast-forming processes.

The evolution of the coastal zone during the *Holocene* is indicated by mature relief features wholly created by sea waves. Particularly abundant material for paleogeographical analysis is found in the aggradation features of the coastal zone. Each usually consists of a series of offshore bars, from whose lengthwise arrangement the development stages of the aggradation feature and of the shore area as a whole can reproduced. In a morphological analysis of offshore bars, the following points should be considered: The direction of an offshore bar under any formation conditions indicates that of the coastline at the given shore section as it was at the time when the bar was formed. When offshore bars of a certain generation are undercut by bars of another generation lying closer to the present outline, the second-generation bars must be of a more recent age. Abrupt disconformity in the direction of offshore bars belonging to different aged systems is a sign of an interruption in the growth of the aggradation feature and of its erosion during the time separating the completion of an older system of offshore bars from the starting of a younger system. Bending of bar terminals and their fan-line spreading are usually oriented in the direction of the growth of the aggradation feature. Free terminals of offshore bars indicate that these had been

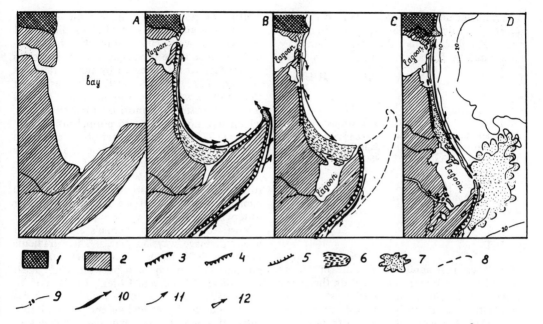

FIGURE 1. An example of reconstruction of development stages of a complex of coastal aggradation features: 1. average altitude mountain massifs broken by erosion and affected by glacier; 2. piedmont plain with monticulate-indented (morainic) relief; 3. active cliffs; 4. extinct cliffs; 5. erosion benches in aggradation features; 6. offshore bars on the surface of aggradation features; 7. wave-resistant surfaces of rubbly material; 8. old coastlines; 9. isobaths; 10. along-coastal detrital flows (thickness of arrows corresponds to relative carrying capacity); 11. drift migration; 12. supply of material to detrital flows due to erosion of main coast and washout of aggradation features.

formed in the conditions of unrestricted water area that existed at the given coastal section in the meantime.

An example of reconstruction of the development stages of a coastal strip on the basis of coastline configuration and the outline of the offshore bars on an aggradation feature are shown in Figure 1.

From the present coastline configuration, the direction of the offshore bars, the existence of an extinct cliff, and the washed-out moraine at sea floor (Fig. 1 D), it can be concluded that the aggradation feature was larger in size and that the washed-out moraine was a peninsula, Reconstruction of the coastal strip changes is shown in Figure 1 A–C.

PAVEL A. KAPLIN

References

Kaplin, P. A., 1973. *Recent History of the World Ocean Coast*. Moscow: Moscow State University, 264p.

Zenkovich, V. P., 1967. *Processes of Coastal Development*. London: Oliver and Boyd, 738p.

Cross-references: *Archeology, Geological Considerations; Barrier Islands; Bars; Classification; Coastal Morphology, Research Methods; Flandrian Transgression; Geomorphic-Cycle Theory; Global Tectonics; Isostatic Adjustments; Raised Beach; Sea Level Changes; Sea Level Curves; Shore Terraces; Spits.* Vol. III: *Paleogeomorphology*.

PALUDAL—See SALT MARSH

PALVÉ

The word *palvé* originated in the Baltic (Lithuanian) language and is widespread in German, Russian, and Lithuanian geomorphological literature. The term refers to a coastal plain of aeolian accumulation (äolische Verebnungsfläche). The palvé plains are widely stretched along the southeastern and southern Baltic coasts. Such plains are developed on marine- and glacial-formed basements. The thickness of the blown sand cover reaches to 10 m and more. In many places intercallations of humic laminae are to be seen—an evidence of blown sand accumulations. The age of the palvé plain formation is not the same in different areas, although the beginning is as early as the late-Atlantic period (Gudelis and Karužaite, 1962).

In certain localities the palvé plains originated as a result of deforestation occurring in coastal areas. Morphologically the palvé plains are represented by flat and hummocky plain types.

VYTAUTAS GUDELIS

References

Gudelis, V., and Karužaite. G., 1962. Grobšto ragas: A landscape study of the Kuršiu nerija spit (in Lithuanian, with summaries in Russian and German), *Geografinis metraštis* 5, 173-196.

Cross-refernces: *Europe, Coastal Morphology; Geographic Terminology; Soviet Union, Coastal Morphology.*

PAMET

Pamet was first applied as a landform term to the underfit or dry stream valleys on the glacial outwash plain along the outer coast of Cape Cod, Massachusetts, by Woodworth and Wigglesworth (1934). These valleys trend from east to west and are deeply incised into the outwash plains, where the valleys tend to be roughly parallel and often occupied along their courses by kettle holes, many water-filled. Pamet, as defined by Woodworth and Wigglesworth (1934), was a "long depression in a thick deposit of stratified gravel and sand . . . resembles a valley formed by running water but its sides and bottom has the forms of mounds and hollows produced by irregularities in the deposition of glacial drift, either ice-laid or water-laid." They further implied that the pamet valleys were formed by "the dissection [of] westward flowing streams."

The term *pamet* comes from the Pamet River on Cape Cod, which occupies one of these former and largest relict stream valleys. The river takes its name from the Pamet tribe of the Algonguin Indian Nation who lived in this area during precolonial times. The lower end of the river is a tidal estuary, Pamet Harbor, into which the Pilgrims sailed in exploration before landing at Plymouth. Pamet, as an Indian word, means "wading place or shallow cove" (Huden, 1962), and therefore probably refers to this estuary.

Formation of these pamets is considered to have occurred during a glacial interstadial, when streams carved the valleys. When the ice returned, debris found its way into these former stream channels (Chamberlain, 1964). The source of the meltwater of the original streams was thus a stagnant ice lobe offshore to the east. Strahler (1966) pointed out that this meltwater valley channel had irregularities from melting of stagnant ice blocks. Along certain pamets, the numerous kettle holes indicate that the valleys were cut before the ice blocks melted, since these proglacial stream valleys do not appear to be deflected by the kettle holes (Fisher, 1972). Oldale (1976) suggests that these valleys were cut not by meltwater streams but by later surface runoff over frozen ground, which was impermeable to normal infiltration. Fisher (1976) pointed out that other features in this area, possibly similar to pamets and having no distinct channel but a line of kettle holes, may indicate a line of subsurface ice. Thus subsurface ice might be responsible for some of the pamets rather than surface flow, meltwater or otherwise.

Most of these relict channels are no longer major drainage ways in places. Their gradients are from 8-20 m/km, almost four times the gradient of the adjacent outwash plain itself (Hartshorn, Oldale, and Koteff, 1967). Most of these valleys are graded to below present sea level, and the lower end or mouth is often a tidal stream or estuary owing to drowning of the mouth of the relict channel by rising sea level. Blackfish Creek, such a drowned pamet—in fact the southernmost one on Cape Cod—was the stream up which the supplies were ferried for Marconi's first wireless radio station in the United States on the marine cliffs of Cape Cod.

The heads of these pamet valleys terminate within the outwash plain or extend completely from the bay to the ocean. Along the ocean, at the sea cliff, the pamet terminate in sags or the *hanging valleys* of Strahler (1966), where the eroded cliffs breach the relict valleys. These gaps, locally called *hollows*, often provide the only means of reaching the beach from the cliff top. In fact, these hollows are identified in Blunt's *American Coast Pilot* for the benefit of shipwrecked sailors; the Massachusetts Humane Society in the late 1700s built emergency shelter huts in seventeen of these hollows (Chamberlain, 1964).

Because of its larger size, the Pamet River pamet, is almost at sea level along the ocean shore, and only a dune ridge separates it from the ocean. N. S. Shaler, well-known early coastal geologist, suggested in 1897 that this Pamet River pamet might thus be the possible site of a Cape Cod canal. A site further west was finally chosen. In 1850-1860 Henry David Thoreau, early philosopher and naturalist, walked this entire Cape Cod coast and observed where the Pamet River breaches the sea cliff ". . . the whole river is steadily driven westward butt-end foremost; fountain-head, channel, and lighthouse, at the mouth, all together (1865, p. 131). These pamet valleys, although localized to the Cape Cod coast, have been important in the history of man's activities in this region from early colonial times to the present.

JOHN J. FISHER

References

Chamberlain, B. B., 1964. *These Fragile Outposts.* Garden City, N.Y.: Natural History Press, 327p.

Fisher, J. J., 1972. *Field Guide to Geology of Cape Cod National Seashore.* Kingston, R.I.: American Association of Stratigraphic Palynologists and University of Rhode Island Dept. Geology, 5th Annual Meeting, 53p.

Fisher, J. J., 1976. Coastal geology and geomorphology of Cape Cod: A mesoscale view, in B. Cameron, ed., *Geology of Southeastern New England.* Boston: New England Intercolligate Geological Conference, Boston Unviersity, 239-255.

Hartshorn, J. H., Oldale, R. N., and Koteff, C., 1967. Preliminary report on the geology of the Cape Cod National Seashore, in O. C. Farquaher, ed., *Economic Geology in Massachusetts.* Amherst: University of Massachusetts Graduate School, 49-58.

Huden, J. C., 1962. *Indian Place Names of New England.* New York: New York Museum of the American Indian, 408p.

Oldale, R. N., 1976. *Notes on the Generalized Geologic Map of Cape Cod.* Washington, D.C.: U. S. Geological Survey, Open File Report 76765, 23p.

Shaler, N. S., 1897. Geology of the Cape Cod District, *U. S. Geol. Survey Ann. Rept.* **18**, 503-593.

Strahler, A. N., 1966. *Cape Cod: A Geologist's View.* Garden City, N.Y.: Natural History Press, 115p.

Thoreau, H. D., 1865. Cape Cod. New York: W. W. Norton and Co., 300p.

Woodworth, J. B., and Wigglesworth, E., 1934. Geography and geology of the region, including Cape Cod, Elizabeth Islands, Nantucket, Martha's Vineyard, Nomans Land Island and Block Island, *Harvard Univ. Mus. Comp. Zoology Mem.* **52**, 328p.

Cross-references: *Cliff Erosion; Fiard, Fjard; Glaciated Coasts; North America, Coastal Morphology; Spits.*

PANHOLE—See SOLUTION AND SOLUTION PAN

PARALIC

Areas of paralic sedimentation (from the Greek *paralia,* meaning seacoast) pertain to the intertongued marine and continental deposits and environments found on the landward side of a coast. These include alluvial, lagoonal, littoral, and shallow neritic environments.

Marine swamps and marshes representing mixed continental and marine environments and their possible subsequent coal formation are often referred to as *paralic swamps* and *paralic coal,* respectively. Development of paralic swamps usually begins in a protected part of the shallow water zone produced by an offshore barrier beach, spit, baymouth bar, emergence of a sea bottom, or submergence of a land area. The sedimentation history of a paralic swamp is generalized as follows. Prior to the building of the barrier, the environment and sediments are of typical marine origin. After the building of the barrier, quiet water prevails behind it and the sediments are generally finer and of a brackish or freshwater origin. Eventually, as the water becomes shallower, a succession of grasses and reed plants follows. The vegetation captures gravels, shells, sands, and muds, and organic matter derived from the land, air, and storm waves. Ultimately the accumulation of sediment builds the swamp above saltwater level, and a freshwater environment ensues. Compaction and settling of sediments may bring a return of saltwater conditions. Under conditions of slow submergence whereby sediment accumulation balances the rise of water level, thick paralic deposition is possible (Twenhofel, 1961).

GARRY D. JONES

Reference

Twenhofel, W. H., 1961. *Treatise on Sedimentation,* 2 vols. New York: Dover Publications, 926p.

Cross-references: *Aktology; Coastal Sedimentary Facies; Euxinic; Sediment Size Classification.* Vol. VI: *Paralic Sedimentary Facies.*

PEAT—See BOGS

PEBBLE

A *pebble* is a small, more or less rounded stone that has been characteristically abraded by the indirect action of water—whether fluvial, lacustrine, or marine—or, less frequently, by ice and blown sand. The size may be a result of initial particle size or attrition of larger material. Rounding is usually more complete than with *boulders* and *cobbles,* while resistance to abrasion is frequently greater than in larger but more locally derived rocks.

As with cobbles and sand, there are various definitions of size. While in sedimentology a pebble usually falls between *granule* and cobble (4-64 mm, -2 to 6 phi; Gary, McAfee, and Wolf, 1972), sometimes granules (2-4 mm, -1 to -2 phi) are included, thus extending the range from 2-64 mm (-1 to -6 phi). In the terminology of soils, a pebble is defined as a rock or mineral fragment having a diameter of 2-20 mm or 2-64 mm.

Sometimes *gravel* is used as a synonym—for example, in *British Standards* (British Standards Institution, 1967). Particularly in Britain, the term shingle is used to describe the size range covered by beach pebbles, but the word lacks precision and cobbles are sometimes also included.

Because of continuing ambiguity, terms such as boulder, cobble, pebble, sand, silt, clay, and gravel need to be defined at the outset of any individual paper or report. In general, however, the Wentworth classification or some modification of it is used in sedimentation studies. Krumbein's phi scale has been widely adopted but is really satisfactory only for particles less than 1 mm in diameter (that is, its original range of positive values).

ALAN PAUL CARR

References

British Standards Institution, 1967. *Methods of Testing Soils for Civil Engineering Purposes, B.S. 1377.* London, 234p.

Gary, M.; McAfee, R., Jr.; and Wolf, C. L., eds., 1972, *Glossary of Geology.* Washington, D.C.: American Geological Institute, 805p.

Cross-references: *Beach Material; Beach Material, Sorting of; Boulder; Clay; Cobble; Gravel; Sand; Sediment Analysis, Statistical Methods; Sediment Size Classification; Shingle and Shingle Beach; Silt; Soils.*

PEELS

In order to preserve for future study the mineralogy, paleontology, and structure of sediments, a thin layer of the sediments, a *peel,* is removed from the surface. Peels are made by impregnating a sediment with a compound that penetrates some small distance and cements the grains together, making an artificial sedimentary rock. The depth of thickness of penetration depends on the viscosity of the binding agent and the permeability of the sediment.

Many binding materials have been used since Voigt (1933) made the first lacquer peels; among them are laquer, gelatin, glue, latex, paraffin, polyester resin, and epoxy. The individual properties of the diverse substances and the time required to harden are given by Bouma (1969) and Klein (1971). Most of the methods require the preparation of a dry, smooth, flat surface over which cheesecloth is spread and affixed. The binding agent is then brushed on. After drying is complete, the peel is removed by placing a board against the cheesecloth and pulling the board, cheesecloth, and sample from the flat surface.

If the surface to be preserved is semiconsolidated or consolidated, then nitrocellulose or acetate peels can be made (McCrone, 1963; Katz and Friedman, 1965). A piece of cellulose acetate is placed over the etched and polished surface, or the peel solution (parlodoin and butyl acetate) is poured over it and left to dry. The result is a thin replica or transparent print of the sample surface. These peels show textural and structural details often not apparent in the original, and can be used to study fossils, size and shape and orientation of the grains, and the structure and mineralogy of carbonates. These peels, moreover, can be photographed or projected directly (Moiola, Clarke, and Phillips, 1969).

BENNO BRENNINKMEYER

References

Bouma, A. H., 1969. *Methods for the Study of Sedimentary Structures.* New York: Wiley Interscience, 458p.

Katz, A., and Friedman, G. M., 1965. The preparation of stained acetate peels for the study of carbonate rocks. *Jour. Sed. Petrology* **35**, 248-249.

Klein, G. deVries, 1971. Peels and impressions, *in* R. E. Carver, ed., *Procedures in Sedimentary Petrology.* New York: Wiley Interscience, 217-250.

McCrone, A. W., 1963. Quick preparation of peel-prints for sedimentary petrography, *Jour. Sed. Petrology* **33**, 228-230.

Moiola, R. J., Clarke, R. T., and Phillips, B. J., 1969. A rapid field method for making peels of unconsolidated sands, *Geol. Soc. America Bull.* **80**, 1385-1386.

Voigt, E., 1933. Die Ubertragung fossiler Wirbeltierleichen auf Zellulose-Filme, eine neue Bergungsmethode fur Wirbeltiers aud der Braunkohle, *Paläont. Zeitschr.* **15**, 72-78.

Cross-references: *Beach Material; Beach Processes; Coastal Morphology, Research Methods; Sample Impregnation.* Vol. VI: *Sedimentological Methods.*

PERCHED BEACH

A *perched beach* has been defined as "a beach or fillet of sand retained above the otherwise normal profile level by a submerged dike" (Coastal Engineering Research Center, 1977). This is accomplished by constructing a long, low embankment, referred to as a *sill* or *dike,* parallel to the shore. Usually the sill is constructed at the mean low water line, although the exact placement varies with the nature of the beach. For example, extensive tidal flats might require a sill to be placed somewhat closer to the mean high water line. Usually the area shoreward of the sill is artificially filled with sediment, but sometimes it is allowed to fill by natural processes (Anon., 1977).

The sill inhibits the seaward movement of beach sediment, thus causing an accumulation of sediment shoreward of the sill. The accumulation results in the raising of the beach profile and a seaward migration of the shoreline, so that the original shoreline is buffered from erosion. Although sills can be constructed from common material, such as concrete, a new economical method using large polyvinylchloride-coated tubes filled with sand shows good potential for use on lower-energy sand beaches.

Perched beaches have several advantages over other erosion control methods such as *bulkheads, groins,* and *riprap.* A perched beach allows waves to break over it and spend their energy, while bulkheads directly oppose wave energy, reflecting it so that the bulkhead is eventually undermined. Boulders making up riprap reflect waves in the same way, causing them to settle into the bottom. Groins interrupt the natural shore drift, causing sediment accumulation on the updrift side and accelerated erosion on the downdrift side. Perched beaches, which are designed to harmonize with natural processes, are more effective in the proper circumstances than erosion control methods that directly oppose natural forces.

EDMUND E. JACOBSEN

References

Anonymous, 1977. Perched beach: Temporary relief from shore erosion? *American Shore and Beach Preservation Assoc. Newsletter,* 2-3.
Coastal Engineering Research Center, 1977. *Shore Protection Manual.* Washington, D.C.: U.S. Army Corps of Engineers, 1264p.

Cross-references: *Beach Processes; Bulkheads; Coastal Engineering; Coastal Engineering, History of; Coastal Erosion; Groins; Protection of Coasts; Sediment Transport.*

PERIGLACIAL EFFECTS

The coasts of present-day arctic regions are characterized by features related to both periglacial and marine processes. Such coasts may be dominated for part or all of the year by the effects of sea ice, which inhibits wave generation and wave action, so that these are typically low-energy coasts. The existence of *permafrost* (Embleton and King, 1975) in the beach material will further limit wave action and will also prevent percolation; thus, unlike coasts outside the periglacial zone, the volumes and velocities of swash and backwash may be roughly equal.

Beach material is relatively unsorted and angular or subangular, since its movement is restricted by the permafrost beneath and often by a cover of frozen sea spray and swash; in addition, wave disturbance will cease when the sea itself freezes. On Devon Island, Canada (lat. 75°N), Owens and McCann (1970) note that the period of effective wave action is as short as 8 weeks a year. Although permafrost limits the free movement of beach material at more than shallow depths, it does not prevent recession of coastlines by *thermoerosion* (Czudek and Demek, 1973) in areas of unconsolidated rocks—the process is one of melting by the relative warmth of the sea when the latter is unfrozen. The most distinctive features of Arctic coasts are associated with the disturbance of the beach by pack ice driven onshore by wind. Pack ice can move boulders 15 m^3 in size, push up ridges of beach debris several meters high, and grind its way inland for tens of meters. In thaw periods, blocks of sea ice may remain on the beach or partially encased in beach debris, giving rise to a variety of melt-out features.

CLIFFORD EMBLETON

References

Czudek, T., and Demek, J., 1973. Die Reliefentwicklung während der Dauerfrostbodendegradation *Československé akademie věd (Prague) Rozpravy* 83, 69p.
Embleton, C., and King, C. A. M., 1975. *Periglacial Geomorphology.* London: Edward Arnold, 203p.
Owens, E. H., and McCann, S. B., 1970. The role of ice in the Arctic beach environment, with special reference to Cape Ricketts, south-west Devon Island, Northwest Territories, Canada, *Am. Jour. Sci.* 268, 397-414

Cross-references: *Antarctic, Coastal Morphology; Arctic, Coastal Morphology; Chattermarks, Frost Riving; Glaciated Coasts; Glaciel; High-Latitude Coasts; Ice along the Shore; Ice-Bordered Coasts; Ice Foot; Kaimoo; Monroes; Mud Volcanoes; Shore Polygons; Stone Packing.* Vol. I: *Sea Ice;* Vol. III: *Ice Thrusting; Periglacial Landscapes.*

PETERSEN SCALE—See SEA CONDITIONS

PHAEOPHYCOPHYTA

The Phaeophycophyta (Phaeophyta) or *brown algae* are a division (phylum) of lower plants whose more than 1500 species are almost exclusively marine in habitat. Their size ranges from microscopic and filamentous to the massive *kelps,* which may surpass 70 m in length. To most people *seaweed* is synonymous with only a few genera of kelps (order Laminariales).

Brown algae are distinguished from the other groups of algae by the nature of their pigments, stored food products, flagellation and motile cells, and cell wall. The typical color of brown algae is due primarily to the accessory pigment fucoxanthin (a xanthophyll) that masks the colors of chlorophyll a and c, β carotene, and other xanthophylls. The products of photosynthesis may be stored as mannitol (a sugar alcohol), the polysaccharide laminarin (composed of 1:3 and 1:6 linked β glucosides), or as fat droplets. Starch is not found in the Phaeophycophyta.

In contrast to the nonmotile adult brown algae, reproductive bodies (gametes and spores) are generally motile and of characteristic appearance. They are usually pear- or bean-shaped with two laterally inserted flagella, unequal in length and type. One flagellum is the whiplash type, while the other is the tinsel type with two rows of fine hairlike processes (mastigonemes) along its length.

The cell wall contains cellulose and alginic acid. The latter may represent more than 35% of the dry weight of some kelps and is important commercially (Chapman, 1970; Levring, Hoppe, and Schmid, 1969; Okazaki, 1971). In 1970, world production of algin reached 12,800 metric tons at an estimated value of 35 million dollars.

The monumental work of Fritsch (1945) is still a basic reference to the morphology (form) and reproduction of these algae, although more recent references should also be consulted, such as the texts of Dawson (1966) and Scagel et al. (1965).

The basic arrangement of the plant consists of a basal rootlike holdfast, a stemlike stipe, and a blade (or blades) of various shapes. Except for some species of *Sargassum*, which may live free-floating, all brown algae must be firmly anchored to a substrate. The habitat in which a particular species is found seems to determine the morphology of the plants that survive. Neushul (1972) discusses the effects of "water motion regions" on morphology.

Basic morphology of the groups of brown algae varies. The kelps include the most highly differentiated of all algae, some with special cells (sieve tubes) that function in translocation of photosynthate to the holdfast or actively growing regions of the plants. More simple forms are filamentous, crustose, saclike, or tubelike. Pneumatocysts (gas bladders) may be present.

Among different groups of the brown algae, sexual reproduction ranges from isogamy (two identical gametes) to oogamy (fusion of a nonmotile egg and a smaller, motile sperm). An alternation of haploid and diploid generations is characteristic of most Phaeophycophyta. The generations may be isomorphic (as in the most primitive, filamentous order Ectocarpales), with both phases morphologically resembling each other, or heteromorphic (e.g., Laminariales) with alternation of a massive, diploid, spore-producing plant with microscopic filamentous gametophytes (gamete-producing plants).

Classification of the Phaeophycophyta has been well covered by Papenfuss (1955), although Scagel (1966) points out alternative systems of classification. As more information becomes available, especially in the field of life history studies (Wynne and Loiseaux, 1976), a rearrangement of the various groups may occur.

Macrocystis is undoubtedly the most studied of all brown algae because it is important commercially and provides a habitat for fish and invertebrates. North (1971) and the various *Proceedings of the International Seaweed Symposium* may be consulted for information on the biology of *Macrocystis*.

Discussions of the physiology and biochemistry of algae, including the brown algae, may be found in Lewin (1962) and Stewart (1974). Phytogeography of the Laminariales (Druehl, 1970) and Fucales (Nizamuddin, 1970) has been discussed, and a general algal phytogeography of the North Atlantic has been proposed (Hoek, 1975). The Japanese literature (with many references to brown algae) has recently been made more accessible (Tokida and Hirose, 1975); Boney (1969) concentrates on British species.

The fossil record does not include many brown algae because most species do not have calcification of cell walls (except *Padina*). Miocene deposits in California have, however, shown several well-preserved genera (Parker and Dawson, 1965).

Reviews of the brown algae by Scagel (1966) and Druehl and Wynne (1971) should be consulted for specific areas of interest. Recent information may be found in these journals that concentrate on algae: *Phycologia, Journal of Phycology, British Phycological Journal, Phykos, Bulletin of the Japanese Society of Phycology, Revue Algologique,* and *Botanica Marina*.

ROBERT B. SETZER

References

Boney, A. D., 1969. *A Biology of Marine Algae*. London: Hutchinson, 216p.

Chapman, V. J., 1970. *Seaweeds and Their Uses*. London: Methuen, 304p.

Dawson, E. Y., 1966. *Marine Botany: An Introduction*. New York: Holt, Rinehart, and Winston, 371p.

Druehl, L. D., 1970. The pattern of Laminariales distribution in the northeast Pacific, *Phycologia* 9, 237-247.

Druehl, L. D., and Wynne, M. J., 1971. Bibliography on phaeophyta, in J. R. Rosowski and B. C. Parker, eds., *Selected Papers in Phycology.* Lincoln: University of Nebraska, 791-796.

Fritsch, F. E., 1945. *The Structure and Reproduction of the Algae,* vol. 2. New York: Cambridge University Press. 939p.

Hoek, C. van den, 1975. Phytogeographic provinces along the coasts of the northern Atlantic Ocean, *Phycologia* 14, 317-330.

Levring, T., Hoppe, H. A., and Schmid, O. J., 1969. *Marine Algae: A Survey of Research and Utilization.* Hamburg: Cram, De Gruyter & Co., 421p.

Lewin, R. A., ed., 1962. *Physiology and Biochemistry of Algae.* New York: Academic Press, 929p.

Neushul, M., 1972. Functional interpretation of benthic marine algal morphology, in I. A. Abbott, and M. Kurogi, eds., *Contributions to the Systematics of Benthic Marine Algae of the North Pacific.* Kobe: Japanese Society of Phycology, 47-73.

Nizamuddin, M., 1970. Phytogeography of the Fucales and their seasonal growth, *Bot. Marina* 13, 131-139.

North, W. J., ed., 1971. The biology of giant kelp beds (Macrocystis) in California, *Nova Hedwigia Beihefte* 32, 1-600.

Okazaki, A., 1971. *Seaweeds and Their Uses in Japan.* Tokyo: Tokai University Press, 165p.

Papenfuss, G. F., 1955. Classification of the algae, in *A Century of Progress in the Natural Sciences 1853-1953.* San Francisco: California Academy of Sciences, 115-124.

Parker B. C., and Dawson, E. Y., 1965. Non-calcareous marine algae from California Miocene deposits, *Nova Hedwigia Beihefte* 10, 273-295.

Scagel, R. F., 1966. The Phaeophyceae in perspective, *Ann. Rev. Ocean. Mar. Biol.* 4, 123-194.

Scagel, R. F.; Bandoni, R. J.; Rouse, G. E.; Schofield, W. B.; Stein, J. R.; and Taylor, T. M. C., 1965. *An Evolutionary Survey of the Plant Kingdom.* Belmont, Calif.: Wadsworth Publishing Co., 658p.

Stewart, W. D. P., ed., 1974. *Algal Physiology and Biochemistry.* Oxford: Blackwell's, 960p.

Tokida, J., and Hirose, H., eds., 1975. *Advance of Phycology in Japan.* The Hague: W. Junk, 355p.

Wynne, M. J., and Loiseaux, S., 1976. Recent advances in life history studies of the Phaeophyta, *Phycologia* 15, 435-452.

Cross-references: *Algal Mats and Stromatolites; Chlorophyta; Chrysophyta; Coastal Flora; Coral Reef Coasts; Cryptophyta; Cyanophyceae; Euglenophyta; Organism-Sediment Relationship; Reefs, Non-Coral; Rhodophycophyta.* Vol. III: *Algae;* Vol. VI: *Algal Reef Sedimentology.*

PHASE DIFFERENCE

Phase difference is defined as the ratio of time of uprush of *swash* (t) to the wave period (T) (Kemp, 1960, 1975). Using the ratio t/T, Kemp classified the wave-beach relationship into three categories with characteristically different flow regimes. For low values of phase difference, the broken wave surges up the beach to the limit of uprush and returns as *backwash* to the breaker point before the succeeding wave has broken (*surge* condition). Flow landward of the breakers is distinctly oscillatory and beaches steep and plane. Increasing wave height (or shorter wave period) and *plunging breakers* result in backwash from one wave, causing interference with the uprush of the next and producing a *transition* flow regime. Further increase in phase difference values greater than unity produces what Kemp termed *flow* conditions in which successive lines of breakers continuously spill water into the inshore zone, producing a corresponding seaward return flow. This situation, Kemp's so-called *surf* condition, is associated with *spilling breakers* on relatively flat beaches.

ALAN PAUL CARR

References

Kemp, P. H., 1960. The relation between wave action and beach profile characteristics, in *Conf. Coastal Eng., 7th, Am. Soc. Civil Engineers, The Hague, Proc.,* 262-276.

Kemp, P. H., 1975. Wave asymmetry in the nearshore zone breaker area, in J. R. Hails and A. P. Carr, eds., *Nearshore Sediment Dynamics and Sedimentation.* Chichester, U.K.: Wiley, 47-68.

Cross-references: *Beach Processes; Nearshore Hydrodynamics and Sedimentation; Wave Energy; Wave Meters; Waves; Wave Statistics.*

PHORONIDA

At first glimpse this organism appears to be one of the many "worms" that are so abundant in marine environments, but upon close inspection the phoronids display unique features that cause them to be classified in their own phylum. Although they have worldwide distribution, only two genera are now recognized, these containing about 12-15 species (Ayala et al., 1974; Marsden, 1959). However, planktonic larval forms have been observed for which no adults have yet been discovered (Silen, 1954). Adult phoronids typically dwell in soft-bottomed, shallow marine environments, where they lie buried within a cylindrical chitinous tube. Their length ranges from several millimeters to several centimeters. The front of the wormlike body bears a crown of cilia-covered tentacles called the *lophophore,* used to gather plankton, upon which the organism feeds. Phoronids are hermaphroditic, having both male and female sex organs. A planktonic larval stage, the *actinotroch,* lasts for several weeks before settling to the ocean bottom. Since phoronids lack a skeleton, the fossil record is poor, and the origin of the phyllum is unclear. They may possibly be represented by various fossil burrows in Late Precambrian rocks. Both the Brachiopoda and Ectoprocta bear certain structural similarities, notably the presence of the lophophore; the Phoronida may represent the ancestral stock from which these organisms developed.

GEORGE MUSTOE

References

Ayala, F. S., Valentine, J. W., Barr, L. G., and Zumwalt, G. S., 1974. Genetic variability in a temperate intertidal phoronid, *Phoronopsis Viridis, Biochem. Genetics* 11, 413–427.

Marsden, J. R., 1959. Phoronidea from the Pacific Coast of North America, *Canadian Jour. Zoology* 37, 87.

Silen, L., 1954. Developmental biology of Phoronidea of the Gullmar Fiord area (west coast of Sweden), *Acta Zool.* 35, 215.

Cross-references: *Annelida; Chactognatha; Coastal Fauna; Organism-Sediment Relationship; Platyhelminthes.*

PHOTIC ZONE

The photic zone is the region of water that is penetrated by light and is therefore characterized by the presence of plant life, notably phytoplankton and larger attached algae.

The depth to which the photic zone extends is determined largely by the density of suspended matter in the water, which in turn depends on weather conditions and substrate stability. Light penetration is described fairly well by $dI/dt = -kI$, where I is incident light, t is depth, and k is the extinction coefficient.

Even very clear water ($k = .03$) absorbs 50% of the light in the first meter and 90% in the upper 20 m. Open oceanic water may have a k of .1 and coastal seawater, a k of .3 and upward (Krebs, 1972).

The component wavelengths of sunlight are differentially absorbed by water and dissolved compounds, and penetrate to greatly varying depths (Fig. 1). This characteristic, together with the total drop in intensity with increasing depth, leads to biotic zonation in the photic region. Because different algae contain varying amounts of photosynthetic pigments, each utilizing specific wavelengths, many have a distinct distribution within the photic zone. Many faunistic distributions likewise follow the algal zonations.

CHARLES R. C. SHEPPARD

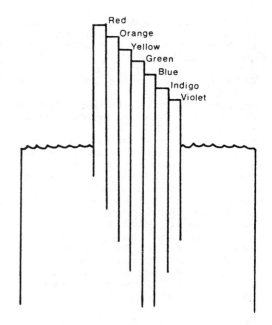

FIGURE 1. Relative light penetration of different wavelengths in seawater.

Reference

Krebs, C. J., 1972. *Ecology: The Experimental Analysis of Distribution and Abundance.* New York: Harper and Row, 694p.

Cross-references: *Biotic Zonation; Coastal Fauna; Coastal Flora; Coastal Waters Habitat; Food Chain; Nearshore Water Characteristics; Productivity.*

PHOTOGRAMMETRY

The value of aerial photographs as a tool in intelligence gathering during military operations has been recognized since the Franco-Prussian and the American Civil wars, when balloon-carried cameras were used (American Society of Photogrammetry, 1975). By World War I, air photography and photointerpretation proved their military significance through the development of improved cameras and airplanes for military use. In the past forty years, peacetime applications of aerial photography have been developed at an increasing rate in many fields, ranging from topographic mapping, natural resources surveys, and city planning to pollution detection. Aerial photographs have also proved an indispensable tool in the study and mapping of shoreline features and changes as well as in coastal engineering and coastal zone management (El-Ashry, 1977).

Field measurements and observations as well as laboratory modeling were, and remain, the major activities for collecting and interpreting data from coastal areas. Nautical charts and topographic maps were used for gathering information about the configuration of the coasts and the changes that take place along them. However, air photographs and other sensors proved valuable in showing the larger features and circulation patterns that do not normally show on photographs taken from ground levels. Over shallow and clear waters, they serve to indicate shoals and channels that exist along many shorelines and provide accurate means for revising and updating maps and nautical charts.

Radar instruments were developed during World War II to obtain images of almost photographic quality during day or night and through dense cloud covers. From these techniques, Side Looking Radar (SLR) emerged as an all-weather, all-time photographic technique to be used from airplanes. This method obtains a picture through a series of lines similar to those on a television screen, and the image can easily be recorded on photographic film.

In the 1960s weather satellite systems NIMBUS (started in 1958) and TIROS as well as manned space flights Gemini and Apollo provided spectacular photography of the earth's surface from altitudes many times above the levels that can be reached by jet airplanes. This was the beginning of the field of *remote sensing*, which is considered an outgrowth of air photographic interpretation. The major advantage of space photographs is that the area of the surface shown in a single exposure is many times larger than that shown on a conventional aerial photograph. Typical coverage of a Gemini space photo, for example, is about 38,500 km^2, whereas typical coverage of a 9 X 9 inch vertical aerial photograph (at a scale of 1:20,000) is only about 20 km^2.

In the 1970s the supply of remotely sensed data about the surface of the earth and the importance of remote sensing greatly increased with the launching on July 23, 1972, of the unmanned ERTS-1 (Earth Resources Technology Satellite-1), now known as LANDSAT-1, which was placed in near-polar orbit, 900 km above the earth. The manned SKYLAB-A in 1973 orbited at 435 km above the surface of the earth.

Types of Aerial Photographs

Two basic types of aerial photographs are generally recognized, based on the orientation of the camera axis relative to the surface of the earth. A *vertical* aerial photograph is obtained when the camera axis is pointing vertically down perpendicular to the surface of the earth. An *oblique* aerial photograph is obtained when the camera axis is at angle considerably away from the vertical. When the inclination is so great that the horizon appears in the picture, such photographs are called *high obliques*. Those with less inclination of the camera axis that do not show the horizon are called *low obliques*.

The type of film used in air photography is one of the factors that determine the tone and sharpness of an object. Four principal types of film are available for aerial photography: black and white, black-and-white infrared, conventional color, and color infrared.

Images on black-and-white photographs appear in different shades of gray, with each tone comparable to the color density of an object as it appears to the human eye. On the other hand, the gray tones on black-and-white infrared photographs do not reflect the object's true color since the infrared film is mainly sensitive to blue-violet and infrared radiation.

Conventional color film is sensitive to all visible radiation, although with variations according to film types; color infrared film is sensitive mainly to green, red, and infrared radiation. Accordingly, colors on color infrared photographs are false colors.

Mechanical scanners are highly sensitive scanning radiometers capable of scanning a scene from an airplane or a satellite and producing an image the same way a picture is shown on a television screen. A *radiometer* is a nonimaging device that measures the absolute energy emitted directly from an object or the amount of solar energy reflected from it. A radiometer may be sensitive to energy from $0.3-14\,\mu$ (ultraviolet to far infrared) depending on the type of sensing element being used (Scherz and Stevens, 1970).

The value of a multispectral scanner stems from the fact that it allows an investigator to view imagery obtained from a different number of narrow band widths. Multispectral scanners utilize anywhere from 3 to 18 or more channels and have the ability to look at all channels simultaneously and pick those that are most useful for a desired discrimination. The ERTS-1 multispectral scanner subsystem (MSS) provided images of the earth's surface in four different bands of the electromagnetic spectrum. Band 4 has the capability of water penetration, and thus is useful in studying underwater features. Band 5 best defines the extent of turbid plumes and sediment laden waters. Band 6 is useful in the differentiation between water and vegetation in marsh areas. Band 7 is useful in coastline mapping because it yields excellent demarcation between land and water.

Applications in Coastal Studies

The literature contains a large number of papers and a few books that emphasize the value of aerial photography (from conventional aircraft and satellite) as a tool in coastal studies. Examples include: Dietz (1947), Coleman (1948), Shepard (1950), Zeigler and Ronne (1957), El-Ashry and Wanless (1967 and 1968), Nichols (1970), Shepard and Wanless (1971), Magoon (1973), and El-Ashry (1977, 1979).

Coastal features in aerial photographs can be identified by their shape, tone, texture, and pattern. Beaches appear as grayish white bands. Changes in tone and texture of a beach can be indicative of local conditions. Answers to questions about the changes may be found in the parent material of the beach as present in rock outcrops, types of vegetation in the area, exposure, prevailing surf and wind conditions, and slope and capillarity of beach material, among others (Coleman, 1948). Beach features exposed above water normally appear lighter in tone than submerged ones. Dried sand and shingle reflect more light than wet ones.

An important area of application of aerial photography is in information gathering for military purposes. During World War II, aerial photographic interpretation of coasts and beaches was of great interest to the military. Coleman (1948) stated that: "A great deal of World War II was fought along enemy-held coastlines. . . . Not one of the thousands of assault troops or landing craft personnel had seen those beaches before. . . . Aerial photography was the greatest, and in some cases the only source of information."

One of the major advantages of aerial photographs is their utilization in mapping projects in any geologic environment. However, in mapping coastlines or in revising older maps, there is always the problem of where to place the shoreline on the photographs. It is generally agreed that the shoreline should be placed at the high water line. However, the exact position of the high water line is not always easy to determine in aerial photographs. The commonly used method is based on a combination of photographic tone, texture, and alignment patterns.

The availability of comprehensive air photography of coastal areas, along with rephotographing of the same areas at intervals of time during the last 40 years, has made it possible to observe changes in beaches, sand dunes, and inlets as well as the effects of severe storms (Fig. 1) and man-made structures, between dates of photography. Published topographic maps and coastal charts, as well as field sketches by coast surveys, permit observation and comparison of changes prior to the availability of aerial photography. From sequential aerial photography it is possible to measure changes resulting from sediment

FIGURE 1. Aerial photographs of Ocracoke Island, North Carolina. Left: after hurricane Helene in October 1958; right: approximately one year later, August 1959 (reproduced by permission of the National Ocean Survey [NOAA], U.S. Department of Commerce).

FIGURE 2. ERTS-1 view of the New York–New Jersey coastal zone. Barrier islands, beaches, tidal inlets, lagoons, tidal deltas, tidal marshes, and spits are clearly visible (NASA photo).

shift during the interval between dates of photography.

In the past decade, weather satellites and manned as well as unmanned satellites and spacecrafts have been placed in orbits about the earth at altitudes many times the height jet planes are capable of reaching. Spectacular photographs were obtained from these flights, from which valuable information regarding the coastal zone has been obtained. Nichols (1970) and Magoon (1973) demonstrated the varied areas of application of space photography. Shoreline features, including barrier islands, beach ridges, dunes, salt marshes, tidal flats, tidal inlets and deltas, are portrayed clearly on space photography (Fig. 2). Changes in shoreline configuration and identification of major areas of erosion or deposition can also be obtained from space photography. Considerable underwater preparation is possible—thus its utilization in water depth determination. Information regarding littoral drift direction can be obtained from certain shoreline features or changes as well as from sediment plumes discharged from river mouths and bays. The possibility of widespread offshore movement of nearshore sediments was first appreciated after space photography became available.

MOHAMED T. EL-ASHRY

References

American Society of Photogrammetry, 1975. *Manual of Remote Sensing,* vol. 1, Falls Church, Va., 27–30.
Coleman, C. G., 1948. Photographic interpretation of coasts and beaches, *Photogramm. Eng.* 14, 463–472.
Dietz, R. S., 1947. Aerial photographs in the geologic study of shore features and processes, *Photogramm. Eng.* 13, 537–545.
El-Ashry, M. T., ed., 1977. *Air Photography and Coastal Problems.* Stroudsburg, Pa.: Dowden, Hutchinson & Ross, 425p.
El-Ashry, M. T., 1979. Use of Appolo-Soyuz photographs in coastal studies, in *Appolo-Soyuz Test Project, Summary, Science Report, Earth Observations and Photography,* vol. 2. Washington, D.C.: National Aeronautics and Space Administration, 531–543.
El-Ashry, M. T., and Wanless, H. R., 1967. Shoreline features and their changes, *Photogramm. Eng.* 33, 184–189.
El-Ashry, M. T., and Wanless, H. R., 1968. Photo interpretation of shoreline changes between Cape Hatteras and Fear, North Carolina, *Marine Geology* 6, 347–379.
Magoon, O. T., 1973. Use of ERTS-1 in coastal studies, in *NASA/Goddard Space Flight Center ERTS-1 Symp. Proc.,* 108–116.
Nichols, M., 1970. Coastal processes from space photography, in *Conf. Coastal Eng., 12th, Am. Soc. Civil Engineers, Proc.,* 641–649.
Scherz, J. P., and Stevens, A. R., 1970. *An Introduction to Remote Sensing for Environmental Monitoring.* Madison, Wis.: University of Wisconsin Remote Sensing Program, Report 1, 80p.
Shepard, F. P., 1950. Photography related to investigation of shore processes, *Photogramm. Eng.* 16, 756–769.
Shepard, F. P., and Wanless, H. R., 1971. *Our Changing Coastlines.* New York: McGraw-Hill Book Co., 579p.
Zeigler, J. M., and Ronne, F. C., 1957. Time lapse photography—Aid to studies of the shorelines, in *Naval Research Rev.,* vol. 4. Washington, D.C.: U.S. Office of Naval Research, 1–6.

Cross-references: *Base-line Studies; Coastal Characteristics, Mapping of; Coastal Morphology, Research Methods; Environmental Impact Studies; Protection of Coasts; Remote Sensing.* Vol. XIII: *Remote Sensing.*

PIER

A *pier* is a structure extending from the shore out into the water, serving as a berthing place for vessels or as a recreational facility (Coastal Engineering Research Center, 1977; Gary, McAfee, and Wolf, 1972). It can be of open construction, as a deck on piles, or of solid-fill construction, as a *mole* (Quinn, 1961). The term is also used synonymously with *jetty* in England and the Great Lakes region of the United States.

In some instances a pier may be roughly parallel to the shore and connected to the shore by a mole or trestle extending perpendicularly from the pier. Such piers are called T-head or L-shaped piers, depending upon the location of the connecting structure.

CHARLES B. CHESTNUTT

References

Coastal Engineering Research Center, 1977. *Shore Protection Manual.* Washington, D.C.: U.S. Army Corps of Engineers, 1264p.
Gary, M.; McAfee, R., Jr.; and Wolf, C., 1972. *Glossary of Geology.* Washington, D.C.: American Geological Institute, 850p.
Quinn, A. F., 1961. *Design and Construction of Ports and Marine Structures.* New York: McGraw-Hill, 531p.

Cross-references: *Coastal Engineering; Jetties; Mole; Protection of Coasts.*

PILE, PILING

A *pile* is a structural member of timber, concrete, or steel that is driven or jetted into the earth or seabed to transmit surface loads to lower levels in the soil mass (Bowles, 1968). This load transfer may occur through friction, end bearing, or a combination of the two. Those piles transmitting a large portion, or all of the load, through friction between the pile and soil are known as *friction piles;* those the points of which lie on a firm soil or rock are called *end-bearing piles.* Piling is a group of piles, sometimes known as the *pile system.* Piling is used when the shallow soil zone has a poor bearing capacity or when settlement problems could occur (Coastal Engineering Research Center, 1973). Piles range in area and length with some of today's offshore piles up to 350 m long consisting of a series of short piles.

JOHN B. HERBICH

References

Bowles, J. E., 1968. *Foundation Analysis and Design.* New York: McGraw-Hill, 659p.
Coastal Engineering Research Center, 1973. *Shore Protection Manual,* Washington, D.C.: U.S. Army Corps of Engineers, 750p.

Cross-reference: *Coastal Engineering.*

PINICAE

The subdivision Pinicae (also known as Coniferophyta) of the division Pinophyta (gymnosperms) includes about 600 living species, as well as a number of extinct forms. Two classes are recognized, the Ginkgoopsida and Pinopsida. The latter group consists of only a single order, Ginkgoales, and the unique characteristics of the ginkgos cause them to be classified by some authors as an independent division (Ginkgophyta). The Pinopsida contains two orders: Pinales (conifers) that bear seeds enclosed in cones, and Taxales (taxads) containing yews and related trees that do not produce female cones. The Pinopsida contains a variety of shrubs and trees, divided into seven families: Voltziaceae (an extinct group), Araucariaceae (Monkee Puzzle family), Cephalotaxaceae (Plum-yew family), Cupressaceae (cypress family), Pinaceae (pine family), Podocarpaceae (podocarpus family), and Taxodiaceae (deciduous cypress family). Other coniferlike plants are known by fossils, though the taxonomy of these organisms is unclear.

The ginkgos are known to have existed as far back as the early Permian, when they were represented by the genus *Trichopitys,* a plant that had striking similarities to some Devonian seed ferns. The modern *Ginkgo* genus achieved worldwide distribution during the Jurassic, but declined rapidly during the Cenozoic. *G. biloba,* the only surviving ginkgo, was preserved from extinction as an ornamental tree in China and Japan. The ginkgo, or maiden hair tree, is characterized by fan-shaped dichotomously veined deciduous leaves. Sexes are separate and, unlike the Pinopsida, the trees do not produce cones. Instead, male trees release pollen from soft stalks. Female trees bear seeds that are surrounded by fleshy prunelike fruit. Because of their unusual resistance to air pollutants, ginkgos are sometimes used as ornamental trees. Male trees are preferred, because the soft, sticky, and foul-smelling fruit produced by the females poses an esthetic problem for pedestrians. The inner seed is nearly odorless, however, and is sometimes sold in Chinese food stores.

The order Taxales includes the families Taxaceae, Cephalotaxaceae, and Podocarpaceae, which have either solitary or pairs of seeds that are fleshy coated or surrounded by a fleshy cupule. The plum-yew family (Podocarpaceae) includes seven genera found mainly in the southern hemisphere, extending as far north as Mexico, Cuba, southern Japan, and China. Cephalotaxaceae are trees and shrubs in eastern Asia, while the Taxaceae (true yews) includes about 20 species of shrubs and small trees found mostly in Asia. Exceptions are *Torreya* (stinking yew) that occurs locally in North America, and *Taxus* (yew) that has widespread distribution.

Members of the Pinopsida typically are freely

branching trees or shrubs with relatively dense wood and clusters of small, narrow leaves ("needles"). Unlike flowering trees (angiosperms), which have broad leaves with complex branching veins, leaves of the conifers have only a single vein in the center. *Tracheids* bring water and nutrients to the leaves and *sieve tubes* transport dissolved sugars produced by photosynthesis to other parts of the plant. Food-producing cells, *mesophyll,* surround this central core, accompanied in most genera by ducts that extrude resin as needed for healing wounds. The outer surface of the leaf is covered by a hard, waxy cuticle. *Stomates* (pores) are usually arranged in longitudinal rows along the leaf. Heavily cutinized leaves allow the conifers to conserve moisture, and they grow better than angiosperm trees in cold regions where roots have difficulty absorbing large quantities of water from the soil. Most conifers retain their leaves throughout the year, thus giving rise to the common name "evergreens." A few varieties such as the larch (*Larix*) and swamp cypress (*Taxodium distichum*) shed their needles in the fall. The ginkgos are also deciduous.

Gymnosperm wood is made up of several structural components. *Bark* protects the outer surface and can be up to 10 cm thick near the base of large trees, becoming thinner on the upper trunk and branches. Beneath the bark lies a layer of conductive tissue, *phloem,* which transports food produced in the leaves to other parts of the plant. The *cambium* layer underlies the phloem. Typically only a few cells thick, the cambium is the region where new growth takes place. Beneath it lies xylem, a conductive tissue that sends water and nutrients absorbed by the roots upward to the leaves. The great central portion of the tree is dead wood; only about 1 percent of the wood of a typical pine consists of living cells (Peterson, 1980).

Like all seed plants, Pinicae show alternation of generation between an asexual (sporophyte) phase and a sexual (gametophyte) phase. The familiar trees represent the sporophyte generation, and produce two types of reproductive structures. In members of the Pinales these are generally hard or papery scales arranged in whorls around a central axis, forming "cones." Small male cones produce pollen grains, which represent the gametophyte generation. Female gametophytes develop inside *ovules* located on the upper surfaces of scales of the larger, woody female cones. These cones develop during the fall and remain dormant through the winter. During late spring pollen is released, usually being transported by wind. At this season, scales of the female cones spread apart to allow pollen to enter; unlike angiosperms which have female gametophytes that are enclosed within ovary tissue, ovules of the conifers are readily accessible. Cones of some species contain a small drop of liquid to trap pollen grains. The male pollen releases sperm that fertilize the female gametophyte, followed by a complicated process of development that results in each zygote producing a number of embryos. Because of competition, usually only a single embryo survives (Strasberger, 1976; Wiggins, 1981). The embryo continues to develop in size and complexity, eventually being released as a seed.

Seeds are shed at the end of summer and lie dormant until spring. Conifer seeds may remain viable up to 30 years as long as they remain within the protective cone, but they deteriorate rapidly after release. Only about 20 percent germinate, and few of the seedlings survive their first season. Death may result from heat, drying, damage by insects, fungal infection, browsing animals, and competition for space. Though all species of conifers bear male cones that produce pollen, not all species produce female cones. Thus, the plum yews and podocarps bear their seeds in olivelike "fruits," and seeds of the true yews are enclosed in a fleshy cup (*aril*). Most conifers are monoecious, bearing both male and female reproductive structures on a single tree, while a few species are dioecious, with separate male and female trees.

Rate of growth varies widely among different species and is strongly affected by environmental conditions. Many species add one growth ring per year, a phenomenon that has become widely used as a method of age-dating. Tree ring dating, or *dendrochronology,* has been widely used during archaeological studies and provided a means of improving the accuracy of carbon-14 radiometric dating. Because the modern atmosphere is contaminated with large amounts of very old carbon released by burning of fossil fuels, more reliable estimates of the natural carbon isotope ratio were made by analyzing prehistoric wood samples whose age could be independently determined by counting rings. Descriptions of dendrochronology methods have been published by Bowers (1961) and Stokes and Smiley (1968).

Ring counts indicate that the oldest living trees are bristlecone pines (*Pinus aristata*) found at high elevations of California and neighboring states. The oldest of these trees are nearly 5000 years in age, though they seldom exceed 10 m in height and have an irregular bushy form. For many years the giant sequoias (*Sequoiadendron giganteum,* previously known as *Sequoia washingtoniana* and *S. wellingtoniana*) were considered to be the oldest forms of life. John Muir described a fallen sequoia trunk tht contained over 4000 rings, but the largest verified figure is generally considered to be 3126 (Bowers,

1961). While giant sequoias are the most massive of trees, having trunk diameters reaching 8 m, the tallest trees are the coast redwoods, *Sequoia sempervirens,* which have heights up to 90 m. (The other extreme of the confier size range is represented by creeping junipers, which do not exceed 30 cm in height.)

The early evolution of the Pinicae is not fully understood, but they probably evolved from the seed ferns (pteridosperms). The earliest known seed plant, *Eospermatopteris textilis,* was discovered in eastern New York in 1869 when a forest of fossil stumps was exposed by stream erosion. Fossils include slender trunks up to 10 m high, having large bulbous bases and crowned by a mass of large pinnate fronds. Some fronds bear small oval seeds, while others show evidence of pollen sacs. Other pteridosperms have since been discovered from many Paleozoic formations, but their taxonomic relationships are confusing. Fossils show many structural characteristics of the true ferns (Polypodiophyta or Filicinae), but their reproductive habits were more similar to the gymnosperms (Campbell, 1939).

Conifers first appeared during the Carboniferous, represented by the Voltziacea family, a group that became extinct in the late Mesozoic. The *Lebachia* genus consisted of straight, slender trees represented by fossil wood, twigs, and cones from the late Carboniferous. A form genus, *Walchia,* is used to describe leaf-bearing twigs whose stomatal character cannot be determined.

Several other extinct groups are thought to be similar to the gymnosperms, but their relationships remain uncertain. The Caytonailes resemble seed ferns, while the Bennettitales appear to be more like the cycadophytes. A third group, Czekanowskiales, has been compared to the ginkgos (Emberger, 1944).

As mentioned, the Ginkgoopsida flourished during the Mesozoic, and seem to be the nearest living relative to a once-important group, the cordiatales. These tall trees bore a spreading crown of branches with large straplike leaves having parallel veins. Petrified wood from Cordaites is very much like that of the living confiers of the Araucaracea family, and the general form of the fossil trees resembles the modern Kauri pine (*Agathis australis*). The pithy stem of Cordaites is similar to Cycadophytes, however, and the leaf pattern shares characteristics with both *Ginkgo* and the pteridosperms. The similarities in reproduction suggest that the ginkgo is the closest modern relative.

The apparent confusion among these characteristics probably results because of parallel evolution. Modern seed plants may not have originated from a single common ancestor, and at least some of the extinct forms may have arisen from divergent evolutionary pathways.

Modern conifer families appeared during the mid-Mesozoic, though a few relict forms originated earlier. The oldest member of the Pinales order was the Araucariacae family, at present represented by only two genera, *Araucaria* and *Agathis,* both found only in the southern hemisphere. *Araucaria excelsa* (Norfolk Island pine) and *A. arancana* (Monkee Puzzle tree) are common ornamental plants, while the Kauri pine (*Agathis*) flourishes in Australia, New Zealand, and the islands of the South Pacific. Originally appearing in the Permian, the Araucariacia were once widespread. Triassic sediments at Petrified Forest National Park in Arizona contain abundant *Araucarioxylon* logs, and other fossils resembling *Araucaria* have been reported from Triassic rocks extending as far as Europe and Greenland.

Modern conifer families flourished throughout the Cenozoic, although since the late Carboniferous or early Permian they have been differentiated into two geographical groups. A northern group is composed of 33 living genera, and a southern group contains 17 genera. Only 7 are found in both hemispheres (Dallimore and Jackson, 1967). Some of these belong to genera that are isolated relicts such as *Araucaria* and *Taxodium,* but even *Pinus* (pine), *Abies* (fir), and *Picea* (spruce) that presently cover vast areas of the northern hemisphere have more limited range now than during the late Tertiary (Hu-Lin, 1953).

Most conifers are best adapted to cool temperate environments, and all species occur in temperate or sub-tropical habitats. Although some species have worldwide distribution, most are confined to particular regions. *Picea, Larix, Pinus, Abies,* and *Juniperus* are abundant in North America and northern Europe. The latter three genera are also found in Latin America, along with *Cypressus* (cypress) and *Taxodium* (swamp cypress). Several genera are unique to South America, such as *Fitzroya, Austrocedrus* (Chilean cedar), *Pilgerodendron,* and *Saxegotheae* (Prince Albert's yew).

The most famous Asiatic conifer may be the dawn redwood (*Metasequoia*), a genus widely distributed as fossils but preserved only as a few thousand living trees in central China. More abundant are such Chinese trees as golden larch (*Pseudolarix*), deciduous cypress (*Glyptostrobus*), and arbor vitae (*Thuja orientalis*). Other oriental varieties include the Japanese umbrella pine (*Sciadopitys*), *Taiwania,* and cow's tail pine (*Cephalotaxus*). New Zealand and Australia provide habitats for many unique conifers, such as the alpine microstrobus (*Microstrobus*), Kauri pine (*Agathis*), *Podocarpus,* cypress pine

(*Callistrus*), Australia cedar (*Austrocedrus*), and tear-drop cedar (*Libocerdrus*). African conifers include the cypress pine (*Widdringtonia*), *Tetraclinus,* and *Cedrus* (true cedar).

Although conifer forests are best known in mountain regions, many species are found only along coasts. The Monterey cypress (*Cypressus macrocarpa*) is found only on rocky headlands on the shores of northern California, where it is often grotesquely bent and gnarled from sea winds. The Torrey pine (*Pinus torreyana*), the rarest of California pines, occurs only at elevations less than 30 m along the coast of San Diego County and Santa Rosa Island. Species that grow only in proximity to the sea are also found on other continents, such as *Pinus pityusa,* which is limited to rocky slopes bordering the Black Sea.

Conifers may also be found in broad belts along the seaward slopes of coastal mountains, where ocean winds provide abundant moisture. Thus, the coastal redwood (*Sequoia sempervirens*) grows along the Pacific coast from northern California to central Oregon, and Sitka spruce (*Picea stichensus*) occurs at low elevations from California to Alaska. The latter species is commonly found as dwarfed trees along sandy beaches of Oregon and California. Yellow cedar (*Chamaecyparis nootkatensis*) once formed extensive coastal forests in Alaska and British Columbia, but the best stands are now found in the Olympic and Cascade Mountains of Washington, where the species occurs at elevations of 600 to 2000 m (Bowers, 1961).

Certain conifers have adapted to brackish coastal swamps. Bald cypress (*Taxodium distichum*) grows best in fresh water swamps but extends into the coastal region of brackish tidelands. The tree occurs along the Atlantic and Gulf Coast Plains from Texas to Virginia, and 90% of the trees grow within 30 m of sea level, though in west Texas the range extends to 500 m.

The Pinicae have been extensively exploited for commercial purposes. The Piñon pine (*Pinus cembroides*) produces seeds that were a staple in the diet of Indians of the southwestern United States, and conifers continue to play a minor role as a source of nutrients: the Balkan pine (*Pinus peuce*) provides pitch used to flavor the Greek wine *retsina,* and cellulose obtained from wood is a nondigestible additive found in certain brands of low-calorie diet bread sold in the United States. Turpentine oil, used as a solvent for paint and varnish, is obtained by steam distillation of pitch, and the residue left after distillation, rosin, is used for sizing paper, making soap and greases, and preparation of violin bows. Dammar resin, or kauri gum, is produced by *Agathis* of Australia and New Zealand. This resin is usually collected as large masses of fossil gum found buried in the soil, and is used as an ingredient in some varnishes. The most important product derived from conifers is lumber, along with wood products used for paper making and the manufacture of rayon. The Pinaceae provide the majority of lumber used in North America, with lesser amounts obtained from the Cypressaceae and Taxodiaceae. In the southern hemisphere the Araucariacea and Podocarpaceae are sources of commercial timber. Extensive logging has markedly reduced populations of some trees, such as red cedar (*Thuja plicata*), and the present system of replanting logged lands has led to many of these species being replaced by faster-growing Douglas fir (*Pseudotsuga menziesii*). Rising lumber prices and more stringent air pollution regulations have led to increasingly efficient utilization of timber products, as indicated by the proliferation of chip board and other products made from timber byproducts. On the global scale, timber harvesting and land clearing still continue to outpace reforestation, however.

LOUIS CUTTER WHEELER*
GEORGE MUSTOE

*Deceased

References

Bowers, N. A., 1961. *Cone-Bearing Trees of the Pacific Coast.* Palo Alto, Calif.: Pacific Books, 169p.

Campbell, H. C., 1939. *The Evolution of Land Plants.* Stanford, Calif.: Stanford University Press, 731p.

Dallimore, W., and Jackson, A. B., 1967. *A Handbook of Coniferae and Ginkgoaceae,* revised by S. C. Harrison. New York: St. Martin's Press, 729p.

Emberger, L., 1944. *Les Plantes Fossiles dan leur Rapports Avec les Végétaux Vivants.* Paris: Maison et Cie, 492p.

Hu-Lin, Li, 1953. Present distribution and habitats of conifers and taxads, *Evolution,* 7, 245-261.

Peterson, R., 1980. *The Pine Tree Book.* New York: Brandywine Press, 144p.

Stokes, M. A., and Smiley, T. L., 1968. *An Introduction to Tree-Ring Dating.* Chicago: University of Chicago Press, 68p.

Strasburger, E., 1976. *Strasburger's Textbook of Botany,* rewritten by D. von Deraffer, W. Schumacher, L. Magdefrau, and F. Ehrendorfer. New York: Longman, 877p.

Wiggins, I. L., 1981. Conifer, in *Encyclopaedia Britannica,* vol. 5. Chicago: Encyclopaedia Britannica Press, 1-9.

Cross-references: *Biotic Zonation; Coastal Ecology, Research Methods; Coastal Flora; Coastal Plant Ecology, United States, History of; Pinophyta.*

PINOPHYTA

The classification of vascular plants has posed problems for botanists because of the difficulty of establishing definitive boundaries between various groups. Criteria used for establishing taxonomic systems include morphological and anatomical characteristics as well as fossil evidence indicating evolutionary relationships. Editors of this volume have chosen to use the system of Cronquist (1971), which separates seed-bearing plants into the divisions Pinophyta (gymnosperms) and Magnoliophyta (angiosperms) (q.v.).

The pinophyta compose an important part of the earth's terrestrial vegetation, although the total number of genera and species is far fewer than the Magnoliophyta. The Pinophyta are also adapted to a smaller range of environmental conditions and have more limited geographic distribution. Members display considerable diversity of form, ranging from vines, creepers, and shrubs to gigantic trees. All are woody perennials, sometimes having life spans as much as several thousand years.

Three subdivisions are included within the Pinophyta: the Gneticae, Cycadicae, and Pinicae. The Gneticae and Cycadicae are represented by a small number of genera that exist as "living fossils," though both divisions were widely represented during the Mesozoic. The Gneticae contains only three living genera: *Wilwitschia,* occurs in the coastal mist deserts of southwest Africa and Angola and consists of a large, bulbous turnip-like body that may be one meter or more in diameter and which bears a pair of broad, flat leaves; *Ephedra* includes a variety of desert shrubs found in the Mediterranian area and in arid parts of Asia and America; and *Gnetum* is found in luxuriant tropical forests of Africa, Asia, and South America, where its members typically occur as lianas that twine or trail on other plants.

The Cycadicae includes two orders, the Cycadales and Bennettitales. The latter group is known only from fossils, its members being dominant elements of Mesozoic floras. The Bennettitales appeared in late Carboniferous or early Permian, reached maximum development during the Jurassic, and became extinct in the late Cretaceous—a pattern of evolution markedly similar to that of the dinosaurs. The Cycadales are morphologically similar, but have unisexual cones instead of the flowerlike organs found in the Bennettitales. Both groups were characterized by large trees having columnar trunks and a crown of feathery leaves, thus resembling palms. Cycadales were abundant during the Mesozoic, but are now represented by only nine genera found in tropical and subtropical regions.

The Pinicae have been widely distributed since the Mesozoic. Two classes are recognized: the Ginkgoopsida (ginkgos), and the Pinopsida. The latter group consists of *conifers,* which bear their seeds in cones, and *taxads* (yews and related trees) that have seeds surrounded by fleshy tissue.

More detailed descriptions of the members of the Pinophyta are presented in the specific entries in this volume.

LOUIS CUTTER WHEELER*
*Deceased
GEORGE MUSTOE

Reference

Cronquist, A., 1971. *Introductory Botany.* New York: Harper and Row, 885p.

Cross-references: *Biotic Zonation; Coastal Ecology, Research Methods; Coastal Flora; Coastal Plant Ecology, United States, History of; Cycadicae; Gneticae; Pinicae.*

PLATFORM—See WAVE-CUT PLATFORM

PLATFORM BEACH

A *platform beach* is a significant body of sand deposited by wave action on and toward the rear of a *wave-cut platform.* Platform beaches abut the base of the backing cliff and extend out across the platform, but rarely to the outer edge. They are unstable, often seasonal in nature, and may be completely removed during storms. They occur on platforms where a *sand ramp* leads to the edge or side of the platform providing a sediment source (Fig. 1). Sand is supplied onto the platform from the one or more sources and spreads laterally along the

FIGURE 1. Platform beach lying behind a broad wave-cut platform at Long Reef, New South Wales, Australia. A sand ramp (lower right) provides the sand source that has been transported for 150 m along the back of the platform. Note the development of regular cusps on the lower beach face (photo: A. D. Short).

back of the platform. The beach may develop some regular beach features such as *cusps* and *berms*. Erosion of the beach is rapid because the impermability of the underlying platform greatly increases the volume of surface runoff as compared with that of a natural sand beach. Platform beaches may be composed of sediment ranging from fine sand to boulders (Bird and Dent, 1966). In the latter case, they are termed *boulder beaches* and are more stable platform features, but significant reworking occurs during storms.

ANDREW D. SHORT

Reference

Bird, E. C. F., and Dent, O. F., 1966. Shore platforms on the south coast of New South Wales, *Australian Geog.* **10,** 71-80.

Cross-references: *Erosion Ramp, Wave Ramp; Raised Beach; Shore Platforms; Wave-Cut Bench; Wave-Cut Platform.* Vol. III: *Platforms–Wave-Cut.*

PLATYHELMINTHES

The *Platyhelminthes* or *flatworms* are dorsoventrally flattened, bilaterally symmetrical, nonmetameric (unsegmented) worms with no coelom, typically no anus, and no circulatory, respiratory, or skeletal system. Nearly all species of all classes are hermaphroditic. All have internal fertilization. There is a simple ladderlike central nervous system with enlarged cerebral ganglia. Several kinds of sensory structures, including eyes, are present in some. All but Acoela and Gnathostomulida have excretory systems of the protonephridial type. Connective tissue of mesenchymal origin fills spaces between internal organs (reduced in Gnathostomulida).

The phylum includes four classes: Turbellaria *(free-living flatworms)* with approximately 3000 species, Gnathostomulida with less than 100, Trematoda *(flukes)* with 8000, and Cestoda *(tapeworms)* with some 3500 species. Riedl (1969) considered Gnathostomulida a separate phylum, and was supported by Sterrer (1972). Platyhelminthes have left no fossils other than possible larval flukes or tapeworms in insects preserved in amber. Durden et al. (1969) suggested that conodonts might be jaws of worms related to Gnathostomulida.

Turbellaria

Although called free-living flatworms, this group includes a number of parasitic, commensal, or otherwise symbiotic species living in or on other aquatic animals. Unlike Trematoda and Cestoda, Turbellaria have the external surface at least partly ciliated, with the possible exception of some Temnocephalida. Turbellaria are found in marine, estuarine, freshwater, and terrestrial habitats.

Among symbiotic forms, *Fecampia* is endoparasitic in marine crustaceans, *Graffila* and *Paravortex* are endocommensal in marine molluscs, and *Micropharynx* is ectocommensal on skates. In tropical and subtropical freshwaters, Temnocephalida are ectocommensal on crustaceans and other hosts.

Free-living freshwater genera include *Dugesia* (the textbook planaria), *Prorhynchus,* and *Stenostomum.* A few species can live in both freshwater and saltwater. The land has been invaded by triclads, some 25 cm long. Most triclads, alloeocoels, rhabdocoels and all members of the orders Acoela and Polycladida are marine.

Pelagic marine forms include the acoel genus *Haplodiscus* and the neotenic polyclad *Graffizoon.* Other acoels such as *Amphiscolops sargassi* and polyclads such as *Gnesioceros sargassicola* inhabit floating "gulfweed," *Sargassum.*

Most marine turbellarians are benthic. The triclad *Procerodes ulvae* is one of the many turbellarians found on attached seaweeds. Many of the larger triclads and polyclads live among shells and stones on hard bottoms. The European acoel *Convoluta roscoffiensis* is one of the many species living on softer bottoms; it places itself on the surface of *intertidal flats* in such a way as to facilitate photosynthesis of the symbiotic algae in its tissues, and may be so abundant as to color large areas green or yellow.

Sandy sea bottoms contain many turbellarians specialized for dwelling in the interstices between sand grains. Notable among these are the kalyptorhynchids, which have on the anterior end sensory bristles and a protrusible proboscis used to capture copepods. Well-known kalyptorhynchids include *Gyratrix, Polycystis,* and *Gnathorhynchus. Gnathorhynchus* has a pair of jawlike hooks in its proboscis, suggesting the more complex armature of Gnathostomulida.

Small turbellarians, which may be less than 3 mm long, ingest protozoans, unicellular algae, and bacteria. Larger forms tend to be more carnivorous, scavenging dead animals and swallowing live crustaceans, molluscs, and worms almost as large as themselves. The polyclad *Cryptophallus magnus* swallows clams. The polyclads *Stylochus frontalis* and *S. ellipticus* prey on oysters. Some turbellarians (for instance, the freshwater *Dugesia*) secrete substances distasteful to fish, and some marine turbellarians seem to be toxic.

Gnathostomulida

The gnathostomes are marine interstitial sand dwellers. They differ from Turbellaria by having only one cilium per epidermal cell, by having uniflagellate or nonflagellate spermatozoa (also found in Acoela), by having a pair of cuticularized jaws with complex musculature, by having the parenchyma or mesenchyme much reduced, and by lacking protonephridia (also like Acoela). In other respects, such as rhabdocoel gut, external ciliation, and hermaphroditism, they resemble rhabdocoel Turbellaria. Gnathostomes were neglected until recent years, but by 1972 some 18 genera and 80 species were known, widely distributed.

Trematoda

The class Trematoda traditionally contains the subclasses (or orders in some classifications) Monogenea (1400 species), ectoparasitic flukes of fishes, mainly; Digenea (6500 species), endoparasitic flukes with complex life cycles involving different stages in two or more hosts; and Aspidocotylea (less than 100 species), endoparasitic in molluscs and most capable of living also in cold-blooded vertebrates without change in form. All are parasitic; free-living larval stages, if any, last only a few hours. Adults have no external cilia. A digestive system is always present in adults.

Most fishes, in the sea or in freshwater, have monogeneans on gills or skin or sometimes in mouth cavity or cloaca. In nature, infestations seldom build up to lethal levels, but aquarium and hatchery fish are often killed by monogeneans, such as *Gyrodactylus*. A few monogeneans (*Polystoma, Polystomoides*) have become parasites in the mouth cavity or urinary bladder of amphibians and terrapins.

There is a high degree of host specificity among Monogenea, and zoologists have used them as clues when studying phylogenetic relationships among fishes. In most cases monogeneans have evolved more slowly than their hosts, so that fishes of a family characteristically bear flukes of one genus.

Members of the small subclass (or order) Aspidocotylea resemble Digenea, but the larva that hatches from an egg develops directly into an adult like its parent, so that the life cycle is monogenetic. The most striking feature is the large ventral attachment organ made up of multiple suckers. Some species of *Aspidogaster* become sexually mature in molluscs, but most aspidocotyleans require a vertebrate final host. Aspidocotyleans are about equally divided between marine and freshwater forms. They are not known to have adverse effects on either vertebrate or invertebrate hosts. The better-known genera have worldwide distributions.

Digenea are represented by 6500 species in marine, freshwater, and terrestrial hosts. They have two (or more) reproducing generations, the adult in the final host (nearly always a vertebrate) and a sporocyst or redia in the first intermediate host (nearly always a mollusc). The sporocyst and redia produce larval offspring by polyembryony, while the hermaphroditic adult reproduces by either biparental or uniparental sexual reproduction with internal fertilization. The sporocyst develops from a ciliated miracidium (the stage that hatches from the egg). Daughter sporocysts or rediae give birth to cercariae (usually tailed swimming forms). Blood fluke cercariae penetrate into the final host and become adults; other cercariae encyst in the open or in a second intermediate host and become metacercariae (most families). Metacercariae become adults when ingested by the final host.

On beaches and in shallow coastal waters most Digenea are parasites of teleost (bony) fishes or of shore and water birds, and use either gastropod or bivalve molluscs as first (and sometimes second) intermediate hosts.

Many sporocyst infections cause sterility of the mollusc host, destroying the gonad. Trematode infections therefore play a role in molluscan population dynamics. In the Galveston area, 1279 of 12,131 individual molluscs examined were found by Wardle (1974) to be parasitized by trematode larvae. Incidence of infection reached 50% in some mollusc species and 10.5% overall.

Cercariae of some blood flukes of lower mammals and birds penetrate human skin and cause a severe dermatitis called swimmer's itch or clam-digger's itch. Such species occur in both freshwater and saltwater.

Cestoda

The Cestoda (3500 species) are commonly called tapeworms, but the smaller of the two subclasses, the Cestodaria, are not strobilated (made up of a chain of replicated individuals called proglottids) like the typical tapeworms. Cestoda are all internal parasites of vertebrates in the adult stage; they are nonciliated, covered with cuticula, have no digestive system and no internal organs other than nervous, excretory and reproductive. With few exceptions, in the tapeworms or Eucestoda that make up most of the class, each proglottid contains a complete male and a complete female reproductive system.

The groups of greatest interest on beaches are

several families of cyclophyllidean tapeworms of shore and water birds that use insects as intermediate hosts, and cestodes of four orders that are adult in elasmobranchs and use various invertebrates and sometimes bony fishes as intermediate hosts. In Wardle's survey of Galveston area Mollusca, cestode larvae were found in 984 of the 12,131 individual molluscs examined; adults of all eight kinds are intestinal parasites of elasmobranch fishes. Cake (1975), in a survey of cestode larvae in molluscs of the eastern Gulf of Mexico, found larvae of 11 types; adults of all parasitize elasmobranchs (mostly rays). Neither Wardle nor Cake reported any adverse effects of cestode infections on mollusc hosts. No adverse effects of tapeworms on elasmobranchs have been reported by anyone, but larvae in teleost fishes sometimes cause sterility and even death.

Other relevant literature includes Cheng (1973), Grassé (1961), Hyman (1951), Kaestner (1967), and Yamaguti (1963, 1971).

SEWELL H. HOPKINS

References

Cake, E. W., Jr., 1975. *Larval and Post-Larval Cestode Parasites of Shallow-Water, Benthic Mollusks of the Gulf of Mexico from the Florida Keys to the Mississipps Sound.* Ph.D. dissertation. Tallahassee, Fla.: Florida State University, 411p.
Cheng, T. C., 1973. *General Parisitology.* New York: Academic Press, 965p.
Durden, C. J.; Rodgers, J.; Kochelson, E.; and Riedl, R., 1969. Gnathostomulida: Is there a fossil record? *Science* 164, 855–856.
Grassé, P., 1961. *Traité de Zoologie Anatomie, Systématique, Biologie,* vol. 4. Paris: Masson, 944p.
Hyman, L. H., 1951. *The Invertebrates: Platyhelminthes and Rhynchocoela,* vol. 2, *The Acoelomate Bilateria.* New York: McGraw-Hill, 550p.
Kaestner, A., 1967. *Invertebrate Zoology,* vol. 1. New York: Wiley Interscience, 597p.
Riedl, R., 1969. Gnathostomulida from America: First record of the new phylum from North America, *Science* 163, 445–452.
Sterrer, W., 1972. Systematics and evolution within the Gnathostomulida, *Syst. Zool.* 21, 151–173.
Wardle, W. J., 1974. *A Survey of the Occurrence, Distribution, and Incidence of Infection of Helminth Parasites of Marine and Estuarine Mollusca from Galveston, Texas.* Ph.D. dissertation. College Station, Tex.: Texas A & M University, 322p.
Yamaguti, S., 1963. *Systema Helminthum,* vol. 4, *Monogenea and Aspidocotylea.* New York: Wiley Interscience, 699p.
Yamaguti, S., 1971. *Synopsis of Digenetic Trematodes of Vertebrates,* vol. 1. Tokyo: Keigaku, 1074p.

Cross-references: *Annelida; Coastal Fauna; Intertidal Flats; Intertidal Mud Habitat; Intertidal Sand Habitat; Nemertina; Tidal Flats.* Vol. VII: *Trace Fossils.*

PLAYA

This word stem of middle Latin *plagia* (Catalan *platja,* Portuguese *praia*), has developed semantically from its meaning of "land slope" (a word that among American geomorphologists designates the flat-floored bottom of an interior desert basin) to that of beach—especially a sandy beach—in the Spanish language of today (Baulig, 1956). (Latin American and Spanish morphologists avoid the first meaning, preferring instead *salar, saladar,* and *bolsón.*) These beaches are generally found at flat coasts, but they may appear either at the foot of certain cliffs in shallow waters or in concavities on high coasts of sheltered spurs (Zenkovich, 1967). On the Mediterranean coasts, owing to the lack of tides, there is scarce difference between strand or backshore and *anteplaya* or foreshore. Storms give rise to *crestas de playa* (beach ridges) (Guilcher, 1958; King, 1972), sometimes separated by furrows and also, in special cases, berms or terraces. Some beaches are often backed by dunes or *médanos,* forming a single or double line (shore dunes and back dunes) when not forming a sheet of more extensive but less ordered dunal morphology.

VINCENC M. ROSSELLÓ

References

Baulig, H., 1956. *Vocabulaire Franco-Anglo-Allemand de Géomorphologie.* Paris: Les Belles Lettres, 229p.
Guilcher, A., 1958. *Coastal and Submarine Morphology.* London: Methuen. 274p.
King, C. A. M., 1972. *Beaches and Coasts.* London: Arnold, 570p.
Zenkovich, V. P., 1967. *Processes of Coastal Development.* Edinburgh: Oliver & Boyd, 738p.

Cross-references: *Beach; Beach Cycles; Beach in Plan View; Beach Material; Beach Processes; Beach Profiles; Beach Ridge; Geographic Terminology; Sand.*

PLEISTOCENE

The *Pleistocene* is the episode (epoch) of geologic time that immediately antedates the present, or *Holocene* (Flint, 1971). Collectively the Pleistocene and Holocene constitute the *Quaternary* Period. (In some usages the Quaternary is elevated to era status, in which case the Pleistocene and Holocene are considered periods.)

The Pleistocene was originally defined in 1839 to encompass strata whose content of fossil molluscs included more than 70% living species. Many modifications of this definition have been proposed. The currently favored definition is to assign the beginning of the Pleistocene as coeval with the onset of deposition of the Calabrian marine deposits in southern Italy (ca. 1,600,000 B.P.; Haq, Berggren, and Van Couvering, 1977) and its termination at the arbitrary but useful date of 10,000 B.P.

The Pleistocene is commonly equated with the glacial epoch, ice age, or Great Ice Age, all of which informally refer to the most recent episode of global-scale glaciation. However, there is now much evidence that indicates that the onset of glacial conditions was not globally synchronous but, rather, was attained in some areas (particularly high latitude and high elevations) in pre-Pleistocene time and was not evidenced in others until the Pleistocene was well advanced. Incorrect equation of the Pleistocene with the glacial epoch is a major reason for the broad range (350,000 to 3,500,000 yrs.; Cooke, 1973) in recent estimates for the duration of a glacially equated Pleistocene.

Many subdivision schemes for the Pleistocene have been offered by workers in various disciplines and regions. Among the most widely applied of these are subdivisions based on glacial advances and retreats, mammalian faunas, magnetic stratigraphy, evidence of man and his culture, and various intertebrate fossil taxa. These are subject to constant modification and revision, and almost all present problems of correlation.

In spite of problems of definition and correlation, the Pleistocene remains a time noteworthy for having witnessed major worldwide climatic fluctuations. These were attested to most prominently in the repeated buildup and degeneration of massive continental glaciers, but also in myriad effects experienced in nonglaciated regions. In the context of this volume, the Pleistocene is noteworthy in having been a time of extensive fluctuations in sea level, which resulted in the development of marine terraces and attendant features along many coasts, as well as in the development of the majority of the world's modern lakes and many of its extinct lakes through such activity as glaciation in the regions so effected and pluviation in many modern arid regions.

RONALD C. FLEMAL

References

Cooke, H. B. S., 1973. Pleistocene chronology: Long or short? *Quaternary Research* **3**, 206-220.
Flint, R. F., 1971. *Glacial and Quaternary Geology*. New York: John Wiley & Sons, 892p.
Haq, B. U.; Berggren, W. A.; and Van Couvering, J. A., 1977. Corrected age of the Pliocene/Pleistocene boundary, *Nature* **269**, 483-488.

Cross-references: *Flandrian Transgression; Holocene; Sea Level Changes.*

POLDER

Polder is a Dutch word originally meaning *silted-up land* or *earthen wall*, and generally used to designate a piece of land reclaimed from the sea or from inland water. It is used for a drained marsh, a reclaimed coastal zone, or a lake dried out by pumping. Drainage of lakes, swamps, and river valleys was carried out earlier in ancient Greece and Italy. Reclamation of land from the sea is of later date, and began during the Middle Ages in Friesland, Holland, Zeeland, and Flanders on the North Sea coast. Dike construction was developed to protect the land from the sea, and in the eleventh century the first real polders were made (Harris, 1957; Wagret, 1968). Gradually the practice spread to other parts of the Netherlands and water-lifting techniques were developed to keep the polders dry or to drain inland lakes. The windmill proved the most effective. From the sixteenth century onward land-reclamation and drainage projects were carried out in many West European countries. Polder dikes now form large parts of the coast of the Wadden Zee and other coastal waters between Belgium and Denmark. Mud flats are usually formed in front of a polder dike, so that new reclamations become profitable.

D. EISMA

References

Harris, L. E., 1957. Land drainage and reclamation, *in* C. J. Singer, E. J. Holmyard, A. R. Hall, and T. I. Williams, eds., *A History of Technology*, vol. 3. Oxford: Clarendon Press, 300-323.
Wagret, P., 1968. *Polderlands.* London: Methuen & Co., 288p.

Cross-references: *Bodden; Coastal Engineering, History of; Europe, Coastal Morphology; Land Reclamation.*

POLLUTANTS

Coastal water bodies constitute about 10% of the total oceanic area and include estuaries, lagoons, and inshore waters. Their properties are strongly influenced by boundaries with the continents and the sea floor. They receive direct injections of continental materials via rivers, terrestrial runoff and drainage, and the atmosphere as well as through domestic and industrial outfalls and ships.

Coastal waters are the sites of high rates of biological activity. The marine primary production of organic material (*photosynthesis*, which is the base of the food chain) takes place predominantly in these waters. The open ocean areas, with a few exceptions, such as some productive equatorial waters, are the marine deserts. Within the coastal zone there are some especially productive waters—upwelling areas—where a coupling of strong offshore winds with prevailing boundary currents bring nutrient-rich deep waters to the surface. Here, with even higher levels of primary productivity, fish stocks are large.

High rates of biological activity in coastal seas affect materials introduced from the continents. Some organisms have a remarkable ability to accumulate substances from seawater, even where the materials are in extremely low concentrations. Thus, organisms living in coastal waters may act as conveyors of man's wastes. They can move substances from surface waters to the sea floor through their death or the discharge of metabolic waste products, or they may return such materials to man in the form of food.

The persistence of chemicals in coastal waters, before removal by sedimentation, degradation, mixing with the open ocean, or harvesting of living organisms, may extend from months to decades. Estuaries exchange their waters with the open ocean in such periods of time.

The open ocean, which constitutes approximately 90% of the world ocean, differs from coastal waters in its time scales and its relationships with the continents and sediments. Whereas time spans describing natural processes in coastal waters extend from months to decades, periods associated with the open ocean are between hundreds and hundreds of millions of years.

Open-ocean sediments accumulate natural debris about a thousand times more slowly than do coastal deposits. Since biological activity is less intense in the surface waters of the open ocean than in those of the coastal ocean, there is a lesser potential for the downward transport of pollutants through biological activities. Most of the materials dispersed to the open-ocean environment through man's activities are still in the water column. Only a few have been taken up by the sediments.

As far as metals are concerned, sludge from a sewage treatment plant and top soil from a forest have much in common. Both are rich in the organic products of bacterial degradation that have the capacity to sequester metals. Erosion of the land results in the burdening of rivers with soil particles. When released to a stream, organic products from sewage sludge or from industrial processes replicate the behavior of top soil with respect to heavy metals. Thus, the organic particles of a stream, whether derived from natural or from man-made sources, are rich in heavy metals and can effectively extract much of the dissolved metal delivered to a stream.

Man's activities clearly have increased the flux of metals delivered to coastal waters by streams, sewer outfalls, and atmosphere. Some of this increase is due to the higher metal concentration of the particles delivered by streams, but most of it is due to the dumping of metal-rich dredge spoils from the harbors of the major cities. The highest metal concentrations of harbor sediments are found near sewer outfalls and outfalls from sewage treatment plants.

Another source of marine pollutants is power plants. Cooling is generally accomplished by locating a power plant near a body of water and passing that water through the power plant. The volume of water required for the cooling and the amount of heat rejected can be tremendous. Planktonic organisms, such as phytoplankton, zooplankton, fish eggs and larvae, the larvae of benthic organisms, and others, are all passed through the condenser cooling system. Some organisms may be affected primarily by chlorination, others by mechanical and pressure damage, still others by temperature increase.

It is estimated that by 1980 about one-third of all power plants in the country will be located on estuaries. Estuaries are nursery grounds for many coastal and oceanic species and are among the most productive ecosystems in the world. There is some concern about the ecological consequences—among other things, the increase in temperature of natural waters.

The ocean has long been considered a reservoir for man's wastes, but it is clear that in many areas of the world the capacity of the coastal ocean for such materials is at or past its limit. The multiple use of the coastal area include besides industrial and domestic waste disposal, recreation, shell fishing, fishing (commercial and sport), waterfowl hunting, supplying of industrial and municipal water, and navigation. Each of these uses has considerable economic value.

(Editor's Note: This article first appeared in the newsletter *Martek Mariner* and is reprinted here with the permission of the editor, Oscar P. Zabarsky.)

MARTEK INSTRUMENTS INC.

Cross-references: *Human Impact; Oil Spills and Pollution; Tar Pollution on Beaches; Thermal Pollution; Waste Disposal.* Vol. XIII: *Pollution, Geochemical Aspects.*

POLYPODIOPHYTA

This division of the plant kingdom contains the ferns and is referred to by some authors as the Pterophyta or Filicophyta. Three extinct orders of "preferns" are known from the

Devonian, as well as five orders of living "true ferns." The latter group is represented by abundant fossils dating back to the Carboniferous period. The great majority of living ferns belong to the Filicales order, divided into 14 or more families containing about 170 genera, with nearly 9000 species (Scagel et al., 1965).

Fern leaves are typically in the form of large, feathery fronds that bear spore cases (sporangia) on the under surfaces. In a few genera spores are borne on specialized stalks believed to be reduced fronds. The stem may be very small, sometimes existing only as an underground rhizome, but in the tropical tree ferns the stem is enlarged to form a tall columnar trunk. Reproduction involves alternation of generations between a sexual gametophyte stage and an asexual sporophyte phase. The gametophyte consists of a tiny green platelike *prothallium* connected to the soil by hairlike roots. The prothallium bears two types of sexual organs, male *antheridia* and female *archegonia*. Unlike flowering plants and conifers, the fertilized germ cell does not develop into a seed; instead, the embryo grows into the familiar fern plant. This asexual form matures and produces spores, which are then released to develop into a new generation of prothallia. Fertilization only occurs when the archegonia is surrounded by a film of water. Thus ferns are most commonly found in moist habitats. Some varieties are capable of vegetative reproduction, such as by fragmentation.

Polypodiophyta are particularly abundant in the wet tropics, and constitute an important part of the flora of rain forests. Many species are herbaceous, but the range of habitats is quite diverse, and ferns occur as mosslike fronds, climbers, creepers, and tree- and shrublike forms. A few of the Filicales and the Marsiliales occur as partially submerged aquatic plants, while the Saviniales are found as small, free-floating plants on ponds and lakes. These aquatic ferns commonly have wide geographic distribution because waterfowl ingest their spores. Most terrestrial Polypodiophyta have only regional distribution, but some species have widely disjunct ranges because of wind dispersal of their spores. For example, the bracken fern (*Polypodium aquilinum*) is an agressive cosmopolitan weed (Smith, 1955). A few ferns live in arid regions, and at least one species is adapted to dry portions of salt marshes. Ferns of warm regions remain evergreen, but in cool temperate localities the plants die down during the winter, though the roots remain alive and produce new foliage in the spring.

LOUIS CUTTER WHEELER*
GEORGE MUSTOE

References

Scagel, R. F.; Bandoni, R. J.; Rouse, G. E.; Schofield, W. B.; Stein, J. R.; and Taylor, T. M. V.. 1965. *An Evolutionary Survey of the Plant Kingdom.* Belmont, Calif.: Wadsworth Publishing Co., 658p.

Smith, G. M., 1955. *Cryptogamic Botany*, vol. 2, *Bryophytes and Pteridophytes.* New York: McGraw-Hill, 399p.

Cross-references: *Biotic Zonation; Coastal Ecology, Research Methods; Coastal Flora.*

*Deceased

PORIFERA

Sponges (phylum Porifera) are sessile aquatic Metazoa of simple organization (Figs. 1, 2). They are filter feeders and pump water by means of characteristic flagellated cells (choanocytes) (Fig. 3), which are usually arranged to form small spherical chambers scattered throughout the body. Water enters a sponge through small pores (ostia) and leaves through larger openings (oscula) that are connected to

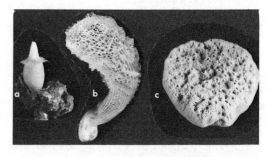

FIGURE 2. Representatives of the major sponge classes: *a. Sycon* sp. (Calcispongea) × 1.7; *b. Euplectella* sp. (Hyalospongea) × 0.3; *c. Spongia* sp. (Demospongea) × 0.4.

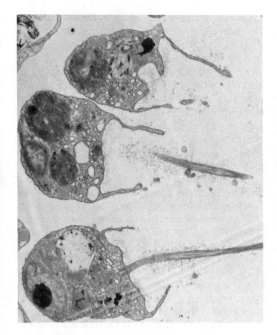

FIGURE 3. Portion of flagellated chamber lined with choanocytes (*Gelliodes* sp., Demospongea) × 7500 (transmission electron photomicrograph).

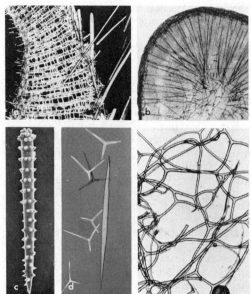

FIGURE 4. Sponge skeletons. *a.* Cross section through the body wall of *Sycon* sp. (Calcispongea); lattice of tri- and tetraradiate calcareous spicules, some large monactins protruding, × 27.5 (polarized light). *b.* Cross section through hemispherical *Stelletta* sp. (Demospongea); radial arrangement of silicious megascleres, × 27.5. *c.* Silicious spicule (acanthosty) of *Agelas* sp. (Demospongea) × 275 (scanning electron micrograph). *d.* Calcareous spicules of *Sycon* sp. (Calcispongea); monactin tri- and tetraradiates, × 68.8 (semipolarized light). *e.* Spongin fibers of *Spongia* sp. (Demospongea).

the choanocyte chambers by an incurrent and excurrent canal system. Most sponges have a skeleton of mineral (calcium carbonate, silicic acid) or organic (*spongin*) nature (Fig. 4), which is of great taxonomic importance. The skeleton supports a body that can be a mere crust a few millimeters in size or an irregular broad mass reaching 2–3 m in diameter. Sponges can assume a variety of shapes, including vases, tubes, and treelike branching rods. Every color of the spectrum, even some fluorescence, can be displayed by shallow-water sponges. The great plasticity in internal organization and the coordination of physiological events support the present view that sponges are individuals rather than colonies. Approximately 5000 species are known. Representatives of the phylum occur in all seas, depths, and climatic zones. The Spongillidae is the largest of the three families that occur in freshwater.

Evolution and Fossil Record

There are similarities between sponge choanocytes and certain choanoflagellates. *Proterospongia* is a colonial choanoflagellate but not acceptable as a transitory stage. Sponges are probably derived from flagellates that were also ancestors to choanoflagellates. Fossil sponges without fused mineral skeletons are generally poorly preserved. The first certain records are Hyalospongea from the Lower Cambrian. Demospongea appeared in the Middle Cambrian Period; Calcispongea not before the Devonian Period. Chert and flint are rocks that may be entirely composed of sponge *spicules*.

Classification

The principal criteria for separating the four classes of Porifera (Demospongea, Hyalospongea, Calcispongea, Sclerospongea) are the mineral composition and symmetry conditions in the skeleton. The largest class and the only one with quantitative importance in recent coastal environments are the Demospongea. The orders and families of this class are distinguished by structural features of the skeleton (major spicule types and spongin fibers) and by certain cytological and embryological characteristics. For the identification of genera and species the type, shape and size of spicules, details of the spongin meshwork, shape and pigmentation of individuals, consistency, surface structures, and arrangement and size of oscula and ostia are of great importance.

PORIFERA

The Role of Sponges in Coastal Environments

Geographic Distribution and Depth. The continental shelves of tropical seas are particularly rich in sponges. The Indo-West Pacific, Central Pacific, and West Indian regions are distinct faunal provinces. The northeastern coasts of South America, the northwestern coasts of Africa, and the Pacific coasts of Central America and California have affinities with the West Indies. The Mediterranean Sea has subtropical character with elements from the Red Sea. The Arctic Ocean has mainly a fauna of boreal Atlantic and Pacific species. The Antarctic is rich in sponges with many endemic species. The upper limit of depth distribution is the lower intertidal. Sponges can not survive desiccation, but some can grow under rocks or algal cover or in tide pools in the littoral zone. They can be abundant in the shallow zone of photophilous algae, but there they are often restricted to areas of low light levels (for example, to caves in the Mediterranean). Highest quantitative values can be obtained in the zone of sciaphilous algae 30–100m.

Habitats. Because all adult sponges are sessile, they depend on the availability of solid substrates. They occur in great abundance on artificial structures—fouling on buoys, pilings, harbor walls. Mangrove roots and sea grass rhizomes are colonized unless they are intertidal. Rocky coasts, coral reefs, and all other hard substrate formations show almost invariably rich sponge populations. Some species connect pieces of gravel, thus creating new habitats in high energy environments. Certain sponges grow on protected soft bottoms if small substrate fragments for initial settlement are available; others can develop root structures for stabilization.

Physicochemical Factors. These always act in combination and cannot be isolated entirely. Temperature influences the geographic and bathymetric distribution, the growth rate of individuals and of spicules, and the time of reproduction. Light controls the distribution within small areas. It favors growth of plants that are competing for substrate but also of algal symbionts that are an important food source for some sponge species. Light gradients in shallow caves can generate population sequences similar to those that occur with depth, only on a smaller scale. Hydrodynamic conditions influence the growth form of sponges and interact with sediments. Fine sediment in stagnant water can have a smothering effect; coarse sediments in agitated environments act as an abrasive. Sufficient water exchange is important also for food transport. The prevailing water movement determines the stability of small substrates. Substrate inclination influences light conditions and exposure to sedimentation. Most sponges are adapted to normal ocean salinity. Only a few (for example, suberitid and clionid) species have invaded brackish water. Silicon is one of the most important mineral substances. Its presence controls the size and number of spicules. Other minerals promote the growth of microorganisms that are needed for food.

Biological Interactions. Under suitable conditions intensive competition for space can develop among sponges and between sponges and other sedentary organisms. Fast-growing incrusting species can be successful competitors by overgrowing and killing their neighbors—other sponges, corals, and bryozoans (Fig. 5). Many cases of epibiosis are also known where sponges grow on other organisms (e.g. on algae, on sponges, on majid, dromiid, and pagurid crustaceans) without harming them, and even protecting them (Fig. 6a). On the other hand, sponges can serve as substrate for settlers like algae, hydroids, zoanthids, barnacles and bryozoans (Fig. 6b). Some species, particularly those with a large interior canal system (for example, of the genera *Hippospongia, Ircinia, Spheciospongia*) can harbor a wealth of endobiotic associates, mostly polychaetes, nematodes, and crustaceans (Fig. 6c). With a few exceptions this relationship is inquilinistic and the sponge host is not harmed. An association of great nutritional advantage to sponges is the symbiosis with bacteria and unicellular algae (Fig. 6d). Bacteria and Cyanophycea occur in great quantities in a variety of Demospongea species. Dinoflagellates (Zooxanthellae) are known only from some clionid species. Preda-

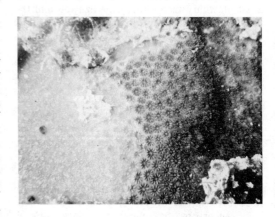

FIGURE 5. Spatial competition. Sponge (*Cliona* sp., left, two oscular openings visible) taking over live coral (*Montastrea* sp., right), × 1.

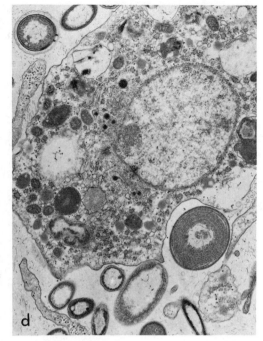

FIGURE 6. Symbiosis. *a.* Crab (*Dromia* sp.) carrying *Chondrosia* sp., × 0.7 (photo: K. Sanved). *b.* Zoanthid (*Parazoanthus* sp.) on *Gelliodes* sp., × 6. *c.* Shrimp (*Synalpheus* sp.) inside *Verongula* sp., × 0.8 (photo: K. Sanved). *d.* Bacteria and unicellular blue-green algae intercellular in *Verongula* sp.; one alga being engulfed by the large sponge cell, some vacuoles contain debris of previously digested symbionts, × 14,000 (transmission electron photomicrograph).

tors sometimes control the growth and distribution of sponge populations. Among these are endobiotic polychaetes and crustaceans, epibiotic mollusks (Nudibranchia) and several associated and nonassociated species of asteroid echinoderms and fishes.

Pollution. Since sponges function as powerful filter pumps, they could potentially purify water from bacteria and other filter organisms. However, they are sensitive to pollutants, and only a few species survive in eutrophicated water.

Commercial Sponges. Softness and absorbency have made the cleaned spongin skeletons of keratose species (genera *Spongia*, *Hippospongia*) a desired article for the household and for art and industry since antiquity. Sponge fishing centers were once located in the eastern Mediterranean, on the west coast of Florida and in the Bahamas. Because of overexploitation, disease, and introduction of plastic products, this industry has steadily declined during the past three decades. Attempts to cultivate sponges under controlled conditions were successful on a small scale but never gained commerical importance.

Bioerosion. A number of sponges, most of them belonging to the family Clionidae, are able to excavate limestone and to live in cavities or galleries formed by that activity. Water exchange is maintained through ostia and oscula, which are located on ectosomal papillae or incrustations at the surface of the substrate. Some species can reach a free, massive "gamma" stage after the available limestone support has been used up. The excavation is accomplished by special cell processes that free calcium carbonate fragments by etching around them. These chips are expelled through the sponge osculum, and form an important constituent of the mud-size fraction of many coastal sediments. An average clionid sponge population can erode as much as 250 gr of calcium carbonate per m^2/year. Excavating sponges cause most of the framework destruction in deep coral reefs (that is, below wave action). Clionid sponges have long been considered a pest in oyster cultures, where they erode the shells of the bivalves until the muscle supports break. Cultures in brackish water are safe; in other areas the sponge growth can be controlled by periodic exposures to low salinity.

Literature

A bibliography of early works, up to 1913, was presented by Vosmaer (1928). Later publications are listed in the continuing *Zoological Record*, published by the Zoological Society of London. The data summarized in the present article are found in greater detail in handbook contributions such as Hentschel (1923), Hyman (1940), Laubenfels (1955), Brien et al. (1973), and Bergquist (1978); and in symposium volumes edited by Fry (1970), Boardman, Cheetham, and Oliver (1973), Harrison and Cowden (1976), and Levi and Boury-Esnault (1979). The most recent summary presentations and some research papers that are particularly relevant to coastal ecology are by Reiswig (1973), Fell (1974), Rützler (1975, 1976, 1978), and Wiedenmayer (1977).

KLAUS RUETZLER

References

Bergquist, P. R., 1978. *Sponges*. Berkeley: University of California Press, 268p.
Boardman, R. S.; Cheetham, A. H.; and Oliver, W. A., Jr., eds., 1973. *Animal Colonies, Development and Function*. Stroudsburg, Pa.: Dowden, Hutchinson & Ross, 603p.
Brien, P.; Lévi, C.; Sarà, M.; Tuzet, O.; and Vacelet, J., 1973. Spongiaires, *in* P. P. Grassé, ed., *Traité de Zoologie*, vol. 3. Paris: Masson et Cie, 716p.
Fell, P. E., 1974. Porifera, *in* A. C. Giese and J. S. Pearse, eds., *Reproduction of Marine Invertebrates*, vol. 1. New York: Academic Press, 1-125.
Fry, W. G., ed., 1970. *The Biology of the Porifera*. London: Academic Press, Zoological Society of London Symp. No. 25, 512p.
Harrison, F. W., and Cowden, R. R., eds., 1976. *Aspects of Sponge Biology*. New York: Academic Press, 354p.
Hentschel, E., 1923. Porifera, *in* W. Kükenthal and T. Krumbach, eds., *Handbuch der Zoologie*, vol. 1. Berlin: Walter de Gruyter and Co., 295-418.
Hyman, L. H., 1940. *The Invertebrates: Protozoa through Ctenophora*. New York: McGraw-Hill, 726p.
Laubenfels, M. W. de, 1955. Porifera, *in* R. C. Moore, ed., *Treatise on Invertebrate Paleontology*, Part E. New York: Geological Society of America, 21-112.
Lévi, C., and Boury-Esnault, N., eds., 1979. *Biologie des Spongiaires*. Paris: Centre National de la Recherche Scientifique, 533p.
Reiswig, H. M., 1973. Population dynamics of three Jamaican Demospongiae, *Bull. Marine Sci.* **23**, 191-226.
Rützler, K., 1975. The role of burrowing sponges in bioerosion, *Oecologia* **19**, 203-216.
Rützler, K., 1976. Ecology of Tunisian commercial sponges, *Tethys* 7, 249-264.
Rützler, K., 1978. Sponges in coral reefs, *in* D. R. Stoddard and R. E. Johannes, eds., *Coral Reefs: Research Methods*. Paris: UNESCO, 299-313.
Vosmaer, G. C. J., 1928. *Bibliography of Sponges, 1551-1913*. Cambridge: Cambridge University Press, 234p.
Wiedenmayer, F., 1977. *Shallow-Water Sponges of the Western Bahamas*. Basel and Stuttgart: Birkhäuser, 287p.

Cross-references: *Algal Mats and Stromatolites; Biotic Zonation; Coastal Fauna; Coastal Waters Habitat; Coral Reef Coasts; Coral Reef Habitat; Mangrove Coasts; Organism-Sediment Relationship.*

POSTGLACIAL REBOUND

Postglacial rebound is a special case of *isostatic adjustment,* which is the vertical displacement of the earth's surface due to removal or buildup of loads on or near the earth's surface. In the general isostatic case, the earth's outer layers are viewed as behaving hydrostatically, as though they were floating on some compensation layer within the earth's *mantle*. Thus, any given column of the earth's outer layers tends to be in equilibrium so that it rises to a height proportionate to its mean density. If this equilibrium is disrupted by load additions or subtractions so that the mean density is altered, a given column will tend to move vertically in proportion to the density change until equilibrium is restored. In practice, only large scale changes (that is, sedimentary buildup of large deltas, water buildup in large reservoirs) have produced measurable displacements.

Postglacial rebound refers specifically to the vertical rise in a land surface caused by the removal of glacial and/or glacial water masses from it, such as occurred with the melting of the late Pleistocene ice sheets (Andrews, 1974, 1975). Rebound from this latest deglaciation has aggregated in excess of 300 m in some areas and continues at maximum rates of up to 1 cm/yr.

In general, if a glacier is in existence long enough to cause the earth's surface beneath it to be depressed into equilibrium, the total amount of rebound to be expected from the melting of the ice is proportionate to the thickness of ice removed, in a ratio nearly equal to the density of ice and the density of the earth at the compensation level (approximately 1:3). Such relationships can be used to estimate ice thicknesses if rebound amounts are known, or reversed to estimate rebound amounts if ice thicknesses can be determined.

Rebound may elevate shoreline features to considerable elevations in areas so affected; marine limits lie at 285 m above present sea level in eastern Hudson Bay and are still rising (Farrand, 1962). Moreover, since rebound amounts differ from area to area, an originally horizontal datum such as a strandline may be warped by differential rebound. Strandlines on the Great Lakes have been warped as much as 100 m, and rebound is not yet complete. Mapping of warped strandlines is useful for estimating the amount of differential rebound which has occurred in a given area.

RONALD C. FLEMAL

References

Andrews, J. T., ed., 1974. *Glacial Isostasy*. Stroudsburg, Pa.: Dowden, Hutchinson & Ross, 491p.
Andrews, J. T., 1975. *Glacial Systems*. North Scituate, Mass.: Duxbury Press, 191p.
Farrand, W. R., 1962. Postglacial uplift in North America, *Geog. Bull.* 17:5-22.

Cross-references: *Emergence and Emerged Shoreline; Holocene; Isostatic Adjustment; Isostatically Warped Coasts; Pleistocene; Present-Day Shoreline Changes, United States; Present-Day Shoreline Changes, Worldwide; Raised Beach; Sea Level Changes;* Vol. III: *Postglacial Isostatic Rebound.*

POTRERO

Potrero is a local name—derived from a Spanish word for "pasture"—for an elongate, grassed, sandy island, one of a chain of six extending for about 23 km near the western edge of the central part of Laguna Madre in south Texas. They are immediately seaward of the great South Texas Sand Sheet or Eolian Plain and surrounded by mud flats.

The potreros range in size from about 3-4 km in length and from about .3-1.2 km in width (see Potrero Lopeno NW, Potrero Cortado, and Yarborough Pass; USGS 7 1/2-minute quadrangles.) They are comprised, as discussed by Price (1958) and Fisk (1959), of parallel stabilized *dune ridges* from less than 1 m to about 9 m in height.

The islands probably formed from, and are currently accreting by, sand blown westward across the Laguna Madre mud flats when the flats are intermittantly drained by seasonal low tides and strong onshore winds (*meteorological tides*). The accumulation of some of the potreros may have started around remnants of previously drowned and eroded eolian topography of the Sand Sheet to the west.

SAUL ARONOW

References

Fisk, H. N., 1959. Padre Island and the Laguna Madre mud flats, coastal south Texas, *in* R. J. Russell, ed., *Proceedings of the Second Coastal Geography Conference*. Washington, D.C.: National Academy of Sciences, National Research Council, 103-151.
Price, J. A., 1958. Sedimentology and Quaternary geomorphology of south Texas, *Gulf Coast Assoc. Geol. Socs. Trans.* 8, 41-75.

Cross-references: *Barrier Islands; Chenier and Chenier Ridge; Coastal Accretion Forms; Coastal Dunes and Eolian Sedimentation; Geographic Terminology.*

PRESENT-DAY SHORELINE CHANGES, UNITED STATES

About 75% of the population of the United States live in states bordering the oceans and the Great Lakes with a total of about 134,400 km of beaches and shorelines. Sediment movement (onshore, offshore, and along the shore) as a result of action by waves causes alternate erosion and accretion of the shorelines. The nature and type of change are generally dependent on weather conditions, the seasons, the direction and intensity of the wave action on the shore, and the type of material that forms the shore. In 1971, the U.S. Army Corps of Engineers reported that 36,900 km, or 24%, of the nation's total shorelines were seriously eroding. Of these, about 4900 km were classified as critical, and 32,000 km were classified as undergoing noncritical yet serious erosion.

Causes of Shoreline Changes

The major causes of present-day shoreline changes in the United States include: storm waves; changes in sea level; and man's activities.

Wave action is the most important factor affecting sediment movement along United States shorelines. Tidal currents are generally negligible in influencing sediment transport in the United States, but may be a major factor in other parts of the world. Owing to changes in the direction of wave approach, sand transport along the shore may be reversed in terms of hours, but most reversals in the direction of longshore drift are seasonal. Such reversals could also be cyclical in nature as a result of major shifts in weather patterns. For example, the unusually cold winter of 1977 in the eastern half of the United States resulted in reduced beach erosion along the mid-Atlantic shore. The major cause seems to have been a shift in weather patterns resulting in fewer severe storms and a reversal in the usual winter longshore transport direction.

Shoreline changes occur all the time as a result

of normal processes during fair weather conditions. However, severe storms and hurricanes can result in changes, in a few hours, that are equivalent to a hundred years of normal shore processes. The great Atlantic coast storm of March 6-8, 1962, caused tremendous damage along the Atlantic coast from New York to North Carolina. Although the winds associated with the storm did not reach hurricane force, it caused serious damage because it lasted through three successive high tides (Kenney, 1962). Offshore bars, barrier beaches, sand dunes, and man-made defenses suffered great damage. A similar storm on December 17, 1970, was estimated to have caused a minimum of 10.1×10^6 cubic yards of sand to move from the beach above MSL along 800 km of ocean front between Cape May, New Jersey, and Race Point, Massachusetts (DeWall, Pritchett, and Galvin, 1977).

The effects of hurricanes and severe storms along the Atlantic and Gulf coasts were studied extensively by the author (El-Ashry and Wanless, 1968; El-Ashry, 1971). These were found to include: erosion of the ocean side of the beaches, washover fan formation on the landward side of the barrier islands or in the lagoons behind them, the opening of new inlets and the closing of others by beach drifting, and cumulative hurricane effects that are most pronounced on low exposed beaches, cape points, and inlets. Material eroded from the beaches during storms may be moved in the direction of dominant longshore currents or offshore to form submerged bars or shoals; finer material may get trapped in strong seaward-moving rip currents and move out onto the continental shelf.

Eustatic rise in sea level can explain some of the long-term and slow recessions of the shorelines. Fairbridge (1961) mentioned that some of the long-term eustatic changes observed involve 100 m (vertical in 200,000 years, which averages 0.5 mm/year; current rise in sea level, measured by tide gauges, is 1.2 mm/year. A rise in sea level of up to 10 mm annually, continuing over many centuries, has been recorded on the coasts of northern Canada and parts of Scandinavia. In discussing the effect of sea-level rise on erosion along a coast, Bruun (1962) assumed that for any rise in sea level the bottom of the nearshore littoral drift zone will be raised simultaneously until it is covered with the same depth of water at the same distance from the (new) shoreline as it was before the rise (see *Bruun Rule*). He concluded that the material needed to raise the bottom comes from the corresponding shore area through the movement of material by rip and diffusion currents. This was experimentally proved by Schwartz (1965) in scale-model and field studies on Cape Cod.

During the past decade, above average precipitation (33 cm above normal) has occurred in the Great Lakes region, resulting in higher lake levels and causing extensive erosion and flooding. Of the 5920 km of United States shoreline on the Great Lakes, erosion is occurring on about 2080 km and another 536 km are being flooded. The increased erosion results from waves being able to break farther shoreward of former beaches and bars and directly against bluffs, in addition to the fact that much eroded material is washed into deeper water rather than being carried along the shore (Hartford and Tanner, 1976).

Man's activities that interfere with normal shore processes and have adverse effects in the coastal zone include: up-stream dams on coastal rivers, resulting in a reduction of sediment contributed to the beaches; jetties for inlet and harbor maintenance and improvements; bulkheads and seawalls to protect property on the upper part of the beach from wave attack; breakwaters for protecting harbors and providing shelter for boats; groins for the interruption of longshore sand movement and accumulation of sand on the shore or retardation of sand loss; and construction in general that results in the destruction of natural dune systems that are a major supply of sediment for the beaches.

Under natural conditions, shorelines develop their own natural defenses to cope with extreme sea conditions. Gradually sloping beaches dissipate most of the wave energy, dunes absorb the energy of storm waves and offer protection to the land behind them, and barrier islands act as a front-line defense affording complete protection of lands along the main coastline. Continued encroachment in the coastal zone with man-made development has in many cases taken place without regard for existing natural defenses. Dunes are leveled and graded to make room for buildings, and barrier islands are modified in many ways for resort and recreational facilities. Where such developments were left unprotected, great damage has resulted from storms, and artificial protection methods were sought.

MOHAMED T. EL-ASHRY

References

Bruun, P., 1962. Sea-level rise as a cause of shore erosion, *Am. Soc. Civil Engineers Proc., Jour. Waterways and Harbors Div.* 88, 117-130.

DeWall, A. E.; Pritchett, P. C.; and Galvin, C. J., 1977. *Beach Changes Caused by the Atlantic Coast Storm of 17 December 1970*. Washington, D.C.: Coastal Engineering Research Center, Tech. Paper 77-1, 80p.

El-Ashry, M. T., 1971. Causes of recent increased erosion along United States shorelines, *Geol. Soc. America Bull.* **82**, 2033-2037.

El-Ashry, M. T., and Wanless, H. R., 1968. Photointerpretation of shoreline changes between Capes Hatteras and Fear, North Carolina, *Marine Geology* **6**, 347-379.

Fairbridge, R. W., 1961. Eustatic changes in sea level, in L. H. Ahrens, ed., *Physics and Chemistry of the Earth*, vol. 4. New York: Pergamon Press, 99-185.

Hartford, F., and Tanner, W. F., 1976. Current Great Lakes shore damage, *Shore and Beach* **44**, 16-19.

Kenney, N. T., 1962. Our changing Atlantic coastline, *Natl. Geog. Mag.* **122**, 860-877.

Schwartz, M. L., 1965. Laboratory study of sea-level rise as a cause of shore erosion, *Jour. Geology* **73**, 528-534.

Cross-references: *Bruun Rule; Climate, Coastal; Coastal Engineering; Coastal Erosion; Human Impact; Hurricane Effects; Present-Day Shoreline Changes, Worldwide; Sea Level Changes, 1900 to Present; Storm Wave Environments; Waves.*

PRESENT-DAY SHORELINE CHANGES, WORLDWIDE

Recent efforts to determine worldwide patterns of shoreline change have been stimulated by growing property losses from such coastal problems as erosion, flooding, and storm surges. To be sure, much of the damage stems directly from the growing intensity of construction and development along coastlines, which are among the most hazardous of earth's environments. But losses in many locations have been inexplicably large, raising speculation that recent changes in geomorphic forces may be partially responsible.

Data Collection Methods

The irregular oscillations of shorelines cause difficulties for anyone attempting to classify coastal dynamics at any scale. The waves of a single hurricane, for example, can remove decades of accretion, changing a growing beach into an eroded one in just a few hours. The many agents of coastal change operate in a sporadic and somewhat unpredictable fashion. For this reason, most coastal classifications have been done with a relatively long time scale in mind, spans of some few thousand years. Their conclusions are based on indirect physical evidence—cliffs imply erosion, while parallel dune ridges mark progradation.

Within the past few decades, however, the availability of modern air photographs, combined with highly accurate maps from the nineteenth century (even earlier for Britain and Denmark) has made possible the accurate determination of recent shoreline changes through the comparison of past shoreline positions with those of the present. Owing to limitations of the data, however, most studies have been determinations of linear changes, with conclusions expressed in *meters of erosion (or deposition)* at sea level. Consequently, these studies might be more accurately termed *measures of shoreline advance or retreat*, rather than the determinations of volumetric erosion or deposition that would result from comparisons of entire shore profiles. The latter technique, although becoming increasingly feasible, has been limited to studies of comparatively small areas.

Extreme Shoreline Changes

The most rapid shoreline prograding occurs along depositional river deltas. Rates of 80 m/year at the southwest pass of the Mississippi delta and 60 m/year at the Irrawaddy in Burma are among the extremes recorded. Volcanic eruptions add many square kilometers periodically to such land areas as Iceland and Hawaii, while earthquake-associated vertical displacements of land surfaces also cause significant additions to coastal lands.

The most rapid erosion, not surprisingly, occurs where loosely consolidated materials are exposed to catastrophic storm or tsunami waves. Rates of cliff erosion of up to 1 m/day and 100 m/year have been recorded in newly deposited volcanic material in the Aleutian Islands, Hawaii, and Iceland, and in glacial drift environments of the Siberian Arctic. Erosion rates of several meters per hour can result when beaches and dunes are assaulted by hurricane waves. At Miami, for example, the beach receded about 150 m between 1884 and 1944, with about 80 m of that occurring during a 1926 hurricane.

Worldwide Trends

Worldwide, slowly eroding beaches and cliffs are the general rule. Verified cases of steadily prograding shore locations are rare and isolated, virtually all confined to river deltas, sandspits, cuspate forelands, or where people have interfered with beach sand movement by building jetties, groins, or breakwaters.

Measurements verify that virtually all cliffed coasts are receding, although many highly resistant cliffs exhibit little change over many decades. Rates of retreat in excess of 1 m/year are considered rapid. The chalk cliffs of the French north coast, for example, vary in average recession rates from 0.1-1.0 m/year, with measurements dating to 1825. Similar rates of retreat have been found to exist along the coastal cliffs of the western United States.

Lower rates have been reported along the northern shore of the Black Sea (up to 0.25 m/year) and the other low-energy coasts.

The continuing erosion of sandy shores is widespread around the world; virtually no extended segment of beached coast is unaffected. Erosion prevails on sparsely settled coasts in Brazil, southern Africa, and Australia; as well as along heavily populated shores in the United States and Britain. Moreover, there are numerous reports, from Ghana, the western United States, eastern Mexico, the Mediterranean, and other locations of presently eroding beaches that had been prograding until recent decades.

This worldwide erosion is too much to be accounted for by recent sea level rise (about 1 mm/year) and is too widespread to be blamed entirely on human interference in the coastal system or on any general increase in rough weather. Russell (1967) hypothesized that large quantities of shore zone sands were brought to their present coastal positions from the sea floor by the actions of a rising sea level during Holocene times. He concluded that the present 5000-year standstill in sea level may have brought about a slight disequilibrium in which normal sand losses are not quite replaced by the remaining sources of sand supply—rivers and cliff erosion.

The continuing erosion of shorelines is a matter of growing concern around the world. Data on recent changes are being compiled by a commission of the International Geographical Union headed by E. C. F. Bird of the University of Melbourne.

There is a large body of literature on this subject. The following is a partial list recommended for further reading: Bird (1973, 1974, 1976), Kidson (1976), Koike (1974), Schwartz (1967), Shepard and Wanless (1972), Steers (1964), and Zenkovich (1958).

JAMES E. STEMBRIDGE, JR.

References

Bird, E. C. F., 1973. *Physiographic Changes on Sandy Shorelines in Victoria within the Past Century.* Melbourne: University of Melbourne, Department of Geography, 31p.

Bird, E. C. F., 1974. Coastal changes in Denmark during the past two centuries, *Aarhus Univ. Lab. fysisk geog. Skr. fysisk geog.* **8**, 21p.

Bird, E. C. F., 1976. *Shoreline Changes during the Past Century: A Preliminary Review.* Melbourne: University of Melbourne, Department of Geography, 54p.

Kidson, C., 1976. *Bibliography 1971-1974, Working Group on Dynamics of Shoreline Erosion.* Aberystwyth, U.K.: University College of Wales, Department of Geography, 127p.

Koike, K., 1974. Preliminary notes on recent changes of sandy shorelines in Japan, *Geog. Rev. Japan* **47**, 719-725.

Russell, R. J., 1967. Aspects of coastal morphology, *Geog. Ann.* **49A**, 299-309.

Schwartz, M. L., 1967. The Bruun theory of sea level rise as a cause of shore erosion, *Jour. Geology* **75**, 76-92.

Shepard, F. P., and Wanless, H. R., 1972. *Our Changing Coastlines.* New York: McGraw-Hill, 579p.

Steers, J. A., 1964. *The Coastline of England and Wales.* London: Cambridge University Press, 750p.

Zenkovich, V. P., 1958. *The Black and Azov Sea Shores.* Moscow: State Geographical Publications, 373p.

Cross-references: *Classification; Coastal Engineering; Coastal Erosion; Coastal Morphology, Research Methods; Human Impact; Hurricane Effects; Present-Day Shoreline Changes, United States; Sea Level Changes, 1900 to Present; Sea Level Curves.*

PRIAPULIDA

The phylum Priapulida includes eight species in six genera of cylindrical, benthonic animals. Enclosed in a chitinous cuticle that is periodically molted, the unsegmented body is divided into an anterior proboscis and a posterior trunk, from which caudal appendages of uncertain function extend in some species (Fig. 1). The eversible spines surrounding the mouth are invaginated at the anterior of the proboscis. The proboscis may also be drawn into the trunk; alternate protrusion and retraction of the proboscis is the major method of locomotion. The superficially annulated trunk holds the fluid forming the hydrostatic skeleton and contains the straight digestive tract, gonads, and excretory organs. The fluid-filled cavity was thought to be bounded by an acellular membrane, indicating that it was a pseudocoel as possessed by the Aschelminthes, including nematodes, rotifers, kinorhynchs; other larval and adult features show similarites to rotifers and kinorhynchs. Shapeero (Barnes, 1975) found that the membrane is cellular, suggesting that the priapulids are coelomates; more work in the developmental biology is necessary to clarify the phylogenetic relationships.

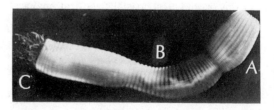

FIGURE 1. A large (120 mm) living specimen of *Priapulus caudatus* in relaxed condition. *A:* proboscis, *B:* abdomen, *C:* caudal appendages (from Van der Land, 1970).

Most priapulids are carniverous, capturing sluggish polychaetes and other slow-moving invertebrates with the barbs around the mouth. Some can tolerate great salinity fluctuations. They live at depths from the intertidal to 7500 m, generally in poorly oxygenated muddy sediments. Priapulids either lie in the mud with their mouths open at the sediment water interface or plow through the sediment in search of prey. The tiny (1-2 mm) *Tubiluchus* is anomalous for it deposit-feeds in coralline sands in warm tropical waters; the larger predaceous priapulids occur in greatest number in high latitudes. In some areas they are common, but are subject to large fluctuation in abundance. Van der Land (1970) suggests that their restriction to high latitudes and seemingly unfavorable substrates is a result of inability to compete successfully in more hospitable environments.

MOLLY FRITZ MILLER

References

Barnes, R. D., 1974. *Invertebrate Zoology*, 3rd ed. Philadelphia: W. B. Saunders Co., 870p.

Van der Land, J., 1970. Systematics, zoogeography, and ecology of the Priapulida, *Leiden Rijksmus. Nat. Hist. Zool. Verh. 112*, 118p.

Cross-references: *Aschelminthes; Biotic Zonation; Coastal Fauna; Coral Reef Habitat; High-Latitude Coasts; Organism-Sediment Relationship.*

PRODUCTION

Production is defined as growth of organic matter per unit of time. It can be subdivided into *primary* and *secondary production.*

Primary Production

Primary production describes the initial formation of organic matter by autotrophs (UNESCO, 1966). In the sea this is accomplished mainly by phytoplankton, macroalgae, bottom-living diatoms and eelgrasses, using inorganic nutrients and solar radiation as energy source. During assimilation, carbon is taken up from carbon dioxide, and this process is used to measure primary production (Raymont, 1980). Plankton samples from different light levels are poured into glass bottles together with a ^{14}C-labelled substance. After submergence to their original depth or in experimental tanks for 6-12 hours, the plankton is filtered and the activity of the incorporated ^{14}C is measured (*gross production*). Dark bottle tests allow the calculation of respiration. Gross production minus respiration is net production. Similar experiments can be conducted with the formation and the uptake of oxygen. Production by small algae can be determined in the same manner; in large algae, growth and degradation must be measured by size or weight changes (Mann, 1972). Production of algae and eelgrasses as well as that of diatoms covering sand or mud flats may exceed phytoplankton production in shallow waters. Primary production ranges from a few milligram under a square meter of sea surface to maximum values of 10g $C/m^{-2}/day^{-1}$, or to 300 g $C/m^{-2}/year^{-1}$ in highly productive upwelling areas (Ryther, 1963). Production in benthic diatoms was found to reach 100g $C/m^{-2}/year^{-1}$ Of the benthic algae, the fucoids derive the highest production rates, with 2 kg $C/m^{-2}/year^{-1}$. Mann (1972) estimated the annual production in the seaweed zone to average 1.75 kg $C/m^{-2}/year^{-1}$. Eelgrass meadows produce about 200 g $C/m^{-2}/year^{-1}$.

Secondary Production

Secondary production, based on available organic matter derived from primary production, is the transfer of organically bound energy. Of this energy, 70-90% is lost between two successive trophic levels. Phytoplankton and algae are fed upon by herbivores, and these in turn are used as food by the organism groups following in the food chain—the carnivores and scavengers. Others, including the bacteria and fungi, live on dead organic matter or on a variety of food sources. The determination of secondary production uses short-term respiration measurements of individuals or of associations on the sea bed (Zeitzschel, 1981). In the long-term surveys, the life cycle of a species is studied with monthly observations on *standing stock,* age, structure, size, and weight of individuals, from which recruitment, growth, mortality, and production can be calculated. Production maxima for net- and microzooplankton under 1 m^2 surface area amount to 100 and 150 g $C/year^{-1}$. In the sediment bacteria, protozoa and meiofauna reach several g $C/m^{-2}/year^{-1}$ each, and the macroinfauna produces up to 500 g $C/m^{-2}/year^{-1}$. Epifaunal associations, especially those with bivalves, exhibit a much higher production. Natural mussel banks produce the highest values, with 1 kg $C/m^{-2}/year^{-1}$.

Production may exceed standing stock (*biomass*), and production per biomass (P/B) within a unit of time is termed turnover rate. The time necessary for one turnover (P = B) is the turnover time. Bacteria have a short turnover time (hours), while that of protozoa ranges from days to weeks, and that of metazoa from weeks and months to years. Turnover times are roughly size-dependent (Fenchel, 1974).

High production rates of organic matter cause high sedimentation rates where currents are not too strong and fine-grained particles settle. The amount of organic matter in sediments is of importance in diagenetic processes.

Productivity is frequently used synonymously with *production*. This should be avoided since productivity is *potential* production—the maximal sustainable production under optimal and (in some publications) experimental conditions (see Ryther, 1963).

HJALMAR THIEL

References

Fenchel, T., 1974. Intrinsic rate of natural increase: The relationship with body size, *Oecologia* **14**, 317–326.

Mann, K. H., 1972. Ecological energetics of the seaweed zone in a marine bay on the Atlantic coast of Canada. II. Productivity of seaweeds, *Marine Biology* **14**, 199–209.

Raymont, J. E. G., 1980. *Plankton and Productivity in the Oceans,* vol. 1, *Phytoplankton,* 2nd. ed. Oxford: Pergamon Press, 489p.

Ryther, J. H., 1963. Geographic variations in productivity, *in* E. Goldberg, ed., *The Seas.* New York: Pergamon Press, 347–380.

UNESCO, 1966. *Determination of Photosynthetic Pigments in Seawater.* Paris: UNESCO Monographs on Oceanographic Methodology 1, 69p.

Zeitzschel, B., 1981. Field experiments on benthic ecosystems, *in* A. R. Longhurst, ed., *Analysis of Marine Ecosystems.* London: Academic Press, 607–625.

Cross-references: *Aquaculture; Benthos; Biotic Zonation; Coastal Ecology, Research Methods; Coastal Fauna; Coastal Flora; Nutrients; Organism-Sediment Relationship; Standing Stock.*

PROFILE OF EQUILIBRIUM

Swell waves cause accretion of the beach so long as material is available from offshore to feed the incoming breakers. When the offshore zone seaward of the breaker line deepens and steepens sufficiently the swell can no longer bring material in, and so the whole profile from berm to offshore becomes reasonably static. This swell-built *summer profile* will exist for most of the year on ocean margins. During storms the beach soon becomes saturated, and backwash increases. The return flow is sediment-laden and travels outward near the sea bed. In combination with strong shoreward drift at the surface, this gives rise to the construction of an *offshore bar,* which continues to grow until its crest is high enough to steepen and break the majority of the incoming storm waves. Once the wave dissipation process develops fully, erosion of the beach almost ceases. Any successive storm in the same season is unlikely to cause denudation of the beach to any further degree, since the offshore bar, perhaps partly returned, is soon reformed to carry out its role of wave dissipator. Although partial replenishment will be affected soon after the abatement of the storm, it may take three or four weeks after the last storm of the season for the hump of the storm profile—*winter profile*—to be removed.

The *profile of equilibrium* is usually referred to as a long-term profile of ocean bed produced by a particular wave climate and type of coastal sediment. Accordingly, there are swell-built summer and storm-built winter profiles of equilibrium. Sitarz (1963) has developed equations, based upon energy principles, for profiles beyond the breaker line to the limit of active bed disturbance. The offshore profile is given by $x = ky^2$, where $k = 0.95/[(s-1)^{1/2} DH^{3/2}]$, and where x is the distance seaward from the breaker line, y is the depth below SWL, s is the ratio density of sediment/(density of seawater), D is the median sediment diameter at some appropriate depth (mm), and H is the wave height just offshore from the breaker zone (m). The width of the surf zone (Z) is given by $Z = BH^{3/2}/[(s-1)^{1/2} D^{1/2}]$, with D of beach face near the SWL. For the swell-built profile the factor B has the value 43.5, while for the storm profile it is 75.0. The cross-sectional area of water in the storm beach surf zone is

$$A = (3.4 g^{1/2} H_{1/3}^2 T_{1/3})/[(s-1)^{1/2} D^{1/2}],$$

where $T_{1/3}$ is the significant wave period, $g = 981$ cm/sec^2, D is the median sediment diameter across the surf zone, and $H_{1/3}$ is the significant wave height.

The offshore bar formed during winter may take some months to be swept completely back to the beach. Not until this is accomplished can the cross section be considered a full-fledged swell profile. Further, the bed is certainly not in equilibrium while material is still actively being removed.

A number of equilibrium profile characteristics have been given recently by Swart (1974). From his extensive study it follows that the exponent 2 in the equation $x = ky^2$ should be replaced by $1.36 \cdot 10^4 D$ (in m). The width Z is determined explicitly by H, D, and the deep-water wave steepness. The equilibrium beach slope at the upper limit of the equilibrium profile increases with increasing particle diameter, while that at the lower limit decreases.

In a first approximation, the equilibrium profile under three-dimensional conditions is equal to that under corresponding two-dimensional conditions. Transport rates and other profile characteristics for three-dimensional cases are also given by Swart (1974).

Equilibrium profiles of coarse material under wave attack have been studied by Hijum (1974).

A contribution to the formation of equilibrium profiles due to sea level rise has been made by Schwartz (1967).

RYSZARD ZEIDLER

References

Hijum, E. van, 1974. Equilibrium profiles of coarse material under wave attack, *in Conf. Coastal Eng., 14th, CERC, Am. Soc. Civil Engineers, Copenhagen, Summaries,* 385-388.

Schwartz, M. L., 1967. The Bruun Theory of sea level rise as a cause of shore erosion, *Jour. Geology* 75, 76-92.

Sitarz, J. A., 1963. Contribution à l'étude de L'evolution des plages à partir de la consistance des profile d'équilibre, *Centre Etude Rech. Oceanog. Trav.,* 10-20.

Swart, D. H., 1974. *Offshore Sediment Transport and Equilibrium Beach Profiles,* Delft Hydr. Lab. Pub. No. 131, 302p.

Cross-references: *Beach Cycles; Beach Processes; Beach Processes, Monitoring of; Beach Profiles; Equilibrium Shoreline; Profiling of Beaches; Sweep Zone.* Vol. I: *Profile of Equilibrium.*

PROFILING OF BEACHES

Beach profiles are lines surveyed *perpendicular* to the shoreline in order to record the form of the beach at a given moment in time. The form of the beach is a reflection of the action of littoral processes on the unconsolidated sediments, and a *time series* of profiles can be used to explain the interactions of the processes and the morphology. Profiles from different environments can be used similarly to indicate the effects of different elements on the materials of the littoral zone.

Profiles are usually surveyed from the low-tide mark to the landward limit of sediment redistribution by storm waves or by eolian processes. Frequently the profiles are continued seaward by use of echo sounders to include the morphology of the adjacent nearshore zone. The data derived from these surveyed lines can be used: to illustrate the morphology of the beach, to record changes in the profile or the plan shape of a beach, or to record volume changes along a section of coast. In all cases where more than one profile is surveyed, the profiles should be tied to the same reference plane, such as a geodetic point or a temporary benchmark, so that direct comparisons between different data sets are possible.

Profiling Techniques

Survey lines are usually run from a fixed point, such as a stake, in the backshore to the low water line or into the surf zone when possible. Profiles should be surveyed at low tide so that the line can be extended as far seaward as possible. A number of techniques of differing accuracies are available for profiling, and the choice of the most suitable method depends on the required accuracy or on budgetary or time constraints.

Optical Method. The most precise surveys require the use of a level. This instrument measures the height difference between two points "by observing a horizontal line of sight on graduated staves held vertically over the points" (Cheetham, 1965). Accuracies greater than ±0.5 mm are possible over a 1000-m line if standard survey procedures are followed. This technique may be time-consuming if such levels of accuracy are required, but the introduction of self-leveling levels now allows rapid setup and adjustment of the instrument.

Visual Method. An alternative technique has been developed specifically for beach profiles. This involves measurement of the height difference between two graduated rods by leveling on the horizon (Emery, 1961). The observer aligns his eye with the top of the seaward rod and the horizon. The distance from the top of his own rod to the intersection with the line of sight is the elevation difference between the two rods (Fig. 1). The line of sight between the observer's eye and the horizon is not exactly horizontal owing to the curvature of the earth, and it is possible to compensate for this by using correction tables. This technique has the advantage of being rapid and requiring no expensive equipment. Accuracies of ±2.0 cm over 500-m lines

FIGURE 1. Pole-and-horizon method of surveying beach profiles. The change in elevation between the two graduated vertical poles is measured by aligning the top of the seaward pole with the horizon and reading the intercept value on the landward pole. In this example the staffs are graduated in 2-cm increments and the elevation difference is -9 cm.

can be obtained by this method. If the horizon is not visible, a builder's spirit level strapped to a rod or pipe held between the survey poles can be used to provide a horizontal line (Schwartz, 1967).

Angular Method. A similar inexpensive method has been developed by Harrison and Boon (1972). This employs an articulated frame that may be used by one man. This technique involves measurement of a slope angle over a fixed distance by use of a protractor and bubble level attached to the frame.

Stake Method. Changes in beach elevation at fixed points can be measured by use of stakes set along a line across the beach. Once the stakes are installed, the heights of the tops are surveyed in to each other, and then the distance from the top of each stake to the beach surface is measured. This method has the advantage of being rapid and accurate and also allows direct comparison of elevation changes at fixed points.

Cliff Profiles. If profiles of the backshore are required that involve survey lines on steep slopes, the measurement of elevation differences can be carried out with a theodolite or tachometer or by use of a clinometer, such as a hand level. These instruments are more practical on steep gradients since they do not require traverses across the slope, though it is necessary to have one person at the top and one at the bottom of the profile to be surveyed.

Subaqueous Profiles. Extension of the beach profile into the adjacent nearshore zone involves use of a boat and some form of depth recorder. The most efficient technique is to use an echo sounder that provides a continuous trace of the bottom as the boat proceeds along a survey line. It is important to tie the nearshore profile to the beach profile so that they both refer to the same datum. This can be achieved by installing a marker in the nearshore zone that indicates the seaward limit of the beach profile and the landward limit of the nearshore profile. It is easier if the beach profile is surveyed at low tide and the nearshore profile is run at high tide. A simple datum correction obtained at the stake allows correction of the depth values to the low water datum. (See also Schwartz, 1967).

Profile Networks. If more than one profile line is to be surveyed on a beach, it is usual to set up in the backshore a line of semipermanent benchmarks or stakes that can be used as a baseline. Single profiles or repetitive profiles can then be surveyed with reference to the same datum plane. Regularly spaced profile lines along a beach can be used to prepare maps of the section as well as provide height information across the beach. If a semipermanent survey network is established, repetitive profiles can then be used to establish a time series of beach maps.

Profiles Through Time

For reconnaissance studies, one profile or set of profiles across a beach provides data on that beach at a given moment in time. More detailed studies may require systematic profiling. Daily, weekly, monthly, or yearly surveys of a profile provide a more representative sample because they include changes in the form of the beach over time. This type of profiling provides data on temporal beach changes on short-term (one tidal cycle) or long-term (seasonal) scales. Thematic profiling is used to study form changes on a beach in relation to a specific event, such as a storm. Profiles before and after a storm reflect the action of changes in the level of wave activity. A set of profiles surveyed at intervals after a storm can be used to show how a beach recovers following erosion (Davis et al., 1972). From a comparison of profiles surveyed through time, it is possible to calculate volume changes between the surveys.

Profiles in Space

A single profile across a beach provides a point sample that may not be representative of the total system. Alongshore variations in morphology can be determined from a series of regularly spaced profiles surveyed with reference to the same datum plane. A single profile is essentially a two-dimensional record of the beach morphology, whereas a set of profiles surveyed on a systematic grid can be used to build a contour map or a three-dimensional model of the section. Beach maps surveyed through time can then be used to determine form changes and to compute volume changes that can be related to variations in the process environment. The accuracy of a contour map drawn from profiles depends on the interval between survey lines and on the number of elevation values along each line. The spacing of data points along and across the beach will vary according to the dimensions of the study area and the scale of features on the beach.

Presentation of Profile Data

Two-dimensional profiles are usually drawn with a *vertical exaggeration* that distorts the scale in order to emphasize elevation changes. The magnitude of the vertical exaggeration ranges between X 4 and X 20, depending on the relationship between the beach width and the local relief. Single profiles may be plotted to provide an indication of beach morphology.

Superimposed profiles can be used to indicate temporal changes at one site or spatial variations at one time. An example of superimposed profiles is given by the *sweep zone* (q.v.). In a more simple manner, superimposed profiles can be used to demonstrate beach changes with reference to a single event (Owens and Rashid, 1976) or to a sequence of variations associated with longer term processes such as poststorm recovery (Davis et al., 1972) or shoreline retreat.

Contour maps or three-dimensional diagrams are used to illustrate morphologic features along the beach and, when more than one data set is available from the same site, can be used to locate areas of erosion or deposition and to compute volume changes. Davis and Fox (1972) have used profile data to construct a four-dimensional model of the beach to illustrate an area-time sequence.

EDWARD H. OWENS

References

Cheetham, G., 1965. *Texbook of Topographical Surveying*. London: Her Majesty's Stationery Office, 388p.

Davis, R. A., Jr., and Fox, W. T., 1972. Four-dimensional model for beach and inner nearshore sedimentation, *Jour. Geology* 80, 484-493.

Davis, R. A., Jr.; Fox, W. T.; Hayes, M. O.; and Boothroyd, J. C., 1972. Comparison of ridge and runnel systems in tidal and non-tidal environments, *Jour. Sed. Petrology* 42, 413-421.

Emery, K. O., 1961. A simple method of measuring beach profiles, *Limnology and Oceanography* 6, 90-93.

Harrison, W., and Boon, J. D., III, 1972. Beach-water-table and beach-profile measuring equipment, *Shore and Beach* 40, 26-33.

Owens, E. H., and Rashid, M. A., 1976. Coastal environments and oil spill residues in Chedabucto Bay, Nova Scotia, *Canadian Jour. Sci.* 7, 908-928.

Schwartz, M. L., 1967. The Bruun Theory of sealevel rise as a cause of shore erosion, *Jour. Geology* 75, 76-92.

Cross-references: *Bars; Beach in Plan View; Beach Processes, Monitoring of; Beach Profiles; Coastal Characteristics, Mapping of; Profile of Equilibrium; Sweep Zone; Time Series.*

PROGRADATION AND PROGRADING SHORELINE

Johnson (1919) wrote, "Just so long as the current aggrades (builds up) the seabottom offshore, the waves will prograde (build forward) the shore. Following Davis we may call any shore which is experiencing such a long-continued advance into the sea, a *prograding shore* and distinguish it from the more usual retreating or *retrograding* shore" (Johnson's italics). (Davis does not appear to have used the word prograde in print prior to Johnson's definition, *aggrade* being used instead). A wave-built beach ridge is the basic structure for a number of prograded land forms: dunes, chenier plain, and spit and barrier beaches (q.v.). The factors controlling the stability or otherwise of these features can be summarized as time, energy, sedimentary supply, organic growth, and sea level change. *Time* allows partial or complete equilibrium between all other factors. *Energy,* in the form of wind-, wave-, and tide-generated currents, controls grain size, direction of sedimentary transport, and speed at which equilibrium is attained. Energy also determines the degree of temporary progradational and retrogradational change, but all other factors being stable, it is the net *rate of sedimentary supply* that determines the net rate of progradation. *Organic growth* is important in fixing wind-blown dune sands on top of the wave-constructed beach ridges, and helps to raise estuarine mud deposits up to high spring tide level.

The effect of a rising sea level on *coastal retrogradation* has been recognized only comparatively recently. The recognition of the converse effect of a falling sea level in promoting coastal progradation is more recent still. Quantitative studies of a number of essentially closed, coastal sand systems in the Auckland region of New Zealand (Schofield, 1975a, 1975b) have shown a linear relationship between degree of sea level fall and amount of coastal progradation. Of equal importance is the fact that mineralogical studies clearly show about 95% or more of the prograded sand has been derived from the sea floor.

J. C. SCHOFIELD

References

Johnson, D. W., 1919. *Shore Processes and Shoreline Development*. New York: Wiley & Sons, 584p.

Schofield, J. C., 1975a. Beach changes in the Hauraki Gulf, 1965-68: Effect of wind, sea-level change, and off-shore dredging, *New Zealand Jour. Geology and Geophysics* 18, 109-128.

Schofield, J. C., 1975b. Sea-level fluctuations cause periodic, post-glacial progradation, South Kaipara Barrier, North Island, New Zealand, *New Zealand Jour. Geology and Geophysics* 18, 295-316.

Cross-references: *Apposition Beach; Beach Processes; Beach Ridge and Beach Ridge Coasts; Chenier and Chenier Plain; Coastal Accretion Forms; Emergence and Emerged Shoreline; Major Beach Features; Retrogradation and Retrograding Shoreline; Sea Level Changes.* Vol. III: *Prograding Shoreline.*

PROTECTION OF COASTS

It has long been known that the best natural protection of the coast is a wide beach (Bruun, 1972; Matthews, 1934). Wherever the beach becomes deficient in sediments, washed away to the bottom by waves and currents or to adjacent stretches of shore by alongshore drift flows, the destruction of the primary coast is stepped up. Coastal destruction proceeds slowly where the submarine slope is gentle and faster where it is steeper because of the differences in the degree of wave deformation, which diminishes the energy of the waves.

Coastal erosion is caused mainly by wave action but also to some extent by currents (tidal and storm surges) (Minikin, 1950). Along a high primary coast, waves undercut the cliff base and force it to retreat. In the surf zone, coarse beach sediments (boulders, pebbles, and gravel) suffer attrition. At a low coast, washout and destruction occur as a result of storm surges, usually during high tide. Occasional floods inflict heavy damage. In cold climates, ice pushing onto the shore is another factor of destruction (Coastal Engineering Research Center, 1973).

A marked increase in coastal destruction has resulted from people moving closer to the shore zone, the building of ports, and the development of recreation facilities. A large part of the shoreline of the United States and many western European countries has been so exploited. Also, the intensified use of water for farming and the dams erected on many rivers have caused a decrease runoff of rivers. Consequently, the coastal areas with beaches composed of alluvial material have been sharply reduced. Destruction is even more evident where beach sediments have been widely used for building purposes (Zenkovich and Kiknadze, 1981).

The natural causes of coastal destruction remain rather obscure. Various authors have attributed it to: a continuing slow rise of the world ocean level, at a rate of 10–15 cm/100 yr; the coming of geomorphologic maturity to most of the globe's coasts, with a resulting decline in abrasion—a major source of beach feeding in many areas; and a heightened overall storminess of the sea, probably owing to climatic changes over the past few decades.

The beginnings of man's attempts to deter coastal destruction date back several centuries. Protective dikes have been built in Holland since the sixteenth century and in England since the seventeenth century. However, in the past few decades both the need for, and the techniques of, coastal protection have changed considerably. Progress in the knowledge of shore zone dynamics, including the underwater belt, as well as in the development of new engineering and building facilities and materials correspondingly has expanded the methods and techniques of coastal protection. Engineering intuition and the experience gained over the centuries have also played major roles. Nevertheless, the trial-and-error approach often has to be resorted to in choosing a protection method and structural design and material.

Methods of coastal protection differ greatly for a pebble beach and a sand beach. The choice depends also on whether there is a unidirectional sediment flow along the part of the shore being protected and on the magnitude of the contraposed sediment migration. A flow may only be formed along a relatively smooth coast. Where the coast is indented, the protection may be local, but on a smooth shore attention must be paid to the likely consequences, over long distances, of any engineering action (Knaps, 1960).

Coastal protection may be passive or active. Passive methods simply protect the existing coast from direct impact of the surf waves and include the building of dikes, rubble mounds, and seawalls of various shapes (straight, concave, with berm blocks at the base, or with reinforced bulkheads); as well as revetments for sloping coasts and above-water or submerged breakwaters built to fence off particularly valuable parts of the coast (Townson, 1974).

Active protection involves the more difficult task of providing conditions for stabilizing or expanding the existing beach width or even of artificial beach accumulation. To achieve such conditions, wise use has long been made of groins, which are a type of variously structured short moles. Series of groins are built in places of an expected natural supply, or artificial accumulation, of beach sediment. Breakwaters too in a certain sense are a means of active protection, since, in their wave shadow, sediments are accumulated, leading to a widening of the beach and a shallowing of the nearshore bottom (Bozhich and Dzhunkovsky, 1949).

Each type of coastal protection has advantages and disadvantages. *Seawalls* assure a strong and safe defense for the adjacent land, but to some extent they always reflect the surf waves. As a result, the interaction of swash and backwash, which maintains the constant beach width, is disturbed. The swash length is shortened, giving more strength to the backwash, which alters the beach profile and narrows the beach.

When the seawall is approached by oblique waves, the velocity of the longshore drift is greatly increased. The longshore water flow at the wall may attain more than 6 m/sec in velocity, with a corresponding increase in its carrying capacity. Consequently, the beach in front of the wall shrinks even more than it does when waves approach it directly (Zhdanov, 1962). The damage can be reduced by a mound of figurate concrete blocks (tetrapods and the like) piled in front, or instead, of the wall. By breaking the wave, the blocks dampen the backwash. Waves grinding at the base of a wall fronting a pebble beach will shatter the whole structure in a short time. Numerous examples are found along the Black Sea, the rate of destruction sometimes being as high as 0.5 m/yr.

Mounds of large boulders or concrete blocks of different shapes are sometimes used to protect a sand beach where there is no cliff. Soviet experience suggests, however, that this can be only a temporary measure. Boulders sink into the sand, and through the openings between the blocks backwash carries sand away into the sea. The shore continues to retreat, and eventually the blocks sink completely below the sediment (Zenkovich, 1958).

Submerged breakwaters are another major means of weakening the wave impact. On tideless seas, the breakwater ridge may be submerged 0.5-0.8 m. Apart from weakening the impact at the shore, these breakwaters produce an additional effect owing to a fan of sediments accumulating at the bottom behind them. The fan usually joins the above-water beach accumulation and adds much to the overall system stability.

Breakwaters of this type do have a shortcoming, though. With oblique waves, a strong longshore current may arise between the breakwater and the shore. Such a current is capable of eroding the bottom as well as of carrying away sand and pebbles. To combat this condition, the breakwater is connected with the shore by concrete traverses spaced at tens or hundreds of meters. Systems of this kind are widespread on the Soviet Black Sea. Accurate special calculations have been developed to assess the appropriate breakwater ridge depth, distance from the shore, and layout of traverses.

Diverse *dikes* are built along low shores to protect them from getting flooded. The area on the outside should be paved with revetments set on concrete plates. The outer slope of the dike should be gentle (1:4, 1:6) to enable swash water to rush up easily. For beach defense, tetrapods or other figurate blocks are sometimes piled at random at the base of a dike. If the area is to be used for recreation, *groins* are built for conserving or widening of the sand beach.

Where the shore land is not particularly valuable and the uprush is not extremely high, simple protection may be made of large (up to 1.5 m in diameter) plastic casings of any length. On site, the casing is filled with sand and cement mortar pumped into it. When filled, the casing fits closely to the rugged features of the bottom surface and quickly sets into a monolith. Strung out along the area to be protected, such defense pieces prove effective under certain conditions (West Germany, the Azov Sea in the USSR). Groins may also be built in this manner.

The design and size of, and the spacing between, groins depend on the average and maximum wave parameters, their angle of approach, sediment composition, nearshore depth, and tidal range. The parameters of groins can be calculated accurately. However, wave-basin model testing is recommended. Along sand shores in Holland and West Germany, for example, groins made up to 200 m long are spaced by one to four groin lengths. Along the pebble beaches of the Caucasus, groins 30 m or so in length, with a spacing of 1 to 1.25 of their length, prove to be effective.

Groins are usually straight in plan and protrude normal to the shoreline. Sometimes they are put at an acute angle to the shoreline, are bent at the middle, or are T-shaped. They may be made up of pilings (usually at sand beaches), concrete blocks, or earth. Metal piles are subject to rapid *corrosion* in saltwater. Wooden structures should not be used in places abounding in boring mollusks and worms. Concrete groins made up of large blocks are laid down on a stone bed and bordered on the outside front and edges with berm plates. Earthen groins are widespread in Holland. These are made of a clay core over which boulders, first small and then larger, are strewn. Cracks between boulders in the surface layer are filled with asphalt. A groin of this kind is smoothly arcuate in section. Groins of rubble mounds or of figurate concrete blocks are practical and safe, but disfigure the landscape and are seldom used near cities or resorts.

Groin-building techniques have developed quickly in many countries. In the USSR there is a convenient design of a groin made of pipelike concrete columns driven into the sea bottom by vibration. These columns are connected by concrete shields inserted into special grooves (Tsaturyan, 1972). Such groins, unlike any other design, suffer virtually no deformation during occasional washouts of bottom and shore.

It is essential that the top of the groin, whatever the design, be elevated over the natural beach surface. Otherwise the sediment will be carried by waves from one pocket into another. Along the Black Sea, groin tops at pebble beaches are raised up to 4.5 m above the still-water level.

Where sediment drifts are heavy, groins are efficient, but they may cause downdrift washouts along the lee side. Many instances can be cited where groins, accumulating sediment drifts, have caused major shore destruction many kilometers downdrift, as along the Black Sea's Caucasian coast southeast of Sochi and south of Ochamchire (pebble beaches). On sand beaches, washouts resulting from this cause have been observed in the Kaliningrad region of the USSR, in Poland, and in East Germany. Similar processes have been investigated along shore in West Germany and England. When the intergroin spaces become filled with sediment and, with the continuing general advance of the shoreline, sediment flow or migration in front of the groin is renewed so that the temporary coast-damaging effect of the groins gradually declines to complete disappearance.

At smoothly embayed coasts, beaches can be created and maintained by a single *jetty* set up opposite the headlands.

Port structures jutting out into the sea (moles, breakwaters) play a major role in both defense and destruction of the nearby coasts (Fig. 1). They obstruct sediment drifts. The downdrift area, deprived of its sediment supply, is destroyed, sometimes over long stretches. As a result of updrift sediment accumulation, the port inlet may become subject to siltation. As a countermeasure and a means of coast protection, *sediment bypassing* using pumps is practiced. The method was first used in the United States and Mexico. At present, it is at work in India (for example, at the Paradip Port) and in other countries.

With advances in dredging and refilling operations, it has become economical to widen or create a beach without any fencelike structures. Although a portion of the sediment is lost with this method, it proves worthwhile at resorts when used for periodic beach refeeding. For such a beach, sediments are usually taken with a dredge from the adjacent shoal bottom. Care must be taken that dredging depths are properly selected so as not to damage the beach. Less frequently, sediments have to be brought in from outside. It is essential for the sediments to be coarser, and at least not finer, than those in the natural beach. Modern equipment can dredge sand from a bottom depth of down to 60 m

FIGURE 1. Polish port along a coast composed of loose Pleistocene sediment. The shoreline position of some seventy years before construction of the moles is indicated by the white dashed line.

and pipeline it to a distance of up to 6–8 km.

Sediment refilling currently has been used in the United States to create and maintain a beach approximately 150 km long. In Germany, a major project of this sort has been implemented on Norderney Island. In the USSR, successful and economically justified beach building has been carried out in several bays on the Black Sea (at Gelendzhik and Planerskoye) and on the Sea of Azov (Albulatov and Pogodin, 1974).

Dunes are a major natural feature, protecting the nearshore land with sand beaches from floods. Dunes are formed naturally at areas of continuous sand supply and predominant shoreward winds. Sometimes a dune barrier has to be protected or created by planting vegetation having strong roots. Brushwood or reed fences, usually laid out in a checkered pattern, are also used. After sand accumulates, the fences are pulled up above the sand surface and the operation is repeated until the desired increase of the dunes is attained (Bülow, 1954).

Rapid storm surges damage the front edge of a dune, but thereby produce a reserve of sand that keeps the beach width constant and prevents the coast from retreating.

Along smooth shores, where local coastal protection may have a harmful effect over long stretches, overall schemes and plans specifying the succession (in time) of coastal protection measures for tens and hundreds of kilometers must be drawn up. This approach ensures the least expense and the best protection of beautiful natural landscapes. Planned coastal protection enables a control of coastal processes to be implemented and the necessary changes to earlier projects to be introduced. Such schemes in the USSR are underway along the Black Sea and the Baltic Sea (Zenkovich and Kiknadze, 1981).

Complex coastal protection combining several techniques is regularly used with the best results, particularly near cities and resorts. For example, near Odessa, where the coast is clayey and abrupt, groins enclosing sand beaches have been built along a stretch of more than 10 km. Submerged breakwaters, with openings for outflowing surge water, have been built between groin heads. Sand has been transported by artificial means from the sea to a distance of about 30 km. Over seven years only about 5% of the sand has been lost (Dodin and Ponomarenko, 1972).

At steep underwater pebble slopes of the Crimea (Fig. 2), groins (some with submerged breakwaters) are also used. Beaches are built on chips of rock transported from elsewhere. The backbeach is bordered by a wall and a quay. At some places groins extend to depths of 8 m.

FIGURE 2. A stretch along the south Crimea coast on the Black Sea reinforced with a groin series and quay-bordered by a seawall. The beach is artificial.

The shore as a whole has been advanced to a width of about 20 m (Chuzhmir, 1968). Both in Odessa and the south Crimea, coastal protection has prevented landslides. Similar methods of coastal protection are in use in the Caucasus. Owing to smaller near-shore depths, groins there may be up to 70 m long (near Sochi).

There are two major facts that must be mentioned with regard to engineering coastal protection. In the surf zone, any structure is subject to destruction and needs regular repair. Over decades, the coastal pattern may change, causing the earlier useful structures to become harmful. Dismantling them may cost no less than repairs. Engineering structures in most cases pollute the water and reduce the attraction for recreation. Therefore, artificially nourished beaches without engineering structures are actually more economical (Grechishchev, Morozev, and Shul'gin, 1972). Over long periods of time and for long stretches, they cost less than engineering protection. Moreover, these beaches help conserve the natural shore in its original form. By planning nourishment or protective structures for long stretches of the shore, it is easier to comply with the rule laid down at the dawn of coastal protection engineering, which requires that any structure should be built along the shoreline without sharp angles or protrusions. The whole pattern of the shore should be *streamlined* so as to prevent a concentration of the destructive energy of waves and storm surges on short areas (Bruun, 1954). Engineers should remember to take into account the specifics of the coast's geomorphology and dynamics when choosing a protection method and the potential effects on adjacent coastal areas.

Further suggested readings are Shul'gin and Zubarenkova (1974), Sokol'nikov, Tsaits, and Khomitsky (1974), and Zhdanov (1963).

VESEVOLOD PAVLOVICH ZENKOVICH

References

Alibulatov, N. A., and Pogodin, N. F., 1974. Dynamics of "free" artificial beach in a sheltered bay (in Russian), *Okeanologiya* 3, 505-511.

Bozhich, P. K., and Dzhunkovsky, N. N., 1949. *Sea Waves and Their Action on Structures and Coasts* (in Russian). Moscow: Mashstroizdat Publishers, 385p.

Bruun, P. M., 1954. *Coast Stability*. Copenhagen: Danish Society of Civil Engineers Press, 7 pamphlets.

Bruun, P. M., 1972. The history and philosophy of coastal protection, *in Conf. Coastal Eng., 13th, Vancouver, Canada, Proc.,* 33-74.

Bülow, K., 1954. *Allgemeine Küstendynamik und Küstenschutz an der südlichen Ostsee*. Berlin: Akademie Verlag, 87p.

Coastal Engineering Research Center, 1973. *Shore Protection Manual*. Washington, D.C.: U.S. Army Corps of Engineers, 750p.

Chuzhmir, A. A., 1968. Theory and practice of complex reinforcement of the coasts of the Crimea and Odessa (in Russian), *in Inzhenernaya zashchita beregov Chernogo marya*. Kiev: Kiev University Press, 16-19.

Dodin, V. V., and Ponomarenko, V. V., 1972. Dynamics of artificial beaches in the conditions of Odessa (in Russian), *in Geologiya poberezhya i dna Chernogo morya No. 6*. Kiev: Kiev University Press, 145-154.

Grechishchev, E. K.: Morozov, L. A.; and Shul'gin, Ya. S., 1972. Artificial sand beaches and organization of their aggradation (in Russian), *in Ukreplenie morskikh beregov*. Moscow: Transport Publishers, 60-68.

Knaps, R. Ya., 1960. On methods of strengthening of sand shores (in Russian), *in Proektirovania i stroitel' stvo beregoukrepitel'nykh sooruzhenii*. Moscow: Transport Publishers, 58-91.

Matthews, E. R., 1934. *Coast Erosion and Protection*. London: C. Griffin & Co., 228p.

Minikin, R. R., 1950. *Wind, Waves, and Maritime Structures*. London: C. Griffin & Co., 216p.

Shul'gin, Ya. S., and Zubarenkova, G. G., 1974. The problems of coast protection and beach preservation along the Black Sea Caucasus part, *Coordinating Hydrotech. Sessions Trans.* (*Leningrad*) 92, 8-70.

Sokol'nikov, Yu. N.; Tsaits, E. S.; and Khomitsky, V. V., 1974. *Coast Protection with Rock Mass Blankets* (in Russian). Kiev: Ukranian Akademy of Sciences, Naukova Dumka Publ., 702p.

Townson, J. M., 1974. *History of Coastal Engineering*. London: Dock and Harbour Authority, 324p.

Tsaturyan, G. A., 1972. Groins on pipe-like columns (in Russian), *in Ukreplenie morskikh beregov*. Moscow: Transport Publishers, 96-109.

Zenkovich, V. P., 1958. *Coasts of the Black Sea and the Azov Sea* (in Russian). Moscow: Geografgiz Publishers, 374p.

Zenkovich, V. P., and Kiknadze, A. G., 1981. The sea coast investigations in Georgia, *in Man and Nature in Geographical Science*. Tbilisi: Georgian S.S.R. Academy of Sciences, 50-72.

Zhdanov, A. M., 1962. Methods of sea coast reinforcement and their present development (in Russian), *Okeanogr. Kom. an SSR Trudy* 10, 113-122.

Zhdanov, A. M., 1963. On basic problems of coast protection at the Black Sea (in Russian), *in Morskie beregoukrepitel'nykh sooruzhenii*. Moscow: Transport Publishers, 5-31.

Cross-references: *Breakwaters; Coastal Engineering; Coastal Engineering, History of; Coastal Engineering, Research Methods; Lakes; Soviet Union, Coastal Morphology.*

PROTOCHORDATA

This group of invertebrate chordates contains several types of organisms generally given the status of individual subphylla within the Chordata. Included are the Urochordata and Cephalochordata and possibly the Enteropeneusta. The basic characteristic shared by these organisms is the presence of a dorsal hollow nerve cord, the *notochord,* during at least part of the life cycle.

Strictly marine, the Urochordata contains 2000 species divided into three classes on the basis of position of the reproductive glands and nature of the gills (Brien and Drach, 1948). Most abundant is the class Ascidacea, the *sea squirts* that live attached to solid substrates. These individuals typically resemble a leathery sack several centimeters or more in height and ranging in hue from red or brown to nearly colorless. The classes Thaliacea and Larvacea are specialized pelagic or free-swimming forms and include only about 100 species. The Urochordata (*tunicates*) have a tadpole-like larva that possesses a notochord. However, the free-swimming larva soon settles to the sea floor and becomes transformed into a barrel-shaped adult in which the notochord is absent. The organism's protective outer wall consists of tunicin, a substance similar to cellulose that is produced by no other animal. Tunicate blood is rich in vanadium and sulfuric acid but contains no respiratory pigment for transporting oxygen. Feeding occurs as the tunicate circulates seawater through a complex internal gill where plankton is trapped by the mucus-covered membranes. Tunicates are hermaphroditic, possessing both ovaries and testes, but also reproduce by budding. Distribution is worldwide, from polar latitudes to the tropics and ranging in depth from the intertidal zone to the deep abyss. Evolutionary status of the Urochordata is unclear, and they may represent either an original chordate ancestor or a product of degenerative evolution at a later stage (Barrington, 1965).

Cephalochordata, the marine lancets, are small fishlike creatures that rarely exceed 5-6 cm in length and are found burrowing in coarse sands in subtemperate and tropic oceans. Two genera are recognized, Amphioxus (=Branchiostoma) and Asymmetron, containing 29 species. Unlike the case in the Urochordata, the notochord is present in the adult stage. Their anatomy generally resembles the higher chordates, but is simpler, including the absence of brain, heart, and capillaries. As in the tunicates, the blood lacks a respiratory pigment. Sexes are separate, and breeding occurs in the early spring. Cephalochordates appear to be a divergent form that arose early in the evolution of the Chordata rather than being an original ancestor. The unusual gill structure, which acts as a filter during feeding, suggests they may have evolved from an early sedentary organism similar to the tunicates.

The Enteropeneusta (acorn worms or tongue worms) include about 80 species of soft-bodied marine animals that burrow in sand or mud (Hickman, 1973). The cylindrical body may be from several centimeters to more than one meter in length. The digestive tract contains an anterior outpouching that resembles a notochord, causing the group to be classified by some zooologists as a separate subphyllum within the Chordata. However, they are often placed as a class within the phylum Hemichordata. Together with the phyla Chaetognatha and Pogonophora, the Hemichordata comprise a problematical group that appears to be related to the Echinodermata as well as to the Chordata. The well-known *graptolites,* abundant as Paleozoic fossils, are sometimes classed within the Hemichordata, although their taxonomic position is uncertain owing to the scarcity of specimens in which anatomical details are preserved.

GEORGE MUSTOE

References

Barrington, E. J. W., 1965. *Biology of Hemichordata and Protochordata.* Edinburgh: Oliver and Boyd, 176p.

Brien, P., and Drach, P., 1948. Procordes, *in* P. P. Grassé, ed., *Traité de Zoologie,* vol. 11. Paris: Masson et Cie, 535-553.

Hickman, C. P., 1973. *The Biology of Invertebrates,* 2d ed. St. Louis: C. V. Mosby Co., 757p.

Cross-references: *Chordata; Coastal Fauna; Organism-Sediment Relationship.*

PROTOZOA

Because of their small size, protozoa are among the least known of the animal inhabitants of coastal environments. Indeed, until about half a century ago, beach sands were thought to be poor in life of any kind. However, in the mid-twenties certain Russian investigators reported that coastal sands sheltered a varied and numerous fauna and flora, and that the former included some highly specialized protozoa. They also introduced the term *psammon* by which these mostly minute organisms are now collectively known. Psammon corresponds to the name *plankton,* long used for the more minute fauna and flora of lakes and seas.

Protozoa, though less numerous than bacteria in coastal environments, occur in beach sands and tidal zones. Hundreds, even thousands, have been counted in a single cubic centimeter of sand collected on the ocean shore. Abundance depends on many factors, among which are the amount of organic material present, moisture at different depths, exposure to wave action and tides, temperature (which may obviously vary greatly with the seasons) and the size of sand grains, since this determines the size of interstices between them. Depth is also an important factor, the number of organisms being greatest at or near the surface.

The most numerous protozoa of beaches and intertidal zones are almost always the Foraminifera (Foraminiferida), especially in the tropics. This is a vast group of amoeboid protozoa (Sarcodina), almost all of which are limited to marine environments. Typically, they have shells of calcium carbonate with many chambers, usually spirally arranged in the same plane, but sometimes almost serially. The size varies greatly even within the same species, since the shell is always started as a single chamber, known as the proloculum. The name is derived from the numerous openings in the partitions separating the chambers, which enable protoplasmic communication to be maintained between them. There may also be foramina in the shell wall itself from which pseudopodia can be extruded. Some Foraminifera are large, and one genus (*Parafusulina*) may be 70 mm long. A giant among all the rest is *Neusina,* sometimes 190 mm in diameter. Foraminifera are a group of great antiquity: fossilized shells occur in large numbers in chalky limestones, such as the famous White Cliffs of Dover, and may even be found in blackboard chalk (Manwell, 1968).

A few Foraminifera are naked or have a very thin chitinous shell (for example, some species of Allogromidae). Although most of these live in freshwater, some are marine, and these are most numerous in tropical waters.

The Foraminifera are a large group. There are at least 50 families (Cushman, 1955), and these are said to include more than 700 genera and an uncertain, though large, number of species.

Amoebae (also Sarcodina) are a far smaller tribe. Although freshwater species are best known, and probably more numerous, some are marine and may occur in, or in the neighborhood of, beach sands. Some can adapt, though slowly, if changed from saltwater to freshwater, or vice versa.

There are also the Labyrinthulidae (order Protomyxida, of the Sarcodina). These organisms are not well known, but some are of real importance as parasites of eelgrass, which grows in shallow water along the ocean shore and affords food and shelter to numerous marine animals. Some years ago a peculiarly destructive variety of *Labyrinthula macrocystis* caused widespread injury to the eelgrass of the United States Atlantic coast, although it also attacks other hosts, including algae. Some of the species in this family also have flagellate stages. (Manwell, 1968).

Oddly enough, true flagellates (Mastigophora) seem relatively uncommon in coastal environments, although they may certainly be found in abundance in the brackish water of coastal wetlands and in tidal pools that are not too ephemeral. These flagellates are all members of the two subclasses, the Phytomastigina (plantlike forms) and the Zoomastigina (animallike forms).

Dinoflagellates make up one of the largest groups in the first two of these subclasses. Most of these are pelagic, but some may always be found in coastal waters; the freshwater and brackish water species may be found in rain and tidal pools, coastal wetlands, and the freshwater bogs that usually abound near coasts. Since almost all contain chlorophyll, they carry on photosynthesis and are of immense importance in the food chain, in common with other green flagellates and algae.

Dinoflagellates are curious forms, with two flagella; one of these is transverse and vibrates in a beltlike groove (the annulus), and the other is directed posteriorly and moves in a longitudinal groove (the sulcus). They are either in a cellulose shell (theca or test), often elaborately marked and bizarre in shape, or naked (*unarmored*). Shelled species are often brown (even though they contain chlorophyll), but the unarmored forms are usually beautifully tinted in a variety of pastel hues.

Among the dinoflagellates often found in coastal waters are the bioluminescent forms belonging to the genera *Gonyaulax* and *Noctiluca*. These can add greatly to the charm of an afterdark boat ride, since the water glows with each dip of an oar. But some species of the two genera have been incriminated as the cause of the *red tides,* which are apparently the result of occasional and sudden bursts of multiplication, triggered by still poorly understood changes in environmental conditions (Hutner and McLaughlin, 1958). When the population density reaches a certain point, exo- or endotoxins are released into the seawater, which kill great numbers of fish. Although these substances are not toxic to molluscs, these animals may nevertheless absorb enough of the toxins to render them poisonous for human consumption and may remain so for some months. Red tides have occurred in many parts of the world.

Other Mastigophora, much less common in saltwater than the dinoflagellates, include many species of the Phytomonadida, small, roundish, green flagellates with from one to a few flagella and a few species of the Euglenoidida, of which an example is *Eutreptia marina*. Like most euglenids, *Eutreptia* has a delicately striated pellicle and orange-red stigma (eye spot); it is spindle-shaped, but can change its shape in typical euglenid fashion.

More abundant in the seas then either of the above are the minute flagellates known as chrysomonads (order Chrysomonadida), which possess one or two flagella and yellow to brown disc-shaped chromotophores. There are also the cryptomonads (order Cryptomonadida). Although some chrysomonads have shells, most lack a cell wall and may therefore put out pseudopods. Cryptomonads, in contrast, possess a definite pellicle and are therefore incapable of changes in body shape; they also have one or two flagella. Some creep rather than swim and are thus more likely to be found in coastal environments. All the smaller flagellates help determine the larger protozoan species that may be present since they serve as food for them.

Among the colorless flagellates (order Zoomastigida) there are also a few species that inhabit saltwater and hence may be encountered along the coast. These include a truly remarkable form called *Multicellular marina* that, as the name suggests, is equipped with an effective locomotor apparatus—40-50 flagella (not cilia) according to Kudo (1966)—but is also amoeboid, apparently just in case. However, the great majority of flagellates in this order are either inhabitants of freshwater or are parasitic.

One of the most authoritative and complete works on coastal organisms is that of Swedmark (1964), although he gives protozoa relatively little attention. But in considering the fauna of such environments, he picks out two main groups of protozoa: the Foraminifera (already briefly considered) and "above all, the Ciliata." Dragesco (1960), a French protozoologist who studied coastal forms, states that there are at least 301 species of interstitial ciliates ("interstitial" meaning that they live in the minute spaces between sand grains), and that these belong to 85 genera and 27 families. Animal groups with species peculiar to marine psammon are listed by Zinn (1969).

Many of these ciliates have adapted themselves to an existence between sand grains of certain types and sizes. Fauré-Fremiet (1951), an eminent French protozoologist, divides such environments into three groups: mesoporal, microporal, and euryporal. These preferred habitats are, respectively: among sand grains 1.8-0.4 mm, 0.4-0.12 mm, and among sand grains of any size. Mesoporal species are from small to medium size, and are free-swimming. Microporal forms tend to have elongate, threadlike bodies, and to glide rather than swim.

Fauré-Fremiet, who had previously studied beach-dwelling protozoa in Europe, also spent a summer investigating such organisms in the neighborhood of the Marine Biological Laboratory at Woods Hole, Massachusetts. There he found many of the ciliates he had previously encountered in Europe as well as a few new ones. Those common to European and New England coasts included 18 mesoporal species and 12 of the microporal sort. One of the three new species illustrates certain characteristics of sand-dwelling ciliate fauna, particularly the ribbonlike body shape that must facilitate movement among the sand grains. A remarkable trait is also the *dorsal brush* of sulfur bacteria, presumably symbiotic, but one can only speculate on what use it is to the host. These bacteria give the host a dark appearance because of the sulfur granules they contain. Fauré-Fremiet also observed that some of sand-dwelling ciliates actually ingested sand grains—one can only wonder why.

However, sand grain size is only one of the important environmental factors. Others include temperature, which varies greatly with the season in temperate regions, and depth. Some species are relatively much less abundant near the surface in winter months. Nutritional factors—primarily the abundance of food organisms, such as diatoms and the smaller flagellates—vary with the season. When such microflora is abundant, predator ciliates are likely to be also abundant.

A characteristic obviously useful in a life among sand grains is the ability to contract, and most interstitial ciliates possess it, although a few cosmopolitan species (for example, *Coleps*) do not.

Unfortunately, most studies of protozoa in coastal environments, other than those on the Foraminifera, have been done on the coasts of Europe, although Fauré-Fremiet has been mentioned. All the evidence thus far indicates that most such species are cosmopolitan. This accords well with the theory of *continental drift,* since most of the protozoa have been around for a very long time.

In general it is still true that much less is known about the ecology of psammophilous protozoa than of those living in other environments. There are almost certainly many still unknown species, and we know little about the life cycles of the known species, and methods of cultivating such species have not yet been devised.

REGINALD D. MANWELL

References

Cushman, J., 1955. *Foraminifera.* Cambridge, Mass.: Harvard University Press, 605p.
Dragesco, J., 1960. Les Ciliés mésopsammiques littoraux (systematiques, morphologie, écologie), *Roscoff Sta. Biol. Trav.,* n.s., **12,** 1-356.
Fauré-Fremiet, E., 1951. The marine sand-dwelling ciliates of Cape Cod, *Biol. Bull.* **100,** 59-70.
Hutner, S. H., and McLaughlin, J. J. A., 1958. Poisonous tides, *Sci. American* **199,** 92-98.
Kudo, R. R., 1966. *Protozoology.* Springfield, Ill.: Thomas, 1174p.
Manwell, R. D., 1968. *Introduction to Protozoology.* New York: Dover, 642p.
Swedmark, B., 1964. The interstitial fauna of marine sand, *Biol. Rev.* **39,** 1-42.
Zinn, D. J., 1969. Psammon, *in* P. Gray, ed., *Encyclopedia of Biological Sciences.* New York: Van Nostrand Reinhold, 775-777.

Cross-references: *Biotic Zonation; Coastal Fauna; Coastal Flora; Euglenophyta; Foraminifera; Intertidal Sand Habitat; Organism-Sediment Relationship.*

PROXIGEAN SPRING TIDES

These are unusually high tides produced by an extreme proximity of the moon to the earth, adding to dynamic conditions associated with the simultaneous occurrence of both perigean and spring tides (see *Tides*). In combination with strong, sustained, onshore winds, such tides are especially conducive to coastal flooding.

Once in each *anomalistic month,* defining the moon's period of revolution around the earth in an elliptical orbit, the satellite reaches its position of closest approach to the earth for that lunation. This position is known as *perigee.* Twice each *synodic* month, at the times of new moon (conjunction) and full moon (opposition), the lunar body also comes into direct alignment in celestial longitude with the earth and sun. Either of these two positions of alignment is called *syzygy.*

The average length of the anomalistic month is 27.554551 days, and that between successive conjunctions or oppositions—the synodic month—is 29.530589 days. Thus, the times of perigee and syzygy do not generally coincide. On those occasions occuring twice in each calendar year when, as the result of mathematical commensurability between the periods involved, these positions occur nearly coincidentally (separation time $\leqslant$ 36 hours), the astronomical alignment is referred to as *perigee-syzygy*.

Because of the increased reinforcement of the gravitational forces of the moon and sun produced by this alignment, together with the closer proximity of the moon to the earth, and other factors created dynamically from the collinearity, tides of considerably greater amplitude and range result. These are known as *perigean spring tides*.

The usual length of time between two successive groupings of perigee-syzygy (in which the individual components are separated by 36 hours or less) is either 6.5 or 7.5 months, depending upon the variable intercommensurability caused by the changing lengths of the synodic and anomalistic months. By contrast, those more exact perigee-syzygy alignments that dynamically create a greater parallax and considerably reduced distance of the moon from the earth occur, on the average, only once in approximately every 2.5 years.

The resultant extremely close approach of the moon to the earth (parallax $\geqslant 61'26.5''$, and separation time between components $\leqslant 10$ hours) is referred to as *proxigee-syzygy*. The corresponding tides, whose amplitudes and ranges are now further amplified by the extreme proximity of the moon to the earth, are designated as *proxigean spring tides*. Those associated with a component-separation $\leqslant 2$ hours and parallax $\geqslant 61'30.7''$ are called *maximum proxigean spring tides*.

Should the occurrence of either of these types of tide be accompanied by strong, persistent, onshore winds, tidal flooding along the coastline is highly probable. Representative examples of more than 100 cases of such major coastal flooding occurring during the past 345 years on both the east and west coasts of North America have been compiled. These were associated with either perigean or proxigean spring tides and the necessary supporting wind conditions. Prominent among these are the cases of severe tidal flooding (and/or strong subsurface currents) experienced near proxigee-syzygy dates during the great mid-Atlantic coastal flooding of March 6-7, 1962 (Cooperman and Rosendal, 1962); in southern California on January 8, 1974 (Wylie, 1979; Wood, 1981a), in La Jolla submarine canyon there on January 29-30, 1979 (Shepard, Sullivan, and Wood, 1981); and from South Carolina to Connecticut on October 24-25, 1980 (Wood, 1981b).

FERGUS J. WOOD

References

Cooperman, A. I., and Rosendal, H. E., 1962. *Great Atlantic Coast Storm, 1962 March 5-9,* U.S. Weather Bur. Climat. Data—Natl. Summary 3, 137p.

Shepard, F. P.; Sullivan, G. G.; and Wood, F. J., 1981. Greatly accelerated currents in submarine canyon head during optimum astronomical tide-producing conditions; astronomical and tidal analysis of unusual currents in a submarine canyon during proxigee-syzygy alignment, *Shore and Beach* **49,** 32-36.

Wood, F. J., 1978. *The Strategic Role of Perigean Spring Tides in Nautical History and North American Coastal Flooding, 1635-1976.* Rockville, Md.: National Oceanic and Atmospheric Administration, 538p.

Wood, F. J., 1981a. When Neptune "sacks" the shores, *Oceans* **14,** 28-29.

Wood, F. J., 1981b. Proxigean spring tides, *Discover* **2,** 6.

Wylie, F. E., 1979. *Tides and the Pull of the Moon.* Brattleboro, Vt.: The Stephen Greene Press, 246p.

Cross-references: *Storm Surge; Tidal Range and Variation; Tidal Type, Variation Worldwide; Tides.* Vol. I: *Tides.*

PUFFING HOLE—See BLOWHOLES

PYRRHOPHYTA

This division of the plant kingdom contains a variety of unicellular microorganisms known as *dinoflagellates,* whose characteristic feature is the presence of two flagella. The longitudinal flagellum trails behind the organism and is used for forward locomotion, while the transverse flagellum encircles the cell and imparts rotational motion. Thus the dinoflagellate moves in a corkscrew path. In many varieties the cell is protected by an armorlike cell wall made up of cellulose plates. Most dinoflagellates are free-living planktonic organisms that depend upon photosynthesis for nutrition. However, some species devour living organisms, and certain varieties have adapted to living as internal or external parasites of unicellular and multicellular organisms. Dinoflagellates are common in marine waters, usually occurring at shallow depths (18-90 m) because they depend on sunlight for photosynthesis. Dinoflagellates also occur in lakes, marshes, estuaries, and even in the snow of polar regions (Chatton, 1952;

Evitt, 1970). They frequently display *bioluminescence,* and are the most important producers of light in ocean waters.

When environmental conditions are especially favorable, a particular species may reproduce at a rapid rate until it overwhelmingly dominates the plankton population, coloring the seawater to form a *red tide.* Highly toxic to fish and molluscs, the toxins excreted by the dinoflagellates may cause death to humans and animals that eat marine organisms. Dinoflagellate *blooms* gave the name to the Red Sea, and may explain why the Nile River "turned to blood" during the plague of Moses. These blooms generally last only a few days before declining owing to dwindling nutrient supply or changing environmental conditions.

Reproduction is usually by asexual cell division, but sexual reproduction has been reported. Some species form a dormant cyst stage during cold periods. These cysts are found as fossils in Mesozoic and Cenozoic sediments. Morphologically similar microfossils of uncertain identity (*acritarchs*) can be traced as far back as the PreCambrian (Sarjeant, 1974).

GEORGE MUSTOE

References

Chatton, E., 1952. Classe des dinoflagellés ou péridiniens, *in* P. P. Grassé, ed., *Traité de Zoologie,* vol. 1. Paris: Masson, 309-390.

Evitt, W. R., 1970. Dinoflagellates: A selective review, *Geoscience & Man* 1, 29-45.

Sarjeant, W. A. S., 1974. *Fossil and Living Dinoflagellates.* New York: Academic Press, 182p.

Cross-references: *Coastal Flora; Pollutants, Protozoa.*

Q

QUARRYING PROCESSES

This process refers to the removal of rocks from a cliff or *wave-cut platform* from their in situ position by wave action. Continued wave attack on a cliff or platform edge exerts tremendous hydraulic pressure and erosive force along lines of weakness, often bedding and joint lines. Gradual widening and deepening of these lines finally permit blocks up to several tons in weight to be shifted and eventually transported from their situ position (Fig. 1). During major storms the blocks are shifted from the wall of the cliff or platform. The blocks are generally moved seaward and downslope, but in some cases, especially in the erosion of *ramparts*, they may be transported toward the rear of the platform. Extremely large blocks quarried in this manner may be subsequently broken into smaller blocks. Removal of the surface blocks releases pressure on the underlying rocks, which leads to a slackening of the rock, permitting them to be more readily attacked. On wave-cut platforms and at cliff bases, quarrying is the major process of erosion, being the prime mechanism of removing blocks lying below the level of saturation. Both Jutson (1954) and Hills (1949) reported channels cut into platforms by wave quarrying, often along lines of weakness, particularly on more exposed sections. This process is therefore concentrated between the upper platform level or level of saturation down to the base of the submarine cliff, which in high-energy regions may be as deep as 10 m. It is best developed on highly jointed or fractured rock. Wentworth (1937), however, found on limestone coasts that quarrying was the major process removing material above the platform, and therefore of cliff erosion.

ANDREW D. SHORT

References

Hills, E. S., 1949. Shore platforms, *Geol. Mag.* **86**, 137-152.

Jutson, J. T., 1954. The shore platform of Lorne, Victoria, and the processes of erosion operating thereon, *Royal Soc. Victoria Proc.* **65**, 125-134.

Wentworth, C. K., 1937. Marine bench-forming processes (abstract), *Hawaiian Acad. Sci. Proc.* **30**, 5.

Cross-references: *Abrasion; Cliff Erosion; Erosion Ramp, Wave Ramp; High-Energy Coast; Hydraulic Action and Wedging; Ramparts; Shore Platforms; Wave-Cut Bench; Wave-Cut Platform; Wave Erosion; Weathering and Erosion, Mechanical.* Vol. III: *Weathering.*

FIGURE 1. Quarrying of large slabs of sandstone from the surface and outer edge of a wave-cut platform at Copacabana Beach, New South Wales, Australia. Note how the quarrying is proceeding along well-defined joint lines in the sandstone. Boulder debris at the back of the platform is derived from both cliff and platform erosion. The platform is 30 m in width (photo: A. D. Short).

QUAY

Quay is frequently used loosely as a synonym for *wharf*, but more precisely refers to the surface upon which railroads and crane tracks are constructed and where the various types of loading appliances operate. In essence, it is the paved surface part of a wharf (Bruun, 1973; Coastal Engineering Research Center, 1977).

CHARLES B. CHESTNUTT

References

Bruun, P., 1973. *Port Engineering.* Houston: Gulf Publishing Co., 436p.

Coastal Engineering Research Center, 1977. *Shore Protection Manual,* Washington, D.C.: U.S. Army Corps of Engineers, 1264p.

Cross-references: *Breakwater; Bulkhead; Coastal Engineering; Protection of Coasts; Wharf.*

QUILLWORTS—See LYCOPODIOPHYTA

R

RADIOCARBON DATING

Radiocarbon dating is a method of *absolute dating* applicable to carbon-containing substances, whether organic or inorganic. Radiocarbon (^{14}C) is formed in the upper atmosphere by primary cosmic ray interaction with atmospheric nitrogen (Libby, 1955). After oxidation to $^{14}CO_2$ and relatively rapid mixing with stable $^{12}CO_2$ and $^{13}CO_2$ in the atmosphere, it is either fixed in the biosphere photosynthetically or dissolved in natural waters (oceanic or meteoric) to become carbonate or bicarbonate ions.

Fixation of $^{14}CO_2$ by one of the above processes constitutes the basis for radiocarbon dating. Living organisms establish an equilibrium level of ^{14}C in their lifetime by one of the following processes: assimilation during photosynthesis, intake of plant-derived foods, or deposition of calcium carbonate. This equilibrium level becomes the base level of ^{14}C activity for an organism upon death, provided no isotopic fractionation or exchange occurs after death. A similar argument holds for inorganic carbon—cave deposits, for instance—except that deposition is synonymous with death for dating purposes. Thus the age of a carbon-containing sample may be calculated by knowing the equilibrium level of ^{14}C activity, the present level, and the rate at which ^{14}C disintegrates.

Age Calculations

Radiocarbon ages are calculated using the following equation: $t = 8033 \ln A_o/A$, where t is the age in years, A is the sample ^{14}C activity, and A_o is the ^{14}C activity. (The ^{14}C activity is actually measured as the $^{14}C/^{12}C$ ratio for a sample, not as the absolute amount of ^{14}C.)

Radiocarbon Dating in Beach and Coastal Environments

Materials generally chosen for dating include charcoal, wood, shells, bone, and peat, with reliability being in the same order (that is, charcoal is the most reliable material for dating, and peat the least). Since beaches are normally high-energy zones where few materials suitable for dating survive, there is seldom a choice of samples. Except for archeological sites where charcoal, shells, and bone may be found, only shells are found in coastal deposits to any extent.

Shell samples must be chosen carefully owing to possible recrystallization, which alters the $^{14}C/^{12}C$ ratio and hence, the calculated or apparent age of a sample. To prevent or at least alleviate such an error, it is advisable to: visually examine all shells in the field for recrystallization—that is, avoid a chalky, weakly consolidated shell; collect only large, solid shells; run 5-10 large shells to date a deposit or site, as opposed to 1 or 2; and obtain mass spectrometric data to ascertain the isotopic ratio of a sample as a check on recrystallization.

THOMAS D. MATHEWS

Reference

Libby, W. F., 1955. *Radiocarbon Dating.* Chicago: University of Chicago Press, 175p.

Cross-references: *Archaeology, Geological Considerations; Archaeology, Methods.* Vol. I: *Radionuclides: Their Application in Oceanography;* Vol. VI: *Radioactivity in Sediments.*

RAISED BEACH

The term *raised beach* has been applied since early in the past century to any former shoreline deposits that today are found emerged, above present high tide level (de la Beche, 1839). Holocene raised beaches are frequently marked by shell accumulations. Natural examples contain mixed species of all sizes. Kitchen middens of shellfish-eating early man are often associated with the raised beaches (Fairbridge, 1976); but the shell mounds, which may contain several billion specimens, are distinguished by monospecific edible forms (also charcoal, tools, and so on). Raised beach shells were noted by early classical philosophers on the coasts of Greece, North Africa, and the Persian Gulf. They were commonly interpreted as evidence of the Noachian Flood. Most raised beaches are marine, but some, as in glacial or pluvial lakes, are lacustrine. An extensive bibliography was summarized by Richards and Fairbridge (1965; with several supplements).

Some authors have felt that the adjective raised implied that the earth's crust had been elevated in the area concerned, and, instead, urge that the term *emerged beach* be used (Zeuner, 1959). Early in the present century, the French geographer de Martonne pointed out that one should say *emergent* (or *submergent*) coasts rather than *uplifted* (or *subsided*), which has a genetic implication.

By custom, however, the label raised beach seems to have been fairly well fixed in the literature for more than two centuries, and it is for geomorphologists to recognize that it carries no genetic implication.

The cause of the emergence of the beach deposit, sometimes also called *uplifted* or *abandoned*, may be the result of at least three causes, separately or in combination: glacio-isostatic uplift, glacioeustatic oscillation, and tectonic uplift. Secondary additional factors include variations in tidal amplitude (for example, in the Bay of Fundy), hydroisostasy, geoidal changes, oceanographic current changes, and others (Chambers, 1848; Geikie, 1881; Charlesworth, 1957; Fairbridge, 1971).

The presence of a raised beach implies the availability of loose, unconsolidated material appropriate for beach building such as at the mouths of rivers, on the shelf offshore, or from exposed cliffs or saprolite, or of soft glacial tills and outwash (moraines, drumlins, eskers, kames, sandurs). Along hard-rock coasts the raised beaches often pass laterally into wave-cut or floating ice-cut rock platforms (Nansen, 1905; Fairbridge, 1971). Collectively these former *water planes* are known as *elevated shorelines* (q.v.) or *strandlines*.

Holocene raised beaches in the 6000 B.P. to present age range are globally widespread in middle to low latitudes, up to 3 m above sea level, where they have been affected only by eustatic and hydroisostatic processes and not by tectonism or glacioisostasy. Tectonically warped Holocene beaches are best known from California, Japan, and Chile, as well as other tectonically active regions; however, within the 6000-year duration of the mid- to late-Holocene epoch (when sea level has been close to the present datum), the amplitude of tectonic displacement has often been less than that of the eustatic influence.

In contrast, the isostatically active regions of Scandinavia, Svalbard, Greenland, Canada (especially Hudson Bay), Patagonia, and Antarctica disclose postglacial raised beaches of great extent and ranging to over 300 m above sea level (Hillaire-Marcel and Fairbridge, 1978; Omoto, 1977). The beaches formed at those times when the eustatic rise rate coincided for a few centuries with the isostatic uplift rate; the specific years were seasons of high wave energy. Negative eustatic oscillations (cool intervals) would be out of phase with the isostatic uplift, and at such times there would be a rapid emergence of the land and corresponding lateral withdrawal of the sea, sometimes tens of kilometers within a few decades, so that no beaches had time to form. An abundant supply of loose sediment is the primary requirement for beach formation; glacial moraines and outwash provide exceptionally favorable sources, even in areas of only modest isostatic uplift such as in northeastern England (Tooley, 1974).

The removal of the ice cover from Scandinavia or North America was not uniform in rate, receding in rapid pulses, alternating with brief readvances sometimes as much as 500 km. The crustal reaction is delayed, however, being rapid at first but decaying exponentially over perhaps 10,000 years. Thus oscillation in the variations in the melt rate are bridged over by the retardation effect, and a rather smooth glacioisostatic rebound curve can be constructed. The raised beach elevations and separations can then be used to construct an approximate eustatic curve rise. The same logical approach can be made in order to solve the eustatic-tectonic problem in the nonglaciated areas.

Pleistocene raised beaches (over 100,000 years old) tend to be destroyed (by leaching, wind erosion, and the like) unless protected by some sort of covering. Charlesworth (1957) refers to them as *infraglacial beaches* (that is, older than the last glacial). There are extensive pre-Wisconsinan raised beaches in southwest Britain (described first by de la Beche, 1839) capped by a protection of periglacial colluvium (locally known as *head*). In the Mediterranean many early Pleistocene raised beaches are protected by a cover of periglacial colluvium of limestone debris that has been secondarily indurated by a carbonate element. In warm latitudes beaches are often contemporaneously cemented to become *beach rock* (q.v.). Ancient beach rocks are often well preserved. On windward coasts in the mid-latitudes, there are widespread calcareous eolianites (fossil dunes) that were carried up during the glacial drops in sea level and then subsequently cemented, thereby forming a protective cover over the underlying raised beach. On aggrading coasts such as those of North America, the raised beach evidence is supplemented by lagoonal and other facies (Doering, 1960; Flint, 1966).

Analyses of the malacological (shell) faunas of eustatically raised beaches shows almost invariably that they corresponded to warmer-than-present phases of earth history. Thus the 6000-year raised beaches of western Scotland (so-called 25-30 foot beaches) contain faunas characteristic of Spain or the Mediterranean. The Tyrrhenian (mid Pleistocene) beaches of the Mediterranean carry elements of a west African (*Senegal*) fauna, for example, *Strombus bubonius* (Zeuner, 1959). In contrast, the late Pleistocene raised beaches of southern Scandinavia and the St. Lawrence valley (*Champlain Sea*) of Quebec and northern New York state contain shells characteristic of extremely cold waters, thus reflecting the presence of melting glacier snouts.

RHODES W. FAIRBRIDGE

References

Chambers R., 1948. *Ancient Sea Margins.* Edinburgh: Chambers, Orr and Co., 388p.
Charlesworth, J. K., 1975. *The Quaternary Era*, 2 vols. London: Edward Arnold, 1700p.
de la Beche, H. B., 1839. *Report on the Geology of Cornwall, Devon, and West Somerset.* London: Longmans, 648p.
Doering, J. A., 1960. Quaternary surface formations of southern part of Atlantic coastal plain, *Jour. Geology* 68, 182-202
Fairbridge, R. W., 1971. Quaternary shoreline problems at INQUA 1969, *Quaternaria* 15, 1-18.
Fairbridge, R. W., 1976. Shellfish-eating preceramic Indians in coastal Brazil, *Science* 191, 353-359.
Flint, R. F., 1966. Comparison of interglacial marine stratigraphy in Virginia, Alaska, and Mediterranean area, *Am. Jour. Sci.* 264, 673-684.
Geikie, J., 1881. *Prehistoric Europe.* London: E. Stanford, 592p.
Hillaire-Marcel, C., and Fairbridge, R. W., 1978. Isostasy and eustasy of Hudson Bay, *Geology* 8, 117-122.
Nansen, F., 1905. Oscillations of shore-lines, *Geog. Jour.* 26, 604-616.
Omoto, K., 1977. Geomorphic development of the Soya coast, East Antarctica, *Tohoku Univ. Sci. Repts.*, ser. 7 (Geog.), 27, 95-148.
Richards, H. G., and Fairbridge, R. W., 1965. *Annotated Bibliography of Quaternary Shorelines (1945-64).* Philadelphia: Academy of Natural Sciences Spec. Pub. 6, 280p. (Supplement for 1965-1969, publ. 1970, n. 10, 240p.; and subsequent issues.)
Tooley, M. J., 1974. Sea-level changes during the last 9000 years in northwest England, *Geog. Jour.* 140, 18-42.
Zeuner, F. E., 1959. *The Pleistocene Period,* 2d ed. London: Hutchinson, 447p.

Cross-references: *Elevated Shoreline (Strandline); Emergence and Emerged Shoreline; Global Tectonics; Isostatic Adjustment; Midden; Postglacial Rebound; Sea Level Curves.*

RAMPARTS

A *rampart* is a band of rock at the seaward edge of a *wave-cut platform* that lies higher than the platform itself (Fig. 1). Ramparts are morphologically placed lines of in situ rock that are more resistant to marine erosion because

FIGURE 1. View seaward across horizontal wave-cut platform, with 1-m-high rampart at seaward edge (to left of person), Culburra, New South Wales, Australia. Note the water-layer weathering on the platform surface (photo: A. D. Short).

their position on the seaward edge of the platform exposes them to greater saturation (Wentworth, 1938), thereby raising the level of saturation and consequently the level at which the overlying rock is removed. For the same reason, ramparts remain more resistant to the secondary processes that denude the backing lower platform. Gill (1971) also attributes their resistance to pedologic mobilization of iron in the original surface, resulting in the formation of a more resistant surface band of which the ramparts are the remnants of the seaward edge. Gill also found ramparts related to *massive* rock, free of jointing, bedding, and the like, and multiple ramparts related to variations in rock hardness.

Ramparts are usually removed by *quarrying processes,* the eroded material, because of its position, often being deposited on the backing platform. Owing to their formation and nature, ramparts are by definition ephemeral features formed during erosion of the primary platform surface subsequent to the recent transgression. Following their removal, they cannot be replaced by the lower backing platforms. As a result, contemporary ramparts tend to be the remains of once extensive features that dominated the outer edge of the primary platform but that have since been removed or denuded where the original platform edge has retreated shoreward.

Ramparts may still form today in situations where a platform is only just beginning to *notch* a cliff owing to changes in relative sea level, relative protection, or energy conditions.

ANDREW D. SHORT

References

Gill, E. D., 1971. Ramparts on shore platforms, *Pacific Geology* 4, 121-133.
Wentworth, C. K., 1938. Marine bench-forming pro-

cesses: Water-level weathering, *Jour. Geomorph.* **1**, 5-32.

Cross-references: *Boulder Barricades; Erosion Ramp, Wave Ramp; Niche, Nick, Nip, Notch; Quarrying Processes; Water-Layer Weathering; Wave-Cut Bench; Wave-Cut Platform; Wave Erosion.*

RASA

Rasa is a Spanish word (from the Latin *radere*, "to scrape or rub out") meaning "ancient and perched rocky surfaces of marine planation" (Guilcher, 1974), with an average breadth of 1-15 km. Rasas can be classified into two types. The first is bordered inwardly by a mountainous escarpment and have been found not only in Spain (Hernández Pacheco, 1950; Nonn, 1966), but in Chile (Paskoff, 1970) and Morocco, with analogies to the Norwegian *strandflat* (q.v.). The stepping of several rasas is often seen. For example, rasas on the Cantabrican coast are frequently bordered by cliffs and form steps on subvertical strata of slate, quartzite, and limestone. The most typical rasa segment is located between the rivers Deva and Sella. Rasa altitudes are variable (10-200 m above sea level), the higher ones being referred to as *sierras planas*.

Rasas ending with a slight break belong to the second type (Guilcher, 1974) and are less extended and less spectacular than those of the first type. One of the best examples of a rasa of the second type is found on quartzite in Cape Peñas (NW Gijón) at 115, 170, and 250 m.

The origins of the rasas are still under study. The first researchers—Barrois, Royo Gómez, Sermet—considered marine action, but about the time of World War I, the lack of evidence for marine erosion supported the reactions of epeirogenists led by Hernández Pacheco. In the 1950s and 1960s such evidence was found (Guilcher, 1974; Asensio Amor, 1970; Nonn, 1966; Mary, 1970), but at the same time the great importance of faulting or flexuring was verified. Erosion and accumulation, some of which may have only continental origins, combine on several rasas; while on lower rasas there are terrestrial features formed prior to wave action. Therefore, as marine features and fauna have often vanished chronology becomes difficult, although late Sicilian or Tyrhenian I and II have been suggested.

VICENÇ M. ROSSELLÓ

References

Asensio Amor, I., 1970. Rasgos geomorfológicos de la zona littoral galaico-astúrica en relación con las oscilaciones glacio-eustáticas, *Estudios Geologicos* **26**, 29-91.

Guilcher, A., 1974. Les rasas: Un problème de morphologie littorale générale, *Annales Géographie* **455**, 1-33.

Hernández Pacheco, F., 1950. Las rasas litorales de la costa cantabrica en su segmento asturiano, *VII Congr. Internat. Geogr. Lisbonne, 1949, Comptes Rendus* **1**, 29-86.

Mary, G., 1970. La rasa cantabrique entre Luarca et Ribadeo (Asturies), *Breviora Geologica Asturica* **14**, 45-48.

Nonn, H., 1966. *Les régions cotières de la Galice (Espagne), étude morphologique* (thèse). Paris: Les Belles Lettres, 591p.

Paskoff, R., 1970. *Recherches géomorphologiques dans le Chili semi-aride* (thèse). Bordeaux: Biscaye Freres, 420p.

Cross-references: *Abrasion, Erosion Ramp, Wave Ramp; Geographic Terminology; Sea Level Changes; Strandflat.*

RAVINEMENT

Ravinement is a disconformity cut by a *transgressing shoreline* (Stamp, 1921; Swift, 1967) as a consequence of the landward translation of the shoreface profile (see *Nearshore Hydrodynamics and Sedimentation*). The disconformity lies within the marine transgressive sequence rather than at its base and generally separates *lagoonal* or *deltaic* from overlying open shelf deposits. Barrier superstructure deposits (*beach, dune*) are generally absent, having been destroyed by the ravinement process.

DONALD J. P. SWIFT

References

Stamp, L. D., 1921. On cycles of sedimentation in the Eocene strata of the Anglo-Franco-Belgian Basin. *Geol. Mag.* **58**, 108-114, 146-157, 194-200.

Swift, D. J. P., 1967. Coastal erosion and transgressive stratigraphy. *Jour. Geology* **76**, 444-456.

Cross-references: *Barrier Islands, Transgressive and Regressive; Beach Processes; Beach Stratigraphy; Bruun Rule; Nearshore Hydrodynamics and Sedimentation; Submergence and Submerged Shoreline.*

RECENT—See HOLOCENE

RECESSION AND RETROGRESSION

The terms *recession* and *retrogression* are synonymous and commonly refer to the permnent landward retreat of a coastline or shoreline. G. K. Gilbert (1884) for instance, speaks of the "landward recession" of the "water margin." D. W. Johnson (1919), following

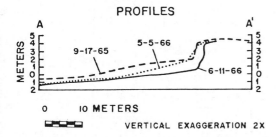

FIGURE 1. Shoreline changes on the northwest coast of Florida, between September 17, 1965, and June 11, 1966 (modified from Warnke, 1969).

W. M. Davis, distinguishes between a *prograding shore,* and a *retrograding shore.* Modern workers speak, for instance, of *shoreline recession* (McCormick, 1973) or *beach retrogression* (Warnke, 1969, 1973). All these terms commonly imply long-term erosional changes (for a discussion of the scale of shore erosion, see Schwartz, 1968).

The erosional processes may proceed slowly and continuously, over a period of weeks, months, years, or may be accentuated during storm episodes. Figure 1 shows such shoreline changes on the northwest coast of Florida: the changes documented for the period between September 17, 1965, and May 5, 1966 (see dashed and dotted profiles, Fig. 1), are mainly changes in the berm profile and minor modifications of the forward-dune apron. These changes do not represent retrogression. The change caused by hurricane Alma (on June 9, 1966), and shown by the profile taken on June 11, 1966 appears to be permanent and represents retrogression. Other examples are given by, among others, Young (1975).

The term recession has also been used to signify the *seaward* translation of a shoreline (see, for instance, L. v. Buch, 1867). Such use, however, is now uncommon.

DETLEF A. WARNKE

References

Buch, L. v., 1967. Reise durch Norwegen und Lappland, *Gesammelte Schriften* 2, 503-504. (An English translation of this article appears in S. L. Mason, ed. K. F. Mather, 1970. *Source Book in Geology, Fourteen Hundred to Nineteen Hundred.* Cambridge, Mass.: Harvard University Press, 702p.)

Gilbert, G. K., 1884. The topographic features of lake shores, *U.S. Geol. Survey Annual Rept.* 5, 75-123.

Johnson, D. W., 1919. *Shore Processes and Shoreline Development.* New York: Columbia University, 584p. (Reproduced in facsimile in 1972 by Hafner, New York.)

McCormick, C. L., 1973. Probable causes of shoreline recession and advance on the south shore of eastern Long Island, *in* D. R. Coates, ed., *Coastal Geomorphology.* Binghamton: State University of New York, 61-71.

Schwartz, M. L., 1968. The scale of shore erosion, *Jour. Geology* 76, 508-517.

Warnke, D. A., 1969. Beach changes at the location of landfall of hurricane Alma, *Southeastern Geology* 10, 189-200.

Warnke, D. A., 1973. Beach changes in a low-energy coastal area in Florida which is in the pathway of rapid urbanization, *in* D. E. Moran, J. E. Slosson, R. O. Stone, C. A. Yelverton, eds., *Geology, Seismicity, and Environmental Impact.* Los Angeles: University Publishers, Assoc. Eng. Geologists Spec. Pub. Oct. 1973, 367-377.

Young, K., 1975. *Geology: The Paradox of Earth and Man.* Boston: Houghton Mifflin, 526p.

Cross-references: *Erosion Ramp, Wave Ramp; Geographic Terminology; Sea Level Changes; Strandflat. Prograding Shoreline; Retrogradation and Retrograding Shoreline.* Vol. III: *Retrograding Shoreline.*

RED ALGAE—See RHODOPHYCOPHYTA

REEFS, NONCORAL

Although hermatypic corals are today's dominant reef builders, several other groups of organisms build potentially wave-resistant structures in intertidal and shallow subtidal environments (for a discussion of reef-related terminology, see Heckel, 1974). All of these organisms may live as solitary individuals, often as suspension-feeding encrusters on coral reefs and other firm substrata. As reef builders, these organisms are apparently excluded competitively by corals from warm, shallow waters of normal marine salinities, and generally able to construct frameworks only in environments that are not favorable to coral growth (Fig. 1). Exclusion of more stenohaline predators is also a factor contributing to construction of noncoral reefs.

Vermetid gastropods cement their shells to hard intertidal surfaces, growing uncoiled calcareous shells resembling worm tubes (Hughes, 1979). In combination with red (coralline) algae, they construct relatively small reefs (100 m wide, a few m thick) on rocky shores lacking coral reef growth, reported from Florida (Shier, 1969), Curacao (Focke, 1977), the eastern Mediterranean, and as local *cup reefs* on Bermudan coral reefs (Safriel, 1975; Dean and Eggleston, 1975).

Reef-building oysters, such as *Crassostrea,* flourish under brackish conditions, and most reefs occur in salinities averaging 15-25 ‰ in tropical and temperate lagoons, bays, and estuaries. These lower salinities exclude many competitors and predators, particularly the

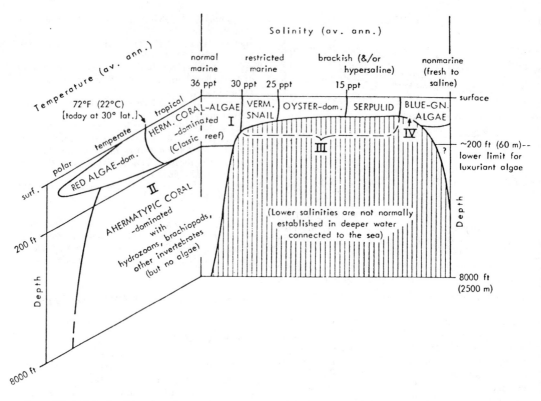

FIGURE 1. Relationship of major reef assemblages to primary physical environmental factors. I, shallow-marine environment; II, deep-marine environment; III, nearshore, reduced-salinity environment; IV, nonmarine environment (from P. H. Heckel, 1974. Carbonate buildups in the geologic record: A review, in L. F. Laporte, ed., *Reefs in Space and Time,* SEPM Spec. Pub. 18, Tulsa, Okla.).

oyster drills of the gastropod family Muricidae (Hedgpeth, 1953; Galtsoff, 1964). Like the vermetids, the oysters cement their shells to hard substrata, including other shells. On the United States Gulf coast, oysters form reefs over 100 m wide and up to 40 km long (Norris, 1953). On a smaller scale, Bahr (1976) described intertidal mounds over 10 m in diameter and 0.7 m high in a Georgia salt marsh-estuary system. Stenzel (1971) discussed oyster reefs and classified them according to their relationship to the shoreline as fringe, string, or patch reefs.

Polychaete worms of the family Serpulidae may produce cemented masses of calcareous tubes in salinities ranging from 4 to 55 ‰ or higher (Heckel, 1974; ten Hove, 1979). Subfossil serpulid reefs found in Laguna Madre, Texas, inhabited waters that were usually hypersaline, but that became nearly fresh after heavy rainfalls. The reefs ranged in size from 8 m in diameter to 45 m in length and rose 1-3 m above the surrounding sediment, built on a pavement of shell debris (Shier, 1969). Serpulid reefs have been reported from Tunis lagoon (Lucas, 1960; Keene, 1980) a sheltered lagoon at Arbear Lock, Eire (Bosence, 1979); hypersaline seepage pools at the Gulf of Elat (Por and Dor, 1975); and in open bays on the Norwegian coast, north of the range of oyster reefs. This last occurrence indicates that serpulids are capable of building frameworks in normal salinities (Heckel, 1974).

The Sabellariidae as well as polychaetes construct tubes of cemented sand grains and shell debris, and may form reefs in the intertidal zone. Wilson (1968) has demonstrated that gregarious behavior in *Sabellaria* depends on larvae recognizing the cement used by adult worms to build their tubes. Kirtly and Tanner (1968) described masses about 0.3 m high and 100 m wide in the Florida intertidal. Sabellariid reefs are widespread, and have been described from France (Mathieu, 1967); England (Wilson, 1971); Delaware Bay, United States (Wells, 1970). Multer and Milliman (1967) have suggested that these agglutinated reefs tend to stabilize the shoreline and retard wave erosion and, by construction of tubes, sort local beach sands. It is likely that other types of noncoral reefs also play an important role in retarding coastal erosion (for example, Focke, 1977) and, when acting as a current baffle, in accumulating sediments.

DAVID JABLONSKI

References

Bahr, L. M., Jr., 1976. Energetic aspects of the intertidal oyster reef community at Sapelo Island, Georgia (USA), *Ecology* 57, 121-131.

Bosence, D. W. J., 1979. The factors leading to aggregation and reef formation in Serpula vermicularis, in G. P. Larwood and B. R. Rosen, eds., *Biology and Systematics of Colonial Organisms.* London: Systematics Association Spec. Vol. 11, 299-318.

Dean, W. E., and Eggleston, J. R., 1975. Comparative anatomy of marine and freshwater algal reefs, Bermuda and central New York, *Geol. Soc. America Bull.* 86, 665-676.

Focke, J. W., 1977. The effect of a potentially reef-building vermetid-coralline algal community on an eroding limestone coast, Curaçao, Netherlands Antilles, *Internat. Coral Reef Symp. Proc.* 1, 239-245.

Galtsoff, P. S., 1964. The American oyster *Crassostrea virginica* Gmelin, *U.S. Dept. Interior Fish. Bull.* 64, 480p.

Heckel, P. H., 1974. Carbonate buildups in the geologic record: A review, in L. F. Laporte, ed., *Reefs in Space and Time.* Tulsa, Okla.: Soc. Econ. Miner. Spec. Pub. 18, 90-154.

Hedgpeth, J. W., 1953. An introduction to the zoogeography of the northwestern Gulf of Mexico, with reference to the invertebrate fauna, *Univ. Texas Inst. Marine Sci. Pub.* 3, 107-224.

Hughes, R. N., 1979. Coloniality in Vermetidae (Gastropoda), in G. P. Larwood and B. R. Rosen, eds., *Biology and Systematics of Colonial Organisms.* London: Systematics Association, 243-253.

Keene, W. C., Jr., 1980. The importance of a reef-forming polychaete, *Mercierella enigmatica* Fauvel, in the oxygen and nutrient dynamics of a hypereutrophic subtropical lagoon, *Eustuarine Coastal Marine Sci.* 11, 167-178.

Kirtley, D. W., and Tanner, W. F., 1968. Sabellariid worms: Builders of a major reef type, *Jour. Sed. Petrology* 38, 73-78.

Lucas, G., 1960. Deux examples actuels de "biolithosores" construits par des annelides, *Soc. Géol. France Bull.,* ser. 7, 1, 385-389.

Mathieu, R., 1967. Le banc des Hermelles de la baie du Mont-Saint-Michel, bioherm à annélides: Sedimentologie, structure et genèse, *Soc. Géol. France Bull.,* sér. 7, 9, 68-78.

Multer, H. G., and Milliman, J. D., 1967. Geologic aspects of sabellarian reefs, southeastern Florida, *Bull. Marine Sci.* 17, 257-267.

Norris, R. M., 1953. Buried oyster reefs in some Texas bays, *Jour. Paleontology* 27, 569-576.

Por, F. D., and Dor, I., 1975. Ecology of the metahaline pool of Di Zahov, Gulf of Elat, with notes on the Siphonocladaces and the typology of nearshore marine pools, *Marine Biology* 29, 37-44.

Safriel, U. N., 1975. The role of vermetid gastropods in the formation of Mediterranean and Atlantic reefs, *Oecologia* 20, 85-101.

Shier, D. E., 1969. Vermetid reefs and coastal development in the Ten Thousand Islands, southwest Florida, *Geol. Soc. America Bull.* 80, 485-508.

Stenzel, H. B., 1971. Oysters, in R. C. Moore, ed., *Treatise on Invertebrate Paleontology,* Part N, *Mollusca* 6, vol. 3. Boulder, Colo.: Geological Society of America and Lawrence, Kans.: University of Kansas, N953-N1224.

ten Hove, H. A., 1979, Different causes of mass occurrences in serpulids, in G. P. Larwood and B. R. Rosen, eds., *Biology and Systematics of Colonial Organisms.* London: Systematics Association, 281-298.

Wells, H. W., 1970. Sabellaria reef masses in Delaware Bay, *Chesapeake Sci.* 11, 258-260.

Wilson, D. P., 1968. The settlement behavior of the larvae of *Sabellaria alveolata* (L.), *Jour. Marine Biol. Assoc. U.K.* 48, 387-435.

Wilson, D. P., 1971. *Sabellaria* colonies at Duckpool, North Cornwall, 1961-1970, *Jour. Marine Biol. Assoc. U.K.* 51, 509-580.

Cross-references: *Algal Mats and Stromatolites; Biogeneous Coasts; Biotic Zonation; Coastal Fauna; Coral Reef Coasts; Coral Reef Habitat; Mollusca.*

REFLECTION—See WAVES

REFRACTION—See WAVES

RELICT SEDIMENT

Relict sediments are inherited from previous phases of sedimentation and are therefore not necessarily in equilibrium with the present environment in which they occur. It has been estimated (Emery, 1968) that approximately 70% of the continental shelves of the world are characterized by a cover of sediment that was deposited when the shelves were largely exposed owing to the low sea levels associated with the glacial maxima of the Pleistocene epoch. At such times the sediments accumulated in a variety of environments (subaerial, littoral, lacustrine, fluvial, lagoonal, paludal, shallow-marine, periglacial) and subsequently became deeply submerged by the postglacial rise in sea level. Thus relict shoreline and channel-fill deposits are common features of many continental shelves.

In some areas of low current velocities, off the mouths of rivers that are now transporting large quantities of fine sediment, relict sediments are being buried beneath a blanket of mud. In many regions, usually shallow and inshore, reworking has modified the character and distribution of relict sediments, in accordance with present hydrodynamic conditions, often forming new sedimentary structures with old grains. Other, truly relict deposits, usually occurring in deeper water, show little modification by the present regime.

Relict glacial and periglacial deposits have been recorded and described from the neritic realm of the higher latitudes of the Northern Hemisphere since the nineteenth century. In recent years, relict carbonate facies have been shown to occur extensively in the coral reef environment, and in some regions they constitute a major part of the sediment cover (Maxwell, 1968a).

The effects of excessive abrasion and corrosion, hematite or limonite staining due to periods of subaerial exposure and weathering, and excessive boring of skeletal debris are among the criteria used to identify relict grains. Mineral components incompatible with the present depositional environment may also indicate the relict nature of the deposit. Strong evidence is supplied by the presence of sedimentary particles derived from species that were favored by previous environmental conditions but that are not characteristic of the present environment. Anomalous structures and textures (modal diameters, sorting, and so on) that are incompatible with the present energy levels and sediment dispersal conditions are also diagnostic of relict sediments. Further, complex grain size distributions may result from the partial adjustment of a relict sediment to the parameters of the present regime and from the mixing of relict grains with Recent sediment (Maxwell, 1968b). However, recognition of relict sediments is difficult where there is similarity between the present and past environments.

G. R. ORME

Emery, K. O., 1968. Relict sediments on continental shelves of the world, *Am. Assoc. Petroleum Geologists Bull.* 52, 445-464.

Maxwell, W. G. H., 1968a. Relict sediments, Queensland continental shelf, *Australian Jour. Sci.* 31, 85-86.

Maxwell, W. G. H., 1968b. *Atlas of the Great Barrier Reef.* Amsterdam: Elsevier, 258p.

Cross-references: *Beach Material; Flandrian Transgression; Nearshore Hydrodynamics and Sedimentation; Sand, Surface Texture; Sediment Analysis, Statistical Methods; Sediment Transport.*

REMOTE SENSING

Remote sensing, the collection of data without direct contact, had its beginning in 1839 when the first photographs were reportedly taken. Black-and-white photographs, first taken from the ground, were eventually recorded from balloons, pigeons, and aircraft, and had marked success in military ventures. During and following World War II, the field of remote sensing—then referred to as *air photo interpretation*—expanded to include color photography and sensors operating outside the visible spectrum. These *new* sensors included photographic sensors (standard color, false color camouflage detection, ultraviolet, and multispectral black-and-white photography) and nonphotographic sensors (thermal infrared, active and passive microwave, laser and luminescence systems, and sonar). Characteristics and capabilities of some of the more important electromagnetic remote sensing systems are given by Estes and Senger (1974).

Remote sensing platforms also changed, becoming more stable, faster moving, and farther away from earth. Low- and medium-altitude aircraft ($<$11,000 m), with cruising speeds of $<$ 600 km/hr, are now complemented by high-altitude ($>$ 35,000 m) supersonic ($>$ 3,000 km/hr) aircraft and manned (Apollo, Skylab, and proposed Space Shuttle) and unmanned satellites (Nimbus, Landsat, and Seasat). More unconventional platforms are helicopters, drones, tethered balloons, and dirigibles.

The application of remote sensing systems to coastal zone research is problem-dependent. Hand-held cameras for black-and-white, color, color infrared, or stereo photos may serve to document field notes, whereas the more sophisticated imaging systems mounted on platforms at various altitudes not only provide a new perspective because of scale but also record spectral patterns (ultraviolet, infrared, and microwave) otherwise not visible to the human eye. In seeming contrast, the best overall remote sensing system, including cost effectiveness, availability, and data retrieval, is a high quality aerial mapping camera flown at altitudes $>$3000 m. In situations where cloud cover makes photography prohibitive or where reflected visible energy does not provide the required information, other systems, such as radar and thermal imagers, are more practical and effective than aerial photography.

Specific applications are many. Conventional aerial photography has long been used for collecting morphometric information of the coastal zone. Photography has been used to gather baseline data on shoreline configuration as well as to map shoreline erosion and deposition over time with sequential photography. Wave refraction, surf zone, wave power spectrum, riptides, sand spits and dunes, subaqueous and subaerial bars, and depth are a few coastal zone features or parameters monitored on aerial photography.

Water depth and underwater topography, for example, have been measured with several sensors utilizing several techniques. Stereo color photography of clear ocean water—maximum transmittance of 0.44–0.54 μm—was employed for mapping to depths of 16 m (Geary, 1967; Specht et al., 1973). The standard photogrammetric techniques used in stereo mapping must be modified to account for two-media transmission (atmosphere and hydrosphere). By incorporating a two-media model, Umbach and Harris (1972) produced bathymetric maps from stereo aerial photography comparable with those from conventional underwater mapping methods. Where turbidity prohibits the use of stereo photography, color change, temperature variation, and wave refraction may serve as a depth indicator (Polcyn, 1969, 1970). Sonar (Sound Navigation and Ranging) operates with acoustic rather than electromagnetic waves and has been utilized successfully in the mapping and locating of submarine features such as faults, submarine bars, and sunken vessels.

Remote sensing data on the physical and chemical properties of water can be collected with varying degrees of accuracy. Absolute measurements of water turbidity have been made from color infrared and multispectral photography. The same is true for temperature measurements from thermal infrared imaging systems where greater than 1°C accuracy is not uncommon and 0.25°C accuracy is occasionally reported. Salinity measurements within 3–5 ‰ from passive microwave data off of the Mississippi River delta were reported by Thomann (1973).

Determining the elemental chemical composition of water is not feasible except on a regional scale, although several techniques and special remote sensing devices exhibit some potential (Stoertz, Hemphill, and Markle, 1969). The Fraunhofer Line Discriminator (FLD) uses solar-stimulated fluorescence to detect certain pollutants, oil slicks, lignin sulfonates, and chlorophyll fluorescence, and to trace certain dyes (especially Rhodamine) useful in monitoring water movement. Digital multispectral scanner Landsat data correlated highly (stepwise linear regression model) with ground measurements of such parameters as temperature, conductivity, secchi depth, and several chemicals (nitrogen, phosphorus, potassium, calcium, and magnesium) (Rogers et al., 1975). Cautious use of such results is warranted since Landsat multispectral scanners record only in the green, red, and near infrared part of the spectrum and since the parameters measured are not always directly related to the reflected energy from the visible and near infrared regions. For example, the study found a high correlation between reflected red light (Band 5 of Landsat) and temperature. This relationship may hold under specific conditions, but it cannot be considered universal since the sensor recorded reflected energy and the target transmitted emitted energy. Basically, if a change in the chemical and/or physical properties of water causes a direct quantifiable change in the visible and/or near infrared reflectance characteristics of water, a cause/effect relationship can be established by using Landsat data; otherwise the results must be considered speculative.

Sea state, a most sought after measurement, has been calculated from sun glitter on aerial and satellite photography (McClain and Strong, 1969). Specular reflection was found to be indirectly related and diffuse reflection directly related to sea state. Active microwave imaging sensors (especially K- and X-band radars) are sensitive to wind-generated gravity-capillary waves, and therefore the intensity of the radar return (tonal value on the imagery) is related to wind speed. Following positive results from side-looking airborne radar (SLAR), an imaging radar is planned for the forthcoming Seasat, and global sea state data are anticipated by 1980. SLAR imaging has also been utilized for distinguishing ocean current, surf zone, shoreline configuration, mangrove vegetation, and other various coastal zone features (Lewis and MacDonald, 1972).

Living aquatic resources are identified, located, and mapped from various types of remotely sensed data. Visual spotting of fish schools and whales from a ship's mast has been operational for centuries. Today spotting is often done from a low- and slow-flying aircraft. On-board sonar and radar are also used to locate fish. Once located, the type of fish is often determined by the size and shape of the school on the video display unit.

Spectral reflectance was used to identify three species of pelagic fish (Loya and Graves, 1968), and whale census taking was facilitated by a specially developed water-penetration film. Light-intensified night-time photography and video displays were successful in the detection and identification of 34 types of fish in the coastal zone of California (Roithmayr, 1971) and of individual fish three meters below the surface (Roithmayr and Whittman, 1972).

By utilizing known preferred environs, fish have been detected by means of remote sensing techniques. Tuna locations off the Oregon coast correlate well with thermal contour maps of surface water derived from airborne thermal infrared scanner data (Pearcy, 1973). Turbidity patterns on Landsat imagery in the Mississippi Sound show a strong relationship with menhaden sightings (Kemmerer and Benigno, 1973).

The mapping of wetlands seems to be best accomplished with color infrared photography from aircraft because of the vegetation enhancement on color infrared photography and the interrelationships between vegetation and wetland environs. Aircraft data seem desirable because of the resolution requirements for identifying species and locating small, isolated wetland enclaves. An inventory of wetlands in Virginia using digital Landsat data and an unsupervised cluster analysis was not successful because a large number of wetland patches were smaller than the one-acre pixel (picture element or resolution cell) of the Landsat multispectral scanner (Penney and Gordon, 1975). Satellite data are functional for meso- and macro-scale studies, such as for monitoring coastal currents, turbidity plumes, and coastal landform changes (erosion and deposition), and for classifying water types (Gordon and Nichols, 1975).

Aircraft remotely sensed data were used by Anderson (1970), Anderson and Carter (1972), and others for monitoring wetland productivity, mapping wetland vegetation, and locating depositional and erosional problem areas and potential dredge sites. Eleven environmental zones in the Mississippi delta area were delineated by Weisblatt (1972) on multispectral aerial photography.

Coastal currents, freshwater springs in coastal environments, and interface mixing have been identified on thermal imagery. Currents and water parameters can be monitored via expendable drogues and over-the-horizon radar (OTH) tracking over an 18,000-km distance. Oil slicks are visible on many types of remote sensing, including ultraviolet, color infrared, thermal infrared, and radar. Coral reefs have been categorized and mapped from Landsat digital data by automatic processing (Smith, Rogers, and Reed, 1975).

For more information relating to remote sensing and its coastal zone applications, the reader is referred to Chapters 19 and 20 of volume 2 of the *Manual of Remote Sensing (Reeves, 1975)*. The *Proceedings of the 10th International Symposium on Remote Sensing of the Environment*, *Remote Sensing and Water Resources Management*, and *Proceedings of a Symposium on Coastal Mapping* are other valuable reference sources.

ANTHONY J. LEWIS

References

American Society of Photogrammetry, 1972. *Proceedings of a Symposium on Coastal Mapping*. Washington, D.C., 319p.

Anderson, R. R., 1970. Spectral reflectance characteristics and automated data reduction techniques which identify wetland and water quality conditions in the Chesapeake Bay, *NASA 3rd Annual Earth Resources Program Review* 3, 53-1–53-29.

Anderson, R. R., and Carter, V., 1972. Wetlands delineation by spectral signature analysis and legal implications, *NASA 4th Annual Earth Resources Program Review*, 3, 78-1–78-9.

Estes, J. E., and Senger, L. W., 1974. *Remote Sensing: Techniques for Environmental Analysis*. Santa Barbara, Calif.: Hamilton Publishing Company, 340p.

Geary, E. L., 1967. Coastal hydrography, *Photogramm. Eng.* 34, 44-50.

Gordon, H. H., and Nichols, M. M., 1975. Skylab MSS vs photography for estuarine water color classification, in *Proceedings of the 10th International Symposium on Remote Sensing of the Environment*. Ann Arbor, Mich.: Environmental Research Institute of Michigan, 421-435.

Kemmerer, A. J., and Benigno, J. A., 1973. Relationships between remotely sensed fisheries distribution information and selected oceanographic parameters in the Mississippi Sound, *Symp. Significant Results from ERTS-1, NASA SP-327* 1, 1685-1695.

Lewis, A. J., and MacDonald, H. C., 1972. Mapping of mangrove and perpendicular-oriented shell reefs in southeastern Panama with side-looking radar, *Photogrammetria* 28, 187-199.

Loya, B. R., and Graves, C. D., 1968. *Fish Identification by Remote Sensing*. Redondo Beach, Calif.: TRW Systems Group Rept. 11435-6001-R0-00.

McClain, E. P., and Strong, A. E., 1969. On anomalous dark patches in satellite-viewed sunglint areas, *Monthly Weather Rev.* 97, 875-884.

Pearcy, W. G., 1973. Albacore oceanography off Oregon, *Fishery Bull.* 71, 489-504.

Penney, M. E., and Gordon, H. H., 1975. Remote sensing of wetlands in Virginia, in *Proceedings of the 10th International Symposium on Remote Sensing of the Environment*. Ann Arbor, Mich.: Environmental Research Institute of Michigan, 495-503.

Polcyn, F. C., 1969. Depth determination by measuring wave surface effects, *NASA 2d Annual Earth Resources Program Review* 3, 52-1–52-17.

Polcyn, F. C., 1970. Measurement of water depth by multispectral ratio techniques, *NASA 3rd Annual Earth Resources Program Review* 3, 61-1–61-ll.

Reeves, R. G., ed., 1975. *Manual of Remote Sensing*, 2 vols. Falls Church, Va.: American Society of Photogrammetry, 2144p.

Rogers, R. H.; Shah, N. J.; McKeon, J. B.; Wilson, C.; and Reid, L., 1975. Application of Landsat to the surveillance and control of eutrophication in Saginaw Bay, in *Proceedings of the 10th International Symposium on Remote Sensing of the Environment*. Ann Arbor, Mich.: Environmental Research Institute of Michigan, 437-446.

Roithmayr, C. M., 1971. Airborne low-light sensor

detects luminescing fish schools at night, *Comm. Fish Rev.* **32**, 42-51.

Roithmayr, C. M., and Whittman, F. P., 1972. Low-light level sensor development for marine resource assessment, *in Proceedings of the 8th Annual Conference and Exposition.* Washington, D.C.: Marine Technology Society, 277-288.

Smith, V. E.; Rogers, R. H.; and Reed, L. E. 1975. Thematic mapping of coral reefs using Landsat data, *in Proceedings of the 10th International Symposium on Remote Sensing of the Environment.* Ann Arbor, Mich.: Environmental Research Institute of Michigan, 585-594.

Specht, M. R., and others, 1973. New color film for water-photography penetration, *Photogramm. Eng.* **39**, 359-369.

Stoertz, G. E.; Hemphill, W. R.; and Markle, D. A., 1969, Airborne fluorometer applicable to marine and estuarine studies, *Marine Tech. Soc. Jour.* **3**, 11-26.

Thomann, G. C., 1973. Remote measurement of salinity in an estuarine environment, *Remote Sensing Environment* **2**, 249-259.

Umbach, M. J., and Harris, W. D., 1972. *Photogrammetric Bathymetry.* Rockville, Md.: NOAA, National Ocean Survey.

Weisblatt, E. A., 1972. *Multispectral Discrimination of Deltaic Environments.* Ph.D. dissertation, Department of Geography, Louisiana State University, 148p.

Cross-references: *Base-line Studies; Coastal Characteristics, Mapping of; Environmental Impact Studies; Photogrammetry.* Vol. XIII: *Remote Sensing.*

REPLENISHMENT—See BEACH NOURISHMENT

RESTINGA

Restinga is a Spanish word for reef or sand bank and was recorded as early as the fifteenth century. It is, perhaps, a derivative of *rock string* and is preserved today to designate a sand bar or barrier (Hoyt, 1967) longitudinally emerged—especially when enclosing an *albufera* (q.v.) or lagoon. When it does enclose an albufera or lagoon (Zenkovich, 1967), the restinga is usually interrupted by one or more inlets (*golas*) connecting the albufera (Rosselló, 1972) with the open sea (Larras, 1964).

VINCENÇ M. ROSSELLÓ

References

Hoyt, J. H., 1967. Barrier island formation, *Geol. Soc. America Bull.* **78**, 1125-1136.

Larras, J., 1964. *Embouchures, Estuaires, Lagunes et Deltas.* Paris: Eyrolles, 171p.

Rosselló, V. M., 1972. Los ríos Júcar y Turia en la génesis de la Albufera de Valencia, *Cuadernos de Geografía* **11**, 7-25.

Zenkovich, V. P., 1967. *Processes of Coastal Development.* Edinburgh: Oliver & Boyd, 738p.

Cross-references: *Accretion Ridge; Albufera; Barrier Islands; Geographic Terminology; Inlets and Inlet Migration; Lagoon and Lagoonal Coasts; Nehrung; Spits.*

RETROGRADATION AND RETROGRADING SHORELINE

Johnson (1919) first used the term *retrograding shoreline* when he distinguished it from a *prograding shoreline.* As with progradation, retrogradation of a shore in loose sediments previously deposited by wave action is a function of time, energy, sedimentary supply, organic growth, and—above all—sea level change. *Time* is needed for equilibrium after any change in the other factors. *Energy* (see Progradation and Prograding Shoreline) increase during storms may produce temporary retrogradation in an area of net progradation or speed up net retrogradation. *Organic growth* on dunes may only retard rather than halt retrogradation. *Sedimentary supply* can be reduced in two ways by man: Dredging from the sea floor may upset the sea floor/sea level equilibrium (see *Sea Level Changes*) and cause coastal erosion; and where longshore drift is important, any man-made structure that halts retrogradation or promotes progradation in one part of the coast will only cause increased erosion of the shore in the down-current direction.

Sea level rise promotes retrogradation for the following reason. In prograded areas—areas of dynamic equilibrium between sea level and sea floor—any sand transported from the shore and placed on the sea floor during a storm is soon returned to the coast providing sea level is static. When sea level is rising, some or all of this storm-removed sand may not be returned to the shore and retrogradation is permanent. Nevertheless, a rising sea level only favors, and need not cause, retrogradation, since other factors—particularly rate of rise as related to sedimentary supply rates—are important. This principle, first recognized by Bruun (1962), is sometimes called the *Bruun Rule* (q.v.), and has been formalized by Schwartz (1968) as $S = f(A_e, T_e)$, where S equals coastal change, A_e equals equilibrium amplitude of sea level change, and T_e equals equilibrium time.

J. C. SCHOFIELD

References

Bruun, P., 1962. Sea-level rise as a cause of shore erosion, *Am. Soc. Civil Engineers Proc., Jour. Waterways and Harbors Div.* 88, 117-130.
Johnson, D. W., 1919. *Shore Processes and Shoreline Development.* New York: John Wiley & Sons, 584p.
Schwartz, M. L., 1968. The scale of shore erosion, *Jour. Geology* 76, 508-517.

Cross-references: *Bruun Rule; Coastal Engineering; Coastal Erosion, Formations; Progradation and Prograding Shoreline; Seal Level Changes; Submergence and Submerged Shoreline.* Vol. III: *Retrograding Shoreline.*

REYNOLD'S NUMBER

Reynold's number (R_e) is defined as $R_e = \rho v L/\eta$, where ρ is the density of water or fluid; η, its viscosity; v, the velocity of the flow; and L, a representative length related to the boundary of the flow. Reynold's number is important in the discrimination between laminar and turbulent flows. In a pipe of diameter L, for small values of R_e (a few thousand; often the boundary value is accepted to be about 2000)— small velocity, small dimensions, and great viscosity—the motion is laminar. The greater the R_e, the more turbulent is the motion.

In laminar motion, the fluid is similar to being layered, without mixing between neighboring layers. In turbulent motion, the concept of *layers* vanishes; here there exists a random motion of the water particles that could be thought of as superimposed on an *averaged* motion. The turbulence involves the amplitude and the frequency of the oscillating motion of the particles, and increases with increasing velocity—rising R_e.

Turbulence occurs not only in pipes but in all natural motions, appearing near the walls of a channel or basin and transmitted to the entire body of water. Oceanic motion is primarily turbulent and this condition can be seen, for example, by assuming that with $L \cong 4$ km and $v = 10$ cm/sec and because $\eta = 0.01$ and $\rho = 1$, then R_e becomes $1 \times 400,000 \times 10/0.01 \cong 4 \times 10^8$, a distinctly turbulent motion. Only the shallowest layer of water near the beach, having little velocity, would not denote turbulence.

In both turbulent and laminar motion, there is a large and swift exchange of substances (dissolved or suspended in the water) and of energy.

R_e may also be expressed as $R_e = \rho v^2/\eta(v/L)$, where ρv^2 represents the inertial force (per unit area) of the flow and $\eta(v/L)$ is the shear stress (viscous forces) per unit area (McCormack and Crane, 1973). In this case, the Reynold's number is the ratio between these two fundamental forces in the motion of water.

FERRUCCIO MOSETTI

Reference

McCormack, P.D., and Crane, L., 1973. *Physical Fluid Dynamics.* New York: Academic Press, 487p.

Cross-references: *Ferrel's Law; Fourier Analysis; Froude Number; Mathematical Analysis; Richardson's Number; Waves.*

RHODOPHYCOPHYTA

The Rhodophycophyta (Rhodophyta), or *red algae*, includes about 4000 species of primarily marine plants. Only about 2% of this total inhabit freshwater (Bourrelly, 1970).

The red algae are characterized as a group by the nature of their pigments, stored food products, and cell wall, and by the lack of any motile stages in the life history. Pigments found in this group are chlorophyll a, possibly chlorophyll d, α and β carotene, several xanthophylls, and the phycobilin pigments allophycocyanin, phycocyanin, and phycoerythrin. It is the latter that is responsible for the red color for which the group is named, although sometimes photodestruction of this pigment or chromatic adaptation will give the plant a greenish or yellowish aspect. Additionally, in its role as an accessory pigment that can absorb shorter wavelengths of light, phycoerythrin allows red algae to live more deeply (to 200 m) than other plants.

Excess photosynthate is stored primarily as floridean starch, a complex carbohydrate. The cell wall is composed of cellulose and often pectin. In the *coralline algae* the outer part of the cell wall is heavily calcified. The complex hydrocolloids agar and carrageenan may also occur in the cell wall and intercellular spaces.

Two classes are usually recognized in the Rhodophycophyta: Bangiophyceae and Florideophyceae. The former includes plants that are relatively simple in construction, while genera in the latter are usually complex. It is best to consider these two classes separately.

In the Bangiophyceae, the morphology may be unicellular, colonial, filamentous or sheetlike. Cell division is primarily intercalary, and pit connections are absent in most. Asexual reproduction by nonmotile spores or fragmentation is common. The occurrence of sexual reproduction is still debated (Dixon, 1973).

The best-studied member of this class is *Porphyra*, the nori or laver of commerce. Its life history involves the photoperiod-controlled alternation of a sheetlike plant (the harvestable *Porphyra*) that grows on rocks or other algae, with a filamentous *Conchocelis* phase inhabiting shells. Miura (1975) discusses *Porphyra* cultivation in Japan.

The Florideophyceae includes most of the genera of red algae. Cell division is apical, and the basic construction is filamentous. The form of all members of this class is determined by the precise arrangement of branched filaments, often so tightly compacted as to obscure the filamentous nature. Pit connections (appearing like pores) occur between adjacent cells of these filaments.

Sexual reproduction is oogamous, involving the transfer of a small, nonmotile sperm to the carpogonium (female gametangium). The nature of the special branch bearing the carpogonium and the postfertilization events are significant in classification, though often difficult to interpret. Life histories in Florideophyceae are triphasic and probably the most complex in the plant kingdom (See Knaggs, 1969, for a review).

Reference should be made to a general text on algae (Dawson, 1966) or cryptogamic botany (Bold, 1973; Scagel et al., 1965) for a more complete discussion of the difference between the classes. Reviews of red algae in general (Dixon, 1973; Hommersand and Searles, 1971) and the classification of the red algae (Papenfuss, 1966) should also be consulted. Fritsch's (1945) classical text on morphology is still valuable, while the serious student must become familiar with Kylin's (1956) study of the genera of red algae. Physiology and biochemistry are covered by Lewin (1962) and Steward (1974).

Several members of the red algae are of significant commercial value. Primarily red algae are used directly as human food (for example, nori, dulse) or extracted to yield the complex polysaccharides agar and carrageenan. These are most valuable because of their properties as emulsifiers and stabilizers. Treatments of the economics of red algae may be found in Chapman (1970), Levring, Hoppe, and Schmid (1969), and Okazaki (1971).

Those red algae of the class Corallinaceae have calcified cell walls and are often found as fossils. Johnson (1961) provides a good treatment of these. Coralline algae actually contribute the bulk of *coral reefs*, the depth of which often exceeds 1000 feet.

Identification of extant marine algae is often difficult because of the plasticity of forms influenced by different environmental factors, and a specialist may need to be consulted for verification.

A specialist should also be consulted about the existence of local floras and their weaknesses. Dawson (1966) lists regional floras for North America; additionally, Abbott and Hollenberg (1976) have provided a flora that will be useful along most of the Pacific coast of North America.

ROBERT B. SETZER

References

Abbott, I. A., and Hollenberg, G. J., 1976. *Marine Algae of California*. Stanford: Stanford University Press, 827p.

Bold, H. C., 1973. *Morphology of Plants*, 3rd ed. New York: Harper and Row, 668p.

Bourrelly, P., 1970. *Les alques d'eau douce*. Tome III, Les alques bleues et rouges, les Eugle niens, Peridiniens et Cryptomonadines. Paris: Ed. N. Boubee & Cie, 512p.

Chapman, V. J., 1970. *Seaweeds and Their Uses*, 2d ed. London: Methuen, 304p.

Dawson, E. Y., 1966. *Marine Botany: An Introduction*. New York: Holt, Rinehart & Winston, 371p.

Dixon, P. S., 1973. *Biology of the Rhodophyta*. New York: Hafner Press, 285p.

Fritsch, F. E., 1945. *The Structure and Reproduction of the Algae*, Vol. II. London: Cambridge University Press, 939p.

Hommersand, M. H., and Searles, R. B., 1971. Bibliography on Rhodophyta, *in* J. R. Rosowski and B. C. Parker, eds., *Selected Papers in Phycology*, Lincoln: University of Nebraska Press, 760-767.

Johnson, J. H., 1961. *Limestone-building Algae and Algal Limestones*. Golden: Colorado School of Mines, 297p.

Knaggs, F. W., 1969. A review of Florideophycidean life histories and of the culture techniques employed in their investigation, *Nova Hedwigia Beihefte* 18, 293-330.

Kylin, H., 1956. *Die Gattungen der Rhodophyceen*. Lund: CWK Gleerups, 673p.

Levring, T., Hoppe, H. A., and Schmid, O. J., 1969. *Marine Algae. A Survey of Research and Utilization*. Hamburg: Cram, De Gruyter & Co., 421p.

Lewin, R. A., ed., 1962. *Physiology and Biochemistry of Algae*, New York: Academic Press, 929p.

Miura, A., 1975. *Porphyra* cultivation in Japan, *in* J. Tokida and H. Hirose, eds., *Advance of Phycology in Japan*. The Hague: W. Junk, 273-304.

Okazaki, A., 1971. *Seaweeds and Their Uses In Japan*. Tokyo; Tokai University Press, 165p.

Papenfuss, G. F., 1966. A review of the present system of classification of the Florideophycidae, *Phycologia* 5, 247-255.

Scagel, R. F.; Bandoni, R. J.; Rouse, G. E.; Schofield, W. B.; Stein, J. R.; and Taylor, T. M. C., 1965. *An Evolutionary Survey of the Plant Kingdom*. Belmont, Calif.: Wadsworth Publishing Co., 658p.

Stewart, W. D. P., ed., 1974. *Algal Physiology and Biochemistry*. Oxford: Blackwell's, 960p.

Cross-references: *Algal Mats and Stromatolities; Chlorophyta; Chrysophyta; Coastal Flora; Coral Reef Coasts; Cryptophyta; Cyanophyceae; Eugleno-*

phyta; Organism-Sediment Relationship; Phaeophycophyta; Reefs, Noncoral, Vol. III: *Algae;* Vol. VI: *Algal Reef Sedimentology.*

RHYTHMIC CUSPATE FORMS

Rhythmic coastal features encompass a wide range of forms, covering considerable variations in size, morphology, and situation. The nomenclature adopted for *shoreline* forms includes beach cusps, having curvature concave to the sea; protuberances (sand waves, shoreline rhythms, giant cusps, nesses) (Dolan and Ferm, 1968); and cuspate spits resulting from lagoon sedimentation. *Submerged* rhythmic coastal forms are classified as crescentic bars, having curvature convex to the sea; multiple bars that are straight and parallel to the coast; and transverse bars (finger bars) that can be straight, at an angle, or transverse to the coast.

Characteristics and Processes

Beach Cusps. Beach cusps are the smallest rhythmic cuspate feature. They form in a variety of materials from coarse shingle to fine sand, but are usually best developed and most common on coarse beaches, often occurring on mixed sand and shingle beaches. The coarser material forms the *horns*, pointing seaward, while the finer accumulates in the *bays*. Records of beach cusps from many parts of the world, collected by Russell and McIntyre (1965), indicate that their *wavelength* falls into two groups, a large mode around 14 m, and a smaller one at about 47 m. They are characteristic of the swash-backwash zone of wave activity, usually forming on the upper foreshore at the high tide level.

A considerable number of different hypotheses have been put forward to explain the generation of beach cusps. Many of them are probably produced by the formation of *edge waves* in the surf and swash zone. Edge waves are free modes of water trapped against a shoaling beach (Bowen and Inman, 1971). The amplitude of the edge wave varies sinusoidally along the shore, and for a zero mode edge wave it decreases exponentially offshore. On steep beaches, which are usually coarse beaches, edge waves are related to the reflection of wave energy from the shore. In these conditions standing waves can be set up, with a longshore component of motion that creates the cusps. The standing waves can either be trapped between headlands on enclosed beaches or be generated along long, straight beaches, sometimes by the process of wave refraction.

Cuspate features are likely to be caused by the finite amplitude of an edge wave at the shoreline, resulting in a sinusoidal run-up pattern along the shore. Erosion will occur at the antinodes, since velocity here is offshore. The wavelength of the cusps will be half that of the edge waves. The cusp points should be opposite rip currents, which form as part of the edge wave circulationn. Edge waves will form only when the relationship between the breaker height and the wave period is suitable; thus the presence of cusps is intermittent where they are created by edge waves. Steep beaches facilitate edge wave formation, since they encourage the reflection of some of the wave energy from the shore. The relationship between the swash and backwash is important in the setting up of edge waves. Field observations on Slapton beach, Devon, revealed edge waves with a length of 32 m, which agreed well with the theoretical value of 34 ± 5 m (Huntley and Bowen, 1975).

Other possible causes of beach cusps include shore drift (Schwartz, 1972), in which sand moves alongshore in pulses, creating waves of sand that could develop into cusps by the action of the swash and backwash. Sheetflood (Gorycki, 1973) has also been suggested as a possible process of cusp formation. The advancing swash forms salients with an element of lateral spread that could develop a circulatory flow resulting in cusp formation.

Protuberances. These features are similar in form to beach cusps, but usually with less pronounced horns. They occur on long, straight, sandy beaches—for example, along the east coast of the United States. Their wavelength is between 150-1000 m, with a value of 500-600 m being common. Their amplitude is usually about 15-25 m between the horns and bays. Coarser sediment occurs on the horns than in the bays. They often migrate alongshore at rates between 100 and 200 m/month.

The protuberances are related to a nearshore cell circulation pattern. The circulation can be set up by wave refraction, resulting in variation of wave height alongshore at the shoreline. Currents flow from zones of high waves to zones of low waves along the shore, meeting in the zone of low waves and flowing seaward as strong rip currents (Komar, 1971; Sonu, 1972). Sometimes the cusp points occur opposite the rips, while at other times or elsewhere the horns appear between the rips; laboratory experiments produced horns opposite the rips. Where waves approach normal to the shore and submarine relief causes suitable refraction, the cell pattern can develop. Where the waves approach the shore at an angle, a meandering system of currents is more likely to develop, and this pattern can also give rise to rhythmic protuberances along the shore. The bays occur where the current diverges offshore; the horns, where the current is directed towards the shore. Oblique wave approach can sometimes result in the formation of skewed, asymmetrical cusps with a rhythmic pattern.

As the protuberances move alongshore, the beach profile changes from concave up in the bays to straight and then convex up as the horn passes. Differences of water level have also been observed over the rhythmic system. The water level was found to be high on the shoals and low in the rip channels, thus helping to maintain the circulation and, with it, the morphological features, including the cuspate shoreline.

Cuspate Spits. Cuspate spits occur in rhythmic sequence in some elongated lagoons. These spits are often sharply pointed and divide the lagoon into almost circular lakes. In other areas the points of the spits interdigitate at either side of the lagoon. The latter pattern is well exemplified on the landward side of Nantucket Island, Massachusetts, while the circular lakes are found in Kosi lagoon on the Zululand coast of southeast Africa. The cuspate spits separate almost circular lakes varying from 1-7 km in diameter. Similar spits in rhythmic sequence occur in the Chukchi Sea, some of which are symmetrical and others asymmetrical. These features are about 1 km apart and tend to increase in symmetry with time as they become adjusted to the pattern of sediment movement under the effect of short, refracted waves in the lagoon.

The features probably form as a result of longshore movement of sediment induced by winds creating oblique waves in the lagoons (Zenkovich, 1959). The system becomes stable as the sediment movement forms symmetrical spits separating circular lakes. Equilibrium is thus achieved between the incident waves and the orientation of the shoreline, reducing the longshore movement and hence achieving stability when the spits and circular lakes between them are fully developed.

Crescentic Submarine Bars. Submarine bars form widely along coasts, especially those with a small tidal range or limited fetch. These bars sometimes develop a crescentic form, with the points of the crescents pointing toward the shore. These features occur, for example, in the Mediterranean, often in enclosed bays. They also occur on open coasts, such as along the Virginia shore, where their wavelengths are often between 400-600 m, with an amplitude of about 50 m. At times two series of crescentic submarine bars occur together, the outer series having the longer wavelength. They occur off the embayed coast of New Brunswick in Kouchibouguac Bay (Greenwood and Davidson-Arnott, 1975), where two systems exist. The inner system changes most rapidly and is often connected to the beach face by giant cusps. It is complex with 2-3 *bars*, which are asymmetrical in profile. Rip currents flow in channels across the bars in the bays, between the horns that are connected to the cusps. The rips are spaced 90-300 m apart, the average being 243 m. The bars vary in pattern both spatially and temporally, being either parallel, transverse, crescentic, or irregular. The outer system is simpler, and the mean wavelength is about 500-600 m, with an amplitude of 30-40 m, the distance offshore being about 250 m. The depth of water over the outer bar is 1.8 m at low tide, with a maximum bar relief of 2.75 m. The outer system is relatively stable, migrating at about 10 m/month alongshore during the summer season.

The crescentic submarine bars have been explained in terms of *edge waves*. The edge waves that create crescentic submarine bars are longer in wavelength than those responsible for the beach cusps. They can be generated either by two incident wave trains or by a single wave train. Standing waves are set up by the reflection of energy from the beach. Edge waves of 30-60 seconds period are needed to form crescentic bars. With a zero-mode edge wave, the drift velocity is offshore, thus tending to build a submarine bar. The bars that form are crescentic, and their length is half that of the edge waves. If the orbital velocities were large enough, there would be a series of crescents equal to the modal number n. The dimensions should be as given in Table 1. The period of wave needed to produce an edge wave of length 1000 m would be of 40-50 seconds for $n = 2$. The crescentic bars would have a wavelength of 500 m, half that of the edge wave. Long-period edge waves can be formed in bays by the creation of standing waves between the enclosing headlands, the wavelength being dependent on the length of the bay. The crescentic forms can result from the operation of the longer-period edge waves. They can also operate on long, straight coasts when *surf beat* is present. These long waves have periods of several minutes. Their reflection from the shore can set up edge waves, which in turn can initiate the crescentic forms.

Multiple Submarine Bars. Series of closely spaced bars lying parallel to the coast have been

TABLE 1. Dimensions of Crescentic Bars

Mode	Position	Bar Length	Distance to Horn	Distance to Bar
$n = 1$		0.5L	0.08L	0.24L
$n = 2$	Inner bar	0.5L	0.046L	0.11L
	Outer bar	0.5L	0.27L	0.51L

described from relatively low energy shorelines, including the north coast of Alaska and Florida, where up to 11 parallel bars may occur. The spacing of the bars usually increases offshore. Those off north Alaska lie within 1 km of the shore, in depths up to 10 m, with 2-3 bars being most common, but 4-5 occur at times. The bars lie outside the normal surf zone, and their spacing increases offshore as the offshore gradient decreases (Short, 1975).

The bars have been explained by the generation of a standing wave, where they are closely spaced and regular, although not all submarine bars are formed in this way. In theory the normal incident wave is partially reflected over water of constant depth. The theory can be extended to a sloping bottom to account for the increase in bar spacing offshore. The standing wave process of multiple bar formation applies where waves do not break. An investigation of the mass transport velocities associated with Stokes bottom boundary layer in waves moving over a gently sloping beach shows that a standing wave could form submarine bars. There will be more bars where the beach slope is small. When the reflection coefficient is between 0.414-1.0 the spacing increases offshore. The predicted positions of the nodal and antinodal points agree with the spacing of the bars, which are probably formed by the reflection of obliquely incident long-period progressive waves from the shore forming standing waves (Lau and Travis, 1973).

Transverse Submarine Bars. Transverse bars are rhythmic features, although they tend to be straight rather than cuspate, running seaward normal to the coast for up to about 3.3 km in length. Formed of sand, they usually have an amplitude of about 0.5-1.0 m and lengths between 80-3300 m. Their spacing also varies from about 10 m apart to values over 2 km. They occur where wave energy is low, sand plentiful, and the beach slope gentle. They have been reported on coasts of Lake Michigan, Brazil, Canada, Denmark, Florida, Japan, and U.S.S.R. They sometimes migrate along the coast, and their angle of attachment varies.

Observations off Florida show that their spacing depends on the inverse of the beach slope and on the square of the drift velocities over the bars (Barcilon and Lau, 1973). Patterns formed by the bars can be complex when two or more sets at different angles intersect and coexist. On the east shore of St. James Island, Florida, the bars had a relief of 20-30 cm, and were covered by 60 cm of water at high tide, but only a few centimetres at low tide. At low tide borelike waves 5-10 cm high and 4 second period refract over the bar crests to form a diagonal pattern of crossing wave crests. The longshore current in the area was 0.25 m/sec, and this current may be necessary to form the bars, although they may be maintained by the wave refraction pattern. The *Froude number* at low tide is calculated to be 0.4, which should result in the formation of dunes in the longshore flow. Thus it is thought that a sustained longshore current may be required to form the transverse bars.

Wave refraction has also been considered as the process that forms the bars by Niedoroda (1973). The wave crests tend to cross over the bar crests, bringing sand onto the crest and maintaining it, while sand is carried seaward along the troughs between the bar crests. The bars are in equilibrium with the pattern of water movement under the refracted waves, but the wave refraction is itself dependent on the relief of the bars. Thus some initiating process is required, and this may well be supplied by the steady longshore current and the dunes associated with it, if the Froude number is suitable for their formation. The longshore current may be tidal, since this would provide the necessary direction and consistency of flow, while refraction would maintain the features once formed.

Conclusion

Rhythmic cuspate and crescentic features occur in a wide range of forms and in many different environments, varying greatly in size and mobility. The two most important features are the concave seaward shoreline cusps, and the convex seaward submarine crescentic bars. A number of different processes are involved in their formation, many of which are interrelated, including wave refraction, longshore currents, rip currents—all of which together produce cellular or meandering currents. These different currents can often be linked as parts of one system induced by the generation of edge waves along the shore. Thus edge waves provide the most satisfactory explanation of many cuspate forms that occur in rhythmic sequence, since one of the fundamental properties of an edge wave is that it varies sinusoidally along the shore. One problem is the feedback between the form and the process. Once formed, the features help to maintain the circulation in equilibrium relative to their form, but their initiation may require special conditions.

CUCHLAINE A. M. KING

References

Barcilon, A. I., and Lau, J. P., 1973. A model for the formation of transverse bars, *Jour. Geophys. Research* 78, 2656-2664.

Bowen, A. J., and Inman, D. L., 1971. Edge waves and crescentic bars, *Jour. Geophys. Research* 76, 8662-8671.

Dolan, R., and Ferm, J. C., 1968. Crescentic landforms along the mid-Atlantic coast, *Science* 159, 627-629.

Gorycki, M. A., 1973. Sheetflood structure and mechanism of beach cusp formation and related phenomena, *Jour. Geology* 81, 109-117.

Greenwood, B., and Davidson-Arnott, R. G. D., 1975. Marine bars and nearshore sedimentary processes, Kouchibouguac Bay, New Brunswick, in J. Hails and A. Carr, eds., *Nearshore Sediment Dynamics and Sedimentation*. London: Wiley, 123-150.

Huntley, D. A., and Bowen, A. J., 1975. Field observations of edge waves and their effect on beach material, *Geol. Soc. Jour.* 131, 69-82.

Komar, P. D., 1971. Nearshore cell circulation and the formation of giant cusps, *Geol. Soc. America Bull.* 82, 2634-2650.

Lau, J. P., and Travis, B., 1973. Slowly varying Stokes waves and submarine longshore bars, *Jour. Geophys. Research* 78, 4489-4497.

Niedoroda, A. W., 1973. Sand bars along low energy beaches: Transverse bars, in D. R. Coates, ed., *Coastal Geomorphology*. Binghamton: State University of New York, 103-113.

Russell, R. J., and McIntyre, W. G., 1965. Beach cusps, *Geol. Soc. America Bull.* 76, 307-320.

Schwartz, M. L., 1972. Theoretical approach to the origin of beach cusps, *Geol. Soc. America Bull.* 83, 1115-1116.

Short, A. D., 1975. Multiple offshore bars and standing waves, *Jour. Geophys. Research*, 80, 3838-3840.

Sonu, C. J., 1972. Field observations of nearshore circulation and meandering currents, *Jour. Geophys. Research*, 77, 3232-3247.

Zenkovich, V. P., 1959. On the genesis of cuspate spits along lagoon shores, *Jour. Geology* 67, 269-277.

Cross-references: *Bars; Beach Cusps; Beach Processes; Cuspate Foreland; Cuspate Spits; Edge Waves; Lagoonal Segmentation; Rip Currents; Ripple Marks; Sediment Transport; Submarine Bar; Surf Beat; Waves.* Vol III: *Littoral Processes;* Vol. VI: *Littoral Processes.*

RIA AND RIA COAST

According to Richthofen's (1886) definition, a *ria* is a funnel- or wedged-shaped drowned valley on coasts that are transverse to the structural grain of the land, and therefore contrasts with the *Dalmatian type* valley of longitudinal folded structure. He adopted ria from the names given to gulfs or wide-mouthed bays, with considerable extensions inland, on the Galician-Asturian coast of northwest Spain (Ria de Vigo, de la Coruña, del Farrol), although in fact this region does not fully meet the structural conditions contained in Richthofen's definition. The deeply embayed coast of southwest Ireland (Bantry and Dingle bays, for example), with its system of major rectilinear subsequent valleys that reach the sea approximately at right angles to the trend of the coast, is considered to be the type *ria coast*.

Various subsequent interpretations of the term have appeared in the geomorphological literature and these have been reviewed by Cotton (1956), who summarizes the descriptions of Penck, Geikie, Gulliver, de Martonne, Davis, Johnson, Jutson, Guilcher, Upson, Birot and Solé Sabarís, to name a few researchers who have referred to this feature. But, as Cotton states: "Some European geomorphologists now treat ria as merely a synonym of drowned valley, without distinction as to kind or size, including some mere creeks in this category. Thus, in a recently published work on coastal geomorphology by Guilcher, 'ria' is defined as any valley system partly invaded by the sea."

De Martonne criticized Richthofen's definition and described *coasts of transverse structure* as the "Finistère type." In fact, he cited the embayed margin of Finistère (western Brittany, France) and the gulfs of the Smyrna coast of Asia Minor as type examples.

Cotton, in his lucid paper, succinctly summarizes the differences of opinion about the exact meaning of rias and rias coasts as follows: "Though it might make for convenience to include all such ria-like bays and gulfs in a broad category of rias, a more conservative and strictly defensible course would be to exclude them as not conforming to Richthofen's definition, and to refer to them instead as *ria-like*. Perhaps it would be in order to think of rias *sensu stricto* ("true" rias) and *sensu lato* ('ria-like')." For example, the drowned valley systems like Botany Bay, Port Jackson (Sydney Harbor), and Broken Bay (Hawkesbury River system), New South Wales, which dissect a terrain of horizontal structure, are well known for their ria-like characteristics.

Rias must also be considered with regard to differential movements as well as to submergence, as exemplified by the rias sensu lato on the coast of northwest Spain.

JOHN R. HAILS

References

Cotton, C. A., 1956. Rias *sensu stricto* and *sensu lato, Geog. Jour.* 122, 360-364.

Richthofen, F. von, 1886. *Führer für Forschungreisende*. Hanover: Jänecke, 734p.

Cross-references: *Estuaries; Geographic Terminology; Submergence and Submerged Shoreline.* Vol. III: *Ria, Rias Coast and Related Forms.*

RICHARDSON'S NUMBER

Richardson's number is involved in the stability of the vertical wind profile, related to the turbulent vertical heat flux.

Richardson's number has several applications in oceanography; in fact wave generation as well as sediment transport are strictly related to the form of the wind profile near the sea surface. However, it is more interesting in meteorology or in problems involving turbulent exchanges between atmosphere and ocean.

Two forms (Roll, 1965) of Richardson's number (R_f and R_i) are related here:

$$R_f = \frac{-gH}{c_p \, T \, \tau(\delta u/\delta z)}$$

where g is the gravity acceleration, H vertical heat flux, $\tau(\delta u/\delta z)$ the shear stress per unit volume, c_p the specific heat of the air at constant pressure, T mean air temperature (absolute) at sea surface; and

$$R_i = g \frac{(\delta \Theta/\delta z)}{T \, (\delta u/\delta z)^2}$$

where $\delta\Theta/\delta z$ is the vertical gradient of potential air temperature and $\delta u/\delta z$ the gradient of mean wind speed.

FERRUCCIO MOSETTI

Reference

Roll, H. U., 1965. *Physics of the Marine Atmosphere.* New York: Academic Press, 426p.

Cross-references: *Ferrel's Law; Fourier Analysis; Mathematical Models; Reynold's Number, Waves; Wind.*

RIDGE AND RUNNEL

Ridges and runnels occur on sandy beaches with a low overall gradient, which often results from a surfeit of sand. The ridges are built up by the waves attempting to establish their equilibrium gradient where the overall gradient is too low. The runnel is deepened by the falling tide. These are common on beaches with a large tidal range and in relatively enclosed seas, such as the North Sea and Irish Sea (King and Williams, 1949). Where waves approach the coast at right angles, the ridges lie parallel to the shore, as at Blackpool; where the waves approach at an angle, the ridges diverge slightly offshore, and on any one profile they appear to move inshore as they are transferred laterally in the direction of the dominant drift. This occurs on the south Lincolnshire coast. The ridges occur mainly in enclosed seas because the *equilibrium gradients* are steeper owing to the shorter waves in these situations. The features have been called *low and ball* by Cornish (1898) and Evans (1940).

CUCHLAINE A. M. KING

References

Cornish, V., 1898. On sea beach and sand banks, *Geog. Jour.* 11, 628-651.

Evans, O. F., 1940. The low and ball of the east shore of Lake Michigan, *Jour. Geology* 48, 476-511.

King, C. A. M., and Williams, W., 1949. The formation and movement of sand bars by wave action, *Geog. Jour.* 113, 70-85.

Cross-references: *Bars; Beach Profiles; Major Beach Features; Ripple Marks.* Vol. III: *Beach;* Vol. VI: *Ripple Marks.*

RILL MARKS

Rill marks are erosional channels found only on the lower foreshore that are formed by the escape of ground water when the tide falls below the water table.

As the tide rises, water enters the sand and flows landward. However, before it can saturate the beach and form a horizontal water table, the tide falls and the water on the seaward side of the beach escapes from the sand, lowering the water table there. Thus, as the tide rises and falls, a wave of tidal period moves landward through the sand. It lags in phase behind the tide, 1hr/20m of beach, diminishes in amplitude in a landward direction (Emery and Foster, 1948), and dies off exponentially to a value of several cm 60 m from the shoreline (Fausak, 1970). This tidal effluent has been estimated at $0.0 \times 29 m^3/m^2/tide$ (Dominick, Wilkins, and Roberts, 1971).

This ground water does not seep out uniformly, but is irregularly spaced owing to grain orientation and permeability inhomogeneities in the sand. Elongate grains align themselves with their long axes to the water movement. Because the swash moves more or less perpendicular to the beach, the grains align themselves at right angles to the beach trend (Curray, 1956) and imbricate themselves in a shoreward direction. Measurements of permeability (Emery, 1960) shows that the lowest value is vertical, owing to horizontal laminae. The greatest permeability is horizontal and parallel to the beach. Therefore only after the intermediate value of permeability perpendicular to the beach has been surpassed will the water escape. The escaping water will scour out the mud or sand over which it flows, forming a transient runnel or rill mark.

Rill marks are initially 2-10 mm wide and about 1mm deep, and are narrowly dendriditic in outline. This pattern may be 1/3 as wide as it is long (Twenhofel, 1932).

BENNO M. BRENNINKMEYER

References

Curray, J. R., 1956. Dimensional grain orientation studies of recent coastal sands, *Am. Assoc. Petroleum Geologists Bull.* 40, 2440-2456.
Dominick, T. F.; Wilkins, B.; and Roberts, H., 1971. Mathematical model for beach groundwater fluctuations, *Water Resources Research,* 7, 1626-1635.
Emery, K. O., 1960. *The Sea off Southern California.* New York: John Wiley & Sons, 365p.
Emery, K. O., and Foster, J. F., 1948. Water table in marine beaches, *Jour. Marine Research* 3, 644-654.
Fausak, L. E., 1970. The Beach Water Table as a Response Variable of the Beach-Ocean-Atmosphere System. Unpublished thesis, Gloucester Point: Virginia Institute of Marine Science, 52p.
Twenhofel, W. H., 1932. *Treatise on Sedimentation.* Baltimore, Md.: The Williams and Williams Co., 926p.

Cross-references: *Backwash Patterns; Beach Processes; Ghyben-Herzberg Ratio; Minor Beach Features; Sediment Transport; Swash Mark; Water Table.*

RIP CURRENT

A *rip current* is a narrow seaward flow of water from the beach to within, or more commonly to outside, the breaker zone. The name was first proposed by Shepard (1936). It now has replaced *undertow,* which had been used for either backwash or rip current. Rip currents are fed by longshore currents, which increase in velocity from about zero midway between two adjacent rips and reach a maximum just after turning seaward. In plan view, rip currents look like a tree with a narrow trunk within the surf zone while the branches or rip head spread out beyond the breaker zone. Inside the surf zone, this return flow may be from one to tens of meters wide; outside the breaker zone, it becomes diffuse and its effect may be noticed for more than a kilometer offshore. Rip currents are best developed on long, gently sloping beaches exposed to large waves. When waves are smaller, the rip currents are weaker and more numerous (McKenzie, 1958). This may be due to the nonlinear terms of the equations of water motion, where the outflowing rip current tends to become narrower with increasing Reynold's number (Arthur, 1962).

The velocity structure within rip currents, however, is not well known. Most authors suggest that the main body of the rip current is located in the upper part of the water column. It has a velocity that ranges from about 1 m/sec to over 5 m/sec in severe storms (Popov, 1956). Only in the breaker zone does the rip current extend from top to bottom, so that it can scour out channels 1-3 m deeper than the adjacent bottom (Cook, 1970). These channels are floored by a rippled lag deposit of coarse sediment and shell fragments with large-scale crossbedding with the foreset laminae sloping seaward.

These currents may extend a considerable distance out to sea. Reimnitz and others (1976) describe 30-100 m wide, 0.5 m deep, seaward widening channels reaching from the surf to 1500 m from shore in water as deep as 30 m. These channels contain 1-1.5 m wavelength ripples whose crests are curved seaward.

When a wave breaks on a beach it imparts momentum to the water and carries it up above mean sea level in the form of surf and swash. When, under the influence of gravity, this returns as backwash toward mean sea level, it often cannot do so, for another wave has broken and more water rushes up the beach. The volume of this water stored above sea level increases as more waves arrive, until at some point on the beach the hydraulic head of the stored water becomes greater than the uprush of the waves and it begins to flow seaward in a rip current.

Rip currents are not randomly distributed along the beach. The spacing between rip currents depends on the variations of wave height along the shore. There are several ways in which this alongshore wave height variation can be produced. The first is *wave refraction,* which, because of irregularities in offshore topography, causes a bending of the waves and concentrate or weaken the wave height along shore. At La Jolla, California, two submarine canyons are close to shore. These cause the waves to bend so that there are always low waves in the lee of the canyons and larger waves on either side. Rip currents are always present in the lee of the canyons, shifting position only slightly as waves arrive from different directions.

On many beaches there are offshore, crescentic or transverse sand bars present within the nearshore zone. These have a tremendous influence on the longshore circulation patterns. Rip currents are frequently found in the roughs or between the bars (Sonu, 1969, 1972a). Even protuberances on shore, such as cusps or bars welded onto the shore, affect the surf zone circulation in that they affect the path of the backwash. Rip currents are frequently found adjacent to cusps (Komar, 1971). These features, however, may owe their existence to the processes discussed in the following section. Once established, these topographic undulations influence the spacing and the direction of rip currents.

If the beach is straight with parallel offshore contours, rip cells may still exist. These may owe their origin to *edge waves,* the intersection of two wave trains of the same period, or wave instabilities.

Incident waves can excite transversal waves (edge waves). Edge waves are surface waves trapped by refraction to the shore having an amplitude maximum at the shoreline and decreasing seaward through several smaller maxima and minima (Bowen, 1969). They can be standing waves with the crests normal to the shoreline and wave length parallel to shore. If these edge waves have the same period as the incoming waves, then their interaction or summation will produce nodes and antinodes that are stable in position. The net result is a consistent variation in breaker height. This will produce a rip current in the areas of low breaker, at every other antinode of the edge waves (Bowen and Inman, 1969). The position of these antinodes can best be observed on the beach itself, where they form the areas of maximum run up (Komar, 1976). Edge waves are common on beaches steeper than 1/10 (Sasaki and Horikawa, 1975). At the nodes of the edge waves, sediment is likely to be deposited, forming a rhythmic offshore topography.

The spacing of the rip currents (Ursell, 1952) is equal to the length of the edge wave

$$\lambda = \frac{gT^2}{2\P} \sin(2n+1)\beta$$

where $2n+1$ has integral values up to $(2n+1)\beta < \P/2$, where β is the slope of the beach. Thus several spacings are possible. The dominant one under a given set of wave conditions is

$$\frac{\chi_b \left(\frac{2\P}{T}\right)^2}{g \tan \beta}$$

where χ_b is the width of the surf zone (Bowen and Inman, 1969).

Sometimes, especially on gently sloping beaches, the spacings between rip currents is larger than that predicted by edge waves (Sonu, 1972b). These larger spacings may be due to a similar interference mechanism formed only by infragravity waves (Munk, 1951; Suhayda, 1974) that have a period of 30 sec–10 min.

The same type of interference pattern of high and low waves may be due to the arrival on shore of waves from two different directions but of the same frequency (Dalrymple, 1975). Then the spacing of the rip currents would be $\lambda = L_o/(\sin \Theta - \sin \phi)$, where Θ and ϕ are the angle of approach of the two wave crests. If the waves differ in frequency by a slight amount, then the rip currents should slowly migrate.

Another theory for the rip current spacing on a smooth coast is the instability theory of Hino (1973, 1975). He argues that even if waves are of the same height everywhere along a coast, they would be unstable and would be changed by even an infinitesimal disturbance, such as minor changes in bottom topography. For any given bottom inhomogeneity, the growth rate of the resultant rip current has a peak value four times the surf zone width. Field data suggest that this is a common spacing on beaches of 1/20–1/40 slopes.

The available data suggest that the spacing of rip currents is proportional to the width of the surf zone. If the surf zone similarity parameter ξ is used (Battjes, 1975), where $\xi = \tan \beta/(H/L_o)$ or approximately the relative depth change across one wave length in the surf zone, then the spacing of rip currents can be summarized by the width of the surf zone that is dependent on wave parameters and beach steepness.

BENNO M. BRENNINKMEYER

References

Arthur, R. S., 1962. A note on the dynamics of rip currents, *Jour. Geophys. Research* 67, 2777-2779.

Battjes, J. A., 1975. Surf similarity, *Conf. Coastal Eng., 14th, Am. Soc. Civil Engineers, Proc.,* 466-480.

Bowen, A. J., 1969. Rip currents, *Jour. Geophys. Research* 74, 5467-5478.

Bowen, A. J., and Inman, D. L., 1969. Rip currents, laboratory, and field observations, *Jour. Geophys. Research* 74, 5479-5490.

Cook, D. O., 1970. The occurrence and geologic work of rip currents off southern California, *Marine Geology* 9, 173-186.

Dalrymple, R. A., 1975. A mechanism for rip current generation on an open coast, *Jour. Geophys. Research* 80, 3485-3487.

Hino, M., 1973. A theory of rip current generation, *Coastal Eng. Japan* 16, 55-60.

Hino, M., 1975. Theory of formation of rip-current and cuspoidal coast, *Conf. Coastal Eng., 14th, Am. Soc. Civil Engineers, Proc.,* 901-919.

Komar, P. D., 1971. Nearshore cell circulation and

formation of giant cusps, *Geol. Soc. America Bull.* **82**, 2643-2650.

Komar, P. D., 1976. *Beach Processes and Sedimentation.* Englewood Cliffs, N.J.: Prentice-Hall, 429p.

McKenzie, P., 1958. Rip-current systems, *Jour. Geology* **66**, 103-113.

Munk, W. H., 1951. Origin and generation of waves. *Conf. Coastal Eng., 1st, Council on Wave Research, Proc.,* 1-5.

Popov, E. A., 1956. On current surges in the coastal zone, *Akad. Nauk SSSR Okeanogr. Komis. Trudy* **1**, 98-104.

Reimnitz, E., and others, 1976. Possible rip current origin for bottom ripple zones to 30 m-depth, *Geology* **4**, 395-400.

Sasaki, T. X., and Horikawa, K., 1975. Nearshore current system on a gently sloping bottom, *Coastal Eng. Japan* **18**, 123-142.

Shepard, F. P., 1936. Undertow, rip tide, or rip current, *Science* **84**, 181-182.

Sonu, C. J., 1969. Collective movement of sediment in littoral environment, *Conf. Coastal Eng., 11th, Am. Soc. Civil Engineers, Proc.,* 373-400.

Sonu, C. J., 1972a. Field observations of nearshore circulation and meandering currents, *Jour. Geophys. Research* **77**, 3232-3247.

Sonu, C. J., 1972b. Comments on paper by A. J. Bowen and D. L. Inman, Edge waves and crescentic bars, *Jour. Geophys. Research* **77**, 6629-6631.

Suhayda, J. N., 1974. Determining nearshore infragravity wave spectra, *in International Symposium on Ocean Wave Measurements and Analysis.* New York: Am. Soc. Civil Engineers, 54-63.

Ursell, F., 1952. Edge waves on a sloping beach, *Royal Soc. London, Proc.* ser. *A* **214**, 79-97.

Cross-references: *Bars; Currents; Edge Waves; Nearshore Hydrodynamics and Sedimentation; Ripple Marks; Sea Puss; Sediment Transport; Tidal Currents; Waves.* Vol. III: *Rip Current.*

RIPPLE MARKS

Ripple marks are generally *depositional* features (formed at a fluid/sediment interface), more or less regular and repetitive, and typically having a spacing greater than about 7 mm, up to a few meters. Most examples occur in coarse silt, sand, or fine gravel.

The term *giant ripple mark* has been applied to certain large depositional features; terms such as *beach cusp, beach pad, bar,* and *dune* represent accumulations that are excluded from the ripple mark category. An early systematic treatment was given by Kindle (1917).

A compact classification of ripple mark types includes the following: elementary ripple marks—wave-formed (water only), current-formed (water; air); flat-topped ripple marks (water only); windrow ridges (water; possibly air); washover crescents (water only); composite ripple marks (see *Rooster Tail*); helical cell ripple marks (parallel with current; water or possibly air); swash zone ripple marks.

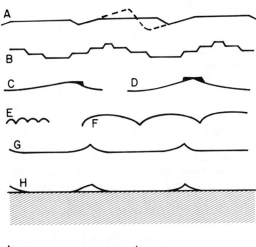

FIGURE 1. Several shallow-water ripple mark varieties: *A.* Flat-topped ripple mark, formed when falling water level (falling tide) permits planing off of previously developed wave or current type ripple marks. *B.* Terraced forms, produced by an irregular rate of fall; each "terrace" represents a pause in the drop of water level. *C.* and *D.* Steep faces on wave-type ripple marks; the "keeled" variety. *E.* Capillary ripple mark, about 2 cm long. *F.* Convex-up ripple mark, much longer than the 2.44 cm. maximum for the capillary variety. The sharp intersections point down. *G.* Concave ripple mark, with widely spaced crests and nearly flat troughs, developed on deep sand. *H.* Incomplete ripple marks, produced in sparse sand over bedrock surface, and having approximately the profile of the concave variety. *I.* The figure-eight, double-reversal, bottommost eddy above wave-type forms. As the water level drops, and the maximum diameter is reduced to the ripple mark wave length, two reversals within a single trough will build a new, smaller crest, as shown in *J.* These two sketches show the same ripple profile at two different moments. If the water level is stabilized, the taller crests may be reduced to match the smaller ones (from Tanner, 1962).

Other varieties, subdivisions, and combinations are sketched (Fig. 1), discussed, and partly explained elsewhere (Tanner, 1960, 1962; Hoyt and Henry, 1963). Allen (1968) devoted an entire book to current ripples, which also treats other types and contains an extensive bibliography. Perhaps ripple marks can be formed by viscous liquids, such as lava, but only air and water varieties are discussed here.

Elementary Types

Wave-formed ripple marks are characterized by good continuity and parallelism and low *ripple index* (the ripple index is the ripple spacing divided by the ripple height). The combination of good parallelism and continuity and intermediate ripple index (> 16) generally identifies wind-formed (current-formed) ripple marks. Ripple marks made by water currents are typically characterized by poor parallelism, poor continuity, and an intermediate ripple index. Wave-formed ripple marks and current-formed ripple marks cannot be distinguished by symmetry or asymmetry. Wave-formed ripple marks may be extremely asymmetrical; the two halves of the bottom orbit (landward motion, seaward motion) are commonly not equal. Symmetry, parallelism, continuity, and other parameters are treated in detail elsewhere (Tanner, 1967).

Wave-formed ripple marks have been photographed in water several kilometers deep; because surface waves do not extend to such depths, these beautifully regular features must have been made by internal waves.

Origin and Decay

Elementary ripple marks are formed initially by a shear mechanism that operates on a smooth sediment surface (Tanner, 1963). The initial shear ridges are low and hard to see; they cause (underwater) instant boundary layer separation, development of horizontal vortexes, and hence modification of the initial form. The modified geometry is what is actually observed.

Ripple marks that are static (no sand motion) tend to be plastered over, partially or completely, by mud (silt and clay), or to be destroyed by wandering bottom organisms (gastropods and echinoids) or by burrowing organisms, the latter especially in the tidal flat. Partial filling of the trough by mud modifies the ripple index, giving misleading or unrealistic values.

In most of the nearshore zone, periods of quiescence do not last long enough for ripple marks to be generally destroyed. In these circumstances, each new wave regime begins to operate on a set of inherited ripple marks, which are then modified to match the new bottom orbital parameters—that is, there is an inherited roughness, which provides for boundary layer separation, and the development of the eddies that shape and maintain common ripple marks. These eddies occupy a thin layer of water below the bottommost wave orbit, typically as thick as two to three ripple mark heights. Within the "bottom eddy layer," eddies are associated in pairs, so that the water motion in any one pair of eddies has the form of a figure eight. There are two or more figure eights under any one bottom orbit.

The Plane Bed

There are basically four regimes for elementary wave-formed ripple marks: static (no motion, relict), simple maintenance (sand moving back and forth across the elementary form), keeled maintenance (a keel or wedge of sand oscillates from one side of the ripple mark crest to the other as the figure eight reverses), and plane-bed motion (ripple marks are eliminated by high-velocity transportation of sand). These four regimes have been listed in approximately ascending order of transport velocity. As the velocity decreases, ripple marks appear again, with certain exceptions, and the history is reversed.

The exceptions occur in extremely shallow water, where the combination of water depth and water velocity is such that a smooth bed is maintained even with falling velocity. In some instances water depth may go to zero without the flow regime ever returning to ripple-making conditions, thus leaving a plane bed. This can be verified by use of the Froude Number, which should stay in the general range of 0.7–1.0 (to provide a smooth or plane bed) until the water depth reaches essentially zero.

Additional Types

Flat-topped ripple marks are generally the product of falling water level, with terminal depths no greater than the final ripple mark height. A related form, also due to falling water level, has large and small ripple marks alternating with each other; the smaller features were built in the troughs of the larger ones as the water level dropped.

Washover crescents are depressions left where curved ripple marks are planed off by falling water and the troughs are filled in here and there by the sand supplied from the former crests.

Convex-up ripple marks, which in some instances have such gentle convexity as to be almost flat on top, are not rare on the tidal flat and in shallow water, but the details of their formation are not known.

Composite ripple marks occur where refraction (such as over transverse bars) provides that a single wave set, having one wave period, crosses itself. Bottom orbit vector addition produces low, imperfectly parallel but essentially symmetrical ripple marks that bisect the acute angle between the intersecting wave crests and that do not parallel either wave crest.

Windrow ridges parallel the long axes of Langmuir cells (helical cells) that are driven in the water by the wind. Such ridges tend to be perfectly straight and parallel, cross the assocciated wave formed ripple marks at angles that do not have to be 90, and have a spacing which is quasi-regular.

Other *helical cell ripple marks* that may be formed by water current flow tend to have much shorter spacing (e.g., 2–20 cm instead of 50 cm or more for the windrow ridges).

Swash zone ripple marks (Tanner, 1965) are commonly almost straight and parallel and parallel with the step at the toe of the beach. The ripple index is extremely high, typically 50–100; hence, symmetry cannot be measured effectively. In at least some cases, swash zone ripple marks are associated with the hydraulic jump (Froude Number crosses 1.0).

Paleogeography

Those wave-formed ripple marks having a spacing of 5 cm or less can ordinarily be attributed to small bodies of water such as lakes. A larger spacing indicates either lake or open ocean waves.

Several factors enter into the relationship between wave-formed ripple mark spacing and fetch (Tanner, 1971), and cannot be treated here. However, a rough empirical relationship, for medium sand, ignoring wind effects, can be written as follows: $f = (40,000)s^{2.5}$, where f is fetch in kilometers, and s is ripple mark spacing in meters. The fetch estimated in this fashion might be small (for example, a spacing of 12 cm = 0.12 m, provides $f = 200$ km, which is not necessarily the width of the water body; ripple marks may be built by waves generated in the nearshore zone of a large ocean).

In studying relict or ancient examples, one must operate without current, wind, or wave data. The problem is indeterminate. However, the possible ranges of values are reasonably small, and interesting estimates can be obtained. Reconstruction of the paleogeography of a Jurassic lake in northwestern New Mexico (Lake Todilto) has been undertaken, using (among other information) ripple mark and grain size data (Tanner, 1974). Water depth was estimated by various means, including limitations placed by the ripple mark data; fetch obtained from the ripple marks was confirmed by field work. With a figure in hand for fetch, wave period and wave height can be estimated for an assumed wind velocity (Murali and Tanner, 1975).

Most ancient ripple marks are elementary forms made by waves, water currents, or wind, in that order, with eolian examples rare (Poole, 1964). Flat-topped ripple marks are fairly common in the lithified section, as are the alternate-size ripple marks formed by falling water level.

Near shore ripple marks provide useful information about the general orientation of the isobaths, but it is not essential that they were precisely parallel with either the ancient wave crest or the isobaths.

Variability

Bottom orbital parameters, under wave action in water, do not change monotonically from deep water toward shore. The oscillations in these parameters (such as bottom orbital diameter and bottom orbital velocity) reflect the interaction of the wave with the bottom slope and the bottom sediment. A banding, more or less parallel with the beach or at least at right angles to the wave orthogonals, is produced; this rough banding, with each component perhaps hundreds of meters wide, controls changes in the ripple mark characteristics from deep water to shallow water.

In extremely shallow water, and particularly intertidal zones, the combination of various agencies (changes in bottom orbital parameters, changes in water level, currents, Langmuir cell circulation, and others) may give rise to a complicated ripple mark pattern.

Water current ripple marks also can be complicated, but for different reasons. In shallow flows (several millimeters to a few tens of centimeters), microbraiding, current lineation, kolk pits, and other features may complicate the ripple mark pattern. In deeper flows, a hierarchy of ripple mark sizes may develop. Wind ripple marks generally have relatively simple patterns.

WILLIAM F. TANNER

References

Allen, J. R. L., 1968. *Current Ripples*. Amsterdam: North-Holland Publishing Co., 433p.

Hoyt, J. H., and Henry, V. J., Jr., 1963. Rhomboid ripple mark, indicator of current direction and environment, *Jour. Sed. Petrology* 33, 604–608.

Kindle, E. M., 1917. Recent and fossil ripple mark, *Geol. Survey Canada Mus. Bull. No. 25*, 1–56.

Murali, R. S., and Tanner, W. F., 1975. Correlation of wave parameters in shallow water, *Zeitschr. Geomorphologie* 19, 479–489.

Poole, F. G., 1964. Paleowinds in the western United States, *in* A. E. M. Nairn, ed., *Problems in Paleoclimatology*, New York: Wiley Interscience, 394–405.

Tanner, W. F., 1960. Shallow water ripple mark varieties, *Jour. Sed. Petrology* 30, 481–485.

Tanner, W. F., 1962. Falling water level ripple marks, *Gulf Coast Assoc. Geol. Socs. Trans.* 12, 295–301.

Tanner, W. F., 1963. Origin and maintenance of ripple marks, *Sedimentology* 2, 307–311.

Tanner, W. F., 1965. High-index ripple marks in the swash zone, *Jour. Sed. Petrology* 35, 968.

Tanner, W. F., 1967. Ripple mark indices and their uses, *Sedimentology* 9, 89–104.

Tanner, W. F., 1971. Numerical estimates of ancient waves, depth, and fetch, *Sedimentology* **16**, 71-88.

Tanner, W. F., 1974. *History of Mesozoic Lakes of Northern New Mexico.* Socorro, N. Mex.: New Mexico Geological Society, Guidebook, 25th Field Conf., Ghost Ranch, vol. 24, 219-223.

Cross-references: *Bars; Beach Cusps; Beach Pad; Bed Forms: Coastal Dunes; Froude Number; Major Beach Features; Minor Beach Features; Nearshore Hydrodynamics and Sedimentation; Organism-Sediment Relationship; Rooster Tail; Tidal Currents; Waves.* Vol. VI: *Ripple Marks.*

RIPRAP

Riprap consists of a layer or mound of stones used for the prevention of erosion, scour, or the crumbling of a structure or embankment. There are two classes: *random riprap* and *hand-placed riprap.*

Randon riprap consists of stones that are usually dumped on location or tossed in by hand. The stones making up random riprap can be of almost any size, but the total height of a random riprap barrier should not be less than 1 m. It is generally considered that 1 m of random riprap is equivalent to .5 m of hand-placed riprap.

Hand-placed riprap consists of stones, small enough to be hand placed, that are laid on a gravel bed that has previously been prepared. Any voids should be filled with smaller stones to achieve a neat compact structure. A hand-placed riprap structure should be at least 45 cm in height (not including base materials).

Stones used as riprap should be hard and durable. They should not undergo breakdown under constant exposure to water, ice or air. Most of the igneous and metamorphic rocks can be used along with many of the limestone and sandstone classes.

For further information see Creager, Justin, and Hinds (1945), Coastal Engineering Research Center (1973), and Saville (1966).

JOHN B. HERBICH
TOM WALTERS

References

Creager, W. P.; Justin, J. D.; and Hinds, J., 1945. *Engineering for Dams.* New York: John Wiley & Sons.

Coastal Engineering Research Center, 1973. *Shore Protection Manual.* Washington, D.C.: U.S. Army Corps of Engineers, 750p.

Saville, T., Jr., 1966. Rock movement in large-scale tests of riprap stability under wave action, *in Conf. Coastal Eng. 10th, Am. Soc. Civil Engineers, Proc.*, 972-976.

Cross-references: *Breakwater; Coastal Engineering; Dike, Dyke; Groin; Protection of Coasts.*

RIVER-DEPOSITION COASTS

River-deposition coasts are low coasts whose principal forms are produced by alluvial sedimentation close to the shoreline.

The expression is used only tacitly by the classic authors. Johnson (1919) attributed alluvial deposition coasts principally to actually stable coasts. Shepard (1948) clearly recognized the alluvial coasts in his classification; subsequently Valentin (1954) established the notion of *advance shoreline by deposition (vorgeruckte Küsten).*

A river deposition coast must usually be an *advance shoreline* like that of Valentin because, the alluvial influence predominates over the marine sedimentation or abrasion if the alluvial forces are major. Then the shoreline advances, at least up to a certain equilibrium state. The following coasts may be included as river deposition forms: *delta coasts, alluvial plain coasts,* and *outwash plain coasts.* Johnson states that these last two are formed where the broad alluvial slope at the base of a mountain range, or the outwash plain in front of a glacier, is built forward into the sea. The principal individual form of the river deposition coast is the delta.

Applying the classification of Zenkovich and Leontiev (Ionin, 1965), these coasts belong to the group formed predominantly under action of factors other than waves, the potamogenous shorelines, with delta shorelines and alluvial plain shorelines. There are various types of deltas and at least two types of alluvial coastal plains.

River-deposition coasts depend on various factors, not only on a stable sea level, as Johnson seems to suggest. Sea level may change slowly and the deltas stop prograding. Low oceanographic energy coasts and rivers with plenty of sedimentary material are more important than the relative movement of sea level, whose changes are slow. Rapid supplies of coarse sediment from torrential arid-region rivers in the north of Chile (Arica) now produce rapidly growing deltas. On the other hand, the postglacial delta of the Bío-Bío River in central Chile was formed, in its upper facies, during an ascending sea level; but because the alluvial supply of detritus was great, even with the sea level rise, it formed slowly.

JOSÉ F. ARAYA-VERGARA

References

Ionin, A. S., 1965. Geomorphology of sea and ocean shorelines, *Soviet Geogr.* 6, 297-298.

Johnson, D. W., 1919. *Shore Processes and Shoreline Development.* New York: John Wiley & Sons, 584p.

Shepard, F. P., 1948. *Submarine Geology.* New York: Harper, 348p.

Valentin, H., 1954. *Die Küsten der Erde*. Gotha: Veb. Geogr. Kartogr. Anstalt, 187p.

Cross references: *Classification; Deltaic Coasts; Deltas; Marine-Deposition Coasts.*

ROCK BEACH—See WAVE-CUT BENCH

ROCK BORERS

Rock borers are plants or animals with the capacity to excavate holes in solid substrates. Sedimentary rocks, including limestone, mudstone, and sandstone are particularly susceptible; boreholes in intertidal and subtidal rocks may be so spaced that a rock sample looks like Emmenthaler cheese, a honeycomb, or sponge. In this weakened state the rock may break or crumble with a slight mechanical disturbance. Of the many hundreds of rock borers, perhaps the most widespread in distribution and best-known groups are clionid sponges, sipunculid and annelid worms, bivalve molluscs (especially pholadids and mytilids), and boring barnacles (*Lithotrya*). Boreholes are excavated by biochemical or biophysical means for food or protection. Much information on wood, shell, and rock borers is available in an annotated bibliography by Clapp and Kenk (1963) and collections of papers edited by Carriker, Smith, and Wilce (1969) and Frey (1975). Although of less economic importance than wood borers like *Teredo*, rock borers can destroy concrete used in coastal structures. They are also locally a source of food.

ROGER F. McLEAN

References

Carriker, M.; Smith, E. H.; and Wilce, R. T., eds., 1969. Penetration of calcium carbonate substrates by lower plants and invertebrates, *Am. Zoologist* 9, 629-1020.

Clapp, W. F., and Kenk, R., 1963. *Marine Borers: An Annotated Bibliography*. Washington, D.C.: U.S. Government Printing Office, 1136p.

Frey, R. W., ed., 1975. *The Study of Trace Fossils*. New York: Springer-Verlag, 562p.

Cross-references: *Alveolar Weathering; Cliff Erosion; Weathering and Erosion, Biologic.*

ROCK PLATFORM—See WAVE-CUT PLATFORM

ROCKY SHORE HABITAT

The unique communities characterized by rocky shores may be found wherever bedrock outcrops into the intertidal zone or wherever (usually in the tropics) biogenic rocks such as coral are formed there. Man has created much additional solid substrate for this community in the form of docks, seawalls, groins, and jetties. Indeed, these structures may harbor the only outposts of the "rocky" fauna and flora to be found on long stretches of sedimentary shore. Some representatives of the hard substrate biota may also be seen on isolated solid objects such as shells, stones, or artifacts in the intertidal or subtidal zone. The following discussion will apply primarily to north temperate shores, although many aspects will be found relevant to tropical situations as well.

In a sense, no two square meters of the rocky intertidal environment are alike, since the biota can be greatly affected by both gross and subtle differences in latitude, elevation, slope, aspect, rainfall and cloud cover regime, ocean currents, roughness of the rocks, exposure to waves, tide range, and whether the adjoining land mass experiences a continental or a maritime climate. Scour by sand and pebbles, ice formation in winter, and battering by drift logs further complicate the picture. But in spite of all the variables, a set of generalizations emerges that seems applicable to solid substrate communities more or less worldwide. In particluar, a pronounced zonation of dominant organisms is everywhere apparent. This zonation has been extensively examined on widely separated coasts in the work of T. A. Stephenson and A. Stephenson (1971).

The highest point on the shore that could be considered marine is the supralittoral fringe or *black zone*, wetted only by the bimonthly spring tides but extended upward to varying degrees by storm surge and salt spray. Only the hardiest organisms can survive the alternation of seawater and rainwater, baking sun and winter cold, that characterize this zone that is neither sea nor land. Lichens, exemplified by *Verrucaria*, and blue-green bacteria, such as *Calothrix*, encased in a slippery, gelatinous coating, encrust the rocks. On northerly shores, a littorinid gastropod with semiterrestrial adaptations will usually be found grazing on these primary producers.

Throughout the intertidal zone of most rocky shores will be found large and small pools of water in natural basins of the rocks containing biotic communities of their own. Small pools in the supralittoral area may harbor only encrusting algae and certain eurytopic copepods that can tolerate the extremes of temperature and salinity to which a tiny body of water, above the waves most of the time, is subject.

Most of the shore profile is contained within the littoral—or truly intertidal—zone, and as one proceeds seaward one encounters successive groups of organisms in sharply defined bands. Sessile animals are almost all filter feeders; motile animals are grazers or predators on other macroscopic animals. Large plants are brown, red, or green algae. Nearly all the sessile animals and many of the rovers reproduce by means of large numbers of planktonic larvae; the plants similarly give rise to motile propagules. Thus the organisms greatly enhance their ability to colonize new substrates—bare rock or the surfaces of other living things. Settling occurs in waves, following the reproductive cycles of the plants and animals. Some larvae delay metamorphosis until they sense that a suitable substrate is present. Others are attracted to the vicinity of established members of their own kind, risking being filtered out of the water column and killed by their planktivorous conspecifics, but benefiting from the fact that the adults may be occupying a location that has been tolerable to the species at least during their lifetimes.

In addition to the large organisms, both bare rock and the surfaces of plants and animals will be covered by a film of bacteria, unicellular algae, and organic detritus. This film seems to enhance the settling of many plant and animal disseminules. Coralline red algae such as *Corallina* and *Lithothamnion* can encrust any surface.

Barnacles such as *Balanus* and *Chthamalus* are almost always present in the upper littoral zone; on some coasts there is an uppermost and a lower band, representing different species. Below the barnacles (on cool temperate shores) will be a band of mussels belonging to *Mytilus* or a closely related genus. Unlike barnacles, which are cemented to the rock, mussles are attached by byssal threads that can be disengaged and then reformed in a slightly different spot, allowing the animals to move very slowly. Mussels require some surface roughness, provided by the rock or sometimes by barnacles, to become established. The bivalves exist in a solid band on some shores. On others they occur in scattered clumps, with barnacles, algae, and other organisms scattered about and considerable bare rock exposed. At other times or places, mussels and barnacles can be found crowded together in the mussel zone. Fucoid algae in varying abundances are characteristic of both the barnacle and mussel zones. In the lower part of that mussel zone, red algae such as *Gigartina* and *Chondrus* grow extensively. On more southern temperate shores (such as the United States east coast of Cape Hatteras), *Mytilus* drops out; its place may be taken by oysters and red or green algae.

One or more species of littorinid gastropods will be found grazing on the tender algae of the intertidal zone, but these snails are ineffective against tougher species such as *Fucus* and *Chondrus*. Other intertidal grazers include chitons and limpets. Some limpets have "homes" on the rock surface to which they return during each low tide; others maintain by their grazing, and defend for their exclusive use, small "pastures" of matlike algae, in which the species composition differs from that on the surrounding rock.

Predators of this zone usually include one or more small muricid whelks, capable of drilling holes in the shells of mussels and other mollusks, and also of feeding on barnacles. These snails can be active only when covered with water, so that prey items low in the interidal zone are at greatest risk. When the tide is high, other predators from the subtidal realm, such as asteroids, crabs, and certain fishes, can attack the intertidal populations. (Asteroids and crabs may remain in the intertidal zone at low tide, but they are inactive and usually hidden under festoons of fucoids.)

Other occupiers of littoral space may include other mollusks, sea anemones, tunicates, sponges, and gooseneck barnacles. The fronds of the various large algae will serve as habitat for small, epiphytic algae, as well as for bryozoans, hydroids, both errant and tubicolous polychaetes, nudibranchs, and many minute crustaceans.

Environmental conditions within littoral rock pools will be determined by the size of the pool and its nearness to the low tide mark. A small, high pool will be almost as harsh a habitat as a pool in the supralittoral zone. Large pools, especially low in the intertidal zone, will have much more benign conditions, since their volume will protect them from sudden changes in temperature or great increases in salinity due to evaporation. Rainwater will float on top of the denser seawater, allowing benthic organisms to live in reasonably normal conditions. These pools often show a well-developed algal flora, including red or brown encrusting forms and sometimes extensive growth of a larger species such as *Enteromorpha* or *Chondrus*. Tide pool animals include most of the same species as are found on the surrounding rocks, although relative abundances may be quite different. Gastropods, asteroids, and crabs need not confine their grazing or predatory activity to high tide. Algal photosynthesis keeps the oxygen content of the pool high during the day, but at night the respiration of all the organisms may deplete the oxygen supply, with an accompanying rise in pH, leading to another set of adverse conditions to which the plants and animals must be adapted.

At the lower edge of the littoral zone will be found a heavy growth of kelps or other large algae, exposed to air at only the lowest tides or not at all. This vegetation shelters a great variety of animal life, including carnivorous and herbivorous gastropods, crabs, lobsters and asteroids, as well as echinoids, which are consumers of large algae. The dense beds of mussels and barnacles characteristic of the intertidal zone are not usually found subtidally.

Obviously the intertidal zone will be much wider on shores with a wide tidal range. On north-facing slopes, not subject to hot sun, each zone will extend higher than on a nearby south-facing slope. Latitudinal effects are most noticeable on the east coast of North America. The area north of Cape Cod is one of cold ocean currents and hard winters, and the faunal affinities are northern. South of Cape Cod there are distinct changes in species composition, with more and more southern elements creeping in, until a tropical community is attained in the Florida Keys. Northern elements continuously drop out; the northern littorines reach their southern limits in the mid-Atlantic region, and *Mytilus* fails to reach maturity south of Cape Hatteras. The overall trend, however, is toward greater diversity as one proceeds southward, as is the case for most community types.

The North American west coast has a more unified biota thanks to the configuration of Pacific Ocean currents. Another distinctive feature of the west coast is the maritime climate it enjoys; cooler summers and milder winters are associated with a more diverse and trophically complex community than that to be found at the same latitude in New England, where summers are hot and winters bitter. However, some west coast shores are so severely battered by drift logs that certain areas are denuded of most macroscopic life.

Any rocky shore far enough north to experience severe winter icing will have an added degree of physical severity for its communities. Finally, scour by sand and pebbles can prevent the colonization of certain surfaces, especially isolated rocks surrounded by finer material.

All these aspects of the physical environment of course have much to do with determining what, if anything, will live on a particular rock; obviously no organism will survive outside its physiological tolerances. But interactions among the plant and animal populations have increasingly been shown to be highly significant in determining community composition. Indeed, the rocky coast has been used as a convenient natural laboratory for the study of biotic interactions. Since most of the organisms are slow-moving or sessile, and live on a two-dimensional surface easily accessible at low tide, they are amenable to experimental manipulation. A number of scientists have taken advantage of these circumstances to document the biotic interactions of the rocky community; a partial listing would include J. H. Connell (1972), P. K. Dayton (1971), J. Lubchenco (1978), J. Lubchenco and B. A. Menge (1978), and R. T. Paine (1974). Techniques for the manipulation of solid substrate populations include complete or partial denudation of the surface, transferral of stones with attached organisms from one part of the shore to another, exclosure or enclosure of certain organisms by means of small cages and fences, and various combinations of these procedures.

The many ingenious experiments devised by the above-mentioned and other workers cannot be described here, but some generalizations that have emerged from their efforts can be noted. Special circumstances of course produce exceptions to almost any generalization, and knowledge continues to be expanded and refined in this active area of research.

Within a particular section of shoreline, the upper distributional limits of any population seem to be set by the physical environment as it determines the organisms' abilities to stand exposure to air, with accompanying drying, heating, or freezing and, for many, deprivation of food and oxygen. Barnacle and mussel bands are arranged in descending order of the abilities of their occupants to survive in air. But when it comes to setting the lower limits for a population, biological interactions often come to the fore. Where there are two bands of barnacles, the uppermost species does not invade the zone of the lower species, because it cannot compete for space with the latter; the barnacles, in turn, cannot compete with *Mytilus*.

Mussels are formidable competitors for space. Owing to their capacity for byssal attachment and reattachment, they are at least potentially able to settle upon, overgrow, and smother other sessile organisms, and sometimes even to entwine and kill such motile animals as limpets. As a mussel clump enlarges, the individuals spread from the center, and already established organisms such as anemones, gooseneck barnacles, and the stipes of large kelps and fucoids may be constricted and cut or crushed, providing another mode by which mussels dominate space. (Barnacles too can deny space to fucoid algae by constricting the stipe; they also may foul the thallus, but if the algae sloughs off, the barnacles go with it.)

The dense mussel zone rarely extends below the low tide mark, and in this case the lower limit tends to be set by predation, usually on the part of asteroids or muricids. On certain shores, there has been shown to be a clear competitive hierarchy among large brown algae of the lower intertidal zone, the competitive outcome depending on degree of wave exposure. In order for algal succession to occur, however, asteroids must be sufficiently abundant to keep the area free of herbivorous echinoids.

Thus physical factors give way to competition, which gives way to predation as the important force controlling community structure, along a gradient from the land to the sea. Predation is most important at the level at which predators can be most effective—that is, low in the intertidal zone and subtidally, where they can spend a maximum amount of time submerged and feeding.

The significant environmental factor in this vertical gradient is exposure to air, severest high on the shore. A horizontal gradient may also be seen in the relative importances of physical factors, competition, and predation as one proceeds from shores subjected to extreme wave action, log battering, and the like to shores enjoying a more benign environment. On the high energy shore, the physical environment controls the community completely and directly. Organisms do colonize the rock surface, but they are obliterated long before any biological interaction can take place. The community is then limited to a few small mussels, barnacles, and algae clinging to survival in cracks and crevices.

Where wave action is a little more moderate, the shore is likely to be as described previously, with bands of barnacles and mussels occupying nearly all the intertidal zone, and the border between the two groups determined by competition for space. Numbers of individuals are high, but diversity is relatively low. Within the zones of abundance of barnacles and mussels, intraspecific competition for space is intense; whole clumps of mussels and barnacles eventually grow too tall or heavy to maintain their hold on the rock. When they slough off, usually in a winter storm, space is cleared and soon recolonized. Within the mussel zone, mussels, barnacles, other invertebrates, and algae may settle, but the mussels will ultimately prevail. Predators such as muricids and asteroids are present, but wave action is strong enough to keep them from effectively controlling community structure above the low tide mark.

Where the shore is more protected, predators have much more effect, especially in the mussel zone. Their inclination to eat the competitive dominants (mussels and barnacles) results in much free space on the rock and therefore more opportunity for competitively inferior invertebrates and algae to colonize. Mussels and barnacles will be present as scattered individuals or patches. They will grow large enough to compete intra- or interspecifically for space only if they manage by luck to escape the attention of predators until they have attained a refuge in size and cannot be successfully attacked.

A situation such as this, with moderate predation on the competitive dominants, appears to maximize diversity; the predator mainly responsible for maintaining the equilibrium has been called a *keystone species*. Cases have been described in the rocky shore environment in which abundant herbivores eating algae have made diversity low whether by eating the dominant species and nearly everything else or by eating all but one subdominant species, which then takes over. Thus diversity can be extremely low in the presence of both minimal and intense predation on the dominant, and high in cases of moderate predation. A unimodal relationship between degree of predation (centered on the competitive dominant), and diversity is suspected by some workers to have wide application.

Further experiments will perhaps clarify how generally these ideas apply within the rocky shore environment and to other community types as well.

MELINDA M. GODFREY

References

Connell, J. H., 1972. Community interactions on marine rocky intertidal shores, *Ann. Rev. Syst. Ecol.* **3**, 169-192.

Dayton, P. K., 1971. Competition, disturbances, and community organization: The provision and subsequent utilization of space in a rocky intertidal community, *Ecol. Mon.* **41**, 351-389.

Lubchenco, J., 1978. Plant species diversity in a marine intertidal community: Importance of herbivore food preference and algal competitive abilities, *Am. Naturalist* **112**, 23-39.

Lubchenco, J., and Menge, B. A., 1978. Community development and persistence in a low rocky intertidal zone, *Ecol. Mon.* **48**, 67-94.

Menge, B. A., 1976. Organization of the New England rocky intertidal community: Role of predation, competition, and environmental heterogeneity, *Ecol. Mon.* **46**, 355-393.

Paine, R. T., 1974. Intertidal community structure: Experimental studies on the relationship between a dominant competitor and its principal predator, *Oecologia* **15**, 93-120.

Stephenson, T. A., and Stephenson, A., 1971. *Life between the Tide Marks on Rocky Shores*. San Francisco: W. H. Freeman, 425p.

Cross-references: *Biotic Zonation; Coastal Fauna; Coastal Flora; Loose Rock and Stone Habitat.*

ROOSTER TAIL

Rooster Tail is a descriptive term long used by engineers to identify a surface spout of spray and/or water generated either by interfering *oscillatory waves* or by *standing waves*. In the latter case, the Froude number exceeds $F = 1.0$, and the water surface curls up locally but in the manner of breaking waves, so that the elevated spout moves for a short distance upstream. If the standing wave does not break or dissipate, water and/or spray may continue to move through the rooster tail. Rooster tails on standing waves are commonly seen in experimental flumes and in canals and inlets where water velocities are sufficiently high to push the Froude number past unity (Kennedy, 1961).

Where two surface waves of the oscillatory type, such as ocean surface waves, cross, the combination of orbital motions may concentrate wave energy enough locally to cause breaking in a very narrow band—and hence the rooster tail (Niedoroda and Tanner, 1970).

Rooster tails due to intersecting surface waves are common along beaches over *transverse bars*. The rooster tail occupies a narrow strip near the center of the bar, moves from the seaward end of the bar toward the beach, and is underlain by *ripple marks* that parallel the rooster tail path rather than either set of intersecting waves.

WILLIAM F. TANNER

References

Kennedy, J. F., 1961. *Stationary Waves and Antidunes in Alluvial Channels*. Pasadena.: California Institute of Tectnology, W. M. Keck Lab. Rept. No. KH-R-2, 146p.

Niedoroda, A. W., and Tanner, W. F., 1970, Preliminary study of transverse bars, *Marine Geology* 9, 41-62.

Cross-references: *Bars; Froude Number; Ripple Marks; Sea Slick; Waves.* Vol I: *Ocean Waves*

S

SABKHA

Sabkha (various spellings include sabkhah, sbabkha, sabkhat, sebkha, and sebcha) is widely used in the Arabic-speaking world to describe a low, flat, usually bare or sparsely vegetated, coastal and/or inland plain in arid regions; it usually has a salt-encrusted surface and is liable to flood (Tricart, 1954; Holm, 1960; Evans et al., 1964; Kinsman, 1969). The term has been adopted by earth scientists to describe a specific type of arid coastal plain where intertidal and supratidal evaporites are being deposited (Shearman, 1966; see references in Ginsburg, 1975).

Coastal sabkhas may be tens of kilometers wide and may extend along the coastline as a relatively unbroken feature for hundreds of kilometers. The surface level of such plains is close to sea level, with occasional beach-dune ridges and dunes forming slightly elevated areas. They are subject to extensive surface flooding from both land and sea. Since this environment is characteristic of areas where evaporation exceeds precipitation, there is an enrichment of salts in the surface. A characteristically blistered puffy surface, sometimes with polygonal patterns and pressure ridges, is usually developed. This surface may, however, be smoothed by floods. Such plains have only an immature soil development and are usually devoid of vegetation, or only have a scant cover of halophytic plants where slightly lower salinity water is available.

Sabkhas are composed mainly of siliciclastic or calcareous sediments, with varying amounts of evaporitic minerals. The sediments consist mainly of sand-grade material admixed with varying proportions of silt and clay grade material; gravel-grade sediments are rare except for the occurrence of skeletal debris that is often concentrated in beach ridges.

Reaction between the host sediment and the interstitial high salinity brines produces various diagenetic minerals within the sediment, often in such quantities that the surface is "jacked-up" or raised.

Sabkhas are formed by coastal progradation (Evans et al., 1964; Evans et al., 1969). Their relatively flat surfaces result from an interplay between marine or aeolian sedimentation, deflation, storm surge erosion, and precipitation of salts from ascending groundwaters. They are best regarded as arid zone analogues of strand plains, such as chenier plains, in less arid regions.

GRAHAM EVANS

References

Evans, G.; Kendall, C. G. St. C.; and Skipwith, Sir Patrick A. d'E., 1964. Origin of the coastal flats, the sabkha of the Trucial Coast, Persian Gulf, *Nature* **202**, 759-761.

Evans, G.; Schmidt, V.; Bush, P. R.; and Nelson, H., 1969. Stratigraphy and geologic history of the sabkha Abu Dhabi, Persian Gulf, *Sedimentology* **12**, 145-159.

Ginsburg, R. N., 1975. *Tidal Deposits: A Casebook of Recent Examples and Fossil Counterparts*. New York: Springer-Verlag, 428p.

Holm, D. A., 1960. Desert geomorphology in the Arabia peninsula, *Science* **132**, 1369-1379.

Kinsman, D. J. J., 1969. Modes of formation, sedimentary associations and diagenetic features of shallow water and supratidal evaporites, *Am. Assoc. Petroleum Geologists Bull.* **53**, 830-840.

Shearman, D. J., 1966. Origin of marine evaporites by diagenesis, *Inst. Miner. Metall. Trans., Section B* **75**, 208-215.

Tricart, J., 1954. Une forme de relief climatique: Les sebkhas, *Rev. Géomorphologie Dynam.* **5**, 97-101.

Cross-references: *Asia, Middle East, Coastal Morphology: Israel and Sinai; Asia, Middle East, Coastal Morphology: Syria, Lebanon, Red Sea, Gulf of Oman, and Persian Gulf; Coastal Dunes and Eolian Sedimentation; Desert Coasts; Intertidal Flats; Salcrete, Tidal Flat.* Vol. III: *Sabkha or Sebkha.*

SALCRETE

Salcrete is a thin, surficial crust of salt cemented, marine beach sediment formed primarily by evaporation of salt spray blown onshore from breaking waves (Yasso, 1966; Gary, McAfee, and Wolf, 1972). Crust thickness varies with duration of wave and wind conditions, and thins with distance from shoreline. After its formation strong winds may create *deflation furrows* where pieces of salcrete are blown away. Depth of deflation furrows is greatest at the berm crest. Salt crusts at the surface of low-lying flat land along the coast of North Africa, termed *sabkha* (q.v.), are formed primarily by evaporation of salt-rich capillary ground water (Fairbridge, 1968). Such crusts should not be confused with salcrete.

WARREN E. YASSO

References

Fairbridge, R. W., 1968. Sabkha or Sebkha, *in Encyclopedia of Geomorphology*. Stroudsburg, Pa.: Dowden, Hutchinson & Ross, 967.

Gary, M.; McAfee, R., Jr.; and Wolf, C. L., eds., 1972. *Glossary of Geology*. Washington, D.C.: American Geological Institute, 805p.

Yasso, W. E., 1966. Heavy mineral concentration and sastrugi-like deflation furrows in a beach salcrete at Rockaway Point, N.Y., *Jour. Sed. Petrology* 36, 836-838.

Cross-references: *Coastal Dunes and Eolian Sedimentation; Salt Weathering; Splash and Spray Zone*. Vol. III: *Induration; Sabkha or Sebkha*; Vol. VI: *Salcrete*.

SALT MARSH

Coastal *salt marshes* are tidal wetlands fringing the land/water interface of many temperate regions. They are part of a much larger marine system, the tidal marsh-estuarine ecosystem, which is recognized as one of the most biologically productive in the world. Chapman (1960, 1977 a, b) describes nine different geographical salt marsh regions throughout the world. In eastern North America, salt marshes occur from Nova Scotia to sourthern Florida, where they are replaced by *mangrove swamps*. They also occur on the Gulf coast. Because of the physiographical nature of the Pacific coast, they are much more limited in their distribution along the western shores of North America.

Geologically, tidal marshes are a relatively recent landform. In the northeastern United States they originated about 3000 years ago, when the rise in sea level slowed sufficiently to favor their development. The origin and dynamics of the Barnstable marshes on Cape Cod have been described by Redfield (1965, 1972). The formation of these marshes has occurred in conjunction with a *barrier beach*, which is apparently a common pattern. Tidal marshes can also develop along river estuaries to form valley marshes. Deposition of silt, sand, and organic sediments is involved in the formation of marshes in either situation. Saltwater cordgrass (*Spartina alterniflora*) becomes established in the intertidal zone and further accelerates sediment accretion. Other species can also become established, and eventually peat deposits may accumulate to a considerable depth (Niering, Warren, and Weymouth, 1977).

The vegetation pattern is influenced by a complex of environmental factors, including frequency and range of tides, salinity, microrelief, substrate, ice scouring, and storms. In addition, historical influences and more recent anthropic factors, including fires, cutting, diking, grazing, and ditching, also influence the distribution of species on salt marshes (Niering and Warren 1977, 1980).

In many marshes a characteristic belting pattern occurs. In southern New England, Miller and Egler (1950) recognize four major zones: the bay to upland sequence. Saltwater cordgrass dominates the intertidal zone; salt meadow cordgrass (*Spartina patens*), the high marsh; black grass (*Juncus gerardi*), a rush, the lower border; and switch grass (*Panicum virgatum*), the upper border. However, within this pattern there is a mosaic of subtypes: stunted *S. alterniflora* pannes, forb areas, algal pannes, and pools. Southward along the Atlantic coast, the width of the tall *S. alterniflora* belt increases dramatically, and *J. gerardi* is replaced by *J. roemerianus*. In southern California, Purer (1942) reports a lower littoral zone with *Spartina foliosa*, a middle littoral dominated by *Salicornia* spp., and an upper littoral with an admixture of *Frankenia grandifolia, Distichlis spicata, Atriplex watsonii*, and *Monanthochloe littoralis*.

The most ubiquitous marsh species is spike grass (*Distichlis spicata*), a typical halophyte, of exceptionally wide distribution in coastal and inland saline wetlands. A diverstiy of forbs lends an especially colorful aspect to certain tidal marshes. Sea lavender (*Limonium carolinianum*), arrow grass (*Triglochin maritima*), gerardia (*Gerardia maritima*), and aster (*Aster tenuifolius*) are among the most common. The most conspicuous shrub associated with the marsh is marsh elder (*Iva frutescens*). The tidal marsh should be viewed as a constantly changing mosaic of vegetation types, continuously being modified by subtle environmental influences. It should also be noted that over a long time frame the overall trend with coastal submergence is toward a landward movement of any bay-to-upland sequence of belts.

As one proceeds up the estuary, the increase in freshwater favors a new spectrum of species such as cattail (*Typha angustifolia*), reed grass (*Phragmites australis*), and sedges (*Scirpus* spp.). Certain of these ultimately replace the typical salt marsh species previously mentioned. *Phragmites* is commonly associated with disturbed sites, especially where the normal saltwater flushing has been arrested.

The dominant animal populations found on tidal marshes can be divided into those occuring in the intertidal zone and those restricted to the higher marsh, which is not inundated by each high tide. The dominant detritus-algae feeders on the high marsh are the marsh snail (*Melampus bidentatus*), amphipods (*Orchestia* spp.), and the isopod *Philoscia vittata*. Among the factors controlling the distribution of the invertebrates are food and protection, as well as elevation and frequency of tidal flooding. Ecologically these marsh invertebrates are important as primary consumers, as enrichers of the grass production that is exported to the estuary, and as a vital part of the estuarine and terrestrial food chain. It is also of interest that most of the dead grass is greatly enriched in amino acids by detritis feeders and, especially, bacteria before being flushed into the estuary. In terms of wildlife productivity, this figure ranges from 2.5-6.5 metric tons (2500-6500 kg). The marsh grasses, mud algae, and phytoplankton of the tidal creeks comprise the three major units of primary production.

The ribbed mussels embedded in the intertidal zone among the tall *S. alterniflora* grasses also play a major role in trapping phosphorus and keeping it in the estuarine system. The physical nature of the tidal marsh-estuarine system with its freshwater/saltwater interface tends also to act as a nutrient trap. In essence the marsh can serve as a primary nutrient source for the highly productive finfish and shellfish resources in the contigous estuarine waters. Marsh creeks also serve as nurseries and spawning grounds for coastal fish and other marine organisms.

Recently it has been documented that coastal wetlands can be important pollution filters. Data from the Tinicum marshes near Philadelphia indicate 50-70% reductions in nitrate and phosphate levels several hours after the waters from sewage and effluent passed over a 202-hectare tidal marsh.

The marshes also play a significant role geologically. They act as sediment accretors or as depositories for sediments, thereby reducing the frequency of dredging for navigation. During severe storms the marshes exhibit resiliency and thereby act as buffers to protect the developed contiguous shoreline.

With man's attraction to the coastal region the salt marshes have suffered greatly from his activities (Clark, 1977). Filling, dredging, ditching, impounding, and draining, as well as polluting, have greatly reduced the total acreage. Some of man's activities have actually obliterated the marshes; others have modified their biotic composition and productivity. Ditching has had a profound effect in reducing invertebrate productivity and modifying the vegetational pattern. Causeways constructed for highways and railroads and the installation of tidal gates restricting tidal flow further modify the vegetation. Considering the ecological functions these liquid assets are performing, every effort should be made to preserve them.

WILLIAM A. NIERING

References

Chapman, V. J., 1960. *Salt Marshes and Salt Deserts of the World*. New York: Wiley Interscience, 392p.

Chapman, V. J., ed., 1977a. *Wet Coastal Ecosystems*. Amsterdam: Elsevier, 440p.

Chapman, V. J., 1977b. *Coastal Vegetation*. New York: Pergamon Press, 292p.

Clark, J. R., ed., 1977. *Coastal Ecosystem Management*. New York: John Wiley & Sons, 928p.

Miller, W. R., and Egler, F. E., 1950. Vegetation of the Wequetequock-Pawcatuck tidal-marshes, Connecticut, *Ecol. Monogr.* 20, 143-172.

Niering, W. A., and Warren, R. S., 1977. Salt marshes, in J. R. Clark, ed., *Coastal Ecosystem Management*. New York: John Wiley & Sons, 607-702.

Niering, W. A., and Warren, R. S., 1980. Vegetation patterns and processes in New England salt marshes, *Bioscience* 30, 301-307.

Niering, W. A.; Warren, R. S.; and Weymouth, C. G., 1977. *Our Dynamic Tidal Marshes: Vegetation Changes as Revealed by Peat Analysis*. New London, Conn.: Connecticut Arboretum Bull. No. 22, 12p.

Purer, E. A., 1942. Plant ecology of the coastal salt marshlands of San Diego County, California, *Ecol. Monogr.* 12, 81-111.

Redfield, A. C., 1965. Ontogeny of a salt marsh estuary, *Science* 147, 50-55.

Redfield, A. C., 1972. Development of a New England salt marsh, *Ecol. Monogr.* 43, 201-237.

Cross-references: *Biotic Zonation; Coastal Fauna; Coastal Flora; Halophytes; Nutrients; Pollutants; Production; Salt Marsh Coasts; Standing Stock; Vegetation Coasts.*

SALT MARSH COASTS

Coastal salt marshes are low-lying meadows of herbaceous vegetation subject to periodic inundations. The development of a marsh occurs during a constructional phase of a coastline— that is, the quantity of sediment deposited exceeds that removed by waves. This results in the progradation of the coastline. Three conditions are most critical to the formation of a marsh. These include large quantities of sediment, low wave energy, and a low coastal gradient. Once sediment deposition reaches a critical height, the mud flats are colonized by *halophytic* plants that aid in continuing the growth of the marsh by the addition of organic material and by filtering sediment as tidal flooding occurs. Plants growing in a salt marsh must be capable of tolerating not only high salinities but also the mechanical stress of tidal inundation and anaerobic soils. These stressful conditions have contributed to limiting the number of species that are found in salt marsh development. Those plants that have adapted, however, display broad geographical distributions and are limited latitudinally by temperature. Nine regional zones of marshes, further subdivided into twelve subzones based on soil characteristics and temperature (Chapman, 1974), have been discerned based on geographic location and species representation:

Arctic Group: dominated by *Puccinellia phryganodes*
Northern European Group: dominated by *Salicornia* spp., *Puccinellia maritima*, *Juncus gerardi*
Mediterranean Group: dominated by *Juncus acutus*, *Arthrocnemum* sp., and *Limonium* sp.
Western Atlantic Group: dominated by *Spartina patens*, *S. alterniflora*, *Distichlis spicata*, and *Juncus gerardi*
Pacific American Group: dominated by *Puccinellia phryganodes* (in northern latitudes); *Salicornia virginica*, and *Spartina foliosa* (in midlatitudes); *Batis maritima*, *Cressa truxillensis*, *Salicornia* spp., and *Monanthochloe littoralis* (in subtropics)
Sino-Japanese Group: primary colonists dominated by *Triglochin maritima*, *Salicornia brachystachya*, or *Limonium japonicum*
Australasian Group: characterized by *Salicornia australis*, *Suaeda nova-zelandiae*, *Triglochin striata*, *Samolus repens*, *Arthrocnemum* sp.
South America Group: characterized by *Spartina* spp., *Distichlis* sp., *Heterostachys* sp., and *Allenrolfea* sp.
Tropical Group: characterized by *Sesuvium portulacastrum* and *Batis maritima*

Salt marshes may also be categorized according to the types of coastlines along which they are found. The marine and terrestrial processes influencing the critical conditions for marsh formation vary in relative dominance in different geographic locations, and result in variations in types and sizes of the marshes. For simplification, marsh coasts can be divided into three types: fringing marshes of estuaries and bays; lagoonal marshes associated with spits and barrier bars; deltaic marshes.

Fringing Marshes

Vast estuarine systems, such as Chesapeake Bay, formed during the last marine transgression are ideal environments for marsh development. Large quantities of sediment were rapidly deposited within the shelter of drowned river and tributary mouths during the rise in sea level. When deposition of sediment reached lower high water, marsh vegetation was able to colonize the flats, forming fringing marshes outlining the estuary shore. Continued sedimentation plus the accumulation of organic plant debris (*peat*) enabled the marsh to extend out into the former bay. As the bay was encroached upon, the marshes also spread inland across former upland areas inundated by the rise in sea level.

The volume of sediment carried by rivers is critical. An excessive amount of sediment relative to the discharge results in rapid sedimentation to elevations above tidal levels conducive to marsh development. This occurred along the southeastern coast of the United States, where during the last rise in sea level the drowned river mouths were filled rapidly, resulting in few estuarine systems of the Chesapeake Bay type. Here marshes are marginal, rapidly being succeeded by transitional and terrestrial vegetation.

Sediment may be derived from marine sources as well as from fluvial sources. This is the case in the Bay of Fundy, where the topographic configuration of the bay funnels incoming tides, increasing them in height as they approach the two heads of the bay. Tides may reach up to 18 m in height and scour the easily erodible rocks which characterize the bay bottom and shore, depositing the material on the flats at the heads of the bays.

Lagoonal Marshes

The growth of spits across rivers provide additional protection for marsh formation. Rivers carrying a large volume of water or that empty along a coast where the longshore drift is interrupted by deep water are able to maintain an opening in the spit. If the discharge is low or the flow ephemeral, the rivers may be closed seasonally, forming a lagoonal environment suitable for marsh development. Examples of the latter are found along the southern coast of California where rainfal averages 36-56 cm/yr. In these situations, there is an additional source of sediment from overwash as the waves move sand over the spit into the lagoon or bay behind it. The overwash, combined with alluvial materials, marine sediment carried by the tides, and organic accumulation result in rapid sedimentation of the marsh.

Spits may develop across irregular coastlines that may not have major river systems providing sediment. Marine processes become the dominant means of marsh development. Erosion of local cliffs, longshore drift, overwash by waves, and runoff from the mainland provide the sediment for deposition. The development of the marsh begins behind the proximal end of the spit and spreads as the spit lengthens downcurrent. Similar to those in estuaries, many marshes also spread inland where upland areas have been inundated. Where a recurved spit forms a number of landward arcuate bends because of changes in wind and wave conditions, the abandoned distal ends provide suitable flats for marsh colonization. This results in ponds between the curved flats. As sediment deposition continues, the marshes encroach on the ponds uniting the flats, Detailed histories of marsh formation in the lee of spits can be found in Chapman (1960) and Redfield (1972).

Barrier bars developed along subsiding coastlines having broad continental shelves also produce marsh environments dominated by marine processes, particularly overwash and tidal dynamics. The marshes develop immediately along the lagoonal side of the bar on flats high enough for colonization. These spread landward as periodic storms wash sand over the bar and into the lagoon, thus providing new flats. Occasionly, with large storms, the bar may be breached, forming a tidal inlet and allowing currents to flow into the lagoon. Although this increase in wave activity may erode portions of the marsh inside the lagoon, it also brings in large quantities of sediment that are deposited in the form of a tidal delta. With time and a return to normal wave conditions, the inlet will close again by longshore drift, leaving the delta to be colonized by plants.

Deltaic Marshes

The third general type of marsh coast is deltaic. This is similar to the others discussed above in that it occurs along a low gradient coast where large quantities of sediment are available, but it is different in that it is characterized by microtidal ranges. The growth of a deltaic marsh is cyclic and is directly related to the sediment input into the river and shifts in the channel. As the river alters its course, one area of the marsh may be deprived of sediment, resulting in a retreat of the shore while a new area is being built, thereby prograding the shoreline seaward. However, the net change for the coastline is a general seaward extension. As progradation of the shoreline occurs, the older marshes become isolated from the tidal effects and are more influenced by the freshwater input from stream runoff and rainfall.

LINDA C. NEWBY

References

Chapman, V. J., 1960. *Salt Marshes and Salt Deserts of the World*. London: Leonard Hill Limited, 392p.

Chapman, V. J., 1974. Salt marshes and salt deserts of the world, *in* R. J. Reimold and W. H. Queen, eds., *Ecology of Halophytes*. New York: Academic Press, 3-19.

Redfield, A. C. 1972. Development of a New England salt marsh, *Ecol. Monogr.* **42**, 201-237.

Cross-references: *Dieback; Coastal Flora; Deltaic Coasts; Halophytes; Lagoon and Lagoonal Coast; Marsh Gas; Salt Marsh; Soils; Spits; Vegetation Coasts.*

SALT WEATHERING

Salt weathering is a geomorphic process resulting in the physical disintegration of rocks or stones and in the *fretting* of their surfaces. It is mainly due to the growth and expansion of various salts crystals. Buildings and building stones can be attacked in a similar way.

The necessary conditions for the process to proceed are a suitable supply of soluble salts and a periodic change in temperature and humidity, so that dissolved salts confined to the pores and fissures of rocks or clasts can undergo a phase change—that is, crystallize and/or hydrate. These are best fulfilled on seacoasts where a large amount of sea spray is frequently produced and subject to drying, and in arid regions supplied by airborne salt particles from rain and dust. Both hot and cold deserts (Yaalon, 1970; Selby, 1971) as well as the seasonally dry Mediterranean coasts (Frenzel, 1965) exhibit significant effects of salt weathering. $NaCl$, Na_2SO_4, and $CaSO_4$ are the most frequent salts involved.

The phenomenon of salt weathering actually includes three separate physical mechanisms, which separately and in combination produce the high stresses required for the disintegration.

Stresses produced by crystallization and *crystal growth* are by far the most significant, and can split large stones (Fig 1). *Hydration* of certain salts (thenardite→mirabillite) and *thermal expansion* of the crystallized salts may also produce large volume changes and stresses that, when repeated frequently enough in confined spaces, may produce, in combination with crystal growth, the granular distintegration of rocks, such as granite, sandstone, or limestone (Evans, 1970; Goudie, 1974).

Where *case hardening* or *desert varnish* protects all or parts of the outer rock surface, the boulders are often hollowed out beneath, frequently leaving only a curved rock shell (Fig. 2). The interior walls of such cavernous rock forms (*tafoni*) are frequently characterized by flaking and scaling, especially in limestones and marls. Fretted surfaces and rock fracturing or cracking are other distinctive and common products of salt weathering.

FIGURE 1. Rock splitting by salt crystal growth in cracks (photo: D. H. Yaalon).

FIGURE 2. Tafoni in granite.

Although laboratory experiments indicate that under optimal conditions the processes are rapid, no field data on their actual rate are available. Coastal terraces of different ages undoubtedly exhibit different degrees of development from which the process rate for each particular combination of environmental conditions could be elucidated. Once elucidated, it could be used, together with other criteria, for relative dating of similar terraces elsewhere.

DAN H. YAALON

References

Evans, I. S., 1970. Salt crystallization and rock weathering: A review, *Rev. Gémorphologie Dynam.* **19**, 153-177.

Frenzel, G., 1965. Studien an mediterranean Tafoni, *Neues Jahrb. Geologie u. Paläontologie Abh.* **122**, 313-323.

Goudie, A., 1974. Further experimental investigations of rock weathering by salt crystallization and other mechanical processes, *Zeitschr. Geomorphologie* **21**, 1-12.

Selby, M. J., 1971. Salt weathering of landforms, and an Antarctic example, *Geogr. Conf., 6th, Christchurch, Proc.*, 30-35.

Yaalon, D. H., 1970. Parallel stone cracking, a weathering process on desert surfaces, *Bucharest Geol. Inst. Tech. Econ. Bull.*, ser. C. (Pedology), No. 18, 107-111.

Cross-references: *Aerosols; Alveolar Weathering; Aquafacts; Muricate Weathering; Tafoni; Weathering and Erosion, Mechanical.*

SAMPLE IMPREGNATION

Sample impregnation provides a means whereby a sediment *core* (q.v.) may be hardened by interstitial-space infilling with a colorless resin. In some cases, sample encasement in a block of resin is an acceptable alternative. An overview of various methods is provided by both Stanley (1971) and Müller (1967).

If impregnation while under water is called for, a technique developed by Brown and Patnode (1953) accomplishes this in two stages: partial hardening *in situ*, followed by thorough curing of the resin in the lab. For cores retrieved and returned to a workroom or laboratory, Ginsburg et al. (1966) describe a step-by-step procedure for drying of the core and introduction of the resin within a vacuum chamber. While the aforementioned methods require specially constructed equipment, Stanley (1971) includes an outline of a much simpler method for partially impregnating half-core cylinders. In all three of those cases, the resulting hardened core may be cut into sections for examination of fabric and textural characteristics.

A slab of core material may be embedded in clear lucite by following instructions given by Shannon and Lord (1967). In using this technique, resin penetrates only slightly under the sediment surface, but it results in a convenient block of lucite within which the sample may be conveniently viewed.

Since the actual details of sample impregnation, utilizing the various methods discussed above, are too lengthy and involved to be included here, the interested reader is referred to the original literature for further advice. As with sediment *peels* (q.v.) sediment impregnation ideally facilitates preservation and observation of the internal characteristics of unconsolidated sediments obtained in the coastal environment.

MAURICE L. SCHWARTZ

References

Brown, W. E., and Patnode, H. W., 1953. Plastic lithification of sands *in situ, Am. Assoc. Petroleum Geologists Bull.* 37, 152-162.
Ginsburg, R. N.; Bernard, H. A.; Moody, R. A; and Daigle, E. E., 1966. The Shell method of impregnating cores of unconsolidated sediments, *Jour. Sed. Petrology* 36, 1118-1125.
Müller, G., 1967. *Methods in Sedimentary Petrology*. New York: Harper, 283p.
Shannon, J. P., and Lord, C. W., 1967. Preservation of unconsolidated sediment cores in plastc, *Jour. Sed. Petrology* 37, 1200-1203.
Stanley, D. J., 1971. Sample impregnation, *in* R. E. Carver, ed., *Procedures in Sedimentary Petrology*. New York: Wiley Interscience, 183-216.

Cross-references: *Beach Material; Cores and Coring Techniques; Minor Beach Features; Peels; Sand, Surface Texture; Sediment Analysis, Statistical Methods.* Vol. VI: *Sedimentological Methods.*

SAND

Sediments are classified by grain size in six major groups: clay, silt, sand, gravel, cobble, and boulder. There exists more than twenty different ways of defining the limits of these scales (Truesdell and Varnes, 1950). Fortunately, only a few are in common usage. Most American workers follow the Udden (1914) scale as modified by Wentworth (1922) and Lane and others (1957). This scale is based on a center of 1 mm and a multiplier or divider of 2. Sand in this so-called Wentworth scale ranges from 0.0625-2 mm, which corresponds to 2^{-4}-2^{+1} mm. This scale has been logarithmically transformed by Krumbein (1934) and later McManus (1963) to obtain the ϕ scale. The phi scale is dimensionless, and makes whole numbers from the Wentworth size divisions. The transformation is: phi = $-\log_2$ (size in mm/ 1 mm). Sand in this classification ranges from +4 to -1 ϕ.

Most Europeans use a scale devised by Atterberg (1905) that has a center of 2 mm and a multiplier or divider of 10. Each of the major grades has two subdivisions, chosen at the closest whole number to the geometric mean of the grade limits. Sand in this scale ranges from 02-2 mm.

Colloquially, sand refers not only to the size of the particles but can also imply a composition, usually quartz. In temperate climates, quartz commonly accounts for 70% of beach sands, but can vary from 0-99%. In hotter climates, calcium carbonate—in the form of shell, coral, algal fragments, and oolites—is often dominant, sometimes up to 100%. Feldspar is usually the next most common detrital mineral. Typically it is from 10-20% of beach sands, but may be more common, particularly near regions of eroding igneous rock. The remainder of beach sands is composed almost completely of heavy minerals (specific gravity greater than 2.87).

The mineralogy of the sand can be used as a natural tracer indicating the source of the sand. Based on inclusions, optical extinctions, and shape of the quartz grains, Krynine (1940, 1946) has been able to determine whether their origin is from plutonic, volcanic, hydrothermal, metamorphic or sedimentary sources. The heavy minerals, however, are more commonly used to determine provenance.

BENNO M. BRENNINKMEYER

References

Atterberg, A., 1905. Die rationelle Klassifikation der Sande und Kiese, *Chem. Zeitung* 29, 195-198.
Krumbein, W. C., 1934. Size frequency distribution of sediments, *Jour. Sed. Petrology* 4, 65-77.
Krynine, P. D., 1940. Petrology and genesis of the Third Bradford Sand, *Pennsylvania State College, Mineral Ind. Expt. Station Bull.* 29, 13-20.
Krynine, P. D., 1946. Microscopic morphology of quartz types, *Anales Segundo Congr. Panamericano de Ing. de Minas y Geol.* 3, 35-49.
Lane, E. W., and others, 1957. Report of the subcommittee on sediment terminology, *Am. Geophys. Union Trans.* 28, 936-938.
McManus, D. A., 1963. A criticism of certain usage of the phi notation, *Jour. Sed. Petrology* 33, 670-674.
Truesdell, P. E., and Varnes, D. J., 1950. Chart correlating various grain-size definitions of sedimentary materials, *U.S. Geological Survey General Mineral Resource Maps*.
Udden, J. A., 1914. Mechanical composition of clastic sediments, *Geol. Soc. America Bull.* 25, 655-744.
Wentworth, C. K., 1922. A scale of grade and class terms in clastic sediments, *Jour. Geology* 30, 377-392.

Cross-references: *Beach Material; Boulder; Clay; Cobble; Entrainment; Gravel; Sand, Surface Texture; Sediment Analysis, Statistical Methods; Sediment Size Classification; Shingle; Silt;*

SAND, SURFACE TEXTURE

The surface texture of sand-sized grains includes those aspects of shape discernible only with the aid of optical or electron microscopy. This surface relief has been studied in an effort to identify textures diagnostic of particular environments of deposition, on the assumption that the agent of transport produces characteristic surface features. The early optical work of Cailleux (1942) created the *classical* morphoscopic method whereby smoothed and polished grains (*émoussés-luisants*) of water-worn origin were distinguished from rounded, frosted grains (*ronds-mats*) of wind-worn origin. Subsequent use of the transmission and scanning electron microscopes, with their greater resolution, has permitted the identification of recurring surface textures, particularly on quartz grains, typical of littoral (beach) and other environments of deposition. An excellent review of early work, the techniques used and results obtained, is contained in Krinsley and Margolis (1971) and Krinsley and Doornkamp (1973).

Quartz grains from the subaqueous littoral environment have as their most characteristic features: V-shaped depressions, 0.5 μm to several microns in size (see Fig. 1a); straight to curved grooves, 1–25 μm in size, occasionally with satellite V-shaped depressions oriented along them (see Fig. 1a); preferentially oriented triangular or rectilinear depressions, 1–1000 μm in size (see Fig. 1b). The first two features are thought to be formed by mechanical fracturing on impact between grains and their orientation, density and size controlled by the direction and energy of impact and the internal structure of the quartz crystal. The third feature exhibits no sign of mechanical breakage and is thought to result from chemical etching of the prismatic and rhombohedral faces of quartz in seawater. The relative frequency of occurrence of these features has been related to the degree of turbulence generated by waves, with mechanical features being predominant on high energy beaches and chemical features on low energy beaches.

Surface textures from coastal dune sands are not yet clearly documented but include: conchoidal breakage patterns not greater than 15 μm in size; and upturned plates consisting of thin parallel to subparallel plates, either continuous or discontinuous and oriented at some angle to the grain surface (see Fig. 1c). The edges are frequently jagged and broken producing a surface relief varying from 0.1 μm to several microns, and the orientation is thought to be controlled by cleavage or defor-

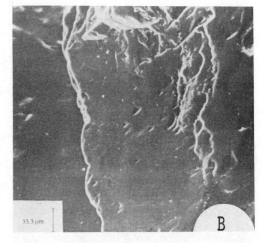

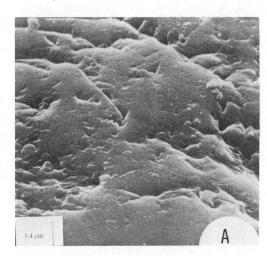

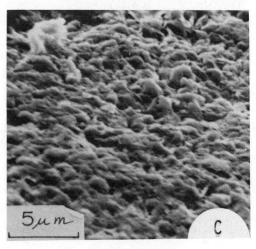

FIGURE 1. Electron micrographs of quartz grain surface textures. *A.* V-shaped depressions and irregularly curved grooves with satellite Vs resulting from impact fracturing in a beach environment; *B.* oriented solution triangles resulting from chemical etching by seawater in a beach environment; *C.* upturned plates formed by grain-to-grain impact under wind transport in a coastal dune environment (*A* and *B* from Krinsley and Doornkamp, 1973; *C* from Nordstrom and Margolis, 1972, by permission of the Society of Economic Paleontologists and Mineralogists).

mation sheeting. Both features reflect mechanical fracturing during wind transport but are not unique to this environment.

A number of problems and limitations relating to the environmental significance of surface textures have been documented (Brown, 1973), and can be summarized: features are not unique to a particular environment; for example, V-shaped impact pits are found, though less frequently, on river sands; features may be inherited from a previous environment; for example, coastal dune sands may carry V-shaped impact pits because of the short distance of transport involved; diagenesis, through solution and reprecipitation of silica, destroys features preventing their use in interpreting paleoenvironments of deposition; the method is still essentially descriptive, and a theoretical basis for the presence or absence of features is not well established.

BRIAN GREENWOOD

References

Brown, J. E., 1973. Depositional histories of sand grains from surface textures, *Nature* **242**, 396-398; **245**, 31-32.
Cailleux, A., 1942. Les actions éoliennes périglaciares en Europe, *Soc. Géol. France Mém*. **46**, 1-176.
Krinsley, D. H., and Doornkamp, J. C., 1973. *Atlas of Quartz Sand Surface Textures*. New York: Cambridge University Press, 91p.
Krinsley, D. H., and Margolis, S. V., 1971. Grain surface texture, in R. E. Carver, ed., *Procedures in Sedimentary Petrology*. New York: Wiley Interscience, 151-180.
Nordstrom, C. E., and Margolis, S. V., 1972. Sedimentary history of central California shelf sands as revealed by scanning electron microscopy, *Jour. Sed. Petrology* **42**, 527-536.

Cross-references: *Beach Material; Beach Material, Sorting of; Sand; Sediment Size Classification; Sediment Transport.* Vol. VI: *Sand Surface Texture;* Vol. XIII: *Electron Microscopy.*

SAND BYPASSING

The *natural* or *mechanical* movement of sand from one side of an inlet or harbor entrance to the other is referred to as *sand bypassing* (Bruun, 1973; Coastal Engineering Research Center, 1977). The littoral drift at most tidal inlets is transferred by either of two processes. Tidal currents move some material from the updrift side of an inlet into the inlet channel on each floodtide and from the inlet channel onto the downdrift side on the ebbtide. Material may also be moved across the inlet along a shoal or offshore bar by obliquely approaching waves. The relative importance of each of these mechanisms depends on the *tidal prism* (q.v.) and the amount of freshwater flow out through the inlet.

Where structures have been built to stabilize an inlet for navigation and to protect vessels from wave action while in the channel, natural sand bypassing is impaired. Mechanical bypassing is frequently needed to maintain the downdrift sand supply. Often the structure (littoral barrier) and the mechanical transfer system are designed as a complete unit. Watts (1965) provides a description of the four basic types of littoral barriers in which sand bypassing systems have been employed. The basic methods of mechanical sand transfer are: land-based dredging plants, floating plants, and land-based vehicles.

CHARLES B. CHESTNUTT

References

Bruun, P., 1973. *Port Engineering*. Houston: Gulf Publishing Co., 436p.
Coastal Engineering Research Center, 1977. *Shore Protection Manual*. Washington D.C.: U.S. Army Corps of Engineers, 1264p.
Watts, G. M., 1966. Trends in sand transfer systems, in *Coastal Engineering Santa Barbara Specialty Conference, Proceedings*. New York: American Society of Civil Engineers, 799-804.

Cross-references: *Coastal Engineering; Inlets and Inlet Migration; Protection of Coasts; Sediment Transport; Tidal Inlets, Channels, and Rivers.*

SAND DUNE HABITAT

The physical environment of coastal dunes is severe in several respects—shifting sand, strong winds, salt spray, and nutrient-poor soils make coastal dunes uninhabitable for most species of plants and animals. Those species that do occur must be adapted to tolerate these conditions. Certain plants, such as American beach grass (*Ammophila breviligulata*), actually grow best when they are subjected to burial by sand. By trapping sand and stabilizing the substrate on which they grow, dune grasses play in important role in the development and maintenance of coatal dunes. This paper will describe the key environmental factors that affect the growth and distribution of dune vegetation.

Much of the information presented here is taken from two coastal ecology texts by the British authors Ranwell (1972) and Chapman (1976). These writers often refer to the extensive work of an earlier investigator, Salisbury (1952). Sand dunes have received considerable attention from British ecologists, who have contributed a great deal to our understanding of dune ecosystems. In the United States, successional changes at the Indiana dunes on Lake Michigan were interpreted in the classic work of Cowles (1899), followed by Olson (1958). Many studies have been conducted on the dunes along the Atlantic coast of the United States, particularly in North Carolina, where Oosting and Billings (1942) and Boyce (1954) focused on the influence of salt spray on coastal vegetation. The major characteristics and processes in the dune environment are described quantitatively in these and other papers. Many unanswered questions, such as the possible role of nitrogen fixation in dunes, are the subject of current research.

The Effects of Wind

Sand Movement The general topography of coastal dunes is often a series of undulating ridges parallel to the shoreline. The *foredunes* adjacent to the beach itself are smaller, less stable, and of more recent origin than the larger *secondary dunes* further inland. Between the dunes are *slacks* or *swales*, which sometimes intersect the water table.

Goldsmith (1978) notes that "dunes occur wherever there is a large supply of sand, wind to move it, and a place to accumulate." Dunes as high as 30 m can be found on parts of Cape Cod, the Outer Banks of North Carolina, the barrier beaches of Oregon and Washington, and the southern shore of Lake Michigan. In many other areas, dunes 1-2 m high are more common. The highest coastal dunes occur where the orientation of the coastline is perpendicular to the strongest prevailing winds, which transport sand landward from the beach and intertidal sand flats.

The rate of aeolian sand deposition is a function of grain size, wind direction, and wind velocity. Coarse sand (600-1100 μ in diameter) is carried in air only by strong winds, while fine particles (< 200 μ) may become airborne in wind speeds greater than 1 m.p.h. (Chapman, 1976). Finer sand particles are carried further inland, resulting in smaller grain size of secondary dune sand. Wind erosion in slacks stops when the moist soil near the water table is reached.

Wind velocity is altered by local topography and the presence of vegetation. Since the velocity is lower on the lee side of a dune, sand is deposited there and stimulates the growth of dune grass (Marshall, 1965). Erosion is retarded by vegetation, while deposition is favored. Olson (1958) demonstrated that experimental plantings of beach grass increased the surface roughness of the ground, enhancing sand deposition.

The maximum vertical accretion rate of dunes is limited by the height, density, and growth rates of dune vegetation, which must be able to survive burial and grow through new deposits. This is the process by which dunes increase in size. The most important dune-forming species in North America are American beach grass (*Ammophila breviligulata*) and its southern counterpart, sea oats (*Uniola paniculata*). Pacific coast dunes are dominated by European beach grass (*Ammophila arenaria*). *Elymus arenaria* occurs on subarctic beaches, while *Ipomea* spp. are important in the subtropics. *Ammophila* spp. can tolerate up to 1 m of sand burial, but show a decrease in density if this rate continues year after year (Ranwell, 1972). Vertical accretion rates on young dunes have been measured at 0.3 m/year in Indiana (Olsen, 1958) and 0.3-0.5 m/year on Cape Cod (Goldsmith, 1978).

Rates of erosion, deposition, and dune migration vary greatly among geographic locations and within a given area. Ranwell (1972) cites dune migration rates of 0-20 m/year. He estimates that in the most mobile areas there may be a change from dune to slack to dune at a given site every 80 years or so. On Shackleford Banks, North Carolina, a migration rate of less than 10 cm/yr has been suggested by Au (1974). Natural blowouts, overwash, grazing, trampling, four-wheel-drive vehicles, sand fences, and artificial plantings will alter the patterns of sand movement in a given area. In vegetated secondary dunes undisturbed by man, sand movement rates are minimal.

Salt Spray Another major effect of wind on coastal dune ecosystems is the aerial transport of salt spray. *Aerosols* of seawater are released by bubbles breaking in the swash of waves, and may be carried 1-2 km inland by wind. Boyce (1954) sampled airborne salt using cheesecloth traps and found the expected relationships between the amount of salt collected, distance from the ocean, and wind speed. The strongest winds on the coast occur during hurricanes, when precipitation may dilute the salt spray that is deposited on plants and soil.

It is well known that salt spray is damaging to plants, and is the most important single factor determining the distribution of seacoast dune species (Oosting and Billings, 1942; Boyce, 1954). Exposed sites, such as the foredune, are colonized by the most salt-tolerant species, while sheltered areas provide a habitat for woody, coastal species and other nonhalophytes.

The chloride ions in salt spray are toxic to plants, but other ions, such as potassium, magnesium, and calcium are important plant nutrients. Because these nutrients must enter the soil before they can be taken up by plants (van der Valk, 1977), their role in plant nutrition will be discussed below in the section on dune soils.

In addition to its effect on sand movement and salt spray, wind causes higher evapotranspiration rates and, at high speeds, may cause mechanical damage (thrashing and sandblasting) to plants growing in exposed areas.

Water Relations

The Water Table Because coastal dunes are almost at sea level, the water table is not deep and may even be visible as ponds or marshes in dune slacks. Since the roots of most plants do not penetrate more than 1 m deep, the water table will be unavailable to plants growing at higher dune elevations. For many species, however, proximity of the water table is an important factor in determining their distribution.

Rainwater percolates through the porous soil of dunes to collect as a dome-shaped freshwater lens. Even within a few feet of the maximum high tide line, the groundwater is fresh rather than saline. (Saline groundwater would drastically reduce the number of species found in dune slacks.) Where dunes near the beach are underlain by *shingle ridges*, the water table below them may fluctuate monthly with the rise and fall of spring and neap tides (Chapman, 1976). In any dune system, seasonal changes occur. The water table is lowest in the summer as a result of greater evaporation, less rain (in most climates), and greater transpiration by plants. The range of water table flucutations may be about 1 m over the course of a year, and 2-3 cm/day due to evapotranspiration (Ranwell, 1972). At any given site, the depth of the water table will increase if sand deposition occurs.

Soil Moisture What is the source of water for plants that are not restricted to dune slacks? Rapid drainage of the water table follows periods of rainfall, and the capillary rise of water through dune soils is insignificant. The amount of water held by soil following drainage is termed the *field capacity*. With approximately 40% of the soil volume as pore space (Salisbury, 1952), dune soils at depths of 10-20 cm have a field capacity of only 3.5-7% (Oosting and Billings, 1942; Salisbury, 1952). This water is available to plants, but could theoretically be depleted by them on warm, sunny days (Chapman, 1976). The field capacity of dry slacks, which are much higher in organic content, is 25-30% (Salisbury, 1952).

Despite high rates of evaportranspiration and low field capacity of the soil, dune plants are able to survive long periods without rain. Their success has been attributed to several factors, including the *internal dew* that condenses in dune soils at night. The daily temperature changes beneath the soil surface are great enough that dew from the humid pore space can amount to 0.9 ml water/100 ml of soil on clear nights, replenishing water lost during the day (Salisbury, 1952). The soil below about 5 cm deep remains relatively moist, although at the surface it may be exceedingly dry. Most maritime climates are characterized by high relative humidity and moderate to high rainfall. Therefore, although coastal dunes appear edaphically xeric, lack of soil moisture does not seem to be a problem for species adapted to this habitat.

Storm Flooding Dunes on barrier beaches are subjected to storm flooding when overwash occurs. Erosion, extensive deposition, and saltwater flooding can result in dramatic changes following a single storm. Overwash channels may be narrow passages between high, stable dune ridges, or may spread through entire low-lying dune fields. Overwash is a recurring phenomenon along the Outer Banks of North Carolina, where *Spartina patens* is the species best adapted to both saltwater flooding and periodic burial (Godfrey and Godfrey, 1974, 1976).

Light and Temperature

The light intensity in dune environments is high because of the great albedo of dune soils and the sparseness of dune vegetation. While high light intensity is required by dune plants for efficient photosynthesis, it has the detrimental effect of increasing transpiration rates. Light is not limiting to the growth of dune plants, even on fog-bound coastlines. In maritime shrublands and forests, the amount of light that filters through the canopy may limit the growth of understory species, which are adapted to lower light intensity.

A more important effect of intense solar radiation reaching the soil surface is that of temperature. At air temperatures of 95-100°F, soil surface temperatures on North Carolina dunes can reach 125-127°F (Oosting and Billings, 1942). High temperatures also occur on leaves, and this heat load can lead to rapid water loss and greater respiration rates. The growth form, leaf shape, and anatomy of many dune species are adapted to minimize heating and water loss. Below the soil surface, temperature descreases rapidly with depth. The maximum daily temperature of soil at a depth of 10 cm can be 30°F cooler than that at the surface (Oosting, 1954). Because sand is a poor conductor of heat, extreme fluctuations in day and night temperatures at the soil surface are common.

The annual temperature range of maritime climates is much less than that at inland locations, owing to the moderating effect of the ocean. Summers are cooler and winters are warmer, and so in this respect the dune environment is less extreme than others in temperate regions.

Dune Soils

Soil Development The soil of most coastal dunes in North America are composed of silica grains, the product of bedrock erosion, followed by transport by rivers and longshore currents to sandy beaches. Other sources of sand include reworked glacial and sedimentary deposits. Shell fragments and limestone break down into calcareous sand, the major constituent of dunes and beaches in southern Florida.

Initially, dune soils contain little organic matter. The organic content of foredune soil in North Carolina is about 0.01%; in the soil of the maritime forest on these dunes, it is about 5% (Au, 1974). Soil development on young dunes proceeds slowly as the soil becomes stabilized by vegetation and organic matter begins to accumulate. The increase in organic content is positively correlated with the water-holding capacity of the soil. Salisbury (1952) measured field capacities of 7%, 33%, 25-30%, and 50% in young dunes, secondary dunes, dry slacks, and the humus of wet slacks, respectively.

Another major difference between young and old dune soil is in pH, which is also correlated with the age and organic content of the soil. The pH of young dune soils is just above neutral (7.4-8.0). As a result of rainfall and rapid drainage, calcium carbonates and other mineral ions are leached from the soil and replaced by hydrogen ions. Thus, in stabilized, siliceous dunes with little or no calcareous input from shell fragments, the soil is highly acidic. The organic acids leached from the litter layer also cause a decrease in soil pH, and even calcareous sand may become acidic where humus has accumulated (Oosting, 1954). At a pH of 3.8 and lower, the soil's cation capacity is almost hydrogen-saturated, a condition that few species of plants are able to tolerate. Members of the heath family (*Ericaceae*), such as bearberry (*Arctostaphylos uva-ursi*) and cranberry and blueberry (*Vaccinium* spp.), thrive on acid soils.

The effects of heavy leaching and low nutrient content often prevent dune soils from developing a substantial humus layer. Soil structure is slightly enhanced over time by *in situ* weathering and the gradual accumulation of fine sand and airborne dust (Olson, 1958).

Soil Salinity Despite the fact that salt spray is continuously deposited on the soil and vegetation of coastal dunes, the amount of salt that accumulates in the soil is extremely low (Au, 1974). Periodic rain and rapid drainage prevent salts from remaining in the root zone at concentrations damaging to plant growth.

Soil Nutrients Certain mineral nutrients, such as calcium, magnesium, potassium, and sodium, are supplied to dune ecosystems in salt spray. In some respects this meteorological input compensates for the lack of nutrients in these highly weathered soils. Art et el. (1974) found that the primary production rate of a maritime forest on Long Island dunes is similar to that of other temperate forests, and that the meteorologic cation input on the coast (excluding Na) is similar to the input from precipitation and weathering bedrock in a New Hampshire watershed. Art et al. even attribute the rapid development of coastal dune forests (200-300 yrs) as compared with the longer time period for Indiana dune forests (several thousand years) to this difference in nutrient supply.

Many studies have shown that the macronutrients nitrogen, phosphorus, and potassium are limiting in coastal dunes (Ranwell, 1972; Chapman, 1976). Fertilization with combined N/P/K stimulates the growth of dune plants, as does nitrogen alone (Willis, 1965). These macronutrients are highly soluble and easily leached from dune soils. Jones and Etherington (1975) measured 0.014% total phosphorus in dune soils and 0.018% in slack soils, which also had higher concentrations of iron and manganese. In general, slack soils are thought to be higher in nutrient content, although waterlogging interferes with nutrient uptake in many species (Jones and Etherington, 1973, 1975).

Measurements of total nitrogen content yield little information on nitrogen availability and rates of nitrogen cycling. Not surprisingly, many species rely on biological nitrogen fixation. The nitrogen-fixing bacterium *Azotobacter* is associated with the rhizosphere of *Ammophila* species (Hassouna and Wareing, 1964). *Myrica* spp. (including bayberry) have root nodules containing endophytic N-fixing bacteria (Morris et al., 1974), as do the legumes and the subtropical "weed" tree *Casuarina*.

Stewart (1967) showed that the nitrogen fixed by blue-green algae (Cyanobacteria) in dune slacks is taken up by mosses and higher plants. Soil pH affects both nitrogen-fixing organisms and the availability of nutrients to plants. Acid conditions tend to depress bacterial populations and favor fungi and Actinomycetes (Ranwell, 1972).

The low nutrient availability of coastal dune

soils influences the productivity, species composition, and rate of successional changes in dune vegetation. In Willis' 1965 fertilization experiments, nutrient applications caused the grasses to predominate over lower-growing dicots and bryophytes. Boyce (1954) showed that the growth response of *Ammophila* to nitrogen applications made the plant less tolerant of salt spray.

ALLISON A. SNOW

References

Art, H. W.; Bormann, F. H.; Voigt, G. K.; and Woodwell, G. M., 1974. Barrier island forest ecosystem: Role of meteorological nutrient inputs, *Science* 184, 60-62.

Au, S., 1974. *Vegetation and Ecological Processes on Shackleford Bank, N.C.* Washington, D.C.: National Park Service Sci. Monogr. Ser. No. 6, 86p.

Boyce, S. G., 1954. The salt spray community, *Ecol. Monogr.* 24, 29-67.

Chapman, V. J., 1976. *Coastal Vegetation*. Oxford: Pergamon Press, 292p.

Cowles, H. C., 1899. The ecological relations of vegetation on the sand dunes of Lake Michigan, *Bot. Gaz.* 27, 95-117, 167-202, 281-308, 361-391.

Godfrey, P. J., and Godfrey, M. M., 1974. The role of overwash and inlet dynamics in the formation of salt marshes on North Carolina barrier islands, *in* R. J. Reimold, ed., *Ecology of Halophytes*. New York: Academic Press, 429-440.

Godfrey, P. J., and Godfrey, M. M., 1976. *Barrier Island Ecology of Cape Lookout National Seashore and Vicinity, N. C.* Washington D. C. National Park Service Sci. Monogr. Ser. No. 9, 160p.

Goldsmith, V., 1978. Coastal dunes, *in* R. A. Davis, Jr., ed., *Coastal Sedimentary Environments*. New York: Springer-Verlag, 171-235.

Hassouna, M. G., and Wareing, P. F., 1964. Possible role of rhizosphere bacteria in the nitrogen nutrition of *Ammophila arenaria, Nature* 202, 467-469.

Jones, R., and Etherington, J. R., 1973. Comparative studies of plant growth and distribution in relation to water-logging. VII. The influence of water table fluctuations in iron and manganese available in dune slack soils, *Jour. Ecology* 61, 107-116.

Jones, R., and Etherington, J. R., 1975. Comparative studies of plant growth and distribution in relation to water-logging. VII. The uptake of phosphorous in dune and dune slack plants, *Jour. Ecology* 63, 109-116.

Marshall, J. K., 1965. *Corynephorous canescens* (L.) P. Beauv. as a model for the Ammophila problem, *Jour. Ecology* 53, 447-463.

Morris, M.; Eveleigh, D. E.; Riggs, S. C.; and Tiffney, W. N., Jr., 1974. Nitrogen fixation in the bayberry (*Myrica pennsylvanica*) and its role in coastal succession, *Am. Jour. Botany* 61, 867-870.

Olson, J. S., 1958. Rates of succession and soil changes on southern Lake Michigan sand dunes, *Bot. Gaz.* 119, 125-170.

Oosting, H. J., 1954. Ecological processes and vegetation of the maritime strand in the southeastern U.S., *Bot. Rev.* 20, 226-262.

Oosting, H. J., and Billings, W. D., 1942. Factors affecting vegetational zonation on coastal dunes, *Ecology* 23, 131-142.

Ranwell, D. S., 1972. *Ecology of Salt Marshes and Sand Dunes*. New York: John Wiley & Sons, 258p.

Salisbury, E. J., 1952. *Dunes and Downs*. London: Bell, 327p.

Stewart, W. D., 1967. Transfer of biologically fixed nitrogen in a sand dune slack region, *Nature* 214, 603-604.

van der Valk, A. G., 1977. Mineral cycling in coastal foredune plant communities at Cape Hatteras National Seashore, *Ecology* 55, 1349-1358.

Willis, A. J., 1965. Braunton Burrows: The effects on the vegetation of the addition of mineral nutrients to the dune soils, *Jour. Ecology* 51, 353-374.

Cross-references: *Aerosols; Biotic Zonation; Climate, Coastal; Coastal Dunes and Eolian Sedimentation; Coastal Flora; Hurricane Effects; Major Beach Features; Splash and Spray Zone; Soils; Washover and Washover Fans.*

SAND WAVES AND LONGSHORE SAND WAVES

Sand waves are large flow-transverse bedforms coupled to oscillatory boundary currents of tidal origin (Allen, 1980). Jordan (1962) summarized the results of observations of sand waves in estuaries, on the shelf, in the English Channel, on ocean banks, and under subarctic ice floes. He found the greatest height attained to be 27 m and wavelengths of 1000 m. The sand waves maintained a height-to-length ratio of between 1:35-1:65. Sand waves are generated and maintained by strong tide- or storm-generated currents flowing over a sandy bed in depths up to 100 m, but generally between 25-30 m. They characteristically lie normal to the dominant current, are asymmetrical in section and capable of migrating in the direction of the current. An extensive field of longshore and shelf sand waves off Virginia is dependent on storm-generated currents (Swift et al., 1972).

Longshore sand waves are large bodies of sand attached to and protruding from a sandy shoreline (Fig. 1). Series of longshore sand waves maintain wavelengths between 100-1000 m, with the protrusion extending between 10-100 m seaward. Their location, formation, persistence, and longshore migration are related to the patterns and placement of offshore bars. Sand waves lie in the lee of offshore bars, or where the bar or sand wave is attached to the shore, they represent the attachment (Swift et al., 1972). Longshore sand waves therefore result from variations in the offshore topography and its effect on the distribution of wave energy along the shore. The offshore bars act to break waves and lower wave energy in their lee permitting shoreline accretion to

FIGURE 1. Longshore sand wave protruding from the beach at Cape Cod Light, Cape Cod. Note people on beach for scale (photo: A. D. Short).

FIGURE 2. Multiple longshore sand waves, also called mega-cusps, forming in the lee of offshore bars and channels, at La Selva Beach, Monterey Bay, California. Waves are arriving from south skewing bars, sand waves, and channels to the north (photo: A. D. Short).

occur; behind channels and areas devoid of bars, greater wave energy reaches the shore, coupled with the channels acting as routes of offshore sediment transport, resulting in shoreline erosion (Fig. 2). The presence and absence of offshore bars therefore generate areas of relative shoreline accretion and erosion, hence longshore sand waves.

Longshore sand waves migrate alongshore in response to longshore migration of the parent bar. They may also originate downdrift of streams or river mouths, which debouch coarse sediment during infrequent floods. In this case the initially large sand wave gradually decreases in size as it migrates alongshore and spreads its sediment along the coast. Along barrier coasts, inlets often permit plurative movement of sediment across their mouths, generating a migrating solitary sand wave on the downdrift side.

The migration and resilience of longshore sand waves are related to their size. In general the larger the sand wave, the slower the migration and greater the resilience. The smaller sand waves, on the order of 100 m in length, are usually associated with bars attached to the shore. They represent sand moving from the bar onto the shore, first as a sand wave and later as a berm. They tend to have a life span in terms of days or at the most weeks. Sonu (1973) termed features of this nature and scale *rhythmic beach topography*. Larger sand waves assoicated with offshore bars are, like the bars, dependent on storms for their formation, maintenance, and migration. They tend to be more resilient features that may persist over several months or years; large sand waves in the Alaskan Arctic on the order of 5-10 km in length and 50-200 m in amplitude, associated with migrating offshore bars, can be traced from the present to the earliest aerial photographs in 1949, and would therefore be considered to have a life span on the order of 100 years (Short, 1975).

ANDREW D. SHORT

References

Allen, J. R. L., 1980. Sand waves: A model of origin and internal structures, *Sed. Geology* **26**, 281-328.

Jordan, G. F., 1962. Large submarine sand waves, *Science* **136**, 839-843.

Short, A. D., 1975. Offshore bars along the Alaskan Arctic coast, *Jour. Geology* **83**, 209-221.

Sonu, C. J., 1973. Three-dimensional beach changes, *Jour. Geology* **81**, 42-64.

Swift, D. J. P.; Holliday, B.; Avignone, N.; and Shideler, G., 1972. Anatomy of a shore face ridge system False Cape, Virginia, *Marine Geology* **12**, 59-84.

Cross-references: *Beach Cusps; Beach Pads; Cuspate Spits; Rhythmic Cuspate Forms; Sediment Transport.* Vol. I: *Subaqueous Sand Dunes.*

SCALE MODELS

The need for model investigations on physical (hydraulic or scale) models is well understood, though sometimes questioned in view of the possibility of using computer-aided mathematical models. The purpose of a scale model is to reproduce certain features of prototype behavior that are of interest. The scale ratios between model and prototype quantities must be chosen according to *model laws*, which are based on a physical understanding of the pertinent phenomena and which result in the desired model similitude. The choice of model laws depends upon the physical process being modeled. All of the scale ratios are usually expressed in terms of one or two scale ratios, which may be chosen independently. On the basis of the understanding of the physical processes involved, all these scale ratios are derived from differential equations of those processes, by methods of dimensional analysis and similarity theory, and by various auxiliary means (e.g., empirical formulas).

A scale model should provide similarity of prototype phenomena over their widest range. Geometric similarity requires that corresponding dimensions maintain the same proportions in the prototype and its model. For kinematic similarity to hold, corresponding distances must be passed by fluid particles during corresponding times. The dynamic similarity brings about the correspondence of forces. Further modeling requirements arise when, instead of a fixed-bed model, movable-bed and sediment motion are considered (mobile-bed model). The sedimentologic similarity requires that coastal processes (erosion, accretion) be reproduced to scale during corresponding times. In the case of waste disposal and thermal discharges, one has to account additionally for the laws of turbulent diffusion and heat transfer, and so on.

Even from this brief outline, it becomes clear that to satisfy the similitude of all factors taking part in numerous coastal processes is impossible. Generally, a decision must be made as to which factors are more important. Let us consider the forces of inertia, gravity, pressure, viscosity, surface tension, and elasticity, controlling the hydrodynamical processes in the prototype and on the fixed-bed model. By dividing the force of inertia by each of the remaining forces, one obtains: V_0^2/Lg, $V_0/(2\Delta p/\rho)$, $V_0 L/\nu$, $V_0/(\varphi/gL)^{1/2}$, and $V_0/(E/g)^{1/2}$, where: V_0 is a characteristic velocity, L = characteristic length, g = acceleration due to gravity, Δp = pressure gradient, ρ = density, ν = kinematic viscosity, φ = surface tension, and E = modulus of elasticity. The above ratios are known as the Froude (Fr^2), Euler, Reynolds, Weber, and Mach numbers, respectively. A properly reproduced scale model should maintain the prototype values of the force ratios—that is, the Froude, Reynolds, and other numbers should be identical under prototype and model conditions. This is practically impossible to accomplish even for two numbers (e.g., from the Froude criterion one has the scale of velocities $n_{V_0} = n_L^{1/2}$ while the Reynolds criterion requires $n_{V_0} = n_L^{-1}$, provided gravity and viscosity are the same in the prototype and model), and thus the choice of the overwhelming criterion (the most important forces) must be made.

When modeling large bodies of water, the length scale must be chosen so as to comply with model space limitations. For large length-scales, n_L applied to depths, the models would be shallow and the surface tension and bottom friction would control the water motion, in contrast to the prototype conditions. Therefore *distorted models* are often used with different horizontal and vertical length scales. To indicate the principal differences, some scales of physical quantities are given below in Table 1 in distorted and undistorted Froudian models (with dominant gravity and inertia forces).

In the fixed-bed models of coastal environment, it is necessary to correctly reproduce waves, currents, tides, and other forms of water motion. They all touch bottom, the friction of which controls their refraction and attenuation. They also interact with each other, thus giving rise to complex velocity fields. The scaling criteria for parameters of waves and currents are generally antagonistic, and a certain compromise can usually be obtained only by careful choice of model roughness. Artifical bottom roughness added for proper reproduction of currents would totally destroy any reasonable wave simulation (note that wave steepness, breaking, and the like are important coastline factors). Thus, for combination models, additional roughness in the form of vertical strips is normally supplied.

In mobile flat-bed models, the upper region, the boundary layer, the bottom configuration, and the sediment motion must be modeled simultaneously. For inshore areas—for example, a littoral drift study—the wave orbital motion normally greatly exceeds the current action, and the governing velocity scale should be determined by the current pattern requirement (Kamphuis, 1974). For offshore areas, for

TABLE 1. Physical Quantities in Froudian Models

Scale	Undistorted Model	Distorted Model
Horizontal Length	n_L	n_L
Vertical Length	n_L	n_Z
Time	$n_L^{1/2}$	$n_Z^{1/2}$
Velocity	$n_L^{1/2}$	$n_L \cdot n_Z^{1/2}$
Flow rate	$n_L^{5/2}$	$n_L^2 \cdot n_Z^{1/2}$

example, around a jack-up platform, where the combined waves and currents move the material, distortions normally do not matter and neither do the exact current patterns, but the shear stress scales must be the same. For mobile beds with bedforms, the total roughness is a combination of grain size roughness and bedform roughness. The grain size component is relatively much more important in the model than in the prototype. Scales are given by Mogridge (1974).

The sediment transport under combined action of waves and currents in mobile-bed models was analyzed by Bijker (1967), who arrived at a set of scales for proper reproduction of current patterns and transport rate distributions. However, scaling of coastal models is still very much an art, as can be indicated by different approaches of various authors (Keulegan, 1966; Bijker, 1967; Yalin, 1971; Migniot, 1973; Kamphuis, 1974). Artificial effects (*scale effects*) of factors present in the model but absent in the prototype must be avoided by careful choice of scales. Also, in the case of two or more important phenomena, a combined approach is used with two or more submodels to reproduce each of the phenomena separately (e.g., near field and far field of thermal discharges; see Stolzenbach and Harleman, 1972).

Proper selection of model sediment is essential. Together with artificial roughness, it is the sediment density that permits one to reproduce various sedimentologic phenomena—for example, threshold of grain motion, formation of equilibrium profile, beach configuration, transport rates. A particular problem of mobile-bed models, still far from being solved satisfactorily, consists in the determination of the sedimentologic time scale. A procedure employed by Migniot (1973) can serve as a basis for approximate computations.

Modeling of heated discharges in the coastal zone is described by Stolzenbach and Harleman (1972), who employed this technique in many practical problems and in addition to their numerical model. Other applications of scale models are described by Keulegan (1966) and elsewhere.

Two interconnected problems arise in hydraulic modeling. The first is the requirement of correct reproduction of prototype conditions, while the second consists in proper interpretation of the data obtained from a scale model. Both problems are aided by model *calibration*, in which various model parameters are modified so as to fully simulate the prototype.

RYSZARD ZEIDLER

References

Bijker, E. W., 1967. *Some Considerations about Scales for Coastal Models with Movable Bed*. Delft: Delft Hydr. Laboratory Pub. No. 50, 150p.

Kamphuis, J. W., 1974. Practical scaling of coastal models, *Conf. Coastal Eng., 14, Copenhagen, Am. Soc. Civil Engineers, Summaries*, 437-440.

Keulegan, G. H., 1966. Model laws for estuary and coastline models, *in* A. T. Ippen, ed., *Estuary and Coastline Hydrodynamics*. New York: McGraw-Hill, 691-710.

Migniot, C., 1973. *Facteurs hydrodynamiques intervenant sur les modeles sedimentologiques*. Laboratoire Central Hydraulique de France, Masions-Alfort.

Mogridge, G. R., 1974. Scale laws for bed forms in laboratory wave models, *Conf. Coastal Eng., 14th, Copenhagen, Am. Soc. Civil Engineers, Summaries*, 441-444.

Stolzenbach, K. D., and Harleman, D. R. F., 1972. *Physical Modeling of Heated Discharge*. Massachusetts Institute of Technology, Ralph M. Parsons Laboratory, Summer Course.

Yalin, M. S., 1971. *The Theory of Hydraulic Models*. London: Macmillan, 266p.

Cross-references: *Beach Processes; Coastal Engineering; Coastal Engineering, Research Methods; Coastal Morphology, Research Methods; Froude Number; Reynold's Number; Wave Refraction Diagrams; Waves*. Vol. III: *Stream Table Construction and Operation*.

SCHIZOMYCETES

The name *Schizomycetes* was first used by Naegli (1857) with reference to *bacteria*. No taxonomic rank was assigned, and the nature and limits of the group were poorly understood at the time. Migula (1900) assigned it the rank of a class. For many years subsequently the Schizomycetes were generally treated as a class of plants.

A fundamental advance in our understanding of the relationships of bacteria was the recognition by Chatton (1937) that there are two basically different general patterns of cellular organization—procaryotic and eucaryotic. This insight has been amply confirmed through the work of Stanier, Doudoroff, and Adelberg (1970) and many others. Some of the differences between these are tabulated in Table 1.

The procaryotes were raised to the rank of kingdom by Murray (1968), to contain the *blue-green algae* and the bacteria. A recent subdivision of this kingdom is that presented in the eighth edition of *Bergey's Manual* (Buchanan and Gibbons, 1974), in which definite taxonomic ranking is deliberately avoided, and the blue-green algae are regarded as a specialized group of bacteria differing from other

TABLE 1. Differences in the Two Patterns of Cellular Organization

Procaryotic cells	Eucaryotic cells
Nucleoplasm not separated from the rest of the cytoplasm by a nuclear membrane.	Membrane bound nucleus present.
DNA a single double-stranded loop not associated with histones, forms 1 linkage group.	DNA in more than 1 loop, intimately associated with histones, forming more than 1 linkage group.
No chromosomes or mitotic spindle formed during cell division.	Cell division marked by condensation of the DNA into discrete chromosomes. Mitotic spindle formed.
No membrane bound cytoplasmic organelles. In photosynthetic and nitrifying bacteria where there is a system of unit membranes, these are derived by invagination of the cell membrane, and in most cases the connection persists.	Cytoplasmic membrane bound organelles (mitochondria and plastids) present, containing their own DNA and replicating as units. Their membranes differ from that of the cell containing them.
Ribosomes small, 70 S type, distributed through the cytoplasm. No endoplasmic reticulum.	Ribosomes larger, 80 S type, arranged on the endoplasmic reticulum.
Cell wall contains the mucopeptide murein, composed of amino sugar muramic acid units.	Cell wall does not contain murein.

photosynthetic bacteria in the nature of their pigment system and photosynthetic mechanism.

Kingdom Procaryotae

Division I. Phototrophic procaryotes
    Class 1. Blue-green photobacteria
    Class 2. Red photobacteria
    Class 3. Green photobacteria
Division II. Procaryotes indifferent to light (Scotobacteria)
    Class 1. Bacteria (17 major groups comprising the Eubacteria, Spirochaetes, Pseudomonads, Actinomycetes and others)
    Class 2. Obligate intracellular bacteria (Rickettsias)
    Class 3. Bacteria without cell walls (Mycoplasmatales)

The class Schizomycetes may or may not prove to be a valuable taxon in this context. As usually conceived, it contains all the procaryotes except the blue-green algae and possibly the Mycoplasmatales.

BAYARD H. McCONNAUGHEY

References

Buchanan, R. E., and Gibbons, N. E., eds., 1974. *Bergey's Manual of Determinative Bacteriology*. Baltimore, Md.: Williams and Wilikins Co., 1246p.

Chatton, E. P. L., 1937. *Titres et Travaux Scientifiques*. Séte: E. Sottano, 405p.

Migula, W., 1900. *System der Bakterien*, vol. 2. Jena: Gustav Fischer, 1068p.

Murray, R. G. E., 1968. Microbial structure as an aid to microbial classification and taxonomy, *Spisy* 43, 249-252.

Naegli, C., 1857. In Caspary, Bericht ueber die Verhandlungen der 33, Versamlung deutscher Naturforsches und Aertze, gehalten in Bonn von 18 bis 24 September 1857, *Bot. Ztg.* 15, 749-776, 784-792.

Stanier, R. Y.; Doudoroff, M.; and Adelberg, E. A., 1970. *The Microbial World*. Englewood Cliffs, N.J.: Prentice-Hall, 873p.

Cross-references *Coastal Flora; Cyanophyceae*.

SCHORRE

This Flemish-Dutch term indicates that part of the beach, and of the delta area of tidal rivers, that is normally emerged land. However, the schorre is regularly submerged at syzygy and equinoctial tides. It is colonized by a vegetation that survives the water invasion. Occasionally a schorre is no longer subject to invasion by the waters—not even exceptionally; it then becomes a natural *polder*.

The schorre results from deposition by the water of suspended material (Francis—Bœuf, 1942). The depositional process is slow, and the materials anchor themselves to the existing bottom during the emergence spans of the bottom at low tide. The gradual piling up of these thin layers of deposits allows the mud bank to evenutally remain above water level during long periods, finally to be covered by water only exceptionally. The schorre has then been built.

ROGER H. CHARLIER

Reference

Francis-Bœuf, C., 1942. Sur la presence du schorre et sa signification dans les estuaires, *Soc. Géol. France Compte Rendu*, 182.

Cross-references: *Deltas; Polder; Slikke; Tidal Flat.*

SCOUR HOLES

Scour holes are areas where removal of underwater bed material has taken place, and are caused by current or wave activity impinging upon a structure-face deflecting wave or current activity upward or downward (Coastal Engineering Research Center, 1973). The downward component causes erosion of bed material at the toe of the structure. Scour holes formed at the base of the structure have dimensions governed by the type of structure face, nature of wave (or current) attack, and resistance of the bed material. Extreme scouring will cause undermining of the structure that may cause structure failure. Prevention of scouring and thus scour hole development requires protection to the seaward toe of the structure utilizing armor stone of adequate size to prevent displacement.

JOHN B. HERBICH
ROBERT E. JENSEN

Reference

Coastal Engineering Research Center, 1973. *Shore Protection Manual*. Washington D.C.: U.S. Army Corps of Engineers, 750p.

Cross-references: *Bulkheads; Coastal Engineering; Coastal Erosion; Colk; Protection of Coasts; Quay; Seawall.*

SCOURING RUSHES—SEE EQUISETOPHYTA

SEA CAVES

Sea caves are deeply hollowed-out sections of sea cliff that occur where agents of coastal weathering and erosion have penetrated far into zones of rock weakness (Zenkovich, 1967). Similar in origin to wave-carved *notches,* sea caves are caused by the hydraulic pressure of ocean waves combined with the abrasive action of wave-hurled rock fragments, and are confined to the bases of highly resistant *sea* cliffs. They differ from notches in that they are highly localized and may be many meters deep. Where waves attacking the sides of a promontory have tunneled caves completely through it, a *sea arch* is produced. In some cases, sea caves and arches are merely limestone solution features or lava tubes modified by sea action.

One of the largest is the Sea Lion Caves near Florence, Oregon, where two sea-level passageways of 400 and 30 m, respectively, join at a one-hectare cavern more than 40 m high. The cave is in a 100-m high sea cliff composed of alternating layers of sediment and highly jointed basalt.

Even though they comprise voids in highly resistant rock, sea caves are to some degree temporary features of coastal erosion, and numerous cases of collapse within this century are known (Bird, 1969; Davies, 1972).

JAMES E. STEMBRIDGE, JR.

References

Bird, E. C. F., 1969, *Coasts.* Cambridge, Mass.: MIT Press, 246p.
Davies, J. L., 1972. *Geographical Variation in Coastal Development.* Edinburgh: Oliver & Boyd, 204p.
Zenkovich, V. P., 1967. *Processes of Coastal Development.* New York: Wiley Interscience, 738p.

Cross-references: *Blowholes; Cliff Erosion; Karst Coasts; Niche, Nick, Nip, Notch; Sea Cliffs; Stack and Chimney.*

SEA CLIFFS

A *sea cliff* is a steep coastal slope created by the erosive power of waves at its base (Bird, 1969; Zenkovich, 1967). The spectacular sea cliffs on the north coast of Molokai, Hawaii, which rise more than one kilometer above the Pacific Ocean, are thought to be the highest in the world.

Some authorities define the term *cliff* to mean *sea cliff,* and so the modifier *sea* may be redundant. *Bluff* and *precipice* are synonyms but, like *cliff,* do not necessarily refer to coastal features. *Escarpment* and *scarp* are usually used in reference to the steep slopes along fault lines. Cliffs and terraces are the most common features of coastal erosion. They are often cited as primary evidence of former shorelines, even when found far from present-day bodies of water.

Sea cliffs result from wave attack, usually

confined to within a few meters of sea level. The wave-carved *notch* is a typical feature resulting from this process. As rock removal continues, the coastal slope eventually becomes oversteepened and its weight causes collapse in the form of rockfalls, landslides, or other mass movement. The resulting talus may serve to temporarily protect the cliff from further erosion, creating a *dead cliff,* one not subject to the oversteepening caused by wave action. A growing cover of vegetation is evidence of stability in such cliffs. Thus, the steepness of a sea cliff generally depends on its resistance and the severity of wave attack, though numerous other factors are involved.

The minimum angle at which a coastal *slope* becomes a *sea* cliff has not been defined, although Davies (1972) divides cliffs into two groups based on the presence or absence of undercutting. Variations in lithology, structure, bedding plane dip, vegetation, climate, wave regime, and beach form create the varieties of form and feature present in sea cliffs. Many such features are discussed elsewhere in this volume.

As erosional features, active sea cliffs are in a state of continuous change, cycles of steepening and collapse. Rates of retreat vary with the erosive power of the sea and the friability of the rock. Knowledge of cliff recession rates is increasing worldwide, and is of special importance along densely populated coasts where damage due to landslides can be extremely costly.

JAMES E. STEMBRIDGE, JR.

References

Bird, E. C. F., 1969. *Coasts.* Cambridge, Mass.: M.I.T. Press, 246p.
Davies, J. L., 1972. *Geographical Variation in Coastal Development.* Edinburgh: Oliver & Boyd, 204p.
Zenkovich, V. P., 1967. *Processes of Coastal Development.* New York: Wiley Interscience, 738p.

Cross-references: *Blowholes; Cliff Erosion; Coastal Bevel; Fossil Cliffs; Karst Coasts; Niche, Nick, Nip, Notch; Slope-Over-Wall Cliffs; Stack and Chimney.*

SEA CONDITIONS

A variety of qualitative and quantitative scales have been developed to provide an estimation of the condition of the ocean surface in terms of roughness and wave height (Schule, 1966). The *Douglas Sea Scale* was introduced in 1921 and subsequently adopted by the International Meteorological Organization (later renamed the World Meteorological Organization) in 1929. Developed by Captain H. P. Douglas of the Royal Navy, the scale is based on a range of descriptive terms from 0 (calm) to 9 (confused) that is partially correlated with the Beaufort Wind Scale (q.v.). A subsequent refinement introduced by the German Captain Petersen in 1939 allowed an estimation of wind force from the appearance of the sea surface. The *Petersen Scale* therefore provides a bridge between the Beaufort Wind Scale and the Douglas Sea Scale. In 1947 the WMO adopted a single ocean wave code that differs slightly from the Douglas Sea Scale. These sea state codes are still used in ships' logs and in marine meteorological forecasts as a descriptive classification of sea conditions (Table 1).

EDWARD H. OWENS

References

Defense Mapping Agency Hydrographic Center, 1977. *American Practical Navigator.* Washington, D.C.: Government Printing Office, H. O. Pub. No. 9, 1386p.
Schule, J. J., Jr., 1966. Sea state, *in* R. W. Fairbridge, ed., *Encyclopedia of Oceanography.* New York: Reinhold, 786-792.

Cross-references: *Beaufort Wind Scale; Climate, Coastal; Waves; Winds.* Vol. I: *Sea State.*

SEA LEVEL CHANGES

Sea level changes (herein considered exclusively from the oceanographer's point of view and studied principally on the basis of continuous sea level records) are the consequences of a considerable number of contributing factors (Lisitzin, 1974). Excluding the disturbing effect of the astronomic bodies upon the sea surface, there remain to be examined the meteorological and hydrographical factors that may affect the sea level to a substantial degree. Among these factors the most significant are atmospheric pressure, wind, and water density. Currents, frequently originated by wind effect, and precipitation and evaporation also must be taken into consideration. Particularly in coastal zones, the sea level may also be influenced by seiches (q.v.) and tsunamis (q.v.). Seismic and volcanic activities, secular (long-term) changes such as vertical movements of the earth's crust, and eustatic factors also affect the relative position between the mean sea level and the benchmark on land.

The sea level and its variations along the coasts of continents and islands are recorded by various kinds of *tidal gauges* or *marigraphs* having differing degrees of accuracy. One of the biggest problems in sea level research on a global scale is, however, the fact that some coastal

TABLE 1. Sea State Codes Corresponding to the Beaufort Wind Scale (after Defense Mapping Agency Hydrographic Center, 1977)

Beaufort Wind Scale	Wind Speed mi/hr	Wind Speed m/sec	Seaman's Term	U.S. Weather Bureau Term	Effects of Wind Observed at Sea (Petersen Scale)	U.S. Navy Hydrographic Office (Douglas Scale) Term and Height of Waves, in Feet	Code	International (W.M.O.) Code Term and Height of Waves, in Feet	Code
0	under 1	0.0–0.2	Calm		Sea like mirror.	Calm, 0	0	Calm, glassy, 0	0
1	1–3	0.3–1.5	Light air	Light	Ripples with appearance of scales; no foam crests.	Smooth, less than 1	1	Rippled, 0–1	1
2	4–7	1.6–3.3	Light breeze	Light	Small wavelets; crests of glassy appearance, not breaking.	Slight, 1–3	2	Smooth, 1–2	2
3	8–12	3.4–5.4	Gentle breeze	Gentle	Large wavelets; crests begin to break; scattered whitecaps.	Moderate, 3–5	3	Slight, 2–4	3
4	13–18	5.5–7.9	Moderate breeze	Moderate	Small waves, becoming longer; numerous whitecaps.			Moderate, 4–8	4
5	19–24	8.0–10.7	Fresh breeze	Fresh	Moderate waves, taking longer form; many whitecaps; some spray.	Rough, 5–8	4	Rough, 8–13	5
6	25–31	10.8–13.8	Strong breeze	Strong	Larger waves forming; whitecaps everywhere; more spray.				
7	32–38	13.9–17.1	Moderate gale	Strong	Sea heaps up; white foam from breaking waves begins to be blown in streaks.	Very rough, 8–12	5	Very rough, 13–20	6
8	39–46	17.2–20.7	Fresh gale	Gale	Moderately high waves of greater length; edges of crests begin to break into spindrift; foam is blown in well-marked streaks.				
9	47–54	20.8–24.4	Strong gale	Gale	High waves; sea begins to roll; dense streaks of foam; spray may reduce visibility	High, 12–20	6	High, 20–30	7
10	55–63	24.5–28.4	Whole gale	Whole gale	Very high waves with overhanging crests; sea takes white appearance as foam is blown in very dense streaks; rolling is heavy and visibility reduced.	Very high, 20–40	7		
11	64–72	28.5–32.6	Storm	Whole gale	Exceptionally high waves; sea covered with white foam patches; visibility still more reduced.	Mountainous, 40 and higher	8	Very high, 30–45	8
12	73–82	32.7–36.9	Hurricane	Hurricane	Air filled with foam; sea completely white with driving spray; visibility greatly reduced.	Confused	9	Phenomenal, over 45	9
13	83–92	37.0–41.4							
14	93–103	41.5–46.1							
15	104–115	46.2–50.9							
16	115–125	51.0–56.0							
17	126–136	56.1–61.2							

areas yield considerable amounts of data while other areas yield little or none.

Factors Affecting Sea Level

Atmospheric Pressure Every change in atmospheric pressure over the surface of the sea, which at least theoretically reacts like a reverse barometer, affects the sea level. The sea level should increase by 1 cm for every decrease of 1 mbar in atmospheric pressure, and vice versa. However, such changes rarely occur because every variation in atmospheric pressure is always associated with changes in the direction and velocity of the wind and it is the wind that usually produces the most pronounced fluctuations in sea level in the coastal regions and in more or less enclosed sea basins.

Wind. Sea level variations may be caused by several markedly different wind-produced phenomena. *Stress* upon the surface of the sea is the most significant effect of wind-water interactions, producing the largest departure of sea level from the predicted height of the tide, or in basins with weak or practically no tidal motion, from the mean sea level. Either transient winds associated with rapidly moving storm formations or steady winds of various strengths can be involved, but the effect of the tangential stress of the wind upon the sea surface always causes an increase or decrease in sea level. The term *storm surge* (q.v.) is generally used in more extreme cases of stress, while the term *set up* of the water surface has been recommended instead by some authors. If the phenomenon is less pronounced, the term *pile up* is adequate, being a substitute for the widely used German term *Windstau*.

The most conspicuous and disastrous storm surges usually arise in rather shallow sea regions. The dreaded storm surges that occur along the Atlantic Ocean and the Gulf of Mexico coasts of the United States are known as *hurricane surges*, while *typhoons* principally ravage the coasts of Japan. The west coast of the British Isles is also frequently subject to the devastating effect of open coast storm surges. Another highly exposed area is the shallow Bay of Bengal, where destructive floods may be accompanied by catastrophic losses of life and property. The North and Baltic seas are included among the more restricted sea basins with pronounced storm surges and the phenomenon has been closely examined in these regions.

The computations needed to investigate complicated cases of storm surges must be performed with the help of hydrodynamic differential equations, which reproduce the motion of the sea as functions of atmospheric pressure, wind stress, and water density, and the use of modern high-speed computers has been invaluable in making these computations.

Waves. Ordinary wave motion of course has a disturbing effect on the sea level; however, its study does not belong in the field of sea level research proper and *waves* (q.v.) are considered in a separate entry in this volume. *Seiches* (q.v.), though, are standing waves (or stationary oscillations) that occur in enclosed or semi-enclosed water basins such as lakes, bays, gulfs, or harbors. The main characteristic of a seiche is its natural period of oscillation that depends on the horizontal and vertical dimensions of the body of water in which it occurs and upon the number of nodes of the standing wave. The phenomenon of a seiche has been connected primarily with standing waves in lakes but has come to describe also cases of stationary sea level oscillation in all other more or less enclosed bodies of water. A *tsunami* (q.v.) is a long surface wave caused by submarine earthquakes, volcanic eruptions on the sea bottom, or, in recent times, by powerful atomic bombs exploded over the oceans. As a consequence of the vertical movements of the sea floor in connection with an earthquake, progressive long waves spread annularly as unbroken ridges in all directions from the epicenter soon after the shock. The height of the wave is not always a function of the magnitude of the earthquake; it also depends upon the inclination of the sea floor in the neighborhood of the coast and the configuration of the coastline, the highest amplitudes being reached in the inner parts of gradually narrowing bays.

Water Density. The influence of the changing water density, and consequently also of its reciprocal term, water volume, upon the sea level has been especially investigated by Pattullo et al. (1955). Density, one of the basic properties of seawater, is a function of temperature, salinity, and pressure. The density of pure water at $4°C$ is equal to unity and reaches its highest value at this temperature. The density of seawater deviates from unity owing to not only the temperature effect but also as the consequence of salinity, which causes the density to increase slightly with an increasing quantity of salt. Seawater density *in situ* is influenced, moreover, by the pressure exerted by higher-lying water layers. In general then, the density of seawater increases as the result of increasing salinity, cooling down and heating up to $4°C$, increasing pressure, evaporation, and ice formation. It decreases when the discharge by rivers, precipitation, and melting of ice results in changing water temperature and a dilution of salinity. When considering water volume and sea level, the situation is reversed.

The effect of water density upon the sea level

was recognized years ago and there have been extensive studies on this subject for the separate regions of the oceans. However, Pattullo et al. (1955) were the first to examine this phenomenon on a worldwide scale, paying special attention to the problem of seasonal variation in sea level in different zonal areas.

Currents. As a consequence of the uneven distribution of heat over the earth's surface, there may arise accentuated differences in atmospheric pressure and water density between the equator and the poles. The gravitational force tends to lessen these deviations, and the final result is a continuous circulation of air and water masses. Atmospheric circulation produced by the deviating structure of air masses is more pronounced than the water circulation in the oceans and seas, particularly since the density of air is considerably less than that of water. Nevertheless, when the velocity of the air currents moving over the water surface reaches a high enough value, the frictional affect at the boundary between the two elements causes the so-called wind or *drift currents*. These currents are of great significance for the circulation in oceans and seas and they exert an additional effect on the distribution of the water masses. This circulation is influenced by the fact that every motion on the rotating earth is subject to the deflecting effect of the earth's rotation, generally known as the *Coriolis Effect* (q.v.).

The Coriolis Effect is manifested by the deflecting of water masses to the right in the Northern Hemisphere and to the left in the Southern Hemisphere, when looking in the direction of the current. In this manner there arise, across the current sea level, differences that increase with increasing current velocity. Since currents frequently move along the shore, these deviations may be of considerable significance in coastal regions. In sounds and narrows with a less variable direction of the current, the Coriolis Effect upon the sea level on the opposite coasts must always be taken into consideration, especially when dealing with different problems connected with the determination of the mean sea level.

Precipitation and Evaporation. Compared with the previously mentioned factors, the effects on changes in sea level of precipitation and evaporation are not as marked and must be considered principally in connection with the seasonal variation in mean sea level and with the regional distribution of mean sea level on a worldwide scale.

Ice. In recent years special attention has been paid to the disturbing effect of ice along the coast and on the sea upon the height of the wind-produced pile up of the sea surface. In some cases, this effect has been proved to be of considerable significance and requires further detailed study.

Earthquakes. There are several known cases where an earthquake has influenced the mean sea level, changing the zero height of the tidal gauge operating in the coastal area of the affected region. The most pronounced recorded case is probably the earthquake in Messina, which occurred in 1908. The zero of the gauge fell not less than 57 cm and continued to fall, although more slowly, during the following years. More recent cases of displacement in connection with earthquakes have been studied in more detail, especially in Japan. Statistical results based on 103 earthquakes in that country showed that the sea level curve representing the monthly averages reached, on the average, a maximum approximately every four months and a minimum every one or two months before the occurrence of the earthquake. In the months immediately following the occurrence, the mean sea level rose in all cases examined.

Secular (Long-term) Changes in Sea Level

The vertical movements of the earth's crust are, according to a theory advanced by Jamieson in 1865, the consequence of the extensive ice masses that covered the northern parts of the earth during the glacial periods and their subsequent melting. These ice masses exerted a tremendous pressure upon the earth's crust, which subsided but began to reoccupy its original position as soon as the ice pressure began to decrease. *Land emergence* has also continued after the disappearance of the ice and it is taking place at the present time as well, although at a diminishing rate. Other theories on recent crustal movements have been advanced, but they are of little significance in connection with sea level investigations.

Land emergence was observed in the Fennoscandia area centuries ago because this phenomenon is highly pronounced in these regions and occurs over a relatively short period of time. Several attempts have been made since the middle of the eighteenth century to evaluate the rate of the continuous decrease in mean sea level in Finland and Sweden. Early estimates were, however, only rough. More exact results could not be achieved until the twentieth century because the necessary computations have to be based on uninterrupted series of sea level observations that should be not only relatively prolonged but also highly accurate. In order to achieve satisfactory results, astronomical tides and various meteorologic and hydrographic disturbing factors must also be taken into consideration.

The continuous decrease in the mean sea level has been established in the northern parts of the British Isles and in Alaska in addition to the evidence of the Fennoscandia land emergence. However, the computations based on sea level records in the the Netherlands have confirmed the well-known fact that the country is in the process of *submergence* and that the sea gains continuously at the expense of the land. This same fact can also be established for a number of sea level recording stations around the world. The possibility must always be kept in mind that the increase in mean sea level can be ascribed to eustatic terms, mainly to the continuous melting of the continental ice in the higher latitudes of the Northern Hemisphere. Differing results have been obtained as to the average yearly quantity of the melting ice masses. In order to give a numerical value, it may be mentioned that the eustatic increase in mean sea level amounts to 1.0–1.1 mm/yr. This correction must always be subtracted from or added to the computed values representing the secular variation in mean sea level when determining the approximate rate of land emergence or submergence.

Seasonal Cycle in Mean Sea Level

Oceanographers interested in the problem of mean sea level and its fluctuations in time and space have long paid considerable attention to the seasonal changes of the height of the water surface in oceans and seas. It has been possible to develop a fairly reliable picture of the average fluctuations in mean sea level during the course of a year given the computed variations of the water volume as a function of changing temperature and salinity and taking into account meteorological factors, especially atmospheric pressure and wind stress. For example, it seems likely that the seasonal heating and cooling of seawater, followed by corresponding changes in the volume of the water column, causes the mean sea level to reach its highest point in the low and middle latitudes of the Northern Hemisphere early in autumn, on the average in September, and in the corresponding zones in the Southern Hemisphere in the early austral autumn, in March. This distribution is characteristic not only of the coast but of the open deep regions of the oceans as well.

The Pacific Ocean has been examined in more detail on the basis of sea level records collected as an important part of the extensive research program elaborated for the International Geophysical Year 1957–58. Consequently, a considerable deviation has been established between the central regions and the bordering areas of this extensive basin, resulting in large part from the effect of the changing seasonal distribution of atmospheric pressure. The deviation reaches its most pronounced states during the summer and winter. The persistent high pressure body situated over the northern parts of the Asiatic continent in winter results in a low atmospheric pressure over the central parts of the Pacific Ocean during this season. Therefore, in December the mean sea level is higher by about 10–15 cm in the middle parts of the basin than along the continental coasts and in June the situation is reversed. In autumn and spring the difference is only weakly pronounced.

Such deviations may also be described as follows. There occurs in the oceans a seasonal oscillation in mean sea level that is characterized by a nodal line running in the vicinity of the equator and that reaches its greatest height departures in September and March. Another oscillation could be established, at least in the Pacific Ocean, with two nodal lines running longitudinally between the coastal and deep-sea regions. The typical feature of this oscillation is that the most pronounced deviations from the mean sea level are noted, on the average, in the months of June and December.

The influence of wind stress upon the sea surface was not considered in the previously described studies, but Gill and Niiler (1973) have paid particular attention to this phenomenon. The results of their numerical investigations clearly indicate that the effect of wind stress is significant but not of decisive character for the problem as a whole, its value lying in the clarification of certain details that improve upon the results of studies based on fewer elements.

The annual fluctuation in the strength of the extensive current systems in the oceans, associated with the seasonal variation of wind stress, is another factor that may contribute to the seasonal variation in mean sea level and its deviation between the coastal areas and the open deep-sea regions on a global scale, Also, the deflecting force of the earth's rotation may be the cause of considerable changes in mean sea level inside and outside of the extensive gyres.

It has been possible to determine that the total quantity of water in the oceans is also subject to seasonal changes. On the basis of all available sea level records covering the period 1957–1962, it can be computed that the amount of water in the oceans increases by 0.59×10^{19} g from March to September, corresponding to an average sea level increase of 1.6 cm. It can also be estimated that roughly 70% of the total water gained is stored in March as snow and ice in the higher latitudes of the

Northern Hemisphere, mainly in Siberia. Approximately 20% of the water gain probably comes from the atmosphere, based on the fact that in the Southern Hemisphere evaporation is considerably higher than precipitation. Also there is a marked seasonal cycle in evaporation and an accentuated moisture transfer across the equator primarily connected with the effect of the Asiatic monsoon. Because there are some additional conditions—such as humidity of the soil, groundwater, and organic material—that may augment the amount of water gained by the oceans in September, the numerical results achieved on the basis of extensive sea level records may be considered reliable as a first approximation.

Regional Distribution of Mean Sea Level

It is common practice among oceanographers, geodesists, and geographers to choose the average surface at a given place as the reference level not only for all depth determinations but also for all measurements of altitude on land. Nevertheless, it is a well-established fact that the mean sea level in the oceans and seas varies considerably from one place to another (Fig. 1), and that the deviations from the average value are of such a magnitude that they cannot be neglected even for practical purposes. For example, the variations of the average heights of sea level in the open deep-sea areas of the oceans situated between 60°N and 60°S have an average range of 270 cm, the lowest mean sea level occurring in the southern parts of the Atlantic Ocean off the Antarctic coast and the highest in the western parts of the Pacific Ocean.

The determination of the regional distribution of mean sea level along the coasts of a continent or across the continent itself is facilitated by the information of fundamental significance yielded by precise geodetic leveling, in addition to sea level records and the numerical computations based on these data. The conformity between oceanographic methods (those utilizing water density data and knowledge of the effects of meteorological factors) and geodetic methods may be illustrated by the following results. The adjustment of first-order leveling in the United States (Braaten and McCombs, 1963) has shown that the water surface at Neah Bay, Washington, stands 71 cm higher than at Portland, Maine, while the height difference in mean sea level between San Francisco, California, and Norfolk, Virginia, is 62 cm and between San Diego, California, and Fernandina, Florida, 65 cm. The averaging of these figures indicates that the Pacific Ocean stands approximately 66 cm higher than the Atlantic Ocean. Oceanographic computations for the open deep-sea areas of the two oceans have resulted in a mean sea level difference of 72 cm. The deviation between the two height differences may, at least to some degree, be ascribed to the fact that the mean sea level is generally lower in the eastern parts of the oceans than in their western parts, the highest values for each particular ocean being reached in the western areas (as mentioned previously pertaining to the Pacific Ocean).

In contrast to the conformity of oceanographic and geodetic results between the eastern and western coasts of the United States,

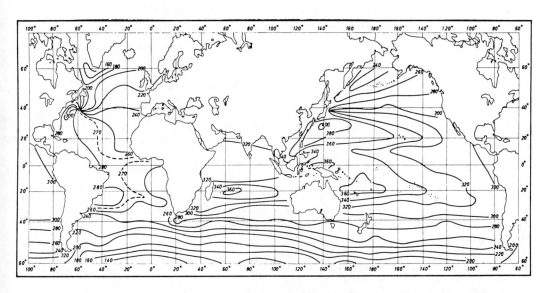

FIGURE 1. The distribution of mean sea level heights (dyn/cm) in the oceans (from Lisitzen, 1965).

the results found along each coastline are not in such accord. According to the adjustment of the first-order leveling, the lowest mean sea level along the Atlantic and Gulf of Florida coasts of the United States is noted at Key West, Florida. The mean sea level increases continuously along the coast toward the north and the rise from Key West to Portland, Maine, amounts to not less than 58 cm. This marked increase in mean sea level from the south to the north is in disagreement with the requirements of the observed flow of the Gulf Stream. Further extensive measurements on land and in the oceans are therefore indispensable in order to solve the problem. A similar unconformity between the geodetic and oceanographic results is characteristic also for the Pacific coast of the United States.

EUGENIE LISITZIN

References

Braaten, N. F., and McCombs, C. F., 1963. *Mean Sea Level Variations as Indicated by a 1963 Adjustment of First Order Leveling in the United States*. Washington, D.C.: Coast and Geodetic Survey, 22p.

Gill, A. E., and Niiler, P. P., 1973. The theory of the seasonal variability in the oceans, *Deep-Sea Research* **20**, 141-177.

Lisitzin, E., 1965. The mean sea level of the world ocean, *Commentationes Physico-Mathematicae 30*, 35p.

Lisitzin, E., 1974. *Sea Level Changes*. Amsterdam: Elsevier, 286p.

Pattullo, J.; Munk, W.; Revelle, R.; and Strong, E., 1955. The seasonal oscillation in sea level, *Jour. Marine Research* **14**, 88-156.

Cross-references: *Coastal Morphology, Research Methods; Coriolis Effect; Emergence and Emerged Shoreline; Hurricane Effects; Sea Level Curves; Seiche; Submergence and Submerged Shoreline; Tides; Tsunamis.*

SEA LEVEL CHANGES, 1900 TO PRESENT

Mean sea level is defined as the mean (arithmetic) height of the open sea surface, measured at hourly intervals for all stages of the tide over at least a nineteen year period (one lunar nodal cycle). *Tide gauges* or *mareographs*, accurately positioned relative to fixed, permanent land points, are used to collect the tide height data. Because of the time asymmetry in tidal cycles, mean tide level is not, strictly speaking, identical to mean sea level. Although one lunar nodal cycle (18.6 years) is necessary to define mean sea level, yearly averages enable the description, interpretation, and evaluation of variations in mean sea level.

Periodic variations, anomalies, and secular trends in mean sea level are the results of numerous interacting factors, variable in duration, efficiency, and extent. Meteorologic, hydrologic, and oceanographic factors—such as air pressure, wind stress, river runoff, and ocean density—are important in the production of periodic variations and anomalies. Isostatic/tectonic and eustatic changes are considered to be the most important factors affecting long-term secular trends in mean sea level (Gutenberg, 1941; Fairbridge, 1961; Fairbridge and Krebs, 1962; Rossiter, 1967; Mörner, 1974). Mörner (1976) argued that small-scale/short-term fluctuations in the geoid itself may be recorded in mareograph data and that geoid changes should be considered of equal importance with isostatic/tectonic and eustatic changes in the evaluation of secular trends.

Gutenberg (1941) and Fairbridge and Krebs (1962) averaged world tide gauge data, omitting those gauges located in areas obviously affected by tectonic/isostatic vertical movements, to compute estimates of 1.1 mm/yr and 1.2 mm/yr, respectively, for the eustatic rise of mean sea level. This technique is not completely satisfactory since mean sea level data represent an interaction between tectonic/isostatic and eustatic factors wherever such data are collected, and averaging does not separate the two. Fairbridge (1961) and Schofield (1970) applied independently determined tectonic/isostatic vertical movement rates to the Amsterdam tide gauge data and computed the eustatic factor at this gauge to be 0.8 mm/yr and 1.0 mm/yr, respectively, for the interval 1682-1930. By applying tectonic/isostatic corrections to tidal gauge data from Amsterdam, Stockholm, and Warnemünde, Mörner (1973) identified a complex, fluctuating, eustatic mean sea level history for the interval 1682-1970 (Table 1). However, of 96 north European tide gauge stations analyzed by Rossiter (1967), only 26 required a quadratic or cubic regression expression, and of these 26 only 8 had sufficiently long periods of record (70 years) to completely rule out accidental correlation with large mean sea level anomalies. In addition, the tendency for mean sea level rise to accelerate between 1880 and 1960 was apparently contradicted by the fact that 23 stations with more than 70

TABLE 1. Major Eustatic Sea Level Changes During the Past 290 Years Based on Tide Gauges in Amsterdam, Stockholm, and Warnemünde (after Mörner, 1973)

Period	Characteristic	Rate of Eustatic Change
1682-1740	stability	zero
1740-1820	fall	0.25 mm/yr
1820-1840	stability	zero
1840-1950	rise	1.1 mm/yr
1950-1970	stability (or fall)	zero (or slight fall)

TABLE 2. Mean Sea Level Trends
(Obtained by Regression Analysis) of Selected Cities
(after Hicks, 1973, and Rossiter, 1967)

Location	Interval	Mean sea level trend, mm/yr
New York	1893–1971	+ 2.83 ± 0.12
Baltimore	1903–1971	+ 3.32 ± 0.15
Key West	1913–1971	+ 2.07 ± 0.18
Galveston	1909–1971	+ 5.73 ± 0.34
Seattle	1899–1971	+ 1.89 ± 0.18
San Francisco	1898–1971	+ 1.95 ± 0.18
San Diego	1906–1971	+ 1.95 ± 0.15
Honolulu	1905–1971	+ 1.62 ± 0.21
Cristobal, C.Z.	1909–1971	+ 1.13 ± 0.15
Trieste, Italy	1905–1962	+ 1.40 ± 0.32
Marseilles, France	1894–1958	+ 1.69 ± 0.15
Cascais, Portugal	1894–1962	+ 1.08 ± 0.22
Brest, France	1894–1961	+ 2.08 ± 0.26
Aberdeen, Scotland	1874–1962	+ 0.78 ± 0.11
Sheerness, England	1874–1959	+ 2.37 ± 0.17
Den Helder, Netherlands	1874–1962	+ 1.45 ± 0.12
Vlissingen, Netherlands	1890–1962	+ 3.04 ± 0.19
Kobenhavn, Denmark	1889–1962	+ 0.23 ± 0.16
Oslo, Norway	1886–1962	− 4.11 ± 0.33
Stockholm, Sweden	1889–1962	− 3.97 ± 0.19
Helsinki, Finland	1879–1962	− 3.15 ± 0.18
Warnemünde, East Germany	1882–1939	+ 1.48 ± 0.20
Swinemünde, Poland	1871–1942	+ 0.75 ± 0.15

years of record, from this same region, did not require nonlinear expressions.

The Permanent Service for Mean Sea Level, Bidston Observatory, Birkenhead, Merseyside L43 7RA, England, is the repository for worldwide sea level data.

FRANK W. STAPOR, JR.

References

Fairbridge, R. W., 1961. Eustatic changes in sea level, in *Physics and Chemistry of the Earth*, vol. 4. New York: Pergamon Press, 99–185.

Fairbridge, R. W., and Krebs, O. A., Jr., 1962. Sea level and the Southern Oscillation, *Royal Astron. Soc. Geophys. Jour.* 6, 532–545.

Gutenberg, B., 1941. Changes in sea level, postglacial uplift, and mobility of the earth's interior, *Geol. Soc. America Bull.* 52, 721–772.

Hicks, S. D., 1973. *Trends and Variability of Yearly Mean Sea Level, 1893-1971.* Washington, D.C.: U.S. Department of Commerce, NOAA Tech. Memo. No. 12, 11p.

Mörner, N.-A., 1973. Eustatic changes during the last 300 years, *Palaeogeography, Palaeoclimatology, Palaeoecology* 12, 1–14.

Mörner, N.-A., 1976. Eustasy and geoid changes, *Jour. Geology* 84, 123–151.

Rossiter, J. R., 1967. An analysis of annual sea level variations in European waters, *Royal Astron. Soc. Geophys. Jour.* 12, 259–299.

Schofield, J. C., 1970. Correlation between sea level and volcanic periodicity of the last millenium, *New Zealand Jour. Geology and Geophys.* 13, 737–741.

Cross-references: *Marigraph and Marigram; Mean Sea Level; Sea Level Changes; Sea Level Curves; Tide Gauge; Tides.*

SEA LEVEL CURVES

The present sea level is a function of the geoid or equipotential surface of the earth's gravity field. The geoid relief amounts to several tens of meters. Further, it is the function of local effects such as the major wind pattern, major currents, and density variations due to salinity and temperature. The past sea level is the function of: global glacial volume changes (glacioeustasy) due to global climatic changes; ocean basin volume changes (tectonoeustasy) due to earth movements; and ocean level distribution changes (geoidoeustasy) due to geoid changes (Mörner, 1976a). Then too, it is influenced by changes in local meteorological, hydrological, and oceanographic factors. The present position of past shorelines is the function of crustal activity: isostasy and tectonism. In some areas, there is a secondary effect from compaction.

A *sea level curve* is a graph with the past sea level depth and/or elevation plotted against time. There are two main types of sea level curves: the *relative* sea level curve that gives the present position of the past sea level changes regardless of the changes in level during the interjacent period; and the *absolute* sea level curve or *eustatic curve* that gives the real ocean level changes unaffected by crustal movements, compaction, and local meteorological, hydrological, and oceanographic changes.

A rise in sea level is termed a *transgression*, and leads to *submergence* of land. A fall in sea level is termed a *regression* and leads to *emergence* of land. A global increase of glaciers (for example, during early ice ages) leads to a eustatic regression, while a global decrease of glaciers (the end of ice ages) leads to a eustatic transgression. In uplifted areas, the amplitude of a eustatic transgression is decreased (and may even change sign), while the amplitude of a eustatic regression is increased. In subsiding areas, the amplitude of a eustatic regression is decreased (and may even change sign) while the amplitude of a eustatic transgression is increased. A relative sea level transgression in an uplifted area and a relative sea level regression in a subsiding area therefore provide minimum figures of the corresponding eustatic (absolute sea level) change.

Figure 1 shows eleven different sea level curves representing various parts of the world. Curves 5, 8, 9, and 11 are relative sea level curves (8–9, subsiding areas; 11, the uplifted

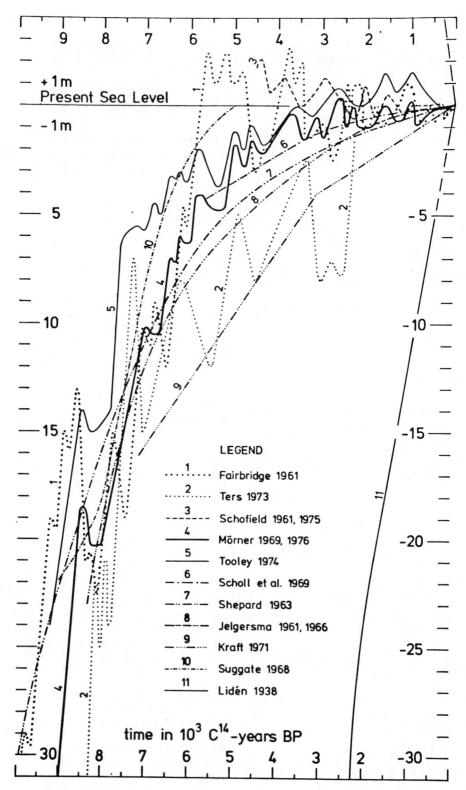

FIGURE 1. Eleven Holocene sea level curves representing different parts of the globe. Curves 1–4, 6–7, and 10 are supposed to be eustatic (absolute sea level) curves. The others are relative sea level curves (with curve 11 being plotted with changed sign).

northern Sweden plotted with changed sign), and the others supposed eustatic curves.

Relative Sea Level Curves

Relative sea level curves or *shore level displacement curves* show the combined effect of eustatic and crustal changes. In areas of strong uplift, such as northern Sweden, the crustal factor is predominating; the shore level displacement curves have therefore often been termed *uplift curves* (Swedish: *Landhöjningskurvor*), though these curves are in fact only *relative* uplift curves. This applies also to the geodetically determined present rate of relative uplift that on maps is termed *present uplift,* though the eustatic effect is not considered. Mörner (1969) expressed the relations, with respect to the present sea level, between shore-level displacement (S), isostatic uplift (I), and eustatic changes (E) in the following equation:

$$S = I - E \text{ (or: } S + E = I),$$

where positive S and I are measured in elevation above present zero (and negative values in depth below) and positive E in depth below present zero (and negative values in elevation above).

Uplifted Areas In uplifted areas, the sea level record is elevated above the present sea level and can therefore be studied in great detail. In the central areas of the Fennoscandian, Canadian, and Scottish glacioisostatic uplift the shore-level displacement curves show more or less continuous relative regressions; in the marginal areas, they show oscillating changes between transgressional and regressional periods. The Fennoscandian region is a classical area for Late Quaternary sea level investigations (Mörner, 1979). However, all pre-radiocarbon studies are marred by dating errors, with the outstanding exception of Lidén's (1938) shore level displacement curve from Angermanland in northern Sweden (Fig. 1, curve 11; plotted with changed sign), which is based on varve-dated emerged delta levels. Mörner (1969) made a detailed sea level study of southern Scandinavia (the marginal area of the isostatic uplift and tilting), and calculated the eustatic factor for the past 13,700 years (Fig. 1, curve 4) by separating the isostatic and eustatic factors from the recorded relative sea level changes (other eustatic curves were found to be incompatible with these changes). A similar method was used by Chappell (1974) to calculate the major eustatic changes during the past 120,000 years from the tectonically rising New Guinea relative sea level records.

Subsiding Areas In subsiding areas, the sea level records are submerged below the present sea level, and must therefore be studied by coring and off-shore investigations. Curve 9 (Kraft, 1971) from the Delaware coast and curve 8 (Jelgersma, 1961, 1966) from the Dutch coasts and the North Sea basin represent relative sea level curves of subsiding areas. The curves show more or less continuous transgressions. In the Netherlands, however, there is stratigraphical evidence of minor regressions and retardation periods. Because the Dutch curve represents a slowly subsiding area (about 0.4 mm/yr), it provides a maximum depth of the corresponding eustatic (absolute sea level) curve.

Absolute, Eustatic, Sea Level Curves

Absolute, eustatic, sea level curves can be established by studies of supposed stable areas (Shepard, 1963), by synthesis of global data (Fairbridge, 1961), or by calculations of the eustatic factor from relative sea level records (Mörner, 1969).

Figure 1 shows at least six supposed eustatic (absolute sea level) curves for the Holocene epoch (the postglacial period of Northern Europe). The curves differ markedly. However, at about 7500 B.P. all curves converge. If one of the curves would have been the true eustatic curve and the others relative sea level curves, one would expect that the curves would diverge farther and farther back in time. According to Mörner (1971a), this convergence therefore indicates that some factor has affected the ocean level distribution in a cyclic manner. This factor was later found to be geodicoeustatic changes (Mörner, 1976a). Because of geoid changes, eustatic (absolute sea level) curves are probably valid only locally or regionally (but not globally as earlier assumed).

It is a well-known fact that the ocean level must have fallen glacioeustatically during the ice ages when water was removed from the oceans and stored in the continental ice caps. The amount of this regression, however, is much debated. The records of the eustatic level during the last glaciation maximum (with a first complex maximum at 20,000-17,200 B.P., and a second maximum at about 14,800 B.P.) vary from -75 m to -125 m (-100 ±25 m). Figure 2 gives three main "eustatic" curves for the last 35,000 years.

The Holocene sea level records (Fig. 1), which we know by far the best, include at least four main types of eustatic curves: oscillation (high-amplitude) curves reaching above the present sea level in mid-Holocene time (curves 1 and 3); oscillating (low-amplitude)

SEA LEVEL CURVES

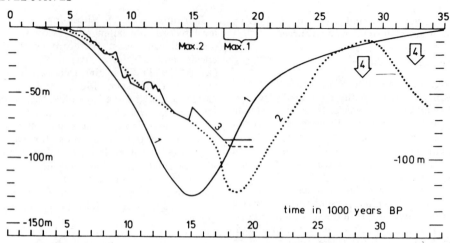

FIGURE 2. Three eustatic curves covering the time of the Late Wisconsin-Weichselian glaciation maxima (max. 1 and max. 2). Time in thousands of years B.P.; depth in meters. (1) the Milliman-Emery curve (1968), (2) the Curray curve (1965), (3) the Mörner curve (1969), and (4) arrows representing climatic-glaciologic indications of a much lower sea level at around 30,000 B.P. (Mörner, 1971b). The differences between the curves have been discussed by Mörner (1971b, 1976a).

curves rising more or less continuously (curve 4); smooth, steadily rising curves (curves 6 and 7); and smooth curves becoming stable at about the present sea level in mid-Holocene time (curve 10).

High-Amplitude Oscillation Curves In 1961 Fairbridge (Fig. 1, curve 1) made a famous synthesis of the global sea level records and arrived at a curve that is characterized by high-amplitude oscillations and mid-Holocene sea levels far above the present sea level.

Ters (1973) made a synthesis of the sea level data from the Atlantic coast of France and arrived at a curve that fluctuates with a high amplitude (Fig. 1, curve 2). It is in drastic disagreement with the well-established Dutch and English relative sea level curves and the northwest European eustatic curve in Fig. 1 (curves 8, 5, and 4, respectively).

Schofield (1961, 1975) established a detailed sea level curve for the past 4500 years from beach ridge systems in the Auckland area of northern New Zealand (Fig. 1, curve 3). The curve is oscillation with a medium to high amplitude and lies above the present sea level.

Low-Amplitude Oscillation Curves Mörner (1969) made a thorough investigation of the shorelines and the shore level displacement in southern Scandinavia over an area of 350 km in the direction of tilting. He tested available eustatic curves and found that none could be added to the shore level displacement curves and give logical isostatic curves. He calculated (Mörner, 1969, 1971c, 1976b) the corresponding eustatic curve, which is a low-amplitude oscillation curve that rises more or less steadily (Fig. 1, curve 4), and argued that this is *the* eustatic curve, at least, for the northwest European/northeast Atlantic region.

Tooley (1974) made a detailed study of the Holocene sea level records in northwestern England and arrived at a relative sea level curve that is a low-amplitude oscillation curve that rises more or less steadily (Fig. 1, curve 5) and exhibits a high similarity to the northwest European eustatic curve (Fig. 1, curve 4).

Smooth, Steadily Rising Curves Scholl (1964) and Scholl, Craighead, and Stuiver, (1969) gave a eustatic curve for the Florida area that is important because it is well dated and comes from a limited area (Fig. 1, curve 6). The curve rises smoothly to the present, but Scholl states that this smoothness does not preclude "high-frequency low-amplitude oscillations," because those are "easily lost or confused with possible hurricane deposits."

Shepard (1963) presented an average curve of nine "relatively stable areas" that rises smoothly and steadily to the present (Fig. 1, curve 7). The curve lies close (even below during the past 2500 years) to Jelgersma's curve of the slowly subsiding Dutch coast (Fig. 1, curve 8).

Smooth and Stable Curves Suggate (1968) presented a eustatic curve for the Christchurch area of New Zealand. The curve is smooth and rises steadily until about 5000 B.P., after which it becomes stable at about the present sea level (Fig. 1, curve 10).

In summary, the disagreements between the eustatic curves in Figure 1 are partly the effect of errors, misinterpretations, crustal movements, and overgeneralizations, and partly the effect of geoid changes as discussed by Mörner (1976a).

NILS-AXEL MÖRNER

References

Chappell, J., 1974. Geology of coral terraces, Huon Peninsula, New Guinea: A study of Quaternary tectonic movements and sea-level changes, *Geol. Soc. America Bull.* **85**, 553-570.

Curray, J. R., 1965. Late Quaternary history continental shelves of the United States, *in* H. W. Wright and D. G. Frye, eds., *The Quaternary of the United States*. New Haven, Conn.: Yale University Press, 723-735.

Fairbridge, R. W., 1961. Eustatic changes in sea level, *in Physics and Chemistry of the Earth*, vol 4. New York: Pergamon Press, 99-185.

Jelgersma, S., 1961. Holocene sea-level changes in the Netherlands, *Geol. Stichting Med.*, ser. C, **6**, 101p.

Jelgersma, S., 1966. Sea-level changes during the last 10,000 years, *in Proceedings of the International Symposium on World Climate 8000 to 0 B.C.* London: Royal Meteorological Society, 54-71.

Kraft, J. C., 1971. Sedimentary facies pattern and geologic history of a Holocene marine transgression, *Geol. Soc. America Bull.* **82**, 2131-2158.

Lidén, R., 1938. Den senkvartara strandförskjutningens förlopp och kronologi i Ångermanland, *Geol. Fören, Stockholm Förh.* **60**, 397-404.

Milliman, J. D., and Emery, K. O., 1968. Sea levels during the past 35,000 years, *Science* **162**, 1121-1123.

Mörner, N.-A., 1969. The Late Quaternary history of the Kattegatt Sea and the Swedish west coast: Deglaciation, shorelevel displacement, chronology, isostasy and eustasy, *Sveriges Geol. Undersökning*, C-640, 487p.

Mörner, N.-A., 1971a. The Holocene eustatic sea level problem, *Geologie en Mijnbouw* **50**, 699-702.

Mörner, N.-A., 1971b. The position of the ocean level during the interstadial at about 30,000 B.P.: A discussion from a climatic-glaciologic point of view, *Canadian Jour. Earth Sci.* **8**, 132-143.

Mörner, N.-A., 1971c. Eustatic changes during the last 20,000 years and a method of separating the isostatic and eustatic factors in an uplifted area, *Palaeogeography, Palaeoclimatology, Palaeoecology* **9**, 153-181.

Mörner, N.-A., 1973. Eustatic changes during the last 300 years, *Palaeogeography, Palaeoclimatology, Palaeoecology* **13**: 1-14.

Mörner, N.-A., 1976a. Eustasy and geoid changes, *Jour. Geology* **84**, 123-151.

Mörner, N.-A., 1976b. Eustatic changes during the last 8,000 years in view of radiocarbon calibration and new information from the Kattegatt region and other northwestern European coastal areas, *Palaeogeography, Palaeoclimatology, Palaeoecology* **19**, 63-85.

Mörner, N.-A., 1979. The Fennoscandian uplift and Late Cenozoic geodynamics: geological evidence, *Geo Journal* **3**, 287-318.

Schofield, J. C., 1961. Sea level fluctuations during the last 4,000 years as recorded by a chenier plain, Firth of Thames, New Zealand, *New Zealand Jour. Geology and Geophys.* **1**, 92-94.

Schofield, J. C., 1975. Sea-level fluctuations cause periodic, post-glacial progradation, south Kaipara Barrier, North Island, New Zealand, *New Zealand Jour. Geology and Geophys.* **18**, 295-316.

Scholl, D. W., 1964. Recent sedimentary record in mangrove swamps and rise in sea level over the southwestern coast of Florida, *Marine Geology* **1**, 344-366.

Scholl, D. W., Craighead, F. C.; and Stuiver, M., 1969. Florida submergence curve revised; its relation to coastal sedimentation rates, *Science* **163**, 562-564.

Shepard, F. P., 1963. Thirty-five thousand years of sea level, *in* T. Clements, ed., *Essays in Marine Geology in Honor of K. O. Emery*. Los Angeles: University of Southern California Press, 1-10.

Suggate, R. P., 1968. Post-glacial sea-level rise in the Christchurch metropolitan area, New Zealand, *Geologie en Mijnbouw* **47**, 291-297.

Ters, M., 1973. Les variations du niveau marin depuis 10000 ans, le long du littoralatlantique francais, *in Le Quaternaire; Géodynamique, Stratigraphie et Environnement*. Paris: Comm. Natl. France INQUA, 114-136.

Tooley, M., 1974. Sea-level changes during the last 9000 years in North-West England, *Geog. Jour.* **140**, 18-42.

Cross-references: *Coastal Morphology, Research Methods; Emergence and Emerged Shoreline; Flandrian Transgression; Global Tectonics; Holocene; Isostatic Adjustment; Isostatic Uplift Coasts; Pleistocene; Present-Day Shoreline Changes; Raised Beach; Sea Level Changes; Sea Level Changes, 1900 to Present; Submergence and Submerged Shoreline.* Vol I: *Coastal Stability.*

SEA PUSS

A *sea puss* is a dangerous longshore or rip current. The term is also loosely used to mean the submerged channel or inlet caused by those currents (Coastal Engineering Research Center, 1977).

Longshore currents flow parallel to the shoreline, and are generated by the longshore current of motion in waves that obliquely approach the shore (Shepard and Inman, 1950). Rip currents are concentrated jets of water that provide an exchange mechanism between the offshore and surf zones. They are most noticeable when long, high waves produce wave setup on the beach. The rise in local sea level causes the water to flow back into the offshore zone. The rip current reaches a maximum velocity during periods of low waves, and therefore the flow is frequently pulsating. However, the irregularity of natural waves does not seem to be an essential feature in generating these currents, because steady rips produced by regular waves have been observed in a wave tank (Arthur, 1962).

CHARLES B. CHESTNUTT

References

Arthur, R. S., 1962. A note on the dynamics of rip currents, *Jour. Geophys. Research* 67, 2777-2779.
Coastal Engineering Research Center, 1977. *Shore Protection Manual.* Washington, D.C.: U.S. Army Corps of Engineers, 1264p.
Shepard, F. P., and Inman, D. L., 1950. Nearshore water circulation related to bottom topography and wave refraction, *Am. Geophys. Union Trans.* 31, 196-212.

Cross-references: *Nearshore Hydrodynamics and Sedimentation; Rip Current.*

SEA SLICK

Sea slicks are visible bands on the sea surface distinguished by reduced surface roughness (Curtin and Mooers, 1975). Their slick appearance results from the local accumulation of natural oils or other impurities that damp capillary waves. Slicks are most conspicuous under calm sea conditions; they are obscured by wind chop. The occurrence of sea slicks as elongate streaks can be caused by several factors (Neumann and Pierson, 1966; Phillips, 1969). Winds blowing over the sea surface may cause surface oils to collect into *wind rows*, producing linear slicks parallel to the wind direction. Slicks are also frequently associated with *internal waves*, and on windless days indicate the presence, length, and direction of propagation of internal waves. Internal wave motion concentrates oils along lines halfway between an internal wave crest and the succeeding trough (Fig. 1). This creates a sequence of long, narrow, quasi-parallel slicks that slowly migrate in the direction of internal wave propagation. The coherence of slicks tends to be enhanced and sustained by an inward force provided by the dissipation of short-period waves along the slick edges.

LYNN D. WRIGHT

References

Curtin, T. B., and Mooers, N. K., 1975. Observations and interpretation of a high-frequency internal wave packet and surface slick pattern, *Jour. Geophys. Research* 80, 872-894.
Neumann, G., and Pierson, W. J., Jr., 1966. *Principles of Physical Oceanography.* Englewood Cliffs, N.J.: Prentice-Hall, 545p.
Phillips, O. M., 1969. *The Dynamics of the Upper Ocean.* Cambridge: Cambridge University Press, 261p.

Cross-references: *Beaufort Wind Scale; Internal Waves; Sea Conditions; Waves; Wind.* Vol. I: *Slicks, Ripples, and Windrows.*

SEA STATE—See SEA CONDITIONS

SEAWALL

A *seawall* is a structure separating land and water areas, primarily to prevent erosion, overtopping by water, and other damage by wave action (Wiegel, 1964). The region of protection is only to the area immediately shoreward of the structure location. A secondary purpose of a seawall is to diminish slumping of coastal sea cliffs that it may front. Seawalls are placed parallel or nearly parallel to the shoreline and are constructed of solid or block concrete, steel sheets, timber, or stone. Seawalls are sometimes constructed as a vertical wall, but this promotes erosion at the toe of the wall (Komar, 1976). More commonly seawalls slope upward toward land to help diminish wave energy and subsequent erosion of materials at the toe of the structure.

JOHN B. HERBICH
ROBERT E. JENSEN

References

Komar, P. D., 1976. *Beach Processes and Sedimentation.* Englewood Cliffs, N.J.: Prentice-Hall, 429p.
Wiegel, R. L., 1964. *Oceanographic Engineering.* Englewood Cliffs, N.J.: Prentice-Hall, 532p.

Cross-references: *Breakwater; Bulkhead; Coastal Engineering; Protection of Coasts; Quay.*

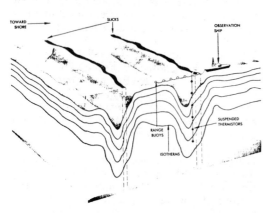

FIGURE 1. Relationship between sea slicks and internal wave structure (U.S. Navy official photograph).

SEDIMENT ANALYSIS, STATISTICAL METHODS

Since Udden's pioneer and classic work on the grain size of wind-blown deposits, just before the turn of the century, sedimentary petrographers have held divergent views about the usefulness of size-frequency analyses to determine sedimentary environments and the transport history of sediments. Despite several notable contributions to the understanding of grain-size distributions during the past two decades, relatively few researchers have attempted to study the relationship between depositional processes and grain-size distributions.

Folk (1966), by referring to more than 150 papers, has reviewed many of the graphical and mathematical techniques that have been proposed for the statistical summary of grain-size data. A cursory glance at the recent literature on the subject of grain-size statistics and environmental interpretation unequivocally shows that there is still disagreement about which mathematical technique to use in routine sediment analysis. Although several papers refer to the defects of both *moment* and *graphical* methods, few geologists in general, or sedimentologists in particular, have evaluated in any detail the problem of random operator error and, to a lesser degree, operator bias. As this writer has mentioned previously (Hails, Seward-Thompson, and Cummings, 1973), the relationship that exists between the accuracy of results and the sieve interval used merits more attention than it has received hitherto.

Graphical and Moment Measures (Statistical Parameters)

Any assessment of the relative significance of graphical and moment measures for distinguishing ancient and modern sedimentary environments must take cognizance in particular of the lack of standard sampling procedures used in the field, the different analytical techniques used in the laboratory, and of course experimental errors.

As Folk (1966) has stated, few quantitative data can be read from graphical methods. Most statistical work is undertaken with cumulative curves, which should always be drawn on percentage probability graph paper. This makes normal curves plot out as straight lines. Of course grain-size parameters can be made directly from the data by computer or by hand without drawing a cumulative curve.

Grain diameter is indicated in terms of the phi (ϕ) scale, which is a log transformation to simplify the arithmetic involved in computing statistical parameters, with transformation from negative to positive at 1 mm.

Many graphical measures of average grain size and several descriptive scales for sorting have been proposed hitherto. These have been reviewed succinctly by Folk (1966), who has also evaluated the significance of *skewness* (or asymmetry) and *kurtosis* (or peakedness)— two parameters that are widely used to measure the non-normality of a distribution. Folk and Ward (1957) developed graphical versions of statistical measures that have been adopted by many sedimentologists, summarized as follows.

$$\text{Mean size } (M_z) = \frac{\phi 16 + \phi 50 + \phi 84}{3}$$

Inclusive graphical standard deviation (σ_1)

$$= \frac{\phi 84 - \phi 16}{4} + \frac{\phi 95 - \phi 5}{6.6}$$

$$\text{Skewness } (SK_1) = \frac{\phi 16 + \phi 84 - 2\phi 50}{2(\phi 84 - \phi 16)} + \frac{\phi 5 + \phi 95 - 2\phi 50}{2(\phi 95 - \phi 5)}$$

$$\text{Kurtosis } (K_G) = \frac{\phi 95 - \phi 5}{2.44 \, (\phi 75 - \phi 25)}$$

Friedman (1961, 1967) has also summarized several distinct statistical methods that have been proposed for interpreting depositional environments from grain-size data. He used the first to fourth moments of a grain-size distribution to distinguish among dune, beach, and river sands. Equations for the calculation of these statistical parameters are as follows.

Mean ($\overline{X}_\phi$), first moment, $= 1/100 \Sigma f m_\phi$ (where f is a grade-size frequency and m_ϕ is the midpoint of each grade size in phi units)

Standard deviation ($\sigma\phi$), second moment,

$$= [\Sigma f(m_\phi - \overline{X}_\phi)^2 / 100]^{1/2}$$

Skewness ($\alpha 3 \phi$), third moment,

$$= (1/100) \alpha \phi^{-3} \Sigma f(m_\phi - \overline{X}_\phi)^3$$

Kurtosis ($\alpha 4 \phi$), fourth moment,

$$= (1/100) \alpha \phi^{-4} \Sigma f(m_\phi - \overline{X}_\phi)^4$$

The method of moments is undoubtedly the best way of obtaining parameters of a frequency distribution, since the entire frequency distribution is used in the determination instead of a few selected percentiles. It is pertinent to mention here that faulty sieves can affect the third and fourth moments (skewness and kurtosis), two particularly sensitive parameters. Also, because the method includes the entire distribution, it is necessary to make some arbitrary assumption about the grain size of the *fines* before a computation is made, if the fine-grain fraction (silt and clay) is not analyzed.

It is generally accepted that sediments follow a log-normal distribution. Spencer (1963) demonstrated that the majority of grain-size frequency curves are actually mixtures of one or more log-normally distributed sediment populations and that mixing of these populations may be indicative of depositional processes.

Significance and Limitations of Statistical Parameters

Researchers who have used statistical parameters are readily aware that such estimators are limited by truncation, bias, and the grouping process involved in sieve analysis (see, for example, references in Folk, 1966; Jones, 1970). Any assessment of their relative significance for distinguishing sedimentary environments must also take cognizance of inadequate sampling techniques, in the field, whether on land or at sea, different analytical methods in the laboratory, and of course experimental errors.

For example, seabed sediments obtained by *grab sampling* can be considerably disturbed by the sampling device being used, and by ship movement or ship-induced disturbance in the water column to depths of about 6 m and 14 m, respectively (Sly, 1969). This can produce an unrepresentative sample that is significant when relating the statistical parameters obtained to the original environment.

The computed values of the higher moments—skewness, and kurtosis—reflect the composition of the tails of a distribution, and for this reason all sample data should ideally be complete and not open-ended. Therefore it may be necessary to obtain weights at the 1.0 ϕ interval by the Andreason Pipette method, rather than lumping the fine-grained fraction as a *residue-pan* fraction. In this way the sample data, from which moment measures are computed, are not open-ended.

Higher moments are extremely sensitive to small changes, and/or inaccuracies in the grain size distribution data, and therefore the values of these moments must be treated with caution unless some numerical limit can be placed on their accuracy by virtue of a careful evaluation of both laboratory and sampling techniques.

It has been found from studies of deeply podzolised Pleistocene barrier sands in New South Wales, Australia, and on the Georgia coastal plain, United States, that both diagenetic and pedogenetic processes may cause aberrations in the third moment measure, skewness.

Thus, as the writer has stated previously (Seward-Thompson and Hails, 1973), there are still a number of aspects that need further clarification and research in sediment analysis and the application of statistical methods to the interpretation of sedimentary environments. These range from the improvement of laboratory and sampling techniques to the statistical manipulation of the data obtained. In particular, more work needs to be undertaken in order to explain the apparent sensitivity of higher moment measures to sedimentary environments and the relationship of present-day sediments to their ancient analogues.

JOHN R. HAILS

References

Folk, R. L., 1966. A review of grain-size parameters, *Sedimentology* 6, 73-93.

Folk, R. L., and Ward, W. C., 1957. Brazos River bar: A study in the significance of grain size parameters. *Jour. Sed. Petrology* 27, 3-26.

Friedman, G. M., 1961. Distinction between dune, beach, and river sands from their textural characistics, *Jour. Sed. Petrology* 31, 514-529.

Friedman, G. M., 1967. Dynamic processes and statistical parameters compared for size frequency distribution of beach and river sands, *Jour. Sed. Petrology* 37, 327-354.

Hails, J. R., 1967. Significance of statistical parameters for distinguishing sedimentary environments in New South Wales, Australia, *Jour. Sed. Petrology* 37, 1059-1069.

Hails, J. R., and Hoyt, J. H., 1969. The significance and limitations of statistical parameters for distinguishing ancient and modern sedimentary environments of the lower Georgia Coastal Plain, *Jour. Sed. Petrology* 39, 559-580.

Hails, J. R.; Seward-Thompson, B.; and Cummings, L., 1973. An appraisal of the significance of sieve intervals in grain size analysis for environmental interpretation, *Jour. Sed. Petrology* 43, 889-893.

Jones, T. A., 1970. Comparison of the descriptors of sediment grain-size distributions, *Jour. Sed. Petrology* 40, 1204-1215.

Seward-Thompson, B., and Hails, J. R., 1973. An appraisal of the computation of statistical parameters in grain size analysis, *Sedimentology* 20, 161-169.

Sly, P. G., 1969. Bottom sediment sampling, in *Proceedings of the 12th Conference on Great Lakes Research*. Buffalo, N.Y.: International Association for Great Lakes Research, 883-898.

Spencer, D. W., 1963. The interpretation of grain-size distribution curves of clastic sediments, *Jour. Sed. Petrology* 33, 180-190.

Cross-references: *Beach Material; Beach Material, Sorting of; Coastal Morphology, Research Methods; Gravel; Moment Measures; Sand; Sediment Size Classification; Silt.* Vol. VI: *Grain Size Parameters, Environmental Interpretations.*

SEDIMENT BUDGET

A volumetric accounting of material eroded and deposited for a given stretch of coast constitutes a *sediment budget* (Stapor, 1973). The

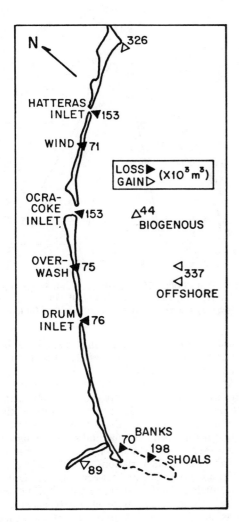

FIGURE 1. Long-term volumetric changes along the North Carolina outer banks between Capes Hatteras and Lookout. Solid arrows show estimated amount of losses to different areas; open arrows show source and amount at input necessary to balance losses (after Pierce, 1969).

quantitative data essential for the determination of such budgets can be obtained in numerous ways: bathymetric/topographic map differencing, maintenance dredging data, measured beach and nearshore changes related to engineering structures, empirical formulas relating measured areal changes with corresponding volumetric transport rates (see Pierce, 1969, for references to specific techniques). The time interval over which the data were collected is of major importance in evaluating the accuracy of the resultant budget; accuracy should increase with increasing time interval as short-term variations and anomalies become less and less significant.

FRANK W. STAPOR, JR.

References

Pierce, J. W., 1969. Sediment budget along a barrier island chain, *Sed. Geology* 3, 5-16.

Stapor, F. W., 1973. History and sand budgets of the barrier island system in the Panama City, Florida, region, *Marine Geology* 14, 277-286.

Cross-references: *A-B-C Model; Coastal Engineering; Coastal Morphology, Research Methods; Sediment Transport; Shore Drift Cell; World Net Sediment Transport.*

SEDIMENT CLASSIFICATION

Clastic sedimentary particles are most commonly classified by grain size (see Sediment Size Classification). Sand and silt may be further modified by the terms (very) coarse, medium, and (very) fine. Gravel is subdivided, in ascending size, into: granules (2-4 mm), pebbles (4-64 mm), cobbles (64-256 mm), and boulders (256 mm).

Folk (1954) devised a terminology for mixtures of gravel, sand, silt, and clay (Fig. 1 and Table 1). *Mud* is a term that includes both silt and clay. *Loam*, a term rarely used outside soil science, is a mixture of approximately equal parts of sand, silt, and clay.

Relevant publications on this topic include Folk (1974), King (1959), Krumbein (1936),

TABLE 1. Terminology of Mixtures of Gravel, Sand, Silt, and Clay

A.	Gravel
B.	Sandy gravel
C.	Muddy sandy gravel
D.	Muddy gravel
E.	Gravelly sand
F.	Gravelly muddy sand
G.	Gravelly mud
H.	Slightly gravelly mud
I.	Slightly gravelly muddy sand
J.	Slightly gravelly sandy mud
K.	Lightly gravelly mud
L.	Sand
M_1.	Silty Sand
M_2.	Muddy sand
M_3.	Clayey sand
N_1.	Sandy silt
N_2.	Sandy mud
N_3.	Sandy clay
O_1.	Silt
O_2.	Mud
O_3.	Clay

Source: R. L. Folk, 1954, The distinction between grain size and mineral composition in sedimentary rock nomenclature, *Jour. Geology* 62, 344-359; © 1954 by the University of Chicago. Reprinted with permission.

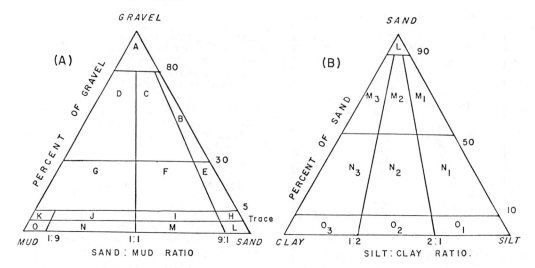

FIGURE 1. Grain-size nomenclature according to Folk (from R. L. Folk, 1954, The distinction between grain size and mineral composition in sedimentary rock nomenclature, *Jour. Geology* **62**, 344-359; © 1954 by the University of Chicago. Reprinted with permission). Letters in ternary diagrams correspond to letters in Table 1.

Udden (1898), Wentworth (1922), and Zenkovich (1967).

JOANNE BOURGEOIS

References

Folk, R. L., 1954. The distinction between grain size and mineral composition in sedimentary rock nomenclature, *Jour. Geology* **62**, 344-359.
Folk, R. L., 1974. *Petrology of Sedimentary Rocks.* Austin, Texas: Hemphills, 182p.
King, C. A. M., 1959. *Beaches and Coasts.* New York: St. Martin's Press, 403p.
Krumbein, W. C., 1936. Applications of logarithmic moments to size-frequency distributions of sediments, *Jour. Sed. Petrology* **6**, 35-47.
Udden, J. A., 1898. The mechanical composition of wind deposits, *Geol. Soc. America Bull.* **25**, 655-744 (also Rock Island, Ill.: Lutheran Augustana Book Concern, Printers, 69 p.).
Wentworth, C. K., 1922. A scale of grade and class terms for clastic sediments, *Jour. Geology* **30**, 377-392.
Zenkovich, V. P., 1967. *Processes of Coastal Development.* New York: Wiley Interscience, 738p.

Cross-references: *Clay; Gravel; Sand; Sediment Analysis, Statistical Methods; Sediment Size Classification; Silt.*

SEDIMENT SIZE CLASSIFICATION

Size classifications are important not only as a means of describing the particles that are incorporated to form sediments but also because of the implications that sizes have on transport and sorting processes and such characteristics as porosity.

Apart from their generality, terms such as silt, sand, or pebbles are not entirely satisfactory as means of description because of the varying interpretation given to them. For example *silt* has both popular and scientific usage, while even within the latter there is a range of definitions. Such variations usually relate to different disciplines or between different countries, such as United States or United Kingdom usage. Additionally, more elaborate classifications are likely to incorporate criteria other than purely dimensional ones and thus be defective in another respect.

Pettijohn (1975) has pointed out three types of size classification: those that are divided geometrically on a decimal basis and in a cyclic fashion; those that are divided geometrically but in a nondecimal and noncyclic manner; and those that are irregular. Examples of each type are the classifications of Hopkins and Atterberg; the *family* of Udden; Wentworth and Lane; and the U.S. Bureau of Soils, respectively.

Two size classifications are particularly widely used, those of Wentworth (1922) and Krumbein (1934). Both are shown in Figure 1. A derivation of the Udden-Wentworth scale, which merely omits the *granules* category, was adopted in 1947 by the National Research Council (Lane et al., 1947).

The Krumbein classification (the so-called Phi Scale) was designed to convey much the same information as, but to simplify the format of, that of the Wentworth-type classifica-

SEDIMENT SIZE CLASSIFICATION

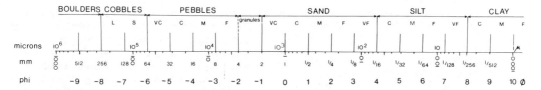

FIGURE 1. Relationship between typical sedimentary particle size classifications (after Pettijohn, 1975).

tion. Krumbein's scale has made statistical manipulation of sediment size data much easier. The Wentworth Scale follows the sequence 1, $\frac{1}{2}$, $\frac{1}{4}$, $\frac{1}{8}$ ... mm, which can be expressed as 2^0, 2^{-1}, 2^{-2}, 2^{-3} ... mm—that is, in $\log_2$ diameter (mm) form. To avoid negative values of the derivative (phi), a minus sign was introduced in front of the $\log_2$ term. The use of the term "diameter" for anything other than a sphere is something of a misnomer. Its actual sense varies but is often regarded as synonymous with intermediate dimension or particle breadth. Wadell (1932) took the diameter to be that of a sphere of equivalent volume to the particle in question.

Subsequently the Phi Scale was extended by other workers to particles greater than 1 mm, and thus negative phi values finally came into existence. While the Phi Scale is valuable within the range for which Krumbein created it and where number of particles tends to be large, it is highly unsatisfactory within pebble size and coarser grades of sediment. This is because an individual particle in, say, the -5ϕ category weighs more than 32,000 times comparable particles in the zero ϕ category, and weight is the customary means of expression of quantity. If absolute number, rather than weight, were to be used in quantifying results, such an objection would no longer apply. In effect, quantification by weight is mixing criteria.

In general, particles between about 0.06 and 4 mm are sieved, while larger sizes may be sieved or the absolute linear dimensions measured. With rectangular mesh sieves, particles are effectively sorted on the relationship between the length of the diagonal of the mesh and the intermediate diameter of the particle. Evidence with pebbles suggests that this is the least relevant dimension in sorting and transport (Carr, 1975).

Within the sand fraction a fall settling tube is sometimes used for size analyses. At one time elutriation was a popular technique. For finer grades of sediment, sedimentation cylinders, pipette analyses, and the proprietary Coulter Counter are widely employed. In the laboratory measurement can also be done optically by the use of scaling devices in association with a lens, microscope, or electron microscope, as appropriate. When grain size is determined by using settling velocity, both a spherical shape and specific gravity of 2.65 are normally assumed, or data are adjusted on this basis.

Typically the information obtained is plotted graphically on logarithmic probability paper with a cumulative weight percentage on the y-axis, and particle size along the x-axis. Size is represented by diameter in millimeters, Phi Scale units and/or sieve mesh numbers. These data are used for statistical calculations either by graphical or mathematical analysis of moment measures (Folk, 1966). The first four moments give mean, standard deviation, skewness of distribution, and kurtosis, respectively. Various sorting coefficients exist to express *average* size, spread, symmetry, and *peakedness*.

Various workers, including Krumbein (1938), have noted that size distribution of clastic sediments appears to be broadly log normal, and this is another reason for the Phi Scale approach. However, this view is not held universally, and even if it is true of the sand fraction where most work has been done, there does appear to be a dearth of particles in, for instance, the 2-4 mm size. This may reflect actual worldwide availability or the mode of origin of clastic material. In work on Chesil Beach, England, the size distribution of pebbles and cobbles found there was largely normal, not log normal (Carr, 1969). In other cases, distributions have been found to be bimodal or polymodal, and such data have been invoked to suggest two or more formative or sorting processes at work (Passega, 1964; Visher, 1969). A similar argument has been adopted in determining environments of deposition.

Size classifications are important both for their intrinsic value and for the information that can be derived statistically from size data. While the Phi Scale is suitable within the sand fraction, a satisfactory universal means of classification is still wanting.

ALAN P. CARR

References

Carr, A. P., 1969. Size grading along a pebble beach: Chesil Beach, England, *Jour. Sed. Petrology* 39, 297-311.

Carr, A. P., 1975. Differential movement of coarse

sediment particles, *Conf. Coastal Eng., 14th, Copenhagen, Am. Soc. Civil Engineers, Proc.* **2**, 851–870.
Folk, R. L., 1966. A review of grain size parameters, *Sedimentology* **6**, 73–93.
Krumbein, W. C., 1934. Size frequency distributions of sediments, *Jour. Sed. Petrology* **4**, 65–77.
Krumbein, W. C., 1938. Size frequency distributions of sediments and the normal phi curve, *Jour. Sed. Petrology* **8**, 84–90.
Lane, E. W.; Brown, C.; Gibson, G. C.; Howard, C. S.; Krumbein, W. C.; Matthes, G. H.; Rubey, W. W.; Trowbridge, A. C.; and Straub, L. G., 1947. Report of the sub-committee on sediment terminology, *Am. Geophys. Union Trans.* **28**, 936–938.
Passega, R., 1964. Grain size representations by C. M. patterns as a geological tool, *Jour. Sed. Petrology* **34**, 830–847.
Pettijohn, F. J., 1975. *Sedimentary Rocks.* New York: Harper and Row, 628p.
Visher, G. S., 1969. Grain size distributions and depositional processes, *Jour. Sed. Petrology* **39**, 1074–1106.
Wadell, H., 1932. Volume, shape, and roundness of rock particles, *Jour. Geology* **40**, 443–451.
Wentworth, C. K., 1922. A scale of grade and class terms for clastic sediments, *Jour. Geology* **30**, 377–392.

Cross-references: *Beach Firmness; Beach Material; Cobbles; Gravel; Sand; Sediment Analysis, Statistical Methods; Sediment Tracers; Sediment Transport; Silt.* Vol. VI: *Phi Scale.*

SEDIMENT TRACERS

Sediment transport (q.v.) presents a serious problem in coastal engineering. Unprotected coasts may have large losses, and ship ways may silt up. It is therefore desirable to trace the movement of the sediment, to find the principal direction of its movement, and to determine the rate of transport. This is usually done in the following way: a representative sample of the sediment, usually between 10–100 kg, is taken from the area of investigation; each grain of the sample is marked; the sample is then deposited where it was originally taken; and the movement of the sample is followed. Different methods of marking the sediment determine the way of tracing the sediment transport. A good tracer must fulfill the following conditions: The marked sediment must have the same properties as the unmarked; the marked sediment must be clearly and easily distinguishable from the unmarked; the tracer must be bound tightly to the sediment so that it cannot be dissolved by water or rubbed off by other grains; and the individual grains must be marked proportional to their volume (mass), not proportional to the surface.

In practice, there are basically two different ways of marking the sediment. The simple (economical) way is by means of *fluorescent tracers.* Each grain of the sediment sample is coated with a fluorescent dye, using water glass (potassium silicate) as cement. The sample is deposited in the sea. At regular intervals a sample is taken from the investigation site, and the number of marked grains in each sample is counted. The fluorescent grains are visible in UV light. The grains are counted either visually or electronically with a photomultiplier. The number of marked grains per mass unit is a measure for the transport rate from the place of deposit to the place of taking the sample. If *ferromagnetic tracers*—for example, Fe_2O_3—are used instead of fluorescent dyes, the method is the same except for detection. The advantages in using fluorescent or ferromagnetic tracers are that they are inexpensive and present no hazards. However, the disadvantages include: inaccurate results because of poor statistics (only a few marked grains per sample); fluorescent tracers do not decay and make later measurements difficult; only the surface can be sampled; no *in situ* measurements can be made; the work is time-consuming; continuous measurements can not be made; and results are obtained only several days after taking the sample.

The more advanced (but expensive) method is the use of *radioactive tracers.* A chemical solution containing a particular isotope is added to a glass smelting. The isotope is radioactivated either before adding it to the glass smelting or afterwards, depending on the facilities available. The radioactive glass is ground to the size of the sediment under investigation. One disadvantage of this method is the fact that the hydrodynamic transport properties of the radioactive glass are different from those of the sediment.

The sediment sample taken from the area of investigation is separated by sieving into various classes of grain size; each grain of each class is coated with a chemical compound containing the γ-emitting radioisotope (Table 1). The coating of the larger size classes is more strongly radioactive, according to the volume of proportional activation.

After depositing the sample in the sea, a probe containing a scintillation counter and mounted on a cart on the bottom of the sea is towed across the site of investigation from behind a ship. The measured γ-ray intensity gives the sediment transport rate.

The use of radioactive tracers is advantageous in that highly accurate results (typical error 1–3%) can be achieved, *in situ* measurements can be taken, the results can be obtained immediately, and real time data processing is possible. On the other hand, the method is expensive and presents environmental hazards. Also, it requires qualified and experienced personnel.

TABLE 1. Frequently Used Radioisotopes

Isotope	Half-life	Energy (MeV)	
^{24}Na	15.0 h	1.37	2.75
^{198}Au	2.7 d	0.41	
^{133}Xe	5.3 d	0.08	
^{140}Ba } ^{140}La	12.8 d	1.6	
^{51}Cr	27.8 d	0.32	
^{46}Sc	84.0 d	0.89	1.12
^{65}Zn	250.0 d	1.12	

Criteria for selection: The half-life should be approximately ¼–½ of the duration of the particular investigation; the γ-radiation should be able to penetrate a possible overlying sediment layer; therefore the γ-energy should be approximately 0.5 MeV or greater.

Scintillation Counter

A scintillation counter consists of a scintillating crystal—for example, NaJ (Tl)—a photomultiplier, an amplifier, a pulse hight analyzer, and a counter.

If a γ-quant traverses matter, it transfers either a fraction of its energy by inelastic scattering (Compton effect) or its full energy by being absorbed (photoeffect) to an electron (Pair production is negligible up to several MeV). The free electron is quickly absorbed, and in a scintillating material a short ($\sim 10^{-6}$ s) light pulse proportional to the electron's energy is emitted. The light pulse excites the photomultiplier cathode, which emits electrons that are subsequently multiplied ($\sim 10^5$) by several stages (~ 10) of the photomultiplier. The pulse height of the amplified electron pulse is analyzed and counted if within the preset thresholds. If the thresholds are set around the total γ-energy, only γ-quants interacting by photoeffect are counted; if the thresholds are set around one-half of the total γ-energy, only γ-quants interacting by Compton effect are counted.

Figure 1 shows the cross section of a typical scintillation counter. The large conical volume at the bottom is filled with air, and defines the opening angle of the scintillation counter. Inside the opening angle, absorption of γ-quants is negligible; outside, they are strongly absorbed ($> 10^4$) by layers of 50 cm of water and 10 cm of lead. A sharply defined opening angle is required to relate the number of counted γ-quants to a radioactive quantity, which, in turn, is related to a certain mass of radioactive sediment.

Qualitative Data Analysis

The measured count rates are plotted throughout the area of investigation, and the points of

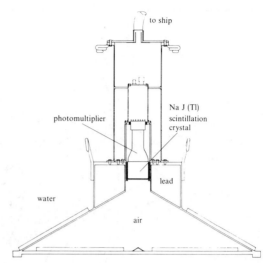

FIGURE 1. Scintillation counter used for radioactive tracer studies.

equal count rates are connected. This gives the iso-count-rate lines. In many cases this measurement is sufficient to indicate the preferred direction(s) of sediment transport (Fig. 2). This simple procedure, however, has a fundamental disadvantage. During the transport process, both the sediment and the sediment tracer are transported, burying each other. In areas of strong sediment transport—and those are usually the interesting ones—the overlying layer may be several feet thick. If fluorescent tracers are applied, the samples are usually taken from the surface; any burying effect remains undetected, which may lead to a substantial underestimate of the transport rate. If radioactive tracers are used, the overlying sediment layer causes absorption of the γ-rays, and again the transport rate will be underestimated.

Quantitative Data Analysis

Fortunately the γ-spectrum is changed in a characteristic way when the γ-quants traverse a material layer. Recall that the energy of γ-quants interacting by photoeffect is completely absorbed, whereas γ-quants interacting by Compton effect lose only a fraction of their energy and are still present after the interaction. Therefore, Compton scattered γ-quants can interact several times by Compton effect, and can penetrate a thicker layer of material— that is, they are less strongly absorbed than γ-quants interacting by photoeffect. This is called *build-up effect*, and leads to a characteristic change of the γ-spectrum. Figure 3 shows a γ-spectrum before (upper curve) and after (lower curve) traversing a sediment layer of 20 cm thickness. Notice that the curves are

SEDIMENT TRACERS

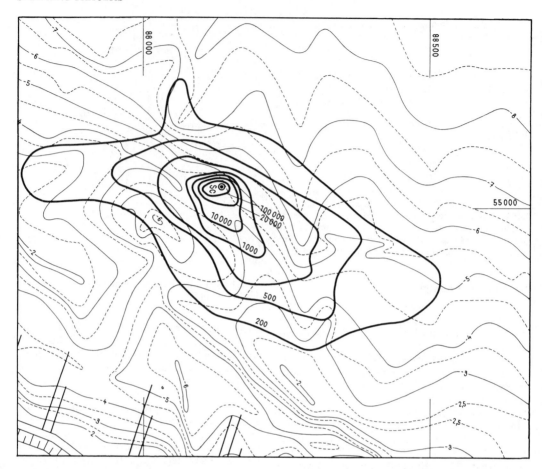

FIGURE 2. Iso-count-rate lines (thick lines). The dot in the center indicates the place where the sediment radioactivated with ^{46}Sc was deposited. The thin lines are depth contours. The coastline is at the bottom— notice the groynes.

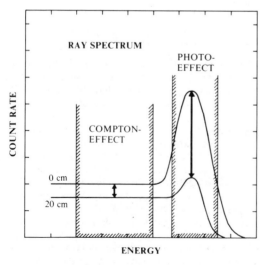

FIGURE 3. The build-up effect.

closer together in the Compton part than in the photo part of the spectrum. The ratio of the count rates in these two parts increases with the thickness of the sediment layer. If that ratio is measured, the thickness of the absorption layer can be estimated by means of a calibration curve and the count rate can be corrected.

This of course gives only a rough estimate since radioactivated and natural sediment are not overlying each other but are mixed in reality. To obtain a more exact correction, one has to actually measure how the radioactivated sediment is distributed within the natural sediment. A probe that measures the distribution function has been developed recently.

Before correcting for self-absorption of the radioactivated/natural sediment mixture, one must subtract the natural background radiation from the measured count rates. The natural background radiation is produced by cosmic rays, the natural sediment itself, the seawater, and the housing of the scintillation counter, and must be determined throughout the site of investigation prior to deposition of the radioactivated sample.

If corrected for self-absorption, the count rate is directly proportional to a certain quantity of sediment. The proportionality factor is calculated by taking into account: the correction for half-life (radioactive materials decay and emit fewer γ-quants at later times); the calibration of the detector system, which means that a certain count rate corresponds to a certain radioactivity (measured in Curie); and the relation of radioactivity to sediment quantity that is given by the known parameters of the deposited radioactive sample. Following this procedure gives the amount of sediment that was transported from the place of deposit to the place of measurement. Integrating the quantities over the entire area of investigation should give roughly the total amount of sediment originally deposited.

Pertinent literature on this subject includes: Courtois and Squzay (1966), Crickmore and Waters (1966), DeVries (1971), Duane (1970), Inman and Chamberlain (1959), Reinhard (1974), Schuster et al. (1970), Schwartz (1966), Smith and Parsons (1965), and Yasso (1966).

RUEDEGER REINHARD

References

Courtois, G., and Squzay, G., 1966. Les méthodes de bilan des taux de comptage de traceurs radioactifs appliquées à la mesure des débits massiques de charriage, *Houille Blanche* 3, 279–290.

Crickmore, M. J., and Waters, C. B., 1966. *The Calibration of Radiation Detectors Used in Sediment Tracer Investigations by the Hydraulics Research Station.* Wallingford, England: Hydraulics Research Station, Rept. No. INT 48, 15p.

DeVries, M., 1971. *On the Applicability of Flourescent Tracers in Sedimentology.* Delft, Netherlands: Delft Hydraulics Laboratory, Pub. No. 94, 19p.

Duane, D. B., 1970. Tracing sand movement in the littoral zone: Progress in the radioisotopic sand tracer (RIST) study, July, 1968–February, 1969, *U.S. Army, Coastal Engineering Research Center, Miscellaneous Paper 4-70*, 1–46.

Inman, D. L., and Chamberlain, T. K., 1959. Tracing beach sand movement with irradiated quartz, *Jour. Geophys. Research* 64, 41–47.

Reinhard, R., 1974. Ein Verfahren zur quantitativen Erfassung von Sandwanderungsvorgängen bei Messungen mit radioadtiven Tracern, *Die Küste* 26, 25–106.

Schuster, S.; Pahlke, H.; and Grimm-Strele, J., 1970. *Messung von Sandbewegungen mit Leitstoffen.* Berlin: Versuchsanstalt für Wasserbau und Schiffbau, Bericht Nr. 530/70, 89p.

Schwartz, M. L., 1966. Fluorescent tracer: Transport in distance and depth in beach sands, *Science* 151, 701–702.

Smith, D. B., and Parsons, T. V., 1965. *Silt Movement Investigation in the Oxcars Spoil Ground, Firth of Forth, Using Radioactive Tracers, 1961 and 1964.* Harwell, England: Hydrology and Coastal Sediment Group, AERE-R 4980, 1–30.

Yasso, W. E., 1966. Formulation and use of fluorescent tracer coatings in sediment transport studies, *Sedimentology* 6, 287–301.

Cross-references: *Beach Material, Sorting of; Coastal Engineering; Coastal Morphology, Research Methods; Protection of Coasts; Sediment Transport; World Net Sediment Transport.*

SEDIMENT TRANSPORT

Beach materials are in almost constant motion under the influence of waves, currents, tides, and winds. All loose particles, whether of clastic or biogenic origin and even trash, such as bottles and cans, are moved and arranged into sedimentary structures. The resulting *accumulation forms,* including beaches, spits, bars, barrier islands, and dunes reflect the origins and methods of transport of their constituent materials. It is convenient to discuss *normal* (perpendicular to the shore) and *alongshore* sediment movements separately because different problems are involved in each movement.

Theory

The capacity of the transporting agents to move sediment is influenced by the size, shape, and mass of the individual particles. Before a particle can move, it must be separated from its companions and started from rest. To do this, wave, current, or wind must overcome the forces of gravity and friction and support the weight of the particle for long enough to impart forward motion. *Suspended sediment* moves with the speed of the water in which it is contained. *Bed material* moves at considerably lower speeds determined by the hydraulic size of the particles and the current speed in the bottom layer. A current has to achieve a *threshold velocity* before particles of a particular size can be started from rest. Movement of sediment in the sea close to the shore depends on the nature of fluid stress on sediment particles under wave action. Bagnold (1963) discussed the shearing of particles over each other and the bed in terms of the limiting friction coefficient T/P usually expressed as the tangent of the *friction angle* ϕ. This signifies the ratio of the shear force or stress T (needed to be exerted over a plane of static contact between two bodies in order to maintain relative movement between them) to the perpendicular force or stress P across that plane. Ippen and Eagleson (1955) diagrammatically ex-

TABLE 1. Critical Velocity Values Taken from Various Sources (cm/sec)

Source		0·1 mm	0·2 mm	0·4 mm	0·6 mm	0·8 mm	1·0 mm
Penck, 1894	*V_H	4·5	6·3	9·0	11·0	12·7	14·2
Eagleson, 1958	†V_C	6·9	9·7	13·7	16·8	19·4	21·7
Larras, 1957	–	5·2	6·2	8·9	11·1	13·4	15·6
Bagnold, 1946	–	11·1	15·2	20·5	24·4	27·7	30·5
Gugnyaev, 1959	V_H	4·7	7·5	10·4	12·7	14·5	16·0
Volkov, 1960	V_H	–	8·0	10·2	12·8	15·3	17·3

Source		2·0 mm	3·0 mm	4·0 mm	5·0 mm	10·0 mm	50·0 mm	T, sec
Penck, 1894	*V_H	20·0	24·4	28·3	31·8	44·8	100·0	–
Eagleson, 1958	†V_C	30·6	37·4	43·4	48·6	68·6	153·2	–
Larras, 1957	–	23·6	27·8	31·5	34·6	47·1	100·6	6·0
Bagnold, 1946	–	41·2	49·0	55·5	61·1	82·5	165·3	6·0
Gugnyaev, 1959	V_H	23·0	28·0	32·0	–	–	–	–
Volkov, 1960	V_H	25·2	29·8	–	–	–	–	–

Source: Zenkovich (1967)
*V_H-commencement of movement of individual particles
†V_C-commencement of mass displacement

pressed the interaction between sediment and the forces acting on it.

Once started, particles move by rolling or sliding, by saltation or in a suspended state as the velocity of water movement increases. According to Airy's law (1845), the linear dimension of particles that roll along the bottom are proportional to the square of the current velocity. His equation for the instant of movement is: $C^2/gd = k([p/po] - 1)$, where d is the diameter of the particle, C is the current velocity, p and po are the densities of the particle and the water, and k is a shape coefficient.

Zenkovich (1967) has summarized the views of many workers on critical velocity values and compared them with wave-tank results obtained in 1960 by Volkov.

Transport Normal to the Shore

Seaward of the Breaker Zone. Both the theory of sediment motion and experimental evidence (Bagnold, 1947) suggest that waves, other than those of small steepness (the ratio of wave height to wave length, H/L), induce a shoreward drift of sediment close to the bed in the area seaward of the breaker line. This ceases close to the breaker line. Cornaglia (1891) had suggested that for each grade of sediment there exists a *null point* or neutral line landward of which the bottom sediment moves landward and seaward of which it moves seaward. Ippen and Eagleson (1955) partially confirmed Cornaglia's hypothesis experimentally. On a model beach of gradient 1 in 15, they demonstrated the reality of both landward and seaward movement. However, they showed that seaward of the null point movement rapidly ceases while landward movement reaches maximum velocities just seaward of the breaker line and then falls rapidly to zero.

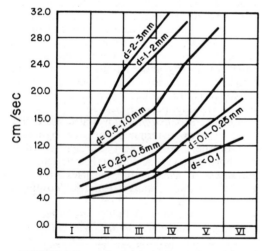

FIGURE 1. Stages in the movement of particles of material from a sandy bottom for various maximum bottom velocities of wave motion (according to Volkov, 1960). I: sand grains at rest. II: agitation of individual grains. III: oscillation without movement from the bottom. IV: movement from the bottom and to and fro movements of individual particles. V: mass dislodgment and displacement. VI: formation of ripple marks (after Zenkovich, 1967).

King (1959) similarly demonstrated movement outside the breaker line on a model beach, and showed how it varies in relation to varying wave parameters, the nature of the bed, and distance from the break point. The depth at which movement in *relation to the shore* begins under wave action is still a matter for debate. Trask (1955), in investigating the movement of sand round rocky promontories on the California coast, found that movement becomes almost nil at depths greater than 18 m. This depth also marked the outer limit of ripple formation and disturbance of the bottom by waves. Kidson and Carr (1959) confirmed their earlier work with pebbles off Orfordness in eastern England, which suggested that in conditions of up to gale-force winds, no movement of significance occurred in depths of greater than 6 m. While Inman (1957) has shown that limiting depths for sand may be greater than those suggested by Trask, it is probable that movement in deeper water may occur only in extreme conditions (Draper, 1967). While waves can theoretically disturb sediment in depths equal to half their wavelength, it is probably that *wave base* is less significant than *surf base*. Dietz (1963) argued that the "active near shore sand wedge" pinches out rapidly a short distance seaward of the break point, and is unrelated to wave base; *surf base* is more significant.

Landward of the Breaker Zone. The *breaker zone* is one of great bed disturbance, and much sediment is put into suspension. The ultimate resting place of sediment particles continuously or intermittently disturbed depends on the net movement of water. There is a tendency for the formation of a *bar* close to the break point where sediment put into suspension comes to rest. Since waves of different wavelengths break in different depths, more than one bar is not unusual. Bars tend to be more common when tidal range is low and the break point for a particular size of waves does not migrate significantly. Early work on bars includes that by de Beaumont, Lehmann, Gilbert, and Cornaglia; more recent work by Keulegan, Shepard and Williams is significant (see references in King, 1972).

On breaking, the motion of the water particles in a wave changes from orbital to linear. That part of the wave to which a landward component is imparted surges up the beach face as *swash* and returns as *backwash*. The swash zone, like the breaker zone, is one of great turbulence with rapid movement of sediment particles in a variety of directions. The net movement is reflected in changes in the beach profile. When short steep waves, according to Johnson (1956), those with a deep water steepness ratio $Ho/Lo > 0.025$, comb down the beach, material from above the water line tends to be deposited below water. When longer *constructive* waves of lower steepenss ratios replace storm waves, the reverse occurs. This is a process known as *cut and fill*.

The shoreward-seaward transport involved in cut and fill tends to cancel out on stable coasts. However, many coasts show gains or losses of material over time that suggest sediment transport not only alongshore but also normal to the shore over a wider band than that involved in cut and fill. Sediment transport through the breaker zone must therefore be considered.

Transport Through the Breaker Zone. Results from field experiments are difficult to obtain in the breaker zone, one of great turbulence. However, Bagnold (1940) has shown in wave tank experiments that for waves of low steepness (*plunging breakers*) distinctly separate circulation cells exist landward and seaward of the breaker zone and that sediment is unable to pass to the nearshore zone from seaward of the breakers. For high steepness values (*spilling breakers*) Bagnold noted a tendency for the two circulation cells to merge and for a large "head" of water to be built up and for sediment entrained by the backwash to be carried seaward through the breakers. The head of water can in certain conditions give rise to *rip currents*.

Little hard evidence exists of the mechanism, if any, by which sediment moves landward across the breaker zone from offshore. Johnson and Eagleson (1966) have suggested a promising line of research into the theory of fluid exchange across the breaker zone in terms of phase differences. The arrival of the swash of one wave may inhibit the effectiveness of the backwash of the previous breaker. Where the two are coincident, little "cross breaker" sediment exchange is likely. Where they are widely separated, complete exchange may be possible. The present writer believes that even if the breaker zone is an effective barrier to landward sediment movement, some replenishment of the nearshore zone from offshore may be possible because of migrations of the breaker zone in terms of variations in the length of waves acting on a coast and in relation to tidal rise and fall. In the first of these connections, it is important to appreciate that more sediment transport may be achieved in the one in 50- or 200-year type of event than in more "normal" conditions. In this sense, field experiments, like model tests, have severe limitations.

Transport Alongshore

Longshore Drift. Waves arriving in the offshore zone from deep water begin to adjust to the bed contours. For those of long wavelength, including many *swell* waves, which "feel bottom" a long way out, this process of *refraction* is complete by the time the swash runs up the beach. For many waves, however, particularly locally generated storm waves, the process is not completed and waves arrive at an angle to the shore (the angle of wave incidence). The oblique upward rush of the swash succeeded by the return of the backwash down the beach, following the path of maximum gradient, results in a longshore movement of sediment known as *beach drift*. In addition, the continuous action of waves arriving at an angle to the shore, also induces a longshore current which carries sediment parallel to the shore, complementing beach drifting. The whole alongshore system, combining the effects of waves and wave induced currents, shown diagrammatically in simplified form by Bird (1968) (Fig. 2), is known as *longshore drift*. Inman and Bagnold (1963) give a model for the transport of sand by longshore currents. The total immersed weight of sediment transported per unit time past a section of beach is given by

$$l_1 = \frac{ps - pfg}{ps} J = K \frac{ECn}{u_0} \frac{\cos \alpha}{\tan \phi} \bar{u}_1$$

where l_1 is the dynamic transport rate, pf is the fluid density; ps is the sand density; J is the sediment discharge in mass per unit time; K is a proportionality factor; E is the mean energy per unit surface area of a wave approaching a long straight beach at an angle α; ECn is the rate of transport of energy available for dissipation; C is the wave phase velocity; Cn is the velocity at which the energy is propagated forward; u_0 is the mean friction velocity; $\tan \phi$ is the intergranular friction coefficient and $\bar{u}_1$ is the longshore current velocity.

Zenkovich (1967) gives the simpler equation $L + m/l = 1 + \sin \alpha/\cos \alpha$ for calculating the

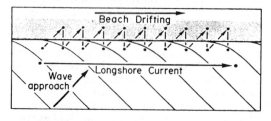

FIGURE 2. Longshore drifting, the flow of coastal sediment produced by wave and current action when waves approach at an angle to the shoreline (from E. C. F. Bird, 1968, *Coasts*. Cambridge, Mass.: MIT Press; © 1968 by MIT Press).

horizontal displacement for beach drifting, as distinct from the current displacement referred to by Inman and Bagnold. In the Zenkovich equation, L is the perpendicular distance, and m is the slant distance moved by the swash up the beach, l is the horizontal displacement and α is the angle of wave approach.

Enormous volumes of sediment are involved in what has been variously termed the longshore *conveyor belt* and the coastal *river of sand*. For example, Watts (1956) estimated longshore drift at Atlantic City to be of the order of 305,500 m³/yr. Where movement of sediment alongshore onto a particular length of coast is exceeded by movement out of it, erosion occurs, as it did at Atlantic City. Aibulatov, Boldyrev, and Zenkovich (1960) have shown that, in the microtidal Black Sea, sediment is moved alongshore not only on the beach but throughout the nearshore zone. Maximum movement tends to take place along nearshore bars in quiet weather and in the breaker zone under stormier conditions.

Counter Drift. The process of longshore drift has come to be better understood in recent years than it was before the late 1950s. This has come about largely as a result of extensive field experiments using sediment marked either with fluorescent dyes, marine paints, or radioactive isotopes or using sub-

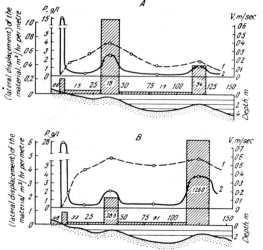

FIGURE 3. Graphs of the rate of displacement of suspended sand in various zones of the submarine slope at Anapa, and of the magnitude of the determining factors (from Aibulatov, 1960). A: in force 2-3 seas. B: in force 4-5 seas; 1: coastal current velocities (V). 2: concentration of suspensions in the bottom layer (P). The shaded rectangles and the figures in them correspond to the flow rate of material m³/hr) in different zones (from Zenkovich, 1967).

stitute sediment particles. The concept of the longshore conveyor belt has been considerably modified, and it is now generally agreed that longshore drift is not unidirectional. Redman (1851–52), for example, thought of sediment moving along the entire south coast of England to accumulate in the cuspate foreland of Dungeness as the "shingle end" of a sediment transport system. Robinson (1955) regarded the double spits across the mouth of many harbors and estuaries on this coast as having been formed as the result of breaching of single spits created by this longshore drift from the west to the east, despite the apparent growth of a number of these spits in the opposite direction. Kidson (1963), however, argued that many spits on the west and east coast of England, as well as on the south, have grown in directions opposite to that generally held to be the drift direction. He introduced the concept of *counter drift,* arguing that the *drift direction* discussed by many workers was the resultant of drift when waves approached from the dominant direction and counter drift when they approached from other directions. Some double spits thus included elements that had grown in the counter drift direction in areas sheltered, often by the major spit, from the dominant direction.

The Cell Concept. More recently still, the cell concept in coastal sediment transport has been developed (Stapor, 1971; Tanner, 1974). This concept argues that along rocky, indented, and embayed coasts the concept of a river of sand is misleading when thought of in terms of longshore transport over hundreds of kilometers. Instead, such coasts, even in conditions of extremely high wave energy, are compartmented in such a way that movement alongshore is restricted to tens rather than hundreds or thousands of kilometers. Trask (1955) has shown that sediment cannot pass rocky promontories that extend into deep water. Even though inlets do not prevent littoral sediment movement for either sand (Bruun and Gerritson, 1959) or pebbles (Kidson, Carr, and Smith, 1958), the idea of the limitation of sediment transfer to cells of varying size is beginning to commend itself to coastal geomorphologists in many parts of the world. The ideal coastal cell has been defined as one area of erosion, one area of deposition, and the transport path between them. While much more work is needed, particularly in the context of interaction between adjacent cells and between transport within the cell and sediment movement in deep water, the concept is one of major significance.

The Role of Tides

There seems little doubt that tidal streams are of great significance in sediment movement on the continental shelves. For example, Stride (1963) has demonstrated that movement in the seas around the southern part of the British Isles is governed by the resultant of the ebb and flood streams. The significance of tides in the nearshore zone, even in areas of considerable tidal range with resulting relatively high velocity flows, is less well understood. On open coasts tidal velocities, except at half tides of spring tides, are rarely high enough to start from rest sediment coarser than medium sand. However, in association with waves or wave-induced longshore currents, tidal streams can move sediment of larger size and many offshore banks, and nearshore bars tend to follow directions dictated by tidal flow. Where there is a difference between ebb and flood velocities, such bed forms as well as their constituent particles migrate in the direction of the resultant. There is some interchange of sediment between nearshore banks and the beach under the influence of tidal flow and wave currents in the channels between them. Generally, however, tidal velocities do not reach values high enough for independent sediment movement until some distance offshore.

Tidal streams become important in relation to sediment movement when they turn and flow into coastal inlets. The alternate ebb and flood streams combined with wave energy play a significant role in the bypassing of inlets by sediment moving alongshore (Kidson, Carr, and Smith, 1958). In addition, the tendency of the ebb and the flood to follow discrete channels in estuaries and their approaches is fundamental in controlling the shapes and locations of banks and channels (Van Veen, 1950).

The Role of Winds

The major influence of winds in coastal sediment transport lies in their role in wave propagation. The direct effects must not, however, be underestimated. Broadly speaking, where winds are predominantly onshore on sandy coasts, dunes develop. Where they are predominantly offshore, dune development is restricted and the nearshore zone tends to be shallow with many sand banks and bars. The selection by winds of the finer sand particles is significant in sediment sorting. Friedman (1961) has shown that beach sands can be distinguished from dune sands on the basis of size distributions. Beach sands tend to be negatively skewed, reflecting a tail of coarser particles, dune sands tend to be positively skewed, implying a tail of finer particles.

While coastal sediment movements follow general patterns that can be predicted from theory, there are as many variations from such patterns as there are coastal situations. The interactions between the processes and coastal morphology can be almost infinitely variable.

CLARENCE KIDSON

References

Aibulatov, N. A.; Boldyrev, V. L.; and Zenkovich, V. P., 1960. Some new data on longshore drifting, *Symp. 21st Internat. Geol. Cong., Akad. Nauk SSSR*, 164-174.

Airy, G. B., 1845. On tides and waves, *in Encyclopedia Metropolitana*, vol. 5. London: B. Fellows, 241-396.

Bagnold, R. A., 1940. Beach formation by waves: Some Model experiments in a wave tank, *Inst. Civil Engineers Jour.* 15, 27-52.

Bagnold, R. A., 1946. Motion of waves in shallow water, *Royal Soc. London Proc.*, ser. A, 187, 1-15.

Bagnold, R. A., 1947. Sand movements by waves; some small-scale experiments with sand of low density, *Inst. Civil Engineers Jour.* 27, 457-469.

Bagnold, R. A., 1963. Mechanics of marine sedimentation, *in* M. N. Hill, ed., *The Sea*, vol. 3. New York: Wiley & Sons, 507-528.

Bird, E. C. F., 1968. *Coasts*. Cambridge, Mass.: M.I.T. Press, 246p.

Bruun, P., and Gerritsen, F., 1959. Natural bypassing of sand at coastal inlets, *Am. Soc. Civil Engineers Proc., Jour. Waterways and Harbour Div.* 85, 75-107.

Cornaglia, P., 1891. *Sul regime della spiagge e sulla regulazione dei porti*. Turin: G. B. Paravia, 569p.

Dietz, R. S., 1963. Wave-base, marine profile of equilibrium, and wave-built terraces: A critical appraisal, *Geol. Soc. America Bull.* 74, 971-990.

Draper, L., 1967. Wave activity at the sea bed around northwestern Europe, *Marine Geology* 5, 133-140.

Eagleson, P. S.; Dean, R. G.; and Peralta, L. A., 1958. The mechanics of the motion of discrete spherical bottom sediment particles due to shoaling waves, *U.S. Army Corps of Engineers Beach Erosion Board Tech. Memo. 104*.

Friedman, G., 1961. Distinction between dune, beach, and river sands from their textural characteristics, *Jour. Sed. Petrology* 35, 900-910.

Gugnyaev, Ya. E., 1959. The modeling of sea drift, *Akad. Nauk SSSR Okeanog. Kom. Trudy* 4.

Inman, D. L., 1957. Wave generated ripples in nearshore sands, *U.S. Army Corps of Engineers Beach Erosion Board Tech. Memo. 100*, 42p.

Inman, D. L., and Bagnold, R. A., 1963. Littoral processes, *in* N. M. Hill, ed., *The Sea*, vol. 3. New York: Wiley & Sons, 529-553.

Ippen, A. T., and Eagleson, P. S., 1955. A study of sediment sorting by waves shoaling on a plane beach, *U.S. Army Corps of Engineers Beach Erosion Board Tech. Memo. 63*, 83p.

Johnson, J. W., 1956. Dynamics of nearshore sediment movement, *Am. Assoc. Petroleum Geologists Bull.* 40, 2211-2232.

Johnson, J. W., and Eagleson, P. S., 1966. Coastal process, *in* A. T. Ippen, ed., *Estuarine and Coastline Hydrodynamics*. New York: McGraw-Hill, 402-492.

Kidson, C., 1963. The growth of sand and shingle spits across estuaries, *Zeitschr. Geomorphologie* 7, 1-22.

Kidson, C., and Carr, A. P., 1959. The movement of shingle over the sea bed close inshore, *Geog. Jour.* 125, 380-389.

Kidson, C.; Carr, A. P.; and Smith, D. B., 1958. Further experiments using radioactive methods to detect the movement of shingle over the sea bed and alongshore, *Geog. Jour.* 124, 210-218.

King, C. A. M., 1959. *Beaches and Coasts*. London: Arnold, 403p.

King, C. A. M., 1972. *Beaches and Coasts*, 2d ed. London: Arnold, 570p.

Larras, J., 1957. *Plages et Cotes de sable*. Paris.

Penck, A., 1894. *Morphologie der Erdoberflache*, 2 vols. Stuttgart.

Redman, J. B., 1851-52. Alluvial formations of the south coast of England, *Inst. Civil Engineers Proc.* 11, 162-184.

Robinson, A. H. W., 1955. The harbour entrances of Poole, Christchurch, and Pagham, *Geog. Jour.* 121, 33-50.

Stapor, F. W., 1971. Sediment budgets on a compartmented low-to-moderate energy coast in northwest Florida, *Marine Geology* 10, M1-M7.

Stride, A. H., 1963. Current-swept sea floors near the southern half of Great Britain, *Geol. Soc. London Quart. Jour.* 119, 175-199.

Tanner, W. F., 1954. The "a-b-c . . ." model, *in* W. F. Tanner, *Sediment Transport in the Nearshore Zone*. Tallahassee, Fla.: Department of Geology, Florida State University, 12-21.

Trask, P. D., 1955. Movement of sand around southern California promontories, *U.S. Army Corps of Engineers Beach Erosion Board Tech. Memo. 76*, 60p.

Van Veen, J., 1950. Eb en vloedschaar systemen in de Nederlands getijwatern, *Koninkl. Nederlandsch Aardrijksk. Genoot. Tijdschr.* 67, 303-327.

Volkov, P. A., 1960. The velocity below which waves do not erode a loose bottom, *Rechnoi Transport No. 6*.

Watts, G. M., 1956. Behavior of beach fill at Ocean City, New Jersey, *U.S. Army Corps of Engineers Beach Erosion Board Tech. Memo. 77*, 33p.

Zenkovich, V. P., 1967. *Processes of Coastal Development*. Edinburgh: Oliver and Boyd, 738p.

Cross-references: *A-B-C Model; Beach Material, Sorting of; Beach Processes; Coastal Dunes and Eolian Sedimentation; Currents; Cut and Fill; Sediment Analysis, Statistical Methods; Sediment Budget; Sediment Tracers; Shore Drift Cell; Tides; Waves; Wind; World Net Sediment Transport*. Vol. III: *Sediment Transport, Fluvial and Marine*.

SEICHES

Seiches are free oscillations of the water contained in a closed basin (lake) or in a semienclosed basin (some seas, gulfs, estuaries). In some cases, the water moving over a continental shelf from a deep ocean to a coastal region can produce a particular kind of seiche (*platform seiche*). Open ocean basins with an irregular bottom can also produce seiches (Dietrich, 1963).

Seiches are stationary waves whose periods are strongly related to the geometrical dimensions of the basin. The periods of the seiches may be found as solutions to hydrodynamic equations, with the boundary conditions as eigenperiods related to the geometry of the basin. A simple rectangular basin with constant depth h and length L, may present a set of free oscillations whose periods are:

$$T_n = \frac{2L}{n\sqrt{gh}} \quad (n = 1, 2, 3 \ldots)$$

A natural basin with irregular form and depth, closed or semienclosed (often with secondary forms such as gulfs and bays), may have different sets of oscillations (of first, second, third order) acting along the length or width of the basin. The maximal amplitudes or the nodes of these different oscillations will appear in different parts of the basin, illustrating the complicated nature of the natural basins. In a complicated natural basin the periods and amplitude distribution characteristics of such seiches are found by solving hydrodynamic equations. Comparison with *marigraphic* (or liminographic) records is useful for testing the validity of the mathematical model.

It is not always simple to define the periods with numerical models. An interesting recently developed method (Dovier, Manca, and Mosetti, 1974; Radach, 1971) consists of forcing the simulated basin (in the computer) by a continuous set of frequencies (or, to be practical, by an almost continuous set) with constant amplitude. This eliminates the improper frequencies among those introduced. The basin is like a filter, so the frequency spectrum at a point in the basin (after the introduction of a set of extreme frequencies) denotes only the proper frequencies, which are those passed through the filter. In this way different amplitudes related to nodal zones are shown at each point. This method of calculation is similar to the optical spectrometry of some substances (emission or absorption spectrum).

Sea or lake seiches are commonly induced by wind, pressure disturbances, and thunderstorms, rarely by earthquakes or by submarine (or sublacustrine) slides. Seiches may have different amplitudes, depending on the intensity of their cause.

Some basin seiches can have larger amplitudes than the regional tides. Seiches are irregular oscillations (in relation to time), while tides are a regular (in time) phenomenon; though seiches and tides can be superimposed, in some cases even interacting. The separation of the different effects can be highly complicated. Prediction of seiches is important in storm surge forecasting, when the seiche effect should be added to the tide or to the rise of sea level produced by wind. The most appreciable effect of seiches is the level variation but a *seiche current* can follow from the level variation.

The period of seiches is variable; from a few minutes to about one day, according to the geometry of the basin. Long-period seiches involve great energy; only the most intense causes (the strongest winds, the greatest pressure anomalies) are able to produce long-period seiches in those basins whose geometries permit seiche formation.

FERRUCCIO MOSETTI

References

Dietrich, G., 1963. *General Oceanography*. New York: Wiley Interscience, 588p.

Dovier, A.; Manca, B.; and Mosetti, F., 1974. I periodi di oscillazione del Golfo di Trieste calcolati con un nuovo metodo, *Riv. Italiana di Geofisica*, 23, 64-70.

Radach, G., 1971. Ermittlung zufallsangeregter Bewegungvorgange fur zwei Modellmeermittels des Hydrodynamisch-numerischen Verfahrens, *Inst. Meersk. Univ. Hamburg, Mitt.* 1-74.

Cross-references: *Forced Waves; Marigram and Marigraph; Mathematical Models; Sea Level Changes; Tides; Waves; Wind.* Vol I: *Seiche*.

SEISMIC SEA WAVE—See TSUNAMI

SHELF SEDIMENTATION

Continental shelf sediments show a pattern of recent (Holocene) sediments, reworked older material, and relatively undisturbed deposits of mainly Pleistocene age (Shepard, 1973). In most areas the continental shelf is dominated by relict features (a drowned topography and relict sediments) dating from one or more periods of low sea level during the Pleistocene, modified by the postglacial transgression and Holocene sedimentation. Thus, about 70% of the sediment on the shelf is classified as relict (old river beds, peat layers, fluviatile, estuarine and former coastal sediments, and, at higher latitudes, glacial and periglacial deposits); a large part of the remainder is considered a mixture, containing reworked as well as recent material. Because of this variable origin, continental shelf sediments usually show a patchy distribution, but definite trends may be present, depending on the coastal configuration, distribution of currents and waves, shelf topography, and the character of the adjacent land area.

The sources of recent sediments include coastal erosion, rivers, glaciers, sea floor erosion, organic production in the sea, and, to a smaller extent, volcanism and wind-blown material. Sands are usually found on open shelves with small relief, off sandy beaches, on shoals and banks, outside narrow bay entrances, and along the shelf edge where tidal currents are important. Muddy bottoms are present off large rivers, in sheltered bays and gulfs, and in depressions on the shelf. Gravel, pebbles, stones, and rock bottoms are found off cliffs, in straits, on elevations on the shelf, and on open shelves with strong currents. Because of Pleistocene glacial influence, gravel and pebbly deposits are most numerous at higher latitudes.

Outside the arctic, where glacial and glaciomarine deposits dominate, ranging from very coarse to very fine, mud is most abundant in areas with high rainfall and high temperatures (the humid tropics), while gravel is most common in areas of low temperature. Sand is abundant in all types of climate. Coral reefs and carbonate sands (calcarenite), consisting of coral debris and remains of other calcareous organisms (bryozoa, calcareous algae, foraminifera, ostracoda, and mollusk shells), occur in the tropics and subtropics. Shell deposits are present on all types of shelves, but are most abundant in areas where the inflow of an organic material is low, as off arid coasts. Diatom deposits (diatomites) occur locally in areas of high diatom production, as on the Namibian shelf and off Antarctica.

Dispersal of material over the shelf and sedimentation are often complicated processes of interaction between surface waves and various types of currents, of which tidal currents are the most important. Only small amounts of material are transported from the shelf into the deep ocean, except where canyons cut the shelf and a direct downward movement through the canyon is present. Sedimentation of fine-grained material on the shelf has a blanketing effect, smoothing the topography. Sandy material is often deposited in the form of banks or ridges rising only a few meters above their surroundings and with a large variety of shape and size. Gravels often occur as banks.

D. EISMA

Reference

Shepard, F. P., 1973. *Submarine Geology*. New York: Harper & Row, 517p.

Cross-references: *Flandrian Transgression; Holocene; Nearshore Hydrodynamics and Sedimentation; Sediment Size Classification*. Vol. VI: *Continental Shelf Sedimentation*.

SHELL PAVEMENT

The term *shell pavement* was introduced to describe laterally extensive accumulations of shell valves within the tidal flat environment (Van Straaten, 1950; Schäfer, 1970). However, Carter (1976) has indicated that shell pavements have a wide environmental distribution, occurring in all places where mixed sand/shell sediments are subjected to selective erosional winnowing, leaving a superficial lag deposit of the coarse shell material. The pavements are common in shallow, low-energy marine environments where fine sediment can be moved by tidal currents, wave action, channel migration, or even the burrowing activity of intertidal fauna, leaving a residual veneer of shell material at or just below the surface. Powers and Kinsman (1953) have recorded similar features in more dynamic wave environments where the shells form a *traction carpet* on the sea bed.

Above the high water line, wind action often removes the finer sediment to a point where a relatively process-resistant deposit is formed (Fig. 1). Deflation rates depend upon the wind speed and direction, sediment size, shape and density, and the moisture content of both air and sediment. The depth to which deflation continues is related to the proportions of sand and shell in the admixture.

The presence of a shell pavement changes the nature of the air or water flow above it, inducing new boundary layer conditions, and also controls the amount of sediment available for transport. In the case of the aeolian shell pavement, adjacent sand dunes may grow rapidly at first, but will eventually become starved of fresh sediment as the pavement extends. In addition, a resistant shell pavement may offer some protection against infrequent erosional events—for example, winter beach berm erosion appears to be substantially reduced if a pavement is present.

RICHARD W. G. CARTER

References

Carter, R. W. G., 1976. Formation, maintenance, and geomorphological significance of an aeolian shell pavement, *Jour. Sed. Petrology* 46, 418-429.

Powers, M. C., and Kinsman, B., 1953. Shell accumulations in underwater sediments and their relation to the thickness of the traction zone, *Jour. Sed. Petrology* 23, 229-234.

FIGURE 1. A shell pavement in western Ireland, with embryo dunes forming on the shell surface (photo: R. W. G. Carter).

Schäfer, W., 1970. *Ecology and Palaeoecology of Marine Environments.* Edinburgh: Oliver and Boyd, 568p.

Van Straaten, L. M. J. U., 1950. Environment of formation and facies of the Dutch Wadden sea sediments, *Koninkl. Nederlandsch Aardrijksk. Genoot. Tijdschr.* 67, 374-381.

Cross-references: *Beach Processes; Coastal Fauna; Deflation Phenomena; Dunes and Eolian Sedimentation; Midden; Nearshore Hydrodynamics and Sedimentation; Sediment Transport.*

SHINGLE AND SHINGLE BEACHES

Shingle is coarse beach material that is on the whole more common in previously glaciated areas, where many of the stones are derived from glacial and fluvioglacial deposits. Shingle includes material in the *pebble* and *cobble* range of the Wentworth and ϕ scales, being, respectively, 4-64 mm and 64-256 mm, and -2.0 to -6.0 and -6.0 to -8.0 ϕ units. Granule-sized particles of 2-4 mm or -1.0 to -2.0 ϕ units are rare as beach material. Marine shingle becomes well rounded, and usually consists of resistant rocks, such as flint, quartzite, granite, and sandstone (Emery, 1955).

On a beach, shingle tends to be moved to the top of the beach, where it occurs mixed with sand, forming a steep bank near the high water level. A typical *shingle beach* has a ridge at the level of the extreme storm wave action and minor ridges at lower elevations marking the levels of lesser storms or lower tides. A terrace commonly occurs at the low tide level. The *gradient* of a shingle beach is steep owing to the high permeability; common slopes can be as much as 1 in 2, and 1 in 5 to 1 in 10.

The most permanent feature of a shingle beach is the storm wave, high level ridge. The orientation of the ridge reflects the direction from which the dominant storm waves come. Chesil Beach in Dorset, one of the three major shingle structures of southern Britain, illustrates a relict shingle beach adjusted to lie normal to the storm waves that come from the direction of maximum fetch from the southwest. The beach is also graded laterally, probably in response to variations in wave energy along its length (Carr and Blackley, 1973). The larger pebbles, about 6 cm in diameter, lie at the southeast end, where the energy is greatest and the beach has been built to 13.3 m above high water level. The other major shingle structures of southern Britain are Dungeness in Kent (Hey, 1967), *a cuspate foreland,* and Orfordness in Suffolk, a *spit* trending southward in the direction of beach drifting.

CUCHLAINE A. M. KING

References

Carr, A. P., and Blackley, M. W. L., 1973. Investigations bearing on the age and development of Chesil Beach, and the associated area, *Inst. British Geographers Trans.* 58, 99-112.

Emery, K. O., 1955. Grain size of marine beach gravel, *Jour. Geology* 63, 39-49.

Hey, R. W., 1967. Sections in the beach-plain deposits of Dungeness, Kent, *Geol. Mag.* 104, 361-370.

Cross-references: *Beach Material; Beach Material, Sorting of; Cobble; Pebble; Sediment Size Classification.* Vol. VI: *Wentworth Scale;* Vol. XIII: *Sieve Analysis.*

SHOALING COEFFICIENT

The *shoaling coefficient* (K_S) is defined as the ratio of wave height (H_i) at a particular point of interest (x_i) to the original or deep water wave height (H_o). As a first approximation, the assumption is made that there exists no loss or gain of energy from the system (*conservation of energy flux*). The simplest case uses Airy linear wave theory and yields:

$$K_S = H_i/H_o = (2 \cosh^2 kh / (\sinh 2kh + 2kh))^{1/2}$$

where k is $2\pi/L$, L is wavelength, and h is still water depth at the point of interest.

The limitations on the validity of linear theory are well known (see Wave Theory, Vol. I of this series), and so attempts to refine the determination of the shoaling coefficient have used other wave theories. LeMéhauté and Webb (1964) computed the shoaling coefficient using third order Stokes' wave equations. Their study showed that K_S is larger than that resulting from linear theory, and that K_S increases with increased H_o. These general results are corroborated by experimental observation.

Koh and LeMéhauté (1966) computed K_S based on the fifth order Stokes' wave equations. The results are similar to those obtained from third order theory, except that relative water depth (h/L) is more restrictive for higher order theory. They conclude that the shoaling coefficient based on third order theory is probably more useful at intermediate depths, and that that based on linear theory is better for shallower depths.

In regions where the offshore slope is steep, conservation of energy flux may be valid; however, the effects of reflection from steep shores may be significant. Reflection has been shown to be significant for slopes greater than 4.5°, or about 1/13.

Conservation of energy flux may not be valid for regions where the offshore slope is gentle. Energy losses at the bottom in shoal water (where depth is less than one-half wavelength) may be significant. Putnam and Johnson (see references in May and Tanner, 1973) developed an equation for the shoaling coefficient based on linear theory and taking into account bottom friction. Putnam also considered the additional effects of percolation into a permeable bottom. Bretschneider and Reid expanded on these previous studies and presented graphical solutions for shoaling coefficient. May and Tanner (1973) independently developed an equation for shoaling coefficient based on linear theory and considering bottom friction and refraction effects. This equation is in the form of a numerical series and was presented as part of a computer program to determine wave energy conditions at the shoreline given specified bathymetry and deep water wave conditions:

$$K_s = H_i/H_o = [(b_o/b_i)(2\cosh^2 kh / (\sinh 2kh + 2kh))]^{1/2}$$

$$- 0.67 \sum_{j=1}^{i} (x_j - x_{j-1})$$

$$[T c_f \sigma^3 H_j^2 / (L_j \pi g \sinh^3(k_j h_j))]$$

where b is wave ray spacing, subscript o refers to deep water, subscript i refers to shoal depth (at the point x_i), T is period, c_f is a friction coefficient (a variable, related to Reynold's number, assumed to have a numerical value on the order of 0.03), σ is angular velocity (equals $2\pi/T$), g is acceleration of gravity, and j is the jth increment along the path of the wave ray.

JAMES P. MAY

References

Koh, R. C. Y., and LeMéhauté, B. J., 1966. *Wave Shoaling.* Pasadena, Calif.: National Engineering Science Co., 22p.

Le Méhauté, B. J., and Webb, L., 1964. Periodic gravity wave over a gentle slope at a third order of approximation, *Conf. Coastal Eng., 9th Am. Soc. Civil Engineers, Proc.,* 23-40.

May, J. P., and Tanner, W. F., 1973. The littoral power gradient and shoreline changes, *in* D. R. Coates, ed., *Coastal Geomorphology.* Binghamton: State University of New York, 43-60.

Cross-references: *Coastal Engineering, Research Methods; Reynold's Number; Wave Energy; Wave Refraction Diagrams; Waves; Wave Statistics.* Vol. I: *Wave Theory.*

SHOESTRING SANDS

The term *shoestring sands* was introduced into the geologic literature by Rich (1923) to refer to oil-bearing sand bodies encapsulated by comparatively impermeable clays and shales and considerably longer than they are wide, thus producing a type of stratigraphic trap (see Levorsen, 1967, for a comprehensive review). Krynine (see summary in Pettijohn, 1975) defined such bodies as having width-thickness ratios of < 5:1. While Rich originally contrasted shoestring sands with those of equidimensional ground plans, Pettijohn finds them to be only one of several types of sandstone bodies, without oil trap connotations, in a classification that is a product of recent, more extensive information on the geometry of sandstone bodies (see review in LeBlanc, 1972). The term is being used currently in a nongenetic, strictly geometric sense.

Shoestring sands range in width from less than 1 km to over 5 km, and extend for several km to over 300 km in length. The longer sands may be discontinuous. Thicknesses are mostly less than 60 m. Almost all known ones consist of siliceous (quartzose) sand; a few have conglomeratic material.

A variety of depositional processes and environmental frameworks have been described as producing shoestring sands, including linear, discontinuous trends of barrier islands, sinuous channel, and point bar deposits of streams meandering in alluvial valleys, simple channel fillings in bedrock valleys, and portions of delta distributaries and bar finger sands. The lack of continuity of some shoestring sands may be due to their original depositional pattern or to their later erosion.

SAUL ARONOW

References

LeBlanc, R. J., 1972. Geometry of sandstone reservoir bodies, *in* T. D. Cook, ed., *Underground Waste Management and Environmental Implications.* Tulsa, Okla.: American Association of Petroleum Geologists Mem. 18, 113-190.
Levorsen, A. I., 1967. *Geology of Petroleum.* San Francisco: W. H. Freeman and Co., 724p.
Pettijohn, F. J., 1975. *Sedimentary Rocks.* New York: Harper & Row, 628p.
Rich, J. L., 1923. Shoestring sands of eastern Kansas, *Am. Assoc. Petroleum Geologists Bull.* 7, 103-113.

Cross-references: *Barrier Islands; Bars; Beach Material; Deltas; Sand.*

SHORE DRIFT—See SEDIMENT TRANSPORT

SHORE DRIFT CELL

A stretch of coast for which identifiable introduced material (from rivers, offshore erosion, coastal erosion) can be traced to a significant depositional site (prograding beaches, spits, beach ridge plains, shoals) constitutes a *shore drift cell.* An overall sand budget probably most precisely identifies shore drift cells (Pierce, 1969; Stapor, 1971, 1973), although systematic changes in grain size and bulk composition as well as natural tracers (distinctive pebbles, heavy minerals) can serve also. The degree of independence among shore drift cells is determined by the efficiency of the interact-

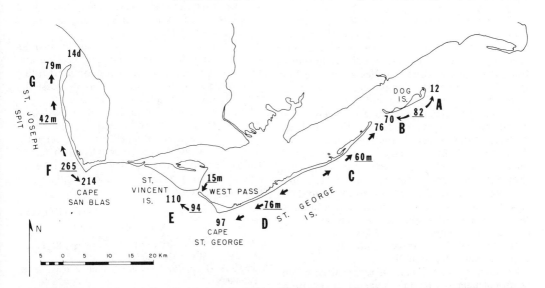

FIGURE 1. Seven shore drift cells (*A* through *C*) identified along the Florida panhandle coast. The numbers are yearly rates of erosion (underlined) and deposition times 1000 cubic meters. Rates followed by the letter *m* are minimum values; those by the letter *d*, from maintenance dredging data. Arrows indicate the transport direction for individual cells (data and diagram after Stapor, 1971).

ing factors creating the various cells. Wave energy, coastal and offshore geology, offshore bathymetry, size, geometry, and geographic spacing of tidal inlets, tidal range, and climatic conditions are perhaps the more important of these factors interacting to create shore drift cells.

Stapor (1971, 1973) identified seven shore drift cells along the Florida panhandle coast, all but two of which experience little if any net exchange of sand (Fig. 1). Pierce (1969) identified two major cells along the North Carolina outer banks, both of which experience net exchange of sand with adjacent cells, offshore regions, and lagoonal areas.

FRANK W. STAPOR, JR.

References

Pierce, J. W., 1969. Sediment budget along a barrier island chain, *Sed. Geology* **3**, 5-16.
Stapor, F. W., 1971. Sediment budgets on a compartmented low-to-moderate energy coast in northwest Florida, *Marine Geology* **10**, M1-M7.
Stapor, F. W., 1973. History and sand budgets of the barrier island system in the Panama City, Florida region, *Marine Geology* **14**, 277-286.

Cross-references: *Beach Processes; Coastal Erosion, Formations; Coastal Characteristics, Mapping of; Coastal Engineering; Coastal Morphology, Research Methods; Sediment Budget; Sediment Transport; World Net Sediment Transport.*

SHOREFACE AND SHOREFACE TERRACE

The term *shoreface* is generally applied to a rather narrow zone of the ocean floor adjacent to the shoreline. The term was originally used by Johnson (1919) to define that portion of the beach zone that lies below the low tide shoreline and is permanently covered by water. With changing wave and tide conditions, beach materials (sand and gravel) actively move within the beach zone including the shoreface. Although the upper boundary between the *foreshore* and the shoreface is rather easily delineated, the lower boundary between the shoreface and the *offshore ramp* is less distinctive. Johnson defined the lower boundary as the break between the rather steeply sloping shoreface and the beginning of the more nearly horizontal offshore surface. W. A. Price (1954) suggested that the shoreface could extend down to a maximum depth of 16.5 m.

Frequently the shoreface is represented by an erosional bench with only a veneer of sediment actively moving across it by oscillitory wave action or longshore currents. However, the outer portion of the shoreface may become a more or less temporary sedimentary wedge known as the *shoreface terrace*. During heavy or prolonged winter storms, foreshore beach material tends to be removed by the stronger waves and deposited offshore as a shoreface terrace. With a change in wave regime, this material will tend to migrate back onto the foreshore beach.

DONALD B. KOWALEWSKY

References

Johnson, D. W., 1919. *Shore Processes and Shoreline Development.* New York: John Wiley & Sons, 584p.
Price, W. A., 1954. Dynamic environments-reconnaissance mapping, geology and geomorphic, of the continental shelf of Gulf of Mexico, *Gulf Coast Assoc. Geol. Socs. Trans.* **4**, 75-107.

Cross-references: *Beach; Erosion Ramp, Wave Ramp; Shore Platforms; Wave-Built Terrace; Wave-Cut Platform; Wave Erosion.*

SHORELINE DEVELOPMENT RATIO

The *shoreline development ratio* (or index) is a number that relates the measured shoreline length of a given lake to the shoreline length of a perfectly circular lake of equal area. The formula for the ratio is $D_L = L/(2\sqrt{\pi A})$, where L is the measured shoreline of a lake and A is the area of the lake. The ratio can be no smaller than 1.0, the ratio for a perfectly circular lake.

The significance of the ratio can be variously interpreted—for example, in terms of origin of the lake basin. Lakes in volcanic craters or caldera, in limestone sinkholes, in meteor craters, or in some glacial kettles may have small ratios; those in oxbows of alluvial plains or in glacial troughs, large ones.

Modifications of the edges of the original depressions may influence the ratio. Erosion and mass wasting of salients and concomitant filling of recesses along the shoreline may decrease the ratio. Deltas filling lakes may increase it.

The ratio may be an indication, if used with caution, of a lake's productivity, as noted by Cole (1975) and Wetzel (1976). A larger ratio may mean a larger zone through which terrestrially derived nutrients may enter a lake, as well as a larger zone for the development of littoral communities.

SAUL ARONOW

References

Cole, G. A., 1975. *Textbook of Limnology.* St. Louis, Mo.: C. V. Mosby Co., 283p.
Wetzel, R. G., 1975. *Limnology.* Philadelphia: W. B. Saunders Co., 743 p.

Cross-references: *Beach Processes; Cuspate Spits; Lagoonal Segmentation; Lakes, Coastal Morphology; Production.*

SHORE PLATFORMS

Shore platforms are the bevel cut by the sea in the hard rock edges of the land. Their morphology is a function of the dynamic relationships between the rocks (passive factor) and the sea (active factor). The properties of the sea and the air above it are a direct result of the climate of the area. This varies greatly, as also does the nature of the rocks, so that shore platforms vary greatly in sizes and shapes. Sea level plays a part in that the platforms are cut during times of small oscillations, generally called stillstands. This is varied by the local tectonic style, because if the area is in active movement, the time available within which to establish a platform is shorter than where no movement occurs. For example, since the mid-Holocene at least, the north coast of New Guinea at Aitape has risen at a rate of about 1m/88 yr, so that the shoreline of 4550 years ago now stands at 51.5 m. Another factor of importance is tidal range, because on a microtidal coast the marine energy is concentrated within a narrow range, whereas on a macrotidal coast the energy is spread over a wide zone with much less platform formation.

No rocky shore is static but in evolution. The degree of platform development depends on the factors named, except in the case of aeolianite and other granular calcareous (*soluble*) rocks. These have a different mode of formation, the platform being cut level from the beginning (Gill, 1973). For this reason no supratidal platforms occur in this lithology.

EDMUND D. GILL

Reference

Gill, E. D., 1973. Rate and mode of retrogradation on rocky coasts in Victoria, Australia, and their relationship to sea level changes, *Boreas* **2,** 143–171.

Cross-references: *Erosion Ramp, Wave Ramp; Wave-Cut Bench; Wave-Cut Platform; Wave Erosion.* Vol. III: *Platform — Wave-Cut.*

SHORE POLYGONS

Three types of *shore polygons* or *shore polygonal patterns* are known. The first type is found on lake shores and shallow lakes (water depth less than 2 m) in periglacial environments. These are *patterned ground,* including sorted circles, polygons, and nets of various forms and sizes, composed of a border of stones and a cell of fine material (Fig. 1). These have been observed in arctic and subarctic

FIGURE 1. Shore sorted polygons at low water level, Lake Nouveau, central subarctic Quebec. Cells are composed of fines and borders of coarse material (from Dionne, 1974).

regions of Canada and Europe (Conrad, 1933; Lundqvist, 1962; Mackay, 1967; Dionne, 1974; Shilts and Dean, 1975). They occur within or outside permafrost regions, but it is not known if those found today outside permafrost regions are relict features. They result from a sorting process that is related to frost action, especially to freeze-thaw cycles in the active layer or in the seasonally frozen layer. They are a valuable indicator of a periglacial environment characterized by a mean annual temperature lower than $-1°C$, but they do not necessary indicate presence of permafrost.

A second type of shore polygonal patterns is found in tidal flats of cold regions (Dionne, 1971). At breakup, mud ridges, 10-20 cm high, 5-25 cm wide, and up to 4 m long, form polygonal patterns (3- to 5-sided polygons) at the surface of mud flats. The central part of the polygons are depressed. Mud ridges form under the *icefoot* by upward injection of soft mud into tidal cracks occurring in the ice. At low tide, the downward pressure of the ice forces the mud to flow into the cracks. At breakup, ice floes carried away by ebb currents leave a polygonal pattern at the surface of tidal flats. This peculiar type of sedimentary structure can provide valuable information to characterize the climatic sedimentary environment.

A third type of shore polygon also occurs in tidal flats of cold regions (Dionne, 1976). In James Bay, polygons (4- to 6-sided) are formed by irregular mud ridges, 40-90 mm high, 50-95 mm wide, and up to 100 cm long (Fig. 2). The central part of the polygons are depressed. Mud ridges result from upward injection of liquified clay, silt, and fine sand from the underlying strata into surficial mud. In early summer (beginning of July), segregation ice or buried shore ice is found in the center of polygons at 30-35 cm depth. Upward injection dikes composed of liquified sediments form into the fissures or spaces separating ice lenses or ice blocks, producing a polygonal pattern at the surface of the mud flat. In James Bay, numerous miniature *mud volcanoes* are associated with this type of shore polygonal pattern, which is a valuable indicator of a cold environment.

JEAN-CLAUDE DIONNE

References

Conrad, V., 1933. Ein Unterwasser-Strukturboden in den Ostalpen, *Gerlands Beitr. Geophysik* **40**, 353-360.

Dionne, J.-C., 1971. Polygonal patterns in muddy tidal flats, *Jour. Sed. Petrology* **41**, 838-839.

Dionne, J.-C., 1974. Cryosols avec triage sur rivages et fond de lacs, Québec central subarctique, *Rev. Géographie Montréal* **28**, 323-342.

Dionne, J.-C., 1976. Miniature mud volcanoes and other injection features in tidal flats, James Bay, Quebec, *Canadian Jour. Earth Sci.* **13**, 422-428.

Lundqvist, J., 1962. Patterned ground and related frost phenomena in Sweden, *Sveriges Geol. Undersökning*, ser. C, **583**, 101p.

Mackay, J. R., 1967. Underwater patterned ground in artificially drained lakes, Garry Island, N.W.T., *Geog. Bull.* **9**, 33-44.

Shilts, W. W., and Dean, W. E., 1975. Permafrost features under arctic lakes, District of Keewatin, Northwest Territories, *Canadian Jour. Earth Sci.* **12**, 649-662.

Cross-references: *Glaciel; High-Latitude Coasts; Ice along the Shore; Ice-Bordered Coasts; Ice Foot, Kaimoo; Monroes; Mud Volcanoes; Stone Packing.*

FIGURE 2. Shore polygons in a mud tidal flat, James Bay, Quebec. Polygons are 30-125 cm long and ridges are up to 10 cm high (from J.-C. Dionne, 1976. Miniature mud volcanoes and other injection features in tidal flats, James Bay, Quebec, *Canadian Jour. Earth Sci.* **13**, 422-428; reproduced by permission of the National Research Council of Canada).

SHORE TERRACE

Complexes of ancient forms of coastal relief, lying above and below present sea level, are referred to as *shore terraces* or *old coastlines*. Shore terraces are mainly shaped either as curves on the surface of submarine coastal slope at the sea floor or as curves on the surface of seashore land; in either case, they are strung out along the present coastline. Shore terraces indicate the old coastline as it was in the past, when sea level lay below or above the present line. Shore terraces derive their origin from eustatic sea level fluctuations or from tectonic land movements, but mainly from the combined effect of both. In a study of shore terraces, it is practically impossible to identify the true cause behind the old coastline's sinking or rising.

A shore terrace may be a rock bench or an aggradation terrace. A rock bench is a former erosive submarine slope adjoining a cliff, formed below the then sea level. Aggradation terraces are former beaches, strings of coastal bars, old aggradation features formed above the respective sea level. This is one reason why rock benches and aggradation terraces of the same age usually differ in hypsometric level.

The surface level of both rock benches and aggradation terraces depends upon the impact of swell pounding against the coast concerned. Thus, depending on wave height, an aggradation terrace, consisting of a series of coastal bars may vary in height from 5-8 m, at open ocean shore, to 1-2 m above the sea edge in closed bays. Cut into the submarine slope, rock benches likewise lie at different depths, from 1-2 to 10-15 m, because of differing amounts of wave action and varying seafloor resistance to erosion. Formation conditions of shore terraces, determined by hydrometeorological factors and lithology of the shore rocks, may vary greatly along the coast, thus affecting the hypsometric level of terraces. These factors are often disregarded by investigators of elevated terraces, who attribute all terrace elevation differences to differential tectonic movements.

Shore terraces are important to the study of coastal paleogeography. They serve as reference points in judging sea level changes in the past and coastal tectonic regimes and in correlating past events in different areas.

Shore terraces frequently appear in sequences of different elevation complexes, which represent general trends in coastal development and show the course of transgressions and regressions during glacial and interglacial periods (for example, North Africa's Mediterranean terraces).

For literature on this topic the reader is referred to Gill (1967), Kaplin (1973), Mii (1963) and Zeuner (1959).

PAVEL A. KAPLIN

References

Gill, E. D., 1967. Criteria for the description of Quaternary shorelines, *Quaternaria* 9, 237-243.
Kaplin, P. A., 1973. *Recent History of the World Ocean Coast* (in Russian). Moscow: Moscow State University, 264p.
Mii, H., 1963. Relation of shore erosion to sea level, *Jour. Marine Geology* 2, 8-17.
Zeuner, F. E., 1959. *The Pleistocene Period.* London: Hutchinson, 447p.

Cross-references: *Isostatic Adjustment; Paleogeography of Coasts; Raised Beach; Sea Level Changes; Wave-Cut Bench; Wave-Cut Platform.* Vol. III: *Terraces—Marine.*

SILT

Varying particle size limits and descriptions have been given, depending upon the particular science and the specific country involved. Thus, in the United States, in sedimentation, *silt* is generally regarded as a partially rounded rock fragment or detrital particle smaller than very fine *sand* and larger than coarse *clay*. In this definition, the diameter falls between 1/256 and 1/16 mm (4-62 μ; 8-4 ϕ). However, in engineering, the emphasis is on the relative lack of plasticity and lack of strength when dried (unlike clay). The diameter is defined as less than 0.074 mm (that is, passing U.S. standard sieve 200). In U.S. soil science, the diameter is currently defined as falling within the range 0.002-0.05 mm (Gary, McAfee, and Wolfe, 1972), effectively the same as British Standards (1967) (0.002-0.06 mm). The International Society of Soil Sciences silt classification covers the range 0.002-0.02 mm—that is, that originally adopted by Atterberg.

Other definitions of silt include: silt-size particles, either in a loose aggregation or in suspension in water; a largely unconsolidated sedimentary deposit of fine-grained clastics.

In popular usage, the term silt is applied to a sediment containing a high proportion of both sand and clay particles. Silt is variable in composition but usually has a high proportion of clay minerals.

ALAN PAUL CARR

References

Gary, M.; McAfee, R.; and Wolfe, C. L., eds., 1972. *Glossary of Geology.* Washington, D.C.: American Geological Institute, 805p.
British Standards Institution, 1967. *Methods of Testing Soils for Civil Engineering Purposes,* B.S. 1377. London, 234p.

Cross-references: *Beach Material; Beach Material, Sorting of; Boulder; Clay; Cobble; Gravel; Pebble; Sand; Sediment Analysis, Statistical Methods; Shingle and Shingle Beach; Sediment Size Classification; Soils.*

SIPUNCULA

Sipunculans are marine, *unsegmented, wormlike animals* first named by Linnaeus in 1767 and elevated to a phylum by Sedgewick in 1898. The body, usually 20-40 cm in length, consists of a narrow eversible introvert, with an anterior group of tentacles surrounding the mouth, and a posterior trunk into which the introvert is retracted (Fig. 1). A large coelomic cavity holds fluid serving as a hydrostatic skeleton and

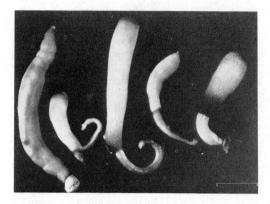

FIGURE 1. Carbonate rock-dwelling sipunculids. Scale 5 mm (from Rice, 1969).

houses a U-shaped intestine that ends in the anus at the anterior of the trunk. The embryological development, the larval stages, and the adult nervous system resemble those of the Annelida, but because of the absence of metamerism during the life history of the Sipuncula most authors (Hyman, 1959; Stephen and Edmonds, 1972) recognize it as a phylum with about 320 species, distinct from, but closely related to, annelids.

Sipunculans can be found from intertidal depths to over 3000 m and in all latitudes. They are particularly common in shallow water and in low latitudes. They burrow in muddy, sandy, or gravelly sediment, live in crevices in rocks and among roots, or inhabit worm tubes or borings in calcareous rocks. Gut contents of skeletal debris, diatoms, and sediment indicate that most sipunculans use the introvert for deposit feeding, but some living in rocks may be ciliary suspension feeders. Those inhabiting soft substrates do not construct permanent burrows, but move through the sediment, disrupting sedimentary laminations. Other sipunculans are thought to bore into calcareous rock, possibly using both chemical secretion and mechanical abrasion, but the exact mechanism has not been delineated. In tropical and subtropical areas, these sipunculans are common in coral debris and beach rock.

MOLLY FRITZ MILLER

References

Hyman, L. H., 1959. *The Invertebrates: Smaller Coelomate Groups,* vol. 5. New York: McGraw-Hill Book Co., 783p.
Rice, M. E., 1969. Possible boring structures of Sipunculids, *Amer. Zool.* **9,** 803-812.
Stephen, A. C., and Edmonds, S. J., 1972. *The Phyla Sipuncula and Echiura.* London: British Museum (Natural History), 528p.

Cross-references: *Annelida; Biotic Zonation; Coastal Fauna; Coral Reef Coasts; Coral Reef Habitat; Echiura; Organism-Sediment Relationship; Platyhelminthes; Weathering and Erosion, Biologic.*

SKERRY, SKERRY GUARD

Extending seaward from many glacially eroded coasts, there is often a zone of low rocky islands (*skerries*) and partially submerged platforms of erosion (*strandflats,* q.v.). The term skerry is derived from a Scandinavian root (Old Norse *sker*) and signifies a rocky islet, sometimes covered by the sea at high water. *Skerry guard* is an Anglicized form of the Norwegian *skjaergård* (Swedish: skärgård) that refers to the area of sea within which the skerries are found (*gård* is a yard or enclosure, not a *guard*). The features are well developed off the Norwegian coast, where the fringe of islands and reefs appears around Stavanger and continues (with breaks) to the North Cape, reaching a width of 50 km off Ranfjord in latitude $66°$. Other typical skerry guard coasts are to be found off western Iceland, western Greenland, and Spitsbergen. The islets represent the postglacial accident of partial submergence of certain strandlat levels, or the partial submergence of glaciated lowland *fjard* coasts (Embleton and King, 1975). The islets bear witness to severe glacial erosion, and although frost weathering in the surf zone has played a part, wave action seems to have only slightly modified the glacial forms.

CLIFFORD EMBLETON

Reference

Embleton, C., and King, C. A. M., 1975. *Glacial Geomorphology.* London: Edward Arnold Ltd., 573p.

Cross-references: *Europe, Coastal Morphology; Fiard, Fjard; Glaciated Coasts; Stack and Chimney; Strandflat.*

SLIKKE

Slikke is a Dutch and Flemish term (Fl.: slijk) designating the part of a beach or of a tidal river bank that is left emerged at low tide and is gradually submerged as the incoming tide rolls in again. Dark gray in color, it is an area of harsh biological conditions. Estuarine slikke is exposed in such Atlantic rivers as the Scheldt (Belgium/Netherlands), and represents large expanses along the Atlantic coast of France south of La Rochelle (Francis-Bœuf, 1947). French geologists have adopted the term and refer to *la slikke.*

Rich in organic matter, this complex contains

70-90% of particles with a diameter below 40 μ. Larger constituents are sand, quartz, and mica. Bourcart and Francis-Bœuf (1939) estimate that the organic proportion is 6% on the average but may reach 15%. Water represents a major fraction.

Bourcart and Francis-Bœuf designated the bonding material under the name of *algon*, an organogenic compound in evolutive stage. Colloidal iron also plays a bonding role, accounting for 3-7% of the mass. Oxygen is either absent or present only in minute quantities, but the thin upper layer is the site of photosynthetic activity during the emergence spans of the slikke.

ROGER H. CHARLIER

References

Bourcart, J., and Francis-Bœuf, C., 1939. Sur la véritable signification des vases sableuses et des sables vaseux, *Acad. Sci. Comptes Rendus,* 209-568.

Francis-Bœuf, C., 1947. Sur la teneur en oxygène dissous du milieu interieur des vases fluvio-marines, *Acad. Sci. Comptes Rendus,* 225-392.

Cross-references: *Schorre; Tangue; Tidal Flat.*

SLOPE-OVER-WALL CLIFFS

The original definition of *slope-over-wall cliffs* was: "If a harder rock is overlain by a softer one the resultant form of cliff is what is known as slope over wall. The hard rock ... will form a more or less vertical cliff, while the softer rock will yield readily ... until an angle of rest is reached and a condition of comparative stability is gained" (Whittaker, 1911). This definition is particularly applicable in southeastern England, where the rocks are horizontal or dip only gently and there is an alternation of resistant and weaker beds. The same coastal form is seen in cliffs of uniform lithology but then has a different origin (see *Coastal Bevel*).

ALAN WOOD

Reference

Whittaker, W., 1911. *3rd and Final Report,* Command Paper 5708. London: Royal Commission on Coastal Erosion, p. 6.

Cross-references: *Coastal Bevel; Fossil Cliffs; Sea Cliffs.*

SOILS

Although there are many different kinds of *soils* showing extreme ranges of development in the coastal zone, they can be grouped in general according to environments of formation and soil properties. They range, for example, from incipient soils of Holocene dune systems and mangrove muds to more strongly developed soils of older Pleistocene beach ridges. Many coastal soils occur in tidal marshes and swamp lands and are related to alluvial and estuarine deposits (Chapman, 1977), while others are developed in pyritic marine clays. Also to be considered are ancient soils, formed independently of littoral processes, that are now encountered on drowned coasts. Thus *paleosols* and *weathering crusts* (q.v., Vol. XII, Pt. 2) are also observed along certain coasts.

Coastal soils forming within the present intertidal belt are essentially *regoliths* because they are subjected to periodic flooding by tidal water, which inhibits the development of a true pedogenetic profile. Soils that are associated with marine marshes of coastal lagoons, estuaries, or deltas reflect mainly the origin and composition of their sedimentary parent materials (Phleger, 1977). The immediate source of sediments may be the land or the sea or both. In many estuaries or lagoonal areas of pronounced relief and high discharge, most of the sediment supplied to marshes is transported and deposited by fluvial processes. Other estuaries on low relief shores, such as those on the U.S. Atlantic Coastal Plain and the borders of the North Sea (for example, The Wash, the Wadden Zee), are, in contrast, being filled by a landward movement of sediment derived from offshore (Meade, 1969). In still other cases, wind-blown silt and sand may be a significant source of sediment, especially in arid climatic regions.

The physical properties of most coastal soils are largely due to the presence of alternating laminae of coarse and fine sediment (Allen, 1965), biogenic reworking (Allen and Curran, 1974), and the effects of decaying plant roots. Soil chemical properties such as pH, Eh, and salinity are affected mainly by the periodicity and duration of submergence by saltwater (Phleger and Bradshaw, 1966; Servant and Favori, 1966).

Study of different coastal types and their associated (sedimentary) materials suggests that coastal soils can be grouped in the following categories: dune sands and stranded beach ridges, organic soils, coastal alluvial soils, acid sulfate soils, coastal paleosols and weathering crusts, and submarine soils.

Dune Soils and Stranded Beach Ridges

Stabilized but extremely young calcareous coastal dune soils normally show no profile development beyond a surface accumulation of organic matter. The sands are usually weakly coherent when moist but loose and unstable when dry. Generally the sands are deep, medium, to coarse textured, and range from pale yellow to brown in color. With increasing age these soils tend to lose their calcareous fraction by *leaching* (q.v., Vol. XII, Pt. 1), or they may develop a more marked profile differentiation (in the so-called Podzols, for example). Holocene dune sands on the present coastline of southwestern Australia, for example, are unlithified calcareous shelly sands (made up of foramineferids and molluscan fragments) with minimal profile development (Semeniuk and Meagher, 1981) that are related to the *Flandrian transgression* and more recent minor sea level fluctuations. Late Pleistocene dune soils overlie a substrate of eolianite, eolian calcarenite (Fairbridge, 1950; Fairbridge and Teichert, 1963) that was cemented by percolating rainwater during pluvial phases. Siliceous sands entirely leached of their original calcareous material occur farther inland on the coastal plain as subdued hillocks and are believed to be Middle Pleistocene in age (McArthur and Bettenay, 1960; Mulcahy and Churchward, 1973). Still farther inland at somewhat higher elevations adjacent to uplands, there are Late Tertiary or Early Pleistocene lateritic soils that must have been influenced by sea levels of earlier cycles (Finkl and Churchward, 1976; Finkl and Fairbridge, 1979).

Like coastal dune systems, the soils of stranded beach ridges form chronosequences where variations in physical and chemical properties occur mainly in response to increasing age with distance from the coast. Being older and showing greater profile differentiation, these soils reflect changes in the coastal environment that are associated with fluctuating sea levels. Some Holocene beach ridge sequences possess more than 100 beach ridges (and minor dunes) formed during the past 6000 years. Studies by Blackburn, Bond, and Clarke (1965) indicate a relative age sequence on the South Australian coast where Pliocene lateritic soils occur on the oldest ridges. Lateritic-podzolic soils are featured on ridges of Middle Pleistocene age, while the Late Pleistocene is marked by Podzols and Terra Rossa soils. All of the former dune systems have been converted to coastal limestones with solution features during successive pluvial periods. In general, thickness of kunkar and red soil increases with distance inland (and thus age). There is also evidence of an eolian accession of a fine-grained soil component having its source on the exposed continental shelf.

Organic Soils

Peats and other organic soils are characteristic features of the coastal zone being associated with interdunal swales, deltaic backwater marshes, intertidal swamps and lakes, mangrove flats, and former estuaries and lagoons that were cut off from the sea by dunes. These soils, especially the deep coastal peat deposits, which may be several meters in thickness, consist largely of the remains of mangroves and associated plants. Along subsiding coasts, even thicker peats may accumulate (for example, in New England). On *mangrove coasts* (q.v.), sedimentary peats and marls, which also contain some mangrove root beds, are often conspicuously interstratified. Dating of such stratification may relate it to minor eustatic oscillations of sea level during the past 6000 years (Fairbridge, 1974). The deep peat deposits of southern Florida are renowned because there are few places of such extent that are so favorable for their convenient study (Davis, 1940). Buried organic soils also often occur in low-lying coastal areas where there are shifting dune sands and changes in coastal stream and distributary courses (Gadel, 1969).

Coastal Alluvial Soils

Young coastal alluvial soils occur mainly at mouths of estuaries and deltas, where fluvium imperceptibly merges with marine deposits due to tidal action. Soils of these areas are usually poorly drained, and at places their salt content is high. They tend, however, to show a great variability in *electrical conductivity* (q.v., Vol. XII, Pt. 1), which depends on the history of flooding (Saini, 1971). Recent deposits of alluvium occurring in marshes and mangrove swamps are usually subjected to daily flooding by seawater and are little affected by soil-forming processes (Horn et al., 1967). Their texture varies according to their distance from the sea; at the coast, most of the sediment is contributed by tidal waters, which pick up their load from outlying bays and deposit it at high tide. The low velocity of the tide is just enough to transport fine particles forming deposits of silt and clay. Sandier textures occur toward the upper reaches of estuaries and along natural levees.

Older alluvial soils of emerged coastal plains tend to be better developed, showing definite profile differentiation. In places, they are associated with alluvial fans that overlie prior soils and occur farther inland adjacent to fringing uplands. Even though these medium-textured soils are comparatively well drained, some are still subject to flooding causing seasonal swamps. Many alluvial soils feature a *gilgai* (q.v., Vol. XII, Pt. 2) microrelief.

Acid Sulfate Soils

Acid sulfate soils (formerly called *cat clays*) develop from reduced pyritic soils by oxidation of iron sulfide to sulfuric acid. These soils are extremely acid, with pH values ranging from 2 or 3–4.5 (Bloomfield and Coulter, 1973). Millions of hectares of actual and potential acid sulfate soils occur on coastal plains in the tropics, but they are less extensive in middle latitudes (Kawalec, 1973).

An explanation of the formation of acid sulfate soils involves a consideration of the reduction-oxidation cycle. Under reducing conditions, sulfates in the marine environment are reduced bacterially to sulfides (for example, pyrite), and on aeration the sulfides are oxidized again into sulfates. Upon exposure, due to a lowering of sea level or the construction of dikes, pyritic sediments become aerated, favoring the establishment of the aerobic bacterium *Thiobacillus ferro-oxidans*, which because of its ability to oxidize ferrous iron, helps transform the former sea deposits into oxidized soils containing jarosite, a process referred to as *ripening* of the marine muds (Pons, 1975). With each oxidation step there is mineral expansion, causing the soils to heave. If the soils should become inundated by freshwater or saltwater for long periods owing to floods, dike breaches, or the inflow of seepage water, the vegetation is killed, with a return to a reducing environment. The anaerobic bacterium *Desulfovibrio desulfuricans* utilizes the dead plant material and sulfate, and helps reduce jarosite to mackinawite (Ivarson and Hallberg, 1976). With age mackinawite turns to pyrite and the cycle is completed (see Vol. IVB: *Soil Mineralogy*).

Coastal Paleosols and Crusts

Deep weathering profiles in a coastal zone indicate that the region has been exposed to tropical and subtropical weathering regimes over a prolonged period of time. Although mature deep weathering mantles are generally associated with older, stabilized landward margins of coastal plains (Hays, 1967), both incipient and mature profiles sometimes develop in close proximity to the coast (Finkl, 1971; McArthur and Bettenay, 1960), suggesting that some laterites, for example, may be forming at the present time. Lateritic and saprolitic paleosols in the coastal zone can be modified by changes in environmental conditions that result in their truncation, burial, or submergence. Laterites near Darwin and Broome (Australia), for example, have been truncated by coastal erosion and now form massive sea cliffs with the present beach developed on the pallid zone (Hays, 1967). Bauxites now extend under the Gulf of Carpentaria, and laterites on the southwestern Australian coast as well as those on the Bataan Peninsula (Philippines) also range below sea level. Deeply weathered basalts at the Giant's Causeway, Northern Ireland, with preglacial profiles, are today preserved at sea level.

The intercalation of sesquioxidic paleosols bebetween limestone layers on the south coast of Puerto Rico (Alexander and Johnson, 1976) provides another interesting example of dramatic changes in the coastal environment, as is the occurrence of a now submerged subaerial weathering crust in Bermuda (Steiner, Harrison, and Mathews, 1973) that is similar to those presently exposed in the Florida Keys (Multer and Hoffmeister, 1968), the Bahamas (Kornicker, 1958), Morocco, the eastern Mediterranean, southern Arabia, South Africa, and Australia. The occurrence of kunkar and eolianite below sea level in the Spencer and St. Vincent gulfs, South Australia, also shows that soil crust formation was contemporary with sea levels lower than present ones (Jessup, 1967). The occurrence of pedogenic relics in present coastal environments is important for interpretations of regional geomorphology, being indicative of climato- and/or tectono-eustatic changes in sea level (Riding and Wright, 1981). These materials are challenging topics for study because of their geotectonic and paleoclimatic implications.

Submarine Soils

The submergence of existing coastal soils or the accumulation of *soil sediments* results in the formation of submarine soils (Deelman, 1972). The major biopedological factors involved in their formation include the activity of benthic organisms (for example, bioturbation) and the accumulation of organic matter. The inundation of terrestrial soils by seawater results in the destruction of pedological features such as *cutans* (see Vol. XII, Pt. 2: Pedological Features). Submergence is also manifested by changes in soil structure and stratification, in the character of organic matter, in the distribution and mineralogy of iron and manganese compounds, and in the mineralogical composition of the clay fraction (Buurman, 1975; Ponnamperuma, 1972). These modifications in soil properties are largely due to changes in Eh and pH conditions, in the composition of interstitial waters, burrowing activities of marine organisms, and microbial activity in the surface layer of the bottom sediment. Other submarine soils result from the accumulation of coastal and lagoonal sediments, with a major contribution by deposition and humification of organic substances (Gadel, Cahet, and Bianchi, 1975). Aquatic soils are most characteristic of low energy environments under a continental influence, where organic and transported soil materials (such as terrestrial clays) tend to accumulate. Soils submerged permanently tend to be developed in gray-ocher muds, muddy ocher sands, and organogenic gravels.

Use and Reclamation

Even though tidal marshes and swamps of saltwater and brackish-water environments have generally been relegated for use as wildlife habitats and recreational areas (Coultas and Calhoun, 1976), many coastal soils have a high agricultural production potential (Odum, 1971). High quality mangal soils, for example, have been successfully used for growing rice in Sierra Leone, for the cultivation of sugar cane in Puerto Rico and Australia, and for growing coconut palms in Cambodia (Walsh, 1977).

Coastal soils the world-over warrant greater appreciation as sensitive indicators of the well-being of a fragile environment that is coming under the increasing pressure of man's use. Human needs may best be served, as indicated by Coover, Bartelli, and Lynn (1975) and Walsh (1977), by managing coastal soils as a commercial resource—that is, they should not be "reclaimed" indiscriminantly or used as dumping grounds. Once a reclamation process of the type used in Holland, for example, has been completed, the diked and drained area is no longer marsh, and the benefits resulting from the subsequent agricultural activity should not be thought of as coming from the use of marsh soils (Queen, 1977). Similarly, economic benefits resulting from commercial or industrial activity on filled sites is in no way dependent on the natural characteristics and processes of coastal soils.

The reclamation of coastal soils requires careful study prior to the drainage of wetlands or the construction of *polders*. The rate and magnitude of changes in the water table, salt content, and pH indicate that coastal flood plain soils are part of a rapidly changing system. Important inputs to the system, such as water, vary so much with time that the soils may never be in a state of equilibrium as regards soil moisture, reaction, and solute concentration (Walker, 1972). Upon drying, many coastal soils are capable of acid and salt production, may contract forming cracks with attendant subsidence, or become hydrophobic and rewet only with difficulty (Bloomfield and Coulter, 1973). Other soils are subject to: periodic erosion and deposition, as well as flooding, which results in acidification when the soil is exposed and oxidized, and deacidification when submerged and reduced.

CHARLES W. FINKL, JNR.

References

Alexander, C. S., and Johnson, D. L., 1976. Some problematic Late Cenozoic Caribbean paleosols and their geotectonic and paleoclimatic implications, *1976 AAPG-SEPM Convention Abstracts,* 38.

Allen, E. A., and Curran, H. A., 1974. Biogenic sedimentary structures produced in lagoon margin and salt marsh environments near Beaufort, North Carolina, *Jour. Sed. Petrology* 44, 538-548.

Allen, J. R. L., 1965. Late Quaternary Niger Delta and adjacent areas: Sedimentary environments and lithofacies, *Am. Assoc. Petroleum Geologists Bull.* 49, 547-600.

Blackburn, G.; Bond, R.; and Clarke, A. R. P., 1965. Soil development associated with stranded beach ridges in south-east South Australia, *C.S.I.R.O. Australia Soils Pub.* 22, 65p.

Bloomfield, C., and Coulter, J. K., 1973. Genesis and management of acid sulphate soils, *Adv. Agron.* 25, 265-326.

Buurman, P., 1975. Submarine soil formation changing fossil terrestrial soils, *Soil Sci.* 119, 24-27.

Chapman, V. J., ed., 1977. *Wet Coastal Ecosystems.* New York: Elsevier, 428p.

Coover, J. R.; Bartelli, L. J.; and Lynn, W. C., 1975. Application of soil taxonomy in tidal areas of the southeastern United States, *Soil Sci. Soc. America Proc.,* 703-706.

Coultas, C. L., and Calhoun, F. G., 1976. Properties of some tidal marsh soils of Florida, *Soil Sci. Soc. America Jour.* 40, 72-76.

Davis, J. H., 1940. The ecology and geologic role of mangroves in Florida, *Carnegie Inst. Washington Pub. 517,* 303-412.

Deelman, J. C., 1972. Recent and fossil submarine soils, *Soil Sci.* 114, 164-170.

Fairbridge, R. W., 1950. The geology and geomorphology of Point Peron, Western Australia, *Royal Soc. Western Australia Jour.* 34, 35-72.

Fairbridge, R. W., 1974. The Holocene sea-level record in southern Florida, *in* P. J. Gleason, ed., *Environments of South Florida: Present and Past.* Miami: Miami Geological Society, Mem. 2, 223-231.

Fairbridge, R. W., and Teichert, C., 1963. Soil horizons and marine bands in the coastal limestones of Western Australia, *Royal Soc. New South Wales Jour.* 86, 68-86.

Finkl, C. W., Jnr., and Churchward, H. M., 1976. Soil stratigraphy in a deeply weathered shield landscape in south-western Australia, *Australian Jour. Soil Research* 14, 109-120.

Finkl, C. W., Jnr., and Fairbridge, R. W., 1979. Paleogeographic evolution of a rifted cratonic margin: S. W. Australia, *Palaeogeography, Palaeoclimatology, Palaeoecology* 26, 221-252.

Gadel, F., 1969. Etude des substances humiques de sédiments lagunaires et mariens, *Vie et Milieu* 20, 221-256.

Gadel, F.; Cahet, G.; and Bianchi, A. J. M., 1975. Submerged soils in the northwestern Mediterranean Sea and the process of humification, *Soil Sci.* 119, 106-112.

Hays, J., 1967. Land surfaces and laterites in the Northern Territory, *in* J. N. Jennings and J. A. Mabbutt, eds., *Landform Studies from Australia and New Guinea.* Canberra: Australian National University Press, 182-210.

Horn, M. E.; Hall, V. L.; Chapman, S. L.; and Wiggins, M. M., 1967. Chemical properties of the coastal alluvial soils of the Republic of Guinea, *Soil Sci. Sco. America Proc.* **31,** 108-114.

Ivarson, K. C., and Hallberg, R. C., 1976. Formation of mackinawite by microbial reduction of jarosite and its application to tidal sediments, *Geoderma* **16,** 1-7.

Jessup, R. W., 1967. Soils and eustatic sea level fluctuations in relation to Quaternary history and correlation in South Australia, *in* R. B. Morrison and E. Wright, eds., *Quaternary Soils; 1965 INQUA Congress Proceedings.* Reno: University of Nevada, 191-204.

Kawalec, A., 1973. World distribution of acid sulphate soils, *in* H. Dost, ed., *Acid Sulphate Soils.* Wageningen: Institute for Land Reclamation and Improvement (INRL), Pub. 18, 292-295.

Kornicker, L. S., 1958. Bahamian limestone crusts, *Gulf Coast Assoc. Geol. Socs. Trans.* **8,** 167-170.

McArthur, W. M., and Bettenay, E., 1960. The development and distribution of the soils of the Swan Coastal Plain, Western Australia, *C.S.I.R.O. Australia Soils Pub.* **16,** 55p.

Meade, R. H., 1969. Landward transport of bottom sediments in estuaries of the Atlantic Coastal plain, *Jour. Sed. Petrology* **39,** 222-234.

Mulcahy, M. J., and Churchward, H. M., 1973. Quaternary environments and soils in Australia, *Soil Sci.* **116,** 156-169.

Multer, H. G., and Hoffmeister, J. E., 1968. Subaerial laminated crusts of the Florida Keys, *Geol. Soc. America Bull.* **79,** 183-192.

Odum, W. E., 1971. Pathways of energy flow in a south Florida estuary, *Univ. Miami, Sea Grant Tech. Bull. No. 7,* 162p.

Phleger, F. B., 1977. Soils of marine marshes, *in* V. J. Chapman, ed., *Wet Coastal Ecosystems.* New York: Elsevier, 69-77.

Phleger, F. B., and Bradshaw, J. S., 1966. Sedimentary environments in a marine marsh, *Science* **154,** 1551-1553.

Ponnamperuma, F. N., 1972. The chemistry of submerged soils, *Adv. Agron.* **24,** 29-96.

Pons, L. J., 1975. Acid sulphate soils, *in* E. Dost, ed., *Acid Sulphate Soils.* Wageningen: International Institute for Land Reclamation and Improvement (INRL), Pub. 18, 3-27.

Queen, W. H., 1977. Human uses of salt marshes, *in* V. J. Chapman, ed., *Wet Coastal Ecosystems.* New York: Elsevier, 363-368.

Riding, R., and Wright, V. P., 1981. Paleosols and tidal-flat/lagoon sequences on a Carboniferous carbonate shelf: sedimentary associations of triple disconformities, *Jour. Sed. Petrology* **51,** 1323-1339.

Saini, G. R., 1971. Chemical and physical properties of coastal alluvial soils of New Brunswick, *Geoderma* **5,** 111-118.

Semeniuk, V., and Meagher, T. D., 1981. Calcrete in Quaternary coastal dunes in southwestern Australia: A capillary-rise phenomenon associated with plants, *Jour. Sed. Petrology* **51,** 47-68.

Servant, J., and Favori, J. C., 1966. Sur les sols salés du littoral Languedoc-Roussillon: essai de classification, *in Transactions of the Conference on Mediterranean Soils.* Madrid: Sociedad Española de Ciencia del Suelo, 1-6.

Steiner, R. P.; Harrison, R. S.; and Mathews, R. K., 1973. Eustatic low stand of sea level between 12,500 and 105,000 B.P.: Evidence from the submergence of Barbados, West Indies, *Geol. Soc. America Bull.* **84,** 63-70.

Walker, P. H., 1972. Seasonal and stratigraphic control in coastal floodplain soils, *Australian Jour. Soil Research* **10,** 127-142.

Walsh, G. E., 1977. Exploitation of mangal, *in* V. J. Chapman, ed., *Wet Coastal Ecosystems.* New York: Elsevier, 347-362.

Cross-references: *Coastal Dunes and Eolian Sedimentation; Coastal Sedimentary Facies; Aeolianite; Flandrian Transgression; Humate; Land Reclamation; Mangrove Coasts; Polder; Sediment Transport.* Vol. IVB: *Soil Mineralogy;* Vol. XII, Pt. 1: *Electrical Conductivity; Leaching; Redox Reactions; Soil Science;* Vol. XII, Pt. 2: *Calcareous Soils; Gilgai; Paleosols; Weathering Crusts.*

SOLUTION AND SOLUTION PAN

Solution pans are small-scale, surface weathering features, and are not a feature of coastal environments exclusively (Bird, 1969; Davies, 1973). Solution pans can be found in most climatic belts, in such diverse rock types as basalt, granite, limestone, or quartz arenite. Their formation can be attributed to essentially *biochemical solution* and accompanied induration. The pans develop from small solution pits or depressions, which range from a few millimeters to several centimeters in diameter. The pans extend laterally in all directions from the original solution pit, retaining a smooth, shallow surface. Large solution pans grow to approximately 2 m in diameter, and frequently have undercut rims.

Solution pans found in coastal environments are frequently present on limestone or lime-cemented rocks. They are found on *shore platforms,* ranging from the *spray zone* down to the low tide limit. As is the case with most coastal, solution features, solution pans are best developed in low wave energy environments. In areas of strong wave energy or where rock tools are available, *abrasion* will be the dominant geomorphic process, and will tend to destroy any solution features that might otherwise be developing. A prime example is the chalk formation of Dover, in which flint nodules and wave action are continuously abrading a smooth intertidal platform.

The solution of limestone by water, in the presence of carbon dioxide, into calcium bicarbonate, is represented by the following equation: $CaCO_3 + H_2O + CO_2 \rightarrow Ca(HCO_3)_2$. Rainwater or groundwater containing dissolved carbon dioxide will cause the solution of limestone, provided the water is not already saturated with calcium bicarbonate and therefore incapable of dissolving more limestone. Although rainfall may account for some solution on limestone coasts, it cannot be responsible for all solution. This is demonstrated by the occurrence of many solution features on essentially arid coasts.

It has been found that diurnal variations in the dissolving capacity of sea water can range from extreme values of pH 6.5 to 10. A nocturnal increase in carbon dioxide can result in two ways. Nocturnal cooling of seawater increases its capacity to take in carbon dioxide, which is more soluble at lower temperatures. Carbon dioxide is also accumulated at night from plants (chiefly *algae*) when the lack of sunlight halts photosynthesis. The increase in acidity, caused by the nocturnal accumulation of carbon dioxide, causes limestone to go into solution as calcium bicarbonate. During the daylight hours carbon dioxide is removed from the system by photosynthesis and an increase in temperature, causing basic conditions to prevail. Calcium bicarbonate is then reprecipitated back to limestone.

The flushing action of waves and tidal currents removes the reprecipitated limestone as the tide fluctuates, leaving the solution feature. The nocturnal solution process, mainly operating within the tidal zone, will continue to proceed until it reaches the level of permanently saturated calcium carbonate, usually near the low tide level.

ROBERT T. SIEGFRIED

References

Bird, E. C. F., 1969. *Coasts*. Cambridge, Mass.: M.I.T. Press, 246p.

Davies, J. L., 1973. *Geographical Variations in Coastal Development*. New York: Hafner, 204p.

Cross-references: *Alveolar Weathering; Erosion Ramp, Wave Ramp; Muricate Weathering; Salt Weathering; Shore Platforms; Shore Terrace; Splash and Spray Zones; Tafone; Weathering and Erosion, Biologic; Weathering and Erosion, Differential; Weathering and Erosion, Mechanical.* Vol. III: *Solution Pits and Pans.*

SOUNDING SANDS

When in motion, some sands emit sounds that are audible to the human ear. Such sands have been described from several parts of the world and have been variously called *roaring, booming, squeaking, singing, musical* and *sounding sands*. The quality of noise differs from place to place and time to time. Early in the fifteenth century the Emperor Baber described sands which sounded like "drums and nagarets"; Thesiger (1959) records moving sands from southern Arabia that sounded like a "low vibrant humming" but that increased in volume until it sounded like an airplane flying low overhead; Miller (1858) describes sands on the Island of Eigg in western Scotland as making a shrill and sonorous sound; and Gibson (1946) likened the sound made by moving sands on a beach on Boston Island in southern Spencer Gulf, South Australia, to that produced by "a whistling kettle just before the water is thoroughly boiling."

The sounds made by moving sands are produced by the impact of sand grains on other sand grains. The quality of sound produced varies according to the velocity, volume, and purity of the sands involved. Not all moving sands are sonorous, but those that do produce sounds are remarkably well sorted and uniform in size. With only one known exception, sounding sands are less than 0.5 mm in diameter, but the sound produced nevertheless varies with the precise size of particles. The noises produced by sands of varied grades from Boston Island differed as given in Table 1. The shape of sand grains is apparently not significant in this regard, nor are the interstitial air spaces, for it has been shown that musical sand retains its acoustical properties when moved *in vacuo*. Neither heating nor cooling is significant. Most sounding sands are of quartz, but carbonate sands from coral islands possess similar qualities. However, experimental work suggests that musical quartz grains have greater elasticity than those that are acoustically dead.

Contamination of musical sand by dust and moisture is significant. Sand that sings on the beach will, if kept dry and dust-free, retain its properties for a number of years and possibly for an indefinite period. If the same sand is exposed to dust and moisture, it quickly loses its acoustic properties but, if washed and dried, becomes acoustically alive again.

Thus, although many of the properties necessary to make singing sands are known,

TABLE 1. Comparison of Sand Size and Sound Produced (after Gibson, 1946)

Size (mm)	Nature of Sound
< 0.251	Poor sound
0.251–0.295	Feeble squeak
0.295–0.353	Good musical sound
0.353–0.422	Higher-pitched sound
> 0.422	Lower pitch

the ultimate cause of the phenomenon is as yet unexplained.

C. R. TWIDALE

References

Gibson, E. S. H., 1946. Singing sand, *Royal Soc. South Australia Trans.* 70, 35-44.
Miller, H., 1858. *The Cruise of the Betsey.* Edinburgh: T. Constable and Co., 486p.
Thesiger, W., 1959. *Arabian Sands.* London: Longmans, 326p.

Cross-references: *Beach Material; Coastal Dunes and Eolian Sedimentation; Sand; Sediment Size Classification.* Vol. III: *Singing Sands.*

SOUTH AMERICA, COASTAL ECOLOGY—
See CENTRAL AND SOUTH AMERICA, COASTAL ECOLOGY

SOUTH AMERICA, COASTAL MORPHOLOGY

Investigation into the coastal geomorphology of Latin America does not have a lengthy history, but there are some indications that interest by Latins and others is turning to this important area (Psuty, 1970). Large portions of the coastal zone remain unstudied in detail, and frequently only general descriptions exist (Dolan et al., 1972; Putnam et al., 1960). Questions about regional correlations of depositional and erosional features must await basic research into topical problems. However, despite the paucity of extensive inquiry, the following interpretation of the coastal geomorphology of South America is presented as a base upon which to build many future layers of information.

Colombia

The Guajira Peninsula is the northernmost arm of the Andean mountain system. The extreme northeastern tip is fronted by an active coral reef. Some evidence has been presented (Anderson, 1927) that there are raised coral platforms and fossil mollusks representing tectonic displacement of the land terminus. Sandy beaches without accompanying coral line the northwestern margin of the peninsula southward to Cape San Juan de Guia, where crystalline cliffs occur with pocket beaches.

South of Santa Clara the coastal zone is dominated by the delta of the Magdalena River. There is a broad arcuate delta in this area with several active distributaries. Great quantities of sediment are transported to the ocean to be reworked. Mangrove-covered mud flats are characteristic of this area, as is beach-ridge and chenier topography. The fluctuating activity of the distributaries is responsible for episodes of accretion and erosion of the shoreline. Beach ridges and cheniers in the delta demonstrate the old shoreline alignments, and their truncated forms give evidence of reorientation of the shoreline in past times.

Near the Colombian-Panamanian border the Atrato delta provides the source of sediments that form the shore features, but there is little change. Mangrove forest, distributary channels, and sand ridges along the distal margins continue to characterize the shoreline.

The Pacific coast of Columbia begins as a high cliffed coast at the Panamanian border. A portion of the flank of the Andes comes to the sea to create steep, vegetation-cloaked precipices plunging directly into the water. South of Cape Corrientes, short drainage systems leading from this steep mountainous ridge have been successful in building a narrow coastal plain. The shoreline consists of these fluvially derived sediments broken by river mouths or stream channels. With the exception of three places where the shoreline is cliffed (West, 1956), there is a narrow sandy beach lying between the ocean and a dense mangrove forest. Large mud flats occur in front of the beaches and in front of the fluvial plain at the foot of the mountains. Near Buenaventura and Tumaco deltaic forms line the coast, and change the landform patterns slightly by filling the zone behind the fringing beach.

Ecuador

The northern coast of Ecuador is drenched with rainfall, and supports a dense vegetation down to the shoreline. However, at the position of the southern margin of the country, the coast is stark and almost devoid of all vegetation. At both extremes, the coast consists of high cliffs fronted by a sandy fringing beach or pocket beaches. The high cliff extends consistently from near the border with Colombia southward to near Santa Elena. Near the latter location the shoreline takes a right angle bend and trends southeastward. Along this portion of the shore Sheppard (1930) describes a series of marine terraces rising steplike to an elevation of 60 m. He interprets these features as evidence of tectonic displacement.

The Gulf of Guayaquil is an estuarine system whose shoreline is fringed by dense stands of mangrove on mud flats. Near the distal margins of the estuary, beach-ridge systems compose several of the exposed points. However, the

Peru

The hyperaridity of coastal Peru is responsible in large part for the specific geomorphologic features found there. The lack of precipitation at present limits the contribution of fluvially transported sediments to the littoral zone, and thus limits the development of beaches. During the Holocene, as sea level was rising and meltwaters were coursing through the river valleys leading to the Pacific Ocean, there were great quantities of sediment discharged to be reworked by the waves and currents. Large sediment volumes were sent into the small estuaries at the river mouths and onto the narrow continental shelf. However, only a few rivers continuously reach the sea at present, and the situation has changed from that of several thousand years ago. The rise of sea level has continuously encroached upon the small estuarine locations, and scattered tectonic displacement has produced further alterations of the shoreline features.

Peru is characterized by the presence of a cliffed shoreline interspersed with pocket beaches and beaches fronting the river mouths. The pocket beaches tend to be located where more easily eroded rock formations of sandstone, shale, or marl are exposed at the shoreline. The rocky promontories are usually composed of crystalline rock units. In a few places the cliffed coasts are cut into colluvial deposits that have moved down spacious interfluves during the Quaternary. At locations where the cliffs are in colluvial gravels and cobbles, a narrow fringing beach usually exists.

South of Pisco there is little or no coastal plain, and the shoreline usually consists of towering cliffs or small crescent-shaped embayments with narrow beaches. From Pisco north to Chiclayo there are a number of rivers that reach the sea and contribute to the development of localized coastal plains. According to Parsons and Psuty (1975), the mouths of the major river valleys have had a similar geomorphic history and a similar assemblege of landforms. Archeologic evidence suggests that the fluvial plains found in these river mouths obtained their present characteristics about 1000-1500 yrs. B.P. At that time the embayments were filled, and the leading margin of the fluvial plain included a classic beach/dune profile. The profile has been migrating inland since that period over a narrow backmarsh that has formed between the beach and the fluvial plain.

Coastal sand dunes reach impressive size along the central portion of the coast. At the southern margins of the valleys, there is frequently a dynamic beach area where sand is being moved inland by the prevailing winds out of the southwest. The coastal dune ridge is frequently enlarged in this area and transformed into longitudinal dunes extending inland. In some instances, barcan dunes break off the sand sheet and migrate independently across the terrace surfaces.

Shoreline displacement is evidenced in many locations of coastal Peru by raised shore platforms and raised beaches. Investigators have identified as many as ten coastal terraces at elevations of up to 250 m at the mouth of the Ica River (Broggi, 1946) and up to 75 m (250 feet) in the Sechura Desert area. However, there is some question whether the displacement is part of regional uplift or local movement. Several detailed inquiries (Parsons and Psuty, 1975; Craig and Psuty, 1968) have shown that the movement is local because the platforms do not extend great distances. Further, flanking terraces in the valleys are Pleistocene depositional features. Most of the major river valleys contain no evidence of post-stillstand uplift (Psuty, 1978).

In northern Peru, Richards and Broecker (1963) have collected marine shells from several terraces that imply emergence of the platforms. A low terrace at 4.5 m (15 ft) has shells collected on its surface dated at 3000 yrs. B.P., whereas a high terrace of 75 m (250 ft) had shells assayed beyond 30,000 yrs. B.P.

Chile

The northern two-thirds of coastal Chile has characteristics similar to those of Peru. The Andean coastal range comes down to the sea, and the cliffed coast is interrupted by small pocket beaches or alluvial embayments where infrequent streams lead down to the shoreline. Numerous terraces appear along the cliffed headlands fronting the foothills. According to Börgel (1967), there are a series of steps reaching to 200 m thought to be terraces of abrasion north and south of Valparaíso. Paskoff (Fuenzalida et al., 1965) believes that the highest terraces found in northern Chile at 250-400 m are probably Pliocene, whereas those below 250 m are likely to be Pleistocene. Numerous investigators (Fuenzalida et al., 1965) have identified the considerable number and variety of terraces found in coastal Chile. It is unlikely that the terraces can be considered wholly as products of eustatic sea level changes, as was suggested in the early investigations. Rather, the lack of uniformity of number, elevation, and kind points to localized tectonic events. However, one terrace at the 85/100 m level does appear to persist throughout much of northern and central coastal Chile. It is considered to have extensive deposits of marine gravel on its surface and also marine mollusks. It is possible that this surface may represent an episode of regional displacement.

Rainfall increases southward in Chile, but there are no major coastal alluvial plains developing. The coast range comes to the sea and provides for only modest embayments. Pocket beaches prevail south from Valparaíso to Chiloé. However, the shores near the mouth of the Bío-Bío River and south of the Mataquito River are well developed with sizable active dunes penetrating inland. Further, there is estuarine development at Valdivia, where a small fluvial plain has accumulated at the mouth of a river and a narrow belt of coastal features bounds its shoreward margin. Frequently some coastal dunes are part of the coastal phenomena, and there is evidence that the coastal unit has been migrating inland over the fluvial surface. The inland shift is most probably the result of a negative sediment budget aided by the slow eustatic rise of the sea.

South of the latitude of Puerto Montt the coastal configuration is dramatically changed. The coastal range becomes broken. The island of Chiloé retains many of the characteristics of the northern shoreline, but to the south the coast is altered by the processes of glaciation. A fiord coast is present south of 43°S. latitude with many channels extending entirely across the crestal portion of the coastal range into the interior passage. Many of the pocket beaches are cobble-strewn and exist only in sheltered areas.

Weischet (1959) reports that southern Chile has several terrace levels, but they are not continuous through the region. They may be dissected remnants of a larger surface or products of local tectonic activity. However, the degree of glaciation is so thorough in this area that it is the characteristics of glaciation that must be considered rather than those of coastal processes.

Argentina

Argentina is characterized primarily by a cliffed shoreline with a narrow beach zone before it. The cliffs vary from only a few meters to the spectacular elevations of greater than 500 m south of Comodoro Rivadavia. Occasional areas of sedimentary accumulations do exist either as beaches fronting the cliffs or as substantial areas of beach ridges and coastal dunes.

The Río de La Plata estuary dominates the northern portion of the Argentinian shoreline. From the mouth of the river to Cabo San Antonio, the shoreline is a tidal mud flat. Wave energies are low, and the fine-grained sediments derived from the fluvial system are not reworked into a beach form. However, Urien (1972) has interpreted beach ridge and chenier forms that were created in the period of 7000-3000 yrs B.P. along the southern shore of the estuary. These features were part of the sand wedge that was being pushed up the continental shelf as sea level was rising, and they developed prior to the silting of the estuary. Urien suggests that tectonic displacement of these forms has raised them to elevations of 9 m, where they and a marine terrace form the margin of Samborombón Bay. From Mar del Plata to Bahía Blanca, the coast consists of a low cliffed shoreline with a narrow beach before it. Occasionally there are large dune fields leading from the beach. This portion of Argentina is an extension of the Pampas coming to the shoreline. South of Bahía Blanca the Negro and Colorado rivers transport considerable quantities of sand to the shoreline, and the beaches are extremely broad. The alluviation is not complete, however, for there are rocky islands and points located between these two river mouths. The Colorado delta is extensive, and there is evidence of coastal aggradation in the form of beach ridges and distributary elongation. The Negro delta is not nearly so extensive; it manages to fill its estuary.

With the exception of well-developed beaches and associated landforms at the Gulf of San Marcos and the Gulf of San Jorge, the southern half of Argentina is a cliffed shoreline. Investigators have identified a number of terraces that have been used to describe tectonic or eustatic displacement. Terraces ranging to 140 m have been noted in Patagonia. A 9 m terrace at Comodoro Rivadavia has been dated at 3000 yrs B.P. (Richards and Broecker, 1963), and has been used to suggest local tectonic movement. Urien (1970) has indicated that a 9 m terrace also exists in Tierra del Fuego along parts of the Río de La Plata estuary and bordering the Paraná delta. However, he cautions against attempting any broad correlations.

From Santa Cruz to the eastern tip of Tierra del Fuego the cliffs are cut into glacial morainic material. Occasional outcrops of bedrock are noted, as are pocket beaches (Etchichury and Remiro, 1967). At Punta Dungeness there is an excellent series of beach ridges created where currents converge at the point. The more protected western margin shows considerable accretionary history, whereas the exposed southeastern flank gives evidence of truncation of the ridges and extensive dune fields extending inland.

Uruguay

Coastal Uruguay is extremely diverse for such a relatively short coastline. The northern area consists of an extension of the barrier island system of southern Brazil. The sand beach continues in Uruguay but narrows and becomes discontinuous, so that it becomes a series of sandy embayments set into the crystalline Brazilian Shield. These embayments are not products of marine processes but, rather, the prior irregular topography of the shield encroached upon by a rising sea. In several places the embayments contain small lagoons behind a sand barrier. This feature is the product of incomplete filling of small estuaries along the margin of the shield (Chebataroff, 1960).

From near Maldonado westward, the shoreline is the margin of the Rıo de La Plata estuary. For nearly this entire length there is a cliffed shoreline with a sand beach lying at its base. Delaney (1963, 1966) suggests that there are terraces cut into this cliff. The estuary is receiving little sand, and it is proposed by Urien (1970) that this sand is the product of local erosion of the bluffs rather than fluvial transport. The exposure of this portion of Uruguay to the southeast would tend to allow greater wave energies to reach this shore and favor the necessary erosion and sorting responsible for sandy beaches, while much of the rest of the estuary is a shallow muddy tidal flat. Occasionally mud flats dominate along the cliffed shoreline.

Brazil

The great size of Brazil allows for considerable diversity of coastal exposure and geomorphologic development. There are three principal portions of the shore. The first is the area influenced by the Amazon River and its sediments; the second is the narrow coastal margin fringing the huge Brazilian Shield, creating an escarpment nearly adjacent to the ocean; the third is the southern area, where considerable quantities of sediments have accumulated to provide a barrier island formation.

The mouth of the Amazon River is a great estuary stretching for nearly 1000 miles. Large quantities of sand and especially silt and clay are discharged by the river and accumulate along the shore margins. From the border with Surinam eastward to the Gulf of San Marcos, the fine-grained sediments blanket the shoreline and are cloaked with mangrove. The Gulf of San Marcos is another, much smaller, estuary that similarly contributes large quantities of fine-grained sediment.

East of the Gulf of San Marcos the shoreline begins to be characterized by sandy beaches lying before low hills. The sand beaches are interspersed with mangrove stands that dominate where local deltaic buildout occurs in association with short drainage systems leading off the eastern margin of the Brazilian Shield. Beginning in Rio Grande do Norte and continuing southward to the coastal margin of Alagoas state, the beach zone is severely attenuated. The dry climate and the short drainage systems limit the transport of sediment to the ocean margin. Further, this portion of Brazil is bordered by a fairly extensive coral reef. Where beach sediments have accumulated, there are also likely to be well-developed dune forms migrating inland over terrace surfaces.

South of Recife the coast is cliffed. The combination of cliffed coast and the presence of coral reef extends for about 300 miles. Near Recife a small promontory has been investigated for evidence of a high sea level. Van Andel and Laborel (1964) have dated a fossil biogenous limestone that is encrusting granite a few meters above modern sea level as having been active 1200 to 3600 yrs B.P. The authors have interpreted these dates to hypothesize that sea level was higher in that time period.

The sandy beach backed by an escarpment begins near the Alagoas-Sergipe border and continues south to Rio Grande do Sul. The beach often broadens in large curvilinear embayments, and there may be local mangrove stands, beach ridges, and deltaic buildout. However, the escarpment dominates the horizon, and the coastal geomorphic features occupy a small niche on the continental margin. In the state of Paraná there is an extensive area of beach ridge development associated with the Maciel River. These ridges appear to resemble cheniers in that they are bounded by clayey deposits rather than forming a broad sandy surface. The beach ridges attain elevations of 10 m in their interior location, and gradually decrease to elevations of 2–3 m near the shore. Bigarella (1965) believes this is further evidence for a fluctuating and generally falling sea level.

The coastal margin of the state of Rio Grande do Sul is distinct from the rest of Brazil—it consists of a classic barrier island-lagoon sequence. Delaney (1963, 1966) describes the geologic origin of the barrier island sequence as occupying a unique position in a geologic depositional basin with sediments being transported into it from several directions. Certainly the positive sediment budget that existed in this area with the changing sea level caused the particular assemblage of broad sandy beach extending along the coast for 640 km and incorporating broad beach ridge systems and large coastal dunes reaching 25 m in elevation. The northern margin of this Holocene coastal plain comes against a terrace surface with elevations of 15 m. It is probably of Pleistocene age, but whether its origin is wave-cut is unknown.

Guyana, Surinam, French Guiana

The coast of what was once referred to as the European Guianas is somewhat similar for its entire length of over 1100 km. Basically, the shoreline is the product of massive quantities of fine-grained sediments discharged by the Amazon River that proceed to drift westward. Some small quantities of sand are contributed by the Amazon and by the smaller streams leading from the Guiana Shield to the shoreline.

The beach and inland coastal geomorphology of the Guianas is characterized by the active development of *cheniers* (q.v.). These ridged features are coarse-grained deposits of sand and shell accumulating on a mud or clay foundation. Intermittent development of these coarser accumulations creates a series of sandy ridges bordered on either side by the fine-grained sediments. At times the shoreline is a broad mud flat rather than a sandy beach. Several investigators (Vann, 1959; Zonnenfeld, 1954; Wells and Coleman, 1978) have described the massive clay waves that migrate along this shoreline and extend far out into the ocean.

There is a type of pattern to the chenier or sand ridge development along this coast. A kind of apex forms near the west bank of the river mouth from which a series of ridges extends. The number and extent of the ridges decrease westward to the point where the coarser sediments no longer are found. Some of the river mouths, such as those of the Marowijne and the Suriname, show evidence of west bank progradation for several kilometers by means of mud flat development interspersed with sandy ridges. The ridge trends also give evidence that the progradation sequence has not been continuous because numerous ridges are truncated and new fulcrums have developed from which a fan-shaped series of ridges has spread.

Much of the mud flat area is colonized by mangrove. These trees line the shores in places as well as occupy the troughs between the ridges. Wave action appears to be attenuated by the extremely turbid waters, and thus the coastal clays are only infrequently disturbed. However, during those infrequent higher wave energies, there is considerable movement of the clay waves. Large units of clay are displaced, and the sand and shell are sorted to accumulate as a beach on top of the clay base. These beaches continue to develop to form the cheniers.

Venezuela

Coastal Venezuela tends to be dominated by the northern terminus of the Andean mountain system. The distal portion of the Andes splits into two north-trending prongs and creates the depression occupied by Lake Maracaibo and the Gulf of Venezuela region between them. The eastern prong makes a sharp bend due east and establishes the northern margin of the country for most of its length. Modifications of the mountainous coastal topography are caused by deltaic development and by breaks in the Andean ridges.

The eastern coast of Venezuela is completely the product of the Orinoco River and its deltaic forms. A fairly large arcuate delta occupies the position from the Gulf of Paria south to the border of Guyana (van Andel, 1967). The southern third of the delta tends to be a series of coalescing distributaries that retain their fluvial forms as they discharge into the Atlantic. However, the northern two-thirds of the delta is lined with cheniers forming a well-developed shoreline. Mangrove forests occur at the frontal lobes of the southern margin of the delta and at the northern margin. However, in the area of chenier development, the mangrove is in the protected troughs between the ridges rather than at the exposed coast.

The Paria Peninsula is bounded on its southern side by sand beaches and fairly broad mud flats. The sand beaches are found in association with small streams that drain the high peninsula and contribute their coarser sediments. The eastern half of the south side of the peninsula is a steep rocky cliffed coast, as is the entire northern portion of this Andean extension. There are some pocket beaches, but not until the Gulf of Barcelona is a well developed beach found. The shore consists of a broad curvilinear beach that has characteristics of barrier island development because several lagoons are formed behind it.

Westward beyond the Barcelona embayment, a cliffed coast appears once again with a number of pocket beaches. A well-developed sandy beach is located at the Triste Gulf, and there is a sizable tombolo connecting the Paraguana Peninsula with the mainland. Along the Gulf of Venezuela most of the shoreline is sheltered from marine processes, and thus the features are of fluvial origin and the shores are marshy or lined with mangrove. However, that portion of the gulf exposed to swell waves from the northeast does have good coastal forms. The exposure to the northeast is also the dominant fetch direction for the tombolo to Paraguaná and the Triste Gulf beaches as well.

Tanner (1970) has shown that the western shore of the Gulf of Venezuela has a barrier island formation with a series of beach ridges prograding seaward over a distance of 7 km. The ridges are low, only about 1 m local relief, and there have been several interruptions in the accumulation sequence. These erosional breaks in beach ridge development are marked by longitudinal and parabolic dunes at the erosional shoreline whose form extends inland over older beach ridges.

NORBERT P. PSUTY
CHIZUKO MIZOBE

References

Anderson, F. M., 1927. Nonmarine tertiary deposits of Colombia, *Geol. Soc. America Bull.* **38**, 591-644.
Bigarella, J. J., 1965. Sand-ridge structures from Paraná coastal plain, *Marine Geology* **3**, 269-278.
Börgel, R., 1967. Correlaciones fluviomarinas en la desembocadura del Rio Choapa, *Informaciones Geográficas* **13-14**, 55-68.
Broggi, J. A., 1946. Las terrazas marinas de la Bahía de San Juan en Ica, *Soc. Geol. Peru Bol.* **19**, 21-33.
Chebataroff, J., 1960. Sedimentacion Platense, *Inst. Estudios Superiores Rev. Año 4, No. 7*, 544-566.
Craig, A. K., and Psuty, N. P., 1968. *The Paracas Papers: Studies in Marine Desert Ecology.* Boca Raton, Fla.: Florida Atlantic University, 196p.
Delaney, P. J. V., 1963. *Quaternary Geologic History of the Coastal Plain of Rio Grande do Sul, Brazil.* Baton Rouge, La.: Louisiana State University Press, Coastal Studies Series No. 7, 63p.
Delaney, P. J. V., 1966. *Geology and Geomorphology of the Coastal Plain of Rio Grande do Sul, Brazil and Northern Uruguay.* Baton Rouge, La.: Louisiana State University Press, Coastal Studies Series No. 15, 58p.
Dolan, R.; Hayden, B.; Hornberger, G.; Zieman, J.; and Vincent, M., 1972. *Classification of the Coastal Environments of the World, Part 1: The Americas.* Charlottesville, Va.: University of Virginia, Department of Environmental Sciences, 163p.
Etchichury, M. C., and Remiro, J. R., 1967. Los sedimentos litorales de la provincia de Santa Cruz entre Dungeness y Punta Desengaño, *Mus. Argentino Cienc. Nat.–Geologia* **6**, 323-376.
Fuenzalida, H.; Cooke, R.; Paskoff, R.; Segerstrom, K.; and Weischet, W., 1965. High stands of Quaternary sea level along the Chilean coast, *in* H. E. Wright, Jr., and D. G. Frey, eds., *International Studies on the Quaternary.* Boulder, Colo.: Geological Society of America, 473-496.
Parsons, J. R., and Psuty, N. P., 1975. Sunken fields and prehispanic subsistence on the Peruvian coast, *Am. Antiquity* **40**, 259-282.
Psuty, N. P., 1970. Contributions to the coastal geomorphology of Latin America, *in* B. Lentneck, R. L. Carmin, and T. L. Martinson, eds., *Geographic Research on Latin America: Benchmark.* Muncie, Ind.: Ball State University, 250-264.
Psuty, N. P., 1978. *Peruvian Shoreline Stability/Instability during Stillstand of Sea Level.* Lagos: International Geographical Union Regional Conference, Commission on Coastal Environments, 19p.
Putnam, W. C.; Axelrod, D. I.; Bailey, H. P.; and McGill, J. T., 1960. *Natural Coastal Environments of the World.* Los Angeles: University of California, Department of Geography, 140p.
Richards, H. G., and Broecker, W., 1963. Emerged Holocene South American shorelines, *Science* **141**, 1044-1045.
Sheppard, G., 1930. Notes on the climate and physiography of southwestern Ecuador, *Geog. Rev.* **20**, 445-453.
Tanner, W. F., 1970. Growth rates of Venezuelan beach ridges, *Sed. Geology* **6**, 215-220.
Urien, C. M., 1970. Les rivages et le plateau continental du Sud du Brésil, de l'Uruguay ey de l'Argentine, *Quaternaria* **12**, 57-69.
Urien, C. M., 1972. Rio de la Plata estuary environments, *Geol. Soc. America Bull.* **133**, 213-234.
Van Andel, Tj. H., 1967. The Orinoco delta, *Jour. Sed. Petrology* **37**, 297-310.
Van Andel, Tj. H., and Laborel, J., 1964. Recent high relative sea level stand near Recife, Brazil, *Science* **145**, 580-581.
Vann, J. H., 1959. *The Physical Geography of the Lower Coastal Plain of the Guiana Coast.* New Orleans: Louisiana State University, Department of Geography and Anthropology, 91p.
Weischet, W., 1959. Geographisches beobachtungen auf einer forschungsreise in Chile, *Erdkunde* **13**, 6-21.
Wells, J. T., and Coleman, J. M., 1978. Longshore transport of mud by waves: Northeastern coast of South America, *Geologie en Mijnbouw* **57**, 353-359.
West, R. C., 1956, Mangrove swamps of the Pacific coast of Colombia, *Assoc. Am. Geographers Annals* **45**, 98-121.
Zonnenfeld, J. I. S., 1954. Waarnemingen langs de kust van Surinam, *Koninkl. Nederlandsch Aardrijksk. Genoot. Deel Tijdschr.* **71**, 18-31.

Cross-references: *Antarctica, Coastal Morphology; Central America, Coastal Morphology; Central and South America, Coastal Ecology; Chenier and Chenier Plains; Coral Reef Coasts; Mangrove Coasts.* Vol. VIII, Pt. 1: *South America.*

SOVIET UNION, COASTAL ECOLOGY

The coasts of the Soviet Union, washed by three oceans, are situated in different landscapes—geographical zones—thus offering a fair opportunity to investigate coastal marine biomes distributed in the region, ranging from the subtropical waters of the northwestern part of the Pacific and the Black Sea to the high latitudes of the Polar basin.

Regional Setting

By modern classification, the coastal waters of the Soviet Union occupy four major biogeographical regions. The coastal Asian waters from the shoals of Tokyo Bay and Bozo peninsula of Honshu, and those in the latitudes of 38°N off the Korean peninsula in the south up to the Bering Strait in the north, belong to the **Pacific boreal region**. (The American coastal waters from the Bering Strait down to the state of Washington belong to this region as well.) This region is divided into two subregions: low-boreal and high-boreal. The border between them is Southern Kuril islands near Iturup and Terpenija Bay (near Sakhalin). The coastal *Pacific low-boreal subregion* includes the northwestern part of the Sea of Japan, the southern part of the Okhotsk Sea (to Cape Terpenija in Sakhalin and the region of Iturup Island in the north), the coastal waters of the South Kuril Islands, of Hokkaido, and the northeastern part of Honshu. However, in the protected bays of the latter area where the water is efficiently heated in summer (up to 25-26°C), the biogeographical compostion of populations has a subtropical character, so that it has much more similarity with the populations of the coastal Pacific-Asian subtropical subregion of the West Pacific region, which lies further to the south.

In unprotected open districts of the coastal low-boreal subregion the summer water temperature does not rise above 16-20°C. In winter the protected bays are covered with solid ice, while in the open districts drift ice seems to be more common. The temperature of surface waters from December until April is generally everywhere below zero. In the majority of protected bays salinity is, as a rule, no lower than 28‰, while in open districts it is about 32-33‰.

No marked differences can be traced between the biogeographical composition of the shoal populations of the open districts and that of the protected districts in coastal Asian waters of the *Pacific high-boreal subregion*. The temperature of surface waters is never above 8-12° C. The winter hydrological regime in coastal Pacific high-boreal waters is similar to that in the low-boreal subregion. In the majority of districts in the high-boreal subregion salinity is about 32-33.5‰.

The **Arctic region** (defined by the distribution of different groups of invertebrates and algae) in the coastal waters of the Soviet Union borders near the Pacific Ocean, near the southern part of Anadyr Bay or Bering Strait, and near the Atlantic Ocean in the eastern part of the Barents Sea (bordering on the Novaya Zemlya Islands), northern Iceland, western Greenland and Newfoundland Island. This region is divided into three provinces: the Eurasian province (Greenland, Barents, Kara and Laptev seas), the Amerasian province (Vostochno-Sibirsk, Chuckchi and Beaufort seas) that is divided near Novosibirskije Islands and 90°W off Arctic Canada, and the estuary-arctic interzonal provinces. This latter province includes the regions of mixed high-arctic and fluvial waters near the large rivers in the Siberian arctic seas, Anadyr Bay, White Sea, and Baltic Sea.

The temperature of high-arctic surface waters remains below zero the entire year, and even at the height of hydrological summer it does not rise above -0.4-0.6°C. In high latitudes (near the northern islands of the archipelago of Severnaya Zemlya, Franz Josef Land, and De Long) the ice cover stays solid the whole year round, while to the south the surface of the sea is free of solid ice cover during 1-1.5 summer months a year. In low-arctic waters (in regions where the high-arctic and boreal waters co-mingle) the temperature- and ice-regime are considerably milder, so that in summer the temperature of surface waters may rise up to +4°. Salinity in surface waters is 30-35‰ in the majority of high-arctic and low-arctic regions. However, in regions of mixed high-arctic and fluvial waters of the Siberian seas (the estuary-arctic water mass) the salinity of 6-24‰ and the rise of temperature up to +2-3°C are more typical during hydrological summer.

A part of the high-boreal province of the **Atlantic boreal region** in the coastal waters of the Soviet Union embraces the southern region of the Barents Sea shoals and the shoals of the White Sea. The water temperature of the shoals in the Barents Sea rises in summer up to +6°-12°C, while in the White Sea it is sometimes +18°C. The western part of the Barents Sea has no solid ice cover even in winter. In summer the eastern parts of the Barents Sea and the White Sea are free of ice cover during a period of 3-4 months. Salinity of surface waters in the majority of regions of the Barents Sea is 33-34.5‰, while in the White Sea it is 24-26‰.

The Black Sea, included in the **Mediterranean-Lusitan subtropical region**, has a surface water salinity of 17-18‰ in its greater part. The surface water temperature in summer is 26-27°C, while in winter, in the southern and eastern part of the sea, it is seldom below +8°C. However, in the northwestern part it is often below 0°C.

Ecology

The animal and plant species, typical of each biogeographical region, compose biogeographical floro-faunistic complexes. The components of each complex have been formed under similar conditions, at the same period of time, and in a similar locality, which fact has determined their distribution borders.

Separate species in the optimal for their genotype conditions, at definite depths and on certain grounds, may accumulate high bioenergetic capacity, forming a basis for the ecosystem; other organisms join the same ecosystems by coadaptation to the leading forms, occupying ecological niches of some other quality.

The coastal ecosystems of subtropical latitudes of the northwestern part of the Pacific Ocean are, by their origin, the most ancient among those we have studied. Paleontological material together with data obtained by the application of the ecological principle of evolutionary reconstructions (Golikov and Tzvetkova, 1972: Golikov, 1974; Scarlato and Kafanov, 1975) suggest the idea that species — having originated in the first half of the Miocene (about 25 million years ago) in the region of central and southern Japanese islands — provided a foundation for the formation of these ecosystems. There are among them such background-forming species as species of the genus *Sargassum, Bivalvia, Crassostrea gigas*, the prosobranchs *Littorina brevicula* and others, that were widely distributed in coastal Asian waters in this remote and milder geological epoch. Later on, when the climate of the northern Pacific became colder, these species largely diminished the areas of their distribution so that they retained their dominant position only in a number of ecosystems in protected bays of the Sea of Japan (efficiently heated in summer time), in the southern part of the Okhotsk Sea, in northern Japan, and in the South Kurils.

Accordingly, at present, species subtropical by origin are determining the bionomy of protected bays, confined to the Asian low-boreal subregion, thus giving an intrazonal effect, so that districts of a definite biogeographical origin are distinctly wedging themselves into another biogeographical region (Golikov and Scarlato, 1967, 1968).

A typical picture of a partially protected bay landscape in the western part of the Sea of Japan can be drawn from the scene of vertical distribution of biocoenoses in the protected districts of Possjet Bay. Near stony and rocky abrasion coasts, between the tide marks (in the littoral), the sea-snails (*Littorina brevicula* and *L.mandshurica*), having the average biomass of 300–400 gr/m^2 in summer, are predominant. At the depth of 1.5–2 m bottom ecosystems are more common. Here the *Sargassum miyabei* predominate and at greater depths is succeeded by ecosystems with the predominance of *S.pallidum* (the biomass of biocoenoses of *Sargassum* algae are about 2000-3000 gr/m^2 at the vegetation period).

In districts in protected bays with warm waters where there are no *Sargassum* algae, there are prospering ecosystems of *Crassostrea gigas*. At the depth of 3–4 m on abrasion coasts, there is often to be found a biocoenosis wherein the bivalves are predominant (*Arca boucardi* and *Crenomytilus grayanus*). Actinias, Bryozoa, and Sponges are leading in protected bays on firm grounds in more deep waters. On gravelly and shingly beaches, with rotten leaves of sea grasses cast ashore in the supralittoral and in the upper horizon of the littoral, the predominance of Talitridae biocoenoses is common enough (in summer their biomass is about 20–70 gr/m^2). On soft, sandy and silted grounds in protected bays at the depth of 0–1 m (sometimes at the depth of 3–4 m) there are ecosystems with *Zostera marina japonica* as the leading species (the biomass is 4000–6000 gr/m^2 at the vegetation period). In this ecosystem the bivalves *Musculus senhousia* are often dominant; in districts without the cover of seagrasses they can form independent ecosystems. At depths of 4–5 m on muddy grounds, ecosystems with predominant subtropical species of polychaets *Spirochaetopterus variopedatus* and sea stars *Patiria pectinifera* are more common. Further down, on silted grounds biocoenoses of large-sized bivalves *Anadara broughtoni* or relatively smaller bivalves, *Macoma incongrua*, are more typical. The biocoenosis of *Zostera nana + Batillaria cumingii* is highly typical of alluvial lagoon beaches with lowered salinity (it is sometimes lower than 8–12‰) on soft and mixed grounds. Sometimes *Batillaria cumingii* forms an ecosystem of its own. If the salinity in estuaries falls to 8–9‰ or lower, mass populations are formed by small-sized prosobranchs *Assiminea lutea* or *Fluviosingula nipponica*.

In remote isolated protected bays of the coastal Pacific low-boreal subregion ecotypic biocoenoses (isolated biocoenoses with one and the same leading species) tending to form associations seem to be a common trait. All the associations of protected bays lie practically within the northern border of the coastal Pacific low-boreal subregion, though they can be found in the open districts of the subtropical subregion of the West Pacific region.

Subtropical by origin, biocoenoses of protected bays of the coastal Pacific low-boreal subregion have many general characteristics in common with Black Sea biocoenoses. The

Black Sea biocoenoses were composed by many species that originated in the first half of the Miocene in reservoirs of the former Tethys Sea. Parallel biocoenoses are f. i. ecosystems formed by Sargassum algae (*Cystoseira barbata*), mussels (*Mytilus galloprovincialis*), oysters (*Ostrea edulis*), and others.

Carnivorous prosobranchs *Rapana venosa* enter into the compositon of ecosystems in a number of protected bays in the Sea of Japan and are common in the Black Sea (the latter species acclimatized itself in the Black Sea nearly 40 years ago in consequence of a chance contamination of its egg capsules by ships).

Associations of protected bays in the coastal Pacific low-boreal subregion form biomes also analogous to those in biocoenoses of the Black Sea (biomes are complexes or aggregations of associations, made up by species of identical or analogous forms, i.e. by species with similar morphofunctional and ecological peculiarities according to Humboldt, who first coined the term "life-form").

The change of climate in the Northern Hemisphere succeeded by a temperature fall in the second half of the Miocene (15-12 million years ago) resulted in the formation of boreal waters in the northern part of the Pacific and in the origin of widely-distributed boreal species abiding in such waters. These species appear to have been formed in the southern part of the Bering Sea; they became widely distributed owing to the fact that at that time there were no marked hydrological distinctions in the northern part of the Pacific stretching to the latitude of about 38-39°N. Later, when the summer hydrological regime of low-boreal waters became strikingly different from the hydrological regime of high-boreal waters, these species either shifted their spawning period or retained their place only in the southern part of the aquatorium under observation in relatively cool waters off open surfy coasts and in the depths.

At present many of the widely distributed Pacific boreal species play the leading part in shelf- and bathial-ecosystems within the whole Pacific boreal region. Among such species forming the biological background to the coastal ecosystems, there are Cirripedia *Chthalamus dalli* (in the supralittoral and the upper horizon of the littoral among rocks and stones with a biomass of about 100-200 gr/m^2); sea-snails *Littorina kurila* (in the upper and central horizon of the littoral in rocks, stones and mixed ground with a biomass of 200 gr/m^2 on the average); sea mussels *Mytilus edulis* (in the central and lower horizon of the littoral in rocks, stones, and mixed ground with a biomass of 300-600 gr/m^2); phaeophyta *Fucus evanescens* (in the central and lower horizon of the littoral among rocks and stones with a biomass of 1000 gr/m^2); a number of prosobranch species of the genus *Collisella*, actinias *Metridium senile fimbriatum* (at 10-30 m among rocks, stones and boulders with a biomass of 1500 gr/m^2); and others.

Biocoenoses with the predominance of these species form associations of considerable meridional extent, which is characteristic of the shoals in the whole Pacific boreal region.

In the course of time, at the second half of the Miocene, both the growing discrepancy in the hydrological regime between the coastal Asian and coastal American waters and the formation of greater depths in the southwestern part of the Bering Sea caused a number of boreal species to differentiate and to develop into coastal Asian and coastal American species, which were formed correspondingly off the Kamchatka and Alaska coasts. A number of these species play the leading part in a series of shoal ecosystems on the coasts of Asia–f. i. phaeophyta *Gloiopeltis capillaris* with a biomass of 40-50 gr/m^2 on firm grounds between tide marks, and the sea grass *Zostera asiatica* with the average biomass of about 3000-4000 gr/m^2 on sandy grounds and some others.

At the end of the Miocene (8-10 million years ago), due to the further fall of temperature, the high-boreal and low-boreal waters of the Pacific acquired firm hydrological distinctions, which in turn resulted in the formation of primary high-boreal species at the southeastern coasts of Kamchatka, and in the development of coastal Pacific low-boreal species in the regions of modern south Sakhalin and Hokkaido.

In comparatively open districts of the coastal Pacific low-boreal subregion near abrasion coasts, on firm grounds, the dominant part is played by such low-boreal, background-forming species as brown algae *Pelvetia wrighti* (found between tide marks with a biomass of 1500-2000 gr/m^2), sea grass *Phyllospadix iwatensis* (found at a depth of 0.5-1 m with a biomass of 4000-5000 gr/m^2), trade brown algae *Laminaria japonica* (in depths of 1-3 m with a biomass of 15,000-20,000 gr/m^2) and *L. chichorioides* (in depths of 12-16 m with a biomass of 300-800 gr/m^2), and bivalves *Crenomytilus grayanus* and *Modiolus difficilis* (in depths of 8-20 m with a biomass of 2000-5000 gr/m^2).

At the coasts with accumulative relief forms, on sandy beaches near bars and spits and between tide marks, Amphipoda and Talitridae generally play the leading part in biocoenoses (biomass of 40-80 gr/m^2), while in the sublittoral to the depth of 2-5 m, edible bivalves *Spisula sachalinensis* and *Mactra sulcataria* (biomass of 2000–3000 gr/m^2) or flat sea urchins *Echinarachnius griseus* and *E. mirabilis* (biomass of 500–800 gr/m^2) are mostly predominant.

Three types of landscape can be distinguished in the shoals of the northwestern part of the Sea of Japan, in the southern part of the Okhotsk Sea, and in the adjoining regions of the Pacific. There are: (1) Landscapes of protected bays wherein the population has a subtropical character and where, at the coasts of abrasion-sculpture relief forms, there developed biomes of *Sargassum* algae, oysters and other organisms specific to firm grounds in subtropical waters. In districts with alluvial relief forms there are formed biomes of sea grasses and associations of heat-loving molluscs *Batillaria cumingii* and *Assiminea lutea* (the latter in case salinity is considerably below the average). (2) There are landscapes of surfy and moderately surfy coasts with prevailing abrasion-sculpture relief forms where the low-boreal and widely distributed boreal species of hard facies are predominant. This landscape type is most common in the Pacific low-boreal subregion. (3) Lastly, there are landscapes with prevailing accumulative relief forms at relatively open coasts, with bivalves burying themselves in the ground and with seagrasses *Zostera asiatica*.

There are also districts of transitory type—ecotons—situated between the above-mentioned types of landscape. The part played by widely distributed boreal species in associations increases with depth and near open coasts.

In the coastal waters of the high boreal subregion the landscape is sharply changed. At stony and rocky coasts there are prevalent in the ecosystems Pacific high-boreal and widely distributed boreal species: associations of *Balanus cariosus* (biomass of 2000–3000 gr/m^2), of *Littorina kurila* (biomass of 600–900 gr/m^2), of *Mytilus edulis* (biomass of 1000–2000 gr/m^2), and of *Fucus evanescens* (biomass of 2000 gr/m^2). Near the position of water at low tide, *Porphyra umbilicalis* and *Rhodymenia stenogona* (biomass of 400–800 gr/m^2) are most typical for the rocky littoral of the Pacific high-boreal subregion.

In the sublittoral with abrasion-sculpture relief forms the biomes of *Laminaria* with leading *Laminaria bongardiana* and *Alaria marginata* are well developed to the depth of 2–3 m with leading *Laminaria digitata* to the depth of 7–8 m (biomass of 2000–4000 gr/m^2).

Gigantic *Alaria fistulosa* (about 20 m high and 1 m wide), forming intermittent associations at depths of 8–12 m and having biomass up to 30,000–40,000 gr/m^2, is a specific component part of this biome in the Pacific high-boreal waters. Further down, at depths of 15–18 m, rocks and stones are usually covered with belts of brown algae *Agarum cribrosum* and *Thalassiophyllum clathrum* and then of red algae with the leading *Ptilota asplenioides* (biomass of 500–800 gr/m^2).

Further, on firm grounds the leading part in bottom biocoenoses is played by adhesive bivalves *Pododesmus macroshisma*, sea urchins of the genus *Strongylocentrotus*, Actinias, *Cucumaria japonica*, and Sponges (biomass of 800–1500 gr/m^2). On sandy grounds at depths of 2–3 m (often down to 100–150 m) there stretches, forming a wide belt, an association of flat sea urchins *Echinarachnius parma* (biomass of 500–800 gr/m^2) characteristic for high-boreal waters of the Pacific. In cool districts and on silted grounds the association of a boreal-arctic species of bivalves *Macoma calcarea* (biomass of 300–400 gr/m^2) is more common.

In districts with fresh water, having salinity of 20–26‰, in secluded bays in the Pacific high-boreal subregion with alluvial relief forms on silted-sandy grounds, there are often found at the lower horizon of the littoral and near ebb-tide associations of *Macoma baltica*, and at depths of 0.5–3 m associations of *Zostera marina marina* (biomass of 2000–4000 gr/m^2). Still lower down there are biocoenoses with the leading ascidia of the *Molgula* genus.

In coastal Asian waters there can be distinguished three types of landscape: (1) landscapes of secluded bays with fresh water and prevailing alluvial relief forms, wherein *Macoma baltica* and *Zostera marina marina* play the leading role in coastal biocoenoses; (2) landscapes of littorals with prevailing abrasion-sculpture relief forms and the predominance of *Laminaria* in the upper regions of the shelf (this landscape type is common enough in the coastal Asian waters of the high-boreal subregion); and (3) landscapes of accumulative sandy-spits, characterized by a general decrease of biomass, by a descent of fucoids to the sublittoral, by a complete lack or by only a sparsity of *Laminaria*, and by the presence of associations of *Echinarachnius parma*.

A vigorous oceanic transgression in the second half of the Pliocene caused a mass migration of a number of representatives of the Pacific flora and fauna into the North Atlantic through the northern part of Canada, which was covered with water. At that time the boreal hydrological regime had already been established in the North Atlantic, and original Atlantic boreal ecosystems were being formed. Along the coasts of America and Europe, immigrants from the Pacific had reformed themselves into special peculiar species (formerly subspecies); they entered these local ecosystems, having made up the greater part of their components. A minority of species were formed along the coasts of Europe from local Tethys and Paratethys genesis of elements; species of the genus *Gammarus* (Amphipoda), *Gadus* (Pisces), and so on.

Many associations of boreal Atlantic shoals are to be found in that part of the high-boreal province of the Atlantic boreal region which is situated along the coasts of the Soviet Union (the southern part of the Barents Sea). On stony and rocky grounds near abrasion beaches, the following vertical distribution of associations seems to be most common. In the supralittoral and the upper horizon of the littoral there are associations of *Balanus balanoides* and *Littorina saxatilis,* often forming joint populations and alternatively playing the leading part in biocoenoses. At the upper floor of the central horizon in the littoral *Fucus vesiculosus* and *Littorina obtusata* are predominant, while at the lower floor of this horizon, the same part is played by *Ascophillum nodosum* and *Mytilus edulis.* Dense populations in the lower horizon of the littoral and in the sublittoral at depths of 2-3 m are often formed by sea mussels. In the lower horizon of the littoral there are often to be found associations of *Fucus serratus, Rhodymenia palmata* and *Littorina littorea.* In the sublittoral at a depth of 8-10 m there are associations of *Alaria esculenta, Laminaria saccharina,* and sometimes *Laminaria digitata.*

Still further down there is a belt of red algae, followed in many places by associations of *Strongylocentrotus droebachiensis* and by *Lithothamnion* algae. On sandy and silted grounds of alluvial relief forms *Macoma baltica* and *Mya arenaria* are more common, while in the sublittoral at a depth of 2-3 m there are *Zostera marina*, and in relatively cool districts *Mya truncata.* Large bivalves, *Arctica islandica,* form a characteristic component of populations on sandy grounds at depths above 20 m. The descent of fucoids down to the sublittoral (*Fucus inflatus*) is most common with accumulative forms of relief on sandy grounds with separate stones, while on silted grounds at depths over 3-4 m the association of *Astarte borealis* is more typical.

Within the high-boreal province of the Atlantic boreal region again three types of submarine landscape can be pointed out corresponding to the general relief character and its population: (1) landscapes of firm facies off abrasion beaches with a well developed belt of *Laminaria* in the sublittoral; (2) landscapes of mixed facies near accumulative beaches with associations of *Fucus inflatus* in the upper sublittoral and with *Astarte borealis* lower down; and (3) landscapes of sandy and silted facies with chiefly alluvial relief forms with the growth of *Zostera marina* in the upper sublittoral.

Many associations of the coastal waters of the North Atlantic are bionomically parallel to the corresponding associations of the high-boreal Pacific waters, or they have proved to be sequel to the same associations. In the Atlantic high-boreal waters the biomass are of the same order, as in the corresponding parallel associations of the Pacific, though they are on the average lower.

A progressive fall of temperature at the end of the Pliocene (about 2.5 million years ago), which had caused the formation of glaciers in Greenland and mountain glaciers in the islands of the North Atlantic and the Bering Sea, followed by an excessive fall of water temperature, put an end to all migrations of boreal population from the Pacific into the Atlantic Ocean. At the same time (or somewhat later) there were formed the boreal-arctic species in the Bering Sea, by origin Pacific (a majority), and in the Barents Sea, by origin Atlantic (a minority). These species distributed themselves widely in the Arctic seas, and in relatively cool waters joined in the compositions of associations of the Pacific and Atlantic boreal regions, or formed in these regions associations of their own. Among these we may count species of bivalves *Macoma calcarea* and *Nuculana pernula* (with a biomass of 300-400 gr/m^2), dominant in many ecosystems, a number of the species of the genus *Astarte, Musculus, Neptunea, Buccinum, Eunephyta* and others.

A vigorous regression of the ocean, which occurred at the beginning of the Pleistocene (1.8-2 million years ago), suddenly restricted or even completely blocked the way for comparatively warm North Atlantic drift to the coasts of Northern Europe. Such may have been one of the most vital factors which evoked the first wave of a formidable glacial period in coastal Atlantic high latitudes; it may have also caused the formation of the high Arctic water mass in the Polar basin. In these waters from boreal-arctic species, there originated the high-arctic species that enriched the already existing associations of boreal-arctic species and more seldom formed associations of their own (f.i. *Laminaria longicruris, Anonyx sarsi, Strongylocentrotus pallidus, Buccinum hydrophanum, Oenopota gigantea,* and so on). The biomasses of associations of high arctic species on the average are seldom above 200-300 gr/m².

In high arctic waters with abrasion sculpture relief forms there can be distinguished two types of landscape. First there are those free from ice during 1-1.5 months of hydrological summer, with moderately developed algal cover of Laminariae (*Phyllaria dermatodea, Alaria esculenta, Laminaria saccharina, L.longicruris*) and red algae (*Phyllophora spp.* and others). The littoral zone in districts devoid of solid ice soldering is lifeless. There is no apparent criopelagic population on the lower surface of the ice cover. Second, there are landscapes of districts that are covered with ice the year round and have no algae. Criolittoral biocoenoses with the predominance of Amphipoda *Gammarus setosa* and Diatomea are developed in the littoral in solid soldering ice, while on the lower surface of pack ice there are criopelagic biocoenoses with the leading Amphipoda *Apherusa glacialis*. The biomass of criopelagic biocoenoses is 20-40 gr/m^2. Associations of bivalves *Hiatella arctica* and *Musculus corrugatus,* Cirripedia *Balanus balanus,* Ascidias, soft corals, Sponges, and Bryozoa are most common in districts where firm facies prevail, while associations of bivalves *Mya truncata, Macoma moesta,* and *Yoldia hyperborea,* as well as Ophiuroidea and Foraminifera, are chiefly distributed in districts with mostly soft, clayey-silted and mixed facies. The average biomass of the population of landscapes free from solid ice-cover in summer is 400-500 gr/m^2, while that of biocoenoses of landscapes covered with ice the whole year is 200-300 gr/m^2.

Near the end of the early Pleistocene regression in the ocean in the hollow formed by that time in the western part of the Okhotsk Sea, there originated the glacial Okhotsk species, which did not widely distribute themselves and have a subordinate position in the majority of Okhotsk Sea associations.

During the period of the next to the last transgression, which followed the early Pleistocene regression, in the zone where the high Arctic Sea waters and comparatively warmer fluvial waters mixed together at the coasts of Siberia, there was formed an estuary-arctic water mass, wherein a corresponding faunistic complex sprang to life.

Due to evribiont peculiarities of estuary-arctic species and to a set up of wide migration connections between the estuary systems of Eurasian reservoirs at the period of transgressions, some of these species have an intrazonal distribution. In landscapes of estuary-arctic waters with prevailing alluvial relief forms and soft clayed-silted, sandy and mixed grounds the associations of Isopoda of the genus *Mesidothea (M.eutomon, M.sibirica, M.sabini)* and the bivalves of the genus *Portlandia* (especially *P. arctica*), Ascidia of the genus *Molgula,* and some others are more typical.

Summary and Conclusions

The distribution of some most characteristic biomes and associations in the coastal waters of the Soviet Union is represented in Table 1. As we can see from the table, biomes of *Sargassum* and *Ostrea* are sepcific for subtropical waters, while biomes of *Fucus* and *Laminaria* are typical for boreal waters (including the Arctic, for the latter).

The arctic region is characterized by specific criolittoral biocoenoses and estuary-arctic associations of *Portlandia arctica*. Thus, from the viewpoint of landscape geography the Arctic region has few original peculiarities, fewer than the boreal region. The same point of view is shared by many biogeographers. The principle of landscape-geographical regioning is, indeed, akin to the same principle in biogeography, since the distribution of many associations is restricted to biogeographical subregions, and the succession of associations according to depth is often typical of biogeographical provinces. This is largely due to the determining influence that physicochemical changes in environment exercise over the formation and distribution of floro-faunistic complexes. Considering that periods of sudden climate changes and of intensive species formation coincide in locality and epoch with periods of intensive orogenesis and with changes in landscape morphology, it is becoming increasingly evident that landscape- and biogeographical regioning have a common character.

It is noteworthy that species having geologically a more ancient origin and having a wide geographical distribution, considerably more often occupy the leading position in ecosystems and form associations than recently formed species and species having a more restricted distribution. This phenomenon may be caused by the fact that species of ancient origin had formerly occupied all optimal ecological niches, while recently formed species only later on in more recent times co-adapted themselves to the former.

It appears that already formed ecosystems can now only rearrange themselves while new ecosystems are not formed, since with relatively stable climate conditions there can be observed no species-forming processes.

The principal peculiarities of ecology and life-distribution in the coastal waters of the Soviet Union given in the present paper are a result of direct observation and of analyzing the quantitative data obtained by the authors by means of quantitative diving methods of hydrobiological research in different landscape-geographical zones (see papers by Golikov and Scarlato in the list of literature).

Some Typical Biomes in Coastal Waters of the USSR Seas.

Biomes (formations) of hard grounds	Pacific Boreal Region				Mediterranean-Luzitanian Region	Atlantic Boreal Region	Arctic Region	
	Asian low boreal subregion		High boreal subregion		Black Sea province	High boreal province	High Arctic province	
	Associations (facia) of landscapes of protected bay in South Primorski Territory (the Sea of Japan)	Associations (facia) of landscapes in open region of the Sea of Japan	Associations (facia) of landscapes of temperately opened region of the Pacific Ocean		Associations (facia) of landscapes of the region of the northern coast		Associations (facia) of landscapes released in summer from solid ice	Associations (facia) of landscapes ice-covered the entire year
1	2	3	4		5	6	7	8
Littorina-Melaraphe	*L. brevicula* *L. mandshurica* (supralittoral–middle horizons of littoral)	*L. kurila* (middle horizon of littoral)	*L. kurila* (supralittoral–middle horizon of littoral)		*M. neritoides* (supralittoral)	*L. saxatilis* (supralittoral, upper horizon of littoral)	—	*Gammarus setosus*
			L. sitchana (lower horizon of littoral)			*L. obtusata* (middle and lower horizons of littoral)		
	L. squalida (lower horizon of littoral, upper sublittoral)	*L. squalida* (lower horizon of littoral, upper sublittoral)	*L. squalida* (lower horizon of littoral, upper sublittoral)			*L. littorea* (lower horizon of littoral, upper sublittoral)		
Balanus-Chthamalus	—	*Ch. dalli* (supralittoral, upper horizon of littoral)	*Ch. dalli* (supralittoral, upper horizon of littoral)		—	*B. balanoides* (supralittoral, upper horizon of littoral)	*B. balanus* (5–30 m)	
			B. cariosus (middle horizon of littoral)					

Some Typical Biomes in Coastal Waters of the USSR Seas. (Continued)

Biomes (formations) of hard grounds	Pacific Boreal Region			Mediterranean-Luzitanian Region	Atlantic Boreal Region	Arctic Region	
	Asian low boreal subregion		High boreal subregion	Black Sea province	High boreal province	High Arctic province	
	Associations (facia) of landscapes of protected bay in South Primorski Territory (the Sea of Japan)	Associations (facia) of landscapes in open region of the Sea of Japan	Associations (facia) of landscapes of temperately opened region of the Pacific Ocean	Associations (facia) of landscapes of the region of the northern coast		Associations (facia) of landscapes released in summer from solid ice	Associations (facia) of landscapes ice-covered the entire year
1	2	3	4	5	6	7	8
Mytilus-Crenomytilus		*M. edulis* (lower and middle horizons of littoral)	*M. edulis* (littoral upper sublittoral)	*M. galloprovincialis* (0–10m)	*M. edulis* (littoral, upper sublittoral)	*Hiatella arctica* (5–30 m)	—
		M. coruscus (0.5–1 m)					
	Cr. grayanus (3–4 m)	*Cr. grayanus* (3–12 m)					
Fucus-Pelvetia-Ascophillum	—	*P. wrighty* *F. evanescens* (littoral)	*F. evanescens* (littoral)	—	*F. vesiculosus* (upper horizon of littoral)	—	—
					A. nodosum (middle horizon of littoral)		
					F. serratus (lower horizon of littoral)		
					F. inflatus (upper sublittoral on mixed grounds)		
Sargassum Cystoseira	*S. miyabei* (0.5–2 m)	—	—	*C. barbata* (0.5–10 m)		—	—
	S. pallidum (2–3 m)						

Ostrea-Crassostrea	Cr. gigas (0.5–3 m)	–	–	O. edulis (0.5–10 m)	–	–
Laminaria-Alaria	–	L. japonica (0.5–3 m)	L. bongardiana (0–5 m); A. marginata (0–3 m); A. fistulosa (6–12 m)	–	A. esculenta (0–3 m); L. sacharina (0–12 m)	A. esculenta (0.5–3 m); L. sacharina (5–30 m)	–
Strongylocentrotus	–	S. nudus (0–3 m); S. intermedius (5–20 m)	S. sachalinense (3–30 m)	–	S. droebachiensis (2–30 m)	S. pallidus (10–50 m); S. golikovi (5–40 m)
Biomes (formations) of soft and mixed grounds						
Zostera-Phyllospadix	Z. nana (upper horizon of littoral); Z. marina japonica (0–4 m)	Z. asiatica (3–15 m); P. iwatensis	Z. marina marina (0.3 m)	–	Z. marina marina (0.5–3 m)	–
Macoma	M. incongrua (1–6 m)	M. lata (15–20 m)	M. baltica (0–1 m); M. calcarea (30–100 m)	–	M. baltica (littoral–1 m); M. calcarea (30–100 m)	M. moesta (10–30 m)
Mya	–	M. japonica (5–10 m)	–	–	M. arenaria (littoral); M. truncata (littoral–3 m)	M. truncata (5–50 m)
Astarte	–	A. borealis (60–120 m)	A. borealis (60–120 m)	–	A. borealis (3–15 m)	A. crenata (20–40 m)

For additional references see Clements (1905, 1916), Clements and Shelford (1939), Deryugin (1915, 1928), Gurjanova et al. (1928), Shelford (1945), Ushakov (1952, 1953), Vorobiev (1949), and Zenkevich (1951, 1956).

A. N. GOLIKOV
O. A. SCARLATO

References

Clements, F. E., and Shelford, V. E., 1939. *Bioecology.* Lincoln, Nebraska: The University Publishing Co., 334p.

Clements, F. E., 1916. *Plant Succession.* Washington, D.C.: Carnegie Institute of Washington, Pub. 242, 512p.

Clements, F. E., and Shelford, V. E., 1939. *Bioecology.* New York: J. Wiley & Sons, 425p.

Deryugin, K. M., 1915. Fauna of the Kola Bay and conditions of its existence, *Acad. Imperiale Sci. (St. Petersburg) Zapiski,* ser. 8, 34, 929p.

Deryugin, K. M., 1928. Fauna of the White Sea and conditions of its existence, *Leningrad Hydrol. Inst. Issled. Morei,* 7-8, 511p.

Golikov, A. N., 1965. Comparative ecological analysis of some marine bottom biocenoses of south Primorski Territory and south Sakhalin and perspectives of their trade mastering: Problems of hydrobiology, *in The First Congress of the All-Union Hydrobiological Society.* Moscow: Nauka, 94-95.

Golikov, A. N., 1966. Ecological peculiarities of coastal marine bottom biocenoses of the south Primorski Territory and south Sakhalin in connection with hydrological regime, *in The Second International Oceanographic Congress.* Moscow: Nauka, 136-137.

Golikov, A. N., 1974. *Hydrobiology and Biogeography of the Shelf of the Temperate and Cold Waters of the Ocean,* Akad. Nauk SSSR Zool. Inst., 145p.

Golikov, A. N., and Averintsev, V. G., 1971. Some regularities of life distribution in upper parts of the shelf of Franz Josef Land archipelago, in *Scientific Session of Zoological Institute, Academy of Sciences, USSR, on Results of Works.* Leningrad: Nauka, 11-12.

Golikov, A. N., and Scarlato, O. A., 1965. Hydrobiological investigations in the Possjet Bay using diving technique, *in Investigations of Marine Fauna in North-Western Part of the Pacific Ocean.*, vol. 3 (11). Leningrad: Nauka, 5-29.

Golikov, A. N., and Scarlato, O. A., 1967. Ecology of bottom biocoenoses in the Possjet Bay (the sea of Japan) and the peculiarities of their distribution in connection with physical and chemical conditions of the habitat, *Helgoländer Wiss. Meeresunters* 15, 193-201.

Golikov, A. N., and Scarlato, O. A., 1968. Vertical and horizontal distribution of biocoenoses in the upper zones of the Japan and Okhotsk seas and their dependence on the hydrological system, *Sarsia,* 34, 109-116.

Golikov, A. N., and Scarlato, O. A., 1970a. Regularities of biocoenoses distribution in upper zones of the shelf in temperate waters which is connected with the character and structure of water masses. Biological process in marine and continental waterbodies, *in Second Congress of the All-Union Hydrobiological Society, Kishinev,* 83-84.

Golikov, A. N., and Scarlato, O. A., 1970b. Hydrological exploration in cold and temperate waters of the USSR with light diving equipment *(SCUBA), Internat. Rev. Gesamten Hydrobiologie* 55, 305-315.

Golikov, A. N., and Tzvetkova, N. L., 1972. The ecological principle of evolutionary reconstruction as illustrated by marine animals, *Marine Biology,* 14, 1-9.

Gurjanova, E.; Zaks, I.; and Ushakov, P., 1928. Littoral of the Kola Bay, *Leningrad Soc. Naturalists Trans* 58, 89-143.

Scarlato, O. A., and Kafanov, A. I., 1975. The main features of the recent marine malacofaunas formation in the shelf of cold and temperate waters of the northern hemisphere on the example of bivalve molluscs, *in All-Union Conference on Biology of Shelf, Abstracts of Papers.* Vladivostok: Institute of Marine Biology for East Scientific Center of the USSR Academy of Sciences, 161-162.

Shelford, V. E., 1945. The relative merits of the life zone and biome concepts, *Wilson Bull.* 57, 248-252.

Ushakov, P. V., 1952. The Chuckchee Sea and its bottom fauna, *in Extreme Northeast of the USSR,* vol. 2, 5-82.

Ushakov, P. V., 1953. *Fauna of the Sea of Okhotsk and Conditions of Its Existence.* Leningrad: Academy of Sciences, USSR, 459p.

Vorobiev, V. P., 1949. *Benthos of the Sea of Azov.* Krymizdat: Simferopol, 193p.

Zenkevich, L. A., 1951. *Seas of the USSR Their Fauna and Flora.* Moscow: Uchpedgiz, 368p.

Zenkevich, L. A., 1956. *Seas of the USSR, Their Fauna and Flora,* 2nd ed. Moscow: Uchpedgiz, 424p.

Cross-references: Asia, Eastern, Coastal Ecology; Coastal Fauna; Coastal Flora; Europe, Coastal Ecology; Ice-Bordered Coasts; North America, Coastal Ecology; Organism-Sediment Relationship; Soviet Union, Coastal Morphology; Standing Stock; Vol. VIII, Part 2: U.S.S.R.

SOVIET UNION, COASTAL MORPHOLOGY

The total length of continental shoreline within the boundaries of the USSR exceeds 70 thousand kilometers, and with islands included the figure surpasses 100 thousand km (Fig.1). Along its entire length the shoreline borders different geological structures, having survived a diverse developmental history during the Tertiary and Quaternary periods. This circumstance determines the enormous variety as well as the frequent alternation of coastal types, and the structureal types of a shore zone as a whole. Structural type incorporates the composition of nearshore bottom areas within the limits of the drift-shifting action of storm waves and the adjacent land strip, which has been constantly

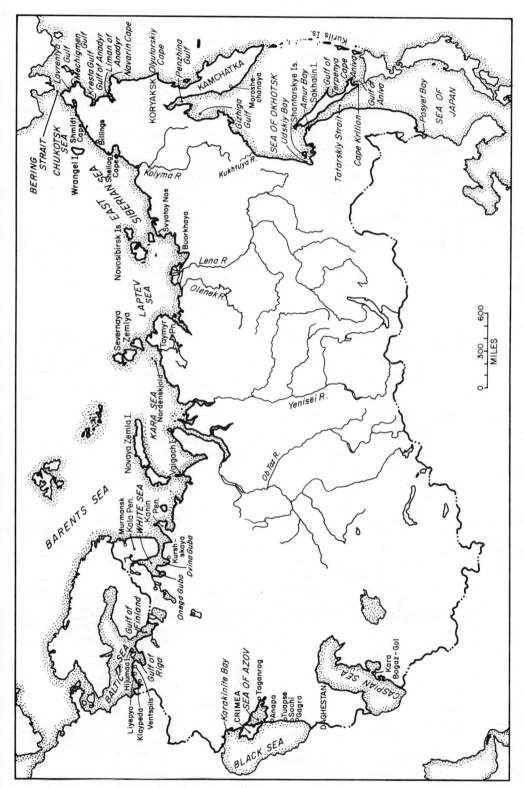

FIGURE 1. U.S.S.R. coastal location map.

influenced by the sea throughout the Holocene epoch as well as in recent times.

The morphology of a sea's coastal zone is determined chiefly by the geological structure of the adjacent land areas, as well as by ancient and recent relative vertical movements. The dynamics of a shore zone are strongly affected by solid coarse-grained river discharge (its amount and composition) and the sea's general hydrodynamic conditions (shores of high and low energy). The positioning and exposure of the given site in relation to the resultant wave regime are also of much importance. Consequently, the description and dividing into districts of the USSR seacoast ought to be based on, and mainly concerned with, the geological structures of different regions and the composition of rocks bordering on the various bodies of water.

White Sea to Baltic Sea

In the northwest of the USSR the Barents and the White seas border a Fennoscandian crystalline shield. During the Late- and Postglacial periods the adjoining land was subjected to isostatic elevation, which has not ceased up to recent days. The main feature of coast in the Murmansk area (ranging from the frontiers with Norway to the funnel of the White Sea) is expressed in a minute, but sharply distinguished, dissection by tectonic fractures of the crystalline mass's borderland. There the shore is cut along the line of an ancient fault and reveals a sawlike contour. Owing to the hardness of the rocks in this area, the shores have suffered little from the seas' erosional effect. This situation ought to be ascribed to, and is attested to, by great depths (over 100 m) at the base of the coast scarps. Waves are reflected from the gneissose-granite *walls*, while a true surf is observed only on small, gently sloping sites. *Fjords*, which in miniature resemble Norwegian ones, are noted in the northwest along the lines of the basic fractures dissecting the crystalline shield. Their heads are gradually filled in by large rivers.

In the southeastern region the seas' primary depths, and those of tectonic depressions, gradually lessen, and the river alluvium is deposited closer to the sea. The fjords shorten and then disappear entirely; their presence is marked by terraced sand sites. The rate of the recent coast elevation is reduced.

In the southern area of the Kola Peninsula long stretches of primary coast are bordered by a sand strip of moraine origin and only at large capes do the gneissose - granite formations contact the sea directly.

The western coast of the White Sea borders land with low, but adequately dismembered, relief. In these areas the shoreline is dissected into narrow bays, of the Swedish *fiard* type, and in front of the coast one notes numerous isles, *skerries*, composed of native rocks or boulders washed out of moraine. The coasts of Murmansk and of the western White Sea area are subjected to rather high tides. However, the tides do not strongly affect coastal structure if one discounts the formation of wide, silty tidal flats in the heads of bays and in sites protected by islands from wave action.

The basic outlines of the eastern coast of the Baltic Sea, within the boundaries of the USSR, have also been determined by glaciation. However, these are mainly features of glacial accumulation. The protrusions of native rocks of Paleozoic age determine only the morphology of the southern shores of the Gulf of Finland and the Estonian SSR (Fig. 2). The Estonian archipelago and the mainland coast of the northwest are subjected to isostatic elevation with a maximal rate of 3 mm yearly (Hijumaa Island).

South of the Gulf of Riga, the Baltic Sea shores are almost entirely levelled. The only large protrusion is observed in the Samland Peninsula. The east Baltic drift flow originates at its abraded shores and now migrates to the mouth of the Riga gulf. In earlier times, the drift had reached its head, the Daugava River mouth, at a range of about 550 km. At a number of sites in the course of its flow the drift is replenished, at the cost of additional nourishment provided by coastal erosion of loose glacial deposits.

In its northern part the east Baltic drift flow has a load of 1 million m^3/yr. It causes a severe drift accumulation at the canals leading to the ports of Klaypeda, Liyepya, and Ventspils. The bulk of the drift flow, up to 500,000 m^3/yr, settles at the Kolkas Rags Cape. Presently about

FIGURE 2. Cliff composed of Ordovician rock undergoing erosion at the Estonian coast, Gulf of Finland.

FIGURE 3. Tidal flats along the eastern shore of the White Sea.

10,000–15,000 m³ of drift material moves past the cape in a southeastern direction.

As the Baltic Sea retreated, barrier islands that fenced off some lagoons were formed along the east coast. The water areas of those lagoons have partly turned into swamps. The biggest of the barrier islands is Kurshskaya bay-bar near the Neman River mouth; it is crowned with dunes up to 60 m high. On the basement of the submarine slope, stony material of the water-worn moraine is exposed almost all the way along.

The southern and eastern coast of the White Sea form another region where the shores are composed mainly of glacial accumulation products and their derivatives. The southern coast is subdivided into two open bays, the Dvina and Onega gubas. The action of waves upon these bay shores is insignificant, owing to the minute wave fetch and the relative shallowness of the sea. In the south of the Dvina Bay one notes the delta of a river bearing the same name and a premouth shallowness.

The eastern shore of the White Sea is in a less protected position, especially in its northern part and within the limits of the Mezen Gulf (Fig. 3). It is strongly abraded for almost its entire length. In the Mezen Gulf the abrasion rate reaches 5 m yearly, being facilitated by the permafrost thawing. The high tides of the gulf run up to 10 m. Strong currents in the funnel area of the White Sea have built up sand ridges stretched out for tens of kilometers, similar to those in the southwest of the North Sea.

The *gorlo* (narrow) of the White Sea gives rise to a drift flow bound southward. Its material has served to build a large spit and sandy island, Mudjug, at the head of the Dvina Gulf.

Kanin Peninsula to Bering Strait

The coast of the Kanin Peninsula of the Barents Sea is of abrasional nature and composed of solid Paleozoic rocks. The coast east of Cheshskaya Guba, and beyond its limits up to the Pechora River mouth, is low and composed of loose material. Here one observes mighty transportation of drifts in a northeastern direction. This flow feeds a series of spits and wave-built isles in outer shallow water areas of the Pechora Gulf.

The coasts of Novaya Zemlya and Vaigach islands, which are a structural continuation of the Ural Mountain ridge, reveal some peculiar features. The Novaya Zemlya coast is characterized by fjord dismemberment. In the northern part of the island many glaciers fall into the sea. Vast strandflats, areas similar to those on Spitsbergen, are widely developed and common. The west coast of the southern island features lobate dismemberment and diagonal structure.

The mainland coast of the Kara Sea's western portion is of a lowland nature. It is abraded everywhere, the rate reaching 7–8 m/yr, for it is composed of loose rocks bound by permafrost. Its long stretches are levelled. The world's most spacious *limans* are situated in front of the Ob Taz and Yenisei river mouths.

Further to the east, the coast of the Taymyr Peninsula is composed of resistant metamorphic rocks and its relief is caused by glacial excavation. Consequently the coast is jagged, with small bays, and the adjoining offshore area abounds in rocky isles (Nordenskjold Archipelago). Characteristic of the Severnaya Zemlya islands coast is a rugged topography of large-scale lobation. With respect to their geomorphology, these shores are insufficiently studied.

To the east of Taymyr the low coast of the Laptev Sea stretches to the Olenek River mouth. It is undergoing intensive destruction under the combined effects of wave abrasion and thermal abrasion. Along the levelled areas drift flows are formed, which build small accumulative forms (spits).

The rectangular protrusion of the multi-branched delta of the Lena River is a major coastal formation. At the present time this delta is in an erosion stage, and its branches are cut through the old alluvium bound by permafrost.

A vast plain borders the sea east of the Lena River's delta. The low permafrost-bound coast (Fig. 4) stretches from the Lena River to the mouth of the Kolyma River. The adjacent water areas of the Laptev and East Siberian seas are exceptionally shallow, so that the bathymmetric line of 20 m is, at some places, 150 km distant from the shore. Not only the land but the sea bottom too are here bound by the permafrost. Along this stretch, cliffs in unconsolidated deposits alternate with blocks and veins of ice, as well as with flat muddy shores.

SOVIET UNION, COASTAL MORPHOLOGY

FIGURE 4. Tundra coast of the Laptev Sea.

Where thermal abrasion occurs the shore retreats at a rate of more than 10 m/yr, both on the continent and on the Novosibirsk Islands. During the last decades some small islets have disappeared because of the destruction by the sea that progressed at a rate of about 50 m/yr. At the same time, thermal destruction of the sea bottom to depths of about 10 m has been taking place.

For long stretches of muddy plains bordering the Laptev Sea there is no fixed shore line, due to the frequent sea-level oscillations under wind action. Driftwood, only recently cast ashore, can be found a few kilometers from the sea. In such places the slope of a frozen, or even an icy, sea bottom is of negligible inclination. Here and there long *arrows* (arrowlike strips) composed of argillaceous material are formed near the shore (these are analogous to barrier islands).

Lacustrine basins, emptied and partly cut off by the sea in the process of abrasion and similar to those in Alaska, occur in the eastern part of the Laptev seacoast. On the whole the coast has a large-scale lobation. Capes composed of bedrock (Buorkhaya, Svyatoy Nos, and some others) present a solid base fastening the major protrusions to the land. Buorkhaya Gulf and Sellyakh Gulf lie between the capes. The total length of this section of the Arctic coast is more than 1000 km. In spite of the high rate of destruction at many places, these shores should be considered low-energy, since the sea in front of them is shallow and only for a short period of time is it free of ice.

East of the Shellag Cape the coastal characteristics vary gradually, the solid rock hills occurring closer to the East Siberian Sea. Therefore, the shoreline becomes wavy, with abrasion sections. The Billings Cape is to be noted as a large-scale accumulative feature, the formation of which was possibly associated with the waves-shadow effect of Wrangel Island.

To the east of the Shmidt Cape there are many narrow lagoons. They are located along the coastal lowland between protrusions of bedrock masses, some being over 30–40 km long. It is in these lagoons that the process of their dividing into a series of roundish lakes was studied (see Lagoonal Segmentation).

Barrier islands along the Chukchi Sea coast are composed mainly of *shingle*. It may be assumed that thermal disintegration of piedmont strata containing rubble material is still occurring on the sea bottom. The rock debris becomes mobile and is moved shoreward by waves to form barriers.

In the eastern part of the Chukotsk Sea the mountain spurs are abraded, and in some places, where particularly resistant rocks are exposed, the coast is of minute (crenulate) dissection.

Bering Strait to Sea of Japan

Along the Bering Strait blocks of bedrock alternate with lowland areas. The loose sediments of the latter are bound by permafrost. Deeply cut gulfs of fjord type appear here (Lavrentya Gulf, Mechigmen Gulf, and others). The southeastern protrusion of the Chukotsk Peninsula is dissected by fjords, the heads of which are almost converging.

At the north and south, the coast of the large Gulf of Anadyr is made up of resistant bedrocks. These areas are deeply dissected highlands. However, due to the shallowness of the gulf, the fjordlike bays are separated by baymouth barriers (Tymna Bay and Kaina-Pilgen). Two large gulfs, namely the Cross (Kresta) Gulf and the Liman of Anadyr, are pre-mouth lowlands submerged under the sea. Long spits have formed at their mouths. The Meechken barrier island is quite remarkable. It is formed of boulder and shingle materials of bottom erosion origin and thus is similar to the previously described barriers of the North Chukotsk coast.

The Koryaksk coast of the Bering Sea (Fig. 5) between the capes of Olyutorskiy and Navarin, is dissected by small fjords that occupy the narrow intermontane depressions. Most of the fjords are fenced off from the sea by barriers.

FIGURE 5. Cliff scarp of the Koryaksk ridge, Bering Strait.

They remain open in the south because the depth of the sea increases in this direction.

From the Chukotsk Peninsula southward the mainland coasts of the Soviet Far East have a shared characteristic with regard to their structure. Almost everywhere they are composed mainly of resistant metamorphic and igneous rocks of Mesozoic and Tertiary ages that underwent intensive folding. Alternating along the mainland margin are fjord areas, now cut off by old faults and almost rectilinear. Lobate dissection into large bays is characteristic of eastern Kamchatka (Fig. 6) and of the northern part of the Sea of Okhotsk. There are numerous alluvial-marine lowlands that depend on the amount of the solid river discharge and on the sea depth near the shore for their formation.

The length of Far East shoreline is enormous, and some parts of that coast deserve special mention. The four large protrusions, of volcanic origin, of eastern Kamchatka make its outline look like gigantic gear teeth. These protrusions have bluff shores up to 800 m high, which are subject to slow abrasion. Between the protrusions lowland areas with many shore ridges—recent aggradation terraces—were formed of coarse alluvial sediments. River mouths are separated from the sea by garlands of spits, as is the case with the Kamchatka River. In the south of the peninsula small fjords similar to those of the Koryaksk coast are developed.

The shore of western Kamchatka is of a different structure. It is composed of loose deposits for a distance of more than 600 km. The low coast is fenced off by a pebbled, sandy barrier behind which, at some places, there are lagoons; while, to a greater extent, it fences off the lower reaches of numerous rivers that are thus deflected for many dozens of kilometers. There are two flows of shore drift along this coast, the northern one having built a large spit near the mouth of the Moroshechnaya River and the southern one having produced a dune accumulation several kilometers long.

At the heads of the Penzhina and Gizhiga gulfs the high tides run up to over 10 m. Wide tidal flats, prevailingly rocky ones, were developed here. Occasionally the sandy–silty low relief of tidal flats is broken by these forms.

A characteristic feature of the northwestern part of the Sea of Okhotsk is an alluvial marine plain of the Kukhtuya and Okhota rivers. Its shores are changeable. The western Okhotsk sea coast is fault-cut and is, on the whole, almost rectilinear. Its monotony is broken only by some bays (Ayan and others) and by far-protruding narrow promontories.

Notable in the southern part of the sea of Okhotsk are the archipelago of the Shantarskye Islands and the large gulfs: Udskiy, Tugurskiy, and Ulbanskiy bays. The nature of the shores changes here within short distances. The wave abrasion is slowed down, but high tides in addition to the frost weathering effect facilitate shore destruction. There are also wide tidal flats here. On the southeast of the Shantarskye Islands the shore assumes a lagoonal character.

The throat of the sound between the Sea of Okhotsk and the Sea of Japan is remarkable. An immense area is formed here by mobile shoals, wide tidal flats, and transient supraaqueous accumulative features. The relief of this region is unstable because of the interaction between the fluvial run-off flows and high tides.

The mainland shores of the Sea of Japan are similar to those of the western part of the Sea of Okhotsk and even of the Murmansk Sea. They stretch in a smooth line, having been cut off along the line of an old fault. At the same time these shores are intermittantly dissected into small bays. Major ingression bays were formed in the south (the Ussuri, Amur, and Posyet bays) and in the middle of this region. The rate of the shore abrasion is greatly re-

FIGURE 6. Abrading coast of the northeastern Kamchatka Peninsula. In this section, marine clay lies between morainal sediments at the bottom and peat at the top.

duced here because of the resistance of the rocks (basalts) and the great depth close to the very base of coastal bluffs. There are no large accumulative features, but small ones are frequently found within the ingression bays.

The Kurils and Sakhalin islands stand apart in the Soviet Far East. The shores of the former are typically volcanic, with sheets of lava and accumulations of volcanic ash strata. The coasts are jagged and very picturesque. Here and there the abrasion is very active. In addition, these islands are subject to *tsunami* wave action, as is the eastern Kamchatka coast.

Sakhalin Island is formed of a complex of Cretaceous and Tertiary rocks of medium or poor resistance, with the exception of some solid protrusions (Krillon and Aniva capes, Shmidt Peninsula). The western shores of Sakhalin are levelled by abrasion and are notable for their wide benches. Drift flows along these shores in the northern direction. Near the head of the Tatarskiy Gulf (strait) this flow has formed a number of *en echelon* spits.

The middle part of the eastern shores bears a lagoonal character (Fig. 7). Wide sand barriers are formed by an intricate complex of several offshore bars of different age. A considerable retreat of the shore has been caused by the modern sea level. The lagoon region in the north borders an abrasion area, along which the sea has truncated a number of rounded basins of large lakes and has separated them by barrier beaches. Small sections of lagoon shore are also developed at the heads of the open gulfs of Aniva and Terpenya (Aranika) near the south end of the island.

The Black and Caspian Seas

Most interesting to Soviet scientists are the shores of the Black Sea (the Sea of Azov included) and the Caspian Sea; they are the best developed shores with respect to industrial exploitation and health-resort utilization, and are therefore the most thoroughly studied. Both of these seas are isolated from the ocean. In the completely detached Caspian Sea considerable autonomous level fluctuations take place owing to river discharge variations. The last phase was represented by the lowering of sea level by 2.5 m during the period of 1929–1956. Since then the sea level has remained nearly stable.

The fall in sea level results in the casting ashore, on a mass scale, of coarse bottom sediments (mainly valves and shell fragments of mollusks, and, on the eastern coast, oolites too). The beaches become wider, and the bottom has been exposed in many places to depths of 10–15 m.

FIGURE 7. East Sakhalin coast. The wide barrier has been formed by two generations of sandy beach ridges. Segmentation of the lagoon is being initiated by the growth of cuspate spits.

FIGURE 8. Erosion of deltaic coast near the mouth of the Kura River, Caspian Sea.

Because of the elongated shape of the Caspian Sea drift flows are developed on its shores, particularly in the west. The eastern side is devoid of any clastics transported by run-off, owing to the arid climate, and the shores are therefore abraded for very long stretches. In the west, to the south of the Kura estuary (Fig. 8) and opposite the termination of the Caucasus range in Daghestan, there are areas of alluvial coastal plains.

With the exception of Apsheron Peninsula and the adjacent territory on the south, the western Caspian Sea shore contours are simple. In the east there are a number of lagoons and spits, though the lagoons have dried up and are covered by a layer of salt. One of them, namely the vast Kara Bogaz-Gol, is the site of chemical mining (salt mines).

The northern Caspian Sea shores are uniformly levelled and low-lying, and are the margins of ancient alluvial plains. The predominating shallowness abets the frequent occurrence of wind and storm-surge piling-up level fluctuations. For this reason, the shoreline is not fixed along great stretches. In this respect, this specific type of shore resembles the Arctic coasts of the Laptev Sea (east of the River Lena) and of the east Siberian Sea.

A major element of the northern Caspian shore is the Volga River delta. Before the construction of artificially impounded reservoirs this dendritic delta was advancing at a rapid rate. In recent years, however, the rate of its advance has slowed down sharply.

The Black Sea is connected with the Mediterranean Sea through the narrow Strait of Bosporus. Its level therefore follows the variations of the oceanic level. The reduced near-surface salinity of the Black Sea results in a sparceness of fauna there. However, shell fragments form a considerable part of the deposits in the northwestern region, and particularly in the Sea of Azov.

A low plateau composed of neogenic, moderately resistant rocks and loesses borders the sea on the northwest, where the shores are even. Still their destruction is going on at a rate of up to 10 m per year and landslides frequently occur on these coasts. The rivers bring practically no beach-forming alluvium to this region. During the Holocene transgression their lower courses were submerged under the sea and formed numerous limans. There are drift flows along these shores, but they are of moderate magnitude—up to 30 thousand m^3/yr. In the area near the Danube River delta and in the western Crimea, extensive lagoons with sand barrier beaches have been formed. In Karkinite Bay there are large spits composed of shelly material; one of which, the Tendra spit, is 70 km long.

The shores of the Sea of Azov (Fig. 9) are of a similar nature. A characteristic feature of this sea is the shelly barrier of the Arabat in its western part, more than 100 km long, and quite a series of shelly-sandy spits along the northern shore. These spits stretch at an angle of about $45°$ with respect to the shoreline, because of the prevalence of northeastern winds blowing parallel to the shore (oblique waves).

A moderately high mountain range, up to 1500 m, rises abruptly at the southern coast of the Crimea. The mountainous upper part is composed of resistant Jurassic limestones, the basement rocks being relatively soft Triassic rocks. Owing to this structure, gigantic landslides and slumping have taken place here, so that the shore is encumbered with limestone rock waste. Due to this circumstance the shoreline is jagged in many places; its crenulate outline being still more emphasized by separate outcrops of igneous rocks, with the shore being abraded very little. For the eastern Crimea, see Figure 10.

The Caucasian shores of the Black Sea present a series of forms that change in a southeasterly direction with the increase in the humidity of climate and in the abundance of rivers. The relatively even coast in the west, between Anapa and Tuapse, presents a continuous line of active cliffs cut in rocks of Cretaceous and Paleogenic flysch. Southward from Sochi the cliffs are not so active and in some places they have disappeared because the amount of alluvium evacuated by the mountain rivers was sufficient for the formation of wide protective pebble beaches. At present the shore in this region is much altered because of man's technical activities.

South of Gagra the alluvium discharge by major rivers is increasing steadily, hence a continuous alluvial marine terrace is gradually being formed along the sea coast.

FIGURE 9. Slumping in cliffs along the Taganrog Bay coast, eastern Sea of Azov.

Along the coasts of the Georgian Republic (300 km) about 4.5 million m³/yr of coarse alluvium is delivered to the sea. However, less than 10% of this amount remains at the shore, the rest being carried down to the deep sea along submarine canyons, the heads of which come near to the shore.

The elongated form of the Black Sea, and the prevailing west winds and waves, determine the existence of drift flows composed mainly of shingle. In some places their capacity is over 100 thousand cubic meters annually. A continuous drift flow that was in existence during the Holocene is now broken into separate sections by the coastal protrusions near river mouths. The sediment flow material was delivered both from the north and south to the Kolkhida lowland, which in Holocene time was a major gulf of the Black Sea. At the present time a sand terrace, a former barrier, stretches along the Kolkhida shores and behind it are swamps. The Kolkhida shores are subject to destruction in spite of the huge amount of alluvium brought here by the rivers (Fig. 11).

For further readings on the Soviet coast see BALTICA (1963-1974), Grigoriev (1966), Leontiev and Khalilov (1965), Orviku (1974), Zenkovich (1958, 1967), and Zenkovich and others (1967).

VSEVOLOD PAVLOVICH ZENKOVICH

FIGURE 10. Karadag Mountain, composed of resistant volcanics, forming the abrupt east Crimea coast. The cliff extends to a considerable depth below sea level.

References

BALTICA, *International Yearbook*, No. 1 (1963)-No. 5 (1974). Vilnius: Lithuanian S.S.R. Academy of Sciences.

FIGURE 11. Erosion of clay substrate and destruction of shore protection facilities in the Kolkhida sector of the Georgian Republic, southeastern Black Sea coast.

Grigoriev, N. F., 1966. *Permafrost Soils of Yakutsk Region Coastal Zone.* Moscow: Nauka, 123p.

Leontiev, O. K., and Khalilov, A. I., 1965. *Natural Conditions of Caspian Sea Coast Development.* Baku: Azerbaijan S.S.R. Academy of Sciences, 206p.

Orviku, K. K., 1974. *Estonian Sea Coasts.* Tallin: Estonian S.S.R. Academy of Sciences, 111p.

Zenkovich, V. P., 1958. *The Coasts of Black and Azov Seas.* Moscow: Geographical Publishing House, 374p.

Zenkovich, V. P., 1967. *Processes of Coastal Development.* London: Oliver and Boyd, 738p.

Zenkovich, V. P., and others, 1967. *The Pacific Coasts.* Moscow: Nauka, 374p.

Cross-references: *Asia, Eastern, Coastal Morphology; Asia, Middle East, Coastal Morphology; Europe, Coastal Morphology; Flandrian Transgression; Holocene; Lagoonal Segmentation; Liman and Liman Coasts; Soviet Union, Coastal Ecology,* Vol. VIII, Part 2, *U.S.S.R.*

SPITS

Spits are quasilinear, subaerial landforms that are caused by longshore deposition of sediment by the prevailing waves and currents along coasts. They may be found anywhere in the world where there exists a net longshore movement of material away from a break in shoreline orientation. A spit represents longshore deposition, and is morphologically defined as extending from land into the offshore zone. Therefore it has exposed seaward and sheltered bayside shores of higher and lower energies, respectively.

The general form of a spit is usually not straight but curved shoreward. This offset is the end product of oblique waves causing the growth to take place at and around the free end of the spit. This tip is often referred to as the *terminous* or *distal* portion, while the attached end is often described as the *root* or *proximal* area (Fig. 1). The distal portion, owing to greater wave refraction of the predominant waves, generally has lower energies and is therefore relatively depositional and wide in form. The root is a higher energy section with less deposition and therefore narrower in plan. Energetically, the exceptions to this description exist when storm waves approach from a direction opposite of the spit growth trend. Spits should be distinguished from other similar forms such as some *bars*, a term to be used only for subaqueous features, and *tombolos*, which have no free end to the depositional structure. While the formational process of spits may sometimes be similar to those of *cuspate forelands*, the latter lack the linearity of form.

While spits have been of interest since ancient

FIGURE 1. Sandy Hook, N.J., a compound, complex barrier spit. Note the recurvature, bayside hooks, and wide distal and narrow proximal areas (photo: J. R. Allen).

times (to cartographers and mariners, for example) systematic inquiry into their taxonomy, genesis, and implications dates back only to the nineteenth century, with the bulk of the research having been since the 1940s. An invaluable reference to spit research is edited by Schwartz (1972) and includes important papers ranging from Gilbert's study of Lake Bonneville shore features, Evans's seminal paper on spit processes, Meistrell's establishment of the spit platform in the laboratory, to King's computer simulation of spit dynamics.

Classification

While there is no established taxonomy of spits, a number of types have been observed in nature and reflect variations in the direction of energy delivery and sediment transport. *Simple spits* are relatively straight forms extending into deeper water from a headland. *Recurved spits* bend slightly shoreward; when reattached, they are sometimes called *looped spits.* Some spits are concave to the sea and are a response to the approach of convexly refracted wave crests. *Hooked spits* have acute recurves extending bayward as a result of occasional but strong opposing wave action. *Compound spits* possess several recurves or hooks from episodic growth. *Complex spits* exhibit at least two of the above. *Bayhead, bayside,* and *baymouth spits* are so named for their respective locations within an embayment. *Winged headlands* exist where diverging currents extend spits in opposite directions off of a headland. *Double spits* converge toward each other from two nearby headlands, often associated with estuaries. *Radial spits* extend into atoll lagoons with overwash deposits and subsequent lagoonal reworking. When an atoll is breached in several places with a spit trailing inward from each remnant, they have been called *hammerhead spits.* Similarily, *trailing* or *tailing spits* extend in the lee of some island although there is no washover contribution to the platform. *Barrier spits* are located in conjunction with barrier island systems where there is a headland attachment. Davies (1973) discusses many spit types.

Dynamics

Because the exact contributions of the processes of construction are locally unique and depend on wind and wave climate, sediment budget, tidal range, and the impacts of man, only relative—not quantitative—comparisons can be given. Present research indicates that beach drifting is more important than longshore currents in the building of spits (Hine, 1979).

Whenever the coastal orientation changes abruptly, momentum losses lead to nearshore and offshore deposition of sediment, resulting in the formation of a submarine embankment. In 1890 Gilbert (in Schwartz, 1972) suggested that inside the breakpoint location, wave action could elevate the deposits into a subaerial position. While this argument has not yet been totally rejected, it appears that the embankment, or *spit platform*, provides a shallow water setting for subaerial extension of the spit by beach drifting, according to the 1942 work of Evans (in Schwartz, 1972), and mentioned by others. In 1966 Meistrell (in Schwartz, 1972) gave a laboratory demonstration to show that the spit platform is a necessary precursor for overwash and beach drift deposition into deep water. Many studies indicate that offshore and nearshore sediments can move above sea level where a subaerial form already exists. It is only under this latter condition that Gilbert's thesis should be accepted today.

The growth of spits often results in a series of *beach ridges* marking former shorelines. These attest to the contributions of overwash and the episodic nature of extension that is caused by variations in the longshore transfer rate. The wave climate controls sediment mobility, and the angular approach causes the recurvature and hooked features. Large tidal variations also appear to enhance the amount of curvature by increasing wave refraction at the distal end and deposition around the tip. The tidal range also influences tidal current velocities—hence sediment movement; the movement appears to be relatively more important on the bayside and distal beaches because of the limited contribution to movement by wave-induced currents. Wind activity is important in dune growth and wave generation. Offshore topography also induces perturbations along the spit of wave energy and sediment movement. In sum, beach drifting seems to be the primary mechanism for spit growth upon the current deposited platform. Along the spit periodic fluctuations in the sediment budget result in cyclic beach erosion and progradation. The subaerial form is further modified by storm-wave washover and eolian redistribution.

The declining energy continuum along the shoreline leads to local differences in spit unit dynamics that also define the total dynamics. Complex segmentation of the shoreline may result from variations in the sediment/energy field due to engineering structures and offshore controls on wave refraction. Bayside beaches are less dynamic than the more exposed seaward beaches. The often irregular bayside shoreline can prevent the development of the continuous alongshore drift system that is characteristic of the oceanside. The low energies result in poorer water quality and organic infilling plus minimal beach recovery following the occasional storm waves, as shown by lag deposits and profile analysis. On the higher energy oceanside, there is a greater rate and extent of change. The distal beaches are broad, with beach ridges and dune fields which reflect their depositional nature. The proximal unit is usually much narrower with the higher wave energies, reflecting its closeness to the erosional headland. This area is susceptible to storm-wave breaching due to the lack of a substantial sediment reserve. Usually the spits' sediments become finer downdrift as a response to the decreasing wave energy/increasing wave refraction toward the tip. Exceptions to this would appear to result from offshore contributions and storm lag deposits. A substantive example of the study of spit dynamics using an analytical approach is given by Allen (1981).

Location

Spits occur within any region of the world, arctic/tropical, ocean/lake, because they are the end member of a longshore sediment transfer cell upon which there is no regional constraint. They are more numerous, however, where there is shallow coastal water and large volumes of movable sediment. Therefore there are relatively few on tectonically advancing crustal plate coasts compared with receding plate coasts. Also they are more common on irregular coasts because these offer more sediment traps for deposition. Their material foundation will reflect the characteristics of local provenance, with shingle spits being more common in the upper latitudes and sandy quartz and carbonate spits in the mid and lower latitudes respectively.

The location also exhibits local controls on the equilibrium states of shape, sediment supply, and wave energy. As such they are usually attached to detrital or erodable headlands, although rivers and offshore shoals may be important contributors to shape and material. Because these are depositional forms, their location also corresponds to low energy segments of coasts, areas where wave rays diverge.

Age and Time

Owing to the relative stillstand of sea level over the past 3000+ years, few spits antedate this age, and most represent recent developments. They are dynamic in their position and shape—locationally transgressive in response to eustatic and isostatic fluctuations and short-term responsive to meteorologic events. Linear growth rates may be rapid. Orford Ness has extended 6.5 km since the early sixteenth century (Steers, 1962), while a spit at the Courantyne River mouth at the border of Guyana and Surinam has been prograding 62 m annually (Hine, 1979). Such growth is not indefinite, however, because of climatic, sea level, and sediment budget changes. Certainly root erosion, thoroughly discussed by Kidson (in Schwartz, 1972), leads to breaching where the inlet may act as a sediment barrier causing downdrift starvation of the beaches. The short-term dynamics and the long-term variations in the depositional/erosional fulcrum and transgression are characteristic of a spit's history.

Land Use Management Problems

Great demands have been placed on many spits for recreational, commercial, and residential land uses. Traditionally the large economic investments have lead to a protectionist philosophy of shoreline stabilization that is clearly at odds with spit dynamics. If the sensitive proximal beaches are bulwarked by engineering structures, the longshore transport rate is decreased, thus merely displacing the erosion problem downdrift. At the terminus, deposition can be a problem by leaving beach accesses and structures (lighthouses) far behind and by infilling navigation channels that are often routed around the spit's end. While recent research has shown the natural state is the healthiest course of coastal management, many spits are characterized by longstanding and intensive use by man with considerable investment. Therefore these spits do need protection where the costs can be justified versus those of replacement or relocation. Many of these spits, because of their dynamics and restricted size, should easily fit into sediment recycling schemes that take sand from the distal weir and reimplace it on the proximal beaches. Effective land use planning for the present as well as future spit users must then take the spit's dynamics into consideration.

JAMES R. ALLEN

References

Allen, J. R., 1981. Theoretical model of shoreline dynamics at Sandy Hook spit, New Jersey, *Northeastern Geology* **3**, 243-251.
Davies, J. L., 1973. *Geographical Variations in Coastal Development.* New York: Hafner, 186p.
Hine, A. C., 1979. Mechanisms of berm development and resulting beach growth along a barrier spit complex, *Sedimentology* **26**, 333-351.

Schwartz, M. L., ed., 1972. *Spits and Bars.* Stroudsburg, Penna.: Dowden, Hutchinson & Ross, 452p.
Steers, J. A., 1962. *The Sea Coast.* London: Collins, 292p.

Cross-references: *Barrier Beaches; Bars; Beach Processes; Coastal Dunes and Eolian Sedimentation; Coastal Engineering; Cuspate Forelands; Cuspate Spits; Global Tectonics; Nehrung; Sediment Transport; Tombolo; Washover and Washover Fans.*

SPLASH AND SPRAY ZONES

The direct contact of a wave with a near vertical cliff face produces *primary abrasion* to a level slightly above that reached by spring tides. This is a zone of extreme physical erosion (Zenkovich, 1967).

Secondary abrasion occurs as the wave breaks down and splashes up on shore. The *splash zone* is the area just above the zone of primary abrasion where physical weathering associated with wave breakdown predominates. This zone also includes a great deal of *chemical weathering* (q.v.), mostly in the form of *salt weathering* (q.v.), and alternate wetting and drying. E. S. Hills has referred to such wetting and drying by sea water as *water-layer weathering* (q.v.) (Davies, 1972).

The *spray zone* is the area of wind-blown sea water above the splash zone that is dominated by chemical weathering. Therefore this zone begins where the strong physical weathering of the *splash zone* ends and extends to where there is no longer any water layer weathering.

Guilcher and Bodere (1975) attribute most of the surficial sculpturing associated with these zones to be due to salt weathering. They also state that the higher the latitude, the less the spray erosion.

PAUL HELLER

References

Davies, J. L., 1972. *Geographical Variation in Coastal Development.* New York: Hafner, 204p.
Guilcher, A., and Bodere, J. Cl., 1975. Formes de corrosion littorale dans les roches volcaniques aux Moyennes et hautes latitudes dans L'Atlantique, *Assoc. Géographes Francais Bull.* **426**, 179-185.
Zenkovich, V. P., 1967. *Processes of Coastal Development.* New York: Wiley Interscience, 739p.

Cross-references: *Alveolar Weathering; Muricate Weathering; Salt Weathering; Solution and Solution Pan; Tafone; Water-Layer Weathering; Weathering and Erosion, Chemical; Weathering and Erosion, Mechanical.*

SPONGE—See PORIFERA

SPOUTING HORN—See BLOWHOLES

SPRING PITS

Spring pits are irregular depressions, of various sizes, created by the upward escape of groundwater. They may occur on land, as is well known, or under water in lakes and seas. Common sizes are one cm across to 10 or more meters across. Smaller pits, developed in loose sand, may be wiped out temporarily by a surge in current or wave activity and when reestablished may have a different location. They should not be confused with *mud* or *sand volcanoes*, which are active during and immediately after an earthquake.

Spring pits in clay or mud may have steep walls. If, as spring activity decreases, a different material (such as sand) is deposited in the pits, unique depositional features, shaped much like a multidip ice cream cone, may be formed. Sand-filled cones in a Cretaceous shale (*Providence formation*) in Pike County, Alabama, are thought to have had this origin (Tanner, 1964). The cones extend downward into the shale. They range up to a maximum of tens of centimeters in diameter at the top, and up to a meter in height. Model studies designed to duplicate these examples showed that in the simplest case a tripartite vortex trail accounted for the scalloped map view at any given horizon prior to infilling. In tiny examples developed in sand, a tripartite trail would not be necessary.

Spring pits developed temporarily on the lower beach face at low tide, because of outflow of water stored in the beach sand, commonly have enough discharge for sand to be transported, building *low-tide deltas* (q.v.); these deltas (or fans) are typically only some tens of centimeters or a few meters from the pits from which the water flowed.

WILLIAM F. TANNER

Reference

Tanner, W. F., 1964. Filled submarine spring vents in Cretaceous rocks of Alabama, *Southeastern Geology* 5, 113–116.

Cross-references: *Beach; Low-Tide Deltas; Minor Beach Features; Monroes; Mud Volcanoes.* Vol. III: *Littoral Processes;* Vol. VI: *Littoral Processes; Primary Sedimentary Structures.*

STANDING STOCK

Associations of organisms and their activity are described as *standing stock* and *production*.

TABLE 1. High Values of Standing Stock (Wet Weight) for Benthos (g/m^2) and for Plankton (g/m^3)

Benthos	g/m^2
bacteria	1
microflora	2
eelgrass	10×10^3
macroalgae	30×10^3
protozoa	1
meiofauna	5
macrofauna	10^3

Plankton	g/m^3
phytoplankton	3
microzooplankton	3
netplankton	1

Standing stock measures the organisms present at the time of observation within a unit area of sea bottom or a unit volume of water. It is given in terms of number of organisms, displacement volume, wet weight (often termed *biomass*), dry weight, plasma volume, organic carbon, chlorophyll, ATP, or energy content. It may characterize such different catergories as individuals, species, size groups, or total samples. The latter may include dead organic matter (detritus, debris), containing organic carbon and energy (Friedrich, 1969). Examples of standing stock in coastal waters are given in Table 1.

The term standing stock is sometimes replaced by *standing crop*. This expression implies an exploitation of the biological system and the harvest of an amount of organic matter, and it is therefore not recommended for plankton and benthos studies. For the comparison of results from different plankton studies, rough conversion factors are available (Hagmeier, 1961), however, their validity should be checked for each set of investigations.

Standing stock is continuously changed by the metabolism of its components. Under conditions of steady state processes, production and decomposition are equal, and standing stock, in terms of weight or organic carbon, does not change measurably. Interaction between organisms, however, alters abundance and age structure of the association (Mann, 1972). Standing stock of autotrophs depends on nutrients, light conditions, temperature, and mixing of water. For heterotrophs, food and oxygen availability are the main factors. Thus standing stock varies through the course of time (days to years) according to physical, chemical, and biological conditions. In higher latitudes, the seasonal weather changes are of importance, regulating nutrient and food availability and finally growth, reproduction, and recruitment of species and populations.

HJALMAR THIEL

References

Friedrich, H., 1969. *Marine Biology: Introduction to Its Problems and Results.* London: Sedgwick & Jackson Ltd., 474p.
Hagmeier, E., 1961. Plankton-Äquivalente, *Kiel Univ. Meeresforsch* 17, 32-47.
Mann, K. H., 1972. Ecological energetics of the seaweed zone in a marine bay on the Atlantic coast of Canada. I. Zonation and biomass of seaweeds, *Marine Biology* 12, 1-10.

Cross-references: *Aquaculture; Benthos; Biotic Zonation; Coastal Ecology, Research Methods; Coastal Fauna; Coastal Flora; Nutrients; Organism-Sediment Relationship; Production.*

STILLSTAND—See SEA LEVEL CURVES

STOKES THEOREM

Stokes Theorem (Li and Lam, 1964) is given by the expression

$$\int_s (n \cdot E) ds = \oint_c q \cdot d\ell \qquad (1)$$

Since the left hand side of this equation is commonly referred to as circulation, Stokes theorem reduces to:

$$\Gamma = \oint_c q \cdot d\ell \qquad (2)$$

The development of this equation is as follows. Referring to Figure 1, consider the surface S bounded by the closed curve c. Let the surface be divided into many small pieces that may be of various sizes but all are approximately rectangular and flat. Construct a local coordinate system (x, y, z), the z-axis being perpendicular to the surface s, the value of $n \cdot E$, from Equation (1), at any point on this surface is approximately E_z (since n and the z-axis are everywhere parallel). Therefore

$$\int_s (n \cdot E ds) = \iint_s E_z dxdy$$
$$= \int_s \left(\frac{v}{x} - \frac{u}{y}\right) dxdy, \qquad (3)$$

where v and u are the components of velocity in the y- and x-direction, respectively. According to Figure 1, Equation (3) can be partitioned as follows:

$$\iint_s \frac{v}{x} dxdy = \int (V_{ab} - V_{cd}) dy, \qquad (4)$$

where V_{ab} and V_{cd} denote the values of velocity along the sides ab and cd, respectively. Similarly,

$$\iint_s \frac{u}{y} dxdy = \int (U_{bc} - U_{da}) dx, \qquad (5)$$

where U_{bc} and U_{da} are the values of velocity along their respective sides. Recombining Equation (3) yields

$$\int_s (n \cdot E) ds = \int_a^b U_{ab} dy - \int_b^c U_{bc} dx$$
$$- \int_c^d U_{cd} dy + \int_d^a U_{da} dx \qquad (6)$$

It can also be seen from this expression that components of the right-hand side of Equation (6) cancel each other for adjacent squares. Thus, when the results of all the line integrals are summed over the entire surface, the right-hand side of Equation (6) reduces to the line integral around the outside of the outer rectangles in surface s and in the limit equal to the line integral taken over the boundary curve s. Hence Equation (1) becomes

$$\int_s (n \cdot E) ds = \int_c q \cdot d\ell,$$

where q is the velocity component along the differential line element $d\ell$.

JOHN B. HERBICH
TOM WALTERS

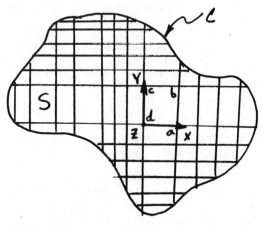

FIGURE 1. Arbitrary closed surface s, bounded by curve c. (Z-axis directly out of paper.)

Reference

Li, W-H., and Lam, S-H., 1964. *Principles of Fluid Mechanics.* Reading, Mass.: Addison-Wesley, 374p.

Cross-references: *Coastal Engineering; Coastal Engineering, Research Methods; Fourier Analysis; Froude Number; Mathematical Models; Reynolds Number.*

STONE PACKING

Vertical packing is the deposition on edge of flat stones or shells forming peculiar pavements without any fine matrix. *Stone packings* occur on *shore platforms* cut into slates, shales, or other thinly bedded sedimentary rocks (Fig. 1) while *shell packings* occur in tidal flats. *Vertical stone packings* are found in marine, lake, and river environments throughout temperate-cold regions of the world, while shell packings have been found in marine environments of temperate and warm regions. In Quebec (St. Lawrence Estuary and Lake St. Jean), flat stones packed on edge form pavements having up to 1000 m^2. Slabs are tightly packed and interlocked, and commonly show flow and radiate structures. At places, vertical flat stones are packed concentrically around boulders or large clasts. Vertical stone and shell packings are produced by wave action (Mii, 1957; Dionne and Laverdière, 1967; Dionne, 1971; Sanderson and Donovan, 1974), rather than by currents (Kostyaev, 1973; Grinnell, 1974) or by periglacial processes (Gregory, 1930; Rozanski, 1943). In temperate-cold regions, frost shattering provides abundant slabs, which are reworked by moderate waves, put on edge, and tightly packed in fractures and small depressions. Vertical flat stone and shell packings have been reproduced experimentally in nature during the summer, giving evidence of the non-periglacial origin of the feature.

JEAN-CLAUDE DIONNE

References

Dionne, J. C., 1971. Vertical packing of flat stones, *Canadian Jour. Earth Sci.* **8**, 1585-1591.

Dionne, J. C., and Laverdière, C., 1967. Sur la mise en place en milieu littoral de cailloux plats posés sur la tranche, *Zeitsch. Geomorphologic* **11**, 262-285.

Gregory, J. W., 1930. Stone polygons beside Loch Lomond, *Geog. Jour.* **74**, 415-418.

Grinnell, R. S., 1974. Vertical orientation of shells on some Florida oyster reefs, *Jour. Sed. Petrology* **44**, 116-122.

Kostyaev, A. G., 1973. Some rare varieties of stone circles, *Biul. Peryglacjalny* **22**, 347-352.

Mii, H., 1957. Peculiar accumulation of drifted shells, *Saito Ho-Onkai Museum (Sendai), Research Bull.* **26**, 17-24.

Rozanski, G., 1943. Stone-centered polygons, *Jour. Geology* **51**, 330-341.

Sanderson, D. J., and Donovan, R. N., 1974. The vertical packing of shells and stones on some recent beaches, *Jour. Sed. Petrology* **44**, 680-688.

Cross-references: *Glaciel; High-Latitude Coasts; Ice along the Shore; Ice-Bordered Coasts; Ice Foot; Kaimoo; Monroes; Mud Volcanoes; Shore Polygons.*

STONE REEF

Stone reefs are bands of beach rock, either semi- or fully submerged, separated from the shoreline by open water or a lagoon. They represent a stage in tropical shoreline development where, following normal beach rock formation below the active beach, coastal retreat has occurred stranding the beach rock as a *stone reef* seaward of the present shoreline or lagoon shore (Fig. 1). Stone reefs often lie

FIGURE 1. Vertical flat stone packings at the surface of a rock shore platform, at Saint Fabien-sur-Mer, lower St. Lawrence Estuary, Quebec (from Dionne, J. C., 1971, Vertical packing of flat stones, *Canadian Jour. Earth Sci.* **8**, 1585-1591, by permission of the National Research Council of Canada).

FIGURE 1. Straight band of beach rock, now stranded as a stone reef 13 km in length at Suape, Pernambuco, Brazil. Note the lagoon shoreline responding to the amount of wave energy allowed through the reef, with the embayment occurring in lee of the lower section of reef. Photo taken at low tide (photo: L. D. Wright).

in a series of a parallel bands, each band representing former beach lines. Where the outer bands occur at successively greater depths, they represent beach rock developed at stillstands in the recent transgression. Stone reefs are therefore restricted to the tropical areas of beach rock formation, in particular where significant coastal retreat has occurred.

Stone reefs are extensive along the northeast coast of Brazil between 5° and 14°S latitude (Fig. 1), where they are termed "recife" (Branner, 1905). They are composed of cemented, horizontally bedded quartz and carbonate sands (Mabesoone, 1964).

ANDREW D. SHORT

References

Branner, J. C., 1905. Stone reefs on the northeast coast of Brazil, *Geol. Soc. America Bull.* **16**, 1-13.

Mabesoone, J. M., 1964. Origin and age of the sandstone reefs of Pernambuco (N.E. Brazil), *Jour. Sed. Petrology* **34**, 715-726.

Cross-references: *Bombora, Bomby; Coral Reef Coasts; Reefs, Noncoral.* Vol. III: *Beachrock.*

STORM BEACH

The product of a storm's impact on a beach environment and its associated erosion is a *storm beach*. The storm beach is characterized

FIGURE 1. Storm beach showing uniformly seaward slope and accumulation of heavy minerals.

by a particular profile configuration and generally by a unique sediment distribution as well.

Storms typically cause rather extensive beach erosion that results in a beach profile nearly uniformly sloping in a seaward direction with a slight concave-upward configuration (Hayes, 1969). The beach is relatively narrow and essentially all foreshore (Fig. 1). This is in contrast to an *accretional beach,* which contains a wide backshore area separated from the foreshore by a prominent berm. Much of the sediment removed from the beach during a storm is deposited just offshore from the beach

FIGURE 2. Trench in storm beach showing thick accumulation of heavy minerals. Note pen for scale.

in the form of a sand bar. During low energy conditions following the storm, the bar moves landward, eventually replenishing much of the sediment lost during the storm.

A particular sediment distribution, especially in the upper part of the beach, is also common to the storm beach. A veneer of heavy minerals forms as a lag concentration of the erosion process. Beaches and adjacent sand dunes, if present, may contain up to 3-4% heavy minerals. These are mineral grains or rock fragments with specific gravity values greater than 2.85.

Included are magnetite, garnet, zircon, rutile, and many others. Concentrations of essentially pure heavy minerals up to 15 cm thick have been accumulated from severe dune erosion (Fig. 2). The presence of a veneer of these materials on the landward portion of the beach is excellent evidence of recent beach erosion even if the profile itself has recovered.

RICHARD A. DAVIS, JR.

Reference

Hayes, M. O., ed., 1969. *Coastal Environments: NE Massachusetts and New Hampshire.* University of Massachusetts, CRG, Contr. No. 1, 462p.

Cross-references: *Beach Cycles; Beach Stratigraphy; Coastal Erosion, Environmental-Geologic Hazard; High-Energy Coasts; Hurricane Effects; Mineral Deposits; Storm Surge; Storm-Wave Environments.*

STORM SURGE

Sustained and strong winds may cause water level changes, called *storm surge*, along the coast. The terms *wind setup* and *wind tide* are used synonomously with storm surge, although wind tide may be either an increase or decrease in water level whereas storm surge and wind setup refer to an increase in water level.

Most often associated with intense onshore moving storms such as hurricanes, this phenomenon actually occurs to a much lesser extent almost continuously. Whereas hurricanes may cause a surge of up to several meters, the regular passage of minor weather systems also has a measureable effect on water level. The magnitude of water level change is dependent on wind velocity and direction, fetch, depth of water and slope of the inner shelf (Coastal Engineering Research Center, 1973). In addition, the configuration of the coast is an important factor, in that embayments may experience a funneling effect with surges being much higher than adjacent areas without an embayment.

The storm surge may be one of the most destructive aspects of a hurricane because of the potential for widespread flooding, especially along heavily populated estuary margins. Although most emphasis is rightfully associated with the increase in water levels, there is much erosion that takes place during the generally rapid decrease in water level that follows the passage of a tropical storm. In barrier-estuarine coastal complexes, such as along the eastern and Gulf coasts of the United States, there is much water that is pushed into coastal bays, lagoons, and estuaries. The release of the force that created the surge permits the great quantity of water to leave these areas rapidly. It moves through inlets of passes created by the storm itself and causes much erosion and seaward transport of sediment (Hayes, 1967).

It is rather easy to separate the various major components of water level fluctuation and to determine the role of each. Standard tide gauges record water level changes, incorporating both the astronomical tide and the wind tide or storm surge. If all fluctuations with a period of 48 hours or less are extracted from the tide gauge record, the result is the astronomical curve, with the residual being the curve of atmospheric related water level changes (Fig. 1). Note that

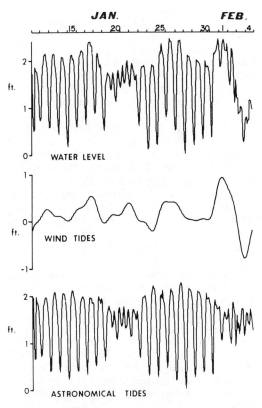

FIGURE 1. Record of the tide gauge at Aransas Pass, Texas, showing the astronomical and atmospheric components for water level variation (Jan.-Feb., 1970).

both positive and negative changes occur, depending on the presence of onshore or offshore moving wind.

RICHARD A. DAVIS, JR.

References

Coastal Engineering Research Center, 1973. *Shore Protection Manual*, vol. 1. Ft. Belvoir, Va.: U.S. Army Corps of Engineers, 750p.
Hayes, M. O., 1967. *Hurricanes as Geological Agents: Case Studies of Hurricanes Carla, 1961 and Cindy, 1963*. Austin, Tex.: University of Texas Bureau of Economic Geology, Rept. No. 61, 54p.

Cross-references: *Beach Cycles; High-Energy Coasts; Hurricane Effects; Storm Beach; Storm-Wave Environments; Tide Gauge*. Vol. II: *Hurricanes*.

STORM-WAVE ENVIRONMENTS

Storm-wave environments is a term introduced by Davies (1964) to describe a coast on which "a significant proportion of waves is generated in local centers by gale-force winds, these being short, high-energy waves of varying direction" (Davies, 1972). The major storm-wave generating areas are along the *Polar Fronts* in both hemispheres (see *Wave Environments*). Characteristically, levels of wave energy are high throughout the year in storm-wave environments, but in the Northern Hemisphere, where the storm belts are geographically less stable owing to the effects of land masses on the general circulation pattern, there is considerable local variability. The storm waves are generated by the dominant westerlies; therefore west-facing coasts have higher wave energy levels than east-facing coasts in the storm belts.

Beach profiles in storm-wave environments exhibit large short-term changes owing to the effects of storm-wave erosion and poststorm deposition. In *plan*, the orientation of beaches is related more to local *fetch* than to the direction of deepwater waves.

A secondary region of storm waves results from the generation of tropical cyclones at the Intertropical Convergence during the period surrounding the summer solstices. These storms are highly variable in frequency, magnitude, and location. They are locally important and can have a dominant effect on the coastal environment—for example, by building *boulder ramparts* in carbonate environments or by the deposition of *washover fans* on barrier island systems.

EDWARD H. OWENS

References

Davies, J. L., 1964. A morphogenic approach to world shorelines, *Zeitschr. Geomorphologie* 8, 127-142.
Davies, J. L., 1972. *Geographical Variation in Coastal Development*. Edinburgh: Oliver and Boyd, 204p.

Cross-references: *Beach in Plan View; Climate, Coastal; High-Latitude Coasts; Waves; Wind; Wind, Waves, and Currents, Direction of*. Vol. III: *Coastal Geomorphology*.

STRAND AND STRANDLINE

The word *strand* is commonly used to describe sandy beaches, or even occasionally offshore sandbanks exposed at low water, on the coasts of Scotland and Ireland. The word, derived from the Old Norse *strond*, has a number of both local and national synonyms particularly in northwest Europe.

The compounded derivative *strandline* is synonymous with *shoreline* or *water line* (Johnson, 1919), but also has two distinct scientific meanings.

First, strandline is used by Pleistocene geologists to cover the range of relict features associated with ancient or abandoned shorelines (Andrews, 1970; Flint, 1971). Careful analysis of such features as beach terraces, deltas, and shingle ridges may allow relative strandline chronologies to be compiled (see, for example, Andrews, 1970) indicating possible extent of both former sea levels and other geomorphic changes.

Second, strandline is used, in the more immediate sense, by ecologists to indicate the zone in which tidal litter is deposited during a period of high water (Ranwell, 1972). Because tidal litter increases stability, nutrient status, and moisture retention, these zones are rapidly colonized by strandline plants (for example, *Atriplex, Elymus, Agropyron,* and *Ammophila*) that act as nucleii for sand deposition and subsequent foredune growth. In this manner the strandline is an important element in controlling coastal progradation.

RICHARD W. G. CARTER

References

Andrews, J. T., 1970. *A Geomorphological Study of Post-Glacial Uplift, with Particular Reference to Arctic Canada*. London: Institute of British Geographers, 156p.
Flint, R. F., 1971. *Glacial and Quaternary Geology*. New York: John Wiley & Sons, 892p.

Johnson, D. W., 1919. *Shore Processes and Shoreline Development.* New York: John Wiley & Sons, 584p.
Ranwell, D. S., 1972. *Ecology of Salt Marshes and Sand Dunes.* London: Chapman and Hall, 258p.

Cross-references: *Beach Processes; Coastal Flora; Geographic Terminology; Sea Level Curves.* Vol. III: *Strandline.*

STRANDFLAT

Strandflats are low, wide bedrock surfaces on land or submerged under shallow coastal seas. On land, they include accordant levels of erosional remnant surfaces that slope gently seaward and offshore. A multitude of *skerries, stacks* and low islands dot their shore zone in subpolar high latitude regions (Andersen, 1965; King, 1972). Steep, high cliffs terminate these surfaces landward. Occasional monadnocks (*nyker,* in Norwegian) rise a few hundred meters above the strandflats. Beyond their smooth transition under the sea, they form broad, submerged rock platforms. Strandflat widths range between 40–65 km along the west Norwegian coast.

Hans Reusch introduced the term in Norway in 1894 (Tietze, 1962). He thought that preglacial marine abrasion played the decisive role in strandflat development. D. W. Johnson (1919) also supported the wave-cut platform theory. Nansen (1922) and Holtedahl (1960) pointed out the importance of frost wedging and glacial erosion, respectively. Ahlmann (1919) believed in the decisive role of subaerial erosion at base level and in later glacial modifications. Today it is recognized that not one single but several erosional factors, as well as postglacial uplift, all contributed to the evolution of present high latitude strandflats of Norway, Spitsbergen, Iceland, Greenland, Antarctica, and possible other regions. In a tropical area, Cotton explained Indian strandflat formation by subaerial weathering and erosion.

ERVIN G. OTVOS

References

Ahlmann, H. W., 1919. Geomorphological studies in Norway, *Geog. Annaler* **1,** 1–148, 193–252.
Andersen, B. G., 1965. The Quaternary of Norway, in K. Rankama, ed., *The Quaternary,* Vol. 1. New York: Wiley Interscience, 91–138.
Holtedahl, O., 1960. Geology of Norway, *Norges Geol. Undersökelse,* No. 208, 540p.
Johnson, D. W., 1919. *Shore Processes and Shoreline Development.* New York: John Wiley & Sons, 584p.
King, C. A. M., 1972. *Beaches and Coasts.* New York: St. Martin's, 570p.

Nansen, F., 1922. The strandflat and isostasy, *Norske Vidensk.-Akad. Oslo Skr., Mat.-Naturv. Kl. 1921, No. 11,* 2 vols., 313p.
Tietze, W., 1962. Ein Beitrag zum geomorphologischen Problem der Strandflate, *Petermanns Geog. Mitt.* **106,** 1–20.

Cross-references: *High-Energy Coast; High Latitude Coasts; Shore Platform; Wave-Cut Bench; Wave-Cut Platform; Wave Erosion; Weathering and Erosion, Mechanical.* Vol. III: *Platforms-Wave-Cut.*

SUBAQUEOUS SAND DUNE–See Vol. 1

SUBMARINE SPRINGS

For at least 3000 years *submarine springs* have been known to man and often utilized as a source of fresh, potable water. Strabo, born in 63 B.C., described its collection in leather bags by boatmen offshore. The same technique can be seen used today in the Persian Gulf. The freshwater bubbles up giving a boiling impression. An offshore spring near Crescent Beach, Florida, is about 15m in diameter, and from a depth of 17m freshwater bubbles up at $42m^3/$sec (1500 cfs or 11,000 gal/second).

Submarine springs are characteristic of limestones affected by karst processes. These are most active when the landscape is more emergent than today, as, for example, when sea level is lowered during a glacial phase of an ice age. Rainwater (which has a low pH) causes solution to take place in the limestone. As it sinks deeper into the rock, it leads to the development of an intricate underground network of pipes and tunnels. The subsurface waters tend to sink until they reach the level of saltwater (which is denser), and thus freshwater is said to float on a saltwater wedge in coastal areas. In sandy or other porous formations, the freshwater/saltwater interface is found at about mean low tide level. In limestones, however, the system of underground karst tunnels facilitates an artesian hydrology, and the water will boil up offshore wherever there is an old solution pipe. From deep aquifers the natural heat flow may augment the hydraulic pressure. In the Bahamas these pipes are often visible in the surface of the coral reefs; they are known as *blue holes, ocean holes,* or *boiling holes.* During the rising tide the water here may be 10°C cooler than elsewhere, because it has generally risen from over 100 m depth. Owing to the extra nutrients brought up, there is often a prolific fish population in or near the pipes.

A small-scale natural replica of this hydraulic phenomenon may be observed on some coral reefs, where small karst pipes are subject to the hydraulic pressure of waves; the result is a *blow hole,* where the water is forced up by the compressed air trapped in the pipe.

The total volume of freshwater transported from the land to the ocean every year by rivers or melting ice is 38,000 km^3/yr, whereas the amount of groundwater outflow is only 1600 km^3/yr, but 1 km^3 of water amounts to 252 $\times$ 10^9 gal, so that the global loss of freshwater through submarine springs is appreciable.

In volcanic areas and, more important, at crustal plate boundaries such as active sea floor spreading centers, there are *hydrothermal springs.* These hot waters carry in solution large amounts of metals, such as iron, manganese, and copper. Normally these solutions are diffused and lost in seawater, but in stagnant basins they may precipitate and accumulate to form potential ore deposits. Many ancient mineral deposits are believed to have formed in this way.

Relevant literature on this subject includes: Bryan (1919), Cooper et al. (1964), Newell and Rigby (1957), and Visher and Mink (1964).

RHODES W. FAIRBRIDGE

References

Bryan, K., 1919. Classification of springs, *Jour. Geoology* 27, 522-561.

Cooper, H. H., Jr.,; Kohout, F. A.; Henry, H. R.; and Glover, R. E., 1964. Sea water in coastal aquifers, *U.S. Geol. Survey Water-Supply Paper 1613-C,* C1-C84.

Newell, N. D., and Rigby, J. K., 1957. Geological studies on the Great Bahama Bank, *Soc. Econ. Paleontologists and Mineralogists Spec. Pub. No. 5,* 15-72.

Visher, F. N., and Mink, J. F., 1964. Ground-water resources in southern Oahu, Hawaii, *U.S. Geol. Survey Water-Supply Paper 1778,* 133p.

Cross-references: *Ghyben-Herzberg Ratio; Water Table.* Vol. I: *Submarine Springs.*

SUBMERGENCE AND SUBMERGED SHORELINE

Submergence of a shoreline may take place because of a rise in sea level or a downward adjustment of the land surface; from a geometrical point of view, the two are indistinguishable. The basic idea, along with that of *emergence,* was known before the turn of the century (see summary in Johnson, 1919).

Shepard (1948), in a refined version of his *coastal classification,* included as examples the following: Ria coasts (drowned river valleys—two varieties); drowned glacial erosion coasts (two varieties); drowned karst; partly drowned deltas; partly submerged moraines, partly submerged drumlins; depressed wave-cut bench coasts; and perhaps others (reef types).

In view of the fact that sea level has both risen and fallen in the past few thousand years (as well as over the past 10^5 years), features of both emergence and submergence can be expected along many shores. Where submergence has been dominant, the coast is either immature (for example, still exhibits nonmarine characteristics, as in Shepard's classification) or mature (now has basically marine characteristics). An example of the latter can be seen in the Chandeleur Islands, off the coast of southeastern Louisiana. This set of discontinuous barrier islands, remnants of an earlier delta rim, is migrating landward and maintaining a geometry that was shaped by marine agencies (primarily waves).

A temporary rise in water level, such as during a hurricane or, in the case of a lake, a wet cycle of perhaps 5-10 years duration, is not generally taken to constitute submergence.

The general effects of submergence depend on whether the waves are depositing material (such as on a transgressing sand beach), or etching out zones of weakness in exposed bedrock, thereby making the coastline more irregular.

WILLIAM F. TANNER

References

Johnson, D. W., 1919. *Shore Processes and Shoreline Development.* New York: John Wiley & Sons, 584p.

Shepard, F. P., 1948. *Submarine Geology.* New York: Harper and Row, 517p.

Cross-references: *Classification; Emergence and Emerged Shoreline; Global Tectonics; Retrogradation and Retrograding Shoreline; Ria and Ria Coast; Sea Level Changes; Sea Level Curves; Tectonic Movements.* Vol. III: *Submerged Shorelines.*

SUCCESSION

In the context of ecology, *succession* implies the observed sequences of changes that take place in any plant community or biotic association. The dependence of animal populations upon plant cover or vegetation logically leads to noncommitant changes in animal populations as the vegetation changes.

The concept of community change, or the fact that plants might so alter the site as to bring about a change in species composition, is more than two centuries old. In 1863 Thoreau observed in central New England that with cutting of a white pine stand, hardwood species tended to dominate the site. He named this process succession. At the turn of the present century one of America's early ecologists, H. C. Cowles (1899), described plant succession on the sand dunes of Lake Michigan. In 1935 A. G. Tansley, an English ecologist, distinguished *autogenic* and *allogenic* succession, the former resulting from the action of the organisms themselves and the latter due to external factors. Ecologists have also used such terms as primary succession referring to the sequence of biotic changes that occur on previously unvegetated sites until a relatively stable (*climax*) or cyclically oscillating system has been created. Such sites include dunes, bare rock outcrops, and open bodies of water. In contrast, secondary succession includes the sequence of vegetation or biotic development on sites previously vegetated but disturbed by storms, fire, drought, or man's activities such as clearing and subsequent abandonment. F. E. Clements (1916, 1928), an American grassland ecologist, formulated many of our early succession and climax concepts. He felt that succession was the embryonic development that ultimately led to a superorganism unit of vegetation, the climax dependent primarily on the regional climate. Early successional diagrams tended to indicate that basic trends in terrestrial systems moved from dry (*xeric*) to more moist (*mesic*) conditions and that hydric (*aquatic*) situations tend to development toward more mesic conditions. Although such trends can occur, strictly unidirectional development—for example, from open water to wetland and ultimately to upland forest—is oversimplified. Seldom does an upland forest actually occupy a former pond or wetland.

The unidirectional sequence was early emphasized by Cowles (1899) in his sand dune studies, with pioneer grasses giving way to cottonwood and pine followed by oak and ultimately a beech-maple forest—the so-called regional climatic climax. However, more recent studies by Olson (1951) documented that an oak forest is as far as the vegetation development will proceed, since oak makes the soil unfavorable (too acidic) for beech-maple seedling establishment. In coastal environments salt spray can also be a significant factor in modifying or arresting vegetation development (Boyce, 1954).

Primary succession on rock outcrops along coastal areas has also been thought of as a unidirectional sequence from lichens to mosses to herbaceous plants to shrubs and ultimately a forest. However, this too is an oversimplification of the process on such rigorous sites, where vegetation change is more typically cyclic owing to periodic droughts that constantly set back or interrupt the development. In fact, lichens still cover some rocks in New England where for thousands of years they have resulted in a relatively stable (climax) rock lichen community. Cutting, forest fires, and postagricultural abandonment of fields all result in a sequence of change that may lead to the ultimate establishment of a forest. Yet the process or manner of development may be highly variable, and in some situations a shrub or thicket community may tend to be relatively stable for a long time (Niering and Goodwin, 1974). With old field abandonment, annuals, perennial grasses, forbs, and a thicket of woody species, both trees and shrubs, tend to be the physiognomic aspects that one may observe over time. There is no so-called preparing the way from one type to the next, as suggested in classical succession. Yet this does not imply that site modification does not occur or that nurse plants may not be involved in some situations. The sequence of individual species invasion, establishment, and differential rate of growth are among the critical factors governing just how a given field or pasture will ultimately develop. If shrub cover becomes dense, it is possible that tree establishment may be suppressed to a point where a shrub community is the relatively stable type for some time. Thus F. E. Egler's relay versus initial vegetation floristic concepts are relevant (Egler, 1954). Relay floristics implies the traditional unidirectional sequence of one physiognomic type following another, whereas initial floristics implies that many species populations, including woody species, become established early in the process, and thereafter much of the pattern observed is a mere differential development of the species present at the start. However, in many cases both initial and relay floristics may be involved. Ecologists are currently reevaluating traditional successional concepts, as revealed by the reviews of Drury and Nisbet (1973) and Connell and Slatyer (1977).

In coastal marshes, belting or vegetation zonation should not be confused with succession. There may be a successional movement of belts in terms of a landward movement of zones with coastal submergence, but periodic tidal inundation serves as a tidal subsidy or equilibrium factor in maintaining the belting pattern. One can predict that saltwater cordgrass (*Spartina alterniflora*) can be succeeded by salt meadow cordgrass (*Spartina patens*) along the developing marsh front. Yet many tidal marshes represent a constantly changing mosaic of vegetation types.

One can only conclude that succession has been oversimplified as an ecological process. Unidirectional stereotyped diagrams are often misleading since they set preconceived goals toward which the development will proceed. This may or may not be the case. Viewed as a dynamic process involving interacting populations of plants and animals operating in a constantly changing environment, cyclic or nonpredictable patterns of development are more common, especially along the coast, where differential hydroperiod, salinity, and wind are stressing the environment.

WILLIAM A. NIERING

References

Boyce, S. G., 1954. The salt spray community, *Ecol. Monogr.* **24**, 29-67.
Clements, F. E., 1916. *Plant Succession.* Washington, D.C.: Carnegie Institute of Washington, Pub. 242, 512p.
Clements, F. E., 1928. *Plant Succession and Indicators.* New York: H. W. Wilson, 453p.
Connell, J. H., and Slatyer, R. O., 1977. Mechanisms of succession in natural communities and their role in community stability and organization, *Am. Naturalist* **3**, 1119-1144.
Cowles, H. C., 1899. *The Ecological Relations of the Vegetation on the Sand Dunes of Lake Michigan.* Chicago: University of Chicago Press, 119p. (See also *Bot. Gaz.* **27**, 95-117, 167-202, 281-308, 361-391, 1899).
Drury, W. H., and Nisbet, I. C. T., 1973. Succession, *Harvard Univ. Arnold Arboretum Jour.* **54**, 331-368.
Egler, F. E., 1954. Vegetation science concept. I. Initial floristic composition, a factor in old-field vegetation development, *Vegetatio* **14**, 412-417.
Niering, W. A., and Goodwin, R. H., 1974. Creation of a relatively stable shrublands with herbicides: Arresting "succession" on rights-of-way and pastureland, *Ecology* **55**, 784-795.
Olson, J. S., 1951. *Vegetation-Substrate Relations in Lake Michigan Sand Dune Development*, Ph.D. thesis. Chicago: University of Chicago, 127p.
Tansley, A. G., 1935. The use and abuse of vegetational concepts and terms, *Ecology* **16**, 284-307.
Thoreau, H. D., 1863. *Excursions.* Boston: Ticknor and Fields, 319p.

Cross-references: *Biotic Zonation; Coastal Fauna; Coastal Flora; Production; Salt Marsh; Standing Stock; Vegetation Coasts.*

SURF BEAT

Surf beat results from the generation of long waves by the high and low waves of two different, interacting wind-wave trains as they approach the coast. Groups of high waves raise the water level temporarily at the shore (King, 1972). Surf beat periodicity is reported to range from 1-5 minutes (Wilson, 1966) to 1-10 minutes (Bates and Jackson, 1980). The amplitude is measured as just a few centimeters.

Surf beat was first identified, and named, by Munk in 1949.

JOSÉ F. ARAYA-VERGARA

References

Bates, R., and Jackson, J. A., eds., 1980. *Glossary of Geology.* Falls Church, Va.: American Geological Institute, 749p.
King, C. A. M., 1972. *Beaches and Coasts.* London: Arnold, 570p.
Munk, W. H., 1949. Surf beats, *Am. Geophys. Union Trans.* **30**, 849-854.
Wilson, B. W., 1966. Seiche, *in* R. W. Fairbridge, ed., *Encyclopedia of Oceanography,* New York: Reinhold, 804-817.

Cross-references: *Bars; Beach Processes; Minor Beach Features; Ripple Marks; Wave Energy; Waves; Wave Statistics.*

SURFING

Surfing is the sport of riding water waves, accomplished using various forms of equipment—surfboards, inflated mats, canoes, sailboats—or no equipment in body surfing except for perhaps fins (Kelly, 1973). The primary objective in surfing is to balance the force of gravity against the advancing wave profile. Surfing is practiced on several types of waves. Ocean swell and sea waves are the most common types that are surfed, but waves made by mechanical means, boat wakes, and tidal bores are also used for surfing. The most desirable form for surfing is to ride a breaking wave in a region between the broken portion and the unbroken portion of the wave crest: in some forms of surfing, nonbroken waves or broken waves are ridden (Walker, 1974).

JAMES R. WALKER

References

Kelly, J., 1973. *Surf Parameters, Final Report, Part II. Social and Historical Dimensions.* University of Hawaii, James K. K. Look Laboratory of Oceanographic Engineering, Tech. Rep. No. 33, 251p.
Walker, J. R., 1974. *Recreational Surf Parameters.* University of Hawaii, James K. K. Look Laboratory of Oceanographic Engineering, Tech. Rep. No. 30, 299p.

Cross-references: *Biotic Zonation; Coastal Fauna; Human Impact; Thalassotherapy; Tourism; Waves.* Vol. I: *Ocean Waves.*

SWEEP ZONE

Sweep zone is a term introduced by Barnes and King (1955) to define the area between the upper and lower lines that join the highest and lowest points on a set of superimposed beach profiles. The sweep zone indicates the vertical envelope within which the beach surface varied during the period of the profiles. This method can be used to illustrate the variability of a beach surface and, if more than one set of profiles surveyed over different time periods is used, can be used to determine seasonal or long-term changes in the beach profiles.

EDWARD H. OWENS

Reference

Barnes, F. A., and King, C. A. M., 1955. Beach changes in Lincolnshire since the 1953 storm surge, *East Midland Geog.* 4, 18-28.

Cross-references: *Beach; Beach Cycles; Beach Processes; Beach Processes, Monitoring of; Coastal Engineering, Research Methods; Coastal Morphology, Research Methods; Profiling of Beaches; Protection of Coasts.*

SWELL AND ITS PROPAGATION

The water waves having the greatest effects on coasts are often those generated by the winds in relatively distant, but severe, storms. These waves are called *swell*, and have longer periods than locally generated waves, typically in the range 8-20 sec.

Swell travels across oceans, as shown by observations at a single station. Since the velocity of water waves in deep water is inversely proportional to their period, the longest waves from a particular storm arrive first and the rate of change of wave period enables the distance of the storm to be deduced. Barber and Ursell (1948) measured swell arriving at the British coast from a storm off Cape Horn 10,000 km away. Snodgrass et al. (1966), by using a line of measuring stations across the Pacific Ocean, showed that the longest waves are not attenuated in deep water, though they are scattered and absorbed by shallow water and coasts. Some of the wave energy they measured had traveled more than halfway round the earth.

Swell is not influenced by the *Coriolis* effect since, although it may take days to cross an ocean, the time scale of motion of the water is that of the wave period, which is much shorter than a day. Thus in the deep oceans, swell travels along a path perpendicular to the wave crests, following a great circle. Kenyon (1971) points out that this result is modified by the presence of ocean currents that refract the swell. He shows that the Antarctic circumpolar current refracts swell sufficiently to account for discrepencies that puzzled Snodgrass et al. (1966) in measurements in Alaska of swell from the Indian Ocean.

Near coasts, swell is influenced by the effect of water depth on wave velocity as soon as it is over the continental shelf. The subsequent *refraction* deflects it from its great circle path, but if the swell has a long travel over these shallower waters, the spherical geometry of the earth's surface should be allowed for. For example, compare an east-west line at latitude λ on a chart using the standard equatorial Mercator projection with the great circle starting at the same point and with the same initial direction. After following both lines for a distance s, the distance between them Δs is given by $\Delta s/s = \frac{1}{2} (s/R) \cot \lambda$, where R is the earth's radius.

D. HOWELL PEREGRINE

References

Barber, N. F., and Ursell, F., 1948. The generation and propagation of ocean waves and swell. I. Wave periods and velocities, *Royal Soc. London Philos. Trans.* **A240**, 527-560.

Kenyon, K., 1971. Wave refraction in ocean currents, *Deep-Sea Research* **18**, 1023-1034.

Snodgrass, F. E.: Groves, G. W.; Hasselmann, K. F.; Miller, G. R.; Munk, W. H.; and Powers, W. H., 1966. Propagation of ocean swell across the Pacific, *Royal Soc. London Philos. Trans.* **A259**, 431-497.

Cross-references: *Wave Climate; Wave Environments; Wave Refraction Diagrams; Waves; Wave Statistics; Wave Theories, Oscillatory and Progressive; Wind, Waves, and Currents, Direction of.* Vol. I: *Wave Theory.*

SWING MARK

At times, wind blowing from various directions will move a fixed plant stem or exposed root tip to inscribe a circular or semicircular furrow in the surrounding loose sand (Fig. 1).

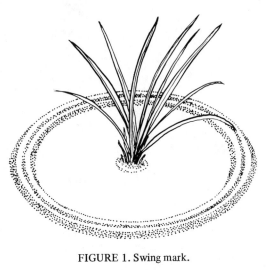

FIGURE 1. Swing mark.

SWING MARK

This rather ephemeral sedimentary structure is called a *swing mark* (Gary, McAfee, and Wolf, 1972.) The radius of the circle is determined by the length of the root or the distance over which the plant stem arcs to touch the ground.

MAURICE L. SCHWARTZ

References

Gary, M.; McAfee; R., Jr.; and Wolf, C. L., 1972. *Glossary of Geology.* Washington, D.C.: American Geological Institute, 805p.

Cross-references: *Coastal Dunes and Eolian Sedimentation; Coastal Flora; Minor Beach Features.*

TAFONE

Originally used in Corsica to describe unusual weathering features, *tafone* (or tafoni, plural) has become established as the generic name for a type of cavernous weathering characterized by the existence of hollows or cavities that range in size from a few centimeters to a meter or more in diameter. Depth is variable, the cavities commonly being nearly hemispherical. The shape and orientation are strongly influenced by the rock fabric, causing the hollows to be elongated in the direction of foliation or bedding planes. Coalescence of the cavities produces mushroomlike shapes, natural arches, and other unusual sculptured forms. Eventually the outcrop surface may be destroyed by this process of expansion. When numerous small cavities occur, the resulting spongelike texture is termed alveolar or honeycomb weathering. Both alveolar weathering and tafone may occur independently, but often the two forms coexist and seem to originate from a similar process. Tafone occurs commonly in granitic rocks and in sandstone, but has also been observed in schist, gneiss, gabbro, and limestone. The rate of development may be rapid, cavities having formed in the surface of stone seawalls built less than a century ago.

The origin of tafoni is not well understood, particularly in view of the wide range of environments in which it appears. Tafoni are abundant in the polar regions of Victoria Land, Antarctica, but occurs also in tropical and subtropical climates. The cavities form in the deserts of Australia, Africa, central Asia, and South America as well as in such humid regions as Hong Kong and the island of Aruba, West Indies. In North America, tafoni occurs in such diverse areas as the deserts of New Mexico, Utah, Arizona, and the coast of Washington state, and at midcontinent locations in Wisconsin and Illinois (Blackwelder, 1929; Bryan, 1928; Mustoe, 1982).

Tafoni often occur in outcrops having a hardened surface layer caused by precipitation of iron hydroxides or other compounds derived during weathering of the interior rock. The cementing action of these oxides produces a resistant rind. Thus any form of erosion that attacks the outcrop in a localized fashion will produce cavities, since penetration of the protective layer leads to rapid destruction of the weaker interior rock. Differential attack of the surface might result from variations in lithology, the presence of fissures or zones of high porosity that allow water to penetrate, or localized attack by organisms such as lichens. Because the formation of the protective rind requires moisture, this explanation is consistent with the observation that tafone seldom occurs in extremely dry environments, being more common along the margins of deserts than in the arid central regions. In many locations tafoni occur mainly as coastal features presumably owing to the existence of a favorable microclimate, since even in extremely arid regions sea winds may cause the shoreline to be relatively humid. The existence of a protective rind is most favored by an environment where some moisture is present but where periods of dryness allow evaporation to occur at the rock surface.

Recently salt weathering has been proposed as an explanation for tafone, particularly in Antarctica (Prebble, 1967; Evans, 1970; Johnson, 1972). In general, the distribution of tafone throughout the world correlates well with environments where salt crystallization occurs. In arid zones salt weathering is concentrated where moisture is retained along the base of cliffs and undersides of boulders, where shadow weathering (tafone) may often be found. Along the coast, cavities are presumed to form by salt crystallization as wave-splash evaporates. Because salt weathering may attack in a relatively selective fashion, being controlled by variations in moisture or exposure to salt spray, tafoni might be formed even on outcrops where a hardened surface layer is absent, though the development of cavities would be enhanced when such a rind is present. Early explanations of tafoni invoking erosion by wind action, temperature fluctuation, or frost wedging have largely been abandoned owing to lack of substantiating evidence.

GEORGE MUSTOE

References

Blackwelder, E., 1929. Cavernous rock surfaces of the desert, *Am. Jour. Sci.* **17**, 393-399.

Bryan, K., 1928. Niches and other cavities in sandstone at Chaco Canyon, New Mexico, *Zeitschr. Geomorphologie* **3**, 125-140.

Evans, I. S., 1970. Salt crystallization and rock weathering: A review. *Rev. Géomorphologie Dynam.* **19**, 153-177.

Johnson, J. H., 1972. Salt weathering processes in the McMurdo Dry Valley regions of South Victoria Land, Antarctica, *New Zealand Jour. Geology and Geophysics* 16, 221-224.

Mustoe, G. E., 1982. Cavernous weathering in the Capitol Reef desert, Utah, *Earth Surface Processes and Landforms* (in press).

Prebble, M. M., 1967. Cavernous weathering in the Taylor Dry Valley, Victoria Land, Antarctica, *Nature* **216**, 1194-1195.

Cross-references: *Alveolar Weathering; Muricate Weathering; Salt Weathering; Solution and Solution Pan; Weathering and Erosion, Biologic; Weathering and Erosion, Chemical; Weathering and Erosion, Differential; Weathering and Erosion, Mechanical.* Vol. III: *Solution Pits and Pans; Tafoni.*

TANGUE

Tangue (Bourcart, Jacquet, and Francis-Boeuf, 1944) is a tidal deposit of calcereous, silty mud occurring in shallow bays along the coast of Brittany. It has the permeability of fine sand and is remarkably coherent (Bourcart and Charlier, 1959). The calcareous fraction (35-75%) comes from the grinding of *coquina*, facilitated by the destructive action on the shell material of mushrooms. The noncalcareous fraction is derived from reworked fluviolacustrine silts, and consists of diatoms and fine quartz grains. Sponge spicules are an additional component. Deposits may be > 15 m thick; only Holocene deposits are known for certain.

JOANNE BOURGEOIS

References

Bourcart, J.; Jacquet, J.; and Francis-Boeuf, C., 1944. Sur la nature du sédiment marin appelé tangue, *Acad. Sci. (Paris) Comptes Rendus* **218**, 469-470.

Bourcart, J., and Charlier, R. H., 1959. The tangue, a "nonconforming" sediment, *Geol. Soc. America Bull.* **70**, 565-568.

Cross-references: *Beach Material, Sorting of; Sediment Classification; Sediment Size Classification; Soils; Tidal Flat.* Vol. VI: *Tangue.*

TAR POLLUTION ON BEACHES

Tar accumulates on beaches as a result of marine oil spills by oil tankers (ballast water and sludge), ships (bilge water), leaking oil terminals, and land-based industries. The oil in the water loses its light fractions due to evaporation and gradually forms tar balls. Winds, currents, and waves push the floating tar balls landward, where sand grains attach to the tar, increasing its specific gravity and preventing it from floating back into the sea. On the beach, the tar balls form successive strips along the wave swash marks. The tar then dries in the sun, shrinks in size, and eventually breaks down into small particles that are blown by winds either landward or seaward.

The increase in oil traffic during the past two decades has made tar pollution on beaches a worldwide phenomenon (Butler, Morris, and Sass, 1973; Saner and Curtis, 1974). Scientific reports are available from India, Africa, Mediterranean coasts, the United States, and the Bahamas, giving quantities ranging from hundreds of grams to several kilograms of tar per meter of running beach.

Tar balls are a major nuisance for bathers. Special machines have been designed to clean beach tar, but they are inefficient as yet. The tar pollution problem may be greatly reduced if harbors provide proper installations for dumping bilge water, sludge, and the like and if laws forbidding marine oil pollution are enforced.

ABRAHAM GOLIK

References

Butler, J. N., Morris, B. F.; and Sass, J., 1973. *Pelagic Tar from Bermuda and the Saragasso Sea.* Bermuda Biological Station for Research, Spec. Pub. No. 10, 346p.

Saner, W. A., and Curtis, M., 1974. Tar ball loadings on Golden Beach, Florida, *in Marine Pollution Monitoring (Petroleum).* Washington, D.C.: National Bureau of Standards, Spec. Pub. No. 409, 79-81.

Cross-references: *Human Impact; Oil Spills and Pollution; Pollutants.* Vol. XIII: *Oil Spill Clean-Up on Coasts.*

TECTONIC MOVEMENTS

Tectonic movements are crustal dislocations of the earth. When vertical, such movements may cause a change in relative sea level as the land moves up or down on a continental scale (*epeirogeny*) or local scale (*orogeny*). Epeirogeny, a form of *diastrophism*, was defined by Gilbert (1890) as producing plateaus and basins on a continental scale. In contrast to this, the term orogeny, adopted in the mid-19th century (Bates and Jackson, 1980), refers to the regional formation of mountains.

Other tectonic movements that have been identified are *continental flexure, post-glacial rebound*, and *marine isostasy*. Continental flexure is the tilting of the edge of the continent that further submerges the continental shelf and raises the terrestrial coastal region, as if pivoted at the shore. During times of continental glaciation, the weight of the ice may depress the surface of the land. Return of the surface to the original level following deglaciation is called post-glacial rebound. And finally, the added weight of the water on the continental shelves during interglacials, causing a downwarp of the coastal region, is a form of marine isostasy. All three of these tectonic movements may cause a relative change of sea level at the shore, evidenced, in part, by submerged or raised terraces.

Many systems of coastal classification (q.v.) incorporate in some way, relative changes of sea level due to tectonic movements.

JOSÉ F. ARAYA-VERGARA

References

Bates, R. L., and Jackson, J. A., eds., 1980. *Glossary of Geology*. Falls Church, Va: American Geological Institute, 749p.

Gilbert, G. K., 1890. *Lake Bonneville*. U.S. Geological Survey, Monograph 1, 438p.

Cross-references: *Classification; Emergence and Emerged Shorelines; Global Tectonics; Isostatic Adjustments; Isostatic Uplift Coasts; Sea Level Changes; Submergence and Submerged Shorelines.*

THALASSOTHERAPY

Little has been published in American periodicals on the subject of seawater therapy. Apparently the most significant and most recent paper on *thalassotherapy* was read at the Thirteenth Annual Session of the American Congress of Physical Therapy, in Philadelphia on September 10, 1934, and subsequently published in the *Archives of Physical Therapy, X Ray, and Radium* (Singer, 1935). Singer briefly reviews the history of thalassotherapy and traces it back, in modern times, to the Margate royal sea-bathing hospital. He credits Barellai of Italy, Perochaud of France, and Benecke of Germany as being the founders of contemporary seawater therapy. His definition of thalassotherapy is, however, somewhat too narrow by today's scope. It is no longer limited to the therapeutic utilization of the maritime climate. Well beyond that utilization is the direct administration of seawater orally and by injection, of the spray of the water, of the pounding effect of the waves, of heated seawater baths, and most recently of the combination of electroacupuncture and seawater therapy. The 400 seashore sanatoria and preventoria, with their 30,000 beds in 1935, have multiplied, especially since World War II.

Even though for a while thermal spas were *fashionable* in the United States, surprisingly there has never been a major interest, similar to the one prevailing in Europe, in tapping the curative properties of sea air and seawater. Belgium, France, Germany, Romania, the Soviet Union, for instance, have several centers for marine cures. Seawater therapy, practically abandoned after World War I, seems to have regained a new vitality. Before the 1920s, seawater was injected intravenously in medical centers in Paris, Brest, Nancy, Reims, and elsewhere, and cured thousands of children of blood and nervous diseases of major importance (Singer, 1935; Larivière, 1958). Baths and jet pressure of seawater have brought about near miraculous cures of nervitis, lumbagos, cellulite, and obesity. During World War II in German concentration camps such as Dachau, criminal physicians used political and ethnic prisoners for seawater experiments. Observers at the subsequent war crimes trials recall that prisoners were forced to absorb large, lethal quantities or received intravenous injections of seawater. The aim was of course to study rescue possibilities for flyers downed at sea. Fortunately, those experiments are but a horrible recollection today yet injections and oral administration of sea water even in minute quantities have serious therapeutic effects.

The ancient Greeks placed considerable faith in the healing power of the sea and appreciated thalassotherapy. Their historian Xenopho, wishing to pictorialize the magnitude of the "march of the 10,000" in Asia Minor, used the expression *"kai kata gai kai kata thalatan"* (they traveled by land and by sea). Both Greek words remained in modern languages, *gai* appears in geography and geology, for instance, while *thalatan* still designates the oceanic environment. Thus we find this last vocable in thalassotherapy: the therapeutic uses of the sea. Euripides said: "The Sea restores man's health" and Plato, cured by a marine treatment, wrote, "The Sea washes all man's ailments." More than twenty centuries later, another historian, Michelet, wrote, "La terre vous supplie de vivre; elle vous offre ce qu'elle a de meilleur la mer, pour vous relever . . ." (Earth begs you to live; it offers you the best she has, the ocean, to perk you up . . .).

Jean-Pierre Range (Larivière, 1958) points out that thalassotherapy rapidly faded away in the Western world imbued with Aristotlian logic, later nurtured by Galileo and Descartes and, more recently, by Pasteur and Claude Bernard.

Rebirth of a Therapy

Whether we endorse seawater cures or not, the recent success, principally in Europe, of *thalassothermal* centers deserves our attention at a time when we are turning to the sea for many uses. Curative properties of seawater have been heralded for a long time, and what we witness today is more a resurgence of interest than an entirely new concept.

The specific therapeutic value of seawater was mentioned in 1750 by Richard Russell, a British physician, who praised it for the treatment of glandular disturbances. The *Proceedings of the French Imperial Academy of Medicine* (February 5, 1856) contain a report stating:

Therapeutics draw good results, every day from the use of sea water and from salt springs; and although its use for baths in tubs is often not as advantageous as when taken in the surf, where the mechanical action of the fluid is added to its chemical action, one can still expect much of this therapeutic application....

In the twentieth century, under the impact of Freudian writings and the psychosomatic philosophical movement, there was a splitting of medical thinking into a traditional scientific approach and what Range (Larivière, 1958) calls enthropological medicine, in which he includes thalassotherapy and acupuncture. Whether these views are accurate is for physicians to debate, but there is a strange coincidence of timing between the renewed interest in marine cures and a growing disenchantment with ways of life in an ever increasingly technological society.

Water, Sun, and Air

Thalassotherapy does not limit itself to, nor is it based only upon, the use of seawater; part of the treatment is the change of life style, the new surroundings, the freeing of the individual from modern life's stresses. It is concomitant with natural *aerosoltherapy* and, during the good seasons, *heliotherapy* as well. Aerosoltherapy is the absorption through the pulmonary parenchyme of the negative marine air ions whose short-lived existence, linked to offshore winds, is limited to a few hundred yards of shore. Woodcock and Blanchard (1957) established that sea air contains several hundred of thousands of tiny salt particles per cubic foot; aerosols made of very small particles that work their way into the deepest parts of pulmonary alveoles and settle on their walls. Since the chemical composition of the aerosol salts is nevertheless the same as that of the sea water bubbles from which they originated, their physiological effect is probably not negligible. As for the heliotherapeutic effects, the high proportion of ultraviolet in seaside sun rays has a recognized effect upon calcium metabolism.

Warm sea water baths facilitate the passage through the skin of such natural oligoelements as calcium, magnesium, manganese, and cobalt, which are defense agents of the human organism able to substitute for, or at least buttress, drugs fighting off viruses and bacterial infections. The biochemical properties of seawater, according to such French physicians as Labeyrie (Larivière, 1958). are successful substitutes for comfort medications, without having side-effects. They have been used successfully against arthrosic lesions and further arthritic degeneration.

A major difference between an actual swim in the ocean and the bath or swimming pool swim is the absence of the initial shock of cold water, which might even be dangerous for older persons. Heated sea water causes a dilatation of cutaneous vessels and under water jet streams have the same beneficial effect as the pounding of the waves against the body and its spraying by sea foam. At less than 80°F (27°C) baths are stimulating, between 82°F (28°C) and 90°F (32°C) they are sedating; they should not exceed 100°F (38°C) except under medical control (Fig. 1).

According to Larivière (1958), hydrostatic pressure acting in concert with salinity, provides relief to muscles and articulations, exerts its action upon the skin's intra- and extracellular fluids, the peripheral veins, and capillary vessels, thereby activating blood circulation, desintoxication, and cleansing of tissues. The blood volume leaving the heart is 24% higher.

In localized action of hypertonic waters osmosis plays a dominant role; concentrated saline water flushes blood through the skin,

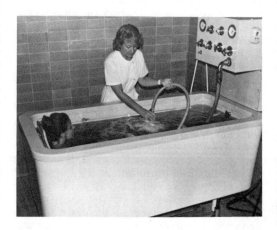

FIGURE 1. Heated saltwater bath at Trouville, France (photo courtesy of Centre Thalassotherapeutique, Trouville-sur-mer).

removes fats from it, and softens deposits and incrustations. Furthermore, a bioelectric action occurs due to the difference of potential between seawater and tissues, and ionic effects modify the electrical charge of the skin.

Mere balneation brings about an increase of 24% of the quantity of blood chased from the heart to the large vessels. The warm seawater bath has a decongestant and smoothening action upon articulations; it is especially recommended for the treatment of chronic evolutive polyarthritis. In the swimming pool, seawater facilitates movements due to its high salt content.

Physiological effects of marine climates are reflected in a slowing down of the rates of breathing and heart beat. The amplitude of the respiratory movement and pulmonary ventilation are increased, and so are hematies in the blood and hemoglobin ratio, while heart contraction is reinforced; the body is better prepared for the beneficial effect of seawater baths due to an increase of cutaneous exchanges. Additional symptoms include a neuro-endocrinic and growth stimulation, and an increase of diuresis and of basal metabolism.

The Role of Algae Algae can reinforce seawater effects, and algae powder has been added to tub water. However, many physiologists contend that the passage through the skin of algae components such as iron, cobalt and others, remains hypothetical.

Though the Romanians publicize marine mud applications the most, German physicians had already used such muds in 1929, having collected them in a removed sector of Wilhelmshaven harbor. French marine cures use an alga jelly mixed with wet sea sand heated in a double boiler. When applied to the body, it slowly releases its heat. While no agreement has been reached on this matter, some believe that there occurs, through the skin, an ionic displacement of marine electrolytes and algal constituants.

Showers and Irrigations Showers of seawater are taken before or after the bath; they exert a dual thermal and mechanical action on vessels and nerve endings of the skin (Fig. 2). Most often a short cold spray is followed by a warm, even shorter spray. The effect must be somewhat similar to the tonifying action of the Finnish sauna.

Overall showers are either direct or interrupted. Localized showers are used for decongestion and antineuralgic treatment. Gynecological irrigation is particularly effective because of seawater's hypertonic action upon mucous tissues and its penetrative ability in remote areas. Finally, nasal irrigations, aerosols, and gargles are widely recommended for sinus troubles and eye-nose-and throat ailments.

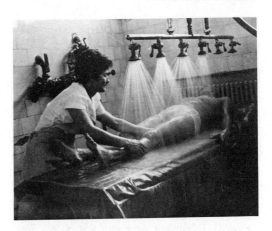

FIGURE 2. Shower massage at Black Sea resort, Mangalia Nord, Romania (photo courtesy of the National Tourist Office, Romania).

The Marine "Cocktail" Quite obviously drinking of sea water is very limited. Nevertheless it is prescribed for gastric problems, anemia, and growth problems. Both German and French pharmaceutical firms market vials of seawater generally tapped some 30 miles offshore at 7 to 10 fathoms. Physicians often prescribe these products for children.

Over the last decade seawater with a salinity reduced to from 1.16 to 0.38 ounce per quart has been use in medicine; optimum concentration seems to be 0.12 ounce per quart. Such water remains rich in magnesium and oligoelements as confirmed through spectrographic analyses. This "living" water, freed of chlorides, closely resembles blood plasma.

Finally, health stores and pharmacies sell bread, crackers, and pasta made by using sea water.

Treatment Centers

If some thalassotherapy stations already have a long history, new centers are being opened and existing ones greatly improved. In Europe, Eastern Germany and France have the largest number of stations. Belgium and Romania also have well known cure centers.

The Quiberon thalassotherapeutic station is located on a very unusual presque isle of Southern Brittany. This spur of land, some 14 km long, protrudes into the ocean and faces Belle-Ile, the largest of France's offshore islands. Its bay is thickly carpeted with sediments used in thalassotherapy, due to the slowing down of oncoming currents. Other treatment centers are located at Trouville-sur-mer, a seaport in the Calvados, a sector of Normandy, northeast of the very fashionable sea-shore resort of Deauville.

FIGURE 3. Saltwater pool with wave effect, Westmorland, Island of Sylt, North Sea, Germany (photo courtesy of Federal Republic of Germany Tourist Office).

Today no less than twenty-two of forty German North Sea and Baltic resorts are designated as marino-therapeutic spas; most of them having developed during the last half century. While surf bathing is still the recommended "cure," the shortness of the season has led to the construction of swimming pools, many with artificially created wave effects (Fig. 3). German resorts also provide silt pack treatments (*pelotherapy*); the silt, washed ashore here, is particularly rich in vegetal and mineral substances, showing some similarity to the content of moor-silt used at inland spas. Physicians have praised the North Sea resorts for treatment of allergies, bronchial asthma, hay fever, eczema, spastic ailments, and upper respiratory diseases.

Occasionally, natural springs are found that possess the desired mineralization. Such a mineral spring was discovered in 1856 in the gardens of the Belgian royal shore residence at Ostend, a few hundred feet from the beach. Today Ostend is a renowned center of both thermalism and thalassotherapy that in a single year may cater to as many as 80,000 curists. It provides closed and open-air seawater swimming pools and *hammans*, the famed hot Turkish baths.

The marine vegetal mud of Ostend is principally made up of compacted peat brought back, to the beach area, in large quantities at strong high tides and then gathered by spa personnel. After having been dried, it is turned into a powder to which is added a certain quantity of marine clay that has been first dried and pulverized. When it is to be used the peloid is mixed with seawater and heated in a double-boiler. Applications are local.

The Black Sea Area Along the Romanian littoral, south of Constantsa, the nearshore land area is dotted with hypersaline lakes. Some of these lakes maintain connections with the Black Sea. The extent of their salinity depends on the salinity of the adjoining sector of the Black Sea, and varies between 1.32 ounces per gallon (10 mg/l), and 2.53 ounces per gallon (19 mg/l); the presence of bromine in their waters confirms their marine origin. According to the season, water temperatures range from 32°F to 81°F (0°C to 27°C), and the pH remains alkaline. Some thirty species of algae have been identified and microfauna is abundant. Water level falls as low as 5.6 inches (14 cm) below that of the nearby Black Sea. The bottom muds, rich in amino-acids and carcyonoids, have a high rate of natural radioactivity. Of these hypersaline lakes, Lake Techirghiol, a Sarmatian (upper Miocene) lake, is one of the most important. These coastal lakes remind one somewhat of the limans found north of Constantsa; shallow and muddy lagoons, or lakes, close to river mouths and protected by bars, spits, or sedimentary accumulations. Some muds are sapropelic, made of plant remains, especially algae, which putrefied in an anaerobic environment (Pricăjan and Opran, 1970).

The Romanians and Soviets have actively developed treatment centers along the Black Sea coast (Fig. 4). These centers cater to thousands of patients from their own countries, but are rapidly drawing increasing numbers of foreign visitors as well.

Mangalia is the southernmost harbor and resort on the Romanian coast. It has attracted seamen since classical times and was actually

FIGURE 4. Thalassotherapy center, Olymp Resort, on the Black Sea coast of Romania (photo courtesy of the National Tourist Office, Romania).

built on the ruins of a 6th century B.C. Doric colony, dominated by the Callatis citadel. Located on a shore plateau, the cliffs overlook a fine sand beach between 170 and 500 feet (50-150 m.) wide. The modern name Mangalia crops up for the first time in writings of Paolo Girgis Ragasa around 1593. Neptun and Eforie are artificial creations just a few kilometers north. These three communities form the Romanian cure center.

The Romanians have greatly enlarged their sea thermal stations ever since they decided to lift, at least in one direction, their iron curtain. The Mangalia-Neptun complex offers a therapy based upon seawater and sapropelic muds use, sulphurous mesothermal springs, and a balmy climate. Conditions are particularly favorable for aero-ionization and insolation; the air is rich in iodine, magnesium, bromine, and, naturally, sodium chloride. The hypertonic seawater contains between 0.22 and 0.25 ounce per pound (14-16 g/kg) or 2.07 ounces per U.S. gallon (15.5 g/l) of mineral substances. The hydrotherapy includes salt baths in lake- and seawater, warmed seawater swimming, subaquatic showerbaths and subaquatic elongations. But equally praised are mud baths and applications of mud poultices.

The mud comes from Lake Techirghiol, containing black, pasty sapropetic mud of extremely fine granular structure and a water retention factor of 10.28 ounces per pound of mud (643.22 g/kg). Compared with roughly 0.23 ounce per pound of minerals in the Black Sea water, Lake Techirghiol is a *"hypersaline"* lake whose water has a mineralization of 10.67 ounces per U.S. gallon (80 g/l) equivalent to 1.27 ounce per pound.

Water and mud properties are maintained by the existence of a highly specific and rich flora and fauna. Concentrated mud extracts are shipped to distant treatment centers. Professor Ana Aslan, a controversial physician outside her own country, maintains that Black Sea thalassotherapy and mud treatments have brought relief to rheumatism and arthritis patients, are potent remedies for geriatric illnesses, and have made spectacular cures of psoriaris and varicose veins.

Predominant salts are sodium chloride, sulphate, and bromide. Peat mud, also used, is found in Mangalia Lake some 150 m from the shore. Bicarbonated, hypotonic, mesothermal (78.8°F, 26°C), radioactive, sulfurous water spouts from springs on and near the beach at Mangalia-Neptun.

Sapropel is an aquatic ooze, rich in carbonaceous or bituminous matter. The sapropelic mud has a plasticity value of 8.81 ounces (250 g) water, a thermal capacity index of 20.99 (metric). Its organic substances content reaches 24.16 ounces per U.S. gallon (181 g/l) and total mineralization in the solid stage is 2.4429 ounces per thousand (69.2647 g ‰). Its classical values are enhanced by bacteriostatic, batericidal, and anti-allergic qualities due to a high content in C, E, B_2 and B_{12} vitamins, nicotinic acid, hormones and organic substances.

Lake Techirghiol is separated from the sea by a 500-foot (150 m) wide strip of land. On the beach, cold bath facilities follow the Egyptian method of open-air treatment.

At the Neptun center seawater inhalations are sometimes coupled with Gerovital H3 and Aslavital, controversial medications, treatments against premature aging and to enhance regeneration of vital functions.

The Eforie-Neptun-Mangalia complex can handle well over 1,000 patients a day. From a strictly thalassotherapeutic viewpoint it provides hot mud applications and packings (Fig. 5), tub baths in highly concentrated saline water at 97°F (37°C) from Lake Techirghiol, tub baths of hot mud (97°F), and indoor pools of hot sea water at 97°F. Open air treatments can be given to 5,000 persons a day. After body warming on the hot sand, they are covered with a thin layer of mud (Fig. 6), then dried in the sun; this then followed by bathing, washing and salt water swimming.

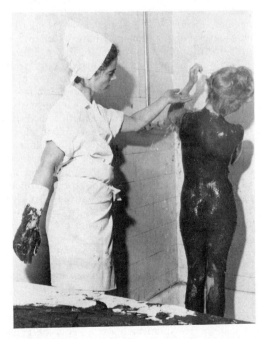

FIGURE 5. Mud therapy, Eforie Nord, Romania (photo supplied by R. Charlier).

FIGURE 6. Patients covered with radioactive mud on sun-bathers' beach area, Eforie Nord, Romania (photo: R. Charlier).

Thalassotherapy and Acupuncture

The Trouville-sur-mer marine health establishment combines thalassotherapy and acupuncture. Both medical techniques act at skin level, both are natural treatments, and both attempt to alleviate the same ailments. Experimental essays have included gold-and-silver needle-mud therapy, swimming pool underwater treatment with needles applied and the passing of a microelectric current through the needles. *Electrothalassopuncture,* a recent Chinese development, seems to have excellent results for the treatment of arthrosic conditions.

The warm seawater bath causes a cutaneous vaso-dilatation and a hyperhemia of the conjunctive sub-cutaneous tissue that permits an increase of ionic and electronic exchanges between the stable and perfect marine environment and the patient's internal milieu whose physicochemical equilibrium is unbalanced through illness. Acupuncture needles cause, at the carefully selected puncture point, a cellular depolarization bringing about biochemical modifications which are transmitted to nerve centra and the ailing organ. Both techniques aim thus at reestablishing a biological, or energetic, equilibrium.

A recent personal communication received from Dr. S. Pavie, formerly of the Faculty of Medicine of the University of Paris, and presently medical director of the Etablissements de Cures Marines, in Trouville-sur-mer, describes electroacupuncture as a treatment particularly in use for rheumatismal algies. The therapist places the acupuncture needles at the algic symptomatic points and then spreads a layer of ground algae, rehydrated with seawater previously heated at $37°C$.

The steel needles thus heated by this algic mud, which is richly ionized, provoke an oxidation phenomenon that appears to be beneficial in rheumatology. This treatment, however, is not comprehensive by itself and is administered in addition to the standard marine cure involving underwater massage, swimming pool re-education, high pressure jet showers, and algae baths.

While the *Trouville Hospital Bulletin* claims favorable reactions for many more ailments, the list is so long one may wonder whether thallasso-acupuncture is indeed that successful. The enumeration includes rheumatism, arthrosis, lumbagos, sciatica, neuralgies and such neuro-vegetative problems as insomnia, palpitations, cellulitis, dyspepsia, and chronical colitis.

Oceanic Medical Complex

Benefits from the sea are not limited in medicine to treatments at cure stations or of minute quantities mixed with tap or distilled water inland. Marine physiology and pharmacology are domains that have hardly been investigated and, yet, offer considerable promise. Cod-liver oil was the forerunner of the synthetic vitamin.

Several human, or even mammallian, organs are difficult to study. There is usually a good chance to find a similar organ in more convenient form in marine animals. Squids, for instance, have nerve fibers in their mantles as thick as pencil heads while individual fibers of living human nerve cells are extremely small. The octopus's brain is sufficiently complex to serve as a model for a human brain (Gruber, 1968).

If the pharmaceutical value of algae has been praised by some, the oceanic drug chest has been opened for many centuries. Iodine, algas, cod-liver oil, and others won their place in the pharmacopeia a long time ago. Extracts of many sea foods can kill viruses and bacteria, while antibiotics can be extracted from phytoplankton and sponges. Research has led to encouraging results in the areas of arteriosclerosis, cardiology, and oncology (Arehart, 1969).

Future Prospects

Widely attended symposia on drugs from the sea have been organized in recent years, but little emphasis has been placed on the healing effects of the sea itself. Though there are agencies concentrating on medical oceanography, marine therapy appears to be a domain that has not yet gained coordinated international attention. Efforts and research remain fairly individual affairs with each station going its own way (Spaulding, 1976).

Often the marine water cure is closely associated with concurrent therapy: air, sun, and mud treatments. The most recent development is, as we underscore here, its combination with acupuncture, and even more recently, electroacupuncture. Whatever the claims of its effectiveness, lacking any information from the mainland Chinese, no assessments of its eventual merits can be made at this time.

Yet medical authorities have recognized remarkable recoveries, and spectacular improvements, of hundreds of patients who followed thalassotherapeutic treatments. This in itself should generate a rekindled interest in the curative powers of the ocean. And, thence, perhaps, the ocean might bring many renewed hope and longer life, in still another way, besides providing food, mineral resources, transportation avenues, energy and possibly habitat.

ROGER H. CHARLIER

References

Arehart, J. L., 1969. Oceanic drug chest, *Sea Frontiers* **15**, 99-107.
Gruber, M., 1968. The healing sea, *Sea Frontiers* **14**, 74-86.
Larivière, A., 1958. *Les Cures Marines*. Paris: Aubier, 155p.
Pricăjan, A., and Opran, C., 1970. La protection et l'exploitation rationelle des lacs et des boues thérapeutiques, in *Géographie des Lacs, Travaux du Colloque National de Limnologie Physique*. Bucharest: State Publishing, 307-311.
Singer, C. I., 1935. Thalassotherapy, *Arch. Phys. Therapy, X Ray, and Radium*, 662-666.
Spaulding, I. A., 1976. *Recreational Adequacy of Beach Activity and Comparative Regulating Influences*. Kingston, R.I.: University of Rhode Island, Marine Tech. Rept. 47, 48p.
Woodcock, D. C., and Blanchard, A. H., 1957. Bubble formation and modification in the sea and its meteorological significance, *Tellus* **9**, 145-158.

Cross-references: *Beachcombing; Coastal Reserves; Europe, Coastal Morphology; Human Impact; Surfing; Tourism; Soviet Union, Coastal Morphology.*

THERMAL POLLUTION

Electricity production, either with nuclear boilers or with conventional fossil-fired boilers, creates waste heat that must be disposed of in the site environment. Because of the growing energy demand and the amount of discharged heat, a shift to atmospheric as well as to sea cooling of water from thermal plants is now required in many countries. Seawater offers good potential as coolant as well as the possibility of large-scale heat rejection with a small environmental impact when due care is taken to avoid local effects in the discharge area. The local hydrographic conditions need to be studied and precautions taken to avoid recirculation of the discharged water. Then too, extensive analytical studies may be required to gain a full understanding of the heat dissipation capability of a site.

Heated water releases are commonly divided into two categories of discharge: *surface* and *submerged*. The first procedure involves the spreading of the lighter, heated water as a jet on the surface of an impoundment. Bottom diffusers ensure the jetlike dispersion of heated water on its way to *naviface*, or submerged level of entrapment. Although the second procedure (submerged discharge) produces smaller surface temperatures above diffusers, which conforms better to the pollution control policies of some countries (for example, some U.S. states permit a $1.5°F$ rise in surface temperature within a radius of 300+ feet), the surface discharge imposes a weaker impact on the environment because of faster heat transfer to the atmosphere and smaller total amount of heat restored in water.

There are five basic processes that contribute to the dispersion of heat in a large receiving body of water. *Jet entrainment, turbulent diffusion*, and *buoyant spreading* are coupled primarily with the mixing of the heated and ambient fluids. *Heat transfer* to the overlying air transfers the thermal energy to the atmosphere. *Interaction* of the initial jet and ambient current generally determines the location of the plume temperature field in relation to the outfall structure (and, equally important, to the intake structure) and receiving water body. In the initial or *near-field* region, temperature changes are governed mostly by the geometry and hydrodynamics of the discharge. Mechanisms that affect the temperature reduction in the near-field region are the dilution and entrainment due to the momentum of the discharge jet and the buoyancy effects due to the temperature difference between the discharge and the receiving water. The temperature distribution in the *far-field* region is controlled by the receiving water. The important properties of the receiving water body are natural thermal stratification; advection; diffusion and dispersion due to tidal currents, wind-driven currents, and wave action; and heat dissipation from the water surface. There is no universally accepted criterion for defining the exact limits of the transitions zone between the near field and the far field. In general, the use of only a near-field model has been adequate to determine whether required temperature standards are being met. However, sometimes (for example, in the implementation of the United States National

Environmental Policy Act) it is necessary to assess the total environmental impact (that is, the total amount of heat disposed of), and therefore the far-field effects also require evaluation.

Three fundamental sets of equations are used for the analysis of the spreading of cooling water in the sea: equations of motion considering eddy viscosity, equations of continuity, and thermodynamic equations including heat budgets and heat exchange between the sea surface and the atmosphere. By solving numerically these simultaneous equations under the boundary conditions concerning factors such as quantity, velocity, and temperature of released cooling water; topography of the coast; location of the outlet; natural structure of temperature in the sea region; meteorological parameters (wind, insolation, air temperature, humidity, cloud cover); characteristics of turbulence in the sea; and maritime conditions (tidal, coastal, and wind-driven currents and waves), one can obtain a detailed distribution of both velocity and temperature in the sea region in front of the outlet for each power station. A number of analytical models have been developed to deal with the near field, far field, and both of them. Summaries and discussion are given by the International Atomic Energy Agency (1974), Policastro (1972), Zeidler (1975), and others. Stolzenbach and Harleman's model (1971) with their computer program can be recommended as one of the most comprehensive near-field models. Typical isotherm patterns as predicted by some models are shown by the International Atomic Energy Agency (1974).

Hydraulic models have been important engineering tools for many years and have been used as a means for developing quantitative estimates of some of the flow quantities—such as velocities, depths, and pressures—to aid in the design of prototype water works. A thorough review of the use of hydraulic models for thermal plume modeling has been prepared by Silberman and Stefan (1970). However, neither hydraulic nor analytical models have been generally compared or verified against prototype field data. With this in mind, the Argonne National Laboratory has undertaken the collection of perhaps the most comprehensive prototype data available for the near field (Frigo and Frye, 1972). Data comparisons and the resulting differences in model predictions are discussed by Policastro (1972).

Temperature affects nearly every physical property of concern in water quality management, including density, viscosity, vapour pressure, surface tension, gas solubility, and diffusion. Dissolved oxygen is perhaps the most important single parameter, since it is essential for sustaining the forms of aquatic life. High temperatures in receiving water bodies, especially if combined with organic pollution loads, may reduce the saturation capacity of oxygen to the extent that fish may not survive. An increase in water temperature may also have a significant effect on chemical reactions, with the rate of reaction being approximately doubled for each $10°$ rise in temperature. However, without adverse biological effects on water quality and direct impact of waste heat on living organisms, the physical and chemical problem of heat disposal would be minimal. Of primary interest is the possible impact on fish populations. Perhaps the most widely used diagram to describe this impact is Brett's (1959) temperature tolerance trapezium. Of course there is no single lethal temperature for an organism. Extensive literature (Coutant, 1971; Policastro, 1972) is devoted to different effects of temperature on aquatic forms, which can be classified as decomposers, detrital feeders, producers, and consumers. For microorganisms, the higher the temperature, the more active a microorganism, unless the temperature or a secondary effect becomes a limiting factor. However, there are only a few studies on complete aquatic ecosystems.

Waste heat can also be used beneficially for aquatic cultivation, desalination, sewage and waste water treatment, urban and district heating, and elimination of ice from water bodies. This aspect should be taken into account together with the detrimental heat impact in siting of power plants in coastal environments.

RYSZARD ZEIDLER

References

Brett, J. R., 1959. *Thermal Requirements of Fish—Biological Problems in Water Pollution*. Cincinnati, Ohio: R. A. Taft Engineering Center, Rept. No. W60-3.

Coutant, C. C., 1971. Thermal pollution—Biological effects, *Jour. Water Pollut. Control. Fed.* 1292.

Frigo, A., and Frye, D., 1972. *Physical Measurements of Thermal Discharges into Lake Michigan*. Argonne, Ill.: Argonne National Laboratory, Center for Environmental Studies.

International Atomic Energy Agency, 1974. *Thermal Discharges at Nuclear Power Stations: Their Management and Environmental Impacts*. Vienna: Tech. Rept. Ser. No. 155.

Policastro, A. J., 1972. State-of-the-art of surface thermal plume modelling for large lakes, *in Argonne National Laboratory 1972 Annual Meeting of the American Institute of Chemical Engineers, Proceedings*. New York.

Silberman, E., and Stefan, H., 1970. *Physical (Hydraulic) Modelling of Heat Dispersion in Large Lakes: A Review of the State-of-the-Art*. Argonne, Ill.: Argonne National Laboratory, ANL/ES-2.

Stolzenbach, K. D., and Harleman, D. R. F., 1971. *An Analytical and Experimental Investigation of Surface Discharge of Heated Water.* Cambridge, Mass.: M.I.T. Ralph M. Parsons Laboratory, Tech. Rept. No. 135.

Zeidler, R., 1975. Hydromechanics of heated- and waste-water discharges into the marine environment, *Oceanologia 5.*

Cross-references: *Coastal Engineering; Dispersion; Human Impact; Nearshore Water Characteristics; Nuclear Power Plant Siting; Pollutants; Production; Waste Disposal.*

THERMAL POWER

Tapping the difference of temperatures between surface and deep water in coastal areas has been a subject of interest ever since d'Arsonval (1881) suggested utilization of the energy thus generated. Early attempts include the schemes of Georges Claude (1930) on the Meuse River (Belgium), in Matanzas Bay (Cuba) and offshore Brazil. The first full-scale attempt, initiated by the French in 1942, led to the construction of an experimental power plant near Abidjan (Ivory Coast) (Beau, 1955). There an *upwelling* of deep, cold water provided excellent conditions even though conduits had to reach to depths of about 400m. The plant was needed because of the construction of chemical and other industries to be located on the coast. The plant did go out of business because of repeated ruptures of the ducts, the large dimensions of the turbines, and the decision to build a conventional plant that at the time could produce cheaper electricity. For bibliographies dealing with ocean thermal power see Charlier and Gordon (1982) and Kim (1981).

The subject has been revived, particularly in the United States, although France still conducts research on a modest scale (Weeden, 1975). Among others, TRW, Hydronautics, Lockheed, Carnegie-Mellon University, the University of Massachusetts, Johns Hopkins University, and the Andersons (Solar Power, Inc.) have developed *thalassothermal* projects; ERDA refers to such plants as OTEC (Ocean Thermal Energy Converter). Schemes propose open and closed cycle arrangements, and most current plans favor the use of an intermediate fuel to reduce turbine size (Charlier, 1978). A plant was tested offshore of Hawaii in 1979 and 1980. The principle of thalassothermal energy is based on the fact that the ocean collects large amounts of solar radiation and stores it as heat energy. This energy can be converted into electrical energy by using surface water (warm) as a heat source and subsurface water (cold) as a heat sink (ERDA, 1977). It is estimated that today such electricity can be produced at competitive prices. From a coastal and nearshore viewpoint, environmental impact is not unfavorable and, except for some probable slight modification of surface water temperature, no major ecological damage is foreseen.

Numerous sites worldwide are suitable. In United States waters, Florida, Hawaii, and the Virgin Islands hold promise. Nigeria, India, Israel, and Australia are also conducting feasibility studies. Several Pacific Ocean island agencies have sollicited support to build OTEC plants. The success of thalassothermal plants depends on circumventing some construction and operation problems that engineers feel can be overcome. Currently, offshore plants are favored over land-based ones. OTEC plants cannot be substituted for other types of electricity production plants but, like *tidal power* (q.v.), they are a supplemental source that should not be left untapped (Avery, 1980). Although experimental plants will be of modest size, plans for the future foresee major schemes combining electricity production with the possible manufacture of ammonia, extraction of hydrogen, and coast industrial development. However, serious consideration should be given to the possibility of constructing small plants, particularly in energy-poor, developing nations, that would prompt regional coastal growth.

ROGER H. CHARLIER

References

Avery, W. H., 1980. Ocean thermal energy conversion, contribution to the energy needs of the United States, *Johns Hopkins APL Tech. Digest* 2, 101-107.

Beau, Ch., 1955. L'état actuel des études et travaux en vue de la construction d'une centrale électrique thermique des mers en Cote d'Ivoire, *Industrie & Travaux d'Outre-Mer III* 17, 222-223.

Charlier, R. H., 1978. Other non-living resources, *in* E. M. Borgese and N. Ginsburg, eds., *Ocean Yearbook*, vol. 1. Chicago: University of Chicago Press, 170-174.

Charlier, R. H., and Gordon, J. R., 1982. *Handbook of Ocean Energies.* New York: Industrial Press (in press).

Claude, G., 1930. Power from the tropical seas, *Mech. Eng.* 52, 1039-1044.

d'Arsonval, A., 1881. Energie de l'échange thermique des mers, *Rev. Scientifique* Sept. 17.

ERDA, 1977. *Ocean Thermal Energy Conversion.* Washington, D.C.: Energy Research and Development Administration, 60p.

Kim, Y. C., ed., 1981. *Bibliography on Ocean Energy Prepared by the Task Committee on Ocean Energy.* New York: American Society of Civil Engineers, Spec. Pub., 39-75.

Weeden, S. L., 1975. Thermal energy from the oceans, *Ocean Industry* 10, 219-228.

Cross-references: *Coastal Engineering; Demography; Human Impact; Thermal Pollution; Tidal Power.*

TIDAL BASIN

A *tidal basin* is a natural or man-made body of water that experiences the periodic rise and fall of water level produced by the creation of a tidal wave within the basin itself or by the introduction of tidal effects from connecting basins. Although minor tidal effects may occur on even the smallest enclosed lakes, the term tidal basin is more properly applied to bodies of water where tidal changes are not obscured by daily changes in weather conditions.

The form, magnitude, and direction of the tide-induced forces depend upon the size, shape, and orientation of the basin as well as the degree of connectivity with other tidal basins. Natural basins are irregular structures, and most may be considered composites of several interconnecting sub-basins, each of which exhibits very different tidal characteristics. Certain large bodies of water, such as the Gulf of Bothnia, may be relatively tideless as a result of their orientation to other tidal basins and to *amphidromic systems,* whereas relatively small *estuaries* may have a considerable tidal range owing to resonance and friction affecting the tidal wave entering from a larger tidal basin (Defant, 1961; Dronkers, 1964).

KARL F. NORDSTROM

References

Defant, A., 1961. *Physical Oceanography.* New York: Pergamon Press, vol. 1, 729p.; vol. 2, 598p.
Dronkers, J. J., 1964. *Tidal Computations in Rivers and Coastal Waters.* Amsterdam: North-Holland Publishing Co., 518p.

Cross-references: *Amphidromic Systems; Estuaries; Tidal Prism; Tidal Range and Variation; Tides.*

TIDAL CURRENTS

Tidal currents have magnitudes up to 6 m/sec in coastal regions. They vary in magnitude in a way similar to tidal height (see *Tidal Type, Variation Worldwide),* but the phase of the current relative to the height may differ from place to place. In nautical circles they are known as *tidal streams* in order to further distinguish them from currents due to other causes such as winds.

In a confined channel the direction of the tidal current must be along the channel. However, if the channel is more than a few kilometers wide, the *Coriolis effect* leads to a cross-channel slope of the water surface with gradient $2\omega \sin\lambda \ U/g$ (ω = angular velocity of the earth, λ = latitude and U = current magnitude). There is also a corresponding variation of velocity across the channel. In the Northern Hemisphere, the maximum current and water elevation is to the right of the current direction (to the left in the Southern Hemisphere). For details, see Proudman (1953), chapter 12.

If a tide travels for 100 km or more in shallow water, a normally symmetric tide will become asymmetrical with the incoming (flood) current being stronger and lasting for a shorter time than the outgoing (ebb) current. This has its most pronounced form when *tidal bores* occur and the flood tide arrives suddenly in a wave, which in the largest bores is several meters high and traveling at several meters per second (Defant, 1961).

In more open seas, the direction of a tidal current usually varies continuously. If only one tidal harmonic is considered, then the velocity vector **U** will describe an ellipse in the velocity plane—for example: $\mathbf{U} = \mathbf{U_i} \cos\Omega t + V_j \sin\Omega t$. However, often more than one tidal harmonic is important and the velocity variation is more complex. The mean value of the current is usually zero except when topographic features cause asymmetries in the flow—for example, flow separation in the lee of a headland. The mean displacement of the water is not necessarily zero, since the distance it travels may be sufficient to carry it into an area with a different current variation.

In all cases, tidal currents vary with depth. They are just like any other turbulent flow in water that is only a few meters deep: the flow diminishes toward the bottom with a logarithmic velocity profile near the bed. In deeper water, density stratification often leads to more varied velocity profiles. The stratification may be due to heat, salt, or suspended sediments. Since tidal flow is essentially unsteady, the direction and magnitude of currents at depth may lag behind the surface current because of bottom friction. Changes with depth occur also due to bottom topography and Coriolis effects (Defant, 1961).

D. HOWELL PEREGRINE

References

Defant, A., 1961 *Physical Oceanography,* vol. 2. New York: Pergamon Press, 598p.
Proudman, J., 1953. *Dynamical Oceanography.* London: Methuen & Co., 409p.

Cross-references: *Currents; Tidal Deltas; Tidal Range and Variation; Tidal Type, Variations Worldwide; Tides; Tide Tables and Charts.* Vol. I: *Oceanography, Inshore.*

TIDAL DELTAS

Tidal deltas at inlets develop as *flood deltas* (landward of inlet) and *ebb deltas* (seaward). The more common delta—the flood delta—may range in shape from irregular sand shoals to the typical deltoid shape (Fig. 1). Flood and ebb tidal currents flowing rapidly through the inlet are dispersed and reduced in the shallow lagoon depositing their sediment load as *delta shoals*. In general, the flood delta is more prominent (Johnson, 1919) with both intertidal and even supratidal forms, while the ebb delta is often subtidal and less prominent.

Lucke (1934) was the first in the United States to study tidal deltas with a complete report on the morphology, sedimentation, and stratigraphy of the flood tidal delta at Barnegat Inlet, New Jersey. Others studying these aspects of flood deltas have been Fisher (1962), Price (1963), El-Ashry and Wanless (1965), Morton and Donaldson (1973), and Hayes and Kana (1976). Ortel (1972, 1975) described ebb delta features and processes, while Pierce (1970) discussed differences of flood tidal deltas and washover fans. Hydraulic and engineering aspects of tidal inlets, as part of the continuing General Investigation of Tidal Inlets (GITI) by the United States Army's Coastal Engineering Research Center and its Waterways Experiment Station, resulted in over a dozen reports, including an annotated bibliography to 1973 of over 1000 references on these and geologic aspects of inlets (Barwis, 1976).

Flood and ebb tidal flow hydraulics develop characteristic delta morphology (Fig. 2) consisting of a flood tidal delta and an ebb tidal delta with flood tidal delta shoals, flood tidal delta channels, and distributaries (Fisher, 1962). The flood tidal deltas (Hayes and Kana, 1976) often have a flood ramp, flood channels, ebb shield, ebb spits, and ebb spillover lobes. Similar morphology of the ebb tidal delta consists of a major ebb channel, channel-margin bars, terminal lobe, marginal flood channels, and swash bars.

Flood tidal deltas should not be confused with *washover fans* (usually temporary in nature), which do not have channels reaching below sea level. In paleogeographic reconstruction, flood tidal deltas should not be confused with *fluvial deltas*. In the former, in addition to morphologic differences, grain size decreases landward (lagoonward); in the latter, grain size decreases seaward.

JOHN J. FISHER
ELIZABETH J. SIMPSON

FIGURE 1. Prominent flood tidal delta at Charlestown Inlet, Rhode Island, exhibiting typical deltoid shape in lagoon. This is a low tidal range coast, and the ebb delta is a minor feature (photo: J. J. Fisher).

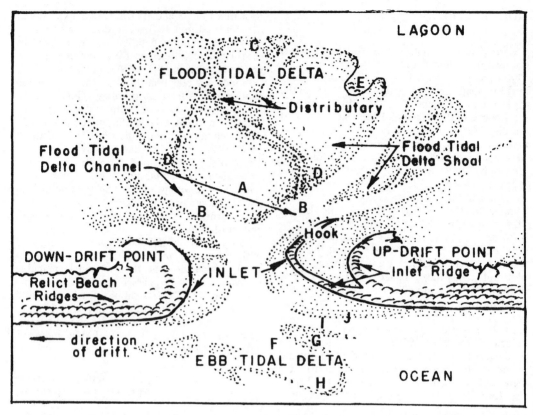

FIGURE 2. Major morphologic features of a typical inlet tidal delta (from Fisher, 1962). Minor features: *A.* flood ramp; *B.* flood channels; *C.* ebb shield; *D.* ebb spit; *E.* ebb spillover lobes; *F.* major ebb channel; *G.* channel-margin bars; *H.* terminal lobe; *I.* marginal flood channels; *J.* swash bars (terminology after Hayes and Kana, 1976).

References

Barwis, J. E., 1976. *Annotated Bibliography on the Geologic, Hydraulic, and Engineering Aspects of Inlets.* Vicksburg, Va.: U.S. Army Corps of Engineers, Waterways Experiment Station, GITI (General Investigation of Tidal Inlets) Rept. 4, 333p.

El-Ashry, M. T., and Wanless, H. R., 1965. Birth and early growth of tidal delta, *Jour. Geology* 73, 404-406.

Fisher, J. J., 1962. *Geomorphic Expression of Former Inlets Along the Outer Banks of North Carolina.* Chapel Hill, N.C.: University of North Carolina, unpub. Master's thesis, 120p.

Hayes, M. O., and Kana, T. W., eds., 1976. *Terrigeneous Clastic Deposition Environments.* Columbia, S.C.: University of South Carolina, Department of Geology, Coastal Research Division, Tech. Rept. 11-CRD, 302p.

Hayes, M. O.; Owens, E. H.; Hubbard, D. K.; and Abele, R. W., 1973. The investigation of form and process in the coastal zone, *in* D. R. Coates, ed., *Coastal Geomorphology.* Binghamton: State University of New York, 11-41.

Johnson, D. W., 1919. *Shore Processes and Shoreline Development.* New York: John Wiley & Sons, 584p.

Lucke, J. B., 1934. A study of Barnegat Inlet, *Shore and Beach* 2, 45-94.

Morton, R. A., and Donaldson, A. C., 1973. Evolution of tidal deltas along a tide dominated shoreline, *Geol. Soc. America Abs. with Programs* 4, 604.

Ortel, G. F., 1972. Sediment transport of estuary entrance shoals and formation of swash platforms, *Jour. Sed. Petrology* 42, 857-863.

Ortel, G. F., 1975. Ebb-tidal deltas of Georgia estuaries, *in* L. E. Cronin, ed., *Estuarine Research, Geology and Engineering,* vol. 2. New York: Academic Press, 267-276.

Pierce, J. W., 1970. Tidal inlets and washover fans, *Jour. Geology* 78, 230-234.

Price, W. A., 1963. Patterns of flow and channeling in tidal inlets, *Jour. Sed. Petrology* 33, 279-290.

Cross-references: *Inlets and Inlet Migration; Tidal Currents; Tidal Flat; Tidal Inlets, Channels, and Rivers; Tides.*

TIDAL FLAT

Tidal flats are depositional surfaces built in a shallow-water marine environment. They are especially well-developed in areas protected from strong wave action. Characteristically, these surfaces are subjected to diurnal or semidiurnal tidal flooding, although the higher parts of tidal flats may be inundated only at spring tides, during storm surge, or wind tide conditions. The flats consist mainly of mud or fine sand, with evaporite deposits being prevalent in arid environments.

The best-suited tidal flats are those bordering the North Sea (Reineck, 1972). In North America, similar tidal flats surround the Bay of Fundy and occur along the New England coast. Such flats are developed in a humid, temperate climate in the lee of barrier islands or sand bars or in sheltered estuaries. Salt marsh plants constitute a significant part of these tidal flats. In tropical climates, most emphasis has been on tidal flats in semiarid regions such as the northwestern Gulf of Mexico (Thompson, 1968), Trucial Coast of the Persian Gulf (Evans et al., 1969), and northwestern Australia (Thom, Wright, and Coleman, 1975). The characteristic features of these flats are the occurrence of evaporites, the broad expanse of surfaces in the high intertidal or even supratidal zone bare of vegetation, and the presence of algal mats. The role of mangroves on tidal flats in tropical regions should also be noted (Thom, Wright, and Coleman, 1975).

The principal environments of tidal flats are distributed laterally into distinct zones reflecting changing dynamics of sedimentation and ecological conditions in relation to the degree of tidal inundation. Sediments become progressively finer away from inlets of channels toward the more landward portion of the tidal flats. Typically, the tidal channels that dissect the flats contain the coarsest sediments, often sandy with bidirectional crossbedding. In macrotidal areas such as the Ord River of Western Australia, well-developed flood-tide megaripples are exposed on these flats at low tide. Adjacent to the channels are low mud flats or channel banks through which pass tidal gullies that feed the higher flats. In temperate latitudes, these high flats are vegetated by salt marsh plants, such as *Spartina*, especially in the zone covered only at high spring tides. However, in tropical semiarid regions, this zone is frequently bare of vegetation, with mangroves and other halophytes concentrated along the margins of dissecting channels and gullies (Thom, Wright, and Coleman, 1975).

The mechanism of sedimentation on tidal flats has been attributed to a settling lag process as elaborated by van Straaten and Kuenen (1958). Tidal waters spread onto the flats through channels and gullies. As the flood tide wanes, the muddy water crossing the flats progressively loses velocity, and mud particles begin to settle. Vertical accretion takes place if ebb currents cannot entrain these particles. Mangroves, salt marsh grasses, or algae may assist in reducing bottom water velocities and thus encourage deposition. However, wind erosion may deflate bare tidal flat surfaces, especially in the dry season, producing *clay dunes* on, or marginal to, the tidal flats. Wave action is also important on many tidal flats—for example, in the deposition of beach ridges on the high flats of the northern Gulf of California (Thompson, 1968).

The principal environments of tidal flats are also represented in vertical section reflecting upward growth since sea level has been close to its present position. Vertical and lateral growth appears to have been initiated during the final stages of the Holocene marine transgression when the supply of tidally supplied muds exceeded the ability of waves and tidal currents to erode and disperse. In many areas, lateral accretion is limited by tidal channel networks, and vertical accretion appears to be dominant (see Thom, Wright, and Coleman, 1975). Transgressive and regressive tidal flat sequences have been documented in several places including the North Sea and Colorado River areas (Reineck, 1972; Thompson, 1968).

BRUCE G. THOM

References

Evans, G.; Schmidt, V.; Bush, P.; and Nelson, H., 1969. Stratigraphy and geologic history of the sabhka, Abu Dhabi, Persian Gulf, *Sedimentology* 12, 145-159.

Reineck, H. E., 1972. Tidal flats, *in* J. K. Rigby and W. K. Hamblin, eds., *Recognition of Ancient Sedimentary Environments*. Tulsa, Okla.: Soc. Econ. Paleont. Miner., Spec. Pub. 16, 146-159.

Thom, B. G.; Wright, L. D.; and Coleman, J. M., 1975. Mangrove ecology and deltaic-estuarine geomorphology: Cambridge Gulf-Ord River, Western Australia, *Jour. Ecology* 63, 203-232.

Thompson, R. W., 1968. Tidal flat sedimentation on the Colorado River Delta, northwestern Gulf of California, *Geol. Soc. America Mem.* 107, 133p.

van Straaten, L. M. J. U., and Kuenen, Ph. H., 1958. Tidal action as a cause of clay accumulation, *Jour. Sed. Petrology* 28, 406-413.

Cross-references: *Algal Mats and Stromatolites; Coastal Flora; Mangrove Coasts; Salt Marsh; Tidal Range and Variation; Tides*. Vol. VI: *Tide Flat Geology*.

TIDAL FLUSHING

Tidal flushing refers to the systematic replacement of water in a bay or estuary as a result of tidal flow. The ocean is assumed to be a sink for water discharged during the ebb and a source of new water carried in by the flood. Ketchum (1951a) broadly defined tidal flushing in terms of an *exchange ratio* (r) representing the fraction of water in a specified location that is replaced during a tidal cycle. The exchange ratio, also called the *flushing rate* or *water renewal rate* is construed as $r = P/P + V$, where P equals the intertidal volume or *tidal prism*, which is the difference between the volumes of water occupying the location at high and low tide; and V equals the low tide volume. In estuaries, the tidal prism is composed of a mixture of river-contributed water and more saline water introduced by the flood tide. The expression assumes that complete mixing occurs within the specified water body and that none of the water discharged on the ebb returns on the flood.

If river water is discharged into an estuary undergoing tidal flushing, a steady-state distribution of salinity results. This steady-state distribution represents a *salt balance* wherein the quantity of salt introduced by more saline flood tide water is balanced exactly by the loss of salt in the mixed water escaping as the tide ebbs. Hydrographic data permit calculation of the exchange ratio using the following expression (Ketchum, 1951b): $r = I/Q$, where I equals the input of river water per tidal cycle, and Q equals the total accumulated river water in the area of interest. The reciprocal of this expression is called the *flushing time*, defined as the time required for half of the water in the area to be replaced by the flushing process. As such, this is the mean residence time of any substance introduced to the water body.

Ketchum (1951a, b) devised a more accurate method for describing tidal flushing in an estuary. This method involves division of the estuary into discrete longitudinal segments whose length equals the tidal excursion or distance traversed by a water particle during a half tidal cycle. The segmentation concept provides for tidal flushing, freshwater influx, and incomplete mixing resulting from salinity stratification.

Tidal flushing is relevant to pollution problems and causes dilution and removal of any substance being continuously discharged into a bay or estuary culminating in a steady state distribution. The distribution of pollutants can be predicted by Ketchum's segmentation method, more sophisticated math modeling, or, alternately, by field simulation using tracer dye.

DAVID O. COOK

References

Ketchum, B. H., 1951a. The exchanges of fresh and salt waters in tidal estuaries, *Jour. Marine Research* 10, 19-38.

Ketchum, B. H., 1951b. The flushing of tidal estuaries, *Sewage and Industrial Wastes* 27, 1288-1295.

Cross-references: *Effluents; Estuaries; Human Impact; Pollutants; Tidal Basin; Tidal Prism; Tidal Range and Variation; Tides; Waste Disposal.*

TIDAL INLETS, CHANNELS, AND RIVERS

Tidal inlets, channels, and *rivers* are dynamically related features characterized by bidirectional tidal flow and by tide-dominated sediment transport. The three features are distinguished from each other primarily by their geometry and by the relative volume of fluvial input. A *tidal inlet* is a short, narrow passage connecting two larger and wider bodies of water. Typically, tidal inlets link barrier lagoons or impounded estuaries with the sea and are maintained largely by tidal currents. The primary components of most tidal inlets include: an *inlet throat*, which is normally the narrowest and deepest part of the inlet and through which tidal currents attain their maximum velocities; a *flood-tidal delta* consisting of material deposited by deceleration of flood-tide currents landward (or lagoonward) of the inlet throat; and an *ebb-tide delta* consisting of material deposited by deceleration of ebb-tide currents seaward of the inlet throat. Tidal inlets may be stable and semipermanent with respect to position and geometry; they may be ephemeral and subject to periodic closure; or they may migrate alongshore.

Tidal channels are of three major types: *intertidal channels*, which drain intertidal flats during low and falling tide; *tidal creeks* with subaerial banks; and *wide-mouthed tidal channels*. Intertidal channels often create complex networks through tidal deltas and within the intertidal flats fronting macrotidal coasts. Tidal creeks are normally sinuous with parallel or subparallel banks and relatively high depth/width ratios. Tidal creek meanders often differ from riverine meanders in that the bends of the former tend to be angular rather than rounded. Wide-mouthed tidal channels are often similar in their upper (landward) regions to tidal creeks; however, their seaward reaches are characterized by relatively low sinuosities, low depth/width ratios and funnel shapes with pronounced seaward widening.

Tidal rivers are fluvial systems that experience

tidal influence from the mouth to a significant distance upstream. In contrast to tidal inlets and tidal channels, which owe their existence largely to tidal currents, the channels of tidal rivers are the downstream continuations of riverine channels. Prominent examples of tidal rivers include the lower courses of the Ord, Chao Phrya, Mekong, Irrawaddy, Parana, and Shattel-Arab.

For further reading on this subject see Bruun and Gerritsen (1960), Hayes (1975), Pestrong (1965), and Wright, Coleman, and Thom (1973).

LYNN D. WRIGHT

References

Bruun, P., and Gerritsen, F., 1960. *Stability of Coastal Inlets*. Amsterdam: North Holland, 123p.

Hayes, M. O., 1975. Morphology of sand accumulations in estuaries, *in* L. E. Cronin, ed., *Estuarine Research*, vol. 2. New York: Academic Press, 3-22.

Pestrong, R., 1965. The development of drainage patterns on tidal marshes, *Stanford Univ. Pubs. Geol. Sci.* **10**, 87p.

Wright, L. D.; Coleman, J. M.; and Thom, B. G., 1973. Processes of channel development in a high-tide-range environment: Cambridge Gulf-Ord River Delta, Western Australia, *Jour. Geology* **81**, 14-51.

Cross-references: *Inlets and Inlet Migration; Tidal Deltas; Tidal Flushing; Tidal Range and Variation; Tides.*

TIDAL POWER

Tidal energy is derived from the inherent force of the earth's rotation. In coastal areas where the sea is no deeper than 100 m, friction dissipates almost all radiated energy as progressive waves. The generating forces are strongest in the open sea (Charlier, 1969). Useable tidal power potential reaches perhaps as much as one trillion kwh/yr or about 3-4% of the projected electricity consumption for the year 2000 (Charlier, 1978).

While numerous tide mills were in operation during the eighteenth century in England, Wales, Brittany, and later on the U.S. east coast, plans to transform this ocean energy into *electrical* energy were formulated in the 1920s; abortive attempts were made in Brittany in 1925 and in the United States in 1933. Only two plants have actually been completed and are in current operation: one near St. Malo on the Rance River (France), and the other on Kislaya Bay (U.S.S.R.) (Bernstein, 1972).

Harnessing tidal power poses no engineering problems (Dalton, 1961). If more plants have not been built, the reasons are to be found in economic considerations (including construction costs) and problems caused by phases of the moon (Lawton, 1972). Since the time of the earlier considerations, transmission systems have improved greatly and reversible blade turbines that permit electricity production at ebb and flow have become available, as has the use of daily and intersyzgial control. In the 1960s construction costs of a tidal power station was 2½ times as high as for an equivalent conventional hydroelectric station, with cofferdams accounting for 30% of total costs. The Soviet Kislaya plant dispensed with cofferdams and was constructed on land in modules that were towed to the site and sunk in place on a prepared foundation, thus reducing construction costs considerably (Bernstein, 1972).

There are many suitable sites for tidal power electrical plants. Among these, the Severn River (Great Britain), the Kimberleys (Australia), Cape Tres Puntas (Argentina), Passamaquoddy Bay (United States and Canada), and the Bay of Fundy (Canada) were the subjects of in-depth studies (Laba, 1964; Lewis, 1963; Wilson, 1965, 1973). The rating of a potential site is based upon tidal range, basin size, dam length required, and proximity to the consumer market as well as the characteristics of the basin and foundation soil, and other geological considerations. Gibrat (1966), who was instrumental in the building of the Rance plant, introduced the coefficient for *site value*, being the ratio of dam length to natural energy; the smaller the coefficient's value, the more desirable the site.

The greatest tidal amplitudes are found in the Bay of Fundy (16 m) and in the Passamaquoddy region (15 m). In North America, interest has also been expressed in the Minas Basin in Nova Scotia and Cook Inlet in Alaska. A Canadian scheme predicted production of 1300 million kwh, and an American plant proposed by General Electric foresaw 1594 million kwh for a plant that would have cost in 1927 less than one-tenth of the current price to build.

The Argentinian potential involves many sites, predominantly the gulfs of San José and San Jorge. The latter alone could provide 10,000 million kwh. The proponents of a Severn River plant near the Bristol Channel estimated 800 million kwh (Wilson 1973). Some fifty sites were examined along 1700 km of coastline in northwestern Australia; resources from Broome to Darwin Harbour could amount to 300 million kwh (Lewis, 1963).

The French Rance River plant (Fig. 1) provides about 500,000 kilowatts. According to the builders, Electricité de France, it has had

FIGURE 1. Rance tidal power station (photo courtesy of Electricité de France).

negligible adverse environmental impact, but it has strongly influenced the coastal region (Lebarbier, 1975). It has allowed for the creation of some new communities, the road across the dam has shortened by about 30 km the distance between St. Malo and Dinard, and the plant itself has become a major tourist attraction while providing industrial potential. The plant is a symbol of innovation and has the advantages of functioning with precise regularity and never running out of fuel.

While, like ocean *thermal power* (q.v.), it probably will never become a replacement for other energy sources, tidal power can be an important supplemental source, particularly in areas otherwise deprived of energy resources. Tidal ranges of at least 9-10 m are considered necessary for an economically viable plant but smaller amplitudes should not automatically disqualify a site. Small schemes could be valuable, particularly in developing nations, to spur on regional growth, as opposed to national and international growth.

ROGER H. CHARLIER

References

Bernstein, L. B. 1972. Kislayaguba experimental tidal power plant and the problem of use of tidal energy, *in* T. J. Gray and O. K. Gashus, eds., *Tidal Power*. New York: Plenum, 215-238.

Charlier, R. H., 1969. Harnessing the energies of the ocean, *Marine Tech. Soc. Jour.* **3**, 13-32.

Charlier, R. H. 1978. Other non-living resources, *in* E. M. Borgese and N. Ginsburg, eds., *Ocean Yearbook*, vol. 1. Chicago: University of Chicago Press, 163-168.

Dalton, F. K. 1961. Tidal electric power generation, *Royal Astronom. Soc. Canada Jour.* **55**, 22-33.

Gibrat, R., 1966. *L'énergie des Marées*. Paris: Presses Universitaires de France, 220p.

Laba, J. T., 1964. Potentials of tidal power on the North Atlantic Coast in Canada and the U.S. *in Conf. Coastal Eng. 9th, Am. Soc. Civil, Engineers, Proc.*, 832-857.

Lawton, F. L., 1972. Economics of tidal power, *in* T. J. Gray and O. K. Gashus, eds., *Tidal Power*. New York: Plenum, 105-130.

Lebarbier, C. H. 1975. Power from tides. The Rance tidal power station, *Naval Engineers Jour.* **87**, 57-71.

Lewis, J. G., 1963. The tidal power resources of the Kimberleys, *Australia Inst. Eng. Jour.* **35**, 333-345.

Wilson, E. M., 1965. The Solway Firth Tidal power project, *Water Power* **17**, 431-440.
Wilson, E. M. 1973. Energy from the sea. Tidal power, *Underwater Jour.*, 175-181.

Cross-references: *Coastal Engineering; Demography; Human Impact; Thermal Power; Tidal Range and Variation; Tides.*

TIDAL PRISM

Tidal prism is the total amount of water flow into or out of an *inlet* with the rise and fall of the *tide*, excluding any freshwater discharges. Tidal prism for any tidal period is the product of the mean of the high- and low-water surface areas of the bays behind the inlet entrance and the tidal range in each segment. Allowance must be made for the time involved in propagation of the tidal wave through the inlet into the basin, since the rise and fall of the tide are not uniform over the entire bay.

Tidal prism has been related to inlet dimensions, particularly area (O'Brien, 1976), and to flow characteristics. The relationship of tidal prism to littoral drift (or wave power) determines inlet stability. Bruun and Gerritsen (1960) indicate that if the ratio of the tidal prism to the quantity of littoral drift delivered per year is in excess of 300, the inlet will have a high degree of stability, whereas ratios of less than 100 reflect a low degree of stability. Tidal prism is also important in maintaining the interior channels behind the inlet. O'Brien (1976) has indicated the importance of locating as much of the tidal prism as possible at the maximum distance from the inlet.

KARL F. NORDSTROM

References

Bruun, P., and Gerritsen, F., 1960. *Stability of Coastal Inlets*. Amsterdam: North-Holland, 123p.
O'Brien, M. P., 1976. *Notes on Tidal Inlets on Sandy Shores*. Vicksburg, Va.: U.S. Army Corps of Engineers, Waterways Experiment Station, GITI (General Investigation of Tidal Inlets) Rept. 5, 26p.

Cross-references: *Boat Basin Design; Inlets and Inlet Migration; Tidal Basin; Tidal Flushing; Tidal Range and Variation; Tides.*

TIDAL RANGE AND VARIATION

Tidal range is the vertical distance through which the tide rises and falls, the difference in water height between *low tide* and *high tide*, or, quite simply, the "size" of tide. The range of tide varies from place to place as well as from time to time (Davies, 1972; Zenkovich, 1967).

Tides vary in size through time because of variations in the distance and direction of the moon and, to a lesser extent, the sun. Greater ranges (*spring tides*) occur when the sun and moon align with the earth in a position called *syzygy* (pronounced *siz-eh-jee*). Greater ranges occur also when the moon is at *perigee*, its closest approach to the earth. The largest tidal ranges, occurring when these factors combine, are called *perigean spring tides*. Perigean spring tides add about 40% to mean tidal ranges.

Tidal size varies from place to place because of differences in both coastal configuration and sea-floor topography. Tides range between 1-2 m on open oceanic coasts, reaching more than 2 m where the semidiurnal component becomes pronounced. Tidal ranges are near zero in almost totally enclosed seas, such as the Mediterranean, Baltic, and Red seas and the Gulf of Mexico.

Extremely high tidal ranges are found in funnel-shaped embayments, where periods of *seiche* and tide nearly coincide. The best example of such a funnel-shaped embayment is the Bay of Fundy, separating New Brunswick from Nova Scotia in eastern Canada, where the tidal range exceeds 15 m at springs, the largest in the world. Other high spring tidal ranges exist at Penzhina Bay in the Sea of Okhotsk, U.S.S.R., 12 m; the Bristol Channel separating England from southern Wales, 12 m; the Gulf of Mezen in the White Sea, U.S.S.R., 10 m; the Bay of St. Michel, France, 9 m; and at the Colorado River delta in the Gulf of California, Mexico, 9 m. Where exceptionally high tidal ranges occur at the mouths of large rivers, *tidal bores* are created as the incoming tide collides with the river current.

Differences in tidal range are important in explaining variations in coastal geomorphology, especially where they widen the zone of wave action or influence inlets (Rosen, 1977). Knowledge of tidal ranges is also important in the prediction of destructive coastal flooding, which is most likely to occur when perigean spring tides combine with *storm surges*, low atmospheric pressure, or *tsunami*.

JAMES E. STEMBRIDGE, JR.

References

Davies, J. L., 1972. *Geographical Variation in Coastal Development*. Edinburgh: Oliver and Boyd, 204p.
Rosen, P. S., 1977. Increasing shoreline erosion rates with decreasing tidal range in the Virginia Chesapeake Bay, *Chesapeake Science* **18** 383-386 (See also *Virginia Institute of Marine Science Contribution No. 735*).
Zenkovich, V. P., 1967. *Processes of Coastal Development*. New York: Wiley Interscience, 738p.

Cross-references: *Storm Surges; Tidal Type, Variation Worldwide; Tides; Tsunami.*

TIDAL TYPE, VARIATION WORLDWIDE

Differences in the relative importance of the diurnal and semidiurnal components of the tide-producing forces lead to the existence of recognizably different tidal types around the coast of the world. The ratio (F) between the aggregate amplitude of the *diurnal tides* and the aggregate amplitude of the *semidiurnal tides* is generally used as a measure of variation.

When $F = 0.0$-0.25, the resultant tide is semidiurnal, with two high and two low waters of approximately equal height in the 24 hours 50 minutes of the tidal day. Such tides are strongly characteristic of Atlantic Ocean coasts and of extensive parts of the Artic. Mixed, predominantly semidiurnal tides ($F = 0.25$-1.5), also produce two clearly defined highs and lows per day, but they are of pronouncedly unequal range. They are especially prevalent in the Pacific and Indian oceans. In mixed, predominantly diurnal tides ($F = 1.5$-3.0), the inequalities of range become even larger, and for part of the lunar cycle only one high tide a day will be observed. In the diurnal tide ($F > 3.0$), the second tide disappears completely. Both these two latter types are relatively rare and are found most notably in relatively enclosed seas such as some of those of eastern Asia.

Tidal type affects such things as the length of the drying period in the intertidal zone and the intensity and duration of tidal currents. The most important contrast is probably between the semidiurnal type and all the other. (Fig. 1).

J. L. DAVIES

Reference

Davies, J. L., 1980. *Geographical Variation in Coastal Development.* London: Longman Group Ltd. 212p.

Cross-references: *Tidal Range and Variation; Tide Guage; Tides; Tide Tables and Charts.*

TIDAL WAVE—See TSUNAMI

TIDE CURVE—See MARIGRAPH AND MARIGRAM

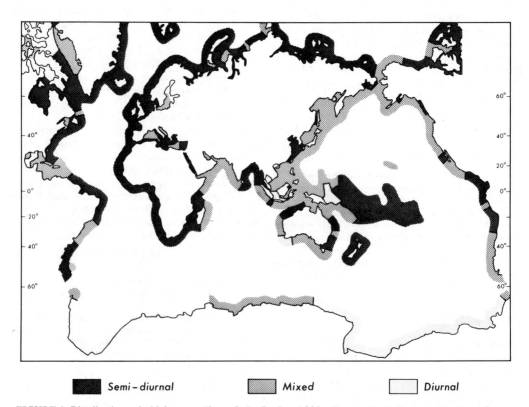

FIGURE 1. Distribution of tidal types (from J. L. Davies, 1980, *Geographical Variation in Coastal Development.* London: Longman Group Ltd.).

TIDE GAUGES

Observing the tides is the greatest bore upon earth, or on the waters, and the greatest exhaustion of a man's patience and trial of his temper. Professor Herschell, 1835.

A *tide gauge* is an instrument for recording simultaneously two measurements: the level of the undisturbed sea referred to some fixed point on land, and the time at which the particular level occurs (Ingham, 1975). The simplest type of tide gauge, which is often termed a *visual tide scale*, consists of a graduated pole set vertically in the sea so that visual observations of sea level can be made. Conventional tide gauges measure variations in sea level inside a cylindrical well that, in most cases, is connected to the open ocean through a circular orifice near the bottom of the well. Several types of automatic recording tide gauge have been developed for a variety of purposes, including hydrographic surveying, harbor operations, observation of mean sea level for connection to land leveling systems (Lisitzin, 1974), positive and negative surge warning systems, scientific analysis of mean sea level, tidal analysis, and prediction.

Attempts have been made recently to develop shore-based instruments using electronic or pneumatic sensors that will operate effectively in open water and eliminate the need for a stilling well. A stepped sensor has been used successfully to provide a direct electrical indication of water level in digital form. Examples of this type of instrumental development are the ICOT (Institute of Oceanographic Sciences, U.K.) digital tide gauge and the one developed at the Institute of Applied Physics at Delft, The Netherlands. Gauges using acoustic or microwave techniques are being developed, but so far their real worth has not been fully tested.

Relatively few offshore gauges, which measure pressure at the seabed to infer the water level, have been used operationally. These are commercially available as a standard instrument, having been developed during the past decade or so at various scientific establishments. Tide gauges have also been developed for use both on and beyond the continental shelf. Those deployed on the shelf to measure sea level, both around a coast and offshore, have important applications in the safe operation and navigation of deep-draught ships with small under-keel clearance, the provision of flood warning, coastal defense systems, and scientific research. The Institute of Oceanographic Sciences, Wormley Laboratory, England, has developed and used a tide gauge for scientific research on the continental shelf so that data from both analog and frequency sensors are recorded in noncomputer compatible format on magnetic tape. There are difficult sensor problems associated with continental shelf gauges, because parameters such as water temperature can change by several degrees over a tidal cycle.

Sites for tide gauges have to be chosen carefully with regard to wave action, *impounding* (a gauge will not record the full tidal range as a result of the effect of impounding), variations in salinity, the strength and direction of tidal streams (a strong stream will cause a change in pressure at the orifice of the well), and occasionally, the availability of power supplies. Although the stilling well is reported to smooth out wave action, its efficiency is suspect when wave action is strong.

There is still doubt about the reliability and frequency response of some tide gauges and therefore the extent to which records from these gauges can be used to study, for example, high-order tidal components, and *seiches* must be critically reviewed.

JOHN R. HAILS

References

Ingham, A., ed., 1975. *Sea Surveying*. New York: John Wiley & Sons, 306p.

Lisitzin, E., 1974. *Sea Level Changes*. Amsterdam: Elsevier, 286p.

Cross-references: *Coastal Morphology, Research Methods; Tidal Range and Variation; Tides; Tide Tables and Charts; Tsunami; Waves.*

TIDE POOLS

Tide pools occur along *cliffed coasts* at and above high water level. They are filled by the incoming tide, swash and spray, and may therefore have upper levels above high water level, and are drained with the falling tide. The pools, whether open directly to the sea or indirectly by a series of interconnected channels, are by definition drained during each ebb, to, at, or near low water level. Because of their location within the intertidal zone and the continued recycling of water, they support an abundance of marine organisms characteristic of more sheltered rocky coast environments (Singletary, 1972).

Tide pools originate in several ways. First, they may enlarge an existing pothole or closed surface pool until it connects directly with the sea. On limestone platforms where pools are perhaps most common, solution plays an important role in enlarging the pools (Fig. 1). Second, they may originate by quarrying

FIGURE 1. Tide pools on a heavily denuded wave-cut platform. The platform is 25 m wide, cut in sandstone-shale at Turimetta Head, New South Wales, Australia. The pools are being generated by a combination of potholing and erosion along joint lines (photo: A. D. Short).

along lines of lithological or structural weakness. In this case, they are usually oriented parallel to the lines. Tide pools are best developed on older, deteriorating platform surfaces where secondary processes have had sufficient time to form and enlarge them.

ANDREW D. SHORT

Reference

Singletary, R. L., 1972. Tide pools: Nature's marine aquaria, *Sea Frontiers* 18, 2-9.

Cross-references: *Biotic Zonation; Cliffed Coasts; Coastal Fauna; Coastal Flora; Rock Borers; Solution and Solution Pan.*

TIDES

The word *tide* refers to the gravitationally induced, periodic rise and fall of the waters of the oceans—and, to a much lesser extent, the waters of the major seas and lakes—with respect to the land. Tides may also occur to a minor degree in the atmosphere and in a very small, although detectable amount in the solid earth. The height of coastal waters, especially in basins and harbors, also may be raised above the average by strong, persistent onshore winds (see *Storm surges*), deep atmospheric low pressure systems, and gravity-type waves produced in the ocean by vertical uplift or subsidence along a geological fault on the sea floor or by other causes (see *Tsunami*). However, the phenomenon produced in the latter case—popularly called a *tidal wave*—bears no relationship to the tide, other than that its disastrous effects (Wylie, 1979) may be superimposed upon the normal, astronomically induced high tide.

Historical and Utilitarian Background

As far back as the late fourth century B.C., a Greek navigator named Pytheas of Massilia, while sailing northward from the Pillars of Hercules (Gibraltar) with an exploring expedition along the west coast of Europe, wrote in his account of the expedition the first record of the effect of the full moon in producing heightened tides along the ocean shores. This was a relationship yet undiscovered in the Mediterranean area, where the tidal range was too small to be noticed or pondered upon by the average, incurious observer. A knowledge of this natural phenomenon was undoubtedly conveyed to Pytheas by the inhabitants of this west European region who, living on the seacoast and noting the influence that the tides had on their everyday activities, had long since become acustomed to the relationship between the lunar cycle and the tides.

Today the influence of the tides upon the daily affairs of mankind is of continuing and ever growing importance. Fishing boats put to sea and return with close observance to the ebb and flow of the tides, while in estuaries, bays, and nearshore waters (discounting other *deepwater* habitats) clam digging, lobster trapping, and oyster tonging are prominent tidewater activities. The shipbuilding industry must rely on the tides for the launching of ships as well as the floating of vessels into and out of basin-, graving-, and dry docks. In coastal waters where shoals, reefs, and submerged hazards abound, a knowledge of the tides is of utmost importance to safe navigation (Sager, 1959). The maneuvering of large vessels into bays and harbors possessing large tidal changes generally requires the services of a pilot having a close familarity with the local tide regime as well as the effects of marine meteorological conditions.

Military amphibious-landing operations and the work of underwater demolition teams are necessarily scheduled to take account of changing tidal conditions and—as in the case of offshore oil-drilling platforms and early-warning radar towers—may be affected by associated tidal currents (q.v.). Dredging and salvaging operations, and shallow-water diving in accompaniment thereof, are markedly influenced by the tides. Swimming, boating, and other oceanic and estuarine sports activities are frequently dependent upon both tides and tidal currents. Marine engineering, harbor and breakwater construction, the onshore laying of communications cables and the generation of tidal power (Gray and Gashus, 1972) are activities closely tied to the times of tidal high and low waters. Biologically, the tides have many effects on marine organisms, from establishing the daily habits and patterns of existence of the small creatures of the tidewater belts to the spawning of the tiny grunion on certain southern California beaches during nighttime spring tides.

Principal Types of Tides

Depending upon the pattern and frequency of occurrence of high and low tides at any one location during the tidal day, four general types of tides are distinguished (Russell and Macmillan, 1970), as follows.

Tides that, on the average and as a function of diurnal inequality, display but one high tide and one low tide in each tidal day are termed *diurnal*. (Examples are found in the northern part of the Gulf of Mexico, along portions of the coasts of Alaska and China, the Phillippine Islands, and in the Java Sea.)

Where two high tides and two low tides occur during each tidal day, with comparatively little diurnal inequality between successive tides of the same phase, the tides are referred to as *semidiurnal*. (Representative cases are those found along the east coast of the United States and the Atlantic coast of Europe.)

A combination of the characteristics of both diurnal and semidiurnal tides may also exist. These are known as *mixed* tides. (Such tides are exemplified notably by those along the Pacific coast of the United States and where the Pacific Ocean joins the Indian Ocean.) Large diurnal inequalities are common in this type of tide. Numerous variations may occur where the diurnal inequality is almost wholly in the high tide, entirely in the low tide, or balanced fairly equally between the two (that is, one high tide and one low tide may each have a greater height than the other). Stylized examples of each of these basic types of tides are illustrated in Figure 1.

Finally, a type of tide exists that consists of two distinct maxima of approximately the same height, with a single small minimum located between them, or a phase distribution may occur in which two tidal minima are separated by a single, relatively small maximum. Either phenomenon is known as a *double tide* or *agger*. An occasional physical merging of tide levels, plus an overlapping of tidal days and effective disappearance of all but one tidal phase is termed a *vanishing tide*.

Tidal Prediction Methods

An aspect of considerable importance in connection with tides is their mathematical predictability for any seaport in the world (see *Tide Tables and Charts; Marigraph and Marigram*). This capability is of course based upon years of observation of tide-producing conditions and variables through many complete 19-year Metonic (lunar) cycles. Tide-prediction data are processed by modern, high-speed electronic computers. The determination of times and heights of both low and high waters is achieved by either harmonic or non-harmonic methods of analysis (Marmer, 1926; Doodson and Warburg, 1941; Godin, 1972). In these predictions, the tidal movements are regarded as consisting of a series of partial tides, and the reduction process involves the combination and resolution of the various components and constituents. In the most common or harmonic solution (Defant, 1961; Dronkers, 1975), the periods of the harmonic terms are derived from the corresponding periods of the astronomical cycles that produce the various tidal effects (Schureman, 1941).

Nature of the Tide-Raising Forces

The popular belief that the moon lifts the waters of the earth to produce the tides is a common misconception. In actuality, lunar gravitation is not nearly strong enough to lift the waters of the earth vertically against the far more powerful force exerted by the gravitational attraction of the earth acting to pull the same waters toward its center. The attraction of the earth upon these surface waters is almost nine million times greater than that of the moon.

The heaping of the ocean waters in the two positions corresponding to high tides is produced predominantly by the horizontal, rather than the vertical, component of the moon's gravitational attraction. This *horizontal*, or so-called *tractive* (drawing) component of the moon's gravitational force tends to cause the ocean waters to flow laterally toward a point directly beneath the moon (this position is termed the sublunar point) and to accumulate there, creating a tidal bulge. This same force produces a simultaneous amassing of water on the side of the earth directly opposite the moon, in a position known as the *antipodal point*. However, longitudinal displacements in the two tidal bulges caused by the simultaneously acting gravitational force of the sun result in the fact that these tidal humps are not always located directly in the sublunar or antipodal positions. In terms of corresponding variations in the times of high water, this influence is responsible for the tidal phenomena of *priming* (acceleration) and *lagging* (retardation).

The Horizontal and Vertical Tide-Producing Components The horizontal or tractive com-

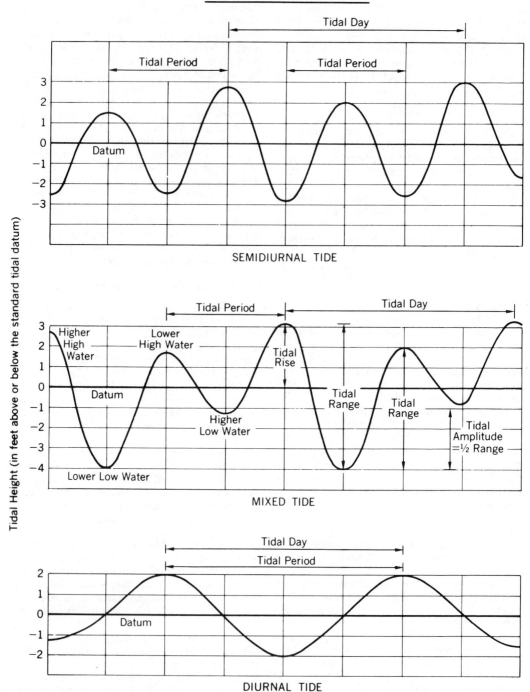

FIGURE 1. The principal types of tides, showing the moon's declinational effect in the production of semidiurnal, mixed, and diurnal tides.

ponent of the moon's gravitational force at any point on the earth's surface occurs along a line that is tangent to the earth's surface at that point (this force accordingly acts perpendicularly to a line connecting that point on the earth's surface with the earth's center). The horizontal component of the moon's gravitational attraction is widely variable in amount at different points on the earth's surface. It is zero at the sublunar point, as well as at a point diametrically opposite it, and at all points along a great circle approximately 90° in between. It is maximum along two circles passing through all points 45° and 135° around the earth in either direction from the sublunar point.

The *vertical component* of the moon's gravitational attraction is that component of force directed away from the center of the earth. The maximum vertical gravitational attraction of the moon on the earth obviously occurs along a line of action connecting the centers of mass of the earth and moon and passing through the sublunar point.

Where θ is the angle between the line connecting the centers of the earth and moon and a line from the center of the earth to the surface point under consideration, m is the mass of the moon, d is the distance from the center of the earth to the center of the moon, and r is the radius of the earth, the horizontal and vertical components of force are given by the expressions:

$$F_h = \frac{3mr}{d^3} \sin \theta \cos \theta$$

$$F_v = \frac{3mr}{d^3} (\cos^2 \theta - 1/3)$$

Subject to the force exerted by the horizontal component of the moon's gravitational attraction, there is a constant tendency for the waters of the earth to flow tangentially across its surface toward the sublunar point and to pile up in the vicinity of this point. However, the presence of the continents and other land masses prevents this free flow of water, and the particles of water do not actually move around the earth beyond the limits of their containing ocean basins. What is actually produced is an ellipsoid-shaped envelope of forces in which the maximum tide-producing force vectors at the surface of the earth are grouped around, and very nearly in the direction of, the major axis of the ellipsoid — one end of which points toward the moon (Fig. 2).

The earth's waters likewise assume the figure of an ellipsoid-shaped fluid envelope, whose major axis passes through the sublunar and antipodal points (with some modification

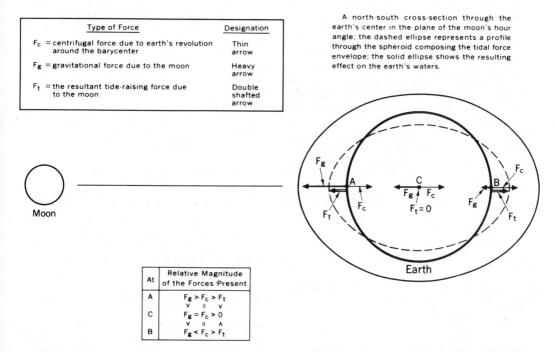

FIGURE 2. The tidal force envelope. The combination of forces of lunar origin producing the tides (a similar complex of forces exists in the earth-sun system).

introduced by the sun's gravitational attraction) and whose minor axis rotated describes a circle everywhere 90° from these two positions. Because zero-magnitude horizontal force vectors exist along this circle, no tendency for amassing of water occurs here. Conversely, with a tendency for withdrawal of water from this region, locations within a narrow band on either side of this circle will experience a low tide. The tidal force vectors (if transferred in three dimensions to the surface of the actual fluid envelope) will converge from this band directly toward the sublunar and antipodal points. In both locations, high tides will be produced.

Direct and Opposite Tides

This simultaneous tide-raising action at two points on opposite sides of the earth, which lie in a straight line joining the centers of mass of the earth and moon (although affected by the position of the sun and certain other resonance and inertial factors), is illustrated in Figure 3, looking down on the northern hemisphere of the earth. Point C represents the center of mass of the earth where, according to Newton's universal law of gravitation, all of the earth's attractive force may be assumed to be concentrated; M similarly represents the center of mass of the moon; the bulge D represents the amassed waters of the *direct tide*, immediately underlying the moon; O represents the *opposite tide*, halfway around the earth from D; S is the sublunar point and S' is the antipodal point; A and B are two points each 45° around the earth, in opposite directions from S; and F and G are similar points, each 135° around the earth from the sublunar point. All points represented in this diagram are considered to be in the plane of the moon's orbit around the earth, as viewed from the north pole of this plane.

The horizontal component of gravitational attraction exerted by the moon strives to draw the waters from all points on the respective hemispheres of the earth toward the points S and S'. In the latter case, the effect of the gravitationally induced motion is similar to the one that would be experienced by tying a rope to the point F in this diagram, now regarded as that of a freely rotating wheel, looping the rope over the wheel hub at C, and pulling on the rope from the point M. The point of maximum force application F will move toward point S' under the action of the force exerted from point M, but will stop when it reaches S'. During this movement, the original maximum horizontal force component applied at F has been converted into a purely vertical force component at S', and no further tractive or drawing action occurs.

Barycentric Motion of the Earth-Moon System

In general considerations of astronomy not involving gravitational perturbations or the tides, it may be assumed that the moon revolves around the center of mass of the earth. Actually, however, the earth and moon revolve together around the center of mass (*barycenter*) of the combined earth-moon system. The barycenter is a point which lies some 1718 km (1068 mi) below the surface of the earth. Although this point is fixed with reference to space rather than with respect to the earth itself, it is always located on the side of the earth turned momentarily toward the moon, along a line connecting the individual centers of mass of the earth and moon.

In Figure 4 as the moon revolves (*1*) (*2*) (*3*) (*4*) (*5*) (*6*) around the barycenter G, the center of mass of the earth describes a very nearly circular orbit, $C_1 C_2 C_3 C_4 C_5 C_6$, around the same point. This motion of the earth's center of mass is entirely independent of the diurnal rotation of the earth and does not in any way affect the latter movement. For dynamic purposes, therefore, the revolutionary motion around the center of mass of the earth-moon system may be thought of as occurring without any accompanying rotation of the earth.

Concept of Revolution without Rotation

Assuming such a system of revolution without rotation, it may be shown graphically (Fig. 5) that during each revolution of the moon every point in the body of the earth revolves around its own discrete center in a nearly circular orbit whose radius is the same as that of the revolution of the center of mass of the earth around the barycenter. In consequence, the centrifugal force exerted on unit masses anywhere on or beneath the earth's surface is the same in both magnitude and outward direction as that produced by the revolution of the earth's center of mass around the barycenter.

As a purely representative example, a point

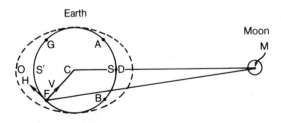

FIGURE 3. Explanation of the direct and opposite tides.

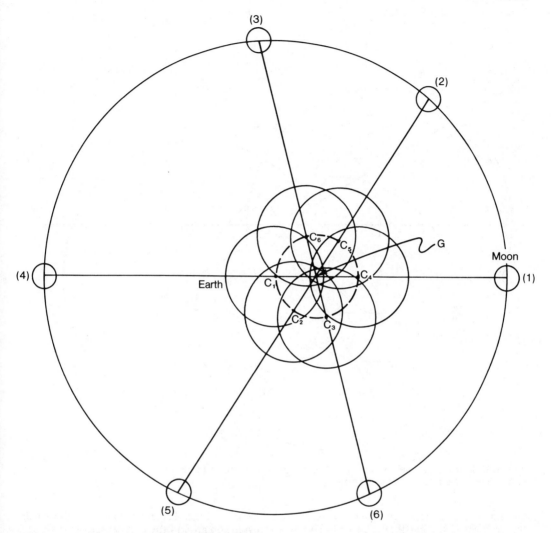

FIGURE 4. Revolution of the earth and moon around the center-of-mass of the earth-moon system.

on the surface of the earth has been selected to show that during the motion of revolution of the earth's center of mass C around the barycenter G (without accompanying rotation) this point P describes an approximate circle around C_p of exactly the same radius as that of the earth's center of mass around the barycenter.

Since the earth is a rigid body, the point P (or any other point selected) must always remain at a constant distance from the center of mass of the earth and since the motion considered is one of revolution without rotation the orientation of the line connecting this point with the center of mass must always remain constant in space. As the center of mass revolves around the barycenter, and C_1 moves to C_2, the point P_1—retaining the same distance from C_2 as well as orientation in space with respect to it that it had with respect to C_1—moves to the position P_2. Similarly, as C_2 moves to C_3, P_2 moves to P_3, and so on. The resulting orbit is represented by the successive points $P_1 P_2 P_3$, and so on.

If the points $P_1 C_1$ are assumed to be joined by a rigid bar whose orientation remains fixed in space, with passage of one end of the bar from P_1 to P_2 to P_3 and so on, the other end describes an orbit around G having the same size and radius as that described by $C_1 C_2 C_3$ and so on. Since the radius of revolution is the same, the centrifugal force exerted on a unit mass will also be the same.

Any other point P_x within the figure of the earth can similarly be shown to describe a nearly circular (actually slightly elliptical) orbit whose radius is equal to $C_x G$ as C_x revolves

TIDES

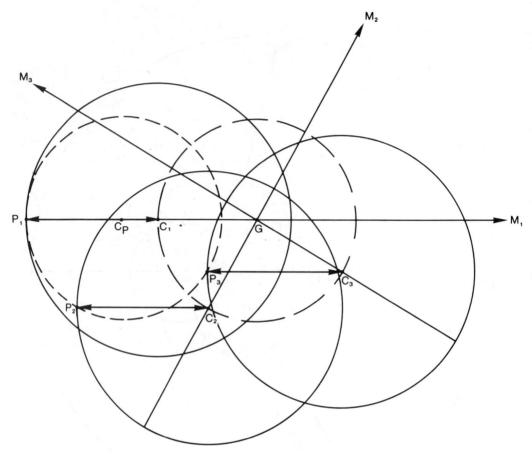

FIGURE 5. Demonstration of the overall equality of centrifugal force produced by the revolution of the earth's center of mass around the barycenter.

around G. For a point anywhere on or below the surface of the earth, the centrifugal force produced by revolution of the earth's center of mass around the barycenter is the same.

The Balance of Gravitational and Centrifugal Forces

The centrifugal force resulting at the earth's center of mass from its revolution around the barycenter is exerted in a direction that is always opposite to that of the moon. This centrifugal force exactly compensates for the gravitational force acting to pull the center of mass of the earth toward the moon. Since these two forces are in equilibrium, the earth does not fall toward the moon, even as the moon is prevented from falling toward the earth by the centrifugal force produced by the revolution of its own center of mass around the barycenter. All other points on the earth are, however, in a state of imbalance between the gravitational force of the moon (and sun) and the centrifugal force produced by the earth's revolution around the barycenter.

Since the centrifugal force caused by revolution around the barycenter is equal at all points on the earth, there exists, at the surface of the earth on the side directly opposite the position of the moon, a centrifugal force component having the same magnitude as that at the earth's center. It is, likewise, directed away from the moon.

In terms of the earth's gravitational effect alone, the attraction for a point on the earth's surface is the same, either on the side directly beneath the moon or on the side directly opposite the moon. The earth's gravity therefore plays no role in the differential forces that produce the tides. These forces are provided by the moon and the sun.

The Differential Tide-Producing Forces

The lunar tide-generating force is defined as the difference between the centrifugal force

832

created by the revolution of the earth around the center of gravity (barycenter) of the earth-moon system and the gravitational attraction of the moon exerted on the earth's surface waters. While the sun is much more massive than the moon, its much greater distance from the earth reduces its tide-raising force to a magnitude only 0.455 that of the moon. However, an important set of tide-generating influences also exists in the earth-sun system.

It is the varying gravitational influence of the moon (and sun) upon different portions of the earth, combined with the constant centrifugal force produced at these same points, that is responsible for the unbalanced force components causing the tides. Providing a slight modification of Newton's universal law of gravitation, the *tide-raising* forces vary directly as the product of the masses of the celestial bodies concerned, but inversely as the *cube* (rather than the square) of the distance between them. Because of its greater distance from the moon, a slightly smaller force (reduced by 3.4%) is exerted on the earth's center of mass compared with that exerted at the sublunar point, and a successively smaller force (reduced again by 3.3%) exists at the antipodal point compared with that acting on the earth's center of mass. These varying values of lunar gravitation constitute the so-called *differential tide-producing force*.

Variable Tidal Range- and Amplitude-Producing Effects

The tide produced in the open ocean by the gravitational actions of the moon and sun is believed to be only about one or two feet high, but where the tidal bulge is thrust up against land masses or channeled into estuaries, its effects are much more noticeable. The effects of coastline configurations and shoaling waters are very important in this connection. The purely astronomically induced amplitudes of the tides will—with a number of specific exceptions existing along certain coastlines—in general be a function of the moon's declination; its phase; its proximity to the earth; the distance of the earth from the sun; the possible coplaner alignments of the earth, moon, and sun; and certain other cyclical factors.

Cyclical Constituents of the Tides

A considerable number of constantly changing astronomical positions, motions, distances, forces, perturbing functions, and alignments may be considered basic to the establishment of the harmonic constituents used in analysis and prediction of the tides. The most important among these factors of influence, together with the special types of tides they produce, and their associated periodic terms, are described below.

Semidiurnal Tides (**Average Period: 12.4167 hours**) In Figure 3 it was seen that as the waters are drawn from adjoining portions of the ocean and tidal bulges are produced at D and O to create high tides there, low tides simultaneously occur along a circumferential belt at all points $90°$ removed from D and O. As the solid portion of the earth rotates beneath the two maxima and two minima of the tidal force envelope thus produced, an alternation of high and low waters takes place. Since, in order for a transit of the moon to occur, any observing station on the surface of the rotating earth must catch up with the motion of the moon revolving in the same direction in its orbit (Fig. 6), this catch-up time is additive to the normal period of the earth's rotation.

The resulting difference, expressed as a time equivalent, makes the period between successive lunar transits across the upper and lower meridians of any place 25 minutes longer than 12 hours, on the average. This period is one-half of the *mean tidal day* of 24 hours, 50.415 minutes, the time required for a point on the earth's surface, rotating from west to east at the equator, to "catch up" to a position directly beneath the moon again after the moon's own orbital motion in one day has carried it some $12.5°$ farther eastward. This interval of $50^m 24.9^s$, the average extension in the mean solar day due to the moon's revolution, is known as the *mean daily lunar retardation*.

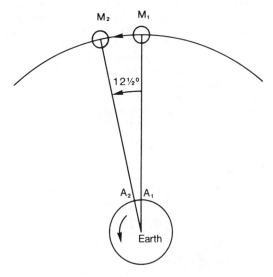

FIGURE 6. Lunar retardation. The necessary catch-up motion in the earth's rotation caused by the moon's orbital revolution.

With certain exceptions to be noted below, a completely open and straight section of ocean coastline should thus be subject, on the average, to two high tides and two low tides during each 24 hours, 50.415 minutes. Such tides, having essentially equal amplitudes in the two highs and two lows of each set, are known as *semidiurnal tides*. The opposite tides at O (Fig. 3), despite the greater distance of this point from the moon, are indistinguishably different in amplitude from those of the direct tides that occur at D.

Diurnal Tides **(Average Period: 24.8333 hours)** Prominent among the astronomical influences having an effect on the tides is that associated with the moon's *declination* (its perpendicular angular distance north or south of the celestial equator) and the changing values of the maximum lunar declination during the course of the year. The direct and opposite tidal bulges tend to follow the moon in a north-south direction as well as east and west. Since the moon may range through a maximum of approximately $57°$ in declination north and south, the positions of the two tidal bulges may vary considerably in latitude over the earth.

Consider the diagram (Fig. 7) in which the position of the moon is indicated by M and the direct and opposite tides produced by its gravitational action are indicated by D and O, respectively. Now consider the effect of recurring high and low tides along any parallel of latitude such as AB. The height of the tides will be determined by the depth of water lying vertically above these points at any time. Because of the unsymmetrical distribution of the overlying tidal bulges with respect to the earth's axis of rotation, the height of the high tide at point A will be considerably less than the corresponding high tide at this same point when it is carried halfway around by the earth's rotation to point B. This occurs, as has been shown approximately $12^h 25.208^m$ later but subject to variations in elapsed time as well, owing to the nonuniform water levels.

The difference in height between two tides of the same phase at upper and lower transits of the moon is known as the *diurnal inequality*, since the increased high water and decreased low water occur only once in each day under the conditions specified. To distinguish between the two unequal high tides and the two unequal low tides in each day, these tides are denoted as *higher high water, higher low water, lower high water*, and *lower low water*, but not necessarily in this order.

Although any situation in which two high tides and two low tides occur each day is, strictly speaking, described as an example of semidiurnal tides, this term is generally reserved for the case in which the diurnal inequality is not too marked and in which successive tides of the same phase are of approximately the same height.

Obviously the diurnal inequality will completely disappear when the moon is in the plane of the earth's equator. Thus twice each month, when the moon in its orbit around the earth crosses the celestial equator, the heights of the respective pairs of high and low waters will be equal at diametrically opposite positions along any parallel of latitude. Tides in which the diurnal inequality almost completely disappears or becomes minimum under these conditions are known as *equatorial tides*. Conversely, *tropic tides* possessing maximum diurnal inequalities occur as twice each lunar month the moon reaches its maximum declination for the month, north or south of the celesital equator. Every 18.613 years, as the moon attains its extreme declination, the diurnal inequality also reaches an extreme value.

Spring Tides **(Average Period: 14.7653 mean solar days)** Of further significance among the various astronomical influences controlling the tides is the effect of the gravitational attractions of the moon and sun acting together to reinforce or to oppose one another as the moon revolves around the earth in its mean synodic period (from conjunction to conjunction) of 29.530589 mean solar days (Fig. 8). When the moon is at new phase (conjunction), it is located in the same meridian plane with, and between, the earth and sun. Although, considered in three dimensions, these three bodies lie in or near the same plane in celestial latitude —the ecliptic—as well only at times of a solar eclipse. At each new moon, therefore (and especially when the moon and sun are at the same time coplaner in either celestial latitude or declination), the gravitational forces of both are combined vectorially to produce a considerably increased tide-raising force.

At full moon (opposition), an average of 14.76529 days later, the moon is situated opposite the earth from the sun, and the

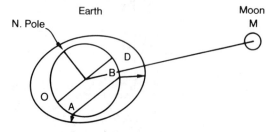

FIGURE 7. Dirunal inequality. The unequal height of the tides at the same latitude on opposite sides of the earth, with the moon at a large declination.

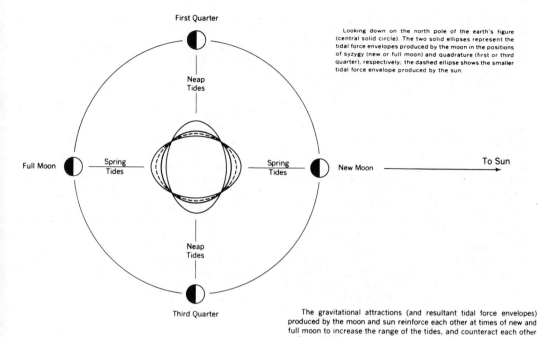

FIGURE 8. The phase inequality, spring and neap tides.

gravitational attractions of the moon and sun are mutually opposed. However, in terms of their tide-producing influences, the respective attractions of the two bodies are still exerted along a common meridian plane passing through the central positions of the direct and opposite tides. The solar and lunar gravitational forces again act as if they were combined. Thus, at both new and full moon (either of which positions of alignment of the sun and moon with the earth is called *syzygy*), the combined gravitational attractions of the moon and sun result in especially high-amplitude tides known as *spring tides*. (This expression derives from the tendency of these tides to "well" or "spring" up, and bears no relation to the season.)

At first- and third-quarter phases (east and west *quadrature*, respectively), the moon is in a position halfway around the earth toward its opposite hemisphere and at right angles to a line joining earth and sun. The gravitational attractions of the moon and sun then are exerted perpendicularly to each other. Thus, although the moon's gravitational attraction is always greater than that of the sun, the sun's gravitational force acts in part to counteract the gravitational pull of the moon on the earth's tidal waters. The resulting high tides are not as high and the low tides are not as low as in the case of either spring tides or the "average"

tides produced when the moon's age is some 3.7, 11.1, 18.5, or 25.8 days in its monthly cycle. As a result, the range or difference in height between successive high and low tides is also not as great. These diminished-amplitude and reduced-range tides are termed *neap tides* (from the Anglo-Saxon word *nep*, meaning "scanty").

If the height of the tides produced by the moon alone is considered as unity (1.0), the addition of the sun's gravitational force at new or full moon can raise spring tides that have 1.455 the amplitude of the tides produced by the moon alone. At first- and third-quarter phases, the counteraction of the moon's gravitational pull by that of the sun can reduce the tides to 0.545 of their value as raised solely by the moon.

The effect of phase inequality may combine with various other astronomical variants to produce tides of extreme amplitude and range, but this phase influence affects only the semidiurnal portions of the tide, since no change in declination of the moon is directly involved.

Perigean Tides (Average Period: 27.5546 mean solar days). Because the moon revolves in a slightly elliptical, rather than a circular, orbit around the earth, with the earth occupying one focus of the ellipse, the distance of the moon from the earth may vary between extreme limits separated by some 50,338 km

(31,278 miles). During an *anomalistic month* of 27.554551 mean solar days, the moon revolves around the earth from *perigee* to *perigee*. At perigee, the point of the moon's closest monthly approach to the earth (widely variable), the moon can come within an absolute minimum distance of about 356,354 km (221,428 miles). At *apogee*, the moon's greatest monthly distance from the earth, the satellite can recede to an extreme maximum distance of 406,692 km (252,706 miles). This represents a change in distance of approximately 13%. The moon's gravitational attraction upon the tides varies inversely as the distance cubed.

Thus, at perigee, the lunar semidiurnal range is increased, and the range of the perigean tides produced may be as much as 20% above the mean range. When the tidal effects of perigee are combined simultaneously with those of syzygy (new moon or full moon), *perigean spring* tides showing an increase in range of 40% or more may be created (Wood, 1978).

Mathematical Development

Evaluation of the Relative Tide-Raising Forces of the Moon and Sun Where m is the moon's mass; M is the earth's mass ($= 1$); D is the distance between the centers of mass of the earth and moon (in terms of the earth's radius, taken as 1); and g is the value of the acceleration of gravity at the earth's surface—from Sir Isaac Newton's universal law of gravitation the moon's gravitational force f_c acting upon the center of mass of the earth is given by

$$f_c = gm/D^2$$

Indicating the earth's radius by $R = 1$ (with D in the same units), then the gravitational force f_s of the moon upon a point at the surface of the earth on the side toward the moon is represented by

$$f_s = gm/(D - R)^2 = gm/(D - 1)^2$$
$$f_s - f_c = gm/(D - 1)^2 - gm/D^2$$
$$= \frac{gm\,[D^2 - (D - 1)^2]}{D^2(D - 1)^2}$$

Thus the actual tide-producing force f_t of the moon opposing the gravity pull of the earth is

$$f_t = \frac{gm(2D - 1)1}{D^2(D - 1)^2}$$

Since the earth's radius is only 3,959/238,857 or 1/60th part of the average distance from the earth to the moon, the difference between D and R is approximately the same as D itself. Therefore $D - R = D - 1 \approx D$ and $2D - R = 2D - 1 \approx 2D$. Substituting these latter values in the immediately preceding equation gives:

$$f_t/g = 2mD/D^2 D^2 = 2m/D^3$$

This demonstrates that the moon's (or sun's) tide-producing force is directly proportional to the mass of either of these attracting bodies and inversely proportional to the cube of its distance from the earth.

Expressed as a fraction of the earth's mass M, the moon's mass $m = 1/81$. Similarly, the average distance D of the center of mass of the moon from the center of mass of the earth, in units of earth radii, is $D = 238,857/3,959 = 60$. Thus the moon's tide-producing force in terms of the earth's gravity at the sublunar point and with the moon at its average distance from the earth is

$$f_t/g = 2(1/81)/60^3 = 1/8,800,000$$

Since the mass of the sun M_s is 332,000 times that of the moon, but its distance D_s from the earth is also 23,400 times that of the moon from the earth, the tide-producing force of the sun compared with the earth's gravitational attraction for the waters on its surface is

$$f_t/g = 2(332,000)/(23,400)^3 = 1/19,300,000$$

And, comparing the sun's tide-producing influence with that of the moon from the preceding computation,

$$1/19,300,000 \div 1/8,800,000 = 5/11$$

Therefore, the sun's tide-producing force is only 5/11ths that of the moon.

Similarly, for a point on the opposite side of the earth from the moon, the moon's tide-producing force f_t' is given by

$$f_t' = f_s' - f_c$$
$$= mg/(D + 1)^2 - mg/D^2$$
$$= mg\left[\frac{D^2 - D^2 - 2D - 1}{D^2(D + 1)^2}\right]$$
$$= -mg\left[\frac{(2D + 1)1}{D^2(D + 1)^2}\right]$$

Assuming $D + 1 \approx D$ and $2D + 1 \approx 2D$, then $f_t' = -2(mg/D^3)$. The minus sign indicates that the force is directed away from the moon, thus demonstrating mathematically that tidal bulges are produced both on the side of the earth S

nearest the moon and on the side of the earth S' turned directly away from the moon.

It may also be seen by a direct comparison of the appropriate values in the above equations that $f_s > f_c > f_s'$.

FERGUS J. WOOD

References

Defant, A., 1961. *Physical Oceanography,* vol. 2. New York: Pergamon Press, 598p.

Doodson, A. T., and Warburg, H. D., 1941. *Admiralty Manual of Tides.* London: His Majesty's Stationery Office, 270p.

Dronkers, J. J., 1975. Tidal theory and computations in V. T. Chow, *Advances in Hydroscience,* vol. 10. New York: Academic Press, 145-230.

Godin, G., 1972. *The Analysis of Tides.* Toronto: University of Toronto Press, 264p.

Gray, T. J., and Gashus, O. K., 1972. *Tidal Power.* New York: Plenum Press, 630p.

Marmer, H. A., 1926. *The Tide.* New York: D. Appleton and Company, 282p.

Russell, R. C. H., and Macmillan, D. H., 1970. *Waves and Tides.* Westport, Conn.: Greenwood Press, 348p.

Sager, G., 1959. *Gezeiten und Schiffahrt.* Leipzig: Fachbuchverlag, 173p.

Schureman, P., 1941. *Manual of Harmonic Analysis and Prediction of Tides.* Washington, D.C.: U.S. Department of Commerce, Coast and Geodetic Survey, Spec. Pub. No. 98, 317p.

Wood, F. J., 1978. The strategic role of perigean spring tides, *in Nautical History and North American Coastal Flooding, 1635-1976.* Rockville, Md.: National Oceanic and Atmospheric Administration, 49-122.

Wylie, F. E., 1979. *Tides and the Pull of the Moon.* Brattleboro, Vt.: The Stephen Green Press, 246p.

Cross-references: *Coastal Engineering; Intertidal Mud Habitats; Intertidal Sand Habitats; Marigraph and Marigram; Proxigean Spring Tides; Storm Surges; Tidal Currents; Tidal Range and Variation; Tidal Type, Variation Worldwide; Tide Tables and Charts; Tsunami.* Vol. I: *Tides;* Vol. II: *Earth Motions.*

TIDE TABLES AND CHARTS

Tides have always been of importance to mariners, and most tidal predictions are given in a form convenient for them. Predictions for the astronomical tide are based on a combination of measurement and theory. No attempt is made to include in tide tables variations due to winds or river flows.

The moon and sun cause the tides, and periods of half a day (semidiurnal), a day (diurnal), a lunar month, and a year are conspicuous in most tidal regimes. A record of tidal height (or current) measured at one station is used, after harmonic analysis, to predict future tides at that place. The same relative configuration of earth, moon and sun recurs closely every nineteen years, and so nineteen years of measurement are ideal for prediction. For practical purpose, predictions made on one year's record are sufficient. This is fortunate since major dredging or land reclamation schemes do affect tides.

The hydrographic departments of the leading maritime authorities compute tide tables for a set of standard ports. They exchange information, so that their publications include tide tables for a wider area. Commercial publishers and other concerns also use the information to produce tables to suit particular groups of users or particular localities.

The main tables give the times of high and low water and their height relative to some datum. The datum is usually the lowest level of the astronomical tide, but may be some other level of relevance to navigation such as the level of the sill of an entrance lock or *mean low water.* Information may also be given for interpolation between tabulated values and for predicting tides at other ports and tidal stations. (These latter are usually called secondary ports, though in some cases they are important ports, and in others not ports at all).

Adequate interpolation between tabulated values is obtained by fitting a sine curve when the rise or fall of the tide is between 5-7 hours. When there are substantial shallow-water influences on the tide, special tables or curves are used to interpolate. This is often the case in northwest Europe (*Admirality Tide Tables,* Vol. 1). The other case where sinusoidal interpolation is inadequate is when tides are diurnal or are mixed semidiurnal and diurnal. In this case, simplified direct prediction using the principal harmonic components may be needed. The harmonic constants and relevant tables for doing this by the "Admiralty method" are given in Vol. 2 of the *Admiralty Tide Tables.* For a general account of this type of approach see Defant (1961), chapter 10.

In many cases, the information needed to predict tides at secondary ports is a simple table of *tidal differences*—the difference in time of the tide and the difference in height of the tide from a standard port. The standard port is usually one fairly close to the secondary port, but occasionally it may be far distant but have similar harmonic components—for example, Galveston, Texas, is the standard port for a large portion of Antarctica. However, in mixed semidiurnal and diurnal tidal regimes, tidal differences are unreliable, and a simplified direct prediction is preferable.

Tidal predictions at other places, offshore, or remote from any tidal station are more difficult.

A short period of observation (ideally a month) is the best basis for prediction. Otherwise estimates of the harmonic constituents must be made from other sources. These include tidal charts, which show the variation of individual harmonic components, such as the moon's semidiurnal component (M_2). Lines of equal amplitude of the component are called *co-range line*, and those of equal phase are called *cotidal* lines. These charts are based largely on coastal observations, so that there is much scope for error. Direct theoretical computations have been made, but still need improvement. Hendershott (1981) gives a review.

Another form of tidal chart is a chart of *tidal streams*—that is *tidal currents*. These are published for those parts of the world where there are significant tidal currents and shipping. Usually they are given for hourly intervals covering half a lunar day. Arrows or lines denote the current direction with figures alongside denoting the usual range of current speeds.

D. HOWELL PEREGRINE

References

Admiralty Tide Tables, 3 vols. Taunton, England: Ministry of Defence, Hydrographic Department.

Defant, A., 1961. *Physical Oceanography*, vol. 2. New York: Pergamon Press, 598p.

Hendershott, M. C., 1981. Long waves and ocean tides, in B. A. Warren and C. Wunsch, *Evolution of Physical Oceanography*. Cambridge, Mass.: M.I.T. Press, 292-341.

Cross-references: *Currents; Tidal Range and Variation; Tidal Type, Variation Worldwide; Tide Gauges; Tides.* Vol. I: *Tides.*

TIME SERIES

The measurement of a parameter at some type of regular interval is called a *time series*. In coastal environments, the most obvious phenomenon producing time series data is the fluctuation of tides. Other phenomena that are commonly studied by time series means are various wave and current parameters, meteorological observations, and sometimes changes in coastal morphology.

One of the primary reasons for collecting data in such a fashion is to analyze it for any trends that might be present. Usually this is best done by quantitative means such as *Fourier analysis* or *harmonic analysis* (Fox and Davis, 1973). In this way relatively long-period trends can be subtracted from a relatively short-term cyclic phenomenon. Again, tides are a good example, with the storm surge or wind tide providing the trend, and the astronomical tides, the regular cyclic variations. A similar treatment can be applied to wave data in order to determine the *surf beat*.

Although the emphasis above has been on dynamic parameters, it is also possible to apply time series techniques in a spatial sense. This is regularly done by geologists investigating a sequence of rock strata in which regularly spaced sampling of the sequence becomes the time series (Miller and Kahn, 1962). A similar application can be utilized in coastal areas where data collection is regularly spaced in a horizontal or vertical (core) manner.

RICHARD A. DAVIS, JR.

References

Fox, W. T., and Davis, R. A., Jr., 1973. Simulation model for storm cycles and beach erosion on Lake Michigan, *Geol. Soc. America Bull.* 84, 1769-1790.

Miller, R. L., and Kahn, J. S., 1962. *Statistical Analysis in the Geological Sciences.* New York: John Wiley & Sons, 483p.

Cross-references: *Computer Simulation; Currents; Fourier Analysis; Storm Surge; Surf Beat; Tides.*

TOMBOLO

A *tombolo* is a beach or beaches that connect an island to the mainland or to another island. This term was first proposed by Gulliver (1899) to "distinguish island-tying bars from those of other kinds." Johnson (1919) notes that the

FIGURE 1. Vertical aerial photograph of St. Peter's Island, Nova Scotia. Note wave refraction patterns (original photograph supplied by the Surveys and Mapping Branch, Department of Energy, Mines, and Resources).

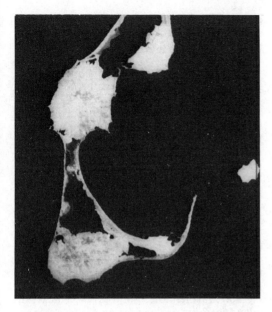

FIGURE 2. Satellite image of the Magdalen Islands, Gulf of St. Lawrence, Canada. The double tombolo connects the two islands to each other. The lagoon enclosed by the barriers is rapidly infilling. Scale approximately 1:1,000,000 (photo: NASA LANDSAT Image, produced by Canada Centre for Remote Sensing).

original Italian word refers to coastal sand dunes and to sand barriers. Tombolo is a descriptive term defining a geomorphic feature that may result from the development of a beach or barrier either in the lee of an island or parallel or oblique to the shoreline.

A *single tombolo* results from the development of a beach deposit or barrier that joins an island to the mainland or to another island. Examples of this are St. Peter's, Nova Scotia (Fig. 1); Chesil Beach, England; and Orbetello, Italy. A *double tombolo* has two barriers that formed separately to enclose a lagoon—for example, Monte Argentario, Italy (Johnson, 1919), and the southern Magdalen Islands, Canada (Fig. 2). The lagoon between the double barrier may be infilling and eventually form a single tombolo. A *multiple* or *complex tombolo* results from the development of a series of beaches or barriers that join a group of islands to the mainland or to each other—for example, Nantucket, Massachusetts, where a series of drowned glacial drumlins have been connected to each other and to the mainland.

EDWARD H. OWENS

References

Gulliver, F. P., 1899. Shoreline topography, *Am. Acad. Arts Sci. Proc.* **34**, 151-258.

Johnson, D. W., 1919. *Shore Processes and Shoreline Development.* New York: John Wiley & Sons, 584p.

Cross-references: *Barrier Beaches; Bars; Beach Processes; Cuspate Spits; Spits.* Vol. III: *Tombolo.*

TOURISM

Coastal recreation includes both tourist and residential use, each of which strains coastal resources (Charlier, 1960). The search for clean air and uncluttered horizons has diverted a huge segment of vacationers to the shores of lakes and seas for relaxation, wildlife study, swimming, waterskiing, boating, or sportfishing (Clark, 1977). In the 1930s, New York public beaches had about 5 million visitors per year; in the 1960s over 61 million made use of them. Such growth led the National Park Service to recommend that at least 15% of the shoreline be reserved for public recreation purposes; one mile of private beach is of use to 100 persons as compared to 5000 for a public one. Yet less than 10% of the entire Atlantic-Gulf coastline is federal or state land (Alexander, 1977; Brahtz, 1972; Cole, 1974). Underwater tourism has been developed in Hawaii, Florida, and the U.S. Virgin Islands, where underwater national parks provide scuba divers with an opportunity to follow marked paths and observe *in situ* aquatic flora and fauna. Undersea tourist facilities (e.g., Aquapolis) were experimented with during the 1975 Okinawa Development Exhibition (Craven, 1977; Fuller, 1972; Nettleton, 1971).

Seashore and water-related outdoor recreational activities have a considerable impact on a shore region's economy (Ketchum, 1972, Latortue, 1979). For example, in the United States it has been estimated that such activities in 1972 accounted for an expenditure of $5.2 billion, projected to rise to $12.8 billion by the year 2000. In particular, marinas and boating constitute a major source of income and employment in coastal areas (Charlier, 1976).

Coastal tourism became fashionable in eighteenth-century Europe when the more sophisticated segment of the population took on the habit of sea bathing (Haulot, 1974; Haulot et al., 1977). Shore resorts flourished rapidly where beaches were readily accessible, esthetically attractive, and pleasant of climate (Bird, 1977; Bryden, 1973). Areas favored by warming marine currents, such as Scandinavia and the Baltic coast, developed rapidly.

Although such areas as the Baltic and Scandinavian coasts are sought-after sites, the most popular shore resorts are located between $40°N$

FIGURE 1. Doctor's Cove Beach, Montego Bay, Jamaica (photo supplied by R. H. Charlier).

and 40°S; in North America beaches line the California coast down to the tip of Baja and southward and in the east, the Atlantic and Gulf coasts and the Caribbean Islands (Fig. 1). In Europe beaches stretch from Denmark to the Algarve, on both sides of the English Channel, and the entire Mediterranean fringe. The Indian and Pacific oceans attract tourists to Hawaii, Tahiti, New Zealand, Madagascar, the Seychelles, the Ryukyus, the Phillippines, Malaysia, and Japan (Matsuishi, 1977).

The surge of development in coastal zones related to tourism can be illustrated by some statistical data from European examples (Aubert, 1973; Ducsik, 1974; Fischer, 1975; Legrain, 1975). Denmark reserves 55% of its 7352 km of shoreline and of its 510 islands (400 of which are inhabited) for tourism. Norway has 21,000 km of coast, 2500 fjords, and some 50,000 islands and islets but climatic conditions limit their use for tourism. Germany has 1000 km of coastline along the North and Baltic seas and limits their use exclusively to tourism except for port activities (e.g., Bremen and Lubeck). The Netherlands developed coastal tourism on 265 km of its 1300-kilometer coastline. Belgium uses 90% of its 67 km of coast along the North Sea for high-density tourism; expansion can only occur landward and current activity is severely challenged by industrial development (Haulot et al., 1977; Vanhove, 1972).

Including coastal islands, Great Britain possesses 15,316 km of coastline, of which 10,162 km are in Scotland. Some 7% (1274 km) of the area is used for tourism activities, but regional density varies considerably. While 50% of island shoreline is used, this proportion declines to 19% in England and Wales, 2% in Scotland, and no use in Northern Ireland. The Irish Republic, however, uses 15% of its 5628 km of littoral for tourism.

France has a coastline that extends over 5650 km, of which roughly 800 km are on Corsica. Coastal occupancy by tourists runs to 52% (Fig. 2); however, this figure is deceptive because only 75% of the coasts are useable for tourism. Based on the 4240 km of coastline that is useable, tourism use climbs to a staggering 90%; furthermore, the tourism belt extends inland for 7 km on the average.

The Iberian Peninsula has witnessed a coastal tourism explosion second to none. In Spain the negative results are cause for serious alarm as a result of environmental and cultural degradation. One may consider that use is maximal: 45% of the 3904 km of its continental coasts and 70% of the 1522 km of insular coasts are used for tourism. Among the islands, the Baleares and the Canaries are prominent. Portugal uses 90% of 850 ha of beaches for tourism.

Although a relative newcomer in appealing to the Western tourist, Yugoslavia (Fig. 3) claims 100% tourism for its 2092 km of coastline and 4024 islands. Of Greece's 15,000 km of shoreline, only 3000 km are suitable for tourism and only 12½% are currently in use.

Once reluctant to open their borders to residents of noncommunist countries, the Black Sea countries such as Romania and

FIGURE 2. Soulac-sur-Mer, Gironde, France (photo supplied by R. H. Charlier).

FIGURE 3. Beach at Dubrovnik, Yugoslavia (photo supplied by R. H. Charlier).

Bulgaria are steadily expanding the tourism activity of their shores. Romania has combined these activities with *thalassotherapy* (q.v.) resorts. Consequently, entire new cities such as Neptun, Jupiter, and Mangalia have arisen in a few short years. The Soviets have developed numerous resorts along their northern and eastern coasts and with Iran have developed shore tourism along the Caspian Sea.

Planning and Needs

World wide efforts to harmonize the growing needs of coastal recreation with economic development and environmental protection often lack coordination, sufficient funding, and even outright sincerity (Brahtz, 1972). Pollution has taken a serious toll of shore recreation facilities. Italy, Greece, Monaco, and Israel prohibit bathing in polluted water, but several countries impelled by economic concerns still allow swimming in coastal waters that contain 50,000–150,000 fecal germs/l.

As taxes increase and more profitable industries move to the coast, recreational activities become less economically viable and often get low priority in planning and management considerations, resulting in the further diminishing of recreational opportunities and public access to the beach.

Private facilities persist but private ownership often engenders further destruction of the beach by intensifying erosion owing to construction that disturbs the environmental balance. Likewise public boardwalks can cause beach erosion: The beach at Coney Island has shrunk considerably, Nice has lost its sandy beach to the Promenade des Anglais, and the fashionable beaches of Belgium's eastern shore, already reduced to a minimal strip, are further threatened by the harbor extension at Zeebruggee and the impending construction of Noordzeepoort (Craven, 1977; Haulot et al., 1977).

Because of the increasing demand for space for leisure activities, thoughtful planning will be required to maintain and even reestablish an adequate balance between tourism and nature (Winslow and Bigler, 1969). Some outstanding achievements to this end can be found at Port Grimaud, France, and in Italy and Spain, while novel ideas for a physical rehabilitation of the seashore are being implemented in other European countries, such as Belgium (Haulot et al., 1977). Vanhove (1972) showed that the absolute capacity of the Belgian coast is 358,000 on the basis of $9m^2$/person of beach. Clearly the continuing pressure of the recreational use of seashore resources will require energetic intervention by ecological engineers (Matsuishi, 1977; Purpura, 1974; Spaulding, 1974; Vines, 1970).

ROGER H. CHARLIER
ARTHUR HAULOT

References

Alexander, L., ed., 1977. *Proceedings of the Conference on the U.S. Coastal Belt.* Kingston, R.I.: University of Rhode Island, Center for Ocean Management Studies, 47p.

Aubert, M., 1973. *Pollution marine et aménagement des rivages.* Nice: Centre d'Etudes et de Recherches Biologiques et Oceanographiques de la Mediterranée, 64p.

Bird, J. B., 1977. Beach changes and recreation planning on the west coast of Barbados, West Indies, *Geog. Polonica* 35, 31–41.

Brahtz, J. F. P., ed., 1972. *Coastal Zone Management: Multiple Use with Conservation.* New York: John Wiley & Sons, 352p.

Bryden, J. M., 1973. *Tourism and Development.* Cambridge: Cambridge University Press, 236p.

Charlier, R. H., 1960. The recreational role of the beach, *Jour. Economic and Social Geog.* 51, 284–285.

Charlier, R. H., 1976. *Economic Oceanography.* Brussels: Uitgaven Vrije Universiteit Brussel, 98–101.

Clark, J. R., 1977. *Coastal Ecosystem Management. A Technical Manual for the Conservation of Coastal Zone Resources.* New York: John Wiley & Sons, 928p.

Cole, B. J., ed., 1974. *Planning for Shoreline and Water Uses.* Kingston, R.I.: University of Rhode Island Marine Advisory Service, 20p.

Craven, J. P., 1977. Use of the ocean space, *in Okinawa, Symposia of Ocean Exp. '75.* Japan Assoc. for the Int. Oc. Expo., 396–399.

Ducsik, D. W., 1974. *Shoreline for the Public, a Handbook of Social, Economic and Legal Considerations Regarding Public Recreational Use of the Nation's Coastal Shoreline.* Springfield, Ill.: National Technical Information Service, 257p.

Fischer, J.-C., 1975. L'integration sito-écologique des aménagements littoraux, Amé*nagement et Nature* 39, 23–24.

Fuller, R. B., 1972. Geoview: floating cities, *World* **1**, 40–41.
Haulot, A., 1974. *Tourisme et Environement.* Verviers: Marabout, 411p.
Haulot, A.; Vanhove, N.; Verheyden, L. R. A.; and Charlier, R. H., 1977. Coastal belt tourism, economic development and environmental impact, *Internat. Jour. Environmental Studies* **10**, 161–172.
Ketchum, B. H., ed., 1972. *The Water's Edge: Critical Problems of the Coastal Zone.* Cambridge, Mass.: M.I.T. Press, 393p.
Latortue, G., 1970. *The Demand for Water-based Recreation in Southwest Puerto Rico.* Mayaguez: Puerto Rico University Water Resources Research Institute, 160p.
Legrain, D., 1975. Le conservatoire du littoral, *Aménagement et Nature* **39**, 10–12.
Matsuishi, H., 1977. General view of coastal development in Japan, *in Okinawa, Symposia of Ocean Expo '75.* Japan Assoc. for the Int. Oc. Expo. 92–99.
Nettleton, A., 1971. Cities in the sea, *Oceans Magazine* **5**, 71–75.
Purpura, J., and Sensabaugh, W. M., 1974. *Coastal Construction Setback Line.* Gainesville, Fla.: University of Florida, Sea-Grant Program, 18p.
Spaulding, I. A., 1974. *Factors Related to Beach Use.* Narrangansett, R.I.: University of Rhode Island, Marine Advisory Service Tech. Rept. 13, 20p.
Vanhove, N., 1972. *Micro- and Macro-ekonomische verantwoording van de toeristische rekreatieparken aan de belgische kust.* Bruges: Westvlaams Ekonomisch Studiebureau, 142p.
Vines, W. R., 1970. *Recreation and Open Space.* West Palm Beach, Fla.: Palm Beach County Area Planning Board, 90p.
Winslow, E., and Bigler, A. B., 1969. Marine recreation: problems, technologies and prospects to 1980, *Ann. Conf. Marine Technol. Soc. Proc.* **5**, 53–54.

Cross-references: *Boat Basin Design; Coastal Engineering; Coastal Reserves; Coastal Zone Management; Demography; Human Impact; Protection of Coasts; Pollutants; Thalassotherapy; Waste Disposal.*

TREND SURFACE ANALYSIS

Trend surface analysis is a mathematical technique for separating a mapable variable into its regional components and local fluctuations. Data consist of three variables: a *dependent variable* (X) that possesses both magnitude and location, and two *independent variables* (U, V) that are the geographic coordinates of the dependent variable measured within an orthogonal grid. Methodology consists of determining the coefficients in a standard mathematical expression of the form $X = f(U, V)$ that provides a best fit to the dependent variable by the criterion of least squares.

Of the several types of mathematical expressions in use, polynomials are the most common. A second variety of expression, double *Fourier* series, have proved useful for fitting surfaces to data that possess wave forms. In the simplest polynomial case, data are fitted to an expression of the form

$$X = a_0 + a_1 U + a_2 V \qquad (1)$$

Equation (1) is the expression for a linear, or first-order, trend surface. It is a simple planar, dipping surface, and has a direct two-dimensional analog in the standard regression line $X = a + bY$.

Polynomial surfaces of order greater than one are obtained by addition of appropriate terms to the right-hand side of Equation (1). The second-order surface has the expression

$$X = a_0 + a_1 U + a_2 V + a_3 U^2 + a_4 U \cdot V + a_5 V^2 \qquad (2)$$

The second-order trend surface may have the form of a dome, basin, ridge, or trough, depending on the values assumed by the coefficients (a_1). The third-order surface is constructed by the addition to Equation (2) of the four terms of positive exponentials involving the product $U^n \cdot V^{3-n}$, the fourth order, by addition to the third of the five terms involving the product $U^n \cdot V^{4-n}$, and so on. The higher the order of the surface, the more complex the surface may be.

Residuals or local fluctuations are determined by calculation of the trend value at each sampling point and subtraction of this from the observed value at the point. Depending on the goals of the study, either the trend surface or the residuals from it may be of primary interest.

Relevant publications on trend surface analysis include those by Davis (1973), Koch and Link (1971), and Lusting (1969).

RONALD C. FLEMAL

References

Davis, J. C., 1973. *Statistics and Data Analysis in Geology.* New York: John Wiley & Sons, 550p.
Koch, G. S., Jr., and Link, R. F., 1971. *Statistical Analysis of Geological Data,* vol. 2. New York: John Wiley & Sons, 438p.
Lusting, L. K., 1969. Trend-surface analysis of the Basin and Range Province, and some geomorphic implications, *U.S. Geol. Survey Prof. Paper 500-D,* 70p.

Cross-references: *Computer Simulation; Fourier Analysis; Mathematical Models; Sediment Analysis, Statistical Methods.*

TSUNAMIS

The name *tsunami* is applied to a long-period gravity wave in the ocean caused by a sudden large displacement of the sea bottom or shores. They are also called *seismic sea waves* and are popularly known as *tidal waves*, but this latter term is a misnomer because tsunami waves are not associated with the astronomical tidal forces of the sun and moon. A tsunami is accompanied by a severe earthquake; however, earthquakes do not cause tsunamis but, rather, both phenomena are caused by the same sudden crustal displacement. The waves of a tsunami have periods of several minutes to several hours. For example, the waves found to occur during the 1964 Alaskan earthquake included periods from two minutes to five hours (Wilson and Thorum, 1968). The potential of these types of waves is of immediate importance for the design of coastal and harbor structures. The speed of the tsunami can be 700–900 km/hr in the open ocean, depending on the water depth. In shallow water, the speed decreases but the wave height increases greatly.

Tsunamis are classified on a magnitude scale much like the earthquake classification scheme. The tsunami magnitude, however, is measured at a coastal area 100 km from the epicentric location. A typical scale prepared by Imamura (1949) for Japanese tsunami is listed in Table 1.

Iida (1958) has found a relationship between tsunami generation and earthquake magnitude. He determined that whether or not a tsunami is generated depends not only upon the energy of the earthquake but also upon the focal depth associated with it. The shallower earthquake is able to transfer more energy into tsunami energy. The relationship was if $M < 0.017D + 6.42$, no tsunami; if $0.017D + 6.42 < M < 0.008D + 7.75$, moderate tsunami; and if $M > 0.008D + 7.75$, disastrous tsunami, where M is the earthquake magnitude on the Richter scale, and D is the focal depth in km. From the above relationship and Table 1, a plausible relationship between earthquake magnitude (M), and tsunami magnitude (m) is $m = 2M - 13.5$. Wilson (1962) extended the work of Iida further to determine the relationship between earthquake magnitude and resulting tsunami wave height. It was found that $\log_{10} H = 0.75M - 5.07$, where H is the average wave height in the coastal vicinity (approximately 100 km) of the earthquake. The actual tsunami height will be more or less than this depending on many factors such as refraction, reflection, convergence, and backflow, and hence will vary considerably along the coast.

Additionally, further away from the local generating zones there will be felt a lesser amount of tsunami energy due to the radial propagation of the energy. The problem of tsunami prediction from actual seismic data was proposed by Adams (1969). This algorithm involved knowledge of earthquake magnitude, water depth at epicenter, and focal depth. However, crustal displacement length and displacement area are not used, and the actual effectiveness of this prediction algorithm has yet to be tested for a large tsunami. Once the tsunami elevation at the coastline has been determined, then it is possible to estimate the wave run-up and inundation.

Reported tsunami run-ups vary considerably from one location to the next owing to various effects such as flow convergence, friction, reflection, and wave refraction. The convergence, reflection, and refraction effects depend on the land topography and bathymetry. Conversely, the frictional effect depends primarily upon the terrain conditions over which the surge propagates. For rough terrains, the run-up is lower than for smooth terrains. Therefore, if a development is made and the terrain is made smoother than the original, higher tsunami run-ups can be expected.

The frequency of tsunamis within the Pacific Ocean has been summarized by the U.S. Army Corps of Engineers (1962), Honolulu District. For 142 years the report has given 41 damaging occurrences in Hawaii, or once every 3.5 years. There have been several locally generated severe tsunamis, one in 1866 and one in December 1975. Other recent severe tsunamis originated from earthquakes off the coast of South America (notably Chile) and the Gulf of Alaska (notably the Aleutian Islands). Table 2 summarizes the tsunamis through May 1960. The maximum elevation that the water surged at

TABLE 1. Tsunami Classification

Height of Tsunami (m)	Potential Damage	Magnitude
0.5	Nil	-1
1.0	Very little	0
2.0	Shore and ship damage, some inland damage	1
4–6	Inland damage and loss of life	2
10–20	Severe destruction over 400 km of coast	3
30	Severe destruction over 500 km of coast	4

Source: Imamura, 1949

TABLE 2. Tsunamis Affecting Hawaii

Date	Source	Wave Period (Minutes)	Damage
Apr. 12, 1819	Chile	11	Slight
Feb. 20, 1835	Chile	–	Severe
Nov. 7, 1837	Chile	28	Very severe
May 17, 1841	Kamchatka	40	Moderate
Apr. 2, 1868	Hawaii	–	Very severe
Aug. 13, 1868	Peru & Bolivia	–	Very severe
July 25, 1869	South America	–	Severe
Aug. 23, 1872	Hawaii	6	Moderate
May 10, 1877	Chile	20	Very severe
Jan. 20, 1878	–	–	Slight
Aug. 27, 1883	–	–	Moderate
June 15, 1896	Japan	–	Slight
Aug. 9, 1901	Japan	–	Slight
Jan. 31, 1906	Colombia	–	Slight
Aug. 16, 1906	Chile	–	Moderate
Oct. 11, 1913	New Guinea	–	Slight
May 26, 1914	New Guinea	–	Slight
May 1, 1917	Kermadec Islands	–	Slight
June 25, 1917	Samoan Islands	–	Slight
Aug. 15, 1918	Philippine Islands	–	Slight
Sept. 7, 1918	Kurile Islands	–	Moderate
Apr. 30, 1919	Tonga Islands	90	Slight
Nov. 11, 1922	Chile	–	Slight
Feb. 3, 1923	Kamchatka	15	Very severe
Apr. 13, 1923	Kamchatka	–	Slight
Nov. 4, 1927	California	–	Slight
Dec. 28, 1927	Kamchatka	–	Slight
Jun. 16, 1928	Mexico	–	Slight
Mar. 6, 1929	Aleutian Islands	–	Slight
Oct. 3, 1931	Solomon Islands	15	Slight
June 3, 1932	Mexico	–	Slight
Mar. 2, 1933	Japan	–	Moderate
Nov. 10, 1938	Alaska	–	Slight
Apr. 6, 1943	Chile	–	Slight
Dec. 7, 1944	Japan	–	Slight
Apr. 1, 1946	Aleutian Islands	15	Very severe
Dec. 20, 1946	Japan	–	Slight
Aug. 21, 1951	Hawaii	–	Slight
Nov. 4, 1952	Kamchatka	38	Moderate
Mar. 9, 1957	Aleutian Islands	18	Moderate
May 23, 1960	Chile	33	Very severe

Source: After U.S. Army Corps of Engineers, 1962

Hilo was 10.7 m for the 1960 tsunami, whereas the water reached 7.6 m in 1946 and 4.3 m in 1957. Water elevations have been higher in other parts of the Hawaiian Islands, depending on topography, orientation, and direction. In 1946, the water surged to 15 m or more at several locations, such as Waikolu Valley, Molokai, and Waipio Valley, Hawaii. In 1868 a locally generated tsunami reached 15–18 m near Punaluu, southeast coast of Hawaii.

The April 1, 1946, tsunami was very severe, and followed an earthquake off Aleutians. The magnitude of the earthquake was 7.4 on the Richter scale. The November 4, 1952, tsunami was moderate, and followed an earthquake off Kamchatka. The magnitude of this earthquake was 8.5 on the Richter scale. The March 9, 1957, tsunami was moderate, and followed an earthquake off the Aleutians. The magnitude of the earthquake was between 8.0 and 8.5 on the Richter scale. The May 23, 1960, tsunami was very severe, and followed an earthquake off the west coast of Chile. The magnitude of the earthquake was 8.5 on the Richter scale.

The April 1, 1946, Alaskan earthquake and the May 23, 1960, Chilean earthquake tsunamis were of opposite earth hemispheric origin.

Other geographical source areas of the Pacific basin apparently offer less tsunami potential for the State of Hawaii.

Most records of tsunami waves are sinusoidal, similar to the much longer-period astronomical tide waves. When tsunami waves approached shallow water and a relatively flat continental slope, the sinusoidal oscillatorial tsunami wave transforms nearly into a solitary wave and then possibly into a breaking wave bore. In case of a very steep continental slope or a vertical seawall, the tsunami might not be a breaking wave but, instead, can be a reflected wave.

Thus, there are two conditions to consider: the reflected wave, and the breaking wave. The second condition requires a knowledge of the type of breaker and whether or not it breaks before the structure or actually breaks direct on the structure. If the tsunami breaks before it reaches the structure, it will travel as a bore and impact its momentum against the structure.

In studying the forces produced by tsunamis, it is desirable to describe the damage produced by select tsunamis as reported by field investegators. There are three types of damage that are produced by tsunamis: inundation resulting in rapid flooding of foundations and grounding or carrying vessels inland; severe damage due to very high water velocities that destroy buildings, strip and erode land and vegetation, and scatter rocks, coral, and debris; and a combination of both the above where the major effect would be flooding as a tidal inundation; however, severe backflows may develop, and erosion and local impact damage may occur. The second type of damage appears to be the most destructive, and occurs when the tsunami transforms itself into a bore, as was the case, in part, in the 1960 Chilean tsunami experienced in Hilo. The damage caused by this tsunami was extensively studied by a team of structural engineers (Matlock, Reese, and Matlock, 1962). In this study a field survey was made of buildings and structures in Hilo that were damaged by only wave forces (minimal debris damage) to determine velocities and forces exerted by tsunamis. Some of their findings were:

All of the evidence gathered in the survey of structural damage indicates that the third wave of the tsunami approached the shoreline of Hilo Harbor as a bore approximately 15 ft. above normal sea level, and swept inland at high velocity as a sheet of water. It was this mass of high-velocity water moving laterally which inflicted the heavy damage to structures on shore. Evidence of structural damage indicates that the height of this high-velocity water ranged from 8 to 12 ft above ground level, and then it traveled at 25 to 40 ft/sec. Analysis of specific structural elements indicates that a lateral force greater than 400 lb/ft^2 and less than 1800 lb/ft^2, with a mean average of approximately 700 lb/ft^2, was exerted by the water.

The force of the water completely demolished all light frame buildings and most heavy timber structures and inflicted varying degrees of damage to structural steel and reinforced concrete structures. Properly designed reinforced concrete construction seemed to withstand the force of the water with little serious damage to the structural frame. However, damage was particularly severe where the fronts of buildings were open or glassed, but where the rear walls were relatively continuous.

The destruction caused by the 1960 tsunami at Hilo was clearly of the third type. However, certain coastal locations do not produce bore-type tsunami waves; hence, consideration must be given to the first and second categories. Thus, a theoretical method can be used for tsunami wave forces and overturning moments, in which case the wave can be treated either as a bore or as a nonbreaking long wave (similar to a tidal-type wave) with a period on the order of 5 minutes to 2 hours.

For a bore impacting a vertical wall, it can be shown by conservation of momentum that

$$\frac{d_w}{H} = \sqrt{1 + 2\left(\frac{U}{\sqrt{gH}}\right)^2} \qquad (1)$$

where H is initial height, U is the velocity of the bore, g is the acceleration due to gravity, and d_w is the maximum height the bore attains after impacting the wall.

The hydrostatic force, F, on a vertical wall of unit width is given by

$$F = 1/2\, \rho g\, d_w^2 \qquad (2)$$

where ρ = the mass density of water.

It then follows that

$$F = 1/2\, \rho g\, H^2 \left[1 + 2\left(\frac{U}{\sqrt{gh}}\right)^2\right] \qquad (3)$$

Note that for $d_w = 3H$,

$$F = 4.5\, \rho g\, H^2 \qquad (4)$$

Studies by Alavi (1964) and Fukui et al. (1963) confirm that $d_w/H = 3$ is a valid approximation for a vertical wall of sufficient large breadth (5–10 times H); hence the force is given by equation (4). For structures of limited breadth, the force becomes

$$F = 1/2\, \rho g\, H^2\, (1 + 4\, C_D) \qquad (5)$$

where F is the force per unit width of structure

and C_D is the characteristic drag coefficient of the structure. For a plate of infinite breadth, $C_D = 2$ and Equation (5) reduces to Equation (3). Values of C_D can be found in various references on fluid mechanics. In the case of the nonbore where it is definitely clear that a tsunami bore will not form because of steep cliffs at the coastline, then the wave will rise unbroken similar to an extraordinarily high tide. Any structure in this path will then be exposed to an elevated hydrostatic pressure. In this case, it will be sufficient to use Equation (2) for determining the force (per unit width) on a vertical structure of infinite breadth.

In this case, the maximum rise on the structure will be the run-up, R, and $d_w = d + R$, where d is the still water depth. Thus, for a vertical structure of large breadth, the total force per unit width will be

$$F = 1/2 \, \rho \, g \, (d + R)^2 \qquad (6)$$

(kg/lineal m of breadth)

When the trough of the wave arrives, $d_w = 0$ and there will be no wave force, and the structure should be designed against the water and soil pressure behind the wall.

The total broadside force is then given by:

$$F_B = B \times F \text{ (kg total)} \qquad (7)$$

where F_B = total broadside force and B = breadth of structure.

The total overturning moment per unit width is given by:

$$M = 1/3 \, F \times d_w \qquad (8)$$

or

$$M_B = B \times M \qquad (9)$$

where M_B = total overturning moment and B = breadth of structure.

A detailed list of references on tsunamis is given by Wiegel (1970). For other literature on the subject the reader is referred to Adams (1967), Bretschneider and Wybro (1976), and Iida, Cox, and Pararas-Carayannis (1967).

CHARLES L. BRETSCHNEIDER

References

Adams, W. M., 1967. *Relative Susceptibility to Wave Inundation at the Hawaiian Islands Determined from Historical Observations.* University of Hawaii, Tech. Rept., U.S. Atomic Energy Commission Ref. Contract No. At(25-1)-235, 13p.

Adams, W. M., 1969. *Prediction of Tsunami Inundation from Current Real-Time Data.* Hawaii Institute of Geophysics, Rept. No. HIG-69-9, 57p.

Alavi, M., 1964. Surge run-up of sloping beach and surge forces on piles, unpublished memo to R. L. Wiegel. University of California, Berkeley, College of Engineering.

Bretschneider, C. L., and Wybro, P. G., Tsunami inundation prediction, in *Conf. Coastal Eng., 15th, Am. Soc. Civil Engineers, Proc.,* 1006-1024.

Fukui, Y.; Nakamura, M.; Shiraishi, H.; and Sasaki, Y., 1963. Hydraulic study of tsunami, *Coastal Eng. Japan* 9, 67-82.

Iida, K., 1958. Magnitude and energy of earthquakes accompanied by tsunami and tsunami energy, *Nagoya Univ. Jour. Earth Sci.* 6, 101-112.

Iida, K.; Cox, D. C.; and Pararas-Carayannis, G., 1967. *Preliminary Catalog of Tsunamis Occurring in the Pacific Ocean.* University of Hawaii, Hawaii Institute of Geophysics, Rept. No. HIG-67-10, 277p.

Imamura, A., 1949. List of tsunami in Japan, *Zisin,* ser. 2, **2**, 23-28.

Matlock, H.; Reese, L. C.; and Matlock, R. B., 1962. *Analysis of Structural Damage from the 1960 Tsunami at Hilo, Hawaii.* Austin, Tex.: University of Texas, Structural Mechanics Research Laboratory, Tech. Rept. to Defense Atomic Support Agency, 1268, 95p.

U.S. Army Corps of Engineers, 1962. The tsunami of May 23, 1960, *Final Post Flood Report, U.S. Army Corps of Engineers,* Honolulu District, 11p.

Wiegel, R. L., ed., 1970. *Earthquake Engineering.* Englewood Cliffs, N.J.: Prentice-Hall, 518p.

Wilson, B. W., 1962. *The Nature of Tsunamis: Their Generation and Dispersion in Water of Finite Depth.* National Engineering Science Co., Tech. Rept. Contract No. CGS-801 (2442), 150p.

Wilson, B. W., and Thorum, A., 1968. *The Tsunami of the Alaska Earthquake, 1964: Engineering Evaluation.* U.S. Army Corps of Engineers Coastal Engineering Research Center, Tech. Memo No. 25, 401p.

Cross-references: *Coastal Engineering; Coastal Erosion, Environmental-Geologic Hazard; Tides; Waves.* Vol. I: *Tsunami.*

U

UNDERTOW—See RIP CURRENT

U. S. NAVAL HYDROGRAPHIC OFFICE SCALE—See SEA CONDITIONS

V

VEGETATION COASTS

Vegetation coasts, occupied by higher plant communities, are found in areas where shore physiography provides shelter from strong wave action, and are best developed where the shore is gently shelving and the tide range large. In certain circumstances, the plants of vegetation coasts can influence physiographic evolution.

Mangrove coasts

Mangrove coasts (q.v.) are best developed in tropical or subtropical zones, but several species occur with salt marsh in the warm temperate zones of both hemispheres. Mangrove shrubs or trees grow on sand, peaty soil, coral, and rock, although they are best developed on muddy substrates. Species or groups of species are frequently arranged in zones parallel to the shoreline in response to factors such as tidal inundation, light, substrate, and salinity. Zonation may be interrupted by the presence of drainage creeks that are characteristic of most mangrove swamps.

The role of mangroves in promoting sedimentation has been a subject of considerable controversy. Coastal progradation due to accretion around the aerial roots of mangroves was reported by Davis (1940), but Van Steenis (1958) claimed that progradation was a prerequisite to mangrove colonization and merely continued after their establishment. Mangroves are thought to modify the rate at which geomorphic processes take place without changing the pattern of landform evolution (Carlton, 1974). However, measurements at several sites on the Australian coast show that *Avicennia* colonization has resulted in reshaping of the intertidal topography and development of a depositional terrace. It is thought that variation in the structure of aerial root systems of mangroves may result in differing physical effects on sedimentation (Bird and Barson, 1982).

Salt Marsh Coasts

Salt marshes (q.v.) occupy similarly protected sites on sand and mudflats in the mid to high latitude regions, and are also found on low energy sectors of open coasts.

Salt marshes may be dominated by marine algae, grasses, rushes, herbs, shrubs, or a mixture of these communities depending on the nature of the substrate, tidal range, climate, and factors of historical plant geography. The muddy, macrotidal, humid temperate marshes are generally floristically the most diverse. Where temperature is a limiting factor, grasses or sedges tend to dominate, but toward the equator, where extreme salinity becomes a limiting factor, succulent shrubs are important. In areas of lowered salinity regime, *Phragmites communis* forms coastal marshland. Species zonation is also observed in salt marshes, especially on coasts with a large tide range. Tidal creeks are also common features of salt marshes, and residual enclosed unvegetated pans may develop.

Measurements made on salt marshes show that they promote accretion of sediments, particularly at mid-marsh level where vegetation cover is dense and tidal inundation frequent. Floristic differences are also thought to affect patterns of accretion; species vary in their efficiency as sediment traps and in the ability of their root systems to withstand erosion.

The vigorous hybrid grass *Spartina townsendii*, which originated in Britain, has been planted in intertidal environments elsewhere in Europe, Australia, and New Zealand, where it spreads rapidly and promotes accretion. It is thus of value for stabilization and reclamation purposes.

MICHELE M. BARSON

References

Bird, E. C. F., and Barson, M. M., 1982. Stability of mangrove ecosystems, *in* B. F. Clough, ed., *Mangrove Ecosystems in Australia; Structure and Management*. Canberra: Australian National University Press (in press).

Carlton, J. M., 1974. Land building and stabilization by mangroves, *Environmental Conservation* 1, 285-294.

Davis, J. H., 1940. The ecological and geological role of mangroves in Florida, *Carnegie Inst. Washington Pub. 517*, 303-412.

Van Steenis, C. C. J., 1958. Rhizophoraceae, *Flora Malesiana*, ser. 1, 5, 429-445.

Cross-references: *Biogenous Coasts; Biotic Zonation; Coastal Ecology, Research Methods; Coastal Flora; Low-Energy Coast; Mangrove Coasts; Nearshore Water Characteristics; Salt Marsh.*

VISOR—See NICHE

VOLCANO COASTS

Volcano coasts are shorelines typified by volcanic structures. Frequently they have a lobate or circular pattern and are associated with modern volcanism. According to Shepard's (1948) criterion, volcano coasts are *primary shorelines* and are structural coasts.

Darwin (1842, in DeMartonne, 1926) was the first to study volcanic islands from the viewpoint of coral reefs. Fouqué (1879, in DeMartonne, 1926) studied the caldera coast of Thira (Aegean Sea) (Fig. 1). Dutton and Hahn studied the Hawaiian and other islands (1883, in DeMartonne, 1926). However, the expression *volcano shoreline* was not used until this century by Johnson (1919).

A clear classification of volcano coasts was made by Guilcher (1954). Using this scheme we may distinguish:

Circular or Lobate Coasts. There are three

FIGURE 1. Caldera coast of Thíra, Greece (photo: M. L. Schwartz).

types: *volcanic peninsulas and lobes*, such as those near Rabaul on New Britain Island where a volcano formed recently in 1937. In this instance volcanoes form peninsulas and calderas form lobes. Jolo Island (Sulu Archipelago, Philippines) shows two stages of volcanism and the far spread distribution of marine terraces (Voss, 1971). The coastal form of Jolo is controlled by volcanism. *Circular or elliptical islands* appear when there is one principal volcano; there are generally other cones that are parasitic on the sides of the main cone. Pata Island is a well-formed elliptic island close to Jolo. Two ancient, central craters are associated with the cone, which is a recent crater. Some islands of this type have steps of terraces on the cone slope, as evidenced by the steps on Pata (Voss, 1971). Lastly, Isla de Pascua is an example of *islands with two or more* volcanoes, and it has three principal volcanoes that form a triangle on the island.

Caldera Islands. Deception Island (South Shetlands, Antarctica) is a typical caldera, represented by a closed, semicircular, natural bay. Violent volcanic explosions destroyed the entire central portion of the original volcano. There remains only a great central depression (caldera), generally the result of collapse as evidenced by subsidence faults. Latent igneous activity is shown in parasitic craters over pyroclastic cones within the caldera.

Many volcanic islands are likewise coral reefs and atolls, and their evolution must be studied as a whole. Different islands have distinct morphologies, depending on the type of volcanic activity and history, as seen in the Sulu Archipelago.

JOSÉ F. ARAYA-VERGARA

References

DeMartonne, E., 1926. *Traité de Geographié Physique,* vol. 2., Paris: Colin, 1057p.

Guilcher, A., 1954. Morphologie Littorale et Sousmarine. Paris: Press Université France, 216p.

Johnson, D. W., 1919. *Shore Processes and Shoreline Development.* New York: John Wiley & Sons, 584p.

Shepard, F. P., 1948. *Submarine Geology.* New York: Harper, 348p.

Voss, F., 1971. Quartärer Vulkanismus und junge Hebungsbewegungen der Insel Jolo im Sulu Archipel der Philippinen, *Zeitschr. Geomorphologie* **15**, 1-11.

Cross-references: *Antarctica, Coastal Morphology: Coral Reef Coasts; Lobate Coasts.* Vol. III: *Coastal Classification.*

W

WADDEN

The *wadden* area, Europe's largest coastal tidal marsh, lies along the North Sea coast of the Netherlands, Germany, and Denmark. *Wad* (Dutch) refers to any tidal marsh, while *Wadden* (Dutch) or *Watten* (German) often refers to these specific North Sea tidal marshes. The Wadden is fronted by the Frisian barrier islands in the west, unsheltered along the German coast in the east, and is partially island-sheltered again along the Danish coast to the north (Fig. 1). Averaging less than 10 km in width and with a tidal range varying from 1.5 m (east) to 4.0 m (west), its four major environments between mainland and barrier island shorelines are the: supratidal, grass-covered salt marshes; high intertidal, unvegetated tidal flats between high tide and midtide; low intertidal marine fauna rich tidal flats; subtidal meandering, branching, tidal channels and sand bars (Van Straaten, 1951).

Numerous sedimentological studies have been conducted on these flats by Postma (1961), Van Straaten (1951), and Reineck (1972) among others (see also reprints in Klein, 1976). In general, the tidal flats off the Dutch coast are less muddy than the tidal flats off the German coast. The high concentration of muds on the tidal flats, beyond that due to decreasing tidal currents, was first thought to be the result of *settling lag* effect, the time necessary for a particle to settle after a velocity decrease. Later it was considered a *scour lag* effect, the difference between the velocity of deposition and greater velocity of erosion (*scour*). More recently it was related to tidal current asymmetry with low current velocity existing for longer periods of time at high tide than at low tide and allowing increased deposition of the finest size particles.

The significance of the Wadden sedimentation environment to the geologic record is that extensive mud deposits occur not only at great

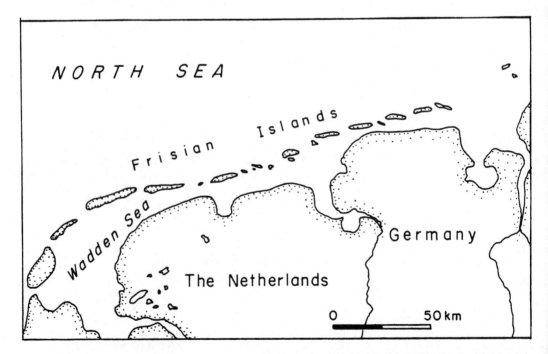

FIGURE 1. Wadden (*watten*) tidal flats of North Sea coast developed behind the Frisian barrier islands of the Dutch, German, and Danish coasts.

depths or far from shore but also at shallow depths and close to shore. Wadden sedimentation also provides recent analogs for petroleum organic source muds, mudstone origin, linear sand bodies, and *laser* (alternating sand and mud) beds. Ecologically, in recent years, the Wadden's tidal *wetlands* have been under various stresses such as industrial pollution, land reclamation, oil drilling, military exercises, and increasing recreational use.

JOHN J. FISHER

References

Klein, G. deV., ed., 1976. *Holocene Tidal Sedimentation.* Stroudsburg, Pa.: Dowden, Hutchinson & Ross, 423p.

Postma, H., 1961. Transport and accumulation of suspended matter in the Dutch Wadden Sea, *Netherlands Jour. Sea Research* 1, 148-190.

Reineck, H.-E., 1972. Tidal flats, in J. D. Rigby and W. K. Hamblin, eds., *Recognition of Ancient Sedimentary Environments.* Tulsa, Okla.: Society of Economic Paleontologists and Mineralogists, 146-159.

Van Straaten, L. M. J. U., 1951. Texture and genesis of Dutch Wadden Sea sediments, *Internat. Congress Sediment., 3rd, Netherlands, Proc.,* 225-244.

Cross-references: *Europe, Coastal Morphology; Geographic Terminology; Nearshore Hydraulics and Sedimentation; Organism-Sediment Relationship; Sediment Transport.*

WASHOVER AND WASHOVER FAN

A *washover,* sometimes referred to as an *overwash* in the literature, is a small deltalike feature or fan that is built in a lagoon or on a salt marsh when the sea breaches a dune ridge and invades the landward side of a barrier during periods of severe storm wave activity and abnormally high tides. Such features, described by Lobeck (1939), are particularly common in areas like the Gulf coast and eastern seaboard of the United States, that experience hurricanes or cyclones. Various washovers have been described by Hayes, Godfrey, Dolan, Kraft, and others (Coates, 1973).

Washover material often provides new substrate for salt-marsh development because it is an ideal surface for colonization by marsh grass. Some washover fans are capped by aeolian deposits that have been subsequently stabilized by vegetation. Usually washover fans can be distinguished from other depositional features by their sedimentary structures, which, in the case of estuarine washover barriers, often display high-angle cross-lamination facing landward, indicating their formation by strong washover currents toward the land.

Thin estuarine washover barriers occur along the broad marshes fringing the submerged Delaware estuary, United States, where they are migrating landward and upward in space and time across the marsh areas.

The effects of hurricanes as geological agents have been examined by Hayes (1967), who studied hurricane washover fan deposits on Padre Island and in adjacent areas after Hurricanes *Carla* (1961) and *Cindy* (1963) crossed the south coast of Texas. As a result of *Carla,* offshore and beach sediments, 9 cm thick, were spread as a washover fan over wind-tidal flats (*algae mat*) sediments by storm-surge flood waters.

JOHN R. HAILS

References

Coates, D. R., ed., 1973. *Coastal Geomorphology.* Binghamton: State University of New York Publications in Geomorphology, 404p.

Hayes, M. O., 1967. Hurricanes as geological agents: Case studies of hurricanes Carla, 1961, and Cindy, 1963, *University of Texas, Bureau of Economic Geology Pub. No. 61,* 56p.

Lobeck, A. K., 1939. *Geomorphology.* New York: McGraw-Hill, 781p.

Cross-references: *Algal Mats and Stromatolites; Barrier Islands; Beach Processes; Beach Stratigraphy; Coastal Dunes; Hurricane Effects.*

WASTE DISPOSAL

Coastal lands and shallow ocean bottoms have been extensively used for *waste disposal* for centuries. The prime example is the filling of wetlands by landfills for solid refuse and garbage, especially in urbanized coastal regions. About one-fifth the area of New York City is filled wetland, and at least half of that area was formerly used for disposal of refuse and garbage. The process continues in many cities, but is usually not well studied and documented.

The coastal ocean is used in many areas to dilute and disperse wastes, especially those from coastal cities or industries. Various wastes, usually dissolved or suspended in water, are discharged into bays and harbors or directly into the coastal ocean from pipelines. Ocean waters dilute the wastes, and currents disperse them. Eventually wastes are carried away from the discharge site to be diluted and dispersed in open ocean waters. When disposal operations are not working properly or have overloaded the capacity of the coastal ocean to disperse or decompose them, waste concentrations may build up to levels where they become noticeable and can make the area unusable for other purposes.

Municipal Waste Water Treatment Sludges

Municipal waste water-treatment-produced *sludges* are discharged to coastal ocean waters in the United States (New York, Philadelphia, Los Angeles) and in the United Kingdom (London, Glasgow, Liverpool). Although produced in large volumes, these wastes contain only about 5% solids, typically low-density particles. The materials tend to be widely dispersed by mixing when discharged and subsequently by tidal currents and wave action. Sludge solid deposits are usually characterized by high carbon and metal contents. They may also be identified by human artifacts (hair, cigarettes, tomato seeds, plastics) and by high bacterial populations coming from human wastes.

Dredged Material Disposal

Dredging, the removal of deposits from navigation channels or other facilities, constitutes one of the largest sources of wastes for disposal in coastal areas (Boyd, 1972). In the early 1970s about 3×10^8 m^3 were dredged each year from United States waterways; about 60% of this volume was disposed of in open waters.

In the New York Bight, approximately 200×10^8 m^3 of waste solids are known to have been deposited between 1888 and 1973. These wastes have filled in part of the Hudson Shelf Valley and formed hills 10–15 m high on the ocean bottom (Williams, 1975).

Industrial Wastes

Several industrial processes produce large volumes of wastes that, because of the amounts involved and the characteristics of the wastes, are often not acceptable for disposal on land and are discharged at sea. Among these are the fermentation wastes from the pharmaceutical industry and from the brewing industry. Such wastes were disposed in the waters of eastern Long Island Sound and in the deep waters north of Puerto Rico. Because of the dominantly organic composition, such wastes are not likely to result in the formation of waste deposits on the ocean bottom, especially if discharged in areas of strong tidal currents or in deep water.

Alumina production from bauxite using the Bayer process and the extraction of titanium dioxide from ilmenite or other minerals leave a residue of extracted ore that has been dumped at sea in many localities. Being soillike materials, these extracted ores are difficult to detect on the bottom from their chemical characteristics. The iron in acid extraction liquids are also discharged at sea. Their oxidation causes the formation of iron-oxide flocs that characteristically discolor the water in the disposal site.

Ash from coal burning has long been disposed at sea. Clinkers and ash are common constituents in sediments in many coastal ocean and estuarine regions. A minor amount of ash from landfill and various industrial products is used, but the bulk of the ash produced must be disposed of, finally going into ocean waters

TABLE 1. Common Wastes (after Gross, 1972)

Wastes	Sources	Major Constituents	Minor Constituents
Dredged wastes	harbor and channel construction and maintenance	sand, shell, gravel, silt, river sediment	sewage solids, industrial wastes
Rubble	construction and demolition	stone, concrete, steel	
Sewage solids	municipal sewage systems and treatment plants	organic matter (50%) alumino-silicates (50%)	industrial wastes
Coal ash	coal combustion, primarily power generation	quartz, mullite	
Fermentation wastes	breweries, distilleries, pharmaceutical industry	organic matter	
Red mud	alumina production from bauxite	bauxite	
Acid-iron waste	T_1O_2 production	ilmenite, sulfuric acid	

either directly or through erosion of ash deposits on the land.

Effects of Waste Disposal

Deep ocean basins have been used for disposal of toxic or radioactive wastes. Concentrated wastes have been sealed in weighted containers and dumped into deep waters, where they will remain out of contact with humans for long periods. Even if the containers leak, the large volume of water above them greatly diminishes the probability of human exposure.

Effects of long, continuous waste disposal operations are often subtle and hard to identify (Ruivo, 1972). The problems are usually compounded by a lack of observations on conditions prior to onset of disposal activities. Phosphates and nitrates, both nutrients necessary for plant growth, are discharged into estuaries and coastal waters by sewage treatment plants, by runoff from fertilized agricultural lands, and by discharges of untreated sewage from areas without sewers or sewage treatment facilities.

Addition of nutrients may increase local productivity, upsetting delicate ecological balances between food production and food consumption. Phytoplankton growth can occur in such abundance that production exceeds consumption by animals. The uneaten phytoplankton sinks to the bottom, and decomposition of this organic matter consumes dissolved oxygen in near-bottom waters. If enough carbon is produced and if renewal of dissolved oxygen is sufficiently slow, then all the dissolved oxygen may be used up during bacterial decomposition so that no organisms except anaerobic bacteria can survive in the bottom waters. Even intermittent depletion of dissolved oxygen can make the area uninhabitable by bottom-dwelling organisms. And undecomposed organic matter can wash up on beaches causing odor or aesthetic problems.

M. GRANT GROSS

References

Boyd, M. B., 1972. *Disposal of Dredge Spoil.* Vicksburg, Va.: U.S. Army Corps of Engineers Water Experiment Station, Tech. Rept. H-72-8, 121p.

Gross, M. G., 1972. Geological aspects of waste solid disposal: New York metropolitan region, *Geol. Soc. America Bull.* 83, 3163-3176.

Ruivo, M., ed., 1972. *Marine Pollution and Sea Life.* London: Fishing News Ltd., 624p.

Williams, S. J., 1975. Anthropogenic filling of the Hudson River (Shelf) Channel, *Geology* 3, 597-600.

Cross-references: *Dredging; Human Impact; Land Reclamation; Nutrients; Pollutants; Thermal Pollution; Tourism.*

WATER LAYER WEATHERING

Water layer weathering is a secondary erosional process occurring on *wave-cut* (shore) *platforms* and *benches* that acts to lower the primary platform surface to approximately low water level. Water layer weathering occurs in pools of free-standing seawater contained on the platform surface, the water level being maintained by waves, spray, and tide action. On exposed coasts, the pools may lie up to 5m above sea level.

Originally termed *water-level weathering* by Wentworth (1938), it referred to the layer of water in the pools. Hills (1949) suggested the term *water layer weathering* to avoid any confusion that the processes are lowering the platform to water level. A number of separate processes are associated with water layering, the ultimate effect of which is to lower the platform level to where it is permanently awash, somewhere above low water level depending on wave energy and exposure.

The most obvious process associated with water layer weathering is the miniaturized wetting and drying that take place at the pool sides and, combined with small scale abrasion, act to widen the pools (Fig. 1). Continued widening and coalescing of the pools have the effect of producing extensive horizontal surfaces, best developed in naturally horizontal strata (Fig. 2). Dipping strata and resistant rock types such as quartzite hinder its development. Warmer temperatures and dry climates assist the process.

More recently water layer weathering has been used to include all weathering processes that operate as a result of wetting and drying, both in pools and on exposed surfaces (Davies, 1973). In this case, the end result remains the same—that is, reducing the original platform surface to a near horizontal surface at about low water level. However, usually before this ultimate platform level can be achieved, *quarrying processes* remove the platform. Only

FIGURE 1. Water layer weathering generating a pitted surface on horizontally bedded, siltstone platform, at Avoca, New South Wales, Australia (photo: A. D. Short).

FIGURE 2. Water layer weathering acting on a flat horizontally bedded sandstone platform at Culburra, New South Wales, Australia. The low relief pools are coalescing within structurally controlled high relief *joint* pools (photo: A. D. Short).

FIGURE 3. Solution acting on a limestone platform, above the level of saturation (to left) and on platform (to right) where pools have developed at several levels, Nanakuli, Oahu, Hawaii (photo: A. D. Short).

where platforms are protected by a lessening in wave energy, usually due to offshore shoaling and beach progradation, can water layer weathering processes complete their function on usually inactive platforms.

Other processes associated with water layer weathering are solution, potholing, and organic burrowing in the pools, and pitting, flaking, and salt crystallization on exposed surfaces (Fig. 3).

ANDREW D. SHORT

References

Davies, J. L., 1973. *Geographical Variation in Coastal Development*. New York: Hafner, 204p.

Hills, E. S., 1949. Shore platforms, *Geol. Mag.* 86, 137-152.

Wentworth, C. K., 1938. Marine bench forming processes: Water-level weathering, *Jour. Geomorph.* 1, 5-32.

Cross-references: *Erosion Ramp, Wave Ramp: Quarrying Processes; Rock Borers; Salt Weathering; Solution and Solution Pan; Wave-Cut Bench; Wave-Cut Platform: Wave Erosion: Weathering and Erosion, Biologic; Weathering and Erosion, Chemical; Weathering and Erosion, Differential; Weathering and Erosion, Mechanical.*

WATER TABLE

There is a definite relationship between the beach *water table* and the stability of the beach. High water tables accelerate foreshore erosion, while low water tables retard erosion and promote accretion.

The interaction between the waves and the beach is regulated by the position and *pressure head* of the beach water table, which, in turn, is normally a function of wave input tidal level and permeability of the sand body.

When the beach water table is low beneath the foreshore (sand is dry), water from the uprush of a wave will infiltrate into the foreshore, trapping sediment transported up the beach face. Just before the termination of the uprush, the velocity of the water drops below the critical *Reynolds* value, and the flow becomes laminar, with sand further dropping out of suspension. With lower volume flow in the backrush, the flow will remain laminar for a greater distance down the beach face, leaving more of the sediment carried up in the uprush (Grant, 1948).

As the tidal wave passes through the beach, there is a rise and fall of the beach water table. There is a lag of about 1-3 hours between cresting of the water table in the beach and cresting of the tide (Harrison et al., 1971). After the tide begins to ebb, the beach water table is still relatively high. This high water table gives rise to an outflow of water at the toe of the beach (resulting in *rill marks* if the foreshore is exposed). As the water flows out of the beach, sediment is lifted and transported down the beachface. Zones of erosion and deposition migrate up and down the foreshore in response to the relative positions of the water table (Duncan, 1964).

Any factor that generates a rise in the water table beneath a beach will be in effect increasing the erodability of the beach. Factors that tend to raise the water table include: disposal of storm drainage onto a beach (natural or manmade), the existence of an *aquaclude* beneath the beach, which can perch the beach water table; or the presence of bulkheads or subsurface barriers in the back beach that prevent the temporary flow of groundwater inland when storm sea level exceeds normal water table levels.

PETER S. ROSEN

References

Duncan, J. R., 1964. The effects of water table and tide cycle on swash-backwash sediment distribution and beach profile development, *Marine Geology* 2, 186-197.

Grant, U.S., 1948. Influence of water table on beach aggradation and degradation, *Jour. Marine Research* 7, 655-660.

Harrison, W.; Boon, J. D., III; Fang, C. S.; and Wang, S. N., 1971. *Investigation of the Water Table in a Tidal Beach.* Gloucester Point, Va.: Virginia Institute of Marine Science, Spec. Sci. Rept. No. 60, 165p.

Cross-references: *Beach Firmness; Beach Processes; Beach Profiles; Ghyben-Herzberg Ratio; Hydrogeology of Coasts; Minor Beach Features; Reynolds Number; Rill Marks; Tides.* Vol. IVA: *Groundwater.*

WAVE-BUILT PLATFORM—See WAVE-BUILT TERRACE

WAVE-BUILT TERRACE

The term *wave-built terrace* originally referred to subaerial deposits of sand heaped along the shore edge of Lake Bonneville by exceptionally high storm waves (Gilbert, 1890). This definition by G. K. Gilbert would more appropriately belong to the term *backshore terrace* as proposed by D. W. Johnson (1919). Gilbert's usage of the term wave-built terrace is considered archaic. The term now follows Johnson's definition of *continental terrace.* Thus, wave-built terrace refers to a "gently sloping coastal surface entirely constructed at the seaward or lakeward edge of a wave-cut platform by sediment brought by rivers or derived from the wave cutting and drifted along the shore or across the platform and deposited in the deeper water beyond" (Gary, McAfee, and Wolf, 1972).

In numerous basic geology texts, *wave base* is used to separate the zones of wave erosion and wave deposition. Above wave base, wave-cut terraces are formed. Below wave base, wave-built terraces are constructed of sediment that can no longer be carried by wave-induced water motion. The concepts of wave-base and wave-built terraces have been challenged by R. S. Dietz (1963). He suggests that wave-built terraces were predicated on the existence of a wave base prior to sufficient bottom sampling and modern geophyscial sub-bottom profiling methods. Dietz states that not only are no wave-built terraces found on land adjacent to uplifted wave-cut terraces, but they have not been encountered in subbottom profiles of continental shelves. Instead of being deposited adjacent to the edge of wave-cut platforms, sediment is deposited as a thin, mobile sheet on the continental shelf and as thick wedges between the continental slope and abyssal floor as continental rise deposits.

DONALD B. KOWALEWSKY

References

Dietz, R. S., 1963. Wave-base, marine profile of equilibrium, and wave-built terraces: A critical appraisal, *Geol. Soc. America Bull.* 74, 971-990.

Gary, M.; McAfee, R., Jr.; and Wolf, C. L., eds., 1972. *Glossary of Geology.* Washington, D.C.: American Geological Institute, 805p.

Gilbert, G. K., 1890. *Lake Bonneville,* U.S. Geological Survey, Monograph No. 1, 438p.

Johnson, D. W., 1919. *Shore Processes and Shoreline Development.* New York: John Wiley & Sons, 584p.

Cross-references: *Beach; Erosion Ramp, Wave Ramp; Marine Erosion Coasts; Shoreface and Shoreface Terrace; Shore Platforms; Wave-Cut Terrace; Wave Erosion.*

WAVE CLIMATE

Wave climate is defined as the distribution of wave height, period, and direction averaged over a period of time for a particular location (Wiegel, 1964) (care must be taken as to the number of years and the particular years for which the wave characteristics are averaged). In deep water offshore, the wave climate is rather invariant and may be similar over large distances. Near shore however, the wave climate depends on the offshore wave climate caused by prevailing winds and storms and on the bottom topography, which tends to modify the waves. An important reason for knowing the wave climate of a region for engineering design calculations is the fact that the wave climate will determine the effect of storms.

Shipboard observation of the wave climate (wave height and direction) can be obtained from the National Technical Information Service, Springfield, VA 22151, in the form of summaries.

JOHN B. HERBICH
TOM WALTERS

Reference

Wiegel, R. L., 1964. *Oceanographical Engineering.* Englewood Cliffs, N.J.: Prentice-Hall, 532p.

Cross-references: *Coastal Engineering; Design Wave;*

Protection of Coasts; Swell and Its Propagation; Wave Environments; Waves; Wave Statistics.

WAVE-CUT BENCH

A *wave-cut bench* is a relatively flat, bench-like section of cliffed coast, lying above the *wave-cut* or *shore platform* (Fig. 1). Benches occur on the more exposed sections of rocky coasts where waves may attack and erode the cliff face on several levels (Jutson, 1951). Actively forming benches are generated by wave attack along lines of weakness above the wave-cut platform. They are well developed in horizontally bedded sandstone-shale cliffs, where the shale is more readily eroded.

Benches may also result from a lowering in wave energy at the cliff face, thus lowering the level of saturation and bench planation. In this case, extension of a younger, lower wave-cut platform erodes the higher, older bench. The reduction in wave energy is often related to offshore shoaling producing greater wave attenuation. Second, a bench may be a relic wave-cut platform from a former higher stillstand, implying that the bench is not a contemporary feature but, rather, a relic wave-cut platform formed during an earlier Pleistocene interglacial and possibly the result of several episodes of coastal denudation (Bird and Dent, 1966).

ANDREW D. SHORT

References

Bird, E. C. F., and Dent, O. F., 1966. Shore platforms on the south coast of New South Wales, *Australian Geographer* 10, 71-80.

FIGURE 1. A relic wave-cut bench (upper center) standing 4 m above the lower, active wave-cut platform (lying just above HWM). Bench and platform cut in sandstone-shale at Warriewood, New South Wales, Australia (photo: A. D. Short).

Jutson, J. T., 1951. The shore platform of Flinders, Victoria, Australia, *Royal Soc. Victoria Proc.* 60, 57-74.

Cross-references: *Cliff Erosion; Erosion Ramp, Wave Ramp; Marine Erosion Coasts; Pleistocene; Quarrying Processes; Ramparts; Shore Platforms; Solution and Solution Pan; Water Layer Weathering; Wave-Cut Platform; Wave Erosion; Weathering and Erosion, Mechanical.* Vol. III: *Platforms—Wave-Cut.*

WAVE-CUT CLIFF—See SEA CLIFF

WAVE-CUT PLATFORM

A *wave-cut platform* is a relatively flat rock platform cut into the base of a sea cliff, usually above high water level, by wave removal of overlying cliff material. The upper platform level or basal cliff level is determined by the level of permanent rock saturation (Bartrum, 1938; Edwards, 1951). Tidal fluctuations, wave action, and spray tend to place this above high water level at the initial planation edge. The actual level is related to wave energy, rock lithology and structure, and tidal range (Fig. 1). As a platform widens, affording greater protection from spray and swash, or extends into lower wave energy conditions, such as into an embayment, it permits a lowering of the level of saturation and hence the level of planation and elevation of the platform. Therefore the platforms tend to be cut deepest at their seaward, most exposed edges in the highest wave energy conditions, and lowest in sheltered, low wave conditions. In addition, *ramparts* are a result of the higher level of planation of the platform edge and subsequent lowering of the level of planation as the platform widens.

The platform is cut above the level of saturation by the action of waves. The prime functions of the waves are to supply moisture for the wetting and drying processes that accelerate weathering of the cliff base (Bird and Dent, 1966), to generate *abrasion* of the cliff base, and to remove all debris from the platform and cliff base. The waves are therefore never the sole agency in cutting the platform. Rather, they are the dominating force of platform formation, assisted by secondary atmospheric and biological processes, in both the delivery and removal of cliff material and denudation of the platform itself.

The width of a wave-cut platform is a function of the rate of cliff retreat versus the rate of platform retreat. Platform retreat is primarily a function of wave energy, and tends to be fastest in areas of wave convergence, especially head-

exist (Jutson, 1940). For this reason some cliffed coasts are devoid of platforms.

ANDREW D. SHORT

References

Bartrum, J. A., 1938. Shore platforms, *Jour. Geomorph.* **1,** 266-268.
Bird, E. C. F., and Dent, O. F., 1966. Shore platforms on the south coast of New South Wales, *Australian Geographer* **10,** 71-80.
Edwards, A. B., 1951. Wave action in shore platform formation, *Geol. Mag.* **88,** 41-49.
Jutson, J. T., 1940. The shore platform of Mt. Martha, Port Phillip Bay, Victoria, Australia, *Royal Soc. Victoria Proc.* **52,** 164-174.

Cross-references: *Cliff Erosion; Erosion Ramp, Wave Ramp; Marine Erosion Coasts; Quarrying Processes; Shore Platforms; Solution and Solution Pan; Splash and Spray Zones; Wave-Cut Bench; Wave Erosion; Water Layer Weathering; Weathering and Erosion, Biologic; Weathering and Erosion, Mechanical*. Vol. III: *Platforms—Wave-Cut.*

FIGURE 1. Active wave-cut platforms in different marine environments and rock lithologies and structures: *A.* sandstone platform 30 m in wide in gently dipping strata at Copacabana, New South Wales, Australia; *B.* limestone platform in coral-algae conglomerate, Makaha, Oahu, Hawaii; *C.* platform in steeply dipping metamorphosed material at Bermagui, New South Wales, Australia (photos: A. D. Short).

WAVE-CUT TERRACE—See WAVE-CUT PLATFORM

WAVE DRAG LAYER

As a natural phenomenon, a *wave drag layer* in the coastal zone is developed only at those locations where beaches exist and the submarine slope is covered by nearshore drift (Shuisky, 1971 *a, b*).

The wave drag layer (WDL) is the third dimension of longshore drift flow, along with the width and the length. All processes of coastal sedimentation take place in the WDL. Deposits of sediments with high heavy-mineral concentrations are formed here as well.

The WDL can be defined as a mobile layer of bottom and beach sediments, the thickness of which is changing, at a constant sea level, as a function of time and location or as a result of repeated drift roiling by waves, or a movement of mobile micro and mezoforms of nearshore zone relief. WDL is a regime phenomenon. It should be differentiated from an *active layer*, —that is, a layer of coastal sediments worked over just once during a given swell.

The thickness of WDL, H_W, is its most important parameter. According to the change of WDL across the profile of coastal zone, five types of WDL can be distinguished (Fig. 1). The distribution of H_W across the profile corresponds to the specific energy distribution of the breakers, Generally H_W has greater value in those locations where longshore drift flows

lands, the rate decreasing into embayments or areas of lower wave energy. Cliff retreat, however, is also dependent on atmospheric and lithological parameters. The rate of retreat of each must therefore be determined separately to calculate the relative platform width. Only if cliff retreat exceeds platform retreat will a platform

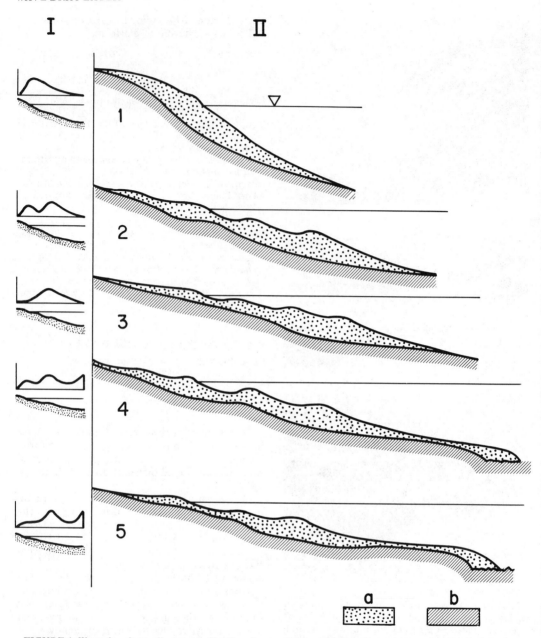

FIGURE 1. Wave drag layer diagram: I. Family of curves characterizing H_w changes across the profile of the coastal zone. II. WDL types in the coastal zone having different steepness of submarine slopes: 1. for sandy shores with steep gradient; 2. for conventionally deeper shores with easy gradient; 3. for sandy shores with easy gradient; 4. for relatively deeper submarine slope where sedimentary layer is joined with a bench; 5. the same for gently sloping submarine slope; a. mobile drift in H_w; b. stationary deposits.

have great capacity. H_W gradually decreases with the decrease of capacity along the flow path.

Separate WDL types can be changed with time within the boundaries of the same region, together with the change of longshore drift regime under the influence of transgressional-regressional fluctuations of sea level. These changes lead either to burying or to washout of recent coastal sediments. Thus, the change of longshore drift regime characterized by the substitution of the second and then the third type of WDL leads to depositing of the coastal sediments. The reverse substitution of WDL type accounts for partial or complete washout of the sediments.

YURII D. SHUISKY

References

Shuisky, Yu. D., 1971a. Wave drag layer placers and their genetic peculiarities, *U.S.S.R. Acad. Sci. Rept.* **196**, 1430-1433.
Shuisky, Yu. D., 1971b. On the notion of coastal placers in connection with their genesis, *Jour. Lithology and Mineral Resources* **2**, 122-130.

Cross-references: *Bars; Beach Processes; Nearshore Hydrodynamics and Sedimentation; Sediment Transport; Wave Work.*

WAVE ENERGY

Waves receive their energy from the wind by means of a complicated transfer whose yield is not yet clearly understood. Energy in the wave is more concentrated than in the wind because the water density is about a thousand times greater than the air density. For this reason a motor utilizing wave power could be more remunerative than a wind motor; there is an important interest in proposals for utilizing *wave energy* (Masuda, 1974).

Energy must be continuously furnished from wind to wave, the wave energy being dissipated by internal friction—moving particles lose their energy against particles at rest. Likewise, the wave can dissipate its energy against floating bodies, since the energy of the wave is concentrated at the surface of the sea. Projects for harbor or beach defense using floating buoys, artificial algaes, or air bubbles injected into the sea are based on this fact (Wiegel, 1969). In nature, analogous effects are produced by floating ice cakes or suspended natural materials (for example, plankton, sand). Since the energy is confined in a thin superficial layer, only this layer acts in erosive processes against the coast or in the transport of the sediments.

In quantitative studies, it is necessary to distinguish between the energy of a single regular wave and the confused sea, corresponding to the actual complicated wave motion. A calculation for the regular single wave can be valid far from the source of the waves, often near a coast. In the open sea, exposed to a gale, the motion is complex and confused; mathematical models for interpreting the motion are not available—only statistical approaches are possible. The energy of a confused sea could be commonly related to the power spectrum of the wave (to the power spectrum of a *wave record*); the integral of the spectrum is proportional to the total energy developed in any particular case.

According to the theory of a single simple wave, the energy of the true wave is proportional, as is well known, to the square of the wave height. In a confused sea, the energy is obviously proportional to the square of the heights of all the waves (integral of the heights square of the waves).

The energy of the waves is both *potential* and *kinetic,* and can be related to the unit area of the sea or to the unit length of the wave crest. The energy yielded in wave motion can be great. The power often reaches more than 1 kW/m^2.

FERRUCCIO MOSETTI

References

Masuda, Y., 1974. Study of wave-activated power generator for island power and land power, *Coll. Internat. Expl. Oceans, 2nd, Bordeaux* **2**, 119p.
Wiegel, R. L., 1964. *Oceanographical Engineering.* Englewood Cliffs, N.J.: Prentice-Hall, 532p.

Cross-references: *Coastal Engineering; Energy Coefficients; Protection of Coasts; Waves; Wind.* Vol. I: *Ocean Waves; Waves as Energy Sources.*

WAVE ENVIRONMENTS

Three generalized *wave environments* may be delineated (Fig. 1) that are related to zonal variation in the input of wind energy to the world ocean (Davies, 1980).

Storm wave environments lie between roughly 40° and 60° of latitude in both hemispheres. They are characterized by a relatively large number of local gale force winds, generating high energy waves. Storms occur with so small a recurrence interval that they tend to dominate the shore regime. Swell waves are of background significance.

Swell environments lie between roughly 40°N and S, and here high energy events are relatively scarce. Fine weather swell and waves generated by the trades tend to dominate the shore regime, and this is only infrequently upset by waves from tropical and temperate cyclones. A subdivision may be made between west coast swell environments, where swell from temperate storm regions is especially important and consistent, and east coast swell environments, where such swell is less in evidence and the trade wind systems are prominent.

Protected sea environments form the third group. Generally, they are coasts of enclosed or partially enclosed seas or coasts in high lati-

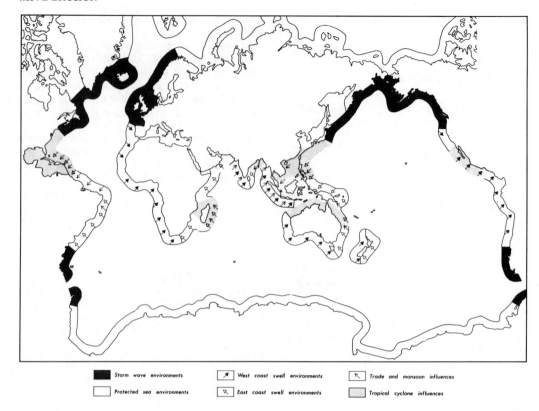

FIGURE 1. Major world wave environments (from Davies, J. L., 1980. *Geographical Variation in Coastal Development*. London: Longman Group Ltd.).

tudes, where seasonal sea ice provides an additional protection against the incidence of high wave energy. Typically, they are characterized by almost continuous low energy levels and infrequent high energy events.

J. L. DAVIES

Reference

Davies, J. L., 1980. *Geographical Variation in Coastal Development*. London: Longman Group Ltd., 212p.

Cross-references: *High-Energy Coast; High-Latitude Coasts; Low-Energy Coast; Low-Latitude Coasts; Mid-Latitude Coasts; Moderate-Energy Coast; Swell and Its Propagation; Waves; Wind; Zero-Energy Coast.*

WAVE EROSION

Erosion of beaches by waves occurs on several different spatial and temporal scales. Every rise and fall in the tide is accompanied by the migration up or down the beach of a scour zone, (Strahler, 1966). This zone, which is found at the tide level, is also the zone where the backwash meets the incoming surge and is an area of turbulence where suspension of sediment is prevalent (Brenninkmeyer, 1975). The depth of the erosion ranges from 3% (King, 1951) to 40% and higher (Williams, 1971) of the breaker height. The breaker zone also migrates across the beach as the tide changes. Inside the breaker zone, sediment movement is the greatest of anywhere within the foreshore-inshore area. In fact, about 50% of the sediment movement takes place within the breaker zone, (Ingle, 1966). The sediment movement is largely by bed load in oscillating horizontal to flat elliptical paths with small alongshore migration during the breaking of each wave (Fig. 1). Acceleration of the shoreward water takes place immediately offshore of the breaking point, while a seaward acceleration occurs shoreward of the breaking wave (Iverson, 1952; Morrison and Crooke, 1953). This action may produce a bar that accretes from both sides. Inside the bar, at the plunge point, a trough may be excavated. This trough may owe its origin to a variety of causes. Depending on the beach slope, the maximum horizontal water velocities occur at 0.09 wave period after the breaking crest has passed (Adeyemo, 1971). Also, if the wave breaks in shallow water, the ascent of air bubbles trapped

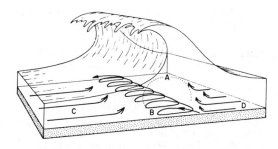

FIGURE 1. Schematic diagram of grain motion beneath a breaking wave. Largest grains present saltate along the bottom at position B. Largest percentage of grains follows horizontal to flat-elliptical-alongshore paths during the collapse of each wave. Finest particles travel in suspension at position A. Grains shoreward (C) and seaward (D) of the breaker zone move toward the plunging wave essentially tracing water-particle motion. Diameter of grains at any position is primarily a function of available wave energy or power (from Ingle, J. C., 1966. *The Movement of Beach Sand.* Amsterdam: Elsevier).

by a breaking wave after penetrating the bottom has a suctional effect sufficient to draw up sand (Aibulatov, 1958, 1961).

When waves grow steeper during storms, generally in winter months, sand is moved from the beach to the offshore. When waves decay after the passage of a storm or under the long swell prevalent during the summer, the offshore deposit gradually moves onshore, and sand is spread out over the beach (Sonu and Van Beek, 1971). Saville (1950) and Scott (1954) have shown that wave steepness is the most important variable in determining the beach slope. Most authors follow Saville, who divides the beach profile into two categories: a storm outline that is produced by waves with a steepness (H/L) greater than 0.03, and a summer profile formed by waves less than 0.025 steep.

During storms, when waves are steep or of short period and arrive at the coast in many wave trains from a great variability of directions, much more water is thrown on the beach. The beach soon becomes saturated, and the water table becomes almost coincident with the beach face. Because of this, less of the swash can percolate through the sand, and so the backwash volume is really equal to the uprush. This downflow drags sediment from the beach face and increases the turbulence within the scour zone. At this mean water line, the groundwater flow returns to the sea. Its upflow at this point tends to provide a quicksand effect, which sets the stage for more rapid removal of sediment by the backwash (Silvester, 1974). In fact, the elutriation due to the groundwater itself can remove the finer-sized sediment (Emery and Foster, 1948).

The backwash carries its load until its velocity is reduced sufficiently; then the eroded sediment is deposited offshore. If the foreshore slope is less than $4°$, this deposit tends to accumulate in a bar (Shepard, 1950). This bar will grow until it is high enough to steepen and force the early breaking of the storm waves. Since waves break farther from shore, most of the erosion of the beach face stops. Erosion usually reaches its peak within several hours after the onset of a storm. Conversely, the first of the winter storms will cause the largest and most spectacular beach erosion, for it tends to build the offshore deposit or bar.

The direction of the wind also has an important effect on the amount of scour and direction of transport of sediment. A strong onshore wind forces the water near the bottom to flow seaward in shallow and deeper water, reversing the normal onshore transport of flat waves, and increases erosion under the breaker zone (King, 1972).

Group comparison of beach profiles before and after storms suggests that erosion on a beach above MSL may range from 5000–240,000 m3/km during a storm (Shuisky, 1969; Everts, 1973). As large as this may seem, it is minor when compared with the typical values of 380,000 m^3/yr moved by longshore currents (Johnson, 1956).

Another important consequence of storm or seasonal changes is the shifting of sand along a beach as a result of a change in wave approach. Beaches tend to align themselves at right angles to the direction of wave approach. If storm or winter directions of wave approach are different from the predominant swell approach, then the longshore direction may change. This is especially noticeable on large crescent beaches, where a wide beach may be present on one side in summer conditions and on the other during the storm months.

BENNO M. BRENNINKMEYER

References

Adeyemo, M. D., 1971. Velocity fields in the wave breaker zone, *Conf. Coastal Eng., 12th, Am. Soc. Civil Engineers, Proc.* **1**, 435-460.

Aibulatov, N. A., 1958. New investigations of longshore migration of sandy sediment in the sea (in Russian), *Akad. Nauk SSSR Okeanograf. Kom. Biul.* **1**, 72-80.

Aibulatov, N. A., 1971. Observations of the longshore sand drift along a shallow coast of accumulation (in Russian), *Akad. Nauk SSSR Inst. Okeanol. Trudy* **53**, 3-18.

Brenninkmeyer, B. M., 1975. Mode and period of sand transport in the surf zone, *Conf. Coastal Eng., 14th, Am. Soc. Civil Engineers, Proc.* **2**, 812-817.

Emery, K. O., and Foster, J. F., 1948. Water table in marine beaches, *Jour. Marine Research* **3**, 644-654.

Everts, C. H., 1973. Beach profile changes in western Long Island, *in* D. R. Coates, ed., *Coastal Geomorphology*. Binghamton: State University of New York, 279-301.
Ingle, J. C., 1966. *The Movement of Beach Sand*. Amsterdam: Elsevier, 221p.
Iverson, H. W., 1952. Waves and breakers in shoaling water, *Conf. Coastal Eng., 3rd, Council on Wave Research, Proc.*, 1-12.
Johnson, J. W., 1956. Dynamics of nearshore sediment movement, *Am. Assoc. Petroleum Geologists Bull.* **40**, 2211-2232.
King, C. A. M., 1951. Depth of disturbance of sand on sea beaches by waves, *Jour. Sed. Petrology* **21**, 1131-1140.
King, C. A. M., 1972. *Beaches and Coasts*. New York: St. Martin's Press, 570p.
Morrison, J. R., and Crooke, R. C., 1953. The mechanics of deep water, shallow water, and breaking waves, *U.S. Army Corps of Engineers, Beach Erosion Board Tech. Memo. 40*, 14p.
Saville, T., 1950. Model studies of sand transport along an indefinitely straight beach, *Am. Geophys. Union Trans.* **31**, 555-556.
Scott, T., 1954. Sand movement by waves, *U.S. Army Corps of Engineers, Beach Erosion Board Tech. Memo. 48*, 37p.
Shepard, F. P., 1950. Beach cycles in southern California, *U.S. Army Corps of Engineers, Beach Erosion Board Tech. Memo. 20*, 26p.
Shuisky, Y. D., 1969. The effect of strong storms on sand beaches of the Baltic eastern shore, *Oceanology* **9**, 388-391.
Silvester, R., 1974. *Coastal Engineering*. Amsterdam: Elsevier, vol. 1, 457p.; vol. 2, 338p.
Sonu, C. J., and Van Beek, J. L., 1971. Systematic beach changes on the Outer Banks, North Carolina, *Jour. Geology* **74**, 247-268.
Strahler, A. N., 1966. Tidal cycle of changes in an equilibrium beach, Sandy Hook, New Jersey, *Jour. Geology* **74**, 247-268.
Williams, A. T., 1971. An analysis of some factors involved in the depth of disturbance of beach sand by waves, *Marine Geology* **11**, 145-158.

Cross-references: *Bars; Beach Cycles; Beach in Plan View; Beach Orientation; Beach Processes; Beach Profiles; Cut and Fill; Depth of Disturbance; Sediment Transport; Storm Beach; Sweep Zone; Time Series; Water Table.*

WAVE METERS

In order to define the w*ave climate* of a given region, it is necessary to measure the main characteristics of sea waves: H (height), T (period), and Θ (direction). Visual observations, made from lighthouses, sea vessels, or by volunteers living on maritime coasts, are of great interest and utility. From this amount of data (covering many years in the past), the estimated wave height is given as the *significant height* (H_s), the average height of the one-third highest waves, which leads to the knowledge of other statistics. However, visual estimates of the wave height have been correlated with actual values (recorded by precise methods), and they are usually overestimated by approximately 25%.

For a more precise knowledge of sea waves, the use of *wave meters* or *wave gauges* is required. Most of these gauges provide discrete or continuous records (paper, film, magnetic tape, punched paper tape) on which the wave profile over time is recorded. In this way it is possible to know H and T in reference to the measuring point.

A simple and easily mounted wave meter consists of an optical device located on the beach and a graduated bar or buoy at sea; the wave oscillation is then visually registered. Other commonly used systems are listed below.

Pile-supported wave gauges consist of a floater within a tube, connected with a recording mechanism, or electrically sensitive parts, and using resistance or capacitance variations caused by the rising or falling of the water. A *step-resistance wave gauge* is an example of this type of gauge (Meyers, 1969).

Pressure-type wave gauges register the variation of pressure induced by the passing wave. Chatou, OSPOS and other Japanese models are some of the available pressure-type wave gauges.

Inverted echo-sounder wave gauges use an ultrasound beam produced by a device placed at the bottom of the sea and reflected by the surface.

Accelerometer buoys record two successive integrations of the acceleration induced by the wave. "Data well" is a well-known Dutch model (Baird and Glodowski, 1971).

Other systems include: remote-sensing techniques (aerial surveys of the sea surface using photography or laser rays), wave gauges for the determination of directional spectrum, and some devices for measuring the *direction* of the wave.

CARLOS MORAIS

References

Baird, W. F., and Glodowski, C. W., 1971. Accelerometer wave recording buoy, *in Offshore Tech. Conf. Proc.*
Meyers, J. J., ed., 1969. *Handbook of Ocean and Underwater Engineering*. New York: McGraw-Hill.

Cross-references: *Coastal Engineering; Coastal Engineering, Research Methods; Design Wave; Wave Climate; Wave Energy; Waves; Wave Statistics.*

WAVE REFRACTION DIAGRAMS

Wave refraction diagrams are a graphical technique used to illustrate and predict the *refrac-*

tion—the bending effect—of waves approaching a shoreline. The illustration is done by means of *orthogonals*, wave rays drawn perpendicular to wave fronts traveling toward the shoreline. An example of one such diagram using the orthogonal method of construction is shown in Figure 1. The orthogonal method is considered the most practical, quickest, and widely used technique at present. Other methods used to construct wave refraction diagrams include the wave front method (Johnson, O'Brien, and Isaacs, 1948) and computer methods (Harrison and Wilson, 1964).

Theory

The theory behind the orthogonal method is based on the work of Arthur, Munk, and Isaacs (1952). The fundamental principle for all numerical and hence graphical procedures for the construction of these diagrams is *Snell's law:* $\sin \alpha_2 = (C_2/C_1) \sin \alpha_1$, where α_1 is the angle of a wave front (which is perpendicular to the orthogonal) with the bottom contour over which the wave is passing; α_2 is a similar angle measured as the wave front passes over the next bottom contour; C_1 is the wave velocity (celerity) at the depth of the first contour; and C_2 is the wave celerity at the second contour.

Refraction occurs as one part of a single wave front moves slightly faster than another part; this change in speed across the face of a wave will tend to change the direction of advance of the wave. A deepwater wave approaching a shoreline will alter its speed as the ratio of water depth to wavelength increases. The effects of the bottom topography on the wave will start at about one-half the wavelength. Development of celerity equations is found in the small amplitude wave theory clearly seen in Ippen (1966). From Snell's law a template is constructed that will show the angular change in α that occurs as an orthogonal passes over a particular contour interval and allows the construction of the direction changed orthogonal. Figure 2 is the template that is necessary in the construction of the diagrams.

Using the orthogonal method, we can make the following assumptions: Wave energy between orthogonals remains constant. The direction of the wave advance is parallel to the orthogonals. Celerity of a wave of a given period at a particular location depends only on the water depth at that location. Changes in sea-floor topography approaching a shoreline are gradual. Waves are long crested, constant period, small amplitude and monochromatic. Other factors that can effect wave refraction, such as currents, winds, and reflected waves, are negligible.

Construction

The basic materials needed to construct the diagrams are bathymetric charts of the area to be studied, the template in Figure 2, and a set of wave tables such as those in *Shore Protection Manual* (1973).

Bathymetric charts are normally any navigation chart showing soundings taken over the study area. The National Ocean Survey (NOS) publishes these charts, which are readily available at any marine supply establishment. Earlier, the United States Coast and Geodetic Survey was responsible for these charts, and today most of the charts in circulation will be from that organization.

The template should be reproduced on a transparent material to a size where the C_2/C_1 scale is approximately 25 cm long.

The Orthogonal Method

To construct the orthogonals, first a tracing paper overlay is placed on the bathymetric chart. Contours are then traced with small irregularities smoothed out. The depth interval between contours is so chosen that the smaller the interval the greater the accuracy of plotting. The shoreline should be traced for reference purposes. Each refraction diagram will represent one direction of wave approach with one deepwater wavelength. The deepwater wavelength $L_o = 1.56\, T^2$, where L_o is in meters and T, the wave period, is in seconds. A set of evenly spaced and parallel orthogonals from a chosen direction of approach is drawn up to the bottom contour closest to $\frac{1}{2} L_o$. The number of orthogonals should be selected to sufficiently cover the shoreline in question when the wave front has terminated on the shore. Close spacing between orthogonals will increase the resolution when estimating wave energy attacking the shore. Deepwater wave periods and hence wavelengths are determined by hindcast study of historical weather charts or from other historical wave climate information. One source for this historical information is the National Oceanographic Data Center, Silver Springs, Maryland.

The next step is to compute the C_2/C_1 ratio values for each contour interval. Table 1 is a sample of the computations used in plotting the

TABLE 1. Computations Used in Plotting Orthogonals (T = 15 sec., L_o = 349 m)

Bottom Contour Depth d (m)	ratio d/L_o	tanh $2\pi d/L$ (in wave tables)	C_1/C_2	C_2/C_1
128.0	0.37	.9825	.98	1.02
110.0	0.31	.9653	1.00	1.00
91.5	0.26	.9611		

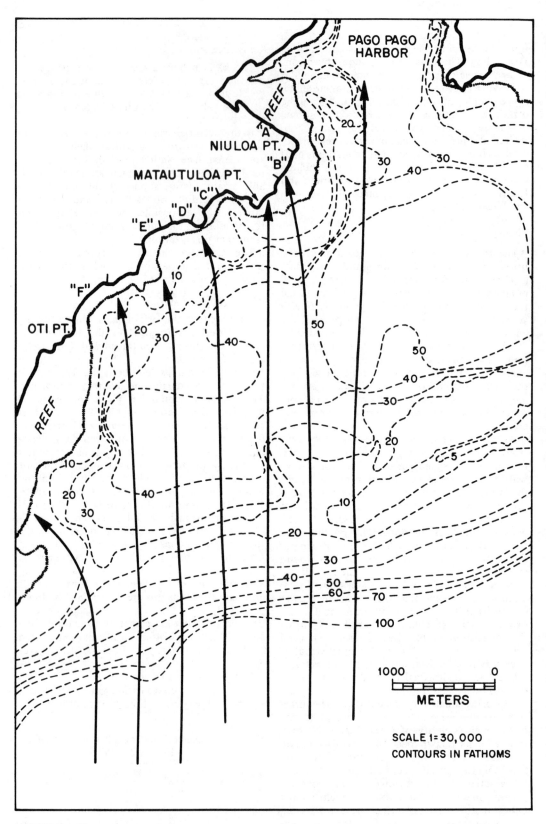

FIGURE 1. Wave refraction diagram using the orthogonal method of construction. The orthogonals represent deepwater waves with an L_o of 349 m travelling north.

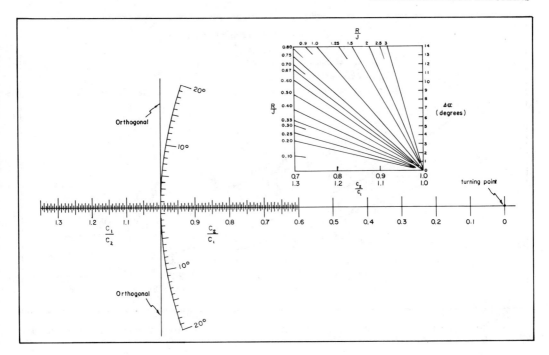

FIGURE 2. Refraction template (from Coastal Engineering Research Center, 1973).

orthogonals of Figure 1. The C_1/C_2 ratio is the quotient of successive terms in column 3.

Once the C_2/C_1 ratio has been calculated for each depth interval, it should be marked on each depth interval. The C_2/C_1 values are used when orthogonals are constructed from shallow to deep water.

In most cases the angle α will be less than 80°. The following procedure should be followed to construct each orthogonal: Sketch a contour contour midway between the first two contours to be crossed, extend the orthogonal to the mid-contour, and construct a tangent to the mid-contour at this point. Lay the line on the template labelled *orthogonal* along the incoming orthogonal with the point marked 1.0 at the intersection of the orthogonal and the mid-contour. Rotate the template about the turning point until the C_1/C_2 value corresponding to the contour interval being crossed intersects the tangent to the mid-contour. The orthogonal line on the chart now lies in the direction of the turned orthogonal on the template. Place a triangle along the base of the template and construct a perpendicular to it so that the intersection of the perpendicular with the incoming orthogonal is midway between the two contours when the distances are measured along the incoming orthogonal and the perpendicular. Note that this point is not necessarily on the mid-contour line. This line represents the turned orghogonal. Repeat the above steps for successive contour intervals. As long as the angle α is less than 80°, the orthogonal method will produce a wave refraction diagram of good validity. In the event that α is greater than 80°, the R/J method of constructing orthogonals must be used.

The R/J Method

In Figure 3, the angle α is 90°. In using this technique, first, the contour interval is divided into a series of rectangles by traverse lines drawn perpendicular to the contours to be crossed. The spacing R of the transverse lines is set as a ratio of the interval spacing J in the vicinity of each newly drawn transverse. In using the R/J method, the celerity ratio used is C_2/C_1 when moving from deeper to shallower water.

Once the R transverses are constructed, refer to Figure 2, where $\Delta\alpha$ is plotted as a function of C_2/C_1 for various values of the R/J ratio. Choose the correct $\Delta\alpha$ for each $R \times J$ rectangle and turn the orthogonal by $\Delta\alpha$ in the center of these rectangles. Once the orthogonal crosses a new contour and the angle α is less than 80°, the R/J method is replaced by the orthogonal method.

Limitations and Significance

Refraction diagrams will provide a reasonably good indication of how bottom topography will

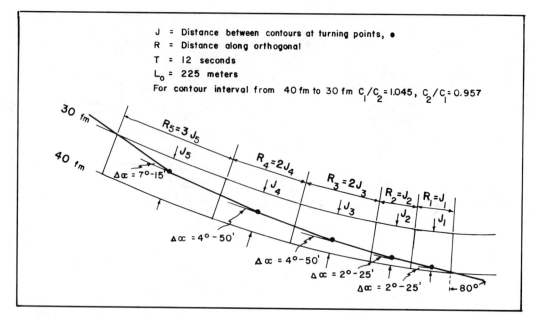

FIGURE 3. Refraction diagram using the R/J method of construction (from Coastal Engineering Research Center, 1973).

affect small amplitude waves approaching a shoreline. The theory assumes a mild sloping topography—that is, slopes of 1 : 10 ratio or flatter. In areas of nonuniform slope with complex bottom features, the accuracy of wave refraction diagrams will decrease. *Model studies* can enhance the results derived from these diagrams, and it is recommended that model tests be run when critical design work depends on the wave energy spectra approaching a shore.

Just as bottom topography is important to orthogonal paths, the study of naturally occurring wave refraction can point out the characteristics of the sea floor. *Submarine ridges* will tend to bring orthogonals closer together, especially if the ridges are extensions of coastal headlands. Wave energy will then be concentrated at these ridges. Conversely *submarine canyons* perpendicular to the shoreline will cause waves to spread out away from the canyon. The slope of the sea floor just offshore of a beach can be estimated by visually studying aerial photographs of wave patterns adjacent to the shoreline.

JOHN A. RIPP

References

Arthur, R. S.; Munk, W. H.; and Isaacs, J. D., 1952. The direct construction of wave rays, *Am. Geophys. Union Trans.* **33**, 855-865.

Coastal Engineering Research Center, 1973. *Shore Protection Manual.* Washington, D.C.: U.S. Army Corp of Engineers, 750p.

Harrison, W., and Wilson, W. W., 1964. *Development of a Method for Numerical Calculation of Wave Refraction TM-6.* Washington, D.C.: U.S. Army Corps of Engineers, Coastal Engineering Research Center, 64p.

Ippen, A. T., ed., 1966. *Estuary and Coastline Hydrodynamics.* New York: McGraw-Hill, 740p.

Johnson, J. W.; O'Brien, M. P.; and Isaacs, J. D., 1948. *Graphical Construction of Wave Refraction Diagrams HO No. 605, TR-2.* Washington, D.C.: U.S. Naval Oceanographic Office, 45p.

Cross-references: *Beach Processes; Coastal Engineering; Coastal Morphology, Research Methods; Scale Models; Wave Environments; Waves; Wave Work.*

WAVES

Shoaling waves are those that have progressed into water depths at which the interaction with the bottom affects the wave motion. This point is generally accepted as occurring where the depth is less than one-half wavelength (that is, the relative depth is less than 0.5; $h/L < 0.5$, where h is still-water depth and L is wavelength). The region between this point and the breaker zone is called the *zone of shoaling waves.*

Deepwater *wave theory* (see Vol. I) is applicable to the shoaling zone only to a limited extent. Linear theory (Airy, 1845) is based on several assumptions that become untenable as water depth decreases: amplitude is negligibly small compared with water depth; fluid is frictionless (that, is, viscosity is zero); motion is irrotational (that is, no vorticity); fluid motions are so small that higher-order terms of the equations of motion can be ignored; and the lower boundary is fixed, horizontal, and impermeable.

Linear wave motion can be described by the velocity potential function:

$$\phi = \frac{HL \cosh kz}{2T \sinh kh} \sin(kx - \sigma t) \quad (1)$$

that gives for the horizontal and vertical orbital velocities, respectively:

$$u = \frac{\delta \phi}{\delta x} = \frac{\pi H}{T \sinh kh} \cosh kz \cos(kx - \sigma t) \quad (2)$$

$$w = \frac{\delta \phi}{\delta z} = \frac{\pi H}{T \sinh kh} \sinh kz \sin(kx - \sigma t) \quad (3)$$

In the above equations the coordinate axes are arranged so that the positive x axis in the direction of wave propagation with origin under the wave crest, and the z axis is vertical with the origin at the bottom, measured positively upward (the still-water surface occurs as $z = h$); H is wave height, T is period, k is wave number $2\pi/L$, σ is angular frequency $2\pi/T$, and t is time.

Stokes (1847) developed a theory for waves of finite height, overcoming the limitations imposed by Airy's first assumption. Stokes' theory is nonlinear and has been solved to higher orders of approximation; hence Airy's fourth assumption is less significant. Airy's theory predicts a more realistic trochoidal surface profile and net transport of water. Both phenomena are in accordance with the natural system. Stokes' equations are more complicated mathematically, and the higher order solutions become unstable as relative depth decreases past 0.1. Koh and LeMéhauté (1966) conclude that the *shoaling coefficient* (q.v.) based on fifth-order solutions to Stokes' theory is less valid than that derived from the third-order equations (Skjelbreia and Hendrickson, 1960).

Cnoidal theory (Korteweg and de Vries, 1895) was developed for waves of steepness (H/L) greater than 0.02 and relative depth less than 0.1, where Stokes' theory begins to fail. Cnoidal theory is seldom used because of its mathematical complexity, being based on the Jacobian elliptic function "cn," rather than the more familiar circular functions. Solutions must be derived graphically or numerically, a limitation that should be less restrictive with the increased availability of digital computers. Cnoidal theory is appealing for it gives reasonably good results and has as depth-limiting cases: the solitary wave in very shallow water and the linear Airy wave in deep water. Thus, it tends toward being *universal* theory.

It is appropriate to test the ability of the various wave theories to predict accurately certain pertinent parameters, considering the many simplifying assumptions that go into their development. Since we are primarily interested in the situation near the bottom, the horizontal component of fluid velocity due to waves (u) at the bottom ($z = 0$) is the significant parameter to test. LeMéhauté, Divoky, and Lin (1968) evaluated twelve different theories on the basis of their ability to predict maximum particle velocities. These experiments were performed in a wave tank under shoal water conditions only ($h/l = 0.1$). They concluded that the best theory for describing the velocity field in shoal water is the cnoidal theory of Keulegan and Patterson (1940). Their data show, however, that an equally accurate predictor of the maximum horizontal component of velocity near the bottom is the simple, linear theory of Airy (1845). This finding was experimentally corroborated by May (1975) using measurements of orbital velocities of natural waves.

The most commonly used theory is that of Airy (1845), which gives the simplest (hence mathematically most attractive) expressions for wave motion. It does a reasonably good job of describing deepwater waves of low height. Airy's theory describes the surface profile as a simple sinusoidal wave form; hence it is obviously inaccurate in most situations. It describes the particle motions as closed orbits, disallowing any net mass transport of water contrary to observation. Measurements in shoal depths indicate that the temporal and spatial distribution of particle orbital velocities disagrees with that predicted by Airy's theory. These limitations do not prevent the theory from being useful and most frequently used. Airy's theory is mathematically simple; hence it is most suitable for inclusion into more complicated mathematical formulations (for example, Longuet-Higgins and Stewart, 1964). Further, this simplicity appeals to the investigator who is interested in obtaining a reasonably good approximation to some wave parameter, such as height, length, and so on, based on a minimum of computation. Airy's theory is known to be least accurate in shallower water, and yet it has been found to provide a reasonably good prediction for the maximum horizontal component of orbital velocity at the bottom (May 1975), even though it does a poor job of predicting the distribution of velocities through space and time.

The group velocity of a wave set represents the velocity at which wave energy is propagated. Rayleigh (1876) showed that in the case of finite depth, the average energy transmitted per unit time (power) per unit width of wave crest is:

$$P = ECn \tag{4}$$

where E is wave energy density (energy per unit surface area):

$$E = \rho g H^2/8 \tag{5}$$

where ρ is fluid density and g is acceleration of gravity. C is wave phase velocity and equals L/T, or:

$$C = [(g/k) \tanh kh]^{1/2} \tag{6}$$

n is the ratio of group velocity to phase velocity:

$$n = (\tfrac{1}{2} + kh/\sinh 2kh) \tag{7}$$

Equation (4) can be written:

$$P = (\rho g H^2/8)\,[(g/k) \tanh kh]^{1/2}\,(\tfrac{1}{2} + kh/\sinh 2kh) \tag{8}$$

In deep water, where kh is large, equation (8) reduces to:

$$P_0 = (\rho g H_0^2/8)(g/k)^{1/2}(1/2) \tag{9}$$

Note that in deep water the phase velocity is twice that of the group velocity (subscript zero indicates deep water condition).

In shallow water, where kh is small, equation (8) reduces to:

$$P_s = (\rho g H_s^2)(gh)^{1/2}(1) \tag{10}$$

In shallow water, the phase velocity is equal to that of the group velocity (subscript s indicates shoal water condition).

If it is assumed that conservation of energy flux is valid for shoaling waves (that is, there is no loss or gain of energy in the wave system) and that the wave period is constant, we can write:

$$P_i = P_0 \tag{11}$$

where P is wave power per unit width of wave crest and i represents some point x_i along the wave ray path. Based on these equations, values for wavelength, height, phase velocity, and group velocity can be computed at any point along the ray path if the water depth, wave period, and initial height are known. The ratio of height to deep water height (H_i/H_0) is called the shoaling coefficient (q.v.).

Wave rays are *refracted* upon entering shoal water as described by *Snell's equation*:

$$C_i/C_0 = \sin a_i/\sin a_0 \tag{12}$$

where the angle a is measured between the wave ray and the normal to the bathymetric contour line (isobath). Equation (6) indicates that C decreased as a wave moves into shoal water depth—that is, as h decreases. Therefore, as a wave enters shoal water, the ray is bent to a position more nearly normal to the bottom isobath but never perfectly normal, for this would require the phase velocity to go to zero.

The character of the refraction pattern is dependent on the form of the bathymetry. There are three general bathymetric types that occur in the shoaling wave zone: isobaths parallel to the shoreline; isobaths concave seaward (as in an embayment); and isobaths convex seaward (as off a headland or foreland). In the first type, where wave rays enter shoal water with an orientation originally normal to the isobath, they remain so. If they approach at some oblique angle, they are refracted toward the shoreline. The effect is that adjacent wave rays are more widely spaced, and thus the wave crest is extended and the energy density diminished. In the second type, the waves are refracted regardless of the original angle a_0. In this case energy density is always *decreased*. In the third type, refraction causes the rays to converge, resulting in an *increase* in energy density. This is the reason that wave heights increase over shoals.

Since E is proportional to H^2 (Equation 5), H varies with the square root of ray spacing:

$$H_i/H_0 = (b_0/b_i)^{1/2} \cdot M \tag{13}$$

where b is the distance between adjacent wave rays, and M contains other variable terms (see *Shoaling Coefficient*).

The wave power per width of crest delivered to the shoreline is given by Equation (4) as:

$$P_b = E_b\,C_b\,n_b \tag{14}$$

where the subscript b refers to the breaker zone. In most regions this cannot be equated with deep-water wave power because the assumption of conservation of energy flux is not valid (that is, Equation 11 does not hold). For example, the presence of *ripple marks* on the bottom provides a clue that the waves are performing work on the bottom in the form of sediment transport. The power consumed in doing this work must be subtracted from that originally possessed by the wave. Other energy losses are caused by frictional effects such as turbulence, bottom drag, and percolation. If these effects can be compensated for, then Equation (14) can be derived.

Most breaker energy is dissipated through turbulence during breaking. However, a portion of that energy is used to transport sediment in the breaker zone. Much of this transport is in the form of oscillatory shore-normal motion, but when there is a breaker angle or a shore-parallel current bias, shore-parallel drift of sediment is effected (see *Sediment Transport*). The wave power that is effective in causing shore-parallel drift is given by (Watts, 1953; Komar, 1971):

$$P_l = 0.5 P_b \sin 2a \qquad (15)$$

where the subscript l refers to a shore-parallel component.

In shallow water, Equation (6) reduces to:

$$C = (gh)^{1/2}, \qquad (16)$$

which signifies that phase speed depends on water depth and not on wavelength (or period), as in deeper water. At this stage, the wave can be considered as a *solitary wave* (Galvin, 1972). As it proceeds up the beach slope, the water depth ahead of the wave is shallower than the depth behind the wave. According to Equation (16), the trailing edge of the wave travels faster than the leading edge, which causes a modification in shape to an asymmetrical form having a steeper leading face and gentler trailing face. At some critical depth, this process results in oversteepening so that the wave form is unstable and topples over forward—that is, the wave breaks. This depth can be related to the wave height (H/h) to form a dimensionless parameter, and is usually assigned a critical value of 0.78 (McCowan, 1894). Other theoretically derived values for the critical H/h range from 0.73–1.03 (Galvin, 1972). Experimental data show an even greater range.

One of the reasons for the variability of the critical H/h is the effect of bottom slope. Furthermore, short period waves that occur on smaller water bodies may not be well characterized as solitary waves, but rather as periodic waves of such steepness (H/L) that Equation (6), instead of Equation (16), is appropriate.

The type of breaker is dependent on wave height, period, and bottom slope (Wiegel, 1964). There are four types of breakers (Galvin, 1972): spilling; plunging; collapsing; surging. This sequence is in order of decreasing wave height for specified period and slope conditions. The same sequence is observed if height and slope are held constant and period is increased or if height and period are held constant and slope is increased. Breaker type is a continuous variable. Any given type grades into the adjacent types as wave or slope conditions change; therefore all types might be observed on a given day along a stretch of beach.

JAMES P. MAY

References

Airy, G. B., 1845. On tides and waves, *Encyclopaedia Metropolitana*, vol. 5 London: B. Fellowes, 241-396.

Galvin, C. J., Jr., 1972. Wave breaking in shallow water, in R. E. Meyer, ed., *Waves on Beaches*. New York: Academic Press, 413-456.

Keulegan, G. H., and Patterson, G. W., 1940. Mathematical theory of irrotational translation waves, *Nat. Bur. Standards Jour. Research* 24, 47-101.

Koh, R. C. Y., and LeMéhauté, B. J., 1966. *Wave Shoaling*. Pasadena, Calif. National Engineering Science Company, 22p.

Komar, P. D., 1971. The mechanics of sand transport on beaches, *Jour. Geophys. Research* 76, 713-721.

Korteweg, D. J., and de Vries, G., 1895. On the change of form of long waves advancing in a rectangular canal, and on a new type of long stationary waves, *Philos. Mag.*, ser. 5, 39, 422-443.

LeMéhauté, B.; Divoky, D.; and Lin, A., 1968. Shallow water waves: A comparison of theories and experiments, *Conf. Coastal Eng.*, 11th, Proc. 1, 86-107.

Longuet-Higgins, M. S., and Stewart, R. W., 1964. Radiation stresses in water waves; a physical discussion, with applications, *Deep-sea Research* 11, 529-562.

McCowan, J., 1894. On the highest wave of permanent type, *Philos. Mag.*, ser. 5, 38, 351-358.

May, J. P., 1975. Ability of various water wave theories to predict maximum orbital velocities in the shoaling wave zone, *Coastal Research Notes* 4, 7-9..

Rayleigh, J. W. S., 1876. On waves, *Philos. Mag.*, ser 5. 1, 257.

Skjelbreia, L., and Hendrickson, J., 1960. Fifth order gravity wave theory, *Conf. Coastal Engineering*, 7th, Proc. 184-196.

Stokes, G. G., 1847. On the theory of oscillatory waves, *Cambridge Philos. Soc. Trans.* 8, 441, and Supplement, *Sci. Papers* 1, 314.

Watts, G. M., 1953. Study of sand movement at South Lake Worth Inlet, Florida, *U.S. Army Corps of Engineers, Beach Erosion Board, Tech. Memo.* 42, 24.

Wiegel, R. L., 1964. *Oceanographical Engineering*. Englewood Cliffs, N.J.: Prentice-Hall, 532p.

Cross references: *Ripple Marks; Sea Conditions; Sediment Transport; Shoaling Coefficient; Swell and Its Propagation; Wave Energy; Wave Environments; Wave Meters; Wave Refraction Diagrams; Wave Shadow; Wave Statistics; Wind.* Vol. I; *Wave Refraction; Wave Theory.*

WAVE SHADOW

Barriers such as islands and breakwaters intercept the normal transmission of wave energy and thus create a sheltered area in the wave lee from which waves are excluded. This sheltered area is called the zone of *wave shadow*. The size and location of the shadow zone can have great importance in the protection of the coastline and coastal structures and in the control of the longshore transport of sand. Further, the wave shadow phenomenon can be used as an aid to navigation as evidenced by the early Micronesians and Polynesians and the early explorer-navigators who used the shadow zone disturbances of Pacific islands for island-to-island travel.

Even though the direct transmission of wave energy into the shadow zone in precluded, leakage of wave energy around the ends of the barrier prevents the area from being completely quiescent. *Refraction* and *diffraction* of the waves produce wave activity in the shadow zone, while variability in the direction of wave approach reduces the size of the wave shadow (Arthur, 1951). Diffraction, which is the transmission of wave energy normal to the direction of wave propagation, and refraction, which is the change in the direction of wave travel with changes in water depth, can be predicted for barriers with simple bottom topography (Weigel, 1964), and such calculations are commonly made during breakwater design. The effect of the variability of wave direction can

FIGURE 1. Aerial photograph of a breakwater at Venice, Santa Monica Bay, California. Wave shadow in lee of breakwater has caused a tombolo to form joining breakwater to the beach (from Inman, D. L., and Frautschy, 1966. Littoral processes and the development of shorelines, *in Coastal Engineering Speciality Conference, 1965*. Santa Barbara: American Society of Civil Engineers, 511-536).

be estimated from wave hindcast information, which gives the percentage of wave energy occurring in sectors of the compass.

When the wave shadow falls directly on the coastline, the amount of wave energy reaching the beach is reduced, and consequently the longshore transport of sand decreases. This causes sand accretion in the shadow zone, and in some cases the barrier and shore eventually become connected by a *tombolo* (Fig. 1). Sand accretion is likely to occur when the distance offshore to the barrier is less than three to six times the length of the barrier normal to the direction of wave propagation (Inman and Frautschy, 1966).

JOHN R. DINGLER

References

Arthur, R. S., 1951. The effect of islands on surface waves, *Scripps Inst. Oceanog. Bull. 6*, 24p.
Inman, D. L., and Frautschy, J. D., 1966. Littoral processes and the development of shorelines, *in Coastal Engineering Speciality Conference, 1965.* Santa Barbara: American Society of Civil Engineers, 511-536.
Wiegel, R. L., 1964. *Oceanographical Engineering.* Englewood Cliffs, N.J.: Prentice-Hall, 532p.

Cross-references: *Coastal Engineering; Sediment Transport; Tombolo; Wave Refraction Diagrams; Waves.*

WAVE STATISTICS

Wind waves can be treated like other random phenomena, although it is difficult to clearly define individual waves in a record. One technique has been to estimate the *mean* of the record and to count only those waves with troughs below and crests above the mean water level (*zero-up-crossing*). The U.S. Army Corps of Engineers Beach Erosion Board adopted a procedure based on the concept that individual wave heights should conform to a Rayleigh distribution function. Given the latter, it is possible to compute the probability of any particular wave and other wave characteristics (for example, *significant wave height*, the average height of the one-third highest waves). It is sometimes assumed that the Rayleigh distribution function is also valid for the square of the period of individual waves. However, a more consistent fit is provided by a log-normal distribution function. From the joint distribution of individual wave heights and periods (Battjes, 1971), it is seen that the highest waves correspond to intermediate periods and the *wave height* increases with an increase in the *wave period* up to the maximum height value and then decreases as wave period increases further. A higher correlation is generally obtained when a parabolic function is assumed to relate wave height with the wave period.

The distribution of instantaneous water elevations is closely approximated by the Gaussian distribution function. However, strongly non-linear effects may appear in shallow water. Therefore the probability density function for the elevations is not Gaussian; it can be expressed through the Hermitte polynomials with the coefficients of skewness and kurtosis (Zeidler and Massel, 1976). The presence of skewness and kurtosis also generates a certain deviation from the distributions mentioned above for the wave period and height; thus the mean wave period is 20-30% greater than the linear value following from the deep-water distribution, but the mean wave height is not so much affected by the nonlinearities.

RYSZARD ZEIDLER

References

Battjes, J. A., 1971. Run-up distributions of waves breaking on slopes, *Am. Soc. Civil Engineers Proc., Jour. Waterways and Harbors Div.* 97, 91-114.
Zeidler, R., and Massel, St., 1976. Coastal diffusion, currents, and waves in the south Baltic, *in Conference of Baltic Oceanographers, 10th.* Göteborg: Fishery Board of Sweden.

Cross-references: *Coastal Engineering; Swell and Its Propagation; Wave Climate; Wave Environments; Waves.*

WAVE THEORIES, OSCILLATORY AND PROGRESSIVE

The wave motion of water is well described by the Navier-Stokes equations. Particular problems arise in solving these equations because of the unknown position of the free surface and complicated boundary conditions. For analytical progress, the fluid motion is often assumed to be inviscid and irrotational. This is a good approximation when waves enter a region of water that was previously at rest and where there is no initial vorticity.

The most widely used method of finding approximate solutions to the equations for inviscid, incompressible, irrotational motion is to *linearize* the free surface boundary conditions and apply them to the undisturbed position of the free surface. The first approximation solution yields sinusoidal free surface; further approximations can be found by a perturbation expansion, an approach first used by Stokes (Peregrine, 1972, p.100). Nearly all the work done with the higher-order Stokes wave approximations has been on uniform wave trains or on slowly varying wave trains (Chu and Mei, 1970).

For $kh = (2\pi h)/L$, where h = depth, L = wavelength, a different approach is needed because of higher wave steepness, higher crests, and flatter troughs. Adequate *finite-amplitude, shallow-water* equations are often derived by making one of the equivalent assumptions: that the pressure is hydrostatic or that the horizontal velocity is uniform with depth. These equations are relatively easy to solve by the method of characteristics for constant depth. They usually become invalid when the water surface steepens so much that the wave breaks or spills shortly before breaking.

If wave amplitude is small compared with depth, the previous equations become inapplicable owing to the effects of the vertical velocity on both the horizontal velocity and the pressure, which give the Boussinesq equations (Peregrine, 1972, p.107). The typical solution is the *solitary-wave*. From comparison with experiments it is found that the Boussinesq equations give reasonable solutions for amplitudes up to about one-half the depth.

The Boussinesq equations have a limiting form: translational periodic waves, known as *cnoidal waves*. The latter result from the Korteweg-de Vries equation.

Some of the wave theories predict pure *oscillatory* motion of water particles, while others provide for simultaneous oscillations and resulting continuous displacement (*mass transport*), either in the direction of propagation or against it, depending on the elevation of a particular water particle. Special application of these theories for *progressive* waves (that is, propagating in one direction) is required in consideration of wave reflection, either complete or partial (see Rundgren, 1958).

The problem of reflected waves over sloping bottom has been considered in a number of studies for various uniform-depth equations mentioned herein (Peregrine, 1972). However, little progress has been made toward solving three-dimensional problems of waves over a variable depth of water. Also, when a wave breaks, the motion becomes exceedingly difficult to describe analytically.

RYSZARD ZEIDLER

References

Chu, V. H., and Mei, C. C., 1970. On slowly varying Stokes waves, *Jour. Fluid Mechanics* **41**, 873-887.

Peregrine, D. H., 1972. Equations for water waves and the approximations behind them, *in* R. E. Meyer, ed., *Waves on Beaches.* New York: Academic Press, 95-121.

Rundgren, L., 1958. Water waves forces, *Stockholm Inst. Hydraulics Bull. No. 54.*

Cross-references: *Coastal Morphology, Research Methods; Swell and Its Propagation; Wave Climate; Waves; Wave Statistics.*

WAVE WORK

Waves are periodic, more or less regular deformations of the interface between two fluids or, in some instances, of an internal layer within a fluid. Ocean surface and lake surface waves are observed on the interface between water and air. Their descriptive parameters are *period* (the time interval between two successive deformations), *height* (twice the amplitude), and *length* (horizontal distance between two successive crests, taken at right angles to the crests). Measurable periods of surface waves generally fall in the range 0.5-20 sec. Longer waves include *internal waves* and *tides.*

Ordinary ocean (and lake) waves are generated by the wind, travel in groups, and have a measurable *celerity* (velocity, length/period). The celerity (phase celerity, referring to each individual wave) is greater than the group velocity. *Standing waves*, which ordinarily are not transported from point to point, are created in special ways, such as by reflection off a seawall or by high Froude effects in a channel (such as a tidal inlet).

Very long waves include internal waves (commonly developed along a water/water interface owing to a thermal or salinity difference) and tides. Very small water surface waves, less than 2.44 cm in wavelength, are capillary waves, generated as part of the process of energy transfer from the wind to the water. Tides and internal waves may be responsible for sediment transport, the former by means of *tidal currents,* and the latter in the form of *ripple marks* in water too deep to be reached by surface wave activity.

Water Orbits

In a general way, the water motion during wave passage is in the form of a circle or ellipse in shallow water. It is not necessary, however, that the ellipse be closed. In shallow nearshore waters it is not closed, and there tends to be a net shoreward drift of water, elevating the water surface and thus giving rise to several kinds of current gyres (those where seaward return is more rapid than landward flow form *rip currents;* those where seaward return is more diffuse and slower than landward flow are associated with *transverse bars*) (Niedoroda and Tanner, 1970). In deep water, orbital circles decrease in size downward to the vanishing point; in shallow water, the orbits are deformed into ellipses, the flattest of which immediately overlie the bottom eddy layer, which consists of a series of eddies combined in pairs to make "figure eights."

"Deep," "intermediate," and "shallow" depths are defined in terms of the wavelength of a given wave. The boundary between deep and intermediate water may be found where the water depth is one-half the wavelength; the boundary between intermediate and shallow water, at 1/20 the wavelength. It is not necessary that a shallow water zone appear for any given wave; storm waves may break in intermediate water.

Limited Fetch

Lakes and ponds differ from the ocean in that the *fetch* (the distance over which the wind blows to generate the waves) is necessarily severely limited. The effective fetch for an ocean wave coming into the surf zone cannot be known for sure. For most lakes, however, only one wave train can be generated at one time, and this process utilizes all of the available fetch. The scatter (high standard deviations, for example, for wave height) observed on most lakes is due to hydrodynamic and bathymetric effects on a single wave group; ocean waves also have scatter due to the mixing of various groups, each of which may have been generated in a different part of the ocean. Further, once an ocean wave is generated, it may coast (without additional wind effects or despite a mild contrary wind) for thousands of kilometers, whereas lake waves rarely coast. Lake waves are almost always forced (driven) by the wind all the way into the surf. Very large lakes may have both coasting waves and mixed wave trains.

Wave steepness is the height divided by the wavelength. Swell (coasting waves) is characterized by a steepness more than 0.030, and by rounded crests and great geometrical regularity. Sea (forced waves) has a steepness less than 0.015, and irregular, peaked crests.

Refraction

When a wave passes from deep to intermediate water, it is said to feel bottom. Energy begins to drain from the wave into bottom effects, perhaps in ripple mark maintenance or some other form of sediment transport. On very gently sloping bathymetry, all of the wave energy may be spent ultimately on the bottom, thus providing that no wave reach the shore (*Zero-Energy Coast*, q.v.). On ordinary bathymetry, however, the energy drain increases — generally not in a linear progression — until a significantly large fraction of the energy is dumped (in the surf zone).

The angle between the wave crest and the local isobath (which at the shoreline is the zero contour) is the wave, or beta, angle. Where beta is zero, the wave approaches the coast in "head-on" fashion. Deep water beta, however, is typically not zero. The process by which the wave crests are swung around, minimizing beta, is *refraction;* this is achieved because one part of the crest reaches intermediate (or shallow) water sooner than other parts and is therefore subjected to retardation due to bottom effects prior to those other parts.

Beta is commonly small, but still not zero, in the surf. This small but finite beta angle gives rise to littoral drift effects, which include a littoral current as well as the longshore transport that develops in the swash zone (Tanner, 1974).

Energy Density

The wave energy density of water surface waves, E, is defined as: $E = \gamma \rho H^2/8$, where γ is the acceleration of gravity, ρ is the mass density of the water and H is the wave height (Eagleson and Dean, 1966). In Si units, $E = 1250\ H^2$. The units are joules/meter^2 (energy or work per unit area).

Wave period is highly conservative for coasting, shoaling waves; except for new waves (multiples) created by partial breaking on an offshore bar or elimination of part of the train by such breaking, the period at the surf is the same as the deep water period. Wave height, however, can be attenuated greatly, even to zero; it is the only variable needed to define wave energy density, which shows an overall landward decrease along all coasts where shoaling occurs. (Wavelength behaves in a manner intermediate between the behavior of period and that of height.)

The change in E per unit distance (R) travelled along the wave ray or path is $\delta E/\delta R$. In deep water under wind stress, this expression is positive (the wave grows); in shallow water for coasting waves, this expression is negative (energy is drained to the bottom, and the wave is attenuated). In shallow lakes, the clearly positive segment of the $\delta E/\delta R$ profile can be short (but is inversely proportional to the wind velocity); the rest of the profile may be slightly positive, negative, essentially zero, or variable. In many instances, it is close to zero. This means that after a certain critical point is reached within a few kilometers of the upwind shore, all new energy added to the waves is drained to the bottom into sediment motion or other activity (such as bending aqueous vegetation or disturbing the subaqueous meadow).

Energy Loss

Seaweed, marsh grasses, and other aqueous vegetation may account for a significant energy drain. Values of $-\delta E/\delta R$ for representative examples include the following: sedge, -0.16; *Spartina*, -0.10; and *Thalassia*, -2.25; obtained by observing wave height attenuation over a measured course, depending on plant density (spacing) and individual bending characteristics.

Values of $-\delta E/\delta R$ in joules/m^2-m vary from a minimum of zero (deep water) to a maximum on the order of 10^2 (storm waves on an open ocean coast). Ordinary fair-weather ocean waves crossing a gentle slope may drain 2.0 joules/m^2-m or more into bottom activity. Severe storm waves on a small, shallow lake can exceed 2.0 but may not reach 10. These changes in energy content are much larger than necessary to move sand or fine gravel.

A widely used expression for wave-driven, bed shear stress (τ_o) has never been developed, because the original definition, applicable to rivers, contains the slope or gradient, a term that is meaningless in wave activity. Recent work by Jonsson and Carlsen (1976) can be extended to an estimate of bed shear stress: $\hat{\tau}_o = (c/u_o)(\delta E/\delta R)$, where c is the wave phase celerity (length/period) and u_o is the bottom orbital velocity. If maximum u_o is calculated (a simple matter in computer simulation), then maximum bed shear stress is estimated at the top of the bottom eddy layer. This last expression should be multiplied by some value close to 2.0 to extrapolate it downward through the bottom eddy layer, in which the geometry of motion requires that local velocities exceed u_o, thus producing τ_b at the bed. The bed value can be divided by 0.81 to give an estimate of the largest grain size (of quartz sand) that can be moved under these conditions (expressed in millimeters).

Transport Efficiency

The total energy drained from a wave train as it crosses a selected area can be compared with the work actually done in moving sediment, in order to obtain a *transport efficiency* for the wave–and–sediment system. This efficiency is represented by the symbol k, defined as: $k = I/P_L$, where I is the immersed weight of the sediment mass moved per unit time ($\gamma \rho p q$), P_L is the littoral component of wave power and a function of both H and β, and k is dimensionless. Numerical values for k, averaged over tens of years, commonly fall in the range 10^{-2} to 10^{-4}, cannot equal or exceed unity and probably rarely if ever equal or exceed 10^{-1} (Entsminger, Banks, and Tanner, 1975). Although the definition given above is correct, it poses certain methodological problems, which can be circumvented in part by a direct (but less obvious) computational method involving the *reduced mass* of the sand ultimately transported (May and Tanner, 1975). In either case, volumes (and hence masses), distances, and rates of transport must be known.

If P_L can be obtained (by computer simulation) and k is known or can be estimated reasonably well, I can be calculated. For the expression I, γ (acceleration of gravity), ρ (mass density of quartz immersed in water) and p (the packing factor = 0.6) are known, and hence q, the quantity of sand delivered (m^3/sec) can be computed. This delivery volume is easy to convert to an annual volume, which can be distributed along the eroding segment to produce a predicted annual recession rate (May and Tanner, 1973).

Studies of wave-versus-sediment interactions, from an energetics point of view, permit the partitioning of wave energy into various categories. One such study (May and Tanner, 1975) for a moderate-energy coastal cell (average breaker height 0.10-0.50 m) showed the following long-term average values: total energy, 10^{15} joules (or 130 joules/sec-m. of beach); NUST (net unidirectional sediment transport, 10^{12} joules (or 0.13 joules/sec-m. of beach); and efficiency, 10^{-3} joules. From these reasonably well established data for the study area, 99.9% of the energy appears to have been used for purposes other than NUST. An unknown part of this fraction must have been spent in back-and-forth sediment shuffling and most of the remainder in other fluid turbulence (there is no submarine meadow in the study area). The average $-\delta E/\delta R$ was close to 0.5, but inshore this parameter can increase to about ten times the average. (See Figure 1.)

Water Level Rise

Lake levels may rise or fall several meters in a few years, for meteorological reasons, and sea level may rise or fall 10-100 m over 5-15 thousand years, thereby transferring the location of the surf zone either landward or seaward. A rise of lake level of one meter in a couple of years provides a new and significantly different bathymetry for incoming waves to cross. Until the bottom geometry has been reshaped, the $\delta E/\delta R$ profile will be radically different, with much smaller numbers at almost every step. As a result the energy delivered in, and close to, the surf zone will be greater (by one or two orders of magnitude) (Berquist and Tanner, 1974), and erosion will be severe and rapid, provided β is not zero. The serious erosion experienced along Great Lakes coasts during the late 1960s and early 1970s was due to such a water level rise.

In due time sand taken offshore from the beach will reconstruct the *equilibrium profile*

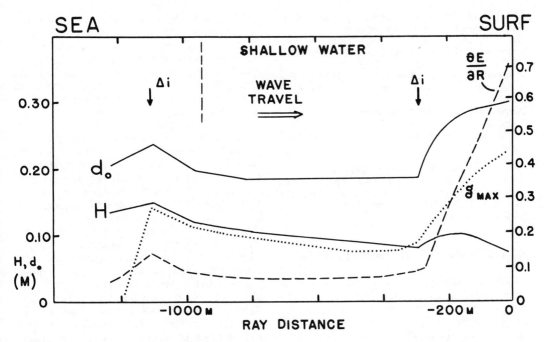

FIGURE 1. Profiles of selected parameters for an ocean wave coasting (from left to right) across a gently sloping bottom to the surf. The period is 5 sec, the deep water height is 1.0 m, and the β angle is zero (head-on approach). Wave height, shown as a solid line (H), is read against the left ordinate in meters; it is roughly representative of wavelength. Bottom orbital diameter, also shown as a solid line (d_o), is likewise read against the left ordinate; it is roughly representative of bottom orbital velocity and bottom orbital acceleration. The change in wave energy density per unit distance traveled along the ray is shown as a dashed line ($\delta E/\delta R$), and is read on the right ordinate in SI units (see text). The maximum quartz particle size that can be moved, based on the estimated maximum bed shear stress, is shown as a dotted line (estimate, g_{max}), and is read on the right ordinate in centimeters. Two important slope changes are shown by the symbol Δi; the middle slope segment is less steep than the other two. Results were obtained by computer simulation, and most of the intermediate depth zone has been omitted (to the left of the diagram). Computational step length along the ray varied from a maximum close to 60 m, near the left side of the diagram, to a minimum of 17.4 m, close to the surf zone. Because the β angle is close to zero, the longshore component of wave power (P_L) is also close to zero, and the computed longshore current is only 0.06 m/sec. However, medium sand should be in motion in all parts of the area closer to the beach than about 1150 m.

at a higher elevation, and erosion rates will gradually return to normal. A later drop in lake level will have a different effect. The change in bathymetry will produce a breaking point farther offshore, with part of the sand being carried into still deeper water, and part being driven back onto the beach.

The profile of $\delta E/\delta R$ differs from the profile of wave height or of any other surface wave parameter along the wave ray, but is similar to (and locally steeper than) profiles of the various bottom orbital parameters (diameter, maximum velocity, maximum acceleration). All of these profiles are relatively simple to obtain by machine methods. They can be used to predict or to clarify much that otherwise is difficult to understand in terms of nearshore sediment transport. They show, for example, that sediment transport must take place in different bands more or less parallel with the isobaths. This suggests that bottom markings may be different from band to band.

WILLIAM F. TANNER

References

Berquist, C. R., and Tanner, W. F., 1974. Analysis of water-level rise effects, *Gulf Coast Assoc. Geol. Socs. Trans.* 24, 255-256.

Eagleson, P. S., and Dean, R. G., 1966. Small amplitude wave theory, *in* A. T. Ippen, ed., *Estuary and Coastline Hydrodynamics.* New York: McGraw-Hill, 1-92.

Entsminger, L. D.; Banks, R. S.; and Tanner, W. F., 1975. The coefficient "k" in the "a-b-c . . ." model, *Geol. Soc. America Abs. with Programs* 7, 486.

Jonsson, I. G., and Carlsen, N. A., 1976. Experimental and theoretical investigations in an oscillary turbulent boundary layer, *Jour. Hydraulic Research* 14, 45-60.

May, J. P., and Tanner, W. F., 1973. The littoral power gradient and shoreline changes, *in* D. R.

Coates, ed., *Coastal Geomorphology*. Binghamton: State University of New York, 43-60.

May, J. P., and Tanner, W. F., 1975. Estimates of net wave work along coasts, *Zeitschr. Geomorphologie* **22**, 1-7.

Niedoroda, A. W., and Tanner, W. F., 1970. Preliminary study of transverse bars, *Marine Geology* **9**, 41-62.

Tanner, W. F., ed., 1974. *Sediment Transport in the Near-Shore Zone*. Tallahassee, Fla.: Florida State University, Geology Department Coastal Research Notes, 147p.

Cross-references: *A-B-C Model; Bars; Bruun Rule; Equilibrium Shoreline; Internal Waves; Lakes, Coastal Morphology; Profile of Equilibrium; Rip Currents; Tidal Currents; Tides; Wave Drag Layer; Waves; Zero-Energy Coast.*

WEATHERING AND EROSION, BIOLOGIC

Coastal erosive processes on rocky and semiconsolidated substrates fall into three categories: *mechanical, chemical,* and *biological* erosion. Of these the last has been viewed globally as least important, although few studies have evaluated the relative magnitude of each erosive process at one site. In a recent study where this was done, Trudgill (1976) found that mechanical and biological factors are clearly the dominant marine erosional processes on limestone of Aldabra atoll, in the Indian Ocean and that chemical solution accounts for only 10% of the erosion that occurs. A great many other studies have demonstrated the local significance of individual or groups of rock-destroying organisms as erosional agents, especially on intertidal platforms and notches, subtidal ledges, and scarps in low and middle latitudes.

Biologic erosion (*bioerosion*) is the erosion of rocks by organisms. A surprisingly large range and number of groups in the plant and animal kingdoms are known erosional agents, including some bacteria, fungi, algae, sponges, worms, crustacea, molluscs, and echinoids. Many of these are discussed in a collection of papers edited by Carriker, Smith, and Wilce (1969). Organisms that actually erode the rock surface are those that graze upon algae or excavate a shallow depression on the rock surface, others penetrate deeply into the rock itself and the extent of boreholes is apparent only when the rock is broken. The first group contributes directly to retreat of the rock surface, while borers presoften the rock, which, once weakened, tends to be more susceptible to attack by other erosional agents. Surficial microscopic algae also weaken the rock surface by bioweathering rather than by direct erosion. Bioerosion is carried out either by biochemical (e.g., secretion of solvents) or biophysical (e.g., mechanical abrasion by action of radula or shell) means. The motivation for erosion is to obtain food, shelter, protection, or a place to live. The nature and extent of rock destruction are determined partly by the type of organism and partly by the characteristics of the substrate: limestones—including coral reefs—are perhaps most susceptible to bioerosion, followed by mudstones and softer sandstones. The harder rocks, like granites and basalt, are rarely eroded.

A large number of terms are used to describe the micromorphological results of bioerosion: pipes, perforations, pits, galleries, tunnels, cavities, burrows, boreholes, among others. Individual structures vary enormously in size, from microscopic to a few centimeters in maximum dimension and in shape. Generally the size and shape of structures are species-specific. The relevance of biogenic structures in paleontology, stratigraphy, sedimentology, and paleoecology, together with equivalent terms in English, German, and French, is given in Frey (1973, 1975). Rates of erosion vary greatly, depending on rock type, exposure, tidal position, and type and number of organisms. On coastal limestones, rates of surface lowering are about 1 mm/yr, but much higher rates—5-20 mm/yr—have been measured where species of sponges, molluscs, and echinoids are present. While these rates appear minor, they are important because organisms of several types usually work together and mechanical removal of weakened rock accompanies biological attack.

ROGER F. McLEAN

References

Carriker, M.; Smith, E. H.; and Wilce, R. T., eds., 1969. Penetration of calcium carbonate substrates by lower plants and invertebrates, *Am. Zoologist* **9**, 629-1020.

Frey, R. W., 1973. Concepts in the study of biogenic sedimentary structures, *Jour. Sed. Petrology* **43**, 6-19.

Frey, R. W., ed., 1975. *The Study of Trace Fossils*. New York: Springer-Verlag, 562p.

Trudgill, S. T., 1976. The marine erosion of limestone on Aldabra Atoll, Indian Ocean, *Zeitschr. Geomorphologie* **26** (suppl.), 164-200.

Cross-references: *Alveolar Weathering; Rock Borers; Weathering and Erosion, Chemical; Weathering and Erosion, Mechanical.*

WEATHERING AND EROSION, CHEMICAL

Marine erosion is a largely mechanical process carried out by action of waves and currents, affecting rock particles that have been *prepared* by chemical weathering (Fairbridge, 1952, 1968). The surfaces of corroded rocks are etched into pits and pinnacles ("marine karren,"

Ley, 1979), as distinct from mechanical abrasion that tends to produce smooth boulders and rounded surfaces (Guilcher, 1958; Tricart, 1959).

Most familiar rock types outcropping in coastal regions contain minerals that are theoretically soluble to some extent in either fresh or saltwater (Birot, 1964). Experiments, however, have shown that solubility rates for most minerals are extremely low, so that in the time frame of the existence of the present coastline (c. 6000 yr.) for most rock types chemical weathering is not an effective preparation for erosion.

Chemical weathering is effective only with respect to a few specific rocks. First and foremost are limestones and dolomites, which contain minerals in the following order of chemical stability: dolomite > calcite > high magnesium calcite > aragonite > gypsum > halite (Revelle and Emery, 1957).

Aerated freshwater, saturated with respect to CO_2 (as happens during heavy rainfall) is several times more reactive with respect to these minerals than is seawater. In fact near-surface seawater is normally supersaturated with Ca^{++} and $CO_3^=$ ions, with respect to both the calcite and aragonite phases (but less so with respect to aragonite); thus limestone is normally stable in near-surface contact with seawater. Because of this differential solubility in fresh and saltwater, low tide level becomes a very distinctive weathering boundary plane (*water-layer weathering*; Wentworth, 1939). This constitutes *littoral* weathering, a special case of the general phenomenon. Below it limestone and dolomite are stable, but above it a *littoral karst* evolves (see *Karst Coasts*).

Certain coastal rocks are a mixture of minerals —e.g., in the calcareous eolianites (lithified dune rocks) that characterized the regressive phases during Quaternary sea level oscillations in intermediate (semi-arid) latitudes; the rock is constituted of a clastic mixture of calcite, high magnesium calcite, aragonite, and (on continental shores) insoluble quartz and heavy mineral grains. In such cases the most soluble mineral is weathered first, and the rock tends to break up mechanically (Bartrum, 1926).

Within the intertidal zone and somewhat below it, the carbonate rocks are also susceptible to attack by boring molluscs, echinoids, sponges, browsing fish, crabs, and gastropods (Hodgkin, 1964). Biochemical weathering is thus amplified in this zone by biomechanical erosion as well as by the usual processes of mechanical wave action (abrasion and hydraulic erosion).

Other rocks besides limestone are also affected by intertidal and spray weathering. These include rocks rich in feldspars—e.g., arkose and graywacke, as well as many volcanic rocks, especially tuffs (Hills, 1949; Edwards, 1958).

Besides intertidal and spray weathering of chemical origin, all the normal soil processes (dominated by bacterially generated carbonic acids). affect rock surfaces above sea level but not below it. Thus mechanical erosion by waves tends to remove soil and regolith previously rotted by soil processes. The result favors production of a *shore platform* (q.v.), which tends to correspond approximately to mean sea level. In favored situations and with resistant rock types formerly active, shore platforms may be found as emerged strandlines—witness of former coastlines.

An associated feature of the littoral weathering zone is *honeycomb weathering* or *alveolar erosion*. This is a process of differential pitting that is primarily a mechanical product caused by salt crystallization (*salt fretting*) but often supplemented by chemical breakdown and biochemical actions.

RHODES W. FAIRBRIDGE

References

Birot, P., 1964. Experiences sur la desagregation des roches en milieu acidi et oxydant. *Zeitschr. Geomorphologie* 5(suppl.), 28-29.

Bartrum, J. A., 1926. Abnormal shore platforms, *Jour. Geology* 34, 793-806.

Edwards, A. B., 1958. Wave-cut platforms at Yampi Sound in the Buccaneer Archipelago, *Royal Soc. West Australia Jour. and Proc.* 41, 17-21.

Fairbridge, R. W., 1952. Marine erosion, *Proc. 7th Pacific Sci. Congr., New Zealand, 1949,* 3, 349-359.

Fairbridge, R. W., ed., 1968. *Encyclopedia of Geomorphology.* New York: Reinhold, 1295p.

Guilcher, A., 1958. Coastal corrosion forms in limestones around the Bay of Biscay, *Scottish Geog. Mag.* 74, 137-149.

Hills, E. S., 1949. Shore platforms, *Geol. Mag.* 86, 137-152.

Hodgkin, E. P., 1964. Rate of erosion of intertidal limestone, *Zeitschr. Geomorphologie,* N.F. 8, 385-392.

Ley, R. G., 1979. The development of marine karren along the Bristol Channel, *Zeitschr. Geomorphologie* 32 (suppl.), 75-89.

Revelle, R., and Emery, K. O., 1957. Chemical erosion of beach rock and exposed reef rock, *U.S. Geol. Survey Prof. Paper 260-T,* 699-709.

Tricart, J., 1959. Problemes geomorphologiques du littoral oriental du Bresil, *Cahiers Oceanog.* 11, 276-308.

Wentworth, C. K., 1939. Marine bench-forming processes: Solution benching, *Jour. Geomorphology* 2, 3-25.

Cross-references: *Alveolar Weathering; Karst Coast; Rock Borers; Splash and Spray Zone; Weathering and Erosion, Biologic; Weathering and Erosion, Mechanical.* Vol. III: *Limestone Coastal Weathering.*

WEATHERING AND EROSION, DIFFERENTIAL

Weathering and erosion occur at varying rates within a given area. The variation is caused by differences in resistance or susceptibility of various rock types to weathering and erosion. The result is an uneven surface, with the more resistant rocks forming protrusions, hills, or ridges and the more susceptible rocks forming indentations or valleys. The scale of *differential weathering and erosion* can vary greatly, as almost all rock masses are affected. On a small scale, different minerals in the rock will vary in resistance so that rock surfaces become pitted or rough from differential weathering. On a larger scale, the resistance or susceptibility of various rock types to weathering and erosion can result in the formation of ridges and valleys (Byrne, 1963; Pipkin and Ploessel, 1973; Ploessel, 1973). This is particularly apparent in tilted sequences of sedimentary rocks. Differential weathering and erosion often result in uneven coastlines (Sunamura, 1973); headlands are formed of resistant rocks, with the bays eroded into less resistant materials (Fig. 1).

MICHAEL R. PLOESSEL

References

Byrne, J. V., 1963. Coastal erosion, northern Oregon, in T. Clements, ed., *Essays in Marine Geology in Honor of K. O. Emery*. Los Angeles: University of Southern California Press, 11-33.

Pipkin, B. W., and Ploessel, M. R., 1973. *Coastal Landslides in Southern California*. Los Angeles: University of Southern California, Department of Geological Sciences Sea Grant Pub., 20p.

Ploessel, M. R., 1973. Engineering geology along the southern California coastline, in D. E. Moran, J. E. Slosson, R. O. Stone, and C. A. Yelverton, eds., *Geology, Seismicity and Environmental Impact*. Los Angeles: Association of Engineering Geologists Spec. Pub., 365-366.

Sunamura, T., 1973. Coastal cliff erosion due to waves. Field investigation and laboratory experiments, *Tokyo Univ. Fac. Engineering* 32, 1-86.

Cross-references: Alveolar Weathering; Muricate Weathering; Salt Weathering; Water Layer Weathering; Weathering and Erosion, Biological; Weathering and Erosion, Chemical; Weathering and Erosion, Mechanical.

FIGURE 1. An uneven coastline at Laguna Beach, California, resulting from differential erosion. Headland consists of hard, resistant andesite. Larger bays are incised into less resistant siltstone, sandstone, and shale. The small pocket cove in the center of the photograph is cut into hydrothermally altered tuff (photo: R. E. Stevenson).

WEATHERING AND EROSION, MECHANICAL

The sea operates as an agent of *mechanical weathering and erosion* in several ways: through hydraulic action and wedging; by abrasion (*corrasion*), using sand, gravel, and larger rock fragments as tools; by attrition of the rock particles themselves during this abrasive action; by salt weathering or fretting; by organisms (bioerosion); and by chemical attack, or corrosion, which weakens the rocks and accelerates erosion. The rates of mechanical weathering and erosion are a function of the exposure of a coast to wave attack, the offshore gradient, and the resistance of the constituent rocks.

Loose materials may be picked up directly by currents and wave action and moved offshore or along the coast. Further, solid rock may be directly eroded by the hydraulic action of breaking waves. When a wave breaks against a cliff, the impact of the water is capable of exerting pressures of greater than 250 kg/m^2 (6000 lb/ft^2). Water is thus driven into cracks in the rock, entrapping bubbles of air. In so doing, the compressed air acts as if a wedge was suddenly driven into the cracks. As the wave recedes, the explosive expansion of the air produces an effective quarrying action, capable of moving large blocks of rock.

Another method of marine erosion is the abrasive or corrasive action of sand, gravel, and large rocks picked up by the waves and driven back and forth against the exposed shore. This action includes dragging (friction), rolling, bouncing (saltation), and the *artillery action* of large fragments, which can shatter solid rock when hurled against a cliff. At Tillamook Rock on the Oregon coast, rocks weighing as much as 297 kg have been picked up during severe storms and thrown through the windows of the lighthouse, 44 m above mean high water. Similarly, the windows of the Dunnet Head

Lighthouse on the north coast of Scotland, more than 100 m above the high water mark, have been broken by rocks propelled by wave action. The abrasive action may produce notching and undercutting of cliffs, setting the stage for mass wasting. Seepage of groundwater, frost action, and wind—all combine with the undercutting to produce cliff recession. The abrasion along the shore works horizontally to form a *wave-cut cliff* and a *wave-cut platform* (q.v.). Differences in the resistance of the rocks being eroded may produce an irregular cliff face.

The impact and rubbing together of the rock fragments during wave action, as well as the grinding of the blocks that fall as the cliff is undercut, cause the tools themselves to wear down to smaller sizes and become rounded.

Salt weathering (q.v.) or fretting takes place when salt crystals grow within pore spaces in rocks; the crystal growth expands the pore and may cause fracturing of the rock. This process is most effective in areas exposed to salt spray and subject to alternative wetting and drying. The chemical and solvent action of seawater may be especially important along limestone coasts. However, even noncalcareous rocks may weather rapidly as a result of the chemical action of seawater. It has been experimentally demonstrated that basalt and other rocks weather 3-14 times faster in saltwater than in fresh water. Thus, the mechanical action of salt fretting and chemical weathering by seawater may act in consort.

Additional factors that may contribute to mechanical weathering in certain areas include large diurnal temperature variations and biological activity. Biological erosion may be important, especially in tropical or subtropical areas having limestone coasts. Boring and browsing organisms erode rocks in the intertidal zone. This bioerosion, combined with the chemical solution of limestone by algae, produces the intricate scalloping known as *phytokarst*.

Relevant literature on mechanical weathering and erosion includes Bromley (1978), Gaillard (1904), Garner (1974), Johnson (1919), King (1972), Longwell, Flint, and Sanders (1969), Sunamura (1977), Thornbury (1966), and Wellman and Wilson (1965).

MICHAEL R. RAMPINO

References

Bromley, R. G., 1978. Bioerosion of Bermuda reefs, *Paleogeography, Paleoclimatology, Paleoecology* **23**, 169-197.

Gaillard, D. B. W., 1904. *Wave Action in Relation to Engineering Structures.* Washington, D.C.: U.S. Army Corps of Engineers, 232p.

Garner, H. F., 1974. *The Origin of Landscapes.* New York: Oxford University Press, 734p.

Johnson, D. W., 1919. *Shore Processes and Shoreline Development.* New York: John Wiley & Sons, 583p.

King, C. A. M., 1972. *Beaches and Coasts.* New York: St. Martin's Press, 570p.

Longwell, C. R.; Flint, R. F.; and Sanders, J. E., 1969. *Physical Geology.* New York: John Wiley & Sons, 685p.

Sunamura, T., 1977. A relationship between wave-induced cliff erosion and erosive force of waves, *Jour. Geology* **85**, 613-618.

Thornbury, W. D., 1966. *Principles of Geomorphology.* New York: John Wiley & Sons, 618p.

Wellman, H. W., and Wilson, A. T., 1965. Salt weathering: A neglected geological erosive agent in coastal arid environments, *Nature* **205**, 1097-1098.

Cross-references *Cliff Erosion; Marine Erosion Coasts; Frost Riving; Hydraulic Action and Wedging; Quarrying Processes; Salt Weathering; Shore Platforms; Wave Erosion; Weathering and Erosion, Biological; Weathering and Erosion, Chemical.*

WHARF

A *wharf* is a marine structure built contiguous with and parallel to the shore of a harbor, river, or canal. (Bruun, 1973; Cornick, 1958; Quinn, 1961). It is generally used for the mooring or tying of vessels during the loading and unloading of cargo or receiving and discharging of passengers.

A wharf is frequently backed up by warehouses, industrial areas, roads, and railroads. Wharves are usually created by extensive fill operations, and act as a retaining wall or bulkhead. The term *quay* is frequently used synonymously.

CHARLES B. CHESTNUTT

References

Bruun, P., 1973. *Port Engineering.* Houston: Gulf Publishing Co., 436p.

Cornick, H. F., 1958. *Dock and Harbor Engineering,* vol. 1. London: Charles Griffin & Co., Ltd., 316p.

Quinn, A. F., 1961. *Design and Construction of Ports and Marine Structures.* New York: McGraw-Hill, 531p.

Cross-references: *Coastal Engineering; Protection of Coasts; Quay.*

WIND

Nearly all of the processes that act upon the coast are either directly or indirectly the result of circulation of the atmosphere (*wind*). Primary among these wind-generated processes are waves and longshore currents. The friction between the moving atmosphere and the water surface causes the formation of *waves*, which in most circumstances are propagated toward the coast. As these waves approach the coast, they commonly do so at an angle with the shoreline, thus generating *longshore currents*.

In addition, sustained and strong onshore wind causes *setup*, also called *storm surge* or *wind tide*. This phenomenon can be of considerable significance in coastal processes, in that it provides a mechanism whereby coastal waves and currents may interact with portions of the beach or coastal bays that are not normally inundated. Coastal winds are also important as direct agents of sediment transport as they interact with the dry sand on the back beach and coastal dunes. Some of the largest dunes in the world are along the coast of Oregon and on the eastern coast of Lake Michigan.

Wind is the horizontal movement of air, typically generated by gradients in barometric pressure. Under such circumstances the wind actually moves normal to the pressure gradient owing to the earth's rotation and *Coriolis effect*. A wind that moves along the isobars so that the pressure gradient is balanced by deflective (Coriolis effect) and centrifugal effects is called a *gradient wind*. Under circumstances of uniform gradient, the centrifugal force is insignificant and the gradient and Coriolis effect are in balance. This produces a *geostrophic wind* (Blair and Fite, 1957).

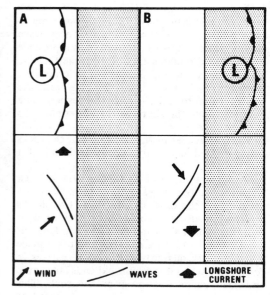

FIGURE 1. Position of cyclonic system (Northern Hemisphere) and orientation of associated coastal processes.

Cyclones

Cyclonic circulation occurs as winds move in a near circular pattern around a region of low pressure. Cyclones have *counterclockwise* circulation in the Northern Hemisphere and a *clockwise* circulation in the Southern Hemisphere (Critchfield, 1974). Throughout the midlatitude belt of westerlies, the passage of these cyclonic low pressure weather systems has a profound effect on coastal processes, particularly on the beach and nearshore environment.

Assuming a north–south-trending coast with the water to the west and the land to the east, the low-pressure system (cyclone) moves west to east approaching the coast. As the system approaches, barometric pressure is falling and the wind blows from the southwest (Fig. 1a). This generates waves, which approach from the southwest, causing a northward-flowing longshore current along the coast. As the center of the cyclone passes over the coast there is an abrupt change in wind direction. Wind on the trailing side of the system is blowing from the northwest, which causes a reversal in wave direction and longshore current (Fig. 1b). This pattern is one that is repeated in a fairly regular cyclic fashion with a period of 5–8 days (Fox and Davis, 1973).

Local Winds

Another important wind phenomenon occurs along the coast and has a diurnal period. This wind is generated by local temperature differences between the land and the water mass. In summer the land warms more than the water during the day. As a result the warm air expands over the land, flows out over the water, and thereby creates a pressure gradient from the water mass to the land. This generates a *sea breeze* that is somewhat of a convential circulation (Fig. 2a).

The opposite situation results during the night, when the land cools more than the adjacent water. The upper air moves landward, is cooled, and a gradient is formed in a seaward direction. This causes a *land breeze* to blow seaward (Fig. 2b); however, it typically is more local and has less speed than the sea breeze. It is uncommon that these local winds have much effect on coastal processes in restricted bodies of water, but where swell waves are absent they may be significant, especially in periods between the passage of low pressure systems when

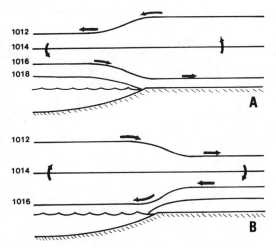

FIGURE 2. Position of isobars and resulting circulation that causes sea breeze (A) and land breeze (B).

regional winds are absent. Breakers approaching a meter in height have been observed along the coasts of Lake Michigan and the Gulf of Carpenteria (Australia). Land breezes may dampen waves as they approach the coast, and in this fashion they have an effect on coastal processes.

RICHARD A. DAVIS, JR.

References

Blair, T. A., and Fite, R. C., 1957. *Weather Elements.* Englewood Cliffs, N.J.: Prentice-Hall, 414p.

Critchfield, H. C., 1974. *General Climatology.* Englewood Cliffs, N.J.: Prentice-Hall, 446p.

Fox, W. T., and Davis, R. A., Jr., 1973. Simulation model for storm cycles and beach erosion on Lake Michigan, *Geol. Soc. America Bull.* 84, 1769-1790.

Cross-references: *Coriolis Effect; Storm Surge; Storm Wave Environments.* Vol. II: *Wind.*

WIND, WAVES, AND CURRENTS, DIRECTION OF

Because of a seemingly contradictory set of rules, confusion often arises in naming the direction of movement of wind, waves, and currents. By tradition, winds (Strahler, 1963, p. 227) and waves (Davies, 1973, p. 42) are designated by the direction they *come from,* while currents (Strahler, 1963, p. 297) are identified by the direction in which they *are going.* For example, *northerly* winds or waves originate in the north and are moving south, while *northerly* currents are flowing from south to north.

Reporting of wind, wave, or current direction may be done in azimuth degrees (45°, 135°, 225°, 315°), compass bearings (N45°E, S45°E, S45°W, N45°W), or compass points (NE, SE, SW, NW) all of these being, respectively, the same four directions.

MAURICE L. SCHWARTZ

References

Davies, J. L., 1973. *Geographical Variation in Coastal Development.* New York: Hafner, 204p.

Strahler, A. N., 1963. *The Earth Sciences.* New York: Harper and Row, 681p.

WORLD NET SEDIMENT TRANSPORT

As a general rule, *littoral drift* may be expected to extend from areas of higher energy toward areas of lower energy. Its incidence is also strongly associated with obliquity of the coast in relation to the direction of approach of the most important sediment-shifting waves. Such relationships are well appreciated at local and regional scales, but the first attempt to look at *sediment transport* on a world scale was that of Silvester (1962), who used map evidence of the distribution of *zetaform bays* to deduce likely directions of net movement (Fig. 1).

Subsequently, similar conclusions were reached by Davies (1972), as a result of mapping world trends in the orientation of constructional shore forms, and were explained in terms of predominant wave directions. Highest aggregate wave energy is generated in the temperate cyclone belts between about 40° and 60° of latitude in both hemispheres. From here swell waves move toward the equatorial zone and are reinforced by the trade winds blowing in similar directions. By extrapolation from what is known at the local scale, net long-term littoral drift systems may be expected to have a strong equatorward component. This trend should be most evident on relatively regular continental coasts closer to the source of big waves and least evident where coasts are not so strongly oblique to wave directions and where seasonal reversals take place because of monsoonal influences. Existing field studies support this deduction. In particular, net equatorward movement is well documented from Southern Hemisphere coasts of Australia, Africa, and South America.

During the Quaternary, when the disposition of the continents and position of the poles was not very different from the present, a long-continued tendency for coastal sediments to

WORLD NET SEDIMENT TRANSPORT

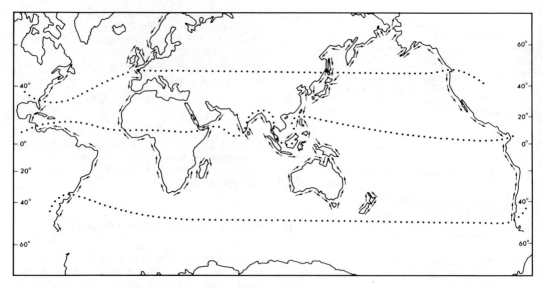

FIGURE 1. Major overall trends of net sediment movement around world coasts (after Silvester, 1962; Davies, 1972). Dotted lines indicate major changes in trend.

move equatorward may have contributed to the presence of large masses of unconsolidated sands in equatorial latitudes.

J. L. DAVIES

References

Davies, J. L., 1972. *Geographical Variation in Coastal Development.* London: Longman, 204p.

Silvester, R., 1962. Sediment movement around the coastline of the world, Paper No. 14 *in Proc. Conf. on Civil Engineering Problems Overseas,* London 289-304

Cross-references: *Beach Processes; Nearshore Hydrodynamics and Sedimentation; Sediment Transport; Wave Environments; Wind; Zetaform Bays.*

Y

YEAST—See FUNGI

Z

ZERO-ENERGY COAST

Zero-energy coast was defined by Tanner (1960) as the coast where the long-term average breaker heights are 3–4 cm or less, the wave energy is zero or almost zero, and there is no significant littoral transport of sand. It is different from other coasts where the effective littoral drift is low though the wave energy is high or moderately high.

The coastline between St. Marks and Tarpon Springs, Florida, is a classic example of the zero-energy coast. The coastal region consists of a broad, shallow, sand-floored marine shelf, tidal marshes, low-lying swamps, rare oyster reefs, and short strips of sandy beaches.

Factors reducing the wave energy, thus causing the zero-energy coast are: location of the coast in the upwind direction; wide-shallow offshore slope (less than 0° 4'); divergence of waves into the large coastal concavity, reducing offshore wave energy; old, submerged beaches, submarine meadows, and marshgrasses; and insignificant supply of new sediment by rivers.

Many ponds, small lakes, and protected open beaches or coasts fall into this category. Other examples are Lake Manitoba (Murali and Tanner, 1975) and the east coast of China between the Yangtze and the Shantung Peninsula (Keulegan and Krumbein, 1949). *Low-energy coasts* and zero-energy coast may be good examples of many interior geosynclinal coasts of the past.

R. S. MURALI

References

Keulegan, G. H., and Krumbein, W. C., 1949. Stable configuration of bottom slope in a shallow sea and its bearing on geological processes, *Am. Geophys. Union Trans.* **30**, 855–861.

Murali, R. S., and Tanner, W. F., 1975. Correlation of wave parameters in shallow lakes, *Zeitschr. Geomorphologie*, N.F., **19**, 479–489.

Tanner, W. F., 1960. Florida coastal classification, *Gulf Coast Assoc. Geol. Socs. Trans.* **10**, 259–266.

Cross-references: *High-Energy Coast; High-Latititude Coasts; Low-Energy Coast; Low-Latitude Coasts; Medium-Energy Coast; Sediment Transport; Wave Energy; Wave Environments; Waves.*

ZETAFORM BAYS

A *zetaform bay* has a shoreline that in plan view is asymmetrical, with a decreasing radius of curvature toward one end. The term *zeta curve* was first applied to such forms by Halligan (1906), who saw a resemblance to the Greek letter ζ (zeta), while Silvester (1960) applied the description *half-heart shape*.

Yasso (1965) studied the morphology of some zetaform bays and found that the shoreline curves corresponded closely to the logarithmic spiral, $r = e^{\Theta \cot a}$, where r = radius, Θ = angle of rotation, and a = angle between radius and tangent (constant for any given logarithmic spiral). These parameters are illustrated in Figure 1.

Zetaform bays are common on swell-dominant coasts where refracted persistent swell approaches obliquely and the coast consists of alternating rocky headlands and sandy bays. In such a situation the beach plan form becomes adjusted to refracted swell so that the fairly straight upcoast section of beach is more or less normal to incoming swell *orthogonals*,

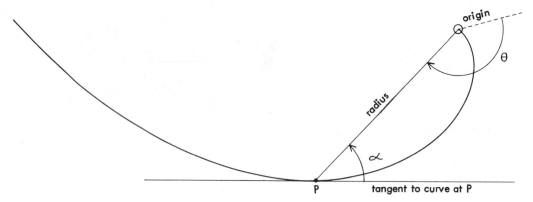

FIGURE 1. Parameters of logarithmic spiral curve.

this straight section being tangential to a curved part having a decreasing radius of curvature downcoast. Examples of zetaform bays include Half Moon Bay, California; Gold Coast, Queensland, Australia; and Mossel Bay, South Africa. (See also Figure 2.)

DAVID M. CHAPMAN

References

Halligan, G. H., 1906. Sand movement on the New South Wales coast, *Linnean Soc. New South Wales Proc.* **31**, 619-640.

Silvester, R., 1960. Stabilization of sedimentary coastlines, *Nature* **188**, 467-469.

Yasso, W., 1965. Plan geometry of headland-bay beaches, *Jour. Geology* **73**, 702-714.

FIGURE 2. A zetaform bay at Pearl Beach, New South Wales (reproduced by courtesy of the Deputy Under Secretary for Lands, New South Wales).

Cross-references: *Beach in Plan View; Coastal Erosion, Formations; Cuspate Spits; Headland Bay Beach.* Vol. III: *Cuspate Foreland or Spit.*

ZINGG SHAPE

It has often been observed and frequently demonstrated that beach gravels are flatter than river or glacial gravels—hence the term *shingle* (q.v.). Flatness is but one shape property, and particle shape, in turn, is but one of four common attributes that collectively make up particle morphology, the others being roundness, rollability, and surface texture. The larger types of sedimentary particles are usually considered in shape studies.

There is an extensive and long-term literature on the definition, origin, and significance of particle shape in both European languages and English. Many shape measures have been suggested, including those of Wadell, Wentworth, Cailleux, Krumbein and Sneed, and Folk; and there has been considerable confusion of terms; Lees (1964) provides a review and bibliography. While Flemming (1965) proposed that 13 parameters are necessary to define the form of a sedimentary particle, others have settled for fewer. Zingg (1935) is an example.

Zingg's classification of shape is based on simple ratios of a particle's three principal axes: a axis or long (L) diameter; b axis or intermediate (I) diameter; and, c axis or short (S) diameter. From these three easily measured lengths, six ratios can be made. Zingg selected the b/a and c/b pair, the indexes ranging from zero to unity. For the conditions $a > b > c$, he defined four shape classes on the basis of a fixed 2/3 ratio. These shapes can be called I, disc; II, sphere; III, blade; and IV, rod (Fig. 1).

The Zingg diagram has since been divided into different zones, using different class ratios. Terms and boundaries have been changed and further classes added. On it have been superimposed sphericity classes and roundness and sphericity values of a smooth ellipsoid. It has been criticized on grounds that the rodlike field (IV) is disproportionately small and that there are too few classes. And yet it is still the most widely used method for designating particle form. Perhaps the fact that investigators are frequently confronted with the need to classify large numbers of particles, and therefore require rapid and simple measurement and analytical techniques, has something to do with its survival. The basic Zingg plot is likely to endure, albeit in modified form.

ROGER F. McLEAN

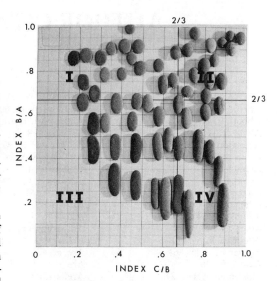

FIGURE 1. Array of pebbles superimposed on Zingg diagram to illustrate shape classes. All pebbles have a *B* axis diameter of 24–32 mm and were collected by David Harrowfield and the author in thirty minutes from the beach on the south bank of the Rakaia River mouth, Canterbury, New Zealand.

References

Flemming, N. C., 1965. Form and function of sedimentary particles, *Jour. Sed. Petrology* **35**, 381-390.

Lees, G., 1964. A new method for determining the angularity of particles, *Sedimentology* **3**, 2-21.

Zingg, Th., 1935. Beitrage zur Schotteranalyse, *Schweizer. Mineralog. u. Petrog. Mitt.* **15**, 39-140.

Cross references: *Beach Material; Cobble; Pebble; Sand, Surface Texture; Sediment Analysis, Statistical Methods; Sediment Size Classification; Shingle and Shingle Beach.*

AUTHOR CITATION INDEX

Abbott, A. T., 172
Abbott, I. A., 687
Abbott, R. T., 555
Abele, R. W., 818
Accerboni, E., 346
Acelrod, D. I., 222
Achituv, Y., 84
Adam, P., 409
Adams, D. A., 309
Adams, J., 537
Adams, J. E., 422
Adams, W. M., 846
Adelberg, E. A., 720
Adeyemo, M. D., 861
Adie, J.R., 46
Agadzanin, A. K., 627
Agassiz, L., 132
Ager, D. V., 179
Agershou, H. A., 250
Ahlmann, H. W., 799
Ahlstrom, E., 245
Ahmad, E., 37, 484, 486
Aibulatov, N. A., 668, 748, 861
Ainsworth, G. C., 440
Airy, G. B., 250, 748, 869
Alavi, M., 846
Alekhin, V. V., 168
Alessi, A. H., 147
Alestalo, J., 478
Alexander, C. S., 31, 222, 235, 762
Alexander, L., 841
Allan, T. D., 102
Allen, E. A., 53, 762
Allen, J. R. L., 31, 120, 366, 397, 431, 549, 567, 697, 717, 762, 792
Allen, P. H., 191
Aller, R. C., 614
Allersma, E., 207
Almagor, G., 97
Altinvilek, H. D., 164
Alvarino, A., 202
American Association of Petroleum Geologists, 423
American Society of Civil Engineers, Task Force on Bed Forms in Alluvial Channel, 164
American Society of Photogrammetry, 639, 684
Amiel, A., 37
Ammar, M. Y., 16
Ampferer, O., 453
Anan, F. S., 531
Andersen, B. G., 389, 799
Anderson, D. T., 491
Anderson, F. M., 770
Anderson, R., 37
Anderson, R. R., 684
Anderson, T. W., 337

Anderson, W. W., 400
Andrews, A. R., 604
Andrews, J. T., 61, 388, 464, 498, 655, 798
Andrews, P. B., 321, 531
Andrews, R. W., 470
Anters, E., 474
Antonakis, J., 608
Anwar, Y. M., 97
Arad, A., 97
Araya, R., 47
Araya-Vergara, J. F., 150, 301, 520
Arber, E. A. N., 231
Are, F., 478
Are, F. E., 61
Arehart, J. L., 813
Arnaud, P. M., 43
Arndorfer, D., 367
Aronow, S., 513
Art, H. W., 716
Arthur, R. S., 694, 734, 866, 871
Arunachalam, B., 486
Aschenbrenner, S. E., 52, 53
Asensio Amor, I., 678
Asgaard, U., 385
Ashraf, A., 32
Atkinson, L. P., 538
Atterberg, A., 710
Au, S., 716
Aubert, M., 841
Aubert de la Rue, E., 200
Aufrère, L., 222
Augelli, J. P., 201
Ault, W. V., 520
Australian UNESCO Committee for Man and the Biosphere (MAB), 328
Averintsev, V. G., 780
Avery, W. H., 815
Avignone, N., 717
Avnimelech, M. A., 97
Awramik, S. M., 354
Ayala, F. S., 636
Ayyad, M. A., 16
Axelrod, D. I., 533, 770

Baaker, W. T., 256
Baarghorn, E. S., 440
Bagnold, R. A., 122, 142, 242, 531, 549, 748
Bahr, L. M., Jr., 681
Bailey, H. P., 222, 230, 533, 770
Baines, B. G. K., 456
Baird, W. F., 862
Baisley, J. B., 264
Baker, P. E., 47
Baker, R. A., 557
Bakler, N., 97

Bakus, G. J., 288
Balasingam, E., 76
Ball, M. M., 2
BALTICA, 788
Bandoni, R. J., 635, 650, 687
Bandy, O. L., 433
Banks, R. S., 875
Banu, A. C., 473
Barash, Al., 85
Barbaza, Y., 185
Barber, N. F., 803
Barbour, M. G., 592
Barcilon, A. I., 690
Bardach, J. E., 49
Bardin, V. I., 47
Barnes, F. A., 803
Barnes, P. W., 61
Barnes, R. D., 40, 64, 202, 332, 555, 659
Barnes, R. S. K., 409
Barr, L. G., 636
Barrell, J., 567
Barrett, B. B., 400
Barrett, D. R. M., 31
Barrington, E. J. W., 669
Barson, M. M., 109, 119, 848
Bartelli, L. J., 762
Bartrum, J. A., 579, 857, 877
Barwis, J. E., 488, 818
Bascom, W. N., 141, 156, 158, 531
Bass, G., 54
Basse, R. Z., 133
Bate, R., 321
Batelle Memorial Institute, 610
Bates, C. C., 367, 513
Bates, M., 200
Bates, R. L., 338, 537, 802, 807
Bathurst, R. G. C., 342
Battistini, R., 16, 31
Battjes, J. A., 694, 871
Baulig, H., 647
Bauman, D., 97
Bayer, A. W., 16
Beard, J. S., 200
Beau, Ch., 815
Beaujeu-Garnier, J., 374
Beaumont, L. E. de, 132, 505
Becker, R. D., 147
Beckinsale, R. P., 295
Bee, A. G., 32
Beeftink, W. G., 409
Behrendt, J. C., 31
Belknap, D. F., 53
Bemmelen, R. W. van, 81
BenAvraham, Z., 97
Ben-Eliahu, M. N., 85
Benigno, J. A., 684
Bennett, I., 109

Benson, W. N., 333
Bentley, C. R., 47
Benton, W., 484
Bentor, Y. K., 97
Berg, R. R., 321
Berggren, W. A., 648
Bergquist, P. R., 654
Bergström, R., 462
Bernard, H. A., 134, 710
Berner, R. A., 614
Bernstein, L. B., 822
Berquist, C. R., 467, 875
Berry, A. J., 75
Berry, L., 31, 81
Berry, R. W., 356
Berryhill, H. L., 601
Bertrand, G., 473
Bessette, D., 289
Bettenay, E., 763
Betzer, P. R., 32
Beutow, D. E., 403
Bianchi, A. J. M., 762
Bicker, R. D., 321
Bierhorst, D. W., 393, 454
Bigarella, J. J., 147, 242, 321, 770
Biggs, H. E. J., 321
Biggs, R. B., 133, 397, 400
Bigler, A. B., 842
Bijker, E. W., 719
Billings, W. D., 171, 310, 716
Birch, M. C., 491
Birch, W. R., 16
Bird, E. C. F., 38, 119, 132, 156, 295, 339, 398, 513, 561, 580, 645, 658, 721, 722, 748, 764, 848, 856, 857
Bird, J. B., 177, 841
Birot, P., 877
Biscoe, C., 31
Black, M., 36
Black, W. L., 562
Blackburn, G., 762
Blackley, M. W. L., 751
Blackman, D., 54
Blackwelder, E., 805
Blair, T. A., 881
Blanchard, A. H., 813
Blatt, H., 321
Bloeser, B., 37
Bloom, A. L., 222, 389, 627
Bloomfield, C., 762
Bluck, B. J., 122
Blumer, M., 245
Boardman, R. S., 654
Bodere, J. Cl., 792
Boersma, J. R., 321, 549
Bold, H. C., 403, 687
Boldyrev, V. L., 748
Boltovskoy, E., 433
Bond, R., 762
Bondesen, E., 560
Boney, A. D., 634
Bonnin, J., 31
Boodle, L. A., 309
Boolootian, R. A., 385
Boon, J. D., III, 663, 855
Boothroyd, J. C., 144, 353, 531, 663

Borchert, H., 423
Bordeau, P. F., 309
Borengasser, M., 230
Börgel, R., 770
Borgese, E. M., 324
Borichansky, L. S., 367
Bormann, F. H., 716
Boschung, H. T., 209
Bosence, D. W. J., 681
Boucart, J., 759, 806
Bouma, A. H., 632
Bourliere, F., 200
Bourrelly, P., 687
Boury-Esnault, N., 654
Bowen, A. J., 138, 142, 148, 158, 272, 337, 386, 531, 567, 580, 691, 694
Bowers, C. E., 33
Bowers, N. A., 643
Bowles, J. E., 640
Bowman, H. H. M., 309
Boyce, S. G., 171, 309, 716, 802
Boyd, M. B., 853
Boysen Jensen, P., 245
Bozeman, R. N., 531
Bozhich, P. K., 668
Braaten, N. F., 728
Bradley, W. C., 156
Bradshaw, J. S., 763
Bradshaw, M. P., 139
Brahtz, J. F., 328, 377, 841
Braitsch, O., 423
Branner, J. C., 796
Brashear, H. R., 568
Braun, M., 37
Breckon, C. J., 592
Breedlove, D. E., 200
Breier, A., 473
Brenninkmeyer, B. M., 337, 353, 861
Bretschneider, C. L., 377, 846
Brett, J. R., 814
Brice, W. C., 102
Bricker, O. P., 223
Bridges, P., 321
Brien, P., 183, 654, 669
Briggs, L. I., 423
Briggs, S. R., 32
British Standards Institution, 328, 455, 632, 757
Broad, A. C., 57
Brochu, M., 177
Broecker, W. S., 389, 770
Broggi, J. A., 770
Bromley, R. G., 385, 879
Brothwell, D., 543
Broussard, M. L. S., 358, 368
Brown, C., 740
Brown, G. W., Jr., 200
Brown, J. E., 712
Brown, L. F., 235
Brown, N. J. F., 520
Brown, R. G., 322
Brown, R. W., 33
Brown, W. E., 710
Brush, B. M., 158, 328
Bruns, E., 533
Bruun, P. M., 142, 150, 180, 256, 267, 377, 488, 567, 615, 656, 668, 674, 686, 712, 748, 821, 823, 879
Bryan, K., 800, 806
Bryden, J. M., 841
Buch, L. v., 679
Buchanan, R. E., 720
Budanov, V. I., 301
Buick, R., 354
Bülow, K., 668
Burdon, J. F., 400
Burg, M. E., 592
Burgess, F. J., 140
Burk, C. A., 453
Burke, K., 453
Burkholder, P. R., 309
Burne, R. V., 560
Burns, G. W., 528
Bury, H., 207
Bush, P. R., 321, 704, 819
Busson, G., 31
Butcher, R. W., 345
Butler, J. N., 806
Butler, P., 309
Butzer, K. W., 31, 185, 543
Buurman, P., 762
Byrne, J. V., 207, 224, 878
Byrne, R. J., 338, 606

Cabot, J., 31
Cacchione, D. A., 492
Caddle, R. D., 3
Cahet, G., 762
Cailleux, A., 38, 47, 524, 712
Cake, E. W., Jr., 647
Calder, D. M., 109
Calhoun, F. G., 762
California Division of Highways, 187
Calvin, M., 37
Cameron, B., 133
Campbell, H. C., 643
Carbonnel, J. P., 81
Carder, K. L., 32
Carey, A. E., 47
Carlisle, R. J., 481
Carlsen, N. A., 875
Carlton, J. M., 848
Carlton, J. T., 40
Carothers, Z. B., 403
Carr, A. P., 147, 148, 158, 295, 739, 748, 751
Carriker, M., 699, 876
Carrington, A. J., 31
Carson, C. E., 615
Carstens, M. R., 164
Carter, G. B., 337
Carter, R. W. G., 356, 543, 750
Carter, T. G., 138
Carter, V., 684
Case, F. N., 568
Cassie, R. M., 574
Casson, L., 54
Castahano, J., 148
Castañares, A. A., 321
Cathles, L. M., III, 498
Cato, I., 462
Challis, D. A., 621

Chamberlain, B. B., 631
Chamberlain, C. J., 355
Chamberlain, T. K., 743
Chamberlin, T. C., 206
Chambers, R., 388, 677
Champion, H. G., 484
Chandy, M., 484
Chaney, C. A., 460
Chapman, D. J., 210, 403
Chapman, S. L., 762
Chapman, V. J., 16, 75, 109, 200, 210, 242, 285, 288, 309, 378, 403, 409, 533, 574, 593, 634, 687, 706, 708, 716, 762
Chappell, J. M. A., 139, 388, 389, 733
Charlesworth, J. K., 388, 428, 537, 677
Charlier, R. H., 806, 815, 822, 841, 842
Charnell, R. L. 568
Chatton, E. P. L., 673, 720
Chebataroff, J., 770
Cheetham, A. H., 654
Cheetham, G., 663
Chen, T. P., 75
Cheng, L., 285, 491, 492
Cheng, T. C., 647
Chew, E., 75
Childress, J. J., 167
Chorley, R. J., 295
Chowlert, G., 31
Chrysler, M. A., 309
Chu, V. H., 872
Chuang, S. H., 75
Chubb, L. J., 627
Churchward, H. M., 762, 763
Chuzhmir, A. A., 668
Clapp, C. H., 338
Clapp, W. F., 699
Clark, J. R., 706, 841
Clark, L., 533
Clark, R. B., 40
Clarke, A. R. P., 762
Clarke, F. E., 264
Clarke, F. W., 473
Clarke, L. D., 109
Clarke, R. T., 632
Claude, G., 815
Clausen, H. B., 464
Claypool, G. E., 538
Clayton, M. N., 109
Clements, F. E., 310, 780, 802
Clifton, H. E., 321, 531
Cloud, P. E., Jr., 142
Cloudsley-Thompson, J., 200
Clovis, J. F., 593
Coastal Engineering Research Center, 145, 159, 164, 179, 183, 187, 250, 256, 259, 267, 378, 388, 425, 438, 445, 456, 499, 551, 605, 633, 640, 668, 674, 698, 712, 721, 734, 798, 866
Coates, D. R., 132, 295, 851
Cochrane, G. R., 579
Cohn, B. P., 481
Coker, W. C., 309

Cole, B. J., 841
Cole, G. A., 754
Coleman, C. G., 639
Coleman, J. M., 61, 321, 358, 368, 398, 518, 559, 568, 770, 819, 821
Collier, A., 400
Collins, B. W., 424
Collins, J. D., 492
Colquhoun, D. J., 133
Conard, H. S., 309
Connell, J. H., 702, 802
Conrad, V., 756
Conybeare, C. E. B., 537
Cook, D. O., 531, 694
Cook, P. J., 207
Cooke, C, W., 132, 187
Cooke, H. B. S., 648
Cooke, R., 770
Cooper, H. H., Jr., 470. 800
Cooper, W. S., 242, 309, 593
Cooperman, A. I., 672
Cooray, P. G., 81
Coover, J. R., 762
Coque, R., 31
Corliss, J. B., 123, 243
Cornaglia, P., 748
Cornick, H. F., 551, 879
Cornish, V., 692
Cotet, P., 473
Cotton, C. A., 222, 231, 424, 445, 558, 579, 691
Couch, E. B., 309
Coultas, C. L., 763
Coulter, J. K., 762
Countryside Commission, (U. K.), 314
Courtois, G., 743
Coutant, C. C., 814
Cowden, R. R., 654
Cowell, P., 139
Cowles, H. C., 309, 716, 802
Cox, D. C., 846
Craig, A. K., 543, 770
Craighead, F. C., 733
Crampton, P. J. S., 321, 568
Crane, L., 426, 438, 682
Craven, J. P., 841
Creager, W. P., 698
Crenshaw, M. A., 167
Crickmore, M. J., 743
Critchfield, H. C., 881
Critchum, P., 495
Cronin, L. E., 489
Cronquist, A., 528, 644
Crook, K. A. W., 537
Crooke, R. C., 531, 862
Crosse, M., 31
Crowell, J. C., 453
Csanady, G. T., 568
Cummings, L., 736
Curran, H. A., 762
Curray, J. R., 132, 321, 429, 550, 567, 568, 627, 693, 733
Curtin, T. B., 734
Curtis, H., 200
Curtis, M., 806
Cushing, S. W., 486

Cushman, J., 671
Czudek, T., 633

Dagodag, W. T., 265
Daigle, E. E., 710
Daily, J. W., 187
Dakin, W. J., 109, 332
Dal Cin, R., 337
Dales, R. P., 40
Dallimore, W., 643
Dalrymple, D. W., 36
Dalrymple, R. A., 142, 694
Dalrymple, R. W., 321
Dalton, F. K., 822
Daly, R. A., 82, 177, 627
Dan, Y., 98
Dana, J. D., 222
Danin, Z., 85
Dansereau, R., 175
Dansgaard, W., 464
D'Appolonia, E., 259
Darcy, H., 470
d'Arsonval, A., 815
Dartnall, A. J., 109
Darwin, C. R., 388, 537, 627
Daubenmire, R., 200
Davidson, C. F., 224
Davidson-Arnott, R. G. D., 138, 321, 691
Davies, D. K., 321
Davies, J. L., 31, 82, 119, 150, 222, 235, 295, 349, 397, 453, 461, 463, 524, 544, 570, 721, 722, 764, 792, 798, 823, 824, 854, 860, 881, 882
Davis, C. A., 309
Davis, D. W., 445
Davis, G. M., 75
Davis, G. R., 37
Davis, J. C., 337, 842
Davis, J. H., 466, 762, 848
Davis, R. A., Jr., 138, 141, 144, 156, 159, 337, 338, 353, 478, 531, 568, 663, 838, 881
Davis, S. J., 604
Davis, W. M., 295, 438, 445, 513, 627
Davy, A. J., 409
Davy, J. B., 309
Dawson, C. E., 209, 400
Dawson, E. W., 621
Dawson, E. Y., 208, 528, 634, 635, 687
Day, J., 40
Dayton, L. B., 43
Dayton, P. K., 43, 702
Dean, B., 400
Dean, R. G., 250, 748, 875
Dean, W. E., 37, 681, 756
de Boer, J., 32
Deelman, J. C., 296, 762
Deevey, E. S., Jr., 175, 474
Defant, A., 816, 837, 838
Defant, F., 230
Defense Mapping Agency Hydrographic Center, 161, 722
De Geer, G., 462, 474

de Heinzelin, J., 537
De Jong, T. M., 592
de la Beche, H. B., 677
Delaney, P. J. V., 200, 770
Delany, A. C., 3
Dell, R. K., 43
Dellow, V., 574
del Moral, R., 593
Demangeon, P., 223
Demarest, D. F., 121, 550
DeMartonne, E., 849
de Matos, J. E., 31
Demek, J., 633
Dent, O. F., 119, 645, 856, 857
Denton, G. H., 464
de Oliveira, J. T., 31
de Raaf, J. F. M., 321
Deryugin, K. M., 780
DeSanto, R. S., 201, 533
Det Norske Veritas, 609
Deuser, W. G., 102
Devoy, R. J. N., 420
de Vries, G., 869
DeVries, M., 743
DeWall, A. E., 656
Dewey, J. F., 31, 453
De Wiest, R. J. M., 446, 470
Dexter, R. W., 309
Dickas, A. B., 264
Dickenson, K. A., 601
Didge, J. D., 345
Dietrich, G., 749
Dietz, R. S., 156, 453, 639, 748, 855
Dill, W. A., 49
Dillon, W. P., 132
Dingle, R. V., 31
Dionne, J.-C., 206, 448, 478, 481, 537, 558, 560, 756, 795
Disraeli, D. J., 593
Divoky, D., 869
Dixon, P. S., 687
Dobson, R. S., 337
Dodin, V. V., 668
Doering, J. A., 677
Doflein, F., 400
Dolan, R., 31, 132, 142, 200, 222, 230, 235, 242, 264, 425, 691, 770
Dominick, T. F., 693
Donaldson, A. C., 818
Donnellan, R. A., 264
Donovan, D. T., 333
Donovan, R. N., 795
Doodson, A. T., 542, 837
Doornkamp, J. C., 712
Dor, I., 681
Dornhelm, R. B., 97
Doudoroff, M., 720
Dougherty, E. C., 67
Dovier, A., 749
Dozier, J., 481
Drach, P., 669
Dragesco, J., 671
Drake, C. L., 102, 453
Draper, L., 748
Dronkers, J. J., 816, 837
Drooger, C. W., 185
Drucker, D. M., 606

Druehl, L. D., 634, 635
Drury, W. H., 802
Duane, D. B., 66, 132, 328, 489, 568, 743
Duarte, G. M., 147, 321
Dubertret, L., 102
Dubois, G., 537
Dubois, R. N., 142, 180
Ducsik, D. W., 841
Dumas, F., 54
Dunbar, M., 57
Duncan, J. R., 156, 353, 550, 855
Dunham, J. W., 460
Dunham, R. J., 186
Dunlop, J., 354
Dunn, A. J., 295
Durden, C. J., 647
Dzhunkovsky, N. N., 668
Dzulinski, S., 122

Eagleson, P. S., 148, 391, 748, 875
Eales, N. B., 85
Eaton, F. H., 309
Eckart, C., 386
Eden, J. J., 250
Edmonds, S. J., 109, 385, 758
Edmondson, W. T., 64
Edwards, A. B., 857, 877
Eggimann, D. W., 32
Eggleston, J. R., 681
Egler, F. E., 310, 706, 802
Eilers, H. P., 593
Einsele, G., 31, 37
Einstein, H. A., 391
Eipper, A. W., 245
Eitam, Y., 97
Ekman, S., 75, 285
Ekman, V. W., 568
El-Ashry, M. T., 639, 656, 657, 818
El Dashlouty, S., 31
Elder, R. B., 400
Eleuterius, L. N., 400
Elliot, D. H., 47
Elliott, T., 138, 321
El Samni, E. A., 391
Eltringham, S. K., 64, 200, 285
Emberger, L., 643
Embleton, C., 427, 428, 435, 447, 633, 758
Emelyanov, E. M., 97
Emery, K. O., 32, 97, 102, 121, 139, 156, 168, 342, 389, 502, 537, 550, 627, 663, 682, 693, 733, 752, 861, 877
Emmel, F. J., 321, 568
Endean, R., 76, 109, 322
Energy Research and Development Administration, 815
Engel, I., 102
Entsminger, L. D., 875
Environmental Protection Agency, 610, 611
Epstein, E., 459
Erol, O., 52
Eronen, M., 388
Estes, J. E., 684
Etchichury, M. C., 770

Etherington, J. R., 716
Ethington, R. L., 141
Ethridge, F. G., 321
Evans, G., 102, 321, 704, 819
Evans, I. S., 38, 709, 806
Evans, O. F., 132, 138, 142, 349, 513, 531, 692
Eveleigh, D. E., 716
Everett, D. K., 368
Everhart, W. H., 245
Everts, C. H., 862
Evitt, W. R., 673
Ewing, G. C., 123, 243

Faegri, K., 474
Fairbridge, R. W., 2, 82, 119, 168, 230, 295, 388, 429, 453, 464, 465, 474, 502, 537, 542, 580, 627, 657, 677, 705, 729, 733, 762, 877
Falcon, N. L., 102
Faller, A. J., 568
Fang, C. S., 855
Farnham, W. F., 285
Farquhar, O. C., 513
Farrand, W. R., 655
Faure-Fremiet, E., 671
Faure-Muret, A., 31
Fausak, L. E., 693
Fauvel, P., 85
Favori, J. C., 763
Fay, P., 354
Felbeck, H., 167
Fell, P. E., 654
Fenchel, T., 167, 495, 660
Fenneman, N. M., 513, 568
Ferm, J. C., 142, 691
Ferre, Y. de, 525
Ferreyra, R., 200
Field, M. E., 132
Filloux, J., 156, 531
Finkl, C. W., Jnr., 762
Finn, A. A., 460
Fischer, A. G., 568
Fischer, J. A., 264
Fischer, J.-C., 841
Fisher, J. J., 132, 180, 264, 438, 445, 631, 818
Fisher, R. V., 520
Fisher, W. L., 133, 235, 321
Fisk, H. N., 2, 124, 368, 601, 655
Fite, R. C., 881
Flemming, N. C., 148, 885
Fletcher, M. J., 491
Flint, R. F., 206, 462, 474, 648, 677, 798, 879
Floodgate, G. D., 604
Flores-Silva, E., 47
Focke, J. W., 681
Fogg, G. E., 354
Folger, D. W., 505
Folk, R. L., 148, 186, 342, 502, 522, 557, 736, 738, 740
Fonda, R. W., 593
Forbes, D., 235
Fosberg, F. R., 75
Foster, A. S., 355

Foster, J. F., 693, 861
Foster, W. A., 492
Fox, F. L., 264
Fox, W. T., 138, 144, 337, 338, 353, 531, 568, 663, 838, 881
Francis-Boeuf, C., 721, 759, 806
Frautschy, J. D., 871
Frazier, D. E., 368
French, H. M., 437
Frenzel, G., 709
Frey, D. G., 474
Frey, R. W., 184, 699, 876
Fried, I., 97
Friedman, G. M., 37, 48, 321, 557, 632, 736, 748
Friedman, J. M., 264
Friedrich, H., 167, 794
Frigo, A., 814
Fritsch, F. E., 403, 635, 687
Fritts, H. C., 465
Frohne, W. D., 593
Fry, W. G., 654
Frye, D., 814
Fuenzalida, H., 770
Fukui, Y., 846
Fuller, H. J., 403
Fuller, R. B., 842
Furon, R., 31

Gadd, P. E., 568
Gadel, F., 762
Gagliano, S. M., 321, 358, 368, 559
Gaillard, D. B. W., 879
Gaillard, D. D., 468
Galem, J. F., 550
Gallagher, J. L., 288
Galloway, W. E., 321, 368
Galtsoff, P. S., 681
Galvin, C. J., Jr., 606, 656, 869
Gammelsrod, T., 568
Ganong, W. F., 309
Gantt, E., 345
Garner, H. F., 879
Garrett, C., 492
Garrison, C. J., 250
Gary, J. S., 614
Gary, M., 33, 123, 176, 328, 445, 455, 470, 503, 551, 558, 559, 632, 640, 705, 757, 804, 855
Gashus, O. K., 837
Gaskell, R. E., 337
Geary, E. L., 684
Gebelein, C. D., 322
Gee, E. R., 486
Geer, R. L., 609
Geike, A., 295
Geikie, J., 677
Gellert, J. F., 301
Genser, H., 31
George, C. J., 97, 102
Gerritsen, F., 488, 748, 821, 823
Ghyben, B. W., 446
Gibbons, N. E., 720
Gibbs, C. F., 604
Gibbs, F. E., 621
Gibbs, R. J., 207
Gibrat, R., 822

Gibson, E. S. H., 765
Gibson, G. C., 740
Gibson, R., 570
Gibson-Hill, C. A., 75, 76
Gierloff-Emden, H. G., 191
Gifford, E. M., Jr., 355
Gilat, E., 85
Gilbert, G. K., 132, 295, 368, 513, 679, 807, 855
Gill, A. E., 728
Gill, E. D., 2, 119, 388, 677, 755, 757
Gill, W. D., 560
Gillespie, M. C., 400
Gillson, J. L., 545
Gimingham, C. H., 519
Ginsburg, N., 324
Ginsburg, R. N., 321, 322, 493, 704, 710
Ginzburg, A., 97
Girdler, R. W., 102
Giresse, P., 31
Gjessing, J., 428
Glasby, J., 109
Gleason, R., 148
Glodowski, C. W., 862
Glover, R. E., 800
Godfrey, M. M., 132, 171, 243, 716
Godfrey, P. J., 132, 171, 243, 593, 716
Godin, G., 837
Gofna, A., 97
Gojdics, M., 403
Goldberg, E. D., 328
Goldberg, R. R., 437
Goldring, R., 321
Goldsmith, V., 97, 243, 321, 337, 338, 606, 716
Golik, A., 97
Golikov, A. N., 780
Gómez-Pompa, A., 201
Gontcharoff, M., 570
Gonzáles-Ferrán, O., 47
Goodman, L. A., 337
Goodman, P. J., 378
Goodspeed, T. H., 201
Goodwin, R. H., 802
Gopinadha Pillai, C. S., 76, 82
Gordon, H. H., 684
Gordon, J. R., 815
Gorgy, S., 85
Gorsline, P. S., 37
Gorycki, M. A., 142, 691
Gosner, K. L., 64
Gosselink, J. G., 516
Gottlieb, E., 85
Goudie, A., 709
Gould, H. R., 120, 207, 368, 602
Gove, P. B., 445
Govorukhin, V. S., 168
Grabau, A. W., 186, 423
Grabert, G. F., 461
Graf, W. H., 391, 397
Grancini, G., 349
Granlund, E., 457
Grant, U. S., 855
Grass, A. J., 391
Grassé, P., 647
Grave, C., 518

Graves, C. D., 684
Gray, J., 140
Gray, T. J., 837
Graybill, F. A., 337
Grechishchev, E. K., 668
Green, C. D., 321
Green, J., 497
Greene, H. G., 478, 500, 550
Greenwood, B., 138, 139, 321, 691
Gregory, A., 61
Gregory, J. W., 427, 453, 795
Grewe, F., 16
Griffin, G. M., 223
Grigoriev, N. F., 789
Grim, R. E., 223
Grimm-Strele, J., 743
Grinnell, R. S., 795
Griswold, A. H., 378
Gross, M. G., 387, 466, 853
Groves, G. W., 522, 803
Gruber, M., 813
Grumman Ecosystems Corp., 264
Gudelis, V., 569, 630
Gugnyaev, Ya. E., 748
Guilcher, A., 31, 82, 102, 150, 156, 388, 420, 447, 531, 647, 678, 792, 849, 877
Guiler, E. R., 201
Gulland, J. A., 76
Gulliver, F. P., 186, 295, 349, 506, 534, 839
Gumbel, E. J., 467
Gunter, G., 209, 400
Gurjanova, E., 780
Gutenberg, B., 542, 729
Gutmann, W. F., 614
Guza, R. T., 142

Haas, G., 85
Haff, K. W., 568
Hageman, B. P., 465
Hagen, G. M., 322
Hagmeier, E., 794
Häikiö, J., 478
Hails, J. R., 120, 132, 148, 158, 295, 328, 467, 736
Halden, B. E., 462
Hall, J. K., 97
Hall, J. R., 538
Hall, V. L., 762
Hallberg, R. C., 763
Hallberg, R. O., 167
Halligan, G. H., 884
Halsey, S. D., 133
Hamelin, L. E., 448
Hammitt, F. G., 187
Han, G. A., 568
Hancock, J. L., 609
Hannau, H. W., 201
Hannon, N. J., 109, 533
Hansen, C. L., 31
Hansen, K., 435
Hansen, W., 541
Hanshaw, B. B., 264
Hanson, H. C., 593
Hantzchel, W., 550
Haq, B. U., 648

Harbaugh, J. W., 337
Hardcastle, P. J., 148
Hardin, G., 328
Hardy, J. R., 337
Hardy, N. E., 164
Harleman, D. R. F., 541, 719, 815
Harpes, J. R., 478
Harris, C. R., 145
Harris, D. L., 509
Harris, I. M., 560
Harris, L. E., 648
Harris, R. L., 328
Harris, S. E., 206
Harris, W. D., 685
Harrison, C. M. H., 245
Harrison, F. W., 654
Harrison, R. S., 763
Harrison, W., 663, 855, 866
Harroy, J. P., 200
Harshburger, J. W., 309
Hartford, F., 657
Hartog, C. den, 528
Hartshorn, J. H., 631
Hassan, F., 31
Hasselmann, K. F., 522, 803
Hassouna, M. G., 716
Hathaway, E. S., 310
Haulot, A., 842
Hawley, A. H., 374
Hay, R. F., 628
Hayden, B. P., 31, 200, 222, 230, 235, 770
Hayden, R., 425
Hayes, D. E., 32
Hayes, M. O., 132, 144, 164, 321, 353, 397, 531, 606, 663, 797, 798, 818, 821, 851
Hays, J., 762
Hecht, A., 97
Heckel, P. H., 681
Hedberg, H. D., 559
Hedgpeth, J. W., 43, 285, 400, 497, 681
Hedley, C., 109, 339
Heezen, B. C., 484, 486
Hellstrom, B., 509
Hemphill, W. R., 685
Hendershott, M. C., 349, 434, 838
Henderson, D. A., 201
Hendrickson, J. A., 250, 869
Hendrickson, J. R., 76
Henry, H. R., 800
Henry, V. J., Jr., 322, 349, 550, 697
Hentschel, E., 654
Herak, M., 502
Herbich, J. B., 33, 379
Herm, D., 37
Hernández Pacheco, F., 678
Heron, S. D., Jr., 133
Herring, J. P., 289
Herron, W. J., 328
Hertweck, G., 614
Hervé, F., 47
Herzberg, B., 446
Hesse, R., 400
Heward, A. P., 322
Hey, R. W., 752

Hickman, C. P., 64, 67, 176, 669
Hicks, S. D., 729
Higgs, E., 543
Hijum, E. van, 661
Hillaire-Marcel, C., 388, 465, 677
Hills, E. S., 119, 224, 393, 674, 854, 877
Hilmy, M. E., 97
Hinde, H. P., 309
Hinds, J., 698
Hine, A. C., 489, 792
Hino, M., 139, 694
Hinze, J. O., 378
Hirose, H., 635
Hjulström, F., 397
Hobbs, C. H., III, 606
Hobday, D. F., 31
Hodgkin, E. P., 580, 877
Hodgson, J. M., 537
Hodgson, W. D., 133, 602
Hoek, C. van den, 635
Hoffman, P., 322
Hoffmeister, J. E., 763
Hogan, C., 32
Holden, J. C., 453
Hollenberg, G. J., 687
Holliday, B., 717
Holm, C. H., 183
Holm, D. A., 704
Holme, N. A., 245
Holmes, C. W., 601
Holtedahl, O., 799
Holthuis, L. B., 85
Homma, M., 531
Hommersand, M. H., 687
Hopley, D., 119, 388
Hoppe, H. A., 635, 687
Hori, N., 31
Horikawa, K., 695
Horn, M. E., 762
Hornberger, G., 200, 222, 230, 235, 770
Horodyski, R., 37
Horowitz, A., 97
Hospers, J., 31
Houbolt, I. I. H. C., 102
House, H. D., 309
Howard, C. S., 740
Howard, J. D., 322, 568
Howard, R. A., 201
Howe, H. V., 207
Hoyt, J. H., 2, 132, 322, 349, 550, 697, 736
Hsu, S.-A., 61, 158, 230
Hsu, T. L., 82
Hubbard, D. K., 489, 818
Hubbard, G. D., 420
Hubbert, M. K., 470
Hubbs, C. L., 201
Huden, J. C., 631
Hughes, G. C., 440
Hughes, R. N., 681
Hu-Lin, Li, 643
Humbert, F. L., 148
Hume, J. D., 61, 478
Humphrys, C. R., 445
Hunter, R. E., 321, 531, 550

Huntley, D. A., 139, 142, 158, 691
Hurlburt, H. E., 541
Hurtig, T., 174
Hussey, K. M., 615
Hutchins, L. W., 285
Hutchinson, G. E., 513
Hutner, S. H., 671
Hydraulics Research Station, 148
Hyman, L. H., 67, 183, 202, 332, 346, 385, 555, 647, 654, 758

Ignatiev, G. M., 627
Iida, K., 846
Imamura, A., 846
Imperial Gazetteer of India, 484
Ingham, A., 454, 825
Ingle, J. C., Jr., 141, 295, 862
Ingle, R. M., 400
Ingram, R. L., 223, 505
Inman, D. L., 31, 82, 97, 123, 132, 138, 142, 148, 156, 158, 222, 224, 243, 272, 328, 337, 368, 386, 453, 531, 537, 580, 691, 694, 734, 743, 748, 871
International Atomic Energy Agency, 814
Ionin, A. S., 698
Ippen, A. T., 148, 250, 295, 328, 379, 748, 866
Irvine, D. E. G., 285
Isaacs, J. D., 866
Isbell, R. F., 109
Iseki, H., 82
Israel Port Authority, 97
Issar, A., 97
Ito, K., 76
Itzhaki, Y., 97
Ivarson, K. C., 763
Iverson, H. W., 862
Iverson, J., 474

Jackson, A. B., 643
Jackson, J. A., 338, 537, 802, 807
Jacquet, J., 806
James, R. W., 250
James, W. R., 337
Jauzein, A., 31
Jefferson, C., 593
Jeffreys, H., 391
Jeffries, R. L., 409, 593
Jelgersma, S., 429, 733
Jennings, J. N., 119
Jessup, J. W., 763
John, B. S., 61, 463
John, C. J., 53
Johnson, A. F., 592
Johnson, D. L., 762
Johnson, D. S., 309
Johnson, D. W., 37, 121, 132, 142, 156, 222, 295, 309, 446, 468, 488, 505, 513, 534, 535, 550, 568, 602, 615, 663, 679, 686, 698, 754, 799, 800, 818, 839, 849, 855, 879
Johnson, J. H., 687, 806
Johnson, J. W., 250, 391, 748, 862, 866

Johnson, S. J., 464
Johnson, T., 57
Johnson, T. W., Jr., 440
Johnson, W. W., 201
Jolliffe, I. P., 271
Jones, C. F., 201
Jones, E. B. G., 440
Jones, H. S., 33
Jones, J. R., 132, 133
Jones, O. A., 76, 109, 322
Jones, R., 716
Jones, T. A., 736
Jonsson, I. G., 875
Jordan, G. F., 717
Joreskog, K. G., 337
Joustra, D. Sj., 256
Juday, C., 513
Judd, F. W., 593
Justin, J. D., 698
Jutson, J. T., 674, 856, 857

Kaczorowski, R. T., 615
Kaestner, A., 385, 647
Kafanov, A. I., 780
Kahn, J. S., 479, 481, 838
Kaiser, K., 295
Kaizuka, S., 82
Kalk, M., 16
Kamphuis, J. W., 164, 719
Kana, T. W., 321, 818
Kaplan, I. R., 538
Kaplin, P. A., 429, 627, 629, 757
Kareh, G. el, 102
Karlén, W., 464
Karlson, R. H., 436
Karužǎite, G., 630
Kashef, A. I., 470
Katz, A., 632
Katsui, Y., 47
Kaufman, A., 97
Kawalec, A., 763
Kayan, I., 52
Kaye, C. A., 580
Kear, D., 627
Kearney, T. H., 309
Keefe, C. W., 516
Keene, W. C., Jr., 681
Keith, J. M., 516
Kelletat, D., 502
Kellogg, F. B., 309
Kelly, J., 802
Kemmerer, A. J., 684
Kemp, A. F., 310
Kemp, P. H., 635
Kendall, C. G. St. C., 704
Kenk, R., 699
Kennedy, J. F., 164, 703
Kenney, N. T., 657
Kenny, R., 109
Kensley, B. F., 31
Kent, P. E., 31
Kenyon, K., 54, 803
Ketchum, B. H., 328, 400, 820, 842
Keulegan, G. H., 513, 719, 869, 883
Khalilov, A. I., 789
Khomitsky, V. V., 668
Kidson, C., 513, 658, 748

Kiener, A., 16
Kiknadze, A. G., 668
Kim, Y. C., 815
Kindle, E. M., 145, 697
King, C. A. M., 37, 38, 124, 133, 139, 141, 142, 148, 156, 158, 243, 295, 301, 337, 338, 349, 353, 374, 425, 427, 428, 435, 447, 468, 513, 542, 615, 633, 647, 692, 738, 748, 758, 799, 802, 803, 862, 879
King, R. H., 423
King, R. J., 109
Kinsman, B., 161, 250, 390, 750
Kinsman, D. J. J., 453, 704
Kirk, R. M., 531, 579
Kirkpatrick, J. B., 109
Kirtley, D. W., 681
Klein, G. de V., 322, 632, 851
Klovan, J. E., 337
Klug, K., 86
Knaggs, F. W., 687
Knapp, R. T., 187
Knaps, R. Ya., 668
Knight, R. J., 321
Knox, G. A., 109
Kobayashi, T., 453
Koch, G. S., Jr., 842
Koch, H., 57
Kochelson, E., 647
Kofoed, J. W., 180
Kofoed, L. H., 495
Koh, R. C. Y., 752, 869
Kohlmeyer, E., 440
Kohlmeyer, J., 440
Kohn, A. J., 76
Kohout, F. A., 800
Koike, K., 658
Kolb, C. R., 368, 602
Komar, P. D., 123, 139, 141, 142, 148, 156, 295, 337, 338, 391, 550, 691, 694, 695, 734, 869
Konai, T., 346
Kornicker, L. S., 763
Korteweg, D. J., 869
Kostyaev, A. G., 795
Kosugi, K., 82
Koteff, C., 631
Kouyoumontzakis, G., 31
Kraft, J. C., 52, 53, 133, 134, 135, 322, 733
Krebs, C. J., 636
Krebs, O. A., Jr., 729
Kreyzig, E., 557
Krinsley, D. H., 712
Krumbein, W. C., 301, 337, 421, 509, 557, 710, 738, 740, 883
Krynine, P. D., 710
Kudo, R. R., 671
Kudryashov, L. V., 168
Kuenen, Ph. H., 50, 82, 103, 142, 502, 560, 580, 819
Kuhl, H., 167
Kuhn, G. G., 602
Kumar, N., 133, 322
Kunselman, C. A., 568
Kurz, H., 310

Kussakin, O. G., 43
Kutina, J., 31
Kwon, H. J., 133
Kylin, H., 687

Laba, J. T., 822
Laborel, J., 770
Lachance, T. P., 32
La Cruz, C., 368
Ladd, H. S., 168, 342, 627
Ladd, J. W., 32
LaFond, E. C., 139, 156
Lakshim, G. J., 400
Lam, S-H., 794
Lambert, A. M., 516
Lanan, G. A., 142
Landes, K. K., 423
Landon, R. E., 103
Landry, A., 374
Lane, E. W., 710, 740
Langford-Smith, T., 120
Lapes, D. N., 33
Laporte, L. F., 322, 680
Lappalainen, A., 495
Latortue, G., 842
Larivière, A., 813
Larras, J., 33, 685, 748
Larson, R. L., 32
Lau, J. P., 690, 691
Laubenfels, M. W. de, 654
Lauff, G. H., 295, 322, 402
Lavelle, J. W., 568
Lavenia, A., 349
Laverdière, C., 478, 795
Lawrence, J. M., 385
Lawson, G. W., 16
Lawson-Kerr, C., 491
Lawton, F. L., 822
Learmonth, A. T. A., 486
Leatherman, S. P., 124, 133, 171, 243
Lebarbier, C. H., 822
LeBlanc, R. J., Sr., 133, 134, 368, 602, 753
Lecointre, G., 32
Lee, H. A., 389
Leedale, G. F., 403
Lees, G., 885
Lees, G. M., 102
Lefond, S. J., 545, 546
Legrain, D., 842
Le Méhauté, B. J., 250, 752, 869
Leontiev, O. K., 505, 518, 534, 535, 558, 627, 789. See also Leont'yev, O. K.
Leont'yev, O. K., 133, 222. See also Leontiev, O. K.
Leopold, L. B., 264
Le Pinchon, X., 32
Lerner-Seggev, R., 85
LeRoy, D. O., 207
Lessing, P., 264
Leu, T., 76
Levedev, V. L., 47
Levenson, E. B., 481
Lévi, C., 654
Levorsen, A. I., 753
Levring, T., 635, 687

AUTHOR CITATION INDEX

Levy, Y., 97
Lewellyn, R., 61
Lewin, R. A., 635, 687
Lewinsohn, Ch., 85
Lewis, A. J., 191, 518, 519, 684
Lewis, I. F., 310
Lewis, J. G., 822
Lewis, J. R., 285, 409
Lewis, T. E., 328
Lewis, W. V., 153
Ley, R. G., 877
Li, W-H., 794
Libby, W. F., 675
Lidén, R., 733
Lin, A., 869
Lindén, A., 462
Linder, D. H., 440
Lindner, M. J., 400
Lindstrom, M., 356
Ling, S. C., 259
Ling, S. W., 76
Lingsma, J. S., 420
Link, R. F., 842
Linthurst, R. A., 288
Lipkin, Y., 85
Lisitzin, E., 728, 825
Liu, P. L. F., 138
Lloyd, F. E., 310
Lloyd, R. M., 322
Lobeck, A. K., 438, 851
Lockman, W. O., 164
Loeblich, A. R., 433
Logan, B. W., 322, 376
Logvinenko, N. V., 322
Loiseaux, S., 635
Løken, O., 178
Lomas, J., 3
Lombard, J., 31
Lonard, R. I., 593
Longinov, V. V., 295
Longuet-Higgins, M. S., 337, 869
Longwell, C. R., 879
Lord, C. W., 710
Lortet, L. C., 473
Lowe, D. R., 123
Lowe, R. L., 158
Lowenstam, H. A., 385
Loya, B. R., 684
Lubchenko, J., 702
Lucas, G., 681
Lucas, J., 495
Lucke, J. B., 488, 818
Ludwick, J. C., 322
Lundqvist, G., 457
Lundqvist, I., 462
Lundqvist, J., 756
Lusczynski, N. J., 470
Lusting, L. K., 842
Lyell, C., 178, 448, 537
Lynd, L. E., 545
Lynn, W. C., 762

Mabesoone, J. M., 796
McAfee, R., Jr., 33, 123, 176, 328, 445, 455, 470, 503, 551, 558, 559, 632, 640, 705, 757, 804, 855
McAleer, J. B., 609

McAllister, R. F., 183
McArthur, W. M., 763
McBride, E., 557
McBride, E. F., 550
McCammon, R. B., 557
McCann, S. B., 61, 235, 478, 479, 481, 633
McClain, E. P., 684
McCombs, C. F., 728
McConnaughey, B. H., 289, 324, 542
McCormack, P. D., 426, 438, 686
McCormick, C. L., 679
McCormick, M. C., 377
McCowan, J., 869
McCrone, A. W., 632
McCullagh, M. J., 164, 337, 338
McCurrach, J., 528
Macdonald, G. A., 172, 520
MacDonald, H. C., 191, 519, 684
MacDonald, K. B., 593
McDougall, I., 627
Macfadyen, W. A., 102, 502
McFarlan, E., Jr., 207, 602
McGee, W. J., 133
McGill, D. A., 102
McGill, J. T., 222, 235, 533, 770
McGinnies, W. G., 201
McGowen, J. H., 321, 322
McGuirt, J. H., 207
McHarg, I. L., 264
MacIntire, W., 531
McIntire, W. G., 120, 133, 142, 691
Macintosh, D. J., 76
McIntyre, A. D., 245
Mackay, J. R., 61, 756
McKee, E. D., 322, 531
McKenzie, D. P., 32
Mackenzie, F. T., 223
McKenzie, P., 695
McKeon, J. B., 684
Mackin, J. G., 400
McKinney, T. F., 568
MacLaren, C., 388
McLarney, W. O., 49
McLaughlin, J. J. A., 671
McLean, R. F., 224, 456, 531
McManus, D. A., 710
McMaster, R. L., 32
Macmillan, D. H., 162, 837
Macnae, W., 16, 76, 109, 288, 533
MacNeil, F. S., 502
McRoy, C. P., 285
Madsen, H., 57
Magoon, O. T., 639
Maillard, C., 102
Major, C. F., Jr., 134
Maldonado, A., 368
Malloch, A. J. C., 3, 409
Manca, B., 346, 434. 749
Mann, K. H., 40, 660, 794
Manners, J. G., 378
Mantell, G. A., 537
Manton, S. M., 64
Manwell, R. D., 671
Maree, B. D., 32
Margolis, S., 712
Mariscal, R. N., 245

Markle, D. A., 685
Markov, K. K., 47
Marmer, H. A., 837
Marsden, J. R., 636
Marsh, D. B., 481
Marsh, G. P., 516
Marsh, W. M., 481
Marshall, J. K., 716
Marshall, P., 103
Martin, A. R. H., 120
Martin, L., 32
Martin, P. S., 465
Martin, W. E., 310, 593
Mary, G., 678
Masch, F. D., 250
Mason, C. C., 557
Mason, D. T., 57
Mason, S. L., 295
Masry, D., 85
Massel, St., 871
Mast, R. F., 550
Masuda, Y., 859
Mather, A. S., 235
Mather, K. F., 295
Mathew, P. M., 337
Mathews, R. K., 763
Mathieu, R., 681
Matlock, H., 846
Matlock, R. B., 846
Matsuishi, H., 842
Matthes, G. H., 740
Matthews, E. R., 256, 468, 668
Matthews, R. J., 389
Matthews, W. H., 328
Mauermeyer, E. M., 53
Maxwell, W. G. H., 76, 120, 339, 342, 682
May, J. P., 1, 506, 752, 869, 875, 876
May, V. J., 205
Mayer, D. A., 568
Mazor, E., 97
Mead, D. W., 470
Meade, R. H., 430, 763
Meagher, T. D., 763
Meckel, L. D., 368
Medeiros, C. A., 31
Medvedev, V. C., 627
Megalitsch, P. A., 64
Mei, C. C., 138, 872
Meighan, C. W., 543
Meigs, P., 376
Meistrell, F. J., 513
Melton, F. A., 602
Menge, B. A., 702
Mergeritz, M., 97
Merling, P., 86
Merrell, J. L., 400
Merriam, D. F., 337
Merrill, F. J. H., 133
Merrill, W. M., 337
Mesolella, K. J., 389
Metcalf, W. G., 368
Meyer, P. D., 141
Meyer, R. E., 295
Meyers, J. J., 862
Meyers, S. P., 440

Michaud, J-L., 374
Michelson, H., 97
Middleton, G. V., 123, 135, 321, 391, 550
Migniot, C., 719
Migula, W., 720
Mii, H., 531, 550, 757, 795
Mikhailov, V. N., 367
Milankovitch, M., 474
Miles, M., 235
Miller, D., 37
Miller, G. R., 522, 803
Miller, H., 765
Miller, L., 32
Miller, M., 574
Miller, M. C., 148, 391
Miller, R. L., 139, 337, 550, 838
Miller, W. R., 310, 706
Milliman, J. D., 32, 322, 342, 681, 733
Mills, E. L., 614
Miltner, F., 53
Minikin, R. R., 668
Mink, J. F., 800
Mitchell, A., 380
Mitchell, A. H., 32
Mitler, P. R., 139
Miura, A., 687
Moebius, K., 285
Mogridge, F. R., 391
Mogridge, G. Rm., 164, 719
Mohamed, A., 97
Moiola, R. J., 632
Molengraaff, G. A. F., 76, 82
Monahan, E. C., 3
Monin, A. S., 379
Monkhouse, F. J., 445
Monteiro Marques, M., 32
Moody, D. W., 568
Moody, R. A., 710
Mooers, N. K., 734
Moore, C. H., 502
Moore, G. W., 500
Moore, H. B., 201, 495, 497
Moore, J. C., 520
Moore, R. C., 64, 179, 322, 519
Morelli, C., 102
Morgan, J. P., 133, 322, 358, 368, 513, 559
Morisawa, M., 158
Morison, J. R., 250
Mörner, N.-A., 465, 729, 733
Morozov, L. A., 668
Morris, B. F., 806
Morris, I., 403
Morris, M., 716
Morris, W. D., 338
Morrison, J. R., 531, 862
Morrison, T. P., 400
Morskaja geomorphologia, 168
Morton, J. E., 555, 574, 621
Morton, R. A., 818
Moruzzi, U. L., 368
Moser, D., 201
Mosetti, F., 346, 349, 434, 749
Moshkovitz, S., 97
Moss, A. J., 123, 148

Moss, F. E., 328
Moul, E. T., 593
Mounteney, N., 420
Muir, R. O., 423
Mukundan, C., 76
Mulcahy, M. J., 763
Müller, G., 710
Muller, R. A., 230
Multer, H. G., 681, 763
Munk, W. H., 337, 492, 522, 695, 728, 802, 803, 866
Murali, R. S., 153, 697, 883
Murray, J., 256
Murray, J. W., 321, 433
Murray, R., 321
Murray, R. G. E., 720
Murray, S. P., 568
Mustoe, G. E., 38, 806
Myers, J. J., 183

Naegli, C., 720
Nagle, J. S., 321
Nakamura, M., 846
Nakano, T., 82
Nance, R. M., 445
Nansen, F., 388, 537, 677, 799
Nathan, Y., 98
National Environmental Policy Act, 264
Naumann, C. F., 568
Naylor, D., 420
Neev, D., 97
Neill, W. T., 201
Neilson, F. M., 164
Nelson, B. W., 322
Nelson, H., 704, 819
Nelson, M., 321
Nelson, T. H., 453
Nesbit, D. W., 310
Nesteroff, W., 102
Nettleton, A., 842
Neuenschwander, L. F., 289
Neumann, A. C., 102, 580
Neumann, D., 492
Neumann, G., 250, 568, 734
Neumann-Mahlklau, P., 560
Neushul, M., 635
Neville, W. E., 560
Newell, N. D., 555, 627, 800
Newman, W. A., 627
Newton, R. S., 531, 568
Nichols, D., 385
Nichols, G. E., 310
Nichols, M. M., 639, 684
Nichols, R. L., 47, 478
Nickerson, G. A., 264
Nicolae, T., 473
Niedoroda, A. W., 139, 691, 703, 876
Nielsen, E., 356
Nielsen, N., 47, 481
Niering, W. A., 171, 706, 802
Niiler, P. P., 728
Nikiforov, L. G., 133, 222, 505, 627
Nilsson, L., 474
Nir, Y., 97, 98
Nisbet, I. C. T., 802
Nizamuddin, M., 635

Noble, S. M., 174
Nonn, H., 374, 678
Noosin, J. A., 82
Nordin, C. F., 123
Nordstrom, C. E., 31, 82, 132, 222, 224, 337, 368, 453, 712
Nordstrom, K. F., 438
Norris, R. M., 356, 681
North Sea Island Study Group, 66
North, W. J., 635
Nortin, C. F., 550
Norton, D., 57
Nummedal, D., 489
Nybakken, J. W., 245

O'Brien, J. J., 541
O'Brien, M. P., 823, 866
Ochoterena, I., 201
Odum, E. P., 507
Odum, W. E., 516, 763
Oertel, G. F., 489
Ogden, J. G., 593
Okazaki, A., 635, 687
Oldale, R. N., 631
Olive, S. L., 441
Oliver, F. W., 47
Oliver, J. S., 43
Oliver, W. A., Jr., 654
Ollier, C. D., 520
Olson, J. S., 310, 716, 802
Olson, T. A., 140
Olsson-Seffer, P., 310
Omoto, K., 677
O'Neill, M. E., 310
Oosting, H. J., 171, 309, 310, 716
Oomkens, E., 368
Opran, C., 473, 813
Oren, O. H., 102
Orford, J. D., 337
Orlov, A. I., 47
Orme, A. R., 16, 31, 32, 133
Ortel, G. F., 818
Ota, Y., 82
Otvos, E. G., 133, 160, 207, 322, 353, 374, 389, 550
Overeem, A. J. A. van, 82
Ovriku, K. K., 789
Owens, E. H., 235, 478, 479, 480, 481, 633, 663, 818

Pahlke, H., 743
Paine, R. T., 43, 702
Palacas, J. G., 467
Pallary, P., 85
Palmer, H. D., 387
Papenfuss, G. F., 635, 687
Pararas-Carayannis, G., 846
Parker, B. C., 403, 635
Parker, R. H., 519
Parrott, B. S., 134
Parsons, J. J., 191
Parsons, J. R., 770
Parsons, T. R., 245
Parsons, T. V., 743
Pascoe, E. H., 486
Paskoff, R., 678, 770
Passega, R., 740

AUTHOR CITATION INDEX

Patnode, H. W., 710
Patterson, G. W., 869
Pattullo, J., 728
Pavoni, N., 579
Peach, B. N., 537
Pearcy, W. G., 684
Pedley, L., 109
Peevey, R. M., 609
Pegrum, R., 420
Penaherrera de Aguila, C., 201
Penck, A., 185, 520, 535, 748
Penfound, W. T., 310
Pennak, R. W., 64, 332, 555
Penney, M. E., 684
Penrose, B., 296
Peralta, L. A., 748
Peregrine, D. H., 872
Perkins, E. J., 285
Perthuisot, J-P., 31
Pessl, F., 478
Pestrong, R., 821
Petersen, C. G. J., 245, 285
Peterson, R., 643
Peterson, R. W., 264
Petrie, G. M., 57
"Petros," 448
Petrov, M. V., 201
Pettijohn, F. J., 120, 148, 186, 301, 322, 421, 455, 537, 557, 740, 753
Pevear, D. R., 223
Pfeiffer, W. J., 288
Philip, G., 98
Phillips, B. J., 632
Phillips, O. M., 492, 734
Phillips, R. C., 285
Phillips, R. L., 321, 531
Philp, R., 37
Phipps, C. V. G., 120
Phleger, C. P., 459
Phleger, F. B., 37, 321, 433, 763
Picard, L., 97
Picha, F., 102
Pickett, T. E., 505
Pickwell, R., 256
Pierce, J. W., 133, 322, 737, 754, 818
Pierson, W. J., Jr., 250, 568, 734
Pilger, C., 355
Pilkey, O. H., 489
Pillay, T. V. R., 49, 76
Pipkin, B. W., 878
Pitman, W. C., 31
Platzman, G. W., 509
Ploessel, M. R., 878
Plough, H. H., 209
Poggie, J. J., Jr., 201
Pogodin, N. F., 668
Polach, H. A., 207
Polcyn, F. C., 684
Policastro, A. J., 814
Pollock, W. H., 3
Pomerancblum, M., 98
Ponnamperuma, F. N., 763
Ponomarenka, V. V., 668
Pons, L. J., 763
Poole, F. G., 697
Pope, E. C., 109
Pope, R. M., 516

Popov, E. A., 695
Por, F. D., 85, 86, 681
Porter, C. L., 528
Porter, D. M., 201
Postma, H., 397, 851
Potter, P. E., 120, 322, 550
Potts, A. S., 437
Powell, E. N., 167
Powers, M. C., 750
Powers, W. H., 522, 803
Pratje, O., 458
Prebble, M. M., 806
Prêcheur, C., 205
Prénant, A., 31
Prescott, G. W., 403
Pričăjan, A., 473, 813
Price, J. A., 655
Price, J. H., 285
Price, W. A., 123, 133, 149, 160, 164, 187, 207, 222, 264, 338, 461, 519, 522, 550, 754, 818
Pritchard, D. W., 397, 400
Pritchett, P. C., 656
Proctor, C. V., Jr., 321
Proudman, J., 816
Prouty, W. F., 187, 602
Psuty, N. P., 543, 770
Public Service Electric and Gas Company, 264
Pugh, K. B., 604
Purdy, E. G., 627
Purer, E. A., 310, 593, 706
Puri, G. S., 484
Purpura, J., 842
Purser, B. H., 102, 322, 376
Putnam, W. C., 222, 446, 533, 770

Queen, W. H., 763
Quennel, A. M., 102
Quinn, A. F., 183, 640, 879

Radach, G., 749
Ragsdale, J. A., 322
Raj, U., 621
Ralph, E. J., 400
Ramalay, F., 310
Ranwell, D. S., 380, 409, 459, 519, 716, 799
Rao, D. B., 509
Rao, R. P., 156
Rao, V. S., 250
Rapp, G., Jr., 52, 53
Rashid, M. A., 663
Ratcliffe, D. A., 409
Rayleigh, J. W. S., 869
Raymont, J. E. G., 324, 660
Rayss, T., 86
Read, J. F., 322
Read, L., 289
Reading, H. G., 32, 322
Reaside, J. D., 333
Redfield, A. C., 706, 708
Redman, J. B., 748
Reed, L. E., 685
Rees, G., 420
Rees, W. J., 332
Reese, L. C., 846

Reeves, R. G., 684
Reid, C., 123, 537
Reid, L., 684
Reidl, R. J., 604
Reimers, H., 525
Reimnitz, E., 61, 478, 602, 695
Reimold, R. J., 288
Reineck, H.-E., 2, 121, 139, 322, 493, 531, 550, 568, 614, 819, 851
Reinhard, R., 743
Remiro, J. R., 770
Reiss, Z., 86
Reiswig, H. M., 654
Resio, D., 31
Reusch, H., 537
Revelle, R., 502, 728, 877
Reyment, R. A., 337
Rex, R. W., 500, 550, 615
Rheinheimer, G., 614
Rhoads, D. C., 385, 614
Rice, M. E., 758
Rich, J. L., 753
Richards, H. G., 32, 82, 677, 770
Richardson, E. V., 123, 164, 550
Richard-Vindard, G., 16
Richter-Bernberg, G., 423
Richthofen, F. von, 222, 691
Riding, R., 763
Riedl, R., 167, 647
Rieger, R. M., 167
Riek, E. F., 492
Rigby, J. K., 800
Riggs, S. C., 716
Riley, Ch. M., 207
Ritchie, W., 235
Rittschof, W. F., 264
Rizvi, A. I. H., 82
Robb, J. M., 31
Roberts, H., 693
Roberts, H. H., 502
Robilliard, G. A., 43, 235
Robinson, A. H. W., 748
Robson, J., 445
Rochford, D. J., 109
Rockwell, F. G., 353
Rodeick, C. A., 478
Roden, G. I., 201
Rodgers, J., 647
Roeder, D. H., 453
Rogers, R. H., 684, 685
Roithmayr, C. M., 684, 685
Roll, H. U., 692
Ron, Z., 98
Rona, P. A., 32
Roney, J. R., 264
Ronne, F. C., 639
Rosen, P. S., 178, 180, 352, 478, 823
Rosenan, E., 98
Rosenberg, E. S., 592
Rosendal, H. E., 672
Rosenqvist, I. Th., 223
Rosowski, J. W., 403
Ross, E. H., 102
Ross, J. R. P., 183
Ross, P., 76
Rosselló, V. M., 33, 185, 685
Rossignol, M., 98

Rossiter, J. R., 542, 729
Rounsfell, G. A., 324, 400
Rouse, G. E., 635, 650, 687
Royal Commission on Coast Erosion, 257
Rozanski, G., 795
Rubey, W. W., 740
Rudowski, S., 478
Ruivo, M., 140, 853
Rundgren, L., 872
Rusnak, G. A., 156, 397, 474
Russell, F. S., 495
Russell, R. C. H., 162, 837
Russell, R. J., 2, 120, 142, 207, 296, 368, 531, 580, 602, 658, 691
Russell-Hunter, W. D., 324, 332
Rust, B. R., 560
Ruttner, F., 507
Rützler, K., 654
Ryan, W. B. F., 31, 98
Ryland, J. S., 183
Ryther, J. H., 49, 660

Saenger, P., 109
Safianov, G. A., 505
Safra, D., 98
Safriel, U. N., 85, 86, 98, 681
Safyanov, G. A., 222
Sager, G., 837
Sahu, B. K., 337
Saini, G. R., 763
St. Denis, M., 66
Salisbury, E. J., 409, 716
Salisbury, F. P., 180
Sall, M., 32
Sallenger, A. H., 606
Salomone, L. A., 264
Salvayre, H., 223
Sanders, J. E., 48, 122, 133, 322, 879
Sanderson, D. J., 795
Saner, W. A., 806
Sanlaville, P., 102
Sarà, M., 654
Sarjeant, W. A. S., 673
Sasaki, T. X., 695
Sasaki, Y., 846
Sasekumar, A., 76
Sass, E., 97
Sass, J., 806
Sater, J. E., 61
Sauer, J. D., 16, 109, 191, 201
Saurin, E., 82
Savage, R. P., 48, 66, 531
Savigear, R. A. G., 231
Saville, T., Jr., 353, 698, 862
Scagel, R. F., 635, 650, 687
Scarlato, O. A., 780
Schäfer, W., 385, 751
Schalk, M., 61, 478
Schaller, F., 64
Schattner, I., 98
Schenk, P. E., 560
Scherz, J. P., 639
Schieferdecker, A. A. G., 445
Schiffman, A., 550
Schild, R., 465
Schlee, J., 31

Schlieper, C., 245
Schmalz, R. F., 423
Schmid, O. J., 635, 687
Schmidt, P., 321
Schmidt, V., 704, 819
Schmidt, W., 186
Schneider, D. E., 57
Schneider, J., 502
Schofield, J. C., 579, 627, 663, 729, 733
Schofield, W., 445
Schofield, W. B., 635, 650, 687
Scholander, P. F., 76
Scholander, S. I., 76
Scholl, D. W., 733
Schopf, J. W., 37, 354
Schou, A., 301
Schreiber, J. F., Jr., 376
Schreiver, W., 602
Schubel, J. R., 397, 398, 402, 430
Schule, J. J., Jr., 722
Schureman, P., 837
Schuster, J., 355
Schuster, S., 743
Schuster, W. H., 76
Schwartz, M. L., 124, 133, 139, 142, 144, 156, 180, 296, 353, 461, 513, 657, 658, 661, 663, 686, 691, 743, 792
Schwartz, R. K., 531
Schwarz, G., 374
Schwarz, H., 37
Schweizer, J. E., 353
Sclater, J. G., 32
Scoffin, T. P., 342
Scott, A. J., 133
Scott, I. D., 513
Scott, J. T., 568
Scott, T., 353, 862
Scrutton, R. A., 31
Searle, A. G., 76
Searles, R. B., 687
Sears, P., 180, 568
Segada-Vianna, F., 175
Segerstrom, K., 770
Seibold, E., 568
Seigh, I. B., 121
Seilacher, A., 184
Selby, M. J., 709
Selley, R. C., 322
Semeniuk, V., 763
Senger, L. W., 684
Sengupta, S., 121, 550
Sensabaugh, W. M., 842
Servant, J., 763
Seshadri, R., 441
Seward-Thompson, B., 736
Sewell, R. B. S., 486
Shachnai, E., 97
Shackleton, N. J., 543
Shah, N. J., 684
Shaler, N. S., 310, 631
Shamir, U. Y., 541
Shannon, J. P., 710
Sharaf el Din, S. H., 32
Sharkov, V. V., 558
Sharp, A. J., 181

Sharp, R. P., 550
Shaver, R. H., 322
Shaw, A. B., 423
Shaw, J., 54
Shea, M. L., 171
Shearman, D. J., 704
Shedlovsky, J. P., 3
Shelford, V. E., 780
Shepard, F. P., 32, 37, 123, 124, 133, 139, 156, 158, 222, 296, 353, 368, 424, 429, 513, 531, 534, 535, 559, 602, 615, 627, 639, 658, 672, 695, 698, 733, 734, 750, 800, 849, 862
Sheppard, G., 770
Sherlock, R. L., 516
Shevchenko, V. Ya., 518
Shideler, G., 717
Shields, A., 391
Shier, D. E., 681
Shilts, W. W., 756
Shinn, E. A., 322
Shiraishi, H., 846
Shirley, M. L., 322
Short, A. D., 61, 478, 480, 500, 550, 691, 717
Shreve, F., 310
Shuisky, Yu. D., 518, 859, 862
Shukri, N. M., 98
Shulenberger, E., 492
Shul'gin, Ya. S., 668
Shulyak, A., 164
Shunk, I. V., 310
Sides, S. L., 593
Sieburth, J., 441
Siesser, W. G., 32
Siever, R., 120, 322
Silberman, E., 814
Silen, L., 636
Silverstein, M. K., 31
Silvester, R., 153, 250, 272, 328, 353, 368, 391, 862, 882, 884
Simons, D. B., 123, 164, 550
Singer, C. I., 813
Singh, I. B., 139, 322, 531, 550
Singletary, R. L., 826
Sitarz, J. A., 661
Sivanesan, A., 378
Skipworth, Sir P. A. d'E., 704
Skjei, R. E., 516
Skjei, S. S., 516
Skjelbreia, L., 250, 869
Skutch, A. F., 309
Slayter, R. O., 802
Sleath, J. F. A., 164
Sloss, L. L., 421
Sly, P. G., 454, 736
Smart, J. S., 368
Smiley, T. L., 643
Smith, A. G., 420
Smith, D. B., 743, 748
Smith, E. H., 699, 876
Smith, F. E., 328
Smith, G. G., 109
Smith, G. M., 393, 650
Smith, J. B., 310
Smith, J. D., 322

AUTHOR CITATION INDEX

Smith, J. E., 328
Smith, J. L., 560
Smith, J. S., 235
Smith, R. I., 40
Smith, V. E., 685
Smith, W. G., 358, 368, 378
Smosna, R. A., 264
Snead, R. E., 82, 201, 533
Sneh, A., 98
Snodgrass, F. E., 328, 522, 803
Snow, L. M., 310
So, C. L., 205
Sokol'nikov, Yu. N., 668
Solovjev, V. F., 558
Somero, G. N., 167
Sonu, C. J., 139, 142, 150, 337, 353, 531, 568, 691, 695, 717, 862
Southard, J. B., 391, 492
Spar, M. S., 98
Sparrow, F. K., Jr., 440
Spate, O. H. K., 486
Spaulding, I. A., 813, 842
Specht, M. M., 109
Specht, M. R., 684
Specht, R. L., 109
Spencer, C. P., 604
Spencer, D. W., 736
Sporne, K. R., 393, 525, 528
Squzay, G., 743
Stamp, L. D., 76, 445, 678
Standley, P. C., 201
Stanier, R. Y., 720
Stanley, D. J., 568, 710
Stapor, F. W., 160, 161, 392, 431, 737, 748, 754
Staub, R., 453
Steele, J. H., 431
Steers, J. A., 133, 420, 428, 456, 658, 792
Stefan, H., 814
Stein, J. R., 635, 650, 687
Steiner, M., 16
Steiner, R. P., 763
Stembridge, J. E., 379
Stenzel, H. B., 681
Stephen, A. C., 385, 758
Stephenson, A., 16, 76, 109, 201, 285, 702
Stephenson, J., 41
Stephenson, T. A., 16, 76, 109, 201, 285, 702
Stephenson, W., 109
Sterrer, W., 647
Sterrett, T. S., 531
Stevens, A. R., 639
Stevens, G. R., 579
Stevenson, R. E., 593
Stewart, F. H., 423
Stewart, R. W., 869
Stewart, W. D. P., 354, 635, 687, 716
Steyermark, J. A., 201
Stickney, R. R., 49
Stoddart, D. R., 76, 82, 342, 621
Stoertz, G. E., 685
Stokes, G. G., 869
Stokes, M. A., 643
Stolzenbach, K. D., 719, 815

Strahler, A. N., 144, 156, 296, 353, 446, 631, 862, 881
Strasburger, E., 643
Straub, L. G., 740
Strickland, C., 445
Strickland, J. D. H., 245
Stride, A. H., 333, 748
Stringfield, V. T., 502
Strong, A. E., 684
Strong, E., 728
Stubblefield, W. L., 568
Stuiver, M., 733
Stunkard, H. W., 542
Suess, E., 82, 222, 388, 420, 453
Suetova, I. A., 47
Sugden, D. E., 61, 463
Suggate, R. P., 733
Sugiyama, T., 82
Suhayda, J. N., 61, 139, 368, 695
Sullivan, G. G., 672
Summerhayes, C. P., 32
Sunamura, T., 225, 878, 879
Sundborg, A., 397
Supan, A., 453
Susman, K. P., 133
Sutherland, J. P., 436
Svenson, H. K., 201
Sverdrup, H. U., 537
Swan, S. B. St. C., 82, 235
Swanson, V. W., 467
Swart, D. H., 661
Swayne, J. C., 445
Swedmark, B., 64, 671
Swift, D. J. P., 133, 180, 322, 489, 568, 678, 717

Tada, F., 82
Tait, R. V., 201, 533
Takahashi, T., 82
Takashi, I., 484, 486
Tammekann, A., 503
Tanner, S. H., 328
Tanner, V., 175, 178
Tanner, W. F., 1, 150, 153, 161, 222, 352, 392, 431, 461, 467, 506, 522, 524, 531, 550, 657, 681, 697, 698, 703, 748, 752, 770, 793, 875, 876, 883
Tansley, A. G., 802
Tappan, H., 433
Tarr, R. S., 448, 537
Task Committee, 259
Tauman, J., 98
Taylor, J. du P., 54
Taylor, N., 310
Taylor, R. B., 61, 235, 479, 481
Taylor, R. J., 57
Taylor, T. M. C., 635, 650, 687
Tebble, N., 86
Tedrow, J. C. F., 296
Teichert, C., 762
ten Hove, H. A., 681
Termer, F., 191
Termier, G., 445
Termier, H., 445
Ters, M., 733
Terwindt, J. H. J., 322

Tharp, M., 484, 486
Thesiger, W., 765
Thom, B. G., 120, 139, 358, 368, 398, 467, 819, 821
Thomann, G. C., 685
Thomas, I. M., 109
Thomas, R., 52
Thompson, R. W., 819
Thompson, W. O., 156, 322, 531
Thoreau, H. D., 631, 802
Thorn, R. B., 257
Thornbury, W. D., 879
Thorpe, S. A., 492
Thorson, G., 167, 285, 495
Thorsted, T. H., 289
Thorum, A., 846
Throckmorton, P., 55
Thurber, D., 389
Tietze, W., 799
Tiffney, W. N., Jr., 716
Tijia, H. D., 82
Tinkler, K. J., 32
Tipper, G. H., 486
Tirey, G. B., 343
Todd, T. W., 2
Tokida, J., 635
Tolman, C. F., 446
Tom, M., 86
Tooley, M. J., 420, 465, 677, 733
Tooms, J. S., 32
Tortonese, E., 86
Tos'ar, H., 98
Townson, J. M., 668
Tracey, J. I., Jr., 168, 342, 627
Tracy, S. M., 310
Transeau, E. N., 310
Tranter, D. J., 76
Trask, P. D., 145, 296, 531, 557, 748
Travis, B., 691
Traylor, M. A., 337
Treadwell, G. T., 460
Trefethen, J. M., 145
Treherne, J. E., 492
Trenhaile, A. S., 337
Tricart, J., 301, 524, 704, 877
Tripp, D. R., 592
Trowbridge, A. C., 740
Trudgill, S. T., 876
Truesdell, P. E., 710
Tsaits, E. S., 668
Tsaturyan, G. A., 668
Tsuriell, D. E., 380
Tsurnamal, M., 86
Tur Caspa, Y., 98
Tuttle, S. D., 156
Tuzet, O., 654
Twenhofel, W. H., 176, 550, 631, 693
Tzvetkova, N. L., 780

Udden, J. A., 423, 710, 738
Ul'st, V. G., 161
Umbach, M. J., 685
Umbgrove, J. H. F., 76, 82
UNESCO, 660
U. K. Hydrographic Office, 205
U. S. Army Corps of Engineers, 141, 264, 343, 846

AUTHOR CITATION INDEX

U. S. Atomic Energy Commission, 264
U. S. Department of the Interior, 264
U. S. Department of Transportation, 460
U. S. Environmental Protection Agency, 264
U. S. Navy Department, 156
U. S. Navy Hydrographic Office, 191
U. S. Nuclear Regulatory Commission, 265
Urien, C. M., 770
Ursell, F., 695, 803
Ushakov, P., 780
Usoroh, E. J., 32
Uziel, J., 98

Vacelet, J., 654
Valentin, H., 82, 222, 339, 447, 542, 699
Valentine, J. W., 636
Van Andel, Tj. H., 770
Van Beek, J. L., 353, 568, 862
Van Couvering, J. A., 648
van Dam, L., 76
Van der Land, J., 659
van der Lingen, G. J., 531
van der Valk, A. G., 3, 716
Vanhove, N., 842
van Lopik, J. R., 368
Vann, J. H., 201, 533, 770
Van Olphen, H., 430
Vanoni, V. A., 391
Van Steenis, C. C. J., 848
van Straaten, L. M. J. U., 133, 322, 493, 550, 751, 819, 851
van Veen, J., 257, 333, 466, 748
Varlet, F., 32
Varnes, D. J., 710
Vaughan, T. W., 627
Vaumas, E. de, 102
Veatch, J. O., 445
Veeh, H. H., 627
Veenstra, H. J., 148
Velain, C., 537
Venkataramiah, A., 400
Venkarathnam, K., 98
Verheyden, L. R. A., 842
Vermeer, D. E., 201
Vermeij, G. J., 495
Verstappen, H. T., 580
Vevers, H. G., 621
Vincent, M. K., 31, 200, 222, 230, 235, 770
Vines, W. R., 842
Visher, F. N., 800
Visher, G. S., 740
Vistelius, A. B., 337
Vita-Finzi, C., 388
Vogl, R. J., 289
Vohra, F. C., 76
Voigt, E., 632
Voigt, G. K., 716
Vold, M. J., 430
Vold, R. D., 430
Volker, A., 76

Volkov, P. A., 748
Von Arx, W. S., 346
Vonder Haar, S. P., 37
von Post, H., 457
von Post, L., 457
Von Zittel, K., 388
Vorobiev, V. P., 780
Voronov, A. G., 627
Voronov, P. S., 47
Vortisch, W., 356
Vosmaer, G. C. J., 654
Voss, F., 849

Wadell, H., 740
Wadia, D. N., 486
Wagner, P. L., 201
Wagner, R. H., 593
Wagret, P., 648
Waisel, Y., 459, 533
Walcott, R. I., 453
Walker, H. J., 61, 296
Walker, J., 296
Walker, J. R., 802
Walker, P. H., 763
Walker, R. A., 516
Walker, R. G., 322
Wall, J. R. D., 82
Walsby, A. E., 354
Walsh, G. E., 76, 763
Walter, H., 16
Walter, M. R., 37, 354
Walters, C. D., 61
Walton, W. C., 470
Walton, W. R., 434
Wang, S. N., 855
Wanless, H. R., 296, 602, 615, 627, 639, 657, 658, 818
Warburg, H. D., 837
Ward, H. B., 64
Ward, W. C., 557, 736
Warden, R. E., 265
Wardl, W. J., 647
Wareing, P. F., 716
Warme, J. E., 593
Warnke, D. A., 679
Warren, R. S., 171, 706
Waterman, L. S., 102
Waters, C. B., 743
Watkins, L. L., 250
Watson, A. F., 593
Watson, E. V., 181
Watson, I., 264
Watson, J. G., 76
Watts, G. M., 712, 748, 869
Wax, C. L., 230
Weaver, C. E., 223
Weaver, J. E., 310
Webb, J. E., 368
Webb, L., 752
Webber, H. J., 310
Weeden, S. L., 815
Wegener, A. L., 420, 453
Weide, A. E., 133
Weisblatt, E. A., 685
Weischet, W., 770
Weise, B. R., 124
Weissbrod, T., 98

Welch, P. S., 507
Weller, G., 57
Wellman, H. W., 879
Wells, B. W., 175, 310
Wells, H. W., 681
Wells, J. T., 770
Wells, J. W., 76, 342, 621, 627
Welsh, B. L., 289
Wendelken, W. J., 380
Wendorf, F., 465
Wentworth, C. K., 580, 674, 677, 710, 738, 740, 854, 877
Werner, F., 31, 568
Werth, E., 427
West, R. C., 201, 533, 770
Wetzel, R. G., 507, 754
Weydert, P., 32
Weyl, R., 191
Weymouth, C. G., 706
Weymouth, F. W., 400
Wheeler, Sir M., 55
Whelan, T., 368, 538
Whipple, G. C., 64
White, D. A., 453
White, S. J., 391
White, W. A., 124
Whitfield, C. J., 310
Whitlock, C. H., 338
Whittaker, R. H., 441
Whittaker, W., 759
Whittman, F. P., 685
Whittow, J. B., 420
Wichdorff, H. von, 569
Wiedenmayer, F., 654
Wiegel, R. L., 250, 344, 390, 456, 580, 734, 846, 855, 859, 869, 871
Wiens, H. J., 628
Wiens, J. W., 76
Wiggins, I. L., 643
Wiggins, M. M., 762
Wigglesworth, E., 631
Wilce, R. T., 699, 876
Wilhelmy, H., 82
Wilkins, B., 693
Wilkinson, B. H., 133
Williams, A. T., 82, 353, 374, 862
Williams, H., 628
Williams, P. F., 560
Williams, P. W., 579
Williams, S. J., 66, 853
Williams, W., 692
Williams, W. T., 378
Williamson, I. A., 560
Williamson, W. C., 121
Willis, A. J., 716
Willis, D. H., 149, 338
Wilson, A. T., 879
Wilson, B. W., 802, 846
Wilson, C., 684
Wilson, D. P., 681
Wilson, E. M., 823
Wilson, L., 230
Wilson, W. W., 866
Winant, C. D., 337, 349
Winkelmolen, A. M., 148
Winslow, E., 842
Winslow, J. H., 428

AUTHOR CITATION INDEX

Wirszubski, A., 86
Wiseman, W. J., Jr., 61, 478, 479, 480, 481, 500, 550
Wissman, H. von, 82
Woldstedt, P., 435
Wolf, C. L., 33, 123, 176, 328, 445, 455, 470, 503, 551, 558, 559, 632, 640, 705, 757, 804, 855
Wolf, S. C., 478
Wolken, J. J., 403
Wollast, R., 223
Womersley, H. B. S., 109
Wood, A., 230, 435
Wood, B. L., 627, 628
Wood, F. J., 672, 837
Wood, J., 537
Woodcock, D. C., 813
Woodford, A. O., 121, 550
Woodhouse, W. W., Jr., 66
Woodland, A. W., 435
Woods Hole Oceanographic Institute, 436
Woods, P. J., 322
Woodwell, G. M., 716
Woodworth, J. B., 631
Wright, A., 337
Wright, E., 474
Wright, H. E., Jr., 465

Wright, L. D., 61, 139, 358, 368, 398, 568, 819, 821
Wright, R., 433
Wright, V. P., 763
Wunderlich, F., 121
Wunsch, C., 492
Wybro, P. G., 846
Wylie, C. R., 337
Wylie, F. E., 672, 837
Wynne, M. J., 635
Wyrtki, K., 1, 76

Yaalon, D. H., 3, 98, 709
Yabe, H., 82
Yaglom, A. M., 379
Yalin, M. S., 164, 719
Yamaguti, S., 647
Yasso, W. E., 153, 461, 705, 743, 884
Yeakel, L. S., 550
Yensen, D., 492
Yensen, E., 492
Yensen, N., 492
Yingst, J. Y., 614
Yonge, C. M., 495
York, H. H., 309
Yoshikawa, T., 82
Young, D. K., 385, 614

Young, K., 679
Youngs, W. D., 245

Zahran, M. A., 16
Zaks, I., 780
Zaremba, R., 171
Zeidler, R., 815, 871
Zeigler, J., 33
Zeigler, J. M., 133, 156, 550, 639
Zeitschel, B., 660
Zenkevich, L. A., 57, 780
Zenkovich, V. P., 98, 123, 133, 139, 143, 156, 168, 222, 243, 272, 296, 301, 349, 352, 456, 505, 506, 534, 535, 568, 629, 647, 658, 668, 685, 691, 721, 722, 738, 748, 789, 792, 823
Zennaro, P., 434
Zeuner, F. E., 388, 537, 677, 757
Zhdanov, A. M., 103, 668
Zieman, J., 200, 222, 230, 235, 770
Zingg, Th., 885
Zinn, D. J., 671
Zonnenfeld, J. I. S., 770
Zubarenkova, G. G., 668
Zumberge, J. H., 513
Zumwalt, G. S., 636

SUBJECT INDEX

Aa, 519
"A-B-C . . ." MODEL, **1**, 151, 152, 506
Aber, 185
Aberdeen, U.K., 541
Abidjan, 26, 815
Abies, 582, 642
Abies balsamea, 582
Abrasion, 271
Abronia latifolia, 589, 590
Abronia maritima, 589, 590
Abu Dhabi, 101
Acacia, 107, 108, 194, 197
Acacia longifolia, 107, 108, 109
Acadian region, North America, 582-583
Acajutla, 190
Acanthaceae, 288
Acanthaster, 617
Acanthocephala, 67, 279
Acanthopterygi, 483
Acanthozostera, 617
Acanthuridae, 72
Acanthus ilicifolius, 621
Acapulco, 597
Accra, 19, 26
Accretion, lateral, 1
ACCRETION RIDGE, 1-2
Accumulation forms, 743
  according to Zenkovich, 214
Acer rubrum, 175, 582, 583, 584
Acer saccharum, 582
Achillea, 582
Achillea millifolium, 581, 582
Acipenseriformes, 283
Acochlidium, 619
Acoela, 645, 646
Acoelorraphae, 198
Acrania, 283
Acrasiomycetes, 439
Acre, 84
Acritarchs, 673
Acropora, 72, 618
Acropora hyacinthus, 618
Acropora hystrix, 618
Acropora reticulata, 618
Acrostichum, 70
Acrostichum aureum, 9, 13, 194
Acrothoracica, 281
Actinias, 774
Actinomycetes, 715, 720
Actinopypga, 617
Actinulida, 331
Actostaphylos alpina, 581
Adelaide, 110, 119
Adenanthos sericea, 108
Adenocystis utricularis, 199
Adhesion warts, 549

Adocia, 618
Adour River, 414
Adriatic Sea, 126, 359, 416, 434, 465, 501, 558
Aedes, 491
Aegean microplate, 19
Aegean Sea, 52, 405, 417, 848
Aegialitis, 108
Aegiceras, 107, 108, 621
Aeodes orbitosa, 12
Aeolian calcarenite, 2
AEOLIANITE, **2**, 755.
  See also Eolianite
Aepophilus bonnaire, 491
Aeposis robini, 491
Aerial photographs, 637
AEROSOLS, **2-3,** 713, 808, 809
Aerosoltherapy, 808
Aerosol transport, 226
Africa, 41, 67, 77, 126, 167, 212, 220, 235, 284, 298, 323, 359, 365, 369, 374, 418, 454, 495, 523, 532, 533, 544, 628, 644, 652, 658, 675, 689, 705, 757, 805, 806, 881
AFRICA, COASTAL ECOLOGY, **3-16**
  Atlantic coast, 9-11
  biotic factors, 7
  climatic factors, 3
  edaphic conditions, 6
  geomorphic factors, 6-7
  hydrologic factors, 4-6
  Indian Ocean coast, 12-16
  Mediterranean coast, 7-9
  Red Sea coast, 16
  southern tip, 11-12
AFRICA, COASTAL MORPHOLOGY, **17-32**
  Atlantic coast, 23-27
  coastal processes, 20-22
  geologic history, 17-20
  Indian Ocean coast, 28-30
  Madagascar, 30
  Mediterranean coast, 22-23
  Red Sea coast, 30-31
  southern tip, 27-28
African Plate, 17, 19, 26, 30, 100
Afzelia cuanzensis, 13
Agadir, 19, 20
Agariciidae, 619
Agarum cribrosum, 774
Agathis, 642
Agathis australis, 642
Agavaceae, 587
Agave, 194
Age calculations. *See*
  RADIOCARBON DATING

Agelas, 650, 651
Agger, 827
Agglomeration, 429
Aglossopsida, 524
Agnatha, 208, 283
Agonis flexuosa, 108
Agoseris apargiodes, 589
Agropoli, 416
Agropyron, 288, 408, 459, 798
Agropyron junceum, 7, 107
Agropyron junciforme, 237, 408
Agrostis alba, 590, 592
Agrostis pallens, 589
Agulhas Bank, 17, 27
Agulhas Current, 4, 11, 12, 21, 28
Agulhas Plateau, 17
Ahmadabad, 484
Aigues Mortes, 415, 419
Aikona, 411
Aira praecox, 591
AIR BREAKWATERS, **32-33**
Air holes, 141
Airy's Law, 744, 752, 867
Aistmarès, 458
Aitape, 755
Aizoon dinteri, 11
Akhziv, 90
Akita, 235
Akkar, 99
Akko, 86, 92, 93, 94, 95
Akovitika, 52
AKTOLOGY, **33**
Alabama, U.S.A., 127, 351, 596
Alagoas, 200, 768
Alaria, 286
Alaria esculenta, 775, 776
Alaria fistulosa, 774
Alaria marginata, 774
Alaska, U.S.A., 55, 57, 59, 60, 61, 224, 236, 351, 352, 374, 389, 397, 427, 461, 462, 464, 477, 479, 505, 521, 526, 527, 543, 554, 590, 592, 593, 594, 599, 600, 615, 643, 690, 726, 772, 784, 803, 821, 827, 843, 844
Alauda arvensis, 407
Albania, 416-417, 558
Albany, Australia, 104, 106, 108, 110
Albemarle Sound, 397
Albenga, 53
Albert Lake, 112
Albizzia amara, 482
Albrohos, 443
ALBUFERA, **33,** 685
Albuhaira, 33
Albunea, 620
Alcidae, 284

SUBJECT INDEX

Alcyonaria, 72, 331
Alcyonium, 42
Alcyonium aurantiacum, 572
Aldabra, 72, 876
Aleutian Islands, 600, 657, 843, 844
Alexander, 83
Alexandretta Bay, 96
Alexandria, 20, 23, 53, 94, 371
Alexandria Basin, 27
Alexandria Lake, 112
Algae
  blue-green, 33, 34, 55, 104, 165, 344, 345, 353, 354, 407, 506, 523, 612, 653, 680, 715, 719-720. *See also* CYANOPHYCEAE
  brown, 104, 105, 165, 195, 207, 208, 286, 404, 571, 573, 616, 633-634, 700-701, 773, 774. *See also* PHAEOPHYCOPHYTA
  Fucalean, 104
  golden brown, 209-210
  green, 83, 165, 195, 207, 208, 286, 341, 506, 573, 612, 617, 618, 700. *See also* CHLOROPHYTA
  red, 9, 83, 104, 105, 165, 195, 199, 207, 208, 286, 344, 404, 571, 612, 618, 679, 680, 686, 687, 700, 774, 775, 776. *See also* RHODOPHYCOPHYTA
  in thalassotherapy, 809
Algal communities, Europe, 404
Algal film, 286
ALGAL MATS AND STROMATOLITES, **33-37,** 502, 819, 851
Algal zonal studies, 306-307
Algarve, 414, 840
Algeria, 7, 8, 9, 22
Algiers, 23
Alicante, 375, 414
Alisio, 25
Al Khums, 23
Alkmaar, 515
Allcoela, 279
Allenrolfea, 707
Allenrolfea occidentalis, 196, 589
Allogromidae, 669
Allogromiina, 431, 432
ALLUVIAL PLAIN SHORELINE, **37,** 214, 216, 698
Almometers, 158
Alnus rugosa, 582
Aloidis gibba, 83
Alpheidae, 617, 620
Alps, 451, 469
Alternathera, 199
Althenia, 286
Altithermal, 474
Alum Bay, 203
ALVEOLAR WEATHERING, **37-38,** 805, 877. *See also* Honeycomb weathering
Amalfi, 416

Amapa, 374
Amaranthaceae, 587
Amazon delta, 360, 398
Amazon River, 206, 211, 359, 370, 396, 768, 769
Ambergris, 141
Amblycera, 490
Amblyrhynchus cristatus, 283
Ambriz, 27
Ambrosia chamissonis, 589, 590
Amelanchier, 582, 583, 584
Amelanchier canadensis, 583
Amelanchier laevis, 583
Ameronothrus lineatus, 56
Ammophila, 169, 170, 288, 303, 379, 408, 583, 584, 585, 715, 716, 798
Ammophila arenaria, 7, 107, 108, 237, 408, 589, 590, 591, 713
Ammophila breviligulata, 237, 582, 589, 712, 713
Amoebae, 670
Amorgós, 417
Ampelisca abdita, 612
Ampelopsis arborea, 585, 586
Ampharete vega, 55
Ampharetidae, 40
Amphibia, 209, 283
Amphibola crenata, 573, 574
Amphidesma australe, 573
Amphidesma ventricosum, 573
Amphidromic point, 38
AMPHIDROMIC SYSTEMS, **38,** 816
Amphineura, 551
Amphinomiidae, 620
Amphioxus, 208, 283, 669
Amphipoda, 281, 611, 773, 775, 776
Amphiroa, 618
Amphiscolops sargassi, 645
Amsterdam, 515, 541, 728
Amundsen Sea, 43, 45
Amur Bay, 785
Anadara, 75, 621
Anadara broughtoni, 772
Anadyr Bay, 771, 784
Anáfi, 417
Anamastrea, 11
Anapa, 746, 787
Anastomosis processes, 299-300
Anatolia, 52
Ancient evolution, coastal morphology, 300
Ancona, 416
Andamans, 72, 81, 482, 483, 484, 485, 486
Anderson's Inlet, 117
Andes Mountains, 197, 211
Andhra, 482, 484
Andira, 195, 198
Andromeda glaucophylla, 175
Andropogon, 194, 305
Andropogon scoparius, 170, 583, 584, 588
Andropogon virginicus, 584, 585
Ándros, 417
Angara Shield, 57

Angel de la Guarda Island, 597
Angelica atropurpurea, 581, 582
Angelica lucida, 589
Angermanland, 731
Angiospermae, 526
Angle of repose, 237, 239
Angola, 4, 9, 19, 21, 27, 374, 454, 644
Anguilliformes, 283
Aniva Cape, 786
ANNELIDA, **39-41,** 62, 280, 385, 758
Annona, 198
Anodontia, 619, 621
Anomalistic month, 671, 672
Anomalocardia subragosa, 518
Anonyx sarsi, 775
Anopheles, 491
Anopheles melas, 9
Anopla, 569
Anoplura, 281, 490, 491
Ansarieh Jebel, 98
Anseriformes, 284
Ansiodactyli, 483
Antarctica 37, 38, 42, 43, 45, 225, 281, 354, 381, 387, 427, 462, 463, 479, 535, 652, 676, 727, 750, 799, 803, 805, 837, 848
ANTARCTICA, COASTAL ECOLOGY, **41-43**
ANTARCTICA, COASTAL MORPHOLOGY, **43-46**
Antarctic Circumpolar Current, 41
Antarctic Convergence Current, 41
Antarctic Sea, 41, 42, 43, 277, 323
Antedon, 282
Anteplaya, 647
Anthephora, 197
Anthocerotes, 181
Anthocerotidae, 181
Anthomyiidae, 490
Anthophyta, 526
Anthospermum littoreum, 13
Anthozoa, 279, 331, 611
Anti-Atlas, 19, 20, 25
Antidunes, 163, 546, 547
Antigona, 619
Antikythera, 53
Antipatharia, 619
Antofagasta, 199, 371
Antrim, 203, 411
Antwerp, 253, 412
Anurida, 281, 491
Anzio, 416
Apalachee Bay, 587
Apennines, 415, 418
Apherusa glacialis, 776
Aphididae, 490
Aphroditidae, 40
Apiaceae, 287
Aplacophora, 551, 553
Aplidium, 617
Apodida, 384
Apophloea sinclairii, 571, 572
Apostomatida, 279
Appalachicola Delta, 366

SUBJECT INDEX

Appalachicola River, 595
Apsheron Peninsula, 787
APPOSITION BEACH, **47.**
    See also BEACH
Aquaclude, 854
AQUACULTURE, **48-49,** 74, 75
AQUAFACTS, **49-50**
Aquitaine Gulf, 414
Arabat, 787
Arabellidae, 39
Arabia, 761, 764
Arabian Plate, 19, 30, 98, 100, 101
Arabian Sea, 359
Arachnida, 61, 64, 280
Arafura Sea, 116
Arakan, 69
Aralia spinosa, 584, 586
Aral Sea, 168, 470
Aranika, 786
Aransas, 585, 797
Araucaria, 642
Araucaria arancana, 642
Araucariaceae, 640, 642, 643
Araucaria excelsa, 642
Araucaria imbricata, 197
Araucarioxylon, 642
Arbacia lixula, 84
Arbear Lock, 680
Arca, 617
Arca boucardi, 772
ARCHAEOLOGY,
    GEOLOGICAL
    CONSIDERATIONS, **50-53**
ARCHAEOLOGY, METHODS,
    53-55
Archanglica, 287
Archaster typicus, 620
Arches, 419, 534
Archiannelida, 39
Archipelago, 443
Archipélago de las Mulatas, 194
Archipelago of Pearls, 190
Architectonica, 619
Arcopagia, 619
Arcs, volcanic, 450
Arctic, 126, 175, 176, 187, 226, 273,
    294, 305, 323, 381, 406, 449,
    462, 463, 464, 474, 475, 590,
    600-601, 606, 657, 824
ARCTIC, COASTAL ECOLOGY,
    55-57
ARCTIC, COASTAL
    MORPHOLOGY, **57-61**
Arctic regions, North America,
    580-581
Arctica islandica, 775
Arctic Ocean, 42, 43, 55, 57, 61,
    225, 364, 449, 451, 480, 513,
    652, 784, 787
Arctic Sea, 56, 228, 277, 364, 775,
    776
Arctophila fulva, 581
Arctostaphylos alpina, 581
Arctostaphylos uva-ursi, 583, 715
Arctotheca nivea, 108
Ardenne, 410
Ardos, 94

Arecaceae, 527, 587
Arenaria peploides, 581
Arenicola marina, 611
Arenicolidae, 40
Arequipa, 371
Argens, 415
Argentina, 176, 198, 199, 200, 462,
    767-768, 821
Arica, 199, 698
Ariciidae, 620
Ariidae, 283
Arizona, U.S.A., 642, 805
Arles, 415
Arm, 442
Armeria, 288, 419
Armeria maritima, 581
Arnhemland, 106, 110
Arno River, 415
Aroe Islands, 443
Aronia melanocarpa, 583
Arrows, 784
Artemia, 55, 280, 288
Artemidactis, 42
Artemisia california, 197
Artemisia caudata, 583
Artemisia stelleriana, 582
Artemisium, 53
Arthrocnemum, 107, 108, 707
Arthrocnemum africanum, 11
Arthrocnemum arbusculum, 107,
    108
Arthrocnemum decumbens, 14
Arthrocnemum glaucum, 9, 16
Arthrocnemum halocnemoides,
    107, 108
Arthrocnemum indicum, 13
Arthrocnemum macrostachyum, 9
Arthrocnemum perenne, 11, 14
Arthrocnemum pillansii, 11
Arthrocnemum subterminale, 589
Arthrophyta, 392
Arthropoda, 61-64, 280
Articulata, 383
ARTIFICIAL ISLANDS, **64-66,**
    465, 514, 606
ARTIFICIAL SHORELINES, **66**
Aruba, 805
Asaphidae, 619
Asaphis, 621
ASCHELMINTHES, **66-67,** 202,
    279, 658
Ascidiacea, 283, 611, 668
Asclepiadaceae, 587
Asclepios, 491
Ascomycetes, 440
Ascomycotina, 439
Ascophyllum, 56, 202, 286, 582
Ascophyllum nodosum, 581, 775
Ascothoracia, 281
Ashdod, 83, 91
Asia, 17, 280, 355, 359, 370, 374,
    454, 532, 533, 621, 640, 644,
    691, 805, 824
ASIA, EASTERN, COASTAL
    ECOLOGY, **67-76**
    coral reefs, 71-73
    economic exploitation, 74-75

mangrove swamp forests, 67-71
rocky shores, 73-74
sandy beaches, 73
ASIA, EASTERN, COASTAL
    MORPHOLOGY, **76-82**
ASIA, MIDDLE EAST,
    COASTAL ECOLOGY, **82-86**
ASIA, MIDDLE EAST,
    COASTAL MORPHOLOGY,
    86-102
ASIA, MIDDLE EAST,
    COASTAL MORPHOLOGY:
    ISRAEL AND SINAI, **86-98**
    bays, 89
    beach sediments, 89-94
    climate, 87
    geographic background, 86-87
    offshore ridges, 89
    rivers, 89
    shoreline, 94-97
    waves, currents, and tides, 87, 89
ASIA, MIDDLE EAST,
    COASTAL MORPHOLOGY:
    SYRIA, LEBANON, RED
    SEA, GULF & OMAN, AND
    PERSIAN GULF, **98-102**
Asia Plate, 76
Asphodelus microcarpus, 8
Aspidochirotida, 384
Aspidocotylea, 646
Aspidogaster, 646
Aspidosiphon, 617
Asplenium, 288
Asplenium platyneuron, 586
Assa River, 416
Assateague Island, 242, 584
Assiminea lutea, 772, 774
Astarte, 775
Astarte borealis, 775
Aster, 287, 419, 581
Asteraceae, 287, 527, 528, 587
Asterina wega, 84
Aster linarifolius, 582, 583
Asterodon, 382
Asteroidea, 381
Aster tenuifolius, 586, 705
Aster tripolium, 491
Astragalus magdalenae, 589
Asturia, 414, 691
Aswan High Dam, 23, 90, 356
Asymmetron, 669
Atacama Desert, 354, 375
Atactodea, 619
Atchafalaya River, 207, 398
Atelostomata, 384
Atergatis floridus, 617
Athens, 371, 417
Atheriniformes, 283
Atlantic City, New Jersey, U.S.A.,
    32, 130, 148, 746
Atlantic Ocean, 57, 188, 210, 332,
    352, 359, 409, 419, 451, 504,
    532, 584, 586, 591, 705, 713,
    724, 727, 771, 775, 824, 839,
    840
Atlit, 89, 95
Atollon, 443

Atolls, 168, 332, 504, 621, 623, 624-626
Atoll systems, 616
Atrato River, 765
Atrina, 619
Atriplex, 199, 288, 581, 798
Atriplex barclayana, 589
Atriplex cinerea, 107, 108
Atriplex hastata, 582, 584, 586
Atriplex leucophylla, 589
Atriplex paludosa, 107
Atriplex patula, 589
Atriplex watsonii, 705
ATTRITION, 103, 271
Auckland, 235, 573, 578, 732
Aulacomya, 574
Aulacomya maoriana, 571, 572
Australasia, 67
Australia, 33, 37, 77, 126, 130, 144, 175, 179, 207, 220, 223, 236, 281, 284, 294, 298, 311-313, 325-327, 329, 332, 333, 354-355, 359-360, 364-366, 369, 370 375, 392, 397, 398, 434, 466, 470, 490-491, 514, 518, 521-523, 532, 544, 621, 642-644, 658, 674, 677, 736, 760-762, 815, 819, 821, 826, 848, 853, 854, 856, 857, 881, 884. *See also* WESTERN AUSTRALIA
AUSTRALIA, COASTAL ECOLOGY **103-110**
 biogeography, 103-104
 intertidal ecology, 104-106
 vegetation, 106-109
AUSTRALIA, COASTAL MORPHOLOGY **110-120**
 biological factors, 117
 climatic factors, 113-115
 geological factors, 110-113
 impact of man's activities, 119
 inherited features, 117-118
 modern dynamics, 118-119
 oceanic factors, 115-117
 shore platforms, 118
Austrocedrus, 642, 643
Austrocochlea concamerata, 104
Austromytilus, 105
Austrosignum, 42
Auvergne, 410
Avalon Peninsula, 581
Aves, 209, 284
Avicennia, 14, 70, 108, 167, 200, 287, 459, 514, 532, 620, 621, 847
Avicennia africana, 9
Avicenniaceae, 532
Avicennia germinans, 171, 588
Avicennia marina, 4, 13, 14, 15, 16, 106, 107, 117, 574
Avicennia nitida, 192
Avicula, 617
Avoca, 853
Avrainvillea, 617
AVULSION, **120**

Axim, 25
Ayan Bay, 785
Áyios Eustrátios Island, 417
Azotobacter, 715
Azov Sea, 516, 665, 667, 786, 787
Azuero Peninsula, 189, 196

Bab el Mandeb, 13, 16, 30
Baccharis halimifolia, 584, 585
Back Bay, 442
Backshore, 140
 features, 530-531
 terrace, 855
Backwash, 745
 marks, 121
 rippleo, 547
BACKWASH PATTERNS, **121,** 155
Backwater, 442
Bacopa monnieri, 586, 588
Bacteria. *See* SCHIZOMYCTES
Bactris, 198
Baculogypsina sphaerulata, 341
Baffin Island, 58, 176, 427
BAGNOLD DISPERSIVE STRESS, **121-123**
Bagnold effect, 122
Bahamas, 33, 185, 313, 466, 501, 532, 653, 761, 799, 806
Bahía Blanca, 228, 767
Bahrain, 101, 371
Baie d'Antongil, 30
Baja, California, U.S.A., 33, 34, 35, 36, 195, 196, 235, 287, 375, 589, 590, 591, 597, 840
Baku, 38
Balanoglossus, 282, 620
Balanoglossus australiensis, 573
Balanophyllia, 619
Balanus, 74, 199, 202, 404, 700
Balanus amphitrite, 11, 84
Balanus balanoides, 56, 775
Balanus balanus, 776
Balanus cariosus, 774
Balanus crenatus, 56
Balanus elizabethae, 11
Balanus nigrescens, 574
Balasore, 485, 486
Balboa, 189
Balearic Islands, 185, 414, 419, 840
Bali, 371
Balikpapan, 81
Ball, 126
Ball and point, 479
Ball and ridge, 529
Ballard Point, 203, 205
BALLAST, **123,** 806
Baltic Sea, 57, 67, 126, 165, 176, 177, 204, 220, 235, 236, 237, 404, 405, 406, 409, 410, 411-412, 419, 434, 444, 448, 449, 458, 461, 502, 503, 535, 541, 569, 629, 667, 724, 771, 782-783, 810, 823, 839, 840
Baltic Shield, 409, 410
Baltimore, Maryland, U.S.A., 371

Banana holes, 501
Bangiophyceae, 686
Bangladesh, 67, 69, 75, 77, 371
Banks, 163, 444. *See also specific names*
 artificial, 388
Banksia, 107
Banksia intergrifolia, 108
Banksia serrata, 109
Banks Peninsula, 575, 578
Banrohu, 483
Bantry Bay, 691
Banyas, 99
Baraka River, 31
Baram River, 81
Barari, 7, 23
Barbados, 371, 501
Barbatia, 617
Barcelona, 371, 415, 769
Bardawil, 91, 92
Bardawil Lagoon, 23, 84, 87, 89, 94, 96
Barents Sea, 56, 359, 771, 775, 782, 783
Barfleur, 413
Barnegat Inlet, 817
Barnstable, 705
Barrages, 514
BARRIER BEACHES, **123,** 124, 125, 663, 705
 chain, 125, 130, 131
 estuaries, 394
Barrier coastal lowlands, North America, 594
Barrier coasts, 216, 593, 596
Barrier complex, 125
BARRIER FLATS, **123-124**
BARRIER ISLAND COASTS **124,** 217
BARRIER ISLANDS, 123, **124-133,** 472, 509, 512, 593
 chain, 130
 classification, 130
 distribution, 126-127
 economic and environmental aspects, 130-132
 lake, 512
 origin, 127-130
 white sand, 594-596
BARRIER ISLANDS, TRANS- ORESSIVE and REGRESSIVE, **133-135**
Barrier reefs, 125, 168, 332, 339, 616, 621, 623
Barriers, 125, 214, 512
 coastal, 458
 drumstick, 127
 primary, 130
Barringtonia, 619
Barringtonia racemosa, 15
Barrow, 477
Bars, **135-139,** 162, 163, 391, 495, 504, 511, 528, 529, 565, 580, 688, 689, 693, 694, 695, 703, 708, 745, 747, 789, 796, 860, 861

SUBJECT INDEX

barrier, 125, 126
emergent, 127, 128
longshore, 125, 135, 511-512
looped, 349
lunate, 271
offshore, 125, 126, 504, 511-512, 660
processes, 300
submarine, 156, 511-512, 689, 690
submergent, 128
swash, 548
tangent, 438
transverse, 136, 138, 150, 703, 872
Barshageudd, 160
Barycentric motion, 830
BASE-LINE STUDIES, **139-140**
Basement rocks, 18
Basidiomycetes, 440
Basidiomycotina, 439
Basin, 442. *See also specific names*
Basra, 101
Bassia muricata, 8
Bassin d'Arcachon, 414
Bass Point, 350
Bass Strait, 105, 110, 112, 333
Bataan, 79, 761
Bat Galim, 92
Bathurst, 9
Bathynormus giganteus, 63
Batillaria cumingii, 772, 774
Batina Plain, 102
Batis maritima, 193, 196, 586, 590, 591, 707
Batrachoidiformes, 283
Batrun, 99
Baumea juncea, 574
Bayche, 521
Bay of Bengal, 67, 74, 359, 484, 724
Bay of Biscay, 408, 415
Bay of Buenaventura, 198
Bay of Fundy, 307, 676, 707, 819, 821, 823
Bay of Islands, 573, 575
Bays, 442-443. *See also specific names;* ZETAFORM BAYS
barriers, 125
bayhead features, 442, 444, 512, 534
baymouth features, 125, 130, 131, 512
lobated, 45
Bdellonemertini, 570
BEACH, **140-141**. *See also specific names;* APPOSITION BEACH; BARRIER BEACHES; BEACH ORIENTATION; FEEDER BEACH; HEADLAND BAY BEACH; MAJOR BEACH FEATURES; MINOR BEACH FEATURES; PERCHED BEACH; PLATFORM BEACH; PROFILING OF BEACHES; RAISED BEACH; SHINGLE AND SHINGLE BEACHES; STORM BEACH; TAR POLLUTION ON BEACHES
accretional, 144, 796
accretion bedding, 1
bayhead, 442, 534
bayside, 442
boulder, 572, 645
concentrates. *See* MINERAL DEPOSITS
coral reefs, 342
crescent, 442, 799
drift, 154, 746. *See also* SEDIMENT TRANSPORT
emerged, 676
equilibrium, 149
erosion control, 257-258
false, 442
flora, 589-590
fossil, 485
gray, 619
headland, 442
hurricane, 455
infraglacial, 676
log-spiral. *See* HEADLAND BAY BEACH
longitudinal, 149
major forms, 155, 156
mining, 325-326
oblique, 149
open, 619
parallel, 149
pocket, 140, 195, 199, 200, 442
reserve, 312
replenishment, 257
retrogression, 679
rhythmic, topography, 717
scarp, 140
sedimentation, 143
summer, 144, 159
systems, 316
transverse, 149
unbalanced, 149
winter, 144, 159
BEACHCOMBING, **141,** 379
BEACH CUSPS, **141-143,** 150, 155, 271, 529, 688, 695. *See also* Cusps
BEACH CYCLES, **143-144,** 155
BEACH FIRMNESS, **144-145,** 496
Beach gravel, 157
BEACH MATERIAL, **145**
BEACH MATERIAL, SORTING OF, **145-148**
BEACH NOURISHMENT, 119, **148-149,** 246, 256, 356, 579
BEACH ORIENTATION, **149**
BEACH PADS, **150,** 695
BEACH IN PLAN VIEW, **150-153**
Beach pools, 443
BEACH PROCESSES, **153-157**
BEACH PROCESSES, MONITORING OF, **157-158.** *See also* Dynamic processes, measurement of
BEACH PROFILES, 140, 143, 157, **159**
BEACH RIDGE PLAIN, **159-160,** 567
BEACH RIDGES AND BEACH RIDGE COASTS, 47, 126, **160-161,** 358, 360, 365, 387, 455, 510, 511, 528, 530, 534, 647, 663, 791
Beach rock, 27, 28, 77, 81, 82, 84, 90, 95, 96, 99, 100, 101, 186, 189, 221, 342, 376, 485, 501, 502, 523, 545, 624, 626, 627, 676, 758, 795, 796
Beak, 444
Bear Island, 56
Beaufort Force, 21
Beaufort Sea, 64, 359, 585, 771
BEAUFORT WIND SCALE, **161,** 722
Bedfordia, 108
BED FORMS, **162-164.** *See also* Plane bed
Bed load, 396
Bed material, 734
Before and after studies, 139
Beheira, 371
Beirut, 99
Belgica antarctica, 281
Belgium, 65, 411, 413, 465, 471, 648, 758, 807, 809, 815, 840, 841
Belize, 192, 194, 501, 597, 650
Belle Ile, 809
Bellingshausen Sea, 45
Bembicium nanum, 105
BENCH, **164,** 500. *See also* WAVE-CUT BENCH
abrasion, 534
Bengal, 371, 482, 483, 484
Benghazl, 20, 23
Benguela, 4, 11, 20, 26, 369
Benicarlo, 415
Benin Gulf, 370
Benioff zones, 450
Bennettitales, 642, 644
BENTHOS, **164-167**
hand substrate, 243
soft substrate, 243-244
Benue Graben, 20, 26
Benue River, 26
Berchemia scandens, 584, 586
Beringia, 589
Bering Sea, 57, 600-601, 773, 775, 784
Bering Strait, 771, 783-784, 784-786
Berlenga Island, 414
Berm, 528
crest, 141
Bermagui, 857
Bermuda, 83, 185, 305, 371, 679, 761
Bernoulli's equation, 249
Beta, 288
Betic Cordillera, 414

Betula, 408, 582
Betula alleghaniensis, 582
Betula nana, 581
Betula papyrifera, 582, 583
Betula populifolia, 583
Bezinkvelden, 515
Biafra (Gulf), 370
Bicrisia, 182
Bigeneria, 432
Bight, 442
Bight of Benin, 4
Bihar, 486
Bijagos Archipelago, 25
Bilbao, 468
Billings Cape, 784
Billiton, 81
Bío-Bío River, 698, 767
Biocoenosis, 273
Bioerosion, 653-654, 876, 878
BIOGENOUS COASTS, **167-168**
Biokarst, 501
Biomass, 167, 244, 405, 659. *See also* Standing stock
BIOTIC ZONATION, **169-171**, 636
Bioturbation, 611-612
Birgo, 281
Birkenhead, 729
Birling Gap, 204
Biscayne Bay, 394, 470
Bittern salts, 421
Bivalvia, 282, 551, 611, 772
Bizette, 23
Blaavands Huk, 411
Blackfish Creek, 630
Blackpool, 692
Black Sea, 126, 127, 235, 289, 294, 359, 370, 420, 445, 471, 472, 473, 512, 516, 643, 658, 665, 666, 667, 746, 771, 772, 773, 786-788, 809, 810, 811, 840
Black zone, 285, 699
Blakeney Point, 203, 455
Blanche Bay, 520
Blanes, 415
Blastomycetes, 439
Blennioidea, 283
Blind Pass, 596
Block Island, 583
BLOWHOLES, **171-172**, 800. *See also* Spouting horn
Blowout, 240, 713
Bluefields, 189, 190
Blue Grotto, 501
Blue holes, 501, 799
Blue Nile, 23, 90
Bluff, 721. *See also* SEA CLIFFS
BOAT BASIN DESIGN, **172-174**
Boat channel, 168, 625, 626
BODDEN, 174, 411
coasts, 217
Bodega Bay, 460, 599
Boehmeria cylindrica, 586
BOGS, **174-175**, 538, 582. *See also* Peat
lakes, 513

pocosin, 175
quaking, 513
Bogue Banks, 586
Bohemia, 410
Boiler. *See* Blowholes
Boiling holes, 799
Bolinas Bay, 460, 599
Bolivar Island, 127
Bolivia, 615, 844
Bombacaceae, 198
Bombax augaticum, 198
Bombay, 371, 482, 483, 484, 485, 486
BOMBORA, BOMBY, **175-176**
Bonaparte Gulf Basin, 112
Boodlea, 490, 617
Bora, 416
Bora-Bora, 627
Boreal region, North America, 581-582
Boreogadus saida, 56
BORES, 176
Borings, 184
Boritis, 7
Borneo, 68, 70, 77, 79, 323, 532
Bornholm, 411
Borrishia frutescens, 585, 586, 588
Borrow, 257
Bosporus, 417
Boston, Massachusetts, U.S.A., 370, 593
Boston Island, Australia, 764
Bostrychia, 9, 11, 572
Bostrychia arbuscula, 571, 572, 574
Bostrychia harveyi, 574
Botany Bay, 691
Bottle chronology, 430
Bottleneck Bay, 442
Bottom scores, 60
Botula, 617
Boubian Island, 101
Boughaz, 87, 90
BOULDER, **176**, 455, 631, 710
BOULDER BARRICADES, 60, **176-178**
Boulder flats, 177
Boulder ramparts, 177, 624, 625, 626, 627, 798
Boulonnais, 413
Bourem, 26
Bourgneuf, 414
Bournemouth, 148, 207, 270
Boussinesq equations, 872
Bouxas, 52
Bowdichia virgiloides, 198
Bowenia, 355
Bowerbankia, 182
Bozo Peninsula, 771
Brachiodontes, 199
Brachiodontes rostratus, 104
Brachiodontes variabilis, 83
BRACHIOPODA, **178-179**, 282, 611, 636
Brachylaena discolor, 13
Branching Bay, 442
Branchiopoda, 280

Branchiostoma, 669
Branchiostoma floridae, 399
Branchiura, 280
Branta bernicla bernicla, 405
Brasilia, 371
Brassicaceae, 527
Brazil, 36, 176, 198, 199, 200, 235, 238, 239, 443, 451, 532, 559, 658, 690, 768-769, 795, 796, 815
Brazilian Shield, 768
Brazos River, 127, 366, 596
Breakers, 153. *See also* WAVES
plunge point, 529
plunging, 635, 745
spilling, 635, 745
Breaker zone, 745
Breakup, 476, 477, 480, 513, 756
BREAKWATERS, **179**, 247, 257, 376, 499, 509, 551, 657, 664, 665, 870
air, 32-33
submerged, 665, 667
hydraulic, 32
Bremen, 412, 840
Bremerhaven, 412
Brest, 413, 807
Bretagne, 185
Breton, 414
Brevoortia, 399
Breydon water, 413
Bridge islet, 444
Brig, brigg, 444
Brighton, 202, 536
Brissopsis, 383
Bristol Channel, 333, 404, 413, 821, 823
British Columbia, 374, 397, 427, 543, 593, 600, 643
British Isles, 291, 292, 405, 408, 413, 462, 724, 726, 747
Brittany, 230, 333, 370, 404, 410, 419, 536, 691, 806, 809, 821
Broads, 413, 443
Broad Sound, 207
Broken Bay, 691
Bromelia, 194
Bromeliaceae, 587
Brooks Range, 601
Broome, 106, 110, 114, 761, 821
Brosimum, 195
Brouwershavense Gat, 243
Broward County, 587
Bruguiera, 14, 70, 75, 108, 167, 287, 532, 620, 621
Bruguiera gymnorrhiza, 13, 14, 15, 16
Brunel, 369, 370
Brunt-Vaisala, 492
BRUUN RULE, **179-181**, 656, 685
BRYOPHYTA, 181
Bryopsida, 181
Bryopsis, 286
BRYOZOA, 41, **181-183**, 282, 611, 617
Buccinum, 775

SUBJECT INDEX

Buccinum hydrophanum, 775
Buenaventura, 765
Buenos Aires, 198, 371
Bugula, 182
Built platform or terrace. *See* WAVE-BUILT TERRACE
Bulbostylis barbata, 108
Bulgaria, 841
BULKHEAD, 66, **183,** 246, 388, 509, 633, 656, 664
Buller. *See* BLOWHOLES
Bumelia tenax, 585
Bunbury, 108
Bund, 444
Bunnies, 207
Buorkhaya, 784
BUOYS, **183-184**
Burdekin River, 114, 116, 359, 365, 366
Burma, 69, 72, 76, 77, 657
Burren, 412
BURROWS AND BORINGS, **184**
Bursera, 194
Burullus Lagoon, 23, 87
Bushehr, 100, 102
Bushland, 197, 198
Büyük Menderes, 52
Bypassing, 247, 270, 712, 747.
 See also SAND BYPASSING
Byron Bay, 581
Byrsonima crassifolia, 195

Caatinga, 198
Cabinda Basin, 20, 26
Cabo Corrientes, 198
Cabo San Antonio, 767
Cabo San Lucas, 590
Cactaceae, 587
Cadaba farinosa, 13
Cadiz, 414
Caecum, 619
Caepidium antarcticum, 199
Caesalpinia, 194
Caesarea, 83, 89, 94
Cagliari, 418
Cairns Bay, 117, 119
Cakile, 108, 288
Cakile edentula, 107, 582, 584, 585, 589
Cakile maritima, 107, 108, 109, 237, 589, 590
CALA and CALA COAST, **185,** 217
Calaglossa, 9
Calamophyta, 392
Calanoida, 280
Calanques, 185, 186, 415, 442
Calappa, 620
CALCARENITE, **185-186**
Calcarina hispida, 341
Calcilutite, 185
Calcirudite, 185
Calcispongea, 650, 651
Calcrete, 376
Calcutta, 371, 486
Caldera Islands, 848

California, U.S.A., 32, 235, 247, 283, 302, 307, 313, 325, 326, 356, 387, 389, 424, 445, 460, 490, 508, 514, 554, 589, 590-592, 598-599, 634, 641, 643, 652, 672, 676, 683, 693, 705, 707, 717, 727, 827, 840, 844, 870, 878, 884
Caligoida, 280
Callao, 371
Callianassa, 620
Callinectes sapidus, 399
Calliopus laevuisculus, 55
Calluna vulgaris, 408
Calocephalus brownii, 107
Caloglossa, 573
Caloglossa leprieurii, 573, 574
Calophyllum, 195, 619
Calophyllum inophyllum, 108
Calothrix, 699
Calothrix fasciculata, 104
Calotte, 46
Calpurnus, 619
Calvados, 413, 809
Calystegia soldanella, 589
Camargue, 415
Cambay, 482, 484, 485, 486
Cambodia, 75, 79, 762
Cambridge, U.K., 413
Cameron Parish, 206
Cameroon, 20, 26, 532
Camissonia cheiranthigolia, 589
Campbell Island, 574
Campeche, 192, 195, 374, 597
Campidano, 418
Campsis radicans, 584, 586
Canaceidae, 490
CANADA, 55, 58, 61, 127, 135, 175, 220, 311, 370, 387, 389, 427, 461, 464, 474, 475, 582, 590, 593, 633, 676, 690, 756, 771, 774, 821, 823, 839
Canadian Archipelago, 57, 58, 479, 580
Canal, 443
 intracoastal, 594
Canale di Lerne, 416
Canal estates, 514
CANALI, **186,** 416
Canaries Current, 4, 9, 25
Canary Islands, 25, 369, 840
Canavalia, 619
Canavalia maritima, 13, 191
Canberra, 371
Canche River, 413
Cannes, 415
Canning Basin, 112
Cantabrica, 678
Canterbury, 578, 885
Canterbury Plains, 575, 578
Cap d'Antibes, 415
Cap Blanc, 3
Cape, 349, 444, 506
Cape d'Agde, 415
Cape Agulhas, 3, 12, 20, 27
Cape d'Ambre, 30

Cape Ann, 593
Cape Canaveral, 587, 594
Cape Carmel, 84
Cape Carvoeiro, 414
Cape Charles, 56
Cape Cod, 130, 131, 175, 235, 236, 237, 238, 313, 384, 438, 447, 460, 465, 582, 583, 584, 594, 630, 631, 656, 701, 705, 713, 717
 National Seashore, 438
Cape Comorin, 482, 484, 485, 486
Cape Corrientes, 597, 765
Cape Cullera, 415
Cape Cuvier, 110, 112
Cape Dra, 23, 25
Cape de Espichel, 414
Cape Fear, 585, 594
Cape Finisterre, 414
Cape Flats, 27
Cape Flattery, 118
Cape Fria, 27
Cape de Gata, 414
Cape Gelidonia, 54
Cape of Good Hope, 355
Cape Gracias a Dios, 190
Cap de la Hague, 413
Cape Hatteras, 125, 130, 141, 237, 242, 262, 272, 313, 398, 465, 594, 595, 601, 700, 701, 737
Cape Henry, 585
Cape Horn, 803
Cape Juby, 20
Cape Leeuwin, 110, 116
Cape Lisburne, 59, 61
Cape Lookout, 594, 737
Cape Lopez, 26
Cape May, 247, 656
Cape Mendocino, 599
Cape de la Nao, 414
Cape Oropesa, 415
Cape Otway, 112
Cape Padaran, 79
Cape Palmas, 17, 25, 26
Cape de Palos, 414
Cape Peñas, 678
Cape Point, 12, 21
Cape Prince of Wales, 601
Cape Province, 4, 11, 19, 27
Cape Ranges, 20, 26
Cape Raso, 414
Cape Ricketts, 475
Cape Rojo, 597
Cape Romain, 585, 594
Cape Romano, 594
Cape Sable, 594
Cape St. Andre, 30
Cape San Blas, 595
Cape San Juan de Guia, 765
Cape San Maria di Leuca, 416
Cape de San Martin, 414
Cape de Santa Antonia, 414
Cape de Sines, 414
Cape Spartel, 23
Cape Three Points, 25, 26
Cape Town, 10, 11, 27

SUBJECT INDEX

Cape Trafalgar, 414
Cape Tres Puntas, 821
Cape Turakirae, 578
Cape Varella, 79
Cape Verde, 371
Cape Verga, 25
Cape Vert, 4, 9, 20, 23, 25
Cape Vidal, 28
Cape York Peninsula, 112, 118
Capitellidae, 40, 620
Capophyllum maschalocarpum, 572
Capparis, 197
Capparis cartilaginea, 13
Caprellida, 281
Capri, 416, 501
Capricorn Group, 117
Carabidae, 490
Carbon 14. *See* RADIOCARBON DATING
Carcharhinus, 15
Cardigan Bay, 412
Cardiidae, 619
Cardioid, 151
Cardisoma, 494
Cardisoma carnifex, 621
Cardium, 96
Cardium edule, 84, 89, 405
Carex, 287, 581, 583, 586
Carex cryptocarpa, 590
Carex glareosa, 581
Carex lyngbyei, 590, 591, 592
Carex macrocephala, 589, 590
Carex maritima, 581
Carex misandra, 581
Carex nardina, 581
Carex obnupta, 591
Carex ramenskii, 55, 581, 590
Carex silicea, 582, 583
Carex ursina, 55, 581, 590
Caribbean Islands, 369, 840
Caribbean Plate, 188
Caribbean region, North America, 586-588
Caribbean Sea, 188, 189, 190, 195, 287, 355, 359, 527
Carissa carandas, 482
Carmel, 86, 92, 93, 94, 95, 599
Carmel Mountain, 86, 87, 89
Carmel Plain, 87
Carnarvon Basin, 112
CAROLINA BAYS, **186-187,** 594, 601, 614
Caroline Islands, 625, 627
Carolinian region, North America, 585-586
Carpentaria Basin, 112
Carpillus maculatus, 617
Carpinus caroliniana, 586
Carpobrotus, 109
Carpobrotus rossii, 107
Carpophanginae, 483
Carpophyllum, 571
Carpophyllum angustifolium, 571
Carr, carrig, 443
Carse, 444
Cartagena, 414

Caryophyllaceae, 527
Caryophyllia, 619
Casablanca, 371
Casamance, 370
Cascade Mountains, 643
Casco Bay, 306
Case hardening, 502, 708
Caspian Sea, 38, 126, 168, 235, 359, 470, 471, 472, 512, 558, 786-788, 841
Cassia fasciculata, 588
Cassididae, 619
Cassidula, 621
Cassiopeia, 331
Cassiope tetragona, 581
Cassytha, 619
Castellammare, 418
Castilla, 195
Casuarina, 13, 69, 107, 109, 715
Casuarinaceae, 107
Casuarina distyla, 109
Casuarina equisitifolia, 108, 109
Casuarina paludosa, 108
Casuarina pusilla, 108
Casuarina stricta, 107
Catalonia, catalunya, 185, 371
Catania, 418
Catastrophism, 291
Catenella nipae, 574
Catomerus polymerus, 104, 105
Cattolica, 416
Caucasian Riviera, 445
Caucasus, 665, 667
Caulacanthus ustulatus, 11
Caulerpa, 74, 105, 617
Caulerpa ashmeadii, 618
Caulerpa peltata, 618
Caulerpa racemosa, 618
Caulerpa scalpelliformis, 83
Caulerpa serrulata, 618
CAUSEWAY, **187**
Cauvery River, 483, 485
Cave. *See* SEA CAVES
CAVITATION, **187-188**
Cayman Graben, 188
Cayo Costa Island, 587
Cayor, 25
Cayos, 188, 189
Cays, 100, 342
Cay sandstone, 485
Caytonailes, 642
Ceará, 198
Cecropia, 194
Cedar Keys, 587, 594
Cedrus, 643
Ceiba, 194
Celebes, 79
Cell, 150-151, 747
Cellana, 572
Cellana capensis, 11
Cellana creticulata, 574
Cellana tramoserica, 104
Celtis, 586
Celtis phillippinesis, 108
Cenchrus, 194, 197
Cenchrus tribuloides, 584, 585

Centaurium spicatum, 9
Centella asiatica, 586
Central America, 192, 323, 374, 495, 504, 532, 593, 597, 621, 650, 652
Central America, coastal ecology. *See* CENTRAL AND SOUTH AMERICA, COASTAL ECOLOGY
CENTRAL AMERICA, COASTAL MORPHOLOGY, **188-191**
Central Sahara Basin, 20
CENTRAL AND SOUTH AMERICA, COASTAL ECOLOGY, **191-201**
Centroceras clavulatum, 11
Cephalocarida, 280
Cephalocerus, 194
Cephalochordata, 208, 283, 668, 669
Cephalopoda, 282, 551, 553, 554-555
Cephalotaxaceae, 640
Cephalotaxus, 642
Ceramium rubrum, 199
Cerastium, 288
Cerastoderma glaucum, 84
Ceratiola ericoides, 585
Ceratopogonidae, 281, 490, 491
Ceratozamia, 355
Cerianthus, 620
Ceriops, 108, 287, 532, 621
Ceriops tagal, 13, 14, 16
Cerithidea, 74, 621
Cerithidea decollata, 11
Cerithium, 74, 619
Cerithium kochi, 84
Cerithium scabridum, 83
Cestoda, 279, 645, 646
Cetacea, 272, 273, 284, 483
Cetraria islandica, 581
Cetraria nivalis, 581
Ceylon, 81, 126, 355, 523, 545
Chactopteridae, 40
Chaetangium corneum, 571
Chaetangium ornatum, 12
Chaetangium saccatum, 12
CHAETOGNATHA, 201-202, 282, 669
Chaetomorpha, 618
Chaetopteridae, 40
Chakan, 168
CHALK COASTS, **202-205.** *See also* Cliffs, chalk
Chama, 74, 616
Chamaecyparis nootkatensis, 643
Chamaecyparis thyoides, 583
Chamaedaphne calyculata, 174
Chamaesipho, 571, 572
Chamaesipho brunnea, 570-572
Chamaesipho columna, 104, 105, 570, 571, 572
Chamaesyce leucophylla, 589
Champerico, 190
Champlain Sea, 677
Chandeleur, 127, 596, 800

Channel Islands, 247, 327, 413
Channels, 443. *See also specific names*
  dieback, 377
  interreef, 340
  sand. *See* SHOESTRING SANDS
  storm-overwash, 316
Chanos, 75, 483
Chao Phraya River, 77, 79, 359, 360, 821
Chaparral, 196, 197
Chapeiro, 443
Char, Chur, 444
Charadriiformes, 284
Charente, 414, 415
Charlestown Inlet, 488, 817
CHART DATUM, **205**
Charybdis, 620
Charybdis longicollis, 84
Chasshowitzka, 587
Chathamiidae, 490
Chatham Island, 532
CHATTERMARKS, **205-206**
Cheilosporum, 618
Chelicerata, 280
Chelonia, 283
Chelonidae, 283
CHENIER AND CHENIER PLAIN, 1, 69, 116, 126, 160, **206-207**, 318-319, 366, 389, 523, 567, 596, 704, 765, 767, 768, 769
Chenier-plain complexes, 318-319
Chenier ridges, 365
Chenolea diffusa, 11, 15
Chenopodiaceae, 107, 287, 407, 527, 528
Chenopodium album, 582
Chenopodium rubrum, 582, 584
Chesapeake Bay, 126, 242, 394, 397, 399, 402, 465, 584, 594, 707
Cheshskaya Guba, 783
Chesil Beach, 146, 147, 156, 312, 408, 413, 455, 530, 739, 751, 839
Chetco River, 592
Chetumal Peninsula, 194
Chiapas, 196, 197
Chiclayo, 766
Chile, 149, 197, 199, 200, 370, 375, 387, 389, 427, 455, 534, 559, 593, 615, 676, 698, 766-767, 843, 844
Chilean mattaral, 197
Chiloé, 199, 520, 767
Chilomonas, 345
Chilopoda, 61
Chimaeras, 283
Chimney, 534. *See also* MARINE EROSION COAST
China, 48, 67, 68, 78, 176, 208, 284, 356, 360, 640, 642, 827, 883
China Sea, 74
CHINE, **207**
Chione stutchburyi, 573

Chiriqui Lagoon, 194
Chironomidae, 281, 490
Chiton, 199
Chittagong, 69, 79
Chlamydospermae, 453
Chlamys, 617
Chloris, 194, 197
Chloris petrea, 585
Chlorodesmis comosa, 618
Chlorohydra, 279
Chlorophyceae, 286, 573
CHLOROPHYTA, 55, **207-208.** *See also* Algae, green
Chnoospora, 616
Chondrichthyes, 283
Chondrosia, 653
Chondrus, 306, 700
Chonotrichida, 279
Chorda, 56
CHORDATA, **208-209**, 283, 668, 669
Chortitza, 254
Christchurch, 207, 732
Christensen Land, 43, 45
Christmas Island, 72, 73, 502
Chroomonas, 345
Chrysobalanus, 198
Chrysobalanus icaco, 191
Chrysocapsa, 210
Chrysomonadida, 670
Chrysomonodales, 210
Chrysophyceae, 210
CHRYSOPHYTA, **209-210**
Chrysops, 490
Chrysopsis falcata, 583
Chrysosphaera, 210
Chthamalus, 10, 74, 84, 199, 616, 621, 700
Chthamalus antennatus, 104, 105
Chthamalus cirratus, 199
Chthalamus dalli, 773
Chthamalus dentatus, 11
Chthamalus malayensis, 106
Chthamalus rhizophorae, 9
Chthamalus stellatus, 83
Chubut, 374
Chukchi Sea, 55, 56, 59, 516, 601, 689, 771, 784
Chukotka, 504
Chukotsk, 784, 785
Chutes and pools, 163
Chytridiomycetes, 439
Ciconiformes, 284
Cicuta maculata, 586
Cidaroida, 384
Ciliata, 278, 670
Circus cyaneus, 284
Cirratulidae, 40
Cirripedia, 280, 436, 611, 776
Cities
  demography, 371
  floating, 327
Civitavecchia, 415, 416
Cladium jamaicense, 588
Cladium ustulatum, 574
Cladocera, 83, 280
Cladonia alpestris, 581

Cladonia lichens, 582
Cladophora, 84, 286, 506, 572
Clape, 415
Clarence River, 578
CLASSIFICATION (coasts), **210-222,** 425-453, 510, 807. *See also* Shorelines
  dynamic morphology, 215-220
  dynamic processes, 220-221
  genetic, 214-215
  morphological, 212-214
  structural, 210-212
Clavularia, 42
CLAY, **222-223,** 710, 757
  cat, 760
  dunes, 819
Clearwater, 587, 594
Cleome strigosa, 13
Clerodendron, 108
Clew Bay, 412
CLIFFED COAST, 176, 212, 217, **223-224,** 825
Cliffed headland, 444
CLIFF EROSION, 224, 267, 271
Cliff line, 444
Cliffs, 444, 721, 759. *See also* CLIFFED COASTS; Cliffed headland; CLIFF EROSION; Cliff line; FOSSIL CLIFFS; SEA CLIFFS
  abandoned, 435
  chalk, 412, 413
  coastal, flora, 288
  dead, 435, 722
  hogs-back, 230
  old, 435
  plunging, 578
  wave-cut. *See* SEA CLIFFS
CLIMATE, COASTAL, **225-230**
  desert coasts, 375
  world coastline, 228
Climate, hydraulic, 562-564
Climatic controls, 225
Climatic optimum, 464, 474. *See also* HYPSITHERMAL
Clints, 99
Cliona, 652
Cliona celata, 617
Clione, 268
Clionidae, 653
Clipperton Island, 623
Clithon, 621
Cloeosiphon, 617
Club mosses. *See* LYCOPODIOPHYTA
Clupeiformes, 283
Cnemidocarpa, 42
Cnidaria, 279, 328
Cnoidal theory, 867
Cnoidal waves, 872
Coagulation, 429
COASTAL BEVEL, **230-231,** 759
COASTAL CHARACTERISTICS, MAPPING OF, **231-235,** 300
Coastal dune and beach vegetation, 191

COASTAL DUNES AND
 EOLIAN SEDIMENTATION,
 235-243. *See also* Dunes
 wind regime, 236
Coastal ecology. *See also*
 AFRICA, ANTARCTICA,
 ARCTIC, ASIA,
 AUSTRALIA, CENTRAL
 AND SOUTH AMERICA,
 EUROPE, INDIA, NEW
 ZEALAND, NORTH
 AMERICA, PACIFIC
 ISLANDS, and SOVIET
 UNION *entries;* COASTAL
 ECOLOGY, RESEARCH
 METHODS
 experimental methods, 244, 245
 lakes, 506-507
COASTAL ECOLOGY,
 RESEARCH METHODS,
 243-245
COASTAL ENGINEERING,
 245-250
 lakes, 508-509
 structures, 245-247
COASTAL ENGINEERING,
 HISTORY OF, **250-257**
COASTAL ENGINEERING,
 RESEARCH METHODS,
 257-259
COASTAL ENVIRONMENTAL
 IMPACT STATEMENTS,
 259-265
Coastal environments, agricultural
 effects on, 465
COASTAL EROSION, **265-267,**
 267-271, 272
COASTAL EROSION,
 ENVIRONMENTAL-
 GEOLOGIC HAZARD,
 267-271
 cost-benefit analysis, 269, 271
COASTAL EROSION,
 FORMATIONS, **271-272**
COASTAL FAUNA, **272-285**
 intertidal regions, 275-276
 shallow seas, 273-274
 systematic composition,
 277-285
 zoogeographical patterns,
 276-277
COASTAL FLORA, **285-289,**
 408-409
 marsh, 589-590
 rocky intertidal zones, 286
Coastal geobotany, 303, 309
Coastal geomorphology research,
 294-295
Coastal morphology, lakes,
 509-513. *See also* AFRICA,
 ANTARCTICA, ARCTIC,
 ASIA, AUSTRALIA,
 CENTRAL AMERICA,
 EUROPE, INDIA, NEW
 ZEALAND, NORTH
 AMERICA, PACIFIC
 ISLANDS, SOUTH

AMERICA, and SOVIET
 UNION *entries;* COASTAL
 MORPHOLOGY *entries*
COASTAL MORPHOLOGY,
 HISTORY OF, **289-296**
COASTAL MORPHOLOGY,
 OCEANOGRAPHIC
 FACTORS, **296-298**
COASTAL MORPHOLOGY,
 RESEARCH METHODS,
 298-301
Coastal paleosols, 761
Coastal plant ecology
 descriptive, 301-303
 experimental, 308
COASTAL PLANT ECOLOGY,
 UNITED STATES,
 HISTORY OF, **301-310**
Coastal protection, 326-327.
 See also PROTECTION
 OF COASTS
COASTAL RESERVES, **310-314**
Coastal sandrock, 466
COASTAL SEDIMENTARY
 FACIES, **314-322**
Coastal slope, 230
Coastal structure design, 258, 259
Coastal waters, 245
COASTAL WATERS HABITAT,
 322-324
Coastal zone, 211
COASTAL ZONE
 MANAGEMENT, 311,
 324-328, 502
 environmental quality, 325
Coasts. *See also main entry list in
 front of book for specific*
 COASTAL *and* COASTS
 entries
 Afro-trailing edge, 212, 213,
 223, 359
 Amero-trailing edge, 212, 213,
 359
 Atlantic-type, 19, 77, 210, 451
 barrier, 216, 593, 596
 bold, 444
 circular, 848
 collision edge, 19, 211, 212, 213,
 223, 359
 consequent cliffed, 534
 consequent plain, 534
 crenulate, 212, 214, 215, 534
 dalmatian, 186, 444
 desert, landforms, 375-376.
 See also DESERT COASTS
 designated, 327
 emergent, 215, 448
 erosion, 534
 exposed, ecology of, 506-507
 fault-line, 216, 424
 fault-scarp, 216, 424
 Finistère type, 691
 fjord, 214, 215, 216, 217, 463
 fold, 216, 424, 444
 glacial, Antarctica, 45
 glaciated, North America,
 593-594

Heritage, 312
 marginal, 213, 223, 359
 Middle Atlantic, 130
 neo-trailing edge, 212, 213, 223,
 359
 neutral, 37, 214
 old, 756
 open, 444
 outwash plain, 214, 615, 698
 Pacific type, 76, 210, 452
 phytogenetic, 167
 primary, 215, 216, 509, 534
 reed, 168
 secondary, 215, 216, 534
 submergent, 448
 terminology, regional, 444
 trailing-edge, 19, 126, 211
 trucial, 101, 819
 types of, 452. *See also*
 CLASSIFICATION
 warped, 498
 zoogenetic, 167, 339
Cobb, 444
COBBLE, **328,** 631, 710, 751
Coccolithaceae, 209, 210
Coccolithophorida, 278
Coccolithus, 210
Coccoloba, 194
Coccoloba uvifera, 191
Cochin, 482
Cochlearia, 287, 288
Cochlearia officinalis, 55, 581, 590
Cocos Islands, 72
Cocos, 619
Cocos nucifera, 526
Cocos Plate, 188
Codakia, 619
Codium, 105, 286
Codium adhaerens, 571, 572, 573
Codium magnum, 207
COELENTERATA, 279, **328-332,**
 346
Coelomycetes, 439
Coelopa frigida, 490
Coelopodidae (coelopidae), 282, 490
Coenobita, 616
Coeruleus, 572
Coffee rock, 466. *See also*
 HUMATE
Cofferdams, 53
Coiba Island, 190, 196
Coimbra, 414
Cold Spring Inlet, 247
Coleoptera, 282, 490, 491
Coleps, 671
COLK, **332-333**
Collembola, 281, 490, 491
Collier Bay, 116
Collioure, 415
Collisella, 773
Colombia, 190, 198, 199, 200, 371,
 559, 765, 844
Colon, 374
Colorado River, Argentina, 767
Colorado Delta, 398
Colorado River, U.S.A., 127, 196,
 207, 365, 366, 597, 819, 823

Colpomenia, 199
Colpomenia capensis, 12
Colubrina, 194
Colubrina asiatica, 13
Columbia, 168, 532
Columbia Glacier, 600
Columbia River, 398, 599
Columna, 572
Colville Delta, 601
Colville River, 60, 359
Comacchio, 416
Comatulidae, 617
Comayagrea Graben, 188
Combretaceae, 287, 532
Comodoro Rivadavia, 767
Comoro Islands, 283
Compositae, 527, 528, 588
Compton effect, 741
COMPUTER APPLICATIONS, 333-337
COMPUTER SIMULATION, 337-338, 874, 875
Conakry, 25
Conavalia lineata, 482
Conchagua, 189
Conchaquita Island, 189
Conchocelis, 687
Concholepas, 199
Cone of depression, 470
Cone penetrometer tests, 258
Cones, 549
Coney Island, 841
Congelifraction, 437
Congo River, 9, 26, 27, 211
Conidae, 619
Coniferophyta, 640
Conifers. See PINICAE
Conioselinum chinense, 589
Connecticut, U.S.A., 302, 307, 447, 672
Conocarpus, 532
Conocarpus erectus, 9, 171, 192
Constantsa, 472, 810
Constitución, 534
Continental drift, 409, 448, 450-451
Continental flexure, 806, 807. See also TECTONIC MOVEMENTS
Continental plate, 126
Continental terrace, 855
Continuity equation, 538, 539
CONTRAPOSED SHORELINE, 338
Conus, 72
Convection hypothesis, 451-452
Convoluta roscoffiensis, 645
Convolvulaceae, 527, 587
Cook Islands, 616, 621, 624, 625, 626
Cook Strait, 571, 572, 821
Coolangatta, 325
Coombe Rock, 204
Coos Bay, 235, 236, 237, 302, 599
Copacabana Beach, 674, 857
Copenhagen, 371

Copepoda, 280
Copernicia, 198
Coquimbo, 371
Coral. See COELENTERATA
Corallina, 104, 286, 571, 572, 573, 700
Corallinaceae, 687
Corallina cuvieri, 105
Corallina officinalis, 105, 570
Coralliobia violacea, 619
Corallocoris marksae, 491
CORAL REEF COASTS, 168, 214, 216, 217, **339**, 504
CORAL REEF HABITAT, **339-342**
Coral Sea, 359
Corange lines, 38
Corcagh, corcass, 443
Cordaites, 642
Cordia, 194
Cordia somaliensis, 13
Cordylophora, 279
Core Banks, 130
Coregonus autumnalis, 56
Coregonus nasus, 56
Coregonus sardinella, 56
Corema conradii, 583
CORERS AND CORING TECHNIQUES, **342-343**, 454
Corfu Island, 417
Corinth, 54
Corinthian Gulf, 123
Corinto, 190
CORIOLIS EFFECT, 38, **343-344**, 347, 348, 425, 426, 725, 803, 816, 880
Corixidae, 490
Corner Inlet, 113, 116, 117, 532
Cornus canadensis, 581
Cornus florida, 586
Cornwall, 412, 413, 541
Coromandel, 482, 483
Corophium, 494, 612
Corpus Christi, 596
Corrasion, 271, 878. See also WEATHERING and EROSION, MECHANICAL
Correa alba, 108, 109
Corrosion, 250, 508, 561, 606, 665, 878. See also WEATHERING and EROSION, MECHANICAL
Corsica, 37, 38, 185, 410, 415, 418, 805, 840
Corunna, 414
Corwin Bluffs, 224
Cosa, 53
Coseguina, 189
Coseguina Volcano, 188
Cossuridae, 40
Costa Brava, 415
Costas de Garraf, 415
Côte de Maures, 415
Costa Rica, 188, 190, 191, 195, 197, 553, 597
Costa del Sol, 415

Cotentin, 413
Cotidal lines, 38
Coto Doñana, 313
Cotula, 199
Cotula corenopifolia, 590, 591
Coulter Counter, 739
County Claire, 521
Courantyne River, 792
Coutances, 413
Cove, 442
Cove Point, 350
Crama, 288
Craniata, 283
Craspedacusta, 279
Crassostrea, 74, 573, 616, 621, 679
Crassostrea cucullata, 11
Crassostrea gigas, 772
Crassostrea glomerata, 571, 573
Crassostrea virginica, 518
Crassula maritima, 11
Crater Lake, 508
Creep, 240
Crenomytilus grayanus, 772, 773
Crescentia, 197
Crescentic rhythmic forms, 271
Cressa truxillensis, 199, 707
Crestas de Playa, 647
Crete, 369, 418
Crevassing, 120
Crimea, 424, 667, 787
Crinoidea, 381, 383, 611
Criques, 190
Cristobal, 189, 191
Crithium, 288
Critical erosion velocity, 396
Crocodilia, 284
Crocodylus niloticus, 14
Crocodylus porosus, 284
Cromer, 202
Cross-dams, 252
Cross Gulf, 784
Crossopterygii, 283
Croton punctatus, 191, 585, 588
Crotone Peninsula, 416
Crown-of-thorns starfish, 331
Crozet Islands, 282
Crozon Peninsula, 413
Crucianella maritima, 7
Cruciferae, 407, 527
Crustacea, 61, 63, 280, 573
Crusts, 761
Crypsis aculeata, 9
Cryptocarpus, 197
Cryptochiton, 617
Cryptomonadida, 670
Cryptophallus magnus, 645
CRYPTOPHYTA, **344-345**
Cryptosula, 182
CTENOPHORA, 279, **345-346**
Ctenostomata, 182
Cuanza River, 4, 9
Cuba, 640, 815
Cubomedusae, 331
Cucumaria, 382, 620
Cucumaria japonica, 774
Cuillins, 411
Culbin, 410, 419

SUBJECT INDEX

Culburra, 677, 854
Culcita, 617
Cul-de-sac, 443
Culex, 491
Culicia, 619
Culicidae, 281, 490, 491
Culicioides, 281, 491
Culver Cliff, 203
Cumacea, 281, 573
Cunene River, 27
Curacao, 679
Curatella americana, 195
CURRENT METERS, 258, **346-347**
CURRENTS, **346-349,** 394, 725, 851, 863, 879, 880. *See also specific names;* CURRENT METERS; RIP CURRENT; Sea Currents; TIDAL CURRENTS; WIND, WAVES, AND CURRENTS, DIRECTION OF
 bedding, 185
 crescents, 121, 141, 548
 inertia, 348, 426
 longshore, 153, 299, 348
 marks, 548
 ripples, 155, 163, 546
 seiche, 749
 turbidity, 347, 428
 wave-induced, 348
Currituck Sound, 235, 237
Cuscuta salina, 590
Cuspate features, 271
CUSPATE FORELAND, 160, **349,** 352, 444, 593, 595, 747, 751, 789
CUSPATE SPITS, **349-352,** 512, 689
Cusps. *See* BEACH CUSPS
Cussonia umbellifera, 15
Cutandra memphitica, 8
CUT AND FILL, 157, 267, **352-353,** 374, 745
Cutans, 761
Cutch, 481, 482, 485
Cutuco, 189
Cuxhaven, 412
Cyamidae, 281
Cyanobacteria, 353, 715
Cyanophora, 345
CYANOPHYCEAE, **353-354,** 483, 652. *See also* Algae, blue-green
Cyanophytes, 33
Cyathomonas, 345
Cycadaceae, 354
Cycadales, 354, 644
CYCADICAE, **354-355,** 644
CYCADOPHYTA, 354-355, 642
Cycas, 355
Cycas circinalis, 355
Cycas revoluta, 355
Cycle of erosion. *See* GEOMORPHIC-CYCLE THEORY
Cyclograpsus punctatus, 11

Cyclones, 326, 880
Cyclopoida, 280
Cyclostomata, 183, 208, 283
Cymatiidae, 619
Cymodocea, 10
Cymodocea nodosa, 405
Cynanchum palustre, 585
Cynodon dactylon, 585
Cyperaceae, 527, 587, 588
Cyperus, 198
Cyperus maritimus, 13
Cyperus papyrus, 14
Cypraea, 72, 74
Cypraea arabica, 617
Cypraea caput-serpentis, 618
Cypraea mauritiana, 618
Cypraeida, 619
Cypressaceae, 643
Cypressus, 642
Cypressus macrocarpa, 643
Cyprinodontoidei, 283
Cyprus, 54
Cyrenaica, 20, 23
Cyrilla racemiflora, 175
Cyrtodaria kurriana, 55
Cystophora, 104, 105, 106
Cystophora intermedia, 105
Cystophora retroflexa, 571
Cystophora scalaris, 572
Cystophora torulosa, 105
Cystoseira, 286
Cystoseira barbata, 773
Czekanowskiales, 642

Dactylia, 617
Daghestan, 787
Dahomey, 20
Dahra Massif, 23
Dalbhum, 486
Dallol Basin, 31
Dalmans, 186
Dalmatia, 419
Dalmatian Valley, 691
Damietta, 23, 87, 90, 92, 94, 371
Dampierian Province, 104
DAMS, EFFECTS OF, 356
Danakil, 31
Danube River, 359, 364, 472, 787
Danzig, 411
Danzig Bay, 411
Darcy's Law, 469
Dardanelles, 417
Dar es Salaam, 20, 28, 29
Darien Gulf, 188
Darling River, 114
Darnah, 23
Darss, 411
Darwin, 116, 761
Darwin Harbour, 821
Dasyhelea, 281, 491
Data Analysis, 333
Daugava River, 782
David, Panama, 197
Dead Sea, 470, 471
Deauville, 809
Decapoda, 281
Deception Island, 848

Decodon verticillatus, 174
Deep Shelf Mixed Assemblage, 42
Deep Shelf Mud Bottom Assemblage, 42
DEFLATION PHENOMENA, 265, **356**
 furrows, 704
Deinocerites, 491
Delagoa Bay, 28
De La Panne, 471
Delaware, U.S.A., 126, 302, 565, 731, 851
Delaware Bay, 50, 51, 126, 396, 397 583, 584, 594, 680
Delaware River, 394
Delesseria, 56
Delft, 825
Delmarva, 51, 130, 584
De Long, 771
DELTAIC COASTS, 212, 213, 214, 216, 217, **356-358,** 593, 698
Deltaic Complexes, 318
Deltaic Forelands, 160
Deltaic Marshes, 708
Deltaic Processes, 299-300
DELTAS, 163, **358-368.** *See also specific river names;* DELTAIC COASTS; ESTUARINE DELTA; LOW-TIDE DELTAS; TIDAL DELTAS
 bird's-foot, 358, 366, 596
 components, 360-361
 compound, 593
 crowfoot, 398
 deformational processes, 364
 ebb, 487, 489, 817, 820
 fan, 318
 flood, 487, 489, 817, 820
 fluvial regime, 360
 lake, 512-513
 "models," 366
 -molding factors, 359
 occurrence of, 359
 ocean current regime, 360
 plain, 364-365
 plan, 412
 receiving basin, 360
 sediments and sedimentary structures, 365-366
 shelf configuration, 360
 shoals, 817
 tectonics, 360
 variability, 366
 wave and tide regimes, 360
DEMOGRAPHY, 369-374
Demospongea, 279, 650, 651, 652
Dendrochirotida, 384
Dendrochronology, 641
Dendrophyllia, 83, 619
Dendropoma, 616
Dendropoma petraeum, 83
Den Helder, 412
Denia, 415
Denmark, 126, 174, 203, 205, 250, 255, 256, 273, 294, 369, 411,

427, 434, 435, 444, 447, 462, 466, 648, 657, 690, 840, 850
Density Structures, 298
DEPTH OF DISTURBANCE, **374**
Dermaptera, 281
Dermochelyidae, 283
Dermochelys coriacea, 73
Deschampsia, 287
Deschampsia caespitosa, 590, 591, 592
Deschampsia flexuosa, 583
DESERT COASTS, **374-376**
Desert Varnish, 708
Desiccation Theory, 387-388
DESIGN WAVE, **376-377**
Desulfovibrio desulfuricans, 761
Deuteromycetes, 440
Deuteromycotina, 439
Deva River, 678
Devil's Lake, 511
Devon, 147, 412, 413, 688
Devon Island, 475, 581, 633
Devonshire, 314
Dhílos, 417
Dhufar, 102
Diadema, 72, 617
Diadematacea, 384
Diamond Coast, 235
Diamond Head, Hawaii, 393
Diapirs, 364
Diara, 444
Diatomea, 776
Dicemida, 542
Dichromena colorata, 586, 588
Dicrodon, 197
Dictyochaceae, 209
Dictyopteris delicatula, 10
Dictyosphaeria, 617, 618
Dictyota, 616, 617
Dicyema, 279
Didemnum, 617
DIEBACK, 4, **377-378**
Diffraction, 249, 250, 333. *See also* WAVES
Diffusion Equation, 539
Digenea, 646
Digitaria erianthia, 13
DIKE, DYKE, **378,** 388, 514, 648, 664, 665
Diluvialism, 290, 291
Dimorphotheca fruticosa, 15
Dinard, 822
Dingle Bay, 691
Dinobryon, 210
Dinoflagellates, 278, 670, 672
bioluminescent, 670, 673
Dinophyceae, 483
Diodia crassifolia, 589
Diodia virginiana, 584, 585
Diodora, 617
Dionaea muscipula, 175
Diospyros rotundiflora, 13
Diospyros virginiana, 584
Diplanthera, 171
Diplopoda, 62
Diptera, 281, 490, 491
Dirt cones, 476

Discomycetes, 439
Discovery, 41
Discovery Bay, 112, 116, 311
Discrimination Analysis, 335
Dismal Swamp, 307
DISPERSION, **378,** 386
Dispersive Pressure, 122
Disphyma australe, 107
Distichlis, 199, 287, 585
Distichlis distichophylla, 107
Distichlis spicata, 171, 191, 199, 582, 586, 589, 590, 591, 592, 705, 707
Distichopora, 618, 619
Distributary Margins, 364
Distributary Networks, 364
Distributional Isopleths, 262
Diurnal Inequality, 834
Djakarta, 371
Djawa, 370, 371
Djeffara, 23
Djursland Peninsula, 203
Dneiper River, 359
Dniester River, 472
Dobrudja, 471, 472
Docks, 699
Doctor's Cove Beach, 840
Dodecanese, 417
Doghole, 442
Dolichopodidae, 490
Dollart Inlet, 412
Donax, 169
Donax cuneatus, 619
Donax faber, 619
Donax pulchellus, 10
Donegal, 411
Dong Hoi, 79
Dor, 84, 89
Dordrecht, 251, 252
Doridacea, 617
Dorset, 203, 269, 333, 408, 412, 751
Dotilla 73, 619
Douarnenez, 413
Douglas Scale, 722. *See also* SEA CONDITIONS
Douro, 414
Dover, 202, 669, 763
Downdrift offset, 605
Drainage, 514-516
Dra River, 25
Dredge, 53, 379
DREDGING, **379,** 386, 465, 466, 826, 837, 852
Drift, of sediments, 154, 746-747. *See also* SEDIMENT TRANSPORT; SHORE DRIFT CELL
Drift Ice, 447
Drift Line, 407-408
Drift Logs, 699, 701
DRIFTWOOD, **379,** 784
Drogues, 257, 258
Dromia, 653
Dromiidae, 617
Droogmakerijen, 515
Dropped Block, 535

Dropstone, 535
Drosanthemum paxianum, 11
Drosera, 174
Drosera anglica, 526
Drowned Valley Lowlands, North America, 594
Drupa morum, 618
Drupa ricinus, 618
Drupa rubus-caesius, 618
Drupa violacea, 618
Dryas integrifolia, 581
Dryomyzidae, 490
Dubai, 101
Dubrovnik, 417, 841
Dugesia, 645
Dumbbell Island, 443
Dune Rock, 523
Dunes, 162, 163, 356. *See also* COASTAL DUNES AND EOLIAN SEDIMENTATION; DUNE STABILIZATION; Dune rock; SAND DUNE HABITAT; Sand dunes
barchan, 239
classification, 263-237
clay, 819
coastal, 235-243, 316, 444, 543
flora, 236, 237-239, 589, 590-591
foredunes, 141, 265
fossil, 676
insemination, 241-242
internal geometry, 238, 240
limestone 2, 240
migrating, 239
precipitation, 239
-ridge coasts, 217
ridges, 655
transverse, formation of, 236, 239, 240
vegetated, 236
washed-out, 163
wind shadow, 141
DUNE STABILIZATION, 119, **379-380**
Dungeness, 254, 270, 408, 455, 747, 751
Dunnet Head, 878
Dupontia fisheri, 55, 581
Durban, 17, 28
Durban Bluff, 28
Durham, U.K., 413
Durnford Point, 21
Durvillea, 571
Durvillea antarctica, 199, 570, 571, 574
Durvillea potatorum, 105
Durvillea willana, 571
Dvina, 783
Dwarka, 483
Dwip, 444
Dy, 456
Dybs, 411
Dye Patches, 257, 258
Dynamic Energy, 218, 220
Dynamic Morphology, 215
Dynamic Processes, Measurement of, 220-221. *See also* BEACH

PROCESSES, MONITORING OF
Dytiscidae, 490

Eager, Eagre. *See* BORES
Earthquakes, 725
Earthquake sea wave. *See* TSUNAMI
East African Rift, 19, 20, 28, 449
East Anglia, 127, 434, 447
East Australian Current, 103
Eastbourne, 202, 204
East Cape, 577
East China Sea, 78, 359
Eastern Highlands, Australia, 110, 112, 113, 114
Eastern Scheldt, 253
East Friesland, 251
East Germany, 809, 810
East Indies, 208
East Pacific Rise, 452
East Siberian Sea, 56
Eatule adentula, 237
Ebro River, 359, 365, 415, 419
Echinacea, 384
Echinarachnius griseus, 774
Echinarachnius mirabilis, 774
Echinarachnius parma, 774
Echinocardium cordatum, 83, 612
Echinodera, 279
ECHINODERMATA, 282, **381-385,** 669
Echinoidea, 381, 383, 384
Echinometra, 383
Echinometra lucunter, 10
Echinometra mathaei, 617, 618
Echinophthiriidae, 490
Echinophyllia, 619
Echinopora, 72, 618
Echinops spinosissimus, 7
Echinothrix calamarius, 617
ECHIURA, 280, **385**
Echiurida, 67
Echiurius, 385
Ecklonia, 104
Ecklonia buccinalis, 12
Ecklonia radiata, 105, 106, 571
Ecocide, 484
Ecological communities
  Atlantic coastal, 191-194
  Pacific coastal, 195-196
Ecological impact, 260
Ecotones, 285
Ectocarpales, 634
Ectoprocta, 67, 179, 181, 636
Ecuador, 197, 198, 532, 765-766
Edentata, 483
EDGE WAVES, 136, 138, 142, 146, 271, 296, 334, **385-386,** 580, 688, 689, 694
Edinburgh, 148, 291
Edwardsia, 42, 620
Eel River, 599
EFFLUENTS, **386-387**
Eforie, 811, 812
Egregia, 286

Egypt, 7, 8, 16, 20, 22, 37, 86, 251, 369, 470
eh-Tineh, 87, 89, 92, 93, 94, 96
Eigg Island, 411, 764
Eighty Mile Beach, 112
Eil, 30
Ekman number, 564
El Arish, 83, 89, 94, 96
El Arish River, 90
Elasmobranchii, 209, 283
Elba, 415
Elbe River, 166, 411, 412
El Bluff, 190
Eleocharis, 199, 287
Eleocharis bella, 592
Eleocharis obtusa, 588
Eleocharis palustris, 592
Eleocharis parvula, 590, 591
ELEVATED SHORELINE, **387-388,** 676
El Ferrol, 414
Ellice, 624, 625
Ellobium, 621
Elminius, 199, 572
Elminius modestus, 571, 573
Elminius plicatus, 570, 571, 572, 574
Elopiformes, 283
Elos, 52
El Salvador, 188, 189, 597
El Segundo Pier, 32
El Tigre Island, 189
Elymus, 169, 170, 408, 582, 798
Elymus arenaria, 237, 408, 581, 713
Elymus mollis, 55, 581, 589, 590
Elymus vancouverensis, 589
Elytrigia, 408
Embankment processes, 300
EMBANKMENTS, **388,** 444
Embayments, 442-443
Emiotocidae, 283
EMERGENCE AND EMERGED SHORELINE, **389,** 445
Empetrum nigrum, 408, 581
Ems River, 252, 397, 412
Encelia california, 197
Encephalartos, 355
Enchylaena tomentosa, 107
Enchytraedae, 280
Enchytreaus albidus, 56
Enclosed soft shores, 620-621
Encounter Bay, 112
Endarachne, 199
ENERGY COEFFICIENTS, **390**
England, 65, 146, 147, 148, 156, 204, 205, 207, 230, 235, 237, 254-255, 267, 268, 269, 270, 281, 294, 312, 313, 314, 327, 333, 377, 387, 397, 405, 406, 413, 419, 420, 434, 442, 530, 536, 541, 639, 664, 666, 676, 680, 729, 732, 739, 745, 747, 759, 821, 823, 825, 839, 840. *See also* Great Britain; United Kingdom

English Channel, 165, 407, 716, 840
Enhalus, 73
Enhalus acoroides, 617
Enhydra lutris, 284
Enopla, 570
Ensenada de Tumaco, 198
Entada scandens, 526
Enterolobium, 194
Enteromorpha, 10, 35, 55, 56, 84, 199, 286, 405, 490, 572, 700
Enteropeneusta, 208, 668, 669
Entophysalis, 35
Entophysalis crustacea, 10
Entoprocta, 67
ENTRAINMENT, **390-391**
  and sediment transport, 240, 241
Environmental baseline studies, 261
Environmental impact
  assessment, 261
  statement, 259-265
Eolianite, 240, 376, 502, 676. *See also* AEOLIANITE
Eolian ripples, 241
Eolian sedimentation, 235-243
Eospermatopteris textilis, 642
Epacridaceae, 107
Ephedra, 453, 454, 644
Ephedrales, 453
Ephedropsida, 453
Ephesus, 52, 289
Ephydra cinerea, 490
Ephydridae, 282, 490
Epifagus virginiana, 583
Epifauna communities, 274
Epigaea repens, 583
Epilobium angustifolium, 581
Epiphytes, 586
Epistomaroides, 432
Equatorial Current, 12
Equilibrium curvature, 152
EQUILIBRIUM SHORELINE, 152, **391-392**
EQUISETOPHYTA, **392-393**
Equisetum, 55, 392, 592
Eragrostis pilosa, 585
Ericaceae, 174, 582, 715
Eridu, 52
Erie Canal, 254
Erigeron pusillus, 585
Eriodictyon, 197
Eriogonum fasciculatum, 197
Eriphia, 617
Erosion
  abatement measures, 269-270
  glacial, 446-447
  indicators, 389
  and sedimentation, 465
EROSION RAMP, WAVE RAMP, **393**
Erratics, 123, 535-537
  glacial, 123
Erythrean, 452
Esbjaerg, 411
Escarpment, 721

913

Escudo de Veraguas Islands, 190
Eskimos, 56, 370
Esperance, 104
Essaouira, 25
Essex, 413
Esterel, 415
Estonia, 177, 502, 503, 782
ESTUARIES, **393-397**. *See also specific names;* ESTUARINE *entrieo;* Estuarine-shelf dependency
ESTUARINE COASTS, **397-398**
ESTUARINE DELTA, 366, **398, 400**
ESTUARINE HABITAT, **398-400**
ESTUARINE SEDIMENTATION, **400-402**
rates, 396
Estuarine-shelf dependency, 399
Étangs, 415, 443
Étangs de Berre, 415
Étangs de Vaccares, 415
Etched pothole. *See* SOLUTION AND SOLUTION PAN
Etesian winds, 99
Ethiopia, 90, 449
Ethiopian Highlands, 23
Etna, 418
Euapta, 382
Eubacteria, 720
Euboea, 417
Eucalyptus, 107, 109
Eucestoda, 646
Eucheuma, 74
Euchone incolor, 612
Euchordata, 208
Eucidaris, 382
Eucla Basin, 112
Euclea natalensis, 13
Euechinoidea, 384
Eugenia, 70
Euglena, 35, 402
Euglenoidida, 670
EUGLENOPHYTA, **402-403**
Eulerian criterion, 346
Euler number, 718
Eulittoral zone, 104, 165
Eumycota, 439
Eunephyta, 775
Eunicella, 619
Eunice viridis, 617
Euniciidae, 39, 40, 617, 620
Euphausiacea, 281
Euphausia crystallorophias, 281
Euphausia superba, 281
Euphorbia ammannioides, 585
Euphorbiaceae, 304, 528, 587
Euphorbia dentata, 585
Euphorbia paralias, 7
Euphorbia polygonifolia, 584, 585
Euphrates River, 101, 465
Euplectella, 650
Eurasia, 19
Eurasian Plate, 17
Euripus, 443

Europe, 127, 128, 176, 186, 203, 228, 236-237, 241, 251, 269, 282, 285, 289, 313, 359, 369, 371, 374, 375, 403-409, 419, 420, 450, 463, 464, 471, 490, 498, 501, 521, 541, 543, 544, 569, 642, 664, 671, 731, 756, 774, 775, 798, 807, 809, 826, 827, 837, 839, 840, 848, 850
EUROPE, COASTAL ECOLOGY, **403-409**
EUROPE, COASTAL MORPHOLOGY, **409-420**
Europoort, 65
Eurypterida, 63, 280
Eurythoe, 620
Eustatic. *See* SEA LEVEL CHANGES
Eutreptia marina, 670
Euxinic, 82, **420-421**
Euzkadi, 371
Evans Head, 466
EVAPORITES, **421-423**
Everglades, 594
Everglades National Park, 313, 586
Evvoia, 417
Exchange ratio, 820
Excirolana latipes, 10
Excoecaria, 108, 621
Excoecaria agallocha, 527
Exfoliation, boulder of, 176
Exmoor National Park, 313
Exsudation. *See* SALT WEATHERING
Eyre Peninsula, 110
Eyrr, 442

Fabaceae, 527, 528, 587, 588
Faceting, 50
Facies
  complexes, 318
  elements, 314
  systems, 316
Factor analysis, 335
Fagus grandifolia, 582, 583
Fairway, 443
Falaise, 444
Falconiformes, 284
Falcon Island, 623
Falkland Plateau, 17, 18
Fano Island, 411
Farafangana, 30
Farallon Plate, 452
Farasan Islands, 100
Farewell Spit, New Zealand, 577
Farilttoes Island, 414
Faro, 414
FAULT COAST, 214, 216, **424**
Favia, 11, 72, 618
Faviidae, 618
Favites, 618
Fecal pellets, 40, 612
Fecampia, 645
FEEDBACK, **424-425**
FEEDER BEACH, **425**

Felixstowe, 541
Fens, 369
Fernandina, 727
Fernando Poo, 26
Ferns. *See* POLYPODIOPHYTA
FERREL'S LAW, **425-426**
Festuca, 55, 288
Festuca littoralis, 107, 108
Festuca rubra, 581, 582, 583, 584, 589, 590, 591
Fetch, 248, 508, 798, 873
FIARD, FJÄRD, 46, **426-427**, 434, 447, 758, 782
Ficus, 194, 197, 198, 199
Ficus hippopotamus, 15
Ficus sycamorus, 15
Ficus tremula, 13
Fidalgo Island, 142
Field capacity, 714
Fifeshire, 410
Fiherenana River, 30
Fiji, 616, 621, 623, 624
Filicales, 650
Filicinae, 642
Filicophyta, 649
Filograna, 617
Fimbria, 619
Fimbristylis castanea, 585, 586, 588
Fimbristylis spadicea, 586
Fingal's Cave, 411
Finland, 410, 502, 725, 782
Finmark, 374
FIORD, FJORD, 46, 186, 394, 410, 426, **427-428**, 434, 435, 447, 543, 575, 593, 600, 767, 782, 784, 840. *See also specific names*
Fiordland, 575
Fire Island, 583
FIRTH, 410, 427, **428,** 569
Firth of Forth, 411
Firth of Lorn, 428
Firth of Tay, 428
Firth of Thames, 207, 625
Fissurella, 199, 617
Fissurella nubecula, 10
Fitzroya, 642
Fitzroya patagonica, 197
Fitzroy River, 114, 117
Fiume, 416
Fjärd. *See* FIARD, FJÄRD
Fjardur, 427
Fjord. *See* FIORD, FJORD
Flabellum, 619
Flamborough, 202, 413
Flanders, 369, 648
FLANDRIAN TRANSGRESSION, 26, 28, 128, 188, 215, 349, **428-429**, 486, 504, 521, 625, 628, 760
Flatworms, 542
Fleet, 442, 443
Flensburg Förde, 434
Fleurieu Peninsula, 110
Flevo Lacus, 515
Flies, 282

Flinders Current, 103
Flindersian Province, 104
Flinders Island, 110
FLOCCULATION, **429-430**
Flores Sea, 67
Florida, U.S.A., 126, 127, 150, 158, 160, 171, 187, 247, 270, 284, 287, 302, 303, 313, 325, 326, 329, 355, 394, 461, 466, 491, 501, 522, 532, 547, 585-587, 594, 595, 596, 653, 679, 680, 690, 701, 705, 715, 727, 728, 732, 753, 754, 760, 761, 799, 815, 839, 883
Florida Peninsula, 586-588
Florideophyceae, 686, 687
FLOTSAM AND JETSAM, 236, 281, 282, 381, **430-431**
Flow, 442
Flowering plants. *See* MAGNOLIOPHYTA
Flow-regime conditions, 162
Flustra, 182, 236
Flustrella, 56
Fluting, 50
Fluvialism, 290
Fluviosingula nipponica, 772
Fly Delta, 366, 398
FOAM MARK, **431**
Foeniculum, 288
Föhrde, 411, 434-435
Folegandros, 417
Fonseca transverse fault, 188
FOOD CHAIN, **431,** 439, 659
Food web, 431
FORAMINIFERA, 164, 278, 339, 341, **431-434,** 612, 617, 618, 669, 670, 776
FORCED WAVES, **434-435,** 533
Forcipulatida, 381
FÖRDE, 427, **434-435,** 447
Forecasting, 299
Foreland, 444
  traveling, 349
Foreshore, 140
  features, 529-530
  step, 140
Formentura, 415
Fort Myers, 587
Fort Walton Beach, 158
Fosse de Cap Breton, 414
FOSSIL CLIFF, 230, **435,** 485
FOULING, 181, **435-436**
FOURIER ANALYSIS, 334, **436-437,** 838, 842. *See also* Harmonic analysis
Foveron Links, 519
Foxe Basin, 176
Fragaria chiloensis, 589
Fragum, 619
France, 37, 65, 186, 202, 235, 254, 294, 313, 369, 387, 397, 413, 414, 415, 420, 444, 445, 471, 520, 536, 680, 691, 732, 758, 807, 808, 809, 815, 821, 823, 840, 841
Frankenia, 196, 199
Frankenia grandifolia, 705

Frankenia palmeri, 589, 590
Franklin County, 160
Franseria chammisonis, 590
Franz Josef Land, 771
Fraser Island, 106, 108, 114, 115, 116, 118, 119, 311, 325
Fraxinus pensylvanica, 586
Freetown, 25
Freeze-up, 475, 479, 480, 481
French Guiana, 769
Fresco, 9, 26
Freshwater/saltwater interface, 468
Fretting, 708
Friesland, 515, 648
Frisches, 411, 458, 569
Frisian Islands, 412, 489, 515, 850
Fromia, 617
FROST RIVING, **437,** 795
Frost shattering, 437, 463, 795
Frost splitting, 437
Frost table, 476
Frost weathering, 758
Frost wedging, 437, 799, 805
FROUDE NUMBER, 163, **437-438,** 546, 547, 690, 696, 697, 703, 718, 872
Fucales, 105, 404, 634
Fucus, 56, 202, 286, 305, 404, 582, 700, 776
Fucus ceranoides, 404
Fucus evanescens, 773, 774
Fucus inflata, 56, 775
Fucus serratus, 775
Fucus vesiculosus, 581, 775
Fuirena simplex, 588
FULCRUM EFFECT, **438**
Fully arisen sea, 249
Funafuti Atoll, 168, 456
Fundulus, 283
FUNGI, **438-441**
Fungia, 72, 618
Fungiidae, 618
Funnel Sea, 442

Gabes, 19, 20
Gabon, 9
Gabon Basin, 20, 26, 451
Gadus, 775
Gafrarium, 73, 621
Gagra, 787
Gahnia, 107
Gahnia filum, 108
Gahnia trifida, 108
Galápagos Islands, 200, 209, 277, 284, 623
Galeolaria caespitosa, 104, 105, 106
Galicia, 185, 410, 414, 419, 691
Galilee, 93, 95
Galium palustre, 581
Galveston, Texas, U.S.A., 242, 588, 646, 647, 837
Galveston Island, 127, 130
Galway Bay, 412
Gambia River, 9
Gammarus, 775
Gammarus setosus, 55, 776

Gammarus wilkitzkii, 56
Ganges-Brahmaputra Delta, 77, 359, 365, 482, 483, 485
Ganges River, 69, 79, 292, 369, 482, 483, 484, 485, 486
Gap, 443
Gapeau, 415
Gargano Peninsula, 416
Garian horst, 20
Gascogny, 235, 236, 370
Gasteromycetes, 439
Gasterosteiformes, 283
Gastrochaena, 617
Gastropoda, 282, 551, 553
Gastrotricha, 67, 279
Gat, gate, 443
Gaultheria procumbens, 583
Gaultheria shallon, 591
Gaussian distribution, 555, 871
Gay Head, 594
Gaylussacia baccata, 583, 584
Gaylussacia dumosa, 174
Gaza, 23, 92
Gazania uniflora, 13
Gdansk, 411
Geest, 411
Gela, 418
Gelappten küsten, 520
Gelendzhik, 667
Gelidium, 573
Gelidium cartaligineum, 12
Gelidium pristoides, 12
Gelidium pusillum, 571, 573
Gelidium reptans, 11
Gelliodes, 651, 653
Geloina, 621
Gelsemium sempervirens, 586
Genoa, 369
Geo, gia, gja, goe, 442
GEOGRAPHIC TERMINOLOGY, **442-445**
Geograpsus, 621
GEOMORPHIC-CYCLE THEORY, **445-446**
Geonemertes, 279
Georgia, U.S.A., 126, 128, 129, 238, 550, 586-587, 594, 680, 736
Georgian republic, U.S.S.R., 788, 789
Georgia Sea Island, 127
Gerardia maritima, 586, 705
Gerardia purpurea, 585
Germany, 126, 174, 254, 273, 397, 411, 412, 434, 435, 447, 471, 666-667, 807, 810, 840, 850
Gerridae, 281, 490, 491
Ghana, 658
GHYBEN-HERZBERG RATIO, 342, **446,** 469
Giant's Causeway, 411, 761
Gibraltar, 414, 826
Gibraltar Point, 237
Giens, 415
Giganta Range, 598
Gigartina, 286, 700
Gigartina alveata, 570, 571

SUBJECT INDEX

Gigartina minima, 11
Gijón, 414, 678
Gilbert Islands, 624, 625, 626
Gilgai, 760
Ginkgo, 640, 642
Ginkgoales, 640
Ginkgo biloba, 640
Ginkgoopsida, 640, 642, 644
Gippsland, 114
Gippsland Lakes, 112, 118, 398
Gironde, 176, 408, 414, 840
Gizhiga, 785
Glacial deposition, 216, 447
GLACIATED COASTS, 212, 213, **446-447**
GLACIEL, **447-448**, 535
Glacio-eustatic. *See* SEA LEVEL CHANGES
Glacioeustatic oscillation, 387
Glacioisostic uplift, 387
Glasgow, 410, 852
Glaux, 287, 288
Glaux maritima, 582, 584, 590, 592
Glehnia littoralis, 589
GLOBAL TECTONICS, 211, 215, **448-453**
Gloiopeltis capillaris, 773
Gloriosa virescens, 13
Glossophora kunthii, 572
Glossopsida, 524
Glottidia, 282
Gloup. *See* BLOWHOLES
Glyceria, 581
Glyceridae, 39, 40, 620
Glycimeris, 96
Glycimeris violacescens, 84, 89
Glyptonotus antarcticus, 42
Glyptostrobus, 642
Gnathorhynchus, 645
Gnathostomata, 384
Gnathostomulida, 67, 279, 645, 646
Gnesioceros sargassicola, 645
Gnetales, 454
GNETICAE, **453-454**, 644
Gnetophyta, 453, 454
Gnetopsida, 454
Gnetum, 454, 644
Goa, 486
Goascoran River, 189
Gobiescoiformes, 283
Godavari River Delta, 366, 482, 483, 485, 486
Goëmen, 370
Goeree Island, 412
Golas, 685
Gold Coast, 884
Golden Hoop, 252
Golfito, 190
Gomphrena canescens, 108
Gondwanaland, 17, 19, 20, 26, 28, 451
Goniadidae, 39, 40
Goniastrea, 11, 72, 618
Goniopora, 72, 618
Gonyaulax, 278, 670
Goodwin Sands, 268
Gordonia lasianthus, 175

Gorge, 486
Gorgonaria, 611
Gorgonia, 619
Gorlo, 783
Gotland, 160
Goulet, 333
Gourliea, 198
Gower Peninsula, 412
GRAB SAMPLERS, 244, **454**, 736
Gracilaria secundata, 574
Grading, 147
Graffila, 645
Graffizoon, 645
Grain movement, physics of, 240, 241
Grain stones, 185
Graminae, 195, 527, 528
Grand Congloué, 53
Grand Forks, 511
Grand Harbor, 418
Grand Isle, 127, 596
Granules, 455, 631, 738, 751
Grapsus, 616
Graptolites, 669
GRAVEL, **455**, 632, 710
GRAVEL RIDGE AND RAMPART, **455-456**
Great Australian Bight, 107, 110, 112
Great Barrier Reef, 72, 104, 106, 110, 112, 115, 117, 313, 332, 339, 571
Great Britain, 220, 311, 312, 313, 369, 406, 407, 410, 444, 447, 455, 465, 603, 632, 657, 658, 676, 751, 803, 821, 840, 848. *See also* England; United Kingdom
Great Escarpment, 17
Great Geba Flat, 25
Great Holland Poulder, 252
Great Lakes, 135, 263, 463, 475, 506, 509, 511, 550, 639, 655, 656, 874
Great Salt Lake, 470, 471, 498
Great Sandy Desert, 112, 375
Greece, 52, 53, 54, 369, 387, 417-418, 419, 434, 648, 675, 840, 841
Greenland, 55, 56, 57, 58, 387, 397, 427, 461, 462, 464, 642, 676, 758, 771, 775, 799
Greifswald, 174
Grevelingen, 253
Grewia glandulosa, 13
Grijalva River, 359, 366
Grindelia cueifolia, 590
Grindelia integrifolia, 590, 592
GROINS, 66, 246, 255, 257, **456**, 499, 509, 518, 633, 656, 657, 664, 665, 699
Dutch, 254
Groningen, 412
Gross production, 659
Groundwater, 468, 470, 523, 547, 548, 549, 727, 861. *See also* WATER TABLE

Grou River, 25
Gruiformes, 284
Grus americana, 284
Guadalquivir, 414
Guadalquivir River, 313
Guadalupe, 196
Guadiana River, 414
Guairamar, 238
Guajira Peninsula, 198, 765
Guam, 355
Guatemala, 188, 189, 190, 196
Guaymas, 597
Guazuma, 194
Guerrero, 197, 235
Guerro Negro, 597, 598
Guettarda speciosa, 108
Guiana, 200, 206, 207
Guiana Shield, 769
Guinea, 3, 4, 6, 19, 20, 21, 22, 25
Guinea Bissau, 25
Guinea Current, 20, 26, 360
Gujarat, 484, 486
Gulf, 442. *See also specific names*
Gulf of Aden, 19, 30, 99
Gulf of Alaska, 228, 452, 600, 843
Gulf of Ancud, 197
Gulf of Aqaba, 30, 100
Gulf of Argolikós, 417
Gulf of Barcelona, 769
Gulf of Bothnia, 228, 406, 462, 463, 816
Gulf of California, 206, 207, 323, 449, 452, 532, 597, 598, 819 823
Gulf of Cambay, 484, 485
Gulf of Carpentaria, 112, 114, 116, 761, 881
Gulf Coast, 126, 278, 394, 797, 839, 840, 851
Gulf of Elat, 680
Gulf of Eufemia, 416
Gulf of Fonseca, 188, 189, 190, 597
Gulf of Fréjus, 415
Gulf of Gabes, 23
Gulf of Gioia, 416
Gulf of Guayaquil, 198, 765
Gulf of Guinea, 4, 20, 359
Gulf of Kalamata, 434
Gulf of Kavalla, 417
Gulf of Lakonikós, 417
Gulf of the Lion, 471
Gulf of Manaar, 481
Gulf of Messiniakós, 417
Gulf of Mexico, 127, 128, 134, 143, 159, 220, 359, 365, 369, 398, 461, 467, 472, 504, 550, 585, 586, 588, 596, 597, 606, 646, 724, 819, 823, 827
Gulf of Morrosquillo, 198
Gulf of Oman, 100-102
Gulf of Orosei, 418
Gulf of Panama, 188, 189, 190, 197, 522, 597
Gulf of Paria, 769
Gulf of Patras, 417
Gulf of Po-Hai, 77, 78

916

Gulf of Policastro, 416
Gulf of Rapallo, 415
Gulf of Rijeka, 416
Gulf of Rosas, 415
Gulf of St. Lawrence, 135, 479, 581, 582, 593, 839
Gulf of Salerno, 416
Gulf of San Jorge, 415, 767
Gulf of San Marcos, 767, 768
Gulf of San Matías, 199
Gulf of San Miguel, 190, 518
Gulf of Siam, 359, 360
Gulf of Sidra, 23
Gulf of Spezia, 415
Gulf Stream, 347, 369, 527, 581, 587, 728
Gulf of Suez, 12, 16, 30, 100, 374
Gulf of Taranto, 416
Gulf of Tehuantepec, 189, 597
Gulf of Thailand, 74
Gulf of Tonkin, 359
Gulf of Trieste, 416
Gulf of Venezuela, 769, 770
Gully, 442, 443
Gunnarea capensis, 12
Gut, 443
Guyana, 374, 769, 792
Guysborough County, 156
Gwettarda, 195
Gymnocarpos decandrum, 8
Gymnodinium, 278
Gymnolaemata, 182, 282
Gymnospermae, 526
Gymnostomatida, 279
Gyodo River, 499
Gyratrix, 645
Gyrodactylus, 646
GYTTJA, **456-457**, 569

Haarlemmermeer, 412
Habitats
  coastal waters, 332-324
  coral reef, 339-342
  estuarine, 398-400
  intertidal, 42
  intertidal mud, 493-495
  intertidal sand, 495-497
  loose rock and stone, 520-522
  rocky shore, 699-702
  sand dune, 712-716
Hadera, 89
HAFF, 174, 411, **458**, 471
Haifa, 83, 87, 89, 94
Haifa Bay, 82, 83, 84, 87, 90, 91, 92, 93, 95
Haiti, 371
Halacaridae, 280
Halammohydra, 619
Halecium, 42
Half-heart shape, 151, 883
Half-Islands, 444
Half Moon Bay, 884
Haliaetus leucocpehalus, 284
Halicore dugong, 483
Halicryptus spinulosis, 55
Halimeda, 339, 341, 617, 618, 624
Halimeda tuna, 83

Halimione, 288
Halimione portulacoides, 406
Haliotis, 617
Halobates, 281, 491
Halocnemon strobilaceum, 8, 16
Halodule, 286, 617
Halokarst, 501, 502
Halomitra, 618
Halopeplis perfoliata, 16
Halophila, 490
Halophila stipulacea, 405
HALOPHYTES, 10, 304, 406, **458-459**, 620, 704-706, 713, 819
Halopteris hordacea, 571
Halosaccion, 286
Halovelia, 491
Halul ou Das Islands, 101
Hamburg, 251, 371, 412
Hammans, 810
Hampshire, 207, 270, 413
Hamra, 91
Hanging valleys, 444, 630
Hantsholm-Bulbjerg, 204
Haplodiscus, 645
Haplopappus sugarrosus, 197
Haplosclerina, 279
Haptophyceae, 210
Hapuku River, 578
HARBORS, **459-460**
  Central America, 190-191
  and estuaries, New Zealand, 572-573
  structures, design of, 172, 173
Haringvliet, 253
Harmatobatidae, 281
Harmonic analysis, 838. *See also* FOURIER ANALYSIS
Harpacticoida, 280
Harpidae, 619
Harwich, 413
Havana, 371
Haven, 442
Havsband, 443
Hawaii, 172, 387, 393, 450, 452, 519, 526, 616, 621, 622, 623, 624, 626, 657, 721, 815, 839, 840, 843, 844, 845, 854, 857
Hawkesbury River, 691
HEADLAND BAY BEACH, **460-461**
Headlands, 444
Hebridean Shield, 409
Hebrides, 370, 410
Heliacus, 619
Helianthemum arenicola, 585
Helice, city of, 123
Helice, 621
Helice crassa, 574
Helichrysum paralium, 108
Heligoland Island, 412
Heliocidaris erythrogramma, 106
Heliopora, 626
Heliotherapy, 808
Heliotropium, 199
Heliotropium curassavicum, 589
Hellevoet, 412

Heloecius, 621
Hemiascomycetes, 439
Hemichordata, 282
Hemiptera, 281, 490, 491
Henricia, 617
Hepaticae, 181
Hepatidae, 181
Heptopsida, 181
Heracleia, 52
Heracleum maximum, 581
Heritage Coast, 312
Heritiera, 70
Heritiera fomes, 482
Heritiera littoralis, 13
Heritiera sundri, 69
Hermatobates, 491
Hermatobatidae, 490
Herodiones, 483
Herpolithon, 618
Heterocentrotus mammillatus, 618
Heteronema, 617
Heteronemertini, 570
Heterostachys, 707
Heterotheca subaxillaris, 585, 588
Hevea brasiliensis, 528
Hexacorallia, 331, 332
Hexactinellida, 279
Hiatella arctica, 776
Hibiscus, 586
Hibiscus moscheutos, 584
Hibiscus tiliaceus, 13, 14, 108
High Atlas, 20
HIGH-ENERGY COAST, 151, 218, **461**, 467
HIGHEST COASTLINE, **461-462**. *See also* Marin Gräns
HIGH-LATITUDE COASTS, **462-463**
High mangrove forest, 198
Hijumaa Island, 782
Hilazon, 94
Hildenbrantia, 572
Hilleh Rud, 102
Hilo, 844, 845
Hindcasting, 249, 257, 299
Hippa, 73
Hippa cubensis, 10
Hippa pacifica, 619
Hippmane mancinella, 191
Hippophae rhamnoides, 237, 408
Hippospongia, 617, 652, 653
Hippotamus amphibius, 14
Hippurus tetraphylla, 581
Hirudinea, 39, 280
Histria, 472
Hoang-Ho Delta, 77
Högsta kustlinjen, 461
Hokkaido, 78, 771, 773
Holderness, 268, 269, 413
Hole, 442
Holland, 65, 369, 371, 397, 408, 411, 412, 419, 648, 664, 665, 762
Hollows, 630
HOLOCENE, **463-465**
Holocephalii, 283

Holoplankton, 323
Holothuria, 617
Holothuria scabra, 620
Holothuroidea, 381, 384, 611
Homaxinella, 42
Homo sapiens, 285
Homosira, 572
Homotrema, 617, 619
Honduras, 188, 189, 190, 191, 194, 195, 597
Honduras Gulf, 188
Honeycombing, 37, 224, 500
Hong Kong, 78, 79, 371, 397, 514, 805
Honkenya, 288
Honkenya peploides, 55, 237, 589
Honshu, 78, 371, 771
Hooghly River, 485
Hooghlyside, 484
Hook, 444
Hope, 442
Hoplonemertini, 279, 570
Hordeum boreale, 592
Hordeum brachyantherum, 590
Hordeum jubatum, 582
Hormosira, 105, 573
Hormosira banksii, 105, 106, 571, 573, 574
Hormuz, 102
Horn, 444
Horn of Africa, 4, 12, 19, 20, 228
Hornsea, 254, 255
Horsens Fiord, 411
Horsetails. *See* EQUISETOPHYTA
Horton Delta, 366
Hot spots, 451
Houtman Abrolhos, 104, 110, 117, 501
Huahine, 627
Huai-Ho River, 78
Huang-Ho River, 78
Hudson Bay, 389, 448, 449, 462, 463, 593, 655, 676
Hudsonia tomentosa, 582, 583, 585
Hudson River, 397
Hudson Strait, 176
Huelva, 414
HUMAN IMPACT, **465-466**
HUMATE, **466-467.** *See also* Coffee rock
Humber River, 255, 413
Humboldt Bay, 591, 599
Humboldt Current, 623
Humphrey spiral, 546
Humus, 456
Hunstanton, 413
Hunter River, 114
Hurghada, 16
HURRICANE EFFECTS, **467**
Hurricane, 326, 797, 800, 851
 beaches, 455
 surges, 724
 tide or wave. *See* STORM SURGE
Hwang Ho River, 356, 359

Hyalospongea, 650, 651
Hydnophora, 618
Hydra, 279
HYDRAULIC ACTION AND WEDGING, **468**
Hydraulic zones, 562, 564
Hydric deficit, 369
Hydrobia, 494
Hydrocharitaceae, 527
Hydroclathrus, 617
Hydrocotyle, 586
Hydrocotyle bonariensis, 585
Hydrodea bossiana, 11
Hydrodynamical equation, 538, 539
HYDROGEOLOGY OF COASTS, **468-470**
Hydrophilidae, 284, 490
Hydrophylax carnosa, 15
Hydrothermal springs, 800
Hydrozoa, 330-331, 611
Hymenaea, 194
Hymenomonas, 210
Hymenomycetes, 439
Hymenoptera, 491
Hymenostomatida, 279
HYPERSALINE COASTAL LAKES, **470-474**
Hyphaene coriacea, 13
Hyphalus insularis, 491
Hyphochytridiomycetes, 439
Hyphomycetes, 439
Hypnea rosea, 11
Hypnea spicifera, 11
Hypogastrura viatica, 491
Hypogastruridae, 490
Hypotrichida, 279
HYPSITHERMAL, **474.** *See also* Climatic optimum
Hyptis, 199

Iberia, 410, 559
Iberian Peninsula, 840
Ica River, 766
Ice. *See also* ICE *entries;* Ice-push; Sea ice; Shelf ice; Shore ice
 effects of, 476-477
 ground, 462
 jams, 513
 nearshore, 477
 pack, 60, 77
 rafting, 123, 475, 477
 ramparts, 509, 513
 ridge, air flow, 226
 terrestrail, 477-478
 wedges, 60
ICE ALONG THE SHORE, **475-478,** 725
ICE-BORDERED COASTS, **479-480**
ICE FOOT, 46, 60, 462, 475, 476, 477, 479, **480-481,** 500, 558, 559, 560, 756
 floes, 535
Iceland, 55, 56, 374, 427, 447, 449, 452, 461, 462, 615, 657, 758, 771, 799

Ice-push, 177, 475, 476, 477
 terrace, 513
Ichthyosaurs, 209
Icy Cape, 352, 479
Idku Lagoon, 23
Ijmuiden, 235
Ijssel Meer, 412, 515
Ikaría, 417
Iles Glorieuses, 30
Iles d' Hyères, 415
Iles aquifolium, 408
Ilex glabra, 584
Ilex opaca, 583, 584, 585
Ilex verticillata, 582
Ilex vomitoria, 585
Illinois, U.S.A., 805
Impact Statements, Coastal Environment, 259-265
Impact Studies, 139
Imperial Valley, 597
Impounding, 825
Incipient Berm, 163
India, 19, 37, 69, 73, 77, 126, 280, 284, 289, 323, 351, 366, 369, 374, 451, 481-484, 484-486, 527, 544, 666, 799, 806, 815
INDIA, COASTAL ECOLOGY, **481-484**
INDIA, COASTAL MORPHOLOGY, **484-486**
Indiana, U.S.A., 713, 715
Indian Ocean, 6, 7, 12, 16, 18, 19, 30, 67, 72, 73, 74, 77, 99, 103, 116, 289, 359, 449, 451, 484, 523, 532, 803, 824, 827, 840, 876
Indian Plate, 76
Indonesia, 69, 74, 75, 77, 79-81, 284, 374, 387, 453, 466, 495
Indonesian Archipelago, 76
Indus Delta, 77, 376
Indus River, 81, 359
Industrial Wastes, 852-853
Inertial Regime, 122
Infauna Communities, 274
Infralittoral Fringe, 165
Infralittoral Zone, 165
Inhaca Island, 14, 28
Inhambane, 14
INLETS, MARINE-LAGOONAL AND MARINE-FLUVIALS, **489**
INLETS AND INLET MIGRATION, 443, **486-488,** 489
 overlapping, 605
Inner Shelf Floor, 562
Inner Shelf Systems, 316
Insecta, 281
INSECTS, 55, 61, 62, 243, 273, **489-492,** 507
Integral Transform, 436
Interdistributary Features, 364-365
Internal Dew, 714
INTERNAL WAVES, 296, 297, 394, **492,** 734, 872
INTERTIDAL FLATS, **493**

INTERTIDAL MUD HABITAT, **493-495**
INTERTIDAL SAND HABITAT, **495-497**
Intertropical Convergence Zone, 3
Inuit, 56
Inverness, 374
Ionia, 52
Ionian Sea, 450, 558
Íos, 417
Ipomoea, 70, 189, 616, 619, 713
Ipomoea bilboa, 13, 482
Ipomoea brasiliensis, 589
Ipomoea pes-caprae, 13, 108, 109, 191
Ipomoea sagittata, 585
Ipomoea stonifera, 585, 588, 589
Ipsopoda, 281
Iran, 101, 387, 841
Iranian Plate, 101
Iraq, 360, 364
Ircinia, 652
Ireland, 203, 370, 387, 410, 411, 412, 443, 447, 452, 521, 544, 680, 691, 751, 761, 798, 840.
  See also United Kingdom
Irian, 374
Iridea, 286, 432
Iridophycus capensis, 12
Irifi, 25
Irish Republic, 840
Irish Sea, 412, 692
Iris setosa, 581
Ironstone, 79
Irrawaddy River, 69, 70, 77, 79, 359, 366, 370, 657, 821
Isactis plana, 104
Isanda cf. *holdsworthiana,* 84
Ischia Island, 416
Ischnocera, 490
Isfod, 480
Isidaea, 199
Island Beach, 303
Islands, 443. See also specific names; ARTIFICIAL ISLANDS
  demography, 370, 371
  floating, 65, 443
  land-tied, 443
Isla de Pascua, 848
Islas de la Bahía, 190
Islas del Istmo, 190
Islas del Rey, 190
Isle of Purbeck, 412
Isle of Wight, 203, 412, 413
Islets, 624-625
Isochrysidales, 210
Isoetales, 525
Isoetes, 525
Isoetopsida, 524, 525
Isognomon, 617
Isonzo River, 416
Isopoda, 611
Isostasy, 449
ISOSTATIC ADJUSTMENT, **497-498**, 508, 654

ISOSTATICALLY WARPED COASTS, **498**
Isostatic Uplift, 406
Isotomidae, 490
Israel, 82, 83, 84, 86, 87, 235, 238, 239, 370, 815, 841
Istanbul, 371
Isthmus of Panama, 188, 189, 190
Istria, 416
Italy, 185, 348, 387, 415-416, 419, 445, 501, 648, 807, 839, 841
Iturup, 771
Iva imbricata, 585
Iva frutescens, 584, 585, 586, 705
Iviza, 415
Ivory Coast, 4, 25, 370, 815
Ivory Coast Basin, 20, 26

Jacksonville, Florida, U.S.A., 585
Jahresringe, 422
Jamaica, 501, 840
James Bay, 559, 756
Jania, 616, 618
Janjira, 485
Japan, 48, 64, 65, 68, 74, 76, 77, 78-81, 179, 235, 283, 294, 311, 313, 370, 387, 389, 466, 499, 514, 532, 543, 640, 676, 687, 690, 724, 725, 772, 840, 844
Jardim Sao Pedro, 239
Jaumea, 287
Jaumea carnosa, 590, 591
Java, 70, 72, 77, 79, 81, 323, 452
Java Sea, 67, 79, 545, 827
Jebel el Akdar, 23
Jeddah, 100
Jersey, 371
Jervis Bay, 112
Jetsam, 430-431
JETTIES, 66, 247, **499**, 509, 639, 656, 657, 666, 699
Jiang Su, 371
Johns Pass, 247
Jolo Island, 848
Jordan River, 471
Joseph Bouaparte Gulf, 114
Jounieh Bay, 99
Juba River, 30
Juncaceae, 527
Juncus, 107, 199, 287, 574, 581, 583, 584, 586
Juncus acuminatus, 584
Juncus acutus, 707
Juncus arcticus, 581
Juncus biglumis, 581
Juncus effusus, 584
Juncus falcutus, 591
Juncus filiformis, 581
Juncus gerardii, 459, 582, 705, 707
Juncus krausii, 14, 107, 108
Juncus lesueurii, 590, 592
Juncus maritimus, 459
Juncus phaeocephalis, 591
Juncus roemerianus, 585, 586, 705
Juncus triglumis, 581
Juniperus, 642

Juniperus communis, 582
Juniperus virginiana, 583, 584, 585, 586
Jupiter, 841
Jutland, 204
Jylland, 411, 434

Kaikoura Mountains, 578, 579
Kaimeni Islands, 417
KAIMOO, 476, 479, 481, **500**, 549
Kaina-Pilgen, 784
Kaitorete Barrier, 577, 578
Kalimantan, 77, 79, 81, 374
Kaliningrad, 666
Kalmia angustifolia, 174, 583
Kamaran Islands, 100
Kamchatka, 76, 77, 374, 424, 772, 785, 786, 844
Kandellia, 287
Kangaroo Island, 110
Kanin Peninsula, 783-784
Kaohsiung Harbor, 75
Kaokoveld, 27
Kapuas River, 81
Kara Bogaz, 471, 472, 787
Karadag Mountain, 788
Karang, 443
Kara Sea, 56, 771, 783
Karimata Islands, 81
Karkinite Bay, 787
Karman Constant, 241
Karnataka, 483, 485
Karrenfeld, 626
Karst, 185, 558
  littoral, 500
KARST COAST, **500-502**, 877
Karun River, 101
Karwar, 482, 485, 486
Kasegaluk Lagoon, 479
Kathiawar, 374, 482, 484, 485, 486
Kattegat, 434
Kauai, 172, 526, 622, 624
Kawhia Harbour, 578
Kayals, 485
Kéa, 417
Kefallinia Island, 417
Kei River, 11, 12, 14
Kenchreai, 54
Kent, 202, 268, 270, 408, 413, 751
Kenya, 7, 12, 13, 19, 21, 29, 313, 366
Kenya Basin, 20, 28
Kerala, 371, 482, 483, 484, 485, 544
Kerguelenella, 199
Kerguelen Islands, 282
Kerguelen Ridge, 45
Kérkira Island, 417
Kermadec Islands, 574, 844
Kess-Kess, 443
Kettles, 476, 549
Keys, 127, 701
  coral, 594
Keystone species, 702
Key West, 487, 728
Khalkidhiki Peninsula, 417
Khios, 417

Khulna, 69
Kiel Bay, 411
Kimberley Ranges, 110, 821
King Island, 110
Kingscliff Cudgen, 326
King Sound, 114, 117, 118
Kinorhyncha, 67, 279
Kirra Beach, 327
Kislaya Bay, 821
Kismayu, 30
Kíthnos, 417
Klaipéda, 458, 569, 782
Klang River, 359, 360, 365
Klint, 444
KLINT COAST, **502-503**
Knolls, 342
Kobe Harbor, 64, 65
Koilu, 374
Kola, Peninsula, 782
Kolkas Rags Cape, 782
Kolkhida, 788, 789
Kolk Pits, 697
Kólpos Korinthiakos, 417
Kolyma River, 783
Königstuhl, 411
Konkan, 482, 486
Korea, 67, 68, 74, 75, 77, 78, 79, 532
Koryaksk, 784, 785
Kosi Bay, 14, 689
Kosteletskya virginica, 584, 586
Kotor, 417
Kotzebue Sound, 601
Kouchibouguac Bay, 689
Krillon Cape, 786
Krishna River, 485, 486
Kristianstad, 411
Krka, 417
Kukhtuya River, 785
Kukpuk Delta, 601
Kura River, 787
Kuria Muria, 102
Kuril Islands, 771, 772, 786, 844
Kurisches Nehrung, 411, 458, 569
Kurkar, 82, 83, 84, 87, 89, 90, 91, 94, 95, 96
Kuroshio Current, 347, 369
Kurshskaya, 783
Kurshyo, 411, 458
Kurtosis, 147, 555, 556, 557, 735, 739, 871
Kusaie, 622, 627
Kuskokwim River, 600
Kuwait, 101
Kwa-Zulu, 13
Kyle, 443
Kyrenia, 54
Kyushu, 79, 332, 499

Labiostrombus, 619
Laboulbeniomycetes, 439
Labrador, 56, 57, 176, 177, 463, 580, 581, 593
Labyrinthuales, 439
Labyrinthula macrocystis, 670
Labyrinthulidae, 670
La Camargue, 313

Laccadive, 484
Lacustrine Deltas, 512-513
Laevicardium, 619
Lagan, 430
Laganum, 620
Lag Deposits, 356
Lagenaria vulgaris, 526
LAGOON AND LAGOONAL COASTS, 341-342, 352, 443-444, **504-505**
  fiord, 504
  hypersaline, 35, 171
  liman, 504
  segmented, 152-153, 352. *See also* LAGOONAL SEGMENTATION
LAGOONAL SEDIMENTATION, **505**
LAGOONAL SEGMENTATION, 352, **505-506,** 784
Lago di Patria, 416
Lagos, 9, 26
Lagrangian criterion, 346
Laguna Beach, 878
Laguna Guerrero Negro, 34, 35, 36
Laguna Madre, 124, 171, 472, 588, 596, 655, 680
Laguna di Marano, 416
Laguna Mormona, 34, 35
Laguna Ojo de Liebre, 34, 35, 36
Laguna Superior, 197
Laguna de Términos, 192, 195, 597
Laguncularia, 287, 532
Laguncularia racemosa, 9, 167, 192
Lagynion, 210
La Jolla, 672, 693
Lake Agassiz, 511
Lake Agighiol, 472
Lake Babadag, 472
Lake Bonneville, 127, 292, 498, 509, 511, 512, 790, 855
Lake District, 412
Lake Eyre, 470
Lake Golovitza, 472
Lake Ladoga, 502
Lake Managua, 188
Lake Manitoba, 883
Lake Maracaibo, 198, 769
Lake Melville, 176, 177
Lake Michigan, 143, 144, 150, 235, 236, 303, 305, 690, 713, 801, 880, 881
Lake Nasser, 23
Lake Nouveau, 755
Lakes, 443-444, 470, 508-509, 510-511, 512-513, 601. *See also specific names;* HYPERSALINE COASTAL LAKES; LAKES *entries*
LAKES, COASTAL ECOLOGY, **506-507**
LAKES, COASTAL ENGINEERING, **508-509**
LAKES, COASTAL MORPHOLOGY, **509-513**
Lake St. Jean, 795
Lake St. Lucia, 4, 14, 15

Lake Sinoe, 472
Lake Siutghiol, 472, 473
Lake Smeica, 472
Lake Tahoe, 508
Lake Techirghiol, 472, 810, 811
Lake Todilto, 697
Lake Victoria, 23
Lake Worth, 270, 326
La Libertad, 190
Lambis, 619
Lamellbranchia, 551, 554
Laminaria, 55, 56, 104, 165, 202, 275, 286, 306, 404, 582, 774, 775, 776
Laminaria bongardiana, 774
Laminaria chichorioides, 773
Laminaria digitata, 774, 775
Laminaria japonica, 773
Laminariales, 633-634
Laminaria longicruris, 775, 776
Laminaria pallida, 12
Laminaria saccharina, 55, 775, 776
Laminaria solidungula, 55
la Morro, 559
Lampra, 42
Lamu, 7, 13
Lanark, 371
Lancashire, 371, 412
Land Breeze, 880, 881
Land Emergence, 725
Landes, 414, 419, 471
LAND RECLAMATION, 465, **513-516,** 648, 837, 851
Landscape Development, 290-291
Langmuir Cells, 696, 697
Languedoc, 471
Laomedia, 621
Lapies, 501
Lapin, 374
Laptev Sea, 56, 359, 771, 783, 784, 787
Larache, 25
Lariidae, 284
Larisa, 417
Larix, 582, 641, 642
Larix laricina, 582
La Rochelle, 414, 758
Larrea, 198
Larrea divaricata, 196
Larvacea, 283, 668
La Selva Beach, 717
Laser, 851
Las Tortugas, 198
Latakia-Jable, 99
Lates calarifer, 483
Lathyrus, 170, 582
Lathyrus japnicus, 581, 582, 584, 589
Lathyrus littoralis, 589, 590
Lathyrus maritima, 55
Latimeria, 283
Latin America, 191, 198, 559, 642
Latrobe Valley, 112
Lau Islands, 624, 625
Launaeo resedifolia, 7
La Unión, 189, 190
Laura Basin, 112

SUBJECT INDEX

Laurasia, 451
Lauwersee, 412
Lavrentya Gulf, 784
Lawai Blowhole, 172
Leathesia, 286, 574
Lebachia, 642
Lebanon, 84, 87, 93, 94, 98, 99, 251, 370
Lechea maritima, 582, 583, 585
Ledum groenlandicum, 175, 581
Lee Shore, 444
Leghorn, 415
Leguminosae, 527, 528
Le Havre, 202
Lemnaceae, 526
Lena River, 61, 359, 783, 787
Lepidophthirus macrorhini, 491
Lepidophyllum, 199
Lepidoptera, 491
Leptocarpus, 574
Leptoconchus, 619
Leptoconops, 281, 491
Leptograpsus, 199
Leptograpsus variegatus, 104
Leptoria, 618
Leptoseris, 619
Leptospermum, 108
Leptospermum juniperimum, 108
Leptospermum laevigatum, 108, 109
Leptospermum myrsinoides, 108
Leptosynapta, 620
Lepturus repens, 13
Lernaepoida, 280
Les Sables d' Olonne, 413, 414
Lessonia nigrescens, 199
Lessonia variegata, 572
Leucopogon parviflorus, 107, 108, 109
Leucothoe axillaris, 175
Leuresthes tenuis, 283
Levantine Basin, 82, 86, 92
Levees, 378
Leven Bank, 30
Liao River, 78
Liao-Tung Peninsula, 78
Liberia, 17, 25, 370
Libocedrus, 643
Libocedrus chilensis, 197
Libya, 7, 20, 22, 23
Libyan Basin, 20
Licata, 418
Lichina confinis, 104, 572, 574
Lido, 130, 442
Ligia, 74
Ligia exotica, 104
Liguria, 371
Ligusticum scoticum, 589
Lilaeopsis occidentalis, 590, 592
Liliopsida, 354, 527
Lima, Peru, 371
Lima, 617
LIMANS AND LIMAN COASTS, 186, 471, 472, **516-518**, 534, 783, 787, 810
Lime Fiord Barriers, 255, 256
Lim Fiord, 411, 434
Limicolae, 483
Limnichidae, 490
Límnos Island, 417
Limoniastrum quyonianum, 8
Limonietum linifolium, 11
Limonietum sebkarum, 8
Limonietum sinuatum, 8
Limonietum spathulatum, 9
Limonium, 406, 419, 585, 707
Limonium axillare, 16
Limonium californiicum, 591
Limonium carolinianum, 584, 586, 705
Limonium commune, 590
Limonium japonicum, 707
Limonium nashii, 582, 588
Limonium pruinosum, 16
Limpopo River, 18, 21
Limulus polyphemus, 280
Linckia, 617
Lincolnshire, 235, 236, 237, 254, 413, 692
LINEAR SHELL REEFS, **518-519**
Lineus corrugatus, 42
Lingula, 179, 282, 620
LINKS, **519**. *See also specific names*
Lioconcha, 619
Liocyma fluctuosa, 55
Liomera, 617
Lipari Islands, 418
Lippia nodiflora, 584, 586
Liquefaction, 172, 173
Liquidambar styraciflua, 584, 586
Lisbon, 371
Lithacrion, 617
Lithodomus, 501
Lithophaga, 72, 617
Lithophyllum, 616, 618
Lithothamnium, 56, 83, 100, 286, 574, 700, 775
Lithotrya, 617, 699
Lithuania, 535
Litorina Transgression, 174
Little Bay, 582
Littoral Community, 506
Littoral Component of Power, 1
Littoral Concrete, 485
LITTORAL CONES, **519-520**
Littoral Drift, 265. *See also* SEDIMENT TRANSPORT
Littoral Fringe, 104
Littoral Population Concentrations, 370-374
Littoral Settlement, 369-370
Littoral Shelf, 510
Littoral Weathering, 877
Littoraria, 615
Littorina, 6, 56, 171, 202, 404, 501
Littorina africana, 11, 12
Littorina angulifera, 9
Littorina brevicula, 772
Littorina irrorata, 494
Littorina knysnaensis, 12
Littorina kurila, 773, 774
Littorina littorea, 775
Littorina mandshurica, 772
Littorina obesa, 11
Littorina obtusata, 775
Littorina praetermissa, 104, 106
Littorina punctata, 10
Littorina saxatilis, 56, 775
Littorina scabra, 621
Littorina unifasciata, 104, 105, 106
Liverpool, 852
Liyepya, 782
Llanos de Almeria, 414
Llansa, 415
Lleyn Peninsula, 412
Loam, 737
LOBATE COASTS, **520,** 848
Lobito, 9, 10, 11, 27
Lobophyllia, 617
Lobophytum, 618
Loch Morar, 410
Lochs, 410
Loculoascomycetes, 439
Lofotens Island, 410
Log Battering, 702
Log Spiral, 151, 460
Log-Spiral Beach. *See* HEADLAND BAY BEACH
Loire River, 413
Lomas, 197
London, 251, 371, 413, 464, 852
Long Island, 126, 130, 254, 302, 306, 398, 594, 715, 852
Long Reef, 175, 644
Lonicera japonica, 586
Lonicera morrowi, 583
Lonicera sempervirens, 586
LOOSE ROCK AND STONE HABITAT, **520-522**
Lopha, 621
Lorient, 413
Los Angeles, California, U.S.A., 371, 598, 852
Lotus, 197
Louisiana, 127, 129, 207, 227, 302, 306, 307, 326, 398, 518, 538, 596, 800
Low and Ball, 135, 156, 692. *See also* RIDGE AND RUNNEL
Low Countries, 369
LOW-ENERGY COAST, 218, **522,** 883
Lower Murray River, 112
LOW-LATITUDE COASTS, **522-524**
LOW-TIDE DELTAS, **524, 793**
Luanda, 20, 27, 374
Luan-H River, 78
Lübeck, 411, 840
Lucinacea, 619
Ludwigia, 586
Luida, 620
Lumbricella lineatus, 56
Lumbrineridae, 39
Lumnitzera, 14, 108, 621
Lumnitzera Racemosa, 13
Lunella, 617
Lupinus littoralis, 590, 591
Luzon, 79

Lycium brevipes, 589
Lycopodiaceae, 524
LYCOPODIOPHYTA, **524-525**
Lycopodiopsida, 524
Lycopodium, 524
Lymington Marsh, 377
Lyngbya, 35
Lyngbya aestuarii, 35
Lythraceae, 288
Lytleton, 572

Ma'agan Mikhael, 86
Maas River, 252
Maba buxifolia, 482
Mablethorpe, 254
Macalister Range, 112
Machaeranthera phyllocephala, 588
Machair, 408
Machilidae, 490
Mach number, 718
Maciel River, 768
Mackenzie River, 61, 359, 601
McMurdo Sound, 42
Macoma baltica, 494, 774, 775
Macoma calcarea, 774, 775
Macoma incongrua, 772
Macoma moesta, 776
Macomona, 573
Macrocystis, 286, 384, 634
Macrocystis angustifolia, 105
Macrocystis pyrifera, 12, 105, 199, 571
Macrometeorology, 227
Macrophthalmus, 621
Macrophytes, 507
Mactra, 169
Mactra corallina, 83
Mactra olorina, 84
Mactra sulcataria, 773
Madagascar, 3, 14, 18, 19, 20, 21, 22, 30, 355, 371, 374, 840
Madeira, 371
Made land. See LAND RECLAMATION
Madras, 484, 485, 486
Madreporaria, 279, 332
Mafia, 29
Magdalena Bay, 597
Magdalena Plain, 196
Magdalena River, 359, 559, 765
Magdalen Islands, 234, 839
Magellan, 520
Magelonidae, 40
Magilidae, 619
Magilus, 619
Magnolia grandiflora, 585
Magnolia virginiana, 175, 584, 585
MAGNOLIOPHYTA, 354, **526-528,** 644
Magnoliopsida, 527
Mahanadi River, 485
Maharashtra, 485, 486
Mahdia, 53
Maianthemum canadense, 583
Maiidae, 617

Maine, U.S.A., 306, 307, 582, 593, 594, 727, 728
MAJOR BEACH FEATURES, **528-531**
Majorca, 185, 414
Makaha, 857
Makassar Strait, 72, 81
Makatea, 443
Makkovik Bay, 177
Makran, 102
Malabar, 483, 486
Malacca Straits, 68, 75, 359, 360
Malacostraca, 281
Malacothrix incana, 589
Malaga, 414
Malagasy fracture zone, 18
Malaspina Glacier, 600
Malay, 72, 443
Malaya, 78, 79, 233, 523
Malay Archipelago, 323
Malay Peninsula, 79, 544, 545
Malaysia, 73, 74, 75, 77, 358, 360, 443, 527, 545, 621, 840
  Peninsular, 69, 70, 73, 75
Maldanidae, 40
Maldive Islands, 72, 443
Maldonado, 768
Mali, 26
Mallomonas, 210
Mallophaga, 281, 490, 491
Malmo, 411
Malta, 371
Maltese Islands, 418
Malvaceae, 587
Malvan, 485
Mamaia, 473
Mammalia, 209, 284
Manaar, 482
Manasquan Inlet, 326
Manawatu, 575
Mandibulata, 280
Mangaia Island, 622, 626, 627
Mangalia, 809, 810, 811, 841
Mangalore, 485
Mangareva, 622, 627
MANGROVE COASTS, 167-168, 216, 217, 272, **532-533,** 760, 847
Mangroves, 189, 207, 361, 593, 620, 819
  Australia, 106-107
  eastern Asia, 67-70, 71
  forest, 74
  swamps, 171, 192, 196, 200, 287, 342, 522, 574, 705
Manila, 371
Manila Bay, 75
Manilkara, 195
Manzala Lagoon, 23, 87
Manzanilla, 597
Mapping, 231-235
Maps, geomorphological, 300-301
Maranhão, 198, 200, 374
Maremma, 186, 443
Maremme, 186
Mareograms, 299, 728
Margate, 807

Marginal Bulge, 449
Marginopora vertebralis, 341
Mariager Fiord, 411
Marie Byrd Land, 45
MARIGRAM AND MARIGRAPH, 434, **533,** 722, 749, 827
Marine abrasion. See WAVE EROSION
Marine communities, coastal flora, 285-286
Marine-cut platform. See Wave-cut platform
MARINE-DEPOSITION COASTS, 216, **534**
MARINE-EROSION COASTS, **534-535**
MARINE ERRACTICS, **535-537**
Marine fouling, 435
Marine fungi, 440
Marine inlet, 489
Marine insects, 489-492
Marine isostasy, 806, 807
Marine karst, 558
Marine park, 313
Marine planation, 291-292
Marine salina, 443
Marine swamp. See PARALIC
Marine terrace, 389. See also WAVE-BUILT TERRACE
Marin Gräns, 461. See also HIGHEST COASTLINE
Maritime Provinces, 581
Marker-and-cell, 249, 250
Markov chains, 336, 338
Marlborough Sound, 575
Mar Menor, 414
Marowijne River, 366, 769
Mar del Plata, 767
Marquesas Islands, 623, 624
Marsala, 418
Marseille, 186, 415
Marshall Islands, 72, 625, 626
Marsh, 443-444, 593. See also specific names; MARSH GAS; SALT MARSH
  fringing, 707
  lagoonal, 707-708
  pontine, 416
  vegetation, 591-592
MARSH GAS, **537-538**
Marsiliales, 650
Martha's Vineyard, Massachusetts, U.S.A., 594
Martinique, 371
Maryland, U.S.A., 126, 242, 350, 863
Maryut Lagoon, 23, 87
Mascaret, 176
Mascate, 102
Masira Island, 102
Massachusetts, U.S.A., 127, 128, 129, 143, 152, 175, 235, 236, 237, 246, 302, 304, 307, 350, 351, 352, 438, 583, 593, 630, 656, 671, 689, 839

Massawa, 31
Mass transport, 265
Mastigomycotina, 439
Mastigophora, 278, 670
Matagorda Island, 123, 127, 130, 588
Matanzas Bay, 815
Mataquito River, 767
MATHEMATICAL MODELS, 259, 347, **538-541**
Matrix Analysis System, 264
Matthiola, 288
Matuta, 620
Maugean Province, 104
Maui, 622, 623
Maupiti, 627
Mauritania, 8, 25, 35, 374
Mauritia, 198
Mauritius, 371
Mayo, 411
Maytenus, 196
Mazarron, 414
Mazatlán, 597
Mean, 147, 555, 556, 739
Meanguera Island, 189
Mean low water spring tide, 205
MEAN SEA LEVEL, **541-542**, 728
Mean Water Level, 249
Mechigmen Gulf, 784
Médano, 444, 647
Medaños, 236
Median, 147, 556
Mediterranean Sea, 7, 8, 16, 30, 33, 52-53, 82-83, 86-87, 90, 94-95, 126, 139, 186, 220, 228, 289, 359, 369, 404-405, 419, 434, 471, 501, 628, 644, 647, 652-653, 658, 676-677, 679, 689, 708, 757, 761, 787, 806, 823, 826, 840
Medjerda Delta, 23
Meechken Island, 784
Megabalanus nigrescens, 105, 106
Megalopolis, 370
Megapontum, 416
Mehetia, 622, 623, 627
Mekong, 370
Mekong River, 77, 79, 358, 359, 366, 821
Melaleuca, 107
Melaleuca ericifolia, 108
Melaleuca lanceolata, 107, 108
Melaleuca squarrosa, 108
Melampus, 171, 619, 621
Melampus bidentatus, 706
Melanesia, 370, 621
Melanophseal, 199
Melarapha, 571
Melbourne, 112, 119, 312, 370, 658
Meliaceae, 288, 532
Melostra, 10
Memecylon edule, 482
Menton, 415
Merca, 30
Mercenaria, 169

Mercierella enigmatica, 84
Meretrix, 73
Mergui Archepelago, 72
Meroplankton, 323
Merostomata, 63, 280
Merseyside, 729
Mertensia maritima, 55, 581, 582, 589
Merulina ampliata, 618
Mesembryanthemum cheilense, 589, 590
Mesembryanthemum edule, 13, 589
Mesembryanthemum salicornioides, 11
Meseta, 414
Mesidothea, 776
Mesidothia eutomon, 776
Mesidothea sabini, 776
Mesidothea sibirica, 776
Mesodinium rubrum, 345
Mesohaline, 470
Mesolithic, 50
Mesopotamia, 52, 101, 376
Mesoveliidae, 281
MESOZOA, 279, **542**
Messenia, 52
Messerschmidia, 619
Messersmidia argentea, 108
Messina, 418, 725
Messina Strait, 348
Metapenaeus stebbing, 84
Metasequoia, 642
Metazoa, 650
Meteorology. *See also* CLIMATE, COASTAL
  micro- and meso-scale, 225-226
  synoptic, 227
Méthana, 417
Metonic cycles, 827
Metopograpsus, 621
Metridium senile fimbriatum, 773
Meuse River, 815
Mexico, 33, 34, 35, 36, 126, 127, 191, 192, 194, 196, 235, 236, 239, 355, 366, 504, 527, 588, 597, 598, 623, 640, 658, 666, 823, 844
Mezen Gulf, 783, 823
Miami, Florida, U.S.A., 130, 254, 470, 587, 594, 657
Michigan, U.S.A., 127, 128
Michigan Basin, 422
Miconia, 195
Microcladia, 286
Microcoleus, 35
Microcoleus chthonoplastes, 35
Microfans, 476
Micronesia, 370, 552
Micronutrients, 603
Micropharynx, 645
Microphiline, 619
MICROSEISMS, **542**
Microstrobus, 642
Mid-Atlantic Ridge, 17
MIDDEN, 117, **542-543**, 552, 675

Middendorfia caprearum, 84
Middle America, 194. *See also* Central America
Middle America Deep Sea Trench, 188
Middle Atlantic Bight, 566
Middle ground, 444
Middleton, 270
Middlothian, 371
MID-LATITUDE COASTS, **543-544**
Migrating humps, 155. *See also* SAND WAVES AND LONGSHORE SAND WAVES
Mikania scandens, 584, 586
Mikhmoret, 83, 84
Míkonos, 417
Miletus, 52
Miliolina, 431, 432, 433
Millepora, 618
Million Dollar Pier, 32
Mílos, 417
Mimosae, 198
Mimusops caffra, 13
Minas Basin, 821
Mindanao, 79
MINERAL DEPOSITS, **544-545**
MINERAL DEPOSITS, MINING OF, **545-546**
Mineral placers, 530. *See also* Placers
Miniacina, 617
Mining, 7, 91, 325, 466. *See also* MINERAL DEPOSITS, MINING OF
Minnesota, U.S.A., 508
MINOR BEACH FEATURES, 155, **546-550**
Minorca, 415
Miohaline, 470
Mirounga leonina, 491
Mirtoan Sea, 417
Misratah, 23
Mississippi, U.S.A., 127, 129, 206, 302, 594, 596
Mississippi Delta, 358, 364, 365, 366, 448, 498, 595, 596, 657
Mississippi River, 126, 129, 211, 292, 294, 302, 358, 359, 360, 363, 364, 365, 378, 396, 397, 398, 498, 518, 559, 588, 594, 683
Mississippi Sound, 127, 129, 208, 399, 684
Mitchell Delta, 398
Mitidja, 23
Mitridae, 619
Miyazaki Port, 499
Mobile, Alabama, U.S.A., 595
Moçamedes, 371
Mode, 147
Model laws, 718
Models, 271, 300, 334, 347, 717-719
  process-response, 334

scale, physical, 259. *See also* SCALE MODELS
MODERATE-ENERGY COAST, 218, **550**
Modiolus, 616, 621
Modiolus demissus, 171
Modiolus difficilis, 773
Modiolus fluviatilis, 573
Modiolus neozelanicus, 570, 571, 572, 574
Mogadisho, 30, 374
Moher, 521
Mold. *See* FUNGI
MOLE, **551**, 639, 664, 666
Molgula, 774, 776
Molgus littoralis, 56
MOLLUSCA, 282, **551-555**, 647
Molokai, 721, 844
Molotschna, 254
Molpadia, 384
Molpadia oolitica, 611
Molpadida, 384
Moluccas, 70
MOMENT MEASURES, **555-557**, 735-736, 739
Monaco, 841
Monanthochloe littoralis, 588, 705, 707
Monastir, 23
Mondego, 414
Monerma cylindrica, 9
Monetaria annulus, 617
Monetaria moneta, 617
Monfalcone, 416
Monmonanthochloe littoralis, 196
MONOCLINAL COAST, 424, **557-558**
Monodonta turbiformis, 83
Monodonta turbinata, 83, 84
Monogenea, 646
Monomoy Island, 237, 238, 239
Monoplacophora, 551, 552, 553
Monostroma, 56
MONROES, 477, **558**
Monrovia, 9, 19
Møns Klint, 203, 205, 411
Montastrea, 652
Monte Argentario, 416, 839
Monte Circeo, 416
Montego Bay, 840
Monte Le Madonie, 418
Monte Nebrodi, 418
Monte Peloritani, 418
Monte di Procida, 416
Monterey Bay, 302, 599, 717
Monterey Peninsula, 302
Montevideo, 199, 371
Montipora, 617, 618
Montipora foliosa, 618
Mont St. Michel, 419
Moorea, 627
Moray Firth, 410
Morbihan, 333, 413, 414
Morecambe Bay, 405
Moreton Island, 311
Morison equation, 249

Morocco, 4, 7, 8, 9, 19, 20, 21, 22, 25, 374, 443, 678, 761
Moroshechnaya River, 785
MORRO, **559**
Morro de Bonifacio, 559
Morro de Puercos, 559
Morro Vermelha, 559
Mosquito Gulf, 188, 190
Mossamedes, 27
Mossel Bay, 884
Motril, 414
Motu, 443
Mount Cameroon, 26
Mount Casius, 96
Mt. Desert Island, 307
Mt. McKinley, 600
Mt. Waialeale, 526
Mousse, 610
Mozambique, 6, 12, 13, 14, 18, 19, 27, 28
Mozambique Basin, 20, 28
Mozambique Channel, 19, 20, 21
Mozambique Current, 4, 12, 28
Mtunzini, 27
Mud, 737
Mud flat, 404-405, 593. *See also* TIDAL FLAT
Mudjug, 783
Mud Lump Islands, 596
MUD LUMPS, 318, 364, **559**
MUD VOLCANOES, 558, **559-560**, 756, 793
Muehlenbeckia complexa, 574
Mugilidae, 283
Muhlenbergia capillaris, 585
Mull, 411, 444
Multicellular marina, 670
MURICATE WEATHERING, **560-561**
Muricidae, 680
Murman Peninsula, 55, 56
Murmansk, 424, 782, 785
Murray River, 114, 397
Musci, 181
Musculus, 775
Musculus corrugatus, 776
Musculus senhousia, 772
Mushroom. *See* FUNGI
Mussidae, 617
Mustang Island, 239, 596
Mustelidae, 284
Mya arenaria, 494, 775
Mya truncata, 775, 776
Mycetozoa, 439
Mycoplasmatales, 720
Myoporum insulare, 108
Myoxocephalus quadricornis, 55, 56
Myriapoda, 64
Myrica, 715
Myrica californica, 591
Myrica cerifera, 175, 584, 585
Myrica gale, 583
Myrica pensylvanica, 237, 582, 583, 584
Myrsiniaceae, 288, 532
Myrtaceae, 107

Myrtle Beach, 594
Mysidacea, 281
Mysis oculata, 56
Mysis relicta, 55
Mytilus, 74, 199, 700, 701
Mytilus crenatus, 12
Mytilus edulis, 55, 56, 199, 571, 572, 573, 574, 773, 774, 775
Mytilus galloprovincialis, 773
Mytilus meridionalis, 12
Mytilus pernus, 10, 12
Myxicola infundibulum, 494
Myxochrysis, 210
Myxomycota, 439
Myxosporidea, 279
Myzostomaria, 39

Nab, nase, ness, naze, nore, 444
Nahariya, 84
Naias marina, 405
Naididae, 280
Namaqualand, 27
Namib Desert, 27, 374
Namibia, 4, 19, 27, 454, 750
Nanakuli, 854
Nancy, 807
Nandidae, 483
Nannorrhops richieana, 482
Nantucket, 152, 350-352, 583, 689, 839
Napali Coast, 624
Naples, Florida, U.S.A., 587
Naples, Italy, 53, 371, 416
Narbonne, 415
Nardoa, 617
Narva River, 502
Nassariidae, 620
Nassarius, 494
Nassarius obsoletus, 612
Natal, 4, 7, 12, 13, 14, 20, 27
Natica, 73, 619
Natrix, 283
Nauru, 168, 502
Nauset Spit, 237
Nautilus, 555
Navarin Cape, 784
Navier-Stokes equations, 250, 871
Naviface, 813
NAVIGABLE WATERS, **562**
Náxos, 417
Nayarit, 197
Nazaré, 414
Neah Bay, 727
Neanuridae, 490
Nearshore features, 528-529
NEARSHORE HYDRO-DYNAMICS AND SEDIMENTATION, **562-568**
NEARSHORE WATER CHARACTERISTICS, **568-569**
Nebaliacea, 281, 573
Neck, 444
Nectonema, 279
Negro River, 767
NEHRUNG, 174, 569

Nekton, 244, 274
Nelson-Golden Bay, 575, 578
Nemalion helminthoides, 104
Neman River, 783
Nemaster, 382
Nemastoma oligarthra, 571
Nematodes, 202, 279
Nematomorpha, 67, 279
Nemertea, 279
NEMERTINA, **569-570**
Nemetoda, 67
Nemunas River, 458
Neo-glacial events, 464
Neogoniolithona notarisi, 83
Neolithic, 50, 52
Neomysis mercedis, 281
Neomysis rayii, 55
Neopalina, 553
Nepean Peninsula, 112
Nephthya, 618
Nephtyidae, 39, 620
Neptune, 811, 841
Neptunea, 775
Neptunism, 290
Nereidae, 39, 40
Nereis, 494
Nereocystis, 286
Neretva River, 417
Nerine cirratulus, 10
Nerita, 74, 571, 616
Nerita atramentosa, 105
Nerita exuvia, 621
Nerita melanotragus, 574
Nerita senegalensis, 10
Neritic waters, 323
Neritidae, 621
Ness, 267, 272, 688
Netanya, 90
Netherlands, 126, 127, 129, 235, 236, 251-254, 270, 333, 378, 395, 412, 420, 465, 466, 489, 514, 515, 648, 726, 731, 758, 825, 840, 850
Neusina, 669
Nevada, U.S.A., 508
New Britain Island, 848
New Brunswick, 302, 374, 582, 689, 823
New Caledonia, 616
Newcastle, Australia, 112, 370
New England, 143, 175, 312, 387, 461, 535, 582, 819
Newfoundland, 57, 452, 581, 582, 771
New Guinea, 387, 389, 443, 501, 520, 731, 755, 844
New Hampshire, U.S.A., 583, 715
New Hebrides, 616
New Jersey, U.S.A., 65, 126, 130, 148, 158, 187, 247, 262, 302, 305, 307, 326, 438, 583, 594, 639, 656, 790, 817
Newlyn, 541
New Mexico, U.S.A., 697, 805
New Orleans, Louisiana, U.S.A., 326, 596

New South Wales, 105, 106, 108, 109, 110, 112, 114, 115, 118, 175, 223, 326, 327, 398, 466, 544, 621, 644, 674, 677, 691, 736, 826, 853, 854, 856, 857, 884
New York, U.S.A., 307, 371, 464, 465, 583, 594, 599, 639, 656, 677, 839, 851, 852
New Zealand, 37, 207, 224, 228, 235, 281, 311, 333, 387, 389, 392, 397, 424, 427, 455, 490, 520, 532, 543, 545, 572, 621, 625, 642, 643, 732, 840, 848, 885
NEW ZEALAND, COASTAL ECOLOGY, **570-575**
NEW ZEALAND, COASTAL MORPHOLOGY, **575-579**
Niadales, 286
Nicaragua, 188, 189, 190, 194, 195, 597
Nice, 415, 841
NICHE, NICK, NIP, NOTCH, 77, 91, 224, 271, 389, 393, 500, **579-580,** 626, 677, 721, 722, 876, 879. *See also* Visor
Nicobar Island, 72, 81
Nicobars, 481, 482, 483, 485, 486
Nicoya Peninsula, 197
Niemen River, 535
Niger Basin, 20, 26
Niger Delta, 20, 25, 26
Nigeria, 370, 532, 815
Niger River, 26, 359, 365, 366
Nile Delta, 23, 83, 90, 92, 93, 356, 358, 366, 376
Nile River, 87, 90, 126, 211, 358, 359, 369, 378, 464, 673
Ninety Mile Beach, 112, 575
Nipa fruticans, 482
Nissum Fiord, 411
Nitraria retusa, 8, 16
Nitraria schoberi, 107, 108
Niue Island, 622, 626
Noctiluca, 278, 670
Nodal point, 38, 438, 489
Nodal zone, 438
NODE AND ANTINODE, **580**
Nodilittorina millegrana, 106
Nodilittorina pyramidalis, 105, 106
Nodolittorina, 74
Noirmoutier, 414
Nome, 601
Nook, 444
Noordzeepoort, 841
Norbotten, 374
Nordenskjold Archipelago, 783
Norderney Island, 667
Norfolk, U.K., 202, 312, 413, 419
Norfolk, Virginia, U.S.A., 304, 585, 727
Normandy, 809
Norrland, 374
North African-Arabian Shield, 409
North America, 17, 135, 176, 188,
208, 211, 221, 231, 241, 256, 275, 277, 290, 306, 352, 359, 370, 375, 377, 379, 384, 399, 447-450, 454, 462-464, 474, 490, 498, 541, 543, 550, 564, 566, 589, 640, 642, 643, 672, 676, 687, 701, 705, 713, 715
NORTH AMERICA, COASTAL ECOLOGY, **580-593**
NORTH AMERICA, COASTAL MORPHOLOGY, **593-602**
North American Plate, 449, 452, 498
North Arcot, 486
North Beveland Island, 412
North Cape, 55, 574, 758
North Carolina, U.S.A., 125, 128, 129, 130, 131, 141, 158, 170, 175, 235, 236, 237, 242, 262, 302, 305, 306, 307, 397, 487, 518, 584, 585, 586, 588, 638, 656, 713, 714, 715, 737, 754
North central region, North America, 583-584
Norh Dakota, U.S.A., 511
Northeast Monsoon Current, 4
Northern Sporades, 417
North Island, 570, 571, 573, 575, 577, 578
Northland, 575
North Manakau Head, 578
North Sea, 65, 126, 165, 237, 254, 255, 407, 411, 412, 413, 419, 448, 465, 471, 515, 566, 606, 648, 692, 724, 731, 759, 783, 810, 819, 840, 850
Northumberland, 413
North Vietnam, 366. *See also* South Vietnam; Vietnam
North West Cape, 114, 116
Norton Sound, 601
Norway, 55, 56, 389, 397, 406, 410, 411, 420, 427, 447, 461, 527, 543, 593, 782, 799, 840
Nothofagus antarctica, 197, 198
Nothofagus betaloides, 197, 198
Nothofagus dombeyi, 197
Nothofagus obliqua, 197
Nototanais, 42
Nourishment, 425. *See also* BEACH NOURISHMENT
artificial, 667
sand, 327
Nova Scotia, 156, 230, 302, 530, 582, 705, 821, 823, 838, 839
Novaya Zemlya, 55, 56, 58, 771, 783
Novosibirskiye Islands, 771, 784
Nuclear power plants, 258, 262, 465, 602-603
NUCLEAR POWER PLANT SITING, **602-603**
Nuculana pernula, 775
Nuda, 345
Nudibranchia, 653
Nullarbor Plain, 112, 375

Null point, 146, 154, 744
Null zone, 281
NUTRIENTS, 323, 433, 568, **603-604**, 715, 799
Nyker, 799
Nypa, 528
Nypa fructicans, 527
Nyssa sylvatica, 583, 584, 586

Oahu, 393, 854, 857
Oaxaca, 197
Ob River, 61, 783
Ocean holes, 799
Ocean ranching, 48
Ocean Shores, Washington, U.S.A., 548
Ocean thermal energy converter, 815
Ochamchire, 666
Ochetostoma, 620
Ochlockonee Bay, 595
Ochromonas, 210
Ocotales, 190
Ocracoke Island, 242, 304, 638
Octocorallia, 331, 332
Octomeris angulosa, 11
Ocypode, 73, 169
Ocypode africana, 10
Ocypode ceratophthalmus, 619
Ocypode cordimanus, 619
Ocypode cursor, 10
Oder River, 411, 458
Odessa, 667
Odonata, 491
Odontaster, 41
Odontaster validus, 42
Odonthalia, 56, 286
Oenopota gigantea, 775
Oenothera drummondi, 588
Oenothera humifusa, 585
OFFSET AND OVERLAP, **605-606**
OFFSHORE PLATFORMS, **606-609**
Ogadir, 374
Ogooue River, 9
Ohio, U.S.A., 396
Ohiwa, 574
Oia Shio, 369
OIL SPILLS AND POLLUTION, **609-611**
Okhota River, 785
Okhotsk Sea, 57, 77, 462, 771-774, 776, 785, 823
Okinawa, 371
Öland Island, 411, 502
Olearia axillaris, 107, 108
Olenek River, 783
Oleron Island, 414
Olifants River, 26, 27
Oligochaeta, 39, 280
Oligohaline, 470
Oligotrichida, 279
Olivella, 619
Olssonia, 432
Olympic Mountains, U.S.A., 643
Olymp Resort, 810

Olyutorskiy Cape, 784
Oman, 102
Omaniidae, 490
Oncorhynchus gorbuscha, 56
Oncorhynchus keta, 56
Onega, 783
Onilahy River, 30
Onisimus litoralis, 55
Onitsha, 26
Ontika, 502
Onuphidae, 39, 40
Onychophora, 67, 280
Oolites, 710
Oomycetes, 439
Oostrombus gibberulus, 619
Opheliidae, 40
Ophiacantha, 41
Ophiactis parva, 83
Ophiarachna, 617
Ophiarthrum, 617
Ophiocomina scolopendrina, 617
Ophioderma longicaudata, 84
Ophiolepis, 617
Ophiomastix, 617
Ophiothrix, 382
Ophiuroidea, 381, 383, 611, 776
Opifex, 491
Oporto, 414
Opuntia, 194
Opuntia drummondii, 585
Opuntia humifusa, 584
Opuntia lindheimeri, 588
Oran, 8
Orange River, 17, 21, 27, 544
Orbetello, 839
Orbignya, 198
Orbiniidae, 40
Orchestia, 706
Orchomena, 42
Orcinus, 284
Ord Delta, 398
Ord Estuary, 333
Ord River, 358, 359, 360, 364, 366, 819, 821
Oreaster, 617
Oregon Inlet, 487
Oregon, U.S.A., 224, 228, 235, 236, 237, 288, 302, 326, 461, 508, 589, 590, 591, 592, 599, 643, 683, 713, 721, 878, 880
Orford, 419, 455
Orfordness, 408, 745, 751, 792
ORGANISM-SEDIMENT RELATIONSHIP, **611-614**
Organ pipes, 501
ORIENTED LAKES, **614-615**
Orinoco River, 200, 359, 769
Orissa, 482
Orthogonals, 149, 299, 863, 865, 883
Orthonectida, 279, 542
Orthoptera, 491
Oryctolagus cuniculus, 408
Osaka, 371
Osbornia, 108
Oscillatoria nigroviridis, 573
Oslo, 370, 389

Oslo Fiord, 411
Osmanthus americanus, 585
Osmussaar, 502, 503
Osteichthyes, 209, 283
Ostend, 810
Osteospermum moniliferum, 13
Ostia, 250
Ostracoda, 280, 573
Ostrea, 776
Ostrea angasi, 573
Ostrea cucullata, 10
Ostrea edulis, 773
Ostrea tulipa, 9
Otago, 333, 571, 575
Othonna sarnosa, 13
Otóque Island, 190
Otway Range, 112
Oum er Rbia River, 25
Outer Banks, 125, 126, 130, 131, 140, 141, 236, 585, 586, 588, 713, 714
OUTWASH PLAIN SHORELINE, 615
Ovalipes, 620
Overhang, 500
Overlap, 605-606
Overwash, 130, 170, 242, 316, 319, 584, 585, 586, 707, 708, 713, 714, 791, 851
Oviressoa, 42
Ovula, 619
Oweniidae, 40
Oxystele tabularis, 11, 12
Oxystele variegata, 12
Oyster grass, 207

Pachycereus, 194
Pachygrapsus transversus, 84
Pachymenia himantophora, 570, 571
Pachyseris, 619
PACIFIC ISLANDS, COASTAL ECOLOGY, **616-621**
PACIFIC ISLANDS, COASTAL MORPHOLOGY, **621-628**
Pacific Ocean, 12, 67, 72, 103, 167, 168, 188, 189, 210, 211, 292, 352, 398, 451, 522, 523, 527, 532, 577, 597, 701, 705, 721, 726, 727, 765, 766, 771, 772, 775, 803, 815, 824, 827, 840, 843
Pacific Plate, 76
Padina, 72, 616, 617, 634
Padre Island, 124, 472, 588, 596, 851
Pads, 150, 391
Pahoehoe, 519
Päite, 503
Pakistan, 76, 77, 81
Pakri, 503
Palaeonemertini, 569
Palamos, 415
Palembang, 70
PALEOGEOGRAPHY OF COASTS, **628-629**
Palermo, 418

SUBJECT INDEX

Palimpsest, 501
Palmae, 527
Palmer Peninsula, 387
Palm Forest, 198
Palos Verdes, 389, 598, 599
Paludal. *See* SALT MARSH
PALVÉ, 569, **629-630**
Palythoa, 618
Palythoa nelliae, 11
PAMET, **630-631**
Pamisos River, 52
Pamlico Sound, 397, 505
Pampas, 198, 767
Panadus, 619, 621
Panama, 188, 189, 190, 191, 194, 195, 196, 197, 198, 518, 559, 595, 597
Panama Canal, 189, 190
Pandanus candelabrum, 9
Pandion haliaëtus, 284
Pangaea, 451
Pangalanes, 30
Panhole. *See* SOLUTION AND SOLUTION PAN
Panicum, 199
Panicum amarulum, 584
Panicum amarum, 584, 585
Panicum marulum, 588
Panicum virgatum, 705
Pannes, 171, 471, 705
Pans, 501, 847
  dieback, 377
Panthalassa, 451
Paphia, 619
Papua, 374
Pará, 374
Paraclunio, 490
Paradip Port, 666
Parafusulina, 669
Paraguaná Peninsula, 200, 769, 770
PARALIC, **631**
Paraná, 235, 359, 767-768, 821
Paraonidae, 40
Paravortex, 645
Parazoanthus, 653
Paria Peninsula, 200
Paris, 807
Parnassia palustris, 581
Páros, 417
Parthenocissus, 585
Parthenocissus quinquefolia, 583, 584, 586
Pas de Calais, 536
Paspalum, 194, 199
Paspalum monostachum, 588
Paspalum vaginatum, 9, 193, 194
Pass, passes, 443, 624
Passamaquoddy Bay, 821
Passeriformes, 284
Passerina rigida, 13
Passiflora incarnata, 586
Passiflora lutea, 586
Patagonia, 199, 228, 375, 387, 462, 543, 676, 767
Patagonia Steppe, 198
Pata Island, 848
Patella, 202, 268

Patella argenvillei, 12
Patella caerulea, 83, 84
Patella cochlear, 12
Patella granularis, 11, 12
Patella lusitanica, 83
Patella safiana, 10
Patella variabilis, 11
Patelloida alticostata, 104
Patiria pectinifera, 772
Patiriella exigua, 490
Patraikós Kólpos, 417
Patterned ground, 478, 755
Pavement, 356, 750
Pavona, 619
Pawan River, 81
Peachia hastata, 494
Peak, 444
Pearl Beach, 884
Pearl Islands, 190
Peat. *See* BOGS
Peat moss, 181
PEBBLE, 455, **631-632,** 738, 751
Pechora River, 359, 783
Pectinia, 619
Pectiniidae, 619
Pedestals, 356
Pedicularis, 581
Pedum, 617
PEELS, **632,** 710
Pelagophycus, 286
Pelecaniformes, 284
Pelecypoda, 551, 552, 553-554
Pella, 52
Pelliciera rhizophporae, 196, 532
Pelopónnisos, 417
Pelotherapy, 810
Pelusaic, 94, 97
Pelusium, 96
Pelvetia, 56, 202, 286
Pelvetia wrighti, 773
Pelvetiopsis, 286
Pemba, 29
Pembrokeshire, 412
  Coast National Park, 313
Pemphigus treherneri, 491
Pemphis acidula, 13
Penaeidae, 620
Penaeus fluviatilis, 400
Penaeus japonicus, 84
Penaeus setiferus, 399
Penepatella kermadecensis, 574
Peniche, 414
Peninsula of Darien, 189
Peninsulas, 444. *See also specific names*
Pennatularia, 611
Pennell Bank Assemblage, 42
Penner, 482
Pensacola, 595
Pentland Firth, 428
Pentostomida, 67
Penzhina, 785, 823
Peoples' Republic of China, 371
Perched aquifer, 502
PERCHED BEACH, **632-633**
Perciformes, 283
Percnon planissimum, 617

Peridinium, 278
Perigee-syzygy, 672
PERIGLACIAL EFFECTS, 633
Periglypta, 619
Perim Island, 100
Periopthalamus, 495
Periopthalmidae, 621
Perischoechinoidea, 384
Peritrichida, 279
Permafrost, 55, 58, 77, 462, 475, 477, 478, 600, 633, 756, 783
Perna, 572
Perna canaliculus, 570, 571, 572
Pernambuco, 795
Peronian Province, 104
Persea borbonia, 175, 584, 585
Persian Gulf, 33, 100-102, 359, 360, 465, 501, 532, 675, 799, 819
Perth, 108
Peru, 197, 199, 200, 235, 239, 298, 369, 375, 766, 844
Peruvian Current, 197
Pesaro, 416
Pescara, 416
Peterborough, 413
Petersen Grab, 273
Petersen scale, 722. *See also* SEA CONDITIONS
Petrobius, 281
Pett level, 254
Peyssonelia, 618
Phacelocarpus, 104
Phaeodermatium, 210
Phaeophyceae, 286, 404
PHAEOPHYCOPHYTA, **633-635.** *See also* Algae, brown
Phaeophyta, 55, 56, 402
Phaeothamnion, 210
Phanerozonida, 381
Phascolosoma, 617
PHASE DIFFERENCE, **635**
Phaseolus, 459
Philadelphia, Pennsylvania, U.S.A., 551, 706, 807, 852
Philanisidae, 490
Philanisus plebeius, 490
Philinoglossa, 619
Philippines, 67, 70, 73, 75, 76, 77, 79, 179, 527, 761, 827, 840, 844, 848
Philoscia vittata, 706
Philoxerus vermicularis, 588
Phi scale, 738, 739, 751
Phoenix farinifera, 482
Phoenix paludosa, 482
Phoenix reclinata, 14
Phoenix spinosa, 9
Phoenix sylvestris, 482
Pholas, 501
PHORONIDA, 179, 282, **635-636**
PHOTIC ZONE, **636**
PHOTOGRAMMETRY, 53, **636-639**
Photo interpretation, 682
Phragmites, 168, 588
Phragmites australis, 14, 15, 16, 168, 705

Phragmites communis, 583, 592, 847
Phrynophiurida, 383
Phuket, 73
Phycomycetes, 439, 440
Phylactolaemata, 181, 282
Phyllaria dermatodea, 776
Phyllodoce coerulea, 581
Phyllodocidae, 40
Phylloglossum, 524
Phyllophora, 56, 776
Phyllospadix, 286
Phyllospadix iwatensis, 773
Phyllospongia, 617
Phyllospora comosa, 105, 106
Physalis, 330, 331
Physalis maritima, 585
Physostomi, 483
Phytokarst, 501, 879
Phytomastigina, 670
Phytomonadida, 670
Phytoplankton, 286, 603
Piano di Catania, 418
Piaster, 382
Piaui, 198, 374
Picea, 642
Picea glauca, 582
Picea mariana, 175, 581, 582
Picea rubens, 582
Picea sitchensis, 591, 643
PIER, 499, 551, **639-640**
Pieris nitida, 175
PILE, PILING, **640**
Pile Up, 724, 725
Pilgerodendron, 642
Pilgerodendron uriferum, 198
Pillau Strait, 569
Pilumnus, 617
Pinaceae, 640
Pinales, 640, 641, 642
Pinctada radiata, 84
Pinguitellina, 619
PINICAE, **640-643,** 644
Pinna, 619
Pinnacles, 534
Pinnipedia, 273, 284
PINOPHYTA, 526, 527, 640, **643-644**
Pinopsida, 640, 644
Pinus, 408, 642
Pinus aristata, 641
Pinus banksiana, 582
Pinus caribaea, 190, 195
Pinus cembroides, 643
Pinus contorta, 591
Pinus elliottii, 586
Pinus palustris, 586
Pinus peuce, 643
Pinus pityusa, 643
Pinus rigida, 582, 583, 584
Pinus strobus, 582
Pinus taeda, 584, 586
Pinus torreyana, 643
Piombino Peninsula, 415
Pipes, 501
Pirenella conica, 84
Pisces, 209, 775

Pisco, 766
Pisonia, 194
Pitar, 619
Pitcairn Island, 622, 627
Pithecolobium, 195, 197
Pits, 501
Placers, 578
  deposits, 325, 545
  mineral, 530
Plage, 442
Plagianthus divaricatus, 574
Plane Bed, 163, 696. *See also* BED FORMS
Planerskoye, 667
Plankton, 244, 274
Plantaginaceae, 287, 527
Plantago, 288, 582, 583
Plantago albicans, 8
Plantago eripoda, 581
Plantago juncoides, 582
Plantago maritima, 581, 582, 590, 591, 592
Plantago oliganthos, 582
Plasmodiophoromycetes, 439
Plate Tectonics, 126, 188, 409, 448-449
PLATFORM BEACH, **644-645**
Platforms (geomorphological), 500, 755. *See also* SHORE PLATFORMS
  abrasion, 1, 2. *See also* WAVE-CUT PLATFORM
  corrosional, 99
  cut. *See* WAVE-CUT PLATFORM
  erosion, 271
  low-tide, 625-626
  spit, 790, 791
  wave-built. *See* WAVE-BUILT TERRACE
Platforms, man-made. *See* OFFSHORE PLATFORMS
Platform seiche, 749
Platygyra, 618
PLATYHELMINTHES, 279, 569, **645-647**
Platynereis australis, 573
PLAYA, **647**
PLECTOGNATHA, 483
Plectomycetes, 439
Pleisiosaurs, 209
PLEISTOCENE, **647-648**
Pleshet, 94, 95, 96
Pleuronectiformes, 283
Pliohaline, 470
Plocamium, 104
Plow Marks, 536
Pluchea, 586
Pluchea purpurascens, 585, 586
Plumbaginaceae, 527, 528, 532
Plum Island, 129
Plutonism, 290
Plymouth, Massachusetts, U.S.A., 246, 630
Poa, 55, 108, 288
Poaceae, 527, 528, 587, 588
Poa douglasii, 589

Poa eminens, 581, 589
Poa macrantha, 590, 591
Pocillopora, 11
Pocillopora damicornis, 618
Pocosin bogs, 175
Podicipediformes, 284
Podocarpaceae, 640, 643
Podocarpus, 642
Pododesmus macroshisma, 774
Pogonophora, 669
Point, points, 444, 512
Point Barrow, 352, 477, 590, 601
Point Conception, 599
Point Franklin, 352
Point Hope, 601
Point Loma, 598
Point Peron, 387
Point Reyes, 313
Point Roberts, 546, 549
Poitevin, 414
Poland, 254, 666
POLDER, 65, 412, 369, 514, **648,** 720, 762
Polinices, 619
POLLUTANTS, 173, 325, 378, 433, 468-469, 472, 484, 569, 609-611, **648-649,** 653, 806, 813-815, 820, 841, 851
Pólwysep Hel, 411
Polychaeta, 39, 280, 611
Polycladida, 279, 645
Polycystis, 645
Polygonaceae, 407
Polygonum paronychia, 589
Polyhaline, 470
Polynesia, 370
Polynices, 73
Polynoidae, 39
Polyodontidae, 39, 40
Polyphyllia, 618
Polyplacophora, 282, 551, 553
POLYPODIOPHYTA, 642, **649-650**
Polypodium aquilinum, 650
Polypodium polypioides, 586
Polystoma, 646
Polystomoides, 646
Pomatoceros, 572, 573
Pomatoceros caeruleus, 571, 572
Pomatoleios crossland, 12
Ponape, 622, 627
Pond culture, 48
Poneroplax, 105
Pontomyia, 490
Pontoporeia affinis, 55
Populus tremuloides, 582
Porichthys, 283
PORIFERA, 279, 611, **650-654**
Porites, 72, 618, 619, 626
Porites andrewsi, 617
Porites fusca, 617
Porites lutea, 617
Poritidae, 617
Po River, 359, 416, 419, 465
Poro, 416
Porolithon, 618
Pororoca, 176

SUBJECT INDEX

Porosity, 738
Porphyra, 199, 286, 490, 571, 686, 687
Porphyra capensis, 12
Porphyra columbina, 105, 571, 574
Porphyra umbilicalis, 199, 774
Port de Bouc, 415
Port Bouet, 26
Port Campbell, 115, 119
Port Durnford Formation, 28
Port Elizabeth, 11
Port Grimaud, 841
Port Hedland, 116
Port Hueneme, 326, 327
Port Jackson, 691
Portland, Australia, 114
Portland, Maine, U.S.A., 727, 728
Portlandia, 776
Portlandia arctica, 776
Port Lavaca, 588
Portobello, 148, 572
Porto Fino, 415
Porto Nova, 238
Port Phillip Bay, 108, 112, 117, 119, 312, 333, 434
Port Said, 89, 92
Port Said Lagoon, 23
Portugal, 185, 414-415, 840
Portuguese man-of-war, 329, 330
Portunus, 620
Port Venres, 415
Posen, 254
Posidonia, 405
posidonia oceanica, 83, 405
Possidonia, 370
Postelsia, 286
POSTGLACIAL REBOUND, **654-655,** 806, 807
Post Point, 38
Posyet Bay, 772, 785
Potamogeton, 507
Potamogetonaceae, 527
Potamogeton filiformis, 405
Potamogeton pectinatus, 405
Potamogeton pusillus, 14
Potentilla, 287
Potentilla anserina, 581
Potentilla egedii 582
Potentilla pacifica, 590, 591, 592
Potentilla tridentata, 581, 582
Potholes, 522
POTRERO, **655**
Potrero Cortado, 655
Potrero Lopeno, 655
Pouteria, 195
Pozzallo, 418
Pramort, 411
Prandtl log height equation, 240
Precipice, 721
PRESENT-DAY SHORELINE CHANGES, UNITED STATES **655-657**
PRESENT-DAY SHORELINE CHANGES, WORLDWIDE, **657-658**
Presque Isle, 444, 809
Pressure Head, 854

Preston Links, 519
PRIAPULIDA, **658-659**
Priapuloidea, 67, 279
Priapulus, 279
Priapulus caudatus, 658
Pribilof Islands, 277
Priene, 52
Primovula, 619
Primula borealuis, 581
Primulaceae, 287
Prince Edward Island, 127
Princess Charlotte Bay, 112
Prince William Sound, 600
Prionospio cirrifera, 55
Priseres, 287
Procaryotae, 720
Procellariformes, 284
Procerodes ulvae, 645
Procida Island, 416
Procyon lotor, 284
PRODUCTION, 400, **659-660,** 793
Productivity, 41, 324
  organic, 568
Profile. *See also* BEACH PROFILES; PROFILE OF EQUILIBRIUM; PROFILING OF BEACHES
  storm, 159, 352
  summer, 352
PROFILE OF EQUILIBRIUM, 131, 155, 179, 445, 504, 529, 564-566, **660-661,** 874
PROFILING OF BEACHES, **661-663**
PROGRADATION AND PROGRADING SHORELINE, **663,** 679, 685
Prokljan Lake, 417
Promontory, 444
Prorhynchus, 645
Prosopis, 194, 197, 198
Prosopis glandulosa, 588
Prostoma, 279
Protected sea environments, 859-860
PROTECTION OF COASTS, 326-327, **664-668**
Proterospongia, 651
Protista, 439
PROTOCHORDATA, 208, **668-669**
Protomyxida, 670
Protothaca grata, 518
PROTOZOA, 278, **669-671**
Protuberances, 272, 688
Provenance, 300
Provence, 185, 186
Province Lands, Cape Cod, 175, 583
  spit, 438
Provincetown, Massachusetts, U.S.A., 175
PROXIGEAN SPRING TIDES, **671-672**
Prunus caroliniana, 585
Prunus maritima, 583
Prunus pensylvanica, 583

Prunus serotina, 583, 584
Prunus virginiana, 583
Prymnesiales, 210
Psammocora, 11
Psammohydra, 619
Psammon, 669
Psammosere, 379, 380
Pseudolarix, 642
Pseudolithophyllum hyperellum, 105
Pseudomonads, 720
Pseudotsuga menziesii, 591, 643
Pteranodon, 209
Pteria occa, 84
Pterobranchia, 208
Pterocarpus, 194, 198
Pterocarpus officinalis, 198
Pterocereus, 194
Pterocladia, 104, 616
Pterophyta, 649
Ptilota, 56
Ptilota asplenioides, 774
Puccinellia, 287, 288, 592
Puccinellia distans, 9
Puccinellia maritima, 582, 707
Puccinellia palustris, 9
Puccinellia paupeala, 590
Puccinellia phryganodes, 55, 581, 590, 707
Puccinellia pumila, 590, 591
Puccinellia trifolia, 590, 592
Puerta Isabela, 190
Puerto de los Alfaques, 415
Puerto Barrios, 190
Puerto Cabeza, 190
Puerto Cortés, 190
Puerto Limón, 191
Puerto Montt, 767
Puerto Motaquia, 190
Puerto Rico, 371, 761, 762, 852
Puerto Somoza, 190
Puerto Vallarta, 597
Puffing Hole. *See* BLOWHOLES
Puget Sound, 38, 592, 600
Pulmonata, 282
Punaluu, 844
Punta Amapala, 189
Punta Arenas, 190
Punta Dungeness, 767
Punta Galero, 597
Punta Mala, 189
Punta Pinas, 189
Purbeck, 203, 204, 205
Puu Hou, 519
Pycnoclines, 298
Pycnogonida, 41, 64, 280
Pyrazus, 621
Pyrenees, 415
Pyrenomycetes, 439
PYRRHOPHYTA, **672-673**
Pyrus melanocarpa, 584
Pythia, 621
Pyura chilensis, 199
Pyura stolonifera, 12, 105, 106

Qatar, 101
Quaternary, pluvial phases, 2

Qeshm, 101, 102
Quarried, 579
QUARRYING PROCESSES, 486, 521, 523, 543-544, **674,** 677, 825, 853
QUAY, **674,** 879
  wall, 183
Québec, 448, 677, 755, 756, 795
Queenscliff, 119
Queensland, 103, 104, 106, 110, 112, 114, 116, 117, 118, 119, 144, 311, 313, 325, 327, 371, 544, 621, 884
Quelpart Island, 532
Quercus alba, 583
Quercus borealis, 582
Quercus falcata, 584
Quercus ilicifolia, 583
Quercus laurifolia, 585
Quercus nigra, 584, 585
Quercus phellos, 585
Quercus stellata, 583
Quercus velutina, 583
Quercus virginiana, 206, 306, 585, 588
Quiberon, 413, 809
Quidnipagus, 621
Quillworts. See LYCOPODIOPHYTA
Quilon, 483
Quimper, 413
Quinqueloculina, 432
Quintana Roo, 374

Rabaul, 520, 848
Race, 443
Race Point, 656
Radar, side looking, 637
Rade de Brest, 333
RADIOCARBON DATING, 389, 463, 641, **675**
Radiolaria, 278
Rafah, 90, 92
Rafted Block, 535
Raglan, 574
Ragusa, 417
Raiatea, 627
RAISED BEACH, **675-677**
Rajang River, 81
Rakaia River, 885
Ralfsia, 286, 572
Rambla, 185
Rameswaram, 483
RAMPARTS, 455-456, 674, **677-678**
Ramps, 271. *See also* EROSION RAMPS, WAVE RAMPS
  abrasion, 118, 393
  slope, 550
Rance River, 821
Randers Fiord, 411
Randia, 194
Randia dumelorum, 482
Ranfjord, 758
Ranunculus flamella, 591
Ranunculus hyperboreus, 581
Ranunculus reptans, 581

Rapa, 619
Rapana venosa, 773
Raphia, 9, 198
Raritan River, 394
Rarotonga Island, 622, 623, 624
RASA, **678**
Ras Asir, 28, 30
Ras Banas, 31
Ras et Bar, 90
Ras et Bassit, 98
Ras Beirut, 98
Ras Chekka, 98
Ras el Hadd, 102
Ras Hafun, 30
Ras ibn Hani, 98
Ras el Khaima, 101
Ras Masendam, 102
Ras Tanura, 101
Rastrelliger, 74
Ratnagiri, 482
RAVINEMENT, **678**
Rayalaseema, 482, 484
Rayleigh, J. W. S., 868, 871
Razelmul Mare, 472
Reach, 442, 444
Recent, *See* HOLOCENE
RECESSION AND RETROGRESSION, **678-679**
Recife, 768, 796
Reclamation, 251, 254, 513, 762
Red Beach Cave, 572
Red Delta, 77
Red River, 77, 79, 359, 366, 370
Red Sea, 4, 7, 8, 13, 16, 19-21, 30, 31, 33, 82, 99, 100, 185, 186, 369, 374, 405, 424, 449, 451-452, 532, 652, 673, 823
Red Tide, 278, 345, 670, 673
Reefs, 443. *See also* CORAL REEF COASTS; CORAL REEF HABITAT; Great Barrier Reef; LINEAR SHELL REEFS; REEFS, NONCORAL; STONE REEF
  barrier, 125, 168, 332, 339, 616, 621, 623
  coral, 71-73, 194, 196, 273, 342 522, 532, 616-617, 623, 687, 848
  crest, 340
  cup, 679
  dead, 617
  environments, 340-341
  flat, 168, 341
  floor, 340
  fragile, 618
  forereef, 340
  fringing, 168, 332, 339, 523, 616, 617, 621, 623
  front, 340
  infralittoral, Asia, Middle East, 83
  lagoonal, 341, 616
  outer, 617-618
  raised, 342, 623, 626-627
  rock, 339, 342
  string, 518

substrates, 339-340
systems, 317
REEFS, NONCORAL, **679-681**
Reflection, 333, 752. *See also* WAVES
Refraction, 333. *See also* WAVES
Reggio, 416
Regression analysis, 335
Reims, 807
Ré Island, 414
RELICT SEDIMENT, 123, **681-682**
REMOTE SENSING, 637, **682-685**
Replenishment, *See* BEACH NOURISHMENT
Reptilia, 209, 283
Reserves, 310-314
Restinga, 27, 33, **685**
Reteporella, 617
RETROGRADATION AND RETROGRADING SHORELINE, **685-686**
Revetments, 66, 246, 388, 509, 664, 665
Revilla Gigedo, 623
REYNOLD'S NUMBER, 390, 437, 564, **686,** 693, 718, 752, 854
Rhabdocoela, 279
Rhagodia baccata, 107, 108
Rhagodia crassifolia, 107
Rhine River, 127, 252, 413, 419
Rhine River Delta, 413
Rhinoclavis, 619
Rhizocephala, 281
Rhizochrysidales, 210
Rhizochrysis, 210
Rhizoclonium, 573
Rhizophora, 14, 70, 74, 108, 117, 167, 198, 200, 287, 482, 495, 532, 620, 621
Rhizophora apiculata, 621
Rhizophoraceae, 287, 532
Rhizophora harrisonii, 9
Rhizophora mangle, 9, 171, 192, 526, 621
Rhizophora mucronata, 13, 14, 16, 621
Rhizophora racemosa, 9
Phizophora stylosa, 621
Rhode Island, U.S.A., 127, 488, 583, 817
Rhododendron viscosum, 174
Rhodomela, 286
Rhodomela lycopoides, 55
Rhodomenia, 56
Rhodophyceae, 286, 404
RHODOPHYCOPHYTA, **686-687.** *See also* Algae, red
Rhodophyta, 55, 207, 574
Rhodymenia, 618
Rhodymenia natalensis, 11
Rhodymenia palmata, 775
Rhodymenia stenogona, 774
Rhombic patterns, 155
Rhombold marks, 548
Rhombold rills, 548
Rhone River, 126, 313, 365, 415, 419

Rhum, 411
Rhus, 197
Rhus copallina, 583, 584
Rhus toxicodendron, 582, 583, 584, 586
Rhus typhina, 583
Rhus vernix, 174
Rhynchocoela, 569
Rhynchosia minima, 588
RHYTHMIC CUSPATE FORMS, **688-691**
Rhythmic Forms, 271
RIA AND RIA COAST, 78, 79, 100, 101, 112, 185, 186, 214, 216, 217, 397, 410, 412, 413, 414, 415, 417, 426, 471, 485, 514, 575, 593, 597, **691**, 800
Ria de la Cornuña, 186, 691
Ria del Farrol, 691
Ria de Vigo, 691
Ribe, 411, 419
Ribes hirtellum, 582
RICHARDSON'S NUMBER, **692**
Ricinus communis, 528
Rickettsias, 720
Ridge embayment, 128
Ridge Plain, 159
RIDGE AND RUNNEL, 136, 156, 162, 163, **692**. See also Low and ball
Ridges. See also ACCRETION RIDGE; BEACH RIDGE PLAIN; BEACH RIDGES AND BEACH RIDGE COASTS; GRAVEL RIDGE AND RAMPART
 circalittoral coralligenous, 83
 pressure, 480
 shingle, 455, 714
 shore, 127
Rif, 20
Riga, 782
Rijeka, 416
RILL MARKS, 141, 155, 549, **692-693**, 854
Rimini, 416
Rincon Island, 514
Ringkobing Fiord, 411
Rio Balsas, 597
Rio Chubut, 199
Rio de la Cienega, 597
Rio Coco, 190
Rio Colorado, 199
Rio Crati, 416
Río Escondido, 190
Rio Grande, 127, 190, 198, 596, 597
Rio Grande do Norte, 198, 200, 768
Rio Grande do Sul, 768
Rio de Janeiro, 370, 371
Río Negro, 199, 374
Río Paraná, 199
Río Patuca, 190
Río de la Plata, 199, 370, 767, 768
Río Prinzapolca, 190
Río de Vigo, 186

RIP CURRENT, 154, 271, 299, 509, 529, 546, 562, 565, 566, 567, 656, 688, 690, **693-695**, 733, 745, 872. See also Undertow
RIPPLE MARKS, 163, 185, 365, 391, 431, 524, 528, 546, **695-698**, 703, 744, 868, 872, 873
Ripples, 141, 162, 163, 495. See also RIPPLE MARKS
 adhesion, 549
 antiripples, 163
 current, 155, 163, 546
 eolian, 241
 granule, 241
 index, 696
 ladder-back, 547
 megaripples, 163, 546
 oscillation, 155
 rhombohedral, 121
 rhomboid, 121, 548
RIPRAP, 633, **698**
Rips, 610
Rithrohaspanopeus harrissii tridentatus, 473
RIVER-DEPOSITION COASTS, 216, **698-699**
River Discharge, 299
Riverhead, 573
River Mouth Systems, 317, 363
Riviera, 415, 444
Riviera di Levante, 415
Riviera di Ponente, 415
Rivularia, 35
Rivularia firma, 104
Rivularia mesenterica, 35
Road and Roadstead, 442
Roatán Island, 190
Robe, 105
Roccella, 416
Rockall Plateau, 410
Rockall Trough, 410
Rock beach. See WAVE-CUT BENCH
ROCK BORERS, 617, **699**
Rock platform. See WAVE-CUT PLATFORM
Rock String, 685
Rocky Mountains, 211
ROCKY SHORE HABITAT, **699-702**
Rogoz, 168
Rogue River, 592
Roll-over, 129
Romania, 472, 473, 807, 809, 810, 811, 812, 840, 841
Rome, 250
Romo Island, 411
ROOSTER TAIL, 695, **702-703**
Rootan Island, 190
Rosaceae, 287, 582
Rosario, 196
Rosario Beach, 142
Rosa rugosa, 583, 584
Rosa virginiana, 237, 582, 584
Rosetta, 7, 23, 87, 90, 94

Rosh-Haniqra, 82, 84, 86, 87, 92, 93, 94
Rossella, 42
Rossellidae, 42
Ross Sea, 41, 42, 43, 45, 284
Rotaliina, 431, 432, 433
Rotifera, 66, 279
Rotten Spots, 377
Rotterdam, 251, 253, 371, 412
Rouad Island, 98
Rubble Mounds, 664
Rubiaceae, 587
Rubus, 582
Rubus chamaemorus, 581
Rubus idaeus, 582
Rubus pubescens, 581
Rügen, 411
Rumani, 92
Rumex salicifolius, 581
Rung-Sat, 75
Runnels, 529, 549, 692
Ruppia, 286, 287
Ruppia maritima, 14, 84, 171, 405
Rutaceae, 587
Ryukyu Islands, 68, 79, 355, 840

Sa de Arrábida, 414
Sabal palmetto, 585, 586
Sabatia stellaris, 585
Sabella pavonina, 494
Sabellaria, 572, 621, 680
Sabellaria kaiparaensis, 571, 572
Sabellariidae, 40, 680
Sabellidae, 40
Sabine Lake, 160
SABKHA, 33, 35-36, 87, 96-97, 101, 221, 376, 387, **704**, 705
Sabtan Island, 79
Saccostrea, 106
Saccrostrea cucullata, 574
Sacoglossa, 618
Sacramento River, 397, 599
Sado, 414
Saduria entomon, 55
Sagavanirktok River Delta, 359, 366
Sagitta, 282
Sagittaria latifolia, 586
Saguenay Fjord, 447
Sagunta, 415
Sahara, 4, 20, 25, 371, 374, 464
Sahel, 4
Sahul Shelf, 67, 68
Saida, 98, 99
Saigon, 75
St. Elias Range, 600
Saint Fabien-Sur-Mer, 795
St. Helena Bay, 374
St. James Island, 690
St. Johns River, 326, 581
St. Joseph Peninsula, 150, 531
St. Kilda Islands, 443
St. Lawrence Island, 505
St. Lawrence River, 176, 535, 536, 677, 795
St. Malo, 821, 822
St. Marks, 883

SUBJECT INDEX

St. Michel Bay, 823
St. Petersburg, Florida, U.S.A., 594
St. Peter's Island, 838, 839
St. Vincent Cape, 414
St. Vincent Gulf, 107, 761
St. Vincent Island, 160
Sakhalin, 76, 77, 78-81, 374, 504, 771, 773, 786
SALCRETE, **704-705**
Salicornia, 68, 107, 108, 171, 199, 287, 407, 585, 705, 707
Salicornia ambigua, 590
Salicornia arabica, 8
Salicornia australiensis, 574
Salicornia australis, 573, 707
Salicornia bigelowii, 584, 586, 590, 591
Salicornia brachystachya, 707
Salicornia europaea, 582
Salicornia fruticosa, 9
Salicornia herbacea, 8, 15
Salicornia herbica, 590
Salicornia pachystachya, 14
Salicornia pacifica, 196
Salicornia perennis, 193
Salicornia guadichaudiana, 199
Salicornia quinqueflora, 107, 108
Salicornia virginica, 584, 586, 588, 589, 590, 591, 592, 707
Salidae, 281, 490
Salinas River, 599
Salinity balance, 539
Salix, 55, 581
Salix hookeriana, 591
Salix waghornei, 581
Salmacina, 619
Salmoniformes, 283
Salsola, 288
Salsola kali, 108, 237, 582, 584, 585
Salt
  balance, 820
  in the coastal zone, 305
  crystallization, 118, 561
  fretting, 877
  pan, 443
  pannes, 171, 582, 586
  particles, 2
  pond, 443
  plugs, 101
  spray, 306, 407, 590, 699, 704, 712, 713, 714, 715, 716, 801
  wedge, 363, 489
  wedge estuary, 394
Saltation, 122, 146, 240, 744
Salt-crust structures, 356
Salting, 444
SALT MARSH, 106-107, 171, 172, 192, 196, 199, 287, 307, 308, 361, 377, 395, 405-407, 523, 538, 574, **705-706,** 819. *See also* Marsh
SALT MARSH COATS, **706-708,** 847-848
SALT WEATHERING, 38, **708-709,** 792, 805, 878, 879
Salvelinus alpinus, 56
Salvia, 197

Salviniales, 650
Salzwasserkarst, 501
Sambian Peninsula, 569
Samborombón Bay, 767
Samland Peninsula, 782
Samoa, 616, 621, 623, 624, 844
Samolus repens, 574, 707
Samothráki Island, 417
SAMPLE IMPREGNATION, **709-710**
Sanaga, 26
San Andreas Fault, 508, 599
San Clemente Island, 424
SAND, 455, **710,** 738, 757, 764-765. *See also* Sand *entries;* SAND *entries;* SHOESTRING SANDS; SOUNDING SANDS
  finger, 559
  indurated, 466
  movement, 713
  nourishment, 327
  quick, 172, 173, 496
SAND, SURFACE TEXTURE, **710-712**
Sand Banks, 125, 163
Sand Bars, 125, 163, 693
SAND BYPASSING, 179, 488, 666, **712.** *See also* Bypassing
Sand Cays, 616
San Diego, California, U.S.A., 599, 643, 727
Sand domes, 155, 548
Sandrock, 466
SAND DUNE HABITAT, **712-716**
Sand dunes, 265, 288, 376, 379, 408
  airflow, 226
  vegetation, 107
Sand flats, 404-405
Sand ramp, 644
Sand reefs, 125
Sand ripples, 241
Sand shadows, 141
Sand storms, 96
Sand volcanoes, 793
SANDWAVES AND LONGSHORE SAND WAVES, 136, 142, 150, 155, 162, 163, 272, 391, 529, 688, **716-717.** *See also,* Migrating humps
Sandy Hook, New Jersey, U.S.A., 438, 583, 790
San Feliu, 415
San Francisco, 246, 302, 371, 460, 599, 727
San Francisco Bay, 237, 280, 281, 333, 394, 397
Sanibel Island, 596
San Joaquin River, 397, 599
San Jorge, 821
San José, 821
San Juan Islands, 600
San Juan del Sur, 190
San Pedro, 26, 598
San Simeon, 599
Santa Barbara, 302, 326

Santa Catarina, 200, 532
Santa Clara, 765
Santa Cruz, 198, 767
Santa Elena, 765
Santa Lucia Range, 599
Santa Monica, 302, 599, 870
Santander, 414
Santa Rosa Island, 643
Santa Rosalia, 598
Santiago, 200
Santiago River, 198
Santo Domingo, 371
São Francisco River, 359, 365, 366
São Paulo, 198, 200, 370
Sapelo Island, 238
Sapropel, 456
Sarasota, Florida, U.S.A., 594
Sarawak, 79, 374
Sarcodina, 278, 431, 669, 670
Sarcophyton, 618
Sardinia, 410, 415
Sargassum, 10, 11, 12, 72, 104, 105, 106, 286, 616, 618, 634, 645, 772, 774, 776
Sargassum cristaefolium, 618
Sargassum fissifolium, 574
Sargassum miyabei, 772
Sargassum pallidum, 772
Sargassum sinclairii, 571
Saronic Gulf, 417
Saros Körfezi, 417
Sarracenia purpurea, 174
Sassafras albidum, 583, 584
Sassandra River, 25
Satellite communities offshore, 327
Saturated flow, 505
Satureja coccineum, 585
Saudi Arabia, 100, 371
Sauna, 809
Savannas, 195
Savarashtra, 374
Saxegotheae, 642
Saxifraga oppositifolia, 581
Saxostrea, 572
Scaevola, 109, 619
Scaevola crassifolia, 108
Scaevola keonigii, 191
Scaevola lobelia, 9
Scaevola plumeri, 589
Scaevola sericea, 108
Scaevola taccada, 13
Scaevola thunbergii, 13
SCALE MODELS, **717-719**
Scammons Lagoons, 597
Scaphopoda, 551, 552, 554
Scarp, 721
Scheelia, 198
Scheldt River, 333, 758
Schepradmolens, 515
Schizodonta, 282
SCHIZOMYCETES, **719-720**
Schizoporella, 182
Schizoporella errata, 84
Schleswig-Holstein, 252, 434, 515
Schmidt Cape, 784, 786
Schokland Island, 253
SCHORRE, **720-721**

Schouwen, 252
Schouwen Island, 412
Sciadopitys, 642
Scintillation counter, 741
Scirpus, 287, 583, 705
Scirpus americanus, 582, 585, 586, 588, 590, 591
Scirpus littoralis, 14
Scirpus maritimus, 193, 574, 590, 591
Scirpus nodosus, 107, 108
Scirpus paludosus, 237
Scirpus robustus, 585
Scleractinia, 279
Sclerospongea, 651
Scolecolepides arcticus, 55
Scolymastra joubini, 42
Scoresby Sound, 427
Scorpaeniformes, 283
Scotland, 148, 234, 284, 370, 387, 397, 406, 408, 410, 412, 420, 428, 442, 443, 468, 519, 676, 764, 798, 840, 879. *See also* UNITED KINGDOM
Scotobacteria, 720
Scour-and-fill, 185
SCOUR HOLES, 332, 495, **721**
Scouring rushes. *See* EQUISETOPHYTA
Scripus, 199
Scrophulariaceae, 587, 588
Scutus, 617
Scylla serrata, 621
Scyphozoa, 279, 331-332
Scytothamnus, 573
Sea (storm waves), 248
Sea arch, 721
Sea banks, 254
Sea breeze, 226, 880, 881
SEA CAVES, 171, 389, 534, 571-572, **721**
SEA CLIFFS, 407, 435, 463, 510, 593, **721-722**, 734. *See also* Cliff; Bluff
SEA CONDITIONS, **722**. *See also* Douglas scale; Peterson scale; Sea state
Sea of Crete, 417
Sea currents, 348
Sea floor mounds, 559
Seaford Head, 202
Sea Grass, 405
Sea ice, 60, 77, 462, 475, 479, 480, 500
Sea Island, 126, 489, 585
Sea of Japan, 78, 179, 332, 771, 772, 773, 774, 784-786
Sea level, 541-542
 factors affecting, 724-725
 indicators, 389
SEA LEVEL CHANGES, 447, 685, **722-728**, 729
SEA LEVEL CHANGES, 1900 TO PRESENT, **728-729**
SEA LEVEL CURVES, **729-733**
Sea Lion Caves, 721
SEA PUSS, **733-734**

SEA SLICK, **734**
Sea state, 226, 376, 683. *See also* SEA CONDITIONS
Sea surface temperature, 298
Seattle, Washington, U.S.A., 600
SEAWALL, 66, 246, 257, 388, 509, 656, 664, 699, **734**
Sea water therapy, 807-813
Sebkha el Melah, 23
Sebou River, 25
Sechura Desert, 766
SEDIMENT ANALYSIS, STATISTICAL METHODS, **735-736**
Sedimentary deposits, 444
Semidentary facies, 134, 314-322
Sedimentary regimes, 566-567
Sedimentation, 562-568. *See also* COASTAL DUNES AND EOLIAN SEDIMENTATION; ESTUARINE SEDIMENTATION; SHELF SEDIMENTATION
 allochthonous, 567
 autochthonous, 566
 and erosion, 465
 lagoonal, 505
SEDIMENT BUDGET, 7, 91, 157, 270, 271, **736-737**
SEDIMENT CLASSIFICATION, **737-740**
Sediment-organism relationship, 611-614
Sediments. *See also* Drift, of sediments; SEDIMENT entries
 anthropogenic, 464
 deltaic, 365-366
 suspended, 743
Sediment size analysis, 555
SEDIMENT SIZE CLASSIFICATION, **738-740**
SEDIMENT TRACERS, **740-743,** 746
SEDIMENT TRANSPORT, 157, 509, 740, **743-748,** 869, 881-882. *See also* Beach, drift; Drift, of sediments; SEDIMENT TRANSPORT; SHORE DRIFT CELL
 along the beach, 154, 155
 normal to the beach, 154
Sedum, 288
SEICHES, 348, 419, 434, 436, 505, 509, 533, 539, 602, 722, 724, **748-749,** 823, 825. *See also,* Shelf seich
Seine River, 397, 413
Seismic sea waves, 843-846. *See also* TSUNAMIS
Sekondi, 26
Selaginella, 524, 525
Selaginellales, 524
Sele River, 416
Sella River, 678
Sellyakh Gulf, 784
Selsey Peninsula, 269

Senecio frigidus, 581
Senecio pseudo-arnica, 581, 589
Senegal, 9, 25, 366, 677
Senegal River, 9, 20, 25, 359, 365, 366
Seoul, 371
Sepia, 554
Septaria, 621
Septifer, 616
Sequoiadendron giganteum, 641
Sequoia sempervirens, 642, 643
Sequoia washingtoniana, 641
Sequoia wellingtoniana, 641
Serenoa repens, 585
Sergipe, 768
Seriatipora hystix, 618
Sérifos, 417
Serpulidae, 40, 680
Sertella, 182
Sertularia, 56
Sesarma, 14, 70, 494, 621
Sesarma catenata, 11
Sesuvium, 588
Sesuvium portulacastrum, 10, 13, 14, 191, 199, 585, 588, 589, 707
Sète, 415
Seto Inland Sea, 78
Setubal, 414
Setup, 296, 724, 733, 880
Seven Sisters, 312
Severnaya Zemlya, 771, 783
Severn River, 176, 419, 821
Sewage effluent, 258, 386
Sewage treatment plants, 258
Seward Peninsula, 601
Seychelles, 19, 72, 840
Shabelle River, 30
Shackleford Banks, 713
Shamal, 101
Shanghai, 371
Shank Painter Pond, 175
Shannon River, 412
Shantarskye Islands, 785
Shantung Peninsula, 78, 883
Sharm, 100
Sharon, 94, 95, 96
Sharon Plain, 89, 93
Shatt-el-Arab River, 101, 359, 360, 364, 366, 821
Shear sorting, 122
Sheetflood wave swash, 142
Shelf ice, 462, 535
SHELF SEDIMENTATION, **749-750**
Shelf seiche, 296, 297
Shellag Cape, 784
Shell midden, 542
Shell mound, 542
Shell packings, 795
SHELL PAVEMENT, **750-751**
Sheringham, 202, 413
Sherm, 185
Sherum, 185, 186
Sheveningen, 235
Shikoku, 79, 332, 499
SHINGLE AND SHINGLE BEACHES, 124, 145, 157,

SUBJECT INDEX

254, 255, 288, 408, 511, 512, 520, 521, 632, 638, 747, **751-752**, 784, 788, 791, 885
Shiqmona, 83, 92
Shoal, 444
  emergent, 128
Shoalhaven Delta, 398
Shoalhaven River, 114, 365
SHOALING COEFFICIENT, **752**, 867, 868
Shoaling waves, 866
Shock marks, 206
SHOESTRING SANDS, **752-753**
Shorea robusta, 482
SHORE DRIFT CELL, 154, **753-754**. *See also* SEDIMENT TRANSPORT
SHOREFACE AND SHOREFACE TERRACE, 562, 565, 567, **754**
Shoreface slope, 550
Shore ice, 141
SHORELINE DEVELOPMENT RATIO, **754**
Shorelines, 262, 798. *See also* ALLUVIAL PLAIN SHORELINE; ARTIFICIAL SHORELINES; CONTRAPOSED SHORELINE; ELEVATED SHORELINE; EMERGENCE AND EMERGEO SHORELINE; OUTWASH PLAIN SHORELINE; PRESENT-DAY SHORELINE CHANGES, UNITED STATES; PRESENT-DAY SHORELINE CHANGES, WORLDWIDE; PROGRADATION AND PROGRADING SHORELINE; RETROGRADATION AND RETROGRADING SHORELINE; SHORELINE DEVELOPMENT RATIO; SUBMERGENCE AND SUBMERGED SHORELINE
  boggy, 513
  broken, 444
  changes, 655, 657
  compound. *See* CLASSIFICATION
  crenulate, 444
  of emergence, 127
  fault, 424
  graded, 438. *See also* EQUILIBRIUM SHORELINE
  marshy, 513
  neutral, 510, 615. *See also* CLASSIFICATION
  primary, 848
  recession, 679
  rhythms, 688
  rocky, 195, 404
  secondary, 510
  spit, 128
  stranded, 389
  of submergence, 509, 615

transgressing, 678
uplifted, 389
SHORE PLATFORMS, 118, 393, 463, 523, **755**, 763, 795, 856, 877
SHORE POLYGONS, **755-756**
Shore ridges, submergent, 127
Shore and shore system, 157
SHORE TERRACE, 628, **756-757**
Shore zone, 211
Sibenik, 417
Siberia, 55, 57, 58, 61, 76, 77, 449, 462, 477, 498, 601, 615, 657, 727, 783, 784, 787
Sicily, 418-419
Sideroxylon diospyroides, 13
Sidon, 53, 94
Sierra de Irta, 415
Sierra Leone, 25, 522, 762
Sierra de Montsia, 415
Sierras Planas, 678
Sierra Vizcaino, 196
Sífnos, 417
Sigalionidae, 39
Si-Kiang Delta, 77
Si-Kiang River, 78, 397
Sikínos, 417
Silene, 288
Silicoflagellaceae, 483
Silicoflagellida, 278
SILT, 710, 738, **757**
Siluriformes, 283
Silver Springs, Maryland, U.S.A., 863
Simulation, 336-337
Sinai, 23, 83, 84, 86, 87
Sinaloa, 197
Singapore, 75, 79, 371
Sinkholes, 595
Sinularia, 618
Sinum, 619
Sinus, 442
Siphonaria, 11, 199, 616
Siphonaria diemenensis, 104
Siphonaria pectinata, 10
Siphonops, 617
Siphonosoma, 620
SIPUNCULA, 67, 280, 385, **757-758**
Sipunculida, 611
Sipunculus, 620
Siracuse, 418
Sirbonian Lagoon, 84
Sirenia, 284, 483
Síros, 417
Sirte, 20
Sjaelland, 411
SKERRY, SKERRY GUARD, 427, 443, 444, 447, **758**, 782, 799
Skewed, 747, 748
Skewness, 147, 555, 556, 557, 735, 739, 871
Skíathos, 417
Skiros Island, 417
Skjaergoard, 410
Skye, 411

Slacks, 408
Slapton Beach, 147, 314, 688
SLIKKE, **758-759**
Slipoff points, 163
SLOPE-OVER-WALL CLIFFS, 230, **759**
Slough, 443
Sluiceway, 443
Smilax, 584
Smilax auriculata, 586
Smilax bona-nox, 586
Smilax laurifolia, 175, 586
Smith Cove, 156, 530
Smyrna, 691
Snell's law, 863, 868
Snow ramps, 61
Sochi, 445, 666, 667, 787
Society Islands, 552, 621, 623, 627
Socotra, 13
Sogne Fiord, 410, 427
SOILS, 509, 519, 527, 757, **759-763**, 877
  and boat-basin dredging, 172
  development, 715
  dune, 715, 759-760
  moisture, 714
  nutrients, 715
  salinity, 715
  submarine, 761-762
Solanderian Province, 104
Soledad Mountain, 598
Solent River, 207
Soletellina, 619
Solidago pauciflosculosa, 585
Solidago sempervirens, 582, 584, 585
Solidago spathulata, 590
Solomons, 616, 844
Solution benching, 580
Solution pits, 99, 389
SOLUTION AND SOLUTION PAN, **763-764**, 825, 854, 876
Somalia, 6, 12, 13, 21, 30, 371, 374
Somali Basin, 20, 28
Somali Current, 4
Somme River, 413
Sonneratia, 14, 108, 532, 620, 621
Sonneratia alba, 13, 14, 167
Sonneratiaceae, 532
Sonora, 195, 196, 197, 228
Sophora tomentosa, 588
Sorrento Peninsula, 416
Sorting, 556, 557, 682, 735, 738, 747. *See also* BEACH MATERIAL, SORTING OF
  coefficients, 147
Soudanese Niger, 26
Soulac-Sur-Mer, 840
Sound, 442
SOUNDING SANDS, **764-765**
Sous River, 25
South Africa, 11, 12, 14, 15, 27, 28, 130, 284, 311, 351, 369, 374, 454, 615, 761, 884
South America, 17, 26, 41, 45, 77, 126, 167, 176, 188, 193, 194,

197, 200, 211, 220-221, 231, 277, 284, 323, 355, 359, 369-370, 375, 454, 461, 495, 520, 532, 577, 642, 644, 652, 805, 844, 881
South America, coastal ecology. See CENTRAL and SOUTH AMERICA, COASTAL ECOLOGY
SOUTH AMERICA, COASTAL MORPHOLOGY, **765-770**
Southampton Water, 406
South Australia, 103, 104, 105, 106, 107, 108, 109, 110, 112, 114, 116, 117, 240, 375, 760, 761, 764
South Carolina, U.S.A., 126, 127, 175, 302, 550, 585, 586, 594, 672
South China Sea, 69, 81, 168, 359
South Equatorial Current, 4
Southern Alps, 577
Southern Bight, 566
Southern Ocean, 20, 30, 103, 115, 451, 521
South Island, 37, 427, 571, 572, 575, 577, 578, 579
South Shetlands, 45, 46, 848
South Texas Sand Sheet, 655
South Victoria Land, 37
South Vietnam, 70, 74. *See also* North Vietnam; Vietnam
South Westland, 575
SOVIET UNION, COASTAL ECOLOGY, **770-780**. *See also* U.S.S.R.
SOVIET UNION, COASTAL MORPHOLOGY, **780-789**. *See also* U.S.S.R.
Spadella, 282
Spain, 186, 313, 414-415, 419, 468, 559, 676, 678, 691, 840, 841
Spartina, 119, 170, 193, 287, 490, 819
Spartina alterniflora, 171, 307, 406, 440, 459, 582, 585, 586, 588, 705, 706, 707, 801
Spartina anglica, 117, 377
Spartina bakeri, 588
Spartina brasillensis, 194
Spartina cynosuroides, 586, 588
Spartina densiflora, 199
Spartina foliosa, 459, 590, 591, 705, 707
Spartina maritima, 9, 11, 406
Spartina patens, 169, 170, 171, 305, 459, 582, 584, 585, 586, 588, 705, 707, 714, 801
Spartina pectinata, 171, 582, 583
Spartina townsendii, 377, 406, 407, 848
Spas, thermal, 807
Spatangus, 620
Spencer Gulf, 107, 761, 764
Spergularia, 288
Spergularia marina, 582, 590, 591
Sperlonga, 416

Sperrgebied, 27
Sphacellaria, 56
Spheciospongia, 652
Sphenisciformes, 284
Sphenopsida, 392
Spilanthes urens, 199
Spinifex, 482
Spinifex hirsutus, 107, 108, 109
Spinifex longifolius, 108
Spinulosida, 381
Spiochaetopterus variopedatus, 772
Spionidae, 40
Spiraea, 583
Spiraea latifolia, 582
Spirochaetes, 720
Spisula, 169
Spisula sachalinensis, 773
SPITS, 512, **789-792**
  barrier, 123-125, 130-131, 790
  complex, 128, 790
  longshore, 127
  trailing, 444, 790
Spitsbergen, 55, 56, 427, 758, 783, 799
Splachnidium rugosum, 104, 105
SPLASH AND SPRAY ZONES, 13, 165, 500-501, 560, 763, **792**, 856, 877
Spondias, 194
Spondylus, 616, 617
Sponge. *See* PORIFERA
Spongia, 650, 651, 653
Spongillidae, 279, 651
Sporobolus virginicus, 11, 13, 107, 108, 191, 193, 199, 585, 588, 589
Sporozoa, 279
Spouting horn, 171. *See also* BLOWHOLES
SPRING PITS, 524, **793**
Spurn Head, 255
Squillidae, 617
Sri Lanka, 313, 355, 371, 545
Stabilization, 379
Stac, 443
Stacks, 419, 534, 799
Staffa, 411
Standard deviation, 147, 557, 735, 739
Standing Crop, 200, 793
STANDING STOCK, 165, 167, 659, **793-794**. *See also* Biomass
Staphylindae, 282, 490
Statistical inference, 334-336
Statistical methods, 735-736
Statistical parameters, 735-736
Stauromedusae, 279, 331
Stavanger, 411, 758
Stavoren, 252
Stellaria, 581
Stellaria humifusa, 55, 581
Stellaria longifolia, 581
Stelletta, 651
Stenolaemata, 183
Stenostomum, 645

Step, 547
  plunge, 140
Sterechinus neumayeri, 42
Stereocaulon, 581
Sterna, 408
Sternaspidae, 40
Stevns Klint, 203, 411
Stichopus, 617
Stikine Flats, Alaska, U.S.A., 592
Stillstand, *See* SEA LEVEL CURVES
Still water level, 249
Stipa, 198, 574
Stipa teretifolia, 107, 108
Stockholm, 53, 370, 463, 728
STOKES THEOREM, 690, 752, **794-795**, 867, 871
Stomatopoda, 281
Stonehaven, 410
STONE PACKING, **795**
STONE REEF, **795-796**
STORM BEACH, 141, 144, **796-797**
  cycle, 143
  profile, 143
Storm flooding, 714
STORM SURGE, 176, 267, 269, 326, 346, 467, 539, 657, 664, 667, 699, 704, 724, 749, 787, **797-798**, 819, 823, 826, 838, 851, 880
STORM-WAVE ENVIRONMENTS, **798**, 859
Storm waves, 220, 248, 467
Stradbrokes Island, 311
Strait, 443. *See also specific names*
Strait of Bab el-Mandeb, 99, 100
Strait of Belle Isle, 581
Strait of Bosporus, 787
Strait of Hormuz, 102
Strait of Juan de Fuca, 599
Strait of Messina, 416
STRANDFLAT, 410, 447, 463, 536, 678, 758, 783, **799**
Strandplain, 319, 444, 704
STRAND AND STRANDLINE, 387-388, 676, 798-799, 877
Streaky Bay, 107, 110
String reefs, 518
Stromatolites, 33-37, 354
Strombidae, 619
Stromboli Island, 418
Strombus, 72, 619
Strombus bubonius, 677
Strongylocentrotus, 382, 774
Strongylocentrotus droebachiensis, 775
Strongylocentrotus pallidus, 775
Strophostyles, 170
Strophostyles helvola, 584, 585
Strudel scour, 477
Struvea, 617
Stylaster, 618, 619
Stylochus, 279
Stylochus ellipticus, 645
Stylochus frontalis, 645
Stylophora, 11

SUBJECT INDEX

Stylophora mordax, 618
Stylophora pistillata, 618
Suaeda, 108, 196, 287
Suaeda californica, 590, 591
Suaeda fruticosa, 9, 11
Suaeda linearis, 584, 588
Suaeda maritima, 11, 582, 584
Suaeda monoica, 13, 16
Suaeda nova-zelandiae, 707
Suaeda vermiculata, 8
Suape, 795
Sub-aerial zone, 299, 300
Subaqueous sand dune. See Vol. I of series
Subduction, 450
Sublittoral border, 165
Submarine canyons, 347, 428, 866
Submarine cones, 347
Submarine fans, 347
SUBMARINE SPRINGS, **799-800**
Submarine zone, 299, 300
SUBMERGENCE AND SUBMERGED SHORELINE, 445, **800**
Subonoba, 42
Subsiding areas, 731
SUCCESSION, 406, 408, **800-802**
Sudan, 16
Suez, 16, 99
Suez Canal, 82, 405
Suffolk, 408, 413, 751
Sulawesi, 73, 79, 81
Sulu Archipelago, 848, 849
Sumatra, 68, 70, 72, 73, 77, 79, 81, 323, 369, 370, 532
Sumbawa Island, 81
Sumerians, 52
Sundaland, 67, 81, 369
Sundanella, 182
Sundarbans, 69, 70, 481
Sunda Sea, 79, 81
Sunda Shelf, 67, 68, 69, 70, 72, 74
Sundri, 482
Sungei, soengei, 443
Supralittoral fringe, 165
Supralittoral zone, 165
Surface creep, 240
Surface induration, 502
Surf base, 155, 745
SURF BEAT, 136, 146, 689, **802**, 838
SURFING, **802**
Surge base, 155
Surinam, 366, 374, 768, 769, 792
Suriname River, 769
Suspended load, 395, 402
Suspended material, 568
Suspension, 240
Susquehanna River, 402
Sussex, 123, 202, 204, 269, 270
Suwannee River, 595
Svalbard, 387, 462, 676
Svyatoy Nos, 784
Swalda, 199
Swash, 745
Swash marks, 141, 155, 431, 548
Sweden, 65, 160, 410, 411, 426,
 444, 447, 462, 464, 502, 725, 731
SWEEP ZONE, 663, 802-803
SWELL AND ITS PROPAGATION, 248, **803**
Swell wave environments, 220, 859
SWING MARK, **803-804**
Swinoujsche, 411
Sycon, 650, 651
Sydney, 118, 370, 371, 514
Sydney Basin, 110
Sydney Harbor, 691
Sylt, 411, 810
Symphyllia, 617
Synalpheus, 653
Synapta, 617
Synodic month, 671, 672
Synura, 210
Syria, 94, 98, 99, 369
Syringodium, 617
Systems analysis, 300, 424-425
Syzygium cordatum, 15
Syzygium guineense, 15
Syzygy, 671, 672, 720, 823, 835, 836

Tabanidae, 282, 490
Tabanus, 490
Tabebuia, 194
Tadpole nests, 155, 547
TAFONE, 3, 708, 709, **805-806**
Taganrog Bay, 788
Tagliamento, 416
Tagus, 414
Tahaa, 627
Tahiti, 627, 840
Tail, 444
Taiwan, 75, 76, 77, 79, 355, 370, 371, 501
Taiwania, 642
Takoradi, 26
Talitridae, 773
Tallinn, 177
Talorchestia quoyana, 573
Tamar Estuary, 117
Tamatave, 30
Tamaulipas, 194, 228
Tambaks, 74
Tamilnadu, 484
Tanacetum camphoratum, 590
Tanacetum douglasii, 589
Tana Delta, 30
Tanaidacea, 281
Tana River, 359, 366
Tanga, 13, 29
Tanganyika, 232
Tangier, 25
TANGUE, **806**
Tanzania, 12, 19, 29, 532
Tapti, 482
Taranaki, 578
Taranto, 416
Tardigrada, 67
Tarfaya Basin, 20, 25
Tarifa, 414
TAR POLLUTION ON BEACHES, **806**
Tarpon Springs, 587, 594, 883
Tarragona, 415
Tarsus, 52
Tartous, 99
Tasmania, 103, 104, 105, 108, 110, 112, 117, 118, 228, 235
Tasman Sea, 103, 577
Tatarskiy Gulf, 786
Tavoliere, 416
Taxaceae, 640
Taxales, 640
Taxodiaceae, 640, 643
Taxodium, 642
Taxodium distichum, 641, 643
Taxus, 640
Tay Estuary, 428
Taymyr Peninsula, 783
Tectarius granosus, 10
Tectarius natalensis, 11
Tectarius rugosus, 106
Tecticornia, 108
Tectona grandis, 482
TECTONIC MOVEMENTS, **806-807**
 uplift, 387
Tectus, 617, 618
Tegula, 199
Tel Aviv, 84, 89, 92
Teleostii, 483
Teliomycetes, 439
Tellinella, 619
Tell, the, 23
Telmaro, 443
Telpher, 253
Temnocephalida, 645
Tenasserim, 69
Tenby Peninsula, 412
Tendra, 787
Tenebrionidae, 490
Tentaculata, 345
Ten Thousand Islands, 532, 594
Tephrosia canescens, 13
Terebellidae, 40
Terebellides stroemi, 55
Terebralia, 621
Terebratalia, 179
Terminalia, 195, 619
Terminal scour, 268
Terminology, geographic, 442-445
Terpen, 369, 412, 515
Terpeniya Bay, 771, 786
Terraces, 721. See also SHOREFACE AND SHOREFACE TERRACE; SHORE TERRACE; WAVE-BUILT TERRACE
 cut. See WAVE-CUT PLATFORM
 ladder, 628
 low-tide, 163
 pendant, 444
 series, 628
 subaqueous, 510
Terracina, 416
Terrestrial maritime communities, flora, 287-288

SUBJECT INDEX

Ter River, 415
Terschelling Island, 313
Tesseropora rosea, 105
Tethya, 617
Tethys Sea, 17, 19, 773
Tetraclinus, 643
Tetraclita serrata, 11
Tetraclita squamosa, 106, 616
Tetraclita vitata, 106
Tetraodontiformes, 283
Tetrapods, 665
Texan region, North America, 588-589
Texas, U.S.A., 33, 123, 124, 126, 129, 134, 140, 160, 171, 233, 234, 235, 237, 238, 239, 242, 284, 326, 397, 399, 472, 518, 531, 585, 596, 643, 655, 680, 797, 837, 851
Textularia, 432
Textulariina, 431, 432, 433
Thailand, 73, 74, 75, 77, 466
Thais, 10, 74
Thais orbita, 104
Thalamita, 620
Thalassina, 171, 286, 617
Thalassina anomala, 621
Thalassiophyllum clathrum, 774
THALASSOTHERAPY, **807-813,** 841
Thaliacea, 283, 668
Thames, New Zealand, 424
Thames River, 397, 412, 413, 419
Thanet, 413
Thásos Island, 417
Thelkeldia diffusa, 108
Thelonotus, 617
Thermal discharge, 262
Thermal effluents, 386
THERMAL POLLUTION, **813-815**
THERMAL POWER, **815,** 822
Thermoclines, 298
Thermoerosion, 633
Thespesia acutiloba, 14
Thespesia populnea, 108
Thigmotrichida, 279
Thiobacillus ferro-oxidans, 761
Thipha, 168
Thipha latifolia, 168
Thíra, 418, 848
Thoracica, 281
Thraustochytriales, 439
Threshold velocity, 743
Thuja occidentalis, 582
Thuja orientalis, 642
Thuja plicata, 643
Thyborøn Barriers, 254
Thymelaea hirsuta, 8
Thysanoptera, 491
Thysanura, 281, 490
Tialpai mossambica, 483
Tiber River, 250, 416
Tibet, 451
TIDAL BASIN, **816**
Tidal bores, 816, 823

Tidal colks, 332
Tidal creek, 443, 820
TIDAL CURRENTS, 299, 348, 750, **816,** 817, 820, 826, 838, 850, 852, 872
TIDAL DELTAS, 317, 395, 402, 444, **817-818,** 820
TIDAL FLAT, 317, 402, **818-819.** *See also* Mud flat
TIDAL FLUSHING, **819-820**
TIDAL INLETS, CHANNELS, AND RIVERS, 317, 486, **820-821**
TIDAL POWER, 815, **821-823,** 826
TIDAL PRISM, 712, 820, **823**
TIDAL RANGE AND VARIATION, **823**
Tidal streams, 816, 838
TIDAL TYPE, VARIATION WORLDWIDE, 816, **824**
Tidal wave. *See* TSUNAMIS
Tide curve. *See* MARIGRAPH AND MARIGRAM
TIDE GAUGES, 541, 722, 728, 797, **825**
TIDE POOLS, **825-826**
TIDES, 296, 299, 541, 577, 671-672, 747, **826-837,** 872. *See also* PROXIGEAN SPRING TIDES; TIDAL *entries;* Tide curve; TIDE *entries*
astronomical, 205, 225, 229, 385, 467
autonomous, 434
diurnal, 824, 834, 837
double, 827
equatorial, 834
induced, 434
meteorological, 655
perigean, 672, 823, 835-836
semidiurnal, 824, 833-834, 837
spring, 823, 834-835
tropic, 834
vanishing, 827
TIDE TABLES AND CHARTS, 827, **837-838**
Tied Island, 443
Tierra del Fuego, 197, 198, 199, 200, 374, 536, 615, 767
Tiger Bay, 27
Tigris River, 101, 465
Tihama, 100
Tilapia, 49
Tillamook, 326, 878
Tillandsia usneoides, 586
TIME SERIES, 333-334, 579, 661, 662, **838**
Timor, 116, 359, 452, 501
Tinicum, 706
Tinnevelly, 486
Tínos, 417
Tintinnida, 278, 279
Tipasa, 23
Tipulidae, 490

Tirso River, 418
Tiryns, 52
Tofieldia borealis, 581
Tokar Delta, 31
Tokyo, 371
Tokyo Bay, 771
Tomales Bay, 599
TOMBOLO, 443, 512, 534, 789, **838-839,** 871
Tonga Islands, 452, 623, 844
Tongue, 444
Tonnidae, 619
Tordera, 415
Toredo, 699
Torres Strait, 333
Torreya, 640
Torrvieja, 414
Tosa Bay, 499
Toulon, 415
TOURISM, 370, 419, **839-842**
Toxopneustes, 617
Trace fossil, 184
Trace metals, 568
Tracers, 740-743
 ferromagnetic, 740
 fluorescent, 740, 746
 radioactive, 740, 746
Traction carpet, 750
Traganum nudatum, 8
Trainasse, 443
Transantarctic Mountains, 43
Transition, 163
Transgressive complexes, 319
Transkei, 14
Transport, 743-748
 longshore, 265, 267, 745
Trapani, 418
Travemünde, 411
Treasure Island, 247
Trema, 194
Tracheids, 641
Trematoda, 279, 645, 646
Trend analysis, 335
Trend maps, 335
TREND SURFACE ANALYSIS, **842**
Trent River, 413
Trichocorixa reticulata, 490
Trichodesmium, 354
Trichomycetes, 439
Trichopitys, 640
Trichoptera, 281, 490
Tricladida, 279
Tridacna, 617
Trifolium wormskjoldii, 590, 591, 592
Triglochin, 287, 581, 583
Triglochin maritimum, 581, 582, 590, 591, 592, 705, 707
Triglochin palustris, 581, 582
Triglochin striata, 707
Trilobita, 280
Trinidad, 200
Triodia, 108
Triplasis purpurea, 585
Triple junction, 449

Tripneustes, 617
Tripoli, 20, 98, 99
Tripoli Archipelago, 98
Tripolitanian Basin, 20
Triste Gulf, 769, 770
Trochodota dendyi, 573
Trochopus, 491
Trochus, 74, 617, 618
Trombidiform mites, 280
Tropic of Cancer, 623
Tropic of Capricorn, 623
Trottoir, 83, 443
Troughs, 135, 593
Trou-sans-Fond, 26
Trouville-sur-mer, 808, 809, 812
Trujillo, 199
Truk, 622, 627
Truro, 235
Tsuga canadensis, 582
TSUNAMIS, 176, 296, 297, 299, 385, 577, 602, 623, 657, 722, 724, 786, 823, 826, **843-846**.
  See also Seismic sea wave
Tuamotu, 627
Tuapse, 787
Tubastrea, 618, 619
Tubificidae, 280
Tubiluchus, 659
Tubipora, 618
Tubulanus annulatus, 570
Tubularia, 42
Tubulipora, 182
Tugela River, 27, 28
Tugurskiy Gulf, 785
Tuléar Bay, 30
Tumaco, 765
Tumbes, 199, 766
Tunicata, 208, 283, 668, 669
Tunis, 374, 680
Tunisia, 7, 8, 20, 22, 23
Tunnel valleys, 434
Turbellaria, 279, 346, 542, 645, 646
Turbidity maxima, 402
Turbinaria, 74, 616, 617
Turbinaria murrayana, 618
Turbinaria turbinata, 618
Turbo, 74, 617
Turbo argyrostoma, 618
Turbo pethalatus, 618
Turimetta Head, 826
Türisalu, 503
Turkey, 52, 54, 96, 419
Turnover rate, 659
Tuscan Archipelago, 416
Tuvalu, 456
Tuxpan, 597
Tweed River, 327
Tymna Bay, 784
Typha, 198, 574, 583, 588
Typha angustifolia, 171, 583, 584, 586, 705
Typha domingensis, 16
Typhoons, 724
Tyre, 53, 94, 98, 99, 251

Uber, 619
Uca, 14, 70, 73, 171, 494, 621

Uca dussumierii, 621
Uca lactea, 621
Uca marionis, 621
Uca vocans, 621
Ucides, 494
Udden-Wentworth scale, 738
Udotea, 617
Udskiy Gulf, 785
Ulbanskiy Bay, 785
Ulmus americana, 586
Ulva, 10, 84, 199, 286, 306, 490
Ulvaceae, 208
Ulva lactuca, 12, 199
Umm el Qawain, 101
Umnagazana River, 532
Undercut, 500
Undertow, 154, 693. *See also* RIP CURRENT
Underwater equipment, archaeology, 53
Uniformitarianism, 290
Uniola, 169, 170, 305, 588
Uniola paniculata, 191, 237, 584, 585, 588, 713
U.S.S.R., 55, 56, 127, 235, 254, 294, 311, 351, 424, 502, 504, 665, 666, 667, 690, 770-780, 780-789, 807, 821, 823. *See also* SOVIET UNION, COASTAL ECOLOGY; SOVIET UNION, COASTAL MORPHOLOGY
United Arab Emirates, 101
United Kingdom, 146, 255, 327, 328, 410-411, 412-413, 466, 520, 738, 852. *See also* England; Great Britain; Ireland; Scotland; Wales
United States, 37, 50, 55, 65, 124, 125, 126-128, 135, 139, 148, 159, 169-171, 175, 228, 231, 236, 237, 239, 242, 254, 256, 259, 268, 292, 294, 301-310, 311-313, 325, 328, 349, 352, 356, 370, 393, 397, 399, 407, 422, 442, 449, 465, 466, 470, 489, 498, 504, 505, 510, 511, 543, 544, 562, 584, 586, 587, 593, 597, 602, 630, 639, 655-657, 658, 664, 666, 667, 670, 680, 688, 700, 705, 707, 712, 713, 724, 727, 728, 736, 738, 757, 759, 774, 797, 806, 807, 813, 815, 817, 821, 827, 839, 851, 852
U.S. Naval Hydrographic Office Scale. *See* SEA CONDITIONS
Upcoast, 444
Upogebia africana, 11
Upolu, 623
Upstart Bay, 518
Upwelling, 298, 323, 347, 649
Ur, 52
Ural Mountains, 783
Urechis, 620
Urechis caupo, 385

Urochordata, 208, 668, 669
Urstromtäler, 434
Urticinopsis, 42
Uruguay, 199, 768
Usedom Island, 411
Ussuri Bay, 785
Utah, U.S.A., 127, 128, 470, 498, 509, 805
Uyea, 622, 623

Vaccinium, 715
Vaccinium angustifolium, 583
Vaccinium arboreum, 586
Vaccinium corymbosum, 174, 583, 584
Vaccinium macrocarpon, 174, 582, 585
Vaccinium ovatum, 591
Vaccinium oxycoccos, 174
Vaccinium uliginosum, 581
Vaccinium vacillans, 583
Vaccinium vitis-idaea, 581
Vaigach Islands, 783
Vaigai, 482
Valdivia, 767
Valencia, 415
Valletta, 418
Valley of Zevulun, 89
Valonia, 617, 618
Valparaíso, 199, 766, 767
Vanderford, 427
Van Diemen Gulf, 114, 116
Vane shear tests, 258
Varanus niloticus, 14
Variance, 555
Varves, 422
Vasa, 53, 54
Vasterbotten, 374
Vasticardium, 619
Vasto, 416
Veerse Gat, 253
VEGETATION COASTS, **847-848**
Vejle Fiord, 411, 434
Veliidae, 281, 490, 491
Vembanad Lake, 485
Veneridae, 619
Venezuela, 198, 200, 228, 769-770
Venice, 126, 130, 346, 369, 416, 419
Venice, California, U.S.A., 870
Ventifacts, 49
Ventspils, 782
Ventura, California, U.S.A., 598
Venus gallina, 83
Veracruz, 194, 195, 235, 236
Verbenaceae, 287, 527, 528
Verkhoyansk Range, 449
Vermetus triquetrus, 83
Vermilion Parish, 206
Verongula, 653
Verrucaria, 404, 699
Vertebrata, 209
Vesteralen Island, 410
Vesuvius, 416
Viburnum dentatun, 583
Victoria, Africa, 2
Victoria, Australia, 103, 105, 107, 108, 109, 110, 112, 113, 114,

SUBJECT INDEX

115, 116, 117, 118, 119, 312, 313, 532
Victoria Land, Antarctica, 805
Victoriella pavida, 84
Vietnam, 75, 77, 79, 370. *See also* North Vietnam; South Vietnam
Vigna, 619
Vigna lutea, 482
Vijzelmolens, 515
Virginia Beach, 246
Virginia, U.S.A., 126, 237, 242, 246, 302, 304, 583, 585, 643, 684, 689, 716, 727
Virginian region, North America, 584-585
Virgin Islands, 313, 815, 839
Viscous regime, 122
Vishakhapatnam, 485
Visor, 389, 500, 501, 579. *See also* NICHE, NICK, NIP, NOTCH
Vitex amboniensis, 13
Vitis aestivalis, 586
Vitis rotundifolia, 584, 586
Vittina, 621
Vizcaino Bay, 597, 598
Vladivostok, 77, 370
Voacanga dregei, 15
Vochysia, 195
VOLCANO COASTS, 214, 216, 520, 616, **848-849**
Volcanoes, 623
Volga River, 359, 787
Volkerak, 253
Volta River, 26
Volturno River, 416
Voltziaceae, 640, 642
Volutidae, 619
Volva, 619
Vostochno-Sibirsk, 771
Vulcano Island, 418

WADDEN, 126, 166, 188, 253, 495, 496, **850-851**
Wadden Zee, 273, 395, 404, 405, 412, 515, 648, 759
Wadi, 89, 91, 96, 98, 101, 375
Wadi el Arish, 23, 89, 91
Wadi el Ghweibba, 16
Waikiki Beach, 623
Waikolu Valley, 844
Waipio Valley, 844
Wairau, 577, 578
Waitaki River, 578
Walcheren Island, 252, 412
Walchia, 642
Wales, 312, 313, 412, 420, 821, 823, 840. *See also* United Kingdom
Walland, 254
Wallis Island, 623
Walmer, 202
Walther's law, 134
Walvis Bay, 11
Walvis Island, 17
Walvis Ridge, 27
Wampum, 552

Warnemünde, 411, 728
Warriewood, 223, 856
Warrnambool, 110
Warrnambool-Port Fairy, 2
Wash, 444
Washington, D.C., U.S.A., 370
Washington, U.S.A., 38, 142, 228, 235-236, 387, 461, 530, 546, 548-549, 589, 590-592, 599, 643, 713, 727, 771, 805
WASHOVER AND WASHOVER FAN, 124, 125, 160, 214, 236, 268, 395, 467, 505, 528, 531, 565, 588, 596, 656, 696, 790, 791, 798, 817, **851**
WASTE DISPOSAL, **851-853**
Waste water
  effluents, 386
  treatment, 852
Water
  density, 724-725
  jet, 53
  —level marks, 549
  line, 798
  orbits, 872-873
  planes, 676
  quality, 173, 258
WATER-LAYER WEATHERING, 224, 393, 523, 544, 677, 792, **853-854,** 877
Waterloo Rock, 466
Watersipora cuculata, 573
WATER TABLE, 353, 408, 462, 469, 470, 510, 514, 548, 713, 714, **854-855,** 861. *See also* Groundwater
WAVE-BUILT TERRACE, 510, 511, **855.** *See also* Marine terrace
WAVE CLIMATE, **855**
WAVE-CUT BENCH, 510, 511, 800, **856**
Wave-cut notch, 501
WAVE-CUT PLATFORM, 224, 536, 579, 623, 644, 674, 677, 799, 826, 853, 855, **856-857,** 879. *See also* Platform, Abrasion platform
WAVE ENERGY, 1, **859,** 863, 883
WAVE DRAG LAYER, **857-859**
Wave energy, zero relative, 522
WAVE ENVIRONMENTS, 798, **859-860**
WAVE EROSION, 216, **860-862**
WAVE METERS, 533, **862**
Wave recorders, pressure-type, 157
WAVE REFRACTION DIAGRAMS, **862-866**
Waves, 247-250, 265, 296, 274, **866-869,** 879-881. *See also* Breakers; DESIGN WAVE; Diffraction; EDGE WAVES; FORCED WAVES; INTERNAL WAVES; Reflection; Refraction;

STORMWAVE ENVIRONMENTS; WAVE *entries;* WIND, WAVES AND CURRENTS, DIRECTION OF
  action, 153, 154, 299
  angle of approach, 1
  base, 155, 510, 745, 855
  front, 618-619
  gauges, 862
  gravity, 437, 492
  height, significant, 862, 871
  intersecting, 142
  long, 438
  ramp, 393. *See also* EROSION RAMP, WAVE RAMP
  refraction, 693
  ripples, 547
  solitary, 176, 869, 872
  standing, 163, 529, 872
  steepness, 248, 425
  train, 467
WAVE SHADOW, **870-871**
WAVE STATISTICS, **871**
WAVE THEORIES, OSCILLATORY AND PROGRESSIVE, 247-250, **871-872**
Wave washing, limit of, 461
WAVE WORK, **872-876**
Weald, 412
Weathering. *See also* MURICATE WEATHERING; SALT WEATHERING; WATER LAYER WEATHERING; WEATHERING AND EROSION *entries*
  frost, 758
  honeycomb, 805, 877. *See also* ALVEOLAR WEATHERING
  littoral, 877
  organic. *See* WEATHERING AND EROSION BIOLOGIC
  physical, 224
  shadow, 805
WEATHERING AND EROSION, BIOLOGIC, 617, **876**
WEATHERING AND EROSION, CHEMICAL, 79, 224, 236, 792, **876-877**
WEATHERING AND EROSION, DIFFERENTIAL, **878**
WEATHERING AND EROSION, MECHANICAL, **878-879.** *See also* Corrasion; Corrosion
Weather shore, 444
Weber number, 718
Weddell Sea, 41, 43, 45
Wedging, 468
Wellington, New Zealand, 424, 572
Welwitschia, 454, 644
Welwitschiales, 454
Welwitschia mirabilis, 454
Wentworth scale, 710, 738, 751
Weser River, 412
West Bay, 269

SUBJECT INDEX

West Beach, 530
West coast region, North America, 589-592
Western Australia, 103, 104, 106, 108, 109, 110, 114, 240, 366, 369, 375, 387, 501, 544, 819
Western Ghats, 481, 486
Westernport Bay, 112, 116, 119
Western Shield, 112
West Flandes, 371
West Friesland, 253
West Germany, 411, 665, 666
West Greenland Current, 55
West Indies, 235, 330, 371, 527, 553, 652, 805
Westmorland, 810
West Palm Beach, 487
Wexford, 412
WHARF, 183, 674, **879**
Whatipu Beach, 578
Whidbey Island, 530, 548
Whitecaps, 226
Whitecliff Bay, 203
White Cliffs of Dover, 669
White Nile, 23
White Sea, 56, 771, 782-783, 823
Wick, 468
Widdringtonia, 643
Wieringen, 253
Wildlife reserves, 313
Wilhelmshaven, 809
Wilson's Promontory, 105, 110, 113, 114, 117, 313
WIND, 265, 724, 747-748, 863, **879-881**
  Beaufort scale, 161
  ripples, 528, 549
  rows, 610, 734
  set-up, 268, 797
  shadow dunes, 141
  tide, 797
WIND, WAVES, AND CURRENTS, DIRECTION OF, **881**
Windstau, 724
Winged headlands, 790
Wisconsin, U.S.A., 508, 805
Withernsea, 255
Wittrockiella salina, 573
Wolds, 251
Wolffia, 526
Wollin Island, 411

Woodland, 50, 51
Woods Hole, Massachusetts, U.S.A., 304, 305, 671
WORLD NET SEDIMENT TRANSPORT, **881-882**. *See also* SEDIMENT TRANSPORT
World shoreline length, 225
Wormley, 825
Wrangel Island, 784

Xanthoxylum clava-hercules, 584, 585, 586
Xenia, 618
Xenostrobus, 105
Xenostrobus pulex, 105
Xerochloa, 108
Xiphophora, 574
Xiphophora chondrophylla, 105, 571, 572
Xiphophora gladiata, 105
Xiphosura, 63, 280
Xylocarpus, 14, 620, 621
Xylocarpus granatum, 13, 14
Xylocarpus moluccensis, 13, 14

Yacyama Islands, 68
Yampi Sound, 116
Yanbo, 100
Yang-tse-Kiang Delta, 77
Yang-tse-Kiang River, 78, 359
Yangtze River, 360, 883
Yarborough Pass, 655
Yavne-Yam, 84
Yeast. *See* FUNGI
Yellow Sea, 77, 359
Yemen, 100
Yenisei River, 61, 783
Yioúra, 417
Yokohama, 371
Yoldia hyperborea, 776
Yoldia limatula, 612
Yorke Peninsula, 110
Yorkshire, 202, 254, 255, 268, 419
Yorkshire Ouse, 413
Yucatán, 127, 191, 192, 194, 195, 501, 589, 597
Yucca, 194
Yucca aloifolia, 585
Yugoslavia, 186, 416-417, 840, 841
Yukon, 374, 477, 600, 601

Zákinthos Island, 417
Zalew, 411, 458
Zambezi, 13
Zambezi Cone, 18
Zambezi Delta, 28
Zambezi River, 13, 18
Zamia, 355
Zandkreek, 253
Zannichellia palustris, 405
Zanzibar, 29
Zatoka Gdansk, 411
Zawn, 442
Zeacumantus lutulentus, 574
Zeebruggee, 841
Zeeland, 252, 648
Zemmour Fault, 20
Zenobia, 175
ZERO-ENERGY COAST, 151, 153, 218, 391, 467, 873, **883**
ZETAFORM BAYS, 523, 881, **883-885**
Zieriksee Promontory, 333
Zihuatanejo, 597
ZINGG SHAPE, **885**
Zire, 98
Zoantharia, 331
Zoanthus, 618
Zonation, 165, 275, 285
  benthos, 165, 166
  biotic, 169-171
Zoobotryon, 282
Zoomastigina, 670
Zoophyta, 328
Zooxanthellae, 617, 652
Zostera, 15, 83, 286, 405, 494, 574
Zostera asiatica, 773, 774
Zostera capensis, 11, 14
Zostera japonica, 772
Zostera marina, 55, 56, 171, 405, 527, 591, 772, 775
Zostera marina marina, 774
Zostera nana, 574, 772
Zostera tasmanica, 574
Zozimus, 617
Zululand, 13, 14, 28, 689
Zuyder Zee, 253, 412, 465, 515, 541
Zwartkops, 11
Zygomycetes, 439
Zygomycotina, 439
Zygophyllum album, 8, 16
Zygophyllum simplex, 11